Jane's

Avionics
2007-2008

Edited by Edward Downs

Twenty-sixth Edition

ISBN-13 978 0 7106 2796 4
ISBN-10 0 7106 2796 3

Printed in Great Britain by Cambridge University Press

Front cover image: *For many, the star of the 2006 Farnborough Airshow was the MiG−29 OVT, equipped with RD−33 engines and Thrust Vectoring Control (TVC) manufactured by the Klimov Scientific Production Association. The combination of two−dimensional TVC and Full Authority Digital Engine Control (FADEC) on the MiG−29 OVT is described by Klimov as 'KliVT', which stands for 'Klimov Thrust Vectoring'. This addition to the MiG−29/35 provides for superior manoeuvrability to the Su−30/37, with no Angle of Attack (AoA) or low speed limitations. While previous displays of thrust vectoring in fighter aircraft have shown limited potential combat value, the ability of the MiG−29 OVT to 'fuselage point' around all of its axes represents a powerful close−in air combat capability. However, the spectacular manoeuvres flown during the Farnborough display was not simply 'seat of the pants' stuff; while the entry parameters were set by the pilot, it was a combination of KliVT and the MiG−29 Flight Control System (FCS) that controlled the aircraft through its various transitions between thrust−borne and wing−borne flight.* (Jane's/Patrick Allen)

Contents

Jane's Avionics website: jav.janes.com

If you find it hard to capture
the small details while
it's stationary,
imagine doing it at 300kts

Today's battlefield is a rugged fast-paced arena, hosting fast moving threats and targets ranging from aircraft to naval vessels. Elop's range of stabilized payloads is specifically ruggedized to withstand the harshest conditions and are customized to be installed on fixed and rotary wing aircraft of all kinds, ships and armored vehicles. CoMPASS IV is Elop's latest generation advanced stabilized system designed for medium and high enforcement observation and surveillance requirements. CoMPASS IV is an optimal solution for fixed and rotary wing aircrafts, marine patrol boats, UAVs, and ground vehicles. Elop's battle proven payloads, integrate technologically advanced multi-sensors including high resolution FLIRs, eye-safe LRFs, state-of-the-art color or b/w TV cameras and automatic trackers. All payloads are highly adaptable to existing radar, ECM, FCS, HMS and other command & control systems. After more than six decades of developing high stability, supreme quality, electro-optical payloads, we are sure that having them on-board, will allow you to safely fire up and hit the road.

ELECTRO-OPTICS **elop** Ltd.

If it's out there - you'll see it

Advanced Technology Park, Kiryat Weizmann, P.O.B 1165 Rehovot 76111, ISRAEL
Tel. 972-8-9386211, Fax. 972-8-9386237, Web-Site: www.el-op.com

How to use: Jane's Avionics

Content

Jane's Avionics describes airborne electronic equipment designed civilian and military aircraft. To be included within this title, equipment should be operated in manned aircraft, characterised by significant electronic content, and be in development, production or, if no longer in production, in widespread service. Equipment which is in service in legacy platforms has been archived in order to maintain a manageable overall number of entries in the Yearbook. This archive is accessible through the Online service. The exception to the above is equipment which, in the opinion of the editor, is significant and therefore relevant to the reader.

Avionics equipment designed specifically for Unmanned Aerial Vehicles (UAVs) is covered in *Jane's Unmanned Aerial Vehicles*, although this distinction is becoming increasingly difficult to make in view of the development of Unmanned Combat Air Vehicles (UCAVs), which may be regarded as combat aircraft operated by a remote, ground-based pilot. Systems that are applicable to both manned and unmanned aircraft will continue to be found in *Jane's Avionics*.

Structure

Supporting the main sections of *Jane's Avionics*, a leading *Analysis* section contains technical overviews covering some of the core technologies to be found in the book, namely, Digital Datalinks, Airborne Radar, Electro-Optic systems and Helmet Mounted Display Systems (HMDS); these are intended to provide useful background information regarding some of the fundamentals of civil and military avionics systems.

The main body of *Jane's Avionics* is divided into six main sections that are easily identifiable to the casual or expert reader alike:

- **Avionic Communication, Navigation and Identification (CNI) Systems** - This section makes use of a commonly used acronym in the aerospace industry, CNI, which covers a diverse range of equipment including aircraft radios and transponders. In order to further aid the reader, this main section is further divided into subsections, covering communications, navigation, combined comm/nav and safety of flight systems.
- **Flight/Mission Management (FM/MM) and Display Systems** - This section includes most equipment which directly interfaces with the crew, such as electronic and analogue displays, Head Up Displays (HUDs) and other Man-Machine Interface (MMI) systems.
- **Airborne Electro-Optic (EO) Systems** - This section contains all synthetic and enhanced vision systems, including Near-, Mid- and Far Infra-Red (IR) systems, such as Aviators' Night Vision Imaging Systems (ANVIS) and Pilot's Night Vision Systems (PNVS).
- **Airborne Radar Systems** - This section includes all types of airborne platform-integrated and podded radar systems.
- **Airborne Electronic Warfare Systems** - This section covers important Electronic Warfare (EW) systems, including RF jammers, Counter Measures Dispensing Systems (CMDS), radar and missile approach warners, Towed Radar Decoys (TRDs) and Electronic Support (ES) systems, such as ELINT, COMINT and SIGINT equipment.
- **Aircraft Control and Monitoring Systems** - The last main section covers Flight Control (FC), autopilot and onboard monitoring and Fault Warning Systems (FWS).

Equipment and systems are listed alphabetically under country of origin, which is also entered alphabetically.

After the six main sections and prior to the various indexes, *Aircraft Cockpits* provides the reader with reference and comparison of modern and legacy cockpit configurations, together with descriptions of important features.

After the main text are three indexes - a contractors index, an alphabetical index and an index that correlates entries to contractors.

The contractors index contains details of the name, address and, where available, the telephone/fax number/website of each contractor represented in the book.

The alphabetical index lists every entry in alphabetical order.

The manufacturers index lists entries alphabetically by contractor.

There is no delineation in *Jane's Avionics* between civil and military systems, reflecting current trends in the utilisation of Commercial Off-The-Shelf (COTS) equipment in military applications wherever possible, thereby minimising acquisition and through-life costs. Noting the global use of the terms 'airborne' and 'aircraft' in the subsection titles, which must be surely taken 'as read' in a book detailing avionic equipment, the reason lies in the integrated nature of the Jane's Defence Equipment portfolio and the online medium - users searching for particular equipment capabilities may not necessarily know of its intended application.

Record structure

Within each main section, entries are arranged alphabetically by country of manufacture. Entries are structured as follows:

- **Title**
- **Type** - describes the group or classification of the equipment by role and/or physical characteristic(s).
- **Description** - contains a technical description and information regarding functionality.
- **Specification** - lists the main technical parameters and physical characteristics.
- **Status** - contains information on development, production and operational status.
- **Contractor** - details prime and relevant sub contractors.

Of particular note to manufacturers, the type headings may not exactly match the detailed function of the system in question - for taxonomy purposes, in order to keep the number of type groupings manageable and members of those grouping at a reasonable and balanced level, it is sometimes necessary to describe a piece of equipment with a more generic type heading. In all such cases I have taken care to ensure that the type description will capture that piece of equipment and all similar.

Images

The reader will find that parts of the book contain colour images in order to more effectively convey the information therein - a large colour image is vital to adequately describe the properties of a full-colour electronic display, for example.

Images are annotated with a seven-digit serial number which uniquely identifies them within Jane's image database.

The widespread use of colour in relevant areas of *Jane's Avionics* continues, with further expansion of the Aircraft Cockpits section and update of the technical overviews to maintain pace with the development of the technologies.

Jane's Libraries

To assist your information gathering and to save you money, Jane's has grouped some related subject matter together to form 'ready-made' libraries, which you can access in whichever way suits you best – online, on CD-ROM, via Jane's EIS or through Jane's Data Service.

The entire contents of each library can be cross-searched, to ensure you find every reference to the subjects you are looking for. All Jane's libraries are updated according to the delivery service you choose and can stand alone or be networked throughout your organisation.

Jane's Defence Equipment Library

Aero-Engines
Air-Launched Weapons
Aircraft Upgrades
All the World's Aircraft
Ammunition Handbook
Armour and Artillery
Armour and Artillery Upgrades
Avionics
C4I Systems
Electro-Optic Systems
Explosive Ordnance Disposal
Fighting Ships
Infantry Weapons
Land-Based Air Defence
Military Communications
Military Vehicles and Logistics
Mines and Mine Clearance
Naval Weapon Systems
Nuclear, Biological and Chemical Defence
Radar and Electronic Warfare Systems
Strategic Weapon Systems
Underwater Warfare Systems
Unmanned Aerial Vehicles and Targets

Jane's Defence Magazines Library

Defence Industry
Defence Weekly
Foreign Report
Intelligence Digest
Intelligence Review
International Defence Review
Islamic Affairs Analyst
Missiles and Rockets
Navy International
Terrorism and Security Monitor

Jane's Market Intelligence Library

Aircraft Component Manufacturers
All the World's Aircraft
Defence Industry
Defence Weekly
Electronic Mission Aircraft
Fighting Ships
Helicopter Markets and Systems
International ABC Aerospace Directory
International Defence Directory
Marine Propulsion
Naval Construction and Retrofit Markets
Police and Homeland Security Equipment
Simulation and Training Systems
Space Systems and Industry
Underwater Security Systems and Technology
World Armies
World Defence Industry

Jane's Security Library

Amphibious and Special Forces
Chemical-Biological Defense Guidebook
Facility Security
Fighting Ships
Intelligence Digest
Intelligence Review
Intelligence Watch Report
Islamic Affairs Analyst
Police and Homeland Security Equipment
Police Review
Terrorism and Security Monitor
Terrorism Watch Report
World Air Forces
World Armies
World Insurgency and Terrorism

Jane's Sentinel Library

Central Africa
Central America and the Caribbean
Central Europe and the Baltic States
China and Northeast Asia
Eastern Mediterranean
North Africa
North America
Oceania
Russia and the CIS
South America
South Asia
Southeast Asia
Southern Africa
The Balkans
The Gulf States
West Africa
Western Europe

Jane's Transport Library

Aero-Engines
Air Traffic Control
Aircraft Component Manufacturers
Aircraft Upgrades
Airport Review
Airports and Handling Agents –
    Central and Latin America (inc. the Caribbean)
    Europe
    Far East, Asia and Australasia
    Middle East and Africa
    United States and Canada
Airports, Equipment and Services
All the World's Aircraft
Avionics
High-Speed Marine Transportation
Marine Propulsion
Merchant Ships
Naval Construction and Retrofit Markets
Simulation and Training Systems
Transport Finance
Urban Transport Systems
World Airlines
World Railways

EDITORIAL AND ADMINISTRATION

Chief Content Officer: Ian Kay, e-mail: ian.kay@janes.com

Group Publishing Director: Sean Howe, e-mail: sean.howe@janes.com

Publisher: Sara Morgan, e-mail: sara.morgan@janes.com

Compiler/Editor: Welcomes information and comments from users who should send material to: Information Collection Jane's Information Group, Sentinel House, 163 Brighton Road, Coulsdon, Surrey CR5 2YH, UK
Tel: (+44 20) 87 00 37 00 Fax: (+44 20) 87 00 39 00
e-mail: yearbook@janes.com

SALES OFFICES

Europe and Africa
Jane's Information Group, Sentinel House, 163 Brighton Road, Coulsdon, Surrey CR5 2YH, UK
Tel: (+44 20) 87 00 37 50 Fax: (+44 20) 87 00 37 51
e-mail: customer.servicesuk@janes.com

North/Central/South America
Jane's Information Group, 110 N Royal Street, Suite 200, Alexandria, Virginia 22314, US
Tel: (+1 703) 683 21 34 Fax: (+1 703) 836 02 97
Tel: (+1 800) 824 07 68 Fax: (+1 800) 836 02 97
e-mail: customer.servicesus@janes.com

Asia
Jane's Information Group, 78 Shenton Way, #10-02, Singapore 079120, Singapore
Tel: (+65) 63 25 08 66 Fax: (+65) 62 26 11 85
e-mail: asiapacific@janes.com

Oceania
Jane's Information Group, PO Box 3502, Rozelle Delivery Centre, New South Wales 2039 Australia
Tel: (+61 2) 85 87 79 00 Fax: (+61 2) 85 87 79 01
e-mail: oceania@janes.com

Middle East
Jane's Information Group, PO Box 502138, Dubai, United Arab Emirates
Tel: (+971 4) 390 23 36 Fax: (+971 4) 390 88 48
e-mail: mideast@janes.com

Japan
Jane's Information Group, Palaceside Building, 5F, 1-1-1, Hitotsubashi, Chiyoda-ku, Tokyo 100-0003, Japan
Tel: (+81 3) 52 18 76 82 Fax: (+81 3) 52 22 12 80
e-mail: japan@janes.com

ADVERTISEMENT SALES OFFICES

(Head Office)
Jane's Information Group
Sentinel House, 163 Brighton Road,
Coulsdon, Surrey CR5 2YH, UK
Tel: (+44 20) 87 00 37 00 Fax: (+44 20) 87 00 38 59/37 44
e-mail: defadsales@janes.com

Janine Boxall, Global Advertising Sales Director,
Tel: (+44 20) 87 00 38 52 Fax: (+44 20) 87 00 38 59/37 44
e-mail: janine.boxall@janes.com

Richard West, Senior Key Accounts Manager
Tel: (+44 1892) 72 55 80 Fax: (+44 1892) 72 55 81
e-mail: richard.west@janes.com

Nicky Eakins, Advertising Sales Manager
Tel: (+44 20) 87 00 38 53 Fax: (+44 20) 87 00 38 59/37 44
e-mail: nicky.eakins@janes.com

Carly Litchfield, Advertising Sales Executive
Tel: (+44 20) 87 00 39 63 Fax: (+44 20) 87 00 37 44
e-mail: carly.litchfield@janes.com

(US/Canada office)
Jane's Information Group
110 N Royal Street, Suite 200,
Alexandria, Virginia 22314, US
Tel: (+1 703) 683 37 00 Fax: (+1 703) 836 55 37
e-mail: defadsales@janes.com

US and Canada
Robert Silverstein, US Advertising Sales Director
Tel: (+1 703) 236 24 10 Fax: (+1 703) 836 55 37
e-mail: robert.silverstein@janes.com

Sean Fitzgerald, Southeast Region Advertising Sales Manager
Tel: (+1 703) 836 24 46 Fax: (+1 703) 836 55 37
e-mail: sean.fitzgerald@janes.com

Linda Hewish, Northeast Region Advertising Sales Manager
Tel: (+1 703) 836 24 13 Fax: (+1 703) 836 55 37
e-mail: linda.hewish@janes.com

Janet Murphy, Central Region Advertising Sales Manager
Tel: (+1 703) 836 31 39 Fax: (+1 703) 836 55 37
e-mail: janet.murphy@janes.com

Richard L Ayer
127 Avenida del Mar, Suite 2A, San Clemente, California 92672, US
Tel: (+1 949) 366 84 55 Fax: (+1 949) 366 92 89
e-mail: ayercomm@earthlink.com

Rest of the World

Australia: *Richard West* (UK Head Office)

Benelux: *Nicky Eakins* (UK Head Office)

Eastern Europe (excl. Poland): MCW Media & Consulting Wehrstedt
Dr Uwe H Wehrstedt
Hagenbreite 9, D-06463 Ermsleben, Germany
Tel: (+49 03) 47 43/620 90 Fax: (+49 03) 47 43/620 91
e-mail: info@Wehrstedt.org

France: Patrice Février
BP 418, 35 avenue MacMahon,
F-75824 Paris Cedex 17, France
Tel: (+33 1) 45 72 33 11 Fax: (+33 1) 45 72 17 95
e-mail: patrice.fevrier@wanadoo.fr

Germany and Austria: *MCW Media & Consulting Wehrstedt*
(see Eastern Europe)

Greece: *Nicky Eakins* (UK Head Office)

Hong Kong: *Carly Litchfield* (UK Head Office)

India: *Carly Litchfield* (UK Head Office)

Iran: Ali Jahangard
Tel: (+98 21) 88 73 59 23
e-mail: ali.jahangard@eidehinfo.com

Israel: *Oreet International Media*
15 Kinneret Street, IL-51201 Bene Berak, Israel
Tel: (+972 3) 570 65 27 Fax: (+972 3) 570 65 27
e-mail: admin@oreet-marcom.com
Defence: Liat Heiblum
e-mail: liat_h@oreet-marcom.com

Italy and Switzerland: *Ediconsult Internazionale Srl*
Piazza Fontane Marose 3, I-16123 Genoa, Italy
Tel: (+39 010) 58 36 84 Fax: (+39 010) 56 65 78
e-mail: genova@ediconsult.com

Japan: *Carly Litchfield* (UK Head Office)

Middle East: *Carly Litchfield* (UK Head Office)

Pakistan: *Carly Litchfield* (UK Head Office)

Poland: *Carly Litchfield* (UK Head Office)

Russia: Anatoly Tomashevich
1/3, appt 108, Zhivopisnaya Str, Moscow, 123103, Russia
Tel/Fax: (+7 495) 942 04 65
e-mail: to-anatoly@tochka.ru

Scandinavia: *Falsten Partnership*
23, Walsingham Road, Hove, East Sussex BN41 2XA, UK
Tel: (+44 1273) 77 10 20 Fax: (+ 44 1273) 77 00 70
e-mail: sales@falsten.com

Singapore: *Richard West* (UK Head Office)

South Africa: *Richard West* (UK Head Office)

South Korea: Kim TaeHae
HARNTER
#213 Ventures Incubating Center, Korea Aviation University
200-1 HwaJeon, GoYang City, GyoungGi-Do, 412 791 Korea
Tel: (+82 (0)) 167 14 11 13
e-mail: taehae_kim@hotmail.com

Spain: Macarena Fernandez
VIA Exclusivas S.L., Virato, 69 - Sotano C, E-28010, Madrid, Spain
(+34 91) 448 76 22 Fax: (+34 91) 446 02 14
e-mail: macarena@viaexclusivas.com

Turkey: *Richard West* (UK Head Office)

ADVERTISING COPY
Kate Gibbs (UK Head Office)
Tel: (+44 20) 87 00 37 42 Fax: (+44 20) 87 00 38 59/37 44
e-mail: kate.gibbs@janes.com

For North America, South America and Caribbean only:
Tel: (+1 703) 683 37 00 Fax: (+1 703) 836 55 37
e-mail: us.ads@janes.com

Alphabetical list of advertisers

EL-OP
Advanced Technology Park, Kiryat Weizmann, PO Box 1165, Rehovot 76111, Israel ....[3]

Executive Overview: Avionics

Civil and military aerospace

During 2006, the ongoing deployment of Automatic Dependant Surveillance - Broadcast (ADS-B) capability increased pace markedly, with growing impetus to see a truly global system within a time frame measured in years rather than decades.

Automatic Dependent Surveillance (ADS) is the process of creating and sending a message, which incorporates the aircraft flight identification, current position and other surveillance information, including routing, speed and temperature/wind vector. This information is used to monitor aircraft separation during long-range oceanic/polar route flying, increase Situational Awareness (SA) for ground controllers and aid their decision making in the management of airspace such as the North Atlantic Track System (NATS). ADS data can be used in co-operation with data from other sources, such as Air Traffic Control Radar Beacon System (ATCRBS), Mode S, Traffic Collision Avoidance System (TCAS) and primary Air Traffic Control (ATC) radar, although the intention is that ADS should become the sole means of ground-based surveillance. The application of this technology most recognised by pilots is ADS-Addressed (ADS-A), which provides a surveillance data report that is sent to a specific addressee over the course of a route through the applicable airspace segment. ADS-A reports are employed in the Future Air Navigation System (FANS) using the Aircraft Communication Addressing and Reporting System (ACARS) as the communication protocol. During oceanic flight operations, including Minimum Navigation Performance Specification/Reduced Vertical Separation Minimum (MNPS/RVSM) airspace over the North Atlantic, a report is downloaded to the relevant air traffic sector (Shanwick or Gander in the case of NAT MNPS/RVSM) at each way point along the route (each 10° of longitude for NAT MNPS/RVSM). As aircraft transit air traffic sectors, the system automatically changes addressee. If the aircraft is also Controller-Pilot Data Link Communication (CPDLC) equipped, then messages can be exchanged, allowing for requests (climb/descend), lateral separation (time at way point), weather reports, etc.

The main difference between ADS-A and ADS-B is that ADS-B aircraft and other vehicles continuously broadcast a message including position, heading, velocity and routing to all receiving ground-based and airborne stations. Other uses may include the position of significant obstacles to flight and/or emergency contingency information. Instead of simply a two-way exchange of information, ADS-B is a broadcast system which facilitates accurate air traffic management over airspace sectors at medium/high level, or terminal areas around major airports for all suitably equipped platforms.

During June 2006, the Federal Aviation Administration (FAA) Joint Resources Council (JRC) approved ADS-B for next-generation surveillance in the National Airspace System (NAS). The JRC approved an initial two-year baseline programme, after which, assuming satisfactory progress, a full programme will proceed. It seems confidence is high, since the FAA has also been reported as planning on ADS-B as a cornerstone of its Next Generation Air Transportation System (NGATS). Accordingly, the programme will "seek to develop a multi-segment, life cycle managed, performance based ADS-B strategy that aligns with NGATS and generates value for the NAS".

Programme activity will develop two services, namely:
- Surveillance Broadcast Services (en route, terminal and surface)
- Traffic / Flight Information Services - Broadcast (TIS-B/FIS-B)

and five applications:
- Enhanced visual acquisition
- Enhanced visual approaches
- Final approach and runway occupancy awareness
- Airport surface Situational Awareness (SA)
- Conflict detection

TIS-B is a service which provides ADS-B equipped aircraft with position reports from secondary surveillance radar on non-ADS-B equipped aircraft. FIS-B transmits graphical National Weather Service (NWS) products, Temporary Flight Restrictions (TFRs), and special use airspace.

Since this initial activity in mid-year, the US programme has gained impetus - dates and schedules have been defined, with Screening Information Request (SIR) to suppliers during November. The SIR requires prospective bidders to describe in detail, including cost estimates, what they intend to offer - in this case, a nationwide network of some 400 ADS-B Ground-Based Transceivers (GBTs). This in turn will allow FAA officials to closely examine each response to ensure it meets all the programme technical requirements and to request further information when necessary. Once all of these questions are answered,

the FAA plans to issue a Request For Offers (RFO) to qualified bidders in March 2007. The FAA has requested USD80 million in its Financial Year 2007 (FY07) budget, in part to work on backup analysis, separation standards, hardware development and activities leading to a Notice of Proposed Rule Making (NPRM) in FY07/08, as nationwide coverage becomes imminent.

So, what does this mean to the aviation community? Rather like the introduction of the Traffic Collision Avoidance System (TCAS) Change 7 amendment, allowing for RVSM operations (1,000 ft vertical separation) with TCAS II, there will most likely be a graded introduction, with various classes of airspace reserved for aircraft suitably equipped for full ADS-B operations. This will force major operators to adopt the system while protecting the general aviation community in the US from the heavy financial burden of mandated acquisition of the system.

For those who might consider a 'stretched' timescale for the NPRM (historically this has been the case), it should be noted that, where TCAS Change 7 was primarily brought about by safety concerns over traffic density across NAT airspace, ADS-B brings with it, as well as enhanced SA and flight safety, an eventual USD1 billion saving through the decommissioning of secondary radar throughout the US - this goal alone should convince all as to the seriousness of the FAA in making this programme work.

Pilot's ADS-B display. Traffic within 30 n miles is shown; note the enhanced information (identification and vector) TIS-B provides over a standard TCAS presentation
1130095

While the US might be leading the ADS-B charge, the rest of the world is not far behind - during May 2006, Airservices Australia completed integration of ADS-B services into the Australian Advanced Air Traffic System (AAATS), deploying ADS-B GBTs across Australia to enhance surveillance throughout non-radar airspace and complement the existing en-route and terminal radar coverage. In addition to the initial Bundaberg ground station, four new sites across Australia are being integrated into the operational system (Longreach, Bourke, Woomera and Esperance). The 23 remaining stations will progressively be integrated during 2007. In total, 56 GBTs will be located at the 28 sites throughout the continent. This implementation is the first wide-scale implementation of ADS-B to be undertaken by an air navigation service provider. The system deployed in Australia is Thales' AS 680 ADS-B Mode S Ground Station, based on its 1,090 MHz Mode S Extended Squitter (1090 ES) datalink.

TAAATS airspace covers 56 million square kilometres and the system controls more than three million air traffic movements per year. While almost full radar coverage exists along the east coast of Australia and the majority of commercial air traffic in Australia is currently under radar coverage, over 90 per cent of Australian airspace is outside radar coverage - the deployment of ADS-B will provide coverage in non-radar airspace and facilitate the provision of secondary radar-like services.

The majority of civilian air transport will undoubtedly be obliged to adopt the system and acquire the necessary hardware, although some companies, notably UPS, are well ahead of schedule in this respect. UPS, in conjunction with Boeing and ACSS, launched an ADS-B integration programme during the 2005 Paris Air Show to improve the

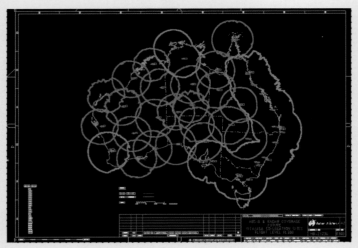

Australian ADS-B coverage. Red shows current secondary radar coverage, green is new ADS-B coverage   1130096

efficiency of the company's operations at its Louisville hub. ACSS software aids UPS pilots to merge at cruising altitude before flying Continuous Descent Approaches (CDA) at fixed intervals, markedly reducing controller workload in busy terminal areas. This 'merging and spacing' application improves the efficiency of flight operations, saving the carrier an estimated USD1 million in fuel during the course of a year, a fact which has not been lost on commercial carriers worldwide in the face of increased aviation fuel costs. Also, the reduction of low altitude holding will reduce environmental impact around airports, particularly in respect of noise around urban areas.

Civil aircraft market

As is usual for *Jane's Avionics*, a brief examination of the fortunes of the major aircraft manufacturers will give some insight regarding the health of their major subcontractors - the avionics manufacturers. Continuing last year's resurgence, Boeing has built on the initial warm reception for its new B787 Dreamliner, with the order books full for some years to come. This has had a knock-on effect for its B777, which has thoroughly trounced the rival Airbus A340 line in terms of orders. Indeed, with the introduction during 2006 of the much-redesigned A350 XWB, it seems that, aside from an Increased Gross Weight (IGW) variant of the A340-600 (380 tonnes MTOW), there will be no further development of the line. Reinforcing the opinion that attention at Airbus has shifted away from the A340, the A350 will be eventually fielded in

245, 265 and 295 tonne variants, exactly overlapping the current A330/340 range. So, with Airbus' gaze firmly on the large and ultra large class, the avionic configuration of the European civil air transport cockpit is set - sidesticks, portrait Active Matrix Liquid Crystal Displays (AMLCDs) and Electronic Flight Bags (EFBs), with all Airbus Fly By Wire (FBW) models providing a Head Up Display (for both pilots) as standard, a feature echoed by the B787.

This final iteration of the A350 XWB cockpit is a great deal more sophisticated than that offered to customers for the original (non-XWB) A350, which closely followed the then near 20-year old configuration found in the A330. However after the move to XWB configuration, plus a redesign of the entire nose section (reportedly incorporating the crew rest module under the floor of the cockpit - hopefully including a toilet!), space was found to accommodate the extra depth required to install the portrait displays of the new system. This decision to introduce a more advanced cockpit was likely spurred on by the competing Boeing B787's standard large format display/HUD layout. Of note is the employment in the A350 XWB of QWERTY keyboards for data input via Keyboard Cursor Control Units (KCCUs); Airbus have made this departure from the traditional ABCD keyboard found in current Multifunction Control and Display Units (MCDUs) in response to the proliferation of personal computers and hence a greater familiarity with the QWERTY keyboard format. Boeing has notably resisted following this lead with the B787 - it retains a more 'traditional' MCDU format with ABCD keyboard, citing "operator preference" as its main driver in retaining the old layout. Presumably, this is the same, a similar reason for their retention of a centrally mounted yoke for the fully Fly By Wire B777; this current operator certainly prefers the greatly increased useable space in the cockpit of a sidestick-equipped Airbus, although having become used to the ABCD layout of current MCDUs, judgement will be reserved on the QWERTY layout until a reasonable amount of exposure to the new system permits a balanced view.

As a footnote, Boeing has not exercised dominance in overall sales for 2006. Despite the much-publicised maladies afflicting Airbus, the traditional 'late charge' by the European manufacturer has seen a significant number of orders for its A320-series narrow bodies, which continues to hold sway in the class. The 5,000th order for A320 family aircraft was reached with the sale of 30 A319 aircraft to US-based customer Spirit Airlines. In the last two years, the A320 family has received orders for around 1,600 aircraft, a significant achievement. Until Boeing can introduce replacement(s) for its ageing middle/lightweight aircraft range, this situation is likely to remain,

The flightdeck of Airbus Industrie's new A350 XWB is the same variation of Thales' Integrated Modular avionics System (IMS) as found in the A380 ultra large airliner (Airbus Industrie)   1128630

although, as has happened to Airbus with the 'leap frogging' of the A330 by the B787, Boeing should be wary of an A320-series replacement waiting in ambush.

Also of note, during 2006 Eurocopter reinforced its claim to be the world's number one helicopter manufacturer with a total of 381 new helicopter deliveries for new military and civil aircraft. As part of the EADS group (which also owns the beleaguered Airbus), Eurocopter accounted for a consolidated turnover of EUR3.8 billion. Compared to 2005, Eurocopter turnover increased by 18 per cent, with orders for 615 new helicopters totalling EUR4.89 billion (increased from EUR3.52 billion in 2005). The company's order backlog by the end of December 2006 was on a historic high, reaching EUR11 billion.

So, as the civil air transport community catches up with the business jet/small commuter aircraft category in recognising the value of HUDs in the cockpit, what of the military's move away from a fixed HUD to a 'virtual HUD' within a Helmet Mounted Display (HMD)? As previously reported in *Jane's Avionics*, while aircraft such as the Eurofighter Typhoon may be regarded as a 'halfway house', retaining both a fixed HUD and integrating a virtual HUD in its associated HMD, the first aircraft to fully embrace the concept of a virtual HUD is the Lockheed Martin F-35 Lightning II. Recent initial flights of the first production-representative aircraft have shown the upper coaming to be devoid of any Up Front Control Panel (UFCP)/HUD combination, although the pilot did not seem to be wearing any form of HMD during these flights. With many fighter aircraft equipped with some form of helmet mounted display, surely the step to a fully-functional HMD to replace the fixed HUD is just a technical formality?

While there has been an increasing number of Helmet Mounted Sight (HMS) systems, such as the Vision Systems International (VSI) Joint Helmet Mounted Cueing System (JHMCS), successfully deployed in a variety of fighter aircraft, including the Lockheed Martin F-16, Boeing F-15 and Boeing F/A-18, these are comparatively simple systems when compared to a full HMD. JHMCS is a day-only, weapon aiming module attached to a modified HGU-55/P shell, which must be interchanged with the complimentary QuadEye™ Night Vision Cueing and Display (NVCD) system for night operations. However, despite these limitations, pilots have reaped the benefits of head-slaved weapon aiming and, perhaps more importantly, vastly increased Situational Awareness (SA) during air combat through the presentation easily interpreted 'spatial' (three dimensional) information, rather than having to 'visualise' information presented on a two-dimensional Multi-Function Display (MFD). Having proven the utility of such systems, it is natural that squadrons are keen to see the introduction of the even more capable HMD, offering full day/night capability with Infra-Red (IR) and Infra-red Imaging (II) sensor overlay and virtual HUD, all in the same helmet.

However, while a single-layer display of off-boresight weapon aiming is a relatively simple task for an HMS, only requiring a pointing accuracy sufficient to cue an air-to-air missile seeker or Laser Designation Pod (LDP) sightline close to the Desired Mean Point of Impact (DMPI), a full HMD must be far more accurate to successfully overlay sensor imagery and symbology. While 1-2 mrad accuracy has been achieved in azimuth and elevation under laboratory conditions, roll accuracy (important in an agile fighter) falls short of this figure. In the cockpit, 3-4 mrad is achievable - JHMCS accuracy is approximately 6 mrad. The F-35 HMD pointing accuracy is reported as approximately 2 mrad, a figure which is assumed to be under laboratory conditions.

A day/night HMD must also satisfy the opposing requirements for symbology. Since display designers (for fixed HUD or HMD) clearly cannot manipulate background luminance by day or night, hence adequate contrast depends on sufficient symbology luminance for the displayed information. In a fighter cockpit, background luminance at high altitude is very high, resulting in a minimum specification for peak image luminance that is as high as possible (typically up to 17,000 cd/m^2 against a background of >10,000 cd/m^2). So far, so good. However, for the same display to be useful at night, it must be capable of dimming to an extremely low level, creating an enormous range of display luminance many orders of magnitude greater than for a day-only HMS. This range and fine adjustment (at lower levels) of display luminance is critical to operational acceptance of a system. In addition, an automatic adjustment of display luminance is essential to avoid increasing pilot workload. Having covered symbology brightness, the HMD must also (accurately) overlay sensor information, either helmet mounted II or aircraft mounted IR, which also must be set at a suitable luminance with respect to both the outside scene and the displayed symbology. The balance between these elements is absolutely pivotal to the effectiveness of the HMD - if the sensor image luminance is too great, it will swamp both the outside scene (which may not be a disadvantage at night if there is no useful information) and elements of the displayed symbology. If the symbology is too bright, detail in the scene may be missed and (at night) the pilot may not achieve an optimum level of dark adaption.

This image of the second flight of Lockheed Martin's F-35 Lightning II AA-1 shows the smooth upper coaming, following the line of the top of the 20 × 8 in Panoramic Cockpit Display (PCD), devoid of any UFCP or HUD. This configuration is indicative that the F-35 will be introduced into service with a 'virtual HUD' displayed within the associated VSI HMD (Lockheed Martin)

1128685

EXECUTIVE OVERVIEW: AVIONICS

Having already mentioned the display of helmet mounted II and aircraft mounted IR onto the pilot's HMD visor, it is worth considering the different mechanisms employed in each case. A recent iteration of the F-35 HMD has seen the central, brow-mounted solid-state low-light TV camera replaced by twin cameras mounted in-line with the eyes on the sides of the helmet, as in the Eurofighter Typhoon Integrated Helmet. Since this display is directly attached to pilot's head, the display, and hence the displayed image, moves with his/her head, requiring no compensation for motion or image displacement. However, displaying an image derived from a fixed Forward Looking Infra-Red (FLIR) or Distributed Aperture System (DAS), as in the F-35, requires a great deal of image processing to maintain an accurate view of the outside scene during rapid head movements. Noticeable display lag, possibly coupled with a smearing of the display as it catches up with the pilot's head, can lead to disorientation as the image display competes with the outside scene or II image.

Lastly, if an HMD is to introduce a Primary Flight Display (PFD) - qualified virtual HUD, then it must be certified (in terms of hardware and software) at the highest level of integrity as a safety-critical system. In this respect, all aspects of the design, from power supply to symbology set(s) must pass rigorous safety requirements. At present much research is still ongoing with respect to the symbology displayed to the pilot (in terms of aircraft orientation and vector) for various head positions - misinterpretation of a pitch/roll 'ladder' when the pilot is looking over his shoulder could be catastrophic! For aircraft which retain a fixed HUD, while these problems are certainly equally pressing, their solution is not, since qualification of the HMD as a PFD need not be undertaken. It could also be argued that the F-35 HMD need not qualify as a PFD, since the pilot could simply utilise his head-down displays - a valid point, but this would surely seriously limit operations in Instrument Meteorological Conditions (IMC)?

The foregoing is clearly only 'the tip of the iceberg' with respect to the technological hurdles facing the operational deployment of a fully-featured HMD, but it is included to illustrate the significant advance a full HMD represents over an HMS. It will therefore come as no surprise that the HMD is arguably the most important piece of the F-35's avionic system and must therefore fulfil its remit if the aircraft is to realise its full potential. As many of the aforementioned issues, at least to the knowledge of the editor, are some way from being resolved, it will be interesting to see if initial deliveries of the aircraft, scheduled for 2012, have full capability in this respect.

The HMD is not the only avionic innovation in the F-35; head-down, the pilot will not be confronted by the now usual trio of 8 × 6 in portrait AMLCDs, complete with an absolute plethora of bezel-mounted softkeys, rather a single full-colour, landscape 20 × 8 in Panoramic Cockpit Display (PCD) with (zoned) touchscreen capability. The contract for the PCD was awarded, during November 2005, to L-3 Communications Display Systems for the display and Display Management Computer (DMC). The display will feature a high-brightness panel for daylight readability, combined with full NVIS compatibility for night operations with Night Vision Equipment (NVE). The display computer is described as highly redundant, as is the power supply; this last point is important when the majority of the tactical capability of the cockpit is vested in a single display. The PCD will significantly increase the total effective screen area available to the pilot, divided into four main zones, providing any combination of tactical, sensor, systems and weapons information as selected by the pilot, who will have the ability to 'customise' zones according to personal preference.

Elsewhere in the military arena, 2006 saw Russian Air Arms beginning the process of renewal and upgrade of their aircraft fleets. At the beginning of 2007, the Russian Air Force's fleet of seven Mi-28Ns had completed 478 sorties as part of trials and needed to undergo a further 185 successful test flights to complete certification. The Air Force received its most recent Mi-28Ns in late December and early January. The Russian Air Force's fleet of large aircraft is set to be boosted with a Russian Ministry of Defence (MoD) order for four Tupolev Tu-214s in command-post/VIP configuration to replace Ilyushin Il-62M and Tu-154M aircraft. Additionally, the air force expects to take delivery of two new Tupolev Tu-160 strategic bombers in the next three years, the first being delivered in February 2007. There is also a programme under way to upgrade the strike, navigation and

The cockpit of an Indian Navy MiG-29KUB. The MiG-29/35 has begun to stimulate interest amongst international customers, latterly with all-Russian avionic display systems rather than European replacements for legacy items. This cockpit features displays from the RPKB avionics design bureau of Ramenskoye, with a IKSh-1M HUD and three MFI-10 multi function displays (Piotr Butowski) 1128686

aiming systems of the 15 in-service Tu-160s. The Ministry of Defence has also cleared the MiG-31BM upgrade programme to be carried out on in-service MiG-31 aircraft after pre-production aircraft completed six months of flight trials. It is anticipated that the MiG-31BM will receive full military certification in the second half of 2007. The MiG-31BM upgrade is designed to provide the aircraft with a multirole combat capability involving improved air-target interception and ground-attack facility. The aircraft incorporates software and antenna array improvements to its N007 Zaslon electronically-scanned radar through the installation of new modules and new software.

Despite the recent lack of orders from the home market, Sukhoi and MiG have continued to do well in the international arena, particularly with customers who would find negotiations difficult with the US. Variants of the Su-30 have been ordered by China, India, Indonesia and most recently, Venezuela, with an order, confirmed during July 2006, for 24 aircraft. The acquisition of the Su-30 by Venezuela, which would represent the most powerful fighter in South America, was undoubtedly a contributory factor in the resurrection of the Brazilian F-X fighter programme in the early part of 2007. The Su-30s are intended to replace Lockheed Martin F-16s in service with the Venezuelan Air Force, which are in low operational usage after the US suspended the sales of spares and upgrades to Venezuela. In early February India committed itself to the purchase of 40 more SU-30 MKI fighter aircraft on top of 190 already delivered or being manufactured.

Elsewhere, having reached agreement with MiG in January 2004 over a USD700 million contract for 16 MiG-29K carrier-based fighters to equip the aircraft carrier *INS Vikramaditya*, formerly Admiral Gorshkov, the Indian Navy is preparing to receive 12 single-seat and four two-seat MiG-29KUB fighters by 2008, with options on a further 30 aircraft before 2015.

Indicating a definite 'sales offensive', RSK MiG has been reported as offering a free MiG-29 upgrade with each sale of a MiG-35. The serious point here is that the market for avionic upgrade of legacy platforms remains extremely strong; with a large number of rugged Russian aircraft designs still in worldwide service, the recent trend of retrofitting of European or Israeli avionics as part of upgrade packages looks set to be reversed as indigenous Russian avionic systems begin to match their Western counterparts, at least in terms of the level of capability demanded by typical upgrades.

Indian Air Force Su-30Ks during a deployment to France. In May 2006, India was negotiating for return of all 18 Su-30Ks to the Irkut production plant for replacement with new Su-30MKIs; Belarus has shown interest in the ex-Indian aircraft 1158626

Podcast

UK Royal Air Force (RAF) Panavia Tornado GR.4 reconnaissance/attack aircraft are currently deployed in Iraq conducting what is termed Non-Traditional Intelligence, Surveillance and Reconnaissance (NTISR) missions in support of coalition troops. US and UK aircraft operating over Iraq are using their electro-optical targeting pods to improve SA for ground troops and detect insurgent threats. Currently, RAF Tornados are equipped with Selex Sensors & Airborne Systems Thermal Imaging and Laser Designating 500 (TIALD 500) pods, with information from currently passed to ground forces using voice communications; in early February TIALD is due to be replaced by Rafael Armament Development Authority Litening III advanced targeting pods, which will also feature a datalink to allow the download of imagery from the pod directly to ground controllers.

In a subsequent request, the UK RAF launched a competition to purchase new third-generation targeting pods for BAE Systems Harrier GR.9 ground-attack aircraft supporting NATO forces in Afghanistan. Late in 2006 the UK MoD issued an invitation to several international companies asking them to bid for a contract which has evolved from an Urgent Operational Requirement (UOR) raised by Royal Air Force (RAF) commanders in theatre. The new pods are needed to improve the ability of RAF and Royal Navy Harrier pilots to find and identify small groups of Taliban insurgents during close-quarter fighting in difficult terrain, according to RAF sources. Similarly to the Tornado GR.4 experience in Iraq, the Thermal Imaging Airborne Laser Designator

A Tornado GR.4 from 41(R) Squadron carrying a Litening III pod on its port underfuselage pylon (Martin Keen) 1184850

(TIALD) pod does not offer the required level of resolution. Also part of the requirement is the Remote Operated Video Enhanced Receiver (ROVER) viewing capability, which is currently available to US Joint Terminal Attack Controllers (JTACs) or forward air controllers to view real-time imagery from RQ/MQ-1 Predator unmanned aerial vehicles or Fairchild-Republic A-10A fighter/attack aircraft equipped with the Litening AT pod. MoD sources said it is expected that the purchase of the pods will be announced early in 2007. It was reported that in early 2006, RAF commanders in Afghanistan requested that a version of the Litening pod, with ROVER capability, be purchased, but the MoD subsequently ruled that a competition should be conducted. An invitation to tender has since been issued to Selex, Rafael and its UK partner Ultra Electronics, Thales, Lockheed Martin, Raytheon and Northrop Grumman.

While it is laudable that the RAF should seek the best equipment for its aircraft and crews, the total procurement for the Typhoon, Tornado GR.4 and Harrier GR.9 will likely be in the order of 24-plus pods, so perhaps it would be prudent to ensure maximum interchangeability between platforms to cater for unserviceability or operational deployment (clearly, if Tornados are deployed and Typhoons are not, then the ability to borrow extra pods is sensible). However, the overruling of in-theatre Harrier commanders to launch a competition may result in an entirely different pod to the Litening III being selected. BAE Systems has already flown a Sniper pod on a Harrier GR.9 as part of its bid with partner Lockheed Martin; Sniper is undoubtedly an excellent design and more recent than Litening III, benefiting from the most recent advances in sensor technology, but it is entirely different to Litening III and would provide for little interchangeability. It is assumed that Northrop Grumman's contender will be its more comparable (to Sniper) Litening AT+, the latest iteration of the line. While there will undoubtedly be small differences in performance between the contenders, for such a small total procurement, surely it makes sense that a Litening AT buy for the Harrier force will maximise the overall capability for all three types?

Lockheed Martin's contender for UK RAF Harrier GR.9s is the Sniper Advanced Targeting Pod (ATP) (Jamie Hunter) 0589835

Corporate news

So, in the civil arena, with order books for new-build aircraft full, coupled with military air arms urgently re-equipping with new sensors and communications equipment more suited to asymmetric operations in Iraq and Afghanistan, we should expect the major avionic manufacturers to have made good progress in 2006 and with excellent prospects for 2007 - and so it has transpired.

Honeywell, aided by its avionic business sector, increased its turnover for 2006 by 13 per cent from USD27.6 billion in 2005 to USD31.4 billion, driven partly by 14 per cent sales growth in the fourth quarter (Q4). Full-year earnings per share (EPS) grew by 31.2 per cent to USD2.52 against USD1.92 in 2005, producing net profits of USD2.08 billion. Sales of Honeywell Aerospace's civil and military equipment and maintenance services increased by six per cent to USD11.12 million, generating a 13 per cent year-on-year increase in segment pre-tax profit to USD1.89 billion. Defence and space sales grew by only three per cent in 2006, however, with stronger growth recorded in the commercial and business aviation sectors. Significant defence contracts in 2006 included a USD36 million, three-year contract with Lockheed Martin to supply F-16 cockpit displays to the Turkish and Greek air forces. Honeywell expects to build on these successes in 2007, with sales forecast to be up five per cent to approximately USD33 billion.

USD	2006	2005	± %
Sales	31.4 bn	27.6 bn	+13
Gross profit	2.79 bn	2.29 bn	+21.8
Net profit	2.08 bn	1.64 bn	+26.8
EPS	2.52	1.92	+31.2

Northrop Grumman, also with a large avionics sector, reported its fourth consecutive year of double-digit growth in EPS, together with double-digit growth in operating margin for FY06. On 25 January 2007, the defence and technology group reported that income from continuing operations for FY06 rose 13 per cent to USD1.6 billion, or USD4.44 per diluted share, from USD1.4 billion, or USD3.83 per diluted share, for FY05. Sales for FY06 stood at USD10.9 billion, "comparable" to those for FY05 - USD10.3 billion. Total segment operating margin for 2006 increased by 16 per cent to USD2.8 billion from USD2.4 billion in 2005, while total operating margin for 2006 increased 12 per cent to USD2.5 billion from USD2.2 billion in 2005. According to the group, the increase was driven by higher operating margin in all four of the company's businesses. The company expects 2007 sales to range between USD31 billion and USD32 billion. Earnings per diluted share from continuing operations are expected to range between USD4.80 and USD5.05, and include estimated net pension income of approximately USD70 million. Net cash provided by operating activities is expected to range between USD2.5 billion and USD2.8 billion in 2007.

However, not everything has been plain sailing for Northrop Grumman, which has a two-year timetable to resolve technical glitches plaguing three international electronics contracts. The company has absorbed earnings charges totalling USD151 million since January 2006 on three programmes, which include a USD61 million write-down in the fourth quarter for the Multirole Electronically Scanned Array (MESA). Last year, Northrop Grumman also reported a USD68 million charge on the mission suite for the Lockheed Martin F-16 Block 60 delivered to the United Arab Emirates. The company was also charged USD25 million to fix errors on the Raytheon Airborne Self-Protection Integrated Suite-2 (ASPIS-2) installed on Greek Air Force F-16s.

Staying over the Atlantic, Rockwell Collins has reported a quote "great start" to its FY07 operations, with Commercial Systems sales leading the way. On 25 January, the US avionics company reported a net income of USD143 million for the first quarter of 2007 (Q107,) ended on 31 December 2006 - an increase of USD39 million, or 38 per cent over its Q106 net income of USD104 million. EPS improved USD0.25, or 42 per cent, to USD0.84 compared to an EPS of USD0.59 last year. EPS growth exceeded the growth rate in net income owing to the favourable effect of the company's share repurchase programme. Q107 sales increased USD112 million, or 13 per cent, to USD993 million compared to sales of USD881 million a year ago, while organic revenue growth stood at USD88 million - a 10 per cent increase. The stronger than planned first quarter financial performance from the commercial systems business, coupled with expectations for continued overall favourable market conditions for the remainder of the fiscal year, have allowed the company the luxury of raising expectations for full year Commercial Systems' sales and operating margins, as well as total company sales and earnings per share. FY07 outlook projects total company sales in the range of USD4.25 billion to USD4.3 billion (previously USD4.2 billion to USD4.3 billion), with EPS in the range of USD3.25 to USD3.35.

Coming back across the Atlantic, BAE Systems has enjoyed an equally encouraging year. As the UK's number one defence manufacturer, the company has performed well by delivering improved corporate performance on extant programmes and addressing the delays which have traditionally plagued them, winning new business, notably the sale of Eurofighter Typhoon to the Kingdom of Saudi Arabia, revising company pension arrangements and, controversially, selling its share in Airbus to EADS. In considering the ramifications of the Airbus sale, it should be noted that considering BAE Systems as a UK company is a rather outmoded view; the company is the seventh largest US supplier, employs some 35,400 employees in the US (compared to 32,400 in the UK), with defence procurement in the US at USD159.1 billion for 2005/2006 (USD13.0 billion in the UK) and 30 per cent of total sales (UK 37 per cent). In terms of avionic-related sector performance, excluding the effect of the sale of Airbus, the Electronics, Intelligence and Support (EI&S) sector accounted for 32 per cent of total sales (up 8 per cent) out of a total of GBP6.4 billion.

For those who viewed the sale of the stake in Airbus as a mistake, it should be noted that BAE Systems, as a truly global player, sees a far more lucrative future in continuing its predation of value-enhancing companies, which will elevate its position in the US market by even greater involvement in the highly lucrative programmes traditionally reserved for US-only companies, while maintaining a streamlined and efficient presence in the domestic and European markets.

Overview

Much as reported last year, strong performances by all of the major players in the global avionics market reflects strong demand in the civil and military sectors for new and upgraded avionics, which directly and cost-effectively enhance economy and operational effectiveness in both camps. 2007 seems set for more of the same.

Edward Downs
February 2007

Acknowledgements

I am most grateful to those manufacturers, contractors, armed services and research establishments who have provided invaluable support in the supply and verification of data presented in *Jane's Avionics*; supplied comments, changes, new information and images are critical to the continued accuracy of this edition. It will come as no surprise that a great deal of work goes on in the background during the hardcopy cycle and my thanks, as usual, go to the tireless staff at Jane's Information Group who guide and assist me through the publishing process; my thanks go to Jacqui Beard, who has dealt with all the information-gathering activities for this edition, and Colin Maslin and those in IT Support, who have maintained my over-worked computer throughout the year!

Seemingly a regular feature of my acknowledgements in recent Yearbooks, I must extend my warmest thanks to my outgoing Content Editor, Laura Kew. After many years at Jane's, she has decided to leave for pastures new - I wish her well in her future endeavours. Replacing Laura as my Content Editor is Angela Velasco, who, in concert with Rebecca Davies, Senior Content Editor for the Air Desk, has been taking an active part in the update cycle this year, freeing valuable time to revise and incorporate new information into the universally well-received Analysis section of Jane's Avionics.

Finally, my thanks go to Sara Morgan, my Publisher, who has provided valuable guidance and support as I further develop *Jane's Avionics* to better serve both the researcher and casual reader.

Edward Downs

The Editor of *Jane's Avionics*, Edward Downs, is a former Royal Air Force pilot, retiring from the Service in 1997 after tours of duty as a frontline Tornado GR1 pilot, (31 Squadron, RAF Bruggen), a Qualified Weapons Instructor (Tornado Weapons Conversion Unit, RAF Honington) and finally as an operational test pilot at the Defence Evaluation and Research Agency (DERA), Boscombe Down. While at Boscombe Down, he was involved in weapons and Electro-Optic systems development for RAF ground attack aircraft, including the Tornado GR4, Harrier GR7 and Jaguar GR3. Currently a civilian pilot with a UK international airline, a Night Electro-Optic flight instructor and director of an aerosystems consultancy specialising in Aviator's Night Vision Imaging System (ANVIS) integration in fixed- and rotary-wing platforms, Edward has accumulated over 3,000 hours in military aircraft and 8,500 commercial hours flying a number of aircraft, including the Airbus A340-300 and -600.

Jane's Users' Charter

This publication is brought to you by Jane's Information Group, a global company with more than 100 years of innovation and an unrivalled reputation for impartiality, accuracy and authority.

Our collection and output of information and images is not dictated by any political or commercial affiliation. Our reportage is undertaken without fear of, or favour from, any government, alliance, state or corporation.

We publish information that is collected overtly from unclassified sources, although much could be regarded as extremely sensitive or not publicly accessible.

Our validation and analysis aims to eradicate misinformation or disinformation as well as factual errors; our objective is always to produce the most accurate and authoritative data.

In the event of any significant inaccuracies, we undertake to draw these to the readers' attention to preserve the highly valued relationship of trust and credibility with our customers worldwide.

If you believe that these policies have been breached by this title, you are invited to contact the editor.

A copy of Jane's Information Group's Code of Conduct for its editorial teams is available from the publisher.

Intelligence and Insight You Can Trust

Glossary of acronyms and abbreviations in Jane's Avionics

The following is a list of acronyms and abbreviations associated with avionic systems that are used in *Jane's Avionics*:

A	Ampère
AAA	Anti-aircraft artillery
AAAA	Advanced architectures for airborne arrays
AAAM	Advanced air-to-air missile
AACC	Alternate airborne command centre
AACMI	Autonomous air combat manoeuvring installation
AACS	Airborne advanced communications system
AADS	Airborne active dipping sonar
AAED	Advanced airborne expendable decoy
AAICP	Air-to-air interrogator control panel
AAIP	Analogue autoland improvement programme
AAR	1. Air-to-air refuelling
	2. Air-to-air receive
AAT	Advanced avionics testbed
AAU	Audio amplifier units
AAW	Anti-air warfare
AAWWS	Airborne adverse weather weapon system
ABC	Automatic brightness control
ABCCC	Airborne battlefield command and control centre
ABE	Air-based electronics
ABICS	Ada-based integrated control system
ABIS	All-bus instrumentation system
ABIU	Avionics bay interface unit
ABL	Airborne laser
ABS	Agile beam system
AC	Alternating current
ACAS	Airborne collision avoidance system
ACARS	Automatic communications and reporting system
ACC	1. Avionics computer control
	2. Axis-controlled carrier
ACCS	Airborne computing and communications system
ACE	1. Actuator control electronics
	2. Advanced communication engine
	3. Autonomous combat (manoeuvres) evaluation
	4. Aircraft common equipment
ACEM	Aerial camera electro-optical magazine
ACI	Armament control indicator
ACIDS	Automated communications and intercom distribution system
ACIS	1. Armament control indicator set
	2. Advanced cabin interphone system
ACLS	Automatic carrier landing system
ACM(I)	Air combat manoeuvring installation
ACMS	1. Aircraft condition monitoring system
	2. Armament control and monitoring system
	3. Avionics control and management system
ACNIP	Auxiliary (or advanced) communication navigation and identification panel
ACP	1. Advanced controller processor
	2. Armament control panel
	3. Audio converter processor
ACP(C)	Automatic communications processor (control)
ACS	Armament control system
ACT	Airborne crew trainer
ACTIVE	Advanced control technology for integrated vehicles
ACU	1. Adaptive control unit
	2. Airborne computer unit
	3. Annotation control unit
	4. Antenna control unit
	5. Audio control unit
	6. Auxiliary control unit
ADAAPS	Aircraft data acquisition, analysis and presentation system
ADACS	Airborne digital automatic collection system
ADAD	Air defence alerting device
ADAS	1. Airborne data acquisition system
	2. Auxiliary data annotation set
ADAU	Auxiliary data acquisition unit
ADC	1. Advanced data controller
	2. Air data computer
ADDS	Advanced digital dispensing system
ADECS	Advanced digital engine control system
ADELE	Alerte détection et localisation des emetteurs
ADELT	Automatically deployed emergency locator transmitter
ADF	Automatic direction-finder

ADI	1. Attitude director indicator
	2. Azimuth display indicator
ADID	Aircraft data interface device
ADIRS	Air data and inertial reference system
ADIRU	Air data inertial reference unit
ADLGP	Advanced datalink for guided platforms
ADM	1. Advanced development model
	2. Air data module
Ad-Me	Advanced metal evaporated
ADOCS	Advanced digital optical control system
ADR	Accident data recorder
ADS	1. Audio distribution system
	2. Automatic dependent surveillance
	3. Airlifter defence systems
ADS-B	Automatic dependant surveillance broadcast
ADS-C	Automatic dependant surveillance contract
ADSU	Air data sensor unit
ADTU	Auxiliary data transfer unit
ADU	1. Air data unit
	2. Annotation display unit
ADVCAP	Advanced capability
ADVICE	Acoustic data vessel identification, classification and explanation
AEC	Automatic exposure control
AEGIS	Airborne early warning/ground environment integration segment
AELS	Airborne electronic library system
AERA	Automated en-route air traffic control
Aero-H	Aeronautical high-gain antenna
Aero-I	Aeronautical intermediate-gain antenna
AESA	Active electronically scanned array
AES	Aeronautical earth station
AESOP	Airborne electro-optical special operations payload
AESS	Aircraft environment surveillance system
AETMS	Airborne electronic terrain-mapping system
AEU	Airborne electronics unit
AEW	Airborne early warning
AF	Audio frequency
AFA	Audio frequency amplifier
AFC	Automatic frequency control
AFCAS	Automatic flight control augmentation system
AFCS	Automatic flight control system
AFDAS	Aircraft fatigue data analysis system
AFDS	Automatic flight director system
AFH	Advanced fibre heater
AFIS	Airborne flight information system
AFMCS	Advanced flight management computer system
AFMS	Automatic flight management system
AFMSS	Air force mission support system
AFSAT	Air force satellite
AFSATCOM	Air force satellite communications
AFTI	Advanced fighter technology integration
AFV	Armoured fighting vehicle
AG	Auto gate
AGA	Auto gate area
AGC	Automatic gain control
AGES	Air-to-ground engagement simulator
AGINT	Advanced GPS inertial navigation technology
AGL	Above ground level
AGPPE	Advanced general purpose processor element
AGRA	Automatic gain ranging amplifier
AGREE	Advisory group on the reliability of electronic equipment (US)
AGTV	Active gated television
AGU	AIRLINK gateway unit
AHEAD	Attitude, heading and rate of turn indicating system
AHIP	Army helicopter improvement programme
AHRS	Attitude and heading reference system
AHS	Attitude heading (reference) system
AI	Air interception
AIBU	Advanced interference blanker Unit
AICS	Airborne integrated communications system
AIDC	Aero industry development centre

AIDS	Aircraft integrated data suite (or system)
AIG	ACAS implementation group
AIM	1. Air intercept missile
	2. Air traffic control beacon/IFF/Mk XII system
AIME	Autonomous integrity monitored extrapolation
AIMES	Avionics integrated maintenance expert system
AIMS	1. Advanced integrated MAD system
	2. Air traffic control radar beacon, IFF, Mk 12 transponder system
	3. Aircraft information management system
	4. Aircraft integrated monitoring system
AINS	1. Aided inertial navigation system
	2. Airborne inertial navigation system
AIP	Australian industrial participation
AIPT	Advanced image processing terminal
AIRSTAR	Airborne surveillance and target acquisition radar
AIT	Advanced intelligence tape
AIU	1. Armament interface unit
	2. Automatic ignition unit
AJB	Audio junction box
AKU	Avionic keyboard unit
ALARMS	Airborne laser radar mine sensor
ALAT	French army light aviation corps
ALB	Airborne LIDAR bathymeter
ALE	Automatic link establishment
ALFS	Airborne low-frequency sonar
ALLTV	All light level television
ALRAD	Airborne laser range-finder and designator
AM	Amplitude modulation
AMAC	Airborne multi-application computer
AMC	Advanced micro-electronics converter
AMCS	Airborne missile control system
AME	1. Amplitude modulation equivalent
	2. Angle measuring equipment
AMHMS	Advanced magnetic helmet-mounted sight
AMICS	Adaptive multidimensional integrated control system
AMIDS	Advanced missile detection system
AMIMU	Advanced multisensor inertial measurement unit
AMLCD	Active matrix liquid crystal display
AMMCS	Airborne multiservice/multimedia communications system
AMP	Advanced modular processor
AMPA	Advanced mission planning aid
AMRAAM	Advanced medium-range air-to-air missile
AMRS	Advanced maintenance recorder system
AMS	1. Airborne maintenance subsystem
	2. Avionics management system
AMSAR	1. Airborne multirole steerable array radar
	2. Airborne multirole multifunction Solid-state active array radar
AMSL	Above mean sea level
AMSS	Aeronautical mobile satellite service
AMU	1. Audio management unit
	2. Auxiliary memory unit
	3. Avionics management unit
ANC	Active noise cancellation
ANDVT	Advanced narrowband digital voice terminal
ANMI	Air navigation multiple indicator
ANSI	American national standards institute
ANVIS	Aviator's night vision imaging system
AOA	1. Airborne optical adjunct
	2. Angle of attack
AOC	Assumption of control (message)
AOCM	Advanced optical countermeasures
AOPT	Advanced optical position transducer
APALS	Autonomous precision approach and landing system
APAR	Advanced phased-array radar
A/PDMC	Aircraft products data management computer
APHIDS	Advanced panoramic helmet interface demonstrator system
APIRS	Aircraft piloting inertial reference system
APR	Auto power reserve
APS	1. Adaptive processor system
	2. Aircraft position sensor
	3. Altitude position sensor
APSP	Advanced programmable signal processor
APU	Auxiliary power unit
AQL	Advanced quick look
ARBS	Angle rate bombing set
ARCADS	Armament control and delivery system
ARDS	Airborne radar demonstrator system
ARI	Azimuth range indicator
ARIA	Advanced range instrumentation aircraft
ARIC	Airborne radio and intercom control
ARIES	Airborne recorder for IRLS and EO sensors
ARINC	Aeronautical radio incorporated
ARJS	Airborne radar jamming system

ARM	Anti-radiation missile
ARPA	Advanced Research Projects Agency (US)
ARPS	Advanced radar processing system
ARSA	Advisable radar service area
ARTCC	Air route traffic control centre
ARTI	Advanced rotorcraft technology integration
ARWS	Advanced radar warning system
AS	Anti-spoofing
ASAC	Airborne surveillance airborne control
ASAP	Airborne shared aperture programme
ASARS	1. Airborne search and rescue system
	2. Advanced synthetic aperture radar system
ASAS	Airborne separation assurance system
ASC	Airborne strain counter
ASCB	1. Aircraft standard communications bus
	2. Avionics standard communications bus
ASCII	American Standard Code for Information Interchange
ASCOT	Aerial survey control tool
ASCTU	Air supply controller/test unit
ASDC	Armament signal data converter
ASE	Autostabilisation equipment
ASETS	Airborne seeker evaluation test set
ASGC	Airborne surveillance ground control
ASI	ARINC standards interface
ASIC	Application specific integrated circuit
ASIT	Adaptive surface interface terminals
ASMD	Anti-ship missile defence
ASMU	Avionics system management unit
ASNI	Ambient sea noise indication
ASP	1. Advanced signal processor
	2. Airborne signal processor
	3. Aircraft systems processor
ASPIS	Advanced self-protection integrated suite
ASPJ	Airborne self-protection jammer
ASPRO	Associative processor
ASR	1. Advanced special receiver
	2. Auto scene reject
ASSG	Acoustic sensor signal generator
AST	Airborne surveillance testbed
ASTAR	Airborne search target attack radar
ASTE	Advanced strategic and tactical expendables
ASTOR	Airborne standoff radar
ASU	Acoustic simulation unit
ASuW	Anti-surface warfare
ASV	Anti-surface vessel
ASVW	Anti-surface vessel warfare
ASW	Anti-submarine warfare
ASWAC	Airborne surveillance warning and control
AT³	Advanced tactical targeting technology
ATA	1. Actual time of arrival
	2. Advanced tactical aircraft
ATAC	1. Applied technology advanced computer
	2. Air transportable acoustic communication
ATAL	Appaeillage de télévision sur aéronef léger
ATARS	Advanced tactical air reconnaissance system
ATC	Air traffic control
ATCRBS	Air traffic control radar beacon system
ATCS	Advanced tactical communications system
ATDS	Airborne tactical data system
ATE	Automatic test equipment
ATF	Advanced tactical fighter
ATFLIR	Advanced targeting forward looking infra-red
ATH	Autonomous terminal homing
ATHS	Automatic target handover system
ATIRCM	Advanced threat infra-red countermeasures
ATIS	Automatic terminal information service
ATLANTIC	Airborne targeting low-altitude navigation thermal Imaging and cueing
ATLIS	Automatic tracking laser illumination system
ATM	1. Air traffic management
	2. Asynchronous transfer mode
ATN	Aeronautical telecommunications network
ATOPS	Advanced transport operations system
ATR	1. Air transport tracking
	2. Automatic target recognition
ATRJ	Advanced threat radar jammer
AVAD	Automatic voice alert device
AVDAS	Airborne video data acquisition system
AVE	Airborne vehicle equipment
AVICS	Air vehicle interface and control system
AVTR	Airborne videotape recorder
AWACS	Airborne warning and control system
AWADS	Adverse weather aerial delivery system
AWARE	Advanced warning of active radar emissions
BBU	Battery back-up unit

BCSV	Bearing compartment scavenge value
BDHI	Bearing, distance and heading indicator
BDI	Bearing distance indicator
BER	BIT error rate
BFDAS	Basic flight data acquisition system
BFO	Beat frequency oscillation
BIT	Built-in test
BITE	Built-in test equipment
BPI	Bits per inch
BPSK	Bi-phase shift keyed
B-RNAV	Basic area navigation
BSIN	Bus system interface unit
BSP	1. Barra side processor
	2. Bright source protection
BT	Bathythermograph
BTH	Beyond the horizon
BTU	Basic terminal unit
BVCU	Bleed valve control unit
BVR	Beyond visual range
C³CM	Command, control and communications countermeasures
C³I	Command, control, communications and intelligence
C⁴I	Command, control, communications, computers and intelligence
C/A	Coarse acquisition
CAA	Civil Aviation Authority (UK)
CAB	Common avionics baseline
CACTCS	Cabin air conditioning and temperature control system
CAD	Computer-aided design
CADC	Central air data computer
CAINS	Carrier aircraft inertial navigation system
CAMBS	Command activated multibeam sonobuoy
CAMEL	Cartridge active miniature electromagnetic
CARA	Combined altitude radar altimeter
CARABAS	Coherent all radio band sensing
CARS/DGS	Contingency airborne reconnaissance system/deployable ground station
CASS	1. Command active sonobuoy system
	2. Consolidated automatic support system
CAT	Category
CBL	Control by light
CBT	Computer-based training
CC	Countermeasures computer
CCD	Charge coupled device
CCG	Communication control group
CCIL	Continuously computed impact line
CCIP	Continuously computed impact point
CCIR	Comité Consultatif International des Radiocommunications
CCRP	1. Continuously computed release point
	2. C4ISR co-operative research programme
CCS	Conformal countermeasures system
CCTWT	Coupled cavity travelling wave tube
CCU	1. Cockpit control unit
	2. Common control unit
	3. Communication control unit
CDC	Cabin display computer
CDI	Course deviation indicator
CDIRRS	Cockpit display of infra-red reconnaissance system
CDM	Collaborative decision making
CDNU	Control and display navigation unit
CDR	Critical design review
CDS	Control and display system
CDTI	Cockpit display of traffic information
CDU	1. Cockpit display unit
	2. Control and display unit
CEAT	Centre d'Essais Aéronautiques de Toulouse
CEC	Co-operative engagement capability
CEDAM	Combined electronic display and map
CEP	Circular error of probability
CEV	Centre d'Essais de Vol
CFAR	Constant false alarm rate
CFD	Chaff and flare dispensing
CFDCU	Chaff and flare dispenser control unit
CFDIU	Centralised fault display interface unit
CFDS	Centralised fault display system
CFIT	Controlled flight into terrain
CG	Centre of gravity
CGCC	Centre of gravity control computer
CHAALS	Communications high-accuracy airborne location system
CHT	Cylinder head temperature
CI	Control Indicator
CIBS-M	Common integrated broadcast service-modules
CI-F	Control indicator - front
CIP	Common integrated processor
CIRCE	Cossor interrogation and reply cryptographic equipment
CIRTEVS	Compact infra-red television system

CI-S	Control indicator - side
CIS	1. Commonwealth of Independent States
	2. Control indicator set
CIT	Combined interrogator/transponder
CITS	Centrally integrated test system
CIU	1. Cockpit interface unit
	2. Control interface unit
CLASS	Coherent laser radar airborne shear sensor
CLDP	Convertible laser designation pod
CLDS	Cockpit laser designation system
CM	COMSEC module
CMD	1. Colour multipurpose display
	2. Countermeasures dispenser
CMDR	Card maintenance data recorder
CMDS	Countermeasures dispensing system
CMF	Central maintenance function
CMLSA	Commercial microwave landing system avionics
CMOS	Complementary metal oxide silicon
CMP	Central maintenance panel
CMRA	Cruise missile radar altimeter
CMRS	Crash/maintenance recorder system
CMS	Computer module system
CMT	Cadmium mercury telluride
CMUP	Conventional mission upgrade programme
CMWS	Common missile warning system
C-Nite	Cobra nite
CN2H	Conduit nuit, second-generation, helicopters
CNC	Communication/navigation equipment controls
CNI	Communications, navigation and identification
CNI-MS	CNI management system
CNIU	Communications/navigation interface unit
CNS	Communications navigation surveillance
CNS/ATM	Communications navigation surveillance/air traffic management
CNS/ATN	Communications navigation surveillance/aeronautical telecommunications network
CODAR	Correlation detection and ranging
COIL	Chemical oxygen iodine laser
COMAC	Cockpit management computer
COMED	Combined map and electronic display
COMINT	Communications intelligence
COMPASS	Compact multipurpose advanced stabilised system
COMSEC	Communications security
COTIM	Compact thermal imaging module
COTS	Commercial-off-the-shelf
CP	Computer processor
CPA	Cabin public address
CPC	Cabin pressure controller
CPCS	Cabin pressure control system
CPDLC	Controller-pilot datalink communications
CPM/P	Command post modem/processor
CPR	Covert penetration radar
CPS	Covert penetration system
CPU	Central processing unit
CPU-F	Control panel unit - front
CPU-S	Control panel unit - side
CPVR	Crash protected video recorder
CR	Countermeasures receiver
CRE	Communications radar exciter
CRPA	Controlled reception pattern antenna
CRT	Cathode ray tube
CS	Communications subsystem
CSA	Control stick assembly
CSAS	Command and stability augmentation system
CSC	1. Communication system controller
	2. Compass system controller
CSCG	Communications system control group
CSD	Common strategic Doppler
CSMU	Crash survivable memory unit
CSS	1. Computer support system
	2. Complimentary satellite system
CSU	1. Communications switching unit
	2. Control status unit
	3. Crew station unit
CSVR	Crash survivable voice recorder
CT	1. Control transmitter
	2. Crew member terminal
CTAS	Chrysler Technologies airborne systems
CTC	Cabin temperature controller
CTS	Central tactical system
CTT/H-R	The commanders' tactical terminal/ hybrid-receive only
CTU	Control terminal unit
CU	Control unit
CUGR	Cargo utility GPS receiver
CV	Aircraft carrier

CVFDR	Cockpit voice and flight data recorder
CVR	Cockpit voice recorder
CVR/DFDR	Cockpit voice recorder/digital flight data recorder
CW	Continuous wave
CWI	Continuous wave illuminator
CWS	Control wheel steering
D	Detectivity
D*	Normalised detectivity
DADC	Digital air data computer
DAFCS	Digital automatic flight control system
DAFD	Digital autopilot/flight director
DAFICS	Digital automatic flight inlet control system
DAI	DCMS audio interface
DAIS	Digital avionics information system
DAIRS	Distributed-architecture infra-red sensor
DAMA	Demand assigned multiple access
DAMS	Drum auxiliary memory sub-unit
DAP	Downlink aircraft parameters
DAPU	Data acquisition and processing unit
DAR	1. Direct access recorder
	2. Drone anti-radar
DARP	Digital audio record and playback
DARPA	Defense Advanced Research Projects Agency (US) (now ARPA)
DARTS	Digital airborne radar threat simulator
DARU	Data acquisition and recording unit
DAS	1. Distributed Aperture System
	2. Defensive Aids Suite
DASH	Display and sight helmet
DASS	Defensive aids subsystem
DAT	Digital audio tape
DAU	1. Data acquisition unit
	2. Digital amplifier unit
dB	decibel(s)
DBC	DCMS bus coupler
DBI	DCMS bus interface
dBm	decibels $\times 10^{-3}$
DC	Direct current
DCC	Digital computer complex
DCI	DCMS crew member interface
D/CM	Diagnostic and condition monitoring
DCMS	Digital communication management system
DCPS	Data collection and processing system
DCSU	Dual crew station unit
DCU	1. Data collection unit
	2. Digital computer unit
DDI	1. Data display indicator
	2. Digital display indicator
DDIC	Digital display indicator control
DDPS	Digital display processing system
DDS	Direct digital synthesis
DDU	Disk drive unit
DDVR	Displayed data video recorder
DEC	Digital electronic control
DECS	Digital engine control system
DECU	Digital engine control unit
DED	Data entry display
DEEC	Digital electronic engine control
DEFCS	Digital electronic flight control system
DEMON	Demodulation of noise
DES	Data encryption standard
DEU	Display electronics unit
DF	Direction-finding
DFAD	Digital feature analysis data
DFDAU	Digital flight data acquisition unit
DFDR	Digital flight data recorder
DFGC	Digital flight guidance computer
DFIR	Deployable flight incident recorder
DFLCC	Digital flight control computer
DFU	Deployable flotation unit
DG	Directional gyro
DGA	Displacement gyro assembly
DGNS	Differential global navigation system
DGPS	Differential GPS
DIANE	Détection identification analyse des nouveaux émetteurs
DICASS	Directional command activated sonobuoy system
DICU	Display interface control unit
DID	Data insertion device
DIFAR	Directional acoustic frequency analysis and recording
DIFM	Digital instantaneous frequency measurement
DII	DCMS interphone interface
DII COE	Defence information infrastructure common operating environment
DIL	Digital integrated logic
DIM	Dispense interface microprocessor

DIRCM	Directional infra-red countermeasures
DITACS	Digital tactical system
DITS	Digital information transfer system
DITU	De-icer timer unit
DIU	Data interface unit
DLMS	Digital land mass system
DLPP	Datalink pre-processor
DLS	Data loader system
DLT	Digital linear tape
DMA	Direct memory access
DMDG	Digital map display generator
DME	Distance measuring equipment
DME-P	Distance measuring equipment - precision
DMG	Digital map generator
DMM	Data management module
DMPI	Desired mean point of impact
DMS	Data multiplexer sub-unit
DMU	1. Data management unit
	2. Digital master unit
DNATS	Day/night airborne thermal sensor
DOA	Direction of arrival
DoD	Department Of Defense (US)
DOLE	Detection of laser emissions
DOLRAM	Detection of laser, radar and millimetric threats
DOP	Digital onboard processor
DOS	Disk operating system
DP	Display processor
DPCM	Digital pulse code modulation
DPG	Data processor group
DP/MC	Display processor/mission computer
DPRAM	Dual-port random access memory
DPS	Data processing system
DPSK	Digital phase shift keying
DPU	Digital processing unit
DRAM	Dynamic random access memory
DRC	Data recording cartridge
DRD	Digital radar display
DRFM	Digital radio frequency memory
DRU	Data retrieval unit
DSCS	Defence satellite communication system
DSDC	Digital signal data converter
DSP	1. Day surveillance payload
	2. Digital signal processing
	3. Digital signal processor
DSS	Data storage set
DSU	1. Data storage unit
	2. Digital switch unit
	3. Dynamic sensor unit
DSUR	Data storage unit receptacle
DTC	Data transfer cartridge
DTD	Data transfer device
DTE	Data transfer equipment
DTED	Digital terrain elevation data
DTF	Digital tape format
DTIU	Data transfer interface unit
DTM	Data transfer module
DTM/D	Digital terrain management and display
DTN	Data transfer network
DTS	1. Data terminal set
	2. Data transfer system
	3. Digital terrain system
DTU	1. Data transfer unit
	2. Display terminal unit
DU	Display unit
DUCK	DCMS universal configuration key
DV/A	Doppler velocimeter/altimeter
DVI	Direct voice input
DVOF	Digital vertical obstruction file
DVRS	Display video recording system
DVS	Doppler velocity sensor
EADI	Electronic attitude director indicator
EAMED	Eventide airborne multipurpose electronic display
EAP	1. Emergency audio panel
	2. Experimental aircraft programme
EAR	Electronically agile radar
EAROM	Electrically alterable read-only memory
EARS	ECI airborne relay system
EASU	Engine analyser and synchrophase Unit
EATCHIP	European Air Traffic Control Harmonisation and Integration Programme
EATMS	European Air Traffic Management System
EAU	Engine analyser unit
EBI	Equivalent background illumination
ECA	Electronic control amplifier
ECAC	European Civil Aviation Conference

ECAM	Electronic centralised aircraft monitor
ECB	Electronic control box
ECCM	Electronic counter-countermeasures
ECDU	Enhanced control and display unit
ECIPS	Electronic combat integrated pylon system
ECL	Emitter coupled logic
ECM	Electronic countermeasures
ECNI	Enhanced communications, navigation and identification
ECOP	Electronic co-pilot
ECP	Engineering change proposal
ECR	Electronic combat and reconnaissance
ECS	Environmental control system
ECU	1. Electronics control unit
	2. Environmental control unit
	3. Exercise control unit
EDAU	Extended data acquisition unit
EDC	Error detection and correction
EDIU	Engine data interface unit
EDM	Engine data multiplexer
EDTS	Expanded data transfer system
EDU	1. Electronic display unit
	2. Engine diagnostic unit
EEC	Electronic engine controls
EEMS	Electrostatic engine monitoring system
EEPROM	Electrically erasable programmable read-only memory
EEZ	Economic exclusion zone
EFB	Electronic flight bag
EFCS	Electronic flight control system
EFCU	Electrical flight control unit
EFDARS	Expansible flight data acquisition and recording system
EFIS	Electronic flight instrumentation system
EFMCS	Enhanced flight management computer system
EGAC	Enhanced general avionics computer
EGI	Embedded GPS-inertial
EGNOS	European Geostationary Navigation Overlay System
EGT	Exhaust gas temperature
EHF	Extra high frequency
EHSI	Electronic horizontal situation indicator
EIA	Electronic Industries Association (US)
EICAS	Engine indication and crew alerting system
EID	Emitter identification
EIS	Electronic instrument system
EIU	1. Electronic interface unit
	2. Engine interface unit
EL	Electroluminescent
ELAC	Elevator and aileron computer
ELB	Emergency locator beacon
ELF	Electronic location finder
ELINT	Electronic intelligence
ELIOS	ELINT identification and operating system
ELIPS	Electronic integrated protection shield
ELMS	Electrical load management system
ELS	1. Electronic library system
	2. Emitter location system
ELT	Emergency locator transmitter
E-MAGR	Enhanced-miniaturised airborne GPS receiver
EM	Electromagnetic
EMC	Electromagnetic compatibility
EMD	Engineering model derivative
EMDU	Enhanced main display unit
EMI	Electromagnetic interference
EMMU	Engine monitor multiplexer unit
EMP	Electromagnetic pulse
EMS	Entry monitor system
EMSC	Engine monitoring system computer
EMTI	Enhanced moving target indicator
EMU	Engine monitoring unit
EMux	Electronic multiplexing
EO	Electro-optic
EOB	Electronic order of battle
EOCM	Electro-optical countermeasures
EOIVS	Electro-optical/infra-red viewing system
EOSS	Electro-optic sensor system
EOVS	Electro-optical viewing system
EPAD	Electrically powered actuator design
EPI	Engine performance indicator
EPLD	Electronic programmable logic device
EPMS	Electrical power management system
EPR	Engine pressure ratio
EPROM	Erasable programmable read-only memory
EPRT	Engine pressure ratio transmitter
EQAR	Expanded quick access recorder
ERAPS	Expandable reliable acoustic path sonobuoy
EROM	Erasable read-only memory
ERP	Effective radiated power

ERS	Electronic resource system
ERWE	Enhanced radar warning equipment
E/SA	Embedded and special application
ESC	Engine supervisory control
ESG	Electrostatically suspended gyro
ESM	Electronic support measures
ESP	1. Expandable system programmer
	2. Expendable signal processor
ESS	Exercise support system
ESU	Electronic storage unit
ETA	Estimated time of arrival
ETE	Estimated time en route
ETIPS	Electrothermal ice protection system
ETMP	Enhanced terrain masked penetration
ETPU	Engine transient pressure unit
EU	Electronics unit
EUROCAE	European Organisation for Civil Aviation Electronics
EVM	Engine vibration monitor
EVS	Enhanced vision Sensor
EW	Electronic warfare
EWAAS	End-state WAAS (wide area augmentation system)
EWMS	Electronic warfare management system
EWMU	EW management unit
EWPI	Electronic warfare prime indicator
EXCAP	Expanded capability
FAA	Federal aviation administration (US)
FAADC²I	Forward area air defence command, control and intelligence
FAC	1. Flight augmentation computer
	2. Forward air controller
FACS	Fully automatic compensation system
FACTS	FLIR augmented cobra TOW sight
FADEC	Full authority digital engine control
FAF	Final approach fix
FAFC	Full authority fuel control
FAM	Final approach mode
FAMIS	Full aircraft management/inertial system
FANS	Future air navigation system
FAP	Final approach
FBL	Fly-by-light
FBS	Fly-by-speech
FBW	Fly-by-wire
FCC	Flight control computer
FCDC	Flight control data concentrator
FCMC	Flight control and monitoring computer
FC/NP	Fire control/navigation panel
FCPC	Flight control primary computer
FCS	Flight control system
FCSC	Flight control secondary computer
FCU	Flight control unit
FDAMS	Flight data acquisition management system
FDAU	Flight data acquisition unit
FDE	Fault detection and exclusion
FDEP	Flight data entry panel
FDIU	Flight data interface unit
FDM	Frequency division multiplex
FDMU	Flight data management unit
FDP	Flight data panel
FDR	Flight data recorder
FDR/FA	Flight data recorder/fault analyser
FDS	Flight director system
FDT	Flight deck terminals
FDU	Flight data unit
FET	Field effect transistor
FEWSG	Fleet electronic warfare support group
FFSP	Full function signal processor
FFT	Fast fourier transform
FGCP	Flight guidance control panel
FINAS	Ferranti inertial nav/attack system
FIR	Far infra-red (band)
FIRAMS	Flight incident recorder and aircraft monitoring system
FIRMU	Flight incident recorder memory unit
FIS	Flight information service
fl	Foot-lambert
FLAG	Four-mode laser gyro
FLAGSHIP	Four-mode laser gyro software hardware implemented partitioning
FLASH	Folding light acoustic systems for helicopters
FLIR	Forward looking infra-red
Flops	Floating point operations per second
FM	Frequency modulated
FMA	Flight mode annunciator
FMC	1. Flight management computer
	2. Forward motion compensation
FMCS	1. Fatigue monitoring and computing system
	2. Flight management computer system

FMCW	Frequency modulated continuous wave	h	hour(s)
FMGC	Flight management and guidance computer	HAC	Anti-tank variant of Tiger helicopter
FMGS	Flight management and guidance system	HACLCS	Harpoon airborne command, launch and control system
FMICW	Frequency modulated interrupted continuous wave	HAD	Hybrid analogue digital
FMS	Flight management system	HADAS	Helmet airborne display and sight
FMU	Flight management unit	HADS	Helicopter air data system
FOAEW	Future organic airborne early warning	HAINS	High-accuracy inertial navigation system
FOC	Full operational capability	HAP	Escort variant of Tiger helicopter
FOM	Figure of merit	HAPS	Helicopter acoustic processing system
FoR	Field of regard	HARM	High-speed anti-radiation missile
FoV	Field of view	HBR	High bit rate
FPD	Flat-panel display	HDD	Head-down display
FPMU	Fuel pump monitoring unit	HDDR	Head-down display recorder
FQIS	Fuel quantity indication system	HDG	Heading
FRPA	Fixed reception pattern antenna	HDS	Hard disk subsystem
FRP	Federal (US) radionavigation plan	HDTV	High-definition TV
FRS	Fighter/reconnaissance/strike	HER	Harsh environment recorder
FSA/CAS	Fuel savings advisory and cockpit avionics system	HERALD	Helicopter equipment for radar and laser detection
FSAS	Fuel saving advisory system	HEU	HUD electronics unit
FSC	Fuel savings computer	HF	High frequency
FSD	Full-scale deflection	HFAC	Helicopter flight advisory computer
FSK	Frequency shift keying	HFDU	High-frequency data unit
FSRS	1. Flight safety recording system	HgCdTe	Mercury cadmium telluride
	2. Frequency selective receiver system	HGS	Holographic guidance system
FSS	Flight service station	HHTI	Hand-held thermal imager
ft	Feet	HHUD	Holographic head-up display
FT	Fault tolerant	HIADC	High-integration air data computer
FT-ADIRS	Fault tolerant air data inertial reference system	HIBIRD	Helicopter identification by infra-red detection
FT-ADIRU	Fault tolerant air data inertial reference unit	HIBRAD	Helicopter identification by radar detection
FTC	Fast time constant	HICU	HIPSS interface control unit
FTI	Fixed target indication	HIDAS	Helicopter integrated defensive aids system
FTIT	Fan turbine inlet temperature	HIDEC	Highly integrated digital engine control
FTRG	Fleet tactical readiness group	HIDSS	Helmet integrated display sight system
G	Giga = 1,000,000,000	HIPAS	High-performance active sonar
GaAs	Gallium arsenide	HIPSS	Helicopter integrated power and switching system
GANS	Global access navigation safety	HIRES	High resolution
GAS	Global positioning adaptive antenna system	HIRF	High-intensity radiated field
GATM	Global air traffic management	HIRNS	Helicopter infra-red navigation system
GATR	Ground air transmit receive	HIRS	Helicopter infra-red system
GATS	GPS-aided targeting system	HISAR	Hughes integrated surveillance and reconnaissance system
GBIB	Ground-based integrity broadcast	HIT	Hughes improved terminal
GCA	Ground collision avoidance	HITMORE	Helicopter installed television monitor recorder
GCAS	Ground collision avoidance system	HLD	Head level display
GCR	Ground clutter reduction	HLWE	Helicopter laser warning equipment
GDE	Graphics differential engine	HMCU	Hydraulic monitoring computer unit
GDP	Graphics drawing processor	HMD	Helmet-mounted display
GEM	GPS embedded module	HMDD	Helmet-mounted display device
GEN-X	Generic expendable	HMDS	Helmet-mounted display system
GES	Ground earth station	HMFU	Hydromechanical fuel unit
GGP	GPS guidance package	HMP	Helmet mounting plate
GHz	Giga hertz	HMSS	Helmet-mounted sighting system
GIC	GPS integrity channel	HMU	Hydromechanical unit
GIC	GPS/WAAS integrity channels (GPS global positioning system) (WAAS wide area augmentation system)	HNVS	Helicopter night vision system
		HOCAS	Hands-on collective and stick
GIG	GPS integration guidelines	HOFIN	Hostile fire indicator
GIS	Geographic information system	HOPS	Helmet optical position sensor
GIT	General interface terminal	HOS	Helitow observation system
GLINT	Gated laser illuminator for night television	HOTAS	Hands-on throttle and stick
GLONASS	Global orbital navigation satellite system	HOWLS	Hostile weapons location system
GMR	Ground mapping radar	HP	High pressure
GMT	Greenwich mean time	HPA	High-power amplifier
GMTI	Ground moving target indication	HPAG	High-power amplifier group
GNC	General navigation computer	HRP	Headset receptacle panel
GNLS	GPS navigation and landing system	HSDB	High-speed databus
GNSS	Global navigation satellite system	HSI	Horizontal situation indicator
GNSSU	Global navigation satellite sensor unit	HSI	Hyperspectral imagery
GP&C	Global positioning and communication	HSIC	High-speed integrated circuit
GPCDU	General purpose control display unit	HSSL	Helicopter self-screening launcher
GPIN	Global positioning laser inertial navigation	HTC	Hover trim control
GPIRS	Global positioning/inertial reference system	HTNS	Helicopter tactical navigation system
GPIRU	Global positioning inertial reference unit	HTTB	High-technology testbed
GPS	Global positioning system	HUD	Head-up display
GPSSU	Global positioning system sensor unit	HUDC	Head-up display computer
GPVI	Graphics processor video interface	HUDWAC	Head-up display and weapon aiming computer
GPWS	Ground proximity warning system	HUMC	Health and usage monitoring computer
GRCS	Guardrail common sensor	HUMS	1. Health and usage monitoring system
GRE	Ground readout equipment		2. Health and usage monitoring and sensing
GRIS	Guardrail common sensor interoperability system	HVPSU	High-voltage power supply unit
GS	Groundspeed	Hz	Hertz
GSDI	Groundspeed and drift indicator	IAC	Integrated avionics computer
GSE	Ground support equipment	IAD	Integrated antenna detector
GSM	GPS sensor module	IAHFR	Improved airborne high-frequency radio
GTAR	GEC Thomson airborne radar	IAHFR/NOE	Improved airborne high-frequency radio nap of the Earth
GTRE	Gas-Turbine Research Establishment (UK)	IAM	Initial approach mode
GUI	Graphics user interface	IAMS	Integrated armament management system

IAPS	Integrated avionics processing system		**IRFIS**	Inertial referenced flight inspection system
IAS	Indicated air speed		**IRIS**	1. Infra-red imaging subsystem
IB	Interconnecting box			2. Integrated radar imaging system
IBS	Integrated broadcast service		**IRLS**	Infra-red linescanner
I/C	Interface and control		**IRMS**	Integrated radio management system
ICAAS	Integrated controls and avionics for air superiority		**IRP**	Interphone receptacle panel
ICAO	International civil air traffic organisation		**IRRS**	Infra-red reconnaissance system
ICAP	Increased capability		**IRS**	Inertial reference system
ICCP	Integrated communications control panel		**IRST(S)**	Infra-red search and track (system)
ICDU	1. Integrated control and display unit		**IRU**	Inertial reference unit
	2. Intelligent control display unit		**IR/UV**	Infra-red/ultraviolet
ICE	Improved combat efficiency		**IRV**	Infra-red vision
ICMS	Integrated countermeasures system		**IRVAT**	Infra-red video automatic tracking
ICNIA	Integrated communications navigation identification avionics		**ISA**	Instruction set architecture
ICNIS	Integrated communications navigation identification set		**ISAHRS**	Improved standard attitude and heading reference system
ICP	Integrated control panel		**ISAR**	Inverse synthetic aperture radar
ICS	1. Intercommunications system (or set)		**ISB**	Independent sideband
	2. Internal countermeasures set		**ISBA**	Inertial sensor-based avionics
ICSM	Integrated conventional stores management		**ISC**	Intercommunications set control
ICSM/GPS	Integrated conventional stores management/global positioning system		**ISDS**	IRCM self-defence system
			ISIS	Integrated strike and interception system
ICU	1. Interface computer unit		**ISLS**	Interrogation side-lobe suppression
	2. Interface converter unit		**ISO**	International standardisation organisation
	3. Interstation control unit		**ISP**	Intelligence, surveillance and reconnaissance
ICW	Interrupted continuous wave		**ISS**	Integrated sensor system
ID	Identification		**ISU**	Intercommunications set control unit
IDACS	Integrated digital audio control system		**ITAR**	Integrated terrain access and retrieval system
IDAP	Integrated defence avionics programme		**I-TOW**	Improved tube-launched optically tracked wire-guided (missile)
IDAS	Integrated design automation system			
IDECM	Integrated defensive electronic countermeasures		**IUMS**	Integrated utilities management system
IDF	Instantaneous direction-finding		**IVMMS**	Integrated vehicle mission management system
IDL	Interoperable datalink		**IVSC**	Integrated vehicle subsystem controls
IDM	Inductive debris monitor		**IVSI**	Instantaneous vertical speed indicator
IDP	Imagery display processor		**IWAAS**	Initial WAAS (wide area augmentation system)
IDS	1. Infra-red detection set		**JAR**	Joint airworthiness requirement
	2. Interdiction Strike		**JASS**	Joint airborne SIGINT system
IEEE	Institute of Electrical and Electronic Engineers (US)		**JAST**	Joint advanced strike technology
IEU	Interface electronics unit		**JDAM**	Joint direct attack munition
IEW	Integrated electronic warfare		**JIAWG**	Joint Integrated Avionics Working Group (US)
IEWCS	Intelligence and electronic warfare common sensor		**Joint STARS**	Joint surveillance and target attack radar system (also JSTARS)
IF	Intermediate frequency			
IFF	Identification friend or foe		**JPALS**	Joint precision approach landing system
IFFCP	Identification friend or foe control panel		**JPATS**	Joint primary aircraft training system
IFM	Instantaneous frequency measurement		**JPO**	Joint Program Office (US)
IFM/SHR	IFM superheterodyne receiver		**JSF**	Joint strike fighter
IFMU	Integrated flight management unit		**JSAF LBSS**	Joint SIGINT avionics family low-band subsystem
IFOG	Interferometric fibre optic gyro		**JTIDS**	Joint tactical information distribution system
IFoV	Instantaneous field of view		**JTRS**	Joint tactical radio system
IFR	Instrument flight rules		**JTW**	Joint targeting workstation
IFTS	Internal FLIR targeting system		**k**	1,000
IGAC	Israeli general avionics computer		**K**	Kelvin
IHADSS	Integrated helmet and display sighting system		**KAPS**	Kollsman auto-schedule pressurisation System
IHAS	Integrated hazard avoidance system		**kbit**	Kilobit
IHDTV	Intensified high-definition television		**KBU**	KeyBoard unit
IHEWS	Integrated helicopter electronic warfare suite		**kbyte**	Kilobyte
IHU	Integrated helmet unit		**KCCU**	Keyboard cursor control unit
IHUMS	Integrated health and usage monitoring system		**kg**	Kilogram
IIDS	Integrated instrumentation display system		**KFD**	Key fill device
IIS	Infra-red imaging system		**kHz**	Kilohertz
IJMS	Interim JTIDS message system		**kips**	Thousand instructions per second
ILS	Instrument landing system		**kops**	Thousand operations per second
IMA	Integrated modular avionics		**kt**	Nautical miles per hour (knot)
IMC	1. Instrument meteorological conditions		**KTD**	Key transfer device
	2. Image motion compensation		**L1**	GPS carrier frequency (1227.6 MHz)
IMSS	Integrated multisensor system		**L2**	GPS carrier frequency (1575.42 MHz)
IMU	Inertial measurement unit		**LAAP**	Low-altitude autopilot
in	inch(es)		**LAAS**	Local area augmentation system
INACP	Integrated navigation aids control panel		**LAASH**	LITEF analytical air data system for helicopters
INEWS	Integrated electronic warfare system		**LAAT**	Laser augmented airborne TOW
INS	Inertial navigation system		**LADGNSS**	Local area differential global navigation satellite system
INU	Inertial navigation unit		**LADGPS**	Local area differential global positioning system
I/O	Input/output		**LAEO**	Low-altitude electro-optical
IOC	1. Initial operational capability		**LAHRS**	Kearfott low-cost altitude heading reference system
	2. Input output computer		**LAIRS**	1. Light aircraft reconnaissance system
I/P	Identification position			2. Loral advanced imaging radar system
IP	Intermediate pressure		**LAMPS**	Light airborne multipurpose system
IPADS	Improved processing and display system		**LAN**	Local area network
IPCS	Intelligent power control system		**LANA**	Low-altitude night attack
IPEC	In-flight passenger entertainment and communications		**LANE**	Low-altitude navigation equipment
IPNVG	Integrated panoramic night vision goggle		**LANTIRN**	Low-altitude navigation and targeting infra-red for night
IPT	Intelligent power terminal		**LASS**	Low-altitude surveillance system
IPU	Interface processor unit		**LASTE**	Low-altitude safety and target enhancement
IR	Infra-red		**LCC**	Leadless ceramic chip-carrier
IRCCD	Infra-red charge coupled device		**LCD**	Liquid crystal display
IRCM	Infra-red countermeasures		**LCF**	Low cycle fatigue

LCINS	Low-cost inertial navigation system		MASPA	Minimum aviation system performance standards
LCLU	Landing control logic unit		MATSS	Mobile aerostat tracking and surveillance system
LCM	Lance-cartouches modulaire		MAW	1. Missile approach warner
LCOS(S)	Lead computing optical sight (system)			2. Mission adaptive wing
LCTAR	Le Centre Thomson d'Applications Radar		MAWS	Missile approach warning system
LCWDS	Low-cost weapon delivery system		mb	Millibar
LDT/SCAM	Laser detector and tracker/strike camera		MBAT	Multibeam array transmitter
LDU	Launcher decoder unit		MBE	Molecular beam epitaxy
LEA	Leurre electromagnetique actif		Mbit	Megabit
LECOS	Light electronic control system		Mbyte	Megabyte
LED	Light-emitting diode		MC	Mission computer
LED-RHA	Light-emitting diode-recording head assembly		MCCP	Main communications control panel
LEP	Laser eye protection		MCDU	Multifunction control and display unit
LESM	Lightweight electronic support measures		MCE	Modular control equipment
LF	Low frequency		Mcops	Million complex operations per second
LGCIU	Landing gear computer and interface unit		MCP	Microchannel plate
LH	Light helicopter		MCT	1. Mercury cadmium telluride
LHN	LITEF helicopter navigation			2. Metal oxide semiconductor controlled thyristor
LHR	LITEF helicopter reference		MCU	1. Management control unit
LICAS	Low-intensity conflict aircraft system			2. Master control unit
LIDAR	Light detection and ranging			3. Missile control unit
LIMAR	Laser imaging and ranging			4. Modular concept unit
LINS	Laser inertial navigation system			5. Multifunction control unit
LIP	Limited installation programme		MCW	Modulated continuous wave
LISCA	Leurre infrarouge à signature et cinématique adaptée		MDF	Mission data file
LITDL	Link 16 interoperable tactical data link		MDG	Map display generator
LIVAR	Laser illuminated viewing and ranging		MD(G)T	Mission data (ground) terminal
LLLTV	Low-light level television		MDI	Multifunction display indicator
LLTV	Low-light television		MDL	Mission data loader
LNA	Low-noise amplifier		MDP	1. Maintenance data panel
LO	Local oscillator			2. Modular display processor
LOA	Line of attack		MDRI	Multipurpose display repeater indicator
LOCUS	Laser obstacle cable unmasking system		MDT	Mission data terminal
LOFAR	Low frequency omnidirectional acoustic frequency Analysis and recording		MDTC/P	Mega data transfer cartridge with processor
			MDTS	Mission data transfer system
LORAN	Long-range aid to navigation		MEA	Minimum en route altitude
LORES	Low resolution		MEC	Modular electronics concept
LOROP	Long-range oblique photography		MEECN	Minimum essential emergency communications network
LOS	Line of sight		MERLIN	Modular ejection-rated low-profile imaging for night
LOSI	Line of sight indicator		MESA	Minimum emergency safe altitude
LP	Low pressure		METR	Multiple emitter targeting receiver
LPC	Linear predictive coding		MFCD	Multifunction colour display
LPCBA	Low-pressure compressor bleed actuator		MFD	Multifunction display
LPD	Low probability of detection		MFD(S)	Multifunction display (system)
LPHUD	Low-profile head-up display		MFHD	Multifunction head-down display
LPI	Low probability of interception		MFMS	Military flight management system
LQA	Link quality analysis		MGR	Miniature GPS receiver
LRCU	Landing roll-out control unit		MHDD	Multifunction head-down display
LRM	Line-replaceable module		MHz	Megahertz
LRMTS	Laser ranger and marked target seeker		MIAMI	Microwave ice accretion measurement instrument
LRU	Line-replaceable unit		MICNS	Modular integrated communications and navigation system
LSB	Lower sideband		Micro-AIDS	Micro-aircraft integrated data system
LSI	Large scale integration		MIDA	Message interchange distributed application
LST	Laser spot tracker		MIDS	Multifunction information distribution system
LTD/R	Laser target designator/ranger		MIGITS	Miniature integrated GPS/INS tactical system
LTR	Loop transfer recovery		MIL-SPEC	Military specification
LVDT	Linear variable differential transformer		MILSTAR	Military strategic and tactical relay
LWA	Laser warning analyser		MIL-STD	Military standard
LWF	Lightweight fighter		MIMU	Multisensor inertial measurement unit
LWIR	Long wave infra-red		min	Minute(s)
LWR	Laser warning receiver		MIPS	Million instructions per second
m	Metre		MIR	Middle infra-red (band)
M	1,000,000 or mega		MIRLS	Miniature infra-red linescan system
M-ADS	Modified automatic dependent surveillance		MIRTS	Modular infra-red transmitting system
MACC	Multi-application control computer		MLI	Mid-life improvement
MACS	Multiple application control system		MLPRF	Modular low-power radio frequency
MAD	Magnetic anomaly detector		MLS	Microwave landing system
MADAR	1. Maintenance analysis, detection and recording		MLU	1. Mid-life update
	2. Malfunction detection, analysis and recording			2. Monitor and logic unit
MADC	1. Micro air data computer		MLV	Memory loader and verifier
	2. Miniature air data computer		mm	millimetre(s)
MADGE	Microwave aircraft digital guidance equipment		MM	Mission manager
MAESTRO	Modular avionics enhancement system targeted for retrofit operations		MMC	Mission management computer
			MMIC	Monolithic microwave integrated circuit
MAG	Micromachined accelerometer gyro		MMLSA	Military microwave landing system avionics
MAGR	Miniature airborne GPS receiver		MMMS	Multimission management system
MAHRU	Microflex attitude and heading reference unit		Mmo	Maximum operating mach number
MAP	Missed approach point		MMP	Maintenance monitoring panel
MAP	Modular airborne processor		MMR	Multimode receiver
MARA	Modular architecture for real-time applications		MMRS	Multimission radar system
MARE	Miniature analogue recording electronics		MMS	Mast-mounted sight
MARS	1. Modular airborne recording system		MMSS/R	Mobile mass storage system model R
	2. Multi-application recorder/ reproducer system		MMW(R)	Millimetric wave (radar)
MASINT	Measurements and signatures intelligence		MNOS	Metal nitride oxide silicon
MASTER	Military aircraft satcoms terminal		MNPS	Minimum navigation performance standards

MNS	Modular navigation system
MOA	Minimum operating altitude
MODAR	Modular aviation radar
MODAS	Modular data acquisition system
MODIR	Modulated infra-red jammer
MONOHUD	Monocular head-up display
MOP	Measure(s) of performance
Mops	Million operations per second
MOPS	Minimum operating performance standard
MOS	Metal oxide silicon
MOSFET	Metal oxide silicon field effect transistor
MOSP	Multimission optronic stabilised payload
MOUT	Military operations in urban terrain
MP	Mission planner
MPA	Maritime patrol aircraft
MPCD	Multipurpose colour display
MPD	Multipurpose display
MPHD	Multipurpose head-down
MPM	Multipurpose modem
MPPU	Multipurpose processor unit
MPS	Mission planning subsystem
MQLF	Mobile quick-look facility
MR	Maritime reconnaissance
MRAALS	Marine remote area approach landing system
MRAAM	Medium-range air-to-air missile
mrad	milliradian
MRLG	Monolithic ring laser gyro
MRT	1. Miniature receive terminal
	2. Multirole turret
MRTD	Minimum resolvable temperature difference
MRTI	Multirole thermal imager
MRTU	Multiple remote terminal unit
MRU	1. Maintenance recorder unit
	2. Mobile reporting unit
ms	millisecond
m/s	metres per second
MSA	Minimum safe altitude
MSAR	Miniature synthetic aperture radar
MSCADC	Miniature standard central air data computer
MSD	1. Map storage display
	2. Mass storage device
MSF	Mission support facility
MSI	Multispectral imagery
MSIP	1. Multi(national) staged improvement programme
	2. Multistaged improvement programme
MSIS	Multisensor stabilised integrated system
MSP	Mission system processor
MSPS	Modular self-protection system
MSR	1. Marconi secure radio
	2. Modular strain recorder
	3. Modular survivable radar
MSS	Maritime surveillance system
MSSP	Multi sensor stabilised payload
MSTAR	Moving and stationary target acquisition and recognition
MSU	1. Maintenance station unit
	2. Mass storage unit
	3. Mode selector unit
MSWS	Multisensor warning system
MTAS	Millimetric target acquisition system
MTBF	Mean Time between failures
MTBR	Mean Time between repairs
MTBUR	Mean Time between unscheduled repairs
MTI	Moving target indicator
MTIS	Modular thermal imaging sight
MTF	Modulation Transfer Function
MTL	Magnetic tape loader
MTTR	Mean time to repair
MTU	Magnetic tape unit
MUST	Multimission UHF Satcom transceiver
Mux	Multiplexer
mV	Millivolt
MVP	Military VME processor
mW	Milliwatt
MWIR	Mid-wave infra-red
MWR	Microwave radiometer
NAS	Navigation attack system
NASA	National Aeronautics and Space Administration (US)
NASH	Navigation and attack system for helicopters
NAT	National air traffic
NATO	North Atlantic Treaty Organisation
nav/com	navigation and communication
NavHARS	Navigation, heading and attitude reference system
NavWASS	Navigation and weapon aiming subsystem
NBC	Nuclear, biological and chemical
NCS	Network control station

NCU	Navigation computer unit
NCW	Network centric warfare
ND	Navigation display
NDB	Non-directional beacon
Nd:YAG	Neodymium/yttrium aluminium garnet
NEACP	National emergency airborne command post
NEP	Noise equivalent power
NETD	Noise equivalent temperature difference
NFoV	Narrow field of view
Ng	Gas generator RPM
NGH	Next generation helmet
Ni/Cd	Nickel/cadmium
NIDJAM	Navigation/identification deception jammer
NIIRS	National imagery interpretability rating scale (US)
NILE	NATO improved link 11
NIR	Near infra-red (band)
NIS	NATO identification system
NITE-OP	Night imaging through electro-optics
NIU	Navigation interface unit
n mile(s)	Nautical mile(s)
NMOS	Negative metal oxide semiconductor
NMS	Navigation management system
NMU	Navigation management unit
NOCUS	North continental US
NOE	Nap of the earth
NOR	Logic circuit usable as either AND/OR
NOTAR	No tail rotor
NOVRAM	Non-volatile random access memory
NPR	No power recovery
NPRM	Notice of proposed rule making
NPU	Navigation processor unit
NQIS	Navigation quality inertial sensor
NRT	Near real-time
ns	Nanoseconds
NSA	National Security Agency (US)
NSIU	Navigation switching interface unit
NSP	Night surveillance payload
NSSL	National Severe Storms Laboratory (US)
NTDS	Naval tactical data system
NTS	Night targeting system
NV	1. Night vision
	2. Non-volatile
NVG	Night vision goggles
NVG/HUD	Night vision goggles/head-up display
NV/HUD	Night vision/head-up display
NVIS	Night vision imaging system
NVM	Non-volatile memory
NWDS	Navigation and weapon delivery system
OADS	Omnidirectional air data system
OAS	Offensive avionics system
OASYS	Obstacle avoidance system
OAT	Outside air temperature
OBEWS	Onboard electronic warfare simulator
OBI	Omni-bearing indicator
OBTEX	Onboard targeting experiments
OCU	Optical control unit
ODIN	Operational data interface
ODS	1. Operational debrief station
	2. Optical disk system
ODU	Optical display unit
OEM	Original equipment manufacturer
OFP	Operational flight programme
OHU	Optical head unit
OIS	Onboard information system
OMMS	Oxygen mask-mounted sight
OMS	1. Onboard maintenance system
	2. Operational management system
OMT	Onboard maintenance terminal
OOOI	Out/off/on/in
OQAR	Optical quick access recorder
OQPSK	Offset quadrature phase shift keyed
ORS	Offensive radar system
OSA	Operational support aircraft
OSC	Optical sensor converter
OSF	Optronique secteur frontal
OSU	Omega/VLF sensor unit
OTAR	Over the air re-keying
OTCIXS	Officer in tactical command information exchange subsystem
OTH(T)	Over the horizon (Target)
OTIS	Optronic tracking and identification system
OTPI	On-top position indicator
OWL	Obstacle warning ladar
OWL/D	Optical warning, location and detection
OWS	Obstacle warning system
P	Precision

P³I	Preplanned product improvement
PA	1. Pilot's associate
	2. Public address
PAC	1. Precision attitude control
	2. Public address set control
PACA/IAGSS	Precision attitude control augmentation/improved air-to-ground sight system
PACIS	Pilot aid and close-in surveillance
PAL	Phase alternation line
PAM	Pulse amplitude modulation
PAMIR	Passive airborne modular infra-red
PAR	Power analyser and recorder
PAS	Performance advisory system
PAT	Pilot access terminal
PAWS	Passive airborne warning system
PBDI	Position, bearing and distance indicator
PBIL	Projected bomb impact line
PC	1. Personal computer
	2. Printed circuit
	3. Pulse compression
	4. Photoconductive
PCB	Printed circuit board
PCM	1. Power converter module
	2. Pulse code modulation
PCMR	Probability of correct message receipt
PCSB	Pulse coded scanning beam
PCU	Pilot's control unit
PDC	Programme development card
PDC/PMM	Programme development card/ performance monitor module
PDES	Pulse Doppler elevation scan
PDNES	Pulse Doppler non-elevation scan
PDR	Programmable digital radio
PDS	Portable data store
PDU	Pilot's display unit
PE	Processing element
PENETRATE	Pass no enhanced navigation with terrain referenced avionics
PEO	Program executive officer
PEP	Peak envelope power
PERP	Peak effective radiated power
PFCS	Primary flight control (or computer) system
PFD	Primary flight display
PFD/ND	Primary flight display/navigation display
PFIS	Portable flight inspection system
PI	Process (or programme) instruction
PICC	Processor interface controller and communication
PIDS	Pylon integrated dispenser station
PILOT	Pod integrated localisation, observation, transmission
PIN	Positive-intrinsic-negative (semiconductor)
PIRATE	Passive infra-red airborne track equipment
PISA	Pilot's infra-red sighting ability
PITS	Passive identification and targeting system
PIU	1. Processor interface unit
	2. Pylon interface unit
PLAID	Precision location and identification
PLB	Personal locator beacon
PLGR	Precision lightweight GPS receiver
PLL	Phase-locked loop
PLRS	Precision location reporting system
PLS	Personnel location system
PLSS	Precision location strike system
PLU	Programme load unit
PMA	Projected map assembly
PMAWS	Passive missile approach warning system
PMF	Processeur militaire Français
PMM	Performance monitor module
PNVG	Panoramic night vision goggle
PNVS	Pilot night vision sensor
PODS	Portable data store
POET	Primed oscillator expendable transponder
PP	Parallel processing
PPI	Plan position indicator
PPM	1. Preprocessor module
	2. Programmable processing modules
pps	Pulse per second
PPS	1. Photovoltaic power system
	2. Precise positioning service (GPS)
PRF	Pulse recurrence frequency
PRI	Pulse repetition interval
PRIDE	Pulse recognition interval de-interleaving
PRNav	Precision area navigation
PROM	Programmable read-only memory
PRU	Performance reference unit
PSC	Performance seeking control
PSCS	Photographic sensor control system
PSI	Pounds per square inch

PSK	Phase shift keyed
PSP	Programmable signal processor
PSU	1. Power supply unit
	2. Power switching unit
PTC	Pack temperature controller
PTM	Pressure transducer module
PTMU	Pressure and temperature measurement unit
PTP	Programmable touch panel
PTT	Press to transmit
PULSE	Precision up-shot laser steerable equipment
PV	photovoltaic
PVD	Paravisual director
PVI	Pilot-vehicle interface
PVS	Pilot's vision system
P/Y	GPS precision code
QAR	Quick access recorder
QDM	Magnetic heading to runway
QEC	Quadrantal error correction
QPSK	Quadrature phase shift keyed
QR	Quadrant receiver
QWIP	Quantum well infra-red photodetector
RA	1. Relay assembly
	2. Resolution advisory
RADC	Rome Air Development Centre (US)
RAE	Royal Aerospace Establishment (UK)
RAF	Royal Air Force (UK)
RAID	Redundant array of independent disks
RAIM	Receiver autonomous integrity monitoring
RAM	Random access memory
RAMS	1. Racal avionics management system
	2. Removable auxiliary memory set
RAMU	Removable auxiliary memory unit
RAN/RAWS	Royal Australian Navy role adaptable weapons system
RAPPORT	Rapid alert programmed power management of radar targets
RA/TA	Resolution advisory/traffic advisory
RA/VSI	Resolution advisory/vertical speed indicator
RC	Resistance capacitance
RCS	Radar cross-section
RCU	1. Remote-control unit
	2. Rudder control unit
RDAS	Reconnaissance data annotation set
RDC	Remote data concentrator
RDI	Radar Doppler à impulsions
RDM	Radar Doppler multifunction
RDP	Range Doppler profile
RDU	Remote Display unit
REACT	Rain echo attenuation compensation technology
REU	Remote electronics unit
REWTS	Responsive electronic warfare training system
RF	Radio frequency
RFA	Royal fleet auxiliary
RFI	Radio frequency interference
RFP	Request for proposals
R/FPU	Recorder/film processor unit
RFTDL	Range-finder target designator laser
RFU	Radio frequency unit
RGB	Red, green, blue
RGS	Recovery guidance system
RHA	Recording head assembly
RHWR	Radar homing and warning receiver
RIMS	Replacement inertial measurement system
RIS	Reconnaissance interface system
RISC	Reduced instruction set computer
RLG	Ring laser gyro
RMI	1. Radio magnetic indicator
	2. Remote magnetic indicator
RMPA	Replacement maritime patrol aircraft
RMR	Remote map reader
RMS	1. Reconnaissance management system
	2. Root mean square
RNav	Area navigation
RNP/ANP	Required navigation performance/ actual navigation performance
RNS	Radar navigation system
RO	Range only
ROC	Read-out circuit
RODS	Ruggedised optical data system
ROM	Read-only memory
ROMAG	Remote map generator
RPA	Rotocraft pilot's associate
RPG	Receiver processor group
rpm	revolutions per minute
RPU	Receiver processor unit
RPV	Remotely piloted vehicle
RRU	Remote readout unit

RSC	Remote switching control	**SEOS**	Stabilised electro-optical system
RSIP	Radar system improvement programme	**SEP**	Spherical error probability
R/T	Receiver/transmitter	**SETS**	Severe environment tape platform
RTA	Receiver-transmitter-antenna	**SEU**	Sight electronics unit
RTCA	Radio Technical Commission For Aeronautics (US)	**SFCC**	Slat flap control computer
RTCA FFSC	Radio Technical Commission For Astronautics' Free Flight Select Committee (US)	**SFDP**	Smart flat-panel display
		SFDR	Standard flight data recorder
RTD	Real-time display	**SGU**	Signal generator unit
RTIC	Real-time information into cockpit	**SHARP**	1. Standard hardware acquisition and reliability programme
RTMM	Removable transport media module		2. Strapdown heading and attitude reference platform
RTT	Radio telemetry theodolite	**SHF**	Super high frequency
RTU	Radio tuning unit	**SHIP**	Software/hardware implemented partitioning
RVDT	Rotating variable differential transformer	**SHR**	Superheterodyne receiver
RVSM	Reduced vertical separation minima	**SHUD**	Smart head-up display
RWR	Radar warning receiver	**SICAS**	Secondary surveillance radar (SSR) improvements and collision avoidance systems
RWS	Range-while-search		
s	second(s)	**SID**	1. Sensor image display
SA	1. Selective availability		2. Standard instrument departure
	2. Situation awareness	**SIF**	Selective identification facility
SAAHS	Stability augmentation and attitude hold system	**SIFF**	Successor identification friend or foe
SAARU	Secondary attitude and air data reference unit	**SIGINT**	Signals intelligence
SA/AS	Selective availability/anti-spoofing	**SIMOP**	Simultaneous operation (of collocated RF sets)
SAC	Strategic air command (US)	**SIMS**	Signal identification mobile system
SACT	Signal acquisition conditioning terminal	**SINCGARS**	Single-channel ground/air radio system
SADANG	Système acoustique d'Atlantique nouvelle génération	**SIPRNET**	Secret Internet protocol router network
SAFCS	Standard automatic flight control system	**SIRFC**	Suite of integrated RF countermeasures
SAFFIRE	Synthetic aperture fully focused imaging radar equipment	**SIS**	Superhet IFM subsystem
SAFIS	Semi-automatic flight inspection system	**SIT**	Silicon intensified target
SAHRS	Standard attitude and heading reference system (Kearfott)	**SITREP**	Situation report
SAIRS	Standardised advanced infra-red sensor	**SIU**	1. Sensor interface unit
SAM	1. Situation awareness mode		2. Sidewinder interface unit
	2. Surface-to-air missile	**SKE**	Station-keeping equipment
SAMIR	Système d'alerte missile infra-rouge	**SLAM**	Standoff land attack missile
SAMSON	Special avionics mission strap-on (pod)	**SLAMMR**	Side-looking airborne modulated multimission radar
SAR	1. Search and rescue	**SLAR**	Side-looking airborne radar
	2. Signal acquisition remote	**SLIR**	Side-looking infra-red
	3. Synthetic aperture radar	**SLOS**	Stabilised long-range observation system
SARPs	Standard and recommended practices	**SMA**	Surface movement advisor
SARSAT	Search and rescue satellite aided tracking	**SMDU**	Strapdown magnetic detector unit
SASS	Small aerostat surveillance system	**SMEU**	Switchable main electronic unit
SAT	1. Situational awareness technology	**SMP**	Stores management processor
	2. Static air temperature	**SMS**	Stores management system
SATCOM	Satellite communications	**SMT**	Surface mount technology
SATURN	Second generation of anti-jam tactical UHF radios for NATO	**SMU**	Systems management unit
		SMWP	Standby master warning panel
SAU	1. Safety and arming unit	**S/N**	Signal/noise ratio stress against number of alternating load cycles to failure
	2. Signal acquisition unit		
SAW	Surface acoustic wave	**SOCUS**	South continental US
SAWS	Silent attack warning system	**SOF**	Special operations force
SBC	Single board computer	**SOJ**	Standoff jamming
SC	Single card	**SOLL**	Special operations low level
SCADC	Standard central air data computer	**SOTAS**	Standoff target acquisition system
SCAR	Sistemi de control de armamento	**SP**	Serial processing
SCAT	Speed command of attitude and thrust	**SPASYN**	Space synchro
SCAT-1	Special category 1	**SPEES**	Système pour l'élévation de l'endommagement structural
SCD	Signal command decoder	**SPEW**	small platform electronic warfare (system)
SC-DDS	Sensor control-data display set	**SPEWS**	Self-protection electronic warfare system
SCDL	Surveillance and control datalink	**SPHERIC**	System for protection of helicopters by radar and infra-red countermeasures
SCI	Serial communication interface		
SCNS	Self-contained navigation system	**SPI**	Short pulse insertion
SCP	Single card processing	**SPILS**	Spin prevention and incidence limiting system
SCS	Sidewinder control system	**SPN/GEANS**	Standard precision navigator/gimballed electrostatic aircraft navigation system
SCT	Single-channel transponder		
SCU	1. Station control unit	**SPO**	System program office (US)
	2. Stores control unit	**SPRITE**	Signal processing in the element
	3. Supplemental control unit	**SPS**	Standard positioning service (GPS)
SDC	1. Signal data computer	**SPT**	Signal processing tools
	2. Signal data converter	**SPU**	1. Signal processing unit
	3. Situation display console		2. Stores power unit
SDCS	Satellite data communications system	**SRA**	1. Shop repair assembly
SDCU	Smoke detection control unit		2. Shop replaceable assembly
SDP	Signal data processor	**SRAM**	Static random access memory
SDS	Satellite data system	**SRC**	Surveillance radar computer
SDU	1. Satellite data unit	**SRFCS**	Self-repairing flight control system
	2. Smart display unit	**SRL**	Systems research laboratories
SEAD	Suppression of enemy air defences	**SRPC**	Strain range pair counter
SEAFAC	Systems engineering avionics facility	**SRS**	1. Sonobuoy reference system
SEC	Spoiler and elevator computer		2. Superhet receiver subsystem
SECU	Spoilers electronic control unit		3. Survival radio set
SELCAL	Selective calling		4. Speed reference system
SEM	Standard electronic module	**SRU**	1. Scanner receiver unit
SEMA	1. Smart electromechanical actuator		2. Shop replaceable unit
	2. Special electronic mission aircraft	**SS**	System status
SEMC	Standard electronic memory cartridge	**SSB**	Single sideband
SENAP	Signal emulation of aero-navigation and landing	**SSCVFDR**	Solid-state combined voice and flight data recorder

SSCVR	Solid-state cockpit voice recorder
SSFDR	Solid-State flight data recorder
SSICA	Stick sensor and interface control assembly
SSID	Solid-state ice detector
SSQAR	Solid-state quick access recorder
SSR	Secondary surveillance radar
SSTI	Stabilised steerable thermal imager
SSU	Sensor surveying unit
STAIRS	Sensor technology for affordable infra-red systems
STANAG	Standard NATO agreement
STAR	1. Signal threat analysis and recognition
	2. Standard terminal arrival route
STARS	1. Small tethered aerostat relocatable system
	2. Stand-off target attack radar system
START	Solid-state angular rate transducer
STC	1. Sensitivity time control
	2. Supplemental type certificate (US)
	3. Swept time constant
STCA	Short-term conflict alert
STEVI	Sperry turbine engine vibration indicator
STIRS	Strapdown inertial reference system
STIS	Stabilised thermal imaging system
STORMS	Stores management system
STP	Status test panel
STR	Sonar transmitter/receiver
STRAP	Straight through repeater antenna performance
STS	Support and test station
STTE	Special-to-type test equipment
STU	Satellite terminal Unit
SUAWACS	Soviet union airborne warning and control system
SUM	Structural usage monitor
SWIP	Systems weapons improvement programme
T²A	Total terrain avionics
TA	Traffic advisory
TACAMO	Take charge and move out
TACAN	Tactical air navigation
TACCO	Tactical co-ordinator
TACDS	Threat adaptive countermeasures dispensing system
TACNAVMOD	Tactical navigation modification
TADIL	Tactical digital information link
TADIXS	Tactical data information exchange subsystem
TADIXS-B	Tactical data information exchange system broadcast
TADS	Target acquisition designation sight
TADS/PNVS	Target acquisition and designator set/ pilot night vision sensor
TADS/PNVS	Targeting air data system/pilot's night vision system
TAFIM	Technical architecture for information management
TAIMS	Three-axis inertial measurement system
TANS	Tactical air navigation system
TAOM	Tactical air operations module
TARPS	Tactical aircraft reconnaissance pod system
TAS	True airspeed
TAT	Total air temperature
TAWS	Terrain awareness and warning system
TBCP	Telebrief control panel
TBMS	Tactical battlefield management subsystem
TC	Thermal cue
TCA	1. Terminal control area
	2. Thermal cueing aid
TCAS	Traffic alert and collision avoidance system
TCCP	Take command control panel
TCS	Television camera set
TDI	Time delay integration
TDM	1. Tactical data modem
	2. Time division multiplex
TDMA	Time division multiple access
TDMS	Tactical data management system
TDS	Tactical data system
TDU	Test display unit
TED	Transferred electron device
TEMS	Turbine engine monitoring system
TEORS	Tactical electro-optical reconnaissance system
TERCOM	Terrain contour matching
TEREC	Tactical electronic reconnaissance
TERPROM	Terrain profile matching
TESS	Threat emitter simulator system
TEVI	Turbine engine vibration indicator
TEWS	Tactical electronic warfare system
TF	Terrain-following
TF/TA²	Terrain-following, terrain-avoidance, threat-avoidance
TFEL	Thin film electroluminescent
TFPRT	Thin film platinum-resistant thermometer
TFR	Terrain-following radar
TFT	Thin film transistor
TFTS	Terrestrial flight telecommunication system

TIALD	Thermal imaging and laser designator
TIBS	Tactical information broadcast service
TICM	Thermal imaging common module(s)
TIFS	Total in-flight simulator
TILS	Tactical instrument landing system
TINS	Thermal imaging navigation set
TIP	Technical improvement programme
TIRRS	Tornado infra-red reconnaissance system
TIS	Traffic information service
TISEO	Target identification system electro-optical
TIU	Time insertion unit
TJS	Tactical jamming system
TLR	Target locating radar
TM	Transverse magnetic
TNR	Tornado nose radar
TODS	Tactical optical disk system
TOF	Trigger-on-failure
TOPMS	Take-off performance monitoring system
TOW	Tube-launched optically tracked wire-guided (missile)
TP	Tactics planner
TPU	Transmitter processing unit
T/R	Transmitter/receiver
TRAAMS	Time reference angle of arrival measurement system
TRAC-A	Total radiation aperture control-antenna
TRANSEC	Transmission security
TRAP	Tactical related applications
TRF	Tuned radio frequency
TRIXS	Tactical reconnaissance intelligence exchange service
TRN	Terrain reference navigation
TRSB	Time reference scanning beam
TRU	Transmitter/receiver unit
TSD	Tactical situation display
TSFC	Thrust specific fuel consumption
TSO	Technical service order (US)
TSPJ	Tornado self-protection jammer
TSS	Target sighting system
TSSAM	Tri-service standoff attack missile
TTC	Tape transport cartridge
TTFF	Time to first fix
TTL	Transistor/transistor logic
TTS	Time to station
TTTS	Tanker transport training system
TTU	Triplex transducer unit
TWMS	Tactical weapons management system
TWS	Track-while-scan
TWT	Travelling wave tube
TV	Television
UAC	Universal avionics computer
UAV	Unmanned aerial vehicle
UCS	Utilities control system
UDL	The universal datalink
UFCD	Up-front control display
UFCP	Up-front control panel
UFD	Up-front display
UFDR	Universal flight data recorder
UHF	Ultra high frequency
UHU	Multipurpose escort/anti-armour variant of Tiger helicopter
UKIRCM	United Kingdom infra-red countermeasures
UK MoD	United Kingdom Ministry of Defence
ULAIDS	Universal locator airborne integrated data system
URR	Ultra reliable radar
USAF	United States Air Force
USB	Upper sideband
USN	United States Navy
USTS	UHF satellite tracking (or terminal) system
UTM	Universal transverse mercator
UV	Ultraviolet
V	Volt
VA	Visual acuity
Vl	Relative velocity of flow on the underside of an aerofoil
VADR	Voice And data recorder
VAMP	VHSIC avionics modular processor
VATS	Video augmented tracking system
VAWS	Voice alarm warning system
VBM	Volatile bulk memory
VCR	Video cassette recorder
VCS	1. Video camera system
	2. Visually coupled system
VDA	Versatile drone autopilot
VDL	VHF (very high frequency) datalink
VDM	Visual display module
VDU	Visual display unit
VEMD	Vehicle and engine management display
VFR	Visual flight rules
VG	Vertical gyro

VGG	Visual graphics generator
V/H	Velocity/height (ratio)
VHF	Very high frequency
VHSIC	Very high-speed integrated circuit
VID	Virtual image display
VIEWS	Vibration indicator engine warning system
VIGIL	Vinten integrated infra-red linescan
VISTA	Variable stability in-flight simulator test aircraft
VITS	Video image tracking system
VLA	1. Vertical line-array
	2. Very large array
VLAD	Vertical line-array DIFAR
VLC	Very low clearance
VLF	Very low frequency
VLLR	Very light laser range-finder
VLSI	Very large scale integration
VMO	Maximum permitted operating speed
VMS	Vehicle management system
VMU	1. Velocity measuring unit
	2. Voice message unit
Vnav	Vertical navigation
VNE	Never to be exceeded speed
Vocodor	Voice encoder/decoder
VOGAD	Voice-operated gain adjustment device
VOR	VHF omnidirectional range
VOR/Loc	VHF omnidirectional range/locator
VORTac	VHF omnidirectional range/tacan
VOS	Voice-operated switch
Vox	Voice keying or activation
VPU	Voice processor unit
VPU(D)	Voice processor unit (with data mode)
Vr	Radial velocity
VRD	Virtual retinal display
Vref	Typical approach speed
VRU	Vertical reference unit
Vs	Stalling speed
VSD	Video symbology display
VSI	Vertical speed indicator
VSRA	Vertical/short take-off and landing research aircraft

V/STOL	Vertical and short take-off and landing
VSVA	Variable stator vane actuator
VSW	Verification software
VSWR	Voltage standing-wave ratio
VTA	Voice terrain advisory
VTAS	Visual target acquisition system
VTM	Voltage-tuneable magnetron
VTO	Volumetric top off
VTOL	Vertical take-off and landing
VTR	Videotape recorder
V/UHF	Very/ultra high frequency
W	Watt
WAAS	Wide area augmentation system
WAC	Weapon aiming computer
WACCS	Warning and caution computer system
WAGS	Windshear alert and guidance system
WAN	Wide area network
WASP	Weasel attack signal processor
WBC	Weight and balance computer
WBS	Weight and balance system
WCCS	Wireless communication and control system
WCMS	Weapons control and management system
WCP	Weapon control panel
WDA	Weather display adapter
WDC	Weapon delivery computer
WDIP	Weapon data input panel
WDNS	Weapon delivery and navigation system
WEAC	Weapons computer
WFoV	Wide field of view
WIU	Weapon interface unit
WNC	Weapons and navigation computer
WPU	Weapon processing unit
WRA	Weapon-replaceable assembly
WSIP	Weapons system improvement programme
WSW	Windshear warning
WSW/RGS	Windshear warning/recovery guidance system
WTC	Windshield temperature controller
YIG	Yttrium indium garnet
ZTC	Zone temperature controller

The Electromagnetic Spectrum

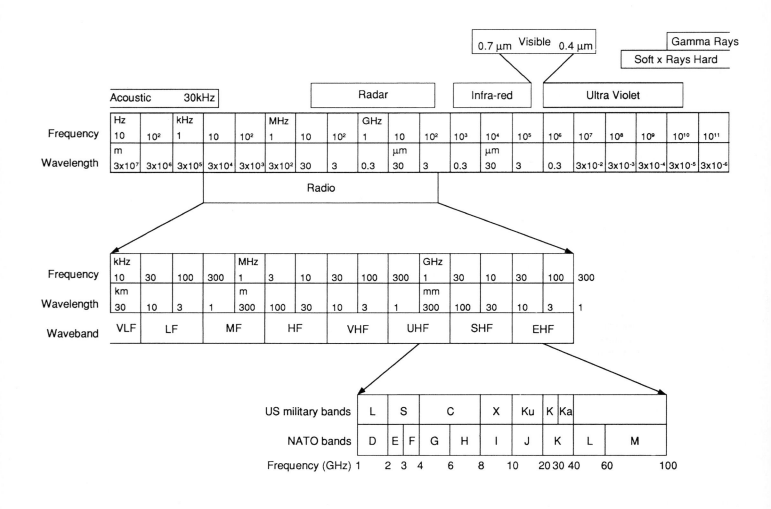

The Joint Electronics Type Designation System (JETDS)

The Joint Electronic Type Designation System (JETDS) is an unclassified US military system designed to identify, by a series of numbers and letters, the use of an equipment and its major components. The system was formerly known as the AN system. A typical example of the system would be:

AN/FPS-5A

In this example the prefix AN indicates that the type number has been assigned in the JETDS system. It does not necessarily mean that the army, navy or air force uses the equipment.

The next letter (F) gives the type of installation where the equipment is used such as fixed, mobile, shipborne and so on. See Table A for a complete listing of installations. The following letter (P) indicates the type of equipment (radar, teletype, radio and so on). See Table B for a complete listing of equipment types. The letter before the dash (S) gives the purpose of the equipment such as transmitting, receiving, detecting and so on. See Table C for a complete listing of equipment purposes. The number after the dash (5) is the model number. The final letter (A) gives notification when a model has been modified but can be interchanged with, or substituted for, the equipment in its original state. If the modified system cannot be interchanged with, or substituted for the original, a new type designation is assigned.

Table A Installation

A	Piloted aircraft
B	Underwater mobile, submarine
C	Air transportable (inactive)
D	Pilotless carrier
F	Fixed, ground
G	Ground, general
K	Amphibious
M	Ground, mobile
P	Pack or portable
S	Water surface craft
T	Ground, transportable
U	General utility
V	Ground, vehicular
W	Water, surface and underwater
Z	Piloted and pilotless vehicle combination

Table B Type of Equipment

A	Infrared
B	Pidgeon (inactive)
C	Carrier (wire)
D	Radiac
E	Nupac (inactive)
F	Photographic
G	Telegraph or teletype
I	Interphone and public address
J	Electromechanical
K	Telemetering
L	Countermeasures
M	Meteorological
N	Sound in air
P	Radar
Q	Sonar and underwater sound
R	Radio
S	Special types or combinations of types
T	Telephone (wire)
V	Visual and visible light
W	Armament
X	Facsimile or television
Y	Data processing

Table C Purpose of Equipment

A	Auxiliary assemblies (inactive)
B	Bombing
C	Communications
D	Direction-finder or reconnaissance/surveillance
E	Ejection or release
G	Fire control or searchlight directing
H	Recording or reproducing
K	Computing
L	Searchlight control (inactive)
M	Maintenance and test assemblies
N	Navigation aids
P	Reproducing (inactive)
Q	Special, or combination of purposes
R	Receiving, passive detecting
S	Detecting, or range and bearing search
T	Transmitting
W	Control (automatic flight control or remote control)
X	Identification and recognition
Y	Surveillance (search, detect and tracking) and control (fire control, air control)

ANALYSIS

Digital datalinks
Airborne radar
Electro-optic systems
Helmet-mounted display systems

DIGITAL DATALINKS

Figure 1 – What price SA? A stealthy F-22 Raptor performs an undetected launch of an AIM-120A at long range, with inevitable results   0581543

Digital Datalinks

Introduction

As every fighter pilot knows, the major factor in determining the outcome of any modern air-to-air engagement is not the agility of his aircraft or the range of his weapons; it is his ability to acquire and maintain better Situational Awareness (SA) than his opponents throughout the combat. But what is SA? In terms of the aforementioned combat scenario, it is simply the pilot's ability maintain a 'mental air picture', including the positions and vectors of all participants, including friendly forces as well as bogeys. In the mêlée of modern air combat, this is not as easy as it might sound; the majority of air-to-air kills to date have been as a result of an unseen attack – perhaps the most severe example of a disparity in SA! While it is fair to say that, as aircraft onboard sensor systems become more sophisticated, this ratio is reducing, the advent of stealth technology (delaying detection by the opponent's radar) and longer range (possibly launched beyond the effective range of onboard sensors), extremely agile (high off-boresight capability, slaved to helmet-mounted aiming systems) air-to-air missiles has, however, once again raised the distinct possibility of being killed by a completely undetected opponent.

How, then, can the modern pilot, equipped with limited-range onboard sensors, hope to maintain an effective 'mental air picture' when many of the major protagonists are out of sensor range? The solution lies in datalink technology – the ability to transmit and receive large amounts of data, including imagery, between aircraft, surface platforms and ground stations, thus 'sharing' sensor data and effectively 'pooling' fragments of an 'air picture' to produce a definitive and accurate version, to the tactical advantage of all participants.

This overview of digital datalinks is intended to provide technical information to potential users of datalink systems, and also to provide useful background information to engineers and technicians working in related areas.

Datalinks

There has long been a requirement for interoperable data communications for fighter aircraft, in order to exchange elements of the Common Operational Picture (COP). Today, the majority of US and NATO allies' fighters still communicate using insecure analogue radios that provide only interactive voice communications. This severely limits their ability to share a wide range of combat data, in addition to voice, over a secure, jam-resistant communications network. Communications systems that include datalink capabilities offer a near-term solution for exchanging digital data over a common network that is continuously and automatically updated. Precise data can be sent faster and more reliably via direct digital (that is, computer-to-computer) communications. In addition, text messages need only a small fraction of the communications resources that interactive voice messages require and can be delivered much more reliably than voice in high-stress combat conditions. Moreover, digital modulation offers

many advantages over analogue modulation, including the ability to send data, encrypt voice or data, and the employment of error detection and correction coding. This increases the reliability and quality of transmissions over channels affected by noise, interference, or fading. Also, depending on the digital modulation scheme used, such systems provide a low probability of detection or resistance to jamming.

Several communications systems have been developed to support datalink communications, or the near-realtime exchange of data among tactical data systems. Each such system is specified by hardware/software characteristics (for example, waveform, modulation, data rates, transmission media, and so on) as well as by message and protocol standards. The most recent is the Joint Tactical Information Distribution System/TActical Digital Information Link-J (JTIDS/TADIL-J) system, which is commonly referred to as Link 16. Link 16 is an encrypted, jam-resistant, nodeless tactical digital datalink network, established by JTIDS-compatible communication terminals that transmit and receive data messages in the TADIL-J message catalogue. Link 16 data communications standards and technology were developed in the US JTIDS program which began in 1975. The first JTIDS terminals, or Class 1 terminals, were large and installed only on AWACS aircraft and at US, UK, and NATO ground-control facilities. Smaller JTIDS terminals (Class 2) were also developed. However, because of their high cost, relatively large size and reliability issues, only a limited number of these terminals were procured to equip US fighters, specifically US Navy F-14Ds and a single squadron of US Air Force F-15Cs. The Multifunction Information Distribution System (MIDS) program was created to put small, lightweight Link 16 terminals on US and allied

fighter aircraft. MIDS is a major international program led by the United States.

The digital datalinks that have been operated throughout NATO prior to the introduction of JTIDS are of three types:

- **Point-to-Point.** Point-to-point connections imply that each participant is linked directly to only one other participant and that the units are, in general, non-mobile
- **Broadcast.** One unit only transmits (broadcasts) its data for all other participants to receive
- **Netted.** Several units exchange data with each other on a flexible network.

Link data can be transmitted over high-grade telephone lines, UHF, HF or SATCOM using the following modes of operation:

- **Simplex.** Information can only be passed one way
- **Duplex.** Participants transmit and receive information simultaneously
- **Semi-duplex.** Participants transmit and receive data but not simultaneously.

Message Formats and Standardisation

Digital Datalinks are capable of passing several thousand, separate, identifiable pieces of data which can be slotted into various categories and subcategories. In tactical operations data can be grouped under the following headings:

- **Picture compilation.** Picture compilation can be subdivided into air, surface, sub-surface and electronic warfare
- **Mission management.** Data used for air and sea co-ordination
- **Command and Control.** Data for the tactical control of assigned weapons or units.

To enable computers within a Tactical Datalink System (TDLS) to communicate and interpret items of information correctly, messages are compiled and transmitted in previously agreed

Figure 2 – Some US Air Force F-15Cs are equipped with JTIDS Class 2 terminals (US Air Force)
0524964

formats. These formats are specific to the datalinks in which they are used; for example, S-series for Link 1 and M-series for Link 11.

DDL System Descriptions and Operational Documents

Major datalink interoperability problems are overcome by the use of NATO Standardisation Agreements (STANAGs). These agreements provide specifications for the automatic data exchange of tactical information between NATO tactical data systems; for example, STANAG 5501 for Link 1 operations. In addition to the STANAGs, operational procedures for all the links currently in service are defined by a group of documents called the Allied Data Publications (ADatPs). Again, these are numbered in accordance with the particular datalink; for example, ADatP 11 for Link 11.

DATALINK SYSTEMS
Link 1

NATO Link 1 was designed in the late 1950s as a point-to-point, digital datalink employed for the automatic data exchange of real-time tactical air defence and control information between land-based air defence and aircraft control units. NATO Link 1 can be used as a one-way (simplex) or two-way (duplex) link.

Operational Characteristics

Communications medium. NATO Link 1 is normally transmitted over landlines (telephone), but can be transmitted over multichannel radios. Multichannel radios may be established within LOS, use troposcatter techniques for over-the-horizon communications, or employ satellite communications. Both ends of the link must use a compatible communications system. Point-to-point (line) encryption devices are not normally used with NATO Link 1, though standards are provided in Annex A of STANAG 5501. US Marine Corps and US Air Force Tactical Data Systems (TDSs) and US Navy buffers use Differential Frequency Shift Keying (DFSK), with transmission design 1316 modulator/demodulator (modems) to interface on NATO Link 1.

Communications architecture. NATO Link 1 units are normally fixed land-based centres that are assigned a geographic area of responsibility called an Area of Operational Interest (AOI). Within the AOI, a Track Production Area (TPA) is established. Information exchange is between two adjacent centres for the purpose of cross-telling tracks that transit from one AOI to another. To prevent TPA overlap, a Track Continuity Area (TCA) is established. A TCA is a belt on either side of the common boundary of two TPAs of adjacent centres, or some other area, positioned to ensure

Figure 3 – Airborne Early Warning (AEW) aircraft, such as this US Air Force E-3A, are integrated into NATO Link 1 (US Air Force) 0524969

that continuity of tracking is established. All tracks within the TCA are automatically transmitted (cross-told) with the exception of tracks being cross-told in the opposite direction and tracks whose cross-telling is inhibited either locally or by request message.

NATO Link 1 communications require a separate, dedicated link between each Radar Unit (RU), which simultaneously transmits and receives data on two channels – one dedicated transmit and one dedicated receive line for duplex operations. When data messages are not being transmitted, each unit transmits a continuous standby signal (idling period) to maintain time synchronisation. Land-based US Air Force and US Marine Corps TDSs have a NATO Link 1 capability and can link directly with a NATO site. US Navy mobile TDSs (ships and aircraft) operate on Link 11, Link 16, and Link 14 and use a ground based buffer to forward data to NATO Link 1 or Link 11B.

Airborne Early Warning (AEW) aircraft are integrated into NATO Link 1 through a data forwarder and have AEW ground target areas designated, in which air defence information is forwarded from a ground air defence site to the AEW aircraft.

Data Encryption. NATO Link 1 data is not normally encrypted and is exchanged between NATO centres over dedicated telephone lines. Due to this limitation, US systems that forward Link 11, Link 11B, or Link 16 data to NATO Link 1 are required

to activate a Special Processing Indicator (SPI) filter to ensure classified data is not compromised when forwarded to NATO Link 1.

Data Rates. NATO Link 1 is operated at a basic speed of 1200 bps with alternate speeds of 600 and 2400 bps using a DFSK modulated serial bit stream with a synchronous transmission mode. Phase shift keying may be used by mutual agreement between two adjacent sites.

Message Standard. NATO Link 1 uses S-series messages and transmits data in pairs as a single 98-bit data message. Messages fall into various functional groups:
- Air surveillance
- Strobe information
- Management
- Test
- Frame filler.

Information Exchange Scheme. NATO Link 1 units exchange positional information on strobes, tracks, and points by transmitting X and Y co-ordinates. Measurements are made with respect to a system co-ordinate centre, agreed by the transmitting site and the receiving site, if authorised by the proper national configuration manager. The transmitting site transmits information referenced to the system co-ordinate centre of the specific Link 1 interface. Normally, a NATO Link 1 site uses the centre of its Data Storage Area (DSA) as the co-ordinate centre.

NATO Link 1 Capabilities

Nation	System	Tactical Data System	Link Capabilities
Belgium	NATO Air Defence Ground Environment	Sector Operations Centre	
France	Système de Traîtement et Représentation des Informations de Défence	AWACS	Link 1, Link 11, AEGIS
Germany	German Air Defence Ground Environment	Sector Operations Centre	Link 1, Link 11, Link 11B, AEGIS
Iceland	Icelandic Air Defence System	Sector Operations Centre	Link 1, Link 11
Italy	System for Point Air Defence		Link 1, Link 11
Norway	NATO Air Defence Ground Environment	Sector Operations Centre	
Netherlands		Sector Operations Centre	Link 1, Link 11
Portugal	Portuguese Air Command and Control System	Sector Operations Centre	
Spain	Spanish Air Defence System	Sector Operations Centre	
UK	Improved United Kingdom Air Defence Ground Environment	Sector Operations Centre	Link 1, Link 11, Link 16, United Kingdom AEGIS
US	Joint Tactical Air Operations	USAF-AOC	TADIL-A, TADIL-B,
		AN/TYQ-23	NATO Link 1 TADIL-A, TADIL-B,
		USAF- CRC/CRE	TADIL-C, ATDL-1, NATO Link 1,
		AN/TYQ-51	TADIL-A, TADIL-B,
		USMC-TACC	NATO Link 1 TADIL-A,
		AN/TYQ-23	TADIL-B, TADIL-C, ATDL-1,
		USMC-TAOC	NATO Link 1 TADIL-A,
		USN- Ship Shore	NATO Link 1
		Ship Buffer	An ADSI is used in the USAFE CAOC at Vicenza, Italy. They receive an IJMS feed from an AF CRC

Figure 4 – NATO Link 1 capabilities and the equipment

Figure 5 – US Navy F-14 Tomcats are fitted with Link 4 to support the Automatic Carrier Landing System (ACLS) (US Navy) 0524971

A DSA is the maximum area within which an automated air defence facility is able to store tracks in its computers. The DSAs of adjacent air defence facilities must overlap one another.

Track Data. Track data is assigned priority for transmission (telling).

NATO Track Numbers (NTNs). NTNs consist of two alphanumeric characters and three octal numbers. The alphanumeric characters are assigned from A, E, G, H, J, K, L and M, and the three-digit octal numbers range from 001 to 777. When tracks with non-NTNs are forwarded to NATO Link 1, the forwarding unit or buffer is normally assigned a TN block using the prefix AA. This TN prefix indicates to the NATO Link 1 units that the track source is from a data forwarder.

System Employment. NATO Link 1 is used by the US NAVY, Marine Corps, Air Force and NATO centres. NATO centres include the NATO Air Defence Ground Environment (NADGE), the United Kingdom Air Defence Ground Environment (UKADGE), the Spanish Air Defence System, the Italian System for Point Air Defence and the German Air Defence Ground Environment sites. NATO Link 1 capabilities and the equipment of each US service system are shown in Figure 4.

NATO Link 1 Operation. Each unit reporting on NATO Link 1 is assigned a unique address. A unit operating on multiple NATO Link 1 interfaces uses the same address for all NATO Link 1 links. Units operating on a single NATO Link 1 are designated RUs. Units forwarding between Link 11, Link 11B, or Link 16 and NATO Link 1 are designated forwarding RUs. Additionally, the USN employs a land-based buffer that forwards between Link 11 or Link 16 and NATO Link 1 or Link 11B. Buffer units forward data to and from NATO Link 1 and are designated FPUs in US terminology. The TN assigned to a track by a NATO unit specifies track origin as the two-character alphabetical code.

Link 4/TADIL-C

Link 4 is a non-secure datalink used for providing vector commands to fighters. There are two variants: Link 4A and Link 4C. Link 4A uses a command-and-response protocol and the principle of Time Division Multiplexing (TDM) to derive apparently simultaneous channels from a given frequency spectrum. It connects two points by assigning a sequence of discrete time intervals to each of the individual channels. Thus, one controller can control multiple aircraft independently on the same frequency. At any given time, a Link 4 unit is transmitting, receiving, or idle on a single point-to-point circuit. Link 4C is a fighter-to-fighter datalink.

Link 4A

Link 4A is a controller-to-aircraft datalink which uses V-series messages and R-series messages to support the Automatic Carrier Landing System (ACLS), Air Traffic Control (ATC), Air Intercept Control (AIC), strike control, the Ground Control Bombing System (GCBS) and the Carrier Aircraft Inertial Navigation System (CAINS).

The system has a limited data throughput, no ECM resistance and can cater for a limited number of participants (up to eight). It was, however, the first datalink to be integrated with fighters to provide a digital C² unit-to-fighter communication path. A 56-bit control message is transmitted every 32 msec, for an effective one-way tactical data rate of 1,750 bps. The effective two-way tactical data rate is slightly better than 3,000 bps.

Link 4 has been fitted to US Navy ships, US Airborne Early Warning and Control (AEWC) aircraft and several fighters, fighter/bombers and supporting platforms. Since NATO E-3s, UK E-3Ds and French E-3Fs are based on the US E-3, they also implement Link 4, although they do not use it operationally. However, the French E-3F does employ the system and the French Ground Environment (GE) has been equipped with a capability to forward Link 4 messages.

Link 4C

Link 4C is a fighter-to-fighter datalink which is intended to complement Link 4A, although the two links do not communicate directly with each other. Link 4C uses F-series messages and provides some measure of ECM resistance. Link 4C is fitted to the F-14 only; the aircraft cannot communicate on Link 4A and 4C simultaneously. Up to four fighters may participate in a single Link 4C net.

V-Series Messages

V.0A	Addressed dummy message
V.0B	Unaddressed dummy message
V.1	Target data exchange message (Navy)
V.2	Aircraft vectoring message
V.3	Vectoring and close control message
V.5	Traffic control message
V.6	Automatic landing control message
V.9	Target information message
V.18	Precision direction final message
V.19	Precision direction initial message
V.31	Inertial navigation system alignment message
V.3121	Strike control message
MCM-1	Test message
MCM-2	Test message
UTM-3A	Universal test message
UTM-3B	Universal test message

R-Series Messages

R.0	Aircraft reply message
R.1	Aircraft reply message
R.3A	Reply message (tactical data)
R.3B	Reply message (position report)
R.3C	Reply Message (target velocity report)

It is planned that Link 16 will assume the role of Link 4A in AIC and ATC operations, and the role of Link 4C in fighter-to-fighter operations. However, Link 16 is not currently capable of replacing Link 4A's ACLS function, and it is likely that controlled aircraft will remain equipped with Link 4A to perform carrier landings.

Message standards are defined in STANAG 5504 while standard operating procedures are laid down in ADatP 4.

Link 11/TADIL-A

Link 11 (also known as TADIL-A in the US) is a medium-speed, NATO-standard, HF/UHF tactical data information link. It uses netted communication techniques and standard message formats for the exchange of digital information between maritime, air- and land-based units. Data is exchanged using the Conventional Link Eleven Waveform (CLEW), over a differential quadrature phase-shift keying modulated datalink, operating at a rate of 1,364 (HF/UHF) or 2,250 (UHF) bps.

Link 11 is designed for operation on High-Frequency (HF) ground wave and thus has a Beyond Line-Of-Sight (BLOS) capability (to a theoretical range of approximately 300 n miles). Link 11 can also operate in the UHF band but is then limited to LOS ranges (approximately 25 n miles surface-to-surface or 150 n miles surface-to-air). As an alternative, satellite and fibre optic links can be used. Units that exchange data via Link 11 are designated Participating Units (PUs) or Forwarding Participating Units (FPUs).

Link 11 is based on 1960s technology and is therefore a relatively slow link that normally operates on a polling system, with a Net Control Station (NCS) polling each participant in turn for their data. In addition to this 'Roll Call' mode, Link 11 may be operated in broadcast modes in which a single data transmission, or a series of single transmissions, is made by one participant. Link 11 is, therefore, a half-duplex link which is secure but not ECM resistant.

Functions

Link 11 is capable of exchanging data and carrying out Command and Control (C²) functions in the following areas:
- Picture compilation, including surface tracks, air tracks, subsurface tracks, special points, ESM/ECM detections and IFF/SIF information
- Track management which includes: identity/ environment difference reports, identity/ environment change reports, data update requests, track alerts, drop track reports, cease reporting orders, track pairing and association, track correlation and pointing
- Transfer of aircraft control and EW control and co-ordination
- Aircraft status, ASW aircraft status and weapon and engagement status
- Command, which includes weapon engagement orders and duties and procedures.

Frequency Spectrum and Range. Link 11 operates over conventional radio transmitters and receivers in the HF 2–30 MHz and UHF 225–400 MHz bands. When operating in the HF band, it is designed to provide omnidirectional gapless coverage in line with other HF transmissions. In the UHF band it has the same range constraints as with all UHF communications. SATCAT airborne relay extends UHF range markedly.

Data Format. Data is transmitted in two 30-bit frames, arranged in a number of pre-determined Link Message Formats (LMFs), with each frame consisting of 24 information bits and six error detection and correction bits. Two such frames, in parallel, constitute a Link 11 message.

Net Speed. The Link Terminal equipment operates at data speeds of 2,250 or 1,364L/1,364S bps, which includes the EDC and Doppler correction bits. The Link terminal equipment incorporates a parity-type error detection and correction system that can detect and sometimes correct errors in the received data frames that have been caused by propagation conditions or by equipment malfunction.

Encryption. The Link data transmissions are encrypted using the United States Navy KG 40 equipment. Link 11 voice communications can be encrypted by a Bid 250.

Roles

A station on a Link 11 net operates in one of the following roles:
- **Net Control Station (NCS).** Only one unit per net can operate as the NCS. The NCS is responsible for controlling, monitoring the frequency and analysing the net. The NCS is the unit that polls the other net participants.
- **Participating Unit (PU).** All other netted units are termed PUs, also known as Pickets.

Figure 6 – UK Royal Air Force Nimrod Maritime Patrol Aircraft (MPA) employ Link 11

Modes of Operation
Link 11 has three modes of operation:

Roll Call. Roll Call is the mode of operation when the NCS interrogates each PU in sequence, with the PUs transmitting their data in response to this call. There are three subtypes of this mode as follows:
- **Full Roll Call.** In Full Roll Call, all units are active on the net and all PUs respond to each call by the NCS with their data
- **Partial Roll Call.** In Partial Roll Call, some units are switched to a radio silence mode (Silence) and do not respond when interrogated by the NCS. If a unit observing radio silence in this way needs to make a report on the Link, he will switch to become an active picket, pass his data in the next cycle when interrogated and then return to Silence
- **Roll Call Broadcast.** In Roll Call Broadcast, all units except the NCS are in Silence. The NCS transmits all its data, but when the other PUs are interrogated sequentially they do not respond. A unit in Silence having raid data to report does so as in Partial Roll Call.

Broadcast. In Broadcast Mode, the NCS broadcasts repeated transmissions of its data. Other PUs are in Silence, are not interrogated and have no capability to report. If the OTC requires a broadcast mode of operation but wishes to retain the capability for some units to make reports when necessary, or to issue/reply to commands, the Roll Call Broadcast mode must be used; however, the possibility of net capture must be considered.

Silence. In Silence Mode, all units are in Radio Silence. If a unit needs to initiate a report, he will do so by making a single transmission (short broadcast) to all units.

Standard Link Messages
Standard Link messages are arranged in a series of numbered and predetermined formats. The messages are of two kinds, basic and amplifying. Basic messages are numbered in the series 1 to 15 (excluding 8), sometimes with a suffix letter (for example, Message 4A). Amplifying messages are all numbered 8, with an additional digit corresponding to the number of the basic message it amplifies (for example, Message 83 amplifies a Message 3). For ease of reference, message designations are abbreviated to M2, M82, M84C, and so on. Figure 3 lists M-series and amplifying messages.

Link 11 Procedures
DataLink Reference Point (DLRP). A DLRP is a geographic position, given in latitude and longitude, from which each PU calculates and reports other System Co-ordinate Centres (SCCs) expressed as X and Y from the DLRP. The DLRP is designated ahead of time and must be manually injected into the system before Link 11 operations can begin. The difference between own position and that of the DLRP is transmitted in the M1 message.

System Co-ordinate Centre (SCC). An SCC is a PU's datum position, to which track reports that follow are referenced. Systems may use a variety of grids and co-ordinate transformation algorithms, but on Link 11, positions are expressed as X and Y from the SCC. Thus, no matter which algorithm is being used, a PU system must convert the position into X and Y co-ordinates from its own established datum before the information can be transmitted in a Link 11 message.

Picture Matching. An essential element in data exchange is the ability for a unit to match data received from other participants with that produced by its own sensors. Various errors, but particularly navigational errors, cause the received data to be displaced. The process applied in picture matching is known as Registration.

Registration. This is used by GE sites to correct an AEW picture so that it is aligned to the GE's track base.

Gridlock. Gridlock is used by mobile units to correct any error between their true position and the picture held by their system. A unit with the best navigational aid or a fixed site is designated the Grid Reference Unit (GRU). Gridlock can be carried out in two ways:
- **Manual.** Provides simple matching of a single track reported by the GRU (M1) and the local radar contact (ICCS report). Only positional correction is applied
- **Automatic.** Compares the positions of up to eight pairs of GRU reported tracks with the equivalent local tracks. The average error of all tracks is taken into consideration and a positional and rotational correction is applied.

Correlation. Correlation is used in Link 11 operations to indicate that two tracks that are currently being reported on the Link are in fact the same contact. Link correlation, which can be either manually or automatically initiated, compares a received track with its own system track, taking into account position, course, speed, height, identity, IFF, and so on. When the checks are complete, the system will either allow or reject a correlation and will also initiate appropriate merging action. Decorrelation will be carried out when the correlated tracks fail the system checks that are carried out on a routine basis after correlation has been initiated.

Track Quality (TQ). TQ is a set of numbers from 0 through to 7, which indicates the system's estimate of the accuracy of a track's reported position. A TQ equaling 0 indicates a non-realtime track and a TQ of 1 to 7 indicates a real-time track with increasing levels of accuracy. The main purpose of TQ is to define who has Reporting Responsibility (R^2) for a track which can be seen by two or more units.

Reporting Responsibility (R^2). Reporting Responsibility is part of the design philosophy of Link 11 which ensures tracks are reported by the unit that has the most accurate and up-to-date information. Since TQ gives a measure of such information, it is used to determine who should have R^2. Therefore, a PU with real-time information on a track will take R^2 from a unit with non-realtime information on the same track. A unit having a TQ of 2 greater than another unit (on the same track) will assume R^2.

Within the UK, Link 11 is employed by the Royal Navy, Royal Marines and Royal Air Force in its Ships, Ship-Shore-Ship Buffers (SSSBs), E-3D AEW, Nimrod MPA, Tactical Air Control Centre (TACC), and so on. Within NATO, Link 11 is primarily used as a maritime datalink.

Link 11B/TADIL-B
Link 11B employs a dedicated, point-to-point, full-duplex digital datalink, using serial transmission frame characteristics and standard message formats transmitted by individual signal elements or binary digits on a time sequential basis. Data is exchanged over a fully automatic, phase-continuous, full-duplex, frequency shift-modulated datalink, operating at a standard rate of 1,200 bps with optional capabilities of 600 and 2,400 bps (or multiples of 1,200 bps; for example, 3,600, 4,800, and so on). Units which exchange data via Link 11B are designated Reporting Units (RUs) or Forwarding Reporting Units (FRUs).

Within the UK, Link 11B will be employed by the TACC, the second generation SSSBs and for ground-to-ground communications with the Iceland Air Defence System (IADS). Within NATO Link 11B is used to integrate Ground Based SAM Command and Control and Fire Distribution

Centres into the Air Defence Ground Environment using CRC SAM Interfaces (CSI). Within the US, and some other NATO Nations (for example, France), Link 11B is used as the primary datalink for ground based TACS (for example, USAF MCE, and USMC TAOC).

Message standards for both Link 11 and Link 11B are defined in STANAG 5511 while standard operating procedures are laid down in ADatP 11.

M-Series Messages

Message	Purpose
M.0	Test message
M.1	Data reference position message
M.81	Data reference position amplify message
M.2/82	Air track position/air position amplify message
M.3/83	Surface track position/surface position amplify message
M.4A thru D	Anti-submarine warfare primary/secondary message
M.84A/C/D	ASW amplification message
M.5/85	Special points position/special points amplify message
M.6A	ECM intercept data message
M.6B	ESM primary message
M.86B	ESM amplify message
M.6C	ESM parametric message
M.86C	ESM parametric amplify message
M.6D	Electronic warfare co-ordination and control message
M.86D	Electronic warfare co-ordination and control amplify message
M.7A/B	Theatre Missile Defence 2
M.87 A/B	Theatre Missile Defence amplify message 2
M.9A0	Information management message (data source and sim)
M.9A1	Information management message (ID difference)
M.9A2	Information management message (change data order)
M.9A3	Information management message (data update request)
M.9A4	Information management message (drop track)
M.9A5	Information management message (track alert)
M.9A6	Information management message (controlling unit report)
M.9A7	Information management message (terminate track alert)
M.9A9	Information management message (IFF management message)
M.9B	Information management message (pairing)
M.9C	Information management message (pointer)
M.9D	Information management message (Link 11 monitor)
M.9E	Information management message (supporting information)
M.9F/89F	Area of probability message
M.9G	Datalink reference point message
M.10A	Aircraft control message
M.11B	Aircraft status message
M.11C	ASW aircraft status message
M.11D	IFF/SIF message
M.11M	Intelligence information message
M.811M	Intelligence amplification message
M.12.31	Timing message
M.13	Worldwide national message
M.14	Weapon/engagement status message
M.15	Command message

Note: M-series messages support both Link 11 and Link 11B

Figure 7 – M-series messages

Army Tactical Datalink-1 (ATDL-1)
Army Tactical Datalink-1 (ATDL-1) is a dedicated, point-to-point, duplex, digital datalink employed to exchange real-time tactical data between a C^2 land unit and a Surface-to-Air Missile (SAM) system, or between multiple SAM systems. Units that exchange data via ATDL-1 are designated Supporting Units (SUs).

Operational Characteristics

Communications Medium. ATDL-1 can be transmitted over landlines (wire or fibre optic cable), single-channel radios, or multichannel radios. Multichannel radios may be established within LOS, use troposcatter techniques for over-the-horizon communications, or employ satellitecommunications. Both ends of the ATDL-1 link must use a compatible communications system. US forces employ two basic communications systems, the Direct Current (DC) digital form and the quasi-analogue form:

- The US Marine Corps Tactical Air Operations Centre (TAOC) and Battery Command Post (BCP), the Army Adaptable Surface Interface Terminal (ASIT) and joint TADIL-A distribution system and the Air Force Control and Reporting Centre (CRC) use the quasi-analogue Transmission Design (TD), which employs TD 1089 or TD 1316 modulator/demodulator (modems)
- US Air Defence Artillery (ADA) C² systems operate using a quasi-analogue form. Transmission and reception of ATDL-1 data, however, is accomplished using DC digital form and bulk encryption. Data conversion is via an air defence interface to render analogue signals to digital (or vice versa) for transmission and receipt over the Mobile Subscriber Equipment (MSE) net. Data transmissions are bulk encrypted by the MSE.

Communications Architecture. SAM units that use ATDL-1 can be linked together to form a chain (round-robin) or use a single forwarding unit as a hub, with multiple units tied through the hub. The round-robin mode permits data to be circulated within three Hawk batteries using ATDL-1 formatted messages. However, this is used only in the absence of higher echelons. In the normal mode, ATDL-1 communications require a separate, dedicated link between each SU that simultaneously transmits and receives data on two channels – one dedicated transmit line and one dedicated receive line. When data messages are not being transmitted, each unit transmits a continuous standby signal to maintain time synchronisation.

Data Encryption. ATDL-1 is normally operated in the secure mode using encryption devices provided by the tactical unit. The configuration of encryption devices at both ends of the ATDL-1 links must be compatible and may be accomplished by means of 'strapping' (physical) or 'initialisation' (electrical). The configuration means used are dependent on the Tactical Data System (TDS) configuration. When MSE communications are used to provide the ATDL-1 link, secure mode operations are accomplished through bulk encryption of data transmissions by the MSE nodes. NATO or allied ADA C² units provide the necessary equipment for data compatibility with US Air Force or US Marine Corps C² facilities. When ATDL-1 is transmitted over landlines (wire or fibre optic cable), single-channel radios or multichannel radios, then KG-30 (analogue) or KG-84 (digital) single-channel data encryption devices are used.

Data Rates. ATDL-1 can be transmitted at 600, 1,200, or 2,400 bps. If the quasi-analogue transmission scheme is used, audio signals are modulated using frequency shift keying.

Message Standard. ATDL-1 uses B-series messages and transmits data in an 81-bit message frame, which consists of a start group, seven data groups and a check group. ATDL-1 messages are defined in MIL-STD-6013.

Co-ordinate Reference Convention. ATDL-1 units exchange positional information on units, tracks, strobes, and points by transmitting an X and a Y co-ordinate, measured from the SU's system co-ordinate centre (SCC). An SU's location is reported to other interfaced units by transmitting its X and Y co-ordinates measured from the DLRP. Once the position of an SU is known, all positional information received from that SU is displayed in relation to the SCC.

Transmission Structure

Test Message. Once the ATDL-1 link has been initialised, a test message is transmitted. The continuing operational status of the link depends on the exchange of the B-series ATDL-1 test message. This is a special message that is transmitted on a periodic basis by both ATDL-1 TDSs. Failure to receive this message for a specific period of time causes the ATDL-1 link to lose its operational status. When the link status drops to less than operational, all data exchange ceases except for the idle pattern (no information signal) and the ATDL-1 test message.

Idling Condition. Once the link is established and no information is to be transmitted, the system automatically assumes an idling condition. The idle signal consists of alternating marks and spaces (1 0 1 0) and is continuously transmitted by each unit on the ATDL-1 link. This signal permits the two modems to remain in time synchronisation and is interrupted only for transmission of a data message, termination of the link by operator action, or equipment malfunction. The modem indicates when a good idle pattern is received. This standby signal, or idle pattern, always starts with a (1) and ends with a (1) bit set to distinguish it from the beginning of a start code.

Data Message. ATDL-1 uses an 81-bit message frame that consists of:
- **Start group** – Nine bits; all set to (0)
- **Data group** – Seven data groups of nine bits, each immediately following the start group, with the first bit of each data group set to (1) and the next eight bits consisting of the B-series message
- **Check group** – One check group completes each transmission frame and comprises one fixed bit, followed by eight check bits.

Contiguous Data Message Grouping. B-series message-transmit rules in MIL-STD-6013 for ATDL-1 require transmission of multiple message sets. These message sets are transmitted as a contiguous set with no requirement for an idle pattern to be transmitted between messages. In a contiguous set, messages are transmitted back-to-back, with no time delay between sets. Contiguous message sets occur with message start codes immediately following the last bit of the previous message's check group. It is therefore possible to have extended interruptions of the 'no information signal' (idle pattern).

Non-Continuous Data Message. A non-continuous message set is a group of multiple messages separated by the insertion of a 'no information signal', rather than transmission of messages back-to-back. When a 'no information signal' (idle pattern) is inserted between messages, it always starts with a (1) bit and ends with a (1) bit.

Unit Address. Each unit reporting on the ATDL-1 link is assigned a unique station address. This address or TN is assigned in an OPTASKLINK message and consists of two alpha characters that represent the station address and a three-digit octal number, 000 to 777. The first alpha character can be from A to N, P and Q. The second alpha character can be from A to H. A unit operating on multiple ATDL-1 links uses the same address for all links. An SU (fire unit) is assigned a Track Number (TN) by its Reporting Unit (RU).

Restrictions. The address AA000 is used to indicate no statement/unknown. AA and QH are illegal values for use as station addresses. If the address of the directly tied unit changes, the datalink requires re-initialisation.

Supporting Units (SUs). Units operating on a single ATDL-1 link are designated SUs. C² units operating on multiple links (ATDL-1 and Link 11B) are designated Forwarding Reporting Units (FRUs). Units operating on Link 11 and/or Link 11B and ATDL-1 are designated Forwarding Participating Units (FPUs). For an FPU, a single two-digit address is used in place of the normal three-digit address for RUs. This restriction is imposed by the Link 11 Data Terminal Set (DTS),

which only allows a two-digit octal address from 01 to 76. The ATDL-1 address used by an FRU or FPU is required to be two alpha characters and three octal digits.

Track Numbers (TNs). An ATDL-1 TN consists of two alphabetical characters and three octal digits. An ATDL-1 TN derives its two alphabetical characters from the station address of the unit initiating the track. A unique three-digit octal number is derived from the TN block assigned to the unit.

Link Initialisation

Full Data Transmit (FDT) mode. Once the operator has designated an ATDL-1 link, it is physically activated. On activation, the ATDL-1 standby signal is transmitted along with the periodic ATDL-1 test message. The ATDL-1 test message contains the datalink address of the transmitting unit, which provides a capability to perform a loop-back check of the communications path, data path, and encryption path. The FDT mode provides for automatic reporting of all tracks and information as soon as the link is operational.

Limited Data Transmit (LDT) mode. When an ATDL-1 SU requests to initialise its ATDL-1 link, it may be directed to enter the LDT mode as a means of resolving dual designations prior to activating its FDT mode. Only one unit is in LDT mode for a single link. LDT mode allows the reception of track reports but inhibits the transmission of local air tracks until an operator or system action is taken to switch to the FDT mode. The unit in the LDT mode will begin transmitting its surveillance tracks after the operator has determined that all necessary correlations are complete, or a maximum time limit has elapsed since the transition to an operational link state.

Link States. ATDL-1 has two link states, normal and round-robin.

System employment. US Air Force, US Marine Corps, NATO, and Allied C2 and SAM units employ ATDL-1.

SURVEILLANCE AND CONTROL DATALINK (SCDL)

The Surveillance and Control Datalink (SCDL) is used between the E-8 JSTARS aircraft and tactical ground stations. The JSTARS aircraft distributes its Moving Target Indicator (MTI) and Synthetic Aperture Radar (SAR) data to Army and Marine Corps Common Ground Stations (AN/TSQ-179 – CGS) via the SCDL. This datalink is half duplex, Ku-band (12.4 to 18 GHz) and uses a robust frequency-hopping waveform, error correction coding and diversity techniques to mitigate jamming. The network employs a Time Division Multiple Access (TDMA) scheme that can, in real time, revise network assignments and priorities with a selectable data rate up to 1.9 Mbps. The SCDL also provides an 'uplink' to the E-8 aircraft for radar service requests from the CGS. The Marine Corps employs the SCDL in the Light Common Ground Station (LCGS) variant, transported on a High-Mobility, Multipurpose Wheeled Vehicle (HMMWV). The US Navy is testing a capability to receive JSTARS MTI radar data via the SCDL and integrate this data with the Global Command and Control System-M (GCCS-M) track database.

Figure 8 – The SCDL is used between the E-8 JSTARS aircraft and tactical ground stations (Northrop Grumman)   0524975

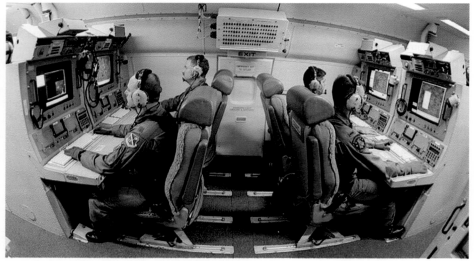

Figure 9 – Console operators on an E-8 JSTARS aircraft (Northrop Grumman)   0524976

Figure 10 – JSTARS tactical picture
(Northrop Grumman)   0524978

Link 14

Link 14 is a 75 bps, non-realtime, broadcast HF Radio Teletype (RATT) link for maritime units, designed to transfer surveillance information from ships with a tactical data processing capability to non-tactical data processing ships. Teletype transmission allows reception over very long ranges. Link 14 provides the capability to broadcast picture compilation and status information for use in units unable to receive Link 11 transmissions, which is then recorded on a teletype or displayed via an Autoplot VDU display. The Link can be HF, VHF or UHF dependent on unit-communication fits.

Link 14 is capable of transmitting the following information:
• Air tracks
• Surface tracks
• Subsurface tracks
• ESM information
• Special points
• Non standard/free text messages.

In addition, Link 14 can be used to broadcast the Fleet Ocean Surveillance Product (FOSP), the High-Interest Target Broadcast (HIT-B) and the Recognised Maritime Picture (RMP).

Each nation within NATO has its own Link 14 transmission formats which are promulgated in ADatP-14. Message protocol is defined in STANAG 5514.

VARIABLE MESSAGE FORMAT (VMF)

VMF is a 'Link 16 family protocol' which employs Link 16 data elements to create variable length messages suitable for near-realtime data exchange in a bandwidth-constrained combat environment. Earlier US Army VMF protocols contained ATDL-1 message elements. The final and Joint Services version is still under development.

VMF is a bit-orientated digital information standard consisting of variable-length messages. VMF is not man-readable. It is designed as a common means of exchanging data between combat units at various organisational levels, with differing requirements for volume and detail of information. VMF provides the user with the flexibility to send only required information on demand. This is particularly significant when operating in a bandwidth-constrained environment. VMF is media-independent and can operate over any digital-capable radio frequency, broadcast, or point-to-point system.

VMF allows the 'embedding' of other message formats (for example, ATDL or PADIL packets, or even an unformatted digital stream) into messages that can be transported over Link 16 networks, returning them at destination to the same format in which they originated. Therefore, the host need not implement the full TADIL-J message catalogue, since VMF converts the user traffic into Link 16 packets and reassembles to the original formats at the receiving terminal. This process incurs a small increase in Link 16 overhead (though it is more efficient where relaying is involved), but adds to the flexibility of terminal use (that is, users can use their own message formats, end-to-end). The JTIDS Class 2M and MIDS terminals are used for VMF.

VMF message processing rules provide for consistent message construct implementations across multiple systems. They include cases for each use of a message and the inter-element conditions within messages for basic processing. Message processing rules also include defaults, service restrictions, expected responses, special considerations and case level minimum implementation. Structured, natural language, using established logical operators, rigorously and unambiguously defines the construction rules for VMF messages.

By design, JVMF protocol eliminates the need to transmit empty, placeholder data packets that would be required in a fixed-format, character-based protocol. This means that only message fields that have valid user data, as defined by a bit known as the field presence indicator, are transmitted. This reduces the transmission load on the attendant hardware. However, the complexities of the variable format place very stringent requirements on the message parsers that encode the user data into properly formatted binary data, and then subsequently decode the binary data back into user data on the receiving end. If one bit is set incorrectly, the entire message is unreadable by the recipient. It is also worthy of note that the specification of the message types and field definitions is expected to evolve continuously.

The messages used to exchange information over a VMF interface are known as the K-series messages. **Kn.m** is the numbering convention for a VMF message, where **n** is the Functional Area Designator (FAD) and **m** is the message number assigned consecutively. Therefore, Kn.1 is the first defined message of all currently defined messages within a functional area. For example, K02.1 is the numbering convention for the Fire Support (FS) functional area 'Check Fire' message. Figure 11 contains the FADs defined in the VMF Technical Interface Design.

K00	Network Control
K01	General Information Exchange
K02	Fire Support Operations
K03	Air Operations
K04	Intelligence Operations
K05	Land Combat Operations
K06	Maritime Operations
K07	Combat Service Support
K08	Special Operations
K09	Joint Task Force Operations Control
K10	Air Defence/Air Space Control

Figure 11 – Functional Area Designators

COMMON DATALINK/TACTICAL COMMON DATALINK

The Common DataLink (CDL), which includes the Tactical Common Datalink capability, is the US DoD primary datalink for unprocessed data. The CDL will support air-to-surface transmission of radar, imagery, video, and other sensor information from manned aircraft and UAVs.

The CDL is a flexible, multipurpose, radio link-based digital communication system that was developed by the US DoD for use in imagery and signals intelligence collection systems. It provides standard waveforms that follow a LOS microwave path (link) and allows both full-duplex and simplex communications between airborne/spaceborne platforms and surface-based terminals. The link consists of an uplink that operates at 200 Kbps and a downlink that operates at 10.71, 137 and 274 Mbps. All links use the C, X and K frequency bands. The uplink is secure and jam resistant. Currently, the downlink is secure only for the 10.71 Mbps rate. However, new platforms are coming into service that will require a secure downlink for the 137/274 Mbps rates. In addition, rates of 548 and 1,096 Mbps will be supported. The CDL system supports air-to-surface, and air-to-satellite (relay/Beyond Line-Of-Sight, BLOS) communications modes.

The term CDL refers to a family of interoperable datalink implementations that offer alternative levels of capabilities for different applications/platforms. Five classes (Class I through Class V) of CDL have been defined:
• Class I – Ground-based applications with airborne platforms operating at speeds up to M2.3 at altitudes up to 24,400 m (80,000 ft)
• Class II – Speeds up to M5.0 and altitudes up to 45,750 m (150,000 ft)
• Class III – Speeds up to M5.0 and altitudes up to 152,500 m (500,000 ft)
• Class IV – Terminals in satellites orbiting at 750 n miles
• Class V – Terminals in relay satellites operating at greater altitudes.

Current Class I terminals include the land-based Modular Interoperable Surface Terminal (MIST) and maritime Common Datalink-Navy (CDL-N), previously known as Common High-Bandwidth Datalink Surface Terminal (CHBDL-ST). The latter supports the Advanced Tactical Air Reconnaissance System (ATARS – see entry) and the Battle Group Passive Horizon Extension System (BGPHES).

The system permits the remote operation and exploitation of sensors carried by CDL-equipped platforms from BLOS locations via satellite. There are two satellite CDL systems in use, Senior

Figure 12 – The CDL supports the transmission of tactical reconnaissance imagery, gathered by aircraft such as this F-18D equipped with the Advanced Tactical Air Reconnaissance System (ATARS)   0524983

Figure 13 – Senior Span, one of the satellite CDL systems, is carried by the U2-R. The Span Airborne Datalink system equipment is carried in the dorsal housing 　　0524986

Span and Senior Spur. Senior Span makes use of a Span Airborne Datalink system operating in the I-band via Defense Satellite Communication System II/III spacecraft. Senior Spur operates in the Ku-band in order to gain extra bandwidth. In August and November 1999, the US Navy demonstrated a phased array antenna that can handle both satcoms and CDL operation. Frequencies utilised were in the EF/S band (2.2–2.3 GHz) and I/X band (7.25–8.4 GHz) for military satcom, plus I/X band (9.7–10.5 GHz) and Ku-band (14.5–15.35 GHz) for CDL and Ku-band (10.95–14.5 GHz) for commercial satcom.

Link 22

Link 22 (also known as the NATO Improvement to Link Eleven (NILE) Programme), is a secure, ECM-resistant, tactical data communications system that will interface US Navy C^2 P-equipped ships and NATO systems. NATO systems will employ a Datalink Processor (DLP) or integrate processing within their host systems to provide real-time data exchange over a continuous 300 n mile range using NILE Communications Equipment (NCE). Link 22 uses fixed frequency or frequency hopping in the HF and/or UHF bands. Architecture can be TDMA or Dynamic TDMA. The US Marine Corps plans to implement Link 22 in the Common Aviation Command and Control System (CAC²S).

Link 22 is designed for the exchange of EW data, air, space, land, surface and subsurface tracks and points with amplifying data in real time, together with the transmission of orders, commands and alerts between C^2 systems.

Link 22 and Link 16 utilise common data structures; Link 22 is fundamentally an upgrade to the Link 11 message standard and is based on Link 16 data elements. The programme goal is to replace Link 11 and enhance Link 16. Programme participants are NATO members: Canada, France, Germany, Italy, Netherlands, the United Kingdom and the United States. Implementation is planned during the period 2002 to 2009, dependent on nation and service.

Link 22 Features

Primarily intended as a maritime (ASW/ASUW) datalink, Link 22 will have the following features:

- Time Division Multiple Access (TDMA) or Dynamic TDMA
- Spread spectrum, frequency-hopping NILE link-level crypto
- Netted architecture and polling protocol (Roll Call)
- A 72-bit word message standard that can carry embedded J-Series Link 16 messages, known as the FJ-series, as well as newly defined Link 22 F-Series messages. (F/J-series messages are 'J' messages with an additional 2 bits of overhead)
- Extended surface wave and long-range sky wave HF
- Ship-ship relay networking, to extend connectivity
- HF data rate 500 to 2,200 bps; UHF data rate 12.6 Kbps (and others)
- Wave form: UHF JTIDS, HF single tone Link 11
- Multiple nets/enhanced networking capabilities. A Link 22 unit may operate up to four networks simultaneously, each on a different media, with any participant on any network able to communicate with any other participant
- Data Terminal Set AN/USQ-125 (Single Tone Link 11 Wave Form), AN/URC-107 or MIDS (JTIDS Wave Form)
- Operating Procedures: STANAG 5522 and ADatP 22.

F-Series Messages

F00.0 NU	Performance
F00.1-0	EW bearing initial
F00.1-1	EW fix initial
F00.1-2	EW position
F00.1-3	EW amplifying
F00.2-0	EW area of probability initial
F00.2-1	EW area of probability
F00.3-0	EW emitter and ECM
F00.3-1	EW frequency
F00.3-2	EW PD/PRF/scan
F00.3-3	EW platform
F00.4-0	EW co-ordination initial
F00.4-1	EW association
F00.4-2	EW co-ordination ECM
F00.4-3	EW co-ordination emission control
F00.7-0	Frequency allocation
F00.7-1	Network media parameters
F00.7-3	Network management order
F00.7-3P	Network management order with parameters
F00.7-5	Radio silent order
F00.7-6	Network status
F007-7	MASN
F00.7-7	Network status
F00.7-7C	MASN create
F00.7-7M	MASN modify
F00.7-10	Key roll-over
F01.0-0	IFF
F01.4-0	Acoustic Brg/Rng resolved
F01.4-1	Acoustic Brg/Rng ambiguous
F01.5-0	Acoustic Brg/Rng amplification
F01.5-1	Acoustic Brg/Rng sensor
F01.5-2	Acoustic Brg/Rng frequency
F01.6-0	Basic command
F01.6-1	Command extension
F01.6-2	Air co-ordination
F02.0-0	Indirect PLI amplification
F02.1-0	PLI IFF
F02.2-0	Air PLI course and speed
F02.2-1	Air PLI additional mission correlator
F02.3-0	Surface PLI course and speed
F02.3-1	Surface PLI mission correlator
F02.4-0	Subsurface PLI course and speed
F02.4-1	Subsurface PLI mission correlator
F02.5-0	Land point PLI continuation
F02.5-1	Land point PLI additional mission correlator
F02.6-0	Land track PLI course/speed
F02.6-1	Land track PLI mission correlator
F02.7-0/7	ANFT TBD
F03.0-0	Reference point initial
F03.0-1	Reference point position
F03.0-2	Reference point course/speed
F03.0-3	Reference point axis
F03.0-4	Reference point segment
F03.0-5	Reference point anti-submarine
F03.0-6	Reference point friend weapon danger area
F03.0-7	Reference point theatre ballistic missile
F03.1-0	Defence emergency point initial
F03.1-1	Emergency point position
F03.4-0	ASW contact information
F03.4-1	ASW contact confirmation
F03.5-0	Land track/point initial
F03.5-1	Land track/point position
F03.5-2	Land non-realtime track
F03.5-3	Land track/point IFF
F1-0	Indirect PLI position
F1-1	PLI position
F2	Air track position
F3	Surface track position
F4-0	Subsurface track position
F4-1	Subsurface track course and speed
F5-0	Air track course and speed
F5-1	Surface track course and speed
F6	EW emergency

Figure 14 – F-series messages

FJ-Series Messages

FJ3.0	Reference point message
FJ3.1	Emergency point message
FJ3.6	Space track message
FJ6.0	Intelligence message
FJ7.0	Track management message
FJ7.1	Data update request message
FJ7.2	Correlation message
FJ7.3	Pointer message
FJ7.4	Track identifier message
FJ7.5	IFF/SIF management message
FJ7.6	Filter management message
FJ7.7	Association message
FJ8.0	Unit designator message
FJ8.1	Mission correlator change message
FJ10.2	Engagement status message
FJ10.3	Handover message
FJ10.5	Controlling unit change message
FJ10.6	Pairing message
FJ13.0	Airfield status message
FJ13.2	Air platform and system status message
FJ13.3	Surface platform and status message
FJ13.4	Subsurface platform and system status message
FJ13.5	Land platform and system status message
FJ15.0	Threat warning message
FJ28.2(0)	Text message

Figure 15 – FJ-series messages

Link 16

Over the last 20 years, the development of Digital Datalinks (DDLs) has concentrated on the exchange of tactical information between C^2 elements. The complexity of the modern battlefield, combined with the requirement to transfer information in 'real time' and the growing electronic warfare (EW) capability of potentially hostile nations, has stretched the capabilities of current NATO DDL systems. Accordingly, a requirement for an Electronic Countermeasures-Resistant Communications System (ERCS), capable of providing secure communications between all elements involved in Anti-Air Warfare/Air Defence (AAW/AD) operations, has emerged. In addition, NATO has a parallel requirement for a Multifunctional Information Distribution System (MIDS). The system adopted to meet these requirements is the Joint Tactical Information Distribution System (JTIDS). JTIDS was developed for US forces as a result of experiences in trying to effect C^2 over joint- and single-service operations in the Vietnam War and is designed to be a high-capacity, secure, flexible, real-time, ECM resistant and survivable Integrated Communication System (ICS).

Several names and acronyms are associated with Link 16. The TActical Digital Information Link (TADIL) is a term used by the US Joint Services. The TADIL designation for Link 16 is TADIL-J. The nomenclature JTIDS refers to the communications component of Link 16, which includes the terminal software, hardware, RF equipments and the waveforms they generate. The NATO term for JTIDS is MIDS. For simplification, the term Link 16 will be used throughout when referring to this system.

Within the US, confusion arises when JTIDS and Link 16 are used interchangeably for the datalink. JTIDS and JTIDS-compliant radio equipment (such as MIDS) are merely the communications element. There also seems to be some confusion concerning the use of the designation TADIL-J; in some quarters, the term is applied to the link itself, while in others, the term applies only to the message formats and protocols (as defined by MIL-STD-6016).

Link 16/TADIL-J/JTIDS/MIDS Specifications

Standard	US Specifications	NATO Specifications
Terminal interface standards (hardware/software)	JTIDS SSS DCB79S4000C	MIDS STANAG 4175
Procedural interface standards (message formats/protocols)	TADIL J MIL-STD-6016	Link 16 STANAG 5516
Standard operating procedures	JMTOP CJCSM 6120.01	ADATP-16

Figure 16 – US/NATO Link 16 specifications

NATO has a different view of this terminology. The TADIL-J messages and protocols become 'Link 16' (STANAG 5516), while the JTIDS communication element becomes 'MIDS' (STANAG 4175). Thus, NATO uses Link 16 in a narrower sense than that used in the US.

There are also differences in standard operating procedures. The US uses the Joint Multi-TADIL Operating Procedures (JMTOP) (Chairman Joint Chiefs of Staff Manual CJCSM 6120.01), whereas NATO uses the Allied Data Publication-16 (ADAP-16). JMTOP has been recommended to NATO for adoption. The different specifications are listed in Figure 16.

Essentially, JTIDS is a netted system, with each participating unit being allocated discrete periods in which to transmit data, with the link medium being UHF radio. When not transmitting data, all units will receive data transmitted from all other JTIDS Users (JUs) which are in LOS. To overcome the range limitations of UHF frequencies, JTIDS can incorporate an automatic relay capability between units that are not within LOS. JUs within a network can exchange data on many aspects of operations, including tracks, unit position and status, command and control messages and secure speech. The following describes the operating system of JTIDS/Link 16 and the planned implementation of JTIDS within NATO.

JTIDS/Interim JTIDS Message Specification.
JTIDS/Interim JTIDS Message Specification (JTIDS/IJMS) was created because the hardware for JTIDS operations was available before the J-series message catalogue; Link 16 uses the J-series message. Therefore, an IJMS was developed to provide an initial JTIDS operating capability, introduced into US Air Force and NATO E3As, the UK ASACS, the GEADGE and the NADGE, using the JTIDS Class 1 terminals. The IJMS message catalogue is based closely on the M-series message catalogue associated with Link 11. However, that is the limit of any similarity between the systems; IJMS has greater capacity

Figure 17 – MIDS Low-Volume Terminal (LVT)
0524979

than Link 11 and is ECM resistant. Furthermore, the surveillance capacity, message catalogue for aircraft control and the limited C² capability of IJMS is not sufficient for the scenarios predicted in the AD sphere of operations over the next 20 years. Moreover, no naval forces implement IJMS.

Despite these shortcomings there is no clear date when IJMS will be phased out.

JTIDS/Link 16. JTIDS/Link 16 uses the J-series message standard, which is designed to make optimum use of the JTIDS architecture and is based firmly on the US TADIL-J message standard. It is worth noting at this stage that JTIDS refers to the overall bearer system, whereas Link 16 is the NATO-approved message specification which is carried by JTIDS. JTIDS encompasses more than just TADIL messages; for instance, it supports free text messages. However, all messages sent via JTIDS share the same link characteristics and timing architecture.

Multifunction Information Distribution System (MIDS). The Multifunction Information Distribution System (MIDS) is designed to meet the NATO operational requirement for a flexible, secure, ECM-resistant, high-capacity

Figure 19 – JTIDS Class 2 terminals have been fitted to UK Royal Air Force Tornado F3 interceptors
(I Black)   0524967

communications system capable of supporting a wide range of functions. The only existing system capable of meeting this requirement is JTIDS, using the Link 16 message standard, and this combination should be referred to as JTIDS/Link 16. However, recently terminal manufacturers have started to refer to JTIDS/Link 16 equipment as MIDS terminals. JTIDS/Link 16 and MIDS are nevertheless the same system.

JTIDS Terminal
History
The first JTIDS terminals introduced were the Class 1 type produced by Hughes and fitted to USAF and NATO E-3s and C² ground sites. They are physically large, support IJMS operations only and have been operational in the UK since 1982. Class 2 terminals were developed to meet the requirement for a smaller Link 16-capable terminal which could be integrated into other platforms, giving greater capacity but with a lesser weight penalty. Class 2 terminals were developed by the then GEC-Marconi Electronic Systems Corporation (GMESC) and Rockwell-Collins. The first Class 2 terminal was delivered in March 1984.

Although the Class 2 terminal has been fitted to the F3, F-14 and F-15, it is considered too large for the next generation of fighter aircraft such as the Eurofighter Typhoon, F-35 JSF and Rafale. Consequently, in 1985 it was decided to develop a Low-Volume Terminal (LVT) which would encompass 1980s technology to provide a smaller, lighter and more maintenance-free terminal. The development of JTIDS terminals to decrease size and cost is a continuing process. Some Class 2 terminals are capable of transmitting and receiving both JTIDS/IJMS and Link 16, and these are known as bilingual terminals. The UK is implementing the JTIDS Class 2 terminal for Link 16 platform integration.

System features
Link 16 consists of the JTIDS waveform, the TADIL-J message standard and the Time Division Multiple Access (TDMA) protocol. The system features of Link 16 are programmable and can be combined to enhance throughput, anti-jam or a specific operational requirement. These features include:

- Spread spectrum techniques
- Variable throughput
- Variable output power
- Unique antenna gains
- Variable capacity
- Message formatting
- Assignable access.

Figure 18 – The MIDS Low-Volume Terminal (LVT) is fitted to the Eurofighter Typhoon
(BAE Systems)   0524981

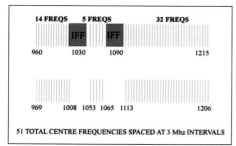

51 TOTAL CENTRE FREQUENCIES SPACED AT 3 Mhz INTERVALS

Figure 20 – Hopping pattern 0525836

Network Participation Groups (NPGs)

In older style datalinks such as Link 11, all participants received all of the data transmitted, regardless of whether it was relevant to their role or information requirements. Link 16 data can be grouped into Network Participation Groups (NPGs) which enable networks to be planned so that the transmission and reception of data can be selectively managed. For example, non-C² units do not need to receive Mission Management Messages exchanged between C² Units. The use of NPGs means that the source does not need to know who the recipient will be, and only the users that need the data will have access to that NPG and will receive all associated messages. NPGs that will be implemented by NATO are described later.

Waveform

The JTIDS terminal produces the Link 16 waveform. This waveform is designed to provide complete communications service to multiple users within a hostile electromagnetic environment. The waveform employs the following techniques to ensure maximum jam resistance:

- Spread spectrum
- Frequency hopping
- Error Detection and Correction (EDAC) coding
- Pulse redundancy
- Pseudo-random noise coding
- Data interleaving
- Automatic data packing
- Inherent relay.

Frequency Band

The Link-16 JTIDS waveform uses pulsed transmissions in the 960–1,215 MHz (C/D) frequency band. The selection of this wide band both enhances Command, Control, and Communications (C³) throughput and complicates enemy jamming strategy. Bandwidth coverage and the jammer power required to overcome the JTIDS jam-resistant waveform will provide sustained connectivity in all projected tactical scenarios.

Spread Spectrum

Spreading information in the frequency/time domain, using pseudo-random code sequences, creates JTIDS pulses. The spreading is implemented via continuous phase shift modulation at a 5 MHz rate, which results in a 3 MHz bandwidth pulse transmitted by the receiver/transmitter or High-Power Amplifier (HPA). Since the spreading code is a continuously changing phase pattern determined by the transmission encryption variable (stored in the Secure Data Unit (SDU)), the transmitted pulse appears as random modulation to interceptors for whom the transmission is not intended. Receiving units that possess an accurate knowledge of system time (in synchronisation) and the right transmission encryption variable are able to detect and decode the signal into five data bits.

Frequency Hopping

During transmission each radiated pulse is pseudo-randomly assigned to one of the 51 authorised centre frequencies. This pseudo-random assignment of frequency is referred to as the hopping pattern. The functional circuit Network Participation Group (NPG), the net number, and the assigned transmission crypto variable determine the hopping pattern. Units operating on different NPGS, crypto variables, or net numbers would have different hopping patterns. This allows multiple transmitters to be within LOS of each other without causing mutual interference.

During any transmission, pulses are evenly distributed among the 51 different frequencies, with no two consecutive pulses falling 'close together'. Since each outgoing pulse is on a different frequency, the instantaneous hopping rate is 1/13 msec or 76,923 hops per second (current military frequency hopping systems employ less than 2,000 hops/sec). Theoretically, up to 20 circuits with different hopping patterns can be co-located before appreciable interference occurs (that is, hopping onto the same frequency). JTIDS, unlike existing military systems, frequency-hops the sync pulses as well as the data pulses, which further complicates enemy jamming strategies.

Pulsed RF

Each pulse is 3 MHz wide after spreading. The pulses are transmitted on centre frequencies that are spaced at 3 MHz intervals on 51 frequencies located in the 960–1,215 MHz range. Frequencies (notches) are reserved in order to preclude adverse effects on Identification Friend or Foe (IFF) systems. The even distribution of pulses (over the 51 frequencies) and the extremely fast hopping rate makes JTIDS appear as random pulses to electronic countermeasures receivers.

Anti-jam

Link 16 maintains good performance in all anticipated scenarios against an aggressive jammer strategy. The waveform and system hardware, couple with flood relays, provide continuous communications services within severe hostile electromagnetic environments. Furthermore, Link 16's use of the C/D band forces jammer assets to be diverted from jamming other communication or radar systems.

Jam Margin

The system's ability to overcome an optimised band-matched jammer is based on the built-in features of the waveform. These include processing gain provided by the spread spectrum techniques (direct sequence coding and frequency hopping), signal-to-noise reduction techniques provided by the double pulse format, Reed Solomon coding, interleaving and jitter.

Electromagnetic Propagation

Link 16 features provide considerable anti-jam margin by employment of techniques that increase the signal level or decrease the jammer level at the receiver. These features allow Link16 to continue to provide connectivity in the predicted operational jamming environment.

Time Division Multiple Access (TDMA)

Link 16 system access is controlled through the TDMA protocol. In this format, opportunities to transmit are apportioned on a timeslot basis. Therefore, users take turns transmitting based on time. Each unit, depending on message format, transmits a burst series of pulses (usually 258 or 444). Each user receives multiple transmit opportunities (timeslots) per frame.

Frame. The basic recurring unit of time for Link 16 is the frame. It is 12 seconds in length and is composed of 1,536 individual access/transmit units called timeslots. Frames occur repeatedly, one following another, as long as the link is operational. Since there are 1,536 timeslots per 12-second frame, each timeslot is 7.8125 msec in length. A 'network' is used to define which functions (surveillance, air control, mission management, and so on) are performed in each timeslot.

Timeslot. The timeslot is the basic unit of access to the JTIDS network. Each platform operating in a Link 16 network is assigned to either transmit or receive in each of the 1,536 timeslots. Timeslots are assigned to both the function they support (surveillance, Voice Group A, and so on) and to the participating platforms (E3D, Tornado, and so on). If a unit is assigned a set of transmit timeslots (for example, every 16th timeslot) and has messages to be sent, it will transmit a series of pulses in each timeslot until all messages have been sent.

Epoch. Every 24-hour day (1,440 minutes) is divided into 112.5 epochs. An epoch is a measure of time equal to 12.8 minutes' (768 seconds) duration. The epoch is then further divided into 98,304 timeslots, each of which are 7.8125 msec in duration.

The timeslots of an epoch are grouped into sets, named A, B and C. Each set contains 32,768 timeslots numbered 0 to 32,767; this number is called the Slot Index (98,304 (the total number of slots in an epoch) divided by 3 equals 32,768). By convention, timeslots within an epoch are named in an alternating pattern. The timeslots of each set are interlaced, or interleaved, with those of the other sets to provide the repetitive sequence:

A-0, B-0, C-0,
A-1, B-1, C-1,
*
*
*
A-32767, B-32767, C-32767

This sequence is repeated for each epoch (see Figure 22).

Transmission Format

The JTIDS terminal accepts TADIL-J words from the host computer and turns them into a stream of pulses. The transmission pulses vary in number depending on the message type Round Trip Timing (RTT), voice (or data), and data packing format selected.

Jitter. Prior to transmitting in an assigned timeslot, the JTIDS terminal waits for a randomised period, referred to as jitter, and then begins to emit pulses. The jitter period is a transmit delay interval. This period of dead time varies pseudo-randomly, and correctly initialised terminals will know the jitter period from timeslot to timeslot. This feature is designed to defeat a jammer that targets its jamming at the beginning of each transmission, which, if there was no jitter,

Frequency Number	(MHz)	Frequency Number	(MHz)	Frequency Number	(MHz)
0	969	17	1,062	34	1,158
1	972	18	1,065	35	1,161
2	975	19	1,113	36	1,164
3	978	20	1,116	37	1,167
4	981	21	1,119	38	1,170
5	984	22	1,122	39	1,173
6	987	23	1,125	40	1,176
7	990	24	1,128	41	1,179
8	993	25	1,131	42	1,182
9	996	26	1,134	43	1,185
10	999	27	1,137	44	1,188
11	1,002	28	1,140	45	1,191
12	1,005	29	1,143	46	1,194
13	1,008	30	1,146	47	1,197
14	1,053	31	1,149	48	1,200
15	1,056	32	1,152	49	1,203
16	1,059	33	1,155	50	1,206

Figure 21 – Centre frequencies spread at 3 MHz intervals in the band 960–1,215 MHz

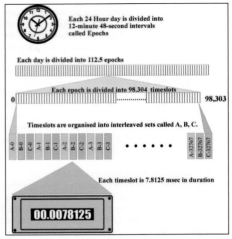

Figure 22 – Time division   0525839

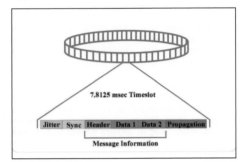

Figure 23 – Jitter   0525848

would take place at the same time within every timeslot. In some timeslot structures, the jitter is deleted at the expense of some ECM resistance to allow more data to be transmitted in that slot. Each transmission begins with synchronisation (sync) pulses, followed by time refinement pulses, a message header, and then data. Transmissions contain 72 pulses for synchronisation, time refinement, and the message header.

Sync. The synchronisation preamble is the first element within the timeslot to be transmitted and therefore received. The receiver knows when a transmission will start, but does not know the propagation time of that transmission from the unit with which it is trying to synchronise. The Synchronisation preamble allows for this and enables the receiver to synchronise with the incoming pulses. Due to the varying propagation time, synchronisation must be achieved on receipt of every message. Sync preambles are always transmitted as 16 double pulse symbol packets.

The time refinement pulses are further used to align the receiving terminals' clock oscillator, and therefore reduce the time-of-arrival uncertainty of the message data that follows. Time refinement consists of four double pulse symbol packets.

Header. The header contains information that allows the terminal to correctly interpret and process the messages that follow in the main body of the timeslot. The header is transmitted as 16 double pulse symbol packets. The following information is contained within the header:
- Source TN of originating unit
- Relay indicator
- Message packing level
- Message format (fixed format or free text)
- Secure Data Unit (see below) information for decryption.

Data Content. Within the timeslot, the principal element that is of interest to the host platform is the data. This data may be fixed format or free text. Free text is primarily used for exchanging voice data. The data content of the standard double pulse timeslot consists of three words. Each fixed format word contains 75 data bits (15 symbol packets). A Link 16 J-series message is made up of one or more words. For example, a message containing a large amount of data such as a full air track report, may take 3 words, whereas a short message, such as a disengage

command, may take only one word. It therefore follows that a standard double pulse timeslot can accommodate a single 3-word message, or one 2-word message and one 1-word message, or three 1-word messages.

Propagation and guard time. The data content is immediately followed by a period to allow for propagation delays. This additional period of dead time allows for the transmitted signal to propagate throughout the network before the next timeslot begins, thereby avoiding interference between successive transmissions. There are two propagation dead times available, normal and extended. Normal range mode allows for a dead time equivalent to 300 n miles of range, while extended range mode allows a dead time equivalent to 500 n miles. However, this does not imply extended terminal range; the system is still constrained to LOS, and the extra period required for the extended mode is subtracted from the maximum jitter time available. This leads to a slight reduction in the anti-jam margin against a smart jammer. Therefore, extended range mode is only selected when absolutely necessary. All terminals operating within the same network must operate in the same range mode.

Fixed format words. As previously discussed, each Link 16 fixed format message consists of a number of 75 bit words. The three types of words used in fixed format messages are the initial word, the extension word and the continuation word. The way in which these words are used in messages is illustrated below.
- **Initial word.** An initial word identifies the message and the number of extension and continuation words that follow to make up the message. It also contains basic data included in the message
- **Extension word.** The extension word contains additional data that is normally included in the message. Extension words need not be included for all messages, but if they are, they must immediately follow the initial word and be in serial order
- **Continuation word.** Continuation words contain amplifying data that may be included in the message. The inclusion of the continuation word can be periodic, be based on the availability of amplifying data, or may only be necessary for some uses of the message. Continuation words are transmitted after the initial word and any extension words.

J3.2 Air Track Message

		Example Contents	Inclusion
J3.2	**Initial Word**	Track Number Force Tell/Emergency Indicator Identity Strength Altitude	Mandatory
J3.2	**Extension Word**	Lat and Long Position Heading Speed	Mandatory
J3.2	**Continuation Word**	IFF Data Time for Non-realtime Tracks	Optional Periodic/On Change

Figure 24 – Example of Words forming a 'J' Series Message

The number of data pulses is dependent on the 'packing structure'. STandard Double Pulse (STDP) and Packed 2 Single Pulse (P2SP) have 186 pulses of data. Packed 2 Double Pulse (P2DP) or Packed 4 Single Pulse (P4SP) have 372 pulses of data. The remaining time within the timeslot is reserved for signal propagation through the air. The time left for propagation allows for a distance between platforms of 300 n miles (500 n miles in extended mode).

Data Packing
The JTIDS terminal can produce a number of different timeslot formats depending on the type of message being supported, the throughput demand and the required anti-jam performance. These timeslot formats differ in the number of data pulses, the amount of EDAC encoding and the amount of redundancy.

When the data rate required is less than 26.9 Kbps,each unit will transmit three TADIL-J words (3 × 70 bits) per timeslot using the STDP format. However, when demand increases, the terminal will pack three additional words (53.8 Kbps each) into each timeslot. STDP is always used whenever possible to insure maximum anti-jam performance. The maximum 'packing' limit is specified in the network design for each NPG.

P2SP and P2DP both allow for the transmission of six TADIL-J words per timeslot while P4SP allows 12 words. Voice transmissions are not TADIL-J data and are always sent using P2SP.

Double Pulse Format
In the double pulse format (STDP and P2DP), each set of three TADIL-J words is sent twice in the same timeslot. Successive pulses, although using identical data, are separately encoded and transmitted on two different frequencies. Thus, receiving units need only receive one of these two pulses to obtain all of the data. This format is preferred for maritime use as it compensates for multipath effects, platform manoeuvring and partial band jamming.

Data Encoding
All data transmitted over Link 16 has Reed-Solomon EDAC coding applied. Link 16 uses a '31-15' Reed-Solomon technique to guarantee the original dataword can be reconstructed even if half of the data is lost in transmission. Each Link 16 word, consisting of fifteen 5-bit characters (70 bits plus 5 parity bits), is expanded into 31 Reed-Solomon characters (155 bits). If a unit receives more than

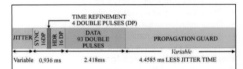

Figure 25 – The Standard Double Pulse (STDP) affords the lowest throughput but the highest level of jam resistance, since the same data is transmitted in both pulses of the slot   0525853

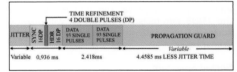

Figure 26 – The Packed-2 Single Pulse (P2SP) doubles the throughput by putting different data in each of the pulses, at the expense of jam resistance offered by redundant pulse transmission   0525857

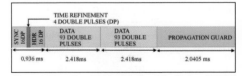

Figure 27 – The Packed-2 Double Pulse (P2DP) keeps the throughput increase from double packing while restoring some of the jam resistance, at a cost of the loss of Jitter   0525858

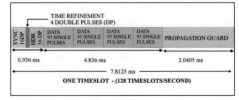

Figure 28 – The highest throughput is offered by the Packed-4 Single Pulse (P4SP) format, which gives up Jitter and pulse-redundancy to get maximum data capacity   0525863

one-half (16) of the transmitted pulses error free, its EDAC mechanism can regenerate the entire message.

When data is relayed, the relay platform 'corrects out' errors in the received message prior to retransmission, which insures that each 'new' transmission of the message is accurate. Messages received with too many erasures and/or errors are discarded prior to message decryption and are not relayed.

It should be noted that voice does not contain Reed-Solomon EDAC and can be relayed with errors.

Round Trip Timing (RTT)

RTT messages do not contain a data stream. They are used to support active fine synchronisation and consist of only the sync, time refinement, and header portions of a message. RTT-Is (RTT Interrogations) are transmitted at the beginning of a timeslot by units attempting to gain or maintain fine sync. When another terminal (which has a better time quality) receives the RTT-I, it generates a response, an RTT-R, which is transmitted in that same timeslot. By analysing the total time from transmission of the RTT-I to receipt of the RTT-R, unit synchronisation can determine timing adjustments required to gain and maintain synchronisation.

Precise Participant Location and Identification (PPLI)

The Precise Participant Location and Identification (PPLI) is the friendly unit reporting message used within Link 16. PPLI data is used to support many link functions, including passive synchronisation and navigation, link participation determination, data forwarding requirements and air control initiation. PPLIs are similar to the M1 messages used by Link 11, but they contain much more data.

The PPLI message contains the following information:

- The geodetic and relative position of the user
- The type of unit (for example, fighter), function (for example, CAP) and identity (implicitly friendly) of the user
- The network status of the JU
- Other miscellaneous information (for example, IFF).

The PPLI message clearly identifies all friendly forces participating in the Link 16 network and establishes the JU position without ambiguity. The PPLI messages are transmitted generally in

NPGs 5 and 6 and occasionally in NPG 27. The PPLI messages are defined by the prefix J2 in the J-series catalogue. The J2 message is subdivided and the message transmitted is dependent on the environment of the user:

- J2.2 – Air PPLI
- J2.3 – Surface maritime PPLI
- J2.4 – Subsurface PPLI
- J2.5 – Land ground point PPLI
- J2.6 – Land ground track PPLI.

A further PPLI message exists, the J2.0 Indirect Interface Unit PPLI message. This is transmitted by a Link 16 JU on behalf of a Link 11 unit which is not operating on Link 16. This message uses information from the Link 11 M1 message, the PPLI equivalent, and allows the position of the Link 11 unit to be reported on the Link 16 network.

Secure Voice

Link 16 supports two secure voice circuits, each with 127 subcircuits. Each JTIDS terminal contains two 16 Kb voice encoders. At 16 Kb, these analogue-to-digital converters provide 'voice recognition' quality communications.

The two voice encoders are referred to as 'Voice Group A' and 'Voice Group B'. Each voice channel receives an exclusive set of timeslots for the digitised data. Thus, they do not interact with each other (that is, Voice A cannot interfere or communicate with Voice B). Voice A is a stacked net and, therefore, the operators can select from 127 (0 to 126) subcircuits. Voice Group B, when available, has another 127 mutually exclusive subcircuits. Voice and voice relay may be unavailable if the feature is not included in the network design.

Unlike data, the digitised voice is not Reed-Solomon error encoded. In the STDP format there are 930 bits packed into each voice timeslot. Without Reed-Solomon, transmission errors are not corrected prior to conversion to audio or prior to retransmission for relay. In practice, 16 Kb voice is understandable with up to 10 percent of the bits in error (that is, a 10 per cent bit error rate). When operating in jamming or in low signal-to-noise environments, relayed voice will become unusable first, followed by garbled direct voice, and finally unrecoverable data.

Throughput

There are several features that combine to determine the system throughput of a Link 16 network. As previously discussed, the packing

format of timeslots varies the instantaneous terminal throughput from 26.9 Kbps to 107.6 Kbps (3 to 12 TADIL-J words). Packing affects 'terminal throughput' as opposed to 'system throughput'. Unlike other communications systems, Link 16 system capacity is not limited by terminal capacity. Many platforms can be transmitting simultaneously using different hopping patterns, thereby increasing the 'system throughput'.

System capacity. The features that increase 'system throughput' are:

- Access
- Stacked nets
- Multinet
- Multiple networks.

In Link 16, up to 20 different hopping patterns can be collocated without mutual interference. The nominal system capacity is 1 Mbps (20 × 54 Kbps).

As more functions operate simultaneously, then system throughput approaches 1 Mbps. Functions that currently enable simultaneous transmissions are voice, air control, fighter-to-fighter and EW/High update rate PPLI. These functions use one or more of the multiple access techniques.

Access

Associated with each JTIDS NPG is an access method. There are four types of access modes: contention (RTT-B), Push-To-Talk (PTT), dedicated, and dedicated with timeslot reuse. These access rights are defined by the network design and how operators use the system.

The contention access method gives all platforms transmit rights to the same set of timeslots. If two or more platforms attempt to transmit simultaneously, platforms only hear the closest transmitter. JTIDS voice uses a form of access called PTT. In PTT, protocols are told to wait until no one is speaking, which eliminates multiple transmitters.

In the dedicated access mode for a given hopping pattern, only one platform is assigned transmit rights to a timeslot, and all others receive. The timeslot, net number and crypto determine the hopping pattern. Dedicated with timeslot reuse is a variation of dedicated access, where the battle group commander selects one platform as the transmitter. Mission management weapons co-ordination uses this method of access, where one platform is selected for sending force orders. All C² units have the capability to transmit, but only the designated unit is allowed to transmit at any given time.

To implement the various access methods, timeslots are designated as transmit, relay transmit, or receive. A platform cannot receive and transmit in the same timeslot, with the exception of the RTTs. Relay transmit timeslots are described below.

Stacked Nets

Link 16 networks use stacked nets to increase system throughput. Stacked nets support the voice, air control, and fighter-to-fighter functions of Link 16. The selection of a particular net from a stacked net function changes the terminal's possessive hopping pattern. For these functions, operators manually select the net number, 0 to 126, on which to participate.

The air control stacked net, each air control platform, and the fighters under its control select unique net numbers. Since nets do not interfere with each other, multiple air control circuits can then operate using the same set of timeslots. Platforms can switch from net to net by simply changing the net number; the terminal does not have to be reinitialised. Similarly, for fighter-to-fighter, multiple independent groups of fighters are supported. JTIDS voice is also a stacked function enabling each channel to support 126 voice circuits.

Multinet

Link 16 datalink functions (surveillance, voice, EW, and so on) are allocated to NPGs; not all users need to participate in every function. Therefore some functions in networks are mutually

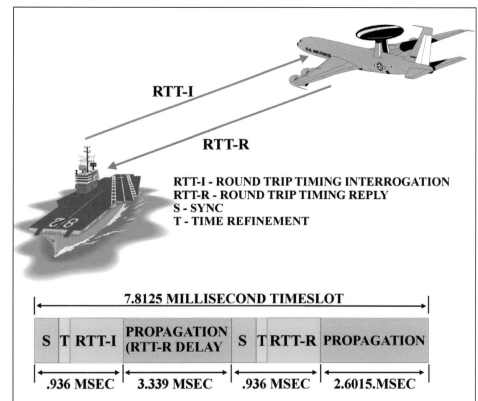

RTT-I

RTT-R

RTT-I - ROUND TRIP TIMING INTERROGATION
RTT-R - ROUND TRIP TIMING REPLY
S - SYNC
T - TIME REFINEMENT

7.8125 MILLISECOND TIMESLOT						
S	T	RTT-I	PROPAGATION (RTT-R DELAY	S	T	RTT-R	PROPAGATION

.936 MSEC	3.339 MSEC	.936 MSEC	2.6015.MSEC

Figure 29 – RTT   0525868

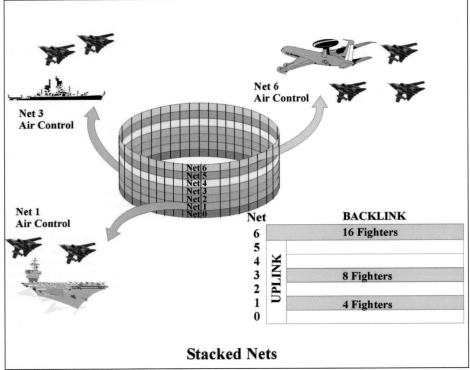

Stacked Nets

Figure 30   0525885

exclusive (for example, EW and high update rate PPLI). Platforms are assigned to one function or the other but cannot participate on both. These mutually exclusive NPGs are 'multinetted'. The same timeslot is used by different platforms for different functions; thus, network-wide throughput increases. For instance, Eurofighter performs high update rate position reporting while ships and E-3Ds perform EW co-ordination. This is implemented by placing each functional group (that is, PPLI-A NPG and EW NPG) on different nets. This multinet arrangement differs from stacked nets in that participants perform different functions and cannot selectively switch from one net (function) to the other.

Multiple Networks
Link 16's ability to separate functions via different hopping patterns can also be used to separate entire networks. Multiple spread spectrum systems operating in the same area maximise the use of the available electromagnetic spectrum. To prevent mutual interference, networks have different time references and should be crypto isolated. There is no information transfer between the networks and platforms cannot operate in more than one network. Switching networks requires reinitialising the terminal. For example; in Net 0, EW can be transmitted, while on Net 1, fighters transmit their high update rate PPLI, and on Net 2 a JSTARS platform transmits its surveillance information. Because these three activities do not conflict in interest, they can be put onto different nets coinciding in time, but separated by frequency hopping pattern.

Relay
Utilising the Link 16 relay feature creates an Extended LOS (E-LOS) datalink. Paired slot flood relays are often used to provide enhanced connectivity and improved anti-jam margin. The PPLI, surveillance, and air control NPGs are often relayed. Relays must be designed into a network since it requires timeslot assignments. Relays double (for one hop) the number of slots required to support an NPG. Multiple functions can be relayed and multiple hops are possible. Link 16 platforms relay all specified NPGs continuously. Prior to retransmission of the relayed data, the relay platform's terminal corrects errors in the message. Relayed voice does not have error coding and is retransmitted with errors. Message originators transmit in paired relay timeslots as well as in the primary timeslot. This eliminates the reception of a relayed message (loop-back) by its originator.

Track Numbers (TNs)
To ensure items reported on Link 16 are uniquely identified, they are assigned unique reference numbers, and whenever an item is transmitted over the link, it is told with that particular reference number. On Link 16, this unique reference number is called a Track Number (TN); TNs are also used for reference points, EW bearings, and so on. A Link 16 TN consists of two alphanumeric characters followed by three numeric characters, for example, AAnnn.

JU Numbers
The Link 16 version of Link 11 Participating Unit (PU) identifiers or Source Track Number (STN) is the JU. JU numbers may be any of the TNs, as none are reserved for this use. Each Link 16 participant is given a unique JU. This identifier acts as the unit identifier (in the PPLI), STN, and the link address.

Track Blocks
Track blocks (similar to Link 11 track blocks) are a range of track numbers set aside for use by surveillance participants.

Passive Operation
For the purposes of Emission Control (EMCON), Link 16 supports passive datalink participation. Units can enter the network passively or select passive operation after active entry. There are two types of passive operation in Link 16, Long Term Transmit Inhibit (LTTI) and Data Silent (DS). Passive units can receive all messages, but transmissions are prohibited or limited. In either passive mode, units continue to perform sync maintenance and navigation passively.

Crypto
Crypto is an inherent function in the JTIDS terminal. Networks are designed such that the JTIDS terminal requires an SDU for operation; thus, all transmissions from the terminal are encrypted. The JTIDS terminal encrypts each timeslot in two ways. Firstly, the waveform transmitted varies depending on the current Transmission Security (TSEC) variable. Secondly, the message itself is encrypted via the Message Security (MSEC) variable. JTIDS Crypto performs NPG and network isolation as well as data encryption.

Link 16 Functions
The functions currently supported by Link 16 include net entry (synchronisation), relative navigation (navigation and gridlock), Blue Force

reporting (identification), Mission Management and Weapons Control (MM/WC), surveillance, air control, secure voice and EW. Link 16 categorises these functions into communities of interest or NPGs. In order to provide interoperability with existing Link 11/Link 4A platforms, Link 16 incorporates a data-forwarding unit that is denoted operationally as the FJUA. Data forwarding enables the transfer of friendly tracks, force orders, and surveillance tracks between Link 11 and Link 16.

Network Participation Groups (NPGs)
Network Participation Groups (NPGs) are groups of J-series messages that combine to carry out a common function, such as surveillance. The function of an NPG is to hold and file messages created within a terminal, and then to release these messages for transmission in the correct TSB as laid down in the network design. Each message is allocated an NPG number; the host terminal then holds those messages within its system, only transmitting those messages in the TSB assigned for that particular NPG. NPGs are defined by the PG index 0 to 31 and can be divided into two groups, those which are used for network maintenance and those used to exchange tactical data. NPGs 1 to 4 are for network maintenance, NPGs 7 to 31 are for tactical data, and NPGs 5, 6, 7, 27 and 30 can be used for both. The capacity of each NPG is determined by:
- The transmission requirements of the units participating in that NPG
- The upper limit of message packing
- The access mode defined in the network design
- The requirement to relay data or voice.
- NPGs implemented in Link 16 networks in the UK are shown in Figure 31.

NPG	Functional Group
1	Initial Entry
2	RTT-A
3	RTT-B
4	Network Management
5	PPLI and Status Group A
6	PPLI and Status Group B
7	Surveillance
8/18	Mission Management/Weapon Co-ordination
9	Control
10	Electronic Warfare
12	Voice Group A
13	Voice Group B
19	Non-C^2 to Non-C^2
21	Spoke
22	Composite A
23	Composite B
27	Non-C^2 to Non-C^2 PPLI Exchange
29	Alphanumeric Messages E3D-E3D
30	IJMS Position and Status
31	Other IJMS Messages

Figure 31 – Network Participation Groups (NPGs)

The functions of the different NPGs are:
NPG 1 – Initial Entry. The initial entry NPG is used for the transmission of the initial entry and network time update messages.
NPG 2 – RTT-A. The RTT NPGs are used to achieve active synchronisation within the net. The RTT-A is used for the transmission of addressed RTT messages.
NPG 3 – RTT-B. The RTT- B NPG is used for the transmission of broadcast RTT messages.
NPG 4 – Network Management. The network management NPG is used to transmit changes to timeslot assignments, provide connectivity status, relay re-allocation and over-the-air-rekeying of crypto-variables.
NPG 5/6 – Precise Participation Location and Identification (PPLI). There are two PPLI NPGs – PPLI A and PPLI B; both are capable of fulfilling the same function:
PPLI A equates to NPG 5
PPLI B equates to NPG 6.
NPG 7 – Surveillance. The surveillance NPG is used by C^2 JUs to report tracks detected on their own sensors and tracks received from their controlled non-C^2 units.

NPG 8 – Mission Management and Weapon Co-ordination (MM/WC). The MM/WC NPG is used by C² JUs to report and receive the status of, and manage the effective use of assets at their disposal. Although this NPG is primarily used for C² to C², it can be monitored by non-C² units.

NPG 9 – Control. The control NPG can be used to fulfil the following functions:

Uplink. Uplink allows a C² unit to send messages to any fighters under its control, which can include commands, vectors and even a co-ordinated air picture specially tailored for the fighters' use

Backlink. Backlink allows fighters to pass messages, including command replies and radar targets, back to their controlling unit

Fighter-to-Fighter. This element of the backlink allows fighters to exchange radar targets they hold among themselves.

It should be noted that passing of fighter radar targets on the backlink and fighter-to-fighter are not separate functions. They result from the same transmission of data by a fighter. However, due to the large variety of messages in the air control function of the system, the US Air Force and Navy separate the same group of messages into NPGs 9 and 19. NPG 9 is used for control messages on the backlink and uplink, and NPG 19 for fighter-to-fighter. US forces do not allow fighter aircraft to contribute to the air picture, whereas UK Tornado F3s will use NPG 9 to pass target information to C² units or to other F3s.

NPG 10 – Electronic Warfare. The EW NPG is used by JUs capable of receiving ECM/ESM information to report that data to other net participants.

NPG 12 – Voice A. The voice A NPG can be used to support voice communications using either 16 Kbps between C² units and their fighters and from fighter-to-fighter, or the 2.4 Kbps ERV between C² units.

NPG 13 – Voice B. The Voice B NPG will be used to support the same functions as NPG 12.

NPG 21 – Spoke. UK non-C² JUs may report ECM spoke data on NPG 21 to other non-C² JUs or to C² JUs.

NPG 22 – Composite A. The composite NPGs are used to make maximum use of limited peacetime timeslot availability. The Composite A NPG, when used, replaces NPGs 4, 5, 6, 8, 9, and 21. Grouping all these NPGs into one PG with one large TSB adds to the flexibility of the peacetime network. For example, a JU may wish to transmit a number of control messages but may not have sufficient timeslots to fulfil this requirement. Therefore, that unit must seek additional timeslots within the net. However, by using the composite NPG, that unit can automatically use any spare timeslots within that Composite A to achieve his task.

NPG 23 – Composite B. The Composite B NPG, when used, groups the functions of NPGs 7 and 10 and provides the same flexibility described above.

NPG 29 – Residual Messages. This is a special NPG, provided to ensure a transmission opportunity for messages which are not assigned to one of the other NPGs. NPG 29 is not generally implemented by the UK, however, the UK E-3D uses NPG 29 to transmit J28.2 (0) alphanumeric free text messages to other E-3s.

NPG 30 – IJMS Position and Status. NPG 30 is used for net synchronisation and for transmitting an IJMS unit's position and status. Although IJMS is generally associated with the Class 1 terminal, Class 2 terminals should be able to synchronise with Class 1 terminals using this NPG.

NPG 31 – IJMS Messages. The IJMS NPG 31 is used for all other IJMS messages not included in NPG 30. Only the bilingual Class 2 terminals will be able to work IJMS in NPG 31.

Synchronisation

As a time-based system, Link 16 requires that each participating JTIDS terminal acquires and maintains an accurate knowledge of system time. A single designated unit, the Network Time Reference (NTR), establishes system time. This role designation can be given to any singular (but only

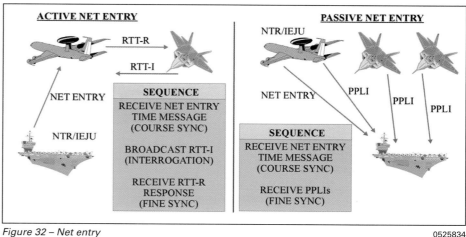

ACTIVE NET ENTRY

RTT-R
RTT-I
NET ENTRY
NTR/IEJU

SEQUENCE
RECEIVE NET ENTRY TIME MESSAGE (COURSE SYNC)

BROADCAST RTT-I (INTERROGATION)

RECEIVE RTT-R RESPONSE (FINE SYNC)

PASSIVE NET ENTRY

NTR/IEJU
NET ENTRY
PPLI
PPLI
PPLI

SEQUENCE
RECEIVE NET ENTRY TIME MESSAGE (COURSE SYNC)

RECEIVE PPLIs (FINE SYNC)

Figure 32 – Net entry   0525834

one) unit. Once operating, this unit will assume fine sync and begin broadcast transmissions of the system time and net entry messages. Once in fine synchronisation, units within LOS of the NTR can be designated as Initial Entry JTIDS Units (IEJUs) to transmit net entry messages for terminals that are beyond LOS of the NTR. Units within LOS of the NTR, or IEJUs, acquire system time during the net entry process.

Net Entry. During initialisation, the operator enters their best estimate of system time. Unless a time hack is obtained from the unit designated as NTR, Greenwich Mean Time (GMT) should be used for the time estimate. During net entry, this time establishes a window during which a terminal 'looks for' the net entry message. Once a message is received, the entering unit starts transmitting RTT messages in order to refine its estimate of system time. When a unit's time is properly refined, 'fine sync' is declared and full link participation is possible. Units in coarse sync can receive all messages or voice, but can only transmit RTTs. RTTs are used to refine time and thus aid in achieving fine sync.

Passive net entry can be achieved through the receipt of multiple PPLI messages. It requires that the passive units (and secondary users) be within LOS of at least three active units with good geometry.

Synchronisation Maintenance. Synchronisation is continuously maintained during link operations through the receipt of system time messages, the exchange of RTTs, and the receipt of PPLIs. Regardless of the method used to establish and maintain synchronisation, all units (except the NTR) develop a 'model' of the NTR internal clock. The accuracy of this model is referred to as the 'time quality'. The higher the time quality (NTR highest = 15), the longer the unit will maintain synchronisation when connectivity is lost. If a platform is positioned so that communications with the NTR and all IEJUs is lost, its clock model estimate will begin to 'drift'. Depending on the total time without connectivity (beyond LOS or jammed), the terminal will fall back to coarse sync, resend RTTs and attempt to resync. If the platform has drifted too far off the NTR clock model, a net entry reset may be required to re-enter the net.

Synchronisation NPGs. Three virtual circuits, called NPGs, support the synchronisation function. These are NPG 1, where the NTR (and any IEJU in fine sync) broadcasts net entry and system time, and NPG 2 or 3, where units perform RTTs. NPGs 2 and 3 differ in the type of access – dedicated or contention, respectively.

Identification

The identification function of Link 16 is performed through the transmission of PPLI messages. These messages are transmitted periodically by each member of the Link 16 network. The data forwarder also transmits 'Indirect PPLIs' (over Link 16) for the Link 11 units within the

battle group. These messages provide complete friendly force position and status reporting.

The Status portion of the identification function includes enough information to eliminate the need for voice status updates. Controllers can monitor the fuel status, engagement status and equipment status of all participating engaged aircraft. Status information includes:
- Platform type
- Voice net number
- Air control net number
- Network status (active, passive)
- Reply status
- Navigation qualities
- Equipment status
- Fuel status.

Identification NPGs. NPGs 5, 6, and 14 support the transmission of PPLIs. NPG 5 is the High Update Rate (HUR) PPLI NPG, used by fighters to provide a report every two seconds. NPG 6 supports all Link 16 participants (including fighters) at a 12-second rate. Only the Data Forwarder transmits on NPG 14. It is used to support the 'Indirect PPLIs' of the Link 11 participants.

Relative Navigation

Relative Navigation (RELNAV), the name of the JTIDS navigation function, is inherent to the TDMA architecture and the synchronisation process. It uses information from received PPLIs and the host navigation system to compute a common Link 16 navigation solution. This navigation solution provides the data registration reference for all information passed over the link. Since all platforms exchange navigation information, those platforms with high-quality fixing (Global Positioning System (GPS)) raise the navigation quality of all other platforms.

The RELNAV process uses a Kalman filter to compute position and velocity solutions in two separate grids, the relative grid (RELGRID (U,V,W)) and the geodetic grid (GEOGRID (latitude, longitude, altitude)). It uses host-provided navigation information and received PPLI navigation information to calculate its position and the estimated accuracy of that position within these two grids. These are then broadcast in the PPLI to all other participants.

Improved position and velocity data are sent to the host platform where it can be used for display, fixing or in-flight alignment purposes. The use of this information and the selection of the grid type are dependent on host platform implementation.

RELNAV accuracy is improved by the availability of high-quality users, good relative motion between sources, good geometry and time in the network.

Relative Grid. The RELGRID is a flat-plane grid, 1,024 sq n mile square. The plane is tangent to the earth at the grid origin. Units estimate position in the U (east), V (north), and W (altitude) co-ordinate system and report in feet from the grid origin estimate. Proper operation of the grid requires that a unit be assigned the role of Navigation Controller (NC) (and optionally, a Secondary Navigation Controller), and that all participating units initialise a common grid origin. The RELGRID is not required for proper link operation and is optional.

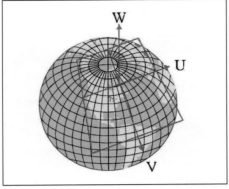

Figure 33 – RELGRID 0525837

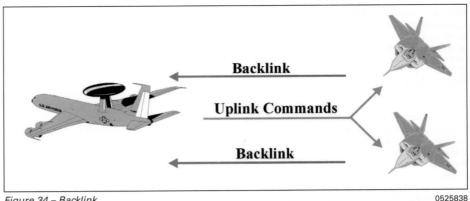

Figure 34 – Backlink 0525838

Geodetic Grid. The GEOGRID uses the standard latitude/longitude/altitude co-ordinate system. Each unit automatically computes its position (and quality) and broadcasts it in its PPLI. The solution is computed by using own navigation data coupled with received PPLI data. This allows units with high quality fix information (for example, GPS) to improve the latitude and longitude estimates of every other participant.

Navigation Resets. Navigation Resets can be initiated by the JTIDS terminal (automatically) or manually by an operator. Automatic resets may occur when the terminal is changing navigation modes or when inconsistent information has been provided. They are transparent to the operator in active sync (a momentary drop of position quality may be seen), but may cause a loss of fine sync when operating passively. Recurring automatic resets can indicate that another user is broadcasting overly optimistic qualities in its PPLI or that problems with own navigation processing exist. Manual resets are performed only to correct unusually poor PPLI track correlation or navigation errors.

Surveillance

Link 16 surveillance is the exchange of track and track management messages. The Blue Force reporting, force orders, and EW messages that are commonly associated with surveillance on Link 11 are performed using Link 16 NPGs (PPLI, MM/WC, EW, and so on).

Surveillance NPG. The surveillance function is supported by NPG 7. All C2 platforms (Ships, E-2Cs and E-3As) are assigned dedicated transmit timeslots during which they broadcast track reports. Track and data management also involves the resolution of data conflicts such as identity, IFF, strength and the management of track numbers. Non-C^2 units are primarily passive members within this function. The surveillance NPG varies in size depending on the projected use of the network. Current networks vary from 2,850 to 5,700 tracks. The number of timeslots assigned to each unit is based on its expected track loading.

Air (Weapons) Control

C^2 units (Ground Environment, Ships and AWACS) exercise control over fighters, providing tracks, mission assignment and vectors. The circuit is functionally divided into an uplink, where the controller broadcasts or addresses messages to all his fighters, and a backlink, where the fighters send track information to the C^2 unit. Fighters automatically backlink all tracks. To initiate air control between a ship/AWACS and a Link 16-equipped fighter, the handshaking process requires the PPLI NPG in conjunction with the air control NPG.

The messages used to support weapons control are used within the Control NPG and are as follows:

J12.0	Mission Assignment
J12.1	Vector
J12.3	Flight Path
J12.4	Controlling Unit Change
J12.5	Target/Track Correlation
J12.6	Target Sorting

Mission assignment. The mission assignment message is used by C^2 units to pass commands to non-C^2 JUs and details the position of the object of the mission, its identity and other supporting information. When the mission assignment information has been assessed by the non-C^2 JUs, they will respond with either a WILCO (WILl COmply), CANTCO (CANnoT COmply) or HAVCO (HAVe COmplied) message to the controlling unit. Mission assignments may be sub-divided into five groups as follows:

- Intercept (attack): the interception and attack of specific targets
- Intercept (non-attack): the interception, but not attack, of specific targets
- Procedural: non-intercept messages such as return to base, refuel, and so on
- Intercept (attack): cancel
- General mission assignments: air raid warnings, weapon control orders.

Vector Messages. The vector message enables a C^2 unit to give steering information to non-C^2 units

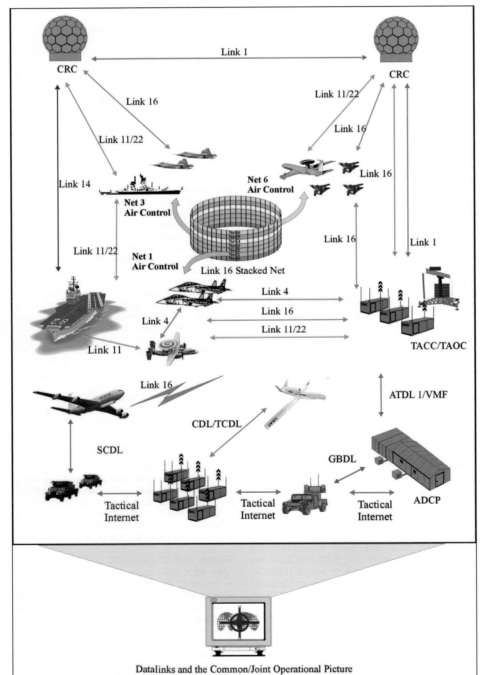

Datalinks and the Common/Joint Operational Picture

Figure 35 – Datalink connectivity 0525835

under its control. This message could consist of an advisory message before the transmission of mission assignment, or a command message to instruct a non-C² JU to assume a specified heading and altitude.

Flight Path. The flight path message enables the controlling C² JU to transmit flight path information to non-C² JUs. A flight path message may consist of up to fifteen individual waypoints.

Controlling Unit Change. The controlling unit change message supports the automated exchange of messages between C² JUs and non-C² JUs for the handover and termination of control of non-C² JUs.

Target/Track Correlation. All target reports received by a C² unit from a controlled non-C² unit go through the correlation process. The C² unit then reports the results of the correlation to its controlled non-C² JUs by means of the target/ track correlation message.

Target Sorting Messages. The target sorting message is a multipurpose message which is used to report targets and to pass mission commands. Non-C² JUs will use the target sorting message to transmit targets detected by their sensors to their controlling unit. These messages will be transmitted on the control NPG approximately every six seconds. If a target sorting message is received by the AEW from a non-C² JU, and no local or crosstold track exists in the AEW database, then a local track will be generated with a NTN/TN with a pre-set TQ. This track is then transmitted on the surveillance NPG

and back on the control NPG to the non-C² JU, if necessary. The AEW can also transmit this track out on Link 11 or IJMS if required.

Mission Management (MM)
Mission management messages are transmitted, primarily by C² JUs, in the MM NPG. Weapons management messages are used by the C² JUs to report, receive the status of and manage the assets under their command. The messages used to support this function are as follows:

J7.3	Pointer
J9.0	Command
J10.2	Engagement Status
J10.3	Handover Message
J10.5	Controlling Unit Report
J10.6	Pairing.

Pointer Message. The pointer message consists of Latitude and Longitude information, used to transmit data on items of specific interest. The pointer message is used to pass the precise position of a particular air target or ground feature.

Command Message. The command message is used by controlling C² JUs to direct other C² JUs in weapons control activities and weapons condition orders. Weapons control activities include directions for the launch aircraft and instructions to return to base. Command messages are also used to transmit changes to alert states.

Engagement Status Message. A non-C² JU reports its engagement status within the target sorting message on the control NPG only; therefore, this message is transmitted only

to those units on the control NPG. In order that a non-C² JU's engagement status can be received by a wider community, the C² JU uses the engagement status message, J10.2, on the management NPG to forward the non-C² JU's status. Wide circulation of engagement status information enables the effective management and utilisation of friendly forces by reducing the number of assets engaging the same target.

Handover Message. The handover message provides the data necessary to perform an automatic digital handover of the control of aircraft between C² JUs.

Controlling Unit Report Message. The controlling unit report message is used by C² JUs, within the air control mission, to report which aircraft they have under their control. This message is reported every eight frames.

Pairing Message. The pairing message is used to indicate a pairing between a friendly track and another track or point.

Secure Voice
JTIDS uses 16 Kbps Continuous Variable Slope Delta (CVSD) digital voice. The voice recognition quality of Link 16 allows operators to respond instantly to voice commands from wingman, air controllers, and so on, without the time-lag introduced by formal identification. Voice-to-data conversion is performed in the JTIDS terminal and does not require host processing.

Voice Utilisation. On aircraft and ships, the JTIDS voice circuits (labelled J1 and J2) are wired into the existing intercom system. Operators at existing

---

J-SERIES MESSAGE CATALOGUE

Network Management		**Weapons Co-ordination and Management**	
J0.0	Initial Entry	J9.0	Command
J0.1	Test	J 10.2	Engagement Status
J0.2	Network Time Update	J10.3	Handover
J0.3	Timeslot Assignment	J10.5	Controlling Unit Report
J0.4	Radio Relay Control	J10.6	Pairing
J0.5	Repromulgation Relay	**Control**	
J0.6	Communication	J12.0	Mission Assignment
J0.7	Timeslot Reallocation	J12.1	Vector
J1.0	Connectivity Interrogation	J 12.2	Precision Aircraft Direction
J1.1	Connectivity Status	J12.3	Flight Path
J 1 .2	Route Establishment	J12.4	Controlling Unit Change
J 1.3	Acknowledgement	J 12.5	Target/Track Correlation
J 1.4	Communicant Status	J 12.6	Target Sorting
J1.5	Net Control Initialisation	J12.7	Target Bearing
J1.6	Needline Participation Assignment Group	**Platform and System Status**	
Precise Participant Location and Identification		J13.0	Airfield Status Message
J2.0	Indirect Interface Unit PPLI	J 13.2	Air Platform and System Status
J2.2	Air PPLI	J13.3	Surface Platform and System Status
J2.3	Surface PPLI	J13.4	Subsurface Platform and System Status
J2.4	Subsurface PPLI	J13.5	Land Platform and System Status
J2.5	Landpoint PPLI	**Electronic Warfare**	
J2.6	Land Track PPLI	J14.0	Parametric Information
Surveillance		J14.2	Electronic Warfare Control/Co-ordination
J3.0	Reference Point		
Threat Warning			
J3.1	Emergency Point	J15.0	Threat Warning
J3.2	Air Track	**National Use**	
J3.3	Surface Track	J28.0	US National 1 (Army)
J3.4	Subsurface Track	J28.1	US National 2 (Navy)
J3.5	Land Point or Track	J28.2	US National 3 (Air Force)
J3.7	Electronic Warfare Product Inform	J28.3	US National 4 (Marines Corps)
Anti-submarine Warfare		J28.4	French National 1
J5.4	Acoustic Bearing and Range	J28.5	French National 2
Intelligence		J28.6	US National 5 (NSA)
J6.0	Intelligence Information	J28.7	UK National
Information Management	J29.0	J30.0	National Use (reserved)
J7.0	Track Management	**Miscellaneous**	
J7.1	Data Update Request	J31.0	Over-the-air Rekeying Management
J7.2	Correlation	J31.1	Over-the-air Rekeying
J7.3	Pointer	J31.7	No Statement
J7.4	Track Identifier		
J7.5	IFF/SIF Management		
J7.6	Filter Management		
J7.7	Association		
J8.0	Unit Designator		
J8.1	Mission Correlator Change		

Figure 36 – J-series message catalogue

consoles can select the channel they wish to access for secure voice communications, but not the Link 16 net number. However, net selection is neither available nor indicated at all access points. Of the 127 selectable stacked nets, platforms can access only one per channel. Furthermore, a platform relays only the net it has selected for use. Accordingly, the selections of nets for use require both internal and external co-ordination.

Voice NPGs. Voice channels A and B are supported by NPGs 12 and 13 respectively. When present, each requires 224 timeslots (P2SP or PDP) into which the digital bit stream from the CVSD encoder is placed. These are unique timeslot blocks and interaction between the channels is not possible. All platforms can transmit in these timeslots and access is controlled by push-to-talk protocol. All units receive when not transmitting. One voice channel typically requires 14.5 per cent of network capacity; relayed voice would require 29 per cent of network capacity.

Electronic Warfare

EW is supported by the Link 16 EW NPG. This function enables detailed reporting of emitters. EW track correlation messages are also available, along with intelligence data and emitter parametric data. When EW platforms receive Link 16, this function can be used to perform target correlation and track association. An EW JU is defined as a JU possessing an organic EW processing capability. EW data may consist of bearing intercepts from ESM equipment, jamming strobe data from active radars, and processed data such as fixes, areas of probability and system tracks. An EW JU can report data on the EW NPG using the J14.0 Parametric Information Message (PIM). The J14.0 message contains either parametric data or emitter evaluation. Within an EW community, a single

unit is nominated as the EW Co-ordinator (EWC) to control the exchange of EW information. The role of the EWC is to designate EW information of tactical interest to the wider surveillance community, and then to transmit that data within the surveillance NPG using the J3.7 EW message.

EW NPG. The EW function is performed on NPG 10. Since only C^2 units operate on this NPG, it is usually multinetted (same timeslots but different net number) with the HUR PPLI NPG (fighters only). Currently, each C^2 platform is assigned four dedicated transmit timeslots.

Spoke Reporting. The J29.0 spoke message is used by non-C^2 JUs to report spoke data on Link 16. The spoke message is transmitted on the control NPG or on the spoke NPG, if one has been defined. A non-C^2 JU can report up to four spokes. However, C^2 JUs cannot transmit spoke messages. The C^2 JUs do not perform any spoke processing, but do display the data received from the controlled non-C^2 JUs.

Fighter-to-Fighter (F/F) Group. Each fighter 'group' operates on its own stacked net (hopping pattern). The net number is operator selectable, allowing the fighters to change groups if necessary. The maximum number of fighters allowed in a group is determined by the F/F option selected.

F/F NPG. The F/F function is supported by NPG 19 (non-C^2 to non-C^2). This is a stacked net NPG and operators enter the net number on which they wish to operate. Fighters wishing to exchange target sorting (track) information must have the same net number selected. C^2 units participating (F/F advisory) in a fighter group must also select that group's net number.

J-Series Messages

Figure 35 defines the J-series messages, many of which have been discussed in the preceding paragraphs:

The Goal – Battlefield Digitisation

Digitisation is the near-realtime transfer of battlefield information between diverse fighting elements, which permits the shared awareness of the tactical situation. Digitisation uses information-age technologies, and the use of digital technology in sensors, intelligence fusion systems, communications systems and smart munitions increases the ability to manage, process, distribute and display C^2 information rapidly and globally.

A major product of digitisation is the Common Operational Picture (COP), mentioned at the beginning of this article. The COP enables commanders to synchronise the actions of air, land, sea, space and special operations forces. In essence, the COP enables common battlefield visualisation. A major element in the provision of sensor data to the COP is datalinks, which additionally provide the means of distributing tactical command decisions. The COP is still only an embryonic idea, as many technological problems, including the management, storage, filtering and dissemination of data, and the ever-increasing bandwidth requirement, need to be overcome before it can become an operational reality. However, datalinks are here today, and provide a sound picture at the tactical level for Command and Control. Figure 25 illustrates how some of the many datalinks contribute to the COP and allow the necessary elements to be fed back, in particular by Link 16, enabling aircraft and other assets to be provided with both Situational Awareness (SA) and tasking.

AIRBORNE RADAR

Introduction

The following overview of the theory and fundamentals of airborne radar encompasses most of today's modern radar systems from simple civilian weather radars to complex Track While Scan (TWS) systems used in combat aircraft. While the theoretical outline is not necessary to gain a sound understanding, it is included for those engineers and/or operators who require an in-depth knowledge of the subject.

PRINCIPLES OF RADAR

Radar, by definition, is an electromagnetic system for the detection of objects, be they ground based or airborne. The term 'radar' was originally coined during the Second World War as an acronym for 'Radio Direction and Ranging'. Nowadays, the acronym is interpreted as 'Radio Detection and Ranging'. The term 'radio' refers to the use of electromagnetic waves with wavelengths in the so-called radio wave portion of the spectrum, which covers a wide range from 10^4 km to 1 cm. Radar systems typically use wavelengths on the order of 10 cm, corresponding to frequencies of about 3 GHz. The detection and ranging part of the acronym is accomplished by timing the delay between transmission of a pulse of radio energy and its subsequent return. If the time delay is Δt, then the range may be determined by the simple formula:

$$R = c\Delta t/2$$

Where $c = 3 \times 10^8$ m/s, the speed of light, a fundamental value and the speed at which all electromagnetic waves are assumed to propagate. The factor of two appears in the denomination since the radar pulse must travel to the target and back to the receiver before detection, a distance of twice the range to the target, R.

The most common radar waveform is a train of narrow, rectangular shaped pulses modulating a sinewave carrier. The common radar carrier modulation, known as the pulse train, is shown in Figure 1, which also defines the common radar parameters.

PW (Pulse Width). PW has units of time and is commonly expressed in µs. PW is the duration of the pulse.

RT (Rest Time). RT is the interval between pulses. It is measured in µs.

PRT (Pulse Repetition Time). PRT has units of time and is commonly expressed in ms. PRT is the interval between the start of one pulse and the start of another. PRT is also equal to the sum, PRT = PW + RT.

PRF (Pulse Repetition Frequency). PRF has units of time and is commonly expressed in Hz (1 Hz = 1/s) or as pulses per second (pps). PRF is the number of pulses transmitted per second and is equal to the inverse of PRT.

RF (Radio Frequency). RF has units of time or Hz and is commonly expressed in GHz or MHz. RF is the frequency of the carrier wave that is being modulated to form the pulse train.

Radar components

A practical radar system requires seven basic components, as illustrated in Figure 2:

Transmitter

The transmitter creates the radio wave to be sent and modulates it to form the pulse train.

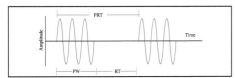
Figure 1 – Basic radar waveform 0528534

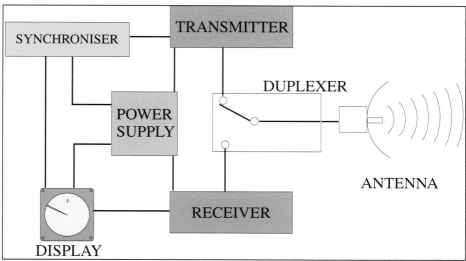
Figure 2 – Components of a basic radar system 0528535

The transmitter must also amplify the signal to a high power level to provide adequate range. The source of the carrier wave could be a magnetron, klystron or Travelling Wave Tube (TWT). Each has its own characteristics and limitations:

Magnetron, klystron and Travelling-Wave Tube (TWT)

The transmitter of an airborne radar system should be efficient, reliable, relatively lightweight and easily maintained. In addition, it requires the wide bandwidth and high power characteristic of radar applications. In MTI, pulse Doppler, and CW applications, the transmitter must generate noise-free, stable transmissions so that extraneous signals from the transmitter do not interfere with the detection of the small Doppler frequency shifts produced by weak moving targets.

The invention of the magnetron in the late 1930s facilitated radar systems that could operate at the higher microwave frequencies. While the magnetron transmitter has certain limitations, it continues to be widely used – generally in low average power applications, such as ship navigation and airborne weather avoidance radars. The magnetron is a power oscillator, which is to say that it self-oscillates when voltage is applied. Other radar transmitters are usually power amplifiers in that they take low-power signals at the input and amplify them to high power at the output. This provides stable high-power signals, as the signals to be radiated can be generated with precision at low power.

The klystron amplifier is capable of extremely high power levels and has good efficiency and stability. The disadvantages of the klystron are that it is usually large and it requires high voltages (about 90 kilovolts for one megawatt of peak power). At low power, the instantaneous bandwidth of the klystron is small, but it is capable of large bandwidth at high peak powers of a few megawatts.

The Travelling-Wave Tube (TWT) is related to the klystron. It has very wide bandwidths at low peak power, but as the peak power levels are increased to those needed for radar, its bandwidth decreases. As peak power increases, the bandwidths of the TWT and the klystron approach one another.

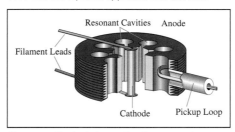

Figure a – Typical elements of a magnetron 0528569

Magnetron

Magnetrons can be manufactured relatively inexpensively because they require so few parts – a cathode, anode, tank circuit, and magnet. A typical magnetron is shown in Figure a.

The cylindrical anode structure of the magnetron contains a number of equally spaced cavity resonators with slots along the anode surface adjacent to the cylindrical cathode. A permanent magnet is usually used to provide the necessary magnetic field. The power output can be coupled through a slot in the cavity or by means of a coupling loop.

The cathode is heated so that it emits electrons. A strong DC voltage is applied between the electrodes, anode (positive) and cathode (negative). The electrons accelerate towards the anode, attracted by the positive voltage. As the velocity of each electron increases, the magnetic field produces an increasingly strong force on the electrons, causing them to follow a curved path that carries them past the openings of the cavities. This produces an oscillating electromagnetic field as the electrons sweep past the cavity openings. The frequency of this radio wave is the resonant frequency of the cavities. As in other types of oscillators, the oscillation originates in random phenomena in the electron space charge and in the cavity resonators. The cavity oscillations produce electric fields that spread outward into the interaction space from the slots in the anode structure, as shown in Figure b. The oscillation propagates from cavity to cavity via the interaction space. Energy is transferred from the radial DC field to the RF field by interaction of the electrons with the RF field. The first orbit of an electron occurs when the RF field across the gap is in a direction to retard the electron. This transfers energy from the electron to the tangential component of the RF field. The electron comes to a stop and is again accelerated by the radial DC field, making an orbit adjacent to the next cavity slot. If the RF field across the next cavity slot has changed phase by 180 degrees, the direction of the RF field is in the same direction as that of the electron, whose rotation is synchronous with the RF field. The electrons are gradually slowed down by their interaction with the travelling wave, and in the process give up energy to the wave before it finally terminates on the anode surface. There is

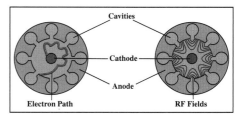

Figure b – Magnetron theory 0528570

a net delivery of energy to the cavity resonators because electrons that absorb energy from the RF field are quickly returned to the cathode. The energy in the rotational component of motion of the electrons in the retarding RF field remains practically unaffected, and the electrons may orbit around the cathode many times.

Klystron

In klystrons, the velocities of electrons emitted from the cathode are modulated to produce a density-modulated electron beam. The principle of operation involved here can be explained in terms of a two-cavity klystron amplifier, the basic structure of which is shown in Figure c.

The first grid next to the cathode controls the number of electrons in the electron beam and focuses it. The voltage between the cathode and the cavity resonators (termed the *buncher* and the *catcher*, which serve as reservoirs of electromagnetic oscillations) is the accelerating potential and is commonly referred to as the beam voltage. This voltage accelerates the DC electron beam to a high velocity before injecting it into the grids of the buncher cavity. The grids of the cavity enable the electrons to pass through, but confine the magnetic fields within the cavity. The space between the grids is referred to as the interaction space, or gap. When the electrons traverse this space, they are subjected to RF potentials at a frequency determined by the resonant frequency of the buncher cavity and the input signal frequency. The amplitude of the RF voltage between the grids is determined by the amplitude of the incoming signal in the case of an amplifier, or by the amplitude of the feedback signal from the second, or catcher, cavity if the klystron is used as an oscillator. Electrons traversing the interaction space when the RF potential on grid 3 is positive with respect to grid 2 are accelerated by the field, while those crossing the gap one half-cycle later are decelerated. In this process, essentially no energy is taken from the buncher cavity, since the average number of electrons slowed down is equal to the average number of electrons speeded up. The decelerated electrons give up energy to the fields inside the cavity, while those that have been accelerated absorb energy from its fields.

Upon leaving the interaction gap, the electrons enter a region called the drift, or bunching, space, in which the electrons that were speeded up overtake the slower-moving ones. This will cause the electrons .to bunch which results in density modulation of the beam, with the electron bunches representing an RF current in the beam. The catcher is located at a point where the bunching is greatest. This cavity is tuned to the same frequency as the input frequency of the input cavity resonator. The power output at the catcher is obtained by slowing down the electron bunches. If an alternating field exists at the output cavity resonator and grid 4 is positive with respect to grid 5, the electron bunches passing through the grids are decelerated, and they deliver energy to the output cavity. The electron bunches induce a RF current on the walls of the cavity identical to that in the beam. At resonance, the oscillation in the output cavity builds up in proper phase to retard the electron bunches. The power output is equal to the difference in the kinetic energy of the electrons averaged before and after passing the interaction gap. The positive electrode, or collector, located beyond the catcher, collects the electrons; it is designed to minimise secondary emission. (Such emission occurs because of the impact of electrons that reach the end wall).

The klystron amplifier can be converted into an oscillator by employing feedback from the output cavity to the input cavity in proper phase and of sufficient amplitude to overcome the losses in the system.

The power levels of klystrons are achieved through the use of very high beam voltages and currents. In simple terms, the output power is given by:

$$P_o = \text{efficiency} \times I_o E_o$$

where I_o and E_o are the beam current and voltage and the efficiency is defined by how well the

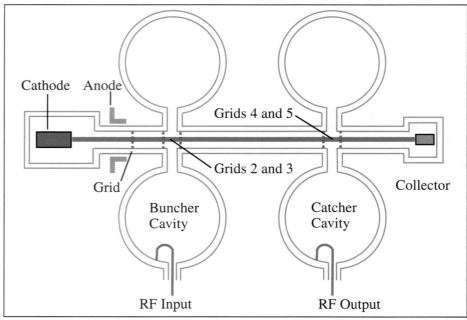

Figure c – A two-cavity klystron   0528568

Cathode Anode
Grids 4 and 5
Grids 2 and 3
Grid
Collector
Buncher Cavity
Catcher Cavity
RF Input
RF Output

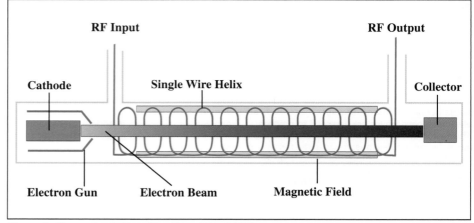

Figure d – Elements of a Travelling-Wave Tube (TWT)   0528571

RF Input
RF Output
Cathode
Single Wire Helix
Collector
Electron Gun
Electron Beam
Magnetic Field

DC power supplied is converted to RF power. For klystrons, the efficiency can be as high as 70 per cent. By collecting the spent electron beam at a potential significantly below that of the cavities, even higher efficiency can be achieved – perhaps as much as 10 to 15 per cent higher.

The power gain of the klystron is dependent on the voltage and current as well as on the number of cavities used. The larger the number of cavities employed, the larger the gain that can be obtained. There is, however, a practical limit imposed by the onset of RF instability.

Travelling-Wave Tubes (TWTs)

TWTs are generally used to amplify microwave signals over broad bandwidths. The main elements of a TWT are an electron gun, a focusing structure that keeps the electrons in a linear path, an RF circuit that causes RF fields to interact with the electron beam and a collector to gather the electrons.

There are two main types of TWTs, differentiated by their RF structures. One employs a slow-wave circuit called a helix for propagating the RF wave for electron-RF field interaction, whereas the other employs a series of staggered cavities, coupled to each other for wave propagation. Each type has different characteristics and finds its use in different applications. The helix TWT is distinct from other electron tubes, as it is the only one that does not use RF cavities. Since cavities have bandwidth limitations, the coupled-cavity TWT also is bandwidth limited to, typically, 10 to 15 per cent.

The helix TWT, however, has no particular bandwidth limitations and, for all practical purposes, an octave bandwidth (100 per cent) is attainable. The basic helix TWT is shown schematically in Figure d.

The electron gun contains a cathode that emits electrons, which are formed into a beam that is injected into the opening of the helix. As a result of the space-charge forces tending to make the electrons diverge radially, a focusing structure is used to keep the beam at a desired diameter by causing diverging electrons to be sent toward the axis of the helix. In this manner, the electron beam is maintained at the desired diameter all along the length of the helix. This is necessary because the electron-RF field interaction takes place continuously over the length of the helix within the helix diameter. In order to achieve this interaction, the diameter and pitch of the helix must be such that the RF wave travelling on the helix wire at the speed of light is slowed down in its axial travel to be synchronous with the velocity of the electrons in the beam. The axial phase velocity of the wave is approximated by multiplying the speed of light by the ratio of the pitch to the circumference of the helix. The axial phase velocity is relatively constant over a wide range of frequencies and this characteristic provides for the large bandwidths of helix TWTs. For typical applications, the electrons travel down the helix axis at about one-tenth the speed of light. The voltage required to impart this velocity to the electrons is on the order of 10,000 V. The RF output power and frequency required determine the actual voltage and current used.

The amplifying action of a TWT is via a continuous interaction between the axial component of the electric field wave travelling down the centre of the helix and the electron beam moving along the axis of the helix at the same time. The electrons are continually slowed down, and their energy is transferred to the wave along the helix. The electrons tend to bunch in regions where the RF field ahead is decelerating and the field behind is accelerating. The interaction between a bunched electron beam and a helix may be viewed in terms of induced currents.

The 'bunches' of electrons induce positive charges on the helix and these charges move in phase with the bunches. If the phase is correct, this current adds to the current associated with the RF wave flowing in the helix and causes the wave to grow. This interaction is continuous along the length of the helix. The wave amplitude growing on the helix in turn causes the electrons to bunch more and the growing bunches of electrons precipitate a continuous exponential growth of the helix wave with distance. Typical gains are in the order of four dB per cm and overall gains are 40 to 50 dB for helix tubes of practical sizes and applications. The DC-to-RF conversion efficiency of TWTs, both helix and coupled-cavity, is similar and is in the range of 20 to 60 per cent, depending on the power level and bandwidth.

Receiver

The receiver is tuned to the range of frequencies being transmitted and provides amplification of the returned signal. In order to provide the greatest range, the receiver must be very sensitive without introducing excessive noise. The ability to discern a received signal from background noise depends on the signal-to-noise ratio (SNR). The background noise is specified by an average value, called the noise-equivalent-power (NEP). This directly equates the noise to a detected power level so that it may be compared to the return. Using these definitions, the criterion for successful detection of a target is:

P_r > (SNR) NEP

where P_r is the power of the return signal. Since this is a significant quantity in determining radar system performance, it is given a unique designation, S_{min}, and is called the *Minimum Signal for Detection*:

S_{min} = (SNR) NEP

Since S_{min}, expressed in Watts, is usually a small number, it has proven useful to define the decibel equivalent, MDS, which stands for *Minimum Discernible Signal*:

MDS = 10 Log (S_{min}/1 mW)

In the receiver, SNR dictates a threshold for detection which determines those returns which will be displayed and those that will not. In theory, if SNR = 1, then only returns with power equal to or greater than the background noise will be displayed. However, noise is, as previously mentioned, a statistical process and varies randomly, with NEP representing the average value of the noise. There will be occasions when the noise exceeds the threshold set in the receiver. This noise will therefore be displayed and appear to be a legitimate target – termed a *false alarm*. If the threshold is set at a high level (coupled with a high SNR), there will be few false alarms, but some actual targets may not be displayed – known as a *miss*. If the threshold is set at a low level, there will be many false alarms, or a high False Alarm Rate (FAR). Some receivers monitor the background and constantly adjust the threshold to maintain a Constant False Alarm Rate, known as CFAR receivers.

Power Supply. The power supply provides the electrical power for all the components. The largest consumer of power is the transmitter, which may require several kW of average power. The actual power transmitted in the pulse may be much greater than 1 kW. The power supply only needs to be able to provide the average amount of power consumed, not the high power level during the actual pulse

transmission. Energy can be stored during the rest time in a capacitor bank. The stored energy is then put into the pulse when transmitted, thus increasing the peak power. The peak power and the average power are related by the Duty Cycle (DC). DC is the fraction of each transmission cycle that the radar is actually transmitting. Referring to the pulse train in Figure 1, the duty cycle can be seen to be:

DC = PW/PRF

Synchroniser

The synchroniser co-ordinates the timing for range determination. It regulates that rate at which pulses are sent (that is, sets PRF) and resets the timing clock for range determination for each pulse. Signals from the synchroniser are sent simultaneously to the transmitter, which sends a new pulse, and to the display, which resets the return sweep.

Duplexer

This is a waveguide switch, which alternately connects the transmitter or receiver to the antenna. Its purpose is to protect the receiver from the high power output of the transmitter. During the transmission of an outgoing pulse, the duplexer will be aligned to the transmitter for the duration of the pulse, PW. After the pulse has been sent, the duplexer will align the antenna to the receiver. When the next pulse is sent, the duplexer will shift back to the transmitter.

Antenna

The antenna takes the radar pulse from the transmitter and radiates it. Ideally the antenna will focus the energy into a well-defined beam which increases the power and permits a determination of the direction of the target. On reception, the aerial collects the energy contained in the reflected signal and delivers it to the receiver. The antenna keeps track of its own orientation which can be accomplished by a synchro-transmitter. Antenna systems which do not physically move but are steered electronically are known as Active Electronically Steered Arrays (AESA) and are described later.

Beam width. The beam-width of an antenna is a measure of the angular extent of the most powerful portion of the radiated energy. A desirable feature of antenna design is low sidelobe radiation, with maximum energy directed into the main lobe. For the purposes of this overview, the main lobe is all angles from the perpendicular where the power is not less than ½ of the peak power, or, in decibels, -3 dB. The beam-width is the range of angles in the main lobe and is usually resolved into a plane of interest, such as the horizontal or vertical plane. The antenna will have a separate horizontal and vertical beam width. For radar

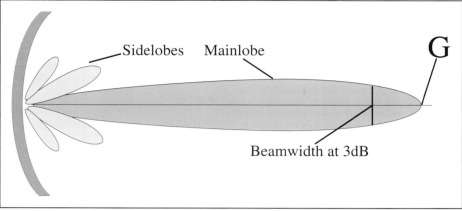

Figure 3 – Antenna mainlobe and sidelobes   0528536

antenna, the beam width can be predicted from the dimension of the a ntenna in the plane of interest by:

$\theta = \lambda/\mathbf{L}$

where:
θ = beam-width in radians
λ = wavelength of the radar
$\mathbf{L}$ = dimension of the antenna, in the direction of interest (that is, width or height).

Gain and directivity. The gain and directivity of the antenna main lobe depend on the dimensions of the antenna in relation to the wavelength (λ) and the efficiency, (η). Gain (**G**) is defined as the ratio between the radiated energy along the radioelectric axis and that radiated by an omnidirectional antenna. Where **S** is equal to the antenna surface area, the gain is as shown in Equation 1.

$$G = 4\pi \frac{S}{\lambda^2} \eta$$

Equation 1   0526774

Display

The display unit may take a variety of forms but in general is designed to present the received information to an operator in a form that satisfies the operational requirements, allows control of the radar functions and facilitates extraction of target data as required. Common display formats are shown in Figure 4.

Most air-to-air radar displays employ a B-Scope type of display, with a controlling radar computer filtering raw radar data to present a synthetic display to pilot, free from clutter and other interference in the picture. Radar information can also be displayed on the pilot's Head Up Display (HUD).

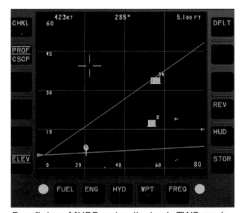

Eurofighter MHDD radar display in TWS mode
0121564

Radar performance

All of the parameters of a basic pulsed radar system will affect the performance in some way. The most common factors affecting radar performance are:

Pulse width

The duration of the pulse and the length of the target along the radial direction determine the duration of the returned pulse. In most cases the length of the return is usually very similar to that of the transmitted pulse. In the display unit, the pulse in terms of time will be converted into a pulse in terms of distance. The range of values from the leading edge to the trailing edge will create some uncertainty in the derived range to the target. In simplest terms, the ability to accurately measure range is determined by the pulse width.

If the uncertainty in measured range is designated the range resolution, R_{RES}, then it must be equal to the distance equivalent of the pulse width, namely:

$$R_{RES} = c\ PW/2$$

Ideally, measurement would be taken from the leading edge of the pulse, as the range can be determined with much finer accuracy. However, it is virtually impossible to create the perfect leading edge. In practice, the ideal pulse will appear as in Figure 5.

To create a perfectly formed pulse with a vertical leading edge would require infinite bandwidth. In fact bandwidth (β) of the transmitter may be equated to the minimum pulse width, PW by:

$$PW = 12\ \beta$$

Therefore, the range can be determined no more accurately than cPW/2 or:

$$R_{RES} = c/4\ \beta$$

High-resolution radar is often referred to as wide-band radar; these can be seen as equivalent statements, with one term referring to the time domain and the other the frequency domain. The duration of the pulse also affects the minimum range at which the radar system can detect a target. The outgoing pulse must physically clear the antenna before the return pulse can be processed. Expressed in terms of time, this event lasts for an interval equal to the pulse width, PW and the minimum displayed range is then:

$$R_{MIN} = c\ PW/2$$

The rear cockpit of an F-16B with prominent radar/HUD repeater in the centre of the coaming (I Black) 0528527

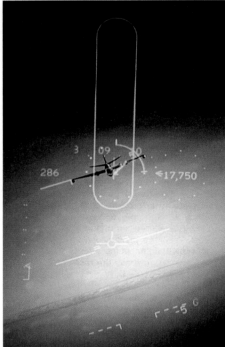

Radar information displayed in the HUD of a Tornado F-3. On the left, another Tornado F-3 is locked with the radar, showing range to target (quarter arc). On the right, a Canberra is the target aircraft, with the HUD display showing the radar search volume and range-to-target (I Black) 0528529

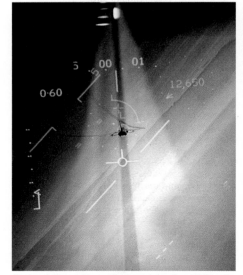

Radar information displayed in the HUD of a Tornado F-3. On the left, another Tornado F-3 is locked with the radar, showing range to target (quarter arc). On the right, a Canberra is the target aircraft, with the HUD display showing the radar search volume and range-to-target (I Black) 0528528

COMMON RADAR DISPLAYS

Plan Position Indicator

PPI (plan position indicator). The A-scan information is converted into brightness and then displayed in the same relative direction as the antenna orientation. The result is a top-down view of the situation where range is the distance from the origin.

A Scope

The most basic display type is called an A-scope (amplitude vs. Time delay). The vertical axis is the strength of the return and the horizontal axis is the time delay, or range. The A-scope provides no information about the direction of the target.

B Scope

The B Scope displays targets on a rectangular plot of range versus amplitude.
The B Scope is widely used in fighter applications.

C Scope

The C Scope displays target elevation angle versus azimuth. Widely used for HUD Displays as it corresponds to the pilots view.

Sector PPI

Sector PPI displays are widely used for weather and mapping applications as they give an undistorted view of the area being scanned.

Patch Map

Patch Maps are commonly used for high-resolution (SAR) ground mapping.

Figure 4 – Common types of radar display presentation 0528537

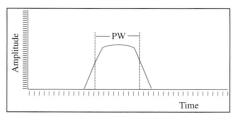

Figure 5 – Radar pulse characteristic　　　0528538

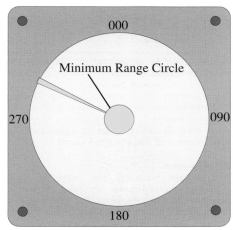

Figure 6 – Minimum range　　　0528539

The minimum range effect can be seen on a Plan Position Indicator (PPI) display as a saturated or blank area around the origin (Figure 6).

Increasing the pulse width while maintaining the other parameters will also affect the duty cycle and therefore the average power of the system. For many systems, it is desirable to keep the average power fixed. In this case, the PRF must be varied simultaneously with PW in order to keep the product PW × PRF the same. For example, if the pulse width is reduced by half, in order to improve the resolution, then the PRF is usually doubled.

Pulse Repetition Frequency (PRF)

The frequency of pulse transmission affects the maximum range that can be displayed. Recall that the synchroniser resets the timing clock as each new pulse is transmitted. Returns from distant targets that do not reach the receiver until after the next pulse has been sent will not be displayed correctly. Since the timing clock has been reset, they will be displayed as if the range where less than actual. If such a condition were allowed, under such circumstances the range information would be ambiguous, with the operator uncertain as to whether the displayed range was correct or less than the actual range to target.

The maximum actual range that can be detected and displayed without ambiguity, or *the maximum unambiguous range*, is the range corresponding to a time interval equal to the pulse repetition time, PRT. Therefore, the maximum unambiguous range is given by:

$$R_{UNAMB} = c\ PRT/2 = c/(2PRF)$$

When a radar is scanning, it is necessary to control the scan rate so that a sufficient number of pulses will be transmitted in any particular direction in order to guarantee reliable detection. If too few pulses are used, then it will more difficult to distinguish false targets from actual ones. False targets may be present in one or

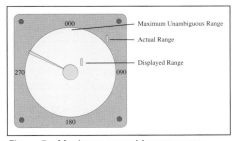

Figure 7 – Maximum unambiguous range

0528540

two pulses; however, to maintain a low false detection rate, the number of pulses transmitted in each direction is normally above ten.

For systems with high pulse repetition rates (frequencies), the radar beam can be repositioned more rapidly and therefore scan more quickly. Conversely, if the PRF is lowered the scan rate needs to be reduced. For simple scans it is easy to quantify the number of pulses that will be returned from any particular target. Let τ represent the *dwell time*, which is the duration that the target remains in the radar's beam during each scan. The number of pulses, N, that the target will be exposed to during the dwell time is:

$$N = \tau PRF$$

We can rearrange this equation to make a requirement on the minimum dwell time for a particular scan:

$$\tau_{min} = N_{min}/PRF$$

So it can be seen that high pulse repetition rates require smaller dwell times. For example, for a continuous circular scan, the dwell time is related to the rotation rate, τ (radians/sec) and the beam-width τ (radians), by:

$$\tau = \theta/\Omega$$

These relationships can be combined, giving the following equation from which the maximum scan rate may be determined for a minimum number of pulses per scan:

$$\Omega_{MAX} = \theta PRF/N$$

Radar frequency

The frequency of the radio carrier wave will also have some effect on how the radar beam propagates. At the low frequency extremes, radar beams will refract in the atmosphere e and can be caught in 'ducts' which can result in long range performance. At the high extreme, the radar beam will behave much like visible light and travel in very straight lines. Very high frequency radar beams will suffer high losses and are not suitable for long range systems. The frequency will also affect the beam-width. For the same antenna size, a low frequency radar will have a larger beam width than a high frequency system. In order to keep the beam width constant, a low frequency radar will therefore require a large antenna.

Theoretical maximum range equation

A radar receiver can detect a target if the return is of sufficient strength. The minimum return signal that can be detected is designated S_{min}, with units of Watts (W). The size and ability of a target to reflect radar energy can be summarised into a single term, σ, known as the Radar Cross Section (RCS), which has units of m². If all of the incident radar energy directed onto the target was reflected equally in all directions, then the radar cross section would be equal to the target's cross-sectional area as seen by the transmitter. In practice, some incident energy is absorbed and reflected energy is not distributed equally in all directions. As a result, RCS is quite difficult to estimate and is normally determined by measurement. A simple model for the radar power that returns to the receiver is given by Equation 2.

$$P_r = \left[P_t G \times \frac{1}{4\pi R^2} \right] \times \left[\sigma \times \frac{1}{4\pi R^2} \right] \times A_e$$

Equation 2　　　0526775

The terms in this equation have been grouped to illustrate the sequence from transmission to collection. The transmitter outputs peak power P_t into the antenna, which focuses it into a beam with gain G. The power gain is similar to the directional gain, G_{dir}, except that it must also include losses from the transmitter to the antenna. These losses can be summarised by the single term for efficiency, ρ. The gain is therefore G = ρ G$_{dir}$.

Physically, radar energy spreads out uniformly in all directions; the power per unit are a must therefore decrease as the area increases. Since the energy is spread out over the surface of a sphere, the factor of $1/4\pi R^2$ accounts for the energy reduction. The radar energy is collected by the surface of the target and reflected. The RCS, σ, accounts for both of these processes. The reflected energy spreads out in an identical manner to the transmitted energy.

The receiving antenna collects the energy proportional to its effective area, known as the aperture, A_e. This term also includes losses in the reception process until the signal reaches the receiver, hence the subscript 'e', meaning 'effective aperture'. The effective aperture is related to the physical aperture, A, by the same efficiency term used in power gain, given the symbol ρ, so that $A_e = \rho A$.

The criterion for detection is simply that the received power, P_r must exceed the minimum, S_{min}. Since the received power decreases with range, the maximum detection range will occur when the received power is equal to the minimum, that is, $P_r = S_{min}$. Solving for range, the equation for the maximum theoretical radar range becomes as shown in Equation 3.

$$R\max = \sqrt[4]{\frac{P_t G \sigma A_e}{(4\pi)^2 S_{min}}}$$

Equation 3　　　0526776

Continuous Wave (CW) radar

Principle of operation. Continuous Wave (CW) radar systems emit electromagnetic radiation at all times. Conventional CW radar cannot measure range because there can be no basis for the measurement of pulse-to-pulse time delay, as opposed to the aforementioned basic radar system, which creates pulses and uses the time interval between transmission and reception to determine target range.

However, CW radar can measure the instantaneous rate-of-change of target range. This is accomplished by a direct measurement of the Doppler shift of the returned signal. The Doppler shift is a change in the frequency of the electromagnetic wave caused by motion of the transmitter, target or both. For example, if the transmitter is moving, the wavelength is reduced by a fraction proportional to the speed it is moving in the direction of propagation. Since the speed of propagation is a constant, the frequency must increase as the wavelength shortens. The net result is an upwards shift in the transmitted frequency, called the Doppler shift.

If the receiver is moving opposite to the direction of propagation, there will an increase in the received frequency. Furthermore, a radar target which is moving will act as both a receiver and transmitter, with a resulting Doppler shift for each. The two effects caused by the motion of the transmitter/receiver and target can be combined into a net shift of the frequency. The amount of shift will depend of the combined speed of the transmitter/receiver and the target along the line between them, termed the Line of Sight (LoS).

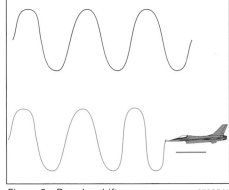

Figure 8 – Doppler shift　　　0528541

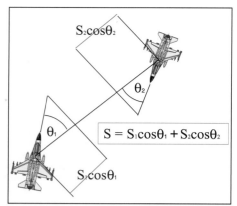

Figure 9 – Relative speed in the mutual Line of Sight (LoS)　0528542

The Doppler shift can be calculated with the transmitter/receiver and target speeds, designated as S_1 and S_2, and the angles between their directions of motion and the LoS, designated θ_1 and θ_2. The combined speed in the LoS is thus:

$$S = S_1 \cos\theta_1 + S_2 \cos\theta_2$$

This speed can also be interpreted as the instantaneous rate of change in the range, or *range rate*. While the problem is confined to two dimensions, the angles also have simple interpretations: θ_1 is the relative bearing to the target. This is the difference between the course of the transmitter/receiver and the true bearing to the target. Assuming that the range rate is known, the shift in returned frequency is:

$$\Delta f = 2s/\lambda$$

where λ is the wavelength of the original signal. As an example, the Doppler shift in an X-band (10 GHz) CW radar will be about 30 Hz for every 1 mph combined speed in the LoS.

CW radar systems are used in military applications where a measurement of the range rate is desired. Of course, range rate can be determined from the basic pulsed radar system by measuring the change in the detected range from pulse to pulse. However, and most usefully, CW systems measure the instantaneous range rate and maintain continuous contact with the target.

Frequency Modulated Continuous Wave (FMCW) radar

Frequency Modulated CW radar systems measure range instead of range rate by frequency modulation of the transmitted wave, which overlays a unique 'time stamp' on the transmitted wave at every instant. By measuring the frequency of the return signal, the time delay between transmission and reception can be determined and therefore the range, by $R = c\Delta t/2$. The amount of frequency modulation must be

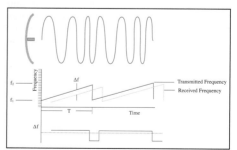

Figure 10 – FMCW theory of operation　0528543

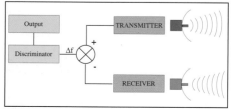

Figure 11 – FMCW system block diagram

0528544

significantly greater than the expected Doppler shift in order to prevent ambiguous results.

The simplest way to modulate the wave is a linear increase in the frequency where the transmitted frequency will change at a constant rate. The FMCW system measures the instantaneous difference between the transmitted and received frequencies, Δf. This difference is directly proportional to the time delay, Δt, which is the time it takes the radar signal to reach the target and return. From this the range can be found using the usual formula, $R = c\Delta t/2$. The time delay can be determined by:

$$\Delta t = T\Delta f/(f_2 - f_1)$$

where:
f_2 = maximum frequency
f_1 = minimum frequency
T = period of sweep from f_1 to f_2
Δf = the difference between transmitted and received.

In order to prevent problems in such a system when the sweep resets the frequency and the frequency difference becomes negative (as shown in the plot of Δf vs. time), the system uses a discriminator to clip off the negative signal, leaving only the positive part, which is directly proportional to the range. The components of an FMCW system are shown in Figure 11.

The foregoing equations can be combined to ascertain the range:

$$R = 2cT\Delta f/(f_2 - f_1)$$

where Δf is the difference between the transmitted and received frequency (when both are from the same sweep, that is, when it is positive).

An FMCW system can also be designed to compare the phase difference between the transmitted and received signals after they have been demodulated. This system does not have to discriminate the negative values of Δf. In both cases however, the maximum unambiguous range is still determined by:

$$R_{unamb} = cT/2$$

FMCW systems are often used for radar altimeters, or in radar proximity fuses for warheads. These systems do not have a minimum range like a pulsed system. However, they are not suitable for long-range detection because the continuous power level they transmit is considerably lower than the peak power of a pulsed system. As previously discussed, the peak and average power in a pulse system were related by the duty cycle,

$$P_{ave} = DC \times P_{peak}$$

For a continuous wave system, the duty cycle is one, or alternatively the peak power is the same as the average power. In pulsed systems the peak power is many times greater than the average.

Radial velocity discrimination

There are many radar applications where it is useful to know both the range and the radial velocity of the target. Since the relative radial velocity is the range rate, measurement of the radial velocity can be used to predict the target's range in the near future. For example, it allows system prediction of the time a target will be inside the effective range of a weapon system.

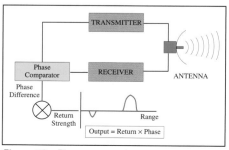

Figure 12 – Phase comparison output　0528545

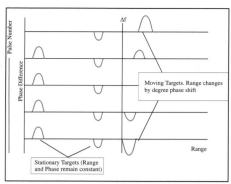

Figure 13 – Five sequential returns from the pulse comparison output　0528546

Radial velocity discrimination can also be used to eliminate potential false targets, such as sea clutter or buildings, from the display. There are three methods used that can give simultaneous measurement of range and range rate.

Differentiation

This system measures the range at fixed intervals and computes the rate of change between the measurements. For example, if a target is at 2,000 m for the first measurement and at 1,990 m for the next measurement made 1 second later, the range rate is –10 m/s. Light Detection And Ranging (LIDAR) systems also use this method. Accuracy is improved by taking several quick measurements and computing the average rate of change; the intervals cannot be too small however, as the target must be able to change range by a significant distance during the measurement interval.

Moving Target Indicator (MTI)

MTI measures changes in the phase of the returned signal to determine motion of the target. In order to measure the phase, a sample of the transmitter pulse is fed into a *phase comparator*, which also samples the return signal. The output of the phase comparator is used to modulate the display information. Returns will be the largest and positive when they are in-phase; they will achieve the largest negative value when out of phase.

When the range to a target is changing, the phase comparison output will vary between its extreme values, as well as moving in range. One full cycle of phase shift is completed as the range changes by one-half wavelength of the radar. This is because the radar signal travels both to and from the target, so that the total distance travelled by the radar pulse changes by a factor of two. For a typical radar wavelength of 3 cm, the phase comparison output will rapidly vary for targets whose range is changing.

Stationary targets have a fixed phase difference which can be exploited to remove them from the display by a cancellation circuit. The MTI processor takes a sample from the phase comparison output and averages it over a few cycles. Moving targets will average to zero, while stationary targets will have non-zero averages. The average signal is then subtracted from the output before it is displayed, thereby cancelling out stationary targets.

Stationary targets, by definition, are those returns that are not changing in range, relative to a fixed transmitter. For moving transmitters, returns from fixed objects on the ground will be changing in range and therefore displayed. Therefore, MTI systems for moving transmitters must provide a modified input to the phase comparator, which includes the phase advance associated with the motion of the transmitter.

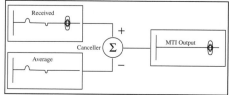

Figure 14 – Cancellation circuit of MTI processor　0528547

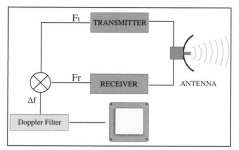

Figure 15 – Pulsed Doppler radar system components 0528548

PD radars are widely utilised in modern Air Interception (AI) radar systems, such as the AN/APG-70 in the Boeing F-15E Strike Eagle 0083123

Pulse Doppler (PD) radar

PD radar systems add additional processing equipment to the basic pulsed radar system. A sample of the transmitted signal is directed to a mixer, which also samples the output from the receiver. The output of the mixer is the Doppler shift, Δf. The Doppler shift is passed to a filter that modifies the display information accordingly.

A common practice in PD systems is to colour code the return information on the PPI display. The Doppler shift is sorted into categories, for example positive, zero and negative, which are then associated with colours. This method is also seen in weather radar images.

Pulsed Doppler radar systems are used in numerous military applications, most notably, Airborne Intercept (AI) radar.

Limitations

MTI and PD radar systems cannot measure velocities above a certain value, known as the first blind speed, or maximum unambiguous speed. For MTI systems, the first blind speed occurs when the change in range between pulses is exactly one-half of the wavelength. At this relative speed, phase change equals 360°, which is the same as 0°, or no phase shift at all. Therefore, a target moving at the first blind speed will appear to be stationary and be cancelled from the radar display. While this may seem to be a major limitation, it should be noted that any relative velocity change, precipitated either by the fighter or the target,

Figure 16 – A pulsed Doppler radar antenna, in this case as part of Thales' RC400 AI radar system 0011895

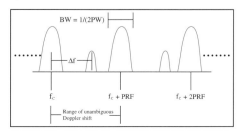

Figure 17 – Spectrum of pulsed Doppler radar 0528549

away from the first blind speed will render the target immediately visible; in the fluid arena of air-to-air combat a long-term condition with a relative velocity equal to the first blind speed is highly unlikely.

Further, in PD radar systems, there will be a main lobe centred on the radar carrier frequency. The bandwidth (BW) of the main lobe will be determined by the pulse width (PW), from the fundamental relationship:

$$BW = 1/(2PW)$$

Since the cycle is repeated at a frequency equal to the PRF, there will be images of the main lobe at multiple intervals of the PRF on either side of the carrier frequency. All of the information from the return will be repeated at intervals of PRF, including the Doppler shifted return. The only useable portion of this spectrum is the interval between the main lobe and the first harmonic, at $f_c + PRF$. Therefore, only Doppler shifts that fall within this range can be measured unambiguously. The condition when the Doppler shift is equal to the PRF defines the maximum unambiguous speed that can be measured. From this the maximum unambiguous speed is:

$$\Delta f = 2s/\lambda = PRF$$
$$S_{unamb} = \lambda PRF/2$$

This is also the same condition described for MTI systems, namely when the target moves one-half wavelength in the period PRT:

$$s_{blind} \times PRT = \lambda/2 \quad s_{blind} = \lambda PRF/2$$

From the spectrum, it is also apparent that the Doppler shift must be larger than the bandwidth of the main lobe in order to be detectable. This defines a minimum detectable speed:

$$\lambda f_{min} = 2s_{min}/\lambda = 1/(4PW)$$
$$s_{min} = \lambda /(8PW)$$

There are two solutions for the problem of ambiguous speed and range measurements. The first is to increase the PRF, which will reduce the maximum unambiguous detection range accordingly. The other is to vary the PRF. Ambiguous returns will vary either in range or velocity, while accurate ones will not; this method can be used to identify conditions where the target range is beyond R_{unamb} or when the target radial speed is beyond s_{unamb}.

High resolution radar
Pulse compression

This is a method that combines the high energy of a long pulse width with the high resolution of a short pulse width. The pulse is frequency modulated, which provides a method to further

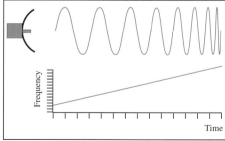

Figure 18 – Pulse compression using frequency modulation 0528550

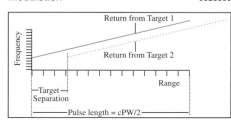

Figure 19 – Overlapping returns separated by frequency 0528551

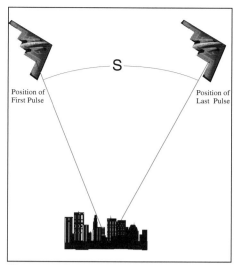

Figure 20 – Synthetic aperture 0528552

resolve targets which may have overlapping returns. Since each part of the pulse has unique frequency, the two returns can be completely separated. The pulse structure is shown in Figure 18.

The receiver is able to separate two or more targets with overlapping returns on the basis of the frequency. Figure 19 illustrates a return showing two targets with separation less than the conventional range resolution.

When the pulse is frequency modulated in this manner, the process of resolving targets on the basis of frequency is called *Post-Detection Pulse Compression* (PDPC). The ability of the receiver to improve the range resolution over that of a conventional system is called the *pulse compression ratio*. For example, a pulse compression ratio of 20:1 means that the system range resolution is reduced by 1/20 of a conventional system. Alternatively, the factor of improvement is given the symbol PCR, which can be used as a number in the range resolution formula, which now becomes:

$$R_{res} = c\ PW/(2\ PCR)$$

This process does not improve the minimum range. The full pulse width still applies to the transmission, which requires the duplexer to remain aligned to the transmitter throughout the pulse. Therefore R_{min} is unaffected.

Synthetic Aperture Radar (SAR)

As previously described, angular resolution is determined by the beamwidth of the antenna. At a given range, R, the ability to resolve objects in the cross-range direction, known as the *cross-range resolution*, is calculated by:

$$\theta R_{cross} = R\theta$$

where θ is the beamwidth, expressed in radians. Mathematically, this value is merely the arc length swept out by the angle θ at radius R. It is also the width of the radar beam at the range R. For example, a 6° beamwidth (0.1 radians) will be 10 m wide at a range of 100 m. For most radar antennas the beamwidth is sufficiently large so that the cross range resolution is fairly large at normal detection ranges. As such, these systems cannot resolve the detail of the objects they detect.

Synthetic Aperture Radar (SAR) uses the motion of the transmitter/receiver to generate a large effective aperture. In order to accomplish this, the system stores several returns taken while the antenna is moving and then reconstructs them as if they came simultaneously. If the transmitter/receiver moves a total distance S during the period of data collection, with several return pulses stored, then the effective aperture upon reconstruction is also S.

This large synthetic aperture creates a very narrow beamwidth which can be calculated by the usual beamwidth formula, substituting the synthetic aperture for the physical antenna

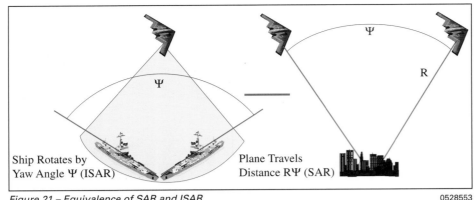

Ship Rotates by
Yaw Angle Ψ (ISAR)

Plane Travels
Distance RΨ (SAR)

Figure 21 – Equivalence of SAR and ISAR   0528553

aperture. The new beamwidth can be used to predict the improved cross-range resolution:

$$\Delta R_{cross} \approx R\, \lambda/S \text{ (SAR)}$$

where:
R = target range
S = distance travelled by the transmitter/receiver during data collection
λ = the wavelength of the radar

The most frequent application of SAR is with satellite radar systems. Since the satellite is travelling at high velocity, the accuracy of these systems can be made very high. Furthermore, if the target is fixed in location, the period for data collection can be made very long without introducing significant error. Therefore satellite SAR is used primarily for the imaging of fixed objects like terrain, cities, military bases, and so on.

Inverse Synthetic Aperture Radar (ISAR)

A large synthetic aperture can be achieved without moving the transmitter/receiver. If the target rotates by a small amount, it has the same effect as if the transmitter/receiver were to travel a distance equal to the arc length at the range R. ISAR systems are typically used for long-range imaging and identification of possible targets. The ISAR platform may be fixed or moving. Figure 21 illustrates this effect for a yaw angle ψ of a ship at range R.

Since the effect is the same for the cross range resolution, a similar calculation substituting the distance Rψ for the aperture may be used.

$$\Delta R_{cross} = R\,(\lambda/R\psi) – \text{the range cancels out to}$$
give:
$$\Delta R_{cross} = \lambda\,/\psi \text{(ISAR)}$$

Note that the result is independent of range.

Phased array radar – Active Electronically Scanned Arrays (AESA)

A radar beam may be formed using a planar array of simple antenna elements (for example, dipole antennas) in, for example, a linear array as illustrated in figure 22.

If all of the antennas are driven coherently, that is, at the same frequency and phase, then the condition of maximum constructive interference will occur in the two directions perpendicular to the array axis. If this system is modified to include a variable phase shift on the input to each element, the condition of maximum constructive interference and, hence, the axis can be changed.

Taking the example of a three element linear array, all fed from a common source with a variable phase shift at each element, in order to change the direction of maximum constructive interference by an angle θ, the phase shifts must be chosen to exactly compensate for the phase

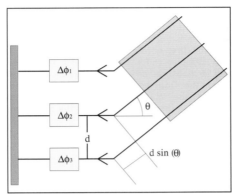

Figure 23 – Path length difference for beam steered off-axis   0528554

shift created by the extra distance travelled, d sin(θ). The condition for the phase shift between adjacent elements can be found from:

$$\Delta\Phi_{adj} = (2\pi/\lambda)\,d\,\sin(\theta)$$

It is essential to assign an appropriate sign to the phase shift. In the example above, the phase of the signal from antenna element 3 must be advanced relative to element 2, therefore the phase shift is positive.

This same principle may be applied to a planar array. In this case, the phase shift will steer the beam both in azimuth and elevation. The phase shifts required may be computed independently and combined algebraically to give the net phase shift required. It is necessary to establish a co-ordinate system for calculation; in Figure 24, it is as follows. When facing the array, the upper left element will have co-ordinates (0,0). The elements will be assigned co-ordinates with the first number representing the elevation and the second the azimuth element number. The general element co-ordinates will be (e,a), where:
e = elevation element number
a = azimuthal element number

When referring to the phase shift at a specific element, it will be in reference to the element (0,0). Using this system, the angles will be positive in the directions indicated.

The phase shift required to steer the beam to an elevation angle φ (defined so that upwards is

The AN/APG-79 AESA radar will enter service in the F/A-18E/F, offering a fivefold increase in reliability over the mechanically scanned AN/APG-73 currently used. Note the slight upward tilt on the array, designed to deflect any incident radar energy away from the likely receiver, thus enhancing the stealth characteristics of the host aircraft   0118181

The upgraded AN/APG-63(V)2 radar for the Boeing F-15 Eagle includes an AESA antenna   0111020

positive) and azimuth angle θ (positive when to the left as seen looking into the array), will be:

$$\Delta\phi_{e,a} = (2\pi/\lambda)[\,e\,d_e\,\sin\phi + a\,d_a\,\sin\theta\,]$$

where d_e and d_a refer to the element spacing in the vertical and horizontal directions respectively.

Radar tracking systems

Tracking systems are used to determine the location or direction of a target on a near-continuous basis. An ideal tracking system would maintain contact and constantly update the target's bearing (azimuth), range and elevation. The output of the tracking system can be sent to a Fire Control System (FCS), which stores the information and derives the target's motion and therefore its future position. Tracking systems not only provide an automatic target-following feature but also determine the target's position with sufficient accuracy for weapons delivery.

Radar servo tracking system. One of the most basic tracking system designs is the servo tracking system. Here, the radar antenna is initially trained

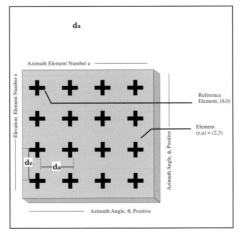

Figure 24 – Planar phased array   0528555

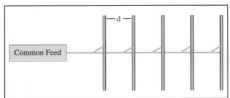

Figure 22 – Linear array of antennas   0528572

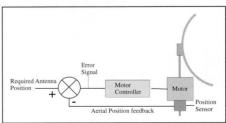

Figure 25 – Servo-mechanism   0528556

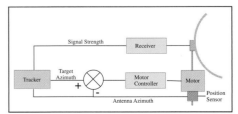

Figure 26 – Servo-tracking mechanism   0528557

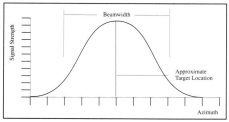

Figure 27 – Return signal strength as beam sweeps over target 0528558

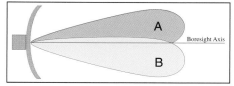

Figure 28 – Dual beams 0528559

on a target after which it automatically remains pointed at the target as it follows target motion. The system provides continuous positional information to the operator and possibly to an FCS. The antenna is rotated by a servomechanism that provides a negative position feedback signal to a controller (Figure 25).

The error signal drives the motor to reposition the antenna until the position feedback indicates the antenna is at the desired azimuth, at which point the error signal is zero and the motor stops. This servomechanism can be combined with a *tracker*, which determines the azimuth to the target, which the system now uses as the input.

Figure 26 illustrates the same servomechanism block diagram, although the input now comes from the tracker. This combination is called a radar *servo-tracking system*. The heart of this system is the tracker, which takes the return signal and positional information and determines the location of the target. There are several ways to do this, with varying degrees of complexity and accuracy.

For example, Figure 27 illustrates the data obtained as the radar sweeps across a target.

The individual returns are shown as vertical lines. As the beam sweeps across the target, a number of returns are generated, whose strength increases until the target is centred in the beam and then falls off. The target location can be

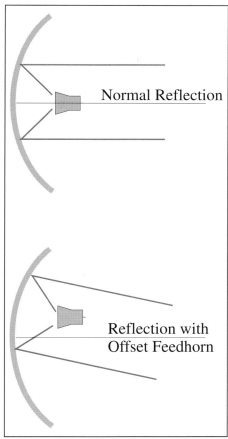

Figure 29 – Offset feed 0528560

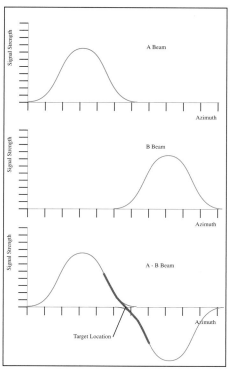

Figure 30 – Constructing the dual beam output 0528561

determined from the point of maximum return strength. However, the region of maximum return is broad and therefore the exact location of the target cannot be determined with high accuracy. While the target can be located within some fraction of the beamwidth, using this system the tracking accuracy would only be about one-quarter of the beamwidth which would be insufficient for weapons delivery.

Dual beam and monopulse systems. Tracking accuracy can be improved by using a dual-beam system. The two beams are offset in angle by a small amount to either side. The centre between the beams is known as the *boresight axis*.

The two beams can be created by a dual-feed system, where the two parallel beams are fed into the reflector slightly to one side or the other. When the beam is reflected, the offset in the feed axis will cause the beam to be reflected at an angle relative to the boresight.

When a dual-beam system scans across a target, the return will be the sum of the two beams. If one of the beams is inverted or out of phase, the result will be a well-defined target location where the difference between the beams is zero.

Since the signal return strength is changing rapidly to either side of the axis, the location of the target can be determined with great accuracy. For a typical radar beam that is 3° wide, a dual-beam

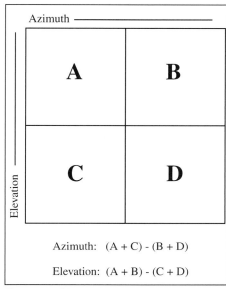

Figure 31 – Monopulse radar 0528562

system could track the target with 0.1° accuracy, which is sufficient for weapons delivery.

The return strength of a dual-beam system varies nearly linearly in the vicinity of the target. It is therefore easy to measure the target location, even if the boresight is not directly on the target. The difference between the target location and the boresight position will be linearly proportional to the return strength as long as the target is not too far off centre. Using the maximum strength method, it is difficult to determine the correct direction to reposition the antenna, since the return strength varies equally to either side of the boresight.

A dual-beam system can also be used for tracking in elevation. A *monopulse* system uses two dual-beam systems, one for elevation and one for azimuth. This requires four beams which are measured in pairs.

Range tracking. Range tracking is accomplished in a similar manner to dual-beam angle tracking. Once the range has been measured, the tracking system attempts to predict the range on the next pulse. This estimate becomes the reference to which the next measurement will be compared. The comparison is made by using two range windows called the *early* and *late range gates*.

The area of the return in each gate is computed by integration. The difference between the area in the early and late gates is proportional to the error in the range estimate. If the two areas are equal, the return is centred directly on the range estimate, and there is no error. If the return has more area in the early gate, the range estimate is too great, and therefore the range error is positive. In the near vicinity of the range estimate, there will be a linear relationship between the range error and the difference in the areas.

If the range estimate is not too far off, the tracking error can be determined and the target range updated. Again, like dual-beam tracking, the range tracking system can measure the target range with greater accuracy than the range resolution of the system, R_{res}, which is determined by the pulse width and possibly the pulse compression ratio.

Track-While-Scan (TWS). In many cases, particularly AI radar applications, it is undesirable to dedicate

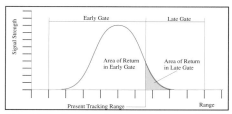

Figure 32 – Range gates 0528563

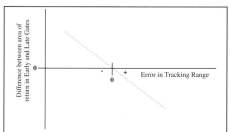

Figure 33 – Range error 0528564

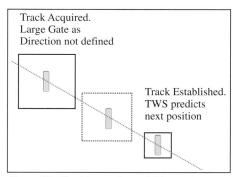

Figure 34 – Tracking and acquisition gates 0528573

the entire radar system to tracking a single target. As discussed, a servo tracking system maintains the antenna pointed in the vicinity of the target at all times which means there can be no search capability when tracking using this method. A track-while-scan (TWS) system maintains the search function, while a computer performs the tracking functions and is capable of automatically tracking many targets simultaneously. Furthermore, TWS systems can also perform a variety of other automated functions, such as collision or close proximity warnings.

TWS systems manage targets using gates in a similar manner to those used in a range tracking system. A TWS system may use range, angle, Doppler and elevation gates in order to sort targets from one another. When a target is first detected, the computer will assign it an *acquisition gate*, which has fixed boundaries of range and bearing (angle), and possibly other parameters, depending on the system. When the radar sweeps by the target again, if the return still falls within the acquisition gate, the computer will initiate a track on the target.

By following the history of the target positions, the course and speed of the target can be found. The combination of range, bearing, course and speed at any one time is known as the target's *solution*. It is used to predict where the target will be at the next observation. Once a solution has been determined, the computer uses a *tracking gate* about the target's next predicted position. If the target falls within the predicted tracking gate, the computer will refine its solution and continue tracking. If the target is not within the tracking gate at the next observation, it will check to see if the target is within a *turning gate*, which surrounds the tracking gate (Figure 35).

The turning gate encompasses all the area that the target could be in since the last observation. If the target is within the turning gate, the computer restarts its computation to obtain the new solution. If the target falls outside of the turning gate, the track will be lost. However, the system will continue to predict tracking gates in case the target reappears. Depending on the system, after a track is lost, the operator may be required to drop the track.

The process of assigning observations with established tracks is known as *correlation*. During each sweep, the system will attempt to correlate all returns with existing tracks. If the return cannot be correlated, it is assigned an acquisition gate, and the process begins again. On some occasions, a new target may fall within an existing tracking gate. The system will attempt to determine which return is the existing target and which is the new target, but may fail to do so correctly. It is common for TWS systems to have difficulty when there are many targets, or when existing tracks cross each other. In the latter case, the computer may exchange the identity

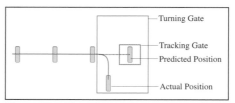

Figure 35 – Use of a turning gate to maintain track on a manoeuvring target   0528565

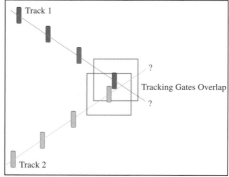

Figure 36 – Crossing tracks   0528566

of the two crossing tracks. In all these cases of mistaken identity, it is incumbent on the operator to intervene and correct the problem.

TWS systems use a *track file* for each established target. The track file contains all of the observations that are correlated with that particular target; for example, range, bearing and time of observation. The track file is given a unique name known as the *track designation*. This may be either a simple number, such as 'track 25', or it may contain other useful information, such as the classification of the target (friendly or hostile). The radar computer may use this information when determining the track and turning gates. Finally, the track file also contains the current solution to the target motion. Some systems maintain a history of solutions which can be useful in determining the pattern of a manoeuvring target.

Phased array tracking. Phased array radar systems can not only electronically steer the beam but can also perform a TWS function. Since the planar array has many independent elements, they need not all be used to form a single beam. In a phased array radar, some beams can be dedicated to search functions, while others perform dedicated tracking functions giving a Search-While-Track (SWT) feature, but with the added benefit of continuous contact on the targets (Figure 37).

Tracking networks. A natural extension of the TWS system is to create a system that shares tracking information between users. All that is required is to transmit the contents of the track file, since it contains all of the observations and the current solution. The sharing of tracking information has been incorporated extensively into modern combat aircraft via datalinks.

RADAR SYSTEMS IN OPERATION

Having looked at the main elements of modern radar systems, we shall now look at an in-service airborne radar system. Modern radars incorporate many of the elements discussed so far, and are complex in their operation. Many different radar modes and methods are combined in order to provide the user with the best possible capability.

AN/AWG-9 radar

The AN/AWG-9 radar is an important milestone in the development of modern air-to-air radars for a number of reasons. The heritage of the AN/AWG-9 can be traced back through the original 1958 Hughes XY-1 Fire-Control System (FCS), which was later designated the AN/ASG-18 for implementation in the cancelled F-108 M3 interceptor. Much of the technology from this system was employed in the design of the AN/AWG-9 radar. Revolutionary for its day, the AWG-9 was fitted to the Northrop Grumman F-14A Tomcat fleet defence fighter and, coupled with the AIM-54 Phoenix long-range air-to-air missile, gave the US Navy an extremely long-range interception capability, far beyond that of current systems. Unsurprisingly, the lineage of the missile is also linked to the radar, since the AIM-54 was developed from the GAR-9, paired with the F-108/ASG-18, and later re-designated AIM-47 for integration in the YF-12A, again with the ASG-18 radar.

Before the introduction of the AN/AWG-9, flight crews were forced to interpret the cluttered analogue radar returns on the screen in an attempt to distinguish real targets from the background noise. The AWG-9 was one of

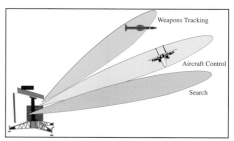

Figure 37 – Multiple beams in a phased array radar   0528567

The F-14A Tomcat is fitted with the AN/AWG-9 radar   0131065

the first radars to feature a signal processor, so excess clutter could be filtered out, leaving the crew with a much clearer picture of the airspace before them.

The AWG-9 comprises two power-supply units, three units which make up the general-purpose computer, five signal processors, four radar control units, three transmitter units, three missile auxiliary units, five units associated with cockpit display, and the antenna. The antenna's slotted, planar array measures 914 mm (36 in) in diameter. On to the antenna is mounted the Identification-Friend-Foe (IFF) system made up of two rows of six dipole arrays. The output power of the AWG-9 is rated at 10.2 kW. The transmitter is capable of generating continuous wave (CW), pulse, and pulse-Doppler (PD) beams.

In addition to providing a clear picture of the airspace before the aircraft, the signal processor provides another benefit to the Tomcat and its mission. The AWG-9 is able to remove the surface clutter of the Earth's surface, providing the radar with a look-down, shoot-down capability against low-flying aircraft. This feature also proves useful in the detection and interception of sea-skimming anti-ship missiles. The AWG-9 is capable of detecting targets as low as 50 ft and as high as 80,000 ft at ranges exceeding 130 miles. At its greatest range and widest azimuth, the radar can sweep a volume of sky measuring more than 170 miles across.

However, the hallmark of the AWG-9 is its ability to track a total of 24 targets at once and make near-simultaneous attacks against any six of them with the AIM-54 Phoenix missile. Like any of today's TWS modes, the radar keeps a file for each target detected. That file includes the target's range and angular position. On each occasion the radar detects a new contact, it compares that contact with the list of files that have already been generated. If the new contact matches the predicted position of one of the filed contacts, the radar assumes that this new contact is in fact the contact on file, and that contact's file is updated accordingly. As a result, it is not necessary for the radar to stay locked on any one target. It is left free to keep track of other targets in different locations while also scanning for new targets.

This same method allows the radar to update multiple airborne AIM-54 Phoenix missiles by transferring each updated file to the appropriate missile. As a result, even though the missile's target may have travelled many miles since the AIM-54 was launched, the Phoenix is always kept updated with the target's new position and the missile adjusts its course accordingly. Prior to the launching of Phoenix missiles, the AWG-9 computer establishes a threat hierarchy for the various contacts. As a result, six AIM-54s can be launched at targets that pose the most danger, thus increasing the survivability rate of the aircraft in a multiple-threat environment.

Additional radar modes for the AN/AWG-9 aid the pilot in searching for and targeting enemy aircraft. Pulse Doppler Search (PDS) provides range and bearing information of contacts to the pilot. Range While Search (RWS) utilises a high Pulse Repetition Frequency (PRF) to gather range and bearing information and, as a result, the radar is able to determine the rate of change in a target's range. Thus, rate of closure can be determined. The primary distinction between RWS and TWS is that in RWS, the AWG-9 is not able to determine a firing solution for the contacts, which are essentially just that – contacts. No tracking files exist which will provide the F-14 with the ability to engage the aircraft. However, RWS is able to scan a larger volume of airspace than TWS. This limitation is necessary so that the antenna is able to provide a two-second scan rate.

A Pulse-Doppler Single-Target Track mode (PDSTT) provides a very reliable method for maintaining lock against hard-manoeuvring targets at ranges up to 128 km (80 miles) from the aircraft. In addition, PDSTT mode has a Jam Angle Track (JAT) function that can provide range, bearing, and speed information for targets that are employing on-board jamming equipment.

The AWG-9 features a separate transmitter for CW illumination. The aircraft carries the semi-active AIM-7 Sparrow missile which requires CW for guidance. The missile homes in on the CW radar energy that is reflecting off the target. Semi-active missiles such as the Sparrow require a steady radar lock by the launching aircraft so that the missile never loses site of the target aircraft while in flight. Should the lock be broken, the missile will be unable to complete an intercept. Active radar guided missiles (such as the AIM-54 Phoenix and AIM-120A AMRAAM) track the target via occasional updates from the launching aircraft until the target is within the range of the missile's own radar unit. The on-board radar then takes over in the terminal phase, guiding its missile the rest of the way to the target. Throughout the entire process, the weapon utilises the radar only periodically, freeing the unit to perform other tasks, such as searching for new threats.

Vertical Scan Lock-on (VSL) mode is an automated mode that scans a 40° vertical sweep with a 4.8° wide beam. This mode is used primarily during a tight-turning engagement, usually at fairly close ranges (approximately 5 miles). In such an engagement, the F-14 would be oriented banked at or near 90°. At such an attitude, the opposing aircraft would be in the vertical plane relative to the aircraft, requiring the radar to scan the airspace ahead of the aircraft's turn. When a target is located, it is automatically locked and a firing solution is presented to the pilot.

Pilot-Rapid-Location (PRL) mode is essentially a boresight mode featuring a 2.3° wide beam that remains stationary. It is aimed by the pilot physically altering the attitude of the aircraft until the beam makes contact with the opposing aircraft. As with VSL, once the beam intercepts the target, which is locked and a firing solution is presented.

AN/AWG-9 maximum ranges
- Pulse search: 117 km (73 miles)
- PSTT: 91 km (56 miles)
- PDS of 5 sq miles target: 213 km (132 miles)
- RWS /TWS: 167 km (104 miles)
- VSL / PRL: 9 km (6 miles)
- CW for AIM-7: 70 km (44 miles)

In addition to all of these active modes, the radar can be slaved to the Infra-Red Search and Track (IRST) system. Using IRST, the pilot can passively search for other aircraft by scanning for their heat-signatures. Once an aircraft is detected and tracked, the pilot can then engage the radar in order to launch an attack.

AN/APG-71
With the introduction of the F-14D, the AWG-9 radar was subject to a total overhaul and re-designated the AN/APG-71. The changes were so complete that only the transmitter, power supply, and backseat Tactical Information Display (TID) remained from the original AWG-9. Unlike the more analogue AWG-9, the APG-71 uses a digital signal-processor to provide better target detection and greater tracking range while being less susceptible to jamming. The speed at which the radar processes information is also improved by a factor of six.

The APG-71 features additional operating modes. Targets can now be identified without the need for visual identification as dictated

by modern Rules-of-Engagement (RoE) using a technique called Non-Co-operative Target Recognition (NCTR), which employs the greater processing capability of the radar to analyse radar returns and compare their characteristics with a threat library. By cross-referencing radar characteristics with range and angle off, the aircraft type can be determined, thus satisfying RoE. This means that threats can be intercepted at greater ranges, increasing the margin of safety for both aircraft and aircraft carrier. The APG-71 also features a Raid Assessment (RA) mode similar to that found on the F/A-18. This mode is used to resolve a situation where a pilot faces an undetermined number of closely spaced targets. Normally the group would appear as one contact on the radar display. Effective up to a range of 55 km (35 miles), RA uses Doppler Beam Sharpening (DBS) to provide increased resolution of the patch of airspace around the group. Typically, the minimum separation between aircraft must be approximately 150 m (500 ft). The APG-71 also features a ground-mapping mode in order to deliver accurate attacks against ground installations.

ELECTRONICALLY SCANNED ARRAYS
Having examined the theory behind the electronically steered radar beam, it is worthwhile investigating current and future applications of this important new technology. The introduction of new AESA radars, such as the AN/APG-63(V)2, has demonstrated a major leap in radar capabilities, including greater flexibility and a broader range of missions that an AESA-based sensor can support. Recognition of this potential has motivated a great deal of research and development on the central building element of an AESA, the T/R module.

Progress in both semiconductor device technology, as well as the integration and packaging of this technology into modules sufficiently small and low cost has made AESAs a practical reality. The advantages of ESAs over mechanically steered antennas have long been recognised, the most important of which is the ability of ESAs to steer, near instantaneously, the radar beam point-to-point within the Field of View (FoV). As discussed, ESAs accomplish this via an electronic phase shifter at each radiating element (or column of radiating elements in the case of electronic steering in only one dimension). Beam steering is accomplished by subjecting the signal at each radiating element to an additional phase shift, such that the wavefront at the antenna face, which results from the coherent addition of the output of each element, propagates in the desired beam direction. Similarly, on receive, the coherent addition of the signal from the individual elements will cause the gain of the antenna to be maximised in the direction selected by the phase shift at each element (beam pointing). Since the electronic phase shifters can be switched very rapidly, beams can be steered over large angles in a small fraction of a Pulse Repetition Interval (PRI). Computing the appropriate phase shift for each element to achieve a particular beam steer direction is carried out by the Beam Steering Controller (BSC), an important subsystem of any ESA. Many current ESAs rely on passive components, typically ferrite phase shifters, at each controlled radiating element (or groups of elements) to achieve the necessary phase shift function. Such arrays are generally referred to as 'passive' ESAs, in contrast to 'Active' ESAs (AESAs), which feature active electronics in the form of a T/R module at each element, which provides the transmit and receive functions in addition to the phase shift required for electronic steering. Although both passive and active ESAs share the aforementioned benefit of near instantaneous beam steering, there are some significant additional advantages that an AESA enjoys.

In describing the key differences between passive and active arrays, a one-dimensional array, such as that found in the JSTARS and ASARS-2 radars, will be assumed for the passive ESA. Passive ESAs feature a centralised, high-power tube transmitter

Later batches of the BAE Systems/Saab Gripen multirole fighter will employ an Active Electronically Scanned Array (AESA) known as NORA (Not Only Radar) 0111267

generating the required RF power. This power is distributed to the radiating element via a power dividing network, incorporating a waveguide to handle the large peak power (typically in the tens of kW range) of the tube transmitter. To generate this power, these transmitters need high-voltage power supplies (tens of kV). At each controlled element (radiating column in this case) a ferrite phase shifter is inserted in the signal path as described previously. On receive, the signal travels the reverse path, through the phase shifter, multiple corporate feeds (for multiple phase centres/antenna beams), to receivers with Low Noise Amplifier (LNA) front ends, and then to Analogue to Digital (A/D) converters whose outputs are fed to the signal processor. Additional components in this signal chain are required to switch the elements between transmit and receive (T/R switches), provide isolation and protect against element mismatches (circulators).

Each of the passive components introduces a loss in the overall system, thus contributing to a degradation of radar sensitivity. In the case of the transmit path, RF power is lost in the power distribution network, phase shifters, circulators and T/R switches. On the return path these losses reduce the signal strength prior to the LNA, thereby reducing the received Signal to Noise Ratio (SNR). In combination, these losses can reduce the overall sensitivity on the order of 10 to 15 dB (a factor of 10 to 30). By integrating the final stage of transmit power generation and the receive amplifier at the radiating element, these losses can be greatly reduced. To accomplish this, T/R modules in AESAs, containing the phase shifter, transmit power amplifier, T/R switch (possibly including a circulator) and the LNA/receiver, are located as close to the radiating element as possible, with the radiating element in some cases being an integral part of the module itself. While every T/R module still requires signal distribution and combining networks, losses due to these elements are effectively insignificant, since they occur prior to the final transmit power amplifier and thus do not affect the final power output. In the receive path, the losses associated with the combining networks (analogue beamformer) occur after the LNA and receiver preamplifier and hence do not reduce the incoming signal strength prior to amplification. The net result is that while not all losses can be eliminated (for example, circulator losses), a reduction in overall system losses in the order of a factor of 10 is feasible in active ESAs compared with passive ESAs. In terms of hardware integration, T/R modules can be manufactured at lower cost, components feature far less bulk and weight and printed circuit boards can be employed in place of waveguide components. In a similar vein, the phase shifter, which is part of the T/R module, is also placed before the power amplifier and after the receive amplifier. This facilitates the use of more solid state integrated circuitry which, compared to ferrite phase shifters, can be switched much more rapidly and with essentially no power penalty.

Just as the RF power generation is distributed over the array, the regulated DC power supplies are similarly distributed, with one supply typically feeding multiple T/R modules. Unlike high power tube transmitters, voltages for T/R modules are low (~10 V); thus power distribution over multiple power supplies close to the T/R modules they supply is necessary to avoid high Ohmic losses and problems with supply regulation. Once again, these supplies can be integrated into the antenna itself to simplify power interfaces.

Another advantage of an AESA over a passive ESA or a mechanical array is the Mean Time Between Failure (MTBF). Since power supplies, final power amplification and input receive amplification are distributed over the entire array, MTBF is significantly higher, typically 10 to 100 times than that of a passive ESA or mechanical array. In terms of the effect on the MTBF of a fighter radar system based on an AESA, such as the APG-63(V)2 over the previous, mechanically scanned, APG-63(V)1 in the F-15, the replacement of the antenna resulted in higher system readiness with attendant significant savings in terms of life cycle cost of the weapon system. The AN/APG-79 AESA radar in the F/A-18E/F Super Hornet will require little or no radar array maintenance (save for battle damage) for the life of the platform.

While all of the above advantages of AESAs have long been recognised by the radar community, the major challenge has been to develop the technology and manufacturing capability to achieve the size, weight, power, and cost characteristics necessary to make AESAs affordable and cost-effective when compared to current radar systems. Key to achieving this aim was the development of the T/R module. Integration of component chips (including digital logic) into a module that can be manufactured using a high degree of automation for both assembly and test, coupled with supporting research into semiconductor chip technology to achieve Low Noise (receive) Amplifiers (LNAs) and high Power (transmit) Amplifiers (PAs), incorporating high power-added efficiency (a key parameter in defining prime power and cooling requirements), have enabled designers to provide a significant leap in operational effectiveness.

With the technology maturing rapidly, AESAs have become the preferred approach for high performance, multifunction radars. The performance advantages offered by AESAs have been particularly important for modern fighter aircraft and have undoubtedly been the major driving force behind the rapid progress made in recent years. An illustration of this progress is the reduction in T/R module size, facilitating the development of compact, high power AESAs suitable for integration into the nose of a fighter, epitomised by the AN/APG-77 in the F/A-22 Raptor. Further reduction in T/R module size, combined with even greater levels of integration, has resulted in further reduction in the depth (and weight) of new AESAs, such as the AN/APG-81 in the F-35 Joint Strike Fighter (JSF). This has been accomplished by 'tiling' – the various functions required for the AESA (power distribution, RF distribution, timing and control) are implemented in a multilayer 'circuit board' which also contains layers for the T/R modules and antenna radiators.

Obviously, a radar antenna composed of an array of 'tiles' can be easily scaled to produce an antenna of almost any size; this characteristic makes AESAs particularly well-suited to a 'family-of-radars' approach, with commensurate gains in operational integration and savings in manufacturing and support costs. This concept is already bearing fruit, exemplified by the development path of the APG-63(V)2 to (V)3 benefiting from the development effort for the APG-79 (see relevant entries). In applying AESA technologies to other platforms, such as the E-10A, it is clear that such a scaling approach undoubtedly brings advantages in terms of cost savings, with the same T/R tile applied to the larger array. This enables ISR platforms to benefit directly from the extensive effort in T/R module development, driven by the relatively much larger market for fighter radars. However, noting the different demands of ISR radars

in terms of power requirements, it is unlikely that T/R modules from a fighter radar would be employed in an ISR radar unmodified. Benefiting from the inherent modular design of the T/R module, replacement of specific chips within the module to more efficiently match differences in power requirements between the two types of applications would be a relatively simple task. Also, since fighter radars generally feature circular apertures, side-looking radars typically are rectangular with high length-to-width ratios (longer antennas are desirable to achieve good GMTI performance), therefore, antenna array (and sub-array) characteristics will be vastly different. However, at higher levels, for example after the (sub)array, a high degree of commonality between fighter and ISR radars is possible. In particular, there is a high level of functional commonality between these classes of radar, with both required to perform surveillance and target tracking. Therefore in parallel with hardware commonality, a modular, open systems architecture for software is critical to reaping the full benefits associated with this 'family of radars' approach.

Such technology growth will benefit all AESA applications, but is of special interest for platforms such as Global Hawk where size, weight, power and cost play a particularly important role in enhancing operational effectiveness. Research in lightweight apertures, conformal arrays, and improved receiver/exciters will help to take full advantage of available platform payload capacity and offer new opportunities for radar and other sensors to be integrated into the platform. While the greater overall power efficiency offered by AESAs is an important factor in increasing the performance of airborne radars, the extent of increases in UAV platforms is limited by the prime power available for the radar. While incremental improvements in T/R module efficiency and in the power efficiency of other components can alleviate the problem somewhat, to achieve a major increase in UAV operational capabilities in general, and radar sensitivity in particular, will require a significant increase in the available prime power. In a similar vein to the fighter community driving T/R development, UAV budgets will drive development work into power generation to bring the benefit of highly efficient generation of prime power at high altitude, technology that would also benefit the fighter community.

CONFORMAL ARRAYS
In considering the further implications of the ability to electronically steer a radar beam, it becomes clear that, with the need for a complicated and aerodynamically inefficient mechanically scanned antenna system abrogated, there is much scope for alternative solutions to be developed. Indeed, with the effective physical

'depth' of the radar reduced to a fraction of mechanical systems, any relatively flat surface could be employed as an alternative antenna position – this brings us to the concept of the 'conformal array'.

Operational requirement?
In explaining the potential advantages of conformal arrays, let us consider an example which also serves to reveal why this technology is currently very well-funded. Air Forces equipped with conventional fighter designs, with nose-mounted Fire-Control Radar (FCR), have developed Beyond Visual Range (BVR) tactics that maximise the Probability of Kill (PK) of their own missile, while also helping to protect them from incoming enemy missiles. The most critical test of these tactics occurs with weapons system (aircraft performance, radar and missile) equivalence and both adversary aircraft ('blue' and 'red') firing their missiles simultaneously.

Assuming on-boresight shots with high launch velocity, if, in supporting their respective missiles, both pilots simply maintain their course and speed after launch, then, unsurprisingly, both aircraft will be hit simultaneously. Current tactics are designed to enable the pilot to manufacture a significant increase in the enemy missile's fly-out time compared to that of his own while gaining valuable time to prepare to attempt to defeat the incoming missile directly using Electronic Counter Measures (ECM) techniques. This tactic, unfortunately, is limited by the need to support his own missile on its way to target. While our pilot would like to ideally 'fire and forget', thereby released to turn through 180° and utilise maximum performance to 'run away' from the incoming missile, to have any chance of achieving a hit, he must update his own missile via radar or datalink. Assuming no third-party assistance, he must therefore maintain the target within his radar scan. With typical radar semi-angles of approximately 60°, this means that his maximum turn is a fraction less than 60°; any more and he will lose lock on his target. Having established a potential 'geometric stalemate', perhaps the key lies in accepting the arrival of the enemy missile and looking to defeat it at end-game through defensive manoeuvre and ECM? However, considering the Electronic Counter-Counter Measures (ECCM) capabilities of modern missiles, this is a very risky tactic; we must still strive to ensure that the missile does not arrive at all. We should still aim to destroy the support mechanism for the incoming missile before it can reach its 'autonomous point' – either fusing for a semi-active missile, or seeker lock-on for a radar-guided Active/IR missile. If this is not achieved, then our 'blue' pilot, having achieved a greater fly-out time, but not enough to deny

The AN/APG-68(V)9 equips Lockheed Martin F-16 Block 50 fighters (Northrop Grumman) 1127356

The RDY-2 radar, which equips Mirage 2000-5 Mk2 and later variants, incorporates a SAR mode for high-resolution ground mapping
1039842

The AN/APY-3 radar in the E-8C is a passive ESA; the US Air Force is planning a radar upgrade to a full AESA (Northrop Grumman) 1127364

The Northrop Grumman MESA L-band (1.215 – 1.4 GHz) surveillance radar is an AESA; an integrated Transmit-Receive (T/R) module with internal phase shift and RF gain controls drives each antenna element (Northrop Grumman) 1121439

the incoming missile a final update, will only momentarily enjoy the satisfaction of seeing his opponent destroyed before suffering a similar fate – a somewhat hollow victory.

While it is not the purpose of this discussion to examine current missile defence techniques, it is apparent that much depends on the pilot to execute his tactics promptly and accurately if he is to have any chance of surviving such an encounter. As is current practice, when designing new fighters, evolutionary technologies are introduced to obviate the need for missile defence by bestowing the new design with a significant 'edge' in combat against current designs.

A more powerful radar, providing much longer range detection and target tracking, which, by necessity must be combined with a longer range missile (a point sometimes overlooked), will give a fighter pilot such an advantage – indeed this development route has been employed for many new fighters over the years. Conversely, stealth technology, which effectively minimises the Radar Cross Section (RCS) of an aircraft, makes it nearly 'invisible' to the enemy FCR; thus an aircraft such as the F/A-22 is capable of approach, launch and achieving a kill before the enemy has even detected its presence! However, while it would seem that stealth technology is the ultimate solution to making a truly superior aircraft, it is extremely expensive to procure and, perhaps more importantly, maintain. Also, to be truly effective, stealth must be combined with other signature suppression technologies (such as Infra Red and visual) without prejudicing aerodynamic capabilities if the aircraft is to be capable of prevailing once detected. This point is exemplified by comparing the F-117 Nighthawk with the F/A-22 Raptor – the Nighthawk was designed with stealth as an *absolute* priority, to the exclusion of performance, whereas the F/A-22 design is more of a compromise, enabling it to retain true dogfighting capabilities. Accordingly, the F-117 is only committed at night and alone, without other, less stealthy platforms to draw attention to its presence. As for the Raptor, its afterburners, despite shielding, introduce some vulnerability to passive Infra-Red Search and Track (IRST) systems. Further, more sophisticated multiwaveband sensors and powerful onboard mission processors, combined with data link information from third-party (air/ground based) sensors can enable enemy fighters to gain valuable Situational Awareness (SA) from the fragments of information the Raptor cannot withhold. So, learning from the US experience, a highly expensive stealthy platform might not match cost effectiveness with perceived threat for many other airforces worldwide. So what other technologies, less expensive and easier to integrate into current and planned fighter designs, are available which can provide a *decisive* advantage in air combat?

Having eschewed total stealth, it is clear that our provisional new fighter will be detected at some stage, so we must return to the core of missile defence – destroy the support mechanism for the enemy missile in order to generate a survivable miss distance. This means being able to utilise as much of the performance of the aircraft (speed) as possible to separate from the enemy missile in flight. Going back to the geometry problem, it is clear that the constraining factor is the semi-angle of the radar – even increasing this by 10° (from 60 to 70°) provides a significant advantage if all other factors remain equal. However, what if this angle could be increased to 90°? Would this bring a greater advantage than, say, designing a faster missile? While the simplistic answer to this question is simply a matter of vector geometry to determine overall fly-out times, much modelling and simulation has been performed since the mid-1990s to gain a full understanding of the tactical implications. The short answer is yes – increasing the radar semi-angle beyond 90° does bring a significant advantage, even over high-speed missiles. So how might this increase be achieved? Even if a mechanically scanned array could achieve such angles, it would encounter interference from the airframe at extreme look-back. AESAs offer the ability to electronically steer the radar beam, thus obviating the need for mechanical scanning, but they also cannot achieve the large semi-angle required. Mechanically scanning an AESA, while seemingly re-introducing many of the problems of earlier designs, is actually a promising idea. Such an array would not require the high-speed mechanical scan of current designs, since the AESA would bring this facility; the mechanical components could move relatively slowly to simply facilitate greater look angles. If the mechanical arrangement were capable of ±60° and the AESA also capable of ±60°, then our theoretical ±120° would be possible; however, if the arrangement were mounted in the nosecone of a fighter, we would still encounter airframe interference.

The optimum antenna solution

In order to solve this interference problem, while retaining the weight and bulk savings from the elimination of any form of mechanical scanning, we need a radar antenna that requires no modification to the host platform's aerodynamic profile (which also brings further benefits in terms of drag and thus performance and fuel efficiency). What if we could utilise the entire skin of the aircraft as an antenna? This, in a somewhat round about way, brings us to the idea of using aircraft-mounted conformal antennas. From the foregoing, for an AESA application, any flat portion of the skin of an aircraft could be utilised for a radar array, either as a subordinate panel to a nose radar, or as part of an aircraft skin-mounted system. In either instance, each array would add its FoR to the overall system swathe, facilitating near 360° radar coverage if sufficient panels are employed. There are, however, a number of options and constraints to the implementation of this technology. While the most simplistic application of sub panel arrays would be to simply graft an array directly onto an available aircraft panel, this would introduce extra weight to the airframe, which may be significant in terms of reducing range/payload. While this may be of limited impact to the integration of two sideways-looking sub panels into a high performance fighter, it may be critical to the range and sensor payload capability of a UAV. In such cases, the radar panel could be integrated into the design at the earliest stage, making the panel part of the load bearing structure of the aircraft, although this would bring additional problems. Many external structures are not truly flat, which would bring the challenge of maintaining fine radar beam control over a continuously curved surface. Further, while traditional radar feature some form of stabilisation, large arrays integrated into external structures, such as wings, would be subject to the full range of aerodynamic loads, with inevitable flexing, twisting and vibration which would require software compensation within the radar processor.

However, while the technological hurdles to achieving a fully fledged conformal radar capability for a range of airframes are many and will be costly to clear, the operational imperatives will drive research. With ever more stringent Rules of Engagement for both air-to-air and particularly air-to-ground delivery, modern weapons now feature smaller warheads (to minimise collateral damage) and far greater accuracy; these weapons demand highly accurate targeting and hence sensor resolution at long range (for platform survivability). If most of the airframe skin (of an ISR platform, UAV or tactical fighter) could be used as a sensor, the size of the aperture would be constrained only by the platform's overall dimensions. A larger aperture, as previously discussed, means greater resolution for GMTI applications. Moreover, if the main sensor is no longer an 'addition' to the airframe or attached as a bulky payload, more fuselage space is freed for defensive ECM/ECCM, weapons or fuel. The aircraft's aerodynamic integrity would be preserved insofar as the array structure will sustain such loads.

In considering the challenge of integrating a conformal array into a curved surface, new algorithms will be needed to control AESA beamforming, adjusting for the added complexity of a non-linear surface. To date, the US Air Force Research Laboratory's (AFRL) Sensors Directorate has been able to map and estimate 'first-order' performance differences between flat planar and curved conformal arrays.

For the technology to work, however, it must be practical – in addition to the above mentioned factors, we must also consider other in-flight events, such as birdstrikes, gust-loads and battle damage. Whatever array technology is introduced in place of the airframe 'skin', it must be affordable, as reliable as the skin it is replacing and capable of easy assembly on the manufacturing line. US research into achieving these goals is moving forward; in 1999, the AFRL Structures Division, in co-operation with a Northrop Grumman research team achieved their first success in a ground test of a prototype antenna. The curved, load-bearing AESA, measuring 1×1 m, was subjected to loads of 280 kg/cm^2, a condition comparable to the in-flight loads on a mid-fuselage panel of a F/A-18. The array structure not only had to survive the loading over a fatigue life cycle, but also perform its surveillance function. The experiment was reported as a success, with the array achieving a 150 per cent design limit load before failure. Northrop Grumman has claimed that the array has survived loads of more than 700 kg/cm^2.

Current research efforts into conformal antennas are in support of UAV applications, which is entirely logical, considering the main advantages of the technology (overall airframe weight, aperture size). However, once the goals of cost and manufacture/supportability have been achieved, there will be transfer of this technology to (probably first in line) fighter platforms (who will greatly benefit) and eventually ISR aircraft (which will then become smaller while maintaining or enlarging the all important aperture size).

ELECTRO-OPTIC SYSTEMS

The Tactical Night Environment

Introduction
This overview of Electro-Optic (EO) systems, covering Night Vision Goggles (NVG) and Image Intensifiers (IIs), together with Forward-Looking Infra-Red (FLIR) systems, will address the physical attributes of electro magnetic radiation, environmental modifiers, the theory of operation of II and FLIR devices and the technical operation of current systems. This will serve the reader in three ways:

- To provide a technical overview to potential users of Night Vision Devices (NVD) and IR systems in the night environment.
- To provide useful background information to engineers and technicians working in the field of night vision as to the operational employment of NVD and FLIR.
- To illustrate the complementary nature of FLIR and NVG for airborne operations.

MILITARY AIR OPERATIONS BY NIGHT
Nowadays, it is taken as read that military air operations, be they ground attack or air-air, both for fixed and rotary wing aircraft, will be carried out continuously, that is, 24-hour operations. Recent conflicts have also shown that, until absolute air supremacy is achieved, the conduct of initial air operations (both offensive and defensive counter-air) are carried out exclusively during the hours of darkness, thereby maximising the tactical effect of night sensors and thereby minimising risk to crews. However, this 'holy grail' of air operations, unrestricted by both light levels and, in parallel, weather, enabling pressure to be kept up on an enemy and allowing no time for reorganisation or resupply, is a comparatively modern concept for offensive counter-air operations. During the late 1970s, the advent of aircraft such as the Panavia Tornado GR1 and General Dynamics F-111, equipped with Terrain Following Radar (TFR), allowing blind flight at extremely low level, gave aircrews the ability to prosecute operations on a 24-hour basis. However, while aircraft equipped with TFR were truly all-weather capable, weapon delivery at night was based purely on the ability of the Ground Mapping Radar (GMR) of the aforementioned aircraft to discriminate the target, which could prove a severe limitation against small, mobile targets. Further, TFR systems, with their inherent need for an integrated autopilot system, are a costly solution for many air forces and are comparatively noisy in electro-magnetic terms when faced with sophisticated air defence systems.

The use of Electro-Optic (EO) sensors offers a cheaper, totally passive solution to night operations. Although only able to offer pilots the capability to operate visually (the ability of such sensors, due to the wavelength of their operation, is severely limited in terms of 'seeing' through significant weather; in contrast, TFR operations, utilising the radar wavelengths, are truly all-weather), Night Vision Goggles (NVGs), operating in the 0.6 to 0.9 µm waveband, complemented by Forward Looking Infra-Red (FLIR), operating in the 8 to 12/14 µm or 3 to 5 µm wavebands, affords pilots the ability to visually acquire and deliver weapons against all types of target, including mobile targets such as armour. The complementary nature of NVGs and FLIR enables pilots to continue operating when one of the sensors is degraded due to atmospheric attenuation (more of which later).

However, operations at night bring their own problems – humans, by definition, are day creatures and are thus not able to readily adapt to night operations. While some claim that EO sensors 'turn night into day', this is not strictly true, to the extent that night flying techniques while utilising EO sensors are sufficiently different to daytime operations that unique training methods are required to maintain proficiency and, perhaps more importantly, safety. Loss of peripheral

Image of a Panavia Tornado IDS taken through night vision goggle equipment　　0110046

vision in traditional NVG designs, degraded depth perception (the ability to determine relative position of objects in the scene) and distance estimation while operating with NVGs conspire to deny pilots much of the raw data they process routinely while manually flying with reference to ground features, thus degrading situational awareness and spatial orientation.

While many air forces equipped with day/night-capable multirole aircraft claim 24-hour capability, the reality is that day and night operations are difficult to mix effectively for a typical fighter squadron, since such squadrons are invariably organised in a 'traditional' manner, with insufficient manpower and support to maintain continuous operations. So, during peacetime, squadrons generally conduct either day flying or night flying during any given period. However, past air operations in the Arabian Gulf and Bosnia/Kosovo have highlighted these deficiencies, resulting in increases in manpower (pilots and groundcrew) and support (airframes and spares) for night attack squadrons sent into theatre.

In addition, dedicated night attack training is made difficult due to the fact that (at least in Europe) it is inevitably carried out close to urban areas, and at times which are generally severely restricted by the need to minimise disturbance to the slumbers of the civilian population.

Many countries see the acquisition of EO sensors as a relatively 'cheap' route to 24-hour operations for their air forces, but it should also be noted that the sensors are really only part of the capability: for instance, effective operational

employment of NVGs is greatly dependent on effective internal and external lighting, both NVIS (Night Vision Imaging System) compatible (that is, filtered, to block the majority of IR radiation which causes the NVG image to 'gain down' and thus degrade the scene viewed by the pilot) and tactical (purely IR lighting visible only with IIs – used to facilitate identification of elements within a formation, or to friendly ground forces equipped with suitable NVGs). If an aircraft is to be modified for EO operations, the modification of the lighting suite should be considered to be as important as the integration of the sensors themselves (although this has certainly not been the case in many instances). In addition, development of specific night tactics and operational procedures, supervised pilot conversion and regular and effective training are also pivotal to the acquisition of a true night capability.

In order to fully appreciate the strengths and limitations of IIs and NVGs, we should first examine some of the characteristics of electromagnetic radiation and its propagation through the atmosphere:

THE ELECTROMAGNETIC SPECTRUM
Characteristics and measurement of light
The energy that all EO devices receive, and the resulting visible image seen by the operator, is all electromagnetic energy. Light, or optical radiation, may be characterised in two ways:

- As particles, called photons
- As waves, which are propagated through a medium, which, for our purposes, is air.

Definition	Photometric Units Name	Unit (SI)
Energy	Luminous energy	Lumen-sec
Energy per unit time (power)	Luminous flux	Lumen
Power input per unit area	Illuminance	Lumen/m² (lux)
Power emitted per unit area	Luminous exitance	Lumen/m²
Power emitted per unit solid angle	Luminous intensity	Candela
Power emitted per unit solid angle per unit projected area	Luminance	Candela/m²

Note: Figure 1 details metric units only. In the US, the term foot-lambert (fl) is used to quantify luminance, where 1 fl = 3.42626 cd/m² and the term foot-candle (fc) to quantify illuminance, where 1 fc = 10.7639 lux.

Figure 1 – Photometric Units (SI)

The particle theory of light describes the emission of light from a source, such as the moon. The intensity of this energy, which is useful in dealing with the amount of reflected light available for IIs, can be measured as the amount of light that strikes an object or surface at some distance from a source. When considering units for the measurement of incident and reflected light, we usually employ photometric units, as opposed to the perhaps more recognised radiometric units. Reflected light, or luminance, is normally expressed in terms of foot-lamberts (fl), while the available light, or illuminance, is expressed in terms of lux, which is lumens per square metre (lm/m²). Other commonly used SI units and their definitions are shown at Figure 1.

The wave theory of light characterises light in terms of wavelength, frequency and velocity, enabling us to fully describe the propagation of light through the air, or an optical system such as the human eye, conforming to the well-known equation $\lambda = e/f$, where e is the speed of light (for our purposes, a constant), λ is the wavelength of the radiation and f is the frequency. Thus wavelength is inversely proportional to frequency.

For a pilot to exploit the night by utilising the available illuminance, then he needs an imaging system which can interpret the visual scene across a broad range of selected wavelengths (near-, mid- or far-IR for the purposes of EO sensors) and present it to the pilot within the visual spectrum. This is exactly what NVGs and FLIR systems are designed to do for night aviators. As we shall see in more detail later, these devices are particularly sensitive to emitted and/or reflected energy in these IR bands. We shall now examine the two main sensors utilised for the exploitation of these IR wavelengths in tactical and paramilitary night operations.

NIGHT VISION GOGGLES AND IMAGE INTENSIFIERS

NVG and the Electromagnetic Spectrum
Many different types of radiation are present in the atmosphere, such as light, heat and radio waves. Just as a radio must be tuned to receive specific frequencies, II devices, such as NVGs, and the human eye, are sensitive to different wavelengths (or bands) of the electromagnetic spectrum. These bands are inherently similar in nature, and as such, can be best related by their position in the entire electromagnetic spectrum, as shown in Figure 2. The optical band, covered by visible light, is a relatively small portion of the entire spectrum. All these radiations obey the same laws of reflection, refraction, diffraction and polarisation, and, as previously mentioned, the velocity of propagation can be considered to be constant. Thus they differ from one another only in their frequency. Visible light is bounded on the short wavelength side by UV radiation and, on the long wavelength side, by IR radiation.

The human visual system, operating in the visible range, and NVG, operating in the visible and near IR range, are dependent on reflected energy, which requires scene illumination. When a scene is irradiated by an external source such as the sun, it is observed primarily by reflected energy. This holds true for near ultra-violet, visible and near IR wavelengths.

SOURCES OF NIGHT ILLUMINATION
There are several natural sources of illumination available at night:
• Residual Sunlight
• Moonlight
• Starlight
• Airglow.

The primary source of daytime illumination is the sun. The illumination produced by the sun on a horizontal surface on the earth depends on its angle above the horizon (elevation angle). However, due to refraction effects, the sun also provides light when it is below the horizon, therefore residual sunlight is also of interest to the night aviator. Significant times are civil twilight, when many tasks in the cockpit can be achieved with unaided vision and there is generally too much residual

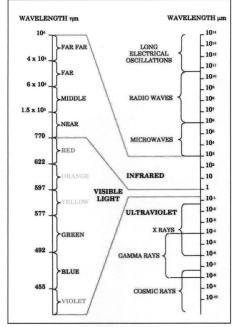

Figure 2 – The Electromagnetic Spectrum

0110032

sunlight for NVG operations, nautical twilight, when the sun's effects steadily decrease to the extent that normal NVG operations are possible and finally, astronomical twilight, when all residual sunlight has disappeared from the night sky and normal NVG operations are possible given other forms of natural or artificial illumination.

After the sun has fully set, the most significant source of light is the moon. The illuminance provided by the moon depends on its elevation and phase, and knowledge of these is important in planning NVG operations.

To the casual observer, the most obvious moon characteristic is that it changes phase regularly. The phase of the moon is described either by its age, or by the percentage of the illuminated disc. For the purposes of night operations, it is useful to consider the four lunar phases and their characteristics:

New Moon. The new moon phase is characterised by very low light levels that limit useful NVG flying conditions.

First Quarter. Relatively good light levels are available for NVG operations with the best period in the early part of the evening.

Full Moon (Second Quarter). This phase is characterised by very high light levels once the moon rises above the horizon.

Third Quarter. Relatively good light levels are available for NVG operations, although optimum lighting periods progress towards the early morning.

After rising, the moon continues to increase its elevation, which is dependant on its declination and the observer's latitude. The former depends on the position of the moon in its orbit, and its plane of motion. This characteristic is important when considering those moon phases where the elevation is critical for providing sufficient light.

However, the moon is not the only source of natural light in the night sky. Starlight also provides illuminance, which is equivalent to about 25 to 30 per cent of the total light available from a moonless night sky. The majority of stars emit a significant amount of radiation between 0.8 and 1.0 µm, which means that a great deal of their output is invisible to the human eye, but falls within the sensitivity band of the latest generation of NVGs, termed Generation III (Gen III) NVGs (0.6 to 0.9 µm). This sensitivity facilitates night operations in starlight only conditions.

Perhaps surprisingly to anyone who has never worn NVGs, the greatest portion (approximately 40 per cent) of natural light in a moonless night sky comes from an effect called airglow, which originates in the upper atmosphere and is caused by solar radiation, which produces emissions from atmospheric particles.

The remainder of night illumination comes from other sources such as aurora, luminous patterns of light which sometimes appear in the night sky at high latitudes.

Having described the main sources of natural night illumination, we should now attempt to illustrate how the modern NVG's response is tuned to respond in the wavelengths that characterise these phenomena. Figure 3 shows approximate curves for the three main contributors to illumination in the night sky, namely starlight, moonlight and airglow, as a function of wavelength. Note that starlight and airglow are mainly present in the near IR band, which means they are largely invisible to the unaided eye, but fall well within the sensitivity of NVGs, as shown by the overlaid response curves for the unaided eye and NVGs.

Having discussed the sources of natural illumination in the night sky, defined the position of their radiations in the overall electromagnetic spectrum and described the tuning of NVGs response to their characteristic wavelengths, it would be useful to consider the practical application of all of this theory!

By considering the positions of the sun and moon, and prevailing weather conditions (cloud cover), a theoretical luminance level can be calculated. Figure 4 shows illuminance levels for various moon phases and cloud conditions.

It should be noted that, while the luminance levels shown in Figure 4 serve as a useful guide to mission planning for NVG operations, they should only be considered as a guide, since the contribution of other factors, particularly that of cultural lighting (described later) to the scene, are not considered. Accordingly, actual light levels encountered may be considerably greater than predicted in urban areas, or drastically less if cloud predictions were in error.

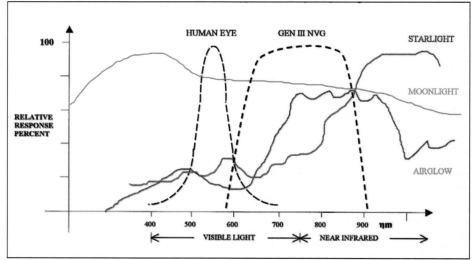

Figure 3 – Response of the Human Eye vs. NVG/Sources of Night Illumination

0110033

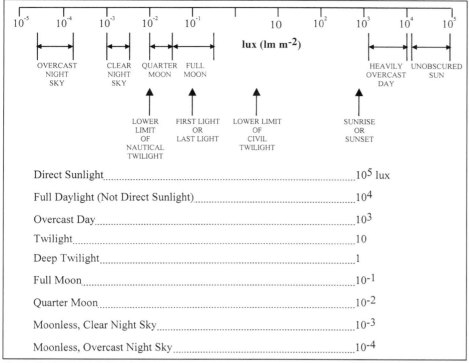

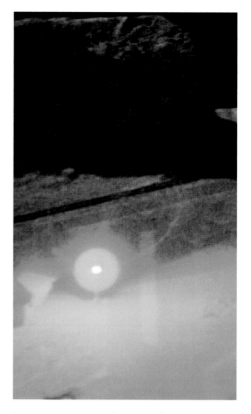

Figure 4 – Luminance Levels   0110034

While it is obvious that accurate information concerning luminance is critical to the safe conduct of night attack sorties, calculation of light levels can seem an extremely complicated proposition for the night aviator; however, mission planning is facilitated by computer programmes, such as the UK Met Office Illumination Program and the US WinANVIS light level prediction program, which calculate the orientation of the sun and moon, factor for prevailing weather (cloud cover) and predict the illumination levels for any point on the earth's surface at any time.

Cultural Lighting

As witnessed during recent conflicts, such as allied operations in Kosovo, there are very many sources of man-made light in the overall night scene. In the absence of natural lighting, for example, when operating under a thick cloud cover, cultural light from urban areas, reflected from the cloud base, can provide significant illumination for NVGs. As a result, low-level operations will often be possible even when there is no natural illumination available, although care should be taken to make use of such lighting without subjecting pilots' NVGs to adverse conditions. If NVGs are pointed directly at brightly lit areas, this may saturate NVGs, causing them to 'gain down' and the wearer will effectively be blinded. If the area of bright light is small point (such as a single street light), then this will be shown as a so-called 'halo' around the light, leading to a loss of contrast in the surrounding image.

Recent conflicts such as the Gulf War and Bosnia/Kosovo have shown that 'blackout' conditions, as practised during WWII, are not guaranteed during modern air warfare and the presence of cultural lighting in the target area should still be considered, particularly for Military Operations in Urban Terrain (MOUT).

ATMOSPHERIC TRANSMISSION

Having considered available sources of illumination for IIs, we should now consider the attenuation mechanisms within the atmosphere. For all of the wavelengths of interest, there are components within the atmosphere that will refract, scatter and absorb radiation. The transmission of electromagnetic energy through the atmosphere is dependent upon the concentration and distribution of those attenuating atmospheric constituents, which is, in turn dependent on meteorological conditions. Therefore, the atmosphere itself is a limiting factor for transmission.

Scattering occurs due to the presence of molecules and particles of matter in the atmosphere (aerosols). Scattering causes attenuation of an incident beam of radiation because in the scattering process the energy is redistributed in all directions of propagation.

Another form of attenuation of radiation is absorption. Water is the primary absorbing molecule for IR and visible radiation. The severity of scattering and absorption is a function of radiation wavelength and particle size, and it is often difficult to determine which is the primary attenuation mechanism. However, for particular wavebands, 'windows' exist where the transmittance of the atmosphere is consistently high, therefore modern EO systems are designed to operate at wavelengths within these windows. Figure 6 shows the three windows within the visible/IR range of wavelengths where atmospheric transmittance is high, together with examples of associated EO systems found within *Jane's Avionics*.

The design of all EO sensors utilises one or more of the windows in the near-, mid- and far-IR regions of the spectrum. Modern NVGs utilise the window that extends from the deep red area of the visible range into the near IR region (0.6 to 3.0 μm). The other two windows are both in the IR portion of the spectrum. One is the 3 to 5 μm band, which is located in the mid-IR region; this band is associated mostly with missile seeker heads, although more recently with modern targeting FLIR systems, such as the AN/ASQ-228 Advanced Targeting FLIR (ATFLIR). The other window is in the 8 to 12 μm band, which lies in the far IR; this band is normally used for navigation and surveillance purposes in systems such as the AN/AAR-50 NAVFLIR and TIALD.

WEATHER AND VISIBILITY

Due to the generally similar operating wavelengths of the unaided eye (0.4 to 0.7 μm) and modern NVGs (0.6 to 0.9 μm), the attenuating effects of weather phenomena on night illuminance may be considered to be largely as would be seen by the naked eye during the day. The following summarises some of the major weather effects and highlights the subtle differences associated with the higher wavelength of operation of NVGs:

Cloud

Clouds are highly variable in their attenuating effects. Water, the main contributory constituent in low-level clouds, can be found in all states, thus making judgement of overall degradation of sensors difficult. Clouds reduce the natural

Figure 5 – The effect of cultural lighting. The above two images, showing the same terrain board scene, were taken through a NVG. The image on the left was taken at low overall illuminance (dark night), while the image on the right was taken under high illumination conditions (half moon). Note the higher intensity of the halo around the cultural light in the scene on the left; under low light levels, NVG gain is high, therefore the light appears bright and obscures the scene detail in the vicinity (the building to the right is almost invisible). Conversely, the halo in the high light level scene on the right is less troublesome to the observer; the higher overall scene illumination drives NVG gain lower, thus the cultural light appears less intense, allowing the observer to see the buildings to the left of the light. The presence of cultural lighting severely degrades the ability of current NVGs to operate effectively, hindering operations in urban areas

0581544

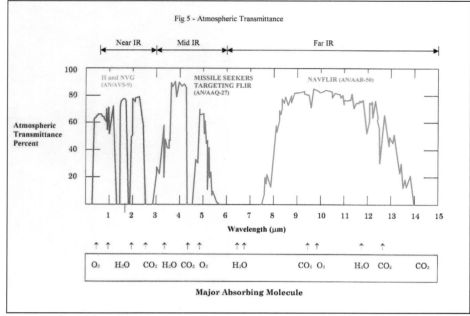

Figure 6 – Atmospheric Transmittance 0110035

night sky illumination to an extent dependent on the amount of cloud coverage and the overall density and/or thickness of the clouds. It follows that a thick, overcast layer of cloud will reduce the ambient light much more than a thin, broken layer.

Fog
The effect of fog on NVGs is similar to that of clouds. In meteorological terms, only their distances from the ground differentiate fog and cloud. When a fog layer is very thin (10 m or so) and it is made up of very small droplets, NVGs operating in the near IR wavelength may be able to see through it. However, it should be noted that encountering fog and poor visibility during flight at low level is a particularly hazardous aspect of NVG operations. When viewing the NVG scene, a decrease in the intensity of ground lights, possibly coupled with increased intensity of halos around ground lights is an indication that moisture in the air is high and that fog may be forming.

Rain
The effect of rain on NVG scene is difficult to predict, since it is usually associated with cloud, which, as discussed, is a major attenuator of the available natural illumination. In isolation, individual droplet size and the overall density of droplets will determine the level of attenuation. Similarly to rain viewed with the unaided eye, heavy rain (large droplet size and high density) obstructs light transmission and the low light levels associated with it can be very noticeable when flying at low level. However, light and medium rain (lower density and much smaller droplet size) may be almost invisible to NVGs.

Snow
Snowflakes are generally large in comparison to the wavelength of visible and near-IR energy; therefore snow, sleet and hail will all significantly attenuate illumination levels. As for the unaided eye, under these conditions the NVG scene will be unusable.

Battlefield Obscuration
The battlefield will yield many obscurants to NVG performance, such as exhaust smoke, fires and debris thrown into the air by explosions. While these effects are inevitably most severely felt at low level, some obscurants, such as oil fires during the Gulf War, can affect medium level operations. While operational planners will tend to route aircraft missions away from areas of intense ground warfare, sometimes aircraft will be tasked into such areas as part of Close Air Support (CAS) or Battlefield Air Interdiction (BAI) missions. In such cases, pilots should be aware that degradation of NVGs (indeed, most aircraft

sensors) may be sudden and severe, with similar actions required to those on encountering a volcanic dust cloud.

INTERPRETATION OF THE NVG SCENE
In this context, contrast is a measure of the luminance difference between two or more surfaces. Assuming a constant level of illumination, contrast as perceived by NVGs is dependent upon the differing reflectivities, or albedos, of each object and on the amount of texture, or roughness, in the scene. Although NVGs rely on reflected light from the scene, just as the naked eye, it should be noted that the II tubes are sensitive mainly in the near IR wavelengths, therefore we should consider the reflectivities, or albedos, of objects in the night scene at those higher wavelengths, to gain a true understanding of the NVG scene.

While it would be impossible to quantify all significant objects which make up the night environment in terms of how they are perceived on NVGs, the following is intended to provide a broad perspective of the NVG scene:

Roads
Generally, roads show up well on NVGs, due to their contrast (large albedo difference) with flanking terrain. Roads that cut through heavily forested areas are also easily recognised if they are visible through the tree canopy. Light coloured concrete roads/motorways are easily identified under most medium to high light conditions, whereas asphalt roads can be difficult to identify because the dark surface absorbs available light.

Water
There is very little contrast between the land and a body of water in low light conditions. As the light level increases, land-water contrast increases and reflected moonlight is easily detected. It is usually reflected moonlight that confirms that a dark patch on the ground, as viewed through NVGs, is actually water!

Fields and Forests
Ploughed fields look dark on NVGs due to the roughness of the surface, which absorbs most of the incident light. Despite their low albedos, cultivated fields provide for a 'patchwork' effect which provides for excellent contrast with surrounding roads and other terrain features.

Trees have different albedos depending on their type (deciduous or coniferous) and the time of the year. Large areas of unbroken forest can mask underlying terrain and generally provide poor contrast, although small areas of forest, divided by open fields, cultural objects and roads contribute to the generally excellent contrast levels found in European countries.

Desert
Open desert, without significant vegetation or cultural objects, produces a very bland image on NVGs due to lack of texture in the terrain resulting in poor contrast in the NVG scene. During bright light conditions, mountain ranges can be easily identified due to intense shadowing, although these shadows can obscure other less significant features which can represent a serious hazard to pilots flying at low level in such terrain.

Snowfields
Snow has an extremely large albedo, reflecting approximately 90 per cent of incident light. As previously discussed, under high levels of illuminance, this can result in too much light entering the NVG, causing a decrease in gain and loss of resolution. However, during typical winter conditions in temperate climates such as Europe, snowfall is usually associated with significant cloud cover, therefore snow cover can be extremely useful increasing illuminance.

However, significant amounts of snow lying on the ground can increase overall light levels to the point where the NVG reduces gain considerably, thus decreasing resolution. This, coupled with drastically reduced contrast associated with complete snow cover can make for subtly hazardous flying conditions at low level. Conversely, and similarly to forest areas, as a snowfield becomes more broken, with roads, fields and trees becoming discernible, contrast levels increase and provide excellent conditions for low level flight on NVGs.

NIGHT VISION GOGGLES (NVGs)
NVGs have been in widespread service for years, but only relatively recently have pilots of fast jet tactical aircraft used NVGs for operational purposes. All NVGs operate on the same basic principles. Image intensifier tubes produce a bright monochromatic (green) EO image of the outside world in light conditions too low for normal vision. Unlike FLIR, however, which uses the far IR area of the electromagnetic spectrum to create an image based on object temperature and emissivity relative to its surroundings, NVGs use the red and near IR area of the spectrum to produce their own unique visible image of the world. This image is based on the relationship between the amount of light present, referred to as illuminance, and the amount of light which is reflected from objects in the scene, referred to as luminance or brightness. Under defined luminance conditions, the NVGs are able to provide navigation and terrain avoidance references for pilots.

Night Vision Devices – Background
Generally, image intensifier technology is expressed in terms of first-, second-, third- and, latterly, so-called fourth-generation systems:

First-generation (Gen I) systems were used in starlight scopes used by snipers in Vietnam. These devices featured gains in the range of 40,000 to 60,000 by means of a three-stage cascade configuration of simple intensifier tubes contained in the scope. First-generation tubes were extremely durable and offered extended life, but were not suitable for aviation applications due to a number of factors, which included blooming (a tendency for the tube to washout if a bright light source appears anywhere in the device FoV), a high voltage requirement and, most significantly, their size and weight.

Second-generation (Gen II) image intensifiers introduced the Micro Channel Plate (MCP). The invention of the MCP was the breakthrough that facilitated miniaturisation of image intensifier tubes to the extent necessary for helmet-mounted systems for use in fast jet aircraft. However, the gain associated with second-generation tubes is somewhat lower than first-generation tubes at 20,000 to 30,000. Second-generation tubes are still somewhat susceptible to blooming when exposed to a bright light source but this tendency is minimised by the function of the MCP. The MCP allows the saturation to be confined to individual channels instead of the entire FoV as was the case in first-generation tubes. This localised

Figure 7 – The binocular section of an AN/AVS-6 (V)1 NVG, with a section of an 18 mm Gen III image intensifier tube and a MCP. Note the relatively small size of the image intensifier tube compared to the monocular and, in turn, the small part of the intensifier tube occupied by the MCP; most of the length of the former is occupied by the fibre optic twister. See also Figure 11, which shows a cutaway section of an II tube in isolation   0533842

Figure 8 – NIGHTBIRD Gen III NVG   0110044

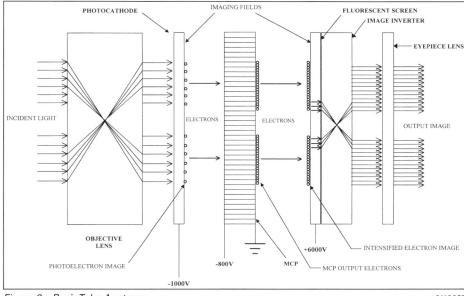

Figure 9 – Basic Tube Anatomy   0110037

saturation appears on the intensified image as a 'halo' effect around any bright light source. Image contrast is drastically reduced inside the halo, and the overlapping of halos from multiple light sources removes scene detail between the lights. Although now superseded by third- (and soon fourth-) generation tubes, it should be noted that second-generation tubes remain in widespread use amongst many ground forces.

Most fast jet aircraft today employ third generation (Gen III) tubes in their NVGs. There are two major differences between second- and third-generation tubes: the S-20 multi-alkali photocathode of second-generation tubes has been replaced by a gallium arsenide photocathode, and a metal oxide film has been applied to the MCP. The gallium arsenide photocathode surpasses the photosensitivity of the multi-alkali photocathode beyond 550 nm. This means that third-generation tubes are far more sensitive in the region where near IR radiation from the night sky is plentiful, and the 800 to 900 nm peak sensitivity range demonstrated by the tube has a five to seven times greater photon rate than in the visible region. This translates directly to increased gain for NVGs utilising these tubes. Also, the addition of the metal oxide film to the MCP in third-generation tubes greatly extends their service life over that of second-generation tubes. This is due to the fact that the life of an intensifier tube is largely a function of the lifetime of the photocathode. End of life for a photocathode is primarily caused by ion bombardment from the MCP as it is struck by electrons from the photocathode. This bombardment results in a gradual loss of photocathode luminous efficiency. The higher the light input, the more ions are generated, and the shorter the life expectancy of the tube. For this reason the NVGs should never be exposed to any bright light. The addition of the metal oxide film to the MCP in third-generation tubes helps to offset this problem. The metal oxide film effectively 'traps' the ions, thus preventing contamination of the photocathode. While the film is almost transparent to electrons, some are lost to the intensification process, thus reducing the effective Signal-to-Noise Ratio (SNR) of the tube. A further disadvantage encountered with the introduction of the metal oxide film is an increase in the bias voltage requirement between the photocathode and the MCP. This, in turn, requires increased spacing between the two components to prevent arcing. Ultimately this spacing results in an increased halo size, as compared to second-generation performance, when viewing bright light sources.

While Gen III image intensifiers from ITT and Northrop Grumman are predominant in the US, the aforementioned disadvantages associated with the technology led to the further development of Gen II devices in Europe. Although the S-25 multi-alkaline photocathode material employed in the latest DEP XR-5 and Photonis HyperGen™ tubes exhibits lower sensitivity than the gallium arsenide used in Gen III tubes, overall system performance is comparable due to a superior SNR (XR-5 and HyperGen™ tubes do not have a metal oxide film). In addition, these tubes feature reduced photocathode-to-MCP spacing and thus benefit from smaller halos.

In addition to photocathode life, and depending on light levels encountered during the life of an image intensifier, another factor which determines the lifetime of the tube is a gradual drop in secondary electron emissions through the MCP.

The most recent evolution of the image intensifier is often termed 'fourth-generation' (Gen IV) tube (although this definition has not been formalised yet), which utilises an autogated power supply, which matches the applied power to the gain of the tube, thus avoiding saturation and possible damage which the tube is exposed to high light levels. This has enabled the manufacture of a gallium arsenide photocathode tube without a metal oxide film ion barrier since autogating provides protection for the MCP; with a decreased gap between the MCP and photocathode, bias voltage requirement is less. These modifications, together with advances in MCP manufacturing techniques enabling smaller pitches to be achieved, has resulted in a tube which offers a significant performance advantage over third-generation tubes in both high and low light conditions.

However, during operational testing of truly 'filmless' Gen IV tubes, difficulties were encountered with ensuring adequate life for the tubes. This, in part, led to the development of a hybrid tube, termed Gen III +, combining a third-generation tube with a thinner (metal oxide) film and the autogating power supply of the Gen IV tube. Claimed to offer near 'Gen IV' performance, combined with tube life comparable to current gallium arsenide-based designs, NVGs based on this technology have been developed and fielded by the US military. Due to the smaller halos and rapid gain response, Gen III + NVGs have shown considerably greater performance under MOUT conditions. Such technologies are, at present, restricted to US operators and a very few foreign customers.

Night Vision Goggles – Theory of Operation

IIs consist of three main components – a photocathode, a Micro Channel Plate (MCP) and a phosphor screen (Figure 9):

Photocathode. Photocathode developments represent the principal difference between Gen II and Gen III intensifier tubes. Gen II equipment utilises a multi-alkali compound, whereas a Gen III tube has a gallium arsenide coated photocathode. As already discussed, the photocathode converts photons into electrons under the influence of an electrical field and these electrons are accelerated into the MCP channels. The photocathode is fitted with a brightness output control, called Bright Source Protection (BSP) circuit, which limits the number of electrons leaving the photocathode by reducing the voltage between the output of the photocathode and the input to the MCP. This function is performed automatically.

Micro Channel Plate. Electrons collide with the sides of the channels, Figure 10, and result in an output of, typically, 1,000 electrons for each electron that enters. Consequently, the current leaving the MCP is 1,000 times higher than that leaving the photocathode. The MCP is extremely thin (approximately one millimetre), therefore IIs using them have become known as 'wafer type' intensifiers. The MCP is encased in an 18 mm tube, although recent advances in MCP technology have facilitated the development of 16 mm tubes without loss of resolution. Within this diameter, there are over 6 million glass tubes, or channels (in the latest US tubes), uniformly tilted at approximately 5 to 6° (see Figure 10) to ensure that input electrons strike the sides of the tubes. The MCP also has a means of protection from high light levels, the Automatic Brightness Control (ABC). As the name suggests, the circuit automatically adjusts MCP voltage to maintain image brightness at a constant level for a wide range of ambient illumination levels.

Phosphor Screen. The phosphor screen converts the magnified number of electrons produced by the MCP back into photons, thus producing an intensified, visible image. The choice of phosphor is governed by a requirement for resolution, eye readability and persistence (of the resultant image). The 'green' image seen through most NVGs is due to the phosphor used; in early US Gen III (OMNIBUS 2/3 contract) goggles, the P-20 phosphor produces a saturated green image, whereas later Gen III (OMNIBUS 4–6 contracts) goggles utilise a P-43 phosphor which produces a green image with a yellow tint.

The final components in the optical train are the image inverter and the eyepiece lens. The fibre-optic image inverter, sometimes referred to as a twister, consists of a bundle of millions of fibre-optic strands, which rotates through 180° to re-orient the intensified image. The image inverter also collimates the image for correct positioning in relation to the wearer's eye.

The eyepiece lens is the last optical component in a NVG. It focuses the visible image onto the retina of the viewer and incorporates a limited dioptre adjustment to allow for some correction for individuals' vision. However, the eyepiece does not correct for astigmatism, therefore the wearer should still wear any prescribed corrective spectacles. Modern Gen III goggles provide for 25 mm of eye relief (the distance between the eye and the eyepiece lens) at 40° Field of View (FoV), enabling spectacles to be worn and facilitating view beneath the goggle (at cockpit instrumentation).

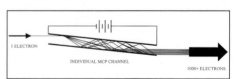

Figure 10 – Individual MCP Channel   0110036

Figure 11 – The fibre optic twister consists of a bundle of millions of strands rotated through 180° in order to re-invert the intensified image to the viewer   0533843

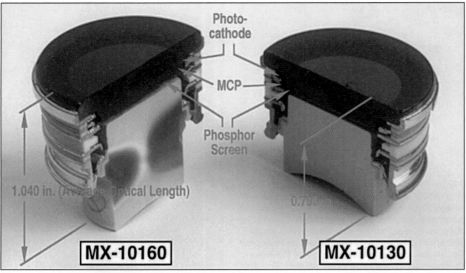

Figure 12 – ITT Night Vision's MX-10160 18 mm (diameter) Image Intensifier (II) tube is utilised in US AN/AVS-6/9 Night Vision Goggles (NVG), while the MX-10130 18 mm II tube is used in AN/PVS-7A/B/D NVG   0130038

Types of NVG

NVGs fall into two main categories – Type 1, or straight-through optics, are the most common type of NVG in use today, characterised by Aviators' Night Vision Imaging System (ANVIS) (USAF) and NIGHTBIRD (UK RAF) and Type 2, or folded optics, characterised by CATSEYE (used in the UK until superseded by the NIGHTBIRD and in the US by the USMC, now superseded by the AN/AVS-9). Figures 13a and 13b show NIGHTBIRD and CATSEYE NVGs, while Figure 14 illustrates the different optical paths of Type 1 and Type 2 NVGs.

The reason for the development of two entirely different designs of NVG can be traced back to the approaches of engineers in the US and UK. In the US, the first NVG designed for aviation use was the binocular AN/AVS-6, which was developed from the biocular AN/PVS-5, an NVG designed for vehicle drivers. The AN/PVS-5 is an enclosed system (that is, it fully covers the face) utilising a single II tube, with the optical path split into two, providing an image for each eye. The AN/AVS-6 is a logical development of that system, utilising two II tubes without any faceplate to allow for viewing around and below the goggle at flight instrumentation.

In the UK however, engineers designed an entirely new type of goggle aimed purely at the fast jet cockpit. The resulting Ferranti (now BAE Systems) CATSEYE NVG featured two separate optical paths, one an intensified image from the II tube and the other a direct view of the outside world. The reason for this approach was to minimise interference

Figure 13a – Nightbird NVG   0110042

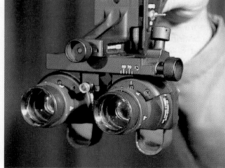

Figure 13b – Catseye NVG   0110043

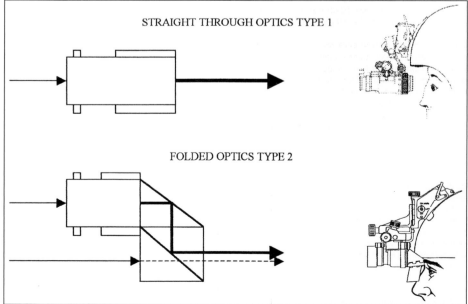

Figure 14 – NIGHTBIRD and CATSEYE goggles characterise different approaches to NVG design   0110038

from the II image when the pilot viewed raster FLIR imagery in the Head-Up Display (HUD). A sensor on the goggle (the sensor can be seen situated between the objectives of the CATSEYE goggle in Figure 13b) removed the II image when the goggle was aligned with the HUD, reinstating the image when the goggle moved out of line. A three-position power switch enabled pilots to select/deselect this Auto Scene Reject (ASR) feature. This system allowed the pilot to view FLIR imagery and/or the outside world without the II image and provided for a much better view underneath the goggle at the aircraft Head-Down instruments and displays. The major drawback to the CATSEYE system was the relatively poor 30° FoV, compared to 40° for the AN/AVS-6, although the system found favour with the USMC, who used the CATSEYE goggle (US designation MXU-810/U) in all their fixed wing fast jet aircraft until only recently replacing it with the AN/AVS-9 Type 1 NVG.

Offering a wider FoV and a generally brighter image, the Type 1 NVG is now the dominant version in service. In the US, the original AN/AVS-6 has been replaced by the AN/AVS-9, which, although largely similar in appearance, offers increased image resolution and brightness. In the UK, the CATSEYE NVG was used briefly in the Harrier aircraft until it was replaced by the Type 1 NIGHTBIRD NVG, which is similar to early marks of AN/AVS-6 in terms of performance.

NVG Adjustment

• The adjustments and controls for Type 1 (NIGHTBIRD) and Type 2 (CATSEYE) NVGs are shown in Figures 15 and 16. Note the similarity in positioning of the major controls.
• Vertical Height Adjustment – This raises and lowers the monocular optical axis with respect to the helmet datum.
• Interpupillary Distance (IPD) – This moves the individual monocular assemblies equal distances from a fixed central datum position. A scale allows pre-setting of the IPD to a known personal setting.
• Tilt Adjustment – The monocular assemblies, which are linked together mechanically, will elevate or depress from the nominal. By convention, forward rotation of the knob tilts the NVG downwards.
• Fore and Aft Adjustment (Eye Relief) – Fore and aft adjustment is carried out by rotating the adjustment knob located at the front between the monocular assemblies.
• Objective Focus – For the CATSEYE NVG, this control focuses the goggle for distance. For the NIGHTBIRD NVG, the objective lens assembly is fixed focus at between 20 m and infinity.
• Fore/Aft Adjustment (dioptre) – Dioptre adjustment is achieved by moving the eyepiece lens assembly axially, relative to the monocular housing. Normally, dioptre adjustments will be required for each operator prior to use. Adjustment to a known personal setting is also possible by means of a graduated scale.

Ejection Safety – Autoseparation

Fast jet aircrew helmets are designed to exacting specifications in terms of mass and balance, to ensure optimum protection during the ejection sequence. However, until the advent of integrated helmets, none were designed from the outset to include the mass of a set of NVGs attached to the front, which obviously upsets the balance and centre of gravity of the helmet assembly, resulting in possible complications for the pilot should the goggles be in place during ejection. As a result, pilots were obliged to takeoff and land with the goggles stowed, since it was judged that the pilot would have insufficient time to manually doff the goggles prior to ejection during these flight phases. This presented some quite serious operational limitations, since airfields would have to maintain full lighting for aircraft movements, thus presenting an easy target for enemy air attack, and pilots would have to allow sufficient time to recover full night vision prior to landing during poor weather conditions.

In order to address the problem of unpremeditated ejection when wearing NVGs, the UK and US both developed autoseparation systems, which

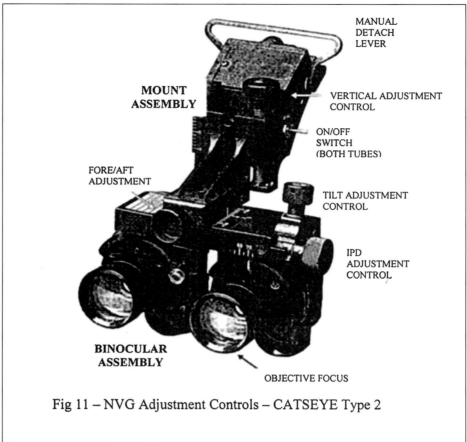

Fig 11 – NVG Adjustment Controls – CATSEYE Type 2

Figure 15 – NVG Adjustment Controls – CATSEYE Type 2   0110039

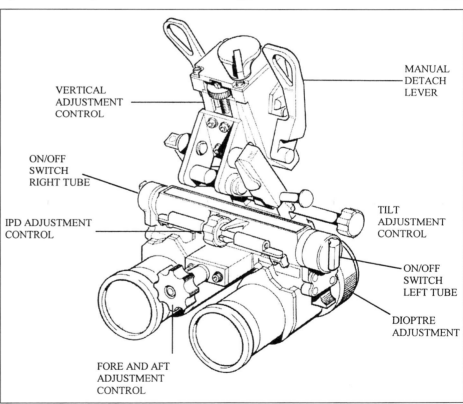

Figure 16 – NVG Adjustment Controls – NIGHTBIRD Type 1   0110040

sensed movement of the ejection seat at the beginning of the sequence and fired a small charge to detach the goggles from the pilot's helmet. While the UK developed and has fielded autoseparation for all of its fast jet aircraft, the US developed a system (the AN/AVS-8), but never fielded the system.

Considering the UK system as fitted to the Martin Baker Mk10 ejection seat as fitted to RAF Tornado aircraft and aircrew using the GEC-Ferranti NIGHTBIRD Gen III NVG, operation of the system is as follows: During unpremeditated ejection, the NVG assembly will be released from the Helmet Mounting Plate (HMP) automatically. A gas motor

assembly, located within the NVG mounting assembly, will actuate the release mechanism for the NVG when power is supplied from the Power Supply/BIT Unit (PS/BIT) (Figure 17). The wiring to carry the power connects from the PS/BIT Unit to the Pilot's Equipment Connector (PEC), then up through a modified connector to the helmet cableform; the connections are passed through on spring loaded contacts at the helmet mounting plate and then to the gas motor.

In terms of operational experience with autoseparation of NVG in the UK, compared to that in the US without such a system, evidence as to the overall utility of NVG autoseparation

Figure 17 – NVG PS/BIT unit as fitted to a Tornado
GR4   0110045

during ejection is rather difficult to identify directly:

The UK Nightbird and AN/AVS-6/9 are different designs, the former being a one-piece system which would probably remain attached to the helmet without autoseparation, thus injuring the pilot. The AN/AVS-6/9 is a two-piece design; to the editor's knowledge, during night ejections, while no US pilot has manually doffed NVGs before ejecting, the binocular section has detached at some stage during the ejection sequence, leaving only the approximately 250 gm mount attached to the brow of the helmet, resulting in no significant injuries directly attributable to NVGs sustained.

Of the few instances of UK ejections while wearing NVGs, in one instance the pilot doffed his goggles before ejecting; in other instances the autoseparation system functioned correctly with no significant injuries directly attributable to NVGs sustained.

So, while there have been no significant injuries sustained by US pilots ejecting with NVGs, this, it would seem is as a result of the physical design of the AN/AVS-6/9, which was originally designed for helicopter use, with the binocular section designed to break away at loads of 10–15 g during crash landing. This has translated into a 'breakaway' capability during ejection from a fighter aircraft. For one-piece NVGs, like the UK Nightbird, the differing physical design precludes such 'de facto' autoseparation.

Thus, despite US experience without a fielded autoseparation system, the UK and other users of one-piece NVGs will require autoseparation if they wish to conduct unrestricted NVG operations.

As a footnote, NVG autoseparation can provide a stumbling block to the upgrade of NVGs – during 2005/2006 the UK was forced to fit new IITs into a number of its venerable Nightbird goggles due to fall-off in performance, possibly accelerated by their employment around well-lit cities during the recent Gulf War. In order to maintain the physical characteristics of the NVG, due to the requirements of aircraft 'release to service' for the autoseparation system, coupled with the absence of any IIT manufacturing capability in UK, 're-tubing' with US Gen III IITs was the only alternative. In the editor's opinion, this was most likely a more expensive and less operationally efficient use of resources than simply learning from the American experience of NVG ejection and procuring AN/AVS-9s, thus enabling the UK to benefit directly from the huge effort the US puts into development and support for their NVGs.

NVG DEVELOPMENTS

As previously mentioned, while II and NVG development has been virtually non-existent in the UK in recent years, with greater emphasis placed on research into uncooled FLIR technologies, the two major II manufacturers in the US, ITT and Northrop Grumman (formerly Litton), have made some significant improvements in II performance as the new 'fourth-generation' (Gen IV) specification becomes defined.

NVG Limitations

Before outlining current and future developments in NVG technology, we should first examine the major limitations, as perceived by pilots, of NVGs. By far the most important limitation of operating with NVGs is the nominal 40° FoV, which seriously degrades situational awareness and forces an exaggerated eye scan to build up a mental picture of immediate NVG scene. Close behind in terms of importance, resolution of the current range of NVGs also presents serious limitations to their tactical employment, with degraded depth perception and distance estimation posing problems for pilots, particularly in the low level arena. While there are other issues to be addressed regarding future operational applications of NVGs, including Laser Eye Protection (LEP) and ejection safety, it was the issues of goggle FoV and resolution which were considered of highest importance in developing future goggles.

Resolution and Field-of-View

In common with most imaging systems, FoV and resolution for NVGs are directly linked; for any goggle/tube pairing, resolution can be increased, but FoV must be proportionally decreased to achieve this, and vice versa. This is often referred to as 'the optical constant'. Therefore, in order to increase resolution for an NVG while maintaining the FoV, the basic resolution of imager (in this case the IIT) must be increased directly. In order to achieve this we should examine the most significant limiting factors to resolution for an IIT.

The principle limiting factors governing IIT resolution are the distance between the individual MCP channels (termed pitch – see Figure 18), the distance between the photocathode and the MCP and the distance between the MCP and phosphor screen. Of these, MCP pitch is the most powerful factor in terms of resolution. In terms of visualising MCP pitch, this can be likened to the number of pixels in a modern flat screen television – the MCP channels each produce a single pixel of information on the phosphor screen as viewed by the pilot; hence, by increasing the number of MCP channels over a given screen area, then a greater number of discrete pixels will be displayed, hence increasing resolution. MCP channel spacing has been reduced from 12 μm in Gen II tubes to 6 μm in Gen III tubes, corresponding to over 6 million channels in an 18 mm OMNIBUS 4 specification tube. This has resulted in improvements in sensitivity up to the current 64 lp/mm of current goggles (Figure 19). As manufacturing techniques improve, the MCP pitch will undoubtedly decrease further (perhaps as low as 4 μm), allowing for greater resolutions (up to 125 lp/mm), and/or a reduction in tube diameter (the latest II tubes are 16 mm in diameter, down from 18 mm, while still maintaining the overall resolution of the larger diameter tube – see PNVG).

However, in the editor's experience, NVG resolution, taken in the context of their use of a navigation sensor and night vision aid, is sufficient for most fixed-wing operations; accordingly, a better use of decreased MCP pitch would be to maintain the resolution at the current 64 lp/mm and simply expand the FoV proportionally. Thus instead of producing a 16 mm tube with the same performance as an 18 mm item, by maintaining the II tube diameter at 18 mm, a greater FoV would be realised.

Panoramic Night Vision Goggle (PNVG)

Having described the ongoing work into improving IIT resolution, it is apparent that trading system resolution for increased FoV is not an option. Therefore, a means of increasing the total FoV without affecting system resolution was sought by manufacturers. As part of a US Small Business Innovative Research (SBIR) programme, the Night Vision Corporation was awarded a contract to investigate Wide FoV (WFoV) NVG technologies and fabricate a proof of concept demonstrator. This resulted in the Panoramic Night Vision Goggle (PNVG) Phase 1 SBIR, which consisted of an arrangement of four II tubes mounted on an aircrew helmet in a Type 2 indirect viewing goggle producing a 100° horizontal by 40° vertical FoV.

The work of Phase 1 SBIR led to Phase II SBIR, which ended in July 1999 with 12 working models of two types of PNVG delivered. PNVG I is a fast jet Type 2 NVG, with four II tubes mounted along the brow of a standard HGU-55P aircrew helmet. The outer tubes are canted outward and the tubes are connected to biaxial eyepieces. The result is a panoramic NVG with a 100 × 40° FoV, with resolution as for the donor AN/AVS-9 tubes. Later prototypes were intended to incorporate an Electro-Luminescent (EL) display for symbology or imagery. The relatively flush fit of the Type 2 arrangement of the system means that the NVG can be retained during ejection.

The PNVG was intended to interface with the helmet and the aircraft in the same way as the Joint Helmet-Mounted Cueing System (JHMCS). Installed on any JHCMS-compatible aircraft, the PNVG was able to display flight data and sensor imagery. The aircraft's avionics and sensors give the pilot cueing information to find targets, and the pilot is able to designate targets with head movements. Infra-red imagery can be displayed on an inset panel using a 640 × 480 EL source.

Another version of the PNVG, designated PNVG II (Figure 20), was designed for pilots of helicopters and transport aircraft. This differed from PNVG I in being a Type 1 system, consisting of the familiar AN/AVS-9 with two further tubes grafted onto the outside of the goggle to provide for the wide horizontal FoV.

The PNVG I underwent operational evaluation at Nellis AFB, using F-15 and F-16 aircrew, in the US during 1999, with greatly increased pilot SA reported. This positive reaction led to the development of a specification for an Integrated Panoramic Night Vision Goggle (IPNVG), with the design aims of resolution equal to or better than the AN/AVS-9, compatibility with standard aircrew spectacles and based around a 16 mm II tube. Other key objectives were:

- Wide FoV
- LEP/Laser Hardening
- Ejection/Crash/Ground Egress Safety
- Fit/Comfort
- Reduced Halo
- Increased Image Quality
- Integrated Symbology/Imagery Display
- Field Supportability
- Reliability/Maintainability/Affordability.

The specification for the IPNVG also demands a minimum 86 × 36° FoV, an LEP visor and II tube hardening; Type 1 and Type 2 variants for army and air force use have been developed by two vendors under competitive tender. During 2003, a production contract was awarded by the US Air Force to Insight Technology, Inc. for a WFOVNVG based on the PNVG II (see separate entry).

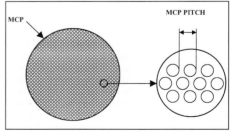

Figure 18 – MCP Pitch   0110041

TYPE	PHOTOCATHODE	MCP PITCH	PHOTO RESPONSE μA/lumen	RESOLUTION lp/mm
Gen II	S-25 Multi Alkali	12 micron	100 – 400	22 – 36
Gen III	GaAs	9 micron (OMNI 3)	600 – 1800	45
		6 micron (OMNI 4)		64
'Gen IV'	GaAs	4 – 6 micron	Up to 2200	>64

Figure 19 – II tube performance improvements by Generation

Figure 20 – A pilot with the 422nd Test and Evaluation Squadron at Nellis Air Force Base, Nevada, wearing IPNVG in an A-10 Thunderbolt II. The baseline version of the system, which has recently entered service in the A-10C, is designated AN/AVS-10 (US Air Force)   1127358

Ongoing test and evaluation of panoramic NVGs has so far not yielded a 'quantum leap' in operational capability – current Type 1 examples have significant issues regarding focus of the outer tubes, with opinion divided as to whether panoramic NVGs should include object focus at all. Further, panoramic NVGs do not feature dioptre adjustment, which is perhaps no loss to the pilot community at all – extensive work in the US has been carried out to determine an optimum 'average' dioptre setting for panoramic NVGs, which will cover the vast majority of wearers (clip-on dioptre adjusters can be employed if the wearer falls outside of this setting). In use, the wearer operates with a panoramic NVG in much the same way as for unaided viewing – the increased peripheral vision is employed to detect objects and relative movement at the edges of the FoV, with head movement bringing objects of interest into the central portion for best viewing and interpretation of the scene, so perhaps the relatively poor focus of the outer channels will be of little practical consequence to future iterations. Overall, it is the editor's opinion that while panoramic NVGs provide a useful increase in (short-range) SA to the pilot, the overall 'quality' of the image (brightness and resolution) is exactly the same as currently fielded systems, so the real-world (not necessarily pilot-perceived) benefits of acquisition must be considered carefully.

From 2006, the baseline configuration of the IPNVG, designated AN/AVS-10, entered service in US A-10C aircraft to enhance the type's effectiveness in night Close Air Support (CAS) operations. Considering the predominance of medium operations, including CAS, being carried out in air operations over Iraq at the time, it is questionable whether a wider horizontal FoV will prove useful in this instance: one of the eternal problems the NVG user encounters is determining the relative ranges of light sources within the FoV – by increasing this, particularly if the airspace is busy, there may well be a great deal of no factor/long-range but nonetheless distracting activity in his/her peripheral vision!

Considering current restrictions on the export of cutting edge NVIS technology, coupled with the vastly increased cost of WFoV systems (last noted at approximately USD65,000 per set), prospective NVG operators must surely question their cost-effectiveness. While a wide FoV is undoubtedly a powerful tool in maintaining SA close to the ground for the helicopter pilot, until there is significant feedback to the contrary from A-10C pilots, there is little evidence to support the view that a WFoV system (in current configuration) will significantly enhance the (correctly trained and supervised) operations of fighter squadrons.

However, we should not dismiss current efforts in the area of enhanced FoV too quickly – there is undoubtedly a 'middle road' in terms of providing the pilot with, arguably, a better compromise in terms of enhanced FoV and usability/cost. Continuing with the previous reasoning concerning the relationship between NVG resolution and FoV, if the MCP pitch is decreased, facilitating a decrease in overall II tube diameter while maintaining resolution, it is therefore reasonable to assume that utilising the same decreased MCP pitch in a full-size 18 mm tube would realise greater FoV than the current 40°. Assuming a linear relationship, this would yield a full resolution of 45° – not a massive increase, but with further decreases in MCP pitch anticipated, this could increase up to the 55 to 60° before NVGs, at least based on IIT technology, have been superseded. It is the editor's opinion that most fighter pilots, confronted with choosing a PNVG, or an 'Advanced' NVG (ANVG) with a 60° full circular FoV, would choose the latter for a number of reasons. Firstly, the increased vertical FoV of the ANVG would be most useful during hard manoeuvring to better clear the forward flightpath. Second, the outer tubes of PNVG restrict the pilot's unaided peripheral vision, which is employed quite extensively during formation manoeuvring and takeoff and landing. Add less weight and bulk for the ANVG, most useful in the tight confines of a fighter cockpit, and assuming less cost (although the quantum may well not be as great as intimated above), then a resounding market for such an ANVG would materialise.

High and Low Light Performance

While increased II tube performance at low light levels has been steadily increased with improvements to Signal-to-Noise Ratio (SNR) resulting from increased photocathode sensitivity, problems with high light level performance have recently come to the fore in recent conflicts involving MOUT. As previously discussed, Gen III tubes utilise a metal oxide film in order to protect the photocathode and extend the life of the goggle. Reduction in thickness of this barrier film, made possible by the integration of an Autogated power supply, improves both high- and low-light performance by directly improving response and SNR.

Halo Size

Halo size is a function of the spacing between the photocathode and the MCP. This gap determines the perceived size of all halos within the FoV, while the intensity of the halo is a function of the overall system gain; hence halos are more intense under overall low light conditions when tube gain is high. While halo size increased in Gen III tubes due to increased bias voltage requirements, thereby forcing an increased distance between the MCP and photocathode, the introduction of new technologies such as autogated power supplies (with a thinner metal oxide film) has enabled this gap to be decreased with resultant reduction in halo size.

The Future

The development of NV technologies for tactical aircraft has become entwined with that of integrated helmets, such as Thales' Topowl helmet, designed for helicopter application, and the Striker display helmet for Eurofighter Typhoon. The Charge Coupled Device (CCD) has matured rapidly, to the extent that it is now being developed as an alternative to current II technologies in terms of night vision. Compact, lightweight and cheaper to produce than traditional II tubes, CCDs unfortunately suffer from comparatively high Equivalent Background Illumination (EBI – a measure of noise in the system) – of the order

of 10^{-3}, compared with around 10^{-7} for II tubes. This reduces effective SNR, and thus low light performance, for current CCDs, to the level of a Gen II tube. However, CCD systems lend themselves to image processing, facilitating the 'cleaning up' of the intensified image, allowing an intensified CCD camera system to drastically reduce halo effects and enhance contrast in the scene; therefore overall performance, as perceived by the wearer, may not lag too far behind. Furthermore, technologies involving Complimentary Metal Oxide Semiconductors (CMOSs), often considered the 'poor relation' to the CCD, may provide another alternative. Exhibiting lower EBI, intensifiers based on this technology may mature in the long term.

More recently, a hybrid II CCD, which couples a traditional IIT with a CCD (or CMOS) camera, shows promise in terms of marrying the performance of current II tubes with the lightness and compact dimensions of a solid-state camera. However, it should also be noted that, in contrast to NVGs based on II tubes, which display the image directly to the observer, CCD/CMOS camera systems are dependent on a video display to convey scene information. Current video standards, 625-line (PAL) and 525-line (NTSC), would therefore be limiting in terms of system performance. The best video performance currently available for cockpit use is of the order of 1,000 lines, which closely matches the vertical pixel count of the latest $1,024 \times 1,024$ pixel hybrid II CCDs utilised in helmets such as the Eurofighter Integrated Helmet. However, while this resolution still lags behind that of a current direct view NVG, with suitable image processing, the subsequent image as presented to the pilot, is, in the opinion of the editor, perfectly acceptable for use in pilotage and navigation.

So, for those air forces and paramilitary organisations considering the purchase of NVGs, the death of the II tube is certainly a long way off; it will be many years before solid-state technology (on its own) provides comparable performance to the II tube in terms of resolution, even assuming that video display technology will be able to keep pace. Other sensor technologies, such as uncooled FLIR, are fundamentally limited in their performance potential, while cooled systems are too bulky for helmet applications, while both remain ultimately limited by video display capability. Recent conflicts, such as the US campaigns in Afghanistan and Iraq, have only served to reinforce the pivotal position of NVGs in any conflict, with the capability to be able to 'see' more and further than your opponent often the deciding factor in any action. As a footnote, this realisation on the part of the US military has led to considerable tightening up of export regulations concerning II technology.

FORWARD-LOOKING INFRA-RED (FLIR) SYSTEMS

FLIR vs. NVG – complimentary sensors

While IR systems are prevalent in all areas of ground, shipborne and airborne operations, both military and civilian, the following will concentrate on the application of FLIR to tactical airborne fixed- and rotary-wing operations.

As previously mentioned, the use of Electro-Optic (EO) sensors offers a totally passive solution to night operations and all comments regarding the implementation of EO sensors, including targeting, human physiology, logistics and crew training apply equally to FLIR and NVGs.

Although the aforementioned NVGs are relatively inexpensive (in military acquisition terms) to integrate into an aircraft cockpit, costs associated with FLIR systems, whether podded or fully integrated, particularly modern navigation/ targeting/Infra-Red Search and Track (IRST) systems (such as PIRATE on the Eurofighter Typhoon – see entry), can be prohibitive. Linked systems, including Helmet-Mounted Displays (HMDs) and laser-guided weapons, should also be considered when budgeting for the overall cost of acquiring an advanced targeting capability.

It cannot be stressed too strongly that NVGs, operating in the 0.6 to 0.9 µm waveband, and FLIR, operating in the 8 to 12 µm and/or 3 to 5 µm wavebands, should be regarded as complementary sensors – as a generalisation, when one sensor is degraded due to atmospheric attenuation, light levels or diurnal effects (to name but a few) the other is likely be operating satisfactorily.

Similarly to the foregoing description of NVG, before embarking on a description of IR sensor systems and FLIR in particular, it is necessary to understand the physical attributes of IR radiation and the utilisation of entirely different waveband of the electromagnetic spectrum to that described previously.

PHYSICAL ATTRIBUTES OF IR RADIATION
IR Systems and the Electromagnetic Spectrum

The Infra-Red (IR) portion of the electromagnetic spectrum can be subdivided into the Near IR (NIR – 0.8 to 1.5 µm, NVGs and target designating laser systems), Middle IR (MIR – 1.5 to 6 µm, IR missile seekers and advanced targeting systems) and Far IR (FIR – 6 to 30 µm, navigation FLIR and IR reconnaissance systems), see Figure 2.

Passive IR systems depend on emitted energy, termed exitance (formerly emittance) from objects in the scene. In contrast to NVIS, ambient illumination is not required for the detection of IR energy. IR detection is an integral part of a modern tactical aircraft's integrated sensor suite, utilised in both air-to-air and air-to-ground roles. Characteristics of passive IR systems with application to combat aircraft operations are:

- IR sensors (MIR and FIR bands) are complementary to NVIS (NIR band).
- IR sensors do not require a source of illumination.
- Conventional combat camouflage techniques are ineffective against IR sensors.

For avionic applications, IR systems are concerned with differential temperature sensing, rather than the absolute level of IR radiation.

Radiometric Terms

In describing the processes involved in the interaction of IR radiation with matter, standard radiometric terms (as opposed to the subtly different photometric terms, as applied to NVGs and image intensifiers) are used. Figure 23 summarises the basic radiometric quantities, definitions, units, and symbols:

Radiation – Descriptive Laws

Any object whose temperature is above absolute zero (0 K or −273°C) will emit electro-magnetic energy; since most objects we are concerned with for the avionic application of FLIR meet this requirement, they therefore radiate EM energy in the IR portion of the spectrum. Laws describing the nature of thermal radiation have been derived by Kirchhoff, Planck, Wien and Stefan-Boltzmann.

Figure 21 – FLIR systems are utilised by night as an aid to navigation as well as targeting　0096321

Figure 22 – FLIR is an important pilotage sensor in the WAH-64D Longbow Apache. Note the TADS (see entry) mounted on the lower forward turret of this UK Army aircraft.　0096029

Kirchoff's Law

Emissivity is described, by convention, in terms of how much infra-red energy an object radiates compared with that of a 'Blackbody'. A 'Blackbody' is defined as an object that absorbs all incident infra-red energy. If part of the incident energy is reflected rather than absorbed, the object is termed a 'Greybody'. Kirchoff's law shows a direct link between absorption and emission and that the emissivity (ε) of an object or, more correctly, a surface can be determined as:

$$\varepsilon = \frac{\text{Total Radiant Exitance of a Greybody}}{\text{Total Radiant Exitance of a Blackbody}}$$

1041550

ε = Total Radiant Exitance of a Greybody

Total Radiant Exitance of a Blackbody where both bodies are at the same temperature.

From the above, the emissivity of a Blackbody is equal to one, since all of the infra-red energy incident upon the object is absorbed and re-radiated. If part of the energy is reflected, emissivity becomes <1 and the object is a Greybody. Different types of surface h ave different emissivities. Typically, a dull, dark surface will absorb and re-radiate most of the incident energy, displaying high emissivity, while a bright shiny surface will reflect much of the energy, displaying low emissivity. Figure 24 shows approximate radiator response curves for a Blackbody, a Greybody and a selective radiator. Figure 25 lists emissivities for some common materials.

Symbol	Term	Meaning	Units (SI)
U	Radiant Energy	Energy transported by electromagnetic radiation	Joule (J)
P	Radiant Flux	Radiant energy transfer per unit time	Watt (W)
M	Radiant Exitance	Radiant flux emitted per unit area of source	Wcm^{-2}
J	Radiant Intensity	Radiant flux per unit solid angle	Wsr^{-1}
N	Radiance	Radiant intensity per unit solid angle per unit area	$Wsr^{-1}cm^{-2}$
H	Irradiance	Radiant flux incident per unit area	Wcm^{-2}
ε	Emissivity	Ratio of radiant exitance of source to that of a blackbody at the same temperature	
a	Absorptance	Ratio of absorbed radiant flux to incident radiant flux	
r	Reflectance	Ratio of reflected radiant flux to incident radiant flux	
t	Transmittance	Ratio of transmitted radiant flux to incident radiant flux	
M	Spectral Radiant Exitance	Radiant exitance per unit wavelength interval at a particular wavelength	$W\mu m^{-1} cm^{-2}$
N	Spectral Radiance	Radiance per unit wavelength	$Wsr^{-1}\mu m^{-1} cm^{-2}$

Figure 23 – Radiometric Terms

Figure 26 – MWIR (3 to 5 µm) image of a warship from a target tracking system   0131070

Material	ε
Highly-Polished silver	0.02
Highly-Polished aluminium	0.08
Polished Copper	0.15
Aluminium Paint	0.55
Polished Brass	0.60
Oxidized Steel	0.70
Bronze Paint	0.80
Gypsum	0.90
Rough Red Brick	0.93
Green or Grey Paint	0.95
Water	0.96

Figure 25 – Emissivity of Common Materials

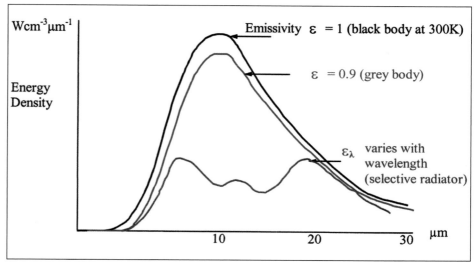

Figure 24 – Radiator response curves   0131027

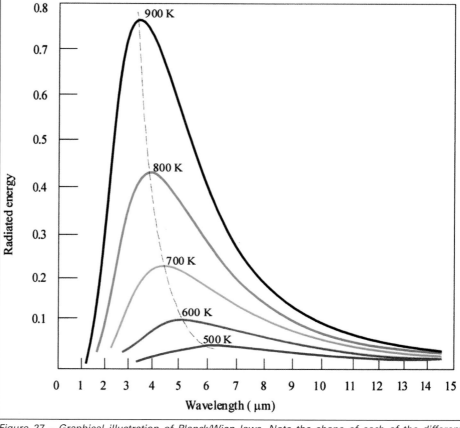

Figure 27 – Graphical illustration of Planck/Wien laws. Note the shape of each of the different temperature curves follows Planck's Law, while Wien's Displacement Law gives the wavelength of peak radiation   0131028

Planck's Blackbody Law

The spectral radiant exitance of an ideal Blackbody source whose absolute temperature is T, can be described by Planck's Blackbody Law which states that the radiation emitted by a Blackbody, per unit surface area per unit wavelength is given by:

$$ M_\lambda = \frac{2hc^2}{\lambda^5} \left[\frac{1}{e^{(hc/\lambda kT)} - 1} \right] $$

0131057

Where:
h = Planck constant
k = Boltzmann constant
c = Speed of light (3×10^{10} cm/sec)
λ = Wavelength of emitted radiation

The effect of a change in temperature may be graphically illustrated by plotting radiated energy versus wavelength at specific blackbody temperatures.

The behaviour is illustrated in Figure 27. Planck's Law gives a distribution that peaks at a certain wavelength, the peak shifts to shorter wavelengths for higher temperatures, and the area under the curve grows rapidly with increasing temperature.

Planck's Law describes the behaviour of Blackbody radiation, but we can derive from this law two other radiation laws that are very useful – the Wien Displacement Law, and Stefan-Boltzmann's Law:

Stefan-Boltzmann's Law

Stefan-Boltzmann's Blackbody Law is expressed by the following formula:

$$ M = \sigma T^4 $$

where:
M = Radiant exitance (W cm^{-2}) of radiating surface
σ = Stefan-Boltzmann constant = 5.67×10^{-12} W cm^{-2}K^{-4}
T = absolute temperature (K)

Applying Kirchoff's Law, we can express the above for a Greybody by:

$$ M = \varepsilon \, \sigma T^4 $$

where:
ε = Emissivity

Wien's Displacement Law

A graphical representation of spectral distribution of energy illustrates that the wavelength of peak radiation decreases as temperature increases. Peak wavelength and temperature are related by Wien's Displacement Law, which states that the wavelength (λ_{max}) at which M_λ is multiplied by the absolute temperature (in K) of the Blackbody is equal to a constant.

$$ \lambda_{max} T = 3 \times 10^7 $$

where:
T = absolute temperature (K)
λ_{max} = wavelength of maximum energy (angstroms, Å)

Rendering the above into wavelength units more commonly used in IR terminology:

$$ \lambda_{max} = \frac{3000}{T} $$

Where λ_{max} is expressed in µm (1 µm = 10^4 Å)

Wien's Law gives the wavelength of the peak of the radiation distribution, while Stefan-Boltzmann's Law gives the total energy being emitted at all wavelengths by the Blackbody (which is the area under a curve of Planck's Law). Thus, Wien's Law explains the shift of the peak to shorter wavelengths as the temperature

increases, while Stefan-Boltzmann's Law explains the growth in the height of the curve as the temperature increases. Notice that this growth is very abrupt, since it varies as the fourth power of the temperature. The value of the peak radiation is a function of the fifth power of temperature.

Therefore, doubling the temperature of a Blackbody will increase total radiation 16-fold and peak radiation 32-fold.

Inverse Square Law (relating radiated power and range from source)

In common with all forms of electromagnetic radiation, IR radiation from a point source emits in all directions, with the power measured at a receiver varying inversely with the square of the range to the source. This is known as the Inverse Square Law. The full equation relating power and distance from the source is as follows:

$$ J = \frac{\sigma T^4}{4\pi d^2} $$

0131058

Figure 28 – Aircraft at slant range d, normal to power source. If received power is J, then if slant range increases to 2d, received power decreases to 14 J   0131029

Thus, it can be seen that doubling the distance (d) will cause the power at the receiver (J) to decrease by a factor of 4 (Figure 28).

Lambert's Law of Cosines (radiated power related to incident angle)

Another factor influencing the amount of power available at the receiver is the angle at which the radiating source is viewed. The amount of power is a function of the cosine of the angle from which the surface is viewed (θ). This is known as Lambert's Law of Cosines and is graphically illustrated in Figure 29. The law is expressed by the following formula:

$$J = \frac{MA \cos \theta}{2\pi d^2}$$

where:
J = radiant intensity received at the detector
M = radiant exitance
A = area of source

IR SOURCES

By definition, all bodies with a temperature above 0 K will radiate energy, and the aforementioned laws will predict the type (wavelength) of radiation that a body will emit (or reflect). For airborne applications, rather than classify IR sources by type of radiation, it is simpler to categorise sources more broadly. Commonly, IR sources are referred to as a target, background, or controlled.

A target source, unsurprisingly, is an object, which may be detected, located, or identified by an IR sensor system. Background sources, such as the sky, are defined as any distribution, pattern or radiation in the external scene, which is not a target (as classified). Of course, what may be a target in one situation may be background in another depending upon the situation. For example, in a ground attack scenario, if the target area includes both troop concentrations and armoured vehicles, if the desired target is the armour, then the troops are part of the background; if the desired target is the troops, then the armour is the background. Controlled sources, such as laser designators, supply the radiant power required for active IR systems (seekers).

Background sources

For all IR sensors, background radiation will be present in the detection system in the form of unwanted noise, which must be reduced or eliminated to obtain optimal performance. Natural IR sources will always be present in the scene, which, from the above definition, constitute background sources. These natural sources may be broadly classified into 2 categories: terrestrial and atmospheric:

Terrestrial sources

Whenever an IR system is looking below the horizon it encounters terrestrial background

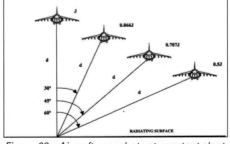

Figure 29 – Aircraft snapshots at constant slant, ranged at various angles to radiating surface
0131030

radiation. Every object in the terrain scene radiates IR energy.

A further source of terrestrial background radiation is reflected sunlight. The sun is a strong IR source and surfaces such as rock, sand, metal, and green foliage (due to the chlorophyll effect) are excellent reflectors of IR energy.

Atmospheric sources

Whenever an IR system is looking above the horizon, it encounters atmospheric background radiation. The prime IR sources in the sky are celestial and atmospheric. The radiation characteristics, as observed within the atmosphere, of celestial sources depend primarily on temperature, coupled with modification by interaction with the atmosphere. Since the density of the atmosphere falls off with altitude, the level of interaction of incident radiation must also decrease with altitude, therefore received radiation must change characteristically according to the altitude of the observer.

The Sun

The Sun may be assumed to be an approximate Blackbody radiator at a temperature of 6,000 K; from Wien's Displacement Law, this yields a radiant energy peak at 0.5 µm, with approximately half of the radiant power in the IR waveband, as shown in Figure 30.

Research has shown that reflected sunlight (from clouds, terrain, and so on) is quite similar to direct sunlight in terms of observed wavelength. Another consideration that must be accounted for is the effect of the earth's atmosphere. As discussed, the characteristics of solar radiation are modified by absorption, scattering, and (to a lesser extent) re-radiation. Broadly speaking, shorter wavelength ultra-violet radiation diminishes significantly and, by comparison, the longer wavelength IR proportionally contributes to more of the total solar power. As altitude is increased, the solar spectral distribution moves towards that of the ideal 6,000 K Blackbody.

The Moon

The next most important celestial source of IR radiation is the moon. The bulk of the energy received from the moon is solar radiation, modified by reflection from the lunar surface and the earth's atmosphere. The moon is also a natural radiating source. During the lunar day its surface is heated to as high as 100° C, or 373 K, and during the lunar night the surface temperature falls to -150° C (123 K). Other celestial sources are very weak point sources and are thus disregarded.

Diurnal effects

Due to the enormous influence of the sun in the overall IR scene, background radiation due to the sky should be considered separately for day and night. The primary differences between night sky background radiation normal to the earth's surface are shown in Figure 31, which shows the spectral distribution of energy for clear night and day skies. At night, the shorter wavelength background radiation, caused by the scattering of sunlight, air molecules, dust, and other particles, disappears. In fact, at night there is a tendency for the surface and the atmosphere to blend with a loss of the horizon, since both are at the same ambient temperature with the same high emissivity. Radiation from the clear night sky approximates that of a Blackbody at 273 K, with the peak intensity occurring at a wavelength of about 10 µm, and, as Figure 46 illustrates, the overall energy level is slightly lower than that of a clear daytime sky.

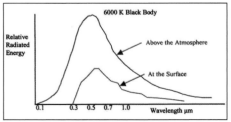

Figure 30 – Spectral distribution of solar radiation
0131031

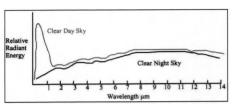

Figure 31 – Spectral energy distribution of background radiation from the sky   0131056

INTERACTION OF RADIATION WITH MATTER

Having described the physical attributes of IR radiation and some of the naturally occurring sources of IR radiation, which contribute to the background scene, we must now consider how such radiations interact with the major attenuating mechanisms encountered within the atmosphere.

Considering the optical region of the electromagnetic spectrum, the most important interactions of radiation with matter, or the action of radiation passing between two differing mediums, is fully described by the processes of emission, reflection, refraction, absorption, diffraction and scattering, outlined in Figure 32.

EMISSION	Molecular energy is changed into radiative energy
REFLECTION	Radiation is turned back when striking a boundary between different media
REFRACTION	Radiation which passes into the second medium (transmission)
ABSORPTION	Radiative energy changes into other form(s) of molecular energy
DIFFRACTION	Spreading of radiation as it passes through an aperture (interference)
SCATTERING	Direction change(s) of radiation due to interference from suspended particles

Figure 32 – Interactions of radiation with matter

From the above, it can be seen that scattering is a form of reflection, therefore, for our purposes, the processes of absorption, reflection and refraction (transmission) can be considered to account for the interactions of all incident radiation in any particular situation. Applying the conservation of energy law, changes occurring due to interaction of radiation with a medium must maintain the total amount of energy present:

a + r + t = 1

The interaction of radiation with matter is a function of the wavelength of the radiation compared to inter-atomic size and spacing of the particle. Further, strong effects are noted when the photon energy of the radiation is approximately equal to one of the quantised energy levels of the matter.

ATMOSPHERIC EFFECTS

The atmosphere is the major attenuating mechanism affecting the performance of FLIR systems. Attenuation takes place by refraction, scattering or absorption. The effect of refraction is almost negligible except for orbital applications where the entire depth of the earth's atmosphere is of concern, so scattering and absorption are the only attenuation mechanisms addressed here. Scattering will be considered first, although absorption is usually the more powerful factor.

Scattering

The size of the incident radiation wavelength compared to the size of the scattering centres suspended in the atmosphere is crucial to the amount of scattering that results. Maximum scattering occurs where the wavelength of incident radiation is approximately the same as the radiation of the scattering particles. There are two general categories of scattering, referred to as molecular scattering and aerosol scattering.

1. **Molecular scattering**. Molecular scattering deals with light scattered by particles, which are much smaller than the wavelength of the incident radiation, which applies to all of the primary molecules in the air (nitrogen, oxygen, water vapour and carbon dioxide). Each particle in the light path behaves as if it were a secondary light source. However, this effect is of more interest to the NVG user (0.6 to 0.9 μm), since molecular scattering becomes negligible at wavelengths greater than approximately one μm (most NAVFLIR operate in the 8 to 12 μm band).

2. **Aerosol scattering**. Aerosol scattering involves large particles such as dust or smog and scatters incident radiation by reflection. This process occurs when the diameter of the particle is greater than or equal to the wavelength of the radiation. This causes attenuation of the radiation by redistributing the energy in all directions. It is the size of atmospheric scatterers (Figure 33), which determines their effect on the FLIR. For example, for a NAVFLIR operating in the 8 to 12 μm band, fog droplets, which have their greatest distribution of radii in the 5 to 15 μm band, produce nearly 100 per cent scattering over typical ranges (see Figure 34). However, small particles (less than 0.5 μm) such as smoke, affect FLIR far less than the visible (and NIR) region. FLIR visibility through fog, light rain, snow, and dust, while severely degraded, is therefore usually superior to normal unaided eye (and NVG) visibility.

Atmospheric Effect	Particle Size (μm)	Remarks
SMOKE	0.2 to 2	
HAZE	0.05 to 0.5	Tiny dust and salt particles
DUST	1 to 10	
FOG AND CLOUD	0.5 to 80	Condensation formed around haze particles Visibility limited to less than 1 km
FUMES	Up to 100	
MIST	50 to 100	Visibility greater than 1 km
DRIZZLE	100 to 500	
RAIN	500 to 5,000	

Figure 33 – Comparative sizes of atmospheric particulates

Absorption

Atmospheric absorption, particularly molecular absorption, is the primary source of atmospheric attenuation. Molecular absorption occurs mainly within several narrow absorption bands, which correspond to the resonant frequency of the molecules concerned.

Absorption occurs when energy contacts the molecule and causes it to oscillate. To return to its ground state, the molecule radiates energy in arbitrary directions. Thus the IR energy is attenuated along its direction of initial travel.

An 8 to 12 μm FLIR is most affected by three molecules – water (H_2O), carbon dioxide (CO_2), and ozone (O_3) as shown in Figure 5 (previous NVG paper). Other minor constituents which also contribute to absorption include nitrous oxide (N_2O), carbon monoxide (CO) and methane (CH_4); however, because of relatively low densities of these constituents, their concentration and variation is considered a constant and does not affect performance appreciably.

1. **Water vapour**. Water vapour is the most important absorbing gas in the atmosphere, and certainly the most variable. Local humidity conditions can easily double the water vapour content in a matter of hours, with a changing weather front. When considering FLIR performance, *absolute* humidity is more relevant than *relative* humidity. (Relative humidity expresses the amount of moisture in the air compared to the maximum amount that could be held at that temperature, while absolute humidity is an actual measure of water vapour in a given area). A dry mid-latitude winter day, with an absolute humidity of 3.5 g/ms, is almost completely transparent to FLIR. However, a wet tropical atmosphere of 20 g/ms becomes an effective wall that severely hinders the operational capability of the FLIR. It is difficult to exactly determine (and thus predict) at what point the FLIR becomes unusable as a navigation or target detection device. Pilots should realise the effects of water vapour on the FLIR, and expect FLIR degradation as absolute humidity rises.

2. **Carbon dioxide**. Carbon dioxide (CO_2) is second in importance to water vapour, in terms of absorption. The vertical distribution of CO_2 in the atmosphere is essentially constant with volume. However, in urban areas the concentration of CO_2 is normally higher.

3. **Ozone**. Ozone (O_3) is rarely encountered at low level, since the vast majority is confined to a layer which is centred approximately 30 km above the earth. However, as highlighted by environmentalists, significant concentrations of reactive O_3 are created in car engines and as a by-product of industry. Large release of fluorocarbons, industrial pollution and heavy concentrations of vehicle exhausts can create concentrations of O_3 near the ground which can attenuate the IR signal and thus limit FLIR performance, particularly in association with anti-cyclonic weather conditions (high pressure) which trap the ozone-producing pollutants in the lower atmosphere.

By convention, the effects of scattering and absorption are combined and expressed in terms of atmospheric transmittance rather than relative absorption. As discussed, since water vapour is the most important of the absorbing gases, absorption transmittance increases with altitude because the less dense and cooler air holds less water vapour.

Cloud

The further effects of water vapour on the total background radiation are apparent when condensation occurs and clouds form. Clouds produce considerable variation in sky background with the greatest effect occurring at wavelengths shorter than 3 μm. This is caused by solar radiation reflected from the cloud surfaces. As previously discussed, short-range IR missiles are sensitive to this wavelength, a fact all too painfully obvious to early fighter pilots who have had trouble locking their missile (AIM-9B/G era of weapons) onto a target against a bright cloud background. Later IR missiles (AIM-9L onwards) incorporate spectral filtering, which eliminates the shorter wavelengths, and spatial filtering, which distinguishes the smaller area of the target from the larger area of the cloud edge.

Terrain/Sky contrast

Terrain produces a higher background-energy distribution than that of the clear sky. This is caused by reflection of sunlight in the short-wavelength region and by natural thermal emission at the longer wavelengths. The absence of sharp thermal discontinuities, breaks in the radiation pattern between an object and its surroundings, complicates the problem of the detection of terrestrial targets. While generally good pictures are obtained, occasionally objects with different emissivities and temperatures produce the same radiant emittance, and thus there is no contrast in the picture. This is known as thermal crossover (discussed later) and results in a loss of picture information.

THERMAL SIGNATURES

Features of the thermal scene

The thermal scene sensed by the FLIR comprises temperature differences, emissivity differences and reflected radiation. The temperature of most objects within the thermal scene will not be far from ambient; however, their IR signal will be different from the background by virtue of their differing abilities to absorb and to release heat (their emissivity). Most objects within the thermal scene obtain their energy from the ultimate IR source, the sun. The temperature of these objects follows a daily cycle in response to daytime solar heating and overnight cooling. However, some objects (such as vehicles and buildings) have internal heating, and will not necessarily follow the cyclic heating and cooling exhibited by the surrounding scene.

Diurnal heating

Objects are heated through the absorption of solar energy. Even during overcast days, some solar radiation is absorbed. Daily solar heating begins at sunrise. After midday, the sun declines and the objects begin to cool. After sunset, the objects cool down to approach the temperature of the surrounding air. This daily two-part heating and cooling cycle is called the diurnal cycle. During the diurnal cycle, individual background and target objects heat and cool at different rates. Large dense objects, such as larger rock formations and, heat and cool slowly, while other objects, such as foliage heat and cool quickly. Heavy, dense objects are said to have 'high thermal mass', while lightweight objects are said to have 'low thermal mass'.

Thermal crossover

There are times during the day and night when the spectral radiances of target objects and the background scene are identical; at these times contrast in the FLIR scene is at a minimum. As a rough guide, the following describes a typical cycle. Just before sunrise, the temperature difference is at a maximum and a good FLIR picture can be expected. After sunrise the entire thermal scene is heated but the temperature of the target material rises more quickly (low thermal mass). Just before noon the temperatures 'crossover' – this is when a poor FLIR picture can

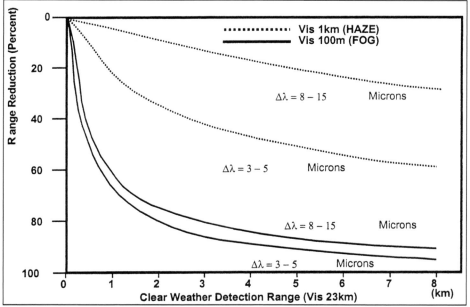

Figure 34 – Haze and fog attenuation

0522495

be expected. During the afternoon the target material reaches a higher temperature than the background, resulting in an improved FLIR picture. Differential cooling overnight causes a second thermal crossover towards midnight, when again, the FLIR picture would be poor.

While the above example gives an approximation of thermal crossover times, exact prediction is normally undertaken by the relevant meteorological authorities, since precise timings vary considerably according to the scene objects and environmental factors.

Weather factors and emissivity

Thermal signatures are strongly dependent upon cloud cover, insulation characteristics and relative aspect between the viewer, scene objects, sun and sky. The cooling rate of scene objects at during the night will depend on their total heat capacity, conductivity, contact with the surrounding air, IR emissivity in the sensor waveband (generally, for NAVFLIR = 8 to 12 µm) and weather conditions (humidity, cloud cover). For example, in a tropical location, when total humidity is high, the sky is cloud covered and the temperature remains roughly constant, the thermal scene does not exhibit a great deal of variance from day to night. However, in a very dry location, such as the desert, where there is no cloud cover, all the thermal radiation in the 8 to 12 µm band will be radiated into space and objects will cool down rapidly. Vegetation which is in close contact with the surrounding air and water surfaces, which have large heat capacities, will radiate fairly evenly during the day and night.

Contrasts between objects in the IR region beyond 4 µm are at a minimum on very cloudy and on rainy days. An object with low emissivity tends to take on the air temperature with a lag determined by thermal mass. An object with high emissivity responds more readily to temperature changes around it (more so if the object has low thermal mass).

Day thermal scene

A common mistake made by aircrew is the assumption that day thermal signatures are broadly similar to those at night. This is not true, with the greatest difference between day and night noted for a sunny day preceding a clear night. This leads us to the (correct) assumption that the main culprit in this marked difference is solar heating; the sun being a much more powerful contributor to the thermal scene than the moon. It therefore follows that differences between the day and night scenes are less for overcast or rainy days and nights.

Targets are heated by the sun on warm days to relatively high temperatures. The outer hull of a vehicle does not always heat evenly due to shadowing. IR reflections during the daytime can also provide spurious hot spots. This combination of uneven heating and reflections can make the recognition of objects in the scene, utilising FLIR alone, during the day difficult, particularly from mid-morning through to mid-afternoon. The many hot background objects make long-range detection and, more importantly, identification of targets difficult, to the extent that, at low level, positive target identification may be perilously close to weapon release.

FLIR image considerations

The FLIR image appears similar to a visual image and, therefore, allows the pilot to use many of his normal cues such as size, shape, shadow, surroundings and texture. However, as described previously, this can be misleading because often the thermal scene does not behave like the visual scene, producing some confusing effects:

1. **Target size and shape.** The true size of a target within the FoV may not be portrayed on the FLIR image, due to 'blooming' at high energy levels. The hotter an object in the scene, the greater the effect. This is true of FLIR sensors employing Signal PRocessing In-The-Element (SPRITE, described later), where a detector element can 'leak', affecting those adjacent and creating 'blooming' at high energy levels.

Figure 35 – Internal heating of targets. Note the bright areas around the wheels of the armoured vehicle, indicating current (or recent) motion of the vehicle. Also, some heating of the hull is apparent in the area of the engine compartment   0111387

Thus, shape may indicate the type of objects, but hot objects often do not appear in their true shape.

2. **Shadow.** A potentially highly confusing feature of the FLIR scene is shadow information. We habitually utilise shadow information to orient the visual scene, so to be confronted by 'thermal' shadows, often covering the same areas as visible shadows, can result in misinterpretation of the FLIR scene. These thermal shadows are caused by the surface temperature being lower in the area shaded from direct radiation, causing a reduced amount of NIR radiation to be reflected from the shaded area. While this may produce great correlation between the visual and thermal scenes when the IR source producing thermal shadows is the same as the light source causing visual shadows (that is, the sun), this may not be the case. Other shadows may indicate activity, such as a thermal shadow from a vehicle that has sheltered the background terrain before moving off; these 'silhouettes' can be detected hours after the vehicle has moved and aid identification of the type concerned. Perhaps even more confusing, an aircraft, engines running, can heat the ground, producing not a shadow, but a thermal 'footprint', before moving off, leaving a 'negative shadow'. Wind shadows can sometimes be noted in the lee of objects on the ground; since the wind does not disturb the surface in the lee of objects, this area may be warmer or cooler than the surrounding area, depending upon whether the breeze is warming or cooling the surface over which it is blowing.

Knowledge of the appearance of various materials in the thermal scene will aid the user in interpreting the FLIR image more readily. As will be discussed later, with most NAVFLIR systems incorporating polarity switching (enabling 'hot' objects in the scene to be portrayed as bright, white hot, or dark, black hot), the following description of materials and textures in the thermal scene will use terms such as 'hot', 'warm', 'cool', and so on; display settings will determine how these objects appear on video:

1. **Grass.** Grass appears very cold on FLIR. Grass is unable to draw heat from the earth because of its poor thermal contact with the ground, and rapidly becomes cold by radiation. For this reason, the air temperature at ground level is usually lower at night than that a few feet above ground, a phenomenon known as 'night inversion'.

2. **Trees.** Trees appear cool to warm on FLIR. This is believed to be associated with the convective warming of the trees by the air, in conjunction with night air temperature inversion, although some heating from the life processes in the tree play a role. Indeed, most vegetation is a strong reflector of IR from the so-called 'chlorophyll effect', making trees, which appear dark at night in the visible spectrum, appear warm on FLIR. During daytime the same leaves appear colder than the ground, because the air temperature at treetop height is cooler than at ground level.

3. **Concrete surfaces.** Concrete surfaces appear quite warm on FLIR because they have high emissivity and are in good thermal contact

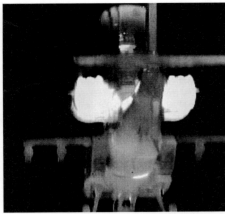

Figure 36 – A FLIR image of an AH-64D Apache attack helicopter seen from the rear aspect. Note the engine exhaust, radome and main rotor bearing (polarity white hot)   0083806

with the earth, which acts as a constant heat source. Pavement retains more of the heat received from the sun during the day, because of its high thermal capacity. In late evening it can appear black due to a loss of heat, which exceeds that of the surrounding area. This is generally true for all types of surfacing, including concrete, asphalt, and blacktop. Runways are a very good example of pavement, and can normally be seen on the FLIR at a much greater range than is possible with the NVGs.

4. **Earth/Soil.** Under normal conditions, soil, which includes various types of earth, sand, and rock, appears quite warm on FLIR. This results from high emissivity, sun heating during the day, and the high heat capacity of the earth.

5. **Water surfaces.** The reflectivity and emissivity of a water surface varies greatly with a change in incidence to the surface. At shallow angles (up to 5°), a calm water surface will reflect most radiation incident on it from the sky or surrounding objects. At steeper angles (20 to 90°), the surface will be almost entirely emissive, radiating its surface temperature. A large swell is normally clearly visible on FLIR because the variations in the incidence of the radiation on the surface will be seen as a change in reflective and emissive radiation from the surface.

6. **Clouds.** Clouds can affect the FLIR scene in many ways. The presence of cloud can block the view of the FLIR scene behind it; in practice, FLIR can see through haze and thin mist, but not fog or cloud. Full cloud cover reduces night-time FLIR contrast between sun-heated surface elements, blanketing the scene and equalising the temperature of included objects. Total cloud cover lasting several days can produce 'washout' of the entire scene.

7. **Wind.** As previously mentioned, wind can a strong environmental modifier of the FLIR scene. A strong wind will reduce temperature differentials within the scene.

8. **Internally heated sources.** The preceding descriptions have assumed that all thermal energy in the FLIR scene arises from solar heating. In this case, the best FLIR image is acquired under clear skies with no wind. However, many targets, such as aircraft, vehicles, buildings and personnel, have internal heat sources, making them much warmer than the background. Further, within targets, significant temperature differences can be observed. Targets such as aircraft, tanks and other vehicles have internal temperature variations which form visible patterns that combine (with other characteristics) to produce a 'unique' and identifiable target signature. Assuming a 'white hot' polarity, the hottest vehicle parts, such as the engine and exhaust (see Figure 36), stand out brightly. Medium temperature objects, such as radomes (heated by the radar), wing leading edges and tank tracks appear medium bright,

while a relatively cool tank hull appears dark. It should be noted that these heated areas might remain significantly warmer than their surroundings for some hours after use. However, it must be stressed that emissivity of the material must be considered in all cases. Special low-emissivity coatings can greatly reduce the radiation from a hot surface of a vehicle or building.

FLIR SENSORS

Infra-red sensors are currently used in a wide variety of military and civilian applications, including night vision for ground, air and naval applications, surveillance, airborne reconnaissance, fire control and missile seekers. Early systems were generally bulky (mainly resulting from cooling requirements), restricting their use to larger vehicles and airborne applications.

Current developments such as uncooled staring arrays of detectors have drastically reduced the complexity, size, weight and cost of IR systems, bringing them into the truly man-portable arena, although such lightweight systems are suitable only for very short-range operation (<1 km), which renders them unsuitable for airborne NAVFLIR and targeting systems.

FLIR developments
Within industry and the military, the terms 'first', 'second' and 'third generation' are used to describe different stages of thermal imager development. While there is broad consensus as to the definition of first- and second-generation systems, argument over the true definition of a third-generation system remains. For the purposes of this paper (and the convention used within *Jane's Avionics*), the following definitions will be used:

Current generations
First generation
- LWIR, 8 to 12 μm
- Two dimensional, electro-mechanically scanned photodetector
- Linear array of single detector elements (60 × 1, 120 × 1, 180 × 1)
- Cooled
- Example sensor: Thermal Imaging Common Module (TICM)
- Example system: AN/AAQ-13 Low Altitude Navigation and Targeting Infra-Red for Night (LANTIRN) system.

Second generation
- LWIR, 8 to 12 μm
- One dimensional scan using several horizontal or vertical strips of detectors
- Larger linear array of multiple elements (480 × 4, 768 × 6, 320 × 256)
- Cadmium Mercury Telluride (CMT) or Quantum Well Infra-red Photodetector (QWIP) technology
- Cooled
- Example sensor: Standard Advanced Dewar Assembly (SADA) II
- Example system: FLIR Systems AN/AAQ-22 SAFIRE.

'Generation 21/2'
- MWIR, 3 to 5 μm
- Staring array
- Example system: Raytheon ATFLIR.

Figure 37 – The Boeing F-15E carries the LANTIRN system. The AN/AAQ-13 Navigation pod is shown on the right chin station – the FLIR window can be seen directly above the Terrain Following Radar

1041601

Figure 38 – Raytheon ATFLIR mounted on the left chin station of an F-15E and imagery taken of the city of St Louis at 22 nm slant range

0581545

Future generations
Generation 2 + +
- LWIR, 8 to 12 μm
  Uncooled staring array (640 × 480 for detection)
- Laser Illuminated Viewing And Ranging (LIVAR) in the SWIR band.

LIVAR
In order to meet pilotage requirements and the demands of current ground and airborne weapon systems, a method of extending the identification range of current IR targeting sensors is required. At present, a pilot or tank gunner can only expect to truly *identify* a target at approximately 4 km, well inside the envelope of his weapon; if we can find a way to extend this range out towards 10 km, then we will then be able to fully exploit the capabilities of weapons such as Laser-Guided Bombs (LGB).

LIVAR technology introduces an active element to IR detection systems. By adding a laser illumination system operating in the 1.55 μm range (eyesafe), a target within the FLIR Narrow FoV (NFoV) can be 'painted' by the laser at near video frame rate. A SWIR gated camera, featuring, say, a nominal 1280 × 1024 array, a 0.2 × 0.2° FoV and a 140 mm aperture, receives the laser returns to provide a visual image to the operator. This system, currently offered as an upgrade to the sensor package of Main Battle Tanks (MBT), facilitates long-range target identification without incurring the expense of integrating a true third-generation dual-waveband sensor.

Third generation
- Large format staring array of detectors (>1,000 × 1,000)
- Dual band (application dependent), combination of LWIR and MWIR (LWIR + LWIR/ LWIR + MWIR)

- On chip processing, no external electronics
- Cooled or uncooled.

Note – while many manufacturers refer to their products as 'Generation 3' IR systems, classification within *Jane's Avionics* will follow the above technically-recognised definitions.

IR SENSOR SYSTEMS
While there are many different types of IR sensor systems for avionic applications, including missile seekers, Infra-Red Search and Track Systems (IRSTSs), IR Reconnaissance Systems (IRRSs) and IR Line Scanners (IRLSs), we shall be concerned primarily with FLIR systems utilised for navigation and targeting in tactical aircraft. Accordingly, discussion of technology and systems will concentrate on those of direct relevance to the employment of FLIR.

Detection versus Imaging
Early generations of IR missile seekers employed a simple detector, which merely looked for a source of energy in a specific waveband (for early stern-aspect only missiles, this was approximately 3 μm). This energy source would establish a seeker head angle, which, via internal electronics, was converted into a lead-pursuit to collision course, with a sensing fuse establishing lethal range. This is an example of a simple IR detector, which was only interested in locating an energy source of a specific waveband with no attempt to form any sort of picture of the target source.

Clearly, for FLIR systems, in order to facilitate pilotage via a visual picture of the IR scene, an imaging system is required which is able to show a true 'picture' of terrain and objects ahead of the aircraft. This means that the system must deliver sufficient spatial resolution and thermal sensitivity if it is to be effective. Insufficient spatial resolution (the ability to discriminate individual

Figure 39 – The Eurofighter Typhoon features the PIRATE IRSTS

1041602

Figure 40 – A FLIR image of a CH-53 helicopter refuelling from a KC-130 Hercules taken from the AN/AAQ-22 SAFIRE mounted on another CH-53. Note the hot turboprop exhausts and blades of the KC-130 (FLIR Systems)

0131091

objects at close proximity to each other at long range) will be manifest as a blurred image on the video screen and inadequate thermal sensitivity (the ability to discriminate small temperature differences) will result in an image which contains very few shades, giving a 'washed out' appearance to the scene.

Having established the requirement for an image forming IR system, we will now consider how such a system may function:

Image forming IR systems fall into two categories: those in which an image is formed directly and that in which optical scanning is employed.

1. **Direct imaging systems**. By definition, a direct imaging system requires a two-dimensional Very Large Array (VLA) of individual detectors, each contributing a very small portion of the overall scene. The array must provide full spectrum coverage over the full dynamic range at high resolution. The performance of such a device would depend on the number of detectors in the array (to cover the whole scene) and their pitch (distance between detector elements). While arrays are certainly becoming ever larger, at present, the best method of providing high quality FLIR imagery is to use an array of cooled quantum detectors which is mechanically scanned.

2. **Scanning systems**. The simplest method of obtaining a thermal image using scanning techniques is to use a single small IR detector to scan all the points in a scene and use the output to modulate a display. The use of a single detector would necessitate scanning in two dimensions. However, if a linear array of detectors were used, scanning could be confined to one dimension. Both single- and twin-axis techniques are used in airborne systems.

FLIR devices generally employ a rectilinear scanning pattern to build up a picture of the forward Field of View (FoV).

FLIR – Functional description
Most thermal imaging devices for airborne FLIR applications are of the opto-mechanically scanned type in which a number of detectors are mounted in the focal plane of an optical system which collects IR energy, focuses and scans it in a prescribed pattern over the detectors. The result is an output electrical current which, after processing, is used to construct a raster television picture.

IR sensor systems for airborne applications incorporate components chosen to optimise system performance for a particular wavelength region, detection range, resolution and so on, depending on application (such as navigation or targeting) and type of source. However, all IR

systems include the following basic components (see Figure 41):
1. A window, which allows IR energy to pass while affording protection to the optics
2. Optics, which collect and focus the IR energy
3. A scanner, which samples IR energy from the scene
4. IR detectors, which convert the IR energy into electrical signals. May be cooled or uncooled
5. Electronics, which amplify the electrical signals
6. A display, which converts the electrical signals into a visual image of the scene.

We will now consider each of these components in turn:

Window
Windows for IR systems are manufactured from IR transmitting materials such as germanium and zinc sulphide. For airborne applications, in addition to the required IR transmissivity, the window requires strength to withstand aerodynamic loads and erosion due to atmospheric particulates, stability to survive aerodynamic heating which can cause opacity and form to minimise drag as a result of its position on the airframe.

Optics
The optics of an IR system produces a focused image of a distant scene while eliminating radiation from sources outside the scene. Filters limit the range of wavelengths passed through to the detector, facilitating an optimal compromise between performance (resolution) and IR envelope (waveband).

Line-of-Sight (LOS) stabilisation is required in order to reduce the image blur caused by platform motion, thereby maintaining sensor image quality. Stabilisation, or isolation, of the LOS from platform motion can be accomplished in three basic ways, open loop, closed loop, and free gyroscope control. With open loop control, the gimbals' angular control servo loops are slaved to a master stabilisation reference, generally the Inertial Reference System (IRS) of the aircraft (or a stand-alone reference for independent or podded systems). This technique is most often utilised for airborne FLIR systems. Closed loop control utilises inertial sensing components (rate gyroscopes and/or accelerometers) mounted directly on the mechanical (or gimbal) element that defines the LOS. These components provide inputs to the gimbal control servos. Free gyroscope control achieves LOS stabilisation by incorporating a high angular momentum wheel as an integral part of the inner gimbal assembly with the spin axis aligned with the LOS.

Scanning
In order to generate a high resolution, real-time visual image of the IR scene, the radiated energy

Figure 42 – Tornado GR4 IR window (shutter closed) (Jane's/E L Downs)   0130870

across the entire scene must be collected, processed and read out in less than 1/25 th sec. This frequency is the minimum required to produce a flicker free image, although most systems double this rate to 50 Hz, in line with common video standards. However, to collect sufficient energy from a region in the scene, a certain amount of time is required in order to discriminate received radiation from electrical noise in the detector. Therefore, to satisfy these conflicting requirements, an array of detectors is required, each staring at a particular region of the scene.

However, as previously mentioned, an extremely large array would be required to cover the entire FoV, so the image is built up by scanning the scene across these detectors to produce a single frame.

A scanning mechanism is therefore required to match the instantaneous FoV of the array to the FoV of the sensor. As discussed, FLIR devices employ a rectilinear scanning pattern to build up a picture of the forward Field of View (FoV). The two most commonly used techniques are:
1. **Serial Processing**, in which a single detector is used in a twin-axis scan to cover a scene line-by-line and element-by-element.
2. **Parallel Processing**, in which a line array of detectors is used in a single-axis scan to cover the scene element by element. Usually, the elevation field of view is covered by the line array and the scanning action covers the azimuth field of view.

Serial Processing (SP) produces a good quality TV-like display but requires very high-speed scanning, which calls for detectors of very short-time constant or high detectivity (described later). Parallel Processing (PP) permits a slower scan rate but requires multiple channels to handle the simultaneous line output signals and signal processing may be complicated by differences in performance of the detectors, which comprise the

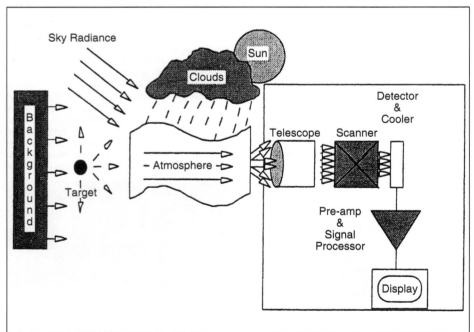

Figure 41 – Simplified schematic of IR sensor system   0522496

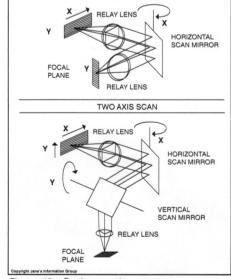

Figure 43 – Basic scanning arrangements   0097652

Figure 44 – TICM II modules   0131072

array. In this case, the detectors are stationary and scanning is achieved by mechanical movement of an optical system.

In first-generation FLIR systems, the whole field was scanned across one row of detectors (row or column). More modern systems use S/P (Serial/Parallel) scanning, incorporating 6 series elements arranged in 8 rows which are scanned to build a picture of the scene in a series of adjacent swathes. This is known as Signal Processing In The Element (SPRITE), which forms a major component of the UK Thermal Imaging Common Module II (TICM II). Incident IR energy is focused on a detector in the form of a strip of Cadmium Mercury Telluride (CMT). This strip is voltage-biased so that the released charge carriers are swept along its length at exactly the same linear speed as the scanning speed. Thus the charge is delayed and integrated to yield the same output current as a long array of discrete elements. This function is termed long-linear Time Delay and Integration (TDI). Eight Sprites give an imaging performance that would otherwise need 50 to 100 individual detectors. The TICM II module currently equips the majority of UK and US combat aircraft.

Detectors

The detector is the heart of an electro-optical sensor; radiation collected by optics and focused onto the focal plane of the detector is transformed from an optical signal to an electrical signal. The strength of the output electrical signal is proportional to the temperature of the corresponding part of the scene. The purpose of this conversion is to provide a signal suitable for electronic processing. While a single detector can only measure a small part of the total scene at any time, a complete picture can be constructed by 'scanning' the scene across a single detector, or by construction of a 'staring array' of multiple detectors covering the entire scene. The success of modern IR sensing systems is largely due to the development of greatly improved detectors capable of sensing IR radiation beyond 3 μm.

Detector devices are of two basic classes, thermal detectors and quantum detectors. Thermal detectors absorb radiant energy, thereby increasing in temperature. The increase in temperature produces changes in the bulk physical properties of the detector material. These changes are then measured as indications of the amount of radiant energy absorbed.

Conversely, quantum detectors do not rely on bulk temperature changes in the detector material. Instead, absorbed radiation causes change in the energy states of electrons within 'wells', which is detected by a read-out circuit.

These technologies will now be examined in greater detail:

Thermal Detectors
Types of photodetector

There are two types of radiation detector commonly utilised in IR sensing systems, Photo-Conductive (PC) and Photo-Voltaic (PV).

A PC radiation detector consists of a piece of semiconductor material with two electrodes attached. The external circuit comprises a voltage source and a current measuring device. Photons entering the semiconductor material produce photo-conductive photo-electrons within the semiconductor material (and

corresponding 'holes' in the bound states of the atoms). Under the influence of the externally applied electric field, the photo-electrons move toward the positive electrode and the 'holes' move toward the negative electrode, thus causing a measurable electrical current to flow in the external circuit. Since the energy required to induce photo-conduction is less than that required to induce photo-emission, photo-conductive detectors can respond to longer wavelength (far IR) radiation.

A PV radiation detector consists of a semiconductor diode with an external circuit with an electrical voltage measuring device. Photons entering the semiconductor material produce photo-electrons and corresponding 'holes'. As a result of the difference in energy levels in P-type and N-type semiconductor materials, the photo-electrons tend to migrate into the N-type material and the holes tend to migrate into the P-type material, thus producing an electric potential difference across the P-N junction. This potential difference is measured by the external voltmeter. The photon energy required for PV detection is essentially the same as that for PC detection, therefore cut off wavelengths are approximately the same.

Intrinsic and extrinsic semiconductors

Photo-Conductive (PC) detectors are most commonly used in IR systems since, as previously mentioned, they are capable of operating out into the far IR region. Two kinds of semiconductors are available – intrinsic semiconductors and extrinsic semiconductors.

Intrinsic semiconductors are made by refining the material to a high degree of purity. These are insulators at a temperature of absolute zero (0 K or –273° C) since the valence band is full and the conduction band is empty. As the temperature is increased, an increasing number of electrons obtain sufficient energy to cross the energy gap between the valence and conduction bands. If the semiconductor is irradiated with photons of the correct energy, electrons can again be excited into the conduction band and this is the photo-conductive process. However, the thermally excited electrons are also a prime source of detector noise; to counter this effect, the detector is cooled to very low temperatures. The characteristics of some commonly used intrinsic detectors are given in Figure 45:

As can be seen in Figure 45, intrinsic detectors suffer from cut-off wavelengths of approximately 4 to 6 μm, denying their use in the longer wavelengths where short-range atmospheric transmissivity is at a maximum. Detector operation at longer wavelengths has been achieved by exploiting the small amount of energy required to ionise impurity atoms in the highly pure host material. The deliberate introduction of impurity atoms into the host material is known as doping, and quite small concentrations of impurity have been found to substantially increase the conductivity of the material, which results from irradiation. This process is called extrinsic detection. It follows that for low levels of incident ionisation energy,

the detector must be kept at a commensurately lower operating temperature if thermal noise is to be kept to an acceptably low level. The characteristics of common extrinsic detectors are given in Figure 46:

Cadmium Mercury Telluride (CMT)

The detector material most often utilised in thermal imaging systems is Cadmium Mercury Telluride (CMT), which provides outstanding performance in the 8 to 14 μm region with a spectral response that extends as low as 2 μm. CMT can be fabricated to optimise its response in either the MWIR or LWIR wavebands.

Quantum Well Infra-red Photodetector (QWIP)

A Quantum Well Infra-red Photodetector (QWIP) is a semiconductor detector employing quantum wells. Simply, a quantum well is a two-dimensional quantum 'box' with characteristics of height and width. When the quantum well is sufficiently deep and narrow, its energy states are quantised (discrete). The potential depth and width of the well can be adjusted so that it holds only two energy states, a ground state near the bottom, and a first excited state near the top. A photon striking the well excites an electron in the ground state to the first excited state, and then an externally applied voltage sweeps the photo-excited electron out, producing a photocurrent. Only photons having energies corresponding to the energy separation between the two states are absorbed, resulting in a detector with a sharp absorption spectrum. Designing a quantum well to detect emissions of a particular wavelength becomes a simple matter of tailoring the potential depth and width of the well to produce two states separated by the desired photon energy, in this case matching the energy of LWIR photons at approximately 8 to 10 μm. A Read-Out Circuit (ROC) detects the excited electrons and the signal amplified and processed to create the image.

QWIP technology is a relatively mature technology, commonly based on structures made from gallium arsenide (GaAs)/aluminium gallium arsenide (AlGaAs). The first stage of fabricating a detector element is to grow the quantum-well structure on the GaAs wafer. The well width is controlled by controlling the GaAs layer thickness; depth is controlled by controlling the Al composition in the barrier layers. Modern crystal-growth methods like Molecular Beam Epitaxy (MBE) fabrication allow the growth of highly uniform and pure crystal layers of semiconductors on large substrate wafers, with control of each layer thickness down to a fraction of a molecular layer. The GaAs/AlGaAs structure allows the quantum well shape to be adjusted over a range wide enough to enable detection at wavelengths longer than about 6 μm.

A typical single-pixel QW 'stack' consists of 50 individual quantum wells, each 5 ζm wide, in the direction of epitaxial growth, sandwiched between AlGaAs layers 35 ζm wide. On either side of the QW structure is a contact layer of GaAs doped with silicon to provide a source of ground-state electrons.

Detector Material	Operating Mode	Cut-off Wavelength (μm)	Operating Temperature (K)
Lead Sulphide (PbS)	PC	4	77
Indium Arsenide (InAs)	PV	4	77
Lead Selenide (PbSc)	PC	6	77
Indium Antimonide (InSb)	PC	5	77

Figure 45 – Characteristics of Common Intrinsic Detectors

Detector Material	Operating Mode	Cut-off Wavelength (μm)	Operating Temperature (K)
Mercury-Doped Germanium (GeHg)	PC	14	27
Cadmium Mercury Telluride (CdHgTe) (CMT)	PV	14	77
Cadmium-Doped Germanium (GeCd)	PC	20	4.2
Antimony-Doped Silicon (SiSb)	PC	23	4.2
Zinc-Doped Germanium (GeZn)	PC	40	4.2

Figure 46 – Characteristics of Common Extrinsic Detectors

Measures Of Performance (MOP)

There are many ways to express the comparative performance of detectors; the following are some of the most commonly used Figures Of Merit (FOM) for IR detectors:

The Noise Equivalent Power (NEP) of an infra-red detector is the amount of electromagnetic radiation which must be received to obtain an output response, which is equal to the noise of the detector. Noise equivalent power can be defined by the equation:

$$NEP = \frac{JA}{V/N}$$

0131059

where:

J = RMS incident energy signal
A = Surface area of the detector
N = RMS noise output of the detector
V = RMS detector output signal

The Detectivity (D) is simply the reciprocal of the NEP. This is a more convenient measure of detector performance since large D values correspond to high sensitivity, whereas small values of NEP would be associated with high sensitivity. Hence, D is expressed by:

$$D = \frac{1}{NEP}$$

0131060

As defined, D is a figure of merit comparable with the noise factor (NF) of electronic amplifiers. However, for perfect detectors NEP varies as $\sqrt{A}$ and $\sqrt{\Delta f}$ – where A is detector area and Δf is the bandwidth over which the noise is measured. Therefore, unless the detector specimen sizes and noise bandwidths were identical, D would not give a true comparison between different detectors. Hence a better measure of Detectivity is given by D*, which is defined as follows:

$$D^* = \frac{D\sqrt{(A,\Delta f)}}{NEP} = \frac{\sqrt{(A,\Delta f)}}{NEP}$$

Thus, D* is D normalised to unit area and unit bandwidth. This is a more convenient MOP for detectors of differing area used in circuits having differing bandwidths.

However, a high D* for a single pixel is not enough to ensure a high-performance imaging array. Pixel-to-pixel nonuniformities produce a spatial fixed pattern noise which must be minimised. A general FOM to describe the performance of a large imaging array is the Noise Equivalent Temperature Difference (NETD), which includes this spatial noise. NETD is the minimum temperature difference across the target that would produce a SNR of unity; thus NETD is a measure of the Minimum Resolvable Temperature Difference (MRTD) in a scene.

CMT vs. QWIP

The use of CMT in current applications includes the QinetiQ STAIRS C (Sensor Technology for Affordable IR Systems, model C), a prototype second-generation LWIR imager that is claimed to provide more than double the detection and recognition performance of the first generation TICM module. STAIRS C includes a scanned long-linear (768 × 8 element) PV CMT array detector, made up of two columns of 384 × 4 detectors. These are vertically offset by half the pitch of the detectors to provide 768 vertical and 1,280 horizontal pixels. With an f/2 optic and cooled to 77 K, NETD will be 0.07 K. The 14-bit digital resolution of STAIRS C typically enables an observer to pick out the structure in trees at ranges of the order of 3 to 4 km.

QWIP technology is employed by a consortium led by FLIR Systems and Saabtech Electronics for the Bill Infra-Red Camera (BIRC), which incorporates a patented optical coupler using a two-dimensional reflection grating, a claimed

Figure 47 – Photograph taken with the STAIRS C demonstration imager. The system features a NETD of 0.07 with an f/2 optic (QinetiQ) 0111402

NETD of 0.016 K with an f1.5 optic and a 38 µm pitch in a 320 × 240 array (640 × 480 has been demonstrated). Operating in conjunction with a two FoV (14.1 and 4.7°) telescope, the system enables detection of a tank at ranges up to 10 km.

Many believe CMT will remain the detector material of choice for the foreseeable future, citing the low quantum efficiency of QWIP and the higher absorption coefficient of CMT, although this may hold true only if advances are made in the areas of dynamic range and SNR (NETD), since the end user may only be interested in the superior performance of current QWIPs in this area (typically 4 ×).

As discussed, while CMT functions in the MWIR range, InSb is producing promising results in detectors designed for the 3 to 5 µm waveband. InSb is used for large arrays under investigation at QinetiQ's Malvern facility. As part of the Albion project, a third generation array of 1,024 × 768 pixels at 26 µm pitch, combined with an f/2 optic, the bare detector has achieved a NETD of 0.01 K (see Figure 48).

Cooling

The proper operating temperature for a detector is determined by studying the effect of temperature on the detector parameters and selecting the temperature that provides the optimum results for the detector under consideration. The major requirements of a FLIR detector cooling system are long operating time, stable temperature, light weight, small size and maximum reliability. Cooled systems, while more sensitive, add complication, weight and expense.

As discussed, quantum detectors require cooling in order that their response to infra-red radiation can be distinguished from electrical noise. The required temperature for detectors made from CMT is 193 K (–80° C) for mid-wave imagers and 80 K (–193° C) for long-wave imagers. Three methods of cooling are used: thermoelectric, Joule-Thomson and Stirling Cycle Engines:

Figure 48 – A prototype Albion MWIR InSb detector array, alongside a British one penny piece for size comparison. QinetiQ Malvern has demonstrated that four 1,024 × 768 arrays with a 26 µm pitch can be grown epitaxially on a 3 in InSb wafer (QinetiQ) 0111405

Thermoelectric coolers

Thermoelectric coolers exploit the so called 'Peltier Effect' by which current flowing across a junction between dissimilar metals causes one metal to heat while the other cools. By arranging stacks of such junctions (in both series and parallel), it is possible with today's materials to sustain temperature differences of about 100° C across a device. Therefore, in an ISA standard atmosphere with a temperature of 15° C, a detector can be cooled to –85° C, low enough to cool a MWIR CMT detector. However, these devices are relatively inefficient (approximately 0.3 per cent) and thus demanding of power.

Joule-Thomson coolers

The Joule-Thomson cooler is an 'open system' device which cools by expansion of a high-pressure gas. By forcing the gas (usually nitrogen) through a narrow nozzle, it rapidly expands and cools. By applying this gas flow to rear of a detector, it will be cooled. This technique can yield rapid cool-down times (a few seconds for missile seekers) and the low temperatures required for LWIR CMT detectors. However, since the gas is effectively dumped, Joule-Thomson coolers require replaceable gas bottles (and thus cannot sustain continuous operation) or a separate compressor unit (adding weight), making them unsuitable for FLIR applications.

Stirling Cycle engine/Dewar

The Stirling Cycle engine is a 'closed cycle' system that cools by mechanical refrigeration. The cooler uses two pistons; one alternately compresses and expands the refrigerant and a second shuttles between two operational positions. The shuttle piston is phased relative to the main compressor piston so that the working fluid is kept in contact

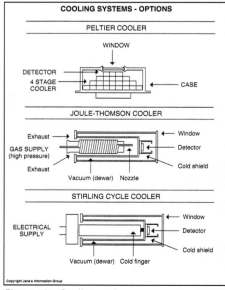

Figure 49 – Cooling options 0522498

with the area to be cooled during the time the compressor piston is allowing the refrigerant gas to expand and cool. During the compression and warming cycle, the shuttle moves to its other extreme position. A regenerator is located in the fluid path between the cold and hot stations and alternately removes heat from and returns heat to the fluid as it is passed from the cold to the hot station. In this manner, a large temperature difference is maintained between the two stations. Excess heat generated in the compression cycle is dissipated through an external heat exchanger. By using gases with very low boiling points (for example, helium), cool-down times of a few minutes are obtained and thermal efficiency is better than thermoelectric devices at some 2 percent. Stirling Engines are utilised widely in LWIR sensors.

Most modern IR sensors utilise a hybrid of the traditional Stirling Engine, combining it with a Dewar. A vacuum Dewar is used to preserve the low temperatures generated by the Stirling Engine and minimise the transfer of heat from the outside environment. The IR detectors are mounted inside the vacuum chamber. IR, which has passed through the optical system, enters the Dewar assembly through a special window and impacts the detector array. The Dewar assembly is mounted directly to the cooler via a 'cold finger' fitted into the Dewar to maintain good thermal contact with the detector.

Split cycle

Some systems utilise what is termed a 'split cycle cooler'; this is essentially a Stirling Engine in two parts, allowing the mechanical parts of the assembly to be isolated from the detector.

Uncooled systems

There is currently considerable research ongoing in the area of high-performance uncooled staring arrays, which offer the benefits of light weight and compact dimensions, when compared with traditional cooled systems. The elimination of scanning and cooling systems also makes them considerably cheaper to produce, bringing such sensors within reach of the dismounted soldier. Programmes such as QinetiQ's STAIRS (Sensor Technology for Affordable Infra-Red Systems) A and B are uncooled second-generation devices intended to establish technology and define common modules for battlefield thermal imagers. Without the benefit of cooling however, current technologies are unable to provide NETDs of better than approximately 0.1 K; cooled systems for avionic applications will thus dominate for the foreseeable future.

Signal processing

As we have seen, infra-red energy is absorbed by the detector(s) and transformed into a very small electrical signal. This very small signal is then used as the input signal for the preamplifiers. There is one pre-amplifier channel for each detector. The signals generated by the detectors are unusable as a composite video signal because of their low level, therefore several stages of amplification are required to produce a workable level.

Aside from the obvious electrical isolation, signal processing circuits must minimise thermal conduction (which will decrease the efficiency of the detectors). Also, SNR is very critical when working with the aforementioned small signals. With a signal of only a few microvolts, even small amounts of noise from the detectors or interconnecting wiring could mask the target signal to the point that no primary output signal can be discerned. Therefore, connecting leads must be long and thin to preserve thermal isolation and must also be physically constrained to avoid generating spurious signals by vibrating. Further, adequate shielding is essential for the detectors, wiring, and preamplifiers and special care must be taken to limit the distance between the detectors and preamplifiers. All of these requirements present a difficult engineering task, which has been addressed to some effect in modern systems by the incorporation of more processing 'on chip', thereby eliminating much of the complicated wiring associated with signal processing.

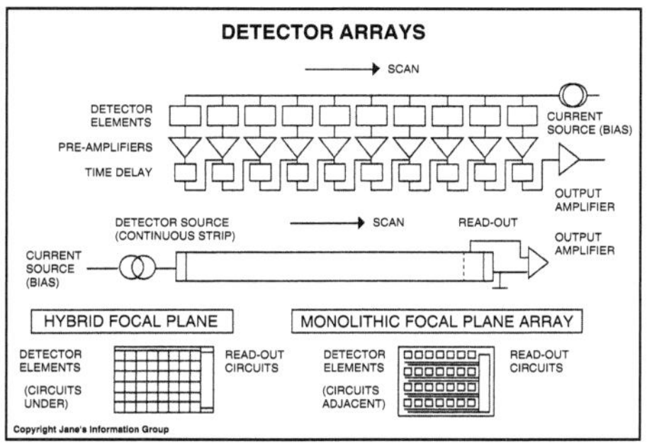
DETECTOR ARRAYS
SCAN
DETECTOR ELEMENTS
PRE-AMPLIFIERS
TIME DELAY
CURRENT SOURCE (BIAS)
OUTPUT AMPLIFIER
DETECTOR SOURCE (CONTINUOUS STRIP)
SCAN
READ-OUT
OUTPUT AMPLIFIER
CURRENT SOURCE (BIAS)
HYBRID FOCAL PLANE
MONOLITHIC FOCAL PLANE ARRAY
DETECTOR ELEMENTS
(CIRCUITS UNDER)
READ-OUT CIRCUITS
DETECTOR ELEMENTS
(CIRCUITS ADJACENT)
READ-OUT CIRCUITS
Copyright Jane's Information Group

Figure 50 – Signal processing 0522497

The output from the preamplifiers provides the input for the post-amplifiers which is sometimes referred to as video processing or video control. There is a separate post-amplifier channel for each pre-amplifier channel. In the post-amplifier, the video signal is amplified to the level necessary to drive the video. The post-amplifier circuit also contains the summing circuitry for video gating, gain control, level control, and video polarity (white hot/black hot, described later). Video on/off time is determined by the gating circuits located in the auxiliary electronics section of the system. As the scan mirror moves across the scene, IR energy will be reflected onto the detectors. At certain points in the scan cycle the mirror will be in a position where IR energy passing through the optics will not be reflected onto the detectors. This part of the scan is referred to as the inactive part of the scan cycle, and that part of the cycle where video is reflected on the detectors is called the active part. The gating circuit gates the video 'off' during the inactive part of the scan.

One of the major tasks for the signal processing circuitry is the automatic scaling of the signal from the detectors. 'Scaling' means matching the thermal contrast in the scene to the visual contrast in the display, that is, the hottest areas of the image are displayed at highest intensity and the coolest areas at lowest intensity (as determined by the selected polarity); intermediate emissivities/temperatures are matched using the available intervening shades. However, if there are large areas of the scanned scene that exhibit very small temperature/emissivity differences (δT), but the total scene also includes 'hot spots' – areas of extreme differences, then, with only a limited number of discrete shades available, the displayed image of the scene would include large areas of little or no contrast – in effect, any thermal contrast has been compressed by the need to display the 'hot spots'. For FLIR systems, a prime example of this problem is the contrast between the sky and the land, which can exhibit a large temperature difference, resulting in the loss of small δTs in objects on the ground. To counter this effect, FLIR systems incorporate systems and controls that will automatically manage thermal 'offset' and 'gain' and provide some manual intervention to the pilot. These features are described later.

Developments in IR imaging

Recent conflicts have demonstrated that future weapon systems must meet the demands of strict Rules of Engagement (RoE) (which require definite target identification, rather than simple detection, prior to weapon release) and enhanced aircraft/aircrew survival (greater standoff at weapon release). Current targeting systems which depend on IR and visual sensors cannot deliver a target image of sufficient reliability to guarantee satisfaction of both of these conflicting requirements, particularly if weather factors are taken into account.

Noting continuing work in the area of MWIR imagers for longer range targeting applications, this technology will only partially satisfy standoff – if aircrews are to be properly protected against more sophisticated and layered air defence systems, they must be afforded much more tactical flexibility than current TV/IR attack systems confer. Realising these rather fundamental limitations to TV/IR acquisition (there is only so much transmissivity that can be 'squeezed' from a standard atmosphere!), designers are expending significant effort in the area of sophisticated processing systems which can deliver true Automatic Target Recognition (ATR). Such systems will enter service with weapons such as the Storm Shadow attack missile, which uses IR imaging to refine the Designated Mean Point of Impact (DMPI) during final approach to target. While such systems do not approach the capability of the human eye/brain combination and are constrained by Line-Of-Attack (LOA) and approach angle during mission planning (making the attack of non-persistent/mobile targets rather more difficult), they show promise in combination with other onboard and off-board EO sensors to achieve true target recognition (and thus attack release) at significant range.

FEATURES OF FLIR SYSTEMS

FLIR systems must be capable of displaying a visual image of the thermal scene to the pilot over a wide range of temperatures. While temperature variations under normal flying conditions will rarely be greater than –50 to + 50° C, manmade objects, such as tank engines and aircraft exhaust nozzles, are at very high temperatures. Depending on mode of operation, a FLIR must be capable of displaying targets and background under all conditions. Having established the requirement for a large dynamic range, we must also consider the display of objects in close proximity with very small δTs – the ability to display a detailed picture of the thermal scene will aid the pilot's Situational Awareness (SA), including terrain perception and object recognition. To achieve this, FLIR needs to resolve δTs of the order of 0.1° C. This feature is the previously mentioned thermal resolution. The following description of elements of a modern fixed FoR, scanning NAVFLIR highlights some operational aspects of interest to aircrew:

Definitions – Area of regard

Having examined the idea of VLAs and described how current systems 'build up' an image by scanning the scene across a smaller array of detectors to produce a single frame, we should define exactly what we mean by the terms Field of View (FoV) and Field of Regard (FoR). For modern FLIR systems, three 'fields' are of interest:

- I_DFoV – the instantaneous FoV of the detectors
- IFoV – the image FoV- that part of the scene which is scanned in a single frame on the display
- FoR – the total area to which the sensor can be pointed (more relevant to targeting systems).

For the purposes of a modern fixed FLIR system, IFoV = FoR.

Gain and offset

Under normal flying conditions, a δT across a land scene may be as little as 10° C; over water, the variation may be only 1 to 2° C. Across the entire FLIR scene (including man-made objects), a total δT of the order of 40° C is typical. A FLIR is designed to 'select' a 'temperature window' from the total range across the scene. The central datum position of this window is the offset. The width of this window, or temperature range about the offset position is called the gain. A FLIR working at a high gain will be using a small temperature window and very small δTs to generate proportionately large signal voltages.

The Auto Gate Area (AGA)

As previously mentioned, the FLIR scene can exhibit large temperature differences, which can depress the contrast of a display with a limited number of discrete shades. The major 'culprit' in most cases is the ground/sky δT.

In modern scanning FLIR systems, a defined 'Auto Gate Area' (AGA) is established, which provides delineation between the ground and sky. For ground attack applications, the FLIR Auto Gate (AG) upper limit is usually set on a line just below the horizon, extending to another line at the lower limit of the FLIR vertical field. The area between the two lines, stretching across the FLIR azimuth FoV is called the AGA. Utilising the aircraft Inertial Reference System (IRS), the area is fully roll-stabilised to maintain the AG upper limit; limited pitch stabilisation against gentle aircraft manoeuvre is also provided. The thermal scene within the AGA is analysed to determine the average temperature and variation. The average temperature is used to drive the offset and the temperature variation is used to drive the gain. The automatic offset function in a modern FLIR can vary the position of the temperature window over a range of approximately 90° C and the auto gain function can reduce the temperature window size to as little as 2° C.

As discussed, the AGA is constrained to lie below the horizon. As the aircraft nose is pitched up and the nominal horizon moves lower in the FoV, the AGA will also be lowered and reduced in size, but only up to a certain point. Beyond this the AGA is frozen until the aircraft nose is lowered. As the aircraft nose is raised further, the AGA will also be raised to include some, or all of the sky, which is thermally uniform (bland). This will cause the offset to reduce (lower average temperature) and the automatic gain to be increased markedly. This will cause any terrain still within the FoR to appear as a bland, unfocussed area. This may be of little importance to the pilot if he is climbing away from the ground to return to base. However, if a sharp pull away from the ground immediately precedes a dive bombing attack, the immediate picture presented to the pilot on lowering the nose (so that now, the entire FoR is terrain) will be devoid of all contrast, making target recognition virtually impossible until the automatic circuits have reacted to reset offset and gain.

Control of gain and offset
It was realised very early in the development of EO operations that pilots were unable to devote adequate time and attention to the control of gain and offset due to workload constraints. Controlling algorithms were developed which facilitated automatic control of these functions to achieve optimum video resolution within the AGA.

For routine flying utilising FLIR, the automatic gain and offset functions are capable of providing optimal video presentation to the pilot, however, there are some circumstances where the automatic function requires operator intervention, such as the transition between flying over a thermally bland sea to coasting in over land. Similarly to the problem of a sharp pull-up followed by a dive back towards the ground, offset and gain will be set to low and maximum respectively (for over-sea flying), making for uncomfortable pilotage on coasting in to an unfocused blur in the distance!

While the solution would seem to be to allow manual control of the gain, this has proved to introduce more problems than solutions, namely the forgetful pilot will most likely interpret forgetting to reset the gain to automatic after intervention as weather deterioration. The solution incorporated into most modern systems involves two facilities, namely gain freeze and 'nudge'.

Gain freeze
Gain freeze, as the name implies, allows the pilot to hold the gain and offset at current levels. By freezing the gain prior to pull-up for subsequent weapons attack, the pilot will ensure that his picture on diving into the target area will be close to optimum for the local conditions. Return to automatic control maybe on a time-out basis, or by re-selection of the relevant switch. Manual selection of gain freeze/unfreeze is a form of full manual control, which may also allow the pilot to shift gain and offset over a (more limited) range of values.

Nudge
A system less workload intensive is called 'nudge', which, unsurprisingly, allows the pilot to adjust the automatic datum around which the system is applying gain. This facility allows the pilot to intervene to adjust his picture without full manual intervention, thus still benefiting from the automatic's ability to adjust settings for changing conditions. Usually, the 'nudge' facility is only available on the gain portion of the system and not on the offset portion.

Polarity (Black/White hot)
A FLIR, like any other monochrome video, creates images which comprise various shades of grey, from fully bright to fully dark. A normal video would, quite logically, show white objects as fully bright and black objects as fully black, with other colours appearing as levels of grey depending on how bright they appear to the video camera. However, a FLIR sensor is not seeing white and black or amounts of light, but is instead seeing a range of temperatures (emissivities). This is displayed to the pilot on a display screen, which necessitates conversion of detected voltages into a video signal. The pilot may therefore select either white hot or black hot via a (hopefully Hands-On Throttle And Stick) polarity switch. This switch activates an additional inverter stage on each of the post-amplifiers, and reverses the polarity of the video signal. Thus, hot infra-red targets can be made to appear as either black or white targets on the display.

Grey scales
The displayed FLIR image is made up of a defined number of brightness levels (corresponding to discrete voltages) which are referred to as the FLIR levels. In practice, while the FLIR image may comprise a large number (over 100) levels, the ability of the pilot to see them is limited by the ability of the head-up or head-down display to show them. One measure of the ability to display these FLIR levels is the number of shades of grey visible to the pilot. Shades of grey refers to the number of increments in luminance that can be resolved on a display. Some FLIR displays provide a 'grey scale' display below the video picture. This comprises a series of squares of increasing darkness, ranging from white to black. There are usually eight of these squares and their purpose is to facilitate adjustment of the FLIR image for optimum contrast and brilliance.

HUD scene reject
While the fixed FoV FLIR generally provides a good quality image for pilotage and targeting, since FoR = FoV = 20° (approximately), the FLIR should not be utilised as the sole reference for manoeuvring flight; NVGs, with a more usable 180° (or more, limited by available head movement) FoR provide for much greater SA, particularly at low level. Also, there will be many occasions when the FLIR image is poor and should not be used for terrain avoidance in the HUD. Accordingly, a FLIR Scene Reject (FSR) switch is usually provided for this purpose. After FSR, HUD symbology remains in view despite the FLIR having been rejected.

Thermal Cueing Aid (TCA)
Bearing in mind that NAVFLIR is aptly named and designed for navigation purposes only, the ability to use the system for targeting should be viewed as something of a bonus. Under good environmental conditions the resolution of the FLIR is adequate for the detection of large targets. The detection of smaller objects such as vehicles and tanks can be more problematical, especially in poor weather or areas of significant thermal contrast. Some FLIRs incorporate Thermal Cueing (TC) software to aid the pilot in the detection of possible targets. The Thermal Cueing Aid (TCA)

Figure 51 – Tornado GR4 head-down display showing FLIR image with HUD overlay in Black Hot (BHOT) (Jane's/E L Downs)   0130866

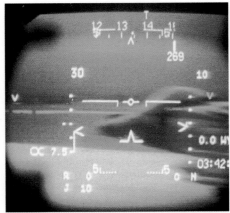

Figure 52 – Harrier GR7 HUD FLIR image with overlaid symbology. Note the TCA caret (V shape just behind the cockpit of the aircraft)   0130867

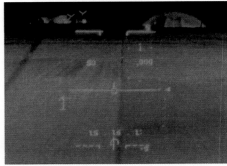

Figure 53 – Picture showing a FLIR display with overlaid HUD symbology and TCAs (upper left) highlighting 'bright spots' within the FoR   0130869

analyses the FLIR scene and detects and tracks 'hot spots', which fit predefined target criteria. These potential targets are shown as carets on the FLIR (see Figures 52 and 53).

THE BAE Systems FIN1010 NAVFLIR
The following short description of the BAE Systems FIN1010 NAVFLIR will highlight several aspects of the foregoing theory in an attempt to reinforce the major aspects involved in integration of a modern FLIR system into a combat aircraft:

UK Harrier GR7 and Tornado GR4 aircraft are equipped with a fixed Field of View (FoV) NAVFLIR, operating in the 8 to 12 µm band. The equipment is the BAE Systems (formerly GEC-Marconi) Modular FLIR, incorporating the UK Thermal Imaging Common Module II (TICM II). The following description will concentrate on the installation in the Harrier GR7 ground attack aircraft:

Harrier GR7 NAVFLIR
The FLIR sensor head has a fixed FoV of 16 × 25° (vertical × horizontal), depressed at an angle of approximately 7° from the aircraft waterline, optimised for lookdown over the nose of the aircraft for low-level attack operations. The FLIR image may be displayed in the Head-Up Display (HUD) or either of the MultiPurpose Colour Displays (MPCDs) at 1:1.

The FLIR is a differential sensing system, measuring signal variation between objects in the scene. These signal variations are expressed as a temperature variation, or δT, between objects (rather than absolute level, as discussed). A δT can arise due to difference in temperature or difference in emissivity. For example, in the 8 to 12 µm region, a 5 per cent emissivity difference around ε = 0.95 is equivalent to approximately a 7° C temperature difference (ISA).

Description
The FLIR is comprised of two main units, the FLIR sensor head and the Electronics Unit (EU). The FLIR sensor head is boresighted to the aircraft and mounted in the upper portion of the nose. The sensor head is mounted at an angle of approximately –7° to the aircraft waterline in order to provide the best possible FoV over the

Figure 54 – The BAE Systems Modular FLIR equips Tornado GR4 and Harrier GR7 aircraft of the UK RAF   0581546

Figure 55 – Harrier GR7 Cockpit showing FLIR image on HUD and (right) MultiPurpose Colour Display (MPCD)   0130868

Figure 56 – The Boeing F - 15E carries the LANTIRN system. The AN/AAQ-13 Navigation pod is shown on the right chin station – the FLIR window can be seen directly above the Terrain Following Radar (TFR)   0131090

aircraft nose, with the FLIR EU mounted directly below.

The FLIR sensor head gathers thermal radiation from the scene by scanning incident IR radiation onto a cryogenically cooled detector array. The IR signals are then converted into electrical signals, which are then amplified and processed prior to transmission to the electronics unit, which converts the electrical signals into a video output to provide a visual image of the IR scene to the pilot.

The IR window is composed of germanium for IR transmissivity, with high efficiency carbon coatings for durability. The IR telescope is situated directly behind the IR window. It provides the magnification necessary for the one-to one registration of the real world and sets the FLIR video FoV. A motor driven scanner assembly scans the IR radiation through the optics, which uses a rotating polygon mirror and an oscillating frame mirror. The reflected energy from the oscillating frame mirror is then focused through a lens assembly onto an eight-element detector array.

The system utilises a Serial/Parallel (S/P) scan, yielding eight lines of raster image per swathe, with image output in a standard 525-line format.

The detector array consists of eight separate detector elements arranged in a vertical line. Each detector element is approximately 70 × 700 μm (height × width), within an array of 910 × 700 μm. Each detector element provides a signal output therefore the detector array provides eight parallel IR video signals.

The detector is a CMT extrinsic semiconductor, which responds to wavelengths in the FIR band (8 to 12 μm). Initial signal processing takes place within the detector elements, using previously discussed SPRITE and TDI to improve sensitivity and increase SNR.

An internal sealed Stirling Cycle cooling engine is used in the Harrier FLIR because it eliminates the logistic support and possible contamination associated with Joule-Thomson coolers, as

Figure 57 – The BAE Systems Modular FLIR configuration for the UK Harrier GR7   0131077

utilised in the AIM-9L via nitrogen air bottles. The Stirling Engine and 'cold finger' are an integral part of the FLIR sensor head.

The detector is insulated from external heat sources by being mounted in a Dewar, with the tip of the cold finger fitted into the Dewar so that it maintains good thermal contact with the detector.

The cooling cycle is a closed loop operation, which is controlled by the cooling engine drive electronics. These electronics sense the cryogenic temperature and control the cooling engine. The detector is thereby maintained at a constant temperature of approximately 80 K or –193° C. The system requires approximately three minutes to achieve operating temperature.

The main function of the FLIR EU is to process the eight parallel IR video signal outputs from the sensor head into a video image for display to the pilot. The EU controls the functions of the scanner assembly and provides the special signal functions required to stabilise and enhance the quality of the FLIR image. An important function of the electronics unit is to provide the automatic control of gain and offset.

An integral subsystem of the electronics unit is the Thermal Cueing Unit (TCU), which utilises the FLIR signal to detect hot spots, which may be potential targets. These hot spots may then be marked with carets and displayed on the FLIR image in order to enhance target detection.

The EU also generates the Auto Gate (AG). The automatic control of gain and offset for the entire image is determined by the thermal conditions in the Auto Gate Area (AGA). The AG is an electronically produced, dynamically controlled area, which utilises aircraft roll, pitch, and height data for its determination. The location and size of the AG changes as flight parameters change. For attitudes up to approximately 7.5° nose-up, the top of the auto gate remains locked just below the horizon; beyond this pitch attitude the auto gate is no longer ground stabilised and will lock in pitch and roll relative to the aircraft. It will resume a relative ground orientation when the pitch attitude returns to less than 7.5° nose-up. There is no indication of the AG position or shape displayed to the pilot.

Within the EU, the TV signal processor combines, processes, and outputs a TV compatible video signal. The TV signal processor also provides several other functions:

- Detection circuitry used to control the automatic gain and offset settings;
- Polarity (black hot/white hot) selection;
- Grey scale generation;
- Graphics injection.

The video signal is amplified with the required transfer characteristics, buffered and synchronising pulses added before being sent to the aircraft display computer.

Operation

The FLIR in the Harrier GR7 is a fully integrated part of the aircraft nav/attack system. Accordingly, the cockpit features Hands-On Throttle And

Stick (HOTAS) control of all of the major FLIR functions, including FSR, Polarity and TC display/ reject. Display of the FLIR scene is via HUD or either MPCD, according to pilot preference and/ or tactical constraints. By day, the FLIR is utilised to aid target recognition, providing indications of thermal activity in the target area of interest. By night, the FLIR is also used as an aid to navigation and for terrain avoidance in the low-level arena, in conjunction with NVGs.

It should be noted that, due to the constraints of the fixed FoV (20°, HUD limited), the FLIR is not normally used as the primary means of terrain avoidance during manoeuvring flight. The NVGs, with a greater than 180° FoR, are used to greater effect in this particular area. For purely navigation and pilotage purposes at low level, the ratio of NVG/FLIR usage is ideally approximately 80:20, although, due to the complementary nature of the sensors, this ratio can virtually reverse if light levels drop significantly.

CONCLUSIONS

The use of FLIR, particularly when partnered with NVGs, provides a powerful operational capability to modern tactical rotary- and fixed-wing aircraft. While the advent of uncooled technologies promises much in terms of reduced weight, complexity and cost, the performance of such systems still falls some way short of modern cooled systems. Thus, for the foreseeable future, cooled IR systems will dominate in airborne applications. However, as arrays become larger and larger, staring arrays, without any form of scanning system, will become more prevalent, enabling reduction of total weight and simplification of the signal-processing portion of the process. These VLAs will also bring tactical benefits in terms of multirole use (similar to Active Electronically Scanned Antennas in radar systems) in dual FLIR/IRST systems.

Panoramic NVGs, only recently introduced into service in US air arms, have yet to prove that the increase in horizontal FoV is worth the extra cost (4 × 16 mm high-performance tubes do not come cheap), bulk, head-carried weight and moment (at least in the case of the fielded Type 1 AN/AVS-10, see entry). Due to the characteristics of the design, an adjustable dioptre, in the sense of current binocular types, is not possible (clip on adjusters are employed to expand the range of dioptre) – which, in the editor's opinion, is not necessarily a bad thing, and those which incorporate objective focus pose new problems to pilots to achieve a consistent and sharp image across the FoV. However, as discussed, this may not be a major problem – how often do we attempt to focus on an object in our periphery? The major problem for the widespread adoption of panoramic NVGs may be in the form of HMDs – as major air forces seek to upgrade pilot SA and tactical effectiveness, the panoramic NVG may find itself 'leapfrogged' in favour of a fully day/night capable helmet system, such as BAE Systems, Stryker/Cobra (see entries and HMDS analysis section).

See Airborne Electro Optic (EO) Systems for related entries in *Jane's Avionics*.

HELMET-MOUNTED DISPLAY SYSTEMS

Helmet-Mounted Display Systems

Introduction
The subject of Helmet-Mounted Display Systems (HMDS), covering all aspects of the projection of flight, system and sensor information into the eyeline of the wearer, as well as the traditional aspects of helmet design, including head protection, sound attenuation, life support, and so on, is an extremely wide subject. For the purposes of *Jane's Avionics* while it might be deemed sensible to therefore concentrate on the purely avionics-related areas of symbology and sensor display techniques, this would misinform the reader regarding the fundamental interactions between the multitude of design trade-offs demanded by the aforementioned elements of HMDS design. Accordingly, the following will address the overall concept of HMDS design and describe each of the major interacting elements in turn. Much of the following will show a logical progression of theories expounded in the Electro Optics analysis paper; accordingly, it is recommended that the reader should peruse that section prior to continuing for the fullest benefit.

HMD – NECESSITY OR LUXURY?
Pilots of the current fighters, used to employing their HUD to keep heads-out as much as possible during the inevitable melee of air combat, will shortly receive a major and operationally highly significant upgrade in the HMDS. The introduction of the Eurofighter Integrated Helmet (production examples are undergoing test) and others such as the Gerfaut HMD for Rafale and the F-35 HMD will bring greatly enhanced capability over currently fielded systems such as the Joint Helmet-Mounted Cueing System (JHMCS). If we consider a fully capable HMDS as a logical extension of the HUD, bringing with it a Primary Flight Display (PFD), weapon aiming, sensor display for pilotage and EW display for enhanced survivability, then by slaving all of this to the pilots eyeline, we can begin to gain some insight as to the enormous benefits to be gained – as long as it works! This last comment, while seemingly flippant, is a very real consideration in bringing a complex system like this to operational acceptance – more of which later. Experience with the HUD in current fast-jet cockpits has led to user acceptance and to the display of a wide range of information. Conversely, there is currently minimal experience of flight with HMDs and hence little is known of the effects of this technology on mission performance. Although the HMD has the benefit of cueing pilots as they move their head, it is unclear what the appropriate selection of stabilising cues should be, for example, attitude information. HUD and HMD displays are distinct from colour Head-Down Displays (HDDs) in a number of ways, and a range of technological, perceptual and cognitive issues must be considered. For those used to employing HUDs with symbology and sensor (for example, Forward-Looking Infra-Red) information, the trade-offs required between viewing a 'virtual' scene versus the real world, when they are overlaid (note I have not used the word 'fused'), is a familiar problem. In listing some of the aspects common to both HUD and HMD, we can begin to build an appreciation of some of the issues:

- Collimated images
- Projection of information against a real world background
- Obscuration of outside scene by display content
- Cluttered displays
- Limitations on writing of symbology
- Harmonisation of 'virtual' and real world scenes
- Monochromatic imagery requiring interpretation.

The HMD has additional issues related to technology, such as image lag with head motion. If all of the above issues can not be solved then there will be a real danger that pilot/mission

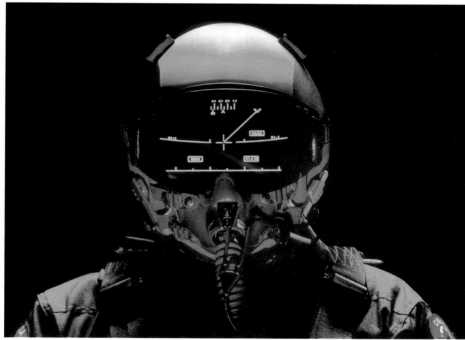

Saabtech HMD (Saabtech)

1128607

performance may be adversely affected. For example, the collimation of the HUD/HMD may cause perceptual problems and the way information is presented can disrupt or reduce the efficiency of cognitive processing. All of these issues need to be considered in the design of optimised HUD/HMD symbology, and their impact must be quantified.

Before we get too carried away with the purely avionic issues pertaining to HMD, let us not forget that the basic helmet has protective duties to perform, and, similarly to the aforementioned view that a flawed avionic implementation will actually decrease pilot effectiveness, the same will be true for a heavy or poorly balanced helmet – this point is arguably the biggest hurdle to an effective design.

So, we have established that the introduction of an HMDS is not a simple undertaking – which means it is expensive; the cost of bringing a fully capable HMDS into service will dwarf any other single element of a fighter aircraft's avionic suite (except, perhaps, the radar system). This fact alone might dissuade an operator from even attempting to acquire such a capability. However, early indications from the employment of 'interim' HMS/D, such as the JHMCS, has shown that the enhanced fighting capability they bring cannot be ignored – recent mock engagements between JHMCS-equipped F/A-18s and Eurofighter Typhoons in the UK illustrated the advantage of high off-boresight Short-Range Air-to-Air Missile (SRAAM) aiming. The advantage of Specific Excess Power (SEP) and high-speed agility of the Typhoon was reportedly utterly negated by the F/A-18 pilots' ability to employ their SRAAMs without limitation in the close combat arena.

That the Tranche 2 upgrade package for Typhoon includes its own HMDS was probably of scant comfort at the time.

However, this is perhaps something we already knew, since helmet cueing devices in fast-jet aircraft were introduced originally for this task and have proved highly effective – it could be argued that, with knowledge of a similar situation, Typhoon pilots would simply remain at longer range and 'wear down' the opposition with Medium Range Air-to-Air Missile (MRAAM) tactics. So our highly expensive HMDS needs to bring something more than just weapon aiming to the table. This it achieves via sensor display – the ability to display a sensor picture (Low-Light TV, Image Intensified, NAVFLIR, IRST) directly into the pilot's sightline bestows greatly enhanced Situational Awareness (SA). This, as every fighter

Initial operational feedback has suggested that the introduction of the Integrated Helmet with Tranche 2 production will finally see the Typhoon perform to its full capability (Eurofighter)

1128622

During recent mock combat sorties against UK Typhoons, US F/A-18 pilots found that their JHMCS enabled them to more than hold their own against their more agile opponents during close combat

1116246

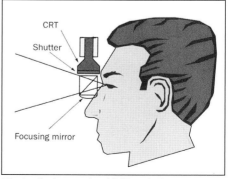

CRT

Shutter

Focusing mirror

The basic theory of an HMD – a helmet-mounted display source generates an image which is positioned in the wearer's sight line

0002734

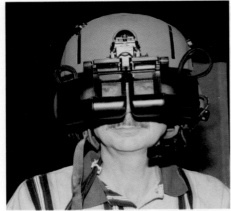

A simple approach to HMD is exemplified by the Kaiser 100 system – note the dual Cathode Ray Tube (CRT) Display Units (DU) mounted below the combiners in front of the eyes   0062063

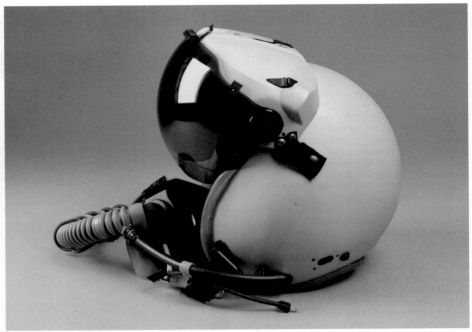

The Joint Helmet-Mounting Cueing System comprises a day-only display module mounted onto a modified HGU-55/P helmet (VSI International)   1128608

pilot knows – is the key to success; he who can quickly attain and, importantly, maintain the most accurate air/ground picture will generally win.

Further, the helicopter community provides a good indication as to the benefits of sensor display; the in-service IHADSS on the AH-64 Apache and more capable systems entering service, such as Topowl, have shown that it is not just weapon aiming where HMDS can make a significant operational contribution.

DEFINITION OF TERMS – HMS OR HMD?

For the purposes of this paper, HMDS will be sub-divided into two categories: Helmet-Mounted Sights (HMS), such as BAE Systems Helmet-Mounted Sight System (HMSS) for the

Jaguar ground attack aircraft, provide the pilot with a reticle to for Line-of-Sight weapon aiming/cueing. More complex Helmet Mounted Displays (HMD), such as BAE Systems Eurofighter Integrated Helmet, incorporate not only weapon aiming/cueing, but also Primary Flight Display (PFD)-qualified flight symbology, together with overlayed sensor (TV, IR or synthetic)

information. Bridging the gap between these are systems, such as the Display And Sight Helmet (DASH)/Joint Helmet Mounted Cueing System (JHMCS) and Topsight®E, which provide some flight symbology in support of weapon aiming. While often referred to as HMD, these will be considered as HMS here. The distinction between HMS and HMD is important when considering the prospective integration of an HMDS into the harsh (high-g/high sightline rate) environment of a fighter aircraft cockpit.

BIODYNAMICS

Having previously mentioned the importance of weight and balance in an aircrew helmet, it is fitting we should begin with a technical appraisal of this subject, termed biodynamics. The most basic (and important) role of any aircrew helmet must be to provide head protection, a role which has remained unchanged since the advent of flight helmets. Further, the prescribed (by use and environment) level of protection must not be adversely affected by the introduction of any enhancements, such as power supplies and display electronics for HMD applications. When we use the phrase 'head protection', it should be noted that, for our purposes, this means not only ensuring the physical well-being of the pilot's head, but also his senses, including protecting the eyes against sunlight and laser devices, and the ears against noisy cockpit environment of some current fighter designs (the BAE Systems Harrier springs to mind in this case).

With this overriding requirement for head protection in mind, it becomes clear that, as we attempt to integrate a display in the pilot's Line of Sight (LoS), with its associated power supply and projection system, such 'additions' must not compromise protection. While there is currently a high level of research and developmental effort in the area of helmet-mounted display technologies, much of this effort is concentrating on packaging – the helmet/display combination should not increase the risk to the wearer's well-being due to excessive weight and/or misaligned/shifted Centre of Mass (C of M). By definition, since integrating a power supply and projection/display system into a helmet will increase mass and, due to the need to position the optical train along the pilot's eye line, shift the overall C of M, it now becomes readily apparent that the core design considerations of a modern HMD are entirely contradictory. These basic design issues of protection, comfort, stability, aural and eye protection are not the only ones involved in HMD design – as we shall see later with the definition and integration of the optical train, basic conflicts between most of the fundamental characteristics of an HMD dog designers throughout the design process.

The Helmet-Mounted Sight System (HMSS) is a simple off-boresight aiming system integrated onto Jaguar aircraft   0106288

GETTING TO GRIPS WITH THE PROBLEM
EVOLVING CAPABILITY

Historically, helicopter aircrew helmets have taken the lead in providing the pilot with a true HMD capability, due, in no small part, to the less harsh environment and lesser manoeuvre capability (roll, *g*), placing less burden on packaging of the HMD and display of sensor information and symbology generation (latency). While introducing helicopter HMD designs, companies have conducted parallel efforts in further refining and developing these designs to be compatible with fighter aircraft (for example, BAE Systems' Striker HMD family ultimately spawned the Eurofighter Integrated Helmet). Current fighters, including US F-15, F-16 and F/A-18, and Russian MiG-29 and Su-27/30 incorporate the JHMCS and Sura respectively. As previously mentioned, these are basic HMS, utilised for weapon aiming and limited SA. As further advances have been made in area of display technology, facilitating tighter/lighter packages, next-generation aircraft, such as the Eurofighter Typhoon, Gripen, Rafale and F-35 Joint Strike Fighter (JSF) all feature fully capable (sensor display and symbology) HMD. Indeed, the JSF HMD will be the sole PFD for the pilot since the aircraft will not have a HUD. The success of this new crop of designs is absolutely pivotal to the overall mission effectiveness of the aircraft as they will be the primary display medium for the enormous amount of tactical and environmental information the latest onboard and 3rd party sensors will 'bring to the party'. Indeed, if the poor (single-seat) pilot has any hope to adequately correlate this avalanche of data, the HMD will be a crucial tool in helping the pilot 'visualise' the three-dimensional volume of airspace around him. To prove this point, pilots currently operating US ANG F-15Cs equipped with the AN/APG-63(V)2 AESA radar and JHMCS have expressed a preference for the JHMCS during night operations over the AN/AVS-9 NVG, citing the immensely enhanced SA gained from a simple scan of the HMD to determine instantaneous to all blue/red aircraft within the (large) scan volume of the radar. In future, the ability to overlay IR and other sensor information onto this basic real-world/tactical picture will aid the pilot even further; The old premise applied by most of to our Instrument Rating Test (IRT), often conducted under a hood to deny any view of the outside world, 'a peep is worth a thousand scans' – springs to mind in this instance! Similarly for future fighter pilots, scanning his MultiFunction Displays (MFD) for critical tactical information will only be if his HMD has failed.

HELMET MASS

For both fixed- and rotary-wing applications, the increase in capability of the HMD comes with an attendant increase in mass of the helmet/display combination, often with an asymmetric C of M. A modern basic fighter helmet, such as the HGU-55/P weighs in at around 1.4 kg; this compares to current monocular (single display source) HMS at around 1.8 kg. While an increase of 400 *g* seems small, consider that this represents an increase in neck-supported mass of 3.6 kg, or a fully featured 15 in laptop computer (as this is probably an apposite metaphor), at a sustained 9 *g* (well within the manoeuvre boundary for the aforementioned Fourth Generation fighters). Consider that the 1.4 kg of the basic helmet serves functions which include head impact protection and mounting for communication systems, hearing protection, eye protective visors and oxygen systems. Accordingly, the weight increase in the transition to HMD must service the needs of weapon aiming, enhanced vision, sensor display, and flight symbology display. Additionally, chemical defence and nuclear flash protection should also be considered, since the HMD may not be compatible with existing integrations. These requirements demand more complex mounting devices on the helmet than presently in place, which, ultimately, will result in further increases in system weight and potentially less than optimal C of M. In the end, however, the limit to this iteration is the ability of the pilot's neck to support this mass/C of M increasing the fatigue rates and neck injury risk in any accident/ejection.

Thales Avionics' Topsight®E 0051713

The fighter cockpit is a harsh environment for HMD, with sustained loads of 9 g *well within the capability of modern Fourth Generation aircraft* (BAE Systems) 1128612

IMAGE ENHANCEMENT DEVICES – LEADING THE WAY TO HMD DESIGN

During the 1980s, the introduction of Night Vision Goggles (NVG) into both rotary- and fixed-wing military operations brought about the first true mass/C of M problem due to the introduction of image enhancement devices to both helicopter and fighter helmets. This brought the mass of helmets on both sides of the Atlantic, up to around 3 kg for a US Army SPH-4 helicopter helmet equipped with AN/AVS-6 NVG and a UK RAF Mk4D equipped with Nightbird NVG. This increased mass contributed to an overall decrease pilot performance due to neck muscle strain and fatigue. These disadvantages, however, were offset by the enhanced visual capability for night flying and increased weapons aiming capability offered by helmet-mounted image intensification devices and other helmet-mounted displays, resulting in positive pilot feedback for the overall system. A note of caution for those pilots still employing NVGs however – despite the increased SA brought about by NVGs, there is still a net *increase* in workload due to these sensors – a point which is often missed and more often the cause of NVG-related mishaps.

In order to permit the use of these 3 kg helmet/NVG systems without overloading the neck in severe helicopter crashes, the US Army Night Vision Laboratory developed a then new type of NVG, incorporating two distinct parts – a helmet-attached mount and a separate binocular section. Connecting the two, a spring-loaded, ball-socket NVG mount permits AN/AVS-6/9 NVG to separate during a crash (or, as was later and perhaps luckily found, an ejection). The 0.6 kg binocular section was designed to break free of the helmet mount at a goggle deceleration of 10 to 15 g (per MIL-A-49425 (CR), 1989). While this two-piece NVG offered a solution to the problem of potential broken necks in the aftermath of a helicopter hard landing/crash, the overall problem of too great a mass placed on the pilot's head remained. In the UK, the Night-Op and subsequent Nightbird NVG were both one-piece designs, weighing in at around 800 g; initially on introduction to service, both types were mandated to be removed before any serious impact or ejection – all very well if the pilot remembers to doff the NVG during the mayhem of an aircraft emergency! While UK military helicopter pilots continue with the basic Night Op/Nightbird system, an autoseparation system, which forced the NVG from the helmet on initiation of the ejection sequence, was introduced for fighter aircraft to alleviate stresses on the pilot's neck during the up to around 30 g applied during ejection with current rocket seats. On the other side of the Atlantic, the Americans developed and tested the AN/AVS-8, equipped with autoseparation, but never fielded the system, possibly as a result

of the 'de facto' autoseparation of the binocular section inherent to the two-piece US design. While it is still preferable to remove the NVG prior to ejecting, to the editor's knowledge, on only one occasion have the crew of an ejection seat-equipped aircraft remembered to doff NVGs prior to 'pulling the handle' – so autoseparation of the NVG is a must. However, depending on mass and the effect of the attached display device on overall C of M, it may be possible to retain the HMD during ejection, but first we must know the practical limit of what the average human neck will withstand in terms of force and moment.

DEFINING THE LIMITS

In the US, as part of initial research into HMDs in Army helicopters, studies sought to actually define a safe limit on flight helmet mass. In 1982, the USAARL proposed a limit of 1.8 kg during the development of the AH-64 Apache IHADSS helmet. The helmet system subsequently developed met this mass limitation while providing the desired platform for the monocular HMD and the required acoustic and impact protection. However, the current SPH-4 plus AN/AVS-6/9 combination used for night operations in all other Army helicopters continues to exceed the proposed 1.8 kg limit by more than 1 kg. Although there are many pilot reports concerning the discomfort of wearing this combination, which correlate closely with the editor's own experience over many hours wearing the Mk4/Nightbird system, the effects on pilot performance of carrying this much mass, coupled with adverse C of M have not yet been adequately quantified. It is the editor's opinion that this degradation, while apparently offset by the aforementioned vastly increased SA bringing a commensurate 'increase' in pilot confidence, is more serious than might be assumed and certainly a background causal factor in many incidents/accidents.

Furthermore, the dynamic consequences of crashing with head-borne masses approximating 3 kg remain largely speculative. Historically, helicopter helmet mass and C of M requirements have been non-existent or vague, with requirements written 'to fit' an extension of an existing in-service design. As a result of much valuable effort conducted in the US, seven parameters were identified to define the mass properties of a helmet system. These include mass, the C of M position along three orthogonal axes, together with the mass moment of inertia about the three respective axes. The resulting co-ordinate system used by the US Army aviation community is based on a head anatomical co-ordinate system, with the x-axis is defined by the intersection of the mid-sagittal and Frankfurt planes with the positive direction anterior of the tragion notch. The y-axis is defined by the intersection of the Frankfurt and frontal planes with the positive y-axis exiting through the left tragion notch. The z-axis is oriented perpendicular to both the x- and y-axes following the right hand rule. According to this research, the rationale for defining helmet mass requirements can be segregated into three areas: Aircrew health, operational effectiveness, and user acceptance.

Aircrew health can be affected by both short- and long-duration head and neck loading. Long duration exposures are the result of total helmet mass and C of M position in normal flight

conditions, including discomfort from a sore or stiff neck after a normal sortie. Most aircrew who have routinely operated with NVGs will readily identify with this problem. The NVG, attached to the brow of the helmet, not only increases the neck-supported mass, but also moves the overall C of M forward, which has two effects for the wearer. The stability of many helmets, whether featuring a foam liner or an internal cradle, is upset, causing the helmet to rotate on the head. This generally forces a tighter fit, which in turn causes discomfort, often resulting in headaches after long missions. Exacerbating this discomfort, the forward C of M places greater stress on the head-supporting neck muscles. For the helicopter community, positioning the battery pack on the rear of the helmet enables better weight distribution by bringing the C of M back along the x-axis. Indeed, the discomfort brought about by the forward C of M is so severe, most helicopter pilots also attach further counter weights to the rear of the helmet to bring the C of M back to its original position, resulting in an even heavier helmet! For the fighter community, such alleviation is not available, primarily due to the presence of the parachute head box of the ejection seat, which would interfere with the battery pack, but also due to ejection restrictions (noting the previous discussion regarding autoseparation) and the greater need to keep total mass to a minimum as a result of the greater dynamic envelope of fighter aircraft. Short duration exposures result from increased *g* over short periods which may be produced by a snap turn in a fighter, adverse environmental effects (turbulence) or indeed a crash. In tandem-seat aircraft, with both occupants wearing NVGs, a high level of crew co-ordination is required since, if the pilot executes a hard turn without informing his navigator/Weapons System Operator (WSO), a severe neck injury to the rear seat occupant could ensue if his/her head is in an 'awkward' position at the beginning of the manoeuvre. However, injuries are generally less severe, with strains and muscle tears being the norm. An aircraft crash is arguably the harshest environment in terms of neck injuries, whether this results in ejection or a crash landing. Ignoring direct injuries to the head/neck as a result of the break-up of the aircraft and/or striking terrain or other objects after ejection, neck injuries are caused by the transfer of energy to the neck from a head impact or as a result of a high g event combined with excessive head supported mass and extreme C of M.

The increase in mass and shift of C of M can also adversely affect operational effectiveness of crews through increased fatigue. A subject close to the hearts of all night operations commanders, controlling the fatigue level in crews is a key factor in maintaining effective sustained airborne operations at night, where crews are not only combating the effects of circadian rhythm but also the increased workload of Electro Optic (EO) operations. Adding the effects of a heavy and unbalanced helmet system and the extent of the problem becomes clear. Aircrew operating with high levels of fatigue are less efficient, exhibit less concentration and are more likely to make mistakes. Accordingly, the operational impact of an unbalanced and heavy HMD would be seriously limited.

As mentioned earlier, an extreme C of M, due to the rotational moment exerted, can physically shift the helmet on the pilot's head. If NVGs are attached, this can move the NVG optical axis away from the pilot's eye axis, thus restricting the observed FoV. The allowable limits of eye position relative to the display, termed the eye box, is generally small for the current crop of HMDs, so such slippage is clearly unacceptable, as the pilot would be effectively blind to information and/or sensor imagery.

Another area which can be affected by head-supported mass is user acceptance. Clearly, even if HMD characteristics are within acceptable weight and balance limits, if the users' opinion is that the system is too heavy or unbalanced, then the system will be the subject of unauthorised 'modification' (note extra weight carried by

Fighter pilots employing NVG cannot take advantage of a counterbalancing battery pack on the back of the helmet due to interference with the head box of the ejection seat. In addition, the extra mass and extended C of M of the 800 g NVG, if not discarded during the ejection sequence, poses an extreme hazard

0049138

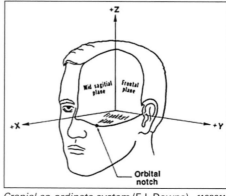

Cranial co-ordinate system (E L Downs)   1128611

helicopter crews on the back of the helmet when utilising NVG) and/or misuse (NVG often mounted too high, allowing pilots of fighter aircraft to view the HUD, with or without FLIR sensor imagery, below the NVG). Hence, the final HMD configuration must be acceptable to the final user prior to fielding to operational units. However, user 'acceptability' is extremely difficult to define since aircrew feedback is, by definition, subjective. On both sides of the Atlantic, there is no empirical data on existing systems which could be utilised to quantify user acceptable mass property limits. In the US, the development of the Comanche HIDSS prompted efforts to develop head-supported weight and C of M requirements, resulting in recommendations for total allowable mass, together with x- and z-axis C of M locations:

The recommended allowable mass requirement is based on neck tensile strength; x-axis C of M location is based on measured biodynamic responses of aviators wearing various helmet mass and C of M combinations; z-axis C of M is based on maintaining a constant moment about the C7/T1 juncture resulting from the helmet mass and vertical C of M position. As can be concluded from the foregoing, a sound understanding of the mechanisms of neck injury are required in order to accurately establish mass limits on HMDs. Drawing on research into neck loadings, US investigations derived two injury mechanisms which are most likely to be affected by the mass properties of HMDs. These are axial tension and forward bending (flexion). While neck extension/compression injury mechanisms were not considered to be affected by HMD mass properties, it should be noted that this research was based in helicopter-based application of HMD – the helicopter crew station (seat/headrest) is very different to that of current fighters, so perhaps this assumption should be further scrutinised as part of fighter HMD design.

Examination of US Army helicopter crash injuries discovered the importance of the C7/T1 juncture – the cervical and upper thoracic vertebra with the highest frequency of fracture was the 7th cervical. The lower thoracic and the lumbar region experienced a higher frequency rate, but these injuries are believed due to compression loadings resulting from high vertical impact loads in pre-crashworthy helicopter seat designs. More recent helicopter designs incorporate various levels of crashworthiness with specific performance levels for crew seats. US helicopter crew seats are typically procured to military performance specifications with a 30 g longitudinal static load requirement and a vertical energy absorption capability (MIL-S-58095(AV), 1986). The 30 g longitudinal requirement is a structural integrity check of the seat and its mounting hardware to provide assurance that the seat will not be ripped

from the floor. The vertical energy absorber is a mechanical device which restricts the vertical crash loads experienced by the occupant. The desired vertical load is an average of 14.5 g over the range of seat stroke. Peak loads of 18.3 g have been measured in anthropomorphic test dummies during seat qualification trials. Therefore the worst case condition would be a seat experiencing 30 g longitudinally and stroking with a peak vertical load of 18.3 g. The resultant from these two load vectors is 35 g directed 31.4° downward from the horizontal. Dynamic tests with rigid seat structures indicated a range of possible 'dynamic overshoot' values (the ratio of measured head or chest acceleration of a test dummy to the input floor or seat acceleration). This increase in acceleration results from harness slack, neck tissue stretch and upper body compression (by contact with the restraint harness), which allows a relative velocity to be created between the occupant and surrounding structure. The dynamic overshoot value is also dependent on when the shoulder strap inertia reel locks (which is activated by occupant motion). A dynamic overshoot value of 1.5 was selected as the magnification of seat acceleration to the head acceleration, based on an average based on dynamic tests of aircrew seats for the UH-60 Black Hawk helicopter.

So, having established a maximum load value, the US researchers needed to define neck strength, such as the ability of the neck to withstand such loads. Utilising operational data, automotive injury data and testing of cadavers, a neck tensile strength threshold of 4,050 N was selected as the maximum limit for US Army helicopter pilots.

The determination of maximum allowable head-supported mass is based on Newton's second law, $F = ma$. This equation is then applied with the neck tensile strength threshold of 4,050 N and the applied acceleration (35g) with the aforementioned dynamic overshoot ratio of 1.5 as a multiplier. The effective mass acting on the C7/T1 juncture is then given by:

$$F = ma$$
$$m = F/a$$
$$m = 4{,}050\ N/[35g \times 1.5 \times 9.81\ m/sec^2]$$
$$m = 7.86\ kg$$

However, the mass acting on the C7/T1 juncture includes the helmet, head and, for the sake of conservatism, the neck. By subtracting the average head and neck mass (4.32 and 1.04 kg respectively) from the above value, we arrive at the allowable helmet mass for the given impact condition –2.5 kg.

The next stage requires the definition of the limiting C of M, based on the maximum allowable helmet mass, and acting as a constant mass moment about the C7/T1 juncture. Applying this technique, it can be seen that, as the

C of M moves downwards, then the maximum allowable helmet mass will increase – this point is pertinent to current NVG users who regularly don helmet/NVG combinations well in excess of the 2.5 kg limit. In selecting a maximum constant mass moment, and thus a spread of limiting C of M for total mass, US researchers were once again forced to examine in-service systems due to a lack of empirical data. The then 'worst case' was found to be the AH-1 cobra helmet, which has a mass of 1.74 kg and a vertical C of M located 5.2 cm above the tragion notch. In order to determine the constant mass moment, the vertical distance between the C7/T1 juncture and the tragion notch was required. A value of 11.94 cm was selected, which represents the 95th percentile female and the 85th percentile male measurement (US). Having gathered the necessary values, the constant mass moment can be defined by:

$$M = md$$

where M is the mass moment, m is the (helmet) mass and d is the (total) distance of the vertical C of M position above the C7/T1 juncture. So:

$$M = [1.74\ kg] \times [11.94 + 5.2\ cm]$$
$$M = 29.8\ kgcm$$

While the above gives a value for the limiting mass moment (based on a worst case system), we can also use this to establish a relationship between the vertical C of M position and mass by rearranging the equation:

$$29.8\ kgcm = [m_{HELMET}] \times [11.94\ cm + z_{HELMET}(cm)]$$

rearranging:

$$z_{HELMET} = [29.8\ kgcm/m_{HELMET}] - 11.94\ cm$$

Plotting this relationship results in a curve defining an allowable area of mass/C of M combinations. The allowable mass is limited to 2.5 kg as determined previously and the vertical C of M position is limited to 5.2 cm, a figure determined by the worst case fielded system – note that these parameters are, strictly, valid only for the conditions described previously, based on helicopter operations. This 'assumed' maximum mass and C of M limit should be re-visited for fighter applications if the impact conditions are found to be significantly different. However, by inspection, it would seem that these figures may be fairly accurate for fighter aircraft equipped with latest-generation rocket powered ejection seats (relatively long burn) and in-service HMD masses between 1.8 and 2.3 kg (Eurofighter HMD without/with night vision cameras).

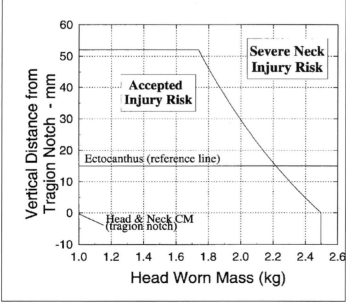

Graph of C of M versus head worn mass, showing the limit of vertical C of M (E L Downs)
1128609

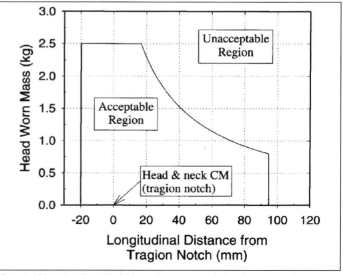

Graph of longitudinal C of M position versus helmet mass. Note the rear limit of –2 cm (from US research which recommended avoiding negative moments) and forward limit of 9.5 cm (arbitrary). Current NVG users might wish to validate the 'acceptable' region for themselves! (E L Downs)
1128610

A US Navy HGU-86 fitted with AN/AVS-6 NVG and a rear-mounted battery pack and counterweight. The approximately 800 g NVG extends the longitudinal C of M forwards, partially offset by the battery pack/counterweight combination (E L Downs)   1128618

Having dealt with total mass and vertical C of M limits, we must now consider longitudinal and lateral C of M. The longitudinal C of M location is believed to have greater effects on aviator fatigue and performance degradation than on crash-induced injury mechanisms. However, this assumption is borne out by current NVG-mounted helmets only inasmuch as the two-piece design of the US AN/AVS-6/9 incorporates a 'breakaway' facility which limits the effect of the forward C of M position (also minimised by the aforementioned battery pack/counterbalance) during crash/ejection. Research conducted in the 1990s, utilising the limiting mass of 2.5 kg, combined with various longitudinal C of M positions (relative to the head C of M), recommended a head supported weight moment of 82.8 ± 22.8 Ncm, measured about the occipital condyles. The work also recommended avoiding a negative moment and set an arbitrary forward limit of 9.5 cm.

In terms of the lateral C of M position, and noting that operational implementation of laterally 'unbalanced' HMDS is relatively scarce (the Apache IHADSS helmet, with monocular HMD attached, is one example), the limit of 1.9 cm off the mid-sagittal plane has not been challenged. In practice, no helicopter incidents/accidents involving neck injuries have had them attributed to lateral C of M. Similarly to the case of NVG, this may be attributed to the breakaway capability of the IHADSS HDU when exposed to contact forces and acceleration induced inertia loads.

HELMET CHARACTERISTICS
As alluded to, the subject of pieces falling off the helmet, known more formally as frangibleness, is a rather 'back door' way of enabling current in-service helmet/NVG combinations to exceed the previously mentioned limits of mass and C of M. The purpose of incorporating frangible (automatically detachable) devices into the helmet assembly is to remove the mass from the helmet, thus reducing the risk of neck injury. Early NVG introduced into US service were originally attached to the then SPH-4 helmet with 'hook and pile' fasteners and elastic tubing. This method did not allow the goggles

to easily or consistently detach during a crash. During ANVIS development, an NVG attachment mechanism was designed with a spring loaded 'ball and socket' engagement which allowed the NVG to separate from the mount when exposed to an 10 to 15 g loading (for more detail see the NVG analysis paper). This mechanism has performed well in US Army helicopter crashes. The IHADSS HDU, which is a monocular CRT display, is also easily attached and detached from the helmet mount. Mounted on the right lower edge of the helmet shell, the HDU also detaches from the helmet during a crash, with empirical evidence showing satisfactory results during actual crashes. Both, the ANVIS and IHADSS HDU detachment mechanisms operate when the device is exposed to crash loads with attendant dynamic inertia loads which exceed the mechanical retaining forces.

In the UK, NVG were first designed for the fixed-wing community – the (then) Ferranti Cats Eyes NVG was a one-piece design, which was fixed to the helmet via a rigid bracket. In order to satisfy ejection requirements, a pyrotechnically activated detachment mechanism (termed Autoseparation) was developed. This device was activated when it received an electrical signal at the beginning of an ejection sequence. This signal activated a squib which performed the mechanical release of the NVG. This Autoseparation system was continued with later Nightbird NVG in UK service. In contrast, the originally helicopter-developed AN/AVS-6 was pressed into service in US fixed-wing aircraft, where the 10 to 15 g detachment loading has served fighter crews well in protecting them in the ejection environment; of all fixed-wing ejections with NVG attached, only one crew (to date) has remembered to doff their NVGs prior to

IHADSS HDU   0111830

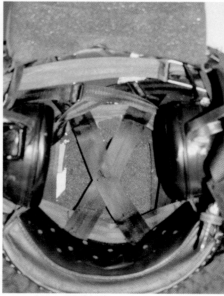

The editor's Mk4D helmet, with neck padding folded out, showing the foam shell lining and internal harness system (E L Downs)   1128614

'pulling the handle' – a small consolation to the UK Tornado crew involved!

In more general terms, pursuing frangibleness in HMD design has drawbacks – it increases complexity of the system and thus technical risk and, in turn, increases overall cost. Dictating when breakaway should occur is complicated (is a rapid onset of 9.5 g a manoeuvre or a crash?) and, for each frangible element, we introduce a portion of it which remains on the helmet, forcing mass up. The trade-off between the amount of weight being removed versus the amount being retained may become negligible without careful consideration. As was discovered during NVG Autoseparation testing, a subject dear to all aircrews' hearts is what happens to components after separation from the helmet – frangible components should not come in contact with 'vulnerable' parts of the wearer!

It should be noted that the foregoing treatment of mass and C of M should not be considered 'gospel', since the definitions are based on limited data from in-service systems and operational research and based heavily on the requirements of helicopter operations. Ongoing HMD research aims to further refine these parameters to encompass the harsher environment of fighter aircraft and expand the amount of empirical/proving data, including increasing the amount of human tolerance data to validate and refine the current mass and C of M limits. This should also extend to sampling a wider range of age/build, since much of the current human characteristics data is based around a relatively young and fit US Army helicopter pilot! As many airforces employ reservist and part-time pilots to supplement short-term shortages in manpower, the effect of age on the ability to withstand injury at the extremes of mass and C of M (noting the current characteristics of in-service HMD and helmet/NVG combinations) should be conducted as a matter of urgency.

It goes without saying that current standards of head impact protection should not be compromised by HMDS. The core attributes of an effective design, which provides a protective outer shell (against penetration and spreads the load over a greater area of the skull), combined with a liner which provides sufficient stopping distance between the shell's outer surface and the skull, are tried and tested. With many HMDS favouring a two-piece (inner and outer) helmet, this requirement can often be more complicated to achieve, with resultant pressure (sic) to perhaps 'relax' the limits to achieve operational effectiveness. While head impact protection limits vary (note UK versus US helmets), the limit for a particular application must be carefully considered. The operational requirements for helicopter aircrew are, by definition, different than those for fighter pilots. By example, the head impact velocity in survivable helicopter crashes has been estimated at 5.97 m/sec, via computer simulation and sled test. However, is this figure applicable to fighter aircraft?

Human head impact tolerance is an area of continuing research. Again referring to research conducted by the USAARL, a threshold of 150 to 175 g (headband region), depending on the impact location, was derived. Considering specifications for other types of helmet (motorcycle, equestrian, fighter aircrew, and so on) reveals a range of thresholds ranging from 200 up to 400 g. The USAARL recommended value is based on the concussion threshold as a result of linear accelerations, not on skull fracture, fatality, or rotational acceleration thresholds. The USAARL recommended value for the earcup and crown regions (150 g) is based on the risk of basilar skull fracture.

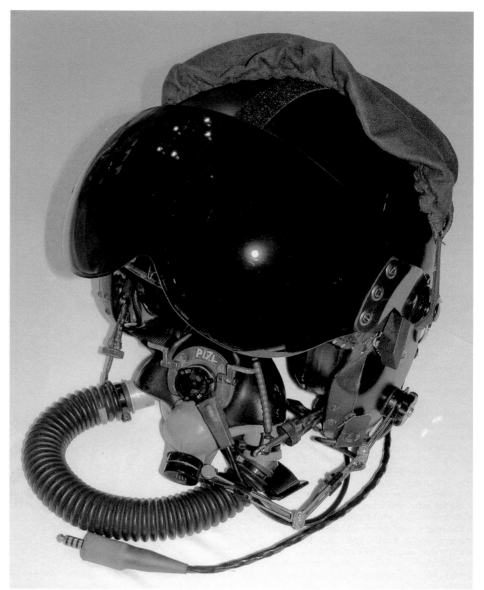

The editor's Mk4D helmet (E L Downs)   1128615

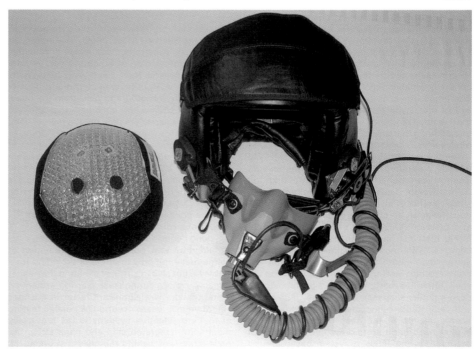

The editor's HGU-55/P with TPL liner (E L Downs)   1128613

For details of the latest updates to *Jane's Avionics* online and to discover the additional information available exclusively to online subscribers please visit
jav.janes.com

The helmet liner, as well as providing for a comfortable and stable fit, must provide a decelerative medium for the skull exposed to an applied impact. Helmet design for impact attenuation is based on the laws of physics. To bring a test head form to a stop requires a deceleration. If the deceleration magnitude is not to exceed 175 g, sufficient stopping distance must be provided in the form of the helmet liner. The required stopping distance is dependent on the acceleration's pulse shape which results from the impact. The three basic pulse shapes are square, triangular, and half sine, which each involve slightly differing approaches to derive the required minimum stopping distance. These derivations are not included for the sake of column space! Suffice to say that the impact mechanisms, as well as their maximum magnitude, should be considered in detail for each operational application of HMD; this includes careful consideration of sources of hazard particular to the cockpit (type specific). Existing data for in-service aircraft and requirements for AEA will provide a baseline for further investigations. The theoretical stopping distance can be used to help determine the required energy liner thickness used in helmet construction.

For those of us present during the first introduction of NVG into fighter aircraft, the realisation that our helmets were (often) woefully fitted and, thus, unstable, came with the first application of g in the aircraft and an immediate loss of NVG image as the entire helmet assembly rotated forwards! In the UK, the solution for most was a sometimes extremely uncomfortable tightening of the inner harness; US pilots fared rather better – the Thermo Plastic Liner (TPL) employed in US fighter helmets, together with some rather ingenious modifications to the basic design (the integrated chin/nape strap in the HGU-55 is a particular favourite of the editor!) made for a better (and therefore more stable) basic fit. Without the benefit of a counterbalancing battery pack/weight set, often tightening the oxygen mask (P-type in the UK and MBU-12 in the US at the time) salvaged some comfort from the vice-like grip of the helmet! However, and particularly under high g loadings, current helmet/NVG combinations still leave a lot to be desired in terms of fit and stability. So much so, if a straw poll of experienced night pilots was conducted, most would agree that they experience significant fatigue during long missions, with opportunities to doff goggles (during air-to-air refuelling for example) often welcomed, despite the almost inevitable (when it is vitally important!) palaver to don them again afterwards. HMD, featuring generally smaller FoV and more stringent eye box limits (for example the allowable limits of eye position with respect to the optical axis of the display – more of which later), place great demands on stability of the helmet, which often means a less comfortable fit.

The primary function of the fitting system is to provide a comfortable, stable fit to the wearer. Comfort can be achieved by distributing the helmet weight across the head, minimising 'hot spots' (points of increased pressure). Stability is dependent on an effective retention mechanism and accurate fit. Over the years, many different fitting methods have been devised for aircrew helmets, with most UK helmets employing a foam liner/internal harness combination or, more recently, variable thickness pads inside the shell. In the US, early SPH-3 and SPH-4 helmets used a similar suspension system to UK helmets, with adjustable nylon cross straps which ran across the top of the head and an adjustable leather head band around the circumference of the head. In the 1980s, a system based around a single-piece Thermo Plastic Liner (TPL) was introduced with the SPH-4B and the HGU-56/P helmets. The TPL system typically consists of 2 to 5 plies of thermoplastic sheets with 1/4 in dimples, enveloped in a cloth cover.

TPL was adopted to improve comfort and to alleviate fitting problems with SPH-4 as a result of longer sorties, the mounting of NVGs and the wider spread of anthropometric cranial

Early aircrew helmets featured a limited number of size options; the introduction of female aircrew brought a requirement to extend the anthropometric range of Aircrew Survival Equipment. With more stringent fit and stability requirements, HMD will force even greater flexibility of fitting and finer adjustment   0049147

dimensions brought with the introduction of female aircrew. Most pilots can obtain an adequate fit with minor adjustments with this system. Another type of fitting system, used in US Air Force and Navy fixed-wing helmets, is based on a custom-fit foam technique. A wax mould which is heated and placed on the head. After it has cooled and solidified, the resulting head cast is used to create a unique helmet liner, matched to the wearer's head shape. While an excellent solution to the problems of stability and comfort, the editor's experience of this system in his own HGU-55/P is mixed – without the instantaneous cushioning effect of the TPL, the helmet can be uncomfortable when carrying the extra weight of AN/AVS-6 NVG, particularly when employing a fighter mount (without the benefit of a counterbalancing battery pack).

Having considered the helmet lining, we should now turn our attention to retention of the helmet – not only in terms of providing for a stable fit, but also in terms of ensuring the helmet remains in place during crash/ejection – retention system failure is a significant factor in mishaps where helmet losses occurred. A notable feature of modern retention systems is the previously mentioned integrated chin and nape strap. The nape strap runs behind the head just under the occipital region. The chin strap runs under the chin, avoiding the areas around and about the trachea. A properly designed retention system will prevent the helmet shell from undergoing excessive forward or rearward rotation when the head is exposed to crash induced acceleration.

Head characteristics will also affect helmet stability. Variation of head shape and the amount of hair (considering female aircrew) under the

helmet may allow the helmet to shift, which will, in turn, shift the FoV of integrated vision devices, introduce hot spots and potentially compromise impact protection zones. For mission execution, helmet stability is critical when helmet-mounted displays or image intensification devices are used. Short-term helmet slippage can distract aircrew during extreme manoeuvring and, in the case of longer term slippage, necessitate in-flight adjustment of the helmet itself.

HELMET COMPATIBILITY

In our efforts to design HMD within these strict limitations of mass and balance, it is important to remember that the resulting system must be operationally viable – key to this is the ability of the system to integrate with existing associated equipment, including Nuclear Biological and Chemical (NBC) equipment, night vision devices, protective masks, ballistic, laser and nuclear flash blindness eye protection. Of these, perhaps compatibility with NBC Aircrew Respirators (AR) poses the greatest challenge. Current designs, such as the UK AR5, employ a rubberised 'hood', which reaches down to shoulder level and provides an air-tight seal, with oxygen mask (including pressure breathing), clear visor and communications built in. The pilot uses his existing helmet, minus oxygen mask, to complete the system. With its associated blower/NBC filter (either hand-carried or aircraft-powered), the system provides NBC protection, but it also brings with it a number of problems. Chief amongst these is the problem of visor mist up during conditions of high activity in the cockpit – loss of visibility to the pilot is critical at the best of times, but if it is also coupled with loss of vital flight/targeting information, then the entire concept of protecting the pilot is surely compromised. If an operationally effective HMDS is to be realised, then *all* aspects compatibility with associated equipment must be examined. Recent US experience with the Combat Edge Pressure Breathing for G (PBG) system, designed to improve a high-performance fighter pilot's tolerance to high G manoeuvres and prevent G – induced Loss Of Consciousness (G-LOC) that occurs when blood pressure within the brain drops below a critical level, serves as a good example. Combat Edge, standard equipment on all US Air Force F-15 and F-16s, functions by sensing when the load on the aircraft increases beyond the operational threshold level of 4 *g*, whereupon the system sends a pressure signal to a regulator which increases the pressure of the oxygen-enriched air that the pilot breathes up to a maximum of 1.2 psi above ambient. During PBG, pressurised breathing gas is simultaneously routed to the Oxygen Mask and a mask tensioning bladder in the rear of the Helmet and a Counterpressure Vest (CV) which inflates to apply pressure to the pilot's chest to balance the breathing pressure supplied to the lungs. Having studiously followed the story so far, the eagle-eyed reader will have spotted the inherent problem with this system for the purposes of HMD – the tensioning bladder will, by definition, shift the helmet on the pilot's head, even if fit and stability are good. When employed with NVG, Combat Edge was found to 'move' the NVG FoV during high-*g* turns and combat manoeuvres; unfortunately, asking a fighter pilot if he wouldn't mind restricting his manoeuvres to less than 4 *g* is not an option!

Linking the physical aspects of HMD to optical considerations is the existing helmet visor system, be it a single- or dual-visor (clear and tinted) arrangement. Visors are normally made from polycarbonate to provide impact protection to the pilot from aircraft parts in the event of crash, wind blast and airborne particulates in the event of ejection, or from projectile fragments in the event of attack. The addition of a tinted visor provides for protection against sun glare. MIL-V-43511C, the specification for aircrew helmet visors (although it is quoted for all types of face protection visors and goggles), specifies two types of visor, Class I or II. Class I visors are clear, having a photopic (daytime) luminous transmittance of ≥85 per cent through the critical area. Class II visors are neutrally tinted, having a photopic luminous transmittance between 12 to 18 per cent, again with respect to the critical area. An exception to the Class II luminous transmittance requirement is the IHU of the IHADSS in the AH-64 Apache. The IHADSS Class II visor has a photopic luminous transmittance between 8 to 12 per cent. This lower range of transmittance is needed to improve visibility of real-time imagery provided on the IHADSS HMD.

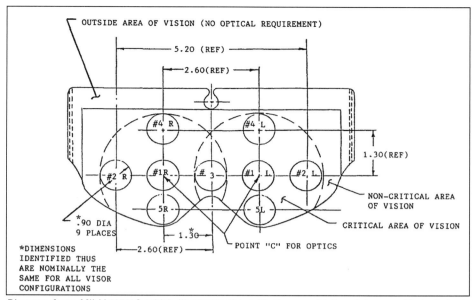

Diagram from MIL-V-43511C which illustrates the critical area of interest for specified photopic luminance transmittance – note this area coincides with the dedicated area for the display of symbology and/or imagery for most HMD (E L Downs)   1128616

This point is extremely pertinent to forthcoming discussion regarding the balance between the pilot's view of the outside scene and his display symbology/imagery. The MIL-V also specifies optical specifications for refractive power, prismatic deviation, distortion, haze, impact resistance, and so on. The test for compliance of impact resistance uses a 0.22 calibre T37 fragment-simulating projectile at an impact velocity between 550 and 560 feet per second.

CURRENT SOLUTIONS
In order to comply with the aforementioned regulatory requirements for impact protection and weight and balance, while delivering a stable foundation for a display system, and maintaining some measure of comfort to the pilot, current HMD almost universally employ a two-piece helmet design. The inner helmet, which may or may not fully enclose the skull, provides an individual fit to the wearer. Employing lasers to exactly model the head, a highly accurate fit can be achieved, which provides for the required level of stability, while affording a relatively high level of comfort.

Having produced a close-fitting and stable inner helmet, this can then be utilised to mount a (non-sized) external/display system, which clamps firmly to the inner. While the inner helmet is generally made lighter by making it non-enclosing, this point should be reviewed according to operational implementation. If operations may be conducted under NBC conditions, then the helmet must integrate with

NBC equipment to ensure that, if the outer/display portion of the helmet is to remain with a specific airframe, pilot protection is not compromised.

The outer helmet must house the display electronics and (possibly) power supplies, all of which mean weight, so in choosing specification of the display system, issues such as whether a monocular (single-eye display, as for IHADSS) or binocular (dual-eye display, such as Topowl) display is required is important, since this will determine directly the number of display sources which must be helmet mounted. A more detailed examination of helmet visual specification will be conducted later.

Bringing the weight and balance issue to a close, current systems such as the Typhoon helmet are within the aforementioned limits of mass and C of M, although it must be (re)emphasised that these limits must be applied with care when applied to the fast jet, rather than their derived helicopter application. Comfort will also be derived from the fact that aircrew routinely operate with NVG attached to their existing helmet, a combination exceeds the weight and C of M limits as previously laid down.

VISUALLY COUPLED SYSTEM (VCS)
One of the enhancements a true HMD, as opposed to an HMS, brings to the pilot is video imagery from a head-mounted or aircraft-mounted sensor, such as Image Intensifier (II) or FLIR imagery. German Tornado ECR pilots are familiar with the limitations to employing a 20° fixed FoV FLIR at low level at night! The ability

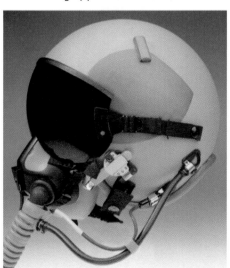

HGU-55/P with Combat Edge modification; note the tube from the pilot's supply tube to the mask tensioning bladder at the rear of the helmet
0089058

A typical internal helmet portion of a fast-jet HMDS. Note that the inner helmet does not enclose the pilot's head – this could have implications under NBC conditions, if the outer (display) portion of the helmet is designed to remain with the airframe (E L Downs)   1128623

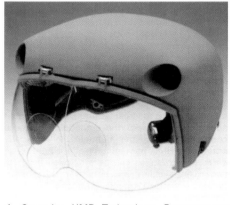

A Crusader HMD Technology Demonstrator (TD) outer (display) module. Note the dual-CRTs mounted inside the shell, projecting onto (possibly dichroic) patches on the visor to produce a fully overlapped FoV (note also the distinct seam down the middle of the visor). Also, embedded in the shell are two sensors, possibly Image Intensifiers, to give the pilot head-mounted night vision (E L Downs)   1128624

A Tornado F2 has been modified with a head-steered FLIR sensor as part of QinetiQ research into a Visually Coupled System (VCS) at Boscombe Down in the UK 0005542

to head-slave sensor information vastly expands the available Field of Regard (FoR), even if the instantaneous FoV remains limited. This is an important point and the subject of hot debate within the night vision community – exactly how much FoV is required? If the sensor is fixed, then FoV = FoR and the answer is simple – as much as possible. If the FoV is moveable and is limited only by physical head movement, then the answer may not be a large as first thought. Once peripheral aspects are considered (hovering is one area which is greatly aided by enhanced peripheral vision), a FoV is generally adequate for most pilotage applications.

While head-mounted sensors, such as NVG, are inherently slaved to the pilot's eyeline, aircraft sensors must be slaved to head motion – the sensor must be 'visually coupled' to the head. While it is acknowledged that eyeline is not simply defined by head motion, for most applications, as we shall see, the small inaccuracies involved in not directly tracking the eye (for example, by assuming the eye is looking directly forward) can be ignored. To accomplish visual coupling, a head tracking system must be incorporated into the HMD. As previously discussed, this definition of sightline provides the capability to aim weapons. Since all pilots use their eyes as their primary sense during flying, it is a natural coincidence that natural psycho-motor skills of the aviator integrate neatly with the operational implementation of HMD).

HELMET TRACKING TECHNIQUES
In a VCS, the pilot's Line-of-Sight (LoS) is continuously monitored, and any change is replicated in the LoS of aircraft mounted, HMD-slaved sensors. The Helmet Tracking System (HTS) detects these LoS changes and sends continuous updates via (ideally) a databus to the aircraft sensor control system. As previously mentioned, tracking systems may detect only head position (head trackers), may detect only eye position (eye trackers), or may be a combination (providing both eye and head position tracking). The level of accuracy delivered by eye trackers would enable the pilot to single out an individual switch on his control panel, until such accuracy is required, head trackers provide enough accuracy to direct pilotage/targeting sensors and weaponry.

Tracking systems with helmet-mounted components must minimise the additional weight, volume, and packaging impacts on the HMD. Tracking components which must be helmet-mounted can be modular (add-on), but integrated approaches allow for the embedding of these components into the helmet shell, thereby optimising HMD packaging. While a number of different techniques can be employed for HTS, two main systems are employed at present, namely, Electro Magnetic (EM) and EO. Early implementations of HTS in helicopters employed either AC or DC EM, incorporating 6 Degrees-of-Freedom (DoF) tracking and delivering high accuracy with low impact on both helmet and aircraft weight, size, and packaging. AC and DC HTS use a similar method. Each uses a transmitter attached to the aircraft and a receiver attached to the helmet. The transmitter fills the cockpit with a magnetic field. Through the measurement of the magnetic field strength at the receiver,

the position and orientation of the HMD can be determined. The major problem with EM HTS is their susceptibility to distortion by conducting metallic objects within the cockpit. This problem can be alleviated by mapping the magnetic field of the cockpit to produce compensation terms, although, as has been found, each cockpit map only applies to its associated airframe – different tail numbers of the same aircraft type can have significantly different EM maps. Other problems with EM trackers have included a limited motion box (volume through which the head can move and the tracker perform effectively), noise, jitter and poor dynamic response. In response to these problems, advances in AC EM HTS have produced more tolerant systems which also overcome many previous problems.

More recently, and particularly for HTS applications in fighter aircraft, the problems associated with EM HTS led to greater interest in EO HTS, which, by definition, are not sensitive to metallic components in the cockpit – useful within the tight confines of many fighter aircraft. There are several approaches to EO HTS: These range from the use of video cameras to infrared beams. The AH-64 Apache uses an EO HTS, which operates using two pairs of lead sulphide photodiodes mounted on the helmet, with two IR sources mounted behind the pilot's seat. The photodiodes continuously calculate their relative position to the sources and, accordingly, the position and orientation of the pilot's LoS. This positional information is sent to the helicopter's TADS/PNVS sensor package to align the sensor sightline. While offering fewer problems in integration than EM systems, EO HTS are not without issues of their own – similarly to metal objects interfering with EM systems, it will come as no surprise that 'stray' light and changes in ambient conditions can cause problems to some aircraft with large expanses of canopy transparency, such as the F-16.

Regardless of the technology, an HTS must provide defined measures of accuracy. System parameters include motion box size, pointing angle accuracy, pointing angle resolution, tracker update rate and jitter. The motion box size defines the linear dimensions of the space volume within which the HTS can accurately maintain a valid LoS. The box is referenced to the design eye position of the cockpit – note all those fighter pilots who sit too high in the seat in an effort to increase their view! It is desirable that the motion box provide angular coverage at least equal to that of normal head movement,

i.e. ±180° in azimuth, ±90 in elevation, and ±45° in roll. By example, the motion box size for the AH-64 IHADSS is 12 in forward, 1.5 in aft, ±5 in laterally, and ±2.5 in vertically from the design eye position.

Pointing angle accuracy, measured within the central portion of the motion box, covers, by convention, ±30° in azimuth and ±70° in elevation, which corresponds to the pointing areas most used by the pilot. Current HTS deliver between 2–4 milliradian (mr) accuracy, limited by the system's pointing resolution. Pointing resolution is simply a function of the number and distribution of sensors which will determine the smallest movement which will register an angular difference. US research indicates that an HTS should be able to resolve changes in head position of at least 1.5 mm along all axes over the full motion box. HTS also need to provide a specified dynamic accuracy, which means the ability of the tracker to measure velocity. Dynamic tracking accuracy (excluding static error) should be less than 30 mr/sec.

Defined the sightline accuracy requirements, however, is only half of the story for a fully-capable HMDS. The system also must be able to display sensor video at a sufficient refresh/update rate that the pilot will see, even under conditions of rapid head movement, a clear, smear-free image, accurately harmonised with the outside scene. Fulfilling this aim, while easily defined in qualitative terms, is both more difficult to define quantitatively and has been far more arduous a task to achieve than engineers first thought. The first limiting factor to consider in achieving the stated aim is the update rate of the HTS – after all, in order to produce an accurate and fast-reacting image, the display 'writing' subsystem must have accurate and timely information concerning pilot sightline, since this will provide the datum for the image display. The next factor is the display update rate; for a simple weapon-aiming HUD, a symbology update rate of 50 Hz (20 msec) was sufficient to accurately aim ballistic weapons. However, the task of writing a limited amount of stroke symbology is simple compared to a 40° circular FoV FLIR scene! If we assume a base display update of 50 to 60 Hz (20 to 16.67 msec) then it follows that the HTS sampling rate must be at least this figure if a coherent display is to be realised. HTS sampling rates of >100 Hz (10 msec) are possible, but excessive rate is nugatory effort, since this will be lost to the pilot, who only sees the display refresh rate. Noting that domestic TV and flat-screen displays operate with 24–16 msec

The Eurofighter Typhoon Integrated Helmet uses an EO HTS; the diodes on the outer helmet can be clearly seen in this image (Eurofighter) 1128625

refresh rates, and seem to offer smooth and crisp high-rate images (objects moving at high speed across a static screen vs. a static image viewed by a rapidly moving screen). Interestingly, however, some of the latest flat panel displays, offering refresh rates of 8 msec (125 Hz) have exhibited 'smearing' of the image during high-speed changes in scene, raising the possibility that there is an optimum display refresh rate.

Variations in head position output due to vibrations, voltage fluctuations, control system instability, and other sources are collectively termed jitter. Techniques to determine the amount of jitter vary from system to system.

To round off tracking techniques, we should address the value of eye trackers – as cockpits (assuming we will still have manned cockpits, of course) become more sophisticated, with greater levels of direct control via systems other than physical switches, the ability to visually select a function by simply looking at it (and perhaps using a Direct Voice Input command) may become a reality for the oft overloaded pilot. Current helmet-tracking systems do provide the level of angular resolution required in this case; also, when viewing or tracking objects in the real world, a combination of head and eye movements is always used, so some means of tracking the eyeline, as opposed to just the 'head line' is required. Normal head and eye coordinated motion begins with the eye executing a saccade towards the object of interest, with velocities and accelerations exceeding those of the associated head motion (a 'flick' of the eye). Consequently, the eye reaches the object well before the completion of the head motion. Eye movements are generally confined to approximately ±20° about the head LoS (which indicates the limit of an effective symbology layout if the pilot is to avoid eye strain). To track this eye/head movement, clearly, a more sophisticated VCS is required which also tracks the eye directly.

However, by increasing the performance of our head + eye tracker, we also introduce a greater resolution requirement for the display, otherwise all this new pointing accuracy would be obviously wasted! Several HMD designs have explored the concept of creating a small inset area of increased resolution, slaved to eye movement, within the overall FoV. The so-called 'high resolution area of interest' is not a new idea, it has been used in simulators for many years to overcome the computational problems of trying to provide high resolution, wide field-of-view imagery in real time. This technique relies on replicating the design of the human eye, with maximum visual acuity within a central high-resolution area (the fovea, a 2° diameter area). This argument will be familiar to NVG users seeking higher resolution, without sacrificing FoV – the 'optical constant' has denied a FoV in excess of 40° for current binocular designs. The last point to note in this short examination of eye trackers is that the aforementioned head rates, which obviously must be tracked accurately by the system, are far exceeded by the eye. Saccadic eye movements, which can be >800°/sec in velocity and >2,000°/sec² in terms of acceleration, present a challenge to designers to produce a high-resolution area of interest which meets all of the aforementioned qualitative requirements.

DISPLAY TECHNOLOGIES
ELECTRO OPTICS
An aircraft sensor, (TV or IR, typically), renders the outside scene with its FoV into an electrical signal for relay via databus to a suitable display source, composed of an image generator (which converts the electrical signal into a visual image), a projector (which outputs the image) and a screen (which displays the resultant image), such as a colour MultiFunction Display (MFD). The concept of an HMD is almost exactly the same – the sensor still outputs the scene within its FoV as an electrical signal, although in this case the image generator, projector and screen are integrated into the pilot's helmet. Since the scene capture and conversion to an electrical output all takes place within the

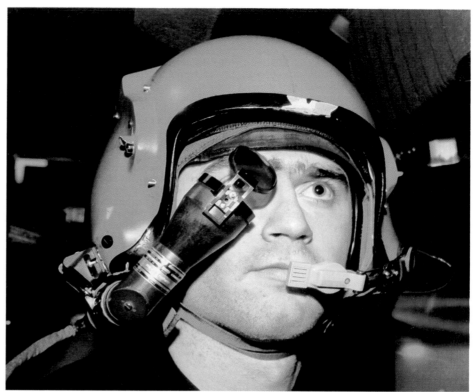

This image of NASA testing of the IHADSS shows the miniature CRT display source and combiner, which adjusts to place the image precisely in the pilot's eyeline (NASA) 1128617

aircraft, as far as an HMD is concerned, the story begins with the image signal arriving at the helmet. By convention, the combination of image generator and projector is termed the image source.

In early HMD designs, a Cathode Ray Tube (CRT), essentially as found in a conventional domestic television, was utilised as the image source. An example of this is the IHADSS, whose image source is a 1 in diameter CRT. While new image generation technologies are now available, CRTs have remained popular due to their relatively low cost, availability, image quality and reliability. However, the inherent drawbacks of CRT technology, which include weight and size (of primary importance to the HMD designer), high power requirement and heat generation, has pushed designers to seek new display technologies. Recently, Flat Panel (FP) technologies have begun to mature and exhibit comparable image quality to CRTs, while offering significantly lower weight and bulk (of primary importance), together with lower power consumption and, hence, heat output.

CATHODE RAY TUBES
A CRT generates an image via a scanning electron beam which strikes (and excites) a phosphor-coated display surface. The electron source, focusing coils, deflection plates and screen are all enclosed within a glass tube whose dimensions are dictated by the dimensions of the display and the necessary distance the electron beam must travel to focus on the entire display surface. Confirmed by the only gradual acceptance of flat panel televisions as a credible rival, CRTs produce a high-quality, full-colour, high-resolution image which is versatile in presenting all types of information. In order to minimise the impact of the weight and bulk of one or two CRTs in an HMD, miniature (<1 in diameter) CRTs were developed.

FLAT PANEL DISPLAYS
The most widely known employed non-emissive display is the Liquid Crystal Display (LCD). They produce images by modulating ambient light, which can be reflected light or transmitted light from a secondary, external source (a backlight). LCDs utilise two sheets of polarising material with a liquid crystal solution between them. An electric current passed through the liquid causes the crystals to align so that light cannot pass through them. Each crystal, therefore, is like a

shutter, either allowing light to pass through or blocking the light. Monochrome LCD images usually appear as blue or dark gray images on top of a grey/white background. Colour LCD displays use two basic techniques for producing colour. The first, termed Passive Matrix (PM), is the less expensive of the two technologies, while the other, Thin Film Transistor (TFT) or Active Matrix (AM), produces colour images that are approaching the same quality as CRT displays, although this technology is more expensive. In PMLCDs, pixels are defined by the intersection of a pair of vertical and horizontal electrodes. Voltages applied to any selected pair causes the LC material at the intersection to respond. AMLCDs employ an array of individual pixels, each controlled by an electronic switch. In TFT display, a TFT and a capacitor are used to switch each LC cell on and off. Monochrome LCDs usually use a backlight consisting of one or more fluorescent lamps, a reflector and a diffuser. In colour LCDs, pixels are composed of three or more colour subpixels. By activating combinations of these subpixels and controlling the transmission through each, a relatively large colour gamut can be achieved. However, the overall resolution capability of a colour LCD is defined by the total number of groups of subpixels – for a display with a maximum horizontal and vertical resolution of X × Y pixels, a monochrome solution will have a threefold better resolution than a colour display utilising groups of three subpixels. For this reason, until overall resolutions increase markedly, a monochromatic display will remain the primary choice for HMD applications.

Backlighting is an important issue with LCDs and even more important for HMD designs based on LCDs. Due to the requirements for good daylight readability (where the background luminance may be as high as 10,000 fL) together with adequate dimming to enable the same display to be useful at night, backlights must feature efficiency (>40 lumens/watt), high luminance (>20,000 fL, a high number due to the poor through-put of LCDs), high luminance uniformity (<20 per cent variation in readability) and a wide dimming range (>5,000:1). These figures should be regarded as a starting point; as related technologies mature, performance of LCDs for fighter HMD applications will improve. Noting the above, it is clear that an integrated backlight has some work to do! Backlight sources include Cold Cathode Fluorescent Tubes (CCFTs) and, more recently, Light Emitting Diodes (LEDs) and Field Emission Displays (FEDs).

LEDs are emissive displays composed of multiple LEDs arranged in various configurations which can range from a single status indicator lamp to large arrays. The individual LEDs operate on the principle of semiconductor physics where electrical energy is converted into light energy by the mechanism of electro-luminescence at the diode junction. Light energy is produced when this junction is forward biased by an applied voltage. LED output is a relatively narrow spectral band (useful for NVIS compatible cockpit applications) and often monochromatic and identified by a dominant wavelength. The colour of the LED is a function of the semiconductor material. LEDs are typically monochromatic, but the use of subminiature LEDs in RGB configurations can provide full colour.

Also emissive, FEDs comprise a matrix of miniature electron sources which emit electrons from the surface of a metallic conductor into a vacuum under the influence of a strong electric field. Light is produced when the electrons strike a phosphor screen. FEDs are driven by addressing a matrix of row and column electrodes. Full grey scale monochrome and full colour displays have been developed.

The stringent requirement for brightness has been a difficult hurdle to overcome, but continuous development in performance has produced significant increases towards the goal of full daylight readability. Arguably, one technology which offers the greatest benefits for HMD applications is the Organic LED (OLED). OLEDs employ a thin-film LED in which the emissive layer is an organic compound. These devices promise to be much less costly to fabricate than traditional LEDs. When the emissive electroluminescent layer is polymeric, OLEDs can be deposited as an array using simple printing methods to create a graphical colour display. OLEDs are available as distributed sources while the inorganic LEDs are point sources of light. One of the benefits of an OLED over the traditional LCDs is that OLEDs do not require a backlight to function, with attendant power saving.

Electro Luminescent (EL) displays feature a layer of phosphor material sandwiched between two layers of a transparent dielectric (insulator) material, activated by an electric field. Pixels are formed by patterning the phosphor into dots. An Active Matrix EL (AMEL), which uses active matrix addressing, can provide reasonably high luminance, contrast and speed. All EL displays are emissive by nature and can be configured as monochrome or full colour. Colour is achieved either by filtering or by patterned phosphors similar to those used in conventional CRTs.

The lower weight, lesser bulk and far lower power requirement of flat panel displays made them a logical alternative to CRTs in HMD applications. Flat panel technologies are classed as emissive or non-emissive. Emissive displays produce their own light, while non-emissive displays operate by the transmission and/or reflection of an external light source.

DISPLAY PERFORMANCE
CRT versus FPD

Traditionally, a number of characteristics define the overall 'quality' of a displayed image. Due to the differing electronic solutions offered by CRTs and FPDs, some of these characteristics will differ, although the core subjective parameters are similar for both:

Luminance
Luminance uniformity
Resolution
Contrast

These are probably the most important image characteristics, but there are others, including the number of grey shades, colour, and Modulation Transfer Function (MTF). Having defined the image characteristics, the performance of a CRT display which delivers the image can be described by:

Display size
Aspect ratio
Phosphor efficiency/persistence
Beam spot size
Number of scan lines/spacing

Interlace ratio
Linearity

Again, there are characteristics, but these are the most important. To complete this audit trail of CRT display performance, we must characterise the background electronics, which should deliver the necessary:

Bandwidth
Dynamic range
Signal to Noise Ratio (SNR)
Frame rate

While all parameters contribute to overall performance, and all should be balanced to produce the desired result, some of them are weighted more than others in their contribution. These include phosphor efficiency and persistence, and electron beam spot size. Adequate luminance and contrast (ratio of luminances in bright and dim areas of the display) require efficient phosphors. Good resolution depends on a small spot size. High speed/dynamic imagery requires short phosphor persistence. Adequate luminance and contrast ratios are a function of anode voltage, beam current, and phosphor luminous efficiency. For HMD applications, anode increasing anode voltage to achieve increased luminance is of interest depending on whether power supplies are mounted on the helmet or off-board; in both cases, a high power source can produce a number of problems in terms of integration and, more importantly, safety. obviously, a high-efficiency phosphor will offset the need for higher driving voltage to achieve a desired luminance output (which will be high for daylight displays). CRT phosphor efficiencies, defined as the ratio of the luminous energy output to the electron beam energy input, range from 1 to 20 per cent. Overall luminance is also affected by beam writing speed, with slower speeds generating higher luminances through greater persistence. Miniature CRTs employed in CRTs have demonstrated luminances of greater than 6,000 fL in stroke mode and greater than 3,000 fL in raster mode.

Contrast is an important Figure Of Merit (FOM) for CRT displays – it describes the ability of the pilot to detect luminance difference between two adjacent areas. In CRTs, the range of contrast available is often expressed using the shades of grey. Shades of grey are luminance steps which differ, by convention, by root two (1.414). The minimum number of shades of grey displayed by a CRT is defined by operational implementation; clearly, if the display has a targeting function, then, depending on maximum sensor range. The required number of shades can be deduced using Johnson's Criteria, which describes the minimum number of cycles (discernable contrast lines) across a target to achieve detection, recognition and identification. However, if the display is to be utilised for pilotage, the shades of grey requirement will be far less (6 to 8).

The selection of a CRT phosphor is based on luminous efficiency, spectral distribution and persistence. Fielded NVGs employ P20/P22 (older goggles) or P43 (later goggles including US AN/AVS-9). The Apache IHADSS also uses P43. As alluded to, the selection of a suitable phosphor depends on application, but also on the response of the eye. The eye's daytime (>1 fL) response peaks at approximately 555 nanometres, while the nighttime response peaks at approximately 507 nanometres (green region of the visible spectrum). While this fact explains why NVG, HUD and HMD feature 'green' symbology/imagery, there is no reason, depending on application, why other phosphors may not be chosen. The persistence of a phosphor, normally defined as the time required for a phosphor's luminance output to fall to 10 per cent of maximum, is the major factor in the dynamic or temporal response of a CRT. As previously mentioned, for an HMD utilised for sensor imagery for pilotage applications, a short persistence/high luminance display is needed to 'keep up' with the dynamic outside scene. The loss of temporal response results in degraded modulation contrast at all spatial frequencies. When modulation contrast degrades below a certain threshold, targets

begin to blend with the background, and the pilot will lose the ability to discriminate targets from the background. For example, problems experienced with the original P1 phosphor (24 msec persistence), employed in early IHADSS for its MTF performance (see later), were solved by replacing it with P43 phosphor (1.3 msec persistence), with only minor degradation in MTF.

Considering the performance of current sensor systems, with 1,024 × 1,024 pixel arrays currently in service, MFD, HUD and HMD CRT sources are most often the limiting factor terms of perceived resolution. CRTs are described in terms of both vertical and horizontal resolution. Horizontal resolution is a function of bandwidth and spot size, whereas vertical resolution is a function of beam diameter. CRT resolution, in common with domestic television sets, is expressed only as the number of raster lines for the display, but for completeness, line width, spot diameter and, importantly, MTF should be included. The essential meaning of MTF is rather simple. Consider a pattern consisting of a sine wave, varying between black and white. At frequencies where the MTF of an imaging system viewing this pattern is 100 per cent, the pattern is unattenuated, for example, it retains full contrast. At the frequency where MTF is 50 per cent, the contrast half its original value, and so on. There are several methods which historically have been used to obtain MTF curves. These include subjective techniques of shrinking raster, line width, and TV limiting resolution and the objective techniques of discrete frequency, half power width, and Fast Fourier Transform (FFT).

Similarly to CRTs, a number of characteristics define the overall 'quality' of an image displayed by a Flat Panel Display (FPD); refining the previous list describing image quality, we find:

Luminance
Luminance uniformity
Resolution
Contrast
Viewing angle
Reflectance

The two added parameters of viewing angle and reflectance are peculiar to the physical attributes of FPDs. To achieve the above, FPD characteristics include:

Total pixels (H × V)
Pixel size
Pixel pitch
Refresh rate (pixel excitation)
Spectral distribution
Chromaticity

This list reflects the difference in the construction of the image by an FPD, although the sets of parameters for each type of display are broadly parallel in their application. Having pointed out some of the subtleties in characteristics between the two main display mechanisms, we will now examine in more detail, some of the ways we can assess the performance of displays for HMD applications:

PERFORMANCE PARAMETERS

In general, the above Measures Of Performance (MOPs) are exactly as would be applied to any visual display system; they can also be utilised for HMD image assessment since the final image is that of the source image modified by the transfer function of the relay optics. However, there are more MOPs which relate to the overall HMDS:

CONTRAST

As previously described, the term contrast refers to the difference in luminance between two adjacent areas of a display. However, the term contrast is often joined by other terms which aim to further indicate the performance of a display. Contrast, contrast ratio and modulation contrast are three of the more common expressions used. Occasionally, display specifications denote luminances according to their relative values and, therefore, labelled as the maximum luminance (Lmax) and minimum luminance (Lmin). However, if the area at one luminance value is much smaller than the area at the second luminance, the luminance of the smaller area sometimes is referred to as the target luminance

(Lt), and the luminance of the larger area is referred to as the background luminance (Lb). Common mathematical expressions for luminance contrast include:

1. C (Contrast) = (Lt – Lb) / Lb for Lt > Lb
2. C = (Lb – Lt) / Lb for Lt < Lb
3. C = (Lmax – Lmin) / Lmin = (Lmax / Lmin) – 1
4. Cr (Contrast Ratio) = Lt / Lb for Lt > Lb
5. Cr = Lb / Lt for Lt < Lb
6. Cr = Lmax / Lmin
7. cm (Contrast Modulation) = (Lmax – Lmin)/ (Lmax + Lmin)
8. CM = (Lt – Lb) / (Lt + Lb)

Mathematical convention orders the equations to ensure that negative contrast value is avoided.

Applying a target (point of interest) luminance of 120 fL, a background (majority of the display area) luminance of 30 fL and inserting these values into (1), (4) and (8) yields:

C = (120 – 30)/30 = 3
Cr = 120/30 = 4
Cm = (120 – 30)/(120 + 30) = 90/150 = 0.6

Noting the previous discussion regarding increasing power to increase contrast and luminance, let us now apply this to the equations by setting Lt to 2,000 fL and Lb to 50:

C = (2,000 – 50)/50 = 39
Cr = 2,000/50 = 40
Cm = (2,000 – 50)/(2,000 + 50) = 0.95

From the above, it can be seen that, as the difference between Lt and Lb grows, then so does the contrast, with C approximating to Cr at large differences. With regard to the modulation value, we can see that the value approaches 100 per cent as the luminance difference approaches infinity. For perfect displays, with no aberrations, these results are what we would expect to see empirically.

As we can deduce from the above, the available contrast depends on the range of luminances a display can produce. The range from minimum to maximum luminance values is termed the dynamic range. For CRT displays the luminance range can be described by plotting the luminance of an area of the display as a function of the voltage on the anode of the CRT. This plot produces a so-called Gamma Curve, which demonstrates the continuous (linear), or analogue, nature of this type of display. Analogue displays are also defined by the number of shades of grey they can show. As previously described, shades of grey are luminance steps which differ by root two (1.414). For example, if the lowest (minimum) luminance value within a scene is 10 fL, then the next square-root-of-two grey shade would be 10 multiplied by 1.414 or 14.14 fL. The next grey shade, if present, would be 14.14 multiplied by 1.414 or 20.0 fL, and so on. For a display with Lmax = 20 and Lmin = 10, the dynamic range could be described as three shades of grey.

For an analogue system, which includes CRTs, the relationship between the number of shades of grey and the contrast ratio is given by:

9. Shades of Grey = [log(Cr)/ log(root 2)] + 1

The added '1' at the end of the equation provides for the first grey shade, which indicates the condition of uniform luminance across the entire display. Under this condition, Lt = Lb or Lmax = Lmin, so the contrast ratio is 1. Hence:

Shades of grey = log(1)/0.1505 + 1
= 0/0.1505 + 1 = 0 + 1 = 1

which shows that a scene of uniform luminance includes only one grey shade. Employing (9) above, we can derive a number of pairings for contrast ratio and shades of grey:

SHADES OF GREY VS. CONTRAST RATIO

Shades of Grey	1	2	3	4	8	16
Contrast Ratio	1.00	1.41	2.00	2.83	11.3	181

Note that the factor of root 2 in equation (9) does not indicate the smallest discernable threshold of the human visual system – the eye can detect luminance differences of far less than this figure. This unit has been used historically for CRTs, with confusion arising as (digital) FPDs are becoming the preferred choice for military and civilian display application. As a generalisation, luminance values for FPDs are not continuously variable as for CRTs, they can take on only certain discrete values. It has therefore become increasingly commonplace to count the total number of possible luminance steps and use this number as a MOP. However, this parameter should not be termed 'shades of grey'. For example, for an LCD capable of showing 16 luminance increments – let us call them 'grey scales' – if this is interpreted as 16 shades of grey, then the contrast ratio, from the table above, would be 181, implying a very large luminance range. However, this statement assumes a difference of root 2 between each grey scale – without examining the luminance value of the lowest and highest levels, this assumption cannot be made, since our original specification only mentioned 16 luminance increments, not their spacing, or whether each of the 16 levels can be discriminated by the eye. To avoid confusion, perhaps shades of grey should only be utilised to define only analogue displays, or perhaps we should discard this system altogether, not only as a result of digital FPDs entering the fray, but also the continuous development of TV/IR targeting systems, such as Northrop Grumman's Litening AT, demanding displays which are capable of showing as many *discernable* shades of grey as possible. For example, Honeywell's colour cockpit display for the AH-64 Apache (see entry) features a 512 × 512 pixel display, maximum luminance of >250 fL, contrast ratio of >5 (high light) and >100 (low light) and 64 levels. Utilising Cr = 5 for the high light condition, applying equation 9:

Shades of grey = log(5)/0.1505 + 1
= 0.6989/0.1505 + 1 = 5.6439

So, the display is capable of showing 5.6439 discrete shades of grey, each separated by root 2. However, as alluded to, if the display only showed just over 5 actual shades, then it would be of only limited use to identify targets (Johnson's criteria specifies 8.0 cycles for identification of a one-dimensional target). The display actually shows 64 levels – whether these can all be discerned is not known, but let us assume they can and that for Lt = 250 fL and Cr = 5, then Lb = 50 fL; so 64 levels between 50 and 250 fL gives an increment of 3.125 fL/level for high light (daylight) conditions.

As we have seen, while shades of grey can be an unreliable MOP for visual displays, in all cases contrast ratio, derived form from the maximum and minimum luminance, is applicable to both CRTs and FPDs. Other contrast MOPs are also used – in many cases adapted to conform to the unique characteristics of FPDs. If the image is formed by the collective turning on or off of an array of pixels, the concept of contrast ratio is redefined to indicate the difference in luminance between a pixel that is fully 'on' and one fully 'off'. The resulting refined equation for 'pixel' contrast ratio is:

Cr = (Luminance of ON pixel)/(Luminance of OFF pixel)

While this MOP is often used, it reflects a contrast condition in the absence of ambient lighting – whether the viewer could discern this contrast in full ambient light depends on the luminance difference. Additional factors may also need to be taken into consideration. An example of this is the dependence of luminance on viewing angle – note most manufacturers' specifications which denote luminance specification within viewing angle bounds.

While we have concentrated so far on the ability to discriminate between two points of interest by differing luminance levels, for images where the background and target have the same luminance levels, discrimination can still be achieved discerned by colour difference (for a colour display!). Colour contrast, in isolation, is less distinct in terms of visual acuity than luminance contrast. Colour can be considered to be composed of three elements – hue, saturation, and brightness. Hue refers to what is normally meant by colour, a subjective measure. Saturation refers to colour purity and is related to the amount of neutral white light that is mixed with the colour. Brightness refers to the perceived intensity of the light. The appearance of colour can be affected greatly by the colour of adjacent areas, especially if one area is surrounded by another. The use of colour in displays increases the information capacity of displays and the natural appearance of the images, enabling faster and more efficient assimilation of information by the pilot.

Colour CRTs use three electron beams to individually excite red, blue, and green phosphors on the face of the display. By using the three primary colours and the continuous control of the intensity of each beam, a CRT display can provide 'full colour' imagery. Likewise, FPDs can be monochrome or colour. Full colour capability is available in all of the aforementioned FP technologies, including LC, EL, LED and FED. Since colour is defined by three distinct elements, colour contrast must address differences in all of these factors, as well as luminance. Other factors, such as the size and shape of the point/area of interest and whether the perceived colour is intrinsic or reflected, also affect the perceived colour differences. As a result, formally defining colour contrast is extremely difficult. In attempting to define colour contrast, we must consider a technique to define colour in other than subjective terms.

CHROMATICITY

'Perception' was, for many years, the main (and highly subjective) tool utilised in defining a colour; for example, the term 'bright red', while often used, will undeniably attractive many different definitions, depending on the observer! However, such loose definitions are unacceptable for aviation authorities, whose interest in maintaining the colour balance of lighting systems for NVG users has been a major driving force. The CIE colour standard is based on imaginary primary colours XYZ, which do not exist physically. They are purely theoretical and independent of device-dependent colour gamuts such as RGB or CMYK. These virtual primary colours have, however, been selected so that all colours which can be perceived by the human eye lie within their colour space. The XYZ system is based on the response curves of the eye's three colour receptors. Since these differ slightly from person to person, CIE defined a 'standard observer', whose spectral response corresponds more or less to the average response of the population. This 'objectifies' the colorimetric determination of colours. The three primary colours of the CIE XYZ reference system call for a spatial model with co-ordinates (X), (Y) and (Z), which is drawn as a chromaticity triangle. To arrive at a two-dimensional diagram (the sail shape), this chromaticity triangle is projected into the red-green plane.

This is only meaningful, however, if appropriate standardisation is performed at the same time which allows the lost value (Z) to be read from the new two-dimensional model. This is achieved by introducing the chromaticity co-ordinates x, y and z, defined as:

x = X / (X + Y + Z)
y = Y / (X + Y + Z)
z = Z / (X + Y + Z)

where x + y + z = 1. The value z of any desired colour can thus be obtained by subtracting the chromaticity co-ordinates x and y from 1:

1 – x – y = z

However, a colour is not defined fully by its chromaticity (x and y). A brightness coefficient also needs to be specified. The eye's response curve for green is standardised in the XYZ system so that it simultaneously reflects the sensation of brightness. A colour is only described in full if it contains the values x and y plus the brightness coefficient Y.

The introduction of the CIE colour system made it possible to transform colour determination from a quality-describing process (as previously mentioned – bright red) into a process, which can be expressed in exact quantitative and numerical terms. In addition to the quantitative judgement it allows, the CIE colour space permits the results of additive colour mixing to be presented in simple form. However, the CIE chromaticity diagram does have some limitations: Brightness is difficult to include and there is a discrepancy between perceived colour differences and the actual spacing of colour in the system.

The preceding discussion has been general in nature, equally applicable to MFDs, HUDs and HMDs. However, see-through displays, such as HUDs and HMDs, by their very nature, introduce additional considerations due to the effect of the outside scene on the displayed image. This effect is very significant during daytime flight when ambient illumination is highest. Ambient scene luminances vary greatly over a 24-hour period. They can range from 0.001 fL under moonless, clear starlight conditions to over 10,000 fL for bright daylight. Luminance levels which qualify as 'daylight' begin at around 300 fL. Luminance values provided by a CRT and its associated optics can be selectively ranged from 100 fL, to provide for nighttime viewing, to over 1,000 fL for daytime viewing. During nighttime, when scene contrast levels are lower and the scene is generally dark, achieving workable display contrast and luminance is not usually problematic. However, the employment of any see-through display during the daytime is a tougher prospect; ambient luminance reaching the eye becomes part of the background luminance and the sum of the CRT and background luminances reaching the eye becomes the target luminance.

Several factors are important when considering HMD luminance and contrast:

For a given ambient luminance, increasing the CRT display luminance increases contrast

For a given CRT display luminance, increasing ambient luminance decreases contrast

For a given set of CRT display and ambient background luminances, the use of a shaded visor over a clear visor increases contrast (rather like introducing nighttime contrast).

Having considered methods for measuring contrast, minimum values should be established. The level of contrast required to perform a task with a display depends on numerous factors. These factors include the type of visual task (target detection, localisation or identification), the viewing environment (ambient light level, interference sources, size and distance of the display from the eye), the nature of the displayed information (symbology and/or imagery) and other display characteristics (screen resolution, sharpness, jitter, colour, pixel geometry and so on). Drawing on research into symbology requirements for electronic displays, for text to be legible on a directly viewed display, it is recommended that the modulation contrast for small characters (between 10 and 20 arc minutes in size) displayed on a monochrome CRT is:

10. $Cm = 0.3 + [\,0.07 \times (20 - S)\,]$

Where S is the vertical size of the character set, in minutes of arc. Consider display characters 17 arcminutes in size. Equation (10) specifies a modulation contrast of 0.5 (contrast ratio of 3 to 1). However, in practice, a modulation value of 0.75 (contrast ratio of 7 to 1) is recommended. So, if the background luminance is 3.3 fL, than the character luminance should be at least 10.0 fL. For displayed video, a minimum of 6 shades of grey is recommended. Note that the preceding recommendations

Computer Standard	Resolution (H × V)	Screen Aspect	Total Pixels
QVGA	320 × 240	4:3	77K
VGA	640 × 480	4:3	307K
SVGA	800 × 600	4:3	480K
XGA	1,024 × 768	4:3	786K
WXGA	1,280 × 768	15:9	983K
SXGA	1,280 × 1,024	5:4	1.3M
SXGA +	1,400 × 1,050	4:3	1.5M
WSXGA	1,600 × 1,024	25:16	1.6M
UXGA	1,600 × 1,200	4:3	1.9M
WUXGA	1,920 × 1,200	16:10	2.3M
QXGA	2,048 × 1,536	4:3	3.1M
WQXGA	2,560 × 1,600	16:10	4.1M
QSXGA	2,560 × 2,048	5:4	5.2M
WQSXGA	3,200 × 2,048	25:16	6.6M
QUXGA	3,200 × 2,400	4:3	7.7M

apply to direct view monochromatic displays. As must be apparent, the definition of minimum contrast values for colour displays are more difficult. In applications where direct view displays are supplemented or replaced by HMD, the task of defining minimum contrast values is further complicated by optical and EO design considerations.

During the development of the HIDSS for the ill-fated RAH-66 Comanche helicopter programme, the HMD specification for contrast was given as >4.66. This contrast value of 4.66 is equivalent to a Cr value of 5.66 which corresponds to 6 shades of grey. For day symbology, the contrast ratio was required to equal or exceed a value of 1.5:1 for a 3,000 fL background and equal to or exceed 7:1 for a background of 100 fL; both values were based on the use of a tinted visor. For nighttime viewing of sensor imagery, a minimum contrast ratio value of 11.2 which corresponds to 8 shades of grey, was required.

RESOLUTION

Resolution, probably the most important parameter in determining the effective image quality of any display, refers to the amount of information (detail) which can be presented to the viewer and, hence, the fidelity of the image. For military applications, the resolution determines the smallest (and/or farthest) target which can be displayed. In HMDs using CRTs as the image source, the CRT's resolution is the limiting resolution of the system. While CRT resolution is usually given as the number of horizontal lines over the display, it is immediately apparent that a 525 line picture on a large format display offers no greater detail than a smaller display with the same number of lines. A more useful indicator of resolution is the raster line width – the smaller the line width, the better the resolution. Currently, 20 µm is a good working figure for miniature CRTs. However, in discrete displays such as FPDs, resolution is given by the number of horizontal by vertical pixels. These numbers depend on the size of the display, pixel size, spacing between pixels, and pixel shape. Typical resolution values for displays are given below; note the propensity of a 4:3 aspect ratio – these have been defined for domestic TV sets. There are more (higher resolution) standards, but for the purposes of HMD, the included resolutions encompass all current and aspirational designs.

Relating the above to what the eye can discern, pixel-related expression of display resolution can be presented in a number of ways:

Note the visual limits above – while not potentially a HMD issue, as displays become more capable in terms of resolution, cockpit designers should take care not to overspecify displays, depending on the distance of the viewer from the screen.

When addressing colour displays, a single colour pixel may consist of several subpixels (RGB). Depending on the subpixel arrangement, the colour pixel count can be different for the horizontal and vertical directions – this may

detract from the maximum possible system performance if a colour display sacrifices resolution to show colour. Accordingly, many current displays are monochromatic in order to maximise all available pixels to yield maximum resolution from associated sensors, which often output higher resolution imagery than can be displayed to the operator. For example, the output of current FLIR sensors is of the order of 1,355 (H) × 960 (V) pixels – far greater than the ability of a 525/625 line CRT or a commonplace VGA FPD to handle. Development of higher resolution displays for military HMD applications continues, with the current fielded limit of 1,280 × 1,024 (SXGA) set to be bettered as manufacturing techniques are perfected.

As displays become more capable, the eye will finally become the limiting factor in the system. From the above table, and utilising a widely accepted measure in the world of NVG users, the eye can detect approximately 1.72 cy/mr (which equates to the commonly accepted 20/20 or 6/6 vision). Ideally, an HMDS should match or exceed this value, although we are some way off this goal to date – the capability of the human eye in terms of FoV, FoR, resolution and colour sensitivity is still the benchmark visual system! To put this figure into perspective, current AN/AVS-9 NVG equipped with the very latest Image Intensifier Tubes (IIT) provide on-axis resolution of 1.3 cy/mr while IHADSS provides 0.57 cy/mr. Research has shown that (depending on application), values between 0.39 and 0.77 cy/mr being acceptable for HMD. Note that these resolutions are maximum values and generally obtained under high light level conditions and utilising high contrast targets for measurements.

However, expressing resolution simply in terms of scan lines or addressable pixels does not convey enough information as to the capability of a display. Another measure, used in conjunction with resolution, is to quantify how modulation is transferred through a display as a function of spatial frequency. A plot of such a transfer is called a MTF curve. The MTF is most often associated with CRT displays and will be familiar to NVG users when attempting to quantify overall system performance. Since any scene theoretically can be resolved into a set of spatial frequencies, it is possible to use a system's MTF to determine image quality. While there a number of ways of obtaining the MTF of a system, the most accepted seems to be the employment of discrete frequency measurement – a time-consuming technique, but arguably the most reliable and repeatable. To apply MTF to describe a system, the response of the system must be uniform through the FoV (homogenous) and in all directions (isotropic), and the response must be independent of input signals. CRT imagery has continuous horizontal sampling (scan) but discrete vertical sampling (lines). While this would imply the requirement for two MTFs, one vertical and one horizontal, to describe a CRT, the horizontal MTF is the one normally specified. A CRT display MTF is defined

DISPLAY RESOLUTION	UNITS	VISUAL LIMIT
Pixels/FoV	Pixels/Degree	60 Pixels/Degree
Pixels/(2 × FoV)	Cycles/Degree	30 Cycles/Degree
(8.74 × Pixels)/FoV	Milliradians/Degree	1.72 Milliradians/Degree
(60 × FoV)/Pixels	Arcminutes/Pixel	1 Arcminute

by scan rate, spot size, phosphor persistence, bandwidth, and luminance.

During the 1990s, there was much heated discussion as to whether MTF is a meaningful measure for FPDs, with MTF for discrete displays expressed as a function of pixel pitch, fill factor and active pixel size.

An important consideration for HMD designers when considering MTF is that traditional MTF curves are static, such as the modulation in the scene is not varying. However, in the aviation environment, relative motion is always present. In addition to the relative target-aircraft motion, when a VCS is employed, sensor gimbal jitter and head motion are present. When motion is present, the temporal characteristics of the scene modulation interact with those of the imaging system (for example, scan rate and phosphor persistence for CRTs) and the transfer of modulation from the scene to the final display image can be degraded. Phosphor persistence is an important display parameter affecting temporal response in CRT displays. Excessive persistence reduces modulation contrast and causes a reduction of grey scale in a dynamic environment where there is relative motion between the target and the imaging system. This may not be important at low spatial frequencies, where there may be multiple grey steps. However, at higher frequencies, where there is perhaps only enough modulation contrast to provide one or two grey steps under static conditions, the loss of even one grey step would represent significant degradation.

As an example, and considering the previously mentioned IHADSS issue with phosphor persistence, the original P1 phosphor was selected to satisfy the high luminance daytime symbology requirement. Initial flight testing revealed an image smearing problem, with pilots reporting tree branches seeming to 'disappear' as they moved their heads in search of obstacles and targets. This was found to be due to the long persistence phosphor coupled with rapid head movements. The CRT phosphor was changed to a shorter persistence (1.2 msec) P43 phosphor. This point illustrates the need to measure a system's dynamic as well as static response. Modulation transfer for a static image can be quite different from that achieved for a dynamic image (with relative velocity).

The degradation in image contrast due to temporal factors is not limited to CRT displays; AMLCDs are frequently used to present moving imagery. Noting previous comments concerning smearing of LCD imagery, the liquid crystal molecules require a finite time to re-orient themselves when the pixel is changing. In a similar fashion to CRT phosphor response, the slow transition between luminance values will degrade modulation transfer in dynamic images on AMLCDs.

The dynamic response of a display and its interaction with other imaging system components is a critical area in the design of HMDs and it is therefore important to be able to measure the dynamic MTF of such systems.

FIELD OF VIEW

For operational effectiveness of an HMDS, display FoV is as important as contrast and resolution. Similarly to NVG, the HMD FoV provides information and (with injected symbology) flight data, with the amount of information (including the outside scene) limited only by the extent of the FoV (note previous comments regarding the presentation of symbology to the eye to prevent eye strain). In principle, the larger the FoV, the more information available, limited only by the maximum field of the human eye (150° × 120° H × V). Considering both eyes together, the binocular FoV measures approximately 200° × 120° (H × V). Current HMDS incorporate far smaller FoV, mainly due to packaging and the optical constant constraining resolution vs. FoV. Again, drawing on lessons from the NVG community, as FoV decreases, head motion must compensate to construct a total image of the outside scene. However, reduced FoV should be considered in

terms of all obstructions (canopy rails, cowling and so on) and operational requirement (is there a requirement for greater FoV?). In current fighter HMD, such as the Eurofighter Integrated Helmet, a fully overlapped binocular 40° FoV is provided – exactly the same as for current NVG, so pilots will be familiar with the nigh vision characteristics of the HMDS. IHADSS provides a 30° (V) × 40° (H) rectangular FoV, presenting an image to the pilot which is equivalent to a 7 ft (diagonal) CRT being viewed from 10 ft away. Although the monocular HDU design obstructs unaided lateral vision to the lower right, the IHADSS provides an unimpeded external view throughout the range of PNVS movement (± 90° azimuth and + 20° to –45° elevation). In common with NVG users, IHADSS users also employ a head scan to 'fill-in' information outside the straight ahead position. An interesting characteristic of this early HMDS, although equally disorienting, occurs when the pilot's head motion exceeds the PNVS range of motion – the IHADSS image suddenly stops, despite continuing head motion!

The IHADSS is designed to present the FLIR FoV in such a manner that the image on the combiner occupies the same area in front of the eye, resulting in unity magnification. However, to achieve this goal, the aviator must position his eye within the exit pupil of the HDU optics. The major determinant of whether this can be achieved is the physical distance between the eye and the edge of the HDU optical barrel. Variations in head and facial anthropometry greatly influence the ability of the aviator to comfortably obtain a full FoV with this physical arrangement. In general, most combiner HMD systems, such as the Knighthelm helicopter HMD, suffer similar physical problems in positioning the CRT-combiner interface; the development of visor-projection techniques, employing the extant blast protection visor as the combiner, has largely eliminated this problem, although this technique is not without its own set of problems.

A number of studies have been conducted in an attempt to understand the role of FoV in pilotage and targeting tasks. A study conducted by the Centre for Night Vision and Electro-Optics, Fort Belvior, Virginia, investigated the tradeoff between FoV and resolution in helicopter operations. In this study, five aviators using binocular simulation goggles, performed terrain flights in an AH-1S Cobra helicopter. Seven combinations of FoV (40° circular to 60° × 75°), resolutions (20/20 to 20/70), and overlap percentages (50 to 100 per cent) were studied. The study reported the lowest and fastest helicopter flights were achieved using the 40° – 20/60 – 100% and 40° – 20/40 – 100% conditions, with the aviators preferring the wider FoV (60°) condition. However, it should be noted that the wider FoV were obtained by reduction of binocular overlap and the results pertain to helicopter operations where nap-of-the-earth flying is significantly aided by wider FoV. Whether an expanded horizontal FoV would be similarly useful in a fast jet HMD is open to question – it is the editor's opinion, drawing on experience of NVG operations, that a wider FoV, especially if no increase in resolution is applied, would be of limited operational use for pilotage and low-level navigation in fighter aircraft. However, conceding the fact there is a great deal of difference between high speed low level flight in a fast jet and hover turning in a helicopter, there is evidence, based on the sensory cues used for slow speed hovering in helicopters, that a wider FoV would greatly benefit helicopter pilots. Also, the FoV required to maintain orientation depends on workload. A small attitude indicator bar (or cue), occupying only a few degrees on the display image, does not provide much information to the peripheral retina, which normally mediates visual information regarding orientation in the environment. Acquiring this orientation information from the central (foveal) vision requires more concentration and renders the pilot susceptible to disorientation should his attention be diverted to other cockpit tasks for even a brief period.

While it is important to provide the pilot with suitable enhanced FoV, it is equally important that he should also be able to perform routine

cockpit tasks without interference from the HMD. It is especially important that caution and warning lights be visible, along with other instruments, in order to be able to perform tasks such as tuning radios and operating other onboard equipment. In an HMD, the available visual field can be impacted by the helmet, image source and the display optics. Visual field can be further reduced when NBC devices and/or oxygen masks are worn. It is therefore obvious that display FoV and available unaided visual field are inversely related.

It is generally accepted that imagery used for pilotage should be one-to-one with the sensor FoV. Magnification can result in disorientation and inaccurate distance and velocity estimations. However, magnification is used in targeting with head-down systems, which has resulted in some research into selective magnification of head-slaved imagery. For HMDs, using high magnification with narrow FoV for targeting would be undesirable because of head jitter and pilot disorientation.

OPTICAL CONSIDERATIONS

Similarly to Catseye and PNVG I, HMD designs are indirect view systems, utilising a beamsplitter (combiner) to present sensor imagery while allowing limited see-through vision of the outside scene. To avoid excessive conflict between the real and virtual scenes, consideration must be made of the luminous and spectral characteristics of the display. A certain percentage of the luminance of the background must be transmitted, however, high ambient background luminances must be attenuated to provide sufficient imagery contrast – a constant problem to HMD designers. This interference may affect the perception of information conveyed by the display imagery and/or the external scene – a problem experienced by pilots viewing a FLIR scene at night through a HUD. Luminance contrast can be reduced and spectral deviations may be introduced due to the combiner's characteristics. To achieve higher contrast, combiners are often designed to attenuate the external background luminance while being highly reflective to the peak wavelength of the monochromatic image source. Reducing combiner transmittance has been shown to be effective in increasing HMD imagery contrast in the Apache, although this technique, by definition, tips the balance of what the pilot sees towards the virtual sensor scene. Another way of increasing contrast is to employ the sun visor to decrease the effects of ambient luminance under high light conditions. While this practice under average luminance conditions is, in effect, an admission that the peak luminance (amongst other factors) of the display is insufficient for the conditions, this is perhaps preferable to the aforementioned modification of the combiner to transmit less of the outside scene to the pilot. While visors used for sun and wind protection usually will be spectrally neutral, some visors are designed to provide protection from directed energy sources (such as lasers) and will have spectrally selective transmittance characteristics.

The exit pupil of a (pupil forming) HMD is the area in space where all the light rays pass; for clarity, it should be noted that an NVG is not an exit pupil forming system, it uses a simple magnifier instead. The exit pupil may be visualised as a two-dimensional hole – to see the full displayed FoV, the eye must be located at the exit pupil. Conversely, if the eye is totally outside of the exit pupil, none of the FoV will be seen. As the viewer moves back from the exit pupil, the perceived FoV will decrease (the eye has an entrance pupil; when the exit pupil of the HMD is larger than the entrance pupil of the eye, the eye position has some latitude while retaining full view of the display FoV. The important feature of an exit pupil forming system is that the final optical path to the exit pupil can be utilised to 'fine tune' the position of the optical train with respect to the eye. The exit pupil has three characteristics – size, shape and location. For maximum flexibility, the exit pupil should be as large as possible. IHADSS has a circular 10 mm diameter exit pupil.

The Guardian HMD has a 16 mm exit pupil. Research has suggested that the exit pupil size should include the eye pupil (approximately 3 mm), an allowance for eye movements that scan across the FoV (approximately 5 mm), and an allowance for helmet slippage (±3 mm). This would set a minimum exit pupil diameter of 14 mm for HMD. Since the exit pupil is the image of an aperture stop in the optical system, the shape of the exit pupil is generally circular and, therefore, its size is given as a diameter. The exit pupil is located at a distance called the optical eye relief, defined as the distance from the last optical element to the exit pupil. Of importance to the fit of HMD is the *physical* distance from the plane of the last physical element to the exit pupil, a distance called the eye relief. This distance should be sufficient to allow use of corrective spectacles and allow for the wide variations in head and facial anthropometry. This has been a continuous problem for IHADSS, where the optical eye relief value (10 mm) is greater than the actual eye clearance distance. This is due to the required diameter of the HDU objective lens and the bulk of the barrel housing. Contact lenses have provided a solution to this particular problem.

HMDs can be classified as monocular, biocular, and binocular. These terms refer to the presentation of the imagery by the HMD: Monocular means the HMD imagery is viewed by a single eye (Guardian HMD), biocular means the HMD provides two visual images from a single sensor (F-35 JSF HMD night vision element), i.e. each eye sees exactly the same image from the same perspective and binocular means the HMD provides two visual images from two sensors displaced in space (Eurofighter Integrated Helmet). A biocular HMD may use one or two image sources, but must have two optical channels. A binocular HMD must have separate image sources (one for each eye) and two optical channels.

The advantages of monocular HMDs include less bulk, lighter weight and lower design costs. Drawbacks include FoV limitations, a small exit pupil, potential for binocular rivalry and eye dominance problems. Reduced FoV results in the need for a more exaggerated head scan. The small exit pupil size requires the display to be very close to the eye and a very stable helmet to avoid the slippage and the eye moving outside the exit pupil. Binocular rivalry causes viewing conflicts between the aided eye viewing the display imagery and the unaided eye viewing the outside world. Most monocular systems project imagery into the right eye (IHADSS, Guardian), which causes problems for left eye dominant pilots, in the form of extreme disorientation. Before the problem was fully understood, the failure rate of experienced pilots during training for the AH-64 during the demanding night/IHADSS phase was a cause for concern. Not merely a function of utilising a monocular system, the problem was exacerbated during periods of high workload, exemplified by night nap-of-the-earth flying with IHADSS. This fact explained that eye dominance itself is not a singularly defined concept and is task dependent. Another monocular problem demonstrated by IHADSS was differential dark adaptation. The problem of one eye fully dark adapted (scotopic viewing), with the other partially adapted due to the bright display (mesopic viewing) can cause a depth illusion for laterally moving objects caused by image delay to the darker adapted eye.

As previously discussed, the reduced FoV of fielded monocular HMDs is their greatest operational disadvantage. Reduced FoV impairs many visual tasks. In HMD designs, the size/diameter of the relay optics limits the available FoV. To provide larger FoVs, designers have adopted a method of partially overlapping the FoVs of two optical channels. This provides for a larger, partially overlapped FoV, consisting of a central binocular region and flanking monocular regions. Overlap can be either convergent or divergent. In a convergent design, the right eye sees the central overlap region and the left monocular region, and the left eye sees the central overlap region and

the right monocular region. In a divergent design, the right eye sees the central overlap region and the right monocular region, and the left eye sees the central overlap region and the left monocular region. The cancelled Comanche HIDSS design features a divergent overlap of approximately 30 percent (based on a 17° overlap region within the 52° horizontal FoV).

Further development of HMD electronics has resulted in lighter and less bulky binocular systems; coupled with a decreasing difference in cost, this has resulted in most modern HMDS being binocular designs. However, binocular designs bring more issues for consideration, including Inter Pupillary Distance (IPD), image registration and luminance balance between the two channels. Problems with these parameters can result in eye strain and fatigue.

Humans view a scene with binocular vision; as a result of the small distance between the eyes, each sees a slightly different scene (try closing one eye at a time and comparing what you see). Typically, an adult male's eyes are approximately 60 mm apart with the individual eyes lines-of-sight converging at the accommodation distance. These slightly differing views of the scene confer depth perception and stereopsis. Biocular HMDs use a single sensor to present the same image of the scene to both eyes, but since the displayed scene is *exactly the same*, there will be no stereopsis. The Comanche HIDSS was a biocular system, but since the Comanche FLIR sensor FoV was larger than the display optics FoV, HIDSS could present approximately two-thirds of the sensor FoV to each eye, resulting in a display FoV that matched the sensor FoV, consisting of two monocular regions and a central region seen by both eyes. NVGs on the other hand, with each sensor aligned with the eyeline, is a simple binocular system providing separate images to the two eyes. However, close-in depth perception and distance estimation is denied since NVGs are focused to nominal infinity, which represents a distance beyond which we can judge distance directly. Topowl is a binocular helicopter HMD which incorporates two II sensors mounted outside, but aligned with, the pilot's eyes. This system has exhibited enhanced stereopsis which, under certain circumstances, has proven disorienting to pilots.

Introduction of partial image overlap to achieve a larger horizontal FoV can bring with it certain additional concerns. The overlapped images must be aligned continuously across the horizontal FoV and the individual images must have matched luminances for a consistent scene when viewed by the pilot. In a fully overlapped binocular system, the full FoV consists of one contiguous binocular region, whereas a partially overlapped FoV consists of three regions, one binocular overlapped region, flanked by two monocular regions. This situation can result in visual fragmentation of the three regions into three phenomenally separate areas, separated by the binocular overlap, with each interpreted in a subtly different manner. This can lead to misinterpretation of the scene. In a fully overlapped binocular system, individual channel problems, such as lower luminance, noise, aberrations etc. will be cancelled out by the brain, which will naturally select the areas of best quality in each channel and combine them to construct the best possible image for the brain to interpret.

Having discussed the influence of colour on image quality and operational effectiveness, it must be emphasised that the majority of fielded HMDs are monochromatic, with the dominant colour of the (CRT) display dictated by the phosphor. While the previous discussion has hinted at the reasons why colour HMDs have been lagging behind, it is worth re-emphasising some of the issues particular to colour displays. The advent of small form FPDs has mitigated a couple of the problems in colour HMD design – power requirements, weight and, unsurprisingly, cost. Additionally, and again as previously described, colour displays require resolution and luminance trade-offs, so the operational case for the implementation of colour has often been

difficult to prove if a monochromatic system of lesser price can provide greater resolution and contrast. The use of colour image sources increases the complexity of the relay optics design since a polychromatic design must be used.

However, all of these hurdles have not decreased the desire of the operators for colour systems, since colour is an intrinsic part of our recognition capability. Since the human views the outside scene in colour, it is natural that the greatest comfort with a system designed to augment or replace unaided viewing should be with colour. Since colour is suppressed at night, NVGs, FLIR and Low Light (LL) TV systems have been effective in their monochromatic guises. Recent research into adding colour to II and FLIR devices for night operations found only limited acceptance. For daytime operations however, colour has the potential to reduce workload and improve performance. However, it should be noted that colour contrast has not been proven to be a substitute for luminance contrast, so a HMD design should not compromise in this respect. Chromatic contrast improves performance as long as luminance contrast has been maximised. If luminance is at a premium in a display system, then noting previous comments regarding the luminous efficiency of the eye (during daylight, the eye is most efficient at 555 nm, or the green wavelength) then other colours will require 'over driving' to make them appear at the same intensity to the viewer.

DEFINING THE STANDARD

Having outlined the major issues concerning the design of HMDs, it will come as no surprise that considerable effort has been made in establishing criteria for their implementation in military aircraft. While a draft standard was produced in the US as early as the mid-90s, the speed of technological development in all areas of HMD design rendered the definition of standards difficult. Typical physical parameters for miniaturised FPDs are 20 × 20 mm, 15 mm depth and <25 g mass. In addition to these (desirable) size and mass characteristics, miniature displays must be adaptable to see-through systems and have sufficient resolution and luminance. Image source size dimensions (approximately 28 mm diagonal) has been loosely dictated by FoV (25° to 50°), eye relief (>25 mm) and exit pupil size (15 mm) requirements. Resolution for FP displays is defined as the highest spatial frequency which can be presented. It is usually expressed as the number of picture elements (pixels) in both the horizontal and vertical directions. An important concern when selecting the resolution of pixelated image sources is to ensure that, when viewed by the eye through the display optics, individual pixels are not resolvable. Based on the human eye's minimum resolution of 1 arcminute, an HMD with a field of view of 40° should not have less than 2,400 pixels in either dimension. Currently, the fielded display limit of 1,280 pixels falls some way short of this requirement. For displays with >20° FoV, this problem can be overcome by defocusing the image source to 'soften' the picture, making it more visually acceptable. For see-through and visor projected HMDs, the image is viewed against the background of the outside scene, which can take on a wide range of luminance values, ranging from direct sunlight (10,000 fL) to a moonless night (0.001 fL). While the low light condition does not present a problem with a suitable dimming circuit, the daylight condition remains a considerable challenge to designers. Miniature display sources currently cannot approach the required luminance output and discernable shades of grey, especially considering losses through the optics which can be as high as 80 per cent.

THE FUTURE

For future HMD applications, there are a number of FP technologies which show promise, including AMLCD, AMEL, OLED and FED, although none of them can as yet be considered as mature as the CRT. Manufacturing techniques, in part due to

Although not an aviation application, this picture of Kaiser Electro-Optics' ProView S035 monocular HMD, incorporating an eMagin full-colour active-matrix OLED microdisplay, shows the potential of small format displays. This one provides SVGA (800 × 600 pixel) resolution over a field of view measuring 32° horizontal by 24° vertical 0532659

the domestic LCD market, are rapidly becoming mature, although AMLCD and AMEL displays do not yet provide sufficient symbology luminance. AMELs additionally suffer from insufficient video luminance. AMLCDs require extremely high backlight luminances and have limited temporal response for presenting the dynamic imagery required for military HMD applications. FEDs are considered by some HMD designers to show the greatest potential for future applications, due to their very low power requirements, wide viewing angle, excellent resolution and high contrast (>100:1). Additionally, FEDs are suited to aviation HMD applications due to their robustness.

Another novel imaging source, with potential for HMDs, is the laser. Laser image generators produce imagery on a screen using the basic scanning method of CRTs. Instead of an electron beam, a laser beam is scanned in two dimensions, with the beam intensity modulated at every pseudo 'pixel' position. If scanned at frequencies of 60 Hz or greater, a flicker-free image is produced. Laser projectors have shown the potential to produce sufficient luminance, colour gamut and colour saturation. It is claimed that a scanning laser-based HMD will be able to provide high resolution, high luminance, and

Planar Systems' Active Matrix Electro Luminescent (AMEL) display, showing the small size. The display surface is the rectangle at the bottom of the flexible connector, on which is an image of an F-15 Eagle fighter aircraft 0002743

monochromatic or colour imagery within the small weight and volume requirements described above. Potential disadvantages to such a system include scanning complexity, susceptibility to degradation in high vibration environments, limited exit pupil size, together with perhaps obvious safety concerns.

EXAMPLE SYSTEM – VIPER

In choosing an example system to bring out the lessons of the preceding analysis, an early fast jet visor-projected system was selected to illustrate the origins of recent developments in the implementation of such systems in aircraft like the Eurofighter Typhoon. Much of the technology introduced in the Eurofighter Integrated Helmet has its origins in the GEC-Marconi Avionics (GMAv) Viper range of HMD.

Viper was conceived as a lightweight modular system, designed to be integrated onto existing aircrew helmets or lightweight helmets designed from the outset for HMD applications. This approach was adopted to minimise acquisition costs for the system by integrating with existing aircrew equipments, including oxygen and communication systems. GMAv also developed a helmet shell designed specifically for the Viper HMD which provided reduced weight and improved fitting techniques optimised for stability and comfort.

Initially a set of basic design parameters were laid down to guide the design towards the desired outcome:

Viper should be compatible with existing service helmets and custom helmets
The interface to the helmet must be simple
One size of Viper display module should fit all sizes of helmet shell
Prototypes should use existing visor hardware
Inter Pupillary Distance (IPD) adjustment must be provided
Total head supported mass less than 2 kg
Monocular 20° FoV
No outside world coloration
Eye relief greater than 30 mm
Exit Pupil at least 15 mm
Minimise use of high risk technologies
Reduce cost of ownership by minimising in service logistics costs.

The Viper HMD was designed for fast jet installations, primarily for daytime operations involving off boresight targeting and missile aiming. It was therefore important that the resultant HMD should not impede the pilots freedom of movement and be lightweight. It should also be usable in all ambient light levels whilst not adversely reducing or distorting the pilots view of the outside world. Finally, it must be compatible with helmet tracking systems.

Go ahead for the HMD was given in April 1993. Various technologies were reviewed, such as lightweight composite materials for structural elements and holographic coatings for optical corrections and high display efficiency.

Viper I HMD 0001461

Although these offered some real advantages, detailed analysis showed some drawbacks. The use of composite material in structures offered good potential, but to ensure an athermalised structure which would have sufficient rigidity and stiffness to act as an optical bench on the helmet, lengthy and expensive analysis would be necessary to evaluate and develop suitable lay ups and thicknesses. The required timescale did not allow time for this activity to be completed so it was decided that prototypes would be built using aluminium alloy for portions of the internal structure.

The optical design concept offered a considerable challenge, having to meet the criteria of low weight, high performance with a large usable exit pupil and adjustable interpupillary setting. The latter two parameters can, to some extent, be a trade off as a larger exit pupil minimises the need for a full range of IPD adjustments. However, a large exit pupil presents two problems, aberration control across its diameter and increased relay lens diameter. These caused considerable problems in configuring the display module around the helmet to minimise protrusions from the existing helmet profile. These problems were solved and the resultant design concept developed for Viper enabled the optical relay system to be rotated about the centre of curvature of the spherical visor, thus allowing pupil positions to be moved without affecting the optical prescription. Experience from flight test of previous HMDs had shown that a usable exit pupil must be in the region of 15 mm to accommodate positioning the helmet on the head, settlement during use and movement brought about by high g manoeuvring.

A collimated display presentation to the pilot was achieved by reflection from the inner surface of the visor. The use of selective dielectric coatings was considered initially as this would allow enhanced display brightness and efficiency. However, a review of customer requirements showed a marked preference for zero real world coloration, so a neutral density coating technique was selected. Calculations of the required display contrast ratio in ambient lighting conditions showed that a real world transmission of 70 per cent could be achieved with a contrast ratio of greater than 1.2:1 in ambient lighting of 10,000 ft Lamberts when using a high efficiency neutral density semi-reflection coating. This approach also conferred the benefit of compatibility with other cockpit instruments including the HUD. HMDs and HUDs normally use a narrow band phosphor such as P53. HUDs in many modern aircraft are holographic so that the use of selective coating on the HMD could result in poor compatibility with the HUD.

In addition to structural integrity, from an optical point of view it was also necessary to meet the requirements of loads induced by the more demanding environment of windblast during ejection. The design could not compromise the protection provided by the helmet. It was therefore important to ensure that the load paths through the display module and helmet were such that loads were transferred through the module without causing damage to frame, optics or helmet. It was therefore important to maintain the stability of the module and its interface to the helmet shell. Aerodynamic shaping of the module and visor combination helped to deflect loads and prevent the helmet shell from ballooning out as it was exposed to the airblast. This reduced the danger of the chin strap releasing and of the helmet subsequently detaching from the pilots head. The Viper display module shell was manufactured from a Kevlar/carbon composite and was carefully blended to the helmet in such a way that the gap between the module shell and the visor was minimised to exclude any airflow from the pilots face. The visors and side arms wrapped around to prevent ballooning. The module shell was secured to the support frames providing additional rigidity in a monocoque form of construction. The interface to the helmet was via a special mount which provided longitudinal stiffness whilst also allowing lateral compliance to accommodate the range of helmet shell sizes available and also to permit flexure of the helmet shell during donning and doffing.

The prototype Viper HMD was designed to fit the HGU-55/P and the HGU-53/P although initial design activity indicated that with only minor adjustment it would fit onto several other helmet types. The projection optic was designed to use a single reflection off the visor surface, the spherical form of which would act as the collimator. The relay optics were positioned around and close to the helmet shell with the collimator operating off-axis. This introduced some complexity into the optical design in terms of coma and astigmatism and also chromatic aberration which was corrected by careful selection of glass types and filtering. A 1/2 in CRT display source was employed with a P53 phosphor. Filter glasses were incorporated into the optical design to remove side bands, allowing the primary peak at 545 nm to pass. The suppression of the other secondary peaks was deemed important as this removed the likelihood of double imaging.

The first prototype of the Viper HMD was assembled in early August 1993, some 5 months after go ahead. The Viper display module was fitted to a standard HGU-53/P helmet with an MBU-12/P oxygen mask. To minimise time scale and cost, a standard spherical aircrew visor was used for the display/blast visor. This had a high efficiency neutral density semi reflective coating applied in front of the pilots right eye and an anti reflection coating applied to the visor outer surface. Both tinted and clear visors were produced, the visor being easily changed without special tools or specialist jigs. The vertical size of the visor limited the uplook angle to 40° for the prototype HMD, with the module shell extended downwards to maintain the aerodynamic form of the HMD. A small bracket was fitted to the module to house a miniature colour camera intended to be used for mission recording and training purposes (see image of Viper I). The CRT was an off-the-shelf unit operating at 8.5 kV final anode voltage. A GMAv HMD drive electronics system provided for full aircraft interfaces, processing, symbol generation and CRT drive. The HMD incorporated an EEPROM which contained setup data for the CRT and optical characteristics of the HMD which were used to initialise the display. Initial testing indicated that the HMD met or exceeded the design specification:

Field of View	>22°
Display Focus	Infinity
Exit Pupil	>20 mm
IPD adjustment	55–75 mm
Real World Transmission	70% (coated area)
Symbol Line Width	1 mR
Optical Resolution	Better than 1.5 cy/mR
Display Contrast Ratio	1.2:1 at 10,000 fL
Total Head	2.1 kg (including
Supported Weight	HGU-53/P and MBU-12/P)
Centre of Gravity (C of G)	Within 0.2 in of head C of G

In parallel with the display testing, Safety Of Flight (SOF) testing was conducted. As the intention was to clear Viper for flight test on the X-31 programme, some testing was carried out by GMAv and some by NASA. Much of the electronics had already been cleared for similar fast jet applications. This included the quick disconnect system which had already undergone extensive testing including ejection and explosive atmosphere tests. Also, the use of the HGU-53/P helmet greatly simplified testing as many parameters could be cleared by similarity. Those parameters affected by the addition of the Viper display module were fully tested. These tests included successful windblast testing up to 450 knots, explosive decompression testing, fit assessment trials, centre of gravity testing, EMC testing, parachute riser trials and valsalva tests. This activity concluded with the Viper HMD cleared for flight test on the Rockwell/iDASA X-31 aircraft in December 1993.

For the X-31 programme, the HMD Electronics Unit (EU) was reprogrammed to provide a novel set of symbology to the pilot to further the aims of the X-31 research, which involved intense manoeuvring and simulated close combat against an F-18 adversary aircraft. The first test flights with Viper were made on the 16 December 1993. These flights were successful and the HMD was cleared for use in the close in combat tactical evaluations to assess the utility of two unique attitude display formats. In general, the fit of the Viper HMD was felt to be comfortable with no hot spots. Helmet weight and centre of gravity received no adverse comment, possibly as it was so close to current flying helmets. The unaided eye FoV was considered excellent, with few viewing restrictions. The uplook angle, although limited to 40° on prototype HMDs, was easily accommodated by head movement. Display resolution, clarity, sunlight readability contrast and brightness were all considered excellent.

The Viper HMD design started in April 1993 and resulted in the flight test of prototype hardware in December 1993. Further development of the basic Viper helmet was undertaken for other applications. The initial 'day' variant of Viper was renamed 'Viper I', and a new wide FoV binocular HMD, designated Viper II, was produced. In parallel, full scale development of a full integrated HMD, called Crusader, was commenced, intended for the Eurofighter 2000 aircraft.

Further work on Viper I was undertaken to improve the HMD performance. The available uplook was increased to 55° by providing a visor with a larger vertical dimension. Minor coating changes enhanced the display efficiency and image quality. Higher performance CRTs were also developed which provided for brighter displays with improved image quality. HMD mass was reduced by replacing the then current aluminium alloy optics support frame with a composite design fully integrated with the module shell.

Viper II was a binocular visor projected HMD derived from Viper I but offering a wider FoV. Viper II was designed for day and night applications such as the display of imagery from

Viper II HMD. Note the spherical visor arrangement 0001460

a head steered FLIR with symbology overlay onto the HMD. The Viper II HMD comprised a display module fitted onto a standard helmet in the same way as Viper I. The design of the optical system, although different, used the same basic principles proven on Viper I. The key design parameters for Viper II were:

Viper II should be compatible with existing service helmets and custom helmets

One size of Viper II display module should fit all sizes of helmet shell

Prototypes should use existing visor hardware

Inter Pupillary Distance (IPD) adjustment of 55 to 75 mm

Total head supported mass less than 2 kg

Binocular 40° FoV with 100% overlap

Real world transmission of >70% in coated area

No outside world coloration

Display contrast ratio of 1.3:1 at 10,000 fL

As manned tactical aircraft seem set to be a thing of the past, perhaps the extensive research into HMDS will bear further fruit in the application of immersive displays for remote manual control of UCAVs
0566900

Exit Pupil greater than 15 mm

Minimal use of high risk technologies

Reduce cost of ownership by minimising in service logistics costs.

Testing of Viper II commenced during 1994. Further variants of Viper, Viper III (Night Viper) and Viper IV (employed during the VISTA trials) were developed, which explored further technologies for HMD applications. Viper and Crusader were pivotal projects in the history of development of the recently fielded Eurofighter Integrated Helmet, the first day/night fully overlapped binocular 40° visor-projected HMDS to enter service in fighter aircraft.

CONCLUSIONS

While this paper has addressed a number of the important aspects of HMD design, this area of avionics is undergoing constant and rapid evolution as designers seek to resolve some of the contradictory characteristics intrinsic to advanced HMDs. Considering that many believe that the current F/A-22 Raptor, F-35 JSF and Eurofighter Typhoon may well represent the last manned advanced fighters, replaced by sophisticated UAVs in the next 20 to 30 years, there is a school of thought that suggests the development of Human Machine Interface (HMI) avionic systems should be curtailed in favour of automated systems. However, if Unmanned Combat Air Vehicles (UCAVs) are to become a reality without the benefit of full Artificial Intelligence (AI), then operator intervention will be required for some years to come. If remote/manual control is to be an effective mechanism to enhance the survivability and combat effectiveness of UCAVs operating at (noting the latency in current datalink systems) line-of-sight ranges, then the operator (né pilot) will require an immersive display with high fidelity and large FoR linked to an onboard sensor suite; it would seem that the HMD might well outlive manned aircraft!

AVIONIC COMMUNICATIONS SYSTEMS

Combined CNI systems
Avionic communications systems
Aircraft navigation systems
Aircraft identification and safety of flight (SOF) systems

COMBINED CNI SYSTEMS

France

TLS-2020 MultiMode Receiver (MMR)

Type
MultiMode Receiver (MMR).

Description
The TLS-2020 MultiMode Receiver is a new VOR/ILS receiver designed for fighters and helicopters and developed to comply with all the requirements regarding Precision and Approach Landing System (PALS). It fully complies with new ICAO – Annex 10 regulations (FM immunity requirements). It also includes a VHF receiver for the DGPS datalink.

The TLS-2020 receiver can optionally be configured with Marker, MLS, DGPS and GPS functions by simply adding modules.

The TLS-2020 MMR utilises the latest technology developed both for the commercial aviation and French military programmes. The TLS-2020 belongs to the TLS-2000 product line.

Specifications
Basic functions: VOR/ILS/VHF receiver for DGPS
Options: Marker, MLS, DGPS and GPS
Dimensions: ¼ ATR short
Weight: <3.3 kg

Thales' TLS-2020 MultiMode Receiver (MMR)
0015353

Power: 28 V DC
Interfaces: MIL-STD-1553B, ARINC 429, analogue interfaces
Frequency bands:
(VOR): 108–118 MHz
(ILS loc): 108–112 MHz
(ILS glide): 329–335 MHz
(Marker): 75 MHz
GPS: 1,575.42 MHz (L1), 1,227.6 MHz (L2)
MLS: 5031–5091 MHz
VHF datalink: 108–118 MHz
C-Band datalink: 5031–5091 MHz
MTBF: 24,000 h AIC
Operating temperature: –40 to +71°C
Environmental:
(EMI/EMC): MIL-STD 461C/462
(Design): MIL-E-5400T Class 2 (Fighter)

Status
In production and in service in Rafale fighter aircraft and NH 90 helicopters. Selected for the Nimrod MRA4 Maritime Patrol Aircraft (MPA).

Contractor
Thales Communications.

TLS-2030 MultiMode Receiver (MMR)

Type
Avionic navigation/communication receiver.

Description
The TLS-2030 MultiMode Receiver is a new VOR/ILS receiver which complies with ARINC 755.

The TLS-2030 is the military version of the TLS-755 MMR that has been developed for commercial aircraft. The VOR function has been added to the functions of the TLS-755.

It offers VOR/ILS capability and a VHF receiver for DGPS functions.

Optionally it can be configured with MLS, DGPS and GPS. The TLS-2030 belongs to the TLS-2000 family.

Thales' TLS-2030 MultiMode Receiver (MMR)
0015354

Specifications
Basic functions: VOR/ILS/VHF receiver for DGPS
Options: MLS, DGPS and GPS
Dimensions: 3 MCU
Weight: 4 kg
Power: 115 V, 400 Hz
Interfaces: ARINC 429

Status
In production.

Contractor
Thales Communications.

Germany

PrimeLine communications and navigation system for business aviation

Type
Avionic navigation/communication receiver.

Description
The Becker PrimeLine communications and navigation system fulfils the need of the business aviation market for a complete dzus rail-mounted family of Cat I avionic products. The basic PrimeLine products are completely housed in single units, requiring no remote boxes. Products in the range include:
• AR 3202 20 W airborne VHF transceiver, AR 3209 10 W airborne VHF transceiver;
• NR 3320/30 VOR/LOC/GS navigation receivers;
• IN 3300 series VOR/ILS navigation indicators, including a 3 in VOR/LOC/GS CDI, a 3 in VOR/LOC CDI, and a 2.25 in VOR/LOC/GS CDI;

• ADF 3500 automatic direction-finder systems with RA 3502 remote control;
• RMI 3337 radio magnetic indicator;
• HSI 421 (4 in) and HSI 8131 (3 in) horizontal situation indicators;
• Audio selector and indicator systems;
• ATC 3401 transponder;
• DVCS 5100 digital voice communication system;
• AirScout Moving map system.

Status
In service.

Contractor
Becker Avionic Systems.

PrimeLine II communication and navigation system

Type
Avionic navigation/communication receiver.

Description
The PrimeLine II is a remotely controlled system consisting of COM, NAV, ADF and ATC systems. It is designed for installations where there is minimum available panel space. It utilises small,

The Becker Avionic Systems PrimeLine communications and navigation system for business aviation
0081496

lightweight, control units, which fit into standard 2¼ in (57 mm) round instrument panel cut-outs. The units are only 2½ in deep.

The connection between the control units and the transceivers is achieved by use of two prefabricated cable harnesses. The interfaces to all COM and NAV modules are standard and universal. This minimises installation time and eases the addition and replacement of individual modules as desired, without the need of additional cables or major changes.

Status
In service.

Contractor
Becker Avionic Systems.

ProfiLine communication and navigation system

Type
Avionic navigation/communication receiver.

Description
The ProfiLine communications and navigation system comprises:
- The CU900-(1) dzus-width control unit with ARINC 429 and ARINC 410 interface for COM transceivers and NAV receivers (optionally NVG-compatible).
- The NR900 VOR/ILS receiver, designed as a retrofit replacement for the Rockwell Collins 51RV-1C, within a ½ ATR short (ARINC 4-4A)

Becker ProfiLine NR900 receiver and CU900-(1) control unit   0005424

housing. It is based on a modern design with integrated BITE, which also tests the HF section. It complies with the new ICAO annex 10 requirements for FM immunity. The NR900 has an ARINC 410, an ARINC 429 control interface, and a built-in marker receiver.

A DGPS receiver and a MIL-STD-1553 interface will be available optionally in the future.

Contractor
Becker Avionic Systems.

United States

Auxiliary Communications, Navigation and Identification Panel for the AV-8B

Type
CNI system controller.

Description
The Auxiliary Communications, Navigation and Identification Panel (ACNIP) is an integral part of the communications, navigation and identification system used on the AV-8B and TAV-8B Harrier.

When interfaced with the other components of the aircraft communications system, the ACNIP performs audio amplification, control inhibit and distribution functions, generates audio warning messages in response to discrete serial and/or analogue inputs, provides code, mode, remote variable load, baseband/diphase, and control zeroing functions for two KY-58 secure communications units and controls functions of the identification system such as zeroing and emergency operation. Additionally, the ACNIP provides logic-controlled push-to-talk switch closures and switch functions for landline telephone communications, control for ground crew communications and a hot mic capability for the operator. BIT circuitry detects 98 per cent of all electrical component failures.

The ACNIP controls a non-volatile EEPROM for the storage of the code and mode operating parameters of the secure speech units. On initial power-up, the ACNIP will update the KY-58 units to the operating modes as selected before power-down.

An LCD module, backlit and with variable illumination level located on the front panel of the ACNIP, provides a visual readout of the functional status of the KY-58 units, the operating mode, code and type of cypher used by each unit being displayed.

Status
In service in AV-8B aircraft.

Contractor
Sanmina-SCI Corporation.

CNS-12™ ACARS/GPS/ADS communication, navigation and surveillance system

Type
Avionic navigation/communication receiver.

Description
The CNS-12™ communication, navigation and surveillance system features a WAAS-capable 12-channel GPS receiver with Receiver Autonomous Integrity Monitoring (RAIM), an optional D8PSK receiver card for differential GPS and aircraft navigation technology to meet the ICAO VDL requirements. The GPS, together with the ARINC VHF Aircraft Communications Addressing and Reporting System (ACARS) two-way datalink, provides timely and accurate Automatic Dependent Surveillance (ADS) position reports from any aircraft via the ACARS network.

The CNS-12™ employs a PCMCIA card capable of storing information on nearly 21,000 airports, navaids and waypoints. The Jeppesen database card contains detailed information on airports, VOR, DME, ILS/DME, VORTAC, TACAN and NDB, as well as en route and terminal intersections. GPS overlay approaches, SIDs and STARs are included on the expandable database card. Additionally, a pilot-defined waypoint database stored in RAM can accommodate up to 500 waypoints. The information is used to create up to 50 flight plans with up to 40 waypoints in each. User flight plans and waypoints can be stored on a programmed PCMCIA card. The active flight plan and stored waypoints can be selected from the internal database, user datacard or Jeppesen NavData card and displayed in a variety of ways. En route, approach and vertical navigation is displayed on the CNS-12™ display.

A second, dedicated D8PSK VHF receiver for differential GPS precision approach information is optional. The receive only module will make precision differential GPS approaches possible as the ground infrastructure becomes available.

The CNS-12™ enables two-way communication for sending and receiving clearances, text messages, system essential messages, weather reports, Out/Off/On/In (OOOI) reports, engine data and other operational messages over the GLOBALink/CNS network. It also automates predeparture clearance and delivery of flight papers. The CNS-12™ communications element relays digital Automatic Terminal Information Services (ATIS) arrival and departure messages for any airport during flight.

The two-way VHF datalink complies with ARINC 745-2 for ADS. The CNS-12™ will automatically transmit data derived from its integrated GPS to air traffic service centres, providing surveillance beyond the radar horizon.

Specifications
Dimensions: 146 × 107.9 × 203.2 mm
Weight: 3.08 kg
Power supply: 14 or 28 V DC, 7 A (max)
Temperature range: –15 to +55° C
Altitude: up to 15,000 ft
Frequency: 108–136.975 MHz
Accuracy:
  (position) 15 m RMS
  (with DGPS) 8 m RMS
  (velocity) 0.1 kt (with DGPS)
  (time) UTC to nearest µs

CNS-12™ ACARS/GPS/ADS communication, navigation and surveillance system   0015387

Status
In service on a variety of commercial aircraft, CNS-12™ links via ARINCs GLOBALink/CNS datalink service and can be integrated with the ADS system.

Contractor
Thales Navigation.

GNS 480 Comm/Nav/GPS system

Type
Avionic navigation/communication receiver.

Description
Garmin claims that the GNS 480 is the first GPS navigator to feature precision guidance via Wide Area Augmentation System (WAAS); previously, Instrument Landing System (ILS) capability/availability was needed for a precision approach in low-visibility conditions. Now, with US FAA approval, the GNS 480 can utilise satellite-based navaids for precise lateral and vertical approach guidance without the need for ground-based assistance, since the system is certified to Gamma-3 requirements which meet the FAA's standards for Localiser-Equivalent Precision with Vertical (LPV) guidance.

The GNS 480 incorporates the following features:
- Glideslope guidance for LNAV/VNAV (Gamma 2) and LPV (Gamma 3) approaches and advisory vertical guidance for other non-precision approaches with vertical path data
- Reversionary approach modes
- Automatic SUSPend key function
- Interfaces to Garmin GTX 330/33/320/32 Mode S/C transponders
- Interface to display traffic data from L-3 Avionics Systems' Skywatch
- NAV page with a full compass rose CDI display
- Dual-GNS 480 flight plan crossfill capability
- Ability to calculate RAIM when out of WAAS coverage
- Automatic frequency checking for ILS, backcourse and VOR approaches
- Notification of navigation to non-WGS 84-compliant waypoints in the active flight plan
- Ability to manually select maximum CDI scaling.

At the heart of the system is a 15-channel WAAS receiver that updates the aircraft's position at a rate of five times per second. The GNS 480 provides oceanic-approved IFR GPS/NAV/COM functionality and ILS/VOR capabilities and includes built-in controls for a remote transponder that can be used to deliver traffic advisories via the US FAA's Traffic Information Service (TIS). With a TIS datalink, the pilot essentially sees what the ATC sees, including trend vectors and altitude data of nearby aircraft.

The GNS 480 integrates with the MX20 MultiFunction Display (MFD). When coupled with the MX20, range changes performed on the GNS 480 are simultaneously reflected on the MFD. When the pilot selects an approach on the GNS 480, the approach plate is automatically depicted on the MX20 with Jeppesen(R) ChartView™. Pilots can also view a vertical profile of the approach using the MX20's split-screen capability.

Working in concert with standard autopilots that accept roll-steering commands, the GNS 480 provides the required guidance to facilitate a coupled approach.

Specifications
Number of channels: 15 (12 GPS and 3 GPS/WAAS/SBAS)
Frequency: 1575.42 MHz L1, C/A code
Sensitivity (acquisition): −116 dBm to −134.5 dBm GPS; −116 dBm to −135.5 dBm WAAS
Sensitivity (drop lock): −144 dBm
Dynamic range: > 20 dB
Lat/Long position accuracy: <1 m RMS typical with WAAS (horizontal/vertical)
Velocity: 1,000 kt maximum (above 60,000 ft)
TTFF (time to first fix): 1.75 min (typical)
Reacquisition: 10 seconds (typical)
Position update interval: 0.2 sec (5 Hz)
1 pps (pulse per second): ±275 nsec of UTC second
Datum: WGS-84

Features
FMS operation: Dedicated buttons and soft-key operation for expanded functionality
VHF COM transceiver: 760 channels and 8-W transmitting power; standby frequency monitoring capabilities
VHF NAV receiver: 200 channels for VOR, localiser and glideslope operation; dual-station tracking feature to cross-check position fixes during approach procedures
Flight planning: Capacity for 50 flight plans with 150 legs
Annunciation: Voice messaging and audio alerting

VHF communications transmitter
Class: 4
Output power: 8 watts minimum carrier at >12 V DC input
Frequency range: 118.000 to 136.975 MHz, 760 channels

VHF communications receiver
Class: D
Frequency range: 118.000 to 136.975 MHz, 760 channels
Sensitivity: 1 μV (2 μV hard) for 6 dB S+N/N with 30% modulation at 1,000 Hz
Selectivity: <6 dB variation at ±7 kHz, >60 dB at ±22 kHz
Squelch control: Automatic squelch with manual override

VOR
Frequency range: 108.00 to 117.95 MHz in 50 kHz increments
Receiver sensitivity: 108 MHz − 115 dBm; 117 MHz − 117 dBm (typical)
Course accuracy
RTCA DO-196 two sigma limit: 3°
GNS 480 performance: less than 0.5° typical

Localiser
Frequency range: 108.00 to 111.95 MHz
Receiver sensitivity: 115 dBm (typical)

Centring error
RTCA DO-195 two sigma limit: 6.6% of full scale
GNS 480 performance: less than 1.0% typical (1.5 mV)

Glideslope
Frequency range: 329.150 to 335.00 MHz
Receiver sensitivity: 95 dBm (typical)

Centring error
RTCA DO-195 two sigma limit: 6.7% of full scale
GNS 480 performance: less than 2.0% typical (3.0 mV)

General
Power: 10 to 36 V DC
Dimensions: 82.55 × 158.75 × 299.72 mm (H × W × D)
Weight: 2.6 kg (unit only); 0.3 kg (mounting tube)
Environmental: DO-106D
Operating temperature: −20 to +55°C
Altitude: Up to 55,000 ft
Interface options: RS232, RS422 and ARINC 429
Aircraft support: STC Approved Model List (AML) for more than 700 aircraft

Status
In production and in service.

Contractor
Garmin International.

GNS 530 Comm/Nav/GPS system

Type
Avionic navigation/communication receiver.

Description
The GARMIN GNS 530 is a 'one-box' WAAS (Wide Area Augmentation System) upgradeable IFR GPS, Comm, VOR, LOC and glide-slope system, with colour moving map. The TSO qualified VHF Comm facility offers a choice of 25 kHz or 8.33 kHz spacing, for a 760 or 2,280 channel configuration, respectively. A Jeppesen database (which can be updated with front-loading data cards) contains all airports, VORs, NDBs, airway intersections, FSS, approach, SIDs/STARs and SUA information.

The GNS 530 features a 5 in colour display which separates land data, terminal areas, route, and approach information for easy scanning and reduced pilot workload, a feature more commonly found in commercial FMS systems.

Specifications
Dimensions: 159 (W) × 117 (H) × 279.4 (D) mm
Weight: 3.9 kg
Power supply: 27.5 V DC
Display: Colour LCD
GPS receiver: PhaseTrac12™ 12-parallel channel
Accuracy: 15 m (position), 0.1 kt (velocity) RMS (steady state)
Database: Jeppesen Americas, International or Worldwide
VHF transceiver: 760-channel (25 kHz spacing) or 2,280-channel (8.33 kHz spacing), 10 W
Interface: ARINC 429, RS 232
Certification: TSO C129a Class A1 (en route, terminal and approach), TSO C37d Class 4 and 6 (transmit), TSO C38d Class C and E (receive), TSO C40c (VOR), TSO C36e (LOC), TSO C34e (GS)

Garmin GNS 480 (Garmin International)   1127914

Garmin GNS 530 Comm/Nav/GPS (Garmin International)   0099754

Status
In production and in service.

Contractor
Garmin International.

KLN 35A GPS/KLX 135A GPS/COMM systems

Type
Avionic navigation/communication receiver.

Description
The panel-mounted KLN 35A GPS incorporates moving map graphics, a high-visibility display and a choice of customised Jeppesen NavData databases – including coverage for America, Atlantic and Pacific areas.

The moving map, useful for providing situational awareness, displays Special-Use Airspace (SUA) boundaries and provides SUA alerting. All three databases contain appropriate Flight Service Station (FSS) and Air Route Traffic Control Center (ARTCC) frequencies, and airport runway data.

The KLN 35A presents this information via an advanced double super-twist nematic LCD, offering improved viewing in direct sunlight and extended side-to-side visibility.

The KLX 135A offers greater capability. Incorporating all the same GPS performance and features, the KLX 135A GPS/COMM also integrates a TSO'd, 760-channel, Very High-Frequency (VHF) communications radio with its navigation functions. A serial RS232 output provides navigation and flight plan data for external moving map displays and there is also an external data port.

A new capability developed specifically for the KLX 135A, QuickTune, allows the pilot to enter the standby COMM frequency directly from the GPS database, saving effort and reducing the chances of making an entry error.

Specifications
KLN35A GPS
Dimensions: 158.7 × 50.8 × 289.1 mm
Weight: 0.94 kg
Power requirements: 11 to 33 V DC

KLX 135A GPS/COMM
Dimensions: 158.7 × 50.8 × 289.1 mm
Weight: 2 kg
Power requirements: 14 V DC (28 V DC with available KA 39 voltage converter)
VHF communications transceiver transmitter power: 5 W (min) (7 W (nominal))
Certification:
  (transmitting) TSO C37d
  (receiving) TSO C38d
  (stuck mic) TSO C128

KLN 35A and KLX 135A
Temperature: −20° to +55° C
Altitude: 35,000 ft
Lighting: 14 V DC or 28 V DC
GPS receiver:
(satellite tracking) up to 12 satellites simultaneously
(antenna) KA 91 or KA 92 active patch antenna

Contractor
Bendix/King.

KX 155 nav/com system

Type
Avionic navigation/communication receiver.

Description
The Bendix/King KX 155 is a self-contained nav/com transceiver which utilises a solid-state, gas discharge digital display and a push-button frequency flip-flop with display of active and standby nav and com frequencies. The main features of the system are:
- 720 or 760 communications frequencies
- 200-channel nav transceiver
- 40-channel glideslope receiver (option)
- built-in audio amplifier (option)
- 14 or 28 V DC power supply option
- 10 W power output
- 25 or 50 kHz com receiver selectivity
- fully TSO qualified.

The KX 155 does not contain its own VOR/LOC converter and so has to be combined with VOR/LOC indicators that contain converters. It can therefore be used with the KI 203/208 VOR/LOC or the KI 204/209 VOR/LOC/GS indicators. The GS function requires the KX155 to have the glideslope receiver option included.

Specifications
Dimensions: 159 × 520 × 258 mm
  (6.25 × 2.05 × 10.16 in)
Weight: 2.16 to 2.5 kg (4.75 to 5.5 lb) depending on options fitted

Electrical
Electrical: 14 or 28 V DC
Power output: ≥10 W
Temperature: −20 to +55°C
Altitude: up to 50,000 ft
Compliance: TSO C34c (GS option), C36c, C37b, C38b, C38c, C40a, RTCA DO-131, −132 (GS option), −153, −156, −157

Status
In service.

Contractor
Bendix/King.

KX 155A nav/com system

Type
Avionic navigation/communication receiver.

Description
The Bendix/King KX 155A nav/com transceiver is a development of the KX 155 (see separate entry), offering the following enhanced or standard fit (where optional on the KX 155) features:
- 760 communications frequencies available with 32 programmable channels
- automatic microphone shut down if activated for more than 33 seconds (stuck microphone)
- bearing-to-station mode and radial-from-station mode
- elapsed time and approach timer
- built in 4 Ω audio amplifier
- internal CDI
- 28 V DC power only
- full backlighting of bezel nomenclature and control
- 2 lines of text information available on the display.

The KX 155A can be used with the KI 203/208 VOR/LOC or the KI 204/209 VOR/LOC/GS indicators, or as part of the KCS 55A compass system with the KI 525A HSI (KN 72 VOR/LOC converter required). When interfaced with the KLN 94 GPS, both the standby Comm and Nav frequencies can be selected and remotely tuned from the GPS database.

Specifications
Dimensions: 159 × 508 × 258 mm
  (6.25 × 2.00 × 10.16 in)
Weight: 1.9 kg incl GS
Frequency range:
Comm receiver: 118.00 to 136.975 MHz in 25 kHz increments
Nav receiver: 108.00 to 117.95 MHz in 50 kHz increments
Glideslope receiver: 329.15 to 335.00 MHz in 150 kHz increments
Electrical: 28 V DC
Power output: 10 W
Temperature: −20 to +55°C
Altitude: up to 50,000 ft
Compliance: TSO C34e, C36e, C37d, C38d, C40a, C40c, RTCA DO-131, −153

Status
In production and in service.

Contractor
Bendix/King.

The Bendix/King KX 155A nav/comm receiver (Bendix/King)   0589141

KX 165 nav/com system

Type
Avionic navigation/communication receiver.

Description
The Bendix/King KX 165 is self-contained nav/com transceiver which utilises a solid-state, gas discharge digital display and a push-button frequency flip-flop with display of active and standby nav and com frequencies. The KX 165 is similar in specification to the KX 155 (see previous entry) but includes a built-in VOR converter and radial display. The main features of the system are:

- 720 or 760 communications frequencies
- 200-channel nav transceiver
- 40-channel glideslope receiver (option)
- built-in VOR converter
- digital display of radial from VOR or VORTAC in lieu of standby NAV frequency
- 14 or 28 V DC power supply option
- 10 W power output
- 25 or 50 kHz com receiver selectivity
- fully TSO qualified.

The KX 165 can be used with the KI 202 VOR/LOC or the KI 206 VOR/LOC/GS indicators.

Specifications
Dimensions: 159 × 520 × 258 mm
  (6.25 × 2.05 × 10.16 in)
Weight: 2.16 to 2.5 kg (4.75 to 5.5 lb) depending on options fitted
Frequency range:
Comm transceiver: 118.00 to 136.975 MHz in 25 kHz increments
Nav receiver: 108.00 to 117.95 MHz in 50 kHz increments
Glideslope receiver: 329.15 to 335.00 MHz in 150 kHz increments
Power requirements:
28 V DC 0.4 A (Receive), 6.0 A (Transmit)
14 V DC 0.7 A (Receive), 8.5 A (Transmit)
Power output: 10 W
Temperature: −20° to +55°C
Altitude: up to 50,000 ft
Compliance: TSO C34c (GS option), C36c, C37b, C38b, C40a, RTCA DO-131, −132 (GS option), −153, −156, −157, −160

Status
In service.

Contractor
Bendix/King.

KX 165A TSO nav/com system

Type
Avionic navigation/communication receiver.

Description
The Bendix/King KX 165A TSO nav/com transceiver is a development of the KX 165 (see separate entry). The unit is fully TSO qualified, with European/JTSO qualification underway. System features include:

- 25 or 8.33 kHz channel spacing
- 32-channel com memory
- internal glideslope
- stuck microphone reset
- radial and bearing display
- elapsed time and approach timer
- composite nav output
- HSI output (left/right/GS/flags)
- internal audio amplifier
- electronic CDI
- 10 W transmitter
- 28 V DC power
- full backlighting of bezel nomenclature and control.

The Bendix/King KX 165A TSO nav/com transceiver   0103876

When interfaced to the KLN 94 GPS, both standby Comm and Nav frequencies can be selected and remotely tuned from the GPS database.

Specifications
Dimensions: 159 × 508 × 258 mm
  (6.25 × 2.00 × 10.16 in)
Weight: 1.81 kg incl GS
Power requirements: 27.5 V DC, 0.6 A (Receive), 6.0 A (Transmit)
Power output: 10 W
Frequency range:
Transceiver: 118.0000 to 136.9916 MHz in 8.33 kHz increments
Receiver: 108.00 to 117.95 MHz in 50 kHz increments
Glideslope receiver: 329.15 to 335.00 MHz in 150 kHz increments
Temperature: −20 to +55°C
Altitude: 50,000 ft
Compliance:
Comm transceiver C37d, C38d
Nav receiver and VOR/LOC converter C40c, C36e
Glideslope receiver C34e
Environmental DO-160c

Status
In production and in service.

Contractor
Bendix/King.

Mk 12 series NAV/COM receivers

Type
Avionic navigation/communication receiver.

Description
The Narco Mk 12 is a VHF communications/navigation radio system designed principally for light and general aviation aircraft. The communications section is a transmitter/receiver covering the VHF band from 118 to 136.975 MHz and encompassing 760 channels. Two channels are preselectable, one for active and the other for standby use. Transmitter power output is nominally 8 W.

The navigation section consists of a VOR/Loc receiver covering the VHF band from 108 to 117.95 MHz providing 200 channels; again, two of these are preselectable for active and standby use.

In each section, new frequencies may be entered into the standby mode at any time and transfer buttons are activated to exchange the selected frequency between active and standby positions.

Navigation section output will drive compatible horizontal situation indicators, area navigation systems, VOR radio magnetic indicators or the system's companion ID 824 VOR/Loc indicator. It will also drive the ID 825 VOR/ILS indicator and, for full ILS capability, a combined 40-channel glide slope receiver, covering the band 329.15 to 335.0 MHz, is available.

Mark 12D+
The TSO'd Mark 12D+ significantly reduces cockpit workloads with its 10 COM frequency storage, digital radial readout in NAV, and full 760 COM channels.

The Mark 12D+ is outwardly similar to the Mark 12D with the exception of a mode selector knob which is located beneath the COM display windows. The new radio retains the standby and active frequency display and flip-flop transfer features, but allows the user to program each of the 10 channels. The frequency selector also doubles as the channel selector.

Active and standby navigation frequencies are displayed in the NAV window. Except during entry of a standby NAV frequency, the right position of the NAV window always displays a digital radial readout from the VORTAC (or dashes if the signal is too weak).

Mark 12 DR
The latest product in the Mark 12 range is the Mark 12 DR which directly replaces the other models in the range. It has similar features but adds a 'Keep alive' active frequency memory option and has minor technical differences.

Specifications
Dimensions:
  (Mark 12 D+) 159 × 64 × 279 mm
  (Mark 12 DR) 156 × 77 × 297 mm
Weight:
  (nav/com with glide slope receiver) 2.0 kg
  (without glide slope receiver) 1.9 kg
  (mounting tray) 0.34 kg
Power: 14 V DC or 28 V DC
Power output: 8 W nominal (transmitter); 10 W minimum (speaker); 50 mW minimum into 300 Ω (headphones)

Narco Avionics Mark 12D/2 NAV/COM receiver   0001342

Narco Avionics Mark 12D+ NAV/COM receiver   0015388

For details of the latest updates to *Jane's Avionics* online and to discover the additional information available exclusively to online subscribers please visit
jav.janes.com

TSO compliance: C37c Class IV (transmitter); C40b (VOR receiver); C36 Class C (localizer); C38c Class C (audio output)

Status

The Mark 12D+ updates the earlier Mark 12 A to D models. Narco also manufacture a Mark 12D NAV/COM 'Direct Cessna Replacement' which replaces the RT308 through to RT328C NAV/COM units.

Contractor

Narco Avionics Inc.

Pro Line 21 CNS

Type

Integrated navigation/communications suite.

Description

Pro Line 21 CNS was designed by Rockwell Collins to meet the demands of current Communications Navigation Surveillance/Air Traffic Management (CNS/ATM) operations. They retain the same form factor as the current Pro Line II and Pro Line 4 products (3.3 in high, 14 in long) but have reduced widths that allow size and weight reductions of up to 50 percent. They are also designed and tested to meet the more stringent DO-160D environmental requirements for helicopters and composite body aircraft. Pro Line 21 CNS maintains the same electrical interfaces as the Pro Line II products they are intended to replace, allowing them to fit into existing installations with a minimum of modification to the existing wiring. These legacy interfaces duplicate the operation of today's products. For new installations the radios provide high-speed I/O buses which interface with the radio interface unit. These interfaces allow the use of new features such as software data loading, enhanced maintenance reporting, digital audio, data communications management and data and audio concentration. Key benefits and features of the system include:

- Data communications management system tailored to business and regional aircraft
- Digital VHF communications transceiver supports Data Mode 2 with future growth to VDL Mode 3 (NEXCOM)
- Weight and size reductions of up to 50 per cent over previous generation products
- Digital audio interfaces for noise-free communications
- Integrated product range facilitates reduced aircraft wiring

- Maintaining legacy interfaces allows for simple replacement of Pro Line II and 400 series products
- Software data loading and enhanced maintenance functions for improved maintenance efficiency.

Pro line 21 CNS components are:

VHF-4000 VHF voice/data communication transceiver

The VHF-4000 provides both voice and data, at 31.5 kbps, to deliver uplinked automated digital messages, flight plan changes and graphical weather depictions. The VHF-4000 includes Modes 2,3 and 4 capability.

ADF-4000/VIR-4000

The ADF-4000 and VIR-4000 include only ADF or VOR/localiser, glideslope and marker beacon functionality in packages that are a half-inch smaller in width than the NAV-4000/4500 configuration.

NAV-4000 navigation receiver

The NAV-4000 includes both the VIR (VOR/ILS/MKR) and ADF functionality in a single Line Replaceable Unit (LRU). The VOR/ILS, marker beacon and ADF receiver modules are the same used in the ADF-4000 and VIR-4000. The NAV-4000 supports CSDB tuning only for the VIR, not the ADF. The NAV-4000 power supply is identical to that in the ADF-4000. The +15 V output is removed for use in the NAV-4500 and VIR-4000. The rear interconnect board is unique to the NAV-4000 in that it includes interconnects for both the VIR and ADF functions. As a result, the NAV-4000 is approximately 0.5 in wider than either the ADF-4000 or the VIR-4000.

The NAV-4000/4500 is packaged in a 2.5 CMU configuration rather than the 2.0 CMU configuration of the ADF-4000 and VIR-4000. The VIR and ADF receivers in the NAV-4000/4500 share a number of interfaces such as tuning inputs, strapping inputs and power inputs. Some of the VIR and ADF legacy signals that are only required for Pro Line II digital system installations have not been included on the NAV-4000/4500 connector.

NAV-4500 navigation receiver

NAV-4500 is identical to the NAV-4000 except it does not include the ADF receiver for installations with a single ADF receiver and an option for a dual ADF receiver.

RIU-4000 CNS manager

The RIU-4000 CNS manager interfaces all the radios with the rest of the avionics system

and also provides high-quality audio for the flight deck as well as digital communication management.

DME-4000 distance measuring unit

The DME-4000 is coupled to the NAV-4000 to provide a comprehensive terrestrially based navigation capability.

SAT-4000 and SAT-6000 satellite communication systems

The SAT-4000 and SAT-6000 systems provide global AERO-I and AERO-H/I services.

HF-900D HF communication system

The HF-900D is available for back-up voice and data communication worldwide, if required.

Specifications
Physical

	NAV-4000/ NAV-4500	ADF-4000/ VIR-4000
Size	2.5 CMU	2.0 CMU
Height	87.4 mm	87.4 mm
Width	63.5 mm	50.8 mm
Length	358.5 mm	358.5 mm
Weight	1.54 kg /1.14 kg	0.95 kg /1.09 kg

Status

In production and in service. The first set of new radios, made available during 2002, were the VHF-4000 VHF voice/data communication receiver, the NAV-4000 navigation receiver, the DME-4000 Distance Measuring Equipment and the RIU CNS manager.

Elliott Aviation introduced an FAA STC-approved avionics flat-panel retrofit for the Hawker 700A which includes the replacement of the exiting instruments with ProLine 21 displays and dual-AHC-3000 Altitude Heading Reference System (AHRS).

The S-80E helicopter, an export version of the CH-53E Super Stallion operated by the Turkish Air Force, incorporates an avionics system based on ProLine 21, including four 152 × 254 mm (6 × 10 in) liquid crystal displays, dual-CDU-9001 Flight Management Systems (FMS), embedded GPS and a data-loading facility.

Contractor

Rockwell Collins.

Pro Line II digital nav/com family

Type

Integrated navigation/communications suite.

Description

Acknowledging the advances in technology, notably in signal processing since the appearance of Pro Line in 1970, Rockwell Collins decided in the late 1970s to develop a replacement. The result was Pro Line II, the first members of which (VHF communications transceiver, VHF navigation transceiver and DME) appeared in January 1983. Other units were added and there now exists a complete range of Pro Line II equipment.

Pro Line II boxes contain analogue/digital and digital/analogue circuits so that individual units of the earlier family can be exchanged on a one-for-one basis without change to the aircraft wiring or racks. Microprocessors within the new units are programmed to accept either analogue or digital frequency tuning arrangements.

The Pro Line II family consists of:

- **ADS-82** air data system
- **AHS-85** attitude/heading reference system
- **APS-65** autopilot
- **APS-85/95** autopilot
- **CTL-22** communication control unit
- **CTL-23** comm/nav control unit
- **CTL-32** navigation control unit
- **CTL-62** automatic direction-finder control unit
- **CTL-92** transponder control unit
- **DME-42** DME receiver
- **EFIS-85/86** electronic flight instrument systems
- **EHSI-74** electronic flight instrument system

Elliot Aviation's cockpit upgrade for the Hawker 700A includes Collins' ProLine 21   0568953

Collins Pro Line II installed in a Falcon 50　　　　0105283

- **IND-42** DME control unit
- **MCS-65** compass system
- **TWR-850** turbulence weather radar system
- **VHF-21/22** VHF communication transceiver
- **VIR-32/33** navigation receiver.

Status
In service in a wide range of regional and business aircraft.

Contractor
Rockwell Collins.

ProLine nav/com family

Type
Integrated navigation/communications suite.

Description
Introduced in 1970, the ProLine series avionics was intended for medium and large general aviation piston and turbine-engined twins. The family was originally designated Low-Profile to emphasise the compact size and form factor of individual units but, in 1975, was renamed Pro Line in recognition of its acceptance by professional pilots in regional airlines, and by the defence forces. Unlike the self-contained members of the Micro Line family, ProLine systems comprise panel-mounted indicators and controls driven by or controlling separate rack-mounted processing and computing boxes. The size and form factor was laid down by Rockwell Collins, there being no industry-wide agreement on packaging for general aviation electronics, in contrast to the highly defined ATR standards governing the characteristics of equipment for airlines. In the military field, ProLine is found on a wide range of aircraft, not only those equivalent in size and performance to general aviation types, but also on attack and surveillance aircraft such as the A-4, F-5E, C-130 and E-3 AWACS.

The ProLine family comprises:
- **346B** audio control/isolation and speaker amplifiers
- **ADF-60** automatic direction-finder
- **ALT-50/55** radio altimeters
- **AP-105** autopilot
- **AP-106A** autopilot
- **APS-80** autopilot
- **BDI-36** bearing/distance indicator
- **CTL** series control heads
- **DME-40** distance measuring equipment
- **FDS-84** flight director system (comprising FD-108 flight director and FIS-70 flight instrument system)

- **FDS-85** flight director system (comprising FD-109 flight director and associated horizontal situation indicator)
- **FPA-80** flight profile advisory system
- **HF-230** HF transceiver
- **PN-101** pictorial navigation system
- **TDR-90** transponder
- **VHF-20A/B** communication transceivers
- **VIR-30A/M and VIR-31A/H** navigation receivers
- **WXR-220/270/300/350** colour weather radars.

Status
In service but no longer in production. Well over 225,000 Pro Line boxes and controls have been manufactured and sold.

Contractor
Rockwell Collins.

Quantum™ line communication and navigation systems

Type
Avionic navigation/communication receiver.

Description
The Quantum™ line of communication and navigation equipment comprises:
- **ALA-52B** radio altimeter
- **DFA-75B** automatic direction-finder
- **DMA-37B** distance measuring equipment
- **MMR** multimode receiver
- **RIA-35B** instrument landing system
- **RTA-44D** VHF data radio
- **RVA-36B** VOR

Quantum™ line features extend reliability over previous designs, with a 30,000 hour MTBF and a guarantee on Mean Time Between Unscheduled Removals. It also offers commonality of numerous parts, assemblies and modules.

Quantum™ line equipment utilises more than 80 per cent of components and software that are common across the line. Quantum™ line features include 486 main processor; common monitor processor in the ILS and radio altimeter; liquid crystal display; packaging to ARINC 650; HIRF/lightning protection; ICAO FM immunity; ETOPS. A common front LCD panel permits easier maintenance with easily understood fault message displays. A flash card is provided for onboard data loading, rapid evaluation of system performance and data recording. A PC-compatible maintenance access port allows for diagnostics and fault isolation on the aircraft. Upgraded mechanical packaging improves protection for electromagnetic interference, reduces heat rise and ensures structural integrity.

Performance has been improved by providing 8.33 kHz channel spacing on the VHF data radio and raising the radio altimeter range to 5,000 ft.

Status
Widely employed on civil aircraft.

Contractor
Honeywell Aerospace, Electronic & Avionics Lighting.

RMA-55B MultiMode Receiver (MMR)

Type
Avionic navigation/communication receiver.

The RMA-55B multimode receiver 2000

0062203

The Quantum™ line is composed of the RIA-35B ILS receiver, DMA-37B DME interrogator, DFA-75B ADF receiver, ALA-52B radio altimeter, RTA-44D VHF Data Radio and MMR (VDR)　　　　0018183

Description

The RMA-55B MultiMode Receiver (MMR) is a member of the Quantum Line of communications and navigational equipment. Building on the modular architecture of the product line, the MMR ensures full compatibility with current and future landing systems.

The RMA-55B meets industry-defined sensor requirements for Category III Instrument Landing Systems (ILS), including requirements for ICAO Annex 10 FM immunity and initially en route/ non-precision Global Positioning System (GPS) approaches. In a single 3 MCU unit, the MMR replaces the functions previously performed by an ARINC 710 receiver and an ARINC 743 GPS receiver. In addition, it provides for incremental growth to Microwave Landing Systems (MLS) and Differential GPS (DGPS) as requirements evolve.

Specifications

Power: 115 V AC
Weight: 5.44 kg (ILS/GPS/MLS)
Dimensions: 324.0 (L) × 90.9 (W) × 194.0 (H) mm

Form factor: 3 MCU per ARINC specification 600
Cooling: forced air per ARINC specification 600
Temperature operation: −15 to +70°C
Warm-up period: stable operating within 1 min after application of power
Frequency selection: serial digital in accordance with ARINC specification 429
Certification: DO-160C

Contractor

Bendix/King.

AVIONIC COMMUNICATIONS SYSTEMS

Canada

AA22 PA controller

Type
Avionic audio management system.

Description
The AA22 series of airborne loudhailer/siren/PA controllers are 28.5 mm high, DZUS mounted controllers for use with the PA110, PA220, PA250 and PA700 remote amplifiers. Features include panel and remote keying, front panel power switch and high and low level inputs.

Status
In production and in service.

Contractor
Northern Airborne Technology Ltd.

Northern Airborne Technologies's AA22-160 PA controller (Northern Airborne Technologies)
1179179

Access/A A710/A711 airborne audio systems

Type
Avionic audio management system.

Description
The Technisonic Access/A system is based on an integrated family of modular, configurable, panel-mounted controls, coupled with external special function units, which produce a complete airborne communications suite. This audio system architecture supports extensive alerting, warning and signalling functions, voice message storage, partitioned crew services, and interface capabilities.

Access/A provides interconnection and interface facilities between airborne crews and their radio systems, as well as providing internal intercom, internal and external paging, airframe alerting, music and entertainment functions. Access/A is designed for applications like emergency medevac, forestry aircraft, search and rescue, customs and emergency services operations.

Access/A A710 and A711 controls are designed to be continuously expanded, relabelled and upgraded. Up to ten crew stations, and up to six crew members per crew station can be supported within a single aircraft. Many special modes and functions such as voice alerting, voice storage, NVG compatibility can be implemented.

Access/A A710/A711 stations support both speaker and headset installations, and provide for internal paging as well as interfaces to commercially available external paging systems.

Specifications
Power: 28 VDC, 350 mA
Dimensions:
(A710 dzus panel) 146 (W) × 47.6 (H) × 154.2 (D) mm
(A711 dzus panel) 146 (W) × 66.7 (H) × 154.2 (D) mm
Weight:
(A710) 1.09 kg
(A711) 1.36 kg
Temperature:
(operating) −40 to +70°C
(survival) −55 to +85°C
Altitude: 25,000 ft
Environmental: DO-160C
Features:
1– 6 users per crew station
direct push-button Tx function
optional voice alerting, with priority
optional voice message storage/replay
six transceiver operation, plus PA
headset and speaker outputs
VOX, PTT or live ICS modes
individual Tx enunciation
split ICS/Rx volume controls
emergency Rx/Tx capability, including boom microphone
changeable front panel legends
configurable ICS loops
simulcast operation
Qualification: RCTA C50c

Status
In wide use.

Contractor
Technisonic Industries Limited.

ADT-200A aeronautical data terminal

Type
Avionic datalink system.

Description
The ADT-200A aeronautical data terminal has been designed principally as an aeronautical transportation fleet management tool, utilising satellite communications. The communications system is based on Inmarsat Standard C communications format and is capable of two-way communications consisting of position reporting, general messaging and the sending of coded messages. The fleet communications control is performed through a network control centre which allows communications with the individual ADT-200A terminals. Position information is derived from a GPS receiver and encoded along with a terminal identification code in the transmit signal.

The ADT-200A is compliant with all applicable mechanical and electrical airborne regulations. The system consists of a compact keyboard/display unit, modular transceiver and antennas.

The keyboard/display unit is easily mounted in the cockpit or elsewhere in the aircraft to alert the operator to incoming messages. It enables the operator to read messages under any lighting condition and to respond via a keyboard.

The transceiver houses the transmitter, GPS position locator and power supply unit. The transceiver provides a link between the operator's keyboard/display unit and the antenna, sends and receives messages and transmits the position of the aircraft as determined by the GPS.

The antennas are rugged, lightweight and weather resistant with a durable aerodynamic cover. They are used for both L-band satellite data communications and for receiving position signals from the GPS system. Once installed, the antennas are automatically aligned with the satellite.

Specifications
Dimensions:
(transceiver) 190.5 × 210.8 × 324.1 mm
(MDS antenna) 25.4 × 127 mm diameter
(GPS antenna) 19.1 × 76.2 mm diameter
(keyboard/display) 267 × 140 × 51 mm
Weight:
(transceiver) 7.5 kg
(MDS antenna) 0.6 kg
(GPS antenna) 0.1 kg
(keyboard/display) 0.9 kg
Power supply: 22–32 V DC
Frequency:
(transmit) 1,626.5–1,660.5 MHz
(receive) 1,530–1,559 MHz
Channel spacing: 5 kHz

Status
In service.

Contractor
EMS Technologies Inc.

AMT-100 aeronautical mobile terminal

Type
Avionic communications system.

Description
The AMT-100 aeronautical mobile terminal provides air-to-ground and ground-to-air voice communications to and from any location on the globe via the Inmarsat satellite system. The unit communicates through aeronautical standard ground stations operated by Inmarsat signatories.

The AMT-100 is an all-in-one system providing a single-voice channel with direct dial capability for automatic global operation. The system, consisting of a transceiver, high-power amplifier and non-obtrusive high-gain antenna, has been designed for corporate aviation aircraft.

The transceiver assembly handles all protocols, performs voice and data modulation and handles frequency conversion and Doppler correction. It also contains the antenna control processor. The high-power amplifier is a reliable solid-state unit designed for a wide range of aircraft installations. The antenna assembly is a high-gain tailfin-mounted steerable antenna which provides full azimuth and elevation coverage without gaps. The antenna may be pointed using aircraft navigation system inputs.

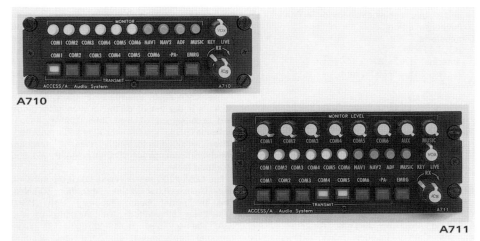

A710

A711

The Technisonic Industries Limited Access/A A710 and A711 airborne audio systems
0062785

The AMT-100 provides voice communications with the public telephone network via the Inmarsat satellite system　　0503990

Specifications
Dimensions:
(transceiver) 381 × 190.5 × 317.5 mm
(amplifier) 127 × 190.5 × 317.5 mm
(diplexer/LNA) 431.8 × 50.8 × 198.1 mm
(antenna drive assembly) 355.6 × 63.5 × 190.5 mm
Weight:
(transceiver) 14.5 kg
(amplifier) 7.7 kg
(diplexer/LNA) 2.72 kg
(antenna drive assembly) 3.17 kg
Power supply: 28 V DC

Status
In service.The first AMT-100-equipped Bombardier Challenger was in service by the end of 1992. Full access approval was gained from Inmarsat in August 1993.

Contractor
EMS Technologies Inc.

AN/ASH-503 voice message system

Type
Avionic voice messaging/aural alerting system.

Description
The AN/ASH-503 voice message system is a single-box device using advanced microprocessor and memory technology to store and reproduce high-quality digitally produced speech messages.

The AN/ASH-503 Micro VMS can store 15 three-word messages or 13 seconds of continuous speech. Each system incorporates extensive self-test facilities. High-quality speech reproduction is ensured by using a 30 kHz digitising sampling rate. Any type of voice or language can be stored and reproduced. Each message can be preceded by an alerting tone and repeated as often as necessary.The messages can be sorted in order of priority so that, in the event of coincidental inputs, the most important message is reproduced first.

Specifications
Dimensions: 64 × 65 × 67 mm
Weight: 0.45 kg
Power supply: 28 V DC, 2.5 W

Status
In service.

Contractor
CMC Electronics Inc.

CMA-2102 high-gain satcom antenna system

Type
Aircraft antenna, satellite communications system.

Description
The CMA-2102 high-gain satcom antenna system is designed to support the Inmarsat® Aero-H and

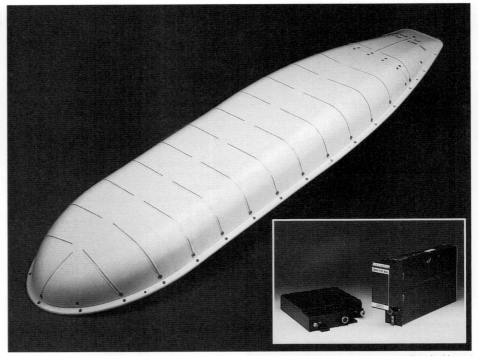

The CMA-2102 high-gain satcom antenna system with (inset) the low-noise amplifier (left) and beam-steering unit (right)　　0018149

Aero-H+ satellite communications services, which provide aircraft with simultaneous two-way, digital voice and real-time data communications capability for flight crew, cabin crew and passengers. It provides a suitable baseline for the forthcoming Communications Navigation Surveillance/Air Traffic Management (CNS/ATM) system.

The CMA-2102 is a single, phased-array, electronically steered, top-mounted, high-gain antenna, with a single Beam Steering Unit (BSU) and Diplexer/Low Noise Amplifier (D/LNA) which provides hemispherical coverage in an extremely reliable installation. Since the CMA-2102 is mounted on the top of the aircraft, its coverage pattern does not suffer from keyholes/blindspots, and installation is greatly simplified. The system conforms to ARINC Characteristic 741 and Inmarsat SDM.

The CMA-2102 covers 360° in azimuth and −5 to +90° in elevation and operates over the frequency range 1,525 to 1,660.5 MHz.

Specifications
Frequency:
(receive) 1525.0 –1559.0 MHz
(transmit) 1626.5 – 1660.0 MHz
Polarisation: right-hand circular; axial ratio <6.0 dB
Multipath rejection: >12.9 dB rejection at 5° elevation
Satellite discrimination: >13 dB over coverage region
Gain: between 12 dBiC and 17 dBiC over 90% of Inmarsat hemisphere; minimum of 9 dBiC over 100% of Inmarsat hemisphere
Axial ratio: less than 6 dB for all steering angles and all frequencies of operation within coverage region
Dimensions:
(BSU) 2 MCU
(antenna) 1,702 × 470 × 126 mm
(D/LNA) 281 × 197 × 50 mm
Weight:
(BSU) 2.7 kg
(antenna) 27.9 kg
(D/LNA) 3 kg
Power consumption:
(BSU) 12 W
(antenna) 45 W
(D/LNA) 12 W

Status
CMA-2102 systems have been selected by many major airlines and VIP/military operators, including American Airlines, Air France, Air Canada, Cathay Pacific, China Airlines, EVA Airways Corporation, Japan Airlines, KLM,

Lufthansa, Quantas, Singapore Airlines, Saudi Arabian Airlines, Swissair, United Airlines and several others. Commissioned on the following aircraft types (TCs and/or STCs): Boeing (707/727/737/747/757/767/777); DC-10; MD-11 and MD-90; and Airbus (A300/319/320/321/330/340). BAE Systems, Canada claims that the CMA-2102 has claimed over 70 per cent of the wide-body aircraft market share.

In the military role, the CMA-2102 has been selected for a variety of tactical helicopters for attack, combat search and rescue and anti-submarine warfare roles. Installations include the AH-1P, Bell CF-UTTH, Denel Rooivalk, Eurocopter Tiger, E-4B, F-27/28, GKN Westland and Agusta Cormorant and P-3C.

During 2004, Airbus Industrie selected a variant of the 2102 satcom antenna, designated 2102LW (Light Weight), as standard Seller Furnished Equipment (SFE) for its A380 ultra-large passenger aircraft.

Contractor
CMC Electronics Inc.

CMA-2200 intermediate-gain satcom antenna system

Type
Satcom antenna system.

Description
The CMA-2200 intermediate-gain satcom antenna system design supports the requirements of the new-generation Inmarsat Aero-I satellite communication service. The Inmarsat-3 satellite provides aircraft with telephony, fax and real-time data communications and has been developed to meet the communication requirements of short-/medium-haul regional and corporate jet operators.

The CMA-2200 Intermediate Gain Antenna (IGA) is a derivative of the proven technology and architecture of the CMA-2102 High-Gain Antenna (HGA).The CMA-2200 is a top-mounted, linear array, electronically steered antenna, which conforms to ARINC Characteristic 761 and Inmarsat SDM.

The CMA-2200 system consists of the antenna and Diplexer/Low Noise Amplifier (D/LNA). The traditional Beam Steering Unit (BSU) has been eliminated, the steering of the IGA being performed directly by the terminal equipment provided by leading manufacturers.

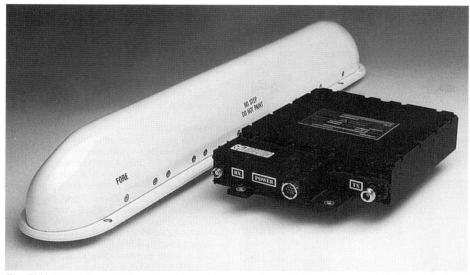

The CMA-2200 intermediate-gain satcom antenna system with the antenna unit (left) and diplexer/low noise amplifier (right) 0018148

Specifications
Coverage: >95% of the Inmarsat hemisphere
Frequency:
(receive) 1,525–1,559 MHz
(transmit) 1,626.5–1,660.5 MHz
Gain: 8.5 dBiC typical; 6.0 dBiC min
Polarisation: right-hand circular; axial ratio <6.0 dB
Multipath rejection: >10 dB at 5 elevation
Satellite discrimination: 16 dB typical; >7 dB over coverage region
Dimensions:
(D/LNA) 280 × 197 × 50 mm
(IGA) 759 × 97 × 109 mm
Weight:
(D/LNA) 3.0 kg
(IGA) 2.7 kg
Power consumption:
(D/LNA) 12 W
(IGA) 7 W

Status
Inmarsat's rigorous AERO-I IGA testing was completed in August 1997. CMA-2200 black-label production shipsets started deliveries in January 1998. The CMA-2200 system is part of the Honeywell/Thales MCS-3000/6000 and MCS-7000 satcom systems.

Contractor
CMC Electronics Inc.

DigiVOX™ and InterVOX™ intercom systems

Type
Avionic internal communications system.

Description
AA83/AA82 InterMUSIC™ stereo intercom
The AA83/AA82 InterMUSIC™ stereo intercom is a high-quality intercom combined with a wideband stereo music processor, incorporating RX audio functions to provide an integrated communication system. Standard features include an ICS Tie Line, Pilot/Split/All Modes, VOX/Live/PTT modes, Music/ICS volume, stereo

Northern Airborne Technology's AA83 stereo intercom (Northern Airborne Technology) 1179180

Northern Airborne Technology's AA85 (Northern Airborne Technology) 1179181

4 + 6 place configurations, ANF™ (Active Noise Filtering), high-quality Surface Mount Technology (SMT), pilot isolate, gradual return of music mute, pilot priority transmit override, patented individual microphone gating and emergency failsafe.

Specifications
Dimensions: 33 × 66 × 142 mm
Weight: 0.32 kg
Power: 11–32 V DC, 550 mA
Displays: 2-colour LED

AA85 InterVOX II™ intercom system
The AA85 InterVOX II™ is a high-quality voice-activated intercom using Northern Airborne Technology's InterVOX™ and ANF™ technology, designed as a direct replacement for the earlier AA80 series intercoms. A single unit provides services for two, four or six place installations, and it provides pilot/crew/all modes to include or isolate other crew and passengers from the services available. The provision of independent microphone circuits reduces noise and other interference.

Specifications
Dimensions: 33 × 66 × 142 mm
Weight: 0.31 kg
Power: 12–30 V DC, 600 mA
Displays: 2-colour LED
Approvals: TSO-C50c

Status
In production and in service.

Contractor
Northern Airborne Technology Ltd.

JS-100A aeronautical satcom system

Type
Avionic communications system.

Description
The JS-100A satcom system has been designed specifically for use on smaller aircraft. The unit has been engineered to minimise impact on airframe and performance.

The system consists of a transceiver assembly, high-power amplifier and antenna assembly. The transceiver assembly handles all protocols, performs voice and data modulation and handles frequency conversion and Doppler correction. It also contains the antenna control processor. The high-power amplifier is a reliable solid-state unit designed for a wide range of aircraft installations. The antenna assembly is a high-gain steerable antenna which provides full azimuth and elevation coverage without gaps. The antenna may be pointed using aircraft navigation system inputs or a signal strength measurement from the transceiver.

Specifications
Dimensions:
(transceiver) 317.5 × 317.5 × 190.5 mm
(high-power amplifier) 127 × 317.5 × 195.6 mm
(antenna) 1,371.6 × 203.2 × 200.7 mm
Weight:
(transceiver) 12.25 kg
(high-power amplifier) 5.44 kg
(antenna) 15.42 kg
Power supply: 115 V AC, 400 Hz or 28 V DC

Status
In service.

Contractor
EMS Technologies Inc.

Multi-user audio controllers

Type
Avionic audio management system.

Description
Northern Airborne Technology Ltd produces an extensive range of multi-user audio controllers in the AMSXX/AAXX series, including:

AMS42/42F
The AMS42 is designed for VFR//OAS/forestry/corporate/fleet use. The AMS42F is specifically intended for forestry application. The AMS42/42F systems are dedicated dual-channel audio controllers with VOX ICS operation and additional internal switch programming; these controls will work in both LH or RH pilot configurations; support is provided for the pilot, co-pilot and five passengers; five transceiver positions are provided. Functions include: COM 1, COM 2, FM 1, FM 2, AUX, PA transceivers, RX and ICS lever controls.

AMS43/43H/43P
The AMS43 series is designed for VFR/IFR/OAS/forestry/corporate/fleet use. All models represent a dedicated configuration of the AA95 series audio controller with LIVE/KEYED/VOX ICS operation; support is provided for the pilot, co-pilot and four passengers; five transceivers and a PA position are provided, as well as six additional switched receivers and two tape inputs; emergency operation and three-level alerting are also featured in these controllers. The AMS43H has special panel legends for Canadian and Australian operators with HF. The AMS43HP has an internal switch to select between PA Key out or Alert 3 in. Functions include: COM 1, COM 2,

Northern Airborne Technology's AMS42F (Northern Airborne Technology) 1179185

Northern Airborne Technology's AMS43 (Northern Airborne Technology) 1179186

Northern Airborne Technology's AMS44
(Northern Airborne Technology)   1179187

Northern Airborne Technology's AMS50
(Northern Airborne Technology)   1179188

Northern Airborne Technology's AA12S
(Northern Airborne Technology)   1179178

AUX, FM 1, FM 2, PA transceivers, NAV 1/2, ADF 1/2, DME/MKR receivers.

AMS44

The AMS44 is designed for police/forestry/charter/corporate/fleet use. It is a dedicated version dual-channel audio controller with VOX ICS operation and front panel Nav Aid selections. It will work in either LH or RH pilot configurations, with support for pilot, co-pilot and five passengers. It includes five transceiver positions, as well as five additional audio sources, together with ICS Tie line and Direct inputs. Functions include: COM 1, COM 2, FM 1, FM 2, AUX, PA transceivers, NAV 1, NAV 2, ADF, DME MKR RX switches.

AMS50

The TSO'd AMS50 was the first general aviation audio control panel to provide an independent transmit capability. Individual selection of transmitters can be activated for the pilot and co-pilot positions. COM 3 can be configured for duplex cellular telephone communication at installation. The controller accepts 3 or 4 COMS and 5 NAV receivers. Panel features include independent microphone gating, Active Noise Filtering (ANF), stereo for all six positions, adjustable music muting, pilot/crew/all modes (optional PTT available for crew positions), an emergency failsafe feature, and pilot/crew isolate. An optional marker beacon receiver is also available. The AMS50 audio panel is similar in size and intended to directly replace the KMA24 H audio controller.

Specifications
Dimensions: 34 × 159 × 165 mm
Weight: 1.0 kg
Power: 11–32 V DC
Display: 2-colour LED
Approval: TSO-C50c (audio selector panel); TSO-C35d Class A (marker beacon)

Northern Airborne Technology's AA95-912
(Northern Airborne Technology)   1179182

Northern Airborne Technology's AA95-913
(Northern Airborne Technology)   1179183

Northern Airborne Technology's AA97-402
(Northern Airborne Technology)   1179184

Northern Airborne Technology's N335-011
(Northern Airborne Technology)   1179191

AA12

The AA12 is designed as a VFR audio controller. It provides stereo audio to two crew and four passenger positions with VOX, LIVE and KEYED ICS modes. Functions include: COM 1, COM 2, COM 3, PA transceivers, NAV, ADF and DME/MKR Nav Aids.

AA92H

The AA92H series are custom dual-channel controllers based on the AMS42 design. Models available include: AA92H-407, AA92H-410, AA92H-411.

AA94

The AA94 series are custom dual-channel controllers based on the AMS44 design. Models available include: AA94-400 (special), AA94-426, AA94-427 (law enforcement).

AA95

The AA95 series are functional equivalents of the AA90 series incorporating a 500 mW headset driver to meet USFS/OAS requirements. Models available include: AA95-512 (police), AA95-824 (police), and a variety of AA-95 emergency medical crew models.

AA96

The AA96 series is again a variant of the AA90 design, customised for tactical (AA96-001) and USFS (AA96-400) operations.

AA97

The AA97 is also a variant of the AA90 design, optimised for law enforcement operations, with individual control of each of six transceivers. Functions include VHF 1/2, TAC 1/2/3/4 and PA transceivers, NAV 1, NAV 2, ADF 1, DME switched Nav Aids.

N335

The N335 is Northern Airborne Technology's 'next generation' dual-user audio panel. It combines new technology with the modular design concept of the N301A family, plus the flexibility and control of the AA97. The N335 supports two operators with either high or low impedance systems and contains a selectable, fully redundant back-up power supply, audio amplifiers, ICS bus, and secondary volume potentiometers. Functions include: NAV 1, NAV 2, DME 1, DME 2, ADF, MKR, DF, VHF 1, VHF 2, UHF, FM1, FM 2, FM3, GSM.

Status

In production and in service in a wide variety of fixed- and rotary-wing platforms.

Contractor

Northern Airborne Technology Ltd.

Single-user audio controllers

Type
Avionic audio management system.

Description
Northern Airborne Technology Ltd produces a range of single-user audio controllers including the AA24, AA25 and N301A.

AA24 and AA25
The AA24 and AA25 provide high performance audio to a single-user in a network with VOX, LIVE and KEYED Inter-Communication System (ICS) capability. The small size and extensive functions make these models ideal for single-seat aircraft. Inputs are fully floating, including a dedicated alerting input, and units can work with either high impedance headsets (models AA24-001 and AA25-001) or low impedance headsets (models AA24-801 and AA25-801).

Both models provide the following services: COM1, COM 2, COM 3, COM 4 + PA, NAV, ADF, DME, MKR and AUX, ICS, RX and VOX controls.

The AA25 also provides independent radio volume adjustment for up to six sources.

Specifications
Dimensions:
AA24 38.1 × 146.05 × 149.35 mm
AA25 66.8 × 146.05 × 149.35 mm
Weight:
AA24 0.5 kg
AA25 0.68 kg
Power: 22–32 V DC, 0.4 A

N301A
The N301A. It is designed for use in new installations where single- or multiple-user networks of 10+ stations are required. User controlled features include split RX and ICS volume controls, a live or 'hot microphone' ICS function switch, an additional RX input for the AUX TX position, and an extra Nav Aid position. Additional internal inputs have been added to improve threat alerting capability. TSO approval pending.

The N301A single-user audio controller updates the earlier A301-6 and N301 systems. Fully floating inputs remove ground-loop or common mode noise. Both high- and low-impedance headset and microphone systems are supported. Microphone outputs can be individually adjustable for each transceiver. The N301A can be utilised in new installations where single or multiple user networks (up to 10 stations) are required.

User controlled features of the N301A include split RX and ICS volume controls, VOX squelch control, an additional RX input for the auxiliary TX position and an extra navigation aid position. Additional internal inputs have been added to improve airframe threat alerting capability. Each panel is self-contained, utilising solid-state modular plug-in construction, sealed switching devices for low noise and gold switch/relay contacts for long life. Other features of the N301A include a CVR output, TX annunciator light, emergency mode switch, and a second connector to permit supplemental inputs with two additional ICS buses to support more complex ICS networks. A VOX circuit and front

AA24 single-user audio controller   0044776

AA25 single-user audio controller   0044777

Northern Airborne Technology's N301A-000
(Northern Airborne Technology) 1179190

panel control on the N301A series provides voice activated ICS.

NAT offers the N301A in standard configurations or customised panel layouts with 5 or 28 V DC standard blue/white lighting or NVG compatible.

Specifications
Dimensions: 66.68 × 146.05 × 93.98 mm
Weight: 0.75 kg
Power: 22–32 V DC, 0.17 A

Status
In production and in service.

Contractor
Northern Airborne Technology Ltd.

Tac/Com™ Tactical FM Communications systems

Type
Avionic communications system, VHF/UHF.

Description
Northern Airborne Technology Ltd has designed a range of Tac/Com tactical FM communications systems in consultation with law enforcement, emergency medical services and forestry agencies. The design aim was to minimise problems associated with complex multiradio installations and multimission aircraft such as panel space, simultaneous operation, inexperienced users, complex operational modes and future upgrades.

The Tac/Com™ family is a series of configurable multiple radio control heads, agile transceivers and supplemental equipment. The modular control heads are capable of operating either Northern Airborne Technology or vendor

Northern Airborne Technology's TH250 control head (Northern Airborne Technology) 1179195

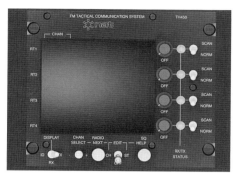

Northern Airborne Technology's TH450
(Northern Airborne Technology) 1179196

radios. The Northern Airborne Technology radios are synthesised agile transceivers covering VHF and UHF bands and providing multiple scanning modes, agile synthetic guard, priority scanning, CTCSS and DPL squelch control, and they are DCS encryption ready. Supplemental equipment includes: encoders/decoders, DTMF keypads, relay-simulcast controllers and antennas.

All Tac/Com NTX series transceivers are harness and tray compatible and all Tac/Com 250/350 series control heads can be retrofitted into existing C722, C962 or C1000 (3 in dzus) openings.

Tac/Com II™ control heads
Tac/Com II control heads provide operators with multi-radio operational capabilities, control of up to four radios, with each radio supporting up to 128 channels. Models available include: TH250 (two-channel), TH350 (three-channel) and TH450 (four-channel). Variants of these options are available to meet customer requirements.

NTX series transceivers
The NTX generation of FM transceivers are the latest evolutionary step within Northern Airborne Technology's series of Tac/Com remote-control airborne radios. Using microprocessor controlled digital technology, the NTX radio is capable of high speed scanning, high volume channel storage, and it is compatible with the existing Tac/Com II control head. Depending on the model, optional features can include a two-channel dedicated real-time guard (USFS compatible), selectable bandwidth filters, and DF outputs.

A major advantage of the NTX transceiver is very compact size: 2.0 in wide (1/4 ATR), 5.1 in high and 10.5 in deep, with a total weight of under 2.5 kg, including undertray.

Specifications
Model NTX138
Frequency coverage: 138.000–173.995 MHz
Channels: 126, plus 2 guard, channel memory
Channel increments: 2.5 kHz
RF power: 1 or 10 W selectable

Northern Airborne Technology's NPX136D
(Northern Airborne Technology) 1179192

Weight: 1.8 kg
Power: 28 V DC
Compliance: USFS

Model NTX066
Frequency coverage: 66.000–87.995 MHz
Channels: 128 channel memory
Channel increments: 2.5 kHz
Weight: 1.8 kg
Power: 28 V DC

Model NTX403
Frequency coverage: 403.000–511.99375 MHz
Channels: 128 channel memory
Channel increments: 6.25/10 kHz
Weight: 1.8 kg
Power: 28 V DC

NPX series FM transceivers
The NPX136D panel-mount P25 VHF transceiver is a stand-alone single-user radio. It provides all the features necessary for communications within the high VHF band. Each of the systems 255 available channels can operate in one of three modes: wide band analogue, narrow band analogue, or digital P25 Phase 1. A SCAN function allows scanning of selected channels. Simplex and semi-duplex operations are available and a Guard receiver is standard with the -070 variant. The aircraft dimmer bus provides control for the panel lighting.

The NPX138N panel-mount FM transceivers are designed as very small size (2 in high), light weight (1.36 kg) stand-alone radios for single-mission users. The NPX138N series are synthesised, frequency-agile, transceivers that provide easy access to all pre-programmed channels and the ability to modify or add channels in the air. Each of the 100 memory channels can store a receive frequency with CTCSS tone, a transmit frequency with CTCSS tone, scan function and alpha-numeric identifier on a 32-character two-line LED display. Transmit power of 1 or 10 W can be selected. Each channel is selectable for wide- (5 kHz deviation) or narrow- (25 kHz deviation) band operation. DTMF encoding and direct keyboard programming can be added using Northern Airborne Technology's DTE12 datapad. Panel lighting can be 28, 14 or 5 V DC controlled, with dimmer. NVG compatible lighting is optional.

Specifications
Model NPX136D
Frequency coverage: 136.000–173.9975 MHz
Channels: 255 channel memory, optional
  2-channel guard
Channel increments: 2.5/6.25 kHz
RF power: 1 or 10 W selectable
Power: 28 V DC

Model NPX138N
Frequency coverage: 138.000–173.995 MHz
Channels: 100 channel memory, optional
  2-channel guard
Channel increments: 5/6.25 kHz
Channel selectable: 25 or 12.5 kHz
  (Model NPX138N)
RF power: 1 or 10 W selectable
Weight: 1.36 kg

Northern Airborne Technology's NPX138
(Northern Airborne Technology) 1179193

Northern Airborne Technology's TH350 (Northern Airborne Technology) 1179194

Power: 28 V DC
Compliance: USFS

Status
In production and in service in a wide variety of fixed- and rotary-wing platforms.

Contractor
Northern Airborne Technology Ltd.

TFM-138 series airborne VHF FM transceivers

Type
Avionic communications system, VHF.

Description
The Technisonic TFM-138 series of airborne VHF FM transceivers utilise frequency synthesis techniques to provide FM communications on every currently available channel within the VHF FM high band. These radios cover the entire band from 138 to 174 MHz in 2.5 kHz increments. Data entry and function control are via a front panel 12-button keypad. Operating frequency and other related data are presented on a 48-character two line LED matrix display, available in green or red.

Technisonic FM transceivers can be operated in 'direct entry' or 'simplex' mode by keying in the operating frequency, or can function without restriction on any split frequency pair within the band. The TFM-138 features 100 preset memory positions, each of which is capable of storing a receive frequency, a transmit frequency, a separate CTCSS tone for each receive and transmit frequency and an alpha numeric identifier for each channel. The TFM-138A and TFM-138B feature similar storage capability, but provide 120 channels of preset memory and offer

Specifications

	TFM-138	TFM-138B
Frequency	138.000 to 174.000 MHz	138.000 to 174.000 MHz
Tuning increments	2.5 kHz	2.5 kHz
Operating mode	F3E simplex or semi-duplex	F3E simplex or semi-duplex
Channel spacing	12.5, 25 or 30 kHz	12.5, 25 or 30 kHz
Memory channels	100	120
Output power	1 W or 10 W	1 W or 10 W
Dimensions	203.2 × 76.2 × 146.05 mm	203.2 × 76.2 × 146.05 mm
Weight	1.4 kg	1.5 kg
Power	28 V DC, 2 A	28 V DC, 2 A
Guard receiver	two channel synthesised	two channel synthesised
Certifications	FCC and DOC type approved; DO-160c	FCC and DOC type approved; DO-160c

the additional capability of allowing for DPL or DCS coded squelch operation. The TFM-138 and TFM-138B allow either 25 kHz wide band or 12.5 kHz narrow band operation on any or all of the preset channels. Data can be entered into any of the preset non-volatile memory positions for both main and guard channels via the front panel keyboard. Data on stored channel settings is instantly available.

Technisonic FM transceivers feature a synthesised two channel guard receiver, a DTMF encoder for signalling during transmit, and a scan function which will scan any or all of the frequencies stored in the preset memory.

The TFM-138 series transceivers are panel mounted (Dzus mounting) and are self-contained in the chassis. The front panel contains controls for main and guard channels including a HI/LO power output switch. High power is 10 W output and low power is 1 W output, which is necessary to comply with marine harbour environment rules. The small size and light weight of the transceivers make them suited to helicopter installations.

The TFM-138 series consists of four models: the TFM-138, TFM-138A, TFM-138B and TFM-138C.

TFM-138A specifications are identical to TFM-138B, with the exception that the TFM-138A does not offer 12.5 kHz operation. TFM-138C specifications as per TFM-138B, but with the addition of voice encryption.

Status
In wide use.

Contractor
Technisonic Industries Limited.

TFM-403 airborne UHF FM transceiver

Type
Avionic communications system, UHF.

Description
The Technisonic Industries TFM-403 UHF FM transceiver utilises frequency synthesis techniques to provide FM communications on currently available channels within the public safety, forestry, government agency, and general service UHF FM band. The TFM-403 covers from 403.000 to 512.000 MHz in 2.5 kHz steps. Data entry and function control are via a front panel, 12-button keypad. Operating frequency and other related data are presented on a 48-character, two line LED matrix display, available in green or red.

The TFM-403 transceiver can be operated in the 'direct entry' or 'simplex' mode by keying in the desired operating frequency, or without restriction on any split frequency pair within the band. There are 120 memory positions, each of which can store a Rx and Tx frequency, a separate CTCSS and/or a DPL/DCS (Digitally Coded Squelch) tone for each receive and transmit frequency, and a nine-character alpha numeric identifier for each channel. Data is entered into the preset memory positions for main and guard channels via the front panel keyboard, or downloaded from a PC. All stored data is available for instant recall.

The TFM-403 transceiver features a synthesised, programmable, two-channel guard receiver, a DTMF encoder for signalling during transmit, display of CTCSS tone frequency as well as EIA identifier and/or display of a DPL/DCS code, and a scan/priority scan function that can scan any or all of the channels stored in preset memory. Additional operator selectable capabilities include a 90 second transmitter time out, a keyboard lockout feature and a direct/repeat function for simplex operation. The TFM-403 is dzus mounted.

The Technisonic Industries Limited TFM-138B airborne VHF FM transceiver 0062786

The Technisonic Industries Limited TFM-403 airborne UHF FM transceiver 0062787

Specifications
Frequency: 403.000– 512.000 MHz
Tuning increments: 2.5 kHz
Operating mode: F3E simplex or semi-duplex
Channel spacing: 20 or 25 kHz
Memory channels: 120
Output power: 1 W or 10 W
Dimensions: 203.2 × 76.2 × 146.05 mm
Weight: 1.4 kg
Power: 28 V DC, 1.3 A (1 W transmit), 2.0 A (10 W transmit)
Temperature range: –45 to +70°C
Altitude: 50,000 ft

Guard receiver: two-channel synthesised
Certifications: FCC- and DOC-type approved

Status
In wide use.

Contractor
Technisonic Industries Limited.

TFM-500 airborne VHF/UHF FM Transceiver

Type
Avionic communications system, VHF/UHF.

Description
The Technisonic TFM-500 airborne VHF/UHF FM transceiver utilises frequency synthesis techniques to provide FM communications on currently available channels within the general radio service VHF FM high band and general service UHF FM band. The VHF module covers from 138.000 to 174.000 MHz in 2.5 kHz increments, while the UHF module covers from 403.000 to 512.000 MHz, also in 2.5 kHz increments. Data entry and function control are via a front panel, 12-button keypad. Operating frequencies and other related data are presented on a 96 character, four line LED matrix display, available in either green or red.

The TFM-500 can be operated in the 'direct entry' or 'simplex' mode by keying in the desired operating frequency. It can also function without

The Technisonic Industries Limited TFM-500 airborne VHF/UHF FM transceiver 0062788

restriction on any split frequency pair within either band, and offers 'cross band' repeat capability. The unit features 400 preset memory positions (200 VHF and 200 UHF) each capable of storing a receive frequency, a transmit frequency, a separate CTCSS tone for each receive and transmit frequency, an alpha numeric identifier for each channel, and the additional ability to provide DPL or DCS coded squelch operation on each channel. The TFM-500 can operate with

either 25 or 12.5 kHz bandwidth on all memory channels. Operating data can be entered via the front panel or from a PC. Data can also be downloaded to a PC.

The TFM-500 includes a synthesised two channel VHF or UHF guard receiver, a DTMF encoder for signalling during transmit, and a scan function which will scan any or all of the frequencies stored in the memory. A remote control head is offered (RC-500) which provides for slaved operation of the main transceiver from a remote location, allowing a second position in the aircraft to exercise frequency control. Both UHF and VHF operating frequency as well as guard frequency can be controlled from the remote position. Active frequency is displayed on both local and remote displays. The TFM-500 is dzus mounted.

Status
In wide use.

Contractor
Technisonic Industries Limited.

Specifications

	VHF module	UHF module
Frequency	138.000 to 174.000 MHz	403.000 to 512.000 MHz
Tuning increments	2.5 kHz	2.5 kHz
Operating mode	F3E simplex or semi-duplex	F3E simplex or semi-duplex
Channel spacing	12.5, 25 or 30 kHz	12.5, 25 or 30 kHz
Memory channels	200	200
Output power	1 W or 10 W	1 W or 10 W
Dimensions	overall TFM-500 dual band transceiver: 203.2 × 95.3 × 146.05 mm	
Weight	1.4 kg	1.4 kg
Power	28 V DC, 2 A	28 V DC, 2 A
Guard receiver	two channel synthesised	two channel synthesised
Certifications	FCC and DOC type approved; DO-160c	FCC and DOC type approved; DO-160c

---

Czech Republic

LUN 3520 LPR 2000 VHF/UHF airborne transceiver system

Type
Avionic communications system, VHF/UHF.

Description
The LUN 3520 VHF/UHF airborne transceiver is a modular system which is suitable for installation in a wide variety of civil/military fixed- and rotary-wing platforms. The technical characteristics of the system comply with current ICAO and NATO standards.

Specifications
Frequency ranges:
  (VHF) 118.000 to 155.975 MHz
  (UHF) 220.000 to 399.975 MHz
Channel spacing:
  (VHF) 8.33 and 25 kHz
  (UHF) 25 kHz
Receiver sensitivity:
  Max 3 µv (S/N = 15 dB
  M = 0.5; 1 kHz)
RF power output:
  (full) 16 W
  (reduced) 6 ±2 W
Spurious rejection: ICAO Annex 10 Part II Chapter 1 Section 2.3.25
Power: 27 V DC, 25 W on receive, 220 W on transmit
Temperature range: – 60 to + 60°C

LUN 3520 components
  LUN 3520.11-8 transmitter/receiver, including guard receiver capabilities at 121.5 and 243 MHz;

LUN 3520.24-8 control box, designed for remote control from a single position, in conjunction with the audio switch box LUN 3520.44, with the ability to select from 20 pre-set channels;

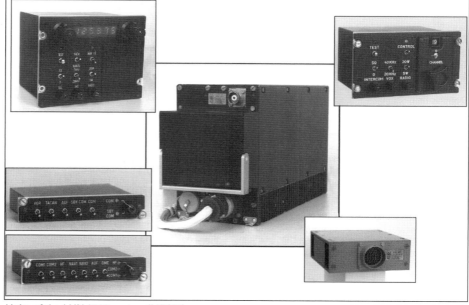

Units of the LUN 3520 system (MESIT) 0079893

LUN 3520.40-8 control box, designed for remote control from a single position, with the ability to tune the frequency in use, and to select from 20 pre-set frequencies;

LUN 3520.41-8 control box, designed for remote control from a single position, with the ability to tune the frequency in use;

LUN 3520.42-8 control box, designed for remote control from two positions, in conjunction with audio switch box LUN 3520.52 or LUN 3520.53; it provides for operation of the aircraft intercom system, as well as control of the radio and its tuned frequency;

LUN 3520.43-8 control box, designed for remote control from two positions, it provides control of the aircraft intercom and other radio signals aboard the aircraft, such as the radio compass, but not control of the radio transmitter;

LUN 3520.44-8 control box, designed for remote control from one location; in conjunction with the control box LUN 3520.24 it is possible to tune the frequency in use;

LUN 3520.45-8 control box, designed for remote control from a single position, with the ability to tune the frequency in use;

LUN 3520.47-8 control box, designed for remote control from two positions, it provides for operation of the aircraft intercom and control of other radio signals aboard the aircraft, but not control of the transmitter;

LUN 3520.48-8 control box, designed for remote control from two positions, it provides for operation of the aircraft intercom and control of other radio signals aboard the aircraft, including control of the transmitter;

LUN 3520.49-8 control box, designed for remote control from two positions, it provides for operation of the aircraft intercom and control of other radio signals aboard the aircraft, but not control of the transmitter;

LUN 3520.50-8 audio switch box, designed for audio control of two transmitters and the aircraft identification system;

LUN 3520.50-8 audio switch box, designed for audio control of the three transmitters and four radio navigation systems, including ADF and DME;

LUN 3520.53-8 audio switch box, designed for audio control of two transmitters, the aircraft identification systems and other navigation devices, including TACAN, VOR and ADF;

LUN 3520.60-8 audio frequency unit to interconnect third and fourth users to the aircraft intercom system; also generates a 500 Hz warning tone.

Status
In production and in service. The LUN 3520 was certified to International Civil Air traffic Organisation (ICAO) standards in 1998.

The system has been widely employed in modernisation of Russian-made aircraft and helicopters to comply with ICAO/NATO communications standards.

Contractor
MESIT pristroje spol, s.r.o.

LUN 3526 VHF Airborne Transceiver

Type
Avionic communications system, VHF.

Description
The LUN 3526 VHF airborne transceiver is a cockpit-installed modular system. Designed to comply with European airspace requirements for 8.33 kHz channel spacing, the system comprises a transceiver module with a power supply unit and cooler, a receiver/synthesiser, a control computer with AF section and a control module. The transceiver modules incorporate multilayer PCBs, fitted using Surface Mount Device (SMD) techniques. Once assembled, the modules are protected by a compact metal cover at the rear of the unit. Connection to the aircraft host is via D-SUB and a 50 Ω BNC for the antenna.

There is a six-digit display and four control knobs on the front of the unit, which facilitate: Frequency setting

Channel setting
Transceiver/intercom volume control
Display brightness control
On/off and squelch
Setting/checking preselected channel frequencies.

Specifications
Frequency range: 117.975 to 137.000 MHz
Programmable channel frequencies: 118.000 to 136.975 MHz
Channel spacing: 8.33 or 25 kHz
Channels: 760 (at 25 kHz) or 2280 (at 8.33 kHz)
Transmitter power: 12 W min (25 kHz); 6 W min (8.33 kHz)
Maximum continuous transmission time: 2 min
Modulation: A3E
Memory channels: 20 channels
Intercom: 2 users
Dimensions: 82.5 × 82.5 × 281.6 mm
Weight: 1.5 kg
Power: 27.5 V DC nominal; 24–29 V DC operating
Reliability: 2,000 h MTBF minimum
Environmental: DO/160D
Compliance: RTCA/DO −186a, −214, and −160D; TSOs -C38d and -C37d, ITU Radio Regulations Chapter II, Article 5, Annex 8

Status
In production and in service.

Contractor
MESIT pristroje spol, s r. o.

LUN 3526 airborne communications transceiver (MESIT)   0079894

Denmark

Aeronautical CAPSAT

Type
Avionic datalink system.

Description
Aeronautical CAPSAT uses British Telecom's Standard Positioning Service (SPS), a worldwide fax, telex and data satellite service. It provides automated Global Positioning System (GPS) position reporting and two-way messaging capability between aircraft and ground controllers located anywhere in the world within Inmarsat coverage.

Aeronautical CAPSAT enables ground controllers to monitor the position of helicopters from a central location. Since helicopters typically operate at low altitudes, the use of satellites provides a more reliable two-way communications link than other methods of transmission.

The system can be programmed to send regular, automatic GPS position reports to a PC connected to the worldwide telephone

and data network. The system's data format is compatible with those of the international Future Air Navigation System (FANS) and Automatic Dependence Surveillance (ADS).

The ground controller also has a satellite connection, making the system ideal for operations in areas where the communications infrastructure is unreliable or non-existent. The accuracy of the satellite GPS is integrated into Aeronautical CAPSAT and the system is geared towards safety and is very user-friendly.

Able to report the aircraft's position automatically every 2 minutes, or on direct request by ground operators, no physical aircrew input is needed. In the event of a forced landing, ground operators will have rapid access to accurate position data that is not limited by geographical position.

Fully two-way, the system can operate certain modes in near-realtime and lets aircrew alter flight plans on the basis of the latest meteorological information. They can also make immediate changes to forward maintenance and refuelling plans and report aircraft position, regardless of where it is in the world.

The system comprises a small aerodynamic fin antenna; an internal transceiver and amplifier subsystem; a compact message terminal with a full QWERTY keyboard and integrated LCD (Liquid Crystal Display); and a mini printer unit for receiving hard copies of messages.

Ground operators connect to the telephone or data network via a modem or when in remote regions, to British Telecom's Goonhilly ground earth station using a portable satcoms unit. The CAPSAT Manager PC software is a full featured fleet and solo aircraft tracking program that lets ground operators monitor flight paths on a moving map display and communicate directly with aircrew, over the satellite link, via their onboard system.

Status
Supplied to the Royal Air Force, the UK Customs and Excise, and to the UK Meteorolgical Research Flight.

Contractor
Thrane & Thrane A/S.

TT-3000M Aero-M single-channel aeronautical satcom

Type
Avionic communications system, satellite.

Description
The TT-3000M Aero-M satcom is the airborne version of the Inmarsat Mini-M system, of which Thrane & Thrane have supplied over 15,000 worldwide.

TT-3000M Aero-M is a single-channel system, which provides voice, secure voice, fax and data transmission capabilities and can accommodate two handsets, for use from the cockpit or passenger cabin.

The TT-3000M system comprises the TT-3068A Inmarsat Aero-M satcom receiver and the TT-5006A antenna (see separate entries for the TT-3068A and the TT-5000.

Specifications
Operating frequencies:
Receive 1525.0 to 1559.0 MHz
Transmit 1626.5 to 1660.5 MHz
Voice: 4.8 Kbps AMBE (3.6 Kbps voice 1.2 Kbps FEC)
Data interface: Serial EIA Standard RS-423
Weight:
Satellite data unit 2.2 kg
Antennae 2.2 kg
Operating temperature: −25 to +55°C

Status
In service.

Contractor
Thrane & Thrane A/S.

TT-3024A Inmarsat-C aeronautical capsat

Type
Avionic communications system, satellite.

Description
The compact TT-3024A aeronautical Inmarsat-C/GPS transceiver is designed for automatic data reporting and message transfer of position reports, performance data and operational mesages on a global basis, from sea level to 55,000 ft and from 70° north to 70° south.

The TT-3024A operates through the established network of Inmarsat and GPS satellites with interconnection to the international telex, fax and packet switched data networks, offering fast and reliable transfer of information, 24 hours per day.

The integrated GPS receiver calculates position, altitude, speed and heading every second, used for automatic Doppler compensation and transfer of position status reports to any predefined air traffic control or other receiver at specified intervals.

Data collecting equipment may be connected to the TT-3024A via an RS-422/423 port, enabling all selected data to be transferred as data reporting packages every 2 minutes or whenever a special event occurs.

Operational messages, weather and flight plan information as well as passenger messages may be transferred to/from any telex or data subscriber via the international networks.

Specifications
General specifications: the TT-3024A receiver meets or exceeds all Inmarsat specifications for the Inmarsat-C aero system and all relevant GPS specifications
Antenna: integrated Inmarsat-C/GPS omnidirectional antenna, RHC polarised
Figure-of-merit (G/T): −23 dB/K at 5° elevation
EIRP: 12 dBW min at 5° elevation
Transmit frequency: 1,626.5–1,646.5 MHz
Receive frequency: Inm-C 1,530.0–1,545.0 MHz, GPS 1,575.42 MHz
Channel spacing: 5 kHz
Modulation: 1,200 symbols/s BPSK
Data rate: 600 bits/s
RX frame length: 8.64 s
TX signalling access mode: Slotted ALOHA
TX message channel: TDMA and FDMA, interleaved code symbol
Position reporting: built-in 5-channel GPS receiver with Lat/Long/alt/speed/track calculation, update rate 1 s
Position accuracy: C/A code with 96 m spherical error probability
Initial stabilisation: 15 min at max Doppler shift correction and GPS almanac update
Solid-state storage: 256 kbyte RAM memory
Onboard message/data interface: RS-422/423, 110–9,600 bps and Centronics parallel
Navigational interface: RS-422/423 V.10 Special for interface to onboard navigational systems
Roll and pitch: min ±25° from level flight
Altitude: MSL-55,000 ft
Airspeed: full Doppler compensation to min 620 kt
Power: floating 10.5–32 V DC, 9.5 W Rx, 80 W Tx
Dimensions:
(antenna) 114 × 274 × 98 mm
(LNA/HPA) 161 × 213.9 × 49.5 mm
(electronics assembly) ¼ ATR short
Weights:
(antenna) 0.75 kg
(LNA/HPA) 2.3 kg
(electronics assembly) 2.5 kg

Status
Full US FAA and Inmarsat approval for aeronautical use.

Contractor
Thrane & Thrane A/S.

TT-3024A Inmarsat-C aeronautical capsat

0022245

TT-3068A Inmarsat Aero-M system

Type
Avionic communications system, satellite.

Description
The TT-3068A Aero-M, together with the TT-5006A antenna (see TT-5000 series Inmarsat Aero-I system entry), are designated the TT-3000M Aero-M system. This system provides true independent and worldwide global interconnection telephone call, fax and data access, with Inmarsat spot-beam operation.

The Aero-M is a single-channel satcom that is operated from the handset or from the two-wire DTMF phone interface.

Integration of the Navigational Reference System (NRS) makes the Aero-M totally independent of aircraft systems.

Specifications
EIRP: 14 dBW
Coverage volume: >85%
Operating frequencies:
receive: 1525.0 to 1559.0 MHz
transmit: 1626.5 to 1660.5 MHz
Channel spacing: 1.25 kHz
Voice: 4.8 kbps AMBE (3.6 kbps voice, 1.2 kbps FEC)
Asynchronous data rate: max 2.4 kbps
Phone 1 interface: RS-485, 4-wire handset
Phone 2 interface: 2-wire 600 Ω CCITT Rec. G. 473, standard DTMF telephone
Fax interface: 2-wire 600 Ω CCITT Rec. G. 473, T.30 Group-III Fax, 2.4 kbps, RJ-11 jack
Data interface: Serial EIA standard RS-423, Hayes compatible, max 19.2 kbps
Printer interface: Serial EIA standard RS-423 max 19.2 kbps
Power: 10.5–32 V DC
Roll and pitch: minimum ±25° from level flight
Altitude: MSL-55,000 ft
Airspeed: Full Doppler compensation to minimum 620 kt groundspeed
Airspeed acceleration: minimum ±1 g
TT-5006A Aero-I satcom antenna:
(dimensions) 560 × 149 × 129 mm
(weight) 0.68 kg

Status
In service.

Contractor
Thrane & Thrane A/S.

TT-5000 series Inmarsat Aero-I system

Type
Avionic communications system, satellite.

Description
The Thrane & Thrane compact, lightweight, low-power TT-5000 series Aero-I system functions as a high-quality multiple-channel communication centre, for telephone calls, fax prints, data transfers or e-mail messages to any destination.

The Aero-I system is designed to be an integrated part of the CNS/ATM system, offering three simultaneous channels (two voice/fax or data channels and one packet data channel). The built-in Cabin Telephone Unit (CTU) provides up to four handsets and two fax/phone/modem ports.

Integration of an optional Navigational Reference System (NRS) utilising 12-channel GPS data and 3-D flux gate technology, makes the TT-5000 series totally independent of aircraft systems. However, aircraft navigational systems can also be connected to create an integrated aircraft system.

Specifications
General specifications: the TT-5000 system meets or exceeds current and proposed Inmarsat specifications for the Inmarsat Aero-I system
Figure-of-merit (G/T): −19 dB/K min

TT-5000 series Inmarsat Aero-I system   0018940

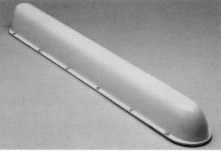

TT-5004A Aero-I satcom antenna   0044782

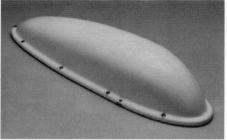

TT-5006A Aero-I satcom antenna   0044783

TT-5002A/B Aero-I satcom antenna   0044781

EIRP: >16.5 dBW total
Coverage volume: >85%
Operating frequencies:
(receive) 1,530.0–1,559.0 MHz
(transmit) 1,626.5–1,660.5 MHz
(GPS) 1,575.42 MHz
Channel spacing: 2.5 kHz
Antenna system: Jet Sat-97 mechanically steered antenna, or CAL ANT-30 electronically steered antenna
Navigational interfaces: stand-alone with NRS (option) or ARINC 429 IRS bus
Altitude: MSL to 55,000 ft
Airspeed: full antenna tracking and Doppler compensation to min Mach 1.0
Dimensions: 559 × 150 × 122 mm

Weight: 2.7 kg
Operating temperature: –25 to +55°C
Power: 28 V DC, 25 W

TT-5033A SDU
Channels: 3 (2 voice/fax/modem data, 1 packet data). Voice 4.8 kbytes/s, fax and modem data 2.4 kbytes/s, packet data 0.6/1.2 kbytes/s
Dimensions: 2 MCU ARINC 600
Weight: 3.6 kg
Power: +28 V DC, 20 W

TT-5010A HPA
Dimensions: 260 × 243 × 74 mm
Weight: 4 kg
Power: +28 V DC, 25–100 W
Output power: 18 W linear

TT-5012A DLNA
Dimensions: 254 × 193 × 50 mm
Weight: 2.2 kg

TT-5620A handset
Display: 2 × 12 character LCD
Key pad: 21 keys
Interface: 4 wire and RS-485
Power: +12 V DC, 0.2 A from SDU
TT-5622A handset
WH-10 AlliedSignal
Antenna options

TT-5002A/B Aero-I satcom antenna
Dimensions: 600 × 336 × 127 mm   (TT-5002A)
222 diameter × 119.5 mm (TT-5002B)
Weight: 4.8 kg (TT-5002A) 3.8 kg (TT-5002B)
Steering: mechanically steered, 2-axis

TT-5004A Aero-I satcom antenna
Dimensions: 886 × 117 × 86 mm
Weight: 2.6 kg
Steering: electronically steered phased array

TT-5006A Aero-I satcom antenna
Dimensions: 560 × 149 × 124 mm
Weight: 1.9 kg
Steering: mechanically steered, 2-axis

TT-5006 OmniPless antenna
TT-5008A NRS antenna
Dimensions: 38 × 243 × 129 mm
Weight: 0.68 kg

Status
In service.

Contractor
Thrane & Thrane A/S.

France

12000 VHF AM/FM radio

Type
Avionic communications system, VHF.

Description
The 12000 is a remote-controlled VHF transceiver designed for air traffic and air-to-ship communications in the 100 to 173 MHz frequency range with a channel spacing of 12.5 kHz. The AM/FM radio can operate with direction-finding equipment and sonobuoys in anti-submarine warfare and is capable of radio relay and voice scrambling. It has basic output powers of 15 and 3 W, with a 6 mW discrete capability. An integrated Guard receiver function is incorporated. The equipment's interface could be either ARINC 410 or 429. The transceiver is suitable for combat aircraft, helicopters and maritime surveillance aircraft. The transceiver can be controlled by the TC 20 control unit.

Specifications
Dimensions:
(transceiver) 197 × 61 × 375 mm
Weight:
(transceiver) 5 kg

The 12000 VHF AM/FM radio   0503993

Power supply: 22.5-31.5 V DC
Frequency:
(transmission) 100–156.975 MHz
(reception) 100–S173.5 MHz

Status
In service with the French Navy and Air Force.

Contractor
Thales Communications.

12100 VHF FM radio

Type
Avionic communications system, VHF.

Description
The 12100 radio is a tactical VHF FM remote-controlled airborne transceiver designed specifically for air-to-ground communications between military aircraft and ground forces.

It covers 30 to 88 MHz. In addition to the standard military FM mode, it provides interoperability with security forces, with a 12.5 kHz step and lower frequency deviation mode.

The Series is capable of KY 58 cyphered voice, external homing and radio delay. Nominal power is 15 W, with a reduced power output of 3 W.

The transceiver can be controlled by the TC 20 control unit.

Specifications
Dimensions:
(transceiver) 200 × 100 × 340 mm
Weight:
(transceiver) 5 kg

Status
In service with the French Air Force and Navy on the AS 355.

Contractor
Thales Communications.

The 12100 VHF FM radio 0503994

3527 HF/SSB radio

Type
Avionic communications system, HF.

Description
The 3527 HF/SSB radio comprises the 3527F transceiver, 3527H remote control, 3596A antenna coupler, 3597A digital preselector and the 3598A FSK modem. The preselector filter enables the simultaneous operation of two 3527H units on the same aircraft, one for transmission and the other for reception.

The fully solid-state transmitter supplies a modulated peak power of 400 W. The transceiver is driven by a digital synthesiser tuned for setting to 280,000 channels spaced at 100 Hz intervals. The antenna coupler provides tuning with wire antennas from 10 to 30 m long. The FSK modem handles messages of 50 to 100 bauds.

The comprehensive fault identification BITE increases maintenance efficiency and the modular construction minimises repair time by fast access to all circuit modules.

Specifications
Power supply: 115 V AC, 400 Hz, 3 phase
Power output: 400 W PEP, 200 W average (SSB)
Modes: USB, data (USB), CW, Link 11
Frequency stability: $\pm 5 \times 10^7$
Number of channels: 280,000
Operational specification: STANAG 5035, AIR 7304

Status
In production.

Contractor
Thales Communications.

Airborne Multiservice/ Multimedia Communication System (AMMCS)

Type
Avionic communications system, multiple waveband.

Description
The Airborne Multiservice Multimedia Communication System (AMMCS) provides voice, computer data, telex and facsimile air-to-ground communications facilities over different media such as satellite, HF, VHF or UHF radio.

Message handling facilities allow editing, retrieving, modifying and encryption/decryption of messages. The AMMCS provides automatic store-and-forward message services including logging, routeing and relaying. It supports a wide variety of message types and formats including telefax, telex, and e-mail. Multiple compression algorithms allow message transmissions to be speeded up. Message routeing is made automatically using different criteria such as urgency, medium availability, cost and priority.

The reliable transmission of long messages, such as high-quality telefax or error-free computer data files, over HF implies some specific modems, datalink protocols and transmission management protocols.

AMMCS integrates the Rockwell MDM-2501, multi-waveform adaptive high-speed modem. MDM-2501 offers 12 different standard waveforms, ensuring interoperability on different media. Among these waveforms is the MIL-STD-188-110A single-tone waveform which allows the efficient transmission of data up to 2,400 bits/s with forward error correction on HF between 300 and 3,000 Hz. When this waveform is used, the automatic multipath compensation provided by this modem overcomes the data rate limitations usually imposed on HF channels.

The AMMCS uses a multimedia FED-STD-1052 link protocol. It can operate over half- or full-duplex channels. This type of robust protocol is mandatory for HF links when propagation conditions may vary drastically and the link can be interrupted for variable periods of time. This protocol supports the automatic adaptability of system parameters to optimise the transmission performance in terms of throughput and bit error rate.

The AMMCS features functions which automatically adapt its parameters during a transmission depending on the type of link, type of service, required performance and propagation characteristic variations to optimise the performance in terms of bit error rate, required transmit power, transmission delay and throughput. These functions automatically select the best HF propagating channel and change the frequency when conditions degrade. The system uses the MIL-STD-188-141A/FED-STD-1045A ALE protocol on HF.

The AMMCS may present different levels of automation. In a fully automated system, direct dialling is offered to aircraft users for voice communications. The system selects the appropriate media and uses the dialling transcription to perform automatic selective call and link establishment procedures. A similar procedure applies to telex, data or faxes queued in the message handling system file for transmission. AMMCS uses the routeing information contained in the message envelopes to establish the appropriate link automatically and to manage the message transmission.

The AMMCS system is integrated into a high-impact magnesium alloy case laptop PC featuring 486/33 MHz Intel processor, up to 500 Mbyte hard disk, high-view angle touch panel colour screen and mouse or trackball interface. A TEMPEST version is available.

Status
In service.

Contractor
Rockwell-Collins France, Blagnac.

Audio-Radio Management System (ARMS)

Type
Avionic audio management system.

Description
The TEAM Audio-Radio Management System (ARMS) is designed for aircraft and helicopter applications. It is used to manage radio communication, radio navigation, the aircraft identification system, and the aircraft intercom system, together with several other functions. The basic ARMS comprises two Audio-Radio Control and Display Units (ARCDU); and one Audio Management Unit (AMU). Optionally, the audio capability can be specified separately, in which case the system would include only two radio control and display units (RCDU), or the system could have more than two ARCDU and audio control panels, located either in the cockpit or elsewhere in the aircraft when fitted in special mission aircraft.

ARMS replaces the standard radio and audio control panels, reducing acquisition, installation, maintenance and logistic costs. ARMS provides immediate access to the main functional capabilities controlled, and the ARCDU is fitted with a high-quality colour LCD active matrix.

ARMS control functions are as follows: radio communication; radio navigation; the intercom system and aircraft identification. Optional functions include datalink; public address; radio auto-tuning; aural warning; SELCAL; GPS or DGPS management. Additionally, ARMS can be integrated with other major avionic systems including the navigation system; anti-collision system; centralised diagnostic system; flight management system; flight data acquisition unit.

Specifications
ARCDU dimensions: 146 (W) × 165 (D) mm × H (Height (H) can be variable, greater than 105 mm, to meet installation requirements)
ARCDU weight: <2 kg
Interfaces: MIL-STD-1553; ARINC 429; other interfaces to meet customer requirements
Compliance: MIL-STD-810C, –461C, –704E, RTCA DO 160C, DO 178B level B, ED-80

Thales' 3527 HF/SSB radio 0503992

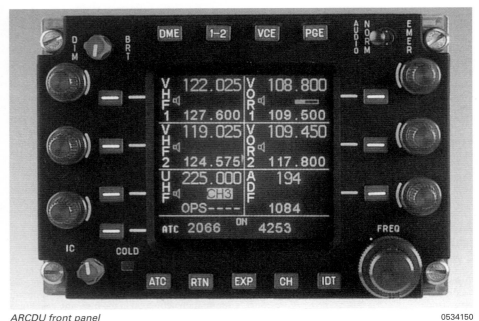

ARCDU front panel

0534150

Status

In service in Embraer Tucano, Bombardier Dash-8-400, Pilatus PC-9(M) and PC-7 Mk II aircraft.

Contractor

TEAM.

Automatic Dependent Surveillance (ADS) unit

Type

Avionic datalink system.

Description

Thales Avionics' Automatic Dependent Surveillance (ADS) unit was introduced in 1993. The reduction of separation distance, and thus the creation of more airspace, depends on establishing the exact position of an aircraft in space and time. To do this, the flight data used by the pilot must be processed and transmitted to a receiving station. The ADS system features the automatic transmission of this information and immediate reporting of data to the airline.

ADS extends surveillance capability to areas not covered by radar. By adding a Secondary Surveillance Radar (SSR), the service can be extended to terminal areas and high-density continental airspace.

ADS provides a number of benefits to both pilots and airlines, including significantly enhanced flight safety, extension of ATC services to oceanic regions and areas not covered by radar, reduced separation minima and highly accurate surveillance protected against interference in high-density airspace.

The Thales approach is two-fold: refining the ADS concept through a standalone computer designed for current aircraft, and developing an embedded ADS function for the avionics suites on future aircraft.

Thales formed an open industry alliance with EADS and Airbus to develop the ADS system in response to initiatives by the European Union, Eurocontrol and European Civil Aviation Authorities. The alliance partners proposed a plan for simultaneous integration and deployment of aircraft capabilities with automated ground air traffic systems to exploit the potential of new technologies enabling co-operation between aircraft and control centres. The plan focused on key areas such as real-time exchange and sharing of flight plan and trajectory data between operations centres, aircraft, airports and ATC centres; datalink-enhanced air-to-air and air-to-ground surveillance to optimise separation and maximise airspace capacity; satellite-based navigation (GPS and Galileo) and automated flight plan data processing supporting fully flexible aircraft routing and dynamic airspace management.

Status

In production and in service.

Contractor

Thales Avionics SA.

BER 8500/8700 transceivers family

Type

Avionic communications system, VHF/UHF.

Description

BER 8500/8700 is a family of transceivers covering the VHF and UHF bands from 100 to 173.975 MHz and 225 to 400 MHz in A3 and A1D (A9) AM modes and in F3 and F1D (F9) FM modes. ECCM versions include Have Quick II embedded operation.

The family consists of the BER 8523 covering VHF and UHF AM/FM with Guard receiver, the BER 8524 VHF and UHF AM/FM with Guard receiver and ECCM, the BER 8751 UHF AM/FM and the BER 8752 UHF AM/FM with ECCM.

The ECCM capability is characterised by TRANSEC provided by embedded Have Quick II circuits; COMSEC provided by an external cypher unit (the family is compatible with KY 58 in both NRZ and diphase operation) and SCP 5000 compatibility through direct connection for external TDP 5000 processor.

BER 8500/8700 transceivers are controlled by one or several units of the ECCM VHF/UHF control systems, including the BCA 1217 main VHF/UHF or UHF control unit, DBC 1317 secondary VHF/UHF or UHF control unit or DMC 1317 and DMCB 1317 loading modules. These modules allow control of the transceivers, display of the mode, frequency and alarm and loading of ECCM parameters. The BER 8523 and BER 8751 can be controlled by the TC20 control unit.

Specifications

Dimensions:
(BER 85xx) 57 × 202.5 × 493 mm
(BER 87xx) 57 × 202.5 × 455 mm
Weight:
(BER 85xx) 8.5 kg
(BER 87xx) 7.5 kg
Power supply: 28 V DC or 115 V AC 400 Hz
(transmit) 180 W
(receive) 35 W
Frequency:
(BER 85xx) 100–173.975 MHz and 225–400 MHz
(BER 87xx) 225–400 MHz

Status

In service with the French Air Force and several other air forces.

Contractor

Thales Communications.

CP3938 audio management system

Type

Avionic audio management system.

Description

The TEAM CP3938 audio management system is a single line replaceable unit, which enables two pilots to operate the following radio services; up to four communications radios; up to four radio navigation systems; and the aircraft emergency warning systems. In addition, it provides inter-communication between: the two pilots; the pilots and up to three passengers.

The CP3938 was designed for the EC120 helicopter, but can be fitted on many other helicopter types. It replaces the TEAM SIB27 (TB27) without any modification to the helicopter wiring.

The front panel is customised, using removable labels, to match the radio communication, radio navigation and warning sets fitted to the aircraft.

The transmission/reception channels and warnings adaptation is customised to the equipment fitted on the aircraft at installation.

Specifications

Power: 22–32 V DC, 12 W
MTBF: 16,000 h
Bandwidth: 300–5,000 Hz
Dimensions (H × W × D): 57 × 136 × 170 mm
Weight: 0.955 kg

Status

In service.

Contractor

TEAM.

The TEAM CP3938 audio management system

0044785

DLP DataLink Processor system for NATO interoperability

Type

Avionic datalink system.

Description

The DLP (DataLink Processor) is an advanced tactical datalink server that supports TADIL-A (Link 11A), TADIL-B (Link 11B) and TADIL-J (Link 16) for NATO interoperability. It receives, transmits, processes and displays tactical data in accordance with STANAG 5511 and/or STANAG 5516 message specification.

The DLP has the capability to simultaneously process data provided by two or more networks (that is Link 11/Link 16, 2 × Link 11).

Depending upon the nature of the application, the DLP can be configured as a tactical datalink front-end processor only, or as a tactical datalink processor together with graphic display and keyboard/trackball.

The DLP provides the necessary external interfaces including NTDS/ATDS, MIL-STD-1553B, Ethernet and serial interfaces.

The DLP product family includes equipment designed for aircraft, ground station (fixed/transportable) and shipboard applications. It uses commercial workstations for benign environments, 19 in ruggedised VME-based equipment for ground and shipboard and ATR VME-based equipment for airborne environments.

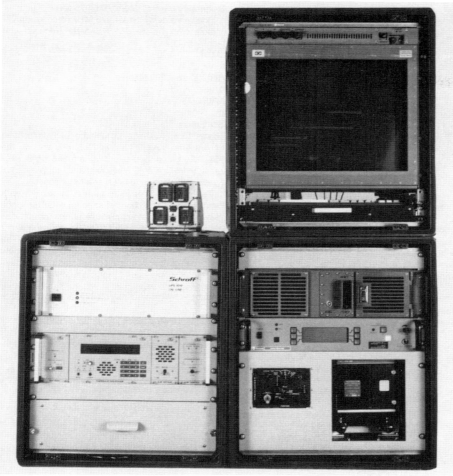

DLP DataLink Processor

0011851

ADLP-100F Airborne DataLink Processor

0011850

Status
In production since 1996. In operation with the French Navy, French Air Force and UK Royal Navy.

ADLP-100F Airborne DataLink Processor for NATO interoperability
The ADLP-100F is part of Rockwell-Collins France DataLink Processor (DLP) product line. It is an advanced tactical datalink processor specifically designed for Link 11 airborne applications including both fixed- and rotary-wing platforms.

The ADLP supports the capability to receive, process, transmit and display tactical data in accordance with STANAG 5511 message specification for NATO interoperability. Depending upon the nature of the application, the ADLP-100F can be installed on board the platform as a front-end processor connected to the onboard mission system through a MIL-STD-1553B interface or equivalent, or as a stand-alone subsystem together with a display and keyboard/trackball.

The ADLP-100F is a VME-based system packaged in an ATR format for easy installation on board the platform. It provides an ATDS interface to a KG-40(A) crypto and a 1553B interface or equivalent to the onboard mission system or sensors.

The ADLP-100F hardware is qualified for both fixed- and rotary-wing aircraft.

Status
Selected by ECF to equip the Super Puma and Panther helicopters with Link 11 for a foreign navy; qualified for the French Navy ATL2 aircraft.

Contractor
Rockwell-Collins France, Blagnac.

ERA-8500 VHF/UHF radio

Type
Avionic communications system, VHF/UHF.

Description
The ERA-8500 transceiver covers the VHF and UHF bands from 100 to 173.975 MHz and 225 to 400 MHz in A3 and A9 AM modes and F1, F3 and F9 FM modes at 25 kHz spacing. It consists of the BER 8500 V/UHF transceiver and the BCA 11XX or BCA 12XX control unit. Its tuning time is compatible with very fast frequency-hopping modes. Power output is 15 W in AM and 20 W in FM.

The ERA-8500 is designed for use by high-performance military aircraft and is offered with a Have Quick II module incorporated in the transceiver. It operates between -55 and +90°C at altitudes up to 100,000 ft.

The ERA-8500 transceiver is coupled with the TDP-5000 radio processing unit to form the SCP-500 integrated ECCM radio communication system.

Specifications
Dimensions:
(BER 8500 V/UHF R/T unit)60 × 202.5 × 493 mm
(BCA 11XX main control unit) 146 × 76 × 173 mm
(BCA 12XX auxiliary control unit) 146 × 47 × 173 mm
Weight:
(BER 8500 V/UHF R/T unit) 8.5 kg
(BCA 11XX main control unit) 1.3 kg
(BCA 12XX auxiliary control unit) 1 kg

The ERA-8500 VHF/UHF radio

0503996

Power supply: 115 V AC, 400 Hz or 28 V DC
(transmit) 180 W
(receive) 35 W

Status
In service.

Contractor
Thales Communications.

ERA-8700 UHF radio

Type
Avionic communications system, UHF.

Description
The ERA-8700 transceiver is limited to the UHF band between 225 and 400 MHz at 25 kHz spacing. It consists of the BER 8700 UHF transceiver and the BCA 11XX or BCA 12XX control unit.

The equipment is available with the Have Quick II frequency-hopping mode.

Specifications
Dimensions:
(BER 8700 UHF R/T unit) 60 × 202.5 × 455 mm
(BCA 11XX main control unit) 146 × 76 × 173 mm
(BCA 12XX auxiliary control unit) 146 × 47 × 173 mm
Weight:
(BER 8700 UHF R/T unit) 7.5 kg
(BCA 11XX main control unit) 1.3 kg
(BCA 12XX auxiliary control unit) 1 kg
Power supply: 115 V AC, 400 Hz or 28 V DC
(transmit) 180 W
(receive) 35 W

Status
In service.

Contractor
Thales Communications.

Thales' ERA-8700 UHF radio

0503995

ETC-40X0F centralised control system

Type
Avionic Communications Control System (CCS).

Description
The ETC-40X0F provides frequency and mode control of any radio communication and radio navigation equipment in the aircraft. It is a fully modular architecture system allowing easy adaptation to the avionics configuration of the aircraft. Main features are tandem seat operation, compatibility with any type of radio com/nav equipment, flexibility, modularity, colour display, easy installation and BITE.

The Rockwell-Collins France ETC-40X0F centralised control system consists of two EDU-40X0F control and display units and the MPU-40X0F bus concentrator (centre)   0011849

ETC-40X0F basic configuration is composed of two EDU-40X0F control and display units and one MPU-40X0F bus concentrator.

ETC-40X0F can control and display ARINC 429 as well as any other type of equipment. ARINC 429-equipped systems are directly linked to the EDU-40X0F, and non-ARINC 429 radios are linked through dedicated interfaced boards, within MPU-40X0F.

The ETC-40X0F allows the operator a global display of all com/nav equipment in the aircraft. For each piece of equipment, essential information is displayed throughout three levels of pages. Top level pages display active and preset frequencies. Technical parameters, such as mode, channel and test, are available on two additional pages. The system provides simultaneous access to all avionics equipment through each of the two EDU-40X0F units.

The MPU-40X0F unit houses the ARINC 429 bus concentrator and interface boards for non-ARINC 429 equipment. ETC-40X0F is also offered with colour LCD technology and embedded GPS.

Specifications
Dimensions:
(EDU-40X0F) 146 × 242 × 105 mm
(MPU-40X0F) 322 × 124 × 193 mm
Weight:
(EDU-40X0F) 3.6 kg
(MPU-40X0F) 4.5 kg
Power supply: 28 V DC, 4 A max
Radio interfaces: 6 ARINC 429, 12 non- ARINC 429
Programmable channels: 16 per radio
Altitude: up to 55,000 ft
NVG compatible

Status
In production for Eurocopter Cougar Mk 1, Mk 2, Fennec, Raytheon Hawker 800 XP and Lockheed Martin C-130 aircraft.

Contractor
Rockwell-Collins France.

EVR 716/750 enhanced VHF Data Radio (VDR)

Type
Avionic communications system.

Description
Thales Avionics' EVR 716/750 (enhanced VHF Radio) is a new VHF AM transceiver designed to comply with 25 and 8.33 kHz channel spacing in the VHF band; it is immune to FM radio signals in accordance with ICAO annex 10.

The EVR 716 offers 25 and 8.33 kHz channel spacing for voice communications and 25 kHz channel spacing data transmission capability with an external modem (MSK/CSMA 2.4 kbytes/s).

The EVR 750 VHF Data Radio (VDR) offers, in addition to the EVR 716, ARINC 750 data modes compatible with the ATN protocols (31.5 kbytes/s) with internal modems.

The EVR 716 complies with ARINC 716 and the EVR 750 with ARINC 750. The two systems share a common hardware configuration and the EVR 716 can be upgraded to the EVR 750 capability by software only.

Specifications
Dimensions: 3 MCU
Weight: 4.8 kg
Power supply: 27.5 V DC
Temperature: –15 to +70°C

Frequency band: 118.0 to 136.975 MHz
Extension down to 111.975 MHz (receive only)
Extension up to 151.975 MHz (receive and transmit)
MTBF: >30,000 h

ARINC specifications
ARINC 716
ARINC 750

Status
Thales' EVR 716-01 VHF Data Radio (VDR) is certified on the Airbus A319/A320/A321 family and on the A340. It is also field-approval certified on B-737/-747-400/-757/-767/-777 aircraft and supplemental type certified on B-737-400 and NG aircraft.

The EVR 716-01 enables airlines to comply with Europe's 8.33 kHz channel spacing requirements, as well as future data communications regulations within the scope of FANS ATN VDL mode A and 2 (Aeronautical Telecommunications Network). The VDR system can easily be upgraded to meet new FANS digital transmission mode requirements.

Operating modes				
Mode ident	Main features		EVR-716	EVR-750
0A	Analogue voice, 25 kHz channel spacing		Compliant	Compliant
0B	Analogue voice, 8.33 kHz channel spacing		Compliant	Compliant
1A	Data MSK-AM 2.4 kbytes/s CSMA non-persistent modem in the (C)MU Standard ACARS mode		Compliant	Compliant
1B	Data MSK-AM 2.4 kbytes/s CSMA P-persistent modem in the EVR ICAO VDL mode 1		Require 750 software upgrade	Compliant
1C	Data MSK-AM 2.4 kbytes/s CSMA non-persistent modem in the EVR ACARS mode for CMU/ATSU interface		Require 750 software upgrade	Compliant
2	Data D8PSK 31.5 kbytes/s CSMA P-persistent modem in the EVR ICAO VDL mode 2		Require 750 software upgrade	Compliant
3*	*Digital voice and data* D8PSK 31.5 kbytes/s TDMA *mode defined in* RTCA DO-224		*Hardware provisions*	*Hardware provisions*
4*	*Data* D8PSK 31.5 kb/s *or* GMSK 19.2 kbytes/s TDMA *mode defined in* STDMA *standard*		*Hardware provisions*	*Hardware provisions*

*This mode is not yet validated and adopted by ICAO

Thales' EVR 716/750 enhanced VHF radio   0001172

The VDR is standard fit on Dash 8 Series 400 regional aircraft.

Thales has delivered more than 3,000 VDRs.

Contractor
Thales Avionics SA.

JETSAT Aero-I satcom

Type
Avionic communications system.

Description
Designed and manufactured by Thales Avionics, the Aero-I satcom system, JETSAT, is marketed and supported worldwide by Thales. It is compliant with Inmarsat SDM, DO160D and DO178B level D, and ARINC 761 (C architecture).

The JETSAT is designed to operate under the Inmarsat 3 satellite spot beams, offering cockpit and cabin voice, fax and data services. This fully integrated system provides five channels (four voice/fax/PC data and one packet data), and complies with CNS/ATM requirements. It comprises one IGA (Intermediate Gain Antenna) and two LRUs (Line Replaceable Units): one HLD (High-power amplifier/Low-noise amplifier/Diplexer) and one SDU (Satellite Data Unit). The JETSAT is a compact, lightweight system offering maximum installation flexibility, with no operational constraint on LRU location (other than the antenna).

Specifications
Dimensions:
2 LRUs each of 4 MCU
(antenna) 650 × 336 × 140 mm
Weight: 18 kg
Power supply: 28 V DC or 115 V AC, 400 Hz
Operating band:
Receive: 1,530 to 1,559 MHz
Transmit: 1,626 to 1,660 MHz
Antenna coverage: >96% above 5° elevation angles
Baud rates: voice 4,800 Bps; fax 2,400 Bps; data 600/1,200/4,800 Bps
Cabin handsets: via external CTU (CEPT E1) or up to 32 handsets with integrated CTU

Status
JETSAT is in production and in service in a number of Falcon and Bombardier Global Express and Challenger aircraft.

Contractor
Thales Avionics SA.

SAVIB69 audio management system

Type
Avionic audio management system.

Description
SAVIB69 audio management system provides control of the aircraft's external and internal communications systems. It includes the ADAV aural warning function with 265 synthetic voice messages. SAVIB69 comprises: one audio management unit CTA 3488 (including voice alert); one audio control panel (BCA 3487) for each cockpit, one fuselage ground connection; and one Telebriefing connection.

Each control panel provides an automatic transmit/receive function on the interphone channel; the ADAV aural warning system memory capacity of 265 messages corresponds to a global alert time of 5 minutes. The system permits the selection of: up to four communication channels with volume control; up to six navigation channels with volume control; four listening channels without volume control; one Telebriefing connection; one double track audio recording system; 24 discrete signals corresponding respectively to one alert.

Each audio control panel transmits the status of its selector switches to the audio management unit via ARINC 429 busses. Each audio management unit incorporates up to two independent audio signal processors. The ADAV mode provides the crew with a prioritised listening monitor on a fixed frequency.

SAVIB69 is connected to the aircraft MIL-STD-1553B databus.

Specifications
BCA
Bandwidth: 300-5,000Hz ±3 dB
Dimensions: 65 (H) × 146 (W) × 110 (D) mm
Weight: 0.7 kg
Power: 17-32 V DC, 10 W
MTBF: 10,000 flight hours

CTA
Bandwidth: 300-5,000Hz ± 3 dB
Dimensions: 194 (H) × 60 (W) × 320 (D) mm
Weight: 3 kg
Power: 17-32 V DC, 20 W
MTBF: 4,500 flight hours

Status
Designed for Rafale, both single-seat and two-seat versions.

Contractor
TEAM.

SCP 5000 VHF/UHF secure radio communications system

Type
Avionic communications system, VHF/UHF, secure.

Description
The SCP 5000 secure radio communications system is designed for military platforms. It provides voice and data transmission services protected against the electronic warfare threat such as jamming, eavesdropping, intrusion and localisation.

The system features pseudo-random fast frequency hopping, time variable unsigned synchronisation dwell, voice and data cyphering and use of the complete UHF band. The Have Quick II mode is available as an option.

The SCP 5000 offers fixed-frequency plain analogue voice, KY 58-compatible fixed-frequency cyphering voice, frequency-hopping plain or cyphered voice, point-to-point data transmission to transmit mission data and inter-weapons system co-operation data and data transmission over a TDMA network to allow participants to exchange tactical information between aircraft or between command and control centres and aircraft.

The SCP 5000 airborne system comprises one or two BER 8500 or 8700 V/UHF or UHF transceivers, a TDP 5000 processor to control the frequency hopping, two BCA 1217 control boxes and an optional BVA 500 frequency display.

The TDP 5000 processor is connected to the aircraft system via the avionics bus or a dedicated link. It manages the frequency-hopping procedures; communicates cyphering/decyphering; manages data transmission; message formatting, link or network management and error detection and correction; and interfaces with the platform for voice digitisation and bus coupling.

Specifications
Dimensions:
(BER 8500) 57 × 202.5 × 493 mm
(BER 8700) 57 × 202.5 × 455 mm
(TDP-5000) 57 × 193.5 × 380 mm
(BCA 1217) 146 × 47.6 × 148 mm
(BVA 500) 41.5 × 41.5 × 125 mm
Weight:
(BER 8500) 8.5 kg
(BER 8700) 7.5 kg
(TDP-5000) 7 kg
(BCA 1217) 1.2 kg
(BVA 500) 0.5 kg
Power supply:
(BER 8500/8700) 28 V DC or 115 V AC, 400 Hz
(TDP-5000, BCA 1217, BVA 500) 28 V DC
Frequency: 100-156 and 225-400 MHz
Channel spacing: 25 kHz

Status
In production for the Mirage 2000 for the French Air Force and the export version Mirage 2000-5.

Contractor
Thales Communications.

SAVIB69 audio management system   0011848

The SCP 5000 has been selected for the French –Air Force Mirage 2000 and the export Mirage 2000–5   0503997

SELCAL Decoder

Type
Avionic communications system, Selective Calling (SELCAL) system.

Description
The purpose of a SELective CALLing System (SELCAL) is to enable exclusive calling of individual aircraft over normal radio communication channels that link the ground station with the specified aircraft. This selective calling avoids continuous monitoring of the communication frequencies by the flight crew.

The TEAM SELCAL aircraft system consists of an Airborne Selective Calling Unit (ASCU) (decoder) SL3924(XX)(XX), a code selection panel BC2065, and visual and audio annunciators. The code selection panel is recommended as part of both the SIB54 (A320) and the SIB73 (A330/340) systems.

The system operates with existing HF and VHF ground-to-air transmitters and receivers. Each SELCAL-equipped aircraft is assigned an identifying four-letter code. The ground communication selects the proper code, sets it up on a control slip and sends the signal. The ground transmitter is then automatically modulated by this coded audio signal. The airborne receiver demodulates the RF signal and the coded audio signal is applied to the appropriate SELCAL decoder.

The decoder is designed to respond only to that for which it is adjusted. When there is a call for the aircraft, the decoder actuates a signal indicator, which may be a lamp, a chime or any combination of such crew alerting devices, and enables swift and reliable answering of the call.

The aircraft decoder is reset by push-button, or automatically when the radio push-to-talk is keyed.

Specifications
Dimensions: 1 MCU per ARINC 600
Weight: 1.13 kg
Power: 27.5 V DC, 200 mA (nominal)
Input
Impedance: 10 k ohms (nominal)
Level: 70 mV to 3.2 V
Tone frequency amplitude difference: less than 10 dB
Tone frequency harmonic distortion: less than 15%
Signal to Noise Ratio (SNR): 6 dB
Tone duration: 1 s (±0.25 s)
Tone spacing: 0.2 s (±0.1 s)
Standards: ARINC 714
Approvals: TSO C59 according to DO 160C and DO 178B level D

Status
In production and in service in a wide variety of airliners, including most Airbus types and a number of Boeing aircraft.

Contractor
TEAM.

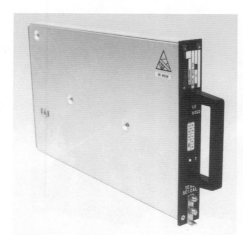

SELCAL airborne selective calling system

0534151

SIB31/43/45/54/66/73/85 audio management systems

Type
Avionic audio management system.

Description
The SIB family of audio management systems provides crew members with an interphone link (in multicrew aircraft) and enables each one to operate the following facilities:
SIB31: radio communications; radio navigation; emergency warning system
SIB43: radio communications; radio navigation; warning system (option: public address)
SIB45: radio communications; radio navigation; emergency warning system
SIB54: radio communications; radio navigation; interphone; public address
SIB66: radio communications; radio navigation; interphone
SIB73: radio communications; radio navigation; interphone; public address; emergency warning system
SIB85 SATCOM: radio communications (including satcom); radio navigation; interphone; public address; emergency warning system.

The systems comprise the following units:
SIB31: junction box BJ1977; three to five main audio control panels CP1976
SIB43: one Remote Control Audio Unit (RCAU); up to three Audio Control Panels (ACP2531); hand microphone, headphones, speakers, jack panel
SIB45: junction box BJ2620A; three to five main audio control panels CP2618A
SIB54: up to five audio control panels ACP2788; 1 × audio management unit AMU2790; 1 × SELCAL code selection panel BC2065
SIB66: up to four audio control panels BC3438; 1 × ground crew panel PC2748; 1 × SELCAL code selection panel BC2065
SIB73: up to five audio control panels ACP2788AC01; 1 × audio management unit AMU3514; 1 × SELCAL code selection panel BC2065; 2 × speakers HP3520
SIB85 SATCOM: up to five audio control panels; ACP2788AF01; 1 × audio management unit AMU3514; 1 × SELCAL code selection panel BC2065; 2 × speakers HP3520.

Facilities provided include:
SIB31: 4 ×Tx/Rx selective tunable; 4 × navigation channels tunable; 3 × listening channels
SIB43: up to six communications channels with volume control; 10 navigation channels; voice recorder; Ground Proximity Warning System (GPWS); Flight Warning Computer (FWC)
SIB45: 6 ×Tx/Rx selective tunable; 8 × navigation channels tunable; 3 × listening channels
SIB54: up to five audio channels; up to 12 navigation channels; one each public address/intercom/cabin channel
SIB66: six audio channels; six navigation channels; 2 × interphone channels; 1 × general call channel
SIB73: up to six audio channels; up to 11 navigation channels; one each public address/intercom/cabin channel

SIB31 audio management system 0011846

SIB43 audio management system 0044786

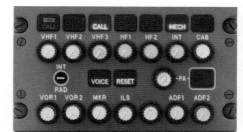

SIB73 audio management system 0011845

SIB85 SATCOM audio management system

0011844

SIB85 SATCOM: up to seven audio channels (including 2 × satcom channels); up to six navigation channels; one each public address/intercom/cabin channels.

Status
Systems have been installed on the following aircraft:
SIB31: Dauphin SA365C; Ecureuil AS355F; Puma SA330J; Super Puma SA332L1 helicopters
SIB43: ATR42, ATR72
SIB45: Super Puma SA332 helicopters
SIB54: Airbus A320 aircraft
SIB66: Transall aircraft of the French Air Force Transport Air Command (COTAM)
SIB73: Airbus A330/340 aircraft
SIB85 SATCOM: Airbus A330/340 aircraft equipped with VOICE SATCOM from the cockpit.

Contractor
TEAM.

SICOP-500 integrated radio communication system

Type
Avionic communications system, VHF/UHF, secure.

Description
SICOP-500 is an integrated ECCM radio communication system designed for air force applications. It operates in the VHF and UHF bands and capabilities include plain data and voice, cipher voice, jam-resistant voice and data, ground-to-air, air-to-air and air-to-ground communications.

The jam-resistant operation is achieved by using frequency-hopping techniques for voice and data transmissions, while error correction is provided for data transmissions. For voice transmissions the ciphering can operate both in jam-resistant and fixed-frequency modes.

The SICOP-500 comprises an ERA-8500 VHF/UHF transceiver, ERA-8700 UHF transceiver (see earlier items) and TDP-500 ECCM radio processor unit. It is designed for use by high-performance military aircraft in a jamming environment. It operates between -55 and +90°C at pressure altitudes up to 96,000 ft.

Specifications
Dimensions:
(BER 8500 V/UHF R/T unit) 57 × 202.5 × 493 mm
(BER 8700 UHF R/T unit) 57 × 202.5 × 455 mm
(BCA 1111 main control unit) 146 × 76 × 173 mm
(BCA 1211 auxiliary control unit) 146 × 47 × 173 mm
(TDP-500 radio processing unit) 57 × 193 × 380 mm
(BCSA-500 radio processing control unit) 146 × 38 × 100 mm
(BMD-500 security unit) 80 × 60 × 160 mm

Thales' SICOP-500 advanced ECCM radio system   0503998

Weight:
(BER 8500 V/UHF R/T unit) 8.5 kg
(BER 8700 UHF R/T unit) 7.5 kg
(BCA 1111 main control unit) 1.3 kg
(BCA 1211 auxiliary control unit) 1 kg
(TDP-500 radio processing unit) 6 kg
(BCSA-500 radio processing control unit) 0.5 kg
(BMD-500 security unit) 0.8 kg

Status
In service on Dassault Mirage 2000 and various other platforms.

Contractor
Thales Communications.

STERN image transmission system

Type
Avionic datalink system.

Description
STERN is a mobile, real time, image transmission system. This digital transmission system implements a new modulation, the Coded Orthogonal Frequency Division Multiplex (CODFM), which facilitates the acquisition of broadcast quality images in environments which are hostile to radio transmissions, such as urban, industrial or mountainous areas.

The compact design of the system enables application in a wide variety of small airborne and ground-based platforms.

Status
In production.

Contractor
SAGEM Défense Sécurité, Optronics and Air Land Systems Division.

TC 20 control unit

Type
Avionic Communications Control System (CCS).

Description
The TC 20 control unit is a multipurpose VHF/UHF control unit which controls all the Thales Communications fixed-frequency transceivers.

It can control two transceivers simultaneously. Programming allows the automatic adaptation of the control unit to the transceiver configuration. Mode controls provided include frequency setting from 30 to 400 MHz, 800 preset channels in four fields, Guard emergency control and Guard receiver, squelch, power and secure

mode control. Preparation of the next frequency or channel to be used is possible on the second line of the display for an instantaneous transfer to the active line by a one-touch button. The unit is Night Vision Imaging System (NVIS) compatible.

Specifications
Dimensions: 57 × 146 × 120 mm
Weight: 1.25 kg
Power supply: 28 V DC

Status
In service.

Contractor
Thales Communications.

Telemir infra-red communication system

Type
Avionic tactical communications system.

The Telemir optical head is fitted on top of the tailfin for infra-red data transfer (Rafale version)   0001173

Description
Telemir uses an infra-red beam for air-to-air, omni-directional air-to-ground, ground-to-air and ground-to-ground communications. The airborne equipment consists of an optical head (mounted on top of the tailfin) and a processing unit. It is extremely difficult to jam.

The system is used by a carrier-based aircraft for the reception of navigational updating and reference data such as attitude, location and speed from the ship's inertial navigation system for aligning its own INS. A new version is now available with a MIL bus 1553 datalink.

Status
In service in French Navy Super Etendard aircraft and integrated into the naval version of Rafale.

Contractor
SAGEM Défense Sécurité, Navigation and Aeronautic Systems Division.

THOMRAD 6000 V/UHF ECCM transceiver series

Type
Avionic communications system, VHF/UHF, secure.

Description
The THOMRAD 6000 product line is designed for voice and data communication and provides a full tri-service interoperability in ECCM mode. It covers the VHF and UHF bands from 30 to 400 MHz and operates in AM and FM modes. All transceivers include the channel spacing at 8.33 kHz in accordance with the new ICAO requirements.

ECCM protection is provided by use of the NATO SATURN mode, or a proprietary embedded fast frequency-hopping and ciphering capability that uses modern digital modulation and synthesiser technology. The provision of embedded COMSEC and TRANSEC modules reduce weight and save installation and integration effort. The THOMRAD 6000 product line features, in addition, the NATO Have Quick II mode and can be associated with external crypto devices.

Data transmission mode is available for point-to-point links or through Time Division Multiple Access protocol. Use of the datalink mode, with fast frequency hopping and error detection and correction procedures, ensures both transmission reliability and minimal detectability. THOMRAD 6000 radios are designed to transmit and receive Link 11 and Link 22.

The THOMRAD 6000 product line includes two airborne versions (TRA 6030 and TRA 6020), ground and shipborne versions (TRG 6030), and a manpack version (TRM 6020). Together they represent a fully integrated and interoperable communication system for all air/ground, air/air and surface/surface communications.

Particular versions are available for specific requirements.

Specifications
Dimensions:
(TRA 6030) 90 × 194 × 320.5 mm
(TRA 6020) 127 × 120 × 183 mm
Weight:
(TRA 6030) 7 kg
(TRA 6020) 5 kg
Power: 28 V DC

Status
All versions in production. Selected for Rafale, Mirage 2000, E2C, C-130, Cougar and Panther.

In March 1999, a variant, designated TRA 6021, was selected by the UK for Royal Air Force Canberra PR.9, Hercules C-130K (C.1/3), Nimrod MR.2 and R.2, and VC-10 aircraft, and for Royal Navy Jetstream T.3 observer trainer aircraft. Aircraft installation by October 1999 was planned. The same equipment has been selected for the British Army WAH-64 Apache attack helicopter development programme, and it was evaluated during 1999 by the UK Royal Navy for its helicopter SATURN programme.

Contractor
Thales Communications.

Thales' Transceivers Airborne TRA 6020 (left) and TRA 6030 (right), of the THOMRAD 6000 V/UHF ECCM transceiver series
0011853

The Thomson-CSF Communications Transceivers Airborne TRA 6020 (left) and TRA 6030 (right), of the THOMRAD 6000 V/UHF ECCM transceiver series
0011854

TRA 2020 VHF/UHF radio

Type
Avionic communications system, VHF/UHF, secure.

Description
The main features of the TRA 2020 are a frequency range of 100 to 400 MHz, Guard receiver, clear or KY 58 secure voice, homing/direction-finding output, discrete power and remote control. The main criteria which guided the design were reliability, maintainability, lightness, low power consumption and the ability to adapt to any military platform.

The TRA 2020 architecture is centred around two printed circuit boards using SMC technology and carrying all power supply, transmission, synthesiser and main receiver functions and optional Guard receiver and control interface functions. The easily adaptable mechanical configuration allows installation on any type of platform. The optional control interface is compatible with MIL-STD-1553 or ARINC 429 busses. In the ARINC 429 versions, it is controlled by the TC 20 control unit.

Specifications
Dimensions: 356 × 57.15 × 193 mm
Weight: <4.5 kg
Power supply: 28 V DC
  (VHF) 118–156 MHz
  (UHF) 225–400 MHz
Channel spacing: 12.5 kHz

Status
In production.

Contractor
Thales Communications.

The TRA 2020 VHF/UHF radio and its associated TC 20 control unit (right)
0011843

TRA 6020 V/UHF ECCM airborne transceiver series

Type
Avionic communications system, VHF/UHF, secure.

Description
The TRA 6020 series of airborne transceivers comprises the TRA 6021, which offers fixed frequency, Have Quick, operation with the NATO SATURN comsec capability, and the TRA 6025 in which the SATURN capability is replaced by a proprietary ECCM algorithm.

The equipment includes an ECCM management system and IDM/EDM compatibility. Ancillaries include a remote control unit, agile filters, multicouplers, and external booster and tuned antennas.

Specifications
Frequency bands:
  30 to 87.975 MHz
  108 to 173.975 MHz
  225 to 399.975 MHz
Channel spacing: 8.33 and 25 kHz
Transmit power:
  (AM) 10 W
  (FM) 15 W
  (LPI) LPI operation available
Modulation: AM, FM, MSK, FSK
Guard channels: 121.5 and 243 MHz
Functions:
  (fixed frequency) clear voice, secure voice, Link 11

TRA 6020 series V/UHF ECCM airborne transceiver   0080280

(ECCM) clear voice, secure voice, point-to-point data with TDMA option
Comsec: external KY 58 – KY 100, or proprietary embedded comsec
Remote control: MIL-STD-1553B, ARINC 429 or RS 422
Power: 28 V DC, 30 W receive only, 150 W transmit
MTBF: 2,000 h
Dimensions: 127 × 124 × 183 mm
Weight: 5 kg

Status
In production and in service.

Contractor
Thales Communications.

TRC 9600 VHF/FM secure radio communication system

Type
Avionic communications system, VHF, secure.

Description
The TRC 9600 is a VHF/FM military airborne transceiver designed for military platforms with a high ECCM protection level ensuring reliable communications in a dense electronic warfare situation. It is the airborne version of the PR4G system, fully interoperable with manpack and vehicular versions and covers the frequency range 30 to 88 MHz at 25 kHz spacing to give 2,320 channels.

The TRC 9600 provides communications protected from interception, direction-finding, jamming, listening-in and spoofing. The lightweight and compact 10 W transceiver embodies frequency hopping, free channel search and high-security digital encryption functions.

Its lightweight, small size and high reliability result from the use of advanced and well-proven technology such as powerful and fast microprocessors of the new HCMOS 68000 family, VLSI circuits, wide use of surface mount components and the use of proximity filters for co-site operation.

Frequencies are generated by a digital synthesiser with an ultra-rapid acquisition time, driven by a high-stability oscillator. Seven channels can be memorised and are stored for more than a year. The output power can be 10, 5 or 0.5 W depending on the requirement. In the analogue fixed-frequency mode, the TRC 9600 is directly interoperable with existing VHF/FM sets and is provided with a noise squelch.

Thales' TRC 9600 VHF/FM secure radio communication system comprises the TRC 9610 transceiver unit (left) and the TRC 9620 control unit (right)   0503999

The system consists of the TRC 9610 transceiver unit, TRC 9620 control unit, shockmount, antenna with logic converting unit and power supply. Accessories include a KY 58 encryption unit, relay cable, aircraft interphone system and fill device.

Specifications
Dimensions:
(control unit) 145 × 76 × 160 mm
(transceiver unit) 125 × 196 × 340 mm
Weight: 8 kg
Power supply: 28 V DC

Status
In production and in service.

Contractor
Thales Communications.

Germany

AR 3202 VHF transceiver

Type
Avionic communications system, VHF.

Description
The single block unit AR 3202 VHF transmitter/receiver is a member of Becker's 3000 Series Prime Line avionic systems which feature microprocessor control. It provides 760 channels in the VHF band which extends from 118 to 136.975 MHz for civil aircraft. This range can be extended to 144 or 152 MHz for military aircraft. The system features a non-volatile memory and solid-state switches which eliminate mechanical contacts, giving increased ruggedness and reliability. Opto-electronic switching is employed for frequency selection. Frequency generation and display are microprocessor-controlled. The display comprises two liquid crystal presentations, one of which indicates the active channel while the other shows a preselected frequency.

The complete system is housed in a single compact unit, which complies with ARINC standards. It does not require the remote boxes, interconnecting cables, or external cooling of older designs.

Additionally the set can be operated using the CU 3202 remote controller. This is particularly useful in a tandem-seat trainer where any frequency selection made in one cockpit is displayed in the other.

The transmitter output power is 20 W and the AR 3202 is suitable for both fixed-wing aircraft and helicopters.

Specifications
Dimensions: 47.5 × 146 × 225 mm
Weight: 1.3 kg
Frequency ranges:
Standard: 118.0–136.975 MHz
Option 1: 118.0–144.0 MHz
Option 2: 118.0–152.0 MHz
Channel spacing: 25 kHz

Power supply: 25–30 V DC; 3.5 A (transmit), 0.24 A (standby). Certified to FAA TSO C37c, TSO C38c and ICAO Annex 10 specifications.

Status
In production and in service.

Contractor
Becker Avionic Systems.

AR 3209 VHF transceiver

Type
Avionic communications system, VHF.

Description
There are two versions of the single block unit AR 3209 transceiver, the AR 3209-(09) 5 W system and the AR 3209-(11) 8 W system.

General features include: storage memory for up to 20 of 760 channels; front panel adjustment of squelch, side tone and intercom; two sunlight-readable LCDs.

The AR 3209 is an ideal retrofit for the Becker COM 2000 and is cable, plug and form/fit compatible.

Specifications
Dimensions: 146 × 47.5 × 229 mm
Weight: 1.2 kg
Radio frequency: 118.0–136.975 MHz

Becker Avionic Systems AR 3209   0001174

Channel spacing: 25 kHz
Memory channels: 20
Power supply: AR 3209-(09): 13.75 V DC
AR 3209-(11) 27.5 V DC
Certifications: JTSO C37d, JTSO C38d, and ICAO Annex 10

Status
In production.

Contractor
Becker Avionic Systems.

AR 4201 VHF-AM transceiver

Type
Avionic communications system, VHF.

Description
The small lightweight AR 4201 offers 760 channels and is certified for use in VFR and IFR equipped aircraft. It is ideal for installation in gliders and home-built and small single-engined aircraft due to its limited power requirement and 57 mm round format.

The equipment has a transmit power of 5 to 7 W. A standby frequency and 99-channel memory are available and can be easily programmed and recalled. The AR 4201 features intercom, panel

Becker Avionic Systems AR4201 VHF-AM transceiver   0098258

lighting, voltage indicator, an RF input, automatic test routines, a serial interface and an optional temperature indication.

Two dynamic and two standard microphones can be connected without any alterations. The display shows the active frequency and either the standby frequency, the memory channel used or the supply voltage and also the external temperature (with the optional temperature sensor).

Using the RS-232 interface, all functions can be remotely operated. With this standard feature, the AR 4201 can be integrated into future Becker Systems.

Specifications
Dimensions: 192 × 60.6 × 60.6 mm
Weight: 0.67 kg
Power supply: 12.4–15.1 V DC;
  (transmit) <2.5 A
  (standby) <0.07 A
Frequency range: 118.0–136.975 MHz.
Channel spacing: 25 kHz
Memories: 99
Certifications: TSO C37d, TSO C38d, and
  ICA Annex 10

Contractor
Becker Avionic Systems.

AS 3100 audio selector and intercommunication system

Type
Avionic audio management system.

Description
A member of the Becker 3000 Series avionic systems, the AS 3100 controls four transmitters and up to six receivers. By addition of an auxiliary unit, a further six receiver units can be added to the audio chain. In different versions the AS 3100 is capable of either voice-operated switch or push-to-talk operation with all other stations, voice filter for ADF and navigation systems, connection of various types of microphone and emergency operation and providing redundancy for the transmitter/receiver operation. The intercom amplifier has a common bus connecting up to six cabin and three cockpit stations. A cockpit voice recorder output is incorporated.

The system is of modular construction and may be tailored to precise customer requirements. It is suitable for fixed-wing aircraft and helicopters. To achieve maximum adaptability, a full range of sub-units has been developed to extend the function of the main AS 3100 controller to a full cabin communication and passenger entertainment system. An intercom amplifier permits communication between passengers and crew in noisy aircraft such as helicopters, and a service station allows communication between the crew and flight attendants as well as a public address facility. A tape player, used in conjunction with the public address amplifier, is the basis of passenger entertainment through headsets or loudspeakers. The public address amplifier includes one mono or stereo amplifier and a double-tone gong which operates when activated by the fasten seat belts or no smoking signs.

An external jack box, which is normally installed in the wheel well or any other readily accessible location, permits communication between cockpits or flight deck and ground crew during starting and departure checks.

All units operate from a 28 V DC supply.

Specifications
Dimensions:
  (main control unit) 38 × 146 × 35 mm
  (auxiliary unit) 29 × 146 × 26 mm
  (cassette player) 57 × 146 × 170 mm
  (service station) 210 × 66 × 115 mm
  (public address amplifier) 129 × 45 × 245 mm
  (external jack box) 117 × 80 × 80 mm
Weight:
  (main control unit) 0.8 kg
  (auxiliary unit) 0.2 kg
  (cassette player) 1 kg
  (service station) 1 kg

The Becker Avionic Systems AS 3100 audio selector and intercommunication system 0098257

  (public address amplifier) 0.8 kg
  (external jack box) 0.6 kg

Status
In production and in service.

Contractor
Becker Avionic Systems.

Digital Voice Communication System DVCS 5100

Type
Avionic Communications Control System (CCS).

Description
The DVCS 5100 is a member of the Becker PrimeLine communication and navigation family; it comprises the Audio Selector Unit (ASU) Remote Electronics Unit REU 5100 ( ) and the Audio Control Unit ACU 5100 ( ). The system can control up to eight radio transceivers, or seven transceivers plus one Public Address (PA) amplifier; it is also capable of monitoring up to eight transceivers and eight navigation receivers, where the volume is individually adjustable, and of monitoring up to six fixed input signals.

The 10 internally generated aural warnings are activated by discrete control lines. While controlling the aircraft intercommunication facilities in 'hot microphone' VOX or PTT modes, the volume can be adjusted independently.

Optical and/or acoustic call and quit functions and separate intercom circuits between the cockpit and passenger cabin are provided.

Automatic switch over to emergency operation is provided in case of failure of the power supply. The system is fitted with a serial interface for programming customer or aircraft requirements; such customising actions need to be performed at the vendor's facility or an approved workshop.

Switch over to SLAVE mode is available for training applications, or in case of partial defect. Full BITE is provided.

Specifications
Audio Control Unit ACU 5100 ( ):
Width: 146.1 mm
Height: 76.2 mm
Depth: 110 mm
Standard: ARINC 8 HE
Mounting: dzus
Weight: 0.6 kg

Remote Electronics Unit REU 5100 ( ):
Length: 320.5 mm
Width: 57.2 mm
Height: 193.5 mm
Standard: 1/4 ATR short
Mounting: ATR Fixture
Weight: 1.8 kg

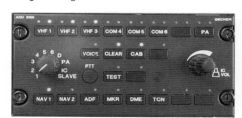

The Becker Avionic Systems digital voice communication system DVCS 5100 0062183

Environmental conditions:
Operating temperature:
  (ACU) –20 to +55° C
  (REU) –40 to +55° C
  (short time) +70° C
Storage temperature: –55 to +85° C
Altitude: = 50 000 ft (15 200 m)
Vibration: S + U
Acceleration: 12 g
Shock: (operational) 6 g/11 ms half-sine wave
  (crash safety) 15 g/11 ms half-sine wave.
Other environmental conditions as per DO-160D.

Applicable documents
RTCA DO-214 audio systems characteristics
EUROCAE/RTCA ED 12B/DO-178B; Level C Software
EUROCAE/RTCA ED 14D/DO-160D environmental conditions
JTSO C50c audio performance
Mil-Spec L 85762A NVG compatibility (Option)

Contractor
Becker Avionic Systems.

FSG 90/FSG 90F VHF/AM transceivers

Type
Avionic communications system, VHF.

Description
The FSG 90/FSG 90F VHF COMM transceivers are designed as dual-mode 25/8.33 kHz channel spacing transceivers, for direct panel mounting into 2.25 in (57 mm) diameter instrument panels.

Frequency coverage is 118.000 to 136.975 MHz, in 760 channels with 25 kHz-only channel spacing, and 2,278 channels in 25/8.33 kHz dual mode operation. For governmental applications an extended frequency range of 118.000 to 149.975 MHz with 3,838 channels in dual mode is available. 99 memory channels are available in 25/8.33 kHz dual mode and another 99 memory channels in 25 kHz-only mode.

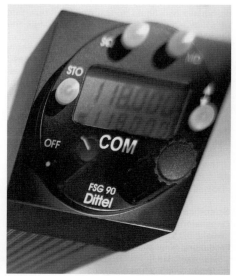

FSG 90 VHF transceiver (Walter Dittel GmbH) 1180333

Three display modes are provided: frequency only, frequency/memory channel, active/standby frequency change.

The system meets EUROCAE ED-23B (airborne), ETS 300 676 (ground) and ETS 300 339 (generic EMI) requirements.

Variants are available with low-profile flat front panels, 10 W power output, 14 V DC supply and band extension for military use, in addition to various specific retrofit adaptors to replace outdated 720 CH radios. Variants are designated FSG90 (6 W output at 50 Ohms), FSG90HH (10 W output at 50 Ohms), FSG90E (extended frequency) and the flat FSG90F.

Specifications

Frequency range: 118.000–136.975 MHz (118.000–149.975 MHz option)
Channel spacing: 25/8.33 kHz dual mode and 25 kHz only
Transmitter power: 6 W into 50 Ohms (FSG 90); 10 W into 50 Ohms (FSG 90H)
Audio power: 8 W
Power supply: 11.0–16.5 V DC
Dimensions (W × H × D): FSG 90: 63 × 58 × 200 mm
FSG 90F 63 × 58 × 230 mm
Weight: FSG 90 0.80 kg
FSG 90F 1.1 kg

Status

Designed for all aircraft and helicopter initial-fit and upgrade applications, as main or standby communications systems. A higher power version, the FSG90H1, producing 10 W RF power at 50 Ohms entered production in the 4th quarter of 1999.

Contractor

Walter Dittel GmbH Luftfahrtgerätewerk.

M3AR VHF/UHF airborne transceiver

Type

Avionic communications system, VHF/UHF, secure.

Description

The R&S® M3AR, the airborne version of the M3xR radio family, is a multiband/multimode radio, designed to provide UHF and VHF, AM and FM, voice and data communication, in fixed frequency or (Electronic Protection Mode/Electronic Counter-Counter Measures (EPM/ECCM) modes with embedded COMSEC and TRANSEC. M3AR transceivers are software defined radios with pre-planned product improvement (P3I) features, which facilitate upgrade to accommodate new developments in communications technologies.

Featuring small size and low weight, the system is capable of providing voice and data communications for a broad range of fixed- and rotary-wing aircraft or UAVs.

The system provides for the use of both NATO EPM/ECCM (Have-Quick I/II and SATURN) and non-NATO (R&S®SECOS) waveforms in all combinations.

The M3AR is available in two form factors (ARC164 for retrofit applications and ARINC600 housings).

The in-cockpit MR6000L is a full form, fit and function replacement for legacy AN/ARC

The MR6000L in-cockpit panel-mount variant of the M3AR 0062218

radio types, with remote control via AN/ARC-164 serial, AN/ARC-164 parallel, ARINC 629 (optional) or RS-485.

The remote-controlled MR6000R incorporates similar features to the in-cockpit variant, plus control of one radio from more than one GB6500 remote-control unit, via serial control RS-485 or ARINC 629 (optional), with remote BITE. Databus versions are available for integration into airborne platforms that incorporate MIL-STD-1553B capability. In addition, emergency control by discrete line is possible.

Specifications

Frequency bands: 108–118 MHz AM receive only; 118-156 MHz AM ATC; 136–174 MHz FM maritime/secure voice; 118–174 MHz FSK; 121.5 MHz Guard channel; 225–400 MHz AM standard/secure voice; 225–400 MHz FM standard/secure voice; 225–400 MHz FSK/ R&S®SECOS; 225–400 MHz FSK/SATURN; 243 MHz Guard channel; 30–88 MHz FM

The M3AR can also be integrated into aircraft equipped with MIL-STD-1553B databus 0062220

Channel spacing: 25 kHz in all bands; 8.33 kHz in VHF ATC band
Tuning capability: 12.5 kHz, 8.33 kHz
Preset channels: 2 × 100 simplex frequencies in standard mode
Modulation: AM, FM, FSK
Power output: AM 10 W; FM 15 W
Frequency accuracy: ≤±1 ppm
EPM/ECCM waveforms: NATO SATURN; NATO Have-Quick I/II, R&S®SECOS
COMSEC: optional embedded NATO (KY58 compatible), embedded R&S®SECOS
ARINC 600 features: Optional embedded NATO COMSEC, Link 11, 20 W AM/30 W FM, integrated pre-/post-selector, sonobuoy command, GMDSS channels, AF wideband

The M6000R remote-control variant of the M3AR 0062219

The MR6000L in-cockpit variant showing construction layout 0062221

output for automatic direction finding and homing

Power supply: MIL-STD-704A
Supply range: 22.5–30 V DC (negative ground)
Emergency operation: 16 V DC
Compliance: MIL-STD-704A, MIL-STD-454, MIL-STD-461/462, MIL-STD-704, MIL-STD-810 and VG 95211, MS25212 among others
Weight: <4.5 kg
Dimensions: ARC 164 or ARINC 600

Status
In production since 2000. In service in Saab Gripen aircraft being delivered to the Royal Swedish Air Force and for export. The M3AR has been selected for the Brazilian Air Force ALX, and Su-30 MK, F-16 and other fighter and transport aircraft and helicopters.

The aforementioned Brazilian Air Force contract is for several hundred systems from 2002, for use onboard the EMB-314 Super Tucano/ALX aircraft as part of the Sistema de Vigilância da Amazónia project.

In late 2002 Rohde & Schwarz announced M3AR orders for German Tiger and NH 90 helicopters. To date, the system has also been delivered to the US and France.

Contractor
Rohde & Schwarz GmbH & Co KG.

M3AR, type XT6923L VHF/UHF SATURN airborne transceiver

Type
Avionic communications system, VHF/UHF, secure.

Description
The R&S® XT6923L is an airborne secure radio communication system designed for clear and encrypted voice and data transmission, in simplex or half-duplex mode in the 100 to 156 MHz (VHF) and 225 to 400 MHz (UHF) frequency ranges. The NATO SATURN Electronic Protection Measures (EPM) waveform, plus embedded crypto functionality (Vinson mode) is used to protect against severe jamming, direction-finding and deception. The airborne transceiver provides downwards interoperable voice modes such as AM and FM in fixed-frequency and NATO Have-Quick I/II ECCM techniques. The system is data capable, in fixed frequency as well as in SATURN mode and complying with NATO Link 11 standards. The Guard receiver monitors 121.5 and 243 MHz international distress frequencies and the 156.525 MHz GMDSS frequency.

Specifications
Frequency range: 108–155 MHz, 225–400 MHz
Channel spacing: VHF 8.33/25 kHz, UHF 25 kHz

RF output power: 20 W/AM, 30 W/FM MSK; low-power mode selectable
Built-in guard receiver: VHF/UHF
Power supply: 28 V DC
NATO EPM/ECCM standards interoperability: Have-Quick I/II, SATURN, embedded COMSEC (Vinson mode)
Operating temperature range: –40 to +71°C
Dimensions: ARINC 600 housing
Weight: <7.5 kg
Compliance: STANAG 4205 (UHF), STANAG 4246 (Have-Quick I/II), STANAG 4372 (SATURN), STANAG 5511 (Link 11).

Status
In production since 2003. Due to enter service in German Tiger helicopters and selected for the NH 90 helicopter.

Contractor
Rohde & Schwarz GmbH & Co KG.

RMU 5000 radio remote-control unit

Type
Avionic Communications Control System (CCS).

Description
The RMU 5000 is a member of the Becker PrimeLine communication and navigation family; it provides control of up to three separate radios from a single compact control head; two variants are available, allowing the customer to select which of the COM/NAV/XPDR [RMU 5000 (1)] or COM/NAV/ADF [RMU 5000 (2)] grouping he wishes to control.

The RMU 5000 provides for three redundant power supply inputs, ensuring that if one bus or power source is lost, all radios remain constantly available and operating. Full redundancy can be obtained through an external panel-mounted switch for IFR operation with dual RMU installation. The front panel and switch knobs are designed to be compatible with NVGs for military and law enforcement applications. The RMU 5000 may be housed in either the standard 160 mm single compact case or in a DZUS-mountable version.

The Becker Avionic Systems RMU 5000 radio remote-control unit 0062184

Specifications
JAR TSO:
Specifications
JAR TSO:
(COM-TX) JTSO-2C37e
(COM-RX) JTSO-2C38e
(VOR) JTSO-2C40c
(LOC) JTSO-C36e
(ADF) JTSO-2C41d
(ATC) JTSO-C74c

Contractor
Becker Avionic Systems.

Series 610 SECOS VHF/UHF airborne transceivers

Type
Avionic communications system, UHF, secure.

Description
R&S® Series 610 SECOS radios are designed for airborne and land-mobile use, featuring integrated COMSEC and TRANSEC, digital encryption, medium-speed frequency hopping with collision-free operation of a large number of networks and interoperability other ground-based or shipboard SECOS radios. The frequency range for ECCM operations is 225 to 400 MHz, with an additional fixed-channel plain language AM option covering the 100 to 156 MHz VHF band.

The equipment consists of transceivers, control units and a key entry device. Control is through a compact cockpit transceiver; there are also remote-controlled versions. A combined UHF ECCM and VHF fixed-channel version is also available.

Specifications
Frequency range: 225 – 400 MHz (100 – 156 MHz option)
Channel spacing: 25 kHz
Operating modes: Fixed frequency, ECCM
ECCM modes: COMSEC, COMSEC +TRANSEC (with or without hailing), voice and data
Power output: 15 W FM, 10 W AM
Power supply: 28 V DC

Status
In production since the early 1990s. In service in F/A-18 aircraft.

Contractor
Rohde & Schwarz GmbH & Co KG.

Series 610 VHF/UHF airborne transceivers

Type
Avionic communications system, VHF/UHF, secure.

Description
Designed for cockpit installation or as remotely controlled equipment, the R&S® Series 610 transceivers provide facilities for radio communication as well as for 16 kb/sec base band data transmission and automatic direction-finding. Models cover the VHF band (XU610/611) and UHF band (XD610/611) in the frequency range from 108 to 400 MHz. All variants feature a transmitter output power of 10 W AM and channel spacing of 25 kHz. Up to 30 channels plus V/UHF guard channel can be preselected.

The equipment's modular design provides a high degree of flexibility in terms of meeting retrofit requirements and individual modules (such as the main receiver or transmitter) have well-defined interfaces to enable them to be replaced without readjustment.

ECCM versions with NATO Have Quick I/II have been introduced according to STANAG 4246.

Status
In production since the early 1990s. In service in German Air Force F-4F and Tornado ECR aircraft.

The Rohde and Schwarz XT6923L VHF/UHF SATURN radio is suitable for a wide variety of military platforms (Rohde and Schwarz) 0587578

The R&S® Series 610 UHF ECCM transceiver is in service in German Air Force Tornado ECR aircraft 0504002

The Rohde and Schwarz/Honeywell XK516D HF voice/data radio 0587577

Contractor
Rohde & Schwarz GmbH & Co KG.

XK516D HF airborne voice/data radio

Type
Avionic communications system, HF, data.

Description
The R&S® XK516D airborne radio, a joint development by Rohde and Schwarz and Honeywell Aerospace, is designed for use in commercial aircraft. The system provides for long-range conventional voice and high-speed data air-to-ground, ground-to-air and air-to-air communications.

The transceiver system consists of the XK516D1 radio and the FK516/517 antenna coupler; modules which provide for high-speed data transfer are fully integrated within the radio.

The equipment is controlled by an integrated test system in which a number of functions are continuously monitored. After a test routine has been triggered, any faulty module will be located and indicated to the crew. BITE results are reported to the onboard CFDS/CMC system via two ARINC 429 serial data interfaces. The system also features built-in interfaces to the Central Maintenance Systems (CMS) of major aircraft manufacturers, including Airbus and Boeing.

The XK516D is designed to meet the requirements of ARINC 719 for voice and ARINC 753/635 for data functions. The integrated data communication capability meets the specifications of ARINC 753 and 635 and high-speed data communication at up to 1,800 bps is provided. The data capability also means that operators can obtain the benefits of ACARS and the Aeronautical Telecommunications Network (ATN) beyond limited VHF coverage.

The FK516 antenna coupler is digitally tuned and features tuning times of (typically) less than 3 seconds; in the 'learn' mode, even shorter tuning times of several hundred milliseconds are achieved. The antenna coupler is designed to tune suppressed shunt/notch antennas that are common on the fin structure of most modern aircraft, while the FK517 antenna coupler is designed to tune the tailcone antennas fitted to older aircraft.

Specifications
RF power: 400 W PEP
RF range: 2–30 MHz
Tuning increments: 100 Hz
Data rate: 150, 300, 600, 1,200, 1,800 bps
Dimensions: 6 MCU
Weight: <12 kg
Power: <1 kW

Status
In production and in service onboard civil long-haul aircraft.

Contractor
Rohde & Schwarz GmbH & Co KG.
Honeywell Aerospace.

XT 3000 VHF/UHF radio

Type
Avionic communications system, VHF/UHF.

Description
The R&S®XT 3000 is a combined VHF/UHF air-to-air and air-to-ground transmitter/receiver covering the frequency ranges 100 to 162 MHz and 225 to 400 MHz, in both FM and AM modes for voice and data. Frequency increments are spaced at 25 kHz intervals, but the channel spacing is switchable by increments of 25, 50 or 100 kHz as required. Transmitter output is 10 W. Up to 28 operational channels, together with the international distress frequencies of 121.5 and 243 MHz, may be preselected on remote-control units. The system incorporates built-in test equipment and inputs for special-to-type and automatic test equipment.

The XT 3000 comprises a single VHF/UHF transmitter/receiver, VHF and UHF amplifiers, two control units for remote operation and a channel/frequency indicator.

Specifications
Modulation: AM/FM
Frequency range: 100–162 MHz; 225–400 MHz; distress frequencies 121.5 MHz, 243 MHz
Channel spacing: 25/50/100 kHz, switchable
Power output: 10 W
Dimensions: ½ ATR short plus two ¼ ATR short units
Weight: 30 kg
Compliance: MIL-E-5400
Environment: MIL-STD-810B
EMC: MIL-STD-461/462/463
Reliability: MIL-STD-781B

Status
In production and in service with the German Air Force since the 1990s.

Contractor
Rohde & Schwarz GmbH & Co KG.

XT622P1 VHF/UHF SATURN airborne transceiver

Type
Avionic communications system, VHF/UHF, secure.

Description
To offer maximum protection for vital communications links against hostile jamming, the R&S® Series 620 VHF/UHF airborne transceivers incorporate the latest second-generation of Anti-jam Tactical UHF Radio for NATO (SATURN) Electronic Protection Measures (EPM).

The XT622P1 SATURN radio covers the UHF frequency band from 225 to 400 MHz and provides clear and encrypted voice transmission/reception in UHF AM, NATO Have Quick I/II and NATO SATURN modes. The VHF frequency band is covered from 108 to 156 MHz for ATC operations.

For voice digitisation, a 16 kbit/sec delta modulator is integrated into the system. Control and data interfaces of the radio are in compliance with MIL-STD-1553B and STANAG 3910 (fibre-optic bus). Other relevant standards include STANAG 4205 (UHF), STANAG 4246 (Have-Quick I/II) and STANAG 4372 (SATURN).

Specifications
Frequency range: 108–155 MHz, 225–400 MHz
Channel spacing: VHF 12.5 kHz, UHF 25 kHz
Preset channels: 24
RF output power: 20 W/AM, 30 W/FM MSK; low-power mode selectable
Built-in Guard receiver: VHF/UHF
Power supply: 28 V DC

The Rohde and Schwarz XT 3000 airborne transceiver system. From left to right (top), UHF power amplifier, transmitter/receiver and VHF power amplifier. Two control units and a remote frequency indicator are shown at the bottom 0504003

Rohde & Schwarz Series 620 SATURN airborne transceiver (Rohde & Schwarz GmbH)   1179132

NATO EPM/ECCM standards interoperability: Have-Quick I/II, SATURN, embedded COMSEC (Vinson mode)
Bus interface: dual-redundant fibre-optic
Serial maintenance interface: RS-232C
Operating temperature range: –40 to +71°C
Dimensions: ⅜ LRU short L-shape
Weight: <7.5 kg

Status
In production and service in Eurofighter Typhoon fighter aircraft.

Contractor
Rohde & Schwarz GmbH & Co KG.

FSG 2T VHF/AM transceiver

Type
Avionic communications system, VHF.

Description
The FSG 2T is a 760-channel panel-mount airborne transceiver covering the 118 to 136.975 MHz range. It has an output power of 5 W and a 20 channel electronic memory. It mounts into a 57 mm instrument panel space. There are also portable cased and base station versions with a built-in rechargeable battery.

FSG 2T VHF transceiver (Walter Dittel GmbH)   1180329

Status
In production and in service.

Contractor
Walter Dittel GmbH Luftfahrtgerätewerk.

India

Audio Management Unit (AMU) 1303A

Type
Avionic audio management system.

Description
AMU 1303A is designed for use in fighter and trainer aircraft. It provides management control of two transceivers, three receivers and seven audio warning signals. Additional features include a telebrief facility and voice-operated switching for hands-free operation.

Specifications
Dimensions:
(junction box) 60 × 175 × 180 mm
(station box) 67 × 146 × 180 mm
Weight:
(junction box) 1.1 kg
(station box) 1.2 kg
Power supply: 27.5 V DC, 1 A
Bandwidth: 300 Hz to 3.5 kHz

Status
In service.

Contractor
Hindustan Aeronautics Ltd, Avionics Division.

Hindustan's audio management unit 1303A
0044790

COM 32XA HF/SSB communication system

Type
Avionic communications system, HF.

Description
The all-solid-state modular construction HF/SSB communication system is designed for air-to-air and air-to-ground communication on 2.06 to 29.999 MHz (COM 326A) or 2.5 to 23.5 MHz (COM 327A). Channel spacing is 100 kHz and the system has seven in-flight programmable preset channels and instant manual selection of any channel.

Hindustan's COM 32XA HF/SSB communication system
0044791

The salient features of the system include solid-state design with software-controlled advanced automatic antenna matching, high-stability frequency synthesiser, high-power solid-state amplifier with protection against high VSWR, high-performance receiver with front panel configurable logic, in-flight programming of channnel and modes and BITE.

Specifications
Dimensions:
(receiver/exciter) 190 × 194 × 320 mm
(power amplifier) 190 × 194 × 320 mm
(ATU for COM 327 and 325) 150 × 180 × 321 mm
(control unit) 146 × 124 × 102 mm
Weight:
(receiver/exciter) 11.5 kg
(power amplifier) 15 kg
(ATU for COM 327 and 325) 7.5 kg
(control unit) 1.5 kg
Power supply:
(COM 326A) 200 V AC, 400 Hz, 3 phase
(COM 325/327/328/329A) 115 V AC, 400 Hz, single phase 27.5 V DC
(receiver) 100 W DC
(transmitter) 850 W AC 100 W DC
Temperature range: –40 to +55°C
Altitude: up to 65,000 ft

Status
The COM 325A is fitted on the Hawker Siddeley 748, COM 326A on the Jaguar, COM 327A on the An-32, COM 328A on the Do 228 and the COM 329A on the I1-38.

Contractor
Hindustan Aeronautics Ltd, Avionics Division.

COM 150A UHF transceiver

Type
Avionic communications system, UHF.

Description
The COM 150A UHF transceiver provides 7,000 channels at 25 kHz spacing between 225 and 399.975 MHz for radio telephony (A3) transmissions. There are four preset channels.

Specifications
Dimensions: 121 × 174 × 294 mm
Weight: 6.5 kg
Power output: 5 W
Temperature range: –55 to +55°C
Altitude: up to 70,000 ft

Hindustan's COM 150A UHF transceiver
0044793

Status
In service.

Contractor
Hindustan Aeronautics Ltd, Avionics Division.

COM 1150A UHF communication system

Type
Avionic communications system, UHF.

Description
The AM transceiver COM 1150A is designed with hybridised circuits for improved reliability, reduced volume and weight, providing A3 communications in the UHF band.

Specifications
Dimensions:
(transceiver unit) 124 × 178 × 250 mm
(control unit) 80 × 80 × 95 mm
Weight:
(transceiver unit) 5 kg
(control unit) 0.3 kg
Operating frequency: 225–399.975 MHz
Channel spacing: 25 kHz
No of preset channels: 10
Type of transmission: A3
Receiver sensitivity: –97.0 dBm

Antenna impedance: 50 Ω nominal
Audio output: 150 mW into 150/600 Ω
Power output: 4 W (nominal)
Microphone: EM type 75 Ω, balanced 6 mV RMS; input microphone signals up to 1 V RMS

Status
In service.

Contractor
Hindustan Aeronautics Ltd, Avionics Division.

INCOM-1210A integrated radio communication system

Type
Avionic communications system, VHF/UHF, secure.

Description
The INCOM-1210A is an airborne, secure, jam-resistant V/UHF communication system designed for air-to-air and air-to-ground voice/data communications. The system incorporates ECCM for tactical communications.

INCOM-1210A is compact and of moderate weight, suitable for all fighter applications and compatible with MIL-STD-1553B and ARINC 429 data transfer.

Specifications
Frequency range: VHF: 108–173.975 MHz (AM, FM); UHF: 225–399.975 MHz (AM, FM, ECCM)
Guard receiver: VHF: 121.5 MHz (AM), 156.8 MHz (FM); UHF: 243 MHz (AM)
ECCM modes: DS, FH and combined FH-DS
Preset channels: 40
Receiver sensitivity: Minimum 10 dB SNR for 2 µV input (conventional voice); BER of 1 × 10⁻⁴ for 4 µV input (data)
Transmitter power: 20 W
Audio output: 50 mW across 150 Ω
Dimensions:
(transceiver unit) 257 × 124 × 318 mm
(control unit) 85 × 210 × 80 mm

Weight:
(transceiver unit) 10 kg
(control unit) 1.2 kg
Power supply: 28 V DC

Status
In service.

Contractor
Hindustan Aeronautics Ltd, Avionics Division.

Intercom – Audio Management Unit (AMU) 1301A

Type
Avionic audio management system.

Description
The Intercom AMU 1301A is designed for use in fighter aircraft, transports and helicopters, with capacity for up to five crew members in a net including call and conference facilities. The system provides management control of five transceivers, three receivers and five audio warning signals. Additional features include VOS for hands-free operation and dual redundancy.

Specifications
Dimensions:
(junction box) 120 × 120 × 250 mm
(station box) 76 × 146 × 210 mm
Weight:
(junction box) 1.4 kg
(station box) 1.5 kg
Power supply: 27.5 V DC, 2 A
Bandwidth: 200 Hz to 2.5 kHz

Status
In service.

Contractor
Hindustan Aeronautics Ltd, Avionics Division.

Hindustan's COM 1150A UHF communication system
0044794

Hindustan's AMU 1301A Intercom
0504279

International

Link 16 JTIDS (Joint Tactical Information Distribution System) airborne datalink terminals

Type
Avionic datalink system.

Description
The Link 16 JTIDS family of airborne datalink terminals evolved from early US Air Force/MITRE and US Navy studies related to Time Division Multiple Access (TDMA) communications and relative navigation. Early contractors included Hughes, ITT, Singer and Rockwell. Development of Class 1 terminals began in 1974 for E-3 AWACS; 70 were delivered. Development of Class 2/2H terminals began in 1990. Low rate initial production began in 1991. Full rate production began in 1995; over 500 Class 2 terminals have been delivered. Developments have continued to

Specifications

	Class 2/2H	MIDS-FDL	MIDS-LVT	SHAR	ATDL
Dimensions: (mm) RF:	396 × 259 × 193	330 × 191 × 194	330 × 191 × 195	318 × 257 × 191	361 × 191 × 194
DDP:	445 × 325 × 193				
Weight (kg):	RF: 26.81, DDP: 33.18	16.36	23.18	20.00	20.45
Power (W):	200	50	200	200	200
Range (nm):	300/500	185	400	400	400
TACAN:	Yes	No	Yes	Option	Option
Voice (kbps):	2.4/9.6	No	2-channel; 2.4/16	2-channel; 2.4/16	2-channel; 4.8/16
MTBF (h):	500	N/A	N/A	>2,000	N/A
Availability:	In production	1999	2001	1999	1999

meet US and NATO requirements, and the overall range of airborne terminals now comprises (see separate entries, as applicable):

1. Joint Tactical Information Distribution System (JTIDS) Class 2/2H terminals. Application: aircraft, ADGE and ship platforms.

2. Multifunctional Information and Distribution System – Fighter DataLink (MIDS-FDL) terminal. Application: F-15 aircraft.

3. Multifunctional Information and Distribution System – Low Volume Terminal (MIDS-LVT). Application: fighter aircraft, ADGE and ship platforms.

4. SHAR Link 16 low-volume datalink terminal (AN/URC-138(V)1(C). Application: fighter aircraft and helicopters.

5. Advanced Tactical DataLink (ATDL). Application: country-unique air/ground terminal.

The main capabilities provided by this family of terminals includes: jam resistant, crypto secured, data and voice, line-of-sight, communication, navigation and identification, with relay for extended range, using TDMA, frequency-hopping transmissions, in the 960 to 1,215 MHz band.

Status
In production or selected for the following platforms: US Air Force: ABCCC, AWACS, B-1B, F-15, F-16, JSTARS, RC-135 Rivet Joint, tankers, transports. US Navy: F-14D, F/A-18, E-2C, EA-6B, S-3. US Marine Corps: AV-8B, F/A-18. NATO: AWACS, Eurofighter Typhoon, Rafale, Sea Harrier F/A-2, Sea King AEW Mk.7, Tornado F-3.

Contractor
BAE Systems North America.
Rockwell Collins.

MCS 3000/6000 aeronautical satellite communications system

Type
Avionic communications system, satellite.

Description
Honeywell and Thales Avionics co-operated for the development and have partnered for the manufacture and marketing of a multichannel satellite communications system for commercial aircraft, fully compatible with ARINC 741 and ARINC 761 for Aero-I.

The Honeywell/Thales MCS 3000/6000 systems provide a three- or six-channel full-duplex voice and data communications capability supporting such functions as Airline Communications And Reporting System (ACARS), Automatic Dependent Surveillance (ADS), and flight deck and passenger telephone and fax communications between an aircraft and the ground.

The MCS 6000 airborne terminal comprises a Satellite Data Unit (SDU), Radio Frequency Unit (RFU) and High-Power Amplifier (HPA), and may be interfaced to a variety of high-gain phased-array antenna subsystems and voice/data communications devices.

The SDU performs the functions of system controller, data modulation and demodulation, data synchronisation and decoding and voice coding/decoding.

The RFU performs the functions of down converting the received L-band (NATO D-band) signals to a lower frequency for input to the digital processing circuits and up converting the modem output signals to the L-band (NATO D-band) transmit frequency for each operational channel. The RFU operates in full-duplex mode,

simultaneously supporting both receive and transmit functions at all times.

The HPA is a linear power amplifier which provides the gain to generate the required output power. The output power of the HPA is under the control of the SDU which receives data from the ground station commanding an increase or decrease in output power to maintain the satellite signal at a satisfactory level. The HPA is available in either an ARINC 741 or ARINC 761 configuration.

The system operates at L-band (NATO D-band) frequencies. Signals are relayed via the Inmarsat space segment satellites, linking in to the ground telecommunications network through a series of dedicated Ground Earth Stations (GES). The satellite and GES networks combine to provide a worldwide communications service.

MCS 3000/6000 systems support both air-to-ground and ground-to-air communications. The systems are capable of supporting 9.6 kbytes/s voice, 4.8 kbytes/s fax, 2.4 kbytes/s PC modem and 10.5 kbytes/s packet data services for ARINC 741 applications; the systems will support 4.8 kbytes/s voice, 2.4 kbytes/s fax, 2.4 kbytes/s PC/modem and 1.2 kbytes/s packet data services for ARINC 761 (Aero-I) applications. Typical applications for these services break down into the areas of passenger services, airline operational and administrative services and air traffic control.

Passenger services include telephone, fascimile, PC and value added data services such as catalogue sales, hire car reservations and duty free sales.

Airline operational and administrative services include the ACARS datalink supporting engineering, operational and cabin management functions.

Air traffic control uses include Automatic Dependent Surveillance aircraft position reporting.

MCS 3000/6000 operates in full accordance with Inmarsat specifications and type approval has been received on all major wide-body aircraft types.

Aero-I is an upgraded capability of the MCS 3000/6000 to allow narrow-body aircraft with intermediate-gain antennas to utilise the new Inmarsat third-generation Aero-I spot-beam services. As an extension to the Aero-I upgrade, existing high-gain MCS equipment can be upgraded, via a 'Service Bulletin'; this takes advantage of 'evolved Aero-H' services, which provide a similar range of passenger global communications for telephone, fax and pc-data, as Aero-I. Using 'evolved Aero-H' in spot-beam coverage allows the operator to benefit from reduced service charges and 4.8 kbytes/s voice CODECs (digital transmission).

Specifications
Dimensions:
(SDU) 6 MCU; weight 10.73 kg
(RFU) 4 MCU; weight 7.68 kg
ARINC 741 (HPA) 8 MCU; weight 12.86 kg (Aero-H)
ARINC 761 (HPA) 4 MCU: weight 7.05 kg (Aero-I)
Power supply: 115 V AC, 400 Hz; or 28 V DC
Output power: 60 W typical
Frequency:
(transmit) 1,626.5–1,660.6 MHz
(receive) 1,530–1,553 MHz

Status
Over 1,600 installations have been completed on all major wide-bodied aircraft types and on many top-of-the-range executive jets.

Integration has been completed with all major antenna, ACARS and passenger telephone equipment vendors.

In March 1996, the Thales/Honeywell team was named as the US government's provider of choice for multichannel SATCOM using the MCS-3000/6000 systems, featuring STU-III secure voice, access to Microsoft-Mail, 9,600 bits/s fax and other capabilities, in conjunction with the Tecom T-4000 High-Gain Antenna System and the Honeywell CM-250 Communications Management Units. The MCS-6000 is also fitted to the Advanced Range Instrumentation Aircraft (ARIA) RC-135s.

System upgrades that became available in May 1998 include new MCS-3000i/+ and MCS-6000i/+ variants which, in addition to supporting Aero-H, are able to support Aero-I and Aero-H+.

The Aero-H+ system (Aero I and H) utilises the same high-gain antenna as Aero-H but, with the capability of using spot-beam satellites, it offers a potentially lower service cost. The Thales/Honeywell Aero-I systems operate in the spot-beams of the new-generation Inmarsat-3 satellites. All services offered with Aero-H are available on Aero-H+ and Aero-I.

The Thales/Honeywell team certified the Canadian Marconi CMA-2200 antenna as its exclusive Aero-I antenna. All FAA certification work for the MCS 3000/6000 Aero-I Boeing 737-800 installation has been completed, and launch customers for the MCS 3000/6000 on the Boeing 737-800 were Hainan Airlines and Royal Air Maroc.

Aero-I also obtained European certification on board the UK Ministry of Agriculture's Fisheries Patrol aircraft – a turboprop Cessna 406.

By mid-1999, Aero H+ had been certified on Airbus A330, Boeing 747-400 and Global Express aircraft.

Contractor
Honeywell Inc, Commercial Aviation Systems.
Thales Avionics.

MCS-7000 aeronautical satellite communications system

Type
Avionic satellite communications system.

Description
Honeywell and the then Racal Avionics (now Thales Avionics) launched the MCS-7000 as their latest-generation enhanced satellite communications system for commercial airliners and business jet aircraft, based on their existing MCS-3000 and MCS-6000 systems. The MCS-7000 provides worldwide continuous multichannel (4 or 7 channels) voice/data communications and the capability of Aero-H, Aero-H+ or Aero-I (spot-beam) services, depending on the High-Power Amplifier (HPA) and antenna configuration. The equipment is fully compliant with ARINC characteristics 741 and 761, Aviation Satellite Communication System, and characteristic 746, Cabin Communications System. The MCS-7000 satcom systems consist of only two units, a 6 MCU-size Satellite Data Unit (SDU) and a 4 MCU-size High-Power Amplifier (HPA) with embedded Beam Steering Unit functionality.

The Aero-H+ system will use the same high-gain antenna as Aero-H but, with the capability of using spot-beam satellites, it offers a potentially lower cost. The Aero-I system operates in spot beams of the new-generation Inmarsat-3 satellites. Spot beams have lower power requirements and therefore smaller, lower-power HPAs and smaller Intermediate-Gain Antennas can be used. The selected antenna is the Canadian Marconi CMA-2200.

All services currently offered with Aero-H systems are available on Aero-H+ and Aero-I

6-channel MCS-6000 system

3-channel MCS-3000 system

Honeywell/Thales MCS 3000/6000 aeronautical satellite communications system showing both the six-channel MCS-6000 system (radio frequency unit (left), satellite data unit (centre) and high-power amplifier (right)); and the three-channel MCS-3000 system (satellite data unit (left) and high-power amplifier (right))

0002115

including cockpit voice (allowing instantaneous communication with operations, maintenance and air traffic control); passenger telephony, passenger fax, news and weather broadcasts; interactive passenger services and a PC data capability.

MCS-7000 is the standard production system, available from June 1999, providing several additional features, including an optional internal PBX phone system with digital handsets. For aircraft which are not equipped with an Inertial Reference System (IRS), a new Signal Conditioning Unit (SCU) supports satcom operation and a proprietary interface supports the imminent Complementary Satellite Systems, including Low Earth Orbit (LEO) and Medium Earth Orbits (MEO). Maximum flexibility is provided for operators to choose Inmarsat services for the cockpit (safety/ATC services) and an alternative service for passenger requirements (voice/fax).

Company data also uses the terminology MCS-7000+ and MCS-7000i when describing this system. The MCS 7000+ system can provide secure voice and data communication with the addition of the CM-250 Communication Gateway Unit (CGU). It is also claimed that, in addition to supporting existing Inmarsat Aero-H, Aero-H+ and Aero-I, the MCS-7000 system has growth to support the Complementary Satellite Systems (CSS).

Status

The MCS-7000 was launched in June 1999, with Hainan Airlines on its Boeing 767 aircraft.

MCS-7000i is also supplied by Honeywell, via its Defense Avionics Systems division, to the US Air Force, together with the CM-950 and CM-950 via communications management units, to meet Global Air Traffic Management (GATM II) upgrade programme requirements.

Contractor

Honeywell Inc, Commercial Aviation Systems.
Thales Avionics SA.

MIDS-LVT Multifunctional Information and Distribution System – Low-Volume Terminal

Type

Avionic datalink system.

Description

After successful completion of international development of the MIDS Low-Volume Terminal, the EUROMIDS consortium has been founded by the European developers Thales, Marconi Selenia, INDRA Sistemas and EADS Deutschland GmbH. EUROMIDS acts as the interface to customers for European production and procurement.

MIDS is the standard NATO interoperable data communications system, implementing Link 16 protocols under STANAGs 4175 and 5516. The MIDS terminal also implements the interim JTIDS message standard, to ensure compatibility with the JTIDS Class 1 terminals which are still in use. It serves a nodeless network with a fast data exchange rate. Also secure voice transmission capability is provided.

While maintaining full interoperability with the family of JTIDS terminals, MIDS is a new-generation design that satisfies a broad range of Link 16 applications. To allow its use in the latest and future aircraft designs, the MIDS terminal has been reduced in size, weight, cost and power consumption to less than half that of the present JTIDS Class 2 terminal. It retains, however, all of the JTIDS capabilities, including three-dimensional receiver coverage, 200 W transmit power, complete TACAN capability, full relative/geodetic navigation, precise self-identification and powerful anti-jam capabilities. MIDS-LVT is configured in a single main terminal box and a standoff remote power supply. It has a MIL-STD-1553 bus interface, allowing standard interoperation with designated aircraft systems, plus a high-speed optical databus (3910) and X.25 and Ethernet interface capabilities.

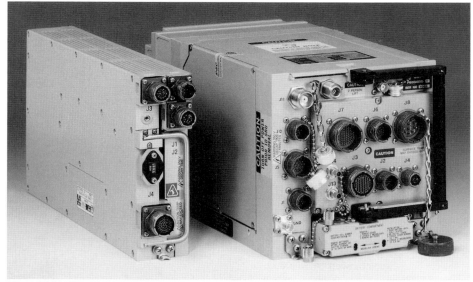

The MIDS – LVT terminal will incorporate the same functional capabilities as the JTIDS Class 2 terminal, with significant decreases in size, weight, power and cost   0044796

An exhaustive Built-In Test (BIT) function eases the failure identification and location. Minimum maintenance effort has to be spent by the modular and interchangeable construction.

The MIDS terminal is designed around multiprocessor architecture, with industry standard VME and IEEE 486 busses offering a truly open architecture. This gives maximum flexibility and additional growth potential. Receiver/transmitter modules incorporate the latest MMIC technology.

Status

MIDS-LVT is a multinational co-operative development programme to design, develop and deliver Link 16 tactical information system terminals that are smaller and lighter than Joint Tactical Information Distribution System (JTIDS) Class 2 terminals. The programme encompasses two US vendors and one European vendor. The US vendors are Data Link Solutions (a limited liability company comprising BAE Advanced Systems and Rockwell Collins) and ViaSat. The European vendor is EUROMIDS, a consortium comprising four companies – Thales, Marconi Selenia, Indra Sistemas SA and EADS Deutschland GmbH. The US and European teams have worked together to develop, test and produce a tactical information system terminal of reduced size, weight and cost, thereby making MIDS-LVT more readily available for use in a wider variety of airborne platforms as well as maritime and ground applications.

As an integral part of the program's acquisition strategy, MIDS-LVT features Open Systems Architecture (OSA) and employs a complex system of performance specifications to allow for continuous competition between MIDS vendors throughout the production phase. This strategy supports the implementation of transatlantic competition after the US and European contractors have completed qualification efforts and established full-rate production capabilities.

There are three variants of the MIDS terminal. MIDS-LVT(1) will be used by US Navy (USN), Marine Corps (USMC) and Air Force (USAF) aircraft as well as by those of European nations. MIDS-LVT(2) will be used by army combat systems such as Patriot for both the US and French Armies. MIDS-LVT(3), also known as the Fighter Data Link (FDL), is already operational and has been installed in USAF F-15C and E aircraft.

European partner nation requirements for MIDS-LVT are contracted through EUROMIDS on a directed sole source basis as set forth in the Programme Memorandum of Understanding (MoU) between the five MIDS nations.

MIDS-LVT platforms include Eurofighter Typhoon, F-15, F-16, F/A-18 and Rafale, with 346 units on order, with a monthly production rate of 16 terminals, through EUROMIDS.

International production is expected to exceed 5,000 terminals. Earlier programme deliveries include 11 MIDS simulators that are being used to integrate the capability into participating aircraft.

Contractor

EUROMIDS, comprising:
EADS Deutschland GmbH.
INDRA Sistemas SA.
Marconi Selenia.
Thales.

Multi Link 2000 M-ADS helicopter satellite communications system

Type

Avionic satellite communications system.

Description

The Multi Link 2000 Modified Automatic Dependent Surveillance (M-ADS) system has been developed by Kongsberg Defence & Aerospace for installation in helicopters operated by Helikopter Service and Norsk Helikopter.

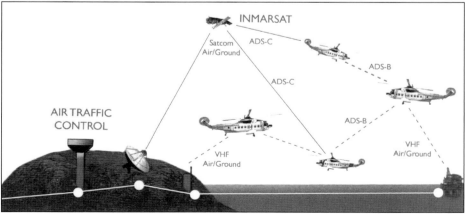

The Multi Link 2000 system concept   0089532

The M-ADS system comprises the following system elements:

1. Airborne equipment consisting of an M-ADS unit; Satellite TRansceiver (STR); low-noise amplifier and antenna, linked to Inmarsat satellite
2. Inmarsat satellite and ground station, linked to ATC centre
3. Air Traffic Control (ATC) centre to set up ADS-Contracts (ADS-C) with the aircraft, so that current position and intent can be displayed.

System functionality includes:

1. ADS – Contract (ADS-C) and ADS – Broadcast (ADS-B).
2. Dual air/ground datalink for safety critical messages: ADS, controller-pilot datalink communication and flight information service.
3. VHF air/ground and air/air communication 9VDL Mode 4/STDMA.
4. Inmarsat aeronautical satellite air/ground communication.
5. ICAO compliant end-to-end data integrity and transfer reliability from built-in Aeronautical Telecommunication Network (ATN) protocols.

The system transmits position, waypoint, groundspeed and other aircraft information from air to ground via the Inmarsat satellite system, providing a means of accurate position reporting. This information, which is automatically transmitted at regular intervals, will provide pinpoint accuracy and keep air traffic controllers up to date with the aircraft's position and intentions.

Specifications
SDU: 2 MCU weighing <4.5 kg
HPA: 2 MCU weighing <4.5 kg
ADSU: 2 MCU weighing <3 kg

Status
The M-ADS system is installed in Eurocopter Super Puma AS332L/LI/L2 and AS365N-2 Dauphin, Bell 214ST and Sikorsky S-61N and S-67 helicopters. Thales Avionics delivers the Satellite TRansceivers (STRs).

Contractor
Kongsberg Defence & Aerospace AS.
Thales Avionics SA.

Satcom conformal antenna subsystem

Type
Avionic satellite communications system.

Description
This Satcom antenna is designed for aeronautical communications in the L band, to meet the specific needs of the airline and general aviation industry. It is compliant with ARINC 741 and meets DO-160C.

The equipment is composed of two side-mounted conformal antennas, two Beam-Steering Units (BSUs) located inside the aircraft and the associated Diplexer/Low-Noise Amplifier (D/LNA) assemblies.

In order to provide a Satcom system with no operational limitations, an antenna with optimised RF characteristics has been designed. A large number of radiating elements arranged in an optimum pattern offers a better combination of high-gain and low-sidelobe levels. The thin profile of the High-Gain Antenna (HGA) results in a negligible drag penalty of less than 0.02 per cent of total aircraft drag and the antenna subsystem is adaptable to all high-gain Satcom avionics subsystems available or under development.

Specifications
Dimensions:
(High-Gain Antenna ×2) 566.4 × 495.3 × 7.6 mm
(Beam-Steering Unit ×2) 342.9 × 261.6 × 88.9 mm
Weight:
(High-Gain Antenna ×2) 7.5 kg
(Beam-Steering Unit ×2) 8.5 kg
(Diplexer/LNA ×2) 3 kg

Status
The antenna has Inmarsat multichannel access approval with no restrictions.

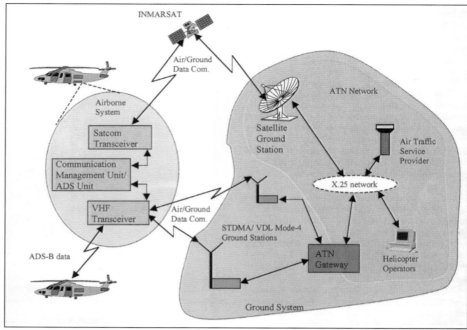

The Multi Link 2000 system equipment   0089531

The antenna system is certified on Airbus A300, A310, A330 and A340; Boeing 707, 737-300, 747-400, 767, MD-11, L-1011 and MD-80 aircraft, and the Falcon 900. More than 400 Satcom antenna systems have been ordered by over 20 major airlines.

Contractor
Honeywell Aerospace, Aerospace Electronic Systems.
Thales Systemes Aeroportes.

SCS-1000 Mini-M aeronautical satcom

Type
Avionic satellite communications system.

Description
The Thales/Honeywell satcom consortium appointed OmniPless (Pty) Ltd of South Africa to be responsible for the exclusive marketing, distribution and product support of the SCS-1000 Mini-M aeronautical satcom worldwide. The OmniPless SCS-1000 Mini-M is a complementary entry system to the MCS 3000/6000/7000 product-line of aeronautical satcom systems manufactured by Thales/Honeywell.

The SCS-1000 Mini-M is based on the Inmarsat land-mobile Mini-M and is intended for small to medium-sized business jet and turboprop operators. Small size offers low cost and flexible installation. The product is type-approved for use with the spot beam capabilities of the latest generation Inmarsat-3 satellites.

The SCS-1000 is a single-channel system, which supports standard (or secure) voice, fax or personal computer data transmission. All antenna steering control sensors, GPS and attitude sensors are integral to the antenna, making them independent of other aircraft systems. Remaining system components comprise an Antenna Control Unit (ACU), Power Supply Unit (PSU), telephone unit and handset.

Contractor
Honeywell Inc, Commercial Aviation Systems.
OmniPless (Pty) Ltd.
Thales Avionics SA.

TRA 6032/XT 621 P1 V/UHF SATURN airborne transceiver

Type
Airborne tactical communications system.

Description
The TRA 6032 (French identity) or XT 621 P1 (German identity) is an airborne secure radio communication system designed for clear and encrypted voice and data transmission, in simplex or half duplex mode in the 100 to 156 MHz and 225 to 400 MHz frequency range.

TRA 6032 uses SATURN as the main transmission security technique which provides the highest level of protection against severe jamming, direction-finding and deception. Communication security is performed with standard NATO crypto devices.

It offers downwards interoperable voice modes such as AM and FM in fixed frequency and first-generation Have Quick I and II ECCM techniques.

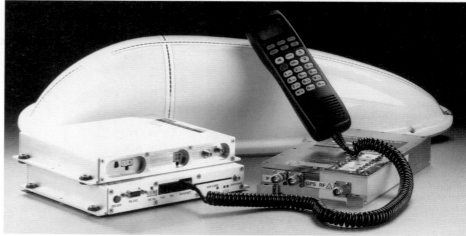

The SCS-1000 Mini-M satcom system   0103885

The TRA 6032 is capable of data transmission, in fixed frequency as well as in SATURN mode and is particularly adapted to Link 11 and NATO Improved Link Eleven (NILE) or Link 22 transmission.

The TRA 6032 is a 3 MCU transceiver compliant with ARINC 600 and is designed for the rotary-wing environment. Its Guard receiver monitors the 121.5 and 243 MHz international distress frequencies and the 156.525 MHz GMDSS frequency.

Specifications
Dimensions: 90 × 194 × 320 mm
Weight: 7.6 kg
Power supply: 28 V DC

Status
In series production and in service in Eurocopter Tiger helicopters. Selected for the NH 90.

Contractor
Thales Communications.
Rohde & Schwarz GmbH & Co KG.

Israel

ARC-740 UHF secure radio

Type
Avionic tactical communications system.

Description
The ARC-740 provides UHF communication for air-to-air, air-to-ground and ground-to-ground use for plain speech and secure speech, using frequency hopping at 10 hops/s and data transmissions at 2.4 kbits/s. The frequency band from 225 to 399.975 MHz is covered by 7,000 channels at 25 kHz steps in both AM and FM modes; any channel can be manually selected. There are also ECCM-protected normal and high-powered modes. There are 99 preset channels and a continuous watch is kept on the Guard frequency.

Specifications
Dimensions:
(transceiver) 250 × 127 × 177 mm
(power amplifier) 190 × 170 × 145 mm
(control box) 143 × 146 × 124 mm
Weight:
(transceiver) 7 kg
(power amplifier) 4.9 kg
(control box) 1.8 kg

Status
In service.

Contractor
Elta Electronics Industries Ltd.

The Elta ARC-740 UHF secure radio 0504005

DL-2000 guided missile datalink

Type
Avionic datalink system.

Description
The DL-2000 is a jam-resistant video command and telemetry link with an operational range beyond 200 km. Interfaces are available for F-4, F-15 and F-16 fighter aircraft and F-111 and B-52 bombers.

Status
In production and in service. Proven with the Israeli Air Force.

Contractor
Rafael Armament Development Authority, Systems Division.

DLV-52 airborne wideband datalink system

Type
Avionic datalink system.

Description
The DLV-52 high-rate datalink is a two-way digital communication system supporting reconnaissance and surveillance missions, providing a full-duplex, near real-time point-to-point link between an airborne platform and a tracking ground station.

The airborne air-to-ground link (downlink) provides a flow of imagery information at data rates of 120 Mbps, along with a low rate channel for telemetry and tracking functions. The ground station link to the airborne platform (uplink) provides a secure link for command and control.

The DLV-52 consists of an Airborne Data Link Segment (ADLS) in the surveillance system (airborne modem, transmitter and receiver; controller and power supply; two (front/rear) selectable gimballed antennas, covering the bottom hemisphere), a Ground Data Link Segment (GDLS) on a mobile trailer (ground station modem, receiver and transmitter; acquisition and tracking subsystem and antennas), together with an optional Ground Control Shelter (GCS) on a trailer or truck (air conditioned shelter including operator console, fibre optic cables, standard long-haul communication equipment, power distribution box, work bench, storage cabinet and power generator).

The ADLS downlink transmits near real-time sensor imagery as well as radar, SIGnals INTelligence (SIGINT) and annotation/telemetry data via a gimballed airborne antenna tracking the ground station. The GDLS signal acquisition and tracking is performed automatically, requiring no intervention by the operator and using a 7 ft dish antenna mounted on a dual axis pedestal. The GDLS uplink provides for command and control of the surveillance system. Sensor imagery and telemetry signals received by the GDLS are transferred to the GCS via fibre optic cables. The GCS functions as a command, control and maintenance centre. It includes the command and control console and a power generator for the GDLS.

Applications for the system include airborne reconnaissance platforms and pods, aerostat-borne surveillance missions and wide band datalink standalone applications for C4I.

Other features and benefits of the DLV-52 include:
• Air-to-ground, near real-time, high-rate data communications
• Very high reliability and availability Line of Sight (LoS) communication
• Air-to-ground imagery data rates – SDH/SONET
• Additional links supporting telemetry and control functions

Specifications
Frequency band: Ku-band (other frequency bands optional)
Data rate
Downlink: Imagery up to 120 Mbps; annotation and telemetry 19.2 kbps
Uplink: Command and control 180 kbps
Bit error rate: 10^{-9} (imagery); 10^{-5} (annotation and telemetry); 10^{-5} (command and control)
Range: Up to 200 n miles
ADLS: Two flat selectable antennas on gimbals
GDLS: 7 ft dish monopulse antenna
Auto-tracking; azimuth 360°; elevation –5° to +85°
Power supply
ADLS: 115 V AC, 3-phase, 400 Hz
GDLS: 220 VAC, 3-phase, 50Hz
Dimensions
ADLS: 540 × 240 × 760 mm (W × H × L)
GDLS, including trailer: 2,500 × 4,000 × 4,700 mm (W × H × L)
GCS: 2,100 × 2,200 × 3,700 mm (W × H × L)
Weight
ADLS: 40 kg
GDLS, including trailer: 5,500 kg
GCS: 650 kg

Status
In production.

Contractor
Rafael Armament Development Authority, Systems Division.

EL/K-1250T VHF/UHF COMINT receiver

Type
Avionic communications system, VHF/UHF.

Description
The compact EL/K-1250T synthesised receiver operating in the 20 to 510 MHz band is used as a building block for larger COMINT or EW systems such as the Arava EW. The unit has four selectable intermediate frequency filters which demodulate AM, FM, continuous wave and single sideband signals. Intermodulation protection is claimed from RF preselection by voltage tracking filters and the fast tuning synthesiser settles within 500 ms between channel changes. Remote digital control operation is possible and the compact dimensions, low weight and power consumption are achieved by extensive use of advanced microcircuit technology.

Specifications
Dimensions: 114 × 193 × 356 mm
Weight: 7 kg
Frequency: 20–510 MHz
Frequency accuracy/stability: ±1 ppm
Synthesiser settling time: 500 µs
IF bandwidth: select 4 of 10, 20, 50, 100, 300, 600, 1,000 kHz
Noise figure:
(20–180 MHz) 12 dB
(180–510 MHz) 11 dB

Status
In production. The unit is in service with Israeli and other armed forces.

Contractor
Elta Electronics Industries Ltd.

HRN-1 airborne data link directional antenna pedestal

Type
Aircraft antenna, Communications, Navigation, Identification (CNI).

Description
The HRN-1 airborne data link directional antenna pedestal provides single-axis pointing capability in a very compact package designed for UAV applications. The system incorporates a precision DC servo-mechanical drive assembly, an integral digital servo-electronic circuit card and associated RF components. The mounted antenna can be pointed to any azimuth angle according to position commands received through an RS-422 serial communications link.

Specifications
RF performance
Operating frequency: 4.4 to 5.1 GHz
Polarisation: Vertical
Pattern (3dB)
(elevation) 35° min
(azimuth) 15° min

Controp Precision Technologies' HRN-1   1031814

Gain: 15.5 dB (C-band)
VSWR: 1.7 max
Power: 50 W CW

Electro mechanical
Velocity: 30°/sec
Azimuth: 360° continuous (rotary joint)
Positional accuracy: 0.7° (1-Sigma)

Physical
Altitude:
    (operating) 35,000 ft
Shock: 15 g, 11 m/sec saw tooth
Power: ±15 V DC, 0.4 A; +5 V DC, 0.5 A; +28 V DV,
    1 A (for −54°C option only)
Temperature:
    (operating) −32 to +71°C
    (option) −54 to +71°C
Weight: 2.3 kg (including C-band antenna)
Options: L, S, X, Ku-band antennas, higher gain
    antennas (18 dB) in C-band

Status
In production and in service in several UAV
applications.

Contractor
Controp Precision Technologies Ltd.

HRN-2 airborne dual directional antenna pedestal

Type
Aircraft antenna, Communications, Navigation,
Identification (CNI).

Description
The HRN-2 airborne dual, independent axis pointing
pedestal provides dual, independent axis pointing
capability in a very compact package designed
for UAV applications. The system incorporates
two precision DC servo-mechanical drive
assemblies, an integral digital servo-electronic
circuit card and associated RF components. Two
mounted antennas can be pointed independently
to any azimuth angle according to position
commands received through an RS-422 serial
communications channel.

Specifications
RF performance
Operating frequency: 4.4 to 5.1 GHz
Polarisation: Vertical
Pattern (3dB)
    (elevation) 35° min
    (azimuth) 15° min
Gain: 15.5 dB (C-band)
VSWR: 1.7 max
Power: 50 W CW

Electro mechanical
Velocity: 30°/sec (upper antenna), 20°/sec (lower
    antenna)
Azimuth: 360° continuous
Positional accuracy: 0.7° (1-Sigma)

Physical
Altitude:
    (operating) 35,000 ft
Shock: 15 g, 11 m/sec
Power: ±15 V DC, 0.8 A; +5 V DC, 0.5 A; +28 V DV,
    1.7 A (for −54°C option only)

Controp Precision Technologies' HRN-2   1031815

Temperature:
    (operating) −32 to +71°C
    (option) −54 to +71°C
Weight: 4.0 kg (including C-band antennas)
Options: L, S, X, Ku-band antennas, higher gain
    antennas (18 dB) in C-band

Status
In production and in service in several UAV
applications.

Contractor
Controp Precision Technologies Ltd.

RAN-1196

Type
Avionic datalink system.

Description
The RAN-1196 airborne tactical data network radio
is a multimode, jam resistant, two-way air-to-air
and air-to-ground digital communication network
transceiver, designed for the tactical environment.
The radio supports multiple networks, providing
an instantaneous, high capacity and secure flow
of data communication. The RAN-1196 seamlessly
integrates with a large variety of associated
equipment via MIL-STD-1553B data bus or an
802.3 Ethernet interface.

The system consists of an applications
transceiver, power amplifier (similar for airborne
and ground terminals), an airborne antenna and
ground station sectorial antennas.

Applications for the system include air-to-air
and air-to-ground communications data
networks, transmission of reconnaissance,
sensors and video information, support for the
Multifunctional Information Distribution System
(MIDS) and communication pod systems.

The main features of the RAN-1196 include:
- Near real-time air-to-air and air-to-ground high
  rate data communications
- Jam resistant, two way digital packet data
  communication network transceiver
- Transparent to IP Networks
- Low Bit Error Rate (BER)
- Provisioning for advanced ad-hoc networking
- Very high reliability and availability of
  communications
- Flexible configuration that can support up to
  40 networks, with up to 50 users per network
- Supports relaying modes for extended range

Specifications
Frequency band: L-band (other frequency bands
    optional)
Jam resistance: Frequency hopping
Range: Air-to-air: 70 n miles (without relay); up to
    140 n miles (with one-hop relay)
Air-to-ground: 200 n miles
Data rate: Up to 3.3 Mbps
RF output power: 50 W

Ground antenna: Sectorial ground antenna,
    instantaneous 360° omni-directional coverage,
    plus zenith
Airborne antenna: Two blade antennas (each 3dBi)
Dimensions
    Transceiver: 118 × 104 × 217 mm (W × H × L)
    Power amplifier: 132 × 85 × 219 mm (W × H × L)
    Ground antenna: 200 × 80 mm (H × Ø)
    Airborne antenna: 45 × 60 × 135 mm (W × H × L)
Weight
    Transceiver: 4 kg
    Power amplifier: 4 kg
    Ground antenna: 165 kg
    Airborne antenna: 300 g
    MBTF: >8,500 hours

Status
In production.

Contractor
Rafael Armament Development Authority,
Systems Division.

RAVNET-300

Type
Avionic communications system, VHF/UHF, data.

Description
RAVNET-300 is a highly integrated voice/data VHF/
UHF airborne tactical radio network, supporting
airborne combat operations. The system is a
multi-waveband, multifunction, jam-resistant
communication system, designed to provide
instantaneous high quality voice and high
capacity data communication in both air-to-air
and air-to-ground modes of operation.

The RAVNET-300 transceiver is based on two
main components – a standard ARC-210 VHF/UHF
radio produced by Rockwell Collins (see separate
entry), combined with a Rafael Applique (directly
attached unit) containing an innovative modem
and network controller. The combined transceiver
incorporates a number of noteworthy features,
including simultaneous listening to three
networks using one radio, extended operation
range, improved voice quality and increased
operational availability.

Major features and benefits of the RAVNET-300
include:
- Advanced multimode UHF and VHF analogue
  AM/FM and digital radio, with backward
  compatibility to existing radio systems
- Advanced ad-hoc networking and automatic
  relay
- Strong Forward Error Correction (FEC)
- Very high voice quality in all modes of
  operation
- Very high voice and data availability facilitated
  by automatic relay, overcoming obstacles to
  Line of Sight (LoS) communications for low-
  flying platforms
- Multiple voice interfaces support including
  standard analogue or E1, G.711 or compressed
  voice, H323/SIP VoIP standards
- Support for Mil-Std-1553B data bus interface
  or optional dedicated control box
- High commonality – the same transceiver is
  used for airborne and ground terminals
- High MTBF

Specifications
Frequency bands: VHF/UHF
Range: 150 n miles (without relay); up to
    300 n miles (with one-hop relay)
Data rate: Up to 36 kbps per user
RF output power: 10–30 W (without external
    power amplifier)
Dimensions: 125 × 170 × 290 mm (W × H × L)
Weight: 7.5 kg
MBTF: >1,500 hours

Status
In production.

Contractor
Rafael Armament Development Authority,
Systems Division.

Italy

AN/ARC-150(V) UHF AM/FM transceiver

Type
Avionic communications system, UHF.

Description
The AN/ARC-150(V) designation represents a family of small, high-performance, lightweight, airborne UHF transceivers manufactured by Selex and based on a Magnavox design, with further contribution from Selex, own research and development. It has produced many versions, including the 10 W panel-mounted ARC-150(V)10, the 10 W remote-controlled ARC-150(V)2 and the 30 W remote-controlled ARC-150(V)8.

Elmer has also developed a series of control panels and frequency/channel repeaters to meet specific installation requirements on different aircraft and helicopters. A feature of this family is 'slice' assembly, which simplifies maintenance and facilitates growth.

A series of mounting adaptors has been developed and produced to allow the basic ARC-150 transceivers to replace older UHF radios such as the AN/ARC-51BX, AN/ARC-109, AN/ARC-52 and AN/ARC-552 without any mechanical or electrical modification to the aircraft.

A version of the ARC-150(V) with ECCM capabilities has also been developed using the frequency-hopping Have Quick technique. This capability is easily implemented by the substitution of the synthesiser slice and by minor changes on the control panel unit. No mechanical or electrical modifications are required on the aircraft. The modified radio retains the normal non-hopping mode.

The ARC-150(V)8 is a development of the basic 30 W ARC-150(V) transceiver incorporating the FM modulation capability by means of an additional slice. This facility makes the unit particularly suitable for use with data, frequency-shift keying and secure voice modems.

It has been demonstrated to be fully compatible with the Vinson KY-58 system both in AM and FM and in the diphase and baseband modes. It has also been successfully used as a main component of an airborne system for UHF satellite communication.

Specifications
Dimensions:
(10 W panel-mounted RT-1136)
146 × 124 × 193 mm
(10 W remote RT-1051) 127 × 120 × 183 mm
(30 W remote RT-1073) 127 × 120 × 291 mm
Weight:
(10 W panel-mounted RT-1136) 4.3 kg
(10 W remote RT-1051) 3.7 kg
(30 W remote RT-1073) 6 kg
Power supply: 28 V DC
Power output: 10 W (30 W for the AN/ARC-150(V)8 model)
Frequency: 225 to 400 MHz
Channel spacing: 25 kHz
Preset channels: 20 using electronic memory (MNOS)
Frequency accuracy: 2 kHz
Guard receiver: 243 MHz
Operating modes: AM voice, ADF, homer, secure voice/data, ECCM

Status
In production and in service with Italian and other armed forces. Over 1,000 units have been delivered and installed on a wide range of aircraft and helicopters including the Panavia Tornado, Aermacchi MB-339, Aeritalia G91Y and F-104S fixed-wing aircraft, and Agusta A 109, Agusta-Bell 212, Agusta-Sikorsky SH-3D and HH-3F and other helicopters.

Contractor
Selex Communications.

CM 105 E encryption equipment unit

Type
Avionic communications system, secure.

Description
The CM 105 E equipment is a digital ciphering device that can be configured for voice or data applications utilising a narrow- or wide-band radio channel (HF or V/UHF). Applications for the system include narrow- and wide-band applications:

Narrow-band applications
Voice mode: The equipment can be connected to a headset. The analogue input signal is converted to a digital signal by an LPC10 Vocoder at 2,400 bps and is transmitted, after encryption, to the radio channel with analogue modulation.
Data mode: Provides for synchronous transmission at 300, 600, 1,200, 2,400 bps. Ciphered data is transmitted to the radio channel in Baseband after analogue modulation.

Wide-band applications
Voice mode: The analogue input signal is converted to a digital signal using CVSDM modulation at 16 kbps and is transmitted, after encryption, to the radio channel in Baseband or Diphase.
Data mode: Provides for synchronous transmission at 16 kbps. Ciphered data is transmitted to the radio channel in Baseband or Diphase.

Plain mode is also available. The working mode and key variable can be selected by the operator, with rapid erasure of all keys provided.

Other features of the cm 105 E include:
- Secure data, secure voice;
- Environmental/EMC/Tempest to military standards;
- Anti-tampering facility;
- Local/remote control, BITE;
- NATO approved Algorithm.

Specifications
Applications: Secure voice and data communications over HF/VHF/UHF radio channels

Narrow-band mode (HF)
Analogue interface (plain and secure): 0 dBm ± 3 dB
Data interface (plain and secure): V.10/V.11 selectable
Baud rate: 300, 600, 1,200, 2,400 bps in synchronous mode
Voice coding: LPC10 at 2,400 bps
Analogue modem: Modem STANAG 4197
Approved algorithm: NATO Saville
Interoperability: NATO approved encryption equipment ANDVT

Wide-band Mode (UHF/VHF)
Analogue interface (plain): 0 dBm ± 3 dB
Data interface: V.10/V.11 selectable
Baud Rate (plain): 300, 600, 1,200, 2,400 bps, 8, 12, 16 kbps in synchronous mode
Baud Rate (secure): 16 kbps in synchronous mode
Voice Coding: CVSDM at 8, 12, 16 kbps

Environmental
Temperature
Operating: –40 to +55°C
Humidity: 95% non-condensing

Electrical
EMI/EMC: According to MIL-STD-461/2
Tempest: According to AMSG 720B

Physical
Dimensions: 448 × 92 × 395 mm (W × H × D)
Weight: ≤5.0 kg
Power: 28 V DC; 25 W

Status
In production and in service.

Contractor
Selex Communications.

SP-1450/N-E intercommunication system

Type
Avionic internal communications system.

Description
The SP-1450/N-E intercommunications system (ICS) is designed for applications onboard fixed- and rotary-wing aircraft; the system will manage all the radio communications and radio navigation equipment, datalinks and internal/external communications facilities, integrating them into a single, Tempest-compliant architecture.

The SP-1450/N-E ICS provides switching and routing functions to configure the aircraft's internal and external communications resources and user access. These functions are supported by control, amplification and interfaces for all the radio communication and intercom facilities installed on the platform.

The system features a modular, flexible design and open, facilitating reconfiguration to customer requirement. Extensive use of fibre optic connections and appropriate wiring layouts are used to reduce crosstalk and EMI, satisfying all Tempest requirements. The SP-1450/N-E ICS is comprised of the following component LRUs:
- Communication Management Unit (CMU), SP-1451/N-E
- Main Station Unit (MSU), SP-1452/N-E
- Secondary Station Unit (SSU), SP-1453/N-E.

CMU
The CMU includes the WTG/DVO (Warning Tone Generator/Direct Voice Output) unit, a digital switching matrix which, under MIL-STD-1553 (or STANAG 3838) control, configures the communications resources by implementing the following functions:
- Control of the operating parameters of the radios, cryptos and modems
- Routing of crypto and modems through external communication equipment
- HF to V/UHF and V/UHF to HF relay (retransmission of the V/UHF channel of information received over the HF channel and vice-versa)
- Public Address function by external audio amplifier
- Management of navigation aids, alarms and alert signals.

The audio signals to/from the communications equipment connected to the CMU is digitised, formatted and routed to the MSU and SSU over a high speed optical bus:

MSU
Each MSU allows the primary crew members (pilots and mission operators) to control the assigned onboard audio sources (radio transceivers, navaids, sonar, ESM, and so on) and to interface with the desired audio channels.

SSU
The SSU allows the secondary crew members (gunners, loadmasters, ground service personnel) to utilise the assigned audio set, to receive external communication channels or to access to intercom facilities.

The CMU is capable of supporting up to 10 MSUs and 5 SSUs and to interface and control up to:
- 12 T/R radio channels (HF and V/UHF transceiver units)
- 8 Crypto equipments (narrow band and/or wideband)
- 25 navaid, plus alarm/alert signals
- 4 Telebrief/ground support channels
- 1 Link 11 modem.

Different System configurations are available for most aircraft types, all using common units and modules.

SP-1450/N-E intercommunication system   0044797

Specifications

Audio interface
Bandwidth: 300 Hz to 3,500 Hz
Audio input (adjustable)
Minimum level at 150 Ω: 100 mVrms
Maximum level at 150 Ω: 2.5 Vrms
Minimum level at 600 Ω: 500 mVrms
Maximum level at 600 Ω: 11 Vrms
Audio Output
Minimum level at 150 Ω: 100 mVrms
Maximum level at 150 Ω: 1.4 Vrms
Minimum level at 600 Ω: 500 mVrms
Maximum level at 600 Ω: 5 Vrms
Headset
Earphone Output mono: At 150/600 W, 250 mW ±10%
Earphone Output stereo: At 300 W + 300 W, 125 mW + 125 mW
Interfaces: MIL-STD-1553, ARINC-429
Power: 28 V DC (MIL-STD-704); 160 W
Operating temperature: –40 to +71°C (without forced air cooling)
Altitude: Up to 70,000 ft
EMC: MIL-STD- 461/462
Tempest: AMSG 784B volume I
Weights
SP-1451/N-E: 25 kg max
CMU: 8 kg
MSU: 1 kg
SSU: 1 kg
Dimensions
MSU: 145.8 × 82 × 140 mm
SSU: 150 × 90 × 80 mm
MTBF: CMU ≥1,500, MSU ≥6,000, SSU ≥6,000 operational hours
MTTR: 15 min

Status

In production and in service. The SP-1450/N-E ICS is installed onboard NH-90 (NFH and TTH) and EH-101 (UK Royal Air Force SH; UK Royal Navy ASW; Italian Navy ASW) helicopters of several NATO countries.

Contractor

Selex Communications.

SP-1527 G Link 11 system

Type
Avionic datalink system.

Description
The SP-1527G has been developed to be integrated into the HF/UHF radio communication systems of the armed forces of several NATO countries in support of Link 11 tactical net operations, in accordance with STANAG 5511. Although primarily intended for installation on board fixed- and rotary-wing aircraft (such as the NH-90 helicopter), the SP-1527G is suitable for use in ground vehicles, small ships and other installations where space is limited. The SP-1527G provides the two fundamental functions of modulation/demodulation of the tactical data, together with net management to support participation in the STANAG 5511 Link 11 network as a Net Control Station (NCS) or as a Picket Unit (PU).

In the transmit mode, the DTS instructs the computer to deliver the data to be transmitted, encodes the data into 30-bit Hamming codewords (24 information bits plus 6 Hamming code bits) and converts the 30-bit codeword into a waveform made up of 15 DPSK modulated audio tones with a frame duration of 13.3 msec, which is applied to the radio transmitter input.

In the receive mode, the reverse process takes place – the DTS demodulates the 15-tone audio waveform at the output of the radio receiver into a Hamming codeword which is decoded to extract the data bits to be sent to the mission computer. All the functions required for Link 11 operation of the SP-1527G are incorporated into a ¼ ATR-short LRU, designated the DTS TD 9411, complete with mounting tray for installation on board the host platform. For specific applications where the SP-1527G operates as a stand-alone system, a dedicated Remote Control Unit (RCU) is available for control and monitoring of the DTS TD 9411 operating functions over a RS-423 interface. When used in integrated avionic system where all control, monitoring and data transfer interconnections are supported by the mission bus, the SP-1527G configuration includes two LRUs – the DTS TD9411 and the Data Link Interface DLI IF 8501, a ⅜ ATR-short unit which interfaces the mission bus protocol to the DTS protocol. The DLI is directly connected to the DTS via an ATDS interface. A KG-40 crypto equipment can be installed between these units. During Link 11 operation over HF, the DTS receives two identical data streams from the radio, one in USB and one in LSB, and implements a diversity selection scheme which minimises the errors. The software-oriented DTS design is based on a multiprocessor architecture, which includes three VLSI processors – one control processor and two slave DSP processors dedicated to digital signal processing for data modulation/demodulation.

In operation, the control processor interprets the received commands and configuration data to task the DSPs for the implementation of the modulation/demodulation algorithms and manages the system protocols and Man-Machine Interface (MMI) functions. The interconnections are supported by internal buses, a feature that allows the addition of hardware and software components to implement new functions. In particular the design of the SP-1527G allows implementation of the following options:
- Link 11 single-tone waveform;
- Extended Link Quality Analysis (LQA) facilities, which provide additional data on link performance, such as sideband power spectrum, fading bandwidth;
- Multipath spread, Doppler offset and average received signal quality for each participating unit;
- Embedded crypto facilities configured as a card of the DLI.

BITE facilities are provided in two different modes, continuous BIT or BIT initiated by external command and include a number of loopback functions to verify correct operation of the DTS, plus a fault isolation routine which identifies failures down to replaceable card level.

Other main features of the system include:

- Single-tone operation (SLEW);
- Interoperability at European and NATO level;
- Advanced design based on new technologies;
- Expandability at hardware and software level for addition of optional functions;
- Two configurations available (DTS TD 9411 plus Remote Control Unit or DTS TD 9411 plus DLI IF 8501) to meet different requirements;
- No mechanical /electrical adjustment required for LRU replacement;
- No scheduled preventive maintenance necessary;
- Error detection and correction code.

Specifications
Data rates: 1,364 bit/sec and 2,250 bit/sec
Tone library: 15 data tones plus Doppler tone
Computer Interface: ATDS or MIL-STD-1553B (via DLI IF 8501)
Preamble: 2 tones (605 Hz and 2,915 Hz), 5 frame duration
Doppler correction: ±75 Hz (acquisition); 3.5 Hz/sec (tracking rate)
Synchronisation: 1 msec max resolution with respect to the 2,915 sync tone acquired during the preamble
Error detection code: Hamming code
Channel bandwidth: 3 kHz nominal

Audio Interface
Impedance: 600 Ω balanced
Audio output: –22 dBm to +2 dBm (adjust)
Audio input: 0 dBm ±3 dB
Keyline Level 1: +6 V DC Tx; 0 V DC Rx
Keyline Level 2: 0 V DC Tx; Open circuit Rx
Power: +28 V DC, 100 W
Diversity reception: USB, LSB, manual or automatic diversity, operator selectable
Temperature: –40 to +70°C (operating); –55 to +90°C (storage)
Altitude: 70,000 ft max
Humidity: Up to 95%
EMC: Per MIL-STD-461/462

Dimensions and weight
DTS TD 9411: 1/4 ATR-short, 5 kg
DLI IF8501: 3/8 ATR-short, 5 kg
MTBF: 3,000 hr
MTTR: 15 min

Status
In production and in service.

Contractor
Selex Communications.

SRT-[X]70/[X] HF transceivers

Type
Avionic communications system, HF, secure.

Description
The SRT-[X]70/[X] series is an advanced family of HF transceivers providing the following features:
- Continuous frequency coverage between 2 to 30 MHz for both transmit and receive
- Voice and data (including Link 11) communication in secure and clear modes
- Comprehensive modulation capabilities (USB, LSB, ISB, AM, CW, FSK, and CPFSK)
- Embedded selcal capability to ARINC 714
- Embedded Automatic Link Establishment (ALE) to MIL-STD-188-141-2A
- Embedded 2,400 bits/s modem

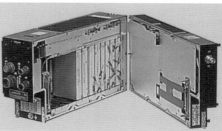

SRT-170/M internal view of receiver/exciter/amplifier unit   0044799

SRT-270/L and SRT-470/L with ATU for loop antenna and notch antenna 0044800

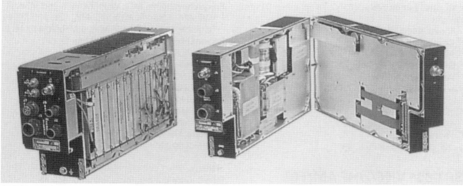

SRT-270/L internal view of 649/L[ ] receiver/exciter 0044801

SRT-170/M (left to right: ATU, control unit, 1/2 ATR receiver/exciter/amplifier) 0044798

- Remote control via MIL-STD-1553B or ARINC 429 databus, or dedicated remote control unit. The SRT-[X]70/[X] family comprises:
- SRT-170/L: 100 W power output transceiver
- SRT-170/M: 175 W power output transceiver
- SRT-270/L: 200 W power output transceiver
- SRT-470/L: 400 W power output transceiver
- SP-648 or SP-648/LA remote control panels
- SP-649/L receiver-exciter, common to all versions
- SP-480/L (100 W), SP-480/L1 (200 W) or SP-484/L (400 W) power amplifiers
- CP-2001/[X] advanced computer-controlled remote multifunction control/display
- SP-1325/L (200/100 W) or SP-1325/L1 (400 W) pre-post selectors
- ATU: antenna couplers to match wire and probe (SP-1127/L for 200/100 W, ATU-1992/LM4 for 400/200 W), notch (ATU-1992/LM for 200/100 W, ATU-1992/LN for 400/200 W) and towel bar antennae (ATU-1992 for 200/100 W, ATU-1992/LN for 400/200 W).

The SRT-170/[M] is a DC power version of the SRT-[170/270]/L variants.

Design characteristics are based on modular digital radio concepts with extensive use of DSP and DDS techniques and digital processors to minimise cost of ownership and optimise flexibility and growth to meet the requirements of STANAG 4444 (ECCM), STANAG 4538 (ARCS) and STANAG 4539 (fixed frequency). The CP-2001[X] provides simultaneous management and control of all on-board communications, intercom systems and navigation systems, as well as providing management of special facilities such as ALE, data modems and satcom;

it features multilayer display techniques and GEN III NVG capability.

Specifications
Frequency range: 2 to 29.9999 MHz
Tuning time: Typical 1s (including ATU); 50 ms on learned channel
Modes of operation: Simplex Rx/Tx on any available channel
Modulation: Clear and secure voice (USB/LSB/AM), CW (USB), Link-11 data (USB,LSB,ISB), TTY (USB), SELCAL, (ARINC-714-6), ALE (8-FSK)
Preset channels: 100 (stored in the control panel)
Power: 115 V AC, 400 Hz 3-phase (SRT-170/M powered by 28 V DC)
Consumption: 400 W typical, for fully configured system
Environmental: MIL-STD-810C (temperature, altitude, vibration, shock, salt, fog)
Temperature: –54 to + 55°C
Altitude: 70,000 ft

Status
The SRT-170/M has been selected for the EH 101 (UK Royal Air Force SH, UK Royal Navy Merlin and Italian Navy ASW versions) and NH 90 (both NFH and TTH versions). The SRT-470 is fitted to Italian Air Force Tornado aircraft. The SRT-470/L has been selected for UK Royal Air Force Nimrod MRA4 and Australian, UK and Italian Air Force C-130J aircraft.

Contractor
Selex Communications.

SRT-170/L, -270/L, -470/L advanced HF airborne transceiver family

Type
Avionic communications system, HF.

Description
Selex Communications has developed an advanced HF/SSB system for fixed-wing aircraft and helicopters. These transceivers provide voice/data radio communications for airborne applications

over the 2 to 30 MHz frequency range, with power output from 100 to 400 W.

The SRT-170 series transceiver systems have been specifically designed for the intense EMI environments found in airborne applications, such as co-located installations requiring simultaneous operation of several radios (voice/data) on the same platform.

In the design and production of the system, a total system development programme was undertaken, including the use of computers and scale model facilities, together with a detailed analysis of development, bonding and mounting problems as they relate to aircraft design and construction.

Extensive flight tests on helicopters and fixed-wing aircraft have demonstrated excellent performance in the most severe environments.

The main features of the system include:
- Efficient thermal design and low power consumption
- Full BITE capability. Interruptive tests of transceiver circuits with an internally generated signal, with BITE signals from the individual LRUs available for use with automatic test equipment
- Direct digital synthesis for spectral purity and fast tuning. Digital Signal Processor (DSP) for superior instantaneous dynamics in receive and transmit modes
- Microprocessor control for flexibility and future upgrade potential (embedded EPM and ARCS modes and high-speed serial modem)
- ALE capability in accordance with MIL-STD-188-141A.

A typical transceiver consists of the following units:
- Receiver/exciter SP-649/L (common to all versions)
- Power amplifier: SP-480/L (100 W), SP-480/L1 (200 W), or SP-484/L (400 W)
- Pre/post-selector SP-1325/L (option)
- Interface. The transceiver can be controlled either through MIL-STD-1553B or ARINC 429, or from a dedicated remote control panel (SP-648/L).

Any of the transceivers can be interfaced with various Antenna Tuning Units (ATUs) suitable for loop, wire and notch antennas. A typical configuration for helicopter application includes an externally mounted, 'silent tuning' ATU (ATU 1992) and a loop antenna. The ATU-1992/LN is the equivalent unit for wire and notch antennas.

Specifications
General
Frequency range: 2 to 29.9999 MHz
Tuning: Automatic at 100 Hz interval
Tuning speed: Less than 1.5 s (less than 50 ms for stored frequencies)
Operating modes: A3J (SSB), A2J (CW), A3(AM)
Power: 115 V AC, 400 Hz, 3-phase

Environmental
(per MIL-E-5400)
Temperature (Normal Operating): –40 to +71°C
Altitude: 70,000 ft
Humidity: Up to 95%

Transmission
RF output: 100/200/400 W PEP
RF power level: Selection of 1/2, 1/4 RF output
Intermodulation: Better than 32 dB below PEP
Harmonic attenuation: Better than 50 dB below PEP
Spurious suppression: Better than 60 dB below PEP
Carrier attenuation in SSB mode: Better than 50 dB below PEP
Duty cycle: 3 min Tx and 3 min Rx in CW, AM and data mode; continuous in SSB voice and Link-11 mode

Reception
Sensitivity for 10 db min ([S+N]/N)
AM: 2.5 µV, 30% modulated by 1 kHz tone
CW/SSB: 0.5 µV
Selectivity
AM: ±3 kHz at –6 dB, 14 kHz at –60 dB
CW/SSB: 300 to 3,050 Hz at –3 dB, 4,400 Hz at –60 dB
Input impedance: 500 ohm

Mechanical

	W (mm)	H (mm)	D (mm)	Weight (kg)	Description
SP-648/L	146	57.1	168	1.1 max	
SP-649/L	90	197.5	320.5	6 max	
SP-480/L	90	197.5	320.5	8 max	200 W PEP 100 W AVG
SP-484/L	257	193.5	320.5	15 max	400 W PEP 400 W AVG
ATU-1992	115	323	318	4 max	Loop antenna
SP-1127/L	120	193.5	320.5	6 max	Wire/probe antenna
ATU-1992/LN	257	193.5	320.5	13 max	Notch antenna 400 W
ATU-1992/LM4	257	193.5	498.5	17 max	Wire antenna 400 W

Status
In production and in service.

Contractor
Selex Communications.

SRT-170/M lightweight HF airborne transceiver

Type
Avionic communications system, HF, secure.

Description
The SRT-170/M is a DC-powered version of the SRT-170/L and is suitable for installation in very severe EMI environments. The system can be controlled via either MIL-STD-1553B, ARINC 429 databus, or from a dedicated control panel (SP-648/L). The bus interface is available as a plug-in module inside the SRT-170/M. Basic configuration is similar to the SRT-170/L, based around the RT-170/M transceiver unit, with associated Antenna Tuning Units (ATUs) performing 'silent tuning':

- ATU-1992, for loop antennas
- SP-1127/L for wire antennas
- ATU-1992/LN for notch antennas.

The Antenna Transceiver Unit (ATU) is able to store up to 256 channels. Antenna tuning data are acquired and stored prior to the mission. When any one of the stored channels is selected during the mission, the stored tuning data are used to speed up the antenna tuning operation, without antenna emission. By using this approach, tuning times of the order of 50 ms are achieved.

The BITE function of the modules is designed to allow facilitate Automatic Test Equipment (ATE) analysis and fault isolation at the maintenance organisation level.

The SRT-170/M interfaces with the following devices:

- Adaptive Communications Processor, for automatic selection of the best operating frequency, based on channel conditions
- Datalink Modem, to set up point-to-point and network links
- Crypto Modem, to operate in the crypto voice narrowband mode.

Specifications
General
Frequency range: 2 to 29.9999 MHz
Tuning speed: Typically 1.5 s (50 ms for stored frequencies by digital ATU)
Operating modes: A3J (SSB), A3(AM), A1(CW), F1 (FSK with internal modem), Data
Preset channels: 100 (SP-648/L)
Power: 28 V DC (MIL-STD-704C), 500 W (fully configured)
Environmental : MIL-STD-810C

Transmission
RF output: 175 W PEP, 85 W AVG
RF power level: Selection of ½, ¼ RF output
Intermodulation: Better than 32 dB below PEP
Harmonic attenuation: Better than 50 dB below PEP
Spurious suppression: Better than 60 dB below PEP
Carrier attenuation in SSB mode: Better than 50 dB below PEP
Duty cycle: 3 min Tx and 3 min Rx in CW, AM and data mode; continuous in SSB voice and Link-11 mode

Reception
Sensitivity for 10 dB min ([S + N]/N)

AM: 2.5 µV 30% modulated by 1 kHz tone
CW/SSB: 0.5 µV
Selectivity
AM: 3 kHz at –6 dB, 14 kHz at –60 dB
CW/SSB: 300 to 3050 Hz at –3 dB, 4400 Hz at –60 dB
Input impedance: 500 Ω

Mechanical
Dimensions
RT-170/M: ½ ATR, short
Control panel: 146 × 57.1 × 178 (W × H × D) mm
Weight
Transceiver: 9 kg
Control panel: 0.9 kg
ATU-1992: 4 kg
SP-1127/L: 6 kg
ATU-1992/LN: 6 kg

Status
In production and in service.

Contractor
Selex Communications.

SRT-651 VHF/UHF AM/FM transceiver

Type
Avionic communications system, VHF/UHF, secure.

Description
The SRT-651 is an all-solid-state, compact, lightweight airborne transceiver covering the 30 to 400 MHz frequency range and using the most advanced techniques in the area of large-scale integration components and microprocessors. The system comprises a CP-1200 control panel, RT-651 receiver/transmitter and optional ID-1151 channel/frequency repeater.

Conceived and designed for airborne use, the SRT-651 is also suitable for a wide range of applications where space and weight are limited. The flexible modular 'slice' design permits easy assembly and disassembly and allows expansion of the operating functions. The RF power rating can be easily upgraded to 30 W by changing the transmitter slice. Different interface standards are available, including an Elmer serial data interface using control panel CP-1200 or MIL-STD-1553B or ARINC 429.

BIT facilities can be implemented by using different plug-in cards for the interface slice. The test result is automatically monitored on the control panel display, which gives a direct identification of the faulty slice.

The SRT-651 has facilities for ECCM including frequency-hopping ('Have Quick' has been implemented) and spread spectrum pseudo-noise modulation. It can be operated with a wide

variety of ancillaries including homer indicators, VHF/UHF direction-finders, UHF emergency beacons, KY-58 secure voice modems in diphase and baseband modes, Elmer SP-1212 ECCM spread spectrum voice and data modem, NATO multitone Link 11 data modem and FSK modems. Various mountings can be provided to retrofit the SRT-651 directly into existing VHF and UHF installations.

Specifications
Dimensions:
(RT-651) 126.7 × 120.6 × 224 mm
(CP-1200) 146 × 57.1 × 150 mm
Weight:
(RT-651) 3.5 kg
(CP-1200) 1.4 kg
Power supply: 28 V DC
(transmit) 180 W max
(receive) 40 W max
Power output:
(AM) 10 W
(FM) 15 W
Frequency: 30–88 MHz VHF/FM, 108–156 MHz VHF/AM, 156–174 MHz VHF/FM, 225–400 MHz UHF AM/FM and UHF-FM 400 MHz to 470 MHz
Channel spacing: 25 kHz in all bands
(12.5/8.33/5/2.5 kHz optional)
Guard receiver: 40.5, 121.5, 156.8 and 243 MHz automatically selected
Preset channels: 99 using electronic memory
Channel change time: 1 msec
Temperature: –54 to 71°C (continuous); –54 to 95°C (storage)
Altitude: Up to 70,000 ft
MTBF: 2,000 h

Status
Fitted to Italian Air Force Tornado aircraft.

In production for Italian, Brazilian and other armed forces and for installation on Alenia/Aermacchi/Embraer AMX aircraft.

The SRT-651/A variant is fitted to Augusta A129 Mangusta helicopters of the Italian Army.

Contractor
Selex Communications.

SRT-651/N V/UHF – AM/FM transceiver

Type
Avionic communications system, VHF/UHF, secure.

Description
The SRT-651/N transceiver is designed for military operation, to provide voice and data communication over the 30 to 400 MHz band (optionally 30 to 470 MHz). It can be configured to provide:

- EPM/jamming resistance through use of frequency hopping techniques including Have Quick and Saturn (export EPM capability by use of EASY II);
- Link 11 operation, in association with a Link 11 data terminal unit;
- secure voice/data operation in association with a wide variety of crypto modems;
- selective call and tone activated repeaters capability;
- CASS/DICASS capability;
- continuous coverage from 30 to 470 MHz;

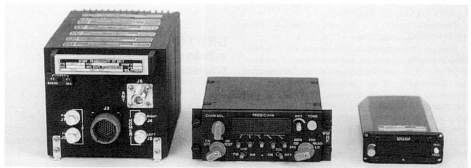

The SRT-651 VHF/UHF transceiver

0504007

- 70 MHz IF standard modem interface for satcom;
- databus control via MIL-STD-1553B or ARINC 429;
- dual system operation.

The SRT-651/N is available in the following configurations:

- standard version: RT-651/N transceiver featuring fixed frequency operation; CP-9000 cockpit control panel;
- databus controlled version: RT-651/N transceiver incorporating a databus controller;
- Have Quick/Saturn/Easy II version: RT-651/N transceiver featuring fixed frequency and frequency hopping operation;
- CP-9000 (Have Quick/Saturn/Easy II) cockpit control panel;
- Have Quick databus controlled version: RT-651/N (Have Quick/Saturn/Easy II) transceiver incorporating databus controller.

Ancillary equipment can be associated with the transceiver to enhance capabilities including:

- homer indicator ID 1351A (homer circuitry is built into the SRT-651/N);
- direction-finders;
- Link 11 data terminal set;
- data modem/FSK modem;
- crypto modem in diphase and baseband modes;
- selcal and Tone Activated Repeater operation via CP-2001.

Operational capabilities include:

- 30 to 88 MHz VHF-FM operation with ground forces;
- transmit inhibitor of 108 to 116 MHz VHF-AM band to avoid interference with VOR or landing systems;
- 118 to 156 MHz VHF-AM operation with ATC;
- 156 to 174 MHz VHF-FM maritime half-duplex operation;
- 225 to 400 MHz UHF-AM/FM air/ground operation;
- 400 to 470 MHz UHF-FM option for compatibility with Tone Activated Repeaters required by law enforcement agencies.

Specifications

Frequency bands:
VHF-FM 30–88 MHz, 156–174 MHz (half duplex)
VHF-AM 108–156 MHz
UHF-AM/FM 225–400 MHz
UHF-FM 400–470 MHz (option)
Preset channels: 99 (with CP-9000 or CP-2001)
Channel spacing: 25 kHz (12.5; 8.33; 5 and 2.5 kHz as options)
Guard channels: 40.5, 121.5, 156.8 and 243 MHz, automatically selected with operating band
Transmit power:
AM 10 W
FM 15 W
Power consumption: 28 V DC; 180 W (Tx) and 40 W (Rx)
MTBF: 2,000 h
Weights: 4.53 kg (transceiver), 1.2 kg (CP-9000), 0.4 kg (frequency indicator, ID-1151), 0.4 kg (mount)
Compliance: MIL-STD-463/810/461/704/1472/462/781/1553, MIL-E-5400

SRT-651/N V/UHF-AM/FM transceiver and control panel   0044802

Status
Selected for the following programmes: C-130J (Australian, UK and Italian air forces); Eurofighter Typhoon; EH 101 (UK Royal Air Force SH, UK Royal Navy Merlin, Italian Navy ASW); NATO AWACS upgrade; NH 90 NFH and TTH versions; Nimrod MRA4, Tornado UK Royal Air Force. Italian armed forces Sicral satellite communications.

Contractor
Selex Communications.

SRT-653 VHF/UHF transceiver

Type
Avionic communications system, VHF/UHF.

Description
Designed to meet a requirement of the Italian Air Force for its version of the Panavia Tornado, the SRT-653 VHF/UHF transceiver comprises the RT-1051 UHF transceiver, SP-1047 VHF transceiver, SP-1203 adapter/power supply, SP-1204 mounting tray, CP-1001 control panel and ID-1150 frequency repeater.

The RT-1051 UHF transceiver belongs to the basic ARC-150(V) family. The SP-1047 VHF transceiver has been derived by Marconi Selenia from this system and maintains its general characteristics.

Specifications
Dimensions:
(ID-1150) 86 × 50 × 159 mm
(SP-1047/RT-1051 on tray) 265 × 155 × 408 mm
(CP-1001) 146 × 95.2 × 165.5 mm
Weight:
(ID-1150) 0.35 kg
(SP-1047 and RT-1051 on tray) 13.5 kg
(CP-1001) 1.6 kg
Power supply: 115 V AC, 400 Hz, 3 phase (transmit) 320 VA max (receive) 140 VA max
Power output: 10 W
Frequency: 108–156 and 225–400 MHz
Channel spacing: 25 kHz
Preset channels: 17 + 2 Guard channels

Status
In service on Italian Air Force Panavia Tornado aircraft.

Contractor
Selex Communications.

The SRT-653 VHF/UHF transceiver is installed on Italian Air Force Tornado aircraft   0504008

Japan

Satellite communication system

Type
Avionic satellite communications system.

Description
Toshiba has developed a satellite communications system which enables air-to-ground international telephone calls from commercial airliners via the Inmarsat satellite service. This was the first type-approved multichannel operation aeronautical satellite communications system in the world. The onboard system consists of radio transceiver equipment for communication with the Inmarsat satellites via a separate antenna and terminal equipment that includes cordless telephones and a control display unit. For passenger use, up to three telephones can be installed in an aircraft and one of the three telephone channels can be used in the cockpit. A 600 bits/s data channel is available for ACARS data communication for automatic dependent surveillance.

Toshiba has developed LSIs to achieve the component miniaturisation necessary for light compact equipment suitable for installation in aircraft.

The radio transceiver consists of a high-power amplifier, radio frequency unit and satellite data unit. The equipment meets ARINC 741 international standards for electronic equipment used in commercial airlines. It can receive and send digital voice signals at 9,600 bits/s, allowing up to three digital voice lines and one 600 bits/s digital data line to be connected to the system.

Specifications
Dimensions:
(high-power amplifier) 8 MCU
(radio frequency unit) 4 MCU
(satellite data unit) 6 MCU

Status
The equipment is FAA Supplemental Type Certified on the Boeing 747-400. The equipment is used by All Nippon Airways.

Contractor
Toshiba Corporation.

Netherlands

Link 11 DataLink Processor (DLP)

Type
Avionic datalink system.

Description
Thales Nederland BV produces its Link 11 DLP to meet the requirements of STANAG 5511 or Opspec 411 for the operation at airborne Link 11 DLP systems in the standard Link 11 environment for all NATO countries and members of the Partners for Peace organisation.

SSP's Link 11 DLP can be fully integrated into an aircraft's command and control system, it can be operated as a stand-alone unit with a simple interface to a host system allowing track data to be passed to/from the Link 11 network. The stand-alone Link 11 DLP is equipped with a Digital Geographic Information System (DGIS), which displays track data on a map or chart of the operational area.

The most common airborne version of the Link 11 DLP is housed in a fully certified. Alternatively airworthy ¼ ATR unit, operating on 28 V DC, but other configurations are available.

Contractor
Thales Nederland BV.

Link 11 datalink system

Type
Avionic datalink system.

Description
Thales Nederland produces lightweight Link 11 systems fully qualified for all airborne environments. The system hardware was developed by Ultra Electronics Limited for a UK programme. All software has been developed by Thales and can be tailored to meet specific requirements.

The DLP has provision for an embedded crypto unit, together with all the necessary controls, resulting in a considerable volume and weight reduction.

The DTS provides the modem and network control functions as defined in MIL-STD-188-203-1A and provides functionality for a unit to act as the Net Control Station (NCS) or 'picket' (active or passive).

The Link 11 application software resident in the DLP enables the exchange of track data, management data, status data and commands between shipborne, airborne and land-based units participating in the network.

The DTS provides the following operational capabilities, both multitone and single-tone operation; automatic or manually selected diversity reception; a modem containing link monitoring software that provides performance and reception quality analysis functions.

The DLP provides all necessary message handling, track correlation, gridlock, filters and conflict management functions. It also provides for exchange of real-time and non-realtime track data, management data such as IFF/SIF data reports and Link 11 management functions.

Specifications
Data terminal set
Interfaces:
(serial) TDS interface to DLP
(control) ARINC 429, MIL-STD-1553B or RS-232C
(audio) 600 ohm balanced interface to HF or UHF transceiver
Dimensions: ¼ ATR short
Weight: 4 kg
Power: 115 V AC, <30 VA
28 V DC, <25 W

Datalink processor
Interfaces:
(serial) ATDS to crypto unit
(control) MIL-STD-1553B
Dimensions: ½ ATR short
Weight: 8.5 kg
Power: 115 V AC, <40 VA
28 V DC, <25 W

Contractor
Thales Nederland BV.

Link-Y Mk 2 datalink

Type
Avionic datalink system.

Description
The Link-Y Mk 2 datalink system exchanges system track data, management data, status data and commands between shipborne, land-based and airborne units participating in the datalink network.

Link-Y Mk 2 allows the establishment of a fleetwide common tactical database through the exchange of real-time and non real-time track data. By exchanging management data, such as IFF/SIF reports, information difference reports and conflict reports, a complete common tactical picture is established.

The exchange of commands and status data, such as force disposition orders, weapon readiness reports and force engagement status reports, is an indispensable tool for engagement planning.

Engagement execution is supported through the exchange of commands, such as weapon doctrine orders, engagement orders and hold-fire orders.

Link-Y Mk 2 allows a maximum of 31 units to participate in the network. Reporting normally takes place using a TDMA method. To each unit, one or more time slots are automatically allocated by the net control station. In radio silence mode all units maintain silence but urgent messages are transmitted by means of a single report.

The Link-Y Mk 2 terminal can be set in the Link-Y Mk 1 mode, resulting in complete interoperability with older Link-Y units which lack the automatic slot allocation and encryption facilities.

The system operates on HF, VHF and UHF frequencies. Effective range is 925 km for HF and line of sight for VHF and UHF.

Specifications
Dimensions:
(Link-Y terminal) 57.2 × 194 × 318 mm
(optional control unit) 146 × 66 × 68 mm
Weight:
(Link-Y terminal) 4 kg
(optional control unit) 0.4 kg
Power supply: 28 V DC
Data rates: 300, 600, 1,200, 2,400 and 4,800 bits/s

Status
In production for several navies.

Contractor
Thales Nederland BV.

Vesta-VC datalink

Type
Avionic datalink system.

Description
The Vesta transponder receiver system can easily be extended with a voice channel datalink. This makes it suitable for data transfer and over-the-horizon targeting for naval vessels or ground control stations by using the helicopter. The target data is received from the helicopter sensors.

The datalink is established via an existing communications voice channel. The Vesta transponder switches the available radio in the helicopter from voice to data, after which the data is transmitted via the voice channel in a frequency shift keyed signal.

A receiver extractor and datalink unit on the base station then converts this data to an 'own position' reference by triangulation of the helicopter position and target data, and interfaces the data handling system (the weapon control system) with the radar displays. Vesta is compatible with all airborne and base system interfaces.

Status
In production for several navies.

Contractor
Thales Nederland BV.

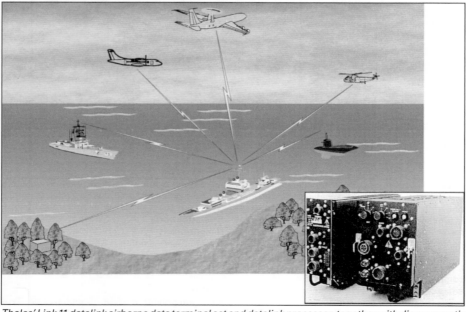

Thales' Link 11 datalink airborne data terminal set and datalink processor, together with diagrammatic concept of operations　　　　0064382

Russian Federation

MS communication and voice warning system

Type
Avionic internal communications system.

Description
The MS communication and voice warning system is designed to meet the internal and external communication requirements of up to three aircrew, including provision of voice warnings.

System components comprise the B27-MS amplifier and switching unit and B7-MS control panels.

System capabilities include:
- intercommunication facilities for up to three crew members and two ground personnel
- control of four radios
- monitoring of seven caution signals and two radio navigation signals
- separate volume control of the intercom and radio facilities
- operation of 256 voice warnings

Specifications
B27-MS
Dimensions: 274 × 126 × 194 mm
Weight: 5 kg
Power: 18 to 31 V DC, <10 W

Status
In production and in service. The MS communication and voice warning system is fitted to the MiG-AT, MiG-29M and YAK-130 aircraft.

Contractor
PRIMA Scientific-Production Enterprise (developer).
Gorky Communications Equipment Plant JSC (GZAS JSC) (manufacturer).

MU-19 multichannel differential Monitoring Unit

Type
Airborne calibration/monitoring system.

Description
The MU-19 multichannel differential monitoring unit provides continuous tracking of all GLONASS and GPS satellites in view, keeping all stored data in memory. It can also be used to monitor the quality of navigation data transmitted via satellites, and to provide differential corrections to satellite data. The MU-19 also monitors the discrepancy between GLONASS and GPS timing data, and determines the discrepancy between the PZ-90 and WGS-84 geodetic co-ordinates at the point of monitoring.

The MU-19 system is controlled via keyboard and display.

Specifications
Measurement accuracy:
(using GLONASS satellites)
pseudo ranges: 1–3 m
pseudo range rates: 1–2 cm/s
(using GPS satellites: (S/A on)
pseudo ranges: 30 cm
pseudo range rates: 30 cm/s
Position accuracy in non-differential mode:
(using GLONASS satellites)
latitude and longitude: better than 5–7 m
altitude: better than 8–10 m
(using GPS satellites)
latitude and longitude: better than 25–35 m
altitude: better than 45–55 m
Interfaces: 2 × RS-232C ports

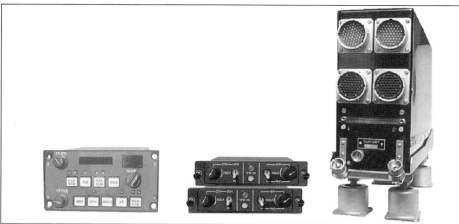
The MS communication and voice warning system 0062794

The SPGU-35 communication system 0062796

Power: 220 V AC; 25 W
Dimensions:
(MU-19 receiver) 345 × 260 × 200 mm
(antenna) 90 × 90 × 40 mm
Mass:
(MU-19 receiver) 6 kg
(antenna) 0.3 kg

Status
Believed in service. Production status unknown.

Contractor
GosNIIAS State Research Institute of Aviation Systems.

SPGU-35 communication system

Type
Avionic internal communications system.

Description
The SPGU-35 communication system is designed to meet the internal and external communication requirements of a three- or four-man crew involved in paramilitary operations such as helicopter casualty evacuation in helicopters such as the Ka-226 and fire fighting operations in the Be-200 amphibian. It also provides for crew monitoring of audio caution and radio navigation systems.

System components comprise the following:
- one B27-35 amplifier and switching unit
- up to three B7-35 control panels
- one or two B7A-35 or B-7B-35 control panels
- one GGO-3 × 2 (two channels × 3 W) address amplifier.

SPGU-35 capabilities include:
- crew communication over up to five air-ground radios
- control of up to 12 radio navigation signals
- monitoring of up to eight caution signals

- passenger address
- communication for up to three crew with ground maintenance personnel
- interface to the voice recorder and automatic test systems
- separate volume controls for different functions
- 6 W audio output power.

Specifications
Dimensions and weight:
(B-27–35 amplifier switching unit) 220 × 160 × 90 mm; 3 kg
(B7-35 control panel) 146 × 80 × 126 mm; 0.8 kg
(B7A-35 and B7B-35 control panels) 146 × 32 × 110 mm; 0.4 kg
(GGO-3 × 2 unit) 131 × 90 × 50 mm; 0.8 kg
Power: 18 to 31 V DC

Status
In service.

Contractor
PRIMA Scientific-Production Enterprise (developer).
Gorky Communications Equipment Plant JSC (GZAS JSC) (manufacturer).

Su-30 MKI communications system

Type
Avionic tactical communications system.

Description
Polyot Research and Production Company produces the communications system for the Su-30 MKI, comprising:
- Simultaneous voice and data communication between air and ground, using HF, VHF and UHF frequencies
- Automatic data transfer of targeting data
- Anti-jamming capabilities

For details of the latest updates to *Jane's Avionics* online and to discover the additional information available exclusively to online subscribers please visit
jav.janes.com

- Two-position control of communications and intercommunication functions
- Emergency frequency monitoring
- Automated BITE.

Specifications
Frequency: 2–18, 100–150, 220–400 MHz
Modes:
HF: AM, SSB
V/UHF: AM, FM, FT
Preset channels:
HF: 20
V/UHF: 20
data: 20
MTBF: 4,700 h

Su-30 MKI communications system (equipment details not divulged)   0044803

Status
Installed in Su-30 MKI aircraft delivered to the Indian Air Force.

Contractor
Polyot Research and Production Company.

South Africa

ACR 500 series air/ground V/UHF transceivers

Type
Avionic communications system, V/UHF, secure.

Description
The ACR 500 series V/UHF transceiver system comprises two airborne transceivers, the ACR 500 V/UHF tactical airborne transceiver and the ACR 520 VHF tactical airborne transceiver; two ground-based transceivers, the GBF 500 UHF power-agile filters transceiver and the GBR 500 V/UHF ground-based transceiver.

The ACR 500 and ACR 520 transceivers employ frequency-hopping and software encryption to ensure communications security, and very high rate direct sequence spread spectrum techniques for 'own probability of detection' communications, and to reduce the probability of interception (ACR 500 only).

High-quality vocoded speech transmission (essentially digitised speech) is used to ensure speech quality.

Other features include bandwidth efficient/4DQPSK data mode, and noise-resistant DSP squelch.

ACR 500 V/UHF transceiver
The ACR 500 is a multimode/multiband transceiver operating over the 30 to 420 MHz range. It also incorporates a fully synthesised auxiliary receiver with analogue and digital modes, to allow simultaneous voice and data reception on two frequencies, to improve situational awareness.

Specifications
(ACR 500 V/UHF transceiver)
Frequency range: 30–420 MHz
Channel spacing: 12.5 kHz; 25 kHz
Modulation formats: AM, FM (WB/NB); SSB (USB); CPFSK Data (binary); Pi/4 DQPSK
RF output, max: 32.4 W PEP AM; 20 W FM, SSB
Collocation: full operation 5% off frequency
ECCM: fast frequency hopping, direct sequence spread spectrum
Encryption: Cat A, built in
Audio capabilities: analogue voice, CVSD, high-quality vocoder
Data capabilities: high-speed, addressable, network function
Auxiliary receiver: built-in
Power: 28 V DC
Dimensions: 124 × 385 × 193 mm
Weight: <9.5 kg

ACR 520 VHF transceiver
The ACR 520 operates in the 30 to 88 MHz portion of the VHF band, its capability for communications with ground forces, using the GBR 500 receiver, and SSB for extended range operation.

Specifications
(ACR 520 VHF transceiver)
Frequency range: 30–88.975 MHz

ACR 500 V/UHF tactical airborne transceiver
0011839

ACR 520 VHF tactical airborne transceiver
0011838

The SAAF has ordered four Super Lynx 300 helicopters, which will be operated by the SAAF's 22 Squadron
0095707

Channel spacing: 12.5 kHz; 25 kHz
Modulation formats: AM, FM (WB/NB); SSB (USB); CPFSK Data (binary); Pi/4 DQPSK
RF output, max: 25 W PEP SSB; 15 W FM
ECCM: fast frequency hopping
Encryption: Cat A, built in
Audio capabilities: analogue voice, CVSD, high-quality vocoder
Data capabilities: high-speed, addressable, network function
Power: 28 V DC
Dimensions: 1⁄2 ATR short
Weight: <6 kg

Status
In production and in service. Selected for South African Air Force (SAAF) Gripen and Hawk 120 aircraft. During 2003, South Africa's Department of Defence (DoD) ordered four AgustaWestland Super Lynx 300 shipborne helicopters equipped with two ACR-500 VHF/UHF radios; the first aircraft is to be delivered by 1 April 2007 and the others will follow

at one-month intervals. The contract is expected to be worth around GBP67 million (USD106.6 million).

Contractor
Reutech Defence Industries (RDI) (Pty) Ltd.

Airborne Communication Management System (ACMS)

Type
Avionic audio management system.

Description
The Airborne Communication Management System (ACMS) is a fully digital system. It is based on the GusBus (LIM), an IYU-T G703 and G704 based serial bus for distribution of audio, data, control and status information. The Gus Control Panel (GCP) and each module in the Communications Management Unit (CMU) contain a LIM.

The CMU is a configurable mechanical housing for a number of independent functional modules, which are individually integrated onto the GusBus. Functional modules provide:
- equipment control and status monitoring; audio management
- data management and networking
- interfaces (audio/analogue/digital/video/discretes)
- GPS (passive frequency hopping and aircraft back-up GPS)
- warning tones and synthetic speech generation
- gateway facilities (MIL-STD-1553B and ARINC 429)
- digital recording (voice and data) and voice command capability.

The GCP provides integrated control, display and audio functions. It also allows direct access to the GusBus for equipment such as radios. The GCP is NVIS compatible and graphics capable.

A fillgun is provided for system application software and mission specific software. Extraction of mission data post-flight is also possible.

A full range of radio navigation and communication products covering the spectrum from 2–420 MHz is available to support the system.

The ACMS is available as a modular concept to meet the needs of all types of aircraft and helicopter, from basic training aircraft to multirole transport and maritime types.

Status
As of early 2006, ACMS is in production and in service in AugustaWestland A109 helicopters of the South African Air Force (SAAF) and Swedish Air Force. In 1998, ACMS was selected for the Saab-BAE Systems Gripen multirole fighter aircraft in international markets.

Contractor
Saab Grintek Communications.

TR 2800 airborne HF transceiver

Type
Avionic communications system, HF.

Description
The TR 2800 is a new-generation 100 W airborne HF transceiver. Its features include: frequency-hopping capabilities for enhanced Electronic Counter Counter-Measures (ECCM) performance, compatible with the TR 250/390 mobile base station HF transceivers; selective calling with channel enhancement and speech enhancement techniques to minimise HF noise and interference.

System control is exercised through the CU2832 control unit and the PA2810 400 W power amplifier is available, as an option, to meet high-power requirements. The CU2832 provides a 2 × 20 line alpha-numeric display. It is NVG compatible and provides links via RS 485 to the radio system and by RS 232 to the fill device.

The system can be configured with three different ATUs to meet aircraft and role requirements:

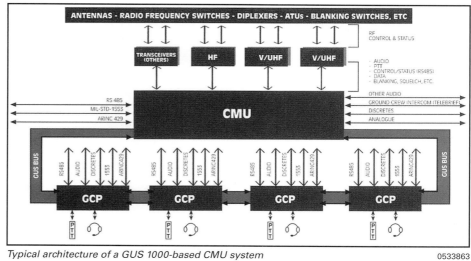
Typical architecture of a GUS 1000-based CMU system 0533863

- The AT2820 100 W, loop, ATU has been developed to provide a high-angle antenna for near vertical incidence skywave propagation, thus eliminating the skip or dead zone, while providing communications in uneven and mountainous terrain up to a distance of 600 km. The loop antenna provides up to 6 dB gain and it is extremely fast tuning as required for the ECCM and ALE operation
- The AT2821 100 W wire ATU provides efficient matching to short wire antennas as installed in fixed-wing aircraft operating over long distances. Capable of storing 99 pretuned frequency settings, the ATU provides fast tuning (typically less than 10 ms) for ECCM and Automatic Link Establishment (ALE) operations
- AT2822 400 W ATU provides fast tuning and high power for ECCM and ALE operations.

The TR 2800 transceiver features advanced electronic techniques to minimise HF noise and interference using a digital voice enhancement technique in accordance with CCIR445-1 and reliable DSP squelch.

Specifications
HF Characteristics
Frequency range: 2 to 29.9999 MHz
Pre-programmable channels: 99
Channel spacing: 100 Hz
Frequency stability: ± 2 in 10^6
Squelch: DSP based on voice detection
Modulation: USB, LSB, AM, CW, AME
COMSEC: Analogue voice encryption
ECCM: Frequency hopping
Data transmission: 75–2400 bps (error corrected)
Interface: RS485 (optional MIL-STD-1553B, ARINC 429)

Dimensions
(transceiver) 1/2 ATR short, inclusive of 100 W power amplifier
(AT2820) 350 (H) × 3000 (L) × 120 (W) mm
(AT2821) 1/2 ATR short
(AT2822) full ATR short

TR 2800 airborne HF transceiver 0011841

CU2832 control unit 0081870

Weight:
(transceiver) <10 kg
(AT2820) <10 kg
(AT2821) <8 kg
(AT2822) <15 kg

Status
As of early 2006, the system is in service in South African Air Force Rooivalk helicopters and upgraded C-130 aircraft.

Contractor
Saab Grintek Communications.

Sweden

AMR 345 dual- and multicommand VHF/UHF airborne radio system

Type
Avionic communications system, VHF/UHF.

Description
The SaabTech AMR 345 dual- and multicommand system is designed for trainers, transport aircraft and helicopters. It consists of the AMR 345 VHF/UHF AM and FM speech and data transceiver and the AMR 349 control units. Performance characteristics are as for the single AMR 345 VHF/UHF transceiver described above.

It features easy handling between instructor and trainee without any priority switches, the latest command being valid, extensive non-volatile reprogrammable memories for up to 1,000 channels and BIT.

The AMR 345 is designed to be operated under the worst flying conditions with a minimum of time and attention required by the pilots. In the trainer installation the trainee has full command over the frequency selection, as in front-line aircraft, and the instructor is fully informed on what the trainee is doing and can at any time take control himself.

The AMR 345 is reprogrammable in the aircraft by means of a fill gun.

In the manual frequency mode a keyboard gives access to any frequency in the VHF or UHF bands. AM or FM is selected by a separate switch on the panel. In the preset mode, a keyboard gives direct access to 1,000 channels and can be used for immediate access to different services at all available airbases. The control features may be adjusted according to customer requirements via software modifications and change of key-tops. One version of the equipment includes VHF and UHF Guard channels.

Specifications
Dimensions:
(transceiver) 76 × 146 × 230 mm
(control unit) 76 × 146 × 40 mm

Weight:
(transceiver) 3.5 kg
(control unit) 0.65 kg
Power supply: 16 to 32 V DC
Power output:
(AM) 10 W
(FM) 15 W
Frequency: 104 to 162 and 223 to 408 MHz
Channel spacing: 25 kHz

Status
Dual- and multicommand radio systems are fitted in the Pilatus PC-7 and PC-9, Raytheon Hawker 125-800, LearJet and Falcon F20.

Contractor
SaabTech AB, Communications Division.

JAS 39 Gripen radio communication system

Type
Avionic communications system, VHF, UHF, secure.

Description
The JAS 39 Gripen communication system is a dual-transceiver installation for simultaneous voice and data communication on the VHF and UHF frequency bands. The system can also operate as a relay link, and there is provision for adding a third Command and Control (C²) data receiver.

The system provides both clear analogue voice and encryption modes, together with ECCM capabilities to ensure jam resistant communication for both voice and data. The Aircraft DataLink (ADL) operates in time sharing mode and enables continuous exchange of data between several aircraft, presented to the pilot via the main computer.

Functions and features included in the system comprise:
- The secure voice function is mechanised by use of a vocoder for narrowband digitising, and by applying encryption and error correction coding. The vocoder is specially designed for operation, including voice recognition, in noisy environments, as found in the Gripen aircraft cockpit
- A ground telecommunication amplifier function is incorporated to provide for analogue (voice) and digital (voice and data) telebrief operation while on the ground to eliminate the need for unnecessary free-space transmissions
- A tone generator and speech synthesiser function is provided to provide warning and alert signals as well as voice messages to the pilot

The AMR 345 VHF/UHF transceiver dual-command version 0504012

JAS 39 Gripen radio communication system 0011837

- A data transfer unit provides for loading: pre-coded frequency channels; crypto and ECCM keys; mission data; time synchronisation; and other requirements for the aircraft avionics system pre-flight briefing and post-flight analysis activities. During 1997, a more modern data transfer unit has been produced with application to other air and ground platforms, which uses 40 Mbytes of memory expansion and additional interfaces including MIL-STD-1553B and Ethernet.

Status
In production for JAS 39 A and B versions, and for the Saab 340 (popularly named Argus), and the S100B airborne radar aircraft. In a modified version, it forms part of the Swedish Air Force Tactical Radio System (TARAS). It is also planned as a baseline for the communication system in the JAS 39 Gripen export system.

Contractor
Saab Tech AB, Communications Division.

United Kingdom

805 series UHF transceivers

Type
Avionic communications system, UHF.

Description
The 805 series is a fully synthesised UHF transceiver covering the frequency range 225 to 399.975 MHz. The state-of-the-art design provides both AM and FM communications, together with a dedicated guard receiver to monitor the 243 MHz distress frequency. When used with the Chelton 715-7 control unit and a suitable UHF antenna, such as the Chelton type 16-1 or any other Chelton multiband antenna, the 805-1 offers a complete AM/FM communications system suitable for connection to the aircraft audio system. The 805-2 variant provides a datalink capability.

The use of software variables in the design permits the implementation of user-specific requirements through simple reprogramming. This ability, coupled with the inbuilt growth facilities, enables the design to be readily configured for a wide variety of applications addressing interfaces such as microphone sensitivity and audio output levels. These

variations are indicated by a suffix to the transceiver part number. A typical system comprises the following units: 805-1 UHF transceiver; 715-7 control unit; 715-7S slave control unit; 27601 mounting tray; optional remote fill gun.

The 715-7 control unit provides control of the transceiver's operating mode and frequency. Selection of the operating frequency may be either by direct entry of the desired frequency or by recalling a stored channel. Up to 100 frequencies can be stored in the transceiver. Each stored channel can be assigned a separate transmit and receive frequency for half-duplex operation and compatibility with satellite communication systems. Alternatively the transceiver can be controlled by a simple command set supplied via an RS-422 databus.

For users who wish to change the stored frequencies on a regular basis a remote fill gun is available. A computer program running on a PC-compatible computer permits the desired frequencies to be assigned to any channel number. The database is then downloaded to the 715-7 controller in a few seconds via an infra-red link using the remote fill gun. This enables

a complete fleet of aircraft to be updated in a matter of minutes without the need to remove any equipment from the aircraft.

For installations requiring dual-control units, a slave control unit, the 715-7S, is offered. When the main control unit is in control, the slave control units acts as a remote readout unit. Control of the system can be passed from the main control unit to the slave control unit and the main control unit will then act as the remote readout unit. Control units can be provided with lighting options, including NVG compatibility, to match cockpit specifications.

The datalink communications transceiver variant, 805-2, is rated to provide a 100 per cent transmit duty cycle.

805 series transceivers are compatible with Chelton 930 Series direction-finders.

Specifications
Transmitter
Power output: 10 W nominal
Frequency response:
(voice) 300 Hz to 3 kHz at +3 dB
(data) 100 Hz to 10 kHz at +3 dB
Sidetone level: 5.0 Vrms into 150 ohms

Chelton 805 series UHF transceivers 0504280

Main receiver
Sensitivity: <3.0 µV for 10 dB (S+N)/N at 30% AM
<1.5 µV for 10 dB SINAD at +3 kHz FM
Audio response:
(AM voice) 300 Hz to 3kHz at +3 dB
(FM voice) 300 Hz to 3 kHz at +3 dB
(adf) 100 Hz to 6 kHz at +3 dB
(AM data) 100 Hz to 10 kHz at +3 dB
(FM data) 100 Hz to 10 kHz at +3 dB

Guard receiver
Sensitivity: <3.0 V for 10 dB (S+N)/N at 30% AM
<1.5 V for 10 dB SINAD at +3 kHz FM

Status
In service.

Contractor
Chelton (Electrostatics) Ltd.

905 series VHF transceivers with integral guard receiver

Type
Avionic communications system, VHF.

Description
The 905-2 is a fully synthesised VHF air and marine band transceiver covering the frequency range 118.000 to 173.975 MHz.

The design provides both AM and FM communications, together with a dedicated guard receiver to monitor the 121.5 MHz distress frequency. When used with the Chelton 715-25 control unit and a suitable VHF antenna such as the Chelton type 21-38-3 or any other Chelton multiband antenna, the 905-2 offers a complete AM/FM communications system suitable for connection to the aircraft audio system. The equipment has been designed to operate with DSC controllers.

Automatic Test Equipment (ATE) employed during the test of the equipment enables calibration data to be stored inside the transceiver ensuring performance can be maintained throughout its life.

The use of software variables in the design permits the implementation of user specific requirements through simple reprogramming. This ability, coupled with the inbuilt growth facilities, enables the design to be readily configured for a wide variety of applications addressing interfaces such as microphone sensitivity and audio output levels. These variations are indicated by a suffix to the transceiver part number. A typical system comprises the following units:
• 905-2 VHF transceiver
• 715-25 control unit
• 21-38-3 antenna
• 27601 mounting tray

The control unit provides control of the transceiver's operating mode and frequency. Selection of the operating frequency may be either by direct entry of the desired frequency or by recalling a stored channel. Each stored channel can be assigned a separate transmit and receive frequency for half-duplex operation. Alternatively, the transceiver can be controlled by a command set supplied via an RS-422 databus.

For installations requiring dual control units, a slave control unit, the 715-25S type, is available. When the main control unit is in control the slave control unit acts as a remote readout unit.

Chelton's 905 series VHF transceivers 0044805

Control of the system can be passed from the main control unit to the slave control unit and the main control unit will then act as the remote readout unit. Control units can be provided with lighting options, including NVIS compatibility, to match cockpit specifications.

905 series transceivers are compatible with Chelton 930 series direction-finders.

Specifications
Transmitter
Power output:
16 W AM/25 W FM nominal
Transmit duty cycle: 25%
Frequency response:
Voice 300 Hz to 2.5 kHz at –3 dB
Data 100 Hz to 10 kHz at –3 dB

Main receiver
Sensitivity:
<3.0 µV for 10 dB (S + N)/N at 30% AM
<1.5 µV for 10 dB SINAD at ±3 kHz FM
Selectivity:
narrowband > ±8.0 kHz at –6 dB
>60 dB at ±25 kHz
wideband > ±15.0 kHz at –6 dB
>60 dB at ±50 kHz
Audio Response:
AM voice 350 Hz to 2.5 kHz at –6 dB
FM voice 350 Hz to 2.5 kHz at –6 dB
ADF 100 Hz to 6 kHz at –3 dB
AM data i/p 100 Hz to 10 kHz at –3 dB
FM data i/p 100 Hz to 10 kHz at –3 dB

Guard receiver
Sensitivity:
<3.0 µV for 10 dB (S + N)/N at 30% AM
<1.5 µV for 10 dB SINAD at ±3 kHz FM
Selectivity:
> ±15.0 kHz at –6 dB
>60 dB at ±50 kHz
Audio Response: 350 Hz to 2.5 kHz at –6 dB

Status
In service.

Contractor
Chelton (Electrostatics) Ltd.

915 series VTAC transceivers

Type
Avionic communications system, VHF.

Description
The 915-1 is a fully synthesised VTAC transceiver covering the frequency range 30.000 to 87.975 MHz.

The design provides FM communications, together with a dedicated AM detector to provide a demodulated output for an associated direction-finding or homing system.

When used with a Chelton 715-22 control unit and a suitable VTAC antenna, such as the Chelton

12-227 pin diode antenna and 7-915PIN227 logic unit, the 915-1 offers a complete FM communications system suitable for connection to the aircraft audio system.

Automatic Test Equipment (ATE) employed during the test of the equipment enables calibration data to be stored inside the transceiver ensuring performance can be maintained throughout its life.

The use of software variables in the design permits the implementation of user specific requirements through simple reprogramming. This ability, coupled with the inbuilt growth facilities enables the design to be readily configured for a wide variety of applications addressing interfaces such as microphone sensitivity and audio output levels. These variations are indicated by a suffix to the transceiver part number. A typical system comprises the following units:
• 915-1 VTAC transceiver
• 715-22 control unit
• 12-227 antenna
• 7-915PIN227 logic unit
• 27601 mounting tray
• Optional remote fill gun.

The 715-22 control unit provides control of the transceiver's operating mode and frequency. Selection of the operating frequency may be either by direct entry of the desired frequency or by recalling a stored channel. Up to 100 frequencies can be stored in the transceiver. Each stored channel can be assigned a separate transmit and receive frequency for half-duplex operation. Alternatively, the transceiver can be controlled by a command set supplied via an RS-422 databus.

For users who wish to change the stored frequencies on a regular basis, a remote fill gun is available. A computer program running on a PC-compatible computer permits the desired frequencies to be assigned to any channel number. The database is then downloaded to the 715-22 controller in a few seconds via an infra-red link using the remote fill gun. This enables a complete fleet of aircraft to be updated in a matter of minutes without the need to remove any equipment from the aircraft.

For installations requiring dual control units, a slave control unit, the 715-25S type, is available. When the main control unit is in control the slave control units act as a remote readout unit. Control of the system can be passed from the main control unit to the slave control unit and the main control unit will then act as the remote readout unit. Control units can be provided with lighting options, including NVG compatibility, to match cockpit specifications.

915 series transceivers are compatible with Chelton 930 series direction-finders.

Specifications
Transmitter
Power output: 15 W nominal
Transmit duty cycle: 25%
Frequency response:
voice: 300 Hz to 2.7 kHz at –3 dB
data: 100 Hz to 10 kHz at –3 dB

Chelton's 915 series VTAC transceivers 30–88 MHz   0044806

Main receiver
Sensitivity:
<1.0 μV for 10 dB SINAD at ±3 kHz FM
<3.0 μV for 10 dB (S + N)/N at 30% AM
Selectivity:
narrowband > ±7.5 kHz at –6 dB
>60 dB at ±25 kHz
wideband > ±15.0 kHz at –6 dB
>60 dB at ±50 kHz
Audio Response:
FM voice 300 Hz to 2.7 kHz at –3 dB
ADF 100 Hz to 6 kHz at –3 dB
AM data 100Hz to 10 kHz at –3 dB

Status
In production and in service.

Contractor
Chelton (Electrostatics) Ltd.

ACCS 3100 audio management unit

Type
Avionic audio management system.

Description
The ACCS 3100 Audio Management Unit (AMU) provides complete integration of audio selection, aural warnings generation and speech recognition functions within a single unit. Selections and control are normally provided via the MIL-STD-1553B avionics databus from an integrated avionics suite, although a dedicated control panel can also be offered.

The AMU implements all the intercommunications functions required on the next generation of military aircraft such as the Eurofighter Typhoon or for avionic upgrades to existing aircraft. Intercommunication is provided between radios, navaids, pilot, co-pilot, ground crew, recorders and the avionics system status outputs which activate aural warnings. Digital signal processing techniques effect noise tracking voice-operated switching for hands-free intercom, and noise filtering and voice-operated gain adjustment devices ensure audio signals routed to transmitters are of optimum clarity and modulation.

Aural warnings include voice warnings as well as attentions and tones, as commanded by external status inputs from dedicated discretes or via a MIL-STD-1553B databus. Voice warnings storage for up to 200 messages is currently provided. Digitised voice output of status and information data can also be activated by direct voice input commands from the aircrew.

Direct voice input uses speech recognition algorithms within the AMU to ensure accurate performance despite ambient cockpit noise and stress-induced voice variations. A programmable 200 word vocabulary, together with a dynamic syntax pointer responsive to avionics system status, gives sufficient flexibility to meet most demanding applications.

The AMU uses a powerful Ada-programmed 68020 processor, running BIT as well as normal operating programs; a non-volatile maintenance memory is available to store fault data. The

AMU databus interface is compatible with MIL-STD-1553B, STANAG 3838 and STANAG 3910 fibre optic databus protocols.

Specifications
Dimensions: 380 × 200 × 124 mm
Weight: 7.7 kg
Power: 100 W
Processing: 30 Mips

Status
In service. Also selected as the communications and audio management unit for the Eurofighter Typhoon.

Contractor
General Dynamics United Kingdom Ltd.

AD120 VHF/AM radio

Type
Avionic communications system, VHF.

Description
Originally designed for civil aviation applications by the King Radio Corporation in the USA, BAE Systems has requalified this VHF/AM system for military roles and, manufacturing under licence, markets the system under the designation AD120. It is installed in most UK military fixed-wing aircraft and helicopters and has also been supplied to a number of overseas customers.

Covering the frequency band 108 to 137 MHz, the AD120 provides channel spacing of 25 kHz. Services provided are double sideband AM voice communication. Power output can be varied between 10 and 20 W. The system is an all-solid-state design of modular construction and is designed for easy installation and maintenance. It is also exceptionally simple to operate. The system's standard remote controller possesses only five controls: an on-off switch, volume control, a test button and two rotary switches for frequency selection.

Tuning is instantaneous. Automatic squelch and gain control eliminate the need for manual adjustment.

The self-test facility may be used during operation as a confidence check.

Specifications
Dimensions:
(transmitter/receiver) 60 × 127 × 315 mm
(controller) 146 × 47 × 90 mm
Weight:
(transmitter/receiver) 2.3 kg
(controller) 0.6 kg

Status
More than 1,200 systems are currently in service.

Contractor
BAE Systems.

AD190 (ARC-340) VHF/FM radio

Type
Avionic communications system, VHF.

Description
The AD190 radio, a company designation for the ARC-340 system, is a tactical VHF/FM equipment designed specifically for air-to-ground communication between military aircraft and ground forces. It is in service with the UK armed forces, in conjunction with Army Clansman VHF equipment, and is in operation in a wide range of fixed-wing aircraft and helicopters in many parts of the world.

The system provides clear, and secure, speech communication, data transmission, automatic rebroadcast for extending the range of tactical communications and homing facilities. The homing mode may be used simultaneously with a communications channel without mutual interference. The range capability of the homing facility is the same as the communications range.

The AD190 covers the VHF band from 30 to 75.975 MHz at selectable channel spacing of either 25 or 50 kHz increments. Tuning is silent and instantaneous and transmitter output power is selectable at either 1 or 20 W.

Multistation collocated operation is possible with a maximum of three systems in the same aircraft. This is subject to the proviso that antennas must be sited at least 3 ft apart and frequency separation of 3.5 per cent for three systems or 3 per cent for two systems must be maintained. The AD190 incorporates diagnostic BITE.

Specifications
Dimensions:
(controller) 146 × 67 × 85 mm
(transmitter/receiver) 385 × 197 × 125 mm
Weight:
(controller) 0.9 kg
(transmitter/receiver) 9.2 kg

The BAE Systems AD120 VHF radio   0504013

Status
In production and in service.

Contractor
BAE Systems.

AD3400 multimode secure radio system

Type
Avionic communications system, VHF, UHF, secure.

Description
The AD3400 provides, in a single transmitter/receiver, coverage of the entire airborne line of sight frequency band from 30 to 400 MHz with channel increments of 25 kHz. Up to 20 channels may be preselected in this range.

In effect, the system provides the equivalent of four radios in a single unit since it covers the following bands and modes: VHF FM for tactical close support, the civil ATC VHF AM band, the civil and maritime VHF FM bands and the military UHF AM and FM bands. Its other major feature is that it provides secure communications for both speech and data transmission. For speech transmission in a secure mode, an encryption unit is used and for data security the appropriate modem is connected to the radio equipment.

Separate, continuously operating Guard receivers are incorporated and homing and ADF facilities are also available with suitable keying and antenna installations.

Construction of the system features 'slice' techniques and Large-Scale Integrated (LSI) circuitry. The system is based on Intel 8085 microprocessors and uses distributed processing. The extensive use of LSI and proven components provides a high measure of reliability. Serviceability is further enhanced by the comprehensive BITE facility which can provide rapid fault diagnosis, particularly since intermittent fault data is stored in a non-volatile memory for maintenance purposes.

The system is convection cooled and protected by a sensing device which progressively reduces power output at high temperatures to prevent transmitter damage. The transmitter is also protected from damage caused by short or open circuits at the transmitter output.

The AD3400 has been designed for ease of installation with particular attention to retrofit requirements in older aircraft in which it is often possible to fit two of these systems in the space formerly occupied by a single radio.

Specifications
Dimensions:
(AA34024 controller) 57 × 146 × 152 mm
(AA34001 transmitter/receiver) 194 × 125 × 256 mm
(AA34601-1 encryption unit)   194 × 58 × 320 mm
Weight:
(AA34024 control unit) 1.3 kg
(AA34001 transmitter/receiver) 6.5 kg
(AA34601-1 encryption unit) 4.5 kg

Status
In production and in service.

Contractor
BAE Systems.

AD3430 UHF datalink radio

Type
Avionic communications system, UHF, data, secure.

Description
The AD3430 is a UHF Link 11 compliant radio which operates in the 225 to 399.975 MHz band. It is tunable in 25 kHz increments and provides transmission and reception of clear and secure speech using amplitude or frequency modulation. Secure speech operation is provided in association

with a BID 250 encryptor. The capability for providing transmission and reception of Link 11 data using frequency modulation is provided in accordance with all the general radio and UHF specific Link 11 radio requirements of MIL-STD-188-203-1A.

The equipment can deliver maximum power outputs of 20 W in AM operation and 60 W in FM operation with a transmit duty cycle of 100 per cent. The primary control of the radio is via an ARINC 429 serial data highway.

Specifications
Dimensions: 365 × 194 × 124 mm
Weight: 9.5 kg

Status
The AD3430 is in production and in service in the UK Royal Navy EH 101 helicopter.

Contractor
BAE Systems.

Advanced cabin audio systems

Type
Avionic audio management system.

Description
Caledonian Airborne Systems' cabin audio system is a stand-alone wall-mounted monaural audio player designed to reduce the workload of the cabin crew and reproduce high quality messages and in-flight entertainment/boarding music.

Specific applications include playback of customised pre-recorded (possibly multilingual) safety briefings, emergency announcements and boarding music.

In contrast to many current cabin audio systems, solid-state digital recording and playback is employed throughout, facilitating weight and volume reduction, vastly reduced programme search time, improved reliability and high quality audio reproduction.

Operationally, each message/briefing/announcement is digitally recorded at Caledonian Airborne Systems. Each programme is organised systematically into arrays or sub-groups depending on the form of the content. For example, all safety briefings would be grouped under the heading BRIEFINGS, emergency announcements under the heading EMERGENCY and so on. Each sub group is represented by a mode button on the control panel. All the contents of each mode can be examined or selected via scroll controls, with the content or title indicated by a backlit alphanumeric display.

The number of buttons, control panel layout and operation mode have been designed to be user friendly.

Specifications
Playback duration: Over 140 minutes
Audio bandwidth: >22 kHz (<0.5% THD)
Output: 300 O output drive to customer PA system
Power: 28 V DC, <1 W
Weight: <1 kg
Dimensions: 155 × 230 mm (footprint)
Approval: CAA approval no. SA 01239

Status
In production and in service.

Contractor
Caledonian Airborne Systems Ltd.

The AD3430 is in production and in service in UK Royal Navy EH 101 helicopters   0099601

Airborne microwave links

Type
Avionic microwave link system.

Description
Skyquest Aviation manufactures a range of microwave link equipment which enables users to send live video and audio data to a ground-based receiving station. Supplied microwave equipment is tuned to the customer's approved frequency.

A typical microwave system comprises the aircraft-mounted equipment (transmitter and antenna), portable- or fixed-ground stations and processing/encryption units.

EX24-10 microwave transmitter
The EX24-10 microwave transmitter has been specifically designed for airborne data transmission and is suitable for both TV outside broadcast and police surveillance applications. The system features a 10 W transmitter with near studio quality video and audio performance, compatible with most existing ground-based receivers. The system can be pre-programmed with up to 16 channels, selectable via a front panel switch. Options include single- or dual-audio channels and video encryption.

Specifications
Frequency range: 2.4–2.7 or 3.4–3.8 GHz
Pre-set channels: Up to 16 over any 200 MHz segment
Operating temperature: –10 to +40°C
Power supply: 28 V DC (–10 to +20%)
Power consumption: 3 A (max); 2.6 A (nominal)
Weight: 3 kg (approximately)
Dimensions: 200 × 123 × 102 mm

RF
Video: Positive/negative
RF output power: >10 W
Pre/de-emphasis: CCIR 405-1 or 725/6

Video
Signal to noise: >55 dB
Video output: 1V p-p composite

Audio
Number of channels: 0,1 or 2
Subcarrier frequency: 6.8 and 7.5 MHz

PRD1000 portable microwave receiver system
The PRD1000 is a portable microwave receiver system consisting of a 10.4 in TFT LCD colour video monitor, integral rechargeable battery and flip-up antenna. The system can process one video and up to two audio channels to outside broadcast quality standards. Designed for rapid deployment helicopter down link, the system is also suited to terrestrial surveillance applications. The PRD1000 is fully compliant with UK Home Office MG46 specifications for microwave video link equipment.

A number of versions of the receiver are available, covering all appropriate frequency bands from 1 to 13 GHz. Each receiver may be factory pre-programmed with up to 10 pre-set channels covering any 300 MHz segment.

The integral battery is capable of over 2 hours of operation; alternatively, the system can be powered and/or charged from an external 12 V DC supply. The flip-up omnidirectional whip antenna can be replaced by an externally mounted or hand-held item. Single- or dual-speakers are specified where appropriate and phono/video outputs are incorporated for connection to external recording equipment.

Specifications
Frequency range: 1–13 GHz
Pre-set channels: Up to 16 over any 300 MHz segment
Operating temperature: 0 to +40°C
Power supply: Internal battery (up to two hours operation); external 12 V DC (–10 to +20%)
Power consumption: <2 A
Weight: 7.5 kg (approximately)
Dimensions: 350 × 270 × 160 mm

RF sensitivity: <–85 dBm
Video: Switchable invert
De-emphasis: CCIR 405-1 or 725/6
Signal to noise: >55 dB
Video output: 1V p-p composite
Luminance linearity: <±5%
Delay inequality: <±20 ns
Differential gain: <±5%
Differential phase: <±3°

Audio
Number of channels: 0,1 or 2 (option)
Subcarrier frequency: 6.8 and 7.5 MHz
Frequency response: (WRT 1 kHz) – 50 Hz to 15 kHz, +2 dB
Output level: 0 dB, 600 Ω balanced
Signal to noise (CCIR 468-4): >50 dB (quasi-peak WRT 0 dBm)

Microwave video/audio encryption unit
A video and audio encryption unit can be added to the microwave link system to ensure the security of data transmitted from the host platform. Utilising line cut and rotate scrambling, a high level of picture concealment and security can be achieved. The system works by adding an encryption unit at the microwave transmitter and a decoder at the receiver. Both units are physically identical.

Specifications
Power: 7–32 V DC
Dimensions: 160 × 90 × 29 mm
Options: Dual-audio scrambling

Status
In production and in service.

Contractor
Skyquest Aviation.

Airborne real-time datalink

Type
Avionic datalink system.

Description
Thales Avionics has integrated its satcom and navigation/mission management products to provide a complete real-time datalink package, which includes both the ground segment and the airborne Line Replaceable Units (LRUs). The airborne LRUs utilised are: the multifunction Control Display Navigation Unit (CDNU) as the operator interface and the Satellite TRanceiver

(STR) as the communications system. The primary function of the CDNU is to provide aircraft navigation and subsystem control. It is also the airborne datalink end system, providing Automatic Dependent Surveillance (ADS) and Controller to Pilot DataLink Communication (CPDLC), which also allows free text messages and pre-formatted messages to be sent to a variety of ground destinations. It displays messages received from the ground and allows the user to reply to those messages. The CDNU also has an Emergency ADS Mode.

ADS reports are initiated from the ground, via an ADS contract, and include: present position, altitude, velocity, and aircraft intent. All ADS contracts are transparent to the pilot.

The STR equipment which supports packet data services – Data 2 (ACARS) and Data 3 (ATN FANS X.25) communications over the worldwide Inmarsat satellite system, comprises two LRUs: the combined Satellite Data Unit (SDU)/Radio Frequency Unit (RFU) and a High Power Amplifier (HPA).

The Inmarsat system utilises a digital format for real-time full-duplex data communications at up to 1,200 bps. Being a low-gain system, the STR operates via a 0 dBic omnidirectional antenna and therefore requires no other navigation input for beam-steering purposes.

The ground terminal allows the operator to initiate ADS contracts with numerous aircraft and displays the requested information on a map and status window. The terminal also allows text messages to be sent to selected aircraft and displays messages received from aircraft. Any aircraft that initiates an Emergency ADS message is highlighted on the map with an audio alert.

Status
In September 1999, the UK Royal Air Force initiated a three-month real-time, helicopter tracking trial of the system, using Search And Rescue (SAR) Sea King Mk 3A helicopters. An automatic call was made by the helicopter system to the Rescue Co-ordination Centre (RCC); the call interval was determined by the RCC, but 60 seconds were used for trial purposes. The call declared helicopter present position, altitude, velocity and operational intent.

The system is an evolutionary development of the Norwegian Modified-Automatic Dependent Surveillance (M-ADS) programme.

Contractor
Thales Avionics SA.

UK Royal Air Force Sea King Mk 3A search and rescue helicopters participating in the airborne real-time datalink trial
0080260

COFDM helicopter digital video downlink

Type
Avionic datalink system.

Description
Available in 2.4, 3.45 and 4.5 GHz versions, Enterprise Control Systems' digital downlink equipment is designed to comply with the UK Home Office MG42C, DVB-T and MPEG-2 specifications and optimised for air-to-ground transmission of two simultaneous video channels from fixed-wing, helicopter or Unmanned Aerial Vehicle (UAV) platforms.

Coded Orthogonal Frequency Digital Multiplexing (CODFM) is a new transmission technology which utilises multiple carriers to provide high-quality, near-realtime video images in environments and situations where conventional analogue video links would be prone to failure through multipath interference. In this application, reflected as well as direct signals can be utilised to provide for an interference free link. Using 1,704 carriers in an 8 MHz bandwidth, the digital downlink provides two video (S-VHS quality), two audio and a GPS data channel simultaneously. The system also features an integrated encryption facility and a high data rate of up to 28 Mbps.

The Encryption Control Panel (ECP) comprises a small panel-mounted housing incorporating a 32 character Liquid Crystal Display (LCD) and a three button keypad. The ECP has been designed specifically for use with the ECS VEM-2.V1.0 COFDM encoder/modulator and associated power amplifier. Three soft keys facilitate user selection and control.

Support equipment includes a base station CODFM receiver with switched sector, long-range antenna and a portable CODFM receiver for incident commanders.

Specifications
Video codec and modulator box
Part number: VEM2-V1.0
Channels: Two Y/C or PAL video, 2 audio and 1 GPS data channel
RF GPS input: TNC female
Video channel data rate: 5–15 Mbps per video channel; 2.4 kbps audio and data
Modulation technique: CODFM
Number of carriers: 2k in 8 MHz
On-air standard: DVB compliant to ISO/IEC 13818
Encryption: Proprietary with separate control unit
Forward error correction: Combination of Reed Solomon, Verterbi interleaving and outer leaving
Power: 28 V DC maximum; 1 Amp
Dimensions: 300 × 175 × 70 mm (dual-video version with integrated GPS receiver)
Weight: 2 kg

Power amplifier
Part numbers: VLT 15-10-6-D2.3 (2.3–2.5 GHz); VLT 15-10-6-D3.5 (3.3–3.6 GHz); VLT 15-10-6-D4.5 (4.4–4.9 GHz)
Output power: 10 W into the antenna
Input: RF @ –7 dBm
Power consumption: 6.0 Amps @ 28 V DC typical
Frequency: 2.3–2.5 GHz, 3.3–3.6 GHz, 4.4–4.9 GHz
Number of RF channels: Up to 16
On-air standard: DVB compliant to ISO/IEC 13818
Dimensions: 260 × 120 × 137 mm
Weight: 5.5 kg

Encryption control unit
Part number: ECU V1.2
Display: 32-character LCD
Controls: 3 button soft key pad
Power: 28 V DC from VEM2-V1.0
Functions: Transmitter on/off and mute; 32 destination codes; transmit mode; channel frequency
Dimensions: 2.25 × 5.75 in (Dzus fitment)
Weight: 300 g

Base station
Power: 12 V DC or mains AC operation
Temperature: –10 to 70°C
Video outputs: PAL, YUV, Y/C, RGB, ASI, MPEG
Range: Up to 50 miles at 2,000 ft agl

Portable receiver
Range: Up to 20 miles
Weight: 5 kg
Battery endurance: 2 h

Status
In production and in service in a number of police helicopter platforms worldwide. The Dutch police were the first to integrate ECS' COFDM transmission technology for their fleet of 8 MD902 helicopters, with approval for installation gained in 2002. In September 2003, the UK Wiltshire Police Air Support Unit commenced operations with the system and during the same year, Norwegian Police EC135 commenced operations with the first encrypted simultaneous uplink and downlink system.

Contractor
Enterprise Control Systems Ltd.

Jaguar-U (BCC 72) UHF radio

Type
Avionic communications system, UHF, secure.

Description
Thales' Jaguar-U system is a range of radio communications equipment operating in the UHF band, available for airborne, shipborne and land use. The airborne version, BCC 72, is designed to operate over the frequency range 225 to 400 MHz and provides fixed frequency or frequency-hopping FM with selectable encryption, and fixed frequency AM compatible with current operational systems. It provides 7,000 channels at 25 kHz spacing, with 30 programmable channels (including a Guard channel on the distress frequency of 243 MHz) with flexibility to select clear, secure, fixed and hopping modes.

Design of Jaguar-U is based on experience gained in the Jaguar-V systems and employs the same method of medium-speed frequency hopping to protect against interception, direction-finding and jamming. To simplify frequency management, the 225 to 400 MHz band has been divided into 13 sub-bands, allowing the co-sited operation of multiple radio systems without interference. However, the radio can also be programmed to use any one of three larger hop bands. In either mode, orthogonal hop sets are available to assist frequency management. Large numbers of nets can operate in the same frequency bands and individual bands can be barred to avoid jamming or to protect other fixed frequency stations. Selective communication can be performed within an individual net or, conversely, a radio can be selectively barred from a net should it be captured. Once the radios have been programmed with hop codes and frequencies they synchronise automatically without the need for time of day input. Communications security is also enhanced by the use of either a built-in encryption unit using a second keystream generator or any external 16 kbits/s system in fixed or frequency-hopping modes.

The airborne BCC 72 system consists of a rack-mounted transceiver and a panel-mounted control/display unit. The transceiver provides output powers of either 10 mW or 15 W on FM and 40 mW or 40 W PEP on AM, output level being selected on the controller. Two control/display units are available, depending on the aircraft requirements. The BCC 584B controller gives full manual selection of any of the 7,000 channels plus selection of the 30 programmable channels, hopping or fixed mode selection, low- or high-power selection and frequency/mode display by light-emitting diodes.

Specifications
Dimensions:
(transceiver) 230 × 90 × 350 mm
(controller BCC 584B) 145 × 66 × 104 mm
Weight:
(transceiver) 6.5 kg
(controller BCC 584B) 0.8 kg

Status
In service.

Contractor
Thales Communications.

Link 11 operational systems

Type
Avionic datalink system.

Description
Aerosystems International has provided Link 11 and Link 11B operational systems into land-based, surface and airborne environments. These systems provide Link 11 message handling, display processing and system control in full compliance with STANAG 5511. The systems are based around Commercial Off-The-Shelf (COTS) software packages developed and supported by Aerosystems for UK and international customers. A variety of COTS hardware and operating system solutions are supported.

Operational systems have been delivered compliant with airborne qualification requirements, including EMC and Tempest.

The main features and benefits of Aerosystems' Link 11 solutions include:
- Operation as a datalink processor in conjunction with a host tactical system
- Operation as a stand-alone Link 11 tactical system
- Available as a software component with a fully defined ICD for direct incorporation into a host system
- Supported on COTS hardware including PC, Sun and Silicon Graphics
- Supported on COTS operating systems including Windows™ NT and Unix
- Proven interoperability with AdatP-11, OPSEC 411.2 and UK Link 11 DTDL SLIRS compliant NATO platforms
- Full correlation and reporting responsibility functions
- Extensive filtering options
- Totes and mini-totes available for control and amplify functions
- Full tactical situation display
- Alerts and warnings
- NATO-standard symbology
- Windows™ and mouse/rollerball-operated HMI
- Configurable for test and training applications
- Full software control of the DTS
- DTS interface via ATDS, NTDS or RS-232.

Status
As of late 2005, in production and in service in UK Royal Air Force Nimrod MR2 Maritime Patrol Aircraft (MPA). Systems have also been delivered to overseas customers.

Contractor
Aerosystems International, Tactical Communications Division.

MCA-6010 Aero SATCOM antenna

Type
Avionic communications system.

Description
The MCA-6010 is a mechanically steered quad helical array, which has been installed at the top of Gulfstream and Challenger vertical stabilisers. The MCA-6010 offers improved coverage over conventional phased-array designs – greater than 95 per cent compared with the minimum

INMARSAT coverage requirements of 75 per cent. This is of particular importance to operators flying high-latitude and Pacific Ocean routes.

The higher gain achievable with a quad helical configuration exceeds the INMARSAT requirements for single- or multichannel operations. The MCA-6010 is capable of supporting in excess of six channels of simultaneous voice, tax or data, even at the edges of INMARSAT satellite coverage.

The system is compatible with all single- and multichannel SATCOM avionics conforming to the requirements of ARINC 741 and the INMARSAT system definition manual.

Other features of the MCA-6010 include:
• High gain >13 dBic
• No beam switching
• Full INMARSAT approval
• Extensive BITE.

The antenna assembly consists of four elements:
• Antenna array Antenna Control Unit (ACU)
• Two-Axis Stabilised Platform (TASP)
• Diplexer/Low Noise Amplifier (D/LNA).

Satellite position relative to the aircraft is computed by the SATCOM avionics and fed to the ACU via an ARINC 429 databus. The ACU converts the data into azimuth and elevation drive signals that are passed to the TASP.

The TASP motors and position sensors steer the antenna array in the direction of the satellite with up to 10°/second of continuous rotation in azimuth from −10 to +90° of elevation to ensure uninterrupted communications during aircraft manoeuvring. The D/LNA provides electrical isolation of the transmit and receive signals.

Specifications
Dimensions:
ACU: 57.7 × 224.0 × 138.0 mm
D/LNA: 42.0 × 195.0 × 280.6 mm
Weight:
ACU: 1.5 kg
D/LNA: 3.2 kg
TASP: 8.6 kg
Power: 28 V DC

Status
In production and in service.

Contractor
Thales Avionics SA.

PTR 1721 V/UHF radio

Type
Avionic communications system, V/UHF.

Description
The PTR 1721, a combined VHF and UHF radio covering the UHF band from 225 to 400 MHz and the VHF band from 100 to 156 MHz, is designed for all types of military aircraft. The ability to communicate on VHF in addition to UHF offers increased flexibility to air forces which, on occasion, may need to operate from civil airfields equipped with VHF facilities only. The PTR 1721 has been chosen for the RAF Panavia Tornado aircraft.

The PTR 1721 offers up to 9,240 channels, 2,240 on VHF and 7,000 on UHF. Channel spacing in the VHF band is at 25 kHz and standard spacing in UHF is at 50 kHz although an optional interval of 25 kHz spacing is available. Frequencies are selectable directly from the remote-control unit which also provides preselection of up to 17 channels with the addition of the UHF and VHF Guards at the international distress frequencies of 243 and 121.5 MHz respectively. Frequency synthesis techniques ensure good frequency stability and channel selection characteristics.

The system is entirely solid-state in construction and, wherever possible, employs conventional technology in the interests of reliability enhancement. The receiver is varactor-tuned and the frequency synthesiser is compared against a single reference oscillator.

The PTR 1721 radio is in service in UK Tornado aircraft   0582982

A sealed case houses the transmitter/receiver unit and access to the modules is gained by removal of the sides of the case, to which the modules themselves are attached. Heat is conducted from the modules via the chassis, which acts as a heatsink, to the sides of the case and hence to external air. Forced-air cooling is directed through the mounting tray installation which also acts as a vibration insulator. Forced-air cooling may be dispensed with in less demanding aircraft environments.

Switching for dual-control operation is external to the system and in two-seat aircraft identical control units are installed at each crew position. Options include an antenna lobe switch for azimuth homing requirements and a UHF antenna switch for automatic direction-finder operation.

Recently, the company has introduced an optional modification which is available for both in-service and new radios. This modification provides a frequency-hopping capability which gives a significant improvement in communications performance when the radio is operated in a jamming environment.

The environmental temperature range extends from −40 to +70°C ambient and the normal operational altitude is to a maximum of 50,000 ft, although operation is possible for short periods at altitudes of up to 70,000 ft.

Specifications
Dimensions:
(controller) 94 × 145 × 175 mm
(transmitter/receiver) 125 × 194 × 339 mm
Weight:
(controller) 1.8 kg
(transmitter/receiver) 11.25 kg

Status
In service in UK Royal Air Force Tornado aircraft.

Contractor
BAE Systems.

PTR 1751 UHF/AM radio

Type
Avionic communications system, UHF.

Description
The PTR 1751 is a lightweight UHF/AM radio designed for all types of military fixed-wing aircraft and helicopters. It provides 7,000 channels in the frequency band 225 to 399.975 MHz

at a channel spacing of 25 kHz and is available with either 10 or 20 W transmitter outputs. Channel spacing of 50 kHz is available as an option.

Both versions comprise a single transmitter/receiver unit with either a manual controller or an optional manual and preset controller. Each controller provides full selection of the range of 7,000 channels together with control of the built-in test functions; the manual/preset unit additionally permits preselection of up to 30 channels, plus Guard frequency, through incorporation of a non-volatile memory store. A remote frequency and channel indicator is also available.

Options include continuous monitoring on the 243 MHz international distress frequency via a separate Guard receiver module which plugs directly into the main transmitter/receiver chassis, an external homing unit and a wideband secure speech facility which requires no additional interface equipment.

Recently the company has introduced an optional modification which is available for both in-service and new radios. This modification provides a frequency-hopping capability which gives a significant improvement in communications performance when the radio is operated in a jamming environment.

The PTR 1751 conforms generally to DEFSTAN-07-55 and operates satisfactorily over a temperature range from −35 to +70°C. It is of all-solid-state modular construction with high reliability as a principal design aim, and a claimed MTBF of 800 hours.

Specifications
Dimensions:
(transmitter/receiver)
(10 W version) ½ ATR short × 160 mm high
(20 W version) ½ ATR medium × 160 mm high
(manual controller) 146 × 48 × 108 mm
(preset controller) 146 × 95 × 108 mm
Weight:
(transmitter/receiver)
(10 W version) 5 kg
(20 W version) 6.7 kg
(preset controller) 1.4 kg
(manual controller) 0.7 kg
(Guard receiver module) 0.3 kg

Status
In service.

Contractor
BAE Systems.

RA690 analogue Communication Control Systems (CCS)

Type
Avionic Communications Control System (CCS).

Description
The RA690 series of CCS equipment provides an all-analogue solution that utilises either distributed facilities or a central amplifier with remote station boxes for crew interface. The RA690 is extensively used in military and civil aircraft worldwide.

The conventional configuration of distributed amplifiers, filters and volume controls is designed using the latest technologies to ensure the most cost effective performance possible. When configured to utilise a central amplifier and remote passive station boxes, the single point audio is routed within the central amplifier, with selection and control implemented via analogue control lines between these major units. Both architectures are built using functional blocks to produce a customised configuration for the application required. The functional blocks include Voice Operated Switches (VOS), attenuators, power supplies, muting and intercom override and interface matching networks.

Facia panel engraving and lighting details (including NVG type), switching and electrical characteristics are configured on each station box type to suit the particular requirements of the user.

Additional flexible RA690 system capability is provided by S690 Audio Selector Panels (ASPs) and D690 switch boxes. The S690 ASP allows an increase in the number of receiver audio inputs which can be accommodated. D690 switch boxes enhance the facilities of the microphone switching, intercom networks and other CCS functions.

Varying combinations of station boxes, ASPs and switch boxes can be installed at each crew position. Where required, the ASP and switch box functions can be integrated within the station box to form a single unit.

The RA690 Series includes a range of general purpose amplifiers and other units. These include specially designed amplifiers to ensure systems conform to Civil Aviation Authority (CAA) specifications 11 and 15 for Cockpit Voice Recorder (CVR) and passenger address systems. Additionally, existing CCS installation performance can be enhanced with use of A6916 units providing noise tracking level control and voice operated switching and a power supply for active noise reduction headsets.

The RA690 range is further complemented by headset connection jack boxes and specially configured audio balance units to ensure correct interface between the CCS and various Tx/Rx equipments.

Status
The RA690 CCS is widely used in military and civil aircraft worldwide. Some 960 noise tracking level control and voice-operated switches have been ordered by the UK Royal Air Force for Sea King helicopters.

Contractor
Thales Defence Ltd.

RA 800 Communications Control Systems (CCS)

Type
Avionic communications system.

Description
The RA800 series of CCS provides full compatibility with aircraft warning and management systems, and is designed to interface with Health and Usage Monitoring Systems (HUMS) to enhance overall maintainability and flight safety.

Two variants of the RA 800 CCS are available, the RA 800 and the RA 800 Light.

Units of Thales' RA690 analogue CCS   0044844

The RA 800 Light offers improved ElectroMagnetic Compatibility (EMC) and reduced weight by use of fibre optic instead of copper routing – a 90 per cent in wiring weight is claimed. The RA 800 Light is a fully digitised system incorporating advanced digital signal processing to ensure high signal quality and integrity, with system flexibility, making it ideal for helicopter installations.

Status
The RA 800 CCS is in service on BAE Systems Hawk aircraft in several countries, and is being delivered to the UK Royal Air Force for fitting on EH 101 and Chinook helicopters following selection for the UK Royal Air Force's Support Helicopter (SH) programme.

In September 1999, the RA 800 CCS was selected to equip the Hawk 115 trainer aircraft of the NATO Flying Training Centre in Canada, and the RA 800 CCS Light system was selected for the Canadian EH 101 Cormorant Search And Rescue (SAR) helicopter programme.

Contractor
Thales Avionics SA.

RA800 digital Communication Control Systems (CCS)

Type
Avionic Communications Control System (CCS).

Description
The RA800 CCS is designed to meet present and foreseeable requirements for all aircraft types. The RA800 provides centralised audio control, interfacing to other aircraft systems and sophisticated intercom networks for the flight crew. It is built on a foundation of intelligent modules, which enables systems to be configured for the particular requirements of the aircraft platform.

The core of the RA800 is the Communication Audio Management Unit (CAMU). All audio routeing is contained within the CAMU: this increases immunity to electromagnetic interference by considerably reducing the volume and complexity of aircraft cabling. Other advantages include ease of installation, weight saving and improved reliability gained through significant cable reduction.

The crew interface to the system is via Audio Control Panels (ACPs) for the control functions and, in some variants, Headset Electronics Units (HEUs) for microphone and telephone audio. Connections form the ACPs and HEUs to the CAMU are via digital links.

A Tempest variant, RA800 'Light', employs Digital Signal Processing of all audio and control functions, together with fibre optic interconnection

between major system elements. The fibre optic connection between the CAMU and the ACP, and the CAMU and the HEU, is a serial datalink. The use of fibre optic cables further improves the immunity to electromagnetic interference hazards and ensures secure TEMPEST operations.

For RA800 equipment configured to use traditional copper wire interfaces, the connection between the ACP and the CAMU is an ARINC 429 bus.

RA800 is microprocessor controlled with centralised or distributed intelligence dependent upon the platform. Where an aircraft is required to undertake a variety of roles, the RA800 system can readily be software reconfigured on board.

The system interfaces to navigation, radio and warning equipment either directly or by means of the aircraft avionics bus (MIL-STD-1553).

Sophisticated BIT is carried out within all units that make up the RA800 CCS system. This data is collated and can be sent to the aircraft maintenance system.

Voice and/or Tone warnings are produced either from an embedded module within the RA800 CAMU or from a stand-alone unit.

Status
Designed for civil and military aircraft applications. In service on BAE Systems Hawk aircraft. Ordered for UK Royal Air Force EH 101 and Chinook helicopters, SAAF C-130B aircraft, Royal Australian Air Force Lead-in-Fighter aircraft, and Canadian SAR helicopters.

Contractor
Thales Acoustics.

Rangeless Airborne Instrumentation Debriefing System (RAIDS)

Type
Avionic datalink system.

Description
RAIDS eliminates the need for deployment to a fixed air combat manoeuvring range, and permits air-to-air and air-to-ground weapon simulation, together with electronic warfare training as required, and safety alerts.

RAIDS is housed in a P4B pod with integral inertial measurement unit (Litton LN-200), GPS receiver (Rockwell Collins GNP-1®) and datalink. No aircraft modifications are required, providing AIM-9 and MIL-STD-1553 interfaces are available. A full debriefing system is provided.

Specifications
Datalink:
(frequency range) 2.2–2.4 GHz
(frequencies available) 32 within 100 MHz

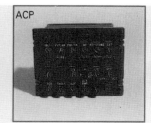

Thales Acoustics' digital RA800 CCS units   0044839

(simultaneous handling) 200 aircraft, without pre-booking
(peak output power) 26–28 W
(range) 60 n miles (120 with relay)
GPS receiver: 10 channel, P/Y or C/A code, 24 state Kalman filter, WAGE
Inertial measurement unit: FOG, 1°/h attitude accuracy
Data storage: 240 MB PCMCIA card, upgradeable
System accuracy (1σ):
(position X and Y) 5 m
(position Z) 8 m
(no drop bomb score) 15 m
(velocity) 0.1 m/s
(pitch/roll/heading) ±1.0°
Ground station:
mission planning
3D debrief
plan, angular and cockpit views
background mapping
digital terrain elevation data
synchronised with aircraft video recordings
DGPS capability
COTS workstation

The BAE Systems rangeless airborne instrumented debriefing system installation   0080284

Status
In service. RAIDS has been chosen for seven air forces. It is flying with the UK Royal Air Force on Tornado F3 aircraft for EW training, with the Fast Jet Operational Evaluation Unit at RAF Coningsby on Harrier GR Mk 7, Jaguar GR3, and Tornado F3 and GR4 aircraft. The system was used as part of fast jet air combat training in the UK North Sea Air Combat Manoeuvring Installation (ACMI) until the area was closed in 2004.

Contractor
BAE Systems.

Scorpio 2000 series datalink systems

Type
Avionic datalink system.

Description
Scorpio 2000 is one of a series of expandable datalink systems which accept host aircraft and designated target global position data from the radar/navigation system databus and translates, encodes and up-converts them for transmission via radio link to a ground station equipped with a ruggedised PC or notebook computer. From the base station's GPS input, the computer predicts range and bearing to the aircraft and target, with data presented on an integrated colour display in either textual or moving map format.

Typically, the radar used with the system is the Honeywell RDR-1500B, and the message data can be secured to meet user requirements. Options include: GPS internal (or external DGPS) interfaces, flat panel display and keyboard control.

Specifications
Radio link:
Frequency: UHF
Output power: To suit customer requirement

Sensitivity: –118 dBm (typical)
Message protocol: High level with data field scrambled
Data rate: 9,600 baud
Interface: ARINC 419
Weight:
Airborne transmitter: 2.4 kg
Base station receiver: 1.2 kg
Base station ruggedised: PC 30 kg
Power: 28 V DC

Status
In production and in service.

Contractor
Caledonian Airborne Systems Ltd.

Series 7-202 HF airborne receiving system

Type
Avionic communications system, HF.

Description
The Series 7-202 HF receiving system comprises a Type 7-202-1 HF receiver and Type 19-197 active antenna. The system is designed to provide worldwide reception of transmissions.

The Type 7-202-1 is a fully synthesised HF receiver capable of receiving transmissions anywhere in the long, medium or short wavebands. The Type 19-197 is a high-efficiency low-noise active antenna providing continuous coverage of the long, medium and short wave and VHF/FM broadcast bands. The antenna is protected against low-level lightning strikes.

The receiving system is fully automatic in operation and, once the desired service has been selected, it will automatically search out and tune to the best available transmission anywhere in the world. A database programmed into the

receiver contains details of the coverage areas of all the available transmitters together with their frequencies and times of transmissions. The receiver, by interrogating the aircraft's navigational computer, obtains latitude and longitude, together with time of day. Using this information, the receiver tunes through available transmissions and, using integrated signal quality measuring circuits and algorithm, determines which transmission offers the best quality and tunes the receiver to the appropriate frequency. The resulting audio is then output for distribution over the aircraft audio system. The integrated microprocessor continuously monitors the quality of the received signal and, should the quality deteriorate for any reason, the radio will automatically retune to any better quality transmission which may be available.

The receiver may be programmed with the worldwide coverage information for two different services which are selected by a single control line. Programming of the receiver's database to define the transmissions available, together with their broadcast times and frequencies, is accomplished with a user-friendly database program incorporating a graphical user interface. This program runs on an IBM PC and, once completed, may be downloaded directly into the receiver, enabling any operator to adapt the receiver to meet the specific needs of passengers and routes. A trace program built into the receiver records the selected frequencies and received signal strengths, thus enabling fine tuning of the database to ensure optimum performance.

Specifications
Weight:
(antenna) 2.5 kg
(receiver) 1.8 kg
Power supply: 28 V DC, 500 mA max
Frequency: 150 kHz to 108 MHz
Channel spacing: 5 kHz and 9 kHz

Status
In service.

Contractor
Chelton (Electrostatics) Ltd.

STR Satellite TRansceiver

Type
Avionic communications system.

Description
The STR Satellite TRansceiver is a single channel satcom system which supports packet data services – Data-2 (ACARS, AFIS, A622 FANS) and Data-3 (ATN FANS, X-25 cabin) communications – over the worldwide Inmarsat satellite system.

Scorpio 2000 series datalink, with Honeywell RDR-1500B radar display   0011835

Thales Avionics' STR Satellite TRansceiver and omnidirectional antenna 0011829

The STR is an Inmarsat Aero 'L' Class 1 system, which meets the integrity and real-time communication requirements necessary to support air traffic services, both today and in the future, while simultaneously supporting operational, administrative and passenger datalink services.

Applications typically supported by the STR are:
- Operational/fleet management services: flight planning data, departure times, ETAs, position reports, manifest and engineering data
- Communications/Navigation Surveillance (CNS) services: FANS applications including: Automatic Dependent Surveillance (ADS), Controller to Pilot DataLink Communications (CPDLC) and Pre-Departure and Oceanic Clearance (PDC/OC) via A622 FANS, and via ATN FANS as a growth option
- Passenger and cabin management data services – X-25 based.

The STR utilises the proven technology of the Racal (now Thales)/Honeywell MCS 3000/6000 multichannel systems, reconfigured into two compact LRUs: a Satellite Data Unit/Radio Frequency Unit (SDU/RFU) and High-Power Amplifier (HPA). The system interfaces to an omnidirectional low-gain antenna and to a variety of cockpit and cabin communication management devices including: ACARS Mus/CMUs, AFIS, FMS, and In-Flight Entertainment (IFE).

Specifications
Data rate: 600, 1,200 bps
Dimensions: each unit; 2 MCU
Weight: each unit; <4.5 kg
Power: 115 V AC, 400 Hz, <275 V A

Status
In production and in service. Selected by Kongsberg Defence & Aerospace for the Modified-Automatic Dependent Surveillance (M-ADS) North Sea System. The STR also forms part of the Thales Avionics airborne real-time datalink fitted to UK Royal Air Force Sea King Mk 3A Search And Rescue (SAR) helicopters.

Contractor
Thales Avionics SA.

T618 Link 11 Data Terminal Set

Type
Avionic datalink system.

Description
The T618 Data Terminal Set (DTS) is a small lightweight unit suitable for helicopter, fixed-wing and other applications. It provides all Link 11 modem and network control functions defined by MIL-STD-188-203-1A, STANAG 5511 and ADat P-11 in picket and network control modes.

As a modem, the DTS converts the Link 11 data into audio tones suitable for transmission over radio circuits. As a network controller, it provides error corrections in USB, LSB and Diversity modes and roll call management functions in the net.

The T618 includes features which are of great importance to Link 11 operators. It provides an outlet to display the performance of every station in the network in real time; the link monitor displays the received signal quality on both upper and lower sidebands. The percentage of time each picket responds to interrogation and

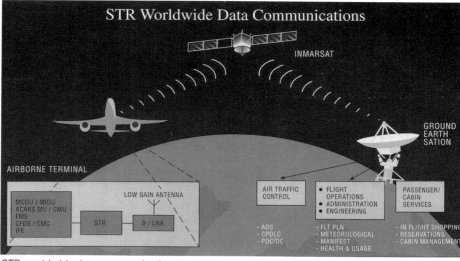

STR worldwide data communications 0011830

Ultra Electronics Link 11 datalink processor (left) and Data Terminal Set (right) 0011828

the analysis of the data being received is also displayed for each picket unit.

For the system maintainer, the T618 contains a BIT which isolates malfunctions to card level. In addition, various loopback modes provide signal paths which help to isolate difficulties in system configuration, including multiple stations. The combination of these controls and the link quality assessment provide a high visibility of the system operation.

The single-tone Link 11 system enhancement is available with the T618. Provided as a switchable option, it enables the DTS to be compatible with both single tone and conventional Link 11 modes.

The DTS is also programmable for operations in combinations of TDMA network protocols and frequency hopping.

The T618 DTS can be upgraded easily to provide NATO Improved Link 11 (NILE) performance.

Specifications
Dimensions: ¼ ATR short
Weight: 3.6 kg
Power supply: 115 V AC, <30 W
Interfaces: serial data ATDS, radio interfaces as defined by MIL-STD-188-203-1A
Computer control: ARINC 429, MIL-STD-1553B, RS-232C as options
Temperature range: −40 to +55°C
Reliability: 7,000 h MTBF (ARW environment)

Status
In serial production. Developed for, and in use on, the EH101 Merlin ASW helicopter; selected for the Nimrod MRA.4 aircraft.

Contractor
Ultra Electronics, Sonar and Communication Systems.

T619 Link 11 datalink processor

Type
Avionic datalink system.

Description
The T619 datalink processor acts as a Link 11 processor. It uses up-to-date technology to provide a small lightweight unit suitable for helicopter, fixed-wing and other applications.

The datalink processor interfaces with both STANAG 5511 and OPSEC 411. It provides database management, track correlation, gridlock, message assembly/disassembly, transmit and receive filtering, operator control and monitoring. All functions are automated where practical to reduce the operator workload to a minimum.

The unit consists of a single processor card and an interface card, both of which are VME-based. A spare slot is available to double processor throughput and memory capability or to provide additional functions such as to integrate a single card cryptographic device. For this purpose the datalink processor has the appropriate qualifications and is equipped with all the required controls on its front panel.

Also available is the A628 speech processor, which allows secure speech with active noise reduction and can be applied to LPC-10 vocoders.

Specifications
Dimensions: ½ ATR short
Weight: <10 kg
Power supply: 115 V AC, 40 Hz, <40 W
Interface: MIL-STD-1553B
Temperature range: −40 to +55°C
Reliability: 6,000 h MTBF (ARW environment)

Status

In serial production. In use on the EH101 Merlin ASW helicopter; selected for the Nimrod MRA.4 aircraft.

Contractor

Ultra Electronics, Sonar and Communication Systems.

Tactical Datalinks for Tanker Aircraft

Type

Avionic datalink system.

Description

Aerosystems International Tactical Communications Division has developed a range of high quality, innovative and cost effective datalink systems for a range of tactical aircraft, with emphasis on Commercial-Off-The-Shelf (COTS) software packages, facilitating short lead-times and precise tailoring of installations to meet customer requirements. One of these systems is an austere JTIDS capability for tanker aircraft, originally developed to meet an Urgent Operational Requirement (UOR) for Royal Air Force TriStar and VC10 tankers.

The operational benefits of JTIDS on Tankers have been proven during conflicts such as Bosnia and Kosovo.

- Tanker and refueling units have accurate information on relative position/location.
- Off load fuel state and Link 16 Precise Participant Location and Identification (PPLI) are transmitted by the Tanker.
- Tanker tow lines can be dynamically changed in response to tactical requirements.
- Tankers can act as relay platforms, transmitting between networks, overcoming line of sight issues.
- Tanker crews have accurate and up-to-date situation awareness, showing precise position of all the friendly and hostile forces, allowing Tankers to operate more effectively and safely.

Aerosystems International offer three installation options: Fully integrated with host aircraft avionics system, stand-alone, as an aircraft upgrade, or as a role fit for specific aircraft missions and roles.

The basic components of the system are:
- JTIDS terminal (which can use existing aircraft Tacan or IFF aerials)
- Processor control interface unit, which acts as the core of the system
- Display unit
- Keyboard
- GPS or aircraft navigation input (a lightweight GPS receiver is an option).

For a fully integrated system, an existing aircraft display unit is used, with processing software integrated with the aircraft computer system and the JTIDS terminal housed in a suitable avionics bay. For a stand-alone system, a dedicated aircraft display, processor and keyboard is provided according to aircraft layout. Navigation information can be either from the aircraft system or via an optional lightweight GPS receiver. For a role fit, the system is crated, which can then be seat-rail mounted and installed when required, typically in less than 2 hours.

The system can interface with a wide variety of terminal types; a MIDS terminal is available.

Specifications

Display (For non integrated versions):
Type: LCD active matrix
Screen dimensions: 211.2 × 158.4 mm
Colours: 262,144
Viewing angle: ±45° horizontal, +10/–30° vertical
Signal input: RGB, VGA or SVGA
Temperature range: –20 to +55°C
EMI/EMC: MIL-STD-461C, FCC-B
Dimensions: 297 × 241 × 140 mm
Power: 28 V DC, or 115 V AC/400 Hz

Processor Control Interface Unit (For non integrated versions):
Video: RGB
Interfaces: MIL-STD 1553B, ARINC-429, RS 422
Temperature range: –25 to +50°C
Cooling: Self-contained fan
EMI/EMC: MIL-STD-461C, FCC-B
Dimensions: 210 (W) × 274 (H) × 442 (D) mm
Weight: 11 kg
Power: 115 V AC/400 Hz

JTIDS Terminal (MIDS):
Interface: MIL-STD 1553B
Cooling: Conductive air
EMI/EMC: MIL-STD-461C, FCC-B

Aerosystems International austere JTIDS installation in a UK Royal Air Force Tristar tanker aircraft. Note the large format display to the right of the co-pilot's seat and control box resting on the flight engineer's table (Aerosystems International)   0099744

Dimensions: 193.5 × 190.5 × 343 mm
Weight: 20.5 kg
Power: 60 Hz AC or 400 Hz MIL-STD 704A
Output: 200 W

Status

The UK Ministry of Defence (MOD) raised an UOR in June 1999 stipulating a requirement to fit up to ten VC10 and TriStar aircraft with the Joint Tactical Information Distribution System (JTIDS). A contract to fit an initial batch of three TriStars (one KC1 and two K1s) and three VC10 MK3/4s was placed with Aerosystems International on the 11th June 1999. The fitting of JTIDS to two RAF TriStar aircraft and one VC10 was accomplished in three months. The first flight test of the system on board the aircraft was successfully completed at RAF Brize Norton in September and the RAF released the JTIDS system for operational service in early October 1999. All installations were complete as of 2002. The system remains in production.

Contractor

Aerosystems International, Tactical Communications Division.

United States

309M-1 Automatic Link Establishment (ALE)

Type

Avionic Communications Control System (CCS).

Description

Rockwell Collins' 309M-1 ALE processor provides for both enhanced communications reliability and operational simplicity. Compliant with new Automatic Link Establishment (ALE) interoperability standards, the 309M-1 includes a host of ALE features, which provide advanced system operational capabilities:

- FED-STD-1045 and MIL-STD-188- 141A (MS-141A) ALE and SELSCAN® interoperability (bilingual)
- Improved communications reliability
- 20 programmable and selectable scan lists
- Sounding and uni- or bi-directional Link Quality Analysis (LQA)
- Special LQA database algorithms reduce sounding requirements (adaptive sounding)
- Listen-before-transmit feature
- Automatic calls accepted while scanning, calling, or sounding
- Unique time-saving automatic foldback channel selection algorithms
- Return-to-Scan automatically via timer, over-the-air command, or manually
- Order wire data message capability via internal data modem – MS-141A AMD

Description	Width[1]	Height[1]	Depth[1]	Weight
ALE processor	124.5 (4.9)	198.1 (7.8)	320.01 (12.6)	5.5 kg (12 lb)
Mount (solid)	132.1 (5.2)	25.4 (1.0)	375.92 (14.8)	1.0 kg (2.0 lb)
Mount (isolated)	132.1 (5.2)	71.1 (2.8)	375.92 (14.8)	1.1 kg (2.3 lb)
Rack mount adapter mm (ft)	482.6 (19)	177.8 (7)	548.62 (21.6)	2.3 kg (5.0 lb)

- Unique conference calling and conversational data mode protocols
- 19 in rack mount adapter option available
- Built-in test (BIT).

The 309M-1 is designed for use with the HF-9000/ARC-217, 718U/ARC-174, ARC-190 and HF-80 systems, using HF-9012D, 514A-13 and 514A-12 controls.

Specifications

Dimensions
Processor: 124.5 × 198.1 × 320 mm
Mount (solid): 132.1 × 25.4 × 375.9 mm
Mount (isolated): 132.1 × 71.1 × 375.9 mm
Rack mount adapter: 482.6 × 177.8 × 548.6 mm

Weights
Processor: 5.5 kg
Mount (solid): 1.0 kg
Mount (isolated): 1.1 kg
Rack mount adapter: 2.3 kg

Status

In production and in service.

Contractor

Rockwell Collins.

7522 series VHF comm radio tuning panel

Type

Avionic Communications Control System (CCS).

Description

The AVTECH 7522 series VHF comm radio tuning panel updates ARINC 500 (analogue) and ARINC 700 (digital) series radios to provide the new 8.33 kHz frequency spacing standard, required in order to meet certain European ATC requirements after 1 January 1999. The new panel has two connectors: a 55-pin connector used exclusively for ARINC 500 series radios, and a 24-pin connector for ARINC 700 series radios. The 500 series radios are tuned with a modified 2 × 5 coding. The 700 series radios are tuned via an ARINC 429 databus. This approach allows airlines with mixed fleets (500 and 700 series radios) to stock only one type of tuning panel. The current range of 760 channels at 25 kHz spacing is increased to 2,280 channels at 8.33 kHz spacing.

Specifications

RTCA DO: 160C
TSO: C37d and C38d

Status

TSO approved September 1997. Selected by a number of airlines to meet new European standards.

FAA STC approval was granted in January 1998 for Rockwell Collins VHF-700B, VHF-900B (digital) and 618M-5 (analogue) radios with AVTECH's 7522-1-2 control panels.

Contractor

AVTECH Corporation.

Advanced Narrowband Digital Voice Terminal (ANDVT)

Type

Airborne communications system.

Description

The Advanced Narrowband Digital Voice Terminal (ANDVT) provides half-duplex secure voice and data communications for a variety of military tactical applications, including shipboard, land-based and airborne. Typical user modes include secure voice, data and signalling, point-to-point and modem processor only.

The CV-3591 Basic Terminal Unit (BTU) provides voice and modem processing by using two similar signal processors. Communications security is achieved by encoding and decoding the digital data to and from the voice processor external data device.

In the standard terminal configuration, the KYV-5 COMSEC Module (CM) is a front panel plug-in unit.

Specifications

Dimensions:
(BTU) 193.8 × 124.7 × 337.8 mm
(CM) 158.75 × 123.95 × 76.2 mm
(MPU/VPU) 157.5 × 124.7 × 75.4 mm
(interface unit) 146 × 69.85 × 285.75 mm
Weight:
(BTU) 9.89 kg
(CM) 1.63 kg
(MPU/VPU) 1.27 kg
(interface unit) 2.13 kg
Temperature range: –46 to +95°C
Altitude: up to 70,000 ft
Reliability: 2,000 h MTBF

Status

In production and in service.

Contractor

Raytheon Company, Space and Airborne Systems.

Aero-I SATCOM satellite communications system

Type

Avionic satellite communications system.

Description

The Honeywell/Thales Aero-I SATCOM satellite communications system provides operators with flexible access to the Inmarsat Aero-I spotbeam service for multichannel voice and data, as well as single-channel data in the continuous global coverage regions.

The three-channel systems consist of a Satellite Data Unit (SDU) and a High-Power Amplifier (HPA). MCS-3000 Aero-I system capability can be expanded to the six-channel MCS-6000 Aero-I version simply by adding a 4-MCU Radio Frequency Unit (RFU). The addition of the Intermediate Gain Antenna and a Diplexer/Low Noise Amplifier completes the system. External beam steering, required by Aero-H, is not necessary with Aero-I, since the new 20W HPAs contain this function.

Status

The Aero-I system complements the established Inmarsat Aero-H service, which has become a standard installation on long haul, wide-body aircraft since its introduction in 1990. All services

currently offered with Aero-H systems are available on Aero-I, including:
- Cockpit voice, allowing instantaneous communication with operations, maintenance and air traffic control
- Passenger telephony
- Passenger fax
- News and weather broadcasts
- Interactive passenger services
- PC data capability.

The Aero-I SATCOM system gained its Supplemental Type Certificate (STC) from the US Federal Aviation Administration on the Next-Generation Boeing 737-800 in November 1998, with launch customers Royal Air Maroc and Hainan Airlines; the system utilises ground stations of the Satellite Aircom Consortium, which were upgraded to provide worldwide coverage for the Aero-I service.

These certifications directly correspond with the Aero-I capabilities of Inmarsat ground earth stations. Inmarsat Aero-I service operates in the spot beams of the new-generation Inmarsat-3 satellites. Since spot beams have lower power requirements, smaller, lower power HPAs can be used and smaller Intermediate Gain Antennas can be installed for operation.

In addition to supporting passenger communications, Aero-I became the first compact SATCOM system to comply with the International Civil Aviation Organisation's (ICAO) Standards and Recommended Practices (SARPS), which form a comprehensive set of specifications for aeronautical satellite communications when used for cockpit and safety of flight applications, such as Air Traffic Control and Future Air Navigation System (FANS) technologies.

Status

In production and in service worldwide.

Contractor

Honeywell Inc, Commercial Aviation Systems.
Thales Avionics SA.

Aircraft intercoms

Type

Avionic internal communications system.

Description

Sigtronics Corporation makes a range of aircraft, panel-mount, intercom systems; -4 and -400 systems indicate 4-place versions; -6 and -600 systems indicate 6-place versions; the -800 is an 8-place variant:

SAS-440/-640 auto squelch panel-mounted intercom series
These intercoms virtually eliminate the need to constantly readjust the squelch during flight. The SAS-440 has radio priority to assure that the only voice heard by air traffic control is that of the crew. SAS-440 supports four headsets. SAS-640 supports six headsets.

SCI-4/SCI-6 dual-squelch voice activated intercoms
Retrofit system for SPA-400/SPA-600 versions, with 'pilot', 'crew' and 'all' functions.

SDB-800 dual-audio panel intercom
Effectively, a dual-SPA-400/-600 installation can support two pilots, and up to eight headsets.

SPA-400/600 intercom series
An industry standard intercom for many years, the SPA-400/600 has radio priority, and a pilot fail-safe feature, which ensures that the pilot will always hear the radios, even if the intercom is set to off.

SPA-400N/-600N intercom series
Specially designed version of the SPA-400/-600 series for very high noise cockpits; helicopters; warbirds and ultralights.

ST-400/-600 stereo intercom series
A full stereo version of the SPA-400/-600 series.

Sigtronics SAS-440 auto squelch panel-mounted intercom 0044831

Sigtronics SDB-800 intercom system 0044832

Sigtronics SPA-400 intercom system 0122440

Sigtronics SCI-S stereo intercom system
0122439

ST-440/-640 stereo intercom series
A full stereo version of the ST-440/-640 series

SCI-S4/-S6 four or six place stereo crew isolate intercom
The SCI-S4 and S6 intercoms have all the features of the SCI-4 and -6 models with the addition of dual-stereo music inputs.

SPA-4S two or four place stereo intercom
The SPA-4S is a voice-activated intercom with inputs for stereo music. It is designed for first time installation or as an upgrade to the SPA-400/600 systems.

Status

In production and in widespread service.

Contractor

Sigtronics Corporation.

AIRLINK antenna system for satcoms

Type

Aircraft antenna, communications.

Description

AIRLINK low- and high-gain antenna systems are designed for use with INMARSAT satellite communications.

The high-gain antenna system uses two conformal, electronically steered, phased-arrays in a side-mounted architecture. This configuration yields superior coverage with minimal aerodynamic drag penalties. The high-gain antenna system is fully approved for multichannel data, voice and data applications.

The low-gain antenna system consists of a single-blade antenna and is used for low-speed data applications. It is ideally suited as a back-up for the high-gain system.

The Ball AIRLINK antenna system showing the conformal array (above) and beam-steering unit (left), diplexer/low-noise amplifier (centre) and high-power amplifier (right) 0504023

Ball has recently introduced the AIRLINK Gateway Unit (AGU) which provides the digital signal processing necessary for operating with INMARSAT's circuit mode data channel. This channel provides users with a host of applications at the 9.6 kbits/s rate and can also provide secure satellite communications.

Specifications
Dimensions:
(antenna array) 407 × 813 × 9.5 mm
(beam-steering unit) 89 × 264 × 343 mm
(diplexer/low-noise amplifier) 51 × 198 × 282 mm
(high-power amplifier) 193 × 257 × 925 mm
(AIRLINK Gateway Unit) 7 MCU
Weight:
(antenna array) 7.1 kg
(beam-steering unit) 8.4 kg
(diplexer/low-noise amplifier) 3 kg
(high-power amplifier) 20 kg
(AIRLINK) Gateway Unit) 13.5 kg
Power supply: 115 V AC, 400 Hz, single phase
Frequency: 1,530–1,559 MHz, 1,626.5–1,660.5 MHz

Status
In service. Selected by United Airlines and British Airways for their Boeing 777 aircraft and by Scandinavian Airline System for its Boeing 767-300s. Also selected by the United States government for its VIP/SAM fleet.

Contractor
Ball Aerospace and Technologies Corp.

AIRSAT™ satcom systems

Type
Avionic communications system, satellite.

Description
Honeywell Aerospace, Electronic & Avionics Lighting has developed a group of AIRSAT satellite communications systems to provide worldwide telephone communication services to aircraft flight deck crew, cabin staff and passengers, via the 66-satellite Iridium network.

AIRSAT 1 satcom system
AIRSAT 1 provides single-channel worldwide telephone services; it comprises a single-channel 3 MCU-sized Iridium Transceiver Unit (ITU), a back-lit digital handset and a low-gain, top-mounted ARINC 761-type blade antenna. AIRSAT 1 is available for voice and/or data communication. The Aircraft Integration Unit (AIU) provides satellite communication through an audio panel that allows pilots to make and receive calls through their headsets and passengers to make calls from the cabin. Data services are available with direct internet connections at around 10 kbps. Using AIRSAT 1 as a data modem, connecting it

The Honeywell Aerospace AIRSAT™ 5 and 8 multichannel satcom system 0044813

to a computer via an RS 232 connection, gives a data throughput of about 2,400 bps.

AIRSAT 5 and 8 multichannel satcom system
AIRSAT 5 and AIRSAT 8 systems provide five-channel and eight-channel worldwide passenger/cockpit voice, fax and data communications; each comprises two 4 MCU-sized processor units, a diplexer and a blade antenna, all ARINC 761 compliant.

Specifications
Radio frequency: 1,616.0–1,626.5 MHz
Modulation: QPSK
Operation: full duplex
Dimensions/weight:
AIRSAT 1: ITU: 3 MCU less than 6.82 kg
AIRSAT 5 and 8: STU-105 (5-channel): 4 MCU/7.28 kg
STU-108 (8-channel): 4 MCU/8.18 kg
HPA-105/108: 4 MCU/9.09 kg
DNLA-105/108: 279.4 × 203.2 × 50.8 mm/2.27 kg
ANT-100: 108 × 279.4 × 120.7 mm/1.36 kg

Status
Marketed under the Bendix/King brand name. AIRSAT 1 was introduced during the second quarter of 1999. AIRSAT 5 and 8 were available from the fourth quarter of 1999.

In July 1999, Lockheed Martin Tactical Aircraft Systems and Honeywell Aerospace, Electronic & Avionics Lighting entered into a co-operative agreement to demonstrate two-way satellite communication on a tactical fighter using the Iridium satellite and the AIRSAT 1 system. This activity is part of the overall US Air Force effort to provide Real-Time Information into the Cockpit (RTIC).

Contractor
Bendix/King.

AM-7189A VHF/FM Improved power amplifier

Type
Avionic communications system, VHF.

Description
Rockwell's Nap of the Earth (NOE) subsystem enables standard VHF/FM or SINCGARS V radios to communicate with parity (range capability) compared to current ground-based VHF/FM radios with improved communications. The solid-state Improved VHF/FM (IFM) subsystem provides an Effective Radiated Power (ERP) of 50 W (nominal), which represents a 10 to 15 dB improvement over current systems, across the tactical communications band of 30 to 87.975 MHz. The IFM is available with MIL-STD-1553B capability.

Specifications
Frequency range: 30–87.975 MHz
Duty cycle: 1 min transmit/5 min receive
Nominal carrier power:
 Low: 3.5–7.0 W
 Normal: 12–19 W
 High: 40–56 W
VSWR: 4:1
Dimensions: 127 × 102 × 422 mm (H × W × D)
Weight: 4.7 kg
Power: 27.5 (±5) V DC, 200 W
Temperature: –40 to +71°C operating
Altitude: 4,500 m (operating)
Cooling: Convection

The Honeywell Aerospace AIRSAT™ 1 satcom system 0044812

Status
The IFM subsystem has been installed in Bell OH-58A/C, UH-1H, AH-1S and Boeing CH-47C/D helicopters. Installations have been planned for the UH-60A, AH-64 and AH-1P helicopters. To date, Rockwell has produced over 5,100 units for the US Army and Navy.

Contractor
Rockwell Collins.

Amplifier Control Indicator (ACI)

Type
Avionic Communications Control System (CCS).

Description
Designed for the US Navy F/A-18 E/F, the Telephonics ACI provides secure communications for onboard radios and the Mutlifunctional Information Distribution System (MIDS). The ACI controls the radio relay, IFF, ILS and cryptographic functions.

The ACI includes a voice activated switch (VOX) for crew interphone communications and extensive built-in testing that automatically checks system circuits and radio lines.

Status
Selected for the US Navy F/A-18 E/F programme, and for the Australian Hornet upgrade programme.

Contractor
Telephonics Corporation, Command Systems Division.

AN/AIC-29(V)1 intercommunication system

Type
Avionic internal communications system.

Description
The AN/AIC-29(V)1 intercommunication system provides secure internal communications between helicopter crew members and direct access between crew members and radios and/or security equipment for external communications. The system can accommodate up to 15 transmit and 18 receive radio channels and provides greater than 100 dB cross-talk isolation between a transmit and any other transmit or receive channel. The system consists of six crew station units, a single maintenance station unit and a communications switching unit. The crew station units are NVG-compatible. The communications switching unit performs switching and mixing of audio channels in accordance with digital data multiplexed from each crew station unit. The switching unit's response to the multiplexed data depends on the communication plan programmed into the system.

The system provides emergency back-up intercom and radio communication selection which bypasses the communications switching unit for audio transmit and receive functions.

System features include the capability for distributing audible alarms to crew stations in response to sensors/switches external to the intercommunication system. The system also provides interface with the MIL-STD-1553 databus, two-way chime call capability and built-in test circuits.

Status
In production and in service with the US Customs Service's fleet of P-3 aircraft for anti-drugs operations. Six shipsets were ordered in September 1999 for installation by Lockheed-Martin Aeronautical Systems in Greenville in 2000.

Contractor
Palomar Products Inc.

AN/AIC-32(V)1 intercommunication system

Type
Avionic internal communications system.

Description
The AN/AIC-32(V)1 intercommunication system provides secure internal communications between crew members and direct access to mission radios and security equipment for external communications. The system's primary units consist of a communication control unit, four flight deck crew station units, five mission area crew station units, eight maintenance station units and one maintenance control unit.

Audio traffic from all station units is controlled by the communication control unit during normal system operations. A back-up operating mode, initiated from selected crew station units, bypasses the communication control unit to provide hard-wired access to a set of predetermined radios. A back-up intercom network, integrated with the call function on all station units, is also provided for emergency intercommunication between crew members.

A test switch on each crew station unit permits preflight verification of audio and lamp indicator circuitry and digital interface with the communication control unit. A public address system, accessed from the flight crew station, allows announcements over the loudspeakers and headsets. Auxiliary control units, located at selected mission area crew station units, expands the total system direct access capability to 30 transmit and 36 receive channels.

Status
In production and in service.

Contractor
Palomar Products Inc.

AN/AIC-34(V)1 and (V)2 intercommunication systems

Type
Avionic internal communications system.

Description
The AN/AIC-34(V)1 secure intercommunication system is a programmable microprocessor-contolled modular audio and digital communication distribution system designed for use on board fixed- and rotary-wing aircraft or in ground-based C³ shelters. The AN/AIC-34(V)1 provides internal communications between crew members as well as crew member access to mission radios and communication security equipment for external communications. Designed in accordance with MIL-E-5400, it is qualified to MIL-STD-810 for air transport, MIL-STD-461 for EMI and NACSIM 5100 for TEMPEST compliance.

The AN/AIC-34(V)1 intercommunication system comprises a Communication Control Unit (CCU), five Crew Station Units (CSU), 16 Control Display Units (CDU), an Emergency Audio Panel (EAP), three Maintenance Station Units (MSU) and 26 jack boxes.

The CCU is a microprocessor-controlled modular audio switching and control unit that provides up to 32 crew stations with access to as many as 48 receive channels and 30 transmit channels. It also accepts up to 70 binary discrete inputs and provides up to 122 switched relay outputs for discrete control and crypto switching.

The CSUs and CDUs are the crew member interfaces to the communication control unit and communication assets. The functions of channel select switches and volume controls on the CSU front panel are firmware dependent and are reconfigurable to meet the needs of specific missions or system requirements. The CDU allows operator assignment of communication assets and displays communication system operational and BIT status on an integral eight-colour CRT display. Both units are capable

of operating with two headsets when operated in a monaural mode or can operate with one headset when operated in a binaural mode. Monaural or binaural operating modes are configured by jumpers at the unit connectors in the aircraft wiring.

The MSU provides intercom access for maintenance personnel and ground crew. Each communication control, control display and maintenance station unit is connected to a jack box. This allows switching of multiple microphone inputs and provides for switching of audio from an auxiliary source or attached unit to various headset or speaker interfaces.

The EAP provides an operator with the capability to select multiple levels of degraded or emergency back-up modes of operation to provide for continued operation in the event of hardware failures or battle damage.

The AN/AIC-34(V)2 secure intercommunication set is a subset of the AN/AIC-34(V)1 system, and is used in airborne military applications where the crew complement is significantly less than that of the AIC-34(V)1 but the requirement to access numerous radio channels and the need for the ability to reconfigure the operating modes of the onboard communication assets are similar. The AN/AIC-34(V)2 intercommunication set is designed to be compliant with the requirements of MIL-E-5400, MIL-STD-810, MIL-STD-461 and NACSIM 5100.

The set comprises a CCU, five CSUs, two CDUs, an EAP, three MSUs and eight jack boxes.

The CCU is the same as that used in the AN/AIC-34(V)1 except that it contains an additional set of relay circuit cards to provide for greater flexibility and expanded switching modes of the communication assets. The firmware resident in this CCU is also unique to the requirements of the AN/AIC-34(V)2.

The CSU, CDU, MSU, EAP and jack boxes are identical to those used in the AN/AIC-34(V)1 except for the CSU front panel switch legends.

Specifications
Dimensions:
(CCU) 497.6 × 257 × 193.7 mm
(CSU) 165 × 146 × 152.4 mm
(CDU) 165.1 × 146 × 152.4 mm
(EAP) 106.7 × 146 × 47.8 mm
(MSU) 78.7 × 160 × 107.9 mm
(jack box) 150.5 × 127.5 × 43.2 mm
Weight:
(CCU) 19.8 kg
(CSU) 2.7 kg
(CDU) 3.6 kg
(EAP) 0.45 kg
(MSU) 1.03 kg
(jack box) 0.34 kg

Status
In service with the US Navy.

Contractor
Palomar Products Inc.

AN/AIC-38(V)1 and AN/AIC-40(V)1 intercommunication systems

Type
Avionic internal communications system.

Description
The AN/AIC-38(V)1 and AN/AIC-40(V)1 secure intercommunication systems are programmable microprocessor-controlled modular audio and digital communication distribution systems designed for use in both fixed- and rotary-wing aircraft or in ground-based C³ shelters. The AN/AIC-38(V)1 and AN/AIC-40(V)1 provide internal communications between crew members as well as access to mission radios and communications security equipment for external communications. Designed in accordance with MIL-E-5400, the systems are qualified to MIL-STD-810D for air transport, MIL-STD-461 for EMI and NACSIM 5100 for TEMPEST.

The systems consist of a Communication Control Unit (CCU), Crew Station Units (CSU), a Digital Switch Unit (DSU) and Emergency Audio Panel (EAP).

The CCU is a microprocessor-controlled modular audio switching and control unit that provides up to 32 crew stations with access to up to 48 receive channels and 30 transmit channels. It also accepts up to 70 binary discrete inputs and provides up to 122 switched relay outputs for discrete control and crypto switching. The communications connectivity plan resident within the CCU may be reconfigured in real time by a higher-order controller via MIL-STD-1553B databus, or by operator control via a CSU.

The CSU provides crew member interface to communication assets. Channel select keyswitches and volume controls on the CSU are firmware controlled and are reconfigurable to meet the needs of specific missions or system requirements.

The DSU, under CCU control, routes eight bidirectional channels of digital data and control lines to data terminals and radios or communication security equipment. Using the EAP, multiple levels of emergency back-up operation are operator selectable for continued operation after hardware failures or battle damage.

Status
In production and in service.

Contractor
Palomar Products Inc.

AN/AIC-39(V)1 intercommunication system

Type
Avionic internal communications system.

Description
The AN/AIC-39(V)1 secure intercommunication system is a programmable microprocessor-controlled modular audio and digital communication distribution system designed for use on board both fixed-wing aircraft and helicopters or in ground-based C³ shelters. The AN/AIC-39(V)1 provides internal communications between crew members as well as crew member access to mission radios and communications security equipment for external communications. Designed in accordance with MIL-E-5400, the AN/AIC-39(V)1 is qualified to MIL-T-5422 and MIL-STD-810D for air transport, MIL-STD-461 for EMI and NACSIM 5100 for TEMPEST.

The system consists of a Communication Control Unit (CCU), Crew Station Units (CSU), Dual Crew Station Units (DCSU), Audio Amplifier Units (AAU), Auxiliary Control Units (ACU), an Emergency Audio Panel (EAP) and secure jack box.

The CCU is a microprocessor-controlled modular audio switching and control unit that provides up to 32 crew stations with access to up to 48 receive channels and 30 transmit channels. It also accepts up to 70 binary discrete inputs and provides up to 122 switched relay outputs for discrete control and crypto switching. The communications connectivity plan resident within the CCU may be reconfigured in real time by a higher-order controller, via MIL-STD-1553B databus, or by operator control via a CSU or DCSU.

The CSU and DCSU provide crew member interface to communication assets. Channel select keyswitched and volume controls on the CSU and DCSU are firmware controlled and are reconfigurable to meet the needs of specific missions or system requirements. DCSUs are identical to CSUs except that the DCSUs are equipped with additional audio circuitry to support a subordinate ACU.

The ACU, used in conjunction with the DCSU, provides limited access to communication assets at remote crew positions. The ACU operates with a subset of the functions provided by the DCSU.

Crew station unit for the AN/AIC-34, AN/AIC-38, AN/AIC-39 and AN/AIC-40 intercommunication sets 0504036

The AAUs provide a binaural headset and microphone interface to the CCU, but have no annunciators or indicators.

The EAP allows operator selection of multiple levels of emergency back-up operation. The back-up capability permits continued operation after hardware failures or battle damage.

Status
In production and in service.

Contractor
Palomar Products Inc.

AN/ARC-22XX HF radios

Type
Avionic communications system, HF.

Description
In June 1999, the UK MoD awarded Rockwell Collins (UK) Ltd a firm-price contract of USD17 million for the replacement of the UK Royal Air Force's E-3D AWACS aircraft HF radiosystems.

The AN/ARC22XX radio features Automatic Link Establishment (ALE), an embedded multimode modem with advanced waveforms, and custom-frequency management/conflict software to enhance mission performance and reduce operator workload.

Each aircraft system comprises triple-fit, 400 W, HF transceivers, controlled over a MIL-STD-1553B serial bush using Control Display Units (CDUs).

Status
In service. All seven UK E-3D aircraft have been upgraded with the AN/ARC-22XX HF radio.

Contractor
Rockwell Collins.

AN/ARC-164 UHF radio

Type
Airborne communications system.

Description
The ARC-164 is the basic member of a family of radio communications equipment and subvariants, each designed for particular applications yet with a high degree of commonality.

The basic ARC-164 covers the UHF band, providing 7,000 channels over the range 225 to 400 MHz in 25 kHz increments. Any 20 channels may be preselected.

A fully solid-state system, the ARC-164 is distinguished by its 'slice' module construction in which a series of modules, connected by a flexible harness, are simply bolted together to form the desired electronic configuration. A typical simple system would comprise transmitter, receiver, guard receiver and synthesiser. The control unit may either form part of this consolidated package or be remotely located. The modular

The ARC-164 (seen on the right-hand side of the cockpit below the main instrument panel) installed in the Hawk aircraft 0504038

approach adopted allows growth capability, extra modules being added as required. A range of optional facilities, such as data transmission, secure speech and ECCM capability, is available by the addition of the appropriate slices.

A number of directly connected or remote-control units are produced for the ARC-164. These include: a simple frequency-selection controller; a 32-channel preset control with LED readout of the selected channel; a 20-channel preset unit with provision for two-cockpit take-control; a microprocessor controller with 400 UHF and VHF, AM or FM, preset channels, liquid crystal channel and frequency readout display and the capability of controlling up to four systems simultaneously. Additional remote frequency/channel indicators are available. Also available is a variety of mounting trays to suit differing installations for new types of aircraft and for the updating of older aircraft equipment.

A remote ARC-164 radio compatible with MIL-STD-1553B databus operation has been developed under contract to the US Army and is in production and service. Panel-mounted radios and certain controls can be furnished with ANVIS Green A lighting compatible with NVG in accordance with MIL-STD-85762. These features can also be obtained by retrofitting appropriate radios and controls. A Low Probability of Intercept and Detection (LPI/LPD) version of the ARC-164, which allows covert UHF communications, is known as StealthComm.

StealthComm waveform features include a hybrid of several LPI/LPD techniques: hybrid direct sequence and frequency hopping; feature suppression modulation techniques; receive sensitivity enhancements; 60 dB of power control.

StealthComm has enhanced data capability from 16 to 80 to 120 kbytes.

All StealthComm features can be incorporated in existing ARC-164 radios through mod kits; or new ARC-164/LPD radios can be installed as direct replacements form-fit for existing ARC-164s.

Specifications
Dimensions:
(transmitter/receiver)
(10 W version) 1/2 ATR × 178 mm
(30 W version) 1/2 ATR × 374 mm
(controller) 1/2 ATR × 83 mm
Weight:
(transmitter/receiver)
(10 W version) 3.7 kg
(30 W version) 6.8 kg
(controller) 2 kg
Frequency: 225-400 MHz
Channels: 7,000
Channel spacing: 25 kHz
Power output: 10 W (standard), 30 W (uprated)
Reliability: 2,000 h MTBF demonstrated

Status
In production and service. More than 65,000 ARC-164s have been produced to date.

The ARC-164 UHF radio 0504037

The system has become standard fit for a wide range of US Air Force aircraft including the F-15 and F-16 fighter aircraft.

The ARC-164 equips the Royal Navy's Sea King helicopters and Sea Harrier aircraft and the Hawk, Jaguar and other Royal Air Force aircraft. Further Hawk aircraft, notably those delivered to Kenya, are also fitted with the ARC-164, as are some Strikemaster strike/trainers.

Contractor
Raytheon Company, Space and Airborne Systems.

AN/ARC-186, ARC-186R and VHF-186R series VHF AM/FM digital transceivers

Type
Avionic communications system, VHF.

Description
The Rockwell Collins ARC-186R and VHF-186R series of VHF transceivers offers digital, form/fit/function replacements for the AN/ARC-186 multiband airborne radios. The Digital Signal Processing (DSP) software-based ARC-186R transceiver provides dual-band VHF operational capability in a single, miniature package. The ARC-186R and VHF-186R transceivers comply with the latest European 8.33 kHz and FM interference ATC standards. Both combine VHF/AM and VHF/FM transmission and reception capabilities into a single compact unit, covering the 30–88 and 108–152 MHz bands. 8.33 kHz spacing will be provided in the 116–137 MHz ATC band. Standard voice and data communications in either band are possible in accordance with international standards and conventions. The radios are panel mounted (RT-1002), MIL-STD-1553 (RT-1001), or RS-422 serial remote (RT-1000). The optional half-size remote control (RC-1000) may be used with a remoted RT-1000 radio.

The pilot may select a channel in either band manually or as a preset without regard to the band. Optionally, the radios may be programmed to restrict access to allowed bands or presets.

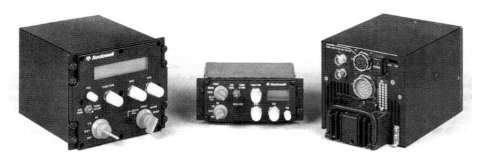

Rockwell Collins ARC-186R series VHF AM/FM digital transceivers   0044848

The ARC-186R features Digital Signal Processing (DSP) technology, allowing the operational capabilities of the radio to be updated or modified by downloading software without disassembling the radio.

The ARC-186R is an all solid-state VHF radio transceiver covering the 30 to 88 MHz and 108 to 152 MHz bands.

The single-piece transceiver is designed for airborne use, either in a panel mounted or remote-controlled configuration. The ARC-186R is a direct form/fit/function retrofit for the proven ARC-186 transceiver.

The ARC-186R will comply with the current European standards for FM overload.

The transmitter utilises a proven, on-channel linear AM modulator for distortion-free, highly intelligible AM transmissions. For FM transmissions, the ARC-186R utilises the new Rockwell Collins Direct Conversion Transmitter (DCT) technique, employing a special vector modulation integrated circuit to synthesise the carrier-frequency FM waveform from the DSP digital inputs. The software-based transmit modes/waveforms/modulations are also field programmable.

The panel mounted ARC-186R (RT-1002) features a digital display (night vision compatible version available) for frequency and radio functions, but retains the proven ARC-186 basic layout, controls and functionality familiar to aircrews worldwide. Similar control layout and functionality minimises operator training needed to transition from the ARC-186 to the ARC-186R. No installation training is required because the ARC-186R directly retrofits the ARC-186.

An optional half-size remote-control head (RC-1000) provides functionality identical to the full-size ARC-186R front panel in half the panel space.

Presets and field programming of options are accessed via a front panel fill port from any RS-232 equipped computer and a special programming cable.

Specifications
Frequency range: 30–87.975 MHz FM
  108-115.99167 MHz AM (receive only)
  116-151.99167 MHz AM
Channel spacing: 25 kHz (30–87.975 MHz)
  8.33, 25 kHz (108-152 MHz) (field programmable)
Emergency presets: 40.5 MHz FM; 121.5 MHz AM
  (customer programmable)
Channel presets: 100 non-volatile (customer programmable)
Carrier power: 10 W min, 16 W max (AM/FM, 52 ohms)
Dimensions (D × W × H): Panel RT
  241.3 × 146.1 × 123.8 mm
  Remote RT 241.3 × 127.0 × 120.7 mm
  Remote control 126.5 × 146.1 × 57.2 mm
Weight: Panel RT 3.29 kg max
  Remote RT 3.18 kg max
  Remote control 1.14 kg

Status
In production and in service.

In December 1999, Rockwell Collins was contracted to upgrade 183 AN/ARC-186 transceivers to the VHF-186R standard for the Royal Netherlands Air Force F-16 aircraft and cougar and Chinook helicopters.

Contractor
Rockwell Collins.

AN/ARC-186(V)/VHF-186 VHF AM/FM radio

Type
Avionic communications system, VHF.

Description
The Rockwell Collins AN/ARC-186(V)/VHF-186 is a tactical VHF AM/FM radio communications system designed for all types of military aircraft.

The basic ARC-186(V) is a solid-state 10 W system of modular construction which provides 4,080 channels at 25 kHz spacing. The 1,760 FM channels are contained in the 30 to 88 MHz band and 2,320 AM channels within the range 108 to 152 MHz. A secure speech facility can be used in both AM and FM modes and the equipment is compatible with either 16 or 18 kbit secure systems in diphase and baseband operation.

Up to 20 channels may be programmed for preselection on the ground or in the air. Preselection is accomplished through incorporation of a non-volatile NMOS memory which continues to retain data in the event of a loss of power supply. Two dedicated channel selector switch positions cover the FM and AM emergency channel frequencies of 40.5 and 121.5 MHz respectively.

Either panel mounting with direct control through an integral controller or remote mounting with an identical control panel presentation is possible. A half-size remote controller which contains the same control functions is also available and a typical configuration in a two-seat aircraft would comprise a full panel mount in the pilot's cockpit with a half-size controller at the crew position. In these dual-control configurations, a manual take-control switch provides full communications control for either crew member.

Conversion from panel to remote control is made by removing the panel controller and replacing it with a plug-in serial control receiver module. A typical conversion is said to require less than 5 minutes. Frequency displays on both types of controller are immune to fade-out during periods of low voltage.

Circuitry of the ARC-186(V) is of modular design. Seven module cards are held in place by the body chassis or card cage and are electrically interconnected by a planar card in which all hard wiring has been virtually eliminated. All radio frequency lines in the interconnecting planar card have been buried to minimise electromagnetic interference. Individual module cards are readily removable and may be replaced in the field to reduce fault finding and repair time.

The Rockwell Collins AN/ARC-186 military VHF system in both direct and remote-control configurations   0504024

Current options for the ARC-186(V) include AM/FM homing facility, but growth capability has been designed into the equipment from the outset and possible future developments could include SELCAL, burst data and target hand-off. One present simple modification, carried out by replacement of the decoder module in the remote transceiver, permits the radio to be directly connected to a MIL-STD-1553 digital databus and it is claimed that the system will be equally compatible with avionics suites of future generation equipment. An additional possibility is the uprating of transmitter output power.

US Air Force testing has demonstrated a MTBF in excess of 9,000 hours.

A principal design objective for the ARC-186 series equipment was that it should be capable of easy retrofit in existing installations. Since the system is considerably smaller than the equipment it is designed to replace, this is accomplished by use of plug-in adaptor trays which permit rapid replacement without disturbance to existing aircraft wiring harnesses.

Specifications
Dimensions:
(remote-mounted transmitter/receiver)
  127 × 165 × 123 mm
(panel-mounted transmitter/receiver)
  146 × 165 × 123 mm
(half-size remote control) 146 × 95 × 57 mm
Weight:
(transmitter/receiver) 2.95 kg
(remote-control) 0.79 kg
(FM homing module) 0.45 kg

Status
Production of the AN/ARC-186(V) has ceased although in-service support is maintained. More than 28,000 sets have been delivered worldwide and it was selected by the US Air Force as the standard equipment for all aircraft requiring VHF AM/FM capability.

Rockwell Collins upgraded 183 AN/ARC-186 transceivers to VHF-186R configuration for the Royal Netherlands Air Force. The upgraded transceivers meet International Civil Aviation Organisation (ICAO) requirements for VHF radios to operate with 8.33 kHz channel separation and FM immunity in European airspace. The upgraded radios are also fitted to the F-16 aircraft, and Cougar and CH-47D helicopters.

Contractor
Rockwell Collins.

AN/ARC-187 UHF radio

Type
Airborne communications system.

Description
The ARC-187 is a further development of the company's ARC-164 US Air Force standard system which has been adapted to meet a US Navy requirement for a low-cost terminal designed to operate with communication satellites, principally the US Navy's FltSatCom. It covers the UHF band from 225 to 400 MHz, in which range it provides

7,000 channels at increments of 25 kHz. Up to 20 channels may be preselected.

Operating modes include AM with a secure speech facility and FM and FSK data transmission in both analogue and digital form. ECCM capability is incorporated internally in the receiver/transmitter.

The system, which is remotely controlled, uses standard ARC-164 'slice' modules including a modified synthesiser section designed for compatibility with communications satellite data rate requirements.

A new control for the ARC-187 has been developed, which provides for compatibility with Satcom and MIL-STD-1553B databus modes of operation. This control incorporates ANVIS Green A lighting in accordance with MIL-STD-85762 for compatibility with Gen III NVG. A Satcom modem for the ARC-187 is under development.

Specifications
Dimensions:
(controller) 132 × 147 × 124 mm
(transmitter/receiver) 440 × 153 × 143 mm
Weight:
(controller) 2 kg
(transmitter/receiver) 7.4 kg
Frequency: 225–400 MHz
Channels: 7,000
Channel spacing: 25 kHz
Power output: 30 W (AM), 100 W (FM/FSK)

Status
In service in US Navy P-3C aircraft.

Contractor
Raytheon Company, Space and Airborne Systems.

AN/ARC-190(V) airborne communication system

Type
Avionic communications system, HF.

Description
Rockwell Collins' AN/ARC-190(V) is a military HF transmitter/receiver designed as a replacement for a number of earlier HF systems in a US Air Force modernisation programme. The system is therefore particularly suited to retrofit applications as well as for installation as original equipment in a wide range of aircraft. The AN/ARC-190 is the mainstay of HF communications in the US Air Force, having been installed in a large variety of fixed- and rotary-wing aircraft, such as the C-130, KC-135, C-141, C-5, C-9, KC-10, B-1, B-52, C-17, F-15, F-16, H-60 and S-2T. As well as retrofit kits for 618T systems, a MIL-STD-1553 system is available. The system can be used in either 1553 or non-1553 applications. A selective calling (SELCAL) AM detector is available in some versions of the ARC-190.

The ARC-190 is automatically tuned in both receive and transmit modes. Built-In Test

Equipment (BITE) and modular construction provide for rapid fault isolation to the box and module level for quick repair. The unit covers the 2 to 30 MHz band, providing 280,000 channels, any 30 of which are preselectable, in incremental steps of 100 kHz. Operational modes include USB, LSB, AME and CW. Data transmission facilities are also available in USB and LSB modes and in these modes the system is able to operate with audio-frequency shift keying or multitone modems.

The system is remotely controlled and dual control of the radio is possible from two crew stations. Serial data control is applied between each of the major units in the system and this is said to render it adaptable to future requirements such as SELCAL and remote frequency management. Transmitter power output is 400 W. The ARC-190 is supplied with AC or DC interface power, as well as MIL-STD-1553 control scheme.

Construction of the ARC-190(V) is all-solid-state. The full system includes an antenna coupler, which ensures compatibility with military cap, shunt, wire, whip or probe antenna systems. An F-1535 bandpass filter unit is available to provide added selectivity and overload protection for improved receiver performance in a strong signal environment. In the transmit mode it provides additional filtering to the exciter RF output. The system operates to pressure altitudes up to 70,000 ft over temperatures ranging from −55 to +71°C.

Rockwell Collins supplies the US Air Force with an automatic communications processor for the ARC-190(V) which automatically selects the optimum HF frequency after the operator has selected the station to be called. The processor also features anti-jam modes. After test and evaluation, production of the processor commenced in 1988.

High Frequency Data Link (HFDL) upgrade
The HFDL mode provides worldwide automated-data communications directly between aircraft systems and the ARINC ground network. HFDL is part of the Aeronautical Telecommunication Network (ATN) and supports a wide variety of data applications. The HFDL configuration of the AN/ARC-190 consists of an RT-1341(V)8 receiver transmitter and a CP-2024(MIL Nomenclature TBD) Automatic Communications Processor (ACP). Upgrade kits are available for both the RT and processor to upgrade prior versions of these equipments to the HFDL configuration.

Specifications
Dimensions:
(controller) 114 × 146 × 67 mm
(transmitter/receiver) 484 × 257 × 194 mm
(antenna coupler) 545 × 211 × 188 mm
(ACP) 198.6 × 121.9 × 495.8 mm
(ACP Controller) 66.5 × 146 × 106.7 mm
Weight:
(controller) 0.68 kg
(transmitter/receiver) 23.60 kg
(antenna coupler) 10.89 kg
(ACP) 9.53 kg
(ACP Controller) 1.81 kg

Status
In service in a wide variety of US Air Force and US Navy aircraft.

Contractor
Rockwell Collins.

AN/ARC-195 VHF radio

Type
Airborne communications system.

Description
The AN/ARC-195 radio may be considered as a variant of the UHF AN/ARC-164, with which it has a component commonality of 93 per cent. It is available in 10 and 30 W output versions. The major differences between the two systems consist of some component value changes and the substitution in the AN/ARC-195 of a synthesiser with a frequency standard appropriate to the VHF section of the radio frequency spectrum.

The AN/ARC-195 covers the VHF band from 116 to 156 MHz, providing 1,750 channels at a frequency separation of 25 kHz. In other respects it is almost identical to its UHF counterpart. Control units are also almost identical and certain units from the AN/ARC-164 range of controllers may be used for combined UHF/VHF operation.

Status
In production and in service.

Contractor
Raytheon Company, Space and Airborne Systems.

AN/ARC-201 VHF/FM transceiver (SINCGARS-V)

Type
Avionic communications system, VHF, secure.

Description
The AN/ARC-201 SINCGARS-V is an airborne VHF/FM frequency-hopping radio and is an all-solid-state equipment for use in helicopters, light observation aircraft and Airborne Command and Control (ACC) aircraft like the C-130. It provides single-channel and frequency-hopping modes. A six-channel non-volatile preset memory is incorporated for single-channel and ECCM modes. An interface and controls for the AM-7189A/ARC 50 W amplifier are incorporated.

The AN/ARC-201 is only available in the MIL-STD-1553B multiplex bus-remote configuration. It is interoperable with the current VHF/FM radios in the single-channel mode, and with the Single Channel Ground and Airborne Radio Subsystem (SINCGARS) VRC-87 to VRC-92 and manpack PRC-119 ground radios in the frequency-hopping mode. Electroluminescent lighting is provided on the front panel, compatible with the use of NVGs.

The radio has module and component commonality with the ground SINCGARS communication equipment and makes extensive use of LSI circuitry and DSP microprocessors for high reliability. A BIT function isolates faults to the module level with 90 per cent confidence. A data-rate adaptor is integrated in the radio to interface with data devices for communication. An automatic single-channel cueing capability in the ECCM modes allows a single-channel user to alert members of an ECCM net. Internal and external COMSEC can be used to provide secure communications in voice and data modes.

In 1998, the AN/ARC-201D radio was introduced as a MIL-STD-1553 bus-controlled unit. The aircraft integrator supplies the audio intercom, antennas, mounting and cabling. The integration of COMSEC and the Data Rate Adaptor combines three line replaceable units into one and reduces overall weight of the aircraft. Additional features such as improved error correction, enhanced data modes (including packet data), more flexible remote control, and GPS capability enable the radio to assume a number of roles supporting future digital battlefield requirements. The

The Rockwell Collins AN/ARC-190(V) HF radio   0504226

The ITT Aerospace/Communications Division RT-1478D/ARC-201D VHF/FM transceiver 0079235

ARC-201D is backward compatible with existing SINCGARS data modes.

Specifications

Dimensions:
(C-11466) 146 × 76 × 132 mm
(RT-1478) 127 × 102 × 259 mm
Weight: (RT-1478) 3.1 kg
Power output: 10 W
Frequency: 30–87.975 MHz
Channel spacing: 25 kHz
Channels: 2,320

Status

In production and in service in a variety of combat helicopters, including WAH-64 Apaches of the British Army.

Contractor

ITT Aerospace/Communications Division.

AN/ARC-210(V) multimode integrated communications system

Type

Avionic communications system, VHF, satellite, data, secure.

Description

The Rockwell Collins AN/ARC-210(V) multimode integrated communications system was derived from the AN/ARC-182 system to provide multimode voice and data communications in either normal or-jam-resistant modes through software reconfiguration. The RT-1556 transceiver is capable of establishing two-way communication links over the 30 to 400 MHz frequency range within tactical aircraft environments. There are currently nine different variants of the transceiver, but a great deal of commonality is retained between these.

The ARC-210 meets the 8.33 kHz European ATC channel spacing requirements.

Rockwell Collins has produced the RT-1794©/ARC-210 which incorporates all the features of the basic ARC-210, plus embedded Satcom/DAMA, COMSEC (KG-84, KY-58, ANDVT, RGV-11), CTIC,-MIL-STD-188-220A JVMFa, Link 4A and is software reprogrammable via the Advanced Memory Loader Verifier (AMLV) feature, also embedded.

The transceiver is the nucleus of the multimode communication system which includes an appliqué for Have Quick, Have Quick II and SINCGARS-V waveforms. In addition, the AN/ARC-210 has been demonstrated to provide Have Quick IIA ECCM and for Link 4A and Link II data communications. The system will also provide Satcom wideband and narrowband operation. Integrated with the AM-7525/-7526 UHF high-power amplifier and MX-11641 low-noise amplifier/diplexer, the ARC-210 provides a flexible satcom terminal and complies with MIL-STD-188-181/182/183 DAMA requirements. Maritime, land-mobile, ATC and ADF are included modes of operation. The system can be controlled by a MIL-STD-1553B databus and includes two types of remote controller for manual operation (C-1189XA/ARC and C-12419A/AERC half-size units with non-embedded DAMA RTs and the full-size C-12561/ARC remote control unit, with RT-1795©/ARC interfaces). A remote indicator and a family of broadband and electronically tunable antennas are also included.

Over time, the system has grown in capability to include Have Quick IIA and SATURN, with further improvements including a SINCGARS Improvement Program (SIP) data rate adaptor, USA ATC DAMA (digital voice) and Differential GPS datalink.

Specifications

Dimensions: 127 × 142.2 × 248.9 mm
Weight: 5.44 kg

Status

Rockwell Collins was placed on contract on 16 August 1995 to incorporate the Secretary of Defence's acquisition streamlining initiative into the ARC-210 communications system.

The ARC-210 is in production and in service in US Navy (F/A-18 C/D/E/F, AH-1W, AV-8B, C-5, C-17, CH-46E, C/MH-53E, KC-130 F/R/T, MV-22, UH-1N, EA-6B), US Air Force, US Marine Corps, US Army and US Air National Guard platforms plus Australian, Canadian, Finnish, Italian, Spanish, Swiss and Taiwanese forces.

A fully functional auxiliary remote control, the C-12571/ARC-210 is available, or the system may be integrated into aircraft Control Display Units (CDUs) via-MIL-STD-1553B databus.

The AN/ARC-210 was also part of the US Air Force C/KC-135 Global Air Traffic Management (GATM) programme. In February 2000, the US Air Force contracted Rockwell Collins to demonstrate complete end-to-end commercial Airline Operational Control (AOC) datalink and Very High Frequency DataLink (VDL) Mode 2 capability on C/KC-135 aircraft as part of the GATM upgrade programme. A VDL Mode 2 enhancement adds a datalink module to the existing Rockwell Collins AN/ARC-210 V/UHF radio and upgrades the CMU-900 software. The datalink module provides VDL Mode 2 functionality and ARINC 750 compliant interfaces to the dual CMU-900s, and provides the necessary hardware to support VDL Mode 3.

In November 2003 Rockwell announced that AN/ARC-210 sales had exceeded USD1 billion. At that time, more than 14,000 radios had been fielded worldwide, equipping over 135 military aircraft types.

Contractor

Rockwell Collins.

AN/ARC-217(V) HF system

Type

Avionic communications system, HF.

Description

The AN/ARC-217(V) (see also HF-9000 entry) is an evolved family of HF systems designed to meet the communications requirements of aircraft ranging from helicopters to military fighters.

The radio was the first Rockwell HF system to use fibre optics. The radio also employs microprocessors, microchip components, digital synthesisers and digital antenna couplers in a fully modular design. Optionally, an embedded or external Automatic Link Establishment (ALE) and embedded ECCM are available.

The ARC-217 system includes the C-12174 series controls, RT-1651 series receiver/transmitters and either the CU-2477 unpressurised or

The Rockwell Collins AN/ARC-210(V) communications system 0504281

The AN/ARC-217(V) HF transceiver is designed for nap of the earth communications. It consists of (left to right): control unit, receiver/transmitter and antenna coupler   0504025

CU-2478/CU-2476 pressurised antenna couplers. All control and status information is transferred via fibre optic cable between the control, receiver/transmitter and antenna coupler. The system can be operated in simplex or half-duplex modes from 2 to 30 MHz in 100 Hz increments.

A non-volatile memory stores operator programmable channels, with each memory channel able to store separate receive and transmit modes and frequencies. Additionally, 249 pre-programmed ITU and six emergency channels are available. The transmitter/receiver employs a direct frequency synthesiser for rapid frequency changes.

Databus control interfaces are integrated into the system, installed in the RT-1651 receiver/transmitter within the existing equipment envelope. MIL-STD-1553B or ARINC 429 databus interfaces provide for full compatibility with advanced cockpit management systems.

HF-9087F
The HF-9087F is the latest variant of the HF-9000 series transmitter/receivers. The HF-9087F is a ruggedised version of the HF-9087D, using Digital Signal Processor (DSP)-based technology to provide highly robust HF voice and data communications. The HF-9087F is designed to surpass the severe environmental requirements of current high-performance fighter aircraft and meets the requirements of MIL-STD-810E (environmental) and MIL-STD-461D (EMI).

Specifications
Weight: 12.7 kg
Power: 175 W; 200 W (option)

Status
In service. Introduced in 1986. By 2003, over 7,000 systems had been installed in more than 60 types of rotary- and fixed-wing aircraft worldwide. The system is also reported as the standard HF airborne system on most large corporate aircraft and has received airworthiness approval in over 20 countries.

Contractor
Rockwell Collins.

AN/ARC-220/URC-100 HF tactical communications system

Type
Avionic communications system, HF, data, secure.

Description
Rockwell Collins' AN/ARC-220/URC-100 is a multifunction, fully Digital Signal Processing (DSP) high-frequency radio intended for airborne applications. Advanced communications features made possible by DSP technology include embedded Automatic Link Establishment (ALE),

Serial Tone Data Modem and Anti-jam (ECCM) functions.

The ARC-220 Advanced HF Aircraft Communications System is suitable for a variety of tactical fixed- and rotary-wing airborne applications. In addition to offering enhanced voice communications capabilities, the ARC-220 is an advanced data communications system capable of providing reliable digital connectivity.

The ARC-220 allows communications on any one of 280,000 frequencies in the High Frequency (HF) band from 2.0000 to 29.9999 MHz. Upper Side Band (USB) voice, Lower Side Band (LSB) voice, Amplitude Modulation Equivalent (AME), Continuous Wave (CW), USB data and LSB data emission operating modes are provided. Communication is possible using either simplex or half-duplex operation. The ARC-220 also includes an integrated data modem, which enables communication in noisy environments where voice communications are not often possible. Up to 25 free text data messages can be pre-programmed via data fill or created/edited in real time. Received data messages can be stored for later viewing and retransmission if desired. Built-in integration with external GPS units allow position data reports to be sent with the push of a button. Long-range wireless e-mail and Internet connectivity is possible when used in conjunction with Rockwell Collins HF Messenger™ software.

In addition to conventional HF communications, the ARC-220 provides MIL-STD-148-141A Automatic Link Establishment (ALE) capability. ALE maintains a database of channel signal conditions and automatically establishes a two-way communication link on the best available frequency. After establishing a link, the operator is alerted and communication can begin. ALE Linking Protection (LP) is also provided, which prevents unauthorised link establishment and ALE jamming. The ARC-220 is also claimed to be the first radio to be fielded featuring

MIL-STD-188-141B Appendix A Alternative Quick Call (AQC) ALE. Rockwell Collins asserts that this advanced protocol/waveform has the potential to reduce ALE over-the-air call time by up to 50 per cent over conventional second-generation ALE.

Electronic Counter-Counter Measures (ECCM) is a frequency method used to combat the effects of communication jammers and direction-finding attempts. The ARC-220 provides ECCM in accordance with MIL-STD-188-148A and CR-CX-0218-001 (Army Enhanced). ALE for both forms of ECCM is provided.

The R/T also supports the modem requirements necessary for the MIL-STD-188-141A ALE, MIL-STD ECCM and Army Enhanced ECCM digital message protocols.

System description
The AN/ARC-220/URC-100 provides three transmit power output settings: low (10 W pep and average), medium (50 W pep and average) and high (175 W pep; 100 W average). Transmit tune time is typically 1 second in manual frequency selection. Transmit tune time is typically 35 ms after a frequency is tuned and tuning data stored.

The AN/ARC-220/URC-100 system is composed of three Line Replaceable Units (LRU) for non-MIL-STD-1553B bus controlled platforms and two LRUs for a MIL-STD-1553B bus controlled platform. The RT-1749/URC receiver transmitter is the heart of the system with the AM-7531/URC power amplifier/coupler providing antenna matching and RF power amplification. The C-12436/URC control allows the user to control system operation.

Radio set control C-12436/URC provides the AN/ARC-220 operator interface. A six-line alphanumeric display provides system status and advisory data. The display includes variable backlighting and is NVG compatible. The radio set control internal power supply operates from the +28 V DC aircraft power.

A radio receiver-transmitter RT-1749/URC provides the electrical interface with other ARC-220 LRUs and associated aircraft systems, such as interphone, GPS and secure voice systems. The receiver-transmitter provides EIA RS-232C and MIL-STD-1553B serial data interface ports to accommodate various system configurations.

The receiver-transmitter uses a microprocessor to perform all ALE, ECCM (optional) and modem functions. The internal power supply operates from +28 V DC aircraft power. An external 6 to 24 V DC power source can be connected to the receiver-transmitter, which will enable retention of clock and data message information when main power is removed from the system.

The RT-1749A/URC version of the receiver transmitter does not have ECCM or ALE LP 3 capability and is intended for foreign military sales.

A power amplifier/coupler AM-7531/URC provides an electrical interface with other ARC-220 LRUs and the antenna system. The PA coupler provides three output power levels: high (100 W average; 175 W PEP), medium (50 W PEP and average) and low (10 W PEP and average). Antenna coupler circuits provide impedance

The AN/ARC-220/URC-100 is an advanced HF aircraft communications system   0504227

matching between the power amplifier output and the transmit antenna which enables maximum power transfer to the antenna. The PA coupler is digitally tuned under control of the receiver-transmitter microprocessor. First time tuning is usually completed in 1 second. The receiver-transmitter stores tuning data for previously tuned frequencies. Stored tuning data reduces future tuning time (to typically 35 ms) when a frequency is used again. The PA coupler internal power supply operates from +28 V DC aircraft power.

Specifications
Frequency: 2.0000 to 29.9999 MHz in 100 Hz steps
Channels: 20 user programmable simplex or half-duplex
20 programmable simplex or half-duplex
20 programmable automatic link establishment (ALE) scan lists
12 programmable ECCM hopsets (with ALE capability)
Modes: USB and LSB voice and data, CW and AME
Power requirements: 28 V DC per MIL-STD-704
Usage: (RT-1749/URC) 50 W
(AM-7531/URC) 450W (transmit)
(C-12436/URC) 50 W
MTBF: 1,000 h min
Fault isolation/detection: 98% fault isolation (LRU level)
Dimensions (W × D × H): (RT-1749/URC) 105.4 × 355.6 × 194.3 mm
(AM-7531/URC) 160.3 × 428.0 × 184.9 mm
(C-12436/URC) 146 × 203.7 × 114.3 mm
Weight: (RT-1749/URC) 5.9 kg
(AM-7531/URC) 8.5 kg
(C-12436/URC) 2.3 kg
NVG compatibility: IAW MIL-L-85762F
Temperature: −40 to +55°C
Altitude: 25,000 ft max
Humidity: 95%
Vibration: 2.5 g

Status
In service with US Army helicopters.

Contractor
Rockwell Collins.

AN/ARC-222 SINCGARS radio

Type
Airborne communications system.

Description
The AN/ARC-222 SINCGARS radio is the replacement for the AN/ARC-186 and is designed for air-to-ground and air-to-air communications. It includes SINCGARS-V capability and covers the frequency ranges 30 to 87.975 MHz in VHF FM, 108 to 151.975 MHz in VHF AM (108 to 115.975 MHz receive only) and 152 to 174 MHz FM for the International Maritime Band.

It consists of a control unit and a receiver/transmitter.

Specifications
Dimensions:
(control unit) 57.4 × 127 × 95.8 mm
(receiver/transmitter) 120.6 × 127 × 174.6 mm
Weight:
(control unit) 0.73 kg
(receiver/transmitter) 4.67 kg
Power supply: 28 V DC, 4.5 A max
Frequency:
30–87.975 MHz (FM)
108–151.975 MHz (AM) (108–115.975 MHz receive only)
152–174 MHz (International Maritime Band)

Status
In production and in service.

Deliveries to the US Air Force began in April 1997. The system was part of the avionics baseline for the C-130J aircraft. Over 500 R/Ts and 250 remote controls had been delivered to multiple customers by mid-1998. Additional capability

to communicate in European 8.33 kHz channel spacing for ATC was added in the MXF-626K version. Other platforms are reported to include A-10, AC-13H, E-8C, EC-130E, F-16, HH-60G, HH-60L, MC-130E, MC-130H, MC-130P, MH-53J, MH-53M, OA-10A and UH-60. Also reportedly installed in Australian P-3C Maritime Patrol Aircraft (MPA).

Contractor
Raytheon Company, Space and Airborne Systems.

AN/ARC-230/HF-121C high-performance radio system

Type
Avionic communications system, HF.

Description
The Rockwell Collins AN/ARC-230 (company designation HF-121C) high-performance radio system is designed for military voice and data HF applications requiring 400W operation. Compliant with the requirements of STANAG 5511/MIL-STD-188-203-1A for Link 11/TADIL A, MIL-STD-188-141A, Appendix A for Automatic Link Establishment (ALE) and MIL-E-5400, MIL-STD-461A, this 400 W PEP and/or average power radio provides maximum performance for airborne applications. The AN/ARC-230 has been optimised for tactical digital data communications and SIMultaneous OPeration (SIMOP) of multiple radio sets with minimum frequency and antenna separation as well as HF-V/UHF SIMOP. The AN/ARC-230/229 transceiver system is specifically configured (firmware programmable) to emulate an AN/ARC-165 and be a form-fit-function replacement for AN/ARC-165/194 transceiver systems on US Air Force E-3B/C AWACS aircraft. For a full description of the system and capabilities overview, see separate entry for Rockwell Collins' HF-121A/B/C HF radios.

Status
In service in US Air Force E-3B/C AWACS aircraft.

Contractor
Rockwell Collins.

AN/ARR-85 Miniature Receiver Terminal (MRT)

Type
Avionic communications system, VLF.

Description
The Miniature Receiver Terminal (MRT) forms part of the Minimum Essential Emergency Communications Network (MEECN) that provides secure VLF links between the National Command Authority and B-52H and B-1B bombers. It includes the Datametrics quarter-page black and white printer.

Status
In service. The MRT was developed for the B-52H and is in service on the B-1B.

Contractor
Rockwell Collins.

AN/ARS-6 (V12)

Type
Avionic Emergency Location System (ELS).

Description
The second generation V12 PLS Receiver/Transmitter (R/T) is a functional replacement for the original AN/ARS-6. The AN/ARS-6 PLS was designed to support the Combat Search and Rescue (CSAR) mission and provides for the covert location of downed pilots carrying the PRC-112 survival radio. The lighter and smaller V12 PLS retains all legacy functionality and is backward compatible with the original AN/ARS-6 (V).

The V12 is interoperable with all standard emergency distress beacons, including the URX-300 GPS based survival radio and ARX-3000 Tactical ELT, together with all US deployed combat survival radios, including the PRC-112 family of radios and the CSEL PRQ 7. Two-way voice has been extended to cover the 225 to 400 MHz band.

A wide band DF antenna from Chelton Electrostatics Inc. provides accurate azimuth measurements at all angles from 110 to 407 MHz, simultaneously monitors 4 channels and decodes the 406 MHz COSPAS-SARSAR embedded GPS position.

The system interfaces with existing PLS installations and wiring, including original hardware (CDU, RDU and ASU LRU's) and software. Additionally, the V12 can be configured as a data bus controlled system, facilitating simple integration into modern databus-equipped platforms.

Specifications
Frequency range: 225–400 MHz (UHF)
Channels: 7000 in 25 kHz steps
Modulation: AM voice, BPSK/OOK (DME), OOK
Temperature: −40° to 55°C (operating)
MBTF: 2843 hours
Receiver:
Type Dual conversion superheterodyne
Sensitivity -113 dBm
Selectivity 70 kHz (IF)
Audio output +17 dBm
Transmitter:
Average power 10 W
Spectrum Nominal 60 dB down at ±1 MHz from the carrier frequency (transpond mode)
Interface: MIL-STD-1553B, ARINC-429 and RS-422
Antenna coverage 360° wide band DF
Dimensions 121.9 × 147.3 × 312.4 mm
DF Antenna diameter 228.6 mm (9 in)
Power 28 V DC; 4.5 A (max), 0.75 A (stby)

Status
In production and in service.

Contractor
Cubic Defense Applications.

AN/ASC-15B communications central

Type
Avionic communications system, V/UHF, HF, secure.

Description
The AN/ASC-15B communications central, referred to as a command console, functions as an airborne and ground command post, providing tactical voice communications in both secure and non-secure modes. This highly mobile communications combat command centre provides NATO and US tri-service forces interoperability during all types of military operations and special missions.

The AN/ASC-15B can be operated from a UH-60A or UH-1H helicopter, or removed and configured for ground operation. It provides HF plus VHF and UHF communications in AM and FM modes, channel scanning of four V/UHF preset channels in each-AN/ARC-182 radio, automatic retransmission in VHF and UHF bands and UHF satellite communications.

The AN/ASC-15B consists of an AN/ARC-174 HF transceiver, three AN/ARC-182 V/UHF transceivers, two AM-7189A IFM power amplifiers, an MX-931B/URC repeater, an AM-7402 Satcom power amplifier and two C-11128 ECCM (HQ) controls.

The ASC-15B has been modified with three ARC-210(V) radios to replace the ARC-182s presently installed. It was given the nomenclature AN/ASC-15C after radio upgrading.

Specifications
Weight: 129.28 kg
Power supply: 28 V DC
Power output:
(2–30 MHz) 100 W PEP
(30–400 MHz) 15 W FM, 10 W AM
Frequency: 2–30 MHz and 30–400 MHz

The AN/ASC-15B command and control console is designed for helicopters such as the UH-60 Black Hawk   0504026

Modes:
(2–30 MHz) HF/SSB, AME and CW
(30–400 MHz) V/UHF AM and FM

Status
In service with the US Army.

Contractor
Rockwell Collins.

AN/AVX-3 Fast Tactical Imagery (FTI) system

Type
Avionic datalink system.

Description
The FTI system is designed to transmit and receive images by Line Of Sight (LOS) in near-realtime (15 sec delay). It replaces the 3/4 in record-only VTR with HI 8 mm 2-hour record and playback in the cockpit. Both the image PhotoTelesis transceiver and the TEAC HI 8 mm VTR are controlled from a small Remote Control Unit (RCU) in the cockpit. The AN/AVX-3 mounts in a standard shock mount for the TEAC V1000 VTR and could be easily adapted to other airborne and ground platforms. A removable PCMCIA SRAM card enables uploading of target and other images as well as maps prior to takeoff. This same card also stores all mission digital images for downloading to PC and laptop after landing. All stored digital images and VTR playback can be viewed on the cockpit display at anytime during the mission. In addition to capturing and transmitting from live video sensors, the AN/AVX-3 will also capture and transmit images from VTR playback.

FTI uses the same datalink fitted to F-14s for the digital Tactical Airborne Reconnaissance Pod System (TARPS). FTI transmissions are made at a burst rate of 15 seconds per image.

Initial application of the system was on the US Navy F-14 aircraft, where it is used to transmit data from the aircraft's integral (long-range TV, HUD, radar) or external sensors, including the LANTIRN (Low Altitude Navigation Targeting Infra-Red for Night) pod, fused with GPS co-ordinates (from the LANTIRN pod receiver). The system can also share imagery with other F-14 aircraft, which can then serve as links to distant users.

FTI was originally intended to transmit targeting data back to a carrier to provide the final approval for an attack. However, upgrades in LANTIRN software, released to squadrons in 1999, allow precise co-ordinates of any Point Of Interest (POI) to be measured. This target information can then be transmitted to aircraft equipped with GPS-guided weapons, effectively allowing F-14 fighters to perform stand-off multispectral reconnaissance or target acquisition for other attackers.

FTI-II
FTI-II is an enhancement of the baseline system (see separate entry).

Specifications
Power: 28 V DC; 87 W
Weight: 11.36 kg
Operating temperature: –40 to +60°C

Status
In service in US Navy F-14 and F-18 aircraft. The FTI system was developed by PhotoTelesis Corporation, under contract to Raytheon, from a design for the US Army's AH-64 Apache helicopter. The US Navy has equipped all of its

176 LANTIRN-equipped F-14 aircraft with the FTI system.

Two F-14 squadrons involved in air operations over Yugoslavia during Operation Allied Force in 1999 were equipped with FTI.

The PhotoTelesis ATR403 image transceiver is now obsolete and has been replaced by the PRISM™ (see separate entry) system.

Contractor
PhotoTelesis Corporation.
Raytheon Company, Space and Airborne Systems.
TEAC America Inc.

AN/AVX-4 Fast Tactical Imagery II (FTI II) system

Type
Avionic datalink system.

Description
The Fast Tactical Imagery II (FTI II) is a follow-on programme to the FTI I programme (see separate entry). FTI II was developed and released for the Navy's F/A-18 Hornet to provide image capture, compression, transmission, reception and display in the cockpit. The system employs a single Photo Reconnaissance Intelligence Strike Module (PRISM™ – see separate entry). Employing FTI II, battlefield commanders can view the battlespace at extended ranges through still frame images provided in near-realtime from airborne assets. Bomb Damage Assessment (BDA) images can be transmitted immediately after a strike to facilitate re-tasking and/or target hand-off to subsequent attacking aircraft. Further, Targets Of Opportunity (TOO) can be transmitted to aircraft on station for immediate attack to provide timely support to ground forces and/or quick reaction to enemy initiatives. At extreme range, the system can 'piggy-back' information across a number of (similarly equipped) aircraft to provide a long-range link.

A removable Compact Flash (CF) card provides for upload and download of target imagery before and after the mission.

Small text messages can also be attached and transmitted with images, or as standalone text files, and then displayed in the cockpit. FTI II can also take full serial control of the associated HI 8 mm video tape recorder for record, rewind, playback and capture of recorded video. For the F/A-18 Hornet, an internal Scan Converter Module automatically detects and down scans 875-line Forward-Looking InfraRed (FLIR) and 675 line-radar video to standard 525-line HI 8 mm recording and image capture format.

Other features of the FTI II system include:
• Burst capture up to six images per second
• Thumbnail display of captured images
• Wavelet and JPEG compression
• Selectable compression ratio
• 'Save to archive' as bit map
• Frame and field capture modes
• Full control of TEAC V80 VTR
• Optional scan converter module
• VTR frame synchronisation during playback
• Capture and transmit from VTR playback
• Serial remote control via RCU403 unit

FTI uses the same datalink fitted to F-14s for the digital Tactical Airborne Reconnaissance Pod System (TARPS). FTI transmissions are made at a burst rate of 15 seconds per image.

The inital application of the FTI system was on the US Navy F-14 aircraft, where it is used to transmit data from the aircraft's integrated (long-range TV, HUD, radar) and external sensors (LANTIRN/GPS).

FTI was originally intended to transmit targeting data back to a carrier to provide the final approval for an attack. However, upgrades in LANTIRN software, released to squadrons in 1999, allow precise co-ordinates of any Point Of Interest (POI) to be measured. This target information can then be transmitted to aircraft equipped with GPS-guided weapons, effectively allowing F-14 fighters to perform stand-off multispectral reconnaissance or target acquisition for other attackers.

Specifications
Power: 28 V DC; 7 W
Weight: 0.9 kg
Operating temperature: −40 to +55°C

Status
In production and in service. F/A-18 E/F Super Hornets from the US Navy's VFA-41, flying from the USS *Abraham Lincoln*, were reported as the first aircraft to operate with FTI-II, during operations over Iraq during 2003.

The FTI II is compatible with all F-14s, all US Navy carriers, US Army AH-64 and OH-58D with AVRIT, US Navy P-3 AIP, UK Jaguar and other Special Forces systems.

Contractor
PhotoTelesis Corporation.
Raytheon Company, Space and Airborne Systems.
TEAC America Inc.

AN/AXQ-14 datalink

Type
Avionic datalink system.

Description
The AN/AXQ-14 is a two-way communication datalink to guide the GBU-15 glide bomb. It provides a video and command link between the command aircraft and the weapon, enabling the systems operator to remain in the control loop while the weapon is being directed to its target. In effect, the datalink permits a command authority similar to a fly-by-wire system, in which the operator can transmit guidance instruction from launch to impact. Alternatively, he may select any one of a number of autonomous weapon control modes, including an override mode which permits target updating or redesignation as required.

The extended weapon control capability conferred by the datalink contributes to weapon system performance in terms of stand-off range and operational utility. Target acquisition is deferred until the weapon, rather than the command aircraft, is closer to the target. Tactically, the aircraft can leave the target zone immediately after launch.

The AN/AXQ-14 system comprises three major elements: a datalink pod mounted on the command aircraft, a datalink control panel used in conjunction with an existing display within the aircraft and a weapon datalink module mounted on the rear of the weapon itself.

The pod is an aerodynamically shaped container mounted on a standard stores carriage strongpoint on the fuselage centreline or on an underwing station, according to aircraft type. It contains four LRUs comprising:
1. An electronics section incorporating all radio frequency generating and receiving equipment, a demultiplexer to decode all aircraft command and pod control signals, an encoder and antenna controls.
2. A phase-scanned array for weapon tracking in normal operation.
3. A forward horn antenna, that provides additional coverage.
4. A mission tape recorder, which maintains a permanent record of weapon video data.

The pod is suitable for high-performance aircraft, is certified for operation at speeds in excess of M1.0 and is also compatible with high- and low-altitude operations. There is said to be no compromise of aircraft performance attributable to carriage of the pod.

Used in conjunction with an existing display system, the aircraft control panel acts as the interface between the weapon system operator and the weapon guidance system. The panel accepts signal inputs from the aircraft as well as from its own controls, and formats these into discrete commands as required via the datalink. Although the panels are tailored to the individual

requirements of the aircraft type and intended customer usage, each unit accepts the standard configurations of the GBU-15/AGM-130 datalink and the pod.

Attached to the aft of the GBU-15/AGM-130 weapon is the ultimate component in the datalink chain, the weapon datalink module. This simultaneously transmits video from the weapon's seeker-head and processes incoming command signals from the aircraft to the weapon. Heading changes during the weapon's flight are effected through discrete command signals. Dual-analogue command channels enable the operator to slew the weapon in pitch and yaw during approach to the target.

Digital techniques are employed in the AN/AXQ-14 system and the transmitter is of all-solid-state construction. The system's electronically phase-scanned antenna array provides the datalink with high-rate tactical manoeuvring capability. A comprehensive range of test equipment is provided, including:
1. A flight checkout unit for testing aircraft cables from the pod connection point.
2. An aircraft simulator unit, which permits functional checks of the control panel.
3. A weapon simulator unit for test of the aircraft pod and isolating faults down to LRU level.

Used together, these two latter units permit full system functional checkout.

Two primary launch modes are envisaged for operation of the GBU-15 weapon and AN/AXQ-14 control combination: low-altitude penetration and high-altitude standoff. It is claimed that use of the datalink has improved weapon delivery accuracy over non-link weaponry in various profiles from airborne platforms such as the US Air Force's B-52, F-4 and F-15 aircraft; and US FMS F-16 aircraft. The system is also said to be compatible with the A-4, A-7 and F/A-18 aircraft. Potential weapon applications include Harpoon, Maverick and cruise missiles.

Status
In production, and in service.

Contractor
Raytheon Company, Space and Airborne Systems.

AN/URQ-33(V) Class 1 JTIDS terminal

Type
Avionic datalink system.

Description
The Joint Tactical Information Distribution System (JTIDS) uses frequency-hopping, spread spectrum, automatic relay and other high-technology techniques to provide data and voice

communications which are highly resistant to jamming.

The AN/URQ-33(V) Class 1 JTIDS terminal is used on board the US Air Force and NATO E-3 AWACS and for a number of ground-based applications for communicating information on command and control, surveillance, intelligence, force status, target assignments, warnings and alerts, weather and logistics.

JTIDS uses a computer-controlled Time Division Multiple Access (TDMA) technique in which information is transmitted in short bursts lasting only a fraction of a second. Bursts are synchronised by computer with bursts from other users, to allow simultaneous transmission on the network without causing interference. The JTIDS burst is spread in frequency, encoded and hopped across a number of frequencies in a split second, making it hard to intercept and almost impossible to jam. The receiver selects pertinent data by means of software filtering.

JTIDS communications are automatically relayed by other terminals. This extends the range beyond the line of sight, and it provides another layer of defence against jammers. It also enables terminals to provide users with position and navigation data, without the need for extra equipment, by means of the highly accurate message time of arrival measurement, which can be converted to range between the transmitter and the receiver.

JTIDS is broadcast in the 960 to 1,215 MHz frequency range. The system consists of a radio set control, transceiver processor unit, high-power amplifier, high-power amplifier power supply, low-power amplifier power supply and antenna coupler. It also includes a general-purpose digital computer, programmed to perform most of the communications tasks and interface with the host platform's computer.

Status
In production and in service in a wide variety of airborne platforms.

Contractor
Raytheon Company, Space and Airborne Systems.

AN/USC-42 (V)3 UHF satcom and line of sight communication set

Type
Avionic communications system, satellite.

Description
The AN/USC-42 (V)3 Miniaturised Demand Assigned Multiple Access (Mini-DAMA) set is a downsized member of the TD-1271 terminal family. It achieves interoperability with the US

The AN/AXQ-14 datalink pod mounted inboard of a GBU-15 glide bomb, under the fuselage of a US Air Force F-15   0504229

Navy's TD-1271B/U multiplexer and AN/WSC-3 and the AFSATCOM system. The Mini-DAMA will function in nine operational modes. Among them is 25 kHz Satcom; here the system will support navy TDMA-1 network operations and non-TDMA communications. On 25 kHz line of sight channels, it will support short-range tactical communications. On 5 kHz UHF Satcom channels, it will interoperate with navy non-TDMA communications, US Air Force DAMA network operations, US Air Force non-TDMA communications and AFSATCOM network operations. Product improvement over the life of the system has included AFSATCOM IIR and Have Quick IIA capabilities.

The US Navy's FLTBDCST, CUDIXS/NAVMACS, SSIXS, OTCIXS, secure voice TACINTEL, TADIXS A and ORESTES will use Mini-DAMA for data exchange. Mini-DAMA modem/receiver/transmitters will come in two configurations: 483 mm rack for ship and shore installations and as a 1 ATR-long package for aircraft.

Principal components include an integrated modem/Receiver/Transmitter (R/T), a separately housed power amplifier and an external Display Entry Panel (DEP). The airborne version will contain a remote operation display/entry panel. Operations will be either half- or full-duplex through the Mini-DAMA embedded radio or through a 70 MHz IF interface to an external receiver/transmitter. The AN/USC-42(V)3-Mini-DAMA is configured for airborne platforms, while the AN/USC-42 (V)2 is the equivalent submarine, ship and shore-based version.

Specifications
Frequency: 225–399.995 MHz
Channel spacing: 5 kHz or multiples of 5 kHz
Power output: 100 W
Temperature range: −32 to +55°C
Reliability: 2,000–4,000 h MTBF

Status
Over 400 Mini-DAMA systems had been installed in US and overseas military forces aircraft, ships, submarines and fixed sites by the beginning of 2002. The system has been reported as part of the US P3 Anti-surface Warfare Program and the Hawkeye 2000 effort.

Contractor
Titan Systems Corporation, Communications and Electronic Warfare Division.

AN/USQ-130(V) MX-512PA Link-11/TADIL-A data terminal

Type
Avionic datalink system.

Description
The AN/USQ-130(V) Link-11/TADIL-A data terminal set is designed to provide all required modem and network control functions in a Link-11/TADIL-A system using either HF or UHF radio equipment. The equipment meets the data terminal set requirements of MIL-STD-188-203-1A and may be operated as a picket or net control station in a TADIL-A net. As a net control station, the AN/USQ-130(V) accepts addresses from the tactical data computer or from a separate control panel.

The equipment provides all the modes of Link-11/TADIL-A systems including net control or picket, high- and low-data rate, net test, net synchronisation, short broadcast, long broadcast and full-duplex (for single station system tests and sidetone verification). Doppler correction circuits which operate independently on both sidebands are operator-selectable. The AN/USQ-130(V) also operates in the Improved Link-11 Waveform (ILEW) mode.

The AN/USQ-130(V) can be externally controlled by a computer over a MIL-STD-188-114, RS-232C-compatible asynchronous control interface, or 1553 databus. The AN/USQ-130(V) may also be controlled from a separate remote-control panel using menus standard to the MX-512P DTS family.

AN/USC-42 (V)3 Mini-DAMA Satcom terminal   0044838

The set is programmable. All modem, network control and link monitoring functions are performed digitally in microprocessors using a modular, multiprocessor architecture. Selection of the conventional Link 11 or ILEW is made over the remote-control interface.

The single-tone waveform for Link-11 provides improved performance in HF Link-11 networks on an SSB HF channel. Single-tone Link-11 uses an eight-phase modulated 1,800 Hz tone. Adaptive equalisation is used to demodulate the signal under the severe multipath conditions typical of HF propagation paths. Error detection and correction codes are used to provide enhanced message throughput.

The AN/USQ-130(V) provides, as an option, a 2,400 bits/s, full-duplex, RS-232C satellite-wireline interface which transmits and receives compatible Link-11 data in digital form. Link-11 data may be sent over satellite, wireline, or other tactical circuits. The AN/USQ-130(V) can be operated in either the digital mode, the conventional mode, or in a mixed mode (gateway), where some pickets operate in the digital transmission mode and some in the conventional HF or UHF mode.

The unit provides, as another option, link quality analysis indicators which include multipath spread, fading bandwidth, net cycle time since last reply, and tone power spectrum for each participating unit in the network. Using these indicators, an operator can troubleshoot equipment failures and configuration set up problems in the net and determine when HF propagation problems require a change of radio frequency. BITE provisions in the AN/USQ-130(V) include loop-back functions which verify operation of the system.

The data terminal set is compatible with ATR short, measures 193 × 57 × 32 mm, and weighs 4.6 kg. It is powered from 24 to 32 V DC and meets certain requirements of MIL-HDBK-217E, MIL-E-5400T, MIL-STD-188 and MIL-STD-1553B.

AN/USQ-130(V) MX-512PA Link-11/TADIL-A data terminal   0002119

Specifications
Dimensions: 1/4 ATR-short:
  Height 19.3 cm
  Width 5.7 cm
  Depth 32.0 cm
Weight: 4.6 kg
Power: 24 to 32 V DC, 28 W
Temperature: −45 to +55°C (operating)
Altitude: Up to 50,000 ft
Compliance: MIL-E-5400T
Reliability: MTBF Over 10,000 h per MIL-HDBK-217F at 50 C A/C

Status
In production and in service.

In November 2004 DRS Technologies announced that it had received new orders to provide tactical datalink and HF data communications systems, including the AN/USQ-130, for US and Allied ground, sea and air forces to support integrated battle space operations. The contracts, with a combined value of USD11.7 million, were awarded by US government, international military and commercial customers.

Contractor
DRS Technologies, Inc.

AN/USQ-140(V)© MIDS LVT-1 anti-jam terminal

Type
Avionic datalink system.

Description
The Harris Corporation, in conjunction with team leader ViaSat Inc and Northrop Grumman's Xetron subsidiary, is one of three MIDS-Low Volume Terminal (MIDS-LVT) manufacturers, the others being Datalink Solutions and EUROMIDS. Utilising ViaSat's Link 16 Enhanced Throughput (LET) concept, the new MIDS LVT-1 produces 10 times higher Link-16 data rates with only a few dB of anti-jam performance degradation. The LET technology will be retrofitted into JTDIDS Class 2 terminals. Embedded modules provide COMSEC and TACAN waveform communications with no added hardware.

The system incorporates terminal modules which are fully interoperable and interchangeable with all other MIDS terminals.

Specifications
Dimensions: 343 × 190.5 × 193.5 mm
Weight: 23 kg (main terminal), 6.5 kg (remote power supply)
Power supply: 115 V AC, 400 Hz 3-phase
MTBF: 1,822 h (shipboard aircraft fighter environment)
Link 16 Message: TADIL-J, IJMS
Cooling: External conductive air
Transmit spectral performance: Greater than −60 dBc in 1030/1090 MHz IFF bands
Output power: 1, 25 or 200 W, plus HPA interface (Option)

Interface: MIL-STD-1553, X.25, Ethernet, STANAG 3910
Data throughput: 115 Kbps (coded), 1.2 Mbps (growth)
Keyfill: DS-101
Voice capability: 2.4 Kbps LPC-10; 16 Kbps CVSD

Status
In production and in service.

The US Space and Naval Warfare Systems Command, on behalf of the Multifunctional Information Distribution System (MIDS) International Program Office, was planning to award up to three contract modifications during February 2004 for Engineering Change Proposal (ECP) modifications to the MIDS LVT. The candidate contractors are DataLink Solutions, ViaSat and EuroMIDS. These ECPs will migrate the LVT to a software architecture that is compatible with the future Joint Tactical Radio System (JTRS), resulting in the MIDS JTRS terminal.

Contractor
ViaSat Inc.
Harris Corporation.
Xetron.

AN/VRC-99 (A) communications system

Type
Avionic communications system, UHF, secure.

Description
The AN/VRC-99 (A) is a programmable, wideband, secure, open architecture communication system. It provides virtual circuit and datagram service guaranteeing reliable, simultaneous, multichannel voice, data, imagery, and video transmission.

The AN/VRC-99 (A) is available in ground vehicular, shipboard and airborne configurations. Digital signal processing provides flexibility and operational simplicity in end-to-end communication connectivity, packet formatting, and packet switching protocols. Use of spread spectrum modulation (LPI/AJ), growth for transmit power controls (LPD), and embedded encryption provides for the security and integrity necessary in modern strategic and tactical networks. The basic system provides coverage from 1,300 to 1,500 MHz.

The AN/VRC-99 (A) can provide wireless LAN service for both Line-Of-Sight (LOS) and Beyond-Line-Of-Sight (BLOS) data and/or voice links. Adaptive routing provides for network survivability in dynamic mobile subscriber applications as required by the Tactical Internet/Warfighter Information Network (WIN) Architecture.

System features include:
1. digitised voice/data/secondary imagery/video
2. user selectable burst rate of 156 kbps to 10 Mbps with growth option for fully adaptive operation*
3. low probability of intercept and antijam (LPI/AJ) utilising specialised spread spectrum techniques
4. low probability of detection (LPD) via transmit power control (ground options)
5. embedded type 1 encryption
6. EKMS/AKMS compliant

BAE Systems North America AN/VRC-99 (A) communications system   0044845

7. forward error correction
8. frequency-hopping mode
9. automatic initialisation/network entry at power on – no user setup required
10. asynchronous/synchronous interfaces; RS-232; RS-422 with X.25 LAPB link layer; Ethernet 802.3
11. TEMPEST-tested COMSEC module with networking efficiency design
12. interoperable with MSE Voice Circuit and Packet Data Switches
13. US Army Technical Architecture (ATA)/Joint Technical Architecture (JTA) compliant.

Specifications
Frequency band: 1,300 to 1,500 MHz (19 channels)
Frequency agility: selection by software controls
Power output: 10 W (optional 50 W with external PA)
Channel access: CDMA/FDMA/TDMA/SDMA/ALOHA/hybrids
Synchronous data: up to 10 Mbps for SDLC/HDLC; 10 Mbps for Ethernet
Asynchronous: up to 100 Kbps
Voice: 4-wire conditioned diphase (4WCDP)
Dimensions: ¾ ATR 193.3 × 190.5 × 318.8 mm
Weight: 11.36 kg
Power: 28 V DC
MTBF: 4,000 hours w/PA

Status
In production and in service.

Contractor
BAE Systems North America, Greenlawn.

ARINC 758 Communications Management Unit (CMU)

Type
Avionic Communications Control System (CCS).

Description
The ARINC 758 CMU is an airborne communications router that supports datalink service access between aircraft datalink applications, such as AOC/AFIS, CPDLC and ADS and their corresponding service providers; it includes an Aircraft Personality Module (APM).

The CMU initially provides an ARINC 724B-compatible datalink router is through which all character-oriented data transmitted to and from the ground-based Aircraft Communications Addressing and Reporting System (ACARS) network. This CMU is designed to ARINC Characteristic 758 and can be upgraded, by software, to an Aeronautical Telecommunications Network (ATN) router when protocols and application infrastructure are available to support CNS/ATM datalink applications.

The initial CMU provides Level 0.1A functional capability, with growth via software updates to Level 2-D, as application functionality becomes available. Level 0.1A functionality is operationally equivalent to an ARINC 724B ACARS management unit. This includes the use of a VHF Data Radio (VDR) operating with ARINC 618 protocols and modulation interfacing to the CMU via an ARINC 429 interface. Level 2-D functionality is operational over the ATN utilising high-speed services and incorporating Air Traffic Services (ATS) applications.

The CMU MMI is provided via an ARINC 739-compatible, Multipurpose Control and Display Unit (MCDU) interface to display datalink status, entry of text messages for transmission and the display of received text messages for review. A hard copy of datalink messages is provided via an interface with an ARINC 740/744 airborne printer.

Access to the ground network is provided via several ACARS air-to-ground sub-networks.

The CMU interfaces to other onboard systems such as the flight management computer and maintenance data computer utilising ARINC 749 interfaces.

The ARINC 758 CMU can be customised to meet operational needs, including future requirements associated with the implementation of the CNS/ATM and ATN systems. The ability to support dual-CMU installations will be available from the third quarter of 2000.

Specifications
Dimensions: 124 (W) × 193.5 (H) × 324 (L) mm
Weight: <5.5 kg

Status
In production.

Contractor
Honeywell Defense Avionics Systems.

ATX-2740(V) airborne datalink system

Type
Avionic datalink system.

Description
The ATX-2740(V) is a lightweight, exportable, ruggedised COTS version of the US Common DataLink (CDL) satisfying STANAG 7085 for digital, point-to-point, datalinks. It provides real-time full duplex sensor data and voice communications between aircraft and ground stations at selectable wide bandwidth data rates.

The ATX-2740(V) system comprises three units: Airborne Modem Assembly (AMA); Radio Frequency Electronics (RFE); one or more fixed or steerable antennas.

L-3 Communications ATX-2740(V) airborne datalink system (left to right) RFE, AMA, antennas   0044822

Specifications

Frequency: X-band or Ku-band (NATO I/J-band or J-band)
Tuning: 5 MHz steps
Data rates:
uplink up to 200 kb/s aggregate data rates (option to 10.71 Mp/s)
downlink selectable aggregate data rates to 274 Mb/s
Dimensions (W × H × D):
AMA 259.1 × 218.4 × 505.5 mm
RFE 241.3 × 101.6 × 431.8 mm
directional antenna 228.6 mm dia
Weight: AMA 15.9 kg
RFE 6.8 kg
antenna 3.2 kg

Contractor

L–3 Communications, Communication Systems – West.

Automatic communications processor ARC-190

Type

Avionic communications processor.

Description

The automatic communications processor and associated ARC-190 400 W airborne radio provide a system that automatically scans multiple frequencies, selects the best frequency on which to make a call and automatically repeats the call until contact is confirmed. The system also provides an anti-jam frequency-hopping capability for effective ECCM.

The CP-2024A Automatic Communications Processor (ACP) and the C-11814/ARC-190(V) Automatic Communication Processor Control (ACPC) operate together as a microprocessor-based remote-control subsystem, which can be added to existing-AN/ARC-190(V) radio systems to automate and simplify HF radio operation. These units are completely interoperable with MIL-STD-188-141A ALE Rockwell Collins Selscan commercial air and ground units and FED-STD-1045.

The ACP combines receive scanning and selective calling under microprocessor control to monitor up to 100 preset channels for incoming ALE calls. Link quality analysis circuits measure and store signal-to-noise and bit error rate characteristics of received ALE signals for use by automatic frequency selection algorithms.

Selective calling addresses and preset channels may be programmed by the user from the ACPC front panel or from a remote ASCII terminal. All presets are stored in non-volatile memory for power-off retention.

The ACP provides frequency control of the associated ARC-190(V) HF radio in order to monitor multiple frequencies by scanning multiple preset channels chosen from a total of up to 100 stored simplex or half-duplex preset channels. Incoming ALE calls are answered automatically and the calling station's address is displayed to the user. Positive squelch is automatically broken whenever contact is established in response to an incoming call or as a result of an outgoing call. The system also provides a standard selective calling (SELCAL) capability when used with the AN/ARC-190 RT-1341(V)6, RT-1341(V)7 or RT-1341(V)8 radios.

Outgoing calls can be initiated on a station-to-station or net broadcast basis.

Specifications

Dimensions:
(ACP) 198.6 × 121.9 × 495.8 mm
(ACPC) 66.5 × 146 × 106.7 mm
Weight:
(ACP) 9.53 kg
(ACPC) 1.81 kg
Power supply: 115 V AC, 400 Hz, 110 W, 28 V DC, 25 W

Status

In service in US C-5, C-20, C-25, C-27, C-130, C-141, KC-10 and VC-135 aircraft.

Contractor

Rockwell Collins.

Automatic Link Establishment (ALE) for HF

Type

Avionic communications processor.

Description

The 309M-1 Automatic Link Establishment (ALE) processor and 514A-12 control set operate together as a microprocessor-based remote-control subsystem which can be added to existing Rockwell Collins HF radio systems to automate and simplify operation.

The 309M-1 ALE processor combines receive scanning and selective calling under microprocessor control to monitor up to 100 preset channels for incoming ALE calls. Link quality analysis circuits measure and store signal-to-noise and bit error rate characteristics of received ALE signals for use by automatic frequency selection algorithms. The ALE processor automatically mutes received audio output from the HF radio while scanning to eliminate distracting HF background noise and irrelevant channel activity.

When an automatic call is placed, the operator selects the preset ALE address of the individual station or net to be contacted and initiates the call. Automatic channel selection algorithms choose the calling channel from the list of channels currently being scanned. Automatic channel selections are made according to the order in which the candidate channels are ranked. The system also features a wire data messaging capability via an MS-141A AMD internal data modem.

Channels scanned, together with the choice of which of the multiple self-addresses are valid at any time, are determined by the scan list or lists selected. Multiple scan lists may be selected simultaneously, resulting in a combined list of channels for scanning purposes. The unique flexibility provided by selectable scan lists allows the 309M-1 to participate in multiple networks simultaneously.

Specifications

Compliance: FED-STD-1045, MIL-STD-188-141A (MS-141A) ALE
Associated equipment: HF-9000, 718U/ARC-174, ARC-190, HF-80
Dimensions:
(309M-1) 198.1 × 124.5 × 320 mm
(514A-12) 66.5 × 146 × 106.7 mm
Weight:
(309M-1) 5.5 kg
(514A-12) 1.36 kg
Power supply: 28 V DC
(309M-1) 30 W
(514A-12) 25 W

Status

In production and in service.

Contractor

Rockwell Collins.

C-10382/A Communication System Control (CSC) set

Type

Avionic Communications Control System (CCS).

Description

The primary function of the Communication System Control (CSC) set is to provide the pilot with integrated, centralised control of data transferring capability, power switching, mode selection, operating frequencies, interconnections and signal flow routes of the aircraft's CNI equipment. The CSC provides for highly efficient operation of communications by integrating these primary CNI systems controls into the aircraft's advanced avionics architecture and also into a single convenient easy-to-operate pilot's control panel.

On the US Navy's F/A-18 aircraft, the controls and displays of the cockpit control panel are engineered to optimise pilot control of the CNI equipment. The control panel is positioned to allow the pilot to keep his eyes focused straight ahead, with only the pertinent information and controls he needs presented in his field of view.

A redundant MIL-STD-1553 multiplex bus provides connection between the CSC and the AN/AYK-14(V) mission computer for the flow of information and control. The mission computer provides CNI control signals, BIT commands and information for the control panel's alphanumeric display. In return, the CSC transmits equipment status, received CNI data, operating options and BIT response to the mission computer. Dedicated serial digital lines interface the control panel with the CSC.

To process data to and from the CSC, mission computer, control panel and CNI equipment, the CSC interfaces serial digital signals, discrete signals, analogue signals, synchro signals, avionic multiplex bus signals and audio signals. As the CSC microcomputer processes at least 1,300 parameters/s, it controls and processes the data required, leaving 40 per cent of real time available for growth.

Status

In service on the US Navy F/A-18 aircraft.

Contractor

Smiths Aerospace.

C-11746(V) communication system control unit

Type

Avionic Communications Control System (CCS).

Description

The C-11746 communication system control unit is designed for secure TEMPEST crew intercommunication and radio transmit and receive control in high-noise airborne applications.

The unit provides individual on/off and receive level control of radios and navigation receivers, voice-operated switching for hands-free intercom control and a remote select capability for HOTAS/HOCAS operation. A single unit can handle five transmit/receive radios, six navigation receivers, four auxiliary inputs and two intercom buses. Units are available with MIL-L-85762A NVG-compatible or standard edge-lit front panels.

Specifications

Dimensions: 66.675 × 127 mm
Weight: 1.18 kg
Power supply: 28 V DC, 7 W
Environmental: MIL-C-58111
TEMPEST: NACSIM 5100

Status

In service on the AH-64 Apache helicopter.

Contractor

Telephonics Corporation, Command Systems Division.

C-1282AG remote-control unit

Type

Avionic satellite communications system.

Description

The C-1282AG Remote-Control Unit (RCU) provides the capability to control and monitor remotely the operation of the RT-1273AG Multimission UHF Satcom Transceiver (MUST) or

The Raytheon Electronic Systems C-1282AG remote-control unit 0504105

the MD-1269A MultiPurpose Modem (MPM) over an asynchronous MIL-STD-188-114A balanced interface. Frequency and mode selection for the AN/ARC-171 and preset selection for the AN/WSC-3 are provided by the RCU via the MPM.

The C-1282AG RCU allows full control of all MUST and MPM radio and modem features, including frequency, data rate, modulation type, modem emulation mode, transmit power level and BIT. Menu and Arrow keys allow the operator to scroll through various menus to view or update radio or modem configurations. The Preset switch allows access to each of the eight radio and modem preset configurations. The volume knob may be used to attenuate plain text receive audio out of the radio or modem.

The C-1282AG RCU has an illuminated panel designed to MIL-P-7788E, which provides a menu-driven keypad/display interface. Display intensity is adjustable and tracks the externally controlled edge-lit panel brightness.

Specifications
Dimensions: 146 × 76.2 × 133.35 mm
Weight: 0.91 kg
Power supply: 19–30 V DC, 0.5 A max
Temperature range: −45 to +55°C
Altitude: up to 30,000 ft

Status
In service.

Contractor
Raytheon Company, Space and Airborne Systems.

CCS-2100 Communications Control Systems

Type
Avionic Communications Control System (CCS).

Description
The Palomar Products' CCS-2100 Communications Control Systems provide intercommunication solutions for secure applications. The secure facilities and extremely high channel-to-channel isolation, provide for multilevel security, private radio channels and private intercom channels.

The integrated, microprocessor-controlled, modular distribution system is designed for real-time adjustment of communication assets to meet the operational requirements of each mission.

The modular design permits systems to be designed to fulfil simple internal/external communication requirements or to provide full command and control systems that include crypto and datalink assignments, clear/secure radio relay, clear/secure simultaneous broadcast, unlimited conference networks and selective dial facilities. In addition, systems can be integrated with a host computer via a MIL-STD-1553B interface or other network links.

A typical CCS-2100 architecture provides a full TEMPEST-compliant communications control capability to flight deck crews and mission personnel that enables them to exercise control over the aircraft's crypto units, radio transceivers, mission receivers and navigation receivers.

Individual control units that comprise the CCS-2100 include Integrated Panel Crew Station Units (IPCSUs); Liquid Crystal Control Display Units (LCCDUs); Crew Station Units (CSUs); a

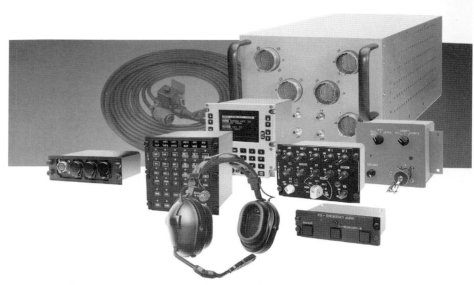

Palomar Products' CCS-2100 communications control system units 0062230

Communication Control Unit (CCU) (full ATR or half ATR sizes available); Maintenance Station Unit (MSU); and Emergency Audio Panel (EAP).

Features include simultaneous clear and secure transmissions; TEMPEST compliance; ElectroMagnetic Pulse (EMP) hardened; voice and data switching with binaural audio; redundant and emergency back-up operation; recorder control/playback.

Status
Palomar Products' CCS-2100 systems are widely used in US military aircraft and in the military aircraft of other nations.

Contractor
Palomar Products Inc.

CDR HF/VHF/UHF DSP Receivers

Type
Avionic communications system, HF/VHF/UHF.

Description
The CDR family of HF/VHF/UHF receivers incorporated digital signal processing microchip technology and is designed to try and solve many of the typical problems of communications, surveillance, and direction-finding on board ships and aircraft. Tuning range is 10 kHz to 3 MHz and 20 MHz to 1200 MHz VHF/UHF, with 1 Hz tuning steps and 3 ms tuning speed. Tuning resolution is 10 Hz.

The receivers are both half-rack versions. The units are intended for local or remote-control operation. Operation is menu-driven and little or no operator training is required. The receivers have digitally tuned preselectors and 250 programmable channels. There are also 51 selectable synthesised IF bandwidths from 1 to 240 kHz. Built-In Test (BIT) constantly troubleshoots the receiver. When a problem is detected, a message on the unit's display screen identifies the module which should be removed and replaced.

Specifications
CDR-3280 HF Receiver
CDR-3580 VHF/UHF Receiver

Cubic Communications' CDR-3580 DSP receiver 0044818

Frequency range: 10 kHz to 30 MHz, 1 Hz tuning resolution 20-1,200 MHz, 10 Hz tuning resolution
Detection modes: ISB, LSB, USB, CW, AM, FM LSB, USB, CW, AM, FM
Sweep and scan: 100 channels per second 100 channels per second
Channels: 250 programmable channels 250 programmable channels
IF bandwidths: 51 standard bandwidths 17 synthesised digital IF filters (1 kHz to 240 kHz)
Remote control: RS-232, RS-422 and IEEE 488 RS-232, RS-422 or IEEE 488
Dimensions: 19 in half-rack, 3.5 in high 19 in half-rack, 3.5 in high

Status
In service.

Contractor
Cubic Defense Applications.

CM-950 Communications Management Unit (CMU)

Type
Avionic Communications Control System (CCS).

Description
In July 1998, Honeywell added the ARINC 758-compliant CM-950 CMU to its WorldNav CNS/ATM product line, designed to support future 'free flight' airborne datalink protocols and applications, as a successor to ACARS management units.

The CM-950 CMU is fully partitioned, allowing airlines to modify user-defined functions without the need for costly and time consuming recertification.

Data loading is available by means of a standard ARINC 615 data loader. For high speed loading, the CMU is equipped with a PCMCIA interface and is provisioned for future Ethernet data loading capability.

The CM-950 CMU provides an easy upgrade path to future airline requirements, and combined with Honeywell's ground-based Airline Maintenance and Operations Support System (AMOSS), it provides a complete end-to-end product for any operator interested in a turnkey solution.

Specifications
Size: 4 MCU
Weight: 5.5 kg
Voltage: 115 V AC 400 Hz or 28 V DC
Power consumption: <40 W
Environmental: DO-160D
Software: DO-178B level C

Honeywell's   CM-950   communications
management unit   0044821

Status

The US Air Force has awarded Honeywell an Indefinite Delivery/Indefinite Quantity (ID/IQ) contract to supply a Communications Management Unit (CMU) and satellite communications system (satcom) to support its Global Air Traffic Management (GATM II) upgrade programme, up to a total quantity of 1,681 MCS-7000I satcom units and 3,554 CM-950 units by 2007.

Contractor

Honeywell Inc, Commercial Aviation Systems.

CMU-900/APM-900 datalink Communications Management Unit/Aircraft Personality Module

Type

Avionic Communications Control System (CCS).

Description

Rockwell Collins' CMU-900 is a datalink communication management unit designed to provide a seamless transition, at minimum cost, as worldwide data link systems moved from ACARS to ATN. The system manages datalink communications over all three air/ground subnetworks (VHF, HF and SATCOM) simultaneously. The CMU-900 supports all aircraft platforms with common hardware and core software and AOC that reduces the configuration management and administrative burden placed upon an airline operating numerous aircraft types and configurations. It was developed and certified to DO160C and DO178B, level D for current ACARS and FANS-1 routing operation, and has now become the industry's first certified ARINC Characteristic 758 compliant unit.

The CMU-900 provides a powerful User Application Development (UAD) capability allowing airlines the option of performing their own datalink non-essential application maintenance and modifications. It minimises software update costs by providing isolated partitions for user-modifiable and non-modifiable, as well as aircraft-specific and aircraft-independent functions. In common with Rockwell Collins' predecessor DLM-700B/C and DLM-900 series of management units, the CMU-900 minimises cost of ownership by maintaining exceptional levels of reliability.

Datalink communication unit/system configurations are dependent upon aircraft type, airline operation and ground system requirements. Customised AOC software configurations are available upon request.

The CMU-900 is also capable of supporting an ARINC 724B backplane wiring configuration, to facilitate installation in existing ACARS provisions.

The APM-900 Aircraft Personality Module complements the CMU-900 in a true ARINC 758 CMU installation. The APM-900 provides a means of permanently storing aircraft-unique parameters necessary for the initialisation and operation of the CMU, including the ICAO address, registration, aircraft type and airline code. The CMU provides power to the APM, program the configuration data upon initial installation and read the configuration data upon power-up.

When combined with the Rockwell Collins VHF, HF and SATCOM (Aero-H and Aero-I) radios, the advanced communications capabilities of the CMU-900 and APM-900 create a comprehensive data communications product line for both forward-fit and retrofit applications.

A Cross-Talk Interface that communicates with a second CMU-900 is planned and will be offered within the timeframe necessary to support applications requiring dual-CMU redundancy.

Specifications

Dimensions: 4 MCU per ARINC 600
Weight: 5.5 kg
Power supply: 115 V AC, 400 Hz, with 28 V DC optional
Power usage: 35 W (nominal); 2.1 W (standby)
Temperature range: –40 to +55°C
Altitude: up to 55,000 ft

Status

In service. The CMU-900 and APM-900 received Type Certification (TC) aboard a Delta Airlines Boeing 737-800 aircraft in May 1999. In addition, the system received Supplementary Type Certification (STC) on the Boeing 767 in June 2001 and full TC at Boeing on the B737 and B767 during the last quarter of 2001, for ACARS over AVLC (AoA) and VDL Mode 2 operation.

The CMU-900 was also selected for the KC-135 and C-17.

Contractor

Rockwell Collins.

COM 810/811 series VHF radios

Type

Avionic communications system, VHF.

Description

Narco's COM 810 and COM 811 models are solid-state microprocessor-controlled communication systems designed principally for light and general aviation aircraft. Each covers the VHF band from 118 to 136.975 MHz in which it provides 760 channels, two of which are preselectable; one is for active and the other for standby use. The 810 and 811 systems are essentially similar except that the former is designed for operation from a 13.75 V DC supply and the latter from a 27.5 V DC supply.

The active and standby frequencies are presented on LED displays which are automatically dimmed during darkness by a built-in photocell circuit. The legend XMT is illuminated when the microphone is keyed for transmission.

New frequencies may be entered in either the active or standby positions when desired; an illuminated arrow indicates which section has been selected for new frequency entry. Frequency selection is completed by use of a concentric tuning control, the outer part of which makes frequency readout changes at the rate of 1 MHz per detent and the inner part providing kHz changes at 25 kHz per detent. Clockwise rotation increases the numerical value of the frequency selection and counter-clockwise rotation decreases it. A transfer switch is used to exchange selected frequencies between active and standby modes.

An optional feature is a connection which enables the last entered frequencies to be retained in the system memory when the radio is inactive. This requires a trickle current of 0.1 mA from the aircraft's battery. If this circuit is not connected, then the radio automatically retunes to the 121.5 MHz internationally designated emergency frequency in the active mode and to the 121.9 MHz ground control frequency in standby mode the next time it is switched on. In the event of a display failure, the radio automatically reverts to these frequencies and may be retuned to the desired channel by counting the detent clicks of the tuning control.

Built-in automatic squelch control, deactivated by use of a pull/test switch, maintains audio silence until a signal is received. Automatic audio-levelling in both transmitter and receiver allows all signals to be heard at the same level regardless of modulation. A built-in 10 W amplifier, provision for multiple audio inputs and intercommunication facilities are also included.

Latest in the range is the TSO'd COM 810+R, which directly replaces all Narco COMs from the COM 11 to the COM 120 and the COM 810/811 and COM 810+/811+.

Specifications

Dimensions:
(transmitter/receiver) 159 × 38 × 279 mm
Weight:
(transmitter/receiver) 1.3 kg
(mounting tray) 0.34 kg
Frequency: 118–136.975 MHz
Channels: 760 incl 10 preset
Power output: 8 W nominal
Modulation: 85% at 1 kHz
TSO compliance: C37b Class IV (transmitter); C38b Class C (receiver)
Audio output: 10 W minimum (speaker); 50 mW minimum into 300 Ω (headphones)

Status

In service.

Contractor

Narco Avionics Inc.

Commanders' Tactical Terminal

Type

Avionic tactical communications system.

Description

Commanders' Tactical Terminal/Hybrid-Receive only (CTT/H-R)
The Commanders' Tactical Terminal/Hybrid-Receive only (CTT/H-R) is a multichannel, multifunction terminal used to receive real-time intelligence reports from a variety of sources. It allows the tactical user to receive data from the Tactical Reconnaissance Intelligence Exchange Service (TRIXS) or the Tactical Information Broadcast Service (TIBS), while simultaneously receiving the Tactical Receive equipment and

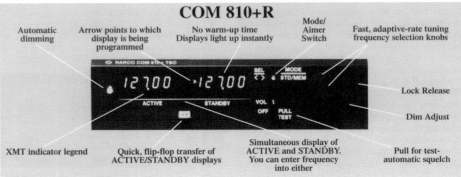

COM 810+R 'Direct Replacement' COM Transceiver   0001186

The Raytheon Electronic Systems commanders' tactical terminal/hybrid – receive only (CTT/H-R)
0044829

related APplications (TRAP) and Tactical Data Information eXchange System-Broadcast (TADIXS-B). This capability provides the user with a significant increase in access to real-time intelligence data and offers the flexibility to utilise whatever resources are available in the theatre.

CTT/H-R is the result of an evolutionary development that began with the full-duplex CTT field terminals and has continued with the development of a CTT/TEREC terminal, a multichannel CONSTANT SOURCE receiver system, a TIBS interface unit and a TIBS receive only unit. The CTT/H-R combines the functions of all these systems into a single receive-only terminal.

The CTT/H-R is packaged into a single unit and uses open architecture VME technology. Some of the key VME modules used include embedded COMSEC modules based on CTIC and Ricebird crypto chips, multiple 680X0 microprocessor modules and a MIL-STD-1553 interface. Additional modules include a single-board UHF receiver/synthesiser and a single-board modem using TMS320 processors. The CTTR/H-R is qualified for installation in fixed-wing aircraft and helicopters, as well as vehicles and ships. It uses Ada software and has a download capability for changes or upgrades.

Specifications
Dimensions: 190.5 × 266.7 × 520.7 mm
Weight: 19.5 kg
Power supply: 100–132 V AC, 47–140 Hz, 180 Wor 200–264 V AC, 47–66 Hz, 180 W
Frequency: 225–400 Hz
Channel spacing: 5 and 25 kHz
Temperature range: –20 to +50°C
Operating modes: TIBS, TRIXS, TDDS (TRAP), TADIXS-B-Tops and OBP
Demodulation: SBPSK or BPSK 2,400, 4,800, 9,600 and 19,200 bps
SOQPSK 4,800 and 19,200 bps
QFSK 32 kbps
Altitude: up to 30,000 ft

Status
In production for aircraft, helicopters, tracked and wheeled vehicles, ships, submarines and ground shelters.

Commanders' Tactical Terminal 3-channel (CCT(3))
The CTT (3) is central to the US Army's Integrated Battlefield Targeting Architecture (IBTA) and a vital link in the US Army's modernisation objectives. It supplies the critical datalink to battle managers, intelligence centres, air defence, fire support and aviation nodes across all service lines.

CTT (3) allows users to exploit all modes and rates of intelligence broadcast networks: Tactical Receive equipment and related APplications (TRAP), TActical Data Information eXchange System-B (TADIXS-B), Tactical Information Broadcast System (TIBS) and Tactical Reconnaissance Intelligence eXchange Service (TRIXS).

It provides tactical intelligence data transmission and network control in the TRIXS

The Raytheon Electronic Systems commanders' tactical terminal 3-channel (CTT(3))   0044846

andTIBS networks; C4I on the move and general situational awareness; support of sensor queuing, Secondary Imagery Dissemination (SID); over-the- horizon targeting and dynamic tasking of forces. CTT (3) enhances survivability through passive tracking, increases response times through early warning, provides data accuracy sufficient for a fire solution and accommodates collateral or SCI traffic.

Configurations available are:
- CTT/H3: full duplex in TIBS, TRIXS and GPL networks while receiving two additional channels of intelligence broadcasts or GPL UHF
- CTT/H-R3: receives three simultaneous channels of intelligence broadcasts or GPL

Specifications
Frequency range: 225 to 400 MHz
Channelisation: 5 and 25 kHz
Operating modes: full- or half-duplex
Primary power: 100–132 V AC, 47–440 Hz, single phase or 200–264 V AC, 47-66 Hz single phase battery-backed key storage and classified software
MTBF: 3,000 hours
Modulation: FM analogue and Secure Voice, FSK,
QFSK (32 kbps), BPSK, SBPSK 975, 300, 1,200, 4,800, 9,600, 19,200 and 38,400 bps)
Encoding/decoding: selectable differential or non-differential, convolutional encoding and Viterbi decoding
Signal acquisition: ±900 Hz offset at 40 Hz per sec
Interfaces: KG-84, KY-57/58, KGR-96, RS-232/422, MIL-STD-1553B, headset, intercom, maintenance
ECCM features: Have Quick II, Adaptive Array Processing (null-steering option), Quadrature Diversity Receiver/Combiner (option)
Transmitter: output power adjustable 11-20 dBW
Dimensions: RRT, RBP, R3, all full ATR tall-long
Weight: RRT, RBP, R3, all 29.55 kg
Cooling: self-contained forced air fans
Temperature: –45 to +55°C operating
–50 to +70°C nonoperating
Altitude: 30,000 ft operating
Vibration/shock: tracked and wheeled vehicles, helicopter, jet and propeller aircraft, shipboard
EMI/EMC: MIL-STD-461

Status
In production. The two-channel versions have been fielded to all US Services.

Contractor
Raytheon Company, Space and Airborne Systems.

DC-COM Model 500 voice-activated panel-mount intercom

Type
Avionic Communications Control System (CCS).

Description
The DC-COM Model 500 voice-activated panel-mount intercom was designed to provide simultaneous dual radio monitor/transmit capability. It provides three modes of operation: COM 1 in which pilot and co-pilot can transmit and receive on COM 1 radios; COM 2 in which the pilot can transmit and receive on COM 1, and the co-pilot can transmit and receive on COM 2; COM 3 in which the pilot and co-pilot can transmit and receive on COM 2 radio.

Facilities provided include: the ability to use up to six aviation headsets; individual VOX circuits for each headset; voice-operated intercom; pilot priority transmission; true stereo capability; fail-safe operation; optional ATC/AUX switches; horizontal or vertical installation.

Specifications
Dimensions: 43.2 × 71.1 × 152.4 mm
Weight: 0.43 kg
Power: 11-30 V DC @ 150 mA maximum
Temperature range: –40 to +70°C
Total headsets: 6
Total output power: 300 mW with 6 headsets

Contractor
David Clark Company Incorporated.

The DC-Com 500   0077760

Digital Audio Control Unit (D-ACU)

Type
Avionic audio management system.

Description
Sanmina-SCI Corporation's digital InterCom System (ICS) is designed for tactical aircraft, such as the C-130J Hercules, V22 Osprey and AH-64D Longbow Apache Helicopter. The Digital-Audio Control Unit (D-ACU) provides digital audio control and distribution, featuring a programmable display panel. Modes of operation and volume levels are controlled through the display unit. Volume levels and mode selection status are displayed in text and graphic formats. A spatial audio capability increases user situational awareness by presenting multiple audio channels in a 3-D format. An open-architecture design allows flexible communication system expansion and upgrades, with extensive use of COTS components yielding efficient life-cycle-costs. A maintenance port allows software modifications through the unit I/O connectors.

System features include:
- Digital Signal Processor logic control
- Programmable audio routing
- Microphone and binaural headset audio
- Three-dimensional headset audio
- Two IEEE 1394 Bus 3 port nodes
- Eight discrete control inputs
- Eight discrete control outputs
- Compatible with Active Noise Reduction (ANR) headsets
- Auxiliary analogue intercom (AIC-10 compatible)
- Maintenance port (RS-232)
- TEMPEST compatible design
- Compatible with A2C2S requirements.

Specifications
Dimensions: 103.4 × 145.8 × 180.8 mm (L × W × H)
Weight: 2.4 kg
Power: 28 V DC, MIL-STD-704D
Display:
Viewing area: 114 × 86 mm at 320 × 240 pixels
Viewing angle: >160°
Temperature: –40 to +54°X
Environmental: MIL-E-4400, Class 1A
MTBF: >6,000 hr

Contractor
Sanmina-SCI Corporation.

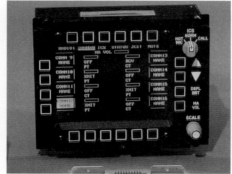

Sanmina D-ACU   0098797

Digital Communication Management System (DCMS)

Type
Avionic Communications Control System (CCS).

Description
The Digital Communication Management System (DCMS) family of intercommunication systems (AN/AIC-37 and AN/AIC-41) is a digital, distributive, TEMPEST certified intercommunication system designed to manage the communications assets of airborne platforms. It integrates and manages all the navigation aids and alert/warning tones, as well as providing for the real-time control and switching of modems and encryption devices between transceivers. Both frequency management and a MIL-STD-1553B interface

Digital Communication Management System (DCMS)   0044836

are offered. In addition, the AN/AIC-37(V) has no single point failure, contains BIT to card level (eliminating the need for intermediate level maintenance) and is field reconfigurable and expandable. A voice-activated switch, capable of operating in high-noise environments, and 25 intercom nets are standard features. Individual volume control for all assets is provided, plus a variety of front panel configurations ranging from push-buttons to a full display.

The AN/AIC-37(V) operates on a redundant 10 mbits/s databus. This permits up to 84 non-blocking channels of communication between crew stations and external radio systems, and a means to interface-with an external computer through a redundant MIL-STD-1553B bus. External voice and data transmissions can be made on encrypted or plain networks. The system can deny access to any crew position not authorised to receive secure information.

When operators communicate, the crew terminal digitises the audio into an allocated data time slot with a destination address to the selected interface. The destination unit then converts the digital information back into baseband audio for application to the specific asset. Similarly, when a radio or remote communication device receives audio, it is digitised and made available to any of the operators that have been enabled to receive the audio. The entire system is digitally reconfigurable, enabling asset assignment based on the operational scenario. All channels within the system can be selected, combined and monitored.

Specifications
Dimensions:
(DCI) 171.5 × 127 × 204.7 mm
(DAI) 152.4 × 127 × 242.8 mm
(DBI) 152.4 × 127 × 242.8 mm
(DII) 123.9 × 127 × 86.1 mm

(DUCK) 19.1 × 80.8 × 132.1 mm
(DBC) 57.1 × 111 × 57.1 mm
Weight:
(DCI) 4.11 kg, (DAI) 4.14 kg
(DBI) 3.64 kg, (DII) 1.02 kg
(DUCK) 0.25 kg, (DBC) 0.34 kg
Power supply: 28 V DC
Temperature range: –54 to +55°C
Altitude: up to 50,000 ft

Status
In production and operational on aircraft such as US Navy EP-3, ES-3 and P-3C, US Air Force E-8 JSTARS and the US President's Air Force 1. Also selected for the Norwegian Air Force P-3 upgrades and the German ATL-1 upgrade.

Contractor
Telephonics Corporation, Command Systems Division.

Digital Control Audio System (DCAS)

Type
Avionic audio management system.

Description
At the heart of each DCAS is the Digital Signal Processor (DSP). DSP circuitry gives flight crews crisp drift-free audio communications with few electronic parts and dramatically increased reliability over analogue systems, extending system flexibility through software add-on features.

The DSP consists of a Remote Electronics Unit (REU), plus an Audio Control Panel (ACP) for each user. In operation, control switch selections are made at the ACP, then multiplexed and sent to the REU. All analogue audio signals are also sent to the REU where they are filtered and converted to digital signals. The resulting digital audio is selected, amplified, filtered and summed in the DSP circuit according to control selections. After processing, the digital audio signal is converted back to analogue for distribution to the user. All processing is done at microprocessor speeds resulting in real-time audio communication with crisp digitally processed quality. Advanced features include a digital control bus between each flight deck ACP, BITE to monitor and report digitally on system operational integrity and advanced data reporting via ARINC 629 and 429 interfaces to other aircraft systems. Jack panels, headsets, microphones and speakers provide audio signal input/output, complete the system.

The digitally controlled audio system for the Boeing 777   0504225

Status

Derivatives of this system are in service in Boeing 777, 767, 757, 737, 717; Embraer EMB 120; Canadair CRJ 100, 200, 700; DeHavilland Dash 8; CASA 235; and IPTN235 aircraft.

Contractor

AVTECH Corporation.

DLink+

Type

Avionic communications processor.

Description

ACSS's communications management system is designed to meet the diverse requirements of aircraft operations and associated ground networks. Dlink+ is a lightweight low cost communications system integrated into a small panel-mount CDU. The system provides extensive digital and analogue Input/Output (I/O), an ACARS processor, internal VDL-Mode 2 radio and optional GPS sensor. The ARINC 429 I/O allows the Communications Management Unit (CMU) access to aircraft information such as engine data, air data, fuel information, and other aircraft parameters.

Dlink+ also provides a growth path with a choice of bi-directional communication media such as SATCOM and airborne telephone and interfaces with host aircraft peripherals, such as ARINC 744 printers and data loaders. System software supports ACARS over AVLC, with growth provisions for Controller-Pilot Data Link Communication/Aeronautical Telecommunications Network (CPDLC/ATN). The optional internal GPS sensor provides dedicated, precise position and time reporting for the CMU.

Dlink+ features a modular design, with separate high-speed Single Board Computers (SBCs) dedicated to each of its major functions. The ACARS SBC is the heart of the system and provides the communication protocol and routing functions. Host aircraft interface is supported y a high-speed dedicated I/O SBC. The VDL-Mode 2 radio is an additional SBC based on the latest digital signal processing technology.

Specifications

Dimensions: 146.1 × 114.3 × 215.9 mm (W × H × D)
Weight: 2.27 kg
Display: Monochrome EL, 9 line display × 24 characters
Environmental: DO-160C; A1/CBA(BMN)XXXXX XBAZAUZAUKXX
Software: DO-178B Level D
Operating temperature: −15 to 550° C
Power: 28 V DC (22–32 V DC range)
Consumption: 20 W (typical); 35 W (maximum)
Processor: Motorola MPC860; 40 MHz; Memory 4 MB Flash, 8 MB SRAM on each microprocessor based PCA
Inputs: ARINC 429 × 5; RS-232/422 × 2/1; discrete × 8; 10 Base-T Ethernet × 1

Status

In production and in service.

Contractor

Aviation Communication & Surveillance Systems (ACSS).
An L-3 Communications & Thales company.

DM C50-17 VHF communication antennas

Type

Aircraft antenna, Communications, Navigation, Identification (CNI).

Description

The DM C50-17 series of VHF communication antennas has incorporated design improvements to enhance corona threshold, corrosion protection and drag characteristics. The DM C50-17 series is low-cost, lightweight and strong blade antennas for use on commercial jet aircraft. Aside from low acquisition costs, further in-service cost savings are claimed due to low installed weight and lower fuel consumption due to the low-drag design. The fuel saving per shipset is claimed to be 160.5 USG per aircraft per year compared to other similar systems. Many of the DM C50-17 series are interchangeable with other types of antennas presently in use.

Specifications

Frequency range: 116 to 156 MHz
VSWR: 2.0:1
Impedance: 50 Ω
Power: 1 kw CW
Polarisation: Vertical
Radiation pattern: Omnidirectional (equivalent to vertical stub)
Weight: 1.5 kg

Status

The DM C50-17 antenna has been selected as original equipment for the Boeing 757, 767 and 777 aircraft. Other models in the DM C50-17 series can also be used on Boeing 707, 727, 737, 747, DC-8, DC-9 and DC-10, Airbus A300 and other commercial aircraft.

Contractor

AIL Systems, Inc.

DM C60 VHF communications antenna

Type

Aircraft antenna, Communications, Navigation, Identification (CNI).

Description

The DM C60 is a vertically polarised VHF communications antenna designed for either top or bottom fuselage mounting. The antenna covers the frequency range of 118 to 137 MHz (−1) or 116 to 152 MHz (−17) for both transmitting and receiving applications. Electrical elements of the antenna are completely sealed by foamed-in-place resin within an outer glass fibre housing. As a result, the antenna is rugged enough for use on both commercial and general aviation aircraft.

Specifications

Frequency range −1: 118 to 137 MHz
Frequency range −3, −4, and −17: 116 to 152 MHz
VSWR: 2.0:1 max
Power: 50 W
Impedance: 50 Ω
Polarisation: Vertical
Radiation pattern: Omnidirectional
Gain: Equivalent to a quarter wave stub
Drag: 0.23 kg (SL at 250 mph)
Side Load Strength: Over 3 psi
Altitude: Up to 35,000 ft
Compliance: TSO C37d and C38d
Weight: 0.73 kg

Status

In production and in service in commercial and business aircraft.

Contractor

AIL Systems, Inc.

DM C70 VHF communications antennas

Type

Aircraft antenna, Communications, Navigation, Identification (CNI).

Description

The DM C70 series of VHF communication antennas are designed for top or bottom installation on high-performance, single, twin and turbo engine fixed- and rotary-wing aircraft. These antennas offer mechanical strength and high-electrical efficiency to provide maximum reliability and full 360/720-channel transceiver operation. The lightweight profile is unobtrusive, resists icing and offers low drag. The DM C70-3 is directly interchangeable with C598501-0104, VF 10-210, CI 109, and CI 121 VHF antennas. The DM C70-4, which supersedes the DM C70-2, is specifically designed for bottom installations where ground clearance does not permit the use of other DM C70 series antennas. The DM C70-4 provides a higher degree of efficiency to assure maximum transceiver performance. The DM C70-6 and DM C70-9 variants (of the DM C70-1/A) have a different mounting hole pattern and connector location. All other characteristics remain the same.

Specifications

Frequency range: 118 to 137 MHz
VSWR:
  DM C70-1/A, −3, −6, −9: 2.0:1
  DM C70-4: 2.5:1
Power: 50 W
Impedance: 50 Ω
Polarisation: Vertical
Radiation pattern: Omnidirectional
Drag: 1.14 kg (25,000 ft at 300 mph)
Speed rating: 400 mph (DM C70-1/A, -4, -6, -9); 250 mph (DM C70-3)
Compliance: TSO C37d and C38d
Weight: 0.34 kg

Status

In production and in service in general aviation aircraft.

Contractor

AIL Systems, Inc.

DM N4-17 VOR/LOC/glideslope antenna

Type

Aircraft antenna, Communications, Navigation, Identification (CNI).

Description

The DM N4-17 VOR/LOC/glideslope antenna is designed for general aviation, commercial, and military aircraft that operate up to M1.0. The DM N4-17 is designed and qualified to provide a low-cost, lightweight, low-drag antenna for modern avionics systems. The antenna is TSO'd and its performance exceeds the environmental specifications of MIL-E-5400 Class 3. The balanced loop design of the DM N4-17 provides for an omnidirectional radiation pattern at the horizon to obtain the maximum signal for standard VOR and area navigation, which in turn provides results in greater reception distance and system performance.

Specifications

Frequency range: 108 to 118 MHz (VOR/LOC); 329 to 335.3 MHz (glideslope)
VSWR: 5:1 max
Gain: 0 ± 2 dB
Impedance: 50 Ω
Polarisation: Horizontal
Radiation pattern:
(VOR/LOC): Omnidirectional
(Glideslope): Forward pointing
Side load: 17 psi
Compliance: TSO C34e, C36e, C40c
Weight: 0.6 kg

Status

In production and in service.

Contractor

AIL Systems, Inc.

DM N4-45 VOR/LOC antenna

Type

Aircraft antenna, Communications, Navigation, Identification (CNI).

Description

The DM N4-45 is a VOR/LOC antenna designed for high-performance aircraft where aerodynamic drag and weight must be minimised. The DM

N4-45 has been designed from the ground up to incorporate latest developments in structural materials, but retaining the electrical performance of a balanced loop VOR/LOC antenna. The DM N4-45 retains the form of the DM N4-15 (see separate entry), while incorporating a new integrated radiating boot material design.

Specifications
Frequency range: 108 to 118 MHz
VSWR: 5:1
Impedance: 50 Ω
Polarisation: Horizontal
Radiation pattern: Omnidirectional
Weight: 2.5 kg

Status
In production and in service in business jet aircraft.

Contractor
AIL Systems, Inc.

DM N25 glideslope antennas

Type
Aircraft antenna, Communications, Navigation, Identification (CNI).

Description
The DM N25 series of glideslope antennas employ a grounded, centre-fed loop as the radiating element, producing symmetrical radiation patterns. The DM N25-2 is a dual-connector version, which effectively isolates the two associated receiving systems providing protection against disablement of both systems in the event of failure in one system. The radiation pattern associated with each DM N25-2 antenna port is identical, so that no coupler dropout problem can occur during automatic landing.

All DM N25 series antennas provide complete lightning protection for the associated receivers. They are suitable for mounting either externally, or within a nose radome. The inherent 180° pattern beam may be modified slightly by the geometry surrounding the installation. The DM N25 antennas are recommended for installations in close proximity to radar dish installations since the antenna sensitivity is relatively unaffected by dish motion.

Specifications
Frequency range: 329 to 335.3 MHz
VSWR:
  DM N25-1: 5:1
  DM N25-2: 3:1
  DM N25-3 (AT-983): 3:1
Gain: 0 dB
Impedance: 50 Ω
Polarisation: Horizontal
Compliance: TSO C34b, MIL-T-5422E
Weight:
  DM N25-1: 0.23 kg
  DM N25-2: 0.39 kg
  DM N25-3: 0.23 kg

Status
In production and in service in air transport and business jet aircraft.

Contractor
AIL Systems, Inc.

DM N27 marker beacon antenna

Type
Aircraft antenna, Communications, Navigation, Identification (CNI).

Description
The DM N27 antenna is a rugged, lightweight antenna for the reception of 75 MHz marker beacon signals. The low-drag design utilises a simple external mounting and requires no cutting of airframe structural members. Protection against moisture ingress is provided by a dielectric foam filled, white polyester-fibreglass radome which is fitted with a metal leading edge for erosion protection. The DM N27 antenna is

designed to meet the performance specifications of antennas AT-134 and AT-536.

Specifications
Frequency range: 74.75 to 75.25 MHz
Centre frequency: 75.0 MHz
VSWR: <1.5:1 (at midband); <5.0:1 (74.8 to 75.2 MHz)
Gain: −9 dB (relative to isotropic)
Impedance: 50 Ω
Polarisation: Horizontal
Compliance: TSO C35a (civilian); MIL-T-5422E military)
Weight: 0.28 kg

Status
In production and in service in air transport and business aircraft.

Contractor
AIL Systems, Inc.

DM N41-1 glideslope antenna

Type
Aircraft antenna, Communications, Navigation, Identification (CNI).

Description
The DM N41-1 is a centre-fed, loop-type antenna which covers the frequency range of 329 to 335.3 MHz for use with glideslope receivers. When the antenna is mounted on a forward-facing metal surface it provides a broad, single radiation lobe. The antenna structure consists of a rugged aluminum die-casting. Featuring high signal output, the DM N41-1 is ideal for use with a coupler for dual-receiver installations.

Specifications
Frequency range: 329 to 335.3 MHz
VSWR: <5:1
Gain: 0 dB
Impedance: 50 Ω
Polarisation: Horizontal
Radiation pattern: Hemispherical
Compliance: TSO C34c
Weight: 0.23 kg

Status
In production and in service in general aviation aircraft.

Contractor
AIL Systems, Inc.

DM N43 marker beacon antennas

Type
Aircraft antenna, Communications, Navigation, Identification (CNI).

Description
The DM N43-1, DM N43-3, and DM N43-4 are low silhouette antennas for use with 75 MHz marker beacon receivers. When properly mounted on the underside of an aircraft, the antennas provide a broad, single-lobed pattern directed downward. The radiation is polarised parallel to the long dimension of the antenna. The DM N43-1 is rugged enough for use on all subsonic aircraft, with all electrical elements of the antenna completely sealed by foam resin within a plastic housing.

Specifications
Frequency range: 74.75 to 75.25 MHz
Centre frequency: 75.0 MHz
VSWR: <5:1; <2:1 (at 75 MHz)
Gain: −10 dB
Impedance: 50 Ω
Polarisation: Horizontal
Radiation pattern: Hemispherical
DC resistance: Grounded
Drag: <0.23 kg (SL at 250 mph)
Compliance: TSO C35a
Weight: 0.45 kg

Status
In production and in service in general aviation aircraft.

Contractor
AIL Systems, Inc.

DM N56-1 VOR antenna

Type
Aircraft antenna, Communications, Navigation, Identification (CNI).

Description
The DM N56-1, designed for the Boeing 757 and 767, represents a concept that can be incorporated in many aircraft designs to provide integral VHF navigation systems performance. The DM N56-1 free standing, balanced-loop array is mounted on the tip of the vertical stabiliser and covered with a structural fiberglass housing forming the fin tip. The electrical design features two-receiver operation, with performance complying to Category III approach requirements. The omnidirectional radiation patterns in the horizontal plane also ensure signal levels are sufficient for R-NAV systems use.

Specifications
Frequency range: 108 to 118 MHz
VSWR: 6:1
Impedance: 50 Ω
Polarisation: Horizontal
Radiation pattern: Omnidirectional
Isolation: 8 dB
Weight: 2.34 kg
Compliance: TSO C40a

Status
In production and in service in Boeing 757 and 767 commercial transport aircraft.

Contractor
AIL Systems, Inc.

DM N102-1-1 GPS antenna

Type
Aircraft antenna, Communications, Navigation, Identification (CNI).

Description
The DM N102-1-1 GPS antenna has been designed to meet ARINC 743A active antenna requirements and takes advantage of the latest in microstrip and microcircuit technology. Among the DM N102-1-1 key features are its lightning protection (direct strike) and enhanced filtering configured to suppress interference from other L-Band Systems such as Inmarsat. DC bias is provided through a RF coaxial connector. The DM N102-1-1 complies with Boeing specification S242T104.

Specifications
Frequency range: 1575.42 ±10 MHz
VSWR: 2.0:1
Gain coverage (min): −1.0 dBic ($0 \leq \theta \leq 75°$);
  −2.5 dBic ($75 \leq \theta \leq 80°$); −4.5 dBic ($80 \leq \theta \leq 85°$);
  −7.5 dBic ($\theta = 90°$ at horizon)
Gain (preamp): 33.0 ±3 dB
Power handling: 1 W
Impedance: 50 Ω
Polarisation: RHCP
Voltage: +12 V DC
Current: 100 mA max
Compliance: TSO C129a, DO-160, ARINC 743A, Boeing S242T104
Weight: 0.11 kg

Status
In production and in service in air transport, business and general aviation aircraft.

Contractor
AIL Systems, Inc.

DM NI50 transponder antennas

Type
Aircraft antenna, Communications, Navigation, Identification (CNI).

Description
Currently in use on business, commercial and military jet aircraft, DM NI50 series antennas are claimed to confer reduced maintenance costs, resulting from unequalled mechanical strength and built-in reliability. Other features of these antennas include:

- Extremely high side load strength to guard against breakage by ground-handling equipment
- Completely sealed construction to prevent failure due to moisture intrusion
- Lightning protection circuits to prevent damage to antenna and safeguard connected electronic equipment

The DM NI50 series antennas, designed to replace DM NI49 series items, are compatible with all standard L-Band equipments due to their very broad bandwidth of 960 to 1,220 MHz. DM NI50 series antennas are claimed to be directly interchangeable with virtually all L-Band blade and flush mounting antennas currently in service on commercial and military aircraft.

Specifications
Frequency range: 960 to 1,220 MHz
VSWR:
  960 to 1,220 MHz: <1.7:1
  1,000 to 1,100 MHz: <1.5:1
Gain: 0 dB (average at horizon)
Power: 3 kw peak (RF); 100 W peak (DC)
Impedance: 50 Ω (RF)
Polarisation: Vertical
Drag: 0.13 kg (SL at 0.8 M)
Side load: >79.5 kg
Speed rating: 400 mph (DM C70-1/A, -4, -6, -9); 250 mph (DM C70-3)
Compliance: TSO C74 and C66a (civil); MIL-T-5422E (military)
Weight: 0.11 kg

Status
In production and in service in air transport and business jet aircraft.

Contractor
AIL Systems, Inc.

DM NI70 DME/ATC transponder antennas

Type
Aircraft antenna, Communications, Navigation, Identification (CNI).

Description
DM NI70 L-Band antennas provide sufficient bandwidth (950 to 1,220 MHz) for use with civil aviation ATC transponders and Distance Measuring Equipment (DME). The single metal element antenna provides omnidirectional, vertically polarised coverage when installed on the top, or bottom, of an aircraft. Its streamlined shape assures minimum aerodynamic drag.

Specifications
Frequency range: 960 to 1,220 MHz
VSWR:
  960 to 1,220 MHz: <1.7:1
  1,000 to 1,100 MHz: <1.5:1
Gain: Equivalent to a quarter wave stub
Power: 2 kw peak
Impedance: 50 Ω (RF)
Polarisation: Vertical
Radiation pattern: Omnidirectional
Drag: 0.23 kg (SL at 250 mph)
Compliance: TSO C74b and C66a
Weight: 0.11 kg

Status
In production and in service in general aviation aircraft.

Contractor
AIL Systems, Inc.

DM PN19 radio altimeter antenna

Type
Aircraft antenna, Communications, Navigation, Identification (CNI).

Description
The DM PN19-1-1 and DM PN19-2-1 employ microstrip technology, developed to fabricate high-quality, low-cost, lightweight, and near conformal antennas. The dielectric insert and radiating elements are recessed into a complete four-sided metal housing. This design feature protects the dielectric surface from the air stream for increased erosion protection in addition to preventing water intrusion and delamination.

Specifications
Frequency range: 4,200 to 4,400 MHz
VSWR: <2.0:1 (4,200 to 4,400 MHz); <1.8:1 (4,275 to 4,325 MHz)
Gain: 10 dBi
Impedance: 50 Ω
Polarisation: Linear
Temperature: –55 to +85°C operating
Altitude: –1,000 to +70,000 ft
Vibration: 10 g

Contractor
AIL Systems, Inc.

Compliance: TSO C87, ARINC 707/552, DO-160c, MIL-E-5400
Weight: 0.14 kg

Status
In production and in service. The DM PN19-1-1 is designed for air transport aircraft; the DM PN19-2-1 is designed for business jet and general aviation aircraft.

Contractor
AIL Systems, Inc.

DSS-100 digital intercom system

Type
Avionic internal communications system.

Description
The DSS-100 uses Commercial-Off-The-Shelf (COTS) components in an open VME architecture, providing a complete solution for systems demanding high-quality digital voice/data switching with combined system management.

Features include a distributive architecture, which can readily be tailored to customer requirements. The system is a secure digital switch with physically separate 'red' and 'black' buses. The red and black bus structure is maintained throughout

POSSIBLE TOUCH PANEL CONFIGURATIONS

Palomar Products' DSS-100 digital intercom system components   0064384

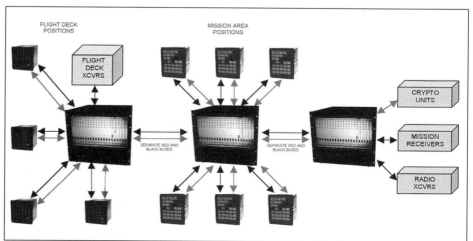

A representative system configuration for the Palomar Products DSS-100 digital intercom system
0064383

all components of the communications suite, including the digital operator positions.

The DSS-100 provides complete management of the communications suite and co-ordinates modes of transceivers, cryptos, data modems and host computers with actions and indications at the operator positions.

The DSS-100 features include up to 2,048 channels; selectable bandwidths; direct signal processing voice algorithms; binaural audio (spatial optional); unlimited conferencing; a variety of colour liquid crystal touchscreen operator display panel styles.

Contractor
Palomar Products Inc.

DVR-IDM

Type
Avionic datalink system.

Description
The Fast Tactical Imagery-Digital Video Recorder-Improved Data Modem (FTI-DVR-IDM) is a multifunction tactical imagery recording and management system which provides broadcast quality Digital Video Recording (DVR), imagery and file transmit/receive (Tx/Rx), Ethernet server, MIL-STD-1553B data recording, messaging and system control in a small integrated unit.

The system has been designed as a modular, three component (DVR only, FTI-DVR or FTI-DVR-IDM) system, featuring open architecture to provide for interoperability with, and support for, both legacy and emerging standards and protocols. The system is fully backwards compatible with all current and prior versions of PhotoTelesis technologies deployed on numerous fixed- and rotary-wing combat aircraft, intelligence platforms (ground/air/vessel), ground targeting/intelligence systems (LVRS) as well as ICE™ imaging and communications software. DVR-IDM provides full functionality while occupying the space currently required for legacy airborne tape recorders. System features include:

• Removable Solid State Memory Module (SSMM) up to 180 Gb
• Erase and declassify functions
• Drop-in replacement for many legacy Hi-8 Video Tape Recorders (VTR)
• Ruggedised for military aircraft
• Optional cockpit Remote Control Unit (RCU)
• Ethernet server functionality with up to 40 Gb of data storage.

The full FTI-DVR-IDM incorporates full Fast Tactical Imagery functionality, including the capture, viewing and Tx/Rx of imagery from onboard sensors. In addition, the system can receive, view, transmit and re-transmit data and images from other ground and airborne sources and utilises proprietary VideoBurst™ technology for multiframe image or video clip capture and transmission.

Specifications
Interoperability: AFAPD, TACFIRE, IDL, MIL-STD-188–220
Radio interfaces: Two primary channels can be configured for analogue (ASK, FSK) or digital (NRZ, ASK, MIL-188–184, synchronous/asynchronous)
Data rate: 16,000 bps (radio channels); 115,000 bps (serial port)
Communication functions: PTAC, TAC02, MIL-188-184 protocols
Two Ethernet 10/100 BaseT ports, two MIL-STD-155 ports
Two fibre channels (option), VMF to Link 16
Two RS-232 channels
One USB 2.0 channel
Two-channel IDM
DVR functions: Up to four channels of record/playback
2–4 hours of high-quality video and audio on all channels
Scalable Motion-JPEG
Frame date and time stamp

Event marker recording and display
Six video inputs supporting RS-170A, NTSC and PAL formats
RS-343 675/875 line scan conversion
Analogue composite and S-Video I/O
PC playback using COTS software
Memory storage: 256 Mb SDRAM; 4 Mb SRAM; 256 Kb EEPROM; 64 Mb FLASH; 180 Gb removable
Power: 28 V DC per MIL-STD-704
Environmental
Temperature: –54 to +71°C
Vibration: Per MIL-STD-810
Dimensions: 177.8 (D) × 147.3 (H) × 152.4 (W) mm
Weight: less than 4.1 kg

Status
In production and in service.

Contractor
Symetrics Industries Inc.

E-SAT 300A satellite receiver

Type
Avionic satellite communications system.

Description
The E-SAT 300A is an all-digital satellite communication system designed for aircraft applications. Airborne telephone calls and data messages are automatically transmitted via the INMARSAT network to ground stations. Two telephone calls can be made simultaneously, while a third channel is available for relaying data messages. The system is designed to accommodate growth to eight simultaneous telephone channels.

The steered high-gain single-helix antenna features small size and weight and minimises the number of LRUs. The design eliminates the keyhole and is mechanically steerable through 360° in azimuth and –30 to +90° in elevation.

Specifications
Dimensions:
(satellite data unit) 9 MCU
(radio frequency unit) 12 MCU
(Class A high-power amplifier) 8 MCU
(antenna control unit) 6 MCU
(system power supply) 8 MCU
(radome) 374.7 × 431.8 × 2,794 mm
Weight: (total system) 99.79 kg
Power: 1.3 kVA
Frequency:
(receive) 1,530–1,559 MHz
(transmit) 1,626.5–1,660.5 MHz
Reliability: 31,500 h MTBF

Status
In service.

Contractor
Raytheon Company, Space and Airborne Systems.

Flexcomm communication systems

Type
Avionic communications system, VHF/UHF.

Description
Wulfberg's Flexcomm systems are synthesised AM/FM communications systems designed specifically for use in fixed- and rotary-wing aircraft. The communications system comprises a control head (C-5000 or C-1000), single-band or multiband transceivers (RT-30, RT-138F, RT-406F and RT-5000) and antennas. All components operate on 28 V DC and have 5 V DC and or 28 V DC lighting.

Flexcomm I transceivers are 10 W FM single-band radios, each covers a particular band: the VHF LO-band of 29.7 to 49.9975 MHz with the RT-30, the VHF HI-band of 138 to 174 MHz with the RT-138F, and the UHF band 406 to 512 MHz with the RT-406F. Options include: a 'guard receiver'; a 'guard receiver' with CTCSS receive tone function; and for special applications, the RT-138F and RT-406F

can be modified for additional sensitivity – this option precludes the use of a 'guard frequency'. Encryption is available using Motorola DVP with a C-1000 or C-5000 control head. The Flexcomm II AM/FM transceiver, the RT-5000, covers all frequencies between 29.7 and 960 MHz.

C-1000 and C-5000 control heads are used to control multiple receivers, thus providing the system with the power to act as a command and control system performing relays, simulcast, repeater and full-duplex operations. Flexcomm I systems include a C-1000 communications management controller and up to three single-band antennas. Flexcomm II systems include a C-5000 communications management controller and any combination of RT-5000 Flexcomm II multiband and Flexcomm I transceivers and antennas.

Status
In production and in service.

Contractor
Chelton Avionics Inc, Wulfsberg Electronics Division.

GA-540 TADIL-A/Link-11 Serial DataLink Translator (SDLT)

Type
Avionic datalink system.

Description
The GA-540 Serial DataLink Translator (SDLT) provides the interface in a TADIL-A/Link-11 system between a serial encryption device and a tactical data system processor. The GA-540 serial interface to the encryption device meets the Airborne Tactical Data System (ATDS) interface requirements of MIL-STD-188-203-1A (Appendix D-2). All data buffering, timing and ATDS handshaking are performed automatically. The GA-540 connects to the tactical data system processor using one of the following interfaces: RS-422/RS-423 conditioned diphase; transformer-coupled conditioned diphase; RS-422/RS-423 synchronous; RS-423 asynchronous; VME bus.

The GA-540 is offered as a 19 in rack-mountable unit, as a 1/4-ATR-short airborne unit, or as a 6U-VME card. The rack-mounted and ATR units include a power supply, BIT indicator, and panel-mounted connectors. The VME card set can be hosted in any VME operating environment.

Specifications
Dimensions:
(size) ¼ ATR-short
(height) 19.3 cm
(width) 5.7 cm
(depth) 32.0 cm
Power:
24 to 32 V DC
15 W
Reliability: 78,000 h

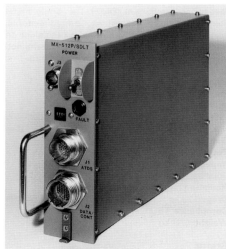

GA-540 TADIL-A/Link-11 Serial DataLink Translator (SDLT)   0002120

Status
In service.

Contractor
DRS Communications.

GPS/telemetry nosecone antenna system

Type
Aircraft antenna, Communications, Navigation, Identification (CNI).

Description
AIL Systems has designed and manufactures a wide variety of AIM 9 pod-mounted and qualified nosecones that provide complete antenna systems, combining both L1 and L2 GPS antennas and various telemetry antennas in one autonomous structure. The antennas operate over narrow bands within the VHF to high L-Band frequency ranges and are flight-qualified on numerous aircraft, including the F-14, F-15, F-16 and F-18. The antennas are housed within a high temperature-capable dielectric nosecone, which provides better radiation pattern coverage than conventional metallic nosecones, without compromising structural integrity. The design is modular in construction, permitting factory substitution of telemetry antennas to allow maximum flexibility as frequency allocations change. Complete systems are available including GPS preamplifiers, GPS antenna switching networks and air-data systems including pitot tubes and air data transducers. Custom configurations are possible to suit project requirements.

Specifications
Frequency range:
GPS: L1, or L1 and L2
Telemetry: 141, 225–400, 433, 512–650, 980, 1350–1390, 1350–1450, 1710–1850 MHz
VSWR: 2.0:1
Impedance: 50 Ω
Polarisation:
GPS: RHCP
Telemetry: LHCP or linear (model dependent)
Radiation pattern:
GPS: Hemispherical
Telemetry: Omnidirectional
Weight: 9.1–11.4 kg
Dimensions: 797.9 × 508.0 mm (L × Ø, across antennas)

Status
In production and in service.

Contractor
AIL Systems, Inc.

Have Quick system

Type
Airborne jam-resistant communications system.

Description
Have Quick provides the user with an effective air-to-air, air-to-ground and ground-to-air jam-resistant UHF voice communication capability for operations in jamming environments.

The Have Quick is an applique system. It consists of an ECCM modification to selected airborne and ground-based radios, which gives them a frequency-hopping capability. Part of the strength of the system comes from the use of channels in an apparently random manner so that no pattern is evident to the external observer. Jamming is consequently more difficult.

The frequency-hopping scheme is implemented by storing a pattern of the frequencies to be used for a given day within every Have Quick radio and utilising this pattern according to the time of day. For every time slot in the day, where each time slot is a small part of a second, there is a specific frequency which must be used

for a given communications net, whether it is transmitting or receiving. This frequency changes pseudorandomly from one time slot to the next. Thus, Have Quick terminals require some means to store the frequency pattern for channel use on a given day and also an accurate clock to control the times at which the pattern is consulted.

The Have Quick radio retains the normal non-hopping mode where it uses any one of the 7,000 channels available in the 225 to 400 MHz UHF communication band. The use and operation of the radio in normal mode is essentially unchanged from present-day procedures.

If jamming is encountered, the Have Quick radios can switch over to the ECCM mode and continue their communication. In order to permit this switch over, the radios must be suitably primed so that they will be synchronised in the ECCM mode; this is usually done before take-off.

The Have Quick system also has a capability termed 'multichannel' or 'break-in' operation. This permits a Have Quick radio to receive two simultaneous transmissions on the same net while avoiding the beat note which typically prevents the listener from understanding either transmission. The technique is implemented automatically in the transmitter where it is recognised whether or not the net is already in use. If so, the transmitter side-steps by one 25 kHz channel. Since the receiver is set to wideband mode, the second signal is received in addition to the original. The multichannel operating capability can be selected by the operator, whenever the system is in ECCM mode.

Status
Have Quick has been defined at two levels of complexity – Have Quick 1 and Have Quick 2. It is in service in a large number of NATO military aircraft, and Have Quick applique systems have been developed by a number of NATO radio manufacturers. AN/ARC-164 is perhaps the most common host radio for the Have Quick system.

Contractor
Raytheon Company, Space and Airborne Systems.

HF datalink

Type
Avionic datalink system.

Description
The HF datalink service is an extension of the VHF Aircraft Communications Addressing and Reporting System (ACARS). It provides an aeronautical data communication link beyond the line of sight limitations of VHF, with the potential for worldwide coverage.

The avionics system consists of an airborne HF Data Unit (HFDU) incorporating an HF modem and a datalink processor which implements the air-ground link and network access protocols. The system makes use of the aircraft's existing HF radio and antenna. Data transmission rates are comparable to those of low data rate Satcom.

The HFDU provides interface between the ACARS MU and HF and VHF radios. In VHF mode, it is transparent to the MU and VHF radios. In the HF mode, there is no operator involvement. The HFDU performs automatic search and selection of HF frequency, controls tuning and keying of the HF radio and employs the HF modem to send or receive ACARS messages. The 4 MCU enclosure provides HF modem and airborne datalink processor, power supply plus backplane and two spare slots for enhancements such as ATN.

Status
In service; certified on Boeing 767-300 airliners.

Contractor
Honeywell Aerospace, Electronic & Avionics Lighting.

HF messenger

Type
Avionic communications system, data.

Description
Rockwell Collins' HF messenger permits personal computers to exchange digital information over a standard HF radio. HF messenger profile software provides all the functions required to send and receive digital data and files between the personal computer and the radio. It also provides all the functions required to send and receive digitised signals over the HF propagation medium.

Normally, computer networks transfer data via the normal ground-based communications infrastructure: telephone lines, fibre optic cables, microwave links and satellite channels. However, for airborne communications, terrestrial-based links are not possible and satellite channels may be too costly or impractical. Under these circumstances, the aircraft HF radio may be used to provide the required connectivity.

HF messenger includes an optional Point to Point Protocol (PPP) type client interface. With this PPP interface, the HF node can be connected to any commercial standard TCP/IP router/remote access server to provide a transparent switched connection over HF – TCP/IP network supporting commercial standard e-mail applications.

HF messenger is based on the protocol defined in the NATO STANAG 5066 (an international public standard for HF data communications developed by NC3A NATO agency). The system can be installed in any commercial PC running Windows NT, 2000 and XP. It can be associated with any external HF modem and radio by loading the appropriate equipment control driver.

Added features for HF radios equipped with ALE
HF messenger provides drivers for use with Rockwell Collins HF radios with embedded ALE functionality. Automatic Link Maintenance (ALM) controls the transmission performance to maintain optimum data rate. It automatically and independently adapts the transmission parameters to any node (data rate, interleaver, frame length) to maximise user data throughput.

HF server
HF messenger defines a profile for a HF server providing, when associated with an external HF modem and radio, a gateway for automatic and transparent HF data interconnection facilities for local and/or remote POP3 or Z-modem.

Multiservice provider
HF Messenger provides connectionless transmissions between several HF node client applications for broadcast, multicast, or point-to-point with ARQ services. Dedicated point-to-point transmissions between two specific client applications (hard links) can also be established. The hard link connected clients can then either use the entire HF connection resources or share those resources with other clients on the same node, providing maximum use of HF resources.

Status
In production and in service.

Contractor
Rockwell Collins.

HF-121A/B/C HF radios

Type
Avionic communications system, HF/data.

Description
The HF-121/A/B/C family of HF radios have been installed in a wide variety of commercial and military aircraft in the United States and in foreign countries. Earlier versions of the HF-121 product line have received military nomenclature designation of AN/ARC-153, -157, -191(V), -207(V), AN/URC-91 and -97(V) by the United States and AN/ARC-512 for other nations.

The HF-121 family can transmit and receive both data and voice signals, providing 280,000 channels covering the full HF-band from 2 to 30 MHz and operating in USB, LSB, ISB and AME modes. HF-121A and B transmitted power output level is selectable at either 100, 500 or 1,000 W peak and/or average. HF-121C transmitted power output is selectable at either 100, 200 or 400 W peak and/or average.

The latest member of the family is the HF-121C (JETDS designation AN/ARC-230 and AN/ARC-2201(V) – see separate entry), a high-performance radio system designed for HF applications requiring voice and data operations, including 400 W PEP/average transmit power. Compatible with the requirements of STANAG 5511/MIL-STD-188-203-IA for Link 11/TADIL A, the HF-121C has been optimised for tactical digital data communications and simultaneous operation (SIMOP) of multiple radio sets with minimum frequency and antenna separation. Embedded MIL-STD-188-141B Automatic Link Establishment (ALE), MIL-STD-188-110A Modem and ARINC 714-6 SELCAL capabilities are also available.

The basic radio set is partitioned into a receiver-exciter with integral pre/post-selector on a mounting shelf and a power amplifier-power supply on a mounting shelf. The receiver-exciter can be used independently of the power amplifier. Serial control via RS-232 or MIL-STD-1553B is available.

The HF-121C HF radio set is composed of the 671Z-3() receiver/exciter, 549E-1 Power Amplifier, 913Z-5 control and the applicable antenna coupler:

671Z-3() receiver/exciter
- The receiver/exciter is composed of a DSP based receiver/exciter-modem, advanced microprocessor control and an agile digitally tuned bandpass filter.

549E-1 power amplifier
- The power amplifier is a solid-state design that provides a power output of 400 W PEP/Average.

913Z-5 control
- The control provides remote control and monitor of the 671Z-3() R/E, the 549E-1 PA and associated coupler using an RS-232 serial interface. In addition to the basic control functions, this control will also provide the control and monitor functions for Automatic Link Establishment (ALE).

Antenna coupler
- The HF-121C HF radio set will interface with any of the standard product line AN/ARC-190 (V) antenna couplers; CU-2275(V)/ARC-190(V) and CU-2314(V). Additional AN/ARC-190 antenna couplers are available for unique platform requirements.

System applications
The HF-121C radio set is compatible with standard Air Transport Rack (ATR) unit dimensioning. Each unit mounts in an equipment tray for ease of removal and maintenance.

The antenna coupler is a pressurised unit, which may be mounted in pressurised or unpressurised areas of the aircraft.

The transmitter is capable of continuous transmission in all modes at ambient temperatures up to +55°C at sea level. Modes such as SSB voice which yield high peak to average power ratios (8 dB or greater) are claimed to show no degradation in power output up to +55°C and 50,000 ft. Short periods of operation up to +71°C are permitted in all modes. For combinations of duty cycle, peak-to-average power ratios, altitude and temperatures that exceed the thermal capability of the transmitter, power output is reduced to a power level at which it can safely operate without equipment damage.

SIMOP operation
The HF-121C radio set is designed to provide high-performance operation in the presence of collocated communication systems. A digitally tuned, four-pole bandpass filter provides filtering in the receive and transmit path to optimise the system's Simultaneous Operation (SIMOP) performance in collocated environments.

To achieve optimum receiver performance in a collocated situation, the level of interfering signals must be reduced to an acceptable level relative to the desired signal strength. A preselector will improve performance in the areas of out-of-band intermodulation distortion, cross-modulation, image rejection, IF rejection, reciprocal mixing and high-voltage protection.

When used in the transmit path, the bandpass filter is useful in reducing the levels of the out-of-pass band broadband noise, spurious outputs and harmonics that are generated in the receiver/exciter-modem.

Link 11
Link 11 is a digital data communication network interconnecting dispersed elements of a task force that provides for the exchange of tactical data among communication systems. The exchange of data uses a 16-tonne composite signal generated by the system's Data Terminal Set (DTS).

The HF-121C radio set has been optimised to provide compatibility with the signal characteristics of Link 11/TADIL A as defined in STANAG 5511/MIL-STD-188-203-1A, Interoperability and Performance Standards for Tactical Digital Information Link. The performance enhancements provided by the DSP technologies used in the 671Z-3( ) R/E enhances the ability of the system to be compatible with Link 11 data transmission and reception.

SELCAL
The HF-121C radio set provides for the demodulation of selective calling waveforms compatible with ARINC 714-6. The SELCAL address is selected from the control unit.

Multimode modem
The HF-121C radio set provides an internal modem compatible with the waveforms of MIL-STD-188-110A and STANAG 4285. The waveforms of MIL-STD-188-110A include the narrow- and wide-shift FSK and the serial (single tone) mode. An RS-232C data port is provided to interface with an external data source.

Automatic Link Establishment (ALE)
The traditional method of operating HF communication systems requires manual frequency-time planning and co-ordination in·addition to skilled radio operators who are needed to perform the frequency selection, frequency monitoring and link establishment. The HF-121C radio set, operating under the control of ALE, will provide automated selective calling, preset channel scanning and real-time channel propagation evaluation to achieve automatic connectivity.

The automatic connectivity of ALE provides improved HF communications reliability with less user training, including: automatic frequency management, automatic link establishment, automatic link confirmation and automatic disconnect. ALE improves the connectivity between communication systems by monitoring multiple frequencies and selecting the best calling frequency. This selection is based on real-time analysis of the quality of the frequency channel. If desired, the ALE system can operate in a 'radio silent' mode, which allows the channel quality to be continuously updated, but which prevents transmissions from the system except by operator intervention.

The sequence of events in an ALE call may be summarised as follows:

The operator selects an individual address to call and then keys the system. The system then automatically;
1. Selects the best preset channel
2. Determines the availability of selected channel
3. Transmits the address data sequence
4. Receives the automatic handshake – link confirmation
5. Breaks the squelch
6. Signals the user to begin communications
7. Monitors system key activity
8. Returns the system to the muted scan mode when communications are completed

The basis for ALE frequency selection is the Link Quality Analysis (LQA) that is continuously performed on the channels within the ALE scan list. The LQA is determined by analysing the signal characteristics (signal-to-noise ratio and delay distortion) of the data signal used in sounding or initiating an ALE call. A database of the LQA values is generated, which ranks the quality of the ALE frequencies associated with each of the addressee's mission directories. The entries in the LQA database are continually updated, with automatic downgrading of old entries.

Status
In production and in service with US air arms.

Contractor
Rockwell Collins.

HF-230 HF radio

Type
Avionic communications system, HF.

Description
The Rockwell Collins HF-230 radio is designed for use in fixed-wing aircraft and helicopters. The system provides 280,000 channels at 100 Hz channel spacing between 2 and 29.9999 MHz. All 176 ITU radio-telephony channels are preprogrammed, giving phone-patch capability over very long ranges wherever this facility is available. Lower sideband operation is possible for international or maritime communications.

The system comprises a TCR-230 transceiver, PWR-230 power amplifier and a range of antenna couplers. A DSA-220 adaptor permits two such systems to be operated in the same aircraft.

The radio features 40 pilot programmable channels and, when selected, channel number and frequency are displayed.

The CTL-230 display unit forms part of the HF-230 and features gas-discharge symbology.

An automatic probe antenna coupler, the PAC-230, is available for helicopter applications.

Specifications
Weight: 11.1 kg
Power output: 100 W PEP
Temperature range: –55 to +70°C
Altitude: up to 55,000 ft (with pressurised antenna coupler)

Status
In production and in widespread service.

Contractor
Rockwell Collins.

The Rockwell Collins HF-230 HF radio and ITU radio-telephony transceiver 0504029

HF-9500 series HF radios

Type
Avionic communications system, HF/data.

Description
Collins' HF-9500 multimode HF system is the Company's latest HF system featuring independent sideband (ISB) and simultaneous operation (SIMOP) capability. Embedded system functionality includes MIL-STD-188-141B Automatic Link Establishment (ALE), MIL-STD-188-110B data modem functionality, ARINC 714-6 SELCAL decoding and compatibility with

the requirements of MIL-STD-188-203-1A for Link 11 (TADIL A) data communications.

The HF-9500 system employs Digital Signal Processing (DSP) techniques and advanced High Efficiency Power Amplifier (HEPA) technology to minimise system size, weight and power consumption.

The HF-9500 includes the HF-9545 Antenna Coupler that tunes most HF airborne antennas (typically) within one second. The learned preset tuning feature of the antenna coupler allows almost instantaneous tuning on frequencies previously tuned by the system. Tuning data is stored in the system's memory for each frequency and channel tuned. This data is recalled and used when that frequency is tuned again to minimize tuning time. The HF-9545 Antenna Coupler also provides an additional third pole of system selectivity which further enhances system RF signal purity, crucial for SIMOP applications.

The versatile control capabilities of the HF-9500 allow it to be installed in a variety of applications. Manual control of the system is accomplished using the HF-9515 Control/Display Unit (CDU) or MIL-STD-1553 data bus. An EIA-232 control interface is also available to provide the system integrator with maximum system integration flexibility.

HF-9500 software can be updated through the radio's MIL-STD-1553B serial control interface, ensuring a cost-effective and renewable growth path to add emerging STANAG waveforms such as STANAG 4444, STANAG 4538, STANAG 5511, and ARINC 635/HFDL.

The HF-9550 Receiver Transmitter internal control software uses the Ada programming language and the software is contained in FLASH reprogrammable memory that is externally reprogrammable, facilitating convenient software field upgrades. Software updates can be accomplished via the MIL-STD-1553B bus utilising the MIL-STD-2217 protocol.

MIL-STD-188-114A modem interfaces provide for integration with virtually all ancillary equipment including the Improved Data Modem (IDM) and Advanced Narrowband Digital Voice Terminal (ANDVT). Global Positioning Satellite (GPS) time and position interfaces allow for position reporting and time acquisition.

Specifications
Dimensions (typical):
(HF-9515 control unit) 146 × 163 × 114 mm
(HF-9550 transmitter/receiver)
257 × 444 × 194 mm
(HF-9545 coupler)129 × 347 × 192 mm
Weight (typical):
(HF-9515 control unit) 1.2 kg
(HF-9550 transmitter/receiver) 20.5 kg
(HF-9545 coupler) 7.7 kg

Status
In production and in service. The HF-9500 radio system is the standard HF replacement radio for the P-3 Orion retrofit upgrade programmes for the US Navy and the Royal Australian Air Force, where it is designated AN/ARQ-57.

Configurations have also been developed for the C-27J, A-310, and EMB-145.

Contractor
Rockwell Collins.

HFS-900D HF Data Radio (HFDR)

Type
Avionic communications system, HF/data.

Description
Collins' HFS-900D HF Data Radio (HFDR) provides operators with a low-cost, long-range data link system for air fleets operating in oceanic, polar, and remote land areas. The HFS-900D is a single, self-contained unit without requirement for an external modem.

The HF Data Radio (HFDR) provides the means to process, transmit and receive data as well as analogue voice. The HFDR transceiver can operate on frequencies spaced 100 Hz apart in the 2 to 30 MHz band. Compatibility with existing ARINC 719 installations is facilitated by the inclusion of Single Side Band (SSB) voice, Amplitude Modulated Equivalent (AME), Continuous Wave (CW), SELective CALIing (SELCAL) and analogue data functions within the HFDR. In addition to providing traditional HF radio functionality per ARINC 753, the HFDR features an internal data modem and controller. Voice transmission is compatible with current SSB HF transceivers. Data transmission is compatible with ground HF transmitting and receiving systems which use conventional HF transceivers and ARINC 635 compliant modems and controllers.

The HFDR uses the bottom three layers of the OSI model which are the physical layer, the link layer, and the network layer. The HFDR interfaces with a CMU in the same manner as other ATN compliant data communications equipment interfaces, that is, through an ISO 8208 DCE.

Key features of the systems are:
- SSB simplex transmission and reception of analogue voice or digital data in a standard channel in the aeronautical HF bands
- Encoding and modulation of digital data, and demodulation and decoding of digital data at user data rates of up to 1800 bps
- Automatic frequency search and link acquisition per the protocols defined in ARINC 635
- Exchange of downlink and uplink data with the ACARS MU or CMU per the protocols defined in ARINC 635
- SELCAL output lines to a selective calling decoder elsewhere in the aircraft.

Specifications
Dimensions:
6 MCU
200 (H) × 195 (W) × 378 (L) mm
Weight: 12.6 kg
Power supply: 115 V AC, 400 Hz, 3 phase
Maximum input power: Tx, 960 W; Rx 170 W
Frequency: 2 to 29.9999 MHz
Channels: 280,000 in 100 Hz increments
Emissions:
(receive)AM, USB/LSB/AME, data, CW
(transmit)USB/LSB/AME, data, CW
Temperature range: –55 to +70°C
Altitude: up to 50,000 ft
Certification: FAA TSOs C31d, C32d; RTCA DOs 160C, 163, 178B; ARINC 429, 600, 604, 635, 719, 753

Status
In service. Certified for Boeing 737, 747, 757, 767, 777, MD-11, MD-90 and Airbus A319, A320, A321, A330, A340 civilian transport aircraft.

Also available are HFDL upgrade kits for Rockwell Collins radios already in service, including 628T-2A, HFS-700 and HFS-900 radios.

Contractor
Rockwell Collins.

ICS-150 intercommunications set

Type
Avionic internal communications system.

Description
The Rockwell Collins ICS-150 intercommunications set is a fully militarised aircraft audio system which provides selectable channels of communications between aircraft crew stations. It also provides communications between each crew station and various transceivers, receivers and warning systems.

This set is particularly applicable for aircraft with multiple crew stations, a variety of communication, navigation and warning receivers or transceivers and stringent requirements for cross-talk isolation, electromagnetic interference and nuclear hardening.

The set delivered to the US Air Force for its B-1B aircraft provides an audio interface for up to eight crew stations, five ground crew/ maintenance stations, eight separate avionics transceivers and 10 receivers.

An InterCommunication Set (ICS) is made up of one central control unit, up to eight crew station units and up to five maintenance station units.

Each crew station unit provides 10 receive-monitor control functions such as on/off and volume and transmit selection control of ICS, plus up to six other transmit functions. The crew station also provides master volume, hot mic, all call and LRU test facilities.

The central control unit is equipped with secure interlock capability to prevent secure communications from being heard on non-secure transmissions.

This system has high channel isolation and provision has been made for the future incorporation of a COMSEC switch capability. The COMSEC switch will permit a single speech encryption device to be switched between several radios.

The ICS-150 was designed to meet all military requirements, from parts utilisation to qualification testing.

A central mixing architecture requires very few interconnect lines between each crew station and the central control unit. Three twisted/shielded pairs of wires are used for microphone audio, headset audio and serial control data.

The ICS-150 central control unit has redundant input regulators and separate line regulators in each module to prevent any single point failure from causing system failure. An additional back-up mode is provided in case there is structural damage to the central control unit. This back-up mode provides for a direct connection between the pilot and co-pilot and two separate transceivers.

Specifications
Dimensions:
(central control unit) 124 × 193 × 497 mm
(crew station unit) 146 × 95 × 112 mm
(maintenance station unit) 117 × 91 × 99 mm
Weight:
(central control unit) 8.3 kg
(crew station unit) 0.9 kg
(maintenance station unit) 0.23 kg
Power supply: 28 V DC

Status
In production and in service. Applications include US Air Force B-1B, B-2 and C-135C, US Army Special Operations Force aircraft and Royal Australian Navy helicopters.

Contractor
Rockwell Collins.

IDC-900 integrated datalink controller

Type
Avionic datalink system.

Description
The IDC-900 uses ARINC 429 to interface with aircraft systems and is ARINC 739 compatible, which enables the Control and Display Unit (CDU) to operate as a Multifunction Control Display and Unit (MCDU) for aircraft functions such as datalink, Satcom and FMS.

The IDC-900 CDU design is based on the ARINC-defined MCDU characteristics which provide for complete access to aircraft systems.

The IDC-900 is capable of serving as a crew interface with ARINC 739 systems. This capability provides an independent control/display unit in the flight deck.

The CDU utilises a large multicolour Active Matrix Liquid Crystal Display (AMLCD) with a viewing area of 4 in by 3 in. Data is shown in six colours (cyan, yellow, magenta, white, green and red), with a display contrast ratio of 20:1 or greater, for viewing angles from 35° left or right and +20 to –40°. Maximum display brightness is 85 fL.

The fixed page format consists of 15 lines of 24 characters each. The page structure is comprised of title line, six label-data line pairs positioned adjacent to left and right line select keys, a data entry line called the scratch pad, and an annunciator line. All manual entries via the CDU keyboard appear on the scratch pad prior to data field entry to allow pilot verification of data. Data may also be copied into the scratch pad from any CDU page. The bottom display line is an annunciator line, reserved for the display of messages requiring crew awareness or action. Actuation of the data entry keys sequentially writes the respective alphanumeric character on the scratch pad, left to right.

Six pairs of line select keys border the display. The line select keys are not dedicated keys, their function being assigned by software depending on which CDU page is displayed. Dedicated keys are provided to facilitate ease of operation by providing most frequently needed functions at the touch of a button.

When the bright/dim key is pushed, it acts as a toggle. The display either brightens or dims and continues as the key is held. When the key is immediately pushed again, the intensity moves in the opposite sense.

Future IDC-900 designs will support use of weather graphics and/or video. Capabilities such as North Atlantic winds, temperatures aloft, composite radar images centred on US Station 1.0, regional radar similarities with tops and movement, regional weather replications, regional infra-red satellites, or regional high-level significant weather can be displayed on the IDC-900. Optional video capability will support standard NTSC video input for display on the IDC-900.

Specifications
Dimensions: 162 × 146 × 1118 mm
Weight: 1.85 kg
Power supply: 115 V AC, 28 V DC
Certification:
FAA TSOs C115b, C129a Class B1
Environmental: DO-160D, DO-170B Level B

Status
In production.

Contractor
Rockwell Collins.

IDM-302

Type
Avionic datalink system.

Description
The design aim of Symetrics' IDM-302 is to act as an international digital communications translator, so that all elements in a multinational force can intercommunicate effectively and efficiently share sensor and tactical data without regard to nationality, language or platform.

By transmitting in a digital format, each participant's platform can receive, process and handle the data in such a manner as to facilitate the operators' specific mission requirements. However, in order that the mission processor can utilise the data, it must first go through a digital translation process to convert the data into a format that the mission processor can use. The IDM is designed to perform that function.

The IDM-302, also known as MD-1295/A, contains four Standard Electronic Module format-E (SEM-E) modules mounted in a chassis: two Digital Signal Processors (DSPs) as modems, a Generic Interface Processor (GIP) for link and message processing, and a power converter for power adaptation for the IDM's internal assemblies. Three additional SEM-E slots are available within the housing for expansion of the functional capability of the IDM. They could include the following modules, which are under development: an internal GPS; a multiband, multimode transceiver module; a video card; a new cryptographic module to address TRAP and TADIXS-B; or additional DSP modules for additional channels.

The DSP module uses a TMS320C30 DSP to modulate and demodulate messages at 33 MHz clock rate/6 Mips. Each DSP can modulate or demodulate two half-duplex channels simultaneously at data rates up to 16 kbps. The DSP module includes all discrete signals required to interface a variety of radios.

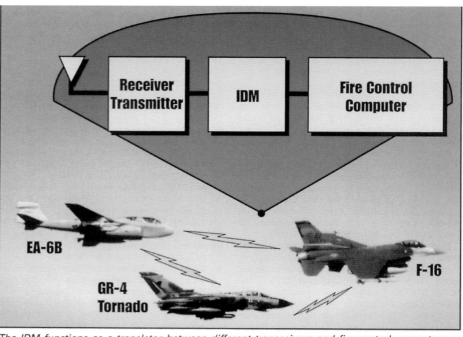

The IDM functions as a translator between different transceivers and fire-control computers on different types of aircraft   0062232

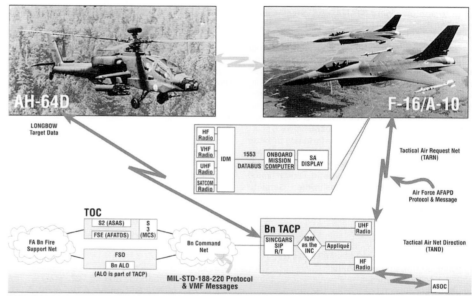

The IDM in typical tactical situation, sharing data between an AH-64D and F-16   0062234

The GIP module uses an Intel 80960MC RISC microprocessor capable of simultaneously processing four half-duplex channels of AFAPD, MTS, TACFIRE or MIL-STD-188-220 messages at 16 MHz clock rate/6 Mips. It also receives and transmits messages and configuration commands via the MIL-STD-1553 databus, processes data and processes configuration commands. It has 6 Mbytes of storage for warehousing messages.

The IDM was designed around the concept that changes in functionality should be accomplished through software changes inside the IDM, and not via more expensive changes to the air/ground platform Operational Flight Program (OFP). In fulfilling this design aim, software 'Radio Look-up Tables' facilitate the easy interconnection to a variety of receiver/transmitters.

Specifications
Interoperability:
AFAPD (US Air Force)
TACFIRE (US Army)
MTS (US Navy/US Marine Corps)
MIL-STD-188-220/VMF (US DoD common)
ATM (commercial)
NITF (imagery)
TADIL-J
IDM radio interfaces:
ARC-164 (Have Quick II/compatible with IIA)
ARC-182
ARC-186
ARC-201 (SINCGARS, with or without DRA182)
ARC-210
ARC-222 (IDM software provides DRA function)
SINCGARS SIP
KY-58 and Whittaker
Radio Look-Up Tables provide flexibility to add other receiver/transmitter software interfaces
Analogue port (CPFSK, duobinary FSK):
(data rate*) 2,400; 1,200; 600; 300; 150; 75 bps
(tone pairs) 2,400/1,200; 2,100/1,300; 1,700/1,300
Digital port (ASK):
(data rate*) 16,000; 9,600; 8,000; 4,800; 2,400;
    1,200; 600; 300; 150; 75 bps
(channel bandwidth) 20 kHz
(input signal level) 0.1 to 25 Vpp
(output signal level) 0.1 to 12 Vpp
KY-58 port:
(data rate) 16,000; 8,000; 1,200; 600; 300; 150; 75 bps
Dimensions: 230.1 (D) × 190.5 (H) × 136.5 (W) mm
Weight: 6.8 kg
MTBF: in excess of 6,200 hours demonstrated
Part number/NSN: 81995-DM001-302/5998-01-
    426-5318
*data rates are limited only by the receiver/transmitter connected; the IDM has achieved 400,000 bps

Status
Under a multiservice programme directed by the US Air Force F-16 System Program Office (SPO), the US Naval Research Laboratory (NRL) developed 25 First Service Units and 200 Low-Rate Initial Production (LRIP) units. Full-rate production

IDM hardware　　　　　　　　　　　　　　　　　　　　0062233

is in progress at Symetrics Industries Inc for a projected total of over 4,000 IDMs. Initially, the MD-1295/B IDM was integrated into F-16C Block 40 aircraft as part of project Sure Strike, enabling Forward Air Controllers (FACs) to electronically transmit GPS-derived target latitude, longitude and elevation data to the pilot as part of the Close Air Support (CAS) mission. Subsequently, the IDM was utilised to provide a datalink capability for the later F-16 Suppression of Enemy Air Defences (SEAD) mission (replacing the F-4G WW aircraft), automatically 'handing off' High-speed Anti-Radar Missile (HARM) targeting data and navigational data directly from one F-16 to another.

Future growth for the system includes an integrated GPS module, Video Interface Module (VIM), Crypto module and a 'programmable waveform' module.

The MD-1295/A IDM is in service with the US Air Force on F-16 Block 40 and Block 50 aircraft. IDMs are also installed on a variety of international and US platforms, including other F-16 aircraft, the Longbow AH-64D and WAH-64, EA-6B, OH-58D, JSTARS, A2C2S, UH-60Q, E-2C, Jaguar GR1/3 (UK) and several UAV platforms.

The IDM supports Suppression of Enemy Air Defenses (SEAD), Close Air Support (CAS), Forward Air Control (FAC), Air Combat, Joint Air Attack Team (JAAT), Fire Support (FS), Intraflight Data Link (IDL), Situational Awareness Data Link (SADL), and Command and Control (C2) missions.

Project Gold Strike
Under a project known as Gold Strike or the rapid targeting system (RTS), a new Video Imagery Module (VIM) was integrated into the MD-1295/B IDM.

The Gold Strike VIM is totally integrated within the IDM, utilising the core system 16 kbyte data transfer capability to transmit, store and receive up to 20 frames of compressed imagery. During trials of the system, imagery sent from the ground to the aircraft required about 40 seconds per frame using a VHF AM/FM secure communications radio. Apart from the length of transmission time and the inability to utilise the IDM for other data transfer functions during image transfer, trials were pronounced a resounding success. The company claims that use of video imagery reduced the nominal target acquisition time from twenty minutes to two minutes. During 1999, as a result of

project Gold Strike, a field modification of 39 Block 40 F-16C aircraft of the 31st Tactical Fighter Wing supporting NATO operations in Bosnia was made.

Contractor
Symetrics Industries Inc.

IDM-501

Type
Avionic datalink system.

Description
Originally designed for UAV applications, the small form factor of the IDM-501/IDM Junior™ facilitates simple physical integration into most fighter/attack aircraft and helicopters. IDM-501 has all the same capabilities of existing IDM systems (such as the IDM-302, see separate entry), including four channels of digital data communications, multiple protocols (including AFAPD, TACFIRE and IDL) and identical aircraft interfaces. The IDM-501 is also available with the Photo Reconnaissance Intelligence Strike Module (PRISM) option for imagery. Physically, the IDM-501 is half the size and weight of the IDM-302.

The IDM-501 can be hosted as a single card in another avionics box and will accommodate MIL-STD-188-220 and VMF protocols.

Specifications
Interoperability: All recognised standards supported
Radio interfaces: Four primary channels can be configured for analogue (ASK, FSK) or digital (NRZ, ASK, synchronous/asynchronous) interface; one multipurpose serial port can be configured for RS-422 or RS-232
Analogue port data rate: 2,400, 1,200 bps
Digital port data rate: 16,000, 2,400, 1,200 bps
Memory storage: 256 Mb SDRAM; 4 Mb SRAM; 256 Kb EEPROM; 32 Mb FLASH
Power: 28 V DC per MIL-STD-704; 13 W
Environmental
Temperature: -54 to +71°C
Vibration: Per MIL-STD-810
MTBF: 5,000 h (forecasted)
Dimensions: 224.0 (D) × 188.5 (H) × 91.2 (W) mm
Weight: 3.8 kg
Part number/NSN: 403145-3/5895-99-612-3735 (without PRISM); 403145-2/5895-99-612-3735 (with PRISM)

Status
In production and in service.

During May 2004, Symetrics announced that it had been awarded a USD1.7 million contract from Aerosystems International Ltd. This was to manufacture and deliver the IDM-501 for integration into the UK Attack Helicopter AH Mk1 Mission Planning Station (MPS) that supports British Army Apache Helicopters.

In October 2005, Symetrics was awarded a further contract from Boeing, valued at more than USD375,000, to manufacture and deliver 15 IDM-501, to support the Turkish Airborne Early Warning and Control (AEWC) Programme, Peace Eagle. The systems will be fitted into a military variant of the Boeing 737-700 aircraft.

Contractor
Symetrics Industries Inc.

Integrated Information System (I²S)

Type
Avionic datalink system.

Description
The aim of the Rockwell Collins I²S is to replace paper with technology in commercial aircraft. This is achieved by automating the transfer of data

The IDM-501 will be integrated into the UK WAH-64D attack helicopter Mission Planning Station (MPS)　　　　　　　0106501

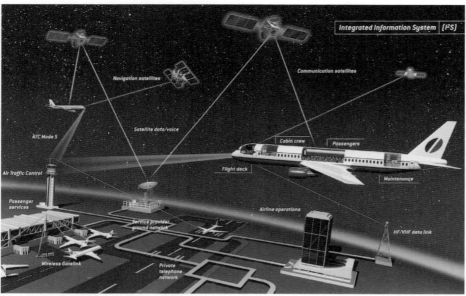

The Rockwell Collins integrated information system concept   0079230

Weight:
(CCU) 13.8 kg, (CNC) 3.14 kg
(ICS) 2.3 kg, (PAC) 0.41 kg
(IRP) 0.23 kg, (HRP) 0.34 kg
Power supply:
(CNC, ICS, PAC) 28 V DC
(ISU, IRP, HRP) 16 V DC
Temperature range: –54 to +71°C
Altitude: up to 50,000 ft
Reliability: 3,300 h MTBF

Status
In production and in service. The IRMS is installed and fully operational on the US Air Force C-17A transport aircraft.

Telephonics also provides the SURE-COMM loadmaster wireless communications system for the US Air Force C-17A, and the company has been contracted by the US Air Force to incorporate Global Air Traffic Management (GATM) requirements into the C-17A communications suite.

Contractor
Telephonics Corporation, Command Systems Division.

---

between the flight deck and the airline's dispatch operations on the ground. Rockwell Collins is currently teamed with Condor Flugdienst, the charter affiliate of Lufthansa, to participate in a year's trial of the concept to link an aircraft-based intranet to airline terminal area databases.

Some of the features of the I²S include:

- Communication via a new high data rate transfer medium on the ground (wireless gatelink), together with HF, VHF and Satcom for air-to-ground links
- A high-frequency datalink between the aircraft and the airline's information system to transmit intranet data including navigation databases, flight plans, graphical weather data and maintenance data
- In-flight entertainment services.

The first products in the I²S product line have successfully completed DO-160D airworthiness testing. These units include, the MAU-2000 (Microwave Airborne Unit) and PMAU-2000 (Pilot's Microwave Airborne Unit), which together facilitate wireless Local Area Netwok (LAN) connectivity onboard the aircraft. They also provide for wireless terminal area gatelink from the aircraft to the airline's Information Technology (IT) system and provide a wirelss LAN environment inside the aircraft.

Contractor
Rockwell Collins.

Integrated Radio Management System (IRMS)

Type
Avionic Communications Control System (CCS).

Description
The Integrated Radio Management System (IRMS) is a communications management system providing control of aircraft radios, radar transponders and intercom. It provides total communications back-up in the event of battle damage through redundant control panels and centralised control units. The system comprises two Communication equipment Control Units (CCU), two Communication/ Navigation equipment Controls (CNC), seven Intercommunication Set control Units (ISU), three Public Address set Controls (PAC), six Interphone Receptacle Panels (IRP) and nine Headset Receptacle Panels (HRP).

The CCU is the centralised 1750 processor-based MIL-STD-1553B bus controller unit which provides control of all audio, digital and analogue signal processing. It interfaces to the CNC, ICS and various radio equipments, via a MIL-STD-1553B interface bus. The CNC is used to tune navigation and communication radios, display radio frequencies

and modes, IFF modes and VOR/ILS course selection. ICS provides microphone, PTT, VOX, audio and volume selection for all audio sources. ISU enables intercom/radio, talk/listen or PA selection. PAC selects speakers and volume levels. IRP and HRP provide microphone preamplification and headset impedance matching.

Specifications
Dimensions:
(CCU ×7) 198 × 191 × 498 mm
(CNC ×2) 95 × 184 × 178 mm
(ICS ×7) 124 × 146 × 152 mm
(ISU ×2) 67 × 146 × 89 mm
(PAC ×3) 38 × 136 × 102 mm
(IRP ×6) 64 × 102 × 38 mm
(HRP ×9) 38 × 127 × 102 mm

Intercommunications set for the V-22 Osprey

Type
Avionic internal communications system.

Description
The full V-22 intercommunications set consists of a Communications Switching Unit (CSU) with up to six Intercom Set Control (ISC) stations, two Audio Frequency Amplifier (AFA) assemblies, a Cabin Public Address (CPA) amplifier and four cabin PA speakers.

It provides simultaneous intercom for multiple crew stations, five channels of radio transmission and reception for each ISC, and reception of four

The Telephonics Integrated Radio Management System (IRMS)   0504043

The intercommunications set for the V-22 Osprey   1120961

composite navaids, five warning tones, one IFF and one radar warning receiver on two ISC panels. All audio switching and mixing functions are software-controlled and interfaces to clear and secure communications equipment are provided. There is also a digital message device I/O channel. The electroluminescent panel provides high-contrast NVG-compatible lighting.

The CSU contains the switching circuitry, logic circuit and a large portion of the audio circuitry required for ICS operation. In addition, it is the interconnect unit for the ICS system components and peripheral devices. The CSU provides impedance matching, audio reporting and push-to-talk functions between the crew station ICS panels and the aircraft radios. Radio selection status is provided to the aircraft control and display subsystem through the CSU.

Each intercommunication set is essentially the individual crew station control panel, on which the frequencies, sources and signal levels are selected for monitoring and/or transmission. The audio frequency amplifier permits the selection of intercom facilities only.

The system is based on a low-power CMOS microprocessor and large-scale integration circuitry. Extensive built-in testing has been designed in for simplified maintenance. Options with the system include a digitised message card, RS-232, ARINC 429 and MIL-STD-1553 interfaces and an incandescent light panel.

Status
In production and in service in the V-22 Osprey.

Contractor
Sanmina-SCI Corporation.

Jet Call

Type
Avionic communications system, Selective Calling (SELCAL) system.

Description
Jet Call is an ARINC ground-to-air selective calling system utilising thumbwheel coding with four buttons in full view. Sixteen available tones provide over 10,000 possible combinations and take about 10 seconds to set. The Jet Call uses no wire jumpers or remote switches. Two or five decoder channels are available to handle up to three VHF comms and two HF transceivers. The Jet Call is TSO'd to FAA C059. It uses all solid-state circuitry and switched capacitor filters with high inherent stability and has low power consumption and easy installation, using standard Mil type D connectors with insertable and removable pins.

Specifications
Dimensions: 127 × 57 × 321 mm
Weight:
　(2 channel) 1.22 kg
　(3 channel) 1.45 kg
Voltage: 28 V DC
Current:
　(2 channel) 320 MA
　(3 channel) 360 MA

Status
As of November 2005, in production and in service.

Contractor
FreeFlight Systems.

Joint STARS Interior Communications System (ICS)

Type
Avionic internal communications system.

Description
The Interior Communications System (ICS) is utilised on the E-8A aircraft as the communication system for the US Air Force/US Army Joint STARS.

The US Air Force/US Army Joint STARS Interior Communications System (ICS)　　0504044

It is a fully distributed digital communications system, supporting 96 full-duplex channels for simultaneous non-blocking secure and non-secure operations. The ICS interfaces with various receiver/transmitters, secure speech devices, communication, navigation and identification receivers via General Interface Terminals (GIT), as well as supporting the interface with all mission crew members and flight deck personnel.

The ICS is fully modular and other crew members and GITs may be added as the system grows or the mission requirements change, with no changes to the existing system hardware. At the present time, there are six Flight Deck Terminals (FDT), 18 Crew member Terminals (CT) and five GITs with five channels each. Each of the mission operator CTs has a single display and keyboard. The keyboard provides radio selection or selection of the mission net, preset conference net, progressive/selective conference net, call and a telephone dialling net. The FDTs are identical to the CTs except that they contain an additional display and control keys for radio frequency and parameter selection. The GITs are digitally configurable for input and output levels, thereby allowing one design to accommodate many different peripheral devices without changing design or adjustments. Each GIT contains the input/output and control for five full-duplex audio channels. Once the system has been configured, all information within all of the units is retained in-non-volatile memory storage.

When operators communicate, the terminals digitise audio in an allocated data time slot with a destination address to the selected interface, which then converts the digital information back into baseband audio for application to the specific asset. Likewise, when a radio or remote communication device receives audio, this audio is digitised and made available to any of the operators that have been enabled to receive the audio. The entire system is digitally reconfigurable, enabling asset assignment based on the operational scenario. All channels within the system can be selected, combined, monitored and individually volume-controlled by any operator.

Battle damage that renders a single CT, FDT or GIT terminal, or multiple terminals, inoperable does not impact on system operation or affect the operation of the other system components. Each of the units is self-contained and utilises microprocessors that transfer function and allow continued operation.

Specifications
Dimensions:
　(CT ×18) 171.5 × 127 × 177.8 mm
　(FDT ×6) 247.7 × 127 × 177.8 mm
　(GIT ×5) 142.7 × 171.5 × 190.5 mm
　(single TAP) 47.8 × 85.1 × 85.1 mm
　(dual TAP) 47.8 × 161.3 × 85.1 mm
Weight:
　(CT) 3.98 kg
　(FDT) 4.45 kg, (GIT) 4.05 kg
　(single TAP) 0.25 kg, (dual TAP) 0.43 kg
Power supply: 28 V DC
Temperature range: –54 to +71°C
Altitude: up to 50,000 ft

Status
In service on the E-8A Joint STARS aircraft.

Contractor
Telephonics Corporation, Command Systems Division.

Joint Tactical Information Distribution System (JTIDS)

Type
Avionic datalink system.

Description
The Joint Tactical Information Distribution System (JTIDS) is a US joint service command and control support system providing secure jam-resistant communications and embedded navigation and identification for land, sea and air platforms. Using Time Division Multiple Access (TDMA) technology, it provides high-capacity networking among diverse airborne and surface users. It allows all stations to share an integrated awareness of the combat situation for friendly forces as well as detected threats in real time, using the tri-service multinational message catalogue TADIL-J. It can thus provide a language-independent method to ensure co-ordinated operations on a multinational battlefield. The data received can be displayed in both symbology and language of the host nation's platform. Thus, data received by a UK system would use English as the language for a given message; an Italian system could display the same message in Italian.

JTIDS operates on 51 frequencies in the 960 to 1,215 MHz frequency band, sharing with Tacan but strictly avoiding the IFF transponder frequencies which are also in the band. The JTIDS TDMA scheme breaks up time into 7.8125 ms time slots and allocates these slots to users, based on projected traffic demand. Users employ their time slots to transmit while listening during all other times, thus enabling relays of opportunity, to maximise the probability of message reception. Participants routinely inject

JTIDS information shown on the colour display of an F-15. The aircraft is shown at the centre of the display, concentric circles indicate distance in nautical miles. Round symbols indicate friendly aircraft, rectangles are unknowns and triangles are enemy aircraft　　0504031

information into the network through their regular broadcast slots without necessarily knowing who needs the information, using a broadcast-oriented architecture. These broadcast slots might include identification and location data, as well as host processor-generated traffic such as target acquisition tracks and platform status information on weapons and fuel status or equipment readiness.

JTIDS is designed to survive the highest levels of enemy radio electronic combat. The links are encrypted with the latest approved crypto devices. Jam resistance is achieved by multiple techniques through fast hopping over all frequencies, direct sequence spread spectrum of the waveform and Reed-Solomon forward error correction. The result of the signal processing gains from this combination of techniques means that the JTIDS omnidirectional radiation pattern can offer as much signal improvement as if it were being transmitted by a highly directional antenna (such as parabolic dish), without the difficulties of beam pointing or the limitations of single path links.

The initial emphasis in the development of JTIDS hardware was on Class 1 terminals developed from 1974. First deliveries took place in 1977 and resulted in terminals on board the Boeing E-3A AWACS aircraft, in ground terminals in the UK Royal Air Force IUKADGE system and in transportable shelters designated as Adaptive Surface Interface Terminals (ASIT) used by the US Army and US Air Force ground control facilities. The terminal is also used in the NATO Air Defense Ground Environment. The Class 1 terminal comprises rack-mounted components occupying almost 0.26 m³ and weighing approximately 192 kg.

Status
The operational fielding of JTIDS began, primarily with navy and army (FAAD) platforms. Multiservice demonstrations, notably the All-Services Combat Identification Evaluation Team (ASCIET), exercises, and combat operations demonstrated the capabilities of Link 16 systems such as JTIDS.

The Class 2 JTIDS terminal for the US Air Force F-15 with the receiver/transmitter (left) and the data processor group (right)   0504228

Exercises and combat experience have proved interoperability among US, UK and French JTIDS equipment. Japanese AWACS aircraft are reported as being equipped with JTIDS capability.

JTIDS Class 2 terminal
The Class 2 terminal, for fighter aircraft installation, began its development early in 1980. It was designed by Singer (now BAE Systems North America) in a leader-follower arrangement, with Rockwell Collins as the subcontractor providing RF design expertise.

The Class 2 terminal, designated by the USA as the AN/URC-107(V)1, AN/URC-107(V)6, AN/URC-107(V)8, and AN/URC-107(V)10, provides the basic JTIDS functions of high-capacity data communications, embedded TACAN functionality, dual-channel/dual-mode integral secure voice, dual-grid relative and geodetic navigation, GPS interface, PPLI self-identification messages, crypto-secure and jam-resistant connectivity and system status monitoring and reporting.

Since JTIDS operates in the TACAN and IFF frequency band, extensive interference protection circuits are built into the equipment to prevent inadvertent interference with those systems. Circuits continuously monitor the JTIDS radio operation and can shut down transmission in the event of any out of specification emission. During peacetime activities, the terminal operation maintains restricted duty cycles, further limiting the potential for interference with civilian facilities. The terminal incorporates SRU/LRU BIT hardware and software, achieving 98 per cent fault detection and 95 per cent fault isolation.

The Class 2 terminal comprises two boxes occupying approximately 0.045 m³ and weighing 57 kg. One box houses the receiver/transmitter circuits; the other is a dual LRU component. The dual LRUs are the digital processor section, standard for all terminals and applications, and the interface unit which is customised for each host platform. The digital processor section carries out all signal processing and digital computing functions, as well as advanced position location and tactical air navigation. The key output of the system is the display of information passed over the JTIDS network, allowing network users to share a common situational awareness. In the fighter terminal, the JTIDS interface feeds the multifunction display, which is capable of selective control by the pilot and automatic display under instructions in preloaded mission files. Selected platforms can receive precise direction from a central control point or can support local decision making based on the shared awareness. In other installations, the JTIDS data can be displayed

on host processor displays, providing real-time depiction of the combat situation.

Specifications
Dimensions:
    (data processing group) 324 × 193 × 485 mm
    (receiver/transmitter) 257 × 193 × 395 mm
Weight:
    (data processing group) 36 kg
    (receiver/transmitter) 25.8 kg
Power supply: 120/208 V AC, 50/60/400 Hz or 240/280 V DC, 1,400 W
Power output:
    (TDMA) 200 W
    (Tacan) 500 W
Data rate: 238 kbits/s
Range: (normal) 557 km, (available) 928 km

Status
The final Class 2 production contract is expected to continue until at least the year 2001.

Terminals are currently being integrated into US Air Force aircraft: F-15, Rivet Joint, Joint STARS, and Airborne Battlefield Command/Control Centers (ABCCC) and the ground-based Modular Control Equipment (MCE). They are also being fitted into the US Navy's F-14 aircraft and the US Marine Corps' Air Defense Command Post (ADCP). They are also operational in many UK Tornado F3 aircraft. Successful testing of Class 2 terminals on US Navy terminals on US Navy submarines has shown the usefulness of JTIDS for the entire fleet.

JTIDS Class 2H terminal
The Class 2H terminal is derived from the Class 2 terminal, to replace the Class 1 terminal. It occupies 0.15 m³ and weighs 155 kg, while giving the additional capabilities of increased throughput (115 kbits/s v 28.8 kbits/s), relative navigation and TACAN functionality, interoperability with TADIL-J message protocols and increased system functionality such as an increased quantity of crypto variables and over-the-air initialisation and rekey.

The Class 2H terminal is the high-power 1,000 W output transceiver version of the basic Class 2 terminal. It currently has three configurations. The first (AN/URC-107(V)4 and AN/URC-107(V)5) is an airborne system with receiver/transmitter, data processor group, high-power amplifier/antenna interface group, control monitor set and power conditioner and crypto loading devices. The other two configurations are for land-based use (AN/URC-107(V)9) and for surface ships (AN/URC-107(V)7). The Class 2H terminal possesses all the functional characteristics of the basic Class 2, and it is totally interoperable.

Specifications
Volume: 0.15 m³
Weight: 155 kg
Power supply: 120/208 V AC, 50/60/400 Hz, 3 phase
Power output:
    (TDMA high power) 950 W
    (TDMA low power) 235 W
    (Tacan) 500 W
Data rate: up to 238 kbits/s
Range: (normal) 557 km, (available) 928 km
Message types: TADIL-J or interim JTIDS message standard
Interface: MIL-STD-1553B

Status
Class 2H terminals are currently being integrated into US Naval surface ships and E-2C Hawkeye aircraft, US Marine Corps Tactical Air Operations Modules (TAOM), and into upgrades to US Air Force E-3 AWACS aircraft. They are also fitted to UK and French AWACS aircraft.

JTIDS Class 2M terminals have been ordered by the US Air Force for JTIDS operations.

JTIDS Class 2M terminal, designated AN/GSQ-240, is a single-box variant of the basic JTIDS terminal, designed for land-based operations.

Contractor
BAE Systems North America.
Rockwell Collins.

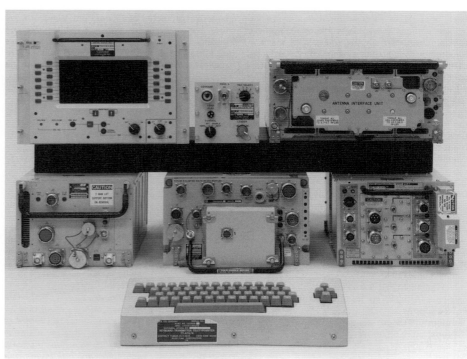

Class 2H JTIDS equipment   0504032

Joint Tactical Radio System (JTRS)

Type
Avionic tactical communications system.

Description
The Joint Tactical Radio System (JTRS) is designed to replace as many as 750,000 radios in between 25 and 30 different families with one radio family comprising some 250,000 multichannel, multiband, software-defined radios. When fielded as a family of related platforms, the system will support joint operations by providing the capability to transmit, receive, bridge and gateway between similar and diverse waveforms and network protocols used within the radio frequency spectrum and across service boundaries. Software waveform selection will define the particular radio's function and reprogramming will determine the 'personality' of the radio at any instant. The radios are based on open software architecture and also designed to integrate with new waveforms and legacy systems. JTRS will handle voice, data and video.

Programme history
The JTRS requirement was initially identified as a formal Mission Needs Statement (MNS) for a Joint Tactical Radio. That statement, approved in August 1997, led to three separate JTRS 'Step 1' contracts in early 1999. The Step 1 awards, to consortia led by Raytheon, Motorola, and Boeing, focused on identifying programme issues and defining the JTRS Software Communications Architecture (SCA).

The resulting Step 2A contract was awarded to a Raytheon-led team that included 11 different companies and academia. Step 2A activities included developing the previously defined SCA and demonstrating the architecture of the various radio platforms.

JTRS Step 2B involved seven contracts, separate from the development of the architecture, to identify unique aspects of various radio technologies and components. Major contract recipients (and general target topics) included:
- Assurance Technology (SINCGARS waveform)
- Boeing (implementation of core framework on multiple packages)
- Harris (man-portable platform issues and battery technology)
- Motorola (now General Dynamics Decision Systems) (commercial aspects of SCA)
- Rockwell Collins (Link-16)
- Racal (now Thales) (hand held radio issues).

There was also a Step 2C contract, awarded to BAE Systems, to produce 220 radios to explore the issues associated with networking, cryptography, and to facilitate greater understanding of the complexities of a multichannel simultaneous

Joint Tactical Radio System (JTRS) Cluster 1 systems (Rockwell Collins)   1129582

operation radio. Testing of these radios began during October/November 2002.

All of these earlier 'Step' contracts paved the way for the 'Cluster 1' (Step 3) contract. The USD475 million contract was awarded during 2002 to a Boeing-led team that included TRW Tactical Systems Division, Rockwell Collins Government Systems Division and BAE Systems Communications, partnered with Harris RF Communications Division. The US Army's Communications Electronics Command (CECOM) announced that the Boeing Company would be the prime system engineering contractor of a multiyear, multimillion-dollar contract to develop the JTRS. Boeing was awarded a USD73,666,000 increment as part of an USD856,539,000 cost plus award fee contract with an estimated total value of USD2,008,116,734 (if all the options are exercised), for development, demonstration, and Low-Rate Initial Production (LRIP) of the Cluster 1 JTRS. Work on this phase of the project is to be completed by 31 January 2008.

Cluster 1 is the first procurement of a number of clusters for the JTRS and consists of US Army, US Air Force Tactical Air Control Party (TACP) and US Marine Corps ground radios, as well as US Army rotary-wing aircraft radios. The deal includes the development of a new wide-band networking waveform. Reports suggest that about 10,000 systems will be produced for US users in Cluster 1. Early operational assessment testing was scheduled for 2004 with LRIP expected to begin in 2005. Full-rate production could begin in 2007 and continue for 12 years. Full-rate production of Cluster 1 could raise the total value of the deal to approximately USD7 billion.

A second JTRS bidding consortium is headed by Raytheon. The company reportedly intends to pursue contracts for future JTRS clusters.

As originally envisioned, future JTRS clusters would include configurations for manpack and hand-held versions (Cluster 2), maritime and fixed-site requirements (Cluster 3), and fixed-wing platforms (Cluster 4).

In September 2003 the Global Grid Product Area Directorate of the Airborne JTRS (AJTRS) Program Office at the US Air Force Materiel Command's Electronic Systems Centre (ESC) indicated its intention to award two parallel contracts, each lasting nine months, for preliminary work that would pave the way for the System Development and Demonstration SDD phase of the AJTRS Cluster 4 effort. This work was intended to focus on preliminary design of a modular, scalable family of radios, collection and analysis of airborne platform requirements and constraints and architectures for the airborne network. The plan was that the ESC would subsequently appoint a single prime system contractor, which would then qualify at least two JTRS production sources, for SDD and LRIP.

In January 2004 the US Air Force and US Navy announced that they had decided to merge their JTRS Cluster 3 and Cluster 4 programmes in order to reduce costs while increasing interoperability. Programme leadership would rotate between the two services, with the Air Force initially taking the lead. A combined request for proposals to conduct the pre-system development and demonstration phase was then being worked up.

At around the same time CECOM and PM WIN-T (Warfighter Information Network-Tactical) issued a request for proposals for Cluster 5 of the JTRS. The resulting contract would address SDD, limited production of the Spiral 1 version and low-rate initial production of Spiral 2.

Spiral 1 would meet critical and urgent needs by modifying commercial-off-the-shelf/non-developmental items and incorporate currently available waveforms. It was intended to involve up to 10,000 examples of a single-channel Hand-Held (HH) radio, and up to 3,000 of a two-channel ManPack (MP). Spiral 2 was tasked to provide expanded capability, include additional waveforms and fully comply with the latest version of the JTRS operational requirements document. Spiral 2 was expected to involve up to 5,000 HH radios (in single- and two-channel variants), up to 9,750 examples of the MPs and up to 5,750 of the Small Form Fit (SFF) implementation.

In August 2004 CECOM awarded a contract worth USD295 million to General Dynamics Decision Systems for SDD of the JTRS Cluster 5 segment. The company had earlier been awarded the Cluster 2 contract from US Special Operations Command. Cluster 5 production could eventually exceed 122,000 radios, with a value in excess of USD1 billion through 2011 if all options are exercised. Cluster 5 team members include Thales Communications Inc, which reportedly expected to receive approximately 20 per cent of the programme value, BAE Systems Communication, Navigation, Identification and Reconnaissance group and Rockwell Collins. Other technology contributors include Agile Communications, Altera, Datasoft, RedZone Robotics, Sarnoff Corp, Tessera, Vanu Inc, General Dynamics Robotic Systems and General Dynamics Advanced Information Systems.

Other manufacturers, inside the USA and outside, are watching JTRS developments with keen interest. The belief is that meeting the overall requirement will necessitate the import of additional technology, expertise and support from non-designated suppliers. As one of these potential participants has pointed out, even the JTRS LRIP is bigger than many countries' total military communications procurement programmes.

Status
A series of Programme Design Reviews (PDRs) began during the latter part of 2002, to include individual system, hardware and software elements. Critical Design Review (CDR) during 2003 aimed to baseline the specific design of the JTRS product. Early Operational Assessment (EOA) testing was scheduled for later in 2004, with Low-Rate Initial Production (LRIP) radios starting to flow in 2005.

During September 2005, Boeing announced that JTRS Cluster 1 radios, for land vehicles and helicopters, will be used at the US Air Force's Joint Expeditionary Force Experiment (JEFX) 2006. This is despite a continuing partial 'stop work' order issued by the army in January 2005 after the service expressed concern about a lack of technical progress in the programme. Boeing will deliver about 40 prototype radios to the army in December 2005 or January 2006 for use in JEFX. The JTRS Cluster 1 radios to be used in JEFX will be pre-EDM (Engineering Development Model) and will not be certified by the National Security Agency (NSA). The 2006 exercise will mark the first time that some FCS systems will participate in JEFX, allowing army and air force planners to experiment on how the army's future land vehicles will work in joint operations.

MIDS JTRS
EuroMIDS is one of the potential bidders for a share in production of the US MIDS JTRS programme, whereby MIDS is being migrated to a JTRS software communications-compliant architecture. The programme objective is for MIDS JTRS to maintain the MIDS LVT (see separate entry) form and fit. The existing terminal amounts to a non-programmable single-channel radio, with dedicated modules for Link 16, two J voice channels, and TACAN, each provided on a shared-time basis. MIDS JTRS is to have full JTRS functionality, incorporating Link 16 and TACAN, all on one channel (with a programmable Link 16 waveform and an integral 200 W PA), plus three additional channels for further JTRS waveforms within the 2 MHz to 2 GHz frequency range. These will have programmable communications security, and provide the terminal with ad-hoc networking and automatic routing/re-broadcasting capabilities.

Phase 1 multivendor independent engineering studies for MIDS JTRS began in July 2002 and were completed in February 2003. The established MIDS vendors conducted four independent efforts, commonly funded by all five MIDS nations (France, Germany, Italy, Spain and US). The programme has since moved on to Phase 2 (multivendor development and qualification), which was further broken down into Phase 2A (specification and preliminary development), commonly funded by all five MIDS nations, and the current Phase

2B (co-operative detailed design), leading to multivendor qualification. This is to be followed by Phase 3 (production), with multiple vendors in competition for production quantities, capitalising on the proven MIDS LVT continuous competition strategy.

It had been planned to progress Phase 2B internationally but security issues arising from the JTRS changeover from hardware to software encryption prevented this. In consequence, the international work share has been reduced from 60 per cent to less than 20 per cent and is being underwritten as an entirely US-funded engineering change programme. However, the MIDS nations are to be offered the Technical Data Pack (TDP) on completion of the effort. The TDP is being prepared under design contracts let to ViaSat and DLS (Datalink Solutions, a joint venture between BAE Systems and Rockwell Collins). Non-US participants include Selex and Rohde & Schwarz working with ViaSat, Indra Sistemas working with BAE, and Thales Communications working with Rockwell Collins.

MIDS JTRS is destined for both the US Air Force (USAF) and US Navy (USN). The Increment 1 common requirements include a JTRS-certified terminal, Link 16 porting, and 'hooks' for both the Airborne Networking Waveform (ANW), being developed as part of the Joint Airborne Network – Tactical Edge (JAN-TE) programme, and for Common Link Integration Processing (CLIP). A Joint USN/USAF initiative, CLIP is a gateway and formatting system that allows additional waveforms to be integrated within the MIDS JTRS box without the need to change the host platform's mission computer, this being achieved by translating all waveforms into Link 16.

The USN's (Level 0) lead platform is to be the F/A-18E/F. Its terminal will have Link 16; in addition it will be tested with the SINCGARS waveform (a laboratory demonstration only, since the USN does not use SINCGARS). The USAF has lately allocated funds for its (Level 1) lead platform, the A-10, which is to have Link 16, CLIP and EPLRS (Enhanced Position Location Reporting System). MIDS JTRS Increment 2, to be fitted aboard the F/A-18C/D and E-2C, will port the ANW and will embody an ANW transceiver, plus CLIP and EMF (Enhanced Modular Functionality).

Contractor
BAE Systems, North America.
Boeing.
Rockwell Collins.
Thales Communications.
Selex.

Joint Tactical Terminal/Common Integrated Broadcast Service-Modules (JTT/CIBS-M)

Type
Avionic datalink system.

Description
JTT/CIBS-M is a high performance, software programmable radio (derivative of the Commanders' Tactical Terminals), providing plug-and-play modular functionality that is backward and forward compatible with the Integrated Broadcast Service (IBS).

JTT/CIBS-M incorporates the Army Technical Architecture (ATA), Joint Technical Architecture (JTA), Defense Information Infrastructure Common Operating Environment (DII COE), Technical Architecture for Information Management (TAFIM) and Joint Vision 2010.

JTT/CIBS-M provides critical datalinks to battle managers, intelligence centres, air defenders, fire support elements and aviation nodes across all services and aboard airborne, sea-going, subsurface and ground mobile mission platforms.

JTT/CIBS-M allow the warfighting Commanders in Chief (CINCs), US Army, US Air Force, US Navy, US Marine Corps, US Special Operations Forces (SOF) and other agency users to exploit the IBS intelligence networks: Tactical

The Raytheon Electronic Systems Joint Tactical Terminal/Common Integrated Broadcast Service-Modules (JTT/CIBS-M)　0044847

Reconnaissance Intelligence eXchange System (TRIXS), Tactical Information Broadcast Service (TIBS), Tactical Related APplications (TRAP) Data Dissemination System (TDDS) and Tactical Data Information eXchange System-B (TADIXS-B-TOPS and OBP).

In addition, JTT/CIBS-M will support the evolving IBS broadcast architecture, including changes to message formats and transmission protocols and use of different portions of the RF spectrum.

JTT/CIBS-M provides an architecture that supports multiple terminal configurations, emerging technology insertion, P3I, and module integration into other terminals and processors.

Specifications
Operating modes: TIBS, TRIXS, TDDS (TRAP), TADIXS-B-TOPS and OBP, 5 and 25 kHz DAMA, SID and GPL
Interfaces: LAN: 10/100 MHz Ethernet, RS-232/422, MIL-STD-188-114, MIL-STD-1553B
Frequency: 225–400 MHz
Operating modes: full- or half-duplex
Transmit power: adjustable 11–20 dBW
ECCM: Have Quick II
EMI/EMC TEMPEST: MIL-STD-461/462, NACSIM 5100
Prime power: 100–132 V AC, 47–400 Hz, single phase, or 200–264 V AC, 47–66 Hz, single phase; battery-backed key storage and classified software
Dimensions: JTT-T/R (12 RX/4 TX; 1 full ATR long and 4X 1/4 ATR long); JTT-R (1 full ATR long)
Weight: 28.2 kg per LRU maximum
Cooling: self-contained or platform-provided forced air
Temperature: –32 to +43°C (operating) –47 to +71°C (nonoperating)
Altitude: 60,000 ft
Vibration/shock: tracked and wheeled vehicles, helicopter, jet and propeller aircraft, shipboard and submarine
MTBF: 6,500 h minimum

Contractor
Raytheon Company, Space and Airborne Systems.

KHF 950 HF radio

Type
Avionic communications system, HF.

Description
The KHF 950 has been designed specifically for fixed-wing aircraft. The system offers a choice of control head to suit the user's requirement: the KCU 1051 with Automatic Link Establishment (ALE), the KCU 951 all-digital remote controller, or the smaller-sized KFS 951. The system uses the remote KAC 952 power amplifier/antenna coupler, and the remote KTR 953 receiver exciter. Both the KAC 952 and KTR 953 are designed to operate up to 55,000 ft in an unpressurised environment when using a grounded antenna.

Frequency coverage is from 2.000 to 29.999 MHz and offers 280,000 frequencies at 100 Hz spacing. The system operates in USB, LSB and AM modes. Transmitter output in each SSB mode is 150 W PEP and 35 W average over the full frequency range.

The KHF 950 employs synthesised frequency generation techniques and uses microprocessor control for easy in-flight operation. The KCU 1051 searches for and finds the best HF frequency to use, while providing the following system features: 100 preset channel memory, fast and simple manual frequency tuning, memory that keeps track of the antenna tuning, data transmission and compliance with US federal ALE standards.

Operation may be either simplex, for normal air traffic or similar communication, or semi-duplex which permits patch-through into public utility telephone circuits. Provision is also made for a SELCAL facility and the dedicated circuits enable continuous SELCAL monitoring to be maintained without having to select the AM mode.

The dzus-mounted KCU 951 control/display for use with the KHF 950 HF radio　0081851

The KAC 952 power amplifier and antenna coupler, and the remote KTR 953 receiver/exciter used in the KHF 950 HF radio   0081852

The KCU 1051 ALE control display unit used in the KHF 950 ALE and KHF 990 ALE HF radios   0081853

A feature of the KHF 950 is its automatic antenna tuning capability, an operation carried out by simply keying the microphone. The system will operate satisfactorily on antennas only 10 ft long. It will also tune to fixed-rod aerials and towel-rail antennas, and can operate from shunt and notch antenna systems.

An optional controller is the KFS 954 which is also panel-mounted and measures only 14.5 cm². This unit contains storage for all 176 ITU maritime radiotelephone channels plus additional preselected simplex air traffic or conventional airborne communication channels, but offers only 19 preset channels. By preprogramming the ITU channels, it is possible for the operator to call any radiotelephone station without having to select the separate transmit and receive channels manually. The operator merely selects the radiotelephone mode and the required channel. The KHF 950 also interfaces with teletype and facsimile systems and a dual-installation equipment allows dual-frequency reception from a single antenna.

Specifications
Weight: 9.16 kg
Temperature:
    (remote units) −55 to +70°C
    (control/display) −20 to +70°C
Power:
    (all units) 27.5 V DC
    (transmit) 16.5 A, 453.7 W peak (total system)
    (receive) 1.5 A, 41 W (total system)
RF power outputs:
    (SSB) 150 W PEP nominal
    (AM) 37.5 W carrier
Certification: FAA TSO C31c and C32c

Status
In production and in service. The system has been widely adopted and has been installed in aircraft such as Bombardier Challenger, Gulfstream G III, Lear 55, Citation III and Falcon 50. The KHF-950 has also been selected by the US Army for helicopters.

Contractor
Bendix/King.

KHF 990 HF radio

Type
Avionic communications system, HF.

Description
The KHF 990 HF radio is a helicopter system which draws on technology used in the company's airborne KHF 950 and marine KMC 95 systems. It provides 280,000 channels in the 2 to 30 MHz HF band at frequency increments of 100 Hz.

KTR 993

KAC 992

The KHF-990 radio for helicopters   0504019

Modulation is in SSB mode and transmitted output power is 150 W PEP.

The KHF 990 has been optimised for helicopter operation. It uses the miniature KFS 594 controller, a KAC 992 combined antenna coupler/probe antenna and a remotely located KTR 993 receiver/exciter/power amplifier. This combination provides a fully capable yet lightweight system. Like the KHF 950, this system has also been updated to offer Automatic Link Establishment (ALE) using the KCU 1051 control display unit.

The KFS 594 controller provides access to 176 permanently programmed ITU marine radiotelephone channels and to 19 programmable channels which may be selected or retuned by the pilot. The KAC 992 is an automatic, digital antenna coupler which is self-contained in the end of a probe antenna system. It may be mounted externally or internally with only the probe portion of the antenna protruding from the aircraft.

The KTR 993 receiver/exciter/power amplifier can be mounted in any convenient location within the helicopter with no restrictions on proximity to the other two units. It meets the TSO requirements for explosion-proof, drip-proof and salt-spray categories.

The KSU 1051 provides automatic selection of the best available frequency, 100 preset channel memory, fast and simple manual frequency tuning, memory that keeps track of antenna tuning, data transmission and compliance with US federal ALE standards.

Specifications
Weight: 9.9 kg
Temperature:
    (remote units) −55 to +70°C
    (control/display) −20 to +70°C
Altitude: 55,000 ft
Power:
    (all units) 27.5 V DC
    (transmit) 16.5 A, 453.7 W peak
    (receive) 1.5 A, 41 W
RF power output:
    (SSB) 150 W PEP nominal
    (AM) 37.5 W carrier
Certification: FAA TSOs C31c and C32c

Status
In production.

Contractor
Bendix/King.

KMA 24 audio control system

Type
Avionic audio management system.

Description
Part of the Silver Crown panel-mount avionics range, the Bendix/King KMA 24 is a compact, lightweight system for the integrated control of a

KFS 594

number of radio communication and navigation systems. Designed principally for the general aviation sector, the solid state KMA 24 is fully TSO qualified.

The KMA 24 can control up to three transmitter/receivers and six receivers, including an internal marker beacon receiver for which it contains an automatically dimmed three-light presentation. Other system features include:

* pushbutton audio selector panel, speaker and headphone isolation amplifiers and a marker beacon receiver
* can be integrated into any 500 ohm output audio system
* different features and combinations of marker light presentation, AUTO, 2 × COM, 2 × NAV, DME, ADF, HF, TEL selections available
* two unswitched inputs for use as altimeter warning and telephone ringer
* outputs for ramp hailer and passenger address or intercom
* isolation amplifiers are provided for headphones and speaker to provide isolation even when the same source is selected
* three isolated 16 ohm resistors provided for transceiver speaker output loads
* 14V or 28 V DC operation

Specifications
Dimensions: 173 × 33 × 159 mm
Weight: 0.77 kg
Temperature: −20° to +55°C
Electrical: 14 or 28 V DC
Compliance: TSO C35d, C50b

Status
In production and in service.

Contractor
Bendix/King.

KMA 24H audio control system

Type
Avionic audio management system.

Description
Part of the Silver Crown panel mount avionics range, the KMA 24H is a variant of the KMA 24 system (see separate entry) where the internal marker beacon facility is replaced by a five-station hot microphone intercom and associated volume control. There are two variants of the KMA 24H, the -50/54 and -70/71. The following summary outlines common system features and differences:

Common Features
* push-button audio selector panel with separate speaker and headphone isolation amplifiers
* can be integrated into any 500 O output audio system

- interphone communication including five microphone audio inputs for up to five intercom stations
- two unswitched inputs for use as altimeter warning and telephone ringer
- speaker outputs for ramp hailer and passenger address or intercom
- PA mute provided to mute passenger background music systems when a microphone is keyed
- 14 or 28 V DC operation
- TSO qualified.

KMA 24H -50/54

- similar to KMA-24 but does not include marker beacon presentation and includes intercom functions/controls
- night vision goggle compatible version available
- designed for use in both single- and dual-audio panel installation
- controls up to three transceivers and six receivers.

KMA 24H -70/71

- similar to KMA-24H -50/54 but includes voice operated intercom, separate alternate action capability and keyed activation of up to five stations
- capability for pilot and co-pilot to isolate out of the intercom system through separate independent amplifiers
- up to five transceivers and five receivers can be controlled in -71 version
- up to four transceivers and six receivers can be controlled in -70 version
- user can select VOX, keyed or hot mic intercom through adjustments of the VOX control on the front panel
- incorporates summing amplifier to combine received audio with the pilots microphone for installations where a voice recorder is necessary.

Specifications
Dimensions: 173 × 33 × 159 mm (–50/54), 165 × 33 × 160 mm (–70/71)
Weight: 0.77 kg
Temperature: –20° to +55°C (+70°C –70/71)
Electrical: 14 or 28 V DC
Compliance: TSO C50b (–50/54), C50c (–70/71)

Status
In production and in service.

Contractor
Bendix/King.

KMA 26 audio control system

Type
Avionic audio management system.

Description
Part of the Silver Crown panel mount avionics range, the Bendix/King KMA 26 is an audio control unit with inputs for up to three transceivers, nine switched audio inputs and four unswitched inputs. Further features of the system are:

- audio amplifier
- six-station intercom with three modes – pilot, crew, and all
- self-contained marker beacon receiver with 3-lamp display
- high and low sensitivity/light test switch for marker lamps
- marker mute button to temporarily suppress marker audio while passing over a beacon
- individual volume controls for crew and passengers
- four separate microphone squelch circuits to reduce background noise
- self-contained mode switching
- emergency (EMG) position on MIC selection switch connects microphone and headphones directly to COMM 1 in the event of a failure
- accepts inputs from two music entertainment sources; entertainment audio is automatically muted when communication audio is detected.

Specifications
Dimensions: 191 × 33 × 159 mm
Weight: 0.77 kg
Altitude: up to 50,000 ft
Compliance: TSO C35d, C50c

Status
In production and in service.

Contractor
Bendix/King.

KMA 28 audio control system

Type
Avionic audio management system.

Description
Part of the Silver Crown panel mount avionics range, the Bendix/King KMA 28 is an audio control unit with five receiver (2 NAV, ADF, DME and AUX) and four transceiver inputs (3 COM and one approved cell phone), together with a six-place stereo intercom. Further features of the system are:

- audio selector panel with 6-place intercom
- 'IntelliVox' feature samples ambient noise and adjusts threshold for clip-free voice communications
- dual high-fidelity stereo inputs
- two split modes (COM 1/2 and 2/1)
- 3-light marker beacon
- stereo entertainment inputs.

Specifications
Dimensions: 173 × 33 × 159 mm
Weight: 0.5 kg
Altitude: up to 50,000 ft
Temperature: –22° to +55°C
Electrical: 11 to 33 V DC, 1.5 A
Compliance: TSO C35d, C50d, RTCA DO-160C, -178B level D, -143, -214

Status
In production and in service.

Contractor
Bendix/King.

KTR 909/909B UHF transceivers

Type
Avionic communications system, UHF.

Description
The all-solid-state KTR 909 transceiver operates in the 225 to 399.975 MHz range in 25 kHz increments and is capable of 10 W of transmitter power. Weighing only 2.09 kg, the compact system includes the KTR 909 remote-mounted transceiver and the Gold Crown III KFS 599A control head. The KFS 599A is offered in two versions: with standard gas discharge display or with ANVIS NVG-compatible display.

The KFS 599A can be tuned by dialling in the desired operating frequency or by selecting any of the 20 user-programmable channels.

The KTR 909 is capable of operating with tandem KFS 599A control heads designated as master and slave, which is ideally suited for training applications. Additionally, one version of the KTR 909 can be tuned via compatible radio management systems, such as the RMS 555 or FMS systems with frequency management.

Dual-monitoring capability allows the KTR 909 to monitor either the main receiver, the Guard receiver or both simultaneously.

The KTR 909B transceiver and KFS 599B control head   0081849

The KTR 909 offers 1,000 Hz tone modulation used in DF operations. In addition, it provides an ADF mode which allows the unit to perform the tuning function for remote ADF systems. This function, when interfaced with peripheral DF equipment, permits an ADF indicator to be used as a bearing indicator for DF operations.

The KTR 909B differs from the KTR 909 in that it incorporates a dedicated guard frequency, monitoring 243 MHz via a dedicated second receiver, while the KTR 909 scans other frequencies when the unit is not in use. The KTR 909B incorporates an ARINC 429 bus interface, and is controlled by the KFS 599B control head.

Specifications
Dimensions:
KTR 909 127 × 44 × 334 mm
KTR 909B 127 × 67 × 277 mm
Weight:
KTR 909 1.59 kg
KTR 909B 2.17 kg
RF power: NTL 8 W (10 W nominal)
Frequency range: 225.000 to 399.975 MHz
Channel spacing: 25 kHz

Status
In production and in service.

Contractor
Bendix/King.

KY 196A/KY 196B and KY 197A VHF transceivers

Type
Avionic communications system, VHF.

Description
During 1987, the KY 196 and 197 transceivers were updated, with the addition of some new features, and redesignated KY 196A and KY 197A. The KY 196A has since been further

The Bendix/King KMA 28 audio control system   0103877

upgraded to the KY 196B standard by the addition of class 4 and class 5 operation to the class 3 standard already available in the KY 196A. The KY 196B is also able to meet the ICAO EUR Region's 8.33 kHz channel spacing requirement providing a total of 2,280 channels in the range 118.000 to 136.990 MHz.

The KY 196A is a compact lightweight panel-mounted transmitter/receiver particularly suitable for light aircraft. It covers the VHF band from 118 to 136.975 MHz in which range 760 channels are provided; channels are selectable at increments of 25 or 50 kHz. One of the features of the update is the expansion of the frequency coverage by 1 MHz at the top end of the range.

A principal feature of the system is that a second frequency, in addition to the one in use, may be stored for immediate selection. Both the operating and standby frequencies are presented on a self-dimming gas discharge display in a window on the front panel. When the standby channel is selected the former operating channel is entered into the standby store. Non-volatile storage, provided by an electrically alterable read-only memory chip, ensures both frequencies remain stored when the power supply is off or disconnected. No separate memory power supply is required.

Solid-state construction is employed throughout. The system is microprocessor-controlled and digital synthesis techniques are used for frequency generation. A MOSFET RF amplifier and mixer stage is used to provide clear signal reception.

Other improvements in the upgrade include a bigger selection knob, lighted push-buttons, pilot programmable lighting and dimming levels and a facility which detects a stuck microphone and stops transmission after 2 minutes.

Specifications
Dimensions: 160.3 × 34.3 × 258.5 mm
Weight: 1.27 kg
Power supply:
(KY 196A/KY 196B) 28 V DC
(KY 197A) 14 V DC
Power output:
(KY 196A/KY 196B) 16 W
(KY 197A) 10 W
Temperature range: −20 to +55°C

Status
In production and in service.

Contractor
Bendix/King.

KY 196B panel-mounted VHF transceiver

Type
Avionic communications system, VHF.

Description
In June 1999, the KY 196B panel-mounted VHF transceiver was announced as an update of its earlier KY 196A model to comply with new European requirements for 8.33 kHz spacing between channels. Transfer from 25 kHz to 8.33 kHz operation is by pull-out of the frequency selection knob, increasing channels available from 760 to 2,280.

The KY 196B is primarily designed for light business jets and turboprop aircraft that fly at high altitude but often have limited space for avionics; it fits a standard 6.3 in 'mark width' opening for panel-mounted avionics. It can also be used as a plug-in replacement for the KY 196A with no modifications to wiring or mounting rack.

AlliedSignal also produces three remote-mounted transceivers with 8.33 kHz channel spacing for larger transport and regional aircraft: the RTA-83A and RTA-83B and the RTA-44D digital voice and data radio.

Contractor
Bendix/King.

LAMPS MK III datalink (Hawk Link)

Type
Avionic datalink system.

Description
The LAMPS MK III datalink system provides full duplex, secure and highly reliable communications between airborne and shipboard platforms. The system, better known as the Hawk Link, was designed specifically to enable the sensors and weapons of the SH-60B Seahawk, LAMPS MK III helicopter to function as integrated subsystems of the US Navy's Light Airborne MultiPurpose System (LAMPS) MK III weapon system. The Hawk Link multiplies SH-60B based processing of airborne sensor information by the power of shipboard processing capabilities through parallel operations, increasing the speed and accuracy of target/threat classification and/or localisation. Airborne sensors become, through the Hawk Link, an extension of the ship's sensor suite.

The Hawk Link system consists of two subsystems, the AN/ARQ-44 Airborne Data Terminal and AN/SRQ-4 Shipboard Data Terminal. The AN/ARQ-44 is functionally responsible for communicating SH-60B Seahawk sensor data to its parent ship over the wide data bandwidth digital downlink and receiving command and control data from the parent ship over the narrower data bandwidth digital uplink. The AN/SRQ-4 is linked with shipboard computers via an NTDS slow interface, hands-off Anti-Ship Surveillance and Targeting (ASST) sensor information (radar video and tactical plot data) to shipboard mission control consoles and hands-off Anti-Submarine Warfare (ASW) sensor information to the ship's sonar signal processing system.

Hawk Link features include: full duplex, digital communication; wide-bandwidth downlink; downlink error correction; high fade margins; up to four wide plus four narrow acoustic channels; communication security (KG-45); full Mil design with online BIT; directional (jam resistant and LPI); narrow-bandwidth uplink; uplink error detection; high-intelligibility voice com; search and track radar video; advanced sensor interfaces, interoperability; LPI upgrades under consideration.

AN/ARQ-44 Airborne Data Terminal Specification
Multiplexer/demultiplexer, TD-1254/ARQ-44, analogue and digital sensor interfaces, 1553A interface, 18 kg.
Radio, receiver/transmitter, RT-1275/ARQ-44, FM FSK modulation, 30 kg.
Directional antenna (2 each), AS-3273/ARQ-44, Azimuth steerable, 3.9 kg each (with radome).

Status
The Hawk Link is operational in the US Navy's SH-60B Seahawk helicopters, and AN/SQQ-89(V) Surface ASW Combat System equipped ships including DDG-51 'Arleigh Burke' Class and DD-963 'Spruance' Class Destroyers, CG-47 'Ticonderoga' Class Cruisers, FFG-7 'Oliver Hazard Perry' Class Frigates. In addition, the Spanish Navy selected the Hawk Link to support its SH-60B Seahawks and LAMPS MK III Frigates.

Contractor
Sierra Research, a division of SierraTech, Inc.

LCR-2000 and LCR-3000 series DSP receivers

Type
Avionic communications system, VHF/UHF.

Description
The Low Cost Receiver (LCR) range of DSP (Digital Signal Processing) receivers comprises: the LCR-2000 single rack-unit (no display or speaker) and LCR-2400 three rack-unit (with display and speaker) LF-HF DSP receivers; the

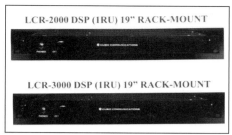

Cubic Communications' LCR DSP receivers

0044819

LCR-2400 and LCR-3400 DSP(3RU) 19 in rack-mounted units share similar enclosures

0062849

LCR-3000 single rack-unit (no display or speaker) and LCR-3400 three rack-unit (with display and speaker) V/UHF DSP receivers.

Specifications
LCR/SMR-2000 and LCR/SMR-2400
Frequency range: 10 kHz to 30 MHz, 1 Hz tuning resolution
Detection modes: LSB, USB, ISB, CW, AM, FM, PM, FSK
Sweep and scan: 100 channels per second
Remote control: RS-232 or RS-422
Dimensions: 19 in rack unit

LCR-3000 and LCR-3400
Fequency range: 20–3,000 MHz, 10 Hz tuning resolution
Detection modes: LSB, USB, CW, AM, FM, PM
Sweep and scan: 100 channels per second
Remote control: RS-232 or RS-422
Dimensions: 19 in rack unit

Status
In production and in service.

Contractor
Cubic Communications Inc.

LG81xx radar altimeter antennae

Type
Aircraft antenna.

Description
Honeywell's LG81 series of radar altimeter antennae (RAAs) have been in continuous service for more than 30 years on air vehicles flown by the US Army, Navy and Air Force as well as NASA and commercial applications. Over 30,000 RAAs in more than 40 configurations have been fielded.

The RAAs are ruggedised, low-profile, conformal micro strip designs, suitable for a wide range of platforms including fighters, bombers, cruise missiles, smart munitions and UAVs. Honeywell produces the US Navy's standard fixed-wing RAA, the US Army's standard helicopter RAA, and numerous specially adapted antennae for unique applications, including antennae for the Space Shuttle and Mars Lander missions and low-RCS designs for covert applications.

Status
In service in all US Navy fixed-wing aircraft, US Army helicopters, F-16, Saab JAS-39 Gripen, Lockheed C-130 aircraft, Tomahawk Cruise Missile and AGM-130 weapon systems and the Space Shuttle.

Contractor
Honeywell Sensor and Guidance Products.

Specifications

RAA Part No.	LG81T	LG81K	LG81AT	LG81V
Dimensions (mm)	89 × 89 × 2.3	108 × 88 × 2.8	146 dia × 2.8	178 × 254 × 5.6
Weight	46 g	122 g	251 g	637 g
Gain	10.0 dBi	10.0 dBi	11.0 dBi	15.0 dBi
E Plane × H Plane Beamwidth	52° × 50°	55° × 50°	42° × 51°	23° × 25°
Application	General	US Navy fixed-wing (standard)	US Army helicopter (standard)	High altitude

TRW's dual-polarisation multi-arm spiral antenna
0044851

Low-observable, conformal, antennas for avionics

Type

Aircraft antenna, Communications, Navigation, Identification (CNI).

Description

For more than 30 years, TRW has developed communications antennas and radio frequency (RF) components. Today, TRW is applying that knowledge to the development of low-observable, conformal aircraft antennas for a wide range of applications in integrated Communications, Navigation and Identification (CNI) avionics, Electronic Warfare (EW) systems and Unmanned Aerial Vehicles (UAVs).

These low-observable, structurally integrated antennas are designed to provide efficient radiation characteristics that meet system performance requirements in avionics. They are capable of receiving and/or transmitting signals from High-Frequency (HF) through millimetre waves and support such functions as CNI, SATellite COMmunications (SATCOM), combat ID and situation awareness.

TRW has pioneered the development of 'Smartskin' antennas. The surfaces, or 'apertures', of these antennas serve as an integral part of an aircraft's outer surface without degrading the low observability of the vehicle. Smartskin antennas provide the bandwidth and efficiency associated with antennas having frequency-independent performance. While flown on a NASA Dryden F/A-18 test aircraft, TRW's VHF 'Endcap' antenna provided complete spherical coverage with a gain improvement of between 15 and 20 dB over the combined existing upper and lower blade antennas.

Until the programme was cancelled, TRW was also studying the application of Smartskin antenna technology to the RAH-66 Comanche helicopter. This extension of the F/A-18 Endcap antenna provides simultaneous VHF and UHF communication from a single, low-observable structurally integrated antenna, meeting mission performance requirements for communication and navigation. This antenna technology enables the Identification of Friend or Foe (IFF), Global Positioning System (GPS) and HF functions.

TRW's CNI antenna designs have been developed for new aircraft such as the F/A-18 E/F.

TRW's situation awareness antennas provide precise direction-finding capability from a single, efficient, extremely broadband, low-observable aperture. This unique antenna has applications to low-observable fighters, bombers, surveillance aircraft and ships in such functions as electronic warfare, Signal intelligence (Sigint), combat ID and other special applications.

TRW has also developed a high-gain, wide-bandwidth phased-array antenna for various applications. This was accomplished by designing, fabricating and testing a low-observable phased array possessing greater than 1,000 elements for simultaneous operation – from S-band through X-band (a five to one bandwidth). This phased-array antenna is highly efficient, since the radar-absorbing material is minimised while the very stringent low-observability requirements are met.

Finally, TRW is developing a similar phased-array antenna operating from C-band through Ka-band applicable to both military low-observable vehicles and commercial vehicles. This design, measuring approximately 3 ft in diameter and having greater than 5,000 elements, will provide military and commercial

multiSatcom functions from a common aperture that eliminates the need for multiple antennas – reducing cost, maintainability and installation issues.

Contractor

TRW Avionics Systems Division.

TRW Smartskin antennas on NASA Dryden F/A-18
0044849

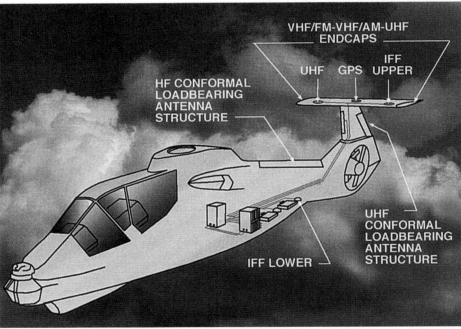

VHF/FM-VHF/AM-UHF ENDCAPS

IFF UPPER

UHF GPS

HF CONFORMAL LOADBEARING ANTENNA STRUCTURE

UHF CONFORMAL LOADBEARING ANTENNA STRUCTURE

IFF LOWER

TRW Smartskin antennas on RAH-66 Comanche
0044850

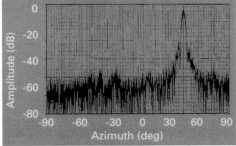

TRW's low-observable, high-gain, wide bandwidth (C- to Ka- band), phased-array antenna and test results
0044852

Mark II Communications Management Unit (CMU)

Type
Avionic Communications Control System (CCS).

Description
The Mark II CMU is designed to meet the needs of current Aircraft Communications Addressing and Reporting System (ACARS) management requirements and Future Air Navigation System (FANS) Air Traffic Management (ATM) regime. The unit can meet the requirements of both older analogue aircraft and newer digitally managed aircraft to meet the complete needs of airlines with a single part number item.

The Mark II CMU is designed around an Intel 486DX processor to provide a flexible format that can be configured by the user to meet specific needs for screen formats, downlink/uplink format and control in accordance with ARINC 429, encryption/decryption and other functions.

Specifications
Dimensions: 4MCU in accordance with ARINC 600
Weight: 5.44 kg
Power: 30 W, 115 V AC or 28 V DC
MTBF: >20,000 h

Contractor
Honeywell Aerospace, Electronic & Avionics Lighting.

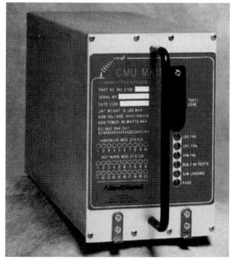

The Honeywell Aerospace, Electronic & Avionics Lighting Mark II communications management unit   0062198

MD-1269A multipurpose modem

Type
Avionic satellite communications system.

Description
The MD-1269A multipurpose modem is an advanced digital signal processor-based, full-duplex 70 MHz modem for UHF Satcom and line of sight communications links. It simplifies existing communications systems and terminals by combining the important modulation modes of six currently used modems into a single unit.

Digital data shaping provides spectral containment and permits 2,400 bps BPSK and 4,800 bps OQPSK operation over a 5 kHz satellite channel. The MD-1269A also includes baseband interfaces with audio equipment, data terminals or encryption devices to provide narrowband AM and FM voice, TADIL A (link 11), TADIL B, TADIL C (Link 4), KY-57/58 (Vinson AM and FM), KY-65/75 (Parkhill), KG-84 and ANDVT compatibility.

The MD-1269A has a standard 70 MHz RF interface and can be used with any 70 MHz compatible receiver/transmitter in the non-DAMA modes. The 70 MHz interface is synthesised to provide 5 kHz channelisation capability with the R/T operation on 25 kHz channels. For DAMA operation, the R/T must be compatible with the stricter requirements

for DAMA operation. The modem is compatible with the US Navy 25 kHz TDMA-1 DAMA and the US Air Force 5 kHz USTS DAMA. All non-DAMA capabilities are preserved in the DAMA version of the MD-1269A for backward compatibility.

The MD-1269A includes synchronous and asynchronous control/data ports for various local or remote-control options. All control and data ports can be configured as balanced or unbalanced MIL-STD-188-114A interfaces to a host terminal controller or C-1282AG control head. The entire modem is contained in a ½ ATR airborne package and is ruggedised to withstand airborne, shipboard or tactical environments.

Specifications
Dimensions: 123.95 × 193.55 × 497.84 mm
Weight: 11.34 kg
Power supply: 115 or 230 V AC, 47–440 Hz, 50 W max
Temperature range: −40 to +55°C
Altitude: up to 30,000 ft

Status
In production and in service.

Contractor
Raytheon Company, Space and Airborne Systems.

MDM-2201 airborne HF data modem

Type
Avionic communications system, HF, data, secure.

Description
The MDM-2201 is a full duplex, multimode single-channel airborne HF data modem contained in a ¼ ATR short package. Operating on 28 V DC power, the MDM-2201 provides up to 4,800 bps synchronous or asynchronous data transmission over half- or full-duplex radio links.

The system provides for data transmission over HF radio channels through MIL-STD-188-110A-compliant single-tone mode, which uses adaptive equalisers to combat multipath and fading, introduced by ionospheric propagation, and employs forward error correction and time interleaving. The MDM-2201 supports applications such as teletype, encryption, digital voice, digital video and adaptive networks. It is interoperable with the MD-1061, MD-522, CV-786, TD-1089 TADIL-B, USC-11 (TE-204), MD-1280, MD-1142 and others. The system's waveform library includes MIL-STD-188-110A 39 tone, MIL-STD-188C 16 tone, fully programmable FSK and multitone FSK.

Remote control and monitoring are performed through the dedicated MCU-2201F control head, RS-232 serial PC interface or MIL-STD-1553B bus controller.

Specifications
Dimensions: 198 × 58 × 334 mm
Weight: 2.7 kg
Power: 28 V DC, 20 W (optional 115 V AC, 400 Hz)
Temperature: −40 to +50°C operating
Reliability: >10,000 h MTBF

Status
In service. The MDM-2201 airborne HF data modem has been selected by the US Army and other customers.

Contractor
Rockwell Collins.

MDM-3001 HF data modem

Type
Avionic communications system, HF, data.

Description
Rockwell Collins MDM-3001 single channel modem is capable of transmitting data over a standard 3 kHz SSB (Single SideBand) at up to 4800 bps. The MDM-3001 provides extensive waveform selection and configurable system parameters

enabling data transmission even in distressed channel conditions. SYSTEM FEATURES Data Transmission Rates of 50 bps to 4800 bps in single HF channel operation. Remote and front panel control JITC certification for MIL-STD-188-110A HF Channel Simulator per CCIR 549-3.

The MDM-3001 is a single channel modem designed for HF data transmission over a standard 3 kHz Single Side Band (SSB) at up to 4,800 bps. Designed to operate with Collins' HF Messenger™ data link (STANAG 5066 protocol) management software, the system features comprehensive Built-In-Test (BIT) and is upgradeable via an RS-232 port.

The MDM-3001 is compliant with STANAG 4285 single tone, and MIL-STD-188-110A single tone waveforms. In addition, 1-, 8- and 16-channel VFCT per MIL-STD-188-342 and STANAG 4481, MIL-STD-188-110A 39-tone, and programmable FSK interoperable with TADIL-B, CV-786, MD-1280 and MD-522 are provided. The system also provides an HF channel simulator per CCIR-549-3.

Specifications
Processor: 28.8 MHz ADSP-21062
Data interface: EIA-232D. RS-422/423, MIL-STD-188-114 via synchronous/asynchronous serial port
Transmit audio: 600 ohm balanced; configurable range −30 dBm to +5 dBm
Receive audio: 600 ohm balanced with dynamic range −30 dBm to +5 dBm (0 dBm nomial)
Dimensions: 44 × 216 × 178 mm (W × H × D)
Weight: 1.6 kg
Power: 85–265 V AC, 47–440 Hz, single phase, <10 W
Temperature:
  (operating) 0 to +50°C
  (storage) −40 to +70°C
Reliability: >12,000 h MTBF (naval, sheltered); >30,000 h MTBF (ground, fixed)

Status
Introduced in 1996. In service with the US Navy, Air Force and Coast Guard.

Contractor
Rockwell Collins.

MDM-700 modem

Type
Avionic datalink system.

Description
The MDM-700 modem is ruggedised for airborne use and provides a relatively inexpensive means of interfacing a computer to a duplex RF link which utilises conventional FM transmitters and receivers. The MDM-700 is a full-duplex interface. However, the modem can be used in the simplex mode without degradation and without special considerations.

The MDM-700 transmit section accepts the RS-232 signal and converts each transition into a half-sine shaped pulse. A positive transition creates a positive pulse and a negative transition creates a negative pulse. Should there be no transition activity, the modem continues to create pulses of the same polarity as the last but at a low frequency. A result of this is that a signal can be AC-coupled; DC response is not required.

The modem receive portion accepts the half-sine pulse train and recreates the original RS-232 data. Since the conversion is one of edge coding, the modem is insensitive to data rate, except for the maximum of 9,600 bps. Clock is not used nor is it required.

Specifications
Weight: 0.23 kg
Power supply: 28 V DC

Status
In service.

Contractor
L-3 Telemetry-East.

MIDS-FDL Multifunctional Information and Distribution System – Fighter Data Link

Type
Avionic datalink system.

Description
BAE Systems North America and Rockwell Collins have entered into a Limited Liability Company (LLC), named Data Link Solutions, to qualify, build and sell Link 16 Multifunctional Information Distribution System (MIDS) Fighter DataLink (FDL) terminals for the US Air Force F-15 aircraft.

FDL brings the advantages of a full Multifunctional Information Distribution (MIDS) system packaged in a smaller, lighter and lower cost Link 16 fighter terminal. Link 16 provides real-time, jam-resistant secure transfer of combat data, voice and relative navigation information between widely dispersed battle elements. Participants gain Situational Awareness (SA) by exchanging digital data over a common communication link that is continuously and automatically updated in near-realtime, reducing the probability of blue-on-blue, duplicated targeting or missed targets. Each participant in the communication link is able to electronically see the battle space, including ownship, assigned targets (by platform), bogeys (potential threats) bandits (confirmed threats).

Sharing the same fundamental Open Architecture of all MIDS-LVT (VME/RS-422, SEM-E), FDL has full Link 16 interoperability (TADIL-J, IJMS), 50 W transmit power and >1 W LPI mode. Dual-antenna transmit and receive provides position, location and identification messages, relative and geodetic navigation and adaptable host interface processing. Growth supports enhanced throughput, video and voice.

Specifications
Performance characteristics:
Pseudo random frequency hopping
51 frequencies – 969 to 1,206 MHz
Frequency hop rate – 13us

238 kbps Data Rate/expandable to 2 Mbps
Jam resistant and crypto secure
128 possible nets
Dimensions: Full ATR form factor
Weight: 24.75 kg (option dependant)
Power supply: 115 V AC, 400 Hz, 350 W average

Status
In production and in service in US Air Force F-15 C/D/E aircraft. MIDS-FDL was the first of the MIDS-LVT (see separate entries) variants to be operationally deployed during air operations over Afghanistan and Iraq. In June 2004, Data Link Solutions announced it had completed deliveries of MIDS-FDL terminals under a fixed-price production contract awarded by the US Navy Space and Naval Warfare Systems Command (SPAWAR) in September 1996. A total of 832 FDL terminals were delivered under this effort. The terminals have been deployed on a number of US Air Force platforms including F-15A/B, C/D and E and KC-135 aircraft as part of the Roll On Beyond Line-of-Sight Equipment (ROBE) programme. In support of the deliveries to F-15s, DLS reached a production rate of 32 terminals per month, the highest output rate of production deliveries of any MIDS supplier. As of mid-2004, DLS was continuing low-rate production of FDL terminals in support of Foreign Military Sales (FMS) and direct commercial sales orders. More than 14 countries have selected and installed Link 16 terminals on 19 platforms.

MIDS-FDL is also part of the MIDS-LVT (Low Volume Terminal) multinational co-operative development programme to design, develop and deliver Link 16 tactical information system terminals that are smaller and lighter than Joint Tactical Information Distribution System (JTIDS) Class 2 terminals. There are three variants of the MIDS terminal. MIDS-LVT(1) will be used by US Navy (USN), Marine Corps (USMC) and Air Force (USAF) aircraft as well as by those of European nations. MIDS-LVT(2) will be used by army combat systems such as Patriot for both the US and French armies. MIDS-LVT(3) is also known as the Fighter DataLink (FDL).

MIDS-FDL Fighter DataLink   0011842

Contractor
Data Link Solutions, comprising:
BAE Systems North America.
Wayne.
Rockwell Collins.

MIDS-LVT Multifunctional Information and Distribution System – Low-Volume Terminal

Type
Avionic datalink system.

Description
Rockwell and BAE Systems, North America have teamed under the banner of Data Link Solutions, to provide the US contribution to the implementation of a full Multifunctional Information Distribution Systems (MIDS) capability in a new, low-cost Link 16 fighter terminal, designated MIDS-LVT (Low Volume Terminal). The system is an extension of proven technologies employed in other Rockwell/ BAE datalink systems, featuring a flexible, open architecture avionic design adhering to commercial standards, coupled with light weight and low acquisition costs.

MIDS-LVT portrays targets, threats and friendly forces on an easy-to-understand, relative Plan Position Indicator (PPI) display, vastly increasing pilot Situation Awareness (SA) and combat effectiveness of aircraft formations.

Specific functional capabilities of MIDS-LVT include:
- Position, location, identification messages
- Relative and geodetic navigation
- Adaptable Host Interface Processing
  In detail, major features of the system include:
- Full Link 16 interoperability
- TADIL-J
- IJMS
- 200 W transmit power and <1 W LPI mode
- Dual-antenna transmit and receive
- Full ATR form factor
- Maximum weight of 50 lb (option dependent)
- 115 VAC, 400 Hz, 350 W average
- Embedded TACAN
- HPA Interface
- Voice Channels
- LPC-10
- CVSD
- MIL-STD-1553 host interfaces
- Open architecture
- VME/RS-422 Busses
- SEM-E cards
- Growth/spare allocation
- High throughput
- Video
- Embedded GPS.

Specifications
Frequency hopping: Pseudo random
Frequencies: 51 frequencies – 969 to 1206 MHz (L-band)
Frequency hop rate: 13 µs (77,800 hops per second)
Data rate: 238 KBps (expandable to 2 MBps)
Security: Jam resistant and crypto secure
Nets: 128

Status
MIDS-LVT is a multinational co-operative development programme to design, develop and deliver Link 16 tactical information system terminals that are smaller and lighter than Joint Tactical Information Distribution System (JTIDS) Class 2 terminals. The programme encompasses two US vendors and one European vendor. The US vendors are Data Link Solutions (a limited liability company comprising BAE Advanced Systems and Rockwell Collins) and ViaSat. The European vendor is EuroMIDS, a consortium comprising four companies – Thales, Finmeccanica, Indra and EADS. The US and European teams have worked together to develop, test and produce a tactical information system terminal of reduced size, weight and cost, thereby making MIDS-LVT more readily available for use in a wider variety

of airborne platforms as well as maritime and ground applications.

As an integral part of the program's acquisition strategy, MIDS-LVT features Open Systems Architecture (OSA) and employs a complex system of performance specifications to allow for continuous competition between MIDS vendors throughout the production phase. This strategy supports the implementation of transatlantic competition after the US and European contractors have completed qualification efforts and established full-rate production capabilities.

There are three variants of the MIDS terminal. MIDS-LVT(1) will be used by US Navy (USN), Marine Corps (USMC) and Air Force (USAF) aircraft as well as by those of European nations. MIDS-LVT(2) will be used by army combat systems such as Patriot for both the US and French Armies. MIDS-LVT(3), also known as the Fighter DataLink (FDL), is already operational and has been installed in USAF F-15C and E aircraft.

During 2002, it was reported that US Defense Department planned to purchase 2,705 MIDS terminals, excluding spares, through FY2011, broken down as: 1,880 MIDS-LVT(1), 97 MIDS-LVT(2) and 728 MIDS-LVT(3) terminals.

Airborne platforms for MIDS-LVT(1) terminals include USN F/A-18 and EA-6B, USAF F-16 and B-2 aircraft, with potential extension to USN P-3, H-60 and MH-60 and USAF B-1, B52 and F/A-22 aircraft. The first US contract split a Lot 1 buy evenly between DataLink Solutions ViaSat. Airborne platforms for MIDS-LVT(3) /FDL include all USAF F-15 aircraft, with 786 units reported ordered, including spares, from Data Link Solutions.

MIDS-LVT Foreign Military Sales (FMS) include Australian F/A-18, Belgian, Norwegian and Danish F-16 (where MIDS-LVT forms part of the M3 upgrade package for Royal Danish Air Force F-16AM/BM) and Swiss F/A-18 aircraft; the contractor was Data Link Solutions.

In October 2003 it was announced that ViaSat had won a USD43.8 million contract for MIDS-LVT(1) and (2) terminals for the US Air Force and Army, with deliveries commencing in October 2004 and continuing through 2006. In November of the same year it was announced that Data Link Solutions had been awarded a USD56.6 million contract for 216 terminals for a number of US and FMS platforms.

In June 2004 it was reported that ViaSat had won a delivery order valued at approximately USD47 million for MIDS from the SPAWAR, San Diego. The order includes MIDS-LVT(1) airborne and MIDS-LVT(2) ground-based terminals, with delivery expected to begin in June 2005 and continue through the first quarter of ViaSat's FY2007.The Lot 5 award includes MIDS-LVT(1) terminals and spares for F/A-18, F-16 and A-10 aircraft, together with MH-60 helicopters.The award also includes a significant number of MIDS-LVT(2) terminals and spares for various US Army and USAF systems.

Contractor
Data Link Solutions, comprising:
BAE Systems, North America.
Rockwell Collins.

Modified Miniature Receiver Terminal (MMRT)

Type
Avionic communications system, VLF/LF.

Description
The Rockwell Collins Modified Miniature Receiver Terminal (MMRT) provides a VLF/LF communication link from the National Command Authority (NCA) to the E-4B National Airborne Operations Centre and the E-6BTACAMO aircraft via the Minimum Essential Emergency Communications Network (MEECN).

The MMRT is an enhanced version of the USAF AN/ARR-85 miniature receiver terminal. It has a new High Data Rate (HIDAR) mode, improved command post operator control capabilities and advanced packaging to enhance MEECN mission performance. Like the ARR-85, the MMRT ensures automatic reception and processing of secure, long-haul Emergency Action Messages (EAM) in benign, hostile and nuclear-stressed environments.

Collins' MMRT is a common VLF/LF receive solution for USAF and USN land, sea and airborne communication platforms. It is a self-contained VLF/LF receiver/demodulator with automatic message processing and embedded cryptographic equipment that allows for on-aircraft updates. It incorporates high-performance, adaptive signal processing, a wide dynamic range front end, programmable operational modes and high-reliability advanced modular design and packaging.

The main benefits of the MMRT and improvements over the MRT are:
- Global interoperability for all MEECN/VERDIN command post platforms
- Enhanced EAM processing
- Nuclear hardened
- Improved anti-jam performance
- Flexible programmability to meet changing mission roles
- Affordable upgrading and capabilities expansion
- Increased operator control and management
- Lower life-cycle support costs
- Improved reliability and lower mean time to repair.

The Collins MMRT provides high-performance EAM reception to globally deployed strategic forces in support of post-cold war scenarios. The architecture includes digital signal processing enhancements to provide high data rate reception and increased anti-jam capabilities. A MIL-STD-1553B interface provides real-time system control. The MMRT uses Ada software and has built-in flexibility for future enhancements.

The main features of the system are:
- Three-channel or single-channel operation
- Interoperability with MEECN modes 15, 9 and 9 MMPM, HIDAR and VERDIN modes 22 and 23
- High dynamic range for simultaneous transmitter operation
- Three-channel TE/TM spatial diversity adaptive combining
- Anti-jam protection via multichannel null steering, NBIS and single-channel, non-linear adaptive processing
- Time diversity, three-channel reception/ message combining
- Automatic dual mode search (one normal/one special)
- Frequency scanning from one to five transmitters (operator programmable), five-day mission plan
- Smaller size and lower power consumption and weight
- Low life-cycle cost through receiver hardware commonality
- Additional nuclear hardening protection.

The MMRT has optional ancillary equipment to enhance operation and adaptability for various mission roles and platforms. They include:
- An antenna coupler for platform antenna interface matching
- A radio set control for mission data set creation and receiver control/status display
- A receiver control panel for receiver on/off and keyfill.

Status
The MMRT is in service on airborne communications platforms of the US forces.

Contractor
Rockwell Collins.

Multifunction/multiband antenna subsystem

Type
Aircraft antenna, Communications, Navigation, Identification (CNI).

Description
The tactical airborne antenna array with multiband and multifunction antennas is integrated into a single aperture structure and configured so that the entire structure can be conformally mounted on the aircraft fuselage. An additional feature is the electronics system which is needed to interface

the aperture to radios and also provides anti-jam adaptive processing. It covers VHF, UHF and L-, S- and C-bands (NATO D, E/F and G/H bands).

The switching electronics is a highly integrated array of RF switches, power dividers and receive amplifiers which provide multi-use selection of each major antenna band. The resource manager is a general purpose processor with interfaces to a radio via mission avionics bus and intercoms. Its function is to manage antenna resources in order to optimise coverage and minimise interference.

Specifications
Weight: 58.1 kg
Reliability: 1,800 h MTBF

Status
In service.

Contractor
Harris Corporation.

OG-188/ARC-96A VLF/LF transmitter

Type
Avionic communications system, VLF/LF.

Description
The OG-188/ARC-96A is a self-contained 100 kw VLF/LF transmitter designed to meet the requirements for the US Air Force World Wide Airborne Command Post mission. Operating over the 17 to 60 kHz frequency range in 10 kHz increments, it provides 100 kw to the dual trailing-wire antenna aboard the EC-135 aircraft.

The OG-188/ARC-69A is an all solid-state transmitter designed to operate in the minimum shift keying, frequency shift keying, continuous shift keying and frequency shift continuous keying modes. The transmitter can be controlled locally or remotely via a standard serial bus. Constant surveillance tuning ensures matching antenna impedances regardless of the environmental effects after deployment.

Specifications
Dimensions: 1,750 × 1,340 × 1,570 mm
Weight: 1,091 kg
Power supply: 115/200 V AC, 400 Hz, 3 phase, 119 kVA, 28 V DC, 8 A

Status
In service.

Contractor
Rockwell Collins.

OK-374/ASC Communication System Control Group (CSCG)

Type
Avionic tactical communications system.

Description
The Communication System Control Group (CSCG) is an airborne communication and

The Communication System Control Group (CSCG) for the LAMPS III SH-60B helicopter
0504045

control system utilised on the LAMPS III SH-60B helicopter. It provides centralised microprocessor-regulated access and control of the various internal and external communication, navigation, voice encryptors and antenna selections via the Relay Assembly (RA) for the four main crew members, while providing for two additional maintenance-only positions utilising the Interconnecting Box (IB).

The CSCG is designed to meet the requirements of NACSIM 5100, MIL-E-5400, MIL-STD-461, MIL-STD-1553A and other applicable standards for airborne communication systems operating in severe helicopter environments. It provides control of communications and intercom configurations based on manual control or serial digital data instructions from an AN/AYK-14 airborne general purpose digital computer. The manual control panel switch settings of the four Remote Switching Controls (RSC) and the Control Indicator (CI) are transferred over a serial digital datalink to the central Audio Converter Processor (ACP). The ACP then performs the required functions such as tuning the radios, selecting audio for presentation to the crew members or selecting ICOM sets.

The CSCG provides integrated control via the CI of the HF radio, two UHF radios, datalink, IFF interrogator and transponder, direction-finder group, voice encryptors and antenna selections. It also provides external aircraft computer control of sonobuoy receivers, sonobuoy command via UHF radio, radio modes and frequencies and other supervisory functions, plus external radio communications access via the two UHF clear or secure radios, the clear or secure HF radio and the secure datalink. Internal ICOM communications are provided over a common ICS net and two separate conference nets: one for the pilot and co-pilot and the other for the sensor operator and the observer. Both PTT and VOX are provided for the ICOM nets and common ICS net access is also provided at two other positions called hoist and maintenance. The system distributes various warning tones such as stabilator warning, radar altimeter warning, helo threat warning and IFF Mode 4.

A manual back-up mode is provided, enabling selected crew members to have access to radios and ICOM in the event of system failure.

Specifications

Dimensions:
(ACP) 185 × 334 × 524 mm
(CI) 372 × 146 × 165 mm
(RSC ×4) 124 × 146 × 165 mm
(RA) 81 × 147 × 112 mm
(IB) 47 × 94 × 54 mm
Weight: (ACP) 22.3 kg
(CI) 6.4 kg, (RSC) 7.9 kg
(RA) 0.9 kg, (IB) 0.2 kg
Power supply: 115/200 V AC, 400 Hz, 3 phase, 200 VA
Temperature range: –40 to +71°C
Altitude: up to 15,000 ft
Reliability: 1,200 h MTBF

Status
In service in the LAMPS III SH-60B helicopter.

Contractor
Telephonics Corporation, Command Systems Division.

Photo Reconnaissance Intelligence Strike Module (PRISM)

Type
Avionic datalink system.

Description
In tandem with development work on the core MD-1295/B Improved Data Modem (IDM-302, see separate entry) as part of a programme known as project 'Gold Strike', a consortium of Symetrics Industries, PhotoTelesis Corporation and

A UK RAF Jaguar GR3A and a USAF F-16C Block 40 trials aircraft on the flightline at the US Naval Air Warfare Center (NAWC), China Lake, California, during Phase 1 of Trial Extendor (QinetiQ)
0111411

Ground engineers prepare a Predator communications relay UAV used during the UK Extendor Phase 2 demonstration in the US during December 2001 (QinetiQ)   0111754

The experimental payload carried by the UAV consisted of UHF radios, a data modem and a payload control processor (QinetiQ)   0118273

The Extendor Phase 2 payload mounted in the rear equipment bay of the Predator UAV. To the right are the two AN/ARC-210(V) transceivers with the PRISM-IDM immediately left of the radios (QinetiQ)   0132690

ARINC Incorporated agreed to work together to develop an alternative approach for integrating a video imagery capability into fast-jet cockpits. Modification of PhotoTelesis' small, hand-held image transceiver, Military MicroRIT (MMT) into a SEM-E module configuration allowed simple integration into the existing MD-1295/B IDM.

The MMT was originally developed under contract with the US Army's Communication Electronics Command (CECOM) to replace the existing outstation computer for the Lightweight Video Reconnaissance System (LVRS). The LVRS allows special forces to transmit and receive tactical imagery to and from ground tactical

operations centres and airborne platforms, such as the US Navy's F-14 Tomcat. In the IDM configuration, the Photo Reconnaissance Intelligence Strike Module (PRISM) is designated as PRISM-equipped IDM or PRISM-IDM. The new video imagery system employs a box-within-a-box approach to the integration. Instead of the more fully integrated approach of the VIM, the PRISM receives only power and electrical connection to the host aircraft's video communications and command buses from the IDM. Essentially, the PRISM acts as a separate unit within an existing aircraft system.

Features of the PRISM include:
- Capability to capture, compress/decompress, and receive/transmit imagery data
- Imagery capture rates of four images per second to show fast sequence images such as missile flyout and weapon impact
- Compatible with existing radios for reduced installation time and cost
- Capture and transmission of images from tape playback
- Accessible FLASH memory, facilitating upload of pre-mission images/maps and post-mission download
- Compatible with fielded systems

US Navy F/A-18E/F aircraft are fitted with PRISM™   0130551

- Control via MIL-STD-1553B or RS232
- Robust data communications using signal averaging and FEC allowing transmission/ reception at longer ranges and reduced transmission times
- Capability to transmit colour or monochrome images
- Relay images over the horizon from one aircraft to another
- Display of thumbnail image directories for quick review
- Programmable image capture rate and time
- Programmable image compression ratios for both Wavelet and JPEG
- Imagery transmission times of under 15 seconds
- Image cropping capability to reduce file size and decrease transmission time
- Single image shot mode
- Abort of image transmission/reception to return radio to voice mode.

Benefits of the integration approach of PRISM include complete interoperability with PhotoTelesis' other ground and airborne imagery products and the ease of integration of the module with the host platform system. The PRISM-IDM requires only a minor modification to the aircraft's Operational Flight Programme (OFP), allowing the transmission of video images over the IDM.

Specifications

Input: NTSC (RS-170 or RS-343), PAL or SECAM
Output: Composite and S-Video
Compression ratio: Variable from 10:1 to 100:1
Image buffer: DRAM-140 images; DISK on CHIP-1400 images
Packaging: Either SEM-E (IDM-302/501) or integrated 6U VME
Power: 28 V DC; 8 W (SEM-E configuration)
MTBF: in excess of 6,200 h demonstrated
Image format: Auto conversion to other formats (JPEG, BMP, NITF and so on) with ICSE S/W
Part number: 402851-01

Status

In production and in service.

During December 2000, the UK RAF conducted trials at the US Navy Air Warfare Center (NAWC), China Lake, California, utilising a Jaguar GR3A fitted with the PRISM-IDM. The trial was successful and resulted in plans to upgrade the RAF fleet of Jaguar GR3As, with possibly Harrier GR7s to follow. This upgrade is one outcome of Project Extendor, a UK Ministry of Defence (MoD)/QinetiQ communications relay Operational Concept Demonstration (OCD) programme, initiated in 1999. It followed a 1998 UK MoD study into the

use of Communications Relay Unmanned Air Vehicles (CRUAVs), which included the provision of low-cost datalinks for tactical air platforms. The baseline configuration is understood to have a radio-equipped UAV, orbiting at 30,000 ft, relaying TACFIRE or AFAPD (Air Force Application Program Development)-format messages from a forward observer or ground station to an Apache helicopter or a Close Air Support (CAS) fast jet aircraft fitted with PRISM-IDM. Ultimately, it is intended to be able to pass situational-awareness information, sourced from Link 16, to a number of IDM-equipped platforms.

Trial Extendor was conducted in three phases, the first of which culminated in the trial at China Lake in December 2000. This involved a British Forward Air Controller (FAC) sending parametric positional data in standard CAS nine-line messages via a Predator UAV to both an RAF Jaguar GR3A and a US Air Force (USAF) F-16C Block 40 aircraft. The Jaguar had been equipped with a four-channel IDM, interfacing with the aircraft weapon computer and HUD via a MIL-STD-1553B databus. Both the ground and air elements had Line Of Sight (LoS) to the UAV but not to each other. Although initially intended to address British objectives, this phase of the trial became a bilateral effort when QinetiQ pooled its resources with the USAF UAV Battlelab at Eglin Air Force Base.

Despite the heavy utilisation of UAVs during operations over Afghanistan, the second phase of the trial was successfully completed late in 2001. As in Phase 1, a Predator was used as a communications relay, this time equipped with a payload processor controlling a Symetrics X302 PRISM-IDM interposed between two ARC-210 UHF radios. The UAV was operated over the China Lake Naval Air Warfare Centre at altitudes up to 21,000 ft, controlled via satellite from the General Atomics Aeronautical Systems, Inc (GA-ASI) test facility at El Mirage. The exercise used the newly configured payload to pass target information transmitted in a TACFIRE digital message format from a FAC via the UAV into the cockpit of two RAF IDM-equipped Jaguar GR3A aircraft. These were holding at low level and out of direct LoS with the FAC. With the aid of a modified AEI Apache Mission Planning Station (MPS), the Jaguars were able to keep track of each other via newly implemented station-keeping messages deliberately mechanised in the AFAPD protocol to give interoperability with the UK WAH-64 Apache attack helicopter (which was to become involved in Extendor Phase 3 during 2002).

According to QinetiQ, the Phase 2 functionality also included network management of the communications traffic via the Predator's onboard

processor. By employing a technique of pre-scripting message forwarding rules, incoming messages could be automatically re-routed to the specified strike assets. Upon pilot acceptance of the tasking message, the target information was then digitally transferred into the navigation/ attack system of the Jaguar, presenting a target designation mark on the Head-Up Display (HUD). To assess the utility of the target mark, the pilot then manoeuvered the aircraft towards the target, visually confirmed the accuracy of the HUD mark against the target and released a practice bomb or a Paveway II Laser-Guided Bomb (LGB). Target designation was provided by a US Marine Corps Cobra AH-1W attack helicopter.

The attacks were imaged by the Predator's Wescam Skyball EO sensor which supplied, via satellite, live video images of weapon impact and target damage.

In addition, in support of the US Air Force's parallel UAV-to-Fighter Imagery Relay (UFIR) Phase II trial, Predator-derived target imagery was passed to the cockpit of a USAF F-16 Block 40 in near-realtime, via satellite communication radios. Under instruction from an operator reviewing the incoming video stream in the ground control station, the Predator's onboard system was also commanded to capture and transmit target video (as still images) directly into the cockpit of a US Navy F-14D equipped with the Phototelesis Fast Tactical Imagery (FTI – see separate entry) package, compatible with PRISM.

During January 2003, QinetiQ announced the successful conclusion of Phase 3 the trial. During a two week period over the Salisbury Plain Training Area and QinetiQ's Larkhill site, the final phase of the trial employed a QuinetiQ BAC 111 aircraft acting as a surrogate for the Predator UAVs used in Phases 1 and 2 of the effort. In order to ensure compatibility of results with the earlier UAV-based phases, the Extendor 3 BAC 111 payload of two UHF radios, an IDM, a payload control processor and a JTIDS Low Volume Terminal (LVT) was controlled from the ground. Phase 3 demonstrated the bi-directional translation of limited tactical data link messages in TACFIRE, AFAPD and Variable Message Format (VMF) into JTIDS/Link 16 and vice versa, thereby allowing a diverse number of UK military assets to operate in a limited Network Enabled Capability (NEC) environment at ranges of up to 556 km. During the flight demonstration, FAC-generated TACFIRE data was transmitted, via the Extendor payload in the BAC 111, into the cockpits of two UK RAF Jaguar strike aircraft, an AH-1 Apache attack helicopter and (utilising the payload translator unit) a UK RAF Tornado F3 fighter and AWACS platform, which were employing JTIDS/Link 16. All of the platforms were out of direct Line-of-Sight (LoS) with the FAC. The trial also demonstrated the transmission of Phoenix UAV-derived target imagery into the cockpit of a suitably modified Strike Attack Operational Evaluation Unit (SAOEU) Jaguar T2 aircraft.

As of 2004, it was reported that PRISM was in service aboard USAF C-130 airlifters and was being evaluated on the F-16C/D and RQ-1 Predator Unmanned Aerial Vehicle (UAV), USN F-14s, F/A-18s and P-3s, together with US Army helicopters. The UK RAF also received 50 systems to equip Jaguar GR3A attack aircraft, with five being supplied to Turkey for its AH-1W attack helicopters. The British Army's upgrade programme for its Lynx battlefield light utility helicopters will include the IDM-501 (see separate entry), as part of its PRISM installation.

The Fast Tactical Imagery – II (FTI-II – see separate entry) system employs a single PRISM™. Employing FTI II, battlefield commanders can view the battlespace at extended ranges through still frame images provided in near-realtime from airborne assets.

Contractor

ARINC Incorporated.
PhotoTelesis Corporation.
Symetrics Industries Inc.

Rockwell Collins HF radios

Type
Avionic communications system, HF/data.

Description
The following table summarises the features of the major products in the Rockwell Collins' family of HF radio products:

Contractor
Rockwell Collins.

Parameter	AN/ARC-190(V)	HF-121C (ARC-230)	HF-9000	HF-9000D	HF-9500	AN/ARC-220 and VRC-(ARC-230)-100
Typical Platform	fixed wing	fixed wing	fixed and rotor wing	fixed and rotor wing	fixed wing	ARC-220: rotor wing VRC-100: vehicular / 2 person lift
DSP based processing	no	yes	no	yes	yes	yes
ALE: MIL-STD-188-141A and FED-STD-1045	yes, via external CP-2024()	yes, embedded DSP	yes, specific version	yes, embedded DSP	yes, embedded DSP	yes, embedded DSP
ALE: MIL-STD-188-141B	growth	growth	no	growth	growth	yes
ALE link protection	yes	yes	no	yes	yes	yes
Quick ALE	growth	growth	no	growth	growth	yes
ECCM: MIL-STD-188-148A	yes, via external CP-2024()	growth	yes, specific ECCM version	growth	growth	yes, embedded DSP
Data modem external MIL-STD-188-110A	yes, data mode	yes, data mode	yes, data mode	yes, data mode	yes, data mode	yes, data mode
Data Modem Internal MIL-STD-188-110A Narrow shift FSK, Wide shift FSK, Serial single tone STANAG 4285	no	yes, embedded DSP	no	yes, embedded DSP	yes, embedded DSP	yes, embedded DSP
SELCAL, ARINC 719 audio via external LRU	yes, dedicated AM audio output	yes, dedicated AM audio output	yes, dedicated AM audio output	yes dedicated AM audio output	no	no
SELCAL, ARINC 714-6 decoder	yes, embedded in CP-2024()	yes, embedded DSP	no	yes, embedded DSP	yes, embedded DSP	growth, embedded DSP
HFDL: ARINC 635/753	yes, embedded in CP-2024()	yes, external HFDU, embedded as growth	yes, external HFDU	yes, external HFDU	yes, external HFDU	yes, external HFDU
FCC Type accepted	yes, specific version	no	yes, specific version	no	no	no
FAA TSO certified	yes, specific version	no	yes, specific version	no	no	no
RF output power: Multitone	400 W Pk	400 W Pk	175/200 W Pk	200 W Pk	400 W Pk	175 W Pk
RF output power: Single tone (RTTY)	400 W Pk and Avg	400 W Pk and Avg	100 W Pk and Avg	100 W Pk and Avg	200 W Pk and Avg	100 W Pk and Avg
Input power required	115 VAC, 3 Ph, 400 Hz	115 VAC, 3 Ph, 400 Hz	+28 VDC	+ 28 VDC	115 VAC, 3 Ph, 400 Hz	+28 VDC
Control interface	discrete CNTL (RS-422), RS-422, and MIL-STD-1553B	discrete CNTL (RS-232), RS-232, and MIL-STD-1553B	discrete CNTL (Fiber Optic), ARINC 429, MIL-STD-1553B	discrete CNTL (Fiber Optic), MIL-STD-1553B	discrete CNTL (RS-422), RS-422, and MIL-STD-1553B	discrete CNTL (RS-422), RS-422, and MIL-STD-1553B
Secure device interface	KY-100, ANDVT	KY-100, ANDVT	KY-100, ANDVT	KY-100, ANDVT	KY-100, ANDVT	KY-100, ANDVT
Environment:	MIL-E-5400T	MIL-E-5400T	MIL-E-5400T	MIL-E-5400T	MIL-E-5400T	MIL-E-5400T
Altitude	Class 1	Class 1	Class 1	Class 1	Class 1	≤25,000 Ft
Temperature	Class 1,	Class 1B	Class 1, 1B (version specific)	Class 1B	Class 1B	Class 1B
Cooling	external, forced air	external, forced air	convection cooled	convection cooled	external, forced air or self cooled with duty cycle	convection cooled
SIMOP Transmit HF to HF	optional, F-1535, F-1602 LRUs	4 pole BPF embedded	optional, 4 pole HF-9060 LRU	growth, optional 4 pole LRU	2 pole BPF embedded	no
SIMOP Transmit HF to VHF/UHF	embedded LP filter in CPLR	embedded LP filter in CPLR	no	no	embedded LP filter in CPLR	embedded LP filter in CPLR
SIMOP receive	optional F-1535, F-1602 LRUs	4 pole BPF embedded	optional 4 pole HF-9060 LRU	growth, optional 4 pole LRU	2 pole BPF embedded	no
Link 11 external modem MIL-STD-188-230-1A	no	compatible, link 11 mode	no	compatible, data mode	compatible, data mode	no
RF emission modes	USB/LSB voice or data, AME, or CW	USB/LSB/ISB voice, data, link 11, or RTTY AME, or CW	USB/LSB voice USB/LSB, data AME, or CW	USB/LSB voice USB/LSB/IS, data AME, or CW	USB/LS voice USB/LSB/ISB, data, AME or CW	USB/LSB voice USB/LSB, data AME, or CW
Duty cycle, 100%, all service conditions	continuous	continuous	continuous, voice	continuous, voice	continuous	2:1, receive: transmit
# of channels	280,000	280,000	280,000	280,000	280,000	280,000
Frequency Range	2.0000 to 29.9999 MHz	2.0000 to 29.9999 MHz	2.0000 to 29.9999 MHz	2.0000 to 29.9999 MHz	2.0000 to 29.9999 MHz	2.0000 to 29.9999 MHz
Syllabic squelch	yes	yes	yes	yes	yes	yes
Rapid tune antenna coupler	yes	yes	yes	yes	yes	yes
BIT to LRU/SRU level	yes	yes	yes	yes	yes	yes
Spare card slot for growth	no	yes	yes	yes	yes	yes
Data fillable	yes	yes	yes	yes	yes	yes
Field software reprogrammable	yes, with HFDL only	no	no	yes	yes	yes

RT-30 VHF/FM transceiver

Type
Avionic communications system, VHF.

Description
The RT-30 transceiver can be operated with either the C-1000 or C-5000 control head. The RT-30 FM transceiver uses a digital frequency synthesiser to provide FM communications over the frequency range 29.7 to 49.99 MHz. There are no band-spread limitations; the receiver can operate at one frequency extreme with the transmitter at the other. Up to 350 preset channels can be programmed into the RT-30 or complete frequency agility is available in the control units.

Fully solid state, it also provides 32 sub-audible CTCSS tones. An available system includes a single channel Guard receiver operating anywhere in the band. Separate audio inputs and outputs are provided for use with external CTCSS tones, tone bursts and DTMF encoders.

Specifications
Dimensions: 111 × 266 × 127 mm
Weight: 3.4 kg
Power output: 10 W continuous
Frequency: 29.7–49.99 MHz
Channels: 20 kHz under Part 90, capable of 10 kHz incremental tuning
Pre-set channels: 350
Temperature range: −40 to +60°C

Status
In production and in service.

Contractor
Chelton Avionics Inc, Wulfsberg Electronics Division.

RT-138F VHF/FM transceiver

Type
Avionic communications system, VHF.

Description
The RT-138F transceiver can be operated with either the C-1000 or C-5000 control head.

The RT-138F uses a digital frequency synthesiser to provide FM communications over the frequency range 138 to 173.975 MHz. There are no band-spread limitations; the receiver can operate at one frequency extreme with the transmitter at the other. Of fully solid-state construction, it also provides 32 subaudible CTCSS tones. A single-channel Guard receiver is available which can operate anywhere in the band.

Separate audio inputs and outputs are provided for use with external CTCSS tones, tone bursts, DTMF encoders, voice scramblers, data and so on.

Specifications
Dimensions: 111 × 266 × 127 mm
Weight: 3.4 kg
Power output: 10 W continuous
Frequency: 138–173.975 MHz
Channels: 25 or 30 kHz under Part 90, capable of 2.5 kHz incremental tuning
Pre-set channels: 350
Temperature range: −40 to +50°C

Status
In production and in service.

Contractor
Chelton Avionics Inc, Wulfsberg Electronics Division.

RT-406F UHF/FM transceiver

Type
Avionic communications system, UHF.

Description
The RT-406F transceiver can be operated with either the C-1000 or C-5000 control head.

The RT-406F uses a digital frequency synthesiser to provide FM communications over the frequency band 406 to 512 MHz. There are no band-spread limitations.

Fully solid state, the system also provides 32 sub-audible CTCSS tones and a single-channel Guard receiver is available operating anywhere in the band.

Separate audio inputs and outputs are provided for use with the external CTCSS tones, tone bursts, DTMF encoders, voice scramblers, data and so on.

Specifications
Dimensions: 111 × 266 × 127 mm
Weight: 3.4 kg
Power output: 10 W continuous
Frequency: 406–512 MHz
Channels: 25 kHz under Part 90, capable of 12.5 kHz incremental tuning
Pre-set channels: 350
Temperature range: −40 to +60°C

Status
In production and in service.

Contractor
Chelton Avionics Inc, Wulfsberg Electronics Division.

RT-1273AG DAMA UHF Satcom transceiver

Type
Avionic satellite communications system.

Description
The RT-1273AG UHF Satcom transceiver is an advanced full-duplex transceiver which upgrades and simplifies existing communications systems by combining modem and transceiver functions into one ATR unit. It incorporates the functionality of the Digital Signal Processor (DSP) based on the MD-1269A multipurpose modem which provides interoperability with all existing and planned UHF satellite communication equipment. The DSP architecture is designed to allow new waveforms to be programmed and incorporated as software updates. This integrated Satcom system reduces cabling, hardware size, weight and power consumption.

The RT-1273AG transceiver section features an integral 100 W transmitter and a highly sensitive receiver for UHF line of sight or satellite communication. Transmitter power output is variable from 1 to 100 W for optimum link performance when using either low- or high-gain Satcom antennas. A high-performance synthesiser provides independent 5 kHz channel selection of receive and transmit frequencies from 225 to 400 MHz for full-duplex operation.

The RT-1273AG modem section is based on the MD-1269A multipurpose modem. The modem section features a DSP-based architecture that offers multiple modulation modes and data rates. DSP-generated AM and FM is provided for low-distortion line of sight communications. Data rates from 75 to 38.4 bits/s are selectable for binary FSK, BPSK, Shaped BPSK, QPSK, Offset QPSK, or Shaped Offset QPSK operation. Doppler frequency offsets up to 1 kHz can be tracked by the modem to allow airborne operation. To maximise data throughput while minimising adjacent channel interference, sidelobes of the transmit signal are significantly reduced by advanced digital signal processing techniques that shape the modulated waveform. This permits 2.4 kbps BPSK and 4.8 kbps OQPSK operation over a 5 kHz satellite channel in compliance with JCS requirements.

The Demand Assigned Multiple Access (DAMA) waveforms are processed internally, allowing the RT-1273AG to participate in either 5 or 25 kHz DAMA networks with no additional equipment. Two simultaneous full-duplex baseband ports are available in the DAMA modes. Provisions for adding embedded COMSEC and digital voice have been designed into the equipment.

The RT-1273AG includes synchronous and asynchronous control/data ports for various local and remote-control options. Both ports can be configured as balanced or unbalanced MIL-STD-188-114A interfaces to a host terminal controller or control head.

Specifications
Dimensions: 256.5 × 193 × 497.8 mm
Weight: 22.68 kg max
Power supply: 115 or 230 V AC, 50/60/400 Hz, single phase
115 V AC, 400 Hz, 3 phase
28 V DC
Frequency: 225–399.995 MHz
Temperature range: −45 to +55°C
Altitude: up to 30,000 ft

Status
In production and in service.

Contractor
Raytheon Company, Space and Airborne Systems.

RT-5000 AM/FM transceiver

Type
Avionic communications system, V/UHF.

Description
The RT-5000 AM/FM transceiver covers the frequency band 29.7 to 960 MHz and can take the place of up to 6 transceivers. When controlled by the C-5000 control head, the RT-5000 can be used in conjunction with other Wulfsberg transceivers to perform cross-band relays between any bands from 29.7 to 960 MHz. The transceiver is tuned using a robust proprietary serial bus designed specifically for the air environment. DF audio output is standard, together with encryption outputs.

Options include an AM/FM guard receiver covering 29.7 to 960 MHz; single channel FM guard receivers; CTCSS and DCS tone squelches are standard on both the main and guard receivers.

New for 2003, the RT-5000 transceiver can perform APCO-25 (P25) digital modulation in addition to Motorola Trunking. Up to two modules covering the VHF, UHF and 800 MHz trunking bands can be incorporated into the transceiver. In addition, numerous encryption algorithms have been built into these modules which are available as upgrades for previous modules.

Specifications
Dimensions: 355.6 × 124.0 × 194.1 mm
Weight: 6.81 kg
Power output: 10 W FM nominal; 15 W AM nominal
Frequency: 29.7–960 MHz
AM/FM 29.7–399.9987 MHz
FM only 400–960 MHz
Channels: 12.5/20/25/30/50 kHz
Tuning: 1.25 kHz steps
IF bandwidths: 12.5 kHz narrow band; 25 kHz standard band; 35 kHz wide band; 70 kHz extra-wide band.
Temperature range: −40 to +60°C

Status
In production and in service.

Contractor
Chelton Avionics Inc, Wulfsberg Electronics Division.

RTA-44D VHF Data Radio (VDR)

Type
Avionic communications system, VHF.

Description
The RTA-44D VHF Data Radio (VDR) is part of the new Quantum Line of communications and navigation equipment.

The VDR, an airborne VHF digital communications transceiver, provides clear voice and data communication among onboard aircraft systems, with other aircraft and with ground-based operations.

The RTA-44D VDR provides standard 8.33 kHz channel capability to fulfil the new European airspace requirement. The unit is fully interchangeable with ARINC 716 communications transceivers, providing retrofit compatibility and satisfying spares pooling requirements.

In an Aviation VHF Packet Communications (AVPAC) system environment, the RTA-44D VDR operates as a simple transceiver with an analogue interface to the ACARS Management Unit (MU) or as a Minimum Shift Keying (MSK) modem. The RTA-44D VDR interfaces with an ARINC 758 Communications Management Unit (CMU Mark 11).

In an Aircraft Communications Addressing and Reporting System (ACARS) environment, the RTA-44D VDR operates as a simple transceiver with an analogue interface to the ACARS Management Unit (MU) or as a Minimum Shift Keying (MSK) modem. The RTA-44D VDR interfaces with an ARINC 724/724B ACARS MU in this environment.

The RTA-44D VDR requires an antenna for RF inputs and outputs, a control head or radio management panel, an audio input source and output sink for analogue voice functions. It may also be connected to a Central Maintenance Computer (CMC) to transfer maintenance data.

In addition to existing ARINC 750 CSMA kbps datalink capability (Mode 2), it features interfaces that allow for Mode-3 TDMA digitised voice capability, soon to be required for operation within US airspace. Processing power and reserve interfaces are built in to provide for the evolving Mode 4 STDMA technology, which may be required to complement Mode-S technology for ADS-B applications.

To ensure the RTA-44D VDR has the functional capability to meet CNS/ATM datalink requirements, the unit contains an Intel 80486SX control processor with 512 kbytes of RAM and 512 kbytes of ROM and two TMS320C31 Digital Signal Processing (DSPs), each with 128 kbytes of RAM.

Current implementation of the ARINC 716 VHF COMM radio and maintenance functions uses only the Main Processor Module (containing the 80486SX processor and a single TMS320C31).

The addition of the second DSP module, combined with the 80486SX and the first DSP provides sufficient power for all foreseeable datalink requirements.

The RTA-44D VDR is designed to the following standards and specifications:
1. ARINC 716 airborne VHF communications transceiver
2. ARINC 750 airborne VHF data radio
3. RTCA DO-186 'Minimum Operational Performance Standards (MOPS) for airborne radio communications equipment operating within the radio frequency range 117.975–137.000 MHz'
4. RTCA DO-207 'MOPS for devices that prevent blocked channels used in two-way radio communications due to unintentional transmissions'
5. EUROCAE ED-23B 'minimum performance specification for airborne VHF communications equipment operating in the frequency range 117.975–137.000 MHz'

The RTA-44D VDR operational modes are ARINC 716 voice, 716 data, and 750 data. The 716 voice and 716 data modes emulate the two operation modes of an ARINC 716 VHF COMM. In the 750 data mode, the RTA-44D VDR bridges ACARS traffic between the CMUNDR 429 bus and the internal 2.4 kbps MSK modem, or bridges AVPAC traffic between the CMU/VDR 429 bus and the 2.4 kbps MSK or the 31.5 kbps D8PSK modems.

Specifications
Power: 27.5 V DC
Weight: 3.76 kg
Dimensions: 319.02 (L) × 90.9 (W) × 194.06 (H) mm
Form factor: 3 MCU per ARINC specification 600

Cooling: forced air per ARINC specification 600
Temperature, operating: –55 to +70°C
Warm-up period: stable operating within 1 minute after application of power
Frequency selection: serial digital in accordance with ARINC specification 429
Certification: O C37d Class 3 & C38d Class C; DO-160C; DO-186; DO-178B; DO-207; ICAO Annex 10 FM Immunity

Contractor
Bendix/King.

RTA-83A VHF tranceiver

Type
Avionic communications system, VHF.

Description
The RTA-83A VHF transceiver was developed to meet the European airspace requirements for 8.33 kHz channel spacing and ICAO Annex 10 FM interference immunity. It meets all ARINC 566A specifications and is fully interchangeable with other ARINC 546 and 566A series transceivers.

The RTA-83A provides VHF voice and data communication between onboard aircraft systems, to other aircraft and to ground-based systems. It operates as a standard double sideband AM analogue voice transceiver. In an ACARS environment, it interfaces to either the ACARS Management Unit (MU) or Communications Management Unit (CMU), providing analogue Minimum Shift Keying (MSK) data capability.

Specifications
Power: 27.5 V DC; receive 1.0 A, transmit: 8.0 A
Weight: 4.5 kg
Form factor: ½ ATR short per ARINC specification 404
Cooling: forced air per ARINC specification 404
Temperature, operating: –55 to +70°C
Warm-up period: stable operation within 1 minute after application of power
Frequency range: 118.000 to 136.992 MHz
Channel spacing: 8.33 kHz or 25 kHz
Frequency selection: 2 out of 5 per ARINC specification 410 or serial digital per ARINC specification 429
Certification: TSO C37d Class 3/5 and C38d Class C/E; O-186; DO-160C

Status
In service.

Contractor
Bendix/King.

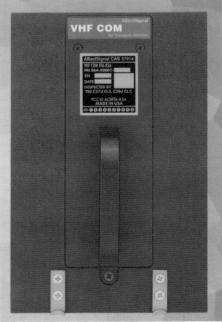

The RTA-83A VHF transceiver 2000 0062204

RTA-83B VHF transceiver

Type
Avionic communications system, VHF.

Description
The RTA-83B VHF transceiver was developed to meet the European airspace requirements for 8.33 kHz channel spacing and ICAO Annex 10 FM interference immunity. It meets all ARINC 716 specifications and is fully interchangeable with the RTA-44A and other ARINC 716 series transceivers.

The RTA-83B provides VHF voice and data communication between onboard aircraft systems, to other aircraft and to ground-based systems. It operates as a standard double sideband AM analogue voice transceiver. In an ACARS environment, it interfaces to the ACARS Management Unit (MU) or Communications Management Unit (CMU), providing analogue Minimum Shift Keying (MSK) data capability.

Specifications
Power: 27.5 V DC; receive: 1.0 A, transmit: 8.0 A
Weight: 4.0 kg
Form factor: 3 MCU per ARINC specification 600
Cooling: forced air per ARINC specification 600
Temperature, operating: –55 to +70°C
Warm-up period: stable operation within 1 minute after application of power
Frequency range: 118.000 to 136.992 MHz
Channel spacing: 8.33 kHz or 25 kHz
Frequency selection: serial digital per ARINC specification 429
Certification: TSO C37d Class 3/5 and C38d Class C/E; DO186; DO160C

Status
In service.

Contractor
Bendix/King.

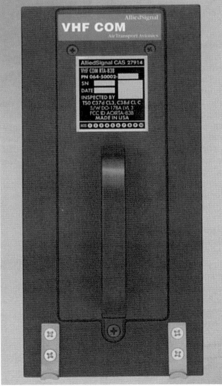

The RTA-83B VHF transceiver 0062205

RTU-4200 series Radio Tuning Units

Type
Avionic Communications Control System (CCS).

Description
Rockwell Collins' RTU-4200 provides for the selection of frequencies, codes, channels,

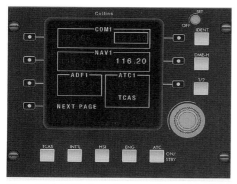

The RTU-4280 has been selected for the Gulfstream V   0504189

operating modes and volume levels for all Collins Pro Line II and for 400 Series CNS sensors and the HF-9000 (see separate entries). Optionally, the system can provide a backup Horizontal Situation Indicator (HSI) display and backup engine data display capability.

Other key benefits are a reduced physical space requirement and the ability to install optional radios without requiring additional controllers. Since all radios are managed by the RTU, one common control scheme and philosophy may be utilised, resulting in reduced pilot workload and training. Consistent operation means that all radios are tuned in the same way using the same set of keystrokes and knob adjustments. Additionally, future expansion of the aircraft radio fit will not change the pilot interaction with the RTU.

Isolation is provided between two installed RTU-4200 units such that failure of one unit will not affect the integrity of the other unit. Dimming of the AMLCD is designed to track other flight deck displays by means of the airplane's master dimming control bus. Additional control within this range is possible by use of the display dimmer on the bezel of the RTU.

Nonvolatile memory is utilised so frequencies and settings will not be lost during power shutdown. At each power-up, all frequencies in use or stored at the last shutdown are again available, eliminating the need to reload often-used frequencies. The RTU-4200 may be integrated with a Flight Management System (FMS) such that tuning of the radios can be performed via the FMS control display unit keyboard. Additionally, the RTU accepts FMS automatic tune requests for the VHF Omnidirectional radio Range/Distance Measuring Equipment (VOR/DME) radios and routes them appropriately. Control of frequencies and Air Traffic Control (ATC) codes is performed on the top level page. This page displays Very High Frequency Communications (VHF COMM), Very

High Frequency Navigation (VHF NAV), Automatic Direction Finder (ADF), ATC and Traffic Alert and Collision Avoidance System (TCAS) information. Pages selected below this top level page will return automatically after 20 seconds of control inactivity.

Radio mode control is performed on the main radio pages. In addition to radio mode selection, these pages provide frequency control, unique operation, self-test initiation and selection of preset frequency channel modes. Radio preset control is performed on dedicated pages. These pages provide preset tune mode selection and numbered channel preset programming. Selecting the preset tune mode will, for that radio, link the top level page tuning to the programmed numbered presets (both channel and frequency will be displayed on the top level page).

Twenty numbered preset channels are provided each for VHF COMM, VHF NAV, ADF and HF COMM radios. Current active frequencies and mode annunciations are provided on all radio preset pages. The RTU has an optional standby HSI page, which displays magnetic or true heading, VOR/ILS selected course and deviation and glideslope deviation. The active COMM, NAV and ATC frequencies/codes are displayed and can be adjusted on this page.

For custom OEM installations, a standby engine display page can be provided. This display presents a simple digital readout of engine parameters (low pressure, high pressure, engine pressure ratio, turbine gas temperature and fuel flow). The user may return to the appropriate top level page via the RETURN selection.

A typical aircraft installation consists of two RTUs to provide the necessary redundancy (in reversion or emergency power operations, the remaining RTU can carry on without any loss of radio tuning function). The 3ATI display provides a clear, bright, sunlight-readable display for both onside and offside views. The RTU-4200 operates concurrently with both Commercial Standard Data Bus (CSDB) and ARINC 429 digital interfaces. The RTU can act as a bilingual translator, formatting FMS ARINC 429 commands for a CSDB radio.

Specifications
Performance
VHF COMM: 118.000–135.975 MHz; 118.000–136.975 MHz, 118.000–151.975 MHz (strap option); 25 kHz or 8.33 kHz spacing
VHF NAV: 108.00–117.95 MHz; 50 kHz spacing
DME: 108.00–117.95 MHz; strap option adds 133.00–135.95 MHz
ADF: 190–1799 kHz; 500 kHz spacing
ATC: 0000–7777
HF COMM: 2.0000–29.9999 MHz; 100 Hz spacing
HF ITU: Channels 401–429, 601–608, 801–837, 1201–1241, 1601–1656, 1801–1815, 2201–2253, 2501–2510
HF Emergency: Channels 1–6

Dimensions: 83 (H) × 114 (W) × 163 (L) mm
Weight: 1.2 kg
Temperature: –20 to +70°C
Altitude: 55,000 ft
Power: 28 V DC, 15 W
Certification: FAA TSOs C34e, C35d, C36e, C37d, C38d, C40c, C41d, C66c, C112, C113, C119a; RCTA DO-160C; EUROCAE ED-14C; software DO-178B

Status
In production and in service.

Contractor
Rockwell Collins.

SADL Situational Awareness DataLink

Type
Avionic datalink system.

Description
The SADL Situational Awareness DataLink integrates air force close air support aircraft with the digitised battlefield via the US Army Enhanced Position Location Reporting System (EPLRS). SADL provides fighter-to-fighter, air-to-ground and ground-to-air data communications that are robust, secure, jam-resistant, and contention free. With its inherent automatic and on-demand position and status reporting for situational awareness, SADL provides an effective solution to the air-to-ground combat identification problem.

Fighter-to-Fighter Operation
The SADL radio is integrated with aircraft avionics over the 1553 multiplex databus, providing the pilot with cockpit displays of data from other SADL-equipped aircraft as well as EPLRS-equipped aircraft and ground units. SADL is capable of fighter-to-fighter network operation without reliance on ground-based EPLRS network control. Fighter positions, radar targets, and ground target positions are shared relative to the fighter's own inertial navigation system or global positioning system. Based on the number of fighters on the network, the data capacity of the network can be customised from the cockpit. Automated fighter-to-fighter relay and adaptive power control capabilities ensure connectivity, jam resistance, and reduced probability of undesired detection.

Air-to-Ground Mode
In the air-to-ground mode, the pilot commands the SADL radio to synchronise with a specific ground network based on encryption keys. The fighter's radio then returns to sharing fighter-to-fighter data while recording ground positions from the ground EPLRS network. EPLRS tracks the fighter and provides the fighter position and altitude for ground-to-air combat identification. At the beginning of an air-to-ground attack, the pilot uses a switch on the control stick to request a view of the five closest friendly EPLRS-netted positions shown with Xs on both the head-up display and the multifunction displays. The pilot decides whether to fire based on the proximity of friendly positions to the target for effective air-to-ground combat ID. Friendly air defence units also have a picture of friendly air tracks to minimise fratricide.

Status
In production and in service. The Joint Interoperability Test Command at Fort Huachuca, in conjunction with the USAF 422nd Test and Evaluation Squadron, successfully conducted gateway interoperability tests connecting the Joint Tactical Information Distribution System (JTIDS) to SADL. Tactical awareness displays were reliably exchanged in both directions between two SADL-equipped F-16s and a JTIDS Class 2-equipped F-15 aircraft. SADL's update rates and message structure permitted exchange of TADIL-J message sets for displays of relative fighter positions, radar targets, aircraft fuel status, weapon loads and additional information.

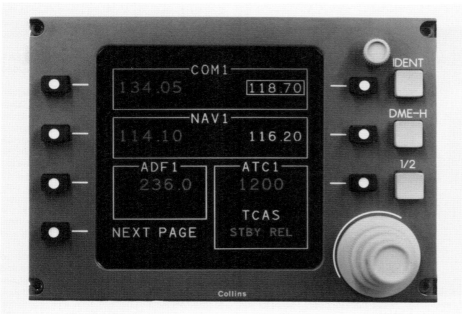

The RTU-4200 radio tuning unit   0105282

In addition to the 1,816 EPLRS radios fielded to divisions of the Contingency Force, EPLRS received a full-scale production award in 1997 for a multiyear 1997–98 buy of 2,100 radios. A further 535 SADL radios were subsequently ordered by the US Air National Guard for F-16 Block 25/30 and A-10 aircraft assigned to the close air-support role.

Contractor
Raytheon Company, Space and Airborne Systems.

SAT-906 Aero-H and -L satellite communications system

Type
Avionic communications system, satellite.

Description
Rockwell Collins' SAT-906 Aero-H and -L satellite communications system provides multichannel voice, facsimile and data capability. The voice channels can be used for cockpit or cabin communications, while the data channel will support low- and high-speed data using INMARSAT Data 2 or Data 3 protocols. The SAT-906 complies with ARINC 741 and 746 characteristics and can be configured for up to six channels through the addition of plug-in modules. The system provides voice channels and CEPT-E1 interface to the passenger telephone system without the need for any additional LRUs. The SAT-906 also provides DTMF capability on two analogue audio lines.

The SAT-906 is capable of low- and high-speed data, cockpit voice and passenger PC, fax and cabin telephone services when mated with a high-gain antenna subsystem.

The SAT-906 consists of three primary components – the SDU-906 Satellite Data Unit, RFU-900 Radio Frequency Unit and the HPA-901A Linear High-Power Amplifier. The SAT-906 operates in conjunction with a High-Gain Antenna System (HGAS), which is offered by several manufacturers. Collins has successfully integrated with all HGAS manufacturers to date using its standardised ARINC 741 interface.

The SDU-906 provides the interface to all other aircraft systems and includes the modems, codecs and protocol support for communication with ground earth stations.

The RFU-900 consists of a wideband L-Band to IF down-converter for receive operation and a wideband IF to L-Band up-converter for transmit operation. The RFU operates in full duplex mode, receiving L-Band signals in the range of 1,530 MHz to 1,559 MHz.

The HPA-901A is a Class A, 60 W, linear, High-Power Amplifier required for simultaneous

multi-channel operation of the INMARSAT Aero and Swift64 services. The added robustness of this amplifier allows operators to operate multiple channels in marginal, edge of coverage, environments. It also operates the Swift64 high speed data services simultaneously.

An optional SIU-900 provides interface between legacy ARINC 561 INS systems and the SDU-906. The SIU-900 converts data from synchro and ARINC 419 interfaces to ARINC 429, in a format required for SATCOM antenna pointing.

The CIU-906 is a companion product to the SAT-906. It provides PBX capability with multiple 2-wire interfaces into the cabin. It also enables the use of STU-III encrypted communications.

The SAT-906 is compatible with all service suppliers that support INMARSAT Data-2, Data-3 and Voice-2 protocols.

The optional HST-900 high-speed data transceiver adds a 64 kbps data service along with up to five channels of voice/facsimile service. The HST-900 utilises the INMARSAT Swift 64 data service, operating in either the circuit data or packet data modes of operation. In the circuit mode of operation, the unit interfaces with the ISDN card of a PC or server, providing a dial up connection to the selected Internet Service Provider (ISP). In the packet mode of operation, individual PCs interface to an Ethernet port of the HST-900 and send/receive data via a shared channel that is continuously available. Billing for the circuit mode is based on minutes of connection time, while billing for the packet mode is based on the number of kilobits transmitted and received.

The optional CDU-2000 multifunction Control Display Unit is available for control of the SAT-906. Alternatively, any ARINC 739-capable multifunction CDU can be used for this purpose.

The SAT-906 interfaces with the INMARSAT constellation of geosynchronous communications satellites that is supported by three consortia of service providers. Ground earth stations located around the globe provide links to the public switched telephone network for bidirectional calling, facsimile and data communications. Aero H, H+, I and L INMARSAT aeronautical protocols are supported by the SAT-906. Future growth capability provisions are in place to support planned upgrades to the system.

In operation, the aircraft earth station can be assigned a common number for all telephones, or an alternate numbering scheme is available to allow direct dialling to a specific aircraft telephone or fax machine. An operator database programme supplied by Rockwell Collins allows easy selection of options using a standard personal computer. This programme prepares files for upload to the system using a standard ARINC 615 data loader, which can be accomplished without removing units from the aircraft.

Specifications
Dimensions (per ARINC 600-8):
(SDU-906) 6 MCU
(RFU-900) 4 MCU
(HPA-901) 8 MCU
(CIU-906) 4 MCU
(HST-900) 2 MCU
Weight:
(SDU-906) 14.5 kg (SDU-906-6); 12.7 kg (SDU-906-4)
(RFU-900) 5.3 kg
(HPA-901) 11.6 kg
(CIU-906) 4.8 kg
(HST-900) 3.85 kg
Temperature:
(SDU-906) – 40 to +70°C
(RFU-900) – 35 to +70°C
(HPA-901) – 55 to +70°C
(CIU-906) – 15 to +55°C
(HST-900) – 40 to +70°C
Altitude: 55,000 ft (HPA-901A, HST-900); 70,000 ft (SDU-906, RFU-900, CIU-906)
Power supply:
(SDU-906) 115 V AC, 400 Hz, ≤180 W
(RFU-900) 115 V AC, 400 Hz, ≤60 W
(HPA-901) 115 V AC, 400 Hz, ≤300 W
(CIU-906) 115 V AC, 400 Hz, ≤100 W
(HST-900) 115 V AC, 400 Hz, ≤34 W

Status
In production and in service on a wide variety of civilian and military aircraft, including the Airbus A300-600R, A310-300, A330, A340, Boeing 737-300, 747-100/200/300/400, 757-300, 767, 777, Dassault Falcon 50 and 900, Gulfstream IV and V and C-135 and C-141.

Contractor
Rockwell Collins.

SAT-2000 Aero-I satellite communications system

Type
Avionic communications system, satellite.

Description
Rockwell Collins' SAT-2000 Aero-I satellite communications system is designed for use with the INMARSAT Aero-I satellite service. The SAT-2000 system has a 25 W High-Power Amplifier (HPA), which will support up to six channels (one data and five voice) within the spot beams. The SAT-2000 takes advantage of the spot beam feature of the Aero-I satellites using a system, which provides the benefits of telephony, fax, and real-time data communications. The 4.8 kbits/sec CODEC supports two-way voice and data traffic. Aero-I is intended for medium- and short-haul missions, but is also suitable for those long-haul missions, which fly mainly within the spot beam coverage. Emergency voice service in the global beam is supported by the SAT-2000 system, (pending INMARSAT approval).

The SAT-2000 system is comprised of the SRT-2000 Satellite Data Unit (SDU) and IGA-2000 Intermediate Gain Antenna (IGA).

The SRT-2000 is a multichannel receiver-transmitter providing both voice and data channels. The SRT-2000 transmits and receives packet-mode data to/from the datalink system (CMU-900), and receives and transmits circuit mode data (voice), in analogue form, to/from the flight crew headphones and microphones (via the audio management unit). Additionally, the SRT-2000 receives and transmits circuit mode data (voice, FAX, or PC MODEM) from passengers via the Cabin Telephone Unit (CTU). Thus, the SRT-2000 combines the functions of the HPA, RF Unit (RFU), Beam Steering Unit (BSU), and Satellite Data Unit (SDU) into a single LRU.

The IGA-2000 IGA allows the SAT-2000 to transmit and receive low-speed ACARS data and high-speed voice traffic efficiently. The diplexer/low-noise amplifier is integrated into the antenna to improve system performance and reduce the number of LRUs. The IGA-2000 IGA is a 6 dBi, top-mounted, electronically steered phased

The Rockwell Collins Satcom 906 communications system showing (left to right) the SDU-906 satellite data unit, the RFU-900 Radio Frequency Unit and the HPA 901 High-Power Amplifier

0001184

array antenna. The antenna complies with the INMARSAT Aeronautical System Definition Manual (SDM) requirements for intermediate gain service. The Aero-I antenna provides BITE information to the SRT-2000 and the SRT-2000 provides beam pointing information to the Aero-I antenna. The smaller size of the intermediate gain antenna in addition to the integration of the diplexer/low-noise amplifier into the antenna, allows SATCOM systems to be installed in smaller aircraft.

The system is also marketed for use in long-range corporate aircraft by Rockwell Collins Business and Regional Systems, where the nomenclature is SATCOM-5000 (see separate entry).

Specifications
Dimensions:
(SRT-2000) 8 MCU per ARINC 600
(IGA-2000)
48.2 × 110.2 × 12.6 mm (W × D × H)
Weight:
(SRT-2000) 15.9 kg
(IGA-2000) 8.2 kg
Power supply:
(SRT-2000) 115 V AC, 400 Hz or 28 V DC; 350 W (max)
(IGA-2000) 60 W
Temperature: –40 to +70°C
Altitude: up to 55,000 ft
Software: DO-178B Level D

Status
In production and in service on a wide variety of civilian and military aircraft, including the Airbus A319, 320 and 321 and Boeing 767-200, 747-400F, DC-10, MD-10, MD-11, KC-135 and C-17.

Contractor
Rockwell Collins.

SAT-2100 satellite communications system

Type
Avionic communications system, satellite.

Description
The Collins SAT-2100 is a compact and lightweight satellite communications system. The SAT-2100 incorporates a High-Power Amplifier (HPA), Radio Frequency Unit (RFU) and Satellite Data Unit (SDU) within an 8 MCU package and provides for multichannel voice, facsimile and low-speed PC data operation. The SAT-2100 also enables up to two simultaneous Swift64 or Swift Broadband channels using its HST-2110 companion. Employing an ARINC 741 electrical interface, the SRT-2100 is compatible with all industry standard high-gain antenna systems. The system also supports STU-III encryption for voice, fax and data using the CIU-906 cabin interface unit.

The SAT-2100 system consists of the SRT-2100 Satellite Receiver/Transmitter, which houses the RFU, SDU and HPA. Novel management software monitors and controls the power between the Aero and Swift64 channels, to ensure safety services integrity.

The optional Collins CIU-906 performs the functions of a private branch exchange for the aircraft. The current model provides up to 30 telephone ports and advanced features such as speed dialling, call conference and seat-to-seat interphone.

The HST-2110 enables the INMARSAT Swift64 service, providing bi-directional 64 kbps packet data and 492 kbps Swift Broadband service enabling services; such as Internet and email access.

Operation
The SAT-2100 operates using the INMARSAT constellation of geosynchronous communications satellites to enable Aero-H, Aero-H+, Swift64 and Swift Broadband services. Ground-earth stations provide links to the public switched telephone network for bidirectional calling, facsimile and data communications. The architecture of the system facilitates continued operation in the event of channel module failure, capable of voice, fax and PC data operation, by removing the failed module from the channel module pool and engaging other modules. A single operational module will support essential data communications such as ACARS or AFIS.

The INMARSAT network is suitable for air traffic services that enable the SAT-2100 to provide access to FANS services for preferred airspace routing and clearances in oceanic regions.

Specifications
Dimensions:
(SRT-2100) 198 × 261 × 323 mm (H × W × L)
(CIU-906)
196 × 126 × 388 mm (H × W × L)
Weight:
(SRT-2100) 13.1 kg
(CIU-906) 5.4 kg
Power supply:
(SRT-2100) 115 V AC, 324– 800 Hz or 28 V DC; 395 W (max)
(CIU-906) 115 V AC 400 Hz or 28 V DC; 100 W (max)
Temperature:
(SRT-2100) – 40 to +70°C
(CIU-906) – 55 to +70°C
Altitude: up to 55,000 ft
Compliance: ED-14C

Status
In production and in service.

Contractor
Rockwell Collins.

SAT-5000 SATCOM

Type
Avionic communications system, satellite.

Description
Collins' SAT-5000 is a small, light satellite communications system that provides the full range of telecommunications services conforming to Inmarsat's Aero-I specifications, delivering multichannel voice, facsimile and PC data capability in the Inmarsat satellites spot beam coverage areas. In addition, low-speed packet data is provided in global beam coverage supporting ACARS/AFIS services as well as FANS applications for preferred routing. The SAT-5000 provides an industry-standard multichannel digital telephone interface (CEPT E-1) and four analogue ports that can be flexibly configured for a variety of cabin setups.

The SAT-5000 consists of the SRT-2000 Satellite Receiver/Transmitter and the IGA-2000 Intermediate Gain Antenna. The SRT-2000 comprises the radio frequency, satellite data unit and high-power amplifier functions in a single 8MCU package. The IGA-2000 is an electronically steered, top-mounted, phased-array antenna that maximizes energy directed to and from the selected satellite regardless of the position and attitude of the aircraft relative to the satellite constellation. The IGA-2000 may be installed or removed without requiring access from inside the fuselage.

The optional CIU-6000 Cabin Interface Unit performs the functions of a private branch exchange for the aircraft. The current model provides ten telephone ports and advanced features such as speed dialing, call conferencing and seat-to-seat interphone. Growth options are provided for encrypted operation and additional telephone ports.

System weight depends on configuration, starting from approximately 25.5 kg plus mounts, cables and cabin equipment. The system requires inertial or GPS/FMS/AHRS data for antenna pointing. The power source is 115 V AC, 400 Hz or 28 V DC and forced air cooling or a mount with fan is required.

The SAT-5000 operates within Inmarsat's Aero I service. Ground-earth stations provide links to the public switched telephone network for bidirectional calling, facsimile and data communications. All current Inmarsat Aero I protocols are supported by SAT-5000, with future growth provisions in place to support planned upgrades to the system, including emergency voice in global beam coverage. The SRT-2000 is capable of operation with up to six channels available in one-channel increments. If one or more modules fail, the system remains capable of voice, fax and PC data operation as long as two modules are operational. A single operational module will support essential data communications such as ACARS or AFIS.

The Inmarsat network is suitable for air traffic services that enable the SAT-5000 to provide access to Communications, Navigation and Surveillance/Air Traffic Management (CNS/ATM) airspace operations for preferred routing and ATC clearances in oceanic regions. The built-in analogue telephone ports are provided with up to 100 speed dial numbers with diagnostics via the telephone unit.

Any ARINC 739-capable multifunction control display unit can be used for controlling and operating the SAT-5000. In operation, the system can be assigned a common number for all telephones, or an alternate numbering scheme is available to allow direct dialling to a specific aircraft telephone or fax machine. An operator database programme supplied by Rockwell Collins allows easy selection of options using a standard personal computer. This programme prepares files for upload to the system using a standard ARINC 615 data loader, which can be accomplished without removing LRU from the aircraft.

Specifications
Performance:
(SRT-2000) Modular, up to six channels; digital voice, fax and data; up to four DTMF (Touch Tone) analogue ports
(CIU-6000) 10– 30 telephone ports; telephone switching and advanced features; interface SRT-2000 digital port
Dimensions:
(SRT-2000) 200 × 261 × 321 mm (H × W × L)
(IGA-2000) 76 × 311 × 711 mm (H × W × L)
(CIU-6000) 197 × 126 × 388 mm (H × W × L)
Weight:
(SRT-2000) 15.4 kg (6 channels)
(IGA-2000) 11.36 kg
(CIU-6000) 5.4 kg
Power supply:
(SRT-2000) 115 V AC, 400 Hz or 28 V DC; 385 VA (max)
(IGA-2000) Supplied by SRT-2000
(CIU-6000) 115 V AC, 400 Hz or 28 V DC; 100 VA (max)
Temperature:
(SRT-2000) –40 to +70°C
(IGA-2000) –55 to +70°C
(CIU-6000) –40 to +70°C
Altitude:
(SRT-2000) 55,000 ft
(IGA-2000) 55,000 ft
(CIU-6000) 70,000 ft

For details of the latest updates to *Jane's Avionics* online and to discover the additional information available exclusively to online subscribers please visit
jav.janes.com

Certification: DO-178B Level D
(SRT-2000) DO-160D; C4PBA[BD] [SB]EXXXXX
AAAAZRRRM [A3E3]XXA; ED-14C
(IGA-2000) DO-160D; A2F2XAC[BD] [SCY]ESFD
FSZXXZC [WYP]H[A4E4] 1AAA; ED-14C
(CIU-6000) DO-160D; [A4]PBAB[SC] EXXXXXZ
[BZ]A[Z]Z [RRR]H[XXZ3] XXXA; ED-14C

Status
In production and in service.

Contractor
Rockwell Collins.

SATCOM HST-2100 high-speed data transceiver

Type
Avionic communications system, satellite.

Description
Collins' SATCOM HST-2100 high-speed data transceiver is the companion to the SAT-6100, enabling Inmarsat's Swift64 high-speed data services. System input/output includes ISDN, Ethernet, Ethernet over ISDN and RS-232. Novel power management software allows the HST-2100 to maintain 'safety services' capability under failure conditions. The integrated high-power amplifier allows for dual simultaneous operation of high-speed and all channels of traditional aeronautical services, while the variable frequency power supply operates in 324 Hz to 800 Hz frequency range.

The system also provides for an upgrade path for next generation services.

Status
In production and in service.

Contractor
Rockwell Collins.

Satellite Data Communications System (SDCS)

Type
Avionic communications system, satellite.

Description
The Satellite Data Communications System (SDCS) uses the INMARSAT network to send and receive information worldwide. Interfaced to Airborne Flight Information System (AFIS), it provides worldwide communication capabilities.

It provides two-way unrestricted message forwarding to another SDCS-equipped aircraft, the Global Data Centre, a fax machine, an auto answer terminal and other service providers such as BASEOPS International, Air Routing International, Jeppesen Dataplan, Universal Weather and Aviation and MEDLINK. It also enables provision of worldwide weather information and flight planning/flight plan filing.

Specifications
Weight:
(HPA/LNA) 2.04 kg
(antenna) 0.4536 kg
(SCU) 2.72 kg

Power supply: 27.5 V DC
(transmit) 4.5 A
(receive) 0.5 A

Status
In service.

Contractor
Bendix/King.

SCDL Surveillance and Control DataLink

Type
Avionic datalink system.

Description
Joint STARS (Surveillance Target Attack Radar System) radar data, comprising wide-area search/moving target indicators, synthetic aperture radar/fixed target indicators, low-reflectivity indicators and sector search information, is broadcast in real time, in continuous communication through Cubic Defense Systems SCDL to the Common Ground Stations (CGS).

SCDL is a time division, multiple access datalink with flexible frequency management capability. The link provides reliable and secure performance in high jamming environments. The link establishes a network consisting of the E-8 aircraft and ground stations.

The main features of the network are: broadcast radar messages to all CGSs within line of sight; CGS access to the aircraft in a time-ordered manner; relay of messages from one CGS to another; automatic acknowledgement of error-free receipt of messages; up to eight multiple networks operating simultaneously and within the same geographical area; all data encrypted.

SCDL uses very wide band fast frequency hopping, coding and data diversity to achieve robust jamming resistance. Novel techniques are used to achieve link acquisition and re-acquisition. The uplink messages are implemented with a unique modulation approach which provides accurate determination of path delay between the E-8 aircraft and the CGS. This provides the capability for a highly reliable, short duration, one-time uplink message to achieve low probability of enemy detection of CGS transmissions.

Airborne Data Terminal (ADT)
The Airborne Data Terminal (ADT) comprises three line replaceable units: an Input/Output Processor (IOP); a transceiver and an RF amplifier. The IOP provides the primary interface with the aircraft through a MIL-STD-1553B databus. It prioritises, encrypts and forwards radar messages for transmission over the datalink, accepts uplink messages, and provides overall control of the datalink, including BIT.

The transceiver contains the downlink transmitter and uplink receiver, which operate with the antenna mounted on the bottom of the E-8 fuselage. All of the datalink functions of modulation, coding, frequency spreading and power amplification are contained in this unit.

The RF amplifier is slaved to the transceiver and provides the capability to transmit and receive data from the top-mounted antenna. The dual antenna installation provides full and continuous ground coverage at all aircraft attitudes, including turns.

Ground Data Terminal (GDT)
The Ground Data Terminal (GDT) comprises four units: the antenna unit; the Lower Control Unit (LCU), which performs data coding, decoding, and datalink timing; the Joint STARS Interface Unit (JSIU), which provides the interface with the CGS using a MIL-STD-1553B databus and provides for encryption and decryption of all data, datalink and BIT control; the AC-AC converter, which converts 50–60 Hz prime power to 400 Hz for use by the datalink.

Narrowband Data Link System (NDLS)
NDLS is used on the UK's Airborne Stand-Off Radar (ASTOR) program. NDLS is virtually identical to the SCDL used by US JSTARS aircraft and provides interoperability between the two aircraft types. NDLS comprises the Air Data Terminal (ADT) – the transceiver and I/O processor in the aircraft, and the Ground Data Terminal (GDT) – the masthead antenna assembly, interface unit and control unit, housed in a vehicle or ground station.

Status
Operational in Joint STARS E-8 aircraft.

Contractor
Cubic Defense Applications.

SELCAL decoders

Type
Avionic communications system, Selective Calling (SELCAL) system.

Description
The purpose of the SELective CALling (SELCAL) system is to permit exclusive calling of individual aircraft over normal radio channels that link the ground station to that aircraft. The system operates with HF and VHF ground-to-air transmitters and receivers and does not interfere with the normal operation of communications except when the SELCAL is performing its calling function.

Each SELCAL-equipped aircraft is assigned a four letter identifier which is used when the ground station wishes to contact it. The SELCAL decoder is designed to respond only to the identifier for which it is set. When the identifier is received by the aircraft, the decoder actuates a signal indicator in the form of a lamp, bell, chime or any combination of these. Typically, the signal will be annunciated on an audio control panel microphone select switch.

The CSD-714 SELCAL decoder is a rack-mounted, 16 tonne decoder designed for the ICAO- and ARINC-standard system. It is designed to surpass the performance requirements of ARINC 714. It may be installed in non-pressurised and non-temperature controlled locations on aircraft up to altitudes of 55,000 ft. The unit is housed in a 1 MCU ARINC 600 package and approved under TSO C59. Code selection is achieved either with installation rack straps or ARINC-compatible code selection panel.

The N1298 decoder is designed to meet ARINC specification 714 and is also housed in an ARINC 600 1 MCU enclosure. Code selection is accomplished by selective jumpers in the mounting rack or via selector switches installed on the aircraft.

The N1335C and N1401C decoders are designed to ARINC specification 531 and are only available in limited quantities. The N1335C is packaged in a ¼ ATR short case and the N401C in a ½ short case as per ARINC specification 404A. Both are two-channel decoders with 16 tonnes per channel.

The CSD-10 is a self-contained SELCAL decoder that may be located anywhere in the cockpit. It is Dzus rail-mounted. Its has an extended operating temperature range (−55 to +55°C) and helicopter vibration category allow installation in uncontrolled environments. Aural and visual signals are internally generated for the aircraft audio system and light annunciators.

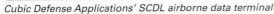

Cubic Defense Applications' SCDL airborne data terminal 0044820

Status
In service.

Contractor
AVTECH Corporation.

SmartComm intelligent frequency management

Type
Avionic Communications Control System (CCS).

Description
SmartComm is an intelligent position-based communications management system. SmartComm organises all database frequencies relative to present position. A standby stack keeps track of the last five frequencies used. The local category list of Centre, Approach, Atis and so on, allows the selection of frequencies by type with up to eight frequencies in each list.

All new Northstar navigators feature SmartComm and include fully operational SmartComm software. Optional 760-channel remote-mounted transceivers instantly tune the frequency selection directly from the Northstar navigator, tuning it into a GPS/comm unit.

Specifications
Nominal voltage: 13.75 V DC
Current (transmit): <2.5 A
Frequency range: 118 to 136.975 MHz
Channels: 760
Weight: 0.55 kg
Dimensions: 70 × 61 × 210 mm

Status
In production.

Contractor
BAE Systems North America.

ST-800S/L series wideband microwave transmitter

Type
Avionic microwave link system.

Description
L-3 Telemetry-East's ST-800S/L series microwave transmitters are designed for highly reliable operation in the severe environmental flight conditions of missiles, space vehicles and aircraft, where size and weight efficiency are critical. These solid-state, crystal stabilised, true FM telemetry transmitters can accommodate various modulation formats such as standard analogue pre-emphasised video, TTL (Transistor Transistor Logic), differential TTL and fully isolated differential TTL (opto-coupled).

Available in 2, 5 and 10 W minimum power output, the ST-800S/L measures 50.8 × 76.2 × 20.3 mm, excluding connectors and weighs 0.2 kg. It operates in the frequency range of 2,200 to 2,400 MHz and 1,435 to 1,540 MHz.

Status
In service.

Contractor
L-3 Telemetry-East.

ST-800S/L series wideband microwave transmitter
0011827

Telephonics STARCOM intercommunication system
0044835

STARCOM intercommunication system

Type
Avionic internal communications system.

Description
STARCOM is a high-intelligibility audio communication system that meets the intercom needs of a wide variety of airborne and ground-based applications. The system is designed to provide secure communications capability in high-noise environments.

The baseline STARCOM system consists of Communications System Controls (CSC) and an Audio Distribution Unit (ADU). It can accommodate five transmit/receive radios, six navigation receivers, two intercom channels and controlled and uncontrolled audio warning signals. Radio and nav receive channels can be individually monitored and controlled for level. Voice-operated switching, with an adjustable threshold for hands-free intercom control, and a remote select capability for HOTAS/HOCAS operation are standard features. Up to 10 CSCs can be interconnected through the ADU. MIL-L-85762A NVG-compatible or standard front panel lighting is available.

Various additions to the baseline system can be installed to support specific applications.

Status
STARCOM has been installed on AH-1W, AH-6,-AS 565 MA, CE-144A, CH-47D ACMS, CH-146, MH-47D, MH-47E, MH-60G, MH-60K, MH-60L, OH-58D and UH-1N helicopters, and C-130 and P-3C aircraft. Also selected for the LCAC upgrade, SH-2G(A), SH-2G(NZ), S-70B (International Sea Hawk), and UH-60Q.

Contractor
Telephonics Corporation, Command Systems Division.

T-300 series airborne UHF transmitter

Type
Avionic telemetry system.

Description
The T-300 series is a subminiature solid-state crystal-stabilised UHF/FM transmitter capable of transmitting wideband telemetry and digital multiplex signals. It is designed for extremely reliable operation in the severe environmental flight conditions associated with missiles, space vehicles or aircraft.

The T-300 operates at 2,200 to 2,400 MHz, with a frequency stability of ±0.002 per cent. Power output is 5 W. The T-300 series meets IRIG-106-93 standards.

Specifications
Dimensions: 63.5 × 38.1 × 19 mm
Power supply: 28 V DC ±4 V
Temperature range: −20 to +70°C

Status
In service.

Contractor
L-3 Telemetry-East.

T-4180 LF-HF Digital Signal Processing (DSP) exciter

Type
Avionic Digital Signal Processor (DSP).

Description
The T-4180 DSP exciter incorporates digital signal processing microchip technology designed to provide greater linearity and spectral purity. It is intended for use with the PA-5050A 1 kw power amplifier, and operates at 1.6 to 30 MHz with 1 Hz tuning resolution.

The T-4180 exciter is available in a half-rack chassis. The unit is capable of local or remote-control operation. In addition, operation is menu-driven and little or no operator training is required. The exciter has a digitally tuned IF filter and 250 programmable channels. Operating modes include LSB, USB, ISB, AM, AM, CW, FSK and FMfax. BITE constantly troubleshoots the receiver and, when a problem is detected, a message on the unit's display screen identifies the module which should be removed and replaced. Remote control via RS-232 or RS-422.

Status
In production and in service.

Contractor
Cubic Communications, Inc.

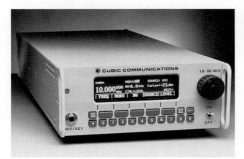

The Cubic Communications T-4180 DSP exciter
0011822

TACAMO II communications system

Type
Avionic communications system, VLF.

Description
The TAke Charge And Move Out (TACAMO) II system provides airborne VLF communications links with the US Navy strategic submarine fleet. The system is a manned communications relay link to strategic forces, normally passing messages one way from the national command to submarines and other strategic forces.

At present, a complete communications centre in the Boeing E-6A TACAMO II aircraft allows simultaneous receive and transmit throughout the frequency range VLF to UHF. The system receives multiple frequency low-level signals, while simultaneously transmitting at high power in a stressed environment. The VLF power amplifier provides amplification of the signal to 200 kW power and automatic tuning of the signal to the dual trailing-wire antenna system. This latter consists of two antennas, one nearly 1,500 m long and the other more than 8,500 m. Only the short wire is charged, the energy reradiating off the longer wire, the length of which varies with the frequency in use. The transmitted signal to the submarine is vertically polarised, with the E-6A aircraft flying in a continuous tight turn. This allows most of the antenna system to hang vertically from the aircraft.

Status
In service on E-6A TACAMO II aircraft of the US Navy.

Contractor
Rockwell Collins.

Tactical common datalink airborne data terminal
0044823

Tactical Common DataLink (TCDL) airborne datalink

Type
Avionic datalink system.

Description
The TCDL programme is a co-operative effort between the US government, L-3 Communications and Rockwell Collins to develop a communications architecture that supports current and future requirements. Its design emphasises Common DataLink (CDL) interoperability, small size, low weight, affordability, and an open, modular and scalable architecture using COTS technology.

The TCDL airborne system is built in a compact, light weight form factor, specifically for UAVs and manned, non-fighter environments. The data rates, modulation techniques, and transmission frequencies are fully interoperable with CDL systems. The programmable modem offers a variety of other waveforms and data rates that can be selected during flight. The Ku-band (NATO J-band) operating frequency is readily changed by replacing the RF converter and the antenna. Options include bulk encryption and programmable uplink and downlink data rates up to 45 Mbps. Full duplex voice and data modes are available. Either omni or directional antennas can be employed and the system can be upgraded to air-to-air operation.

Specifications
Frequency: uplink 15.15–15.35 GHz
    downlink 14.40–14.83 GHz
    optional X-band (NATO I/J-band)
    tuning 5 MHz steps
Data rates: variable up to 45 Mbps; CDL interoperable at 10.71 Mbps; full duplex symmetric or asymmetric
RF power: 2 W (higher power optional)
Bit error rate: 10^{-6} with COMSEC (10^{-8} without COMSEC)
MTBF: >3,800 h
Interface: RS-422
Dimensions (H × W × D): link interface assembly: 76.2 × 171.5 × 254.0 mm; microwave modem assembly: 76.2 × 304.8 × 254.0 mm
Weight: 7.05 kg (including omni antenna)

Status
In production.

Contractor
L-3 Communications, Communication Systems – West.
Rockwell Collins.

TCOMSS-2000 + DAS Telephonics COmmunications Management System and Digital Audio System

Type
Avionic Communications Control System (CCS).

Description
TCOMSS-2000 Telephonics Communications Management System
The Telephonics TCOMSS-2000 is a fibre optic, open architecture, VME-based system that uses COTS hardware to provide secure digital audio and full digital control from all operator positions. TCOMSS-2000 operates on a 100 Mbps fibre optic network that supports full-duplex, non-blocking communication/management of up to 738 audio sources (crew members, external transceivers, navigation aids and encryption devices). Selectable bandwidths support audio and digital data distribution from narrow to wideband signals. System features include: unlimited conferencing, individual volume control, binaural (dichotic) audio, SIMOP, VOX, radio relay, frequency management, data transmission and real-time modem and encryption device switching. Because TCOMSS-2000 is modular, the system can be upgraded with options: auditory localisation (3-D audio), multilevel security, combined or separate red/black busses as well as interfaces with FDDI, ATM, Fibre Channel, MIL-STD-1553, RS-232, and RS-422.

TCOMSS-2000 consists of three module types: the Audio Control Subsystem (ACS);

Telephonics TCOMSS-2000 communication management system   0044837

Audio Control Panel (ACP); and the jack box. The ACS provides the main interface for the TCOMSS-2000. It links the fibre optic rings with operators and communications assets. A plug-in module interfaces with the platform databus. ACS modules can be arranged in single or multiple configurations to support varied platforms.

The ACP is the interface between the operator and the system. Using an RS-422 interface, the ACP enables the operator to manage all assets assigned to a particular crew position. This includes: independent radio transmit and receive selection and control for all assigned radios; programmable intercom/conference net selection and control; VOX; selectable dichotic capability; clear/secure selection; radio relay; and crypto and modem control.

The jack box is available for certain applications, typically larger platforms, to provide digital audio out to the headset jack. The jack box provides the proper controls and interfaces to support an airborne system. In smaller TCOMSS-2000 applications, the ACS incorporates the jack box functions.

When operators communicate, the audio is digitised and placed into an allocated time slot with a destination address within the ACS. The ACS then routes the audio to the appropriate address; either another jack box for interphone audio or for conversion to analogue for radio transmission. Similarly, when a radio or remote communication device receives audio, it is digitised and made available to any of the operators that have been enabled to receive the audio. The entire system is digitally reconfigurable, enabling asset assignment based on the operational scenario. All channels within the system can be selected, combined and monitored.

DAS Digital Audio System
DAS is to be a modified version of TCOMSS-2000 to meet the requirements of the NATO AWACS E-3A Mid-Term Modernisation Programme.

Specifications
Dimensions:
ACS 190.5 × 194.0 × 376.7 mm
ACP 146.0 × 171.4 × 127.0 mm
J-Box 165.1 × 50.8 × 127.0 mm
Weight:
ACS 9.98 kg
ACP 2.04 kg
J-Box .91 kg

Status
TCOMSS-2000 has been selected for the UK Nimrod MRA4 programme. Telephonics Corporation has also been selected by Lockheed Martin Federal Systems to provide a variant of the TCOMSS system for the US Navy SH-60R multimission helicopter and the CH-60 utility helicopter common cockpit programme.

DAS has been contracted by Alcatel Bell Space and Defense of Antwerp to Telephonics

Corporation. Both companies will, together, conduct an engineering and manufacturing development phase to provide three laboratory systems and one flight system. The production phase will be for 17 aircraft and two simulator systems.

Contractor
Telephonics Corporation, Command Systems Division.

TeleLink TL-608 datalink system

Type
Avionic datalink system.

Description
The TeleLink TL-608 digital datalink system meets the needs of the corporate or regional airline pilot by providing a variety of features and versatile connectivity options in a small package. The system exceeds current datalink requirements providing VHF ACARS, SATCOM and airborne phone links. TeleLink has also been designed to comply with Automatic Dependent Surveillance (ADS) functions of the Communications, Navigation and Surveillance/Air Traffic Management (CNS/ATM) programs of the Future Air Navigation System (FANS).

Although derived from ACARS, the TL-608 is designed to take advantage of emerging datalink technologies. Modular hardware and

software design ensures the system is versatile and expandable and open architecture software provides the operator with a choice of service providers and communications media.

Specifications
Dimensions: 1 MCU: 25.4 × 194 × 386.8 mm
Weight: 1.8 kg
Power: 28 V, 10 W
Inputs: ARINC 429 receivers, RS-232/422 receivers, discretes, VHF modem, telephone modem
Outputs: ARINC 429 transmitters, RS-232/422 transmitters, discretes, VHF modem, telephone modem
Interfaces: FMS, CDU, DAU, printer, dataloader, maintenance terminal, laptop PC
Communications media: VHF radio, airborne telephone, Satcom, Mode S
Environmental compliance: DO-160C

Status
Selected by Bombardier Inc as the standard option Communications Management Unit (CMU) for the Global Express aircraft, where it will interface with the Honeywell flight management system and Satcom to provide data and voice coverage worldwide.

Contractor
Teledyne Controls, Business and Commuter Avionics.

TeleLink® helicopter datalink

Type
Avionic datalink system.

Description
The TeleLink® datalink system is a digital communication link for airborne networks. At a minimum, a network consists of a TeleLink®

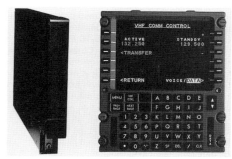

TeleLink helicopter datalink   0011820

The TL-608 is the standard-fit CMU for the Bombardier Global Express aircraft (E L Downs)   0524722

Router Management Unit (RMU), a Control Display Unit (CDU) and a VHF radio. The Teledyne TeleLink® helicopter datalink supports a variety of bidirectional communication media, including: ACARS, satellite and telephone. Capabilities include: flight operations data exchange; GPS-based position reporting; uplink/downlink of FDR/HUMS data; custom messaging. Interfaces include: ARINC 429, 739, CSDB and RS-232/422 I/O.

Contractor
Teledyne Controls, Business and Commuter Avionics.

UniLink air-to-ground two-way datalink

Type
Avionic datalink system.

Description
UniLink is designed to be interfaced with and controlled through Universal Avionics Systems Corporation's colour flat-panel Flight Management Systems, which include the UNS-1B plus, UNS-1C, UNS-1Csp, and UNS-1D. The UniLink menu software integrates with the UNS FMS and provides access for sending and receiving data and graphics. Flight plans can be uplinked through UniLink and loaded directly into the FMS. Position reports and other data from the FMS can be automatically downlinked.

UniLink has been designed to support several communications media including VHF, telephony and Satcom. Other media, such as HF, will be added as they become supported by the Aeronautical Telecommunications Network (ATN).

UniLink has been designed to support all ACARS message types, including triggered events such as OOOI (Out, Off, On, In) and planned interface with Digital Flight Data Acquisition Units (DFDAUs).

The UniLink module is available as a model UL-600 housed in a 1-MCU sized unit, supporting single-, dual- or triple-FMS installations; it will also be available as a separate PCB for the UNC-1C, UNS-1Csp, and UNS-1D FMS.

Universal Avionics Systems Corporation claims that the combined UNS-1 and UniLink suite fulfils the evolving Communications/Navigation/ Surveillance (CNS) routing and communication requirements in the future Aircraft Traffic Management (ATM) system.

Specifications
(UL-600)
Dimensions: 1-MCU
Weight: 1.47 kg

UniLink air-to-ground two-way datalink

0011857

Power: 28 V DC, 5 W
Memory: 2 Mbytes FLASH memory; 1 Mbyte SRAM; 32-bit controller
Interfaces: VHF modern: 1 input/1 output; telephony: 1 input/1 output; ARINC 429: 8 input/3 output; RS-422/232: 8 input/8 output; RS-232 diagnostics/load port: 1 input/1 output; discretes: 16 input/16 output; configuration module 1 input/2 output

Status
In production and in service.

Contractor
Universal Avionics Systems Corporation.

VCS 40 VHF communication system

Type
Avionic communications system, VHF.

Description
The VCS 40 VHF communication system is an all-digital transceiver employing microprocessor-controlled circuitry. It incorporates a full complement of Honeywell Aerospace III performance features including 20 W solid-state transmitter, continuous transmit capability at reduced power, white-on-black dichroic liquid crystal frequency display, 760-channel operation at 25 kHz spacing over 118 to 136.975 MHz, optional 1,360 channel extension up to 151.975 MHz, built-in SELCAL and ACARS capability and automatic self-test and diagnostics.

The VCS 40 consists of the VC 401B digital transceiver and the panel-mounted CD 402B control/display unit.

Specifications
Dimensions:
(transceiver) 104 × 101.6 × 333.6 mm
(control/display unit) 63.5 × 80 × 63.5 mm
Weight:
(transceiver) 2.78 kg
(control/display unit) 0.27 kg
Power supply: 18– 33 V DC
(transmit) 4.5 A
(receive) 0.5 A
Altitude: up to 55,000 ft

Status
In service.

Contractor
Honeywell Aerospace, Electronic & Avionics Lighting.

VHF DataLink System (DLS)

Type
Avionic datalink system.

Description
At Airshow China '98 Honeywell and Aviation Industries of China (AVIC) signed a Memorandum of Understanding to support Communication, Navigation Surveillance /Air Traffic Management (CNS/ATM) in China. As part of the agreement, Honeywell agreed to develop a VHF DLS for Chinese government aircraft. The resultant system includes a new integrated Communications Management Unit and VHF Data Radio (CMU/VDR).

The complete Honeywell VHF DLS comprises the Honeywell integrated CMU/VDR, Honeywell/ Trimble HT9100 GPS navigator and an ARINC 740/744-compliant ACARS printer.

Status
The programme is in three stages: demonstration of the system on three aircraft; installation on a large number of Chinese government transport aircraft; repackaging for smaller aircraft for the remainder of the government aircraft fleet. This final stage is to be completed by 2005.

Contractor
Honeywell Inc, Commercial Aviation Systems.

VHF-21/22/422 Pro Line II VHF Radios

Type
Avionic communications system, VHF.

Description
Designed primarily for general aviation aircraft of all types, the Collins VHF-21/22/422 transmitter receivers are remotely controlled, rack-mounted sets with 20 W transmitter output. They are available in four versions. The A equipment covering the VHF band from 118 to 136.975 MHz and the B variant from 118 to 151.975 MHz. These A and B units utilise 25 kHz channel spacing, and versions with broader receiver bandwidths are available. The C equipment covers the VHF band from 118 to 136.975 MHz and the D variant, the 118 to 151.975 MHz band. The C and D variants provide both 8.33 and 25 kHz channel spacing. Versions are also available for use in installations with GPS receivers.

The radios use digital synthesis frequency generation techniques and are of all solid-state construction. They provide automatic carrier and phase-noise squelch and automatic gain control and are designed to drive cabin audio systems of all types. Principal attractions are low weight, compactness and the low power consumption of 6.5 A during transmission. Consequently they require no forced-air supply, and electronic section cooling is carried out by a combination of heatsink and convective air flow. ICAO Annex 10 FM immunity requirements (DO186) are met.

The VHF-21 can directly replace the earlier VHF-20 series radios. The VHF-21/22 may be controlled by a serial digital signal from a CSDB control, such as the CTL-22. The VHF-422 may be controlled by a serial digital signal from an ARINC 429 control such as the RTU– 4200. It meets DO160C HIRF requirements (Category R) and lighting requirements (level 3).

Either hard or soft mounting may be used and all connections are made through a single connector on the rear of the casing.

Over their production run, several performance improvements have been incorporated into these VHF communications transceivers. A reduction has been made in comm-to-comm interference; this was achieved through the design of a wide dynamic range receiver with sufficient Automatic Gain Control (AGC) to ensure linearity for extremely large desired signals. Only desired signals will affect the AGC; whereas, the first mixer must have enough dynamic range to handle strong interfering signals and small desired signals simultaneously. Intermediate frequencies were selected to minimise any spurious responses.

A reduction has been made in potential for comm-to-nav interference. Improvements in the VHF comm were made through the selection of synthesiser frequencies. The extended-range receiver uses high-side injection below 138 MHz and low-side injection above 138 MHz. The synthesiser will not generate any spurious signals within the nav frequency band into the receiver mixer. An improved phase-noise squelch has been incorporated into the radio that is immune to impulse noise and receiver gain variations. The receiver also includes an override squelch that assures reception of offset carrier signals. The improved squelch and an extremely effective AGC provide excellent reception regardless of altitude or distance from the station. A broad dynamic range ensures that signals as low as three microvolts will be received clearly.

Specifications
Dimensions:
3/8 ATR short/dwarf
85   (H) × 95   (W) × 356   (L) mm   (VHF-22);
85 (H) × 97 (W) × 354 (L) mm (VHF-422)
Weight: 2.1 kg
Temperature: –55 to +70°C
Altitude: 70,000 ft
Power: 28 V DC, 0.6 A receive, 5.0 A transmit
Certification: FAA TSOs C37c, C38c; EUROCAE ED-23, -24,-14A; DO-160A

Status
In production and in service.

Contractor
Rockwell Collins.

VHF-900 series transceivers

Type
Avionic communications system, VHF.

Description
The VHF-900 series radios retain the analogue voice capabilities of the AM Collins VHF-700, while adding both digitised voice and data capabilities, utilising a new Phase-Shift Keyed (PSK) signal in space. The VHF-900 series is designed to be backward compatible with the Collins VHF-700 series (see separate entry).

VHF-900 radios are designed to provide varied functionality and can be upgraded, via service bulletins, from the basic VHF-900 to the advanced, high-speed VHF-920 data radio. The VHF-900B radios provide 8.33 kHz channel operation plus Mode A ACARS operation with built-in modem. The VHF-900B may then be upgraded to the VDL Mode 2 capable VHF-920, which can be utilised in either ARINC 724B or ARINC 758 datalink installations. Software changes are facilitated via a data loader interface for all software read/write operations.

The VHF-920 features the following user benefits:
Provides protection against High-Intensity Radiated Fields (HIRFs) and lightning strike
No additional cooling system requirement
Software is documented to DO-178A (VHF-920 only)
Fully partitioned for Airbus and Boeing Built-In Test Equipment (BITE) software
Internal ACARS modem (ARINC 750 MODE A and Mode 2 Data Link)
No manual adjustments, tuning controlled by an internal processor

Specifications
(VHF-900B)
Dimensions: 3 MCU per ARINC 600
Weight: 4.1 kg
Temperature: –55 to +71°C
Altitude: 50,000 ft
Power: 27.5 V DC, 1 A receive mode, 7.5 A transmit mode
Certification: ARINC 716-10/750-3, 724B, 748, 600-7, 429-17, 604-1, 615-3, 618, 631; TSOs C37d, C38d Class 1, C128; DOs 160D, 186A, 178B, 207, 224A; EUROCAE ED-23B, -24, -12B, -14D, -67

Status
In service in a variety of civilian transport aircraft, including Boeing 737, 747, 757, 767, 777, MD-11, MD-90 and Airbus A319, 320, 321, 330, 340.

Contractor
Rockwell Collins.

VLF/LF High-Power Transmit Set (HPTS)

Type
Avionic communications system, LF/VLF.

Description
The HPTS system consists of a Very Low Frequency/Low Frequency (VLF/LF) 200 kW solid-state power amplifier and dual-trailing wire antenna system. It is designed to improve the reliability of systems that provide survivable communications links from the US Navy's E-6A TACAMO aircraft to the US strategic forces.

Status
In production and in service in US Navy E-6A TACAMO aircraft.

Contractor
Rockwell Collins.

VP-116 voice encryption device

Type
Avionic voice encryption system.

Description
Rockwell Collins' VP-116 voice encryption device is fully compatible with narrowband communications channels, including UHF, VHF, HF-SSB and telephone. The VP-116 is an upgrade of the earlier VP-110 system and fully backward compatible. The main features of the system are:
• Time and frequency division voice encryption
• 10^{19} encryption key variables
• Excellent voice quality and voice recognition
• Reliable in-band FSK synchronisation
• Storage of eight key variables
• Extensive self-test capability
• Clear/secure and local/remote operation
• Over-The-Air Re-keying (OTAR) of key variables
• Dual-algorithm capability
• Night Vision Goggle (NVG) compatibility option
• Embedded version available
• Exportable version available

High voice privacy is provided by a patented combination of time division and frequency scrambling. The scrambling process is implemented entirely with digital techniques and is controlled by a proprietary non-linear algorithm. This algorithm uses a 64-bit key variable, which provides 10^{19} different combinations. Up to eight key variables may be stored in the VP-116 at one time, allowing flexibility of key management. The multiple key variables may be used for different time periods or for different nets.

The VP-116 has analogue audio interfaces for easy integration into a wide variety of communications systems. Transmit audio, receive audio and push-to-talk key are the only three signals from the communications system that need to be routed through the VP-116. In the clear mode, both transmit and receive audio signals bypass the voice privacy circuits to provide clear voice transmission and reception. In the privacy mode, if a private signal is not being processed, any clear signals received will automatically bypass the voice privacy circuits to provide clear voice reception.

Specifications
Dimensions: 67 (H) × 146 (W) × 205 (D) mm
Weight: 0.8 kg
Temperature: –20 to +55°C
Power: 12 to 32 V DC, <6 W
Channel requirement: bandwidth of 400 to 2,800 Hz
Frequency response: 300 to 3,000 Hz with a midband null
Operation: half duplex

Status
In production and in service.

Contractor
Rockwell Collins.

VXI-3000 series DSP receivers

Type
Avionic Digital Signal Processor (DSP).

Description
The VXI-3000 range of DSP (Digital Signal Processing) receivers comprises: the VXI-3250A HF DSP receiver; VXI-3550A DSP VHF/UHF DSP receiver; VXI-3570A VHF/UHF DSP receiver. Supporting systems comprise: the VXI-4400 DF processor; VXI-6180 HF distribution unit; VXI-6581 LO distribution unit; VXI-8110 reference frequency source.

Specifications
VXI-3250A
Frequency range: 10 kHz to 30 MHz, 1 Hz tuning resolution

Cubic Defense Applications' VXI DSP receivers
0044817

Detection modes: LSB, USB, ISB, CW, AM, FM
Sweep and scan: 100 channels per second
Programmable channels: 250
Standard bandwidths: 51 bandwidths from 100 Hz to 16 kHz
Optional interface: C40

VXI-3550A
Frequency range: 20–1,200 MHz
Detection modes: LSB, USB, CW, AM, FM
Sweep and scan: 100 channels per second
Standard bandwidths: 18 bandwidths from 1 to 240 kHz
Optional interface: C40
VXI-3570 A
Frequency range: 20–3,000 MHz, 10 Hz tuning resolution
Detection modes: LSB, USB, CW, AM, FM
Sweep and scan: 100 channels per second
Programmable channels: 250
Standard bandwidths: 18 bandwidths from 1 to 240 kHz
Optional interface: C40

Status
In production and in service.

Contractor
Cubic Defense Applications.

Wireless Communications and Control System (WCCS)

Type
Avionic communications system, UHF.

Description
The Wireless Communications and Control System (WCCS) is a wireless FM communications system which provides voice-operated hands-free full-duplex party line operation for up to six cordless headset users. The system provides highly intelligible communications in a 115 dBSPL environment through the use of a special enhanced noise-cancelling microphone and high-noise attenuation headset. Communication between users occurs within an aircraft through the use of an internally installed leaky coaxial line antenna or externally through a UHF blade antenna. Additional aircraft communications flexibility is provided by a hard-wired two-way audio interface to the aircraft interphone system. In addition to the communications capability which the system provides, it also supports a wireless hand-held remote Control Transmitter (CT) which is used to control cargo and retrieval winches.

The WCCS comprises a Receiver/Transmitter radio (RT), remote-control cargo winch controller, headset rack, up to six cordless headsets and the battery charger.

The RT, which is housed within the headset rack, is the repeater which provides full-duplex operation between users. Remote operation is achieved by using the RT in the portable battery-powered mode. Direct communications between cordless headset users can be accomplished by operating the system in the half-duplex mode, independent of the RT. The simultaneous use of multiple WCCS systems within the same operating range of each other without cross-talk interference is achieved by the use of seven frequency groups and 32-tone squelch codes.

The CT provides variable speed bidirectional cargo winch capability and two single-speed bidirectional retrieval winch capabilities.

Specifications
Dimensions:
(RT) 191 × 203 × 152 mm
(CT) 217 × 84 × 121 mm
(headset) 254 × 226 × 216 mm
(charger) 255 × 229 × 84 mm
(rack) 1,118 × 483 (stowed), 978 (extended) × 178 mm
Weight:
(RT) 2 kg
(CT) 0.55 kg, (headset) 1.23 kg
(charger) 1.9 kg, (rack) 7.73 kg
Power supply: 28 V DC, 28 W
Frequency: 410-420 MHz

Temperature range: –20 to +71°C
Altitude: up to 50,000 ft
Reliability: 5,165 h MTBF

Status
In production and in service in the US Air Force C-17A transport aircraft.

Contractor
Telephonics Corporation, Command Systems Division.

XK 516 HF radio system

Type
Avionic communications system, HF.

Description
Honeywell's XK 516 HF radio system has been designed to eliminate the traditional problems encountered with long-range HF communications, namely link integrity and quality of reception/transmission. In addition to improving voice performance, the XK 516 facilitates the shift from voice to data as the primary means of long-range air-to-ground communication, thereby reducing time and crew workload. Other features and benefits of the system include lower acquisition and message costs than satcom, automatic selection of optimal frequency and data rates, and automatic error correction technology.

Status
In production.

Contractor
Honeywell Aerospace, Electronic & Avionics Lighting.

AIRCRAFT NAVIGATION SYSTEMS

Canada

Allstar and Superstar Wide Area Augmentation System – Differential GPS (WAAS-DGPS) receivers and Smart Antenna

Type
Global Positioning System (GPS) receivers.

Description
The Allstar and Superstar receivers are low-cost Global Positioning System (GPS) receiving units, designed to track the US Federal Aviation Administration (FAA) Wide Area Augmentation System (WAAS) signal, to provide levels of accuracy similar to those of the Differential GPS (DGPS) without the need for extra beacons or receivers.

Simultaneously launched with the aforementioned receivers, BAE Systems Canada Inc introduced a new line of complimentary Smart Antennas. There is an RS-232 seven-pin version and an RS-422 12-pin connector version to optimise compatibility with other host systems. The RS-422 version is available with a 12-pin connector mounted either on the chassis of the Smart Antenna or at the end of a 1 ft cable extending from a central 1 in wide aperture.

The Smart Antenna contains the high-performance 12-channel all-in-view Superstar GPS receiver. It measures 115 (Ø) × 90 (H) mm, and consumes only 1.8 W.

Status
In production and in service. The Allstar receiver was launched in September 1999, the Smart Antenna in October 1999, and the Superstar receiver at the end of 1999.

Contractor
CMC Electronics Inc.

AN/APN-208 and AN/APN-221 Doppler navigation systems

Type
Avionic navigation sensor, Doppler.

Description
The AN/APN-221 Doppler set is supplied for night/adverse weather search and rescue helicopters. This system was derived directly from the AN/APN-208(V) helicopter Doppler navigation system developed to meet the requirements of

ASW and search and rescue helicopters, where versatility in operation and interface options are essential. A recent addition has been the integration of GPS inputs into the navigation solution to provide enhanced performance.

A specific application of the APN-221 Doppler navigation set, which was developed under the sponsorship of the US Air Force Systems Command, Aeronautical Systems Division, is the Pave Low III/Sikorsky HH-53 medium-lift helicopter. The operational requirements of this aircraft called for an extension of the AN/APN-208(V)'s capability to include guidance in poor weather or darkness through the use of advanced computer techniques and cockpit displays. Improvements in the AN/APN-221 have since been incorporated into the AN/APN-208.

The four-unit APN-221 comprises a four-beam lightweight antenna, ATR short signal data converter with 16-bit microcomputer, 6-line by 12-character control and display unit and a steering/hover indicator. The displays are also available in a form that is compatible with Generation III ANVIS night vision goggles. It provides three pilot-selectable navigation co-ordinate systems, with automatic conversion from one to another. These are latitude/longitude, worldwide alphanumeric UTM and arbitrary grid. Data can be stored for up to 75 mission waypoints, 10 targets of opportunity or 25 library waypoints (including Tacan beacon locations). There are also three pilot-selectable search patterns with automatic turning point computation and navigation/guidance outputs: creeping line, expanding square and sector. The system provides aircraft velocity outputs to the flight control system, enabling coupled hover manoeuvres and automatic approaches to the hover to be conducted.

The transmitter/receiver/antenna uses a Gunn diode RF source to generate four Janus configuration beams, and FMCW modulation gives high accuracy and immunity from carrier noise, precipitation, surface spray and reflections from nearby objects such as airframe structure and sling loads.

Specifications
Dimensions:
(antenna) 439 × 439 × 113 mm
(signal data converter) 194 × 198 × 319 mm
(control/display unit) 146 × 114 × 165 mm
(steering/hover indicator) 83 × 83 × 127 mm

Weight:
(antenna) 5.59 kg
(signal data converter) 9.45 kg
(control/display unit) 3.54 kg
(steering/hover indicator) 1.14 kg
Transmission: 4 beam Janus, time-shared (200 ms/cycle), 3 × 6.7° beamwidth from Gunn diode, 13.325 GHz with FMCW modulation optimised for flight envelope
Velocity range:
(forward) –50 to 300 kt
(lateral) 100 kt
(vertical) ±5,000 ft/min
Accuracy:
(forward) 0.3% speed along velocity vector ±0.2 kt
(lateral) 0.32% speed along velocity vector ±0.2 kt
(vertical) 0.2% speed along velocity vector ±20 ft/min
Inputs: pitch/roll attitude and heading from AHARS or INS. Various optional interfaces with map displays, Tacan, sonar, radar and air data computer systems.
Microcomputer:
(architecture) word addressed, 16-bit microprocessor

Status
The AN/APN-208 is in service with the armed forces of many NATO and other countries. The AN/APN-221 is in service for the US Air Force Sikorsky Pave Low III HH-53 helicopter. Variants include CMA 806A/B and CMA 708C. Over 500 systems in service.

Contractor
CMC Electronics Inc.

CMA-2012 Doppler navigation sensor (AN/ASN-507)

Type
Avionic navigation sensor, Doppler.

Description
The CMA-2012 is a single-LRU Doppler navigation sensor designed for both rotary- and fixed-wing aircraft applications where navigation aiding, back-up navigation and hover are of primary importance. Based on CMC's experience in the design and development of Doppler radar systems and related technologies, the design parameters of the CMA-2012 were set to achieve significant reductions in size, weight and cost compared to current systems, while enhancing performance capabilities and improving reliability.

With the CMA-2012, CMC has achieved these improvements by implementing innovative design features, including: digital signal processing for real-time signal analysis; optimised hover-hold mode for precision hover with drift rates significantly less than 1 m/minute; tactical modes include silent, horizontal beam cutoff and EW equipment compatible intermittent track.

Now available in a commercial format suitable for para-military helicopters in Search and Rescue (SAR) operations such as Heli-Dyne Systems, Sentinel Inc and Bell-412.

Specifications
Velocities (land):
Vx (forward) –50 to 250 kt ±0.3% Vt
Vy (lateral) –100 to 100 kt ±0.3% Vt
Vz (vertical) ±5,000 ft/min ±0.3% Vt
$Vt^2 = Vx^2 + Vy^2 + Vz^2$
Altitude: 2 to 15,000 ft
Power: 28 V DC/45 W max
Cooling: convection
Weight: 5.5 kg
Environment: MIL-E-5400 Class 1A/DO-160C
Reliability: 6,700 h MTBF
Dimensions: 372.6 × 345.3 × 49.5 mm
Standard interfaces: MIL-STD-1553B, ARINC-429

The AN/APN-208(V) Doppler system

0503888

CMA-2012 Doppler navigation sensor   0015338

Status

The CMA-2012 has been selected for five models of scout/anti-tank helicopters, as well as a variety of other tactical helicopters in North and South America, Europe, Asia and Africa.

A derivative, the CMA-2012W, has been supplied to GKN Westland Agusta for fitment to 15 Cormorant (SAR) helicopters ordered by the Canadian Armed Forces and 17 helicopters ordered for the Italian Navy's maritime patrol helicopter programme.

Contractor

CMC Electronics Inc.

CMA-3012 Global Navigation Satellite Sensor Unit (GNSSU)

Type

Global Positioning System (GPS) sensor.

Description

The CMA-3012 GNSSU meets the requirements of primary means oceanic/remote area operation as specified by FAA order 8110.60. The Sureflight software package is available to complement all CMC GPS receivers for primary means navigation flight planning and dispatch. Key characteristics include:

1. Twelve simultaneous channels, all of which can be used for continuous satellite tracking, and any two of which are assignable as GPS/WAAS Integrity Channels (GIC)
2. Comprehensive end-to-end receiver Built-In Test (BIT)
3. Carrier phase tracking
4. Differential GPS (SCAT 1) functionality
5. Full compliance with TSO-129A B1/C1 and RTCA DO-208 sensor requirements
6. Growth provisions for WAAS.

Specifications

Receiver: 12 parallel channels
Frequency: L1, 1,575.42 MHz, C/A code
Time to first fix: 95% confidence 75 s max
Time to reacquisition: 5 s max
Accuracy:

CMA-3012 GPS sensor unit   0015339

(horizontal position) 22.5 m, 95%, S/A off
(differential) <2.4 m, 95% (optional)
(altitude) 30 m, 95% S/A off
(velocity) 0.1 kt, 95%, S/A off
Position update: once per second (5 times per second optional)
Dimensions: 66 × 216 × 241 mm
Weight: 2.55 kg
Input power: 18 to 36 V DC, 20 W max
Temperature: −55° to +70°C
Altitude: 55,000 ft
MTBF: 65,000 h
Inputs: 8 ARINC 429, 1 RS-232
Outputs: 3 ARINC 429, 1 RS-232; 1 28 V valid discrete; three 1 Hz time marks
Conformity: ARINC 743A, 429-12; RTCA DOs -160C/D, -208, -217 (optional), -178B Level C, −229 (optional)
Certification: TSOs C129A, C145

Status

In service. FAA approved as primary means of oceanic/remote area navigation. Installed with Thales RNav 2 system on helicopters operating over the North Sea.

Since 1993, over 5,000 units have been installed in numerous air transport and general aviation aircraft types.

Contractor

CMC Electronics Inc.

CMA-4024 GNSSA aviation precision approach GPS receiver module

Type

Avionic navigation sensor, Global Positioning System (GPS).

Description

The CMA-4024 GPS receiver module is designed for incorporation into all avionics applications, while meeting the majority of size constraints, and allowing growth to full functionality with SBAS/WAAS and GBAS/LAAS.

The CMA-4024 has been designed to meet all requirements for en route primary navigation as specified in FAA order 8110.60, TSO C-129a, and TSO-C145 for primary means, oceanic and continental en-route navigation, with growth to terminal, Non-Precision Approach (NPA), and Category 1 Precision Approach (PA). The system can accommodate these added functions by software load through an ARINC 429 input or serial data port. The growth path for GBAS/LAAS Category 2 and 3b is defined with the appropriate software load and a drop-in fault monitor.

Major features and benefits of the system include:

• 24-channel Narrow Correlator® tracking technology receiver which can be used for continuous GPS and SBAS/WAAS satellite tracking
• Two fully independent L1 RF input channels
• GPS and growth for SBAS/WAAS carrier phase tracking
• Numerous inputs and outputs available to support all required aircraft interfaces for complex applications
• 60,000 h MTBF
• Full Fault Detection & Exclusion (FDE)

CMA-4024 GNSSA (CMC Electronics)   1034678

• Predictive Receiver Autonomous Integrity Monitor (RAIM)
• Automatic pressure altimeter calibration and usage
• Comprehensive end-to-end receiver Built-In Test (BIT).

The CMA-4024 is the result of a collaborative effort with NovAtel Inc. for RF front-end and Narrow Correlator® tracking technology.

Specifications

Receiver
Type: 2 active antenna ports with 2 GPS L1 RF channels, 24 parallel Narrow Correlator® digital processing channels
Frequency: L1, 1575.42 MHz, C/A code
Acquisition sensitivity: −121 dBm at 37 dB Hz C/No
Tracking sensitivity: −124 dBm at 34 dB Hz C/No
Time to First Fix (TTF): <75 sec maximum, 95% confidence
Horizontal position accuracy: 15 m, 95%, S/A off
Differential: Better than 1.0 m, 95%
Altitude accuracy: 20 m, 95% S/A off
Velocity accuracy: 0.5 kt, 95%, S/A off
Position update: 1 Hz, 10 Hz optional

Software
Language: Ada
Level: DO-178B Level A (design); DO-178B Level B (certified)
Processor: 64-bit Pentium™ compatible

Physical
Dimensions: 168 × 102 × 14 mm
Weight: <0.23 kg
Power: +3.3, +5.0, ±12.0 V DC (or ±14.0 V DC)
Consumption: 12 W maximum; 10 W typical
MTBF: 60,000 h

Interface
Input: ARINC 429 (9); RS-422/232 (4); discrete (11)
Output: ARINC 429 (5 independent); RS-422/232 (4); discrete (2); time marks (3 at 1 Hz)

Environmental
Temperature: −55 to +85°C
Altitude: 55,000 ft
Humidity: DO-160D Cat C
HIRF: 200 V/m (when properly enclosed)

Status

In production.

Contractor

CMC Electronics Inc.

China

9416 air data computer

Type

Avionic navigation sensor, Air Data Computer (ADC).

Description

The 9416 Air Data Computer (ADC) was designed for the K-8 Karakorum trainer. It features an Intel 8086 processor and combines

with CAIC's 9414 preselector and 9415 altimeter into an air data system. An interface provides outputs to other aircraft systems, including gunsight, warning computer and Fight Data Recorder (FDR).

Specifications

Weight: 4.0 kg
Dimensions: 225 × 124 × 194 mm
Power: 27 V DC; 30 W

Environmental: GJB-150-86
Reliability: 3,000 hr MTBF

Status

In service.

Contractor

Chengdu Aero-Instrument Corporation (CAIC).

ADS-4 air data sensor subsystem

Type
Avionic navigation sensor, Air Data Computer (ADC).

Description
The ADS-4 air data sensor subsystem receives static/dynamic pressure signals from the aircraft and provides independent three-channel analogue output to the Flight Control System (FCS). The system utilises an Intel 80C31 microprocessor and large-/medium-scale Integrated Circuit (IC), coupled with Complementary Metal Oxide Semiconductor (CMOS) circuitry for reduced power consumption.

Specifications
Weight: <2.3 kg
Dimensions: 206 × 133 × 120 mm
Power: 15 V DC; 2.8 W
Output parameters:

Parameter	Range	Accuracy	Analogue Output
Static Pressure	0.680 – 110.4 KPa	0.1%	−10 to +10 V
Dynamic Pressure	0.000 – 160.0 KPa	0.1%	−10 to +10 V

Status
In service.

Contractor
Chengdu Aero-Instrument Corporation (CAIC).

GGK-2C Air Data Computer (ADC)

Type
Avionic navigation sensor, Air Data Computer (ADC).

Description
The GGK-2C is a small, lightweight Air Data Computer (ADC) of modular design, featuring low power consumption and extensive self-test and fault warning. Utilising a silicon pressure sensor, the system provides air data parameters via ARINC 429, RS-422 or RS-232 interfaces.

Specifications
Weight: <5.5 kg
Dimensions: 128 × 62 × 59 mm
Power: 28 V DC; <5 W
Temperature: −55 to +70°C

Status
In service.

Contractor
Chengdu Aero-Instrument Corporation (CAIC).

MD-90 Central Air Data Computer (CADC)

Type
Avionic navigation sensor, Air Data Computer (ADC).

Description
The MD-90 Central Air Data Computer (CADC) has been co-developed by CAIC and Honeywell for the MD-90 aircraft and is compatible with Honeywell's HG280 Digital Air Data Computer (DADC). The system provides precision output of parameters for the following systems in the host aircraft:
• Flight Control System (FCS)
• Flight displays
• Inertial Navigation System (INS)
• Flight Data Recorder (FDR)
• ATC transponder
• Ground Proximity Warning System (GPWS)
• Flight Management System (FMS)
• Windshear warning system
• Autopilot

Other features of the MD-90 include:
• Modular construction – 4 electronic sensor modules
• Built-In Test (BIT), including memory failure
• 32-bit Motorola MC68332 microprocessor and large-scale IC
• Reduced power requirement CMOS circuitry
• EEPROM memory
• C software language, Level I (DO-178A).

Specifications
Weight: <6.6 kg
Dimensions: ½ ATR
Power: 115 V AC, 400 Hz single phase; <30 W
Reliability: Service life 20 years/75,000 flight hours

Status
In service.

Contractor
Chengdu Aero-Instrument Corporation (CAIC).

SS/SC-1, -1G, -2, -4, -5, -10 air data computers

Type
Avionic navigation sensor, Air Data Computer (ADC).

Description
The Chengdhu Aero-Instrument Corporation's SS/SC series of air data computers are all configured for use in fighter aircraft of the J-7 type and trainers such as the K-8 Karakorum. They take data from pressure, temperature and attitude sensors, process it using Intel 8086 series processors and pass the data to the navigation, weapons management and flight control systems, and to other Chengdu flight displays.

Common features and benefits of the SS/SC series include:
• Intel 8086 series microprocessor and large-/medium-scale Integrated Circuit (IC)
• Modular structure
• Vibrating cylinder transducer to enhance accuracy and reliability
• Built-In Test (BIT) capability
• Multiple signal interface capability.

Additionally, the SS/SC-2 includes an interface with Chengdhu's ZG-2 altitude indicator (see separate entry), the SS/SC-5 features only six electronic subassemblies within a modular COTS hardware/software design and the SS/SC-10 incorporates extensive redundancy of operation in a modular design utilising the C-computer language.

The SS/SC-5 has been designed for the Chinese K-8 trainer. The SS/SC-½ and 10 were designed for indigenous fighter aircraft applications.

Specifications
SS/SC-1
Weight: <11 kg
Dimensions: 314 × 240 × 210 mm
Power: 115 V AC, 400 Hz single-phase, 70 VA; 36 V AC, 400 Hz, 0.5 VA; 27 V DC
Environmental: HB 76-76

SS/SC-1G
Weight: <6.3 kg
Dimensions: 240 × 180 × 194 mm
Power: 115 V AC, 400 Hz single-phase, 55 VA

SS/SC-2
Weight: <7.5 kg
Dimensions: 305 × 195 × 190 mm
Power: 115 V AC, 400 Hz single-phase, 70 VA
Environmental: HB 76-76

SS/SC-5
Weight: 6.0 kg
Dimensions: 294 × 124 × 194.5 mm
Power: 115 V AC, 400 Hz single-phase, 60 VA
Environmental: GJB-150-86
Reliability: 1,500 h MTBF
Service life: 1,000 hours/15 years

SS/SC-10
Weight: <6.0 kg
Dimensions: 318 × 124 × 194 mm
Power: 115 V AC, 400 Hz single-phase, 60 VA
Environmental: GJB-150-86
Reliability: 1,000 h MTBF

Status
In service.

Contractor
Chengdu Aero-Instrument Corporation (CAIC).

XSC-1 Air Data Computer (ADC)

Type
Avionic navigation sensor, Air Data Computer (ADC).

Description
The XSC-1 Air Data Computer (ADC) is a microprocessor-controlled, modular system, which,

SS/SC-5 air data computer and K-8 Karakorum trainer   0002361

in combination with a suitable pitot/static system, 102 AUIF total temperature sensor, altimeter, Air Speed Indicator (ASI), Vertical Speed Indictor (VSI) and altitude preselector, is designed to fulfil the functions of a complete aircraft Air Data System (ADS). In addition to digital/analogue outputs for all main aircraft parameters, the XSC-1 features extensive Built-In Test (BIT) and provides outputs for parameter exceedance (overspeed), parallel attitude encoding and fault warning.

Specifications
Weight: <7.0 kg
Dimensions: ½ ATR Short – 324 × 124 × 194.5 mm
Power: 115 V AC, 400 Hz; <70 VA
Temperature: −55 to +70°C
Reliability: 6,000 hr MTBF
Compliance: TSO-C8d, -C10b, -C43c, -C88a, -C106; RTCA/DO-160C, -178B Level B

Status
In service.

Contractor
Chengdu Aero-Instrument Corporation (CAIC).

XSC-3 miniaturised Air Data Computer (ADC)

Type
Avionic navigation sensor, Air Data Computer (ADC).

Description
Functionally, the XSC-3 Air Data Computer (ADC) senses dynamic and static pressure and resolves and calculates outputs to various aircraft systems. Designed to replace the KAD280 Air Data System (ADS) in the Y7L aircraft, the small and lightweight XSC-3 features full self-test capability and provides output via ARINC 429, RS-422 or RS-232 interface.

Specifications
Weight: <1.5 kg
Dimensions: 138 × 112 × 96 mm
Power: 28 V DC; <200 mA
Temperature: −55 to +70°C
Reliability: 6,600 hr MTBF

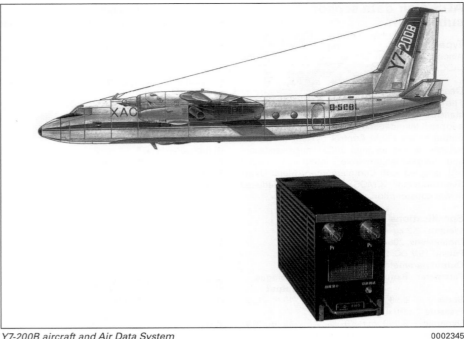

Y7-200B aircraft and Air Data System   0002345

Status
In service.

Contractor
Chengdu Aero-Instrument Corporation (CAIC).

Y7-200B Air Data System (ADS)

Type
Avionic navigation sensor, Air Data System (ADS).

Description
The Y7-200B ADS is designed for the Y7-200B aircraft, and it features the following components: 8903 air data computer incorporating a microcomputer core and large-scale IC; 8904 altimeter; 8905 servo-airspeed indicator; 8908 vertical speed indicator; 89008 altitude preselector/altitude alerter; total air temperature/static temperature/true airspeed indicator.

The Y7-200B ADS provides outputs to the FMS, EFIS and navigation systems, and aural warnings/ground proximity warnings to the pilot.

Other features of the system include:
• Modular architecture
• Built-In Test (BIT) capability
• Vibrating cylinder sensor featuring high reliability and stability.

Status
In service.

Contractor
Chengdu Aero-Instrument Corporation (CAIC).

France

51RV-4/5DF VOR/ILS receivers

Type
Aircraft navigation system.

Description
The 51RV-4DF VOR/ILS receiver forms one part of a flight inspection system, providing test signals and measurement data for flight inspection of VOR localiser and glide slope installations. It is a direct retrofit for ARINC 547 VOR/ILS receivers.

All parameters that need to be accessed are available on the front panel and no wiring harness modification is required. The 51RV-4DF features digital bearing output filtering of the basic bearing information, remote control by two out of five binary control, remote control by serial information in ARINC 429 binary broadcast form, dual conversion of 200 VOR/Loc channels and 40 glide slope channels, dual-instrumentation circuits with internal comparison and level monitoring and BITE.

The 51RV-5DF is identical to the 51RV-4DF except that it meets ICAO FM immunity requirement.

Status
In production for numerous airport and military calibration authorities worldwide.

Contractor
Rockwell-Collins France, Blagnac.

The Rockwell Collins France 51RV-4/5 DF VOR/ILS receiver
0015342

ADC 31XX Air Data Computers

Type
Avionic navigation sensor, Air Data Computer (ADC).

Description
The ADC 31XX is a new line of air data computers for military applications. They are based on a new generation of subminiature pressure sensors (P 90 sensors), which use solid-state, vibrating beam resonators and modern microprocessor technology to measure static and total pressure and impact temperature detected by the aircraft probes.

The modular design and alternative packaging arrangements available make it easy to adapt the system to a number of primary flight needs (AFCS and fly-by-wire) and secondary or back-up requirements (navigation and displays).

Specifications
Dimensions: 121 × 140 × 131 mm
Weight: <1.2 kg
Power: 28 V DC, 10 W (maximum)
Interfaces: MIL-STD-1553B, RS-422, ARINC 429 low and high speed
MTBF: 20,000 h
Performance:
(static and total ± 2 to 5×10^{-4} of full scale pressure)
(impact temperature) ± 1°C
Environment: MIL-STD-810E, 461E and AIR 7306 standards

Status
In production.

Contractor
Thales Avionics SA.

ADU 3000/3008 Air Data Units

Type
Avionic navigation sensor, Air Data Computer (ADC).

Description
The ADU 3000 is a new line of air data measurement units, based on a new generation of micro-machined subminiature pressure sensors that produce a small, highly reliable sensor unit. Different types of sensor can be used for differing applications:

The P 90 sensor is a vibrating beam system that is used for high measurement ranges. Two of these sensors are used in the ADU 3008 (one for pitot channel and one for the static channel). This version is optimised to the needs of transport aircraft – both civil and military.

The P 92 sensor uses piezo-resistive gauges for measurement in a dual installation. One dual-P92 sensor is used in the ADU 3000 to measure static and differential pressure (total static). This unit is optimised for low measurement ranges and is intended for helicopter applications.

In addition to pressure measurements on pitot and static channels, the ADU 3000/3008 is used to perform computations related to air data parameters, including barometric corrections.

Specifications
Dimensions:
(ADU 3000) 150 × 130 × 50 mm
(ADU 3008) 200 × 130 × 50 mm
Weight:
(ADU 3000) 0.8 kg
(ADU 3008) 1.1 kg
Power:
(ADU 3000) 28 V DC, 5 W
(ADU 3008) 28 V DC, 7.5 W
MTBF:
(military) 15,000 to 30,000 h
(civil) 60,000 to 80,000 h
Environment as per RTCA DO 160C or MIL-STD-810C
Interfaces:
(ADU 3000) 1 × ARINC 429, 100 kHz
(ADU 3008) 5 × ARINC 429, 12.5 kHz

Status
In production for several helicopters (including EC 135, EC 155, Super Puma and Gazelle), transport aircraft (including C-130, Dash 8-400, KC-135 and Airbus family) and fighter aircraft (including Rafale, Alpha Jet, Hawk and the Mirage family).

Contractor
Thales Avionics SA.

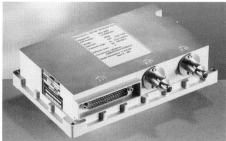

Thales' ADU 3008　　　　0079255

AHV-9 and AHV-9T radio altimeter

Type
Avionic navigation sensor.

Description
The AHV-9 was designed for the Mirage 2000–5; the AHV-9T is the designation for the Panavia Tornado variant.

Low- and high-altitude bands can be used, the variable output being available from a microprocessor-based receiver unit. The system features CMOS integrated electronics, stripline technology antenna and built-in fault detection capability.

Specifications
Dimensions:
(indicator) 61 × 61 × 158 mm
(transmitter/receiver unit 109 × 154 × 324 mm)
(antenna) 78 × 88 × 33 mm
Weight:
(indicator) 0.9 kg
(transmitter/receiver unit 4.5 kg)
(antenna) 0.2 kg
Power supply: 115 V AC, 400 Hz, 70 VA
Altitude: up to 50,000 ft
Accuracy: 1 ft ±2% of altitude
RF power: 60 MW (FM/continuous wave)
Outputs: 5 digital, 3 analogue

Status
The AHV-9T is in service in German Panavia Tornados.

The AHV-9 radio altimeter, and the NC 12 Tacan receiver are fitted to Mirage 2000–5 and 2000–9 aircraft.

Contractor
Thales Communications.

AHV-12 radio altimeter

Type
Avionic navigation sensor.

Description
Designed for use on the Dassault Mirage 2000 and Atlantique 2, the AHV-12 radio altimeter features wide altitude range, high-accuracy and high-integrity levels.

Specifications
Typical installation
Dimensions:
(antenna) circular, rectangular or small size
(transmitter/receiver 193 × 90 × 315 mm unit)
(indicators) ARINC 429 (digital), ARINC 552 (analogue)
Weight:
(transmitter/receiver 5 kg unit)
Power supply: 115 V AC, 400 Hz, 45 VA
Altitude: up to 70,000 ft
Accuracy: 1 ft ±1% of altitude

Status
In service in variants of the Mirage 2000, Atlantique 2 and US reconnaissance aircraft.

Contractor
Thales Communications.

Thales' AHV-12 digital radio altimeter　　0503703

AHV-16 radio altimeter

Type
Avionic navigation sensor.

Description
The AHV-16 microprocessor-based radio altimeter has been built with a reprogrammable memory and is designed for commuter and military aircraft. It has both digital and analogue

outputs to ARINC 552 and ARINC 429 standards. It can interface with existing digital and analogue avionic equipment and with electromechanical instruments as well as electronic flight instrument systems.

Specifications
Dimensions:
(indicator) 3 ATI
(transmitter/receiver unit) 230 × 90 × 90 mm
Weight:
(indicator) 1.2 kg
(transmitter/receiver unit) 2.0 kg
Power supply: 28 V DC;
(R/T unit) 15 W
(indicator) 8 W
Accuracy: 1 ft ±2% of altitude
Reliability: >5,000 h MTBF

Status
In service in transport aircraft and helicopters.

Contractor
Thales Communications.

AHV-17 digital radar altimeter

Type
Avionic navigation sensor.

Description
The AHV-17 is specifically designed for Rafale. It has very low probability of intercept due to an RF management system which adjusts output power as a function of altitude and enhanced receiver sensitivity. The AHV-17 has excellent jamming detection capability, which enables it to have superior resistance even against advanced jammers.

The AHV-17's modular construction, based on a series of integrated tests, allows effective diagnosis of failures and simple replacement without the need to replace or adjust the defective module. Modules can be interchanged independently.

Specifications
Volume: <3 litres
Weight: <3 kg
Power supply: 28 V DC to MIL-STD-740D, 40 W
Frequency: 4.2–4.4 GHz
Altitude: up to 30,000 ft
Accuracy: 3 ft or 1% of altitude
Reliability: >5,000 h MTBF

Status
In production and in service in Rafale multi-role fighter aircraft.

Contractor
Thales Communications.

AHV-18 compact radio altimeter

Type
Avionic navigation sensor.

Description
The AHV-18 is a multimode, modular radar altimeter, which can be adapted for use with fixed- and rotary-wing aircraft, missiles, RPVs and flying weapons.

It has comprehensive ECCM capabilities using power management and jamming detection modes.

Specifications
Dimensions:
(indicator) 3 ATI
(transmitter/receiver unit) 124 × 81 × 81 mm
(antenna) 105 × 90 × 38 mm
Weight:
(indicator) 1.2 kg
(transmitter/receiver unit) 1.2 kg
(antenna) 0.13 kg
Altitude: up to 5,000 ft
Accuracy: 1 ft ±2% of altitude
Reliability: >5,000 h MTBF

Status
In production and in service in Agusta A 109 helicopters and the Penguin Mk 2 anti-ship missile.

Contractor
Thales Communications.

APIRS F200 Aircraft Piloting Inertial Reference strapdown Sensor

Type
Avionic navigation sensor, Inertial Reference System (IRS).

Description
APIRS is SAGEM's latest AHRS. It utilises advanced fibre optic gyros and silicon accelerometers to achieve a rugged design making it suitable for both commercial and military fixed- and rotary-wing aircraft.

APIRS provides attitude, heading, angular rates and linear acceleration data to the automatic flight control system and cockpit displays via ARINC 429 digital databuses. Optional interfaces include: the Honeywell ASCB bus, analogue, and three-wire synchro outputs.

APIRS is well suited to be a replacement for mechanical AHRS and vertical and directional gyros in older avionics systems.

Specifications
Dimensions: 289 × 105 × 114 mm
Weight: 3.2 kg (AHRU); 0.25 kg (MSU)
Power supply: 28 V DC, <30 W
MTBF: AHRU >7,000 hrs; MSU >50,000 hrs
Interface
Input: ARINC 429 digital bus (2 each, low speed), discretes for system control, RS 232 for maintenance access
Output: ARINC 429 digital bus (4 each, high speed), discretes for system status
Performance (2Ó)
  Attitude: 0.5°
  Heading: 1.5°
  DG Mode: 5° per hour
  Angular rates: 0.1° per second
  Acceleration: 5 mg
Qualification: TSO C4c, C5e, C6d; RTCA DO 160 C (including HIRF and lightning), DO 178 B (level A, flight critical)

Status
In production; selected for the following programmes:
- Eurocopter EC 135, 145, 155, 165
- NH 90 NATO helicopter
- Hindustan ALH helicopter
- DASH 8 – 400
- DASH 8 100/200 and 300 upgrade
- Dassault Super Etendard upgrade
- CASA C295

Contractor
SAGEM Défense Securité, Navigation and Aeronautic Systems Division.

Cirus attitude and heading reference system

Type
Aircraft navigation system.

Description
Cirus is a hybrid inertial attitude and heading reference system. Its basic design involves the use of inertial/magnetic heading and air data hybridisation techniques. The system generates heading, attitude, angular velocities, accelerations, true airspeed, pressure altitude and temperature.

As an option, it can also deliver position and groundspeed data in conjunction with a Doppler navigation radar or a GPS receiver. The groundspeed vector supplied by the GPS system, with its medium-term accuracy, can be combined with the angular and linear speed data generated by Cirus, with its excellent short-term accuracy, to form a versatile high-performance AHRS/GPS coupling system.

Specifications
Dimensions: 260 × 150 × 150 mm
Weight: 5 kg
Power supply: 28 V DC, 50 W
Alignment time: <1 min

Status
In production and in service. The Cirus system is fitted to the Eurocopter Super Puma Mk 2 helicopter, French Air Force C-160 Transall (retrofit Cirus 1600 and Totem 200) and Alizé aircraft (Cirus 1500).

Contractor
Thales Avionics SA.

Embedded GPS receiver

Type
Avionic navigation sensor, Global Positioning System (GPS).

Description
The SAGEM GPS receiver is an 'all-in-view' dual-frequency receiver, with Precise Positioning Service (PPS capability). The receiver is designed to ensure INS/GPS tight-coupling, with special attention given to data latency and time tagging.

Comprehensive integrity monitoring is performed, including a RAIM algorithm, in order to ensure the consistency of satellite signals.

Proprietary ASICs and signal processing provide: miniaturisation and reliability; fast acquisition and high resistance to Electronic Counter Measures (ECM).

The SAGEM GPS receiver is also available in Standard Positioning Service (SPS) version. Both PPS and SPS versions are capable of differential positioning through the use of standard connections.

Specifications
Positioning accuracy: in accordance with international standards

Status
Current applications include French Air Force Mirage F1 fighters (French Air Force), NH 90 helicopters (French Army and French Navy), and various export customers.

Contractor
SAGEM Défense Securité, Navigation and Aeronautic Systems Division, Paris.

Gemini 10 navigation and mission management computer

Type
Aircraft navigation system.

Description
The Gemini 10 computer contains a Jeppesen database and, in addition to carrying out conventional FMS functions, also performs tactical military functions. Connected to the VH100-T HUD, it supports missions such as assault landings, airdrops and tactical low-altitude flights.

The computer's compact design is due to its monobloc concept, with the navigation computer and control and display unit integrated in a single item of equipment. This layout offers installation advantages of lower weight, smaller volume and higher reliability, together with easier maintenance and lower power consumption in operation. The Gemini 10's dual architecture helps increase the probability of mission success by allowing reconfiguration on the validated FMS, in case of failure of one of the two units.

The software comprises 150,000 lines in Ada language, conforming to the requirements of DO178, and allows the display of approximately 150 pages of information. The display offers 12 lines of 20 characters each and five variable label keys with functions depending on the mission.

Status
In service in the C-160 Transall.

Contractor
Thales Avionics SA.

MultiMode Receiver (MMR) TLS 755

Type
Aircraft navigation system.

Description
The Thales Avionics MultiMode Receiver (MMR) is a new landing and precision approach sensor that provides navigation, en-route GPS, ILS, MLS and GLS functions in one LRU. In addition, the MMR provides position, time and velocity in a permanent GPS navigation mode. A modular design ensures full scalability that allows the unit to keep pace with applications including the Satellite Based Augmentation System (SBAS) and Ground Based Augmentation System (GBAS).

The receiver has been designed according to ARINC 755 specifications for dual or triplex installations providing Cat III operations. The MMR is immune to FM radio signals in accordance with ICAO annex 10.

The MMR offers a high degree of integrity and reliability due to the digital technology used in the design.

The TLS 755 receivers are ARINC 755 compliant and integrate an ARINC 743 GPS sensor that includes Wide Area Augmentation System (WAAS), Local Area Augmentation System (LAAS) Cat I to Cat III B, and GLONASS capabilities. The system is also FANS compliant.

Specifications
Dimensions: 3 MCU
Frequency bands:
  108–112 MHz for ILS localiser
  329–335 MHz for ILS glideslope

The APIRS Attitude Heading Reference System (AHRS)   0001322

108–118 MHz for GLS VHF datalink
5 GHz for MLS
Channel spacing: 50 kHz for datalink (compliant 25 kHz)
Modulation: AM
DPSK 10 kbps/15kbps
D8PSK 31.5 kbps
Weight: <5.4 kg
Power supply: 115 V AC, 400 Hz single phase
Temperature range: −55 to +70°C
Compliances:
ILS ARINC 710-9
MLS ARINC 727
GPS ARINC 743A
MMR ARINC 755
MTBF: >30,000 flying hours

Status
In production. TSO qualified for all Airbus and Boeing aircraft; certified on Airbus A319/A320/A330/A340 and Boeing 737NG/747–400/777. Selected by over 30 airlines, with more than 2,800 units, reportedly ordered by the end of 1999.

Contractor
Thales Avionics SA.

NC 12 airborne TACAN interrogator

Type
Aircraft navigation system, radio aids.

Description
The NC 12 is a very small and lightweight TACAN interrogator which fully complies with STANAG 5034, the improved MIL-STD-291. It is an all-digital system fitted with ARINC 429, ARINC 582 or MIL-STD-1553B standard outputs. It provides the pilot with digital distance and bearing relative to TACAN beacons on the NR 13 control box indicator. The NC 12 interrogator can be used either on new aircraft or for retrofit applications.

In air-to-air mode the pilot is supplied with distance to another aircraft fitted with an airborne beacon. A warning light indicates the presence of any jamming of the bearing and distance information.

The NC 12 interrogator is also fully adapted to advanced radio navigation through the provision of W and Z channels for DME-P compatibility.

Specifications
Dimensions:
(NR 13 control box) 144 × 63 × 57 mm
(NC 12 transmitter/receiver)
318 × 91 × 193.5 mm
Weight:
(NR 13 control box) 0.7 kg
(NC 12 transmitter/receiver) 5.5 kg
Power supply: 28 V DC, 1.4 A, 40 W

Thales' NC 12 airborne Tacan interrogator
0015352

Transmission power: 350 W
Frequency:
(transmit) 1,025–1,150 MHz
(receive) 962–1,213 MHz
Number of channels: 126 X channels, 126 Y channels
Accuracy: 0.1 n miles and 1°
Altitude: up to 100,000 ft

Status
In production and in service in Rafale, Mirage 2000–5 and Mirage 2000–9, Mirage F1-CT, Super Puma, Dauphin, UH-60, CH-47 and Tucano aircraft.

Contractor
Thales Communications.

RDN 85-B Doppler velocity sensor

Type
Avionic navigation sensor, Doppler.

Description
Thales' RDN 85-B is a single-box Doppler velocity sensor designed for use in helicopters, said to have good operating characteristics over calm seas. The radar interfaces with an ARINC 429 digital databus and along and across track velocities are also transmitted as DC signals for display by a hover meter and for coupling to an autopilot. The system incorporates BITE for in-flight and ground system checkout.

The company claims that the RDN-85-B provides a covert navigation capability which is insensitive to active countermeasures.

Specifications
Dimensions: 437 × 437 × 170 mm
Weight: <10 kg
Power supply: 28 V DC, 30 W
Frequency: 13.325 ±20 MHz
Velocity range: −50 to +350 kt
Altitude: up to 20,000 ft
Accuracy: 0.15% of velocity or 0.12 kt
Transmitter: Gunn diode oscillator
Reliability: >6,400 h MTBF

Status
In production for the French Navy export search and rescue Puma and Dauphin helicopters, French Army Super Pumas and the Fennec helicopter for Singapore. More than 450 units have been ordered.

Contractor
Thales Systemes Aeroportes SA.

RDN 2000 Doppler velocity sensor

Type
Avionic navigation sensor, Doppler.

Description
The RDN 2000 J-band Doppler velocity sensor has been designed for light, medium and heavy helicopters, where autonomous navigation and assistance for the pilot is needed for missions of all types over land and sea. It employs FM/CW techniques and digital signal processing to provide accurate ground velocity data. The lightweight single unit includes an antenna, transmitter/receiver, power supply and signal processor.

Coupled with a computer and an AHRS or INS, the RDN 2000 is claimed to provide a stealthy autonomous navigation system which is insensitive to countermeasures. In addition, it can operate with a local jammer.

The system provides high performance when flying over land or sea, especially over calm water and it has an automatic land/sea transition capability. This allows autopilot operation irrespective of flying conditions.

RDN 2000 Doppler velocity sensor 0010925

Specifications
Dimensions: 437 × 240 × 80 mm
Weight: 4.2 kg
Power supply: 28 V DC, 25 W
Frequency: J-band (10–20 GHz)
Accuracy (95%):
0.15% of Vt or 0.1 m/s (along track)
0.22 % of Vt or 0.1 m/s (across track)
0.15% of Vt or 0.2 m/s (vertical)
Reliability: 8,640 h MTBF

Status
The latest in Thales' range of Doppler velocity sensors for civil and military helicopters; it is claimed that the RDN 2000 radar is particularly suitable for all-weather maritime missions. Installed in the Eurocopter Cougar helicopter, first deliveries began in 1998 for France and Saudi Arabia.

Contractor
Thales Systemes Aeroportes SA.

SIGMA ring laser gyro inertial navigation systems

Type
Avionic navigation sensor, Inertial Reference System (IRS).

Description
The SIGMA family of inertial navigation systems implements a combination of high-performance ring laser gyro sensors, accelerometers and a multichannel GPS receiver. Such systems are intended for use in aircraft equipped with a multiplexed databus.

SIGMA systems offer the benefits of tight hybridisation between ring laser gyros, accelerometers and a GPS receiver supported by a multisensor Kalman filter for both alignment and navigation. The synergy between these three elements brings the following advantages: reduction in size, weight and power consumption through the integration of inertial and GPS functions; short alignment time; sensor performance and integrity monitoring for GPS and INS; automatic in-flight calibration of inertial sensors; long-term stability of inertial performance, and higher resistance to jamming with improved dynamic behaviour of the GPS.

SIGMA 95L (Light) inertial navigation system
0015349

All versions provide: aircraft position, velocity and attitude information; computation of navigation and steering information to waypoints; position updating by navigation fixes; terrain reference updating; some versions combine the navigation function and weapon delivery computations, consisting of ballistics, determination of release point, ripple spacing of weapons, safety pull-up information; head-up display information for target acquisition and commands for the blind release of weapons; attack modes; air data computations and multiplex bus control.

SIGMA 95MF inertial nav/attack system

The SIGMA 95MF (MultiFunction) uses three high-accuracy ring laser gyros and three accelerometers and is fitted with an embedded GPS receiver. SIGMA 95MF is the heart of MAESTRO, SAGEM's avionics system. SIGMA 95MF is a highly integrated system which provides a combination of high-performance, hybrid inertial navigation, attack computations and mission management tailored to advanced multirole combat or tactical transport aircraft.

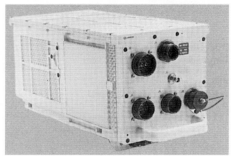

SIGMA 95MF (MultiFunction) inertial nav/attack system 0015347

Specifications
Dimensions: 209 × 200 × 406 mm
Weight: 17 kg
Power supply: 28 V DC, <90 W (115 V AC/400 Hz optional)
Accuracy: 0.6 n mile/h (inertial mode)

Status
In service as part of the Pakistan Air Force Mirage III upgrade.

SIGMA 95N inertial navigation system
The SIGMA 95N (Navigation) uses three high-accuracy ring laser gyros and three accelerometers. It performs hybrid inertial/ GPS navigation. It interfaces with the aircraft avionics systems through a MIL-STD-1553B multiplexed databus. The versatility of its interfaces enables the SIGMA 95N system to be integrated easily with a wide range of carriers, avionic configurations and therefore operational situations.

Specifications
Dimensions: 209 × 200 × 385 mm
Weight: <15 kg
Power supply: 28 V DC, <45 W
Accuracy: 0.6 n mile/h (inertial mode)

Status
In production for the French Air Force/French Navy Rafale aircraft (under the programme name RL-90), for the Cougar helicopter and for several export customers, including the Egyptian Air Force for its Mirage 5 fighters.

SIGMA 95L inertial navigation system
The SIGMA 95L (Light) is a compact inertial reference system fitted with a SAGEM-embedded GPS. It has been designed for applications on helicopters and fixed-wing aircraft requiring a lightweight, small size, high-performance navigation system.

SIGMA 95L can function as a self-contained attitude and heading reference system or as a full inertial navigation system. It performs hybrid inertial/GPS navigation. It interfaces with the aircraft avionics system through a MIL-STD-1553 multiplexed databus and ARINC 429.

Specifications
Dimensions: 180 × 125 × 280 mm
Weight: <8.5 kg
Power supply: 28 V DC, <35 W
Accuracy: 1 n mile/h (inertial mode)

Status
Off-the-shelf production. A (SAPHIR) version of SIGMA 95L, which includes an integrated GPS receiver, is in production for the NH 90 helicopter programme, where it will provide the main navigation data and inertial references for the fly-by-wire system. The SIGMA 95L is also qualified on the Mi-17 and Mi-24 helicopters.

Contractor
SAGEM Défense Securité, Navigation and Aeronautic Systems Division.

SIGMA 95N (Navigation) inertial navigation system (SAGEM) 1146549

Stratus and Totem 3000 flight systems

Type
Avionic navigation sensor.

Description
Thales Avionics produces three types of Ring Laser Gyros (RLG):
1. A three-axis monolithic PIXYZ®, 14 cm path length, 0.1°/h in run stability
2. A three-axis monolithic PIXYZ®, 22 cm path length, 0.001°/h in run stability
3. A single axis, 33 cm path length, 0.001°/h in run stability.

With this family of RLG, Thales produces several inertial reference units for a large range of applications.

Stratus
Stratus is an attitude and heading reference system that uses the RLG PIXYZ® 14 cm path length. It is used as a basic reference system for attitude and hybrid navigation (with Doppler radar and/or air data) on military helicopters, advanced trainers and missile systems.

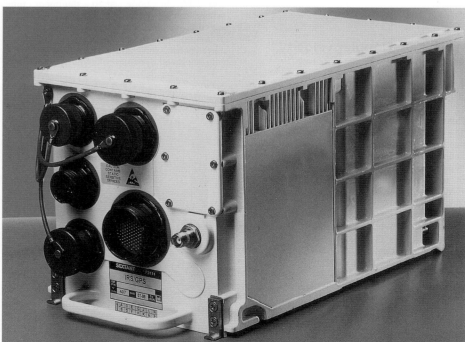

Thales' Totem 3000 navigation and flight management system 0079262

Totem 3000 navigation and flight management system   0001317

Stratus attitude and heading reference system   0001315

Specifications
Attitude: 0.2°
Magnetic heading: 0.3°
Position: 1% of distance (with Doppler)
Weight: 5.2 kg
Power supply: 115 V AC, 45 VA
Interfaces: ARINC 429 and MIL-STD-1553B
Environment: as per MIL-STD-810D

Status
Selected for the Tiger and Rooivalk helicopters.

Totem 3000
Totem 3000 is an inertial/GPS navigation system for high-performance military aircraft. It utilises the PIXYZ® 22 cm path length RLG and the Thales Topstar® 1000 GPS receiver board. The use of these two highly integrated and high-performance subassemblies leads to significant improvement of cost, size, and reliability compared with conventional single axis RLG.

All the functions are in compliance with SNU 84-1 and STANAG 4294

Specifications
Attitudes/heading: 0.05°
Position: 0.5 n miles/h CEP (inertial mode)
   21 m (95%)(inertial/GPS PPS mode)
Velocity: 0.7 m/s RMS (inertial mode)
   0.1 m/s (95%)(inertial/GPS PPS mode)
Reliability: 5,000 hr MTBF
Dimensions: 177.8 × 177.8 × 279.4 mm
Weight: 11 kg including GPS
Power consumption: 50 W including GPS
Dual voltage: 115 V AC/400 Hz; 28 V DC
Interfaces: standard MIL-STD-1553 B and ARINC 429
Environment: as per MIL-STD-810 D

Status
Inertial RLG and GPS systems have been installed on many platforms including: Mirage 2000 and F1, MiG-AT and MiG 21, Alpha Jet, C130, Tiger, Rafale, Iryda I-22, Rooivalk, Ariane, Exocet, Scalp, EG/Storm Shadow.

Contractor
Thales Avionics SA.

TLS-2040 MultiMode Receiver (MMR)

Type
Aircraft navigation system.

Description
The TLS-2040 MultiMode Receiver (MMR) is a new VOR/ILS receiver designed to replace the existing ARINC 547 VOR/ILS receivers. It fully complies with ICAO Annex 10 FM immunity requirements and includes a VHF receiver for the DGPS datalink. It is form-and-fit exchangeable with existing ARINC 547 receivers.

The TLS-2040 receiver can optionally be configured with Marker, MLS, DGPS and GPS functions by simply adding modules.

The TLS-2040 utilises the latest technology developed for commercial aviation and military programmes. The TLS-2040 belongs to the TLS-2000 product line.

Specifications
Basic functions: VOR/ILS/VHF receiver for DGPS
Options: Marker, MLS, DGPS and GPS
Dimensions: 1/2 ATR short

Weight: <4 kg
Power: 28 V DC
Interfaces: As defined by ARINC 547, MIL-STD-1553B, ARINC 429

Status
In service.

Contractor
Thales Communications.

Topstar® family of GPS receivers

Type
Aircraft navigation system.

Description
Topstar® is a family of multichannel continuous tracking sensors. They provide time, three-dimensional position, and speed information at a data rate of up to 10 Hz.

They provide data for speeds up to 800 kt, and for high-acceleration applications. They provide accuracy in compliance with international standards. Additional functions such as DGPS, relative navigation, combined GPS/GLONASS processing, and RAIM are offered as options.

Four variants for aircraft application are available:

Topstar® 100
Topstar® 100 is a stand-alone sensor for high-dynamic military applications in aircraft, helicopters and missiles.

Specifications
SPS or PPS operation
8 parallel channels
Dynamic operation: 1,500 m/s – 10 g
Accuracy: as per STANAG 4294
Dimensions: 130 × 240 × 80 mm
Weight: 2 kg
Power supply: 28 V DC, 20 W
Interfaces: MIL-STD-1553B, ARINC 429, RS-422
Environmental: MIL-STD 810-C

Status
Topstar® 100 is in production and in service in Mirage F1, Mirage 2000 and Rafale aircraft, Cougar, Rooivalk and Tiger helicopters and the Apache missile.

Topstar® 100-2
Topstar® 100-2 is a new-generation, stand-alone receiver.

Specifications
SPS or PPS operation
20 parallel channels
Dynamic operation: 1,500 m/s – 10 g
Accuracy: as per STANAG 4294
Dimensions: 221.5 × 162 × 68 mm
Weight: 1.6 kg
Power supply: 28 V DC, 20 W
Interfaces: MIL-STD-1553B + PTTI, ARINC 429 + PTTI, other by request
Environmental: MIL-STD 810-D

Status
Topstar® 100-2 is in production for French Transall aircraft and Tiger helicopters.

Topstar® 1000
Topstar® 1000 is a new version of Topstar® 100, designed as a PCB module to be integrated in host equipment like IRS. Topstar® 1000 interfaces with host computers through a dual-port RAM and RS-422 databus. It has growth provision for GLONASS reception and processing.

Specifications
SPS or PPS operation
20 parallel channels
Dynamic operation: 1,500 m/s – 9 g
Accuracy: as per STANAG 4294
Interfaces: Dual port RAM, RS-422
Environmental: MIL-STD 810

Topstar® 100   0001320

The Thales GPS Topstar® 200 A with antenna   0079263

Dimensions: 150 × 150 × 19 mm
Weight: 0.75 kg
Power supply: 5 V DC, 15 V DC, 15 W

Status
In production. Topstar® 1000 is embedded in TOTEM 3000 IRS, which has been selected for several retrofit programmes, including Alpha Jet, C-130, and MiG-21. Topstar® 1000 has also been selected for the Storm Shadow/SCALP EG and other missile programmes.

Topstar® 200 A
Topstar® 200 A is an ARINC 743A-compliant stand-alone sensor for civil aviation.

Specifications
C/A (L1) + W.A.A.S. capability
15 parallel channels
Dynamic operation: 2,000 kt, 3 *g*
Accuracy: as per ARINC 743A, RAIM GIC and differential RAIM; and SCAT 1c RTCA DO 217 optional capabilities
Dimensions: ARINC 743A 2MCU format or ARINC 743A alternate format (190 × 240 × 63 mm)
Weight: 2 kg
Power supply: 28 V DC, 15 W
Interfaces: as per ARINC 743A
Environment: as per RTCA DO 160c
    TSO C 129 C1 label

Topstar® 2020
Topstar® 2020 is a version of Topstar® 200, designed as a PCB to be integrated in host equipment such as the IRS or multimode receiver MMR.

Status
Selected for the Boeing MD-82 and for the Airbus family.

Topstar® 250
Topstar® 250 is a development of Topstar® 200 that has been upgraded with: GIC software capabilities; SBAS message processing; augmented PVT for CAT I precision approach; and augmented integrity. It is designed to be connected through and ARINC 429 link to a ruggedised PC acting as a control and display unit. It processes acquisition and tracking of GPS L1 C/A (eight channels) as well as two channels of GEO signals. It operates either in absolute or differential mode to obtain the best performance for accuracy and availability. It ensures integrity based on RAIM or SBAS algorithms. Using the

PC control and display unit, Topstar® 250 is able to perform off-line analysis using raw data collection as well as real-time analysis.

Status
Topstar® 250 has been developed to support the validation of the EGNOS System Test Bed (ESTB) and to perform demonstration to the Aeronautical Users Community in real operational conditions of the experimental service provided by ESTB.

Contractor
Thales Avionics SA.

Type 130 Air Data Computer (ADC)

Type
Avionic navigation sensor, Air Data Computer (ADC).

Description
The modular design Type 130 air data computer is intended for use on new-generation aircraft and for the retrofit of fighter aircraft, and has already been chosen for various Mirage retrofits.

The ADC comprises two pressure sensors for measuring static and total pressures; a CPU board including arithmetic unit; memories and sensor measurement circuits; an input/output board for all analogue and digital multiplexed bus interfacing; and a power supply board. A large number of built-in tests is available and permanent monitoring of all functional circuits takes place during flight, the results being expressed as maintenance words stored in a protected memory or dispatched on the digital lines.

Pressure sensors are Thales' vibrating quartz blade Type 51.

Specifications
Dimensions: 2 MCU
Weight: 4.6 kg
Power supply: 115 V AC, 400 Hz, single phase, 30 VA
Reliability: 10,000 h MTBF

Status
In production for retrofit of Mirage aircraft.

Contractor
Thales Avionics SA.

Type 300 Air Data Unit (ADU)

Type
Avionic navigation sensor, Air Data Computer (ADC).

Description
The Type 300 ADU is designed to measure static and differential (or pitot) pressures and impact temperature. From this data, the ADU computes the true airspeed and provides air data parameters through an ARINC 429 bus. Analogue outputs are available as an option.

The unit is ideally suited to provide the primary reference in helicopters and for retrofits in military aircraft.

Specifications
Weight: 1.2 kg

Status
In production and in service for Super Puma, Tiger and Rooivalk helicopters.

Contractor
Thales Avionics SA.

Uliss inertial navigation and nav/attack systems

Type
Avionic navigation sensor, Inertial Reference System (IRS).

Description
Uliss modular systems all employ high-accuracy inertial components consisting of two dynamically tuned gyros and three dry accelerometers, a microprocessor-controlled computer working at 1 Mops with EPROM memory and highly integrated and hybrid circuits. As an option, Uliss can be equipped with an embedded GPS receiver for high-performance INS/GPS coupling and a higher-performance RISC processor using Ada language.

Uliss systems fall into two categories, in both of which the main inertial navigation unit is contained within a 3⁄4 ATR short case. In the first category are navigation versions with a position accuracy of better than 1 n mile/h. In the second category the navigation function is combined with the computation necessary for weapon delivery. Alignment time is 90 seconds for stored heading and 5 to 10 minutes for self-contained gyrocompassing. Standard interfaces permit the systems to be linked to other equipment via MIL-STD-1553B databusses or ARINC serial data lines. A failure detection system can detect faults at module level with a high level of confidence, isolate them and signal their presence on a magnetic annunciator without external test equipment.

More than 80 per cent of the components and subassemblies are common to all members of the Uliss family, the principal differences being in specific interfaces and computation functions.

In all, close to 2,000 Uliss units are in operation on 25 different types of aircraft.

Uliss 45 inertial navigation system
The Uliss 45 system has been optimised for high accuracy in long-range navigation and certain other special applications. Its interfaces comply with ARINC 561 and embody significant flexibility, for example, in order to communicate with two DME or Tacan receivers, with Kalman filtering for better accuracy. The equipment is used on long-range transport aircraft and as an accurate position and velocity reference for flight development purposes.

Specifications
Dimensions:
(navigator) 420 × 194 × 191 mm
(control/display unit) 209 × 114 × 127 mm
Weight:
(navigator) 16 kg
(control/display unit) 3 kg
Power supply: 115 V AC, 400 Hz, 250 VA
Accuracy: 1 n mile/h CEP for flights of up to 10 h duration.

Status

In service in Transall C-160 and Boeing KC-135 tankers of the French Air Force, and ATL-2 aircraft of the French Navy.

Uliss 52 inertial navigation system

The Uliss 52 navigation system is designed for high-performance combat aircraft. The program is based on Ada. Avionics information and commands are distributed by a digital multiplexed databus. It comprises three units: a UNI 52 inertial navigator, a PCN 52 control/display box and a PSM 52 mode selector fitted with an automatic insertion module to allow information such as flight plan, system data and maintenance information to be fed in.

Specifications

Dimensions:
(navigator) 386 × 194 × 191 mm
(control/display unit) 208 × 114 × 127 mm

Weight:
(navigator) 15 kg
(control/display unit) 3 kg
Power supply: 115 V AC, 400 Hz, 3 phase, 250 VA
Accuracy: 1 n mile/h CEP

Status

In production for French Air Force Dassault Mirage 2000DA, N and D aircraft and the export version Mirage 2000-5 and retrofit of Mirage 2000-5F of the French Air Force.

Uliss 92 inertial nav/attack system

The Uliss 92 combines in a single box all the functions of an inertial navigation and fire-control system. It is based on Ada programming and comprises three units: UNA 92 inertial/attack box, PCN 92 control/display unit and PSM 92 mode selector.

Specifications

Dimensions:
(navigator) 386 × 194 × 191 mm
(control/display unit) 216 × 116 × 153 mm
(mode selector) 151 × 41 × 135 mm
Weight:
(navigator) 16 kg
(control/display unit) 3.5 kg
(mode selector) 1 kg
Power supply: 115 V AC, 400 Hz, 3 phase, 220 VA
Accuracy: 1 n mile CEP position, 5 mrad weapon delivery

Status

In production for various Mirage upgrade programmes.

Contractor

SAGEM SA, Defence and Security Division.

Germany

3300 series VHF navigation systems

Type

Aircraft navigation system, radio aids.

Description

The series 3300 navigation systems have been designed for use in fixed- or rotary-wing aircraft, and have been certified for high-altitude operation. The series includes a variety of VOR/ILS receivers, indicators, special application converters and a glide slope receiver, which can be combined to tailor systems to meet the exact requirements of specific installations. Interfaces are provided for most types of electromechanical or electronic HSI, RMI and CDI displays.

Complete VOR/LOC receiver, VOR/LOC converter and glide slope receiver systems are housed in a single compact unit, which complies with ARINC standards. They do not require the individual control boxes and remote receiver units with interconnecting cables, or external forced-air cooling, which are typical of older navigation receiver designs in this performance class.

Clear LCD displays, which are easily read in the brightest sunlight show both active and standby frequencies. Remote tuning via databus can be provided, and tandem tuning heads are available for dual-cockpit installations. Built-in test functions for the displays and for the proper operation of the receivers and converters are standard features.

The NR 3300 navigation receivers are certified to the stringent requirements of all applicable FAA TSO, LBA, ICAO and RTCA specifications.

Specifications

Dimensions: 146 × 47.5 × 225 mm (excluding clearance for connectors and cables)
Weight:
(NR 3300-(1) (3)) 1 kg
(NR 330-(2) (4)) 1.2 kg
Power: 25-30 V DC; 26 V AC, 400 Hz, 10 mA for RMI converter
Temperature: −20 to +55°C operating
Altitude: >50,000 ft

Frequency range:
(VOR/LOC) 108.00–117.95 MHz (200 channels at 50 kHz spacing)
(glideslope) 329.15–335.00 MHz (40 channels at 150 kHz spacing)
Accuracy: <2° bearing error
Certification: TSOs C35d Class A, C40b, C34d Class D, C36d Class D; RTCA DOs -153A, -131A Class D, -132A Class D, -160A, 178A

Status

In service and production.

Contractor

Becker Avionic Systems.

ADF 3500 system

Type

Aircraft navigation system, radio aids.

Description

In addition to the standard frequency range of 190 to 1,799.5 kHz, the ADF 3500 also receives the international maritime distress frequency of 2,182 ±5 kHz, making it suitable for search

and rescue, and other offshore operations. It is certified for high-altitude operation in turbine-powered aircraft.

The complete system is housed in a single compact unit, which complies with ARINC standards. It does not require remote boxes, interconnecting cables, or external forced-air cooling.

Clear LCD displays, which are easily read in the brightest sunlight show both active and standby frequencies. Preselection of a standby frequency enables instant switch over to a second station for position fixing by cross bearings. A low-profile combined sense and loop antenna is suitable for mounting on high-speed aircraft. Becker RMI converters are available to enable the system to drive most types of indicators.

The ADF 3500 systems are certified to the requirements of applicable FAA TSO, JTSO, RTCA, EUROCEA and FTZ specifications.

Specifications

System model numbers	System components
ADF 3502-(1)	Standard version
	AD 3502 receiver
	AN 3500 antenna
	ID 3502 indicator

NR3300 and Indicator   0581495

RA 3502 ADF receiver system   0001323

ADF 3503-(1)	For RMI with standard synchro input AD 3502 receiver AN 3500 antenna AC 3503-(1) converter
ADF 3504-(1)	For RMI with standard 2.5 V DC Sin/Cos input AD 3502 receiver AN 3500 antenna AC 3504-(1) converter
ADF 3504-(2)	For RMI with standard 5–10 V DC Sin/Cos input AD 3502 receiver AN 3500 antenna AC 3504-(2) converter
CU 3502-(01)	Tandem Control Unit can be added to the system to enable dual control operation

Dimensions
AD 3502 receiver 146 × 47.5 × 245 mm
Weight 1.0 kg
ID 3502 indicator 82.55 × 82.55 × 135 mm
Weight 0.5 kg
AN 3500 antenna 190 × 54 × 330 mm
Weight 1.7 kg
Supply voltage: 25–30 V DC
(20 V DC in emergency)
26 V AC, 400 Hz, for AC 3503-(1) converter
Power
consumption: At 27.5 V DC, without panel lights
ADF 3502 650 mA
ADF 3503 1.15 A
ADF 3504 0.6 A
Frequency
Ranges: 190–1,799.5 kHz and
2,182 kHz ± 5 kHz
Channels
Spacing: 500 Hz
Bearing
accuracy: <3° at 70 µV
190–850 kHz
<8° at 70 µV
>850 kHz

Contractor
Becker Avionic Systems.

AeroNav® Integrated Navigation System (INS)

Type
Aircraft navigation system.

Description
AeroNav® is a TSO C-129a certified GSP navigation system. It provides a 12-channel GPS receiver, integrated with an Inertial Measurement Unit (IMU) for improved navigation performance and Aircraft Autonomous Integrity Monitoring (AAIM). Flight management software with Jeppesen database supports the crew during mission planning and in flight. The system comprises a Navigation and Communication Unit (NCU) and a Control and Display Unit (CDU), which is NVIS compatible. AeroNav®'s open interface architecture facilitates integration with moving map systems, such as the DaimlerChrysler Aerospace DKG-3/4 (see separate entry) and special mission equipment (FLIR) or communications equipment (Satcom/GSM).

AeroNav® is certified for B-RNAV and IFR non-precision approaches.

Specifications
Dimensions:
NCU 214 (H) × 126 (W) × 360 (D) mm (½ ATR short)
CDU 66 (H) × 146 (W) × 185 (D) mm
Antenna diameter 88.9 mm, height 14.5 mm
Weight:
NCU 5.9 kg
CDU 1.4 kg
Antenna 0.15 kg

Status
In production and in service.

Contractor
Aerodata Flugmesstechnik GmbH.

Air data computer/air data transducer for the JAS 39 Gripen

Type
Avionic navigation sensor, Air Data Computer (ADC).

Description
The air data computer/air data transducer for the Saab JAS 39 Gripen combine the primary air data computer with an air data transducer channel as back-up in a single electronics box. Both functions are fully independent of one another and are electrically decoupled. The air data computer contains two high-precision pressure sensors which convert the static pressure and the total pressure into an electronic frequency signal. The signal is digitised and fed to a 16-bit microprocessor. Using software algorithms, the classic air data are computed, taking into account further parameters such as barometric altitude correction and total air temperature. Air data such as speed, altitude and Mach number are passed to aircraft systems via MIL-STD-1553B interfaces.

The other part of the unit contains two additional identical pressure sensors which operate in the same way as in the air data computer. In contrast to the computer, however, the final data are not computed. Instead the static and total pressures are passed to the flight control computer via an ARINC 429 interface. Due to the complete separation of the functions of the air data computer and air data transducer, the air data transducer has its own power supply unit, a separate processor and a separate software programme. In the flight control computer, the pressure values supplied by the air data transducer are used to calculate the air data, taking into consideration the barometric altitude correction and total air temperature, and the result compared with the values from the air data computer.

Both channels of the Nord-Micro air data computer and air data transducer have a highly developed self-testing capability and are able to check their own correct function or to provide information on any faults which may have occurred. In the event of an error, the faulty function is detected. The use of hybrids, gate arrays and LCC circuits enable the units to be housed in a 4 MCU box.

Status
In service on the Saab JAS 39 Gripen.

JAS 39 Gripen air data computer/air data transducer 0015274

Contractor
Nord-Micro AG & Co. OHG.

Air data systems

Type
Avionic navigation sensor, Air Data Computer (ADC).

Description
Nord-Micro AG & Co. OHG manufactures a range of air data systems including Air Data Computers (ADC) and engine air intake control computers for military subsonic and supersonic aircraft. Modular design concepts are used, together with ASICS and hybrid circuits to produce lightweight, compact, low power consumption systems. For critical data, redundant dissimilar algorithms are used to avoid single point failure and improve fault tolerance. Extensive built-in test and health monitoring functions are also included.

Tornado air intake control system
The Tornado air intake control system comprises two digital air intake control units and one pilot's panel. It controls the engine intake ramp position via an electro-hydraulic actuator.

JAS39 Gripen air data computer/air data transducers (ADC/ADT)
The ADC/ADT is configured as two separate computers in one common housing. They communicate with the fly-by-wire flight control computer via ARINC 429 as well as MIL-STD-1553B databus interfaces.

Eurofighter Typhoon ADCs
Two ADCs are used for calibration of the Air Data System on board the first three Eurofighter Typhoon development aircraft.

Status
In service.

Contractor
Nord-Micro AG & Co OHG.

AeroNav® Integrated Navigation System (INS)
0531966

Tornado air intake control system
0018246

Inertial Measurement Unit (IMU) for Eurofighter Typhoon

Type
Avionic navigation sensor, Inertial Reference System (IRS).

Description
The IMU is part of the Eurofighter Typhoon quadruplex fly-by-wire flight control system. It has a strapdown design and is housed in a single box which is internally separated into four channels.

Measurement of aircraft body angular rates and linear accelerations and calculation of attitude and heading angles are accomplished by the IMU. It performs autonomous heading alignment at system start up. To provide damping of air data disturbances and to provide a back-up source of air data. The IMU computes angle of attack, sideslip, TAS and altitude data based on inertial measurement and augmented by signals from the air data sensors, when available.

The IMU contains four dual-axis dynamically tuned LITEF gyroscopes and eight single-axis LITEF pendulum accelerometers mounted in a skewed-axis orientation. This combination of accelerometers and gyroscopes is about half the number of sensors used in conventional systems architecture and therefore represents a considerable weight saving.

Status
In production and in service in the Eurofighter Typhoon.

Contractor
LITEF GmbH.

LCR-92 µAHRS Attitude and Heading Reference System

Type
Avionic navigation sensor, Attitude Heading Reference System (AHRS).

Description
The LCR-92 is an extremely small and light strapdown reference system using LITEF fibre optic gyros. It provides pitch, roll, magnetic heading information and angular rates around the aircraft body axes, replacing conventional vertical/directional gyro installations in one single box.

The system features full compatibility with modern digital cockpit instruments with an ARINC 429 databus. Analogue outputs for use with conventional mechanical indicators are available as an option.

Specifications
Dimensions: 278 × 102 × 128 mm
Weight: 2.1 kg
Power supply: 28 V DC, <25 W
Accuracy (2σ): heading 1°.
Reliability: >6,700 h MTBF
Compliance: RTCA/DO-160

Status
In production and in service in the Raytheon T-6A Texan II JPATS and the Pilatus PC-12 aircraft, and in the Sikorsky S-76 helicopter.

Contractor
LITEF GmbH.

LCR-93 µAHRS Attitude and Heading Reference System

Type
Avionic navigation sensor, Attitude Heading Reference System (AHRS).

Description
The LCR-93 is based on the LCR-92 (see separate entry). It has the same housing and dimensions and offers the same interface configurations. The mechanical layout and connectors ensure mechanical plug-in interchangeability with the LCR-92.

In addition to the LCR-92 capabilities, the LCR-93 standard version provides body axis referenced accelerations as well as inertial altitude and inertial vertical speed. The LCR-93 performance is increased by using air data augmentation.

The LCR-93V combines GPS and air data inputs to provide the necessary outputs to Head-Up Display (HUD) with a speed vector indication. This version of the system gained TSO approval in early 1998. Data optimisation to enhance the performance of the system is achieved by merging low update rate GPS data with high-speed values of inertial angular rates and accelerations in a Kalman Filter (KF). During a customer flight test programme, the performance of the LCR-93V was reported to be extremely good. It is described as an ideal solution for trainer and business aircraft that have a HUD but no INS or IRS.

Specifications
Dimensions: 278 × 102 × 128 mm
Weight: 2.2 kg (2.5 kg with synchro board)
Accuracies (95%):

	Basic	Normal	
Attitude	0.3°	0.3°	static
	1.0°	0.5°	dynamic
Heading	1.0′	1.0′	static
	2.0′	2.0′	dynamic
DG mode	<5°/h	<5°/h	
Angular rates	0.1°/s	0.1°/s	1% max
Acceleration	5 mg	5 mg	1% max

Status
In production and in service in the Cessna Excel and Pilatus PC-7 Mk II and PC-9 aircraft.

Contractor
LITEF GmbH.

LCR-98, VG/DG replacement system

Type
Avionic navigation sensor, Attitude Heading Reference System (AHRS).

Description
The LCR-98 is a strapdown Attitude and Heading Reference System (AHRS), specifically designed to replace Vertical and Directional Gyros (VG/DG) in aircraft like the Dassault Falcon 20, Canadair Challenger 601 and Gulfstream GII and GIII.

Featuring Fibre Optic Gyro (FOG) technology, the LCR-98 offers improved aircraft attitude and heading performance, together with higher reliability and lower cost of ownership.

The LCR-98 is designed as form, fit and function replacement of the VG-311, additionally incorporating the directional gyro function as a substitute for C9-C11 and MHRS-type gyros, using existing aircraft wiring and re-routeing the DG harness.

Operating the LCR-98 is virtually transparent to the pilot but with a tenfold increase in Mean Time Between Failure (MTBF) over mechanical VG/DGs.

Specifications
Dimensions: 219 (L) × 216 (W) × 155 (H) mm
Weight (max): 6 kg
Installation: fits into VG-311 cradle, using existing AC wiring and connectors
Power: 115 V AC, 400 Hz, 70 W with full external load
MTBF: >12,000 h
Outputs:
(synchro) pitch, roll, 2 × headings
(analogue) 50 and 200 mV AC pitch and roll
(discretes) attitude and heading warning and interlocks
Accuracy:
Attitude: 0.5° static, 1.0° dynamic, (95%)
Heading:
(slaved) 0.75° (rms)
(DG) 5.0°/h (95%)
Qualification:
RTCA/DO-160C including HIRF and lightning, RTCA/DO-178 A, Level 1 for flight critical software
Certification: TSO C4c, C5e and C6d

LCR-92 µAHRS Attitude and Heading Reference System

0001330

For details of the latest updates to *Jane's Avionics* online and to discover the additional information available exclusively to online subscribers please visit
jav.janes.com

Status

Certified in 1998. Offered for the Dassault Falcon 20 and Gulfstream GII/GIII aircraft. Future systems will incorporate Global Navigation Satellite System (GNSS) data. These hybrid navigation systems will combine the advantages of inertial sensors – high dynamics, reliability and jamming resistance in the short term – with the accuracy and long-term stability of GNSS. Flight trials with a LITEF prototype have shown that hybrid navigation using differential GPS data supports Category I precision approaches, even with GNSS data outages of up to 90 seconds.

Contractor

LITEF GmbH.

MT-Ultra moving map system

Type

Aircraft navigation system.

Description

The MT-Ultra moving map display is a single-box unit, designed for installation in the cockpit in a standard 6¼ in radio component space. It includes a highly integrated navigation computer and display system, which presents navigational data to the pilot on a full VGA resolution, sunlight readable, screen.

The MT-Ultra moving map display includes a comprehensive worldwide navigational database, supplied by MOVING-TERRAIN, with a 12-channel GPS receiver system. Control is by single alphanumeric keys located around the edge of the display, to minimise workload in high stress conditions. Features that can be selected by the pilot in flight include: zoom in/zoom out, display orientation, instant display of current position, flight planning and replanning facilities.

Data loading and updating is by CD-ROM, and the system can be integrated with other onboard GPS receiver systems if required. A quick release capability is provided to optimise the installation, operation and flight planning options.

Specifications

Dimensions: 158 (W) × 150 (H) × 48 (D) mm
Weight: 1.15 kg
Power: 12–28 V DC, 16 W
Screen: TFT colour display 6.5 in diagonal
Resolution: 640 × 480 pixels
Viewing angle: >60°

Status

In production and in service.

Contractor

MOVING-TERRAIN® Air Navigation Systems GmbH.

NR 3300 series VHF navigation receivers

Type

Aircraft navigation system, radio aids.

Description

The NR 3320-(02)-(01) and NR 3330-(02)-(01) single-block units are new navigation receivers designed as retrofit systems for the Becker NAV 2000 series receivers with which they are pin compatible without mechanical modification.

The NR 3320-(02)-XXX provides a composite navigation signal with glide slope receiver capability, while the NR 3330-(02)-XXX provides its composite navigation output without the glide slope.

Active and standby (preset) frequency read out is given on two sunlight-readable LCDs.

Specifications

Dimensions: 126 × 46 × 186 mm
Weight: 1.2 kg
Power supply: 13.5 V AC/27.5 V DC
NAV Receiver:
Frequency range: 108–117.950 MHz
Channels: 200
Channel spacing: 50 kHz
Memory channels: 20
Glide slope Receiver:
Frequency range: 329.150–335 MHz
Channels: 40
Channel spacing: 150 kHz

Contractor

Becker Avionic Systems.

Remotely-controlled NAV 5300 VOR/ILS navigation systems

Type

Aircraft navigation system, radio aids.

Description

The NAV 5300 systems are part of the Becker Compact Line of equipments that can be customised at installation to customer requirements. They consist of the CU 5301 (control unit) and the remotely-controlled RN 33XX navigation receiver.

The system is ideally suited to installations where minimum panel space is to be used. They utilise a small, lightweight CU 5301 control unit, which fits into a standard 57 mm round instrument panel cut-out, and is only 63.5 mm deep. Lightweight remote receivers, with VOR/LOC or VOR/ILS beacon capabilities are mated with this control unit, to complete the systems.

The receivers can be installed at any convenient place in the aircraft. The CU 5301 control unit uses a clear, high-contrast, double line LCD display, which is readable under all lighting conditions, even bright sunlight.

Both active and standby frequencies are displayed, and can be transferred by a single stroke of the 'flip-flop' button. Up to 20 preset frequencies can be entered from the front panel, and stored in non-volatile memory. Parallel outputs are provided for automatic DME

channelling. The steering signals and flag drive outputs are compatible with most commonly used CDI, HSI, flight director, and autopilot systems.

All systems will be JTSO certified for either VFR or IFR use in all types of fixed-wing and rotary-wing aircraft, and comply with ICAO requirements for VHF radios.

The RN33XX receiver systems can be combined with other Becker PrimeLine avionics systems, such as VHF-COM, ADF and ATC transponders which have similar control units. The remote receiver units can be controlled by other types of CDU or FMS.

Specifications

RN 3330-(1): VOR/LOC receiver and converter
RN 3320-(1): VOR/LOC receiver and converter and GS receiver
CU 5301: control unit
RM 3300-( ): converters to drive RMI
Dimensions:
CU 5301 61.3 × 61.3 × 62 mm
RN 3320/30 134 × 50 × 243 mm
RM 3300-( ) 134 × 50 × 214 mm
Operation voltage:
10 to 32 V DC 26 V AC, 400 Hz, 10 mA (RMI converter)
Power consumption:
At 28 V DC.
without panel lights
RN 3330-(1) 250 mA
RN 3329-(1) 330 mA
Frequency ranges:
VOR/LOC 108.00–117.95 MHz
200 channels
Glide slope 329.15–335.00 MHz
40 channels
Channel spacing
VOR/LOC 50 kHz
Glide slope 150 kHz
Maximum altitude: 50,000 ft
Control interface: RS 422, serial line
Weight:
(RN 3320/3330): 0.81 kg/0.78 kg
(CU 5301): 0.26 kg

Status

In production and in service.

Contractor

Becker Avionic Systems.

NAV 5300 VOR/ILS navigation system   0001325

India

GNS-642 Global Navigation System

Type
Avionic Global Positioning System (GPS).

Description
The GNS-642 system tracks all satellites in view to provide complete positional and navigation data. It is able to handle up to 100 waypoints and 10 routes.

Specifications
Receiver: 5 channel, L1 frequency
Sensitivity: –160 dBW, SNR 10 dB
Horizontal position accuracy: 25 m (RMS)
Vertical position accuracy: 50 m (RMS)
Velocity accuracy: 0.15 kt (RMS)
Output data: position, time, bearing, course and distance to waypoint, ground speed, ETA
Routes: 10 routes, 10 waypoints per route
Output display: 3 lines, 16 characters LED
Dimensions and weight:
(receiver) 124 × 120 × 120 mm; weight <1.5 kg
(display unit) 146 × 56.5 × 90 mm;
weight <0.5 kg
Power supply: 27.5 V DC, 20 W
Interfaces: RS-232 with NMEA-0183

Status
In production and in service.

Contractor
Hindustan Aeronautics Ltd, Avionics Division.

Hindustan's GNS-642 Global Navigation System 0002441

RAM-700A radio altimeter

Type
Avionic navigation sensor.

Description
The RAM-700A radio altimeter features all-solid-state modular construction and operates on a frequency of 4.2 to 4.4 GHz. It provides indication of height over terrain up to 5,000 ft, to an accuracy of ±2 ft up to 100 ft and four per cent above 100 ft.

Specifications
Dimensions: 124 × 97.2 × 366 mm
Weight: <4 kg
Power supply: 115 V AC, 400 Hz, single phase, 80 VA

Status
In service.

Contractor
Hindustan Aeronautics Ltd, Avionics Division.

LCA cockpit 0052370

Hindustan's RAM-700A radio altimeter 0044968

RAM 1701A radio altimeter

Type
Avionic navigation sensor.

Description
The RAM 1701A developed by HAL, Hyderabad for the LCA programme, is an upgraded version of the existing RAM-700A, incorporating state-of-the-art technology. It can be interfaced to an onboard mission computer. The equipment is modular in design, highly reliable and small in size.

The RAM 1701A radio altimeter is being developed for the Indian LCA 0132902

Specifications
Transmitter frequency: 4,200–4,400 MHz
Power: <35 W
Power supply: 22 V DC to 29 V DC
Altitude range: 0–1,500 m
Altitude accuracy: 0–30 m (±1 m +3%)
30 to 1,500 m (±4%)
Pitch limit: ±20°
Roll limit: ±25°
Dimensions:
(without mounting tray) 123 × 175 × 176 mm
(with mounting tray) 132 × 230 × 194 mm
Weight:
(without mounting tray) 3.5 kg
(with mounting tray) 4 kg

Hindustan's RAM 1701A radio altimeter 0044969

Status
Radio altimeters manufactured by Hindustan Aeronautics Ltd are fitted on An-32, MiG-21, Jaguar, Dornier aircraft and on Cheetah and ALH helicopters.

Contractor
Hindustan Aeronautics Ltd, Avionics Division.

International

CMA-3012/3212 GPS sensor units

Type
Global Positioning System (GPS) sensors.

Description
The CMA-3012 and CMA-3212 are designed to meet the requirements as the sole means of navigation for oceanic/remote area operation as specified by FAA order 8110.60. The Sureflight software package is available to complement all CMC Electronics' GPS receivers for primary means navigation flight planning and dispatch. Since 1993, over 5,000 units have been installed in numerous air transport, helicopters and business/commuter aircraft types. Key characteristics for such applications include 12 simultaneous receive channels, all of which can be used for continuous satellite tracking and any two of which are assignable as GPS/WAAS Integrity Channels (GIC); comprehensive end-to-end receiver BITE; carrier phase tracking; differential GPS (SCAT 1) functionality; WAAS functionality and growth provision for LAAS.

Inputs consist of eight ARINC 429 and an RS-232. Outputs consist of three ARINC 429, three 1 kHz time marks and RS-232 and 28 V valid discrete.

Specifications
Dimensions:
(CMA-3012) 66 × 216 × 240 mm
(CMA-3212) 199.6 × 57.1 × 388.6 mm
Weight:
(CMA-3012) 2.55 kg
(CMA-3212) 3.6 kg
Power supply: 18–36 V DC, 20 W (max), 16 W (typical)
Frequency: 1,575.42 MHz, C/A code
BITE: Continuous coverage, >95% fault detection
Accuracy (95%):
22.5 m (horizontal)
30 m (vertical)
0.1 kt (velocity)
Position update frequency: 1 Hz (5 Hz optional)
Reliability: 65,000 h MTBF
Temperature: −55 to +70°C
Altitude: Up to 55,000 ft (16,500 m)
Compliance: ARINC 743A, 429-12; DO-160C/D, -208, -217 (optional), -178B level B/level C, -229 (optional)
Certification: TSO-C129A, -C145

Status
In production and in service. TSO-C129 B1/C1 was received in mid-1995, and SCAT 1 implementation was completed in mid-1996. The system has been certified for Primary Means operation to FAA N8110.57. Ground preflight software (SureFlight) is also available.

Contractor
CMC Electronics Inc.
Honeywell Inc, Commercial Aviation Systems.

Global Navigation Satellite Sensor Unit (GNSSU)

Type
Aircraft navigation system, Global Positioning System (GPS).

Description
Honeywell and Canadian Marconi Company are collaborating in the production of the Global Navigation Satellite Sensor Unit (GNSSU). The 12-channel GNSSU tracks all GPS satellites in view to provide better than 25 m position accuracies. Up to 11 satellites may be in view at one time but only four are required to give a position solution. It offers sensor computational errors of no more than 1.5 m, receiver autonomous integrity monitor and ARINC 743 design. The GNSSU provides accurate worldwide oceanic, enroute and approach navigation, simplified pilot interface, time and position for Automatic Dependent Surveillance (ADS) and Inertial Reference System (IRS) integration.

Increased capabilities result from combining the GNSS data in an 18-state Kalman filter inside the laser IRU. With this, the inertial error of 2 n miles/h is bounded by the accurate GNSS satellite measurements. There is continued system accuracy during periods of less than four satellites and continued integrity with less than five satellites. The high-frequency inertial sensor measurements integrated with the high-accuracy GNSS measurements provide the optimum navigation solution. The GNSS integration in the IRS eliminates the possibility of an added 25 m track error that could occur with blending of GNSS and IRS in the FMS. The GNS/IRS is designed to enable calibration of the inertial sensors after sole source GNSS certification.

Specifications
Dimensions: 63.5 × 215.9 × 241.3 mm
Weight: 3.18 kg
Accuracy:
(position) 25 m
(velocity) 1 kt
(time) 2 ms
Reliability: 55,000 h MTBF predicted
Temperature: −55 to +70°C

Status
The GNSSU has been selected as the standard option on the Boeing 777 and over 1500 have been certified and installed in various airliners and business aircraft. It was TSO'd by the FAA in January 1994. Southwest Airlines ordered the GNSSU for its Boeing 737-700 aircraft. The sensor is being developed to incorporate GLONASS and WAAS once the satellites are installed. It is also being used to evaluate the capabilities of differential GNSS through the FAA and NASA differential evaluations.

Contractor
Honeywell Inc, Commercial Aviation Systems.
Canadian Marconi Company.

H-764G Embedded GPS/INS (EGI)

Type
Aircraft navigation system, integrated.

Description
The H-764G EGI is a small, low weight, inertial navigation system with embedded GPS. It is based on a 32-bit Intel 80960 microprocessor, MIL-STD-1553B databus, and four RS-422 buses on one card. It is programmed in Ada and provides a triple navigation solution (pure inertial, GPS-only, and blended GPS/INS), using a Collins GEM™ GPS receiver module.

The H-764G EGI comprises the following main modules: radar altimeter, remote air data, GPS, inertial sensor, stability augmentation, processor, ARINC and discrete interfaces, and BAE Systems Terprom (optional).

Specifications
Dimensions: 177.8 × 177.8 × 248.9 mm
Weight: 8.4 kg
Pure inertial performance:
(Position) <0.8 n mile/h CEP
(Velocity) <1.0 m/s rms
Blended GPS/INS performance:
Position accuracy (SEP) <16 m
Velocity accuracy (rms) 0.1 m/s
Align time:
(Gyrocompass) 4 min
(Stored heading) 30 s
(in air align) 4 min
Power: 28 V DC, 40 W
Interfaces: MIL-STD-1553B (RT or BC) and four RS-422 serial databus (optional) Synchro, analogue, discrete

Status
Installed on about 50 aircraft and helicopter types, with a total production of over 4,000 units. The H-764G EGI was selected for the Tornado GR4/4A and F3 upgrades, with the TERPROM option, in which installation it is also known as the Tornado LINS 764GT.

Contractor
Honeywell Military Avionics.
BAE Systems.
Honeywell Regelsysteme GmbH.

H-764G Embedded GPS/INS (EGI)

0054178

SLS 2000 Satellite Landing System

Type
Aircraft landing system, Differential Global Positioning System (DGPS).

Description
In January 1995, Honeywell and Pelorus Navigation Systems Inc of Calgary, Canada, teamed to develop and manufacture a Differential GPS (DGPS) ground reference station called the Satellite Landing System (SLS). The SLS allows for improved all-weather operations that reduce delays and operating costs while maintaining high integrity and safety. It can provide Special Cat I (SCAT I) capability to all runway ends within a 30 n mile radius, making it more cost effective than traditional landing aids that are limited to one runway end. It ensures local airport control and enhances satellite coverage and it can provide the flexibility to design approaches that minimise flight time and meet noise abatement objectives.

SLS will be Cat I and II capable with growth to Cat III, and SLS will enable variable geometry precision approaches and departures.

SLS-1000 and SLS-2000
The SLS ground station is available in two configurations, SLS-1000 and SLS-2000. Both systems comprise three major subsystems: ground reference station, Remote Satellite Measurement Units (RSMUs) and VHF (Very High Frequency) datalink transmitter.

The SLS-1000 unit is a fail-safe system that is designed continuously to perform self tests to determine its 'health'. If it detects a problem, it will notify the operator and any aircraft in the area that it is not capable of sending accurate data.

The SLS-2000 unit is a fail-operational system that is not affected by single component failures. The system operator is notified of a component failure and can call for service while the unit continues to operate.

The SLS-1000 and SLS-2000 are self-calibrating. The systems are designed with performance monitors, which eliminate the need for periodic flight checks. Both systems come with a fault-tolerant power supply and battery back-up to ensure continuity of service in the event of a power outage.

Airborne equipment complement
Although Honeywell is involved with design of the ground-based element of SLS-2000, Honeywell alone is designing and manufacturing the airborne element.

For the forward fit market, Honeywell and Pelorus are working with industry and aircraft manufacturers to design the optimum solution for future aircraft.

In the retrofit market, most aircraft will need some combination of the following systems: Flight Management System (FMS) or SLS controller; VHF datalink receiver; Differential Global Navigation Satellite Sensor Unit (DGNSSU); Analogue Interface Unit (AIU).

Maintaining the high integrity of the system is critical to certification and safety. To do this, the Honeywell DGNSSU, designed to ARINC 743 standards, calculates and directs the flight controls. The DO 178B level B, critical-level software, ensures that the system will perform the calculations correctly.

Annunciations and specific interface requirements are resolved on an aircraft-by-aircraft basis.

The FMS or SLS controller tunes the VDL-500 to the SLS ground station datalink frequency at a given airport. The range error corrections and path points received by the VDL-500 are sent to the DGNSSU.

The DGNSSU makes the necessary corrections and calculates the approach path.

The approach path, based on SLS position information, is transmitted to the flight controls as an ILS lookalike signal. The approach is flown by the flight controls using this input.

Status
In September 1998, a Continental Airlines MD-80 flew the inaugural flight of the SLS 2000 system using it for precision landings at commercial airports.

Contractor
Honeywell Inc, Commercial Aviation Systems.
Pelorus Navigation Systems Inc.

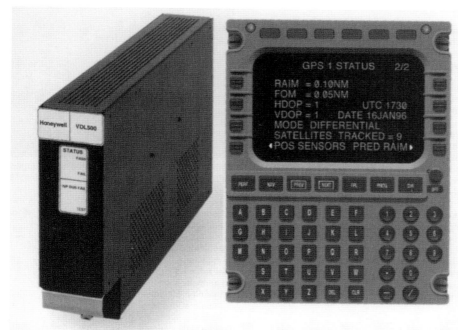

VDL-500 Flight Management System (FMS)   0001404

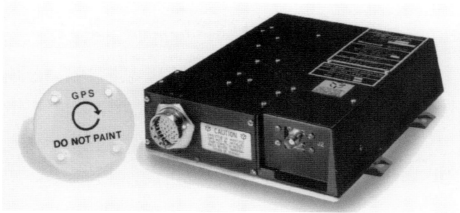

Antenna Differential Global Navigation Satellite Sensor Unit (DGNSSU)   0001405

Timearc 6 GPS navigation management system

Type
Avionic navigation sensor, Global Positioning System (GPS).

Description
The Timearc 6 GPS navigation management system provides accurate position information from an advanced six-channel GPS receiver designed with the latest technology for military and commercial applications. The Jeppesen® NavData card contains information on all airports, VORs, NDBs and intersections in Europe or North America. The three-line 16-character LED alphanumeric display, which is extra bright and adjustable in intensity, is easy to read and is equipped with a left/right cursor indexer.

The Timearc 6 uses GPS satellites for position calculation and navigation and provides worldwide continuous and precise navigation data, offering operation under dynamic conditions up to 10 *g*,

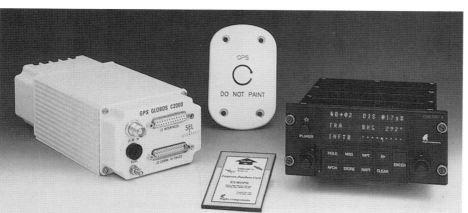

The Timearc 6 GPS navigation management system showing (left to right) the core module, (rear) the antenna with preamplifier, (front) the Jeppesen® NavData card and the control and display unit   0504176

time to first fix of less than 1 minute and optional precise position data with differential GPS.

The associated mission planning system provides the ability to augment the Jeppesen® NavData card with user-defined waypoints and individual mission data. In addition, besides Universal Transversal Mercator (UTM), positions can be displayed in various local grids, including British National Grid and Swiss Grid to compare directly with local maps. Optionally, the Control and Display Unit (CDU) is equipped with an ARINC 429 interface, and the CDU features SID, STAR, and approaches capability.

Specifications
Dimensions: 66 × 146 × 177 mm
Weight: 1.37 kg
Power supply: 10–40 V DC, 0.26 A at 28 V
Temperature range: –20 to +55°C
Altitude: up to 35,000 ft
Accuracy – 3-D position (95%):
(SPS not degraded) 51 m
(SPS with selective availability) 174 m
(DGPS without selective availability) 10 m
(DGPS with selective availability) 20 m

Status
In service.

Contractor
Aerodata Flugmesstechnik GmbH.
Flight Components AG.
Thales Avionics SA.

Israel

BP GPS

Type
Avionic navigation sensor, Global Positioning System (GPS).

Description
The BP GPS board is a parallel-tracking, twelve-channel, C/A code GPS receiver, designed for demanding commercial and military applications.

The system provides accurate three-dimensional position, velocity and time. It also provides raw data measurements for easy integration with inertial systems.

Customisation of the communication protocol, form factor, environmental conditions and dynamic performance are available on request.

Specifications
Type: L1, C/A code, 12-channel, all-in-view GPS
Dimensions: 119 × 122 × 11 mm
Weight: 0.11 kg
Dynamics
Velocity: 515 m/sec
Acceleration: 4 g
Jerk: 4 g/sec
Positional accuracy: 15 m RMS
Velocity accuracy: 0.3 m/sec
Update rate: 1 Hz (navigation)
Time To First Fix (TTFF): 20 sec typical (with ephemeris); 50 sec typical (without ephemeris)
Power: 5 V DC, 2.5 W (average)
Interface: Port 1: RS-422 (host); Port 2: RS-232 (DGPS, data monitor, status messages)
Baud rate: 9,600 to 38,400 bps
Environmental: MIL-STD-810D

Status
In production and in service.

Contractor
BAE Systems ROKAR International Ltd.

EL/K-7200 DME/P-N airborne interrogator

Type
Aircraft navigation system, radio aids, DME.

Description
Elta has been awarded a contract by the US Department of Transportation to develop DME/P-N airborne interrogators for MLS. Elta's EL/K-7200 Precision (P) and Normal (N) DME operates as a subsystem of the MLS, providing precision range and range rate information to the aircraft.

DME/P ground transponders will soon be available for installation in MLS equipment at selected major airports. In its N mode, the DME is compatible with existing transponders. Each DME/P-N includes an interrogator and controller unit for installation in the cockpit.

The EL/K-7200 utilises large-scale integration components and advanced mass production technologies. It rapidly acquires normal and precision ground transponders and remains locked-on during all flight manoeuvres including initial approach, final approach and missed approach through the implementation of proprietary algorithms and adaptive digital filters.

Status
In production and in service. FAA qualified.

Contractor
Elta Electronics Industries Ltd.

The Elta EL/K-7200 DME/P-N interrogator
0503889

GPS NAVPOD-NT rugged receiver

Type
Avionic Global Positioning System (GPS).

Description
The Rokar GPS NAVPOD-NT is an 'all-in-view', parallel-tracking, 12-channel GPS receiver, designed for demanding pod and combat aircraft applications.

It is a fast reacquisition C/A code receiver, which processes navigation data signals transmitted from all satellites in view. In addition to regular navigation data, the NAVPOD-NT also outputs pseudo ranges, delta ranges, system time, measurement quality and other data required for coupling with Inertial Navigation Systems (INSs).

The Rokar GPS NAVPOD high-dynamics, rugged GPS receiver
0015360

The system is capable of operating in differential navigation mode, and of receiving/producing differential GPS corrections.

Specifications
Dimensions: 177.8 × 94.5 × 30.5 mm
Weight: 0.68 kg
Dynamics
Velocity: 800 m/sec
Acceleration: 10 g
Jerk: 4 g/sec
Positional accuracy: 15 m RMS; 2–5 m RMS differential
Velocity accuracy: 0.1 m/sec; 0.05 m/sec RMS differential
Update rate: 1 Hz (navigation)
Time To First Fix (TTFF): 30 sec typical after warm-up (with ephemeris); 2–3 min (without ephemeris)
Reacquisition: >5 sec typical
Power: 18–50 V DC, 6 W (average)
Interface: Port 1: RS-232 (CDU, data monitor); Port 2: RS-422 (host computer interface)
Baud rate: 9,600–38,400 bps
Environmental: MIL-STD-810D/E

Status
In production and in service.

Contractor
BAE Systems ROKAR International Ltd.

GPS SWIFT high-velocity, high-acceleration receiver

Type
Avionic Global Positioning System (GPS).

Description
The Rokar GPS SWIFT is an 'all-in-view', parallel-tracking, 12-channel GPS receiver. The GPS SWIFT is intended for use in platforms that travel at high velocity and high accelerations.

The GPS SWIFT provides accurate position, velocity, acceleration and acceleration-rate (jerk), as well as time and other GPS data. The system is capable of operating in differential navigation mode, and of receiving/producing differential GPS corrections. It is also designed for tight integration with other sensors such as Inertial Navigating Systems (INS).

Key features are high-velocity (2,000 m/s) and high-acceleration (15 g) operation; provision of navigation data including acceleration and jerk; INS integration.

BP GPS (BAE Systems ROKAR)   0561209

The Rokar GPS SWIFT high-velocity, high-acceleration, GPS receiver   0015361

Contractor
BAE Systems ROKAR International Ltd.

NT4RLG GPS

Type
Avionic navigation sensor, Global Positioning System (GPS).

Description
The NT4RLG GPS board is a parallel-tracking, twelve-channel, C/A code GPS receiver, designed for demanding commercial and military applications.

The system provides accurate three-dimensional position, velocity and time. It also provides raw data measurements for easy integration with inertial systems.

Customisation of the communication protocol, form factor, environmental conditions and dynamic performance are available on request.

Other features of the NT4RLG include inertial system compatibility (TNL-16, RNAV, TRF-90M), operation in a high-vibration environment and fast acquisition/re-acquisition of signals.

Specifications
Type: L1, C/A code, 12-channel GPS
Dimensions: 120.2 × 122.7 × 11 mm
Weight: 0.12 kg
Dynamics (higher dynamics by request)
Velocity: 800 m/sec
Acceleration: 6 g
Jerk: 4 g/sec
Positional accuracy: 15 m RMS
Velocity accuracy: 0.1 m/sec
Update rate: 1 Hz (navigation)
Power: 5 V DC, 2.5 W (average)
Power backup: 3.6 V DC, <15 µA
Interface: Port 1: RS-422 (host); Port 2: RS-232 or RS-422 (DGPS, data monitor, status messages)
Baud rate: 9,600 to 38,400 bps
Environmental: MIL-STD-810D

Status
In production and in service.

Contractor
BAE Systems ROKAR International Ltd.

TNL-16G INS/GPS

Type
Avionic navigation sensor, Inertial Reference System (IRS).

Description
Tamam's TNL-16G is a compact lightweight unit that integrates a Ring Laser Gyro (RLG) – based Inertial Navigation System (INS) with GPS to provide a fully integrated INS/GPS navigation solution, along with raw inertial data for control and stabilisation applications, including SAR motion compensation and payload targeting.

Applications include, aircraft, helicopters, UAVs and guided weapons.

Specifications
Dimensions: 231 × 136 × 153 mm (D-type connector); 247 × 165 × 153 mm (D38999 connector)
Weight: <5.4 kg (D-type connector); <5.6 kg (D38999 connector)
Power: 28 V DC; 35 W
Interfaces: RS-422 or MIL-STD-1553
Accuracies After 5 min of manoeuvring with GPS (C/A) updates

	Standard	Optional
Position	20 m RMS	20 m RMS
Altitude:	25 m RMS	25 m RMS
Velocity:	0.3 m/s RMS	0.3 m/s RMS
Attitude:	0.05° RMS	0.04° RMS
Azimuth:	0.15° (1 sigma)	0.1° (1 sigma)

Gyroscopes channel

	Standard	Optional
Dynamic range:	400°/sec	600°/sec
Short term stability:	0.3°/hr (1 sigma)	0.1°/hr (1 sigma)

Accelerometers channel

	Standard	Optional
Dynamic range:	25 G	40 G
Bias, long term:	1.0 mG	0.2 mG
Bias, short term:	0.5 mG	0.1 mG
Scale factor:	1,000 ppm	100 ppm

Parameters:

	Standard	Optional
Temperature:	–40 to +71° C	
Linear acceleration:	25 G	40 G
Shock:	20 G, 11 msec	
Vibration:	6 g RMS (20–2,000 Hz)	

Status
In production.

Contractor
Israel Aerospace Industries Ltd, Tamam Division.

Italy

ANV-141 VOR/ILS/MKR airborne navigation set

Type
Aircraft navigation and landing system, radio aids, VOR/ILS.

Description
The ANV-141 is a compact, all solid-state VHF navigation system, capable of receiving and processing VOR and ILS localiser, glideslope and marker beacon signals. The system provides 200 VOR/LOC channels from 108.00 to 117.95 MHz, with manual and automatic instrumentation outputs, and 20 paired glideslope channels from 329.5 to 335.0 MHz.

The receiver is packaged in a ½ ATR short case and is designed to be mounted in either hard- or shock-mountings, as required; the ANV-141 exceeds TSO specifications and fully meets military requirements.

A standard 2-of-5 frequency selector tunes the system, and the receiver may be adapted to different tuning control codes by changing plug-in circuit cards. Quick access to the separate plug-in modules is provided by a removable cover which can be raised to allow full access for servicing. A test connector is built into the unit to reduce fault-finding and rectification times.

The ANV-141 can be supplied in the following configurations:
- Manual VOR
- Manual VOR with glideslope
- Manual VOR with marker beacon
- Manual VOR with glideslope and marker beacon
- Automatic VOR
- Automatic VOR with glideslope
- Automatic VOR with marker beacon

- Automatic VOR with glideslope and marker beacon.

Specifications
Dimensions: 1/2 ATR short
Weight: <5.5 kg
Power: 27.5 V DC, 42 W; 26 V AC, 20 VA (max)
Altitude: Up to 50,000 ft
Temperature: –55 to +55°C (continuous); up to 70°C (for up to 30 minutes)
Design specifications: TSO C34C, C35C, C36C, C40a

VOR/LOC receiver
Frequency range: 108.00–117.95 MHz
Channels: 200, in 50 kHz increments
Sensitivity: VOR/LOC – 3 µV emf for flag out of view; VOR/LOC – aural 3 µV for 6dB ((S + N)/N)
Selectivity: ±17 kHz (min) at 6 dB; ±35 kHz (max) at 60 dB
AGC: As specified by ARINC-579 and -578
Frequency stability: 0.001% from –55 to +55°C

Glideslope receiver
Frequency range: 328.6–335.4 MHz
Channels: 40, in 150 kHz increments
Sensitivity: 10 µV emf for flag out of view
Selectivity: ±18 kHz (min) at 0 dB; ±45 kHz (max) at 50 dB
AGC: As specified by ARINC-578
Frequency stability: 0.001% from –55 to +55°C

Marker Beacon
Frequency: 75 MHz
Sensitivity: 200 µV emf for high sensitivity; pilot control to 1500 µV emf
Selectivity: ±10 kHz (min) at 3 dB; ±50 kHz (max) at 60 dB
AGC: Output varies not more than 6 dB from 0.2 to 50 mV

Status
In production and in service.

Contractor
Marconi Selenia SpA.

ANV-211 P-DME interrogator

Type
Aircraft navigation system, radio aids, DME.

Description
The ANV-211 is a Distance Measuring Equipment (DME) interrogator, capable of receiving accurate distance information from DME/P transponders and standard DME/N transponders. Performance of the DME/N interrogator is enhanced by the addition of the DME/P IA mode.

When used with other systems, such as VOR ILS and Microwave Landing System (MLS), the ANV-211 meets the needs of a variety of users in the en route, approach, landing, missed approach and departure phases of flight. The ANV-211 P-DME is an integral element of the Marconi Communications MLS.

Angular information is provided by the MLS receiver ANV-201 or by the MultiMode Receiver (MMR) ANV-241. These units provide highly accurate 3-D positioning, capable of elaborate terminal area navigation by following precise (curved and segmented) flight paths from the Approach Area (AA) through to touchdown and roll-out.

The ANV-211 interrogator has been designed around a powerful microprocessor and modular structure to provide maximum growth capability; the system meets the requirements specified in RTCA DO-189.

The main features of the system are:
- Self-contained system
- Accurate 3-D position
- Full terminal area navigation
- Flexible, modular architecture, employing state-of-the-art design and technology
- High reliability
- BITE.

The ANV-211 comprises two Line-Replaceable Units (LRUs), the interrogator unit and the control panel. An optional MLS/DME Combined Control Panel (CCP) is also available.

The ANV-211 can be remotely controlled through standard MIL-STD-1553B or ARINC 429 interfaces.

Specifications
Coverage limits:
  Yaw: 360°
  Pitch: –40 to +25° WRT horizontal plane
P-DME characteristics: RTCA DO-189

Status
In production and in service.

Contractor
Marconi Selenia SpA.

ANV-301 Doppler Navigation System (DNS)

Type
Avionic navigation sensor, Doppler.

Description
The ANV-301 DNS provides a self-contained worldwide precision navigation capability, utilising Doppler-derived measurement of aircraft velocity and external input of aircraft attitude and heading. The ANV-301 design approach allows system functions to be tailored to specific customer requirements at minimum cost, and the electrical interface to be tailored for specific aircraft avionics and subsystems for additional enhancement of system capability.

The system comprises four LRUs: the Receiver/ Transmitter Radar (RTR), Signal Data Converter (SDC), Control Display Unit (CDU) and Pilot Steering Indicator (PSI).

The RTR generates, radiates and detects microwave energy directed to, and back scattered from, the surface of the earth. Doppler frequency shift information for each of four radiated beams is contained in the signals transferred to the SDC unit for further processing. The RTR contains radiating elements, solid-state microwave components and electronic signal processing modules as required to perform this specific function. The Signal Data Converter (SDC) is housed in a 3/4 ATR short cabinet and comprises the following modules:
- COSMO (Computer System Module) Computer
- Power supply
- Embedded Doppler velocity sensor
- Embedded GPS sensor (optional)
- Interfaces

The combination of COSMO computer and sufficient memory endows the SDC with full navigational computing capabilities, including integration of sensors such as INS and GPS to make accurate positional and velocity estimations.

Individual beam frequency shifts are extracted from the Doppler spectra received from the RTR using a single, time-shared IF channel and a cross-correlation frequency tracking loop to compute the aircraft body-referenced velocity components. Utilising aircraft attitude information (pitch and roll), the measured aircraft-body-referenced velocities are transformed into orthogonal, stabilised, earth-referenced aircraft velocity components. These components are then converted into serial digital and analogue formats in the SDC interface modules for transmission to the AFCS, ADI, Hover Indicators, external displays and other external aircraft subsystems. The velocity components, in combination with aircraft heading information and pilot-entered

initial position and waypoint destination data, are also used by the COSMO computer to generate a complete range of navigation data and guidance/ steering outputs for the PSI (or for the HSI, Hover Indicator, and/or EFIS/MFD for the versions with no PSI) and the cockpit displays.

The SDC can host several customer-dedicated interfaces that allow interfacing to various types of aircraft equipment, such as TAS computer, VOR/DME, Tacan, Loran, Radar, FLIR, and others.

The CDU provides the primary ANV-301 man/ machine interface. It contains the circuitry to interface with the SDC and provides operator selection of DNS operating modes, and the entry and display of navigation and mission-related information. The CDU comprises a single CRT display which is capable of displaying alphanumeric text and other symbols. A numeric keyboard and specific function keys (all self-illuminated) are the primary means of data entry to the DNS.

Three pilot-selectable navigation co-ordinate systems with automatic co-ordinate conversion are available: latitude/longitude, worldwide alphanumeric UTM and arbitrary grid.

Specifications
Dimensions:
  SDC: (3/4 ATR) with GPS embedded, 378.5 × 190.5 × 197 mm
  CDU: 217 × 146 × 114 mm
  RTR (helicopter): 445 × 445 × 130 mm
  RTR (fixed-wing): 652 × 425 × 112 mm
  PSI: 127 × 83 × 83 mm
Weights:
  SDC: 9.5 kg
  CDU: 3.6 kg
  RTR (helicopter): 5.8 kg
  RTR (fixed-wing): 8.2 kg
  PSI: 1.3 kg
Power supplies
  Main power supply: 115 V AC, 400 Hz, 90 W
  Synchro ref: 115 or 26 V AC, 400 Hz
On/off relay: 28 V DC, 4 W
Internal lighting: 28 V DC, 10 W or 5 V AC, 400 Hz, 1 A
Temperature range: –40 to +71°C
Cooling: Convection, heat sink dissipation
Align time:
(gyrocompass) 4 min
(memorised) 30 s
Accuracy:
(position) 1 n mile/h CEP
(velocity) 0.8 m/s RMS
(attitude) 0.05° RMS
(heading) 0.1° RMS
Reliability: >900 h (system)

Status
In production and in service.

Contractor
Marconi Selenia SpA.

ANV-351 Doppler Velocity Sensor (DVS)

Type
Avionic navigation sensor, Doppler.

Description
The ANV-351 Doppler velocity sensor provides precision measurement of helicopter velocity components. The ANV-351 is a single-unit sensor specifically designed for integrated avionic systems and weighing less than 7 kg. It comprises two main modules: the receiver/transmitter radar and the signal data converter.

The receiver/transmitter radar module is of fixed design and all-solid-state construction. It comprises separate four-beam transmit/receive antennas, utilising printed circuit planar-array technology. The module is available in two versions, optimised for either ASW or nap of the earth operation.

The signal data converter module comprises power supply, Doppler signature processing and MIL-STD-1553B interface.

Status
Installed on the A129 Mongoose attack helicopter.

Contractor
Marconi Selenia SpA.

ANV-353 Doppler Velocity Sensor (DVS)

Type
Avionic navigation sensor, Doppler.

Description
The ANV-353 DVS provides precision measurement of helicopter velocity components. It is a single-unit low-power sensor specifically designed for integrated avionic systems and weighing less than 5.2 kg. It comprises two modules: a receiver/transmitter radar and signal data converter.

The receiver/transmitter radar module is of fixed design and all-solid-state construction. It comprises separate four-beam transmit/ receive antennas utilising printed circuit planar-array technology. The redundant fourth beam provides high accuracy during extreme attitude manoeuvres. This module contains all RF signal generation and processing functions and receives its power input, modulation and timing signal from the signal data converter.

The signal data converter comprises power supply Doppler digital signal processing and discrete and ARINC 429 interfaces. An MIL-STD-1553B interface is an option. Individual beam frequency shifts are extracted from the spectrum received from the receiver/transmitter radar module, utilising a single time-shared IF channel and digital signal processing.

The extensive use of advanced components allows controllable RF transmitted power and automatic land/sea transition.

Specifications
Velocity range: Vx –25 to +150 m/s; Vy –50 to +50 m/s; Vz –25 to +25 m/s
Altitude range: 7,000 m
Acquisition and tracking: Fully automatic within 2 s, no warm-up required; 3-beam tracking capability with continuous validity/ reasonableness check
Beam geometry: 4-beam Janus configuration, time-shared
RF power: 50 mW radiated at 13.325 GHz
Modulation: FM/CW, optimised for flight envelope
Power: 28 V DC, 30 W (max)
Environmental: MIL-STD-5400 Class 1A, MIL-STD-810E
Cooling: Convection, heat sink dissapation
Dimensions: 365 × 437 × 60 mm
Weight: 5.2 kg maximum

Status
The ANV-353 is an improved variant of the ANV-351 Doppler Velocity Sensor, originally designed by Marconi Selenia Communication for installation on the Agusta A129 Attack Helicopter. The ANV-353 was developed for the NH 90 helicopter programme and selected for installation on other platforms.

Contractor
Marconi Selenia SpA.

ANV-801 multifunction computer navigation system

Type
Aircraft navigation system, Global Positioning System (GPS).

Description
The ANV-801 is a multisensor computer navigation system that can be configured to form the core element in flight management systems for rotary- and fixed-wing aircraft applications. The ANV-801 Multifunction Computer Display

Unit (MCDU) replaces multiple-control layouts with a single, easy-to-read unit, allowing the concentration of flight management information and data presentation. Versatile design, modular hardware and software architecture make it easy to reconfigure the ANV-801 to meet customer requirements.

The ANV-801 display is based on Active Matrix LCD (AMLCD) or LED technologies. The MCDU can host different CPU types (Intel® 16-bit or RISC 32-bit microprocessor based) with corresponding speed and computing power, according to customer requirements.

The ANV-801 integrates external navigation sensors (such as INS, airspeed sensors, Doppler velocity sensors, and so on) into sophisticated navigation packages providing a high-accuracy computation capability, including navigation information for the crew and steering output to flight director and autopilot. It is also available with a built-in GPS receiver sensor, providing self-contained navigation capabilities and a built-in modem for input/output datalink. The ANV-801 also interfaces with a wide range of analogue and digital equipments using ARINC 429, MIL-STD-1553B, serial and modem digital interfaces and analogue interfaces, including nav/comm (VOR, DME, Tacan), landing (ILS, MLS, MMR) and IFF equipments. The ANV-801 also interfaces with a wide range of analogue and/or digital displays and indicators and provides output to EFIS, HSI, ADI, and Hover Indicators.

Principal features of the system are:
- 3D/4D full navigational computations
- Embedded GPS
- Navigation/communication subsystems management
- Display management
- Built-in test
- Highly reliable
- Autonomous navigation
- Pilot-selectable latitude/longitude, UTM/MGRS, grid co-ordinates
- British/metric units
- Avionic equipment management
- NVIS compatible

Specifications
Display: 8 lines of 16 characters (LED); 14 lines of 24 characters (AMLCD)
Screen size: 90 × 90 mm
Keyboard: 12 line keys, 38 alphanumeric keys, 7 control keys, 9 function keys
Dimensions: 185 × 146 × 181 mm
Weight: 4.5 kg (with embedded GPS)
Power: 28 V DC or 115 V AC, 400 Hz, <50 W
HMI: Operator controls designed in accordance with MIL-STD-1472

Contractor
Marconi Selenia SpA.

LISA 200 Fibre Optic Gyro (FOG) Attitude Heading Reference System (AHRS)

Type
Avionic navigation sensor, Inertial Reference System (IRS).

Description
The LISA 200 FOG AHRS, developed from a FOG Inertial Measurement Unit (IMU), is intended for new installation and retrofit into fixed- and rotary-wing aircraft. The system was designed to satisfy navigation and flight control subsystem requirements for highly accurate platform state vector information.

Navigation management functions are integrated into the system, with embedded or stand alone GPS available as an option. Typical fitment is a dual installation, with autopilot inner loop integration.

The LISA 200 is claimed to be the smallest and lightest FOG AHRS currently in production – the system is housed in a single Line Replaceable Unit (LRU) which fits into the footprint of a single

LISA-200 (Litton Italia)   1198697

vertical gyro – this facilitates a space-effective upgrade from legacy AHRS or federated systems of displacement/rate gyros.

A Digital Control Panel (DCP) is also available, providing for enhanced display and control capabilities, greater management and steering functions and improved maintainability through diagnostic and BIT functionality.

Specifications
Dimensions: 190.5 × 101.6 × 114.3 mm
Weight: 2 kg
Power: 115 V AC or 28 V DC; 40 W
Alignment: 30 sec (ground); <120 sec (airborne)
MTBF: >10,000 hours
Compliance: DO-178 Level A; MIL-STD-810; DO-160
Outputs: Digital ARINC-429, MIL-STD-1553B, RS-422/485 (heading, attitude, orthogonal accelerations, rates, speeds)
Syncro ARINC 407 (heading, pitch, roll, attitude; interlock for autopilot)
Analogue (yaw rate, validities)
Accuracy:
(heading) 0.8°
(attitude) 0.3°
Navigation: <1% distance travelled (Doppler/AHRS navigation)

Status
In production and in service. LISA 200 is certified for both military and commercial aircraft. The system is in service aboard several aircraft types, including the AgustaWestland EH-101 and US Army UHZ-60 helicopters.

Contractor
Lital SpA.

Lisa-4000 Attitude Heading Reference System Unit

Type
Avionic navigation sensor, Inertial Reference System (IRS).

Description
The Lisa-4000 performs the basic functions of flight reference and navigation. It provides heading and attitude for display and autopilot, and high-speed, low-latency, anti-aliasing filtered body angular rate and linear acceleration data for autopilot inner loop stability augmentation.

Lisa-4000 (Litton Italia SpA)   1198698

Navigation is performed by coupling the Lisa-4000 with GPS, Doppler radar and air data systems. The Lisa-4000 also outputs high-speed inertial velocity data required for accurate weapon release. It therefore provides all the inertial parameters required for autopilot, navigation, display and weapon delivery.

The Lisa-4000H version is a full-aided inertial navigator with the option of a MIL-STD-1553B or ARINC 429 interface. It can also be supplied with additional synchro outputs to drive back-up cockpit instruments.

Specifications
Dimensions: 1/2 ATR short
Weight: 6.2 kg
Power supply: 28 V DC, 90 W unregulated
Cooling: 0 to 71°C
Accuracy (RMS):
(heading) 0.2° + compass
(pitch and roll) 0.2°
(body rates-PQR) 0.15°/s, 200 Hz, 2.5 ms latency
(body rates-XYZ) 5 m g, 200 Hz, 2.5 ms latency
(horizontal velocity) 2 m/s
(vertical velocity) 1 m/s
(position) 1% of distance travelled
(altitude) 100 ft + 1%
Inputs: GPS, Doppler velocities, barometric altitude, true airspeed, compass, up to 100 waypoints
Outputs: bearing to destination, distance to go, time to go, windspeed and direction, groundspeed

Status
The Lisa-4000B is in production for the A 129 attack helicopter. The Lisa-4000D is operating on the UK Royal Navy EH 101 helicopter. The civil qualified Lisa-4000E is for the EH 101 civil variant. The GPS coupled Lisa-4000G is in production for an RPV. The Lisa-4000H is in production for Italian Navy AB-212, AB-412, SH-30 helicopters, and the EB version on Romanian MiG-21s.

Contractor
Lital SpA.

LN-93EF ring laser gyro inertial navigation system

Type
Avionic navigation sensor, Inertial Reference System (IRS).

Description
The LN-93EF is based on the LN-93 ring laser gyro standard navigator built by Northrop Grumman (formerly Litton) to SNU-84 standards. However, the EF version, modified to Eurofighter Typhoon requirements, includes upgraded electronics, a fibre optic databus, and Ada software. The result is a system with the same accuracy and maturity but with a 30 per cent reduction in size and weight than the SNU-84 system.

Specifications
Dimensions: ¾ ATR (length 360 mm)
Weight: 16 kg
Power supply: 115 V AC, 400 Hz, 85 W
28 V DC back-up
Temperature range: –40 to +71°C
Align time:
(gyrocompass) 4 min
(memorised) 30 s
Accuracy:
(position) 1 n mile/h CEP
(velocity) 0.8 m/s RMS
(attitude) 0.05° RMS
(heading) 0.1° RMS

Status
In production for the Eurofighter Typhoon.

Contractor
Lital SpA.

LOAM Laser Obstacle Avoidance System

Type
Aircraft navigation sensor, Obstacle Avoidance System (OAS).

Description
LOAM is described as a 'navigational aid' for rotary-wing platforms, specifically designed to detect potentially dangerous obstacles placed in or nearby the flight path and to warn the crew in sufficient time to instigate effective avoiding manoeuvres.

LOAM works as a Laser Radar (LADAR), using an eye-safe laser, which periodically scans the area around the flight path and, on the basis of the returned echo analysis and utilising an innovative software algorithm, it identifies possible obstacles and provides the crew with the relevant information and warnings.

The use of a laser beam, instead of a RF emission, to scan for obstacles offers the following advantages:
- Sufficient return power even for obstacles with non-perpendicular incidence
- A very narrow scan beam facilitates a high resolution system

LOAM is capable of classifying the following classes of obstacle:
- **Wires.** Thin obstacles such as, telephone cables and electrical cables (up to 5 mm diameter, electrified or not).
- **Trees.** Vertical obstacles such as trees, poles and pylons.

- **Structures.** Extended obstacles such as bridges and buildings.

Obstacle detection is achieved by laser scan, echo detection and detected echo analysis. LOAM performs echo detection through an analogue process, comprising optical-electrical conversion, signal pre-amplification and threshold comparison. Signal pre-amplification is performed by an automatic gain-controlled amplifier that increases system sensitivity as the elapsed time from the laser emission increases, to allow for lower return signal power from more distant obstacles. Furthermore, the threshold level may be adjusted to compensate for background conditions. This is claimed to eliminate the probability of false echo detection due to atmospheric back-scatter near the laser beam output and optimise system sensitivity for every operational atmospheric condition. LOAM performs echo analysis to determine the presence of possible obstacles and to determine their geometrical characteristics and position. To do this, the system employs two sequential analysis processes, local analysis and global analysis. The local analysis process analyses the single echoes in order to determine range, angular co-ordinates and characteristics of the obstacle portion generating them.

Specifications
Horizontal FoV: 40°
Vertical FoV: 30°
FoV steering: ±20° both in horizontal and in vertical
Maximum detection range: 2,000 m
Minimum detection range: 50 m
Obstacle detection probability: >99.5%
False alarm rate: <1 per 2 hours
Qualification: MIL-STD-810E and MIL-STD-461C
Interface: EIA RS-422 and MIL-STD-1553B
Power supply: 28 V DC, <300 W (MIL-STD-704E)
Laser: Eye-safe according to STANAG 3606 LAS class I

Sensor Head Unit
Dimensions: 320 × 239 × 419 mm
Weight: 24 kg

Control Panel Unit
Dimensions: 146 × 38.1 × 155 mm
Weight: 0.5 kg

Warning Unit
Dimensions: 110.2 × 25.4 × 75 mm
Weight: 0.35 kg

Display Unit
Dimensions: 4 × 5 ATI
Weight: 1.7 kg

Status
In production and in service. LOAM has been selected for the Italian NH90 TTH and Danish EH101 helicopters and the UK Condor 2 Programme.

Contractor
Marconi Selenia SpA.

Russian Federation

705-6 attitude heading reference system

Type
Avionic navigation sensor, Attitude Heading Reference System (AHRS).

Description
The 705-6 attitude heading reference system is part of the inertial navigation system of the MiG-29. It provides the following data outputs: gyroscopic heading, pitch and roll angles, and absolute orthogonal accelerations.

Specifications
Normal erection: 15 min
Fast erection: 3 min
Errors:
(normal erection) 0.15°
(fast erection) 0.7°

Status
Fitted to MiG-29 aircraft.

Contractor
Ramensky Instrument Engineering Plant.

A-744 radio-navigation receiver

Type
Aircraft navigation system, Global Positioning System (GPS).

Description
The A-744 is an integrated GLONASS/GPS navigation receiver that determines position, time, velocity (horizontal and vertical components) in real time. It is designed to be used either independently, or as part of an integrated aircraft system. It can be commanded to GPS only mode, or to GLONASS-GLONASS/GPS mode.

Specifications
Dimensions: 281 × 191 × 64 mm
Weight: 2.7 kg
Power supply: 27 V DC, 18 W
Antenna: active, non-protruding
Interfaces: GOST 18977-79 PTM 1495-75 (ARINC 743A, 429) multiplex channel GOST.26765.52-87 (MIL-STD-1553B)

Number of channels: 6 (universal)
Operating frequency:
GLONASS, EA code: 1598-1616 MHz (F1)
GPS, C/A code: 1575.42 MHz (L1)
Output data: position, velocity, time, quasi-range and Doppler shift measurement
Position error (2e) m:
GLONASS: 45 horizontal/70 vertical
GPS (SA including GLONASS/GPS): 100 horizontal/160 vertical
DGPS/Diff.GLONASS: 6 horizontal/10 vertical
Time to first fix: less than 150 s
Dynamic characteristics: (not more than)
(velocity) 500 m/s
(acceleration) 40 m/s^2
(acceleration increment) 30 m/s^2/s
Temperature limits:
(operation) −50 to +50°C
(storage) −55 to +85°C
Antenna dimensions/weight:
active antenna: 110 mm diameter, 41 mm height, 0.31 kg
passive antenna: 80 × 80 mm, 33 mm height, 0.27 kg

705-6 attitude heading reference system
0018236

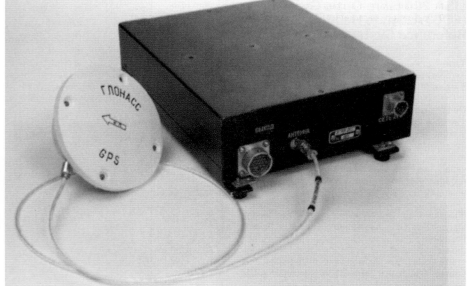

A-744 radio-navigation receiver
0005425

Status
In service.

Contractor
Leninetz Holding Company.

BINS-85 strap-down Inertial Navigation System (INS)

Type
Avionic navigation sensor, Inertial Reference System (IRS).

Description
The BINS-85 INS is designed for civil aircraft, with the ability to compute up to 43 navigation parameters, including geographical co-ordinates, groundspeed and its components, true, magnetic and gyroscopic headings, pitch and roll angles, angular rates and linear accelerations along the axes of the aircraft. System modes are align, navigation, test and Attitude Heading Reference (AHRS).

Specifications
Accuracies:
Position: <3.7 km/flight h
Groundspeed and components: <4 m/s
True/Gyroscopic heading: ≤0.4° in 10 h of operation
Magnetic heading: 2°
Pitch and roll angles: 0.1°
Angular rate: 0.1°/s
Linear acceleration: 0.01 *g*
Status ready time:
(at 0 to +55°C) ≤10 min
(at –20 to 0°C) ≤18 min
Temperature range: –20 to +55°C
Digital outputs: 43 parameters
Interface: GOST 18977–79, ARINC 429
Power: 115 V AC, 400 Hz, 200 VA (max) or 27 V DC, 180 W
Mean time between failures: 5,000 h
Dimensions: 319 × 322 × 194 mm
Weight: 20 kg
Certification: Class I

Status
BINS-85 is installed as the main flight and navigation data system in Tu-204, Tu-214, Tu-334 and Il-96-300 aircraft.

Contractor
Aviapribor Corporation.

BINS-90 Inertial Navigation System

Type
Aircraft navigation system.

Description
The BINS-90 INS is designed for passenger and freight aircraft, with the ability to compute up to 40 navigation parameters. The system has been designed around three LG-1 laser gyros and three AK-6 quartz accelerometers. The BINS-90 can be coupled to a GPS/GLONASS receiver, which dramatically improves the overall accuracy of the system navigation solution.

BINS-90 inertial navigation system 0103861

LG-1 Laser Gyro 0103862

Specifications
Accuracies:
Position: <3.7 km/flight hour (200 m with GPS/GLONASS support)
Speed: <4 m/s (0.2 m/s with GPS/GLONASS support)
Status ready time:
<10 min (autonomous mode)
<3 min (with correction)
Temperature range: –60 to +60°C
Altitude: up to 12,500 m (41,000 ft)
Angular speeds (all axes): up to 200°/s
Digital outputs: 40 parameters
Power: 115 V AC, 400 Hz, 200 VA (max) 27 V DC, <80 W
Mean time between failures: 5,000 h
Dimensions: 194 × 194 × 470 mm
Weight: 16 kg (max)

Contractor
Aviapribor Corporation.

BINS-TVG strapdown Inertial Navigation System (INS)

Type
Avionic navigation sensor, Inertial Reference System (IRS).

Description
The Ramenskoye Design Company BINS-TVG strapdown INS is manufactured as a monoblock, which comprises three accelerometers and three hemispherical resonance gyroscopes rigidly coupled with the monoblock casing. The system electronics utilise large-scale integrated circuits. The monoblock assembly can be integrated with a dual-mode GPS/GLONASS navigation receiver, or supplied separately.

The BINS-TVG system is intended for civil and military application on aircraft and helicopters.

Specifications

	INS alone	with GPS/GLONASS
Accuracies (2 sigma):		
(co-ordinates)	1.85 km/h	20 m
(velocity)	0.6 m/s	0.1 m/s (3 sigma)
(roll/pitch)	0.05°	0.05°
(heading)	0.07°/h	0.07°/h
Warm-up time	3 min	3 min
Data output format	ARINC 429	ARINC 429
MTBF	5,000 h	5,000 h
Dimensions	7 MCU	8 MCU
Weight	14 kg	15 kg

Status
In production and in service.

Contractor
Ramenskoye Design Company AO RPKB.

BINS-TWG strapdown inertial navigation system

Type
Avionic navigation sensor, Inertial Reference System (IRS).

Description
The BINS-TWG strapdown inertial navigation system is designed to provide flight and navigation data for civil and military aircraft and helicopters. It is constructed as a monoblock comprising an inertial sensor unit consisting of three accelerometers and three hemispherical resonance gyros. The unit includes integral GPS/GLONASS capability.

Status
In production and in service.

Contractor
Ramenskoye Design Company AO RPKB.
Ramensky Instrument Engineering Plant JSC.

CH-3301 GLONASS/GPS airborne receiver

Type
Aircraft navigation system, Global Positioning System (GPS).

Description
NAVIS receivers are able to process signals from both GLONASS and GPS satellites, thereby increasing accuracy and reliability by comparison with receivers that process only one source of data.

Specifications
Memory: 500 waypoints, 50 routes
Display: alphanumeric LCD display; 20 symbols on each of 2 lines
I/O: 2 input/output ports RS-232C; 1 output channel in ARINC 429 format; 1 analogue barometric input
Antenna: to ARINC 743A
Dimensions: 384 × 159 × 51 mm
Weight: 2.4 kg
Power: 27 V DC; 115 V AC, 400 Hz
Temperature:
(receiver) –20 to +50°C
(antenna) –55 to +55°C

Status
In service. Production status unknown.

Contractor
NAVIS.

GINS-3 Gravimetric Inertial Navigation System

Type
Aircraft navigation system.

Description
GINS-3 is designed for two applications: stand-alone navigation of aircraft; and aerogravimetry. System components are the gravimetric inertial unit, radio altimeter, air data system, satellite navigation system, electronics unit, and computer.

GINS-3 gravimetric inertial navigation unit 0015366

Specifications

Navigation errors:
(position co-ordinates, independent of flight time and distance) 0.2 to 10 km
(ground and vertical speed components) 0.1 to 1 m/s
(altitude above sea level) 1 to 5 cm
Aerogravimetry error of gravitational anomaly in free air: up to 0.7 mGal
Scales of maps produced: 1/200,000 and more
Power: 27 V DC, 1,000 VA (max)
Weight: up to 120 kg

Contractor

Aviapribor Corporation.

HG 1150BE01 strapdown Inertial Reference Unit (IRU)

Type

Avionic navigation sensor, Inertial Reference System (IRS).

Description

The HG 1150BE01 contains three ring laser gyros and three pendulous accelerometers of Honeywell design in a strapdown configuration to provide primary attitude, heading, body angular rates, body linear accelerations, velocity and position information. Outputs conform to ARINC 704 standards.

Status

In service.

Contractor

Ramenskoye Design Company AO RPKB.
Ramensky Instrument Engineering Plant JSC.

I-21 inertial navigation system

Type

Avionic navigation sensor, Inertial Reference System (IRS).

Description

The I-21 inertial navigation system is designed to autonomously determine flight/navigation data, and provide it to the aircraft system for generation of steering commands between defined waypoints. It conforms to ARINC 61.

Specifications

Navigational errors:
(geographical co-ordinates (in 10 h)) 37 km (max)
(ground and speed components) 12.6 km/h
(true heading) 0.2+0.025°/°C
(angles of roll/pitch, gyro heading) 0.1°
Readiness time: <15 min at 20°C
MTBF: 1,200 h
Power: 500 VA (1,500 VA (max))
Weight: <50 kg

The Ramensky Instrument Engineering Plant I-42-1L strapdown integrated inertial navigation system   0079228

Status

Fitted to An-124, An-224, Il-62, Il-76, Tu-154, Tu-160 aircraft

Contractor

Aviapribor Corporation.
Ramenskoye Design Company AO RPKB.

I-42-1L strapdown integrated inertial navigation system

Type

Avionic navigation sensor, Inertial Reference System (IRS).

Description

The I-42-1L integrated inertial navigation system comprises an inertial unit based on triple GL-1 laser gyros and triple A-L2 accelerometers, together with a proprietary combined GPS-GLONASS/NAVSTAR unit as per ARINC 743A.

Specifications

Accuracy in GPS mode:
(position): 50–200 m
(speed): 5 km/h
(pitch and roll): 0.04°
(true heading): 0.4°
Power: 27 V DC and 115 V AC, 400 Hz

Status

Fitted to civil aircraft including Il-76 and Tu-204.

Contractor

Ramensky Instrument Engineering Plant.

I-42-1S strapdown inertial navigation system

Type

Avionic navigation sensor, Inertial Reference System (IRS).

Description

The I-42-1S strapdown inertial navigation system is designed to provide flight and navigation

parameters as per ARINC 429, as part of an integrated flight and navigation system. The I-42-1S comprises triple KM-11-1A laser gyros and triple A-L2 accelerometers.

Specifications

Accuracy:
(present position): 3.7 km/h
(ground speed components): 22.5 km/h
(true heading): 0.4°
(roll and pitch): 0.1°
Weight: 43 kg
Power: 115 V AC, 400 Hz

Status

In production and in service.

Contractor

Ramensky Instrument Engineering Plant.

INS-80 and INS-97 Inertial Navigation Systems

Type

Avionic navigation sensor, Inertial Reference System (IRS).

Description

The INS-80 system is made as a single unit, comprising gyrostabilised platform, platform electronic units, a computer and I/O devices.

Two free dynamically tuned gyroscopes and three accelerometers are mounted on the gyrostabilised platform. The electronic units are mounted on separate PCBs.

The INS-97 system also includes an integral Satellite Navigation System (SNS), plus antenna.

Optionally, the INS-80 and INS-97 systems can include magnetic heading measurement and display.

Specifications

Position error: 3.7 km/h
Velocity error: 2.0 m/s
Roll, pitch and heading error: 0.1°
Readiness time: 1 to 10 min
Volume: 18 dm³
Weight: 19 kg

Status

In production and in service in Su-30 aircraft.

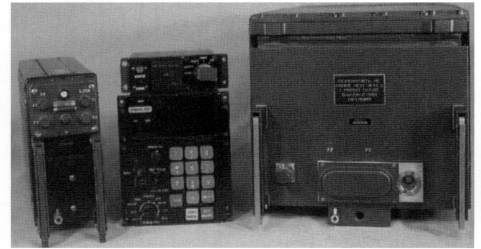

I-21 inertial navigation system   0015367

INS-80 inertial navigation system   0015368

The INS-80 inertial navigation system is fitted in the Su-30 fighter aircraft   0075879

Contractor
Ramenskoye Design Company AO RPKB.
Ramensky Instrument Engineering Plant JSC.

Integrated navigation and control system

Type
Aircraft navigation system, Global Positioning System (GPS).

Description
This integrated navigation and control system provides the following outputs:
1. Vehicle attitude, angular velocities and linear accelerations;
2. Flight parameters such as altitude, vertical speed, angles of attack and slip angles;
3. Speeds and co-ordinates using GPS/GLONASS;
4. Automatic and flight director control of the aircraft;
5. Fatigue monitoring, warning of critical flight conditions, voice and aural warnings;
6. Flat screen display (2) of flight, navigation, radar, weather, map and aircraft data.
   A multiprocessor computer integrates data from all relevant sensor systems.

Specifications
Position error: 80 m
Velocity error: 0.3 m/s
Attitude error:
   roll, pitch: 0.5°
   heading: 1.0°
Readiness time: 10 min
Temperature range: +60 to –60°C
Linear acceleration: 5 *g*
Weight: 28 kg
Interface: ARINC 429

Contractor
Ramenskoye Design Company AO RPKB.

NS BKV-95 integrated Navigation System

Type
Aircraft navigation system, integrated.

Description
The NS BKV-95 integrated navigation system performs three functions: strapdown attitude and heading reference system, GPS/GLONASS receiver, navigation computer. The system is provided with control displays to customer specification.

Specifications
Dimensions: 388 × 124 ×194 mm (4 MCU)
Weight: 8 kg
Power: 27 V DC
Outputs: GOST 18977-79 Sup 3; (ARINC 429)
Accuracy (2 Sigma):
(position) 100 m
(groundspeed) 2 km/h
(roll/pitch angles) 0.25°
(magnetic heading) 1°

Status
Suitable for all types of civil aircraft.

Contractor
Ramenskoye Design Company AO RPKB.
Ramensky Instrument Engineering Plant JSC.

NS BKV-95 integrated navigation system
0015369

NV-1 navigation computer

Type
Aircraft navigation system, control/interface.

Description
The NV-1 navigation computer is designed to provide a flight management interface for the pilot, including flight/navigation systems, radio aids and satellite navigation systems (GPS and/ or GLONASS).

Specifications
Processor: 80486DX4-100
Interface:
ARINC-429
RS-232 (2 channels)
GPS (24 channels)
Electrical: 27 V DC
Power: 60 W (max)
Dimensions: 318 × 194 × 90.5 mm
MTBF: 10,000 h

Contractor
Aviapribor Corporation.

NV-1 navigation computer   0103869

PVD-40 Multifunction air data probe

Type
Avionic flight/navigation sensor, temperature/ pressure.

Description
The PVD-40 multifunction air data probe provides local pitot and static pressure information. If two probes are mounted on either side of the aircraft fuselage, Angle of Attack (AoA) and angle of sideslip can also be derived.

Specifications
Measurement channels
Static pressure: 1
Total pressure: 1

PVD-40 multifunction air data probe   1026863

Integrated navigation and control system   0051596

Operating range
Altitude (m): 0 to 15,000
Velocity (M): 0 to 0.9
AoA: ±20°
Physical
Power: 115 V AC; 400 Hz
Consumption (probe heating): =570 W
Weight: 0.95 kg

Status
In production and in service.

Contractor
AeroPribor-Voskhod Joint Stock Company.

SBKV-2V strapdown Attitude and Heading Reference System (AHRS)

Type
Avionic navigation sensor, Attitude Heading Reference System (AHRS).

Description
The SBKV-2V strapdown Attitude and Heading Reference System (AHRS) is designed for application on civil and military fixed- and rotary-wing aircraft, or as part of their flight and navigation systems, or as a standalone unit. Pre-planned product improvements include provision of a second processor for special purposes, including GPS, SAR and other improvements to navigation and traffic management capability.

Specifications
Accuracy (2σ):
(roll and pitch angles): 0.25°
(magnetic heading): 1.0°
(gyroscopic heading): 0.5°/h
Dimensions: 413 × 130 × 200 mm
Weight: 9 kg
Power: 27 V DC, 80 W; 115 V AC, 60 VA

Contractor
Ramenskoye Design Company AO RPKB.

SNS-2 satellite navigation receiver

Type
Aircraft navigation system, Global Positioning System (GPS).

Description
The SNS-2 is a 24-channel (12 GLONASS, 12 GPS) satellite navigation receiver which computes and outputs the following object movement and positional data:
- Latitude and longitude
- Altitude
- Groundspeed and vertical speed components
- True and magnetic track angles
- UTC time.

SNS-2 is an integrated system, capable of receiving and processing signals from all visible GPS and/or GLONASS constellations. The system is capable of automatic measurement of raw data, almanac and ephemeris download, without preliminary initialisation. These features enable a reduction in selective availability for GPS and satisfactory precision, integrity and availability characteristics for the demands of typical Required Navigation Performance (RNP).

SNS-2 has built-in Receiver Autonomous Integrity Monitoring (RAIM) functionality, utilising data from all tracked satellites and differential capability when RTCM-104 data is available; the system fulfils the technical requirements for onboard GNSS equipment as a primary navigation aid, adopted by the Russian Federal Aviation Service in August 1999. SNS-2 can be installed on all types of aircraft.

Specifications
Output data
Position (95%):
≤35 m
Groundspeed: 15 cm/s
UTC: 0.01 s
Timestamp: 70 ηs
Track: True: 10 arc min; magnetic: 1.5°
Number of receiving channels: 24 (12 GLONASS, 12 GPS)
Output: WGS-84 geodetic system or user specified
Interface: GOST 18977-79/ARINC 429
Physical
Antenna frequency: 1,575.42 to1,609.0 ±18.0 MHz
Weight: 3.4 kg (antenna, antenna amplifier and satellite receiver)
Power: 27 V DC; 10 W
Readiness time: 2.5 min

Status
The SNS-2 is KT-34-01-certified as a B1/C1 class system.

Contractor
Aviapribor Corporation.

SNS-3 satellite navigation receiver

Type
Aircraft navigation system, Global Positioning System (GPS).

Description
The SNS-3 is a 24-channel (12 GLONASS, 12 GPS) satellite receiver and aircraft navigation system which computes and outputs the following aircraft navigational and positional data:
- Latitude and longitude
- Altitude
- Groundspeed and vertical speed
- True and magnetic track angles
- UTC time
- Lateral deviation and rate from desired track
- Time and speed requirement to next waypoint (Time Early/Late –TEL)
- Flight plan data
- Reference data.

In common with the SNS-2, SNS-3 is an integrated system, capable of receiving and processing signals from all visible GPS and/or GLONASS constellations. The system is capable of automatic measurement of raw data, almanac and ephemeris download, without preliminary initialisation. These features enable a reduction in selective availability for GPS and satisfactory precision, integrity and availability characteristics for the demands of typical Required Navigation Performance (RNP).

SNS-3 has built-in Receiver Autonomous Integrity Monitoring (RAIM) functionality, utilising data from all tracked satellites and differential capability when RTCM-104 data is available; the system fulfils the technical requirements for onboard GNSS equipment as a primary navigation aid, adopted by the Russian Federal Aviation Service in August 1999.

SNS-3 also able to accept navigation data from other aircraft systems, such as AHRS, ADS or other guidance systems; in the case of satellite information loss, the system is able to maintain a navigational solution for up to 15 minutes without reacquisition. The system features an alphanumeric display and controls to allow for the display and control of all system modes. SNS-3 can be installed on all types of aircraft, both fixed- and rotary-wing.

Specifications
Output data
Position (95%):
≤35 m (0.01 arc min indication discreteness)
Altitude (95%):
≤35 m (1 m indication discreteness)
Groundspeed: 15 cm/s (1 km/h indication discreteness)
UTC time: 0.01 s (1 s indication discreteness)
True track angle: 10 arc min
Magnetic track angle: 1.5° (0.1° indication discreteness)
DTG: ≤35 m
Track deviation: ≤35 m
Timestamp: 70 ηs
Interface: GOST 1877-79/ARINC 429
Input: 24 receiving channels (12 GLONASS, 12 GPS)
Output: WGS-84 geodetic system of user specified
Physical
Antenna frequency: 1,575.42 to 1,609.0 ±18.0 MHz
Weight: 4 kg (antenna unit, antenna amplifier, satellite receiver)
Power: 27 V DC; 10 W
Readiness time: 2.5 min
Navigation database: 9999 navigation waypoints (100 routes, SIDs and STARs, airport data and intersection points)

Status
The SNS-3 is designed for civilian fixed-wing and helicopter applications. The system is KT-34-01-certified as an A1/C1 class system.

Contractor
Aviapribor Corporation.

SNS-2 satellite navigation receiver showing the antenna unit, amplifier unit and receiver 0103871

SNS-3 satellite navigation receiver

0103872

Specifications

	SVS 85	SVS 2Ts-U	SVS 2Ts-U2	SVS 96
Inputs				
Static pressure	11.55–107.6 kPa	1.07–133.3 kPa	1.07–133.3 kPa	11.55–107.6 kPa
Total pressure	11.55–135.4 kPa	6.66–279.9 kPa	6.66–279.9 kPa	11.55–135.4 kPa
Stagnation temperature	−60 to +99°C	−60 to +350°C	−75 to +400°C	−60 to +99°C
Local angle of attack	−60 to +60°	−60 to +60°	−60 to +60°	−60 to +60°
QFE and QNH	57.7–107.4 kPa	70.1–107.4 kPa	70.1–107.4 kPa	57.7–107.4 kPa
Outputs				
True altitude range	−500 to 15,000 m	−500 to 30,000 m	−500 to 30,000 m	−500 to 15,000 m
True altitude error	–	–	–	±5 m (to 0 m) to ±25 m (to 15,000 m)
Vertical speed range	−100 to 100 m/s	−500 to 500 m/s	−500 to 500 m/s	−100 to 100 m/s
Vertical speed error	–	–	–	±0.15 m/s or 5%
Indicated airspeed range	–	–	–	50 to 800 km/h
Indicated airspeed error	–	–	–	±10 km/h (to 50 km/h) to ±4.5 km/h (to 800 km/h)
True airspeed range	–	–	–	80 to 1100 km/h
True airspeed error	–	–	–	±18 km/h (to 80 km/h) to ±8 km/h (to 1100 km/h)
Mach no range	0.1–1.0	0.2–3.0	0.2–3.0	0.1–1.0
Mach no error	–	–	–	±0.015 (at M0.1)
				±0.005 (at M0.6)
				±0.003 (at M0.8)
				±0.01 (at M1.0)
Stagnation temperature range	−60 to +99°C	−60 to +350°C	−75 to +400°C	−60 to +99°C
Stagnation temperature error	–	–	–	±0.5°C
Outside temperature range	−99 to +60°C	−60 to +60°C	−75 to +60°C	−99 to +60°C
Outside temperature error	–	–	–	±1.0°C
Angle of attack range	−60 to +60°	−60 to +60°	−60 to +60°	−60 to +60°
Angle of attack error	–	–	–	±0.25°
Angle of slip range	–	−30 to +30°	−30 to +30°	−30 to +30°
Angle of slip error	–	–	–	±0.25°
Power	115 V AC, 400 Hz, 50 VA	115 V AC, 400 Hz, 50 VA	115 V AC, 400 Hz, 50 VA	27 V DC <20 VA
Weight	6.5 kg	7.0 kg	4.5 kg	3.5 kg
Interface	ARINC 706	Bipolar code	Bipolar code	ARINC 706
Dimensions	4 MCU	4 MCU	3 MCU	1 MCU

SVS series of digital air data computers

Type
Avionic navigation sensor, Air Data Computer (ADC).

Description
The SVS series of digital air data computers provide measured and computed data on: true altitude; vertical speed; Mach number; stagnation temperature; outside temperature; angle of attack; and angle of slip. Inputs include static pressure, total pressure, stagnation temperature, local angle of attack, local angle of slip, QFE and QNH.

Status
Systems are fitted to all types of Russian military and civil aircraft.

SVS series of digital air data computers 1026864

SVS series of digital air data computers

1026865

Contractor
AeroPribor-Voskhod Joint Stock Company.

SVS-V1 helicopter air data computer system

Type
Avionic navigation sensor, Air Data Computer (ADC).

Description
The SVS-V1 air data computer is designed for measurement of altitude and speed data on

SVS-V1 helicopter air data computer system

0018240

helicopters. It comprises two line-replaceable units: computer, airspeed vector transmitter.

Specifications
Airspeed longitudinal component: −90 to +450 km/h, (±3.5–5 km/h)
Airspeed lateral component: 0–90 km/h, (±3.5 km/h)
Indicated airspeed: 30–450 km/h (±2–6 km/h)
Vertical baroinertial speed: −30 to +30 m/s, ±0.3 ms
Pressure equivalent altitude: −500 to +7,000 m, ±6 m
Outside air temperature: −60 to +60°C, ±2°C
Weight: 6 kg
Power: 30 VA

Contractor
Aviapribor Corporation.

Sweden

NINS/NILS autonomous navigation and landing system

Type
Aircraft navigation and landing system, multimode.

Description
Saab NINS is a total navigation solution, including modular hardware and software based on digital geographical databases: its main characteristics are that it utilises inertial or Doppler navigation system, radar or laser altimeter and air data system; and that it provides high-realtime position and velocity accuracy, plus error awareness.

Saab NILS is a landing system that takes full advantage of the NINS concept by creating a glidepath based on NINS navigation data. Only onboard sensors are utilised, which reduces cost and increases flexibility for operations from austere or dispersed bases with no ground equipment.

The system offers an alternative navigation solution that is fully realtime and independent of ground- or space-based equipment. The software package can be integrated both with aircraft and cruise missiles, with selected hardware and software features as an 'add-on' package for existing platforms or in new designs where the full benefit of Saab NINS can be utilised. Key features of the system include high position and velocity accuracy; fully autonomous operation; awareness of surrounding terrain.

Saab TERNAV is a non-linear terrain-referenced navigation algorithm that has been developed over 20 years and achieves a performance level comparable to military GPS. The system is operational in Swedish Air Force AJS 37 Viggens and in full-scale development for deployment in the JAS 39 Gripen.

A high-precision continuous position fix can be derived from the basic sensor data Inertial Navigation System (INS), Air Data Computer (ADC) and Radar Altimeter (RA) and from the Geographical Information System (GIS) database.

The GIS database and server is where all terrain databases, including those for landing, are stored and prepared for real-time access. TERNAV is one application that uses GIS, together with others such as passive terrain-following and ground-proximity warning.

The data fusion function handles all incoming and outgoing data streams from sensors and system functions. It is the decision-maker in the system and feeds data to all the other functions as well as to the cockpit displays.

At the heart of the system is a Kalman filter for optimal processing of all incoming data streams. The filter utilises all navigation information in the system to estimate the position, velocity and attitude of the platform.

The pilot's standard cockpit displays show information output from the data fusion function, including navigation, approach information and ground-proximity warnings.

Integrity monitoring is a diagnostic system for failure detection and exclusion of failed-sensor signals. With support sensors (for example GPS), it is possible to achieve integrity monitoring at an even higher level, allowing for graceful degradation while still supporting many system functions.

Specifications
(dependent on sensors and terrain characteristic)
Horizontal position error: 5–50 m (CEP 50%)
Vertical position error: 2–4 m (1 sigma)
Horizontal velocity error: 0.05–0.5 m/s (CEP 50%)
Vertical velocity error: 0.02–0.2 m/s (1 sigma)
(with negligible degradation for 3 min over water)
Options:
ground proximity warning
terrain following
virtual reality input to the HUD
passive target ranging
integrity monitoring with GPS
mid-air proximity warning with LINCS

Status
NINS is operational in Swedish Air Force AJS 37 Viggen aircraft and is in full-scale development for the JAS 39 Gripen aircraft.

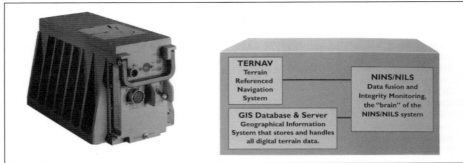

NINS/NILS system hardware (left) and computer architecture (right)　　0051597

NINS/NILS display concept on HUD and head-down displays　　0051598

TERNAV navigation　　0116564

Contractor
Saab Bofors Dynamics AB.

Switzerland

AC32 Digital Air Data Computer (DADC)

Type
Avionic navigation sensor, Air Data Computer (ADC).

Description
The AC32 Digital Air Data Computer (DADC) integrates solid-state pressure sensors for static and pitot pressure, measuring barometric altitude, airspeed and temperature in the atmosphere. The AD32 is Reduced Vertical Separation Minima (RVSM) compliant and provides up to 2 × 16 SSEC curves.

The computed air data parameters are transmitted via the ARINC-429 interface. Two transmit channels and two receive channels are available; an ICAO encoded altitude output is

available as an option. The AC32 is a modular design, with low power consumption (less than 10 W) and low weight (1 kg), optimised for state-of-the-art avionics applications. Extensive Built-In-Test (BIT) assures safe operation and a universal power supply, designed for 14 V DC or 28 V DC facilitates platform integration.

Employing an RS 232 maintenance interface, the AC32 can easily be configured for different applications, ranging from business aviation up to regional aircraft, transports and helicopters. The AC32 DADC meets or exceeds the requirements of the FAA Technical Standard Order (TSO) RVSM accuracy requirements.

Other features and benefits of the system include:
• Solid state pressure sensors
• Continuous BIT
• BIT failure memory.

Specifications
Signal inputs
Primary power: 14/28 V DC
Emergency power: 14/28 V DC
ARINC-429: 2 receive channels (option CSDB, RS 422)
TAT probe: 500 Ω (option 50 Ω) at 0°C

Signal outputs
Encoded altitude: ICAO/TSO C-88a
Warning flag valid: GND/28 V DC
ARINC-429: 2 transmit channels (option CSDB, RS 422)
Pressure altitude: –1,000 to +53,000 ft
Baro corrected altitude: –1,000 to +53,000 ft
Altitude rate: 0 to 20,000 ft/min
Indicated airspeed: 0/40 to 450 kt
Calibrated airspeed: 0/40 to 450 kt
True airspeed: 0/100 to 599 kt

Max allowable airspeed: VMO 150 to 450 kt
Mach number: 0.200 to 0.999 M
Total air temperature: −60 to +99°C
Static air temperature: −99 to +60°C
Baro setting: QNH 20.67 to 31.00 in Hg (700 to 1,050 mbar)

Compliance: FAA TSO-C10b, -C88a, -C106, RTCA/DO-178B Level A, DO-160D
Operating temperature: −20 to +70°C
Reliability: MTBF 20,000 hr (estimated)

Status
Introduced during the first quarter of 2004. Fully certified.

Contractor
Revue Thommen AG.

Turkey

LN100G EGI Embedded GPS Inertial navigation system

Type
Avionic navigation sensor, Inertial Reference System (IRS).

Description
The LN-100G EGI system is an advanced technology unit that includes a sensor assembly with three Zero-lock™ Laser Gyros (ZLGs), an A-4 accelerometer triad and, five electronics assemblies together with two spare card slots, all installed in a very small, lightweight package.

The core LN-100G INS/GPS is optimised for individual applications by appropriate additions of I/O cards and other modules installed in the spare card slots, by modification of the software for different I/O messages and formats, by modified mode control and tuning of the Kalman filter. The LN-100G is an open architecture and hardware/software flexible unit which can be adapted to various air platforms including rotary-wing, fixed-wing and unmanned air vehicles, without the need to change vehicle architecture.

The LN-100G provides three simultaneous navigation solutions; hybrid GPS/INS, free inertial and GPS only. The LN-100G optimally combines GPS and INS features to provide enhanced position, velocity, attitude and pointing performance, as well as improved acquisition and anti-jam capabilities. The embedded GPS module is an L1/L2 CA/P(Y) code unit capable of accepting RF (or IF) inputs from the GPS antenna system. Stand-alone GPS PVT data and stand-alone INS data are provided for integrity and fault monitoring purposes.

Specifications
Laser gyro features: (18 cm nondithered, Zero-Lock™ Laser Gyro and miniature accelerometer technology)
Accuracy:
(position) <10 m CEP
(velocity) <0.01 m/s
Dimensions: 280 × 180 × 180 mm
Weight: 9.8 kg
Interfaces: MIL-STD-1553B/ARINC/discrete databuses

LN100G EGI (Aselsan)   1128605

Status
As of November 2005, in production and in service in a number of platforms.

Contractor
Aselsan Inc, Microelectronics, Guidance and Electro-Optics Division.

Ukraine

Inertial Measurement Unit (IMU)

Type
Avionic navigation sensor, Inertial Reference System (IRS).

Description
The Central Design Office (CDO) Arsenal claims over 20 years' experience in development of monobloc laser for airborne navigation purposes. Its latest platform-independent IMU product, designed for helicopter and missile applications, uses a three-axis laser gyro and pendulous accelerometers. The laser technology used employs a high-stability, two-frequency He-Ne linear laser; work has also been done with CO_2 lasers.

Specifications
Laser gyro:
angular rates measured: ±90°/sec
zero drift (1 sigma): 0.05–0.1°/h
relative error of scale factor: 10^{-4}–10^{-5}

Accelerometer:
Linear accelerations measured: ±100 m/s²
zero drift (1 sigma): 10^{-3} m/s²
relative error of scale factor: 10^{-4}
Precision readiness time: 60 s
Volume: 10 dm³
Weight: 12 kg
Power consumption: 70 W
Operating temperature limits: −20 to +50°C

Status
In service. Production status unknown.

Contractor
Arsenal Central Design Office.

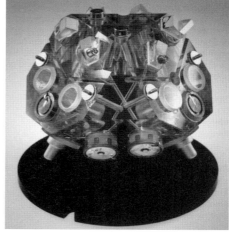

Arsenal CDO Ring Laser Gyro Inertial Measurement Unit (Arsenal)   0097083

United Kingdom

AD2770 TACAN

Type
Aircraft navigation system, radio aids, TACAN.

Description
The AD2770 TACAN navigation and homing aid is suitable for all types of aircraft. This system is used on the majority of front-line aircraft in service with UK forces. It provides range and bearing information from any selected ground TACAN station or from any suitably equipped aircraft and is available in a number of forms offering outputs in digital ARINC 429 or analogue form or a combination of the two. Output signals may be provided to drive range/bearing

or deviation indicators on a pilot's panel or to interface directly with a computer.

The system has full 252 channel X and Y mode capability and operates up to 300 n miles range with a range accuracy of better than 0.1 n mile. Bearing accuracy is said to be better than 0.7° on normally strong input signals. It comprises two units: a transmitter/receiver hard-mounted in the avionics bay and a panel-mounted remote-control unit. A switching unit, also installed in the avionics bay is required when two antennas are fitted. A mounting tray with a cooling air blower is necessary if a cooling air supply is not available from the aircraft's own air-conditioning system.

The transmitter/receiver section with a digital interface only is contained in a 3/4 ATR short case

with a front doghouse. Versions with analogue outputs are accommodated in a case of longer dimensions to house the additional circuitry. Signal processing circuitry is largely digital in the interests of system reliability, and continuous integrity monitoring techniques eliminate the risk of erroneous outputs.

Range and bearing analysis and output formats are prepared in a general purpose computer module called the analyser. This allows both range and bearing signal processing to use the same circuitry, with a consequent reduction in the number of components. The range system uses a parallel search method, said to be unique, which by making use of all signal returns achieves a very rapid lock on.

The AD2770 system operates in three modes: receive (giving bearing information only); transmit and receive; and air-to-air (providing range information only). Transmitter frequency range is from 1,025 to 1,150 MHz with an output of 2.5 kw peak pulse power. Receiver frequency coverage is from 962 to 1,213 MHz. The tracking speed range is from 0 to 2,500 kt.

The design of the AD2770 system is flexible both electrically and mechanically and alternative configurations with appropriate form factors, output characteristics and mechanical and electrical interfaces can be provided for new aircraft or for retrofit.

Specifications
Dimensions:
(control unit) 57 × 146 × 83 mm
(transmitter/receiver standard unit) 194 × 191 × 380 mm
(antenna switch) 69 × 130 × 56 mm
Weight:
(control unit) 0.45 kg
(transmitter/receiver standard unit) 14 kg
(antenna switch) 0.25 kg

Status
In service in the Tornado GR. Mk 1 and F3 and Nimrod MR. Mk 2 aircraft.

Contractor
Selex Sensors and Airborne Systems.

AD2780 TACAN

Type
Aircraft navigation system, radio aids, TACAN.

Description
A follow-on from the AD2770 series supplied for the Panavia Tornado and other front-line types, the AD2780 is also proposed for military applications. The system provides slant range and relative bearing to a standard Tacan station, range rate (which approximates to groundspeed when not used in conjunction with the company's area navigation system), time to go to waypoint or station, ARINC 429 serial data output and 252 channels in X and Y modes. Outputs of range are available in digital format to ARINC 429 and in analogue to dial and pointer displays.

Specifications
Dimensions: 127 × 153 × 318 mm
Weight: 3.5 kg
Frequency:
(transmitter) 1,025–1,150 MHz
(receiver) 962–1,213 MHz
Range rate output: 0–999 kt with accuracies ±15 kt for 0–300 kt, and ±5% for 300–999 kt
Time to station output: 0–99 min
Tracking speed: 0–1,900 kt, 0–20°/s
Memory:
(range) 10 s
(bearing) 4 s

Status
In production and service in the British Army Gazelle and Royal Netherlands Air Force Eurocopter BO 105 helicopters and UK Royal Air Force Shorts S312 Tucano trainers.

Contractor
Selex Sensors and Airborne Systems.

Aircraft sensors

Type
Avionic flight/navigation sensor, temperature/pressure.

Description
Weston Aerospace manufactures a range of aircraft sensors:

Temperature Sensors
Thermocouple probes and systems, Resistance Temperature Devices (RTDs), thermistors,

harnesses and junction boxes. Gas turbine engine applications include measuring: inlet air temperature, compressor stage temperatures, TGT/EGT, turbine inlet temperature as well as providing front and rear overheat detection. Airframe applications include measuring outside air temperature, cargo bay and cabin temperatures and hydraulic fluid temperatures. Other applications include industrial and marine gas turbine engines, environmental control systems, fuel and braking systems.

Speed and Torque Sensors
Speed probes for gas turbine shaft speed, rotor speed and torque probes for output shaft torque measurement. Standard to the product range are probes that operate at temperatures up to 500°C at the tip and with over 820°C body temperatures. Transformer speed sensors generate a high, multiple-channel voltage output, even when operating in very small spaces and at high temperatures.

Pressure Transducers
Pressure sensors for high-accuracy, high-stability, absolute air pressure measurement applications such as Air Data Computers (ADC) and Full Authority Digital Engine Control (FADEC) systems. Due to their high accuracy, proven reliability in both civil and military applications and exceptionally low drift, Weston Aerospace pressure transducers are used in airborne and ground-based RVSM applications.

Status
In production and in service.

Contractor
Weston Aerospace, an operating unit of Esterline Sensors.

CDU/IN/GPS

Type
Aircraft navigation system.

Description
Thales Avionics is responsible for full integration of an Inertial Navigation and Global Positioning System (IN/GPS) and a Control Display Unit (CDU) to interface with the Central Tactical System of the UK Nimrod MRA4 aircraft. The CDU concerned is a derivative of the AMS 2000 Control Display and Navigation Unit, made by Thales Avionics, and the combined unit is to be known as the CDU/IN/GPS.

Status
In development for Nimrod MRA4.

Contractor
Thales Avionics SA.

Digital air data computers for civil aircraft

Type
Avionic navigation sensor, Air Data Computer (ADC).

Description
A range of digital air data computers has been produced for both turboprop and civil jet transports. These computers meet the requirements of ARINC 706 specification, with outputs in both ARINC 429 and analogue format. They incorporate high-performance vibrating cylinder sensors and CMOS technology to minimise weight, space and power requirements.

Extensive BITE facilities are available, including a 10-flight memory to facilitate onboard checks and bench diagnostic routines.

This range of computers is demonstrating extremely high reliability on the Boeing 737 with an MTBF in excess of 30,000 flying hours.

Specifications
Dimensions: ? or ½ ATR
Weight: 4–5.9 kg
Power: 20 VA

Status
In service on the Boeing 737-300, -400 and -500, the BAE Systems ATP and Boeing 727 upgrade.

Contractor
Smiths Aerospace.

Doppler velocity sensors

Type
Airborne navigation sensor.

Description
There are five models in Thales' (formerly Racal) Doppler velocity sensor family: the Doppler 71 and 72 antenna units for helicopters and fixed-wing aircraft respectively; the Doppler 80 for helicopters and light aircraft; and Doppler 91 and 92 antenna units for helicopters and fixed-wing aircraft respectively.

The Doppler 91 and 92 units use the best features of the previous models and incorporates microprocessor technology and new manufacturing techniques. These units use waveguide antennas and varactor multiplier transmitters for high accuracy at greater altitudes.

The Doppler 80, with printed antennas Gunn diode RF source and switched beams, is for low-level helicopters operations where low weight is particularly important.

Antenna units for helicopters have a speed range of –50 to +300 kt forward and 100 kt laterally. The fixed-wing sensors have a corresponding speed range of –50 to +1,000 kt and 200 kt laterally. Velocity data can be provided in either analogue or ARINC 429 digital format and MIL-STD-1553 databus may be specified. The microwave signals produced by the Doppler 90 series are specially tailored to reduce errors created by heavy rain, snow and hail.

The transmission characteristics of the Doppler 91 and 92 units can be remotely controlled. A low-power stealth mode can be selected for minimum detectability, or the transmitter switched off if the beam goes above the horizon, for example when the aircraft is banking. They transmit information to other aircraft systems in ARINC 429 serial digital data form, but the MIL-STD-1553B remote terminal format can also be supplied.

The Doppler 91 can be supplied in a configuration that is optimised for rotary wing over-water operations, where transition to hover and auto-hover are autopilot functions dependent upon the maximum amount of continuously available velocity data.

Specifications
Dimensions:
(Doppler 71/72) 406 × 406 × 127 mm
(Doppler 80) 356 × 381 × 80 mm
(Doppler 91/92) 358 × 391 × 118 mm
Weight:
(Doppler 71/72) 16.5 kg
(Doppler 80) 8.6 kg
(Doppler 91/92) 11 kg
Power supply:
(Doppler 71/72) 115 V AC, 400 Hz
(Doppler 80 and 91/92) 28 V DC

Status
In production and in service. The Doppler 91 has been selected for the EH 101, Sea King, and Lynx helicopters. The Doppler 71 is installed on UK Royal Air Force Aerospatiale/Westland Puma, Sikorsky/Agusta SH-3D, Westland Sea King and UK Royal Navy and British Army Lynx helicopters and Belgian Air Force Sea King Mk 48s. Over 3,500 Doppler 71s are in service.

Contractor
Thales Avionics SA.

FIN 1000 series inertial navigation systems

Type
Avionic navigation sensor, Inertial Reference System (IRS).

Description
FIN 1000 is the designation of a family of inertial navigation systems based on the BAE Systems gimballed inertial platform that use floated rate integrating gyros and precision force feedback accelerometers. The group includes particular systems optimised for long-term high accuracy and for rapid reaction alignment. Versions include:

FIN 1010
Developed for the Panavia Tornado, the FIN 1010 has all-digital interfaces and an accuracy of better than 1 nm/h. Another version of the Tornado system, with analogue interfaces and comprehensive route navigation, is fitted to the Mitsubishi F-1 advanced trainer.

FIN 1012
Fitted to UK Royal Air Force Nimrod MR 2 aircraft, the FIN 1012 is optimised for long-term accuracy.

FIN 1031
The FIN 1031 Navigation Heading and Attitude Reference System (NavHARS) is fitted in UK Royal Navy Sea Harrier FA2 aircraft. The inertial platform is stabilised by twin two-axis ruggedised oscillogyros developed by BAE Systems. NavHARS utilises other aircraft sensors, which provide Doppler radar velocities, true airspeed and flux valve magnetic heading.

A 2-minute alignment can be achieved on land or sea with further refinements when the aircraft is airborne.

The FIN 1031B NavHARS has been fitted as part of the mid-life update installation in the UK Royal Navy Sea Harrier FA2. It is similar to the FIN 1031 system but has two independent dual-redundant MIL-STD-1553B databusses. The addition of a twin MIL-STD-1553B databus is primarily to facilitate interfacing with the Blue Vixen radar, AMRAAM and revised avionics such as the new bus control interface unit.

Specifications
Inertial platform unit
Dimensions: 212 × 215 × 332 mm
Weight: 11.9 kg

Processor unit
Dimensions: 261 × 199 × 381 mm
Weight: 13.64 kg

Control/display unit
Dimensions: 147 × 152 × 139 mm
Weight: 2.5 kg
Power supply: 200 V AC, 400 Hz, 3 phase
28 V DC for switching and lighting

FIN 1064
FIN 1064 is an integrated navigation and attack system, providing inertial navigation and a wide range of weapon delivery modes. The system has been in service since the early 1980s in the UK Royal Air Force Jaguar ground attack/reconnaissance aircraft, and is also fitted to Jaguars of the Omani and Ecuadorian air forces. Navigation data is provided by a FIN 1000 series gimballed inertial platform. The system was fitted in a mid-term upgrade of the aircraft systems and provides a suite of analogue and digital interfaces to the aircraft sensors and control/display systems; it also provides the capability to upgrade the system software at LRU level using a portable, solid-state, programme loader. Weapon aiming computation is provided for both air-to-air and air-to-ground modes. More recently, FIN 1064 has been updated to incorporate a MIL-STD-1553B databus, integration of inertial and GPS navigation data, and integration with the TIALD (Thermal Imaging and Laser Designator) pod. The mission and weapon-delivery capability of the system has been significantly enhanced by in-service software upgrades, in line with the expanding role of the Jaguar aircraft.

FIN 1075
The FIN 1075 inertial navigation system was selected for the Harrier GR5 and GR7 and is currently in service with the UK Royal Air Force. The system is form, fit and function interchangeable with the AN/ASN-130 navigation system in the US Marine Corps AV-8B. The inertial platform interfaces with the Harrier GR7 databus, avoiding the need for a dedicated control/display unit.

The inertial platform can be aligned on land, at sea or in flight. The ground alignment mode has a wander azimuth and does not require an initial heading input.

Specifications
Dimensions: 193 × 286 × 356 mm
Weight: 20 kg
Reaction time: <3 min
Gyrocompass align: 7 min for 0.8 nm CEP
Accuracy:
  (navigation) 0.8 nm CEP
  (heading) ±0.1°
  (attitude) ±0.1°
Reliability: 1,500 h MTBF

FIN 1075G
The FIN 1075G is a variant of the FIN 1075 which is designed to operate in conjunction with a stand-alone GPS receiver to provide a continuous, precision navigation solution under high dynamic conditions with significant periods of GPS outage.

FIN 1075G has been evaluated at Boscombe Down and has been used extensively in overseas Harrier GR7 operations.

The same system is also capable of a GPS-aided moving base alignment and is currently being assessed to provide the Harrier GR7 with an 'at sea' alignment capability.

Status
In service.

Contractor
Selex Sensors and Airborne Systems.

FIN 3110 GTI

Type
Avionic navigation sensor, Inertial Reference System (IRS).

Description
The FIN 3110 GTI (GPS, Terrain, Inertial) is a variant of the FIN 3110 navigation system designed for standoff missiles and aircraft.

The system is based on a Ring Laser Gyroscope (RLG) inertial system capable of autonomous operation to 0.8 n mile/h and includes two sophisticated Kalman filters for the provision of horizontal and vertical integrated navigation solutions using additional sensor data. In addition to the basic inertial system the unit has the capacity to be equipped simultaneously with an embedded military GPS receiver, a Digital Terrain System (DTS) and special-to-type analogue and/or digital interfaces.

The GPS receiver can be either military code or civil C/A code.

The DTS module provides both terrain referenced navigation and terrain following functions and is comprised of: terrain elevation data storage memory (EEPROM); mission specific digital terrain elevation data; data processing hardware; terrain-referenced navigation update algorithms; and terrain-following algorithms providing steering commands.

A feature of the FIN 3110 GTI is the direct use of GPS and TRN measurement data in the same integrated navigation Kalman filter to give improved performance and increased robustness through improved sensor cross-mounting.

The overall size of the FIN 3110 GTI is the same as the basic FIN 3110.

Contractor
Selex Sensors and Airborne Systems.

FIN 3110G ring laser gyro INS/GPS

Type
Avionic navigation sensor, Inertial Reference System (IRS).

Description
The FIN 3110 INS/GPS is designed to meet the requirements of military aircraft, helicopter, self-propelled howitzers, artillery, land vehicles and marine craft.

The FIN 3110 is an Integrated Navigation System (INS) consisting of a ring laser gyro inertial sensor and an embedded Global Positioning System (GPS) receiver module. This system is capable of providing precise and continuous outputs of navigation heading and attitude data to the weapon and flight control systems.

The FIN 3110 is a small (177.8 × 177.8 × 279.4 mm), lightweight (10.5 kg) unit consuming just 55 W from a 28 V DC power source. The INS and GPS are closely integrated in a Kalman filter, which combines the high position accuracy of the GPS receiver with the angular rates and linear acceleration of the INS on a 1553B databus. Other interfaces could be available if required. The applications software is written in Ada and is executed on a Motorola 68040 processor.

The FIN 3110 was the culmination of 15 years of research and development of ring laser gyro technology. The advent of the single module GPS receiver and processor in 1993 allowed the integrated navigation system to be manufactured for the UK MoD and export programmes.

Specifications
Alignment times:
(gyrocompass) 4 min
(rapid reaction) 30 s
Inertial performance:
(position) <0.8 n mile/h CEP
(velocity) (N,E) <2.5 ft/s RMS
(velocity) (vertical) <2 ft/s RMS
GPS performance:
(position) (spherical error) <16 m SEP
(velocity) (per axis) <0.1 m/s RMS
MTBF: >5,000 h
Power: 28 V DC, <55 W
Dimensions: 177.8 × 177.8 × 279.4 mm
Weight: 10.5 kg
Environmental requirements:
MIL-E-5400T;
Temperature altitude operation to Class 2 (optionally Class 2X for forced cooling);
Tested to MIL-STD-810E (temp, altitude, vibration), MIL-STD-461C (EMC), MIL-STD-704E (power supply)
Interfaces:
1 or 2 dual-redundant MIL-STD-1553B (RT or bus control); PTTI/Have Quick; RF for GPS antenna; Dual RS-422 instrumentation; ARINC 429; Synchro/analogue; Panlink

Status
In production.

Contractor
Selex Sensors and Airborne Systems.

Helicopter Air Data System (HADS)

Type
Avionic navigation sensor, Air Data System (ADS).

Description
In 1979, BAE Systems designed and developed a low airspeed Helicopter Air Data System (HADS) for the AH-1S Cobra helicopter.

The system provides full three-axis, prime accuracy, air data information by utilising the Airspeed And Direction Sensor (AADS). A variant, known as the High-Integration Air Data Computer (HIADC), has been developed. It provides three-axis air data information on modern digital busses such as MIL-STD-1553B.

The major parts of the system are shown below:

Airspeed And Direction Sensor (AADS)

The AADS probe is mounted externally to the aircraft below the rotor and swivels to align with the local airflow. The AADS contains pitot and static pressure ports for measuring the magnitude of the local airspeed vector and a pair of resolvers for determining its direction. This enables the system to provide data for: rotor downwash velocity; ground effect; forward, rearward and lateral airspeed (to zero kt); vertical airspeed; wind direction, drift and lift margin (when integrated into an avionics suite); enhanced pilot awareness.

Installation of the system produced major improvements for operational and flight test environments in: fire control; low airspeed, low altitude manoeuvres.

HIADC Interface

The HIADC interface integrates with the AADS by measuring its probe angle, air temperature and pitot and static pressures. Air data calculations are performed and the resultant parameters are made available on digital databusses.

Specifications
Airspeed and direction sensor
Dimensions: 97 × 317 × 246 mm
Weight: 1.1 kg

High integration air data computer
Dimensions: 140 × 102 × 83 mm
Weight: 1.2 kg

Status

Over 1,500 systems have been supplied for attack helicopters including Bell AH-1S Cobra, Agusta A-129 and Super Puma. A major order was received in 1997 for HADS in support of US Army AH-64D Longbow and British Army WAH-64 Apache helicopters.

Contractor

Selex Sensors and Airborne Systems.

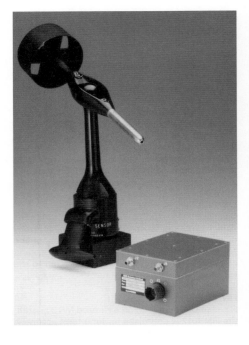

Helicopter Air Data System (HADS)
0001285

High-Integration Air Data Computer (HIADC)

Type
Avionic navigation sensor, Air Data Computer (ADC).

Description
The High-Integration Air Data Computer (HIADC) utilises advanced production techniques and miniaturised air data transducers. It features low individual component count and power consumption to provide an accurate, highly reliable, compact air data computer to meet the growing market demand for distributed pressure sensing devices.

The flexible configuration concept of HIADC offers the ability to satisfy applications for both fixed-wing aircraft and helicopters.

HIADC typically interfaces to the aircraft total air temperature and angle of attack sensors; measures pitot, static and differential pressures and, having corrected for systematic error characteristics, computes a full range of accurate air data parameters. Data is digitally distributed to various aircraft systems in either ARINC 429, RS-422 or MIL-STD-1553B output formats.

HIADC has been designed as a fit-and-forget air data system, which virtually eliminates the requirements for maintenance and provides significant life cycle cost benefits.

Specifications
Dimensions: 140 × 82 × 102 mm (max)
Weight: 1.2 kg
Power supply: 28 V DC, 6 W
Reliability: >25,000 h MTBF

Status
In production for numerous fixed- and rotary-wing programmes worldwide.

Contractor
Selex Sensors and Airborne Systems.

High-Integration Air Data Computer (HIADC) 0002346

LINS 300 Laser Inertial Navigation System

Type
Avionic navigation sensor, Inertial Reference System (IRS).

Description
Selex (formerly BAE Systems Avionics) manufactures a family of LINS 300 SNU 84-1 inertial navigation equipment which meets the requirements of many different types of aircraft. LINS 300 provides vehicle acceleration, velocity, position, heading and attitude. Outputs are available in MIL-STD-1553B, ARINC 429 and synchro formats. For the EH 101 Merlin, LINS can align when airborne using a Kalman filter to interface with a Doppler velocity sensor or GPS; this also enables shipborne alignments. An ARINC 429 interface and Doppler/GPS air alignment mode are unique among SNU 84-1 format systems.

LINS 300-10 is the baseline SNU-84 system used on fixed-wing aircraft and helicopters. LINS

300-20 is the baseline ARINC 429 system with air alignment.

Specifications
Dimensions: 460 × 191 × 194 mm
Weight: 20 kg
Alignment time:
(gyrocompass) 8 min
(stored heading) 1.5 min
Accuracy:
(position) 0.8 n mile/h CEP
(velocity) (horizontal) 2.5 ft/s
(vertical) 2 ft/s
(rates) 400°/s

Status
In 1990, LINS 300 was selected as baseline equipment for the Hawk 100 and 200 series aircraft and in 1995 for UK Royal Air Force variants of the EH 101 helicopter. The system is currently in production and in service.

Contractor
Selex Sensors and Airborne Systems.

PA9052SM GPS receiver

Type
Avionic navigation sensor, Global Positioning System (GPS).

Description
The PA9052SM GPS receiver provides full Precise Positioning Service (PPS) capabilities and includes

all the interfaces expected of a military GPS receiver including MIL-STD-1553B, ARINC 429/575, PTTI and RS-422. Standard Positioning Service (SPS) variants are also available. As well as generating a basic navigation solution, the PA9052SM can provide area navigation facilities and is suitable for use in an integrated navigation system.

BAE Systems is also providing a five/six-channel GPS receiver module for the Eurofighter

Typhoon project. This module also provides full PPS capability and is suitable for use in embedded applications.

Specifications
Dimensions:
(PA9052 receiver) 216 × 194 × 90 mm
(PA9915 antenna) 89 mm diameter
Weight: 4.3 kg

PA9000 Series of GPS equipment showing the airborne GPS receiver, antenna, preamplifier and control/display unit module, together with the naval receiver, control/display unit and antenna
0504179

Generic Air Data Unit (GADU) ADC 0005416

Status
Deliveries of GPS receivers to customers worldwide started in 1989. They were selected for the development phase of the AH-64D Apache Longbow helicopter and are currently in service on UK Royal Air Force Tornado, Jaguar, Harrier and Nimrod aircraft.

Contractor
Selex Sensors and Airborne Systems.

PA9360 GPS modules

Type
Avionic navigation sensor, Global Positioning System (GPS).

Description
The PA9360 family of GPS modules provides a flexible solution for applications requiring an embedded GPS receiver capability. The PA9361 module is the first in the family and is designed for applications in the airborne, naval and land environments. Advanced ECL and VLSI ASIC technology has been used to achieve full six-channel, dual-frequency capability within a single module.

The PA9361 module uses software from the PA9000 Series of military GPS receivers and is suitable for embedded use in a range of navigation equipments. Installation and support are simplified by the incorporation of a PPS-SM device which ensures that the module is unclassified, even when loaded with encryption keys, and the ability to power feed a remote preamplifier via the single-cable RF input. Performance is maximised by the availability of 10 Hz GPS measurements and full navigation capability in high dynamic environments, even in the unaided mode. Extensive BITE and a high inherent reliability combine to lower the customer's support tests.

Specifications
Dimensions: 150 × 150 × 25 mm
Weight: 1.4 kg
Power supply: 5 V DC, 15 V DC or battery
Temperature range: –54 to +71°C

Accuracy:
(position) 16 m (SEP)
(velocity) 0.2 m/s (95%)
(time) 100 ns (1σ)
Reliability: >10,000 h MTBF calculated

Status
In service.

Contractor
Selex Sensors and Airborne Systems.

Primary Air Data Computers (ADC)

Type
Avionic navigation sensor, Air Data Computer (ADC).

Description
Penny & Giles Aerospace produces primary accuracy ADCs in both miniature and ARINC 565 standard form factors. The Generic Air Data Unit (GADU) miniature ADC is a compact lightweight unit with integrated pitot/static solid-state sensors in a rugged enclosure. The GADU is available with either ARINC 429 or MIL-STD-1553 databus interfaces. A height lock function is provided on the GADU and is therefore particularly suited to rotary-wing applications.

Specifications
Miniature primary ADC GADU D60350
Dimensions: 89 × 132 × 172 mm
Weight: 1.3 kg
Power: 28 V DC, 10 W peak
Standards: RTCA DO-160C/178B
Interfaces: ARINC 429/575, airspeed and height lock, MIL-STD-1553 (optional)
MTBF: 15,000 flight hours

Status
Fitted to a wide range of aircraft and helicopters including: Sikorsky S70/S76, Westland Lynx/Sea King, Eurocopter Dauphin, MBB BK117 helicopters; and Boeing 727/737, Lockheed Martin C-130 aircraft. Applications include: primary air

data system; rotary- and fixed-wing versions for RSVM retrofit; FMS; TCAS; GPWS; windshear detection.

Contractor
Penny & Giles Aerospace Ltd.

RNav 2 navigation management system

Type
Aircraft navigation system.

Description
Certificated by the UK Civil Aviation Authority in 1984, and now TSO C129 A/C115 B compliant, the RNav 2 navigation management system is in widespread use by helicopter operators supporting the North Sea oil industry and search and rescue, corporate and special mission operators worldwide. It is also in use in military rotary- and fixed-wing aircraft where the operating environment does not justify the expense of extended military environmental specifications.

The equipment comprises a Control and Display Unit (CDU) and a Navigation Computer Unit (NCU), and can accept inputs from GPS, Decca, VOR/DME, Loran C and Doppler. Four separate navigation plots are maintained within the system, any of which may be selected for guidance. In the latest systems any navigation plot with temporarily invalid sensor input is able to revert to Doppler updating. With certain sensor combinations a total of five input sensors is therefore possible.

The computer provides a variety of display and guidance outputs. Analogue outputs are available for the retrofit installation while digital databusses permit interfacing with EFIS and digital AFCS. Inputs of fuel flow can be accepted from a variety of flowmeter types. Fuel computations include range, endurance, fuel remaining over each waypoint and at destination and estimates of these values when a helicopter is in the hover.

A choice of CDU types is available with letters first or numbers first alphanumeric keys, Gen II or Gen III NVG capability and compatibility with Health and Usage Monitoring Systems (HUMS).

NCU hardware variants provide for interfacing with a variety of VOR/DME types, HSIs and RMIs. Software variants offer options such as grid navigation, vertical navigation, transition down and compatibility with the

The display for the Thales Avionics RNav 2 area navigation system 0503890

Eurocopter Super Puma Mk II integrated flight data system.

RNav 2 is operating worldwide using GPS as one of its input sensors. It will interface with GPS receivers which conform to ARINC 743. It forms the core of search and rescue systems certified by a growing number of airworthiness authorities for IFR use in helicopters such as the Eurocopter Super Puma and the Sikorsky S-76. It is also in demand by fishery and environmental protection agencies in a variety of fixed-wing aircraft.

Specifications
Dimensions:
(control/display unit) 146 × 114 × 208 mm
(navigation computer unit) 124 × 184 × 324 mm
Weight:
(control/display unit) 2.7 kg
(navigation computer unit) 5.5 kg

Status
In production and in service. Employed in large numbers for offshore oil support. Military users include the UK MoD, United Arab Emirates AB-412 SAR and Royal Norwegian Air Force Sea King Mk 43 update.

Also fitted to fixed-wing aircraft such as Dornier 228s of the UK Ministry of Agriculture, Food and Fisheries, Cessna Caravan IIs of the Scottish Department of Agriculture, Food and Fisheries, Dutch police Turbine Islanders and British Army Defenders.

Contractor
Thales Avionics SA.

RNS 252 navigation system

Type
Aircraft navigation system.

Description
RNS 252 is a single-unit panel-mounted navigation computer which can accept inputs from Doppler and one additional sensor such as GPS or Loran C. Sensor control is exercised through the computer keyboard and sensor data is accessed on the computer's dot matrix display.

The system accommodates 200 waypoints which are numbered but may also carry a five-character ident. Any or all waypoints may be vectored if required. Waypoints may be loaded manually through the keyboard or automatically through a data transfer device. A variety of steering modes is available. Steering guidance is available in the basic form of steering arrows on the computer display, but outputs are available for driving instrumentation and both AC and DC analogue autopilots. A weather radar output provides a navigational overlay for use with digital colour radars.

RNS 252 is compatible with Thales' (formerly Racal Avionics) Type 70, 80 and 90 Doppler sensors and with GPS sensors which conform to ARINC 743.

Specifications
Dimensions:
(standard version) 124 × 146 × 201 mm
(Supertans version) 161 × 146 × 240 mm
Weight:
(standard version) 3.5 kg
(Supertans version) 4.2 kg
Power supply: 28 V DC, 40 W

Status
In production. In service in fixed- and rotary-wing aircraft worldwide including British Antarctic Survey Twin Otters, Indonesian Army BO 105s, Norsk Luftambulanse BK 117s and Royal Moroccan Air Force Puma and Gazelle helicopters. Versions with GPS are in service in UK Royal Air Force Chinook and Puma, UK Royal Navy Sea King and British Army Lynx helicopters.

Contractor
Thales Avionics SA.

SAHIS Standby Attitude, Heading and rate of turn Indicating System

Type
Avionic navigation sensor, Attitude Heading Reference System (AHRS).

Description
SAHIS is a Standby Attitude, Heading and rate of turn Indicating System designed primarily for use as an emergency back-up in the event of failure of aircraft generated power and/or loss of primary aircraft attitude and heading information.

The system consists of a three-axis spherical indicator, mounted in the aircraft's instrument panel, and a gyro unit, mounted remotely. The gyro unit contains a vertical gyro for attitude reference and a directional gyro for heading reference.

The indicator displays pitch, roll and heading information by means of a sphere moving behind a fixed aircraft symbol. There is a roll scale below the sphere and slip and rate of turn indications below the roll scale.

All signals and power inputs to the indicator (except slip) are derived from or via the gyro unit.

The system can operate in either of two modes whereby heading information is derived from the aircraft's inertial navigation system or from the unit's own directional gyro.

Specifications
Freedom:
(heading) unlimited
(roll) unlimited
(pitch) ±85°
Accuracy:
(heading) ±1°
(roll) ±1°
(pitch) ±1°
Rate of turn: up to 380°/min
Dimensions:
(indicator) 81 × 203 × 81 mm
(gyro unit) 110 × 265 × 170 mm

Standby Attitude, Heading and rate of turn Indicating System 0018194

Weight:
(indicator) 1.8 kg
(gyro unit) 5.6 kg

Status
SAHIS is in service on UK Royal Air Force Harrier GR7 and T Mk 10 aircraft.

Contractor
Ferranti Technologies Limited.

Secondary Air Data Computers (ADC)

Type
Avionic navigation sensor, Air Data Computer (ADC).

Description
The Digitas and TAS/Plus ADCs provide a wide range of outputs including: altitude; airspeed; airspeed rate; baro corrected altitude; static air temperature; true airspeed, Mach and total air temperature. Depending on the model, these outputs are available on various avionic interfaces, including ARINC 429/575, MIL-STD-1553, RS 232/422 and analogue.

Specifications
Miniature primary ADC GADU D60350
Dimensions: 89 × 132 × 160 mm
Weight: 1.1 kg
Power: 28 V DC, 10 W peak
Standards: RTCA DO-160B/178B
Interfaces: ARINC 429/575, airspeed
Optional interfaces: MIL-STD-1553
MTBF: 15,000 flight hours
Applications: Secondary air data source, rotary- and fixed-wing flight data recording, HUMS

Secondary ADC TAS/Plus 90004
Dimensions: 3/8 ATR short
Weight: 2.04 kg
Power: 28 V DC, 10 W peak
Standards: RTCA DO-160C/178B, TSO-C106
Interfaces: ARINC 429/575, RS-232/-422, altitude rate/airspeed
Optional interfaces: Analogue altitude rate and Mach
MTBF: 10,000 flight hours
Applications: Business and commuter aircraft GPS, FMS, TCAS, GPWS, flight data recording data display

Status
Fitted to a wide range of rotary- and fixed-wing aircraft.

Contractor
Penny & Giles Aerospace Ltd.

Penny & Giles Aerospace Digitas ADC 0005417

Secondary air data sources

Type
Avionic navigation sensor, Air Data System (ADS).

Description
The analogue Air Data Module (ADM) has been especially designed for GPWS (Ground Proximity Warning System) applications requiring a low-cost source of secondary air data information via an analogue channel. The ADM can be mounted in a wide variety of locations to aid retrofit of a GPWS capability.

The TP91 range of altitude and airspeed transducers provides a passive, lightweight, low-cost solution for applications such as flight data recording where a secondary or independent source of altitude or airspeed data is required.

Specifications
Air Data Module D60286
Dimensions: 89 × 132 × 159 mm
Weight: 1 kg
Power: 28 V DC, 5 W peak
Standards: RTCA DO-160C
Applications: GPWS

Altitude and Airspeed Transducer TP91
Dimensions: 83 × 90 × 82 mm
Weight: 0.8 kg
Power: 12 V DC, 4 W peak
Standards: TSO C51a
Applications: Flight data recording

Status
In service.

Contractor
Penny & Giles Aerospace Ltd.

Submetre accuracy Differential GPS (DGPS)

Type
Avionic navigation sensor, Global Positioning System (GPS).

Description
Skyquest Aviation supplies a range of airborne role equipment specifically designed to enhance the capability of surveillance aircraft. Products include GPS-based moving maps, differential GPS decoders, microwave tracking systems, NVG equipment and services, ruggedised video recorders, and a range of advanced aircraft monitors.

Skyquest Aviation is the airborne distributor for the OmniSTAR differential GPS system. This system gives 100 per cent coverage of the UK and the rest of Europe. The position accuracy of the OmniSTAR makes the application ideal for use with moving map systems and FLIR camera pointing devices.

With worldwide satellite coverage, the differential correction signal is broadcast from six geostationary satellites receiving information from 70 reference stations and three network control centres. This reference data is checked for integrity and reliability and is then uplinked to a geostationary satellite that distributes the data over its footprint, providing submetre accuracy.

The DGPS system comprises one OmniSTAR patch-type antenna, which is a combined GPS and DGPS unit, and a small receiver box which houses the GPS receiver. The OmniSTAR airborne antenna is accepted for mounting on aircraft and complies with TSO C-129, RTCA DO 1160c environmental and ARINC 743 specifications.

Virtual Base Station (VBS)
The VBS mathematically weighs each reference station as a function of its relative position and does this every time reference information is received from the satellite. This optimised correction information is constantly updated, making this service ideal for moving applications.

This service uses a worldwide network of reference stations to provide differential corrections by spot beam satellites. All GPS corrections are monitored for quality and precision by a round-the-clock control centre in Aberdeen, Scotland. All reference stations have dual data connections to their network control centre. The European service uses two uplinks (primary and back up).

Specifications
Dimensions:
(Antenna) 119.5 × 76.0 × 19.0 mm
(Receiver unit) 107 × 45 × 200 mm
Accuracy:
(Standard service) 0.9 m in X and Y, and 1.8 m in height
(High performance) <0.1 m in X and Y, and 0.2 m in height

Status
In production and in service.

Contractor
Skyquest Aviation.

Supertans integrated Doppler/ GPS navigation system

Type
Aircraft navigation system.

Description
The Supertans integrated Doppler/GPS navigation system is a replacement for the TANS series of equipments. Based on the RNS 252 (see previous item), it combines a six-channel GPS receiver with any of the Racal family of Dopplers. The package has been designed as a drop-in replacement for current TANS equipments.

Supertans uses all existing connectors and cables, with only minimal additional cabling required for the GPS installation. The system maintains and enhances all the present capabilities of the TANS/Doppler system while adding the precise navigational accuracy of GPS.

Specifications
Dimensions:
(Supertans) 240 × 161 × 146 mm
(GPS receiver) 368 × 197 × 57 mm
(antenna) 102 × 10 × 95 mm
Weight:
(Supertans) 4.2 kg
(GPS receiver) 2.75 kg
(antenna) 0.5 kg
Power supply: 28 V DC, 40 W

Status
In service in UK Royal Air Force Chinook, UK Royal Navy Sea King and British Army Lynx AH. Mk 7 helicopters.

Contractor
Thales Avionics SA.

Terrain Reference Navigation (TRN)

Type
Aircraft navigation system, Terrain Reference Navigation (TRN).

Description
Terrain elevation data can be exploited to achieve automatic micro-navigation and covert terrain-following of the necessary high integrity. The micro-navigation employs modern radar altimeters incorporating variable output power and spread spectrum techniques to provide low probability of detection. In Spartan, the radar altimeter is used to map the vertical cross-section of the terrain beneath the aircraft for matching within the database. This Terrain Referenced Navigation (TRN) produces fixes every 2 seconds. The fix matching algorithm is robust, recovering quickly from the larger inertial errors likely to be met after a prolonged water crossing or from any local errors in the terrain data.

The TRN fixes are used to Kalman filter the INS outputs and thus, even with medium-grade INS, the aircraft position is known with extremely good accuracy and high confidence. With this knowledge, the database can be scanned over the projected flight path to predict the ground profile ahead. The TF algorithm is therefore able to define and demand a totally safe kinematic flight path that ensures the closest maintenance of clearance level, even in manoeuvring flight. The sudden unmasking of close terrain and ballooning over hill crests that can occur with conventional TF radars is avoided. The system also monitors achieved and forecast ground clearances against those required and gives ground proximity warning where necessary. Additionally, obstacles are included in the database and their location and indicated height projected on to the pilot's HUD or helmet-mounted display for obstacle cueing.

Specifications
Dimensions: ¾ ATR short
Weight: 15 kg
Power: 200 W
Coverage: 230,000 sq miles
Database programming: 13 min
Environmental: MIL-STD-810, MIL-STD-461 and 462
Reliability: 3,000 h MTBF

Digital colour map
Advances in techniques for lossless compression, storage and recovery of digital data, combined with developments in cockpit colour displays, have been exploited to produce two- and three-dimensional digital map presentations with wide area coverage. Topographical and cultural data can be stored in pixel or vector formats and recalled for colour video display as realistic reproductions of paper maps or as terrain representations. Additional switchable colour palettes are available to suit different applications, for example to suit cockpit lighting conditions for night operations. Depending upon how the database is compiled, the resulting display can be decluttered of unwanted detail. The viewing area can be rotated in orientation, swiftly and smoothly manoeuvred about the stored area and viewed at any one of the standard aeronautical map scales or with a zoom capability. Additionally, on a sortie-to-sortie basis a mission routing and intelligence overlay can be generated and loaded into the database for use in flight.

The addition of terrain elevation information to the database generates a wide range of enhanced tactical head-down display options to overlay the basic map. For example, dynamic relative height shading provides terrain-avoidance assistance, while ground-to-air intervisibility displays, with ground threat positions and effective ranges, aid threat-avoidance manoeuvring. Conversely, dynamic air-to-ground intervisibility displays demonstrate the achievement of terrain-masking and provide a pseudo radar display to ensure that any mapping radar transmissions are initiated only when the target or fix point is in radar view. Head-up display enhancement is achieved by the generation and precisely placed display of ridge lines to give added pilot confidence in poor visibility or when limited to flat contrast FLIR pictures.

Status
In service in Harrier II, Jaguar, Tornado and C-130J aircraft.

Contractor
Selex Sensors and Airborne Systems.

United States

12-Channel Miniature Precision Lightweight GPS Receiver (PLGR) Engine (MPE)

Type
Avionic navigation sensor, Global Positioning System (GPS).

Description
Collins' 12-channel Miniature Precision Lightweight GPS Receiver (PLGR) Engine (MPE) is a small, lightweight, 2nd-Generation Global Positioning System (GPS) receiver, which provides precise positioning capabilities for military navigation, communications, timing tracking and designating systems.

The MPE is based upon the NightHawk 12-channel signal processor, Phoenix RF front end, ACE fast acquisition engine hardware and mature GPS receiver software; the system is designed to meet a wide variety of applications.

The MPE has been designed to use the same mating connectors and mounting footprint as the Rockwell Collins 5-channel MPE-I, while providing increased functionality.

The MPE transmits position, velocity and timing information via both RS-232 and CMOS serial interfaces. MPE is derived directly from PLGR, MPE-I and PLGR II, therefore it is compatible with existing PLGR, MPE-I and PLGR II integration protocols. The main features of the system are:
- PLGR-II performance in a small size for embedded applications
- 12-channel parallel, P/Y-code GPS receiver
- Selective Availability/Anti-Spoofing (SAAS)
- Direct-Y acquisition
- Military anti-jam capability
- 499 Waypoints
- Military Grid Reference System (MGRS), Universal Transverse Mercator (UTM), Universal Stereo Polar graphic (UPS), ECEF, BNG, ITMG
- Timing data: 1 pps (pulse per second), HaveQuick (HQ)

Specifications
Frequency:
L1/L2 dual-frequency tracking
L1 – C/A, P/Y
L2 – P/Y
Dynamics:
Velocity 1,200 m/s max
Acceleration 9 g max
Time accuracy: 100 ns
Positional accuracy:
SDGPS <2 m CEP
WAGE <4 m CEP
PPS <12 m CEP
Velocity accuracy: 0.03 m/sec RMS (steady rate)
MTBF: >40,000 h
Dimensions: 106 × 68 × 16 mm
Weight: 71 gm (max)
Power supply: +5 and +3.3 V DC, 1.2 W (typical)
Temperature range: –40 to +85°C

Status
In service. See also MPE-HS high-sensitivity SPS GPS receiver.

Contractor
Rockwell Collins.

AA-300 radio altimeter

Type
Avionic navigation sensor.

Description
The AA-300 radio altimeter consists of RA-315 and RA-335 indicators, RT-300 transmitter/receiver and AT-220, -221 or -222 antenna.

The RT-300 transmitter/receiver is a solid-state unit offered in three optional configurations for different outputs.

The RA-315 indicator has a servo-controlled pointer display of radio altitude up to 2,500 ft. Below 500 ft the scale is expanded to enhance readability. There is an adjustable decision height bug and an amber decision height warning lamp. The RA-335 is similar to the RA-315 but is configured for helicopters, having a range of 0 to 1,500 ft. Below 200 ft the scale is expanded to improve readability.

Specifications
Dimensions:
(RA-315 and -335) 3 ATI × 1,143 mm
(RT-300) 104 × 116 × 281 mm
(AT-220) 63 × 159 × 142 mm
Weight:
(RA-315 and -335) 0.7 kg
(RT-300) 2 kg
(AT-220) 0.3 kg
Power supply: 21–32 V DC, 0.5 A
Accuracy:
(RT-300) 0–100 ft ± 3 ft, 100–500 ft ± 3%, 500–2,500 ft ± 4%
(RA-315) 0–100 ft ± 5 ft, 100–500 ft ± 5%, 500–2,500 ft ± 7%
(RA-335) 0–100 ft ± 5 ft, 100–500 ft ± 5%, 500–1,500 ft ± 7%

Status
In production.

Contractor
Honeywell Inc, Commercial Electronic Systems.

ADS-87A Air Data System (ADS)

Type
Avionic navigation sensor, Air Data Computer (ADC).

Description
The ADS-87A Air Data System (ADS), designed for business jets and regional transport aircraft, senses, computes and outputs all parameters associated with aircraft movement through the atmosphere. System performance meets Reduced Vertical Separation Minimum (RVSM) requirements. The core of the ADS-87A is the ADC-87A Air Data Computer, which provides digital outputs for use by associated systems such as Collins' APS-65/80 Autopilot System, AHS-85 Attitude Heading Reference System, ADS-85 Air Data System instruments, EFIS-85/86 Electronic Flight Instrument System. The ADC-87A is a form, fit and functional replacement for the ADC-80( ) (see separate entries).

The ADC-87A ADC analyses the parameters by processing pitot, static and temperature information along with pre-programmed aircraft data. A configuration memory module unique to the aircraft model type contains this data. The pressure transducer in the DC is a piezoresistive solid-state pressure transducer. Such sensors reduce the power input required and the weight of the ADC and increase it's reliability.

The sensor element itself is a tiny silicon chip with piezoresistive elements diffused into its surface. Pressure applied across the chip strains the piezoresistive elements and the resistance of each element changes. The sensor converts these changes to a digital signal which is then processed by a computer. The computer calculates the absolute static pressure and differential pressure between the static and pitot ports of the aircraft and converts them to the familiar air data quantities such as altitude and airspeed.

The ADC-87A uses a completely digital approach, digitising pressure and temperature information from the sensors, then computing and transmitting the data throughout the system. Transmissions to other on-board systems are also digital. This digital approach confers a number of advantages:

- Reduced size and weight;
- Reduced power consumption;
- Continuous monitoring of all functions and internal diagnostics;
- Greater system integrity;
- Simplified installation.

The ADC can be tailored to fit specific customer requirements. It receives pneumatic inputs from the pitot/static system and electrical signals from the aircraft outside temperature probe and uses these to compute altitude, airspeed, Mach and temperature parameters. In turn, this information provides outputs for the flight control system, navigation system, aircraft control system and ATC transponder, as well as for the primary air data instruments. The Collins ADS far exceeds the FAA Technical Standard Order (TSO) accuracy requirements. Features of the ADS-87A include:
- Monitoring of all essential functions;
- On-board fault isolation;
- Solid-state piezoresistive pressure transducers;
- Solid-state synchronization circuits;
- Internal diagnostics and nonvolatile memory.
Computed and available parameters from the ADC-87A are:
- Mach;
- Indicated Air Speed (IAS);
- True Air Speed (TAS);
- Vertical Speed (VS);
- Vmo/Mmo;
- Static Air Temperature (SAT);
- Total Air Temperature (TAT);
- Preselected altitude;
- Overspeed warning;
- ADC valid contacts;
- Uncorrected pressure altitude;
- Barometric corrected pressure altitude.

Air Data Instruments
A set of air data displays is available for use as part of the ADS:

ALI-80A Encoding Altimeter. The ALI-80A decodes the binary, self-clocking Manchester data transmitted by the central ADC, corrects for a value (in ft) equal to the baroset display and displays actual corrected altitude of the aircraft on the circular pointer/dial scale and numeric counter drum readout.

PRE-80 Preselector/Alerter. The PRE-80 provides altitude preselect mode with the flight director system. It accepts baro-corrected altitude data from the ALI-80A and provides selected altitude display and altitude alert operation.

ASI-80D Airspeed Indicator. The ASI-80D has a dual-pointer display of indicated airspeed and maximum allowable airspeed. Scale markings are customized for the particular aircraft, including colored arcs for gear, flap, stall speeds and others. This instrument also has a mechanical bug and push-to-test function.

MSI-80F Mach Indicator. The MSI-80F has a dual-pointer display of indicated airspeed and maximum allowable airspeed driven by independent servos. The Mach airspeed indicator contains a servoed reference bug. Upon engaging airspeed hold, the bug slews to match the airspeed pointer; new airspeed hold commands can be set via the bug knob or the slew control of the autopilot.

VSI-80A Vertical Speed Indicator. The VSI-80A displays the actual vertical speed of the aircraft on the circular pointer/dial scale. The pilot can set a vertical speed marker (bug) by manual activation of the knob located in the lower right hand corner the instrument.

TAI-80A True Airspeed Indicator. The TAI-80A receives, decodes and displays Total Air Temperature (TAT), Static Air Temperature (SAT) and True Air Speed (TAS). The display has a pair of three-digit, seven-bar incandescent readouts simultaneously displaying SAT and TAS. A momentary push-for-TAT button changes SAT display to TAT.

Specifications
Dimensions:
½ ATR short/low
85 (H) × 125 (W) × 360 (L) mm
Weight: 2.7 kg
Temperature: –55 to +70°C
Altitude: 55,000 ft
Power: 28 V DC, 20 W
Certification: FAA TSOs C2c, C8b, C10b, C43a, C46a, C52a, C95; DO-160B; EUROCAE ED-14B

Status
In production and in service.

Contractor
Rockwell Collins.

ADF-462 Pro Line II Automatic Direction-Finder (ADF)

Type
Aircraft navigation system, radio aids.

Description
The ADF-462 is an all-digital ADF operating from 190 to 1,799.5 kHz together with 2,179 to 2,182 kHz, in 500 Hz tuning increments. Two antennas are available: the ANT-462A for single system installations or the ANT-462B for dual system installations. The ADF-462 is compatible only with CSDB or ARINC 429 controls. Outputs are also serial digital and are provided in both CSDB or ARINC 429 characteristics. In addition, a DC sine/cosine output is available for interfacing with conventional RMIs.

Specifications
Dimensions:
(receiver) ³/₈ ATR short/dwarf
(ANT-462A) 432 × 219 × 42 mm
(ANT-462B) 605 × 272 × 28 mm
Weight:
(receiver) 1.7 kg
(ANT-462A) 1.4 kg
(ANT-462B) 2.3 kg
Temperature:
(receiver) –55 to +70°C
(ANT- 462A, -462B): –65 to +71°C
Altitude: 70,000 ft (all units)
Power:
(receiver) 28 V DC, 0.6 A
(ANT-462A, -462B) powered by receiver
FAA TSO:
(receiver) C41d (Class A)
EUROCAE:
(receiver) ED-51
(receiver and ANT-462) ED-14B
Environmental:
(receiver and ANT-462) DO-160B

Typical system	Part number
ADF-462	622-7382-101
ANT-462A	622-7383-001
CTL-62	622-6522-008

Status
In production and service. Designed for business and regional turboprop and jet aircraft and helicopters.

Contractor
Rockwell Collins.

ADF-60A Pro Line Automatic Direction-Finder (ADF)

Type
Aircraft navigation system, radio aids.

Description
The ADF-60A is a lightweight ADF that combines digital tuning with a single crystal frequency synthesizer to eliminate the need to zero-beat tune the ADF receiver. Quadrantal error correction is performed within the receiver and it has integrated sense and loop antennas in one unit similar in size to a conventional loop antenna. Two antennas are available: the ANT-60A for single-system installations or the ANT-60B for dual-system installations. Automatic band switching eases frequency selection over the 190.0 to 1,749.5 kHz frequency range. Tuning is in 0.5 kHz steps with a capture range of ±0.25 kHz.

When used with a CAD-62, the ADF-60A can be controlled with a Pro Line II CTL-62.

Specifications
Dimensions:
(ADF-60A) ³/₈ ATR short/dwarf
(ANT-60A) 419 × 216 × 43 mm
(ANT-60B) 603 × 269 × 27 mm
Weight:
(ADF-60A) 1.9 kg
(ANT-60A) 1.4 kg
(ANT-60B) 2.3 kg
Environmental:
Altitude: Up to 60,000 ft
Temperature:
(ADF-60A) -54 to +55°C
(ANT-60A) -65 to +71°C
(ANT-60B) -65 to +71°C
Power: 28 V DC; 0.6 A

Typical system	Part number
ADF-60A Receiver	622-2362-001
ANT-60A Antenna	622-2363-001
CTL-62 Controller	622-6522-008
CAD-62 Adapter	622-6590-002

Status
In production and in service.

Contractor
Rockwell Collins.

ADF-700 Automatic Direction-Finder (ADF)

Type
Aircraft navigation system, radio aids.

Description
Designed in accordance with ARINC 712, the ADF-700 automatic direction-finder was introduced in 1980. It is based on experience gained with the earlier DF-203 and DF-206 receivers. These technical advances were pioneered by Rockwell Collins and incorporated into draft ARINC characteristic 712, which also provided for ARINC 429 interfaces and an integral loop/sense antenna. Centralised fault monitoring is now incorporated into the design consistent with ARINC 604 characteristics.

In addition to mechanical and reliability improvements introduced by the new characteristic, a significant performance improvement came with the change from analogue to digital technology. In the ADF-700 all bearing signal baseband processing is performed digitally. The reference and bearing signals are converted into digital form using a 12-bit CMOS analogue-to-digital converter and these are subsequently handled by an Intel 8086 16-bit microprocessor. The ARINC 429 input/output functions are performed by an Intel 8049 processor in conjunction with a Rockwell Collins universal asynchronous transmitter/receiver.

The advent of third-generation microprocessors permits the introduction of digital ADF signal processing to improve accuracy and reliability, a significant advantage being the elimination of mechanical adjustments. The system incorporates improved self-test capabilities as a result of using the 8086 microprocessor to control a test sequence incorporating a digital test signal synthesis. The same power supply subassembly is also used in the VOR-700 and ILS-700, leading to reductions in spares inventories and maintenance costs.

Specifications
Dimensions: 2 MCU per ARINC 600, 604
Weight: 3.4 kg
Power supply: 115 V AC, 380 to 420 Hz (1 sigma)
Frequency: 190 to 1,750 kHz
Channel spacing: 0.5 kHz

Modes: ANT-aural receiver, ADF navigation, CW/MCW
Tuning: ARINC 429 dual serial bus
Accuracy: better than 0.9° with ARINC 712 antenna in 35 µV/m field, exclusive of antenna error
Compliance: ARINC 712, 600, 429; TSO C41c; DO-142, -160A; ED14A, EUROCAE

Status
In production and in service in a wide range of commercial aircraft, including Boeing 747, 757/767, Airbus A310, 319/320/321 330/340 and Fokker 100.

Contractor
Rockwell Collins.

ADS-85/86/850 digital air data systems

Type
Avionic navigation sensor, Air Data System (ADS).

Description
The ADS-85/86/850 series is similar to the obsolete ADS-82 except that it uses a solid-state piezo-resistive sensor. The ADC-85 is a form, fit and functional replacement for the ADC-81/82. It interfaces with the Collins electromechanical air data instruments and has an ARINC 429 bus for auxiliary systems. The ADS-86 interfaces with Collins CRT air data instruments. The ADS-850 contains an ARINC 429 bus for flight control and attitude and heading communications.

The ADS-85/86 Air Data System is designed to meet the performance requirements for business jets and regional aircraft and is currently certified to comply with the Reduced Vertical Separation Minimum (RVSM) requirements. The system is all-digital from the Air Data Computer (ADC) to the airdata displays on the instruments. It provides digital outputs for use by the APS-65/85 Autopilot, AHS-85 Attitude Heading Reference System, EFIS-85/86 Electronic Flight Instrument System and other interfacing aircraft systems. It provides for optimal performance of the flight guidance system in all vertical modes and can also display the following air data parameters essential to flight: indicated airspeed, true airspeed, vertical speed, Vmo/Mmo, static air temperature, total air temperature, preselected altitude, overspeed warning contacts, ADC valid contacts, uncorrected pressure altitude and barometric corrected altitude.

Specifications
Dimensions:
½ ATR short/low
85 (H) × 125 (W) × 360 (L) mm
Weight: 2.65 kg
Power: 28 V DC, 20 W
Temperature: –55 to +70°C
Altitude: 55,000 ft
Certification: FAA TSOs C2c, C8b, C10b, C43a, C46a, C52a, C95; DO-160b; EUROCAE ED-14B

Status
In production and service.

Contractor
Rockwell Collins.

ADS-3000 Air Data System (ADS)

Type
Avionic navigation sensor, Air Data System (ADS).

Description
The ADS-3000 Air Data System (ADS) senses, computes and displays all parameters associated with aircraft movement through the atmosphere. It meets the requirements of high-performance business jets and regional aircraft. The ADS-3000 can be certified to comply with Reduced Vertical Separation Minimum (RVSM) requirements.

The system is completely digital, from the Air Data Computer (ADC) to the air data displays on the Primary Flight Display (PFD) and Multifunction Display (MFD). The ADS-3000 comprises the

ADC-3000, which provides ARINC 429 digital outputs from the piezoresistive pressure transducer for use by the AHS-3000 Attitude Heading Reference System (AHRS). The ADC computes parameters to be output by processing pitot, static and temperature information along with pre-programmed aircraft data. Aircraft specific data is uploaded to the ADC using an ARINC 615 data loader. Output parameters are:

- Indicated/Calibrated Air Speed (IAS/CAS);
- True Air Speed (TAS);
- Vertical Speed (VS);
- Vmo/Mmo;
- Static Air Temperature (SAT);
- Total Air Temperature (TAT);
- Pressure altitude (corrected for static source errors);
- Barometric corrected altitude*;
- Mach;
- Total pressure;
- Static pressure;
- Impact pressure;
- International Standard Atmosphere (ISA) Delta Temperature.
  * – Requires external control to provide barometric pressure.

The outputs from the system are also sent to flight data recorders, navigation computers, altitude transponders, Ground Proximity Warning Systems (GPWSs) and auxiliary aircraft subsystems. Each output is capable of driving up to 10 electronic flight instrument loads. The ADS-3000 is standard equipment on many business, regional and air transport aircraft.

Specifications
Dimensions: 89 (H) × 89 (W) × 203 (L) mm
Weight: 53 kg (ADC-3000)
Temperature: –55 to +70°C
Altitude: 55,000 ft
Power: 28 V DC, 8 W nominal
Interfaces: six ARINC 429 12.5 kBd buses, one ARINC 429 100 kBd bus
Environmental: DO-160D
Software: DO-178B Level A
FAA TSO: C-106

Status
In production and in service on a wide variety of aircraft.

Contractor
Rockwell Collins.

Advanced Air Data System (AADS)

Type
Avionic navigation sensor, Air Data Computer (ADC).

Description
Honeywell's AADS introduces a new range of air data systems, designed to meet Reduced Vertical Separation Minimum (RVSM) requirements. Latest products in the range include the AZ-252 Air Data Computer (ADC); AZ-960 ADC; AM-250 barometric altimeter; and integrated multifunction probe.

Status
In August 1999, an installation comprising dual AZ-960 ADCs, dual BA-250 barometric altimeters and an AL-801 altitude alerter was certified for RVSM operation in Gulfstream II and IIB aircraft.

Contractor
Honeywell Inc, Commercial Aviation Systems.

Advanced GNS/IRS integrated navigation system

Type
Aircraft navigation system, integrated.

Description
Honeywell has integrated laser inertial functions with the Global Navigation Satellite System (GNSS). The advanced GNS/IRS offers high reliability, long life, low power consumption, small size, light weight, fast alignment, full performance for alignment up to 78° latitude and improved BITE.

The advanced 4 MCU IRS is 60 per cent smaller, 40 per cent lighter and uses 50 per cent of the power of Honeywell's 10 MCU systems, while meeting the same performance specifications. The heart of the system is the Ring Laser Gyro (RLG). The advanced IRS is fully provisioned for GNSS integration. The blending of these two systems into one navigation solution offers precise navigation accuracies.

The GNSS unit will continuously track all satellites in view to give accuracies of 25 m or better with selective availability switched off. The GNSS unit offers growth potential for Wide Area Augmentation System (WAAS) and Local Area Augmentation System (LAAS) as well as the capability for differential GPS. Further enhancements under development include capabilities to support precision approaches to unimproved runways, automatic dependent surveillance and sole-means navigation.

Specifications
Dimensions: 124.5 × 317.5 × 193 mm
Weight: 12.25 kg
Accuracy:
(navigation)(IRS) 2 n miles/h,
(GNS/IRS hybrid) 25 m

(velocity) (IRS) 12 kt
(GNS/IRS hybrid) 0.3 kt
(attitude) 0.1°
(heading) 0.4°

Contractor
Honeywell Inc, Commercial Electronic Systems.

ADZ air data system

Type
Avionic navigation sensor, Air Data System (ADS).

Description
The ADZ air data system uses an AZ-241 or -242 computer, depending on whether the aircraft is a turboprop or a jet. These are now complemented in production with the AZ-600, for turboprops, and AZ-800, for jets, digital air data systems. All systems use the Honeywell patented vibrating diaphragm pressure sensor to provide outputs for altitude, airspeed, vertical speed, true airspeed, true air temperature and total air temperature information.

If required, a single computer can handle a dual-flight director installation. Consequently, no matter which director is driving the autopilot the full complement of modes is available.

Specifications
AZ-241 (turboprop)
Dimensions: 193 × 79 × 361 mm
Weight: 3.9 kg

AZ-242 (jet)
Dimensions: 193 × 124 × 361 mm
Weight: 5.2 kg

AZ-600 (turboprop)
Dimensions: 193 × 92 × 362 mm
Weight: 3.7 kg

AZ-800 (jet)
Dimensions: 193 × 92 × 362 mm
Weight: 4.08 kg

Status
In production for a wide range of executive turboprop and jet aircraft.

Contractor
Honeywell Inc, Commercial Electronic Systems.

AHS-3000A/S Attitude Heading Reference System (AHRS)

Type
Avionic navigation sensor, Attitude Heading Reference System (AHRS).

Description
The AHS-3000A/S measures aircraft pitch, roll and heading Euler angles for use by the flight deck displays, flight control system, flight management system and other avionics equipment. In addition, high quality body rate, Euler rate and linear acceleration outputs are provided for flight control systems.

The AHS-3000A consists of the AHC-3000A Attitude Heading Computer, FDU-3000 Flux Detector Unit (FDU) and ECU-3000 External Compensation Unit. The AHS-3000S consists of the AHC-3000S Attitude Heading Computer, ECU-3000 External Compensation Unit and a synchro type FDU detector such as the 323A-2G.

The AHS-3000A/S incorporates a solid-state sensor cluster located near the rear of the unit for maximum stability in a high-dynamic environment. The Inertial Measurement Unit (IMU) card receives the low-level analogue signals from the cluster and converts them into a digital output, facilitating superior performance to alternative analogue demodulation techniques. The output is processed utilising proprietary Rockwell Collins algorithms to compute pitch, roll

The Honeywell Inc, Commercial Aviation Systems' advanced air data systems, comprising AZ-252 ADC (top right), AZ-960 ADC (top left), AM-250 barometric altimeter (bottom right) and integrated multifunction probe (bottom left)
0081857

Collins' AHS-3000A (Rockwell Collins)
0131022

and heading Euler angles, along with acceleration information. Data is output via four high-speed ARINC 429 devices. The AHC-3000A/S also outputs this data in synchro and analogue forms. The FDU-3000 is a gimballed two-axis magnetic sensor that detects the horizontal component of the earth's magnetic field. The excitation signal to the FDU-3000 is provided by the AHC-3000A. The FDU-3000 and associated AHC-3000A circuits generate outputs proportional to the sine and cosine of the aircraft magnetic heading angle. Magnetic heading is derived and used to slave the computed heading angle in the AHC-3000A.

The 323A-2G Flux Detector Unit (FDU) is a gimballed magnetic sensor that detects the horizontal component on the earth's magnetic field. The excitation signal to the FDU is provided by the AHC-3000S. The FDU and the associated AHC-3000S circuits generate three outputs that indicate the aircraft magnetic heading. This magnetic heading is used to slave the gyrostabilized heading in the AHC-3000S. The FDU, along with its compensation data, provides an accurate heading reference. The compensation data is used to reduce the hard and soft iron errors and flux detector misalignment. The compensation data is aircraft specific and is stored in the ECU-3000.

When operated in the FDU compensation mode, the AHC-3000A/S determines and stores the compensation data as the aircraft is positioned at eight heading steps. If an index correction is required, the AHC-3000A/S determines and stores the correction during the last step of the compensation procedure while the aircraft is aligned to a known heading. In addition to storing FDU compensation and platform levelling data, the ECU-3000 provides for configuration of various aspects of the AHS-3000A/S, including scale factors for some analogue inputs and outputs, source of air data information and length of time the AHC-3000A/S will operate on backup power. The AHC-3000A/S

is capable of receiving input from ARINC 429, Manchester or analogue air data sources. The AHC-3000A/S determines which air data source it is receiving through the data set stored in the ECU-3000 and performs the corresponding air data interface function. Another ARINC 429 input port is reserved for future GPS data interface that will be used to supplement air data input. Other input sources are primary and battery power sources, strut switch logic, orientation straps, mode select logic, slew command logic, SDI straps, data loader and maintenance interfaces.

Specifications
Data valid for all units unless otherwise specified
Temperature: –55 to +70°C
Altitude: 55,000 ft
Power: 28 V DC, 25 W nominal (AHC-3000A/S)
Performance:
(pitch) ±90° range; ±0.5° steady level flight; ±1.0° manoeuvring*
(roll) ±180° range; ±0.5° steady level flight; ±1.0° manoeuvring*
(heading) ±180° range; ±1.0 steady level flight; ±2.0° manoeuvring*
(body rates) ±880° per second
* – with TAS input
Acceleration: ±15 *g*
FAA TSO: C6d, C4c
Environmental: DO-160D
Software: DO-178B Level A (AHC-3000A)
Dimensions
AHC-3000A/S: 339 × 64 × 127 mm (L × W × H)
FDU-3000: 122 × 122 × 61 mm (L × W × H)
ECU-3000: 70 × 51 × 36 mm (L × W × H)
323A-2G: 121 × 121 × 71 mm (L × W × H)
Weight
AHC-3000A/S: 2.04 kg
FDU-3000: 0.41 kg
ECU-3000: 0.09 kg
323A-2G: 0.7 kg

Status
Designed for regional and business aircraft, the AHS-3000 is in production and in service.

Contractor
Rockwell Collins.

AHZ-800 Attitude Heading Reference System (AHRS)

Type
Avionic navigation sensor, Attitude Heading Reference System (AHRS).

Description
The AHZ-800 is the next generation Attitude Heading Reference System (AHRS) designed for high performance and high reliability while attaining lower power dissipation and reduced size and weight. This is accomplished through the use of advanced manufacturing techniques, such as very large-scale integration and application specific integrated circuits, and fibre optic rate-sensing advanced sensor technology.

Honeywell has developed a practical interferometric fibre optic gyro sensor that replaces the heavier less reliable spinning iron rate-sensors used in the conventional AHRS. The advent of the interferometric fibre optic gyro sensor makes the AHZ-800 a truly solid-state device. The AHZ-800 is a 4 MCU package that outputs attitude, heading and rate data on ARINC 429 and ASCB digital buses. The attitude and heading source approaches the performance of an inertial reference system, but at a significantly lower cost.

Status
The AHZ-800 is in service with the Dornier 328 regional airliner.

Contractor
Honeywell Inc, Commercial Electronic Systems.

Air Data Inertial Reference System (ADIRS)

Type
Avionic navigation sensor, Air Data Inertial Reference System (ADIRS).

Description
The ARINC 738 Air Data Inertial Reference System (ADIRS) is an air data computer combined with an Inertial Reference System (IRS). Together with the Pegasus flight management system and TCAS 2000 collision avoidance system, the ADIRS forms the WorldNav™ system.

The ADIRS provides complete ARINC inertial reference system outputs including primary attitude and heading, body rates, acceleration, groundspeed, velocity and aircraft position and ARINC 706 air data outputs, which include altitude, true airspeed, Mach number, air temperature and angle of attack.

Each ADIRS is equipped with three air data inertial reference units, one control display unit and eight air data modules mounted remotely, adjacent to the pitot and static pressure sensors.

The air data reference electronics and the laser gyro inertial reference units are packaged in a 10 MCU box or 4 MCU box and require a nominal power of 109 W. The unit meets the functional requirements of ARINC 738 and the environmental requirements of DO-160b. On the A320, aircraft maintenance is simplified by extensive reporting of ADIRS LRU Operational Status to the centralised fault data system.

The air data module requires a nominal power of 1.8 W. The unit meets the environmental requirements of DO-160b and features a solid-state pressure transducer.

The control display unit is packaged in accordance with ARINC 738 and requires a nominal power of 5 W exclusive of warning lights. The unit meets the functional requirements of

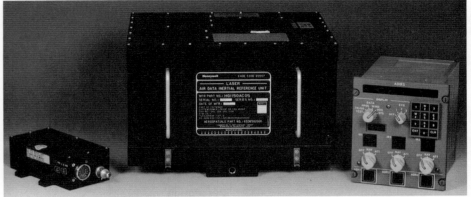

The Honeywell ADIRS air data inertial reference system showing (left) the air data module, (centre) the integrated air data/inertial reference unit and (right) the integrated control display unit 0581497

ARINC 738 and the environmental requirements of DO-160b. It features a liquid crystal display.

Specifications
Dimensions:
(ADIRU) 322.6 × 322.6 × 193 mm
(ADM) 50.8 × 76.2 × 152.4 mm
(CDU) 170.2 × 146 × 152.4 mm
Weight:
(ADIRU) 19.5 kg (10 MCU box)
   or 10 kg (4 MCU box)
(ADM) 0.63 kg
(CDU) 2.27 kg

Status
ADIRS equips Airbus A319, A320, A330 and A340 aircraft.

Fault Tolerant Air Data Inertial Reference System (FT-ADIRS)
The Fault Tolerant Air Data Inertial Reference System (FT-ADIRS) consists of a Fault Tolerant Air Data Inertial Reference Unit (FT-ADIRU), a Secondary Attitude and Air data Reference Unit (SAARU) and six Air Data Modules (ADMs). The FT-ADIRU provides attitude and heading data for inertial navigation as well as air data computations. The SAARU provides a back-up source of attitude and air data computations. The ADMs provide both the FT-ADIRU and SAARU with three redundant sources of static and pitot pressure data.

Status
The FT-ADIRS is standard on the Boeing 777 aircraft.

Contractor
Honeywell Inc, Commercial Electronic Systems.

ALA-52A radio altimeter

Type
Avionic navigation sensor.

Description
The ALA-52A radio altimeter is a lightweight solid-state digital low-range unit which utilises a simplified microprocessor-based design.

The ALA-52's capabilities are achieved by an advanced microprocessor which handles all data computations, including the application of correction factors for aircraft installation delay; the control of tracking filter gain bandwidth characteristics; collection and processing of the beat frequency count representing altitude information; the output of the altitude data for display via the ARINC 429 interface. Flag logic and monitor levels are also controlled by the microprocessor to reference criteria defined in the firmware.

As an added confidence factor, the ALA-52A utilises a second microprocessor of differing design architecture, to compute and verify altitude information by independent comparison.

One of the ALA-52A's other major advantages is its ability to perform continuous automatic self-calibration. By utilising a continuous feedback loop comprising the transmitter, quartz bulk-wave-delay device, a crystal reference and the

modulator, the unit not only monitors the slope of the transmission but also maintains proper calibration. It complies with ICAO Annex 10.

Specifications
Weight: 4.54 kg
Altitude: up to 50,000 ft

Status
In production.

Contractor
Honeywell Aerospace, Electronic & Avionics Lighting.

ALT-50/55 Pro Line radio altimeters

Type
Avionic navigation sensor.

Description
The ALT-50 and ALT-55 Pro Line radio altimeters, the first with a range of 0 to 2,000 ft, the other 0 to 2,500 ft, have been designed for business aircraft. Both types provide decision height annunciation for Cat II landings. The decision height annunciators can be set at any desired altitude and both instruments can interface with high-performance flight directors and autopilots. The DRI-55 indicator is offered with a numeric readout of radar altitude and decision height.

Dual ANT-52 antennas are included.

Specifications
Dimensions:
(transmitter/receiver) 3/8 ATR short dwarf
(indicators) 77 × 77 × 152 mm
(antenna) 25 × 25 × 20 mm
Weight:
(transmitter/receiver) 2.54 kg
(indicator) 0.7 kg
(antenna) 0.1 kg
Power supply: 28 V DC

Status
In production and in service.

Contractor
Rockwell Collins.

AN/APN-217 Doppler Radar Navigation System

Type
Airborne navigation system.

Description
The AN/APN-217 Radar Navigation System (RNS) is a lightweight, low-power, self-contained Doppler radar velocity sensor. The RNS unit detects and processes Doppler-shifted frequency returns from continuous wave, time-multiplexed radar beams to determine three orthogonal velocities in aircraft heading (Vh), drift (Vd) and vertical (Vz) co-ordinates. The AN/ARN-217 is

available in -217, -217(V)2, -217(V)3, -217(V)5 and -217(V)6 models and can accommodate output formats of ARINC or MIL-STD-1553 and provides DC analogue voltages for driving hover indicators and automated flight control systems. Applications encompass helicopters and medium-performance fixed-wing aircraft.

Specifications
Dimensions: 164 × 423 × 408 mm
Weight: 12.72 kg
Power supply: 28 V DC, 48 W
Velocity range:
(heading) −40 to 350 kt
(drift) −100 to 100 kt
(vertical) −4,500 to 4,500 ft/min
Accuracy:
(heading) <0.3% over land, <0.4% over water
(drift and vertical) <0.2%
Reliability: 15,257 h MTBF

Status
The AN/APN-217 is the standard Doppler for US Navy and US Marine Corps aircraft.

Contractor
Northrop Grumman Corporation, Navigation Systems Division.

AN/APN-218 Doppler Velocity Sensor (DVS)

Type
Avionic navigation sensor, Doppler.

Description
The AN/APN-218 Doppler Velocity Sensor (DVS) is a reliable, high-performance, nuclear-hardened Doppler radar for fixed-wing aircraft. Usually referred to as the Common Strategic Doppler (CSD), it is combined with a GroundSpeed Drift Indicator (GSDI) to process digital velocity data from the radar and display the groundspeed and drift angle. The DVS is an LRU consisting of 10 Shop Replaceable Units (SRUs). The GSDI is also an LRU and consists of three SRUs.

The AN/APN-218 continuously measures three velocities: heading (Vh), drift (Vd) and vertical (Vz). These provide accurate in-flight data to navigation equipment. It can output data for all three co-ordinates from the microcontroller and process it into either MIL-STD-1553A or ARINC 575 formats.

Specifications
Dimensions:
(sensor) 636 × 708 × 168 mm
(GSDI) 146 × 76 × 155 mm
(CDU) 146 × 152 × 165 mm
Weight:
(sensor) 33.18 kg
(GSDI) 1.5 kg
(CDU) 3.9 kg
Power supply: 115 V AC, 400 Hz, 10 VA
Velocity range:
(heading) 96–1,800 kt
(drift) ±200 kt
(altitude) up to 70,000 ft
Reliability: >9,740 h MTBF

Status
In production. AN/APN-218 is in service in US Air Force B-52, C-130, KC-135 and MC-130 aircraft, and on the C-130J.

Contractor
Northrop Grumman Corporation, Navigation Systems Division.

AN/ARN-136A(V) TACAN

Type
Aircraft navigation system, radio aids, TACAN.

Description
The AN/ARN-136A(V) lightweight airborne TACAN is a remotely controlled system utilising

large-scale integrated circuits and CMOS technology. The system consists of the RT-1321/ARN-136A radio receiver/transmitter ranging unit and the CP-1398/ARN-136 azimuth computer with bearing unit. The units contain no moving parts and through exclusive use of CMOS circuits, diodes and LSI chips are rated for continuous operation at +70°.

The units meet or exceed all FAA TSO C66a requirements for high-altitude operation up to 70,000 ft. The TACAN system has been designed to work with all TACAN or VORTAC stations meeting MIL-STD-291B and FAA selection order for the US National Aviation Standard 1010.55 and ICAO Annex 10.

The AN/ARN-136A(V) (ruggedised EMI lightweight airborne TACAN) has been designed to meet ElectroMagnetic Compatibility (EMC) requirements of MIL-STD-461B Methods CE03, CS01, CS02, CS03, CS04, CS05, CS06, RE02, RS02 and RS03. In addition, RTCA Do-160B A2E1/A/MNO/XXXXXX2BABA and induced signal susceptibility category A and Z have been met, together with Shock and Random Vibration per MIL-STD-810C Method 514.2 procedure 1A and Method 516.2 procedure I30Gs. This enhancement is in use by the US Navy.

The AN/ARN-136A(V) system utilises an ID-2218/ARN-136 range indicator which displays distances up to 399 n miles, groundspeed up to 999 kt and time to TACAN station up to 99 minutes. To minimise the number of wires between the cockpit and avionics bay, where the remote boxes are installed, the AN/ARN-136A(V) TACAN incorporates a three-wire serial databus that is used for both range and tuning.

The radio receiver/transmitter utilises two LSI circuits to provide digital computations of distance, groundspeed and time to station. All tuning is performed electronically, using a digital frequency synthesiser. The unit provides 63 channels of X mode air-to-air ranging. Channels are selected by the control head (P/N7801-9000-2).

The bearing unit accepts receiver video, pulse-pair decoding information and DC power from the unit. The bearing unit derives aircraft bearing angle, with respect to local magnetic north, to or from the VORTAC or TACAN ground beacon.

The derived bearing information is then converted into several bearing data formats to enable proper interfacing with a variety of aircraft instruments. The unit is designed with two multilayer printed circuit cards, divided so that one card derives the bearing signal digitally and the second interfaces with the aircraft instruments. The bearing unit provides a bearing accuracy of ±0.5°.

Specifications
Dimensions:
(receiver/transmitter) 63.5 × 133.3 × 298.4 mm
(azimuth computer) 63.5 × 133.3 × 298.4 mm
Weight: 3.13 kg
Power supply: 11–33 V DC, 15 W

Contractor
Sierra Research, a division of SierraTech, Inc.

AN/ARN-144(V) VOR/ILS receiver

Type
Aircraft navigation system, radio aids.

Description
The ARN-144(V) VOR/ILS receiver is said to be the first such system to be compatible with the MIL-STD-1553B digital databus. All the standard VOR, localiser, glide slope and ILS beacon facilities are available with 160 VOR channels and 40 localiser/glide slope channels being selectable at 50 kHz spacing.

A number of configurations is produced to meet specific military applications. The R-5094/ARN-514, a version of the ARN-144, is standard on F/A-18. A number of different control panels are also available.

Specifications
Dimensions: 104 × 127 × 304 mm
Weight: 3.6 kg
Power supply: 28 V DC, 25 W

Status
In production and in service.

Contractor
Rockwell Collins.

AN/ARN-147(V) VOR/ILS receiver

Type
Aircraft navigation system, radio aids.

Description
The AN/ARN-147(V) is fully compatible with the MIL-STD-1553B databus. This receiver combines all VOR/ILS functions such as VOR and ILS localiser, glide slope and marker beacon in one compact system.

In late 1997, Rockwell Collins released both a new model of the AN/ARN-147(V) and a modification kit for existing receivers that provides FM interference immunity in accordance with ICAO Annex 10.

All-solid-state modular construction makes the ARN-147 a reliable receiver for either new or retrofit applications on fixed- and rotary-wing aircraft.

Rotor modulation suppression circuitry is pin-selectable in the ARN-147 for reliable operation in rotary-wing aircraft. FAA split-channel requirements are met by providing 50 kHz spacing for 160 VOR and 40 localiser/glide slope channels.

Digital and analogue outputs are compatible with the latest high-performance flight control systems, digital indicators and analogue instruments. In addition, high- and low-level deviation and flag outputs for VOR, Loc and glide slope are provided.

The ARN-147 meets the following US military standards and specifications: MIL-E-5400 Class II environment; MIL-STD-810 vibration, including gunfire vibration; MIL-STD-461/462 electromagnetic interference, and MIL-STD-704 power characteristics. An FM immunity upgrade is now available.

Specifications
Dimensions: 104 × 127 × 304 mm
Weight:
(with MIL-STD-1553B) 3.6 kg
(without MIL-STD-1553B) 3.4 kg
Power supply:
28 V DC, 25 W with MIL-STD-1553B
28 V DC, 20 W without MIL-STD-1553B

Status
In service with US air arms in a range of aircraft including the C-130, C-5, C-17, C-141, CH-53, T-38, T-45, UH-1H, UH-60 and V-22.

Contractor
Rockwell Collins.

AN/ARN-149(V) (DF-206A) Automatic Direction-Finder (ADF)

Type
Aircraft navigation system, radio aids.

Description
The AN/ARN-149(V) was the first low-frequency automatic direction-finder to provide an internal field upgradeable MIL-STD-1553B digital multiplex bus capability. This system, consisting of a receiver, control, antenna and mount, provides a low-frequency automatic direction-finding function in a lightweight easily installed set. The all-solid-state receiver eliminates all moving parts such as goniometers, synchros and mechanical

tuners. Quadrantal Error Correction (QEC) is set by aircraft connector strapping, eliminating corrector modules and airframe specific internal adjustments. The antenna combines the loop and sense antennas and preamplifiers in one compact housing, eliminating expensive sense panels, couplers and special prefabricated cable assemblies. A dual version of the same antenna is also available in one aerodynamic package in either white or black colours. The receiver is controlled by a four-wire serial bus from the control. The system meets MIL-E-5400 and has an MTBF of 4,000 hours.

Specifications
Dimensions:
(receiver) 79 × 127 × 279 mm
(control) 146 × 57 × 96 mm
(antenna) 216 × 43 × 419 mm
Weight:
(receiver) 2.5 kg with MIL-STD-1553B capability
(control) 0.7 kg
(antenna) 1.4 kg
Power supply:
26 V AC, 7.8 W
28 V DC, 18 W

Status
In service.

Contractor
Rockwell Collins.

AN/ARN-153(V) TACAN

Type
Aircraft navigation system, radio aids, TACAN.

Description
The AN/ARN-153(V) is the replacement for the ARN-118, which is now out of production. Unlike its predecessor, a typical installation consists of four not five LRUs: one or two antennas, a cockpit control (same as the ARN-118), the receiver/transmitter and a mounting tray.

Key features of the system are digital outputs for both distance and bearing (with optional analogue outputs), input and output MIL-STD-1553 bus capability, microprocessor-based design, dual-antenna ports, inverse bearing capability in certain applications with the 938Y-1 antenna, pilot-selectable air-to-air range ratio capability, 390 n miles range, signal-controlled search, solid-state 500 W transmitter, compatibility with ARINC 568, 582 and 429, enhanced BIT and W and Z channels for MLS compatibility.

The AN/ARN-153(V) has four basic modes of operation: receive, transmit/receive, air-to-air receive and air-to-air transmit/receive. When used in conjunction with the optional 938Y-1 rotating antenna and an optional control unit, the AN/ARN-153(V) can also provide bearing in certain applications to an air-to-air TACAN if it can transmit unmodulated squitter, as well as bearing to any DME-only ground station.

Specifications
Dimensions:
(control) 144.8 × 76.2 × 99.1 mm
(receiver/transmitter) 104.6 × 172.2 × 304.8 mm
(mounting tray) 114.3 × 50.8 × 312.4 mm
Weight:
(control) 0.9 kg
(receiver/transmitter) 6.49 kg
(mounting tray) 0.54 kg
Power supply: 28 V DC, 1.5 A nominal
Frequency:
(transmitter) 1,025–1,150 MHz
(receiver) 962–1,213 MHz
Number of channels: 126 X and 126 Y Provision for W and Z
Range: up to 390 n miles
Accuracy:
(distance) (digital) ±0.1 n mile,
(analogue) ±0.2 n miles,
(bearing) (digital) ±0.5°,
(analogue) ±1.5°

Status

In widespread service. The AN/ARN-153 has been installed in cargo, fighter, bomber, rotary wing and training aircraft of the US Air Force and Navy and also with other air forces worldwide.

Contractor

Rockwell Collins.

AN/ARN-154(V) TACAN

Type

Aircraft navigation system, radio aids, TACAN.

Description

The AN/ARN-154(V) lightweight airborne TACAN is a remotely controlled multistation tracking system utilising large-scale integrated circuits and CMOS technology. The system consists of the RT-1634 receiver/transmitter, ID-2472 indicator unit, MT-6734 TACAN control unit mounting base and antenna.

The AN/ARN-154(V) has the capability to track up to four ground stations simultaneously in range and two in bearing. Tracking velocity is up to 1,800 kt. It incorporates a sine/cosine bearing output, along with an ARINC 547 CDI interface and ARINC 547/579 low- and high-level flags. Bearing information is pilot selectable from either of the two tracking channels. The system has both X and Y mode MIL-STD-291B air-to-air ranging, plus antenna switching to allow the aircraft to be configured for dual antennas.

The AN/ARN-154(V) provides an ARINC 568 digital range output from either of the two tracking channels for display on remote EFIS or HSI displays.

The RT-1643 is offered in a number of configurations to interface to most navigation flight management systems. The AN/ARN-154(V) can be used as a pilot-controlled positioning system or as a blind navigation sensor controlled by one or more long-range navigation systems.

An ARINC 429 version is available as an option. This version will output range on three separate channels and bearing on two. The input tuning is standard ARINC 429 bus. This unit is ideal for updating VLF/Omega or inertial navigation systems in helicopters and fixed-wing aircraft, or as a standalone TACAN system.

The AN/ARN-154(V) system can utilise an ID-2472 indicator unit which displays distance, TACAN radial or bearing, decoded station ident, groundspeed and time to station. Either or both TACAN stations may be displayed simultaneously. A second ID-2472 can be installed to provide independent cockpit instrumentation.

Specifications

Dimensions:
(display unit) 3ATI × 175.3 mm
(receiver/transmitter) 290.1 × 87.4 × 127.8 mm
(control units) 50.8–56.9 × 71.1–146 mm
Weight:
(display unit) 0.48 kg
(receiver/transmitter) 2.81 kg
(mounting base) 0.2 kg
(control units) 0.57–0.91 kg
Power supply: 18–32 V DC, 1.25 A (max)

The AN/ASN-128 Doppler navigation system: from left, velocity sensor (transmitter/receiver), control/display unit and signal data converter   0503900

Frequency:
(receive) 962–1,213 MHz
(transmit) 1,025–1,150 MHz
Range: up to 400 n miles
Accuracy:
(range) ±0.1 n mile
(bearing) ±0.5°
Temperature range: −54 to +71°C
Altitude: up to 70,000 ft

Status

In service.

Contractor

L-3 Communications Aviation Recorders.

AN/ASN-128B Doppler/GPS navigation system

Type

Avionic navigation sensor, Doppler/Global Positioning System (GPS).

Description

The AN/ASN-128B is the US Army's standard lightweight helicopter airborne Doppler/GPS navigator and comprises three units: a receiver/transmitter/antenna, signal data converter and computer/display unit. A steering hover indicator can also be included as an option. With inputs from heading and vertical references, the system provides aircraft velocity, present position and steering information from ground level to above 10,000 ft.

BAE SYSTEMS, under contract to the US Army, embedded a military code GPS receiver into the AN/ASN-128 GPS receiver. The PCY code receiver was integrated into the signal data converter and the computer display unit software was modified to display both Doppler navigation and GPS data. Over 2,100 AN/ASN-128 systems were modified and are currently installed in UH-60A/L Black Hawk and CH-47D Chinook helicopters. These modifications are applicable to most other AN/ASN-128 systems.

BAE Systems North America has also developed a field kit enabling the AN/ASN-128 Doppler navigator to interface with a stand-alone

Trimble GPS receiver, for continuous update of Doppler-derived present position with valid GPS data.

Specifications

Volume: 20,724 cm³
Weight: 13.61 kg
(hover indicator) 0.9 kg
Propagation: 4-beam configuration operating FM/continuous wave transmissions in K-band. Beam shaping eliminates the need for a land/sea switch. The single transmit/receive antenna uses full aperture in both modes to minimise beamwidth and reduce fluctuation noise
Number of waypoints: >100
Self-test: localisation of faults at LRU level by BITE
Reliability:
(complete system) >2,100 h MTBF

Status

In production. Many thousands of AN/ASN-128 and AN/ASN-128B sets have been manufactured.

The AN/ASN-128B is in service in the Sikorsky UH-60A/L and CH-47D Chinook. The AN/ASN-128 is in service with US Army Bell AH-1F and Boeing AH-64A helicopters; also in Royal Australian Air Force Bell UH-1H, UH-60 and CH-47D, Hellenic Air Force UH-1H, Jordanian Army AH-1S, Pakistan AH-1S, South Korean CH-47, UH-60 and AH-1S, Spanish Army Eurocopter BO 105 and CH-47B, Taiwan Army CH-47B and Turkish UH-60. The system, with hover indicator, has been provided for the German Army Eurocopter BO 105, PAH-1, UH-1D and VBH helicopters. Also built under licence in Japan for the JASDF AH-1S and CH-47D. In use in Austria, Bahrain, Brunei, China, Denmark, Dubai, Egypt, Greece, Netherlands, Singapore, Spain, Taiwan and Thailand. Version kit deliveries started in the second quarter of 1996.

Contractor

BAE Systems North America, Communication, Navigation, Identification and Reconnaissance Division.

AN/ASN-131 SPN/GEANS precision inertial system

Type

Avionic navigation sensor, Inertial Reference System (IRS).

Description

The Standard Precision Navigator/Gimballed Electrostatic Aircraft Navigation System (SPN/GEANS), designated AN/ASN-131, was developed primarily under sponsorship of the US Air Force Avionics Laboratory at Wright-Patterson Air Force Base, Ohio.

The basic SPN/GEANS system consists of an Inertial Measurement Unit (IMU), an Interface Electronics Unit (IEU) and a complete software library. For a stand-alone system capability, these two units are supplemented with a Digital Computer Unit (DCU). The IMU contains, in addition to the Velocity Measuring Unit (VMU) and two ESGs, temperature control electronics, accelerometer pulse rebalance and V readout

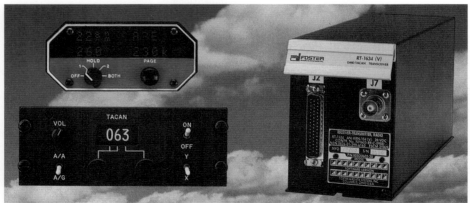

The AN/ARN-154(V) TACAN showing (top left) the ID-2472 display unit, (bottom left) a control unit and (right) the RT-1634 receiver/transmitter   0504180

electronics, precision timing reference, gimbal control electronics and Built-In Test Equipment (BITE) functions, as well as serial digital databus communication electronics. The IEU provides power conversion, control and sequencing electronics, additional BITE circuits and a common serial databus interface with other units of the inertial system and other subsystems.

The ESG has only one moving part, a suspended hollow beryllium ball, which is combined with two optical pick-offs to give error and timing signals. These in turn are used to drive the IMU platform gimbals to maintain a stable reference base for the accelerometers. Three highly accurate, single-axis accelerometers (contrasted with the two-degree-of-freedom ESGs) are used within the VMU which is mounted on the stable platform inner element. These accelerometers, oriented in an orthogonal triad configuration, measure accelerations directly and provide the incremental velocity pulses to the computer which uses them in its software algorithms to calculate velocity and position parameters.

The ESG system requires little or no reliance on other navigation aids for most aircraft applications and thus can be described as self-contained.

In addition to the traditional function of position determination or basic navigation, the higher accuracy outputs of velocity and attitude data have opened up a new realm of possibilities for stabilisation and/or motion compensation for other non-inertial sensors. These include high-precision radars, sonars, lasers, optical and electro-optical devices.

SPN/GEANS is being deployed throughout the entire US Air Force B-52 strategic aircraft and F-117A stealth fighter fleets. Widespread use is also expected for long-range reconnaissance and patrol missions, specialised cargo and transport usage in both military and civil applications and tactical military aircraft.

Status
In service with the US Air Force on the B-52 bomber and F-117A stealth fighter fleets.

Contractor
Honeywell Sensor and Guidance Products.

AN/ASN-137 Doppler navigation system

Type
Avionic navigation sensor, Doppler.

Description
The AN/ASN-137 is the multiplexed version of the AN/ASN-128. It is compatible with the MIL-STD-1553B databus and has ARINC 575 or 429 outputs.

The AN/ASN-137 uses the same receiver/transmitter/antenna as the ASN-128 and has an optional control/display unit of the same size, with the same front panel as the ASN-128. Additional features of the AN/ASN-137 include hover bias correction for precision hovering, 12-point magnetic deviation entry and the

addition of latitude and longitude and UTM grid zone outputs in MIL-STD-1553 output.

Specifications
Weight: 11.7 kg
Power supply: 28 V DC, 87 W
Reliability: 2,800 h MTBF

Status
In production and in service. The AN/ASN-137 is installed on the US Army's Bell OH-58D and the US Navy's Sikorsky VH-3D and VH-60 helicopters. It is also currently installed on AH-64A helicopters, on the US Army MH-47E and MH-60K helicopters and the US Air Force MH-60G Pave Hawk helicopter.

Contractor
BAE Systems North America, Communication, Navigation, Identification and Reconnaissance Division.

AN/ASN-139 (LN-92) carrier-based aircraft Inertial Navigation System (INS)

Type
Avionic navigation sensor, Inertial Reference System (IRS).

Description
Northrop Grumman's LN-92 is being produced as the US Navy AN/ASN-139 Carrier Aircraft Inertial Navigation System (CAINS II) ring laser gyro INS as a form, fit and function replacement for the AN/ASN-130A. It is designed for operation by high-performance carrier-based aircraft. The accuracy specifications are better than 1 n mile/h CEP and 3 ft/s velocity with a 4 minute reaction time.

The gyro in the LN-92 is the LG-9028 28 cm ring laser gyro used in conjunction with the A-4 accelerometer triad. This system incorporates MIL-STD-1750 processors for navigation and signal data processing. The reliability of the unit permits the navy to employ a two-level maintenance concept.

Specifications
Dimensions: 427 × 287 × 190 mm
Weight: 21.46 kg
Reliability: 2,200 h guaranteed MTBF (with 7,634 h MTBF achieved)
Power: 141 V A (.85 pF) or 120 W DC

Status
In production and in service with US Navy AV-8B, C-2A, E-2C, EA-6B, F-14D and F/A-18 aircraft. The system was also selected by Finland, Kuwait, Malaysia and Switzerland for their F/A-18s and was flight-tested in the UK Royal Air Force Harrier GR Mk.7 during carrier suitability trials for the aircraft.

Contractor
Northrop Grumman Corporation, Navigation Systems Division.

AN/ASN-141 (LN-93) inertial navigation unit

Type
Avionic navigation sensor, Inertial Reference System (IRS).

Description
The AN/ASN-141, designated LN-93 by Northrop Grumman (formerly Litton), is the sensing and data processing device that was chosen by the US Air Force as the basis for the standard inertial navigation system on the A-10 and F-16.

The unit contains a P-1000 platform, stabilised by two G-1200 gyros and mounting three A-1000 accelerometers, and an LC-4516C general-purpose computer.

Particular emphasis has been placed on reliability. This is accomplished partly by the use of large/medium-scale integrated circuits and hybrid components, allowing a significant parts count reduction.

There are three versions of the LN-93. The LN-93 INU has achieved a reliability of 3,755 hours mean time between failures and an accuracy better than the 0.8 nm per hour specified. The enhanced accuracy LN93 INU has enhanced accuracy performance in GC and EIA modes and has a direct tie in with GPS. The improved accuracy is due to enhanced OFP/CAL software with no hardware changes. The LN-93G GPS/INS has an embedded GPS module and operates 5 channels P-code. It retains the functions of the enhanced LN-93 and provides bounded GPS/INS and individual GPS and INS functions.

Specifications
Dimensions:
(inertial navigation unit) 191 × 193 × 460 mm
(control/display unit) 146 × 152 × 185 mm
Weight:
(inertial navigation unit) 17.36 kg
(control/display unit) 3.73 kg
Power:
(starting) 550 W
(running) 180 W
Accuracy (position, velocity):
(gyrocompass) 0.8 n mile/h CEP, 2.5 ft/s
(stored heading) 3 n miles/h CEP, 3 ft/s
Align time:
(gyrocompass) 8 min at 21°C
(stored heading) 1.5 min at 21°C
LC-4516C computer: 16-bit single or 32-bit double precision, 65,536 words, semiconductor RAM/ROM or EPROM 24 k words total
Reliability: 740 h MTBF specified goal over full military environment

Status
In service in A-10 and F-16 aircraft. The system is also in service in the Brazilian/Italian AMX, Japan Air Self-Defence Force F-4EJ, Thai Air Force F-4 and F-16s in Bahrain, Egypt, Korea and Turkey. Over 3,000 AN/ASN-141s have been ordered.

Contractor
Northrop Grumman Corporation, Navigation Systems Division.

AN/ASN-142/143/145 (LR-80) strapdown Attitude and Heading Reference System (AHRS)

Type
Avionic navigation sensor, Attitude Heading Reference System (AHRS).

Description
The AN/ASN-142/143/145, designated LR-80 by Northrop Grumman (formerly Litton), was developed as a flexible system to meet the growing operational requirement for an all-attitude, dynamically accurate attitude reference for air, land and sea applications. It is a lightweight self-contained LRU that provides heading, pitch and roll and angular rate reference information for a wide variety of military applications

The AN/ASN-139 INS is installed in the Northrop Grumman F-14D 0504181

including cockpit display, sensor stabilisation, fire control and autopilot reference.

The equipment senses vehicle angular rates and translational accelerations by means of two proprietary G-7 gyroscopes and three A-4 accelerometers.

Specifications
Dimensions: 194 × 192 × 259 mm
Weight: 8.1 kg
Power supply: 28 V DC, 60 W nominal
 115 V AC, 60 W (620 W warm-up)
Accuracy:
(heading) 0.5° RMS
(pitch and roll) <0.25° RMS
(angular rate) 0.25°/s RMS

Status
In service. Over 1,000 systems have been delivered to several programmes for US and overseas services, primarily for the Boeing AH-64 Apache, Bell OH-58D Kiowa Warrior and Sikorsky UH-60.

Contractor
Northrop Grumman Corporation, Navigation Systems Division.

AN/ASN-157 Doppler navigation set

Type
Avionic navigation sensor, Doppler.

Description
The AN/ASN-157 is a Doppler navigation set in a single LRU, emulating the two-unit AN/ASN-137. As well as navigation data, the system provides velocity data in all three axes, including computations for hovering and translational flight in helicopters.

Specifications
Dimensions: 370 × 342 × 57 mm
Weight: 5.7 kg
Reliability: 7,300 h MTBF

Status
In production, and in service. Installed on the Agusta A 109 helicopter for the Belgian Army, the CH-47D glass cockpit Chinook cargo helicopter for the Netherlands Air Force and the AH-64D Longbow Apache attack helicopter for the US Army. Selected for the WAH-64D Longbow Apache attack helicopter for the British Army.

Contractor
BAE Systems North America, Communication, Navigation, Identification and Reconnaissance Division.

APR-4000 GPS approach sensor

Type
Aircraft landing system, (Differential) Global Positioning System ((D)GPS).

Description
The APR-4000 combines essential navigation functions with critical landing functions into a single, integrated GPS sensor that supports en route, terminal area, precision approach and non-precision approach operations. The sensor uses uplinked final approach waypoints and differential corrections of GPS satellite signals to determine the aircraft's precise lateral and vertical position in relation to the associated approach path. The information presented to the flight crew is ILS look-alike deviation data.

The APR-4000 GPS approach sensor utilises a 12-channel GPS receiver and incorporates Receiver Autonomous Integrity Monitoring (RAIM) and predictive RAIM that improves fault tolerance and the integrity of the computed solution by verifying and anticipating the availability of GPS signals throughout the flight plan. The sensor is engineered to provide growth to local area augmentation system specifications for Cat I and Cat II GPS

precision approaches, as well as growth to regional area augmentation systems such as WAAS.

Status
Under development.

Contractor
Rockwell Collins.

AZ-960 Advanced Air Data Computer (AADC)

Type
Avionic navigation sensor, Air Data Computer (ADC).

Description
The AZ-960 Advanced Air Data Computer (AADC), designed for the Gulfstream II and IIB, is one of Honeywell's latest products in its range of air data systems. The AZ-960 AADC is fully compatible with Honeywell's legacy SP-50G and SPZ-800 autopilot systems. It enables operators to comply fully with the requirements of operations within Reduced Vertical Separation Minima (RVSM) airspace.

Status
In production.

Contractor
Honeywell Inc, Commercial Electronic Systems.

Boeing 777 Air Data Module (ADM)

Type
Avionic navigation sensor, Air Data Computer (ADC).

Description
The Boeing 777 Air Data Module (ADM) is a pressure transducer with a single pressure input port and an ARINC 629 bus output that transmits linearised, digital serial pressure data. Each Boeing 777 aircraft is configured with six ADM units: three with pitot pressure sensors and three with static pressure sensors. The ADM transmits ARINC 629 digital serial air data to the Air Data Inertial Reference System (ADIRS) and the Secondary Attitude Air data Reference Unit (SAARU).

Specifications
Dimensions: 88.9 × 63.3 × 171.4 mm
Weight: 1 kg (max)
Power supply: 28 V DC, 10.6 W (max)
Temperature range: −15 to +70°C
Accuracy:
(100–147.4 mb) ± 0.25 mb
(147.4–175.3 mb) <± 0.3 mb
(175.3–1,400 mb) ± 0.3 mb
Reliability: 50,00 h MTBF predicted

Status
In production and in service on the Boeing 777.

Contractor
Honeywell Inc, Commercial Electronic Systems.

CN-1655/ASN (LN-94) ring laser gyro INU

Type
Avionic navigation sensor, Inertial Reference System (IRS).

Description
The CN-1655/ASN, designated LN-94 by Northrop Grumman (formerly Litton), is a variant of the CN-1656/ASN (LN-93), itself developed as a ring laser gyro form, fit and function replacement for the LN-31 conventional gyro INS previously fitted in the F-15. The system uses the same inertial assembly and most of the electronic circuits of the LN-93, but is packaged in the LN-31 chassis and has input/output circuitry that is specific to the F-15.

There are two versions of the LN-94. The LN-94 second-generation F-15 INU has an achieved reliability of 3,097 hours mean time between failures and an accuracy better than the 0.8 nm specified. More than 600 have been delivered in four overseas countries. The LN-94 Enhanced Accuracy INU has a direct tie-in with GPS. The improved accuracy is obtained through enhanced OFP/CAL software and there are no hardware changes. It is interchangeable with the first-generation AN/ASN-109.

Status
Retrofit equipment for all US Air Force F-15A, B, C and D aircraft, as well as for Israel and Japan. The LN-94 is also qualified for the F-15E.

Contractor
Northrop Grumman Corporation, Navigation Systems Division.

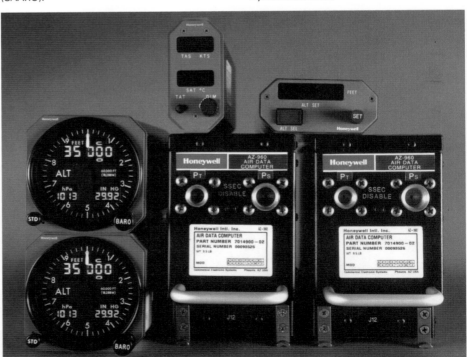

Two of Honeywell's AZ-960 advanced air data computers, shown with other elements of a complete RVSM package: left to right, dual BA-250 altimeters, a DS-125 true airspeed/total temperature indicator and an AL-861 altitude preselect controller 0106206

CN-1656/ASN (LN-93) ring laser gyro Inertial Navigation Unit (INU)

Type
Avionic navigation sensor, Inertial Reference System (IRS).

Description
The CN-1656/ASN, designated LN-93 by Northrop Grumman (formerly Litton), is the US Air Force standard ring laser gyro INU. This unit is now in production, employing 28 cm path length ring laser gyros. Following a contract awarded in 1985, units have been widely fitted to US Air Force aircraft and to aircraft in many other countries.

The LN-93G is form, fit and functionally interchangeable with the LN-39 and LN-93 standard INUs, as well as any other systems that are in compliance with SNU-84-1 used in numerous types of US and foreign tactical aircraft. The LN-93G ring laser gyro INU/GPS is in a single box which contains the US Air Force standard navigation unit integrated with a Collins militarised six-channel P code GPS receiver module.

Specifications
Dimensions: 521 × 204 × 193 mm
Weight: 22 kg
Reliability: 3,755 h MTBF (LN-93 second-generation standard INU)
Accuracy: 1.5 km/h CEP

Status
In production. The LN-93G is in development and a large number of units has been ordered for US fighter aircraft.

Contractor
Northrop Grumman Corporation, Navigation Systems Division.

DFA-75A ADF receiver

Type
Aircraft navigation system, radio aids.

Description
The DFA-75A utilises advanced LSI and microprocessor designs to achieve a greater level of dependability, accuracy and performance. While it uses advanced techniques to deliver performance advantages over its predecessors, its design is based on proven and efficient components which increase reliability and reduce weight and complexity.

The DFA-75A meets or exceeds all ARINC 712 characteristics for form, fit and function. One of the ways it meets the requirement for improved installations is by using a combined loop/sense antenna and digital interface with the receiver. Another major improvement is its ability to deliver more stable bearings in the presence of thunderstorms and during severe ionospheric conditions. This is achieved through a scheme of signal modulation, coherent demodulation and adaptive digital filtering from the antenna.

The DFA-75A ADF receiver　　　　0503894

The DFA-75A also provides automatic bearing adjustments to compensate for circular and quadrantal errors. Adjusted bearing data is smoothed by a second order digital filter.

Microprocessor monitoring permits a higher level of fault isolation diagnostics to be performed within the unit during BIT. The flight fault memory provides a non-volatile memory with the capacity to store 10 faults in each of 64 individual flight segments. It complies with ICAO Annex 10.

Specifications
Weight: 4.175 kg
Frequency: 190–1,750 kHz
Channel spacing: 0.5 kHz

Status
In production.

Contractor
Honeywell Aerospace, Electronic & Avionics Lighting.

DigiData fuel/airdata system

Type
Avionic navigation sensor, Air Data Computer (ADC).

Description
DigiData provides fuel management, navigation and airdata functions including:
E-6B data: Pressure; altitude; density altitude; outside and true air temperature; wind aloft; wind component (speed and direction); indicated and true airspeed; ground speed; Mach; and instantaneous vertical speed.
Fuel flow: Left and right fuel flow; fuel used; fuel remaining; fuel to and reserves at destination; range; cruise efficiency (nautical mileage); and endurance.
Navigation: Heading; ground track; and magnetic variation.

DigiData is connected to aircraft systems to measure Indicated AirSpeed (IAS), pressure altitude, Outside Air Temperature (OAT), heading and fuel flow. From these raw data, TAS and temperature data are calculated by the microprocessor and combined with data from the navigation receiver to calculate automatically wind aloft, fuel needed to destination, specific range and density altitude.

Pressure altitude data is automatically used by most GPS receiver manufacturers to substitute for the fourth satellite range if it is not in view to provide 24-hour, three-dimensional position accuracy with only three satellites, without manual entry of altitude.

The engine fuel flow interface makes the fuel management data totally dynamic, basing it on real-time, winds aloft readings.

With an external Shadin ARINC 429 converter, DigiData can also drive EFIS displays and multisensor navigational management systems.

Specifications
Dimensions: 3.125 (round) × 6½ in (deep) (79 (round) × 165 mm (deep))
Power: 9–35 V DC, 1 W nominal
Weight:
(indicator) 623.7 g
(transducer) 453.6 g
Flow rate: up to 450 gal/h
Max usable fuel: 1,800 gal
Accuracy: ±1%
Operating temperature: −20 to +55°C
Input: pitot pressure, static pressure, outside air temperature, heading synchro, fuel flow
Output: IAS, TAS, ground speed, Mach, P.ALT, D.ALT, OAT, TAT, wind aloft, wind component, fuel flow, fuel used, fuel remaining, cruise efficiency, endurance, left fuel used, right fuel used, IVS, heading, track, magnetic variation
Output format: RS-422/RS-232
ARINC 429 (optional)
Compatible receivers: ARNAV (R-15, R-30, R-40, R-50, STAR 5000, FMS 5000, FMS 7000); Trimble (2000, 2000A, 2100, 3000, 3100); AlliedSignal (KLN90, KLN90A, KLN88); II Morrow (604, 612, 614, 618); Northstar; Garmin; Thales

Status
In production and in service.

Contractor
Shadin Co Inc.

Digital air data computer

Type
Avionic navigation sensor, Air Data Computer (ADC).

Description
Honeywell produces an all-digital ARINC 706 standard air data computer for Airbus Industrie A310 and Boeing 757 and 767 airliners. These all-solid-state systems use large- or medium-scale integrated technology, CMOS master transmitter/receiver chips and a Z8000 microprocessor. Built-in test sequences run continuously in flight to provide a 95 per cent fault-detection capability, with a 99 per cent probability of correct fault identification. Test failure information is stored in the non-volatile memory for post-flight analysis. External sensor failures can also be detected and signalled visually. The system can accommodate a 20 per cent increase in output parameters, more than 50 per cent computational growth and more than 100 per cent memory growth. The unit incorporates Honeywell patented transducers and has ARINC 429 transmitter and receiver interfaces.

Specifications
Dimensions: 4 MCU
Weight: 5.7 kg
Power: 25 W
Accuracy:
(airspeed) ±2 kt at 100 kt, ±1 kt at 450 kt
(altitude) ±15 ft at sea level, ±80 ft at 50,000 ft
(Mach) ±0.003 at 0.8-0.9 in the range 25,000–45,000 ft
Failure memory: 6 failures/10 flights
Reliability: 15,000 h MTBF

Status
In service.

Contractor
Honeywell Inc, Commercial Electronic Systems.

Digital Air Data Computer (DADC)

Type
Avionic navigation sensor, Air Data Computer (ADC).

Description
Astronautics' Digital Air Data Computers are microprocessor-based, using precision solid-state vibrating quartz pressure transducers, with analogue potentiometers and synchro outputs as well as dual-redundant serial digital databusses.

Built-in test equipment allows a high degree of self-diagnosis and the computer has considerable growth potential. Astronautics' DADC family have been designed as a standard air data computer both for retrofit and new aircraft programmes.
DADC-104

The DADC-104 is a fully computerised air data computer, following the DADC-102 developed for the Mirage and the DADC-103 developed for the Kfir.

The DADC-104 is designed for retrofit application into the A-4, delivering high computational power, high accuracy and high reliability (MTBF of over 4,000 hours).

DADC-104 inputs are: static and total pressure from the pitot tube, outside air temperature and several discretes such as BIT initiate. From these inputs the system calculates the required outputs, in both analogue and digital form. An integral BIT checks continuously the circuitry. Any detected failure or performance degradation is reported into the aircraft system through special status output. Special options providing additional I/O and/or different accuracies are available.

The DADC-104 can be supplied in suitable mechanical/electrical configurations for retrofit purposes. Other features and benefits of the system include:

- FFF to AXC 666 (installed on A-4F and upwards)
- Can perform functions of CPU-66 (encoded altitude) and autopilot air data sensor
- High accuracy
- Temperature compensation
- Completely solid state, including digital pressure transducers
- Digital and/or analogue outputs
- BIT
- Modular design
- Highly maintainable
- High reliability.

The DADC-104 calculates Calibrated Air Speed (CAS), True Air Speed (TAS), Mach number, pressure altitude, static free air temperature.

Specifications
Power: 24–32 V DC, 115 V AC
Power consumption: 20 W
Static pressure: 0–32 in Hg
Total pressure: 1–75 in Hg
OAT: –70 to 210°C
Altitude: –1,500 to 60,000 ft
True Air Speed: 70 to 1,200 kt
Calibrated Air Speed: 50 to 880 kt
Mach number: 0.1 to 1.2
Environment: per MIL 5400 Class II
Dimensions: 99.1 × 193 × 264.2 mm
Weight: <5 kg

DADC-105
The DADC-105, designed specifically for the F-16, features the same high reliability as the DADC-104 (MTBF >4,000 h).

DADC-105 inputs are: static and total pressure from the pitot tube, Angle of Attack (AoA), altimeter barometric setting, outside air temperature and several discretes such as altitude hold, flaps setting and BIT initiate. From these inputs, the DADC-105 calculates the required outputs using proprietary algorithms.

The interface with the avionics equipment is through a MIL-STD-1553 MUX BUS as well as several analogue and discrete signals. An continuous integral BIT checks the DADC-105 circuitry. Any detected failure or performance degradation is reported to the aircraft systems through dedicated status words, and valid discretes.

The system calculates: True Air Speed (TAS), Calibrated Air Speed (CAS), Mach Number, baro-corrected pressure altitude, pressure altitude, static free air temperature, air density, static pressure over total pressure, true Angle of Attack (AoA).

Specifications
Power: 115 V AC, 400 Hz
Power consumption: 40 W
Static pressure: 0–32 Hg
Total pressure: 1–90 in Hg
OAT: –70 to +210°C

The Astronautics DADC-105 digital air data computer   0503867

Altitude: –1,500 to 80,000 ft (accuracy 20 ft or 0.2% Hp)
True Air Speed: 70 to 1,700 kt (accuracy 2 kt or as per flight envelope)
Free air temperature: –100 to +150°C
Altitude rate: +6,000 fpm (accuracy 40 fpm)
AoA: –5 to +40° (accuracy 0.25°)
Environment: per MIL-E-5400 Class II
Dimensions: 160 × 177.8 × 340.4 mm
Weight: <7.73 kg

DADC-107
The DADC-107 is a complete Form/Fit/Function (FFF) replacement for the existing Central Air Data Computer (CADC) installed on F-4 aircraft.

The high accuracy of the DADC-107 outputs enhances the performance of the Weapon Delivery System (WDS) in air-to-air and air-to-ground modes of operation. A compact compensating module, which is an integral part of the DADC, executes static pressure compensation for the cockpit indicators.

Other features and benefits of the system include:

- FFF to CADC P/N 32-87094 which is installed on F-4 aircraft
- Temperature compensation
- Completely solid state including pressure transducers
- Analogue, synchro and discrete I/O
- MIL-STD-1553B serial multiplex bus interface available
- Built-In-Test (BIT)

The DADC-107 calculates: impact/static/corrected static pressure, altitude/altitude rate/altitude hold, True/Calibrated Air Speed (TAS/CAS), Mach number, true Angle of Attack (AoA) and other outputs as requested for particular applications.

Specifications
Power: 115 V AC, 400Hz, 1 Phase
Power consumption: 80 VA
Synchro excitation: 26 V AC, 400Hz, 1 Phase
Static pressure: 0.815 in Hg to 31.5792 in Hg
Total pressure: 0.3 in Hg to 105 in Hg
Angle of Attack: –12.5 to +50°
Total temperature: –40 to +200°C
Altitude: –1,500 to 80,000 ft
Mach number: 0.1 to 3.0
True Air Speed: 70 kt to 1,740 kt
Indicated Air Speed: 50 kt to 1,000 kt
Altitude hold: –500 to +500 ft
Altitude rate: ±96,000 ft/min
Mach hold: –0.1 to +0.1
Mach rate: ±2.5 /min
Static Air Temperature: –100 to +50°C
Impact pressure: 0.120 in Hg to 73.5 in Hg
Altitude, barometric: –1,500 to 80,000 ft
Corrected static pressure: 0.815 in Hg to 31.57 in Hg
True Angle of Attack: –15 to +30°
Environment: per MIL 5400 Class II (no cold air required)
Dimensions: 425.5 × 194.1 × 304.8 mm
Weight: 20.46 kg max

Status
In production and operational in A-4, F-4 and F-16 aircraft.

Contractor
Astronautics Corporation of America.

Digital air data computer for the AV-8B

Type
Avionic navigation sensor, Air Data Computer (ADC).

Description
In common with all Honeywell Engine Systems and Accessories' air data computers, this model uses a quartz pressure transducer. The unit is similar in construction to that used on the US Air Force A-10 aircraft. It features a non-volatile memory for the retention of fault messages, even if power is turned off.

Specifications
Dimensions: 129 × 193 × 355 mm
Weight: 7.2 kg
Inputs: 10
Outputs: 78

Status
In service on the US Marine Corps AV-8B.

Contractor
Honeywell Engine Systems & Accessories.

Digital air data computer for the B-1B

Type
Avionic navigation sensor, Air Data Computer (ADC).

Description
This computer was developed for the US Air Force B-1B and features a digital MOS large-scale integrated circuit processor and digital pressure transducers. It has an altitude reporting facility and the built-in self-test system provides continuous failure monitoring.

Specifications
Dimensions: 158 × 195 × 500 mm
Weight: 12.6 kg
Inputs: 19
Outputs: 92

Status
Developed for the US Air Force B-1A and in service in the B-1B.

Contractor
Honeywell Engine Systems & Accessories.

Digital Terrain System (DTS)

Type
Aircraft navigation system, Terrain Reference Navigation (TRN).

Description
As an enhancement to the present F-16 Data Transfer Equipment, Smiths Industries (formerly Orbital Fairchild Defense) has integrated BAE Systems' TERPROM® algorithm with a Data Transfer Cartridge mass memory and a high-performance processor. The resulting Digital Terrain System (DTS) capability is contained within the form factor of the original Data Transfer Cartridge (DTC). The Data Transfer Unit (DTU), the cockpit-resident, intelligent receptacle, remains unmodified resulting in straightforward retrofit or insertion into existing platforms.

The Mega Data Transfer Cartridge with Processor (MDTC/P) performs all existing data transfer functions of mission data load and in-flight data recording on a non-interfering basis while concurrently running DTS algorithms. The cartridge interface has been preserved to ensure compatibility with existing ground support equipment and mission planning systems. The original DTC and the MDTC/P cartridges can be interchanged on aircraft without affecting normal data transfer operations.

The BAE Systems' TERPROM® terrain correlation algorithm has been successfully implemented in numerous aircraft, including the F-16, Jaguar and Harrier. Its unique attribute is its ability to locate precisely the aircraft with respect to its onboard terrain database under adverse conditions, thus providing stealthy, all-weather night operation.

The MDTC/P hosts the Digital Terrain Elevation Data (DTED) database in a non-volatile, solid-state mass memory array. This memory was developed and qualified by Smiths Industries as part of continuing memory system development. It is actively employed on multiple, severe environment, military platforms and requires no special conditions for operation or application.

The cartridge-embedded processor works in conjunction with the cartridge mass memory providing an efficient method of processing DTS algorithms. Direct processor access to the terrain database obviates the transmission of bulk data over the MIL-STD-1553 databus simplifying system integration, reducing bus bandwidth and simplifying bus control.

Through the precise location of the aircraft with respect to terrain and knowledge of the real-time kinematic model of the aircraft, the TERPROM® algorithm can compute ground intercept and provide warning cues to the pilot in advance of entering a dynamically unrecoverable state.

Specifications
Dimensions:
DTU – 178 × 127 × 113 mm
MDTC/P – 191 × 127 × 113 mm
Weight:
DTU – 3.0 kg
MDTC/P – 1.6 kg
Power: 115 V AC, 400 Hz, single phase, 32 W
Cartridge data capacity: Typical configuration uses 72 Mbytes. Growth to over 1 Gbyte
MDTC/P software: coded in Ada
System interface: dual redundant, MIL-STD-1553B
Reliability (MTBF): DTS system – 12,000 h.

Status
In production and in service. During the third quarter of 2001, the USAF awarded Smiths a contract valued at more than USD16 million to supply the DTS, together with the MDTC/P and related ground support, for all F-16 aircraft, including Active, Guard and Reserve Units. The procuring agency for the contract is based at Hill AFB, Utah; systems are manufactured at Smiths' Germantown, Maryland, facility.

Contractor
Smiths Aerospace.

DMA-37A DME interrogator

Type
Aircraft navigation system, radio aids, DME.

Description
The DMA-37A is a fast-scan DME interrogator which meets all ARINC 709 characteristics. The design utilises advanced all-digital technologies to provide pilots with accurate and reliable DME and Tacan slant range information which can be transmitted via the ARINC 429 bus for use by both the visual display instruments and AFCS.

The DMA-37A features a centralised microprocessor system which offers superior signal processing capabilities and handles all unit control monitoring functions. Channel and control information is received through one of two selectable ARINC 429 frequency/function data input ports.

The primary processor automatically decodes the Morse code signal received from the channel

The DMA-37A DME interrogator   0503895

selected for Ident both for single channel timing and when in the multiscan mode. A dedicated second microprocessor translates the logic signal representing the dots and dashes into alphanumeric characters before being formatted into ARINC 429 for transmission on the output ports by the main CPU.

The unit's microprocessor also helps to simplify maintenance through a rigorous self-monitoring and self-diagnostic routine. To further reduce down time, the fault memory and BITE are interfaced with the central fault display system and access to strategic points within the unit is provided via the automatic test equipment connector. It complies with ICAO Annex 10.

Specifications
Weight: 5.85 kg
Frequency:
(transmitter) 1,025–1,150 MHz
(receiver) 962–1,213 MHz
Channel spacing: 1 MHz

Status
In production.

Contractor
Honeywell Aerospace, Electronic & Avionics Lighting.

DME-42/442 Pro Line II DME systems

Type
Avionic navigation sensor, radio aids.

Description
The DME-42/442 are all-digital DME systems which can provide complete information on up to three DME stations using a single transceiver. They were designed for business or commuter aircraft where only a single DME facility is needed, but where additional information is useful. Station information is presented on the associated IND-42 display.

The systems can provide data to show 'distance to station' up to 300 n miles, time to station up to 120 minutes and groundspeed up to 999 kt. They also decode the station identifier and provide for display.

The DME-442 is compatible only with CSDB or ARINC 429 controls and displays. The DME-42 can directly replace the earlier DME-40 system.

Specifications
Dimensions:
(R/T) ½ ATR short/dwarf
(IND-42) 42 × 86 mm
Weight:
(R/T) 2.4 kg
(IND-42) 0.41 kg
Power supply: 28 V DC, 0.7 A, plus 0.3 A for display
Altitude: up to 70,000 ft

Status
In production and in service.

Contractor
Rockwell Collins.

DME-900 Distance Measuring Equipment (DME)

Type
Avionic navigation sensor, radio aids.

Description
Collins' DME-900 Distance Measuring Equipment (DME) has been designed to provide highly accurate distance data outputs in all modes of operation. Error sources in the interrogator are identified and reduced to the lowest level. A highly efficient parallel ranging technique allows the DME-900 to perform less than 15 interrogations, half the allowable maximum average. Only two output devices are required to achieve a 700 W nominal power output. The synthesiser/driver

operates at L-band transmit channel frequencies, significantly reducing the parts count and circuit complexity. In addition, optimum filtering methods reduce naturally occurring noise contained in the DME/Tacan signal.

A comprehensive self-test monitor is capable of determining the operational status of the system to better than 99 per cent confidence level.

The DME-900 Receiver meets all essential level HIRF requirements, provides improved power interrupt capability, and enhanced BITE interface for Boeing, Airbus, and Douglas aircraft.

Specifications
Dimensions: 4 MCU
Weight: 6 kg
Power supply: 115 V AC, 380 to 420 Hz, 35 W (max)
Temperature: –40 to +70°C operating
Altitude: 50,000 ft
Frequency:
(transmit) 962 – 1,150 MHz
(receive) 962 – 1,213 MHz
Channels: 252 in 100 kHz increments
Range: 0 to 320 nm
Accuracy: ±0.1 nm
Certification: FAA TSO-C66b; RTCA DOs -151A, -160C and 178A; EUROCAE EDs -57, 12A and 14C; and ARINC 604, 600, 429 and 709.

Status
In production and in service. Certified on: Boeing 747, 757, 767, 777 and 737X; Airbus A319, 320, 321, 330 and 340; and MD-11 and MD-90.

Contractor
Rockwell Collins.

DMS-44 Distance Measuring System

Type
Aircraft navigation system, radio aids, DME.

Description
The DMS-44 is a digital, solid-state, dual-transmitter, distance-scanning measuring system which can receive three stations simultaneously. It is a lightweight all-digital system consisting of the DM-441 transmitter/receiver and the SD-442 sector display.

The all-solid-state transmitter provides the capability simultaneously to scan stations for Nav 1 and Nav 2 and a third station which is transparent to the pilot for computation. It utilises two separate microprocessors and a video processor. The master processor provides signal processing, control computations and analogue range information. The slave processor performs the digital input/output generation, including the input of frequency-tuning interfaces. The range processor can lock on to data in less than 200 ms and provides an LSB accuracy of better than 0.01 n miles. The system is available with either front or rear connector mounts.

The SD-442A panel-mounted selector/display unit provides full-time display of distance and groundspeed to the selected VORTac. The active Nav is annunciated below the distance readout.

The DMS-44 employs full-time self-monitoring of key circuits such as the synthesiser, receiver, transmitter, power supply and master processor.

Specifications
Dimensions:
(selector/display) 82.6 × 39.4 × 63.5 mm
(front mount transmitter/receiver) 127 × 101.6× 279.4 mm
(rear mount transmitter/receiver) 127 × 101.6× 320.6 mm
Weight:
(selector/display) 0.204 kg
(front mount transmitter/receiver) 2.77 kg
(rear mount transmitter/receiver) 3.22 kg
Power supply: 18–33 V DC, 1.2 A

Contractor
Honeywell Aerospace, Electronic & Avionics Lighting.

FMS navigation systems

Type
Aircraft navigation system, Global Positioning System (GPS).

Description
The Rockwell Collins FMS is a family of satellite-based precision navigation systems designed for all phases of flight, including take-off, en route, approach and landing. It is grouped into series, to match the mission of the aircraft and caters for long-range corporate jets, medium to light jets and turboprops and 30- to 100-seat regional airline passenger aircraft.

At the core of each FMS series is a Rockwell Collins GPS sensor. The capabilities of a flight management system are incorporated within the navigation system. The avionics GPS engine features 12-channel satellite tracking, including those comprising the Wide Area Augmentation System, providing satellite coverage sufficient to meet the FAA-required navigation performance and precision approach criteria; a new software code to ensure certification to DO-178B Level A; Receiver Autonomous Integrity Monitoring (RAIM) and predictive RAIM, and differential correction computation to qualify the receiver for Cats I and II.

The FMS series provides the capability to calculate position and navigation information accurately anywhere in the world; provide VNav guidance and comprehensive pilot annunciations; fly complete airways, SIDs, STARs and approach legs, and predict fuel consumption. FMS also significantly reduces pilot workload by automating many calculations and functions. Information is presented on multifunction display pages, allowing easy, efficient movement through the system. Pilots can remain eyes-up during all phases of flight, resulting in enhanced situational awareness and greater control. Flight planning is simplified to the point where even the most complicated procedures can be generated with a few keystrokes. FMS is a component system of the technology required for the planned 'free flight' regime in which aircraft en route separations will be reduced.

FMS 3000
The FMS 3000 is designed for light jet and turboprop aircraft. It provides the capability for en route, terminal and non-precision approach and lateral/vertical navigation, and includes the growth potential to support precision approach computations.

The 3000 series installation consists of the CDU-3000 Control/Display Unit, FMC-3000 Flight Management Computer (FMC) and FMS GPS sensor. Updated navigation information may be uploaded in the field and 100 flight plans, with 100 waypoints per plan, can be stored in the navigation database.

FMS 3000 also provides access to ground-based messaging, weather data and flight plan transmission through the Airborne Flight Information System (AFIS). Free form messages can be sent to and from the CDU. Frequently used messages can be formatted and stored.

FMS 4000
The FMS 4000 meets the needs of regional airline aircraft, providing worldwide GPS navigation, simplifying cockpit management and enhancing flight crew and aircraft performance. In addition to GPS, long-range navigational inputs such as IRS/AHRS, VLF/Omega, VOR and DME can be used to provide accurate determination of aircraft position. Coupled with flight management technology, the system provides the integrity, flexibility and operational capabilities required by regional airline operators.

FMS 4000 provides the capability for preflight and general flight planning. Complete lateral and vertical navigation is available. Radio sensor management is also available. Two flight plans are available: the primary flight plan is used for active guidance, while a secondary flight plan can be stored as an alternative and activated if needed. Full time and fuel management performance are provided, with fuel prediction available. Up to 1,000 standard company routes, each with up to 100 waypoints, can be stored in the system's database. The database contains navaids, waypoints, non-directional beacons, airports, airport reference points and runway thresholds. Flight planning includes the capability to execute SIDs, STARs, airways and holding patterns.

Once a flight plan is activated, it is presented graphically on a MultiFunction Display (MFD). The MFD also displays navaids, intersections, airports, terminal waypoints and non-directional beacons. In addition, the MFD can present complete operational status, progress and flight plan summary text.

FMS 4000 comprises three LRUs: the CDU-4100 Control/Display Unit, FMC-4200 Flight Management Computer (FMC) and GPS-4000 GPS sensor. The FMC-4200 is packaged as a line-replaceable module, housed in the Collins Integrated Avionics Processing System (IAPS). Pro Line 4 integration allows the FMS 4000 to perform self-diagnostic functions and informs the Pro Line 4 maintenance diagnostic computer of internal faults.

The FMS 4000 has been certified for use on the Saab 2000 aircraft.

FMS 5000
Designed specifically around the needs of medium to light jet and turboprop aircraft, the FMS 5000 centralises control of EFIS, navigation, weather radar, TCAS and radio management in one conventional location to maximise cockpit space. It includes features such as the ability to calculate accurately position and navigation information anywhere in the world, provide VNav guidance and comprehensive pilot annunciations, fly complete airways, SIDs, STARs and approach legs, predict fuel consumption, and offer the right combination of performance, efficiency and ease of use required to perform in complex and demanding operating environments.

FMS 5000 combines GPS technology with flight management capability to provide satellite-based precision navigation systems for take-off, en route, terminal area, approach and landing operations. Other long-range navigational inputs include IRS/AHRS, VLF/Omega, VOR and DME. These may be used to provide the most accurate determination of aircraft position. Vertical navigation is available in each phase of flight. Included is the ability to execute a vertical direct-to, generating a vertical path from the aircraft's current position to a designated altitude.

FMS 5000 also provides access to ground-based messaging, weather data and flight plan transmission through the Airborne Flight Information System (AFIS). Free-form messages can be sent to and from the CDU. Frequently used messages can be formatted and stored.

FMS 5000 comprises three LRUs: the CDU-5000 Control/Display Unit, FMC-5000 Flight Management Computer (FMC) and AVSAT GPS sensor. The FMC is packaged as a line-replaceable module, housed in the Collins Integrated Avionics Processing System (IAPS). Pro Line 4 integration allows the FMS 5000 to perform self-diagnostic functions and informs the Pro Line 4 maintenance diagnostic computer of internal faults.

FMS 6000
The FMS 6000 is designed for long-range business aircraft. It features the ability to calculate position and navigation information accurately anywhere in the world, provide VNav guidance and comprehensive pilot annunciations, fly complete airways, SIDs, STARs and approach legs and manage fuel consumption.

FMS 6000 comprises three LRUs: the CDU-6000 or CDU-6100 Control/Display Unit, FMC-6000 Flight Management Computer and GPS-4000 sensor. It can be configured for single, dual or triple operation by installing the desired number of FMC modules and the same number of control/display units. It is designed to be integrated with the Pro Line 4 avionics system.

FMS 6000 is designed for worldwide navigation, including polar navigation. Long-range inputs in addition to GPS can be used, including IRS/AHRS, VLF/Omega, VOR and DME to provide the most accurate determination of position. It contains a global database of current information used for both flight planning and navigation.

Global Navigation Landing Unit (GNLU)
To meet the requirements of the Air Traffic Management (ATM) system, Rockwell Collins is offering the GNLU system (see separate entry on GLU/GNLU multimode receivers). The system defined by GNLU is of a single navigation unit integrating all the GNSS-based (Global Navigation Satellite Systems) en route, terminal and landing system capabilities required to operate in a global Communications Navigation Surveillance/Air Traffic Management (CNS/ATM) environment.

The GNLU contains an advanced GPS receiver (FAA order 8110.60 compliant) with the required integrity to support precision GNSS-based Landing System (GLS) approach certification to Cat I and II levels.

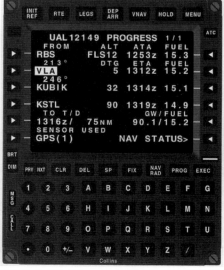

The FMS 3000 CDU   0581498

GNLU multifunction control/display unit 0015375

GNLU-900 GNSS   0015376

Because the role of the GPS sensor is so important, Rockwell Collins has designed and developed a specialised GPS engine for use in air transport system applications. Optional MLS modules, developed in conjunction with DaimlerChrysler Aerospace, and optional ILS modules can be integrated within the GNLU to provide classic aircraft with analogue interfaces. This is an efficient update path for compliance with new landing sensor requirements, such as FM immunity.

As part of the total Rockwell Collins FMS package, the GNLU adds substantially more than GPS receiver inputs to existing flight management capabilities; GNLU is the primary processing and control unit of the GPS/FMS system. The FMS supports complete lateral navigation and provides vertical guidance on approach.

The complete system comprises the cockpit-mounted Multifunction Control/Display Unit (MCDU), the remote 4-MCU sized GNLU, and a GPS antenna.

The GNLU is structured to add the interfaces and processing needed for precision GNSS approaches, using an integral uplink data receiver to simplify installation. The GNLU navigation computer has the volume and processing power to support the Wide Area Augmentation System (WAAS) and Local Area Differential GNSS (LADGNSS) receiver and modem capability; all essential WAAS hardware is provided in the baseline configuration. The system also provides ACARS and ARINC 622 datalink interfaces and is provisioned to interface with ATN compliant architectures.

Specifications
(GNLU-900 GNSS)
GPS performance:
(channels) 12-channel, all-in-view tracking
(frequency) 1,575.42 MHz (L1) C/A code
Accuracy:
(horizontal) 23 m (4.5 m with DGPS); 1.5 m/s (velocity)
(altitude) 30 m (6.0 m with DGPS); 1.2 m/s (velocity)
Dimensions:
4 MCU per ARINC 600
128 (H) × 200 (W) × 369 (L) mm
Weight: 7.3 kg
Power: 28 V DC, 80 W
Temperature: –20 to +55°C operating
Altitude: 25,000 ft
Certification:
FAA TSO: C36e, C34e, C129B1, C129a
RTCA: DO-192, DO-195, DO-160D, DO-178B
EUROCAE: ED-72A, ED-46A, ED-47A, ED-12A, EC-14C
ARINC: 604, 600, 429, 710, 711, 547, 578, 579

Status
Rockwell Collins' FMS are in production and in service in a wide range of business and regional aircraft.

Contractor
Rockwell Collins.

Force™ 5 Module

Type
Avionic navigation sensor, Global Positioning System (GPS).

Description
The Force™ 5 module is an all-in-view, dual-frequency receiver providing both Standard Positioning Service (SPS) and Precise Positioning Service (PPS) capability. Designed in accordance with the GPS Receiver Application Module (GRAM) guidelines, the Force™ 5 module provides a high-performance GPS navigation input to open architecture military applications.

Force™ 5 addresses both Global Air Traffic Management (GATM) and Navigation Warfare (NAVWAR) requirements. Its SPS/PPS Firewall and Switch ensure SPS operation in civil airspace, with Fault Detection and Exclusion (FDE) and Step Detector functions implemented

in accordance with RTCA/DO-229B to ensure the integrity of the GPS solution.

In the PPS mode, the Force™ 5 module corrects for Selective Availability (SA) and uses Anti-Spoof (A-S), augmented with FDE, to protect against satellite/system anomalies and spoofing. Direct Y-Code acquisition enables continuous operation in harsh jamming environments. Advanced digital signal processing and software algorithms, as well as a high-performance Oven Controlled Crystal Oscillator (OCXO), provide precise measurement data even under high dynamics and vibration. Force™ 5 is easily integrated as a standalone GPS receiver or coupled with inertial or Doppler navigation systems, providing both parallel and serial host control interfaces for maximum flexibility in retrofit applications. To support more tightly integrated solutions, Line of Sight (LoS) measurement data, including 10 Hz delta range and carrier phase, is output to the host. The Force™ 5 (PPS) module has been granted the NAVSTAR Global Positioning System Joint Program Office (JPO) security approval. Adherence to GRAM Guidelines reduces system life-cycle cost compared with existing proprietary solutions. Upgrade to SAASM capability is provided by Trimble's Force™ 5 GS (GRAM + SAASM) module, which provides for a backward compatible path to existing Force™ 5 users.

Specifications
Receiver: 12-channel continuous tracking, L1/L2 frequency, C/A, P(Y) code; SPS/PPS Firewall and Switch for operation in civil airspace
Update rate: 1 Hz
Timing: HAVE QUICK IIA timing output or 1 PPS I/O; GPS time mark output
Crypto key input: Compatible with KYK-13, KOI-18, CYZ-10
Security: GPS JPO security approved/unclassified when keyed
Accuracy – autonomous GPS:
(position 3-D) 16 m SEP
(velocity) 0.1 m/s RMS per axis
(time) 100 ns RMS (1 sigma WRT UTC)
Accuracy – Differential GPS (DGPS):
(position 3-D) 5 m SEP
(velocity) 0.1 m/s RMS per axis
Altitude: –701 to +22,860 m (–2,300 to +75,000 ft) AMSL
Dynamic envelope: ±1,200 m/sec per axis; ±100 m/sec^2 per axis
Dimensions:
(receiver) 149 (W) × 145 (D) × 15 (H) mm
Weight: 0.425 kg
Compliance: GRAM-S compliant per GPS-GRAM-001A and TNL-GPS-GRAM-S-001; SPS differential GPS per RTCM-104 Version 2.1; PPS differential GPS per ICD-GPS-155; ICD-GPS-225; CZE-93-105A
Certification: TSO-C129a Class B1/C1

Contractor
Trimble Navigation.

FreeFlight 2101 I/O Approach Plus airborne IFR GPS navigation system

Type
Aircraft navigation system, Global Positioning System (GPS).

Description
The FreeFlight 2101 I/O Approach Plus is the full input/output version of the company's GPS Dzus rail-mount navigator series. Certified to TSO C-129(A1), it may be used for supplemental en route, terminal and approach IFR operations.

The 2101 I/O Approach Plus is also approved for 8110.60 GPS as Primary Means of Navigation for Oceanic Operations and BRNAV European Airspace Operations.

The FreeFlight 2101 I/O Approach Plus minimises pilot workload during approaches. It automatically selects all the required navigation functions and automatically presents and removes all messages.

At the heart of the FreeFlight 2101 I/O Approach Plus is an advanced 12-channel GPS receiver with a defined upgrade path to WAAS capability. It continuously tracks all satellites in view, determining and displaying a new position four times per second and measuring speed to better than a tenth of a knot. It calculates and displays position, bearing and distance to waypoint, ground speed and ground track and estimated time en route. It includes an electronic Course Deviation Indicator (CDI) and a patented track angle error (TKE) display. The 2101 I/O Approach Plus presents the data on a high-contrast LED display. A Night Vision Goggles (NVG) compatible variant is available, which uses specially made glass, filters and polarisers.

The 2101 I/O Approach Plus continuously monitors its accuracy via Receiver Autonomous Integrity Monitoring (RAIM). It also predicts RAIM conditions for the approach arrival time.

The FreeFlight 2101 I/O Approach Plus is easy to install. Installation requires only four external annunciators: Message (MSG), Waypoint (WPT), Approach (APR) and Hold (HLD). Expensive resolvers are not required.

The 2101 I/O Approach Plus supports CDIs, HSIs, altimeters, RMIs, fuel flow gauges and many other flight instruments.

When integrated with an air data computer, the FreeFlight 2101 I/O Approach Plus displays true air speed, density altitude and pressure altitude and calculates and displays winds aloft and applies current wind ETE and ETA calculations. It also automatically sequences through up to 40 waypoints in a flight plan, shows the nearest airport, plans vertical descent profiles and performs a variety of other functions.

The FreeFlight 2101 I/O Approach Plus incorporates Jeppesen's NavData card – an avionics database that includes airports, approaches, SIDS, STARS, VORs, NDBs, intersections, airspace boundaries and more.

A Precision aRea Navigation (PRNav) version will be available in late 2000. The military version is known as the Cargo Utility GPS Receiver (CUGR).

Specifications
Type: 12-channel receiver, L1 frequency, C/A code, continuous all-in-view tracking
Acquisition time: 1.5–3.5 min
Position update rate: 1 time/s
Dynamics: 800 kt (4 *g* tracking)
Accuracy:
(position) 15 m RMS
(velocity) 0.1 kt steady-state
(altitude) 35 m RMS (msl)
Computation range: Great Circle: 0–999 n miles
Distance resolution: 0–9.99 n miles: in 0.01 n mile increments; 10–99.9 n miles in 1.0 n mile increments
GPS antenna: Omnidirectional flat microstrip with integral preamp. TNC connector
Display:
LED: 2 lines of 20 characters each High-intensity, orange alphanumeric
Dimensions: 146.05 × 196.85 × 76.2 mm

FreeFlight 2101 I/O Approach Plus airborne IFR GPS navigation system 0015391

Weight: 1.26 kg
Power: 28 V DC, negative ground, 0.5 A
at 28 V DC, 14 W

Status
In production and in service.

Contractor
FreeFlight Systems.

GEM II/III/IV/V embedded GPS modules

Type
Avionic navigation sensor, Global Positioning System (GPS).

Description
The GPS Embedded Module (GEM) family represents a flexible approach to modular GPS. By offering a family of modules that share the same form factor and interface, the user is offered the ability to select various GPS performance parameters without the added expense of additional integration. To simplify the basic software integration, the GEM family's primary interface is via Dual Port RAM utilising the ICD-GPS-059 format. In addition, HAVE QUICK, KYK-13, RS-422 and 1 pulse/s input/output are also included as standard interfaces.

The GEM series offers full P/Y code performance in a standard module; current applications include manned and unmanned airborne platforms, ground vehicles and missiles. The ability to operate either as a stand-alone GPS receiver or integrated with Inertial or Doppler sensors, Mission Computers and Flight Management systems provides maximum flexibility for platform installation.

GEM II
The GEM II receiver offers full military performance to Miniature Airborne GPS Receiver (MAGR) specifications with an IF interface for retrofit of RCVR 3A, RCVR UH and RCVR OH installations.

GEM III
The GEM III receiver also offers full military performance to MAGR specifications (in the PPS configuration) with an RF interface direct to an antenna. This receiver also supports an RF preamplifier for longer cable runs while still utilising a single cable interface to the antenna.

GEM IV
The GEM IV was developed as a System Specification Equivalent (SSE) replacement for the GEM III and is form, fit and functionally compatible. GEM IV is designed to be a backward compatible replacement of Rockwell Collins' other GEM series modules and allows for an evolutionary path to GPS Receiver Application Module (GRAM) performance. Incorporating the Nighthawk Digital Signal Processor, GEM IV is a 12 channel receiver with a five channel interface, coupled with a dual-frequency L1/L2, P(Y)-code GPS receiver with Selective Availability and Anti-Spoofing (SAAS) capabilities.

GEM V
The GEM V includes all the features of the GEM IV plus Rockwell's third-generation Selective Availability Anti-Spoofing Security Mode (SAASM) with Key Data Processing (KDP-2). The GEM V incorporates all the latest GPS functionality including improved security, ARIM/FDE, Fast Direct Y Code acquisition and the SPS/PPS selection.

Status
The system was selected for the US DoD's first embedded GPS and inertial programme (GINA) for the Navy T-45 trainer aircraft, as well as for the tri-service Embedded GPS/IVS (EGI) programme. The GEM IV now offers a Fast Acquisition Direct Y Capability, which utilises Rockwell Collins'

Specifications
Performance

	C/A code without SA	C/A code with SA
Position accuracy	10 m horizontal	100 m horizontal
Velocity accuracy		
3D	0.2 m/s	0.7 m/s
Time accuracy	70 ns	270 ns

Physical

	GEM II	GEM III	GEM IV/V
Number of channels	5	5	12
P/Y code capable	Yes	Yes	Yes
Frequency	L1/L2	L1/L2	L1/L2
Antenna interface	IF	RF	RF
Power (typical), W	8.9	6.4	6.5
Input	±5 V, −15, +21.4 V DC	±5 V DC	+5 V DC
Temperature range	−54 to +85°C	−54 to +85°C	−54 to +85°C
DPRAM I/0	ICD-059	ICD-059	ICD-GEM-DPRAM or ICD-BDDP-155 (IV) ICD-GEM-155 (V)
Dimensions, mm	145.35 × 144.78 × 14.73	145.35 × 144.78 × 14.73	145.35 × 144.78 × 14.73
Weight	0.45 kg	0.45 kg	0.45 kg

Dynamic conditions: 1,200 m/s, (min) satellite signal power level, and 31 dB jammer signal level. SA degradation is assumed at the current nominal SPS level of 100 m.

High Accuracy, Low Power Time Source (HAL) and the Acquisition Correlation Engine (ACE). The ACE Digital Signal Processor interfaces with the Nighthawk ACE Digital Signal Processor and provides a large correlation bank in order to enhance the acquisition. The HAL device improves the powered-down time-keeping accuracy by more than two orders of magnitude. Combined, these two components provide superior Direct Y acquisition performance to counter the vulnerabilities of C/A code threats. Planned Enhancements to GEM IV include the following:

- all in view tracking and navigation
- DO-229 RAIM/FDE
- Carrier Phase measurements
- Standard Positioning Service (SPS)
- ICD-GPS-155 compatibility

GEM receivers are standard fit in several inertial navigation systems as well as Flight Management (FM) and Mission Computer(MC) systems. GEM II and III in service but no longer in production. GEM IV and V in production and in service.

Contractor
Rockwell Collins.

GPS 155XL TSO Global Positioning System

Type
Aircraft navigation system, Global Positioning System (GPS).

Description
The GPS 155XL TSO is a panel-mounted receiver which replaces the superseded GPS 150 and GPS 155 TSO (see previous items). The unit is IFR approach certified under TSO C129a A1 and includes a very high-definition moving map. The moving map on the GPS 155XL TSO is displayed via the familiar Garmin DSTN (Double SuperTwist Nematic) LCD yellow-on-black display found in the VFR-only GPS 150XL and GNC 250XL.

A photocell in the display automatically controls the intensity of the backlight and will reverse the display from black-on-yellow to yellow-on-black for maximum contrast in daylight or night time viewing. Users may also make these adjustments manually.

The database in the GPS 155XL TSO includes non-precision approaches for US airports, all published SIDs and STARs, 1,000 user-defined waypoints and 20 reversible routes with up to 31 waypoints in each. It contains an extensive, updatable, Jeppesen database providing detailed information on airports, VORs, NDBs, intersections, comms frequencies, runways, FSS and MSAs. In Emergency Search mode, the system lists the nine nearest airports, VORs, NDBs, or user-defined waypoints, together with the two nearest FSS with frequencies and the two nearest ARTCC frequencies. The unit also includes a built-in Ni/Cd battery back-up, capable of providing emergency power in the event of aircraft electrical power failure.

The GPS 155XL TSO can fully interface with flight controls, EFIS, HSI, moving map, altitude encoder, fuel management and other aircraft systems.

Specifications
Dimensions: 159 × 147 × 51 mm
Weight: 0.77 kg
Power supply: 10 to 33 V DC or 115 to 230 V AC (with optional AC adapter)
Display: 80 × 240 DSTN display
GPS receiver: PhaseTrac12™ 12-parallel channel
Performance:
Accuracy: 15 m (position) RMS, 0.1 kt (velocity) RMS (steady state)
Update rate: 1 Hz, continuous
Acquisition time: 15 s (warm), 45 s (cold)
Database: Jeppesen Americas or International
Interface: ARINC 429, RS 232
Certification: TSO C129a Class A1 (en route, terminal and approach)

Contractor
Garmin International.

Garmin GPS 155XL TSO Global Positioning System (Garmin International)

0105002

GPS 165 TSO Global Positioning System

Type
Aircraft navigation system, Global Positioning System (GPS).

Description
The GPS 165 TSO is the dzus-rail configuration of the GPS 155 TSO (see earlier item). The GPS 165 TSO tracks and uses up to eight satellites continuously, providing 1 second updates with horizon-to-horizon coverage.

System features include 1,000 user-defined waypoints, 20 reversible routes with up to 31 waypoints each, single key 'direct to' operation, Jeppesen NavData stored on an updatable removable data card, Auto Search search and rescue grid activation function, optional remote battery pack back-up providing 2 hours of operation in the event of aircraft power failure and interfaces for ARINC 429, RS-232 and RS-422.

In April 1995, the GPS 165 TSO achieved FAA C129 A1 certification for IFR approach operations, and for BRNAV operations in 1998.

Specifications
Dimensions: 146 × 57 × 144 mm
Weight: 0.97 kg
Display: High intensity (600 fL) dot matrix fluorescent
Power supply: 10–33 V DC, 115–230 V AC with battery charger
Battery life: up to 2 h
Temperature range: –30 to +70°C
GPS receiver: MultiTrac8™
Accuracy: 1–5 m (position), 0.1 kt (velocity) RMS (steady state)

Status
The GPS 165 TSO is in production and in service in a wide range of general aviation aircraft.

Contractor
Garmin International.

GPS 400 Global Positioning System

Type
Aircraft navigation system, Global Positioning System (GPS).

Description
The GARMIN GPS 400 combines a global positioning system with a full-colour moving map for enhanced situational awareness. The map features a built-in all-land database that shows cities, roads/highways, railways, rivers, lakes and coastlines, in addition to a Jeppesen database. Thanks to a high-contrast colour display, the information can be easily read from wide viewing angles even in direct sunlight. Like the GNC 420, the GPS 400 is TSO C129a certified for a non-precision approach.

Specifications
Dimensions: 159 (W) × 67.3 (H) × 279.4 (D) mm
Weight: 3 kg
Power supply: 11–33 V DC

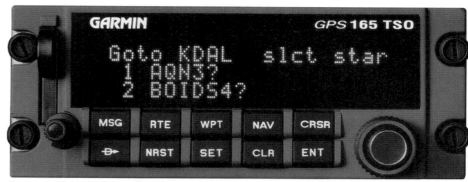

Garmin GPS 165 TSO GPS (Garmin International)   0099752

Display: Colour LCD
GPS receiver: PhaseTrac12™ 12-parallel channel
Accuracy: 15 m (1.5 m with differential correction) (position), 0.1 kt (velocity) RMS (steady state)
Database: Jeppesen Americas, International or Worldwide
Interface: ARINC 429, RS 232
Certification: TSO C129a Class A1 (en route, terminal and approach)

Status
In production.

Contractor
Garmin International.

GPS 500 Global Positioning System

Type
Aircraft navigation system, Global Positioning System (GPS).

Description
The GPS 500 can simultaneously display approach information together with weather and traffic data, in relation to present position, on a large, colour moving map display. The high-contrast display enables the pilot to easily interpret presented information from wide viewing angles and in direct sunlight. The colour moving map features a built-in cultural database that shows cities, highways, railroads, rivers, lakes, coastlines, together with a complete Jeppesen database. The Jeppesen database (that can be updated with a front-loading data card) contains all airports, VORs, NDBs, airway intersections, FSS, Approach, STARs and SUA information.

The GPS 500 is TSO C129a certified for GPS non-precision approaches. The system incorporates growth potential, with defined paths to WAAS compatibility, and TAWS capability. Pilots can utilise the GPS 500 as an MFD, especially when it is coupled with traffic, lightning detection, and weather interfaces, such as Ryan TCAD, TIS from the Garmin GTX 330 Mode S transponder, or Goodrich SKYWATCH™ and STORMSCOPE® WX 500 (see separate entries).

Specifications
Dimensions: 159 (W) × 117 (H) × 279.4 (D) mm
Weight: 3.1 kg
Temperature: –20 to +55°C (operating)
Altitude: –1,500 to 50,000 ft
Velocity: 999 kt (max)
Acceleration: 6 g (max)
Power supply: 11–33 V DC
Display: Colour LCD
Planning features: True airspeed, density altitude, winds aloft, RAIM availability, sunrise/sunset times, trip and fuel planning, vertical navigation (VNAV)
Navigational features: Pilot-defined course selection and waypoint hold, closest point of flight plan, departure and arrival frequencies, approach navigation using published approach procedures stored on NavData card, terminal navigation using DPs/STARs from NavData card
Map datums: 124
Plotting scales: 1/10 to 3000 nm with push-button zoom control
Waypoints: 1,000, user defined
Flight plans: 20 reversible routes of up to 31 waypoints each
GPS receiver: PhaseTrac12™ 12-parallel channel
Accuracy: 15 m (1.5 m with differential correction) (position), 0.1 kt (velocity) RMS (steady state)
Database: Jeppesen Americas, International or Worldwide
Interface: ARINC 429, RS 232, CDI/HSI, RMI (digital: clock/data), superflag out, altitude (serial: Icarus, Shadin-Rosetta, encoded Gillham/Greycode), fuel sensor, fuel/air data, BFG WX 500 StormScope™, BFG SKY 497 SkyWatch™, Ryan 9900B TCAD
Certification: TSO C129a Class A1 (en route, terminal and approach)

Jeppesen database
Coverage: Americas, international or worldwide
Airports: Identifier, city/state, country, facility name, lat/long, elevation, fuel service, control, approach information
VORs: Identifier, city/state, country, facility name, lat/long, frequency, co-located DME/TACAN, magnetic variation, weather broadcast
NDBs: Identifier, city/state, country, facility name, lat/long, frequency, weather broadcast
Airway intersections: Identifier, country, lat/long, nearest VOR
Frequencies: Approach, arrival, control area, departure, Class B, Class C, TMA, TRSA (with sector, altitude, and text usage), ASOS, ATIS, AWOS, centre, clearance delivery, tower, ground, unicom, pre-taxi, localiser, and ILS
Runways: Designation, length, width, surface, lighting, pilot-controlled lighting frequency
FSS: Identifier, reference VOR, frequency, usage
ARTCC: Identifier, frequency, usage
MSA: Minimum safe altitude along and in proximity to active flight plan
Approaches: Non-precision and precision approaches throughout the database coverage
STARs: Contains all pilot/nav DPs/STARs
Airspace: Class B & C with sectors, international CTA & TMA with sectors, all SUAs, including MOAs, prohibited and restricted areas with controlling agency and airport

Garmin GPS 400 global positioning system (Garmin International)   0099756

Garmin's GPS 500 global positioning system has been TAWS certified (Garmin International)

1127912

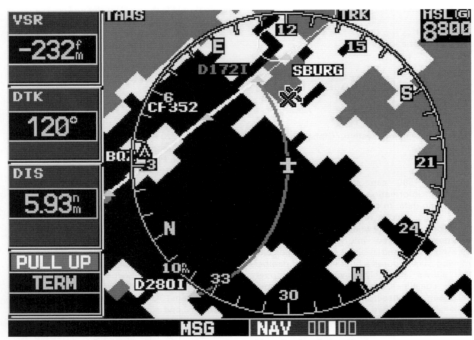

GPS 500 terrain warning display, showing range and bearing to obstructions
(Garmin International)

1127913

Status
During December 2004, Garmin announced that it had received FAA certification to add TSO C151b Class-B Terrain Awareness and Warning System (TAWS) functionality to the GPS 500.

The addition of TAWS enables the GPS 500 to graphically display the surrounding terrain and obstacles in bright yellow and red, relative to the aircraft's current altitude. Yellow is used to depict conflicts 1,000 to 100 feet below the aircraft. Red is used to depict conflicts 100 vertical feet below the aircraft's current altitude and above. Working in tandem with the aircraft's audio system, the GPS 500 provides audible cautions and warnings to alert the pilot of a possible terrain and obstacle conflict. Audible and graphical alerts include forward-looking terrain avoidance, imminent terrain impact, premature descent during approach, altitude loss after takeoff, 500-foot callout and excessive rate of descent.

The US FAA mandated that all US-registered, turbine-powered aircraft with six or more passenger seats be equipped with certified TAWS technology by the end of March 2005.

Contractor
Garmin International.

GPS-505 Global Positioning System receiver

Type
Avionic navigation sensor, Global Positioning System (GPS).

Description
The GPS-505 five-channel GPS receiver incorporates continuous parallel satellite tracking to deliver a 3-D position solution accurate to 15 m horizontally and 27 m vertically, together with speed data accurate to 0.1 kt. Update rate is 1 Hz. The system is associated with the FMS 5000 Flight Management System (see separate entry), also produced by Arnav Systems.

Status
In service.

Contractor
Arnav Systems Inc.

GPS Navstar navigation systems

Type
Aircraft navigation system, Global Positioning System (GPS).

Description
SCI produces two- and five-channel GPS receivers in multiple configurations; the two-channel UH receiver for helicopters, the five-channel 3A receiver for aircraft and the five-channel 3S for shipboard use. The receivers are fully qualified for operation on highly dynamic platforms in a high-jamming environment. As multichannel receivers, they are capable of processing signals from multiple satellites simultaneously, expediting initial acquisition times. The receivers are capable of providing accuracies of better than 16 m SEP in position, 0.1 m/s in velocity and 100 ns in time, even in the presence of the DoD selective availability and anti-spoofing environment.

Specifications
Dimensions:
(3A receiver) 193 × 191 × 484.8 mm
(UH receiver) 193.6 × 191.3 × 382.6 mm
Weight:
(3A receiver) 18.2 kg
(UH receiver) 11.5 kg
Power supply: 120 V AC, 400 Hz
160 V DC
Interfaces:
(3A receiver) ARINC 429 and 575, MIL-STD-1553, PTTI, KYK-13
(UH receiver) ARINC 561, 572, 575, 582, Have Quick, RS-422, KYK-13
Reliability:
(3A receiver) >1,350 h MTBF
(UH receiver) >160 h MTBF

Status
In service.

Contractor
Sanmina-SCI Corporation.

GPS-1000 sensor

Type
Avionic navigation sensor, Global Positioning System (GPS).

Description
The GPS-1000 sensor was specifically designed to interface with the UNS Flight Management and Navigation Management Systems with no additional control or display required. It is a long-range navigation sensor that features a 12-channel engine which tracks all satellites in view simultaneously and includes real-time and predictive Receiver Autonomous Integrity Monitoring (RAIM) capabilities. It features high acquisition and tracking sensitivity, carrier phase smoothing and position updates every second and is local area differential GPS upgradeable.

The GPS-1000 can be certified for en route operations, including primary navigation means for remote/oceanic airspace, terminal operations and non-precision approaches.

Specifications
Dimensions: 194.1 × 25.1 × 386.8 mm
Weight: 1.6 kg
Power supply: 18–32 V DC, 0.255 A at 28 V
Certification: TSO C-129 Class B1/C1

Status
In production and in service.

Contractor
Universal Avionics Systems Corporation.

GPS-1200 sensor

Type
Avionic navigation sensor, Global Positioning System (GPS).

Description

The GPS-1200 long-range navigation sensor utilises a more highly advanced 12-channel GPS engine than the GPS-1000. It features real-time and predictive Receiver Autonomous Integrity Monitoring (RAIM), satellite fault detection and exclusion capability and carrier phase tracking, and is Wide Area Augmentation System (WAAS) ready. It also includes local area differential GPS signal integration capability. Besides en route and terminal operations, the GPS-1200 can be certified as one of the two required long-range sensors for NAT MNPS navigation across the North Atlantic. It can also be certified for non-precision approaches and incorporates DO-178B critical level software which will be required for precision approach certification in the future.

Specifications

Dimensions: 194.1 × 25.1 × 386.6 mm
Weight: 1.6 kg
Power supply: 18–32 V DC, 0.255 A at 28 V
Certification: TSO C-129 Class B1/C1

Status

In production and in service.

Contractor

Universal Avionics Systems Corporation.

GPS-4000A

Type

Avionic navigation sensor, Global Positioning System (GPS).

Description

The Collins GPS-4000A enables aircraft equipped with flight management capability to perform GPS-based en route, terminal area and non-precision approach navigation, as well as primary means of oceanic/remote operations.

The GPS-4000A is part of the continued development of the Collins AVSAT satellite-based communication and navigation system, providing operators with the advanced capabilities required to operate within the evolving CNS/ATM environment.

The 2 MCU GPS-4000A sensor processes the transmissions of up to 12 GPS satellites simultaneously, calculating navigation solutions based on information from all satellites in view. A minimum of four satellites with acceptable geometry, or three satellites plus calibrated barometric altitude, are required to calculate navigation solutions.

The heart of the GPS-4000A is the GPS processor. The design leverages Rockwell Collins' 20 years of experience in the design and production of both military and commercial systems. The GPS processor is an evolution of the precision lightweight GPS receiver, the US military-standard receiver produced by the Company. The processor accepts 12 GPS channels and outputs Earth-Centred, Earth-Fixed (ECEF) position, velocity and GPS time, once a second to the applications processor and provides a time mark discrete coincident with GPS seconds. The engine also provides RAIM, using all satellites in view for both the navigation solution and RAIM. Altitude aiding enhances the availability of the navigation solution and RAIM, and accepts external aircraft velocity to aid its tracking loops. Re-acquisition after power off is facilitated by self-contained non-volatile memory of satellite almanac, last calculated position, velocity, time and, by input of external aircraft position, velocity and time. The RF front end feeds signal processing Application Specific Integrated Circuits (ASICs) to provide a GPS receiver with 12-course acquisition code channels. The ASICs share addresses and data with the microprocessor, RAM and ROM to form the digital subsystem.

The applications processor uses Position, Velocity and Time (PVT) information from the GPS processor and tailors it to provide interface to other aircraft systems. These functions include:

- Dead reckoning in the event of a loss of satellite signals using velocity and heading data from other aircraft systems; maintaining aircraft position aids in rapid re-acquisition of the satellite(s)
- Latitude/longitude conversion from the ECEF co-ordinate system to the WGS-84 World Geodetic System, the required co-ordinate system for navigation
- ARINC 429 aircraft systems input/output interfaces ARINC 624 built-in-test equipment/onboard maintenance system functions
- ARINC 615 software data-loading management.

The data-loading function and the maintenance diagnostic computer interface function are classified as non-essential (they have no involvement in the calculations or transmission of the PVT solution), therefore these functions are partitioned from the rest of the software to ensure that no failures in them can prevent a higher-level software function from operating satisfactorily. Maintenance software is common with other Collins Pro Line 4 products. The maintenance function supports interface with the aircraft maintenance diagnostic computer and storage and retrieval of fault data. A separate maintenance output bus provides diagnostic data on the test bench to facilitate troubleshooting of the system.

ADA language object-oriented design techniques are used for all software modules of the GPS-4000A. Software has been developed in accordance with RTCA DO-178B, Level A (Critical) but verified to Level B (Essential) for the system. This software will be upgraded to Level A (Critical) when the modification to upgrade to the GPS-4000S with SBAS capability is added. Software is physically partitioned within the unit to separate the essential system and signal processing software from the non-essential maintenance functions.

The GPS-4000A GPS sensor is designed to provide basic GPS position service for use in the air traffic control environment to support primary means navigation for en route (including oceanic and remote areas), terminal and Non-Precision Approach (NPA). Within the next few years, significant enhancements to various national airspace segments with the availability of Space Based Augmentation System (SBAS), will allow the use of GPS as a primary means of navigation and near Category 1 approaches. In North America, this service is known as Wide Area Augmentation System (WAAS), in Europe it is known as European Geosationary Navigation Overlay Service (EGNOS).

Specifications

Dimensions:
2 MCU per ARINC 600
200 (H) × 62 (W) × 369 (L) mm
Weight: 2.9 kg
Temperature: –40 to +70°C
Altitude: 55,000 ft
Power: 28 V DC, 11 W maximum
Performance:
(navigation outputs) three ARINC 429 digital buses
(time mark output) three separately buffered outputs to synchronise the GPS outputs
(accuracy) 95% – normal manoeuvres
(horizontal position) 130 ft
(horizontal velocity) 1 kt
(vertical velocity) 200 ft/minute
(time to first fix) <75 sec with initialisation
Certification: FAA TSO C129 Class B1; EUROCAE ED-14C; Environmental EO-160C

Status

The Collins GPS-4000A Global Positioning System sensor has been certified for use on the Beechjet 400A, Canadair RJ, Learjet 60 and Saab 2000.

Contractor

Rockwell Collins.

H-423 ring Laser Inertial Navigation System (LINS)

Type

Avionic navigation sensor, Inertial Reference System (IRS).

Description

The H-423 Laser Inertial Navigation System (LINS) was developed according to the US Air Force SNU-84-1 specification which was the RLG version, an update to ENAC77-1. The system is a self-contained unit comprising three Honeywell GG1342 RLGs, three solid-state Sundstrand QA2000 accelerometers and associated electronics along with a dual-redundant MIL-STD-1553B databus and a complete MIL-STD-1750A navigation processing package. It also contains built-in test circuitry to achieve a greater than 95 per cent fault detection.

Under the US Air Force contract, Honeywell is guaranteeing the H-423 will achieve 2,000 hours MTBF in a fighter/helicopter environment and 4,000 hours MTBF in a transport environment. Honeywell claims the H-423 has the highest reliability and maintainability and lowest life-cycle costs ever achieved by a military aircraft inertial navigation system, allowing the US Air Force to go from a three-level to two-level maintenance system. These performance and reliability advantages have been the primary drivers for the US Defense Services now adopting RLG technology for all future aircraft inertial navigation systems.

Specifications

Dimensions: 459.7 × 193 × 200 mm
Weight: 22 kg
Power supply: AC 140 VA, DC 125 W
Interface: dual 1553B digital databus
Accuracy: <0.8 n miles/h with full performance from 22 stored heading alignment
Specification: SNU-84-1, FNU-85-1

Status

In August 1985, the US Air Force selected the H-423 as the standard inertial navigation system for its C-130, a number of its fixed- and rotary-wing aircraft. Up to the end of 1989 more than 1,000 military systems had been delivered.

In November 1990, the system was selected for Royal Australian Air Force F-111 aircraft in a USD5 million contract and for US Air Force F-16 upgrade programmes. Since then, H-423 systems have been selected for upgrades of Belgian and Danish F-16s and for the Swedish JAS 39, Taiwanese IDF and the Indian Light Combat Aircraft.

In September 1992, the H-423E, a high-accuracy version of the H-423, was selected to upgrade the F-117A fleet. The high standard of accuracy is achieved through the use of an enhanced specialised version of the H-423 standard software.

Contractor

Honeywell Sensor and Guidance Products.

H-770 ring laser gyro inertial navigation system

Type

Avionic navigation sensor, Inertial Reference System (IRS).

Description

Honeywell developed the H-770 ring laser gyro inertial navigation system for the F-15.

The H-770 uses the same ring laser gyro and accelerometers as the H-423 (see earlier entry), but contains two separate databusses for the F-15 applications. The system can determine for itself in which variant of the F-15 it has been installed, and perform accordingly. The H-770 has a reliability 20 times better than the AN/ASN-109 system that it is replacing in the F-15.

Specifications

Dimensions: 381 × 213 × 330 mm
Weight: 26.8 kg
Power required: AC 140 VA, DC 130 W
Accuracy: 0.4 n mile/h CEP
Alignment:
(gyrocompassed) 4 min
(stored heading) 30 s
Reliability: 2,000 h MTBF

Status

Honeywell supplied units for the upgrade of the US Air Force fleet of F-15A, B, C, D and E variants.

Contractor

Honeywell Sensor and Guidance Products.

HG1140 multirole air data computer family

Type

Avionic navigation sensor, Air Data Computer (ADC).

Description

Honeywell's HG1140 air data product family features a broad range of configurations from single-channel air data transducers up to complete air data computers. The product line is based on Honeywell's solid-state silicon sensors and designed to provide primary instrument quality air data information that is compliant with the most stringent Reduced Vertical Separation Minima (RVSM). With flexible input/output capabilities, configurations are easily customised to meet varying interface requirements ranging from Mil Standard 1553 to ARINC 429. The dual use nature of the product line is ideal for both commercial and military aircraft, missile and unmanned vehicle applications.

Specifications

Dimensions: 152.4 × 111.76 × 50.8 mm
Weight: 0.91 kg
Power: 28 V DC, 6 W
MTBF: 100,000+ h (AIC); 50,000 h (AUF)
Long-term stability: 0.015% F.S. over 20 years
Interface options: ARINC 429, RS422, RS485
Optional I/O: Analogue, discrete, synchro MIL-STD-1553
Qualification: RTCA DO-160 and MIL-STD-810E

HG1141 multifunction air data computer

The HG1141 is a derivative of the HG1140 air data computer. It contains a multifunction databus so that the same configuration can be used to satisfy the Gripen aircraft's dual air data system requirements. The HG1141 provides altitude and airspeed information as well as nine other parameters in the Gripen installation.

HG1140 multirole air data computer 0015280

Status

Fully qualified to both RTCA-DO-160 and MIL-STD 810E requirements. Fielded applications include USAF T-38C Talon Upgrade, F/A-18 E/F, JAS-39 Gripen, Airbus A319–A340, Tactical Tomahawk Cruise Missile, E-6B and the Gulf Stream IV–V.

Contractor

Honeywell Sensor and Guidance Products.

IEC 9001 GPS Navigation and Landing System (GNLS)

Type

Aircraft landing system, Differential Global Positioning System (DGPS).

Description

The IEC 9001 GPS Navigation and Landing System (GNLS) is a stand-alone, coarse acquisition (C/A) code, differential GPS-based system intended for installation in Flight Management System (FMS) – equipped commercial transport aircraft.

For en route navigation, the 9001 GNLS provides GPS position in accordance with ARINC 743A format to the flight management computer. For the approach mode, it calculates and provides glideslope and localiser deviations to the Electronic Flight Instrumentation System (EFIS) and Flight Control Computer (FCC) based on the self-contained approach navigation database and preselected FMS approach path.

The system inputs/outputs conform to ARINC 429 low-speed and high-speed data as defined in ARINC 710-9 for the ILS receiver. The 9001 supports en route, terminal, and approach navigation accuracy and integrity requirements.

The 9001 GNLS can be enhanced with a kinematic upgrade to enable CAT-II/IIIb landings. It can also be upgraded to accept and process GLONASS signals.

Specifications

Navigational signal: L1 C/A code (SPS)
Receiver: 12 hardware channels
Time to first fix: <60 s
Dynamics: 900 kt max velocity
Accuracy:
100 m 2σ rms (selective availability on)
5 m utilising differential GPS corrections
Antenna: active conformal meeting ARINC 743A
Environment: fully compliant to environmental specifications of DO160c
Dimensions: 2 MCU
Weight: 2.9 kg
Power: 45 W at 28 V DC

Status

In service.

Contractor

L-3/Interstate Electronics Corporation.

IEC 9001 GPS navigation and landing system
0051615

Integrated Global Positioning/ Inertial Reference System (GPIRS)

Type

Aircraft Navigation Systems, Global Positioning System (GPS).

Description

The integrated Global Positioning/Inertial Reference System (GPIRS) combines the best of global positioning and inertial reference systems to provide very accurate worldwide navigation.

The inertial reference system is upgraded to an integrated GPIRS by adding GPS processing software to the inertial reference unit and coupling it to an ARINC 743 GPS sensor unit.

The GNSSU (Global Navigation Satellite Sensor Unit) is a remote-mounted unit that provides all the functions necessary for either integrated or stand-alone configurations. All existing Honeywell IRUs can be modified into GPIRUs which are one-way interchangeable with existing IRUs.

The GPS sensor unit receives satellite data using a 12-channel design. Data are received from all satellites in view, with updates once per second. The all-digital multiple correlator design allows for satellite tracking during periods of low signal-to-noise. This enables the receiver to track satellites to a 0° elevation angle.

Specifications

Dimensions:
(IRS) 317.5 × 320 × 198.1 mm
(GPSSU) 190.5 × 215.9 × 55.9 mm
Weight:
(IRS) 19 kg
(GPSSU) 2.27 kg
Time to first solution:
(IRS) 10 min
(GPSSU) 4 min typical
Accuracy:
(position) (IRS) 2 n miles/h 95%
(GPSSU) 25 m
(velocity) (IRS) 8 kt
(GPSSU) 1.8 kt
(time) (GPSSU) 350 ns
Reliability:
(IRS) 5,000 h MTBF
(GPSSU) 20,000 h MTBF

Status

Selected as standard equipment by BAE Systems for its smaller air transport aircraft and in a dual fit for the Boeing MD 90-30.

Contractor

Honeywell Inc, Commercial Electronic Systems.

KLN 89/KLN 89B GPS navigation systems

Type

Aircraft navigation system, Global Positioning System (GPS).

Description

Designed to match the Bendix/King Silver Crown series of avionics, the KLN 89 is a panel-mounted, eight-channel, GPS-based navigation system with a database that can be updated by the pilot. The KLN 89B adds IFR-certified en route, terminal, and non-precision GPS approach capability. A basic system comprises panel-mounted unit, altitude input, and KA 92 antenna. Among additional components that may be added to increase capabilities are an external Course Deviation Indicator (CDI) or HSI; RMI, some Shadin or ARNAV fuel management systems; several external moving maps, and certain Shadin air data systems.

The KLN 89 and 89B incorporate Jeppesen NavData™ databases which include room for

KLN 89/89B GPS navigation systems 0001335

up to 500 user-defined waypoints. The databases may be updated via an IBM-compatible PC and flight information is displayed on a simplified four-line moving map display.

Specifications
Dimensions: 160.3 × 50.8 × 272.3 mm
Weight: 1.16 kg
Power: 11 to 33 V DC at 2.5 A
TSO (KLN 89B only): C129 Class A1
Temperature range: −40 to +55°C
Altitude range: up to 35,000 ft

Contractor
Bendix/King.

KLN 90B approach-certified GPS navigation system

Type
Aircraft navigation system, Global Positioning System (GPS).

Description
Together with the capability to perform non-precision GPS approaches, the KLN 90B navigation system offers an easy-to-read CRT map display and a comprehensive Jeppesen database. Designed to meet the FAA's C129 A1 specifications, the KLN 90B features an improved eight-channel parallel GPS receiver for even more reliable satellite tracking. Other enhancements include a more pilot-friendly interface and an expanded database, complete with SID and STAR waypoints and approaches.

Providing all the benefits of satellite-derived input – worldwide coverage, a high degree of accuracy and immunity to atmospheric disturbance – the KLN 90B harnesses the power of GPS to make non-precision approaches easier for the pilot.

Navigation pages were designed specifically for use during approaches, to provide the greatest amount of information with the least amount of effort. In fact, most necessary and en route information is presented concisely and logically on one large, easy-to-read screen. This page also incorporates a moving map, to show progress throughout the approach. Additional approach-specific pages are readily accessed, significantly reducing pilot workload during this critical phase of flight.

KLN 90B database information includes airport data (identifier, name, runway length/surface/lighting, instrument approach availability, customs, types of fuel, oxygen, landing fees, and other data comparable to that found in an airport services guide, and a runway diagram for airports providing published runway threshold co-ordinates); communications frequencies (ATIS, clearance, ground control, tower, CTAF, advisory and various VFR frequencies); VOR information; NDB information; intersection information (data about low altitude, high altitude, approach and SID/STAR intersections, with outer markers and compass locators; ARTCC and FIR boundaries and frequencies); flight service station frequencies and locations; minimum safe altitudes; special-use airspace.

The KLN 90B has sufficient internal memory capacity for up to 26 pilot-programmable flight plans, each with up to 30 waypoints, along with 250 more waypoints in a user-defined database. The database can be updated using a PC, via the internet or with exchangeable cartridges that plug in to the back of the unit. During flight, the KLN

KLN 90B approach-certified GPS navigation system 0131551

90B's flight calculator can calculate true airspeed, pressure and density altitude and actual winds (including headwind and tailwind components), and will make conversions such as degrees celcius to fahrenheit or knots to mph. The unit will also give the pilot up to 10 minutes warning, via a 'Message' prompt, if the current ground track will impinge on any special use airspace, including Class B and C airspace, CTAs, TMAs, and so on.

Specifications
Dimensions: 160.3 × 50.8 × 334 mm
Weight: 2.66 kg
Temperature range: −40 to +70°C
Altitude range: up to 50,000 ft
Power inputs: 11 to 33 V DC at 2.5 A (max)
TSO: C129 A1

Contractor
Bendix/King.

KLN 94 colour GPS receiver

Type
Aircraft navigation system, Global Positioning System (GPS).

Description
The KLN 94 colour GPS navigator/moving-map receiver and display is a new product that is compatible with other products in the Silver Crown and Silver Crown Plus range. Designed to save installation cost and space while upgrading to a colour display and replacing both GPS and Loran receivers, it is a direct plug-in replacement for the KLN 89B. It includes a comprehensive aeronautical database including airports, VORs, NDBs, intersections and special-use airspace. It provides IFR GPS capability (TSO C129a A1) for en-route, terminal and non-precision approaches; vector to final approach; range/map capability; special procedures capability; and a quick-tune interface to the KX 155A NAV/COMM.

The KLN 94's built-in flight calculator will handle a number of manual input computations including true airspeed, pressure altitude, density altitude and actual winds aloft (including headwind and tailwind components). It has an alarm timer and can calculate sunset/sunrise times for any day and location. The navigation database can be kept current by exchanging the front-mounted data card, connecting the unit to a compatible PC and updating with a 3.5 in floppy disk, or by connecting to the internet through a PC.

Specifications
Dimensions: 160.3 (W) × 50.8 (H) × 272.3 (D) mm
Weight: 1.63 kg
Power: 11–33 V DC at 3.0 A max
Temperature: −20 to +55°C
Altitude: 35,000 ft
Certification: TSO C129a; DO-160D

Contractor
Bendix/King.

KLN 900 approach-certified GPS navigation system

Type
Aircraft navigation system, Global Positioning System (GPS).

Description
The KLN 900 approach-certified GPS navigation system meets Basic aRea NAVigation (B-RNAV) requirements for European Civil Aviation Conference (ECAC) airspace when used with AlliedSignal's PreFlight™ software version 2.0 or later. Versions of the KLN 900 will also meet primary means oceanic/remote operation requirements.

The KLN 900 is designed for customers who wish to upgrade the navigation capability of their aircraft, replace outdated systems such as LORAN and RNAV equipment, install an affordable back-up or lower-cost alternative to a Flight Management System (FMS).

The KLN 900 is a dzus-mount, eight-channel unit. It can be IFR certified for en-route, terminal and approach GPS-based navigation. The pilot is able to load his database and flight plan data with a front panel diskette. The unit can be upgraded to provide a precision approach capability. It features a monochrome Cathode Ray Tube (CRT) display, and it can handle a wide variety of analogue and digital inputs and outputs.

A basic installation comprises a dzus-mounted KLN 900 unit, database cartridge, and KA 92 antenna. Among the additional components that can be connected directly to increase the KLN 900's capability are an external Course Deviation Indicator (CDI); a Horizontal Situation Indicator (HSI); Radio Magnetic Indicator (RMI); fuel management system; moving-map display; and air data system.

Two database options provide aeronautical information for the entire world. Both databases contain worldwide information on VORs, NDBs, intersections and minimum safe altitudes. For their primary areas, each database contains public use and military airfields with runways at least 1,000 ft in length along with relevant aeronautical information. Outside their primary coverage areas, the databases contain airports with a hard surface runway at least 3,000 ft in length.

Specifications
Dimensions: 146 (W) × 95 (H) × 240 (L) mm
Weight: 2.0 kg
Power: 11 to 33 V DC, 3.3 A
TSO: C129 A1
Temperature: −40 to +70°C
Altitude: 50,000 ft
Certification: TSO C129 A1

Contractor
Bendix/King.

The KLN 900 approach-certified GPS navigation system 0062193

KN 62A/63/64 digital DMEs

Type
Aircraft navigation system, radio aids, DME.

Description
The compact KN 62A digital DME is a fully self-contained 200-channel, panel-mounted system in the Silver Crown range. It features an all-solid-state transmitter and four LSI chips and requires only 33 mm of panel height.

The KN 62A can be channelled remotely through almost any navigation equipment receiver, or tuned

The KLN 94 colour GPS receiver 0062192

directly with its own frequency selection knobs. Dual channelling makes two DME frequencies available at all times. The self-dimming digital display provides information on simultaneous DME distance, groundspeed and time-to-station, or DME distance and internally selected frequency.

The KN 62A meets the US FAA's TSO performance and environmental standards; the KN 64 is also panel-mounted and offers similar performance but is not TSO'd.

The KN 63 is mounted remotely and includes a KDI 572 Master Indicator display. It is TSO'd, and can be integrated with the KNS 81 integrated Nav/RNav to increase sophistication.

Specifications
Weight: 1.18 kg (KN 62A/64); 1.63 kg (KN 63)
Power supply: 11 – 33 V DC, 15 W (KN 62A/64); 11 – 33 V DC, 17 W (KN 63)
Output power:
(KN 62A) 50 W peak pulsed power minimum
(KN 64) 35 W peak pulsed power minimum
Maximum display range: 389 nm
Range accuracy:
(0 – 99.9 nm) ±1 nm or ±14% whichever is greater
(99.9 – 389 nm) ±1 nm
Groundspeed accuracy: ±1 kt or ±1% whichever is greater
Time-to-station: ±1 minute
Memory time: 11 to 15 s
Altitude: 50,000 ft
Certification: (KN 62A only) TSO C66a, DO-160

Contractor
Bendix/King.

KN-4065 Improved Standard Attitude Heading Reference System (ISAHRS)

Type
Avionic navigation sensor, Attitude Heading Reference System (AHRS).

Description
The KN-4065 Improved Standard Attitude Heading Reference System (ISAHRS) provides both attitude heading reference and full navigation outputs. ISAHRS features Monolithic Ring Laser Gyro (MRLG) technology for high reliability, low cost and weight and low power comsumption.

Items in this family of Kearfott ring laser gyro systems include:
The Improved Standard Attitude Heading Reference System (ISAHRS);
The Modular Azimuth Position System (MAPS);
The Low cost Attitude Heading Reference System (LAHRS);
The Attitude Motion Sensor Set (AMSS).

Specifications
Dimensions: 279.4 × 177.8 × 177.8 mm
Weight: 9.3 kg

The Kearfott Guidance & Navigation Corporation KN-4065 improved standard attitude heading reference system. Note the KN-4060 series and KN-4068GC ring laser gyro inertial navigation systems utilise the same box enclosure 0062845

MTBF: 6,750 h
Outputs: MIL-STD-1553B; Magnetic Azimuth Detector (MAD) interface; Mode Control Unit (MCU) interface; synchro outputs for pitch, roll and heading
Performance:
(heading) 0.5° (1σ) RMS (8 min alignment time)
(pitch/roll) <0.1° (1σ) RMS
(dynamic range) 400°/s max (all axes) ± 10 g acceleration
(back-up navigation) 2 n miles/h

Status
The KN-4065 has been certified by the US Department of Defense for use in military aircraft and is in service in several types of US Navy aircraft.

Contractor
Kearfott Guidance & Navigation Corporation.

KN-4068GC Ring Laser Gyro/ Inertial Navigation System (RLG/INS) with Embedded GPS Receiver (EGR)

Type
Avionic navigation sensor, Inertial Reference System (IRS).

Description
The KN-4068GC system is designed for fixed- and rotary-wing aircraft. It contains RLG technology and a C/A code embedded GPS receiver for enhanced, closely coupled navigation performance, and it accepts other navigation aiding inputs via MIL-STD-1553 or RS-422 databuses or analogue navigation aids (flux valve). During GPS availability, blended INS/GPS performance can provide position accuracies of 300 ft.

Specifications
Dimensions/weight: as for KN-4060 and KN-4065
GPS receiver:
(operating frequency): L1
(antispoof/enhanced antijam): C/A code
(channels): 5

Status
In service.

Contractor
Kearfott Guidance & Navigation Corporation.

KN-4070 Monolithic Ring Laser Gyro/Inertial Navigation System (MRLG/INS)

Type
Avionic navigation sensor, Inertial Reference System (IRS).

The Kearfott Guidance & Navigation Corporation KN-4070 MRLG/INS 0062844

Description
The KN-4070 is designed for aircraft and UAV operation. It is an MRLG-based INS, that is designed to operate in conjunction with an embedded P(Y) or C/A code GPS receiver for enhanced navigation performance and faster satellite acquisition. The KN-4070 provides navigation and autopilot functions in digital formats including MIL-STD-1553B, RS-422 and RS-232.

Specifications
Dimensions: 231.1 × 137.2 × 152.4 mm
Weight: <5 kg
Power: 28 V DC, 35 W
GPS receiver:
(operating frequencies): L1/L2
(antispoof/enhanced antijam) P(Y), C/A code
MTBF: >6,000 h

Status
In service.

Contractor
Kearfott Guidance & Navigation Corporation.

KN-4071 Attitude Heading Reference System (AHRS)

Type
Avionic navigation sensor, Attitude Heading Reference System (AHRS).

Description
Kearfott's KN-4071 AHRS uses a patented Monolithic Ring Laser Gyro (MRLG) and a triad of Kearfott MOD VIIA force rebalance accelerometers as the inertial reference. The system provides analogue outputs of heading and attitude for cockpit, flight director or weapon computer input.

The Kearfott Guidance & Navigation Corporation KN-4071 attitude heading reference system 0062846

The unit includes a self-contained GPS card to improve accuracy of altitude and heading over standard electromechanical AHRS. Growth to full navigation capability is optional.

The AHRS comprises three LRUs: the KN-4072, a Signal Conversion Unit (SCU) and the Mode Select Unit (MSU).

Specifications
Compass accuracy: +/–0.7° typical
Slaved mode:
(heading): +/–0.5° of MAD
(attitude): +/–0.25°
Gyro mode:
(heading): +/–0.5°/h
(attitude): +/–0.25°
Enhanced mode:
(heading): +/–0.25°
(attitude): +/–0.15°
Analogue outputs: four heading; three pitch; three roll
Optional digital outputs: MIL-STD-1553; RS-422, RS-232
MTBF: 2,000 h

Status
In service.

Contractor
Kearfott Guidance & Navigation Corporation.

KN-4072 Digital Attitude Heading Reference System (AHRS) GPS/INS

Type
Avionic navigation sensor, Global Positioning System (GPS).

Description
The KN-4072 digital AHRS GPS/INS is designed for a wide variety of aircraft, UAV and missile applications. It features Kearfott's Monolithic Ring Laser Gyro (MRLG) operating in concert with an embedded P(Y) or C/A code GPS receiver for enhanced navigation performance and faster satellite acquisition. The KN-4072 provides navigation, heading, attitude, velocity, position, together with angle and velocity rate data and autopilot functions in digital formats including MIL-STD-1553B, RS-422 and RS-232.

Specifications
Dimensions: 231.2 × 137.2 × 152.4 mm
Weight: < 5 kg
Power: 28 V DC, 35 W
GPS receiver:
(operating frequencies) L1/L2
(antispoof/enhanced antijam) P(Y), C/A code
MTBF: > 6,000 h

Status
In service.

Contractor
Kearfott Guidance & Navigation Corporation.

The Kearfott Guidance & Navigation Corporation KN-4072 digital AHRS GPS/INS   0062843

KNS 81 Silver Crown integrated navigation system

Type
Aircraft navigation system, radio aids.

Description
A development of the KNS 80, the KNS 81 integrated navigation system can accommodate up to nine waypoints and embodies a number of new features. A remotely mounted DME enables a KDI 572 indicator to be positioned directly in front of the pilot, providing simultaneous digital readouts of distance, groundspeed and time to either a VORTac station or an RNav waypoint. There is simultaneous display of bearing, distance and frequency waypoint details on the KNS 81 panel for easy programming and updating of navigation information. A radial push-button permits a rapid bearing check to a chosen VORTac or RNav waypoint, displayed on the DME panel indicator in place of groundspeed and time to next waypoint. A check push-button permits a rapid cross-check of bearing and distance from the VORTac without disturbing other navigation instrument settings. A radio magnetic indicator output to a KI 229 or KNI 582 indicator gives an accurate bearing to a selected RNav waypoint or VORTac station.

Specifications
Dimensions: 160.3 × 5.1 × 291.2 mm
Weight: 2 kg
Power supply: 11 to 33 V DC, 15 W
Number of waypoints: 9
Distance to next waypoint: 199.9 n miles in 0.1 n mile increments; 0.1° angle increments

Contractor
Bendix/King.

KR 87 ADF system

Type
Aircraft navigation system, radio aids, ADF.

Description
The basic KR 87 system includes the KR 87 receiver, KI 227 indicator with manual compass card, KA 44B combined loop and sense antenna, and mounting racks and connectors. The KR 87 complies with TSO C41c and operates in the 200 to 1,799 kHz range with ADF, ANT, and BFO tuning modes. The all-solid-state receiver operates on DC voltages between 11 and 33 V and draws only 12 W of power. Electronic timers in the KR 87 provide aids to flight management as a flight or elapsed timer. The KI 227-01 slaved indicator can be installed as an option. It takes a heading input from the standard KCS 55A compass system to drive the compass card and display magnetic heading.

The KR 87 features a crystal filter for better long-range reception, coherent detection circuitry to lock on to weak stations, electronic tuning using a microprocessor and an LSI single-crystal digital frequency synthesiser circuit, EAROM non-volatile storage of frequencies during shutdown or power interruptions and a fold-out modular construction for easy access to all circuits and components. The display is self-dimming.

Specifications
Dimensions: 160 × 35 × 286 mm (W × H × D)
Weight: 1.45 kg
Power: 11 – 33 V DC, 12 W
Temperature: –20 to +55°C
Altitude: 50,000 ft
Frequency range: 200 – 1,799 kHz in 1 kHz increments
Certification: TSO C41c

Contractor
Bendix/King.

KTU 709 TACAN

Type
Aircraft navigation system, radio aids, TACAN.

Description
Particularly suited to corporate aircraft, where weight, cost and power requirements are notably important, the KTU 709 Tacan transmitter/receiver is based on Honeywell Aerospace's extensive DME experience and the extensive use of Large Scale Integration (LSI) technology; the system specifically is based on the new-generation KDM 706 distance measuring equipment. This Tacan provides bearing, slant range, range rate and time to station or waypoint information to a KDI 572 control/indicator unit. Transistors provide a 250 W peak-to-peak output for a typical range of 250 n miles and the system covers 252 channels; all tuning is done electronically, using a digital frequency synthesiser designed around a Honeywell Aerospace LSI chip.

Specifications
Dimensions: 76.2 × 127.0 × 260.4 mm
Weight: 2.6 kg
Frequency:
(transmit) 1,025 to 1,150 MHz
(receive) 962 to 1,213 MHz
Number of channels: 250
Reliability: 2,000 h MTBF design

Contractor
Bendix/King.

LANTIRN system

Type
Airborne navigation and laser designation system.

Description
The Low-Altitude Navigation and Targeting Infra-Red for Night (LANTIRN) system consists of two pods: the AN/AAQ-13 navigation pod and the AN/AAQ-14 targeting pod.

The LANTIRN system enables fast jet aircraft, such as the F-15E and F-16C/D, to penetrate hostile airspace at extremely low altitude and high speed, acquire their targets and deliver weapons round the clock. The initial application was the two-seat F-15E fielded by the US Air Force in 1989. LANTIRN also equips the F-16C/D. Compatibility of the LANTIRN system with single-seat navigation and targeting operations has been demonstrated on the F-16. In 1995, the US Navy selected the LANTIRN targeting pod for the F-14 Precision Strike Programme.

The system
The LANTIRN system consists of two separate sets of equipment each contained within its own pod, suitable for underwing or underfuselage attachment. Either or both pods may be carried, depending on mission requirements. This option enhances flexibility and reduces support demands. The equipment within the pods is supplied by a number of manufacturers, but overall responsibility rests with Lockheed Martin as the prime contractor.

The AN/AAQ-13 navigation pod contains a wide Field of View (FoV) (21 × 28°) single cadmium mercury telluride array FLIR unit and a J-band terrain-following radar, together with the associated power supply and pod control computer. FLIR imagery from the pod is displayed on a wide field of view holographic head-up display developed by BAE Systems (formerly Marconi Electronic Systems). This provides the pilot with night vision for safe flight at low level. The Raytheon Electronic Systems Ku-band terrain-following radar permits operation at very low altitudes with en route weather penetration and blind let down capability.

This Lockheed Martin F-16D carries both navigation and targeting pods of the Lockheed Martin LANTIRN system under the air intake on dedicated pylons   0111720

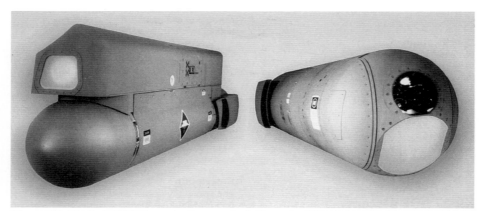

The AN/AAQ-13 LANTIRN navigation pod (left) and AN/AAQ-14 targeting pod (right) (Lockheed Martin)   1044577

Both the F-15E and F-16 have fully automatic terrain-following with inputs from the navigation pods to the digital flight control system in the aircraft. Additionally, terrain-following may be accomplished manually by means of d irective symbology presented to the pilot on the HUD.

The AN/AAQ-14 targeting pod contains a stabilised wide and narrow fields of view targeting FLIR and a Litton Systems Inc, Laser Systems Division laser designator/range-finder in a rotating nose section whose total Field of Regard (FoR), limited only by pod characteristics and airframe configuration, is of the order of ±150°. The centre section houses the Maverick hand-off unit boresight correlator, a multimode tracker, pod-control computer, power supply and provision for Automatic Target Recognition (ATR) equipment. In certain derivatives, global positioning and inertial navigation systems are also integrated into the AN/AAQ-14. The targeting pod interfaces with the aircraft controls and displays, including the Stores Management System (SMS), to permit low-level day and night manual target acquisition and weapon delivery of guided and unguided weapons. It may be configured as a laser designator-only pod, for use with laser-guided munitions and conventional weapons, by deleting the Maverick hand-off subsystem. This configuration of the LANTIRN targeting pod has been redesignated as Sharpshooter.

Both pods have environmental control units in their tail sections to ensure that their systems will function satisfactorily over a wide range of temperatures and flight conditions. Aircraft interfaces include a MIL-STD-1553 multiplex databus, video channels and power supplies.

Service ground support employs typical three-level maintenance: organisational, where no special test equipment is required; intermediate, which employs automatic test equipment, and depot servicing.

In operational use, the pilot will acquire a target using the aircraft (or pod mounted) nav/attack system to cue the line-of-sight (LOS) of the AN/AAQ-14 to the desired area.

Using one of his head-down displays, the pilot will initially utilise wide FoV to confirm the target area, acquire his Desired Mean Point of Impact (DMPI), and then switch to narrow field of view to further refine and positively identify the correct aim-point. The pilot will then engage the automatic target tracker. After the tracker is locked on, the pilot is able to manoeuvre his aircraft, limited only by the need to maintain line of sight between the pod and the target, taking care not to interrupt the LOS with his own airframe. With a stable target lock, the subsequent actions of the pilot are dictated by the type of weapon and attack profile: If the pilot is employing unguided weapons, he may simply wish to employ the laser range-finder to facilitate a highly accurate release solution. If laser guided weapons are to be used, the pilot is able to designate the target for his own weapons, or those of another aircraft. For a Maverick missile, the pod automatically hands the target off to the missile for launch with pilot consent.

Specifications
AN/AAQ-13 Navigation Pod
Dimensions: 1,985 × 310 mm
Weight: 205 kg
FLIR FoV: 21 × 28°

AN/AAQ-14 Targeting Pod
Dimensions: 2,510 × 381 mm
Weight: 241 kg
FLIR Total FoR: ±150°
FLIR WFoV: 6 × 6°
FLIR NFoV: 1.7 × 1.7°

Status
In service on the US Air Force F-15E and F-16C/D and with several other air forces. A modified version of the AN/AAQ-14 targeting pod (as fitted to F-15E and F-16C/D), has been supplied to the US Navy for fitment to upgraded F-14 aircraft. It features a GPS/inertial navigation subsystem, which, coupled with the navy's limited requirement for automatic terrain following, obviates the need for a dedicated navigation pod. Production continues with the US Navy ordering a further 26 targeting pods in June 1998 for F-14 aircraft.

The first production navigation pod was formally accepted by the US Air Force in April 1987, with the final item in the contract delivered in March 1992. US Air Force targeting pod deliveries were completed in April 1994.

As of January 1999, Lockheed Martin had received orders for 763 navigation pods, 579 targeting pods, 12 Pathfinder pods and 184 Sharpshooter pods.

The Taiwanese Air Force integrated LANTIRN into its F-16 fighters under a USD106.2 million contract which was completed with delivery of 28 Sharpshooter/Pathfinder (AN/AAQ-14/AN/AAQ-20) targeting/navigation pods in April 2001.

LANTIRN and its derivatives (Pathfinder/Sharpshooter) are either operational or on order with the US Air Force, US Navy and other air forces, including Bahrain, Denmark, Egypt, Greece, Israel, Netherlands, Saudi Arabia, Singapore, South Korea and Turkey.

In April 2000, Lockheed Martin was contracted by the US Air Force for Engineering, Manufacture and Development (EMD) of a modified version of LANTIRN. The modification integrates a radiometer and digital recorder into LANTIRN to enhance bomb impact assessment capability of the system. This EMD contract was completed during 2001.

The latest version of the LANTIRN system, designated LANTIRN ER (see separate entry), incorporates a large number of enhancements and improvements to the basic system.

Contractor
Lockheed Martin Missiles and Fire Control.

Laser Inertial Reference System (IRS)

Type
Avionic navigation sensor, Inertial Reference System (IRS).

Description
The world's first production ring laser gyro Inertial Reference System (IRS) was chosen by Boeing as part of the avionics package common to both the 767 and 757. It was also selected for the Boeing 737-300/400/500, MD-80 and MD-11, and the Airbus A320, A330 and A340, Fokker 100, and BAE Systems 146-300.

The strapdown configuration is so called because the gyrostabilised platform of current conventional inertial navigation and attitude reference systems is replaced by three ring laser gyro units mounted rigidly to the aircraft and at right angles to one another. The laser gyro detects and measures angular rates of motion by measuring the frequency difference between two contrarotating laser beams made to circulate (hence the term ring) in a triangular cavity by mirrors. When the units are at rest the distances travelled by each beam are the same, as are the frequencies. When the unit rotates, one path lengthens while the other shortens and so a frequency difference is established proportional to the rate of rotation of the unit. The difference is measured and processed digitally in ARINC 704 format as aircraft attitude in pitch, roll and yaw.

Since the accelerometers are mounted rigidly in the box, their signals are related to aircraft axes and have to be processed to convert them to the external inertial reference frame necessary to provide navigation and flight control information and guidance.

A strapdown system has no moving parts to wear, fail or become misaligned; no gimbals, torque motors, spin-motors, slip-rings, or resolvers, and no scheduled maintenance, realignment or recalibration requirements are anticipated. A typical installation comprises three inertial reference units (containing the sensing and computing elements) and a display unit. The Honeywell laser device is contained within a low expansion, triangular glass block, with a 34 cm path length. It has demonstrated a MTBF of 20,000 hours during more than 50 million flight hours.

Specifications
Dimensions: 4 MCU or 10 MCU
Weight:
(4 MCU) 12.24 kg
(10 MCU) 19.5 kg
Power:
(4 MCU) 44 W
(10 MCU) 86 W
Outputs: primary attitude information to displays and Automatic Flight Control Systems (AFCS), linear accelerations, velocity vector and angular rates to AFCS, wind shear detection and energy management, magnetic heading for displays and AFCS and long-range navigation data

ARINC 704
Accuracy (10 h flight – 95% probability):
(position) 2 n miles/h
(velocity) 12 kt
Self-test: BIT (initiated and continuous) detects 95% of failures with 95% confidence level
Reliability: >5,000 h MTBF predicted

Status
In production for Boeing 767, 757 and 737-300/400/500, MD-11 and MD-80 Series aircraft; and Airbus A300-600, A310, A320, A330 and A340.

Contractor
Honeywell Inc, Commercial Electronic Systems.

Laseref II inertial reference system

Type
Avionic navigation sensor, Inertial Reference System (IRS).

Description
Laseref II performs the same functions as Laseref and uses the same inertial sensor assembly, but in addition is designed to interface with new-generation digital avionics, such as flight management systems, using the Aircraft Standard Communications Bus (ASCB). The Laseref II may also interface with aircraft having the ARINC 429 digital databus. Laseref II has been selected as the standard factory IRS installation on the Dassault Falcon 900, Gulfstream IV and Canadair CL-601-3A.

GPIRU configurations are available which contain additional electronics that blend inertial data with satellite information obtained from the Honeywell GPS sensor unit.

Specifications
Dimensions:
(inertial reference unit) 322 × 324 × 193 mm
(mode select unit) 146 × 38 × 62.5 mm
Weight:
(inertial reference unit) 21.1 kg
(mode select unit) 0.45 kg
(inertial sensor display unit) 2.3 kg
Power supply:
(total) 115 V AC, 400 Hz, 137 W or 28 V DC

Status
In production. Laseref II is standard in a dual configuration on the Gulfstream IV, Dassault Falcon 900 and Canadair Challenger 601. Other applications include the Raytheon Hawker 800, Cessna Citation III and de Havilland Dash 8.

Contractor
Honeywell Inc, Commercial Electronic Systems.

Laseref III inertial reference system

Type
Avionic navigation sensor, Inertial Reference System (IRS).

Description
The Laseref III all-digital laser system is 60 percent smaller, 45 per cent lighter and uses 50 per cent less power than its predecessor, the Laseref II. The heart of the Laseref III IRU is a smaller ring laser gyro sensor. In addition to the new sensor, Honeywell is utilising surface mount technology, very large-scale integration, application specific integrated circuits, more powerful and faster processing and enhanced software in the new system. The Laseref III IRU includes integrated GPS processing which further enhances position and velocity data with a hybrid blending of raw inertial and satellite data.

The Laseref III IRU is pin-for-pin compatible with the Laseref II IRU and can be installed in the latter's 10 MCU tray using a mechanical adapter. It operates with the same mode-select unit and optional Lasertrak navigation display unit as existing systems.

Specifications
Dimensions: 124.5 × 320 × 198.1 mm
Weight: 11.79 kg
Align time: 2.5–10 min
Accuracy:
(navigation) 2 n miles/h
(velocity) 12 kt
(attitude) 0.1°
(heading) 0.4°

Status
In production since September 1991. The Laseref III has been selected for the Dornier 328 regional airliner and the upgrade of the Dassault Falcon 2000. Certified on the Raytheon Hawker 1000 in a dual configuration.

Contractor
Honeywell Inc, Commercial Electronic Systems.

Laseref SM inertial reference system

Type
Avionic navigation sensor, Inertial Reference System (IRS).

Description
The Laseref SM inertial reference system comprises the Mode Select Unit (MSU) and the Inertial Reference Unit (IRU). The MSU selects the IRU mode of operation and displays operational status messages on the annunciator panel. The IRU contains the laser inertial components, a processor and associated electronics and BIT. The INU is designated to add a GPS navigation processor card to incorporate an integrated GPS receiver.

By configuring the Laseref SM with an appropriate flight management system, a comprehensive special mission system is created. The combination of these units results in the ability to perform missions requiring special patterns, automatic camera control and computed air release point, in addition to standard flight management functions.

Laseref SM interfaces provide required data to a variety of special mission equipment. Standard ARINC interfaces output essential flight data to digital and analogue flight instruments, autopilots, radars, sensors and other special devices. Additional outputs are provided for special interface requirements.

The Laseref SM IRU is designed to add, as a growth option, an integrated GPS receiver. By adding one card to the IRU, a GPS PreProcessor Module (PPM) and an antenna, the standard Laseref SM becomes a fully integrated inertial/GPS system. The PPM receives satellite data using a two-channel fast sequencing design. Data is received from all satellites in view, up to a maximum of eight. The low signal-to-noise/fast sequencing design allows useful satellite tracking to 0° elevation angle with rapid acquisition of satellite data. Pseudo range and pseudo range rate data is transmitted from the PPM to the IRU where the added GPS navigation processor card processes the pure GPS solution. This card also contains the Kalman filter that blends the GPS and inertial data to provide the GPS hybrid solution.

Specifications
Dimensions:
(inertial) 317.5 × 320 × 198.1 mm
(GPS) 152.4 × 177.8 × 50.8 mm
Weight:
(inertial) 21.32 kg
(GPS) 1.36 kg
Reaction time: 2.5–10 min
Accuracy:
(position) (inertial) 0.8 n miles/h CEP
(GPS) 25 m SEP
(velocity) (inertial) 10 ft/s
(GPS) 0.1 m/s
(time) (GPS) 350 ns
Reliability:
(inertial) 5,000 h MTBF
(GPS) 20,000 h MTBF

Contractor
Honeywell Inc, Commercial Electronic Systems.

LCS-850 Loran C sensor

Type
Avionic navigation sensor, radio aids.

Description
The LCS 850 Loran C sensor has been specifically designed to interface with UNS flight management and navigation management systems with no additional control or display required. Using a true multiple station solution, the LCS-850 provides accurate latitude/longitude position information to the UNS management systems. It tracks up to eight stations in a multiGRI position solution, automatically selecting the best signals from multiple chains. Position is calculated using a minimum of three stations on one GRI or two stations on each of two GRIs. Notch filters reduce interference and provide increased performance in weak Loran signal areas.

The LCS-850 meets the rigorous standards of TSO C-60b and also meets both system and accuracy requirements for RNav operations in the US National Airspace System and has NAT MNPS airspace approval.

Specifications
Dimensions: 194.1 × 25.1 × 386.6 mm
Weight: 1.6 kg
Power supply: 18–32 V DC, 0.225 A at 28 V

Status
In service.

Contractor
Universal Avionics Systems Corporation.

LN-100 advanced navigation system

Type
Avionic navigation sensor, Inertial Reference System (IRS).

Description
In October 1989, the then Litton Industries (now part of Northrop Grumman Corporation) was awarded a contract to develop a low-cost inertial system featuring low power, weight and volume. The resulting LN-100 employs non-dithered Northrop Grumman Zero-Lock™ Laser Gyro (ZLG™) and A-4 accelerometer instrument technologies, together with a 22-state Kalman Filter (KF), which integrates Doppler velocity and GPS position and velocity inputs, to serve as a dual-flight control reference and high-accuracy navigation system for the AH-64D Longbow Apache programme.

In April 1991, the LN-100 INU was selected by Boeing to provide the Navigation Quality Inertial Sensor (NQIS) for the (later cancelled) RAH-66 Comanche helicopter inertial navigation system (LN-100C). The LN-100G Global positioning Inertial Navigation System (GINS) is the Inertial Reference System (IRS) in the Lockheed Martin F/A-22 Raptor.

Specifications
Gyro: non-dithered 18 cm ZeroLock™ laser gyro
Processor: 80960 XA or PowerPC 603 processor, software in Ada
Dimensions: 241.3 × 177.8 × 177.8 mm
Weight: 8.6 kg
Power supply: 28 V DC, 25 W
Accuracy: better than 0.8 n mile/h
Reliability: 5,000 h MTBF
Interface: 2 dual-redundant MIL-STD-1553B Databus, RS-422/485 busses and ARINC 429

Status
In production and in service in a variety of aircraft, including the AH-64D Longbow Apache helicopter.

Contractor
Northrop Grumman Corporation, Navigation Systems Division.

LN-100G navigation system

Type
Avionic navigation sensor, Inertial Reference System (IRS).

Description
The LN-100G has evolved from the proven LN-100 product line. All LN-100 systems use common hardware and software elements,

affording economies of scale from high-rate production. A number of major US DoD avionic programmes have contracted for LN-100G Global positioning Inertial Navigation System (GINS), including the F/A-22 Raptor. The LN-100G, with embedded GPS receiver, was selected by the US EGI tri-service programme office for the F/A-18 and EA-6B upgrades, by the US Navy for the T-45A Cockpit 21 programme and by Boeing for the Japanese 767 AWACS.

The LN-100G provides three simultaneous navigation solutions; hybrid GPS/INS, free-inertial and GPS only. The processing power of the LN-100 product line is the 32-bit PowerPC Motorola microprocessor; the software is Ada.

The LN-100G optimally combines GPS and INS features to provide enhanced position, velocity, attitude and pointing performance as well as improved acquisition and anti-jam capabilities.

The addition of a GPS receiver package on a single circuit card provides a complete hybrid navigation unit with very low size and volume. Two spare card slots allow for the addition of analogue I/O modules, ARINC interfaces, Low Probability of Interception (LPI) radar altimeter, air data and other expansion modules.

The embedded GPS module is an L1/L2 CA/P(Y) code unit capable of accepting RF (or IF) inputs from the GPS antenna system. Stand-alone GPS PVT data and stand-alone INS data are provided for integrity and fault monitoring purposes.

Specifications

Performance

Position	Inertial	GPS/Inertial
4 min gyrocompass align	0.8/0.6 nm/hr	10 m CEP
4 plus 4 min EIA[1] align	0.5 nm/hr	10 m CEP
After loss of GPS	N/A	120 m/20 min
Velocity per axis	2.5 ft/sec (RMS)	0.015 m/sec (RMS)
Attitude (pitch/ roll/azimuth)	0.05° (RMS)	0.02° (RMS)

[1] EIA = Enhanced Interrupted Align – after initial align, aircraft may taxi to heading change >70° to continue align.

Outputs
Digital: 2-dual MIL-STD-1553B, RS-422, ARINC 429
Analogue: Optional 3-wire or 2-wire synchro (pitch/roll/heading)
Options: Range/bearing

Operating
Acceleration: 16g (all axes)
Attitude: Unlimited (all axes)
Roll/pitch azimuth rate: 400°/sec
Roll/pitch azimuth acceleration: >1,500°/sec²
Altitude: –2,100 to 70,000 ft Class 2 (2X optional)
Temperature: –54 to 71°C

Vibration (random): 8.1g RMS performance
Vibration (sine): ±5g sine, 5 to 2,000 Hz
Shock: 21g, 40 msec
Environment: Per MIL-STD-810C
Acoustic noise: 140 dB

Physical
INU
Dimensions: 279 × 178 × 178 mm (L × W × H)
Weight: 9.8 kg

Optional mount
Dimensions: 348 × 183 × 19 mm
Weight: 1.1 kg
Power: 28 V DC; 37.5 W
Cooling: Convection
MTBF: 14,400 h (AIC)

Status
In production and in service. The LN-100/LN-100G is designed for both new aircraft and retrofit applications.

During May 2004 Northrop Grumman was awarded a contract from Lockheed Martin to supply 86 Global Positioning Inertial Navigation Systems (GINS), plus spares, for US Air Force F/A-22 Raptor advanced fighter aircraft.

During August 2005, Northrop Grumman announced a further contract for up to 108 additional GINS for US Air Force F/A-22 Raptors. 53 GINS, and 50 to 55 options, will be delivered under lot 5 and lot 6 of an approximately USD9 million contract over the next two years. Assembly and testing of the inertial navigation systems will take place in Northrop Grumman's Salt Lake City, Utah, facility.

LN-100 technology has been applied to aircraft, Unmanned Aerial Vehicles (UAVs), launch vehicles, missiles, fighters, helicopters and unmanned underwater vehicles. The LN-100G has been selected by more than 70 customers worldwide.

Contractor
Northrop Grumman Corporation, Navigation Systems Division.

LN-200 Inertial Measurement Unit (IMU)

Type
Avionic navigation sensor, Inertial Reference System (IRS).

Description
The Northrop Grumman LN-200 family of inertial equipment uses Fibre Optic Gyros (FOGs) and silicon accelerometers (SiAc's) for measurement of vehicle angular rate and linear acceleration and satisfies tactical missile and guided projectile guidance requirements and aircraft flight control systems. The LN-200 is a small, lightweight sensor, featuring three solid-state Fibre Optic Gyros (FOG) and three solid-state silicon accelerometers, providing for low cost and high reliability.

The system has a wide variety of applications. Customers have purchased the LN-200 for space stabilisation, missile guidance, radar/EO/FLIR stabilisation (including the LANTIRN system – see separate entry), motion compensation, UAV guidance and control, camera/mapping and as the central IMU for higher order integrated systems. The LN-200 is hermetically sealed and contains no moving parts or gaseous cavities, ensuring long, reliable shelf and usage life.

The LN-200 is available in a number of functional and data rate configurations.

Status
The LN-200 has been in production since 1994 and is in service in a wide variety of platforms, including the Predator and Global Hawk Unmanned Aerial Vehicles (UAVs), the CH-46 helicopter and the Aermacchi MB-339 jet trainer.

The LN-100 is installed in a variety of aircraft, including the AH-64D Longbow Apache helicopter

0504182

Specifications

Physical		
	Weight	<0.75 kg
	Size	890 × 850 mm (diameter × height)
	Power	12 W steady state (nominal)
	Cooling	Conduction to mounting plate
Activation Time		0.8 s (5 s to full accuracy)

Performance – Gyro (1 Sigma)		
	Bias Repeatability	1°/h to 10°/h
	Scale Factor Stability	100 ppm
	Bias Variation	0.35°/h with 100 s correlation time
	Nonorthogonality	20 arcsec
	Bandwidth	>500 Hz

Performance – Accelerometer (1 Sigma)		
	Bias Repeatability	200 µg to 1 mg
	Scale Factor Stability	300 ppm
	Bias Variation	50 µg with 60 s correlation time
	Nonorthogonality	20 arcsec
	Bandwidth	100 Hz

Operating Range		
	Angular rate	±1,000°/s
	Angular acceleration	±100,000°/s^2
	Acceleration	±40 g
	Velocity quantisation	0.00169 fps
	Angular attitude	Unlimited
Reliability (predicted)		32,995 h MTBF (30°C missile launch environment)
Input/Output		RS-485 Serial Databus (SDLC)
Data Latency		<1 ms

Environmental		
	Temperature	–54 to +85°C operating (85°C intermittent)
	Vibration	11.9 g rms – performance
		17.9 g rms – endurance
	Shock	90 g, 6 ms terminal sawtooth

Contractor

Northrop Grumman Corporation, Navigation Systems Division.

LN-250 INS/GPS system

Type

Avionic navigation sensor, Inertial Reference System (IRS).

Description

Northrop Grumman Navigation Systems' LN-250 integrated GPS/inertial system provides accurate navigation and motion compensation to all classes of military fixed- and rotary-wing aircraft.

In April 1997 Northrop Grumman was awarded the GPS Guidance Package (GGP) contract, a DARPA sponsored program. The goal was to develop a family of low-cost, modular, miniature GPS-based systems capable of supporting a broad range of DoD air, land and marine platforms, together with strike weapons of all types. Evolving from these GGP activities, Northrop's LN-200 series inertial sensors are lightweight and feature small form factors. The range includes the LN-250 (military air), LN-260 (multi-application) and LN-270 (land) to provide accurate performance in smaller, lighter, more reliable and less expensive configurations than currently available competitive systems.

The LN-250 is a completely integrated navigation system with a Selective Availability/Anti-Spoofing Module (SAASM) compliant embedded Global Positioning System (GPS) receiver. The fully integrated, tightly coupled GPS inertial design provides superior performance relative to other embedded INS/GPS systems. Modular, Open System (OS) architecture allows the LN-250 to be easily adapted to new applications, new system requirements and to improve performance of mission equipment and flight-control systems.

The LN-250 Inertial Navigation System (INS) employs low-noise Fibre Optic Gyros (FOG), in service in the LN-251, which eliminate self-induced acceleration and velocity noise experienced by dithered laser gyro systems, providing for reduced target location errors and improved synthetic aperture radar imagery for equipped aircraft.

Status

All variants in production.

During January 2006, Northrop announced that the LN-260 had been selected by the US Air Force for avionics upgrade of its F-16 aircraft. Earlier flight test activity had confirmed significant improvements in navigation and targeting capabilities over the F-16's current navigation/attack system.

During November 2006, Northrop Grumman's Integrated Systems Division ordered 48 LN-260 systems from its own Navigation Systems Division, as part of a navigation/radar display upgrade for the US Navy's fleet of F-5N/F adversary fighters. Delivery of the kits is scheduled to begin in 2007 and will be complete in 2008.

Contractor

Northrop Grumman Corporation, Navigation Systems Division.

LN-251 INS/GPS system

Type

Avionic navigation sensor, Inertial Reference System (IRS).

Description

The LN-251 is an integrated navigation system with an embedded 12-channel, all-in-view, Selective Availability/Anti-Spoofing Module (SAASM), P(Y) code GPS. The fully integrated, tightly coupled GPS inertial design provides superior positioning performance relative to both

Northrop Grumman LN-251 INS/GPS (Northrop Grumman)   1127369

old style GPS receiver-only updating or earlier embedded INS/GPS systems. The modular open system architecture provides for easy adaptation to other applications and evolving requirements.

Contractor

Northrop Grumman Corporation, Navigation Systems Division.

LN-300 Inertial Measurement Unit (IMU)

Type

Avionic navigation sensor, Inertial Reference System (IRS).

Description

The LN-300 Series IMU incorporates MEMS (MicroElectroMechanical Systems) technology, which has enabled Northrop Grumman to achieve dramatic reductions in the size, weight, power and cost of tactical- and munition-grade inertial navigation systems. The Northrop Grumman LN-300 family of Inertial Measurement Units extends the application of MEMS technology with the success of the SiAc™ Silicon Accelerometer. More than 15,000 units have been delivered. Recently, DARPA chose Northrop Grumman to develop a MEMS-based SiGy™ Silicon Gyro for use in the MEMS INS/G Program that will couple an INS with a GPS receiver.

Status

In production.

Contractor

Northrop Grumman Corporation, Navigation Systems Division.

LTN-101E GNADIRU

Type

Avionic navigation sensor, Air Data Inertial Reference System (ADIRS).

Description

The LTN-101E incorporates the latest enhancements to the LTN-101 Global Navigation Air Data Inertial Reference Unit (GNADIRU) product line. Designed and built for performance, reliability, low weight and economical operation, the LTN-101E is the world's first commercially certified navigation-grade fibre optic inertial sensor. The LTN-101E incorporates cutting edge Fibre Optic Gyro (FOG) and Autonomous Integrity Monitored Extrapolation (AIME®) technologies.

Major features and benefits of the LTN-101E include:

- Very long FOG life
- MTBF of over 50,000 operating hours
- Single LTN-101E GNADIRU part number in Airbus (A300 – A380) aircraft
- Intermix capable with LTN-101 GNADIRUs in Airbus (A300 – A340) aircraft
- Integrates GPS/IRS/Air Data
- Supplies Air Data computation (Airbus A318 – A380)
- ARINC 704/738 compatible without modification
- Reduced weight and power consumption over similar systems
- Adapter tray retrofit (non-4 MCU provisions)
- Provides LNAV advisory, will support growth for precision approaches
- Advanced Built-In Test (BIT)

The LTN-101E introduces new technologies to reduce cost and to increase performance and reliability. FOG sensors provide a significant improvement in reliability and ruggedness over Ring Laser Gyros (RLG), while maintaining critical inertial navigation performance. FOG sensors eliminate life issues that are typical of classic RLGs. A single fibre optic coil has a life expectancy in excess of 3.5 million hours. In addition, the integrated Silicon Accelerometer

(SiAc™), a state-of-the-art rugged MEMS device, produces dependable and precise velocity measurement.

Derived from the successful LTN-101 FLAGSHIP® (see separate entry), the LTN-101E similarly manages major system functions such as GPS, Inertial Reference (IR) and Air Data (AD) in modules, ranks them by criticality and isolates them from all other modules in both execution and memory access. This approach eliminates corruption of one function by another and inhibits functional loss due to a fault.

AIME® technology allows commercial aircraft to achieve sole means of navigation with GPS accuracy. AIME® is a patented software algorithm that integrates GPS and IRS to compensate for the inherent integrity problems of GPS. With AIME®, the LTN-101E supports RNP 0.1 navigation accuracy requirements worldwide.

The LTN-101E is designed to work with or without forced air cooling, in all operational ambient temperatures.

Specifications
Dimensions: 4 MCU
Weight: 7.7 kg
Performance
(Accuracy) 15 m, 95% of the time (with AIME®)
(Availability) 100% (24 satellites)
(Alignment) 10 min or less
(Initialisation) GPS/FMS
(Integrity limit) 0.2 nm or less
(Coverage) worldwide, 24 h a day
(Reliability) >50,000 operating hours MTBF

Input/Output

Type	Number
AFDX bi-directional Ethernet bus	2
Discrete input (open/ground)	77
Digital input ARINC 429 Lo	15
Digital input ARINC 429 Hi	9
GPS time mark	2 (RS-422A)
Analogue input 26 V AC resolver	2
Digital output ARINC 704/429 Lo	8
Digital output ARINC 704/429 Hi	8
Discrete output (open/ground)	13
Discrete output (28 V/open)	5
RS-422 Cal/Test	1

Status
The LTN-101E is in development for certification on the entire Airbus family of aircraft, including the A380.

During January 2006, Northrop announced the selection of the LTN-101E GNADIRU by Lufthansa for 10 Airbus A380 aircraft. Lufthansa will operate its first four aircraft during the summer of 2008.

Contractor
Northrop Grumman Corporation, Navigation Systems Division.

LTN-101 FLAGSHIP® global positioning, air data, inertial reference system

Type
Avionic navigation sensor, Air Data Inertial Reference System (ADIRS).

Description
The Four-mode Laser Gyro (FLAG) Software/Hardware Implemented Partitioning (SHIP) system is designed for a wide range of applications. It can be used in single, dual or triple installations as an Inertial Reference System (IRS), combined IRS and Global Positioning System (GPS), Air Data Inertial Reference System (ADIRS) or ADIRS and GPS. At switch on, the system automatically recognises aircraft type and configures itself for either an ARINC 704 or 738 installation. An adaptor tray allows FLAGSHIP®'s 4 MCU ADIRU to fit into a 10 MCU rack without system or rack modification.

FLAGSHIP® integrates navigational functions and offers reductions in size, weight and power by eliminating the need for external air data computers and their interconnections. This also

leads to savings in spares and maintenance and yields significant increases in reliability.

The GPS-IRS integration is an ideal combination because the two functions are highly complementary. The self-contained IRS contributes to GPS dynamic performance by facilitating satellite acquisition and tracking. The GPS in turn supplies ultra-precise position and velocity to the IRS, whose inertially derived position and velocity accuracies degrade with time. The GPS also makes possible inertial alignment during taxi or flight and furnishes long-term correction data which is used to calibrate the IRS inertial sensors.

The FLAGSHIP® system consists of the LTN-101 Air Data Inertial Reference Unit (ADIRU), Thales Air Data Module (ADM), Global Positioning System Sensor Unit (GPSSU), Mode Select Unit (MSU) and Control Display Unit (CDU). The ADIRU contains the inertial instrument package and performs all system computations with the exception of GPS sensor calculations. Critical air data and inertial reference functions are hardware partitioned to facilitate fault containment. The four-mode laser gyro requires no dithering and is free of conventional ring laser gyro lock-in and other dithering associated errors. The Litton A-4 accelerometer triad completes the sensor package. Surface mount devices and ASIC contribute to an ADIRU 60 per cent smaller than other ARINC 738 systems. Designed for a variety of environments, the ADIRU will operate up to 18 hours without cooling air.

The ADMs interface air data sensors with the ADIRU. Using an aneroid capsule and resonating quartz blade sensor, the ADM converts static and dynamic pressure into electrical signals. These signals are temperature corrected, converted to ARINC 429 format and transmitted to the ADIRU on a digital databus.

The stand-alone GPSSU is a Third-Generation Northrop Grumman design. The eight-channel continuous tracking receiver can employ Autonomous Integrity Monitored Extrapolation (AIME™) Technology providing the means of achieving sole means of navigation with GPS accuracy. AIME software compensates for GPS's inherent deficiencies by integrating GPS and IRS in a patented algorithm. This means the system can provide 0.3 nm accuracy without aids such as WAAS or LAAS. GPS usability is maximised by early acquisition of low-elevation satellites and minimal loss of satellite reception during aircraft manoeuvres. GPS outputs of position, velocity, time and raw satellite data are supplied to the ADIRU and other avionics. Both ARINC 743 configurations are offered: a 2 MCU avionic bay-located unit using an antenna with an internal preamplifier or a remote GPS sensor designed for installation near a passive antenna.

The MSU is a switching device used to apply power, annunciate system operating modes and indicate when the system is running on battery power.

The optional CDU, offered as a flight management computer back-up, provides a keyboard for initialisation data and a data display for auxiliary readout. Rotary switches select individual system modes. Push-buttons and annunciators allow inertial and air data output databusses to be turned off by the operator and indicate faults.

Detailed module BIT history is stored on each module. This includes the identity of the aircraft, ADIRU and other modules and LRUs in the system. System BIT history is stored on the computer module, along with its own history. If this module is replaced, system history is transferred to the new module.

Specifications
Dimensions:
(ADIRU) 4 MCU
(GPSSU) (rack mount) 2 MCU
(remote) 64 × 216 × 241 mm
(ADM) 145 × 97 × 53 mm
(MSU) 89 × 146 × 76 mm
(CDU) 171 × 146 × 152 mm

LTN-101 FLAGSHIP® showing system units; Air Data Inertial Reference Unit (ADIRU) – ARINC 738 Air Data Module (ADM) – ARINC 738 Global Navigation System Sensor Unit (GNSSU) – ARINC 743 Mode Select Unit (MSU) – Control Display Unit (CDU)
0015385

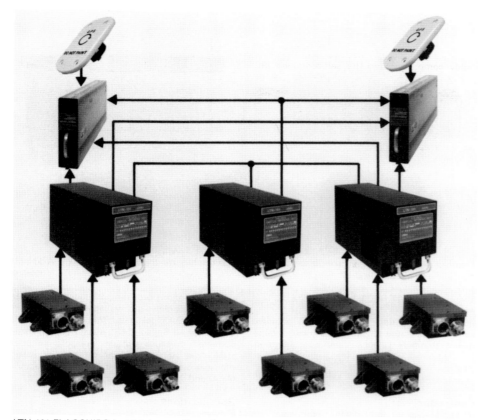

LTN-101 FLAGSHIP® in a triplex configuration

0015386

Weight:
(ADIRU) 12.3 kg
(GPSSU) 3.6 kg
(ADM) 2.4 kg
(MSU) 2.7 kg
(CDU) 10 kg
Performance (with AIME™)
(Accuracy) 100 m, 95% of the time
(Integrity limit) 0.3 nm or less
(Coverage) worldwide, 24 h a day
(Reliability) >15,000 operating hours MTBF

Status
In service in a wide range of wide- and narrow-body civilian transport aircraft, including Airbus A319, A320, A321, A330 and A340 and Canadair RJ-100 and RJ-700 aircraft. Also selected for the Ilyushin Il-96M, Tu-204-200, Saab 2000, CL-604 and An-28.

Superseded by the LTN-101E GNADIRU which features Fibre Optic Gyro (FOG) technology (see separate entry).

Contractor
Northrop Grumman Corporation, Navigation Systems Division.

LTN-2001 Global Positioning System (GPS) sensor unit

Type
Avionic navigation sensor, Global Positioning System (GPS).

Description
The LTN-2001 C/A code GPS provides continuous, worldwide, precision three-dimensional navigation data and offers a means to upgrade and enhance the performance of ARINC 561, 599, 704 and 738 navigation systems. The eight-channel continuous tracking receiver is a mature third-generation Litton design. It maximises GPS usability and features advanced integrity monitoring to deliver performance superior both to sequencing type receivers and those with fewer channels.

The system tracks low-elevation satellites as they rise above the horizon and continues to track them until they disappear, minimises loss of satellite reception during aircraft manoeuvres,

exceeds the ARINC 743 external interference specification and resists multipath reception caused by terrain features and aircraft surfaces. The LTN-2001 integrates with upgradeable navigation systems and provides in-flight alignment, continuous sensor calibration and error bounding capability for inertial navigation systems, plus GPS positional accuracy and worldwide capability.

The LTN-2001 provides GPS outputs of position, velocity, altitude, time, pseudo-range and delta range to inertial navigation systems, flight management systems, smart CDUs and other avionics via ARINC 429 high- and low-speed databusses. An RS-422 serial output is available as an option. Both ARINC 743 configurations are available: a 2 MCU avionic bay-located unit and antenna with internal preamplifier or a remote sensor unit designed for installation within 10 ft of a passive antenna.

Specifications
Dimensions:
(rack) 2 MCU
(remote) 63.3 × 215.9 × 241.3 mm
Weight: 3.63 kg
Power: 15 W
Accuracy (2 DRMS): 100 m
Antenna: active or passive conformal
Compliance: ARINC 743, DO160C, TSO-C129C3 (Mark I), TSO-C129C2 (Mark II)

Status
LTN-2001 Mark I is certified on Airbus A319, A320, A330 and A340, and Boeing 757 and 767. LTN-2001 Mark II is certified on Boeing 737 and 777.

Contractor
Northrop Grumman Corporation, Navigation Systems Division.

LTR-97 fibre optic gyro system

Type
Avionic navigation sensor, Attitude Heading Reference System (AHRS).

Description
The LTR-97 is a Vertical Gyro/Directional Gyro (VG/DG) replacement system which

uses Fibre Optic Gyro (FOG) technology and Application Specific Integrated Circuits (ASIC) in a strapdown configuration to provide superior aircraft attitude and heading and full supportability.

Ideal for retrofit, the LTR-97 offers users of aircraft such as the Boeing 727, 737, DC-9 and MD-8X a cost-effective, advanced technology replacement for their mechanical VG/DG systems with up to 10 times the Mean Time Between Failure (MTBF). The LTR-97 is designed as a plug and play replacement for earlier systems, with no requirements for aircraft wiring changes, or crew training.

A high-speed processor calculates attitude and heading by integrating these rate signals from the FOG and the roll and pitch level sensors. Coupling to the magnetic flux valve is provided through the existing slaving system. Slaved magnetic heading is output on two synchros with automatic switching to free DG mode if the slaving signal is lost.

LTR-97 Lightweight Heading and Attitude System (LHAS)
The LTR-97 LHAS uses FOG technology in a strapdown configuration to provide improved attitude and heading data.

Designed for retrofit applications, the LTR-97 LHAS provides a replacement system for VG/DG mechanical systems on classic aircraft such as the A-300, B-747, DC-10 and L-1011.

Based on Litton's LTR-97 VG/DG replacement system, the LTR-97 LHAS provides up to 10 times the MTBF of mechanical VG/DG systems. The LTR-97 LHAS uses existing aircraft wiring and can be installed in about 20 minutes.

Specifications
Weight: 9.09 kg
Dimensions: per ARINC 561: 256.54 (W) × 269.24 (H) × 508.0 (L) mm
Power: 115 V AC, 100 W
MTBF: >10,000 h
Attitude accuracy:
(static) 0.5°
(dynamic) 1.0°
Heading accuracy: DG mode, dynamic: 1.3°/h
Outputs:
(synchro) 3 pitch, 3 roll, 1 heading
(analogue) 50/200 mV AC pitch and roll
(discretes) ATT and HDG warnings
Qualifications: RTCA DO-160C including HIRF and lightning: DO-178A, level 1 for critical software
Certifications: TSO-C4c, -C5e, -C6d

Status
In production and in service. The LTR-97 LHAS is ARINC 561 compatible.

Specifications
Weight: 5 kg
Dimensions: 256.5 (W) × 154.9 (H) × 248.9 (L) mm
Power: 115 V AC, 64 W (115 VA) – full load
Reliability: MTBF >12,000 flight h
Accuracies:
Attitude (static) 0.5°
(dynamic) 1.0°
Heading:
(slaved mode) dynamic 0.75°
(DG mode) dynamic 1.3°/h (excluding earth rate)
Outputs:
(synchro) Pitch, roll, 2 headings
(analogue) 50/200 mV AC pitch and roll
(discretes) ATT and HDG warns, 6 interlocks
6° roll discrete and interlock
Qualifications:
(RTCA D) 160 C including HIRF and lightning DO-178A, Level 1 for flight critical software
Certifications: TSO C4c, C5e, C6d

Status
Superseded by the LTR-97 LHAS.

Contractor
Northrop Grumman Corporation, Navigation Systems Division.

MD-90 Central Air Data Computer (CADC)

Type
Avionic navigation sensor, Air Data Computer (ADC).

Description
The MD-90 Central Air Data Computer (CADC) is a derivative of the HG280D80. The CADC provides both ARINC 429 digital serial air data outputs and ARINC digital serial and DC analogue air data outputs. Two unique features of the CADC Trigger-On-Failure (TOF) BIT and an internal temperature probe operate in unison to record BIT failures along with corresponding CADC internal temperature data. BIT failure records and other CADC features can be accessed through the front panel connector via the RS-232 bus. Similar to its predecessor, the CADC uses two vibrating cylinder transducers for sensing pitot and static pressure inputs. The vibrating cylinder transducers can be substituted with silicon pressure sensors. Inputs are also provided from a total air temperature probe, baroset signals and programme discretes. Jointly developed by Honeywell and the Chengdu Aero Instruments Corporation.

Specifications
Dimensions: 1/2 ATR long
Weight: 5.35 kg
Power supply: 115 V AC, 400 Hz
Temperature range: −55 to +75°C
Accuracy:
(altitude) ±15 ft at sea level, ±80 ft at 50,000 ft
(airspeed) ±5 kt at 60 kt, ±1 kt at 450 kt
Reliability: 13,000 h MTBF predicted

Contractor
Honeywell Inc, Commercial Electronic Systems.

Micro Air Data Computer (MADC)

Type
Avionic navigation sensor, Air Data Computer (ADC).

Description
The AZ-840 Micro Air Data Computer (MADC) provides both ARINC 429 and bidirectional Avionics Standard Communications Bus (ASCB) digital serial interfaces for the Honeywell Primus 2000, flight control and navigation systems. The MADC uses extensive surface-mount technology for minimising unit size and weight. Full air data outputs and miscellaneous inputs/outputs are provided, including ATC digitiser and input/output discretes for switching functions and aircraft identification selection. Extensive BITE monitors MADC operation and records faults on non-volatile memory for fault analysis on the ground.

The MADC supports Reduced Vertical Separation Minimum (RVSM) requirements.

Specifications
Dimensions: 106.7 × 149.9 × 157.5 mm
Weight: 2.04 kg
Power supply: 28 V DC, 16 W (max)
Accuracy:
(altitude) ±20 ft at sea level, ±150 ft at 60,000 ft
(airspeed) ±2 kt at 100 kt, ±4 kt at 400 kt
Reliability: 6,000 h MTBF predicted

Status
Versions of the MADC are used on the Citation III, Dornier 328 and Raytheon Hawker 1000.

Contractor
Honeywell Inc, Commercial Electronic Systems.

Military air data computers

Type
Avionic navigation sensor, Air Data Computer (ADC).

Description
Honeywell military digital air data computers are in production and in service in the following

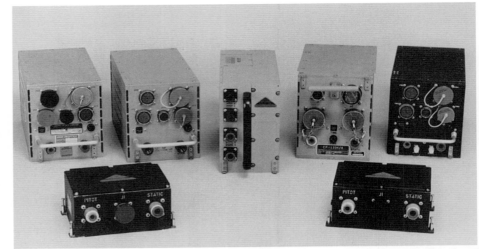

Pictured (left to right, rear) are the central air data computers for the F-15, F-16, C-17, F/A-18 and KC-135R and (front) the B-52   0503869

aircraft: B-52, C-17, C-130, F-16, F-117, F/A-18, FSX and KC-135R.

The latest air data computers use patented solid-state pressure sensors and are said to be so reliable that they will never require recalibration.

Status
In production and service.

Contractor
Honeywell Sensor and Guidance Products.

Miniature Air Data Computer (MADC)

Type
Avionic navigation sensor, Air Data Computer (ADC).

Description
The Miniature Air Data Computer (MADC) -108 is the latest version of Astronautics' Digital Air Data Computers (DADC) for modern aircraft (see separate entry). The MADC-108 features new concepts in air data computer design and performance. The system is fully computerised and uses only solid-state devices resulting in increased computational power, higher accuracy and very high MTBF (>5,000 h).

Other features of the system include:
- Miniature design
- Temperature compensation
- All solid state, including pressure transducers
- Digital and/or analogue outputs
- Modular design.

MADC-108 inputs are: static and total pressure from the pitot tube, Angle of Attack (AoA) sensor readings, altimeter barometric setting, outside air temperature and several discretes such as 'Weight-On-Wheels' (WOW), aircraft configuration and BIT initiate. From these inputs, the MADC-108 calculates the required outputs, using proprietary algorithms.

The interface with the avionics equipment is through MIL-STD-1553 MUX BUS as well as several analogue and discrete signals, including auto flaps control. An integral BIT checks continuously the MADC-108's circuitry. Any detected failure or performance degradation is reported to the aircraft systems through dedicated status words, and failure discretes.

Astronautics' MADC-108 (Astronautics)   0568540

Special options providing additional I/O and/or different accuracies are available.

The MADC-108 calculates: True Air Speed (TAS), Calibrated Air Speed (CAS), Mach number, baro-corrected pressure altitude, pressure altitude, static free air temperature, air density, static pressure over total pressure, true Angle of Attack (AoA) and automatic flaps control.

Specifications
Power: 115 V AC, 400 Hz
Power consumption: 40 W
Static pressure: 0–32 in Hg
Total pressure: 1–90 in Hg
OAT: −70 to 210°C
AoA: −12 to +41°
Altitude: −1,500 to 80,000 ft
Altitude rate: ±6,000 fpm
True Air Speed: 70 to 1,700 kt
Calibrated Air Speed: 50 to 1,000 kt
Mach number: 0.1 to 3.0
Free air temperature: −100 to +150°C
Environment: per MIL 5400 Class II
Dimensions: 190.5 × 99.1 × 238.8 mm
Weight: <4.55 kg

Status
In production and operational in F-5, AT-3, L-159 and other aircraft.

Contractor
Astronautics Corporation of America.

Miniature Air Data Computer (MADC)

Type
Avionic navigation sensor, Air Data Computer (ADC).

Description
The simplification of the interface to MIL-STD-1553B with only a few analogue inputs has made possible the Miniature Air Data Computer (MADC).

Inputs to the baseline MADC are indicated static and total pressure, total temperature (50 or 500 ohm probe), self-test and identification discrete signals with commands via the MIL-STD-1553B databus. Analogue inputs of barometric correction and angle of attack may also be provided.

In addition to the digital databus and altitude reporting code outputs, the MADC has provision for a dual-synchro drive compatible with AAU-19/A, AAU-34/A and AAU-37/A altimeters. These synchro outputs can also be software-programmed to provide other air data functions such as Mach number or true airspeed. Other miscellaneous analogue and discrete outputs tailored to a particular application can be provided. The pressure transducers are self-contained plug-in modules using fused quartz RC sensors.

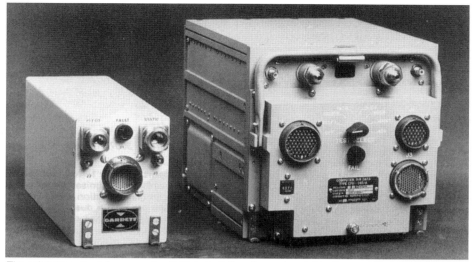

The miniature air data computer (left) is half the size and weight of a standard central air data computer (right) 0503866

Alternative mounting arrangements enhance the flexibility offered by this small package. A front panel Go/No-Go fault annunciator signals the result of the built-in test. Test points are furnished in the front panel input/output connector for fault isolation at the intermediate and depot levels of maintenance.

Specifications
Dimensions: 127 × 108 × 255 mm
Weight: 3 kg
Power: 15 W (max)

Contractor
Honeywell Engine Systems & Accessories.

Model 2100 Doppler Velocimeter/Altimeter (DV/A)

Type
Avionic navigation sensor, Doppler.

Description
The DV/A combines both an altimeter and a Doppler velocity sensor into one compact package. The DV/A consists of two subassemblies: the antenna assembly and the Doppler processor board. The antenna assembly is mounted in the vehicle structure. The Doppler processor board is mounted in a navigation system box. The DV/A measures vehicle velocities and slant range. The output format to the navigation computer is in RS-422.

Specifications
Dimensions:
(antenna assembly) 170 × 386 × 52.1 mm
(Doppler processor board) 10 × 235 × 140 mm
Power: 27 W
Temperature range: −32 to +71°C

Contractor
Northrop Grumman Corporation, Navigation Systems Division.

MPE-HS high-sensitivity SPS/GPS receiver

Type
Avionic navigation sensor, Global Positioning System (GPS).

Description
The MPE-HS high-sensitivity SPS GPS receiver is the latest in Rockwell Collins' range of embedded GPS receivers, providing both embedded PPS (SAASM) and SPS receivers.

For backwards compatibility, the MPE-HS has been designed to provide full compatibility for current users of the embedded MPE I family of PPS and SPS/GPS receivers, utilising the same 80 pin I/O connector and MMCX RF connector.

Every effort has also been made to ensure that the signal pin-out of the I/O connector remains compatible with previous versions of the MPE family and also with the current MPE-S (GB-GRAM) SAASM based receivers. The physical size of the receiver is identical to the MPE-S (GB-GRAM), allowing direct replacement and exchange between the two receivers.

Like all previous MPE receivers, the MPE-HS (SPS) receiver uses the GPS-ICD-153 interface, providing integrators and users with an internationally adopted and maintained ICD.

The MPE-HS (SPS) receiver tracks the L1 GPS satellite frequency; when turned off, an on board time source runs continuously to allow rapid re-acquisition of the GPS satellite constellation once the receiver is switched back on.

The system interfaces via power and data connectors (Berg P/N 87471-640) and RF input connector (Huber Suhner P/N 85MMCX-50-0-1/111).

Hardware interfaces include three independent serial data ports (full duplex):
- Two low power CMOS serial data ports
- One standard RS-232 serial data port (NMEA-0183 data output; 1 PPS input; 1 PPS output; L1 active RF antenna port, 3.3 V DC; HAVEQUICK (SS-110990 and ICD-GPS-060A compliant)

Specifications
Frequency:
L1/L2 dual-frequency tracking
L1 – C/A, P/Y
L2 – P/Y
Dynamics:
Velocity: 515 m/sec or 1,800 m
Acceleration: 4 *g* max
Time accuracy: 100 ns
Positional accuracy: <12 m CEP
Velocity accuracy: 0.05 m/sec RMS (steady rate)
Acquisition time: TTFF <40 sec; TTSF <10 sec
MTBF: >40,000 h
Dimensions: 86 × 62 × 16 mm
Weight: 35 g (nominal)
Power supply: +3.3 V DC, 600 mW (operating); +3.0 to +6.0 V DC, 1 mW (keep alive)
Temperature range: −40 to +85°C (operating)

Status
In production and in service.

Contractor
Rockwell Collins.

MX8000 series military GPS receivers

Type
Avionic Global Positioning System (GPS).

Description
The MX8000 series are single card, high-performance military GPS receivers

designed for embedded applications. The MX8100 provides six parallel continuous tracking, switchable channels, each independently tracking code and carrier measurements for smooth navigation results, high accuracy and greater data availability. Designed to the requirements of GPS-EGR-600, both hardware and software interfaces provide quick and easy mating to military inertial instruments so that initial communication is quickly established.

MX8000 receivers feature a flexible modular architecture, intended for growth and adaptation to changing military requirements. High computer throughput, memory margins and software modularity allow new requirements to be added with minimum impact on the weapon system. A field reprogramming feature allows mission software to be changed as required to meet specific tactical requirements. The receivers can be cold-keyed using standard techniques. Receiver default condition is the Y-code, allowing direct P/Y code acquisition in time-critical applications.

Specifications
Dimensions: 149.4 × 148.6 × 16.5 mm
Weight: 0.27 kg
Power supply: 5 V DC, 7.5 W
Accuracy:
(position) 15 m SEP, 2–3 m with DGPS
(velocity) 0.1 m/s
(time) 100 ns
Temperature range: −40 to +85°C

Status
In service.

Contractor
Raytheon Company, Space and Airborne Systems.

NAV 122D and NAV 122D/GPS self-contained NAV receiver indicators

Type
Aircraft navigation system, Global Positioning System (GPS).

Description
The NAV 122D and NAV 122D/GPS self-contained NAV receiver indicators are being relaunched as technology updates of the earlier discontinued 3 in NAV instruments carrying the NAV 122 nomenclature. The new NAV 122D and NAV 122D/GPS instruments will be form/fit replacements for all earlier models from the NAV 122 back to the NAV 12 (except that replacement of the NAV 12 model will require an interconnect cable).

The new NAV 122D includes self-contained VOR/Localiser/Glideslope receiver and converter, with VOR/Glideslope indicator, DME channelling and autopilot interface. There is no internal marker receiver or lights. It utilises, as a self-contained item, the following elements: 200 channel NAV receiver, VOR, Localiser and Glideslope Indicator, with Marker Beacon Lights.

The new NAV 122D/GPS optional version additionally includes the resolver required to interface with GPS, where VOR-style, 'Left/Right' indication is required for GPS IFR approval.

Narco Avionics NAV 122D self-contained NAV receiver indicator 0015389

Specifications
Size: 27.6 (L) × 8.3 (H) × 8.3 (W) mm
Weight: 1.1 kg
Voltage: 14 V DC
VOR/Loc receiver:
(frequency range) 108.00–117.95 MHz
(TSO compliance) VOR C40c, Localizer C36e
   Class C
Glideslope receiver:
(frequency range) 329.150–335.00 MHz
(TSO compliance) C34e Class D
External interfaces:
(DME channelling) 2 out of 5
(autopilot) left/right, to/from, up/down,
   glideslope flag
(Marker lamps) inputs for amber, white, blue
   lamps

Status
In service.

Contractor
Narco Avionics Inc.

Reduced Vertical Separation Minimum (RVSM) air data system

Type
Avionic navigation sensor, Air Data System (ADS).

Description
The ISS RVSM system is designed to be fitted to the widest possible number of aircraft types as a retrofit or new-build system. In addition to satisfying RVSM requirements, retrofitted aircraft receive an upgraded air data system.

The system is a triplex installation, integrating three high-precision altitude sensing devices which continuously monitor and compare three sources of computed pressure altitude. Compared to conventional RVSM solutions using dual Central Air Data Computers (CADC), the triplex system ensures that an aircraft will not be excluded from RVSM airspace in the event of a single air data component malfunction.

The ISS system provides full triplex operation and, when selected with the ISS Digital Air Data Computer (DADC), it provides the following functions: visual windshear warning, true airspeed display, angle of attack display, stall warning, and take-off monitor. It also warns of excessive deviation, monitors CADC performance, provides an altitude alerting function with minimum descent altitude mode for preselect and voice warnings, and exceeds North Atlantic corridor altitude reference requirements.

ISS claims that the advantage of its system is that the altimeters and airspeed indicators, in their self-sensing mode, act in combination as independent air data computers. In contrast to conventional systems, the pilot using an ISS RVSM system knows where a failure or deviation has occurred and still has two remaining systems to satisfy RVSM requirements.

In normal operation, the altimeters and airspeed indicators function as repeaters displaying information from the DADC's air data sensors. They are, however, also integrated with the aircraft air data sensors and are equipped with their own internal transducers. They constantly compare the DADC signal with their self-sensed information and, if a difference is detected, the altimeter and airspeed indicator automatically switch into their standby mode and display self-sensed information instead of that from the DADC.

System components comprise:
one DADC, two dual-mode solid-state altitude indicators, two dual-mode solid-state airspeed indicators, one multifunction preselect altitude alerter.

The system is fully compliant with North Atlantic RVSM, Free Flight, and Europe 2002 requirements.

Status
Over 500 systems installed on 13 different programmes by summer 1999. Customers include regional and business aircraft, Cargo Airline Association operators and the US Navy, for its C-9 aircraft.

Contractor
ISS Innovative Solutions & Support Inc.

RINU(G) navigation system

Type
Avionic navigation sensor, Inertial Reference System (IRS).

Description
Northrop Grumman Navigation Systems Division was contracted by the US Navy to provide replacement navigation systems for all of the US Navy's P-3 maritime and C-130 patrol and cargo-type aircraft.

The systems replaced legacy units installed in the aircraft since the late 1970s. Two systems are installed in each aircraft.

The new systems, called RINU(G), combine in a single unit the continuous, long-term accuracy of the latest laser gyroscope navigation technology with the geographical precision of a global positioning system satellite signal receiver.

The P-3/C-130 navigation units will be based on Northrop Grumman's production LN-100G system (see separate entry).

Status
In production.

Contractor
Northrop Grumman Corporation, Navigation Systems Division.

RMS 555 radio management system

Type
Aircraft navigation system, control/interface.

Description
The RMS 555 offers high-technology features and performance in a compact RMU 556 control/display unit, driven by a ¼ ATR dwarf KDA 557 data adaptor. Utilising a 3.6 in diagonal CRT display, the RMU 556 provides a multitude of different pages, including the normal active frequency, memory and diagnostic pages.

Line item push-buttons provide quick access to each of the frequencies to be selected. Positive detent concentric knobs are used to provide secure input of frequency, channel or codes.

In a dual-RMS installation, the pilot and co-pilot can each tune all radios in the system, including the cross-side radios. Each RMS is capable of handling 17 different pieces of equipment via an ARINC 429 databus, including three VHF COMM, dual-NAV, dual-TACAN, dual-ADF, dual-DME, dual-ATCRB or Mode S transponders, dual-MLS and -TCAS.

Standby frequencies for COMM, NAV and ADF allow for flip-flop tuning. The memory capability of the RMS 555 system also allows the pilot to store up to 20 pilot-programmable frequencies for each VHF COMM, 10 for each ADF and 10 for each TACAN.

Specifications
Dimensions:
(data adaptor) 99 × 85.9 × 231 mm
(control/display unit) 152.4 × 62.6 × 231 mm
(configuration module) 53.3 × 40.5 × 38.6 mm
Weight:
(data adaptor) 1.87 kg
(control/display unit) 1.46 kg
(configuration module) 0.046 kg
Power supply: 28 V DC, 1.25 A

Status
In service.

Contractor
Honeywell Aerospace, Electronic & Avionics Lighting.

Secondary Attitude and Air data Reference Unit (SAARU)

Type
Avionic navigation sensor, Attitude Heading Reference System (AHRS).

Description
The SAARU, a fail-safe and highly reliable device, operates as a secondary system to the fault-tolerant air data inertial reference unit (see earlier item).

SAARU measures the aircraft's linear and rotational motions and computes air data measurements to provide fail-safe secondary attitude and air data reference information. The 10 MCU device also provides digital attitude and air data reference information to the cockpit LCD standby displays.

Status
Selected for the Boeing 777 aircraft.

Contractor
Honeywell Inc, Commercial Electronic Systems.

Standard Central Air Data Computers (SCADC)

Type
Avionic navigation sensor, Air Data Computer (ADC).

Description
Honeywell Engine Systems & Accessories has completed contracts to supply SCADC to

The ISS reduced vertical separation minimum air data system airspeed and altimeter displays

0064371

Lockheed Martin for the C-5B Galaxy and to Northrop Grumman for the C-2A Greyhound carrier onboard delivery aircraft.

The SCADC configuration uses a non-volatile memory for storing air data and subsystem fault information. There is a high standard of fault detection and high reliability ensures compatibility with a two-level maintenance strategy.

Specifications

CPU-140/A for the C-2A
Dimensions: 143 × 140 × 242 mm
Weight: 9.4 kg
Power supply: 115 V AC, 400 Hz, 60 W

CPU-141/A for the C-5A and C-5B
Dimensions: 1/2 ATR
Weight: 14.4 kg
Power supply: 115 V AC, 400 Hz, 72 W

Status

In service on US Air Force C-5B Galaxy and US Navy C-2A aircraft.

Contractor

Honeywell Engine Systems & Accessories.

STAR 5000 panel mount GPS

Type

Avionic navigation sensor, Global Positioning System (GPS).

Description

The STAR 5000 blends GPS technology with the Arnav R-50i IFR-certified Loran. The panel mount GPS receiver delivers five-channel continuous and parallel tracking of all satellites in view. Its onboard network interfaces with an array of avionics including analogue instruments, fuel computers, air data computers, moving maps, autopilots, CDI/HSI and annunciators.

A two-line 40-character sunlight-readable LED display shows bearing, distance, groundspeed, course deviation, waypoint identification, altitude and track. Waypoints can be located by identification, city, local proximity or proximity to other waypoints. The system uses a 40,000 waypoint Jeppesen NavData card, known as the Gold Card because of the wealth of information. Jeppesen NavData is offered for three geographical areas. The North America card coverage extends from Alaska to Central America. The International card covers all areas outside North America. Worldwide navigation combining the data on the other two cards is available on a single World card.

Airports, VORs, NDBs, terminal and en route intersections are stored on the pilot updatable NavData card. Special use airspace includes floors and ceilings and the card provides TCA, ARSA, ATA, MOA, restricted, prohibited and alert area penetration warnings. Coupled to an optional Mode C encoder interface, pressure altitude with a pilot input barometric correction is displayed. VNav becomes automatic with known present altitude and known waypoint elevations. An altitude hold advisory will notify the pilot of an altitude deviation.

Standard features of the STAR 5000 include nearest airport, VOR, intersection search, true airspeed, wind and fuel calculations, MSA/MESA, 150 waypoint flight planning and 300 user waypoints and the pilot's choice of the 40,000 waypoint North America or International NavData cards. The STAR 5000 has UTM and military grid reference co-ordinates as a standard feature. Options include the altitude encoder interface, GridNav mission management software, ARINC 429 interface and an air data computer.

Contractor

Arnav Systems Inc.

TA-12S SAASM-based GPS receiver

Type

Avionic navigation sensor, Global Positioning System (GPS).

Description

Trimble's TA-12S is a 12-channel, Precise Positioning Service (PPS) GPS receiver that uses both the L1 and L2 frequencies and outputs data over multiple ARINC 429 and RS-422/232 channels. The TA-12S is an upgraded version of the TASMAN™ ARINC TA-12 (see separate entry), incorporating new architecture and design features, including carrier phase tracking, to enhance GPS navigation performance.

The TA-12S meets Navigation Warfare (NAVWAR) and Selective Availability Anti-Spoofing Module (SAASM) requirements for US and authorised Allied military aircraft with FAA certification to TSO-129a for operation in US and international civilian airspace.

The TA-12S takes advantage of modular adaptive legacy software by reusing modules proven on other Trimble military products. The TA-12S integrates with existing or new Flight Management Systems (FMS) for certified IFR operations and/or PPS GPS capabilities.

The SAASM is a single, tamper-resistant Multi-Chip Module (MCM) that incorporates all required GPS Selective Availability (SA) and Anti-Spoofing (A-S) functionality. New SAASM features include Black Keys (Red Keys continue to be accepted), Over-The-Air-Re-Keying (OTAR), signal authentication and contingency recovery operation. The TA-12S receiver corrects for SA, incorporates Receiver Autonomous Integrity Monitoring (RAIM) and Fault Detection Exclusion (FDE) with Step Detection. An offline predictive RAIM programme supports remote/oceanic capability.

The TA-12S receiver uses an all-in-view tracking design over a wide range of dynamic conditions and velocities. Housed in a lightweight, sealed aluminium case, the receiver is designed to withstand extreme temperature and severe vibration. A Trimble L1/L2 low-profile antenna completes the integration package for airborne platforms.

The system is WAGE/WAAS/EGNOS upgradeable.

Specifications

Receiver: 12-channel, L1/L2 frequency, P(Y) code (PPS)
Antenna: Low-profile with preamplifier, L1/L2, TSO
Time To First Fix (TTFF): 1.0 min (nominal)
Update rate: 1 Hz
Timing: HAVE QUICK IIA timing output or 1 PPS I/O; GPS time mark output
Crypto key input: Compatible with KYK-13, KOI-18, CYZ-10
Security: GPS JPO security approved/unclassified when keyed
Accuracy – autonomous GPS:
(position 3-D) 16 m SEP
(velocity) 0.2 m/s RMS
(time) 100 ns RMS
Dynamic envelope: 0–400 m/sec (0–800 kts)
Dimensions:
(receiver) 127 (W) × 241 (D) × 53 (H) mm
(antenna) 89 (D) × 18 (H) mm
Weight:
(receiver) 1.5 kg
(antenna) 0.2 kg
Power supply: 20 to 32 V DC, <15 W
Certification: TSO-C129a Class B1/C1; JAA BRNAV compliant; ARINC 743; RTCA DO-229B; RTCO DO-160D; FAA N8110.60

Status

In production and in service.

Contractor

Trimble Navigation.

TASMAN ARINC TA-12 GPS sensor

Type

Avionic navigation sensor, Global Positioning System (GPS).

Description

Trimble's TASMAN ARINC 12 (TA-12) is a 12-channel, Precise Positioning Service (PPS) GPS receiver that uses both the L1 and L2 frequencies. It corrects for Selective Availability (S/A), and incorporates Receiver Autonomous Integrity Monitoring (RAIM), Fault Detection Exclusion (FDE), and it protects against spoofing. It has interface capabilities to dual-FMS/IRS and air data installations using ARINC 743 standards. It meets US government certification requirements in both PPS and SPS modes, including FDE in accordance with DO-229. It provides users with an upgrade path to WAAS and SAASM. The TA-12 receiver uses an all-in-view tracking design over a wide range of dynamic conditions and velocities. Initialisation is not required and a 3-D fix is calculated nominally within 1 minute. Position and velocity updates occur every second. Trimble's TSO-qualified L1/L2 low-profile antenna provides excellent satellite visibility for airborne platforms.

It provides positioning accuracies of better than 16 m and also supplies velocity and time for navigation and tracking applications.

Specifications

Receiver: 12-channel, L1/L2 frequency, P(Y) code (PPS)
Antenna: Low-profile with preamplifier, L1/L2, TSO
Time To First Fix (TTFF): 1.0 min (nominal)
Update rate: 1 Hz
Timing: HAVE QUICK IIA timing output or 1 PPS I/O; GPS time mark output
Crypto key input: Compatible with KYK-13, KOI-18, CYZ-10
Security: GPS JPO security approved/unclassified when keyed
Accuracy – autonomous GPS:
(position 3-D) 16 m SEP
(velocity) 0.2 m/s RMS
(time) 100 ns RMS
Dynamic envelope: 0–400 m/sec (0–800 kts)
Dimensions:
(receiver) 127 (W) × 207 (D) × 56 (H) mm
(antenna) 89 (D) × 18 (H) mm
Weight:
(receiver) 1.3 kg
(antenna) 0.2 kg
Power supply: 20 to 32 V DC, <15 W
Interface:
 ARINC 429: 8 receiver channels; 3 transmit channels; dual-FMS; dual-IRS, dual-ADS
 RS-422: 2 bi-directional at variable baud rate
 RS-232: 1 bi-directional at 9,600 baud
Certification: TSO-C129a Class B1/C1; JAA BRNAV compliant; ARINC 743; RTCA DO-229; RTCO DO-160D; FAA N8110.60

Status

In September 1999, TASMAN®-ARINC-TA-12 received US Federal Aviation Administration (FAA) certification C-129a Class B1/C1, providing standalone P(Y) code GPS sensor capability,

The TASMAN®-ARINC-TA-12 GPS sensor
0080281

fulfilling both military Precision Positioning Service (PPS) and commercial Standard Positioning Service (SPS) requirements.

Installed in the US Air Force One aircraft.

Contractor
Trimble Navigation.

UNS 764-2 GPS/Omega/VLF sensor

Type
Avionic navigation sensor, radio aids.

Description
The UNS 764-2 is a combination GPS/Omega/VLF long-range navigation sensor. It incorporates a 12-channel GPS receiver that features real-time and predictive Receiver Autonomous Integrity Monitoring (RAIM), satellite fault detection and exclusion capability, carrier phase tracking. It is Wide Area Augmentation System (WAAS) ready.

The Omega/VLF receiver features five independent channels which track all available Omega and up to eight VLF stations simultaneously.

Specifications
Dimensions: 194.1 × 56.9 × 387.6 mm
Weight: 3.6 kg
Power supply: 28 V DC, 30 W nominal
Certification: TSO C-129 Class B1/C1, C-94a

Status
In service.

Contractor
Universal Avionics Systems Corporation.

UNS-764 Omega/VLF sensor

Type
Avionic navigation sensor, radio aids.

Description
The UNS-764 Omega/VLF sensor was designed to operate in conjunction with the Universal's UNS-1 flight management and navigation management systems, working either alone or in combination with inertial, GPS or Loran C sensors. The system comprises an antenna and receiver, the control/ display unit of the UNS-1 system providing the necessary management functions. Gate-array and surface-mount technologies combine to reduce the size and weight of the UNS-764 system over previous Omega/VLF sensors. Operation is automatic, control being exercised via the UNS-1. Five independent receive channels enable all available Omega, and up to eight VLF, stations to be monitored simultaneously.

Specifications
Dimensions: 194.1 × 56.9 × 387.6 mm
Weight: 2.9 kg
Power supply: 28 V DC, 20 W (max)

Status
In service.

Contractor
Universal Avionics Systems Corporation.

UNS-RRS radio reference sensor

Type
Avionic navigation sensor, radio aids.

Description
The UNS-RRS radio reference sensor has been designed to provide DME, VOR and TACAN radio data for the UNS-1 flight and navigation management system. It is remotely tuned through the UNS-1 CDU via an ARINC 429 databus and each RRS can support two UNS-1 systems.

Specifications
Dimensions: 194.1 × 56.9 × 387.6 mm

Weight:
(receiver) 352 kg
Power supply: 28 V DC, 1 A (max)

Status
In service. Further status unknown.

Contractor
Universal Avionics Systems Corporation.

VFR GPS-60

Type
Aircraft navigation system, Global Positioning System (GPS).

Description
The 12-channel, parallel-tracking GPS-60 provides all-weather, worldwide navigation. The VFR-only GPS-60 offers the same features and accuracy found in the M3 IFR GPS (except GPS approaches), at a VFR price. The GPS-60 provides immediate access to flight information, such as distance and bearing to destination; ground speed; NDBs, VORs, Victor and Jet airways, and intersections; winds aloft, and the Class B and Class C airspaces.

The Northstar GPS-60 will interface directly with modern CDIs, HSIs, autopilots and moving maps.

Every Northstar GPS-60 comes complete with a user-updateable FliteCard containing a comprehensive Jeppesen database of over 8,000 US public airports. With Northstar's SmartComm intelligent frequency software, the GPS-60 becomes a GPS/Comm. The SmartComm enables the GPS-60 to organise, tune and display the 40 closest frequencies relative to calculated position. Upgrading to the SmartComm is a designed capability for aircraft with limited panel space or in the need of an extra communication system. Alternative databases available for the GPS-60 include North America, International, and helicopter, all of which include private airports in the United States in addition to public-use airports.

Specifications
Type: L1 frequency, C/A code (SPS)
Multichannel: continuous tracking
Navigation accuracy:
15 m RMS (30 m 2DRMS)
100 m 2DRMS with S/A activated
Navigation update rate: 1 s
Time to first fix: 1 min (typical)
Operating modes:
2D Nav, 3 satellites visible
3D Nav, 4 or more satellites
Automatic cold start: neither time nor position input required
Annunciator output: warn, parallel offset, waypoint alert, VFR
Serial ports: RS-422/RS-485
Power: 10–36 V DC at 14 W
Size: 159 × 298 × 51 mm
Weight: 2 kg

GPS antenna
Type: low-profile patch with integral L1 preamplifier
Length: 87 mm
Width: 56 mm
Weight: 0.14 kg
Mounting: surface-mounts to top of aircraft; requires 12.5 mm cut out in aircraft skin

Status
In production and in service.

Contractor
BAE Systems North America.

The Northstar VFR GPS-60   0051612

VIR-32/33/432/433 Pro Line II navigation receivers

Type
Aircraft navigation system, radio aids.

Description
The VIR-32/33/432/433 are digital VOR/ILS navigation receivers designed for business or commuter aircraft. The VIR-32 can directly replace the VIR-30A or be installed in an all-digital aircraft. Several versions of the VIR-32 are available, but all are identical in format, the variations being selected by making the appropriate wiring connections. The system can receive 200 VOR/localisers and the associated 40 glide slope channels.

The system features a single crystal and simplified frequency conversion for enhanced performance. Digital automatic frequency control centres the received signal precisely in the passband for protection against interference. The VIR-32/33 is controlled by either a wire-saving digital signal from the Pro Line II CTL-32 or by an analogue two-out-of-five control. The VIR-432/433 is controlled by either an ARINC 429 signal from a controller or by a CTL-32. Utilising a CTL-32, the radio's self-diagnostic tests are enabled by the self-test button and displayed on the control.

A significant benefit of the system is that a single version offers a variety of options, depending on how the radio is wired during installation: By example, for helicopter applications, one strap option filters the output to reduce the effect of rotor modulation. Different filtering parameters are activated by connecting power to different pins of the VIR. The radio's microprocessor detects which pin is receiving the power. It uses that information to include or not include rotor mod filtering in the signal processing. For operators of mixed rotary- and fixed-wing aircraft, this option enables the same navigation receiver to be employed for both, resulting in significant economies in logistic support.

Another strap option affects how the instruments respond to self test. Instead of the normal down/ right command, a strap option added during installation provides an up/left indication. A strap option may be used to permit the glideslope deviation bar and flag to remain in view when tuned to a VHF omnidirectional radio range.

Thus, once installed, a single VIR can easily transfer from aircraft to aircraft and automatically produce the operational characteristics required for each aircraft.

Specifications
Dimensions:
3/8 ATR short, dwarf
85 (H) × 96 (W) × 355 (L) mm
Weight: 2 kg
Power supply: 28 V DC, 1.4 A
Altitude: up to 70,000 ft
Temperature: −55 to +70°C
Accuracy:
(VOR deviation) ±0.75°
(VOR bearing) ±0.75°
Certification:
(FAA TSO) C34d, C35d, C36d, C40b
(EUROCAE) ED- 22A, -46, -47, 14A
(environmental) D0-160A
Software:
(VIR-32) DO-178
(VIR-33) DO-178A Level 1
(VIR-432) DO-178
(VIR-433) DO-178A Level 1

Status
In production and in service.

Contractor
Rockwell Collins.

VNS-41 VHF navigation system

Type
Aircraft navigation system, radio aids.

Description
The VNS-41 navigation system is a digital VOR/ ILS receiver and processor system providing

VOR, localiser, glideslope and marker beacon reception. It is a lightweight system consisting of the VN-411 receiver and the CD-412 panel-mounted control display unit.

The VN-411 receiver contains the VOR/Loc, glideslope and marker beacon receiver and processors. It employs full-time self-test monitoring of the key internal circuits such as the power supply, synthesisers and automatic gain control and all receivers and converter circuits. Advanced signal processing provides VOR accuracy within 1°, as well as steady navigation signals.

The CD-412 control display unit provides for a dual-frequency readout, one active and one standby. The frequencies are alternated via the frequency transfer button. The CD-412 displays 200 channels with 50 kHz spacing from 108 to 117.95 MHz. It also features a non-volatile frequency memory which retains the last frequency used, eliminating the possibility of frequency loss due to power interruptions. The CD-412 can display digitally either the bearing or radial to the selected VOR in the standby window.

Specifications
Dimensions:
(control display unit) 63.5 × 79.4 × 63.5 mm
(front connector receiver) 101.6 × 101.6 × 279.4 mm
(rear mount receiver) 101.6 × 101.6 × 320.5 mm
Weight:
(control display unit) 0.27 kg
(front connector receiver) 2.04 kg
(rear mount receiver) 2.81 kg
Power supply: 18–33 V DC, 0.8 A
Frequency:
(navigation receiver) 108–117.95 MHz
(glideslope receiver) 329.15–334 MHz
Channel spacing:
(navigation receiver) 50 kHz
(glideslope receiver) 150 kHz
Number of channels:
(navigation receiver) (VOR) 160, (Loc) 40
(glideslope receiver) 40

Status
In production.

Contractor
Honeywell Aerospace, Electronic & Avionics Lighting.

VOR-700A VHF Omnidirectional Range/marker beacon receiver

Type
Aircraft navigation system, radio aids.

Description
The Collins VOR-700A VHF omnidirectional range/marker beacon receiver is the result of combining digital technology with proven operational and design experience derived from the Collins 5IRV-2, 5IRV-4 and 51Z-4 series VORs.

All bearing signal baseband processing in the VOR-700A is accomplished digitally. The 30 Hz reference and variable signals are converted into digital form using a 12-bit CMOS analogue-to-digital converter and are handled thereafter by an Intel® 8086 16-bit microprocessor. The ARINC 429 input/output functions are performed by an Intel® 8048 microprocessor in conjunction with a Collins universal asynchronous transmitter/receiver large-scale integration-based circuit. Additional functions, such as self-test, auto-calibration and monitoring, are also conducted digitally.

This digital processing improves the accuracy of measuring bearings by reducing the effects of temperature variation and aging. Implementation of 30 Hz bandpass filters in firmware, compared with previous analogue methods, permits improved tracking of ground station modulation frequency variations. It also contributes to increased navigation sensitivity and better rejection of undesired components in the modulation of received signals.

Reliability of the VOR-700A is also greatly increased over earlier systems by the low parts count (40% fewer than previous systems), advanced circuit design and overall quality of components. The design is straightforward, accommodating ground station and environmental anomalies.

Facilitating rapid fault detection and minimising repair time are part of the improved test capabilities of the VOR-700A. Self-test is provided by using microprocessors to control a test sequence using digital test signal synthesis. LRU and card testing are greatly simplified by a carefully organised modular design which results in easily tested functional modules. The system is fully compliant with ARINC 711.

Specifications
Dimensions: 3 MCU per ARINC 600
Weight: 4.5 kg
Power supply: 115 V AC, 400 Hz, 30 VA
Frequency:
(VOR) 108-117.95 MHz
(marker beacon) 75 MHz
Channel spacing:
(VOR) 50 kHz

Status
In production and in service.

Contractor
Rockwell Collins.

VOR-900 VHF Omnidirectional Range/marker beacon receiver

Type
Aircraft navigation system, radio aids.

Description
Collins' VOR-900 is the next generation of Collins VOR radio family. The VOR-900 meets ARINC-700 form, fit and function characteristics, interfaces with other aircraft systems via a serial ARINC 429 databus, meets the racking and cooling requirements of ARINC 600 and is compliant with environmental and software requirements. The VOR-900 can be retrofitted into existing aircraft and interchanged with series 700 VOR units (subject to approval).

The system meets FM immunity requirements as stated in ICAO Annex 10, for 1995 installation and 1999 operational compliance.

The VOR-900 contains partitioned, comprehensive end-to-end self-test that will diagnose and isolate system problems to an individual LRU fault or a fault existing in connected peripherals (such as control panels, antennas, and so on).

In addition to ARINC 604 Built-In Test Equipment (BITE), which connects with aircraft fault maintenance systems, the VOR-900 features a simple and rugged LED-BITE display on the front panel. This allows confirmation of fault status within the equipment bay of older aircraft, which do not include onboard fault maintenance systems.

The VOR-900 features 30 Hz bandpass filters in firm ware, as opposed to analogue techniques, providing for improved tracking of ground station modulation frequency variations, also contributing to increased navigation sensitivity and better rejection of undesired components of received signal modulation. Functional test is provided by employing microprocessors to control a test sequence using digital test signal synthesis. LRU and card testing are simplified by an organised modular design that results in easily tested functional modes.

Specifications
Dimensions: 3 MCU per ARINC 600
Weight: 4.08 kg
Power supply: 115 V AC, 400 Hz; 0.25 Amp
Frequency:
(Marker Beacon) 75 MHz
(VOR) 108-117.95 MHz
Channel spacing:
(VOR) 50 kHz

Sensitivity:
(Marker Beacon) Aural, 1500 µV for low sensitivity, 200 µV for high sensitivity
(VOR) Aural: –99 dBm for 6 dB (s+n)/n; NAV: –99 dBm for valid flag
Selectivity:
(Marker Beacon) ±10kHz at 2 dB, ±50 kHz at 60 dB
(VOR) ±15 kHz at 6 dB, ±33.0 kHz at 60 dB
Altitude: 50,000 ft
Temperature range: –55 to +70°C
Compliance: FAA TSO C40c, C35d; RTCA DO-143, DO-196, DO-160C, DO-178A; EUROCAE ED-22B, ED-12A, ED-14C; ARINC 604, 600, 429, 711

Status
In production and in service in a wide variety of transport aircraft, including the Boeing 737, 747, 757, 767, 777 MD-11 and MD-90, Airbus A319, A320, A321, A330 and A340.

Contractor
Rockwell Collins.

VRS-3000 solid-state Vertical Reference System (VRS)

Type
Avionic navigation sensor, Inertial Reference System (IRS).

Description
The VRS-3000 Vertical Reference System was designed as a replacement for conventional spinning mass vertical gyroscopes. The system utilises solid-state rate and level sensors and provides traditional ARINC 407-synchro information for pitch and roll attitude. The VRS-3000 also provides stable and consistent attitude, body rates, inertial pitch and roll rates and acceleration outputs in ARINC 429 digital format. It can provide attitude information to drive primary or standby flight displays and autopilots.

The VRS-3000 also performs an internal system monitoring routine. This helps provide optimum levels of reliability and system confidence further enhanced by L-3's Standard Product Reliability Acceptance Test Program. This program includes burn-in under environmental operating conditions prior to final acceptance.

The design of the VRS-3000 uses proven solid-state components which are currently in use in the Electronic Standby Instrument System (ESIS) model GH-3000, also developed and manufactured by Goodrich. The VRS-3000 can be installed in fixed-wing aircraft, helicopters, drones, and remotely piloted vehicles. The system meets or exceeds TSO-C4c, weighs only 3.5 pounds, and was designed for low power consumption.

Specifications
Dimensions: 102 × 95 × 156 mm
Weight: 1.6 kg
Power supply: 115 V AC, 400 Hz, single phase

Status
In service.

Contractor
L-3 Communications, Avionics Systems.

VRS-3000 solid-state Vertical Reference System
0098572

AIRCRAFT IDENTIFICATION AND SAFETY OF FLIGHT (SOF) SYSTEMS

Canada

AN/URT-43 recorder locator system

Type
Avionic Emergency Locator Transmitter (ELT).

Description
The DRS Flight Safety & Communications AN/URT-43 recorder locator system is an automatically deployable emergency beacon for fixed-wing aircraft that enables rapid location and rescue of survivors. Optional features include 406 MHz COSPAS/SARSAT capability and a solid state flight data and voice recorder. The AN/URT-43 system comprises: a composite airfoil, which encloses the radio transmitter, transmitter battery and flight recorder (if installed); a mounting tray including the release mechanism; crash detection sensors (frangible switches); hydrostatic switches; and a cockpit control unit for the testing of the system that can also be used for manual deployment. If a flight recorder is installed, the system will include an Aircraft Monitoring Unit (AMU), which provides the interface between the aircraft sensor, voice link and the recorder.

The beacon meets the international requirements for 121.5 and 243 MHz transmitters and is COSPAS/SARSAT compatible. The signal is detectable up to approximately 80n miles (depending on receiver altitude and terrain) and provides direction-finding capability at 50n miles. The system can be configured as a single- or-dual-frequency transmitter with a selection of 121.5, 243 and 406 MHz.

Specifications
Airworthiness certification:
TSO C-91a, C-123a, C-124a
EUROCAE ED-55, ED-56, ED-56A
RTCA DO-183, ED-62
CAA specifications 11, 16, 18

Cockpit Voice Recorder (CVR)
Audio channels conform to ED-56A functional requirements:
    1 channel 150 to 6,000Hz bandwidth;
    3 channel 150 to 3,500Hz bandwidth;
    1h storage for each channel

Flight Data Recorder (FDR)
Conforms to ED-55 requirements;
25 hours of data storage

Storage capability
4-channel audio
Flight data
ATC data link, HUMS and auxiliary data expandable

Recording time
1 hour per channel
25 hours

Interface options
Flexible flight data acquisition options:
Direct from databus: 8 channel ARINC 429 or 2 dual-redundant MIL-STD-1553
analogue acquisition suite: discretes, analogue, synchro, frequency, thermocouple
Industry standard ARINC 573/717 interface

Emergency Locator Transmitter (ELT)
Operates on any combination of any two of 121.5, 243 and COSPAS/SARSAT-compatible 406 MHz
Battery packs designed to TSO-C97

Environmental
System designed to military specifications
EMC suitable for composite aircraft
Compliance: RTCA DO-160C, MIL-E-5400, MIL-STD-810E

Weight
Total system: 17.5 kg
ELT only: 12.5 kg
Airfoil dimensions: 607 × 758.5 × 111 mm
Aircraft Monitoring Unit (AMU): ½ ATR
Power: 28 V DC, 30 W max

Status
In production and in service.

Contractor
DRS Technologies, Inc.

Emergency Avionics Systems (EAS)

Type
Avionic Emergency Locator System (ELS).

Description
DRS Technologies' EAS family of products integrates the functionality of a Flight Data Acquisition Unit (FDAU), Flight Data Recorder (FDR), Cockpit Voice Recorder (CVR) and Emergency Locator Transmitter (ELT), into a deployable beacon offering highly versatile survival and recovery capability. In a deployable system, the beacon increases the survivability of flight and voice data by avoiding the intense destructive forces which occur during a crash by automatically deploying from the airframe away from the accident site.

The beacon, which contains the solid-state crash-protected data recorder memory and ELT, immediately initiates transmission of search and rescue distress signals, providing immediate location of the accident site, thereby assisting early recovery of survivors, protection of valuable flight/voice data and reduced search and rescue and recovery costs. The beacon floats on water indefinitely in the event of an over-water incident. The beacon transmits at a frequency of 121.5 and 243 MHz. As an option, a 406 MHz beacon can be chosen. This allows identification and messaging information to be added to the distress signal to significantly improve localisation of the downed aircraft via the COSPAS/SARSAT system.

The EAS3000 can be configured as a Combi-Cockpit Voice/Flight Data Recorder (CVR/FDR) or optionally a dedicated FDR or CVR. In each configuration the beacon contains the Emergency Locator Transmitter (ELT). An ELT-only configuration (ELB3000) is available for customers who do not require CVR/FDR capability. The ELB3000 can be upgraded with CVR/FDR capability when required. A Cockpit Control Unit (CCU) enables preflight checks to be performed on EAS3000 integrity. The CCU also allows manual deployment of the Beacon Airfoil Unit (BAU).

By combining the recorder and locator functions, installation, operation and maintenance costs are reduced. In addition, the EAS3000 offers significant weight advantages over conventional installations of multiple, fixed onboard systems.

The EAS3000 provides full data acquisition capability enclosed in a ½ ATR short Data Acquisition Unit (DAU) accommodating various inputs, including: ARINC-429; MIL-STD-1553; analogue; discrete; thermocouple; synchro; and RVDT. The DAU contains all sensor interfaces and processor electronics for data inputs plus audio input for the CVR. Provision is made for receiving HUMS data.

Specifications
Cockpit voice recorder
Functionally compliant to ED-56A
4 audio channels, 1h storage per channel
Bulk erase option, ARINC 757 compatible

Flight data recorder
Compliant to ED-55
Selection of data acquisition options MIL-STD-1553 bus; ARINC-429; direct to sensor interface suite: analogue, synchro, discrete, pulse, thermocouple; 25h storage; multiple data storage configurations; ARINC 747 compatible

Recorder system
Solid-state
ED-56A compliant record controls
Download and playback maintenance system allows: FDR data analysis; audo reply: BITE system

Emergency locator transmitter
DO-183 compliant
Operates on 121.5 MHz and 243 MHz
Option for 406 MHz
COSPAS/SARSAT compatible
Automatically deployable
BITE system

Compliance
Transport Canada/Civil Aviation Authority (CAA) appliance approval, including TSOs: C123 (CVR); C124 (FDR); and C91a (ELT)

AN/URT-43 recorder locator system

0051342

EAS3000 (left), cockpit control unit (centre), and data acquisition unit (right)   0015298

Status
Deployable emergency avionics systems are currently installed on the following fixed-wing aircraft and helicopters: C-5, C-9B, C-135, CC-115, CC-130, CP-140, E-3A AWACS, E-4A, E-6A, F/A-18, P-3 (A, B and C variants), T-43A, and Tornado aircraft; 212/412, CH-46, CH-47, Dauphin, EH 101, H-3, HH-1H, HH-1N, Lynx, Puma, S-61, Super Lynx and Super Puma helicopters.

The EAS3000 is fitted on all variants of the EH 101, the Kaman SH-2G for Australia, and the UK Royal Navy Sea King helicopter.

EAS-3000F
The new EAS-3000F is a modular, deployable beacon system that incorporates advanced technology in a single crash-survivable unit. It was developed specifically as a flight safety system for fixed-wing applications, incorporating the latest technology in deployable aircraft monitoring and data acquisition systems. The system integrates a cockpit voice and flight data recorder with a crash-survivable emergency locator beacon for fast recovery of flight data and an increased success rate for search and rescue teams.

The EAS-3000F is released automatically from the aircraft's outer surface during an incident and immediately emits a locator beacon for recovery by search and rescue teams. In water, the system floats indefinitely. On land, it escapes the intense destructive forces that occur during a crash by separating from the aircraft at the time of impact. The recovered data, which provide detailed information of the events during an incident, are utilised for accident investigation, training, aircraft and avionics design and manufacture, and flight safety procedure development.

The design of the EAS-3000F system is based on more than 30 years of experience in deployable data storage technology for emergency beacons, cockpit voice recorders and flight recording systems. DRS has supplied over 4,000 similar units worldwide for combat jet, military and commercial transport and rotary-wing aircraft. The recovery rate of these systems exceeds 95 per cent of the incidents reported. The retrieval rate of data in recovered systems has been 100 per cent.

Status
The EAS-3000F has been selected by the Canadian Department of National Defence to upgrade the capabilities of the CP-140 long-range maritime patrol aircraft.

Contractor
DRS Technologies, Inc.

Czech Republic

SO-69/ICAO airborne transponder

Type
Aircraft transponder.

Description
The SO-69/ICAO airborne transponder is a reconstructed and updated version of the Soviet SO-69 system. It complies with ICAO design standards. The encoder unit S-ICAO and control unit O-ICAO can be modified to support IFF Mk10 modes 1 and 2.

The new transponder can easily be retrofitted in aircraft equipped with the SO-69 transponder. It weighs less than the existing SO-69 equipment and consumes less power.

Test equipment designated ZZ-69/ICAO is available to support the system.

Specifications
Dimensions/Weight:
Transmitter-receiver unit PV-ICAO: 412 × 64 × 160 mm/5.5 kg
Encoder unit S-ICAO: 436 × 86 × 160 mm/2.3 kg
Control unit O-ICAO: 71.5/98 × 112 × 59 mm/0.35 kg
Communication unit BK-ICAO: 217 × 133 × 59 mm/1.1 kg
Frame S0-69: 436 × 155 × 216 mm/2.4 kg
Receiver frequency: 1,030 MHz (bandwidth 6 MHz)
Transmitter frequency: 1,090 MHz
Transmitter power: 250 to 500 W
Modes: A, A/C, B
Power supply: 27 V DC and 115 V AC/400 Hz

Status
The SO-69/ICAO transponder is installed in MiG-23, MiG-29, Su-22 and Su-25 aircraft.

SO-69/ICAO airborne transponder   0002198

Contractor
Elektrotechnika-Tesla Kolin a.s.

France

COSPAS-SARSAT emergency locator transmitters – A06 range

Type
Avionic Emergency Locator System (ELS).

Description
The A06 range of COSPAS-SARSAT emergency locator transmitters includes the following types:
1 For ED-62 requirements; A06-A06V1 models;
2 For TSO requirements; A06T model;
3 For TSO and ED-62 requirements; A06V2-606 models.

The A06 range is packaged in an emergency distress orange container. The transmitter includes a three-postion switch; red high-intensity light and buzzer; antenna connector; three-frequency antenna; remote control (whip, or aircraft skin antenna). Activation is automatic by G-switch.

Transmissions are on 406 MHz COSPAS-SARSAT frequency at 5 W power, and on the 121.5 and 243 MHz international distress frequencies, at 100 mW power.

COSPAS-SARSAT emergency locator transmitters – A06 range   0015340

COSPAS-SARSAT emergency locator transmitters – A06 range   0015341

A remote-control unit option is available for Airbus cockpit fit and as a custom option for other aircraft types. Transmitter weight is 1.3 kg.

An updated variant has been developed for Lufthansa that provides direct identification of the aircraft; aircraft identification data is entered via a new programming module developed for the purpose.

Specifications
Power supply: Lithium batteries providing greater than 48 h at −20°C
Operating temperature: −20 to +55°C
Beacon weight: 1.3 kg
Bracket weight:
(AO6, A06T): 1.3 kg
(AO6V2): 1.6 kg

status
In production and in service.

Contractor
ELTA.

DF-301E direction-finder

Type
Avionic Direction-Finding (DF) system.

Description
The DF-301E (military designation OA-8697 or OA-8697A) is a solid-state direction-finder which utilises digital electronic circuitry to achieve improvements in bearing accuracy, acquisition speed and stability. Operating in the UHF and/or VHF frequency range of 100 to 400 MHz, automatic direction-finding capability is provided within one unit, giving cost, space and weight savings.

Used in conjunction with associated receiver and bearing indicator, the electronically commutated antenna provides relative bearing information to the UHF/VHF signal source transmitter. In the airborne environment, the DF-301E is used for course navigation, or to determine the relative direction of another transmitting aircraft. The unit meets MIL-E-5400 Class 2 requirements for environmental conditions. The unit weighs 3.4 kg and measures 88 × 134 mm.

Status
In production. Over 8,000 units are in operation.

Contractor
Rockwell-Collins France Blagnac.

The Rockwell-Collins France DF-301 EF direction finder   0015343

DF-430F tactical direction-finder

Type
Avionic Direction-Finding (DF) system.

Description
The DF-430F is Rockwell-Collins France's new generation of automatic tactical V/UHF direction-finder.

The Rockwell-Collins France DF-430F tactical direction-finder showing (left to right) the BC-430F control display unit, the RPU-430F receiver processing unit, and the ANT-430F DF antenna   0015344

The DF-430F features an embedded synthesised receiver that covers the 30 to 400 MHz frequency band, thus enabling the DF-430F to function as a full stand-alone tactical direction-finder.

The DF-430F has been designed for simple installation on any type of aircraft (fixed- or rotary-wing). It provides rapid and accurate bearing acquisition.

The DF-430F is a three LRUs system including one DF Antenna (ANT-430F); one Receiver and Processing Unit (RPU-430F); one Control and Display Unit (BC-430F), as an option.

DF-430F main characteristics are, 30 to 400 MHz frequency range; AM or PM antenna modulation according to frequency band; fully solid-state antenna; fast bearing acquisition (50 ms burst); high bearing accuracy; dead reckoning capabilities; remote-control capability via ARINC 429 or MIL-STD-1553B interface; easy DF-301E mechanical retrofit; watertight package; and flush-mount installation.

Specifications
Typical range (up to the line of sight):
30–88 MHz: 100 n miles (20 W)
100–400 MHz: 100 n miles (5 W) ·
Bearing Accuracy
<3° forward/backward axis
<5° for other bearings
Dimensions;
ANT-430F diameter: 278.5 mm; height: 107 mm

RPU-430F: ARINC 600 ¼ ATR short
BC-430F: width: 146 mm; depth: 150 mm; height: 95.5 mm
Weight:
ANT-430F: 2.6 kg
RPU-430F: 3.2 kg, ARINC 600 ¼ ATR short
BC-430F: 0.8 kg

Status
In production and in service in the NH 90 multinational helicopter programme.

Contractor
Rockwell-Collins France, Blagnac.

ELT 90 series Emergency Locator Transmitters (ELTs)

Type
Avionic Emergency Locator System (ELS).

Description
Satori manufactures ELTs:
- **ELT 90-ELT 92** a bi-frequency (121.5/243 MHz) automatic fixed/portable equipment conforming with EUROCAE ED-62/RTCA DO-182
- **ELT 96-406** a tri-frequency (121.5/243/406 MHz) automatic fixed/portable equipment conforming with EUROCAE ED-62/RTCA DO-182

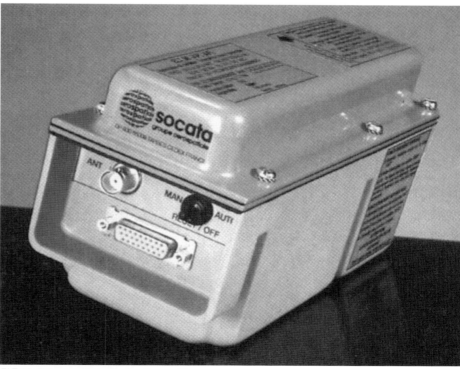

ELT 96-406 emergency locater transmitter   0001194

- **ELT 96S-406** a tri-frequency (121.5/243/406 MHz) automatic fixed/portable equipment conforming with EUROCAE ED-62/RTCA DO-204.

Specifications
Dimensions: 83 × 103 × 215 mm
Weight: 1.52 kg
Output power: ELT 90-ELT 92: 100 mW PERP 48 h
ELT 96 (both models): 5 W PERP 48 h.

Status
In production and in service.

Contractor
Satori SA.

Ground Collision Avoidance System (GCAS)

Type
Ground Proximity Warning System/Terrain Awareness Warning System (GPWS/TAWS).

Description
Designed and manufactured by Thales, the GCAS is jointly marketed and supported worldwide by Teledyne Controls and Thales Avionics SA. It is being marketed to meet airline requirements for Ground Proximity Warning Systems (GPWSs) and Enhanced Ground Proximity Warning Systems (EGPWSs), as well as Terrain Awareness Warning System (TAWS) requirements.

GCAS incorporates a worldwide digital terrain database (like EGPWS) and an alert algorithm that includes aircraft performance potential by incorporating attitude and aircraft type-specific and configuration data (for example, true aircraft flap settings, actual aircraft weight and actual engine performance capabilities). Audio and visual caution and warning messages are similar to those found in GPWS. GCAS is designed as a near form/fit replacement for existing GPWS/EGPWS systems and to be linked to aircraft navigation or GPS data. GCAS reverts to basic GPWS mode if a reliable navigation signal is lost. The digital GCAS can be operated without a display, a feature intended to encourage retrofits on older aircraft, although it is intended to provide display data on appropriate weather radar or EFIS screens, using ARINC 453 protocols. An upgraded 3-D display for future aircraft, such as the Airbus A380, is planned.

GCAS calculates aircraft time to clear obstacles, rather than time to impact. In making its calculations, GCAS demands no more than 0.5 g manoeuvre, 75 per cent airframe load limit and 90 per cent take-off power.

Modes of operation include: Collision Prediction and Alerting (CPA); true look-ahead capability based on predicted aircraft flight path and knowledge of terrain environment; advanced and reliable ground hazard warning and caution; and alert computation based on aircraft climb capability.

Basic modes of operation include: (ICAO Annex 6) modes 1 and 2 (back-up mode); modes 3, 4 and 5; call-outs and bank angle modes.

Specitfications
Dimensions: 2 MCU
Weight: 2.8 kg
Power: 28 V DC, 25 W or 115 V AC

Status
The system commenced test and evaluation at the French Flight Test Centre in an Airbus A300 simulator during 1995 and on board a Dassault Falcon 2000 in 1997. Initial certification in the Falcon 2000 was in 1998 under European Joint Airworthiness Authority (JAA) rules, followed by US FAA certification. The system became available for commercial air transport users during 2000. GCAS is designed to meet the future FAA NPRM and ICAO Amendment of Annex 6 for TAWS and be compliant with TSO C-151.

Contractor
Thales Systemes Aeroportes SA.

KANNAD 406/121 series automatic Emergency Locator Transmitters (ELTs)

Type
Avionic Emergency Locator System (ELS).

Description
Martec manufactures a range of ELTs for land, sea and air application. The KANNAD 406 series provide services covering the three international distress frequencies: 406.025/121.5/243 MHz. The company also produces the KANNAD 121 AF, a two-frequency (121.5/243 MHz) unit for light aircraft applications.

General aviation ELTs
KANNAD 406 AF
The KANNAD 406 AF automatic fixed Cospas-Sarsat ELT is designed to be installed near the tail of the aircraft, and to be connected to an outside antenna; it is a three-frequency (406.025/121.5/243 MHz) unit; overall dimensions: 181 × 107 × 93 mm; weight: 1.18 kg; six-year battery replacement life. A remote-control panel (on option) located in the cockpit allows manual activation and self-test of operational parameters. A buzzer mounted in the ELT warns the pilot should an activation occur. Programming can be done automatically via the ELT front panel, to define nationality and registration marking and ELT serial number up to 4096, aircraft ICAO 24 bit address, aircraft serial number. Activation can be initiated by: automatic G-switch, manual operation, or remotely (from the cockpit). The unit is certified to EUROCAE ED62 and FAA TSO-C91a and JTSO-2C126/TSO-C126.

KANNAD 406 AF-H
The KANNAD 406 AF-H automatic fixed Cospas-Sarsat ELT is designed for flat installation onboard helicopters (rather than the normal 45° installation). It is a three-frequency (406.025/121.5/243 MHz) unit; overall dimensions: 181 × 107 × 93 mm; weight: 1.19 kg; six-year battery replacement life. A remote-control panel (on option) located in the cockpit allows manual activation and self-test of operational parameters. A buzzer mounted in the ELT warns the pilot should an activation occur. Programming can be done automatically via the ELT front panel, to define nationality and registration marking, aircraft designation and ELT serial number up to 4096, aircraft ICAO 24 bit address, aircraft serial number. Activation can be initiated by: automatic G-switch, manual operation, or remotely (from the cockpit). The unit is certified to EUROCAE ED62 and FAA TSO-C91a and JTSO-2C126/TSO-C126.

KANNAD 406 AF ELT   0051582

KANNAD 121 AF
The KANNAD 121 AF automatic fixed ELT is a two-frequency (121.5/243 MHz) unit designed for light aircraft, for installation near the tail, connected to an external antenna; it has a Morse capability; overall dimensions: 181 × 107 × 93 mm; weight: 1.18 kg; six year battery replacement life. A remote-control panel (on option) located in the cockpit allows manual activation and self-test of operational parameters. A buzzer mounted in the ELT warns the pilot should an activation occur. Programming can be done automatically via the ELT front panel, to define aircraft tail number in Morse code. Activation can be initiated by: automatic G-switch, manual operation, or remotely (from the cockpit). The unit is certified to EUROCAE ED62 and FAA TSO-91a and JTSO-2C126/TSO-C126 capability.

Air transport aviation ELTs
KANNAD 406 AP
KANNAD 406 AP automatic portable Cospas-Sarsat ELT is designed to be installed near the tail of the aircraft and to be connected to an external antenna, when removed and connected to the auxiliary antenna, the KANNAD 406 AP becomes a survival beacon. It is a three-frequency (406.025/121.5/243 MHz) unit; overall dimensions: 285 × 107 × 93 mm; weight: 1.290 kg; six-year battery replacement life. A GPS or ARINC 429 interface can be added to load the position of the aircraft in the ELT. In case of activation, the position is transmitted to the Cospas-Sarsat LEO and GEO satellites, together with aircraft identification. A remote-control panel (on option) located in the cockpit allows manual activation and self-test of operational parameters. A buzzer mounted in the ELT warns the pilot should an activation occur. Programming can be done automatically via the ELT front panel, to define nationality and registration marking, aircraft designator and ELT serial number up to 4096,

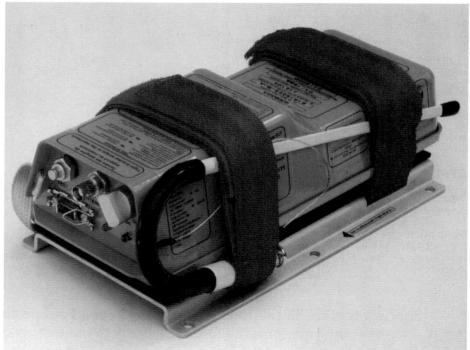

KANNAD 406 ATP ELT   0051583

aircraft ICAO 24 bit address, aircraft serial number, activation can be initiated by: automatic G-switch, manual operation, or remotely (from the cockpit). Certified to EUROCAE ED62 and FAA TSO-C91a/JTSO-2C126/TSO-C126 capability; it also complies with JAR-OPS 1.820 regulation.

KANNAD 406 ATP

The KANNAD 406 ATP automatic Cospas-Sarsat ELT is designed to be installed near the tail of the aircraft and to be connected to an external antenna, when removed and connected to the auxiliary antenna, the KANNAD 406 AP becomes a survival beacon. It is a three-frequency (406.025/121.5/243 MHz) unit; overall dimensions: 300 × 150 × 95 mm; weight: 2.155 kg; six-year battery replacement life. In case of activation, the position is transmitted to the Cospas-Sarsat LEO and GEO satellites, together with aircraft identification. A remote-control panel (on option) located in the cockpit allows manual activation and self-test of operational parameters. A buzzer mounted in the ELT warns the pilot should an activation occur. Programming can be done automatically via the ELT front panel, to define nationality and registration marking, aircraft designator and ELT serial number up to 4096, aircraft ICAO 24 bit address, aircraft serial number. Activation can be initiated by: automatic G-switch, manual operation, or remotely (from the cockpit). The unit is certified to EUROCAE ED62 and FAA TSO-C91a/JTSO-2C126/TSO-C126 capability.

Remote-control units

SERPE-IESM offers two remote-control units for cockpit control of the KANNAD 406 ELT units: RC 200 for fitment in light aircraft and general aviation types; RC 300 for use in air transport aircraft.

Antennas

Rod and blade antennas manufactured by Rayan (Chelton Group) and a blade antenna manufactured by Sensor Systems are offered with the KANNAD 406 ELTs; rod antennas are used for low speed operation, blade antennas for high speed operation.

Status

In service.

Contractor

Martec.

MDF-124F(V2) direction-finder

Type

Avionic Direction-Finding (DF) system.

Description

Recent miniaturisation of technology together with the experience gained, especially in maritime surveillance operation, have led Rockwell-Collins France to develop and release a new version of MDF-124F called MDF-124F(V2).

MDF-124(V2) is a full stand-alone SAR and tactical direction-finder; its embedded synthesised receiver allows it to operate the V/UHF 100–407 MHz frequency range without any external receiver.

MDF-124F(V2) comprises: MDF-124F(V2): antenna and receiver unit; BC-124F(V2): control and display unit.

MDF-124F(V2) features the following direction-finding capabilities: direct access of any of the three international distress frequencies or auto-scanning/alert on these frequencies; direct access to ARGOS channel; direct access to VHF-FM maritime channels 16 and 70; direct access to manual/preset user's frequencies in DF Tactical mode; combined tactical/auto alert SAR modes allowing DF operation on one user's tactical manual/preset channel while still monitoring the three international distress frequencies. Other characteristics are identical to the MDF-124F.

Specifications

Dimension, weight, interfaces and power supply are the same as for the MDF-124F.

Status

In production for Australia, Belgium, China and Japan.

Contractor

Rockwell-Collins France, Blagnac.

NRAI-7(.)/SC10(.) Identification Friend-or-Foe (IFF) transponder

Type

Identification Friend-or-Foe (IFF) system.

Description

The NRAI-7 is a solid-state Mk XII diversity transponder which inhibits replies to interrogator sidelobe transmissions and automatically codes special replies to provide assistance in the position identification of particular aircraft and in emergencies. The diversity function is provided by a dual receiver with inputs connected to upper and lower antennas, a system for comparison of received signals and an antenna switch that directs the response to the antenna that has received the strongest interrogation signal. This allows more accurate identification, particularly during aircraft manoeuvres which can blanket or interrupt signals. The pilot may also insert codes such as radio failure alert and warning of hijackers aboard. It is available in one- or two-box housing (see NRAI-9A). The single-box version is claimed to be one of the smallest transponders in the world. A naval version is also available.

Specifications

Peak power: 500 W
Sensitivity: −77 dBm
Frequency: 1,030 MHz (receive); 1,090 MHz (transmit)
Modes available: 1, 2, 3A/C and Mode 4 capability
No of codes: 32 (Mode 1); 4,096 (Modes 2 and 3/A); 2,048 (Mode C)
Dimensions: 130 × 127 × 145 mm
Weight: 3 kg

Thales' NRAI-7(.)/SC10(.) IFF transponder
0503704

Status

NRAI-7 transponders are reported to have been installed on Mirage 2000, Mirage F-1, Mirage III, Mirage IV, AS 332 and AS 335, C130 Transall and various other aircraft as well as on board several ships of the French and other navies. It has also been integrated in various Polish aircraft, including MiG fighters. Over 2,000 systems have already been delivered to the French defence forces and those of other countries.

Contractor

Thales Communications.

NRAI-9(.)/SC15(.) Identification Friend-or-Foe (IFF) transponder

Type

Identification Friend-or-Foe (IFF) system.

Description

The NRAI-9A is essentially a two-box version of the NRAI-7 which incorporates a number of improvements. These include the elimination of sidelobe response, and automatic special-code referral, with positive identification permitting a ground operator to locate a particular aircraft. Special emergency codes may also be employed, chosen by the pilot, such as radio failure. Dual-receiver channels connected to upper and lower antenna and comparison circuits provide a diversity function.

Specifications

Frequency: 1,030 MHz (receive); 1,090 MHz (transmit)
Power output: 500 W peak
Modes available: 1, 2, 3A/C and Mode 4 capability
Codes: 32 (Mode 1); 4,096 (Modes 2 and 3/A); 2,048 (Mode C)
Dimensions: 58 × 193 × 361 mm (transmitter/receiver); 127 × 130 × 80 mm (control unit)
Weight: 2.5 kg (transmitter/receiver); 1.4 kg (control unit)

Status

In service. The NRAI-9 is believed to be fitted to ATL2 maritime patrol aircraft.

Contractor

Thales Communications.

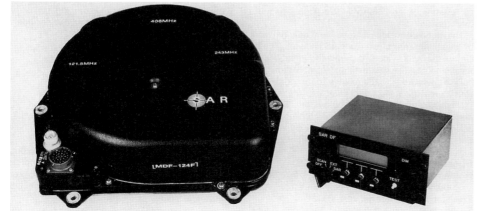

MDF-124F/MDF-124F(V2) direction-finder, showing (left) the antenna/receiver unit and (right) the control unit
0001311

Thales' NRAI-9(.)/SC15(.) IFF transponder
0503705

NRAI-11/IDEE 1 Mk XII interrogator-decoder

Type
Identification Friend-or-Foe (IFF) system.

Description
The airborne NRAI-11/IDEE 1 interrogator-decoder is used to identify a friendly target and determine its range and azimuth. The interrogator-decoder and its control box are integrated in the IFF system of Dassault Mirage 2000 aircraft. The system includes all the functions needed for IFF air-to-air identification and includes an encoder, transmitter, RF switch, two receivers, analogue processing unit, defruiter, passive decoder, evaluator, extractor, staggering circuit and automatic self-test.

Volume and power consumption have been substantially reduced through the use of a solid-state transmitter, low-power integrated and monolithic high-density circuits, hybrid and custom LSI circuits and switching power supply.

Specifications
Dimensions: 124 × 80 × 194 mm
Weight: 12 kg
Power supply: 115 V AC, 400 Hz, 3 phase, 150 VA
Transmission power: 1 kW peak
Receiver sensitivity: −79 dBm between 1,087 and 1,093 MHz
Operating modes: 1, 2, 3/A and Mode 4 capability

Status
In production and in service on the Dassault Mirage 2000.

Contractor
Thales Communications.

Thales' NRAI-11/IDEE 1 Mk XII interrogator-decoder is in service in the Mirage 2000 0504112

RSC-125F personnel locator system

Type
Avionic Direction-Finding (DF) system.

Description
RSC-125F is a personnel locator system which provides combat SAR platforms with the advanced capabilities required for the very demanding combat SAR mission.

This equipment has been developed to recover downed pilots or aircrew members using personal locator beacons.

RSC-125F is a full stand-alone system, with embedded receiver, which can be installed on both fixed- and rotary-wing platforms. Different interfaces give RSC-125F great flexibility and allow it to be integrated within ARINC-429 or MIL-STD-1553B architectures.

RSC-125F also includes tactical DF capabilities in the 30 to 400 MHz band and civil SAR frequencies, when operating all International distress frequencies.

RSC-125F is a three LRUs system including one DF antenna (ANT-430F); one Localisation Processor and Transmitter unit (LPT-125F); one control and display unit (BC-125F) (optional).

The main features of RSC-125F are compatibility with existing AN/PRC-112 and AN/PRC-434 Personal Locator Beacons (PLB); civil SAR (121.5–243 MHz-406.025 MHz COSPAS-SARSAT, VHF/FM maritime channels 16 and 70) and Argos capabilities; tactical DF in the 30 to 400 MHz band (25 kHz step); automatic scanning function on distress frequencies; remote-control capabilities through ARINC 429 or MIL-STD-1553B interfaces; NVG compatible; high-accuracy bearing and distance information on beacon; computed bearing during gap between the interrogations; easy flush-mount installation.

Specifications
Dimensions:
ANT-430F: diameter: 278.5 mm, height: 107 mm
LPT-125F: ARINC-600 1/4 ATR short
BC-125F: 146 × 66.6 × 150 mm
Weight:
ANT-430F: 2.6 kg
LPT-125F: 4.5 kg (ARINC 600 1/4 ATR short)
BC-125F: 0.8 kg

Status
In production and in service in French military aircraft.

Contractor
Rockwell-Collins France, Blagnac.

Spectrum Airborne Surveillance (SAS) system

Type
Avionic Direction Finding (DF) system.

Description
The Thales SAS system is designed to intercept and localise sophisticated radio transmitters throughout the civil and military radio spectrum. The system's radio direction-finders cover frequencies from 20 kHz to 3,000 MHz, while its surveillance capability covers 300 kHz to 3,000 MHz.

Using modular architecture, the all-digital SAS system comprises a network of fixed antennas mounted within the flight deck; an interception/direction-finding antenna system mounted externally on the aircraft; one or more Windows NT™-based workstations, and recording devices. The technical analysis facilities of the SAS system are designed to enable it to identify the transmissions intercepted.

Thales has developed an entirely new network of patch antennas for the SAS system. Thales claims that this system is particularly suitable for localising radio transmissions from commercial cellular networks, that it is easy to integrate, and that it provides a high degree of stealth.

Designed as an open-ended system, SAS can also integrate and merge other data from optical and infra-red sensors and records all mission data for debriefing and further investigation.

Specifications
Frequency range: 20–3,000 MHz (DF); 300 kHz to 3,000 MHz (monitoring)
Precision of DF antenna: <1.5° RMS
Scanning rate: 40-2,000 MHz
Demodulation: A1A, A1B, A2A, A2B, A3E, J2A, J2B, J3E, J7B, F1A, F1B, F1C, F3E, F7B, H3E, R3E, B8E

RSC-125F Personnel locator system, showing BC-125F (left) LPT-125F (centre) and ANT-430F (right)
0001310

Operational sensitivity: 3 µV/m typical (10 mV at 180 km)
Dimensions: 3U rack standard 19 in + compact PC; 134 × 485 × 520 mm
Weight: 45 kg
Power supply: 115 V/230 V AC
Power consumption: 300 W

Status
The system has been reported installed in at least one F406 Vigilant aircraft.

Contractor
Thales Communications.

TCAS 2000 receiver transmitter unit

Type
Traffic Alert and Collision Avoidance System/Aircraft Collision Avoidance System (TCAS/ACAS).

Description
TCAS 2000 is intended for business, commuter and large commercial passenger aircraft, with further military applications. It is available in 6-MCU and 4-MCU sizes and will accept either 115 V AC or 28 V DC power inputs and operate in unpressurised environments. The RT-950 (6-MCU) and RT-951 (4-MCU) units contain dual-microprocessors to implement the surveillance and collision avoidance functions. They then drive the displays that give instructions to the flight crew.

TCAS 2000 has increased display range up to 100 nm to meet CNS/ATM requirements with passive surveillance up to 120 nm. Variable EFIS display ranges are 5, 10, 20, 40 to 100 nm. It can track up to 50 aircraft (24 within 5 nm) that have up to 1,200 kt closing speed and up to 10,000 ft per minute vertical speed and it can provide escape manoeuvre co-ordination. This co-ordination includes normal manoeuvres (climb or descend with vertical speed limits) and enhanced manoeuvres (increased rate of climb or descent and reversed direction of climb or descent). It has air-to-ground datalink, a traffic advisory/resolution advisory data recorder and a level IV transponder, upgradeable to level V.

Specifications
Dimensions: 194.6 × 124.5 × 401.1 mm
Weight: 6.6 kg
Power:
115 V AC, 400 Hz or 28 V DC
55 W standby, 65 W nominal, 100 W maximum

Status
TCAS 2000 is in service with the Airbus family of aircraft, Boeing aircraft (including McDonnell-Douglas airframes) and Fokker aircraft. It is also in use by aircraft of the US Marines, US Navy and USAF.

Contractor
Thales Avionics SA.

TCAS Resolution Advisory/ Traffic Advisory (RA/TA) VSI

Type
Traffic Alert and Collision Avoidance System/ Aircraft Collision Avoidance System (TCAS/ ACAS).

Description
Thales Avionics' Resolution Advisory/Traffic Advisory (RA/TA) VSI combines the vertical speed and Traffic Alert and Collision Avoidance System (TCAS) information on a 3-ATI instrument using an active matrix-type full-colour display. The instrument is capable of acquiring digital, analogue and pneumatic signals, enabling Thales

to propose a unique part number for a given airline. The RA/TA VSI is Thales' first application of its flat-panel instrument programme. This programme aims to replace all the 3 ATI and 4 ATI electromechanical instruments.

Status
In production and in service. Over 50 airlines and aircraft manufacturers of the air transport, regional and business aircraft segments have chosen the 3-ATI Vertical Speed Indicator/Traffic Alert and Collision Avoidance System (VSI/TCAS) liquid crystal flat-panel instrument.

Contractor
Thales Avionics SA.

Thales' TCAS RA/TA VSI is shown at the top left of this group
0504197

TSB 2500 Combined IFF Interrogator and Transponder (CIT)

Type
Identification Friend-or-Foe (IFF) system.

Description
The TSB 2500 consists of two LRUs: the Combined Interrogator/Transponder (CIT) and the Antenna Control Unit (ACU) or the Antenna Adaptor Unit (AAU). The system is compatible with various types of electronically (with ACU) or mechanically (with AAU) scanned antennas. The TSB 2500 is available in both interrogator and interrogator/transponder versions and is a highly flexible system capable of meeting the needs of many military platforms. It is modular in both design and operation.

It is a Mk XII and Mode S level 2/3 transponder and a Mk XII interrogator. It also incorporates provisions for the integration of future mode 5 new-generation IFF, for both transponder and interrogator functions.

The CIT integrates all transmission, reception, signal, and data processing functions required by an interrogator/transponder. It interfaces with the host platform via a MIL-STD-1553B databus.

The ACU controls electronically scanned antennas. The AAU functions as a booster and an RF front end. The packaging of ACU and AAU in separate boxes allows their installation close to the antenna system providing minimum RF losses in the cables.

The TSB 2500 can interface with any NSA Mode 4 crypto computer or secure crypto computer (for non-NATO applications) that is interface compliant with STANAG 4193. It can also be fitted with a dual KIT/KIR appliqué crypto computer on the front panel.

Specifications
Interrogator
Power: >32 dBW
Frequency: 1.030 ±0.2 MHz
Operating modes: 1,2,3/A,C,4 (Mode S upgradable)

Transponder
Power: 500 W (±2 dB)
Frequency: 1.090 ±0.5 MHz
Operating modes: 1,2,3/A,4,S (Mode S upgradable)
Dimensions
(CIT) 228.6 × 157.2 × 193.5 mm
(ACU) 230 × 115 × 105 mm
(AAU) 32 × 193 × 290 mm
Weights
(CIT) <10 kg
(ACU) 5.5 kg
(AAU) <4 kg

Status
In production and in service in Rafale and Erieye aircraft and the NH 90 helicopter.

Contractor
Thales Communications.

TSC 2000 and TSC 2050 IFF Mk XII/Mode S diversity transponders

Type
Identification Friend-or-Foe (IFF) system.

Description
The TSC 2000 and 2050 IFF systems are true diversity Mk XII and Mode S level 3,

TCAS-compatible transponders, that are fully compliant with ICAO Annex 10, STANAG 4193 and DOD-AIMS-65-100B standards. Moreover they include all provisions for the integration of Mode 5 for a new-generation IFF, in accordance with NATO recommendations, and with any NSA Mode 4 crypto computer or secure crypto computer (for non-NATO applications) that is interface compliant with STANAG 4193. Optionally it can be provided with mechanical adaptor to fit the crypto computer as an appliqué.

The TSC 2000 was co-developed by the then Thomson-CSF Communications and DaimlerChrysler Aerospace AG (now Thales Communications and EADS Deutschland).

The TSC 2050 has been designed to be shelf mounted in non-pressurised zones and is intended for both advanced aircraft or for retrofit purposes. It is provided with a MIL-STD-1553B bus interface, or can be operated through a control box.

Specifications
Dimensions: 136 × 124 × 212 mm
Weight: <5.2 kg
Modes: 1,2,3/A,C,4 and S level 3

Status
In production. The TSC 2000 has been selected for both the Franco-German Tiger helicopter and for the TTH (Tactical Transport Helicopter) and NFH (NATO Frigate Helicopter) versions of the NH 90 by Eurocopter Deutschland. The TSC 2050 has been selected by the UK for the Nimrod maritime patrol aircraft and for global retrofit of Romanian aircraft.

Contractor
Thales Communications.

TSX 2500 series of IFF interrogators

Type
Identification Friend-or-Foe (IFF) system.

Description
The TSX 2500 has been ordered to equip the Rafale aircraft and the Erieye AEW&C system.

A derivative version has been chosen by Agusta to equip the NATO Frigate Helicopter (NFH) version of the European NH 90 military transport helicopter, where it will operate with the European Navy Radar (ENR), a mechanically scanned radar developed by the European consortium comprising Thales, EADS and Galileo Avionica SpA.

The latter contract follows the award from Eurocopter Deutschland for the supply of TSC 2000 Mode S IFF transponders for the TTH (Tactical Transport Helicopter) and NFH versions of the NH 90.

Contractor
Thales Communications.

The TSX 2500 IFF interrogator equips the NH 90 NFH helicopter
0080279

Germany

ATC 3401 Mode A/C transponder

Type
Aircraft transponder.

Description
The single-block unit ATC 3401 is certified in accordance with FAA TSO C47c, Class 1A, for the highest level of unrestricted service and can report altitudes up to 62,700 ft.

The ATC 3401 transponder displays 4,096 identification codes in its left window and can display its operating mode or flight level reporting altitude in its right window. A stored VFR code, such as 1200, can be quickly recalled with a VFR push-button. Provisions are included to enable IDENT to be controlled from a button on the control stick and for power to be turned on and off from an external control, if required.

The complete system is housed in a single compact unit, which complies with ARINC standards.

It does not require remote boxes, interconnecting cables, or external forced-air cooling.

To facilitate dual-transponder installations, provisions are included for automatic transfer of one transponder to the standby mode, when the other transponder is selected for normal mode operation. This prevents both transponders transmitting at the same time.

The ATC 3401 Mode A/C transponder is a member of the Becker PrimeLine family of avionics equipments.

Specifications
Dimensions: 146 × 47.5 × 217 mm (not including clearance for connectors and cables)
Weight: 1.2 kg
Power supply: 10-32 V DC; at 28 V DC, 0.8 A normal, 0.3 A standby; at 14 V DC, 1.5 A normal, 0.5 A standby
Transmitter frequency: 1,090 MHz ±0.3 MHz
Receiver frequency: 1,030 MHz ±0.2 MHz
Modes: A, A + C
Power output: 250 W minimum

The Becker Avionic Systems ATC 3401 Mode A/C transponder   0081497

Altitude: 35,000 ft unpressurised, unlimited when pressurised
Operating temperature: –20 to +55°C
Environmental: DO-160C/ED-14L
Software: DO-178A/ED-12A, level c

Status
In service.

Contractor
Becker Avionic Systems.

SAR-ADF 517 Search And Rescue – Aircraft Direction-Finder

Type
Avionic Direction Finding (DF) system.

Description
Becker Avionic Systems' new SAR-ADF 517 is able to locate beacons transmitting on the 406.025 MHz COSPAS/SARSAT emergency frequency, as well as the VHF (121.5 MHz) and UHF (243 MHz) SAR frequencies.

The 406.025 MHz frequency transmits a 450 ms digital pulse every 50 s. Traditional ADF systems rely on a continuous swept tone from the beacon to ensure reliable homing, but the digital pulse from the 406.025 MHz beacon is too brief for the ADG to secure lock. Accordingly, Becker changed its rescue navigation aids to the new SAR-ADF 517 version, which is equipped with the following frequencies: 121.5 MHz, plus an adjustable training frequency near 121.5 MHz; Channel 16 for sea rescue; 243 MHz, plus an adjustable training frequency near 243 MHz; 406.025 MHz, plus a training frequency.

In the 406.025 MHz mode, either 121.5 or 243 MHz can be selected to cover automatically the 49.5 s lapse time between two digital pulses.

The hardware for the system comprises an 80 mm diameter control and display unit connected to a remote antenna mounted on the underside of the aircraft.

Specifications
Method of bearing: Doppler principle (frequency of rotation 3 kHz, cw/ccw)
Accuracy: ±5° rms
Received frequency:
(VHF) 121.500 MHz, 123.100 MHz
(channel 16 marine band) 156.8 MHz
(UHF) 243.000 MHz, 243.500 MHz
(COSPAS/SARSAT) 406.025 MHz, 406.028 MHz
Polarisation: vertical
Polarisation error: <5° at 60° vectorial field rotation
Cone of confusion: approximately 30° measured to the vertical
ELT identification: by direction of audio sweep tone, frequency range 300 to 1,600 Hz and repetition rate 250 to 500 ms
Weights:
(display unit) 0.25 kg
(DF antenna) 2.0 kg
Dimensions:
(display unit) 82 (W) × 82 (H) × 35 (D) mm
(DF antenna) 270 (diameter) × 185 (L) mm
Power: 12–32 V DC, 400 mA

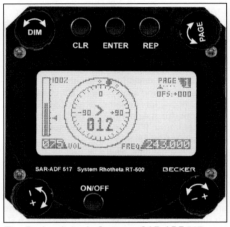

The Becker Avionic Systems SAR-ADF 517   0062185

Status
In service.

Contractor
Becker Avionic Systems.

STR 700 IFF transponder

Type
Identification Friend-or-Foe (IFF) system.

Description
The STR 700 IFF transponder is used by nearly all German armed forces aircraft. It features high reliability, small dimensions, low weight and diversity of operations.

The STR 700 meets AIMS specifications. It consists of two basic units: the receiver/transmitter and the control unit with logic section. The control unit has interfaces for the Mode 4 decoder/coder facility and for an altitude encoder. The STR 700 has been specifically designed for applications where diversity operation is required and is provided with two receiving channels. It can also, however, be equipped with a single receiving channel only. The receiver/transmitter of modular design is accommodated in a standard 1/2 ATR short case for both models. Usually the control unit consists

of the actual control section, with switches and lamps and of the logic section for decoding the interrogation signals and for coding the reply signals. If the available space in the cockpit is too narrow, the logic section can be accommodated separately from the control section.

Reliable identification of the target aircraft depends largely on the correct action of the transponder. Consequently the STR 700 transponder is provided with a large number of test circuits, to reveal the functional condition at any time automatically by internal interrogations. For system checking on the ground, the receiver/transmitter of the two-channel version is provided with 12 LEDs to indicate proper performance of the individual subassemblies.

Specifications
Dimensions:
(receiver/transmitter) 124 × 193 × 382 mm
(control unit/logic section) 146 × 134 × 155 mm

Weight:
(receiver/transmitter) 9.8 kg (single channel)
10.8 kg (two channel)
(control unit/logic section) 3.2 kg
Power supply: 16–32 V DC, up to 70 W
Frequency:
(receive) 1,030 ±1.5 MHz
(transmit) 1,090 ±3 MHz

Status
In production and service in aircraft of the German armed forces.

Contractor
EADS Deutschland, Systems and Defence Electronics, Airborne Systems.

India

400AM IFF transponder

Type
Identification Friend-or-Foe (IFF) system.

Description
The 400AM IFF Mk X operates in Modes 1, 2 and 3A/C, with 4,096 codes to full ICAO standards.

Specifications
Power supply: 27.5 V DC (nominal)
115 V AC, 400 Hz, 90 W (option)
Temperature range: −55 to +55°C
Altitude:
(pressurised) up to 70,000 ft
(unpressurised) up to 40,000 ft

Status
In service.

Contractor
Hindustan Aeronautics Ltd, Avionics Division.

Hindustan's 400 AM IFF transponder   0044966

405A IFF transponder

Type
Airborne multimode radar.

Description
The 405A IFF transponder is all-solid-state and of modular construction. It provides automatic replies to appropriate ground or airborne interrogators operating on the Mk X IFF system, transmitting on 1,090 MHz and receiving on 1,030 MHz. It operates in Modes 1, 2, 3A and 3C on the full 4,096 codes.

Specifications
Dimensions:
(transponder) 122 × 201 × 201 mm
(control unit) 146 × 81.5 × 70 mm
Weight:
(transponder) 9 kg
(control unit) 0.6 kg
(mounting tray) 1.4 kg
Power supply: 115 V AC, 400 Hz, 150 VA (max)

Status
In service.

Contractor
Hindustan Aeronautics Ltd, Avionics Division.

Hindustan's 405A IFF transponder   0044967

ARC 1610A automatic direction-finder

Type
Avionic Direction-Finding (DF) system.

Description
The ARC 1610A automatic direction-finder provides bearing information of the known ground beacons operating in medium frequency along with code reception to identify the selected ground beacons.

The ADF system ARC 1610A is a compact lightweight system which provides accurate, stable bearing information with a high degree of reliability and maintainability using hybrid technology. The system is compatible with ARINC 570.

Specifications
Frequency range: 190–1,700 kHz
Frequency indicator: LED display
Modes of operation: ADF and ANT
Bearing accuracy: ±2°
Hunting: ±2°
Bearing resolution: 6 s
Power input: 27.5 V DC, 2 A
Preset channels: 10
Audio output: 100 mW across 600 Ω
Dimensions:
(receiver) 250 × 195 × 90 mm
(controller) 163 × 14.6 × 66 mm

Hindustan's ARC 1610A automatic direction-finder   0051591

Weight:
(receiver) 4 kg
(controller) 1 kg

Status
In production and in service.

Contractor
Hindustan Aeronautics Ltd, Avionics Division.

IFF 1410A transponder

Type
Identification Friend-or-Foe (IFF) system.

Description
The IFF 1410A transponder system has been designed for operation with a stand-alone control unit, or from a centralised controller through MIL-STD-1553B bus. The equipment uses all-solid-state technology hybrid modules to achieve high

Hindustan's IFF 1410A transponder   0011893

reliability and ease of maintenance. Full operation in accordance with ICAO Annex 10 is provided. Extensive self-test diagnostic features help in identifying the faults to functional level. A secure mode of operation is available as an option.

Specifications
Operation modes: 1, 2, 3/A, C, Secure
Codes: 4,096 in Modes 1, 2, 3/A
2,048 in Mode C
Indentification facility
Military emergency facility
Power supply:
(AC) 108–118 V, 400 Hz, single phase, 0.6 A (max)
(DC) 22–32 V, 0.5 A (max)
Dimensions: 310 × 100 × 194 mm (18 mm extra height with mounting tray)
Weight:
(without mounting tray) 6.3 kg
(with mounting tray) 7.3 kg

Status
The IFF-1410A is in service on a number of aircraft types, including the MiG-21 BIS, Su-30, ALH, Jaguar and IJT.

Contractor
Hindustan Aeronautics Ltd, Avionics Division.

UHS 190A UHF homing system

Type
Avionic VHF/UHF homing system.

Description
The UHS 190A UHF homing system for fixed-wing aircraft and helicopters is used in

conjunction with the normal communications transceiver for locating ground transmitters and personal rescue beacons. It operates in the frequency range 225 to 399.975 MHz with two UHF antennas. The system provides an accuracy of ±5° for homing. The system has the capability to drive two homing indicators.

The UHS 190A consists of a UHF homing adaptor for receiving input signals from the two UHF antennas, homing controller for processing the ADF audio signal from the V/UHF communication set and a homing indicator for indicating the relative bearing of the ground station.

Specifications
Dimensions:
(adaptor) 146 × 98 × 32 mm
(controller) 146 × 51 × 105 mm
(indicator) 10 × 100 × 50 mm
Weight:
(adaptor) 0.3 kg
(controller) 0.55 kg
(indicator) 0.45 kg
Power supply: 22–31 V DC

Status
The UHS 190A is fitted on Indian Navy Sea King, Chetak, Il-38 and Dornier aircraft.

Contractor
Hindustan Aeronautics Ltd, Avionics Division.

International

IFF for Eurofighter Typhoon

Type
Identification Friend-or-Foe (IFF) system.

Description
EADS Deutschland, Marconi Selenia (Italy), Raytheon Systems Limited (UK) and INDRA Sistemas (Spain) have teamed to design, develop and produce the IFF interrogator and IFF Mode S transponder for the Eurofighter Typhoon. The system is compatible with IFF Mk XII, as defined by STANAG 4193.

EADS is also offering a derivative system for the NH 90-NFH helicopter.

Specifications
IFF Mode S transponder
Dimensions: 1/2 ATR intermediate length
Weight: 9 kg
Characteristics:
Mk XII (mode 1,2,3/A,C,4) compliant with STANAG 4193
Mode S
MIL-STD-1553B
Embedded cryptographic module
ADA software
Automatic Code Change (ACC)

IFF for the Eurofighter Typhoon (EADS) 0594528

Growth to Next-Generation IFF (NGIFF)
BIT modes (P-BIT, C-BIT, I-BIT)

IFF interrogator
Dimensions: 1/2 ATR intermediate length
Weight: 11.4 kg
Characteristics:
Mk XII (mode 1,2,3/A,C,4) compliant with STANAG 4193
MIL-STD-1553B
Embedded cryptographic module
Monopulse processing
ADA software
Automatic Code Change (ACC)
BIT modes (P-BIT, C-BIT, I-BIT)

Status
In production and in service in the Eurofighter Typhoon.

Contractor
EADS Deutschland GmbH.
INDRA Sistemas SA.
Marconi Selenia.
Raytheon Systems Limited.

NGIFF (New-Generation IFF)

Type
Identification Friend-or-Foe (IFF) system.

Description
The identification system used by NATO has long ceased to satisfy military requirements. Already in the mid-1980s a standard agreement, STANAG 4162, for a new identification system designated NIS (NATO Identification System) was produced. Latterly, STANAG 4162 has been incorporated into STANAG 4193, ratified by more than eight nations, and promulgated. A further complication has been the civil aviation authorities' request for inclusion of Mode S functionality in military aircraft. Therefore, independent of NGIFF, transponders must include

Mode S functionality from 01 January 2003 for new-build aircraft, and from 01 January 2005 for retrofit application. These measures are to comply with Air Information Circular 13/92.

Status
France and Germany have decided to co-operate in this effort and to split the programme into two phases with the following content:
• Phase 1: Development and procurement of MK X, XII, Mode S and NGIFF upgradeable transponders
• Phase 2: Definition of Mode S NGIFF upgradeable interrogators; Crypto, NGIFF modules for transponder and frequency supportability.

EADS Deutschland GmbH and Thales have agreed to create a combined development company (EISYS GmbH) for performance of the agreed programme.

The German version of the advanced IFF transponder (STR2000) will be installed in the Tornados of the German Air Force as well as in Tiger helicopters and transport aircraft. The French version (TSC 2000) will be used for various French helicopters and transport aircraft with first installation on Tiger and Cougar Resco.

Contractor
EISYS GmbH, a combined development company of:
EADS Deutschland GmbH.
Thales.

SIT 421 (MM/UPX-709) transponder

Type
Identification Friend-or-Foe (IFF) system.

Description
The SIT 421 (MM/UPX-709) is a single-box airborne IFF transponder suitable for fitting in

fixed-wing aircraft or helicopters. It operates in Modes 1, 2, 3/A, 4 and C. The receiver/transmitter includes a 500 W solid-state transmitter, dual-channel receiver and RF interface module. The first of these comprises a delay line oscillator, modulator, driver and power amplifier.

The controls for operation of the transponder, code and mode selection and so on, are mounted on the front of the equipment (which is designed for cockpit mounting) but versions are produced in which remote-control facilities are provided.

Specifications
Weight: 3.5 kg
Frequency:
(receiver) 1,030 MHz
(transmitter) 1,090 MHz
Sensitivity: –77 dBm (adjustable 69–77)
Dynamic range: 55 dB
Output power: 27 ±3 dBW at 1% duty cycle

Status
In service.

Contractor
BAE Systems North America.
MID SpA.

SIT 432 (AN/APX-104(V)) interrogator

Type
Identification Friend-or-Foe (IFF) system.

Description
The SIT 432 (AN/APX-104(V)) is a lightweight airborne IFF interrogator equipment suitable for installation on helicopters or fixed-wing aircraft to provide air-to-air and air-to-ship identification facilities.

The receiver/transmitter module contains a 1,200 W transmitter, a dual-channel receiver and an RF interface module. The receiver operates at 1,090 MHz and is of a dual-channel type which, in conjunction with a dual-channel antenna, provides for receiver sidelobe suppression.

The design employs surface acoustic wave technology in the local oscillator to obtain a reliable, simple design with good stability and no field alignment requirements. The transmitter is solid-state. It accepts coded video pulse trains from an external source and the internally generated Mode 4 ISLS pulse converts the coded

video pulse trains into radio frequency pulse groups for transmission as IFF interrogation.

Specifications
Weight: 6.5 kg
Frequency:
(receiver) 1,090 ±0.2 MHz
(transmitter) 1,030 ±0.2 MHz
Sensitivity: –83 dBm
Output power: not less than 1,200 W
Duty cycle: 1% (max)
Dynamic range: 50 dB

Status
In service.

Contractor
BAE Systems North America.
MID SpA.

STR 2000 IFF transponder

Type
Identification Friend-or-Foe (IFF) system.

Description
The STR 2000 Mode-S ACAS compatible IFF-transponder provides full capability with Mode S Level 3 as defined in ICAO, Annex 10, Volume IV – July 1998, and STANAG 4193 part IV, including SI-Code processing and Extended Squitter (ES) transmission. The system also provides full MkXII capability in compliance with STANAG 4193.

The equipment consists of a transponder and a dedicated control and display unit (CADU).

The transponder provides 3 bi-directional ARINC 429 ports and one ARINC 429 input port together with a software configurable MIL-STD-1553B databus. In addition, it interfaces via a dedicated ARINC 429 bus with an onboard ACAS system. Together with the CADU or a Flight Management System (FMS), the system provides for all simple ACAS control functions (Test, On/Off, TA, TA/RA).

The STR 2000 can be operated via either the dedicated CADU or via any of the following bus systems: MIL-STD-1553 B, ARINC 429, RS 485 (as used by the CADU). A noteworthy feature of the STR 2000 is its modular design; the unit can be easily adapted to any platform by replacement of the rear connector module. All interfaces are programmable by a special configuration parameter file, which is stored in the non-volatile memory of the transponder.

EADS, as the lead German contractor for the Next Generation IFF (NGIFF) Programme, has integrated all feasible provisions for Mode 5 into the design of the STR 2000 (such as connectors, interfaces and slots for two NGIFF modules) to ensure that minimum modification to aircraft cabling and equipment will be required for future upgrade.

Extensive Built In Test (BIT) eliminates the need for scheduled preventive maintenance. Temporary fault conditions are trapped and stored in a non-volatile memory, which is accessed by the Automatic Test System (ATS). The BIT results, including elapsed time and Go/NoGo indications, are available on the CADU front panel or on the data bus used for remote control.

The STR 2000 is completely solid state, making extensive use of modern microprocessors and DSPs, powerful ASIC and FPGA technology, while the mechanical arrangement allows operation in non-pressurised equipment bays at up to 100,000 ft altitude.

Specifications
Modes:
(ATC) Mode A, C, S Level 3
(IFF) Mode 1,2,3 and 4
Transmitter frequency: 1,090 ± 0.5 MHz
Transmitter peak power: 500 W (typical)
Receiver frequency: 1,030 MHz (centre), 10 MHz bandwidth
Power: 28V DC (18–32V range); 90W (transponder) max, 15 W (CADU) max
Dimensions:
(Transponder) 1/2 ATR short (318 × 124 × 194 mm)
(CADU) 146 × 133.3 × 63 mm
Weight:
(Transponder) 8.0 kg
(CADU) 2.0 kg
MTBF:
(Transponder) >2,000 h
(CADU) >6,000 h
Temperature: –54 to +71°C

Status
In production for several NATO countries. Applications include fixed-wing aircraft of the German Air Force (STR 2000) and French Air Force (TSC 2000), NATO E-3A AWACS and TCA, and Eurocopter.

Contractor
EADS Deutschland, Defence and Security Systems Division.
Thales Communications.

Israel

ASARS Airborne Search And Rescue System

Type
Avionic Direction Finding (DF) system.

Description
The Tadiran Spectralink Airborne Search and Rescue System is used in the rescue of downed aircrew by an airborne platform, special forces pick up, or drop zone marking. It facilitates rapid rescue under adverse conditions and is designed for use both in combat and peacetime. Both covert and standard beacon mode operation are possible, as well as voice communication.

ASARS consists of the ARS-700 airborne system and the PRC-434A radio communicator. The system utilises advanced range measurement technology, enabling accurate and quick survivor location in bad weather and difficult or hostile terrain.

First pass pick up is made possible by the accurate azimuth and range measurements provided by the lightweight airborne system. A beacon mode supplements the rescue radio's special transponder capabilities, and downed aircrew survival is enhanced by the fallback

voice communication capability on every channel.

Short burst-type interrogations (600 ms) and special modulation techniques eliminate the dangers of repeated or continuous transmissions being picked up by enemy forces. Ten preprogrammed frequency channels, out of 3,000 operating frequencies, further enhance security. Up to a million call codes allocated to every radio transponder enable selective interrogation, minimising the number of transmissions. Several survivors can be handled by one airborne system, with each one clearly identified. If the selective call code of the transponder radio is unknown, it can be automatically retrieved by the airborne system via secure algorithms, enabling the successful completion of the rescue mission.

The transponder radio is activated automatically upon bail-out, and it can be easily operated with one hand by a disabled survivor. The radio is exceptionally rugged, compact and lightweight, and simple to use. It is powered by a durable lithium battery that provides a long operating life.

The transponder radio is compatible with both the Cubic AN/ARS-6V and the Tadiran ARS-700, and operates over the 225 to 300 MHz frequency range, with an effective range

of 200 km. The beacon mode and default voice capabilities are compatible with other rescue systems and techniques.

ASARS can be installed in all types of fixed-wing aircraft and helicopters, without the need for any aircraft modification. It is easy both to install and remove; re-installation takes less than 30 minutes. The compact control unit mounts in the standard radio slot on the instrument panel. System parameters are easily programmed using a standard PC interface.

DF accuracy is ±2.5° RMS. Range measurement accuracy is 50 m.

The Tadiran Spectralink ARS 700 airborne search and rescue system　　0504006

ASARS with thermal beacon

The latest variant of ASARS integrates a thermal beacon into the system, facilitating optical recognition of survivors' position from a typical range of 5 km. The system includes all the features of earlier models, with the addition of a MWIR/LWIR waveband thermal beacon (NIR is an option), which provides a greater than 90° radiation divergence, greatly enhancing the chances of a first-pass pick-up for rescue aircraft. The system also features a remote operation capability, enhacing covertness for Combat Survival And Rescue (CSAR) operations.

Specifications
ARS 700 airborne system
Frequency coverage: 225–300 MHz
Effective range: 200 km
Final approach accuracy: 20 m
RF channels: 3,000
Range measurement accuracy: 50 m
DF accuracy: 5° rms
Interrogation cycle: 600 ms
Communication: 2-way voice on every channel
Dimensions:
(avionics unit) 190 × 195 × 380 mm
(control display unit) 100 × 127 × 130 mm
(antenna switching unit) 50 × 150 × 150 mm
(remote display unit) 76 × 76 × 50 mm
Weight: 18 kg
Power supply: 28 V DC, 10 A
Power output: 20 W peak (20 W, 2 W, 0.4 W PEP)

PRC 434 radio transponder
Dimensions: 186 × 74 × 38 mm
Weight: 0.9 kg
Power supply:
(lithium battery) 15 h at 1:10 T/R ratio
Power output: 1 W RMS

Status
Selected by the French, Israeli, Italian, Thai and Turkish air forces and the US Navy.

Contractor
Tadiran Spectralink Ltd.

ASARS-G Airborne Search And Rescue System with GPS and relay

Type
Avionic Direction Finding (DF) system.

Description
ASARS-G maintains all the existing facilities of ASARS; in addition, it includes: a complete survivor position locating system, with GPS navigation receiver, DF and full navigation support. Designated ARS-700G; it includes a secure data channel relay capability; voice channel reporting on both VHF and UHF; remote activation of embedded GPS features by the airborne units (ARS-700G and ARS-434R); a simple operator interface that turns the PRC-434 rescue radio into a complete airborne relay system, designated ARS-434R.

Specifications
(for survivor rescue radio PRC-434G)
Frequency range: 220 – 410 MHz, and 121.5 MHz
Channels: 7,000
Programmed channels: 10 + 2 guard
Position accuracy: GPS P-code or C/A code
Modulation:
(voice or swept tone) AM
(narrowband data) FSK
(transponder) OOK, PSK for ASARS
Activation: manual or automatic upon ejection
Remote operation: beacon mode; navigaton mode; data transfer mode
Transmit power:
(beacon in UHF) 2 W peak
(voice in UHF) 0.75 W CW
(121.5 MHz) 0.1 W peak
adaptable power management for data transfer

Status
In production and in service. Fielded by the Israeli Air Force. The system is NATO-certified and has been purchased by the air forces system was employed by French and Spanish forces during operations in Bosnia.

The system was sold to the Thai Air Force in 1999 for use on two types of helicopter. Believed to be in use with other South East Asian customers.

Contractor
Tadiran Spectralink Ltd.

Italy

SIT 421 family of Identification Friend or Foe (IFF) transponders

Type
Identification Friend-or-Foe (IFF) systems.

Description
The SIT 421 family of airborne IFF transponders comprises the following models:
1. SIT 421 cockpit-mounted set;
2. SIT 421T remotely controlled set, including SIT 901 control unit for full capability (or C-6280A(P) APX control box providing reduced capability);
3. SIT 421T-1553 bus-controlled set.

Both the SIT 421 and SIR 421T models have also been adopted by the Italian Navy for frigate and hydrofoil use, where it is given the nomenclature MM/UPX-709.

The SIT 421 provides operation in Modes 1, 2, 3/A, 4 and C. For Mode C operation, the transponder operates with an external pressure/altitude digital converter.

The receiver-transceiver comprises a 500 W solid-state transmitter, a dual-channel receiver and an RF interface module. The solid-state transmitter includes a delay line oscillator, a modulator, a driver and a power amplifier.

The dual-channel receiver handles space diversity operation to ensure reliable transponder response to the received interrogation. The receiver may also be set for operation with one antenna only (single-channel operation).

The signal processor consists of a video processor, a decoder, a coder and a video interface with the KIT-1A-/TSEC for Mode 4 operation.

The SIT 421 includes an open board slot to accommodate anti-jam circuits and other improvement circuits, as growth options.

SIT 421T/SIT 901 control unit
The SIT 421T directly interfaces with the SIT 901 control unit for full operational capability, or a C-6280A(P) APX control box for reduced operational capability.

SIT 421T-1553
The SIT 421T-1553 fulfils the same functions as the SIT 421T, but it is bus controlled via a MIL-STD-1553B bus. Being computer controlled, it includes automatic functions, including automatic code change.

Specifications
Environmental: MIL-STD-810B
EMC: MIL-STD-461, -462, -463
Rx frequency: 1,030 ±0.5 MHz
Sensitivity: –77 dBm
Bandwidth at –6 dB: 7 to 10 MHz

Dynamic range: 55 dB minimum
Tx frequency: 1,090 MHz
Frequency stability: ±3 MHz
Output power: 27 ±3 dB at 1% duty cycle

The SIT 421T-1553 bus-controlled IFF ransponder
0062225

The SIT 421 IFF cockpit-mounted transponder
0062223

The SIT 421T remotely-controlled IFF transponder
0062224

The SIT 901 IFF control unit
0062226

Dimensions:
(SIT 421) 146 (W) × 134 (H) × 190 (D) mm
(SIT 421T) 136.5 (W) × 136.5 (H) × 213 (D) mm
(SIT 901) 146 (W) × 136 (H) × 43 (D) mm
(SIT 421T-1553) 146.8 (W) × 136.5 (H) × 219.8 (D) mm

Weight:
(SIT 421T) 4.1 kg
(SIT 901) 0.8 kg
(SIT 421T-1553) 5 kg

Power:
(SIT 421) 28 V DC, 50 W
(SIT 421T) 28 V DC, 50 W
(SIT 421T-1553) 28 V DC, 65 W

Status
In production and in service.

Contractor
Marconi Selenia SpA.

SIT 423 IFF transponder

Type
Identification Friend-or-Foe (IFF) system.

Description
The SIT 423 IFF transponder system forms part of the Aircraft Attack and Identification Sub-system (AAIS). The transponder operates in accordance with STANAG 4193, providing interoperability with other IFF-equipped platforms of NATO nations and in accordance with the ATC requirements of ICAO Annex 10.

The primary function of the IFF transponder is to accept, decode and reply to interrogations coming from an IFF interrogator. It is capable of operating in the SSR/IFF/ATC scenarios. When receiving correct interrogations requiring a reply, the appropriate reply signal is generated and radiated via one of two separate antennas. Interrogations which do not require a reply are processed and the corresponding transactions are started.

In addition to its main IFF/ATC functions, the SIT 423 performs internal self monitoring

and fault location by means of comprehensive Built-In-Test (BIT) circuitry. The main features of the transponder are:

- Interrogation, recognition and reply generation for Modes 1, 2, 3/A, 4, and C
- Individual enabling of each operational mode
- Mode S operational level 1,2 or 3
- Manual reply code for Mode 2
- Manual or Automatic Code Change (ACC) of Modes 1 and 3/A
- Interrogator Side Lobe Suppression (ISLS) recognition
- Modular construction with plug/socket connections on all modules
- Comprehensive BIT facilities to enable local/remote fault finding to module level
- NGIFF provision
- Crypto appliqué.

Specifications
Rx frequency: 1,030 ± 0.5 MHz
Tx frequency: 1,090 ± 0.5 MHz
Frequency stability: ±3 MHz
Power: 28 V DC; 120 W (max)

Status
In production and in service.

The UPX-719 transponder   0503709

Contractor
Marconi Selenia SpA.

UPX-719 transponder

Type
Identification Friend-or-Foe (IFF) system.

Description
The UPX-719 transponder forms a part of the SMA Intra (interrogator/transponder) system which links ships with their co-operating helicopters. The UPX-719 is the airborne transponder; the UPX-718 is the shipborne interrogation part of the system. The shipborne transmission to the airborne transponder can be integrated with the main radar. Ships can identify up to 10 helicopters and vice versa, due to the characteristics of the coding system. Range is over 100 km for helicopters operating at 1,500 ft.

Status
In production for Italian and other navies.

Contractor
Selex Sensors and Airborne Systems, Radar Systems Business Unit.

Japan

Direction-finding and receiving system

Type
Avionic Direction-Finding (DF) system.

Description
The direction-finding and receiving system is designed to detect SHF frequencies between 6.4 and 7.1 GHz used for video transmissions from ground stations and employs a directional antenna to track the transmission automatically. The system operates up to a maximum aircraft speed of 155 kt and a maximum altitude of 20,000 ft as an emergency communications system.

The system consists of separate direction-finding and receiving antennas, a DF signal processor, DF control panel and monitor on which is displayed the angle of ground transmissions relative to the aircraft to an accuracy of ±5°. One or more optional antenna directing systems may be added, to enable the aircraft to relay transmissions.

Status
In service.

Contractor
Tokimec Inc.

The Tokimec direction-finding and receiving system with the receiving and DF antennas mounted underneath the ventral surface of an S-76 helicopter just aft of the nosewheel. An antenna directing system is also mounted on each side of the cabin   0504010

Netherlands

Vesta transponder

Type
Aircraft transponder.

Description
Vesta is a landing and identification aid primarily intended for ship-based helicopters. It consists of two parts: the helicopter transponder and the

ship receiver. Vesta enables accurate display and tracking of friendly helicopters on the radar display, even in heavy clutter environments. The transponder principle is based on a radar-triggered

VHF reply. Operation is possible with any synchronised surveillance radar (either shipborne or shore-based) in the 1 to 10 GHz band.

For every radar pulse the transponder receives, a VHF reply pulse is transmitted, followed by a code pulse. Up to five helicopters can be identified by means of preselected codes (extension up to 64 is possible). The return signal is received and processed by the Vesta receiver in the ship, which identifies and decodes the transponder reply. Unwanted VHF reply pulses are rejected by digital filters controlled by the allocated radar on the ship.

The Vesta helicopter system consists of a fully solid-state transponder, a control unit, two radar pick-up antennas and a VHF transmitting antenna.

The two radar pick-up antennas are used to guarantee a combined sensitivity pattern which is virtually omnidirectional. The control unit has only two switches; one for sensitivity selection and power on/off and the other for code selection.

Specifications

Dimensions:
(transponder) 167 × 85 × 194 mm
(control unit) 146 × 66 × 68 mm
(radar pick-up antenna) 45 × 106 × 48 mm
(VHF transmitting antenna) 50 × 254 × 123 mm
Weight: 3 kg
Power supply: 28 V DC, 14 W
Frequency:
(transmitter) VHF A-band
(receiver) 1–10 GHz
Transmitter peak power: 10 W
Range: 0–230 km
Pulse duration: 2.2 μs nominal
Number of helicopter codes: 5 standard
(optional extension up to 64)

Thales' Vesta transponder system   0503711

Status
Fitted to several types of helicopters. Over 55 Vesta airborne transponders have been delivered to several navies.

Contractor
Thales Nederland BV.

Russian Federation

IFF 6201R/6202R and 6231R/6232R systems

Type
Identification Friend-or-Foe (IFF) system.

Description
RadioPribor has manufactured airborne radio and navigation equipment for installation on all types of civilian and military aircraft and helicopters produced in Russia and the other RFAS countries.

The IFF responders 6201R and 6202R are fitted on all military and civil aircraft and helicopters. The model 6202R differs from the 6201R in that it incorporates additional signal amplifiers for use on heavy aircraft that have long SHF cable runs.

In aircraft equipped with radar, the 6231R or 6232R interrogators are installed to provide the interrogation function.

The combined system provides three modes of operation: general identification modes; individual identification modes providing 84 interrogation slots and 100,000 reply slots; identification of objects in distress (distress mode with interrogation/alarm modes without interrogation).

Status
Widely deployed on civil and military aircraft and helicopters manufactured in Russia and the RFAS countries.

Contractor
Production Association RadioPribor.

RadioPribor's 6201R IFF responder   0010928

SPPZ ground proximity and warning systems

Type
Ground Proximity Warning System/Terrain Awareness Warning System (GPWS/TAWS).

Description
The SPPZ ground proximity warning systems can be used in all types of passenger and transport

SPPZ ground proximity warning systems   0011883

RadioPribor's 6231R IFF interrogator   0010929

aircraft equipped with flight navigation systems that have digital information exchange. There are two models: the SPPZ-85 and SPPZ-2.

The SPPZ systems compute data obtained from the following systems to produce their warnings: the radio altimeter; air data computer system; ILS or MLS receiver; onboard inertial navigation system; landing gear and flap sensors. The SPPZ-2 system also utilises data from the flight management system and flight control system.

Warning data provided by the systems is as follows: excessive sink rate; excessive terrain closure rate; negative climb rate after take-off or missed approach; insufficient terrain clearance at landing with wrong configuration; inadvertent descent below glide slope; excessive difference in absolute altitude and pressure height; inadvertent flight into dangerous windshear (SPPZ-2 only).

Specifications
Outputs: 2 discrete analogue; 40 voice; ARINC 429
Power: 115 V AC, 400 Hz, 20 VA (SPPZ-85), 25 VA (SPPZ-2)

Dimensions: 2 MCU
Weight: 3.5 kg

Status
Fitted to: An-70, Il-96, Il-114, Tu-204 and Tu-334 aircraft.

Contractor
AeroPribor-Voskhod Joint Stock Company.

South Africa

PT-1000 IFF transponder

Type
Identification Friend-or-Foe (IFF) system.

Description
The PT-1000 provides Mk XII transponder capability and is suitable for airborne and marine applications. The transponder supports South Africa's national secure mode, which is available as a country-specific export version.

The transponder provides full diversity (dual antenna) decoding and replies to modes 1,2,3/A,C and the secure mode, according to STANAG 4193.

The system consists of a tray-mounted transponder and an optional internal cryptographic module and panel-mounted Control/Display Unit (CDU).

Control and status interfacing to the transponder is via ARINC 429.

Specifications
Dimensions: 3/8 ATR (transponder)
5½ in panel, 127 × 113 × 43 mm (CDU)
Weight: <5 kg inclusive of the crypto module
Power supply: 28 V DC, to MIL-STD-704
Power consumption: 35 W nominal, 55 W max

Status
In service.

Contractor
Tellumat (Pty) Limited.

PT-2000 IFF/Mode S transponder

Type
Identification Friend-or-Foe (IFF) system.

Description
The PT-2000 IFF/Mode S transponder is a compact (3/8 ATR) unit, with externally accessible secure mode cryptographic computer within the volume. It offers the following capabilities: full mode1, 2,

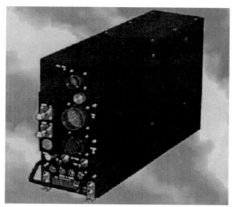

The PT-2000 IFF/Mode S transponder   0044985

3/A and C interrogation/reply functionality; secure mode operation (supplied with Mode 4 capability, which can include a customised national secure mode); the ability to be upgraded to the 'Successor IFF' system; Automatic Code Changing (ACC) facility for secure mode; Mode S Extended Length Message (ELM) capability by software upgrade; software/firmware upgrade and access to maintenance data via front panel connector. The PT-2000 operates off the aircraft 28 V DC supply, and it requires no special cooling. Control and data interfacing via dual-redundant MIL-STD-1553B or ARINC-429 digital databus; TCAS interface via ARINC-429 link.

Specifications
Dimensions: 3/8 ATR short (including cryptographic module)
Weight: <6.5 kg
Power: 28 V DC, 60 W maximum
MTBF: 4,000 h (calculated)

Status
In service.

Contractor
Tellumat (Pty) Limited.

XBT-2000 X-band radar transponder

Type
Aircraft transponder.

Description
The XBT-2000 is a radar transponder providing encoded replies to interrogations from airborne or shipborne X-band (NATO I-band) weather radars operating in the weather band (9,200 to 9,500 MHz). The transponder provides compatibility with the DO-172 radar beacon mode standard (encoded replies), as well as fully independent digital tuning of transmit and receive frequencies for alignment with specific radars. The transponder is suitable for man-pack deployment or may be mast mounted. Applications include demarcation of remote runways or drop zones as well as helicopter decks. Optional packaging is available for airborne use.

Specifications
Dimensions: 172 × 215 × 64 mm
Weight: <5 kg, including rechargeable battery pack and antenna
Power supply: 10.5 to 28 V external, or integral Ni/Cd battery
Power consumption: 2.5 W nominal, 5 W max
Frequency range: receiver and transmitter independently tunable between 8,500 and 9,500 Hz beacon mode 9,375 MHz receive, 9,310 MHz transmit
Set-up: stored in non-volatile memory after configuration via PC RS-232 terminal interface

Status
In service.

Contractor
Tellumat (Pty) Limited.

United Kingdom

7 series homing systems

Type
Avionic V/UHF homing system.

Description
The 7 series is a versatile building block homing system, capable of providing both broadband and emergency guard channel homing. In its simplest form, a 7 series installation comprises a homing indicator, an antenna feed unit, an antenna system and the aircraft receiver. When an independent self-contained emergency guard channel homing system is required, the 7 series uses a Chelton 7–28 Series two-channel receiver in place of the aircraft receiver. It is also possible to interface both the aircraft receiver and the 7–28 Series receiver with the 7 series, thus providing broadband plus guard

channel homing. The 7 series interfaces with all AM receivers, including the ARC116, ARC159, ARC164, ARC182, ARC186, PTR1751 and VHF20, and also with those FM equipments having AM facilities.

The 7 series comprises a number of sub-units which include:

Series 7-24 homing indicator units. These are self-contained homing directors incorporating all necessary electronic control circuits, voltage regulators and computer reference amplifiers, packaged within an 89 mm housing. The indicators have a variable sensitivity control located on the front panel and are edge-lit.

Series 7-27 indicator feed units. These comprise the electronic circuitry otherwise packaged within the 7-24 homing indicator units and are intended

to feed existing aircraft navigation indicators or flight directors.

Series 7-25 antenna feed units. These contain the necessary antenna phasing and switching circuits to couple antennas into the homing system. Unlike the sub-units listed above, the choice of antenna feed unit is dependent upon the frequency band required.

Series 7-28 emergency guard receivers. These small receivers are designed to provide a completely self-contained homing system at VHF and UHF distress frequencies. The 7-28-31 series offers selection or scanning of up to six preset frequencies.

Series 7-60 complete homing adaptors. These are combinations of Series 7-27 indicator feed units and Series 7-25 antenna feed units packaged in one small module.

Specifications
Weight: less than 1 kg
Power: 28 V DC, 500 mA max

Status
In production and in service. More than 750 systems are in service with military and commercial operators worldwide.

Contractor
Chelton (Electrostatics) Ltd.

503 series Emergency Location Transmitters (ELTs)

Type
Avionic Emergency Locator System (ELS).

Description
The 503 ELT system provides full-frequency coverage, including 121.5, 243 and 406.025 MHz with optional inclusion of transmission of last known GPS co-ordinates. It is compatible with the majority of search and rescue systems including the Cospass/Sarsat satellite-based locator system. It is activated either manually by cockpit remote control or automatically by a bi-directional *g*-switch module. The ELT has fixed or portable mounting options.

Specifications
Frequency: 121.5, 243.0 and 406.025 MHz
Peak Effective Radiated Power (PERP): 0.1 W at 121.5/243.0 MHz; 5.0 W at 406.025 MHz
Transmission duration: 24 h (min) at 5W PERP; 48h (min) at 0.1 W PERP
Repetition rate: 520 ms every 50 s
Dimensions: 255 × 105 × 45 mm
Weight: 1.3 kg

Battery life: 6 years
Approval: JTSO C91a, C126 and Cospas/Sarsat

503-15 series ELT interface unit
The 503-15 series ELT interface unit provides last known GPS co-ordinates on ELT, ADELT or CPI. The ELT interface unit is a stand-alone unit, which utilises an RS-232 interface to continually update the ELT, ADELT or CPI. Other interfaces are available optionally. The freedom of a stand-alone unit allows full integration with the aircraft's flight management computer, while retaining the ability to program the ELT/ADELT or CPI via a preset 24-bit aircraft address without requiring specialist maintenance operations at equipment exchange.

Status
As of late 2005, in production and in service.

Contractor
H R Smith.

930 series direction-finding system

Type
Avionic Direction-Finding (DF) system.

Description
The 930 series V/UHF direction-finding antennas enable a standard AM communications receiver to be converted into a 360° direction-finding system. These antennas are suitable for installation on fixed- or rotary-wing aircraft as well as maritime, land-mobile and ground installations. With the adaptor plate, P/N 23468, the 931-1 antenna is essentially a drop-in replacement for the DF301E direction-finder. The 930-1 is a replacement for the AN18 direction-finder.

A typical system interconnection utilises an existing transmitter/receiver. The RF changeover relay enables the transmitter/receiver to be switched between the direction-finding antenna and the normal communications antenna. The action of the direction-finding antenna is to modulate the received RF signal, the sense and depth of this modulation being related to the direction of arrival of the received signal. The receiver demodulates the modulation and passes the resulting audio back to the direction-finding antenna for processing. After processing in the antenna unit, the resulting bearing is available on an ARINC 407 output to drive a synchro indicator. An optional ARINC 429 or RS-422 databus output capable of driving an electronic display can also be specified.

The System 930 has been specifically designed to be compatible with direction-finding on pulsed tone personal locator beams (PLBs), including SARSAT beacons, when used with SARSAT-compatible receivers such as the Chelton 7-28-406 or 7-28-31 series (see separate entry).

Status
In production and in service. 930 series of direction-finding antennas has already been installed on the following platforms: Agusta AB412; BAE Systems (Operations) Limited Nimrod; Bell 212, 214ST, 412SP and CH146; Cessna 416 and Caravan; DHC Dash 8; ECD UH1D; ECF Puma and Super Puma; Learjet; Britten Norman Islander; PZL Mielec AN28; Shrike Commander; and Westland Lynx.

Contractor
Chelton (Electrostatics) Ltd.

935 series direction finding systems

Type
Avionic Direction Finding (DF) system.

Description
The 935-1 tactical direction finder is a standalone direction-finding system with an integral synthesised receiver covering the frequency

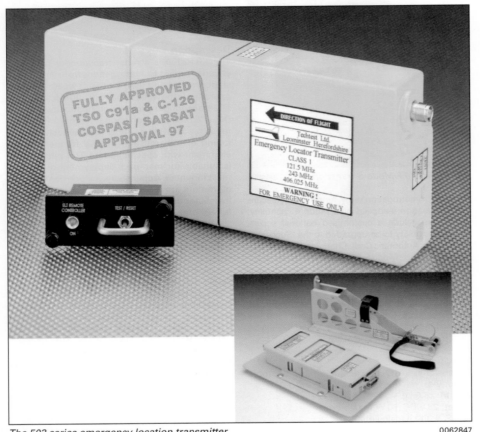

The 503 series emergency location transmitter   0062847

The 503-15 series emergency location transmitter interface unit   0051603

range 30 to 470 MHz, together with 5 Guard receivers to monitor pre-defined distress frequencies.

When used with Chelton's Type 835-1 Personal Survival Radio (PSR) interrogator and a UHF antenna (for example, Type 16-1), the system is capable of operating with PSRs such as the AN/PRC-112 and PRC-434 to provide range and bearing information. In addition, the system can provide bearing co-ordinates from an embedded GPS in the URX3000 PSR.

All variants operate in the 30 to 470 MHz waveband, tuned in 5.0, 6.25 or 8.33 kHz steps; the 935-1 incorporates a MIL-STD-1553 interface, the 935-2 features ARINC-429 and the 935-3 includes rear-mounted direct connectors.

The Guard receivers may be tuned to the following frequency ranges:

Receiver	Waveband	Frequency range
Guard RX1	VHF	121.2–123.2 MHz
Guard RX2	Maritime	155.6–158.5 MHz
Guard RX3	UHF	241.2–245.1 MHz
Guard RX4	GMDSS DSC	156.525 MHz (CH70)
Guard RX5	COSPAS-SARSAT	406.0–406.1 MHz

The VHF, Maritime and UHF Guard receivers can each be programmed with main and auxiliary frequencies, allowing distress monitoring on the main frequencies, with training carried out on the auxiliary frequencies.

The GMDSS DSC Guard receiver is pre-programmed to the VHF DSC Channel 70. The system may be programmed on the MIL-STD-1553 databus to monitor and report distress alerts, all-ships calls, Selective Calls and either distress or urgency categories. Full reporting of the vessel MMSI, nature of distress and GPS co-ordinates is available to the user.

The COSPAS-SARSAT Guard receiver monitors the 406 MHz COSPAS-SARSAT frequencies and provides a decoded output from the beacon including the 15-digit HEX ID, country of registration and any GPS co-ordinates if available.

Specifications

Antenna type: Modified annular slot with a cardioid receiving pattern
Mounting method: In line with airframe, normal or inverted
Accuracy: Better than 5° RMS (dependant on installation)
Weight: 3.8 kg max
Power: 16 to 31.5 V DC, 2.0 A max
Temperature altitude: EUROCAE ED-14C/RTCA DO-160C Section 4 Cat B2 (25,000 ft)
Temperature variation: EUROCAE ED-14C/RTCA DO-160C Section 5 Cat A external to aircraft
Waterproofing: EUROCAE ED-14C/RTCA DO-160C Section 10 Cat R

Status

In production and in service.

Contractor

Chelton (Electrostatics) Ltd.

ADELT Automatically Deployed Emergency Locator Transmitter (ELT)

Type

Avionic Emergency Locator System (ELS).

Description

The ADELT system comprises the beacon, carrier and ejection mechanism, deployment battery and control panel. The beacon is contained within the carrier and mounted externally at a suitable location. The control panel provides for testing of system integrity, system arming, crew activation, and deployment confirmation. Various remote activation devices may be incorporated in the installation design, such as frangible, float, inertia, saline, or hydrostatic switches. At least one sensor should couple with the deployment battery to avoid dependency on the aircraft electrical supply. When operated, whether by crew or remote sensor activation, a small cartridge triggers the release of two coiled

The ADELT CPT-606   0581496

springs, which eject the beacon safely away from the aircraft. Once upright and in the water, the beacon will automatically transmit homing signals on 121.5/243 MHz, while the transponder reacts to aircraft or ship's radar, pinpointing its position on the search vessel's radar screen.

The ADELT system comprises the basic CPT-600, with in-service CPT-606 and -609 variants, together with the latest derivative, the CPT-900, which also integrates additional features introduced to satisfy Automatic Fixed (AF) ELT requirements. The CPT-606 incorporates a homing transmitter on 121.5/243 MHz and a radar transponder on 9 GHz, compatible with aircraft or ship's radar. The CPT-609 adds a satellite transmitter operating on 406 MHz. The Class 2 CPT-900 (part number 070-0900-001) is a direct replacement for the CPT-606/609, providing 121.5, 243 and 406 MHz transmissions in accordance with COSPAS/SARSAT C/S T.001 and EUROCAE ED-62. The beacon is designed to comply with UK Civil Aviation Authority (CAA) requirements for aircraft operating over sea and land and satisfy Scale KKK (i) and (ii) of the amended Schedule 4 of the UK Air Navigation Order (ANO) 2000 and JAR Ops 3.820. Installation of the CPT-900 requires minimal carrier and control panel modification over current -606/609 configurations.

The beacon is programmed with the country code and the aircraft registration marking or radio call sign. When the beacon is activated by deployment into water, the satellite transmitter transmits its programmed information as a burst of coded signals to the orbiting COSPAS/SARSAT satellites receiving on 406 MHz. The message is stored by the satellite and downloaded to the nearest local user terminal ground station. Here the signal is processed to obtain latitude and longitude of the aircraft in distress and its identity. This in turn is passed to a mission control centre which routes the information to the rescue co-ordination centre nearest the incident, from where search and rescue forces will be sent. Accuracy of the 406 MHz signal is to within 1 to 2 km, although testing has proved accuracies of 0.2 km can be achieved.

Once in the general vicinity of the beacon, SAR teams can locate the scene with the aid of the 121.5/243 MHz homing transmitter and the 9 GHz radar transponder which will guide SAR forces to within 10 m of the beacon.

The ADELT CPT-609 can be field programmable, greatly reducing the cost and impact of changing aircraft identity. The CPT-609 can be reprogrammed at remote locations without the need for costly and inconvenient disassembly for replacement of EPROMS.

Specifications
CPT-609

Satellite transmitter: 406.025 MHz
Homing transmitter: 121.5 and 243 MHz

Radar transponder: 9 GHz
Size: 203 mm maximum diameter, 635 mm length excluding antenna
Weight: 5 kg
Operating duration: 48 h continuous
Battery: 15 V lithium
Service life: 3 years

Status

The CPT-606 is in service in Aerospatiale SA332, SA365, Bell 212, 214ST, Sikorsky S61, S76 and MBB Bo 155 and 117 helicopters.
The CPT-609 is in production and in service.
The CPT-900 is in production as a direct form and fit replacement for the CPT-606 and -609 systems.

Contractor

Caledonian Airborne Systems Ltd.

ARI 5983 I-band transponder

Type

Identification Friend-or-Foe (IFF) system.

Description

The ARI 5983 I-band transponder provides a means of locating, identifying and providing navigational assistance to aircraft outside normal radar coverage and range. It is interrogated by a primary radar and gives an edge-of-band response. The response codes are selected on a simple control unit from which a comprehensive BIT routine can be initiated. The control unit also allows either manual or electronic switching between the transponder's two antennas to ensure optimum coverage.

The transponder receives interrogation signals, via the antenna, from pulse radars at any frequency in two bands 100 MHz wide. When interrogated, the transponder will respond with either a single-RF pulse, which provides enhancement of the radar return, or a coded group of up to six pulses, as selected on the control unit, which allows identification. There are 16 different reply codes available.

The transmitted power can be reduced by approximately 11 dB via another switch on the control unit. The transponder output will automatically be suppressed during the operation of other I-band equipment in the aircraft. Similarly, a pulse is supplied by the transponder to allow suppression of other equipment in the aircraft operating in the same frequency band when the transmitter is operating. The BIT self-test facility generates an interrogate signal which is fed into the transponder input. A green LED on the control unit indicates correct transponder operation.

Options include double- or multiple-pulse interrogation to minimise false triggering when several I-band radars are transmitting in the same area.

Both the transponder and control unit are fully NATO codified.

Specifications

Dimensions:
(transponder) 160 × 217 × 87 mm
(control unit) 147 × 117 × 48 mm
Weight:
(transponder) 2.7 kg
(control unit) 0.45 kg
Power supply: 28 V DC, 40 W (max)
Frequency:
(receive) 9,190–9,290 MHz and 9,360–9,460 MHz
(transmit) 9,310 ±7 MHz
Bandwidth: ±50 MHz
Sensitivity: –93 dBW
Output power: 135 W (min) to 300 W (max) peak
Pulse duration: 0.45 μs ±0.1 μs
Reply code: 6-pulse code, 16 settings
single-pulse reply capability
Pulse spacing: 2.9 μs nominal
Duty cycle: 0.005 (max)

Status

In service with the UK Royal Navy on the EH 101 Merlin, Lynx, Sea King and Sea Harrier F/A-2. Also exported for Dauphin and Lynx helicopters and the Do-228.

Contractor

M/A COM Ltd.

BE594 SARFIND Personal Locator Beacon (PLB) airborne decoder unit

Type
Avionic Emergency Location System (ELS).

Description
SARFIND is the cockpit decoder unit used in conjunction with the SARBE 8 and SARBE G2R Personal Locator Beacons (PLBs). SARF1ND decodes and displays the identity and GPS co-ordinates of up to 64 PLBs simultaneously. It is available as an integrated Cockpit Display Unit (CDU) or as a fully portable carry on/carry off unit and integrates with the aircraft radio receiver tuned into the mission operating frequencies. In excess of 18 hours operation is provided by an independent power supply.

The M/A COM ARI 5983 IFF transponder 0504116

During operation, the NVIS-compatible display provides beacon identity, beacon position (latitude and longitude), elapsed time since last beacon transmission and time of GPS acquisition by beacon. When a successful decode is achieved, an alarm sounds alerting the aircrew to the detection of SAR/CSAR transmissions.

Status
The BE594 SARFIND is widely deployed with several air forces worldwide in both fixed- and rotary-wing applications, including the Indian Air Force, British Army and several Special Forces (SF) units.

Contractor
Signature Industries Ltd.

Cossor Interrogation and Reply Cryptographic Equipment (CIRCE)

Type
Identification Friend-or-Foe (IFF) system.

Description
Raytheon Systems Limited supplies a range of modules for all types of IFF applications to facilitate a nationally secure IFF operation in Secure mode which is similar to, but not interoperable with, NATO Mode 4. The modules can be added to any IFF system that is compatible with the NATO Mk XII standard (STANAG 4193).

Raytheon Systems Limited CIRCE – IFF cryptographic unit 0011871

Two key components of CIRCE are the programmer and the fill gun. The programmer is used to hold the key variable necessary to operate the CIRCE units. It is held in a secure location and the electronic fill gun is then used to transfer the key variable data from the programmer to the various IFF systems. The key variable data can be changed whenever required using the fill gun.

Specifications
Dimensions: 123 × 50 × 129 mm
Weight: 1.42 kg

Status
In service with three non-NATO countries.

Contractor
Raytheon Systems Limited.

IFF 4500 interrogator

Type
Identification Friend-or-Foe (IFF) system.

Description
IFF 4500, which is a monopulse interrogator, is suitable for a wide variety of applications. Tactical fighter and maritime patrol aircraft surveillance platforms are already projected.

As well as operating in Modes 1, 2, 3A, C and 4, IFF 4500 has built-in growth potential for upgrade to include Mode 5 system, as specified in STANAG 4193 Part V. Automatic Code Change (ACC) and Mode S can also be added as customer options.

Control of the interrogator from the host primary radar can be via a MIL-STD-1553B databus or via a discrete control alternative. Target reports are fed again via a MIL-STD-1553B databus for display alongside the target information generated by the host primary radar.

Mode 4 or CIRCE cryptograhic units are built in to the IFF 4500, but these 'add-ons' can be removed for IFF Mk10A only applications.

Specifications
Dimensions: ½ ATR Short
Cooling: Convection
Peak output power: 2 kw ±2 dB

Raytheon Systems Limited IFF 4500 interrogator unit 0122513

Status

In service in UK Royal Air Force Tornado F3 aircraft.

Contractor

Raytheon Systems Limited.

IFF 4700 series transponders

Type

Identification Friend-or-Foe (IFF) systems.

Description

There are two versions of the Raytheon Systems Limited 4700 Series of IFF transponders which are designed to meet the NATO IFF Mk XII specification (NATO STANAG 4193); both versions use a common set of modules so they can be supported by a common spares inventory.

IFF 4720

The IFF 4720 series transponder is designed for a wide range of applications, and is a plug-in replacement for the IFF 2720 system. The IFF 4720 has full dual-redundant decoders on each receiver channel to provide optimum anti-jamming performance and the system can be operated by a dedicated remote-control panel, or via a MIL-STD-1553 databus. Full standard coverage of Modes 1, 2, 3/A, 4 and C is offered and the IFF 4720 can operate up to 70,000 ft altitude.

Specifications

Dimensions: 90 × 194 × 314 mm
Weight: 4.5 kg

Status

No longer in production. The IFF 4720 is in service in the BAE Systems Hawk 100 and 200; also fitted to some MiG-29 aircraft and several Naval IFF applications.

IFF 4760

The IFF 4760 transponder is a NATO Mk XII-compatible remotely controlled IFF, designed also to act as a MIL-STD-1553 bus controller. The unit provides an alternative form factor to that of the IFF 4720. All 4700 series transponders also offer tighter frequency tolerances, all-solid-state construction

Raytheon Systems Limited IFF 4700 series transponders, showing: IFF 4720 series transponder (top), IFF 4760 transponder (bottom left), IFF 4770 control unit (bottom right)   0044973

and VLSI processing to enhance performance; operation is identical to the IFF 4720.

Specifications

Dimensions: 136 × 136 × 213 mm
Weight: 4.5 kg

Status

No longer in production. In service with over 250 sets purchased by the UK MoD.

IFF 4770

The IFF 4770 control unit is designed to operate with IFF 4720 and IFF 4760 transponders. It is an NVG-compatible package which is intended to occupy minimum cockpit space, while at the same time having good ergonomics.

Status

No longer in production. In service in CN-235, Hawk 100, Merlin and MiG-29.

Contractor

Raytheon Systems Limited.

IFF 4800 series transponders

Type

Identification Friend-or-Foe (IFF) systems.

Description

As well as operating in Modes 1, 2, 3A, C and 4, the IFF 4800 operates in Mode S Level 3 which is the latest civil aviation selective address SSR system. By using Mode S, military aircraft will be able to continue to use civil airspace when conventional SSR modes are phased out in the early years of the next century.

Additionally, IFF 4800 incorporates the following facilities:

- Interface for Mode S Air Link Data processor
- Airborne Collision Avoidance System (ACAS) interface
- GPS position reporting interface
- Built-in Mode 4 or CIRCE cryptographic computer
- Provision for the incorporation for growth to Mode 5 as specified in STANAG 4193 Part V
- Automatic code change on Modes 1 and 3A.

Raytheon Systems Limited IFF 4700 series transponder, showing: IFF 4720 series transponder (top), IFF 4760 transponder (bottom left), IFF 4770 control unit (bottom right) 0044975

Raytheon Systems Limited IFF 4700 series transponder, showing: IFF 4720 series transponder (top), IFF 4740 transponder (top right), IFF 4770 control unit (bottom right)
    0044976

Raytheon Systems Limited IFF 4810 transponder (right), together with the IFF 4870 control and display unit   0122514

The IFF 4870 control unit is used together with the IFF 4810 transponder. For unified avionics control systems, a MIL-STD-1553D databus version of IFF 4830 transponder is available.

Specifications
Dimensions: 1 ATR short
Weight: 8.5 kg
Peak output power: 500 W ±2 dB
Cooling: convection cooled

Status
In production. As of mid-2006, the IFF 4800 was in service in Royal Australian Air Force Hawk Lead-In Fighter (LIF) and UK Royal Air Force Tornado F3 aircraft, as well as aboard over 40 different aircraft and ships as part of the UK's Successor IFF (SIFF) programme.

Contractor
Raytheon Systems Limited.

ML3500 radar transponder

Type
Identification Friend-or-Foe (IFF) system.

Description
To supplement the more sophisticated ARI5983 I-band transponder, M/A COM introduced the ML3500 radar transponder. The ML3500 provides an edge-of-band response when interrogated by an I-band radar, and is suited to enhancing the radar echoing area of small air and surface targets. Although in essence an active corner reflector, the transponder's response can be simply coded to aid target identification and, being edge-of-band, reduces primary plot clutter on radars with tunable receivers.

The ML3500 is lightweight and fully weatherproof.

Specifications
Dimensions: 120 × 170 × 55 mm
Weight: 1 kg
Power supply: 12 or 24 V DC, 10 W
Frequency:
(receiver) 9,000–9,600 MHz
(transmitter) 9,200–9,400 MHz (factory set)
Sensitivity: –43 dBm (min)
Interrogate pulsewidth: 0.15 to 1.5 µs
Stability: ± 10 MHz
Pulse duration: 0.2 to 1 µs
Reply code: 5 output pulses – one for range mark, four customer settable identifiers
Duty cycle: 1% (max)
Temperature range: –20 to +50°C

Status
In service.

Contractor
M/A COM Ltd.

PA6150 airborne IFF/SSR transponder

Type
Identification Friend-or-Foe (IFF) system.

Description
The PA6150 Mk XII IFF and Mode S SSR transponder is a remote terminal on a MIL-STD-1553B databus. A second, autonomous version uses a separate transponder control and display unit interfaced to the transponder LRI over RS-422A serial interfaces.

The transponder system is specified to meet the requirements of STANAG 4193, Part 1 for Mk XII IFF operation and ICAO Annex 10, Vol 1, Part 1, 4th Edition incorporating Amendment 68 for Mode S Level 3 operation. For the Mode S datalink, it will support Communications Capability Levels 1 to 3. Reply diversity operation is based on received signal level and timing at the two independent receiver channels. Provision is also being included for expansion to NG IFF at a later date.

The transponder system will accept the following Mk XII IFF interrogations and challenges: Modes 1, 2, 3A, 4 and C; automatic code changes for Modes 1 and 3A. Mode 2 will be set by switches mounted on the transponder LRI which are accessible when the LRI is installed in the aircraft.

For Mode S, the transponder will accept the following interrogations/challenges: Intermode A/C/S all-call; intermode A/C only all-call; short air-to-air surveillance; altitude request surveillance; identity request surveillance; Mode S only all-call; comm-A, altitude request; comm-A, identity request and comm-C, ELM.

The transponder system will have comprehensive BIT facilities, including power-up BIT, continuous BIT and initiated BIT.

Specifications
Dimensions:
(transponder) 194 × 124 × 318 mm
(control unit) 95 × 146 × 100 mm
Weight:
(transponder) 9.5 kg
(control unit) 2.2 kg
Power supply: 28 V DC
Frequency:
(transmitter) 1,090 ± 0.5 MHz
(receiver) 1,030 MHz

Status
In service.

Contractor
Selex Sensors and Airborne Systems.

PTR283 Mk1/PVS1280 Mk II IFF interrogators

Type
Identification Friend-or-Foe (IFF) system.

Description
The PTR283 and PVS1280 interrogator systems have been designed to meet the requirements of in-flight secondary radar interrogators. The transmitter/receiver uses pulsed oscillator techniques employing automatic frequency control and a logarithmic receiver using silicon integrated circuits. The equipment interrogates on Modes 1, 2 and 3A, the pulses driving the modulator being generated by the encoder/decoder.

The PTR283 Mk1 equipment consists of a lightweight D-band transmitter/receiver unit, an associated encoder/decoder and a control unit. Other units associated with the system are an antenna switch, dual-antenna system and L-trace radar display.

The PVS1280 Mk II system consists of a D-band lightweight transmitter/receiver and encoder/decoder, which offers the facility of active decoding and defruiting. This equipment is designed to integrate into an airborne primary radar system. Control of the PVS1280 is performed by the radar controller. It offers ISLS operation and an Interrogation SideLobe Suppression (SLS) switch is available to enable the transmitted power to be distributed equally to the two antennas and the two antennas to be fed alternately in phase and anti-phase for ISLS operation.

Specifications
Frequency:
(transmit) 1,030 ± 0.5 MHz
(receive) 1,090 ± 0.2 MHz
Pulse length: 0.8 ± 0.2 µs
Duty cycle: 0.11%
Decoder: 496 codes

Status
In service.

Contractor
Selex Sensors and Airborne Systems.

PTR446A transponder

Type
Identification Friend-or-Foe (IFF) system.

Description
The PTR446A lightweight transponder identifies aircraft in response to secondary radar interrogation and covers civil and military modes. Emphasis in design was placed on reliability combined with small size and low weight and these qualities have been achieved by the use of specially designed micro-electronic circuits. A digital shift register replaces conventional delay lines in the decoder/encoder circuits, so providing time delays independent of temperature. Integrated circuits are used for the logic and video processing circuits and the logarithmic response intermediate frequency amplifier. Decoder, encoder and associated switches in the control unit reduce the number of interconnecting wires to five and substantially cut down the installation weight. The transmitter/receiver houses the pulse selection and power-supply modules. Three-pulse sidelobe suppression is incorporated.

Two control units are available for use with the transmitter/receiver. The smaller of the two is the PV447, of which there are six versions with the following capabilities:
PV447 – Mode 1 or 3A/B and Mode C or off
PV447A – Mode 1 or 2 and Mode C or off
PV447B – Mode 2 or 3A/B and Mode C or off
PV447C – Mode 1 or 3A and Mode 2 or off
PV447D – Mode A or B and Mode C or off
PV447E – Mode A/B or off and Mode C or off.

An alternative to the PV447 is the PV1447 control unit, which meets the requirements of NATO STANAG 4193 for IFF Mk XA and provides Modes 1, 2, 3/A and C. Automatic code changing is provided on Modes 1 and 3A with storage capacity for 48 codes in each mode. Manual code entry for these modes is via a front-panel keypad. Mode 2 codes are entered through screwdriver adjusted switches reached through the top cover of the unit.

The transponder can be used with either control unit without modification to the transmitter/receiver. Comprehensive self-test is incorporated in all units.

Specifications
Dimensions:
(transponder) 57 × 127 × 254 mm
(PV447 control unit) 146 × 57 × 102 mm
(PV1447 control unit) 146 × 95 × 165.1 mm
Weight:
(transponder) 1.7 kg
(PV447) 0.48 kg
(PV1447) 1.6 kg
Power output: 24.7 dBW
Pulse rate: 1,200 replies/s, each containing up to 14 reply pulses
Triggering sensitivity: –72 to –80 dBm

Status
In production and in service with helicopters of the UK Royal Navy, Army and Royal Air Force, and on the BAE Systems Hawk aircraft.

Contractor
Selex Sensors and Airborne Systems.

SAR homing systems

Type
Avionic V/UHF homing system.

Description
Series 406
Fully compatible with the latest 406.025 MHz COSPAS/SARSAT emergency locator beacons, the 406 system is a complete self-contained unit interfacing with a single pair of antennas to provide 'left/right' steering information against a transmission source.

Designed to monitor four distress frequencies: 121.5, 156.8, 243.0 and 406.025 MHz, the unit processes the information and displays it on an analogue indicator. Adjacent-frequency test modes are provided.

RAF Rescue helicopter with Series 406 Homer and Series 500 personal locator beacon   0001180

Series 406-053 NVIS-compatible homing system   0001181

Specifications
Radio frequencies: 121.5, 156.8, 243.0, 406.025 MHz
Dimensions: 146 × 66.6 × 155 mm
Weight: 0.9 kg

Series 406 derivatives
406-1 Homer: The initial system covers 121.5, 156.8, 243.0 MHz. Fitted to both rotary- and fixed-wing aircraft.

406-2 Homer: Identical in function to the 406-1 Homer, with the addition of 406.025 MHz capability. Both standard and night vision compatible; additional remote indicators optional.

406-3 Homer with extended frequency range capability: With a remote controller can be extended over the whole 100–400 MHz band, while retaining the 406-2 Homer features. Can be interfaced to ARINC 429 or 1553 databus.

406-053 Full NVIS-compatible homing system: Four frequency capability: 121.5, 156.8, 243.0, 406.025 MHz, plus additional frequencies at ±1 MHz of each distress frequency. 5 V or 28 V

Green NVIS; choice of white or red lighting for non-NVIS units.

Series 407
The series 407 remote indicators are for use where operators need a lightweight indicator either as a secondary instrument for navigators' use or where panel space precludes use of the 406 series. The 407 provides a visual indication of the 'left/right' steering information to track a transmission source, the directional data is indicated only when a valid 'homing' signal is received from the 406 homing unit.

Specifications
Dimensions: 86 × 48 × 123 mm
Weight: 0.3 kg

Status
As of 2005, in production and in service.

Contractor
H R Smith.

TERPROM®/TERPROM®II

Type
Ground Proximity Warning/Terrain Awareness System (GPWS/TAWS).

Description
TERPROM® is a digital terrain system developed by BAE Systems. Its capabilities include high accuracy, drift-free navigation, predictive ground proximity warning system (GPWS), obstruction warning and cueing, database Terrain Following (TF) and enhanced weapon aiming. The system is entirely passive and produces no forward emissions.

TERPROM® uses standard digital terrain elevation data and vertical obstruction information stored in on board computer mass memory. A Kalman filter compares this data with information from the radar altimeter, allowing in-flight calibration of errors in the aircraft inertial navigation system.

Terrain referenced navigation accuracies of 30 m CEP (horizontal) and 3 m LEP (vertical) are achievable over terrain roughnesses as low as 2 per cent. System robustness is such that accurate navigation can be maintained for long periods when flying over water or during total radio silence.

The recently launched TERPROM®II product has dramatically improved navigation performance which, combined with the use of optional GPS velocity data, eliminates any nuisance ground proximity warnings during tactical flying.

TERPROM® generates 'predictive' ground proximity warnings which are not reliant on current radar altimeter inputs but are based on the output from the navigation capability. Warnings may therefore be provided even when the aircraft is approaching the terrain inverted.

The predictive algorithm is tuned to match the performance of each specific aircraft platform. This, together with the high accuracy and terrain referenced nature of the system's navigation solution, allows timely and accurate warnings to be provided even when the aircraft is being flown aggressively as low as 250 ft above ground level.

TERPROM® provides precursory directional warnings of man made obstructions (for example radio masts) in the aircraft's flight path. This allows the pilot to visually locate and fly around rather than over obstructions, reducing his chance of being detected.

TERPROM® generates terrain following signals, which may be displayed to the pilot. Unlike conventional terrain following radar systems, TERPROM® achieves total freedom of manoeuvre by the use of its onboard map. Through its constant awareness of the shape of the terrain beyond the immediate horizon, TERPROM® enables aircraft to hug contours more closely, reducing exposure to attack.

TERPROM® passive ranging allows undetected approach and accurate weapon aiming accuracy. With its knowledge of terrain elevation and true aircraft height, TERPROM® enables height above target, and hence impact location, to be computed continuously and accurately.

A single shot look aside ranging mode is also available, providing a 'passive laser range-finder' capability.

Inputs to TERPROM® include the: inertial navigation system data; baro/inertial altitude; radar altimeter; Pilot selections (TF and GPW heights); GPS data (optional).

Outputs from TERPROM® are as follows: inertial navigation system corrections; ground proximity warnings; obstruction warnings and cues; terrain following marker; ranging data.

TERPROM®/TERPROM® II is available in the: BAE Systems Modular Navigation System; Honeywell H764G embedded INS/GPS; OSC Fairchild Defence Mega Data Transfer Cartridge with Processor (MDTC/P); AYK14 and Open Systems mission computers from Computing Devices.

Status

TERPROM® was originally developed by BAE Systems for low flying fast jet aircraft, and has been selected for, or is in service on, Eurofighter Typhoon, F-16, Harrier GR. Mk 7, Jaguar, Mirage 2000 and Tornado GR. Mk 4.

The transport aircraft variant of TERPROM, known as the Terrain Awareness Warning System (TAWS), was selected for the US Air Force's C-17 fleet.

A version of TERPROM for rotary-wing aircraft has been developed, and was successfully trialled by the British Army in September 1999 on a Lynx helicopter. The purpose of the trial was primarily to demonstrate TERPROM's capabilities on helicopters flying transit and terrain missions.

Dynamic terrain advisory cues combine with a map of the surrounding terrain enable the pilot to assess his course of action while remaining at low level, with terrain cues being provided down to 100 ft.

Contractor

BAE Systems.

United States

AN/APX-( )MAT IFF transponder

Type

Identification Friend-or-Foe (IFF) system.

Description

The AN/APX-( )MAT contains, in a single unit, all Mk XII, Mode S and cryptographic IFF transponder functions with either a MIL-STD-1553B multiplex bus or a discrete control panel interface. It operates on Modes 1, 2, 3A, C, 4 and S. It is compatible with AIMS and STANAG requirements and uses advanced microwave packaging techniques to minimise space and weight.

Features include a KIV-2 COMSEC appliqué that uses an electronic key fill loader to eliminate requirements for a computer and mechanical code loader. Mode S Level 1 is provided as a standard feature, with optional growth to Levels 2 and 3.

Specifications

Dimensions: 136.5 × 136.5 × 212.6 mm
Weight: 4.55 kg
Frequency:
(receive) 1,030 ± 0.5 MHz
(transmit) 1,090 ± 0.5 MHz
Power output: 27 dBW ± 2 dB at 1.2% duty cycle

Status

In service.

Contractor

Northrop Grumman Corporation, Navigation Systems Division.

AN/APX-100 IFF/Mode S transponder

Type

Aircraft transponder.

Description

The AN/APX-100 IFF mode S transponder is a panel-mounted IFF transponder using microminiature technology in both digital and RF circuitry. It is in production for a number of US military aircraft. The system is a completely solid-state, modular constructed equipment with a complete dual-channel diversity system, comprehensive BIT, digital coding and encoding and a high anti-jamming capability. Two antennas form part of the equipment and the diversity system receives signals from each and switches the transmitter output to the antenna which received the stronger signal. This is designed to cure the problems of poor coverage with a single antenna.

The transmitter is all-solid-state with a 500 W peak power output obtained from four parallel microwave transistors. Two additional transistors complete the transmitter oscillator and driver stages. The diversity system provides improved antenna coverage and allows improvement in performance in overloaded, jamming and multipath environments. Automatic overload control and anti-jamming features are also incorporated.

For aircraft with very limited cockpit space, an equipment bay-mounted configuration, the RT-1157/APX-100(V), is available. In addition, the RT-1471/APX-100(V), a databus version operating in accordance with MIL-STD-155B, is available.

The latest in the APX-100 family includes Mk XII and embedded Mode 4 integrated crypto, full Mode S Level 3 and GPS position reporting for advanced air traffic control systems and TCAS systems. There are also NVG-compatible units available which operate at low light levels.

Specifications

Dimensions: 136.5 × 136.5 × 212.7 mm
Weight: 4.53 kg
Frequency:
(transmitter) 1,090 ± 0.5 MHz
(receiver) 1,030 ± 0.5 MHz
Peak power: 500 W ± 3 dB

Status

The APX-100 replaced the AN/APX-72 as the standard US military ATC transponder. Over 13,000 units have been built to date.

In production for all new US Navy, Army and Air Force aircraft including the AF-1, AH-1S/T, AH-64, AV-8B, C-5, C-12, C-17, C-20, C-21, C-23A, CH-47, C-130, EC-2C, EC-130, F-14D, F-18, F-22, HH-60, HH-65, LAMPS, MH-47, MH-60, OH-58, OV-10, RAH-66, SH-60, T-45A, UH-60, V-22 and VC-6.

The USA is delivering 113 APX-100 Mk XII airborne transponders to Hungary under a USD12.7 million contract.

The equipment is also being produced under licence in Japan by Toyocom.

Contractor

Honeywell Aerospace, Sensor and Guidance Products.

AN/APX-101(V) IFF transponder

Type

Identification Friend-or-Foe (IFF) system.

Description

The AN/APX-101(V) diversity IFF transponder is all-solid-state and consists entirely of replaceable modules. The transponder is housed in a single LRU and is mounted in the airframe without the use of shock-mounts. Crystal-controlled pulsewidth, discrimination, decoding and encoding ensure accurate response to interrogations. BIT circuits monitor the critical parameters of the transponder and provide an immediate status indication.

The transponder operates in Mk XII Modes 1, 2, 3A, C and 4. It receives RF interrogations from upper and lower antenna systems which give near spherical reception coverage. The APX-101 has a separate receiver and video processing channel for each antenna. Normally the transmitter is selected for transmission on the lower antenna, which is usually seen by ground interrogators. The video processing circuits measure the strength of the received messages, process the stronger signal, and direct the transmitter reply to the antenna that received the stronger signal. It then decodes the interrogation into the proper mode, encodes the selected reply and transmits the coded RF reply through the correct antenna.

Specifications

Dimensions: 152.4 × 127.3 × 278 mm
Weight: 6.53 kg
Power supply: 28 V DC, 65 W
Frequency:
(receiver) 1,030 MHz
(transmitter) 1,090 ± 1.5 MHz
Temperature range: −54 to +70°C
Altitude: up to 100,000 ft

Status

Standard equipment on A-10, E-3A, F-5E, F-15 and F-16 aircraft. Over 3,000 units have been ordered.

Contractor

Northrop Grumman Corporation, Navigation Systems Division.

AN/APX-103 Identification Friend or Foe (IFF) interrogator

Type

Identification Friend-or-Foe (IFF) system.

Description

The AN/APX-103 IFF interrogator system was developed for the Boeing E-3 Sentry AWACS, which is operational with the US Air Force,

The Honeywell Aerospace AN/APX-100 IFF transponder   0001212

The Boeing E-3 Sentry AWACS is equipped with the Telephonics AN/APX-103 IFF interrogator, with the antenna mounted back-to-back with the primary radar antenna in the rotating radome   0503737

The AN/APX-109(V)3 Mk XII IFF combined interrogator/transponder   0044986

NATO, UK, France, Japan and Saudi Arabia. The system utilises the features and functions of the co-operative beacon system commonly known as Mark XII Identification Friend or Foe (IFF) and is used to selectively locate and identify transponder-equipped civil and military aircraft. For military operations it can identify friendly aircraft while simultaneously providing instantaneous range, azimuth, altitude and identification in high-density target environments within the surveillance volume.

This information is used in conjunction with the AWACS onboard radar to perform air traffic control and Airborne Early Warning and Control (AEW+C) missions.

The standard configuration consists of two redundant receiver transmitters and a signal processor. The AN/APX-103 is configured as a Digital Beam Split (DBS) system while the later AN/APX-103B and AN/APX-103C utilise advanced monopulse processing, target detection and code processing algorithms.

Status
AN/APX-103 IFF interrogators are currently in service aboard all E-3 and E-767 AWACS aircraft. Two major upgrades for the system are in the process of being deployed. The first is the incorporation of a Mode S interrogation capability as part of the NATO mid-term upgrade programme (nomenclature to be assigned). The second is a reliability upgrade of the receiver transmitter, utilising a new high-powered, solid-state transmitter, low-noise receiver and low-voltage power supply.

Contractor
Telephonics Corporation, Command Systems Division.

AN/APX-108 IFF transponder

Type
Identification Friend-or-Foe (IFF) system.

Description
The AN/APX-108 is a single unit, reduced size Mk XII diversity IFF transponder used on the SR-71 and B-2. Diversity indicates that the transponder has an upper and lower antenna located on the aircraft. Using advanced microwave packaging, high-speed analogue/digital converters, advanced digital signal processing and CMOS LSI gate array circuitry, the component count, volume, weight and power dissipation of the AN/APX-108 has been minimised. It operates in Modes 1, 2, 3A, C and 4 and was designed for the next-generation Mode 5 IFF. It supports Mode S Level 3 and ADS-B with growth to Level 4.

Features include a COMSEC appliqué that uses electronic key fill to eliminate the requirement for a KIT-1A computer. For interface flexibility, the AN/APX-108 can be interfaced with either a MIL-STD-1553B multiplex bus or a C-6280 control box. The unit supports TCAS via an ARINC 429 or 1553B databus.

Specifications
Dimensions: 152.4 × 162.5 × 229 mm
Weight: 6.4 kg
Power supply: 28 V DC, 55 W nominal
Frequency:
(receiver) 1,030 ± 0.5 MHz
(transmitter) 1,090 ± 0.5 MHz
Output power: 500 W min at 1% duty cycle
Temperature range: –40 to +71°C
Altitude: up to 70,000 ft
Reliability: >4,000 h MTBF

Status
In production and in service.

Contractor
Northrop Grumman Corporation, Navigation Systems Division.

AN/APX-109(V)3 Mk XII IFF combined interrogator/transponder

Type
Identification Friend-or-Foe (IFF) system.

Description
The AN/APX-109(V)3 combined interrogator/transponder integrates into a single package functions previously housed in many units. The system is designed for use in all aircraft from helicopters to high performance fighters. The interrogator subsystem uses an electronically scanned antenna to output target reports to the 1553 Bus Controller for display on the radar display. The information for each target includes: Mode of reply (1, 2, 3/A, C or secured); reply code (for Modes 1, 2, and 3/A); altitude (Mode C); and position (range, azimuth, elevation).

The AN/APX-109(V)3 represents a major advance in IFF system design. AIMS and STANAG requirements are both provided in hardware design and system implementation utilises a fully integrated system approach, making practical use of advanced technologies.

The AN/APX-109(V)3 is an efficient solid-state design that reduces the component count, weight and volume of the IFF system significantly. Employing advanced signal processing, the AN/APX-109(V)3 offers: a prioritised four-channel operation; adaptive thresholding of the received video to improve performance in high-noise and jamming environments; and a monopulse processing of received target video for enhanced azimuth accuracy.

Specifications
Dimensions: 152 × 213 × 368 mm
Weight: 15.5 kg (including COMSEC appliqué)
Power supply: 28 V DC, 150 W
Frequency:
(receive) 1,030 ± 0.2 MHz
(transmit) 1,090 ± 0.2 MHz
Temperature range: –40 to +71°C
Altitude: up to 70,000 ft
Reliability: >2,000 h MTBF

Status
In production and in service.

Contractor
Northrop Grumman Corporation, Navigation Systems Division.

AN/APX-111(V) Combined Interrogator/Transponder (CIT)

Type
Identification Friend-or-Foe (IFF) system.

Description
The AN/APX-111(V) CIT consists of an interrogator, a transponder and two associated cryptographic computers in a single, small, lightweight unit. For retrofit applications the CIT can replace six or seven boxes making up the existing interrogator and transponder, yielding reduced weight and freeing space for other avionics equipment.

The AN/APX-111(V)'s architecture enables it to support Mk X (SIF), Mk XII (Mode 4) or custom crypto systems. The crypto module is integral but removable from the front panel. Mode S transponder capability and growth to Mk XV (NIS) are also part of the system architecture.

The CIT meets international standards for IFF and ATC including US-DoD AIMS 65-1000B and NATO STANAG 4193. AN/APX-111(V) utilises miniature low-profile fuselage-mounted electronically scanned antenna arrays. It provides full interrogator range capability without the need for external amplifiers. The modular design is all-solid-state with a unique approach to thermally efficient cooling. A MIL-STD-1750 processor and MIL-STD-1553B databus are included.

The latest IFF techniques are provided, including digital target reports for a clear operator display, monopulse processing for accurate target

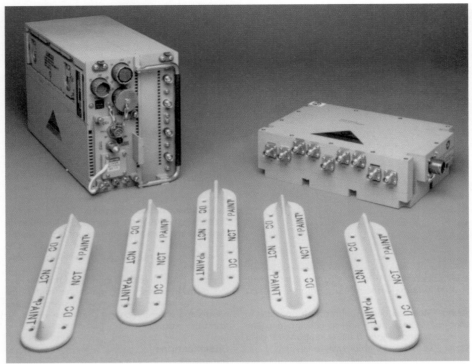

AN/APX-111 (V) combined interrogator/transponder   0044984

azimuth, a statistical reply evaluator for high-confidence identification, a defruiter for dense environments and cryptographic coding for security. Both continuous and operator-initiated built-in tests are provided, with reporting via the MIL-STD-1553B databus.

Specifications
Combined interrogator/transponder RT-1679(V):
    208 × 146 × 320 mm
Beam forming network C-12222(V): 178 × 249 × 76 mm
Fuselage-mounted antenna elements AS-4267(V):
    29 × 55 × 237 mm
Weights:
RT-1679(V): 15 kg
C-12222(V): 5.4 kg
AS-4267(V): 0.2 kg

Status
Fitted in F/A-18 FMS and F-16 MLU programmes, and Greek and Turkish air force F-16 aircraft.

Contractor
BAE Systems North America Greenlawn.

AN/APX-113(V) Combined Interrogator/Transponder (CIT)

Type
Identification Friend-or-Foe (IFF) system.

Description
The AN/APX-113(V) Combined Interrogator/Transponder (CIT) is a complete Mk XII identification system which includes crypto computers. It consists of one unit and incorporates growth for the next generation of IFF and combat aircraft identification equipment. The AN/APX-113(V) provides both interrogation and IFF responses on IFF Mode XII Modes 1, 2, 3/A, 4, C and S Level 3.

The multiple antenna configurations feature electronic or mechanical scan. The system features Ada software and a MIL-STD-1553 bus interface.

Specifications
Dimensions:
(combined interrogator/transponder)
    209.8 × 152.4 × 368.3 mm
(beam-forming network)
    165.1 × 212.9 × 101.6 mm
(fuselage-mounted antenna elements)
    39.4 × 82.6 × 332.7 mm
(lower interrogator antenna) 15.2 × 431.8 × 355.6 m
Weight:
(combined interrogator/transponder) 14.52 kg
(beam-forming network) 4.54 kg
(fuselage-mounted antenna elements) 0.23 kg
Power supply: 28 V DC, 200 W
Range: 1 km
Coverage:
(azimuth) ±60°
(elevation) ±60°

Accuracy:
(range) 500 ft
(azimuth) ±2°
In-beam targets: 32
Reliability: 1,600 h MTBF

Status
Developed specifically for F-16 Falcon. Fitted to F-16 Block 16 A/B MLU, Block 20 A/B. Also fitted to ASW/surveillance helicopters and the Japanese FS-X fighter.

The US Air Force intends to retrofit the AN/APX-113 to its F-16 Block 50/52 aircraft. This installation is designated the Advanced Identification Friend or Foe (AIFF) system. AIFF flight trials commenced in 2001, with aircraft installation starting in July 2002.

BAE Systems (Rochester, UK) was selected to supply Successor Identification Friend or Foe (SIFF) systems for retrofit on UK Royal Navy Sea Harrier F/A Mk 2 aircraft, required to upgrade IFF Mk Xa to Mk XII Modes 1,2 3/A, C and S level 2. Development work, based on the AN/APX-113 system, was completed during 2000, with full-scale production beginning later that year.

Contractor
BAE Systems North America, Greenlawn.

AN/AYD-23 GPWS: Ground Proximity Warning System

Type
Ground Proximity Warning System/Terrain Awareness Warning System (GPWS/TAWS).

Description
Adaptable to all classes of rotary-wing aircraft, GPWS can be installed as a ¼ ATR (short) avionics package interfaced with discrete sensors, or via the ARINC 429/MIL-STD-1553B databusses. It can also be embedded in computers already on board the aircraft.

The GPWS continuously monitors important aircraft parameters in real time:
- Attitude (roll and pitch) relationship to terrain (altitude and closure rate acceleration)
- Terrain type and terrain trends (rising and falling)
- Validity and reliability of sensor inputs
- Response capabilities of the pilot and the aircraft.

These inputs are used to determine the dynamic state of the aircraft. From this data, the GPWS algorithm computes the following:
- The predicted altitude loss due to pilot response time
- The predicted altitude loss due to roll recovery
- The predicted altitude loss during the actual recovery
- The predicted altitude loss due to terrain change.

The sum of these altitude loss quantities is used by the GPWS to compute the need for warnings to be issued.

The GPWS is designed to use existing sensors, including radar altimeter, barometric altimeter, attitude gyros, accelerometers, air data sensors. Growth capability to use forward-looking warning radar systems and windshear data is included in the design.

The AN/APX-113(V) showing the combined interrogator/transponder (left), the lower interrogator antenna (centre, rear), the fuselage-mounted antenna elements (centre, front) and the beam-forming network (right)   0504120

Cubic Defense Applications' AN/AYD-23 ground-proximity warning system   0018926

The GPWS algorithm continually assesses the validity of input data and is optimised for highly dynamic tactical flight environments.

Status
In service in Canadian Forces C-130 transport aircraft and on US Navy H-46 and H-53 helicopters.

Contractor
Cubic Defense Applications.

AT 155 ATC transponder

Type
Aircraft transponder.

Description
The Narco AT 155 transponder is a self-contained panel-mounted unit which meets the FAA's C74c Class 1A TSO specification for ATC transponders. Featuring full 4,096 code capability, the AT 155 is compatible with leading encoding altimeters and blind encoders such as the Narco AR 850 Altitude Reporter.

The AT 155 is both a transmitter and receiver which responds automatically to ground-based radar interrogation. The transponder reply is displayed on ATC radar displays as two short parallel lines which 'fill in' when the system squawks ident.

Specifications
Dimensions: $159 \times 45 \times 286$ mm
Weight: 0.68 kg (panel unit); 0.28 kg (antenna and cable)
Power supply: 13.75 V DC, 1.2 A or 27.5 V DC, 530 mA
Power output: 250 W nominal
Frequency:
(transmit) 1,090 MHz $\pm$ 3 MHz
(receive) 1,030 MHz
Altitude reporting: 100 ft increments to 30,700 ft (with altitude encoder)
Sensitivity: -72 dBm minimum
Range: 200 miles, line-of-sight
Compliance: FAA TSO C74c Class 1A approved to 30,000 ft in pressurised and non-pressurised fixed- and rotary-wing aircraft

Status
In service. The AT 155 is a direct replacement for all Narco AT 50, AT 50A and AT 150 transponders.

Contractor
Narco Avionics Inc.

Configurable Integrated Surveillance System (CISS)

Type
Integrated Surveillance System (ISS).

Description
Collins' CISS-2100 Configurable Integrated Surveillance System is in development for the forthcoming Boeing B787 Dreamliner. The system combines the company's WXR-2100 multiscan weather radar (see separate entry) with a Traffic Collision Avoidance System (TCAS), a Mode S transponder and a (Honeywell) Terrain Awareness and Warning System (TAWS). This drastically cuts both the volume and total weight of the combined system versus separate units, while improving overall reliability.

All of the aforementioned elements fit into an 8 MCU enclosure, replacing a total of 20 MCU for the separate elements.

Status
Type certification for the system on the B787 is expected in 2008.

Contractor
Rockwell Collins.

Crash-Survivable Memory Unit (CSMU)

Type
Flight/mission recording system.

Description
The L-3 Communications/Electrodynamics, Inc. (L-3 EDI) Crash-Survivable Memory Unit (CSMU) receives vehicle, subsystem and environmental parameters, via various communication interfaces, and stores them (in uncompressed form) in EEPROM solid-state memory. The system protects the acquired data from damage and loss to levels exceeding EUROCAE ED-55. The CSMU can be readily adapted to any aircraft. A Flight Data Acquisition Unit suitable for operation with the CSMU is also available.

Other features of the CSMU include:

- 512 KB EEPROM, expandable to Gigabytes of flash memory
- 100,000 memory write cycles
- Power fail protection, memory management, and error detection
- Interface: RS-422 with optional MIL-STD-1553B, 10 Base-T Ethernet, IEEE1394A/B, Universal Serial Bus, microphone pre-amp

- Crash data extraction methods and equipment available
- Simple communications interface
- Less than one bit error per million
- BIT: power-on, commanded, periodic
- RS-422 test bus
- Underwater locator beacon with replaceable battery
- Suitable for data, audio and/or video recording

The CSMU mounts in a crash-survivable and accessible area of the vehicle (usually in the tail section), and may be up to 100 feet from its data source. It contains a micro-controller, communication interfaces, power conditioner, crash survival memory and an externally mounted acoustic beacon.

The system is capable of circular loop rewrites between 20 minutes and 25 hours for up to 17,000 operating hours. The memory format handles periodic and aperiodic parameters and events, with read-after- write and CRC error-detection. All records are independent and self-documenting as stored.

Unit BIT is performed on power-up or on external command, with BIT status available on the communication buses.

Downloading can be performed after an incident or at any time. Card-level and chip-level data extraction equipment and test equipment are also available. The unit is not field-repairable due to its unique construction.

Specifications
Performance
Temperature: -40 to $+71°C$
Shock: 5,100 g/5–8 msec, six axis
Penetration: 10 ft drop of 500 lb weight, six axis, 0.05 in^2
Crush: 20,000 lb, all orthogonal and major diagonal axes
Fire: $1,100°C$ for 60 min
Seawater immersion: 20,000 ft, 3 days
Fluids: Fuel, glycol, hydraulic, fire extinguishing; 48 hr
Dimensions: $76.2 \times 114.3 \times 165.1$ mm
Weight: 2.95 kg (<2.27 kg for titanium version)
Power: $+9$ to $+15$ V DC, 150 mA; 28 V DC
MTBF: 30,000 hr (MIL-HDBK-217E)
Life: 17,000 operating hours; 30 years useful
Cooling: Convection

Status
In production.

Contractor
L-3 Communications, Electrodynamics, Inc.

CPI-406 deployable Emergency Locator Transmitter (ELT)

Type
Avionic Emergency Locator System (ELS).

Description
The CPI-406 121.5 MHz/406 MHz deployable Emergency Locator Transmitter (ELT) meets or exceeds current and planned international ELT requirements for automatically deployable systems on civil and military helicopters. The electronics are packaged within a crash-survivable Beacon Aerofoil Unit (BAU) that is qualified and certified to CAA, TCA and FAA standards.

The BAU separates from the aircraft at the onset of an incident, thereby escaping the effects of the crash. The ELT immediately transmits a distress signal, including the host aircraft identification number, to the international COSPAS-SARSAT network. When fitted with a Global Positioning System (GPS) encoding option, the BAU also transmits the last known latitude and longitude of the aircraft to aid in search and rescue efforts. In addition, the BAU floats indefinitely, thereby enhancing survivability and localisation for both aircraft and crew over water.

A direct upgrade for the CPI-113 system, the CPI-406 is the latest product from the EAS3000 family of emergency avionics equipment that includes deployable cockpit voice and flight data recorders, emergency locator transmitters, and aircraft monitoring systems.

The main features and benefits of the CPI-406 are:

- Meets rigorous standards for crash survivability mandated for automatically deployable ELTs
- 406 MHz GPS encoding provides rapid location and identification
- Built-In-Test (BIT) continuously monitors system performance
- Field reprogrammable to all COSPAS-SARSAT protocols
- Meets requirements for EMC, lightning and High Intensity Radiated Fields (HIRF) protection.

Specifications
BAU
Dimensions: 305×90 mm (Ø $\times$ L)
Weight: 1.55 kg
Cockpit Control Unit (CCU)
Dimensions: $122 \times 146 \times 57$ mm
Weight: 700 g

Release controller
Dimensions: $102 \times 152 \times 84$ mm
Weight: 1.36 kg
Transmitters
121.5 and 406.028 MHz
50 mW at 121.5 MHz PERP
5 W at 406 MHz PERP
>48-hour endurance at $-40°C$ (121.5 MHz)
>24-hour endurance at $-40°C$ (406 MHz)
Optional frequencies available
Battery life: 7 years
Transmitter activation: Automatic when unit deploys
Transmitted information: Aircraft ID, last known position
Positional information interface: ARINC-429, RS-422
Crash survivability: 80 ft/sec impact; high temperature fire; salt and aviation fluids immersion; floats indefinitely
Compliance: Fully meets requirements of DO-160C for helicopters, including composite platforms. Airworthiness certification pending. System approval to performance requirements of TSO C-91A, C-126, TCA, FAA, CAA, JAARS, DO-204, -183, ED-62A, CAA specification 16 and COSPAS-SARSAT C/S T.001, C/S T.007

Status
In production and in service.

Contractor
DRS Technologies, Inc.

Deployable Flight Incident Recorder Set (DFIRS) 2100

Type
Avionic Emergency Locator Transmitter (ELT).

Description
The DFIRS 2100 is a combined Flight Data Recorder (FDR), Cockpit Voice Recorder (CVR) and Emergency Locator Transmitter (ELT), designed to provide instantaneous alert and accurate location/ identification of a downed aircraft, assist in the speedy rescue of survivors, aid in the recovery of the aircraft and assure timely recovery of vital CVR/FDR data for accident investigation.

In the event of an accident, the deployment of the Beacon Aerofoil Unit (BAU) is automatically triggered by on-board sensors, thereby allowing the BAU to clear the immediate area of the impact site. The ELT immediately transmits a distress signal, including aircraft identification number, to the COSPAS-SARSAT network. With the Global Positioning System (GPS) encoding option, the BAU also transmits the last known latitude and longitude of the aircraft to aid in search and rescue efforts. The BAU is able to float indefinitely, enhancing survivability and localisation of crew and aircraft over water.

Advanced Commercial Off-The-Shelf (COTS) technology is employed to provide a highly-reliable, solid-state, lightweight and operationally effective combined FDR/CVR/ELT system in a deployable, crash survivable aerofoil that can be installed on helicopters, fixed-wing and fast-jet aircraft.

Further features of the DFIRS 2100 inlcude:
- COSPAS-SARSAT-certified 121.5/406 MHz ELT (optional frequencies available)
- 406 MHz digital signal provides instantaneous alert and identification
- Meets current and planned CVR and FDR requirements
- Meets high EMC, HIRF and lightning specifications for composite aircraft
- Integrated CVR and FDR data acquisition and processing
- Optional HUMS data collection and processing.

Specifications
Certification: TSO C-91A, -123A, -124A, EUROCAE ED-55, -56, -56A, RTCA DO-183, CAA specifications 11, 16, 18; DO-160D (aerofoil)

Cockpit Voice Recorder
Storage: 1 h storage for each of 4 channels (2 h optional)
Audio channels: 3 channels 150 Hz to 3.5 kHz, 1 channel 150 Hz to 6 kHz bandwidth
Interface: 8-channel ARINC-429 or 2 dual-redundant MIL-STD-1553B
Analogue acquisition: Discrete, analogue, synchro, frequency, thermocouple, fuel flow

Flight Data Recorder
Storage: 25 h

Emergency Locator Transmitter
Frequencies: 121.5/243 MHz; 121.5/406 MHz
Power: 50 mW/50 mW PERP, >48 hrs at –40°C (121.5/243 MHz); 50 mW/5 W PERP, >24 h at –40°C (121.5/406 MHz)
Compliance: DO-183, -204, ED-62, TSO C-97, -147

Ground support equipment
Flight data playback
Data readout in engineering units/audio/ graphical replay (option)
Supports data retrieval directly from the beacon or data interface unit

Physical
Weight: 7 kg

BAU dimensions: 330 × 460 × 76 mm
Data interface unit dimensions: ½ ATR short (320 × 124 × 190 mm)
Power requirements: 28 V DC; 30 W maximum

Status
The DFIRS 2100 is in production and in service on a variety of fixed-wing aircraft.

Contractor
DRS Technologies, Inc.

Deployable Flight Incident Recorder Set (DFIRS)

Type
Avionic Emergency Locator Transmitter (ELT).

Description
The DFIRS is an integrated Flight Data Recorder (FDR) and 243 MHz Emergency Locator Transmitter (ELT) installed as standard equipment on F/A-18C, D, E and F aircraft. The DFIRS' built-in ELT provides immediate alert of a downed aircraft, supporting the prompt location of the crash site/ aircraft crew and recovery of the crash-protected flight recorder module.

In the event of an accident, the deployment of the DFIRS is triggered automatically by an impact sensor or through operation of the ejection seat system. Activation by either of these means initiates the deployment mechanism that releases the DFIRS into the aircraft slipstream where it 'flies' away from the aircraft. The rugged mechanical design of the aerofoil allows it to withstand severe impacts with ground or water, and it will float indefinitely until recovered. The internal ELT is activated simultaneously by the release of the aerofoil. Ground support equipment provided with the DFIRS can be easily connected to the flight recorder to immediately download and analyse the internal data.

The advanced capabilities of the DFIRS are designed to meet the requirements of current and future high-performance jet aircraft. Optional upgrades and interfaces ensure compatibility with the latest avionics suites and associated standards, including future implementations of Firewire IEEE-1394b.

Major features of the system include:
- Automatically deployable FDR/ELT
- MIL-STD-1553 FDR data acquisition
- 243 MHz ELT with optional 121.5 MHz/406 MHz
- GPS position encoding of latitude and longitude with 406 MHz
- 72-hour ELT endurance
- Meets extreme F/A-18 environmental and EMC requirements
- Proven in-service reliability
- Proven crash survivability and recoverability
- Low maintenance requirements
- Full continuous Built-In-Test (BIT).

Specifications
Flight Data Recorder
Interface: MIL-STD-1553
Storage capacity: 256 KByte, optional 2 Gbyte
Recording time: 30 minutes, optional 25 hours

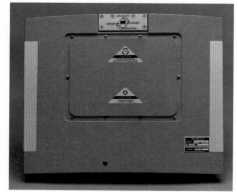

Deployable Flight Incident Recorder Set (DFIRS)   0051317

Emergency Locator Transmitter
Design standard: RTCA DO-183
Transmit frequency: 243 MHz
Output power: 70 mW PERP
Battery endurance: 72 h at 25°C
Battery design standard: NAVSEA S9310-AQ-SAF-010
Antenna pattern: Omni-directional

Physical
Aerofoil dimensions: 330 × 406 × 76 mm
Aerofoil weight: 2.6 kg
Data bus interface unit dimensions: 260 × 194 × 44.5 mm
Data bus interface unit weight: 2.2 kg
Power: 28 V DC, 5 W

Environmental
Compliance: MIL-STD-810, MIL-STD-461
Operating temperature: –55 to +100°C

Upgrade options
Dual frequency 121.5/406 MHz ELT with aircraft position encoding
Allows immediate notification of an aircraft crash through international COSPAS-SARSAT.

Enhanced data storage capability
Additional memory can provide at least 25 hours FDR record time compatible with modern FDR standards and the capability to store other types of data.

High-speed data bus
Ethernet, IEEE-1394b Firewire and fibrechannel can be interfaced to allow greater volumes of data to be stored and retrieved from the DFIRS memory.

Additional data acquisition and storage
Additional data can be acquired directly from analogue sensors to complement the existing list of FDR parameters.

Audio and video data storage
Integration of existing aircraft systems into the DFIRS facilitates storage of audio/video data in a deployable, crash-protected memory.

Status
The DFIRS is installed in US Navy F/A-18 C/D/E/F variants. The system is also employed in more than 600 types worldwide, including the C-130 Hercules, P-3C Orion and German Air Force and Navy Tornado PA-200 aircraft.

Contractor
DRS Technologies, Inc.

DFS-43 direction-finder system

Type
Avionic Direction-Finding (DF) system.

Description
The DFS-43 is an all-digital automatic direction-finding system. It is a lightweight system consisting of the DF-431 receiver, CD-432 control display unit and AT-434 combined loop and sense antenna.

The DF-431 receiver provides accurate reception of en route NDBs, locator outer markers and commercial AM broadcast stations. It is available with front- or rear-mounted connectors.

The CD-432 control display unit provides for a dual-frequency readout, with one active and one standby. Frequencies are alternated via the frequency transfer button. The unit provides frequency control from 190 to 1,860 kHz, with half or whole kHz incremental spacing. It also features a non-volatile frequency memory which retains the last frequency used, eliminating the possibility of frequency loss due to power interruptions.

The DFS-43 automatic direction-finder employs full time monitoring and self-testing of key functions such as the power supply, synthesiser lock, receiver lock and signal processing.

The AT-434 combined loop/sense antenna system is designed specifically for use with the DFS-43 system.

Specifications

Dimensions:
(antenna) 146 × 152.4 × 332.7 mm
(control display unit) 63.5 × 79.4 × 63.5 mm
(front mount receiver) 101.6 × 101.6 × 279.4 mm
(rear mount receiver) 101.6 × 101.6 × 320.5 mm
Weight:
(antenna) 1.72 kg
(control display unit) 0.27 kg
(front mount receiver) 2.36 kg
(rear mount receiver) 2.36 kg
Power supply: 18–33 V DC, 0.6 A at 28 V

Contractor

Honeywell Aerospace, Electronic & Avionics Lighting.

Direction-finding system

Type

Airborne Direction Finding (DF) system.

Description

Received RF signals between 20–1,500 MHz route to distribution and switching circuitry contained in a modular RF unit. This unit provides 14 reconfigurable HF/VHF/UHF RF inputs and is designed to be cascadable, allowing for additional antenna elements or arrays, simply by adding other RF units. The RF unit also includes a switchable dual-channel HF upconverter for coverage of the 2–20 MHz frequency range.

The IF outputs of the two Nanomin receivers are each routed into coherently clocked A/D converters and then into a two-channel DDC. The DDC digitally fine tunes to the specified frequency, generates in-phase and quadrature components of each channel, down-converts the desired portion of the IF bandwidth to baseband, and digitally filters the signal with linear phase Finite Impulse Response (FIR) filters. The DDC also performs basic preprocessing and controls data flow over the VME bus to the Motorola 68040 processor.

Switchable IF filters operate digitally within the DDC. The set of IF bandwidths can be changed easily by downloading a different set of coefficients. The digital decimation filter uses FIR filter designs that exhibit ideal phase characteristics as well as excellent roll-off features. This digital demodulation concept offers unique flexibility for tailoring the DF receiver to specific user needs.

Accurate angle of arrival measurements can be obtained for both data and voice signals using virtually any type of modulation, including AM, FM, continuous wave, single sideband, and independent sideband. Precision results on digital signals can also be achieved using frequency-shift keying or pulse-shift keying modulation.

The DF processor is a single board computer using the Motorola 68040 processor, with its floating point math co-processor running at 25 MHz. Along with 4 Mbyte of random access memory (RAM), a 16 Mbyte electrically erasable programmable read-only memory (EEPROM) board is included for non-volatile storage of calibration data. Interface boards are included for Ethernet, IEEE-488 and navigation systems (ARINC and Synchros).

Angle-of-arrival and position information feed into the DF processor for line-of-bearing (LOB) calculation.

The system handles pulse-type signals with pulse-widths down to 0.4 microsecond and with duty cycles as low as 0.0005.

Specifications

Frequency range: 2–1,500 MHz
DF Accuracy: <1°RMS estimated for fine DF using typical airborne VHF/UHF arrays
DF Sensitivity: –116 dBm (10 dB S/N in 6.4 kHz bandwidth)
DF Output: LOB, quality factor, frequency, time, in/out of geosort limits, emitter location (calculated in workstation)
Options:
(search) 100 freq/s (software upgrade only)
(classification) AM, FM, FDM, FSK, PSK
(copy) digital demod-AM, FM, SSB

Status

In production and in service.

Contractor

Raytheon Company, Space and Airborne Systems.

Display Switching Unit (DSU) for Terrain Awareness and Warning Systems (TAWS)

Type

Ground Proximity Warning/Terrain Awareness System (GPWS/TAWS).

Description

AVTECH's DSU is designed to be a low-cost alternative to replacing the entire display system in aircraft implementing TAWS capability for the first time. AVTECH's DSU is designed to interface with Honeywell's Enhanced Ground Proximity Warning System (EGPWS) terrain presentation on Honeywell UDI Weather Radar Displays (Primus models 200, 300, 300SL, 400 440, 450, 500, 650, 660, 700, 800, 870, 880, 90 and AVQ30). It is also designed to interface with Collins indicators WXR 220, 270 and 300.

The DSU switches up to three sources of display information and facilitates overlays:
1. EGPWS data is in an ARINC 708A/453 bus format; the DSU reformats the data into UDI format;
2. Traffic alert Collision Avoidance System (TCAS); the DSU passes UDI data;
3. Data/navigation/lightning data; the DSU passes through UDI data.

The hardware is qualified to DO-160D and the software to DO-178B. The MTBF is predicted to be 60,000 hours and the DSU is described as 'fail-safe' to the extent that it passes through UDI data, and does not interface with audible warnings and therefore does not deactivate when a failure occurs.

Specifications

Weight: <1.36 kg
Dimensions: 152 × 254 × 63 mm
Power: 28 V DC
TSO/JTSO: C105
ARINC: 413A, 429, 453, 708A

Status

AVTECH received FAA TSO C105 approval for the DSU during the second quarter of 2000.

Contractor

AVTECH Corporation.

ED-55 Solid-State Flight Data Recorder (SSFDR)

Type

Aircraft Accident Data Recorder (ADR)/Cockpit Voice Recorder (CVR).

Description

The ED-55 Solid-State Flight Data Recorder (SSFDR) is an all-solid-state implementation of a crash survivable flight data recorder. It conforms to ARINC 573, ARINC 717 and ARINC 747. The unit eliminates all moving parts and uses solid-state flash memory as the recording medium. The SSFDR comprises three shop replaceable units: the crash survivable memory, power supply and interface and control assembly.

Memory capacity allows for the last 25 hours of flight data to be stored at an input rate of 64 words/s or 128 words/s. Data is stored without the use of any data compression algorithm, which provides for maximum reliability and integrity of critical flight information. An underwater locator beacon is available as an option.

Specifications

Dimensions: ½ ATR long or short
Weight: 8.2 kg
Temperature range: –55 to +70°C
Reliability: >15,000 h MTBF

Status

In production and on order for Airbus aircraft, including the A319 and A320.

The first prototype of a combined SSFDR/SSCVR unit was shipped to Eurocopter for testing in the EC155 medium-lift helicopter in July 1998.

In June 1999, CASA selected the ED-55 SSFDR, as part of an extensive Honeywell Aerospace avionics package, for its new-production C-295 and CN-235 aircraft, also for retrofit installation into existing CN-235 aircraft.

Contractor

Honeywell Aerospace, Electronic & Avionics Lighting.

ED-56A Solid-State Cockpit Voice Recorder (SSCVR)

Type

Aircraft Accident Data Recorder (ADR)/Cockpit Voice Recorder (CVR).

Description

The ED-56A Solid-State Cockpit Voice Recorder (SSCVR) is ARINC 757 compliant. It uses industry

The Raytheon Electronic Systems direction-finding system   0001273

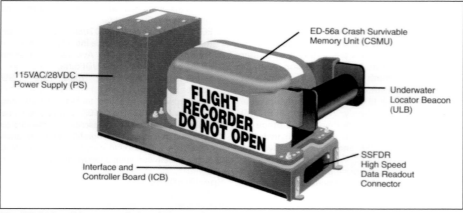

Labels on figure:
- ED-56a Crash Survivable Memory Unit (CSMU)
- 115VAC/28VDC Power Supply (PS)
- FLIGHT RECORDER DO NOT OPEN
- Underwater Locator Beacon (ULB)
- SSFDR High Speed Data Readout Connector
- Interface and Controller Board (ICB)

The ED-56A solid-state cockpit voice recorder   0081864

standard audio compression in both 30 minute and 2 hour models. Interface provisions include: one wideband and three narrowband audio channels; dedicated rotor tachometer input; FSK and GMT time recording; two ARINC 429 interfaces reserved for future ATC datalink messages; and maintenance data.

The ED-56A utilises FLASH EPROM memory and a patent Crash-Survival Memory Unit (CSMU). It is designed to replace older tape-based units.

Specifications
Dimensions: ARINC 404 ½ ATR short
Weight: <7.3 kg
Power: 35 W max

Status
In production and on order for Airbus aircraft, including the A319 and A320.

The first prototype of a combined SSFDR/SSCVR unit was shipped to Eurocopter for testing in the EC155 medium-lift helicopter in July 1998.

In June 1999, CASA selected the ED-56A SSCVR for its new-build C-295 and CN-235 aircraft, as part of a major avionics award to Honeywell Aerospace, Electronic & Avionics Lighting, which also includes retrofit application to its CN-235 aircraft.

Contractor
Honeywell Aerospace, Electronic & Avionics Lighting.

ELT-200HM Emergency Locator Transmitter

Type
Avionic Emergency Locator System (ELS).

Description
The ELT-200HM is designed for helicopter application, it can be mounted horizontally, and weighs only 0.87 kg. Dimensions are only 64 × 69.9 × 158.8 mm.

The ELT-200HM incorporates a 6-axis 'G' switch and a lithium battery pack that gives a four-year replacement life.

The FAA granted TSO C91A approval for the ELT-200HM in July 1999, with shipments beginning in September 1999.

Status
In service.

Contractor
Artex Aircraft Supplies Inc.

ELT 406 series Emergency Locator Transmitters (ELTs)

Type
Avionic Emergency Locator System (ELS).

Description
Artex ELT 406 series Emergency Locator Transmitters (ELTs) are designed to integrate with the Artex ELT to NAV interface, which receives position data from the aircraft navigation system via an ARINC 429 bus. Position information is then converted to ELT format and stored in memory for use upon ELT activation. Upon activation, the ELT transmits the latitude and longitude of the aircraft as part of the digital signal to the COSPAS/SARSAT satellite system.

The C406-1 is a basic 3-frequency (121.5/243 and 406.025 MHz) system, developed for aircraft owners and operators who prefer to use a single rather than a dual-input antenna. As a result of the antenna having one connector that carries all three emergency frequencies, the ELT only requires a single coaxial cable for operation.

The C406-1HM also features a 6-axis 'G' switch, designed for helicopter operations, which allows activation at six separate angles. The C406-1HM can be equipped with either a high-speed rod (aircraft operation up to 350 kt) or blade antenna.

The C406-2 and C406-2HM are similar to the -1 and -1HM, except that they feature a dual-input antenna and are shipped with a high-speed blade antenna which is approved for aircraft operating speeds up to Mach 1.

Type specific solutions include the B406-4 and G406-4. The B406-4 is designed to interface with Boeing's Master Caution System and is approved for use in all models of 737, 747, 757, 767 and 777 aircraft. The G406-4 is designed as a low-cost, General Aviation (GA) solution, featuring a whip antenna approved for speeds up to 200 kt. Both of these systems offer similar capabilities to the C406-1/2.

The latest addition to the range is the C406-N, applicable to both fixed-wing and helicopter platforms. The C406-N features a self-contained NAV interface which eliminates the requirement for a separate ELT to NAV interface box.

Status
In production and in service.

Contractor
Artex Aircraft Supplies Inc.

ELT Emergency Locator Transmitters (ELTs)

Type
Avionic Emergency Locator System (ELS).

Description
Artex Aircraft Supplies Inc manufactures a series of FAA-approved emergency locator transmitters that meet TSO C91a and TSO C126. The ELTs are designed for installation in fixed- or rotary-wing aircraft. The ELTs for fixed-wing aircraft activate via a single-axis 'G' switch and those for helicopters via a 6-axis 'G' switch module.

Artex also manufactures an ELT-to-NAV interface for use with the ELT 406-1 and -2 systems (see separate entries). This interface allows the ELT to couple via ARINC 429 to GPS, Loran or FMC thus broadcasting latitude/longitude as part of the 406 MHz digital message. The overall message includes: type of aircraft, owner, emergency contact, country code, serial number of ELT, ELT manufacturer, latitude/longitude position. Models available include the following:

Specifications
All models provide 50 mW output at 121.5/243 MHz for 50 hours. In addition, the ELT 110-406 systems operate at 5 W on 406.025 MHz for 24 hours.

Status
In production and in service. Widely fitted by fixed- and rotary-wing aircraft manufacturers.

ELT to NAV interface and ELT 406   0051606

Contractor
Artex Aircraft Supplies Inc.

Model number specifications	ELT-200/ 200HM	ELT 110-4	ELT 110-6	ELT-100HM	ELT-200HM	ELT 110-406NAV	ELT C406-1/ C406-2	ELT C406-N/ C406-HM
Application	General aviation, fixed-wing, helicopters	General aviation, Commercial aviation	General, Military & Commercial aviation	Helicopters, general, commercial & military	Helicopters, general, commercial & military	General, corporate helicopters military & commercial aviation	General corporate, helicopters military & commercial aviation	General corporate, helicopters military & commercial aviation
Frequency	121.5/ 243.0 MHz	121.5/ 243.0 MHz	121.5/ 243.0 MHz	121.5/ 243.0 MHz	121.5/ 243.0 MHz	121.5/243/ 406.025 MHz	121.5/243/ 406.025 MHz	121.5/243/ 406.028 MHz
Approvals	TSO C91a	TSO C91a	TSO C91a	TSO C91a	TSO C91a	TSO C126	TSO C126	TSO C126 ETSO 2C/26
Interface capability	N/A	N/A	N/A	GPS/Loran	GPS/Loran	GPS, interface ARINC 429, most RS 232	GPS, interface ARINC 429, most RS 232	Self-contained NAV interface

Emergency Locator Transmitter (ELT) Model 95200000-21

Type
Avionic Emergency Locator System (ELS).

Description
The Model 952-21 ELT is a triple-frequency unit containing a 121.5/243 MHz homing transmitter and a 406.025 MHz satellite transmitter.

The 406.025 MHz transmitter contains a unique digital coded message that can be received by polar orbiting satellites forming part of the COSPAS/SARSAT system.

Status
In service in US Air Force C/KC-135 aircraft.

Contractor
Northrop Grumman Corporation, Component Technologies, Poly-Scientific.

Enhanced Airborne Collision Avoidance System (ACAS II/ TCAS II)

Type
Traffic Alert and Collision Avoidance System/ Aircraft Collision Avoidance System (TCAS/ ACAS).

Description
Enhanced ACAS II is a development of the Honeywell Aerospace CAS 81 TCAS II, designed for commercial air transport aircraft. CAS 81 can be readily upgraded to the Enhanced ACAS II standard.

Enhanced ACAS II meets the full range of ACAS II and Mode S regulatory requirements and, in addition, incorporates increased surveillance range options, better bearing accuracy, improved reliability and an advanced communication datalink.

ACAS II is represented by the TCAS II design requirements, upgraded to the Change 7 software standard.

The Enhanced ACAS II modular architecture assures easy upgrades to future industry-defined systems such as the Airborne Separation Assurance System (ASAS) and Automatic Dependent Surveillance-Broadcast (ADS-B).

Enhanced ACAS II and Mode S is a vital component of the Communications, Navigation, Surveillance/Air Traffic Management (CNS/ATM) concept, embodying the Future Aeronautical Navigation System (FANS) concept and the military GANS/GATM requirements. Honeywell Aerospace also produces a derivative military variant of the Enhanced ACAS II concept, designated Enhanced-TCAS (E-TCAS).

Additional growth provisions include integration with the Enhanced Ground Proximity Warning System (EGPWS) technologies to provide enhanced situational awareness.

Significant capabilities of the Enhanced ACAS II and Mode S transponder system include:
1. ADS-B squitter capabilities, including flight ID, velocity and GPS position.
2. 100 n mile ADS-B extended surveillance range.
3. 40 n mile active interrogation range.
4. Hybrid surveillance.
5. Internal event recording.
6. Cockpit Display of Traffic Information (CDTI).
7. Display range exceeding 100 n miles.
8. Enhanced datalinking functions.
9. Growth capability to meet future requirements.
10. Compliance with relevant ARINC and ICAO requirements.

System components can be selected to meet specific requirements, but include the following, derived from the CAS 67A and CAS 81A ranges:
1. IVA-81A Traffic Advisory/Vertical Speed Indicator (TA/VSI).
2. IVA-81B Resolution Advisory/Vertical Speed Indicator (RA/VSI).
3. IVA-81C and IVA-81D TA/VSIs.
4. PPI-4B colour indicator.
5. CTA 81 ATC/TCAS control panels.
6. Enhanced TRA 67 A Mode S transponder.
7. MST 67A transponder.
8. Enhanced TPA 81A ACAS/TCAS processor unit.
9. KFS 578A controller
10. ANT 81A antenna.

Status
Honeywell Aerospace has produced civil variants configured with both ARINC and non-ARINC form factors. It also produces the military variant designated E-TCAS. Items of equipment from these three separate development lines can be mixed and matched to meet specific customer requirements.

One product line being marketed specifically is designated Enhanced CAS 67A ACAS II for the corporate and regional aviation markets. CAS 81A can be upgraded for the air transport market, and E-TCAS has been contracted by the US Air Force for designated transport and tanker aircraft types.

Contractor
Honeywell Aerospace, Electronic & Avionics Lighting.

Enhanced Ground Proximity Warning System (EGPWS)

Type
Ground Proximity Warning/Terrain Awareness System (GPWS/TAWS).

Description
The EGPWS includes all traditional GPWS functions, but also utilises a proprietary worldwide terrain database. Referencing host aircraft location from the main navigation system, the EGPWS can display nearby terrain and provides aural warnings approximately 60 seconds in advance of a terrain encounter, compared with 10 seconds for a traditional GPWS.

EGPWS provides the following advancements on the GPWS:
1. Look-ahead alerting algorithms.
2. Multiple radio altimeter inputs.
3. Significant reduction in unwanted warnings.
4. Landing-short alerting algorithms.
5. Optionally, an embedded 12-channel GPS receiver can be incorporated to provide an upgrade capability to aircraft that have no existing GPS or FMS output available.

The following variants of the EGPWS are available:
1. The Mk V air transport version for aircraft with digital data interfaces. It has a worldwide database including runways longer than 3,500 ft and man-made obstacles in North America.
2. The Mk VI regional aircraft version, which is similar to the Mk V but smaller and lighter and intended for installation in turboprop aircraft with appropriate analogue avionics. It has a regional terrain database (rather than global) and lacks windshear capability.
3. The Mk VII air transport version is similar to the Mk V, but for aircraft that have analogue data interfaces. Its worldwide database includes runways longer than 3,500 ft and man-made obstacles in North America.
4. The Mk VIII also uses analogue inputs and includes an integrated 75 × 75 mm Terrain Awareness and Display System (TADS), that can be retrofitted into the space occupied by an existing altimeter. It has a worldwide terrain database including runways over 2,000 ft and man-made obstacles in North America.

TADS is also available for the Mk V, VI and VII variants, where it can be fitted as a replacement for an ARINC 453 altimeter. TADS displays threat terrain on a Navigation Display (ND), or on Wx radar display for non-EFIS aircraft, utilising the existing ARINC 453 databus. TADS gives the pilot visual detection of flight paths that could lead to conflict with precipitous terrain; it provides timely alerts with a visual assessment of terrain situation; it provides synchronised aural alert messages; it utilises present and projected aircraft position with built-in terrain databases.

The following modes are available with the EGPWS:

Mode 1
excessive descent rate alert/warning. Mode 1 provides pilots with alert/warning for high-descent rates into terrain. Typical warning time exceeds 20 seconds.

Mode 1 increases alert/warning for sink rate near the runway threshold. The pilot receives

The Enhanced CAS 67A ACAS II installation comprises a TPU 67A processor, MST 67A transponder, two IVA 81D displays, a KFS 578A controller and one or two ANT 67A antennas 0081847

The future CNS/ATM environment will feature the use of extended ADS-B range and advanced datalink systems using GPS, ATN and Mode S technologies that the Enhanced ACAS II is designed to fulfil. 0081846

a timely alert for rapidly building sink rates exceeding 1,000 ft/min near the runway (earlier-generation GPWS computers provided little, if any, warning).

Mode 1 reduces nuisance alerts when visually repositioning down on the glideslope. The pilot has an additional margin to safely reposition the aircraft.

Mode 2

excessive closure rate to terrain warning. The same technology breakthroughs, that allow for the display of threatening terrain, permit advanced, virtual 'look ahead' warnings. 'Caution-Terrain' and 'Terrain Ahead' alerts based on position data and terrain database precede the GPWS warning.

Warning for excessive closure rate to terrain has been dramatically increased by using the speed of the aircraft to expand the warning envelope.

When compared with systems which do not have speed enhancement, warning times are typically doubled for inadvertent flight into mountainous terrain during initial approach or descent.

Terrain alert precedes the *pull-up* and will continue after a *pull-up* alert until 300 ft of barometric altitude has been gained to help encourage the pilot to climb and avoid any following terrain.

This warning mode has not only improved warning time, but has minimised the chance of unwanted warnings.

On final approach, the decreased speed and landing configuration of the aircraft desensitise the warning envelope to allow the aircraft to land at airports situated on terrain without nuisance warnings.

Mode 3

alert to descent after take-off. Mode 3 alerts the pilots to an inadvertent descent into terrain after take-off or missed approach. The alert is given after significant barometric altitude loss has occurred and allows considerable margin for third segment acceleration and flap retraction even under engine-out conditions.

After the aircraft has climbed to a safe altitude, based on climb performance and time, this mode is automatically switched out and replaced by an alert/warning floor below the aircraft based on speed and aircraft configuration.

During initial climb, additional warning protection is provided against a shallow, accelerating climb into rising terrain.

Possible nuisances from premature mode switching caused by terrain undulations or faulty radio altimeter unlock are eliminated.

Mode 4

insufficient terrain clearance. Mode 4 will warn the pilots of insufficient terrain clearance during climbout, cruise, initial descent or approach. This warning mode is especially valuable when the aircraft's flight path, relative to terrain, is insufficient to develop excessive closure rate or descent rate warnings.

Warning times have been essentially doubled over earlier generations of GPWS computers by automatically expanding the warning envelope as the speed of the aircraft increases. Similar to Mode 2, the warning envelope on approach gradually collapses as speed decreases, the landing gear is lowered and the flaps selected. The wording of the warning is also changed to correlate with the phase of flight and the actual cause of alarm. Warning time as compared to the earlier generations of GPWS is greatly improved for situations such as initial descent or approach where landing gear is extended for additional drag. Speed will automatically expand the warning envelope, providing additional warning time.

Terrain Clearance Floor uses navigation position data coupled with an airport/runway database to provide valuable protection in the landing configuration.

Mode 5

alert to inadvertent descent below glideslope. Mode 5 is automatically armed when the pilot selects an ILS frequency and selects gear down. The warning envelope contains two boundaries: a 'soft' alerting region and a 'hard' alert region. Both boundaries are a function of glideslope deviation.

When the aircraft penetrates the 'soft' alerting region, the audio level of glideslope voice is 6 dB below other GPWS computer voices. The voice repetition rate increases as deviation below glideslope increases. If the aircraft subsequently enters the 'hard' alerting region, the *glideslope* voice audio level increases to equal that of other voices. Below 150 ft of radio altitude, the amount of glideslope deviation required to produce an alert is increased to reduce nuisance alerts which could be caused by close proximity to the glideslope transmitter.

Mode 5 can be inhibited by pressing either of the cockpit *below G/S* lights to permit deliberate descent below the glideslope in order to utilise the full runway under certain conditions. Any other warning always has priority over a *glideslope* alert. Possible nuisances from false back course glideslope signals are automatically eliminated.

Mode 6

altitude callouts and excessive bank angle alert. Altitude callouts, which have become a popular feature with many airlines on new aircraft, are available in the system. They are pin-selectable at the rear connector and there are 32 menus available for the purpose of increasing altitude awareness on final approach.

Tones are also available. Automatic audio level increase is available when windshield rain removal is in use. Honeywell Aerospace recommends the use of a few automatic callouts near the runway and a 'smart' callout, which would rarely be heard, for most ILS landings.

Bank angle can be used to alert crews of excessive roll angles. The bank angle limit tightens from 40° at 150 ft AGL to 10° at 30 ft AGL to help alert the crew on landing of excessive roll corrections which might result in wingtip or engine damage. *Bank angle* is also useful to help alert the pilot of severe overbanking which might occur from momentary disorientation during initial climb-out.

Mode 7

Windshear detection and annunciation. Visual and aural windshear warnings are given for windshears that significantly degrade the performance of the aircraft. Optional visual and aural caution alerts can be given for increasing performance windshears that may be a signature of microbursts.

The detection level is automatically varied by outside air temperature and lapse rate, change in flight path, relationship to glideslope, height above ground, bank angle and approach stall margin. This advances the alert/warning time and improves the margin against unwanted alerts or warnings.

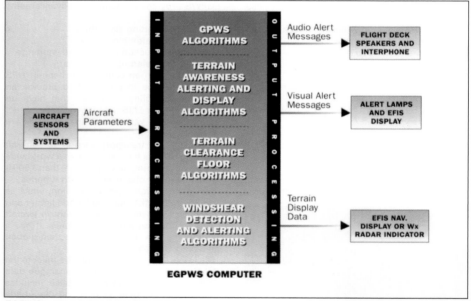

EGPWS typical system configuration   0044979

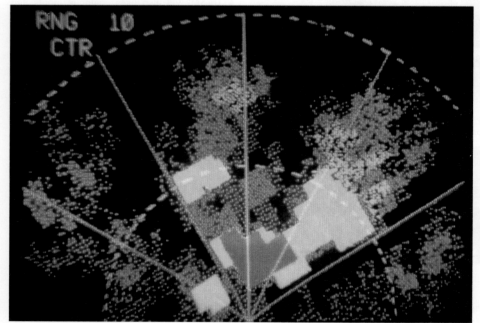

EGPWS terrain display   0044978

The final mode capability is 'alert envelope modulation'. This permits aircraft position data to be used to reduce nuisance warnings at particular problem airports, by modifying algorithms for specific local terrain peculiarities.

EGPWS is a form/fit replacement for the existing Mk V and MK VII GPWS, but some additional cable work is required.

Specifications
Dimensions: 2 MCU
Weight: 3.2 kg
MTBF: 15,000 h

Status
In production. Selected by a large number of air transport operators for fitment to Boeing and Airbus commercial passenger aircraft.

In December 1997, major US air carriers announced a voluntary programme to install EGPWS, in a six year programme up to 2003, affecting about 4,500 aircraft operated by US major and national airlines. US regional airlines were not part of this voluntary programme, nor were charter carriers.

This initiative pre-empted a planned US FAA notice of proposed rule making for the fitment of EGPWS to US long-haul and most regional aircraft.

The Mk V, Mk VI, Mk VII and Mk VIII EGPWS are in production. Military versions of the Mk V and Mk VII are also available.

The EGPWS has been integrated with the RDR-4B forward-looking windshear radar; the combined system has been successfully flight tested and certified on the Boeing 777.

In June 1999, Honeywell Aerospace announced that it had joined with Sikorsky Aircraft to engineer a new version of the EGPWS for helicopters. The system incorporated several features specifically designed for helicopters including a digital moving map, which shows the helicopter's height above terrain with colour or shading contrasts, with wires and other obstructions overlayed on the map and conspicuously marked. The helicopter EGPWS also features high-resolution terrain data at locations away from airports to cater for the requirement to operate close to the ground at distant locations. Flight tests were conducted using an S-76 helicopter from Sikorsky, with certification achieved by 2001.

By June 1999, more than 6,000 EGPWS had been ordered.

In addition to the system's widespread use in commercial aircraft, the system is also standard fitment on many business jets, including the Bombardier Global Express, Dassault Falcon 900EX and Gulfstream V and optional on other business aircraft such as the Cessna Citation X and Excel, Hawker 800XP and Learjet 45.

Contractor
Bendix/King.

Enhanced TRA 67A transponder

Type
Aircraft transponder.

Description
The Enhanced TRA 67A is a new enhanced Mode S transponder designed to meet new European requirements and other anticipated regulations worldwide. It meets Change 7 and Downlinked Aircraft Parameters (DAPs) requirements adopted for European and other nations by the International Civil Aviation Authority (ICAO). It also performs Automated Dependent Surveillance-Broadcast (ADS-B) functions in anticipation of future needs.

In compliance with the DAPs requirement, the Enhanced TRA 67A will broadcast such data as roll angle, groundspeed, true airspeed, magnetic heading and indicated airspeed. Change 7 is an enhancement to the software of Terrain alert and Collision Avoidance Systems (TCAS) and related systems, which has been mandated in Europe. A Mode S transponder is an essential part of a TCAS.

The Enhanced TRA 67A will help facilitate ADS-B by broadcasting digital information such as aircraft location (from the airplane's global positioning system), aircraft altitude, velocity and acceleration. This information, when received by other aircraft that are also equipped with ADS-B, can be used to extend the range and enhance the accuracy of traffic information provided by the TCAS.

Status
The US Federal Aviation Authority (FAA) granted Technical Standard Order (TSO) approval for the Enhanced TRA 67A in June 1999.

Contractor
Honeywell Aerospace, Electronic & Avionics Lighting.

ETCAS – Enhanced Traffic alert and Collision Avoidance System

Type
Traffic Alert and Collision Avoidance System/ Aircraft Collision Avoidance System (TCAS/ ACAS).

Description
Honeywell has designed and developed an Enhanced Traffic Alert & Collision Avoidance System (ETCAS), to provide military aircraft operators with an extended surveillance range and the capability to coordinate formation flying in addition to standard TCAS operations. ETCAS is the only system providing this capability currently certified by the FAA and international aviation authorities. The FAA, US Air Force and international authorities have granted ETCAS the frequency approvals necessary to ensure smooth integration into military operations and prevent frequency overlap with commercial aviation operations.

ETCAS provides two modes of operation. The basic mode is ACAS II which is the same as TCAS II with Change 7.0 software and is RVSM compatible. In addition to the standard TCAS functions of situational awareness, traffic alert and resolution advisories, the ETCAS provides a formation mode. This mode allows aircraft operators to locate, identify, rendezvous with and formate with aircraft equipped with a variety of identification systems, including Identification Friend and Foe (IFF), Modes 1, 2, 3 and 4, Mode A, Mode C and Modes S transponder equipped private, commercial and military aircraft. In formation mode, the ETCAS performs interrogation surveillance to a range of 40 nautical miles with azimuth coverage of 360° and up to 12,700 feet above and below the aircraft. The system can mark and identify formation members and rendezvous aircraft by displaying the aircraft's Mode A code next to its symbol, clearly distinguishing it from other traffic within ETCAS range. In addition to the Mode A codes, the system provides unique symbology for formation members. The symbols remain until deselected manually or until the ETCAS is commanded to switch to normal TCAS operation. ETCAS information can be displayed on a variety of cockpit instruments, including radar displays, navigation displays, electronic flight displays and traffic alert/vertical speed indicators (depending on aircraft type and cockpit configuration). While functioning in formation mode, ETCAS continues

Honeywell's ETCAS has been selected for the USAF transport fleet (Lockheed Martin)   0103882

to monitor the airspace and maintain standard TCAS II collision avoidance capability.

Utilising IFF/Mode-S transponders, the ETCAS provides the operator with continuous TCAS surveillance and protection while maintaining concurrent IFF reply capability. Honeywell will introduce ETCAS with increased surveillance range capability in 2001. Utilising hybrid surveillance technology, ETCAS will track like-equipped aircraft to 100 n miles. The company claims this increased range surveillance will be achieved without increasing transmission power. This feature will be available as an upgrade to existing ETCAS systems.

The US Air Force and Lockheed Martin have let a contract to replace the current station-keeping equipment for providing tanker rescue/rendezvous mission and refuelling capability with a modified version of ETCAS.

Status
ETCAS is standard equipment on US Air Force transport aircraft designated as AN/APN-244. It is also fitted on transport aircraft of the following air forces: Australian; Austrian; Belgian; Brazilian; Canadian; Hellenic; Italian; Israeli; Norwegian; Dutch; Republic of Korea; Saudi Arabian; Spanish; Turkish; United Kingdom; and with the US Coast Guard and Navy. In all, there are about 2,000 aircraft already fitted with ETCAS and several hundred more are anticipated by the company.

Contractor
Honeywell Aerospace, Electronic & Avionics Lighting.

Exceedance Warning System

Type
Aircraft limit warning system.

Description
Safe Flight's Exceedance Warning System employs a collective shaker to provide tactile warning when aircraft operating limits are being approached, improving safety and maximising the available operating envelope.

During high-workload manoeuvres, such as external load manoeuvring in a confined area or hovering on Night Vision Goggles (NVGs), pilots are able to maintain 'eyes out' (of the cockpit) while maintaining a 'feel' for the allowable manoeuvre envelope.

The collective shaker is designed to simultaneously monitor several engine/mast limits, including transmission torque, exhaust gas temperature, engine gas producer speed and rotor speed. If the shaker activates, the pilot's natural reaction to lower the collective will be correct for any of the aforementioned problems.

An associated pedal shaker enhances pilot Situational Awareness (SA) during out-of-ground effect hovering, high crosswind or high density altitude operations.

Status
In production. Installed on the TH-57 helicopter.

Contractor
Safe Flight Instrument Corporation.

Flight Data Recorder Processor (FDRP)

Type
Flight/mission recording system.

Description
L-3 Communications, Electrodynamics' Flight Data Recorder Processor (FDRP) receives key vehicle, subsystem and environmental parameters from up to six MIL-STD-1553 buses, storing formatted records in an external memory unit such as the company's own Crash Survivable Memory Unit (CSMU – see separate entry). The FDRP formats records for incident and mishap investigation,

including 15-minute and 12-hour periodic and 12-hour aperiodic circular buffers, plus a 44-event circular buffer.

Other features of the system include:
- Remote Terminal interface for device control and data acquisition
- Up to 2 KB per second storage rate
- Error-checking on all storage operations
- Uploadable programme memory
- Extraction over RS-422 or MIL-STD-1553
- All records independent and self-documenting
- BIT: power-on, commanded, periodic
- Configurable for new applications
- Cooling ducts for optional forced-air
- GSE available for test and download

The FDRP LRU is a standard ARINC size 6 unit suitable for mounting in an avionics rack. It contains a 16-bit microprocessor, dual-redundant Remote Terminal interface, bus monitor interfaces for five MIL-STD-1553 buses and a power supply. The bus monitoring function receives and edits bus traffic on all buses simultaneously, saving user-specified messages at prescribed rates while also detecting exceedance events. Unit Built In Test (BIT) is performed on power-up, on command and continuously. Downloading of flight data can be performed at any time over the MIL-STD-1553 bus or a dedicated RS-422 channel. The modular design of the FDRP facilitates quick fault detection, isolation and repair under field conditions.

Specifications
Dimensions: 190.5 × 193.5 × 193.5 (H × W × D)
Weight: 7.84 kg
Power: 28 V DC; 28 W
MTBF: 15,000 hr
MTTR: 1.25 hr
Compliance: MIL-E-5400, MIL-STD-1553B, MIL-STD-704, MIL-STD-810, MIL-STD-461/462

Status
In production and in service.
Developed the B-2 Spirit; the unit can be readily adapted to suit other platforms.

Contractor
L-3 Communications, Electrodynamics, Inc.

GPWS Mk V Digital Ground-Proximity Warning System

Type
Ground Proximity Warning/Terrain Awareness System (GPWS/TAWS).

Description
The Mk V digital ground-proximity warning system computer is designed for service with aircraft equipped with ARINC 700 avionics. It provides the flight crew with back-up warning for seven potentially dangerous situations including windshear conditions if connected via ARINC 429 to outputs from an AHRS or INS. The EGPWS computer meets TSOs C-115a, C-92c and C-117a.

Alerts and warnings are provided by steady or flashing visual indications and by audible warnings. Each audio warning is also annunciated to identify the particular situation such as excessive descent rate, excessive closure rate to terrain, significant altitude loss after take-off, insufficient terrain clearance, excessive descent below glideslope, altitude call-outs and windshear detection.

Windshear detection and annunciation are provided by the Mk V computer. When the computer detects an impending windshear situation, an optional amber light is turned on in the cockpit. If the aircraft experiences further windshear severity, a red warning light is displayed along with a voice message 'windshear' repeated three times. The windshear alert function takes priority over other GPWS alerts.

Other alerts are repeated twice. If the aircraft's performance continues to degrade, the message is repeated. A particular advantage of the variety of voice alerts is that its operationally orientated

The Mk V ground proximity warning system
0044980

warnings permit confirmation by cross-checking of the panel instruments. Diagnosis of flight warnings can thus be quickly carried out and corrected. The speed of the ground-proximity warning system envelopes has been increased, providing longer warning times.

The system contains a number of features to assist in test maintenance and repair procedures. These include a non-volatile memory which stores both steady-state or intermittent faults occurring over the last 10 flight sectors and which can be erased only when the unit is removed from the aircraft for bench work. The accepted test procedure is programmed within the computer and a simple test fixture is all that is required to re-address computer output data back into the computer itself. An alphanumeric display on the front of the unit can be used to isolate faults and indicate specific LRUs which require replacement. Faults can be isolated on the bench to board level.

The GPWS complies with ARINC 600 standards and its subcomponents are grouped by circuit function on plug-in/fold-out removable printed circuit boards with easily removed captive hardware. Latitude and longitude are used to modify warning boundaries at certain locations to reduce nuisance probability, or increase available warning time.

Specifications
Dimensions: 2 MCU
Weight: 3.2 kg
Power supply: 115 V AC, 400 Hz, 16 W; and 28 V DC
MTBF: 30,000 h

Status
In production and in service on a wide range of digital avionic air transport and corporate aircraft.

Contractor
Bendix/King.

GPWS Mk VI Ground-Proximity Warning System

Type
Avionic navigation sensor.

Description
The Mk VI ground-proximity warning system is designed for regional, commuter, corporate and business turbojet and turboprop aircraft. It operates in six modes: excessive descent rate alert and warning; excessive closure rate to terrain; alert to descent after take-off; alert to insufficient terrain clearance; alert to inadvertent descent below glideslope; and altitude call-outs and bank angle alert. The Mk VI system is said to cost 40 per cent of the earlier Mk II GPWS.

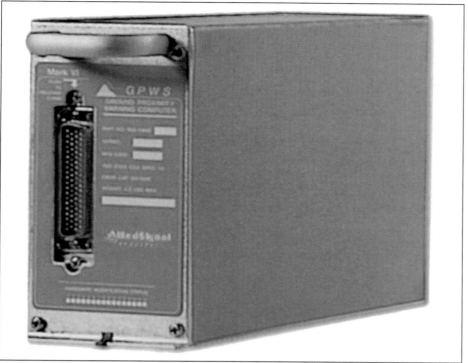

The Mk VI ground-proximity warning system 0044981

The system has been refined to delete unwanted warnings by reducing glideslope and terrain clearance floor limits to trigger warnings at altitudes down to 750 ft above ground level on approach or to 925 ft above ground level with ILS acquired.

Manual functions have been added to minimise the chance of false warnings due to flapless landings or other operational modes, when landing at airports with unique terrain features or in the event of incompatible terrain clearance during approach and departure procedures.

Specifications
Dimensions: 152.4 × 76.2 × 260.4 mm
Weight: 1.9 kg
Power supply: 28 V DC

Status
In production and in service.

Contractor
Bendix/King.

GPWS Mk VII Ground-Proximity Warning System

Type
Ground Proximity Warning/Terrain Awareness System (GPWS/TAWS).

Description
The Mk VII warning computer is designed for all existing and new aircraft with ARINC 500 analogue avionics as a replacement for the Mk I and Mk II ARINC 594 ground-proximity warning computers. The improved Ground-Proximity Warning System (GPWS) dynamics provide the advantages of increased warning times, prioritisation of aural warnings and reduction in nuisance warnings in the cockpit, while implementing a cost-effective windshear warning system. The computer has a common part number used across a wide range of aircraft types, minimising the investment in spares.

The Mk VII warning computer meets the requirements of FAA AC 25-12 for windshear detection and alerting using ARINC 429 inputs from an IRS/AHRS or synchro inputs from a bi-axial accelerometer. The computer can be powered by 28 V DC or 115 V AC and meets TSOs C-115a, C-92c and C-117a.

The GPWS features ground-proximity warning and glideslope alerting; altitude call-out menus; bank angle alerting; reduced audio cockpit clutter; improved take-off monitoring for noise abatement procedures; automatic adjustment of warning modes for ILS approaches; windshear detection and annunciation; optional windshear recovery guidance; verbal annunciation of system faults and front replaceable software modules. It fits into existing GPWS rack space.

Specifications
Dimensions: 1/4 ATR short
Weight: 2.72 kg (max)
Power supply: 115 V AC, 400 Hz, single phase, 15.7 W nominal; and 28 V DC
Environmental: DO-160B
Reliability: 15,000 h MTBF

Status
In production and in service.

Contractor
Bendix/King.

GTX 327 digital transponder

Type
Aircraft transponder.

Description
The GARMIN GTX 327 is a solid-state Mode C digital transponder, which features a 200 W transmitter, a DSTN Liquid Crystal Display, a numeric keypad and a dedicated VFR button, which facilitates rapid selection of 1200/VFR squawking. The GTX 327 also offers timing and display functions such as flight time and count-up and count-down timers, as well as current pressure altitude.

Specifications
Dimensions: 159 (W) × 42 (H) × 208 (D) mm
Weight: 0.95 kg
Power supply: 11–33 V DC, 15 W
Display: DSTN LCD

Garmin GTX 327IFF transponder (Garmin International) 0099755

Transmitter frequency: 1090 MHz
Transmitter power: 200 W nominal
Receiver frequency: 1030 MHz
Mode A capability: 4096 ident codes
Mode C capability: 100 ft increments from –1,000 to 63,000 ft
Certification: TSO C74 Class 1A, certified to 50,000 ft

Status
In production.

Contractor
Garmin International.

GTX 330/330D Mode S transponder

Type
Aircraft transponder.

Description
The Garmin GTX™ 330/330D are Mode S, IFR-certified transponders with datalink capability including local traffic updates and expandability to accommodate future services such as ADS-B.

Garmin GTX 330 transponder (Garmin International) 1127934

The transponders are based on the Garmin GTX 327 transponder and use solid-state receiver and transmitter technology to increase efficiency. They feature a DSTN LCD and offer several timing and display functions: flight time, count-up and count-down timers, and current pressure altitude. Other features include remote ident and auto standby, plus altitude monitor with voice alerting.

The GTX 330D contains all of the features of the GTX 330 but adds antenna diversity for improved visibility to TCAS-equipped aircraft flying at higher altitudes. Both models meet the requirements of Level 2 Mode-S to satisfy the upcoming European Mode-S mandate for Elementary Surveillance.

Specifications
Dimensions: 159 (W) × 42 (H) × 286 (D) mm
Weight: 1.9 kg
Power: 11 to 33 V DC, 27 W typical
Temperature: –20 to +55°C
Altitude: 55,000 ft
Certification: C112

Status
During December 2002, Garmin received a FAA Supplemental Type Certificate (STC) for the GTX 330 enabled to receive data via the Traffic Information Service (TIS).

The Comm A/B protocol within the GTX 330 facilitates TIS datalink functionality. TIS delivers greater traffic awareness to pilots in the United States by transmitting ATC traffic information to aircraft equipped with TIS-enabled Mode S transponders. The data is then displayed on the partner Garmin GNS 430/530.

Contractor
Garmin International.

IHAS 5000/IHAS 8000 Integrated Hazard Avoidance System

Type
Integrated Hazard Avoidance System (IHAS).

Description
The IHAS systems are designed to improve situational awareness and safety.

IHAS 5000 is designed for general aviation aircraft, particularly for piston-powered aircraft not fitted with weather radar. Integrating the four major airborne safety systems – position awareness, weather avoidance, traffic advisories and terrain awareness/warning – the IHAS 5000 presents safety and situational awareness information on a large colour display.

The IHAS 5000 system comprises the following items of equipment:
1. The KDR 510 datalink radio to deliver textual information and colour weather graphics over a high-speed, real-time, VHF DataLink (VDL) mode 2 datalink.

2. The KT 73 datalink Mode S transponder. This system receives traffic information uplinked from Air Traffic Control (ATC) ground stations, and presents a picture of nearby traffic on the KMD 850 MultiFunction Display. At present this service is only available in the USA.
3. The KMD 850 MFD is a full-colour MFD, which displays data on a 5 in diagonal screen. The KMD 850 MFD includes intuitive controls and an extensive aeronautical and cartographic database, to present detailed obstacle and terrain elevation shading data.

An optional addition to the IHAS 5000 system is offered by the KMH 880 unit. KMH 880 is a fully integrated terrain avoidance and traffic system that combines an active traffic sensor capability and Enhanced Ground-Proximity Warning System (EGPWS) to deliver traffic advisories of the same type as those provided to airlines by the TCAS I system.

IHAS 8000 is designed for business aviation users. It is the same as the IHAS 5000 except that the IHAS 8000 supports weather radar.

Specifications
KMD 550
Dimensions: 158 (W) × 101 (H) × 254 (D) mm
Weight: 2.8 kg max with rack
Screen size: 127 mm (5 in) diagonal
Power: 10 to 33 V DC, 25 W
Temperature range: –20 to +70°C
Certification: TSO C113, C110a
Environmental: DO 160D
Altitude: 35,000 ft

KMH 880
Dimensions: 115 (W) × 178 (H) × 351 (D) mm
Weight:
(display) 4.1 kg
(antenna) 0.5 kg
Power: 22–30 V DC
Temperature range: –55 to +70°C
Certification: TSO c147 Class A and C151a Class B
Altitude: 51,500 ft unpressurised

Status
In production.

Contractor
Bendix/King.

KXP 756 Gold Crown III solid-state transponder

Type
Aircraft transponder.

Description
The KXP 756 is a third-generation system incorporating modern avionics techniques such as large-scale integrated circuitry, microprocessor

data programming and all-solid-state transmitter design to provide reliability and simplicity of operation. It operates on Modes A, B and C up to 70,000 ft and can reply on any one of 4,096 preselected codes. Information is provided on one or two 2¼ in square gas discharge digital displays which are automatically adjusted in brightness by a photocell, for maximum visibility under all light conditions. The system is controlled by two concentric knobs – one for mode selection and the other for code selection – and incorporates identification, VFR code and self-test functions.

Specifications
Dimensions:
(control unit) 146.7 × 21.9 × 23.5 mm
(remote unit) 298.45 × 50.8 × 134 mm
Weight:
(control unit) 0.31 kg
(remote unit) 1.727 kg
Power supply: 11 to 13 V DC
Altitude: up to 60,000 ft
Frequency:
(receive) 1,030 MHz
(transmit) 1,090 MHz

Status
In service.

Contractor
Bendix/King.

LandMark™ TAWS Terrain Awareness and Warning System

Type
Ground Proximity Warning/Terrain Awareness System (GPWS/TAWS).

Description
Providing a continuous colour overview of the surrounding terrain, the LandMark™ TAWS Class B system offers predictive warning functions using present position from a GPS receiver, altitude information and aircraft configuration. The system then compares that information to a terrain and obstacle database. Both aural and visual warnings are issued whenever potential Controlled-Flight-Into-Terrain (CFIT) situations arise.

The LandMark™ system can be displayed on any TAWS-compatible ARINC 453 EFIS system or radar indicator, or, when coupled with the Goodrich RGC250 Radar Graphics Computer, terrain can also be viewed on an aircraft's existing compatible weather radar indicator. The RGC250 also offers several other advantages such as improved terrain contouring, runway and obstacle depictions, traffic data and lightning display information, all shown through the radar indicator display.

As a forward-looking terrain avoidance system, the LandMark™ TAWS Class B system has various alerting modes. The forward-looking terrain avoidance mode includes warnings for reduced required terrain clearance and imminent terrain impact when comparing the flightpath with upcoming terrain. The premature descent alert occurs when the aircraft is significantly below the normal final approach path into the nearest runway. The Ground Proximity Warning System

IHAS 5000 integrated hazard warning system   0081845

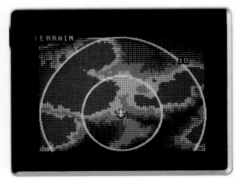

LandMark™ TAWS display   0098573

(GPWS) alert modes include: an excessive descent rate alert when the aircraft is descending too fast for the current height above the terrain, regardless of landing gear and flap configuration; a descending to 500 ft alert when descending to within 500 ft of the terrain, during en route mode, or within 500 ft of the nearest runway threshold during terminal or approach mode; a negative climb rate alert after takeoff or missed approach after the aircraft reaches 50 ft but before reaching 700 ft above the runway.

Certified to TSO-C151a Class B and DO-178B Level C, the LandMark™ System operates at altitudes up to 55,000 ft.

Specifications
Weight: 2.01 kg
Power supply: 28 V DC
Temperature range: –55 to +70°C

Contractor
L-3 Communications, Avionics Systems.

Military Aircraft Surveillance System (MASS)

Type
Aircraft transponder.

Description
The Military Airborne Surveillance System (MASS™ is based on the XS-950SI transponder system (see separate entry). The ACSS Cooperative Member functionality, a subset of MASS, is a software change to the TCAS 2000 or T²CAS™ computer and the Mode S/IFF Transponder that provides a system capable of supporting unique military flight requirements such as formation flight and rendezvous operations in commercial airspace. When Cooperative Flight Mode is not selected, the system will perform normally as a TCAS/ACAS II collision avoidance system.

The use of ADS-B and adaptive power level interrogation techniques is designed to prevent interference with the civil ATC environment. The system will include cooperative member identification using Flight ID and Mode S address, unique cockpit display characteristics for cooperative members and the ability to select an aircraft for display of specific aircraft data.

MASS is upgradeable to T²CAS™ by adding Thales Avionics' Ground Collision Avoidance Module/TAWS system to the TCAS 2000 LRU.

Specifications
TCAS R/T
Dimensions: 124.5 × 193 × 386.1 mm
Weight: 6.68 kg
Cooling: None required
Interface: ARINC 407/429/575/718
Operating altitude: Sea Level to 55,000 ft
Operating temperature: –55 to 70°C
Power: 28 V DC; 115 V AC, 400 Hz
XS-950SI transponder: See entry

Optional CDTI display
Bezel dimensions: 122.3 × 158.8 mm
Image size: 6.3 in diagonal
Weight: 2.95 kg
Case size: ARINC 708A

Interface
TCAS: ARINC 429/735 & 735A
TAWS: ARINC 762/453
FMS: ARINC 702 (growth)
DME: ARINC 568 (growth)
INS: ARINC 561 (growth)
Weather radar: ARINC 708A, Collins WRX-700X, Honeywell RDR-4A/RDR-4B

Status
In production.

Contractor
Aviation Communication & Surveillance Systems (ACSS).
An L-3 Communications & Thales company.

Model A100S Solid-State Cockpit Voice Recorder (SSCVR)

Type
Aircraft Accident Data Recorder (ADR)/Cockpit Voice Recorder (CVR).

Description
The Model A100S Solid-State Cockpit Voice Recorder (SSCVR) is available for new installations or as a direct replacement for existing ARINC 557 CVRs without wiring changes to the aircraft. Existing Fairchild users installing the Model A100S SSCVR do not require new control units, microphones, or changes to existing operating procedures.

While mounted on the aircraft, recorded data cannot be extracted from the SSCVR. This protection is necessary to protect the rights of pilots and other flight crew members. The model A100S SSCVR meets the following specifications in full compliance with worldwide regulatory requirements, including EUROCAE (ED56) and CAA specifications together with FAA ISO-C123.

Specifications
Dimensions: 1/2 ATR short ARINC 404
Weight: 7.3 kg
Power supply: 115 V AC, 12 W

Contractor
L-3 Communications, Aviation Recorders.

SSCVR open   0001306

Model A200S Solid-State Cockpit Voice Recorder (SSCVR)

Type
Aircraft Accident Data Recorder (ADR)/Cockpit Voice Recorder (CVR).

Description
The Model A200S solid-state cockpit voice recorder provides 2 hours of recording, with the last 30 minutes being redundant in a very high-quality format. This unit meets EUROCAE ED-56A and TSO C123a. Readout and copying of data is available instantly by using a hand-held downloading device (digital audio playback unit) without the need to go through a long computer conversion process. The solid-state characteristics of the Model A200S eliminate periodic maintenance requirements, and greatly increase reliability. It meets the severe vibration and environmental requirements of DO-160C without the need for a vibration-mounted tray. A built-in test function includes a tone generator that superimposes the signal on the audio

The Model A200S solid-state cockpit voice recorder   0504269

channel to ensure the unit is functioning correctly. In addition, the unit has a continuous BITE that monitors the entire memory and power supply to ensure that the unit is operating to the highest standards. The A200S accepts four channels of audio including pilot, co-pilot, area microphone and third crew member, PA system and data/timebase source via 429. The A200S includes a front-mounted underwater locator beacon.

With the A200S, audio retrieval and playback is instantly available by using a simple hand-held device without requiring complex computer conversion. However, when mounted on the aircraft, recorded audio data cannot be extracted from the recorder, to protect the rights of pilots and other flight crew.

Specifications
Dimensions: ARINC 404 1/2 ATR short
Weight: 8.44 kg
Power supply: 115 V AC, 400 Hz or 28 V DC, 12 W (max)
Frequency response:
150–3,500 Hz (three inputs)
150–6,000 Hz (one input)

Contractor
L-3 Communications, Aviation Recorders.

Model FA2100 solid-state recorder family

Type
Aircraft Accident Data Recorder (ADR)/Cockpit Voice Recorder (CVR).

Description
Basic family
Model FA2100 was originally designed by Fairchild to meet industry demands for lighter weight and higher reliability. It now includes a number of standard configurations (seven are described), together with custom configurations. Models immediately available are shown in the accompanying photograph.

Solid-State Cockpit Voice Recorder (SSCVR)
Model FA2100 SSCVR was introduced to meet the needs of commercial, regional, business and government fleets. The FA2100 meets the requirements of EUROCAE ED-56A, FAA TSO-C123a RTCA DO-160C and ARINC 757. Readout and copying (both in analogue and digital forms) data is available instantly by using a hand-held downloading device (Portable Interface – PI) without the need to go through a long computer conversion process. The solid-state characteristics of the Model FA2100 eliminate periodic maintenance requirements and greatly increase reliability. It meets the requirements of DO-160C without the need for a vibration-mounted tray. A built-in test function includes a tone generator that superimposes the signal on the audio channel to ensure the unit is functioning correctly. In addition, the unit has a continuous BITE that monitors the entire memory and power supply to ensure that the unit is operating to the highest standards. The FA2100 accepts four channels of audio including pilot, co-pilot, area microphone and third crew member, PA system, data/timebase source via RS-429 interface; rotor speed encoding is available for helicopter installations. The FA2100 includes a front-mounted underwater locator beacon.

With the FA2100, audio retrieval and playback is instantly available in real time using a simple hand-held device without requiring complex computer conversion. However, when mounted on the aircraft, recorded audio data cannot be extracted from the recorder, to protect the rights of the pilots and other flight crew.

Specifications
Dimensions: ARINC 404 1/2 ATR short – footprint– height is 139.7 mm
Weight: <4.36 kg
Power supply: 115 V AC, 400 Hz or 28 V DC, 9 W (max)

The seven standard models of the FA2100 family of solid-state aviation recorders. (Top row, left to right) CVR-helicopter, HUMS configuration (programme unique); CVR-helicopter configuration (programme unique); CVR, 1/2 ATR short, GA model; combined FDR and CVR, 1/2 ATR short. (Centre) FDR, 1/2 ATR. (Bottom Row, left to right) CVR, 1/2 ATR short; FDR, 1/2 ATR short   0051331

Frequency response:
(three inputs) 150–3,500 Hz
(one input) 150–6,000 Hz

Solid-State Flight Data Recorder (SSFDR)
Model FA2100 SSFDR was the second generation of solid-state Flight Data Recorders (FDR) designed by Fairchild. An onboard data retrieval capability eliminates the need to remove the unit from the aircraft to play back flight data. The recorder meets or exceeds all requirements and characteristics of ARINC 573/717/747, EUROCAE ED-55, TSO-C124a and RTCA DO-160C. The FA2100 solid-state design also eliminates the need for a vibration-mounted tray.

With its microprocessor-based architecture, the Model FA2100 incorporates the intelligence to perform sophisticated self-test, as well as provide fault-tolerant features. The recorder can accommodate parameter storage requirements of 64, 128 or 256 words/s with a minimum recording time of 25 hours. Future expansion of memory to allow 512 words/s is planned. A high-speed data dump can be accomplished in approximately 2 minutes. The Model FA2100 includes a front-mounted underwater locator beacon.

Specifications
Dimensions: ARINC 404, 1/2 ATR – footprint – height is 139.7 mm. This flight recorder is available in both long and short versions
Weight: <4.36 kg
Power supply: 115 V AC, 400 Hz or 28 V DC, 9 W (max)

Modular Airborne Data Recording/Acquisition System (MADRAS)
The Model FA2100 combination recorder combines the features of the SSCVR with the SSFDR to provide one combined recorder with both features. The recorder meets the characteristics of TSO-C123a/C124a, EUROCAE ED-56A/ED55, ARINC 757 and RTCA DO-160C. Readout and copying of data is available instantly by using a hand-held downloading device (Portable Interface – PI) for both flight and audio data. The solid-state characteristics of the FA2100 eliminate periodic maintenance requirements and greatly increase reliability. It meets the requirements of DA-160C without the need for a vibration-mounted tray. A built-in test function includes a tone generator that superimposes the signal on the audio channel to ensure the unit is functioning correctly. In addition, the unit has a continuous BITE that

monitors the entire memory and power supply to ensure that the unit is operating to the highest standards. The FA2100 accepts four channels of audio including pilot, co-pilot, area microphone and third crew member, and PA system. The flight data is accepted via an ARINC 573/717 data stream from an onboard acquisition unit.

With the FA2100 combination recorder, the audio and data retrieval and playback is instantly available by using a simple hand-held device. However, when mounted on the aircraft, recorded audio data cannot be extracted from the recorder, to protect the rights of pilots and other flight crew.

Specifications
Dimensions: ARINC 404 1/2 ATR short – footprint– height is 139.7 mm
Weight: <4.36 kg
Power supply: 115 V AC, 400 Hz or 28 V DC, 9 W (max)
Frequency response:
(three inputs) 150–3,500 Hz
(one input) 150–6,000 Hz
Recording duration:
30 min of voice/25 h of data
60 min of voice/10 h of data

Status
In service.

Contractor
L-3 Communications, Aviation Recorders.

Model F1000 Solid-State Flight Data Recorder SSFDR

Type
Aircraft Accident Data Recorder (ADR)/Cockpit Voice Recorder (CVR).

Description
The Model F1000 SSFDR was the first production flight data recorder to provide solid-state operation. An onboard data retrieval capability eliminates the need to remove the unit from the aircraft to recover flight data. The recorder meets or exceeds all requirements and characteristics of ARINC 542A, ARINC 573/717, ARINC 747, EUROCAE ED-55, TSO C124a and RTCA DO-160C. The F1000's solid-state design also eliminates the need for a vibration-mounted tray.

With its microprocessor-based architecture, the Model F1000 incorporates the intelligence to perform sophisticated self-tests, as well as provide fault-tolerant features. The F1000 can receive data from an ARINC 573/717 flight data acquisition unit or process raw data in the ARINC 542A mode by accepting analogue signals in synchro, DC, pneumatic, frequency and discrete formats. The recorder can accommodate parameter storage requirements of 32, 64 or 128 words/s with a recording time of 25 hours. A high-speed data dump can be accomplished in 1 minute. The Model F1000 includes a front-mounted underwater locator beacon.

Specifications
Dimensions: ARINC 404 1/2 ATR long
Weight: 10.2 kg
Power supply: 115 V AC, 400 Hz or 28 V DC, 26 W (max)
Reliability: 15,000 h MTBF

Contractor
L-3 Communications, Aviation Recorders.

MOdular Non-volatile Solid-STate Recorder (MONSSTR®)

Type
Aircraft Accident Data Recorder (ADR)/Cockpit Voice Recorder (CVR).

Description
Harsh environments encountered in many military and aerospace programs often preclude the use of magnetic tape or disk for instrumentation data recording and desired data recording rates often exceed the capability of current magnetic recording techniques. The MONSSTR® system overcomes all of the limitations associated with magnetic recording technology by utilising high-density, non-volatile flash memory in innovative system architecture, enclosed in modular, ruggedised packaging. Featuring sustained data rates of up to 1 Gbyte/sec, storage capacities over 1,640 Gb, MONSSTR® incorporates many features only possible with solid-state, random-access devices.

The basic Model 7000 or 6000 MONSSTR® recorder configuration consists of an enclosure with a power module, one controller and one memory canister. An optional second controller may be included to provide full dual-recorder operation. A data multiplexer may be installed in the secondary controller slot to combine multiple analogue and digital data channels into a single composite data stream for input to the primary controller.

Model 5000 recorders accommodate one controller plus an optional multiplexer board. These recorders do not employ removable canisters, utilising up to 90.1 Gbytes of non-removable memory and contain fixed, non-removable memory.

Power modules are available for use with AC or DC power sources. Standard power modules available for laboratory units are 115 V AC/60 Hz and 230 V AC/50 Hz. Standard power modules for ruggedised units are 115 V AC/400 Hz, 28 and 270 V DC.

Approximately 2 per cent additional memory is held in reserve for internal 'housekeeping' and error recovery. The MONSSTR® system architecture is designed to accommodate higher memory densities, as they become available.

Other system features and benefits include:
- Generic parallel ECL, SCSI or Fibre Channel user equipment interfaces
- High-speed digital cassette tape recorder emulation for DCRsi™, VLDS™ and ID-1
- Full-duplex user equipment data ports support simultaneous read and write
- Read-after-write with re-write error detection and correction
- Hot-swap canisters install in laboratory or rugged systems
- Optional multiple-channel input multiplexer for PCM and MIL-STD-1553

- Recorders may be daisy-chained for increased capacity
- Extensive Built-In-Test (BIT), diagnostic and status reporting software.

Data rates

The maximum sustained data rate of a MONSSTR® recorder is determined by the speed of the internal bus, the mode of operation and the I/O controllers installed. Model 7000 recorders have an internal bus capable of a sustained data rate of up to 256 Mb/sec. Model 6000 internal buses are capable of sustained data rates of up to 128 Mb/sec, while Model 5000 buses are capable of sustained data rates of up to 64 Mb/sec.

The user equipment interface data transfer rate of a MONSSTR® recorder may not always be equal to the maximum data rate of the internal bus. If one controller is active and operating in basic read or write mode, the maximum user interface transfer rate will be equal to the maximum sustained data rate of the internal bus or the user equipment, whichever is less. However, if the controller is operating in record mode with read-after-write enabled, the maximum net throughput will be one-half of the maximum data rate of the internal bus.

I/O controllers

MONSSTR® I/O controller design is organised into three 'layers' of internal hardware and firmware: a physical layer, a logical layer and virtual layer. The physical layer interfaces to the system bus and performs basic memory read and writes. The logical layer performs error detection, error correction and memory management. The virtual layer defines the 'personality' of the controller for the user equipment interface. All MONSSTR® I/O controllers are identical at the physical and logical layers.

Models 7121, 6121 and 5121

Model 21 controllers feature a generic parallel I/O user equipment interface with maximum data transfer rates equal to that of the internal system bus. The user equipment interface is either differential ECL or differential PECL, and may be configured for 8-bit (all models), 16-bit (7000 and 6000) or 32-bit (7000 only) parallel data transfers. On-board time-code reader-generators are used to append time-stamps to user data. Host control is asserted by means of a serial control port that may be configured for standard RS-232 or RS-422 operation.

Models 7122, 6122 and 5122

Model 22 controllers are SCSI standard target devices. User equipment interface data rates of up to 80 Mb/sec (Ultra SCSI mode) are supported. The user equipment interface may be configured for either single-ended TTL or differential LVD operation. No separate control port is necessary; system control is asserted through the use of standard SCSI command protocol via the SCSI bus.

Models 7123, 6123 and 5123

Model 23 controllers emulate the Ampex DCRsi™ tape recorder. Data rates of up to 64 Mb/sec are attainable across the user equipment interface. In read-after-write mode, up to 128 Mb/sec of internal bus bandwidth is used (64 Mb/sec write and 64 Mb/sec read). Model 7000 recorders can accommodate two Model 7123 controllers, each operating in read-after-write mode at the full 240 Mb/sec data rate of the DCRsi.

Models 7124, 6124 and 5124

Model 24 controllers feature ANSI-standard Fibre Channel fibre optic interfaces. Bit-serial I/O is supported at user equipment data transfer rates equal to that of the internal system bus on the Model 5000 version and at the maximum Fibre Channel data rate on Model 6000 and 7000 versions. The user equipment interface features ANSI-standard Fibre Channel hardware. On-board IRIG time-code reader/generators are used to append time-stamps to user data. No separate control port is necessary; system control is asserted via the Fibre Channel interface.

Models 7125, 6125 and 5125

Model 25 controllers feature Enertec™ ID-1 emulation. This interface is based on 8-bit ECL logic and supports the command protocol for ID-1 tape recorders. Host control is accomplished through a separate serial connection.

Multiplexers

A Model 7000 or 6000 multichannel input multiplexer may be installed in the secondary controller slot of the Model 7000 or 6000 MONSSTR®, respectively. Model 7031 and 6031 multiplexers feature eight PCM inputs with bit rates up to 20 Mb/sec each and two analogue inputs suitable for IRIG time code or voice. The format of the data recorded from 7031 and 6031 multiplexers is compatible with the IRIG 106-98 ARMOR standard. The Model 7032 and 6032 multiplexers format the data from up to eight PCM channels, eight MIL-STD-1553 channels, one time code channel, and one voice channel into IRIG-107 Format 1 packets. A Model 5032 packet multiplexer with two PCM channels, two 1553 channels, one time code channel, and one voice channel may be installed in the multiplexer slot of a Model 5000 MONSSTR®.

Read-after-write

All MONSSTR(r) controllers incorporate data buffer memory that preserves user input data until it has been successfully stored in the memory canister. When read-after-write mode is enabled, the controller will read each data word after it is written to verify that it was written correctly. This feature may be enabled or disabled by the user equipment. Since two bus transfers are required for each word recorded in this mode of operation, the maximum net throughput is limited to one-half of the maximum internal bus data rate of the MONSSTR® recorder.

Special memory operations

Special memory operations may be enabled or disabled by the user equipment. Enabling erase-before-write mode allows the MONSSTR® recorder to operate in circular buffer mode. Once the memory is filled, oldest data is erased and re-written with new data. When 'dub' operation is enabled, the recorder will write user input data to two canisters simultaneously, resulting in two 'copies' of the data. A daisy-chain mode allows multiple recorders to be connected together such that when the memory in one is full, the next one in the chain will begin recording.

Hot-swap memory canisters

The power circuits in Models 7000 and 6000 recorders feature separate power buses to each memory canister, allowing one or more canisters to be powered down without affecting ongoing operations with other canisters in multicanister recorders. This feature makes memory capacity virtually limitless, since full canisters may be replaced without interrupting recording.

Momentary power outage

MONSSTR® power modules incorporate sufficient capacitance to support full-up operation of the recorder during power losses of up to 50 milliseconds. Loss of power for more than 50 milliseconds will result in a clean power-down sequence with no loss of previously recorded data.

Status

During November 2001, Calculex announced that they had signed a contract with Zeiss Optronik GmbH for the supply of its MONSSTR® solid-state recorder for Spanish Air Force F-18 RecceLite reconnaissance pods (see separate entry). The recorder being delivered to Zeiss is the MONSSTR® 5000Z, fitted with 12.8 Gbyte of memory and running at 64 Mbyte/sec bandwidth.

MONSSTR® recorders are in service with a wide variety of military test and research programmes, including AIM-9X missile testing with the 46th Test Wing, Eglin AFB, U-2 flight test at the Lockheed Martin Skunk Works, Tornado EW testing at EADS, Tornado flight test at QuinetiQ, Boscombe Down, F/A-22 flight test and Global Hawk reconnaissance system testing at Northrop Grumman.

Contractor

Calculex, Inc.

MST 67A Mode S transponder

Type

Aircraft transponder.

Description

A third-generation Mode S transponder, the compact MST 67A offers all the capabilities of heavier airline-type units in a much smaller package. It incorporates a number of patent pending features, including such advances as a 16-bit microprocessor, programmable gate array digital signal processing and SAW technology. Fully TSO'd, the remote-mounted MST 67A is equipped with standard ARINC 400 series connectors.

A choice of control heads allows the MST 67A to fit virtually any corporate or regional airliner class cockpit. Featuring a photocell-equipped gas discharge display, the KFS 578A control unit supplies ARINC 429 data to all versions of the system. The KFS 578A can also serve as the aircraft's TCAS controller. For aircraft already equipped with a dzus-mount transponder control panel, the CTA 81A is available as a drop-in replacement. Fully compatible with ARINC 718 and ARINC 735 and providing many of the same interfacing and control functions as the KFS 578A, the CTA 81A features a high-contrast liquid crystal display.

In its non-diversity version, the transponder uses a bottom antenna only, for operators who do not anticipate installing TCAS in the aircraft but wish to ensure compliance with ATC reporting standards. A non-diversity MST 67A is fully compliant with air-to-ground/ground-to-air datalink applications.

The diversity version of the MST 67A uses inputs from two antennas, mounted top and bottom of the aircraft. Required for TCAS/ACAS operations, the diversity option provides the aircraft with air-to-air datalink communications capability.

Enhanced BIT features constant monitor transponder status. A bidirectional interface between the transponder and the control unit also enhances diagnostic capabilities. With the test mode selected on the control panel, internally diagnosed problems can be viewed in real time and information stored for as many as the nine previous flights, in non-volatile memory, can be reviewed.

Specifications

Dimensions:
(MST 67A transponder) 57.2 × 381 × 193.8 mm
(KFS 578A control unit) 53.1 × 57.2 × 187.5 mm
(CTA 81A control panel) 146.1 × 57.2 × 119 mm

Weight:
(MST 67A transponder) 3.86 kg
(KFS 578A control unit) 0.45 kg
(CTA 81A control panel) 0.82 kg

Specifications

Model	CSR-5100	CSR-6101	CSR-6102	CSR-7101	CSR-7102	CSR-7103
Capacity	90.1 Gbyte	137.3 Gbyte	274.7 Gbyte	546.6 Gbyte	1.09 Tbyte	1.64 Tbyte
Size (mm)	114.3 × 124.5 × 307.3	193.0 × 157.5 × 327.7	193.0 × 215.9 × 327.7	284.5 × 210.8 × 327.7	284.5 × 330.2 × 327.7	284.5 × 431.8 × 327.7
Weight (kg)	8.18	11.36	15.00	18.18	22.73	27.27
Bandwidth (Mbps)	512	1,024	1,024	2,048	2,048	2,048
Power (W)	42	58	58	75	75	75

Status
The MST 67A transponder forms part of Honeywell Aerospace's line of TCAS/ACAS systems.

Contractor
Honeywell Aerospace, Electronic & Avionics Lighting.

RCZ-852 Mode-S transponder

Type
Aircraft transponder.

Description
Aviation Communication & Surveillance Systems' (ACSS) RCZ-852 Mode S transponder offers a small, light package that is optimised for regional and corporate aircraft and helicopter airline applications.

The RCZ-852 is a stand-alone implementation of current Mode S technology integrated into the radio suites of more than 2,500 regional and business aircraft. The RCZ-852 is designed to perform in high vibration environments in turbojets, fixed wing propeller aircraft and helicopters.

When compared with legacy ARINC Mode S transponders, the RCZ-852 is smaller, lighter and requires less power. The transponder incorporates a TCAS II interface and is designed to be compatible with all TCAS I and TCAS III systems conforming to ARINC 718/735 characteristics.

In order to assure high reliability and extended service life, Highly Accelerated Life Testing (HALT) was performed on all engineering models of the RCZ-852. Each production RCZ-852 unit is subjected to Highly Accelerated Stress Screening (HASS). Extensive BIT (Built-In Test) ensures that approximately 90 per cent of the circuitry is tested during system operation.

The RCZ-852 Transponder has all of the required functionality for ICAO ACAS II mandate compliance, European Elementary and Enhanced Mode S Surveillance Downlink of Aircraft Parameters (DAPs), and ADS-B 1090ES extended squitter, as defined in the Aeronautical Information Circulars (AICs) and ICAO SARPS.

The Traffic Alert and Collision System (TCAS) incorporates a Diversity function with top and bottom antenna ports. Diversity provides reliable RF communication links between both ground-based and airborne interrogators.

The RCZ-852 transponder is an ICAO Level 3 system with growth to Level 4. Level 3 means that it will transmit and receive standard length (112 bit) datalink messages for 'COMM A' and 'COMM B' and receive 16-segment extended length datalink messages for 'COMM C'.

Specifications
Dimensions: 104.1 × 86.4 × 355.9 mm (W × D × H)
Weight: 2.27kg
Cooling: None required
Certification
Environmental: DO-160B
TSO: C-112
Software: DO-178B Level B
ADS-B capability: 1090ES Extended Squitter RTCA/DO-260 for 1090 MOPS ADS-B equipment
Interface: ARINC 407/429/575/718
Operating altitude: Sea Level to 70,000 ft
Operating temperature: –55 to 70° C
Power: 28 V DC
Consumption: 30 W (standby); 55 W (maximum)

ACSS' RCZ-852 Mode S transponder is designed for regional and business aircraft and helicopters (ACSS)   1128673

Status
In production and in service.

Contractor
Aviation Communication & Surveillance Systems (ACSS).
An L-3 Communications & Thales company

Recovery Guidance System (RGS)

Type
Windshear detection system.

Description
Safe Flight offers recovery guidance, an optional enhancement of the basic windshear warning system. With the combined WindShear Warning/Recovery Guidance System (WSW/RGS), as soon as the warning occurs continuously, computed pitch guidance for recovery is displayed on the flight director command bars. The system also provides pitch guidance for take-off and go-around on the same instrument, to maintain pilot familiarity with use of the system and promote confidence in it. The pilot does not have to change his flight scan or depart from accustomed procedures during the crucial emergency escape manoeuvre. The company maintains that, by following the command bars, the best possible climb profile to maximise the performance capabilities of the aircraft will be achieved.

The system is armed automatically, even if the flight director is turned off, by the windshear warning system's alert output, but only becomes operative, displaying pitch guidance for recovery, when the pilot activates the GA switch. System logic may be programmed to perform these switching and display functions automatically.

Safe Flight's RGS displays pitch attitudes up to, but not in excess of, the stick shaker target. However, the RGS can be programmed to display stick shaker target information on the ADI slow/fast indicator alongside the pitch guidance display on the command bars. With this optional function, when the pilot activates the GA switch for recovery guidance, the slow/fast indicator changes from a speed mode to shaker mode with the slow bar representing shaker target. The pilot then has a continuous visual indication of his margin to stick shaker. Internal system monitoring and a self-test function ensure system reliability.

Status
As of March 2006, in production and in service. Recovery guidance is available in combination with Safe Flight's windshear warning system (see separate entry) in a single 3/4 ATR box, or where Safe Flight's Speed Command of Attitude and Thrust (SCAT) system is desired (or already installed), through tie-in of the windshear warning and SCAT system computers.

Contractor
Safe Flight Instrument Corporation.

RT Crash-Survivable Memory Unit (RT-CSMU)

Type
Flight/mission recording system.

Description
The L-3 Communications, Electrodynamics' RT Crash-Survivable Memory Unit (RT-CSMU) receives vehicle, subsystem and environmental parameters from a MIL-STD-1553B data bus, and stores these parameters, in uncompressed format, in Electrically Erasable PROM (EEPROM) solid state memory. It protects the data records from damage or complete loss, including crash protection exceeding FAA-TSO-C51a requirements. L-3/EDI also supplies a Flight Data Recorder (FDR) suitable for recording with the RT-CSMU.

Other features of the RT-CSMU include:
- 0.5 to 2 Mb EEPROM storage capacity
- 100,000 memory write cycles (minimum)
- Power fail protection, memory management and error detection
- RT-to-RT transfer available
- Ready-to-record in 5 seconds from turn-on
- Can be driven from any processor
- MIL-STD-1553B Notice 2 protocol
- Extraction per B-1, B-2, T-45 methods and equipment
- Less than one bit error per million
- BIT: power-on, commanded, periodic
- RS-232 test bus
- Ground strap for composite vehicles

The RT-CSMU mounts in a survivable and accessible area of the vehicle, up to 100 feet from the recorder. It contains a microcontroller, MIL-STD-1553 and RS-232 interfaces, power conditioner and 0.5 to 2 Mb of EEPROM memory. The system is capable of circular loop rewrites, between 20 minutes and 12 hours per cycle, for up to 17,000 operating hours. The memory format handles periodic and aperiodic parameters and events, with read-after-write and CRC error-detecting codes. All records are independent and self-documenting as stored. Card-level and chip-level data extraction equipment are also available. The unit is not field-repairable due to its unique construction.

Specifications
Performance
Shock: 3,400 g/5–8 msec, six axis
Penetration: 10 ft drop of 500 lb weight, six axis, 0.05 in²
Crush: 5,000 lb for 5 min, six axes
Fire: 1,100°C for 30 min
Seawater immersion: 1,500 ft, 14 days
Fluids: Fuel, glycol, hydraulic, fire extinguishing; 24 hr
Dimensions: 96.5 × 146.1 × 137.2 mm
Weight: 2.91 kg (2 Mb version)
Power: +16 to +40 V DC, 250 mA
MTBF: 30,000 hr (MIL-HDBK-217E)
Life: 17,000 operating hours; 30 years useful
Cooling: Convection
Compliance: MIL-E-5400, MIL-STD-883, MIL-STD-2000, MIL-S-19500, MIL-C-38999G, MIL-STD-461/462, MIL-STD-810

Status
In production. The RT-CSMU was developed for the F/A-22 aircraft but can be readily adapted to other platforms.

Contractor
L-3 Communications, Electrodynamics, Inc.

SKYWATCH® and SKYWATCH® HP Traffic Advisory System

Type
Traffic Alert and Collision Avoidance System/Aircraft Collision Avoidance System (TCAS/ACAS).

Description
SKYWATCH® is an active surveillance system that operates as an air-to-air or ground-to-air interrogation device. It is derived from the TCAS 791 system as a more affordable alternative to full TCAS systems for helicopter operators and general aviation aircraft.

When replies to SKYWATCH® Mode C-type transponder interrogations are received, the responding aircraft's range, bearing, relative altitude and closure rate are computed to a fixed position and traffic conflicts within 11 n miles are predicted. Visual targets are displayed using TCAS-like symbology, with aural traffic alerts.

The multifunction capability of the SKYWATCH® system makes it possible to share a 3 ATI cathode ray tube display with late-model WX-1000 Stormscope weather mapping systems. SKYWATCH® and Stormscope display functions are selected via a remote

SKYWATCH(r) transmitter receiver computer, control display unit and low-profile directional antenna

0018064

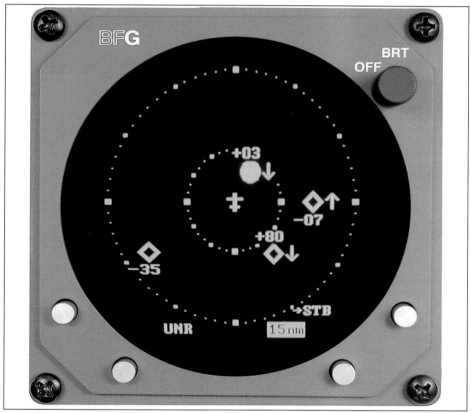

SKYWATCH® HP display unit

0103878

panel-mounted switching device. When operating in the Stormscope mode, the control display unit will temporarily switch to the SKYWATCH® view if an intruder aircraft is detected which poses an immediate collision threat.

SKYWATCH® HP is a development of SKYWATCH®, with a 35 n mile surveillance and display range depending on the selected display interface. The increased power of the system also adds an effective closure rate of 1,200 kts, allowing aircraft at speeds of up to 600 kts to effectively track each other from greater distances. The system tracks up to 35 targets simultaneously, displaying the eight that are the most threatening. SKYWATCH® HP also has enhanced display options, such as an ARINC 429 EFIS output, and the ability to display traffic information on a wide variety of MultiFunction Displays (MFDs) and weather radar indicators using the RGC250 radar graphics computer (see separate entry). The SKYWATCH® HP system can also be installed as a TAS or TCAS I using a conforming display and antenna. Like SKYWATCH®, the new SKYWATCH® HP also has the ability to share a 3in ATI CRT display with the Stormscope® WX-1000 Weather Mapping System.

SKYWATCH® HP is ADS-B equipped in anticipation of the Future Air Navigation System (FANS) requirements.

Specifications
Tracking capability:
(SKYWATCH®) up to 30 targets
(SKYWATCH® HP) up to 35 targets
Range accuracy: ±0.05 n miles (typical)
Bearing accuracy: 5° RMS (typical)
Altitude resolution:
(SKYWATCH®) ±200 ft
(SKYWATCH® HP) ±100 ft
TRC receiver/transmitter:
Dimensions: ARINC standard 404A 3⁄8 ATR short
Weight: 4.06 kg
Power required: 11 to 34 V DC
Multifunction control display unit:
(display) raster scan CRT
(resolution) 256 × 256 pixels
(dimensions) 3 ATI × 209.3 mm
(weight) 1.03 kg
NY164 L-band (NATO D-band) directional antenna:
(dimensions) 279.4 × 158.8 × 35.6 mm
(weight) 1.04 kg

Status
In production.

Contractor
L-3 Communications, Avionics Systems.

Solid-State CVR-30A Cockpit Voice Recorder

Type
Aircraft Accident Data Recorder (ADR)/Cockpit Voice Recorder (CVR).

Description
The SSCVR-30A solid-state cockpit voice recorder provides 30 minutes of data. It accepts four channels of cockpit audio, converts the audio to digital format and stores the data in solid-state non-volatile flash memory. A fifth channel digitally records helicopter rotor speed. The recorder is fully ARINC 557 compatible.

Two control units are available. One conforms to ARINC 557 specifications, the other is a slimline unit designed for installation where cockpit panel space is at a premium and ARINC 557 specifications are not required. Both utilise an LED display for signal level.

Specifications
Dimensions:
(recorder) 1⁄2 ATR short
(ARINC control unit) 57.2 × 148 × 92.1 mm
(slimline control unit) 38.1 × 146 × 98.6 mm
Weight:
(recorder) 10.5 kg
(ARINC control unit) 0.5 kg
(slimline control unit) 0.34 kg
Power supply: 27.5 V DC, 1 A
115 V AC, 400 Hz, 0.2 A option
Reliability: 18,000 h MTBF

Status
In production.

Contractor
Universal Avionics Systems Corporation.

Solid-State CVR-30B/120 Cockpit Voice Recorders

Type
Aircraft Accident Data Recorder (ADR)/Cockpit Voice Recorder (CVR).

Description
The SSCVR-30B/120 solid-state cockpit voice recorders are the latest Universal Avionics Systems Corporation solid-state CVRs. They feature the same capabilities as the SSCVR-30A and are fully ARINC 757/557 compatible. The SSCVR-30B records 30 minutes of data; the SSCVR-120 records 120 minutes. The recorders accept four channels of cockpit audio, convert the audio to digital format and store the data in solid-state, non-volatile flash memory, together with helicopter rotor speed and time.

Specifications
Dimensions: 1⁄2 ATR short
Weight: 5.85 kg
Power supply: 27.5 V DC, 1 A
115 V AC, 400 Hz, 0.2 A optional
Reliability: 30,000 h MTBF

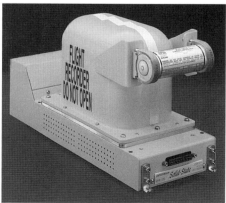

Solid-state CVR-30B/120 cockpit voice recorder

0015337

Status
US FAA certified TSO 123a and ED56A.

Contractor
Universal Avionics Systems Corporation.

Solid-State Data Recorder (SSDR)

Type
Aircraft Accident Data Recorder (ADR)/Cockpit Voice Recorder (CVR).

Description
In September 1999, Metrum-Datatape announced their first solid-state data recorder system. Through a distribution agreement with SEAKR Engineering, Metrum-Datatape is offering the Tape And Rigid-disk Replacement System (TARRS) under the Metrum-Datatape name and SSDR model designator. The SSDR is specifically designed for applications in harsh environments where high speed recording is necessary and data integrity vital.

SSDR features include:
- 96 Gbyte capacity in a single enclosure
- DCRsi compatibility
- Non-volatile, ECC-protected, solid-state memory
- Modularity and expandability
- 100MB/s offload data transfer
- Data rates in excess of 30 Mbytes/s completely solid state.

Status
Available.

Contractor
Metrum-Datatape Inc.

SRVIVR

Type
Flight/mission recording system.

Description
The L-3 Communications, Electrodynamics' next-generation SRVIVR combined voice and data recorder features a soft-core processor utilising a common power supply and flash memory. The system incorporates a flexible architecture, facilitating configuration of the unit for specific fixed-wing or helicopter applications. SRVIVR is capable of monitoring multiple analogue and discrete inputs and is available with a number of interfaces, including 10/100 Base-T Ethernet, MIL-STD-1553B, IEEE-1394b, ARINC 429 and RS-422/485.

The system is designed to EUROCAE ED-112 requirements.

Status
In production.

Contractor
L-3 Communications, Electrodynamics, Inc.

Standard Flight Data Recorder (SFDR) system

Type
Aircraft Accident Data Recorder (ADR)/Cockpit Voice Recorder (CVR).

Description
Smiths' solid-state Standard Flight Data Recorder (SFDR) system has been developed under a US tri-service specification. The solid-state SFDR can replace older oscillograph or tape recorders as well as add more comprehensive aircraft monitoring functions. It was initially introduced by the US Air Force on the Lockheed F-16 fighter.

The Series SFDR consists of the Signal Acquisition Unit (SAU) with auxiliary memory

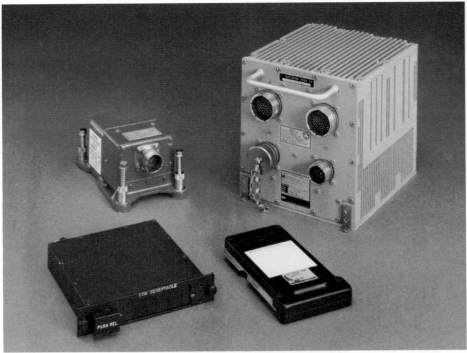

The Smiths Industries standard flight data recorder   0503886

unit, the Crash Survivable Memory Unit (CSMU), the optional Flight Data Panel (FDP) and the Removable Auxiliary Memory Set (RAMS) consisting of an intelligent Data Transfer Interface Unit (DTIU) or a simple Data Transfer Module Receptacle (DTMR) and Data Transfer Module (DTM). The CSMU is also being used on Sweden's Saab JAS 39 fighter.

The SAU is capable of receiving a combination of over 600 discrete, analogue and digital multiplex bus parameters which are converted into digital information, data compressed and stored within the SAU's auxiliary memory unit, RAMS data transfer module and/or CSMU. Daily monitoring and recording of general airframe, avionics and engine health structural loads and engine low-cycle fatigue is typical. All conversion and data management functions are managed witihn the SAU as well as data monitoring and customer designated alerts. The SAU also supports rapid download to ground data logging systems and to graphical replay and analysis systems.

Selected flight data parameters are sent to a compact armoured and insulated CSMU to ensure data recovery following a flight incident. The CSMU is designed to withstand stringent mishap conditions including temperatures of up to 1,100°C, mechanical shock of 3,400 g and penetration. The CSMU non-volatile memory has a life expectancy of 60,000 hours which, together with the built-in test facility, permits it to be mounted in inaccessible regions of the airframe.

Routine maintenance data is retrieved from the aircraft by means of flight line Ground Readout Equipment (GRE) or an installed RAMS. Retrieval data is then transferred to a Data Recovery and Playback Evaluation System for data decompression, analysis and implementation. The STS is developing and loading Operational Flight Program (OFP) data and providing maintenance support for the SFDR.

Specifications
SAU
Dimensions: 157 × 178 × 184 mm
Weight: 6.51 kg
Memory:
(program) 572 kbytes
(scratchpad) 512 kbytes
(non-volatile) 16 kbytes
(auxiliary) up to 3 Mbytes
Environmental: MIL-E-5400 Class II
Power: <50 W

CSMU
Dimensions: 76 × 76 × 117 mm
Weight: 1.58 kg

Memory: 64 k or 256 kbytes
Min recording time:
(attack fighter/trainer aircraft)
15 min active flight, 60 min normal flight
(transport aircraft) 25 h
Environmental:
(impact) 3,400 g for 6 ms
(penetration) 500 lb for 15 ms
(static crush) 5,000 lb for 5 min
(fire) 1,100°C flame for 30 min, or equivalent oven test
(fluid immersion) 48 h at 1,500 ft
Power: <4 W
Locator: acoustic beacon or crash position indicator available

Status
In service on both military and commercial fixed- and rotary-wing aircraft.

Contractor
Smiths Aerospace.

TAWS+

Type
Ground Proximity Warning/Terrain Awareness System (GPWS/TAWS).

Description
The 2 MCU TAWS+ Terrain Awareness Warning System (TAWS) is a component of ACSS' T²CAS™ system (see separate entry), intended for applications where the TCAS component of the fully integrated system is not required. TAWS+ incorporates Terrain Advisory Line (TAL) and Engine Out (EO) functionalities. These advanced features encompass terrain and airport databases. Employing standard aircraft climb rate within its calculations, TAWS+ can be installed with an optional GPS card which provides for greater Situational Awareness (SA) for the crew.

TAWS+ includes the TAL, which extends the conventional TAWS cautionary terrain segment, showing 30° on either side of the aircraft flight path and out as far as two minutes in front of the aircraft, providing crew with an enhanced indication of potential hazards in the current trajectory.

An enhanced variant, designated 'TAWS+performance', builds on the standard system's capabilities by basing alerts on the host aircraft's actual ability to climb. Sophisticated modelling of aircraft climb capabilities provide for further improved SA during critical EO climb-out.

Specifications
Dimensions: 2 MCU
Weight: 3.86 kg
Cooling: Passive
Environmental: DO-160D
Software: DO-178B Level B
Operating altitude: Sea Level to 55,000 ft
Operating temperature: –55 to 70° C
Power: 28 V DC
Consumption: 32 W (nominal)

Status
In production and in service.

Contractor
Aviation Communication & Surveillance Systems (ACSS).
An L-3 Communications & Thales company.

T²CAS™

Type
Ground Proximity Warning/Terrain Awareness System (GPWS/TAWS)/Traffic Alert and Collision Avoidance System/Aircraft Collision Avoidance System (TCAS/ACAS).

Description
Aviation Communication & Surveillance Systems' (ACSS) T²CAS™ traffic and terrain collision avoidance system integrates aircraft performance-based Terrain Avoidance Warning System (TAWS) capability into the TCAS 2000 Line Replaceable Unit (LRU). A Compact Flash Card slot allows for easy loading of terrain and airport databases and other operational software. T²CAS™ is also backward-compatible for operators using ACSS' first-generation TCAS II.

Integrating optional GPS and Windshear functionalities, T²CAS™ provides a unique combination of capabilities. Performance-based T²CAS™ can be applied to any air transport, regional, business or military aircraft to avoid Controlled Flight Into Terrain (CFIT), endowing pilots with more time to clear traffic and terrain with an increased manoeuvre margin.

In contrast to existing TAWS products, T²CAS™ provides alerts predicated on actual host aircraft performance data, thereby minimising nuisance conflict warnings and alerts. For example, if an aircraft suffers an Engine Failure After Take-Off (EFATO), current EGPWS systems may be triggered by the decreased climb performance of the aircraft, whereas T²CAS™ will factor in this decreased performance while accurately alerting the pilot of any potential avoidance manoeuvres. Further benefits of the system include a smaller, lighter system (TAWS can be added to TCAS 2000 without need for an additional LRU), simplified maintenance and reduced spares requirements.

For users of existing GPWS/EGPWS systems, T²CAS™ offers the familiar basic E/GPWS modes, with additional aircraft performance-based capability:

Basic Modes
- Modes 1 and 2 (back-up mode)
- Modes 3, 4 and 5
- Callouts and Bank Angle modes (back-up mode).

Collision prediction and alerting mode
- True 'look ahead' capability in real time, based on predicted aircraft flight path and knowledge of the terrain environment
- Reliable ground hazard warning and caution
- Alert computation according to instantaneous aircraft climb performance
- Eliminates nuisance warnings.

Situational terrain display
In addition to caution and pull-up alerts, T²CAS™ provides two unique display features, designated Alert Line and Avoid Terrain.

System flexibility is facilitated via an updateable worldwide terrain, airport and optional obstacle database.

System options
In addition to TCAS and TAWS functionality, T²CAS™ is also available with optional Reactive Windshear detection and Navigation Quality (NQ) GPS.

System description
T²CAS™ incorporates partitioned modular functions for both TCAS and TAWS functions. TAWS, as required by FAA TSO C151a and other similar standards is provided by an autonomous modular function within the system. This function, termed the Ground Collision Avoidance Module (GCAM), provides fully predictive alerting against CFIT accidents. In addition to its predictive warning capabilities, the GCAM also includes all reactive Ground Proximity Warning System (GPWS) capabilities. The T²CAS™ GCAM supports both analogue and digital aircraft installations in the same unit model, and includes options for Reactive Windshear detection, Obstacle Alerting and, where necessary, an embedded GPS function.

TCAS, also known as the Aircraft Collision Avoidance System (ACAS), required by TSO C119b and other similar standards is provided by fully autonomous functions within the T²CAS™ platform. This part of the system is identical to ACSS' latest generation TCAS product, the TCAS 2000 (see separate entry). As such, it meets all of the latest regulatory requirements for TCAS II and ACAS II products, while providing the necessary growth path for anticipated future requirements. These growth paths include TCAS Change 8 and the new generation of hybrid surveillance capabilities such as ADS-B.

The GCAM predictive warning functions are provided by an advanced Collision Prediction & Alerting (CPA) function, which encompasses terrain and airport databases, as well as modelling of the aircraft climb capability. The CPA computation predicts terrain hazard situations, and generates aural, visual and graphical-display alerts.

The GCAM reactive warning functions include all warning modes required by TSO C151a and DO-161a. GCAM reactive modes include Mode 1 through 5, altitude and bank angle callouts as standard. In addition, the GCAM provides a reactive windshear warning mode.

GCAM's Collision Prediction and Alerting (CPA) provides the primary means for CFIT protection utilising terrain avoidance algorithms. CPA and the Terrain Hazard Display (THD) provide predictive CFIT protection throughout all flight phases from takeoff to landing.

In order to minimise warning confusion and nuisance distractions, Modes 1 and 2 are normally suppressed by the CPA function, and are only active as a back-up to the CPA function. All remaining reactive modes automatically engage during appropriate phases of landing and takeoff.

The CPA function predicts potential terrain conflicts from two primary computational paths:
- Terrain environment determination
- Aircraft flight path prediction

Terrain environment determination correlates precise aircraft position information provided by GPS and/or inertial sources with internally stored worldwide digital terrain and airports databases.

Aircraft flight path prediction utilises current aircraft flight track parameters to project the aircraft flight path more than 2 minutes ahead of the aircraft. A unique feature of the GCAM CPA is the utilisation of a climb rate model to project actual escape capability for the aircraft as part of its warning calculations. This feature is central to GCAM's ability to provide greater warning, thus more time, to pilots for terrain avoidance.

Conventional TAWS vs T²CAS™
Conventional TAWS typically utilises a generic CFIT prediction envelope that is shaped to fit the majority of aircraft. This approach provides an adequate TAWS capability, but does not always provide the protection necessary to avoid the CFIT situation. In some instances of extreme terrain encounter, the existing configuration may not allow the aircraft to achieve the 'fixed' climb assumption built into the generic envelope model. In this case a warning will be provided, but with no assurance of avoiding the CFIT. In contrast, the GCAM prediction envelope is always based on

ACSS T²CAS in 4 MCU form factor (ACSS)

1185933

climb capability of the aircraft model it is installed on. Climb models for the various aircraft types are stored in a database within the GCAM. As a result, the GCAM is always able to predict if the aircraft can avoid the CFIT by a normal 'Pull Up' recovery manoeuvre. In instances where a Pull-Up recovery will be insufficient, the GCAM provides a unique warning- "Avoid Terrain", to immediately notify the crew that the normal recovery will not work, giving the crew time to review the situation and react properly (by a combined pull-up with turn) to avoid the CFIT.

Conventional TAWS allows reactive GPWS alerts and predictive TAWS alerts to be active at the same time. This approach introduces some potential for crew confusion. Crews may be unable to differentiate GPWS nuisance warnings from real predictive warnings and/or may confuse a real GPWS reactive warning with a TAWS predictive warning and thus delay reaction to the potential CFIT.

The GCAM implementation relies on its CPA mode predictive protection whenever sufficient information accuracy exists to make a reliable prediction. While CPA is active, the GCAM suppresses Modes 1 and 2, which are the primary source of GPWS nuisance warnings. When CPA is unable to compute a reliable prediction, the system reverts to the GPWS modes. As a result, the GCAM is able to give the crew a clear understanding of the warning origin so they are fully prepared to react accordingly.

CFIT Detection
The GCAM Collision Prediction and Alerting (CPA) function continuously correlates the projected flight path more than 2 minutes ahead of the aircraft against an internal Terrain Elevation Database (TED). Alerts are generated whenever the CPA computations show that the projected flight path might intersect the correlated terrain elevations underlying that flight path. The CPA computations include two dynamically independent conflict detection algorithms (sensors): a leading caution clearance sensor, and a following warning clearance sensor.

In the horizontal plane, the envelope, or coverage area, of the terrain conflict detection uses a narrow field of view during straight and level flight, diverging at 1.5° to either side of the flight path as it extends outward in front of the aircraft. This narrow view ensures that terrain off either side of the flight path does not initiate unnecessary warnings.

Once the aircraft initiates a turn, the CPA utilises the aircraft turning rate to extrapolate terrain conflict detection over the full terrain area underlying the projected turn between the present aircraft track and the track that is projected by the turning rate up to 90°.

The CPA terrain conflict sensors provide terrain correlation against a world wide 3 nm resolution database of the enroute airspace. When an aircraft enters the airport terminal area airspace, the database resolution increases to 0.5 nm, supporting seamless adaptation of the predictive CFIT protection during low altitude tactical manoeuvres without generating nuisance warnings. The GCAM CPA continues to utilise the same terrain conflict detection rules with >2 min look-ahead to predict potential CFIT threats in the terminal area.

In the vertical plane, the CPA caution envelope extends along the flight path, including flight path angle, from a distance of 20 seconds in front of the aircraft to a distance of 132 seconds in front of the aircraft. 20 seconds provides a comfortable 'time-to-react' point for a crew to initiate a normal climb recovery. The envelope follows the climb capability gradient for the aircraft model; the climb gradient is determined from the current capability of the aircraft including weight, baro altitude, air temperature, aircraft configuration, and climb data for the aircraft model. A 'Caution Terrain' alert is generated when the CPA caution envelope intersects the elevation data representing the terrain underlying the flight path.

The CPA warning envelope is similar in construction to the caution envelope, except that it follows the climb capability gradient from a point 8 seconds ahead of the aircraft to a point 120 seconds ahead of the aircraft. 8 seconds represents the maximum expected time for the crew to react and initiate a recovery manoeuvre. When the CPA warning envelope detects a terrain threat, it is able to generate two types of warning alert. If the CPA calculation determines that the aircraft is able to climb over the terrain, a 'Pull Up' alert will be generated. Since the CPA calculation includes flight path angle and climb capability, the system is also able to determine when the aircraft does not have sufficient climb capability to clear the terrain with a standard vertical recovery manoeuvre. An 'Avoid Terrain' alert will be generated in this instance.

As the aircraft moves closer to the terrain, the GCAM automatically refines its vertical height precision. This is accomplished through consolidation between computed height and measured height. This approach avoids use of baro altitude (which would introduce inherent measurement and temperature errors) and accounts for elevation registration errors between actual terrain and the terrain database.

When well above the terrain, the GCAM CPA utilises computed height in its terrain clearance calculations. As the aircraft nears the terrain, better height precision is available from the radio altimeter system and the GCAM begins consolidating measured height (radio altitude) into its clearance calculation. Below 500 ft, the calculation is made solely using measured radio altitude height.

As the aircraft enters the approach sector, the GCAM begins to measure the aircraft approach relative to an imaginary 'Convergence Envelope', extending from the runway threshold. The Convergence Envelope defines the boundaries for a 'safe' arrival at the runway threshold. CPA alerts are suppressed as long as the calculations indicate that the aircraft can converge (a safe non-CFIT approach is predicted) on the runway. If the calculation shows the aircraft will not converge properly, the CPA is enabled and will provide any appropriate alert should a CFIT potential exist.

The Convergence Envelope has the approximate shape of a tunnel with the smallest part being a landing box that approximately overlays the runway threshold. The bottom and top of the tunnel envelope accommodate normal and steep approaches, while the sides are defined to accommodate straight-on and curved approaches. The aircraft may approach and enter the envelope from any direction and convergence will be calculated accordingly.

Terrain Hazard Display (THD)
In addition to generating aural and visual alerts, the GCAM also provides two Terrain Hazard Display (THD) graphical display outputs for additional situation awareness. The THD image may be presented on the EFIS display, a weather radar display, or another compatible display within the primary field of view. The THD is available during all phases of flight and offers unique features not available in other THD implementations.

The display shows terrain that is above the aircraft Reference Altitude (RA) in increasing density of yellow dot pattern, and terrain below the RA in decreasing density of green. Terrain that is 2,000 ft or more below the RA is shown in black. It is worth noting that the THD does not utilise red terrain depiction – since no crew action is required in this instance and also to avoid confusion with the display format used for a full warning alert.

The THD RA is adjusted relative to the aircraft flight path angle for 30 seconds in front of the aircraft. This approach ensures that the near term terrain situation depicted on the display reflects the real terrain situation that the aircraft will encounter.

A unique feature of the GCAM THD is the 'Caution Alert Line'. Conventional TAWS situation displays show terrain relative to the current aircraft altitude, but give no idea of where an alert will occur. The Caution Alert Line provides the crew with an indication of where the Caution Alert will occur should they continue with the current flight path. The length and direction of the alert line mirrors the cautionary terrain segment within 30° to either side of the aircraft flight path, out to 4 min in front of the aircraft. Once the CPA caution clearance sensor initiates an alert, the alert line is suppressed and the section of terrain associated with the caution alert is highlighted on the THD display in a solid yellow (caution) colour. The alert colouration remains as long as the associated condition exists. The caution colouration area may change according to flight path adjustments. The THD is presented in a 'track up' manner with range selectable in 5 nm increments up to 320 nm. Display images are updated at 0.5 Hz.

When the GCAM CPA warning sensor detects a potential CFIT conflict, it is able to determine if the aircraft has sufficient climb capability to clear the terrain by a standard vertical recovery manoeuvre. If the CPA determines that the aircraft is able to clear the terrain with such a manoeuvre, a 'Pull-Up' alert is initiated and the section of terrain associated with the alert is highlighted on the THD display in a solid red (warning) colour.

When the GCAM CPA warning sensor detects that the aircraft does not have sufficient climb capability to clear the terrain by a standard vertical recovery manoeuvre, it generates a uniquely identifiable 'Avoid Terrain alert'. The Avoid Terrain alert initiates a THD presentation that is unique to the GCAM, which depicts the terrain area that cannot be cleared with a red/black pattern.

The 'Avoid Terrain' feature provided by the CPA is particularly valuable when operating in terminal airspace surrounding Mountainous Area Airports (MAA). MAAs where high terrain exists within 6 nm deserve special treatment, since a turn in the wrong direction can quickly lead into a situation where the crew responding to a

The ACSS product family. From the left, the RT-950 control unit (4 MCU), T²CAS computer unit and AT 910 antenna, XS-950 datalink and RT-950 control unit (6 MCU), CDU and XS-950 SI transponder (ACSS)
1185934

'Pull-Up' warning have no chance of achieving the climb rate necessary to avoid the terrain. The 'Avoid Terrain' alert immediately warns the crew of this danger.

The GCAM THD meets all TSO C151a, Class A, requirements for a terrain situation display. The GCAM implementation recognises most CFIT threats occur in an environment where it is desirable to maximise the crew's 'head-up' Situation Awareness (SA). The GCAM accomplishes this by a CPA that generates consistently intuitive aural alerts that ensure minimum reliance on the THD, then by choosing THD display formats that provide the quickest possible assimilation of the relevant threat information.

GCAM Database

The Terrain Elevation Database (TED) provides a world-wide digital terrain map used by the GCAM for CFIT prediction. Within terminal area airspace, the GCAM draws additional airport/runway information from an Airport Database. Both of these databases are resident within the GCAM memory.

The TED is a world-wide database of all terrain elevations according to WGS84 datum. Terrain elevations covering enroute areas are stored as a 3 nm grid, where the elevation value stored represents the highest point within the grid. Grid density is increased to 0.5 nm in the terminal areas surrounding airports. In some MAA cases where the terrain surrounding the airport is particularly severe, the TED stores the terrain elevations at 0.25 nm grid.

The Terminal Area coverage typically extends 21 nm from the Airport Reference Point (ARP). Where necessary, MAA coverage is provided to a range of 6 nm from ARP. These airports can also be provided with extended terminal area coverage (0.5 nm grid) to 30 nm from ARP. The GCAM includes all airports having runways greater than 3500 feet.

Global Positioning Module

The T²CAS™ may be optionally equipped with an embedded GPS module. This feature is primarily intended for aircraft that are not already equipped with an accurate source of positioning information for the GCAM. A full-featured navigation-capable GPS has been purposely selected for the T²CAS™. This approach enables the flight crew to have access to the same accurate position information as is being supplied for the GCAM alerting functions. The GPS module meets TSO C129a, and is capable of being upgraded for WASS, LASS, SBAS and GBAS, as these become part of the airspace infrastructure.

Connection to an active GPS antenna is all that is required to achieve full-featured GPS functionality for the GCAM. The GPS position information is output on standard bus interfaces for use by external equipment.

Displays

T²CAS™ may be optionally supplied with a 3 or 5 ATI colour display. These displays are primarily intended for those aircraft that are not equipped with a compatible EFIS display or weather radar display. The 5 ATI unit is a MultiFunction Display (MFD), which enables a wide range of terrain, weather radar, and traffic display needs to be consolidated on a single display LRU. The 3 ATI unit is a dedicated THD display, primarily intended for aircraft that are not equipped with a compatible EFIS or weather radar display and have limited space available in the forward instrument panel area. Both displays utilise flight instrument quality Active Matrix Liquid Crystal Display (AMLCD) technology, providing readability equal to modern primary flight displays. The LCD's high-density pixel arrangement with anti-aliasing provides crisp presentation of graphical terrain and textual information. The maximum luminance provided by the display ensures readability in direct sunlight conditions, while the dimming range and built-in dimming features ensure adequate lighting control under all flight deck lighting conditions.

Aircraft Personality Module (APM)

T²CAS™ utilises an ARINC standard APM for storage of aircraft-specific configuration information. The APM is a programmable memory device tethered to the T²CAS™ mounting tray and electrically connected to the T²CAS™ via the tray connector. The configuration information is automatically uploaded to the T²CAS™ during power up and upon request.

Status

T²CAS™ was introduced in Paris by ACSS on 19 June 2001, with customer systems to be available during the latter part of 2002. During February 2002, Northwest Airlines selected the system for its fleet of Airbus A330 aircraft with options for the rest of their fleet. Further orders for the system include FedEx Express for its fleet of Fokker 27, Virgin Express for its Boeing 737 and Aeromexico for its Boeing 757, 767 and MD-80 aircraft. Aeromexico also selected the optional Windshear function and GPS.

During October 2002, the US Customs Service purchased 16 ship-sets of T²CAS™ and the Mode S/IFF (Identification Friend or Foe) transponder for P-3 Orion aircraft on order from Raytheon Aerospace. The equipment will be installed in both the standard and Airborne Early Warning variants of the Customs Service's P-3 fleet.

First flight of T²CAS™ was made during October 2002 on board a King Air C90. The first flight of the aircraft included preliminary testing of the TAWS function of T²CAS™.

Development flight test was completed during December 2002 with the FAA test pilot was quoted as "...pleased with the stability and maturity of the design."

On February 12, 2003, ACSS received Technical Standard Order Authorisation (TSOA) from the Federal Aviation Administration (FAA).

During March 2003, Mesaba, a regional carrier of Northwest Airlines, selected T²CAS™ for installation in its fleet of BAE AVRO-85 aircraft. During May 2003, ACSS began work on the integration of T²CAS™ on the ATR aircraft with the support of ATR to develop a Supplemental Type Certificate (STC) for T²CAS™ on the ATR 42/72 family of aircraft. Also during May, ACSS signed an agreement with Airbus whereby T²CAS™ would be proposed to airlines as a Supplier Furnished Equipment (SFE) option on all-new Airbus long-range, single-aisle and wide body aircraft.

ACSS received its first STC for T²CAS™ for Mesaba BAE AVRO-85 aircraft in November 2003.

In May 2004, the system was selected for the Piaggio P-180, with STC completed in October. In June of that year a further five STCs were announced for Aeromexico (seven B757 and five B767), Air Atlanta Icelandic (four B757), European Air Transport (35 B757) and Virgin Express (nine B737).

During March 2006, ACSS announced that China Sichuan had selected T²CAS for installation on eight new Airbus aircraft. China Sichuan ordered a combination of Airbus A319s and A320s that were scheduled to begin delivering later that year. This order represented the first order for the combined terrain and traffic avoidance system from China.

As of mid-2006, T²CAS was being proposed to airlines as a supplier furnished equipment (SFE) option on all Airbus long range, single aisle and widebody aircraft, and is available as a retrofit solution for existing Airbus and many Boeing aircraft, ATR42/72 and some military aircraft.

More than 1,500 units have been ordered since the initial product launch.

T²CAS customers include Aero California, Aeromexico, Air Asia, Air Atlanta Icelandic, Air Berlin, Air Cariabas, Air Europa, Air Mauritius, Air Niki, Air One, Air Wisconsin, Allegiant Air, American Eagle, AVITEX, Bangkok Airways, China Sichuan, CIT Group, CMC, ComLux, Debis Air, Druk Air, EAT, Estafetta, Etihad, Excel Airways, Falcon Air, Federal Express, German Wings, Gulf Air, Harmony, Hawker Pacific, Icelandic, Independence Air, Islandsflug, Mesaba, NAS, Northwest Airlines, Norwegian Air Shuttle, ONUR Air, QATAR, Raytheon, Royal Brunei, Royal Flight Oman, Silk Air, Sky Airlines, Sukhoi, Swift Aviation, TAROM, TG Aviation, the US Air Force, Virgin Express and the US Customs Service.

Contractor

Aviation Communication & Surveillance Systems (ACSS).
An L-3 Communications & Thales company.

TCAS II Traffic alert and Collision Avoidance Systems

Type

Traffic Alert and Collision Avoidance System/Aircraft Collision Avoidance System (TCAS/ACAS).

Description

Rockwell Collins Business and Regional Systems currently produces the TCAS-94 and TCAS 4000, TCAS II systems. Both systems are suitable for installation in business jets and turboprops, as well as for airliners.

TCAS-94

TCAS-94 tracks up to 150 targets simultaneously and displays as many as 30 at once, with the ability to switch between a short-range display in high-traffic areas and an extended long-range display. An integral part of the system is the advanced TDR-94D Mode S transponder, specifically designed to be lighter in weight and smaller in size than ARINC-compliant transponders used on larger transport aircraft.

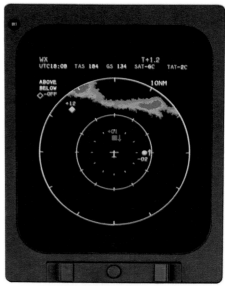

Pro Line 4 MFD format with weather and traffic
0044991

Units of the Rockwell Collins TCAS systems   0018065

MFD-85C with navigation, weather and traffic
0044994

Designed to complement specific cockpit requirements and different operator preferences, the Rockwell Collins TCAS-94 presents a variety of display configurations, whether the aircraft is equipped with electromechanical instrumentation, an Electronic Flight Instrument System (EFIS), or advanced Pro Line 4 or Pro Line 21 avionics.

Other features and benefits of the system include:
* Enhances crew situational awareness of traffic situation
* Provides added safety by detecting and displaying potential collision threats
* Computes avoidance manoeuvres and provides aural and visual commands
* Co-ordinates collision avoidance manoeuvring with other TCAS II equipped aircraft
* Provides a wide variety of display and control options
* Incorporates extensive I/O capability with analogue and digital system interfaces
* Interfaces With single- or dual-Mode S transponders including Collins TDR
* TTR-921 incorporates TCAS Change 7 as a standard feature
* Provides data load feature for on-aircraft software revisions
* Provides internal diagnostics that allow accurate isolation of failed unit online and provides fault history to aid shop maintenance ACAS II compliant.

TCAS-94 comprises the following major items or equipment:

R/T processor: TTR-921
Top directional antenna: TRE-920
TCAS bottom antenna:
  (omni) 237 Z-1
  (directional, optional) TRE-920
Mode S R/T (diversity): TDR-94D
Mode S antennas (2): ANT-42
Control:
  (TCAS) CTL-92T
  (Mode S) CTL-92
  (combined) TTC-920G
Displays:
Traffic Advisory (TA): alternatives
  (EFIS (MFD)) MFD-85C
  EFD-4076
  EFD-4077
Resolution Advisory (RA):
  (EFIS (PFD)) EFD-4076
  EFD-4077
Resolution Advisory – alternative Kollsman VSI/RAL:
  (pneumatic, ARINC 575, VSI, 28 V DC) 47174-003
  (pneumatic, Manchester, VSI, 28 V DC) 47174-004
  (ARINC 429, VSI, 28 V DC) 47174-005
  (Combined VSI/RA/Traffic) (optional) TVI-920D
Power: 28 V DC; 115 V AC, 400 Hz

TCAS-4000
New hardware and software upgrades developed for TCAS-4000 support the addition of Automatic Dependent Surveillance-Broadcast (ADS-B) capability to the existing TCAS/Mode S collision avoidance function. These software upgrades and performance enhancements – defined under the FAA's latest Change 7.0 software standards – set the standard for next-generation, GPS-referenced air traffic management concepts, and ultimately the direct routing efficiencies of 'free-flight'.

With ADS-B, each aircraft uses its transponder to periodically broadcast its identity, altitude and location as defined by GPS co-ordinates. On request from air traffic control, or other TCAS-equipped aircraft, the ADS-B system will transmit additional data such as heading, next FMS waypoint and vertical speed.

Key features of the TCAS-4000 system include:
* DO-185A, Change 7.0 software, providing new verbal commands and display symbology, plus enhanced target surveillance in high-density areas above Fl 180;
* aural and visual warnings 15 to 35 seconds before convergence;
* increased display range greater than 100 n miles;
* tracking of up to 150 aircraft targets;
* operation at up to 1,200 kt closing speed;
* enhanced escape/co-ordination manoeuvres;
* Comm D/Level 4 datalink for Mode S transponder;
* universal fit with analogue or digital interfaces;
* compact 4-MCU R/T design (optional 6 MCU standard unit available); AC/DC power in single unit.

The heart of the TCAS 4000 system is the 4 MCU TTR-4000 TCAS II Receiver/Transmitter, which incorporates all radar surveillance and computer processing functions. In addition, the standard TCAS 4000 system consists of top-and bottom-mounted TCAS antennas, Mode S transponder, L-band antennas, a TCAS/Mode S control panel and cockpit displays. Rockwell Collins offers a variety of TCAS cockpit display options, from integrated EFIS displays, including Pro Line 4 and Pro Line 21 avionics, to individual TCAS instruments.

Major TCAS II Line Replaceable Units (LRUs)
TDR-94/94D Mode S transponders
The TDR-94 provides non-diversity operation, datalink capability.

The TDR-94D provides full-diversity operation: two receivers for top- and bottom-mounted antennas, receiver selection based on better signal, required in TCAS II/IV to ensure data exchange between two aircraft for manoeuvre co-ordination; TCAS interface compatibility; datalink capability.

Certified to Class 2A and Class 3A air-ground datalink. The system supports DAAPS, ADS-B extended squitters and ACAS requirements.

TVI-920D VSI/RA/TA display
The TVI-920D integrates TCAS II RA and TA displays with a conventional vertical speed indicator display using colour LCD flat-panel technology. There are three display modes: vertical speed,

Pro Line 4 PFD format with resolution advisory
0044992

MFD-85C with traffic   0044993

TVI-920D TCAS VSI/RA/TA indicator   0044996

TVI-920D vertical speed indicator 0044995

TAs and proximate traffic; vertical speed, RAs, TAs and proximate traffic; vertical speed, pop-up mode – traffic information displayed only when a TA/RA exists.

TVI-920D is a direct replacement for existing 3 ATI VSIs, and they can integrate with all common VSI sources – pneumatic, analogue and digital.

Status
In widespread service. Collins TCAS II systems have been selected by over 100 major airlines and regional carriers.

Contractor
Rockwell Collins.

TCAS 2000 Traffic Alert and Collision Avoidance System

Type
Traffic Alert and Collision Avoidance System/Aircraft Collision Avoidance System (TCAS/ACAS).

Description
TCAS 2000 is derived from the earlier (Honeywell) TCAS II system. Compared with the earlier TCAS II, it is smaller and lighter, offers double the range and increases the computer capacity by 350 per cent. TCAS 2000 generates advisory information on targets up to 160 km away and can provide this information to other TCAS II-equipped aircraft to co-ordinate manoeuvres.

TCAS 2000 provides for standard TCAS II surveillance up to 32 km for ATCRBS-(Mode A/C) (Air Traffic Control Radar Beacon System) equipped aircraft and up to 64 km for Mode S-equipped aircraft. As an option, TCAS 2000 can provide for extended range surveillance of up to 160 km for Mode S-equipped aircraft. TCAS 2000 is designed to handle closure rates of up to 1,200 kt and vertical rates of 10,000 ft/min. TCAS 2000

computes range, relative altitude, and bearing of nearby transponder-equipped aircraft and visually and aurally alerts pilots of potential collisions, recommending the least disruptive vertical manoeuvre for safe separation. Warning of potential collisions occur at least 20 to 30 seconds before predicted convergence, with more warning at higher altitudes.

A typical TCAS 2000 system consists of a computer unit, Mode S transponder, control panel, resolution and traffic advisory displays, and antennas.

The computer unit performs airspace surveillance, intruder tracking, traffic display, threat assessment, collision threat resolution and TCAS co-ordination. It uses data from airframe and other systems to change performance parameters for varying altitudes and aircraft configurations. Collision avoidance algorithms are used to determine whether a track aircraft is a threat and, if so, the best avoidance manoeuvre.

The Mode S transponder is designed to meet the demands of modern air traffic control. It performs the functions of existing Mode A and Mode C transponders and provides data exchange between TCAS-equipped aircraft. It also communicates with ground-based Mode S sensors, which set TCAS sensitivity levels based on traffic density. The transponder can transmit and receive on either the top or the bottom aerial to optimise signal strength and reduce interference.

The associated control panel selects and controls all TCAS elements including the computer, Mode S transponder, displays and conventional ATCRBS or second Mode S transponder. It includes a transponder failure lamp and four-character LED display for transponder codes, which are set with concentric rotary switches. A variety of displays may be used for TCAS information. The Traffic Advisory (TA) and Resolution Advisory (RA) may be displayed on a colour flat panel display, which integrates vertical speed indication (VSI/TRA). The display is packaged in a 3 ATI-sized indicator. The TCAS also interfaces with EFIS systems to display traffic and resolution advisory information in an integrated display format.

The colours used in TCAS are amber for alert, red for resolution advisory, and blue for non-hazardous traffic. The AT-910 directional antenna features electronic sidelobe suppression and amplitude ratio tracking. The low-profile four-element antennas are mounted on the top and bottom of the fuselage and are capable of transmitting in four selectable directions and receiving omnidirectionally. The antenna transmits at 1,030 MHz and receives at 1,090 MHz.

TCAS 2000 is available in two sizes to meet most upgrade needs, as well as to forward fit a wide variety of aircraft. It is available in both 6 MCU (RT-950) and 4 MCU (RT-951) packages. Both versions offer 28 V DC power connections,

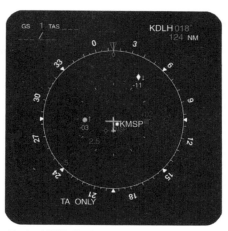

Typical EFIS display

Flat panel VSI/TRA display

TCAS 2000 displays: typical EFIS (above) and VSI/TRA (below) 0018069

while the 6 MCU version also offers a 115 V AC connection. The 6 MCU version is fully compatible in form, fit and function with legacy Honeywell TCAS II installations, while the 4 MCU version only requires the smaller tray for compatibility.

TCAS 2000 is also applicable to military platforms. In this application the system is associated with a new version of the XS-950 Mode S transponder which includes IFF (Identification Friend or Foe) capability, known as the XS-950 SI.

Specifications
System
Processor unit: RT-950 or RT-951
Mode S transponder: RCZ-852 (Business & Regional aircraft)
XS-950 (Air Transport Data Link)
XS-950SI (Military Mode S/IFF)
Antenna: AT 910
Dimensions: 4 or 6 MCU
Weight: 6.66 kg (28 V DC; 4 MCU)
7.16 kg (28 V DC or 115 V AC; 6 MCU)
Cooling: Internal fan (4 MCU); Per ARINC 600/404 (6 MCU)
Certification
Environmental: DO-160C
TSO: C-119b
Software: DO-178B Level B
ADS-B capability: RTCA/DO-260A 1090 MOPS; extended range operations
ADS-B receiver availability: >95%
Operating altitude: Sea Level to 55,000 ft
Operating temperature: –55 to 70°C
Power: 28 V DC or 115 V AC
Consumption: 65 W (nominal)
Max range: 80 nm to meet future Communications, Navigation Surveillance/Air Traffic Management (CNS/ATM) requirements
Display ranges: 5, 10, 20, 40 and 80 n miles
Tracks: 50 aircraft tracks (24 within 5 n miles)
Closing speed: 1,200 kt max

RT-950 TCAS computer unit (6 MCU) (left) and RT-951 TCAS computer unit (4 MCU) (right) 0018068

Vertical rate: 10,000 ft/min max
Normal escape manoeuvres: climb or descend rates; vertical speed limits
Enhanced escape manoeuvres: increased climb or descend; reversed direction of climb or descend

Status

In production and in service. Over 6,000 TCAS 2000 computer units have been delivered to almost 200 operators worldwide since the product's debut in 1997. The TCAS 2000 is offered by most OEMs, including Airbus, Boeing, Bombardier, Cessna, Dassault, Embraer, Gulfstream, Raytheon, Antonov, Ilyushin, and Tupolev. Many airline operators, including American Eagle, British Airways, Federal Express, KLM, Northwest Airlines, Qantas, SAS, Saudia and Swiss Airlines have specified the system for their aircraft. Military operators, including the US Navy, and US Air Force have selected TCAS 2000.

Contractor

Aviation Communication & Surveillance Systems (ACSS).
An L-3 Communications & Thales company.

TCAS 3000

Type

Ground Proximity Warning/Terrain Awareness System (GPWS/TAWS)/Traffic Alert and Collision Avoidance System/Aircraft Collision Avoidance System (TCAS/ACAS).

Description

Aviation Communication & Surveillance Systems' (ACSS) next-generation TCAS 3000 traffic and terrain collision avoidance system is derived from the current TCAS 2000 system and incorporates further advances over the T²CAS system (see separate entries).

TCAS 3000 provides pilots with high accuracy bearing determination (2° RMS vs. the TSO specification of 9°) and extended range surveillance (up to 80 nm active, 100+ nm passive). The system employs enhanced Air Traffic Control Radar Beacon System (ATCRBS) range, altitude and bearing-tracker algorithms, developed by ACSS, to provide greater than 99 per cent track probabilities, while reducing false tracks to less than 0.5 per cent.

TCAS 3000 is completely backward compatible with the TCAS 2000, while offering significant savings in operating costs through its reduced weight and power dissipation. Other features include an internal data recorder with external PC-based data analysis tool and the ability to perform on-board software upgrades.

TCAS 3000 is hosted on ACSS's Surveillance Processor, which delivers additional enhancements in reliability and computing power to handle demanding applications, including the eventual implementation of Change 8 and Airborne Separation Assurance System (ASAS) upgrades. Software-only upgrades allow for ADS-B functions such as Surface Area Movement Management (SAMM) and Cockpit Display of Traffic Information (CDTI) Flight Equipment Rules (CFER).

Other features and benefits of TCAS 3000 include:
- Patented amplitude monopulse antenna with high bearing accuracy
- Superior intruder tracking performance to track up to 400 aircraft
- Extended range
- Significant growth capacity for additional functions:
- Terrain Awareness Warning System (TAWS)
- Mode S Transponder
- Merging & Spacing
- Time and Space partitioned operating system
- Simple operational software upgrades via ARINC 615 Data Loader, Compact Flash or Ethernet

SafeRoute™

SafeRoute™ is a set of Automatic Dependent Surveillance-Broadcast (ADS-B) software applications designed to improve the safety and efficiency of flight operations. SafeRoute's capabilities include Merging & Spacing and Surface Area Movement Management (SAMM); these functions are available individually or as a set.

The SAMM application shows the aircraft's location on an airfield and its proximity to other aircraft on the airfield. The software module leverages other technologies already installed onboard the aircraft to display diagrams of runways, taxiways, gates and airport infrastructure and gives pilots the positions of other aircraft and alerts them to potential runway incursions.

Merging & Spacing enables aircraft to display information that guides merging manoeuvres and spacing behind other aircraft during flight arrivals to bust airports. The resulting improved Situational Awareness (SA) allows pilots to maintain proper sequencing, with output data providing instructions for crews to speed up or slow down to keep a consistent interval between their aircraft and others on approach. These improvements permit operators to keep engines near idle during descent and perform Continuous Descent Approaches (CDA), where demanded, facilitating enable reductions in noise and emissions at lower levels and in the vicinity of large conurbations.

SafeRoute is platform-independent software (a DO-178B Level C system) that can be hosted on ACSS's Surveillance Processor. Its computing power and physical space allows for additional software modules such as TCAS, TAWS, and Mode S Transponder.

Specifications
System
Processor unit: Surveillance Processor
Mode S transponder: RCZ-852 (Business & Regional aircraft)
XS-950 (Air Transport Data Link)
XS-950SI (Military Mode S/IFF)
Antenna: AT 910
Dimensions: 4 or 6 MCU
Weight: 6.28 kg (28 V DC; 4 MCU)
6.97 kg (28 V DC or 115 V AC; 4 MCU)
7.30 kg (28 V DC or 115 V AC; 6 MCU)
Cooling: Internal fan (4 MCU); Per ARINC 600/404 (6 MCU)
Certification
Environmental: DO-160C
TSO: C-119b
Software: DO-178B Level B
ADS-B capability: RTCA/DO-260A 1090 MOPS; extended range operations
ADS-B receiver availability: > 95%
Operating altitude: Sea Level to 55,000 ft
Operating temperature: –55 to 70°C
Power: 28 V DC or 115 V AC
Consumption: 65 W (nominal)

Status

In production and in service. TCAS 3000 was first installed in the Dassault Falcon 7X; the system is currently flying on all three 7X test aircraft.

TCAS 3000 hosts SafeRoute™, ACSS's proprietary software, first unveiled at the Paris Air Show in June 2005.

The launch customer for TCAS 3000 and SafeRoute™ is UPS, which plans to incorporate the system across its entire fleet of some 350 aircraft, to include the new Airbus A380-F cargo aircraft.

Contractor

Aviation Communication & Surveillance Systems (ACSS).
An L-3 Communications & Thales company.

TDR-90 transponder

Type

Aircraft transponder.

Description

The TDR-90 is an air traffic control Mode A and C transponder with 4,096 codes and an altitude reporting capability of up to 126,000 ft when used with an encoding altimeter. It is a remotely controlled system designed primarily for general aviation.

The system has a transmitter output power of 325 W nominal (250 W minimum) on a frequency of 1,090 MHz. Positive sidelobe suppression is incorporated in order to provide a cleaner paint on the interrogator's trace. Two-way mutual suppression avoids interference with DME. Another feature is a strip-line duplexer to control receiver front-end noise while retaining high sensitivity and frequency stability regardless of antenna matching.

A built-in test facility for both the transmitter and receiver functions is included. Test signals are injected at just above the minimum sensitivity level to ensure that receiver, decoder, encoder and transmitter are functioning correctly.

The system's CTL-92 control unit has two-knob code selection, ident, self-test, standby and altitude reporting on/off controls. An optional system selection switch can also be incorporated for use in dual installations. The display is of the gas-discharge type. The TDR-90 electronic unit can, however, interface with most conventional transponder controllers as well as the CTL-92 unit.

Specifications
Dimensions: ¼ ATR short
Weight: 1.59 kg

Status

In production by Rockwell Collins Business and Regional Systems, and in service.

Contractor

Rockwell Collins.

TEC-60i combined interrogator/transponder

Type

Identification Friend-or-Foe (IFF) system.

Description

The airborne IFF TEC-60i combined interrogator/transponder, when functioning as an interrogator, challenges and identifies co-operative active targets. These targets are suitably equipped with compatible transponders and operate within effective IFF range. The unit also functions as a transponder responding to valid interrogations. The unit is a compact, lightweight system installed in the equipment bay, with a control panel mounted in the aircraft cockpit. Peak output power is 1 kW.

Status

In production and in service.

Contractor

Northrop Grumman Corporation, Navigation Systems Division.

Terrain Awareness and Warning System (TAWS)

Type

Ground Proximity Warning/Terrain Awareness System (GPWS/TAWS).

Description

Universal's TAWS provides terrain situational awareness relative to current and predicted aircraft position. This 'look ahead' capability is displayed in three views: Plan, Profile and 3-D perspective. Each view includes the display of the flight plan and flight path intent in conjunction with a detailed display of the surrounding terrain. Full system display benefits are realised using video-capable devices such as Universal's Flat Panel Integrated Displays (FPIDs), its FMS CDU and its new touchscreen cockpit terminal. ARINC 708/708A outputs of terrain in plan view only are available for weather radar displays.

Based upon information from the FMS, the Air Data Computer (ADC), radio altimeter and ILS, the TAWS system is able to determine the aircraft's state and intent, and to provide warnings and alerts well in advance of potential hazards that could result in a Controlled Flight Into Terrain (CFIT). Warnings are provided if any part of the entered flight plan were to pose a threat.

The TAWS system also provides alerts in accordance with standard Ground Proximity Warning System (GPWS) functionality modes. A selectable option is available for bank angle alerting as well.

The TAWS computer is housed in a 2 MCU-sized LRU weighing about 3.2 kg. The worldwide terrain database, stored in flash memory, contains 30 arc sec elevation data with up to 6 arc sec data at mountainous airports. TAWS data is presented as part of Universal's System-1, network of integrated flight management products.

Specifications
Power: 28 V DC at 1.75 A nominal
Environmental: DO-160D
Performance standards:
DO-161A Airborne Ground Proximity Warning Equipment
TSO-C151a Terrain Awareness and Warning System
TSO-C92c Airborne Ground Proximity Warning Equipment

Status
In production and in service.

Contractor
Universal Avionics Systems Corporation.

TPR-900 ATCRBS/Mode S transponder

Type
Aircraft transponder.

Description
Operating with the Air Traffic Control Radar Beacon System (ATCRBS) Mode A and C interrogators, as well as Mode S, the TPR-900 transponder is also compatible with ARINC 735 TCAS systems. It has interfaces for dual Gilham, synchro, ARINC 429 and ARINC 575 input interface ports and is compatible with several types of barometric altimeters.

The solid-state TPR-900 transmitter meets FAA requirements for Class 4 (Comm D) datalink. Up to four segments of downlinked extended length messages can be sent by the transmitter at nominally 450 W power with expansion to 16 segments for Level 5 operation. Reported altitude, discrete address, maximum airspeed, sensitivity control, TCAS control data and Mode S ground station identification are provided as outputs from the system.

The TPR-900 operates with a diversity selection function using two receivers to improve air-to-air surveillance for TCAS. When a signal is received at the two antennas located on top and bottom of the aircraft, diversity selection determines which provides the stronger interrogation signal. The proper reply, depending on the type of interrogation, is then transmitted through the most efficient transmission antenna. Transponder operation is unaffected by aircraft position in relation to other aircraft or ground stations.

The system conforms to the requirements of DO-181A Change 1 and is designed to fulfil all new environmental requirements. It has protection for Cat K lightning and Cat U High-Intensity Radiated Fields (HIRF). In addition, the TPR-900 is fully functional for up to 200 ms of power interruptions.

Specifications
Dimensions: 125 × 193 × 325 mm
Weight: 5.6 kg
Power supply: 115 V AC, 400 Hz, 45 VA
Frequency: 1,090 MHz
Temperature range: –40 to +70°C

Status
The TPR-901 is an update of the TPR-900. Production of the TPR-900 ceased in September 2002; product remains supported.

Contractor
Rockwell Collins.

TPR-901 ATCRBS/Mode S transponder

Type
Aircraft transponder.

Description
The TPR-901 is an ARINC 718, 4 MCU transponder, which operates with older air traffic control radar beacon interrogators in Mode A and Mode C, and with newer sensors that operate with discretely addressed interrogations in Mode S. The transponder is also compliant with ICAO standards and recommended practices for airborne collision avoidance systems (ACAS II, also known as TCAS Change 7.0). Interfaces are provided for dual Gillham encoding altimeters, as well as synchro, ARINC 429 and ARINC 575 air data systems. Diversity antenna ports for reliable air-to-air TCAS surveillance are standard.

The all-solid-state TPR-901 meets all US FAA requirements for Comm A, B and C digital datalink and ICAO Level 3 operations. Growth is provided to support Comm D, Levels 4 and 5 in future. Growth provisions are also included for direct connection to GPS and FMS facilities for extended squitter operation in Mode S, and identification and altitude reporting to support Automatic Dependent Surveillance – Broadcast (ADS-B).

The TPR-901 has a data-loading interface for on-aircraft software updates. Provision is made for transmission of flight identification or radio call sign to meet future European regulatory requirements, and Downlink of Aircraft Parameter Sets (DAPS), a requirement for future CNS/ATM (Communications Navigation Surveillance/Air Traffic Management) operations. DAPS includes datalink of magnetic heading, airspeed, roll angle, track angle rate of change, vertical rate, track angle and ground speed for ATC computations of flight trajectory and air traffic management.

The TPR-901 is the latest Rockwell Collins transponder and is an update of the TPR-720 and TPR-900.

The TPR-901 is qualified to the latest environmental and software verification specifications, including high-intensity radiated fields, lightning and power interruptions to 200 ms, without upset or damage.

Specifications
Dimensions: 125 × 193 × 325 mm
Weight: 6.3 kg
Power: 115 V AC, 400 Hz, 45 VA
Peak power output: 250 to 631 W each pulse
Temperature: –40 to +40°C operating
Certification:
ARINC 718-4A, 735A, 706-1, 600, 407-1, 604, 565, 572-1, 575-3, 429;
RTCA DO -160D, 178B, 181A, 181C;
FAA TSO C112

Status
In service.

Contractor
Rockwell Collins.

TPR 2060 transponder

Type
Aircraft transponder.

Description
The TPR 2060 is a lightweight, compact air traffic control transponder designed for light aircraft and general aviation. It responds automatically to Mode A and Mode C interrogations and, with a suitable encoding altimeter input, will transmit aircraft altitude information with the normal reply pulses. A Mode B capability is optionally available for use in areas employing Mode B interrogation.

The TPR 2060 features special DME suppression circuitry to prevent interference between the transponder and DME installations when the antennas for the two systems are sited in close proximity. The system also permits transmission of a special identification pulse for a 20 second period by an ident button on the front panel. A reply lamp remains lit during this time to reassure the user that the transponder is identing.

Self-test facilities are incorporated. During self-test operation, the unit's coding and decoding circuits are exercised in the same manner as they would be during actual radar interrogation. The unit, which may be panel, console or roof mounted, is in a single case and is of large-scale integrated circuit-type construction.

Specifications
Dimensions: 45 × 160 × 215 mm
Weight: 1.18 kg

Status
In service.

Contractor
Bendix/King.

TRS-42 ATC transponder system

Type
Aircraft transponder.

Description
The TRS-42 ATC transponder system is a digital 325 W solid-state transponder for positive identification in the ATC environment. It consists of the TR-421 transmitter/receiver and the CD-422 control display unit.

The TR-421 transmitter/receiver solid-state design gives 4,096 codes of operation, plus Modes A, B and Mode C altitude reporting when connected to an encoding altimeter. The unit utilises a single chip microprocessor which ensures code data validity and display. To increase system reliability the TR-421 utilises a dual-transmitter design. Under normal operating conditions the dual transmitters work together to provide a full 325 W of power. If one of the transmitters fails, the unit would continue to function, although at a reduced power capability. This feature is especially important in single transponder installations, where the loss of the transmitter would leave no identification capability.

The CD-422 control display unit has the capability to control a dual-transponder installation via a single control head. The selection is made by simply pressing the selector button on the front panel. In a single transponder installation, this button is not provided. The CD-422 also provides an annunciation of the letters ID whenever the transponder replies to an interrogation. When the mode selector is in the VFR position, the active transponder is channelled to the VFR 1200 code. This code may be preprogrammed according to other international VFR codes.

The TRS-422 provides full-time self-testing along with a pilot-selectable TEST mode. The self-testing monitors all key circuits such as the transmitter, receiver, encoder, decoder, video processor and central processor.

Specifications
Dimensions:
(control display unit) 63.5 × 79.38 × 63.5 mm
(front connector transmitter/receiver) 10.16 × 10.16 × 27.94 mm
(rear connector transmitter/receiver) 10.16 × 10.16 × 32.05 mm
Weight:
(control display unit) 0.27 kg
(front connector transmitter/receiver) 2.31 kg
(rear connector transmitter/receiver) 2.73 kg
Power supply: 18–33 V DC, 0.9 A nominal

Contractor
Bendix/King.

Voice And Data Recorder (VADR®)

Type
Aircraft Accident Data Recorder (ADR)/Cockpit Voice Recorder (CVR).

Description
The dual-use Voice And Data Recorder (VADR®) combines the functions of a Cockpit Voice Recorder (CVR) and Flight Data Recorder (FDR) into a highly reliable, lightweight package, available for military and civil applications. The compact recorder system can be hard mounted in virtually any location or orientation, affording the original equipment manufacturer or avionics integrator increased installation flexibility. The VADR® utilises solid-state memory technology, offering increased reliability and low power consumption. The crash-protected memory packaging techniques are based on the US Air Force Standard Flight Data Recorder (SFDR) Crash Survivable Memory Unit (CSMU).

The VADR® provides data collection and mishap recording of audio data and aircraft flight and system parameters to support post-incident analysis. It can be configured as a TSO-C124a Flight Data Recorder (FDR), a TSO-C123a Cockpit Voice Recorder (CVR) or an ARINC 757 combined FDR/CVR. This family of recorders meets the survivability requirements of EUROCAE ED-55 for FDRs and ED-56/56A for CVRs for both ejectable and non-ejectable recorders. The VADR® offers mounting provisions for an underwater locator beacon.

Additional configurations of the VADR® include an ARINC 404 adaptor tray for form, fit and function replacement of existing recorder systems, a CVR/FDR combining voice and MIL-STD-1553, ARINC 429 and ARINC 717/747 serial bus interfaces, and a high-capacity data only FDR for MIL-STD-1553, ARINC 429, ARINC 717/747 or RS-422 serial data recording.

The VADR® is available in a number of models offering audio capacities from one channel × 30 minutes to four channels × 120 minutes. Data capacities from 2 to 30 hours are available depending on application.

Specifications
Dimensions vary with model
Weight varies from: 3–3.5 kg
Audio frequency response:
  3 channels 150–3,500 Hz
  1 channel 150–6,000 Hz

Status
In service on a wide variety of fixed- and rotary-wing aircraft, including: US presidential helicopters, AV-8B, B1-B, ALX Super Tucano, AH-64A, Beech 300, C-2, C-130J, CH-47D, CL-60Y, EC-135, F/A-18C/D, F-111, HH-60J, HH-65A, JPATS, Lear 65, MH-47E, MH-60K, OH-58D, OH-X, RAH-66, SH-60J, UH-60, UP-3, US-1A, VH-3, VP-3, VH-60 and WAH-64 aircraft.

Contractor
Smiths Aerospace.

Windshear systems

Type
Windshear detection system.

Description
Honeywell windshear systems provide detection, alert and guidance in a single unit with two levels of detection – 'caution' and 'warning'. During take-off and approach, the most critical phases of flight, the systems offer an angle of attack reference on the flight director. The ADI also gives an immediate pitch cue to help in exiting a shear. With extensive filtering and an automatic compensation for aircraft manoeuvres and configuration changes, the windshear systems integrate safety with reliability.

Key windshear features enhance the value of the system. The low installation cost is complemented by nearly universal compatibility. Modification of existing aircraft systems is not required and self-contained sensors reduce the proliferation of system configurations. Pin-programmable for multi-aircraft application, the windshear system meets FAA reliability requirements with a 99.9 per cent availability rating and an undetected failure ratio of .00001.

Honeywell windshear systems are available in two configurations: as a stand-alone unit installed in a 3⁄8 ATR short box or as a system integrated with the advanced flight management computer system or flight control computer for new airliners.

Specifications
Dimensions: 3⁄8 ATR short
Weight: 6.8 kg
Power: 22 W
Reliability: 20,000 h MTBF

Status
In production and standard fit on Avro 146/RJ, Boeing 727, 737 and 747, Fokker 100 and F28 aircraft, Lockheed Martin L-1011, Boeing DC-8, DC-9, MD-11, MD-80 and MD-90 aircraft.

Contractor
Honeywell Inc, Commercial Electronic Systems.

Windshear warning system

Type
Windshear detection system.

Description
Safe Flight's airborne windshear warning system provides a voice alert to the crew of high-performance aircraft at the start of an encounter with hazardous low-level windshear. The system is operative during take-off and approach and is a computer-based device which, using conventional sensing elements, resolves the two orthogonal components of a wind gradient with altitude and provides a threshold alert that an aircraft is encountering a potentially hazardous situation. The vectors concerned are horizontal windshear and down-draught drift angle.

Horizontal windshear is derived by subtracting groundspeed acceleration from airspeed rate. The latter term is obtained by passing airspeed analogue data from the airspeed indicator or the air data computer through a high-pass filter. Longitudinal acceleration is sensed by a computer integral accelerometer, the output of which has been summed with a pitch attitude reference gyro to correct for the acceleration component due to pitch. A correction circuit is employed to cancel any errors due to prolonged acceleration. This circuit has a 'dead band', equivalent to 0.2° of pitch, which prevents correction for airspeed rates of less than 0.1 kt/s. Summed acceleration and pitch signals are fed through a low-pass filter, the output from which is summed with the airspeed rate signal to give horizontal windshear.

The vertical computation for downdraught drift angle is developed through the comparison of measured normal acceleration with calculated glide path manoeuvring load. Flight path angle is determined by subtracting the pitch attitude signal from an angle of attack signal sensed by the stall warning flow sensor. This is fed to a high-pass filter and from there to a multiplier to which the airspeed signal has been applied. Thus, the flight path angle rate, corrected for airspeed, provides the computed manoeuvring load term. This is compared in a summing junction with the output of a normal computer integral accelerometer and the failure of the two values to match is the indication of acceleration due to downdraught. The acceleration, when integrated, is the vertical wind velocity and is further divided by the airspeed signal to compute the downdraught angle.

The outputs of both horizontal and vertical channels are determined solely by the atmospheric conditions and ignore manoeuvres that do not increase the total energy of the aircraft. Windshear correctly compensated by increased engine thrust shows no change in airspeed in the presence of an inertial acceleration as thrust is applied. If, however, the shear goes uncorrected, an acceleration or deceleration becomes apparent. Similarly, in the case of the vertical component, a vertical displacement compensated by the crew shows a positive flight path angle rate in a downdraught with less than the computed incremental normal acceleration. Correspondingly, in a downdraught for which the drift angle is allowed to develop, a negative angle rate at a near constant 1 g results in the same computation and output. This is important to the crew as it eliminates the possibility that their actions in anticipating or countering windshear might well mask the condition as far as the warning system is concerned.

Both downdraught drift angle and horizontal windshear signals are combined and the resulting output fed through a low-pass filter to the system computer. This provides two output signals: a discrete alert and, through a voice generator, an audio alert. Warning output is set at a threshold of –3 kt/s for horizontal shear and –0.15 rad downdraught drift angle, or for any combination of the two components which, acting together, would provide an equivalent signal level. According to Safe Flight, any wind condition requiring additional thrust equivalent to 0.15 g to maintain glide path and airspeed will result in a non-stabilised approach.

A crossover network is employed to sense zero crossovers of the combined windshear warning signal and this is sampled every 25 seconds. If the warning signal does not pass through a band close to zero, the network automatically provides failure indication, alerting the crew to the fact that the unit is inoperative. A self-test function activated by the pilot is also built into the system.

Smiths' family of Voice And Data Recorders (VADR®s). From left to right: Model 3255A IDARS (formerly Model 3255 VADR™), Model 3255B IDARS (formerly Model 3253 VADR™), Model 3253A VADR™ and Model 3253C VADR™   0015334

Model 3253 VADR™
ARINC 404 1/2 ATR Adapter

Model 3255 VADR™
ARINC 404 1/2 ATR

Model 3253C VADR™

Model 3253A VADR™

The company has now entered into a licensing agreement with Boeing Commercial Airplanes for the further exploitation of windshear warning technology. This agreement provides for the licensing to Boeing of Safe Flight's existing and pending patents and proprietary data in the areas of windshear detection, alert and escape guidance.

The agreement will facilitate the incorporation of windshear warning capabilities on new models of Boeing commercial aircraft. It is anticipated that the technology will be an added feature on aircraft including the 737-300, 757, 767 and 747-400 models. A version will also be offered for retrofitting on Boeing aircraft currently in service.

Specifications
Dimensions: ¼ ATR
Weight: 2.72 kg

Status
As of March 2006, in production and service. The system has been certified for the Cessna Citation III, Falcon Jet Falcon 50, and Raytheon Hawker 800.

Contractor
Safe Flight Instrument Corporation.

XS-950/XS-950SI Mode S ATDL (Air Transport DataLink) transponder

Type
Aircraft transponder.

Description
XS-950
The XS-950 transponder was designed for the air transport market and meets all ARINC 718 requirements. The XS-950 implements all currently defined Mode S functions with provision for future growth. Current Mode S transponders are used in conjunction with TCAS and ATCRBS to identify and track aircraft position, including altitude. This system transmits and receives digital messages between aircraft and air traffic control. The datalink provides positive and confirmed communications more efficiently than current voice systems.

ACSS' XS-950 is an ICAO Level 4 Mode S transponder, capable of both uplink and downlink extended length messaging (COMM-C/D). The XS-950 meets or exceeds the latest requirements of ICAO Standards and Recommended Practices (SARPs) Annex 10 and RTCA DO-181C Minimum Operational Performance Standards (MOPS) for Mode S equipment and ARINC 718A Mode S characteristics.

The XS-950 has all of the required functionality for ICAO ACAS II mandate compliance, European Elementary and Enhanced Mode S Surveillance Downlink of Aircraft Parameters (DAPs), and

ACSS' XS-950 transponder (ACSS)   1128674

ADS-B 1090ES extended squitter, as currently defined in the Aeronautical Information Circulars (AICs) and ICAO SARPS.

The XS-950 has been designed for reliable performance and ease of maintenance. Highly Accelerated Life Testing (HALT) was performed on all engineering models of the XS-950. Each production XS-950 unit is subjected to Highly Accelerated Stress Screening (HASS) before being shipped to customers to ensure the highest product quality and to prevent failures from ever reaching the end user.

Upgrades to the XS-950 are performed through onboard software loads, providing operators with significant cost savings and operational flexibility.

The XS-950 Mode S Transponder is designed to be upgradeable to ICAO Level 5 datalink or pending US transponder security modifications (also known as the Hijack Mode), with software-only changes.

XS-950SI
Derived from the XS-950 and including all of its core capabilities, the XS-950SI incorporates IFF (Identification Friend-or-Foe), enabling military aircraft to operate within civil airspace and to meet requirements for reduced separation, while providing for IFF reporting required in military operational airspace. All current military IFF functionality is provided in the XS-950 SI. Future growth in the XS-950 SI will parallel changes in the commercial sector, thus affording the military future cost savings. The XS-950 SI can be controlled with a multifunction control display over an ARINC 429 or MIL-STD-1553 bus, or by a special purpose control panel that provides control of the Mode S functions, the IFF function and the TCAS.

The XS-950 SI can perform all the functions of the existing Air Traffic Control Radar Beacon System (ATCRBS), including Selective Identification Features (SIF), Modes 3/A and C operation. The XS-950 SI transponder also meets military IFF Mode 1, 2 and 4 requirements, and it can transmit

ACSS' XS-950SI transponder (ACSS)   1128675

and receive extended-length Mode S digital messages. Growth capability to support CNS/ATM (Communications Navigation Surveillance/Air Traffic Management) Mode-S Level 5 is provided.

Other features of the XS-950 SI include US FAA-specified antenna diversity for simultaneous operation with both top and bottom antennas, interfaces to IFF Mode 4 crypto computer (KIT-C), DoD AIMS 97–1000 certification, Supports Modes 1, 2, 3/A, C, 4 and a demonstrated MTBF in excess of 8,000 hours.

Transmit power is 795 W maximum peak pulse, 316 W minimum and 400 W nominal.

Specifications
Dimensions
(XS-950): 124.5 × 194 × 325 mm (4 MCU)
(XS-950SI): 124.5 × 193 × 386 mm (4 MCU)
Weight: 5.23 kg (both)
Cooling: ARINC 600 DO-160C
Certification
Environmental: DO-160C
TSO/JTSO: C-112/2C-112
Software: DO-178B Level B
ADS-B capability: 1090ES Extended Squitter RTCA/DO-260 for 1090 MOPS ADS-B equipment
Interface: ARINC 407/429/575/718
Operating altitude: Sea Level to 70,000 ft
Operating temperature: –55 to 70°C
Power: 28 V DC; 115 V AC, 400 Hz
Consumption: 40 W (standby); 85 W (maximum)

Status
In production and in service. Over 4,000 XS-950 Mode S transponders have been delivered to more than 100 operators worldwide since the introduction of the system in 1996. Aircraft manufacturers Airbus, Boeing, Antonov, Ilyushin, and Tupolev specify the XS-950, with many airline operators such as British Airways, Crossair, Federal Express, Iberia, KLM, Northwest Airlines, Qantas, Saudi Arabian Airlines, Swiss and UPS having selected the XS-950 for their aircraft.

Contractor
Aviation Communication & Surveillance Systems (ACSS).
An L-3 Communications & Thales company.

FLIGHT/MISSION MANAGEMENT (FM/MM) AND DISPLAY SYSTEMS

Belgium

Cockpit Head-Down Display (CHDD)

Type
Multi Function Display (MFD).

Description
The CHDD-2000 family is a new line of advanced avionics displays that accept both Digital Video Inputs (DVI), together with most currently available analogue video inputs, such as Composite, RGB and STANAG. The use of advanced electronics allows for minimum depth, low power consumption and advanced video processing. The CHDD-2000 line is compatible with ARINC 817/818 and features an optional infra-red touch screen. All variants are also fully compatible with Barco's PU-2000 processor.

CHDD variants can be certified with DO-178B and DO-254 software up to level A and the display can be used as Primary Flight Display (PFD), in combination with an external symbol generator, in most types of aircraft.

CHDD-254
The CHDD-254 is an advanced 5 × 4 in avionics displays that accepts both Digital Video Inputs (DVI) and most of the currently available analogue video inputs (Composite, RGB, STANAG).

The display features a sunlight-readable Active Matrix Liquid Crystal Display (AMLCD). This technology enables the CHDD-254 to provide high brightness and contrast, coupled with low power consumption. The use of advanced electronics allows for a compact footprint (maximum 4.5 in depth).

CHD-268
The CHDD-268 is an advanced, compact 6 × 8 in avionics display that accepts both Digital Video Inputs (DVI) and most of the currently available analogue video inputs.

The display features a sunlight-readable AMLCD and LED backlight technology, high brightness and contrast combined with a wide viewing angle of 80° in all directions. The unit can be used in portrait as well as landscape mode.

Barco anticipates that the CHDD-268 will meet ARINC 817 and 818 digital video standards.

Status
In production and in service. Barco CHDDs have been selected by Boeing Monrovia for

The Barco CDMS   0593659

the BO-105 and by Pilatus Aircraft for the PC-21 advanced training aircraft.

Contractor
Barco.

Control Display and Management System (CDMS)

Type
Control and Display Unit (CDU).

Description
Barco's Control Display and Management System (CDMS) is a compact, lightweight system, featuring a fully sunlight-readable 3 × 4 in full colour AMLCD. This technology provides high brightness and good contrast combined with low power consumption. Customised keyboard panels can be made available, facilitating the installation of the CDMS into various avionics architectures for both new and retrofit aircraft. The CDMS is designed with passive cooling, providing high reliability and maintaining a low-life cycle cost, even in harsh operating environments. The CDMS is available either with MIL-STD-1553 or ARINC 429 interface; the -1553 compliant unit is configured as a remote terminal, whereas the ARINC 429 option is configured as an ARINC 739A MCDU. Both versions have been FAA/JAA/TSO-certified. NVIS-A or NVIS-B compatibility can be specified by the customer.

The main features of the CDMS are:
- 3 × 4 in colour screen
- 320 × 234 standard resolution
- NVG compatibility option
- Extensive Built-In Testing (BIT)
- Modular expandable architecture
- Compact design
- Several interface options
- Different levels of processing power available
- Accepts buyer furnished software
- Software loadable through connector
- Remote processing unit option for limited space applications.

Specifications
Display: 3 × 4 in AMLCD (Silicon TFT)
Panel active area: 127 mm diagonal
Screen:
320 × 234 pixels (QVGA)
264,144 colours, 64 greyscales
Viewing angles: Horizontal: ±45°, Vertical: +45/−20°
Colour temperature >5,000 K

The Barco CHDD-268   1195516

White uniformity >30%
anti reflection multilayer coating MIL-C-14806
Brightness: 550 cd/m2 / 160 fL white surface
Contrast ratio: >100:1 at dark environment; >7:1 at 10,000 fC
Power supply: 28 V DC, MIL-STD-704A and STANAG 1008, RTCA-DO 160D
Power consumption: Operating 60 W, cold start 80 W
Cooling: Convection
Computer programme language: Ada
Interface:
  Option 1: 1553 Remote Terminal Unit: dual-redundant MIL-STD-1553B, discrete board with a minimum of 16 discrete I/O
  Option 2: ARINC 739A MCDU: ARINC 429 board, discrete board with a minimum of 16 discrete I/O
MTTR: Less than 0.5 h
Environmental: Military: MIL-STD-810E; civil: DO160D
High temperature: +55°C operational, +71°C (30 min)
Low temperature: –40°C, warm-up time 5 min, operational 10 min (full specs)
EMI/EMC: MIL-STD-461C/462D
MTBF: 10,000 h (AIC), excluding backlight
Weight: 4.5 kg
Options:
  Retrofit market configuration with remote processing unit
  High-end power engine for FMS or mission applications
  Built-in mass memory
  PCMCIA dataloader

Status
The system has been selected by Eurocopter for the Tiger helicopter, by Honeywell Defense Avionics Systems for the C-27J cockpit upgrade programme and by NIIAO for Tupolev, Ilyushin and Beriev aircraft.

Contractor
Barco.

MultiFunction Display (MFD)

Type
MultiFunction Display (MFD).

Description
Barco's MultiFunction Displays (MFDs) utilise Active Matrix Liquid Crystal Display (AMLCD) technology, providing superior optical

Barco's MFD-6.8/1 multifunction primary flight instrumentation display   0593661

Barco's MFD-5.4/1 multifunction primary flight instrumentation display   0593660

performance in terms of display luminance, contrast and reflectivity. Graphics and video from a number of sources can be displayed in single or overlay mode, with optional NVIS compatibility. The MFD product family uses the same architecture as the CHDD range (see separate entry), enabling full interchangeability of all option boards.

The main features of MFD are:
- AMLCD colour screen
- 640 × 480 VGA resolution
- NVG compatibility option
- Extensive Built-In Test (BIT)
- Modular expandable architecture
- Compact design
- Several communications options
- Processing options
- Accepts buyer-furnished software
- Software loadable through connector
- Remote processing unit option for limited space applications
- Full greyscale anti-aliasing while maintaining a 60 Hz data update rate, providing very fluid ADI/HSI operation under worst case screen loading conditions
- Forced air cooling, giving high reliability and low life cycle cost.

Specifications
Screen:
264,144 colours, 64 greyscales
Colour temperature >5,000 K
White uniformity >30%
anti reflection multilayer coating MIL-C-14806
NVIS compatibility MIL-L-85762A (optional)
Brightness: 685 cd/m2 / 200 fL white surface
Contrast ratio: >100:1 at dark environment; >7:1 at 10,000 fC
Power supply: 28 V DC, MIL-STD-704A
Software: DO178B
MTTR: Less than 0.5 h
Environmental: Military: MIL-STD-810E; civil: DO160D
High temperature: +55°C operational, +71°C (30 min)
Low temperature: –40°C (operational), warm-up time 5 min, operational 10 min (full specs)
EMI/EMC: MIL-STD-461C/462D
MTBF: 10,000 h (AIC), excluding backlight

MFD-5.4/1
Display: 5 × 4 in AMLCD (Silicon TFT)
Panel active area: 130.6 × 97.8 mm
Screen:
  Viewing angles: Horizontal: ±60°, Vertical: +35/–55°

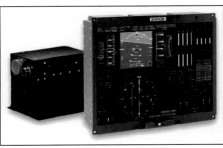

Barco's MFD-12.9   0593662

Power consumption: Operating 60 W, cold start 90 W
Cooling: Forced air via cold wall
Input/output:
  Option boards: Dual-redundant MIL-STD-1553B bus, ARINC 429 (16 Rx/8 Tx), Discrete board with 24 discrete I/O, ARINC 708-453 weather radar, analogue video input board
Interface: RS-170, STANAG 3350B/C PAL/NTSC
Weight: 4 kg

MFD-6.8/1
Display: 8 × 6 in AMLCD (Silicon TFT)
Panel active area: 211.2 × 158.4 mm
Screen:
  Viewing angles: Horizontal: +35/–10°, Vertical: ±45°
Power consumption: Operating 70 W, cold start 150 W
Cooling: Forced air via cold wall
Input/output:
Option boards: Dual-redundant MIL-STD-1553B bus, ARINC 429 (16 Rx/8Tx), Discrete board with 24 discrete I/O, ARINC 708-453 weather radar
Weight: 7.0 kg (single box), 8.5 kg (split box)
Options:
  Extended viewing angle: Horizontal ±65°, Vertical: ±65°
  Landscape mode option

MFD-12.9/1
Display: 12 × 9 in AMLCD (Silicon TFT)
Panel active area: 304.1 × 228.1 mm
Screen:
  Viewing angles: Horizontal: ±60°, Vertical: ±40°
Power consumption: Operating 100 W
Input/output:
  Option boards: Dual-redundant MIL-STD-1553B bus, ARINC 429 (16 Rx/8 Tx), Discrete board with 32 discrete I/O, ARINC 708-453 weather radar, analogue input board, video processing board
Interface: RS-170, STANAG 3350B/C PAL/NTSC, PC input
Weight: <9.0 kg

Status
The MFD-6.8/1 has been selected by Kotlin Novator for the Antonov AN-124 heavy lift aircraft, NIIAO for Tupolev, Ilyushin and Beriev aircraft and Kelowna Flightcraft Ltd. for the T-33 Silver Star. The MFD-5.4/1 has been selected by Eurocopter for the Super Puma helicopter.

Contractor
Barco.

Canada

AWACS Mission Data Recorder (MDR)

Type
Flight/mission recording system.

Description
General Dynamics Canada has developed the Mission Data Recorder (MDR) for the 17 aircraft of the NATO AWACS fleet. The MDR takes all the mission data and the audio communications of

the aircraft, time tags it, records it and facilitates simultaneous replay and record for an entire eight hour mission. Recording time can be extended by changing the storage modules in flight.

Other features and benefits of the system include:
- Compact size – 19 in rack mounted
- COTS, 6U-VME based Open Systems Architecture (OSA)
- Built in Test (BIT)

GD Canada also offers variations of the MDR for applications such as Maritime Patrol Aircraft (MP), surveillance, command and control and other military, paramilitary and civilian applications.

Status
In production and in service.

Contractor
General Dynamics Canada, Maritime Systems.

AWACS Situation Display Console (SDC)

Type
Control and Display Unit (CDU).

Description
General Dynamics Canada provides the Situation Display Console (SDC) for the NATO AWACS modernisation program. The SDC utilises a 20.1 in Active Matrix Liquid Crystal Display (AMLCD), COTS processor and embedded graphics engine and Graphical User Interface (GUI) in an open architecture configuration. This technology is also applicable to maritime patrol aircraft, airborne command and control functions, and similar tactical requirements.

Other SDC features include:
- Programmable, colour LCD, quick-action pushbuttons
- On-screen soft key switches
- Standard QWERTY keyboard
- Sensor-mission system interfaces
- Removable memory device (PMCIA)
- Oxygen regulator compatible

Status
In production and in service.

Contractor
General Dynamics Canada, Maritime Systems.

CMA-900 GPS navigation/Flight Management System (FMS)

Type
Flight Management System (FMS).

Description
The CMA-900 provides full-performance GPS navigation and extensive flight management features, including company routes and worldwide ARINC 424 navigation databases, SID/STAR navigation, GPS instrument approaches, offset tracks, and autopilot-coupled holding patterns and procedure turns. As a multisensor navigator, the CMA-900 integrates other approved navigation sensors, and offers several navigation modes including GPS, DME/DME, DME/VOR, Omega/VLF and optional INS/IRS. This enables the system to provide seamless navigation for all phases of flight with maximum accuracy, integrity and availability. Other features include RNP/ANP, RTA and fuel management functions, and auto tuning of navigation and communications radios. Pre-planned product improvement to provide Vertical Navigation (VNAV) and other Flight Management System (FMS) features was a design criterion.

The modular CMA-900 is authorised to TSO-C129, Class A1 and fully compliant with both the required and desired performance requirements of the FAA Notice N8110.57. Designed for operation in the CNS/ATM environment, the CMA-900 incorporates growth capacity for datalink and ATN compatibilities, as well as local and wide-area augmentation differential GPS (LAAS/WAAS), and GIC for precision approaches and landings. The CMA-900 provides FANS-1 equivalent datalink

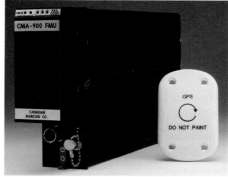

The CMA-900 FMU and GPS antenna   0051579

functions, with Airline Operation Communication (AOC) and Automatic Dependant Surveillance (ADS) datalink software currently certified and Controller Pilot Data Link Communication (CPDLC) to follow. These features make the CMA-900 compatible with any of Eurocontrol's possible P-RNAV requirements.

The CMA-900 incorporates a 12-channel GPS Sensor Module (GSM), which is directly derived from the GPS sensor unit CMA-3012. This system has been selected for the Boeing 777, 737 and Airbus 330/340 series, and has provisions to meet all RTCA requirements for SCAT-I landing equipment. The DGNS capability includes pseudo-ILS approaches and hardware provisions for ARINC 429 or RS-232 interfaces with the local and wide-area datalink receivers and a CMA-2014 Multipurpose Control and Display Unit (MCDU). This is a 14-line 24 character fully ARINC 739-compatible colour AMLCD display.

The technology used in the GSM includes integrity-related features such as proprietary algorithm for high-performance, real-time, satellite Fault-Detection and Exclusion (FDE) including Receiver Autonomous Integrity Monitoring (RAIM), and software written in Ada to critical category standards. Performance is provided by high-dynamic, all-in-view reception of up to 12 channels (including Inmarsat integrity overlay) and carrier phase tracking. The all-in-view tracking and very fast acquisition capabilities of the CMA-900 will be particularly important, in the future, to offset any satellite blanking by the aircraft during terminal-area and precision approach manoeuvres. For aircraft currently equipped with older DME receivers, the DME/DME mode can be implemented using either an ARINC 709 scanning DME, or a digital DME sensor dedicated to the CMA-900.

The CMA-900 operating procedures conform to airline practices, including keyboard and page layouts, full scratchpad, line-select and menu functions. This approach results in fleet-wide commonality in crew interface and a very flexible system with the ability to meet the FANS CNS/ATM environment without keyboard redesign. It also ensures excellent transfer of training as pilots progress through a mixed fleet of aircraft.

The CMA-900 is equipped with the optimum suite of analogue and digital interfaces for retrofit applications, including outputs to conventional and Electronic Flight Instruments (EFIS), suitably modified weather radars, autopilot and flight director systems, and inputs from other navigation sensors, and both analogue and digital air data and heading systems. Data loader interfaces are in accordance with ARINC 615.

Other capabilities include full ARINC 739 compatibility and ARINC 429 file transfer protocols, which allow the FMU to interface with a number of aircraft systems, including ACARS and other ARINC 739-compatible CDUs. The CMA-900 also supports ARINC 429 fuel-computer inputs for fuel management functions in conjunction with operator inputs of fuel on board, and so on.

The CMA-900 Multifunction control display unit
0051580

Specifications
Dimensions:
(FMU) 2 MCU – 200 × 57 × 324 mm (excluding front extension)
(MCDU) ARINC 739a – 171 × 146 × 143 mm
(GA/AEU) 19 × 74 × 119 mm
Weight:
(FMU) 3.7 kg
(MCDU) 2.73 kg
(GA/AEU) 0.5 kg
Power supply:
(FMU) 28 V DC, 38 W max
(MCDU) 28 V DC, 35 W max
GPS channels: 12 channels, L1 band, C/A-code, 1575.42 MHz
Accuracy:
(GPS measurement) 33 ft
(horizontal) 100 ft
(vertical) 130 ft
(groundspeed) 1.5 kt
(horizontal velocity) 1 kt
(vertical velocity) 200 ft per min
(track angle) 0.5°
(time) 2 µs
Signal acquisition: 55 sec to initialised first fix (typical)
Reliability:
(FMU) 10,000 h MTBF
(MCDU) 8,000 h MTBF
Certification:
(FMU) TSOs C113, C115b C129a; RTCA DOs -160C, -178B Level C
(MCDU) RTCA DOs -160C, -178B Level C

Status
In production and in service in a wide variety of aircraft including B-707, B-727, B-737-200, B-747-200/200F/300, Dash 8, DC9-30, DC9-50, DC-10-30, EMB-110, Airbus A300-B4, Gulfstream G2, MD-81, MD-82, -83, -87 P-95, Learjet 35 and C-130.

AOC and ADS datalink software was certified in 2003, with CPDLC planned for certification in late 2004.

The CMA-900 is particularly suited to the upgrade of B747-200 Classic aircraft, with integration of all inertial navigation systems commonly found on the aircraft. While allowing for considerable simplification of cockpit instrumentation, the pilot interface has been specifically designed for high commonality with the B747-400 FMS to minimise cross-fleet training requirements. The existing approach VNAV capability is being extended to performance-based VNAV for other flight phases and will be made available as a future software service bulletin.

Contractor
CMC Electronics Inc.

CMA-3000 Flight Management System (FMS)

Type
Flight Management System (FMS).

Description
The CMA-3000 FMS is an evolutionary step in the company's navigation and GPS products. The CMA-3000 is a single component hybrid made up of the development of CMA-900 GPS/FMS and CMA-2014 Mk III MultiControl Display Unit. The system is integrated into a single control display unit with the CMA-900's multisensor navigation and optional external GPS sensor functions. The system is designed to provide helicopters, regional and commuter aircraft with a versatile GPS-based flight management system housed in one line replaceable unit, featuring a multipurpose control and display function; a full-function embedded navigator and a radio management system.

The basic CMA-3000 features eight ARINC 429 inputs and three ARINC 429 outputs, eight discrete inputs, four discrete outputs and one RS-422 I/O port which makes it suitable for most helicopter applications.

CMC Electronics' CMA-3000 is capable of functioning as a Flight Management System (FMS), Radio Management System (RMS) and Control and Display Unit (CDU)
(CMC Electronics)   1129581

For retrofit into aircraft with analogue interfaces, the CMA-3000 can be complemented with an external analogue adaptor unit. This configuration provides a cost-effective solution while still preserving a common pilot interface.

The CMA-3000 is designed for multisensor RNAV operation during the worldwide transition to GPS primary means navigation. Since certification approvals have already been obtained for the CMA-900, the CMA-3000 GPS/FMS will support installation approval for GPS primary means oceanic/remote operations. In addition, the CMA-3000 will be authorised for GPS-based instrument approaches under TSO-C129 Class A1. This full-function navigation system can also be used with a wide variety of navigation sensors.

The CMA-3000 uses state-of-the-art colour Active-Matrix Liquid Crystal Display (AMLCD) technology. It features a 5 in diagonal display screen with 14 lines of 24 characters each. The system includes a full alphanumeric keyboard; 12 line-select keys; 15 function keys; and nine dedicated annunciators.

The CMA-3000 display and back-lighting are sunlight readable, and compliance with MIL-L-85762A for Class B NVIS operation is available as an option.

The system's operating procedures conform to current airline practices of 'glass cockpit' aircraft, including keyboard and page layouts and full scratchpad, line-select and menu functions. The CMA-3000 conformity allows fleet-wide commonality in the pilot interface. All navigation and radio tuning functions utilise the same scratchpad/line-select crew interface philosophy.

Waypoint, navigation and guidance information is generated in both geographic and track-related reference frames. This information is clearly visible to the pilot on the unit's display. In addition, the system will output the information to flight instruments and autopilot/flight director systems. Its capabilities include:

- Flight planning and route creation, selection and modification
- Complete oceanic, en-route, terminal and non-precision approach navigation and guidance
- GPS instrument approaches
- Continuous and manually initiated predictive FDE
- Outputs to Electronic Flight Instrument Systems (EFISs), and digital autopilot and flight director systems
- Direct-to and leg/course intercept navigation, holding patterns, DME arcs, procedure turns, and offset tracks
- Automatic leg change with fly-by and fly-over leg transitions
- Time and fuel management, including Required Time of Arrival (RTA) computation and display
- Required and Actual Navigation Performance (RNP/ANP) computation and display
- Search pattern navigation

- ARINC 615-3 dataloading capability for software maintenance and database update
- Sensor status information display
- Navigation and communication radio tuning
- Digital map display interface to support route exchange and positioning
- Kalman Filter Integration of GPS/AHRS (INS) (option)
- Compliance with all relevant RTCA and TSO requirements
- Operation in severe (100 V/m) High-Intensity Radiated Field (HIRF) environments.

Specifications
Type: AMLCD
Screen size: 4×3 in (101×76 mm)
Resolution: 320×234 (V $\times$ H) RGB (QVGA)
Viewing angle: $\pm45°$ horizontal, $+10/-30°$ vertical
Contrast ratio: 2:1 daylight (minimum); 20:1 night
Luminance: 0.2 to 120 fl
NVIS compatibility: Per MIL-L-85762A
Alphanumeric data: 14 lines of 24 characters (including scratchpad)
Dimensions: $172 \times 146 \times 162$ (including front bezel) mm (H $\times$ W $\times$ D)
Weight: 3.0 kg
Power requirements: 28 V DC, 40 W maximum (operating), 70 W maximum (warm up)
Reliability: 8,500 h airborne MTBF (rotary wing, predicted)
Interfaces: ARINC 429 (9 inputs, 2 outputs); RS-422 (1 receive port, 1 transmit port); discrete (9 inputs, 4 outputs)
Processor: AMD 486 DX4/4
Software: DO 178B Level C
Certification: TSOs C113, C115b, C129a Class B1C1 (with optional embedded GPS receiver); RTCA DOs -160D

Status
In service.

Contractor
CMC Electronics Inc.

CMA-9000 Flight and Radio Management System (FMS/RMS)

Type
Flight Management System (FMS).

Description
The CMA-9000 advanced flight and radio management system features an embedded datalink capability and was designed for civilian and military fixed- and rotary-wing applications. The system incorporates a sunlight-readable colour Active Matrix Liquid Crystal Display (AMLCD) and flight management functionality which is fully compliant with RNP-1 (with growth to RNP-RNAV) specifications. The embedded datalink is FANS-1/A compatible (AFN, ADS and CPDLC). The CMA-9000 provides an interface to navigation radios and sensors, communications radios, digital moving map displays and other mission equipment.

The Multifunction Control and Display Unit (MCDU) is high-reliability design, with comprehensive Input/Output (I/O) for retrofit as well as new-build installations, and incorporates extensive Built-In Test (BIT) for ease of maintenance.

The CMA-9000 is also available as a Radio Management System (RMS) only.

Specifications
Dimensions: ARINC 739a – $171 \times 146 \times 143$ mm
Weight: 3.6 kg
Power supply: 28 V DC, 60 W (nominal); 90 W (warm up)
Reliability: 12,400 h MTBF (fixed-wing, predicted); 8,500 h MTBF (rotary-wing, predicted)
Certification: TSOs C113, C115b C129a Class A1 C129a Class B1C1; RTCA DO-160D
Display: AMLCD, 8 colours
Screen size: 4×3 in (101×76 mm)
Resolution: 960 horizontal $\times$ 234 vertical RGB

Alphanumeric data: 14 lines of 24 characters (including scratchpad)
Viewing angle: $\pm45°$ (H); $+10/-30°$ (V)
Contrast ratio: 2:1 (day); 20:1 (night)
Luminance: 0.3 to 120 fl (standard and NVIS compatible displays)
NVIS Compliance: MIL-L-85762A NVIS Green + Saturn Yellow
Interface: MIL-STD-1553B, 16 ARINC 429 input; 8 ARINC 429 output; 3 RS-422 transceive port; 1 RS-422 transmit port; 16 unbalanced and 1 balanced discrete inputs
Processor: AMD K6-2E
Software: DO 178B Level C

Status
In production. Suitable for transport aircraft and helicopter applications.

Contractor
CMC Electronics Inc.

Integrated Weapons Delivery System (IWDS)

Type
Mission Management System (MMS).

Description
CMC Electronics' Integrated Weapons Delivery System (IWDS) is an end-to-end, pilot consent-through-weapons-release solution for stores management. Available for both rotary- and fixed-wing aircraft, the IWDS features CMC Electronics' HeliHawk or SparrowHawk™ Head-Up Display (HUD) for navigation and weapons aiming, the FV-4000 Mission Computer (MC) for multidisplay graphics generation and I/O mission processing, an Armament Interface Unit (AIU) for external or internal weapons installations, MultiFunction Displays (MFDs) for the display of critical and non-critical data and a fully customisable Weapons Control Panel (WCP).

IWDS features include passive range finding using high-resolution Digital Terrain Elevation Data (DTED – Level 2 accuracy), Terrain Awareness Warning, an embedded digital map, mission-loadable tactical data and maps, external sensor/EVS display, multi-AIU and multipylon control capability and an embedded ACMI.

The IWDS compact PCI form factor, open architecture design facilitates future growth and upgrade.

Specifications
See individual entries for SparrowHawk™, HeliHawk™, FV-4000
MultiFunction Display (MFD)
Full colour, high resolution
Sunlight readable (330 ftL)
SVGA or XGA resolution
NVIS compatible
4×5; 5×5, 5×7 and 6×8 in sizes.

Weapons Control Panel (WCP)
Customisable control interface for aircraft role and weapon type
NVIS compatible
5/28 V DC
Standard Dzus rail size.

Status
In production.

Contractor
CMC Electronics Inc.

Mission Data Management System (DMS)

Type
Mission Management System (MMS).

Description
General Dynamics Canada's DMS is a flexible mission data handling system designed to support

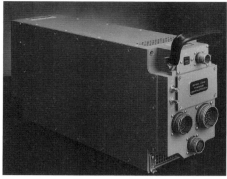

The Mission Computer (MC) for the Data Management System (DMS)   0504109

The DMS has been specified for the Canadian CP-140 Aurora Maritime Patrol Aircraft (MPA)
0593393

maritime and littoral operations by providing highly-integrated airspace, surface and subsurface surveillance. The DMS provides all mission-critical tactical processing, database management, control and display functions to control, monitor and provide for data fusion of a range of multi-spectral sensor and communications systems:

- Flight navigation system
- Internal/external communications including Tactical Data Links (TDL)
- Radar (conventional and SAR)
- Electronic Support Measures (ESM)
- Electro-Optic (EO) sensors (II, TV, IR)
- Sonobuoy acoustic processing
- Dipping Acoustic Sonar
- Magnetic Anomaly Detector (MAD)
- Pre-Flight Data Insertion Package (PDIP)
- Mission Data Recorder (MDR)

The DMS integrates individual subsystem interfaces without requiring modification to the OEM subsystem equipment. All integration functionality is contained within the DMS, allowing the procurement of optimal sub-equipments to suit a specific customer's performance and budget requirements and operational preferences without incurring high integration costs.

The DMS software functionality provides advanced surveillance mission capability including:

- Geographical Tactical Situation Display (TSD)
- Moving Map, Coastline and Bathymetry underlays on the TSD
- Mil-STD-2525 Tactical Symbol Set (TSS)
- Automated sensor contact and target tracking using non-linear Kalman filtering
- Multiple concurrent video-in-window display of sensor data
- Extensive mission system Built-in-Test (BITE) capability to rapidly localise subsystem failures and increase overall mission system availability
- Embedded hi-fidelity tactical operator mission training capability

The system features a modular, scaleable architecture comprising Mission Computers (MCs), Operator Workstation Processors (OWPs) and Tactical Consoles (TCs). The DMS is based on proven hardware Open Architecture principles and COTS 6U-VME form-factor CCAs and enclosures. The Mission Computer (MC) supports a wide variety of industry and military standard digital interfaces to provide integration (control/monitoring and data exchange) with the individual OEM mission subsystems/sensors. The MC provides centralised database management and fusion of all subsystem contact and track data. The MC acts as a Server, providing the OWP Clients with tactical data 'on demand'. Each Operator Workstation Processor (OWP) can support two operators working independently at a shared TC. Each Tactical Console (TC) provides a large, colour, high-resolution, flat-panel tactical display, a touch-sensitive electro-luminescent panel Integrated Control Panel (ICP), QWERTY-keyboard and mouse and/or joy-stick.

The DMS can be scaled-down and reconfigured to satisfy individual customer's performance requirements, mission-tailored sensor suites, operational preferences and budget constraints. The system can be packaged in a variety of hardware configurations including a single air-cooled VME enclosure, PC or laptop form-factor for special mission aircraft applications.

Specifications
Dimensions:
(MC) 132 × 262 × 546 mm
(ICP) 615 × 101 × 316 mm
Weight:
(MC) 15.4 kg
(ICP) 12.4 kg

Status
General Dynamics Canada is currently under contract to supply the DMS to the Canadian Forces for their CP-140 Aurora Maritime Patrol Aircraft (MPA) upgrade programme (7 tactical operators and a cockpit workstation, with a hardware configuration of 1 MC and 4 OWPs each packaged as air-cooled VME enclosures) and new CH-148 Cyclone maritime helicopter (2 tactical operators and a cockpit workstation, with a hardware configuration of 1 MC and 2 OWPs packaged in a single conduction-cooled VME enclosure).

Contractor
General Dynamics Canada.

PilotView™ Electronic Flight Bag (EFB)

Type
Electronic Flight Bag (EFB), Class 2.

Description
The PilotView™ Electronic Flight Bag (EFB), an avionics grade Class 2 EFB per FAA AC 120-76A, consists of two Line Replaceable Units (LRUs) – a lightweight, compact and self-contained Electronic Display and processing Unit (EDU) and a complimentary power and Expansion Module Unit (EMU). The EDU features a high quality, fully dimmable XGA display providing good readability in all light conditions. A 'film-on-glass' touch-sensitive screen works in concert with backlit line select keys arranged around the display bezel, providing for quick access to specific core functions and facilitating navigation around menu-driven, custom or Windows™-based applications, including electronic charts, performance calculations, checklists, e-docs, moving map, weather applications, video display and wireless communications. The EDU is easily positioned in the cockpit using a latching mount which utilises a single hot-pluggable connector to interface with the EMU for aircraft power and addition expansion capabilities.

Other features and benefits of the system include:

- Windows™ environment with simplified menu structure for intuitive pilot interface
- 8.4 in XGA display
- One touch video key enables pilot to quickly toggle between current application and video imagery
- Integrated pilot/co-pilot configuration for dual-installed systems – enables toggling between each display
- Integrated sliding FMS-style keyboard, remains hidden in the EDU when not in use
- Battery back-up.

Specifications
Display: XGA (1024 × 768 pixels)
Dimmable range: 1 to 800 nits
Interface: ARINC-429, RS-232, video in (RS-170/NTSC), Ethernet, Wireless (802.11a/b/g), PCMCIA, USB 2.0
Dimensions: 215.9 × 154.9 × 38.1 mm
Weight: Less than 1.59 kg

Status
In production. Applicable to commercial and business aircraft.

Contractor
CMC Electronics, Inc.

China

8906 Vertical Speed Indicator (VSI)

Type
Flight instrumentation.

Description
The 8906 Vertical Speed Indicator (VSI) is designed for the Y7-200B aircraft. The unit receives ARINC 429 digital signals and displays vertical speed changes to the pilot. The 8906 VSI also incorporates Built-In Test (BIT) capability.

Specifications
Range: –6,000 to +6,000 ft/min
Resolution: ±30 ft/min (0 to 1,000 ft/min); ±2% (1,000 to 6,000 ft/min)
Weight: 1.1 kg
Dimensions: 235.4 × 85.85 × 86.85 mm
Power: 28 V DC; 8 W
Environmental: DO-160B

Status
In service.

Contractor
Chengdu Aero-Instrument Corporation (CAIC).

BGZ-1 precision combined altimeter

Type
Flight instrumentation.

Description
The BGZ-1 precision combined altimeter incorporates barometric altitude measurement, altitude preslection, altitude alerting and parallel altitude reporting. The unit interfaces with a transponder to provide ATC altitude information. Other features of the system include:

- Internal high-accuracy pitot and static pressure sensor for altitude measurement and Static Source Error Correction (SSEC)
- Modular architecture
- Microprocessor control
- Digital servo-indicating
- Integral self-test and alerting.

Specifications
Range: −1,600 to +89,900 ft (±40 ft or 0.4%)
Digitised altitude: −1,000 to 89,900 ft
Pressure setting: 15.95 to 32.11 in Hg (±0.01 in Hg)
Weight: 2.3 kg
Power: 28 V DC; 10 W
Environmental: DO-160B

Status
In service.

Contractor
Chengdu Aero-Instrument Corporation (CAIC).

ZG-2 barometric altimeter

Type
Flight instrumentation.

Description
The ZG-2 barometric altimeter is a modular design, incorporating an ARINC 429 interface to receive signals from a compatible Digital Air Data Computer (DADC). The unit features microprocessor control, a digital servo-pointer and Built-In Test (BIT) capability.

Specifications
Range: −1,000 to +28,000 m (±15 m)
Pressure setting: 4 to 806 mm Hg (±0.5 mm Hg)
Weight: 1.6 kg
Power: 27 V DC; 12 W
Environmental: GJB 150-86

Status
In service.

Contractor
Chengdu Aero-Instrument Corporation (CAIC).

ZKW-1 TAT/SAT/TAS indicator

Type
Flight instrumentation.

Description
The ZKW-1 Total Air Temperature (TAT)/Static Air Temperature (SAT)/True Air Speed (TAS) indicator is a modular, microprocessor-controlled design which incorporates an ARINC 429 interface to receive serial digital information from an Air Data Computer (ADC) or other equipments to provide a digital Light-Emitting Diode (LED) display of output parameters. The system also performs self-test and provides failure warning capability.

Specifications
Range:
TAT −60 to +99°C (resolution 1°C)
SAT −99 to +60°C (resolution 1°C)
TAS 0 to 599 kts (resolution 1 kt)
Display resolution: ±0.1% of display range
Weight: <0.7 kg
Dimensions: 160 × 80.6 × 37 mm
Power: 28 V DC; <12 VA
Temperature: −40 to +70°C
Compliance: Certification TSO-C43c, AS8005; environment RTCA DO-160c; software DO-178b (Level C)

Status
In service.

Contractor
Chengdu Aero-Instrument Corporation (CAIC).

Czech Republic

Airspeed indicators

Type
Flight instrumentation.

Description
Mikrotechna Praha a.s. manufactures a range of 3 in airspeed indicators, including the models: LUN 1106, LUN 1106.XX-8, LUN 1107.XX-8, LUN 1113.XX-8, LUN 1114; LUN 1115, LUN 1116, LUN 1117, and UL 20 series. All are metal pressure-capsule instruments. All series utilise pitot-static pressure. In general, options are available within the following specifications:

Specifications
Range: 0–350 kt, or 0–400 mph, or 0–600 km/h
Lighting: none; 5 V DC; 14 V DC; 27 V DC
Case: ARINC 408 3 ATI; round MS-33638 (AS); round MS-33638 (AS) short
Dial layout: single- or dual-scale and range marked to customer requirements
Approvals: TSO C2d
Temperature range: −30 to +55°C except LUN1107 which is −45 to +60°C
Altitude range: −1,000 to +35,000 ft

LUN 1114 airspeed indicator   0018214

Weight:
(LUN 1116) 0.35 kg
(LUN 1117) 04 kg
(LUN 1106,1115) 0.5 kg
(LUN 1107) 0.55 kg

Leading details of particular models:
LUN 1114
LUN 1114 is calibrated 0–300 kt, and indicates maximum allowable airspeed; it is TSO-C46a, and RTCA/DO 160C approved.

UL 20 series
The UL 20 series are designed specifically for use in ultralight aircraft. Both 3 and 2 in versions are available.

Status
In production and in service.

Contractor
Mikrotechna Praha a.s.

Altimeters

Type
Flight instrumentation.

Description
Mikrotechna Praha a.s. manufactures a range of five altimeters, including the models LUN 1120, 1127, 1128, 1129 and UL 10 series. Altimeters can be manually adjusted to variances in barometric pressure; temperature is automatically compensated by a bi-metallic element.

UL 10 series
The UL 10 series is designed for sports and ultralight aircraft. Both two and three in versions are available, with scales calibrated in inches Hg and millibars.

LUN 1120 encoding altimeter
The LUN 1120 barometric counter pointer altimeter has three moving drums, which

UL 10-10 altimeter   0018213

The LUN 1120 encoding altimeter   0051656

show tens of thousands, thousands and hundreds of ft. The pointer indicates 1,000 ft per revolution on a scale calibrated at intervals of 20 ft. The counter pointer display is actuated by a high-performance mechanism with a high-stability capsule. A built-in vibrator minimises friction and optimises accuracy. All instruments can be customised to meet specific requirements.

LUN 1129 2 in encoding altimeter   0097192

Specifications

Operating range: −1,000 to +35,000 ft, or −1,000 to 50,000 ft, or −1,000 to +65,000 ft
Barometric scale: 950 to 1,050 mbar (28.1 to 31.0 in Hg) and 920 to 1,050 mbar (27.2 to 31.0 in Hg)
Dimensions: ARINC 408 3ATI, round MS33549 (AS)
Dial layout: mbar, in Hg
Output: encoding output according to ICAO
Lighting: 5 V DC, 14 VDC or 28 V DC in colour white, NVIS 28 V DC
Qualification: TSO C10b, C88a, RTCA/DO-160C

LUN 1129 2 in encoding altimeter

The LUN 1129 counter/pointer display is actuated by a high-performance mechanism with a high-stability beryllium capsule. A built-in vibrator minimises friction and enhances accuracy. According to use, the LUN 1129 is supplied with or without an encoder.

Specifications

Operating range: −1,000 to +35,000 ft, or −1,000 to +50,000 ft
Barometric scale: 950 to 1,050 mbar
Dimensions: ARINC 408 2ATI
Dial layout: mbar, in Hg
Output: encoding output according to ICAO
Lighting: 5 V DC, 14 V DC or 28 V DC in colour white, NVIS 28 V DC
Qualification: TSO C10b, C88a, RTCA/DO-160C

Specifications

Range: −1,000 to +20,000 ft; −1,000 to +35,000 ft; −1,000 to +50,000 ft and −1,000 to +65,000 ft (LUN 1120 only)
Lighting: none; 5 V DC, 14 V DC; 27 V DC
Case: ARINC 408 3 ATI; round MS 33549 (AS)
Dial layout: mbar, in Hg; mbar, in Hg
Approvals: TSO C10b

Status

In production and in service.

Contractor

Mikrotechna Praha a.s.

Barometric altimeter LUN 1128, encoding altimeter LUN 1127

Type

Flight instrumentation.

Description

The LUN 1127 and 1128 barometric altimeters are three-pointer, metal-capsule instruments.

Specifications

Weight: 0.5 kg
Operating range: −1,000 to +35,000 ft, or −1,000 to +20,000 ft
Min increment: 20 ft
Barometric scale: 946–1,050 mbar
Qualifications: TSO-C10b, RTCA/DO-160C, TSO C88a

Status

In production and in service.

Contractor

Mikrotechna Praha a.s.

Barometric altimeter LUN 1127.XXXX   0002357

Cockpit Voice and Flight Data Recorder (CVFDR)

Type

Flight/mission recording system.

Description

The CVFDR is the newest product in the SPEEL FDR/CVR line, incorporating the latest recording technologies and improved performance over earlier models. The system can be easily configured to act as a data and voice recorder, or in data only or voice only modes. The CVFDR is extensively qualified and suitable for installation in both civilian and military aircraft.

CVFDR features and functions include:

- Flight data recording incoming from measuring unit via serial link
- Recording of two audio channels:
- Flight crew service calls
- Air-to-ground communication
- Air-to-air communication
- One audio channel recording with background noise of aircraft cockpit
- Digital audio and data recording/transfer to ground equipment
- Voice recording 'online' replay during CVFDR operation
- BIT of CVFDR circuits after start and during recording

SPEEL supplies dedicated Windows™-based PANDA software along with its recorders.

Specifications

Dimensions: 232 × 124 × 122 mm
Weight: 7 kg (max)
Power: 27.5 V DC (11.0–33.0 V DC), 4 W (max)
Recording medium: solid-state flash memory
Data recording capacity: 64 Mb (36 hours of data flow at 512 B/sec)
Voice recording capacity: 140 min each channel
Interface: Two RS422 (9600 Bd, 921,600 Bd); one RS232 (9600 Bd, 115,200 Bd); one LVDS (max 1,500 kBps)
MTBF: 10,000 flight hours
Compliance: ED-55, ED-56A, TSO-C123A, -C124A, DO160D
Operating temperature −55 to +60°C

Contractor

SPEEL PRAHA Ltd.

Combined airspeed indicator with machmeter LUN 1170.XX-8

Type

Flight instrumentation.

Description

The LUN 1170.XX-8 can be used for simultaneous measurement of the indicated and true airspeed, including Mach number.

Combined airspeed indicator with machmeter LUN 1170.XX-8   0051657

It can indicate one mach number and up to three IAS values, and can be equipped with white or NVIS integrated lighting.

Specifications

Ranges:
Indicated airspeed: 100–1,200 km/h
True airspeed: 300–1,200 km/h
Mach number: 0.5–0.9
Ceiling: 15 km
Dimensions: ARINC 408 3ATI, square MS, round MS
Weight: 0.9 kg
Power supply: 5 V or 28 V DC

Status

In production and in service.

Contractor

Mikrotechna Praha a.s.

CVR-M1 Cockpit Voice Recorder

Type

Flight/mission recording system.

Description

The CVR-M1 cockpit voice recorder is an all solid-state unit, designed to be a direct replacement for the MARS-BM mechanical tape-based recorder equipping Sokol PZL W-3A helicopters. The aircraft wiring remains unchanged, only the mechanical fixings are replaced.

The CVR-M1 comprises two individual modules: the connection and communication module; and the solid-state crash-protected memory unit. These modules are fixed together to form a single line replaceable unit. The stored voice data can be replayed using the CVR-M1 or downloaded and replayed using a standard PC, equipped with multimedia capability and the voice replay software supplied by SPEEL PRAHA Ltd.

The CVR-M1 has two audio channels containing flight crew service calls, air-to-ground and air-to-air communications. There is also an audio channel which records background noise in the cockpit. Each channel has 140 minutes recording capacity.

Specifications

Dimensions: 232 × 124 × 122 mm
Weight: 7 kg max
Power: 27.5 V DC (11.0–33.0 V DC), 4 W (max)

SPEEL's CVR-M1 Cockpit Voice Recorder   0051318

Recording medium: solid-state flash memory
Recorded audio channels: 2 crew microphones
(100–3,500 Hz); 1 area microphone (100–6,800 Hz)
Recorded time channel: 1
Recording capacity: 48 Mbyte – corresponds to
the last flight hour
Interface: RS-485/10 Mbps
MTBF: 10,000 h
Compliance: MIL-STD-810E, TSO C124, ED-56A
Operating temperature: –55 to +60°C

Contractor
SPEEL PRAHA Ltd.

FDR 139 Flight Data Recorder

Type
Flight/mission recording system.

Description
The FDR 139 flight data recording system
is designed for the acquisition, processing
and storage of flight parameters for the Aero
Vodochody L-139 light attack/training aircraft.

The FDR 139 consists of three modules:
A solid-state memory module, which stores
measured parameters in case of aircraft crash
A measuring unit module, which processes
analogue and digital parameters including two
ARINC 429 serial bus lines
A power supply module

SPEEL provides proprietary Windows™ based
software, called PANDA, which, combined with
the appropriate GEE and/or GSU configuration,
provides for full data download and evaluation
via RS 422/RS 232 interface; selected records or
the entire FDR memory can be downloaded by
user selection. Data may be transferred to/from
the recorder without interrupting any ongoing
recording action.

Specifications
Dimensions: 258 × 117 × 262 mm
Weight: 8.8 kg max
Power: 16 to 32 V DC, 12 W (max)
Recording capacity: 2 Mb (Approximately 8 flight
hours)
MTBF: 10,000 flight hours
Compliance: TCO-C51a
Operating temperature: –55 to +60°C

Contractor
SPEEL PRAHA Ltd.

*The Aero Vodochody L-139 is fitted with the
FDR 139* 0554712

Flight Data Recorders (FDRs)

Type
Flight/mission recording system.

Description
SPEEL PRAHA has developed a family of
solid-state FDRs that are designed to be direct

Specifications

	FDR39	FDR59	FDR139
Dimensions	258 × 279 × 117 mm	260 × 121 × 243 mm (TM, TME)	267 × 185 × 134 mm
	292 × 185 × 134 mm (BH, BL, BAN)	258 × 262 × 117 mm	
Weight	8 kg (max)	7 kg (max)	8.8 kg (max)
Power	16–32 V DC, <9 W	16–32 V DC, <15 W	16–32 V DC, <12 W
Interfaces	RS232C; RS422	RS232C; RS422	RS232C; RS422
MTBF	10,000 h	10,000 h	10,000 h

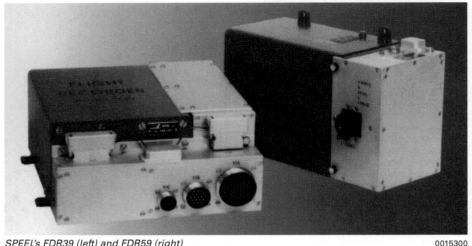

SPEEL's FDR39 (left) and FDR59 (right) 0015300

replacements for the mechanical crash recorders
in Russian aircraft, where SARP-12, TESTER U3
(2T-3M), TESTER U3L (M2T-3), BUR-1 (ZBN1) and
PARES systems are used.

SPEEL's FDR recorders include the FDR39, 59
and 139 series. They are supported by the PMU-F
Portable Memory Unit and Ground Support Unit
(GSU) for downloading recorded information
and by the Ground Evaluation Equipment (GEE)
for evaluating data. Information is downloaded
through an RS232C interface to the GSU at a rate
of 1 hour of flight information in approximately 2
minutes. An RS422 interface transfers the same
amount of information to the PMU-F in about 10
seconds.

FDR39 series solid-state flight data recorders
The FDR39H is a direct replacement of the SARP-
12DM mechanical crash recorders in Mi-17 and
Mi-24 helicopters. The FDR39HG, which also
records GPS system parameters, replaces SARP-
12G recorders.

The FDR39HL system replaces the SARP-12P
recorders in the L410 UVP aircraft. The FDR39HLG
includes GPS recording capability.

The FDR39TM recorder is designed for the L39
aircraft replacing the SARP-12G recorder. The
FDR39TME can be used to replace the SARP-12G
on other aircraft types. In this case, fire signalling
is wired as a negative value instead of a positive
value as on the FDR39TM.

FDR59 series solid-state flight data recorders
The FDR59A recorder is designed for use in L59T
aircraft and can also be used as a replacement of
the cassette tape recorder from the PARES 59E
system. When connected to the MU 59E unit, the
FDR59A is an integral part of the PARES 59M-
KOMPLET ground evaluation system for L59T
aircraft.

The FDR59BH recorder is intended for recording
SOKOL helicopter systems data. It can be used
for the direct replacement of the ZBN-1 magnetic
recorder from the BUR-1 system. The solid-state
memory is divided into two parts, operation
and standby, both of which store information
simultaneously.

The FDR59BL is intended for use in the L410
UVPE aircraft. Memory is again divided in two
and it can also be used to replace the ZBN-1
referred to above.

The FDR59BAN system is designed for use in
aircraft equipped with the MSRP-12-96 system
and is an improved version of the FDR59BL. It
replaces the LPM recording block of the MSRP-
12-96 and can also be fitted with a crash beacon.

FDR139 flight data recorder
The FDR139 recorder is designed for use in the
L139 aircraft and consists of three modules: the
solid-state memory module; the measuring unit
module for processing the analogue and digital
data inlcuding two ARINC 429 serial bus lines;
and a power supply module.

Contractor
SPEEL PRAHA Ltd.

Gyroscopic horizon LUN 1241

Type
Flight instrumentation.

Description
The LUN 1241 gyroscopic horizon is a direct-
indicating instrument, which indicates pitch and
roll angle. The main display indicates angles
within ±20° with greater values of pitch, up to
±70°, being shown digitally in an indication
window. The attitude gyro is provided with a

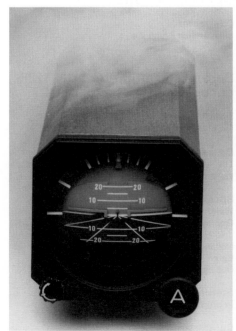

Gyroscopic horizon LUN 1241 0001352

caging gear, an aircraft symbol (with position adjustable between +10 and −7° pitch attitude) and with a failure warning flag. Internal lighting is provided, including NVIS.

Specifications
Dimensions: ARINC 3 ATI round
Weight: 1.6 kg
Power supply: gyro: 27 V DC, or 14 V DC; lighting: 5 V DC/AC, 14 V DC/AC, or 28 V DC/AC
Range: 360° roll and pitch (with controlled precession at more than ±70° pitch angle)
Accuracy: ±0.5°, or ±1.0°
Qualifications: TSO C4c, RTCA/DO -160C and -160B
Temperature range: −30 to +55°C
Altitude: −1,000 to +50,000 ft

Status
In production and in service.

Contractor
Mikrotechna Praha a.s.

Vertical speed indicators

Type
Flight instrumentation.

Description
Mikrotechna Praha a.s. manufactures a number of vertical speed indicators including the LUN

Vertical speed indicator LUN 1184   0051658

1144, 1149, 1184 and UL 30 series. They are metal-pressure capsule instruments.

UL 30 series
The UL 30 series of vertical speed indicators is specifically designed for sport aircraft; both 3 and 2 in variants are available, with scales calibrated in ft/min or m/s.

Specifications
Ranges: 0-2,000; 0-3,000; 0-4,000; 0-6,000 ft/min
Dimensions: ARINC 3 ATI, or MS 33549 (AS) round; LUN 1184 ARINC 2ATI

Vertical speed indicator LUN 1149   0018215

Weight: 0.6 kg
Power: 5 or 27 V DC
Approvals: TSO C8b or C8d, RTCA/DO 160C

Contractor
Mikrotechna Praha a.s.

France

2084-XR mission computer

Type
Aircraft central computer.

Description
The 2084-XR is a general purpose digital computer designed to operate with high reliability under extremely severe environmental conditions.

Using enhanced technology such as VHSIC, ASICs and CMOS battery back-up RAMs, it provides 1.2 Mips throughput with a high performance/volume ratio.

Within a ½ ATR case, the 2084-XR computer comprises: an arithmetic unit; a memory of up to 512,000 18-bit words; a comprehensive input/output system (including the simultaneous management of two multiplexed buses); and a power supply.

Status
The 2084-XR computer is in service in most versions of the Dassault Mirage 2000 aircraft (two computers per aircraft).

Contractor
Thales Systemes Aeroportes SA.

AFDAU/ACMS/QAR

Type
Flight/mission recording system.

Description
The ED 35XX Auxiliary Flight Data Acquisition Unit/Aircraft Condition Monitoring System/Quick-Access Recorder (AFDAU/ACMS/QAR) has been designed for installations with weight or space constraints. It accepts the mandatory crash parameters, meets the requirements of ARINC-573 and can acquire existing ARINC-573 data frame sent from another FDAU or DAU onboard the aircraft. The system can also accept additional analogue, discrete and digital inputs from ARINC-429 databusses. After processing, a new ARINC-573 data frame is transferred to a crash-protected recorder. The system can be expanded by adding a microprocessor and associated electronics to the same box to monitor engine and flight parameters. A printer and quick access recorder may be connected to store maintenance data.

Information can be replayed on the ground through a PCWindows™-based software application to replay QAR (AGS) or ACMS (GSE) data.

A new option that utilises a removable PCMCIA card offers flight data storage equivalent to use of external recorders such as QAR/DAR.

Specifications
Dimensions: 1/2 ATR short
Weight: 4.2 kg
Power: 30 VA

Status
In production for the ATR 42 and 72, Embraer 120 and ERJ 135/145.

Contractor
SAGEM Défense Securité, Navigation and Aeronautic Systems Division.

AFDS 95-1 and -2 Automatic Flight Director System

Type
Avionic flight guidance/autopilot system.

Description
The AFDS 95-1 and -2, designed for VFR and IFR operation with light and medium helicopters, are modular systems consisting of a basic two-to four-axis autopilot computer, flight director computer and associated equipment. In the event of a flight director computer failure, the system is reconfigured in its basic mode, which allows the pilot to complete the mission.

By using inputs from sensors, the autopilot computer provides references for automatic control relative to attitude and heading. The basic autopilot functions are cyclic and yaw damping, long-term pitch and roll attitude hold and long-term heading hold in cruise and hover. The fourth axis facility is intended for operators needing transition and hover capability. The autopilot computer drives the electromechanical actuators and, the trim and artificial feel actuators. The digital, panel-mounted flight director interfaces with the navigation equipment and permits the selection of operating modes, allowing en route and approach applications, search and rescue, or anti-submarine warfare operations. It also provides co-ordination of the collective, yaw and cyclic coupling and longitudinal and lateral speed control functions when using Doppler signal data.

The AFDS 95-1 and -2 flight control systems are transparent, therefore the pilot's control inputs are detected and the attitude hold terms removed to keep the autopilot from resisting pilot control in manoeuvring flight. The AFDS 95-1 system

The AFDS 95-2 automatic flight display system   0079260

is designed for analogue operation, while the AFDS 95-2 is for the digital environment. The AFDS 95-1 system is FAA certified in VFR and single-pilot IFR configurations.

Status
The AFDS 95-1 system is in service on the B 222, AB 212, A 109 K2 helicopters. The AFDS 95-2 is in service on the A 109 M and Bell Model 427 light twin-turbine helicopter operating in the single-pilot IFR configuration.

Contractor
Thales Avionics SA.

AFDS 2000 automatic flight control/director system

Type
Avionic flight guidance/autopilot system.

Description
Thales' AFDS 2000 is an IFR and VFR autopilot system designed for light to heavy helicopters. It has modular architecture which allows it to be configured for a range of requirements from simple stabilisation augmentation to the higher modes necessary for flight director functions and mission assistance. The AFDS 2000 is a fully digital system capable of 3-axis or 4-axis configuration and is compliant with civil regulations FAR 27 and 29. In addition it is a dual-channel system providing redundancy and reversionary features.

AFDS 2000 features two flight control modules (FCM), a set of 3/4 trim actuators with dual-command computation, an autopilot control panel, an actuator position display and a dual set of primary sensors: Attitude and Heading Reference System (AHRS) and Air Data Unit (ADU).

Automatic flight control functions include: Stabilisation Augmentation System (SAS); attitude and heading hold; and side-slip control and turn co-ordination through cyclic and yaw collective compensation.

Flight director functions include: heading select; baro altitude and baro altitude acquisition modes; radar altitude, Indicated Air Speed (IAS) and Vertical Speed (VS) hold; long- and short-range navigation; search and rescue; and approach modes including back course, glideslope and Go Around (GA).

Status
In production.

Contractor
Thales Avionics SA.

AHV-2100 digital radar altimeter

Type
Flight instrumentation, radar altimeter.

Description
The AHV-2100 fully digital radar altimeter has been designed to meet the latest helicopter, trainer and transport aircraft requirements. It improves the FM/CW radio-altimeter technique previously used with digital signal processing and the latest Radio Frequency (RF) technology. This provides enhanced accuracy, integrity, immunity to multipath, and reduces power consumption.

The AHV-2100 has a dual-processing chain and DO-178 qualified software. In addition, power management of the RF output reduces the probability of interception at low altitude over water and the combination of a narrow receiver bandwidth with high-performance digital signal processing provides resistance to jamming.

Specifications
Dimensions: 110 × 90 × 190 mm
Weight: <2.2 kg

Power supply: 28 V DC
Altitude: up to 5,000 ft
Modulation: FM/CW
Frequency range: 4,200 to 4,400 MHz
Accuracy: 3 ft ±2%
Reliability: 5,000 h MTBF
Integrity: up to 10^{-9} (option)
Temperature range: −45° to +71°C operating
EMI/EMC: MIL-STD-461C
Design: MIL-E-5400T
Interfaces: ARINC 429, dual-redundant MIL-STD-1553B or analogue

Status
In service.

Contractor
Thales Communications.

AHV-2900 digital radar altimeter

Type
Flight instrumentation, radar altimeter.

Description
The AHV-2900 is a fully digital radar altimeter designed to meet the latest fighter requirements and stringent requirements for specific aircraft like MPA. It has modular architecture and has MIL-STD-1553B and ARINC 429 interfaces. The AHV-2900 uses digital signal processing and the latest Radio Frequency (RF) technology to provide enhanced accuracy, integrity, immunity to multipath, ECCM performance and reduced power consumption.

The AHV-2900 features: Low Probability of Interception (LPI); high dependability levels with degraded modes; reduced acquisition time; and enhanced tracking performance even in rapid manoeuvring.

Specifications
Modulation: FM/CW
Frequency range: 4,200–4,400 MHz
Accuracy: 3 ft ±1%
Altitude range: up to 30,000 ft
Power: 28 V DC, <35 W
Interfaces: Dual-redundant MIL-STD-1553B and ARINC 429 outputs

Status
In service.

Contractor
Thales Communications.

Airbus new widebody avionics system

Type
Integrated avionics system.

Description
The avionics system for the Airbus Industrie A350 long-range twin and A380 ultra-large airliner is a development of Thales Avionics' Integrated Modular avionics System (IMS – see separate entry). The cockpit follows the current Airbus 'common cockpit' philosophy, sharing overall layout (displays and controls arrangement, dark cockpit, symbology/colour coding) non-back driven thrust levers and side stick controllers. To pilots qualifed on current A340 variants, aside from the introduction of new displays and controls, the most noticeable change in cockpit configuration is the fitment of larger displays has necessitated the displacement of the Electronic Centralised Aircraft Monitor (ECAM) Control Panel (CP) from the front of the centre pedestal to the rear, adjacent to the new Radio Management Panels (RMPs). However, while cockpit layout is largely unchanged, a number of new and enhanced systems have been added for the A350/380:

- EFIS portrait-format displays providing Primary Flight Display (PFD), Navigation Display (ND) and Vertical Display (VD)
- Interactive MultiFunction Display (MFD) accessed via Keyboard Cursor Control Units (KCCUs)
- New Flight Management System (FMS) and Air Traffic Control (ATC) communication interfaces
- Electronic Flight Bag (EFB) as part of the Onboard Information System (OIS)
- Enhanced Electronic Centralised Aircraft Monitor (ECAM)
- New thrust indication system.

The FMS for the A350/380 will be Honeywell's Next-Generation Flight Management System, which will integrate terrain guidance and on-ground (airport) navigation. FMS functionality will follow Airbus' 'second-generation' system, driving a new Control and Display System (CDS) consisting of eight new-generation 7.25 × 9.25 in (6 × 8 in usable screen area) AMLCDs (1 × PFD, 1 × ND in front of each pilot, 2 × ECAM, vertically stacked in the centre of front panel and centre pedestal and 2 × MFD screens flanking lower ECAM. The MFDs are combined with 2 KCCUs; this arrangement replaces the current Multipurpose Control and Display Unit (MCDU) arrangement on current A330/340 aircraft. Displays are video capable and information is fully interchangeable between displays in case of Display Unit (DU) failure. The KCCUs allow for functional redundancy between the Cursor Control Devices (CCDs – one for each pilot, mounted on the centre pedestal) and keyboards (one for each pilot, embedded into each pullout tray table).

The 'trackball' CCD within each KCCU allows the pilot to move the cursor between the associated MFD and ND to highlight and/or 'reveal' expanded information regarding airports, including target runway exit for the automatic 'brake to exit' functionality to be introduced with the A380.

Baseline communications/navigation equipment includes Collins' VHF-920 and HF-900 data radios, multimode receiver, VOR-900 omnirange/marker beacon receiver, DME-900 and ADF-900 (see separate entries). Thales D-HUD (see separate entry) will be certified by Airbus and will be available as an option for single- or dual-fitment. The OIS will integrate database information with operator's in-house software packages and enable flight planning and documentation updates without incurring aircraft down time. OIS pilot interface is via keyboard and large format display for each pilot. The system will also control other ancillary operations including passenger credit card validation, Internet access and In-Flight Entertainment (IFE) and provide near-realtime information to ground crews at base. The OIS also functions as an EFB, which provides aircraft performance calculation, weight and balance (load sheet) computation and provides access to onboard documentation, such as Configuration Deviation List (CDL), Minimum Equipment List (MEL), operational documentation (FCOM) and navigational and approach and airport charts.

System operation
In operation, all Airbus-current Multifunction Control and Display Unit (MCDU) FMS functions (long-term, navigational/time related) will be accomplished with the KCCU. Each KCCU consists of a keypad and trackball system, controlling a cursor which can be invoked on most DUs (except the PFDs). KCCU controls are handed (right for the left seat occupant and vice versa) and include cursor navigation keys, screen cursor select, functional shortcuts, QWERTY and numerical pads, trackball and action/accept key.

While fairly commonplace in business jet aircraft, the provision of a VD is a first for a large commercial airliner. The VD is displayed on the lower portion of the ND and will improve crew Situational Awareness (SA) with respect to safety altitude, significant terrain directly ahead of the aircraft and the relative position of the planned FMS vertical profile. In conjunction with a suitably capable Vertical Profile (VP) weather radar, such as Honeywell's RDR 4000 (see separate entry), as specified for the A380, pilots will also be able to see the vertical extent, as well as lateral extent, of weather ahead and thus make more timely and accurate avoidance decisions.

Airbus A380 takes off at the 2006 Singapore Air Show (Airbus Industrie) 1128634

CM1 position in the A380 showing the Onboard Information System (OIS) display (behind side stick) and keyboard embedded in the tray table (Airbus Industrie) 1044238

The cockpit of the A380 features eight 6 × 8 in portrait format AMLCDs while retaining the familiar Airbus common cockpit layout, facilitating rapid and relatively inexpensive cross training for pilots of other Airbus wide-bodied airliners (Thales) 0578892

The Keyboard Cursor Control Unit (KCCU) (Airbus Industrie) 1044239

The RDR 4000 is part of the A380 Aircraft Environment Surveillance System (AESS), which also includes Terrain Awareness and Warning, Airborne Collision Avoidance and Mode-S ATC Transponder Systems (TAWS, ACAS, ATC). AESS control is via a dedicated CP located at the rear centre of the central pedestal (radar controls, TCAS, TAWS functions), a SURV MFD page (cockpit preparation and CP backup functions) and display control functions on each pilot's EFIS control panel (WX, TERR, TCAS display) on the glare shield. It is likely that the A350 will feature a similar system.

Three new RMPs with integrated AMLCDs are provided (one for each pilot on centre pedestal and one backup on upper right side of overhead panel) which provide for control of all VHF, HF and SATCOM radios, together with radio navigation (VOR, ADF) and ATC transponder.

Instead of more traditional and, perhaps less easily interpreted, Engine Pressure Ratio (EPR) or first-stage shaft speed (N1) indications for selected and available thrust, the new avionics system employs the Airbus Cockpit Universal Thrust Emulator (ACUTE) system, which, regardless of engine limiting parameters, provides a single and much more easily understood measure of thrust: maximum available thrust in any regime is indicated as 100 per cent (thrust levers at TOGA setting); at any time, actual thrust is shown as a percentage of this figure. Accordingly, pilots will instinctively know how much power is in reserve at any time without reference to the limiting value.

A computer-generated image of an Airbus A350-900 (Airbus Industrie) 1128633

The cockpit configuration for the A350 long-range twin, announced in early 2006, exactly mirrors that of the A380, except the provision of two thrust levers and associated controls. Retaining a common cockpit philosophy facilitates a common type rating with the A330 and Cross-Cockpit Qualification (CCQ) from the A350 to A380 with seven days of differences training (Airbus Industrie) 1128630

Status

In production and in service in development and early production A380 aircraft. Announced in early 2006 for the A350 after an extensive re-design of the nose section of the aircraft facilitated the extra depth required to accommodate the increased depth of the portrait displays.

Contractor

Airbus Industrie.
Thales Avionics SA.

All-attitude indicators

Type
Flight instrumentation.

Description
SAGEM has developed a range of spherical indicators on which attitude information is displayed to give the pilot heading, roll and pitch information on a single dial without freedom limits around the three axes.

Some versions are fitted with command bars to display signal information from navigation aids or landing systems. A helicopter version is designated 11-2 and features a manual pitch setting mode of ±10°. The various features of the series are:

Specifications

Type	810	811	816
Roll	Yes	Yes	Yes
Pitch	Yes	Yes	Yes
Heading	Yes	Yes	Yes
ILS/VOR/Tacan	Yes	No	Option
To-from	Yes	No	*
Beacon	Yes	No	Option
Sideslip	Yes	Yes	Yes
Failure detection	Yes	Yes	Yes
Warning flag	Yes	Yes	Yes

*separate unit, on option

Type 810 and 811
Dimensions: 97 × 97 × 203 mm
Power supply: 26 or 115 V AC, 400 Hz
  28 V DC

Type 816
Dimensions: 81 × 81 × 208 mm
Power supply: 26 V AC, 400 Hz
  28 V DC

Status
In production. Four types of instruments are available and equip BAE Systems Harrier GR. Mk 7, Dassault Mirage F1 and 2000, Sepecat Jaguar and Dassault Super Etendard, French Navy Westland Lynx helicopters and Soko Galeb, VTI/CIAR Orao and FMA Pucara aircraft.

Contractor
SAGEM SA, Defence and Security Division.

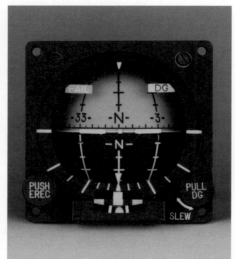

SFIM IS 816-1 attitude indicator   0503929

AP 505 autopilot

Type
Avionic flight guidance/autopilot system.

Description
The AP 505 has been designed for M2.0+ combat aircraft. Pilots of the Dassault Mirage F1, which is equipped with the system, claim to be satisfied with its reliability and ease of use. The system can be switched on before take-off, and engaged or disengaged by a handgrip trigger on the control

column. In basic mode it maintains the longitudinal attitude, held when the pilot releases the stick trigger, and either the heading or bank angle, depending on whether re-engagement is effected at a bank angle of less or more than 10°. Autopilot modes provide automatic flight at a preselected altitude, heading or VOR/Tacan/ILS bearing. Limits on attitude hold facilities are ±40° in pitch and ±60° in roll.

Specifications
Dimensions:
  (system computer) 190 × 202 × 522 mm
  (function selector unit) 132 × 142 × 27 mm
  (heading selector unit) 99 × 35 × 80 mm
Weight: 14.9 kg
Power supply:
  200 V AC, 400 Hz, 3 phase, <100 VA
  26 V AC, 400 Hz, single phase, <6 VA
  28 V DC, <15 W

Status
In service in Dassault Mirage F1 aircraft. More than 600 sets have been built and supplied to eight countries.

Contractor
Thales Avionics SA.

AP 605 autopilot

Type
Avionic flight guidance/autopilot system.

Description
The AP 605 is a digital flight control system developed for the Dassault Mirage 2000 supersonic combat aircraft. The Series 7000 computer allows future extensions to the basic autopilot, providing new modes suited to the particular missions flown by the aircraft. Flight control signals, which pass between the autopilot and the Mirage 2000 fly-by-wire flight control system, use a digibus serial data transmission system.

The AP 605 provides semi-transparent control, meaning that it is engaged or disengaged simply by releasing or taking hold of the control stick, which has a trigger switch in the handgrip. There is a high degree of internal monitoring, and computer design and organisation have been configured

to reduce onboard maintenance. In basic mode the autopilot will hold pitch angle to any value within the range ±40°, or bank angle in the range ±60°. Additionally, there are altitude-capture and hold modes, preset altitude acquisition and fully automatic approach capability down to 200 ft. The Mirage 2000N version autopilot includes a terrain-following mode.

The Mirage 2000D version includes coupling with the air-to-ground fire-control system and the ability to provide very low-altitude capture and hold over the sea.

Specifications
Dimensions:
  (system computer) 194 × 124 × 496 mm
  (control unit) 24 × 146 × 115 mm
Weight: 12.7 kg
Power supply:
  200 V AC, 400 Hz, 3 phase, <100 VA
  26 V AC, 400 Hz, single phase, <1 VA
  28 V DC, <30 W

Status
In service in Dassault Mirage 2000 aircraft.

Contractor
Thales Avionics SA.

AP 705 autopilot

Type
Avionic flight guidance/autopilot system.

Description
The AP 705, equipping the Atlantique 2 ASW aircraft, is designed to provide precise flight path control and a high level of safety at very low altitudes above the sea. The system can hold a course while maintaining altitude. These levels of performance are made possible by the quality of inertial data available on the aircraft and by the versatility of the microprocessor-based computer. Automatic built-in testing enables operation of the system and its safety devices to be checked before take-off and also facilitates onboard maintenance.

The three-axis autopilot includes pitch trim and can drive a flight director system. Autopilot modes permit pressure altitude hold, glideslope

The AP 605 digital autopilot is installed in the Mirage 2000   0113491

AP 705 autopilot   0503906

beam tracking (in Cat I weather minima) and radio altitude hold over the sea at very low levels in reduced visibility. Lateral modes provide for holding the heading or course at the time of engagement, heading hold and homing or tracking on radio navaids, or navigation waypoints.

Specifications

Dimensions:
  (system computer) 384 × 256 × 194 mm
  (control unit) 200 × 164 × 67 mm
  (servo-actuator) 185 × 183 × 101 mm
Weight: 23 kg
Power supply:
  200 V AC, 400 Hz, 3 phase, <50 VA
  28 V AC, <150 VA

Status

In production for the Dassault Atlantique 2 maritime patrol aircraft.

Contractor

Thales Avionics SA.

APFD 800 Autopilot/Flight Director

Type

Avionic flight guidance/autopilot system.

Description

The APFD 800 all-digital autopilot/flight director has been developed for a wide range of aircraft,

The APFD 800 autopilot/flight director has been selected for the upgrade of French Navy Alizé aircraft
0503907

from light jet or prop fighters to medium transport, ASW or SAR aircraft.

It has a modular architecture and all the subassemblies are designed for easy installation on existing aircraft as part of the upgrade operation. It offers basic and higher modes, with many specific functions in the horizontal and vertical planes. The system is designed for use down to very low altitude, with or without autothrottle. It can be coupled with radio navigation, INS or the flight management system and is proposed with brushless DC motor actuators.

The APFD 800 provides semi-transparent control of the aircraft, as it can be instinctively engaged or disengaged hands-on-stick.

Specifications

Dimensions:
  (computer) 4 MCU
  (control unit) 146 × 57 × 140 mm
  (mode selector) 38 × 140 × 146 mm
Weight: 7 kg total
Power supply: 28 V DC, 73 W

Status

Selected by the French Navy for upgrade of its Alizé maritime patrol aircraft; it is also fitted to Nimrod aircraft.

Contractor

Thales Avionics SA.

ATC-TCAS II control panel

Type

Control and Display Unit (CDU).

Description

The ATC-TCAS II control panel combines the independent control of two mode 'S' transponders with the LCD display of ATC code, and the selection of the TCAS II modes. It equips the A319/320/321 and A330/340 aircraft families. A version, which is certified on A300-B4/300-600/310 including TCAS test activation, is available for general cases of TCAS II retrofit installations.

Status

In production.

Contractor

Thales Avionics SA.

Thales' ATC-TCAS control panel
0001355

Attitude director indicators

Type

Aircraft instrumentation.

Description

Thales Avionics produces a wide range of attitude director indicators for both civil and military use.

The current range provides a wide choice of instruments and features flight director command bars, failure flag indicators and warning lights for decision height alerting. Some instruments include a basic yaw indicator for additional assistance in asymmetric flight.

Status

In service in Transall C-160, Dassault Aviation Atlantique 2, Aerospatiale Nord 262 aircraft and

Eurocopter SA 341 and SA 342 helicopters. Civil-standard attitude director indicators are fitted in Airbus Industrie's A300/A310.

Contractor

Thales Avionics SA.

Autoflight system for the A330 and A340

Type

Avionic flight guidance/autopilot system.

Description

The autoflight system for the Airbus A330 and A340 is composed of two Flight Management Guidance and Envelope Computers (FMGECs), a Flight Control Unit (FCU) and three Multipurpose Control and Display Units (MCDUs).

The FCU is used for short-term control of the aircraft autopilot and to select the display modes.

The MCDU is used to make long term changes to the aircraft flight plan (E L Downs)
0137240

The FCU is used to make short-term changes to the aircraft flight path (E L Downs)
0137238

The Rafale cockpit includes a wide angle holographic HUD (Dassault)   0528488

The FCU also includes controls for each pilot's navigation display, including display range and format rotary knobs, ADF/VOR beacon display toggle switches and Flight Director (FD) and Instrument Landing System (ILS) push-buttons (E L Downs)   0137239

The MCDUs are installed on the centre pedestal in the cockpit, beside each pilot, and a back-up/ACARS unit centrally at the rear of the console. They are used for long-term control of the aircraft, initialisation of the fuel management system, and to provide the interface between the aircrew and the maintenance system, ACARS, IRS and GPS.

The autoflight system can operate under autocontrol, using references computed by the flight management guidance and envelope computers on the basis of data selected by the aircrew through the MCDUs, or under manual control.

Status
In production for the Airbus A330 and A340.

Contractor
Thales Avionics SA.

B 39 autocommand autopilot

Type
Avionic flight guidance/autopilot system.

Description
Introduced into French Air Force Dassault Mirage III interceptors since 1975 to replace older autocommand systems, the B 39 is available for retrofit to other versions of the same aircraft type. It is a relatively simple system, but uses modern integrated circuit techniques to confer benefits in terms of performance, safety and

maintenance standards. The new autocommand computer is physically interchangeable with the original equipment.

Autopilot functions are reduced to attitude and altitude hold modes. Attitude hold includes short-term capability, stability augmentation and uniform artificial feel load against load factor irrespective of flight conditions.

Specifications
Dimensions: 264 × 200 × 140 mm
Weight: 6.5 kg
Power supply: 200 V AC, 400 Hz, 3 phase, <25 VA
 28 V DC, <1 A

Status
In service in Dassault Mirage III aircraft.

Contractor
Thales Avionics SA.

CTH 3022 head-up display

Type
Head-Up Display (HUD).

Description
The Collimateur Tete Haute (CTH) 3022 wide angle Head Up Display (HUD) for the Rafale aircraft is a holographic 30 × 22° Field-of-View (FoV) HUD that displays computer-generated symbology and FLIR imagery. An associated touch-screen/

display enables rapid assimilation of aircraft system parameters and mission data. A CCD camera records HUD imagery.

Status
Developed for the Rafale aircraft.

Contractor
Thales Avionics SA.

DDVR Digital Data and Voice Recorder

Type
Flight/mission recording system.

Description
Thales has developed a range of new-generation flight data recorders featuring a crash-survivable Flash memory to replace the conventional on-board tape recorders.

These new recorders are available in commercial and military versions. They can be equipped with an acoustic localisation beacon and are compatible with the ground support equipment associated with conventional recorders.

Increased recording capacity and easy data extraction allow the new recorders to be used for flight analysis and for monitoring essential aircraft parameters.

Specifications
Volume: <5 dm³
Weight: <7 kg
Power supply: 28 V DC
Memory capacity: up to 256 Mbytes
Compliance: EUROCAE ED-55 and ED-56a

Status
In production and in service. The DDVR offers the advantages of an all-digital system. It is certified to the latest survivability standard TSO-C124a.

A top range military version is installed on Mirage F1, 2000, 2000-D, 2000-5 and Rafale aircraft. On Mirage F1 and Mirage 2000 aircraft, the DDVR is interchangeable, both electrically and mechanically, with earlier tape recorders.

A commercial version is certified on Airbus and Boeing 737, 757 and 767 aircraft, as well as MD-90 and IPTN 250 aircraft. It has been selected by a large number of airlines worldwide.

Contractor
Thales Systemes Aeroportes SA.

DFDAU-ACMS Digital Flight Data Acquisition Unit – Aircraft Condition Monitoring System

Type
Flight/mission recording system.

Description
SAGEM produces a range of DFDAU-ACMS units tailored to the requirements of the following aircraft types: A300-600/A310; B737/B757/B767;

Dassault Mirage III   0065981

Specifications

Interfaces:

(DFDR/SSFDR recorder)	ARINC-717
(MCDU)	ARINC-739
(On-board printer)	ARINC-740 or -744
(QAR/DAR recorder)	ARINC-591
(ACARS MU)	ARINC-724 or -724B
(ADL/PDL)	ARINC-615

Acquisition	A320	A300-600/A310	B737/B757/B767	MD 80 series/MD 90
Dimensions	3 MCU	6 MCU	6 MCU	6 MCU
analogues	–	48	48	30
discretes	20	100	135	99
(ARINC-429 busses)	55	56	56	33

Weight: up to 8.5 kg
Power: 115 V AC, 400 Hz, <90 VA

MD 80 series/MD 90. All units are designed to acquire engine and aircraft data reports, including aircraft life data, weather, and other parameters by user programmable requests.

Status

In production and in service.

Contractor

SAGEM Défense Securité, Navigation and Aeronautic Systems Division.

A300-600/A310 common unit DFDAU-ACMS
0015268

B737/B757/B767 common unit DFDAU-ACMS
0015269

MD 80 series/MD 90 DFDAU-ACMS   0015270

Digital Head-Up Display System (D-HUDS)

Type

Head-Up Display (HUD).

Description

Part of Thales Avionics' TopFlight product family, D-HUDS, incorporating Liquid Crystal Display (LCD) technology, offers complete standard flight instrumentation and guidance for all flight phases, together with enhanced graphics quality, limitless symbology combinations and video-image support including new Enhanced Vision System/Synthetic Vision System/Surface Guidance System (EVS/SVS/SGS) capabilities. The modular system features a new HUD computer that facilitates higher integration standards, simplified display controls and increased contrast and luminosity for the overall system. Notably, Thales Avionics also manufactures the Active Matrix LCD (AMLCD) glass in-house.

The D-HUDS system consists of three Line Replaceable Units (LRUs):

HUD COMPUTER (HUDC)

The HUDC receives and processes aircraft equipment and sensor data, develops and generates the display graphics and monitors the integrity and performance of the system.

HUD PROJECTOR UNIT (HPU)

The HPU provides a high-brightness display, combining a backlight module. It also contains a relay lens assembly for projecting graphics to the HCU.

HUD COMBINER UNIT (HCU)

The HCU is an optical element that collimates the image projected from the HPU so that the symbology overlays the real world (conformal). The HCU mechanism allows three positions combiner positions (operational, stowed and breakaway). The HCU may also incorporate an option control panel for display brightness adjustment. The system utilises a single display control versus individual raster/stroke controls in traditional CRT-based HUDs.

For airliner applications, the HPU is installed in the overhead cockpit panel and stows behind the pilot's head when not in use. When required, it moves automatically along the mounting tray to the operating position. The combiner glass is collapsible and can move forward in a crash situation.

The HUDC is located in the electronic bay. It is connected on one side to the display management computer and on the other to the HPU.

Cockpit integration is straightforward, with D-HUDS providing for coherent operations with existing Head-Down Displays (HDDs), including HUD cautions/warnings and FDR recording. Installation is also simplified by the employment of electronic boresighting.

LCD, instead of the traditional Cathode Ray Tube (CRT), technology provides for increased reliability, image luminosity, combined with lower volume, weight and power consumption. It also provides for greater performance, including higher symbology and video brightness and greater contrast.

LCD technology also offers additional graphic capabilities (haloing, priorities, line thickness, grey level). Associated graphic software tools, based on a LCD renderer, are native and permit a high level of flexibility and reactivity with early specification validation including short loop corrections.

The employment of D-HUD in Air Transport applications brings a number of benefits to pilots:

- Faster judgement during critical flight phases
- Immediate and continuous assessment of aircraft trajectory/vector in poor visibility
- Wide field of view in high crosswind conditions
- True EVS integration (no impact on Field of View, image load and brightness)
- Natural integration with SVS/SGS, as a synthetic system.

Specifications

Dimensions: ARINC 600 – 6 MCU
Weight: 23 kg
Power supply: 28 V DC
I/O: ARINC 429, AFDX, digital video, RS-422, RS-232, discrete
Power consumption: 160 W
Field of view: 35 × 26°
Resolution: SXGA
Brightness: 10,000 cd/m2
Contrast: 1:3 (symbology and video)
Controls: Auto/manual brightness; auto/manual declutter
Compliance: A 764, DO 160 D, DO 178 B level A

Status

D-HUDS is available as either a single- or dual-installation and is suitable for Air Transport, Business Jet and regional aircraft cockpits. Thales has installed the system in an Airbus A340–600, with plans for A320 and A380 installations as part of its plan for certification for most Airbus models during 2006.

Contractor

Thales Avionics SA.

DS2100 Ultra-compact airborne digital cartridge recorder

Type

Flight/mission recording system.

Description

The Enertec DS2100 ultra-compact airborne digital cartridge recorder is designed for applications such as acoustic/sonar acquisition on board maritime patrol aircraft and helicopters, and data acquisition and processing for flight tests, UAVs and mission recording.

The DS2100 features a 30 Mbit/s sustained data rate capability, read-while-write capability and simultaneous acquisition of time code (for time damping) with audio and digital annotation. Data download/playback uses IT-standard hardware and software. It has removable cartridge storage capacity of up to 56 Gb per solid-state cartridge and 40 Gb per hard disk cartridge, giving 6 hours of recording at 15 Mbits/s.

The DS2100 has the following interface types: 8-bit parallel, 4 PCM, Ethernet 100 Base T, 16 audio and 2 video.

Specifications

Dimensions: 270 × 136 × 180 (L × W × H)
Weight: 5.5 kg (including 0.5 kg cartridge)
Power: 28 V DC, MIL-STD-704D; 60 W (typical), 70 W (in operation with heaters on)
Temperature: –40 to +50°C (operating)
Altitude: 0 to 70,000 ft
Cooling: Built-in fans
MTBF: >4,000 h
Compliance: MIL-STD-461C

Status

In production and in service. Installed in French Air Force Mirage 2000 aircraft in support of various onboard test programmes. Installed in European NH90 helicopters as the Mission Acoustic Recorder.

Contractor
Enertec.

DS4100 compact airborne digital cartridge recorder

Type
Flight/mission recording system.

Description
The DS4100 is a ruggedised family of products based on commercially available and proven technology.

Design applications are listed as:

- electro-optical, infra-red and synthetic aperture radar (SAR) reconnaissance imagery data acquisition and processing
- acoustic and sonar data acquisition and processing
- system and flight test data acquisition and processing.

Storage capacity per cartridge is 288 Gb. The sustained data rate is 320 Mbits/s (480 Mbits/s burst). Interface flexibility includes a basic 8-bit parallel + clock capability and an optional multichannel analogue/digital/databus capability, including data acquisition and formatting. There is simultaneous acquisition of time code (for time stamping), audio and digital annotation. Data integrity is better than 10–13 and there is a computer-compatible Cartridge Replay Unit available for ground processing.

Specifications
Record time:
(at 240 Mbps): 160 min
(at 80 Mbps): 8 h
Dimensions: 195 (H) × 320 (W) × 254 (D)
Weight: 15 kg, including 4 kg removable storage cartridge
Power: 28 V DC, <120 W (160 W with pre-heaters)

Status
In production and in service. Installed in French Mirage 2000 aircraft in support of various onboard test programmes. Installed in UK RAF Sentry AWACS aircraft as the Mission Audio Recorder (MAR). Installed in other military surveillance/intelligence platforms as a multichannel wideband signal recorder. In the UK, the then Defence Evaluation and Research Agency (DERA, now QinetiQ) evaluated the system during 2000.

Contractor
Enertec.

The Enertec DS4100 compact airborne digital cartridge recorder   0129337

DV 6411 series airborne digital recorders

Type
Flight/mission recording system.

Description
The DV 6411 airborne digital tape recorder is designed for fighter environments but is also space shuttle- and submarine-qualified. It has record data rates of between 10 and 240 Mbits/s with corresponding recording times of 9 hours 48 minutes, at 10 Mbits/s, and 24 minutes at 240 Mbits/s.

DV 6410 series airborne digital recorder used in the central pod of a Mirage 2000   0504083

The DV 6411 has a storage capacity of 350 Gbits on a 19 mm D1-M cassette in a format compatible with ANSI ID-1. It also provides for two auxiliary channels and a full remote-control interface.

Specifications
Dimensions:
(record-reproduce unit) 314 × 240 × 520 mm
(power supply unit) 314 × 140 × 302 mm
Weight:
(record-reproduce unit) 35 kg
(power supply unit) 12 kg
Power supply: 28 V DC < 350 W
Temperature range: –40 to +55°C
Altitude: up to 50,000 ft
Acceleration, 6 half-axes:
(operating) 9 g
(structural) 10 g
Electromagnetic compatibility: MIL-STD-461C
Functional modes: Stand-by; Stop, Read, Write, Pause, Read-after-write, Fast Forward and Reverse search at 31.5 and 300 ips

Status
In service for various airborne, shipborne and space applications. The DV 6411 is in service in the UK Royal Air Force Reconnaissance Airborne Pod for Tornado (RAPTOR) reconnaissance pod. RAPTOR integrates a day-night medium level stand-off imaging capability in Tornado GR. Mk 1A and Tornado GR. Mk 4 aircraft.

Contractor
Enertec.

DV 6421 ruggedised rackmount digital recorders

Type
Flight/mission recording system.

Description
The DV 6421 series of ruggedised rack-mounted digital recorders suitable for large aircraft

consists of the DV 6421 and DV 6221. The recorders are intended for use in: IR/EO/radar imagery acquisition and processing; intelligence data gathering; acoustic/sonar acquisition and processing; and telemetry, test and evaluation. Data rates are 10 to 240 Mbits/s for the DV 6421 and 5 to 120 Mbits/s for the DV 6221, when driven by the users clock (Mode A) and 0 to 240 Mbits/s for the DV 6421 and 0 to 120 Mbits/s for the DV 6221 (Mode B), when used with plug-in buffer memory. Recording time varies between 44 hours at 5 Mbits/s and 55 minutes at 240 Mbits/s.

The DV 6421 is the rack-mounted version of the DV 6410. The basic characteristics are common to both versions, but the rack-mounted version accepts D1 medium or large cassettes and has a storage capacity of 790 Gbits on a D1-L cassette.

A number of standard computer interfaces are available such as HIPPI, SCSI2 and VME64. A higher rate version, the DV 6820, provides 520 Mbits/s throughput rate capability.

Specifications
Dimensions: 483 × 311 × 593 mm
Weight: 65 kg
Power supply: 28 V DC or 115/220 V AC, 50/400 Hz, single phase, <450 W
Temperature range: –20 to +45°C
Altitude: up to 10,000 ft
Tape format: per ANSI ID-1 (x3.175) with one main digital channel and two auxiliary channels.
Functional modes:
Standby, Stop, Write, Read, Pause;
Read-after-write;
Fast Forward and Reverse Search @ 31.5 ips and 300 ips
Acceleration: horizontal 3 g, vertical 4.5 g (operating)
EMI-EMC: MIL-STD-461C

Status
In service in a number of airborne, shipborne and ground platforms. The DV6221 recorder has been selected as the mission recorder of the acoustic sub-system of the Royal Air Force Nimrod MRA4 maritime patrol aircraft.

Contractor
Enertec.

Electronic Flight Instrument System (EFIS) for the A320

Type
Avionic display system.

Description
In August 1984, the then Sextant (now Thales Avionics) was selected to develop and build

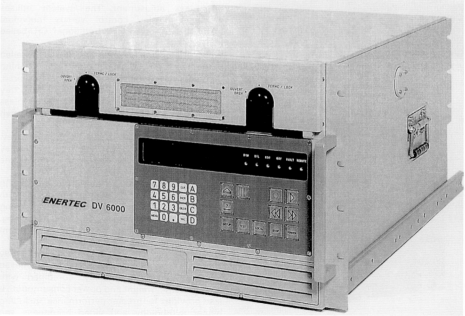

The Enertec DV6421E ruggedised rackmount digital recorder   0051321

the electronic flight instrument system for the Airbus Industrie A320 airliner. The system for this aircraft is an integration of the EFIS and ECAM CRT suites installed on the earlier A310, although with a number of important differences.

As with the A310 there are six CRT displays, but their disposition on the instrument panels is different. While on the earlier aircraft the flight director and navigation director are stacked vertically with the two ECAM instruments side by side on the centre panel, on the A320 the pilots' CRTs are ranged side by side, with the ECAM displays situated one above the other. The CRTs themselves are larger, 184 × 184 mm (7.25 × 7.25 in), compared with 159 × 159 mm (6.25 × 6.25 in), so that more symbology can be displayed without congestion. Each CRT can show at least seven colours. The upper ECAM, for the first time, shows primary engine parameters, replacing the 10 conventional electromechanical dial instruments on the A310. The lower ECAM will normally display systems information, such as electrical or hydraulic circuits, to show the location of faults. Confidence in the electronic display generated by A310 experience has resulted in the decision to delete most of the traditional electromechanical dial and pointer instruments, retaining just a few as a back-up in the most important functions.

As with the A310 system, all six CRTs are identical and interchangeable. However, whereas the A310 employs five symbol generators to drive them, the A320 uses only three and, with more complex functions, they are called display management computers (DMCs). All three are identical. The CRTs and computers are grouped within very advanced architecture, permitting extensive redundancy. In the event of a CRT, sensor or computer failure the system automatically reconfigures itself, with top priority given to flight commands and engine warning. At the same time there has been a significant weight decrease from 153 to 115 kg. As with the A310 system, elimination of software errors in the pilot's flight director instruments is accomplished by the use of two independent development teams. However, unlike the earlier equipment, software is implemented in high-order language.

Status
In production and in service in Airbus A320 and derivatives worldwide.

Contractor
Thales Avionics SA.

Electronic Flight Instrument System for the A310/A300–600 (EFIS)

Type
Avionic display system.

Description
The electronic flight instrument display designed for the A310 uses six identical 6.25 × 6.25 in (159 × 159 mm) shadow-mask colour CRTs. Each pilot has on the flight panel in front of him a pair of EFIS instruments: a primary flight display CRT which replaces the conventional electromechanical attitude director indicator and a navigation display CRT which supplants the earlier horizontal situation indicator and weather radar display. Each pair of CRTs is driven by a single symbol generator and a third system acts as a hot standby. Also associated with each EFIS pair is a control/display unit. Two more displays on the centre panel represent the Electronic Centralised Aircraft Monitor (ECAM) display, presenting information on aircraft systems in any phase of flight and schematic diagrams of the hydraulic or electrical systems, for example, to supervise their operation. One of them is normally reserved for warnings, the other for systems. The ECAM system operates independently of the pilot's EFIS instruments, being driven by two symbol generators (one operational, the other a

Airbus A310 flight deck   0503928

hot spare) and a single control unit mounted on the throttle box. All six CRTs are interchangeable, reducing the number of spares required. Frame repetition frequency is 70 Hz.

Thales has continued to refine the system. Colour stability and visibility of the displays in high-ambient lighting conditions have been demonstrated, the latter at 100,000 lux instead of the 85,000 lux specified. Nine colours are used on the primary flight display, seven are employed for symbology to avoid confusion and blue and brown represent the sky and earth. A uniformly smooth sky unmarred by the raster scan was achieved by slightly out of focus imaging and has proved particularly acceptable to the eye. The three-dimensional effect familiar to pilots who have used electromechanical primary flight displays has been achieved by masking certain symbols when they would normally be occulted by others mounted further forward in the instruments. Thales and Airbus Industrie have thereby devised a presentation which is instantly recognisable as that of a typical primary flight display, but one that can also utilise the vast quantity of newly available digital data circulating within an aircraft on the databus. Airbus Industrie and Thales have added the following information to the periphery of the instrument: a moving speed scale along the left-hand edge with its standard symbols generated by computer; a selected pressure altitude readout along the right-hand edge; an autopilot and autothrottle mode annunciator along the top edge; and radio altimeter heights along the lower edge of the instrument.

For the navigation display, classic compass card symbology has been retained, again in conjunction with the flexibility that digital sensing and processing affords. Its use is innovative in that it creates new symbology to suit each phase of flight, in the form of an electronic map display with five distance scales marking the course to be followed with radio waypoints and their identifying codes. On selection by the pilot, a three-colour weather map can be superimposed on the navigation map, resulting in the saving of panel space that would be required by a separate weather CRT and the more comprehensible integration of weather and navigation information. The warning and system status CRTs on the centre panel show information previously unavailable to flight crews. In addition to special alarms, failure warning is given by presenting the crew with synoptic displays of failed systems, indications of vital actions to be followed to meet the difficulty and the effects on other systems.

Different software is used for each flight director to eliminate common-mode faults. Two separate software teams developed the two sets of software in two different locations. Software was implemented in machine code.

Status
In service in Airbus A310 and A300–600 aircraft.

Contractor
Thales Avionics SA.

Electronic instrument system for the A330/A340

Type
Avionic display system.

Description
In 1989, the then Sextant (now Thales Avionics) was selected to develop the electronic instrument system for the Airbus A330 and A340. The system is an integration of an Electronic Flight Instrumentation System (EFIS) and an Electronic Centralised Aircraft Monitor (ECAM). The EFIS integrates the flight director and the navigation director with a weather radar display. The upper ECAM shows primary engine parameters, displays aircraft status messages (gear down, flap/slat position, speedbrakes, ground spoilers armed, ACARS message, seat belt/no smoking signs, and so on) to the pilot. The lower ECAM displays 14 system pages: cruise, engine, bleed, pressurisation, conditioning, hydraulics, Auxiliary Power Unit (APU), Circuit Breakers (CB), electrical AC, electrical DC, wheels, doors, fuel and flight controls. System pages are automatically displayed when required by system operation (for example, the wheel page is displayed on gear selection), or under abnormal or emergency conditions when a failure is detected by the system monitor.

The system uses six identical 184 × 184 mm (7.25 × 7.25 in) shadow-mask colour CRTs and three display management computers. In the

EFIS Display Management Computer (DMC) switching, PFD/ND brightness and PFD/ND display switching controls on the left glareshield   0126975

The electronic instrument system for the A330 and A340 includes six shadow-mask colour CRTs. Differences between the cockpits of the two aircraft are restricted to the thrust lever and overhead panel, reflecting the differing number of engines and flight systems architecture   0126990

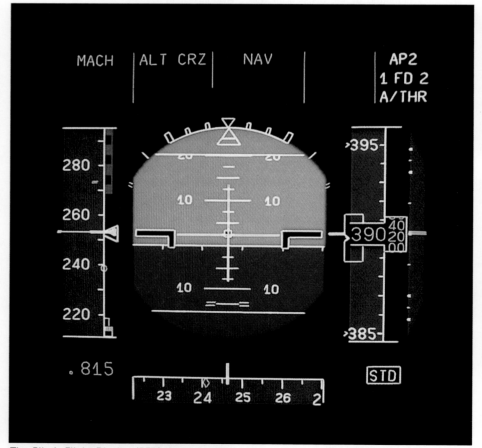

The Pilot's Flight Display (PFD). Note the Flight Mode Annunciator (FMA) along the top of the display; this shows that the aircraft is currently flying to a constant Mach number and cruising at the Flight Control Unit (FCU)-selected altitude (in this case, 39,000 ft). At the far right, the FMA shows Autopilot (AP), Flight Director (FD) and Autothrust (A/THR) engagement status (AP2, both FDs and A/THR engaged)   0126971

event of a CRT, sensor or computer failure, the system is automatically reconfigured, with top priority given to flight instrumentation, engine status information and warnings.

ECAM engine systems page   0126979

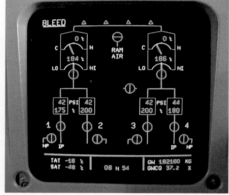

ECAM engine bleed page   0126980

ECAM cabin pressurisation page   0126981

ECAM cruise page, showing engine, pressurisation, conditioning, aircraft weight and balance information, together with outside air conditions   0126978

Status

In production and in service in A330/340 aircraft worldwide.

Contractor

Thales Avionics SA.

Specifications
Weight: 10 kg
Display dimensions: 184.1 × 184.1 × 355.6 mm

The Pilot's Navigation Display (ND), in 80 nm ARC mode, showing Ground Speed (GS), True Airspeed (TAS) and wind vector at the top left of the display, together with next waypoint identification, range, bearing and arrival time (top right), selected ADF and VOR beacons (bottom left and right), predicted track line (green) and radar tilt angle (−1.5°, lower right)   0126972

The Pilot's ND in 160 nm NAV mode, showing waypoints in the area of the aircraft. Note that waypoints associated with navigation aids (VOR or ADF) are shown with differing symbols   0126973

ECAM AC electrical page   0126982

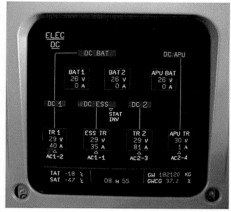

ECAM DC electrical page   0126983

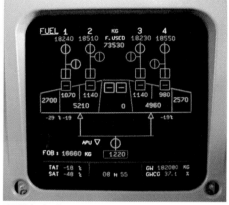

ECAM fuel page, showing automatic fuel transfer forward from the tail tank into the inner wing tanks   0126985

ECAM hydraulics page   0126984

ECAM selector on the forward part of the centre console of the aircraft. Individual pushbuttons control display of each system page, or repeated pressing of the 'ALL' button facilitates scrolling   0126987

ECAM flight controls page, showing computer (PRIM/SEC) status, aileron, elevator and rudder position, together with status of supplying hydraulic systems (G, B and Y)   0126986

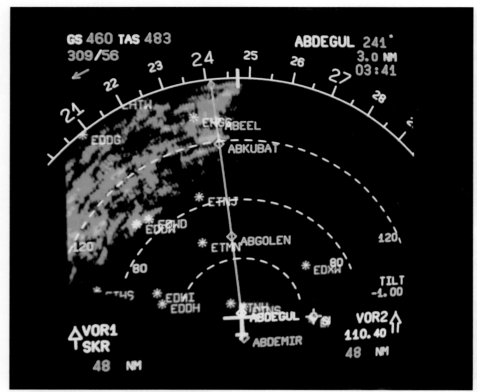

The Pilot's ND in 160 nm ARC mode, showing a depressed radar picture of the northern part of the German-Dutch coastline

0126974

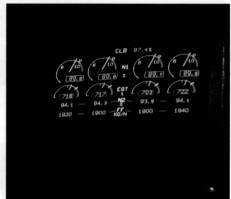

Upper ECAM Display Unit (DU), showing thrust lever detent (CLB setting, maximum N1 95.5 per cent), engine parameters, total Fuel on Board (FOB) and slat/flap position (clean wing in this case). Note the partitioned area below for aircraft status information   0126989

In case of ECAM Display Unit (DU) failure, the primary engine parameters page can be displayed on either ND, selected via the FCU. However, under these conditions, ECAM status messages are not available   0126988

The Flight Control Unit (FCU), showing (left to right) speed, aircraft heading, autopilot/autothrust engagement, altitude and descent rate/angle controls
0126977

Flight Control Unit (FCU) navigation selectors. The top row of buttons control display of overlay information such as airports (ARPT), ADF beacons, Non-Directional Beacon (NDB), VOR/DME (VOR.D), waypoints (WPT) and flightplan vertical/horizontal constraints (CSTR). Range scale and display format controls are below these, with navigation beacon selectors at the bottom. Altimeter setting and PFD flight director (FD) and ILS (LS) selectors are on the right
0126976

Extended-storage Quick Access Recorder (EQAR)

Type
Flight/mission recording system.

Description
Designed and manufactured by Thales, the Extended-storage Quick Access Recorder (EQAR) is used as a quick access recorder and/or digital ACMS recorder. Based on rewritable optical disk technology, the EQAR, which is fully interchangeable with existing magnetic-tape recorders, provides increased storage capacity, enabling more parameters to be recorded at a higher rate. Recorded data can be accessed directly from DOS files read through a PC.

Designed to be fitted on civil aircraft, the EQAR system provides full-time recording throughout the flight, with maximum data integrity.

Specifications
Dimensions: ARINC 404/600 formats
Input data rate: 64–512 words/s
Storage capacity: 128, 256, or 540 Mbytes
Environmental: DO160C

Status
Certified and operated on most commercial aircraft including those manufactured by Airbus and Boeing, EQAR has been selected by more than 30 airlines and operates worldwide with more than 500 units ordered.

Contractor
Thales Systemes Aeroportes SA.

Thales' EQAR optical disk quick access recorder   0051320

FCD 34 colour displays

Type
MultiFunction Display (MFD).

Description
The FCD 34 is a high-resolution, shadow-mask, full-colour multimode CRT display, using video scanning or stroke and raster images to show information on flight and engine control, navigation and weapon status. With an 89 × 114 mm (3.5 × 4.5 in) screen, the FCD 34 compact units adapt to the instrument panels of combat or training aircraft, either as original equipment or as a retrofit.

Status
The FCD 34 is in production and in service in French Air Force Mirage 2000D and the Mirage 2000-5 and has been selected for Mirage 2000 Strike export versions.

Contractor
Thales Avionics SA.

FDAU-ACMS Flight Data Acquisition Unit – Aircraft Condition Monitoring System

Type
Flight/mission recording system.

Description
The SAGEM FDAU-ACMS is designed as a complete package to monitor the mandatory

The Mirage 2000-5 cockpit includes two FCD 34 colour CRT displays   0044545

aircraft data parameters, together with data required for engine maintenance, on the ATR42/ATR72 commuter aircraft. The complete system comprises: the DFAU (Flight Data Acquisition Unit); the FDEP (Flight Data Entry Panel); and a three-axis accelerometer (TAA).

The FDEP displays: data, time and flight number; together with engine maintenance data.

The FDAU performs: acquisition of mandatory parameters; generation of a data frame to a DFDR/SSFDR; generation of the same parameters to the QAR and generation of time code. The system also incorporates BITE.

The ACMS function monitors on-condition engine and flight parameters. It includes all reporting functions via the display memory terminal, disk and ACARS.

Specifications
Interfaces:
(DFDR/SSFDR recorder) ARINC 573
(QAR recorder) ARINC 591
  objective torque output
(DMT (Display Memory Terminal)) via RS-232
(ACARS MU) ARINC 724
Dimensions:
(FDAU) 1/2 short ATR (127.5 × 199 × 318 mm)
(FDEP) 146 × 66.6 × 114.3 mm
(TAA) 101.6 × 91.7 × 63.5 mm

Weight:
(FDAU) 5 kg
(FDEP) 0.7 kg
(TAA) 0.4 kg
Power:
(FDAU) 28 V DC, 60 W (with FDEP)
(TAA) 28 V DC, 3.5 W

Status
In production and in service.

Contractor
SAGEM Défense Securité, Navigation and Aeronautic Systems Division.

FDIU Flight Data Interface Unit

Type
Flight/mission recording system.

Description
The FDIU is designed as a common unit for the A319/A320/A321/A330/A340 aircraft. Its function is to interface with a flight data recorder to ensure recording of all mandatory aircraft parameters. It acquires both discretes and ARINC 429 bus data, and interfaces to: the DFDR/SSFDR – ARINC 717; the QAR – ARINC 591; the CFDS/OMS – ABD 0018 or 0048.

ATR42/ATR72 FDAU-ACMS   0015272

A319/A320/A321/A330/A340 FDIU   0015273

SAGEM's FDIU is in service in the Airbus A321   0103889

Specifications
Dimensions: 2 MCU
Weight: <4 kg
Power supply: 115 V AC, 400 Hz, <40 VA

Status
In production and in service in a wide range of Airbus aircraft.

Contractor
SAGEM Défense Securité, Navigation and Aeronautic Systems Division.

FDS-90 Flight Director System

Type
Avionic flight guidance/autopilot system.

Description
The FDS-90 system comprises an attitude director indicator and a navigation coupler/computer unit.

The H140 attitude director indicator can operate autonomously using self-contained gyros and power inverters. It uses a ball-type real-world display and has a three-cue command capability. There are annunciators for go-around, decision height and flight director mode monitor. The gyro can be caged.

The B152 coupler/computer is a small, panel-mounted unit, which interfaces the attitude director indicator to the navigation equipment. It may have up to 11 push-buttons that permit selection of different flight director operating modes:

HDG
captures and tracks the heading selected on the horizontal situation indicator

NAV or V/L
captures and tracks VOR and ILS localiser beam for short-range navigation or FMS data for long-range navigation. It has the ability to interface with Tacan, Loran or Doppler sensors, depending upon the mission

BC
tracks the back course localiser

BALT
maintains the baro-altitude existing at the time of selection

GS
captures and tracks an ILS glideslope beam

VS
maintains the vertical speed that exists at the time of engagement

IAS
maintains the airspeed that exists at the time of engagement

In addition to the various functions incorporated in the FDS-90, that are more particularly concerned with en route and approach applications at cruise speeds, the search and rescue and anti-submarine warfare coupler functions are specific to hover mode and to the various transition phases. These include:

APP 1
from the initial conditions of height above 200 ft and IAS higher than 60 kt it provides the simultaneous commands for a descent to 150 ft/min and deceleration to 60 kt

APP 2
from the final APP 1 conditions it provides the commands required to control the helicopter's descent to the height selected by the pilot and deceleration to hover

HOV
holds the zero lateral and longitudinal groundspeeds provided by the Doppler radar or the zero sonar cable angles, depending on the submode selected by the pilot

CLB
from the initial conditions of height above 200 ft and IAS higher than 60 kt it provides the commands for a climb at 300 ft/min and acceleration up to 60 kt.

The SAR and ASW coupler is connected to the ADI for manual use in the FD mode or to the basic stabilisation system (Ministab, AFDS 95-1 or any other AFCS) for automatic operation.

Specifications
Dimensions:
(attitude director indicator) 4 ATI
(nav coupler/computer) 3 ATI standard case
Weight:
(attitude director indicator) 2.5 kg
(nav coupler/computer) 1.3 kg
Power supply:
28 V DC, <20 W (attitude director indicator)
28 V DC, <15 W (nav coupler/computer)
Environmental: DOO 160 C, TSO'd

Status
In production and in service, on S-61 (HH3-F) and B206.

Contractor
Thales Avionics SA.

Full-format printers

Type
Avionic information management system.

Description
Thales Avionics C12349 full-format printers are either standard or optional flight-deck equipment on the Boeing 777, 767, 747, 717, MD-11 and MD-90, and on the Airbus A330 and A340.

The system utilises thermal printing technology, and features high resolution, a high-capacity buffer, and a graphical interface capability and associated high-speed datalink.

Thales Avionics is the sole supplier for the flight deck printers of both Boeing 777 and Airbus A330/A340 families. These aircraft were designed to accommodate a full-format printer installed in the centre pedestal which supports both B4 (US standard) and A4 (EU standard) paper sizes. The printers are designed to accommodate all current and future onboard and maintenance printing demands of systems including the Aircraft Communication and Reporting System (ACARS), Aircraft Information Management System (AIMS), Electronic Library System (ELS), Future Air Navigation System (FANS), Satcom, Onboard Information Network System (OINS) and LCD EFIS.

The C12349 A & C series printers incorporate a number of common features:
- Full graphics capability and high resolution (300 dpi)
- Multiport capability (centronics, RS232; Ethernet TCP/IP, ARINC 429 HS & LS)
- Parallel processing

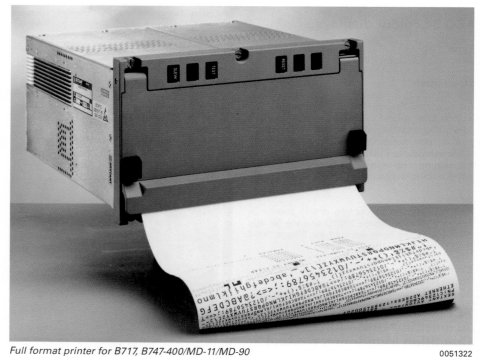

Full format printer for B717, B747-400/MD-11/MD-90   0051322

- Large buffer memory (16 Mb)
- High reliability (MTBF > 15,000 hr).

Fully compliant with ARINC 744A, these full-format printers are also installed in the cabin and interface with the In-Flight Entertainment (IFE) systems from major suppliers. They offer an open design that supports new developments, including Ethernet TCP/IP. Cabin systems have been installed in B777, B767, MD-11, A330 and A340 aircraft.

Status

In production. In February 2001, Continental Airlines selected Thales Avionics full-format printer for all 34 of its Boeing 767-200 and 767-400 airliners.

Contractor

Thales Avionics SA.

Gyro horizons

Type

Aircraft instrumentation.

Description

Thales gyro horizons have been installed on virtually all commercial transport aircraft. Today, approximately 150 airlines and 30 air forces use Thales gyro horizons as stand-by attitude reference instruments.

The Series H3XX gyro horizons are 3 ATI-sized and weigh about 1.5 kg. They make extensive use of modern alloys, have built-in static inverters and gyro speed monitors and Sextant-patented simplified erection and anti-spin devices. The basic versions include H301 (drum display), H321 (sphere display) and H341 (sphere display with ILS capability). 2 ATI and 4 ATI versions are also in production.

Specifications

Power supply:
(Gyroscope) 28 V DC
(Lighting) 5 V/28 V DC or AC
Power consumption:
(Gyroscope) starting, <1.5 A; operating, <0.3 A
(Lighting) 5 V, <0.5 A; 28 V, <0.2 A
Freedom in roll: Total
Freedom in pitch: Total through controlled precession
Erection rate: 3°/min
Rotor speed: >23,000 rpm
Run-up time: <3 min
Run-down time: >9 min

Status

In production and in service.

Contractor

Thales Avionics SA.

Head-Up Flight Display Systems (HFDSs)

Type

Head-Up Display (HUD).

Description

Thales' Head-Up Flight Display System (HFDS) is a multipurpose HUD system that can be fitted to any cockpit, whether analogue or digital.

The system uses total energy and flight path vector symbols to provide assistance to take-off and landing and to improve safety at landing minima. The HFDS has been developed as a series:

Series 100, called Integrated HFDS
Series 200, called Manual HFDS
Series 300, called Hybrid HFDS

The Integrated HFDS is intended for aircraft already designed with a 'Fail Op' autopilot.

The Manual HFDS is intended for aircraft with 'fail passive' autopilots in order to reach Cat IIIa minima on manual landing with 50 ft decision height. Bombardier offers this system on Global Express aircraft.

The Hybrid HFDS is currently in operation on Alitalia MD-82s to enhance the landing capabilities to Cat IIIb standard with respective decision heights of 35 ft and 20 ft. The system allows a 75 m Runway Visual Range (RVR) for take-off.

Display symbology is matched to the series type and operational configuration.

The system is also able to support Enhanced Vision System (EVS) procedures.

Thales Avionics' Head-Up Flight Display System (HFDS) for Cat IIIb operation, shown installed on Aerospatiale's B737-300   0001450

HFDS shipset   0001452

The HFDS comprises 4 LRUs. The projector located in the cockpit above the pilot's head contains a relay lens assembly for projecting symbology to the display combiner and electronic drivers. The HUD combiner is a holographic optical element that allows collimation of the image so that the symbology overlays the real world even in conditions of strong crosswind. The HUD computer receives and processes aircraft equipment and sensor data and generates the display symbology. It also monitors the integrity and performance of the system. The control panel controls the system.

Specifications

Weight: 32.7 kg for complete system including trays
Power consumption: 160 W
Environment: Complies with DO160C
Software category: DO178B level A
Field of view: 40° (lateral) by 26° (vertical)

Status

In production. Selected for the Bombardier Global Express business aircraft as part of the Bombardier Enhanced Vision System (BEVS). First flight is expected in the first quarter of 2003, with qualification testing to begin in the second quarter of 2004, leading to final aircraft certification in early 2005.

The system is also part of the C-130 Topdeck® avionics upgrade, integrated by Marshall Aerospace for South African Air Force C-130 aircraft.

Contractor

Thales Avionics SA.

IMS Integrated Modular avionics Systems

Type

Avionics system.

Description

Thales Avionics has developed a new IMS family of integrated and modular avionics systems to offer regional and business aircraft operators an integrated system for new-build and upgrade aircraft.

The IMS family capabilities are based on innovative avionics technologies – open design and a wide range of functions and performance – adapted to each market segment's specific needs. Thales is developing different versions to provide avionics systems that enhance the intrinsic qualities of regional and business aircraft and ensure long-term customer and product support to regional jets, business aircraft and new generations of airliners.

The first application is the IMS-100 on the Bombardier de Havilland Dash 8-400 regional turboprop, certified in 1999.

Typically, IMS configurations comprise: a new radio-communications and navigation unit; smart LCD units; fibre optic gyro AHRS; FMS; and a centralised diagnostics and maintenance interfacing system.

The cockpit of the Bombardier de Havilland Dash 8-400 incorporating Thales' IMS   0079257

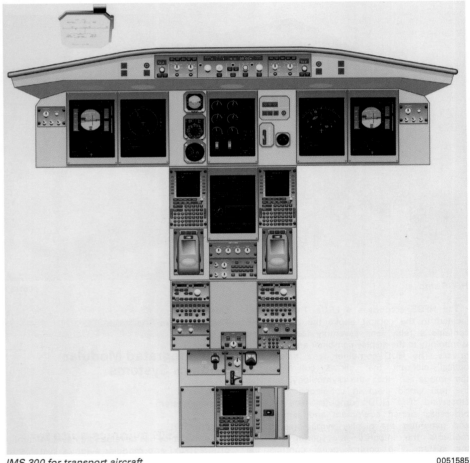

IMS 300 for transport aircraft   0051585

Status
In production and in service in a variety of aircraft.

Contractor
Thales Avionics SA.

Integrated Electronic Standby Instrument (IESI)

Type
Aircraft instrumentation.

Description
Thales Avionics' solid-state Integrated Electronic Standby Instrument (IESI) is part of the company's TopFlight Line and is designed to replace conventional pneumatic altimeter, airspeed and gyroscopic horizon indicators with a single LRU

standard 3 ATI box. It is fitted with a colour, high-definition active matrix LCD and is intended for use in civil aircraft (commercial transport, commuter and business-type helicopters).

The IESI incorporates design growth capability to display: SSEC, ILS (localiser and glide path), back course, slip-heading, dual-baroset scales and Mach number to meet customer requirements.

Two configurations are available: a single LRU configuration to display data from its three internal functions, together with two remote and interchangeable Thales air data modules; a single LRU configuration dedicated to attitude, which acts as a standby horizon indicator.

Specifications
Dimensions: 3 ATI case according to ARINC 408A standard
Weight: 1.9 kg

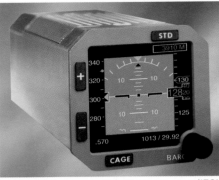

Integrated Electronic Standby Instrument (IESI)
0051659

Altitude: according to TSO C10c (AS 8009A)
Power supply: 28 V DC (emergency bus)
Certification: TSO-C10c (altitude), -C2d (airspeed), -C4c (attitude), -C95 (machmeter), and -C113 (airborne multipurpose electronic displays)
Viewing area: 61 × 61 mm
Resolution: 200 dots per inch
Brightness: >100 fL for civil applications
Dimming range: >2,000/1
Input/Output: 5 standard ARINC 429 buses and 16 discretes
Attitude: +/–90° (pitch), +/–180° (roll) with an accuracy of +/–0.5°
MTBF: >15,000 h
Product design: per RTCA DO-160D standard (civil applications), DO-178B standard

Status
In production since late 1997. Currently in service with the Airbus family, Boeing civil aircraft, the Bombadier Aerospace and Embraer families, Casa C-295 and Fairchild Dornier 728 jet and derivatives.

Contractor
Thales Avionics SA.

LCD engine indicator

Type
Aircraft instrumentation.

Description
The LCD engine indicator uses a 4 ATI size liquid crystal display screen with the same technology as Thales' 3 ATI VSI/TCAS instrument. The basic engine instrument comprises three screens which display data from the FADEC system, secondary sensors, fuel management computer and APU.

Status
In production and in service in Dassault Falcon 2000 and Falcon 50 EX business jets.

Contractor
Thales Avionics SA.

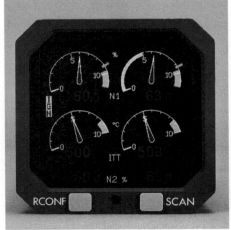

Thales' flat-panel instrument engine control display   0581502

MAESTRO nav/attack system

Type
Mission Management System (MMS).

Description
SAGEM has designed the Modular Avionics Enhancement System Targeted for Retrofit Operations (MAESTRO) with flexibility so it can be easily tailored to any customer's specific requirements and offer capabilities at par with those of current front-line fighters.

MAESTRO provides for full day/night and all weather operations and includes the following SAGEM components (see separate entries):

- DYNAMIC mission computers, interfaced via MIL-STD-1553/1760 databus
- SIGMA Embedded GPS-Inertial (EGI)
- MERCATOR high-capacity data transfer and digital moving map
- IRIS or MATIS wide Field of View (FoV) Forward Looking Infra Red (FLIR)
- Full colour glass cockpit, including Head Up Display (HUD) and MultiFunction Displays (MFDs)
- Full Hands On throttle And Stick (HOTAS)
- MARS Mission Planning System (MPS)
- GIPSY cartographic workshop.

The system provides for air-to-ground and air-to-air fire control in conjunction with a multimode Pulse Doppler (PD) radar and/or Laser Range-Finder (LRF). In addition, a full EW suite comprising radar warning, missile launch detection, chaff and flare self-protection and/or jamming systems is available to complete a full tactical suite for the upgrade of legacy combat aircraft.

The SAGEM MAESTRO modular avionics enhancement system is designed for retrofit applications, including the Mirage 5 (SAGEM)

1146546

Status
In production. Applications include the upgrade of Belgian Air Force Mirage 5, Chilean Air Force Mirage 5 and C-101 combat/trainer aircraft, Indian Air Force Jaguar, PZL Irdya Polish trainer/attack aircraft, Pakistan Air Force Mirage III and 5 and Egyptian Air Force Mirage 5 aircraft.

Contractor
SAGEM Défense Securité, Navigation and Aeronautic Systems Division.

ME 4110 airborne instrumentation recorder

Type
Flight/mission recording system.

Description
The ME 4110 is a lightweight and compact recorder designed for use during the flight testing of military fighters and other aircraft. It uses 10 in reels of 1 in-wide magnetic tape.

Specifications
Dimensions: 400 – 285 – 160 mm
Weight: 20 kg
Power supply: 22–32 V DC, 200 W
Recording time: 7 min – 16 h
Number of tracks: 14 or 28
Recording modes:
FM (up to 500 kHz)
direct (up to 2 MHz)
PCM (up to 4 Mb/s)

Status
In service.

Contractor
Enertec.

ME 4115 airborne recorder/reproducer

Type
Flight/mission recording system.

Description
The ME 4115 is an advanced, microprocessor-controlled recorder designed for use during the testing of aircraft, ships or vehicles or as an ELINT, ASW or reconnaissance mission recorder and can be configured for any application. It uses 15 in reels of 1 in wide magnetic tape and has built-in error correction electronics and read/write heads to give 14 or 28 tracks IB, WB or DD magnetic heads.

Depending on the disposition of the electronics chassis relative to the tape deck, several standard versions are proposed: 14 or 28 track record/reproduce either for anti-vibration mount or 19 in bay installation, tape deck separated from the electronics for integration on board small fighters and 28 track record-only and monitoring.

This latest configuration is fully compliant with MIL-STD-1610 and STANAG 4238 Annex B analogue acoustic recording standards.

The ME 4115 accepts a wide range of high data rate digital formats up to 200 Mbits/s including standard interfaces for MIL-STD-1553 bus monitoring. It has been selected as a flight test recorder by most airframe manufacturers for programmes such as the Alenia/Embraer AMX, Northrop Grumman F-14 and Airbus A340.

Specifications
Dimensions: 581 – 394 – 176 mm
Weight: 35 to 45 kg
Power supply: 22–32 V DC or 115 V AC, 300 W
Recording speed: 4.75–394 cm/s
Number of tracks: 14 or 28
Recording modes:
FM (up to 1 MHz)
direct (up to 4 MHz)
PCM (up to 8 Mb/s)
Cumulative data rate: over 200 Mbits/s

Status
In service on French and Italian Atlantique maritime patrol aircraft. The ME 4115 is operated by the US forces in several airborne and shipborne programmes as the AN/USH-33(V)2. More than 900 units have been produced.

Contractor
Enertec.

MEGHAS® avionics suite for helicopters

Type
Avionics system.

Description
During the latter part of the 1990's, Thales Avionics and SAGEM developed a new generation of avionics systems for helicopters – MEGHAS®. Based on an integrated family of equipment, the new avionics system was designed to satisfy the specific requirements of new helicopters up to the 6 tonne class.

MEGHAS® handles all main helicopter functions: autopilot, guidance and navigation, engine and vehicle control, onboard maintenance, radio communications and radio navigation.

By reducing the crew's workload, MEGHAS® allows helicopter crews to accomplish their missions (EMS, SAR, surveillance and offshore transport) quickly and efficiently. The system's modular design facilitates adding new mission functionality, as well as providing weight savings, lower cost of ownership and higher reliability. Fewer parts and proven technologies provide a significant increase in MTBF compared with conventional systems.

The flexibility of the MEGHAS® architecture also means that it supports all flight configurations (VFR, IFR, single-pilot and dual-pilot), in compliance with international regulations (FAR, JAR 27 or 29).

The major components in the MEGHAS® avionics system include:

- Central Panel Display System (CPDS). The CPDS comprises the Vehicle and Engine Management Display (VEMD) for single-engine helicopters, used in conjunction with the Caution and Advisory Display (CAD) for twin-engine helicopters.

The VEMD comprises two active matrix LCDs. It replaces up to a dozen conventional indicators and allows immediate verification of both vehicle and engine parameters. The design complies with all High-Intensity Radiated Field (HIRF) and lightning requirements (FAR 29 category A).

The CAD uses a single active matrix LCD. In conjunction with the VEMD, it manages all fuel functions and warnings. The CAD also operates as a back-up for main engine parameters in case of VEMD failure.

- Flight Control Display System (FCDS). Thales has developed two 'smart' displays for the FCDS, featuring very high resolution, multifunction colour LCDs; the SMD45H (4 × 5 in LCD) and the SMD68 (6×8 in LCD), with integrated NVG cooling compatibility and video system. These smart displays show both primary flight and navigation information. For specific mission functions, including FLIR, map generator and weather radar, the video functions in these displays allow display of this information using control units on the SMD45H and SMD68CVN.

- Automatic Flight Control System (AFCS). The AFCS is designed and built by SAGEM. It includes a SAGEM attitude and heading reference system (APIRS) and a Sextant ADS 3000 air data system. Use of a removable memory module means that aircraft installation is easy and only a single calibration flight is required.

The AFCS can be upgraded, changing it from a simple stability augmentation system to a complete four-axis autopilot system coupled to a GPS receiver for auto-approach.

- Health and Usage Monitoring System (HUMS). The HUMS system is designed by SAGEM and uses the Miscellaneous Flight Data Acquisition Unit (MFDAU).

- Navigation functions, including a GPS receiver and SAR (Search And Rescue) capability.

- Centralised Radio Control (CRC), comprising radio communications, radio-navigation and identification control systems.

Status

In production and in service. As of early 2004, 1,200 MEGHAS® systems had been delivered. The system has been certified for the Eurocopter EC 120, EC 130, EC 135, EC 145 (BK 117), EC 155 and Ecureuil B3 helicopters.

The EC 120 uses the VEMD, certified in mid-1997, with initial deliveries starting at the end of that year, at the rate of about 100 units per annum. The Ecureuil B3 modernisation programme uses the VEMD to replace all instruments on the central engine and vehicle control console. The installation was certified in late 1997 and deliveries started early in 1998, at a rate of about 60 VEMDs per year.

In August 1998, MEGHAS® was certified on the EC 135, configured as a Central Panel Display System (CPDS), comprising the VEMD, CAD and FCDS. Deliveries at the rate of 40 systems per year started in October 1998. The basic FCDS can be upgraded to include FLIR, map generator and weather radar functions, handled through the SMD68CVN display.

The EC 155, with CPDS, FCDS and AFCS, was certified for IFR dual-pilot operation

The Eurocopter EC 155 cockpit MEGHAS® Installation comprises CPDS, FCDS and AFCS (Eurocopter) 0580569

MEGHAS® in the EC 145 (Eurocopter) 0580570

MEGHAS® new-generation avionics suite for helicopters 0051586

in March 1999. Deliveries of production equipment started at the end of 1998. Single-pilot certification with a four-axis dual-AFCS was achieved in 2000. The basic FCDS can be upgraded in a similar manner to the EC 135 installation.

Thales Avionics and Eurocopter are teamed to provide support for the MEGHAS® system to operators worldwide.

Contractor
Thales Avionics SA.

MERCATOR digital map generator

Type
Avionic Digital Map System (DMS).

Description
The MERCATOR Digital Map Generator (DMG) provides pilots of modern aircraft and helicopters with presentation of a colour moving map display.

To improve flight safety and operational effectiveness, the MERCATOR digital map displays terrain data in colour referenced to aircraft altitude. MERCATOR offers up to 2 Gbyte capacity, corresponding to a very large geographical area at various scales.

Key features of the DMG include a north-up or track-up map/DTED display; real-time smooth rotation, pan/scroll and zoom; the ability to drive up to two multifunction colour displays; up to 2 Gbyte mass storage cartridge, allowing 1,700,000 km² (1:100k) on line map data; standard MIL-STD-1553B/RS-422 databus interface; simultaneous functions of mass memory server on MIL-STD-1553B (mission planning data) and digital map generation.

SAGEM Mercator, onboard digital map generator   0015345

MFD 54 is in service in the Mirage 2000-9 (Dassault Aviation)   0118342

Specifications
Dimensions: 95.2 – 127 – 222 mm
Weight: 2.7 kg
Power: 28 V DC, 17 W

Status
In production and in service. MERCATOR forms part of the MAESTRO nav/attack system upgrade (see separate entry).

Contractor
SAGEM Défense Securité, Navigation and Aeronautic Systems Division.

MFD 55 colour display

Type
MultiFunction Display (MFD).

Description
The MFD 55 is a high-resolution, multifunction display using a 5 – 5 in active matrix LCD, providing for high performance in all light conditions and full Night Vision Imaging System (NVIS) Class B compatibility according to MIL-L-85762A.

The MFD 55 can be provided in dump or as a smart display with embedded bus interface and graphics processor.

Specifications
Display area: 5 – 5 in
Resolution: 120 pixels per inch; stripe
Viewing angles: ±45° horizontal; –10 to +30° vertical

Contrast ratio: 30
Video interface: RGB; STANAG 3350
Interface: MIL-STD-1553; RS-422, -485
NVIS compatibility: Per MIL-L-85762A
MTBF: 1,000 h
Dimensions: 160 – 160 – 265 mm (H – W – D)
Weight: 6 kg
Power: 28 V DC and 115 V AC; 100 W
MTBF: 3,500 h

Status
The MFD 55 is in production and in service in Dassault Mirage 2000-5 and 2000-9 multirole fighter aircraft. Also selected for Sukhoi Su-30, LCA and Hawk.

Contractor
Thales Avionics SA.

MFD55/MFD66/MFD88 liquid crystal multifunction displays

Type
MultiFunction Display (MFD).

Description
The MFD55/MFD66/MFD88 multifunction displays are full-colour multimode displays using 125 – 125 mm (5 – 5 in), 155.7 – 155.7 mm (6.13 – 6.13 in) and 203.2 – 203.2 mm (8 – 8 in) active matrix liquid crystal displays in the MFD55, MFD66 and MFD88 respectively to provide good visibility under any ambient light conditions.

The displays are high-resolution units enabling display of any kind of video image overlaid with synthetic symbols. The units are fitted with surrounding soft keys.

Specifications
Dimensions:
(MFD55) 160 – 160 – 232 mm
(MFD66) 196 – 192 – 238 mm
Weight: <7 kg
Power supply: 115 V AC, 400 Hz, or 28 V DC, 130 W

Status
The MFD55 is in production for the LCA, Mirage 2000 and Sukhoi Su-30. The MFD66 is in

The SAGEM Mercator map display shows colour-coded altitude data   0051581

SMD 88 smart multifunction display   0079265

Thales' range of active matrix liquid crystal displays   0079264

The Nadir 1000 navigation and mission management system for helicopters   0015351

production for the Tiger and Rooivalk helicopters and for Jaguar, Nimrod MRA4 and Sukhoi Su-30 MKI. The MFD88 is utilised in a five-display fit in the NH 90 prototype helicopters.

Contractor
Thales Avionics SA.

Military real-time Local Area Networks (LAN)

Type
Avionic network system.

Description
The DIGIBUS GAM-T-101 became a French tri-service standard multiplex databus in 1982. It is used in many military systems (Mirage F1, Mirage 2000, Atlantique 2, missiles, submarines, navy ships and land-based applications).

STANAG 3910 is a dual-speed version of the MIL-STD-1553B which is widely used in numerous onboard military systems. This new databus has been adopted for the EFA and Rafale aircraft. It can also be used for the modernisation of any MIL-STD-1553-based system.

Status
As of April 2004, in production and in service. Over 41,000 remote terminal bus interface units and 3,800 bus controllers have been produced.

Contractor
Thales Systemes Aeroportes SA.

Mini-ESPAR 2

Type
Flight/mission recording system.

Description
Designed by SAGEM, the Mini-ESPAR 2 has been developed for fitting as a crash recorder on helicopters or military aircraft. The Mini ESPAR 2 can acquire the basic information to cover the crash requirement.

Fitted with an entirely electronic memory, this new-generation recorder is considerably more reliable, and requires less scheduled servicing than earlier models, because of the absence of moving parts; it provides appreciable savings in weight and volume. These savings are achieved by use of a small-size static memory, made possible by the application of a specific data storage algorithm adapted to all types of aircraft and capable of recording up to 25 hours of flight data, and voice recording of up to 1 hour, in commercial aircraft. It has improved environmental protection against crush, pressure, temperature and corrosion.

During data retrieval, a recovery algorithm supplies the expected value of each parameter for comparison, to speed data analysis. A further advantage is the ability to re-read flight data without removal of the recorder. In this case, recorded data is transmitted over a high-speed line to a static data retrieval set which is then connected to the workstation.

The Mini-ESPAR 2 can be adapted easily to any military or small civilian aircraft. It complies with TSO C124 and EUROCAE ED-55.

SAGEM Mini-ESPAR 2   0015302

Specifications
Acquisition capacity: (basic) 3 synchros, 15 analogues, 12 discretes, 3 frequencies (options) 1 or 2 mixed audio channels, serial bus ARINC 429, or serial bus MIL-STD-1553B, or 2 pressure transducers
Recording duration:
(parameters) up to 24 h
(audio) up to 1 h
Frame rate: 64 or 128 words/s
Average parameter sample rate:
0.25-16 Hz, programmable
Dimensions: 183 – 148 – 273 mm
Weight: 9 kg
Power supply:
28 V DC or 115 V AC
Options: voice recording, ARINC 429 recording

Status
In production and in service.

Contractor
SAGEM, Aerospace and Defence Division.

Nadir 1000 integrated navigation and mission management system

Type
Aircraft mission computer.

Description
The Nadir 1000 is an integrated navigation and mission management computer developed to provide mission assistance for military and civil light helicopters over both land and sea. The Nadir 1000 is designed to provide multisensor navigation from Doppler, radio navigation and GPS, for flight management, navigation and mission management, and system links for SAR, hover and ASM roles.

The Nadir 1000 weighs 3 kg in the basic version and consists of two modules: a front panel module and a processing/power supply module. Optionally, analogue/digital and input/output modules can be added.

Status
In production. Selected for several Eurocopter export programmes. More than 100 systems delivered for Fennec and Super Puma helicopters.

Contractor
Thales Avionics SA.

Nadir Mk 2 navigation/mission management system

Type
Aircraft mission computer.

Description
The Nadir Mk 2 is a multipurpose processing and display system that can store details of up to 100 waypoints and the characteristics of up to 100 VOR/DME stations. It is not limited to Doppler but can operate with many other navigation sensors. Nadir Mk 2 comprises a four MCU central processing unit and a general purpose control/display unit. It can provide the following:
1. Navigation management functions based on Doppler, VOR/DME, Tacan, Omega/VLF, GPS, heading sensor, inertial sensor and air data inputs.
2. Flight management functions, with guidance for optimum cruise conditions, fuel and weight management and engine monitoring; air data computations involving speed, altitude and outside air temperature.
3. Interface functions with autopilot, radar and navigation indicators.

Nadir Mk 2 may be used in a dual-system configuration, in which one computer is responsible for navigation and flight management, while the second deals with weapons and aircraft management.

Status
In production. Nadir Mk 2 has been selected for 30 Eurocopter helicopter programmes and for fixed-wing aircraft operations.

Contractor
Thales Avionics SA.

Para-visual display

Type
Aircraft instrumentation.

Description
The para-visual display has been developed as optional equipment for the A320. Mounted on the glareshield, the instrument guides the pilot along the runway centreline during roll-out in poor visibility. Guidance is provided by vertical white and black strips, which move to the left or right, according to the deviation from runway centreline.

The display uses liquid crystal and is within the pilot's field of view when looking outside. It is connected to the flight guidance system and to the flight warning computer of the aircraft, or equivalent sources. The system comprises the autoland warning and the picture generator.

Specifications
Dimensions: 150 – 155 – 36 mm
Weight: 1 kg
Interface: ARINC 429 HS discrete inputs (autoland warning control)
Angle of view: 60° under day or night conditions

Status
Optional equipment for the Airbus A320.

Contractor
Thales Avionics SA.

Rafale aircraft displays

Type
Avionic display system.

Description
Thales Avionics has developed avionics systems for the Rafale multirole fighter comprising:

- Two 5 × 5 in SLCD55 colour, multifunction side-mounted displays, with touch-sensitive screens, to provide the pilot with information on aircraft systems
- TMC 2020 head-level display, providing colour, high-resolution synthetic images collimated for infinity, designed to enhance the pilot's situation awareness
- VEH 3022 (CTH 3022) holographic multimode Head-Up Display (HUD), providing 'short-term' information (symbology, FLIR images) in a wide 30 × 22° Field of View (FoV)

These displays incorporate the following interface technologies:

- SLCD liquid crystal display has its own integral symbol generator, a MIL-STD-1553B bus and a touch-sensitive screen
- TMC 2020 uses LCD technology and an optical device to collimate images to infinity
- VEH 3022 incorporates developments in holography to provide a wider field of view

Completing the man/machine interfaces for Rafale are a control and display unit, combined standby instrument and engine and fuel indicator.

Two computers manage the overall display system:

- The Modular Data Processing Unit (MDPU), developed by Thales, gathers all data processing in the aircraft (symbol generation for the HUD and HLD units, digital mapping functions, mission data, flight data and tactical situation) into one unit
- An alternative symbol generator for the standby instruments.

Thales Avionics is also working on a voice recognition and control system, intended to equip a future version of Rafale. Also in development for Rafale is the CET flight path computer. The CET prefigures the 'electronic co-pilot' which will allow combat aircraft pilots to concentrate on the tactical situation, rather than flight control management.

Thales also supplies the following air data and system management equipment: four multifunction probes; three UMPT33 air data systems; one Topstar® precision positioning GPS receiver; one static magnetometer; a data concentrator to control interface with the control and display unit; and sensors which monitor internal pressure, temperature and other parameters.

Status
In production and in service in Rafale multirole fighters. Thales' Topsight® was replaced by Sagem's Gerfaut as the HMDS of choice for French Air Force and Navy Rafales in 2003.

Contractor
Thales Avionics SA.

Rooivalk helicopter avionics system

Type
Avionics system.

Description
In August 1994, the then Sextant (now Thales Avionics) signed a contract with the South African avionics integrator, ATE Pty Ltd, to supply

Cockpit of the Rafale aircraft (Thales Avionics)   0051660

key avionics equipment for the Rooivalk attack helicopter being developed and manufactured by South African aircraft manufacturer, Denel Aviation.

Equipment supplied by Thales comprises the liquid crystal displays, helmet-mounted sight/displays, ring laser gyro navigation system with embedded GPS, standby instruments and pilot handgrips.

The basic avionics of the Rooivalk are off-the-shelf items, all previously qualified on the Tiger combat helicopter. Thales provides a complete, redundant navigation system, including two inertial navigation units, each based on a PIXYZ® three-axis ring laser gyro, air data equipment, two magnetometers and a GPS receiver. The navigation system is built around the inertial navigation unit, which functions as inertial sensor, navigation computer and controller for other peripherals.

The Rooivalk is equipped with Thales' MultiFunction liquid crystal Display (MFD 66). The 160 × 160 mm (6.3 × 6.3 in) displays offer excellent readability and viewing angles under any lighting conditions, even in direct sunlight. They are part of a complete family of LCDs developed by Thales for both civil and military applications, including rotary-wing aircraft, and are already chosen for a number of programmes, including Tiger, NH 90 and EC135.

Thales Topowl® Helmet-Mounted Sight/Display (HMS/D – see separate entry), developed as part of development contracts for the Tiger programme, also equips the Rooivalk. Topowl® is a biocular, wide field-of-view, day-/night-capable sight/display system. It provides for the projection of flight and weapon aiming data, together with onboard sensor imagery, onto the helmet visor. Topowl® also features helmet-mounted night vision via two Image Intensifiers

Line-up of eight Rooivalk in service with No.16 Squadron of the South African Air Force (Denel)   0580475

Two Thales MFD 66 displays in the front weapons operator position of a SAAF Rooivalk attack helicopter 0097525

Pilot's cockpit of a SAAF 16 Squadron Rooivalk 0556144

(II) mounted either side of, and aligned with, the pilot's eyeline and projected onto the visor.

Status
All 12 Rooivalks ordered by the South African Air Force (SAAF) were delivered by March 2004.

Contractor
Thales Avionics SA.

Smart Head-Up Display (SHUD)

Type
Head-Up Display (HUD).

Description
The SHUD is a fully integrated head-up display with a 24° circular field of view. Connected to the MIL-STD-1553B databus, the SHUD is able to compute and draw symbology and to present simultaneously FLIR video and flight information. The SHUD also incorporates sufficient processing power for weapon system computations. It is particularly well adapted to combat aircraft upgrading.

Status
Selected by Lockheed Martin for the Argentine Air Force A-4M avionics upgrade programme. SHUD has also been associated with the Mikoyan MiG-AT trainer, Spanish Air Force Mirage F1 modernisation and Iryda I-22 trainers of the Polish Air Force. Over 100 SHUDs have been produced.

Thales' SHUD Smart Head-Up Display for the MiG-AT 0051711

Contractor
Thales Avionics SA.

SMD 45 H liquid crystal smart multifunction display

Type
Multifunction cockpit display.

Description
The SMD 45 H was specially designed for helicopter applications and is currently proposed for multiple light helicopter upgrade programmes.

The SMD 45 H is a 'Smart' full-colour multifunction display using a 4 × 5 in Active Matrix Liquid Crystal Display (AMLCD) providing excellent viewability under any aircraft cockpit ambient light conditions.

The SMD 45 H integrates in one panel-mounted LRU all necessary functions required for stand-alone operation such as systems bus interface, data processor and graphics generator.

The SMD 45 H high-resolution display also provides imagery from various aircraft sensors with synthetic flight symbology overlay. As one of the key elements of the MEGHAS® glass cockpit avionics suite, the SMD 45 H can display FLIR, Primary Flight Display (PFD), Navigation Display (ND), moving map and weather radar information.

Specifications
Active screen size: 4 × 5 in
Resolution: 640 × 480 pixels
Brightness and contrast: >150 fL in white
Aircraft interface: ARINC 429 bus (6 in, 4 out) or MIL-STD-1553B bus (optional)
Power: 28 V DC, 50 W

SMD 45 H liquid crystal smart multifunction display 0051661

Overall dimensions:
(front panel) 185 × 134 × 23 mm (without push-buttons)
(case size) 185 × 201 × 166 mm
Weight: 2.8 kg
Video interface: 1 input STANAG 3 350 C
Environmental: DO 160 C
NVG-compatible

Status
Fitted on BK 117C2, EC 120, EC 135, EC 155 and Ecureuil B3 helicopters.

Contractor
Thales Avionics SA.

SMD 66 integrated multifunction display

Type
Multifunction cockpit display.

Description
The SMD 66 is a multipurpose electronic display developed for use on helicopter and fixed-wing aircraft flight decks It integrates in one panel-mounted LRU all necessary functions required for stand-alone operation such as aircraft systems bus interface, data processor and graphics generator. The full-colour high-resolution shadow-mask fully integrated display is capable of providing stroke and raster image for primary flight, navigation and tactical displays and systems, and engine monitoring. The image measures 152 × 152 mm (6 × 6 in). The display features brightness automatic setting up to 8,000 ft candles, to improve legibility, and is compatible with night vision goggles.

Specifications
Dimensions: 210 × 193 × 235 mm
Weight: 7 kg
Power supply: 28 V DC, 80 W
Resolution: 512 × 512 pixels
Brightness and contrast: >150 fL in white
Aircraft interface: ARINC 429, RS485
Environment: MIL-E-5400
Video interface:
2 colours video inputs Stanag 3350 B;
2 image formats: 1/1 or 3/4

Status
The SMD 66 was originally developed for the Tiger program and is now proposed for other applications. It is in production for the Super Puma Mk 2 and is also supplied for the Indian Air Force Jaguar upgrade programme.

Contractor
Thales Avionics SA.

Thales' SMD 66 integrated multifunction display 1041442

SMD 68 CVN liquid crystal multifunction display

Type
Multifunction cockpit display.

Description
The SMD 68 H is a 'smart' full-colour multifunction display designed for helicopter applications. It uses a 6 – 8 in Active Matrix Liquid Crystal Display (AMLCD) providing excellent viewability under all aircraft cockpit ambient light conditions.

The SMD 68 H integrates all the necessary functions required for stand-alone operation (including databus interface, data processor and graphics generator) in one panel-mounted LRU.

As an option, the SMD 68 H can also display imagery from aircraft sensors such as FLIR, map, weather radar together with synthetic flight symbology overlay.

Specifications
Active screen size: 6 × 8 in
Resolution: 1,344 × 1,008 pixels
Brightness and contrast: 0.4 to 315 cd/m²
Aircraft interface: ARINC 429
Power: 28 V DC, 130 W
Dimensions: 192 (W) × 267 (H) × 202 (D) mm
Weight: 6.5 kg
Cooling: internal fan
Video interface: (STANAG 3 350 C or B) 1 input
Environment: DO 160 C
Reliability: > 5,000 flight hours
Options: filtered NVG compatibility

Status
Selected for BK 117C2, C-130, Casa C 295, Dash 8-400, EC 135 and EC 155.

Contractor
Thales Avionics SA.

Thales' SMD 68 H liquid crystal smart multifunction display   0051662

SSCVR Solid-State Cockpit Voice Recorder

Type
Flight/mission recording system.

Description
The SSCVR has been designed in accordance with ARINC 757, TSO-C123 and ED 56(A) by SFIM Industries, Thales and TEAM, to meet the requirements of all commercial aircraft for cockpit voice recording systems. The system has no

SAGEM/TEAM Solid-State Cockpit Voice Recorder (SSCVR)   0015303

moving parts and uses compressed voice storage to provide up to 2 hours of recording time.

Specifications
Recording attributes:
(method) non-sequential adaptive differential pulse code modulation and code exciter linear parameter modelling
(medium) flash memory modules
(inputs) 3 crew microphones; one area microphone; GMT (ARINC 429 format); rotor speed; flight data recorder time marker
(time) 30, 60 or 120 minutes
Dimensions: 1/2 ATR short – ARINC 404
Weight: 8.5 kg
Power:
(DC version) 28 V DC, 0.8 A
(AC version) 115 V AC, 400 Hz, 0.26A

Status
In service.

Contractor
SAGEM, Aerospace and Defence Division.
TEAM.

T100 and T200 weapons sight for helicopters

Type
Head-Up Display (HUD).

Description
The T100 and T200 electromechanical head-up weapons sights show a collimated reticle which can be moved between –10 and +7° in elevation and ±6° in azimuth, angles compatible with the requirements for air-to-air and air-to-ground weapons launch. The device is mounted on the canopy frame of the helicopter, weighs 2 kg and has a field of view of 7.5°.

The sighting system is based on a modular concept. The basic component is the T100 single- or T200 dual-axis monocular sight head which can be used for both day and night weapon firing in conjunction with third-generation microchannel NVG. Sight recording by CCD camera has been validated for both training and operational firing and is available as an option. The physical features of the sight head, such as its low weight, small size, simple and strong high-performance optical system of Angenieux lens, diode array on micro-electronic support, give it the capacity to produce a remarkably high-quality image, both by day and night.

The second main component of the sighting system is a control unit combining command and calculation functions. This unit includes the necessary controls for moving the sight head,

Thales' T100 helicopter head-up display   0018160

Thales' T200 helicopter head-up display   0018159

symbology animation controls, firing tables stored in its internal memory, weapon firing computation system for guns and air-to-air missiles and sighting telescope interface. The links with the missile system and telescope are designed in particular for export requirements and include coupling with Sextant's Nadir computer for target designation. The system is equipped with a built-in automatic testing facility, with status information being displayed on the sight head.

To meet the weaponry requirements of the French Army Gazelles, which are equipped with guns and Mistral missiles, Thales Avionics proposed a multipurpose sighting and fire-control system comprising the above-mentioned components that have already been qualified for the Gazelle. With this system, the pilot can fire the various weapons, including rockets, without any manipulation other than manual selection of functions on the front panel of the control unit.

Status
In production. The system has been installed on a wide variety of helicopters including the Boeing 500/530, Eurocopter BO 105 and BK 117, Sikorsky S-76 and Black Hawk, Bell 206, 406 and 412, Westland WS 70. It has also been selected to equip the Gazelle helicopters of the French Army Light Aviation Corps and the Ecureuil helicopters of the French Air Force, and by Eurocopter for export versions of the Dauphin, Ecureuil, Gazelle and Puma.

Contractor
Thales Avionics SA.

TMV 1451 electronic head-up display for commercial aircraft

Type
Head-Up Display (HUD).

Description
The TMV 1451 Head-Up Display (HUD) comprises the Optical Head Unit (OHU) and the Head-Up Display Computer (HUDC). The OHU is installed in the glareshield panel either behind the glareshield front panel or in the operational or pull-out position with the combiner appearing in the forward field of vision of the pilot. The HUDC is located in the electronics bay. It is linked on one side to the display management computer and on the other side to the OHU itself. The HUD's control panel is included in the OHU.

The OHU is an electronic head-up display system designed specifically to be installed in the glareshield of commercial aircraft and particularly in the Airbus A320. The head-up display, linked to the existing aircraft systems, can be used for roll-out guidance, visual approach guidance and monitoring of automatic approach and flare. The OHU is linked to the HUDC, which generates the symbology to be displayed, according to the flight phase.

The OHU consists of optical lenses and combiner which present collimated symbology to the pilot superimposed on the outside world, a miniature high-brightness CRT, an automatic brightness control which adjusts the symbol brightness to the required level and a control panel. Of limited volume, the OHU is designed and installed in such a way that the pilot's lower field of vision is not interrupted.

The head-up display computer is designed around VLSI circuits already in use in the display systems of A310 and A320 aircraft. Input parameters received from two distinct channels are monitored so that false information is instantaneously detected. The HUDC includes BITE. The symbol generator function of the computer is capable of driving two optical head-up displays.

Specifications
Weight:
(OHU) 10.5 kg
(HUDC) 5 kg
Power supply: 115 V AC, 400 Hz
Field of view: 24 × 15°
Reliability:
(OHU) 5,000 h MTBF
(HUDC) 13,000 h MTBF

Status
In production.

Contractor
Thales Avionics SA.

TMV 544 forward view repeater display

Type
Avionic display system.

Description
Thales Avionics has developed what it terms a 'forward view repeater', which reproduces the pilot's field of view for the benefit of rear-seat occupants. The TMV 544 was developed in response to conclusions that some shortfalls in the instruction of pilots in advanced training aircraft resulted from the instructor's inability to scan the view directly ahead because it was obscured by the pupil's ejection seat headrest. The system is based on a video camera that films the head-up display and outside world from the front compartment and reproduces it on a television monitor in the rear compartment.

Status
In production.

Contractor
Thales Avionics SA.

TMV 980A head-up/head-down display

Type
Head-Up Display (HUD).

Description
The TMV 980A is an integrated head-up/head-down display system for the Dassault Mirage 2000 multirole fighter. It comprises a digital computer and processor to generate the display symbology and help with flight and weapon aiming computation and three display units: a VE 130 CRT head-up display, a VMC 180 interactive multifunction head-down colour display and a VCM 65 complementary monochrome CRT for electronic support measures information.

The head-up display has a high-resolution, high-brightness CRT with a collimating optical system based on a 130 mm lens providing a wide total field of view. The instantaneous binocular field of view is increased in elevation by the use of a twin-glass combiner, which transmits 80 per cent of the light incident upon it. Automatic brightness control, with manual adjustment, permits symbols to be read in an ambient illumination of 100,000 lx. The system provides continuous computation of tracer line in the air-to-air mode and impact and release points in the air-to-ground mode.

The main head-down display presents, in red, green and amber on a 127 × 127 mm CRT radar display, information such as a radar map, synthetic tactical situation and range scales, raster images from television or FLIR sensors and tactical data from the system itself or from an external source.

A helmet-mounted sight may be integrated into the TMV 980A unit to improve target discrimination and off-boresight target designation. Another option is the substitution of the VE 130 head-up display by a VEM 130 system.

Specifications
Weight:
(electronic unit) 9 kg
(head-up display) 13 kg
(VMC 180 head-down display) 14 kg
(VCM 65) 4 kg

Status
In service in the Dassault Mirage 2000. No longer in production.

Contractor
Thales Avionics SA.

Topdeck® avionics for military transport aircraft

Type
Avionics system.

Description
Thales markets the Topdeck® concept as a modular upgrade for military transport and special mission aircraft. Principal components include:
1. Holographic Head-up Flight Display Systems (HFDS)
2. LCD 68 full-colour 6 × 8 in LCDs for Primary Flight Display (PFD), Navigation Display (ND),

The SAAF C-130BZ upgrade programme includes Topdeck® (28 Squadron SAAF)   1123787

The Topdeck® avionics suite equips Polish Air Force C-295M transport aircraft
(Grzegorz Holdanowicz)   0578757

The Venezuelan Air Force C-130H upgrade includes Topdeck® 0563261

Engine Indication and Crew Alerting System (EICAS), weather radar display
3. Multifunction Control Display Units (MCDU)
4. Totem 3000 INS/GPS
5. Gemini 3000 mission management system
6. AFPD 800 autopilot and flight director

Using ARINC-429 and MIL-STD-1553 databus architecture, Topdeck® can interface directly with a wide variety of civil/military equipment, including weather radars, terrain-awareness warning systems, traffic-alert and collision-avoidance systems, IFF, Mode-S transponders and radio/communications systems.

The flight management system integrates an array of sensors, including inertial navigation, global positioning system and air data units, to deliver high-accuracy positional information. The digital autopilot is coupled to the navigation system, greatly expanding the aircraft's operational capabilities.

Status
Selected for C-130 upgrades for South African and Venezuelan Air Force C-130 upgrade programmes and for the CASA CN-235-300 and CN-295 military transport aircraft.

The SAAF upgrade programme has suffered delays – the last of nine aircraft (two ex-US Air Force C-130Bs received via the Excess Defence Articles [EDA] programme and seven SAAF C-130BZs) is due to return to service in late 2006, some three years behind the original schedule. The upgrade enables the SAAF to keep its early 1960s vintage C-130BZs in service through to 2015. In addition to Topdeck®, the SAAF upgrade includes a Saab Avitronics ESM system, including a missile-approach warning system with five sensors, and new radios and related equipment.

Contractor
Thales Avionics SA.

Topflight® avionics suite

Type
Avionics system.

Description
The Topflight® avionics suite is based on avionics designed for Rafale, and it has been designed for basic or advanced training aircraft, as well as for combat aircraft such as the Mirage F1 or Su-22, either as original equipment or for retrofit. It enhances aircraft operational capabilities in a complex environment by

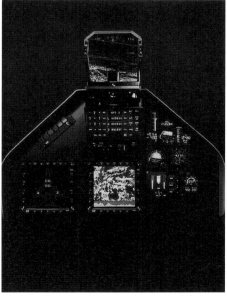

The Topflight® avionics suite for training and combat aircraft 0504175

supporting the pilot in all navigation tasks and in both air-to-air and air-to-surface missions.

Topflight® incorporates advanced technologies, such as holography, liquid crystal displays, laser gyros, head position direction, new-generation computers, GPS, digital mapping and voice command. These are employed in a helmet-mounted display, a key part of the man/machine interface tailored to night mission capability; multimode smart head-up display; multifunction liquid crystal displays; laser gyro navigation system; digital autopilot, and the Precise Position Service (PPS) version of GPS. Topflight® is built around a modular mission computer and symbol generator.

The compact design, incorporating smart liquid crystal displays, gives Topflight® a multiple mounting capability and the modular computer provides the performance levels needed for multirole aircraft.

Status
In production. Selected for the modernisation of Spain's Mirage F1, Argentina's A-4 and India's MiG-21 aircraft. It is also in production for the MiG-AT upgraded-avionics programme, in which Topflight® comprises:
- A SHUD wide field-of-view multimode head-up display, which also controls MIL-STD-1553B databus management
- Two reconfigurable centralised control panels
- Five multifunction colour LCD units (two for the front seat and three for the rear seat) in a 5×5 in (MFD 55) format
- Standby instruments
- A mission management computer and symbol generator which provide weapon systems simulation
- A Totem 3000 laser gyro Inverted Navigation System (INS)
- A Topstar® GPS/GLONASS stand-alone receiver
- A UMPT air data system.

A derivative system called Smart Topflight® has been developed for the Iryda advanced trainer built by PZL Mielec. In this configuration Smart Topflight® comprises:
- A SHUD for piloting, air-to-air and air-to-ground fire-control and MIL-STD-1553B databus management
- A Totem 3000 laser gyro INS for navigation, databus management in back-up modes and system interfaces
- Two configurable centralised control panels
- One or more multifunction colour LCD (SMD 54) to copy the SHUD to the back seat
- Standby instruments
- A UMPT 30 air data system.

Contractor
Thales Avionics SA.

VEM 130 combined head-up and head-level displays

Type
Head-Up Display (HUD).

Description
The VEM 130 is a derivative of the VE 130, which includes an additional capability for raster FLIR image presentation at head level. This display gives a field of view of $14 \times 10°$ and the green phosphor raster display can be set at 525, 625 or 675 lines, 50 or 60 Hz.

Specifications
VEM 130
Dimensions: $645 \times 378 \times 145$ mm (plus combiner glass)
Weight: 23 kg

Status
The VEM 130 is in production and in service in the Dassault Mirage 2000–5.

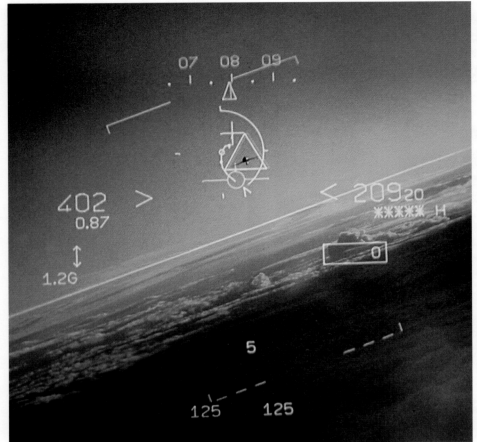

Photograph taken through the VEM 130 Head-Up Display (HUD) of a Mirage 2000 during an air-to-air training engagement. Airspeed/Mach number are shown to the left, while barometric altitude is shown on the right, with aircraft heading scale at the top of the display. The target Mirage is enclosed by a magic IR missile lock (triangle), with air-to-air guns format selected, showing aircraft gun boresight (cross), range to target (arc) and a Continuously Computed Impact Line (CCIL), emanating from the aircraft velocity vector (aircraft symbol) and showing instantaneous bullet impact position (small circle on CCIL). Rounds remaining for each gun appears at the bottom of the display (I Black)   0132089

Contractor
Thales Avionics SA.

VEMD Vehicle and Engine Management Display

Type
Cockpit display.

Description
The VEMD is designed for both single- and twin-engined helicopters. It is interfaced to engine and vehicle sensors and linkable to data concentrators as well as FADEC.

It displays the information on two 127 mm (5 in) diagonal displays. This allows several display modes: normal operation mode (first limitation indication, vehicle information), reversionary mode performance, health monitoring flight report (overlimit detection) and maintenance modes. The dual architecture both in terms of processing and displays, the high level of failure detection and the presence of reversionary modes provide a high level of availability.

The VEMD displays the information on a twin-active matrix colour LCD display unit with a wide viewing angle and provides high readability in any ambient lighting conditions (compatible with filtered NVG). In normal operation the upper matrix displays engine information and the lower one, vehicle information. If one channel fails, the main information is displayed on the remaining display.

The VEMD can also display health monitoring information such as engine cycle count, engine power check, BIT results. This information is stored in a non-volatile memory and is readable on the display in maintenance mode.

Each channel of the dual architecture of the VEMD (display, processing, power supply) performs the acquisition of each parameter and cross-checks its consistency with the other channel. This

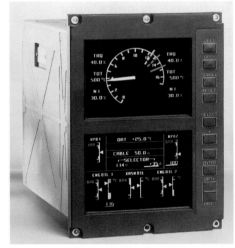

Thales' vehicle and engine management display   0079261

monitoring activity, added to the self-test of each channel, provides a high level of failure detection and low level of erroneous data display.

The VEMD has dual processing and display systems interfaced to engine and vehicle sensors: thermocouples (TOT); frequencies (N1, N2, NR); voltages (torque, fuel quantity, oil pressure, pressure, gearbox oil pressure); resistors (oil temperature, OAT, gearbox oil temperature); voltages (voltmeter, ammeter) pressure sensor as well as numeric interfaces for FADEC (RS-422, ARINC 429, EIA 485). It generates the power supply required for the sensors as well as discrete outputs.

Specifications
Useful screen size: (3 × 4 in) × 2
Lighting in day conditions: 0.4 to 100 Cd/mL
Dimensions: 156 × 212 × 205 mm

Max weight: 3 kg
Power consumption: 44 W
Power supply: 28 V DC
MTBF: 5,000 h
    No cooling required

Status
Fitted on BK 117C2, EC 120, Ecureuil B3, EC 135, EC 155.

Contractor
Thales Avionics SA.

VS1500 ultra-compact airborne multichannel video digital recorder

Type
Flight/mission recording system.

Description
The VS1500 is a mission video and data recorder/data transfer unit for use in fighter, helicopter, UAV, and ground vehicle environments. It is an ultra-compact, standalone, ruggedised unit, giving simultaneous acquisition and recording of four video sources (PAL or NTSC), four audio channels and four MIL-STD-1553B aircraft databus on a single, removable, digital data cartridge. The system provides more than 5 hours' recording time, depending on the selected configuration. It offers up to 4-aircraft synchronised video and mission playback on a single commercial PC with networking capabilities. The system also offers ACMI capability.

Specifications
Memory: Solid-state flash memory cartridge
Dimensions: 120 (H) × 148 (W) × 161 (D) mm
Weight: 3.5 kg, incl 0.5 kg removable storage cartridge
Video acquisition: coherent filtering into D1 or Half-D1 image size
    max aggregate video acquisition rate:
    4 – 25 images/s in PAL
    4 – 30 images/s in NTSC
Video and audio compression: ISO MPEG2 compression algorithm
Interfaces: RS-422A, MIL-STD-1553B, Ethernet 100BaseT
Power: 28 V DC, <20 W typical

Status
In production and in service. Selected by BAE Systems for the Hawk Mk 128. In operational service on the US Air Force A-10 fleet.

Contractor
Enertec.

The Enertec VS1500 ultra-compact airborne multichannel digital video recorder   1140451

VS2100 compact airborne multichannel video digital recorder

Type
Flight/mission recording system.

Description
The VS2100 compact airborne multichannel video digital recorder is a standalone, ruggedised unit, giving simultaneous

The Enertec VS2100 compact airborne multichannel video digital recorder 0129335

acquisition and recording of four video sources (PAL or NTSC), four audio channels and aircraft databus on a single, removable, digital data cartridge. The system provides more than five hours' record time, depending on the selected configuration.

Specifications
Memory: Solid-state flash memory cartridge (optional HDD)
Dimensions: 270 (H) × 136 (W) × 180 (D) mm
Weight: 5.5 kg, incl 0.5 kg removable storage cartridge

Video acquisition:
coherent filtering into D1 or Half-D1 image size max aggregate video acquisition rate:
4 × 25 images/s in PAL
4 × 30 images/s in NTSC
Video and audio compression: ISO MPEG2 compression algorithm
Interfaces: RS-422A, MIL-STD-1553B, Ethernet 100BaseT
Power: 28 V DC, <50 W typical (70 W with pre-heaters)

Status
In production and in service. The VS2100 system has been trialled on Mirage 2000, F-15, F-16 and CT-4 aircraft, and it is in undisclosed French service. Like the DS4100, the system is also reported to have been evaluated by QinetiQ (formerly the Defence Evaluation Research Agency) in the UK.

Contractor
Enertec.

VS2200 compact airborne multichannel video digital recorder

Type
Flight/mission recording system.

Description
The VS2200 compact airborne multichannel video digital recorder is a standalone, ruggedised unit, giving simultaneous acquisition and recording of up to eight video sources (PAL or NTSC), four audio channels and up to eight MIL-STD-1553 aircraft databus on a single, removable, digital data cartridge. Applications for the system include:

pilot training and debriefing; reconnaissance and surveillance; battle damage assessment; and mission uploading. The system provides more than five hours' record time with optional HDD technology increasing this figure to over ten hours. The VS2200 provides a direct form-fit-function replacement for legacy analogue video tape recorders. It offers up to 4-aircraft synchronised video and mission playback on a single commercial PC with networking capabilities.

Specifications
Memory: Solid-state flash memory cartridge (optional HDD)
Dimensions: 114 (H) × 243 (W) × 330 (D) mm
Weight: 8 kg, incl 1.5 kg removable storage cartridge
Video acquisition: coherent filtering into D1 or Half-D1 image size
max aggregate video acquisition rate:
8 × 25 images/s in PAL
8 × 30 images/s in NTSC
Optional built-in scan converter for STANAG 3350A acquisition
Video and audio compression: ISO MPEG2 compression algorithm
Interfaces: RS-422A, MIL-STD-1553B, Ethernet 100BaseT
Power: 28 V DC, <50 W typical

Status
In production and in service. The VS2200 system has been trialled on F-15 and F-16 aircraft, and has been selected for the US Air Force F-117 Nighthawk and the Eurocopter NH 90 helicopter. In operational service on the Indian Air Force Light Combat Aircraft (LCA) and the Canadian Forces CP-140 Aurora.

Contractor
Enertec.

Germany

DATaRec-A4 acoustic recording system

Type
Flight/mission recording system.

Description
The DATaRec-A4 is a complete portable four-channel acoustic/analogue recording system. The system provides connectors, preamplifiers and power supplies for Bruel and Kjäer microphones. With 24-bit digital signal processing and filtering, exceptional phase error specification allows accurate correlation of data between channels. Weighting filters can be selected and, with pre-emphasis, the dynamic range will be extended to 130 dB.

A front panel LCD provides bar graph and alphanumeric displays of input/output voltage, power and spectra. The IRIG generator/decoder allows synchronisation to the internal or external IRIG source, accurately time-stamping all data. Additional data that may be recorded includes voice, RS-232 and opto-coupled rpm and speed data.

With eight different channel configurations, a patented self-calibration system, rugged but lightweight chassis and battery operation, the A4 may be used wherever portability is of importance.

Specifications
Dimensions: 270 × 88 × 265 mm
Weight: 6 kg (incl battery)
Power supply: battery operated for up to 2 h
10–36 V DC or 100–240 V AC (with adaptor)

Status
In production and in service.

Contractor
Heim Data Systems, Inc.

Display Video Recording System (DVRS)

Type
Flight/mission recording system.

Description
The Display Video Recording System (DVRS) has been designed by Bavaria Keytronic Technologie for the German Air Force Tornado. It is also suitable for army, navy and commercial applications. The system can handle up to seven

video inputs with a resolution of 400 TVL/PH each, plus audio channels, event and weapon release markers, databusses, radar and FLIR.

The system consists of a HUD camera, electronics unit, videotape recorder with up to 180 minutes recording tape and a control panel.

The HUD camera consists of the video sensor head with optical components and CCD video colour sensor module, and an electronics assembly consisting of power supplies, control logic, BIT circuitry and video amplifiers. HUD symbology is combined with the picture of the

The display video recording system in service in German and Italian air force Tornado aircraft, showing the electronics unit (top), video recorder (right), control panel (bottom), HUD camera (left) 0051323

outside world and transmitted in colour PAL format via the multiplexing unit to the VTR.

The electronics unit consists of an analogue-to-digital converter unit, multiplex circuitry, microprocessor, control BIT circuitry and power supplies.

The Hi-8 mm VTR is based on a TEAC V 80 recorder. The microprocessor-controlled panel gives the operator complete up-to-date information on the system, including remaining recording time. The control panel consists of the mode selection switch, video input, selection control, alphanumeric display and amplifiers.

The system provides up to 180 minutes recording time on each cassette which can be changed easily during flight.

Specifications
Dimensions:
(HUD camera) 190 × 45 × 210 mm
(electronics unit) 120 × 325 × 170 mm
(videotape recorder) 107 × 139 × 156 mm
(control panel) 55 × 145 × 127 mm

Status
In service in German and Italian Air Force Tornado aircraft.

Contractor
Bavaria Keytronic Technologie GmbH.

DKG 3 and DKG 4 digital map display system for helicopters

Type
Avionic Digital Map System (DMS).

Description
The Systems and Defence Electronics business unit of EADS, at Dornier GmbH in Friedrichshafen, has developed the cockpit map display systems, DKG 3 and DKG 4, destined for use in aircraft and helicopters respectively. These compact electronic devices replace conventional printed maps.

DKG 3
The digital map display system DKG 3, in the form of kneeboard equipment, supports the navigation and communication of helicopter crews by displaying colour maps in various scales, the helicopter's own position, flight route and other flight and mission-related information.

In navigation mode, the current position of the air vehicle on the map is real-time controlled by an onboard navigation system, such as GPS. The map is automatically oriented in flight direction. This specific feature of the DKG 3 reduces the operator's response time especially when flying at low altitude.

Map and flight planning data are stored in PCMCIA format on two memory cards which are inserted in lateral slots in the DKG 3. By pressing a button, the operator can switch easily between the stored maps and, additionally, with a zoom function, magnify or scale down the selected map detail.

With the DKG 3, flight planning can be autonomously performed or changed either on the ground or in the helicopter.

Another primary task of the ground station is providing the logistics for the digital map data. The Geogrid Map Preparation Software (MAPS) generates all the necessary cartographic data adaptations. This cartographic data, which can be stored on a CD-ROM, is loaded, interactively assembled and written on the DKG 3 memory cards. The complete area of the Federal Republic of Germany, with scales of 1:200,000 (survey map of Germany) and 1:500,000 (ICAO map of Germany) can be stored on such a data carrier (mini hard disk with 170 Mbyte, PCMCIA). Currently, mini-hard disks are available with capacities of more than 1 Gbyte.

The DKG 3 has a coloured, 10.4 in diagonal Liquid Crystal Display (LCD). Its removable cover provides antiglare against direct sunlight and reflections. With the addition of a filter, DKG 3 can be made NVIS compatible.

The DKG 3 is used under VFR conditions. As the device does not interfere with flight guidance and control, it is categorised as a 'non-flight-safety-critical' system. The map display system is already in service with the German Federal Border Guard, police helicopter squadrons, the air rescue service of Germany's automobile association (ADAC) and industrial companies. A first lot is in operation with the medium-lift CH-53 helicopters of the German armed forces.

The DKG 3 was specifically designed for retrofitting to existing helicopters.

DKG 4
The DKG 4 was designed for fixed cockpit installation, connected to a large-format map display (typically 6 × 8 in) and acting as the tactical centre for operating the mission suite of the helicopter.

System facilities and modes of operation include:
- Planning and navigation
- Real-time heading-up presentation of true moving map
- Various map scales and continuous zoom-in/zoom-out functions
- Memory cards for map and mission data storage and transfer
- Tactical symbology
- Graphical data communication.

The DKG 4 has interfaces with the following aircraft systems and sensors:
- The navigation system
- Tactical radio
- External sensors, such as camera, FLIR, Radio Direction Finder (RDF)
- The HELLAS Obstacle Warning System (OWS)
- A remote-control system and helicopter displays.

Specifications
Dimensions: 180 × 146 × 76 mm
Power: 28 V DC, 20 W
Data storage:
(mission data) typically 32 MB flash
(map data) up to 1 Gb flash, and more than 1 Gb hard drive.

Status
In service with various German police services. DKG 4 is installed in new EC 135 helicopters of the Federal German Police helicopter squadrons of Bavaria, Saxony and Mecklenburg-Vorpommern, and in MD902 and EC 155 helicopters of the Federal German Police in Baden-Württemberg. DKG 4 is standard fit for all EC 135 and EC 155 helicopters of the Federal German Border Guard.

Contractor
EADS Deutschland.

Flight Control Unit (FCU) for the Airbus A319/A320/A321

Type
Control and Display Unit (CDU).

Description
The FCU for the Airbus A319, A320 and A321 represents a smart, multipurpose control and display unit, interfacing between the pilot and the autoflight and electronic flight instrumentation systems. Installation in the glareshield and separate controls for pilot and co-pilot for

communicating with the primary instrumentation system enable the crew to work head-up. The FCU allows the pilot to engage autopilot, flight director and autothrust systems, and to set flight altitude, speed and course.

The FCU consists of two independent computers with automatic switchover to ensure redundant signal processing. All data exchange between FCU and external systems is done via a discrete interface and ARINC 429 serial datalink. Contrast and illumination of the displays and panels are adaptable to extreme environmental conditions, from direct exposure to sunlight at high altitude to the special requirements of night approaches.

Application of specifically designed LCDs to display set values allows wide viewing angles with high contrast and sunlight readability, even with the pilots wearing polarised sunglasses. Set values are introduced into the FCU using optical encoders. These are incremental opto-electronic devices with a notched input and push-pull capability.

LED keys for the selection of operating modes are based on push-buttons with redundant lighting. Each push-button shows a green confirmation bar and a white illuminated legend. Misreading of non-illuminated bars in full sunlight is prevented by integrated layers of optical filter coating. The push-buttons have double poles to ensure precise operation and repeatable tactile properties.

Status
In production and in service in Airbus A319, A320 and A321 narrow body airliners.

Contractor
Bodenseewerk Gerätetechnik GmbH (BGT).

Flight Safety Recording System (FSRS)

Type
Flight/mission recording system.

Description
The Flight Safety Recording System (FSRS) provides information for analysis of accidents and severe crashes, training of pilots, flight attendants and ground crew and surveillance of passenger and cargo compartments and exterior equipment. It consists of cameras, an electronic control unit and a Hi-8 mm format video recorder.

The cameras may be black and white or colour cameras installed inside or outside the aircraft. The cameras may be installed in the cockpit to record instruments and actions, in the passenger/cargo compartment to observe and record the situation and outside the aircraft to observe and record equipment and the outside world. They may be designed for low-light level operation, with infra-red illlumination. The electronic control unit multiplexes all video signals, controls the recording and has an optional interface to GPS and aircraft busses to record navigation and other aircraft data.

Status
FSRS is designed for installation in rotary- and fixed-wing aircraft and can be tailored to different applications or customer requirements.

Contractor
Bavaria Keytronic Technologie GmbH.

Flight control unit for the Airbus A319, A320 and A321

0581501

IN 3300 series VOR/LOC/GS indicators

Type
Flight instrumentation.

Description
The IN 3300 series indicators are precision engineered instruments which display VOR, localiser and glide slope deviation information for en route navigation and approaches.

Either 5, 14 or 28 V lighting is provided and non-reflective glass is used to ensure reliable readability under all operating conditions.

The model IN 3300-(10) contains a course selector and display, rectilinear VOR/LOC and glide slope cross-pointers and warning flags, and a TO-FROM indicator. The IN-3300-(3) also contains a built-in marker beacon receiver and automatically photocell-dimmed indicator lamps for airway marker, outer marker and inner marker beacon indications.

IN 3300 series indicators can be combined with Becker's NR 3320/30 navigation receivers to form a dependable, lightweight, easily installed VHF navigation system.

The IN 3300 series indicators are approved for operation to 50,000 ft, and are certified to the rigorous requirements of all applicable FAA and RTCA specifications.

Specifications
Dimensions: 82.55 × 82.55 × 130 mm (excluding clearance for connectors and cables)
Weight:
(IN 3300-(3)) 850 g
(IN 330-(10)) 820 g

Status
In production and in service.

Contractor
Becker Avionic Systems.

IN 3300-(3)/-(5)/-(6)   0001328

IN 3360-(2)-B compact VOR/ LOC/GS indicator

Type
Flight instrumentation.

Description
The IN 3360-(2)-B indicator is a precision engineered instrument which displays VOR, localiser and glide slope deviation information for en route navigation and approaches.

The indicator contains a course selector and display, VOR/LOC and glide slope cross-pointers,

IN 3360- (2)-B VOR/LOC/GS indicator   0001329

warning flags, and a TO-FROM indicator. It provides all steering information needed for IFR flying, and is ideal for use in aircraft with limited panel space, or for use as part of an 'emergency bus' avionics package for large aircraft.

The IN 3360-(2)-B indicators are approved for operation to 40,000 ft, and are certified to the requirements of TSO C52a.

Specifications
Dimensions: 60 × 60 × 110 mm
Weight: 0.40 kg

Status
In production and in service.

Contractor
Becker Avionic Systems.

Main computer for the Tornado

Type
Aircraft mission computer.

Description
LITEF GmbH upgraded the main computer in German Air Force Panavia Tornado IDS aircraft to meet the demands of more complex mission requirements and the integration of 'smart' air-ground weapons. It is based on a multiprocessor architecture which still uses the original LR-1432F digital airborne computer but adds one or more high-performance 68040 processor modules. This new structure enables execution of all legacy (and bespoke) software, as well as the newly introduced (industry standard) Ada software running on the 68040 processors.

The computer is equipped with various input/output interfaces including a high number of Panavia interfaces, one/two MIL-STD-1553 bus systems, discretes and special type interfaces.

The increase in computing power has facilitated the introduction of enhanced graphical abilities to the Tornado system.

Status
In service.

Contractor
LITEF GmbH.

LITEF main computer for the Panavia Tornado
0504158

Multiple Dislocated Flight Data Recorder system (MDFDR)

Type
Flight/mission recording system.

Description
The new-generation ultra lightweight Multiple Dislocated Flight Data Recorder (MDFDR) is part of an error tolerant modular solid-state crash and cockpit voice recording system which comprises up to four MDFDR recorders providing sufficient recording capacity. Due to the modularity, small dimensions and standardised interfaces, this recording system is suited for installation into helicopters and general aviation aircraft.

The MDFDR system consists of a Multiple Recorder Data Acquisition Unit (MRDAU), one

Multiple Dislocated Flight Data Recorder 0015305

MRDAU Multiple Recorder Data Acquisition Unit   0015306

to four MDFDR, Hand-Held Terminal (HHT) and ground station, designed to collect onboard data (audio, analogue, frequency, mil-bus, HDU and discrete inputs). It filters data to avoid exceedance and logistic relevant data and stores the resulting different data groups in the FDR and SSDC.

After flight, the data can be checked on board by the HHT or can be transferred via the HHT or SSDC to the ground station for detailed checkout.

Specifications
Dimensions:
(steel ball) 55 mm diameter
Data interface: RS-485 up to 1 Mbit/s (asynchronous datalink)
Power: 8–12 V DC, 1 W
Temperature range:
(operating) –55 to +85°C
(storage) –55 to +85°C
Weight:
(1 MDFDR) 0.35 kg
(MRDAU) 1.5 kg
Memory capacity: 4-32 Mbyte, 1 to 8 tracks, min track size 2 Mbyte

Multiple Recorder Data Acquisition Unit (MRDAU)
The MRDAU controls and supplies up to four MDFDR. The application specific programmable data interfaces of the MRDAU provide flexibility to adapt the MDFDR system to various helicopters and general aviation aircraft. For maintenance recording functions, the MRDAU has an optional data carrier interface which can store up to 170 Mbyte of flight data.

Optional data carrier   0015307

Hand-held terminal   0015308

Specifications
Dimensions:
(front panel) 146 × 76.2 × 6 mm
(case) 180 × 127 × 73 mm
Weight: 1.5 kg (excl data carrier)
Power: 28 V DC, 25 W
Temperature range:
(onboard equipment, operating) –55 to + 71°C

Hand-Held Terminal (HHT) (readout unit)
The HHT is a powerful battery-powered product to read out the MDFDR and the MRDAU. It is capable of 7 to 10 hours of continuous use. It provides a large 640 × 480 pixels, easy to read backlit display, with VGA graphics. The HHT is IBM PC/AT compatible, it is possible to connect an external keyboard, mouse, printer, monitor and floppy disk drive. Optional plug-in modules are available for SCSI, PCMCIA and other ISA bus interfaces.

Status
In production and in service.

Contractor
EADS, Defence and Civil Systems.

Navigation and tactical information systems

Type
Avionic Digital Map System (DMS).

Description
EADS Systems & Defence Electronics Germany, at Dornier GmbH in Friedrichshafen, has developed a family of digital map display systems for helicopter and fast-jet applications. These compact electronic devices replace conventional film-based map drives with a digital map display, combined with a wide range of map overlay information for added Situational Awareness (SA).

DMG EuroGrid
Facilitating enhanced aircrew SA in tactical situations, the EuroGrid digital map not only supports conventional navigation displays with geographic maps/plates of any area of the world but also provides for important mission planning and real-time tactical data, with information exchange by data link; the system stores aircraft sensor (TV /TI) information for post sortie debrief.

The digital map system forms part of the overall aircraft avionics system. DMG EuroGrid consists of two Line Replaceable Units (LRUs): the EuroGrid Digital Map Generator (DMG) and the Mission Data Transfer System (MDTS), which provides two independent display information channels with superimposed maps, graphic overlays, symbols and map correlated video image representations. The moving map portion of the system orients map information with the direction of flight. The System is capable of storing charts for en-route (topographical, Low Flying Chart) and large-scale maps (which may be used for Point-of-Interest search and Initial Point to Target runs), annotated with actual flight path, flight and tactical data. The System exchanges tactical information via the MDTS or by HF, VHF, FM radios and/or Link 16 data transmission.

System features and modes of operation include:
• Onboard planning
• Multiple map scales
• Continuous zoom-in/zoom-out
• Tactical overlay presentation
• Elevation processing
• Terrain profile presentation.
Interfaces are provided for:
• Graphical data
• Host aircraft navigation system
• Tactical radio
• Solid state data carrier for map and mission data storage and transfer
• Aircraft sensors (TV, FLIR)
• Obstacle Warning System (OWS).

Specifications
Dimensions: Available in ARINC 600 and L-Shape versions, 6 MCU
Power: 115 V AC, 400 Hz, 170 VA
Weight: 13.5 kg
Data storage:
Internal memory (map and mission data) up to 768 Mb
Solid state data carrier (map and mission data) currently standard 80 to 160 m bytes, up to 4 g bytes addressable
Video interface: STANAG 3350 class A and B
Image resolutions: 672 × 672 and 512 × 512 pixels
Interface: MIL-STD-1553B, STANAG 3838 and/or fibre optic STANAG 3910 (Eurofighter Typhoon)

Status
In production. In service in Eurofighter Typhoon aircraft and Eurocopter Tiger/NH 90 helicopters.

Contractor
EADS Deutschland.

Smart Flight Data Recording System (Smart FDRS)

Type
Flight/mission recording system.

Description
The new-generation, lightweight Smart Flight Data Recorder System (Smart FDRS) is part of a multiple-redundant, modular solid-state crash and cockpit voice recording system which comprises up to four recorders. Due to the modularity, dimensions and standardised interfaces used, this recording system is suited for installation into helicopters and general aviation aircraft.

The Smart FDRS consists of a Smart Data Acquisition Unit (Smart DAU), one to four Smart FDRs, a Hand-Held Computer/Terminal (HHC/HHT) and a ground station. It is designed to collect onboard data (audio, analogue, frequency, military bus, HDU and discrete inputs). Incident

and event data as well as logistic relevant data are stored as different data groups in the FDRS and Solid State Data Carrier (SSDC).

After flight, the data can be checked on board by the HHC/HHT or can be transferred via the HHC/HHT or SSDC to the ground station for detailed checkout.

Specifications
Dimensions: (metal ball) 55 mm diameter
Data interface: RS-485 up to 1 Mbits/s (asynchronous datalink)
Power: 8-12 V DC, 1 W
Temperature range: (onboard equipment, operating) –55 to +85°C
Mass: (1 Smart FDR) 0.35 kg, (Smart DAU) 1.5 kg
Memory capacity: 4-32 Mbytes, 1 to 8 tracks, min track size 2 Mbytes

Smart Data Acquisition Unit (Smart DAU)
The Smart DAU controls and supplies up to four Smart FDRSs. The application-specific programmable data interfaces of the Smart DAU enable the Smart FDRS to adapt to various helicopters and general aviation aircraft. For maintenance recording functions, the Smart DAU has an optional data carrier interface which can store more than 640 Mbytes of flight data.

Specifications
Dimensions: (front panel) 146 × 762 × 6 mm, (case) 180 × 127 × 73 mm
Mass: 1.5 kg (without data carrier)
Power: 28 V DC, 25 W
Temperature range: (onboard equipment, operating) –55 to + 71°CHand-Held Computer/Terminal (HHC/HHT) readout unit
The HHC/HHT is a battery-powered product to read out the Smart FDRS and the Smart DAU. It is capable of 7 to 10 hours of continuous use. It has a 640 × 480 pixel backlit display with VGA graphics. The HHC/HHT is IBM PC/AT compatible, and it is possible to connect an external keyboard, mouse, printer, monitor and floppy disk drive. Optional plug-in modules are available for SCSI, PCMCIA and other ISA bus interfaces.

Status
In production and in service.

Contractor
EADS Deutschland.

TV tabular display unit

Type
MultiFunction Display (MFD).

Description
The TV tabular display is capable of displaying various sensors such as a TV camera, FLIR,

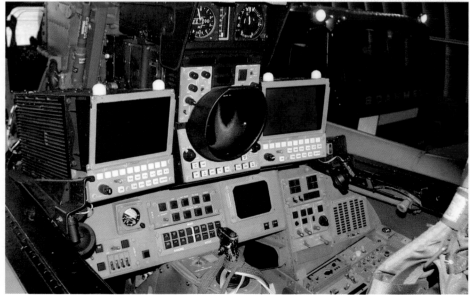

TV tabular displays fitted in the rear cockpit of a UK Royal Air Force Tornado GR4 (E L Downs)
0103860

low-light level TV and reconnaissance cameras and alphanumeric characters. By pressing related buttons on the keyboard the operator can select the preferred sensor out of three which is then internally mixed and displayed with alphanumerics overlaid. The display information on the Display Unit (DU) is also output for recording purposes for future reference or for post-mission analysis.

The DU also generates permanent data output by two 32-bit data words that are transferred to the main computer for selection and control of a computer program. The data words are output in both true and complement form and clocked out by a continuous train of 64 kHz data synch pulses which is also presented in both true and complement form.

The display unit brightness is controlled by an automatic brightness control circuit which guarantees readability of the displayed information during changing light levels, from darkness up to 10⁵ lux ambient illumination. In addition to this, the operator can, within certain limits, override the automatic control system by manually setting a brightness and contrast level.

The TV tab DU consists of modules interchangeable between like assemblies which guarantees good serviceability and low maintenance costs.

Specifications
Dimensions: 450 × 210 × 210 mm
Weight: 11.6 kg (max)
Power supply: 200 V AC, 400 Hz, 3 phase, 80 VA
   28 V DC

Status
In service in early versions of the Panavia Tornado.

Contractor
ESW-Extel Systems Wedel.

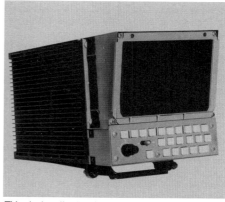

TV tabular display for the Panavia Tornado
0503931

International

AH-64D Longbow Apache HUMS

Type
Flight/mission recording system.

Description
Created jointly by Smiths Aerospace, Strategic Technology Systems, Inc, and the Boeing Company, the AH-64D Longbow Apache HUMS provides comprehensive health and usage monitoring. The diagnostics for avionics, rotors, transmission, engines and airframe are integrated with the helicopter's data management and display systems. It includes exceedance monitoring and cockpit voice and flight data recording.

The system comprises the Expanded Maintenance Data Recorder (EMDR); MultiPurpose Display (MPD); PCMCIA card receptacle and portable ground station.

System features include displays, warnings and pilot initiated functions fully integrated with the Longbow data management system covering 1,800 fault codes; data download via PCMCIA card or MIL-STD-1553B databus to portable PC ground station; ground station for HUMS data management and maintenance. EMDR features include data recording based on the Longbow Integrated Maintenance Support System (LIMSS); Stewart Hughes HUMS; 28 vibration inputs; eight tachometer inputs; blade tracker (day/night); voice recorder; 80 Mbyte crash-survivable memory; up to 160 Mbyte non-crash survivable memory; MIL-STD-1553B databus terminal; RS-422, RS-232, RS-485 serial communications.

Status
In service.

Contractor
Smiths Aerospace.
Strategic Technology Systems, Inc.
The Boeing Company.

The Strategic Technology Systems maintenance data recorder forms part of the AH-64 Longbow Apache HUMS   0051324

Cockpit Displays for the A400M Military Transport

Type
Avionic display system.

Description
The Airbus Military Company was formed in January 1999 to manage the A400M military transport aircraft project, formerly known as the FLA (Future Large Aircraft). Reformed in April 2003 as Airbus Military SL, the industrial partners involved in the programme are BAE Systems, EADS (comprising Aerospatiale Matra, CASA and DASA), FLABEL and TAI. Airbus is the majority shareholder in the company, bringing to the project valuable experience of building large commercial transport aircraft and managing complex international industrial programmes.

During November 2004, Thales Avionics SA was awarded a contract to supply key elements of the A400M avionic system, including:

FMS400 Flight Management System (FMS)
The FMS400 has been designed around a new software architecture and set up to work in combination with a digital map display that is powered by an EADS Defence Electronics-supplied Digital Map Generator (DMG). The system incorporates a mix of civilian (the A400M, in common with all strategic airlifters, is expected to spend 80 per cent of its flying time on civilian airways) and military functionalities (to support typical military airdrop/short field tactical supply missions).

Integrated Modular Avionics (IMA)
The A400M IMA solution largely re-uses the concept for the A380 ultra-large civilian transport, tailored to the specific requirements of the A400M. Thales Avionics (Meudon-la-Forêt, France) will be designing, developing and producing this system in partnership with Diehl Avionik Systeme (Überlingen and Frankfurt/Main, Germany).

Digital Head-Up Display System (D-HUDS)
D-HUDS incorporates Liquid Crystal Display (LCD) technology, which provides outstanding levels of system integration, flexibility of operation and high-quality flight information.

Control and Display System (CDS).
CDS is also derived from the A380 flight deck configuration and which features nine large, portrait format (6 × 8 in) multifunction, interactive

The A400M Class 2 (internal lighting) cockpit mock up in baseline configuration. Much of the forward panel is A380-derived, with the addition of a Control/Display Unit (CDU) on the centre pedestal for the third seat observer's position. In this case, the displays show the rear view of fighter aircraft during air-to-air refuelling operations. (Airbus Military SL)   1159543 'point-and-click' LCD screens, being developed at Thales' cockpit display systems facility in Bordeaux-Le Haillan, France.

The CDS is claimed to be the first use of interactive screens on a frontline military transport aircraft. Located on the central console, three control systems with mouse/keyboard interfaces will give pilots full and intuitive access to navigation functions, particularly important during critical flight phases.

With the two D-HUDS fully integrated the CDS, the A400M cockpit will feature 11 displays in total, with the HUD system certified to function as the Primary Flight Display (PFD) for both normal flight and tactical operations. Whilst, in common with Airbus' large civilian transports, the cockpit of the A400M is designed for two-pilot operation, a third crew member position will be incorporated into the cockpit to assist in critical mission management. In addition, a loadmaster station will interface with the main cockpit systems and provide for intercommunication.

There will also be standby instruments provided by Thales' plant in Vendôme, France.

The cockpit layout of the A400M will be based on the familiar and proven layout used in Airbus' large commercial aircraft, such as the A340 and A380, with addition of military-specific functionality, such as mission management, threat detection and radar mapping. The cockpit features side stick controllers linked to the fully digital Fly-By-Wire (FBW) system, with monitoring of aircraft systems and fault warning carried out by an Electronic Centralised Aircraft Monitor (ECAM). Cockpit seating and HUD installation will cater for both pilots wearing Night Vision Goggles (NVGs).

Status

The A400M is under development. The Thales suite of flight deck avionics is worth approximately EUR1 million (USD1.22 million) per shipset, giving the programme a value of at least EUR180 million to Thales. The systems will already be onboard during the first flight, scheduled for 2007.

During September 2006, Airbus Military announced the achievement of the sixth contractual milestone of the A400M programme, involving the acceptance by the customer of the Class Two A400M cockpit mock-up. Milestone 6 was particularly significant since it was designated as one of the six 'critical milestones' whereby the programme as a whole could be rejected by the customer on the grounds of non-compliance. Other critical milestones include the start of final production, the first flight and certification and delivery of the first aircraft to a customer.

The Class Two mock-up is an exact replica of the aircraft's operational cockpit designed to test and verify a number of important parameters in the overall cockpit environment. Using this facility, validation of the detailed cockpit design and effectiveness of the Man-Machine Interface (MMI) will be undertaken. Parameters to be studied include crew comfort and mobility, ergonomics, Field of View/Field of Regard (FoV/FoR) inside and outside the cockpit, ventilation and ingress and egress.

Some of the most important tests to be carried out in the Class Two cockpit mock-up are those designed to verify the interior light levels of the cockpit under all conditions, including the use of Aviator's Night Vision Imaging Systems (ANVIS).

A formal presentation of the cockpit mock-up was made to customer representatives from the Organisation Conjointe de Coopération en matière d'ARmement (OCCAR) on September 20th 2006 at the Airbus site at Blagnac. OCCAR is the European procurement agency acting as a single interface between the manufacturer and the national customers.

Certification of the overall cockpit is the responsibility of Airbus (supported by Thales), with most of the work scheduled to take place during 2008-09, with first delivery at the end of 2009.

Contractor

Airbus Military Company.
Thales Avionics SA.

Eurocopter Tiger combat helicopter avionics system

Type

Integrated avionics system.

Description

The avionics system for the Eurocopter Tiger combat helicopter is under development by a consortium of European manufacturers. The core avionics system consists of a bus/display system, com radio suite, autonomous navigation system, full ECM suite and AFCS, with connectivity via a redundant MIL-STD-1553B data highway.

TELDIX GmbH, Sextant and VDO-Luftfahrtgeräte Werk have together developed the five onboard computers: the ACSG (Armament Computer Symbol Generator); MCSG (Mission Computer Symbol Generator); BCSG (Bus Computer Symbol Generator); RTU (Remote Terminal Unit); and the CDD mission data concentrator.

The navigation system, by Thales Avionics, Teldix and EADS Deutschland, is fully redundant, including two Thales PIXYZ three-axis ring laser gyro units, two air data computers, two magnetic sensors, one Teldix/BAE Canada CMA 2012 Doppler radar, a radio altimeter and GPS – these sensors also provide signals for flight management, control and guidance; integrated duplex AFCS by Nord-Micro and Thales Avionics; computers by Thales Avionics, Diehl Avionics and Litef.

Cockpit displays include: two colour liquid crystal flight displays per cockpit (showing all flight, aircraft status and onboard system information) by Sextant and VDO-Luft and a central Control and Display Unit (CDU) (Rohde & Schwarz GmbH and Sextant) for avionics system control; a digital map system by Dornier and VDO-Luft (incorporating NH 90's Eurogrid map generation system).

The BAE Systems Knighthelm (see separate entry) fully integrated day/night helmet has been selected for German Tigers; French Tigres will have the similar Thales Avionics Topowl (see separate entry) helmet-mounted sights, with integrated night vision (image intensifiers), FLIR video and synthetic raster symbology.

The major component in the role equipment for Tiger is the Euromep (European mission equipment package), which includes the SATEL (Aerospatiale Matra/Thales Optronics/Eltro consortium) Condor 2 Pilot Vision Subsystem (PVS), air-to-air subsystem (Stinger or Mistral), mast-mounted sight and missile subsystem. Euromep Standard B avionics first flew in February 1995 (PT5); Standard C testing began in late 1997 (PT3R). The PVS has a 40 × 30° instantaneous field of view Thermal Imaging (TI) sensor, steered by a helmet tracker, feeding both crew helmet displays with sensor imagery, flight symbology and weapon aiming. A mast-mounted sight and associated gunner sight electronics, the gunner's heads-in target acquisition display and ATGW 3 subsystem are connected to the main aircraft system via a separate data highway (a HOT 3 missile subsystem also available), and controlled by a Sextant armament control panel

Cockpit of the Eurocopter Tiger combat helicopter. Note the two large displays showing aircraft status (left) and flight (right) information, and the CDU (lower left) showing nav/com information
0015358

Mast-mounted TV/FLIR/LRF (Jane's/Günter Endres)   0114860

and fire-control computer. Identification Friend or Foe (IFF) duties are performed by Thales' TSC 2000 IFF Mk 12.

Depending on the Tiger variant, combat equipment includes a SFIM/TRT STRIX gyrostabilised roof-mounted sight (with IRCCD IR channel) above the rear cockpit, incorporating direct view optics with folding sight tube, television and IR channels and a Laser Range-Finder/Designator (LRF/D). Electronic Countermeasures (ECM) equipment, again depending on variant, includes the EADS Deutschland C-model EW suite (as in NH 90), which includes a Laser Warning Receiver (LWR), Missile Approach Warner (MAW) and Thales EW processor and Radar Warning Receiver (RWR), coupled with chaff and flare dispensers and an optional IR jammer.

Status

In production and in service. Eurocopter estimate a total order of 240 Tigers between France and Germany. Australia selected Tiger for its AIR 87 requirement on 14 August 2001, with an order

for 22 helicopters. Interest has been expressed by Spain, with an anticipated 25 Tigers to be delivered from 2010. Exports of 200 Tigers thought possible.

Contractor

Thales Avionics.
EADS Deutschland.

TELDIX GmbH.
Nord-Micro AG & Co OHG.
Rohde & Schwarz GmbH & Co kg.
Diehl Avionics.
BAE Systems, Avionics.

Flight Management System (FMS) for Airbus

Type
Flight Management System (FMS).

Description
Thales Avionics and Smiths Aerospace are teamed to supply the Flight Management System (FMS) for the Airbus A318/319/320/321 and A330/340 families of aircraft. The system is designed to meet airlines' current and future requirements for forward- and retrofit of Communication/Navigation/Surveillance with Air Traffic Management (CNS/ATM) capability.

The FMS is common to the A318/319/320/321 and A330/340 aircraft families, thereby enhancing flightcrew training across individual airframes. The system incorporates the latest generation hardware and software, with a single aircraft interface for the Flight Management and Guidance Computer (FMGC) (A318/319/320/321 aircraft) or the Flight Management, Guidance and Envelope Computer (FMGEC) (A330/340 aircraft), as installed.

The FMS is based on two FM boards (integrated in the FMGC/FMGEC), each of which features its own high-speed 32 bit processor:

- The Flight Management Processor board (FMP) which processes the 'core FM' functions (navigation, flight plan management, trajectory

The A320 flight management and guidance computer   0079259

Close-up of UHT mast-mounted TV/FLIR/laser ranger sight (Jane's/Günter Endres)   0506400

The A320 flight management and guidance control system MCDU   0079258

predictions, lateral and vertical guidance, performance advisory and so on)

- The Interface and Display Processor board (IDP), which groups and gathers all the interface functions (MCDU display and control, EFIS display, processing of the inputs/outputs, datalink processing, interfaces with the FG(E) – Auto Flight Control and so on).

The FMS includes a colour Multipurpose Control and Display Unit (MCDU), common to the A318/319/320/321 and A330/340 aircraft families and featuring an active matrix LCD flat-panel display. It provides the following operational features:

- A temporary flight plan that allows systematic assessment of flight plan revisions. The temporary flight plan will provide full predictions for all lateral and vertical revisions and will allow multiple flight plan revisions. A graphical display of each revision is displayed on both the Capt and FO's navigation displays, and full predictions are provided on the MCDU. The crew can evaluate the economical implications of any re-route before taking a decision to accept it.
- A Required Time of Arrival (RTA) function that uses a high-fidelity aircraft performance algorithm to improve accuracy. The algorithm ensures that the path calculated accurately represents the path that the aircraft will actually fly when coupled to the autopilot in NAV mode.
- Increased flexibility and commonality across the fleet for datalink applications, using tables that can be loaded with ACARS MU and ATSU (protocol messages, prompts and triggers are airline customised)
- Increased flexibility provided by loadable and cross-loadable databases, including the 5 Mb navigation database. The flight management module is provided with hardware capability for greater navigation memory.

Status
In production and in service.

Contractor
Thales Avionics SA.
Smiths Aerospace.

Head-Up Display (HUD) for the Eurofighter Typhoon

Type
Head-Up Display (HUD).

Description
BAE Systems, leading a consortium with TELDIX, Galileo Avionica and INDRA Sistemas, is responsible for the Head-Up Display (HUD) for the Eurofighter Typhoon. The HUD forms part of the aircraft displays and controls subsystem and provides the primary display of flight information to the pilot.

The Pilot's Display Unit (PDU) uses diffractive optics featuring a single-element holographic combiner which consists of two glass elements bonded to produce a flat parallel-sided assembly.

Eurofighter Typhoon head-up display
0503964

Eurofighter Typhoon HUD, shown fitted into a developmental cockpit (note conventional flight instruments replacing left-hand multifunction display). Note the datalink display below the combiner and the Up-Front Control Panel (UFCP) under the left-hand glareshield (BAE Systems) 0121554

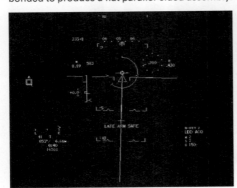

Air-to-ground attack HUD symbology for the Eurofighter Typhoon. Note weapon inventory, (bottom right), timing (bottom left) and safety status (script 'LATE ARM SAFE') information, together with target symbol (triangle) and Bomb Fall Line (centre) 0121551

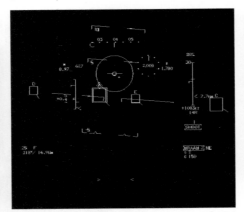

Multiple target designation boxes shown in air-to-air attack symbology for the Eurofighter Typhoon. Note the shoot cue (lower right), weapon fly-out (right), aircraft weapon and in-flight inventory and status (bottom right) 0121552

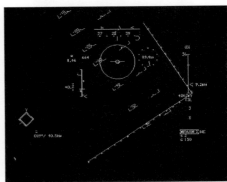

Air-to-air attack HUD symbology for the Eurofighter Typhoon. This post-firing snapshot shows the target designation box (lower left) and weapon fly-out (right) information, together with radar gimbal limits (solid boundary) showing maximum manoeuvre potential while the missile is supported by the aircraft radar (aircraft velocity vector must remain within the gimbal boundary during this phase) 0121553

An optically-powered hologram is recorded on photosensitised gelatine on the spherical interface sandwiched in the assembly and acts as the collimating combiner. The resulting advanced optical system, manufactured using computer-generated holographic techniques, provides new levels of display capability. Additionally, the uncluttered simplicity of the combiner support structure allows virtually a clear out-of-cockpit field of regard. The total field of view is 30° azimuth by 25° elevation and the instantaneous field of view is 30 × 20. The display operates in three modes; cursive, raster and raster/cursive.

The optical module brightness levels are optimised to operate in a very high ambient light environment, whilst minimising solar reflection and maximising outside world transmission and display uniformity from within the large eye motion box.

Associated with the PDU is a HUD control panel, developed by TELDIX, which is attached to the aft face of the unit and incorporates Light-Emitting Diode (LED) technology for multifunctional displays.

Status
In production and in service in the Eurofighter Typhoon.

Contractor
BAE Systems Avionics.
Galileo Avionica, Avionic Systems and Equipment Division.
INDRA Sistemas.
TELDIX GmbH.

Helicopter Flight Data Recording/Health and Usage Monitoring System (FDR/HUMS)

Type
Flight/mission recording system.

Description
The HUMS part of the FDR/HUMS system combines vibration and usage monitoring of critical dynamic power-train components with techniques such as chip detection, rotor track and balance, engine power assurance, cycle counting, exceedance monitoring and oil analysis. Integrated with the FDR, the FDR/HUMS supports helicopter operations, safety and maintenance – both civil and military.

The system comprises a data retrieval unit, sensors and data sources (to customer requirements), crash survivable FDR, and a ground station for fleet data storage.

Status
FDR/HUMS in the forms of North Sea HUMS, EuroHUMS™ and AHUMS™ are in service in North America, Europe, Australia and Southeast Asia. The HUMS element of the system is being supplied by Smiths Industries Aerospace. FDR/HUMS installations have been produced for Bell 412/212 and Sikorsky S-76 helicopters. Applications for these systems include North Sea HUMS: S61N, AS332 Mk 1, BV234 and S76; EuroHUMS: AS332 Mk 1, AS332 Mk 2, AS532 Mk 1 and AS532 Mk 2; AHUMS: Bell 412, CH47D and S-76.

Contractor
Teledyne Controls.
Smiths Industries Aerospace.

MCR500 solid-state combined cockpit voice/flight data recorder

Type
Flight/mission recording system.

Description
The MCR500 is a ruggedised version of the BAE Systems SCR500 solid-state combined cockpit

Helicopter FDR/HUMS equipment   0015310

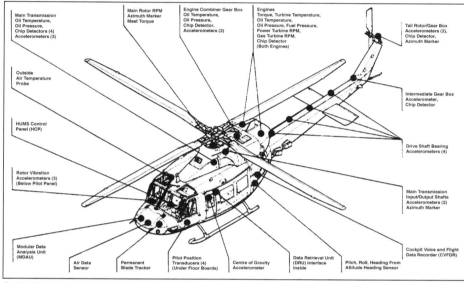
Bell 412 helicopter installation FDR/HUMS   0015311

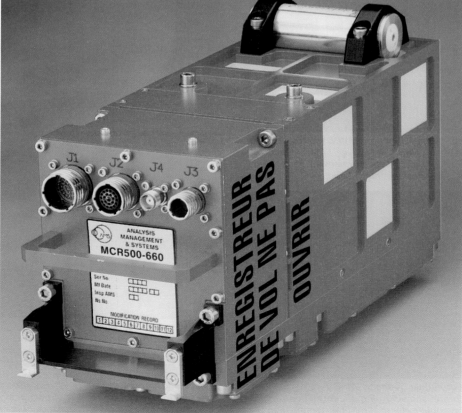
The MCR500 solid-state combined cockpit voice/flight data recorder   0062745

voice/flight data recorder, specifically targeted at fighter aircraft application. The system combines the latest digital memory and material technology with special packaging to produce what is claimed by the Company to be the most rugged CVR/FDR available, able to withstand the extreme temperature and vibration environment in the tail section of a modern fighter aircraft.

In addition, the unit provides expansion capability to integrate a custom Flight Data Acquisition (FDA) interface, thus achieving a single-box solution. The MCR500 is fully compliant with EUROCAE ED-55 and ED-56A and requires no routine maintenance.

The MCR500 is expandable to incorporate customised sensor interfaces for flight data acquisition. The system also includes a PC-based downloader and ground support equipment.

The MCR500 was jointly developed by AMS and BAE Systems.

Specifications

Recording:
(audio) 4 digital audio channels of 60 min duration
(digital data) 64/128 words/s, 10 h duration
Dimensions: 124 (W) × 360 (L) × 188 (H) (including ULB) mm
Weight: 8 kg (including ULB)
Power: 28 V DC, 12 W

Status
Selected for the Royal Australian Air Force Hawk Mk127 Lead-In-Fighter (LIF).

Contractor
Aerospace Monitoring and Systems (Pty) Ltd.
BAE Systems.

Multifunction Head-Down Display (MHDD) for Eurofighter Typhoon

Type
Multifunction cockpit display.

Description
Three Multifunction Head-Down Displays (MHDDs) are installed in the cockpit of the Eurofighter Typhoon, with six in the two-seat trainer version. The MHDD provides flight, tactical situation and sensor data, as well as vital onboard systems information, on a 158.75 × 158.75 mm (6.25 × 6.25 in) usable screen area Active Matrix Liquid Crystal Display (AMLCD). This combines high brightness and resolution to give full legibility in full sunlight of raster scan images overlaid with fine stroke graphics. Normal pilot interface with the aircraft avionics system is via Direct Voice Input (DVI – see separate entry) and Hands-On Throttle And Stick (HOTAS) via a free cursor, slewable across all MHDDs and through the Head-Up Display (HUD); this combined system is designated as V-TAS. Secondary interface is provided by 17 programmable push-button key displays on each MHDD bezel. Display formats are either manually selected by DVI or automatically by the aircraft system, according to flight phase and/or sensor selection.

Smiths Aerospace is the project leader and design authority for the programme.

MHDDs in a UK Royal Air Force Eurofighter Typhoon prior to engine start. Note the extensive onboard systems information on the left MHDD (E L Downs)   0589479

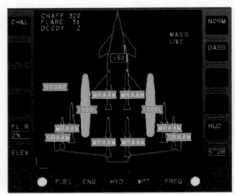

Eurofighter Typhoon MHDD stores inventory format. Note defensive ECM expendables (top left) and Master Armament Safety Switch (MASS) position information   0121559

Eurofighter Typhoon MHDD radio navigation format, showing full-rose information in a similar layout to a classical HIS/RMI combination   0121560

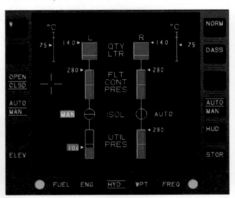

Eurofighter Typhoon MHDD hydraulic system status page, showing reservoir quantity, and system pressure for the split flight controls and utilities circuits   0121561

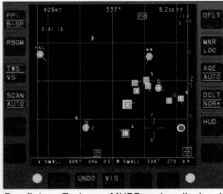

Eurofighter Typhoon MHDD radar display in elevation (ELEV) format. Note the aircraft at bottom left of the screen, with the radar swathe shown as two continuous lines. The free cursor (controlled via HOTAS), slewable across all displays and through the HUD, is also shown (upper left). Range scale (nm) is shown along the bottom and height scale (thousands of feet) is shown up the left side   0121563

Eurofighter Typhoon MHDD engines status page, showing nozzle area (AJ), high- and low-stage rotation speeds (NH and NL), together with intake (INTK and turbine outlet temperatures (TBT)). Also shown are failure conditions (orange and red captions in the centre of the display)   0121562

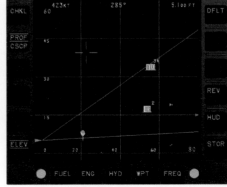

Eurofighter Typhoon MHDD radar display in Track-While-Scan (TWS) mode. Note the free cursor (lower right), used for target designation, and tagged track information with expanded detail shown at the bottom of the display   0121564

Specifications

AMLCD display area: 158.75 × 158.75 mm
(6.25 × 6.25 in)
RGB pixels: 1,024 × 1,024
Viewing angles:
(horizontal) ±35°
(vertical) −5 to +35°
Contrast: >30:1 throughout the viewing
envelope
AMLCD dimming range: >30,000:1
NVIS compatibility: NVIS Class B
Power: 165 W
Weight: <9 kg

Status

In production and in service in the Eurofighter
Typhoon.

Contractor

Diehl Avionik Systeme GmbH.
Galileo Avionica SpA, Avionic Systems and
Equipment Division.
Smiths Aerospace.

Multiplexed Airborne Video Recording System (AVRS)

Type

Flight/mission recording system.

Description

DRS Hadland Ltd, a subsidiary of DRS Technologies
Inc of the US, has teamed with Bavaria Keytronic
Technologie GmbH (BKT) of Germany to produce
the AVRS for mission performance evaluation.

The AVRS provides up to eight channels of
time multiplexed display recording for military
aircraft systems, including Head-Up-Display
(HUD), primary radar, Radar Homing and Warning
Receiver (RHWR), TV tab displays and audio.

Radar and RHWR scan converters are built-in.
It features a colour Charge-Coupled Device (CCD)
HUD camera with 400 line resolution. Full manual
and automatic control of multiplexing sequence
and timing is provided.

The AVRS comprises four units, annotated 1, 2,
3 and 4 on the photograph:

1. The electronics unit houses the primary radar
   and RHWR equipment scan converters, the
   eight-channel multiplexer for signals at up to
   9 MHz each, power supplies, weapon event
   mark generators, BIT circuitry and output to
   the recorder.
2. The videotape recorder is a Hi-8 mm airborne
   videotape recorder that records up to eight
   video inputs, two audio channels and event
   marks for up to four hours.
3. The HUD camera is a 400 TV line colour CCD
   unit, which records the pilot's view and HUD
   symbology via the combining prism or off
   the combining glass; electronic HUD symbols
   input is also available.
4. The control panel provides mode selection,
   manual or automatic control of multiplexing
   sequences, complete AVRS status, BIT and
   elapsed time functions.

The Multiplexed Airborne Video Analysis
System (MAVAS) is used for analysis and replay
of AVRS data. MAVAS is produced by DRS
Hadland Ltd, using software analysis techniques
developed by the then Defence Evaluation
Research Agency (DERA, now QinetiQ) at Malvern
in the UK.

Status

The AVRS was selected to upgrade all of the
German and Italian Tornado aircraft, and a variant
developed by BKT and DRS Hadland Ltd were
installed as a fit and form upgrade on 31 of the
UK Tornado GR1 aircraft. It is also fitted on Jaguar
GR3/T4 upgraded aircraft of the Sultan of Oman's
Air Force.

The MAVAS replay system has been designed
to support a wider range of video recorder
systems than purely AVRS, and it is understood
that it can support all current UK Royal Air Force
aircraft video recorder systems, except the
Tornado F3.

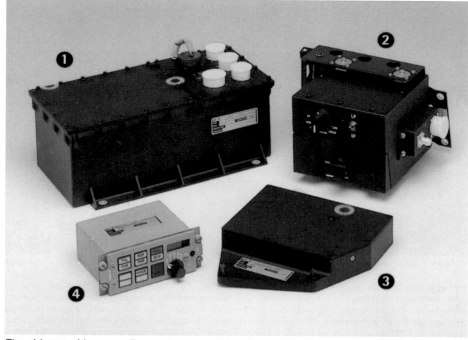

The airborne video recording system comprising: **1** the electronics unit; **2** the video tape recorder; **3** the HUD camera; **4** The control panel
0064378

Contractor

DRS Hadland Ltd.
Bavaria Keytronic Technologie GmbH.
QinetiQ.

New Flight Management (FM) for Airbus

Type

Flight Management System (FMS).

Description

In association with Smiths Aerospace, Thales
Avionics has developed a Second-Generation
Flight Management System (FMS), called New
FM, for the Airbus A318/319/320/321 and A330/340
families of aircraft.

New FM is an advanced flight management
system, compliant with Airbus specification
and meeting with airlines' forward fit as well as
retrofit needs.

Thales supplies a new MultiControl and
Display Unit (MCDU), incorporating a colour
Active Matrix Liquid Crystal Display (AMLCD),
which offers a 15 per cent increase in viewing
area and reduces parallax errors by 50 per cent.
Reliability has been boosted beyond 10,000
hours over current CRT technology, and weight
and power consumption have been decreased by
50 per cent.

The FM Module of the system consists of two
powerful processors (Motorola 68040 with 60
Mhz, Clock), one for the Interface and Display
Processing (IDP) and one for the core Flight
Management Processing (FMP). This design
has demonstrated superior system response
to key-press operation with quick return of
complete flight plan predictions. The use of
independent FMP and IDP processors together
with a unique system architecture minimises
system resets. A 5 Mb Navigation data base
capacity is provided.

New FM features include:

- Multiple-revision temporary flight plan with
  full prediction and UNDO function
- Improved DIR TO interface
- Display of all flight plan transitions on
  Navigation Display (ND)

In more detail, New FM enables single or
multiple lateral and vertical revisions; each time
a revision is introduced, the crew is provided
with a graphical display of the temporary track
on both NDs, as well as full associated flight time
predictions.

A Once UNDO function allows the crew to
delete the last entered revision. The crew is

able to review and decide the operational and
economical impact of a possible revision while
the aircraft remains guided along the original
active flight plan.

New FM includes full FANS A functionality as
a baseline. With ample spare throughput and
memory provided to cope with the demands
of Communication/Navigation/Surveillance/Air
Traffic Management (CNS/ATM) duties, New
FM architecture provides a natural division of
hardware and software resources, reducing
retest when changes evolve to CNS/ATM
requirements.

These features will enable all Airbus operators
to meet the demands of an evolving FANS or
non-FANS environment.

Status

In production and in service. During October
2002, a Frontier Airlines Airbus 319 made its
maiden flight with the New FM installed. The
entire Frontier Airlines fleet is undergoing retrofit
with the new equipment. The new FM entered
service in Alitalia's five Airbus 319s later that
year, with five of that company's A321s to follow.
Volume entry into service began during 2003.

Contractor

Thales Avionics SA.
Smiths Aerospace.

Timearc Visualizer multifunction digital display system

Type

MultiFunction Display (MFD).

Description

The Timearc Visualizer multifunction digital
display system is a versatile display system for
enhanced situational awareness and mission
management; it displays moving map, video
inputs, electronic library and waypoint-find
options.

A large, flat panel, full-colour, high-resolution
active matrix (TFT) LCD driven by the Digital Map
Display Generator (LRU) is used to combine and use
mapping data from multiple raster data sources.

A special Smart Point Track Stick, integrated
into the flat panel LCD, allows the user to scroll
over the entire activated map and to pinpoint
any desired position. A special window shows
the digital readout of such a position in latitude/
longitude, or any other selected grid system. A
pinpointed position can also be used for
an instant transfer into the GPS Navigation

Timearc Visualizer   0018196

Management System or for manually-initiated direct navigation steering.

The Timearc Visualizer does not include a GPS engine, but most stand-alone GPS navigation systems or GPS sensors can be interfaced with the Visualizer, provided they possess a free serial interface either RS-232/422/485 or ARINC 429.

Special purpose software available includes an integrated Electronic Library System (ELS) and 'Waypoint' or 'Street Find' options. The Visualizer version 3.0 displays moving maps and video in real time. The active matrix LCD is available in two models: the standard LCD, and a high-performance LCD for optimal sunlight readability.

Specifications
Colour Display (AMLCD)
Dimensions: 280 × 199 × 52 mm
Resolution: VGA (640 × 480); SVGA (800 × 600) on request
Colours: 256
Weight: approx 1.9 kg
Electrical:
interfaces: keyboard
Display:
screen size: 214 mm
Operating temperature:
(Standard) 0 to +55°C
(Optional) −25 to +70°C
Operating altitude: cockpit

Data Generator (DMD/LGM)
Dimensions: 96 × 163 × 318 mm
Weight: approx 3.0 kg
Electrical:
input voltage: 10–40 V DC
power consumption: 25 W (max)
interfaces: RS-232/422, ARINC 429 (optional)
Operating teperature:
(Standard) −15 to +55°C
(Optional) −25 to +70°C
Operating altitude: up to 35,000 ft

Status
Widely used by police forces, search-and-rescue organisations, forestry/oil/gas industries in helicopter and fixed-wing aircraft installations.

Contractor
Flight Components AG.
Dallas Avionics.

VH 100/130 Head-Up Display (HUD)

Type
Head-Up Display (HUD).

Description
Thales Avionics and Hamilton Sundstrand have produced a family of head-up displays as a part of advanced weapon control systems in helicopters, in particular for the air-to-ground firing of Stinger or Matra Mistral missiles.

The VH 100/130 is designed to provide comprehensive navigation and weapon aiming information to the helicopter crew, while being small enough to obscure the crew's vision to the minimum. In weapon aiming it can offer air-to-air rockets and guns and air-to-ground guns and missile symbology and for navigation the display is compatible with NVG. The system comprises a pilot's display unit, an electronics unit and a control panel. The total field of view is 20°.

The system consists of a Pilot Display Unit (PDU), Electronic Unit (EU) and Optional Control Unit (OCU).

The PDU utilises a miniature CRT, which can generate symbology visible in a high-brightness environment Optical lenses and a combiner are used to to present superimposed symbology over the external scene.

The EU provides power supply and signal processing for the CRT, generation of synthetic symbology, electrical interfacing with weapons system sensors, airframe and fire-control computations and aircraft sensor signal processing.

The CU controls weapons mode selection, symbol luminance adjustment, firing distance selection and weapon elevation offset.

The VH 100/130 has been flight-test fitted to a Gazelle in France, OH-58 and Bell 406 in the USA, and a BO 105 in Germany.

Specifications
Weight:
(pilot display unit) 2.99 kg
(electronics unit) 1.99 kg

Power supply: 28 V DC, 2.5 A
Field of view: 20°
Reliability: >3,600 h MTBF

Status
In service in US Army OH-58 helicopters. Twin VH 100/130 HUDs have been installed in the C-160 Transall upgrade. The VH100/130 has been selected for the Eurocopter Tiger.

Contractor
Hamilton Sundstrand Corporation.
Thales Avionics SA.

Visual Guidance System (VGS)

Type
Head-Up Display (HUD).

Description
BAE Systems has utilised its C-17 and combat aircraft head-up display technology to develop the Visual Guidance System (VGS), previously termed the HUD 2020 (see separate entry) and HUD 2022. The HUD 2020 system is for corporate jet operators and the HUD 2022 for air transport aircraft. In these developments BAE Systems has teamed with Honeywell Commercial Aviation Systems.

The VGS displays essential, stroke and raster, flight information to the pilot in his forward field of view, thus providing:
1. Improved situational awareness
2. Navigation data and precise flight path guidance
3. Warning alerts, including TCAS, TAWS, weather radar and windshear warnings
4. Improved assurance during ground operations on the runway and taxiways in adverse conditions.

The VGS comprises 4 electronic units and a mounting tray:
1. Over-Head Unit (OHU). The OHU contains the CRT, drive circuitry and optical system to project the image on to the combiner assembly. It has a 30 × 25° field of view
2. Combiner Assembly (CA): The CA consists of a lightwieght glass combiner, an ambient light sensor and an integral control panel. The combiner overlays the critical flight symbology projected by the OHU onto the real world image
3. Display Guidance Computer (DGC). The DGC receives data from the various aircraft systems and uses this data to generate symbology and command guidance, which is then supplied to the OHU

VGS cockpit installation   0080277

4. HUD Annunciator Panel (HAP). The HAP is located at the first officer's station and enables the 'pilot not flying' to monitor the status of the VGS

5. The mounting tray carries the OHU and CA in a customised tray above the pilot's position.

Status

The VGS was granted US FAA full Supplemental Type Certification (STC) for use on the Boeing 737-800 in October 1999. UK Civil Aviation Authority (CAA) approval for use of the VGS on all UK registered Boeing 737-600/-700/-800 aircraft was granted in January 2000. Approvals cover VGS use in Cat IIIa landings with the auto-throttle on or off. VGS is also approved for single engine Cat III approach. The VGS can be used to fly the aircraft manually to a decision height of 50 ft when Runway Visual Range (RVR) is as low as 600 ft and for take-off guidance at RVR of 300 ft. The US FAA STC work was accomplished with American Airlines, which has selected the system for its fleet of 737-800 aircraft.

Development work on an Enhanced Visual Guidance System (EVGS) is underway, and BAE Systems has successfully conducted proof of concept trials to interface VGS with both millimetric wave radar and infra-red sensors.

Contractor

BAE Systems.
Honeywell Inc, Commercial Aviation Systems.

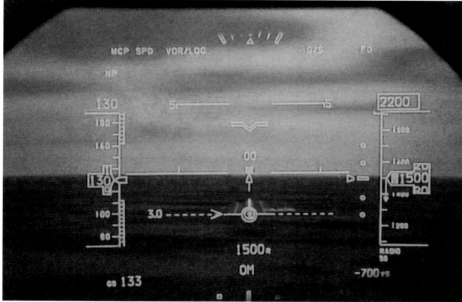

VGS display – aircraft shown on a nominal 3° glideslope aligned with runway with wings level

0081733

Photograph of the VGS display during approach – the aircraft is making a level turn towards the runway centreline, while the pilot waits for the runway threshold to cut the –3° dotted bar before overlaying the aircraft velocity vector (aircraft symbol) to commence descent (BAE Systems)

0114859

Israel

Data Acquisition System (DAS)

Type

Flight/mission recording system.

Description

The Data Acquisition System (DAS) presents new and improved capabilities in data gathering and data acquisition. The two systems that comprise DAS are the VADR® (Voice and Data Recorder), a mishap acquisition and recording unit, and the DAU (Data Acquisition Unit), an I/O device in charge of data routeing.

DAS is being developed jointly by SI Grand Rapids and RADA. The VADR® is a COTS product by SI, while both companies are developing the DAU.

Recent advances in storage density and packaging of digital memory devices have made it possible for SI-IMS to manufacture the VADR® that provides extensive multifunction acquisition and processing of flight and mishap data.

The main features of the VADR® include:

- **Proven reliability:** The system is based on the SFDR and CSFDR systems, proven in actual aircraft mishaps and incidents
- **Multiple recording capabilities:** The VADR® provides recording of up to four channels of aircraft audio information, as well as separate digital flight data, to support post-flight maintenance and incident analysis
- **High performance:** The VADR® provides similar capacities while achieving better performance, higher reliability, lower weight and smaller size compared with current systems
- **Flexible design:** Several chassis configurations have been developed to meet civil and military market application requirements.

The DAU provides extended data acquisition and processing capabilities. A configurable set-up file, without changing the OFP of the aircraft, efficiently defines the acquisition parameters and processing algorithms. DAU interfaces include MIL-STD-1553B, serial communication channels (RS-232, RS-422), analogue channels (AC, DC, strain gauges), discrete, synchro and resistive sensor interfaces.

Features of the DAU include:

- **Non-volatile memory storage:** The acquired data is recorded within the DAU on 16 Mb of Flash Memory
- **Downloadable data:** The data on the Flash Memory can be downloaded through a serial connection to a laptop. This data can be used for different tasks such as strain growth monitoring, DEEC (engine) maintenance and flight debriefing
- **Data acquisition:** The DAU can implement two data acquisition approaches: non-configurable or configurable at the user level set-up
- **Analogue-to-digital signal conversion:** The analogue signals acquired are converted into a digital form
- **Data reduction algorithms:** These filter redundant flight information, and process the data while recording it
- **History-event buffer:** The DAU is capable of Master Data Recording triggered events such as SMS, MFL, release and WOW changes.

Additional events can be defined according to the specific user requirements.

Status
In production and in service.

Contractor
RADA Electronic Industries Ltd.
SI, Grand Rapids.

Digital Video Debriefing System (DVDS)

Type
Flight/mission recording system.

Description
Elbit Systems' DVDS is an advanced PC-based application for producing highly effective, comprehensive real time and post-flight analysis and debriefing of video and audio.

Using in-flight aircraft video (multiple source) and audio (both internal and external), either transmitted or recorded during the flight, the DVDS enables the reproduction and display of multiple synchronised video sources (HUD, MFCD, HMD) and audio. The DVDS displays significant flight parameters and events (for example, weapon release or envelope violation) to facilitate focused, event-driven debrief.

Status
In production.

Contractor
Elbit Systems Ltd.

Digital Video Recorder (DVR)

Type
Flight/mission recording system.

Description
Elbit Systems' DVR is described as a state-of-the-art airborne synchronised video, audio and data recorder which employs a solid-state memory cartridge. The compact, lightweight unit enables simultaneous recording of four video, two audio and two data channels on a single cartridge.

The DVR is designed for high-capacity recording of multiple video, audio and data sources on fixed- and rotary-wing aircraft and UAVs. Based on Commercial Off-the-Shelf (COTS) components, the DVR complies with stringent airborne data recording and mission debriefing requirements. It can operate as a stand-alone system or integrate with existing aircraft avionic systems as an integral part of a cockpit upgrade programme.

Status
In production.

Contractor
Elbit Systems Ltd.

Elbit Systems' DVR (Elbit Systems)   0569012

D-Map Digital Map System (DMS)

Type
Avionic digital map system.

Description
D-Map is an advanced Digital Map System (DMS) which overlays all navigational and mission data, military or civilian, to present the crew with a clear and concise tactical aid. D-Map embraces the paperless cockpit concept and incorporates a retrievable mass memory which stores all mission-critical data in different scales, together with satellite charts, digital terrain, Jeppesen information, scanned images and tactical databases.

The system features open architecture, including standard VME64 interface, PowerPC processor, VxWorks/integrity real-time operating system, commercial interfaces and form factors, Digital Signal Processing (DSP) and multiprocessor design.

The DMS presents all required tactical and navigational information, thereby greatly enhancing Situational Awareness (SA); waypoints, high ground, obstacles, power lines, threats and other details are indicated in the flight plan. Warnings are issued when power lines or no-fly zones conflict with the aircraft flight path. Overlay control allows the crew to select desired information according to mission needs. The DMS may be combined with a Ground Collision Avoidance and Warning System (GCAWS).

Other benefits and features of the D-Map system include:
• Low power consumption
• Four available slots for 6U VME cards
• 50 msec hold up time.

Specifications
Dimensions: 381 × 165 × 200 mm (L × W × D)
Weight: 10 kg (including RMU)
Power supply: 115 V AC, 3-phase, 400 Hz
Power consumption: <70 W
MTBF: 4,000 hours
Interface: MIL-STD-1553B, ARINC 429, VGA, CCIR, RS-170, STANAG 3350 C and B, RS-422, Discrete, Fast Ethernet, SCSI-2, RS-232

Status
In production and in service. D-Map was selected by REGA, a Swiss air ambulance service, for its A 109 EMS helicopter and by the Israeli Air Force for the CH-53 2000 helicopter modernisation programme.

D-Map has also been selected for the Lockheed F-16, Eurocopter Cougar and the V-22 Osprey.

In July 1999, D-Map was selected by the Swiss Air Force for its Super Puma Mk 1 helicopter fleet of 12 aircraft.

Contractor
Elbit Systems Ltd.

FPD-1 flat panel display monitor

Type
Cockpit display.

Description
The FPD-1 is a ruggedised, high-brightness flat panel colour display monitor developed for airborne applications in fixed- and rotary-wing platforms. The unit features a 640 × 480 pixel, 1,000 cd/m², 10.4 in Liquid Crystal Display (LCD) compliant with the requirements of MIL-STD-810.

Specifications
Type: AMLCD (TFT)
Display: 10.4 in (264 mm) diagonal (other sizes available)

Controp Precision Technologies' FPD-1   1031812

Resolution: 640 × 480 pixels (800 × 600 option)
Colours: 256,000
Brightness: 1,000 cd/m²
Response: 20 m/sec
Contrast: 1:150
Viewing angle: ±70° (H), +70 to −40° (V)

Physical
EMI/EMI: MIL-STD 461C Cat CE03, RE02
Altitude:
(operating) 15,000 ft
(storage) 35,000 ft
Shock: 20 g, 11 m/sec saw tooth
Power: 18 to 32 V DC, <45 W (including backlight)
Temperature:
(operating) +55°C
(storage) −25 to +70°C
Dimensions: 300 × 230 × 100 (W × D × H) mm
Weight: 5.3 kg (max)

Interface
Video input: Composite PAL/NTSC
Video output: Option for one buffered output
Options: Serial communication links, video input Y/C, VGA input

Status
In production.

Contractor
Controp Precision Technologies Ltd.

FPD-2 flat panel display monitor

Type
Cockpit display.

Description
The FPD-2 is a ruggedised, high-brightness flat panel colour display monitor developed for airborne applications in fixed- and rotary-wing platforms. The unit features a 800 × 600 pixel, 1,400 cd/m², 8.4 in Liquid Crystal Display (LCD) compliant with the requirements of MIL-STD-810.

Specifications
Type: AMLCD (TFT)
Display: 8.4 in (214 mm) diagonal (other sizes available)
Resolution: 800 × 600 pixels
Colours: 262,000
Brightness: 1,400 cd/m²
Response: 50 m/sec
Contrast: 1:250
Viewing angle: ±60° (H), +60 to −45° (V)

Physical
EMI/EMI: MIL-STD 461C Cat CE03, RE02
Altitude:
(operating) 12,000 ft
Shock: 20 g, 11 m/sec saw tooth
Power: 18 to 32 V DC, <40 W (including backlight)
Temperature:
(operating) −5 to +45°C
(storage) −20 to +60°C
Dimensions: 220 × 165 × 148 (W × H × D) mm
Weight: 3.7 kg (max)

Interface
Video input: Composite PAL/NTSC
Video output: Option for one buffered output
Computer input: Super VGA (option)

Controp Precision Technologies' FPD-2   1031813

Options: Serial communication links, video input Y/C, VGA input, high-altitude operation (20,000 ft), customised mounting

Status
In production.

Contractor
Controp Precision Technologies Ltd.

Multi Function Displays (MFD)

Type
MultiFunction Display (MFD).

Description
4 × 4 in AMLCD Colour Multi Function Display (CMFD)
The 4 × 4 in CMFD was specifically designed for the F-16 aircraft, as a direct replacement for existing F-16 displays. The CMFD functions as a Primary Flight Display (PFD) and incorporates additional functions over F-16 legacy displays that can be useful in other applications. The display features high brightness and contrast, excellent resolution and readability, with a 8:1 contrast ratio in full sunlight with a brightness of 250 fl.

Including full NVIS compatibility, the CMFD interfaces with the aircraft main computer via RS-170 video and RS-422 serial communication interface.

Specifications
Resolution: 480 × 480 colour groups over display area
Contrast ratio: 8:1 at 10,000 fc diffuse and 2000 fl specular ambient illumination
Display brightness: 250 fl (white)
Dimming range: 10,000:1
Grey shades per colour: 256
Viewing angles: ±45° horizontal, +30°/–10° vertical
Power consumption: 50 W max (without heaters)
MTBF: >4,000 h in AIF environment
Video interface
Analogue: RGB, RS-170, CCIR, RS-343 (875 line), VGA
Digital: LVDS (two channels available)
Serial data link: two RS-422
NVIS-compatibility: MIL-STD-85762A Class B and Class C
Aspect ratio: 1:1 or 3:4 (control via serial link)
Controls: 4 rocker controls; 20 on-screen selections
Power supply: 28 V DC or 115 V 400 Hz 3-phase per MIL-STD-704
Panel lighting: 0 to 5 V AC, 400 Hz
Cooling: natural convection and radiation (no fans)
Environmental: MIL-STD-810
EMI/RFI: MIL-STD-461B

Status
In production for Israeli Air Force F-16 aircraft.

4 × 5 in Multi Function Colour Display
The Multi Function Colour Display (MFCD) is designed to serve as an Electronic Flight Indicator (EFI) and processing and displaying all indications and cautions regarding engine performance, while also providing for navigation data backup to the aircraft mission computer.

Based on AMLCD technology, the MFCD features excellent sunlight readability and brightness, with a contrast ratio greater than 7:1 (direct sunlight with 200 fl brightness). The display is fully

NVIS-compatible, with a wide dimming range providing for comfortable viewing by day and night.

The display-computer interface is via an LVDS video channel and RS-422 serial communication line. The display keyboard provides keypads and switches, brightness control and ON/OFF.

Specifications
Resolution: 480 × 640 colour pixels
Contrast ratio: 7:1 at 10,000 fc diffuse and 2000 fl specular ambient illumination
Display brightness: 200 fl (white)
Grey shades per colour: 64 (6 bit)
Processing: three PowerPC processors
Interface: 18 analogue inputs, 61 input discretes, 15 output discretes, 3 data bus channels, 3 ARINC-429 channels, Y/C video output, 7 RS-232 channels, 3 Ethernet channels
NVIS-compatibility: MIL-STD-85762A Class B and Class C
Power supply: 28 V DC
Temperature: –40 to +65°C
Cooling: natural convection and radiation (no fans)
Environmental: MIL-STD-810
EMI/RFI: MIL-STD-461B, -461C
Dimensions: 146 × 210 × 179 mm
Weight: 5.3 kg

5 × 7 in high-performance AMLCD Multifunction Colour Display Unit
Designed for the OH-58 helicopter, the advanced 5 × 7 in (8.4 in diagonal) AMLCD Multifunction Colour Display Unit (MCDU) is employed as the Primary Flight Display for the aircraft. The display is in Landscape configuration, delivering high-resolution SVGA full colour grey scale images, providing a 7:1 contrast ratio in full sunlight and 220 fl brightness. The display is also fully NVIS-compatible for night operations. The display video interfaces are via analogue lines or LVDS digital lines.

The MCDU is capable of interfacing automatically to multivideo standards. Display interface is via RS-422 or RS-232 serial communication for control and status report.

Specifications
Resolution: SVGA 600 × 800 stripe colour groups
Contrast ratio: 7:1 at 10,000 fc diffuse and 2000 fl specular ambient illumination
Display brightness: 240 fl (white)
Grey shades per colour: 64 (6 bit)
Video interface: 2 sets of analogue RGB or one RGB and 3 monochrome
Video standards: RS-170, CCIR, RS-343
Communication interface: serial link per RS-422 or RS-232
Panel lighting: 0 to 5 V DC or AC, 400 Hz
NVIS-compatibility: MIL-STD-85762A Class B and Class C
Power supply: 28 V DC per MIL-STD-704A
Temperature: MIL-STD-5400R Class 1
EMI/RFI: MIL-STD-461
Weight: 4.1 kg

Status
In service in OH-58 helicopters of the Israeli Air Force.

6 × 6 in high-resolution AMLCD Multi Function Colour Display
The 6 × 6 in (display area 6.13 × 6.13 in) AMLCD is designed to function as a Primary Flight Display, offering the high brightness and contrast combined with excellent resolution and day/night readability. The unit features an 8:1

contrast ratio in full sunlight with a brightness of 225 fl.

Fully NVIS-compatible, the display interfaces with the aircraft main computer via RS-170 video and RS-422 serial communication interfaces.

Specifications
Resolution: 512 × 512 colour groups over display area; 512 × 1,024 green pixels for extra video resolution
Contrast ratio: 8:1 at 10,000 fc diffuse and 2000 fl specular ambient illumination
Display brightness: 225 fl (white)
Grey shades per colour: 128
Viewing angle: ±65° horizontal, +15 to –5° vertical
Dimming range: 20,000:1
Video interface: RGB, RS-170, CCIR (option), LVDS
Communication interface: serial link per RS-422
Panel lighting: 50 to 115 V AC, 400 Hz
NVIS-compatibility: MIL-STD-85762A Class B
Power supply: 115 V AC, 400 Hz
Cooling: Internal fan
Temperature: –40 to +55°C (+71°C intermittent)
EMI/RFI: MIL-STD-461B

Status
Elbit/EFW 6 × 6 in flat-panel AMLCDs were specified for US Marine Corps MV-22Bs from FY99, followed by USN HV-22s. AMLCDs are also specified for production USAF CV-22Bs.

6 × 8 in high-performance helicopter AMLCD Multi Function Colour Display
The 6 × 8 in (10.4 in diagonal) AMLCD is designed as a Primary Flight Display offering the high brightness, contrast resolution and readability demanded in the extremely open helicopter cockpit environment.

Other features include a 7:1 contrast ratio in full sunlight with brightness greater than 230 fl and an extremely wide dimming range to ensure consistent and even illumination of the display down to 0.1 fl. The display is also fully compatible with Class A/B NVGs. Interface with the aircraft computer is via analogue video and RS-422 serial communication channel. The display is available in both Landscape and Portrait configurations with optional keyboard layouts.

Specifications
Resolution: SVGA 800 × 600 colour groups over display area
Contrast ratio: 7:1 at 10,000 fc diffuse and 2000 fl specular ambient illumination
Display brightness: 230 fl (white)
Grey shades per colour: 64 (6 bit)
Video interface: 2 sets of analogue RGB or one RGB and 3 monochrome
Video standards: EIA-RS-170, EIA-RS-343 (875 line), CCIR, LVDS (two-channel, option)
Communication interface: serial link per RS-422
Panel lighting: 0 to 5 V DC or AC, 400 Hz
NVIS-compatibility: MIL-STD-85762A Class A/B
Power supply: 28 V DC per MIL-STD-704A
Temperature: –40 to +55°C
Cooling: natural convection or radiation (no fans)
EMI/RFI: MIL-STD-461C
Weight: 4 kg

Status
In production and in service.

Contractor
Elbit Systems Ltd.

Italy

ANV-201 Microwave Landing System (MLS) airborne receiver

Type
Aircraft landing system, Microwave Landing System (MLS).

Description
The ANV-201 MLS airborne receiver is a precision angular position sensor that is capable of operating with any MLS ground equipment facility which transmits the standard ICAO signal format.

The system receives the C-band signal radiated by the ground stations and processes the angle and data functions to generate the offset-corrected 2-D angular position information, which is then sent to peripheral avionics (AFCS, EFIS) in both analogue and digital formats.

The ANV-201 may be operated in two modes: an automatic mode, where an ILS-like straight-in flightpath is followed, and a manual mode, where the pilot has freedom in selecting different glideslope angles and offset azimuth radials for obstacle clearance or tactical reasons.

The system comprises two Line-Replaceable Units (LRUs); the MLS receiver unit and the Control Panel (CP). The CP allows the selection of any magnetic course angle, in 1° increments, and any glideslope angle from 0.0 to 25.5° in 0.1° increments. It also allows selection of any of the 200 available C-band channels.

The ANV-201 has been designed around a powerful microprocessor and modular architecture to provide for maximum growth capability. Special attention has been devoted to building a comprehensive self-test capability into the interrogator. By injecting an internally-synthesised MLS signal ahead of the critical C-band components, and analysing the corresponding processed data, a complete beginning-to-end self-test is achieved, thus providing a high degree of confidence regarding the receiver's operational status.

Specifications
Frequency range
Transmit: 1,025 – 1,150 MHz
Receive: 962 – 1,215 MHz
Coverage limits
Yaw: 360° (dual-antenna installation)
 Pitch: Dependent on antenna mounting
 Range: 20 nm/min
Dimensions: (3/8 ATR Short), 319 × 90.5 × 194 mm
Weight: 6.5 kg
Power supply: 28 V DC, 50 W
Temperature range: –55 to +70° C (operating)
Frequency range: 5031.0 to 5090.7 MHz
Channels: 200 (channel 500 to 699)
Channel spacing: 300 kHz
Signal frequency error tolerance: ±12 kHz (maximum)
Sensitivity: –102 dBm
Compliance: MIL-STD-461, -462, -810D

Status
Selected for the Panavia Tornado, Eurofighter Typhoon and C-130J aircraft and the NH-90 helicopter.

Contractor
Marconi Selenia SpA.

ANV-241 airborne precision landing Multi-Mode Receiver (MMR)

Type
Aircraft landing system, multimode.

Description
The ANV-241 is an integrated Multi-Mode Landing System (MMLS), providing a worldwide precision approach and landing capability. The single-box system operates in five principal modes:

- Protected Instrument Landing System (ILS); ICAO FM-compliant
- Microwave Landing System (MLS)
- Differential GPS (DGPS) – LAAS/WAAS, 12-channel GRAM, C/A and P/Y modes
- Embedded VHF Data Link Receiver (DLR) modem for LAAS
- VOR (for retrofit upgrades)

The ANV-241 can accept DME/N/P, Tacan or GPS ranging sources and can be remotely controlled to transmit/receive data through standard MIL-STD-1553B or ARINC 429 interfaces (analogue interfaces are also provided).

The system is ICAO compatible for civil and military interoperability.

Specifications
Performance
 MLS performance: RTCA DO-177
 ILS performance: RTCA DO-195, -192, -143 (compliant with new FM interference requirements)
Dimensions: 127 × 95 × 258 mm (MIL-E-5400, Class 2)
Weight: 4.5 kg (with embedded GPS)
Power: 28 V DC
BIT coverage: Comprehensive (>95%)
MTBF: >10,500 hours
MTTR: <30 min

Status
Selected for Panavia Tornado, Eurofighter Typhoon and C–130J aircraft and the NH-90 helicopter.

Contractor
Marconi Selenia SpA.

Crash survivable memory unit

Type
Flight/mission recording system.

Description
Logic developed a Crash Survivable Memory Unit (CSMU) in partnership with DRS for use in the Aermacchi M346. It features an RS422 interface operating at 112,600 bit/second for flight parameters. It has two audio channels that offer two hours' flight recording and information can be monitored and downloaded by a USB-type interface. The unit meets the requirements of EUROCAE ED-55 and ED-56A.

Status
In production and in service in development Aermacchi M346 aircraft.

Contractor
Logic SpA.

Get Home Display (GHD)

Type
Cockpit display.

Description
The Logic Get Home Display (GHD) is a 3 in ATI display made using Active Matrix Liquid Crystal Display (AMLCD) technology. It has two independent MIL-STD-1553 RT interfaces and two independent MIL-STD-1553 RT databuses or, as an alternative, and internal attitude sensor. It generates graphic symbols, includes an anti-aliasing function and features on-screen menu selection using a rotating encoder and integrated push button. The display is compatible with NVGs.

Status
In production and in service in development Aermacchi M346 aircraft.

Contractor
Logic SpA.

Head-Up Display (HUD) for the AMX

Type
Head-Up Display (HUD).

Description
Galileo Avionica produces the Type 35 HUD for the AMX. OMI designed the Pilot's Display

Unit (PDU) and Galileo was responsible for the Symbol Generator Unit (SGU). The symbols generated can be changed by software and a display recorder, using either tape or film, can be fitted to the display.

Specifications
Dimensions:
(PDU) 136 × 350 × 650 mm

(SGU) 1/2 ATR short
Weight:
(PDU) 13.75 kg
(SGU) 8.5 kg
Power supply: 115 V AC, 400 Hz

Status
In production and in service in the AMX.

Contractor
Selex Sensors and Airborne Systems, Radar Systems Business Unit.

Maritime Patrol Mission System (MPMS)

Type
Mission Management System (MMS).

Description
The Selex (formerly Galileo Avionica) Maritime Patrol Mission System (MPMS) is based on a company-developed core system Mission Management System (MMS) integrated with various COTS sensors, (including: 360° search radar, SLAR radar, IR/UV scanners, day/night electro-optical systems, ESM, communications and datalink) and a flexible man-machine interface, with software-based tactical solutions and pre-programmed patterns.

The MMS is based on a MIL-STD-1553B databus for data exchange and MPMS integration. It features software controlled video-routing,

The Galileo Avionica head-up display for the AMX

0503968

digital maps, data fusion, communications and datalink to/from fixed and mobile stations. Its open architecture supports system expansion for increased mission requirements and upgrades.

The MPMS can support a wide range of maritime missions, including: Exclusive Economic Zone (EEZ) operations, search and rescue, vessel identification, coastal surveillance, environmental pollution detection, and Anti-Submarine Warfare (ASW) (with additional sensors).

The modular design enables Alenia Difesa to offer a multi console design on larger regional air transport types, and a down-sized configuration on small fixed-wing and rotary-wing aircraft.

Status
Two customers have selected the MPMS for installation on ATR 42 aircraft.

Contractor
Selex Sensors and Airborne Systems, Radar Systems Business Unit.

MISCO Vehicle Computer for Aermacchi M346

Type
Aircraft mission computer.

Description
The Logic MISCO vehicle computer is a fully redundant system in two sections. Each hardware platform uses Power PC architecture operating at up to 1,000 MIPS. It features 24 Mbytes of FLASH memory and 64 Mytes of RAM, two independent MIL-STD-1553 bus controllers and an I/O interface for up to 21 aircraft systems.

The main functions are: back-up mission computer; avionics and general systems interface for caution and warning presentation; strain counter; acoustic alarm generation; maintenance data recording and processing; crash survivable memory management; and engine Health and Usage Monitoring System (HUMS).

Status
In production and in service in development Aermacchi M346 aircraft.

Contractor
Logic SpA.

Mission Symbol Generator Unit (MSGU)

Type
Avionic display processor.

Description
The Mission Symbol Generator Unit (MSGU) is a special configuration of the MCS computer line

The Maritime Patrol Mission System (MPMS) multi console configuration.　　0041391

dedicated to the generation of tactical symbols and the routing of STANAG 3350B video lines. The MSGU can be adapted for different applications. For the NFH NH-90 helicopter, the system is capable of unix/deunix up to six standard video inputs, to be distributed (in any order) on up to 12 displays. Each output can be mixed with internally generated symbols, offering up to six different Raster channels.

Windowing and direct cursor interface capability is offered, independently, on each of the 12 displays. The addition of up to two Stroke channels is also possible.

Status
In production and in service.

Contractor
Selex Sensors and Airborne Systems, Radar Systems Business Unit.

PDU-433 Head-Up Display (HUD)

Type
Head-Up Display (HUD).

Description
The PDU-433 is a Wide Field of View (WFoV) high-performance Head-Up Display (HUD), with Raster/cursive capabilities under all conditions

from full sunlight to NVG operations. The PDU-433 is a single unit with an integrated Up-Front Control Panel (UFCP) and incorporates a colour video camera for mission recording.

The system has been designed for easy adaptation to several fighter applications, including the upgrade of existing aircraft types. The UFCP provides integrated controls for avionic system interface, including nav/attack system initialisation, mission data loading and management, system mode selection and data display. The main features of the system are:
- High brightness/high accuracy (0.5 mrad boresight);
- Cursive, raster and cursive on raster;
- Integrated colour camera;
- NVIS compatible, Class B.
  The UFCP provides integrated controls for:
- Initialisation and manual mission data loading;
- Data control and management;
- Navigation and weapon aiming functions/ modes selections and data presentation;
- Radio mode selection.

Status
In production and in service.

Contractor
Selex Sensors and Airborne Systems, Radar Systems Business Unit.

Japan

Central warning display

Type
Aircraft centralised warning system.

Description
The central warning display has been developed for commercial aircraft. It is designed to work with

two master warning lights which can effectively alert pilots and identify potential problems or hazards.

The indication panel is a liquid crystal display which can indicate a maximum of 10 warning items in red, amber or green. When more than 10 items occur at the same time, up to 50 additional warnings are listed by a scrolling function.

The LCD has a high back-light unit consisting of halogen lamps, allowing the manually or automatically dimmable display to be read under sunlight conditions. The red and amber warning legends can be tailored to customer requirements by a software change. Dual redundancy is incorporated in the power supply, input interface circuit, digital computer and LCD display.

For details of the latest updates to *Jane's Avionics* online and to discover the additional information available exclusively to online subscribers please visit
jav.janes.com

Koito central warning display for commercial aircraft   0581503

The range of HUDs made by Shimadzu Corporation   0504277

Specifications
Dimensions: 134 × 125 × 250 mm
Weight: 2.5 kg
Display size: 55 × 75 mm
Power supply: 16–30 V DC, 1.3 A at 28 V DC
Temperature range: –40 to +70°C
Altitude: –100 to 25,000 ft
Reliability: >5,000 h MTBF

Status
In service.

Contractor
Koito Manufacturing Company Ltd.

Crash Protected Video Recorder (CPVR)

Type
Flight/mission recording system.

Description
The Crash Protected Video Recorder (CPVR) is designed to provide a record of the glass cockpit data presented to the flight crew. This is a natural follow-on from the standard flight data recorder. Knowledge gained over many years in the design and manufacture of crash recorders has been utilised in the design of a prototype CPVR.

The Hi-8 mm video standard has been chosen as it is lightweight, reliable and has a bandwidth approaching 4 MHz. The CPVR provides recording times of 4 hours for NTSC and 3 hours for PAL. These two video standards, plus RS-170, provide 400 TV lines resolution. The system also has a single audio channel with a frequency response of 100 Hz to 10 kHz.

The CPVR will form part of a system utilising closed-circuit television cameras located in suitable positions both within and outside the aircraft. Camera positions will be chosen to provide the aircrew with real-time viewing of internal and external features on a suitable display. This follows recommendations from the UK Aircraft Accident Investigation Branch. The number of cameras will depend upon specific requirements, with the chosen outputs being recorded on the CPVR to enable in-depth analysis to be carried out.

Specifications
Dimensions: ¾ ATR
Power supply: 28 V DC, 15 W
Environmental: MIL-STD-810D, MIL-STD-461B, MIL-STD-462

Contractor
TEAC Corporation.

FMPD-10 EFIS package

Type
Electronic Flight Instrumentation System (EFIS).

Description
FMPD-10 is an Electronic Flight Instrument System (EFIS) package both for commercial and military aircraft. The standard package consists of four smart type displays, two display interface processors and two display control panels. The displays show EHSI, EAI, EHSI/EAI composite graphics, WXR graphics, waypoint map, and fuel information on 3.7 × 3.7 in AMLCD glasses, and also provides hover mode data for helicopter application.

Status
As of 2004, the FMPD-10 was in production and in service.

Contractor
Tokyo Aircraft Instrument Co Ltd.

Head-Up Displays (HUDs)

Type
Head-Up Display (HUD).

Description
The Head-Up Display (HUD) provides information such as altitude, airspeed, magnetic heading, attitude, angle of attack and aiming reticle to the pilot during the mission. The information is presented as symbology with a collimated image overlaid in the pilot's view. The HUD consists of the display unit and a symbol generator.

Status
In production for, and in service with, Japanese Self-Defense Force F-2, AH-1S, F-4EJ Kai, F-15J/DJ, T-4 and US-1A aircraft. Also in limited production for the CCV-T2 and C-1 QSOL experimental aircraft.

Contractor
Shimadzu Corporation.

LK-35 series standby altimeter

Type
Flight instrumentation.

Description
The standby altimeter is an exceptionally accurate pressure-actuated instrument with a –1,000 to 50,000 ft range and P/N LK-35-1 and P/N LK-35-9 provide two resolved outputs of barometric setting.

A three-drum counter supported by ball bearings provides digital indication in tens of thousands, thousands and hundreds of feet, supplemented by a pointer which indicates altitude in hundreds of feet.

A knob on front of the altimeter provides the means for setting barometric pressure which is shown in inches of mercury and mb on two sets of four-digit counters.

Specifications
Altitude range: –1,000 to +50,000 ft
   (LK-35-8) –2,000 to +50,000 ft
Barometric range: 22–31.02 inHg
   (LK-35-7/LK-35-8) 22–31.99 inHg
   (LK-35-9) 22–30.99 inHg
Accuracy certifications: FAA/TSO-10b
Lighting: white (unfiltered)
Electrical input:
   (vibrator) 28 V DC
   (lighting) 5 V AC or DC
   (resolved) 26 V, 400 Hz
Weight:
   (LK-35-1, -9) 1.45 kg
   (LK-35-2, -4, -5, -7) 1.3 kg
   (LK-35-8) 1.5 kg

Status
As of 2004, in production and in service in B737-400, B757, B767 and DC-9 commercial transport aircraft.

Contractor
Tokyo Aircraft Instrument Co Ltd.

LK-35 series standby altimeter   1041441

FMPD-10 EFIS package   0001358

Netherlands

OTA series cockpit cameras

Type
Flight/Mission recording system.

Description
The OTA series is a range of black and white, or colour, cameras designed for the HUDs of helicopters and fixed-wing aircraft as part of a mission recording system. Used with an airborne magnetic-tape recorder, they superimpose and record HUD symbology on real-time video pictures together with flight data. The OTA cameras enable in-flight permanent recording of all the visual information received by the pilot from take-off to landing. The high-sensitivity cameras feature an automatic exposure control to provide fast adjustment to sudden changes in brightness levels. An in-flight replay capability is provided together with display options for the co-pilot.

Current cameras in the OTA range include the OTA-222 black/white HUD camera; and the OTA-1320 Red/Green/Blue (RGB) colour HUD camera, in service in the Rafale multirole fighter.

Specifications
Dimensions: 75 × 65 × 95 mm
Weight: 0.65 kg

OTA 222 B/W HUD camera   0001296

Field of view: 30° horizontal × 22.5° vertical
Line of sight: <1 mrad
CCD sensor: 756 horizontal × 575 vertical pixels
Depth of focus: 3 m to infinity
Spectral range: 0.4–0.7 µm
Bandwidth: 5.5 MHz
Dynamic range: 3–100,000 lux
Video signal: CCIR/STANAG 3350B EIA/RS-70
Synchronisation: int/ext input (RS-422)

OTA-1320 RGB HUD camera selected for Rafale
0015312

Power: 12–60 V DC, 4 W
Temperature range: −15 to +55°C
Reliability: (operating flight hours) 5,000 h MTBF

Status
OTA cameras are in production and in service in Dassault Mirage 2000-5 and Rafale fighter aircraft.

Contractor
Thales Munitronics BV.

Norway

EE 235 solid-state recorder

Type
Flight/mission recording system.

Description
The EIDEL Eidsvoll Electronics AS EE 235 solid-state recorder is designed for use by the aerospace, military and industrial markets, and is qualified for both aircraft and missile use. Its prime purpose is telemetry data storage and features include 256 to 8,192 Mbyte data storage in non-volatile memory; serial data recording at up to 15 Mbits/s; an input data buffer for burst data; extended storage by interconnecting any number of recorders; configurable PCM encoder module for analogue and digital data collection; direct, or control box operation. The system can be configured to user requirements by adding memory and encoder modules as required.

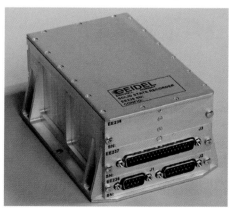

EIDEL EE 235-M3 solid-state recorder, showing EE 236 power module, EE 237 controller module, and EE 238 memory module   0051325

The standard box (M2, M3, M4) is used for data storage of 32 to 1,792 Mbytes. Special systems have capabilities up to 8,192 Mbytes.

Supporting modules include: the EE 240 PCM encoder module, and the sensor signal conditioning modules.

Specifications
Dimensions:
(EE 235-M2) 105 × 143 × 49 mm
(EE 235-M3) 105 × 143 × 74 mm
(EE-235-M4) 105 × 143 × 99 mm
Weight:
(EE 235-M2) 0.8 kg
(EE 235-M3) 1.2 kg
(EE-235-M4) 1.6 kg
Input bit rate: 0–15 MHz
Playback rate: 375 kHz; 1,500 kHz, 3 MHz or external
Input: NRZ-L with CP. TTL and RS-422 level
Output: NRZ-L, CP, BiØ-L. TTL and RS-422 level

Status
Widely used in the aerospace, military and industrial markets.

Contractor
EIDEL Eidsvoll Electronics AS.

EE 255 Solid-State Recorder (SSR)

Type
Flight/mission recording system.

Description
The EIDEL Eidsvoll Electronics AS EE 255 Solid-State Recorder (SSR) is derived from the EE235 and EE240 systems. Its prime purpose is telemetry data storage and features include: 5, 9, 13 or 17 Gbyte data storage in non-volatile memory; simultaneous recording of four channels at up to 30 Mbps each; an input data buffer for burst data; fast playback and 100 Mbps data transfer to remote computer via USB-2 interface; configurable PCM encoder module for analogue and digital data collection; and direct or remote operation via RS232 port.

Specifications
Dimensions:
(EE 255-M3) 105 × 143 × 74 mm
(EE-255-M4) 105 × 143 × 99 mm
Weight:
(EE 255-M3) 1.4 kg
(EE-255-M4) 1.6 kg
Performance: At 10 MHz bit rate input, SSR with 5 Gbyte memory can store 1 h of data
Playback rate: Clock frequencies 1–20,000 kHz
Power: 28 (19–36) V DC or 12 (10–18) V DC
Shock: 50 g, ½ sine, 11 ms, 3-axis
Temperature: −20 to +60°C (operating), extended range available
Vibration: 20 g, 20–2000 Hz, 3-axis

Status
Widely used in the aerospace, military and industrial markets.

Contractor
EIDEL Eidsvoll Electronics AS.

Poland

Solid-State Quick Access Recorder (SSQAR) family

Type
Flight/mission recording system.

Description
The ATM company has designed the Solid-State Quick Access Recorder (SSQAR) with contactless data transmission between recorder and cartridge. This solution with a RAM-based removable cartridge has been developed to give the user high-quality data recording and extremely fast data replay with no pins to damage, nor tape to be mangled. The ATM-SSQAR family is also easy and flexible to operate. Its flexibility allows

Examples of recording time (hours)

Type of cartridge	Recording system			
	ARINC 717			MSRP-64
	64 words/s	128 words/s	256 words/s	
ATM-MC5/30	34	17	8	60
ATM-MC5/70	80	40	20	140
ATM-MC5/70	120	60	30	220

ATM's solid-state quick access recorder

0015314

a wide range of different applications with most major recording systems installed on board civil and military aircraft as well as helicopters and gliders.

Programmable and able to make online analyses, the ATM-SSQAR family can display and export warning signals indicating previously programmed parameter exceedences.

To be able to meet all customer requirements ATM has designed a family of recorders consisting of: ATM-QR3 (modular, most sophisticated type); and ATM-QR4 (modular, advanced type).

Versatility is the main feature of the ATM-SSQAR family. In a single unit it combines the functions of quick access recorder with built-in real-time clock; data acquisition unit; and data management unit (programmable reports depending on installed options).

Recording time depends on the cartridge size and data format. The table below shows examples of recording time for different size ATM cartridges. Values in the table are estimated for the most popular recording standards.

Data replay from the cartridge is performed by the ATM-RD3 reader, and complete operation using a PC-compatible computer, takes up to 3 minutes, for 30 flight hours (for 64 words/s transmission speed).

The basic components include built-in real-time clock; display for recorder status and online programmable flight data analysis (ACMS); data output for external equipment; interface for entry panel; test connector for data monitoring; and data dump.

The optional components: Depending on SSQAR type, the following optional modules are available:

ATM-QR3

Optional internal full-size modules: QR3DC (interface operating two ARINC 429 busses); QR3FT (programmable vibration spectrum analyser); QR3PE (module operating as a DAU).

Optional small size module: QR3AR (ARINC 573/717 bus interface).

ATM-QR4

Optional internal full-size modules: QR4DC (interface operating two ARINC 429 busses); QR4PE (module operating as a DAU).

Optional small size module: QR4AR (ARINC 573/717 bus interface).

Optional external modules

ATM-RT allows connection to any recorder of the ATM-SSQAR family to MSRP-64 and 256 data

recording systems. ATM-DP analogue multiplexer provides the ability to process additional input signals (not available for QR2).

Specifications
Dimensions: $317 \times 57 \times 193$ mm
Power supply: 115 V AC, 40 Hz, 25 W or 27 V DC, 20 W

Status
In service on the Boeing 737 and 767, Airbus A310, Antonov An-28PT, ATR 72, Ilyushin Il-62M and -76 and Tupolev Tu-134/154M/204 and Swift S-1 glider. ATM won a USD250,000 contract to fit SSQARs to all LOT Polish Airlines Boeing 737 and 767 and ATR 72 aircraft. Also fitted to military aircraft: I-22; PZL-130TC; Su-22M4; TS-11; W-3 helicopter.

Contractor
ATM Inc.

Romania

SAIMS data acquisition and recording system

Type
Flight/mission recording system.

Description
SAIMS is designed for the acquisition and recording of the main flight parameters of all aircraft presently equipped with the SARP-12 recording systems: IAR-93, IAR-99, L-39, Mi-8, Mi-17, MiG-21, and MiG-23. It is also available for initial fit to other aircraft types.

SAIMS comprises: the data acquisition unit (UAD); protected recorder (IP); portable data copying equipment (EPPD); ground data processing equipment (ESPD) and ancillary ground equipment.

Specifications
Signals sampled:
7 potentiometric signals
3 synchro signals, 26 V/400 Hz or 36 V/400 Hz
3 frequency type signals, 7–77 Hz or 36 V/400 Hz
1 DC voltage signal (0–32 V DC)
14 discrete digital signals
Sample rate: 1–10 Hz
Resolution: 8 bits
Memory capacity: 512 kbytes, 3.5 flight hours

The SAIMS data acquisition and recording system

0051326

Dimensions:
UAD $178 \times 113 \times 77$ mm
IP $215 \times 108 \times 151$ mm
Weight:
UAD 2.5 kg
IP 7 kg
EPPD 7 kg
ESPD 2.5 kg

Temperature limits: –55 to +70°C
Supply voltage: 27 V DC

Status
In service.

Contractor
SC Aerofina SA.

Russian Federation

Advanced Moving Map System (AMMS)

Type
Avionic Digital Map System (DMS).

Description
The AMMS combines Flight Management System (FMS) and navigation display functions

to support complex Search And Rescue (SAR) and special mission roles.

The AMMS includes an independent source of navigation information (GPS or GPS + GLONASS), together with its own processor and data storage system, a full colour high-resolution display, and both digital and analogue input/output ports.

Route planning is performed directly onto the AMMS, which simplifies this operation

and reduces preparation time. All types of maps are displayed, including: aeronautical, topographic and marine; they can be used separately, or in combination. Planning takes into account wind and current where SAR operations are planned. Automatic planning of five SAR routes can be accommodated, taking into account aircraft performance data.

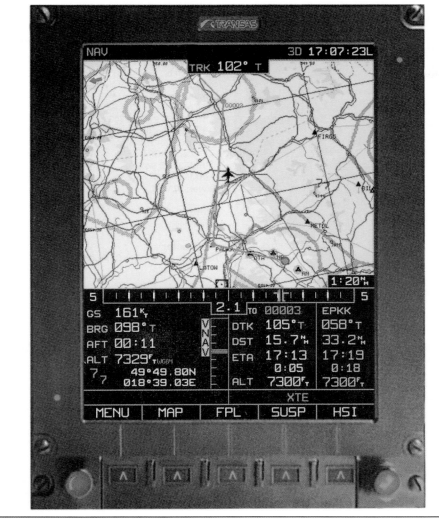

The Transas Aviation Limited advanced moving map system navigation display format   0062790

Terrain alerts and ground proximity time is determined on the basis of vector charts, mathematical models of the terrestrial surface and precise co-ordinates from the satellite navigation system.

The AMMS is designed to accommodate the Automatic Dependent Surveillance – Broadcast (ADS-B) concept. It is connected to the SoTDMA/VDL-4 transponder to provide ADS-V functions. Symbols meet TCAS and RTCA/DO-243 requirements. Information on air traffic is displayed in combination with electronic maps, route and GPWS information. Datalinks can be established for the exchange of data with ATC services and other aircraft. The AMMS can be used as a terminal for INMARSAT satcom.

Functions of the AMMS include:
- determination of position, track, speed and time via GPS or GPS + GLONASS
- horizontal and vertical navigation worldwide
- integration with the navigation and FMS
- provision of warnings if the track departs from the flight plan or FMS demands
- provision of Terrain Awareness and Warning System (TAWS) data
- provision of general aircraft warnings
- cartographic support for all navigation functions, AMMS supports Jeppesen NavData®databases, and claims that its own database The Transas Electronic Chart Database is the second largest in the world, comprising over 5,500 marine and topographic vector digitised charts
- substitution of electromechanical instruments
- cross telling of data to the second pilot's flight director
- automatic saving of data for debriefing purposes.

Status
In service.

Contractor
Transas Marine Limited, Transas Aviation Limited.

Airspeed and altitude indicators

Type
Flight instrumentation.

Description
Aviapribor Corporation produces a range of airspeed and altitude indicators that conform with ARINC 429 data interface requirements. The photograph shows some representative examples; the specifications provide representative data.

Specifications

	VMC altimeters	USC airspeed indicators
Measurement range	0–10,000 m	80–800 km/h
Error	±10 to ±30 m	5.5 to 10 km/h
Dimensions	86 × 86 × 250 mm	86 × 86 × 230 mm
Weight	2.0 kg	1.5 kg

Status
In production and in service.

Contractor
Aviapribor Corporation.

Aviapribor USC airspeed and VMC altitude indicators   0018205

Antonov-70 avionic system

Type
Integrated avionics system.

Description
AviaPribor is responsible for integration of the whole avionic suite for the An-70 aircraft. AviaPribor also developed a number of the onboard systems, including:
- mission management system
- flight control system
- fly-by-wire system
- strapdown inertial system
- satellite navigation system
- standby instruments
- MIL-STD-1553 bus control system.

The airborne information system of the An-70 aircraft is designed by the Leninetz Holding Company. It comprises: centralised firmware, unified information acquisition, processing and output to the flight crew and ground-based maintenance personnel of data on the status of the aircraft functional systems. It is designed to reduce the cost for maintenance and to permit aircraft operation at poorly equipped airfields for 30 days or 200 flight hours. It comprises two subsystems: the Information and Warning Subsystem (IWS), and the Monitoring and Maintenance Subsystem (MMS).

The IWS provides:
- reception of parametric and warning information from functional systems via digital datalinks and equipment sensors
- preprocessing of received information
- continuous monitoring of aircraft functional systems' technical status and crew actions
- generation and transfer of parametric information and monitoring results to the electronic display system, monitoring and maintenance subsystem and emergency warning system.

The MMS provides:
- reception of parametric and warning information, generated by IWS and functional systems
- information processing
- presentation of parametric information and processing results on the displays
- recording of information in operation recorder in automatic and manual modes
- output of information to the printer in automatic and manual modes
- manual input of changed-over constant and other auxiliary information from the control panel
- generation and record of monitoring results in non-volatile memory
- presentation of information stored in non-volatile memory and operation recorder on the MMS display
- initiation of aircraft systems' test through the MMS control panel.

'MONITORING' is the basic operating mode of the system, in which the monitoring of the aircraft systems' technical status and crew action, as well as the record of parametric information and monitoring results are provided.

Flight crew and maintenance personnel receive the following information:
- operating information on the electronic display system, emergency warning system's panel and MMS screen (it is used by the crew in-flight and ground-based services when

The Antonov-70 flight deck, showing 6 multifunctional colour CRT displays   0089534

The Antonov- An 70 aircraft   0089533

controlling the monitored systems and for evaluating their status)

- report information, recorded on paper tape (it is used on the ground for post-flight evaluation of the monitored systems' technical status and the crew actions, as well as for detected failure unit removal)
- report information, recorded in the operation recorder, with pre- and post-histories of errors in the aircraft systems, and information about the aircraft systems failures, stored in non-volatile memory.

'TESTS' is a mode in which the IMS and MMS technical status monitoring and their failure location up to a line-replaceable unit are performed.

'OR OUTPUT' is a mode in which the readout and analysis of information, recorded in the Operation Recorder (OR), are performed directly on board the aircraft. The information processing and its subsequent displaying and printing in the form of digital information and plots are envisaged in this mode.

'FAILURES' is a mode in which the displaying and printing of the reference information, stored in non-volatile memory, are produced on the ground.

Other elements of the avionic system, including the navigation system, flight data system, computing systems and display system (6 multifunctional colour CRTs) were developed and manufactured by Elektroavtomatika.

Specifications

Information and warning subsystem:
(input analogue signals) 560
(input digital signals) 800
(ARINC-429 input channels) 20
(No of types of channels) (analogue) 12; (digital) 2 (multiplex exchange channels) 8
(total no of input parameters) 8,000
(output digital signals, +27 V) 32
(No of generated messages) 2,000
Presentation of parametric and warning information on 5 displays of the electronic display system

(ROM capacity) 300 kbyte
(main memory capacity) 80 kbyte
Monitoring and maintenance subsystem:
Ground-based display:
(screen diagonal) 23 cm
(No of lines) 28
(No of displayed character types) 121
Operation recorder:
(recorded data capacity) 300 Mbyte
(frame capacity, 16-bit word) 1,024
(continuous recording time) 24 h
accelerated playback at readout
(capacity of information, recorded in non-volatile memory) 100 kbyte
(No of input parameters) 8,000
(No of generated messages) 8,000
(ROM capacity) 600 kbyte
(Main memory capacity) 40 kbyte

Status
In service in An-70 transport aircraft.

Contractor
Aviapribor Corporation.
Elektroavtomatika OKB.
Leninetz Holding Company.

ARINC 429 series of electromechanical indicators

Type
Flight instrumention.

Description
The Ramenskoye Design Company has designed a series of flight and navigation instruments as per ARINC 429. They include a Flight Command Indicator (FCI), Horizontal Situation Indicator (HSI) and two Radio Magnetic Indicators (RMI-3 and RMI-5).

Specifications
Number of input lines:
(FCI and HSI) 6

(RMI-3) 7
(RMI-5) 5
Number of output lines:
(FCI and HSI) 1
Number of indicated parameters:
(FCI and HSI) 9
(RMI-3) 5
(RMI-5) 3
Dimensions:
(FCI and HSI) 120 × 120 × 240 mm
(RMI-3) 85 × 120 × 240 mm
(RMI-5) 85 × 95 × 200 mm
Weight:
(FCI and HSI) 5 kg
(RMI-3) 3.5 kg
(RMI-5) 2 kg
Power: 27 V DC

Status
In service.

Contractor
Ramenskoye Design Company AO RPKB.

EFIS-85 electronic flight instrument system

Type
Electronic Flight Instrumentation System (EFIS).

Description
EFIS-85 is an integrated navigation, flight information and cathode ray tube display system. It comprises four displays, three symbol generators and two control panels. The panels are interchangeable and may function as Pilot Flight Displays (PFDs) or Navigation Displays (NDs).

PFDs display attitude, altitude and speed information, flight director and heading information; it also displays glide slope and localiser data, and information from automated systems.

NDs display navigation data, radio navigation data, weather radar and TCAS data.

Each symbol generator can support the operation of both ND and PFD for one pilot or both pilots' displays can be fed from the same symbol generator.

All units have built-in-test facilities, together with redundancy of image generation and display options.

Specifications
Display
Dimensions: 203 × 230 × 356 mm
Weight: 17.3 kg
Power: 250 W

Symbol generator
Dimensions: 194 × 127 × 324 mm
Weight: 7.0 kg
Power: 120 W

Control panel
Dimensions: 146 × 176 × 200 mm
Weight: 2.5 kg
Power: 25 W

Status
In service in Il-96-300 and Tu-204 airliners.

Contractor
NIIAO Institute of Aircraft Equipment.
Ulyanovsk Instrument Design Office.

FMS-85 flight management system

Type
Flight Management System (FMS).

Description
As part of the standard avionics system for the Il-96-300 and Tu-204 aircraft, FMS-85 provides the following function: generation of information

and control signals for 4-D navigation and control of en route flight, SID/STAR procedures, data on navigation aids, optimisation of fuel use.

The FMS-85 comprises the following LRUs: TsVM 80-40001 flight management computer; POOI-85M multifunction control and display unit (MCDU). The MCDU is used to input navigation and flight control data to the central computer and their output to the EFIS; the control of the FMS-85 modes of operation, and control of the I42-1S long-range navigation and satellite navigation systems.

Specifications
RAM capacity: 19 kwords
Non-volatile memory capacity: 236 kwords
ROM capacity: 128 kwords
Inputs;
72 channels; 32 discretes
Outputs: 10 channels; 8 discretes
Power: 115 V AC, 400 Hz, 190 VA; 27 V DC, 2.5 W
Weight: 15 kg

Status
In service in II-96-300 and Tu-204 airliners.

Contractor
NIIAO Institute of Aircraft Equipment.

Head-Up Display (HUD)

Type
Head-Up Display (HUD).

Description
The Electroautomatika HUD is designed for the generation of collimated symbolic flight/navigation and special information with a green colour image, superimposed on the scene outside the cockpit. The HUD comprises: the WCS-3 wide-angle collimating system; projection CRT control unit; built-in symbology information generation module; built-in computer module; built-in video camera for CRT image transfer.

The HUD provides the following functional capabilities: operation with MIL-STD-1553 bus systems; display of TV images exchanged in accordance with STANAG 3350B; wide-angle holographic data presentation; use of a light emitting backlight.

Specifications
Total field of view:
(vertical) 20°
(azimuthal) 30°
Instantaneous field-of-view:
(at 400 mm) not less than 20 × 30°
(at 500 mm) not less than 18 × 24°
Max viewing brightness: 30,000 cd/m²
HUD display error not greater than:
(10° diameter field of view) 7 minutes
(20° diameter field-of-view) 10 minutes
(at wider angles) 15 minutes

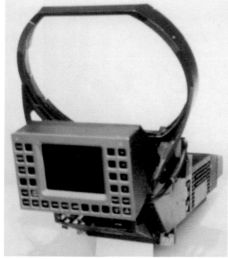

The Electroautomatika head-up display 0062241

TV image display characteristics:
(refresh frequency) 50 Hz
(lines) 625
(active lines) 575
(scan) left to right, top to bottom
(display levels) not less than 8 for up to 50,000 lx; not less than 5 for 75,000 lx
(image brightness) automatic or manual
(front panel controls) symbology brightness; TV brightness; TV contrast; reticule brightness; BIT control
Operating temperature: –40 to +55°C
Weight: 20 kg max
Power: less than 80 W

Status
In production and in service.

Contractor
Electroautomatika OKB.

IM-3, IM-5, IM-6, IGM multifunction displays

Type
MultiFunction Display (MFD).

Description
Aviapribor produces a number of multifunction displays to provide: flight/navigation information; radar data; EFIS/EICAS/EIS advisory, caution, warning and status data, and aural warnings.

Specifications

	IM-3	IM-5	IM-6	IGM
Display technology	CRT stroke	CRT raster	CRT raster	AMLCD matrix
Useful screen size	159 × 159 mm	159 × 159 mm	83 × 107 mm	101 × 101 mm
Pixels pitch	0.15/0.3 mm	0.3/0.3 mm	0.2/0.3 mm	0.2 mm
Refresh rate	50/80 Hz	40/80 Hz	40/80 Hz	80 Hz
Colours displayed	8	8 (16)	16	8 (16)
Allowable ambient illumination	70,000 lx	60,000 lx	75,000 lx	75,000 lx
Viewing angle	±53°	±53°	±53°	±40°
Dimensions	203 × 230 × 356 mm	203 × 222 × 356 mm	130 × 145 × 390 mm	130 × 130 × 230 mm
Weight	17.5 kg	12.5 kg	6 kg	4 kg
Power	200 VA	100 VA	60 VA	90 VA

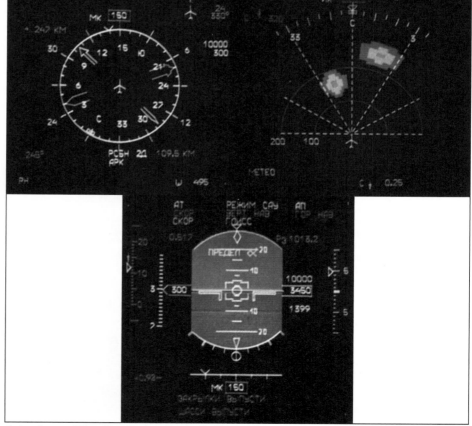

Aviapribor multifunction displays for the Tu-204 and II-96 aircraft 0018203

Status
The displays illustrated are fitted to the Tu-204 and II-96 twin and four turbofan airliners. They are also suitable for other aircraft and helicopters.

Contractor
Aviapribor Corporation.

IRM-1 radio compass indicator

Type
Flight instrumentation.

Description
The IRM-1 radio compass indicator is designed to display data derived from the MKS-1 compass and the ADF system.

Specifications
Compliance: ENLGS, p.82.1 g
Interfaces: MKS-1 compass; KR87 digital ADF receiver (AlliedSignal)
Heading, selected course and ADF bearings: 360°
Error: ±1.5° heading; ±2.0° ADF bearing
Power: 27 V DC
Temperature range: –20 to +55°C
Dimensions: 85 × 85 × 150 mm
Weight: 1.1 kg

MKS-1 compact compass system
The MKS-1 compact compass system has two operating modes; gyro and manual slaving; automatic magnetic slaving. The system

IRM-1 radio compass indicator   0018202

comprises: GK heading gyro: KU-1 slaving unit; ID-6-1 magnetic field sensor.

Specifications
Compliance: AP-23; ENLG-S, p.82.1.d
Warm-up time: 2 min
Gyromagnetic heading measurement error: 2°
Gyro drift: 0.4°/min
Power: 27 V DC
Operating temperature range:
 (GK-1) −55 to +55°C
 (KU-1) −20 to +55°C
 (ID-6-1) −60 to +150°C
Dimensions:
 (GK-1) 115 × 200 ×145 mm
 (KU-1) 55.4 × 55.4 × 94 mm
 (ID-6-1) 102 × 60 mm
Weight:
 (GK-1) 2.2 kg
 (KU-1) 0.35 kg
 (ID-6-1) 0.6 kg

Contractor
Aviapribor Corporation.

KARAT integrated monitoring and flight data recording system

Type
Flight/mission recording system.

Description
The KARAT system comprises KARAT-B the airborne monitoring and recording system and KARAT-N a portable protected computer ground data processing and analysis system.

KARAT-B comprises two line-replaceable units: Flight Data Acquisition and processing Unit (FDAU) and MultiPurpose Flight Data Recorder (MPFDR).

KARAT-B performs the following functions in the FDAU, acquisition, recording and processing of flight information from sensors and onboard systems; in-flight monitoring of onboard equipment and display on aircraft displays of relevant information; determination of maintenance requirements in the MPFDR, crash-protected storage and protection of recorded data; readout to KARAT-N of recorded data via high-speed datalink using RS-232 or RS-422 protocols.

KARAT-B employs an open architecture design that makes it suitable for all types of military and civil aircraft and helicopters. Data is protected in accordance with TSO C 124 requirements.

KARAT-N provides readout, processing and analysis of KARAT-B data; preparation of KARAT-B software and data files.

Specifications
MPFDR
Dimensions: 120 × 143 × 320 mm
Weight: 10 kg
Power: 27 V DC
Memory: up to 64 Mbytes
Inputs: ARINC 717 (1 channel); ARINC 429 (1 channel)
Net interface: IOLA
Crash protection (100% at):
(impact shock) 3,400 g for 6.5 ms
(fire) 1,100°C for 30 min

KARAT-B MPFDR (top right), KARAT-B FDAU (top left), and KARAT-N (bottom)   0015315

(deep sea pressure) 20,000 ft for 1 day; 10 ft for 30 days
(pierce) 500 lb from 10 ft with 6.35 mm steel penetration pin
(static crush) 23 kN

Status
KARAT is installed in the upgraded MiG-21 supplied to the Indian Air Force. KARAT replaces the BASK and Tester U3L systems of the original MiG-21.

Contractor
GosNIIAS State Research Institute of Aviation Systems.
JSC Pribor Design Bureau Aviaavtomatika.

ИМ-68 multifunction display unit

Type
MultiFunction Display (MFD).

Description
Aviapribor's ИМ-68 multifunction display unit is based on a full colour LCD and is designed to receive, process and present flight/navigation data, alert messages and TV-sensor data for aircraft and helicopters.

Specifications
Display: 211 × 158 mm full colour LCD
Dimensions: 200 × 270 × 290 mm

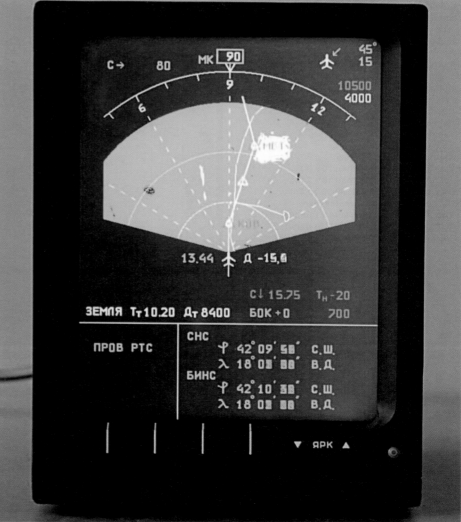

Aviapribor's ИМ-68 multifunction display unit   0103867

Weight: 9 kg
Electrical: 115 V AC, 400 Hz
Power consumption: 110 W (230 W with heating)
Interface: ARINC-429, RS-232, MIL-STD-1553B
Compliance: DO-160
MTBF: ≥ 5,000 h

Status
In production and in service.

Contractor
Aviapribor Corporation.

ИM-7, ИM-8 LCD multifunction displays

Type
MultiFunction Display (MFD).

Description
Aviapribor has developed two new, full colour LCD multifunction displays designed to interface with modern navigation/FMS systems and display: flight/navigation, radar, EFIS/EICAS/EIS, caution, warning and aircraft/system status data, and aural warnings.

Specifications

	ИM-7	ИM-8
Display	Colour LCD	Colour LCD
Useful screen size	101 × 101 mm	159 × 159 mm
Interface	ARINC-429, ARINC-708, RS-422	ARINC-429, ARINC-708
Compliance	DO-160	DO-160
MTBF	10,000 h	10,000 h
Dimensions	130 × 130 × 250 mm	203 × 230 × 280 mm
Weight	5 kg	8 kg
Electrical	27 V DC	115 V AC
Power consumption	50 W 100 W with heating	100 W 230 W with heating

Status
In service.

Contractor
Aviapribor Corporation.

MFI multifunction Active Matrix Liquid Crystal Displays (AMLCDs)

Type
MultiFunction Display (MFD).

Description
The Ramenskoye Design Company produces a range of MFI AMLCDs designed for presentation of raster enhanced graphic TV and other image data. The display specified below forms the display part of the Ramenskoye Design Company's SINUS integrated navigation, flight

MFI active matrix liquid crystal display   0018198

Ramenskoye Design Company MFI-9 and MFI-10 AMLCDs   0054302

Ramenskoye Design Company MFI-10 display installed in Mi-35 helicopter   0075930

management and display system. Other MFI displays, designated MFI-2, MFI-3, MFI-9 and MFI-10 are also specified below.

Specifications
SINUS MFI display
Display size: 130 × 130 mm
Pixels: 864 × 864
Resolution: 0.15 × 0.15 mm
Viewing angles:
(horizontal) ±45°
(vertical) +35 to –10°
Power: 115 V AC, 400 Hz
Dimensions: 205 × 205 × 205 mm
Weight: <7 kg

MFI-2
Display size: 130 × 100 mm
Pixels: 640 × 480
Resolution: 0.15 × 0.15 mm
Viewing angles:
(horizontal) ±60°
(vertical) +55 to –35°
Brightness: 300 cd/m²
Colours: ≥4,096
Power: 115 V AC/400 Hz or +27 V DC
Dimensions: 220 × 280 × 280 mm
Weight: ≤10 kg

MFI-3
Display size: 211 × 159 mm
Pixels: 640 × 480
Resolution: 0.15 × 0.15 mm
Viewing angles:
(horizontal) ±60°
(vertical) +55 to –35°
Brightness: 300 cd/m²
Colours: >4,096
Power: 115 V AC/400 Hz or +27 V DC
Dimensions: 220 × 280 × 280 mm
Weight: ≤10 kg

MFI-9
Display size: 170 × 130 mm
Pixels: 640 × 480
Resolution: 0.27 × 0.27 mm
Viewing angles:
(horizontal) ±50°
(vertical) +30 to –50°
Brightness: 200 cd/m²
Colours: >4,096
Power: 115 V AC/400 Hz or +27 V DC
Dimensions: 210 × 230 × 180 mm
Weight: ≤7 kg

MFI-10
Display size: 211 × 159 mm
Pixels: 640 × 480
Resolution: 0.33 × 0.33 mm
Viewing angles:
(horizontal) ±60°
(vertical) +55 to –35°
Brightness: 500 cd/m²
Colours: 4,096
Power: 115 V AC/400 Hz or +27 V DC
Dimensions: 220 × 280 × 180 mm
Weight: ≤8 kg

Status
Latest variant in production and most variants remain in widespread service.

Contractor
Ramenskoye Design Company AO RPKB.

MFI-68 multifunction display

Type
MultiFunction Display (MFD).

Description
The MFI-68 is designed for equipping newly developed and upgraded aircraft. It presents data from onboard systems and sensors in the form of colour symbol – graphic information, as well as TV data. It incorporates 22 multifunctional control keys.

Specifications
Screen type: colour liquid-crystal panel CS8362
Screen size: 211 × 158 mm
Pixels: 640 × 480
Pixel type: vertical band RGB
Number of bits per colour subpixel: 6
Greyscale: 64
Image contrast: 80:1
Viewing angles:
(horizontal) ±60°
(vertical) –5 to +35°
Interfaces:
ARINC 3 input channels, 1 output channel
GOST 26675.52-87 (MIL-STD-1553B) 1
GOST 7845-79 (STANAG 3350B) 1
Power supply: +27 V, 115 V AC/400 Hz
Power: <60 W (<180 W with heaters)
Operation temperature: –40 to +85°C
Dimensions: 257 × 204 × 242 mm
Weight: <6.8 kg

Status
In production. Included in many cockpit upgrades of Russian combat aircraft and helicopters. The MFI-68 forms part of Stage II of IAPO's Su-30KNM

glass cockpit upgrade, due to commence in the 3rd quarter of 2002. A total of seven displays will be installed, three in the front cockpit and four in the rear. Also included in cockpit upgrades of the MiG-29 and -31, and seen fitted to development versions of the Mi-24 attack helicopter.

Contractor
Russkaya Avionica Joint Design Bureau.

MIKBO series of compact integrated avionic systems

Type
Integrated avionics system.

Description
The MIKBO series of compact integrated avionic systems is designed and produced by the Aircraft Instrument-making Establishment MIKBOTRON, which is part of NIIAO, the Institute of Aircraft Equipment.

The MIKBO series is specially designed for use in superlight aircraft. Included in the series are the following systems:

MIKBO-2
MIKBO-2 is in development for use in superlight single-engined aircraft operating in VFR conditions in the daytime, under visual meteorological conditions in the local area and during route and regional transit flights. MIKBO-2 comprises: magnetic compass MKB-90; VHF radio; microcomputer; AMLCD; ASI; engine speed indicator; cylinder head temperature sensor; integrated instrument panel. Total weight <3 kg; dimensions 350 × 150 × 150 mm; power 12 V DC <5 W.

MIKBO-21
MIKBO-21 is designed for superlight single-engined aircraft operating in VFR conditions in daytime under visual meteorological conditions in the local area. MIKBO-21 comprises: ASI YC-150; barometric altimeter CD-10; variometer BP-5; magnetic compass KI-13; cylinder head temperature sensor; and engine shaft speed indicator. Dimensions 300 × 180 × 200 mm; power 12 V DC, 1.5 W.

MIKBO-22
MIKBO-22 is in preproduction for use in superlight single-engined aircraft operating in VFR conditions in the daytime, under visual meteorological conditions in the local area and during route and regional transit flights. MIKBO-22 comprises: barometric altimeter BD-10; variometer BP-5; magnetic compass MKB-90; ASI; cylinder head temperature sensor/indicator; engine shaft speed indicator; speed indicator, VHF radio. Weight <5 kg; dimensions 350 × 230 × 200 mm; power 12 or 27 V DC, 13 W.

MIKBO-23
MIKBO-23 is in preproduction for use in superlight twin-engined aircraft operating in VFR conditions in the daytime, under visual meteorological conditions in the local area and during route and regional transit flights. MIKBO-23 comprises: barometric altimeter BD-10; variometer BP-5; magnetic compass MKB-90; ASI; cylinder head temperature sensor/indicator; engine shaft speed indicator; speed indicator, slip indicator; main rotor speed indicator (MIKBO-23AJ only); VHF radio 'ptakha'. Weight 6 kg; power 12 or 27 V DC, <13 W.

MIKBO-32
MIKBO-32 is in design for use in superlight and light primary training aircraft certified to AP-23 or FAR-23 operating in VFR conditions from unequipped airfields. MIKBO-32 includes: a multifunction instrument panel with two flat panel displays and databus integrating: VHF radio, ADF, radio altimeter, heading computer, engine and air data sensors, magnetic compass, and CVR/FDR. Weight <30 kg; power <300 W.

MIKBO-43
MIKBO-43 is in design for use in light multipurpose aircraft, including amphibians, operating in IFR conditions from unequipped airfields in all geographical areas, as well as off water in wave heights up to 0.5 m. MIKBO-43 includes the same equipment as MIKBO – 32, plus: multifunction weather radar and emergency COSPAS-SARSAT radio beacon. Weight <75 kg; power <700 W.

Status
In service.

Contractor
NIIAO Institute of Aircraft Equipment.

Multifunction Control and Display Unit (MCDU)

Type
Control and Display Unit (CDU).

Description
Aviapribor's Multifunction Control and Display Unit (MCDU) provides all the required functionality for pilot interaction with a modern Flight Management System (FMS), including route, waypoint and radio aid management and airport standard approach and departure procedures. The unit features a full colour Active Matrix Liquid Crystal Display (AMLCD) and facilities for ARINC-429 and RS-232 interfaces.

Specifications
Display: 100 × 100 mm full colour AMLCD
512 × 512 colour triads
1,024 × 768 pixels
Viewing angles ± 45° horizontal, + 35° to –10° vertical (CR ≥10:1)

Dimensions: 146 × 228 × 266 mm
Weight: 7 kg
Power consumption: 60 W
Interface: ARINC-429, RS-232
MTBF: ≥10,000 hrs

Status
In production and in service.

Contractor
Aviapribor Corporation.

MultiFunction Control Displays (MFCDs)

Type
Control and Display Unit (CDU).

Description
Ramenskoye Design Company produces a range of MFCDs, designated PS-2, PS-3 and PS-5.

Specifications
PS-2 MFCD
Prompt panel: 4 lines with 21 symbols
Data panel: 10 lines with 21 symbols
External lighting: up to 61,000 lx
Screen colour: green
Brightness adjustment: yes
Number of keys: 25
On-screen touch keys: 15
Power: 115 V AC, 400 Hz, 100 VA
Dimensions:
(front panel) 170 × 200 × 85 mm
(electronic panel) 220 × 25 × 130 mm
Weight: 7 kg

PS-3 MFCD
Prompt panel: 2 lines with 21 symbols

Aviapribor's Multifunction Control and Display Unit (MCDU)   0103863

Data panel: 8 lines with 21 symbols
External lighting: up to 61,000 lx
Screen colour: green
Built-in processor: yes
Brightness adjustment: yes
Number of keys: 36 (8 multifunctional)
Power: 115 V AC, 400 Hz, 50 VA (2-phase)
Dimensions: 147 × 102 × 278 mm
Weight: 4.5 kg

PS-5 MFCD
Prompt panel: 2 lines with 21 symbols
Data panel: 10 lines with 21 symbols
External lighting: up to 61,000 lx
Screen colour: green
Number of keys: 33 (10 multifunctional)
Displayed frame storage: up to 256
Power: 115 V AC, 400 Hz, 50 VA
Dimensions:
(front panel) 170 × 200 × 85 mm
(electronic panel) 220 × 25 × 130 mm
Weight: 8.5 kg

Status
In service in a number of Russian aircraft.

Contractor
Ramenskoye Design Company AO RPKB.

PKP-72 and PKP-77 flight directors and PNP-72 compass

Type
Flight instrumentation.

Description
The PKP-72 and PKP-77 flight director indicators and the PNP-72 compass are analogue instruments that have found wide application in Russian aircraft.

Specifications
	PKP-72	PKP-77	PNP-72
Roll error	0.7°	0.7°	
Pitch error	0.7°	0.7°	
Yaw error		1.0°	
Azimuth			0.5 to 1.5 km

Status
In service.

Contractor
Ramensky Instrument Engineering Plant.

PPKR-SVS combined standby flight instrument

Type
Flight instrumentation.

Description
The PPKR-SVS combined standby flight instrument displays fully corrected (pressure, position, instrument, and so on) values for pressure altitude, vertical speed, indicated airspeed and Mach number and provides encoded output to other aircraft systems. Data is displayed on a full colour Active Matrix Liquid Crystal Display (AMLCD) for enhanced readability and interpretation. Different display modes are selected via buttons on the

PPKR-SVS combined standby flight instrument (Aeropribor) 0561214

Ramenskoye Design Company Multifunction Control Displays 0054303

PKP-72 and PKP-77 flight directors and PNP-72 compass 0018197

Specifications
Parameter	Range	Allowable error within range
Static pressure, hPa	115.5 to 1,075	±0.7
Total pressure, hPa	115.5 to 1,150	±1.7
Atmospheric pressure at aerodrome level, hPa	577 to 1,075	±0.5
True altitude (QNH), m	-503	±4
	3048	±6
	9144	±12
	15,240	±24
Relative height (QFE), m	3048	±6
	9144	±12
	15,240	±24
Vertical speed (m/s)	75	±3.5
	0	±0.15
Indicated airspeed (m/s)	111	±9.3
	185	±3.7
	832	±1.85
Mach number	0.1	±0.015
	0.6	±0.005
	0.7 at 6,100 < H < 12,300	±0.004
	0.8 at 7,625 < H < 12,300	±0.003
	0.9 at 7,625 < H < 12,300	±0.003
	0.95 at 7,625 < H < 12,300	±0.004
	1.0	±0.01
Power, V DC	2 sources, 27 each	
Interface	ARINC 429, 706	
Dimensions (mm)	85 × 85 × 225	
Mass (kg)	1.5	

instrument face. Altitude data is displayed in either metres or feet, with warning of deviation from flight level and flight below 1,000 m.

Measured and derived air data is transmitted to aircraft systems via three independent channels as 32-digit serial code according to GOST 18977-79, RTM 1495-75 (Amendment 3) and ARINC 429. Instrument accuracy meets the requirements of MVS-300 and RVSM.

Status
In production and in service.

Contractor
AeroPribor-Voskhod Joint Stock Company.

PV-95 multifunctional control and display unit

Type
Control and Display Unit (CDU).

Description
The PV-95 display computer unit is designed for display of real-time piloting and navigation data; it comprises a CISC architecture processor module (MV58); channel multiplexing module (MO 52) to GOST 26765.52-87 and MIL-STD-1553B standards; channel module (MD 51) to GOST 18977-79 and ARINC 429 bus

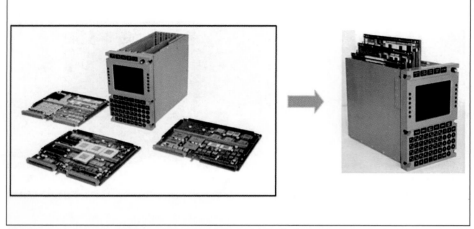

The Electroautomatika PV-95 multifunctional control and display unit 0062188

standards; graphics control module; secondary voltage module (MN 95). The PV-95 system can be readily configured to customer requirements.

Specifications
Screen: monochrome EL flat panel
Active display area: 96 × 77 mm
Resolution: 320 × 256 pixels
Viewing angle: not less than 160°
Display:
(alphanumeric) 14 lines of 24 characters
(graphics) vectors and arcs
Maximum ambient lighting: not less than 75,000 lx
Keys: 75 keys with night backlight
Processor: CISC, 32 bit
Instruction system: Electronica MPK-1839
Clock frequency: 10 Hz
Operands: 8, 16, 32, 64
Internal transfer rate: 5 Mips
Memory:
(RAM) 128 Kbytes
(ROM) up to 4,096 Mbytes
Programmable timers: up to 4
Input/output:
(ARINC 429) 18/6
(discrete) 16/8
(MIL-STD-1553B) 2
BIT: yes
Dimensions:
(overall) 123 × 224 × 255 mm
(front panel) 146 × 228.6 × 23.4 mm
Power: +27 V DC, 50 W

Status
In production and in service.

Contractor
Electroautomatika OKB.

RMI-3 Radio Magnetic Indicator (RMI)

Type
Flight instrumentation.

Description
The RMI-3 radio magnetic indicator displays magnetic heading, relative bearing from two radio beacons, distance from two radio beacons. The display is software programmable. Communication interfaces are digital, as per ARINC 429.

Specifications
Accuracy:
(bearing) ±1.5°
(distance) ±0.1 km
Dimensions: 85 × 120 × 240 mm
Weight: 3.0 kg
Power: 27 V DC

Status
Fitted to the Antonov An-125 Ruslan.

The Ramensky Instrument Engineering Plant RMI-3 radio magnetic indicator 0079227

Contractor
Ramensky Instrument Engineering Plant.

Standby Horizontal Situation Indicator (HSI) INP-RD

Type
Flight instrumentation.

Description
The Ramensky Instrument Engineering Plant standby horizontal situation indicator displays aircraft position, heading and pitch with respect to the radio beacon and four cardinal points. It comprises a sealed electromechanical unit and the control and monitor unit (BUK-14).

Specifications
Accuracy:
(bearing) 1.0°
(distance 0–25 km) ±1 km
(distance 25–999 km) ±3 km
Slaving rate of pointers:
(current heading) 30°/s
(radio beacon bearing) 60°/s
(selected heading) 40°/s
(distance) 15 km/s
Readiness time: 1 minute (maximum)
Weight: 2.3 kg

Status
Fitted to Su-30 military aircraft, as well as civil aircraft.

The Ramensky Instrument Engineering Plant standby horizontal situation indicator 0080655

Contractor
Ramensky Instrument Engineering Plant.

STRIZH gyrostabilised magnetic compass

Type
Flight instrumentation.

Description
The STRIZH gyrostabilised magnetic compass is designed for small civil aircraft. It updates and improves on the capability of earlier products available from the Ramensky Instrument Engineering Plant.

Specifications
Gyromagnetic heading accuracy (2σ): 1.0°
Readiness time: 3 min
Weight: 3.5 kg
Operating conditions:
(flight speed) up to 1,000 km/h
(flight altitude) up to 15,000 m
(axial angular velocity) up to 200°/s
(linear acceleration) up to 4 g
(heading angles) 0-360°
(roll angles) 0–180°
(pitch angles) 0–90°
(latitude) 0–75°

Status
In production and in service.

Contractor
Ramensky Instrument Engineering Plant.

The Ramensky Instrument Engineering Plant STRIZH gyrostabilised magnetic compass
0079232

VBE-SVS electronic barometric altimeter

Type
Flight instrumentation.

Description
The VBE-SVS electronic altimeter is a new-generation instrument that combines the functions of air data computer, altitude indicator and flight level deviation alert in one unit.

The Active Matrix Liquid Crystal Display (AMLCD) indicates the current value of relative height (feet or metres), local barometric pressure at ground level, assigned flight level, warning of deviation from flight level by 60 to 150 m (flashing display frame), warning of deviation from flight level by more than 150 m (illuminated display frame) and warning of flight below 1,000 m, all in a format familiar to pilots used to traditional electro-mechanical instruments. The display is adaptable to customer requirement.

Specifications
VBE-SVS and VBE-SVS-TsM
Inputs

Static pressure, hPa	115.5 to 1,074
Total pressure, hPa	115.1 to 1,150
Stagnation temperature, °C	−60 to 99

Outputs	Range	Permitted error
True altitude (QNH), m	−503 to 15,240	±4.6 (at -503 m), ±24.4 (at 15,240 m)
Relative height (QFE), m	0 to 15,240	±6.1 (at 0 m), ±24.4 (at 15,240 m)
Rate of climb, m/s	±102	0.15 or 5%
Indicated airspeed (CAS), m/s	55.5 to 832	±9.3 (at 55.5 m/s), 1.85 (at 832 m/s)
True airspeed (TAS), m/s	185 to 1,108	±7.4
Outside Air Temperature (OAT), °C	−99 to +60	±1
Total Air Temperature (TAT), °C	−60 to +99	±0.5
Pitot pressure (Pp), hPa	115.5 to 1,150	±1.7
Static pressure (Ps), hPa	115.5 to 1,074	±0.7
Mach number	0.1 to 1.0	
Flight level	0 to 15,000	
Power, V	27	
Interface	ARINC 429, 706	
Dimensions, mm	85 × 85 × 235	
Mass, kg	1.6	

Displayed parameters of true altitude or relative height are corrected for all static and dynamic errors, with the system storing a number of corrective constants to allow for probe positioning and installation in a wide variety of different aircraft.

Vertical speed data and assigned flight level are exchanged with other onboard systems by 32-bit serial data code as per ARINC 429. Flight level deviations (>150 m) and power-on state/fail-free operation/built-in test data are also reported routinely.

The latest addition to the family is the VBE-SVS-TsM, which features a colour AMLCD and specification similar to the -SVS, cleared for RVSM operations.

Status
In production and in service.

Contractor
AeroPribor-Voskhod Joint Stock Company.

VBE-SVS electronic barometric altimeter
0062742

VBE-SVS-TsM colour AMLCD altimeter (Aeropribor)
1026866

VBM mechanical barometric altimeters

Type
Flight instrumentation.

Description
AeroPribor-Voskhod VBM mechanical barometric altimeters are designed on a unified constructional base, and they have a built-in vibrator for reducing friction in the mechanism. Altitude is displayed by a counter calibrated in kilometres. The altimeters are produced in three separate colour options: integral red; red/white; and white dial lighting.

Status
Widely used.

Contractor
AeroPribor-Voskhod Joint Stock Company.

AeroProbor-Voskhod Joint Stock Company VBM mechanical barometric altimeters
0018208

Specifications

	VBM-1	VBM-1A	VBM-2/VBM-2F	VBM-3	VBM-R
Altitude range	−500 to 10,000 m	−500 to 5,000 m	−500 to 15,000 m	0 to 30,000 m	−500 to 30,000 m
Measurement error, at altitude (m)					
0	± 10	± 10	± 10	± 10	± 10
5,000	± 15	± 15	± 20	± 35	± 55
9,000	± 30	± 40	± 50	± 75	
12,000		± 55	± 60	± 100	
19,000		± 300			
Atmospheric pressure range	700–1,080 hPa	700–1,080 hPa	700–1,080 hPa	700–1,080 hPa	525–810 hPa
Dimensions	85 × 85 × 190 mm	85 × 85 × 190 mm	85 × 85 × 240 mm	95 × 85 × 190 mm	65 × 65 × 160 mm
Weight	1.2 kg	1.2 kg	1.4 kg	1.6 kg	1.0 kg

Power (all systems): 27 V DC, 1.5 W; 5.5 V AC, 2.2 VA

Video Recorder System (VRS)

Type
Avionic display system, Head-Up Display (HUD), video camera.

Description
The Electroautomatika VRS is designed to record data displayed on the Head-Up Display (HUD), as well as other data deriving from TV cameras. The VRS is a replacement system for the FKP-EU cine-camera system installed in MiG-29 and Su-27 aircraft.

The VRS comprises a memory unit with a replaceable non-volatile memory video cassette; miniature TV camera; logic and power unit; ground replay unit based on an IBM PC.

Specifications
Cassette memory: 1,024 video frames
Video data recording rate:
8 exposures/s with 2 s delay;
8 exposures/s without delay;
1 exposure/s;
exposures linked to synchronising pulses.
TV camera resolution: 50 lines minimum
Video data scanning standard: 50 interlacing fields

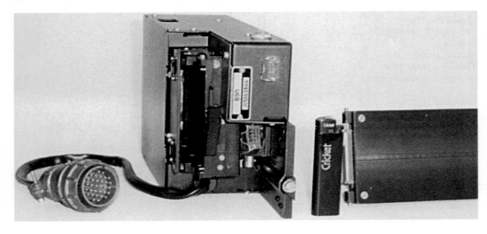

Dynamic range: 0.1 to 10,000 lx
Operating temperature range: −40 to +50°C
Video storage time: 30 days
MTBF: 80,000 cycles minimum
Power: +27 V DC, 25 W
Weight: 4.5 kg

Status
In production and in service.

Contractor
Electroautomatika OKB.

The Electroautomatika video recorder system, showing memory unit, camera and logic/power unit
0062228

South Africa

Display systems

Type
MultiFunction Display (MFD).

Description
The Avitronics family of high-performance Active Matrix Liquid Crystal Display (AMLCD) colour display systems includes multifunction colour displays and multifunction keyboards that can be configured according to customer requirements.

The MultiFunction Colour Display (MFCD) is built around a 5 in diagonal, full colour, AMLCD. The display bezel hosts the custom function keys as well as the brightness and HMI control requirements. The MFCD is provided with onboard programming capabilities and extensive Built-in-Test control circuitry.

Advanced Thin Film Transistor (TFT) AMLCD technology was selected to enhance the visual quality of the display and greater utilisation of colour and graphics. The display technology provides high brightness, excellent contrast and low power consumption.

Software design enables user definable font-set, symbol characters including special bitmap symbols and back-lighting curves that can be down loaded via the external connector by the customer. Display modes include pre-defined pages/symbols that may be displayed in overlay modes.

Further features and benefits of Avitronics' display systems include:
- Ruggedised displays
- Fully sunlight readable
- Wide viewing angles
- NVIS-compatible
- Modular architecture design
- Several communication options
- Automatic and manual brightness control
- Intelligent or non-intelligent displays
- Medium- to very high-performance graphics processing capabilities
- Software/hardware include JTAG debugging capabilities to reduced integration and maintenance effort
- Fast 32-bit data processing electronics
- Software font-set, symbol characters including bitmap symbols and pre-defined pages are loadable through an external connector
- Programmable user definable back-lighting curves
- Modern designed software protocol (or custom design)
- Comprehensive set of graphic commands
- Custom designed bezels to meet client Man-Machine Interface (MMI) requirements
- Reduced bezel thickness to improve MMI characteristics
- MultiFunction Keyboards (MFK), incorporating alpha-numeric and special function keys with tactile feedback and backlighting.

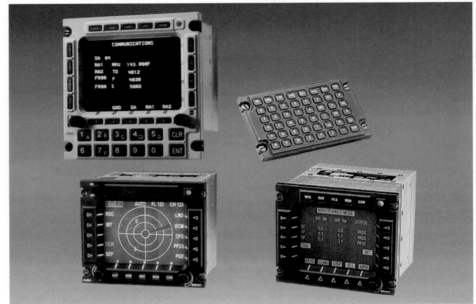

Avitronics multifunction display systems
0593628

Specifications
Power requirements: 28 V DC (accepts 18–32 V)
Data interfaces: MIL-STD-1553B, RS-422, RS-485, ARINC 429
Environmental:
MIL-STD-810E
MIL-STD-461C
DO160C
Software design: RTCA DO178B

Status
In production and in service.

Contractor
Avitronics.

Health and Usage Monitoring Systems (HUMS)

Type
Flight/mission recording system.

Description
AMS Health and Usage Monitoring Systems (HUMS) is a modular system comprising data acquisition processing unit; control and display; Flight Data Recorder (FDR) and Cockpit Voice Recorder (CVR); data transfer device; and ground replay station.

The system covers the full functional spectrum, including flight data recording, cockpit voice recording, engine health and usage monitoring, performance trending, structural fatigue monitoring, limits and exceedance monitoring, vibration monitoring and data storage and transfer.

The data acquisition unit is based on the AM3000 harsh environment avionics computer. This features IEEE 1296 Multibus II, MIL-STD-1398D SEM-E modules, Intel 80960 MC 25 MHz CPU, MIL-STD-704D power supply, multiples of 16 analogue and 32 discrete inputs, extensive BIT capability, 1/2 ATR short ARINC mounting or customised packaging and convection cooling.

The digital flight data recorder has a solid-state memory array as the recording medium, crash protection in accordance with EUROCAE ED-55/56 and is a 1/2 ATR short ARINC mounting unit.

The data extraction computer is a PC-based hardware platform that can be interfaced with the Global Logistics Information System. It operates in a Microsoft Windows-based environment and can function as a data transfer device as well as a powerful analysis station.

Status
Variants of the system have been engineered for the Rooivalk attack helicopter HUMS, and the SMR 95 engine HUMS on Cheetah fighter aircraft.

AMS has also developed a version for the BAE Systems Hawk Mk 100 series jet trainer aircraft that will feature airframe fatigue monitoring; Engine Life Recording (ELR) for the Adour Mk 871 engines; FDR/CVR in accordance with ED55/ED56A (BASE SCR500-660); avionics equipment maintenance recording; flight line maintenance support station; fleet airframe

BAE Systems Hawk Mk 100 and Aerospace Monitoring and Systems' Health and Usage Monitoring System (HUMS)   0015316

usage management system. This system has been selected for the Hawk 100 aircraft being delivered to the Royal Australian Air Force and the NATO Flying Training Centre (NFTC) in Canada. First flight of the Hawk aircraft fitted with this AMS was in April 2000. This variant of the AMS HUMS could also be retrofitted to older Hawk aircraft.

Contractor
Aerospace Monitoring and Systems (Pty) Ltd.

Mission computer

Type
Aircraft mission computer.

Description
ATE's new mission computer is designed to meet the requirements of modern integrated avionics systems. Advanced RISC architecture and modular design provides for a high-performance generic mission computer.

A distributed processor architecture with intelligent Input/Output (I/O) peripherals allows the application software to run independently from the I/O system software. Real-time colour symbology generation/video mixing, as well as HUD stroke symbology generation, is managed by high-performance intelligent symbol generation modules. I/O interfaces include MIL-STD-1553B BC/RT, synchro, analogue, serial and discrete. A user-friendly PC-based applications software development environment is fully implemented and validated. An offline PC-based graphics generation tool allows pilot evaluation of symbology before downloading to the mission computer. Extensive BIT ensures failure detection of between 95 and 98 per cent.

Specifications
Dimensions: 330 × 129 × 197 mm
Weight: 7 kg
Power consumption: 70 W
Reliability: 4,000 h MTBF

Status
Reported to be in production and in service in the Rooivalk attack helicopter.

Contractor
Advanced Technologies & Engineering (Pty) Ltd.

Navigation and Weapons Delivery System (NWDS)

Type
Mission Management System (MMS).

Description
ATE's new mission computer (see separate entry), together with an ATE stores management computer and in-house designed and developed software, forms the core of the Navigation and Weapons Delivery System (NWDS). The system has been applied to both fighter and helicopter platforms, integrating with onboard navigation sensors, systems and weapons via MIL-STD-1553B and -1760 data busses. Two variants of the system are in production:

NWDS for BAE Systems LIFT Hawk
In February 2000, ATE was contracted by BAE Systems for the design, development, integration and production of the NWDS for 24 Hawk Lead-In Fighter Trainer (LIFT) aircraft for the South African Air Force (SAAF). ATE has also been contracted to carry out the NWDS test flight programme to demonstrate the performance of the system.

The NWDS will include an INS with integrated GPS, a glass cockpit with three MultiFunction Colour Displays (MFCDs), a smart HUD and UFCP per pilot station. Subsystems will include a communications management system and self-protection EW suite.

NWDS for Rooivalk helicopter
During 1994, ATE was contracted by Denel Aviation for the design, development, integration and production of the NWDS for 12 Rooivalk combat support helicopters. The system features a fully integrated glass cockpit to facilitate day/night navigation and the detection, identification and engagement of ground and airborne targets. Subsystems, integrated via MIL-STD-1553B/-1760 include an Attitude Heading and Reference System (AHRS) with integrated GPS, the aircraft flight control system, a full communications management system, an EW protection suite and Health Usage and Monitoring System (HUMS – see separate entry).

System control and management is exercised by 2 ATE mission computers and 2 ATE weapons and armament computers supported by proprietary software. Weapons system integration includes a 20 mm turreted gun, 68 mm and 70 mm air-to-ground rockets, Mokopa anti-tank missiles and air-to-air missiles.

Status
The first NWDS has been delivered to BAE Systems for integration into the first flight test Hawk LIFT aircraft.

Twelve NWDS have been delivered for Rooivalk helicopters. The system is in service with the SAAF.

Contractor
Advanced Technologies & Engineering (Pty) Ltd.

The ATE mission computer
0504161

A Rooivalk development aircraft, fitted with the NWDS, fires a Mokopa anti-tank missile during trials (Denel/Kentron)   0083239

Spain

NAT-5 tactical navigation system

Type
Mission Management System (MMS).

Description
The NAT-5 is a computer-based navigation system for airborne ASW tactics. It comprises a control unit, central processor unit and display unit, and accepts information from any or all of Doppler, true airspeed, AHRS, GPS, sonar and radar.

The display unit shows updated positional information plus seven predefined contacts in latitude and longitude, X-Y co-ordinates or any of four different polar co-ordinate systems, with graphic scales from 4.5 × 3 up to 288 × 192 n miles.

The control unit is a microprocessor-based control panel with a keyboard including function, numeric and control keys and a joystick for quick data entry. The control unit allows easy system operation by means of up-down menus on the screen.

Graphics operation is provided by means of a powerful graphics processor with up to 100 kbytes of stored standard search patterns and a cartridge, located in the control unit, with up to 512 kbytes for predefined maps, tactics and tactical scenarios. The cartridge can also store system airborne data for debriefing purposes.

The system provides a serial output for a sensor operator control keyboard and interfaces for cockpit instruments such as the BDHI.

The operational software and the interface electronics are designed in a modular form to accommodate each future change with a minimum impact on the configuration of the system.

Optionally, the system includes ground support equipment and a navigation sensors simulator for training, testing and maintenance facilities.

The ground support equipment has a special software utility which allows the user to create his own tactics and maps for loading into the cartridge.

Specifications
Dimensions:
(control unit) 135 × 146 × 181 mm
(central processor unit) 194 × 124 × 320 mm
(display unit) 265 × 325 × 333 mm
Weight:
(control unit) 2.5 kg
(central processor unit) 10 kg
(display unit) 17 kg

Status
In service in Spanish Navy Sikorsky SH-3D/G Sea King and SH-3D Sea King AEW helicopters.

Contractor
INDRA Sistemas SA.

Sweden

Digital Map System (DMS)

Type
Avionic Digital Map System (DMS).

Description
The Saab DMS is a versatile and comprehensive digital map system for fixed- and rotary-wing aircraft. The Digital Map System (DMS) stores a comprehensive database and generates a one or two channel, digital real-time moving map which can include vector maps, raster maps, Jeppesen data and charts and overlays with flight plans, tactical symbology and other symbology. Terrain awareness and obstacle warning modes as well as still image display (aerial photography) are available. The map and overlay information is selectable in a number of different layers and combinations. North-up/track-up, selectable own Present Position (PP), zoom, Way Point (WP) handling and other features are included. The system is controlled over a MIL-STD-1553B data bus or Ethernet. The map data is stored in a removable Mass Memory Cartridge (MMC). A ground station is available for memory loading and pre-processing of map data.

Specifications
Selectable map scales: 1:10,000, 1:50,000, 1:250,000, 1:500,000, 1:1,000,000, 1:2,000,000
Independent channels: 1 or 2
Storage capacity: Up to 15 Gb
Mechanical: 5 MCU ARINC 600 (194 × 157 × 324 mm)
Interfaces: Digital/analogue colour video output (1 or 2 to XGA resolution), RS-422, ARINC 429 (optional), RS-422, MIL-STD-1553B and Fast Ethernet

Status
In production.

Contractor
SaabTech AB.

DiRECT Digital mission RECording and data Transfer system

Type
Flight/mission recording system.

Description
DiRECT is a universal digital mission recorder for recording and replay of multichannel video, audio and data onboard fixed- and rotary-wing aircraft. It features up to 4 channels of video and audio recording and data from MIL-STD-1553B and Ethernet. The system employs MPEG-2 compression, with adaptable compression rate, with recording and replay controlled by commands over the databus; storage is via a removable Mass Memory Cartridge (MMC). The solid-state version is recommended for the harsh environment of modern combat aircraft, or a hard disk can be supplied for other applications. A variety of information can be transferred to and from the aircraft using the MMC. A PC-Card can be included as a secondary memory to store back-up information, such as maintenance data.

Specifications
Input video: CCIR/PAL, NTSC, RS-170, special
Video recording: 2 to 4 channels, MPEG-2, 1–2 Gb/h per channel
Audio recording: 2 to 4 channels, 64 kbit/sec per channel, μ-law coded
Storage capacity: up to 2 h, utilising all channels simultaneously
Memory: non-volatile, solid-state or hard disk
Coding: MPEG/Layer 2 audio coding
Power: 28 V DC; 175 W (4 channels)
Environment: full military specification
Dimensions: 5 MCU ARINC 600 (194 × 157 × 324 mm)
Weight: 8 kg (including MMC)
Interfaces: Fibre channel, RS-170, RS-422, ARINC 429 (optional), RS-422, MIL-STD-1553B and Fast Ethernet

Status
Prototypes in service in the JAS 39 Gripen fighter (integrated into the display processor) and the F/A-18. Under development for other platforms, including rotary-wing applications. Demonstration units are reported to be available for test flights on demand.

Contractor
SaabTech AB.

Distributed Integrated Modular Avionics (DIMA)

Type
Integrated avionics system.

Description
DIMA combines the advantages of a traditional, federated avionics concept and those of an Integrated Modular Avionics (IMA) system, by being a physically distributed but functionally integrated system. The concept is based on reusable building blocks with viable and standardised interfaces. The concept facilitates a high degree of operational flexibility and expansion capability. The concept supports a layered software architecture by providing hardware independent application software services, making it possible to allocate processing capability more quickly (processor in close proximity to system or sensor) with no performance degradation due to the overload of dedicated processors. The overall effect is that of a digital network, which provides the following advantages:

- Improved capability to re-configure the system architecture
- Improved capability to add, upgrade, or change functionality
- Reduced wiring between units
- Improved electromagnetic compatibility
- Improved transmission capability.

The physical distribution of processing capability also brings advantages in resisting battle damage more effectively, with processing power almost instantaneously re-distributed to maintain operational capability after the loss of processors due to damage or destruction.

Status
In production.

Contractor
SaabTech AB.

EP-17 display system for the JAS 39 Gripen

Type
Avionic display system.

Description
The cockpit of the Saab JAS 39 Gripen features advanced electronic information presentation on four display units: three head-down displays and one head-up display. All four displays are computer-controlled, permitting the presentation to be tailored to every type of mission and flight mode. It also provides considerable flexibility and redundancy under emergency conditions.

The three head-down units are the Flight Data Display (FDD), the Horizontal Situation Display (HSD) and the MultiSensor Display (MSD). The FDD provides the pilot with flight, systems and weapons data as well as HUD back-up. The HSD employs a digital moving map showing geographical features and obstacles, such as radio towers and masts, with the tactical situation superimposed giving excellent situation awareness. Map scale and displayed information are selected manually and automatically to suit different phases of a mission. The MSD presents a computer processed radar picture in air-to-air and air-to-surface modes. The digital map can be overlaid on the radar picture. The flight data display and the multisensor display can also present imagery from other sources, such as IR or TV sensors.

The EP-17 has a display processor, divided into two functional chains each driving two displays and communicating with other aircraft systems by means of MIL-STD-1553B databusses. Each functional chain contains an Ericsson MACS Power PC. High-performance graphics processors provide excellent functionality and dynamics to the graphics. The head-down displays employ raster-generated symbology, while stroke writing is used for the head-up display.

The Lot 1 and 2 version of the Gripen aircraft employs three 152 × 120 mm (6 × 4.7 in) monochrome CRT multifunction head-down displays. The displays are dimensionally identical (the same part number). They use raster to present information and feature three selectable video systems: 525, 675 and 875 lines. Conventional mechanical standby instruments are used as back-ups for some of the more critical parameters. The head-up display uses diffraction optics and presents a large bright picture to the pilot. The field-of-view is 20° × 28°. The normal display mode is stroke, but raster generated imagery can also be presented.

The Swedish Air Force Lot 3 and the export version of the Gripen aircraft employs three 158 × 211 mm (6.2 × 8.3 in) flat panel multifunction colour liquid crystal displays. The displays are dimensionally identical (the same part number). They feature high brightness, high-resolution (SVGA 600 × 800 colour pixels) liquid crystal glass. The displays are NVG compatible. The instrument panel has no space left for back-up instruments. Consequently, back-up functions are included in the multifunction displays. Each display interfaces redundant data sensors and can be supplied from the aircraft battery.

Display processing features full software control in computers and symbol generators, anti-aliasing, full-colour, digital map storage and presentation in several scales and radar scan conversion. The Lot 3 version includes all basic functionality in loadable software to ensure affordable upgrades and future growth.

The multifunction colour display also comes in a smart version designated MFID 68. This display will be used in the rear seat to facilitate special operator display functions. This smart display is ideal for retrofit and has been integrated into the JA 37 Viggen.

Recording capabilities included, with processing in the display processor, are multiplexed sensor video, MIL-STD-1553B bus data and audio using a Hi-8 mm videocassette recorder. The Lot 3 version of the system features a solid-state digital mass memory (DiRECT) instead of the videotape. Several sensor systems sources, video information, bus data and audio will be digitally mixed on the mission recorder. Standard MPEG-2 coding is employed for the imagery. Both systems feature an onboard replay facility for quick evaluation.

Status

The Swedish Air Force has ordered 204 Gripen aircraft in three lots, of which approximately 110 had been delivered by December 2001. In June 1997, the then Saab Avionics (now SaabTech) was contracted for development and production of the Lot 3 version featuring upgraded display processor, colour displays, head-up display and digital recording.

The first delivery of upgraded aircraft was completed during 2002/2003.

A variant of the EP-17 display, the MFID 68, has been retrofitted into two squadrons of JA 37 Viggen aircraft as part of the Viggen Mod D upgrade programme. Each JA 37 Viggen aircraft

This illustration shows the cockpit display suite for the JAS 39 Gripen, comprising a wide-angle HUD and three large format 6 × 8 in head-down displays
0131025

The cockpit of a Swedish Air Force (SwAF) JAS 39C Gripen. Note the high-brightness 6 × 8 in multifunction displays (Gripen International)
0589482

has only one multifunction head-down display, while the Gripen has three.

Contractor
SaabTech AB.

MFID 68 MultiFunction Integrated Display

Type
MultiFunction Display (MFD).

Description
Originally developed for the Saab JAS 39 Gripen, the MFID 68 is a fully integrated, multimode liquid crystal colour display which may be applied to a wide range of aircraft cockpits.

It features a 158 × 211 mm (6.2 × 8.3 in) display surface with a high-resolution NVIS-compliant screen with 20 push-buttons and manual/automatic brightness/contrast controls.

The display comprises computer, interfaces and graphics generator with a digital map. Standard interfaces are used to connect the system to the host aircraft avionics suite, with facilities to implement user-defined functions through software.

The MFID 68 is designed to provide a high-performance interface to the pilot. It is included as part of the EP-17 display system for Batch 3 and export variants of the Saab Gripen, and has also been included in the mid-life upgrade of the Swedish Air Force JA 37 Viggen.

Specifications
Display size: 158 × 211 mm (6.2 × 8.3 in)
Resolution: 600 × 800 pixels
Display processing: Power PC with internal VME bus 16 Mb memory, Ada or Pascal D80 and high-performance graphics with embedded digital map

Status
In service in the JAS 39 Gripen and JA 37 Viggen.

Contractor
SaabTech AB.

MFID 68 MultiFunction Integrated Display
0051668

Mission computers

Type
Aircraft mission computer.

Description
Saab produces a range of avionic mission computers which provide processing, mass memory capability and communication interfaces. The computers are compatible with the ADA software development environment. The use of VHDL design tools makes simulation of a complete board possible and ensures correct functionality; it also facilitates redesign of the electronic functions, either to meet specific customer requirements, or during upgrades to replace legacy/obsolete components. The following functions are available:
- Intel or PowerPC processors
- Multiprocessor capability
- Graphic processing capability
- Shared memory
- Mass memory, up to 320 Mb
- Video interface, digital or analogue
- Interface to MIL-STD-1553B, Ethernet, RS-485, RS-422, RS-232

Specifications
Dimensions: Modular ARINC 600 chassis (other sizes available)
Weight: <10 kg
Vibration:
0.04 g²/Hz functional
0.1 g²/Hz gunfire
Expansion: Slots for six circuit-card assemblies, equivalent to 12 double-Europe PCBs
Cooling: Convection, forced-air or liquid cooling
Power: 28 V DC and/or 115 V AC, 400 Hz
Consumption: <80 W

Status
In production.

Contractor
SaabTech AB.

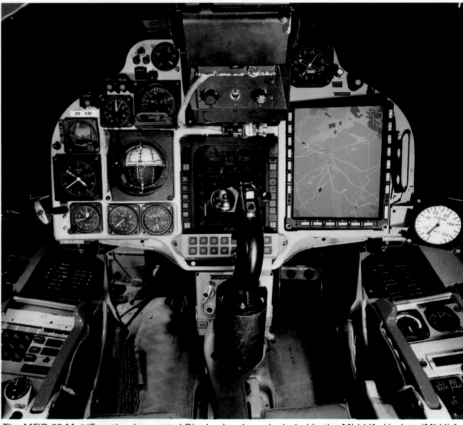

The MFID 68 MultiFunction Integrated Display has been included in the Mid-Life Update (MLU) for the JA 37 Viggen
0131026

Switzerland

AD20 electronic standby altimeter

Type
Aircraft instrumentation.

Description
The electronic Standby Altimeter AD20 from Revue Thommen indicates barometric altitude and can be a direct replacement for mechanical standby altimeters. It measures static pressure by an integrated solid-state pressure sensor and displays the computed altitude by an analogue pointer and a digital Liquid Crystal Display. The computed altitude is transmitted via ARINC-429 serial databus. The Altimeter AD20 has a range of −1,000 to 50,000 ft and can be switched to metric scale indication. The barometric scale can be displayed either in inches of mercury (in Hg) or in millibars (mb). The electronic Standby Altimeter AD20 is designed to operate from 28 V DC primary power. In case of power loss the internal emergency battery power source supplies

the instrument automatically. Lighting is controlled by an external 5 V DC voltage. The lighting colour is available in white or NVIS green. In a dual-D20 installation the baro correction setting (QNH) can be synchronised on both AD20 via ARINC-29. Discrete outputs are available as an option. The RS-232 serial interface provides the communication for maintenance and parameterisation. The test button activates the Built-In-Test (BIT) procedure. The Electronic Standby Altimeter AD20 from Revue Thommen is certified to FAA TSO-C10b.

Other features and benefits of the AD20 include:
- Push-to-test and continuous Built-In Test (BIT)
- BIT failure memory
- LCD display/stepper motor driven pointer
- NVIS green or white lighting
- Built-in emergency battery module (optional).

Specifications
Signal inputs
Primary power: 28 V DC
Lighting power: 5 V DC (28 V DC option)

Emergency power signal: weight off wheels
ARINC-429: 2 receive channels (option)

Signal outputs
ARINC-429: 1 transmit channel
ARINC-429 parameters: Pressure altitude, barometric corrected altitude and altitude rate
Altitude range: −1,000 to 50,000 ft (−300 to 16,000 m)
Baro setting: 20.67 to 31.01 in Hg (700 to 1,050 hPa)
Compliance: FAA TSO-C10b, RTCA/DO-178B Level A, -160D
Operating temperature: 20 to 70°C
MTBF: >12,000 hr (estimated)

Status
In production.

Contractor
Revue Thommen AG.

AD32 RVSM air data display

Type
Aircraft instrumentation.

Description
The AD32 air data display integrates solid-state pressure sensors for static and pitot pressure, measuring barometric altitude, airspeed and temperature in the atmosphere. The AD32 is Reduced Vertical Separation Minima (RVSM) compliant and provides up to 2×16 SSEC curves. The corrected altitude is displayed on a high-contrast LCD in digital format and by a stepper motor driven pointer. The computed air data parameters can be transmitted via the ARINC-429 interface. Two transmit channels and two receive channels are available. The altimeter can be set to operate as a self-sensing standby instrument or to display altitude from an external Air Data Computer source. The AD32 has an integrated altitude alerter and the barometric pressure setting knob has a push-to-reset function. The scale setting allows the selection of English and metric operation. An ICAO encoded altitude output is available as an option. The AD32 is a modular design, with low power consumption (less than 10 W) and low weight (1.25 kg), optimised for state-of-the-art avionics applications. Extensive Built-In-Test (BIT) assures safe operation.

Employing an RS 232 maintenance interface, the AD32 can easily be configured for different applications, ranging from business aviation up to regional aircraft, transports and helicopters. The AD32 air data display meets or exceeds the requirements of the FAA Technical Standard Order (TSO) RVSM accuracy requirements.

Other features and benefits of the system include:
- Solid-state pressure sensors
- Push-to-test and continuous BIT
- BIT failure memory
- ARINC-408, 3-ATI with IMI ring pointers.

Specifications
Signal inputs
Primary power: 28 V DC (option 26 V AC, 400 Hz)
Emergency power: 28 V DC
Lighting power: 28 V DC (option 5 V DC)
Alert lamp power: 28 V DC (option 5 V DC)
Interface: ARINC-429 (2 receive channels)
TAT probe: 500 Ω (option 50 Ω) @ 0°C

Signal Outputs
Encoded altitude: ICAO/TSO C-88a
Visual/aural alert: Relais outputs
Warning flag valid: GND/28 V DC
Baro potentiometer: Option
Pressure altitude: –1,000 to +53,000 ft
Baro corrected altitude: –1,000 to +53,000 ft
Altitude rate: 0 to 20,000 ft/min
Indicated airspeed: 0/40 to 450 kt
Calibrated airspeed: 0/40 to 450 kt
True airspeed: 0/100 to 599 kt
Max allowable airspeed: VMO 150 to 450 kt
Mach number: 0.200 to 0.999 m
Total air temperature: –60 to +99°C
Static air temperature: –99 to +60°C
Baro setting: QNH 20.67 to 31.00 in Hg (700 to 1,050 mbar)
Altitude scale error
–1,000 to 20,000 ft ± 10 ft
20,000 to 29,000 ft ± 20 ft
29,000 to 41,000 ft ± 30 ft
41,000 to 53,000 ft ± 50 ft
Compliance: FAA TSO-C10b, -C88a, -C106, RTCA/ DO-178B Level A, DO-160D
Operating temperature: –20 to +70°C
Reliability: MTBF 20,000 hr (estimated)

Status
In production and in service.

Contractor
Revue Thommen AG.

Altimeters (3 in), types 3A and 3H

Type
Aircraft instrumentation.

Description
Thommen 3A and 3H 3 in altimeters are barometric instruments with the 3H incorporating an optical encoder to provide height reporting output. The counter-pointer display is actuated by a high-performance mechanism with a high stability beryllium capsule. A built-in vibrator minimises friction and optimises accuracy in use. Temperature is compensated by a bimetallic element.

The counter has three moving drums, which show tens of thousands, thousands and hundreds of feet. The pointer indicates 1,000 ft per revolution, on a scale calibrated at intervals of 20 ft. Warning flags are displayed on the counter between 10,000 and 0 ft respectively, below 0 ft. Altimeters can be supplied with measuring ranges of 1,000 to 20,000, 35,000 or 50,000 ft and 300 to 6,000, 10,000 or 15,000 m.

The barometric-counter indicates the pressure as four digits as single baro in mbar or in Hg or as dual baro in mbar/in Hg. Adjustment for QNH/QFE is provided by a knob on the front, built-in stops are included.

The optical electronic system supplies altitude data in digital form, ICAO coded (Gillham) for SSR pressure altitude transmission in 100 ft increments, to standard accuracy TSO-C88.

Encoding Altimeters are provided with a code ON/OFF flag alarm, which indicates any encoding power malfunction.

An electrical indication of pressure (3H67/3H66 only) can be provided by potentiometer geared to the baro-scale mechanism.

The altimeters are installed in lightweight aluminium ARINC, square and round cases.

Lighting is offered in different voltages and different colours. All instruments are available with NVG-compatible lighting.

Status
In production and in service. The 3 in encoding altimeters are installed as primary or standby instruments in civil and military aircraft.

Contractor
Revue Thommen AG.

Altimeters, types 3A and 3H, 2 in

Type
Aircraft instrumentation.

Description
Thommen 2 in altimeters are barometric instruments. The counter-pointer display is actuated by a high-performance mechanism with a high-stability beryllium capsule. A built-in vibrator minimises friction and optimises accuracy in use. Temperature is compensated by a bimetallic element.

The instruments are split into many type designations in order to meet individual customer requirements and are installed in lightweight aluminium ARINC square and round cases.

The counter has two moving drums, which show tens of thousands and thousands of feet. Three fixed zeros are printed on the display. The pointer indicates 1,000 feet per revolution on a scale calibrated at intervals of 20 ft. Warning flags are displayed on the counter between 10,000 and 0 ft respectively below 0 ft. Altimeters can be provided with measuring ranges of 1,000 to 20,000, 35,000 or 50,000 ft, and also of 300 to 6,000, 10,000 or 15,000 m.

The barometric counter indicates the pressure in mbar or in Hg, or in both units of measurement. The display uses a single four-digit scale when indicating barometric pressure in mbar or in Hg, and a dual scale when indicating pressure in mbar/in Hg. QNH/QFE adjustment is provided by a knob on the front face of the instrument. It has built-in end stops.

An electrical indication of pressure is provided on the 3H models by a potentiometer geared to the baroscale mechanism.

The altimeters are installed in lightweight aluminium ARINC, square and round cases.

Lighting is offered in different voltages and different colours. All instruments are available with NVG-compatible lighting.

Status
The 2 in altimeters are installed as primary or standby instruments in civil and military aircraft.

Contractor
Revue Thommen AG.

Clocks and chronographs
Mechanical 8-day aircraft clocks, Type. B13, B15; Mechanical 8-day aircraft clocks with integral revolving bezel, Type B17, B18

Type
Aircraft instrumentation.

Description
Revue Thommen's range of aircraft clocks and chronographs have a quartz-controlled movements with 8 steps or 1 step per second, providing for high accuracy and reliability. All instruments are customer-specific (see type designation) and are mounted to facilitate easy servicing. There is standard 3-function (B13, B17) push button operation of start/stop/reset-to-zero in a fixed sequence, or (B15, B18) dual-mode operating knob with a 2-function push sequence (start/stop) and an independent fly-back-to-zero operation; fly-back-to-zero is possible while the counter is running. The integral revolving bezel models (B17, B18 only) is used as a second elapsed time function. Completely integrated in the case, the revolving bezel is set by a knob at the lower right hand corner. Thommen offers an optional continuously running second display that cannot be stopped, neither by operation of the elapsed-time function nor by normal clock-setting; however, this function can be completely stopped for to-the-second synchronisation with a time signal. The complete range of aircraft chronographs is available with

Altimeters (3 in), type 3A and 3H   0018188

Altimeters, type 3A and 3H, 2 in   0018189

elapsed time ranges of 12 minutes, 60 minutes and 12 hours and supplied in round case or semi-ARINC cases.

Integral lighting is offered in a range of different voltages and colours, with NVIS-compatibility optional. All instruments with lighting are equipped with antireflective-coated glasses according to MIL-C-14806.

Type designation
- B13- Chronograph, 3 Functions (3F)
- B15- Chronograph, 2 Functions (2F)
- B17- Chronograph, 3 Functions with integral revolving bezel
- B18- Chronograph, 2 Functions with integral revolving bezel
- 941- movement 3F, with 60 min elapsed time range
- 942- with 12 h elapsed time range
- 944- with 12 min elapsed time range
- 945- movement 2F, with 60 min elapsed time range
- 946- with 12 h elapsed time range
- 948- with 12 min elapsed time range
- Movement 3F, with continuous running second
- 951- with 60 min elapsed time range
- 952- with 12 h elapsed time range
- 954- with 12 min elapsed time range
- Movement 2F, with continuous running second
- 955- with 60 min elapsed time range
- 956- with 12 h elapsed time range
- 958- with 12 min elapsed time range
- .20- Round case, black
- .30- Semi-ARINC case, black
- .40- Semi-ARINC case, grey
- .01- Dial markings fluorescent white (DTD 573)
- .02- Dial markings lusterless white
- .03- Dial markings fluorescent yellow
- .04- Dial markings luminescent green (Tritium)
- .05- Dial markings day glow yellow (Saturne)
- .07- Dial markings long-time afterglowing luminescent green (DIN 67510)
- .00.0- Without lighting
- .05- Lighting 5 V AC/DC
- .28- Lighting 28 V AC/DC
- .1- Lighting white only (MIL-L-27160)
- .2- Lighting red and white
- .3- Lighting red only (MIL-L-25467)
- .4- Lighting NVIS green (MIL-L-85762).

Specifications
Movement: Chronograph with 3 functions (B13, B17); chronograph with 2 functions (B15, B18)
Number of jewels: 15
Minimum running time: 8 days
Elapsed time range: 60 min
Operating temperature: –35 to +55°C

Status
In production and in service in a wide range of civilian and military aircraft.

Contractor
Revue Thommen AG.

Mach/Airspeed Indicators (MAI), Type 5

Type
Aircraft instrumentation.

Description
Thommen Type 5 Mach Airspeed Indicators (MAI) are pneumatically-operated instruments deriving indicated airspeed and Mach information from the pitot-static sources, which conform to MS 33649. They can be supplied in 2 in or 3 in cases. The combined displacements of the airspeed pointer and the MACH-disk determine the MACH number, which is indicated simultaneously with airspeed by the airspeed pointer. The control relays for the outside warning system (Vmo and Mmo) are actuated by optical detection devices. There are many type designations in order to meet specific customer requirements. Instruments can be supplied with various measuring ranges from 40 to 650 kt, M0.3 to 1.8 and from 70 to 1,200 km/h, M0.3 to 1.8.

Mach/airspeed indicators, type 5, 3 in   0018187

The Mach airspeed indicators are installed in lightweight aluminium ARINC and square cases.

Lighting is offered in different voltages and different colours. All instruments are available in NVG-compatible lighting.

Status
In production and in service. The 3 in MAI are installed as primary or standby instruments in civil and military aircraft.

Contractor
Revue Thommen AG.

MD32 maximum allowable airspeed/digital Machindicator

Type
Aircraft instrumentation.

Description
The MD32 maximum allowable airspeed/digital Mach indicator provides computed/indicated airspeed and Mach number. It has airspeed and max allowable airspeed pointers, a digital Mach display and a fast/slow analogue output. The MD32 is a self-contained, integrally lighted 3 in electronic instrument which can be installed as primary instrument meeting the requirements of (J)TSO-C46a. The MD32 is designed to be a direct replacement for older electromechanical instruments and, as such, meets form, fit, function of such instruments to guarantee easy retrofit. A type-specific adapter cable facilitates different connector pin assignments. The MD32 features a micro-controller and incorporates absolute and differential solid-state pressure sensors for static and pitot pressure with the following operating ranges:
- Airspeed 0/60 to 420 kts (0 mark)
- Max allowable airspeed, VMO/MMO, calibrated for specific aircraft type
- Mach number 0.300 to 0.999
- Altitude -1,000 to +53,000 ft.

The instrument is designed to operate from a 26 V/400 Hz primary power source. Lighting is controlled by an external 5 V DC voltage. A command bug sets the null for the airspeed fast/slow analogue output and a test switch activates a Built-In Test (BIT) procedure. An RS-232 serial interface provides for maintenance purposes. Discrete outputs are available as an option.

Specifications
Compliance: FAA TSO-C46a, RTCA/DO-178B Level A, -160D
Operating temperature: 20 to 70°C
MTBF: 20,000 hrs (estimated)

Signal inputs
Primary power: 14/28 V DC
Lighting power: 5 V DC

Signal Outputs
Fast/slow airspeed error: ±2 V DC
Warning flag valid: GND/28 V DC

Status
In production.

Contractor
Revue Thommen AG.

PC-21 Avionic training system

Type
Avionics system.

Description
The Pilatus PC-21 is a highly advanced twin-tandem-seat turboprop aircraft, designed to provide customised, straight-through training for pilots of modern multirole fast jet aircraft. The PC-21 was conceived in 1998 to fulfil a projected requirement for an aircraft which could provide a realistic military glass cockpit environment for pilots during the earliest stages of their training, while retaining the inherent lower costs of a turboprop compared with the more traditional lightweight jet aircraft.

The very 'clean' glass-cockpit environment of the aircraft closely mirrors that of many of the most modern fighter aircraft. The inclusion of a Head-Up Display (HUD), Up-Front Control Panel (UFCP), Hands-On Throttle and Stick (HOTAS) and full-size portrait MultiFunction Displays (MFDs) provides for a familiar environment to pilots of aircraft such as the F-18, AV-8B, F-16 and Gripen, who would find themselves intuitively able to locate and operate the majority of the controls on first inspection. This familiarity extends to system moding, which has been designed to illustrate the ability of the PC-21 training system to provide moding and switching capable of mirroring that of any required frontline type.

In more detail, each pilot is presented with a custom-designed UFCP, which is used as the primary means for input/manipulation of information into the aircraft's mission/navigation/attack system. The front cockpit features a CMC Electronics HUD based on the Sparrow Hawk™ design (see separate entry), while the rear cockpit has a full-colour HUD repeater display showing the HUD camera view, with an overlay of the HUD symbology. Three Barco portrait 8 × 6 in Active Matrix Liquid Crystal Displays (AMLCDs) are arranged across the cockpit. The central display is a dedicated Primary Flying Display (PFD) including Horizontal Situation Indicator (HSI) information, with the two flanking displays showing navigation, mission and tactical data, as selected via bezel-mounted and/or HOTAS/UFCP buttons. In addition, two Meggitt Avionics' 3 ATI secondary displays flank each UFCP, displaying PFD, systems and engine data. All cockpit displays are NVIS Class B compatible.

The PC-21 navigation/attack system is based around a Laser Inertial Navigation Sensor (LINS), coupled with GPS and Kalman Filter (KF). Systems are linked by ARINC 429 and MIL-STD-1553B interfaces. System moding includes navigation, air-to-ground and air-to-air weapon aiming. System software was designed utilising the Virtual Avionics Prototyping System (VAPS), which enables fast and efficient changes to be made at the customer's request.

PC-21 Training system
Designed from the outset primarily as an advanced training aircraft, the PC-21 incorporates a sophisticated training system aimed at the requirements of basic and LIF students, pilots and navigators. The main features of the system include:
- Customisable software/emulation – the employment of the VAPS system for software design results in a very important ability for the PC-21 – system software can be quickly and easily (cheaply) changed to customer requirements
- Simulated weapons training – the aircraft includes a full simulated weapons capability, with an extensive onboard and customisable inventory, including freefall bombs, missiles and gun. In addition, the system can be fitted with an internal datalink system
- Training modes – aircraft displays, navigation sensor performance/availability, system moding and simulated weapons status (misfire/hang-up) can all be modified by the instructor to increase training effectiveness for each phase of training, according to requirement

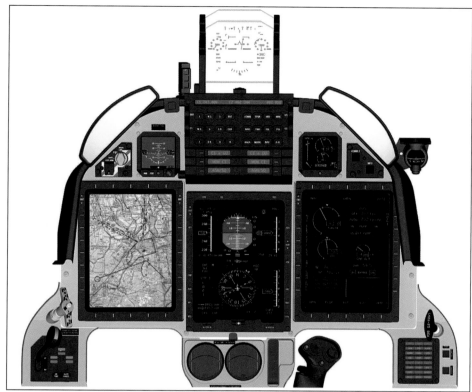

The Pilatus PC-21 front cockpit features a CMC Electronics Sparrow Hawk ™ -derived HUD, proprietary UFCP, three Barco 8 × 6 in AMLCD MFDs and two Meggitt Avionics Secondary Flight Display System (SFDS) AMLCDs (Pilatus) 0524721

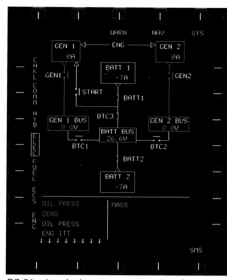

PC-21 electrical system page showing status prior to engine start (Pilatus) 0524720

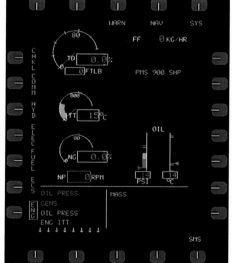 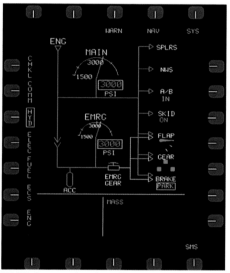

VAPS Screen showing MFD systems hydraulic (HYD) page (left) and engine (ENG) page. Note the lower area of each portrait display is utilised for status and system messages; on the ENG page, further messages for display are indicated by the multiple arrows. The Power Management System (PMS) shows engine output limited to 900 shp. Comparison of the ENG and HYD pages clearly shows the use of colour to indicate system status; in this case, normal indications are green, caution indications are amber and warning indications are red. Note also that Master Armament Safety Switch (MASS) status is indicated on all systems pages. All of these features ensure rapid and intuitive assimilation of essential data, utilising standard conventions for format and layout (Pilatus) 0524719

VAPS Screen showing MFD systems hydraulic (HYD) page (left) and engine (ENG) page. Note the lower area of each portrait display is utilised for status and system messages; on the ENG page, further messages for display are indicated by the multiple arrows. The Power Management System (PMS) shows engine output limited to 900 shp. Comparison of the ENG and HYD pages clearly shows the use of colour to indicate system status; in this case, normal indications are green, caution indications are amber and warning indications are red. Note also that Master Armament Safety Switch (MASS) status is indicated on all systems pages. All of these features ensure rapid and intuitive assimilation of essential data, utilising standard conventions for format and layout (Pilatus) 0524718

- Cockpit decoupling – a switch, located in the front cockpit, can be set to identify the instructor's position. From this position, the instructor will then have access to a cockpit

'de-couple' button, allowing him to separate his cockpit from that of the student. This feature allows the instructor to access training modes and sensor data not available to the trainee

- Mission Planning System (MPS)/simulator integration – in common with modern multirole aircraft, the PC-21 includes an MPS as an integral part of the mission avionics system, facilitating rapid mission planning on the ground, with aircraft loading via data disk. The VAPS software that provides the aircraft avionic capabilities is also utilised in the ground-based and simulator training environments, ensuring that both the aircraft and ground-based training systems always function to the same software release standard

- Mission/Health and Usage Monitoring System (HUMS) data recording – the mission planning data disk also acts as the data storage medium for mission recording. Recording is automatic from power-on to engine shutdown. The system records the PFD, MFDs, and HUD camera (in colour). All MFD pages are recorded, whether displayed or not, to enable a full mission playback for debriefing. Provision for event marking is also provided. A separate solid-state medium is utilised for the recording of engineering data for the HUMS.

Status

PC-21 mission/avionic system definition was carried out in 1999 to identify hardware and software requirements for the training system. Over the same period, initial Proof of Concept (POC) work was carried out on a converted PC-7 Mk II airframe, with an increased power engine, new wing and partial avionics, consisting of a HUD, UFCP and MFD system. The first complete PC-21 aircraft, HB-HZA, was unveiled in April 2002. Pilatus has since reported considerable interest in the PC-21 training system from a number of air forces worldwide, although, as of the second quarter of 2004, no orders have been reported.

Contractor

Pilatus Aircraft Ltd.

Vertical Speed Indicator (VSI), Type 4A

Type

Aircraft instrumentation.

Description

Thommen VSIs operate from the static pressure system using connectors that conform to MS 33649. Type 4A58 VSIs are supplied in a 2 in case and Type 4A16 in a 3 in case. A pointer indicates vertical speed on a fixed dial calibrated in/ft per minute (FPM) or metres per second (m/s). Zero adjustment, used to compensate for internal stresses in the mechanism, is operated by turning the whole mechanism frame in relation to the case/dial assembly. VSIs can be supplied with many options to meet customer requirements and can have measuring ranges up to ±20,000 FPM or ±100 m/s.

The Vertical Speed Indicators are installed in lightweight aluminium ARINC, square or round cases.

Lighting is offered in different voltages and different colours. All instruments are available in NVIS-compatible lighting.

Status
In production and in service. The 3 in Vertical Speed Indicators are installed as primary instruments in civil and military aircraft.

Contractor
Revue Thommen AG.

Vertical Speed Indicator (VSI) Type 4A16, 3 in
0018186

Turkey

CDU-900 flight management system – Control Display Unit

Type
Control and Display Unit (CDU).

Description
The ASELSAN CDU-900 is a cockpit control and display unit for fixed-wing and rotary-wing applications. The CDU-900 provides processing and interface control for all flight management functions, including navigation guidance, flight management, communication and navigation systems management, status monitoring and Built-In-Test (BIT) capabilities.

The CDU-900 serves as the primary avionics computer for communications control, navigation/guidance, equipment status monitoring and MIL-STD-1553 data bus control. The system provides the main aircrew interface for flight management and INS or GPS/INS navigation/management in a reliable, low risk/cost design.

CDU-900 (Aselsan)
1128604

CDU-900 operations are executed using a full alphanumeric keypad, arrow keys, function keys and eight line select keys on the front panel of the unit. The CDU-900 performs guidance computations, data management and subsystem control functions in the FMS-800 system.

Specifications
Display: 220 × 170 pixels, NVIS compatible
Dimensions: 181 × 146 × 168 mm
(7.125 × 5.75 × 6.6 in)
Keyboard: full alphanumeric with 7 generic key functions
Cooling: convection
Interface: Dual MIL-STD-1553B databus, RS-232, ARINC 429, discretes
Weight: 4.5 kg (10 lb)
Power: 28 V DC, 34 W (max)
Compliance: DO-160C, MIL-STD-810D, MIL-STD-461D

Status
As of November 2005, in production and in service installed in a variety of platforms.

Contractor
Aselsan Inc, Microelectronics, Guidance and Electro-Optics Division.

MFD-268E Multi Function Display

Type
MultiFunction Display (MFD).

Description
The ASELSAN MFD-268E MultiFunction Display is a colour flat panel display using Active Matrix Liquid Crystal Display (AMLCD) technology, with an active display area of 6 × 8 in. The display contains internal graphics, video and input/output processing capabilities to support the generation of display formats, using data obtained from system interfaces to the MFD. The MFD-268E is capable of displaying video, graphics, video with

graphics overlay, split screen video/graphics and split screen graphics/graphics.

The ASELSAN MFD-268E has analogue and digital input/output interfaces and a MIL-STD-1553 interface. The -1553 interface allows operational software modification without removal of the display from the aircraft. The MFD is bus programmable in the Ada computing language for customer display format generation.

Specifications
Display type: Colour AMLCD
Display area: 152 × 203 mm (6 × 8 in)
Resolution: 1,024 × 768
NVIS compatibility: MIL-L-8562A, Class B
Dimensions: 208 × 259 × 254 mm (8.2 × 10.2 × 10 in)
Video input: digital, monochrome and colour
Interface: dual MIL-STD-1553B databus, RS-232, ARINC 429, analogue, synchro and discrete
Weight: 9.3 kg (20.5 lb)
Power: 22.5 to 32 V DC operational in accordance with MIL-STD-704A

Status
As of November 2005, in production and in service installed in various platforms.

Contractor
Aselsan Inc, Microelectronics, Guidance and Electro-Optics Division.

MFD-268E showing Primary Flight Display (PFD) and Horizontal Situation Indicator (HSI) information (Aselsan)
1128603

Ukraine

SKI-77 Head-Up-Display (HUD)

Type
Head-Up Display (HUD).

Description
The SKI-77 HUD is designed for use by transport aircraft and helicopters. It provides a collimated display of sensor data, together with appropriate

computing capacity. It includes automatic and manual control of image brightness to match background illumination conditions, together with automatic changes in data formatting to match the different stages of the mission profile.

Specifications
Instantaneous field of view: >20° (V) × 30° (H)
Exit pupil size: 45 × 90 mm

Data display colour: green monochromatic
Weight: <21.5 kg
Power: 115/200 V AC, 400 Hz, <300 W 27 V DC, <5 W 6 V AC, 400 Hz, <5 VA
Temperature limits: – 60 to +60° C

Status
In service in the An-70 tactical transport aircraft.

Contractor
Arsenal Central Design Bureau.

*The Antonov An-70
tactical transport
(Paul Jackson)*
0525782

United Kingdom

5ATI Integrated Display Unit (IDU)

Type
Electronic Flight Instrumentation System (EFIS).

Description
Using Active Matrix Liquid Crystal Display (AMLCD) technology combined with built-in analogue signal and digital databus interfacing, Smiths Aerospace has incorporated EFIS functionality into a single 5ATI instrument. The depth of the unit is compatible with electromechanical ADI/HSI indicators and provides a simple upgrade path for many classic aircraft, with no separate symbol generator required. Other features and benefits of the unit include compatibility with modern weather radar, TAWS, GPS and FMS. EFIS and engine display variants are available.

Specifications
Weight: 3 kg
Dimensions: 5ATI
Usable screen area: 4 × 4 in
Power: 26 V AC or 28 V DC, 35 W
Brightness: 100 fl
Range: 2000:1
Reliability: >15,000 h MTBF
Interface: RS 422, ARINC 429/568, analogue, discrete, digital encoder, analogue input, synchro, ARINC 708 WXR, TAWS
Qualification: DO-160C
Software: DO-178B
Compliance: TSOc3d, 4c, 6d, 52a, 113

Status
The 5ATI integrated display unit is compatible with a wide range of commercial and military platforms, both fixed- and rotary-wing:
- Certified for use in Boeing 747 Classic aircraft as an EFIS
- Specified and flying on the Agusta-Westland Super Lynx 300 helicopter as an Electronic Power Systems (EPSs) instrument
- Flying in Boeing 707 retrofit applications as an EFIS
- Suitable for other platforms such as C-130, Boeing 737 and MD-80 aircraft.

Contractor
Smiths Aerospace.

300 RNA series Horizontal Situation Indicators (HSI)

Type
Flight instrumentation.

Description
The 300 RNA range of Horizontal Situation Indicators (HSIs) is intended for civil and military fixed-wing aircraft and helicopters. Each instrument consists of a mainframe, synchro frame and electronics. Large-scale integrated circuits are used for signal processing and synchros are used to drive the various displays.

EHSI VOR/ILS Mode

EHSI – VOR/ILS Mode (Centred)

EHSI Map Mode

EHSI (Plan) Mode

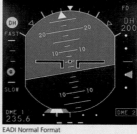
EADI Normal Format

EHSI Map Mode (Centred)

The Smiths Industries 5ATI display upgrade for B747-100/-200/-300 aircraft   *0051686*

Smiths' 330 series HSI is used in the Panavia Tornado GR1 (left of the map display) (E L Downs)
0131117

The instruments can be used in Nav, Tac, App or ADF modes. The range display is a four-digit electronic module at the top of the instrument face reading up to 999 n miles. Another numeric display is used to indicate the setting of the command track pointer. A knob is provided for selecting the relevant runway QDM on the command track pointer. These instruments provide complete ILS information as well as Tacan and ADF displays.

Status

304/305 RNA HSIs are used in the BAE Systems Nimrod and 748 aircraft as integral parts of the SFS6 flight systems. The 309 RNA HSI equips Royal Navy Westland Sea King helicopters and Royal Air Force BAE Systems Hawk trainers. The 306/307 RNA is used in the SEPECAT Jaguar and BAE Systems Strikemaster. The 330 series HSI is used in the Royal Air Force's Panavia Tornado aircraft and Chinook helicopters.

Contractor

Smiths Aerospace.

0732 KEL series low-cycle fatigue counter

Type

Flight/mission recording system.

Description

The 0732 KEL series calculates cumulative low-cycle fatigue and records the number of engine starts and engine hours. It also provides information on exceedances, banding of speeds, voltages and thermocouples, spool-up and spool-down times and snapshots of input data. The 0730 series records a maximum of 18 aircraft or engine input analogue parameters including up to six speed, six voltage, two thermocouple and four discrete parameters. These are all configurable in range and sensitivity to customer requirements. It provides outputs via RS-232 link to a data transfer device, printer or maintenance computer, and current and voltage sources for aircraft sensor excitation. It also provides an output to an integral LED display for manual interrogation of life usage exceedance fault codes.

Specifications

Dimensions: ¼ ATR dwarf
Weight: 2.5 kg
Speeds: pulse probe or tachogenerator 10 Hz to 25 kHz
Voltage: range selectable 0 to 0.5 V, 0 to 5 V, 0 to 40 V FSD
Thermocouple: 0 to 1,000°C Ch/Al
Discretes: 0 to 5 V, 0 to 28 V nominal

Status

In service with Swiss Air Force, the UK Royal Air Force Red Arrows aerobatic team and the Sultan of Oman's Air Force BAE Systems Hawks and Swiss Air Force Eurocopter Super Pumas. Selected for UK Royal Air Force Hawk aircraft and Eurocopter Cougar helicopters.

Contractor

Smiths Aerospace.

0826 KEL health and usage monitor

Type

Flight/mission recording system.

Description

The 0826 KEL health and usage monitor provides snapshots of input data, exceedance monitoring and incident monitoring. It also calculates cumulative low-cycle fatigue. It records up to eight speed, 32 voltage, 16 discrete and four vibration aircraft and engine parameters. Outputs are provided via RS-422 link to a data transfer device, printer or maintenance computer and the system has an integral LED display for manual interrogation of input parameter values.

Specifications

Dimensions: ³/₈ ATR
Weight: 5 kg
Speed: pulse probe 10 Hz to 4 kHz
Voltage: 0–5 V FSD
Discretes: 0–28 V nominal
Vibration: 20–750 Hz buffered

Status

In service in the BAE Systems 146.

Contractor

Smiths Aerospace.

0829 KEL health and usage monitor

Type

Flight/mission recording system.

Description

The 0829 KEL health and usage monitor provides snapshots of input data, exceedance monitoring and incident monitoring. It has capacity available to incorporate LCF counting. There are 30 aircraft and engine input parameters available, including up to eight speed, 15 voltage and seven discrete parameters, and the system incorporates an ARINC 429 databus. Outputs are via RS-422 link to a data transfer device, printer or maintenance computer and there is an integral LED for manual interrogation of input parameter values.

Specifications

Dimensions: ³/₈ ATR
Weight: 5 kg
Speed: pulse probe 10 Hz to 4 kHz
Voltage: 0–5 V FSD
Discretes: 0–28 V nominal
Vibration: 20–750 Hz buffered

Status

In service with the BAE Systems ATP.

Contractor

Smiths Aerospace.

2000 series Solid-State Combined Voice and Flight Data Recorder (SSCVFDR)

Type

Flight/mission recording system.

Description

The 2000 series Solid-State Combined Voice and Flight Data Recorder (SSCVFDR), integrates both cockpit voice and flight data recording functions into a single unit. The recorder complies with the latest EUROCAE ED-55 and ED-56A and the FAA TSO-C123/C124 requirements for airborne recording equipment. Several options are available on the unit, such as extended audio duration, onboard maintenance system interface or integral area microphone preamplifier. The baseline unit records a minimum of 25 hours of flight data, at 128 words/s, with 30 minutes' voice on four channels and can interface to any ARINC 573 or ARINC 717 digital flight data acquisition unit or form part of a health and usage monitoring system. Continuous self-test, front panel interface, for in situ replay and diagnostics checks, and an underwater locator beacon are standard features.

The SSCVFDR can be hard mounted in the aircraft without shockmounts, as there are no moving parts. Because both data and audio recording is achieved in the combined crash-protected unit, the SSCVFDR is particularly suitable for airframes where weight and space are restricted. Advanced design techniques and manufacturing processes, coupled with rigorous environmental testing, mean high reliability and no periodic maintenance.

Specifications

Dimensions: 124 × 320 × 194 mm, ½ ATR Short
Weight: <9 kg
Power supply: 115 V AC, 400 Hz or 28 V DC, 20 W (max)
Temperature range: –55 to +70°C
Altitude: –15,000 to +55,000 ft
Environmental: DO-160C
Reliability: 15,000 h MTBF

Status

In service with military and civil operators.

Contractor

Penny & Giles Aerospace Ltd.

2100 series MultiPurpose Colour Display (MPCD)

Type

MultiFunction Display (MFD).

Description

The Type 2100 MultiPurpose Colour Display (MPCD) meets the stringent requirements found in the bubble-canopy cockpits of combat aircraft. It provides a high-resolution full-colour display in all light conditions from low-level night to high-altitude day; special filters ensure full NVG-compatibility. The ruggedised CRT uses a stretched shadow-mask and a periodic focus electron gun to achieve very high screen brightness without loss of picture quality.

2000 series solid-state combined voice and flight data recorder

0005418

The 2100 series multipurpose colour display
0503946

The MPCD presents data or imaging in stroke, raster or raster with stroke in frame flyback on a usable screen area of 5 × 5 in (127 × 127 mm). Stroke symbology is displayed as the output of an external graphics generator. Raster modes, in a variety of line standards, come from a range of sources such as electro-optical sensors, radar and video map generators. The hybrid mode enables the precise overlay of stroke symbology on to the raster display.

Operator control is exercised by 20 momentary action keys. Display control is via four rocker switches. Display brightness is determined by the operator and automatically compensates for ambient light measured by sensors on the front bezel.

Specifications
Dimensions: 170 × 180 × 440 mm
Weight: 10.9 kg
Power supply: 200 V AC, 400 Hz, 3 phase, 180 VA
Usable screen area: 5 × 5 in (127 × 127 mm)

Status
In production with over 1,500 units delivered. The 2100 series complies with all relevant UK, NATO and US military standards and specifications. The unit is fitted in the US Marine Corps AV-8B Harrier II, US Navy/US Marine Corps F/A-18 C/D Hornet and Royal Air Force Harrier GR. Mk 7.

A version of the MPCD with a 6 × 6 in (152 × 152 mm) usable screen area has been developed for the Eurofighter Typhoon. Other variants are under development or in production for the Tornado GR. Mk 4, Royal Air Force Jaguar and Italian/Brazilian AMX.

Contractor
Smiths Aerospace.

Accident Data Recorders (ADRs)

Type
Flight/mission recording system.

Description
The family of accident data recorders has been developed by Penny & Giles around a recycling mechanism using plastic-based magnetic recording tape. The track configuration is variable to meet customer needs, with up to eight tracks being currently available in various combinations of voice and data. These ADRs are fully protected according to all the major civil and military airworthiness requirements. Versions are currently in production for the Harrier GR Mk.7, Tucano and the EH 101 Merlin.

Harrier GR. Mk 7
This unit has four data and two voice tracks arranged in two channels, each having two data and one voice track. Recording duration is 2 hours.

Tucano
This has two data channels of two tracks each and three voice tracks which are recorded all the time the recorder is operating. Recording duration is 2 hours for data and 1 hour for voice.

EH 101 helicopter ADR
This ADR is in production.

Status
In service on the Harrier GR Mk.7, Tucano and in production for the EH 101 helicopter.

Contractor
Penny & Giles Aerospace Ltd.

ACCS 3300 digital map generator

Type
Avionic Digital Map System (DMS).

Description
GD UK has designed rugged map display equipment intended for installation in harsh airborne environments. The ACCS 3300 digital map generator has been selected for the UK Royal Navy Merlin helicopter. This design has a scalable architecture in order to meet a range of performance criteria, from static north-up maps to full rotating tracking maps with overlays. This will allow presentation of both raster and true vector data with efficient man/machine interface to facilitate zoom and a list of additional functions. The growth capability enables additional functions such as intervisibility, two- and three-dimensional displays. Map data is optimally compressed and stored to provide the operator with control of the map, to enable display of key map and overlay information.

The complexity of overlays can include text with sensor data, providing the ability to perform sensor and map fusion and object correlation presented on a range of low- to high-resolution displays.

Map database expertise has been achieved in compressing raster data and, more significantly where a structured dataset is required, in automatically generating vector data from a raster map using its own Unix-compatible knowledge system. The advantage of utilising vector map data is its easy interpretation, enhancement and attribution. The vector dataset and its associated overlays are more easily maintained and updated than raster.

Specifications
Dimensions: ½ ATR
Power supply: 115 V AC, 400 Hz, 110 W
28 V DC optional
Outputs: RS-170, RGB 625-line CCIRR
Environmental: MIL-STD-810E
Reliability: >2,000 h MTBF

Status
In service in the UK Royal Navy EH 101 Merlin helicopter.

Contractor
General Dynamics United Kingdom Ltd.

The ACCS 3300 digital map generator for the EH 101 Merlin helicopter with a Barco display
0504178

Active Matrix Liquid Crystal Displays (AMLCDs)

Type
Cockpit display.

Description
Smiths' flat panel Active Matrix Liquid Crystal Displays (AMLCDs) are compact, lightweight, high-resolution colour displays, designed to

Smiths Aerospace's 3 ATI airspeed indicator display (left), engine instrument display (centre), and airspeed indicator (right)
0001367

Smiths Aerospace's 3 ATI airspeed indicator display (left), engine instrument display (centre), and airspeed indicator (right)
0001368

meet military and commercial requirements, for both new-build and retrofit aircraft and helicopter installations. Smiths has production contracts for AMLCDs in all its display facilities, both in the UK and USA.

3 ATI flat panel in struments
Selected for the JPATS aircraft, 3 ATI flat panel instruments provide a cost-effective, state-of-the-art means of providing graphics in a high-visibility display. Features include construction based on five modules (one dispay card and four flexible I/O cards); up to 16 colours at 130 ft Lamberts illumination, reconfiguration by pilot-operated switch, ARINC 429 interface, NVIS option. Applications include air data instruments, engine instruments, system status displays and standby displays.

These displays have been selected by Raytheon for the JPATS aircraft; they are also used for the V-22 standby flight display and for the F/A-18 E/F engine fuel display. Civil applications of the 3 ATI display include the Boeing 717.

Specifications
Dimensions: 8.1 × 8.1 × 19.0 mm
Weight: 1.58 kg
Power: 28 V DC, 2 A maximum
Temperature: –20 to +70°C
Reliability: >15,000 h MTBF

5 ATI flat panel instruments
Smiths' 5 ATI design provides a fully self-contained solid-state instrument, capable of directly interfacing to either analogue or digital sensors and presenting primary flight and/or engine information using graphics symbology. The 5 ATI flat panel design uses standard mountings, with no additional requirements for remote symbol generators or cooling air – they are, therefore, suited to upgrade and retrofit EFIS requirements, as well as new-build installations. Features include self-contained interface and graphics processing; analogue and/or digital ARINC 429 or MIL-STD-1553B interfaces; EADI and EHSI formats; ADI/HSI and

Smiths Aerospace's 3 ATI airspeed indicator display (left), engine instrument display (centre), and airspeed indicator (right) 0001369

Smiths Aerospace's 5 ATI attitude director (left) and map/weather display (right) 0001365

engine instrument formats; video option (for example HUD), GPS/CNS/ATM compatibility; passive cooling.

Applications include retrofit and OEM installations, standby displays, message displays, video monitor displays.

These displays have been selected for the Royal Australian Air Force Hawk aircraft. Each twin-cockpit aircraft will have up to six identical display units to show flight, navigation, weapon and system symbology, plus digital map and sensor displays. Fitted to some Boeing 747 aircraft.

Multifunction Control Display Unit (MCDU)

Smiths' flat panel MCDU employs a large full-colour AMLCD which complements modern military and civil cockpits. Features include: flexible reconfiguration of front panel keys; passive cooling; alternative ARINC 429, MIL-STD-1553, RS-422 data-busses. Applications include GPS; CNS/ATM; ACARS; weather radar; FMS; Satcom/radio control; video, cockpit and maintenance displays.

Smiths Aerospace's 5 ATI attitude director (left) and map/weather display (right) 0001366

Smiths Aerospace's multifunction control display unit 0001364

Selected by the UK MoD for the Nimrod MRA.4 aircraft upgrade.

5 × 6 in Electronic Display Unit (EDU)

Smiths' 5 × 6 in Electronic Display Unit (EDU) presents engine and utility system parameters on a multicolour, flat panel AMLCD. Selected for the BAE Systems' Hawk and the Eurofighter Typhoon.

MultiFunction Glareshield Display (MFGD)

Smiths' MultiFunction Glareshield Display (MFGD) is designed for glareshield installation to provide the pilot with continuous peripheral awareness, whether in head-up or head-down attitude. The unit utilises full-colour AMLCD and is suited to the display of tactical messages in the CNS/ATM environment. Boeing has selected this display as a Para Visual Director (PVD) for the B777.

Contractor

Smiths Aerospace.

AD1990 radar altimeter

Type

Flight instrumentation, radar altimeter.

Description

The AD1990 radar altimeter is a covert radar altimeter, which was designed to meet the UK Royal Air Force's needs in the 1990s. The AD1990 radar altimeter directly replaced the initial-fit altimeter in the UK Royal Air Force Tornado aircraft, using all existing fixtures and fittings.

The digital signal processing techniques incorporated in the receiver allow the extraction and simultaneous tracking of height both above the ground and above obstacles such as trees. These two outputs enable the pilot to operate more safely when flying at low level and are also used by the Terrain Reference Navigation (TRN) system to enhance overall navigation performance. The AD1990's fast dynamic response time eliminates the need for groundspeed compensation of height data within the TRN system. Inherent in the signal processing technique is the ability to identify and reject unwanted signals from underslung stores and landing gear, a traditional problem for radar altimeters. Reliable operation is obtained from its maximum operating altitude of 5,000 ft down to ground level.

An important innovation, at the time, was that the altimeter remains covert in operation, rendering it virtually undetectable by the enemy. Such Low Probability of Intercept (LPI) is achieved by spreading the transmitted signal over a very wide bandwidth through the application of pseudo-random phase modulation and adaptive power tailoring which, in addition, gives a high resistance to jamming. AD1990 can also be applied to other modern military aircraft where the ability to remain undetected is the key to mission success.

In addition to analogue height output, the system can be configured in either Panavia or MIL-STD-1553B interfaces.

Specifications

Dimensions: 109 × 154 × 318 mm
Weight: 5.25 kg
Power supply: 28 V DC, 55 W (max)
Frequency: 4.3 GHz
Range:
(height) 0–5,000 ft
(speed) 0–800 kt
(pitch) 0 to ±60°
(roll) 0 to ±60°
Accuracy: ±3 ft or ±3%, whichever is greater
Temperature: −55 to +90°C

Status

In service in Panavia Tornado aircraft.

Contractor

Selex Sensors and Airborne Systems.

AE3000FL series wideband analogue S-VHS data recorders

Type

Flight/mission recording system.

Description

The AE3000FL is a family of 8, 12 and 18 MHz S-VHS analogue recorder/reproducers designed specifically for critical ELINT, SIGINT, ASW and telemetry applications.

There are single- and multichannel variants of the system within each bandwidth range. Other input configurations are available to customer requirements. Some models in the range can be switched between single- and multichannel operation for pre- and post-detection recording from FM, linear and logarithmic receivers. Analogue data can be reproduced in both analogue and 8-bit digital formats.

The AE3000FL series is noted for its excellent reliability, even under the most arduous field conditions; each system is totally self-contained, with power supply and record/reproduce capability, together with all necessary channel multiplexing/demultiplexing systems, if applicable.

Options include a built-in IRIG timecode generator/reader (with Avalon extensions for event marking and year recording) and AV mounting hardware to enable the AE3000FL to be installed into any surveillance aircraft.

The AE3000FL requires no routine calibration and offers a typical MTBF of over 8,000 hours (including scanner). All recorders are backed by a 5,000 hour/3-year warranty.

Status

In production and in service. During the fourth quarter of 2000, Raytheon selected the AE3170FL wideband recorder for a major US Air Force airborne surveillance platform upgrade. The AE3170 features alternative two-channel recording, enabling the operator to quickly change from single-channel 12 MHz recording to one channel at 8 MHz, with a separate 4 MHz channel.

Contractor

Avalon Electronics Ltd.

Avalon Electronics AE3000FL series wideband analogue S-VHS data recorder (Avalon Electronics) 1047895

AE8000 high-performance disk recorders

Type
Flight/mission recording system.

Description
Avalon's third-generation AE8000 series wideband data recorders provide for high-quality SIGINT, ASW and telemetry data recording applications.

The AE8000 series are offered in a range of compact, portable and high-capacity/scaleable configurations; variants recording conventional baseband signals are available, while the latest models have built-in IF/baseband and baseband/ IF converters.

Some baseband models are designed as direct replacements for legacy 6, 8, 12, 16 and 25 MHz cassette tape recorders, while others extend the recording range to 100 to 200 MHz. The flexible IF/baseband interface uses advanced digitisation and decimation techniques to down-convert standard IFs in the 70 to 160 MHz range to a base bandwidth of 70 MHz, or lower if required. These DSP-based converters provide improved Spurious Free Dynamic Range (SFDR) specifications compared to traditional analogue down-converters, permitting low amplitude signals, which would previously have been lost in tape noise, to be easily resolved. Data interfaces for LVDS digital and single/dual channel OC3 telecommunications applications are available.

Many AE8000 variants offer true read-after-write capabilities, whereby previously recorded data can be replayed without interrupting the recording process. Similarly, the LOOP recording feature permits extended mission recording while passages of interest are tagged for transcription to the optional on-board AIT-3 tape streamer.

Other recent enhancements to the AE8000 family include comprehensive recorder control via an Ethernet 10/100 port and simple digital data extraction directly from the AE8000 disk crate to a workstation or computer network via the recorder's 80 Mb/sec SCSI port.

Status
In production and in service.

Contractor
Avalon Electronics Ltd.

Airborne digital video recorder

Type
Flight/mission recording system.

Description
Skyquest Aviation produces the DVCR2640 Digital Video Recorder with an associated dual-deck controller to enable automatic cascade recording (switches one machine to the next when tape is low on first machine), or parallel recording. The compact panel-mounted design of the unit allows the housing of a complete video recorder where previously only a control panel would be fitted. Two recorders can be set up in-line so that, using the optional multideck controller, one tape deck will automatically switch to the other when tape is running low.

One hour of video equates to 90,000 (PAL)/108,000 (NTSC) frames of full resolution stills, each of which may be selected, stored, displayed, downloaded to a PC or printed individually. Post-mission analysis is facilitated by the ability to select a single evidential frame (or sequence at 25/30 frames/s (PAL/NTSC) rather than relying on an event capture or replay from an analogue recorder. The system utilises miniature DV tapes (73 × 51 × 16 mm in a case).

Features and benefits of the system include:
- Vertical or horizontal mounting
- Front panel and multiple remote controls
- 24/28 V DC or 12 V DC option
- Dust, shock and vibration resistant
- Anti-condensation heaters

Avalon Electronics' AE8200 compact 25 MHz SIGINT disk recorder   1047896

- Built-in 'perfect' slow-motion replay
- Multideck controller controls 2 decks to provide Cascade or Instant Replay modes
- 5 min end-of-tape warning.

In IR mode, one deck will always record the output from the onboard camera system. The other deck can be rewound and replayed at any time without affecting the evidential recording on the first deck. Logic lockouts prevent the observer overwriting the evidential recording. The IR mode is extremely useful during chases, where firearms, drugs or evidence is thrown from a vehicle or ship. The evidential recorder continues to record the output from the onboard camera, while the observer can rewind and replay the tape in flight. In Cascade mode, the two VCRs act sequentially, with the second tape automatically starting as the first tape ends.

Specifications
Dimensions: 146 × 76 × 170 mm
Weight: 0.99 kg
Recording system: 2 head, helical scanning
Audio recording system: rotary head PCM
Video signal: PAL colour, CCIR, (NTSC option)
Tape speed: SP 18.83 mm/sec
Recording/playback time: 84 min (nominal dependant upon tape length, 60, 80 and 90 min tapes)

Status
In production and in service.

Contractor
Skyquest Aviation.

Airborne Video Solid state Recorder (AVSR™)

Type
Flight/mission recording system.

Description
The F/A-22's Data Transfer Equipment (DTE) with Mass Memory and Video Recorder (DMVR) uses solid-state flash memory to capture aircraft audio and video data. The F/A-22's Operational Video Debrief System (OVDS™) uses the digitally recorded data and a standard Personal Computer (PC) for post-mission analysis and post-flight operational debriefing – with instant and synchronised access to specific events.

The Airborne Video Solid-state Recorder (AVSR™), a drop-in replacement for single- and triple-deck videotape recorders, is a development of the DMVR. Using the same solid-state memory cartridge and ground station as the F/A-22, the design of AVSR™ has no moving parts and provides a cost-effective solution for the operational and maintenance problems that are typically experienced by tape recorder users.

The AVSR™ consists of an aircraft resident receptacle called the Airborne Video Receptacle (AVR) and a removable Video Data Cartridge (VDC). The AVR uses the F-22's Commercial Off-The-Shelf (COTS) solid-state compression technology to compress and store up to four channels of video data and one channel of audio data to the VDC. On the ground, a standard PC is used to debrief the mission. The PC interfaces to the VDC via a COTS Micro Cartridge Interface Device (Micro CID™) which allows the VDC to appear as a SCSI device to the PC. A MultiCID™ is also available to allow multiple VDCs to be played back simultaneously.

Status
In production for the F/A-22 Raptor.

Contractor
Smiths Aerospace.

Airborne Video Solid-state Recorder – Precision Attack (AVSR™ – PA)

Type
Flight/mission recording system.

Description
The Smiths Aerospace Airborne Video Solid-state Recorder – Precision Attack (AVSR™ – PA), a development of the AVSR™, was developed specifically to satisfy the high-resolution video recording and AVI playback requirements of the LANTIRN Bomb Impact Assessment (BIA) programme. The AVSR™ – PA's small form-factor and digital interface make it ideally suited for podded applications where weight, power and volume are crucial requirements. The system utilises the same solid-state memory cartridge used by the AVSR™ and the F/A-22's Data Transfer Equipment with Mass Memory and Video Recorder (DMVR).

The AVSR™ – PA consists of an aircraft-resident Airborne Video Receptacle (AVR) and a removable Video Data Cartridge (VDC). The AVR uses Commercial-Off-The-Shelf (COTS) solid-state compression technology to compress and store one channel of digital video data to the VDC. When on the ground, a standard PC is used to debrief the mission. The PC interfaces to the VDC via a COTS Micro Cartridge Interface Device (MicroCID™). The MicroCID™ allows the VDC to appear as a SCSI device to the PC.

The AVR serves as the receptacle for the VDC and interfaces to the video source via a HotLink digital interface. The video data is compressed using the JPEG algorithm prior to storage in the removable VDC. A MIL-STD-1553 interface is used to command the AVSR™ – PA as well as to provide BIT status and to support binary data recording. Current configurations allow memory densities of up to 30 Gbytes in the VDC using COTS Flash memory techniques.

Specifications

Dimensions:
(AVR) 228 ×146 × 114 mm
(VDC) 190 × 119 × 41 mm
Weight: 4.5 kg
Input power: 115 V AC, 400 Hz, 31 W typical
Temperature: −40 to +71°C operating

Status
In production.

Contractor
Smiths Aerospace.

Aircraft moving coil indicators

Type
Flight instrumentation.

Description
Weston Aerospace manufactures a range of moving coil indicators for the aircraft industry, in varying case sizes from 1 to 2 in, of modular construction, with integral lighting. Indicators are available to display parameters such as: electrical parameters, temperature, pressure, position, torque and speed. Options for indication of one, two or three parameters on each instrument are available.

Specifications
(typical)
Accuracy:
direct reading (mA or V)
±1.5% to ±4% full scale deflection
indirect reading (pressure, temperature) ±4.5% full-scale deflection
Display: black dial/white markings as standard; colour as required
Lighting: 5 and 28 V options to comply with MIL-L-27160

Status
In production and in service.

Contractor
Weston Aerospace, an operating unit of Esterline Sensors.

AMS 2000 multifunction Control Display Navigation Unit (CDNU)

Type
Control and Display Unit (CDU).

Description
The AMS 2000 multifunction CDNU is a flexible navigation and management system with embedded P(Y) code GPS that is easily configurable to suit the customer's requirements and the host airframe. Typically Search and Rescue (SAR) specific functions are added for aircraft/helicopters, having a SAR role.

This CDNU is configured as a navigation computer using GPS and, if available, a combination of sensors including IN, Doppler and Air Data. From this baseline, the CDNU capability may be increased to meet additional customer requirements – thanks to the modular approach adopted for hardware and software. The CDNU can provide a full mission management facility in single or dual configuration using MIL-STD-1553B and ARINC 429 databus. The CDNU can store up to 50 pre-programmable routes, each with up to 200 waypoints and can host a worldwide navigation database. SAR patterns that are pre-programmed include creeping line ahead, expanding box, spiral and sector searches.

The AMS 2000 CDNU provides centralised control and display of the chosen avionics suite. Data and instructions are inserted manually via the keyboard or Data Transfer Device (DTD). The DTD is used for flight planning and for post-flight data retrieval. A customised mission planning station is available.

The CDNU software is written in Ada. The non-volatile Flash memory is reprogrammable via the DTD. The CDNU's large memory capability allows a wide range of equipment interfaces to be supported – in addition to navigation sensors.

The AMS 2000 multifunction CDNU is compatible with GNSS and provides a control and display function for the Thales Avionics Satellite TRansceiver (STR) system. The CDNU may be used as a cockpit mission system or a tactical system, without any additional interfacing hardware being required.

Specifications
Size: 161.43 × 145.5 × 216.5 mm
Weight: 5 kg (max)

Weston Aerospace aircraft moving coil indicators 0018185

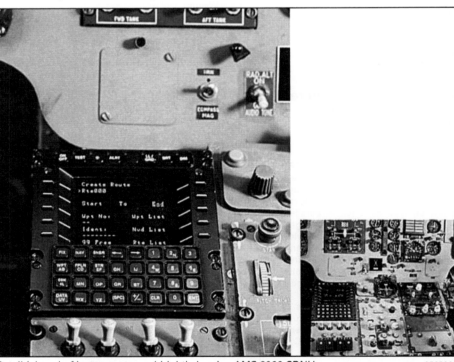

Detail (above) of instrument panel (right) showing AMS 2000 CDNU 0569000

Power: 28 V DC, 45 W (max – for full module complement), typically 36 W
Cooling: convection – no forced air required
Temperature range:
(operating) +55 to +71°C
(storage) –55° to +90°C
Environmental: MIL-STD-810E; MIL-STD-461C/D RTCA/DO-160C; DEF-STAN 5941
Memory: 4 Mbyte Flash PROM
I/O: MIL-STD-1553B; ARINC 419/429 (Hi/Lo) RS-232, -422; DC, AC, synchro, discretes
GPS: embedded C/A or P(Y) 5 channel to STANAG 4294. Includes Have Quick, BID 250 and KYK 13 interface

Status

Selected by the UK MoD for: Lynx HMA Mk.8, Merlin HM Mk.1, Merlin SH Mk.3, Nimrod MR2 Mk.2, Nimrod MRA 4, and Sea King Mk.3, 4, 5 and 6.

During the latter part of 2001, also selected for the British army's fleet of Lynx Mk9 helicopters as a direct replacement for Thales' RNS 252 and MAGR GPS navigation equipment. Trials and development work began at QinetiQ, Boscombe Down, in late 2001.

Contractor

Thales Avionics SA.

ASN-900 series tactical processing systems

Type

Mission Management System (MMS).

Description

The ASN-900 series tactical processing systems provide facilities to correlate and process data for display from the wide variety of sensors and navigation systems installed in modern maritime patrol aircraft, ASW/ASV helicopters and AEW aircraft. The system enables the tactical co-ordinator to display data in an easily assimilated form and assist in the solution of complex navigation, intercept and attack problems. ASN-902 and 924 systems, which are based on AQS-902/920 hardware, have a flexible design, which makes them readily adaptable for use as the central element in an integrated mission management system. It can replace the variety of individual sensor system control and display units with common integrated units, providing flexibility of operation. Standard ARINC 419, 429 and MIL-STD-1553B data interfaces allow installation as original equipment or as a retrofit.

Systems integration considerably improves the efficiency and flexibility of any mission avionics suite, minimises the weight of combinations of multiple sensors and simplifies logistic and training problems with common control units and multipurpose displays. Selex Sensors offers various levels of systems integration, extending to a totally integrated mission management system.

The ASN-902 system, in service in Sea King Mk 42B helicopters, has a monochrome display,

whereas the ASN-924, has a colour display to enhance the information presented to the tactical co-ordinator.

Status

The ASN-902 tactical processing system is in service in UK Sea King Mk 42B helicopters.

Contractor

Selex Sensors and Airborne Systems.

Attitude indicators – FH series

Type

Flight instrumentation.

Description

Attitude indicators of the FH series are electrically driven gyroscopic instruments which display aircraft attitude in two axes by a spherical-type presentation.

The product range encompasses instruments with 2, 3, and 4 in displays, all with a variety of colour and lighting options and a choice of AC or DC electrical input. Options include Night Vision Imaging System (NVIS) compatibility and an integrated slip indicator.

All instruments are hermetically sealed and feature alternative panel angle options, manual caging and automatic gyro control during accelerated flight.

Status

In service in a wide variety of fixed- and rotary-wing aircraft for both military and commercial applications. FH series instruments are specified as primary or standby instrumentation in many aircraft including Jaguar, Tornado, Nimrod, C-130, Hawk, Chinook, Super Lynx, Bell 412 and Eurocopter AS 350.

Contractor

Ferranti Technologies Limited.

Card Quick Access Recorder (CQAR)

Type

Flight/mission recording system.

Description

The Card Quick Access Recorder (CQAR) combines the functions of a high-capacity Quick Access Recorder (QAR), Maintenance Data Recorder (MDR) and Data Transfer Unit (DTU) in one cockpit-mounted unit.

The system enables data transfer to or from an onboard system while simultaneously accepting and recording data from an onboard acquisition system. The CQAR records to standard high-capacity PCMCIA Type II cards.

Specifications

Data inputs: RS232, ARINC 573/717, MIL-STD-1553B
Recording duration: 25–200 hr
Weight: <1 kg
Power: 28 V DC
Qualification: RTCA DO160D, DO178B level C

Status

In production.

Contractor

Meggitt Avionics.

CNI/control and display unit

Type

Control and Display Unit (CDU).

Description

Developed for Hawk 100/200 aircraft, the CNI/control and display unit provides the common point for the control and data information display for the various radio, navigation and IFF systems. The unit has the primary role of controlling and monitoring the radio systems via a MIL-STD-1553B databus. It has a secondary role, upon request, to perform a reversionary bus controller function without affecting its primary role. The unit can, however, be reconfigured for other applications through software changes via an RS-424 serial datalink.

Status

In production and in service.

Contractor

Selex Sensors and Airborne Systems.

The CNI/control and display unit has been developed for the Hawk 100/200 0581505

Cockpit Display System (CDS)

Type

Avionic display system.

Description

The integrated Cockpit Display System (CDS) shows flight, navigation, aircraft systems and mission systems information on eight smart, full-colour Active Matrix Liquid Crystal Displays (AMLCDs). The advanced Man-Machine Interface (MMI) is further enhanced by a Smart Alerting (SA) facility. The SA utility features:

- Automatic selection of appropriate engine/transmission limits in response to changes in aircraft status
- Extensive built-in monitoring of flight and engine limits with automatic highlighting of abnormal conditions
- Cross-monitoring of both sensor input data and displayed symbology.

The integrated mission avionics suite in a Sea King Mk 42B 0503688

Ferranti Technologies' attitude indicators FH22, FH30, FH32 and FH40 (left to right) 0018193

If an individual unit fails, the crew can rapidly redistribute the failed unit's functions among the remaining units, thus ensuring all necessary information required for continued safe operation of the aircraft will still remain available to the crew. The Integrated Display Units (IDUs) have integral I/O and graphics generation and feature a modular design, so the display head is scalable from 5 × 5 in to 10 × 8 in and beyond. The system architecture is built around the ARINC 429 databus, but MIL-STD-1553B could also be used, according to customer preference.

The CDS consists of smart IDUs (6.25 × 6.25 in), 5ATI Electronic Power Systems Instruments (EPSIs), and 3ATI Standby Air Data Instrument (SADI) or Integrated Standby Instrument Systems (ISISs).

Status
The first application of the CDS is in the Super Lynx 300 helicopter, consisting of 4 × IDU, 2 × 5ATI EPSI and 2 × 3ATI Standby Flight Instruments (SFI).

Contractor
Smiths Aerospace.

Cockpit Voice Recorder CVR-90

Type
Flight/mission recording system.

Description
The CVR-90 is compliant with both FAA TSO C-123 and EUROCAE ED-56A survivability and environmental requirements. Voice recording on four channels is provided with a capacity of 30 minutes per channel. The CVR supports ARINC 557 and 757 compatible avionics.

Specifications
Dimensions: ½ ATR short
Colour: International orange
Operating temperature: −55 to + 70°C
Non-operating temperature: −55 to +85°C
Power: 28 V DC or 115 V AC, 400 Hz less than 15 W
Weight: 8.2 kg
MTBF: 20,000 h
Features: built-in test; fail safe erase through double electrical interlock

Contractor
Penny & Giles Aerospace Ltd.

CPT-186 airborne MultiFunction Display (MFD)

Type
MultiFunction Display (MFD).

Description
Caledonian Airborne Systems' range of advanced, smart MFDs incorporate instrument digitiser, video digitiser and symbology generator capabilities for applications associated with demanding civil and military display requirements. Live video landscape to portrait rotation can be accomplished through an integrated Versatile Video Module (VVM).

Features and benefits of the CPT-186 smart MFD include:
Engine, flight instrument, radar, moving map and armament status display modes;
Integrated symbol generator;
Integrated multiple synchro and aircraft instrument digitiser module;
RS-170, CCIR, STANAG 3350B or Computer Graphics video inputs;
ARINC 419/429, MIL-STD-1553, analogue and discrete interfaces;
High resolution active matrix flat panel display with high brightness wide range CCFL backlight;
User interface by programmable softkeys and rotary encoders;

Caledonian Airborne CPT-186 smart MFD, showing (left to right) flight, engine and radar information (Caledonian Airborne Systems)   0528518

Digitised flight data and engine trend monitoring data download option;
Available in standard 8.4 in and 10.4 in versions.

Specifications
Display resolution: 800 × 600 pixels in a 214 mm (8.4 in) diagonal
Colours: 262,144
Video standards: RGB computer graphics (up to SVGA) with separate H & V syncs, composite, STANAG 3350, CCIR/PAL & RS-170/NTSC
Luminance: 1,300 cd/m² minimum
Dimming range: 200:1
Power: 28 V DC to MIL-STD-704, 56 W
Lighting: 5 or 28 V, AC or DC
Weight: 3.0 kg
Dimensions: 210 × 167 × 74 mm (behind panel)
Interface: 18 programmable softkeys, 4 predefined hardkeys, 5 rotary encoders
I/O: ARINC 419/429, MIL-STD-1553, discretes, analogue, synchro
Synchro: Sensitivity to 50 μ rad, excitation frequency from 300 to 600 Hz, 26 to 115 V AC, bandwidth to 200 Hz
Environmental: DO-160D
Temperature: Civil or military temperature ranges
Cooling: Internal forced air
Options
Display resolution: 640 × 480 (VGA) or 1024 × 768 pixels (XVGA)
Display size: 262 mm (10.4 in)
Dimming range: 1,000:1
Video standards: Computer Graphics (up to XVGA), various radar video standards
Interface: ARINC 629 or EIA-RS422
NVIS compatibility: NVIS Type I & II Class B to MIL-L-85762
Multi symbology: Multiple aircraft instrument symbology can be uploaded on a single smart MFD for pilot training

Status
In production and in service.

Contractor
Caledonian Airborne Systems Ltd.

CUGV-1020-01 VGA to video scan converter

Type
Avionic display processor.

Description
The CUGV-1020 is a compact, ruggedised VGA to video signal converter, which converts a computer-generated graphics input (VGA, SVGA, XGA, and so on) into a 525/625 line video signal. Intelligent Auto Scaling accepts RGBHV signals from a computer or Digital Map Generator (DMG) and converts it to a standard video format for display or recording utilising PAL/NTSC or STANAG 3350 signals.

Features of the system include:
• Small, lightweight design
• Automatic detection of incoming resolutions up to 1,600 × 1,200

• Autoset – automatic sizing of incoming signal to exactly fit video display screen
• Computer signal loop-through – allows normal use of the local computer while scan converter is operating
• Simultaneous output in Composite, S-Video (Y/C) and RGB.

Specifications
Graphic signal input
Graphic signal type: RGBHV
Termination: Auto-terminating into 75 Ω
RGB level range: 0.5–2.0 Vp-p
Scan rate detection: Automatic
Maximum resolution: 1,600 × 1,200; 1024 × 768 maximum without dropped lines
Maximum vertical refresh rate:
640 × 480: 210 Hz
800 × 600: 170 Hz
1,024 × 768: 130 Hz
1,280 × 1,024: 100 Hz
1,600 × 1,200: 85 Hz
Maximum horizontal refresh rate: 100 KHz
Computer compatibility: PC, Macintosh

Video Outputs
Television standards: NTSC, PAL-B/G (selectable)
Impedance: 75 Ω
Composite video: 2
S-Video (Y/C): 2
RGBS: 1
Graphic loop-through: 1

General
Size and position: Automatic via AutoSet or Manual
Underscan/overscan: User-definable presets
Image freeze: One video frame
Settings memory: Non-volatile
Zoom/pan range: 2× zoom; 100% pan coverage
Flicker filtering: Selectable 2 or 4 line
Horizontal filtering: Full digital
Conversion technology: Proprietary
Colour resolution: 24-bit (16.8 million colours)
Power: 28 V DC, 15 W

Status
In production.

Contractor
Skyquest Aviation.

Datalink Control Display Unit (DCDU)

Type
Control and Display Unit (CDU).

Description
The Airbus Industrie Future Air Navigation System (FANS) for A330 and A340 aircraft incorporates the Smiths Aerospace Datalink Control Display Unit (DCDU). The DCDU will also be available for retrofit to other Airbus types sharing the common cockpit. The DCDU leverages the company's AMLCD technology, incorporating a 4 × 3 in, full colour, landscape display. Two DCDUs are fitted to each aircraft, providing a datalink messaging interface between the crew and Air Traffic Control (ATC) ground stations. During Reduced

Vertical Separation Minima/Minimum Navigation Performance Standards (RVSM/MNPS) operations across the North Atlantic, the DCDU transmits automatically generated position reports from the aircraft Flight Management and Guidance System (FMGS) and is used to relay requests for lateral and vertical deviation from the flight plan. The unit also receives and displays incoming messages and instructions from ATC.

The system is offered as a retrofit item to current aircraft, avoiding costly modifications to existing display systems.

Status
Certified on the Airbus A330/340 during 2002.

Contractor
Smiths Aerospace.

DC electromagnetic Helmet Tracking System (HTS)

Type
Airborne Electromagnetic Helmet Tracking System (HTS).

Description
The electromagnetic Helmet Tracking System (HTS) has been designed for both fixed- and rotary-wing applications and determines the helmet position and orientation by measuring magnetic fields which it generates in the cockpit. This data can be used to position aircraft sensors and weapon seekers and for referencing Helmet Mounted Display (HMD) systems.

The system has been developed to overcome many of the disadvantages of competing electromagnetic tracker systems while maintaining simplicity of installation. A number of prototype systems have been delivered to US and European customers as part of fully integrated HMD systems.

A small transmitter containing three orthogonal coils is rigidly mounted within the cockpit. The coils are pulsed with currents, which generate known magnetic fields. A small magnetic sensor, again containing three orthogonal coils, is rigidly mounted to the helmet. These coils sense the direction and magnitude of the generated magnetic fields and, using this data, a processor calculates the helmet position and orientation. The pulsed nature of the transmitted signal leads to this type of system being termed a 'DC Electromagnetic Tracker' to differentiate it from the older 'AC electromagnetic' systems.

The key functional elements are the transmitter, the sensor and an electronics module, which is normally installed within the HMD system electronics unit.

This system provides full spherical coverage within the aircraft cockpit offering high accuracy and good dynamic performance. It does not affect and is not affected by other cockpit functions and is significantly less susceptible to cockpit metal than traditional systems, requiring a single generic mapping for all aircraft of a particular type.

Status
In service.

Contractor
Selex Sensors and Airborne Systems.

Digital Map Generator (DMG)

Type
Avionic Digital Map System (DMS).

Description
Selex' Digital Map Generator (DMG) has been developed from COMED/CEDAM experience, to meet the mission requirements of the next generation of combat aircraft. Housed in a ¾-ATR box, the DMG is suitable not only for these new aircraft but also for a large number of retrofit applications.

The DMG hardware and software can handle both true digital maps and digitised chart information. The digital database can be held in solid-state or in optical disk. Both of these options have been developed. Features on the presentation can be selectively displayed or erased. Hazards such as terrain and obstacles above aircraft altitude can be made to stand out in contrasting colour. Safe areas occasioned by terrain-masking can be depicted and areas may be viewed from different angles and altitudes. Data updating is rapid and simple. The map is displayed on a colour electronic display, which incorporates a multifunction keyboard and electronically generated symbology.

The map can be displayed either north up or track up, with the aircraft present position centred or de-centred on the display. Multiple map scales can be accommodated and zoom and declutter facilities are available. Map stabilised overlays for routes and navigational information can be displayed.

Status
In production and in service in UK Royal Air Force Jaguar GR. Mk 1B, Tornado GR. Mk 4 and Lockheed Martin C-130J aircraft.

Contractor
Selex Sensors and Airborne Systems.

Digital Solid-State Recorder (DSSR)

Type
Flight/mission recording system.

Description
General Dynamics' DSSR is a high-performance, flexible, cost effective solid-state storage solution to meet the requirements of many data acquisition and processing applications.

The DSSR, designed to replace the GD UK RMS 3000 (see separate entry) on the UK Royal Air Force Tornado GR1A, utilises many of the latest approaches to open systems and Commercial-Off-The-Shelf (COTS) design.

The DSSR provides a solid-state recording and system control functionality which can be configured for a wide range of airborne, ground based and naval applications.

The DSSR is suitable for the storage of sensor and support data from a wide range of sensors such as infra-red, electro-optical, multispectral, sonar, radar, Synthetic Aperture Radar (SAR) and other sensor types.

Optional features include: simultaneous data compression/decompression; variable ratio compression/decompression; real-time display functionality; flexible data storage options (STANAG 7023, NITFS, raw data); a detachable high storage capacity memory brick (currently up to 25 Gbytes).

Status
In production and in service.

Contractor
General Dynamics United Kingdom Ltd.

Display Processor/Mission Computer

Type
Aircraft mission computer.

Description
The wholly modular Display Processor/Mission Computer (DP/MC) can be configured to meet

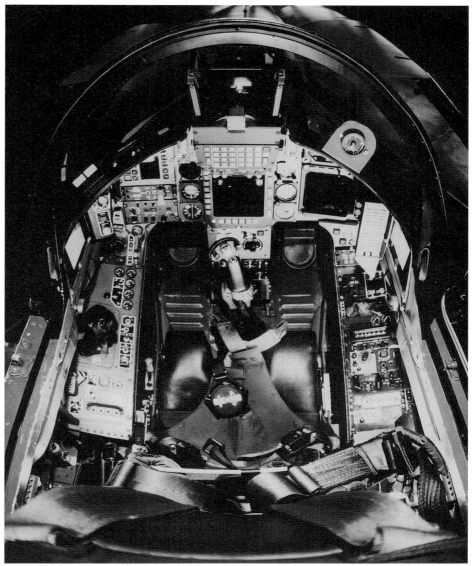

The front cockpit of the Hawk 100/200 with the Smiths HUD, associated Data Entry Panel (DEP) and colour multipurpose display immediately below   0504203

customer requirements. It is based on an extensive library of standard electronic cards, power supplies and ATR short cases. The standardised internal databus allows the unit, and hence the system, to be updated or expanded by replacing or inserting the appropriate cards. This flexibility enables improved components to be incorporated as technology evolves, including faster and more numerous microprocessors and, for example, the adoption of surface-mounted components.

The unit meets military specifications and is suitable for computing and signal generation applications for head-up displays, weapon aiming, navigation, mission and system control and electronic flight instruments.

Various high-speed 32-bit and 16-bit processor cards are available in the library of modules. Other modules provide global memory, raster and/or stroke display generation for either head-up or head-down displays and extensive analogue, discrete and serial input/outputs. MIL-STD-1553B interfaces are also available as remote terminals and/or bus controllers.

Specifications
Dimensions: ¼ ATR to 1¼ ATR
Example dimensions ¾ ATR short
   190 × 193 × 384 mm
Weight: (¾ ATR short) 9 kg

Status
In production. Latest applications include the display electronic unit for the T-45A Goshawk, bus interface control unit, head-up display and weapon aiming computer and DP/MC for the Hawk 100 and 200, electronic unit for the F-5E avionics update and symbol generator unit for the EH 101 Merlin helicopter.

Contractor
Smiths Aerospace.

Display processors/graphics generators

Type
Avionic display processor.

Description
The advent of glass cockpits demands that the display symbol generators have a capability which matches display performance and operational requirements. Selex (formerly BAE Systems Avionics) has adopted a modular approach to graphics generation and can supply boxes which drive either a single display surface in monochrome or a full display suite of multifunction colour displays and a dual-mode head-up display.

A variety of interfaces is offered which can be modified, as necessary, to integrate with existing or new aircraft systems. As well as dedicated display generation, the boxes can have weapon aiming and mission computer functions added easily.

Status
In production for the UK Royal Air Force Tornado GR. Mk 4.

Contractor
Selex Sensors and Airborne Systems.

Displays and Mission Computer (DMC)

Type
Aircraft mission computer.

Description
Smiths has developed a modular DMC, which can be configured to match the exact system requirements of each application by selecting modules from an extensive library of standard electronic cards, power supplies and ATR short cases.

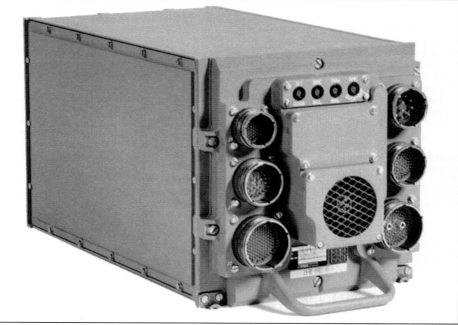

Displays and Mission Computer DMC   0015266

The standardised internal databus allows the system to be updated or expanded by inserting appropriate cards as technology evolves.

The size of the DMC depends on the number of modules needed to meet the operational requirements; ½ ATR, ¾ ATR, and 1 ATR sizes are available.

Electrical connections are via a rear panel which can be adapted to different connector configurations including: DPX and ARINC 600. The panel is detached to provide access to the backplane. Cases are cooled to ARINC 404A/600 specifications.

Status
The DMC is standard equipment on the next-generation BAE Systems Hawk for the Royal Australian Air Force lead-in fighter programme, and for a number of other programmes, including the Boeing AV-8B Harrier II and T-45 Goshawk.

Contractor
Smiths Aerospace.

EICAS/EIDS Engine Instrument Crew Alerting System/Engine Instrument Display System

Type
Engine Instrument Crew Alerting System (EICAS).

Description
The EICAS/EIDS instruments use the latest technology to record engine conditions and

Engine Instrument Crew Alerting System (EICAS)   0002457

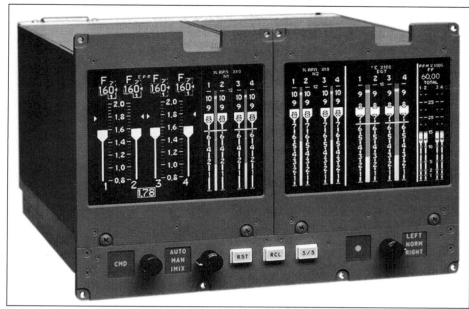

Engine Instrument Display System (EIDS) 0002458

provide indications on AMLCDs (Active Matrix Liquid Crystal Displays) that immediately alert flight crews to abnormal values.

Specifications
Dimensions: 134 × 135 × 242 mm
Weight: 3.4 kg
Power: 18–32 V DC, 50 W (nominal), 70 W (maximum)
Display area: 112 × 84 mm
Grey levels: 64

Contractor
Penny & Giles Aerospace Ltd.

Engine displays

Type
Aircraft instrumentation.

Description
Engine Display Units (EDU)
Two redundant EDUs display all necessary information for the safe operation of turbine engines, including fuel quantity and usage

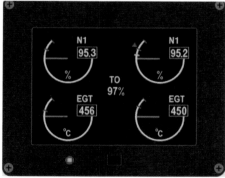

Meggitt Avionics Engine Display Units 0051670

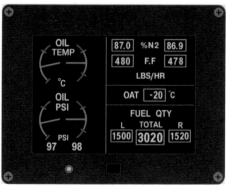

Meggitt Avionics Engine Display Units 0051671

calculations. Engine parameters are normally displayed in both analogue and digital format using colour to highlight exceedance conditions. Displayed parameters include: Torque (TQ); Inter Turbine Temperature (ITT); power turbine rpm (NP); gas generator rpm per cent of maximum (NG); VACuum suction (VAC); oil temperature/pressure; fuel quantity; Fuel Flow (FF); Fuel At Destination (FAD); Time To Destination (TTD); and Outside Air Temperature (OAT). Each EDU also has a selectable reversionary mode where all the data from both screens is compressed and displayed on one screen.

Data Acquisition Unit (DAU)
Meggitt Avionics produces a range of DAUs to acquire engine and fuel data, which is transmitted to cockpit displays via ARINC 429 databusses.

Sterling indicators
The Sterling range of general purpose 2 in indicators use interchangeable personality modules to interface with a wide range of sensors. High reliability, accuracy and low maintenance is guaranteed by a single moving part design. The

Meggitt Avionics Sterling indicator 0051672

Meggitt Avionics Light Off Detectors 0051673

pointer is mounted directly to a precision stepper motor shaft. Use of a microprocessor allows customised dial faces and non-volatile storage of up to 45 hours of flight data. A solid-state digital display can be added to the dial face.

Light Off Detectors (LOD)
For monitoring afterburner ignition on military aircraft, the LOD uses an ultra-violet sensitive Geiger-Muller tube that views the afterburner flame through a port in its liner. Critical attributes of this sensor are immunity to sunlight and rapid detection times despite exposure to afterburner temperatures and pressures.

Contractor
Meggitt Avionics.

Engine Instrumentation Display System (EIDS)

Type
Engine Instrument Crew Alerting System (EICAS).

Description
The EIDS is a direct replacement for legacy electromechanical instruments, providing significant benefits in terms of weight reduction, lower power consumption, increased reliability and reduced cost of ownership. Additionally, retrofit of the EIDS releases valuable panel space for potential display upgrade requirements. Developed for the Lockheed C-130J and P-3 Orion aircraft, the EIDS presents primary parameters in a pointer/counter format, with secondary parameters in a digital format, providing an easily interpreted display that requires minimum crew difference training.

To ensure optimum integrity, three processing channels per engine are provided. Each of three primary parameters is processed by a different channel, along with each of three engine oil parameters. Remaining parameters are apportioned, one to each channel.

Other benefits and features of the system include:
- Two passively cooled, microprocessor-based LRUs per system
- System reliability >20 times better than existing indicators
- COTS-based design
- Comprehensive BITE
- Improved MTBF/MTBUR ratio
- Lower life-cycle costs/low spares holding requirement
- Ease of maintenance
- All electronic
- Improved engine monitoring.

Specifications
Weight: 6.95 kg, each LRU
Power: 28 V DC, 44 W at maximum brightness
Reliability: 7,000 h operating each LRU

Status
In service in Lockheed C-130J and P-3 Orion aircraft.

Contractor
Smiths Aerospace.

Engine monitor panel

Type
Cockpit display.

Description
The engine monitor panel for the Hawk 100 and 200 displays all necessary fuel and engine information as well as indications of any associated malfunction. The display technology combines the latest LCD technology and LED arrays to provide a flat panel which can be mounted in any cockpit position and still maintain high readability over all ambient light conditions.

The engine monitor panel for the BAE Systems Hawk 100 and 200 0581506

The display technology is adaptable to cockpit environments where integration of analogue instruments is required to allow more use of cockpit space.

The panel incorporates a fuel 'bingo' facility which allows the pilot to set minimum fuel levels at which an audio warning will be generated. An option to this panel is the incorporation of engine low-cycle fatigue recording for post-flight evaluation.

Status
In production.

Contractor
Selex Sensors and Airborne Systems.

Eurofighter Dedicated Warnings Panel (DWP)

Type
Aircraft centralised warning system.

Description
The DWP offers a high-intensity LED display which is NVG-compatible. The DWP integrates an Electronic Unit (EU) with a Display Unit (DU). The display has programmable legends which are controllable over a dual-redundant MIL-STD-1553B avionics databus. Internal redundancy is incorporated to ensure continued safe operation in the event of defects and dedicated discrete inputs are also incorporated to ensure correct warning displays should the databus sources fail.

Aircraft status, health and fault data received over the databus is categorised by the internal Ada software for display on a simple menu; amber and red characters are available, with a dedicated display area for catastrophic warnings. An interface is provided to the audio management unit to give the aircrew co-ordinated visual and audible or digitised voice warnings of any new conditions.

The DWP provides the following facilities:

1. Three single colour warning captions for displaying high integrity or catastrophic warnings
2. Two-colour display of system warning captions
3. A control for the paging of warning captions
4. Two attention getter drives
5. Storage of up to 256 system warnings
6. Six audio warning discretes
7. Two modes of operation, normal and reversionary
8. Day, dusk, night luminance modes
9. Dual-channel architecture interface with the Cockpit Interface Unit (CIU) and the Communication and Audio Management Unit (CAMU).

Specifications
Dimensions:
(display unit) 118 × 156 × 70 mm
(electronics unit) 100 × 220 × 237 mm
Weight:
(display unit) 1.7 kg
(electronics unit) 3.95 kg
Power supply: 28 V DC, 270 W (max)

Status
Operational on all Eurofighter Typhoon development aircraft. Production deliveries continuing.

Contractor
General Dynamics United Kingdom Ltd.

Eurofighter Typhoon cockpit Video and Voice Recorder (VVR)

Type
Flight/mission recording system.

Description
General Dynamics' Eurofighter Typhoon cockpit Video and Voice Recorder (VVR) integrates a Cockpit and Video Interface (CVI) with an Off-The-Shelf (OTS) TEAC V-80AB-F Hi-8 mm Airborne Video Tape Recorder (AVTR).

A video demultiplexing system allows post mission replay of recorded (video, MIL-STD-1553B mission data) data on four standard video monitors using a range of search facilities.

The Eurofighter Typhoon VVR is capable of recording colour and/or monochrome video, audio data and the indication of the occurrence of event marking signals.

The Eurofighter Typhoon VVR interfaces with: two Computer Symbol Generators (CSGs) for the recording of video; the Computing Devices' Communications and Audio Management Unit (CAMU) for the recording of audio; dual redundant MIL-STD-1553B terminals to accept event marking, input selection and mode control signals and to output status data; two discrete

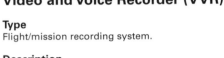

Eurofighter Typhoon cockpit Video and Voice Recorder (VVR) 0051343

output VVR status signals; one discrete input signal to facilitate manual mode control.

The two CSGs output multiplexed video display data to the VVR in RGB format. The minimum record time is 90 minutes.

The VCI comprises four functional blocks: Control Mode and Monitoring module (CMM); Video, Audio and Data module (VAD); Power Supply Module (PSM); Case Electrical Assembly (CEA) including motherboard and Elapsed Time Indicator (ETI).

The CMM module incorporates a microcontroller, MIL-STD-1553B chipset, control interface, data encoder and test bus interface. The VAD module incorporates the discrete interfaces, video interfacing and channel controls and audio conditioning. The PSM accepts +28 V power from the aircraft power systems.

Status
In production and in service in the Eurofighter Typhoon.

Contractor
General Dynamics United Kingdom Ltd.

EuroNav IV

Type
Avionic Digital Map System (DMS).

Description
EuroNav IV is a sophisticated moving map system designed to integrate with other onboard mission equipment and sensors, including Forward-Looking Infra-Red (FLIR) systems, thus easing operator workload in applications such as police surveillance. The system display can be viewed via existing aircraft CRT/AMLCD or via a dedicated display; typically, the operator and pilot each have a display. Core to the EuroNav IV system is a comprehensive mapping database which can hold up to 16 Gbyte of both digital and scanned paper map data of designated geographical areas. Jeppesen flight data is included in the system, with the operator able to select the level of such information required, from none to full IFR information, depending on circumstances. Linked to this mapping information is a multimode database which can display various levels of operational information overlaid on the mapping data.

The EuroNav IV system consists of a monitor (CRT or LCD), monitor switching panel, system controller and processor. Typically, the map picture is sent to two monitors, one of which is also utilised for FLIR imagery and a secondary (LCD) monitor. The monitor switch panel facilitates user control over picture brightness and so on, while the system controller gives full access to all modes and controls. The CPU houses the computer section, which hosts GPS and mapping information. A 3 in floppy drive is included for downloading flight information.

EuroNav IV incorporates many features which serve to ease operator workload and facilitate rapid and highly accurate navigation and mensuration of Points Of Interest (POI):

Display digital data over paper maps
This feature enables the observer to keep familiar 1:50,000 Ordnance Survey map data on screen, while employing a digital map data layer

The programmable aircraft warnings panel has been selected for the Eurofighter Typhoon 0503878

underneath to highlight and name roads. In Camera Pointing mode, the road name in the centre of the observers (camera) video screen is named automatically.

Camera pointing/slaving

When linked to an onboard video camera (FLIR or CCD), the system can display a marker on the map showing where the camera is pointing. This is achieved by correlation of data emanating from the Differential GPS (DGPS), gyrostabilised camera system and the EuroNav digital compass and pitch/roll sensor. Conversely, the Line of Sight (LoS) of the video camera can be slaved to the EuroNav POI.

Appending of digital imagery

Digital pictures can be captured from the onboard camera system and stored as waypoints or attached to operational task data. Pictures can also be stored at higher resolution for downloading on return to base and colour printing.

Messaging

Text messages can be sent via existing microwave equipment or, optionally, via Eurolocator, securely over VHF or AM radios.

Real-time positional information via enhanced moving map

After system power-up, current position is shown on the map display overlaid onto a colour 1:50,000 Ordnance Survey map (or other default map). As the aircraft moves so does the map, allowing the observer to report his position or to plan a route ahead immediately. As the system is also linked to a digital map, a small box in the left-hand corner of the screen shows the village or town currently being over flown and the next road to be crossed. This information box is constantly and automatically updated as the aircraft moves, thus enabling the operator to maintain a timely and accurate commentary to assisting ground units, especially at night.

Navigation and in-flight planning

EuroNav IV provides other operator-selectable features to facilitate rapid re-planning of the flight. The system allows input of several forms of data including National Grid, Lat/Long, stored user waypoints, navigational aids or even road names and individual house names, depending on the mapping information available (in the UK, all of this information is available). Once set up the system displays the flight plan on the screen and enables the observer to pass precise navigational data to the pilot.

Forward viewing

In order to familiarise himself with area of interest as the aircraft approaches, the operator can zoom ahead and look at the area in greater detail. Using UK Ordnance Survey digital Land-Line Mapping data, the operator can zoom down to house boundary level and see house numbers/names and street layouts. The system also allows for secondary flight plans to be readied for action upon completion of the task in hand.

Recording Points of Interest

At any point in the flight the observer can select an event flag marker which places a marker on the map which can later be reviewed or steered to. EuroNav IV is able to mark several areas with different coloured flags which can then be tagged with a name or identification number. These colour-coded flags can be designated according to operator preference, denoting no-fly areas, photographic requests or co-ordination points.

Flight recording

EuroNav IV records every flight in both vertical and plan view.

Emergency mode

The EuroNav IV facilitates rapid response to emergency aircraft situations – a single button press presents the operator with a number of the closest landing sites and airports, with the nearest landing site highlighted. After accepting the system's suggestion, a flight plan direct to

the site is presented. A number of user-specific landing sites can be programmed into the system. If the aircraft is used in the medical role, a full list of hospitals can be entered. Users can also store photographs in the system showing the approved landing site and this information is stored as a link to site with any other special information.

EuroLocator

EuroLocator is a GPS transmitter which is fitted inside a police vehicle on the ground. The locator sends a signal to the aircraft using standard VHF radios or GSM datalinks. The position of the police car is shown on the map in the aircraft enabling the aircraft to make a direct flight plan.

Accuracy

The basic system offers GPS levels of accuracy, which, in the UK, is generally better than 30 m. If greater accuracy is required, a DGPS can be employed, via a local DGPS transmitter, or as a subcarrier on FM radio. Skyquest offer a DGPS package utilising the DGPS signal from FM radio providing accuracy down to 1 m.

Other features and benefits of EuroNav IV include:

- Instant start up in all environmental conditions
- Flight plan directly to:
- Address
- Waypoint
- Ordnance Survey National Grid Reference
- Postcode
- House name/number
- Track Up/North Up orientation of map data, with operator manual orientation in hover
- Aircraft symbol on map showing Present Position (PP) for slow orbit and hover operations
- Digital compass back-up to GPS under slow-flying conditions
- Automatic powerline proximity warning
- Search patterns linked to FLIR FoV to ensure efficient area coverage
- Set up macro keys for commonly used functions
- Removable hard drives (shock proof to 100 g)

Status

In production and in service with UK police air units.

Contractor

Skyquest Aviation.

F-16 HUD video camera

Type

Flight/mission recording system.

Description

This single unit high-resolution colour camera was developed to fit the F-16 C/D HUD. It replaces the existing two-unit monochrome camera.

The latest technology solid-state sensor provides high resolution, excellent sensitivity and accurate colour. Alignment accuracy is high for mounting the camera forward of the HUD. The 525-line/60 Hz NTSC or 625-line/50 Hz PAL configurations are available and the output can be Y/C or composite video.

Status

In production.

Contractor

Selex Sensors and Airborne Systems.

F-16 HUD video camera   0001299

FD4500 Series Head-Up Displays (HUDs)

Type

Head-Up Display (HUD).

Description

The FD4500 Series head-up display system, originally intended for the installation in light attack aircraft and trainers, was designed from the outset to be a dual-mode cursive and raster display with a modular architecture, allowing the mechanical outline to be optimised to fit most aircraft installations.

The display gives a 24° total field of view with a large instantaneous field of view available from the 4.5 in (114 mm) exit lens. P53 phosphor is used on the CRT, giving a very bright display. The upfront control panel (UFCP) gives complete control over the rest of the system and a colour TV camera fitted to record the pilot's view.

The interface unit or Head-up display Electronics Unit (HEU) is a ¾ ATR box containing multiple analogue and discrete synchro interfaces, 1553B R/T or bus control, and high-performance processors, symbol generators and graphics processors. Full weapon aiming, mission computations and head-up and head-down display symbol generation are available.

The Type 4510 has both cursive and raster displays, the latter being selected by a switch on the UFCP. The 4500 and 4510 are physically identical and were based on the optical and symbol generation technology previously used on the COMED system. The extensive production runs of COMED had removed any design problems, while the weapon aiming, interfacing and air data techniques were derived from a series of complete weapon systems designed and built by BAE Systems.

The 4510 head-up display provides a full suite of navigation symbology with steering and location cues available at all times and generates automatic or selected weapon aiming symbology for all known air-to-air and air-to-ground weapons. The control panel allows the pilot to set up the navigation system, select and display modes during flights and then change to a raster display when mission sensors, such as FLIR, necessitate.

The HUD is an integral part of various total system options at present being considered in the worldwide retrofit market. The unit can, because of its compact size, meet the installation requirements of many types of attack aircraft.

Specifications

HUD Display Unit (DU)
FoV: 24°
Refresh rate: 50 Hz
Weight: 11.8 kg
Power supply: Stand-alone PSU

HUD Electronics Unit (EU)
Computer: Power PC 750
Memory: 48 Mb
Weight: 13.6 kg
Dimensions: 3/4 ATR Short
Power supply: 115 V AC, 400 Hz, 150 VA

The cockpit of a UK Royal Air Force Jaguar GR1B features the FD4500 series HUD   0528212

Status

4500 Series HUD systems have undergone rigorous flight trials on the Buccaneer and Harrier Nightbird aircraft at the Defence Evaluation Research Agency, Farnborough. A UK Royal Air Force Jaguar was fitted out with a BAE Systems system and flew early in 1989. Currently in service in UK Royal Air Force Jaguar aircraft, and A-4, C-101, F-5, Mirage III and Mirage 5 aircraft retrofit applications.

Contractor

Selex Sensors and Airborne Systems.

Flat Panel AMLCD

Type

MultiFunction Display (MFD).

Description

This 61/4 in (159 mm) MultiPurpose Display (MPD) is a Line-Replaceable Unit (LRU) which accepts analogue video information, and formats and displays a full-colour or monochrome image on a high-resolution Active Matrix Liquid Crystal Display (AMLCD). The unit is designed to meet military specification and is capable of displaying video, text and graphics, and is fully compatible with NVG. A graphics processor is available for overlaying symbology if required.

The AMLCD array comprises 1,024 × 1,024 individual pixels arranged as 512 × 512 colour groups. Each group contains one red, one blue and two green pixels. The MPD operates in three modes: day, night and green. Day and night modes both provide a full-colour display which is compatible with Class B NVIS equipment. Display brightness is automatically reduced in night mode. In green mode, the red and blue pixels within each colour group are switched off, with the resulting greater density of green pixels enabling a higher-resolution monochrome display which is compatible with Class A NVG.

Specifications

Display:
6¼ × 6¼ in (159 × 159 mm) colour AMLCD
Dimming Range 30,000:1
525 line/60 Hz, or 625 line/50 Hz video standards
1:1 or 4:3 aspect ratio
Operating Modes:
Day – 1.0 to 210 ftL (white)
Night – 0.01 to 5 ftL (white)

Power Requirements:
Input – 3 phase, 115 V, 400 Hz AC or 28 V DC
Consumption – Typically 100 W
Weight: 6 kg
Dimensions (H × W × D): 15.9 × 215.9 × 184.2 mm (81/2 × 81/2 × 71/4 in)
MTBF: In excess of 4,000 h

Contractor

Smiths Aerospace.

Flight Data Recorder FDR-91

Type

Flight/mission recording system.

Description

FDR-91 is compliant with FAA TSO C-124 and EUROCAE ED-55 survivability and environmental requirements. A data recording capacity of 25 hours is provided. The FDR supports ARINC 573, 717 and 747 compatible avionics.

Specifications

Dimensions: 1/2 ATR long
Colour: International orange
Operating temperature: –55 to + 70°C
Non-operating temperature: –55 to +85°C
Power: 28 V DC or 115 V AC, 400 Hz less than 15 W
Weight: 8.4 kg
MTBF: 20,000 h
Features: built-in test; Harvard bi-phase input 64/128 bits/s; high-speed download

Contractor

Penny & Giles Aerospace Ltd.

Flight deck warning system

Type

Aircraft centralised warning system.

Description

The flight deck warning system is a microprocessor-based system suitable for a wide range of aircraft types from small business and commuter aircraft to large airliners. It comprises an electronic monitoring system Central Warning Panel (CWP) and audio warning system.

The electronic monitoring system monitors aircraft systems and sensors providing appropriate alerts to the flight crew via the CWP.

Use of a microprocessor results in a system of compact size, low weight and high flexibility. It enables alert signals to be prioritised and reduces spurious alerts through the use of time delays. Changes to the system can be made simply by altering the software. An important feature of the central warning electronics is the self-test facility which can be activated during the preflight phase and is in continual operation during flight. When a failure is detected in the unit, a warning is given to the pilot on the CWP and BITE indicators isolate the fault to board level. A failure of the microprocessor does not render the system inoperable as warnings will still be indicated on the flight deck.

The CWP is a dedicated display incorporating hidden legend annunciators. It is constructed from identical display modules which can be assembled to suit the application.

Entirely separate from the central warning system, the audible warning system monitors critical aircraft functions and can provide up to 10 warning tones to the flight crew headphones. This high-integrity system has duplicated outputs and dual power supplies to minimise loss of warning under fault conditions.

Status

In service on the Avro RJ.

Contractor

Ultra Electronics Ltd, Controls Division.

Flight displays

Type

Cockpit display.

Description

Primary Flight Display (PFD)
The PFD displays all the information that is needed to fly the aircraft. It receives data from various external sensors in order to display pitch and roll attitude, rate of climb, airspeed, heading, ILS (loc/GS), side slip and barometric setting. If the PFD fails, all this data will be presented on the navigation display.

Navigation Display (ND)
The ND displays all of the required navigation modes; compass rose (HSI), MAP and ARC. Data is received from a number of systems including GPS, radio receivers (VOR and ADF), and the heading system. Outputs for the autopilot are provided.

Smiths Industries Aerospace Flat Panel AMLCD   0087828

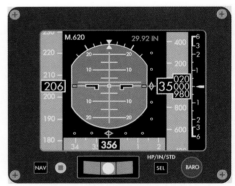

Meggitt Avionics Primary Flight Display 0051674

Meggitt Avionics Navigation Display   0051675

Meggitt Avionics Secondary Flight Display System™
0051676

Meggitt Avionics Secondary Navigation Display™
0051677

Mode, heading and course selection controls are on the bezel of the navigation display.

Secondary Flight Display System™ (SFDS)
The SFDS meets the requirement for standby flight information – attitude, altitude and airspeed – replacing two or three conventional electromechanical cockpit standby instruments with a single 3ATI cockpit instrument. The system comprises the Secondary Flight Display, which provides on a colour Active Matrix Liquid Crystal Display (AMLCD), flight information in a similar and compatible format to modern CRT or AMLCD primary flight displays. Within the SFDS are solid-state inertial sensors and microprocessor systems to measure aircraft pitch and bank attitudes. A second Line Replaceable Unit (LRU), the Air Data Unit (ADU) is connected to the appropriate pitot and static pressure ports and includes solid-state silicon pressure sensors, and a microprocessor to provide digital air data (altitude, indicated air speed and Mach No) to the display.

The ADU can be installed conveniently close to the pitot/static ports to minimise the problems of long pneumatic pipes such as dynamic lags and leaks and eliminates the need to route pipes to the instrument panel.

The air data computations include corrections for known errors in the static pressure source and thus provide air data measurements as accurate as those on the primary flight displays.

The Mark II SFDS is being introduced in early 2000. It provides the following enhancements: video quality graphics; 2.5 million colours to match primary display characteristics (with customer selectable colour options); higher update rates; higher resolution; a heading option; and full anti-aliasing.

The SFDS has been qualified to: TSO-C4c for bank and pitch instruments; TSO-C113 for airborne multipurpose electronic displays; and TSO-C106 for air data computers.

Secondary Navigation Display™ (SND)
The solid-state 3ATI SND cockpit instrument interfaces via ARINC 429 with ADF, VOR and DME radios together with the magnetic flux gate to provide navigation and heading data on a very high-quality AMLCD. It has an integral light sensor and lighting control can be matched to the main cockpit dimmer.

Series 35000 electronic clock
Utilising a high quality Liquid Crystal Display (LCD), this 3ATI clock is entirely solid-state. In GPS mode, the clock receives its time signal via GPS to give the highest accuracy. For aircraft without GPS, or if GPS is not available, the clock operates in Manual mode using its integral high accuracy internal time standard to display time and date. Universal Time Co-ordinate (UTC) is received from the GPS on ARINC 429. Available outputs for other aircraft systems include time and date in either GPS or Manual mode.

Status
The Primary Flight Display (PFD) and Navigation Display (ND) form part of the New Piper Malibu Meridian primary flight display system.

The Secondary Flight Display System™ (SFDS) has been selected by British Airways for retrofit in its B737-300 and B737-400 aircraft and the Mark II version is being supplied from March 2000. The Meggitt SFDS has also been selected for the following aircraft: the Lockheed Martin JSF demonstrator; the Pilatus PC-21 advanced turboprop trainer; the US Air Force C-130/C141B and U-2 upgrade programmes and the US Air Force C-5 fleet; the South African Air Force C-130 upgrade; the Royal Australian Air Force Hawk Lead-In Fighter (LIF) programme; and export upgrades to the Chinook CH-47SD helicopter (initial customers being the Royal Air Force in the UK and the Republic of Singapore Air Force). Meggitt has built over 1,000 systems to date.

The Secondary Navigation Display™ (SND) is in production for Gulfstream IVSP and Gulfstream V aircraft.

Contractor
Meggitt Avionics.

Flight Management Computer System (FMCS)

Type
Flight Management System (FMS).

Description
Conforming to the full ARINC 702 specification, and a standard option on the Airbus A310 and A300-600 aircraft, this Flight Management Computer System (FMCS) is the prime interface between crew and aircraft and enables optimum performance to be achieved from take off to final approach. Main functions include flight planning, navigation, performance optimisation, flight guidance (with coupling to autopilot and autothrottle) and display processing. The operational procedures create a working routine which is easy to implement and is similar for all phases of flight, optimising the factors affecting flight profile to give greater economy of fuel consumption, flight time and aircrew workload.

The system design is based on a parallel multiprocessing arrangement of microprocessors within the flight management computer unit. This technique permits high-processing capability and gives the flexibility to accommodate future expansion of functions and procedures. Two sets of dual 16-bit microprocessors – one dedicated to navigation, the other to performance functions – provide overall throughput of over 1 Mops. Additional microprocessors are dedicated to input/output and database control functions. A bubble memory provides 256 kwords of memory for navigation and performance database storage. There is provision for up to 56 discrete inputs and 16 discrete outputs, plus 32 input and 12 output ARINC 429 channels. The system contains its own built-in test routines which constantly monitor system operations and fault detection.

The crew interface is with the control/display unit which has a 14 lines by 24 character CRT format. The bottom line can be used for scratchpad entries. A full alphanumeric keyboard is provided, together with function keys and 12-line select keys adjacent the CRT. Self-contained built-in test provides a cued step-by-step test of all push-buttons, annunciators and the CRT display. For routine operations, most of the information is defaulted from the navigation database, requiring a minimum of manually entered data.

The Enhanced Flight Management Computer System (EFMCS) supersedes the FMCS. This provides one million words of EEPROM memory for navigation and performance database storage, replacing the 256 kwords of bubble memory in the FMCS. Further reliability and functional improvement are provided, including the facility for interfacing to an ARINC 615 high-speed data loader. The EFMCS is in the preproduction stage.

Singapore Airlines was the launch customer for the second-generation enhanced management computer for six new build Airbus A310 and retrofit of the existing fleet of 15 A310s. The enhanced system offers four times the current database capacity; an essential feature for extended route operations and improved performance and reliability.

Meggitt Avionics series 35000 electronic clock
0051678

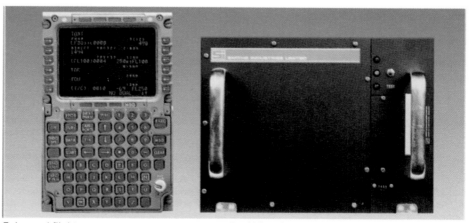

Enhanced flight management computer system for Airbus A300-600 and A310
0051635

Specifications
Dimensions:
(computer unit) 8 MCU
(control/display unit) 267 × 229 × 146 mm
Weight:
(computer unit) 12.7 kg
(control/display unit) 6.3 kg
Power:
(computer unit) 200 W
(control/display unit) 87 W

Status
In production. Airline customers include Kuwait Airways, Saudi Arabian Airlines, Air France, Sabena, Nigeria Airways, Air India, Cyprus Airways, Air Nuigini, Singapore Airlines, Air Algerie and Monarch Airlines. Smiths' FMCS has also been chosen for the Boeing E-6A of the US Navy.

Contractor
Smiths Aerospace.

FV-0110 and FV-0300 FlightVu external airborne video cameras

Type
Flight/mission recording system.

Description
FlightVu FV-0110 and FV-0300 cameras are designed to be fitted to the exterior of commercial aircraft to provide flight-deck crew with an exterior view of the aircraft and its surroundings, both in flight and on the ground. Its role is to enhance safety and security, providing the crew with the opportunity to assess security incidents, to assist them in ground handling or to check for exterior damage, fire or other occurrences. Used with the AD Aerospace FlightVu Video Data Recorder (FVDR), it can also provide instant playback of any hazardous incident to aid evaluation of actions.

The video picture can also be made available to the passengers, as part of the in-flight entertainment, throughout all phases of flight. FlightVu camera data can also be integrated with moving map displays and other in-flight entertainment or advertising.

The cameras are hermetically sealed and are fitted into a 'camera pocket' which is permanently fixed to the exterior of the aircraft. Once initial set up of the pocket has been achieved, no further alignment is required and the rotational alignment of the camera is repeatable. The low-profile, aerodynamically shaped optical window ensures viewing through all weather conditions. Automatic de-icing is provided with an internally fitted 10 W kapton heater and there is nitrogen purging for mist-free operation.

The FV-110 is a monochrome camera and it is interchangeable with the colour FV-0300.

Specifications
Field of view: 103.6 × 76.5° maximum (3.5 mm lens) to 14.6 × 11.0° minimum (25 mm lens) in 6 options
Sensor: 1/2 in CCD technology
Dimensions: 129.8 mm long, 57.9 mm diameter
Weight: 0.8 kg
Power: 28 V DC, 3.5 W
Altitude: to 50,000 ft
Temperature: −55 to +45°C
Environmental: RTCA Do160C

Contractor
AD Aerospace Ltd.

AD Aerospace FlightVu external airborne video camera   0051008

FV-0205 FlightVu internal airborne video camera

Type
Flight/mission recording system.

Description
The FV-0205 is a colour miniature CCD video camera for general aerospace use. It is designed to be fitted within the aircraft in a pressurised, heated area. It is panel mounted and is viewed through the panel. A 'pin-hole' option allows a covert installation and there are a variety of lens options to suit most requirements.

Specifications
Field of view: 73 × 56° (2.6 mm lens) to 8.9 × 61° (25 mm lens) in 6 options; 83 × 61° (3.7 mm pin hole)
Sensor: 1/4 in CCD technology (PAL or NTSC)
Dimensions:
  camera: 45 × 47 × 31 mm
  power supply unit: 120 × 80 × 35 mm
Weight:
  camera: 0.15 kg
  power supply unit: 0.25 kg
Power: maximum 3 W from 6 V supply
Altitude: to 15,000 ft (cabin altitude)
Temperature: −15 to +70°C operating

Contractor
AD Aerospace Ltd.

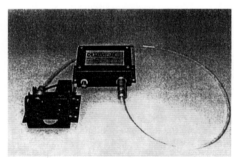

AD Aerospace FlightVu internal airborne video camera   0051009

Genesis SR rugged multiplatform computers

Type
Aircraft mission computer.

Description
The Genesis SR (Short Rack) is an adaptable, rugged, rackmount computer designed for installation in wheeled or tracked vehicles, military aircraft, surface ships and submarines. Typical applications include tactical communications systems, mission planning, combat and logistics support.

The unit can accommodate different processor engines (Intel Pentium™, Power PC™, Compaq Alpha™, Sun UltraSPARC™), it has configurable I/O panels, and can be operated from interchangeable AC, DC or AC/DC power supplies. Optionally it can be supplied to Tempest BTR 01/210.

The DRS Rugged Systems' Genesis SR rugged multiplatform computer   0062797

Specifications
CPU options: Intel Pentium™, Power PC™, Compaq Alpha™, Sun UltraSPARC™
Internal drives: Will support 'hot swap' and disk monitoring (RAID)
Input/Output: Configurable I/O panel (optionally: 104-key hinged removable keyboard with integrated trackball)
Display: DRS Rugged Systems' FPR rugged flat panels from 12 to 20 in
Power: 110–230 V AC (option 400 Hz and 19-32 V DC)
Dimensions: height 4U, depth 400 mm
Weight: 19.5 kg

Status
A variant system, designated Genesis Ultra, is being supplied as the Link 11 driver for UK Royal Air Force Nimrod MR2 Mk2 aircraft.

Contractor
DRS Rugged Systems (Europe) Ltd.

Glareshield displays for Eurofighter Typhoon

Type
Cockpit display.

Description
Smiths has the project and technical leadership for development of the Eurofighter Typhoon's right-hand glareshield displays. This contains a set of high-intensity LED displays used for standby engine and aircraft attitude information. These displays, which present analogue information generated digitally, have automatic compensation for changing ambient light levels.

Status
In production.

Contractor
Smiths Aerospace.

Harrier GR. Mk 7 video recording system

Type
Flight/mission recording system.

Description
The video recording system for the Night Attack Harrier GR Mk.7 was developed as an upgrade to the GR Mk.5 system, to incorporate recording of the FLIR video. Using two sealed video recorders, one for 'head-up' recording and one for 'head-down' recording, the system records colour HUD camera, FLIR and dual-mode tracker. Growth is built in to enable recording of all head-down displays at a later date.

Status
In service in BAE Systems Harrier GR Mk. 7 aircraft.

Harrier GR Mk.7 video recording system 0001303

Contractor
Selex Sensors and Airborne Systems.

Head-Up Display (HUD) for the C-17

Type
Head-Up Display (HUD).

Description
The HUD for the C-17 is claimed to be the world's first HUD designed as a critical flight instrument. The unit has a single box, the electronics being built into the optical unit rather than being separate. It has a 30° azimuth by 24° elevation field of view and has twin integral MIL-STD-1750A processors. Two HUDs are fitted to the aircraft, one each for pilot and co-pilot. When not required, the HUD combiner can be folded away below the line of sight.

Specifications
Power supply: 115 V AC, 400 Hz, single phase, 100 W
Reliability: 5,000 h MTBF

Status
In production.

Contractor
Selex Sensors and Airborne Systems.

Head-Up Display (HUD) for the F/A-22

Type
Head-Up Display (HUD).

Description
BAE Systems (now Selex) developed the HUD system for the Lockheed/Boeing F/A-22 Raptor. The main feature of the HUD is its single flat combiner, which combines excellent brightness and contrast with a large head motion box, enabling the pilot to rapidly assimilate information during extreme combat situations.

The Smart HUD is a critical flight instrument, Primary Flight Display (PFD) qualified, with the display processors and drivers in a single LRU. It uses diffractive optics featuring a single element holographic combiner, which consists of two glass elements bonded to produce a flat parallel-sided assembly. An optically powered hologram is recorded on photosensitised gelatine on the spherical interface sandwiched in the assembly and acts as the collimating combiner. The resulting advanced optical system, manufactured using computer-generated holographic techniques, provides new levels of display capability. Additionally, the uncluttered simplicity of the combiner support structure allows virtually a clear out-of-cockpit field of regard. The total field of view is 30° azimuth by

20° elevation and the instantaneous field of view is 24 × 20°.

The optical module brightness levels are optimised to operate in a very high ambient light environment, while minimising solar reflection and maximising outside world transmission and display uniformity from within the large eye motion box.

Associated with the PDU is a complex HUD control panel attached to the aft face of the unit which incorporates LED technology and a colour camera system.

Status
In production and in service in the F/A-22 Raptor.

Contractor
Selex Sensors and Airborne Systems.

Head-Up Display (HUD) for the F-16C/D

Type
Head-Up Display (HUD).

Description
In March 1983, the then BAE Systems Avionics announced a USD50 million order to begin production of a new, wide-angle non-holographic head-up display for the US Air Force F-16C/D fighter programme. This head-up display was based on development work undertaken for the US Air Force's Advanced Fighter Technology Integration (AFTI) programme.

The head-up display provides electronically generated symbols thrown up on a total field of view of 25° which was much wider than that attained with previous head-up displays. The instantaneous field of view (that is, the field seen by the pilot without moving his head) is 21° in azimuth and 13.5° vertical. The system uses the same electronics unit as that developed for the LANTIRN system. The symbology and raster scan pictures are particularly suited to guidance and target acquisition at night or in poor weather. The system was claimed to represent the first applications of MIL-STD-1750A processor architecture, MIL-STD-1553B digital data transmission and Jovial 73 MIL-STD-1589B high-order language.

Specifications
Pilot's display unit
Dimensions: 635 × 163 × 170 mm
Weight: 21.6 kg
Power: 98 W (including 25 W for the standby sight)
Predicted MTBF: >2,000 h
Electronics unit
Dimensions: 337 × 180 × 191 mm
Weight: 14.1 kg
Addressable memory: 64 kwords (48 kbytes EPROM, 16 kbytes RAM)
Predicted MTBF: >1,000 h

Status
Still in current production for the US Air Force F-16C/D, with approximately 6,000 units delivered.

In September 1999, BAE Systems (now Selex) was awarded a contract for development of a Commercial-Off-The-Shelf (COTS) replacement Head-Up Display Electronics Unit (HUD EU) for F-16 aircraft. The contract covers development, flight qualification and demonstration within 20 months. The new HUD EU will incorporate a MIL-STD-1750 processor emulator and other COTS components that will allow continued use of existing Operational Flight Programme (OFP) software without modification. The lighter form, fit, function replacement EU incorporates a PowerPC microprocessor and enhanced symbol generator that offers significant improvements in reliability, maintainability and support costs. The new EU will also provide significant room for future growth and functionality expansion using vacant module slots and the existing

The C-17A Globemaster is equipped with dual-HUDs 0022134

F-16Cs and Ds are fitted with Selex' wide-angle head-up displays   0581523

for operation in severe environment applications such as helicopters.

The recorder, including interfaces to a typical ASW mission system, is contained in a 16 MCU volume envelope and has a total weight of 30 kg. The recorder is ideally suited to mission recording, as data can be recorded at any rate up to 107 Mbits/s and, because of the incremental tape motion, uses the minimum tape consistent with recording the output of the mission system. One cassette holds 385 Gbits of data, which gives over 8 hours' recording of a typical helicopter ASW mission.

The associated replay system can replay the data at up to 240 Mbits/s, and will reconstitute with high integrity an exact replica of the input signals.

Specifications
Max input data rate: 107 Mbits/s
Record time per cassette: 1 h at 107 Mbits/s; 8 h at 13 Mbits/s
Tape speed: incremental at 5.31 ips
Bit error rate (corrected): better than 1 in 10^7
Total storage per cassette: 380 Gbits
Dimensions:
(Tape Transport Unit (TTU) 11 MCU
(Data Acquisition Unit (DAU) 5 MCU
Weight: 30 kg
Power: 115 V AC, 400 Hz,
   less than 350 VA

Status
Fully qualified and in production for the UK Royal Navy EH 101 Merlin helicopter, also selected for the Nimrod MR2 Mk.2 maritime patrol aircraft upgrade.

power supply. In addition, the new HUD EU will be capable of rapidly updating software over a MIL-STD-1553 data bus in seconds, rather than the 4–6 hours required with the present system.

Contractor
Selex Sensors and Airborne Systems.

Head-Up Display (HUD) for the night attack AV-8B Harrier

Type
Head-Up Display (HUD).

Description
Refractive optics are used on the head-up display for the AV-8B night attack HUD to give a wide instantaneous field of view of 20° horizontally and 16° vertically. The large diameter collimating lens has been truncated on the fore and aft edges to save weight and to place the lens nearer the pilot's eyes in order to achieve this performance.

Conventional stroke symbology can be overlaid during the raster flyback period on a raster picture derived from electro-optic sensors such as FLIR. Brightness of the two displays can be controlled independently.

Status
In service in the Royal Air Force Harrier GR Mk.7 and night attack variants of the AV-8B.

Contractor
Smiths Aerospace.

High-capacity digital data recorder

Type
Flight/mission recording system.

Description
The high-capacity digital data recorder utilises the Ampex DCRSi technology, and is ruggedised

The cockpit of a RAF Harrier GR Mk.7 showing the Smiths Industries HUD (Jane's/E L Downs)
0116562

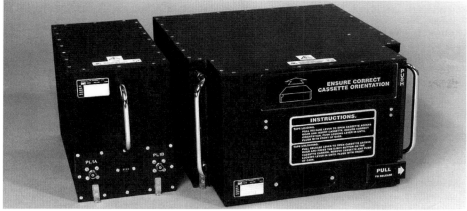

High-capacity digital data recorder showing the Tape Transport Unit (TTU) (right) and the Data Acquisition Unit (DAU) (left)   0001304

Contractor
Honeywell (Normalair-Garrett Ltd).

High-speed Solid-State Recorder (HSSR)

Type
Flight/mission recording system.

Description
The HSSR provides solid-state, fast-access mass memory for applications previously requiring expensive high-speed tape recorders. Each HSSR configuration includes at least one Data Recorder Cartridge (DRC) with non-volatile FLASH memory in configurable memory capacities. This flexibility affords the user the ability to select only the amount of memory required, currently from 8 to 100 Gbytes per DRC. Transferring recorded digital information into ground exploitation systems for evaluation, dissemination or archiving is achieved by either removing DRCs or direct download. After the DRC is erased and initialised, the ruggedised, reusable media is returned to the aircraft for reuse.

Specifications
Dimensions:
(one DRC) 150 × 135 × 337 mm
(two DRCs) 222 × 343 × 337 mm
Weight: 27.2 kg for 2 DRCs
Interfaces: RS-422, RS-232 or MIL-STD-1553
Temperature: −40 to +71°C
Reliability: MTBF 26,000 h (ground), >4,500 h (airborne)
Altitude: 70,000 ft operating

Status
In production.

Contractor
Smiths Aerospace.

In-Step digital technology step-motor aircraft engine instruments

Type
Flight instrumentation.

Description
B&D Instruments and Avionics Inc, a subsidiary of Penny & Giles Aerospace Ltd, has designed microprocessor-controlled micro-stepping motors to reduce the maintenance cost of aircraft engine instruments.

Both 2 in MS style instruments and 2 in ATI style instruments are available, to display the

In-Step exhaust gas temperature (EGT) instrument   0002456

following engine values: torque, propeller speed (Np), gas generator speed (Ng), propeller oil pressure/temperature, engine oil pressure/temperature and exhaust gas temperature.

Contractor
Penny & Giles Aerospace Ltd.

Integrated Mission Equipment System (IMES)

Type
Mission Management System (MMS).

Description
The first IMES to be installed on a non-military aircraft was delivered in August 1998 for installation into a UK Police BK117 aircraft.

The IMES integrates various role equipment together to enhance mission capability and ease observer workload in highly demanding situations. In this particular installation, the IMES has also been fitted with a Voice Recognition System (VRS) to enable voice control of selected mission equipment.

IMES is designed to provide simple integration of various items of role equipment manufactured by different suppliers. The system provides a loop-through facility that allows the equipment to be operated in the normal manner as individual components, or by a central control system using one piece of existing role equipment as the 'Master Control.'

On the BK117 the following equipment is integrated through the IMES system:
* EuroNav III Task Management and Moving Map System ('Master Control')
* FLIR Systems 400 Dual Sensor Camera System
* Video Recorder
* OCTEC Autotracker with Scene Lock and upgraded vehicle tracking mode
* NATS radio systems
* CRT Monitor
* Skyquest 12.1 in LCD monitor
* Searchlight.

This IMES system is fitted with a voice recognition unit that translates voice input commands from the observer to digital control commands for each item of equipment. The main purpose of direct voice control is to allow the observer to concentrate on a task and undertake equipment control without having to look at or touch the role equipment in question.

During all control phases, the observer is presented with messages on his FLIR video screen telling him what the IMES has understood and is doing. The messages stay on the video monitor for five seconds, and individual control of any unit can still be implemented manually at any time.

The Direct Voice Control (DVC) system is similar in technology to that being used in the new Eurofighter Typhoon programme. Voice command recordings have been supervised and implemented at DERA, Farnborough by its speech research unit. At present, the voice recognition system is specific to individual voices, but a new voice recognition card from Octec will allow the system to recognise commands after a brief pre-flight introduction, making the system much more flexible.

IMES is capable of sending observers voice and/or video screen messages that are generated from the EuroNav moving map or other role equipment.

In-Step digital technology step-motor aircraft engine instruments   0002455

For example the EuroNav can be progammed with microwave link receiving sites, and by using its ground terrain mapping and interface to the radalt, can calculate the required height above ground that the aircraft needs to be to enable successful transmission of microwave signals.

When the microwave link is switched ON, IMES tells the EuroNav to switch on its microwave height analysis feature, and if the aircraft is too low, IMES can send a voice and/or video text message to advise the observer of the potential problem.

The functionality of voice feedback can be used for a number of features including powerline proximity warnings, and video tape ending message.

IMES is totally integrated with both the FLIR and map screens and can send data to both screens. Currently under trial is a system that orientates digital mapping information to the FLIR picture and sends digital road name information to the FLIR screen as an overlay onto the FLIR video image. This enables the observer to easily identify the road he is viewing on the FLIR screen without having to look at the map screen individually.

A new generation 14.5 in flat panel, high brightness video screen that allows suitable size picture-in-picture viewing is used. Typically a video image taking up the main screen with an inset moving map picture. At the touch of a button the two images reverse.

The EuroNav moving map system has a tracker feature whereby the observer simply types in a radial being given by the Tracker device in the aircraft. This enables the Tracker information to be displayed graphically on the map screen. This function is being automated.

Contractor
Octec Ltd.
Skyquest Aviation.

Integrated Standby Instrument System (ISIS) – commercial

Type
Flight instrumentation.

Description
Using AMLCD technology, combined with a built-in Air Data Computer (ADC) and Attitude Reference Aerospace System (ARAS), the Integrated Standby Instrument System (ISIS) incorporates all standby functionality, including attitude, altitude and airspeed within a single 3ATI instrument.

There are four variants of the ISIS: fully self-sensing attitude and air data inputs; self-sensing attitude with external air data input; self-sensing air data with external attitude input; and a repeater using external inputs.

Specifications
Weight: 1.7 kg
Dimensions: 3ATI
Power: 28 V DC, 12 W

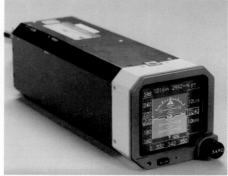

Smiths' Integrated Standby Instrument System (ISIS) 0051687

Brightness Range: 2000:1
Reliability: >15,000 h MTBF
Useable Screen Area: 61 × 61 mm
Sensors: Solid-state rate sensors, accelerometers and pressure transducers
Interfaces: Pneumatic connections for Pitot and Static, external brightness control (option), ARINC 429 (4 receive/2 transmit channels), 12 discrete inputs
Qualification: DO-160C
Software: DO-178B
Certification: TSO C2d airspeed, C3d turn and slip, C4c bank and pitch, C10b altimeter, C113 airborne multipurpose electronic displays, C34e ILS glideslope receiving equipment, C36e ILS localiser receiving equipment, C95 machmeters, C8d vertical speed (planned)

Status
Supplied for MD90, Boeing 717, Bombardier CRJ 200, Learjet 60 and Learjet 45 aircraft.

Contractor
Smiths Aerospace.

Integrated Standby Instrument System (ISIS) Agile

Type
Electronic Flight Instrumentation System (EFIS).

Description
Using AMLCD technology, combined with a built-in Air Data Computer (ADC) and Attitude Reference Aerospace System (ARAS), the Integrated Standby Instrument System (ISIS) Agile is designed to provide all standby functionality, including attitude, altitude and airspeed within a single 3ATI instrument. The system also provides a full range of supporting display functions, including slip, Instrument Landing System (ILS), Mach No, VMO/MMO speed warnings, heading and vertical speed

Key benefits of ISIS Agile include:
- Bright, full-colour, sunlight-readable AMLCD compatible with primary instruments
- Static Source Error Correction (SSEC) and baro-corrected altitude display
- One LRU replaces up to three instruments
- Consumes 40 per cent less power and up to eight times more reliable than traditional standby systems
- Built-in Test (BIT) with maintenance page
- Pin programming allows single part number across different aircraft types
- Requires no active cooling
- Compatible with existing electromechanical instruments, pneumatic connections and panel cutouts
- Automatic fast align on power-up, and manual alignment capability
- MIL-STD-1553B or ARINC 429 interface
- Latest generation processor with anti-aliased raster graphics generation
- Upgraded attitude sensor bay suitable for installation in highly manoeuvrable fighter aircraft.
- NVIS compatible display and bezel.

The ISIS Agile is available in the following variants:
- Fully self-sensing attitude and air data
- Self-sensing attitude, external air data input
- Self-sensing air data, external attitude input
- Repeater using external inputs.

Specifications
Dimensions: 76 × 76 × 229 mm
Useable screen size: 61 × 61 mm
Weight: 1.7 kg
Power: 28 V DC, 20 W
Brightness range: 2000:1
Reliability: >9,000 h MTBF in a military environment
Interfaces: 4 ARINC 429 Receive channels; 2 ARINC 429 Transmit channels; Dual/redundant

MIL-STD-1553; Duplex RS-422; 12 discrete inputs
Qualification:
(Environmental) MIL-STD-810B
(Power) MIL-STD-704A
(EMC) MIL-STD-461D

Status
Selected for F-16 Block 60, Eurocopter Super Puma Mk II, C130 AMP, Embraer ALX and HJT-36.

Contractor
Smiths Aerospace.

Integrated Video System (IVS)

Type
Flight/mission recording system.

Description
IVS is a fully integrated airborne video system which provides distribution for a number of video sources, extensive image processing, the addition of colour symbology, tracking of a number of targets and distribution of video to workstations around the aircraft.

The equipment is packaged to meet airborne environmental and EMC standards. The units include several standard Octec assemblies, including the full range of Octec image processing cards, as follows:
1. VSG 30; an image-processing system.
2. VDB 30, a 16 × 16 video routing switcher, which can be programmed to select any of the 16 video input signals. 8 TTL inputs can also be programmed independently from 8 TTL inputs.
3. VSG 15, a compact (PMC) video routing switcher providing four video outputs, which can be programmed from four video input signals. It also provides a monochrome symbology facility.
4. IMP 15, a PMC card offering a range of complimentary facilities similar to that of the VSG 30, but in a PMC format.

Both the IMP 15 and VSG 15 can be hosted as a mezzanine board on any board capable of hosting PMC modules.

The Integrated Video System is a system tailored to meet individual airborne requirements using a range of standard modules.

Specifications
Mechanical: aluminium alloy chassis, covers and panels
Dimensions (H × W × D): 133 × 486.2 × 263 mm
Weight: 6 kg
Power: dependent on specific application

Status
In production for a NATO ATR 42 maritime patrol aircraft.

Contractor
Octec Ltd.

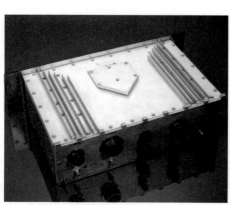

The Integrated Video System (IVS) 0097315

LED engine and system displays

Type
Cockpit display.

Description
Smiths Aerospace's solid-state instrument systems include primary engine displays, aircraft system displays and annunciator panels, and are direct replacements for electromechanical instruments. They are ideal for both new-build and existing aircraft, where they can be retrofitted by easy change of panels and without the need for mechanical or electrical modifications.

Each display system presents information in formats familiar to the aircrew, who thus require minimal training in their use. These displays provide significant benefits in terms of weight, power consumption, reliability, ease of maintenance and cost of ownership.

Primary engine display
The primary engine and aircraft displays are each contained in a single unit which, for ease of maintenance, comprises three modules: a display and associated dimmers, a printed circuit board assembly and a power supply module. Dial markings and legends associated with the displayed parameters are printed on a glass lens which is held in place over the LED displays. Both units are housed in conveniently sealed, lightweight cases with connectors mounted on the rear. Interconnections between the modules and rear connectors are by flexible tape wiring.

For optimum integrity the primary engine display incorporates a separate processor for each engine parameter. Four power supplies are provided to ensure system integrity. The power supplies are configured as two independent pairs so that a failure will result only in partial loss of the parameters of one engine.

Secondary or system display
In the secondary or system display the required reliability is achieved by using multiple processors and power supplies with the parameters to be processed suitably arranged between them.

Brightness is controlled automatically to achieve comfortable viewing over the whole range of ambient lighting conditions on the flight deck, from total darkness to direct sunlight.

Comprehensive BITE facilities are a feature of all Smiths Industries LED display systems. On power-up, a test sequence is initiated automatically. Test routines can be initiated by either the pilot or the maintenance crew.

Annunciator panel
The overhead annunciator panel is a solid-state replacement for existing discrete annunciator panels. Capable of displaying up to 120 cautionary messages, six warnings and 15 advisory indications, it can be installed in any MD-80 series aircraft without the need for electrical or mechanical modifications. On new-build aircraft in particular it makes possible considerable savings in wiring, connectors and weight. Caution messages are displayed on two dot matrix LED panels. Each panel can display six simultaneous cautions of up to 20 characters in length. Both message displays can be scrolled up or down so that a total of 120 different cautions can be viewed.

Eight push-button switches with integral filament lamps operate in conjunction with the two LED message panels. Each switch is associated with a particular aircraft system. Whenever a caution input is received the corresponding lamp is lit so that a system caution is annunciated even if both LED message panels are full.

Normally, cautions received from all aircraft systems are displayed on the message panels in either chronological or priority order as required. If the pilot prefers, he can use the push-button to select one particular aircraft system so that only cautions associated with that system will be displayed.

The two panels are arranged to display a block of 12 consecutive messages. If a fault occurs on one panel then all messages are automatically made available on the remaining panel while the faulty panel remains blank.

A new development of the annunciator panel is the master warning and caution system selected for the MD-90. This system, comprising a separate display and control unit, identifies and presents more than 200 messages.

Status
Fitted in Boeing 737, MD-80 and MD-90, and BAE Systems 146 and Jetstream 41 aircraft. Also available as a developed system for C-130 and P-3 aircraft retrofit.

Engine Instrumentation Display System (EIDS)
Smiths Aerospace has developed, in conjunction with the UK Royal Air Force, a new Engine Instrumentation Display System (EIDS) for the C-130K using LED technology. It has also been selected for use in the US Coast Guard P-3 Orion aircraft.

The EIDS is a direct replacement for the existing 32 electromechanical instruments of existing C-130s and P-3s. The new technology provides significant benefits in terms of weight, power consumption, reliability and cost of ownership. The retrofitting of the EIDS also releases valuable panel space for potential display upgrade programmes.

The EIDS presents primary parameters in a pointer/counter format with secondary parameters in a numeric format to provide one easily read display requiring minimum crew re-training. Three processing channels per engine are provided. Each of three primary parameters is processed by a different channel, along with each of the three engine oil parameters. The remaining three parameters are apportioned, one to each channel. The system provides reliability at least 20 times better than existing indicators. It is a COTS design based on more than 2,000 systems with over 10 million hours in service on commercial aircraft.

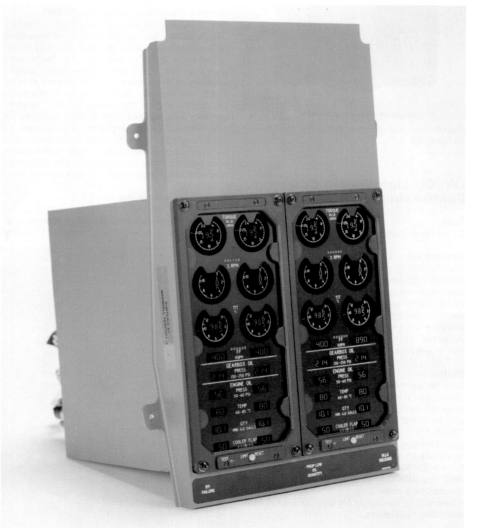

Smiths Aerospace's LED engine and system display   0581510

Smiths Aerospace's Engine Instrumentation Display System (EIDS) for C-130 and P-3 aircraft   0051689

Specifications
Weight: 6.95 kg, each LRU
Power: 28 V DC, 44 W at maximum brightness
Reliability: 7,000 h operating each LRU

Status
Under development as part of a UK MoD contract.

Contractor
Smiths Aerospace.

LED standby engine display panel

Type
Cockpit display.

Description
Smiths Aerospace supplies a standby engine indicator display panel for the Boeing 757 and 767.

The display of engine parameters is by a light emitting diode in a seven-bar format and incorporates four numeric readouts for each engine – N_1 and N_2, EGT and EPR. Eight engine displays are housed in a 3 ATI case.

Status
In production and in service in Boeing 757 and 767 aircraft.

Contractor
Smiths Aerospace.

Standby engine display panel for the Boeing 757 and 767   0503938

LWCCU-LightWeight Common Control Unit

Type
Control and Display Unit (CDU).

Description
The LWCCU is based on the AMS 2000 CDNU (Control Display and Navigation Unit). The LWCCU is the main pilot/co-pilot interface with the aircraft's mission, tactical, navigation and communications systems. It provides significant airframe weight saving (9.8 kg), as well as increased reliability and maintainability over earlier systems.

Status
Selected by the UK MoD for the UK Royal Navy Merlin Mk 1 helicopters and the UK Royal Air Force EH 101 support helicopters. Four LWCCUs will be fitted in each of the Royal Navy's 44 Merlin Mk 1 helicopters: two for the pilot/co-pilot and one each for the mission equipment operators in the cabin. Each of the Royal Air Force's 22 Merlin Mk 3 helicopters will carry only one LWCCU, in the cabin, for use in a maintenance role.

Selected by the Italian MoD for its EH 101 ASW and utility helicopters.

Contractor
Thales Avionics SA.

Thales Avionics will install its LightWeight Common Control Unit into UK Royal Navy Merlin and UK Royal Air Force EH 101 support helicopters   0015374

Maintenance Data Recorder (MDR)

Type
Flight/mission recording system.

Description
The Maintenance Data Recorder (MDR) is a crash-survivable data recorder designed specifically for use on military aircraft. It interfaces with a MIL-STD-1553B databus or RS-422 to store relevant data in crash-protected non-volatile memory. The MDR is a key element of the AH-64D Apache Longbow Integrated Maintenance and Support System (IMSS).

Memory Management Processor (MMP)
The MMP allows user-defined memory segmentation for various classes of data (that is, maintenance, safety, digital voice, and so on) in separate areas of memory. For the Apache Longbow, the MDR segments the crash protected memory to score separately for classes of data: fixed maintenance data is data independent of flight length, for example, tail number; variable maintenance data is used primarily for maintenance or training and is dependent on flight length and characteristics; safety data would be used for incident/accident investigation and is also dependent on flight time and characteristics; and cockpit voice data which is digital audio data.

Cockpit voice recording
Up to two channels of voice recording are provided. Voice digitising utilises an international standard algorithm which meets the performance requirements of ED-56A. The number of channels recorded is a software configurable parameter. The stored voice data can be downloaded and replayed using a standard PC equipped with multimedia capability and voice replay software supplied by Smiths.

Crash-Survivable Memory Unit (CSMU)
The CSMU meets all crash survivability requirements of ED-55 and ED-56. It is expandable up to 640 Mb of memory.

Specifications
Dimensions: $203 \times 165 \times 132$ mm (L × W × H)
Weight: 4.64 kg
Power: 6 W
External connector: 55 Pin. D38999
EMI: MIL-STD-461A
Environmental: MIL-STD-810
Temperature: –40 to +71°C

Status
In production.

Contractor
Smiths Aerospace.

MED 2060 series monochrome head-down displays

Type
MultiFunction Display (MFD).

Description
MED 2060 monochrome head-down displays are small high-brightness raster presentation devices designed for use where space is at a premium. Units are currently available with screen diagonals ranging from 105 to 280 mm. Different aspect ratios are selectable and 525- or 625-line variants are available. The displays can be supplied with or without a passive contrast enhancement filter matched to the CRT phosphor. The filter can be either bonded to the CRT or mounted away from its face to provide optimum visibility in specific conditions.

The display can be used as full multifunction displays for FLIR, radar or stores status, or as simple head-up display repeaters in the rear cockpit of tandem-seat trainers.

Additionally, it has facilities to switch automatically between 1:1 and 4:3 aspect ratios, catering for a range of presentations from maps and engine/systems status data to the display of video from the pilot's head-up display and FLIR night vision system. The unit is also fully compatible with night vision goggles.

Specifications
Weight: 7.5 kg
Power supply: 115 V AC, 400 Hz or 28 V DC, 50 W

MED 2060 series monochrome head-down displays   0581504

Status

The Type MED 2060 series is in service in Sea Harrier FA2 (left and right displays), Hawk 100/200, a variety of Mirage III and 5 aircraft, Danish SAR S-61 helicopters. The display was also utilised for an Asian A-4 Skyhawk retrofit contract.

Contractor

Selex Sensors and Airborne Systems.

Meggitt Avionics' new Generation Integrated Cockpit (MAGIC®)

Type

Avionic system.

Description

MAGIC® is aimed at smaller aircraft and is a suite of displays and sensor systems that utilises LCD and solid-state sensor technology. The system has the following main components:

Two Primary Flight Displays (PFDs), which display all information needed to fly the aircraft.

Two Navigation Displays (NDs), which display all the required navigation modes: compass rose (HSI), MAP and ARC.

Two redundant Engine Display Units, which display information necessary for the safe operation of turbine engines including fuel quantity and usage calculations.

A Data Acquisition Unit (DAU), which converts analogue signals from engine sensors to digital format for display by the EDUs.

An Air Data Attitude Heading Reference System (ADAHRS), which provides ARINC 429 outputs of altitude, speed, rate of climb or descent, pressures, and pitch and roll angles and rates. ADAHRS output recorded by the DAU can be used for engine condition trend monitoring.

Status

The first MAGIC® system was delivered to The New Piper Aircraft company in June 1999 for use on its Malibu Meridian aircraft. Equipment certification followed later that year. Production deliveries under way.

Contractor

Meggitt Avionics.

MOdular Data Acquisition System (MODAS)

Type

Flight/mission recording system.

Description

The MOdular Data Acquisition System (MODAS) has been designed to gather and record any type of signal to be found on an aircraft during flight trials. A large range of signal input types are supported by both the Mk I and Mk II systems. Inputs include strain gauges, voltages, synchros, thermocouples, tachos, ARINC 429, MIL-STD-1553B and RS-232.

MODAS Mk I

MODAS Mk I has up to 4,096 input channels and is expandable to meet specific requirements at sampling rates of up to 128,000 samples/s. Eight separate sampling programs can be selected in flight. A pulse coded modulation digital technique is employed for recording data in either simultaneous multitrack, serial streams or IRIG-106 format. A comprehensive ground replay facility is also available and MODAS can interface with a telemetry system.

MODAS Mk II

The Mk II version of MODAS is less than half the size of the earlier model and is four times as fast. The PA 3101 processor and recorder interface unit can accept data from eight acquisition units, or up to 64 with an expansion

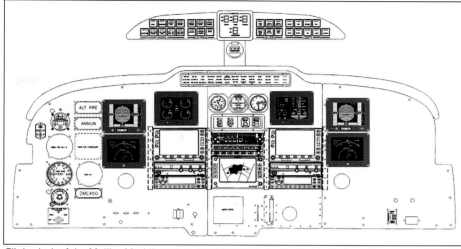
Flight deck of the Malibu Meridian, showing Meggitt Avionics flight display units　　0051683

unit. The PA3120 general purpose acquisition unit houses up to eight interfaces and two MIL-STD-1553B databusses can be monitored with the PA 3130 databus acquisition unit. Data acquisition is up to 512 k parameter samples/s.

Specifications
MODAS Mk I
Dimensions:
(acquisition/processing unit) 1/2 ATR
(control unit) 146 × 191 × 151 mm
(monitor unit) 146 × 191 × 166 mm
(small recorder) 3/4 ATR short
(large recorder) 533 × 360 × 200 mm
Weight:
(acquisition/processing unit) 10 kg
(control unit) 3 kg
(monitor unit) 3 kg
(small recorder) 13 kg
(large recorder) 39 kg
Power supply: 115 V AC, 400 Hz, 3 phase or 28 V DC

MODAS Mk II
Dimensions:
(PA 3101, 3110 and 3130)　135 × 230 × 210 mm
(PA 3120 control and monitor unit) 66 × 146 × 132 mm
Weight:
(PA 3101, 3110, 3130) 5 kg
(PA 3120) 0.9 kg
Power supply: 28 V DC, 60 W (max)

Status

In service in a wide range of UK and European aircraft.

Contractor

Selex Sensors and Airborne Systems.

MultiPurpose Flight Recorder (MPFR)

Type

Flight/mission recording system.

Description

The Penny & Giles Aerospace MPFR functions as both a Cockpit Voice Recorder (CVR) and Flight Data Recorder (FDR), and it uses new materials for impact and fire protection to halve the weight of earlier generation recorders. Hardware and software variants offer wide flexibility in both functionality and recording duration.

Specifications
Cockpit Voice Recorder (CVR)
Duration: four channels × 30 minutes
option for up to 4 × 120 minutes
ATC recording: ARINC 429 digital ATC recording (low- or high-speed)
Audio bandwidth: 150 to 3,500 Hz (3 × voice channels)
150 to 6,000 Hz (area microphone channel)
Playback: real-time (read-after-write) summed output of all audio channels
high-speed digital audio recovery via Ethernet into PC or ground station

Flight Data Recorder (FDR)
Duration: 25 h at up to 256 words/s
shorter duration at up to 4,096 words/s

The Penny & Giles Aerospace multipurpose flight recorder　　0062791

Data acquisition: ARINC 573 or MIL-STD-1553B from data acquisition unit
Replay: ARINC 573 read-after-write
Ethernet to PC high speed
wideband datalink high speed
Fault reporting: ARINC 429 interface to OMS

Common to all configurations
Location beacon: standard Ultrasonic Locator Beacon (ULB)
Power: 28 V DC, 20 W
Cooling: convection, no forced air
Dimensions: ARINC 404A, 1/2 ATR short: 124 × 320 × 194 mm (W × D × H), or miniature outline: 115 × 230 × 86 mm (W × D × H) excluding ULB
MTBF:
(fixed-wing) 15,000 h
(rotary-wing) 10,000 h

Optional features
Wireless datalink system: interface
PC-card copy facility: integrated

Optional equipment
Cockpit-mounted area microphone: optional
Control unit with integral microphone: optional
Portable replay equipment: optional
Remote pre-amplifier and area microphone: optional

Status
Selected for the Bell 609 helicopter.

Contractor
Penny & Giles Aerospace Ltd.

Night Vision Goggle Image Recording System (NVG IRS)

Type
Flight/mission recording system.

Description
The Night Vision Goggle Image Recording System (NVG IRS) is designed to be compatible with most traditional circular eyepiece pilot's NVGs, recording the image viewed by the left or right eye (or both, utilising two systems). The system comprises a NVG camera, which is mounted on a simple collar arrangement and clamps onto the goggle eyepiece and a compact digital video recorder, which is carried on the person, either in a combat vest or a pouch. The camera and recorder are connected by a flexible cable which is routed around the pilot's helmet, under the stole of the Life Saving Jacket (LSJ) and into the recorder. For ease of donning and doffing, connectors are incorporated at natural break positions and a quick-release break can be integrated to cater for NVG autoseparation systems.

The NVG camera can be specified to capture the whole of a typical 40° Field of View (FoV) at medium resolution, or the central portion (approximately 25°) at higher resolution, in a colour or monochromatic format. Digital video is recorded on a standard format mini-DV tape.

The NVG IRS is an invaluable aid to effective NVG conversion and continuation training for all fixed- and rotary-wing users, enabling close monitoring of pilot scan techniques and sensor management. The system is also useful in police and paramilitary applications, where, as a complimentary sensor to Infra-Red (IR) search and track systems, it can provide supporting sensor information and backup capability in case of IR sensor degradation (crossover/high absolute humidity).

The advantages of the NVG IRS include:
• Extremely lightweight, with very small increase in moment for typical 800 g NVGs
• Inexpensive when compared with other systems
• Compact design with negligible interference with peripheral vision
• Compatible with autoseparation systems
• No modification to NVGs required

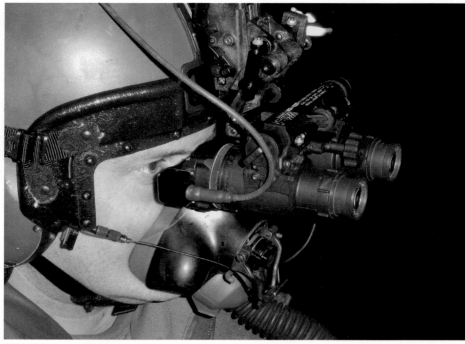

NVG IRS shown fitted to the right eyepiece of a Nightbird NVG. Note the pilot is still able to fit his Face Protection Visor (FPV) and retains his peripheral vision (E L Downs)   0561212

• No aircraft modification required.

The NVG IRS has been tested on current Gen III NVGs, with high-quality imagery recorded down to starlight levels. Integration with most types of circular eyepiece NVGs is accomplished by simply varying the diameter of the attachment collar.

Specifications
Weight:
NVG camera: 78 g
DV recorder: 460 g (no battery); lithium cell weight 76 to 225 g (dependent on capacity)
Power supply:
NVG camera: 9 V DC
DV recorder: Onboard rechargeable lithium cell
Power supply endurance:
NVG camera: In excess of 20 h on single PP6 9 V cell
DV recorder: Up to 9 h (depending on lithium cell capacity)
Recording capacity: Up to 1.5 h on a standard mini-DV tape
Resolution: 380 horizontal lines
Low-light capability: Host NVG-dependant

Status
Variants of the NVG IRS are compatible with AN/AVS-6(V)2/3, AN/AVS-9, Night-Op, Nightbird and CN2H NVGs. The system is in production and in service with military and paramilitary NVG users.

During July 2005, a higher resolution, 600-line camera was announced, with all other physical characteristics remaining the same.

Contractor
Aerosystems Consultants.

Open system computers

Type
Aircraft mission computer.

Description
General Dynamics UK produces an advance range of open system computers, featuring Commercial-Off-The-Shelf (COTS) technology, to provide integrated avionics systems for military aircraft. These systems provide high performance processing capabilities with the additional benefits of rapid, low-risk development, low life-cycle cost, obsolescence protection and significantly increased functional integration.

Advances in electronic design and packaging technology have allowed the functions previously implemented by an entire computer LRU to be compressed to single modules, providing higher integration of functions and the ability to expand the capability of the aircraft avionics suite. The open systems approach enhances this capability by utilisation of open standards for

The old mission computer for the Harrier GR. Mk 7 (left) and its replacement open systems computer (right)   0064381

both hardware and software module interfaces, producing re-usable designs and providing rapid access to rapidly emerging COTS technologies.

This approach provides for scalable avionics systems solutions by re-use in pre-defined modular functions, allowing rapid technology insertion through module replacement strategy, which mitigates risk of obsolescence and provides leading edge performance.

Typical functions which can be integrated in an open systems mission computer include:

- Stores management
- Mission processing
- Ground Proximity Warning System (GPWS)
- Digital maps/situational awareness
- Navigation
- Advanced graphics generation
- Electronic Warfare (EW) processing
- Advanced data modems/tactical datalinks
- Audio management
- Health & Usage Monitoring (HUM)

Typically, such integrated avionics systems combine previously discrete elements of the avionics suite, including navigation, stores management and mission processing, into a single computer using open systems architecture. This provides the user with additional functionality with significantly lower cost, weight and space requirements.

Status
General Dynamics provides its open system architecture computer for the UK Harrier GR. Mk 7 aircraft, while General Dynamics Information Systems, part of the US parent General Dynamics Company, provides a similar solution for the US AV-8B, FA-18 and F-15 aircraft.

Contractor
General Dynamics United Kingdom Ltd.

Optical Quick Access Recorder (OQAR)

Type
Flight/mission recording system.

Description
Penny and Giles new high-capacity Optical Quick Access Recorder (OQAR) enables airlines to transfer data quickly from the flight line to the Flight Operations Quality Assurance (FOQA) system. The compact, ruggedised unit features a fault tolerant design with extensive internal diagnostics and provides up to 40 hours of recorded data at 1,024 words/s and 640 hours of data at 64 words/s. This data allows an airline to conduct the following types of analysis: Engine Condition Trend Monitoring (ECTM); flight operation analysis; aircraft and autopilot performance; engine exceedance reporting; automatic landing analysis and engine derate measurement.

The Optical Quick Access Recorder is designed to provide all the necessary data for effective Flight Operations Quality Assurance (FOQA) programmes. The OQAR can be programmed to record over 2,000 separate parameters, including data from the APU, brake system, and navigation computer which is typically not recorded by other flight data recorders. Airlines that implement FOQA programmes can select which parameters to record, in order to increase safety while reducing maintenance costs, to identify problems and trends more easily and to optimise crew training.

These new high-capacity data recorders document operational characteristics of the aircraft's entire flight, including take-offs and landings. Fully ruggedised, they record even when subjected to vibration or shock, so transient or abnormal events such as hard landings or fast rotations will not be missed. The unit is designed so the rewritable magneto-optical disk can be quickly removed from the aircraft during a level A check, at weekly intervals or even during a normal gate turn. As part of the FOQA programme, the flight data is then downloaded to a ground station where

Optical Quick Access Recorder (OQAR) 0005419

after de-identification, analysis for significant operational patterns can be conducted. Additional ARINC 429 inputs are provided for supplementary TCAS (Traffic Collision and Avoidance System) and GPS (Global Positioning System) recording, or for use as part of a FANS (Future Air Navigation System).

Status
The OQAR units are available in configurations for all new Airbus and Boeing aircraft; and are fully compatible with all digital bus aircraft including ARINC 573/717, and ACMS. They feature embedded control software, which can be upgraded while in service use. A front panel RS-232 port allows for self-diagnostic checks, relay of the stored data, or configuration changes. The recorders use industry standard, 3.5 in magneto-optical disks with MS-DOS-compatible format, which are ISO 10090 compliant.

Contractor
Penny & Giles Aerospace Ltd.

PA3700 Integrated Health and Usage Monitoring System (IHUMS)

Type
Flight/mission recording system.

Description
The PA3700 Integrated flight data recording Health and Usage Monitoring System (IHUMS) has many aircraft applications. It has been designed to fulfil the functions of a Digital Flight Data Recorder (DFDR), a Cockpit Voice Recorder (CVR) and a Health and Usage Monitoring System (HUMS). The system architecture has been optimised to provide a minimum hardware solution by the combination of these functions. Each airborne system comprises a Data Acquisition and Processing Unit (DAPU), Cockpit Voice and Flight Data Recorder (CVFDR), Card Maintenance Data Recorder (CMDR) and Control and Monitor Unit (CMU). In addition to this core system, other options are available, including a control and display unit, pilot interface panel, cockpit warning panel and quick access recorder.

The PA3701 DAPU contains all the conditioning circuitry necessary to sample and accurately monitor a wide range of different types of electrical inputs for subsequent recording, measurement or processing. The mandatory data output interfaces to a standard ARINC 573/747

Flight Data Recorder (FDR), and a standard ARINC quick access recorder. Selected mandatory data, together with raw and partially processed HUM data, is also fed to a ruggedised maintenance data recorder. Data can be displayed on the optional control and display unit as required.

There are several versions of the crash-protected recorder available to customers with differing requirements. For instance, one CVFDR provides 5 hours of continuous digital recording at a data rate of 128 12-bit words/s and 1 hour of voice on each of three separate tracks. Another FDR provides 8 to 10 hours of continuous digital data recording at a data rate of 128 12-bit words/s and is used in conjunction with a separate CVR.

The CMDR was designed to meet ARINC 615 as a high-speed data loader. The recording medium is a 2 Mbyte SRAM card. The data interface is bidirectional using RS-232C protocol. The unit is rugged, compact and the recording medium easily transportable, which makes it ideal for use as a card maintenance data recorder. The bidirectional interface to the DAPU enables upload of documentary data from a card as well as download of raw and preprocessed data for maintenance purposes.

The downloaded airborne DAPU data is supported by a comprehensive ground replay and analysis computer system.

Specifications
Dimensions:
(PA3701 DAPU) ½ ATR short
(CVFDR) 115.6 × 173.5 × 440.5 mm
(CMDR) 38.1 × 133.35 × 132.05 mm
Weight:
(PA3701) 6.2 kg
(CVFDR) 8.5 kg
(CMDR) 0.5 kg
Power supply: 28 V DC, 35 W (max)

Status
The system is fitted to over 120 helicopters worldwide. Systems are currently available for 14 aircraft types or variants covering the AS332L, AS332L1, AS365N, AAS365N2, Bell 212, Bell 214ST, Bell 412, S61, S76A+, S76A++, S76C, S76C+, Sea King.

The Mk II IHUMS has been selected by Sikorsky for the S-76C and S-92 production helicopters. The system architecture has been optimised to provide a minimum hardware solution and comprises a DAPU, CVFDR, CMDR and CMU.

Contractor
Selex Sensors and Airborne Systems.

Units of the IHUMS II equipment showing (left to right) the DAPU, CMU, CVFDR and CMDR 0504167

PA3800 series flight data acquisition units

Type
Flight/mission recording system.

Description
The PA3800 series, of flight data acquisition units, was developed to meet FAA and CAA requirements for flight data recording applicable from 1991. The unit samples data from a variety of input signals, which may be analogue, digital or discrete. The information is sampled in a programmable predetermined sequence and assembled into a digital data stream in a format compatible with any standard ARINC 573/717/747 digital flight data recorder. There are four or five PCB positions available for expansion of the system, perhaps taking the form of extra signal conditioners, or the unit may be expanded into an integrated microprocessor-based monitoring system to provide engine or airframe usage monitoring.

The PA3810 is for accident data acquisition, with four-card expansion available. The PA3820 is for helicopter health and usage monitoring, including accident data. The PA3830 is for fixed-wing aircraft for the Aircraft Integrated Monitoring System (AIMS), including accident data.

The latest development of the system has been designed to meet US FAA rules for August 2002, which require that a minimum of 88 parameters be recorded. This model has been selected for the Avro RJ series Regional Jet aircraft.

Specifications
Dimensions:
(PA3810) 3⁄8 ATR short case to ARINC 404A
Weight: 4 kg typical
Power supply: 115 V AC or 28 V DC, 15 W (max)

Status
In full production for both military and civil applications on fixed-wing aircraft such as the de Havilland Dash 8, Cessna Citation and Bombardier Challenger, and on Bell and Eurocopter helicopters.

Contractor
Selex Sensors and Airborne Systems.

PA5000 series radar altimeters

Type
Flight instrumentation, radar altimeter.

Description
The PA5000 is a software-controlled altimeter operating in the J-band using advanced

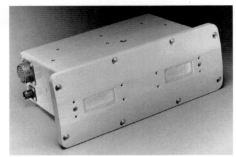

The PA5000 radar altimeter 0504054

microwave and signal processing surface mount VLSI integration techniques.

The transmit and receive antennas are both included within the unit outline and thus the PA5000 is fuselage mounted, requiring only a single fuselage cutout. RF feeders are not required. Operating in the J-band, the PA5000 series radar altimeter provides a covert system and gives precision accuracy and resolution for all high-performance fixed-wing, helicopter, UMA and RPV applications. In helicopter applications, excellent hover and nap of the earth performances are available.

PA5000 series radar altimeters offer height ranges up to 5,000 ft with 2 per cent accuracy, are designed to be compatible with most types of aircraft digital and analogue interfaces and are qualified to MIL-STD-461C and MIL-STD-810D.

Specifications
Dimensions: 218 × 76 × 138.6 mm
Weight: 3.2 kg
Power Supply: 19–32 V DC, 25 W (max) at 28 V
Altitude: 0–5,000 ft options
Accuracy: ±2% height + 2 ft
Reliability: 5,000 h MTBF

Status
Installed in Agusta/Sikorsky HH-3F and the NH500 helicopters.

Contractor
Selex Sensors and Airborne Systems.

PA5200 series radar altimeters

Type
Flight instrumentation, radar altimeter.

Description
The PA5200 series radar altimeter operates in the mid-J-band using microwave Field Effect Transistor (FET) technology. Software-controlled

signal processing techniques are used to enable reliable performance to be achieved up to 5,000 ft with a transmitter power of only 0.5 W. Surface-mount technology is used to give a low-volume, high-reliability package which includes the antenna.

The PA5200 radar altimeter uses a dual leading-edge tracker to ensure tracking of the nearest object. Continuous automatic monitoring of the system ensures high reliability down to ground level.

Specifications
Dimensions: 225 × 76 × 117 mm
Weight: 3 kg
Power supply: 28 V DC, 26 W (max)
Altitude: up to 5,000 ft (can be extended)
Accuracy: ±3% height +3 ft
Temperature range: −40 to +50°C
Reliability: 5,000 h MTBF

Status
In service.

Contractor
Selex Sensors and Airborne Systems.

PA5495 radar altimeter

Type
Flight instrumentation, radar altimeter.

Description
The PA5495 radar altimeter operates in mid-J-band using microwave Field Effect Transistor (FET) technology. Software-controlled signal processing techniques are used to enable reliable performance to be achieved up to 5,000 ft with a transmitter power of only 1 W. Surface-mount technology is used to give low volume. Separate antennas are provided to be compatible with existing C-band installations.

The altimeter uses a dual leading-edge tracker to ensure tracking of the nearest object. Continuous automatic monitoring of the system ensures high reliability with accurate height indication right down to ground level.

Specifications
Dimensions: 140 × 220 × 85 mm
Weight: 4 kg
Power supply: 28 V DC, 26 W
Altitude: 0–5,000 ft (can be extended)
Accuracy: ±3% height + 3 ft
Temperature range: −40 to +70°C
Reliability: 5,000 h MTBF

Status
In service.

Contractor
Selex Sensors and Airborne Systems.

Portrait format intelligent MultiFunction Displays (MFDs)

Type
MultiFunction Display (MFD).

Description
Caledonian Airborne Systems produces a range of portrait format intelligent MFDs designed for both civilian and military applications. Display capabilities include flight instrumentation, moving map and radar/tactical information, coupled with internal character generation, custom overlay and portrait to landscape live video rotation.

Other features and benefits of the MFDs include:
RS-170, CCIR, STANAG 3350B or Computer Graphics video inputs;
ARINC 419/429, MIL-STD-1553, analogue and discrete interfaces;
High resolution active matrix flat panel display with high brightness wide range CCFL backlight;

Caledonian Airborne portrait MFD (Caledonian Airborne Systems)   0528519

User interface by programmable softkeys and rotary encoders;
Available in standard 8.4 in and 10.4 in versions.

Specifications
Display resolution: 800 × 600 pixels in a 264 mm (10.4 in) diagonal
Colours: 262,144
Video standards: RGB computer graphics (up to SVGA) with separate H & V syncs, composite, STANAG 3350, CCIR/PAL & RS-170/NTSC
Luminance: 1,000 cd/m² minimum
Dimming range: 200:1
Power: 28 V DC to MIL-STD-704
Lighting: 5 or 28 V, AC or DC
Weight: 3.5 kg
Dimensions: 257 × 212 × 50 mm (behind panel)
Interface: 32 programmable softkeys, 5 predefined hardkeys, rotary encoders
I/O: ARINC 419/429, discretes, analogue, synchro, EIA-RS422
Temperature: Civil or military temperature ranges
Cooling: Internal forced air

Options
Display resolution: 640 × 480 (VGA) or 1,024 × 768 pixels (XVGA)
Display size: 213 mm (8.4 in)
Dimming range: 1,000:1
Video standards: Computer Graphics (up to XVGA), various radar video standards
Interface: ARINC 629 or MIL-STD-1553
NVIS compatibility: NVIS Type I & II Class B to MIL-L-85762

Status
In production and in service.

Contractor
Caledonian Airborne Systems Ltd.

PRS3500A data/voice accident recorder

Type
Flight/mission recording system.

Description
The PRS3500A consists of the BAE Systems PRS3501A data acquisition unit and the Penny & Giles D50330 accident data recorder.

The system provides 2 hours' continuous recording time, taking in up to 50 parameters at 240 words/s. The small size of the equipment makes it suitable for aircraft with limited cockpit space and digital transmission reduces the size and weight of wiring looms while increasing data integrity. Recorded data can be extracted in 6 minutes using a portable transfer device. The system comprises two units: the flight data recorder and a data acquisition unit.

The PRS 3501A data acquisition unit, and Penny & Giles D50330 accident data recorder   0504270

Specifications
Dimensions:
(data acquisition unit) ½ ATR short
(flight data recorder) 115 × 172 × 440 mm
Weight:
(data acquisition unit) 4.5 kg
(flight data recorder) 8.3 kg

Status
In production and in service in UK Royal Air Force Harrier GR Mk. 7 aircraft.

Contractor
Selex Sensors and Airborne Systems.
Penny & Giles Aerospace Ltd.

PV1820C structural usage monitoring system

Type
Flight/mission recording system.

Description
Designed to monitor airframe fatigue by continuously measuring structural loads during flight, the PV1820C is based on the company's earlier PVS1820 engine usage life monitoring system. Data and the computed results are recorded on a cassette-loaded quick access recorder for ground replay and analysis. The system comprises a PV1820C structural monitoring unit, a PV1819C control unit, quick access recorder Type 1207-003 and a transient suppression unit Type PV1845. Microprocessor operation simplifies the measurement of stress by enabling the output from a number of strain gauges (typically 16) to be monitored along with other flight parameters on a cassette recorder or solid-state data card.

Specifications
Dimensions:
(PV1820C) 94 × 194 × 394 mm
(PV1819C) 146 × 76 × 194 mm
(1207-003) 146 × 51.6 × 169 mm
(PV1845) 95 × 53 × 190 mm
Weight:
(PV1820C) 5.44 kg
(PV1819C) 1.59 kg
(1207-003) 1.77 kg
(PV1845) 1.4 kg

Sampling rate: Each parameter is defined by 10 bits of a 12-bit word, the 11th bit being a compression flag and the 12th being reserved for parity. The sampling rate for each quantity is programmable in binary steps from 1 to 128 times/s
Data compression: programmable 16:1, 8:1 or 4:1
Self-test: comprehensive automatic self-test and fault diagnosis is built in

Status
In service.

Contractor
Selex Sensors and Airborne Systems.

PV1954 flight data acquisition unit

Type
Flight/mission recording system.

Description
The PV1954 fulfils all of the requirements for a 32 parameter Flight Data Recorder (FDR) and provides expansion capability for additional maintenance monitoring.

Conditioning circuits enable the PV1954 to sample data from a wide variety of input signals. These inputs may be analogue, digital or discrete. The information is sampled in a predetermined sequence and assembled into a digital data stream in a format compatible with any standard ARINC 573/717/747 Digital FDR (DFDR).

A programmable read-only memory controls the input signal sampling sequence. This method of control permits the user to define the content of all the words in the frame, except synchronisation words.

The PV1954 provides an output of 64 12-bit data words/s in Harvard bi-phase format to the DFDR. As an option, this data rate may be increased to 128 or 256 words/s.

An auxiliary data output in RZ format is provided for use with an optional quick access recorder, which operates at 64 words/s. Optionally, it is possible to increase the data rate to 128 or 256 words/s.

A time synchronisation output in frequency shift key format is provided to synchronise the DFDR to the cockpit voice recorder.

Specifications
Power input 28 V DC nominal, 15 W (max)
Dimensions: ½ ATR short case to ARINC 404A
Weight: 5 kg typical

Status
In service on Gulfstream IV, BAE Systems 146 Series 2 and ATP aircraft.

Contractor
Selex Sensors and Airborne Systems.

Quick access recorder

Type
Flight/mission recording system.

Description
The quick access recorder is an avionic data logger of modular construction designed to provide rapid access to aircraft performance data recorded in flight. The media used is an industry standard ¼ in magnetic tape cartridge, giving low-cost data storage.

The unit fulfils the quick access recorder requirement in a civil aircraft integrated monitoring system; it also provides automatic read-after-write error detection and correction and an onboard replay facility with GMT search. The quick access recorder is microprocessor controlled and conforms to ARINC 591. It has a built-in power supply, with battery back-up for power drop-out immunity.

Specifications
Dimensions: 124 × 320 × 194 mm
Weight: <7 kg
Power supply: 115 V AC, 400 Hz, 70 VA peak
Reliability: 4,500 h MTBF

Status
Selected by a number of major airlines for use on a wide variety of aircraft.

Contractor
Penny & Giles Aerospace Ltd.

Radar converter – RDR 1500

Type
Avionic display processor.

Description
The RAIU-1000 radar signal converter is designed for use with Bendix/King/Telephonics RDR 1500 series radar units. The converter facilitates replacement of legacy radar indicators, in particular the IN-1502H, with a modern, high-resolution, multifunction Liquid Crystal Display (LCD) such as one of Skyquest's own range (see separate entry), thereby allowing colour video and FLIR images to be viewed in addition to radar information. When the system is employed with a Skyquest LCD, dedicated buttons allow the operator to select Radar, Map (DMG), Forward-Looking Infra-Red (FLIR), video, and Picture In Picture (PIP). Should the operator need to maintain radar information on the screen, he can choose to overlay video as a window on the display. The size and position of this window can also be controlled. At any time an integrated video Freeze Frame (FF) facility can be employed to create a still image from the video stream, which can then be zoomed for further examination of a specific Point Of Interest (POI).

The RAIU-1000 uses existing aircraft wiring and can be employed when a more capable radar display is required, or perhaps more usually, when an aircraft upgrade programme places demands on available cockpit space, requiring a single display unit for multiple sensor/graphic inputs.

Specifications
Dimensions: 140 × 20 × 120 mm (W × D × H)
Weight: <1 kg
Altitude: 25,000 ft
Temperature: –30 to +70°C (standby); –25 to +55°C (operating); –30 to +80°C (storage)
Power: 5 V DC from radar interface unit; 12 V DC from host display

Status
In production.

Contractor
Skyquest Aviation.

RMS 3000 reconnaissance management system

Type
Mission Management System (MMS).

Description
The RMS 3000 Series combines scan conversion and image processing of electro-optic sensor imagery with in-flight display of the terrain overflown. The RMS 3000 Series system is fitted to UK Royal Air Force Tornado GR. Mk 1A and Mk 4A reconnaissance aircraft and provides for the simultaneous recording and display of the imagery from multiple infra-red sensors. Other system features include real-time and near-realtime display of imagery, a wide range of rolling and updating display modes, display facilities including slow speed replay, magnification and processing to optimise

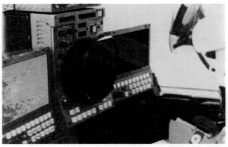

Infra-red images produced by the GD UK RMS 3000, displayed in the rear cockpit of a Tornado GR. Mk 1A   0581492

the displayed image and aspect correction and rectilinearisation of displayed imagery to facilitate in-flight exploitation.

The system design is optimised to ensure that the ground exploitation time is minimised. Relevant features include combined video and aircraft data annotation recording on videotape, in-flight editing of event imagery and rapid removal of video cassettes from the aircraft on return to base.

General Dynamics UK is developing a Digital Solid State Recorder (DSSR) LRI to replace the existing analogue videotape recorder. The DSSR will provide fast data access with greater system reliability.

Status
In service in UK Royal Air Force and Royal Saudi Air Force Tornado GR. Mk 1/4A aircraft as part of the Tornado Infra-Red Reconnaissance System (TIRRS).

Contractor
General Dynamics United Kingdom Ltd.

RMS 4000 reconnaissance management system

Type
Mission Management System (MMS).

Description
The RMS 4000 Series is an all-digital system which is designed to be common across several platforms. Key features which have been used to enhance the system performance and increase the mission success rate are use of Ada high-order language, flexible modular design with high configuration commonality across platforms, high reliability and high level of functional redundancy to ensure no single defect can cause mission critical failure, design for two-level maintenance and growth potential to accommodate new sensors, image and target processes.

The RMS 4000 provides a selection of sensors (EO, IR and SAR) with built-in growth potential, image data management, compression and decompression and routeing, multiple display formats for the selected sensor, recall of imagery for review and edit prior to data transmission, datalinking of key target imagery to provide real-time record and availability of archive data as soon as the aircraft lands.

GD UK is integrating Solid State Recording (SSR) technology into the RMS 4000 to provide customers with significant system cost reductions and performance enhancements.

The RMS 4000 is available for nose-pallet and pod-mounted installation.

Status
In production. Variants of the RMS 4000 are in production (under contracts to Lockheed Martin) for the US Marine Corps F-18 Advanced Tactical Airborne Reconnaissance System (ATARS – see separate entry).

Contractor
General Dynamics United Kingdom Ltd.

Ruggedised flat panel LCD aircraft monitors

Type
Cockpit display.

Description
Skyquest Aviation makes a range of LCD aircraft monitors in 4, 6.5, 8.4, 10.4, 12.1, 15, 18 and 20.1 in sizes.

The AVDU-1010 4 in display is the smallest and most simple display of the range, designed primarily as a secondary viewing device. The display can be fitted in a cockpit or mounted on a small bracket to be positioned conveniently in small spaces. The unit is powered by 28 V DC, with video input via composite PAL or NTSC.

The high-brightness multifunction 6.5 in display is designed for viewing high-quality video images and VGA/SVGA computer signals, making it ideal for applications which require data and TV (or Forward-Looking Infra-Red) compatibility. A key feature of the 6.5 in display is a highly flexible approach to video compatibility, offering all possible combinations of both video protocol and video standard inputs.

The AVDU-2100 is a compact multifunction 8.4 in LCD display providing a wide range of video and VGA compatibility, combined with user-defined input to suit most applications. Two variants of the display are available, one with incorporated soft control keys and one without. Standard display features include Picture In Picture (PIP), video freeze frame, video zoom and full range dimming. The buttons are backlit with customised NVIS-compatible lighting. The user can dim the backlighting to suit the prevailing environment by using the simple set up menu. A rotary brightness knob gives the operator fine control over the brightness of the screen, with special attention paid to precise control of the dimmer ranges, where NVIS-users in particular demand highly accurate and stable levels. The display can also be wired to enable the host platform 'master' cockpit lighting controller to dim and brighten the display in line with other avionics in the cockpit. An on-screen menu can be called up which controls resizing and repositioning of the PIP window and a number of other user-defined options. Optionally, the display can be fitted with 'Skyquest ClearVision' – a specially designed contrast module that enhances vision when mist or haze is present. On selection of the function, a window appears, centralised, with the display and the image within the window enhanced using proprietary algorithms.

The 10.4 in XGA AVDU-264X is available in three variants, a basic display (AVDU-2641), a display incorporating soft keys (AVDU-2640) and a display incorporating soft keys and touch screen overlay (AVDU-2648). While most of the basic features of the 10.4 in unit are similar to those for the 8.4 in AVDU-2100, the touch screen overlay of the AVDU-2648 provides on-screen menus controlled either by stylus or bare/gloved finger. Utilising resistive touchscreen technology, the display offers reliable and accurate operation under the demanding conditions of airborne operations.

The 12.1 in monitor feature set follows the 8.4 and 10.4 in displays, with full PAL or NTSC capability and incorporates a high-brightness, sunlight-readable display with switch-selectable NVIS compatibility as an option. Colour/greyscale capability allows the display to be used with dual-sensor camera systems. The usual multimode video inputs accept all video formats together with VGA/SVGA signals, with instant switching between inputs.

The 15 in XGA AVDU-382X is available in two variants, a basic display (AVDU-3820) and a display incorporating soft keys (AVDU-3822). All of the basic features of the 15 in unit are similar to those for the 10.4 in AVDU-2640 and 8.4 in AVDU-2100.

Completing the line-up of LCD displays, Skyquest's 18 and 20.1 in multifunction video displays include all of the functionality of the smaller displays with soft keys and on-screen touch overlay optional. Featuring all aluminium construction to aerospace standards and fast video response for smear-free video presentation, the displays can be employed in landscape or portrait orientation.

Specifications
AVDU-1010
Viewing area: 4 in diagonal
Dimensions: 146 × 104 × 50 mm
Weight: <1 kg
Connection: 2 × BNC (isolated)
Video: PAL 50 Hz/NTSC 60 Hz
Viewing angle: ±45° horizontal; +10 to –30° vertical
Contrast ratio: >100:1 (at optimum viewing angle)
Brightness: >250 cd/m² (maximum); <7 cd/m² (minimum)
Resolution: 480 × 234 pixels
Pixel pitch: 0.171 × 0.264 mm (W × H)
Pixel configuration: RGB vertical stripe
Grey scales: 64
Colour capability: 6 bits/colour; 256,000 colours
Temperature range: 0 to +60°C (operating); –30 to+80°C (storage)
Power: 28 V DC; 15 W

Multi function 6.5 in display
Viewing area: 6.5 in diagonal
Dimensions: 200 × 166 × 94 mm
Weight: <2.6 kg
Connection:
  Video: CVBS, 75 Ohm, IV p-p; 1 × BNC
  Component video: Y/C MIL 38999
  VGA/SVGA: Mode 3, 12 RGB with separate MIL 38999; H and V syncs
Video: PAL 50 Hz/NTSC 60 Hz
Viewing angle: ±45° horizontal; +15 to –35° vertical
Contrast ratio: >60:1
Brightness: >870 cd/m², 1,000 cd/m² (typical); >60 cd/m² (maximum), 0 cd/m² (minimum) in NVIS compatible mode
Resolution: 640 × 480 pixels
Grey scales: 64
Colour capability: 6 bits/colour; 262,144 colours
Temperature range: –10 to +60°C (operating); –25 to +70°C (storage)
Power: 28 V DC; ≤28 W

AVDU-2100
Viewing area: 8.4 in diagonal
Dimensions: 210 × 180 × 77 mm
Weight: 2.8 kg
Connection: 3 × BNC (isolated)
Video: PAL 50 Hz/NTSC 60 Hz Composite, S-Video (Y/C) and Component video (RGB/Y, Cr, Cb) or STANAG 3350/B/C
Graphic input: VGA, SVGA, XGA, SXGA and UXGA
Viewing angle: ±85° horizontal; ±85° vertical
Contrast ratio: >300:1 (at optimum viewing angle)
Brightness: >350 cd/m² (maximum); <7 cd/m² (minimum) XGA; Option 1,400 cd/m² SVGA
Resolution: 1,024 × 768 pixels
Pixel pitch: 0.1665 × 0.1665 mm (W × H)
Pixel configuration: RGB vertical stripe
Grey scales: 64

Skyquest 12.1 in display showing raster imagery from external sensor   0123259

Skyquest 12.1 (left) and Skyquest 6.5 in displays in map mode (north-up), with TV monitor insert (upper left)   0123258

Colour capability: 6 bits/colour; 256,000 colours
Temperature range: –20 to +55°C (operating); –25 to +80°C (storage)
Power: 28 V DC; 45 W

AVDU-2640
Viewing area: 10.4 in diagonal
Dimensions: 250 × 183 × 63 mm (273 × 225 × 63 for AVDU-2648)
Weight: 2.8 kg (<3.2 kg for AVDU-2648)
Connection: 4 × BNC (isolated)
Video: PAL 50 Hz/NTSC 60 Hz Composite, S-Video (Y/C) and Component video (RGB/Y, Cr, Cb) or STANAG 3350/B/C
Graphic input: VGA, SVGA, XGA, SXGA and UXGA
Viewing angle: ±55° horizontal; +40 to –35° vertical (±45° horizontal; +15 to –35° vertical AVDU-2648)
Contrast ratio: >200:1
Brightness: >800 cd/m² (maximum); <7 cd/m² (minimum) XGA; Option 1,200 cd/m² SVGA (>870/<5 cd/m² AVDU-2648)
Resolution: 1,024 × 768 pixels
Pixel pitch: 0.2055 × 0.2055 mm (W × H)
Pixel configuration: RGB vertical stripe
Grey scales: 64
Colour capability: 6 bits/colour; 256,000 colours
Temperature range: –25 to +55°C (operating); –30 to +80°C (storage)
Power: 28 V DC; 50 W (maximum)
Touch screen (AVDU-2648)
Positional accuracy: <2 mm
Touch activation force: <113 g

12.1 in multifunction display
Viewing area: 12.1 in diagonal

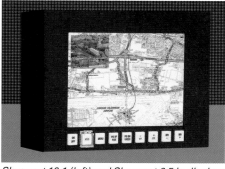

Skyquest 12.1 (left) and Skyquest 6.5 in displays in map mode (north-up), with TV monitor insert (upper left)   0123257

Dimensions: 300 × 259 × 68 mm
Contrast ratio: >120:1
Brightness: >870 cd/m² (sunlight readable)
Resolution: full PAL (scaled) in video mode; SVGA in graphics mode; 1,024 × 768 pixels
TV lines: 730
Grey scales: 64
Colour capability: 6 bits/colour; 262,144 colours
Temperature range: –10 to +60°C (operating)
Power: 28 V DC (built-in)

AVDU-3820
Viewing area: 15 in diagonal
Dimensions: 385.5 × 310 × 62 mm
Weight: <6 kg
Connection: 4 × BNC (isolated)
Video: PAL 50 Hz/NTSC 60 Hz Composite, S-Video (Y/C) and Component video (RGB/Y, Cr, Cb) or STANAG 3350/B/C
Graphic input: VGA, SVGA, XGA, SXGA and UXGA
Viewing angle: ±60° horizontal; ±60° vertical
Contrast ratio: >200:1 (300:1 typical)
Brightness: >800 cd/m² (maximum); <7 cd/m² (minimum) XGA
Resolution: 1,024 × 768 pixels
Pixel pitch: 0.297 × 0.297 mm (W × H)
Pixel configuration: RGB vertical stripe
Grey scales: 256
Colour capability: 8 bits/colour; 16.2 million colours
Temperature range: –25 to +55°C (operating); –30 to +80°C (storage)
Power: 28 V DC

Status
In production and in service.

Contractor
Skyquest Aviation.

Skyquest 12.1 in high-brightness display used by UK police in an AS355 helicopter   0051685

Ruggedised SVHS video recorder

Type
Flight/mission recording system.

Description
Skyquest Aviation produces the AVCR 1030, a ruggedised Super VHS video recorder which has been designed to complement modern high-performance FLIR and TV cameras. The system offers high-resolution, high-quality video and audio recording and playback. Features and benefits of the system include:

- Specially designed DZUS mounted remote control offering full functionality including playback mode
- Confirmation lights on remote control unit to show recording status
- System automatically forwards to clear tape after reviewing pre-recorded footage, ensuring valuable evidence is not destroyed
- Custom-designed anti-vibration tray
- Interference-free freeze-frame capability
- Horizontal resolution of >400 lines
- Automatic video head cleaning
- 4 channel audio recording (selectable)
- Highly reliable and durable tape transport system
- Compact, space-saving design
- Y/C and composite video inputs
- Available in PAL and NTSC versions.

Specifications
General
Power: 28 V DC, 35 W
Operating temperature: +5 to +40°C
Operating humidity: 35 to 80%
Weight: approximately 7.4 kg
Dimensions: 270 × 131.5 × 365.5 mm (H × W × D)

System
Television format: CCIR standard, PAL colour signal (625 lines, 50 fields); EIA standard, NTSC colour signal (525 lines, 60 fields)
Video recording system: Two rotary heads, helical scanning system
Modulation System:
Luminance FM azimuth recording
Colour signal converted subcarrier phase shift recording
Audio track: 2 tracks (hi-fi audio), 2 tracks (normal)

Tape Transport
Tape format: S-VHS/VHS tape
Tape speed: PAL 23.39 mm/s; NTSC 33.35 mm/s (15/16 ips)
Recording time: Dependant on tape: PAL 3 h max; NTSC 2 h max

Video
Input: S-Video In (4P), Line In (BNC)
Output: S-Video Out (4P), Line Out (BNC),
Signal to noise ratio: VHS >75 dB (B/W); PAL >45 dB (colour); NTSC >45 dB (colour)

Audio
Input: Line In (phono), Mic (3.5 mm phono)
Output: Line Out (phono), Audio Monitor (phono), Headphone (3.5 mm phone)
Dynamic range 90 dB (hi-fi audio)

Status
In production and in service.

Contractor
Skyquest Aviation.

SCR 200 flight data/voice recorder

Type
Flight/mission recording system.

Description
The SCR 200 consists of a crash-protected accident data recorder and a data acquisition

Royal Saudi Air Force Tornado GR1 aircraft are fitted with SCR 200 flight data recorders   0103896

unit. It has been designed for the Tornado. The accident data recorder comprises a continuous single-spool tape transport with four recording tracks. Recording rate is 128 12-bit data words/s. Total recording time is 111 minutes data and 37 minutes voice. Audio information is recorded on one track simultaneously with data on the other three.

Specifications
Dimensions:
(ADR) 425 × 172 × 81 mm
(DAU) 383 × 128 × 200 mm
Weight:
(ADR) 7.5 kg
(DAU) 7 kg
Power supply:
115 V AC, 400 Hz, 25 W
28 V DC

Status
In production. Over 750 systems are in service on UK Royal Air Force, Royal Saudi Air Force and Italian Air Force Tornado aircraft.

Contractor
Selex Sensors and Airborne Systems.

SCR 300 data/voice recorder

Type
Flight/mission recording system.

Description
The SCR 300 flight data recording system is designed for use on military aircraft, where space is at a premium. It is believed to be the smallest and lightest tape combined CVR/ADR system currently available.

The main function of the SCR 300 is the preservation of a flight audio and data record in the event of an accident, but it can also be used for maintenance monitoring. The system consists of a crash-protected Accident Data Recorder (ADR) and a Data Acquisition Unit (DAU).

The ADR is mechanically and thermally protected to recent FAA and CAA standards and has provided 100 per cent data and voice retrieval. The system comprises a continuous single-spool tape transport with six 30 minute tracks of information, with tracks assigned to audio and data as required. The recording rate is 128 12-bit data words/s, giving a total recording time of 3 hours. The ADR is fitted with a sonar locating beacon.

The DAU collects information from signal sources in a variety of digital, analogue and discrete forms, samples each at an appropriate rate and converts them into a stream of serial digital data which it then transmits to the ADR. The number of parameters monitored depends

on the application. A typical system has more than 40 digital, analogue, audio and discrete signal inputs. Uniquely, the DAU stores up to 64 different aircraft frame formats within the processor. This feature greatly eases logistics in mixed aircraft fleets, where the single standard DAU is totally interchangeable among numerous aircraft types.

Specifications
Dimensions:
(ADR) 115 × 150 × 250 mm
(DAU) 89 × 125 × 200 mm
Weight:
(ADR) 6 kg
(DAU) 4 kg
Power supply: 28 V DC, 40 W

Status
Available to order and in service worldwide on over 200 Jaguar aircraft, Harrier, Sea Harrier, Hawk 60/100/200, Andover, BAC 111 and Raytheon Hawker 800 fixed-wing aircraft and Chinook, Lynx, Sea King, Gazelle, Scout and Wessex helicopters.

Contractor
Selex Sensors and Airborne Systems.

SCR 500 solid-state cockpit voice and flight data recorders

Type
Flight/mission recording system.

Description
The SCR 500 combines the latest digital recording and flash memory techniques, to provide a lightweight cost-effective range of Cockpit Voice Recorder/Digital Flight Data Recorders (CVR/DFDRs). The SCR 500 range complies with the latest requirements of EUROCAE -55/56A, the latest TSO C123A/C124A for increased fire protection and ARINC 757/747, with recording duration to meet both current and future requirements.

The SCR 500-030 and SCR 500-120 CVRs respectively have 30 and 120 minutes' audio recording duration. The combined CVR/FDRs have a range recording duration from 30 minutes audio/10 hours data to 2 hours audio/25 hours data at 128 words/s or 50 hours at 64 words/s data rates. All SCR 500 recorders have instant real-time audio and data playback.

The SCR 500-630 and SCR 500-660 solid-state combined flight data recorders provide 30 minutes or 1 hour of four-channel audio duration, and 20/10 hours duration of 64/128 digital data words per second

The SCR 500-1530 and SCR 500-1560 solid-state CVR/FDR combined recorders provide

Seven variants of the SCR 500 solid-state cockpit voice and flight data recorders, shown in front are the SCR 500 cockpit control panels (miniature and standard size options).   0504267

Alternative configurations of the Series 3150 colour video cameras   0015319

30 minutes or 1 hour, four-channel cockpit voice and 25 hour digital data. They are designed to provide a long duration audio/digital recorder in a single box, for large helicopters.

The SCR 500-1620 solid-state CVR/FDR combined recorder provides a full 2 hours, four-channel cockpit voice and 25 hour digital data. It is designed to replace both CVRs and FDRs in large passenger aircraft giving the safety benefits of dual redundancy. Utilising an SCR 500 Combi in place of a CVR enables regional aircraft operators to implement cost-effective FOQA procedures in conjunction with the latest analysis software package.

All models can be supplied with control panels (miniature and standard size options) and an underwater beacon. The audio provision includes four analogue channels recorded digitally and one digital data channel plus timebase. Test equipment, designed around common PC architecture, provides Downloading/Testing and Replay of all the SCR 500 family of recorders. A hand-held data downloader is also available.

Specifications
Dimensions: ½ ATR short
Weight: <7 kg for all versions
Power supply: 115 V AC, 400 Hz
28 V DC, 12 W (nominal)
Control panel dimensions:
miniature 28.6(H) × 146(W) × 70.5(D) mm
standard 57.2(H) × 146(W) × 70.5(D) mm

Status
Certified to CAA ED 55/56A, and FAA TSO C123/C124, and compliant with TSO C123A/C124A. In production and in service.

The recorders are being fitted into several major fixed-wing and helicopter programmes: prime fit for Sikorsky S92 and S76C+ (IHUMS) programmes; prime fit for Avro RJ series; prime fit for Royal Australian Air Force Hawk LIF; prime fit for Atlas Rooivalk attack helicopter; prime fit for UK MoD DHFS Bell 412s; selected for UK Royal Air Force BAE 146 and Raytheon Hawker 800 fleets.

Contractor
Selex Sensors and Airborne Systems.

Series 2768 Super VHS video cassette recorder

Type
Flight/mission recording system.

Description
The small lightweight Series 2768 helical scan video recorders are suitable for recording both video and data signals in hostile environments and alternative case designs enable the recorder to be located in either cockpit or equipment bay.

The recorder employs a Super VHS-C/VHS-C cassette to achieve a minimal size. Cassettes may be replayed on the ground by means of a mechanical adaptor in standard Super VHS-C and VHS video equipment, resulting in low-cost and readily available playback systems. Vinten video recorders are capable of producing recordings through high *g* aircraft manoeuvres, high vibration or gunfire.

Series 2768 video cassette recorder   0015318

In operation the Series 2768 recorder is sealed against sand, dust and water and contains a conditioning heater for low-temperature operation.

A significant feature of the Series 2768 is the incorporation of anti-vibration/shockmountings within the case of the recorder. This enables the recorder to be hard-mounted to the aircraft in almost any orientation and requires no sway space. The Series 2768 recorder may be used for unusually formatted signals such as those from linescan sensors.

The Airborne Recorder for IRLS and EO Sensors (ARIES) Video Cassette Recorder (VCR) is one of the Vinten Series 2768 recorders which are designed specifically for airborne environments. The recorder is available in both cockpit and bay-mounting versions.

The VCR is designed for recording linescan formatted video imagery derived from IRLS or EO sensors. It can be fitted internally or in a pod in manned aircraft or in RPVs, drones and UAVs.

Specifications
Dimensions: 125 × 152 × 226 mm
Weight: 3.5 kg
Recording system: rotary helical scan
Video signal system: PAL colour/CCIR monochrome 625-lines or NTSC colour/EIA monochrome 525-lines, plus other formats for non-standard signals
Recording time: up to 2 h
Event marker: visual and audio
Recording bandwidth: up to 4.8 MHz

Status
In production and in service.

Contractor
Thales Optronics (Vinten) Ltd.

Series 3150 colour video camera

Type
Flight/Mission recording system.

Description
The Series 3150 colour video camera is a modular CCD camera specifically designed for airborne recording applications.

The camera is designed to provide high-resolution colour images and is readily installed as either original equipment or as an aircraft upgrading. There are two versions of the camera: the low-profile and the universal. The latter has the ability to separate the lens and sensor from the remainder of the camera by a flying lead. This new concept for an airborne colour camera enables the ideal positioning of the camera lens in a location where there may otherwise be limited room for the complete camera. Previously the only alternatives were either to produce a special camera of dedicated design or to provide an optical periscope and suffer the resulting reduction of light to the sensor.

The camera is able to produce television pictures over a very wide illumination range and operate continuously in very demanding environmental conditions.

For cockpit installation the Series 3150 camera's standard modules are flexible enough in configuration to be orientated to suit most HUD and gunsight mounting requirements. This approach offers both an optical and cost-effective solution. The equipment can be used in association with Vinten airborne video recorders such as the Series 2768.

Specifications
Dimensions:
(camera head) 105 × 56 × 24 mm
Weight:
(camera head) 0.75 kg
Power supply: 28 V DC, 18 W nominal when used with Series 2768 VCR
Camera head sensor: solid-state imager
Field of view: 25 × 19° or 20 × 15°
Refresh rate: 50 or 60 Hz
Dynamic range: 5–170,000 lx

Status
In production and in service.

Contractor
Thales Optronics (Vinten) Ltd.

Solid-state air data instruments

Type
Flight instrumentation.

Description
Altimeters
The range of solid-state 3ATI altimeters is based on silicon pressure transducers and custom Liquid Crystal Displays (LCDs) with unique features to meet requirements for primary flight displays. Configurations are available for civil and military applications and as primary, primary with reversion to standby, standby or repeater variants. The altimeters have altitude reporting and alerting output options and a digital output of altitude is available to an ARINC 429 databus.

Specifications
Weight: <1.1 kg
Altitude range: −2,000 to 99,900 ft
Reliability: 15,000 h operating MTBF
Compliance: DO 160c; TSO C10b

Meggitt Avionics 3 ATI altimeter   0051679

Meggitt Avionics 3 ATI altitude alerter   0051680

Altitude alerter

The Meggitt Avionics altitude alert unit is a solid-state design, utilising a high quality Liquid Crystal Display (LCD) housed in a half 3ATI case.

The pre-selected altitude is displayed on the five digit, seven segment, LCD by adjusting the rotary knob on the front of the unit.

The visual amber altitude alert warning is provided both on this unit and also provides an output to illuminate an amber warning on the Meggitt Avionics primary altimeter via ARINC 429 databus interface. The altitude alert unit monitors the displayed altitude on the primary altimeter (via ARINC 429 databus) and compares that with the altitude set on the alert unit.

Airspeed and Mach/airspeed indicators

With integral pitot and static silicon pressure transducers, microprocessor technology and Liquid Crystal Displays (LCDs), these units provide high quality display of airspeed or Mach/airspeed complete with all markers and bugs. Repeater variants are also available.

Air data and attitude sensors

Meggitt Avionics produces a range of solid-state air data and attitude sensors to provide inputs, via ARINC 429 databuses, to their display systems including: an Air Data Unit (ADU) to provide accurate altitude and airspeed data; and Attitude Reference Unit (ARU) to provide attitude data; and ADAHRS an Air Data Attitude Heading Reference System, which outputs pitch and roll attitude, pitch, roll and yaw rates, altitude, rate of change of altitude (IVSI), airspeed and heading.

Reduced Vertical Separation Minimum (RVSM) system

The RVSM system utilises the above sensor and display units (as depicted in the diagram below) to meet the requirements of full RVSM compliance with a system that is easily installed, and offers low procurement cost. It reduces errors by utilising the latest sensing and processing technology, combined with improvements resulting from

Meggitt Avionics 3 ATI airspeed and Mach/airspeed indicator   0051681

advantages achieved by locating the Air Data Units adjacent to the pitot and static probes.

Status

The RVSM system was selected by the Royal Air Force in the UK for its VC-10 aircraft to meet NATS RVSM requirements.

Standby air data and inertial sensing devices are being supplied to Rockwell Collins for incorporation into the US Marine Corps UH-1Y and AH-1Z helicopter upgrade programmes.

Contractor

Meggitt Avionics.

Solid-State Quick Access Recorder (SSQAR) D51555

Type

Flight/mission recording system.

Description

The Solid-State Quick Access Recorder (SSQAR) D51555 has been developed for light turbine-engined helicopters and turbofan corporate aircraft for offline trend analysis, aircraft performance monitoring and maintenance recording. It is a ruggedised, lightweight and environmentally sealed data recorder which uses industry standard solid-state memory cards. The unit acquires navigational data, typically from a GPS receiver, as well as engine and airframe parameters from various ARINC 429/573/717, RS-422, analogue and discrete sources. These inputs are sampled and decoded as necessary and the acquired data

The Solid-State Quick Access Recorder D51555   0504168

is time-stamped with a GMT value before being written to flash memory cards. Three card slots are provided, each capable of addressing up to 32 Mbytes of memory, subject to card availability of higher-capacity cards in the future.

Specifications

Power supply: 28 V DC nominal
Interfaces: ARINC 573, ARINC 429 (2), RS-422 (2), analogue (4), discrete inputs (16), run control, discrete outputs
Reliability: 15,000 h MTBF

Contractor

Penny & Giles Aerospace Ltd.

Standby Master Warning Panel (SMWP)

Type

Aircraft centralised warning system.

Description

The Standby Master Warning Panel (SMWP) provides the pilot with essential red primary warning indications in the event of any failure of the standard warning system. It is designed for use in glass cockpits, so the one-piece display matches the appearance of glass instruments.

The display panel has 12 red warning captions, lit by LEDs, which flash until acknowledged. Different captions can be included, to satisfy customer requirements, up to 16 characters in length. Brightness control, to switch between day sunlight-readable and night illumination levels, is achieved by a Dim push-switch mounted directly under the display.

Situated adjacent the Dim switch is a rotary mode switch to enable selection of either automatic or manual operation and test. The test position checks more than 97 per cent of the circuit.

In view of its essential role the SMWP is made fault tolerant, so that no single fault will cause the loss of more than one warning. Even in the event of power supply failure, the unit will still operate in a degraded mode.

Specifications

Dimensions: 87.8 × 100.3 × 152 mm
Weight: 0.73 kg
Power supply: 28 V DC, <22 W (max)
Temperature range: –15 to +55°C
Altitude: up to 45,000 ft
Reliability: >100,000 h MTBF

Contractor

Page Aerospace Ltd.

Standby warning panel   0504202

Thales Avionics Management System (RAMS)

Type

Avionics system.

Description

Thales Avionics' RAMS is a family of avionics management systems that can be configured to

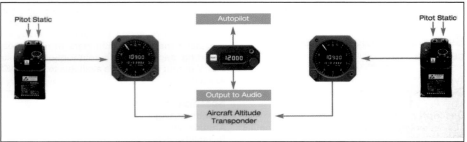

Meggitt Avionics reduced vertical separation minimum system   0051682

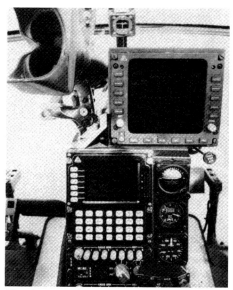

The integrated crew station on the Boeing 530MG helicopter with the RAMS multifunction display (top) and control/display unit   0503910

meet the operational requirements of military and commercial operators for helicopters and fixed-wing aircraft. The purpose of the equipment is to reduce the aircrew workload and so enhance flight safety and mission effectiveness.

The RAMS family starts with the basic RAMS 1000 navigation management system. The RAMS 3000 is similar, but configured around a MIL-STD-1553B databus. The RAMS 2000 is the non-MIL-STD-1553B equivalent of the RAMS 4000.

The RAMS family has a flexible hardware and software architecture for the user-friendly control and display of avionics systems, as well as accomplishing a variety of interfacing and processing tasks. Simplex and duplex system configurations can be provided, depending on the level of redundancy, interfacing and processing required. Management of the Communication, Navigation and Identification (CNI) subsystems can be provided, as well as mission management for the sensors and subsystems specific to the operational role of the aircraft. Other functions provided are performance management, including fuel management and Health and Usage Monitoring and Sensing (HUMS) of engine and transmission parameters.

The interface between the aircrew and the avionics systems is one or more Control and Display Units (CDUs) with alphanumeric keyboard and dedicated function keys; the CDUs are night vision goggle compatible. Associated with the CDU are one or more Processor Interface Units (PIU), commonly called the mission computer, with a hardware and software configuration specific to the operational requirement.

MIL-STD-1553B is the primary interface between the PIU, CDU and other compatible sensors and subsystems, with the bus control provided by the PIU; other interfacing can be analogue or digital, with a variety of industry standards such as ARINC 429, RS-232 and RS-422. A pocket book-sized Data Transfer Device (DTD) with solid-state memory is used to enter information and instructions from a DTD ground loader into the various battery-supported RAMS databases maintained in the PIU.

A variety of liquid crystal display remote indicators, monochrome and colour raster and stroke displays with associated symbol generators, stiff stick controllers and auxiliary control panels are available to make up any required configuration.

A Thales Avionics RAMS 4000 forms the heart of the UK Royal Navy Lynx Central Tactical System (CTS), with two specially extended control and display units and a central tactical situation display. Additional equipment includes a dual-stores management unit for weapons control and two processor interface units, linking RAMS to existing navigation, communications and other systems. A data transfer device allows mission data to be loaded quickly into the system before flight. The high flexibility and wide range of enhancements of RAMS allows it to be tailored for present day and future requirements.

Specifications
Type 5401 processor interface unit
Dimensions: 129 × 194 × 414 mm
Weight: 5.9 kg (depending on number of interface modules)

Type 5407 control and display unit
Dimensions: 146 × 229 × 248 mm
Weight: 4.5 kg

Type 5404/5 series data transfer device with receptacle
Dimensions: 146 × 38 × 171 mm
Weight: 0.6 kg

Status
In service. Superseded by Thales' AMS 2000 CDNU (see separate entry). Notable applications of RAMS are the Central Tactical System (CTS) for the UK Royal Navy Lynx HAS Mk 8 helicopter, the tactical data system in the Royal Danish Navy Lynx, and the Boeing 530MG Defender multirole helicopter.

RAMS has also been fitted in Royal Swedish Air Force Super Puma rescue helicopters and a RAMS version with multifunction displays has undergone trials on German Army PAH-1 anti-tank helicopters. It has been configured for use in maritime patrol fixed-wing aircraft and helicopters.

Contractor
Thales Avionics SA.

Tornado Advanced Radar Display Information System (TARDIS)

Type
Multi Function Display (MFD).

Description
Part of the UK Device Software Optimisation (DSO) programme, which aims to implement an open-architecture approach to avionics, the employment of Wind River Systems' General Purpose Platform, VxWorks Edition software in the Tornado Advanced Radar Display Information System (TARDIS) will facilitate faster and more cost-effective development of advanced software for the system.

TARDIS replaces the 1970s-vintage Combined Radar and Projected Map Display (CRPMD) in the rear cockpit with a new radar and map processor on a flat-panel display. Using Wind River's subscription-based licensing model and integrated platform, BAE Systems developed the necessary advanced software.

The VxWorks software provides an underlying operating system to control the radar display, inserting a software foundation between the existing hardware and the display application software to create a ruggedised software platform. The system takes all the input signals (from the radar, map-projector and inertial navigation systems) to present them in the moving-map display.

The use of such an open-architecture software platform will enable the system to be enhanced and upgraded as required during the remaining service life of the Tornado GR4.

Status
BAE Systems Customer Solutions & Support was awarded the TARDIS contract, worth GBP82 million (USD141 million), in December 2003,

BAE Systems and Wind River Systems, Inc. are collaborating on the Tornado Advanced Radar Display Information System (TARDIS) (BAE Systems)   1184363

following earlier UK Ministry of Defence-funded development.

During June 2005, Wind River Systems, Inc. announced that BAE Systems selected the General Purpose Platform, VxWorks Edition for use in the Tornado Advanced Radar Display Information System (TARDIS). The TARDIS will replace existing radar projected map displays used in the Tornado GR4 aircraft and is scheduled to go into service with the Royal Air Force by the end of 2006.

The software was licensed directly to BAE Systems Avionics at Edinburgh South Gyle, then responsible for developing the TARDIS digital map display. The map display element subsequently moved to BAE Systems Platform Solutions at Rochester.

Contractor
BAE Systems.
Wind River Systems, Inc.

TARDIS has been developed for the UK Tornado GR4 (Jamie Hunter)   1184360

Turn and slip indicators – FTS20 series

Type
Flight instrumentation.

Description
The Ferranti Technologies' FTS20 series turn and slip indicators are single degree of freedom gyroscopes that operate from a 28 V DC nominal supply.

The indicators are housed in a 56 mm diameter case with a 60 mm square mounting flange. The instrument is available with optional integral lighting and Night Vision Imaging System (NVIS) compatibility.

The Ferranti Technologies' FTS20 series turn and slip indicators   0062850

The rate of turn pointer is driven by a rate gyroscope which is powered by an alternating current supply. Aircraft slip is indicated by a ball-in-tube inclinometer. The alternating current for the rate gyroscope is derived from an integral inverter circuit. The direct current supply is monitored by a warning flag which operates when voltage drops below 14 V. Versions are available which provide voltage outputs for differing rates of turn.

Specifications
Scale reading: zero; rate 1 (180°/min); rate 2 (360°/min)
Scale colours: white on black
Altitude rating: 10,700 m (35,000 ft)
Dimensions:
  Case diameter: 56 mm
  Flange size: 60 mm square

Case length (excluding connector);
148 – 172 mm
Mass: 0.7 – 0.75 kg

Contractor
Ferranti Technologies Limited.

Two- and three-terminal light modules

Type
Cockpit annunciator.

Description
Designed for use in single or stacked multichannel configuration, the filament lamp indicator/annunciator modules are a development of a module used extensively in aircraft central warning systems. The modules, available in two- and three-terminal arrangements, provide a low-cost, lightweight and compact alternative to other devices.

Both types of modules incorporate a clip-on caption frame within which a wide range of sunlight-readable blank or engraved caption screens may be contained. The three-terminal module may be used with two screens, one for each lamp, for systems where dual lamping is not essential. In these instances a lamp-holder separator screen may be fitted in a groove provided. Various alternative mounting arrangements facilitate single or multiple panel or stacked installation.

Specifications
Dimensions: 28 × 10.9 × 29 mm
Weight: 0.018 kg
Temperature range: –40 to +55°C
Altitude: up to 60,000 ft

Contractor
Page Aerospace Ltd.

Page Aerospace two- and three-terminal light modules   0581508

Type 1502 Head-Up Display (HUD)

Type
Head-Up Display (HUD).

Description
The Type 1502 HUD is of modular construction. The unit is highly reliable and is easy to maintain; cost, volume and weight have been reduced without any compromise over optical accuracy and capability. The Type 1502 is suitable for installation in many new build attack aircraft or for retrofit, particularly where space or volume is limited.

The total field of view is 25°, achieved by using a 140 mm diameter exit lens, truncated fore and aft, and dual combiner glasses. Stroke symbology, raster video imagery or hybrid formats can be displayed. A video camera and an electronic variable standby sight can be incorporated as options. A customised data entry panel is an integral part of the HUD and enables the pilot to control the aircraft's nav/attack system and HUD mode. The Type 1502 is precision hard-mounted on the aircraft and requires no on-aircraft harmonisation.

Specifications
Dimensions (including combiners):
  489 × 171 × 301 mm
Weight: 11.3 kg

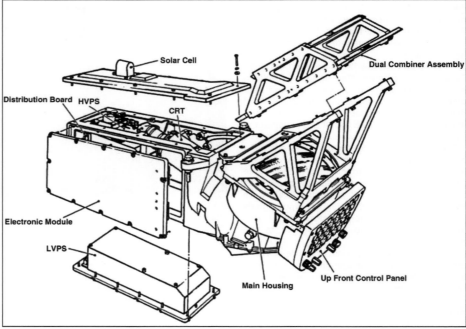

A schematic of the Type 1502 head-up display, showing the major components of the system   0533859

Power supply: 28 V DC, 120 W (excluding smart UFCP); 5.4 V DC, 9 W
Cooling: Natural convection
Reliability: >2000 h MTBF

Optical
Total FoV: 25°
Instantaneous FoV: 21 × 17° (azimuth × elevation)
Exit lens aperture: 140 mm diameter, truncated to 114 mm in the fore and aft plane
Photopic transmission: 70% through dual combiner
Deviation error: <1 mR

Display
Colour: Green P53 phosphor
Boresight accuracy: Better than 1 mR
Symbol position accuracy:
    Boresight: 1 mR
    6° deflection: 2 mR
Reaction time: Display within 30 sec, full performance within 2 min

Stroke
Luminance: >5000 cd/m²
Visibility: Background >35,000 cd/m²; contrast ratio >1.2
Line width: 0.8 mR at 400 cd/m²
Write speed: >16,800°/s

Raster
Format: 525 line 30/60 Hz; 625 line 25/50 Hz
Nominal size: 24 × 18°
Peak brightness: 250 cd/m² minimum (at nominal size)

Stroke in flyback
Write speed: >100,000°/s
Luminance: >840 cd/m²

Status
In production and in service in the F-5E/F, Hawk 100, 200 and LIF.

Contractor
Smiths Aerospace.

Type 3000 integrated colour display

Type
MultiFunction Display (MFD).

Description
The Type 3000 is a self-contained colour multipurpose display designed to be integrated into aircraft navigation and attack systems. All primary flight and navigation data is displayed, replacing conventional attitude and horizontal situation indicators, plus a route map representation. In addition, flight plan, system status including fuel information and engine data, radar warning receiver and maintenance information can be displayed under the control of the 17 soft keys around the front panel.

The unit combines a graphics generator, deflection amplifiers, high-voltage power supplies and a full shadow-mask CRT into a single display unit. The display data and control functions are transmitted via the aircraft MIL-STD-1553B databus. The latest large-scale integration and hybrid packaging techniques have resulted in a compact self-contained unit.

Specifications
Dimensions: 162 × 162 × 317 mm
Weight: 11 kg
Power supply: 115 V AC, 400 Hz, 3 phase, 200 VA
Usable screen area: 127 × 127 mm
Reliability: 1,500 h MTBF

Status
In production for the BAE Systems Hawk 100 and 200.

Contractor
Smiths Aerospace.

Smiths' Type 3000 display is installed in BAE Systems Hawk 200 aircraft (BAE Systems)   0569532

Type 6000 colour video camera

Type
Flight/mission recording system.

Description
The Series 6000 range of colour video cameras are a range of single interchangeable Line Replaceable Units (LRUs), available in a wide range of mechanical packages depending on desired application. Each unit is available in either PAL 625 line/50 Hz or NTSC 525 line/60 Hz formats. The cameras are designed for use in harsh airborne environments, inside or outside the host aircraft and are suitable for HUD, cockpit and instrument recording, external airframe and cargo bay monitoring.

Specifications
TV standard: PAL 625 line/50 Hz or NTSC 525 line/60 Hz
Video output: Y/C composite colour video
Resolution: >550 TV lines horizontal; >350 TV lines vertical
Aspect ratio: 4:3
Automatic scene illuminance range: 170,000 lux
SNR: >46 dB RMS at 1,000 lux
Lens FoV: 25.33° × 19°, or 20° × 15° (H × V; both V ±4%); other options available
Synchronisation: Free running unless externally synchronised
Power: MIL-STD-704A 28 V DC; 5 W (typical) (115 V AC versions available)
Environmental: Per MIL-STD-810C
EMC: Per MIL-STD-461A for Class A1 equipment
Weight: Less than 0.45 kg (excluding mounting assembly and electrical connectors)

Status
In production and in service.

Contractor
Thales Optronics Ltd.

Type 6051 video conversion unit

Type
Flight/Mission recording system.

Description
The Vinten Type 6051 video conversion unit is designed as a 525-line/60 Hz to 625-line/50 Hz video converter for use in combat aircraft. It enables 525-line/60 Hz video sources to be interconnected to 625-line/50 Hz video

Type 6051 video conversion unit   0015320

switching, display and recording equipment and is appropriate for use where multiple video standards exist on an aircraft. The converted video output can be synchronised to another onboard 625-line/50 Hz source.

Status
In production.

Contractor
Thales Optronics (Vinten) Ltd.

Type 6225 colour HUD video camera

Type
Flight/mission recording system.

Description
The Type 6225, part of the Series 6000 range of colour video cameras, is specifically designed for HUD recording, applicable to wide range of fighter, trainer and other airborne platforms. The Type 6225 features a modular architecture to facilitate customisation of installation according to requirements.

Specifications
TV standard: NTSC 525 line/60 Hz (PAL versions available)
Video output: Y/C composite colour video (digital output available)
Resolution: >550 TV lines horizontal; >350 TV lines vertical
Aspect ratio: 4:3
Automatic scene illuminance range: 170,000 lux
SNR: >46 dB RMS at 1,000 lux
Synchronisation: Free running unless externally synchronised
Power: MIL-STD-704A 3-phase, 115 V AC (28 V DC versions available)
Environmental: Per MIL-STD-810
EMC: Per MIL-STD-704A

Status
In production and in service.

Contractor
Thales Optronics Ltd.

Type 9000 Head-Up Display (HUD)

Type
Head-Up Display (HUD).

Description
To cater for larger attack/fighter aircraft or those aircraft with less installation constraints, the Type 9000 wide-angle conventional HUD uses a 6.5 in (165 mm) exit lens. It has self-contained high- and low-voltage power supplies and is capable of cursive and raster display with cursive-in-raster flyback. High-brightness combiners give daylight viewability of the raster FLIR image. A high-resolution colour camera is fitted to record the pilot's view.

The HUD is driven by a Computer Symbol Generator (CSG), which offers full mission computing and faster/cursive symbol generation capability.

Type 9000 HUD 0001457

Type 9000 HUD installed in a Tornado GR. Mk 4A (E L Downs) 0106203

The CSG can be fitted with multiple processor cards to ensure all customer systems moding, control and operational requirements can be easily accommodated. The CSG is fully programmable using the ADA software language.

The CSG provides video routeing and mixing of external video with internally generated symbology and can operate as either a bus controller or an R/T on the aircraft 1553 B databus.

The CSG is of modular construction enabling various combination of symbol generation and aircraft interface to be readily configurable. A typical CSG application utilises 12 cards in a 19 slot box providing expansion capability to add extra functions to the unit. Functions which may be added include digital mapping, terrain following and intelligent ground proximity warning.

Status
In production and in service in the UK Royal Air Force Tornado GR. Mk 4.

Contractor
Selex Sensors and Airborne Systems.

Voice, tone and display warning systems

Type
Aircraft centralised warning system.

Description
Page Aerospace designs and manufactures a wide range of alerting systems for civil and military aircraft. These include centralised and decentralised warning systems, synthesised programmable tone and voice systems, flight mode annunciators, attention alerting devices and indicator modules.

The company has developed a screen finishing process which results in high levels of sunlight legibility.

For night flying operations, Night Vision Imaging System (NVIS) compatible screens and bezels for displays and instruments are available. These screens effectively filter out the infra-red content from cockpit illumination systems, thereby minimising interference with Night Vision Goggles (NVGs).

Other areas of research and development cover high-quality synthesis of audio tones and voice. Units are currently being evaluated by the Defence Evaluation Research Agency, Farnborough and aircraft manufacturers in order to assess the optimum man/machine interface.

Status
Systems are now used in the central warning panel installations of the Saab 340, BAE Systems 146 and ATP and Raytheon Hawker 800 aircraft.

Contractor
Page Aerospace Ltd.

VRDV-4000-01 flash disk video recorder

Type
Flight/mission recording system.

Description
The VRDV-4000 is a lightweight, solid-state, DZUS-mounted video recorder which employs the latest in Flash Disk recording technology to deliver full PAL resolution recording in a small ruggedised system. The system utilises broadcast standard MPEG-2 video compression to record onto a removable flash card. The system does not require anti-vibration mounts, with flash cards easily removed from the unit and plugged into any standard computer PCMCIA compact flash reader for subsequent video exploitation via standard Windows™-based media players. Features of the VRDV-4000 include:

- Full PAL quality video recording
- 25 minutes recording per Gbyte
- 8 and 12 Gbyte flash disks available
- Accepts composite video, Y/C and RGB inputs
- Audio recording
- Date and time stamp facility
- Easy archiving of video via computer
- NVIS-compatible controls
- Dual-recording deck option (VRDV-4010-01)
- All functions can be controlled via any AVDU series LCD video display (see separate entry)
- Optional remote control.

Specifications
Interface: PCMCIA compact flash via adapter
Video standard: PAL (NTSC under development)
Video compression: MPEG-2
Video input/output levels: 1 Vpp 75
Digitising resolution (PAL): 8 bit Y/UV, 720 pixels × 576 lines

Audio
Number of channels: 2 (stereo)
Compression: MPEG1 level 2
Input level: Mic/line (adjustable)
Mic bias: 3.3 V
Power-up to record interval: <2 seconds

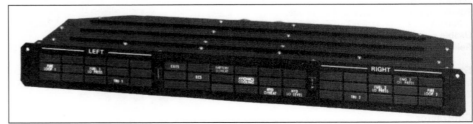

Page central warning panel for the BAe ATP 0503934

System
Power: 28 V DC, 15 W
Temperature: −10 to +50°C (operating); −20 to +70°C (standby); −25 to +80°C (storage)
Weight: approximately 800 g (approximately)
Dimensions: 155 × 146 × 38 mm

Status
In production.

Contractor
Skyquest Aviation.

VU2010 airborne display

Type
Cockpit display.

Description
The VU2010 display is a colour raster display used in UK Royal Air Force and French Air

The VU2010 colour monitor   0503937

Force E-3 Sentry aircraft. The display is a high-resolution unit with a 508 mm (20 in) diagonal tube and a 0.26 mm pitch shadow-mask. The high

performance is obtained by the use of digitally controlled dynamic convergence circuits. Similarly, digitally controlled dynamic focus and active colour purity controls are also incorporated to provide outstanding picture quality. The VU2010 is fitted with a magnetic shield, flashover protection and implosive protection.

Specifications
Dimensions: 464 × 384 × 510 mm
Weight: 53.5 kg
Power supply: 115 V AC, 400 Hz, 3 phase
Reliability: 2,780 h MTBF as MIL-H-217D

Status
In production and in service in UK Royal Air Force, NATO and French Air Force E-3 Sentry AWACS aircraft. Believed ordered by the Japanese government for their Boeing 767-200 AWACS aircraft.

Contractor
Thales Defence Ltd.

United States

2180 series mini-Control Display Unit (CDU)

Type
Control and Display Unit (CDU).

Description
The Smiths Aerospace 2180 series mini-Control Display Unit is a general application CDU which features a seven-colour raster display and has an RS-422 interface. The unit has a full alphanumeric keyboard and a unique colour display which is small in size and offers high resolution, making it ideal for use with navigation and communication systems including GNSS.

CRT units are to be replaced by colour AMLCD 4 in diagonal units (see 2880 series CCDU).

Specifications
Dimensions: 95 × 146 × 203 mm
Weight: 2.2 kg

Status
In production.

Contractor
Smiths Aerospace.

The Smiths Aerospace miniature control and display unit   0503958

2584 series colour flat panel Multipurpose Control Display Unit (MCDU)

Type
Control and Display Unit (CDU).

Description
Smiths Aerospace 2584 series flat panel MCDU is an ARINC 739A compatible control/display unit. The MCDU is designed as an interface unit to support a number of aircraft applications on varying types and is standard

Smiths Aerospace 2584 series MCDU features an Active Matrix Liquid Crystal Display (AMLCD)
0051705

production on Boeing 737 next generation airplanes.

The broad utility of the MCDU permits applications as a control/display unit for the Global Positioning System (GPS), ACARS, Flight Management System (FMS), Digital Flight Recorder (DFR), or Satcom. The unit communicates using ARINC 739 protocol.

The MCDU can receive software upload data via a standard ARINC 615 airborne data loader to accommodate modifications to the operational flight programme.

Specifications
Input/output: ARINC 739 compatible
 7 input ports (hi/low speed)
 2 output ports (hi/low speed)
 discretes 4 input and 1 output
 ARINC 429 compatible
Display: full colour AMLCD display
Brightness: 95fL standard (automatic brightness control)
Resolution: 648 × 532 colour elements
 14 rows of 24 characters
Active display size: 96.8 (W) × 79.5 (H) mm

Control display unit for:
 FMS
 Digital Flight Recorder
 NAV radio tuning
 Datalink
 ACARS
 Satcom
Dimensions: 228.6 (H) × 146.1 (W) × 284.5 (D) mm
Weight: 4.1 kg
Power: 115 V AC, 400 Hz
Keyboard: 3 keyboard FANS options available

Contractor
Smiths Aerospace.

2600/2610 series electronic chronometers

Type
Flight instrumentation.

Description
The 2600/2610 series electronic chronometers utilise two dichroic LCDs to display GMT, chronograph, elapsed time and day and date. Smiths digital clocks utilise a temperature compensated crystal oscillator for extreme accuracy. Trickle current from the aircraft's hot bus maintains functionality with bus power off.

The model 2610 outputs GMT/UTC and Day/Date to the ARINC 429 databus. The model 2620, which has been selected for the Boeing 777, features ARINC 429 input and output, and GPS and ASIC technology.

Specifications
Temperature range: −40 to +85°C
Accuracy: better than 1 second per day
Reliability: >37,000 h MTBF
Power consumption: 0.67 W standby, 2.24 W operational

Smiths Aerospace's 2620 digital clock   0581522

Smiths Aerospace's 2620 digital clock shown installed on an Airbus A340-300 (below standby Horizontal Situation Indicator) (E L Downs)   0116563

Status

In production. Smiths digital clocks are standard fit on the Airbus A320 and A340, BAE Systems Avro 85/100, Boeing 727, 737, 747, 757, 767 and 777 series, MD-80, MD-11 and MD-90, and Fokker 50/70 and 100, IPTN N-250.

Contractor

Smiths Aerospace.

2619 series GPS digital chronometer

Type

Flight instrumentation.

Description

The 2619 series GPS digital chronometer provides display of primary time of day and date reference for the flight crew as well as a chronograph and elapsed time function. Three transilluminated digital liquid crystal displays provide elapsed time, UTC (day/date) and chronograph time.

In primary mode the 2619 series GPS clock receives and processes an external ARINC 429 GPS time source and provides an ARINC 429 output signal of time and date for use by other aircraft systems. The GPS time continually synchronises the internal time base. Should

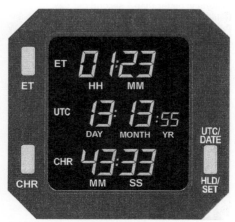

Smiths Aerospace's 2619 series GPS digital chronometer   0051706

the ARINC 429 GPS input fail the clock will automatically use its internal time base.

The clock provides three time keeping modes, UTC mode (00:00 to 23:59), ET (00:00 to 99:59) and CHR (000 to 999) function, and manual mode.

The elapsed time is controlled by a push-button switch which activates the start of ET time, holds CHR time and resets CHR time.

An additional feature of the 2619 series GPS clock is the date display mode. In this mode the clock will display the day, month and year. The date display and output signals are automatically adjusted to achieve an update commensurate with the update of the UTC display. In addition, bissextile (leap) year correction for the date display output signal is accomplished automatically by the clock.

Status

In production.

Contractor

Smiths Aerospace.

2850 series alerting altimeter – 3ATI AMLCD

Type

Flight instrumentation.

Description

The 2850 series alerting altimeter is designed to be installed as: a Reduced Vertical Separation Minima (RVSM) update instrument; as a replacement for electromechanical servo altimeters; or as a multifunction unit.

The primary altimeter displays barometrically corrected altitude in feet on a large AMLCD display. An additional altitude readout in metres is added by pushing a bezel mounted switch. Baro correction counters display in both inches of mercury and millibars. A 'quick set' feature sets 29.92/1,013 baro settings at the push of a button. The unit accepts ARINC 429 digital signals from all standard Digital Air Data Computers (DADC).

The unit incorporates an integral altitude alert function. A bezel push-button mode switch permits selection of alert set mode or the baro set mode. Based on the mode selected, the rotary knob sets the alert altitude or baro setting. No separate altitude alerter unit is required.

The 'altitude alert' setting appears on the flat panel display below the dual barometric

Smiths Aerospace's 2850 series alerting altimeter showing 'alert set' mode   0051707

displays. Alert annunciation is performed both on the display using a 'yellow' caution alert, and through remote aural and visual alerts, which are activated by discrete output from the altimeter. Formats for the display retain flexibility due to software control.

Additional mode/features options include a baro corrected output to other aircraft systems, NVIS B lighting and a settable MDA (Minimum Descent Altitude) bug.

Contractor

Smiths Aerospace.

2880 series AMLCD Compact-Control and Display Unit (CCDU)

Type

Control and Display Unit (CDU).

Description

The 2880 series Compact-Control and Display Unit (CCDU) features a 4 in diagonal colour AMLCD with a wide viewing angle. The 2880 series CCDU is graphic-capable and replaces the 2180 series Mini-CDU (CRT model).

The CCDU employs a full alphanumeric backlit keyboard with either electroluminescent or incandescent lighting. The display is sunlight readable and an automatic brightness feedback loop is provided. The base version of the Mini-CDU is housed in a deep case, 88.9 (H) × 146.1 (W) × 200.7 (D) mm.

The Mini-CDU provides a flexible interface unit for a number of applications including navigation systems, flight control systems and communication systems. NVIS-B capability is an option.

Status

The 2880 series CCDU complements Smiths Aerospace's AMLCD MCDU, multifunction 3ATI electronic display, 5ATI primary flight display and datalink DCDU products.

Contractor

Smiths Aerospace.

Smiths Aerospace's 2880 series AMLCD Compact-Control and Display Unit (CCDU)   0051708

2882 Series AMLCD Multifunction Control Display Unit (MCDU)

Type
Control and Display Unit (CDU).

Description
The 2882 Series flat-panel AMLCD Multifunction Control Display Unit (MCDU) provides a cockpit interface unit with an enhanced display technology, updated ARINC 739 format, advanced electronics and flexible system interface. The unit is designed to facilitate the integration of numerous functions such as GPS, RNav, ACARS, performance advisories and airport maps, and to help conserve space on the flight deck by the elimination of dedicated control heads.

The flat-panel MCDU provides a display of alphanumeric and graphics information, together with a keyboard for crew selection of modes and data entry. A set of dedicated function keys provides the crew with immediate access to particular pages and a set of dedicated alphanumeric keys allows the pilot to enter data. The ARINC 429 digital data enters through the rear connector and is processed by the graphics system processor module where the data is decoded and then displayed on the active matrix LCD. The unit contains standard RS-422 high-speed input and output. An optional MIL-STD-1553 interface unit is available as are NVIS-B and low-temperature operation options.

Specifications
Dimensions: 147 × 183 × 170 mm
Weight: 3.1 kg
Display: 142 mm diagonal
14-lines by 28-characters
Operating temperature: −20 to +55°C
Certification: TSO-C113; DO-160C, D078B Level B
Power: 28 V DC, 1.5 A maximum
Inputs: 7 ARINC 429, 1 RS-422, 4 discrete
Outputs: 1 ARINC 429, 1 discrete

Status
In production and in service. Current platforms include: A300/310, ATR-42/72, BAe 146/RJ-85, B-727, B-747, C-130, C-141, DC-8, DC-10, E-6, King Air, MD-80, Nimrod MRA 4, VC-25 Air Force One and Two.

Contractor
Smiths Aerospace.

Smiths Aerospace's 2882 MCDU   0581520

2888 series AMLCD Cockpit Display Unit (CDU)

Type
Control and Display Unit (CDU).

Description
The 2888 series CDU provides a cockpit control unit that interfaces via a primary RS-422 port with the Flight Management System (FMS). Further interfaces on RS-422 include graphics file transfer, graphical weather and telelink. RGB and NTSC video inputs are also available.

The system is based on open architecture PC-based microprocessor technology. Software is developed to DO-178B, and hardware to DO-160D.

Specifications
Dimensions: 147 × 180 × 178 mm
Weight: 3.1 kg
Display: 142 mm diagonal
Operating temperature: −20 to +55°C
Power: 28 V DC, 1.5 A maximum
Compliance: TSO-C113, DO-160D
Inputs: 2 video (RGB/NTSC), 1 RS-422 interface, 5 discrete
Outputs: 12 discrete, optional ARINC 429 interface

Status
In production.

Contractor
Smiths Aerospace.

2889 series Mode Select Panel (MSP)

Type
Control and Display Unit (CDU).

Description
The Smiths Aerospace 2889 series Mode Select Panel (MSP) provides modern display technology with traditional flight control unit functions in a compact package. This solid-state unit is designed for use as a pilot interface control unit with high reliability and flexible page formats, including graphics if desired.

The launch programme for the 2889 series MSP is an intermediate sized helicopter which will be entering service in 1999.

The 2889 series MSP is a derivative of the Multifunction Control Display Unit (MCDU) product line which is in service on both military and civil programmes.

The 2889 series MSP provides pilot/operator keyboard and display interfaces to: Flight Control Computer (FCC); Aircraft System Computer (ASC); Automatic Flight Control System (AFCS); and in an optional configuration can provide a back-up Engine Instrument Display (EID).

Smiths Aerospace's 2889 series Mode Select Panel   0051709

The pilot accesses functions and modes using the high-contrast lighted keyboard. The pilot receives information from the AMLCD display, which is used to display text information. The 2889 series MSP communicates with the FCC, ASC or AFCS utilising an ARINC 429 digital data bus.

The 2889 series MSP is able to receive external software upload data for updating the operational flight programme. NVIS-compatible display capability is optional.

Specifications
Dimensions: 127 × 156 × 274 mm
Display: 87 mm diagonal
Temperature range: −40 to +55°C
Power: 28 C DC, 1.5 A maximum
Compliance: DO-160C, DO-178B Level B
Inputs: 4 ARINC 429, 10 discretes
Outputs: 1 ARINC 429, 12 discretes

Status
In production.

Contractor
Smiths Aerospace.

3 ATI Engine Performance Indicator (EPI)

Type
Cockpit display.

Description
Astronautics' 3 ATI Engine Performance Indicator (EPI) is a low power, sunlight-readable, NVIS-compatible Active Matrix Liquid Crystal Display (AMLCD). The unit features modular, solid-state construction, making it adaptable to a wide variety of I/O options. The main functions of the EPI are to display propulsion and fuel system parameters; the programmable nature of the system facilitates flexible display configurations. When two displays are installed, a cross-talk channel provides for complete redundancy of operation.

Specifications
Type: High-resolution, AMLCD colour display
Viewable area: 2.3 × 2.3 in (58.4 × 58.4 mm)
Resolution: 256 × 256
Brightness: 150 fl
Processor: Pentium™
Interface: Discrete, analogue, RS-422
Viewing angle: ±45° horizontal, +30 to −10° vertical
Temperature: −40 to +70°C
Power: 28 V DC; 30 W nominal (70 W with heater)
Weight: 2.27 kg
MTBF: 10,000 h

Status
In production and in service.

Contractor
Astronautics Corporation of America.

31400 pneumatic altimeter

Type
Flight instrumentation.

Description
The Kollsman 31400 pneumatic altimeter is a 2 in ATI indicator designed to meet the requirements of TSO C10b. The altimeter is used as a standby indicator for aircraft equipped with EFIS. The indicator measures and displays

accurate pressure altitude. A single pointer affixed to a handstaff makes one revolution per 1,000 ft and is read against a dial graduated in 20 ft increments. The handstaff simultaneously drives a counter-drum in synchronism with the pointer.

The instrument provides a digital display with a baro correction range of 28.1 to 31 in mercury or 950 to 1,050 millibars. Operating range of the instrument is –1,000 to +50,000 ft. The indicator receives inputs of static pressure. It is integrally lit with 5 V white or blue white lighting.

Status
The indicator is in production and is used in a number of aircraft including the de Havilland Dash 8.

Contractor
Kollsman Inc.

44929 digital pressure altimeter

Type
Flight instrumentation.

Description
The Kollsman 44929 is a 3 in ATI altimeter meeting the requirements of TSO C10b and TSO C88a. This digital altimeter senses atmospheric pressure changes, displays altitude with high accuracy, and generates an altitude reporting signal. The altimeter is designed specifically for aircraft that do not require static pressure source correction.

The instrument is driven solely by atmospheric pressure acting on dual aneroid diaphragms. No electrical power is required to operate the pneumatic mechanism. The encoding feature of the altimeter is provided by means of an optical encoder which uses light emitting diode light sources and phototransistor detectors. A code disc is driven by the main shaft of the altimeter mechanism and rotates between the light sources and photo detector array to provide the encoded information to the transponder.

Operating range of the instrument is –1,000 to +50,000 ft.

Status
In production and used on many types of subsonic aircraft.

Contractor
Kollsman Inc.

47174-( ) Resolution Advisory/ Vertical Speed Indicator (RA/VSI)

Type
Flight instrumentation.

Description
The 47174-( ) Resolution Advisory/Vertical Speed Indicator (RA/VSI) was designed to meet wide applications for Traffic alert and Collision Avoidance Systems (TCAS) installations. It is fully compliant with the TCAS guidelines set out in ARINC 735 and will interface with all TCAS equipment.

A single resolution advisory arc is composed of 52 bi-colour surface-mounted LEDs. These LEDs are individually addressed by an embedded microcontroller and provide a high-resolution and flexible RA display.

The modular design of the instrument satisfies all VSI interfaces. These VSI inputs include an internal pneumatic sensor, ARINC 429, ARINC 565, ARINC 575 and Manchester code. Options are available to accommodate specific VSI interfaces and 28 V DC aircraft power. This flexibility allows an operator to utilise a single indicator that will be universal throughout the whole fleet.

The Kollsman resolution advisory/vertical speed indicator 0581515

Specifications
Power supply: 115 V AC, 400 Hz or 28 V DC

Status
The Kollsman RA/VSI is installed in AMR Eagle Saab 340 aircraft.

Contractor
Kollsman Inc.

51Z-4 marker beacon receiver

Type
Aircraft landing system, Instrument Landing System (ILS).

Description
The Rockwell Collins 51Z-4 marker beacon receiver automatically provides aural and visual indication of passage over airway and instrument landing system marker beacons. The system is approved for Category II approaches in a number of Rockwell Collins' all-weather avionic system certifications. Operating at a frequency of 75 MHz, the receiver sensitivity can be varied between two preadjusted levels through a cockpit Hi-Lo switch. The Hi position is used to gain early indication of a marker beacon, the Lo position is then subsequently used closer to the beacon for a sharper position fix. Alternatively, the receiver sensitivity is continuously variable from the cabin or flight-deck by means of a potentiometer.

The system is of all-solid-state construction and contains triple-tuned circuitry for the rejection of spurious signals generated by television and FM broadcast transmitters. It is designed for three-lamp indication but may be easily modified for single-lamp operation by removal of a resistor and wiring the three lamp outputs together. In either type of operation, outputs can operate two sets of indicator lamps in parallel.

An optional self-test facility causes the internal generation of 3,000, 1,300 and 400 Hz marker signals which are detected in sequence by the receiver. The indicators light in order and corresponding aural tones are also generated.

The unit is suited to retrofit installation since it is mechanically and electrically interchangeable with a number of other marker beacon receivers, including the 51Z-2 and 51Z-3 units. An associated marker beacon antenna, the 37X-2 system, is also available. Designed for operation with the 51Z-4 and other compatible receivers, the antenna is plastic-filled and sealed to reduce the effects of precipitation static. This unit, which weighs less than 0.45 kg, can be mounted without cutting into the airframe structure and has negligible drag.

Specifications
Dimensions: ¼ ATR short low
Weight:
(without self-test option) 1.36 kg
(with self-test option) 1.47 kg

Status
In service.

Contractor
Rockwell Collins.

6 × 8 in EFIS/FMS

Type
Integrated avionics system.

Description
Astronautics' 6 × 8 in EFIS/FMS suite is based on their 6 × 8 in multifunction colour displays and EFIs (see separate entries), combined with display controllers, Multi-purpose Control and Display Units (MCDUs), Flight Management (FM) computers and a data loader. The main functions of the system include Primary Flight Display (PFD), navigation, FMS/Map/TCAS/CWX/TAWS/ EGPWS overlays, CNS/ATM growth capabilities, video (optional) and Cockpit Display of Traffic Information (CDTI/ADS-B).

Other notable features of the system are:
- Proven interfaces
- Multiple configuration options
- NVIS compatibility (optional)
- Certified to DO-160D, DO-178B, Level A, HIRF 200 V/m.

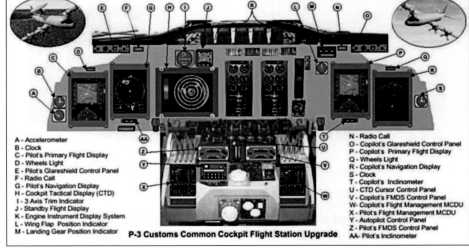

Astronautics' 6 × 8 in EFIS/FMS for the US Customs Service P-3 AEW/Slick Programme (Astronautics) 0568541

Additional system elements include autopilot/flight director, Air Data Computer(s), digital navigation radios, Terrain Avoidance Warning System(s) (TAWS/EGPWS), Colour Weather Radar (CWR), Traffic Collision and Avoidance System (TCAS) and digital radar altimeter(s).

Specifications

Useable display area: 6 × 8 in portrait mode (152.4 × 203.2 mm)
Resolution: 800 × 600 and 1,024 × 768 configurations
FMS: Dual or triple configuration with embedded GPS
MCDU: Dual or triple ARINC-739 configuration
Controls: Two or four EFIS display controllers
Power: 28 V DC
Interface: ARINC-429, -702, -453, -561 (option), -407(synchro, option), RS-422, analogue/discrete, video (optional)

Status

Product applications include P-3 and Boeing 747 series upgrades.

Contractor

Astronautics Corporation of America.

6 × 8 in Electronic Flight Instrument (EFI)

Type

Electronic Flight Instrumentation System (EFIS).

Description

Astronautics' 6 × 8 in Electronic Flight Instrument (EFI) is an AMLCD available as a smart Primary Flight/Navigation Display (PF/ND). The unit can also display sensor, colour weather radar, TCAS, EGPWS, and map data.

Other features of the unit include:

- High brightness, high contrast
- Contrast enhancement filter
- Bezel incorporating multifunction keys and course/heading knobs
- Modular, ruggedised construction
- Built-In Test (BIT)
- ARINC-429 primary interface
- ARINC-708/453 CWX interface (optional)
- Synchro/analogue/discrete interface (optional)
- NVIS compatibility (optional)
- Selectable brightness modes: day, night (auto optional).

Specifications

Useable display area: 6.2 × 8.3 in (157.5 × 210.8 mm)
Resolution: 768 × 1,024
Display luminance: 200 fl (white)
Contrast ratio (high ambient): 10,000 fc diffuse/2,000 fl specular – 5:1 normal to display, 3:1 over FoV
Contrast ration (low ambient): >10:1 normal to display

Viewing angle: >40°
Dimensions: 198.1 × 274.3 × 165.1 mm
Weight: <6.8 kg
Power: 28 V DC (per MIL-STD-704); 65 W (average), 115 W (with heaters)
EMI/EMC: Per DO-160D
Interface: RS-422, Discrete, ARINC-708/453 EGPWS/CWX (optional), ARINC-429 (16 in/4 out)

Status

Product applications include fighters and advanced training aircraft.

Contractor

Astronautics Corporation of America.

ACA Attitude Director Indicators (ADIs)

Type

Flight instrumentation.

Description

ACA 135070 ADI
Operating on the F-5 and F-16, the 3 in (76 mm), two-axis ACA 135070 ADI provides aircraft attitude information. In the flight director mode it also provides pitch and roll steering information. Glide slope information is presented on a separate pointer. When the flight director is not used, the roll steering pointer displays localiser information. The unit is military qualified and hermetically sealed, and is available in a Night Vision Imaging System (NVIS) compatible configuration.

Status

In service in the F-5 and F-16 of the Israel Defence Force.

ACA 129060 ADI
Presently used on the Black Hawk UH-60A, the 5 in (127 mm) ACA 129060 hermetically sealed military qualified ADI provides command as well as raw data information. The command pitch, roll and collective information is provided by Astronautics 3 cue flight director computer. Eyebrow annunciator lights provide go-around, decision height and marker beacon status. Glide slope, localiser and turn and slip information is also provided.

A similar unit used in the SH-60B Sea Hawk – the ACA 126370 – has demonstrated a MTBF in excess of 5,000 hours.

Status

In service in the Sikorsky UH-60A Black Hawk.

ACA 131070 ADI
The ACA 131070 4 in (102 mm), two-axis unit is fitted in the Cobra AH-1S. It has cyclic pitch and roll and collective pitch steering pointers. In addition, glide slope, localiser and turn and slip

information is presented. The hermetically sealed unit is military qualified.

Status

In service in the AH-1S Cobra helicopter.

ACA 135160 ADI
Presently being built for the T-45, the ACA 135160 is a 4 in (102 mm), three-axis ADI. It provides pitch, roll, heading and turn and slip information. Localiser and glide slope information is also provided. The instrument is military qualified.

ACA 137100 ADI
The ACA 137100 3 in (76 mm), three-axis ADI is presently in use on the F-16 and the AMX. It provides pitch, roll and heading as well as turn and slip information. Vertical and horizontal steering pointers provide glide slope and localiser information. The servos are failure monitored and the unit is military qualified and hermetically sealed; it is also available in a Night Vision Imaging System (NVIS) compatible configuration.

Status

In service in the F-16 and Alenia/Embraer AMX.

Contractor

Astronautics Corporation of America.

ACA Horizontal Situation Indicators (HSIs)

Type

Flight instrumentation.

Description

4 in (102 mm) HSI for the Hawk
British Aerospace is the customer for the 4 in (102 mm) HSI which is fitted to export versions of its Hawk light strike/trainer aircraft. Major features include a course bar indicator, to/from indicator, glide slope pointer, two bearing pointers, course set knob, course selection window and digital readout of range which is compatible with ARINC 568 digital input. The instrument meets full MIL-SPEC standards.

Status

In service in the BAE Systems (Operations) Hawk.

ACA 113515 HSI for helicopters
The ACA 113515 HSI has proved to be a popular instrument for helicopters and has course bar indicator, to/from flag, glide slope pointer, two bearing pointers and course set knob with associated course selection window.

Status

In service on Bell 212, 214, 412, Sikorsky S-76, S-61, Eurocopter Super Puma and Agusta AB 212, AB 412 helicopters.

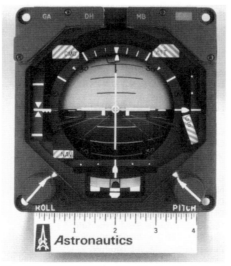

Astronautics' 6 × 8 in EFI (Astronautics) 0568545

The ACA 129060 ADI as fitted in the Black Hawk UH-60A helicopter 0581512

The Astronautics 4 in (102 mm) horizontal situation indicator for the BAE System (Operations) Hawk 0581513

ACA 123790 HSI

The ACA 123790 is a 4 in (102 mm) HSI to military specifications fitted in the Bell AH-1S Cobra helicopter. It has similar features to other members of the company's family of instruments, including a range readout compatible with ARINC 582.

The instrument will also accept direct digital input from Doppler navigation.

Status

In service in the Bell AH-1S helicopter.

ACA 126370 HSI for the SH-60B

The ACA 126370 HSI is a 5 in (127 mm) instrument featuring standard ARINC 407 inputs and outputs with digital interface and extensive built-in test equipment. It was designed for the US Navy Sikorsky Sea Hawk SH-60B LAMPS helicopter programme and is currently in production and operational. An MTBF in excess of 5,000 hours has been demonstrated from operational service.

Status

In service in the US Navy Sikorsky SH-60B LAMPS helicopter.

ACA 126460 HSI

A 3 in (76 mm) HSI, the ACA 126460 has been supplied for the F-5 and F-16 as well as the B-1B. A slightly different version is used on the AV-8C and a type suitable for use in simulators is also available. The ACA 126460 is also available in a Night Vision Imaging System (NVIS) compatible configuration.

Status

Operational in the F-5, F-16 and B-1B.

ACA 130500 HSI

Similar in presentation to other company HSIs, the 3 in (76 mm) ACA 130500 is supplied for the US Marine Corps AV-8B. The instrument will accept direct digital input from a Tacan.

Status

In service in the US Marine Corps AV-8B Harrier.

Contractor

Astronautics Corporation of America.

Actiview

Type

Cockpit display.

Description

The Actiview product line provides ruggedised, flat panel displays to meet a multitude of system architectures and installation requirements. Available with screen sizes ranging from 4 × 3 in to 21.3 in diagonal, each may be acquired as stand-alone or as integrated smart displays, and are available as COTS, rugged, full military specification or customised.

Features include embedded graphics generation, sensor video processing, multiple line rate video inputs and a full complement of interfaces.

Specifications

Actiview 500
Display area: 4 × 3 in
Resolution: 500 × 380; 120 RGB CGPI
Luminance: 0.03 to 280+ fL
Colour: 8-bit colour display
Grey shades: 256 levels
Viewing angle: ±60° horizontal and vertical
Contrast ratio: >8:1 over entire viewing cone
NVIS compatibility: Gen III, Class A/B
Dimensions: 4 × 5 ATI chassis
Weight: 3.18 kg
Power: 28 V DC (5 V DC bezel); 56 W (maximum)
Cooling: Internal fan
Reliability: 7,000 h MTBF
Temperature: –54 to +71°C (operating)

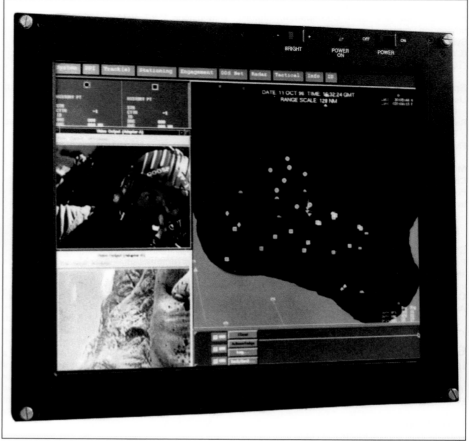

L-3 Actiview 21i flat-panel AMLCD　　　　　　0054304

Altitude: 70,000 ft
EMI/EMC: MIL-STD-461/463
Standard interface: MIL-STD-1553B
Optional interface: ARINC 429/419/575; ARINC 453/708A (2); ARINC 561/568 (2); RS-232/422/485 (2); discrete 30/8 (configurable); analogue 40/7

Actiview 550
Display area: 4.4 × 3.3 in
Resolution: 500 × 400; 120 RGB CGPI
Luminance: 0.03 to 280+ fL
Colour: 8-bit colour display
Grey shades: 256 levels
Viewing angle: ±60° horizontal and vertical
Contrast ratio: >8:1 over entire viewing cone
NVIS compatibility: Gen III, Class A/B
Dimensions: 4 × 5 ATI chassis; 7 in depth
Weight: 3.18 kg
Power: 28 V DC (5 V DC bezel); 56 W (maximum)
Cooling: Internal fan
Reliability: 7,000 h MTBF
Temperature: –54 to +71°C (operating)
Altitude: 70,000 ft
EMI/EMC: MIL-STD-461/463
Video interface: RS-170/RGB-G; RS-343 (875 line); digital
Signal interface: MIL-STD-1553B; ARINC 429/419/575 (6/2); ARINC 453/708A (2); ARINC 561/568 (2); RS-232/422/485 (2); discrete 30/8 (configurable); analogue 40/7

Actiview 5 × 5
Display area: 5 × 5 in
Resolution: 960 × 960; 192 PPI
Luminance: 3 to 250 fL (day); 0.02 to 4.0 fL (night)
Colour: Monochrome display
Grey shades: 256 levels
Viewing angle: ±30° horizontal; +20/–10° and vertical
Contrast ratio: >8:1 over entire viewing cone
NVIS compatibility: Gen III, Class A
Dimensions
Display head: 177.8 × 177.8 × 132.1 mm
Video electronics SEM-E module: 151.1 × 168.9 × 184.2 mm
Power supply SEM-E module: 151.1 × 168.9 × 48.5 mm
Weight: 3.73 kg (display only)

Power: 115 V AC, 400 Hz (MIL-STD-704A)
Cooling: Forced air (display); Conduction (video/power units)
Reliability: 7,000 h MTBF
Temperature: –31 to +55°C (operating)
Altitude: 20,000 ft
Video interface: RS-330 (525 line); RS-343 (875 line)
Serial data interface: RS-232C

Actiview 640
Display area: 5.1 × 3.8 in
Resolution: 125 RGB CGPI
Luminance: 0.03 to 280+ fL
Colour: 6-bit colour display
Grey shades: 64 levels
Viewing angle: ±60° horizontal and vertical
Contrast ratio: >8:1 over entire viewing cone
NVIS compatibility: Gen III, Class A/B
Dimensions: 154.9 × 129.5 × 177.8 mm (W × H × D)
Weight: 3.18 kg
Power: 28 V DC (5 V DC bezel); 56 W (maximum)
Cooling: Internal fan
Reliability: 7,000 h MTBF
Temperature: –54 to +71°C (operating)
Altitude: 70,000 ft
EMI/EMC: MIL-STD-461/463
Video interface: RS-170/RGB-G; RS-343 (875 line); digital
Signal interface: MIL-STD-1553B; ARINC 429/419/575 (6/2); ARINC 453/708A (2); ARINC 561/568 (2); RS-232/422/485 (2); discrete 30/8 (configurable); analogue 40/7

Actiview 840P
Display area: 5.0 × 6.7 in
Resolution: 800 × 600; 119 CGPI (SVGA)
Luminance: 0.03 to 280+ fL
Colour: 8-bit colour display
Grey shades: 256 levels
Viewing angle: ±60° horizontal and vertical
Contrast ratio: >8:1 over entire viewing cone
NVIS compatibility: Gen III, Class A/B
Dimensions: 160.0 × 208.3 × 236.2 mm (W × H × D)
Weight: 3.64 kg
Power: 28 V DC (5 V DC bezel); 88 W (maximum)
Cooling: Internal fan
Reliability: 7,000 h MTBF
Temperature: –54 to +71°C (operating)

L-3 Actiview 550   0125131

Altitude: 70,000 ft
EMI/EMC: MIL-STD-461/462
Video interface: RS-170/RGB-G; RS-343 (875
  line); digital
Signal interface: MIL-STD-1553B; ARINC
  429/419/575 (6/2); ARINC 453/708A (2); ARINC
  561/568 (2); RS-232/422/485 (2); discrete 30/8
  (configurable); analogue 40/7

Actiview 104L

Display area: 8.3 × 6.2 in
Resolution: 1,024 × 768; 125 GPI (XGA)
Luminance: 0.03 to 280+ fL
Colour: 8-bit colour display
Grey shades: 256 levels
Viewing angle: ±60° horizontal and vertical
Contrast ratio: >8:1 over entire viewing cone
NVIS compatibility: Gen III, Class A/B
Dimensions: 297.2 × 223.5 × 165.1 mm (W × H × D)
Weight: 3.64 – 5.45 kg
Power: 28 V DC (5 V DC bezel); 80 W (maximum)
Cooling: Internal fan
Reliability: 7,000 h MTBF
Temperature: -54 to +71°C (operating)
Altitude: 70,000 ft
EMI/EMC: MIL-STD-461/463
Video interface: RS-170/RGB-G; RS-343 (875 line);
  digital
Signal interface: MIL-STD-1553B; ARINC
  429/419/575 (6/2); ARINC 453/708A (2); ARINC
  561/568 (2); RS-232/422/485 (2); discrete 30/8
  (configurable); analogue 40/7

Actiview 104P

Display area: 6.3 × 8.4 in
Resolution: 1,024 × 768; 125 GPI (XGA)
Luminance: 0.03 to 280+ fL
Colour: 8-bit colour display
Grey shades: 256 levels
Viewing angle: ±60° horizontal and vertical
Contrast ratio: >8:1 over entire viewing cone
NVIS compatibility: Gen III, Class A/B
Dimensions: 198.1 × 254.0 × 147.3 mm (W × H × D)
Weight: 3.64 kg
Power: 28 V DC (5 V DC bezel); 88 W (maximum)
Cooling: Internal fan
Reliability: 7,000 h MTBF
Temperature: -54 to +71°C (operating)
Altitude: 70,000 ft
EMI/EMC: MIL-STD-461/463
Video interface: RS-170/RGB-G; RS-343 (875 line);
  digital
Signal interface: MIL-STD-1553B; ARINC
  429/419/575 (6/2); ARINC 453/708A (2); ARINC
  561/568 (2); RS-232/422/485 (2); discrete 30/8
  (configurable); analogue 40/7

Actiview 21i

Display area: 17.01 × 12.76 in (21.3 in diagonal)
Resolution: 1,600 × 1,200 RGB Stripe
Luminance: 40-45 fL (nominal); 50 fL (maximum)
Colour: 16.7 million colours (8-bit)
Grey shades: 256 levels (minimum)
Viewing angle: ±80° horizontal and vertical
Dimensions: 485.6 × 419.1 × 96.5 mm (W × H × D)
Weight: <10 kg
Power: 110 V AC, 60 Hz; 85 W
Cooling: Convection
Reliability: 20,000 h MTBF (predicted)
Temperature: -0 to +50°C (operating)
Altitude: 10,000 ft (operating); 20,000 ft
  (non-operating)
EMI/EMC: MIL-STD-461/463
Video interface: Non-interlaced; Composite; DVI;
  Sync

Status

In production and in service.

Over time, L-3 displays have been utilised in a number of cockpit upgrade programmes:

C-130H Electronic Flight Indicator (EFI)

Over 1000 EFIs have been manufactured for new and retrofit C-130H aircraft. The C-130H EFI provides attitude, heading and pitch/roll/yaw information, as well as radar video, air data and flight director information. These indicators are fully self-contained AMLCD-based display instruments with extensive I/O processing and internally generated graphic presentations of data. A wide variety of signals, including discrete, analogue, ARINC, synchro, serial and resolvers, are processed in real time.C-130J primary flight displays

Contracted in 1996, L-3 designed and delivered hardware for the Colour Multifunction Display Units (CMDUs), the Avionics Management Units (AMUs) and the Communication/Navigation Breaker Panel (CNBP) for C-130J production aircraft. The units interface with the aircraft mission computer via MIL-STD-1553B databus and provide both video reproduction and internally generated graphic presentations of all cockpit flight conditions. The CMDU is a full-colour, NVIS-B portrait 6 × 8 in AMLCD; the AMU and CNBP are green monochrome NVIS-A 4 × 3 in displays with keys for cockpit display and communications control. In addition, L-3 was also selected to provide the C-130J MultiFunction Colour Display (MFCD), which functions as the cargo area interface to the mission computer. The MFCD provides information to the Load Master, computing airdrop data and controlling automated airdrops.

S-3B EFI displays

L-3 supplied 5.5 in colour Electronic Flight Instrument (EFI) displays for the S-3B Viking. There are four EFI displays installed in the cockpit, replacing the older electromechanical instruments, providing for increased reliability and software programmability to support further capability enhancements. The displays provide S-3B aircrew with horizontal and vertical situation data for navigation. L-3 reportedly delivered 390 EFI displays.

U-2 RAMP MultiFunction Displays (MFDs)

L-3 provided the 10.4 in integrated MFDs for the U-2 Reconnaissance Avionics Maintainability Programme (RAMP). Up to three MFDs are installed in the cockpit as part of the RAMP to manage mission-critical and aircraft performance information, and to enhance the pilot's Situational Awareness (SA).

C-27J Spartan

Lockheed Martin and Alenia Marconi selected L-3 to provide the cockpit display suite for the C-27J Spartan. The AMLCD suite consists of a Colour Multifunction Display Unit (CMDU), Single Avionics Management Unit (SAMU) and the Communications Navigation Radio Panel (CNRP). The CMDU is a 10.4 in display, while the SAMU and CNRP are 4 × 3 in displays. Each display suite includes 5 CMDUs, 2 SAMUs and 1 CNRP.

A/MH-6 Little Bird

L-3's 5 × 4 in Electronic Engine Displays (EEDs) are installed in US Army A/MH-6 Special Operations Aircraft. The 5 × 4 in FED provides two display formats for over eight engine parameters. The EED uses a digital signal processor approach to convert the discrete, analogue and synchro signals into digital format. Software is then used to process the information and formatting on the display screen.

JAS-39 Gripen multifunction displays

L-3 supplied colour multifunction, high resolution, 6 × 8 in AMLCDs for the JAS-39 Gripen fighter aircraft. The Gripen cockpit consists of three multifunction displays – a Flight Data Display (FDD), a Horizontal Situation Display (HSD) and a MultiSensor Display (MSD). The FDD presents all flight information.

The HSD displays all tactical information, including classification friend-or-foe, targets, threats, obstacles and guidance information superimposed on a digital electronic map. The MSD is used for radar modes or FLIR imagery. Primary flight information can be routed to any one of the three displays.

Jaguar upgrade programme

L-3 was contracted by BAE Systems to provide cockpit displays for Jaguar aircraft operated by the UK Royal Air Force and the Royal Air Force of Oman. Under the contract, L-3 provided a colour multifunction, high-resolution, 6 × 8 in AMLCD. The display is utilised for primary flight information, TIALD video, FLIR and map information.

Spanish Air Force F-5B MFD

L-3 supplied 6 × 8 in MFDs and 3 × 4 in Electronic Engine Displays (EEDs) for F-5B aircraft operated by the Spanish Air Force. Israel Aircraft Industries was responsible for the integration of the displays.

Joint Strike Fighter concept/demonstration aircraft

L-3 Display Systems provided MFDs for the JSF concept/demonstration aircraft, which incorporated two 8 × 6 in colour, high-resolution, AMLC MFDs.

Contractor

L-3 Communications, Display Systems.

ADAS-7000 Airborne Data Acquisition System

Type

Flight/mission recording system.

Description

L-3 Telemetry-East manufactures a wide range of telemetry and data recording systems for missile and air vehicle flight testing including the ADAS-7000, originally designed for the IAI Lavi flight test programme. This provides a master/slave recording system for aircraft use, using advanced signal conditioning and encoding hardware. A PMU-700 Series III program master unit is at the heart of the ADAS-7000; this is used as a central encoder/controller of the whole system and communicates on a MIL-STD-1553 digital data highway with a number of slave units. Over 5,000 channels can be monitored and recorded.

Contractor

L-3 Telemetry-East.

ADI-330/331 self-contained Attitude Director Indicator

Type

Flight instrumentation.

Description

The ADI-330/331 is a 3-ATI compact, self-contained Attitude Direction Indicator which features glide slope and localiser cross-pointers and an integral inclinometer for slip and skid indication. The ADI-330 accepts analogue glide slope/localiser signals through an ARINC 500 interface, while the ADI-331 accepts ARINC 429 digital glide slope/localiser signals. Direct mechanical linkage eliminates electrical servo response lag. Ideally suited for large corporate and transport aircraft, the ADI-330/331 provides accurate pitch and roll information under all normal conditions. When used in conjunction with an emergency power supply, it also serves as an standby attitude reference.

Electrical erection automatically provides 20°/min fast erect during initial power-up, eliminating the need for manual caging or uncaging during preflight checks. The unit also includes a bezel-mounted push-button for fast erect on demand.

The face presentation of the ADI-330 has been designed to integrate into cockpits which use EFIS electronic displays. The style, colours and integral (blue/white) incandescent internal lighting of the instrument have been chosen to provide a consistent and familiar attitude information presentation in modern cockpits.

Designed for 28 V DC, the unit provides useful attitude information down to 18 V and features power failure monitoring, with 9 minutes of usable attitude information after complete power loss. Estimated MTBF for the ADI-330 is 4,500 operating hours and it is TSO C4c/C4d qualified.

Status
In production and in service in a wide range of corporate and transport aircraft.

Contractor
L-3 Communications, Avionics Systems.

VOR (left) and ILS (right) displays on the ADI-335 standby attitude and navigation indicator   0503955

ADI-332/333 self-contained Attitude Director Indicator

Type
Flight instrumentation.

Description
The ADI-332/333 is a 3 in (76.2 mm) compact, completely self-contained attitude director indicator featuring glide slope and localiser cross-pointers with back course, integral inclinometer for slip and skid indications and mechanical erection with manual caging. The ADI-332 accepts an analogue glide slope/localiser signal via an ARINC 500 interface; the ADI-333 has an ARINC 429 digital interface. Direct mechanical linkage eliminates electrical servo response lag.

The ADI-332/333 features 9 minutes of usable attitude information after complete power loss. It has a selectable switch for BC/ILS/OFF modes, 18 to 30 V DC power supply, blue/white internal lighting, Power Off warning and glide slope/localiser signal validity flags and self-contained lateral and fore and aft acceleration compensation. The design is compatible with EFIS displays and the instrument is qualified to TSO C4c.

Contractor
L-3 Communications, Avionics Systems.

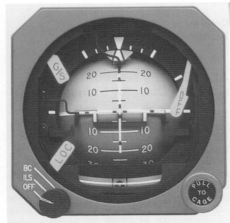

ADI-332/333 attitude director indicator   0504209

ADI-335 standby attitude and navigation indicator

Type
Flight instrumentation.

Description
The ADI-335 standby attitude and navigation indicator provides roll and pitch attitude

information by electromechanical means. In addition, the instrument contains en route VOR/DME and precision approach ILS displays, OBS (with auto-centering) and an eight character LED display for Mode, Radial, To/From and Selected Course presentation. Both the navigation modes contain appropriate validity flags and the indicator will interface with an ARINC 429 bus.

The ADI-335 is qualified to TSOs C4c, C52a, C36e, C40c and C66 and in accordance with DO-160C.

Contractor
L-3 Communications, Avionics Systems.

ADI-350 Attitude Director Indicator

Type
Flight instrumentation.

Description
The ADI-350 is a self-contained gyro indicator with flight director needles, rate of turn and slip indication. Synchro pick-offs are provided for remote indicators, radar stabilisation and flight control functions. Inputs are needed to operate the flight director needles and the rate of turn indicator.

The ADI-350 is compatible with NVGs and designed to military standards.

Specifications
Dimensions: 82.04 × 82.04 × 228.60 mm
Weight: 2.72 kg
Power supply: 115/208 V AC, 400 Hz, 3 phase or 28 V DC

Status
In production. Applications include the Northrop Grumman A-6E, KA-6D and EA-6B, Boeing F/A-18, Canadair CP-140, Rockwell B-1B and Bell OH-58D. The ADI-350V was selected for installation as part of the upgrade programme for the F-14D Super Tomcat.

Contractor
L-3 Communications, Avionics Systems.

Advanced Display Core Processor (ADCP)

Type
Avionic display processor.

Description
The ADCP is a commercially-based processor, designed for the US Air Force F-15E, and sponsored by the US Defense Department's Commercial Operations and Support Savings Initiative (COSSI).

The ADCP is designed to replace two existing line replaceable units: the MultiPurpose Display Processor (MPDP) and a VHSIC (very high-speed integrated circuit) Central Computer (VCC).

Software for the ADCP has been supplied by Virtual Prototypes Inc of Montreal, Canada.

Status
In service.

Contractor
Honeywell Defense Avionics Systems.

Advanced Recorder (AR) series Solid-State Cockpit Voice Recorder and Flight Data Recorders (SSCVR/FDRs)

Type
Flight/mission recording system.

Description
The AR series of recorders is designed for use in business and general aviation aircraft and helicopters. The series comprises the following products:
- AR Cockpit Voice Recorder (CVR)
- AR Flight Data Recorder (FDR)
- AR Combined cockpit voice recorder/flight data recorder (AR Combi)

The newest technology, including solid-state memory, has been used in the AR series to minimise weight, reduce installation cost and delete maintenance. The new units weigh only 4 kg, require no mounting tray and need no maintenance.

The AR CVR simultaneously records up to four channels of audio for 30 or 120 minutes.

The AR FDR can record at data rates of 64, 128 or 256 words/s for 25 hours.

The AR Combi provides both AR CVR and AR FDR capabilities in one unit.

Each system provides as appropriate:
- One area audio channel and two or three crew audio channels
- One ARINC 717 or ARINC 429 data input
- Greenwich Mean Time (GMT) input
- Rotor tachometer input
- Two spare ARINC 429 inputs (reserved for CNS/ATM)
- Maintenance and status outputs.

In order to reduce weight and cockpit panel space, the microphone preamplifier is contained within the microphone unit, eliminating the need for a control panel.

Data downloading is by a hand-held download unit, connected via an RS-422 interface. Playback is on the Playback and Test Station (PATS), which can be hosted on a PC-based workstation using Windows® and Airraft Data Recovery and Analysis Software (ADRAS®).

The Advanced Recorder series of solid-state cockpit voice and flight data recorders   0081866

Specifications
Dimensions: 232.4 (L) × 149.1 (W) × 142.2 (H)
Weight: 4 kg (incl underwater locating beam)
Power: 8 W nominal, 10 W max

Contractor
Honeywell Aerospace, Electronic & Avionics Lighting.

AFD-3000/5000 Adaptive Flight Display

Type
Electronic flight instrument.

Description
Rockwell Collins' AFD-3000/5000 Adaptive Flight Display (AFD) family comprises the 8 × 10 in AFD-3010 and 10 × 12 in AFD-5220 integrated display units.

The AFD-3000/5000 provides for Electronic Flight Instrument System (EFIS) and Engine Indication and Crew Alerting System (EICAS) display.

An EFIS arrangement typically contains four large colour displays and two display control panels:
- PFD Primary Flight Display (×2)
- MFD MultiFunction Display (×2)
- DCP-3000 (DCP-5020) Display Control Panel (×2)

The PFD displays attitude, lateral navigation/compass, flight control and primary air data (altitude/airspeed/vertical speed) functions, receiving databus inputs from input/output concentrators, Integrated Avionics Processing System (IAPS), Traffic Alert Collision Avoidance System (TCAS) receiver/transmitter, Inertial Reference Unit(s) and Air Data System(s). The PFD provides a data bus output to the IAPS and secondary PFD. The MFD displays lateral navigation/compass, radar, TCAS, flight management (map/summary) and diagnostic information. The MFD also provides a reversion backup for the Captain's PFD or the EICAS if that display fails. The MFD also receives input from weather radar, EICAS control panel, EICAS Data Concentrator Units (DCUs), Maintenance Diagnostic Computer (MDC) and Flight Management Computer (FMC). The MFD provides a data bus output to the IAPS and DCP. The left

DCP-3000 (DCP-5020) controls the left side PFD and MFD displays. This panel selects display formats and left side lateral navigation parameter/sources. The DCP receives data bus inputs from the left side PFD, MFD and weather radar panel. The DCP switch data and the weather radar control panel data are output to the on-side and cross-side PFDs and MFDs. The left side display reversion switch allows the pilot to select PFD or EICAS backup display on the pilot MFD. Remote reversion switches allow the Captain or Co-pilot to select either Attitude Heading Reference System (AHRS), DCP and air data source inputs. The Co-pilot PFD/MFD combination offers mirrored functionality to that of the Captain.

In terms of EIS/EICAS functions, the AFD-3000/5000 displays support a range of engine indicating systems. Entry level business jets have an EIS functionality that displays the critical information for the engines. Larger business jets tend to have a full EICAS implemented.

Specifications
Display size: 10 in (AFD-3010) or 13.3 in (AFD-5220) (diagonal)
Resolution:
(AFD-3010) 1024 × 768 × 3 (128 ppi)
(AFD-5220) 1024 × 768 × 3 (96 ppi)
Active area:
(AFD-3010) 6 ×8 in
(AFD-5220) 7.85 × 10.51 in
Brightness::
(AFD-3010) 0.1-100 fL
(AFD-5220) 0.05-100 fL
Dimensions:
(AFD-3010) 253 × 208 × 246 mm (H × W × L)
(AFD-5220) 318 × 250 × 271 mm (H × W × L)
(DCP-3000) 215 × 48 × 155 mm (H × W × L)
(DCP-5020) 48 × 215 × 154 mm (H × W × L)
Weight:
(AFD-3010) 5.8 kg
(AFD-5220) 8.26 kg
(DCP-3000) 0.72 kg
(DCP-5020) 0.81 kg
Power:
(AFD-3010) 28 V DC, 125 W (max), 391 W (max) with heaters
(AFD-5220) 28 V DC, 185 W (max), 440 W (max) with heaters
(DCP-3000) 28 V DC, 3.5 W (nominal), 9.75 W (max)
(DCP-5020) 28 V DC, 9.262 W (max)

Temperature: –20 to +55°C (AFD-3010 and AFD-5220); –20 to +70°C (DCP-3000 and DCP-5020)
Altitude: up to 55,000 ft
Certification:
(AFD-3010) FAA TSO C2d, C3d, C4c, C6d, C9c, C10b, C34e, C35d, C36e, C40c, C41d, C43c, C44b, C46a, C47, C49b, C52b, C55, C63c, C66c, C87, C92c, C95, C101, C104, C105, C110a, C113, C115b, C117a, C129a; DO-160C, -178B
(AFD-5220) FAA TSO C113; DO-160D, -178B
(DCP-3000):
DCP-5020):

Status
The AFD-3000/5000 is in production and in service. The AFD-3000 is certified on the Cessna CJ1/CJ1+, Cessna CJ2/CJ2+, Premier I, Hawker 800XP, King Air 250/300, Cessna CJ3 and Piaggio Avanti light business aircraft. The AFD-5000 is certified on the Bombardier Challenger 300 and Gulfstream G150 business jets.

Contractor
Rockwell Collins.

Aft Seat HUD Monitor (ASHM)

Type
Cockpit display.

Description
Astronautics' Aft Seat HUD Monitor (ASHM) is a compact, high resolution, high brightness, lightweight, ruggedised monochrome raster display, designed specifically for F-16 applications. The ASHM can be used as a training device as well as an advanced weapon system display for presentation of LANTIRN/FLIR video.

The CRT display unit features a modular architecture; each module can be replaced independently without adjustments to the display (including the CRT assembly replacement). Integral BIT continually checks the display circuitry, providing go/no go indications and immediate failure isolation. It also has an internal video test pattern generator.

The ASHM is available with and without NVIS filtering and a colour variant can also be provided for advanced applications.

Specifications
Screen size: 7.83 in diagonal (198.9 mm)
Brightness: 300 fl
Contrast ratio: 7:1
Video bandwidth: 20 to 22 MHz
Scan/line rate: 525 line/30 Hz
Aspect ratio: 4:3
Phosphor type: P-43
Power: 60 W (max)

Status
In production and in service in F-16B, D and F-16 Block 60 aircraft.

Contractor
Astronautics Corporation of America.

AI-803/804 2 in standby gyro horizon

Type
Flight instrumentation.

Description
The AI-804 series was designed specifically as a standby reference indicator and provides up to 9 minutes of usable attitude information after complete loss of electrical power. It eliminates the need for electronic components associated with remote systems using direct mechanical linkages. It has 360° freedom in roll and also in pitch with controlled precession. The AI-804 complies with TSO C4c and is used on fixed-and rotary-wing

aircraft in both military and commercial applications.

The AI-803 military gyro is built to military standards and is compatible with NVGs.

Specifications
Dimensions: 2 ATI
Weight: 1.1 kg
Power supply: 115 V AC, 400 Hz or 28 V DC

Status
In production and widely deployed in civil and military aircraft.

Contractor
L-3 Communications, Avionics Systems.

AIM 205 Series 3 in directional gyros

Type
Flight instrumentation.

Description
Low cost and lightweight characterise the AIM 205 Series of 3 in directional gyros. They are precise flight instruments with a low drift rate, designed to provide the pilot with a constant azimuth reference free from the instability inherent in magnetic compasses. A high-speed electric rotor, mounted in a universal gimbal system, maintains angular momentum to overcome normal bearing friction and establishes a gyroscopically stable datum reference relative to space.

The AIM 205 series complies with TSOs C5c and C5e.

Specifications
Dimensions: 200.7 mm long
Weight:
(205-1) 1.13 kg
(205-2) 1.32 kg
Power supply: 14 V DC, 28 V DC
  or 115 V AC, 400 Hz, single phase
Temperature range: −30 to +50°C
Environmental: DO-160A

Contractor
L-3 Communications, Avionics Systems.

AIM 1100 3 in self-contained attitude indicator

Type
Flight instrumentation.

Description
The AIM 1100 is designed as a form, fit and function replacement for the AIM Model 305. It is characterised by its lightweight and rugged

design. Engineering enhancements include a proprietary bearing design improving rotor life and an improved pointer bar resulting in better performance and reliability under vibration. It has rear-mounting provisions for minimal panel protrusion and can be installed in shock-mounted panels from 0 to 20°.

It is designed for helicopters and general aviation aircraft operating in high duty cycle environments. It is available in 14 or 28 V DC form with an optional slip/skid indicator and fixed or trimmable pitch aircraft symbol.

The AIM 1100 is certified to FAA TSO C4c, RTCA DO-160C, Section 8.0, Vibration Curves S and P.

Contractor
L-3 Communications, Avionics Systems.

AIMS Airplane Information Management System

Type
Avionic information management system.

Description
Honeywell's AIMS architecture, for the B-777 is an advanced avionics system designed to meet very high requirements for functionality, maintainability and dispatch reliability.

The AIMS consists of dual integrated cabinets that contain all of the central processing and input/output hardware to perform the following functions: flight management; displays; navigation; central maintenance; airplane condition monitoring; flight deck communications; thrust management; digital flight data; engine data interface; data conversion gateway.

By eliminating the need for separate Line Replaceable Units (LRUs) for each subsystem – each with its own power supply, processor, chassis, operating system, utility software, input/output ports and built-in test – the AIMS concept saves significant weight, space and power consumption on board the airplane while improving overall system reliability and maintainability.

Displays
The 777's flight deck features six 'D' size (8 × 8 in) Active Matrix Liquid Crystal Displays (AMLCDs) in a horizontal layout similar to the 747-400. These advanced technology screens display primary flight, navigation and engine information with automatic reversion capability.

The flight management function for the Boeing 777 takes advantage of Honeywell's mature systems developed for the 757, 767 and 747-400 aircraft, while at the same time providing greatly improved performance, functional capability and growth potential for the Future Air Navigation System (FANS).

The system includes three passively cooled Multifunction Control Display Units (MCDUs) featuring full-colour flat panel LCDs. The MCDU

controls the flight management function and is capable of controlling other ARINC subsystems.

The Honeywell CMF consolidates the maintenance activity of virtually all systems on the B-777. A primary objective is to reduce maintenance by improving fault isolation and detection capability, thus reducing the number of spares needed as well as minimising dispatch delays and cancellations.

The B-777's ACMF enables the flight crew to make informed decisions about the state of the aircraft. The system continuously monitors engines and aircraft systems and generates reports that can be customised to individual airline needs.

The DCMF/FDCF is the intermediate system of the airborne datalink infrastructure. It serves as the router between various airborne applications, the printer, the onboard Local Area Network devices and ground-based datalink applications via various air-ground subnetworks (such as VHF radio and SATCOM).

It also functions as the crew interface to the datalink system by means of cockpit displays, the CDU and cursor control device. It supports airline programmable ACARS applications.

The Boeing 777 AIMS provides capability for the airlines to customise various aircraft systems such as the Airplane Condition Monitoring Function (ACMF) as well as for flight management, flight deck and data communications, central maintenance and the electronic checklists. This capability is enabled by the AIMS philosophy of fault containment and software partitioning.

Status
In service on Boeing 777 aircraft.

Contractor
Honeywell Inc, Commercial Electronic Systems.

Airborne 19 in colour display

Type
Cockpit display.

Description
Astronautics has designed and produced a large colour display for airborne use. This display is a full militarised high-resolution colour monitor. It has a usable diagonal screen size of 19 in (482.6 mm) and a pitch dot size of 0.31 mm, a feature which contributes to the exceptional resolution. It is equipped with extensive self-test diagnostics for fault isolation and reporting.

The colour display has numerous applications such as for airborne command, warning and control centres, JTIDS operations and control centres and ASW operations. It can easily be adapted to other environments.

Specifications
Dimensions: 706 × 610 × 533 mm
Raster CRT: 280 × 349 mm
Contrast rate: adjustable from 10:1 to 20:1
Frame rate: non-interlaced 60 Hz
Three-phase power distribution: 200 W

Status
In production. Qualified for C-130 aircraft and surface ships and submarines.

Contractor
Astronautics Corporation of America.

Airborne Digital Imaging System (ADIS)

Type
Flight/mission recording system.

Description
The DRS ADIS is a high-resolution, electronically shuttered, solid-state, digital imaging system in a very robust high g package. The system is

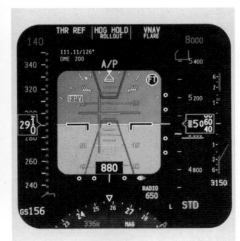

AIM 1100 3 in self-contained attitude indicator
0018174

AIMS Primary flight display   0001436

Airborne Digital Imaging System (ADIS)   0051339

suitable for many applications, including air superiority fighters and aircraft flight test instrumentation.

ADIS features the latest technology, including a new image sensor and internal memory module. The camera contains an ultra-high-density DRAM memory module and a new high-speed mass storage 8 Gbyte hard drive allows transfer of large amounts of digital imagery from the cameras' DRAM for storage and safe keeping.

Windows® application-based software, provides a PC access to camera set ups and the operation of up to 64 cameras. A preview mode provides an easy graphical display to review the overall set up, including pre- and post-trigger settings, plus the number of images to be captured by each camera.

DRS's ADIS is designed for high-speed imaging applications and time lapse, since it is capable of capturing one picture on demand or one pulse-one picture at a synchronous rate of up to 1,000 Hz.

With continuous Image Storing Recording (ISR) of up to 2,000 pictures using a first-in first-out principal, the ability to grab and lock the images to memory upon command is easily accomplished.

A standard feature of the system is Record On Command (ROC) and Burst Record On Command (BROC).

Recalling images from large data files is very fast, given the amount of identification information recorded with each picture. Up to four different camera recordings can be recalled simultaneously from the database and played

together synchronously to analyse different views at the same time.

Specifications
Resolution: 512 × 512 pixels
Image: 24 bit colour or monochrome
Grey levels: 256 (8 bit)
Picture rates: 512 × 512, 30 to 1,000 pps (pictures per second); frame rate profiling (programmable picture rate on a frame-by-frame basis)
Recording time: 1 s at 1,000 pps (1,000 pictures)
Image download rate: 132 Mbits/s
Picture download time: 1 picture in 16.5 ms, 1,000 pictures in <40 s
Dimensions: 87 (H) × 66 (W) × 188 (D) mm
Weight: 2.2 kg
Power: 28 V DC, 22 W

Contractor
DRS Photronics, Inc.

Airborne Flight Information System (AFIS)

Type
Flight Management System (FMS).

Description
Honeywell Aerospace has expanded the capabilities of its GNS-1000 flight management system and GNS-500A Series 4 and Series 5 navigation systems into a full airborne flight information system, with comprehensive facilities for flight plan creation before flight

and amendment in flight. The Airborne Flight Information System (AFIS) is also available for the GNS-X flight management system.

Before flight the pilot can access the Global Data Center via a computer, to obtain flight planning, wind and en route weather information. This information is recorded on a mini-computer disk, which is later loaded into the AFIS system via an onboard data transfer unit. During flight, the comparison of planned and actual flight plan information can be viewed at any time on a control and display unit.

During flight, the air-to-ground and ground-to-air datalink can be used to obtain additional wind and weather information, send and receive flight-related messages and update flight plans. Inflight datalink is through VHF radio or an optional satellite link.

Onboard elements of the AFIS comprise:
1. A Data Management Unit (DMU), which includes receiver/transmitter;
2. A Data Transfer Unit (DTU);
3. A single-unit High-Power Amplifier/Lower-Noise Amplifier (HPA/LNA);
4. An antenna;
5. An optional Satellite Data Communications System (SDCS). An SDCS utilises the INMARSAT satellite network to extend the service to virtually anywhere in the world.
6. An optional graphics Remote Processor Unit (RPU). An RPU allows the AFIS DMU to receive graphical weather information, which is transmitted through VHF datalink every 5 minutes, and helps the pilot to make effective routing decisions.

Specifications
Dimensions:
(DMU) 195 × 124 × 385 mm
(Honeywell/Racal SATCOM) 29 × 30 × 60 mm
Weight:
(DMU) 5.89 kg
(SATCOM) 10.91 kg
Power consumption:
(DMU) 28 V DC; 7 A max (transmit); 2 A (receive)
(SATCOM) 115 V AC, 400 Hz; or 28 V DC

Status
Fully operational.

Contractor
Honeywell Aerospace, Electronic & Avionics Lighting.

Airborne MultiFunction Colour Displays (MFCDs)

Type
MultiFunction Display (MFD).

Description
Astronautics produce 4, 5 and 6 in Multi Function Colour Displays (MFCDs), designed to display video, mission, navigation, sensor and aircraft data in high-performance aircraft. Common display features include:
- Video only or smart display with internal graphics generation and processing
- Colour Active Matrix Liquid Crystal Display (AMLCD)
- High brightness, high contrast
- Brightness control
- Contrast enhancement filter
- NVIS compatibility
- Selectable brightness modes: day, night
- Modular, ruggedised construction
- Designed to military standards
- Built-In Test (BIT)
- Accepts RGB & Monochrome video
- 256 shades of grey.

4 in Multi Function Colour Display (MFCD)
This MFCD Unit is an Active Matrix Liquid Crystal Display (AMLCD) with a useable screen size of 4.24 × 4.24 in (107.7 × 107.7 mm) and

incorporating 20 multifunction keys around the bezel, designed primarily for video display of navigation, sensor, mission and aircraft systems status information. The full colour (RGB) and monochrome capabilities also allow the MFCD to be utilised as a digital map display. Embedded processing and graphics generation options are also available. The 4 in MFCD is manufactured for the F-16, F-5, Kfir, MiG-21 and other advanced fighter and trainer aircraft.

The 4 in display is also available as a MFCD and remote Electronics Unit (EU) two-box system designed for use in aircraft requiring a display with a minimum depth (5.5 in), such as the OV-10.

5 in Multi Function Colour Display (MFCD)
This MFCD is an AMLCD with a useable screen size of 5 × 5 in (127 × 127 mm), incorporates 20 multifunction keys around the bezel and can also be provided as a smart system. The full colour (RGB) and monochrome video capabilities of the display can be utilised to display sensor and digital map data. The 5 in MFCD is in service in the Tornado, A-4, Dauphin and other aircraft.

5 in Pilot Head Down Display (PHDD)
The PHDD is an AMLCD with the same useable screen size and multifunction keys as the 5 in MFCD, designed for use as a Pilot Head Down

Display capable of displaying mission data, navigation, sensor, weather radar and other information in high-performance aircraft. In addition to the internal processing, video over/underlay, freeze frame, and video switching functions of the system, the PHDD can also be utilised to display sensor and digital map data and is equipped with a MIL-STD-1553 bus.

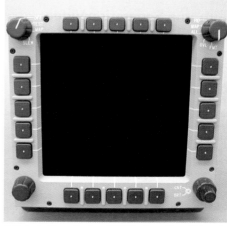

Astronautics 5 in PHDD for the Tornado (Astronautics)   0568551

Specifications

	MFCD (4 in)	MFCD F-16 (4 in)	MFCD (5 in)	PHDD (5 in)	MFCD (6 in)	MFCD (6 × 8 in)	MFCD (8 × 10 in)
Dimensions (W × H × D)	127.0 × 127.0 × 254.0 mm	127.0 × 127.0 × 304.8 mm	160.0 × 160.0 × 254.0 mm	160.0 × 160.0 × 254.0 mm	190.5 × 190.5 × 254.0 mm	198.1 × 274.3 × 165.1 mm	393.7 × 279.4 × 101.6 mm
Weight	<4.55 kg	<4.55 kg	<5.9 kg	<5.9 kg	<6.36 kg	<6.82 kg	9.09 kg
Power	28 V DC/50 W	28 V DC/50 W	28 V DC/60 W	28 V DC/60 W	28 V DC/70 W	28 V DC/65 W max 115 W	115 V AC/65 W 28 V DC/20 W
	MIL-STD-704D	MIL-STD-704D	MIL-STD-704D	MIL-STD-704D	MIL-STD-704D	MIL-STD-704D	
Display size	4.0 × 4.0 in	4.0 × 4.0 in	5.0 × 5.0 in	5.0 × 5.0 in	6.25 × 6.25 in	6.2 × 8.3 in	8.0 × 10.0 in
Luminance	200 to 0.04 fl	230 to 0.04 fl	200 to 0.04 fl	300 to 0.04 fl	200 to 0.04 fl	200 fl (white)	200 to 0.05 fl
Optional	250 fl	250 fl	300+ fl	300+ fl	300+ fl		
Contrast ratio	8:1 in 10,000 fl	8:1 in 10,000 fl	8:1 in 10,000 fl	8:1 in 10,000 fl	8:1 in 10,000 fl	5:1 normal 3:1 over FoV	4:1 (high light) 10:1 (low light)
Colours	256 shades of grey, 16.7 million colours	256 shades of grey, 16.7 million colours	256 shades of grey, 16.7 million colours	256 shades of grey, 16.7 million colours	256 shades of grey, 16.7 million colours		64 shades of grey, 256,000 colours
Temperature	−40 to +55°C	−40 to +55°C	−40 to +55°C	−40 to +55°C	−40 to +55°C	−40 to +55°C	−20 to +55°C
Altitude	0 to 31,000 ft	0 to 31,000 ft					
Interface	RS-170, RS-170A RGB, RS-422, Discrete	RS-170, RS-170A RGB, RS-422	RS-170 or STANAG, RS-485, Discrete	RS-170 or STANAG, Analogue, Discrete, MIL-STD-1553, Ethernet	RS-170 or STANAG, RS-485, Discrete	RS-422, Discrete, ARINC-708/453, EGPWS/CWX, ARINC-429, Mono or RGB video (option)	MIL-STD-1553, ARINC-429, Ethernet, RS-232/ 422/485, Discrete, Analogue
MTBF	4,000+ h	4,000+ h	4,000 h	4,000 h	4,000 h		2,700 h

The 5 in MFCD is in service in the Tornado, A-4, Dauphin and other aircraft.

6 in Multi Function Colour Display (MFCD)

The 6 in MFCD incorporates the same functionality and capability as the 4 and 5 in MFCDs with a useable screen size of 6.25 × 6.25 in (158.7 × 158.7 mm). The 6 in MFCD is in service in Boeing 737 (ARINC D form factor), CH-53, Tornado, K-8 and other aircraft.

6 × 8 in Multi Function Colour Display (MFCD)

The 6 × 8 in (152.4 × 203.2 mm) AMLCD is designed for mission video or EFIS applications, displaying FLIR, map or camera video, as well as optional primary flight, navigation, sensor, colour weather radar, TCAS, EGPWS, and map data. Applications for the display includes helicopters, fighter and transport aircraft.

8 × 10 in Multi Function Colour Display (MFCD)

The 8 × 10 in (203.2 × 254.0 mm) display is a landscape mode display designed for use as a high-resolution mission display for sensor and/or navigation video for special operations and Maritime Patrol Aircraft (MPA). The main functions of the display include graphics processor, graphics overlay on video, touch screen and a non-volatile memory. The large 12.1 in (307.3 mm) diagonal unit features a sunlight readable, 100 pixels/inch display.

Status

Astronautics MFCDs are in production and in service in a large number and wide variety of platforms world-wide.

Contractor

Astronautics Corporation of America.

Airborne Separation Video System (ASVS)

Type

Flight/mission recording system.

Description

The Airborne Separation Video System (ASVS) is the only airborne-qualified high-speed digital imaging system used for weapons separation testing. The ASVS maximises the capabilities of digital imaging by facilitating recording of multiple events during a single mission. The ASVS approach ensures the control, timing, and synchronisation of multiple cameras is accomplished accurately, while providing a central storage module for the imaging data.

From a single user interface, the ASVS can support up to 32 cameras, a single central storage device and an optional telemetry device. The system is comprised of one or more cameras, a MultiCamera System Controller (MSC) and a Cockpit Control Unit (CCU). The camera(s) capture(s) multiple event imagery in the harsh environment of high-performance aircraft at highspeed of up to 500 frames per second (fps) at 1,024 × 1,024 pixel resolution. The MSC provides for ASVS trigger, control, synchronisation, IRIG timing, and mass storage for up to 32 cameras. The removable 20 Gigabyte (Gb) hard drive stores over 27,000 one-megapixel images that are downloaded to a computer via 10/100 Base-T Ethernet or Hotlink directly onto the MSC hard drive. The CCU provides the optional capability to monitor, control, and update ASVS mission data from the aircraft cockpit.

ASVS is designed to replace traditional 16 mm and 35 mm film in both the airborne and ground-based test environments.

Specifications

Camera
Dimensions:
Camera: 87 × 66 × 188 mm (H × W × D)
MSC: 151 × 178 × 308 mm (H × W × L)
CCU: 102 × 146 × 149 mm (H × W × L)
Weight:
Camera: 2.5 kg
MSC: 6.2 kg
CCU: 2.3 kg
Power: 28 V DC or 115 V AC, 400 Hz
Consumption:
25 W (camera); 28 W (MSC); 14 W (CCU)
Temperature range: Operational −55 to +55°C; non-operational −55 to +71°C
EMI/RFI: MIL-STD-461-CE102, CS101, CS114, CS115, RE102, RS103
Sensor: 10-bit CMOS sensor, 1024 × 1024 pixels
Resolution: 1024 × 1024 pixels at 500 fps
Pixel bit depth: 8-bit monochrome and 24-bit colour
Light sensitivity:
colour, 11-110,000 lux, monochrome, <1-100,000 lux
Anti-blooming control:
1,000 × saturation signal
Exposure time: Electronically shuttered: 1/30, 1/100, 1/500, 1/1,000, 1/2,500, 1/5,000, 1/10,000, 1/20,000 (customisable)
Automatic exposure mode: Operator selectable, adjustment up to 1 F-stop per frame
Dynamic range: 54 dB
Grey level: 256 (8-bit)
Recording time: 2 s at 500 pps
Trigger Mode: Delayed trigger 0–30 s, 10 ms; pre-trigger store 0 to memory size; post-trigger store 0 to memory size
Image download rate: 132 Mbps
Picture download time: 1 picture in 16.5 ms, 1,000 pictures in <40 s

Additional Features:
Gain control, multiple-camera synchronisation independent of picture rate, live preview, remote power on/off, self-test

System Performance
Number of cameras supported: 32
Total number of images stored per flight: Over 27,000 one megapixel images (20 Gb removable hard drive)
Programming:
Pre-flight via Windows™-based software
Synchronisation:
Images synchronised to aircraft IRIG time code source

Status

In production and in service since 1997.

Contractor

DRS Data and Imaging Systems, Inc.

AM-250 barometric altimeter

Type

Flight instrumentation.

Description

The AM-250 integrates Honeywell's silicon pressure sensor with an indicator produced by Ametek Aerospace Products Inc. It meets Reduced Vertical Separation Minimum (RVSM) requirements.

In addition to full analogue and digital displays, the AM-250 has two displays to indicate barometric offset in hectoPascals or inches/millimetres of mercury. A rotary encoder and push-button switch allow the entry of barometric offset and auto barometric set. The altimeter is internally lit.

The AM-250 is available in three versions: one provides basic ARINC 429 output labels; another supports True AirSpeed (TAS) for FMS/GPS systems; the third provides an altitude preselect function. The unit can be configured to display height in feet or metres. It includes annunciators for altitude deviation and failure warning.

Specifications

Altitude range: −1,999 to +60,000 ft
Altitude display – accuracy: ±2 ft
Altitude display – pointer: ±10 ft
Barometric setting range: 16.00 to 30.00 InHg
Barometric setting accuracy: ±1 ft
TSO: C10b, C88a, C106c
Dimensions: 3 ATI, length 170.2 mm
Weight: 1.36 kg
Power:
(meter) 28 V DC, 12 W
(lighting) 5 V DC, 2.75 W
Interface: ARINC 429
Output: RS 232

The AM-250 barometric altimeter   0062206

Status

In production for CitationJet, Ayres Loadmaster and Lear 60.

Contractor

Honeywell Inc, Commercial Aviation Systems. Ametek Aerospace Products Inc.

AMLCD 3 ATI HSI/ADI

Type
Cockpit display.

Description

Smiths Aerospace has been involved in Active Matrix Liquid Crystal Display (AMLCD) technology since its inception in the aerospace industry. Various display sizes have been integrated into a number of AMLCD products. Due to the multifunction nature of AMLCDs, a variety of applications has been developed, including digital triple torque, airspeed, altitude, vertical speed, collision avoidance, engine parameters and aerodynamic surface or control position indication presented via MIL-STD-1553B, ARINC 429 digital bus or integrated interfaces.

Specifications

Qualifications: DO-160D and DO-178B
Interfaces: synchro interface added in 2000 to facilitate retrofit to older civil and military aircraft.

Status

In production for Cessna Citation Jet 1 and Citation Jet 2 aircraft.

Contractor

Smiths Aerospace.

AMLCD MultiFunction Displays (MFDs)

Type
Multifunction cockpit display.

Description

CMC Electronics' Active Matrix Liquid Crystal Display (AMLCD) MultiFunction Displays (MFDs) feature full anti-aliased graphic symbology combined with good sunlight readability and full NVIS Class B compatibility. Combined with an appropriate mission processor, multiple pages, for display of navigation, mission and other data, are selectable using soft keys on the bezel. Sensor information, such as radar, Forward-Looking Infra-Red (FLIR) or other real-time video can be presented along with symbology and other graphics which can be overlaid as part of the presentation.

The MFDs may also be configured as a video monitor with RGB or LVDS inputs from the mission processor or as a smart MFD with direct interfaces to the aircraft avionics over standard busses such as MIL-STD-1553B or ARINC-429. The open architecture design accommodates upgrades to meet changing customer needs.

CMC Electronics' MFDs (CMC Electronics)
1030039

Displays are currently available in 4 × 5 (portrait or landscape), 5 × 5, 5 × 7 (portrait) and 6 × 8 (portrait) in screen sizes, although the display architecture will accommodate larger AMLCD screen sizes if required.

Specifications
Colour resolution:
   4 × 5 in: 480 × 640 colour pixels
   5 × 5 in: 640 × 640 colour pixels
   5 × 7 in: 600 × 800 colour pixels
   6 × 8 in: 600 × 800 colour pixels
Horizontal viewing angle: ±45°
Colours: 64 level/colour grey scales (256,000 colours)
Brightness: >300 fl
Contrast ratio: Dark room ambient 100:1 (typical), 80:1 (minimum); sunlight 5,000 lux ambient 12:1 (typical)
NVIS compatibility: compliant as per MIL-L-85762A, Class B
Interface: MIL-STD-1553B (smart only), ARINC-429 (smart only), discrete, analogue, RS-422
Interface to mission processor: RGB analogue
Contrast ratio:
Sunlight: 12:1 typical (5,000 lux ambient)
Night: 100:1 typical, 80:1 minimum (dark room)
Environmental: RTCA DO-160C

Status
In production and in service.

Contractor
CMC Electronics Inc.

AMLCD MultiPurpose Displays (MPDs)

Type
MultiFunction Display (MFD).

Description

4 × 4 in Colour AMLCD MultiPurpose Display (MPD)
The 4 × 4 in MPD offers a complete range of formats, and is ideal for new build and retrofit requirements in fighter aircraft. High resolution, low reflectivity and excellent chromaticity is provided. VAPS software is programmable to meet customer requirements, with AMLCD technology. Simultaneous video and graphics capability is offered, with 15 grey shades (option 64), MIL-STD-1553B interfaces, NTSC output, and NVIS Class B compliance.

Specifications
Dimensions: 141 × 141 × 275 mm
Weight: 4.8 kg
Power Supply: 28 V, 95 W

6.25 × 6.25 in Colour AMLCD MultiPurpose Display (MPD)
The 6.25 × 6.25 AMLCD has been developed for the US Army Longbow Apache helicopter, where the display system comprises four 6.25 × 6.25 colour AMLCD MPDs and two Colour Display Processors (CDPs). The CDPs can be populated selectively to a maximum of four completely independent channels, using a MIL-STD-1553 bus if desired. Full feature digital map system can be embedded within the CDP. The MPD provides 64 grey shades (growth to 256). Resolution is 512 × 512 colour pixels in Quad RGGB Colour Pixel arrangement. The MPD is NVIS Class B compliant, and tested with Class A goggles in mono green mode.

Specifications
Dimensions: 216 × 216 × 184 mm
Weight: 5.8 kg
Power supply: 115 V 3-phase 400 Hz, 115 W

4 × 4 in colour AMLCD multipurpose display
0018182

Four of the Honeywell Aerospace 6 × 8 in AMLCD units (two pilot and two co-pilot) in the US Air Force C-141 cockpit configuration
0018181

Status
In full production for the US Army Longbow Apaches and standard for all Netherlands and UK Longbow Apaches. A total or more than 4,000 MPDs will be required for the Longbow Apache programme including remanufacture of all US Army AH-64s.

6 × 8 in AMLCD Display Unit
The 6 × 8 in AMLCD has been designed for use on the C-141 aircraft upgrade, where it acts as either a primary (ADI, HSI, Airspeed, Altitude display) or secondary (Heading, Waypoint, Weather Map) flight display. A high performance graphics processor is used with a 486DX2 general purpose processor; the Virtual Application Prototyping System (VAPS) permits the drawing and simulation of formats in near-realtime for format changes. Resolution is 480 × 640 colour pixels (RGGB Quad) with 80 colour groups/in. The display is NVIS Class B compliant, with NVIS Class A option. Fitment of 6 × 8 in AMLCDs forms part of the US Air Force C-130 and C-141 integrated avionics upgrade programme. The 6 × 8 in display is also used in the Spanish Air Force C-130 modernisation programme.

Specifications
Dimensions: 196 × 246 × 136 mm
Weight: 8.2 kg
Power supply: 28 V, 117 W

8 × 10 in AMLCD unit
This is currently Honeywell Aerospace's largest ruggedised military and avionic crew station AMLCD MPD. It has been supplied for the US Army's Rotary Pilots' Associate (RPA) next-generation cockpit.

Specifications
Grey shades: 64
Resolution: 1,024 × 768 (XGA)
NVG: NVIS Class B (option)
Dimensions: 292 × 241 × 76 mm
Weight: 6.8 kg
Power: 28 V DC, 75 W

Contractor
Honeywell Aerospace, Defense Avionics Systems.

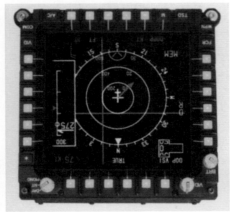

6.25 × 6.25 in colour AMLCD multipurpose display   0018180

8 × 10 in AMLCD display unit   0018184

AMLCDs

Type
Cockpit display.

Description
Planar manufactures a range of AMLCD displays for military, industrial and commercial applications. They include ElectroLuminescent (EL) flat panel displays, miniature Active Matrix ElectroLuminescent (AMEL), 1,000 lines per in units suitable for Head-Mounted Displays, Active Matrix Liquid Crystal Displays (AMLCD), as well as full colour CAT-based avionic systems.

Planar 11 × 8 in landscape orientation flat panel display is an amorphous silicon AMLCD full colour display for use in environmentally demanding applications that require maximum readability, rugged construction, and a wide temperature range.

Specifications
Model AM5: 5.0 × 5.0 in
Active display area: 127 × 127 mm
Pixel configuration: RGGB Quad
Display resolution:
Full colour 480 × 480
Grey levels 255

Viewing angles (20:1 contrast ratio):
horizontal ±57°
vertical +40°; –55°
NVIS compatibility (optional): NVIS-B MIL-STD-L-85762 Class B
Model AM6: 6.25 × 6.25 in
Active display area: 159 × 159 mm
Pixel configuration: RGGB Quad
Display resolution:
Full colour 512 × 512
Grey levels 255
Viewing angles (20:1 contrast ratio):
horizontal ±57°
vertical +40°; –54°
NVIS compatibility (optional): NVIS-B MIL-STD-L-85762 Class B
Model: 6.0 × 8.0-Q/6.0 × 8.0-S/8.0 × 6.0-S in
Active display area:
157 × 211/157 × 211/211 × 157 mm
Pixel configuration: RGGB Quad (Q model); RGGB Quad (S model)
Display resolution:
Full colour 600 × 800
Monochrome 1,200 × 1,600
Grey levels 64/256
Viewing angles:
horizontal >±45°
vertical >+35°; –15°

Planar AMLCD   0001392

B-2 cockpit with Planar displays   0003322

NVIS compatibility (optional): NVIS-B MIL-STD-
L-85762 Class B
Model: 11.0 × 8.0-Q
Active display area: 276 × 200 mm
Pixel configuration: RGGB Quad
Display resolution:
Full colour 1,536 × 1,120
Monochrome 3,072 × 2,240

Status

In December 1999, Planar Systems Inc launched a
3 ATI AMLCD, suitable for both civil and military
flight-deck use in Attitude Direction Indicator
(ADI), Horizontal Situation Indicator (HSI) and
engine data applications.

Contractor

Planar Advance Inc.
Planar Systems Inc.

AMS 2000 Avionics Management System

Type

Flight management and cockpit display system.

Description

Honeywell's AMS 2000 is a fully integrated
modular avionics concept/system designed to
meet the needs of military tanker/transport/
maritime patrol aircraft. It integrates flight
management systems, flight controls, cockpit
instruments and cockpit controls. Modular
architecture is built around the system processor
with growth to TCAS, MLS, HUD, digital map,
NVIS-compatibility, DGPS and CNS/ATM.

Status

No longer in production. In service with C-130
and P-3 aircraft of various countries.

Contractor

Honeywell Defense Avionics Systems.

AMS-5000 Avionics Management System

Type

Avionic Management System (AMS).

Description

Designed specifically for intermediate-range
business aircraft, the AMS-5000 Avionics
Management System provides centralised
avionics management, including control of
EFIS modes, weather radar, TCAS, radio tuning
and flight management, in one convenient
location. The system offers consistent, intuitive
menu-driven operation that facilitates training,

*The AMS-5000 Avionics Management System
offers full AFIS capability*   0504187

minimises entry errors and simplifies avionics
management. A typical system would comprise:
a Flight Management Computer (FMC-5000);
a Control Display Unit (CDU-5000); a Display
Control Panel (DCP-5000); and a Database Unit
(DBU-4100). GPS functionality can be added
using a Global Positioning Sensor and GPS
antenna.

An integral part of the Pro Line 4 system, the
system takes advantage of Pro Line 4 multifunction
displays to provide head-up operation, with
navigation maps and tables and other avionics
information shown on the MFD. Integration of the
AMS-5000 as part of the Pro Line 4 system reduces
weight and size and simplifies interconnections.
Flight management computers are packaged
as line-replaceable modules, centrally housed
in the Collins integrated avionics processing
system. Pro Line 4 integration allows the system
to perform diagnostic functions, identifying any
malfunctioning avionics unit and displaying the
information on the MFD. The system also offers
complete LRU fault history and detailed analysis
of system status.

The menu-driven operation of the AMS-5000
simplifies flight crew route planning. Flight plans
may be entered waypoint by waypoint or by jet
or victor airway designators. The system allows
storage for up to 100 pilot-defined routes, each
with 100 waypoints. En route, terminal area and
non-precision approach navigation as well as
primary means navigation in oceanic and remote
areas are provided by the system database, and
new waypoints may be added within a flight
plan or appended. The system can automatically
execute non-precision GPS approaches, GPS
overlay approaches and multisensor RNAV and
VOR approaches including missed approach
guidance. When editing a flight plan, pressing
the line select key next to the displayed waypoint
automatically brings up the edit page on the CDU
for consistent operation. Pilot-defined waypoints
can be created by plotting a path on the system's
present position map using the joystick, and
direct-to is also available for straight point to
point routeing.

VNav capability is available with the advanced
Collins system, fully integrated with the autopilot,
navigation and air data systems for smooth
operation. The AMS-5000 is available with a full
alpha keyboard to accommodate a wide range of
applications.

The AMS-5000 is also available with AFIS
compatibility, offering flight crews access
to flight planning, weather information,
SIGMET advisories and messaging capabilities.
Additionally, the AMS-5000 provides
straightforward fuel management, computing
fuel used, fuel remaining and endurance at
the present fuel flow. The system's worldwide
database is contained on disks that are updated
every 28 days, which contain global VHF
navigation and airport information.

Specifications

Dimensions:
(FMC-5000) 43 (H) × 222 (W) × 154 (L) mm
(CDU-5000) 162 (H) × 146 (W) ×118 (L) mm
Weight:
(FMC-5000) 0.86 kg
(CDU-5000) 1.82 kg
Altitude: 55,000 ft
Temperature:
(FMC-5000) –55 to +70°C
(CDU-5000) –20 to +70°C
FAA TSO: C115b, C129a Class B1
Environmental: DO-160C
EUROCAE: ED-14C

Status

In service. FAA Supplemental Type Certification
on Beechjet 400A awarded in December 1996
and on Starship in October 1997. FAA TSO C129
Class B1 and Technical Order 8110.60 criteria
for IFR operation to make non-precision
approaches with GPS as primary means of
navigation.

Contractor

Rockwell Collins.

AN/AQH-13 acoustic data recorder

Type

Flight/mission recording system.

Description

The AN/AQH-13 Acoustic Data Recorder is a
form/fit/function replacement for the AN/AQH-
4A unit. Based on DRS's line of Digital Cassette
Mission Recorders (DCMRs) and Common
Recorder Architecture (CRA) technologies, the
VME-base design of the AN/AQH-13 facilitates
straightforward, cost-effective configuration for
other applications.

The system can be controlled remotely (via
RS-232 interface) from a local control panel or by
an identical panel up to 50 ft away. Several special-
purpose control interfaces also are available,
including Ethernet (100 Base T), Proteus digital
bus and MIL-STD-1553. The analogue interface
is capable of recording 36 signal channels, 32
primary signal and 4 auxiliary channels. Digital
interface options, with the ability to record up to
16 digital channels and 16 acoustic channels, are
due for release.

The base unit supports four recording
configurations:

- 32 channels, 20 kHz analogue bandwidth and
  12 bits of resolution (default configuration)
- 24 channels, 20 kHz analogue bandwidth and
  16 bits of resolution
- 16 channels, 40 kHz analogue bandwidth and
  12 bits of resolution
- 12 channels, 40 kHz analogue bandwidth and
  16 bits of resolution

The AN/AQH-13 also supports five channels
of replay: three for fixed analogue channels,
one dedicated to IRIG-B time code, and one
functioning as an analogue monitor channel
to replay recorded signals. The system comes
MIL-qualified and configured to NATO STANAG
format. Custom configurations are also
available.

Other features and benefits of the system
include:

- Recording of 36 analogue channels: 32 primary
  signal, 4 auxiliary
- 20/40 kHz analogue bandwidth
- Rotary helical scan cassette
- Industry standard DTFTM media format
- In excess of 4 h recording time
- 42 Gbyte tape capacity
- Interface options: SCSI-2, analogue, RS-232,
  RS-422, Ethernet, MIL-STD 1553
- VME-based
- Operates from 28 V MIL-STD 704A compliant
  power
- Built-In Test (BIT).

Specifications

Analogue Inputs:
Differential Input: 2.0 V RMS
Input Impedance: 100K ohms
Sampling Method: Delta Sigma
Resolution: 16 or 12 bits
Sampling Frequency: 50 or 100 ksps
Distortion: <0.1% SNR >84 dB broadband
Dynamic Range: 80dB
Cross Talk: < –75 dB
Analogue Outputs:
Single Ended or Differential
Output Voltage: 2.5 V RMS
Output Impedance: 600 ohms
Timing: Less than 250 µs skew; Resolution 16 or
12 bits; Distortion <0.1%
SNR: >84 dB broadband
Dynamic Range: 80 dB
Cross Talk: <–75 dB
Dimensions: 393.7 (W) × 375.9 (D) ×
469.9 (H) mm
Weight: <36.4 kg
Power: 28 V DC, 350 W
Environmental:
Operating Temperature: –20 to +50°C
EMI/EMC per MIL-STD 461C for Class A1b
equipment

Contractor

DRS Technologies, Inc.

AN/ARA-63 microwave landing system

Type
Aircraft landing system, Microwave Landing System (MLS).

Description
The ARA-63 is the airborne portion of the US Navy's standard aircraft approach control system for landing on aircraft carriers and equips aircraft such as the A-6E, F-14A, S-3A and F/A-18. Known as a radio receiving decoder set, it works in conjunction with the AN/SPN-41 on board ships and the AN/TRN-28 transmitters at naval air stations. Pulse-coded microwave transmissions are received by the ARA-63, decoded and displayed on a standard cross-bars indicator in the cockpit. There are 20 channels in the range 14.688 to 15.512 GHz. The system has three LRUs: receiver, pulse decoder and control unit.

Status
In service in US Navy aircraft such as the A-6E, F-14A, S-3A and F/A-18 and also overseas. Over 3,000 units have been produced. Development and early production was by Telephonics. Follow-on production by Stewart-Warner.

Contractor
Telephonics Corporation, Command Systems Division.

The AN/ASN-165 radar navigational data display set showing (left to right) the radar data converter, navigator indicator/control panel and pilot indicator   0504100

AN/ASN-150(V) central tactical system

Type
Flight/mission recording system.

Description
The AN/ASN-150(V) central tactical stem controls and displays navigation, communication and armament data for a wide variety of fixed-wing aircraft and helicopters. It includes two dual-redundant tactical data processors, communications and armament system controllers, multifunction displays and various display and intercom system control panels. Operator interface with the system is provided by four control/display units, each with an alphanumeric keyboard and LED display.

The system interfaces with other avionics by a dual-redundant MIL-STD-1553B databus and by discrete interconnections.

The display system features a high-resolution multifunction display with either 6 × 9 or 19 in diagonal sizes. Up to three live video windows can be presented including radar, acoustic data, FLIR, monochrome or video camera. Window size is under operator control and can present comprehensive graphic, symbol and moving map presentations for tactical data transfer to other land, sea or air platforms.

Specifications
Dimensions:
(tactical data processor) 257 × 193 × 387 mm
(communications system controller)
    356 × 178 × 443 mm

(armament system controller) 191 × 152 × 262 mm
(multifunction display) 267 × 211 × 381 mm
(control/display unit) 116 × 170 × 178 mm
(intercom system) 127 × 38 × 46 mm
(display and control panel) 86 × 147 × 76 mm
Weight: 123 kg
Reliability: 250 h MTBF

Status
In production and in service with US Navy HH-60H and SH-60F, US Naval Reserve SH-2G and US Coast Guard HH-60J. The system is also in use internationally in S-70 variants procured or selected by Greece, Kuwait, China and Thailand. The central tactical system has been modified for use in fixed-wing aircraft such as the China S-2T.

Contractor
Northrop Grumman Corporation, Navigation Systems Division.

AN/ASN-165 radar navigation data display set

Type
Avionic display system.

Description
The AN/ASN-165 is a stand-alone radar indicator system that upgrades the displays on older radars. It replaces the display subsystem on ground mapping, weather avoidance and navigation radars such as the APN-59, APS-133 and APQ-122. The aim is to improve reliability, maintainability and operational performance without replacing the entire radar. The system is designed as a drop-in replacement for the radar displays with little modification to existing structures and cabling in the aircraft. The radar's antenna and receiver/transmitter subsystems remain unchanged.

The AN/ASN-165 consists of a radar data converter, pilot indicator and navigator indicator/control panel. The radar data converter contains two independent digital scan converters that convert the radar data into a high-resolution raster video signal. Navigational data is overlayed on the radar imagery and displayed on the pilot and navigator colour indicators. Aircraft navigational systems are interfaced through a standard ARINC serial bus or dual-redundant MIL-STD-1553B databus.

Radar ground map, terrain-avoidance and weather imagery are typical displays. The imagery is displayed with 16 levels of shading using standard RS-170 or RS-343 video format. Multiple colours are used to indicate different weather intensities. The standard video format also permits recording of the radar imagery for mission review and training.

With the navigational interface, aircraft data, such as true heading, groundspeed and track angle error, is displayed on both the navigator's and the pilot's display. The aircraft's present position is also displayed and the data is used to calculate the position of ground targets identified by a movable cursor that can track a fixed point on the ground. The navigational interface also provides the option of stabilising the radar display to true north.

Flight plan waypoints and navigational aids can be displayed, depending on the navigation system used. Additionally, data from EW systems, such as detected threats and threat zones, can be overlayed on the radar image.

Status
The AN/ASN-165 is installed on the WC-130, C-130 and US Air National Guard KC-135. Variants of the system are used on the MC-130E, HC-130P/N, C-141B, RC-135 and other Special Operations and reconnaissance aircraft.

Contractor
Systems Research Laboratories.

AN/ASQ-212 mission processing system

Type
Mission Management System (MMS).

Description
The AN/ASQ-212 system consists of the CP-2044 computer and several interconnection devices which comprise a form, fit and function replacement for the AN/ASQ-114 computer, data analysis logic units and the signal data converter. The extended memory upgrade of the AN/ASQ-114 computer is transferred to the CP-2044 and used both for global and secondary memory. The CP-2044 incorporates Motorola 68030 processors to provide a throughput ranging from 10 to 25 Mips, which is 30 times greater than the current system in the P-3C Update I/III

Units comprising the AN/ASN-150(V) central tactical system   0504190

aircraft at a fraction of the current size, weight and power requirements. In the full Update III Ada implementation, less than 50 per cent of the CP-2044 minimum throughput and memory capacity is utilised.

The CP-2044 VME bus open architecture can be configured with additional processing, memory and input/output modules to meet the requirements of new subsystems such as GPS and Satcom, and of processing intensive functions such as sensor post-processing and data fusion.

Initially designed for retrofit into P-3C Update I/III aircraft, the AN/ASQ-212 can be easily tailored to the requirements of other P-3C configurations as well as new aircraft.

Status
The AN/ASQ-212 system was developed for the US Navy P-3C aircraft under a two-phase programme that began in September 1989. The first production systems were installed in US Navy test aircraft and training facilities beginning in May 1993, at a rate of four systems per month. Also in service with P-3C export customers.

Contractor
Lockheed Martin Management & Data Systems.

AN/AYK-14(V) standard airborne computer

Type
Aircraft central computer.

Description
The AN/AYK-14(V) is a high-performance general purpose computer with both 16- and 32-bit processing elements. It consists of a family of interchangeable processor, memory, power, enclosure and input/output modules that can be configured to meet specific price, performance and functionality needs for a wide range of applications. The AN/AYK-14(V) computer offers high performance and high reliability with a low life cycle cost, while meeting airborne MIL-E-5400, shipboard MIL-E-16400 and land MIL-E-4158 environments. It is the US Navy's standard airborne computer and is currently being used on a wide variety of military platforms by the US and its allies.

The 16-bit version of the AN/AYK-14(V) is currently in its third generation, which is known as VHSIC AN/AYK-14(V). It provides performance of up to 20 Mips or more within a single enclosure and has a single memory addressing capacity of 16 Mbytes. The VHSIC computer is fully compatible with software written for either of the earlier generations. The instruction set is compatible with that of the AN/UYK-44 and AN/UYK-20 and is supported by the MTASS software development environment. Other 16-bit modules include 128 kbyte core memory, MIL-STD-1553A/B, NTDS, RS-232, Proteus, discrete Input/Output (I/O) and application specific I/O modules. On the Boeing F/A-18, two 16-bit AN/AYK-14(V) computers are installed in a dual-redundant system configuration. One computer serves as the navigation and engine controller, while the other functions in a mission management role, handling and processing weapons and target information. Each computer can accommodate up to 16 Mbytes of memory and provides up to 20 Mips of processing throughput.

The 32-bit version of the AN/AYK-14(V), also known as the Advanced AYK-14, is based on commercial RISC technology. Up to 225 Mips of processor performance and 180 Mbytes of memory capacity are currently available within a single enclosure. The use of commercial open system backplane and processor standards provide built-in performance and capability growth potential that can increase the performance of commercial technologies. Like the 16-bit AN/AYK-14(V) processing elements, the Advanced AYK-14 also provides

The F/A-18 Hornet avionics system incorporates two AN/AYK-14(V) mission computers (Boeing)
0131126

modular processing and I/O components that can be configured to meet specific price and performance requirements of a range of applications. Currently, a MIPS R5271 RISC computing module processor, other processor modules, and SCSI, MIL-STD-1553A/B, RS-422, Proteus and discrete I/O modules are available.

A complete, commercially supported, Ada development environment is available for commercial workstation Sun and RS-6000 networks, providing full development and real-time debugging support. A real-time operating system with POSIX-compatible services, priority management capabilities and Rate-Monatonic scheduling support completes the Advanced AYK-14's capabilities. Ada compilation, debugging and runtime can be obtained from multiple commercial suppliers. Current implementation utilises the Rational V ADS-Advanced commercial product. Device drivers and hardware support software are available from General Dynamics Information Systems.

GD3000 Advanced Mission Computer
The General Dynamics Information Systems (GDIS) Advanced Mission Computer (AMC) is the next generation open systems processor which is deployed on a variety of platforms. The GD3000 AMC is a leading edge, flexible and rugged processing product family, which can be readily configured to meet the needs of modern military systems, from benign laboratory to harsh avionics environments. The AMC is an integrated information processing system, providing complete hardware and software solutions. It is built upon a well-defined open systems architecture allowing for rapid insertion of emerging technologies; GDIS supplies system design and integration services to ensure a precise fit to the requirements of each specific user platform.

The GD3000 AMC is a set of digital computer hardware and software that performs general purpose, I/O, video, voice and graphics processing. Communication is over multiple buses, including 1553, Fibre Channel and Local PCI and all modules integrate in an industry standard 6U VME backplane. The I/O configuration may be tailored via use of PMC mezzanine modules. The design is scaleable and expandable, with a clear and built-in path for technology upgrades and insertion. An Ethernet interface is provided to support software development and maintenance of the system.

The AMC can be used efficiently in a wide range of applications, ranging from embedded module functions to full-scale multicomputer configurations and operates reliably in extreme airborne, ground-based and shipboard conditions. The system has application in display and mission processing, information and stores management.

Specifications
Dimensions: 194 × 257 × 356 mm typical
Weight: 11 to 16 kg

Power: 50 to 300 W typical
Computer type: binary, fixed or floating point
Word length: 16- or 32-bit with double precision and floating point
Typical speed: 500 Kips to 225 Mips
Max addressing: 16 Mbytes (16-bit) to 4 Gbytes (32-bit)
Input/output options: discretes, MIL-STD-1553A or B (dual-redundant buses), NTDS fast, slow, ANEW and serial, 6 MHz Manchester, RS-232C, 85323 Proteus, RS-422, SCSI, RS-485, TM-bus

Status
The AN/AYK-14 is in production and selected for the Boeing F/A-18 (which has two AYK-14 mission computers), Sikorsky SH-60B Seahawk LAMPS Mk III helicopter, Northrop Grumman E-2C Hawkeye, Boeing/BAE Systems AV-8B, Northrop Grumman EA-6B Prowler, F-14D, EP-3E, ES-3A, Joint STARS, Lockheed Martin P-3C Orion and embedded modules in the AN/ALQ-149, US Air Force TAOM/MCE system. The Advanced AYK-14 is used on the Bell/Boeing V-22 Osprey in a lower cost configuration that provides 45 Mips throughput and 12 Mbytes of memory on a single board computer. The GD3000 AMC is part of the F/A-18 E/F, F-15E, AV-8B and T-45 development programs.

A modified version, designated ACCS 2500, is used on the UK Royal Air Force Harrier GR. Mk 7 aircraft (see entry under Computing Devices in the UK part of this section).

Contractor
General Dynamics Information Systems.

AN/USC-45 Airborne Battlefield Command Control Centre (ABCCC III)

Type
Airborne battle management system.

Description
The AN/USC-45 Airborne Battlefield Command and Control Centre (ABCCC III) is an airborne node in the US Air Force Tactical Air Control System. It maximises the efficient use of fighter forces and other air resources by gathering real-time battlefield data from forward areas and by managing aircraft in air-to-ground operations.

Designed for the EC-130E, the ABCCC III self-contained capsule houses 15 automated workstations, allowing the battle staff to manage the tactical air assets conducting over 150 sorties/h effectively. During each 10 hour mission of the EC-130E, the ABCCC III provides a communications link to higher headquarters and co-ordinates forces in the battle area, updating aircraft on their way to the target and collecting their in-flight reports on the way out. Automated capabilities allow the battle-staff to analyse continuing combat quickly and direct offensive air support toward fast developing targets.

The ABCCC III consists of four major airborne subsystems: The Communications Subsystem

The EC-130E ABCCC aircraft features fuselage-mounted heat exchanger pods, a dorsal antenna array and forward-facing HF probes under the outer wing sections   0084529

The interior of the Capsule Subsystem   0084528

(CS), the Tactical Battle Management Subsystem (TBMS), the Airborne Maintenance Subsystem (AMS) and the Capsule Subsystem (CS). There is also a major subsystem on the ground. The Mission Planning Subsystem (MPS) provides the tactical database used by the airborne maintenance technician for system initialisation. On mission completion the MPS can also be used for post-flight playback and analysis of mission data.

Once airborne, the CS and its Automated Communications and Intercom Distribution System (ACIDS) controls and secures all communications between the capsule and forward units, other aircraft and rear bases. Communications within the capsule and between the capsule and the flight crew are also handled by ACIDS.

The TBMS provides comprehensive battlefield management capabilities for up to 12 operators stationed at individual battle-staff consoles. Fast accurate access to communications, as well as tactical and map databases, enhances operator effectiveness.

The AMS provides diagnostic and fault isolation to detect hardware or software malfunctions during a mission. System initialisation and control are also integrated into AMS functions.

The Capsule Subsystem consists of the facilities providing physical, environmental and life support for operating the airborne subsystems.

Contractor
Lockheed Martin Management & Data Systems.

AN/UYH-15 recorder-reproducer set, sound

Type
Flight/mission recording system.

Description
The AN/UYH-15 is a voice recording and reproduction system that uses modern digital speech processing technology. It is a compact

system that allows operators to monitor, record and instantly recall any recorded message. Designed primarily for use with signal acquisition systems as the standard replacement for the US Army AN/UNH-17A analogue cassette recorder, the AN/UYH-15 is ideally suited for all real-time voice transcription and analysis applications. The AN/UYH-15 may be controlled by either a host computer or by one or two control display panels.

The AN/UYH-15 can record six analogue input channels simultaneously. During recording, each operator can either monitor input or play back recorded files. Operator commentary can be recorded and time-correlated to a given signal. The signal and its related commentary can be combined for output to aid in analysis.

Each recorded signal is digitally sampled and compressed before being stored on the AN/UYH-15's hard disk, which makes it possible to store 6 hours of voice input. The voice compression algorithm offers proven performance in the noisy military environment, as well as high-quality reproduction independent of the signal being reproduced.

Specifications
Dimensions:
(chassis) 133.4 × 482.6 × 412.8 mm
(control/display panel) 50.8 × 228.6 × 152.4 mm
Weight:
21.32 kg (with 2 control/display panels)
16.78 kg (without panels)
Power supply: 105–130 V AC or 208–240 V AC, single phase, 47–400 Hz or 22–30 V DC, 80 W typical
Audio channels: 6 input, 2 output
Bandwidth: 300 Hz – 4.4 kHz
Capacity: 6 h
Recording media: 170 Mbyte formatted hard disk
Interfaces: RS-232C, IEEE-488
Environmental: MIL-STD-810D, MIL-STD-461/462, TEMPEST

Contractor
General Dynamics Information Systems.

AN/UYQ-70 advanced display systems

Type
MultiFunction Display (MFD).

Description
The AN/UYQ-70 (Q-70) integrates Commercial Off-The-Shelf (COTS) technology and components with various mission-critical, real-time requirements which can be configured to meet a wide range of configurations and mission requirements. The Q-70 is one of the first standard combat system elements implemented with an open-system architecture.

This flexibility and openness allows the Q-70 family of systems to provide highly flexible computing and display infrastructures for new combat system development or the retrofit of existing systems. The Q-70 family has been developed by Lockheed Martin, DRS Technologies and Raytheon.

A variant of the Q-70 family has been selected by the US Navy and will be deployed in the E-2C Hawkeye Airborne Early Warning (AEW) aircraft.

The Q-70 workstation is based on an open-system architecture which uses the latest available COTS components for mission-critical functions. A ruggedised tactical display and computing system, the Q-70 supports the common operating environment that is being implemented in surface, subsurface, land and airborne platforms. Transition to the Q-70's open operating system does not require that legacy architectures be abandoned to achieve the high level of modernisation necessary for older platforms to effectively participate in the latest warfare applications. This design approach leads to savings in logistic support, reduced cross-training between different legacy display systems, and more flexibility in crew staffing assignments.

Processing for the Q-70 is based on the Hewlett Packard 743 and successor processors. These processors were selected for their performance and compatibility, will support open-system evolutionary products that are being deployed in all branches of the US armed services, and have a strong evolutionary path that features a UNIX-compatible and real-time software base.

The Q-70 programme provides full software compatibility between on-board systems, as well as cross-platform compatibility for shipboard and airborne applications. Q-70 supports growth in native mode applications, using the full benefit of technology advances available to the programme, and also supports back-fit by providing emulation capabilities of older workstation and display architectures.

Q-70 is specifically designed to provide enclosures that are ruggedised, with strong emphasis on protecting COTS hardware in the mission critical environments. The system has been integrated in surface, subsurface and airborne environments and can be adapted to other land-based operations.

Contractor
Lockheed Martin.
DRS Electronic Systems, Inc.
Raytheon.

Angle of attack computer/ indicator

Type
Flight instrumentation.

Description
The angle of attack computer/indicator provides the pilot with a continuous display of aircraft lift information on a decimal scale, with 1.0 representing the stall. The display is valid regardless of bank angle, aircraft weight or wing configurations.

The computer/indicator face is scaled red below 1.1 Vs, amber from 1.1 Vs to 1.3 Vs and black from 1.3 Vs to maximum speed. It also features a settable bug and index slaved to each other, which can be set for a desired airspeed target between 1.2 Vs and 1.5 Vs. Centring and maintaining the angle of attack pointer within the index will result in the selected speed target.

The system computer drives the ADI fast/slow pointer, a 2 in round angle of attack indicator and Safe Flight's speed indexer lights. The panel-mounted angle of attack indicator displays aircraft lift information. An approach reference is provided and the display is valid for all flap positions.

Status

In production and in service. The system has been certified on the Raytheon Hawker 800.

Contractor

Safe Flight Instrument Corporation.

Argus moving map displays

Type

Avionic Digital Map System (DMS).

Description

Argus moving map displays present all vital navigation and position information on an instantly readable display in front of the pilot. By minimising head-down time and giving the pilot more time to manage the aircraft, the Argus makes flying safer. The display can interface with, and receive its navigation data from most GPS, FMS and Loran C navigation systems. Approvals have also been obtained for the display of Ryan TCAD Model 9900B+ TCAS data and BFGoodrich Stormscope® WX-500 weather mapping data. Eventide also makes available hardware adaptors which allow Argus units to display information from older Stormscope models, including the WX-10, 10A, 11 and 1000E (with 429 EFIS option).

The Argus 3000 is designed for VFR use in light single- and twin-engined aircraft. Like the Argus 5000, it interfaces with most popular GPS, FMS and Loran C receivers. Its database contains over 11,000 landing facilities, 6,500 navaids and every special use airspace, including TCAs and ARSAs.

The Argus 5000's comprehensive, versatile moving map display shows all TCAs, ARSAs, navaids and landing facilities. At the touch of a button, the information submode presents detailed information about any on-screen facility from its own field-replaceable database. The Argus 5000 also provides a convenient digital readout of bearing or radial and distance to any selected facility.

The larger Argus 7000 provides all the information contained in the Argus 5000 on a bigger display. The Argus 7000 fits in a similar tray to the Argus 3000 and 5000, but has 2.3 times the screen area of the standard size Argus models. It can be updated without the need to remove it from the instrument panel.

A North American database, containing navigation information for airports, navaids and special use airspace within Canada, United States, Mexico, Central America and the Caribbean is available for use with the Argus 7000/CE, 7000, 5000/CE and 5000. An international database that gives the same worldwide coverage is available for all Argus maps.

The CE and Enhanced monochrome Argus maps are internet compatible for database updates and software upgrades at significant savings.

The Argus family of moving maps offers Enhanced software at the production level, or as an upgrade to existing maps. Enhanced software provides a variety of features: low cost internet database/software upgradeability; full flight planning capabilities; internal demo mode; terminal mode; name/location database search capabilities; Victor and Jet airways; weather display capabilities; collision avoidance display capabilities.

Argus moving map displays showing (left to right) the 3000, 7000 and 5000 models

0503899

With the RMI adaptor, the Argus moving map display can indicate up to two ADF or VOR pointers, as with a traditional RMI display, and also it provides a digital bearing. This mode is approved for ADF or VOR approaches.

The RMI adaptor can be remotely mounted or plugged directly into the back of the Argus. An RMI/ARINC adaptor incorporates a data converter that allows Argus to accept ARINC 419 and ARINC 429 data from flight management systems, inertial management systems, and VLF Omega systems.

Dual-adapter software is now available to allow the Argus to be used with two different data sources, for example, to switch between a TCAD display and a weather display. Standard Argus models have two input ports, one for GPS or other navigation source with the second available for an additional input that must be selected at installation. The new software allows two external sources to be supplied to the Argus with a switch to allow the pilot to select which is displayed.

The Argus 5000/CE and 7000/CE units display colour screen graphics and add several new hardware enhancements including flight recording, an internal barometric pressure sensor (pressurised cockpits require an adaptor to enable this feature), a rotary encoder for easy data entry and convenient database updates.

Advanced display technology allows the Argus 5000/CE and 7000/CE to present graphics in red, green and yellow, while retaining the sharp, bright, sunlight-readable qualities of the monochrome Argus models (which continue to be available). With these new colour Argus models, the pilot can select a colour scheme to display a variety of graphic data – including the ability to 'colour code' Class B and Class C airspace.

The built-in barometric pressure sensor gives these Argus models new capabilities which provide the pilot with greater situational awareness; the aircraft's current altitude is taken into account when displaying restricted areas.

Flight recording is another new feature. The Argus will record current latitude, longitude, time, date, altitude, groundspeed and track and heading in its non-volatile memory for up to 10 hours of flight, which can be 'played back' in real or compressed time on the Argus screen.

With the Argus 5000/CE and 7000/CE, updating the database is easier and less expensive. Updates are available over the Internet, on floppy disk and on PC (PCMCIA) Cards.

Argus can interface with, and receive its navigation data from, most GPS, FMS and Loran C navigation systems. The units are available with a US, North American, and international database. All databases contain navigation information for airports, navaids and special use airspace. The Argus 7000, 7000/CE, 5000, and 5000/CE are TSOd and can be IFR approved. The Argus 3000 is TSOd as well, when equipped with the international database.

Specifications

Dimensions:
(Argus 3000/5000/5000/CE) 81.3 × 81.3 × 269.4 m
(Argus 7000/7000/CE) 81.3 × 121.9 × 273 mm
Weight:
(Argus 3000/5000) 1.6 kg
(Argus 5000/CE) 1.5 kg
(Argus 7000) 2 kg
(Argus 7000/CE) 1.8 kg
Power supply: 11–33 V DC, 15 W

(RMI Adaptor)
Dimensions: 79.4 × 79.4 × 69.34 mm
Weight:
(RMI) 0.37 kg
(RMI/ARINC) 0.41 kg
Power supply: 11–33 V DC, 2–3 W
Environmental: RTCA-DO-160B

Contractor

Eventide Avionics.

ATD-800-II airborne tape system

Type

Flight/mission recording system.

Description

The ATD-800-II ruggedised tape deck is a low-cost, high-performance digital data record/reproduce tape media subsystem suitable for use in the harsh environmental conditions usually associated with flight test applications.

All components of the ATD-800-II are contained in a ruggedised chassis with internal shockmounts to physically isolate all devices from externally imposed vibration. The chassis is sealed from the potential contamination of the surrounding atmosphere and an internal temperature control system is provided to maintain temperature within operating limits. The recording system is the industry recognised DLT-4000 cartridge tape mechanism by Quantam. The front panel of the ATD-800-II contains a comprehensive set of operating status indicators.

The ATD-800-II is designed to interface directly to a standard SCSI-2 controller, such as is readily available on computer systems, as well as on the L-3 Telemetry-East MiniARMOR-700 data multiplexer/demultiplexer. When used in conjunction with the MiniARMOR-700, the ATD-800-II supports recording of multiple combinations of serial and parallel data sources. For example, several channels of PCM may be combined with time, voice, MIL-STD-1553, parallel and digital and analogue inputs. Playback of the data may be accomplished by direct connection of the SCSI-2 interface to a computer system or by using a playback configuration of the MiniARMOR-700. The latter approach permits coherent reconstruction of the original data streams.

Key parameters and features include record/reproduce rates of 1.5 Mbytes/s sustained, 5.0 Mbytes/s burst, 20 Gbytes data storage capacity per cartridge, non-compressed, BER less than one error in 10E17 bits, record time of 3.6 hours at maximum rate, single-ended or differential SCSI-2 interface, tape dubbing software (requires record and playback unit) high-speed access to stored data, remote control option. Records PCM, 1553, voice, time.

Specifications

Dimensions: 387 × 173 × 136 mm
Weight: 6.8 kg
Operating temperature range: –20 to +50°C

Contractor

L-3 Telemetry-East.

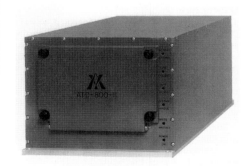

Series II ATD-800 digital data tape recorder

0002444

Attitude Indicators (AIs) and Attitude Director Indicators (ADIs)

Type

Flight instrumentation.

Description

Astronautics' Attitude Indicators (AIs) and Attitude Director Indicators (ADIs) provide the pilot with a pitch and roll attitude (two-axis) display, or pitch, roll and azimuth (three-axis) display, respectively. Added functions are provided in various combinations such as rate of turn and slip indicators, glideslope and localiser pointers, warning flags and command steering bars, to provide flight director or other computer steering information. Basic inputs to the instruments are obtained from the external gyros, radio receivers and flight director or navigation/weapon delivery computers. Astronautics manufactures over 50 different types of AIs and ADIs. These indicators are available in 3, 4, 5 and 6 in sizes with either two or three axes and are installed in a wide variety of military and commercial aircraft. The ADI is available as an independent instrument or as part of an Astronautics Flight Director System (FDS – see separate entry) consisting of an ADI, flight director computer and a Horizontal Situation Indicator (HSI).

The key benefits and features of Astronautics' AI/ADI are:

- Low-power DC servo loops
- 360° roll, pitch and azimuth display capabilities
- Expanded scale pitch displays
- Manual roll and pitch trim
- Localiser pointer
- Flight director command bars
- Collective pitch command bars
- Glideslope pointers and flag
- Integral slip indicators
- Integral rate of turn pointer
- Mode annunciators
- Rising runway presentation
- Special spheroid markings and colours
- Integral electronic buffers for low level signal inputs
- Internal lighting: 5V, 28V MIL Spec white, red or green
- NVIS-compatibility
- Built-In Test (BIT) circuitry.

Status
In production and in service in F-15, F-16, AH-64, UH-60, Hawk, A-10, T-45, F-5, T-38, P-3C, F-16, EF-111, B-212, B-412 and other aircraft.

Contractor
Astronautics Corporation of America.

Avionics Control and Management System for the CH-47D (ACMS)

Type
Flight management and cockpit display system.

Description
The Avionics Control and Management System (ACMS) for the CH-47D helicopter provides the primary interface for the pilot and co-pilot to aircraft pilotage, mission and air vehicle systems. Although some components are essential to more than one function, the ACMS can be divided into the following general functional areas: pilotage subsystem, mission subsystem and air vehicle subsystem.

The pilotage subsystem provides control and display of aircraft sensor data, selection of navigation sources, and other functions essential to flying the helicopter. The EFIS, a primary element of the pilotage subsystem, replaces traditional mechanical gauges, indicators and control heads with electronically generated displays, allowing the operator to select and tailor presentations to provide the optimum level of information during each mission phase.

The mission subsystem provides mission planning utilities, storage and retrieval of navigation reference points, and control of communication and navigation radio lists. Pre-mission planning is normally accomplished on a ground-based workstation and downloaded to the aircraft via a data transfer cartridge. The air vehicle subsystem allows continuous

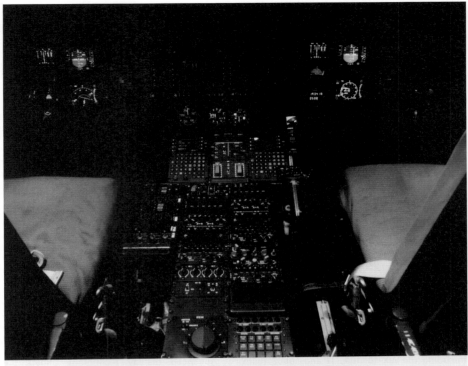

Royal Netherlands Air Force (RNLAF) CH-47 upgraded cockpit

Royal Netherlands Air Force (RNLAF) CH-47 upgraded cockpit 0504273

performance monitoring of all major aircraft systems including warning/caution/advisory annunciations and the capability to interface to the health and usage monitoring system.

The air vehicle subsystem monitors engines and transmissions, fuel and hydraulic systems, flight controls and actuators, the electrical system, and other aircraft systems and equipment. Critical information about the engines and rotors is continuously available on torque/cruise/RPM displays.

System design incorporates redundancy to minimise vulnerability to failures and damage and to enhance the accuracy of navigation equipment. The ACMS interfaces with other aircraft equipment via two dual-redundant MIL-STD-1553B multiplex databusses, special purpose high- and low-speed ARINC 429 buses, and aircraft discrete, synchro and proportional analogue I/O interfaces. The ACMS also receives video signals from the weather radar system for display in the cockpit. A digital map function provides real-time tracking of helicopter position overlaid on paper chart or digital terrain-elevation data displays. Processing and subsystem interfaces are managed by four integrated system processors (two mission, two air vehicle). The primary crew interface is supported by four multifunction displays (two mission, two air vehicle) and two control/display units. Additional information is displayed on two electronic flight instrumentation systems, each comprising of an electronic attitude/direction indicator and an electronic horizontal situation indicator. Two optical display assemblies, which mount onto the pilots' night vision goggles, provide flight symbology during night-time operation. Navigation redundancy is accomplished through use of a Kalman filter to blend navigation sensor data into the most accurate solution of aircraft position and motion, even during transient periods or loss of individual navigation sensor inputs.

The ACMS provides comprehensive integration and control of mission management, air vehicle management and pilotage subsystems. Its flexible architecture will accommodate growth well into the next century. Central computers provide control and data handling capability for a wide variety of aircraft systems and equipment. This control and computation ability decreases crew

workload and enhances navigation accuracy and mission safety. Overall, the ACMS provides the pilot and co-pilot with added functional capabilities, which contribute to a more effective and efficient mission.

Status
In service with the Royal Netherlands Air Force. Two additional export customers have also placed contracts.

Contractor
Honeywell Defense Avionics Systems.

BDI-300A digital Bearing Distance Indicator (BDI)

Type
Flight instrumentation.

Description
The BDI-300A digital Bearing Distance Indicator (BDI) system is designed to complement an integrated cockpit display with traditional VOR/ADF RMI display of bearing, DME distance and compass information. The BDI-303A may be installed as a stand-alone dual-digital RMI, or as part of a system with standby compass when the FXU-803A flux detector and the FXU-803A magnetic compensator are included. System features include:

- Standby compass mode via direct flux detector input
- System status flags and annunciators
- Dual DME distance displays
- Small size and light weight.

Specifications
Dimensions: 106.7 × 86.4 × 190.5 mm
Weight: 1.4 kg
Temperature range: –15 to +55°C
Altitude: up to 15,000 ft
TSO: C-6a, C-66b

Status
The BDI-300A is fitted to the Gulfstream IV and Bombardier Challenger aircraft.

Contractor
Northrop Grumman Corporation, Component Technologies, Poly-Scientific.

BDI-302 digital Bearing Distance Indicator (BDI)

Type
Flight instrumentation.

Description
The BDI-302, a dual-switched RMI with dual-digital DME display, is designed to interface directly with new-generation avionics, transmitting ARINC 429 or Collins CSDB, RS-422 format. The indicator is a 3×4 in (76.2×102.6 mm) form factor.

The indicator is compatible with any high-performance fixed-wing aircraft or helicopter.

Specifications
Dimensions: $106.7 \times 86.4 \times 177.8$ mm
Weight: 1.4 kg
Temperature range: −15 to +70°C
Altitude: up to 15,000 ft
TSO: C-6d, C-66b
Software: DO-178A Level 1
Environmental: DO-160C

Status
The BDI-302 is fitted on the Dassault Falcon 2000 and the Raytheon Hawker 1000 aircraft.

Contractor
Northrop Grumman Corporation, Component Technologies, Poly-Scientific.

The BDI-302 digital bearing/distance/heading indicator
0504205

C-17 Mission computer/display unit

Type
Aircraft mission computer/avionic display system, MultiFunction Display (MFD).

Description
The C-17 computer/display unit, in conjunction with the C-17 multifunction control panel (see separate entry) and other keyboard units, acts as the single point of control for all primary navigation system modes and sensors to control sensors, display sensor and mission computer data, and for input/output control of the mission computer. Display data is available in either 5×7 in (127×177.8 mm) or 7×9 in (177.8×228.6 mm) alpha and numeric forms.

Specifications
Dimensions: $122 \times 146 \times 238$ mm
Weight: 4.32 kg
Power supply: 115 V AC, 400 Hz, 84 W
Reliability: >10,000 h MTBF

Status
In production for the C-17 aircraft.

Contractor
Northrop Grumman Corporation, Guidance and Control Systems Division.

C-17 Multifunction control panel

Type
Aircraft system control, display systems.

Description
The C-17 multifunction control panel serves as the control for data displayed on the multifunction displays and the HUDs. It is used to command the operating modes and display formats, as well as for range selection. It also controls and displays the altitude set function.

Specifications
Dimensions: $160 \times 146 \times 194$ mm
Weight: 3.98 kg
Power supply: 115 V AC, 400 Hz, 35.8 W
Reliability: >10,000 h MTBF

Status
In production for the C-17 aircraft.

Contractor
Northrop Grumman Corporation, Guidance and Control Systems Division.

The multifunction control panel for the C-17
0581516

C-17 Warning And Caution Computer System (WCC)

Type
Aircraft centralised warning system.

Description
Developed for use on the US Air Force C-17 transport aircraft, Northrop Grumman's Warning and Caution Computer system (WCC) accepts discrete aircraft data and data from the MIL-STD-1553B bus connecting specialised peripheral LRUs. The WCC is capable of processing a maximum of 420 discrete inputs although it is currently configured to accept a maximum of 385. It contains dual-redundant MIL-STD-1553B bus communication ports and can function as the bus controller or a remote terminal with automatic switchover occurring in the event of a failure. The WCC processes signals from all sources and manipulates the inputs via Boolean logic to provide warning messages on the warning panel or by lighting annunciator lamps.

Specifications
Dimensions: $190 \times 188 \times 315$ mm
Weight: 8.09 kg
Power supply: 28 V DC, 34 W
Reliability: 20,611 h MTBF

Status
In production for US Air Force C-17 aircraft.

Contractor
Northrop Grumman Electronic Systems, Navigation Systems Division.

C-130 air data and multifunction engine/oil/fuel display system

Type
MultiFunction Display (MFD).

Description
The ISS air data and multifunction engine/oil/fuel system is a quadruple-redundant solid-state system display. The air data system uses four part numbers to replace 19 part units and eliminates 11 Line Replaceable Units (LRUs). Instead of 32 electromechanical engine instruments composed of eight part numbers, the ISS solid-state system consists of 24 indicators composed of just two part numbers (Type I and Type II instrument displays).

System accuracy and redundancy fulfils requirements for RVSM operation in the North Atlantic Free Flight area and metre/foot switching facilities provides for operation in Eastern Europe and other metric airspace areas.

Air data system units are:
One Central Air Data Computer (CADC), which replaces the existing electromechanical True Air Speed (TAS) computer and 12 autopilot and airspeed sensors. It communicates via a dual

MIL-STD-1553B databus and provides height accuracy of 10 ft at 50,000 ft.

Three solid-state airspeed indicators, which provide accuracy of ± 3 kt at > 100 kt. Optional features include: windshear alert with a trend display; take-off monitor; angle of attack display.

Three solid-state altimeters, which are RVSM compliant with accuracy of 30 ft at 30,000 ft.

Three voice annunciating combined altitude alerting units, which provide voice and tone warnings.

The Type I engine display automatically changes function to serve as an RPM indicator, torque indicator, temperature indicator and fuel-flow indicator. The Type II indicator displays five independent oil parameters including pressure,

C-130 aircraft cockpit showing updated ISS multifunction engine/oil/fuel displays
0081871

temperature, quantity and cooler flap settings; the same unit can also display hydraulic indications.

Status
Selected by Lockheed Martin for the C-130H aircraft. The system is catalogued in the US Air Force inventory. It is available for retrofit to older C-130 aircraft.

Contractor
ISS Innovative Solutions & Support Inc.

CCU-800 Cockpit Control Unit

Type
Control and Display Unit (CDU).

Description
The CCU-800 Cockpit Control Unit is designed for use in flight test, trials and certification work. It converts Pulse Code Modulation (PCM) data generated by trials equipment into a display form that allows the pilot to instantaneously verify flight parameters required for safety of flight or to confirm adequate completion of the planned mission.

The CCU-800 is a rugged, sunlight readable, full colour (4,096 colours), high resolution (640 × 480 pixels) 6.4 in liquid crystal display. A microprocessor and PCM demodulator allows third party processor controlled equipment (recorders and control systems) to be controlled during flight, and to present the pilot with real-time data on the status of planned tasks.

The CCU-800 comprises three main items: the CSP-800 Cockpit Switch Panel, the CDP-800 Cockpit Display Panel and the CEU-800 Cockpit Electronics Unit. The CCU-800 is suitable for civil and military use, and is currently in use on a major military flight test programme. Dimensions of the CCU-800 are: 269.2 (H) × 257.3 (W) × 196.9 (D) mm. It meets MIL-STD-810E and MIL-STD-416D.

Contractor
L-3 Telemetry-East.

The L-3 Telemetry-East CCU-800 Cockpit Control Unit   0051693

CD-820 Control Display unit

Type
Control and Display Unit (CDU).

Description
The CD-820 Flight Management System (FMS) CD-820 control display unit is a replacement for the CD-800, CD-810 and CD-815 units.

CD-820 has a large, full colour, Active Matrix Liquid Crystal Display (AMLCD) with an actual writing area of 13.8 in (320 × 240 pixels). CD-820 also interfaces with Honeywell cabin entertainment systems, such as OneView and Airshow.

When used with Honeywell's FMS version 6.0 software, the CD-820 offers additional features such as weather, when interfaced with Teledyne's Telelink, as well as with Honeywell's EGPWS. Single-button selection of video, graphics and Air

The Honeywell Inc, Commercial Aviation Systems CD-820   0081859

Traffic Control (ATC) offers the pilot easy access to camera video, weather and CNS/ATM ATC data.

An additional feature of the CD-820 is incorporation of a new 'fn' key, offering operators an easy way to correct entries into the flight management system, altering flight plans or making immediate updates to ATC requirements.

Specifications
Dimensions: 181 × 146 × 152 mm
Weight: 2.7 kg
Power: 28 V DC
Compliances: TSO C113, DO-160d (environmental), DO-178B level C (software)
Temperature range: −45 to +70°C

Status
In production.

Contractor
Honeywell Inc, Commercial Aviation Systems.

CDU-900 Control Display Unit

Type
Control and Display Unit (CDU).

Description
The CDU-900 is an upgrade of the previous CDU-800 series CDU/CDNU, providing enhanced capabilities. Utilising an Intel 80486 microprocessor and through large-scale integration, the CDU-900 provides powerful built-in processing capability and expansion capacity for embedded functions or interfaces to external equipment.

An embedded military P(Y) code GPS receiver/processor is the first expansion module for the CDU-900. The embedded receiver consists of a GPS Embedded Module (GEM) GEM II or GEM III receiver/processor and accompanying adapters designed to fill two of the three available expansion slots in the CDU-900. Additional expansion modules planned for the CDU-900 include processor and memory modules, various digital, analogue or discrete interface modules for non-MIL-STD-1553B applications and modem/

The Rockwell Collins CDU-900   0581517

datalink modules. The standard and expansion capabilities of the CDU-900 make it suitable for the integration and control of avionics in both fixed-wing aircraft and helicopters.

Specifications
Dimensions: 181 (H) × 146 (W) × 165 (D) mm
Weight: 4.54 kg maximum
Power: 28 V DC, 34 W maximum
Memory: 4 MB flash EEPROM, 1 MB RAM
I/O card: two dual MIL-STD-1553B, ARINC 429 (two independent channels), 16 discrete inputs, 4 discrete outputs
Certification: MIL-E-5400, Class 1A; MIL-STD-461 EMI

Status
The CDU-900 with embedded GPS is integrated with the FMS-800 flight management system in the US Air Force C-5, C-9, E-4B and KC-10. The CDU-900, less the embedded GPS, is integrated in the US Air Force B-1B and KC-135. The CDU-900 is also being fitted to the US Army CH-47D, as part of the upgrade to CH-47F standard.

Latest variant of the CDU-900 family is the CDU-900G which provides a complete capability to meet US DoD GPS Integration Guidelines

(GIG) for stand-alone GPS navigation, including RNAV and airways.

The CDU-900 is also manufactured in Turkey, by Aselsan Inc.

Contractor
Rockwell Collins.

CMA-2082D Avionics Management Systems (AMS)

Type
Avionic Management System (AMS).

Description
The CMA-2082D Avionics Management System (AMS) is a self-contained, high-performance Multifunction Control and Display Unit (MCDU), with onboard mission processing. It integrates navigation sensors and radios, communications radios, displays and other mission avionics with aircraft avionics. Integrated avionics are centrally managed and controlled, reducing pilot workload, and improving pilot situational awareness and mission effectiveness.

The AMLCD VGA display offers a large 4 × 4 in (102 × 102 mm) viewable display area, with a 480 × 480 pixel array and a highly reliable and fault redundant solid-state backlight. Viewing angle is up to ±45°. The system supports both military subsystems through MIL-STD-1553B interfaces and commercial subsystems through ARINC-429 interfaces, as well as other non-standard interfaces. The system can act as a bus controller or remote terminal on a MIL-STD-1553B bus. The MCDU incorporates a powerful Central Processing Unit (CPU) based on an Intel 80486DX or Power PC 8260 with extensive onboard memory. The high brightness colour display is suitable for text, full graphics, and real-time video. The multiprocessor architecture with a dedicated applications processor is well suited for third party software development. Abundant interfaces and spare cards provide ease of integration. An extensive application software library from numerous past programmes reduces programme costs, risks, and schedule.

The CMA-2082D incorporates a comprehensive keyboard comprising 53 data keys, 12 soft (line) keys and two rocker keys. Typical applications for the CMA-2082D include the control of Communication/Navigation/Identification (CNI)

The CMA-2082D avionics management system
1030034

systems, FLIR/radar operating modes, Aircraft Survivability Equipment (ASE), weapons systems and digital maps.

Specifications
Physical
Weight: 5 kg
Power: 28 V DC; 75 W (115 W with heater)
Cooling: Conduction

Signal interface
Standard: MIL-STD-1553B, ARINC-429 (2 out, 4 in), RS-422 × 2
Optional: RS-170A, colour video, analogue or discrete I/O, video output, STANAG 3350, Class B
Spare card slots: 2

Operator interface (keyboard)
Data keys: 32
Rocker keys: 2
Soft keys: 12
Annunciators: 4
Integral lighting: 5 V AC/DC, externally supplied

Display
Type: Full-colour, AMLCD
Pixel array: 480 × 480 RGB × 64 grey shades
Viewing angle: ±45° horizontal, +20 to –10° vertical at CR = 10: 1
Contrast Ratio: 5:1 at 10,000 fc (at optimum viewing angle)
Display luminance: 0.10 to 180 fl
NVIS compatibility: MIL-L-85762A Type 1 Class B

Processor/Software
Applications: Intel 80486DX or Power PC 8260
Graphics/symbol generation: Texas Instruments TMS 34020
Serial interface: Intel 386EX
Applications memory: 4 Mb flash, 2 Mb static RAM
Programme language: ANSI C/Ada95 (application S/W may be written by customer)

Environmental/EMI
Temperature: –40 to +55°C continuous, +70°C for 30 minutes
Humidity: MIL-STD-810 (aggravated)
Vibration: MIL-STD-810 (fixed and rotary wing)
Altitude: to 50,000 ft
EMI/EMC: MIL-STD-461C, Class A1b
RTCA: DO-160C Category C2

Status
In production. The CMA-2082D has been reported to be in service with the German military.

Contractor
CMC Electronics Inc.

CMA-2082M Flight Management Systems (FMS)

Type
Flight Management System (FMS).

Description
The CMA-2082M Flight Management System (FMS) is a self-contained, high-performance Multifunction Control and Display Unit (MCDU), with onboard mission processing. It integrates navigation sensors and radios, communications radios, displays and other mission avionics with aircraft avionics.

The NVIS-compatible flat panel Thin Film ElectroLuminescent (TFEL) display features a large 3 × 5 in (76 × 127 mm) viewable display area, with a 192 × 320 pixel array. Viewing angle is up to ±150°. The system interfaces through a dual-redundant MIL-STD-1553B databus with an optional

ARINC-429 interface for civilian and commercial applications. The MCDU incorporates a powerful Central Processing Unit (CPU) based on 64-bit MPC 8260 with extensive onboard memory.

The CMA-2082D incorporates a comprehensive keyboard comprising 10 soft, 65 independent alphanumeric/mode keys and two rocker keys. Typical applications for the CMA-2082M include the control of Communication/Navigation/Identification (CNI) systems, FLIR/radar operating modes, weapons systems and digital maps.

Specifications
Physical
Dimensions: 146 × 277 ×198 mm
Weight: 5.44 kg
Power: 28 V DC; 45 W (60 W max)
Cooling: Convection/radiation

Signal interface
Standard: MIL-STD-1553B as bus controller, back-up bus controller or remote terminal; RS-422 (BIT)
Optional: ARINC-429, analogue or discrete I/O, non-standard digital, Ethernet, custom
Spare card slots: 4, each providing 154.8 cm²; ARINC-429

Operator interface (keyboard)
Alphanumeric keys: 65
Rocker keys: 2
Soft keys: 10
Annunciators: 3 incandescent, NVIS compatible
Integral lighting: 5 V AC/DC LED, NVIS compatible, externally supplied

Display
Type: Flat panel Thin Film Electro Luminescent (TFEL)
Resolution: 64 lines/in
Display capacity: 20 lines of 21 characters
Colour: NVIS yellow, peak at 575 nm
Viewing angle: ±150° all axes, ±45° at bezel edges

CMC Electronics' CMA-2082M Flight Management System (FMS) (CMC Electronics)
1034676

Sunlight readability: Readable in 10,000 fc incident
MTBF: >5,000 operating hours (MIL-HDBK-217E)
NVIS compatibility: MIL-STD-3009 Type 1 Class A

Processor/Software
Applications: 64-bit MPC 8260
Graphics/symbol generation: Intel 82786 graphics co-processor
Memory (on CPU card): 6 Mb static RAM, 2 Mb UV RAM (programmable via MIL-STD-1553B), 32 Mb flash EPROM; (additional via card slot)
Programme language: Ada, C, Assembler
Software: Capable of downloading application software via MIL-STD-1553B or RS-422; other software to customer spec or by CMC Electronics

Environmental/EMI
Compliance: MIL-STD-461, -462, -5400T Class 1a

Status
In production. The CMA-2082M has been selected for the US Army for its fleet of U-60M helicopters.

Contractor
CMC Electronics Inc.

CMS-80 cockpit management system

Type
Avionic display system.

Description
The CMS-80 cockpit management system family consists of hardware and software building blocks. The baseline CMS-80 consists of a full complement of Rockwell Collins avionics plus selected equipments from other manufacturers.

The CMS-80 unclutters the cockpit by removing individual controls which normally crowd the panel. These are replaced with centrally located control and display units which use a standard screen layout and human interface to control all equipment. System interconnection is via dual-redundant MIL-STD-1553B multiplex cables. The CMS-80 can also provide mission computing and navigation integration.

A CMS-80 option is weapons and sight integration and control. This feature has been implemented on the B-406CS Combat Scout demonstrator and the MD-530 NOTAR helicopter. A single keystroke shifts the CDU from avionics to weapons control and all options for guns, rockets and missiles can be selected and managed. Weapons selection and firing can also be accomplished from the handgrips, allowing the pilot to keep his attention focused outside the cockpit.

Status
In service. The US Army selected the CMS-80 for Special Operations (SO) and Special Electronic Mission Aircraft (SEMA). In addition, the CMS-80 was selected for the AH-64A Apache and for three US Navy/Marine Corps helicopters including the AH-1W, CH-46 and UH-1N.

Various versions of the system have been installed in Sikorsky HH-65A and HU-25A helicopters, together with US Coast Guard Lockheed C-130 aircraft. It was also selected for US Air Force A-10A aircraft.

Historically, in December 1981, Delco Electronics ordered the CMS-80 under a USD15 million contract to support its commitment to provide 300 sets of Fuel Saving Advisory System (FSAS) equipment to the US Air Force, part of a programme to upgrade the fleet of Boeing KC-135s. Deliveries for this application began in January 1983 and ended in 1986. The FSAS system on the KC-135 is expected to save between 2 and 4 per cent of the fuel used by advising the crew of the most efficient speed, engine pressure ratio, altitude and descent profile.

In 1988, the system provided the baseline for the next-generation FMS-800 flight management

system for the German Air Force C-160 Transall autonomous navigation system. In this application the CDU provides crew interface for communications, navigation and IFF, as well as flight management functions.

Contractor
Rockwell Collins.

Cockpit 21 for T-45C Goshawk

Type
Avionic display system.

Description
Cockpit 21 was developed to replace the T-45A's analogue displays with digital displays similar to those found in the US Navy's F/A-18, AV-8B Harrier II and other advanced carrier-based jets. Aircraft equipped with Cockpit 21 are designated T-45C. Since students transitioning to these aircraft from Cockpit 21 will have already mastered cockpit information management skills and situational awareness, they can concentrate on the primary mission of learning how to perform key tactical manoeuvres.

Smiths Industries Aerospace has performed the full systems integration on Cockpit 21 and supplies the Head-Up Display. Cockpit 21 comprises a Display Processor Unit (DPU), Pilot-Display Unit (PDU), and Data Entry Panel (DEP). The DEP drives five display surfaces including the PDU and four raster Head-Down Displays (HDD) (monochrome multifunction displays sourced by Elbit Systems Ltd). The new cockpit provides navigation, weapons delivery, aircraft performance and communication data to both stations in the two-seat cockpit. The cockpit also includes a Global Positioning System/Inertial Navigation Assembly and a multiplex databus

that will allow expansion of cockpit capabilities to accommodate changing training requirements.

Status
The first Cockpit 21 aircraft was delivered in October 1997 for testing of the production configuration at US Naval Air Station Patuxent River. The second T-45C delivered in December 1997 went to US Naval Air Station Meridian for Training Wing. US Navy plans call for the existing 72 T-45A Goshawk aircraft with analogue cockpits to be upgraded to the T-45C Cockpit 21 digital configuration. Smiths Industries will provide Boeing with 103 new digital Cockpit 21 assemblies by 2004. Retrofit kits for 84 T-45As are planned.

Contractor
The Boeing Company, Information and Electronic Systems Division.

Cockpit 4000

Type
Avionic display system.

Description
CMC Electronics' Cockpit 4000, as applied to the Raytheon T-6B advanced turboprop training aircraft, is anchored by two FV-4000 mission computers (see separate entry), each employing a 500 Mhz G4 PowerPC processor and compact PCI/PMC modules. The SparrowHawk™ Head Up Display (HUD), including rear cockpit HUD repeater and camera, together with the six Multi Function Displays (MFDs) in the front and rear cockpits of the T-6B are driven directly from the FV-4000 computers, thereby providing a powerful and flexible architecture for the centralised display and control of navigation and mission

Cockpit 21 in the T-45C Goshawk   0018164

The Cockpit 21 system is to be retrofitted to US Navy T-45A aircraft   0527074

Raytheon Texan T-6B turboprop trainer (Jane's/Patrick Allen)   0583312

CMC Electronics' Cockpit 4000 in the Raytheon T-6B turboprop training aircraft
(Jane's/Patrick Allen)   0583313

data, including the Primary Flight Displays (PFDs), digital map, stores management, and the Engine Indication and Crew Alerting System (EICAS). Control of all navigation, mission and communication functions is via Up-Front Control Panels (UFCPs) and Hands-On Throttle And Stick (HOTAS) inceptors, providing a highly realistic training environment for pilots training for the current crop of multirole fighter aircraft.

The FV-4000 open architecture mission computer is available in 3 and 5 MCU versions and incorporates built-in video processing, HUD symbol generator and driver and interfaces for avionics subsystems. The MFDs are available in 4 × 5, 5 × 7 and 6 × 8 in formats.

Integrated subsystems include INS/GPS, communication radios, Air Data Computer (ADC), Armament Interface Unit (AIU) and front/rear cockpit tandem controls.

Other system features and benefits include:
- Support for MIL-STD-1553, ARINC-429 and Ethernet buses
- RS-422, analogue, discrete and synchro interfaces
- Redundant system architecture
- Proven Operational Flight Programmes (OFPs)
- Full EICAS
- 4-D navigation
- Communications
- Weapons delivery and training
- Simulated air target
- No-Drop Bombing System (NDBS)
- Stores Management System (SMS)
- Digital map
- 'Dumb' MFDs driven by mission computer (fewer LRUs)
- Functionality added by software
- Built-in processing and interface growth
- CNS/ATM compatibility.

As an open architecture mission system, Cockpit 4000 provides a powerful training platform that includes simulated weapon delivery capabilities and can easily accommodate upgrades such as

embedded Air Combat Manoeuvring Installation (ACMI) with virtual radar and Electronic Warfare (EW) simulation, thus facilitating growth and protecting against system obsolescence.

Specifications
See relevant entry for system component.

Status
In production for the Raytheon T-6B. Under their terms of the agreement with Raytheon Aircraft, CMC Electronics will deliver avionics suites consisting of FV-4000 mission computers, a SparrowHawk™ Head-Up Display (HUD) and repeater, stores management and Multi Function

Displays (MFDs). A variant of Cockpit 4000, including three 5 × 7 in and HUD, is also specified as part of Aermacchi's M-311 basic/advanced trainer.

Contractor
CMC Electronics Inc.

Cockpit display system for the F-15E

Type
Avionic display system.

Description
Kaiser Electronics produces the head-up/head-down displays for the F-15E aircraft. The front cockpit has two 6 × 6 in (152 × 152 mm) CRT and one 5 × 5 in (125 × 125 mm) Active Matrix Liquid Crystal Display (AMLCD) head-down MultiFunction Displays (MFDs) and one holographic, Wide Field of View (WFoV) Head-Up Display (HUD), while the rear cockpit has two 6 × 6 in (152 × 152 mm) CRT and two 5 × 5 in AMLCD displays. The larger displays are used for tactical displays (radar, datalink, sensor video), with the smaller displays configured for navigation and pilotage displays (AI, HSI, systems).

By presenting both high-resolution, high-contrast video and fine line, fast-writing stroke symbology, the CRT MFD has the capability to fulfil a wide range of requirements and is compatible with almost any system configuration. High-resolution video from E-O sensors, radar and missiles in either 525- or 875-line rates is fully readable in the high ambient light of the tactical cockpit. Pure stroke modes provide the high information content necessary for such displays as tactical situation displays and JTIDS. The hybrid stroke during retrace mode allows high-resolution stroke symbology to be placed on top of raster displays. A stroke converter has been added to the AMLCD MFD to convert the analogue stroke signals into digital format suitable for display.

The MFD has nine plug-in subassemblies, all replaceable without harmonisation. It is fully equipped with continuous and initiated BIT. Software cueing on the display surface provides in-flight programmability through bezel-mounted push-buttons.

The HUD employs both raster scan and stroke written symbology to accommodate the FLIR imagery from the Lockheed Martin LANTIRN system and the Raytheon Electronic Systems AN/APG-70 radar (see separate entries). The Kaiser WFoV HUD employs a holographic single combiner glass, permitting a smaller and lighter support structure with less obscuration of forward view.

The holograms for this HUD are made by an associate company, Kaiser Optical Systems Inc.

Rear cockpit of a Boeing F-15E, showing two centrally mounted 6 × 6 in CRT displays flanked by two 5 × 5 in AMLCD displays (E L Downs)   1129580

Specifications

6 × 6 in MFD
Dimensions: 345.9 × 190.5 × 196.8 mm
Weight: 10.42 kg
Power supply: 115 V AC, 400 Hz, 3 phase, 180 W
Display: 152 × 152 mm
Resolution: 100 line pairs/in
Reliability: 3,000 h MTBF

Status

In service on the F-15E.

Contractor

Rockwell Collins Kaiser Electronics.

Cockpit Displays for the AH-64D Apache

Type

Avionic display system.

Description

The Honeywell 159 × 159 mm (6.25 × 6.25 in) mid-sized Active Matrix Liquid Crystal Displays (AMLCDs), while designed originally for the AH-64D Longbow Apache, are applicable to all tactical combat helicopters.

In the AH-64D, the display suite comprises four colour MultiPurpose Displays (MPDs) and two COlour display Processors (COPs). The MPD acts as a video monitor and all graphics processing is provided by the COP. This 512 × 512 resolution display is fully NVIS compliant and uses a patented dimming approach to achieve >4,000: 1 dimming on a standard hot-cathode, serpentine lamp.

The MPD is capable of displaying graphics, video, or both simultaneously. A Digital Video Interface (DVI) facilitates real-time presentation of mission-critical targeting/pilotage FLIR and digital map.

Specifications

Dimensions: 216 × 216 × 184 mm (W × H × D)
Weight: 6.1 kg
Viewable area: 159 × 159 mm (6.25 × 6.25 in)
Resolution: 512 × 512 pixels
Luminance (white): >250 fL
Contrast ratio:
(high ambient) >5: 1
(low ambient) >100:1
Viewing angle:
(horizontal) ±25°
(vertical) +15 to +30°
NVIS compatibility: Class B
Video levels: 64 per primary colour
Video interface: 330 Mbps serial digital, redundant
Video processing: external
Interface: RS422, redundant
Operating temperature: –40 to +71°C
Cooling: integral fan
MTBF: 9,900 h
Power: 115/300 W (without/with heater)

Status

In production and in service with the AH-64D Longbow Apache attack helicopter.

Contractor

Honeywell, Defence Avionics Systems.

Cockpit Multifunction Displays and Display Electronics Units (MFDs/DEUs) for V-22 Osprey aircraft

Type

MultiFunction Display (MFD).

Description

The MFDs/DEUs are ruggedised, full colour, 6 × 6 in, beam index CRT cockpit displays and display electronics units for the V-22 Osprey aircraft. The suite comprises four Multifunction Displays (MFDs) and two Display Electronics Units (DEUs) per aircraft. The displays are NVG-compatible and provide graphics/symbology overlays on

The front (weapons systems operator) cockpit of the WAH-64D also incorporates two AMLCD displays (on the right display, three people can be clearly seen on the FLIR picture). The left display is obscured by the weapon aiming yoke (E L Downs)   0126993

The pilot's cockpit of a UK WAH-64D Longbow Apache. Note the two AMLCD displays (E L Downs)   0126992

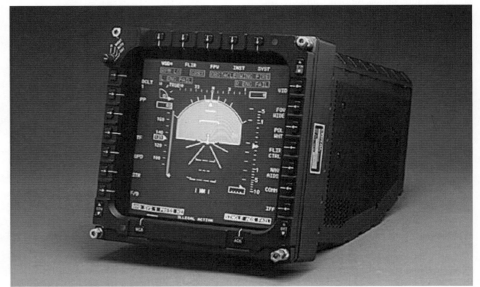

L-3 multifunction display for the V-22 Osprey aircraft   0018167

L-3 display electronics unit for the V-22 Osprey aircraft   0018166

Specifications

	Model 104	Model 550	Model 570	Model 640
Dimensions	8.3 × 6.2 in	4.4 × 3.3 in	4.5 × 3.5 in	5.1 × 3.8 in
Design purposes	split screen HSI, ADI, video engine data warnings	HSI, ADI, video	EFI, HSI, ADI	HSI, ADI, video
Grey scales	64	256		64
Pixels	640 × 480 standard 800 × 600 available 800 × 600 available	640 × 480	640 × 480	640 × 480
Cooling	fan-forced cold wall construction	passive	passive	fan-forced cold wall construction
MTBF	7,500 h	8,200 h	8,000 h	8,200 h

Model 104 AMLCD, MFD, video display unit   0018176

digital map, FLIR and radar sensor video inputs. There are three independent channels of high-performance display processing.

Status
Boeing EMD contract awarded 1994; LRIP contract awarded 1996.

Contractor
L-3 Communications, Display Systems.

Cockpit Voice Recorder (CVR) audio mixer

Type
Flight/mission recording system.

Description
The Cockpit Voice Recorder (CVR) audio mixer is designed to comply with FAA regulations mandating cockpit voice recorders for all multi-engined turbine-powered aircraft that require two crew members and seat six or more passengers. The CVR audio mixer sums and routes audio signals from microphones, headphones and speakers to the cockpit voice recorder. It also provides hot microphone biasing, adjustable channels for balancing audio levels and paired pins to make installation easy and economical. Both single- and dual-station units are available. The systems are qualified to TSO C50c requirements.

Status
In service in various business aircraft.

Contractor
AVTECH Corporation.

Colour AMLCD multifunction indicators

Type
Multi Function Display (MFD).

Description
Universal Instrument Corporation manufactures a range of AMLCD, full-colour, NVIS-compliant, MultiFunction Display (MFD), military-qualified indicators; leading specifications are outlined in the table:

Status
In production and in service in military aircraft. In February 1998, Universal Instrument Corporation, as a subsidiary of Universal Avionics Systems Corporation (UASC), entered into a business alliance agreement whereby UASC will have exclusive rights to market flat-panel integrated displays manufactured by Universal Instrument Corporation to corporate and commercial aviation. The displays will be integrated with Universal flight management systems to produce a new line of flat-panel flight displays (4 × 5, 5ATI, 5 × 6 and 8 × 10 in sizes) for a new-generation of avionic suites. The design is intended for installation in new and retrofit applications, displaying primary flight and navigation data, multifunction data and engine data.

Portuguese Air Force C-130H cockpit upgraded with six Universal Instrument Corporation AMLCD indicators   0018175

Contractor
Universal Instrument Corporation.

Colour Cockpit TV Sensor (CCTVS)

Type
Flight/Mission recording system.

Description
BAE Systems North America CCTVS updates the company's Cockpit TV System (CTVS), of which over 16,000 have already been sold to 28 countries and over 30 different aircraft types.

The CCTVS is an exact form, fit and function replacement for the current monochrome Heads-Up Display (HUD) camera.

CCTVS cameras provide higher-resolution colour imaging sensor and recording capability, with expanded low light night performance.

Specifications
Pointing accuracy: ±0.56 mrad (at factory)
Light levels: 0.05–16,000 fL (both ALC and AEEC)
Resolution: >470 TVL/PH(); >350 TVL/PH(v)
MTBF: 32,000 h

F/A-18 low-light colour camera configuration   0015331

Status

In production and in service in F-14D, F-15E, F/A-18 (3 per aircraft), MB-339 and T-45 aircraft.

Contractor

BAE Systems North America.

Colour MultiFunction Display (MFD) for C-17

Type

MultiFunction Display (MFD).

Description

Honeywell's 6 × 6 in (152 × 152 mm) colour MultiFunction Display (MFD) is a high-resolution, shadow-mask CRT display developed for the C-17 military air transport. Honeywell's newest family of full-colour, military-qualified displays includes a number of advanced technologies to provide significant performance advantages. Superior vibration tolerance, sunlight viewability and high-resolution graphics are inherent in the design of the display.

The MFD can be used to present primary flight and navigation information, colour weather radar, digital map information, engine instrumentation and tactical display formats. The unit presents stroke, raster or hybrid formats in 16 colours. Raster images are driven from sensor inputs with an RS-170 or RS-343 interface.

Honeywell's MFD has been designed as a stand-alone 'smart' unit. The display contains a MIL-STD-1750A processor to convert stored modes into formats using aircraft parameters received over the MIL-STD-1553B bus. An internal vector generator is used to draw the formats on the screen. The unit also includes an expanded built-in test capability and can report failures to the host computer via the MIL-STD-1553B bus.

A separate display processor is not necessary when linking the 'smart' MFD directly to the aircraft's mission computer via the 1553B bus. This configuration reduces system weight, reduces potential interface problems and increases system reliability.

Specifications

Dimensions: 203 × 203 × 381 mm
Weight: 16.8 kg
Resolution: 799 × 820 pixels

Status

The new MFD will be introduced into the 71st of 120 production C-17s. MFD production in excess of 200 units was completed during 2003.

Contractor

Honeywell Defense Avionics Systems.

Colour video HUD cameras

Type

Flight/mission recording system.

Description

Photo-Sonics manufactures a family of colour video HUD (Head-Up Display) cameras for use on the A-4M, A-10 and F-16C/D aircraft; detail specification changes match the cameras to each aircraft type. All cameras utilise a mix of MIL-STD and COTS components; they incorporate extensive filtering to provide noise-free, high-quality imagery, with automatic exposure control. The camera assembly, comprising periscope and colour camera, is mounted on the HUD to record what the pilot is viewing and the symbology on the HUD. Data is output to an onboard video cassette recorder.

Specifications

Horizontal resolution: (NTSC) 470 TV lines (PAL) 460 TV lines
Picture elements: (NTSC) 768 (H) × 494 (V) (PAL) 752 (H) × 582 (V)
Power: 115 V AC 47–440 Hz
Weight: 1.9 kg

Optical specifications

	Lens, focal length	FOV degrees		FOV milliradians	
		H	V	H	V
A-4M	16.0 mm, f1.4	22.6	17.1	395	298
A-10	25.0 mm, f2.5	14.6	11.0	255	191
	16.0 mm, f1.4	22.6	17.1	395	298
	15.0 mm, f1.4	24.1	18.2	420	317
F-16C/D	16.2 mm, f1.4	22.4	17.1	390	294.1

Status

In service.

Contractor

Photo-Sonics Inc.

Command and control display system

Type

Airborne battle management system.

Description

The command and control display system provides the man/machine interface between the mission crew and the sensors and communications systems in the Boeing E-3A AWACS. It maintains the display database, filters and positions the selected data and generates all graphics, alphanumerics and sensor target reports with real-time responses. Situation Display Consoles (SDCs), each comprising a 19 in (482.6 mm) CRT MIL-SPEC colour monitor, data entry, filtering and control panels, trackerball and keyboard with associated processing electronics, provide the mission crew with all display and control features required to carry out surveillance, weapons direction and battle staff functions. Data Display Indicators (DDIs) with monochrome monitors support the communications, maintenance and data processing functions of the mission crew. The E-3A has 14 SDCs and 2 DDIs.

The SDC presents the appropriate colour pictorial representation of the situation required to support the function assigned to the SDC by the operator. Using the high electro-optical qualities of the NDI monitor to achieve high legibility of dense data presentations under all operating conditions, the pictorial contents range from individual symbols indicating only sensor type and target positions to a combination of symbols and tabular notes that display such information as target type, speed, direction of flight, bearing, mission and altitude. Supporting tabular data, also in colour, is presented in the lower 20 per cent of the display surface. From this data, and from background pictorial information such as maps, landmarks and unsafe areas, the SDC operator can determine the appropriate responses to developing situations. The mission crew can also configure the SDCs in flight to serve as battle staff, surveillance or weapons consoles.

Specifications

Power supply: 115 V AC, 400 Hz, 3 phase
Temperature range: –54 to +55°C
Altitude: up to 40,000 ft
Environmental: MIL-E-5400 Class 1

The E-3A AWACS command and control display system has 14 SDCs and 2 DDIs   0581514

Status

In service in US Air Force Boeing E-3A AWACS and Royal Saudi Air Force aircraft, and through a technology transfer on the NATO E-3 AWACS. Selected for the Japanese Air Self-Defense Force Boeing 767 AWACS fleet.

Contractor

BAE Systems North America, Advanced Systems, Greenlawn.

Common Airborne Instrumentation System (CAIS)

Type

Flight/mission recording system.

Description

The CAIS was developed under the auspices of the US DoD to promote standardisation, commonality, and interoperability for flight testing.

The central characteristic of CAIS is a common suite of equipment used across service boundaries and in any airframe or weapon system testing.

CAIS products comprise airborne equipment items, that can be configured to meet project requirements, and comprehensive ground facilities to support the airborne effort:

- MDAUs: Miniature Data Acquisition Units to support both analogue and digital data acquisition
- PMU-700-C5: Programmable Master Controller Unit
- PBC-800: Programmable Bus Controller
- PCU-800C: Programmable Conditioning Unit
- MPC-800C: Miniature Programmable Conditioner
- MiniARMOR-700: High-Speed Multiplexer
- ATD-800-II: Airborne Tape Deck
- CCU-800: Cockpit Control Unit
- GSU-800: Ground Support Unit
- CBE-850: CAIS Bus Emulator
- Lab ARMOR-715: High-Speed Demultiplexer

CCU-800 cockpit control unit   0002348

PMU-700-C5 programmable master controller unit   0002347

Representative CAIS system   0002349

Status

CAIS is used in the F/A-18E/F and F/A-22 aircraft programmes. L-3 Telemetry-East is the instrumentation system integrator for the F/A-22 programme, and is responsible for delivering a complete turnkey system.

L-3 Telemetry-East was contracted by Boeing Aircraft, Missile Systems Division to provide flight test instrumentation for its failed X-32 Joint Strike Fighter (JSF) concept demonstration programme during 2002.

Contractor

L-3 Telemetry-East.

Control Display System (CDS) for OH-58D Kiowa Warrior

Type

Control and Display Unit (CDU).

Description

Honeywell's Control Display System (CDS) provides embedded controls and displays for communication (ETICS), navigation, engine/power-train, Mast-Mounted Sight (MMS) and weapons, as well as Rotorcraft Map System (RMS – digital map) and Video Image cross-Link (VIXL).

The ETICS (Embedded Tactical Information Control System) (see Communications section),

is part of the Honeywell CDS, and it embeds integrated Task Force XXI Variable Message Format (VMF) capabilities for command and control, fire support and situational awareness.

The rotorcraft mapping system provides Kiowa Warrior crews with situational awareness by translating data from ETICS and the Honeywell Embedded Global positioning system/Inertial navigation system (EGI) into icons that report the aircraft's position in relation to the terrain and other air and ground vehicles, both friend and foe.

The VIXL transmits near-realtime reconnaissance video images from the Kiowa Warriors mast-mounted sight over standard combat radio links to other Kiowa Warriors or ground command and control centres. In addition, Honeywell is also providing the SINCGARS/SIP (Single-Channel Ground-Air Radio System/System Improvement Program) radio, and EGI system as part of the overall Kiowa Warrior upgrade. The display system itself comprises two monochrome multifunction displays (one for each pilot), and the radio frequency display.

Status

In service.

Contractor

Honeywell Defense Avionics Systems.

The Kiowa Warrior cockpit, showing the Honeywell Defense Avionics Systems multifunction displays, and the radio frequency display   0018170

Core Integrated Processor (CIP)

Type

Aircraft central computer.

Description

BAE Systems' Core Integrated Processor (CIP) for the C-17A Globemaster III is the central computer that controls all of the aircraft's avionics systems. The computer provides all existing mission computer functions and has growth capability to interface with, and provide data management for aircraft controls, displays and sensor data.

The CIP is a high-speed general purpose computer designed to meet the real-time processing requirements of the C-17 application. The baseline CIP is housed in a single Line Replaceable Unit (LRU) containing four subassemblies. The LRU is a full ATR cross-section, 362 mm in length, providing 10 Versa Module Eurocard (VME) module slots. The MIPS R4400 was selected as the Central Processing Unit (CPU) based on off-the-shelf multisource availability, performance and the error detection and correction capability. The VME backplane is used as the intermodule communication channel providing up to 80 Mbytes/s bus bandwidth. A fundamental requirement for the CIP design is open systems such as VME, POSIX, UNIX and VxWorks for both hardware and software. The entire software approach promotes open architectures which feature industry standards and off-the-shelf solutions for operating systems, kernels, programming languages and software development environments.

Two units will be installed on each C-17, replacing the existing three unit mission computer. The CIP will be retrofitted on existing C-17s and will be installed on all future C-17 aircraft.

Status

In production and in service on the Boeing C-17.

Contractor

BAE Systems Controls.

C-17 and CIP   0001287

CP-1516/ASQ (ATHS) and CP-2228/ASQ (ATHS II) Automatic Target Hand-off Systems

Type

Aircraft mission computer.

Description

Rockwell Collins' Tactical Data Manager (TDM) family of battlefield mission computers is used in conjunction with standard communication transceivers to provide a digital communications network. The C^3I network enables command and firing element crews to manage resources and exchange target and other mission essential information, using a short data burst rather than voice communications. Data-burst transmissions minimise the possibility of jamming and lessen the probability of detection, while increasing the transfer rate of accurate battle information.

The Rockwell Collins CP-1516/ASQ automatic target hand-off system and control/display unit
0504027

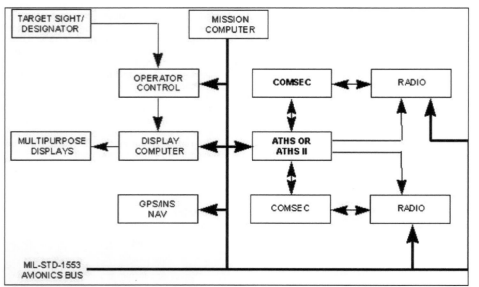

ATHS/ATHS II TDM interface diagram
0105280

avionics, including digitally generated map displays, are completely compatible with the CP-1516.

The computer within the CP-1516 incorporates 8 kbytes of RAM, 2 kbytes of EPROM and 196 kwords of program memory. The system is compatible with MIL-STD-1553A and B databusses.

CP-2228/ASQ Tactical Data Modem (ATHS II)

The CP-2228/ASQ Tactical Data Modem (ATHS II, also known as the TDM-200) is an upgraded and improved version of the CP-1516/ASQ ATHS. ATHS II was developed to provide additional capabilities to meet the more stringent environmental requirements of fighter and close air support aircraft, while also meeting the needs for future datalink applications.

ATHS II is capable of transmitting and receiving FSK from baud rates of 75 to 1,220 and digital data from 75 to 16,000 bits/s. This higher frequency operation dramatically reduces transmission time, thus making it more difficult to detect and jam. It has four ports and up to four modems, which can simultaneously transmit or receive messages.

The computer within ATHS II is a Z80180 with 256 kwords program memory with potential growth to 1 Mwords. The computer also incorporates 64 kbyte RAM and 8 kbyte non-volatile memory. Mission data and operational flight program data may be programmed via the MIL-STD-1553 databus or a digital data loader.

Over 750 ATHSs are installed on various platforms which include the AH-64, OH-58D, JOH-58, ANG F-16, MH-60 A/K, MH-47 D/ E, RAN SH-70B and Belgium Aeromobility A-109 aircraft.

The newest member of the TDM family, the ATHS II entered production in 1997. It is fit and form backward compatible with earlier CP-1516/ASQ models.

ATHS II standard configuration consists of two modems. Up to two additional modems can be added in prewired card slots.

Status

ATHS is in production and in service on a wide variety of US and foreign aircraft.

Contractor

Rockwell Collins.

CP-2108A (3007A) data controller/Mission Computer (MC)

Type

Mission Management System (MMS).

Description

The CP-2108A (3007A) data controller/Mission Computer (MC) was designed to satisfy the combined requirements of the MC-130E Combat Talon 1, the AC-130H gunship and other programmes. Using a single part number MC suitable for multiple missions greatly reduces spares and logistics support. Level 1 test equipment requirements are also minimised by the use of extensive BIT for fault detection and isolation.

The computer employs a dual-CPU architecture. Each processor consists of a high-speed MIL-STD-1750A CPU and a dedicated local memory. The amount of local memory can be varied to suit the application.

The MC has 512 kwords of installed memory, 2.6 Mips of processing capacity and extensive I/O capability high accuracy. There is built-in memory growth to the million word limit of MIL-STD-1750A, multiplexer growth to permit a third dual-channel of MIL-STD-1553B and additional analogue and digital I/O. The unit is mechanised with several blank circuit cards that can be populated with either currently available card designs or new I/O. The 3007A MC is therefore suitable for a variety of other applications.

A plug-in CPU replacement has been developed by Marconi North America which allows one

ATHS and ATHS II Features Comparison and Specifications

Operational features	CP-1516/ASQ (ATHS)	CP-2228/ASQ (ATHS II)
Modulation format	FSK	Same plus digital baseband/diphase, and DCT FSK tone sets
Transmission rates	75, 150, 300, 600, 1200 b/s	Same plus 5k, 8k, 9.6k, and 16 kb/s digital
Radio interfaces	4 ports, 1 modem	4 ports, 2 to 4 modems[1]
TEMPEST	Yes	Yes
Host vehicle interface	MIL-STD-1553B	Same plus MIL-STD-1553A
Programming language	PLM	C
Reprogramming method	Depot, via card edge	On-aircraft, via 1553 bus
Physical characteristics		
Power	35 W	17 W
Weight	4.54 kg	4.54 kg max
Size	137.2 × 167.6 × 203.2 mm	Same
Mounting	Hard mount	Same
Cooling	Convection	Same
Spare card slots	None	Four (If 4 modems implemented)

The TDM units are designed for use with the Rockwell Collins' CMS-80 and FMS-800 Avionics Flight Management Systems, which are capable of integrating with complete avionics systems, including COMM/NAV and weapons control.

CP-1516/ASQ Automatic Target Hand-off System (ATHS)

The CP-1516/ASQ Automatic Target Hand-off System (ATHS) is a battlefield mission management system. It is used in conjunction with a control and display unit and up to four standard HF, VHF or UHF radios to provide a tactical Command Control, Communications and Information (C³I) network. The digital communication network can provide for stores management, target handovers and other similar functions to be passed to airborne, artillery and ground forces.

The CP-1516 features a recall capability for 12 previously received messages and allows the transmitting of preformatted messages or free-text messages using an alphanumeric keyboard. Non-volatile memory in the unit retains all critical information in the event of a power loss. In addition, the CP-1516 maintains the current status of up to 10 active airborne missions and two preplanned missions.

Various control/display unit options are available for data entry and display. The CP-1516 is fully compatible with the AH-64 Apache data entry panel and TADS/PNVS display, the Bell OH-58D control/display and mast-mounted sight display and combat helicopter control/display unit.

Modern electronic battlefield systems including SINCGARS, E-PLRS/JTIDS hybrid (PJH), Tacfire communications, COMSEC and all MIL-STD-1553

or both of the MIL-STD-1750A processors to be replaced by a 25 Mips 32-bit RISC processor. This increases total available throughput to over 50 Mipsfor applications such as digital terrain and digital map database systems, knowledge-based system health monitoring, threat correlation and so on. A corresponding increase in memory to 68 Mbytes can also be supported. A 1/2 ATR variant of the 3007A has also been developed which uses the same dual-CPU configuration.

To assist in developing and debugging software on the 3007A, a Computer Support System (CSS) has been developed. The CSS includes mainframe-based flight software development tools and utilises a micro-VAX host computer for real-time programme debug and validation. An IEEE interface is provided for memory loading and verification.

Specifications
Dimensions: 241 × 330 × 478 mm
Weight: 35.5 kg
Power supply: MIL-STD-704, 356 W (+130 W blower)
CPU: dual MIL-STD-1750A
Throughput: 2.6 Mips (DAIS), growth to 6 Mips
Memory: 512 k × 16 words with growth to 2 million words
Input/output: two MIL-STD-1553B dual-mux channels, approximately 100 discrete, various digital/DC, digital/AC, synchro/digital, digital/ synchro, digital resolver and serial digital channels
Environmental: MIL-E-5400 Class 1A Category III

Status
In production and in service in US Air Force AC-130H gunship and MC-130E Combat Talon I aircraft.

Contractor
BAE Systems North America.

Data Nav V navigation/checklist display system

Type
Multi Function Display (MFD).

Description
Honeywell's Data Nav V turns any Honeywell colour weather radar screen into a versatile en route navigation map or aircraft checklist. It allows pilots to select navigation waypoints and instantly create new waypoints by positioning an electronic designator 'bug' anywhere on the radar screen.

The Data Nav systems are designed to add navigation map and/or checklist display capability to all Honeywell colour weather radar indicators. The latest version, Data Nav V, features an ARINC 429 interface with a variety of Flight Management System (FMS) and long-range navigation systems, enhanced on-screen information and simplified pilot operation. Operators can compile their own aircraft-specific checklists on a personal computer and load them quickly and easily into the Data Nav computer, using the new optional Honeywell Standard Checklist software program.

The ARINC 429 interface links the Data Nav V system, not only with the Honeywell FMZ-2000 FMS system, but those of other makes that provide ARINC 429 outputs, as well as with certain other long-range navigation systems. The FMS or long-range navigation system computer will lay out a flight plan, convert it to ARINC 429 digital data format and transmit it to the Data Nav V computer, which generates navigation map, course, heading, latitude-longitude position and waypoint symbology for display on the Honeywell weather indicator. In addition, Data Nav V, using FMS or nav system inputs, will display VORTAC locations, track lines, projected course, distance to waypoints and ground speed.

En route, the navigation display creates a moving picture of the aircraft's position as preselected courses are flown. Combined with the radar weather display, it gives a complete visual presentation of selected course and weather conditions ahead.

Data Nav V presents a full menu - up to 200 custom preprogrammed pages of normal and emergency preflight procedures and cockpit checklists, together with performance data, operating notes and other selected data.

The Data Nav V Electronic Checklist simplifies the routine checklist procedure and permits quick retrieval of emergency checklists, reducing the chance of errors and oversight by giving positive indication of a checked procedure. Because the system remembers checklist position, a pilot may instantly switch back and forth between weather/map display and the previously selected checklist.

Data Nav V provides nav map and checklist capabilities for both EFIS and non-EFIS equipped aircraft.

It can be used with a two or four-tube EFIS system, where the Honeywell radar indicator serves as an additional display to enhance overall system capability.

The Data Nav V system is available in three configurations, with panel-mounted controllers for selecting nav map with weather only, weather and checklists only, or either map/ weather or checklist displays. Data Nav V is a direct replacement for the Honeywell Data Nav III system, when using the most common FMS interfaces.

Status
Data Nav I, II and III no longer in production. Replaced with Data Nav V, in production.

Contractor
Honeywell Inc, Commercial Electronic Systems.

Data storage and retrieval unit

Type
Flight/mission recording system.

Description
The radiation-hardened data storage and retrieval unit has evolved from the optical disk digital memory unit developed for night attack aircraft. The 300 Mbyte unit is small and lightweight, offering a significant size advantage over larger magnetic storage systems. The unit reduces the risk of loss or compromise of vital information and makes storage of information on board more efficient.

In the Northrop B-2, two rewritable data storage and retrieval units are located in the cockpit console, providing the crew with immediate access to the mission database and the ability to record mission data and performance information in flight.

Status
In service on US Air force B-2 aircraft.

Contractor
Honeywell Defence Avionics Systems.

DC-1590 series magnetic Digital Compass

Type
Flight instrumentation.

Description
The DC-1590 series magnetic digital compass provides encoded heading data. The display is in red, sunlight-readable, 0.5 in (13 mm) high, three-digit numerals and features leading zero suppression. Brightness is adjustable and, to prevent display jitter, display rate is also fully adjustable. A test switch is provided to test all digit segments. Display readings are in 1° increments. No warm-up time is required.

The compass sensor unit is a highquality liquid-filled instrument, fully illuminated for night viewing. All-solid-state to ensure reliable performance, it can be installed up to 38 m from the display. A set of East/West offset switches, mounted internally, provides unlimited compensation for magnetic variation. This remains in operation permanently until readjusted, allowing compensated operation to give true heading values. At the same time, a set of East/West offset push-button switches, mounted on the display,allows input compensation for variation changes.

The DC-1590 may be used to drive remote displays, autopilots, plotters and other electronic navigation equipment through a binary encoded decimal output connector.

Specifications
Weight: 4.5–6.8 kg depending on model

Status
In production and in service.

Contractor
Arc Industries Inc.

DC-2200TM series magnetic Digital Compass

Type
Flight instrumentation.

Description
The DC-2200TM series magnetic digital compass is a twin display system. One display shows true heading, while the other shows magnetic heading. The operator can at any time add or remove any amount of variation for course correction.

Status
In production and in service.

Contractor
Arc Industries Inc.

DCRsi™ 75 digital cartridge recording system

Type
Flight/mission recording system.

Description
The DCRsi 75 is an inexpensive 75 Mbit/s DCRsi variant designed to fill an important niche between 'top-end' S-VHS units and Ampex's existing 107 and 240 Mbit/s DCRsi models. This recording system is targeted particularly at the growing need, particularly within the anti-submarine warfare, airborne instrumentation and telemetry areas, for a severe-environment recorder with a lower rate capability and proportionately lower cost. DCRsi 75 retains the field-proven DCRsi transverse scan recording footprint for full crossplay compatibility with other models in the range. The scanner assembly contains six azimuth record/reproduce heads each with a typical life exceeding 3,000 head-to-tape hours.

The DCRsi features 50 Gbytes of user storage per cartridge and continuously variable record/reproduce data rates of 0 to 9.375 Mbytes per second. There is 18 Mbytes of internal I/O buffer and a programmable buffer near full/empty flag that allows accurate control of burst operations over 15 Mbytes per second. The unit has RS232 and RS422 control.

Status
Selected for the mission recorders for the Merlin EH 101 maritime reconnaissance helicopter.

Contractor
Ampex Corporation Data Systems Division.

The DCRsi™ 75 digital cartridge recording system   0051319

DCRsi™ 107/107R Digital Cartridge Recording system

Type
Flight/mission recording system.

Description
The DCRsi 107/107R rack-mount and modular ruggedised systems are 1 in transverse scan, rotary digital recorders capable of recording and reproducing at any user data rate from 0 to 13.4 Mbytes/s (0-107 Mbits/s). This capability can be sustained for over 1 hour in one tape cartridge with a total storage capacity of 48 Gbytes which is equal to four 14 in tape reels recorded on a conventional 28 track HDDR.

Capitalising on improvements in integrated circuit density and the use of ASICs has resulted in a DCRsi 107 system that is smaller, lighter and has 50 per cent less power consumption than the DCRsi system it replaces. It provides a computer-friendly mass storage data peripheral to any air, sea or land platform.

The DCRsi 107 features a format and interface that is compatible with the DCRsi and the DCRsi 240, constant packing density, 96 Mbit internal I/O buffer and data block, time code addressing and searching and automatic playback alignment. Data transfer can be continuous, in bursts or changing. A ruggedised version, the DCRsi 107R, is designed for hostile environments. The system has RS-232 and RS-422 control interfaces.

The DCRsi 107 is available in both single module, 19 in rack-mount and modular ruggedised configurations. The ruggedised DCRsi 107R is available either as a two module record only or as a three module record/reproduce configuration.

Specifications
Dimensions:
(tape transport module) 373.4 × 274.3 × 175.3 mm
  (rec/rep electronics module)
  360.7 × 485.1 × 152.4 mm
  (optional AC input power module)
  254 × 152 × 76.2 mm
Weight:
(tape transport module) 14.75 kg
(record electronics module) 15.88 kg
(reproduce electronics module) 15.42 kg
(optional AC power module) 4.54 kg

Power supply 28 V DC
210 W (record only)
430 W (record/reproduce)
Temperature:
–30 to +50°C (operating)
–54 to +70°C (non-operating without tape)
Altitude up to 50,000 ft operating

Status
In production for German Air Force Tornado PA200 reconnaissance pods and utilised in the Eurofighter Typhoon development programme.

Contractor
Ampex Corporation Data Systems Division.

DCRsi™ 120 Digital Cartridge Recording system

Type
Flight/mission recording system.

Description
The DCRsi 120 recorder is equipped with a fully integrated internal memory buffer which provides total isolation between the user interface and the instantaneous timing demands of the tape transport. Via this front-end buffer, the DCRsi recorder will unconditionally follow the user's

DCRsi™ 120 digital cartridge recording system   0015321

data clock – any data rate from zero up to 120 Mbits/s can be recorded or played back either continuously, in bursts, or while fluctuating at any slew rate. Instant-on record capability provides for event capture without time lag associated with tape speed lock-up. No operator adjustments are required as the data rate changes. The DCRsi recorder completely emulates a solid-state FIFO memory. The system is a slave to user interface equipment, thus simplifying the task facing the data and control interface designer.

The DCRsi 120 recorder utilises a field-proven transverse-scanning tape transport design, featuring mechanical simplicity, compact size and a short, co-planar tape path. The front loading/unloading of the tape cartridge is done instantly, simply and reliably because there are no elevators or other complex mechanisms used to transport the cartridge. The compact transverse scanner assembly contains six azimuth record/reproduce heads. The scanner assembly is self-contained and is easily replaceable in the field with a typical head life exceeding 3,000 head-to-tape hours.

The DCRsi 120 recorder is available in both a two-module and a single-unit rack-mount configuration.

The rugged 50 Gbyte-capacity DCRsi tape cartridge is made of fibre-reinforced polycarbonate. It is flame retardant, UV resistant, non-toxic, with high impact strength. Tape durability is rated at more than 200 passes. The DCRsi 120 has continuously variable record/reproduce data rates between 0 and 15 Mbytes per second and has 18 Mbytes of internal I/O buffer. A programmable buffer near full/empty flag allows accurate control of burst operations over 15 Mbytes per second. The unit has RS232 and RS422 control interfaces.

Contractor
Ampex Corporation Data Systems Division.

DCRsi™ 240 Digital Cartridge Recording system

Type
Flight/mission recording system.

Description
The DCRsi 240 rack-mount and modular system is a 1 in transverse scan, rotary digital recorder capable of recording and reproducing at any user rate from 0 to 30 Mbytes/s (0–240 Mbits/s). The byte parallel data interface used in the DCRsi, DCRsi 107 and DCRsi 240 consists of eight parallel data lines, one common clock and one enable signal. This interface compatibility, along with the common format on tape, results in full tape interchange among all three DCRsi models. Tapes recorded on the DCRsi 240 can be played on DCRsi and DCRsi 107, and vice-versa. The 240 Mbits/s transfer rate capability is accomplished by increasing the number of record/playback heads from 6 to 12 and adding a second data channel, while retaining the single channel I/O architecture intact. Data is now recorded or played back on two heads simultaneously. The data from each of the two heads is processed by separate data channels at 120 Mbits/s each and the two data streams are combined in the single 72 Mbyte data buffer which yields a sustained total throughput of 240 Mbits/s with a peak

The DCRsi™ 240 digital cartridge recording system   0051329

rate of up to 300 Mbits/s. This capability can be sustained over the entire tape cartridge, resulting in a storage capacity of 50 Gbytes which is equal to four 14 in tape reels recorded on conventional 28 track HDDRs.

Capitalising on improvements in integrated circuitry density and the use of ASICs has resulted in a DCRsi 240 system that is smaller, lighter and consumes less power than the previous DCRsi systems. The DCRsi 240 provides the power of a computer-friendly mass storage data peripheral to any air, sea or land platform.

The DCRsi 240 features constant packing density providing 50 Gbytes user storage per cartridge regardless of data rate, data block and time code addressing and searching and automatic playback alignment. Data transfer can be continuous, in bursts or changing. The two-module ruggedised DCRsi 240 is designed for hostile environments. The system is configured for RS-232 and RS-422 control interfaces.

Specifications

Dimensions:
(tape transport module)
373.4 × 274.3 × 175.3 mm
(rec/rep electronics module) 388.6 × 317.5 × 195.6 mm
(cartridge) 266.7 × 165.1 × 41.9 mm
(optional AC power modules) 254 × 152.4 × 76.2 mm
Weight:
(tape transport module) 14.74 kg
(rec/rep electronics module) 14.06 kg
(cartridge) 1.13 kg
(optional AC power module) 4.54 kg
Power supply: 28 V DC, 450 W typical
Temperature range:
–30 to +50°C (operating)
–54 to +70°C (non-operating without tape)
Altitude: up to 50,000 ft operating

Contractor
Ampex Corporation Data Systems Division.

DCRsi Clip-On™ 1 Gbit/s airborne imagery recorder

Type
Flight/mission recording system.

Description
The Ampex 1 Gbit/s airborne ultra-high rate recorder, known as the DCRsi Clip-On™, offers a 1 Gbit/s snap-shot imaging capability with instant access to cached data.

Clip-On is designed to be used in conjunction with an Ampex DCRsi™ digital cartridge recorder in an airborne image gathering role. Typically, data acquired at 240 Mbits/s or lower will be recorded on cache and tape simultaneously using the high-rate DCRsi 240. Data streams faster than 240 Mbit/s will be stored in cache and then automatically backed up to tape when

the cache is nearly full or on a command from the operator. The baseline solid-state storage capacity is 5 Gbytes although larger memory sizes can be specified.

The system combines the benefits of extremely fast solid-state memory and permanent non-volatile tape storage in one inexpensive, fully integrated package. It has major operational advantages for airborne tactical reconnaissance since up to 5 Gbytes of cached imagery are always available for immediate access, giving the operator the ability to prioritise targets and downlink images almost instantaneously. The Clip-On system has already been selected for a number of classified programmes in the USA and is compatible with the whole range of Ampex ruggedised recorders including DCRsi 75, DCRsi 107 and DCRsi 240.

Status
Launched June 1997. In production and in service.

Contractor
Ampex Corporation Data Systems Division.

Digital map

Type
Avionic Digital Map System (DMS).

Description
CMC Electronics' digital map is a platform independent, open source data, moving map application that offers a significant increase in navigation and mission situational awareness, reducing pilot workload and improving overall flight safety and performance. The Digital Map can be embedded in a processor system, such as the FV-4000 Mission Computer (see separate entry), for display on any available aircraft multifunction display. Use of open source formats, available from numerous international, commercial and government sources, eliminates reliance on a single-source data supplier. Solid-state memory storage, for navigation charts, digital terrain maps, satellite images, Jeppesen-style databases and tactical data, is required.

In concept, the digital map is made up of a number of 'layers' that make up the final image shown on the display:

Background map
Open-source background maps provide operator-loadable raster images of navigation charts, tactical maps, satellite images and terrain maps.

Tactical overlay
The tactical overlay displays flight plan, waypoint, navaid and tactical data. Declutter capability and display selection allows the flight crew to easily set the display as desired for the current mission leg. The tactical overlay uses flight plans from aircraft FMS, tactical database(s) from mission

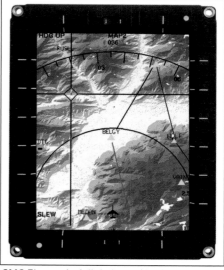

CMC Electronics' digital map (CMC Electronics)
1030042

a planning station, waypoints and navaids from commercial database(s) (for example, Jeppesen) and moving objects from aircraft radar.

Reference overlay
The reference overlay provides navigational references, including full compass mode, partial compass mode, grid mode (lat/long or UTM grid). The digital map can display heading-up modes and North-up modes.

Terrain awareness overlay
The digital map provides terrain awareness features, including terrain height warnings and intervisibility between aircraft position and selected ground positions, to provide enhanced safety and improved approach-to-target capability.

Specifications
Processor: Multiple PowerPC™ 500 MHz G4 with up to 512 Mb of memory each
Memory: 14 Gb mass memory
Operating system: ARINC-653 compatible
Graphics: High-resolution graphics (up to 1,600 × 1,200) generation for MFD drivers, RGB, DVI, LVDS. Full 2D and 3D OpenGL support with real-time video overlay
Dimensions: 3 MCU (96.5 × 198.1 × 381 mm)
Weight: 8 kg
Power: 28 V DC (MIL-STD-704 and DO-160D), or 115 V AC 400 Hz 1- or 3-phase (MIL-STD-704); 100 W
Connector: ARINC-600
Environmental: MIL-STD-810, -461, -462, DO-160D
Interface: MIL-STD-1553/1760, ARINC-429, fibre channel, I2C, RS-422/485, Ethernet, discrete and analogue

Status
In production.

Contractor
CMC Electronics Inc.

Digital map system

Type
Avionic Digital Map System (DMS).

Description
Using the technology and experience gained from the US Navy's AV-8B and F/A-18 night attack programme as a baseline, Honeywell has developed an advanced digital imaging system.

The capabilities of the base system have been expanded to include terrain reference navigation, ground collision avoidance, perspective view, terrain-following computations, threat intervisibility and sensor blending for advanced avionics systems.

Digitised aeronautical charts and Defense Mapping Agency Digital LandMass System

The Ampex Clip-On™ system offers a 1 Gbit/s snap-shot image recording and instant replay capability for airborne reconnaissance applications
0002443

(DLMS) Level 1 and 2 terrain and feature data are stored and recalled from the digital memory unit to generate a full-colour moving map which provides precise navigation and route selection to and from the point of target acquisition and weapons delivery.

The digital memory unit is a stand-alone mass memory unit designed to provide increased storage to military systems requiring high reliability and quick access times. The unit contains a flight-proven fully militarised optical disk capable of storing 520 Mbytes of data. Information is written to the disk by a laser diode in the head assembly. A beam of light is focused on the medium to produce a hole. During the read operation, the hole is recognised as a bit of data. Unlike magnetic media, which can be scratched by a head crash, data stored on the optical disk is written to a thin alloy encased in a protective substrate and a clear protective cover. Digitised reconnaissance photos, radar data, Landsat data, emergency procedures, let-down and approach plates, mission data and flight plans can be stored with digitised aeronautical charts and DLMS data for a variety of missions.

The digital map computer performs all the airborne map generation. The configuration on board the US Navy AV-8B night attack aircraft contains 11 circuit cards, a motherboard, a power supply and additional space to accommodate functions such as terrain reference navigation, terrain-following, ground collision avoidance, target hand-off, in-flight route planning, ridgelines, perspective view, sensor blending and threat intervisibility. The computer receives operational parameters, which include aircraft state vector, map scale, zoom factor, north up/track up, database type, declutter select/deselect and video output mode, from the mission computer via a MIL-STD-1553B multiplex bus. Data requests are then sent to the digital memory unit via a fibre optic link.

DLMS elevation data is compared to absolute elevation limits or limits relative to the aircraft altitude, resulting in a clearance band display. Colour is assigned, based on the band and sun angle shading. Variably spaced contour lines can be added. The overlay selection includes linear, area and point feature data such as roads, rivers and towers in the DLMS mode, and flight path, waypoint symbols, text and threat symbols in the DLMS and digitised chart modes. The video generator combines the background scene memory with the overlay memory to provide a composite video output. The system supports two red/green/blue video outputs and two monochrome video outputs at either 525/60 Hz or 625/50 Hz line rates.

Processing is distributed across several high-performance microprocessors rather than a single unit. Data flow is pipelined through the digital map computer and parallel processing is performed wherever possible. The use of parallel processing provides a level of fault tolerance, since a failure in the map generation does not prohibit the graphics overlay operation.

Status
Production deliveries for the US Marine Corps AV-8B began in June 1989. Honeywell is also under contract to provide the digital video mapping system for the V-22 Osprey tilt-rotor aircraft.

Contractor
Honeywell Defence Avionics Systems.

Digital RMI/DME indicators

Type
Flight instrumentation.

Description
Designed to interface with both the current ARINC 700 sensors and the older-generation analogue avionics, this new series of combined Radio Magnetic Indicators and Distance Measuring Equipment (RMI/DME) forms a family of instruments that share many common features

including electrical, servo, thermal, packaging and lighting methods. Use of a patent digital encoder/driver module, common to each instrument and each channel within it, permits packaging techniques which greatly improve the effectiveness of the heat transfer arrangements. The encoder/driver module consists of an 11-bit digital encoder, DC motor and associated drive circuits, all under microprocessor control. Each of the six channels is isolated from its neighbour for integrity and each has its own single chip microcomputer. Each channel monitors its own faults and activates its own flag or shutter and output signal.

The members of the digital RMI/DME instrument family comprise:

RMI-733A radio magnetic indicator: a compact lightweight VOR/ADF selectable three-servo instrument. A version designated RMI-733A is a three-servo instrument designed to display information from an ADF receiver.

RDMI-743 radio distance magnetic indicator: this instrument features liquid crystal DME readouts for better reliability and readability. The three-servo unit is selectable to either VOR/ADF or VOR only.

RDMI-743A radio distance magnetic indicator: with a magnetic wheel DME display, the RDMI-743A is a four-servo VOR instrument.

Status
All in production.

Contractor
Rockwell Collins.

Display and control system for the F/A-22

Type
Avionic display system.

Description
BAE Systems North America is responsible for the display suite for the F/A-22 Raptor multirole combat aircraft. The F/A-22 display and control system consists of one primary MultiFunction Display (MFD), three secondary MFDs and two upfront displays. BAE Systems also provides the Graphics Processor Video Interface (GPVI), Airborne Video Tape Recorder (AVTR) and Operational Debrief Station (ODS).

The primary MFD presents a situation display of the air and ground situation.

The secondary MFDs present attack and defensive data, together with stores management information. The upfront display presents CNI data, critical flight data and warnings.

The GPVI resides in the Common Integrated Processor (CIP) and hosts software that generates the tactical situation displays. It both generates display formats and mixes these formats with video from other aircraft sources. The GPVI includes the i960MX processor and two graphics ASICs developed by BAE Systems. The Graphics Drawing Processor (GDP) ASIC is the high-performance anti-aliasing drawing processor. Using subpixel addressing and advanced

anti-aliasing algorithms, the GDP generates complex display formats at better than a 30 Hz rate. The anti-aliasing provides for smooth display dynamics, without artifacts such as stair-stepped lines.

The video merge ASIC controls the merging of external video with the locally generated format and transmits the result over the fibre optic interface to the MFDs. The MFDs also utilise the GDP ASIC to enable local display generation. Thus the MFDs can either display video from the CIP or locally generated formats derived from information received over dual MIL-STD-1553 interfaces.

The GDP is a single 3,000 Mops ASIC which accomplishes on-chip translation, rotation, scaling, anti-aliasing drawing, filling and micro-positioning to better than one-sixteenth of a pixel for smooth instrument movement on an active matrix LCD medium. F/A-22 systems engineering trade-offs have resulted in design advances such as dual-brightness sensors for automatic compensation for cockpit ambient light level, expanded luminance ranging from 220 ft lamberts down to 0.1 ft lamberts for enhanced night operations with minimum canopy reflections, lamps designed to outlast the life of the aircraft and comprehensive BIT for improved safety and maintainability.

Specifications
Dimensions:
(primary MFD) 257.8 × 257.8 × 190.5 mm
(secondary MFD) 213.4 × 213.4 × 190.5 mm
(up-front display) 156.2 × 130.8 × 218.4 mm
Weight:
(primary MFD) 7.03 kg
(secondary MFD) 6.12 kg
(up-front display) 3.3 kg
Display size:
(primary MFD) 203.2 × 203.2 mm
(secondary MFD) 158.7 × 158.7 mm
(up-front display) 76.2 × 101.6 mm

Status
Entering service in US Air Force F/A-22 Raptor aircraft.

Contractor
BAE Systems North America, Information and Electronic Warfare Systems.

DMS-1000 Digital Management System

Type
Flight/mission recording system.

Description
The DMS-1000 includes Ampex's proprietary CSRT management system that allows capture, storage, retrieval and transmission of high-speed sensor and image data at up to 1.12 Gbits/s.

The DMS-1000 is a scalable high-speed Solid-State Memory System (SSMS) that is situated between the input/output of the source and the Ampex DCRsi™ or DIS™ recording system. The SSMS allows the input of extremely high-speed data into the data capture system without the added cost of multiple recorders.

The F/A-22 display and control system consists of (left to right) the upfront display, secondary MFD and primary MFD

0581518

DMS-1000 Digital Management System 0051344

The CSRT management system allows the operator to direct data to the recording system, into non-volatile or volatile memory, recall data for immediate display to a monitor or aircraft HUD, or direct it to a datalink system for transmission to other tactical platforms.

Specifications
Performance:
transfer rate: 0 to 1.12 Gbyte/s (140 Mbyte/s) sustained
capacity: up to 10 Gbytes in 2 Gbyte increments
memory technology: 2 Gbyte memory boards can be either high-speed DRAM or non-volatile FLASH or a mix
Interface:
data interface: 8, 16 or 32 bits, differential ECL
recorder interface: connects directly to DCRsi family of airborne recorders
control: RS-232
timecode interface: IRIG-B
Dimensions: 269.9 × 257.2 × 317.5 mm
Weight: 13.64 kg
Power: 28 V DC, 200 W

Status
Used for testing the Tactical Aircraft Reconnaissance Pod System (TARPS – see separate entry).

Contractor
Ampex Corporation Data Systems Division.

DPM-800E PCM encoder

Type
Flight/mission recording system.

Description
The DPM-800E PCM encoder is a fully programmable high-performance 8- to 12-bit resolution data system for acquiring conditioned signals in severe airborne applications where flexibility and reliability are the primary requirements.

Features of the DPM-800E include EEPROM programmable via parallel IF port, 2 Mbit operation, eight programmable bit rates, user programmable for format, bits per word and gain/offset, filtered data outputs, single-ended or differential analogue inputs, single-ended discrete bilevel inputs and subcommutation and supercommutation.

The DPM-800E can be configured as a stand-alone, master or remote unit. Up to seven DPM-800Es can be configured in a master/slave or cluster configuration via the L-3 Telemetry-East standard 10-wire differential interface. This 10-wire interface can be utilised with other L-3 Telemetry-East data acquisition products such as MIL-STD-1553 bus monitors, signal conditioning units and master controllers for large distributed systems.

Specifications
Dimensions: 88.9 × 82.5 × 118.1 mm
Weight: 0.79 kg
Temperature range: −35 to +85°C

Contractor
L-3 Telemetry-East.

EDZ-605/805 electronic flight instrument systems

Type
Electronic Flight Instrumentation System (EFIS).

Description
The EDZ-605 and EDZ-805 electronic flight instruments are intended for corporate aircraft and regional airliners. The EDZ-605 is a 5 in (127 mm) system, while the EDZ-805 is a 6 in (152 mm) system. The products are the result of a new symbol generator and software changes. The generator has increased stroke writing and memory capabilities, while software updating has provided a variety of cosmetic improvements to the display.

The Electronic Attitude Director Indicator (EADI) features an enlarged sphere presentation with linear pitch tape, improved single-cue aircraft symbol and stroke filled single-cue command bar. A digital T-bar airspeed and/or angle of attack presentation has also been added. Other improvements include a stroke filled roll pointer, larger roll indices, shorter horizon indices and colour reversal on glide slope, angle of attack, rate of turn and expanded localiser scales. In addition, the 5 in (127 mm) EADI is now available in a truncated sphere presentation.

The Electronic Horizontal Situation Indicator (EHSI) features a stroke filled lubber line, colour reversal on the glide slope scale and weather radar mode annunciation.

The ED-605/ED-805 are compatible with most analogue systems and all digital systems.

Status
In production. Chosen for many business aircraft.

Contractor
Honeywell Inc, Commercial Electronic Systems.

The Honeywell EHSI for corporate aircraft and regional airliners 0503954

EDZ-705 Electronic Flight Instrument System (EFIS)

Type
Electronic Flight Instrumentation System (EFIS).

Description
The EDZ-705 Electronic Flight Instrument System (EFIS) is tailored to meet the needs of specific helicopter applications from executive use to search and rescue missions. The system provides the ability to display standard search patterns as provided by various navigation sources, collective cue symbology which allows the pilot to follow flight director collective cue demands, hover display symbology which allows the pilot to track and maintain target location in relation to the helicopter position in a search and rescue environment and four-axis helicopter flight director mode annunciation.

Since panel space is at a minimum, the EDZ-705 offers 5 × 5 in (127 × 127 mm) displays.

Contractor
Honeywell Inc, Commercial Electronic Systems.

EFD 4000

Type
Avionic display system.

Description
Collins' EFD-4000 Electronic Flight Display (EFD) product family includes the 7 × 6 in EFD-4076 and 7 × 7 in EFD-4077 integrated display units. The EFD-4000 utilises very large scale integration packaging techniques to combine the symbol generator and the CRT into a single display unit.

The EFD-4000 operational definition can be broken down into two distinct functions, an Electronic Flight Instrumentation System (EFIS) and an Engine Indication and Crew Alerting System (EICAS):

EFIS
The EFIS typically contains four large colour displays and two display control panels. Line Replaceable Units (LRUs) are:
• Pilot Primary Flight Display (PFD)
• Pilot MultiFunction Display (MFD)
• Pilot DCP-4000 Display Control Panel (DCP)
• Co-pilot PFD
• Co-pilot MFD
• Co-pilot DCP-4000.

The pilot PFD displays attitude, lateral navigation/compass, flight control and primary air data (altitude/airspeed/vertical speed) functions. The PFD receives data bus inputs from any/all of input/output concentrators, Integrated Avionics Processing System (IAPS), Traffic Alert Collision Avoidance System (TCAS) receiver/transmitter, Inertial Reference Unit (IRU) and Air Data System. The PFD provides a data bus output to the IAPS and DCP. The MFD displays lateral navigation/compass, radar, TCAS, flight management (map/summary) and diagnostic information. The MFD also provides a reversion backup for its associated PFD or the EICAS, if that display fails. The MFD receives the same data bus inputs that are applied to the associated PFD. It also receives input buses from weather radar, EICAS control panel, EICAS Data Concentrator Units (DCUs), Maintenance Diagnostic Computer (MDC) and Flight Management Computer (FMC). The MFD provides a data bus output to the IAPS and DCP. The DCP-4000 controls the associated PFD and MFD. This panel selects display formats and lateral navigation parameter/sources. The DCP receives data bus inputs from the associated PFD, MFD and weather radar panel. The DCP switch data and the weather radar control panel data are output to the on-side and cross-side PFDs and MFDs. A display reversion switch allows the pilot to select PFD or EICAS back-up display on the MFD. Remote reversion switches allow the selection of either Attitude Heading Reference System (AHRS), DCP and air data source inputs. The remote maintenance switch selects the maintenance display on MFD 1 or MFD 2.

EICAS
The EICAS typically contains two large colour displays, EICAS control panels, two or three EICAS DCUs and a lamp driver unit. LRUs for the EICAS are:
• Pilot primary EICAS display ED-1
• Co-pilot secondary display ED-2
• ECP-4000 EICAS control panel
• Pilot DCU-4000 EICAS
• Co-pilot DCU-4000 EICAS
• Optional third DCU-4000 EICAS
• DCP LDU-4000 Lamp Driver Unit (LDU)
• Pilot ERU-4000 EICAS Routing Unit (ERU)
• Co-pilot ERU-4000.

The primary EICAS (ED-1) displays the primary engine indication instruments and crew alerting messages. The primary display is a fixed format of engine, oil, fuel and gear data. Each ED receives data bus inputs from the EICAS DCUs and EICAS control panel and provides a data bus output to the IAPS data concentrators. The secondary EICAS (ED-2) displays various data pages and serves as a backup to the primary display. EICAS pages are selected using the ECP-4000. A remote EICAS reversion switch allows the pilot or co-pilot to select either display. Selecting one display blanks the other tube and allows data pages to be selected on the remaining display

Specifications
EFD

	EFD-4076	EFD-4077
Dimensions	204 × 157 × 350 mm (H × W × L)	202 × 189 × 337 mm (H × W × L)
Weight	8.1 kg	8.1 kg
Power	28 V DC; 130 W	28 V DC; 130 W
TSO	C2d, C3d, C4c, C6d, C8c, C10b, C34e, C35d, C36e, C40c, C41d, C43a, C44a, C46a, C47, C49a, C52a, C63c, C66b, C87, C94a, C95, C101, C104, C105, C113, C115	C3d, C4c, C6d, C8c, C10b, C34e, C35d, C36e, C40c, C41d, C43b, C44a, C46a, C47, C49a, C52a, C63c, C66c, C87, C95, C104, C110a, C113, C115a, C119a
Altitude	55,000 ft	55,000 ft
Temperature	−20 to +55°C	−20 to +55°C
Cooling	Forced air	Forced air
Qualification	DO-160C, ED-14C, DO-178A Level 1 and 2	DO-160C, ED-14C, DO-178B Level A and C

DCP/DCU

	DCP-4000	DCU-4000/4002	ECP-4000
Dimensions	48 × 146 × 184 mm (H × W × L)	194 × 191 × 356 mm (H × W × L)	67 × 146 × 165 mm (H × W × L)
Weight	0.86 kg	8.41 kg	0.68 kg
Power	28 V DC; 7.5 W	28 V DC; 53 W	28 V DC; 13 W
TSO	C6c, C34e, C36e, C40c, C41d, C52a, C63c, C66b, C104, C105, C113, C115	TC on airplane (PMA)	C43a, C44a, C47, C49a, C113
Altitude	55,000 ft	55,000 ft	55,000 ft
Temperature	-55 to +70°C	−20 to +70°C	−20 to +70°C
Cooling	Convection	Convection	Convection
Qualification	DO-160C, [A2F1]-BB[BMS] E1XXXXXZ [BZ] AZAWA[Z3Z3] XX, ED-14C, DO-178A Level 2	DO-160C, [A2F2]-BB[CLM] E1XXXXXZ [BZ] AZZUZ [Z3Z3]XA, ED-14C, DCU-4000 DO-178B Level A DCU-4002 DO 178B Level B	DO-160C, [A1F1]-CB [BMS] E1XXXXXA AZAYA[Z3Z3]XX ED-14C, DO-178A Level 1 and 2

ERU/LDU

	ERU-4000	LDU-4000	ECP-4003
Dimensions	289 × 318 × 40 mm (H × W × L)	146 × 146 × 179 mm (H × W × L)	67 × 146 × 165 mm (H × W × L)
Weight	2.3 kg	2.2 kg	0.77 kg
Power	Passive	28 V DC; 21 W	28 V DC; 13 W
TSO	TC on airplane (PMA)	C43a, C44a,	C47, C49a, C113
Altitude	55,000 ft	55,000 ft	55,000 ft
Temperature	−55 to +70°C	−20 to +70°C	−20 to +70°C
Cooling	Convection	Convection	Convection
Qualification	DO-160C, A2F2]-BB [CLM] XXXXXXXXXX XXXXXX, ED-14C	DO-160C, [A2F1]-BB [BMS] E1XXXXXZ[BZ] DO-178A Level 2	DO-160C, [A2F1]-BB [BMS] E1XXXXXZ[BZ] AZAWA[Z3Z3]XX, ED-14C, DO-178A Level 1

using the ECP panel. The DCU-4000 collects and formats aircraft data for display on the EICAS and performs flight data acquisition functions for the flight data recorder. Crew alerting logic is processed in the DCU, which receives high- and low-speed ARINC 429 buses, analogue inputs and discrete inputs from the engines and other aircraft systems. The data inputs are concentrated and processed for transmission on ARINC 429 buses. The DCU outputs engine data to the displays; maintenance, diagnostic and aircraft data to the IAPS data concentrators, annunciator lamp data to the LDU, aircraft system data to the flight data recorder and datalink management unit.

The optional DCU-4000 performs the same functions as described above. The LDU-4000 is a dual-channel LDU, capable of driving up to 120 annunciator lamps. Identical channels 1 and 2 receive digital buses from all the DCUs, with the outputs from each side tied together (OR logic). The ERU-4000 are effectively 'junction boxes' for the DCUs, splitting each input signal to three output pins. The pilot ERU routes left-side aircraft data and the co-pilot ERU routes right-side aircraft data.

Status
In production and in service.

Contractor
Rockwell Collins.

EFIS-84 Electronic Flight Instrument System

Type
Electronic Flight Instrumentation System (EFIS).

Description
The EFIS-84 is a 4 in electronic flight instrument system offering full capability and high reliability at an affordable cost. The EFIS-84 combines an EADI and EHSI with a centre panel MFD having the same capabilities as the Collins EFIS-85 and EFIS-86 systems. It can be configured in two- to five-tube systems depending on the flight deck size and application. The system components are: DSP-84 Display Select Panel; DPU-84 Display Processor Unit; EFD-84 Electronic Flight Display; MFD-85B Multifunction Display; and MPU-84 Multifunction Processor Unit.

The system is designed for applications ranging from helicopters to regional airliners and corporate aircraft. Retrofit of existing 4 in electromechanical systems with the EFIS-84 offers significant operational, reliability and cost of ownership benefits.

The EFIS-84 offers bright clear attitude and navigation information in easy-to-interpret formats. Attitude is displayed in full sky presentation which provides an attitude area more than twice as large as that of electromechanical instruments. A race-track attitude display, with a circular attitude depiction similar to the familiar electromechanical ADI, may also be selected as an installation option. A customer choice of V-bar or cross pointer steering commands is also offered.

A full selection of modes is offered on the EHSI, including the traditional compass rose, arc and map. Significant operational advantages are offered by the extensive map capability, particularly to pilots operating in terminal areas. A key benefit of the EFIS-84 is the ability to

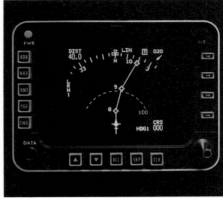

The Rockwell Collins multifunction display offered as an EFIS-84 option   0503945

display weather radar information integrated with the navigation map. The MFD provides additional navigation, weather avoidance, data management and reversionary functions. The system is compatible with the Collins WXR-350 or the advanced TWR-850 turbulence weather radar.

System flexibility allows interface with a variety of analogue and digital aircraft sensors, as well as a broad range of long-range navigation options. The optional multifunction display offered as part of the EFIS-84 provides additional system capability, including expanded map display and checklists.

Specifications
(DSP-84 and EFD-84)
Weight:
(DSP) 1.31 kg
(EFD) 2.3 kg
Temperature: −20 to +55°C
Altitude: 50,000 ft
Power: 28 V DC; 5.5 W (DSP), 55 W (EFD)
Certification:
(DSP) FAA TSOs C6c, C52a; DO-160A, EUROCAE ED-14A
(EFD) FAA TSOs C3b, C4c, C34c, C36c, C40b, C52a; DO-160A; ED-14A

(MFD-85B, DPU-84 and MPU-84
Weight:
(MFD) 4.36 kg
(DPU) 6.4 k
(MPU) 7.7 kg
Temperature: −55 to +70°C
Altitude: 55,000 ft
Power: 28 V DC; 42 W (MFD), 66 W (DPU), 143 W MPU
Certification:
(MFD) FAA TSOs C6c, C63b; DO-160A, EUROCAE ED-14A
(DPU and MPU) FAA TSOs C3b, C4c, C6c, C34c, C36c, C40b, C52a; DO-160A; EUROCAE ED-14A

Status
In service. Certified on the Beech 1900D regional airline aircraft.

Contractor
Rockwell Collins.

EFIS-85/86B Electronic Flight Instrument System

Type
Electronic Flight Instrumentation System (EFIS).

Description
The EFIS-85B and EFIS-86B Electronic Flight Instrument Systems combine Electronic Attitude Director Indicator/Electronic Horizontal Situation Indicators (EADI/EHSI) with a centre panel MultiFunction Display (MFD) to provide a fully integrated flight deck display system. The EADI has a 50 per cent larger earth/sky representation for terminal and en route formats than conventional 5-inch electromechanical instruments. The EHSI combines multiple

Specifications

	EFD-85	DSP-85B	DPU-85N	MFD-85B	MPU-85N
Dimensions	122 × 132 × 249 mm (H × W × L)	48 × 146 × 203 mm (H × W × L)	195 × 129 × 372 mm (H × W × L)	115 × 153 × 320 mm (H × W × L)	195 × 196 × 372 mm (H × W × L)
Weight	3.0 kg	1.36 kg	6.09 kg	4.36 kg	7.64 kg
Power	28 V DC; 55 W; 5 V AC/DC, 1.2 W 28 V DC, 1.4 W (lighting)	28 V DC; 5.5 W; 5 V AC/DC, 5.5 W 28 V DC, 8.4 W (lighting)	28 V DC; 66 W	28 V DC; 41.3 W; 5 V AC/DC, 4.6 W, 28 V DC, 6.1 W (lighting)	28 V DC; 88 W
TSO	C3b, C4c, C6c, C34c, C36c, C40b, C52a	C6c, C52a	C3b, C4c, C6c, C34c, C36c, C40b, C52a	C6c, C63b	C3b, C4c, C6c, C34c, C36c, C40b, C52a
Altitude	55,000 ft	55,000 ft	55,000 ft	55,000 ft	55,000 ft
Temperature	−20 to +55°C	20 to +55°C	−20 to +70°C	−20 to +55°C	−20 to +70°C
Qualification	DO-160A D1/A/KS/ XXXXXXAZAAA, ED-14A	DO-160A D1/A/KS/ XXXXXXAZAAA, ED-14A	DO-160A F2/A/MNO/ XXXXXXZAAAA, ED-14A	DO-160A D1/A/KS/ XXXXXXAAAAA, ED-14A	DO-160A F2/A/MNO/ XXXXXXZAAAA, ED-14A

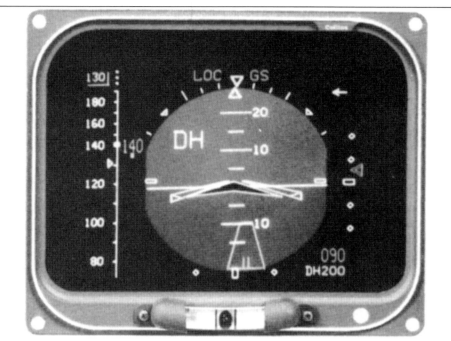

Collins' 5 × 6 in EADI with primary airspeed display (above) and weather imagery (below)   0503947

display formats with weather radar and built-in NAV source selection for simplified co-ordination of all navigation and weather avoidance tasks. The MFD augments the EADI/EHSI displays by providing the pilot and co-pilot with additional navigation, weather avoidance, data management and reversionary functions.

Both stroke and raster scanning techniques are used to provide a large area colour background for weather and distinct alphanumeric/symbology information with high brightness.

The system comprises the following elements:

CHP-86B/85D Course Heading Panel The CHP-86B/85D provides course and heading select functions, selection of time-to-go, ground speed, elapsed time and wind.

DCP-85G Display Control Panel The DCP-85G provides EHSI display sensor control, EHSI display mode control, EADI and EHSI display dimming and EADI radio altitude decision height set capability.

DSP-85B Display Select Panel An alternative to the DCP-85G and one CHP-86B for flight deck where reduced panel space is required. Requires external dimming and stopwatch controls.

DPU-85N Display Processor Unit The DPU-85N provides the necessary deflection and video signals for up to two EFD-85s. Contains aircraft systems I/O interface and comparator warning functions.

DPU-86N Display Processor Unit The DPU-86N partners the EFD-86 as above.

EFD-85 Electronic Flight Display The EFD-85 is a multicolour 5 × 5 in CRT for the display of EADI or EHSI information.

EFD-86 Electronic Flight Display The EFD-86 is a multicolour 5 × 6 in CRT for the display of EADI or EHSI information.

MFD-85B Multifunction Display The MFD-85B is a multicolour CRT instrument display of weather radar, pictorial navigation map, page data, waypoint definition, display of multiple waypoints and remote data.

MPU-85N Multifunction Processor Unit The MPU-85N provides the necessary deflection and video signals for the MFD.

MPU-86N Multifunction Processor Unit As above. The EADI provides conventional ADI information, together with airspeed, airspeed trend and multisource vertical and lateral deviation information. Three EHSI modes can be selected: full compass rose (as in a conventional HSI), expanded sector display or sector display with weather radar paints.

Status
In production and in service.

The system has been certified on the following aircraft in Table 1:

Contractor
Rockwell Collins.

Table 1:

EFIS-85	EFIS-86
BAe 125-700A/800A	BAe 125-800A
Beech 200/300/400	
Dassault Falcon 10/20/50/100/200	Dassault Falcon 20/20F/50/200
de Havilland DHC-6	
Dornier DO-228-200	
Gulfstream G-1159A	Gulfstream G-1159/1159A/1159B
Lear 35/55	
IAI 1124A/1125	IAI 1125
McDonnell Douglas DC-8-72	McDonnell Douglas DC-8 Series
Sabreliner NA-265-40	

Canadair CL-600-2A12
Boeing 727-27C

EFIS-85/86C Electronic Flight Instrument System

Type
Electronic Flight Instrumentation System (EFIS).

Description
With similar capabilities to Collins' EFIS-85/86B, the EFIS-85/86C is available in two-, three-, four- and five-tube configurations with 5 × 5 or 5 × 6 in format CRTs. Compatible with Collins' FCS-85 series and AHRS-85, a full EFIS-85/86C system comprises dual EADI/EHSI with one set of equipment for each pilot, together with a shared MultiFunction Display (MFD) and mode controls. The EADI and EHSI CRTs can display all the information traditionally associated with electromechanical flight director instruments as well as weather radar patterns, navigation maps, performance data and navigation waypoints. The EFD-85 can show conventional HSI information on a circular scale or it can expand just the forward sector of the display and show weather maps.

The EFIS-85 CRT incorporates a three-gun assembly, a shadow-mask, a faceplate with phosphor coating and a glass envelope to enclose the elements. The in-line electron gun assembly provides improved convergence and mechanical rigidity and the high-resolution shadow-mask gives four to six times better resolution than that of a domestic television set because the phosphor dots are so much closer together. The displays use both stroke and raster writing. The high-intensity stroke writing of symbols, in conjunction with contrast enhancement filters, enables displays to be read even in full sunlight. Primary colours are red, blue and green, with easy synthesis of several derivative colours, including white.

The EFIS-85/86C system comprises the following elements:

CHP-85C/86C Course Heading Panel The CHP-85C/86C provides course and heading select functions, selection of time-to-go, ground speed, elapsed time and true airspeed.

DCP-85F/H Display Control Panel The DCP-85F/H provides for navigation sensor selection, EHSI format, bearing pointer selection and the display of weather preset course from a second navigation source or course transfer function on the EHSI. In addition, EADI display dimming and radio altitude decision height set capability is provided.

DSP-85 Display Select Panel An alternative to the DCP-85F/H and one CHP-85C/86C for flight deck where reduced panel space is required. Requires external dimming and stopwatch controls.

DPU-85R Display Processor Unit The DPU-85R provides the necessary deflection and video signals for up to two EFD-85s. Contains aircraft systems I/O interface and comparator warning functions.

DPU-86R Display Processor Unit The DPU-86R partners the EFD-86 as above.

EFD-85 Electronic Flight Display The EFD-85 is a multicolour 5 × 5 in CRT for the display of EADI or EHSI information.

EFD-86 Electronic Flight Display The EFD-86 is a multicolour 5 × 6 in CRT for the display of EADI or EHSI information.

MFD-85B Multifunction Display The MFD-85B is a multicolour CRT instrument display of weather radar, pictorial navigation map, page data, waypoint definition, display of multiple waypoints and remote data.

MPU-85R Multifunction Processor Unit The MPU-85R provides the necessary deflection and video signals for the MFD.

MPU-86R Multifunction Processor Unit As above.

The EADI display incorporates the following information (with associated sensor input):
- Aircraft attitude and flight control system steering commands
- VHF Omnidirectional Radio Range (VOR), tactical air navigation system, Area Navigation (RNAV), Lateral Navigation (LNAV), Microwave Landing System (MLS) azimuth or localiser, deviation
- Glideslope, MLS glidepath or preselected altitude deviation
- Flight control system mode enunciation
- Autopilot engage enunciation
- Attitude source enunciation
- Marker beacon enunciation
- Radio altitude
- Decision height set and enunciation
- Fast-slow deviation or angle of attack deviation
- Indicated airspeed
- Commanded airspeed
- Airspeed trend vector
- Preselected altitude alert
- Excessive Instrument Landing System (ILS) deviation (Category II configurations).

The EHSI display has two basic modes: full compass rose and compass sector. The full compass mode presents compass rose similar to existing electromechanical instruments. Information available in this mode includes the following (with defined sensors inputs):
- Compass heading
- Selected heading
- Selected VOR, localiser, MLS azimuth or LNAV course and deviation (including enunciation of deviation type)
- Navigation source enunciation
- Digital selected course/desired track readout
- Excessive ILS deviation (when Category II configured)
- Heading source annunciation
- ILS and heading comparator warnings (ILS only if Category II configured)
- To/from information
- Back course localiser enunciation
- Distance to station/waypoint
- Glideslope or Vertical Navigation (VNAV) deviation
- Ground speed, time-to-go, elapsed time or true airspeed
- Course information and source information from a second navigation source
- Weather radar target alert
- Waypoint alert when LNAV or RNAV is the navigation source
- Bearing pointer driven by VOR, LNAV or ADF source.

Status
In production and in service.

The system has been certified on the following aircraft as shown in Table 1:

Contractor
Rockwell Collins.

Table 1

EFIS-85	EFIS-86
BAe 125-700A/800A	BAe 125-800A
Beech 200/300/400	
Dassault Falcon 10/20/50/100/200	Dassault Falcon 20/20F/50/200
de Havilland DHC-6	
Dornier DO-228-200	
Gulfstream G-1159A	Gulfstream G-1159/1159A/1159B
Lear 35/55	
IAI 1124A/1125	IAI 1125
McDonnell Douglas DC-8-72	McDonnell Douglas DC-8 Series
Sabreliner NA-265-40	

Canadair CL-600-2A12
Boeing 727-27C

EFS 40/EFS 50 Electronic Flight instrument Systems

Type
Electronic Flight Instrumentation System (EFIS).

Description
The EFS 40 is a 101.6 mm (4 in) electronic flight instrument system designed for operators of light jet and turboprop aircraft and turbine helicopters. The EFS 50 is a 127 mm (5 in) system. They are modular systems comprising an Electronic Attitude Director Indicator (EADI) and Electronic Horizontal Situation Indicator (EHSI). The package can be installed in two-, three-, four- or five-tube configurations.

The system's basic building block is the EHI. This unit incorporates a navigation mapping capability and modular design with interfaces for analogue and digital inputs. Added functions include a weather map display and joystick controller, plus a wide range of sensor input interfaces. The EHI is available with either a combination control/display unit or display with separate control panel.

The interchangeable ED 461/462 in the EFS 40 measures 106 × 106 × 238.8 mm and weighs 2.275 kg. The ED 551A in the EFS 50 measures 117.3 × 133.4 × 238.8 mm and weighs 3.76 kg. The remote SG 465 EADI/EHSI symbol generator unit fits into a 3/8 ATR short box. Control panels are offered as a built-in mode controller in the ED 461 or as the remote CP 468 mode controller for the ED 462 display unit. Standard interfaces are provided for most of the avionics, flying control and long-range navigation systems in current use in business aviation aircraft.

Specifications

	EFD-85	DSP-85C	DPU-85R	MFD-85C	MPU-85R
Dimensions	122 × 132 × 249 mm (H × W × L)	48 × 146 × 203 mm (H × W × L)	195 × 129 × 372 mm (H × W × L)	115 × 153 × 314 mm (H × W × L)	195 × 196 × 372 mm (H × W × L)
Weight	3.0 kg	1.31 kg	6.2 kg	4.35 kg	7.7 kg
Power	28 V DC; 55 W; 5 V AC/DC, 1.2 W 28 V DC, 1.4 W (lighting)	28 V DC; 5.5 W; 5 V AC/DC, 6 W 28 V DC, 5.6 W (lighting)	28 V DC; 66 W	28 V DC; 41.3 W; 5 V AC/DC, 4.6 W, 28 V DC, 6.2 W (lighting)	28 V DC; 88 W
TSO	C3b, C4c, C6c, C34c, C36c, C40b, C52a	C6c, C52a	C3b, C4c, C6c, C34c, C36c, C40b, C52a	C6c, C63b	C3b, C4c, C6c, C34c, C36c, C40b, C52a
Altitude	55,000 ft	55,000 ft	55,000 ft	55,000 ft	55,000 ft
Temperature	−20 to +55°C	−20 to +55°C	−20 to +70°C	−20 to +55°C	−20 to +55°C
Qualification	DO-160A D1/A/KS/ XXXXXXAZAAA, ED-14A	DO-160A D1/B/KS/ XXXXXXAZAAA, ED-14A	DO-160A F2/A/MNO/ XXXXXXAZAAA, ED-14A	DO-160A D1/A/KS/ XXXXXXAZAAA, ED-14A	DO-160A F2/A/MNO/ XXXXXXAZAAA, ED-14A

Specifications
Altitude: 55,000 ft
TSO compliance: TSO C113, RTCA DO 160B
Temperature: −40 to +55°C
Power: 27.5 ±0.5 V DC

Status
In service.

Contractor
Bendix/King.

EHI 40 and 50 Electronic Horizontal Situation Indicator (EHSI)

Type
Electronic flight instrument.

Description
The EHI 40 and 50 provide bright multicoloured displays driven by dedicated stroke and raster technique. They include a built-in or separate mode controller, navigation mapping capability and modular design with interfaces for analogue and digital inputs. The EHI 40/50 has four formats in 360° mode: standard HSI compass rose, nav map, nav with weather and an optional DG-only mode. In ARC mode, display formats include: standard HSI, nav map, nav map with weather, and standard HSI with weather. Added functions include joystick interface and a wide range of sensor input interfaces. The system consists of a control display unit, symbol generator and associated navigation sensors. The EHI 40 is available with either a combination control display unit or traditional display with separate control panel. The EHI 50 requires a separate control panel.

During 1990, a companion attitude director indicator was introduced. The complete 4 in EFIS is designated EFS 40. The system is also available in a 5 in format, designated the EFS 50.

Specifications
Dimensions: (display unit) 106 × 106 × 238.8 mm (symbol generator) 193.5 × 57.2 × 320.5 mm
Weight:
(display unit) 2.275 kg
(symbol generator) 3.64 kg
Power supply: 28 V DC

Status
In service.

Contractor
Bendix/King.

EHSI-74 Electronic Flight Instrument System (EFIS)

Type
Electronic Flight Instrumentation System (EFIS).

Description
EHSI-74 is for the general aviation single- and light twin-turboprop market and comprises a 4 × 4 in (102 × 102 mm) EHSI designed to work in conjunction with the company's ADI-84 attitude director indicator and an information display/radar navigation centre, the Collins IND-270 CRT, which is used in conjunction with the company's WXR-270 weather radar. The system is completed by a DCP-270 display control panel.

The IND-270 CRT can present up to 128 pages of easily accessed pilot programmable text and navigation information from a Collins LRN-85 long-range navigation system. The equipment is programmed by a portable data reader that can load the system with performance tables such as cruise/consumption, emergency checklists and other alphanumeric information. Chapter by chapter indexing facilitates input and retrieval. In addition to the information stored within the DCP-270, many pages of data are available from the LRN-85.

The EHSI-74 no longer requires a US STC as do other EFIS in most installations.

Status
In production and in service. Certified on the Commander 690, King Air 200, Cessna 441, King Air F90, Bonanza, Beech 100, Mitsubishi MU-2 and dual installation for the Learjet Model 35A. Recently the system was installed in the Chinese Y-7.

Contractor
Rockwell Collins.

EHSI-3000 Electronic Horizontal Situation Indicator (EHSI)

Type
Electronic flight instrument.

Description
The design technology for the EHSI-3000 is based on the GH-3000 Electronic Standby Instrument System (ESIS), which was certified in 1997 for display of attitude, air data and navigation data. The EHSI-3000 is designed to replace ageing analogue electro-mechanical HSIs in new or converted aircraft equipped with MIL-STD-1553B databus architecture.

With the inclusion of a graphics processor and anti-aliasing graphics, the EHSI-3000 provides an easily read display and high reliability. It incorporates an Active Matrix Liquid Crystal Display (AMLCD) that is visible in direct sunlight, as well as a dark cockpit, and is NVE compatible. A built-in photocell on the front bezel, when used in conjunction with the setting of the aircraft databus, automatically adapts to ambient lighting conditions.

Designed to replace older electromechanical horizontal situation indicators, the EHSI-3000 provides distance, bearing, and course information from onboard INS and TACAN in a 3ATI display.

Specifications
Weight: < 1.4 kg
Power: 28 V DC at 1.5 A; 26 V AC, 400 Hz, 0–4.5 V AC/DAC bus
Environmental: MIL-STD-461, 462, 810

Status
The EHSI-3000 was selected by Lockheed Martin Tactical Aircraft Systems (Fort Worth) for the US Air Force F-16 Block 40/50 Common Configuration Implementation Program (CCIP). The program is upgrading approximately 750 aircraft. In this application, the EHSI-3000 will interface, via the MIL-STD-1553 databus, to the new upgraded TACAN chosen for the CCIP, as well as other existing analogue equipment.

Contractor
L-3 Communications, Avionics Systems.

EHSI-3000 Electronic Horizontal Situation Indicator (EHSI)   0098574

EHSI-4000 Electronic Horizontal Situation Indicator (EHSI)

Type
Electronic flight instrument.

Description
Derived from the HSI-3000, the EHSI-4000 (Electronic Horizontal Situation Indicator) was designed to replace older electromechanical HSIs and developed specifically for the requirements of business aircraft environments. The EHSI-4000 allows for pilot-commanded heading and course inputs on a single flat panel 3 ATI Active Matrix Liquid Crystal Display. The unit features basic functions for selection among four modes: ILS, NAV, ILS/NAV, VOR, DME, and FMS.

Specifications
Weight: <1.4 kg
Power: 28 V DC at 1.5 A; 26 V AC, 400 Hz, 0–4.5 V AC/DAC bus
Environmental: MIL-STD-461, 462, 810

Status
The EHSI-4000 has been selected for the Cessna Citation Sovereign programme.

EHSI-4000 electronic horizontal situation indicator 0103880

Contractor
L-3 Communications, Avionics Systems.

EIS-3000/EICAS-5000

Type
Engine Instrument Crew Alerting System (EICAS).

Description
The EIS-3000 Engine Indication System is designed to provide engine indication in an integrated flight deck. The information displayed includes torque, interstage turbine temperature, high and low pressure gas generator RPM, fuel flow, oil temperature, oil pressure and any other primary system indication that requires full-time display for aircraft operation.

The EICAS-5000 Engine Indication and Crew Alerting System (EICAS) fully integrates engine indication, crew alerting, synoptic display of aircraft systems and processing of maintenance data into the flight deck.

The crew alerting feature provides CAS messaging, flight deck lamp control and aural alerting. Aural alerts can include voice and synthesised tones.

A typical EIS-3000 utilises two large colour Adaptive Flight Displays (AFDs), a Display Control Panel (DCP) and 2 to 4 EIS Data Concentrator Units (DCUs). Installations with Full Authority Digital Electronic Control (FADEC) and non-FADEC engines are supported.

A typical EICAS-5000 utilises two large colour AFDs, an EICAS Control Panel (ECP), one to three EICAS DCUs, one Remote Data Concentrator (RDC) and a Lamp Driver Unit (LDU). The following line replaceable units can be used to cofigure an EICAS scaled to the specific aircraft installation:
- Two AFDs
- ECP
- 1-3 DCUs (five DCU families allow scaling to any application)
- 1-2 RDCs (two RDC families allow scaling to any application)
- LDU.

Large-format AFDs allow integration of EICAS, EFIS and Situational Awareness (traffic, weather, and so on) information on a single display unit. Two units are usually necessary to insure availability of engine information and meet certification requirements.

A typical EICAS display format would present the primary engine indication, primary system indication and crew alerting messages in fixed regions on the display. Display of this information is usually required throughout all phases of flight and includes engine, fuel, oil and pressurisation parameters.

A secondary region of the display would be used to display secondary system indication and synoptic displays. This information is considered supplementary to the primary information above and different displays may be required during certain phases of flight. Examples of secondary system indication include landing gear and surface position displays. Examples of synoptic displays include:
- Hydraulic system
- AC electrical system
- DC electrical system
- Flight Control System (FCS)
- Environmental Control System (ECS)
- Fuel system
- Anti-ice system
- Engine systems
- Door position.

The EIS-3000 and EICAS-5000 use manual and automatic page control and are generally tailored to the specific airframe application. Automatic page control is typically triggered by an event that requires immediate display of a particular EIS/EICAS parameter. Manual page control is performed with the ECP. Reversion of EIS/EICAS between displays is implemented in accordance with the airframe design, aircraft operational philosophy and flight deck operational philosophy.

The DCU transforms raw aircraft system data into EIS/EICAS information with the following functions:
- **Data Acquisition.** The DCU is connected to the engines and the aircraft systems that are monitored with the EIS/EICAS. The interfaces to these systems include analogue sensors,

discrete sensors and digital data (ARINC 429 high speed and low speed). The raw data is transformed into engineering units that can be routed to other systems and processed into display data.
- **Engine Indication Processing.** The DCU processes engine signals from FADEC and non-FADEC engines for EIS/EICAS display. The engine data is transformed and transmitted to the engine displays.
- **Crew Alert Processing.** The DCU processes aircraft system signals and generates CAS messages alerts, aural alerts and lamp enable signals that are transmitted to the displays, audio system and lamp driver unit.
- **Synoptic Data Processing.** The DCU processes aircraft system signals and generates synoptic display information that is transmitted to the displays.
- **Flight Data Recorder Processing.** The DCU acquires, formats and transmits aircraft system data to the FDR.
- **Maintenance Data Processing.** The DCU acquires, analyses, formats and transmits aircraft system data to the maintenance and diagnostic computer.
- **Data Routing.** The DCU performs direct routing of data between aircraft systems, concentrates ARINC 429 serial data onto general use output busses and transmits transformed aircraft system data to other aircraft systems.

The RDC is used to increase the quantity of aircraft signals that can be processed and provides remote connectivity to aircraft systems in order to reduce the aircraft wiring and weight. It provides a low latency pull through of analogue, discrete and digital signals onto general use ARINC 429 output busses.

The LDU is a dual-channel unit capable of driving up to 120 annunciator lamps. Channels one and two receive digital buses from all the DCUs. The busses contain lamp activation words from the DCUs. Channels one and two are identical and the outputs from each side are tied together (OR logic). If one channel lamp sink fails, the other channel lamp sink will provide the annunciator function. The LDU monitors the lamp sinks to verify correct function and outputs the lamp sink states on a digital bus to the DCUs.

Status
In production and in service.

Contractor
Rockwell Collins.

Electronic Flight Bag/Pilot Information Display (EFB/PID)

Type
Electronic Flight Bag (EFB).

Description
Astronautics' Electronic Flight Bag/Pilot Information Display (EFB/PID) has been certified for multiple applications in all phases of flight. The system brings benefits in terms of organisation and cost savings to the cockpit environment by facilitating a 'paperless cockpit', improving the efficiency of technical information updating, reducing training and operation costs, reducing potential crew errors in checklists and procedures, reduces the potential for runway incursion (via a detailed airport mapping facility) and, as a Class 3 EFB, the PID also enables support of dynamic charts, airport maps, and communications (CPDLC).

Other features and benefits of the system include:
- Available as Class 2 or 3 EFB
- 10.4 or 8.4 in display sizes available
- TSO certified hardware
- Certifiable Linux™ Operating System (OS)
- Accepts Linux™ and/or Windows™ application programmes
- Available with single- or dual-processor and hard drive

- Programmable and dedicated bezel keys, touch screen, and external keyboard.

  Functionality of the EFB/PID includes network connection, ADS-B, TCAS, maintenance/flight manuals, video surveillance and aviation weather information (with optional weather application).

Specifications
Processor: 400 MHz Celeron™
Useable screen area: 6 × 8 in (152.4 × 203.2 mm) or 5 × 7 in (127 × 177.8 mm)
Resolution: 1,024 × 768 or 800 × 600
Luminance: 150 fl
Interface: ARINC-429, -453, RS-422, video, Ethernet
Dimensions: 2 MCU
Power: 28 V DC
Battery: External, 3½ h
Software: Multiple commercially available and in-house airline-developed programmes can be hosted

Status
In production. Applicable to, and qualified for, commercial aircraft and military transports.

Contractor
Astronautics Corporation of America.

Electronic Flight Instrument (EFI)

Type
Electronic flight instrument.

Description
Astronautics' 5 ATI Electronic Flight Instrument (EFI) is certified to display TAWS/EGPWS data and can be installed as a direct replacement for electromechanical HSIs and ADIs on transport aircraft.

The EFI can function as an ADI or HSI with several functional overlays. Designed around an Active Matrix Liquid Crystal Display (AMLCD), the instrument is capable of receiving and processing ARINC-429/702 data from RNAV/FMS/GPS and TCAS/CDTI systems, analogue/ARINC-429 data from radios (VOR, DME, ADF) and ARINC-453/708 from Colour Weather Radar (CWR) and Enhanced Ground Proximity Warning Systems. The EFI also has RS-422 (cross-talk), analogue, synchro, discrete, and differential resolver interface capabilities.

The EFI has been certified to TSO requirements and as 747 EFIS to function as and ADI, HSI, combined ADI/HSI, CWR, FMS/Map, EGPWS/TAWS/TCAS display and CDTI/ADS-B display (optional).

Other features of the system include:
- Sunlight readable
- Remote Control Panel (RCP)
- Modular, rugged construction
- Certified to DO-160D, DO-178B, Level A
- HIRF (200V/M), lightning.

Specifications
Useable screen area: 4.24 × 4.24 in (107.7 × 107.7 mm)
Resolution: 524 × 524
Display luminance: 150 fl (200 fl option)
Viewing angle: ±60°
Processor: Pentium™
Interfaces: ARINC 429, 453, 561, 407 (synchro), RS422, analogue, discrete, video (optional)
Dimensions: 5 ATI
Weight: <3.64 kg
Power: 28 V DC, 45 W (average without heaters)
Control panels: Multiple configuration options
Reliability: 15,000 h MTBF
Temperature:
(operating) –40 to +55°C (+71°C intermittent operation)
(storage) –55 to +85°C
Certification: TSOs C113, C119a, C4c, C63c, C6d; DO-178B Level A
Interfaces: ARINC 561, 419/429/702, 708/453; analogue; discrete; ARINC 407-1 Synchro; RS422

Status
In production. Product applications include the A-109 and S-61 helicopters, Gulfstream III, Boeing 747-200/-300, 747SP, DC-10 and 727, OV-10 and CP-140 Aurora aircraft.

Contractor
Astronautics Corporation of America.

Electronic Warfare Display (EWD)

Type
Cockpit display.

Description
Designed for the extreme environment of fighter aircraft applications, Astronautics' 3 in Electronic Warfare Display (EWD) is a compact, high-resolution, high-brightness monochrome raster CRT display with a useful screen area of 2.3 in square.

The EWD is also capable of displaying electro-optical weapons and sensor data.

The display features a modular architecture; each module can be replaced independently without further adjustments or calibration. An integral BIT continuously checks the display circuitry, providing failure indication and permits immediate failure isolation.

Other features and benefits of the EWD include:
- Designed for military aircraft and helicopter environments
- High brightness and contrast ratio
- Integral contrast enhancement filter
- High resolution
- NVIS-compatible
- Self-contained and extremely compact
- Low weight and power consumption
- Cooling by convection
- High reliability and maintainability.

Specifications
Screen size: 2.7 in diagonal (68.6 mm)
Filter: Narrow band, matched to P-43 phosphor type with anti-reflective coating
Brightness: 150 fl
Contrast ratio: 10:1 (minimum)
Shades of grey: 7 (minimum)
Resolution: 5.5 mR
Video inputs: Stroke
Video impedance: 75 Ω input
Power: 28 V DC (MIL-STD-70D); 35 W nominal, 45 W (max)
Dimensions: 3 × 3 in (76.2 × 76.2 mm)
Weight: 2.27 kg
Temperature: –20 to +55°C
Altitude: 0–30,000 ft operating, 70,000 ft non-operating

Status
In production and reported in service in F-4 aircraft.

Contractor
Astronautics Corporation of America.

Engine Instrument and Caution Advisory System (EICAS) display

Type
Engine Instrument Crew Alerting System (EICAS).

Description
Astronautics' EICAS is a VGA-compatible, low-power, sunlight readable, and NVG-compatible Active Matrix Liquid Crystal Display (AMLCD). Modular, solid-state construction and programmable symbols make the display readily adaptable to a wide range of input/output applications. Large non-volatile memory storage capacity with easy direct downloading simplifies the transfer of engine data to maintenance records. A separate serial test port

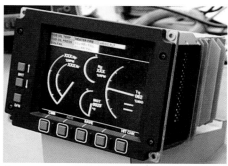

Astronautics engine instrument and caution advisory system display 0568536

can be used to amend parameters to meet user requirements. When two displays are installed, a cross-talk channel provides complete redundancy.

Specifications
Processor: Pentium
Viewing area: 5.1 × 3.8 in (129.5 × 96.5 mm)
Resolution: 640 × 480 pixels
Viewing angle:
(horizontal) ±45°
(vertical) +30 to –10°
Brightness: 150 fL
Multifunction keys: 5 programmable soft keys, plus 2 for brightness and 1 for day/night
MTBF: 10,000 h
Dimensions: 203.2 × 139.7 × 203.2 mm
(190.5 × 152.4 × 203.2 mm with PCMIA)
Weight: 3.64 kg (maximum)
Power: 28 V DC; 40 W nominal
(100 W with heater)
Certifications: DO-16-C, DO-178B Level A
TSO: C113, C43b, 47a

Status
In service, applications include A-119, Ayers Loadmaster and UH-1H.

Contractor
Astronautics Corporation of America.

Engine Performance Indicator (EPI)

Type
Cockpit display.

Description
The Astronautics family of Engine Performance Indicators (EPIs) are Active Matrix Liquid Crystal Displays (AMLCDs), monitoring and displaying five critical engine parameters: Engine RPM, low Turbine Inlet Temperature, Fuel Flow, Oil Pressure and Nozzle Position. It features a contrast ratio of 15:1 during the day and 10:1 at night, and has a viewing angle of ±75°. The system is designed to achieve full dual redundancy, be of minimum weight, volume and power, be easily maintainable and have high reliability. The EPI also functions as an EICAS display for use in aircraft with and without FADECs.

The EPI is designed to operate over a temperature range of –55 to +85°C and within an altitude band from sea level up to 70,000 ft. It meets the requirements of MIL-E-5400T and MIL-STD-810C.

Status
In production for LC-200 and UH-1 aircraft.

Contractor
Astronautics Corporation of America.

Engine performance indicators

Type
Cockpit Display

Description
Honeywell Aerospace makes a series of engine performance indicators for business, commuter

and military aircraft. The indicators provide a continuous analogue display of critical engine variables, such as fan and core rotation speeds, turbine temperature and fuel flow. Configurations with circular or square cross-section are available.

Solid-state circuitry is employed throughout. The indicators are unaffected by the large fluctuations in electrical supply during engine starts.

Specifications
Power supply:
12–34 V DC, 120 mA nominal, 450 mA (max)
5 V DC, 250 mA for lighting
Input signals:
(rotation) monopole pulse
(turbine temperature) chromel-alumel thermocouple per NBS Monograph 125
(fuel flow) second harmonic selsyn mass flow transmitter, 115 V AC 400 Hz reference
Accuracy:
(rpm) (0 to +55° C ambient) ±0.25% at 100% rpm
(−30 to +70°C ambient) ±0.5% at 100% rpm
(temperature) (0 to +55°C ambient) ±5°C at 900°C
(−30 to +70°C ambient) ±10°C at 900°C
(fuel flow) (0 to +55°C ambient) ±1% of full-scale
(−30 to +70°C ambient) ±2% of full-scale
Rpm range: 0 to 110%
Turbine temperature range: +100 to +1,000°C
Fuel flow range: 0 to 2,300 lb/h
Response: 3 s full-scale slew
Temperature range:
(operating) −30 to +55°C
(short term) −30 to +70°C
(ambient extreme) −65 to +70°C
Shock: 6 g for 0.011 s
Vibration: 5 to 3,000 Hz, 0.02 in amplitude, limited to ±1.5 g
Humidity: 95%, +70°C for 6 h, cool to +38°C for 18 h

Status
In production. Typical applications are the Raytheon Hawker 125–700, Gates Learjet 35/36 and 55, Lockheed Martin JetStar 2 and CASA CN-235.

Contractor
Honeywell Aerospace, Electronic & Avionics Lighting.

Enhanced Main Display Unit for the E-2C

Type
Multi Function Display (MFD).

Description
The Enhanced Main Display Unit (EMDU) is the primary information display for the crew of the E-2C Hawkeye AEW aircraft. Each aircraft carries a complement of three EMDUs. The full-colour EMDU displays more than 2,000 radar tracks on a 279.4 × 279.4 mm (11 × 11 in) screen. The EMDU also allows the operators to overlay a map of the search area over their track files and display up to three separate windows containing additional information.

The EMDU utilises beam index CRT technology to provide sharp high-resolution colour images. In addition to the CRT, the EMDU contains processing subsystems that perform graphics generation, radar scan conversion and data input/output functions. The EMDU performs all display oriented processing

internally, thus offloading that requirement from the E-2C's L304 central computer.

Specifications
Weight: 65.77 kg
Resolution: 900 × 900 pixels
Reliability: 1,100 h MTBF

Status
In service in US Navy E-2C aircraft.

Contractor
L-3 Communications, Display Systems.

E-Scope radar repeater display for the Tornado IDS

Type
Cockpit display.

Description
The primary function of the E-scope display is to portray video signals from the Raytheon Systems Company Terrain-Following Radar (TFR) in the Panavia Tornado IDS. In the terrain-following mode it shows sectional representation of the terrain ahead of the aircraft, on which a hyperbolic curve, termed the Zero Command Line (ZCL), is overlaid. This ZCL is calculated to yield no requirement for aircraft manoeuvre (either by command to the autopilot or via Flight Director (FD) in the HUD) when it precisely grazes the terrain contour. If the terrain contour should break through the ZCL (rising ground ahead of the aircraft flight path), the aircraft will be commanded to climb. If the terrain contour falls away from the ZCL (falling ground ahead of the aircraft flight path), the aircraft will be commanded to descend until the ZCL once again just grazes the terrain contour. The allowed distance between the ZCL and the terrain contour is dictated by the pilot-selected ride height. The severity of aircraft manoeuvre to achieve the selected height (adherence of the aircraft flight path to the terrain profile) is controlled by the ride level (again, pilot-selected, hard/medium/soft).

In German Air Force Tornado aircraft, the E-Scope display also acts as a secondary radar repeater display, allowing the pilot to view a small-scale repeat of the navigator's radar display. The display may be checked via an operator-initiated test pattern, with correct data input from the TFR confirmed by a flashing cursor at the top of the display in terrain-following mode.

Status
In service in the Panavia Tornado IDS.

Contractor
Astronautics Corporation of America.

Eventide Airborne Multipurpose Electronic Display (EAMED)

Type
Cockpit display.

Description
The Eventide Airborne Multipurpose Electronic Display (EAMED) is a powerful compact digital avionics development platform. It is a self-contained 16-bit avionics computer with high memory capacity which includes a CPU board, power supply, video board and CRT and a comprehensive set of input/output peripheral connections, including RS-232C and RS-422. EAMED displays both text and graphics on a high-resolution sunlight-readable screen. The unit fits in a standard panel cutout and hardware meets the DO-160B specification.

EAMED has a wide range of applications, including onboard communications, data acquisition, reconfigurable indicators and interactive situation displays.

Specifications
Dimensions: 76 × 76 × 266.7 mm
Weight: 1.6 kg
Power supply: 11–33 V DC, 15 W

Contractor
Eventide Avionics.

F-16 modular mission computer

Type
Aircraft mission computer.

Description
The computer system for the F-16 provides multiple processing functions in a single chassis. It replaces three present computers in the earlier configuration and provides processor power to support the addition of capabilities such as FLIR and digital terrain functions. SEM-E format modules provide data processing, avionic display, power supply and bus interface for the computer. Various aircraft system functions are

The L-3 enhanced main display unit for the E-2C aircraft   0504096

E-Scope display mounted to the left of the HUD in a UK RAF Tornado GR.1 (E L Downs)   0109930

defined by the application software operating on the computer. The primary computing module processors are based on the R3000 32-bit RISC instruction set architecture. Each processor module provides 16 VAX/Mips throughput and has 1 Mbyte of on-module memory. The computer, as configured for the F-16 with 16 digital modules installed, weighs 17.78 kg.

Status
In service in F-16 aircraft worldwide.

Contractor
Raytheon Electronic Systems.

FDS-255 Flight Display System

Type
Avionic display system.

Description
The Rockwell Collins Flight Display System (FDS-255) is made up of MFD-255, 5ATI colour flat-panel multifunction displays offering an extremely wide viewing angle. The FDS-255 is a key element of the Rockwell Collins Flight2 Mission Management Systems. MFD-255s can replace existing instruments, combine functions of existing instruments, or provide the entire display system (FDS-255) in a new or retrofit application. The active matrix liquid crystal design and proven Rockwell display circuitry give the user a highly reliable and easily maintainable flight instrument. FDS-255 is programmed in Ada. User specific display formats can be developed and downloaded over the ARINC 429 bus.

Featuring a 'smart head' design, the FDS-255 contains internal graphics generation, analogue and digital input/output and weather radar. In a dual installation, reversionary colour, full-motion video is provided with a video card slot.

The FDS-255 video with 64 grey scales will display FLIR, digital map and TCAS; colour, full-motion video is provided via an optional video card. The wide viewing angle permits easy cross-cockpit viewing.

Because of the highly integrated display design, Rockwell is able to provide 17.7 sq in of usable viewing area in a 5 ATI format. The FDS-255 has eight standard modes, but can easily be tailored to each customer's specific application.

Flight/navigation operating modes include: ADI, HSI, PFD, ARC, radar-map, windshear, map and hover. TSO versions are also available.

Specifications
Dimensions: 129 × 129 × 222 mm
Weight: 3.6 kg (max)
Power supply: 28 V DC, 50 W typical
(See also Flight2 Systems entry)
Viewing angle: ±65° horizontal, +40°/–20° vertical

Interfaces: ARINC 429 and 708, analogue, synchro and discrete interfaces
Reliability: >11,000 h MTBF
Temperature: –40 to +55°C operating

Status
In production for the US Air Force C/KC-135 Pacer Crag and US Navy P-3. Deliveries began in 1996. Also in production for C-12, C-130, KC-10, MH-60, P-3, VC-10 and various helicopter and commercial aircraft applications.

Contractor
Rockwell Collins.

FDS-2000 Flight Display System

Type
Avionic display system.

Description
Collins' FDS-2000 Flight Display System provides the aircraft interface, data processing, display processing and display control to replace existing cockpit electromechanical indicators. The FDS-2000 replaces both the Attitude Director Indicator (ADI) and the Horizontal Situation Indicator (HSI), with control provided by the Display Select Panels (DSP). The FDS-2000 provides flight crews with primary flight display instrumentation incorporating attitude, heading, flight director guidance and other features available depending on the capabilities of installed avionics.

FDS-2000 system components include Adaptive Flight Displays (AFDs), a Display Control Panel (DCP) and Data Concentrator Unit (DCU). The AFD-2000 is an ARINC 5ATI flat-panel colour indicator. Adaptable to multiple display configurations, the AFD-2000 can be configured as an Attitude Direction Indicator (ADI) or Horizontal Situation Indicator (HSI) flight director instrument. The system's Primary Flight Display (PFD) format combines both ADI and HSI on a single display, with graphics display support for ground proximity warning, weather, FMS map, TCAS and other advanced information. The use of active matrix liquid crystal display technology provides sharp, bright displays that deliver improved daytime and night-time

readability. The DCP-2000 provides the primary interface and integration management between the pilot and the displays. The flight deck-mounted control panel permits the operator to select course and heading, sequence through active navigation source data, choose display formats, select range and initiate diagnostic/built-in-test of the display.

The main features of FDS-2000 are as follows:
- Maximum active display surface available in 5ATI format
- Fits into standard 5 in electromechanical installations
- Full complement of flight instrument software provides ADI, HSI and PFD (combined ADI/HSI formats)
- High-resolution active matrix LCD
- Designed for optimum flight deck viewing
- Capable of displaying EGPWS, FMS, TCAS and weather radar data
- Provides a platform for future digital ARINC 429 I/O growth.

Specifications
Dimensions: (ADF-2000A) 129 (H) × 129 (W) × 230 (L) mm
(DCP-2000) 76 (H) × 146 (W) × 178 (L) mm
(DCU-2000) 76 (H) × 124 (W) × 279 (L) mm
Weight: (ADF-2000A) 3.1 kg
(DCP-2000) 0.54 kg
(DCU-2000) 1.9 kg
Temperature: (ADF-2000A) –40 to +55°C
(DCP-2000) –20 to +70°C
(DCU-2000) –55 to +70°C
Altitude: 55,000 ft (ADF-2000A, DCP-2000); 35,000 ft (DCU-2000)
Certification:
(ADF-2000A) FAA TSO C3d, C4c, C5e, C6d, 34e, C35d, C36e, C40c, C41d, C52b, C66c, C87, C92c, C113, C115b, C117a, C129a; DO-160C; [A2F1]-BA(B'BMS) E1XXX XXZ[BZ] AZZWZ[A3E3]XX; ED-14C; DO-178B Levels A and C
(DCP-2000) FAA TSO C52b, C113; DO-160D; AIA(BB)XXXX FXZAABZYZA2XX ; ED-14D; DO-178B Levels A and C
(DCU-2000) DO-160C; C2/BB/CLM/E1 XXXXXZ[BZ] AZARA[A3E3]; ED-14C; DO-178B Levels A and C
Power: (FMC-6000) 28 V DC; 50 W
(CDU-6000) 28 V DC; 10 W
(DBU-4100) 28 V DC; 14 W

FDS-255 Flight Display System   0131536

Collins' FDS-2000 display system (Rockwell Collins)   0131023

For details of the latest updates to *Jane's Avionics* online and to discover the additional information available exclusively to online subscribers please visit
jav.janes.com

Collins' FDS-2000 AMLCD is claimed to utilise 50 per cent less power, weight and space than conventional aircraft instrumentation 0105290

Status

In production and in service. Certified on the Gulfstream II, IIB, III, Bombardier Challenger 600 and Cessna Citation 650 aircraft.

Contractor

Rockwell Collins.

Flat Panel Integrated Displays (FPIDs)

Type

Electronic Flight Instrumentation System (EFIS).

Description

Universal Avionics produces a range of colour active-matrix liquid-crystal FPID. Current models are: MFD-640; Electronic Flight Instrument, EFI 550, 600, 640 and 890; and Electronic Engine Instrument EEI-640. Details of representative models are shown below.

Electronic Flight Instrument EFI-640

The EFI-640 has a 6.4 in diagonal display for EADI and EHSI applications which are not supported on the MFDs. Likewise, only EFIs support the display of warnings and annunciations and have cross cockpit and same-side reversionary capabilities. The EFI has a full anti-aliased graphics capability and supports both analogue and digital interfaces. It provides horizontal viewing angles of ±60° and electronically adjustable vertical viewing angle, with +45 to −10° vertical coverage, relative to the normal. Sunlight readability with greater than 10,000:1 dimming range is provided.

Specifications

Bezel dimensions: 131.5 × 156.8 mm
Display size: 98.1 × 130.6 mm
Overall dimensions: 5 × 6ATI
Weight: 3.86 kg
MTBF: >7,500 h
Power: 28 V DC, 90 W
Supported functions: HSI, ADI, navigation, map, flight plan, weather radar, TCAS, ADS-B, traffic

The Universal Avionic Systems' EFI-640 colour active-matrix liquid-crystal flat panel integrated display 0062208

The Universal Avionics Systems' MFD-640 colour active-matrix liquid-crystal flat panel integrated display 0062209

information service, cross cockpit and same-side reversionary, warnings and annunciators.

MultiFunction Display MFD-640

The MFD-640 is designed for the display of moving maps, weather radar, terrain awareness, TCAS and other data. It is similar to the EFI-640 in construction and optical capabilities but, unlike the EFI-640, it can display terrain avoidance, EGPWS information and UniLink images.

Specifications

Bezel dimensions: 131.5 × 156.8 mm
Display size: 98.1 × 130.6 mm
Overall dimensions: 5 × 6ATI
Weight: 3.86 kg
MTBF: >7,500 h
Power: 28 V DC, 90 W
Supported functions: Navigation, map, flight plan, weather radar, TAWS, EGPWS, TCAS, ADS-B, UniLink messages, traffic information service

Electronic Engine Instrument EEI-640

The EEI-640 displays provide a flat panel electronic engine instrument and aircraft systems display suite designed specifically for the Falcon 20 with 731 engines. The suite comprises three 5×6 ATI FPIDs stacked in a vertical column in the centre of the instrument panel.

Contractor

Universal Avionics Systems Corporation.

Flight Data Acquisition Management System (FDAMS)

Type

Flight/mission recording system.

Description

The Flight Data Acquisition Management System (FDAMS), combines the functionality of an ARINC 717 Digital Flight Data Acquisition Unit (DFDAU) with the real-time monitoring and troubleshooting capability of a Data Management Module (DMM) in a single 6 MCU box. It is a powerful computing machine with 2 Mbytes of solid-state mass memory.

The FDAMS provides a standard set of 64 maintenance, operational and special algorithms, four of which can be customised for troubleshooting. The airline can also enable or disable any of the 64 algorithms directly by using ground support software.

The ARINC 717 DFDAU provides the mandatory ARINC 573 flight data to the flight recorder. With its seven databases, the DFDAU meets the FAA 88-1, CAA and ICAO requirements for all newly built aircraft. To accommodate any future aircraft modifications easily, Sundstrand's databases can be reloaded from a floppy disk using an ARINC 615 data loader.

The DMM monitors a predefined set of parameters in the aircraft. It then collects and downlinks both exceedance and routine data via a cockpit printer, ACARS or data loader. The data is processed to provide both maintenance and flight operations reports necessary for monitoring the

engines and other aircraft systems. Using an IBM-compatible PC, the readout programs provide information in numerical or graphical format.

Contractor

Honeywell Aerospace, Electronic & Avionics Lighting.

Flight Management Systems

Type

Flight Management System (FMS).

Description

Rockwell Collins' FMS family, including the FMS-3000, -4200, -5000 and -6000, are satellite-based precision navigation systems designed for all phases of flight, including take-off, en route, approach and landing. While offering broadly similar capabilities, each variant is designed to fulfil the needs of specific aircraft groups in terms of capabilities and available flightdeck space.

At the core of each FMS series is a Rockwell Collins GPS sensor. The capabilities of a full FMS are incorporated within the navigation system. The avionics GPS engine features 12-channel satellite tracking, including those comprising the Wide Area Augmentation System, providing satellite coverage sufficient to meet the FAA-required navigation performance and precision approach criteria; a new software code to ensure certification to DO-178B Level A; Receiver Autonomous Integrity Monitoring (RAIM) and predictive RAIM, and differential correction computation to qualify the receiver for Cats I and II.

The FMS series provides the capability to calculate position and navigation information accurately anywhere in the world; provide VNav guidance and comprehensive pilot annunciations; fly complete airways, SIDs, STARs and approach legs, and predict fuel consumption. FMS also significantly reduces pilot workload by automating many calculations and functions. Information is presented on multifunction display pages, allowing easy, efficient movement through the system. Pilots can remain eyes-up during all phases of flight, resulting in enhanced situational awareness and greater control. Flight planning is simplified to the point where even the most complicated procedures can be generated with a few keystrokes.

Specifications

See system entries.

Status

In production and in service.

Contractor

Rockwell Collins.

Flight Management System (FMS)

Type

Flight Management System (FMS).

Description

The Flight Management System (FMS) is used for commercial air transport applications, in single or dual configuration for the Boeing 737 and in a triple configuration for the Ilyushin Il-96. The system was designed to meet ICAO airspace requirements for Communication Navigation and Surveillance (CNS) for Air Traffic Management (ATM). This includes communications with an adaptable datalink to allow each airline to customise its airline communications; navigation with the latest in required navigational performance/actual navigation performance methodology certified to use GPS and surveillance with the building blocks to implement functions such as ATC clearance entry and ADS reporting.

The flight management computer system has a single high-power 32-bit microprocessor

The Smiths Aerospace flight management system is available in single or (as shown here) in dual configurations   0504191

and contains options for 4, 8 or 16 Mbytes of on-aircraft loadable EEPROM for storage of the operational flight program and navigation database. Up to two million bytes of memory is dedicated to the operator's navigation database and can be structured according to the customer's specification. A wide range of options is available from a simple listing of navaids and airports to a detailed coverage of the airline's operating routes, SIDs, STARs and gate assignments. Smiths Aerospace offers airlines a choice of using Jeppesen, Racal or Swissair for the navigation database update service.

The crew interface is via a CRT control and display panel on which 14 by 24 character lines of information can be displayed. Aircraft lateral and vertical profile data is presented and critical information is highlighted by reverse video presentation. The bottom line of the CRT is used as a scratchpad for crew entries.

The system can be coupled to the autopilot and autothrottle for automatic profile tracking and energy management. Fuel savings in the range 4 to 14 per cent are predicted in normal operations.

As discussed, the system, as applied to Boeing 737 variants, is a dual- or single-unit installation. The dual unit manages all performance, navigation and approach functions, as well as improving dispatch availability. The dual-unit format can be installed in place of the current single unit. The computer is a passively cooled 4 MCU form factor that is half the weight, uses one fifth the power and is over five times faster than competitive flight management systems. The system has spare memory and processor throughput to accommodate the functions of the CNS/ATM implementation programme and the Global Air Traffic Management (GATM) requirements of the US Department of Defense.

Specifications
Dimensions:
(display) 267 × 146 × 229 mm
(computer) 320 × 257 × 193 mm (4 MCU)
Weight:
(display) 8.2 kg
(computer) 7.7 kg
Power required:
(total) 75 W
Reliability: 18,000 h MTBF (predicted)

Status
In production for, and in service on, Boeing 737-300, -400, -500, -600, -700 and -800 aircraft, in single or dual configurations.

This FMS has also been selected for Airbus aircraft in co-operation with Sextant. Elsewhere, a triple configuration has been selected for the Ilyushin IL-96M, and in US military service the system has been selected for the US Air Force VC-25 and US Navy E-6 aircraft.

Contractor
Smiths Aerospace.

Flight2

Type
Integrated avionics system.

Description
Rockwell Collins' Flight2 provides fully-integrated and militarised cockpit/upgrade to meet the demands of Global Air Traffic Management (GATM) mandates which require integrated Communication/Navigation/Surveillance (CNS) systems and major flight deck operational changes to traditional military aircraft. These can be custom-configured for new and retrofit applications to include controls and displays, information/data processing and communication, navigation and/or safety and surveillance systems.

Flight2 is an open-architecture system, consisting of a high degree of COTS and Non-Developmental Item (NDI) equipment to ensure civil airspace interoperability, while meeting specialised military mission demands. Users can change military mission functions without affecting FAA-certified civil airspace capabilities, and vice versa. The system provides a powerful ethernet-driven Local Area Network (LAN), providing for technological growth, which can be scaled up for complex mission requirements or scaled down for simpler aircraft integrations.

As the primary integrating component of the Flight2 architecture, the Integrated Processing Centre (IPC), a dual-thread design which works with plug-in, hot-swappable, Line Replaceable Modules (LRMs), provides an adaptable core processing capability. The IPC isolates civil and military functionality.

In addition to providing COTS digital communication functions, the system supports military datalink systems, such as Link 16, Improved Data Modem (IDM) and Automatic Link Establishment (ALE), which grant military users access to global, secure C^2 in a Joint Tactical Radio System (JTRS)-compliant format.

The Flight2 system includes a fully commercial, militarised Flight Management Function (FMF) that provides integrated, seamless, en route navigation (combining FAA commercial flight management requirements and four-dimensional Required Navigation Performance (RNP) down to 0.15 nm) and precision approach, enhanced by Wide and Local Area Augmentation Systems (WAAS/LAAS).

Tactical displays merge complex data and perform the processing necessary to display multispectral imagery, digital terrain maps and digitised charts. Vector graphic overlay of tactical symbology provides the air surveillance situation and combined air- and ground-threat picture. In-theatre C^2 and the display of real-time intelligence are provided via 'windowed' text messages. The system also adds overlays of weather, radar, ground-mapping and target/threat data for the tactical arena. An integrated Head-Up Display (HUD) enables Category III approaches and tactical low-level missions operations.

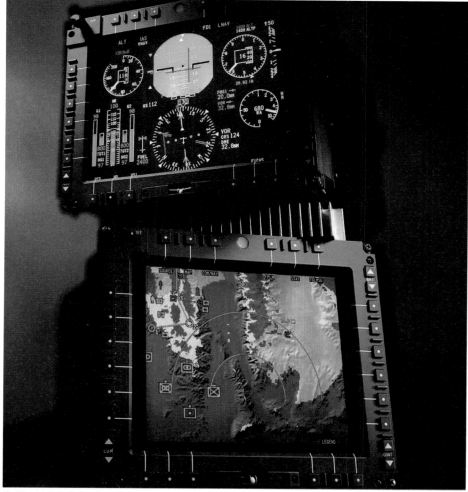

Collins' Flight2 displays (Rockwell Collins)   0534157

Collins' common cockpit architecture for the Sikorsky S-92, being delivered to the Turkish Army (Rockwell Collins) 0534158

Status

Flight2 is the core element of the PACER CRAG C/KC-135 avionics upgrade for the US Air Force. The system also forms the basis of the Rockwell Collins common cockpit architecture for the S-70 and S-92 rotorcraft. This new cockpit design is being delivered to the Turkish Army and features 6 × 8 in AMLCDs, dual FMS, weather radar with flight path overlay, and Forward Looking Infra-Red (FLIR).

Contractor

Rockwell Collins.

FlightMax Flight Situation Displays

Type

MultiFunction Display (MFD).

Description

Avidyne Corporation has developed a new range of fully FAA-certified Flight Situation Displays (FSD), designed to interface with a wide range of Flight Management (FM), sensor and display systems. The FlightMax series of displays offer the following common features:

- Vector-graphic moving map with colour-contoured terrain base (when interfaced with a suitable GPS/FMS)
- BF Goodrich Stormscope® WX 500 Lightning interface
- IFR & VFR charts (Jeppesen NavData for North America with optional worldwide NavData), including Terminal Area Charts (TACs)

- Traffic collision avoidance information (Ryan TCAD™ (Traffic and Collision Alert Device), BF Goodrich Skywatch™ Traffic Advisory System (TAS), or conventional Traffic Alert and Collision Avoidance Systems (TCAS I). Traffic data is displayed on a full-size, full-color display with standard traffic symbology, providing for increased situational awareness for the pilot.

The open-architecture design of the system will also accommodate enhancements including datalink graphical weather via Echo Flight or AirCell, and long-range traffic awareness via ADS-B (Automatic Dependent Surveillance Broadcast) and TIS (Traffic Information System).

Avidyne has integrated three new radar-enhancing features into the FlightMax 850, 750 and 650. *BeamView*, *TiltView* and *AutoTilt* provide additional information on the radar display, which help pilots more easily track and interpret weather radar data. Existing FlightMax 740 systems can also be upgraded to these new features:

- *BeamView* displays a sweeping 'flashlight beam' which corresponds to the beam width of the radar antenna. By displaying the radar beam width as a function of range, pilots can more easily discern targets that are separated by a distance of less than the beam width and which appear on the screen as a single target
- *TiltView* computes the relative altitude of the radar beam at each range, and displays this information adjacent to each range arc on the display scale in thousands of feet. At a glance, *TiltView* gives pilots the relative altitude of any

displayed target, and eliminates the need for cumbersome radar tilt calculations. *TiltView* allows pilots to easily determine aircraft altitude relative to terrain or cloud cell tops

- *AutoTilt* automatically adjusts the antenna tilt angle to compensate for changes in aircraft altitude and selected range, which eliminates much of the workload associated with normal tilt management.

FlightMax 450

The FlightMax 450 is designed to provide a low cost situational awareness capability for non-radar aircraft.

FlightMax 650

Tailored for aircraft with RDR 130/150/160 radar, the FlightMax 650 replaces existing radar indicators and provides all the functionality of the existing radar system, with the added benefits of the FlightMax family.

FlightMax 750

Primarily targeted at turbo props, cabin class twins and radar-equipped high performance singles, the FlightMax 750 includes interfaces for Bendix/King RDR 2000 and RDS 81/82NP radar systems. The system can replace most new and used Bendix/King radar indicators including support for Vertical Profile™.

FlightMax 850

The FlightMax 850 is designed to replace Collins WXR 250/270/270A/300 and Bendix RDR 1100/1200/1300 radar indicators which are found primarily in corporate aircraft, such as the Bombardier Aerospace Learjet, Cessna Citation and Beech King Air, as well as many regional airliners. The system also provides an optional interface for Honeywell's entire family of Enhanced Ground Proximity Warning Systems (EGPWS), providing a versatile Terrain Awareness and Warning System (TAWS) display solution. As an added feature, the FlightMax 850 provides traffic and terrain alerts as text messages with one-button pop-up access to the respective traffic or terrain warning display, allowing the pilot complete avoidance protection, even when viewing other displays.

Specifications

Display:
5 inch diagonal, colour Active-Matrix LCD
Sunlight readable (150fL)
500:1 dimming range,
320 × 234 pixels, 65,536 colours
Size:
FSD (W × D × H) 15.88 × 30.48 × 10.16 cm
Optional CD-ROM Data Loader (W × D × H)
15.88 × 25.40 × 3.18
Weight:
FSD >3.64 kg
CD-ROM Data Loader 0.68 kg
Power: 11–35 V DC, 5.0 Amp @ 14 V
GPS Interface: RS-232 or GAMA 429 high and low speed (optional on FlightMax 450)
Cooling: Forced Air Required
Operating altitude: Up to 25,000 ft. (cabin pressure altitude)
Operating temperature: –20 to +55°C
TSO compliance: TSO C113, TSO C110a, TSO C63c, TSO C118, TSO C147

Status

Avidyne received FAA Technical Standard Order (TSO) approval of its FlightMax 850, FlightMax 750, FlightMax 650, and FlightMax 450 Flight Situation Displays (FSD) during the third quarter of 2000. This is the fifth generation of Avidyne's planned series of certified products, which address situational awareness and safety for business and commercial aircraft operators. This certification includes seven new radar interfaces, with full overlay capability, including terrain and water base maps. Other interfaces include the enhanced versions of the Honeywell Mark V, Mark VI, Mark VII, Mark VIII, and KGP 560 GA-EGPWS, and utilise Honeywell's KCPB picture bus for optimum display clarity and resolution.

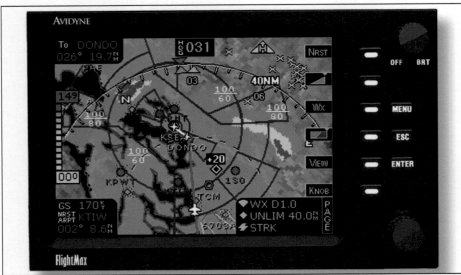

Avidyne FlightMax 850 showing customisable data blocks to top and bottom left, with weather, terrain and traffic collision avoidance data overlaid on the moving map 0099736

Avidyne FlightMax 850 showing current weather data obtained via AirCell datalink interface overlaid on the moving map   0099737

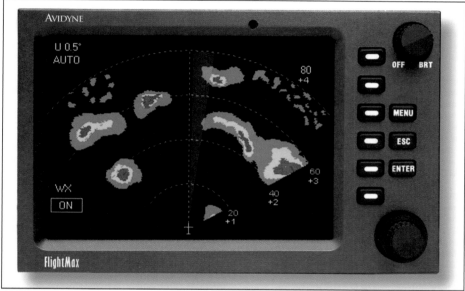

Avidyne FlightMax 850 showing BeamView, TiltView and AutoTilt radar enhancements   0099738

Simultaneously, Rockwell Collins announced an agreement to offer the FlightMax 850 FSD as an additional solution for operators of turbine-powered and entry-level jet aircraft. FlightMax will provide an integrated display solution for Collins Traffic Alert and Collision Avoidance Systems (TCAS II) and Collins weather radar systems, including legacy analog systems.

FlightMax FSDs are in production and in service on a wide range of corporate aircraft and regional airliners.

Contractor
Avidyne Corporation.

FMS-800 flight management system

Type
Flight Management System (FMS).

Description
The FMS-800 flight management system integrates the functions of communications, navigation and IFF control, GPS/INS navigation, flight instruments and controls, autopilot, stores and radar. It is intended for use in transport, tanker, trainer and utility aircraft. The system automates many of the functions normally carried out by the navigator. It also simplifies the complex tasks of the pilot and co-pilot, permitting them to concentrate on mission planning and execution.

The FMS-800 outputs dynamic data to the flight instruments and automatic flight control system using MIL-STD-1553B, ARINC 429/561 or analogue synchro signals. The system is compatible with existing mechanical flight instruments and analogue autopilots as well as digital systems such as the Rockwell Collins CDU-900 Control Display Unit, FDS-255 flight display system and APS-85 autopilot. The FMS-800 integrates GPS/INS navigation using a 12 state Kalman filter for airborne alignment and continued high accuracy. GPS and INS stand-alone navigation are also provided.

Workload is reduced and flight accuracy is improved by the automated processing features of the FMS-800. These features include tactical airdrops, intercepts, raster towlines, orbit/rendezvous, search patterns, VNav, FMS non-precision approach, coupled flight director/autopilot guidance and speed commands for precise time of arrival. The FMS-800 also includes embedded dynamic simulation software for ground mission rehearsal and training.

The standard system consists of dual-control display units, dual mission computers which also provide for the interface of non-MIL-STD-1553B avionics, dual-remote readout units and a data transfer/loader unit.

The FMS-800 flight management system is supported by a Windows 95 Mission Planning Station (MPS), which inputs data to the aircraft system via the DR-200 Airborne PC Card Receptacle. The system allows for integration of aircraft mission planning data with data from Jeppesen and Aeronautical Flight Information File (DAFIF) sources.

Status
The FMS-800 is operational in many international aircraft, including the German C-160 Transall, Canadian, Royal Jordanian and Royal Danish Air Force C-130, and UK RAF VC-10 aircraft. US Air Force and US Army platforms using the FMS-800 include the B-1, C-9, C-12, E-3, E-8, KC-10 and KC-135 (PACER CRAG) aircraft. The FMS-800 has received FAA TSO approvals for installation on US Army C-12 and US Air Force C-9 and KC-10 aircraft.

Contractor
Rockwell Collins, Government Systems.

FMS-3000 Flight Management System

Type
Flight Management System (FMS).

Description
The Collins FMS-3000 Flight Management System provides advanced flight planning and navigation capability designed specifically for the short-range business aircraft. The system comprises the FMS-3000 and GPS-4000A GPS sensor. The FMS-3000 features a number of advanced capabilities, including full flight phase, multiple waypoint vertical navigation, time/fuel planning and predictions based on aircraft flight manual data, database Departure Procedures (DPs) and Standard Terminal Arrival Routes (STARs) and integrated Electronic Flight Instrument System (EFIS) and radar control.

The FMS-3000 provides integrated multisensor navigation, flight maintenance and execution, sensor control, display/radar control, Adaptive Flight Display (AFD) map support and steering/pitch commands to the flight control system. The system is capable of using available combinations of GPS, Distance Measuring Equipment (DME), VHF Omnidirectional radio Range (VOR)/DME, Inertial Reference System (IRS) and dead reckoning data to provide en route, terminal and non-precision approach navigation guidance.

The FMS-3000 comprises three Line Replaceable Units (LRUs) – the FMC-3000 Flight

The FMS-800 flight management system is fitted in the E-8C Joint STARS aircraft   0504186

The FMS 3000 CDU   0581498

Management Computer (FMC), CDU-3000 Control and Display Unit (CDU) and DBU-4100 Data Base Unit (DBU). The FMC is packaged in the Pro Line Integrated Avionics Processing System (IAPS), reducing avionics weight and size and simplifying the avionics interconnect. This integration allows the FMS-3000 to perform self-diagnostic functions and inform the system of internal faults for efficient line replaceable module fault identification and reduced support costs.

The FMS-3000 supports single- or dual-operation. Dual-installation offers fully synchronised operation of the lateral and vertical flight plans. Each of the systems compute position solutions and the FMC modules monitor each other's position solution and issue a warning when a conflict exists. The CDU-3000 uses a scratchpad entry structure that has proven superior to prompting schemes and is widely accepted in both business and airline operating environments.

The FMS-3000 makes optimal use of the available sensor input, weighting GPS to the highest possible degree, to continuously determine accurate aircraft position and velocity. Sensor usage is automatically managed by the FMC and requires no pilot interaction, including automatic position initialisation.

The system enables the pilot to enter waypoints into the flight plan and create additional flight plan legs (100 maximum). Two types of waypoints may be entered into the flight plan, waypoints stored in the navigation database and pilot-defined waypoints (for example, place/bearing/distance). Up to 100 pilot-defined waypoints may be created. As waypoints are entered into the flight plan, the FMS automatically computes and displays the intervening great circle paths and distances.

Once a flight plan is activated, it is presented graphically on an AFD, enabling constant monitoring of aircraft position. Maps may be oriented heading-up, with present position near the bottom centre of the map, or north-up, with the map centred on a specific waypoint. Reference data such as navaids, intersections, airports, terminal waypoints and non-directional beacons may also be displayed. The map may be configured to display altitude restrictions and predicted estimated time of arrival for each waypoint.

As DPs, STARs and approaches are assigned and entered in the flight plan, the corresponding procedural tracks are automatically flown by the FMS. The system gives access to the terminal and approach procedures and transitions via a top-level dedicated function key on the CDU. Once inserted into the flight plan, these procedures, which include all ARINC 424-specified leg types, can be modified in response to real-time air traffic instructions.

The system is designed to automatically execute non-precision GPS approaches, GPS overlay approaches and multisensor RNAV and VOR approaches, as well as to provide missed approach guidance. The navigation database contains the approach definition and corresponding procedures, including the Missed Approach Procedure (MAP). For added flexibility in the heavy workload environment of the terminal area, the system accommodates the use of ATC radar vectors, which may be issued prior to or during an approach.

Full Flight Regime VNAV is provided by the system for each phase of flight – climb, cruise and descent. VNAV ensures altitude and speed constraints at waypoints are met, speed limits at altitudes are honoured and the vertical flight profile, as specified by the pilot, is followed. The system will automatically command the autopilot to sequence modes and set target speeds and target altitudes to ensure the flight plan requirements are met within the constraints of the preselected altitude setting. By integrating the vertical navigation with the autopilot, the pilot has full command of the normal autopilot modes (pitch, flight level change, vertical speed and altitude hold) while maintaining VNAV in an active state. If the pilot commands the aircraft to violate a VNAV constraint, the appropriate alerts are provided. During the various phases of flight, the VNAV follows the script of the flight plan. It levels the aircraft once the preselected altitude is captured and begins descent at a planned location. VNAV cruise mode commands the autopilot to capture and maintain a desired cruise altitude. During the descent mode, the VNAV computes a geographical path to each waypoint and provides guidance relative to that path, ensuring that the descent altitude constraints are honoured. As the approach segment and corresponding procedures are entered, vertical navigation is fully integrated, providing smooth transitions and easing pilot workload.

A basic feature of the FMS-3000 is to provide a measured performance mode based on current fuel flow and ground speed. Parameters available to the pilot include fuel consumption, endurance available currently and at the destination, reserve fuel data and aircraft weight data.

Specifications
Dimensions:
(FMC-3000) 222 (H) × 154 (W) × 43 (L) mm
(CDU-3000) 118 (H) × 146 (W) × 162 (L) mm
Weight:
(FMC-3000) 0.86 kg
(CDU-3000) 1.82 kg
Temperature:
(FMC-3000) –55 to +70°C
(CDU-3000) –20 to +70°C
Altitude: 55,000 ft
Power:
(FMC-3000) 28 V DC; 1.25 A (max)
(CDU-3000):
(IAPS) 5 V DC; 5 A (max)
12 V DC; 0.75 A (max)
–12 V DC; 1.25 A (max)
Certification:
(FMC-3000) FAA TSOs C115b, C129a Class B1; DO-160C; EUROCAE ED-14C
(CDU-3000) FAATSOs C115b; DO-160C; EUROCAE ED-14C

Status
In production and in service. The FMS-3000 is certified on the Premier I and C-12 aircraft.

Contractor
Rockwell Collins.

FMS-4200 Flight Management System

Type
Flight Management System (FMS).

Description
The FMS-4200 is designed to meet the demands of the regional airline environment, providing worldwide GPS navigation, simplifying cockpit management and enhancing flight crew and aircraft performance. The system is available in single- or dual-installations; when installed in a dual-configuration, the system allows synchronised or independent operation. In addition to GPS, long-range navigational inputs such as IRS/AHRS, VLF/Omega, VOR and DME can be used to provide accurate determination of aircraft position. Coupled with flight management technology, the system provides the integrity, flexibility and operational capabilities required by regional airline operators.

FMS-4200 comprises three LRUs: the CDU-4100 Control and Display Unit (CDU), FMC-4200 Flight Management Computer (FMC) and GPS-4000A GPS sensor. The CDU-4100 incorporates an ARINC-style keyboard, 12 line select keys, 22 function keys and a full alphanumeric keyset. The full-size keyboard provides direct access to functions used most often and allows full control of the FMS. The scratch pad-based design allows for easy movement of information from point to point. The CDU-4100 has the optional capability to function as a Multifunctional Control Display Unit (MCDU) controller. If selected, the CDU controls ARINC 439 systems such as ACARS, AFIS™ and Satcom. The FMC-4200 is packaged as a line-replaceable module, housed in the Pro Line 4 Integrated Avionics Processing System (IAPS). The computer provides 16 Mbytes of memory to support the FMS logic, full performance information, company and pilot routes and 10 Mb of navigation database memory. The DBU-4100 Database Loading Unit (DLU) is utilised every 28 days to load the navigation database into the computer. The DBU-4100 is available as an installed unit or as a stand-alone unit. Installation of the Collins GPS-4000A Global Positioning System sensor provides GPS navigation en route and allows the use of GPS Non-Precision Approaches (NPAs). The GPS-4000A can be upgraded to the APR-4000 system that provides GPS precision approach capability.

The FMS-4200 is designed for all phases of flight – takeoff, en route, approach and landing. Complete lateral navigation and vertical guidance is provided, including the ability to fly complete airways, DPs, STARs, parallel offsets and approach legs. Up to 1,000 standard company routes, each with up to 100 waypoints, can be stored with the navigation database. Comprehensive flight plan progress information is provided on both the CDU and MFD. Full predictive fuel and time management is included for both the primary and secondary flight plans as well as the route to an alternate airport. Engine thrust management provides climb and cruise N1 information to the flight deck.

VHF Omnidirectional Radio Range (VOR), Distance Measuring Equipment (DME), Inertial Reference System (IRS), Attitude Heading Reference System (AHRS), Instrument Landing System (ILS) and GPS sensor management is available. The FMS-4200 integrates with onboard equipments to determine the most accurate navigation solution. Crew-selectable automatic and manual radio tuning modes enhance pilot control of the navigation process. In addition, sensors can be readily selected or deselected for use in the FMS navigation solution. A number of features help maximise the operational effectiveness of the system: Insertion of a waypoint into the flight plan and direct waypoint routing are provided with shortcuts. Flight plan edits can be reviewed prior to being accepted or cancelled. Parallel offsets, route to alternate, direct routing, waypoint insertion and up to four holding patterns can be reviewed prior to activation. The flight plan can be reviewed as ATC level language (FPLN) or as waypoints (LEGS). A second flight plan or route to an alternate destination can be inserted for alternate use. Advisory vertical information provides cues to aid aircraft vertical profile management. Vertical information, consisting of top of climb, top of descent, vertical constraints and descent slope are displayed on the CDU and MFD. When climbing or descending, advisory vertical cues are provided on the Primary Flight Display (PFD), including glideslope, vertical speed target and next vertical constraint.

The dedicated Pro Line 4 MFD provides a north-up planning map and heading-up progress map. The north-up planning map is available to assist with flight plan generation and flight plan

Dual FMS-4200 flight management systems in the Canadair Regional Jet aircraft   0504188

Specifications

Dimensions:
(FMC-4200) 233 (H) × 154 (W) × 52 (L) mm
(CDU-4100) 229 (H) × 146 (W) × 262 (L) mm
Weight:
(FMC-4200) 1.04 kg
(CDU-4100) 4.55 kg
Temperature:
(FMC-4200) –55 to +70°C
(CDU-4100) –20 to +70°C
Altitude: 55,000 ft
Certification:
(FMC-4200) FAA TSOs C115a, C129 Class B1;
    DO-160C; EUROCAE ED-14C
(CDU-4100) FAA TSOs C94a, C113, C115;
    DO-160C; EUROCAE ED-14C

Status

In production and in service. The FMS-4200 has been certified for use on the Saab 2000 and Bombardier CRJ-100/200/700 aircraft.

Contractor

Rockwell Collins.

FMS-5000 Flight Management System

Type

Flight Management System (FMS).

Description

Collins' FMS-5000 FMS provides advanced flight planning and navigation capability designed specifically for long-range business aircraft. The system, which comprises the FMS-5000 and GPS-4000A GPS sensor, provides for operational flexibility and enhances situational awareness by offering such capabilities as full-flight phase, multiple waypoint vertical navigation, time/fuel planning and predictions based on the airplane's flight manual data and database Departure Procedures (DPs), Standard Terminal Arrival Routes (STARs) and approaches. The GPS-4000A provides the required level of integrity and monitoring and supports growth to precision approach certification.

The FMS-5000 provides integrated multisensor navigation, flight maintenance and execution, sensor control, MFD map support and roll/pitch steering commands to the host aircraft flight control system. The FMS is capable of using available combinations of GPS, DME, VOR, IRS and dead reckoning data to provide en route, terminal and non-precision approach navigation guidance.

The system is composed of three Line Replaceable Units (LRUs), the FMC-6000 Flight Management Computer (FMC), CDU-6000/6100 Control and Display Unit (CDU) and DBU-4100 Data Base Unit. The FMC is packaged into the Pro Line Integrated Avionics Processing System (IAPS), reducing avionics weight and size and simplifying interconnection. This integration allows the FMS-5000 to perform self-diagnostic functions and inform the system of internal faults for efficient Line Replaceable Unit (LRU), fault identification and reduced support costs.

The FMS-5000 can be specified in single- or dual-configurations by configuring the IAPS with the desired number of FMC modules and installing the same number of CDUs. Dual-installation offers fully synchronised operation of lateral and vertical flight plans. Each of the systems compute position solutions; the two active FMS units monitor each other's position solution and issue a warning when a conflict exists.

The system provides navigation situation information and steering guidance, making optimal use of the available sensor input and weighting GPS to the highest possible degree in order to continuously determine accurate aeroplane position and velocity. Sensor usage is automatically managed by the FMS and requires no pilot interaction, including automatic position initialisation, although pilot override can be

CDU for FMS 4200   0105286

evaluation. The progress display allows the flight crew to monitor the flight plan progression. High-and-low level navaids, airports, terminal waypoints, Non-Directional (radio) Beacon (NDB) intersections can also be displayed on the progress map. Text pages are provided to allow progress, navigation sensor status, position summary and Long-Range Navigation (LRN) status information to be displayed. The progress

map can also be integrated with a text window that displays relevant progress and advisory VNAV information. The combination of text and map on a single display allows for accurate monitoring of the flight while continuing to display map data. Situational Awareness (SA) is further enhanced by displaying Traffic Alert and Collision Avoidance System (TCAS) or weather radar overlays.

Collins' FMS-5000 is certified on the Bombardier Challenger 300 (Rockwell Collins)   1129579

implemented at any time. The pilot can enter waypoints into the flight plan and create additional flight plan legs (100 maximum). Two types of waypoints may be entered into the flight plan: waypoints stored in the navigation database and pilot-defined waypoints (such as place/bearing/ distance). Up to 100 pilot-defined waypoints may be created. As waypoints are entered into the flight plan, the FMS automatically computes and displays the intervening great circle paths and distances. A pilot-defined database of up to 100 routes can be stored in the system memory.

If DPs, STARs and approaches are assigned and entered in the flight plan, the corresponding procedural tracks are automatically flown by the FMS. The system gives access to the terminal and approach procedures and transitions via a top-level dedicated function key on the CDU. Once inserted into the flight plan, these procedures, which include all ARINC 424-specified leg types, can be modified in response to real-time air traffic instructions.

The FMS is designed to automatically execute non-precision GPS approaches, GPS overlay approaches and multisensor Area Navigation (RNAV) and VOR approaches, as well as to provide missed-approach guidance. Additional GPS-based approach types that may be selected by the system include ILS, localiser, localiser back course (BAC), Localiser Directional Aid (LDA), Simplified Directional Facility (SDF), Tactical Air Navigation System (TACAN), Non-Directional Radio Beacon (NDB) and Long-Range Navigation (LORAN) approaches. The FMS database contains the approach definition and corresponding procedures, including the missed-approach procedures.

The system provides multiple waypoint Vertical Navigation (VNAV) for each phase of flight – climb, cruise and descent. The VNAV ensures altitude and speed constraints at waypoints are met, speed limits at altitudes are honoured, and the vertical flight profile, as specified by the pilot, is followed. The system will automatically command the autopilot to sequence modes and set target speeds and target altitudes to ensure the flight plan requirements are met within the constraints of the preselected altitude setting. By integrating the VNAV with the autopilot, the pilot has full command of the normal autopilot modes (pitch, flight level change, vertical speed and altitude hold) while maintaining VNAV in an active state.

With the addition of aircraft flight manual Vspeed tables in the FMS-5000 database, target speed calculations are conducted within the FMC for takeoff and approach. A single key-stroke transfers the calculated Vspeeds to the Primary Flight Display (PFD).

Specifications
Dimensions:
(FMC-5000): 53 (H) × 222 (W) × 151 (L) mm
(CDU-5200): 131 (H) × 162 (W) × 146 (L) mm
Weight:
(FMC-5000): 0.86 kg
(CDU-5200): 1.81 kg
Temperature:
(FMC-5000): −55 to +70°C
(CDU-5200): −20 to +70°C
Altitude: 55,000 ft
Certification:
(FMC-5000) FAA TSOs C115b, C129a Class B1;
  DO-160C; EUROCAE ED-14C
(CDU-5200) DO-160C; EUROCAE ED-14C

Status
In production and in service. The FMS-5000 is certified on the Bombardier Challenger 300.

Contractor
Rockwell Collins.

FMS 5000 Flight Management System

Type
Flight Management System (FMS).

Description
The FMS 5000 consists of a multichain master independent Loran-capable of tracking two Loran chains and up to 12 ground stations simultaneously. All world Loran chains are available, including the US mid-continent NOCUS/ SOCUS chains. The FMS 5000 automatically selects the strongest master and secondary Loran stations, providing hands-off operation.

Three remote GPS receiver options are available for the FMS 5000. A five-, six- or 12-channel tracking receiver can be supplied with the FMS 5000, or added in the future. Together, the Loran and GPS receivers continuously scan up to 12 Loran ground stations and all satellites in view. The FMS 5000 displays the navigation solution with instant exchange of sensor positioning, providing hands-off Loran/GPS operation.

The FMS 5000 interfaces with analogue instruments, fuel computers, air data computers, moving maps, autopilots, CDI/HSI and annunciators. An optional ARINC 429 interface allows coupling to EFIS systems.

A two-line 40-character sunlight-readable LED display shows bearing, distance, groundspeed, CDI, waypoint identity, altitude and track.

Waypoints can be located by identity, city, local proximity or adjacent waypoints.

The FMS 5000 uses a Jeppesen NavData card containing 40,000 waypoints. The North America coverage extends from Alaska through Central America. The International card contains all areas outside North America. Worldwide navigation, combining North America and International data, is available on a single World card. Airports, VORs, NDBs, terminal and en route intersections are stored on the crew updatable NavData card. Special use airspace includes floors and ceilings, TCAs, ARSAs, ATAs, MOAs and restricted, prohibited and alert penetration warnings.

Coupled to an optional Mode C encoder interface, pressure altitude with a pilot input barometric correction is displayed. VNav becomes automatic with known present altitude and known waypoint elevations. An altitude hold advisory will notify the pilot of altitude deviation.

Status
In production and in service.

Contractor
Arnav Systems Inc.

FMS 7000 Flight Management System

Type
Flight Management System (FMS).

Description
The FMS 7000 is a small dzus-mounted Flight Management System (FMS) for business and commuter aircraft. It was designed to provide the pilot with a comprehensive primary or separate secondary navigation tool and database facility. It can be configured with a Loran sensor and a high-precision GPS sensor. The FMS 7000 uses sensor assessments of signal quality to provide the best position information.

A standard Jeppesen North American or International database with over 60,000 aviation facilities is contained on a high-capacity NavData card that can be easily updated. Each card provides all worldwide port-of-entry airports with runway lengths in excess of 6,000 ft, plus all hard-surfaced airports, facilities, frequencies and navigation information for the North American or International geographic areas. An optional Jeppesen World NavData card provides comprehensive worldwide navigation on a single card.

The FMS 7000 automatically builds SID and STAR route waypoints, transition routes, crossing altitudes, arrival routes and arrival frequencies. Jeppesen terminal navigation is also included on the World NavData card.

The Arnav advanced communication network, Arnet, and the ARINC Standards Interface (ASI) have been developed specifically for the FMS 7000. Arnet is a high-speed serial communication device that facilitates integration of other avionics such as autopilot, air data and fuel computers. ASI allows the FMS 7000 to communicate with virtually all cockpit systems, including analogue and digital format DME, EFIS, HSI and flight directors. ARINC 429, ARINC 419, six-wire ARINC 568, synchro, MIL-STD-1553 and several DME controls or data formats are supported.

Cross-talk capability and system redundancy is optional on the FMS 7000. Tandem control/ display units allow the co-pilot to view all flight, fuel and air data information and perform waypoint search routines.

The Arnav GridNav mission management system program is a software option for Loran, GPS and combination FMS. The GridNav option allows the pilot to plot and fly a grid pattern of user-specified dimensions using a minimum number of waypoints and with simplified navigation programming. An event trigger provides a precise timed pulse to a camera shutter or target drop release.

Sensor options for the FMS 7000 include Loran only, GPS only or both Loran and GPS. The Loran-based system automatically selects the proper Loran chains and stations based on aircraft location. The FMS tracks up to 12 Loran stations and automatically selects the signal geometry that provides the best fix. Loran is approved for IFR flight and permits the pilot to file IFR direct when within the boundaries of Loran coverage.

The GPS-based receiver is differential ready and conforms to TSO C-129 certification. The FMS 7000 may be populated with either a five-channel or 12-channel GPS receiver. Both work in any weather and are not subject to precipitation static, low-frequency thunderstorm emissions or any other weather interference. Through a powerful microprocessor, the FMS 7000 tracks all satellites in view and then selects the satellite geometry to acquire the most precise fix. Signal acquisition and tracking are continuous throughout all dynamics of flight. Initial time to fix a reliable position is less than 1 minute and position is updated every second.

Both Loran and GPS sensor options can be installed in the FMS 7000 to provide multisensor blended mode navigation. The combined sensor output can be used for IFR direct navigation when operating within Loran coverage areas and when TSO is achieved. The FMS has TSO C-60b for IFR en route navigation and terminal navigation, TSO C-115a multisensor certifications and Loran approach certification.

Specifications
Dimensions:
(Loran/GPS LRU) 1/4 ATR short
(CDU) 57.1 × 139.7 × 127 mm
Weight:
(Loran/GPS LRU) 1.82 kg
(CDU) 0.68 kg
(Loran/GPS antennas) 0.64 kg
Power supply: 11–35 V DC, 15 W (max)
Temperature range:
(LRU) –55 to +70°
(CDU) –20 to +70°C
Altitude: up to 55,000 ft
Certification: TSOs C-60b, C-115a, C-44a, C-106
Environmental: Do-160c Cats A1/D2, MIL-STD-167

Status
The FMS 7000 is installed in a wide range of civil and military aircraft and helicopters.

Contractor
Arnav Systems Inc.

FMS-6000 Flight Management System

Type
Flight Management System (FMS).

Description
The Collins FMS-6000 FMS provides advanced flight planning and navigation capability designed specifically for long-range business aircraft. FMS-6000 system is designed for worldwide navigation, including polar navigation and comprises the FMS-6000 FMS and GPS-4000A GPS sensor. In common with the FMS-5000, the FMS-6000 extends operational flexibility and enhances situational awareness by providing full-flight phase, multiple waypoint vertical navigation, time/fuel planning and predictions based on the airplane's flight manual data and database DPs, STARs and approaches.

The system is composed of three LRUs, the FMC-6000 FMC, CDU-6000/6100 CDU and DBU-4100 DBU; the FMC is packaged into the Pro Line Integrated Avionics Processing System (IAPS), reducing avionics weight and size and simplifying the avionics interconnection.

The FMS-6000 can be defined for single-, dual-, or triple-operation by configuring the IAPS with the desired number of FMC modules and installing the same number of CDUs. Dual-installation offers fully synchronised operation of the lateral and vertical flight plans. When a triple configuration is installed, the third FMS functions in a standalone manner as a hot spare. Each of the three systems compute position solutions; the two active FMS units monitor each other's position solution and issue a warning when a conflict exists.

Operation and capabilities of the FMS-6000 are broadly similar to the FMS-5000.

Aircraft performance predictions are based on pre-stored data from the aircraft flight manual. Performance data is stored for all flight phases, and a fuel flow correction factor may be applied to customise the performance predictions for each aircraft. Performance predictions account for predicted winds and temperatures, and may be entered for climb, cruise and descent, including individual cruise legs-by pilot entry or by AFIS™ flight plan transfer.

Specifications
Dimensions:
(FMC-6000) 43 (H) × 222 (W) × 154 (L) mm
(CDU-6000) 187 (H) × 146 (W) × 169 (L) mm
(DBU-4100) 92 (H) × 146 (W) × 169 (L) mm
Weight:
(FMC-6000) 0.86 kg
(CDU-6000) 4.04 kg
(DBU-4100) 1.2 kg
Temperature:
(FMC-6000) –55 to +70°C
(CDU-6000) –20 to +70°C
(DBU-4100) –20 to +55°C
Altitude: 55,000 ft (all units)
Certification:
(FMC-6000) FAA TSOs C115b, C129a Class B1; DO-160C; EUROCAE ED-14C
(CDU-6000) DO-160C; EUROCAE ED-14C
(DBU-4100) DO-160C; A2F2-BB[BMN]
E1XXXXXZ [BZ] AZAUA[Z3Z3]XX; ED-14C
Power:
(FMC-6000) N/A
(CDU-6000) 28 V DC; 1.25 A (max)
(DBU-4100) 28 V DC; 0.25 A (standby), 0.375 A (transfer)

Status
The FMS-6000 is certified on the Canadair Challenger 604 (FMS-6000), Falcon 2000 (FMS-6100), Falcon 50EX (FMS-6100), Falcon 20 (FMS-6100), Gulfstream G200 (FMS-6100) and Raytheon Hawker 800XP (FMS-6000).

Contractor
Rockwell Collins.

FMZ-2000 Flight Management System (FMS)

Type
Flight Management System (FMS).

Description
The FMZ-2000 is Honeywell's current generation Flight Management System, designed for applications in corporate and regional airliners. The system is available in two configurations: a stand-alone navigation computer, the NZ-2000, or a circuit card in the Honeywell IC-800 integrated avionics computer. Both configurations are compatible with the Avionics Standard Communication Bus (ASCB) and ARINC 429 system architecture. They include similar operational benefits and feature a full colour Control and Display Unit (CDU) and associated DL-900 data loader (see separate entry).

Both FMZ-2000 configurations provide a premium user interface including an easy-to-read keypad with select keys and a colour display. In addition, the system provides a worldwide navigation database including Standard Instrument Departures (SIDs), Standard Terminal Arrival Routes (STARs) and approach procedures. Multiple sensor inputs, including GPS, IRS, and DME/DME positioning are accommodated. The system provides vertical navigation in climb and descent, multiple holding patterns, and a non-precision approach capability with high precision waypoints.

The system also includes, Honeywell's patented algorithm, SmartPerf™. SmartPerf™; effectively 'learns' aircraft-specific performance. The SmartPerf™ function enables the navigation computer to learn the performance of a specific aircraft, thus affording operators access to performance calculations previously available only to aircraft equipped with a dedicated performance management computer. Using the aircraft's performance database, and both entered and sensed atmospheric data, SmartPerf™ provides performance calculations of time, fuel and predicted altitude at all waypoints; time, fuel, and distance to top of climb and top of descent; maximum endurance targets; optimum cruise altitude; time and distance to bottom of step climb; figure of merit indicating accuracy of fuel calculations and predictions for stored flight plans. Pilots have the option of manually entering altitude and calibrated airspeed or Mach speed constraints.

The FMZ-2000 also performs coupled vertical guidance to multiple, three-dimensional waypoints. Using multiple waypoint vertical navigation allows operators to enter altitudes for each waypoint, both climbing and descending. From this data, the navigation computer displays precise altitude crossings of all predefined waypoints and a complete vertical profile, including top of climb and top of descent on compatible Electronic Flight Instrument Systems (EFIS).

When equipped with Global Positioning System (GPS) sensors and receiving valid GPS data, the FMZ-2000 will navigate entirely by GPS and still maintain input from all available sensors. This affords operators the accuracy of GPS with the integrity and safety provided by a multiple sensor system.

Honeywell's FMZ-2000 flight management system with DL-900 data loader, Control Display Unit and NZ-2000 Navigation Computer
0103883

The latest upgrade, known as FMZ-2000 Version 5.X, is available for both Primus 2000 and NZ-2000 based systems. Version 5.X has gained TSO approval for the Citation X, ATR 42 and Gulfstream V. The Version 5.X upgrade features a Pentium processor together with a larger 16 Mbyte database memory, providing additional capacity for future expansion to meet CNS/ATM requirements. Software improvements to be offered include: parallel database loading, terminal area speed targets, SLS/LAAS compatibility, multiple flight plans on one data loader disk and support of 8.33 kHz communications tuning for European airspace. Take-Off and Landing Data (TOLD) software enhancements, which reduce pilot workload by automating the computation of take-off and landing data, can also be selected by the customer for specific aircraft types.

To make training even easier, Honeywell has introduced optional Personal Computer (PC)-based training for the FMZ-2000. This product will allow pilots and maintenance personnel to accomplish self-paced, interactive training using a CD-ROM-based instruction set. Training is designed in modules and can now be accomplished anywhere the customer has access to a suitably equipped personal computer.

Status
TSO'd in a number of aircraft versions, the stand-alone system is currently standard equipment on the Bombardier Global Express, Gulfstream IV, Dassault Falcon 900B, and Raytheon Hawker 800XP. It is optional on the Dassault Falcon 2000 and the Falcon 50EX, Cessna Citation VII, and Embraer 145. it is also STC'd in a number of retrofit applications including the Challenger 600/601, Hawker 800/1000, Boeing 727, Cessna Citation V and Beechjet 400. The integrated system is standard equipment on the Gulfstream V, Cessna Citation X, Dassault Falcon 900EX and Dornier 328.

Contractor
Honeywell Inc, Commercial Electronic Systems.

Fuel Savings Advisory and Cockpit Avionics System (FSA/CAS)

Type
Avionics system.

Description
Delco provides the Fuel Savings Advisory and Cockpit Avionics System (FSA/CAS) for the upgrading of the current Boeing C-135 and KC-135 fleet of about 700 aircraft.

In this system, a CRT control and display unit replaces one of the standard inertial navigation system units to simplify crew management tasks; in particular, all basic data for flight use can be entered at this one point. In addition to the usual engine and navigation information, a new fuel management panel and centre of gravity display have been installed for inputs of fuel state, disposition and usage. This latter function is invaluable in tanker operations and ensures efficient fuel allocation and usage. The panel is linked to other avionic systems by a MIL-STD-1553B databus.

The FSA provides commands to the pilots during climb, cruise and descent using flight path optimisation algorithms and flight manual data for lift, drag and thrust. An additional mode computes all required take-off and landing parameters based on crew inputs for present conditions.

Many KC-135s are being refitted with GE/SNECMA CFM56 engines, which provide extra thrust over earlier types. FSA/CAS is integrated with this engine, offering more economic operation.

Specifications
Dimensions:
(FSA computer) 1/2 ATR long
(control/display unit) 146 × 181 mm
(bus controller) 207 × 131 mm
(fuel panel) 457 × 123 mm
(fuel management computer) 235 × 249 mm
(remote display unit) 83 × 83 mm
Weight:
(FSA computer) 11.9 kg
(control/display unit) 4.5 kg
(bus controller) 4.1 kg
(fuel panel) 12.5 kg
(fuel management computer) 21.8 kg
(remote display unit) 0.95 kg
Power:
(FSA computer) 95 W
(control/display unit) 44 W
(bus controller) 34 W
(fuel panel) 92.8 W
(fuel management computer) 139.5 W
(remote display unit) 10 W

Status
In service on US Air Force Boeing C-135 and KC-135 aircraft.

Contractor
General Dynamics.

FV-2000 Head-Up Display (HUD)

Type
Head-Up Display (HUD).

Description
FV-2000 HUD
The basic element of the FV-2000 HUD system is the FV-2000 Mission Display Processor (MDP). As a HUD system's computational core, it is designed to interface with virtually any combination of analogue or digital equipment. The 40 MHz 68030 microprocessor generates both flight and military displays.

In civilian application, the FV-2000 HUD utilises an FV-2000 overhead optical unit which provides a 30° field of view. The FV-2000 HUD display capability includes aircraft velocity vector, an accelerate/speed cue and a conformal pitch scale, incorporating both expanded (conformal) and compressed (non-conformal) scales. The system has been certified as a primary flight display and can be retrofitted in aircraft with either electromechanical instruments or EFIS. In military application, the FV-2000 MDP may be combined with CMC Electronics' Sparrow Hawk™ HUD

(see separate entry) to provide a low-cost display capability for training and light attack aircraft.

The HUD system comprises the optical unit, combiner, control panel and HUD computer and weighs 9.98 to 14.25 kg. Each FV-2000 HUD is specifically designed for seamless integration into each aircraft type for which it is certified. The control panel is mounted in the cockpit pedestal or panel. Other features include TCAS, GPWS, runway overlay, raster capability and an RS-170 video port. The processor is certified for use outside the pressure hull. The FV-2000 contains self-diagnostics through the control panel and maintenance pages are accessible through an RS-232 port.

All CMC Electronics HUDs employ holographic combiners to offer maximum transmissivity and contrast ratio.

The FV-2000 system is currently certified and flying on various types of business aircraft, including all Beech King Airs, the Bell 230, Cessna Citation 550, Dassault Falcon 50, Gulfstream III and IV and the Lear 55.

FV-2000E HUD
The FV-2000E HUD is designed to meet Enhanced Vision System (EVS) display requirements. It comprises the FV-2000 optical unit and the FV-2000E Head-Up Display Symbol Generator (HSG).

The current FV-2000 system already incorporates design elements to accept Enhanced Vision Sensor (EVS) technologies including FLIR and Millimetre Wave Radar (MMWR). As new EVS sensors become available, existing systems can be updated by replacing the FV-2000 remote-mounted computer with the FV-2000E computer. The FV-2000's overhead optical unit already has raster capability to display EVS information.

The FV-2000E combines enhanced display generation modules from the FV-3000 (see separate entry) with the less expensive computing engine of the FV-2000. It is designed for users that do not need the processing power and flexibility of the FV-3000 but still require a HUD symbol generator.

The FV-2000E facilitates approach during Category II or III weather conditions to a Category I runway, although, as of this update, it is not known whether the system has been certified for < Cat I approaches.

Status
The FV-2000 is US FAA certified for Falcon 50, Gulfstream III and IV, Learjet 55, Citation 550,

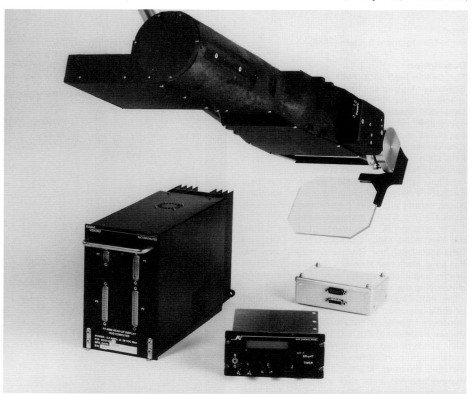

FV-2000 HUD components (top to bottom, clockwise order): straight-mounted overhead optical unit, grid amp, control panel and HUD computer
0001464

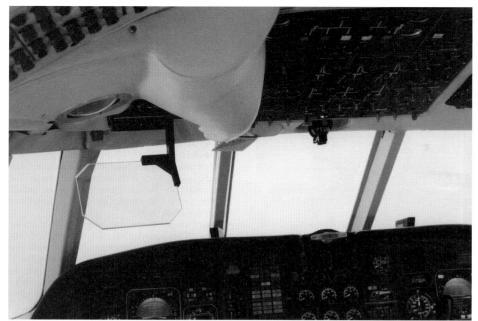

The FV-2000 HUD installed in a Falcon 50   0001465

CMC Electronics' FV-4000 is installed in the Pilatus PC-21   0127158

all Beech King Air models and the Bell 230 helicopter.

Contractor
CMC Electronics Inc.

FV-3000 Modular Mission Display Processor (MMDP)

Type
Avionic display processor.

Description
CMC Electronics' FV-3000 Modular Mission Display Processor (MMDP) has been utilised in the retrofit of upgraded avionics packages into legacy fighter and light attack aircraft. The system can drive Head-Up Displays (HUDs), Helmet-Mounted Displays (HMDs) and MultiFunction Displays (MFDs), while providing spare power for mission processing.

Containing two 50 MHz 68040 microprocessors and a dual-redundant MIL-STD-1553B databus, the FV-3000 generates both flight and military displays. It has the capability to input RS-170 video, synchronise it with stroke-generated graphics and output the combined image for display on a HUD. The FV-3000 also performs mission processing for air-air and air-to-ground weapon delivery and navigation, exercises control over the main system and primary MIL-STD-1553B databus, together with switch processing functions for Up-Front Control Panel (UFCP) and HOTAS.

Designed for the harsh environment of fighter applications, the FV-3000 MMDP is compliant with MIL-STD-810E, and incorporates capability to detect and isolate faults through a combination of operational and intermediate-level testing.

Specifications
Dimensions: 152.4 × 228.6 × 406.4 mm
Weight: 7.3 kg
Power supply: +28 V DC MIL-STD-704D 200 W
MTBF: 3000 h
Interfaces: MIL-STD-1553B; ARINC 429

Status
Superseded by the FV-4000 mission computer (see separate entry). The FV-3000 MMDP forms part of the US Navy F-14B integrated cockpit technologies upgrade, with the Sparrow Hawk™ HUD, providing for significantly improved Mean Time Between Failures (MTBF).

The Lockheed Martin X-35 concept demonstration aircraft utilises the FV-3000 MMDP, combined with CMC Electronics' Night Hawk HUD. The aircraft was selected during the fourth quarter of 2001 as the new Joint Strike Fighter (JSF) to equip the USAF, UK Royal Air Force and

other participating nations. The FV-3000 has been integrated into the Kfir, PC-9 Slovenian Air Force; Czech L-159 and 139; Aero L-59.

Contractor
CMC Electronics Inc.

FV-4000 mission computer

Type
Aircraft mission computer.

Description
The FV-4000 mission computer is a compact, modular, open architecture design which has been developed from its predecessor, the FV-3000 Modular Mission Display Processor (see separate entry). The FV-4000 features conduction-cooled Compact PCI/PMC technology, enabling it to utilise a wide variety of off-the-shelf modules, thus simplifying upgrade and allowing the computer to support new processing and peripheral elements as they become available.

The FV-4000 incorporates one or more 500 MHz Power PC G4 processors, each with up to 512 Mb of memory for processing data in real time and at a high refresh rate. Processing is supported by optional modules which facilitate generation of high-resolution graphical displays or interface with any avionics bus or Input/Output (I/O) signal used with military or civilian systems. CMC Electronics offers five compact PCI cards for the FV-4000: a 14 Gb mass memory card, a video switching module, a 3-D labs Permedia 3 raster

symbol generator, a HUD/HMD stroke/raster graphics generator, and a general interface card.

The FV-4000 is available in 3 MCU and 5 MCU sizes, depending on specification.

Specifications
Dimensions:
 3 MCU: 96.5 × 198.1 × 381.0 mm (3.8 × 7.8 × 15 in)
 5 MCU: 157.5 × 198.1 × 381.0 mm (6.2 × 7.8 × 15 in)
Weight:
 3 MCU: 8.0 kg (17.6 lb)
 5 MCU: 9.3 kg (20.5 lb)
Power supply: 28 V DC per MIL-STD-704 and DO-160D; 115 V AC 400 Hz, 1 or 3 phase per MIL-STD-704
Power consumption: 250 W
Connector: ARINC 600
Interface: MIL-STD-1553B, ARINC-429, RS-422, fibre channel, Firewire, Ethernet, PCMCIA, discrete, RGB/LVDS video, HUD/HMD driver
Environmental: MIL-STD-810; MIL-STD-461/462; DO-160D

Status
In production and in service. Combined with CMC Electronics' Sparrow Hawk™ HUD (see separate entry), the FV-4000 is installed in the Pilatus PC-21 and Korean Aerospace Industries (KAI) KT-1C advanced turboprop trainers.

Contractor
CMC Electronics Inc.

G1000 integrated avionics system

Type
Integrated avionics system.

Description
The G1000 integrated avionics system utilises Liquid Crystal Display (LCD) technology to bring large-format displays with extremely wide viewing angles to the General Aviation (GA) cockpit. The XGA 1,024 × 768 pixel/16 million colour display is powered by an Xscale microprocessor and features a high-performance graphics accelerator for superior 3-D rendering. Physically, the display is 51 mm deep and weighs approximately 3 kg.

CMC Electronics' FV-4000 Mission Computer in 3 MCU and 5 MCU sizes (CMC Electronics)   1030037

Garmin's G1000 system installed in a Diamond DA40 general aviation aircraft. This installation consists of two 10×8 landscape displays with PFD on the left and MFD on the right (Garmin International) 1127380

Garmin G1000 in Cessna 172 Skyhawk 1127381

Major System Components include:

- A 10 in Primary Flight Display (PFD) and 10 or 15 in MultiFunction Display (MFD) with XGA (1024 × 768) resolution, wide viewing angles and clear sunlight readability
- Dual-integrated radio modules to provide WAAS-capable, IFR oceanic-approved GPS, VHF navigation with ILS and VHF communication with 16 W transceivers and 8.33 kHz channel spacing
- Digital audio systems
- Integrated Mode S transponders with Traffic Information Service (TIS) and optional diversity
- Integrated solid-state Attitude and Heading Reference Systems (AHRS)
- Integrated RVSM-compatible digital Air Data Computers (ADCs)
- Integrated four-colour digital weather radar
- Integrated Class-B Terrain Awareness and Warning System (TAWS) with worldwide terrain and obstacle database
- Integrated Automatic Flight Control System (AFCS).

The heart of the G1000 integrated avionics system is its Sequoia Instruments Attitude and Heading Reference System (AHRS), which employs 3-D GPS, 3-D magnetometry (measurement of changes in the Earth's magnetic field), and 3-D air data information to compute the aircraft's attitude. All three components are combined to compute the attitude, but only two are necessary for a full solution at any time, providing a degree of redundancy. A benefit of this system, considering the target audience for the G1000, is that it does not require the aircraft to be stationary during initialisation and requires only 45 seconds to align, even when airborne.

The G1000 is a fully scalable design, with three-display configurations available to provide a dual-PFD/single MFD and upwards to meet the demands of twin-turbine business jets or scaled down for installation in single- or twin-engine pistons. Key to this level of system flexibility is the G1000's ability to integrate with other displays and host aircraft systems:

- Available with two- or three-axis, fail-passive flight control system
- Ethernet data-bus connectivity
- Selectable PFD flight view presentations
- Moving-map MFD with engine/fuel gauge cluster, checklist capability
- Modular rack-mounted LRUs
- Full reversionary display capability
- Integrated IrDA interface, allowing future upload of flight-relevant data from Garmin handheld PDA.

The physical architecture of the G1000 avionics system and extensive use of modern data-bus technology reduces overall weight and simplifies installation and maintenance. The major system components can be inserted into a unique system rack and fitted for the aircraft. These Line-Replaceable Units (LRUs) are architecturally integrated with sensors throughout the aircraft and transfer data seamlessly from these sensors to the PFD/MFD.

As a Primary Flight Display (PFD), the G1000 is capable of replacing all conventional flight instruments and presents a modern integrated display with altitude, airspeed and vertical speed in a tape format, with trend lines, to provide a similar level of predictive information to modern glass cockpits found in air transport aircraft. A pilot-selectable, thumbnail-size moving map can be displayed elsewhere on the screen and/or a window showing flight-plan waypoints. Across the top of the display are a number of pilot-configurable windows that depict typical GPS information, such as next waypoint, time, distance, and desired track. To the upper left of the display are VHF navigation frequencies, controlled by a knob on that side. Below that knob is a control for the heading bug; a push syncs the bug to the current heading. Lower left is the altitude knob for setting altitude bugs. To the upper right are the communications frequencies and their control knobs. Navigation/. communication functionality follows established products such as the Garmin GNS 430/530. There is a joystick for moving the cursor around the moving map, with buttons below for setting up flight plans, instrument approaches and so on.

Soft keys across the bottom of the display control menus and settings.

In dual-display configurations, a second G1000 can be employed as a MultiFunction Display (MFD), for to control of navigation and other sensor information, such as weather information and Terrain Awareness and Warning Systems (TAWS). In an integrated PFD/MFD combination, controls on each can perform functions on the other. For example, a change to a communications frequency on one display will be repeated on the other – an Ethernet bus allows every component to interface with every other component.

All of this combines to provide the GA pilot with a vastly greater level of Situational Awareness (SA), by presenting aircraft performance, navigation, weather, terrain, and traffic information digitally on large-format displays:

- Flight instrumentation: wide horizon; three-axis flight dynamics; air speed; altitude; vertical speed; Horizontal Situation Indicator (HSI) with 360°, arc and perspective modes
- Engine performance: all engine operational parameters for piston, turboprop and turbofan engines; engine trend data; exceedance monitoring
- Weather: weather radar; satellite weather datalink (via Weather Works and XM Radio); lightning-detection interfaces
- Terrain: Class-B TAWS; detailed topographic mapping
- Traffic: TIS data, using Garmin's Mode S technology; Traffic Avoidance System (TAS) interfaces; future expansion to accommodate emerging traffic-awareness technologies

The digital presentation of data eliminates the need to scan multiple instruments, freeing the pilot to devote more time to flying the aircraft and provides for vastly greater SA – a important point, particularly in the continental US, where GA traffic is often intermixed with heavy air transport aircraft.

Status
In production and in service in a number of GA aircraft, including the Cessna Citation Mustang, Diamond DA40/42 and Cessna 172 Skyhawk.

Contractor
Garmin International.

GH-3000, 3001 and 3100 Electronic Standby Instrument Systems (ESIS)

Type
Electronic flight instrument.

Description
The GH series of Electronic Standby Instrument Systems (ESIS) combines the important flight cues – attitude, altitude, airspeed , heading and navigation – in one AMLCD screen. The ESIS GH series of instruments fit into a single 3-ATI opening.

The GH-3000 model utilises the remote-mounted ADC-3000 air data computer for the display of airspeed and altitude information. In common with all the ESIS models, solid-state internal sensors remove the requirement for a separate spinning mass gyro.

The GH-3001 model was designed for military platforms, both fixed- and rotary-wing, and is certified to higher EMI levels. Customised tape readouts allow the addition of a VSI display and the unit is compatible with NVGs.

The GH-3100 model has an integrated air data sensor card with pitot static connections directly to the unit thus eliminating the need for a remote-mounted ADC. The ADC module adds less than 1 in to the length of the system. The GH-3100 can also interface with an ARINC 429 AHRS or INS.

Specifications
Dimensions: 3 ATI
Weight: 1.59 kg
Power: 28 V DC
Certification: DO-160C

GH-3000 electronic standby instrument system   0103879

Status

TSO certification for conformance to the following standards granted in March 1997: TSO-C2d/C4c/C10b/C34e/C36e/C113. Available for production delivery. Certified in Challenger 604, Falcon 50 Gulfstream IV and V, Raytheon Hawker 800XP and Eurocopter AS 365N2.

Selected by the US Army Special Forces for its MH-47E and MH-60K helicopters, and by Bell Helicopter for the 609 Tiltrotor programme.

Contractor

L-3 Communications, Avionics Systems.

Global Star 2100 Flight Management System (FMS)

Type

Flight Management System (FMS).

Description

Global Star 2100 provides the capability to meet the requirements of the Future Air Navigation System (FANS) and Aeronautical Telecommunications Network (ATN) systems. Through the ATN network, the onboard ATN router will connect the aircraft with the ground via VHF, satcom, HF and Mode S datalinks with the Automatic Dependent Surveillance (ADS) position reporting and Air Traffic Control (ATC) conflict resolution system. Global Star 2100 is also compatible, through ARINC 739 connections, with conventional AFIS and ACARS systems.

The Global Star 2100 system combines a navigation computer, multichannel IFR-approach-certified GPS receiver, navigation database, flight management system functionality and a Multifunction Control and Display Unit (MCDU), with 5.5 in display into a single compact unit. A two-box installation can also be provided in which all functions except the MCDU are integrated into a separate LRU that can be located in the avionics bay.

The Global Star 2100A is a derivative system, designed for installations where cockpit and avionics bay space is limited; it is available in

both single-box and two-box variants, but with a 4 in display.

The internal GPS receiver provides real-time and predictive Receiver Autonomous Integrity Monitoring (RAIM) and Fault Detection and Exclusion (FDE). In dual configurations, Global Star 2100 satisfies the requirements for primary means of navigation using GPS alone.

Global Star 2100 provides exceptional flight planning capabilities, with complete SID/STAR procedures, airways, en-route manoeuvres, and 'direct-to' capabilities. Operators are able to store up to 200 flight plans with as many as 100 waypoints, stored in non-volatile flash memory.

Specifications

Display: full-colour LCD flat panel
Display diagonal: 5.5 in
Lines: 12 lines × 24 characters
Dimensions: 180.8 (H) × 146.05 (W) × 199.4 (L) mm
Weight: 3.64 kg
Inputs:
(analogue): fuel flow, air data, heading, VOR/DME
(digital): air data, heading, VOR/DME, weather radar (joystick), radio frequencies
Outputs:
(analogue): HSI course and bearing, crosstrack and vertical deviation, to/from, autopilot steering, annunciators
(digital): EFIS/flight director, autopilot, radio tuning

Status

The Global Star 2100 system is reported to be standard fit in the Cessna Citation Excel, de Havilland DHC-8 Series 400, and Avro RJ, and to be an option on the Bombardier Learjet 45.

Contractor

Honeywell Aerospace, Electronic & Avionics Lighting.

Global Star 2100 flight management system multifunction control display unit   0081855

GLU/GNLU-900 series MultiMode Receivers (MMR)

Type
Aircraft landing system, Multi Mode Receiver (MMR).

Description
The Rockwell Collins 900 series MMR provides two or more landing system standards. Both the ILS and GNSS functions are basic to the landing receiver, while the MLS and GLS functions are categorised as options. The GLU-900 series is designed for use in digital aircraft, while the GNLU-900 series is designed for 'classic' analogue types. Retrofit with the GNLU-900 series MMR permits 'classic' aircraft to operate in accordance with European BRNAV requirements.

The MLS design provides ILS lookalike interfaces to the existing aircraft autopilot and display systems and accommodates requirements for both dual-dual and triplex auto-land architectures. It also offers the accuracy, reliability and integrity required for critical performance in Cat III landings.

The MMR supports existing all-weather landing capability (GLU (ILS) and the GNLU (ILS/VOR)) and capabilities can be expanded as the industry moves toward the future GLS.

The pilot interface to the system is mechanised through a multipurpose control display unit which uses standard ARINC operational philosophy. This allows interoperability training benefits to mixed fleet operators, and future integration of the datalink control function.

The GPS flight management processor and navigation database software can be loaded on the flight line, simplifying system growth.

Specifications
GNLU-900
Dimensions: 4 MCU per ARINC 600
Weight: 7.3 kg approx
Power: 29 V DC
TSO: C129 B1
Qualification: RTCA DO-160 C, DO-178 B

GLU-900
Dimensions: 3 MCU
Weight: 3.86 kg
Power: 115 V AC, 400 Hz
TSO: -C36e, -C34e, (A2D2 YBA (BCL) EIXXXXXZEAEZYZLXX
General: DO-192; DO-195; DO-160C; FCC part 15, EUROCAE ED-46; ED-47A

GPS performance
Channels: 12-channel, all-in-view tracking
Frequency: 1,575.42 MHz (L1) C/A code transmissions
Sensitivity:
(acquisition) –121 dBm
(tracking) –125 dBm
Accuracy (95%):
(horizontal position) 23 m (4.5 DGPS)
(altitude) 30 m (6.0 m DGPS)
(horizontal velocity) 1.5 m/s
(vertical velocity) 1.2 m/s
(time) 100 ns
Time to fix first (95%): 75 s max with valid initialisation; 10 min, max without valid initialisation

MLS performance
Channels: 200, per ICAO Annex 10
Antenna connections: 3 (2 passive, 1 active)
Datalink frequency (optional): C-band

ILS performance
Frequency range: LOC 108.10 to 11.95 MHz; GS 329.15 to 335.0 MHz
Channel spacing: LOC 50 kHz; GS 150 kHz
Navigation outputs: ARINC 429

Status
The GLU-900 series MMR has been certified on a wide range of Airbus and Boeing late-model aircraft.

The GNLU-900 series MMR has been certified for a wide range of retrofit applications including Boeing-737, -747, -757, -767, -777 aircraft and Airbus Industrie A330/A340 and Avro RJ jetliners.

A derivative variant, the GNLU-910 GPS/FMS MMR, has been certified by the European Joint Aviation Authority for use on A300 aircraft, together with the FPI-955 LCD flat panel display as part of a system upgrade to meet European Basic Area Navigation (BRNAV) route requirements, and to provide Global Positioning System (GPS)-based landing approaches. The GNLU-910 GPS/FMS MMR can be integrated with existing analogue sensors.

The GNLU-920 MMR, which will have VHF omnidirectional range and ILS capabilities, is being fitted, as a double GNLU-920 unit fit, into nine US Air Force RC-135 aircraft to meet Global Access Navigation Safety/Global Air Traffic Management (GANS/GATM) requirements, and to comply with ICAO Annex 10 requirements for FM immunity. Aircraft fit was due to begin in late 1999 and to continue throughout 2000.

Yet another variant, the GNLU-945, forms part of the US Air Force's C-5 and KC-135 Global Air Traffic Management (GATM) programme. Each C-5 aircraft will be equipped with two GNLU-945 MMRs, and each MMR will include the following capabilities: VOR, ILS, and MLS providing for Cat III operations. Deliveries are scheduled to begin in 2002 and continue until 2004.

Many thousands of units have been delivered and are on order.

Contractor
Rockwell Collins.

GNS-XL/GNS-XLS Flight Management Systems (FMS)

Type
Flight Management System (FMS).

Description
The GNS-XL and GNS-XLS systems are derived from the earlier GNS-X series of Global Wulfsberg systems. Both are compact, single-box, flight management systems that incorporate eight-channel GPS, and feature flat-panel liquid crystal displays. Both provide control of aircraft navigation sensors, communication, radio and fuel management. With full analogue/digital interfaces, they are well suited for both new programmes, and upgrade/retrofit requirements.

The receiver incorporates RAIM for enhanced reliability, and to meet FAA TSO C129 Class A1/B1/C1 requirements, enabling the operator to make GPS-derived IFR approaches as well as en route and terminal navigation. Both systems include Fault Detection and Exclusion (FDE), enabling them to be used as primary means of navigation during transoceanic and remote area operations. Position and velocity data is accepted from internal and external sensors, using a special navigation filter to generate a composite system position. As well as the built-in GPS receiver, the system includes a VORTAC Position

GNS-XLS flight management system multifunction control display unit 0081854

Unit (VPU) processor, which automatically selects the best available DME/DME and VOR/DME measurements from VOR/DME, VORTAC, TACAN and ILS DME units. External interfaces include those for VLF/Omega, Inertial Reference Systems (IRS), or Inertial Navigation System (INS).

Data communication interfaces include onboard aircraft systems such as the air data computer, compass system, altitude pre-selector, EFIS, weather radar, autopilot, communication and navigation radios and the fuel flow system, together with a full AFIS (Airborne Flight Information System), and optional Satellite Data Communications (SDC) connection.

The GNS-XLS features a full flight planning capability for both area navigation (RNAV) and vertical navigation (VNAV), and it simplifies frequency management.

The GNS-XL and GNS-XLS systems have growth potential for Wide Area Augmentation System (WAAS) and Differential GPS.

Specifications
GNS-XL
Display: full-colour 5 1/2 in diagonal display
Dimensions: 181 × 146 × 200 mm
Weight: 3.64 kg
Inputs:
(analogue) fuel flow, air data, heading, VOR/DME
(digital) air data, heading, VOR/DME, weather radar, EFIS, radio frequencies
Outputs:
(analogue) HSI course and bearing, XTK and vertical deviation, to/from, autopilot steering, annunciators
(digital) EFIS/flight director, autopilot, radio tuning

GNS-XLS
Display: full-colour 4 in diagonal
Lines: 10 lines × 22 characters
Dimensions: 114 (H) × 146 (W) × 165 (D) mm
Weight: 3.18 kg
Inputs: analogue and digital: fuel flow, air data, heading, weather radar, EFIS, VOR/DME/DME
Outputs: analogue and digital: EFIS/flight director, autopilot, radio tuning

Status
GNS-XL is certified by the US FAA to TSO C-129 Class A1; it can be upgraded for compatibility with FANS. It has been standard equipment on Citation Ultra business jets from January 1997.

GNS-XLS was certified for Fault Detection and Exclusion (FDE) by the US FAA in November 1996. Selected by BAE Systems Asset Management as a navigation upgrade for the BAe 146 fleet.

Contractor
Bendix/King.

GRA-2000 Low Probability of Intercept (LPI) altimeter

Type
Flight instrumentation, altimeter.

Description
The GRA-2000 LPI altimeter, has been selected by the US Joint Services Program Office to replace the AN/APN-194, -171, -209 and -232 series altimeters on the majority of tactical jet, helicopter and transport aircraft employed by the US Department of Defense.

The design is based on a very simple durable design employing a single I/F down-convert and specialised algorithms to provide exceptional stealth and jam resistance; it is an outgrowth of the AD-1990 SARA system being procured for Tornado aircraft in the UK.

Specifications
Dimensions: 185.4 × 97.3 × 77.5 mm
Weight: <3.2 kg
Reliability: 8,000 + h
Accuracy: ±2 ft or 2% (1σ)
Performance: 0 to 35,000 ft AGL

Marconi CNI Division GRA-2000 LPI altimeter   0002221

The wide Field-of-View Head-Up Display in the JAS 39 Gripen   1129578

Status

In development against the F/A-18E/F AN/APN-194 configuration.

Contractor

BAE Systems North America, Greenlawn.

Head-Up Display (HUD) for the AH-1W

Type

Head-Up Display (HUD).

Description

For the HUD in the Bell AH-1W, both the control/display subsystem and Full-Function Signal Processor (FFSP) are produced by Kaiser. The HUD is identical to that of the US Army's Bell AH-1S, which Kaiser also supplies. In contrast, the FFSP is an entirely new design, which features dual 68000 processors and software designed to DoD-STD-1679A requirements. The processor is programmable and includes the capabilities for vectors, circles, arcs and rotation.

Status

In production and in service in US Marine Corps AH-1W Cobra attack helicopters.

The rear cockpit of a US Marine Corps AH-1W SuperCobra helicopter, showing the pilot's HUD   0567398

A side view of the front fuselage of a US Marine Corps AH-1W SuperCobra helicopter, showing the pilot's HUD (rear cockpit)   0567399

Contractor

Rockwell Collins Kaiser Electronics.

Head-Up Display (HUD) for the JAS 39 Gripen

Type

Head-Up Display (HUD).

Description

The holographic wide-angle HUD selected for Swedish Air Force JAS 39 Gripen aircraft incorporates diffraction-optics technology -the advantages of this technology are claimed to be a key factor in providing the aircraft with its enhanced capability in the air-to-air, air-to-ground and reconnaissance roles.

Diffractive-optics HUDs have two principal advantages. Firstly, by comparison with conventional systems, they have a much wider field of view, typically $30 \times 20°$ compared with $20 \times 15°$, and are thus more suited to the new generation of combat aircraft in which weapon aiming symbology can make large angles with the flight vector. The wide field will also be useful in night operations to display data from electro-optical sensors such as FLIR. Secondly, the combiner glass on which the symbology is superimposed on the outside world acts as a mirror reflecting only a narrow band of light. The transmission index is about 85 per cent compared with 50 to 70 per cent for refractive HUDs. The symbology is also bright enough to stand out in direct sunlight without having to operate the CRT at such high-power levels that its life is shortened.

The system also provides resistance to glare, reflections and spurious sun images. The latter is particularly important, since bright sunlight can create hot spots on the display that prevent the pilot from seeing the symbology. The design, based on proprietary technology using holography and lasers, employs a single-combiner glass, eliminating the bulky support structure necessary in HUDs that support two or more.

Status

In production and in service in the JAS 39 Gripen.

Contractor

Rockwell Collins Kaiser Electronics.

Helicopter Electronic Flight Instrument (EFI)

Type

Electronic Flight Instrumentation System (EFIS).

Description

Astronautics' 5 ATI Electronic Flight Instrument (EFI) is certified as a direct replacement for electromechanical HSIs and ADIs in helicopters.

The EFI can function as an ADI or HSI with several functional overlays. Designed around an Active Matrix Liquid Crystal Display (AMLCD), the instrument is capable of receiving and processing ARINC-429/702 data from RNAV/FMS/GPS and TCAS/CDTI systems, analogue/ARINC-429 data from radios (VOR, DME, ADF) and ARINC-453/708 from Colour Weather Radar (CWR) and Enhanced Ground Proximity Warning Systems. The EFI also has RS-422 (cross-talk), analogue, synchro, discrete, and differential resolver interface capabilities.

The EFI has been certified for use on civilian and military helicopters, and fixed-wing aircraft. The unit is certified to function as an ADI, HSI,

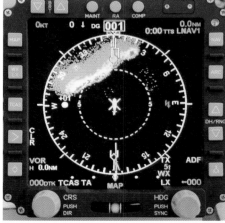

Astronautics' helicopter EFI, as installed in the A109 (Astronautics)   0568543

combined ADI/HSI, CWR, FMS/Map, EGPWS/TAWS (optional) display, TCAS display, CDTI/ADS-B display (optional) and storm scope (optional).

Other features of the system include:
- Sunlight readable
- Remote Control Panel (RCP)
- Modular, rugged construction
- Certified to DO-160D, DO-178B, Level A
- HIRF (200V/M), lightning.

Specifications
Useable screen area: 4.24 × 4.24 in (107.7 × 107.7 mm)
Resolution: 524 × 524
Display luminance: 150 fl (200 fl option)
Viewing angle: ±60°
Processor: Pentium™
Interfaces: ARINC-429, -453, -561, -407 (synchro), RS-422, analogue, discrete, video (optional)
Dimensions: 5 ATI
Weight: <3.64 kg
Power: 28 V DC or 26 V AC
Control panels: Multiple configuration options

Status
In production and in service. Applications include the A-109, Super Puma and S-61 helicopters.

Contractor
Astronautics Corporation of America.

HeliHawk™ Head-Up Display (HUD)

Type
Head-Up Display (HUD).

Description
Derived from the Sparrow Hawk™ (see separate entry), CMC Electronics' HeliHawk™ is a compact, lightweight, single-combiner, raster-capable Head-Up Display (HUD) designed for helicopter applications.

HeliHawk™ provides a stoke or stroke-on-raster display with an electronic boresight capability and 25° total Field of View (FoV). An automatic control adjusts brightness in stroke raster mode and brightness and contrast in raster mode. An optional colour HUD camera and video recording system can be integrated with the system. The fully adjustable overhead combiner stows in a raided position, flat against the bottom of the HUD, enabling simplified cockpit entry/exit without risk of damage to the combiner.

In common with the Sparrow Hawk™, at a total system weight of 8 kg, HeliHawk™ is one of the lightest HUD systems available (claimed 50 per cent lighter than competing systems). A wide range of interfaces are available, from analogue and synchro to ARINC-429 to MIL-STD-1553B. HeliHawk™ features extensive BIT, with most faults diagnosed on-aircraft. Internal data logging and continuous in-flight testing keeps track of system health.

Specifications
Display:
Single combiner
Transmissivity 80%
P-53 phosphor

CMC Electronics' HeliHawk™ HUD
(CMC Electronics)   1030038

5 in aperture
Stroke only or stroke-on-raster
Symbology brightness adjustable 0-3,000 fl
25° circular FoV (IFoV 24° × 15.8° H × V)
24° × 15.8° instantaneous FoV
135,000°/sec maximum draw rate
<1% image distortion
Accuracy 0.0 to 2.0 mr
Contrast Ratio 1.3 to 1 against 10,000 fl ambient
Weight: 8 kg (17.6 lb)
Power supply: 28 V DC per MIL-STD-704D
Power consumption: 60 W per MIL-STD-704E
MTBF: 7,000 h
Environmental: MIL-STD-810E per MIL-5400 Class 1, BS.3G.100
EMI: MIL-STD-461D, -462D, DEF STAN 59–41/RTCA DO-160D

Status
In production.

Contractor
CMC Electronics Inc.

HG7500/HG8500 Series radar altimeters

Type
Flight instrumentation, radar altimeter.

Description
HG7500 configurations available include analogue and/or digital altitude, altitude trips and ARINC 552A. The JG107X height indicator interfaces with the HG7500 and HG8500.

The HG8500 is a form, fit and function replacement for the HG7500. It has a solid-state transmitter and gallium arsenide receiver, and has been qualified to very stringent environmental and EMI requirements.

Options available on the HG8500 include transmitter power management, low-altitude performance in poor antenna installations and unique outputs to meet existing field applications.

Specifications
Dimensions: 137 × 83 × 83 mm
Weight: 1.3 kg
Power supply: 28 V DC, 16 W
Altitude: (HG7502/HG8502) 0–2,500 ft
(HG7505/HG8505) 0–5,000 ft
(HG7508/HG8508) 0–8,000 ft
Accuracy: ±3 ft ±3% analogue altitude
±3 ft ±1% digital altitude

Status
In service on various helicopters and commercial airliners.

Contractor
Honeywell Inc, Sensor and Guidance Products.

HG7700 Series radar altimeters

Type
Flight instrumentation, radar altimeter.

Description
The HG7700 is specifically designed for the low-cost high-performance requirements of tactical

manned and unmanned vehicles. The HG7700 features automatic test capability, noise immune tracker and low power consumption. Available options include digital output, transmitter power management and integrated antenna.

Specifications
Dimensions: 194 × 95 × 72 mm
Weight: 1.55 kg
Power supply: 28 V DC, 12 W
Altitude:
(HG7702) 0–2,500 ft
(HG7705) 0–5,000 ft
Accuracy: ±5 ft ±3%

Status
In production for fighter aircraft, as well as RPVs and tactical weapon systems. The HG7705 is fitted to the Saab JAS 39 Gripen.

Contractor
Honeywell Sensor and Guidance Products.

HG7800 radar altimeter module

Type
Flight instrumentation, radar altimeter.

Description
Honeywell has introduced the new, low-cost HG7800 altimeter module. A unique approach, achieved through the application of ASIC and MIMIC chip technology, delivers the entire radar altimeter on a single, SEM-E short-circuit card. The single-card altimeter is optimised for use as an embedded function and can be added to an available slot within an existing system, such as Honeywell's H-764-G Embedded GPS/INS (EGI) or other integrated systems. The HG7800 is flying in US Navy AH-12 Cobra and UH12 helicopters, USAF T -33, T -38, F-5 fixed-wing aircraft and in a number of US Navy unmanned air vehicles.

HG 7800 options include 5,000 and 8,000 ft variants, and all versions are available as stand-alone LRUs with their own chassis. The HG7800 can be customised for unmanned air vehicles, fixed-wing and rotary-winged platforms.

Specifications
Dimensions: SEM-E Short (5.88 × 6.06 × 0.50 in)
Weight: 0.34 kg
Altitude accuracy: ±2 ft or ±2%
Altitude range: 0 to 8,000 ft
PRF: 35 kHz, PRN coded
Pulsewidth: 30 to 256 ns
Frequency: FCC approved, 4.3 GHz
Transmit power: power managed, 1.0 W peak
Power supply: +5/–15 V DC
Interface: RS 422/485

Status
In service in US Navy and Air Force aircraft.

Contractor
Honeywell Sensor and Guidance Products.

The Honeywell HG7500 radar altimeter
                                    0503730

HG7808 radar altimeter module

Type
Flight instrumentation, radar altimeter.

Description
The HG7808 provides an entire low-cost radar altimeter on a single (SEM-E short) circuit card. The module is highly integrated through the use of ASIC technology and optimised for use as an embedded function added to open/expansion card slots of existing systems such as Honeywell's H-764-G Embedded GPS/INS (EGI) or other integrated systems. The standard circuit card configuration and flexible I/O facilitate integration of the module into other customer-defined installations/applications. The system is also available as a stand-alone LRU with its own chassis.

Specifications
Dimensions: SEM-E Short (5.88 × 6.06 in)
Weight: <0.45 kg
Altitude accuracy: ±2 ft or ±2%
Altitude range: 0 to 8,000 ft
Manoeuvre angles: ±45° up to 5,000 ft, ±30° up to 8,000 ft (nominal antennae, typical terrain roughness, −3 dB cable loss)
Frequency: C-band, 4.3 GHz
RF power: power managed, +30 dBm peak

Status
Honeywell's HG7808 has been selected for the US Air Force T-38 Avionics Upgrade program, various F-5 upgrades worldwide, AH-1W, H1 and the tactical Tomahawk Cruise missile.

Contractor
Honeywell Sensor and Guidance Products.

HG8500 Series radar altimeters

Type
Flight instrumentation, radar altimeter.

Description
The HG8500 is specifically designed for the low-cost, high-performance tactical requirements of manned and unmanned vehicles. The result of Honeywell's independent development efforts, the HG8500 series is a solid-state altimeter utilising Honeywell MMIC technology. Available models feature analogue or digital, or both analogue and digital output, with altitude ranges of 0 to 1,000 ft or 0 to 10,000 ft. The HG8500 offers pulsed leading edge tracking with power management for low detectability. The HG8500 series altimeters are installed in several US Navy missiles and target vehicles.

Specifications
Dimensions: 86 × 86 × 142 mm (3.4 × 3.4 × 5.6 in)
Weight: 1.4 kg
Altitude accuracy: ±2 ft or ±2% (dependent on specific model)
Altitude range: 0 to 10,000 ft (0 to 2,000 ft and 0 to 5,000 ft most common)
Frequency: 4.3 GHz
Track rate: 2,000 ft/s (minimum)
PRF: 25 kHz
Power: +28 (±4.0) V DC, 16 W Max

Status
Operational on US navy land attack missiles and remotely controlled target vehicles.

Contractor
Honeywell Sensor and Guidance Products.

HG9550 LPI radar altimeter system

Type
Flight instrumentation, radar altimeter.

Description
Honeywell claims that the HG9550 represents a quantum leap in altimeter capabilities over its

Honeywell HG9550 LPI radar altimeter system
0018070

earlier products in reliability, covertness, size, weight and cost.

The system was developed by Honeywell and jointly qualified with the US Air Force. The HG9550 was designed as a form, fit and function replacement for MIL-STD-1553B versions of the US Air Force Combined Altitude Radar Altimeter (CARA).

The HG9550 is designed to provide the high accuracy and programmable features of pulsed altimeter designs with the sensitivity and Low Probability of Intercept (LPI) advantages of coherent designs; with less than 1 W of power it is virtually undetectable. A microprocessor permits system characteristics such as track rate and ECCM response to be varied as a function of real-time inputs, or to be preprogrammed according to mission requirements. It is designed to be an element of GCAS systems.

Specifications
Dimensions: 90 × 60 × 222 mm
Weight: 4.43 kg
Frequency: 4.3 GHz
Range and track rate: 0 to 50,000 ft ±2,000 ft/s (minimum)
Altitude accuracy:
(analogue) ±4 ft or ±4% (whichever is greater)
(digital) ±2 ft (0 to 100 ft); ±2% (100 to 50,000 ft)
Manoeuvre angles: ±60° up to 3,000 ft; ±45° up to 5,000 ft; ±10° up to 50,000 ft
RF power: power managed: controlled at 10 dB above track threshold, with less than 1 W transmit power
Input power: 28 V DC (MIL-STD-704), 28 W nominal (35 W maximum)
Programmable features: track rate, ECCM response, sensitivity, altitude range, output formats
LPI features: frequency agility, power management, jittered code and PRF

Status
The HG9550 is an off-the-shelf, fully qualified system, currently in production for US Air Force C-130J and C-17 Globemaster, UK C-130J, Argentine A-4 upgrade, Boeing Joint Strike Fighter and the Lockheed Martin Joint Strike Fighter aircraft.

During 2000, the HG9550 completed an extensive flight test qualification for all blocks of F-16 and was ordered by Lockheed Martin for Block 60 and subsequent production F-16 aircraft during the second quarter of 2001.

Contractor
Honeywell Sensor and Guidance Products.

HGS® Head-Up Guidance System

Type
Head-Up Display (HUD).

Description
Rockwell Collins Flight Dynamics Head-up Guidance System (HGS) is aimed at improving the safety of Cat I operations and providing a landing system capable of operating to less than Cat I weather minima as an economic alternative to Cat III automatic landing systems.

The HGS comprises four main components, an Over Head Unit (OHU), integrating a CRT and lens assembly, a combiner containing the holographic element, a head up computer and a pilot's control panel:

The combiner presents the pilot a view, focused at optical infinity, of HGS symbology overlaid onto the outside scene. The combiner is a single-wavelength mirror, transmissive to all wavelengths of light except the green of the symbology projection, which it reflects back into the pilot's eyeline. Symbology is displayed conformally (true angular representation of symbology onto the outside scene). The combiner can also display conformal IR scene video from the Enhanced Vision System (EVS). A brightness control on the combiner allows the pilot to set the desired intensity of the projected symbology with regard to the outside scene. After this adjustment, an automatic brightness control then adjusts the display relative to changes in the ambient brightness without further pilot input.

The Overhead Unit (OHU) projects symbology onto the combiner. It contains the light source for the display and the lens array that focuses the image to be viewed by the pilot at optical infinity, allowing the pilot to maintain a constant focus when switching from viewing symbology to the outside scene and hence avoiding eye strain under prolonged operations.

The HGS Control Panel (HCP) facilitates selection of display modes according to phase of flight. The HGS can also make use of an existing Multi-purpose Control and Display Unit (MCDU).

Depending on capabilities certified on each specific aircraft type, the HGS Computer provides combinations of symbology for display on the combiner. The computer accepts inputs from a number of sensors on board the aircraft, including air data, attitude/heading, navigation receivers, Flight Management System (FMS) and others.

Along with basic flight information such as airspeed, altitude, course and heading, the HGS displays inertial flight path and acceleration, providing the sensitivity and accuracy required for Cat III operations. The safety of routing Cat I operations is also improved, as industry studies have demonstrated that projected flight path and precise energy management greatly improves the pilot's situational awareness and aircraft control, particularly in difficult or unexpected conditions.

The combiner provides the pilot with a full 30 × 24° field of view, a feature especially useful in high-cross-wind conditions. By comparison, military fighter HUDs typically have 20 × 15° fields of view. The holographic technology improves both the reflectivity of the projected symbology and the transmissivity of real-world details as seen by the pilot.

HGS® combiner showing landing symbology, including runway centreline and touchdown point
0116558

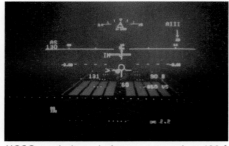

HGS® symbology during an approach at 100 ft in less than category 1 weather conditions. Note the runway centreline (dotted) is aligned with the aircraft actual track (arrow), with a stabilised approach of approximately 3.5°   0116561

The HGS computer provides symbology to assist the pilot in flying an optimal escape from a windshear event and for righting the aeroplane in case of an unusual attitude upset. The ability to follow an extremely precise flightpath facilitates a stable and accurate visual or non precision approach. Depending on certification, the HGS computer can provide ILS approach guidance to allow the pilot to hand-fly an ILS approach to Cat IIIa landing (50 feet DH, 600 feet RVR), at a 1×10-9 level of integrity (equivalent to an autoland or fly-by-wire system). On more recent certifications, the HGS can also provide Low-Visibility Take-Off guidance (LVTO) allowing take-off in visibilities as low as 300 ft RVR on qualifying runways.

In the US, the HGS facilitates approved Cat III operators to obtain lower landing minima on Cat II and on qualifying Cat I runways. FAA Order 8400.13A describes this procedure in detail. The video display capability of the HGS can also be used to display Synthetic Vision information in future iterations of the system.

Flight Dynamics is evaluating the use of the HGS to obtain even lower take-off and landing minima. The commercial transport industry has also expressed interest in combining a fail-passive autoland system with the fail passive HGS to achieve Cat IIIb capability. This hybrid landing system should be certifiable to 300 ft runway visual range and would combine the benefits of an automatic landing system with the projected flight path and head-up advantages of the HGS.

Specifications

Dimensions:
(HCP) 66.8 × 146.1 × 173.2 mm (H × W × L)
(Combiner) 274.1 × 354.3 × 198.6 mm (H × W × L)
(Computer) 198.1 × 256.8 × 387.6 mm (H × W × L)
(OHU) 167.6 × 256.8 × 584.2 mm (H × W × L)
Weight:
(HCP) 1.35 kg
(Combiner) 4.05 kg
(Computer) 11.7 kg
(OHU) 14.4 kg

Status

In July 1984, Flight Dynamics received FAA approval for its HGS® for manually flown Cat IIIa approaches down to a runway minimum range of 700 ft and decision height of 50 ft in the Boeing 727. This was the first manual system certified by the FAA for the demanding low-visibility environment. In August 1987, the HGS was also

HGS® symbology in cruise. Note the aircraft velocity vector (winged symbol) wind arrow (top right) and heading rose (bottom)   0116559

HGS® installation in Dassault Falcon 2000   0116560

certified for windshear detection and recovery guidance and, in 1990, was approved by the FAA for low-visibility take-offs down to 300 ft RVR.

Alaska Airlines, the first carrier to equip its 23-aircraft Boeing 727 fleet with the HGS, received FAA operational approval to operate in revenue service down to 50 ft decision height in late 1988, conducting the first manually-flown Cat IIIa landing with passengers on board in October 1989.

Flight Dynamics claims that its line of head-up guidance systems has received about 60 per cent of the Boeing 737 New Generation orders. More widely, it has sold to a large number of airlines, corporate users, and military transport operators, flight training companies and NASA. Aircraft equipped with the HGS® include the Challenger CL604, Falcon 900 EX, Falcon 2000, Saab 2000, Boeing 727, 737 and BBJ, Bombardier Q100/200/300 and 400, Bombardier Canadair Regional Jet CRJ 100/200 and 700, Embraer ERJ 145, Dornier 328 and the Lockheed Martin C-130J.

The designation HGS-2350 has been used for the installation in the Easyjet B737-300, and HGS-2850 for the Falcon 2000 installation. It is reported that about 50 per cent of all Falcon 2000 buyers are selecting the HGS, together with about 80 per cent of Falcon 900EX customers.

The most recent in the series of Flight Dynamics' HGS is the -4000 series, which incorporates enhanced features aimed at improving pilot situational awareness, including rapid recovery from unusual attitudes; these features are available for retrofit to earlier HGS models. HGS-4000 is installed in Delta Airline's fleet of Boeing 737-300 and 737-800 aircraft. During the first quarter of 2001, Delta announced a further order for the HGS for its fleet of 120 MD-88s, with options to install the system in the remainder of the company's aircraft, including the Boeing 757, 767 and 777. HGS-4100 was reported as having completed flight testing installed in a Bombardier Dash 8 Q400 during the second quarter of 2002.

During September 2006, Rockwell Collins announced that the HGS-4200 had been certified by the European Aviation Safety Agency (EASA), Federal Aviation Administration (FAA) and Transport Canada (TC) for use on Bombardier CRJ 705 and Bombardier CRJ 900 aircraft. The company worked with Bombardier and Air Canada Jazz to achieve this certification. Air Canada Jazz provided the flight test aircraft used to conduct certification flights, including 40 HGS approaches into several Canadian and US airfields. Bombardier provided a CRJ 900 simulator and related support for the certification effort. The certification programme required 335 simulated HGS approaches to be flown in a wide variety of environmental conditions.

Also during September 2006, Lufthansa CityLine (CLH) announced that it would be equipping its fleet of Bombardier CRJ 900 aircraft with the HGS-4200, to be installed on new aircraft as they are delivered. Lufthansa CityLine currently operates a fleet of 36 CRJ 200s and 20 CRJ 700s. All are equipped with HGS systems manufactured by Rockwell Collins. Lufthansa became the first airline to take advantage of the low-visibility takeoff capability on a CRJ by taking off in 100 metre visibility at Milan's Malpensa airport in January 2005.

Contractor

Rockwell Collins Flight Dynamics.

High-speed camera 16 mm-1PL

Type
Flight/mission recording system.

Description
The Photo-Sonics 16 mm-1PL camera provides variable speeds from 10 to 500 frames/s. The 16 mm-1PL features interchangeable magazines of 200, 400 and 1,200 ft and synchronous phase-lock plug-in operation. Utilising the magazine load concept, the 16 mm-1PL can be loaded within seconds without disturbing the camera, the optical alignment or the electrical connection.

The camera features automatic exposure control, and timing pulse marks at 10, 100 and 1,000/s. The 70 series miniaturised Film Data Recording System (FDRS) can be attached to the side of the 16 mm-1PL camera, and it can record numerical and BCD information on 16, 35 and 70 mm film on every frame up to 1,000 frames per second.

Specifications
Shutter angle: 9 settings varying from 7.5 to 160°, set by external knob
Power: 28 V DC or 115 V AC
Weights:
camera body: 2.73 kg
magazines: 200 ft, 2.27 kg; 400 ft, 3.64 kg; 1,200 ft, 6.82 kg

Status
Fitted into the wing tip of Canadian Forces CF-18 (CF-188) aircraft.

Contractor
Photo-Sonics Inc.

The Photo-Sonics 16 mm-1PL high-speed camera with film data recorder attached 0051026

Horizontal Situation Indicators (HSIs)

Type
Flight instrumentation.

Description
Astronautics Corporation of America manufactures more than 140 different types of Horizontal Situation Indicators (HSIs), installed in many military and civilian fixed- and rotary-wing aircraft.

The HSI range provides functionality and indication for heading, range (DME), elapsed time, glideslope pointer, bearing pointers, glideslope deviation flag, slaved course and slaved heading.

Key benefits and features of Astronautics' HSIs include:
- 3, 4 and 5 in sizes
- Single- or dual-bearing pointers
- Remote or manual heading and course pointers
- Digital or analogue interface to navigation sources
- DME readout – digital or analogue inputs
- Digital course readouts
- Bootstrapped azimuth outputs
- Low-power DC servo loops
- Flight director and/or autopilot compatible outputs
- Built-In Test (BIT) circuitry
- Qualified to MIL specs
- Glideslope
- Power warning indicator
- Colour-coded pointers
- To/from arrows

- Failure warning flags
- NVIS-compatible
- Simple front- or rear-mounted installation
- Special functions to meet specific requirements.

Status

In production and in service in F-16, F-5, F-15, EF-111, T-45, C-130, P-3C, AV-8, UH-60, AH-64, Super Puma, A 109, S-61, B-412, B-212, CH-47 and other military and commercial aircraft.

Contractor

Astronautics Corporation of America.

Horizontal Situation Indicators (HSIs)

Type

Flight instrumentation.

Description

The HSI-415 is a 3 in (76.2 mm) form factor indicator capable of functioning as a Horizontal Situation Indicator (HSI) or switchable to perform as an RMI. In the HSI mode, the unit displays aircraft heading, bearing from either ARINC synchro or DC sine/cosine source, manually selected course, VOR/Loc deviation, VOR to/from indication and glide slope deviation. In the RMI mode, the course pointer is continuously motorised to a position that will centre the VOR deviation bar and indicate to the station. With the simultaneous display of ADF bearing, the HSI-415 provides all the functions of an RMI.

The unit is compatible with any high-performance fixed-wing aircraft or helicopter.

The HSI-421 is a 4 ATI form factor instrument which displays aircraft heading, VOR, ILS, ADF, DME, Tacan or long-range navigation information. Several variations of the basic system, configured to operator's requirements, are in service.

The latest variant is the HSI-423, which adds a second bearing pointer. The unit is compatible with any medium- or high-performance fixed-wing aircraft or helicopter. HSI-423 features include:

- Dual servo-bearing pointer display
- Dedicated autopilot outputs
- Isolated bootstrap
- Available with side mount D-sub connection or rear-mounted MS circular style
- Available in US. Night Vision Goggle (NVG) compatible.

Specifications

(HSI-415)
Dimensions: 3 ATI form factor per ARINC 408
Length: 228.6 mm (9 in) (max)
Weight: 1.59 kg (max)
Power supply: 27.5 V DC, 1.5 A (max)
Lighting: Internal, 5 or 28 V, white
Temperature range: −30 to +70°C
Altitude: −1,000 to 55,000 ft
TSO: C6d, C34e, C36e, C40c
Environmental: DO-160B

Status

The HSI-415 is fitted on the Learjet 31A, Sikorsky S-76 and Agusta A 109.

The HSI-421 is no longer in production. A variant of this indicator is fitted to UK Royal Air Force Tucano aircraft.

The HSI-423 is currently fitted to the Agusta A 109 and Eurocopter BK-117 helicopters.

Contractor

Northrop Grumman Corporation, Component Technologies, Poly-Scientific.

ICDS 2000 Integrated Cockpit Display System

Type

Avionic display system.

Description

Arnav Systems' Integrated Cockpit Display System (ICDS) is an Electronic Flight Information

ICDS 2000 MFD 0526162

System (EFIS) designed specifically for general aviation aircraft. The ICDS system uses proprietary symbol computers, displays and software from the ICDS 2000 and MFD 5000 systems (see separate entry). Each ICDS systems display includes a colour LCD display with wide viewing angle; software controls each display independently, so ICDS system display formats can be readily and rapidly developed for each new aircraft type and incorporated with the existing display hardware to achieve maximum installation flexibility.

ICDS system Control Display Units (CDUs) are used to provide a basic integrated cockpit display system. Two ICDS 2000 system displays are used to provide for the three primary software functions. Additional CDUs may be added to provide for system redundancy or displays for the copilot. The displays support the Electronic Flight Instrument System (EFIS), the Engine Indication Crew Advisory System (EICAS) and a general purpose MultiFunction Display (MFD). These three functions are integrated to provide the pilot with a complete picture of the aircraft operating environment, systems status and communications. Each display incorporates a computer system composed of a compatible Signal Generator and a CDU. These hardware elements support the software required to perform the three primary screen functions. The software in each system performs full-time automatic cross-checking of the other systems, and flags any discrepancy between the units. Where only two displays are used, they perform the basic software functions for the Primary Flight Display (PFD) and MFD. In this configuration, The PFD is displayed on a dedicated system CDU while the MFD performs its own functions and also those of the EICAS display. Each Computer Display Unit (CDU) can be replicated for large aircraft aircrew stations or for tandem cockpit aircraft.

The ICDS 2000 PFD brings together all of the essential information for flying the aircraft on a single display. The PFD provides an integrated display for attitude, altitude, heading, vertical speed, airspeed and selected navigation information. The PFD performs continuous diagnostic testing to quickly identify a failure in the system inputs or the display. In the event of a PFD failure the affected function can be switched to the MFD.

Remote mounted solid-state attitude and directional gyros supply primary attitude and heading information. Two independently powered Attitude Heading Reference Systems (AHRS) may be used for redundancy. Attitude is transmitted via a digital bus which sends signals that indicate pitch, roll and heading to the PFD-35 Digital Air Data System (D-ADS) AHRS computer. Each PFD-35 computer converts the digital

information into a common RS-232 data stream. The serial data is then combined with the air data information from the D-ADS and transmitted to the multiple display systems for use with the PFD and NAV functions. Backup attitude and heading data is supplied by a second AHRS, or a set of electrically driven attitude and directional gyros. Alternatively, if only backup roll and altitude is required with no pitch information, then a remote or panel mounted electric turn and bank can be used for roll information. The PFD will indicate if pitch or roll is not available and the pilot can then revert to needle, ball and airspeed flying. The advantage in this mode is that all roll, altitude and airspeed information is still presented on the PFD. The Annunciator system indicates the absence of pitch or roll information, as appropriate, when the primary attitude gyros are not available.

The EADI presents pitch and roll information derived from the AHRS subsystem. When pitch exceeds ±30°, V-pointers appear to indicate the direction of the horizon. The EADI has two modes of operation, Enroute and Approach. The Enroute mode presents a decluttered image for the pilot. In Approach mode, glideslope information is presented in vertical and horizontal guidance windows on the right and bottom sides of the EADI. Magenta balls within the guidance windows indicate the pilot's manoeuvring requirements to capture the desired approach course and glideslope. APP, VNAV, and LNAV flag indicators are controlled by the signals received by the Navigation Management System (NMS).

The Flight Director (FD) integrates input from several sources to provide guidance cueing in both vertical and lateral directions. Vertical modes are controlled by pilot selection of altitude, or vertical guidance supplied by the NMS. Lateral modes include localiser or VOR navigation input supplied by the NMS. The FD has two modes, Pilot-select VNAV or Approach mode. In pilot-select VNAV FD, the pilot enters a target altitude and the FD will give climb or decent cues based upon a comparison of present altitude and target altitude. In Approach FD mode,

ICDS 2000 PFD 0526160

ICDS 2000 installed in a Cirrus SR 20 general aviation aircraft 0526161

the FD automatically locks to the glideslope signals provided by the NMS. The guidance may be derived from either Localiser/Glideslope information, or from computed GPS guidance, as chosen by the pilot on the NMS.

The turn-bank indicator is a graphical display of rate-of-turn only. Turn-rate data is provided by the AHRS subsystem, and is based on a two-minute, 360° turn (three degrees per second). No pitch information is provided from the turn-bank indicator.

The PFD-35 D-ADS, is connected to the aircraft pitot and static ports, heading system and OAT probe to directly sense indicated airspeed, pressure altitude, and outside air temperature. This raw data is collected and combined with the associated AHRS outputs. All of this data is then transmitted to the multiple display systems for use with the PFD and NAV functions. The air data can also be transmitted to the GPS navigators for calculation of winds aloft. When air data is combined with the optional fuel flow input, the navigator can also calculate specific fuel consumption and remaining range.

The system categorises warnings before displaying them to the pilot. System Status messages are located in the left lower section of the EADI and display alert information related to system failures, airframe configuration and other less critical failures. Flight Profile messages, which are displayed in the lower right section of the EADI, display information related to altitude, airspeed, unusual attitudes and other flight-related warnings. Engine data is constantly monitored by the Engine Trend Monitor function of the MFD. If any engine parameter is at 'CAUTION' or 'EXCEED' levels, a message is displayed in the Engine Message Window located at the top left side of the EADI. The EICAS display shows engine and airframe data and lights annunciators in the event of out-of-range conditions. The engine display can be shown full screen on a dedicated EICAS display or 'windowed' onto the MFD moving map display.

ICDS 2000

The ICDS 2000 PFD is an EFIS designed for experimental aircraft. The large, bright display includes engine power indicators, EHSI, Flight Director (FD) and VOR/GPS/ILS-coupled navigation displays, including Glide Slope. The PFD provides predictive data in terms of position, energy and time and also displays airframe configuration data. It provides a wide range of fault detection alerts, and full integration with available ICDS 2000 MFD and engine systems.

The ICDS 2000 MFD integrates applications such as WxLink™ and Stormscope® (see separate entry), together with GPWS, air traffic and engine information, facilitating enhanced pilot Situational Awareness (SA) under high-workload scenarios.

Specifications
Display: 10.4 in (diagonal) colour AMLCD
Dimensions: 206 × 292 × 111 mm (H × W × D)
Weight: 4.0 kg
Voltage: 10–35 V DC (unregulated, positive polarity)
Power: 50 W
Operating temperature: −20 to +70°C
Input/output: 7 RS-232 (4 convert to RS-422)

Status
In production and in service. The ICDS 2000 is available for piston and turbine powered experimental aircraft. The system is also aimed at new aircraft or production aircraft that are being redesigned to new type certificates. It is anticipated that type certification for the ICDS system will be performed in conjunction with certification of the host aircraft. The company claims that development risk for certification in this manner is low because the ICDS system consists of modular tested components that are available off the shelf yet are easily adaptable to the airframe and engine used.

Contractor
Arnav Systems Inc.

IEC 9002/9002M Flight Management Systems (FMS)

Type
Flight Management System (FMS).

Description
The IEC 9002/9002M FMS provides complete LNav and approach capabilities, and accurate navigation in all phases of flight using the 12-channel GPS receiver, which accepts differential GPS corrections. It features a control display unit with a 4 in diagonal, sunlight-readable, high-resolution, 16-colour LCD and can be enhanced with a kinematic upgrade to provide Cat. III landing accuracy. The unit provides rapid GPS satellite acquisition (time to first fix 2 minutes). It is equipped with Jeppesen worldwide navigation database including SIDs, STARs, GPS instrument approaches, airports with runways greater than 4,000 ft, high- and low-altitude airways, intersections, VHF navaids and NDBs, and provides for 400 flight plans (company routes or pilot-defined) of 100 waypoints each; 2,000 pilot-defined waypoints.

IEC 9002/9002M is (S)CAT-I DGPS compatible and GLONASS and WAAS upgradable. It is a full-featured GPS flight management system intended for installation in all types of aircraft. It accepts differential GPS corrections required for (S)CAT-I operations. Accuracies in the 0.5 m range, allowing Cat. III operations, are attained when the IEC 9002 is enhanced with its kinematic upgrade and used with the compatible Model 8000 DGPS ground station.

Specifications
Navigational signal: GPS L1 C/A code (SPS)
Receiver type: 12 GPS hardware channels
Time to first fix: 2 min
Dynamics: 900 kt (max) velocity

Accuracy:
100 m 2σ rms (selective availability on)
5 m using differential GPS correction

Status
The IEC 9002/9002M has been certified by the FAA to TSO C129 Class A1, aboard the B737-200, L-100/C-130, C-12 D/J, UC-12B/F/M and T-44A. The unit is capable of Direct-Y acquisition in heavy jamming environments. Standard features include: an FAA TSO certified P (Y) code input capability; GRAM upgrade capability without TSO impact; anti-jam performance; ARINC 739 compliance; 50 per cent throughput reserve for GATM/FANS functional growth.

The 9002/9002M has been selected by several airlines, and over 100 units have been delivered.

Contractor
L-3/Interstate Electronics Corporation.

IFR/VFR avionics stacks

Type
Avionics system.

Description
Garmin has integrated its products into IFR and VFR avionics stacks comprising the following units:
IFR avionics stack
- GMA 340 TSO'd audio panel
- Dual GNC 300 TSO IFR-certified GPS/Comm
- GTX 320 Class 1A transponder (no longer in production. Replaced by GTX 327, 330, 330D)
- MD 41 annunciator
- GI 102A course deviation indicator
- GI 106A course deviation indicator with glideslope
VFR avionics stack
- GMA 340 TSO'd audio panel
- GNC 250XL GPS/Comm with moving map
- GPS 150XL GPS/Comm with moving map
- GTX 320 Class 1A transponder (no longer in production. Replaced by GTX 327, 330, 330D)
- GI 102A course deviation indicator

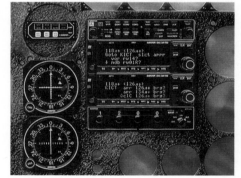

Garmin IFR stack 0015379

L-3/Interstate Electronics Corporation IEC 9002/9002M flight management system 0018230

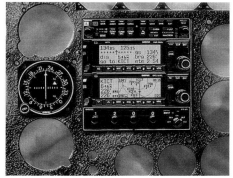

Garmin VFR stack   0015380

Contractor
Garmin International.

ILS-900 Instrument Landing System receiver

Type
Aircraft landing system, Instrument Landing System (ILS).

Description
Collins' ILS-900 is the next generation of Collins ILS sensors. The system meets ARINC-700 form, fit and function characteristics, interfaces with other aircraft systems via a serial ARINC 429 databus, meets the racking and cooling requirements of ARINC 600, and is compliant with environmental and software requirements. The ILS-900 can be retrofitted into existing aircraft and used interchangeably with -700 ILS systems (subject to approval) and meets FM immunity requirements as stated in ICAO Annex 10, for 1995 installation and 1999 operational compliance.

The ILS-900 contains partitioned, comprehensive end-to-end self-test that will diagnose and isolate a system problem to an individual LRU fault or a fault existing in connected peripherals (such as control panels, antennas, and so on). In addition to ARINC 604 BITE, which connects with aircraft fault maintenance systems (Airbus A320/330/340; Boeing B747-400/777/MD-11), the ILS-900 has a simple and rugged LED-BITE display on the front panel. This allows confirmation of fault status within the equipment bay of older aircraft which do not incorporate fault maintenance systems.

The ILS-900 is a high-integrity ILS receiver with software verified to DO-178A Level-1 (critical). This system performs to CAT III fly-by-wire failure and monitor requirements in dual and triplex autopilot installations. This receiver has been specifically designed for CAT III Dual Autopilot operations that place heavy reliance on internal monitoring within the ILS receiver. A high degree of monitoring integrity has been included in the ILS-900. System reliability has been vastly improved by the incorporation of digital technology. Further system features include audio mute in cruise mode and automatic Morse decoding. Absolute partitioning between BIT and the primary monitoring deviation circuitry has been achieved by using four separate processors.

The architecture uses flash memory technology which allows easy code change of BITE protocols to facilitate installation in different aircraft types without reverification of the software.

Specifications
Dimensions: 3 MCU
Weight: 3.86 kg
Power supply: 115 V AC, 400 Hz, single phase
Frequency:
(localiser) 108.1 to 111.95 MHz
(glide slope) 329.15-335 MHz
Channel spacing:
(localiser) 50 kHz
(glide slope) 150 kHz
Temperature range: –55 to +71°C

Status
In production and in service.

Contractor
Rockwell Collins.

Integrated cockpit display for the F/A-18C/D

Type
MultiFunction Displays (MFD).

Description
The integrated cockpit display in the F/A-18C/D includes the Multifunction Display Indicator (MDI) and Multipurpose Display Repeater Indicator (MDRI).

The F/A-18C/D 5 × 5 in (127 × 127 mm) NVG-compatible colour MDI and symbol generator includes an integral display processor. The processor consists of three Motorola 68020 microprocessors that handle input/output via a MIL-STD-1553 bus and provide display functions to the mission computer. The display functions include a variety of alphanumeric symbols, geometric shapes and large format macros such as a compass rose or pitch ladder. Three fully independent symbol generators are included, allowing the MDI to drive four additional displays with stroke or video symbology, or both.

The MDI suports a variety of formats, including 525-line raster in either 1:1 or 4:3 aspect ratio, 675-line square format, arc scan and wedge for display of high-resolution ground mapping radar and 875- or 1,224-line formats for sensor video. The display generators can handle three independent simultaneous video outputs and can overlay stroke-in-retrace on all of them.

Kroma liquid crystal shutter technology is incorporated into these displays, allowing stroke symbols to be presented in red, yellow or green. The use of colour improves pilot situation awareness by providing quick differentiation between friend, enemy and unknown in a cluttered tactical situation without sacrificing resolution and brightness. Colour signals are available externally so that each MDI can drive similar remote colour displays, as in the F/A-18 night attack aircraft.

Originally designed for the night attack version of the F/A-18, the 5 × 5 in (127 × 127 mm) MDRI raster/stroke display provides night vision-

compatible Kroma colour stroke and monochrome raster displays in a wide variety of formats.

While air-to-air displays are presented in colour stroke, with red, green and yellow symbology available, air-to-ground views are matched to the AN/APG-65 radar by means of special arc-scan, DBS patch and rotating raster forms which preserve in the display the inherently high resolution available from that radar. Sensor video, such as FLIR, can also be accommodated with normal rasters in 25-, 675- or 875-line format, in either 1:1 or 4:3 aspect ratios. All of these rasters may be augmented by sharp, clear, fine line stroke symbology written in retrace.

Dependent on an external symbol generator for raster sweeps and stroke symbols, the resultant small size and 'cathedral' top allow installation high in the instrument panel either on the left or right sides or, as in the case of the F/A-18, on both sides. Push-buttons around the display surface are software cued to provide in-flight programming.

Specifications
Dimensions:
(MDI) 658.4 × 170.2 × 179.1 mm
(MDRI) 397 × 170.2 × 179.1 mm
Weight:
(MDI) 18 kg
(MDRI 9.7 kg
Power supply: 115 V AC, 400 Hz, 3 phase
Display: 127 × 127 mm
Resolution: 120 lines/in
Reliability:
(MDI) 1,500 h MTBF
(MDRI) 2,600 h MTBF

Status
In service on the US Navy and US Marines F/A-18C/D aircraft and on export F/A-18C/D aircraft.

Contractor
Rockwell Collins Kaiser Electronics.

Integrated coloured moving map

Type
Avionic Digital Map System (DMS).

Description
The helicopter integrated coloured moving map is a high-performance low-cost system. It comprises the Astronautics modular display processor and multifunction display and provides the pilot with a coloured moving map, night and all-weather capability, nap of the earth capability and workload reduction.

The coloured moving map is stored in the modular display processor and has multiple layers including topography, tactical data, navigation data, flight guidance and obstacles. It has track up and north up modes and a zoom capability from 1:50,000 to 1:1 million.

Options include a helmet-mounted display, data gathering from sensors such as FLIR or CCD and Hands On Collective And Stick (HOCAS).

Contractor
Astronautics Corporation of America.

Integrated Core Processor (ICP)

Type
Aircraft central computer.

Description
Raytheon Electronic Systems was selected by Lockheed Martin Tactical Aircraft Systems to

provide the ICP for its two X-35 JSF Joint Strike Fighter demonstrator aircraft.

The ICP is the central computer system for the JSF, including all the embedded computing elements for multiple subsystems. It provides the digital processing resources for sensors, communications, electronic warfare, guidance and control and cockpit displays. The ICP implements an

open system architecture that maximises the use of commercially supported products and standards, to define, implement and support an affordable ICP that uses open-system concepts to enable seamless incorporation of new technologies.

As the successful candidate, the Lockheed Martin F-35 JSF will utilise a developed version of the Raytheon ICP.

Lockheed Martin1's X-35C Naval Version of the JSF utilises Raytheon's ICP (Lockheed Martin)
0105289

Integrated Mission Display Processor (IMDP)
0051290

Contractor

Raytheon Company, Space and Airborne Systems.

Integrated display system for the Boeing 747-400

Type

Avionic display system.

Description

Rockwell Collins has developed an integrated electronic colour display system for the Boeing 747-400. A Rockwell Collins digital flight control and central maintenance computer (CMC) also are standard on the aircraft. The display system features six 8 in (210 mm) square colour CRT displays. Any one of these units can be selected to show either EFIS or EICAS data.

The new technology gives the 747-400 only 38 per cent of the cockpit lights, gauges and switches compared with older 747 aircraft and leads to better aircraft availability and significant reductions in crew workload.

The IDS-7000 integrated display system utilises the third-generation hardware design first certified in the EFIS-1000 system. Each display unit contains all drive and symbol generation electronics necessary to perform any of the display tasks. Each pilot has a Primary Flight Display (PFD) and Navigation Display (ND) mounted in a side-by-side configuration. All primary air data and heading information is included on the PFD displays. EICAS functions are provided on two displays

mounted in an over/under configuration in the centre panel.

The display system also includes three electronics interface units (EIUs) which function as data conversion and collection for EICAS displays, message processing, snapshot information recording, exceedance data recording and CMC-7000 central maintenance computer data collection. The central maintenance computer, in conjunction with the display units, provides ground maintenance crews with maintenance page displays. The display units, EIUs and CMCs are all capable of being programmed on the aircraft without removal of the equipment.

Status

In production for the Boeing 747-400.

Contractor

Rockwell Collins.

Integrated Mission Display Processor (IMDP)

Type

Aircraft mission computer.

Description

The Integrated Mission Display Processor (IMDP) is an open system architecture based on VME64. It implements reliable and low-cost elements in a configurable military processor. The IMDP is designed to be applicable to airborne, shipboard and ground vehicle applications. The IMDP is

configurable, using common modules, to provide mission management, display management, cockpit interfaces, digital map, weapons management and other functions. The IMDP provides a fully qualified, low-cost approach for new development or upgrades to existing systems.

Architecture

- VME64 – Commercial standard architecture with a large supplier base, proven for cost-effective implementation of real-time military processing applications.

Interfaces

- MIL-STD-1553 – Single or multiple interfaces, configurable as bus controller, remote terminal, bus monitor, or any combination.
- Fibre channel – Fully compliant bidirectional interface, configurable as point-to-point or arbitrated loop.
- Others – RS-422, RS-232, ARINC-429, SCSI, Ethernet, and so on.

Processing

- Power PC – Fully integrated with and support for either Power PC 603 or 604.
- Display processor – Single board provides video switching and outputs for two displays.
- Other microprocessor options – Supports full line of Motorola, Intel, and Silicon Graphics microprocessors.
- Special purpose processing – Variety of special purpose processors and functions (software) to support both graphical and numerical applications.

Memory

- Mass memory module – Existing solid-state (Flash EEPROM) mass memory options (single board with up to 2 GBytes)
- Embedded memory – Existing processing boards contain up to 64 Mbytes of RAM and 8 Mbytes of Flash.
- Memory interfaces – Provides standard interface for other embedded memory boards (VME or PCI disk drives, and so on) or external memories.

Displays

- Single board display processors. Each board provides multiple RS-170 RGB and monochrome inputs, two independent outputs, graphics rendering, and overlays. Configurable for single or multiple display processing boards.

Digital map

- Existing single board map processor. Provides fully anti-aliased, RS-170 RGB, map display, threat overlay (line-of-sight), multiple scales, continuous zoom, and 3-D graphics.

Real-time operating system

- VxWorks – A commercially available program which is supported throughout the VME community for real-time uni- or multi-processing applications.
- Software architecture – Integrated with VxWorks to provide ease of software development or porting and integration of existing applications.

The Rockwell Collins displays and flight control system developed for the Boeing 747-400   0503949

Software development
- PC-based commercial-off-the-shelf (COTS) tools – use of standard processors operating system and languages provides a fully COTS PC-based toolset.

Specifications
Operating environment:
Designed and qualified for military airborne, shipboard and ground vehicle environments.
Power:
28 V DC, 270 V DC, or 115 V AC, less than 120 W.
Dimensions:
¾ ATR without cooling fan, ¾ ATR long with cooling fan.
Weight: <13.64 kg

Contractor
Smiths Aerospace, Germantown.

J.E.T. directional gyro systems

Type
Flight instrumentation.

Description
L-3 Communications Avionic Systems' J.E.T. brand Directional Gyros (DGs) are highly reliability multimode devices, with instantaneous slaving to eliminate start-up delays associated with conventional directional gyros. The DG provides three operational modes: slaved DG, free DG and compass. The microprocessor-controlled gyro permits such features as extensive built-in test and instantaneous slaving.

Standard features include two isolated synchro outputs and autopilot interlock.

Specifications
Dimensions: 155 × 139 × 167 mm
Weight: 3.36 kg
Power supply: 18 to 32.2 V DC

Status
In production. The J.E.T. DGs are in service in UK Royal Air Force Tucano trainers.

Contractor
L-3 Communications, Avionics Systems.

KC-135 Global Air Traffic Management (GATM)

Type
Integrated avionics system.

Description
In October 1999, the US Air Force selected Rockwell Collins to upgrade 544 C/KC-135 aircraft to meet GATM requirements. Rockwell Collins' solution is based on the Collins Flight2 System. This open-architecture system improves situational awareness by integrating flight operations with navigation and guidance functions. It also provides the interactive services required for the future GATM environment.

Flight2 GATM functions allow military aircraft to comply with worldwide civil digital airspace initiatives at a suitable level to meet mission needs. Aircraft solutions can vary from full GATM functionality for high-flying transport aircraft to a reduced functionality fit for low/slow flying helicopters. Flight2 GATM capabilities include:
- Controller/Pilot Data Link Communication (CPDLC) – VHF and SATCOM
- 8.33 KHz frequency spacing
- Reduced Vertical Separation Minimum (RVSM) altitude accuracy
- Protected ILS FM immunity
- Required Time of Arrival
- Required Navigation Performance (RNP) accuracy for current and future requirements.

The C/KC-135 upgrade will provide a COTS solution based on the PACER CRAG avionics baseline, while ensuring unrestricted access to civil airspace and providing enhanced military mission functionality. Rockwell Collins avionics supporting the GATM programme include: the Integrated Processing Centre (IPC), CMU-900 for datalinks; the SAT-2000 Satcom radio; datalink modifications to the AN/ARC-190 HF and AN/ARC-210 V/UHF radios; and the GNLU-955 MultiMode Receiver (MMR).

Status
The initial contract award for USD39.2 million was made for Engineering Manufacturing Development (EMD). This phase included design, development, integration and testing of the completed GATM suite, which comprises the Collins Flight2 System and was scheduled for completion in 2002. A follow-on option for prototype A and B modification kits was issued in December 1999, for completion by the end of August 2003. The installation of 540 production kits and 20 simulators is planned, for completion by 2011.

Raytheon Electronic Systems will perform aircraft installations, under subcontract to Rockwell Collins, at its facility in Greenville, Texas.

Contractor
Rockwell Collins.

KI 825 Electronic Horizontal Situation Indicator (EHSI)

Type
Cockpit display.

Description
The Bendix/King KI 825 Electronic Horizontal Situation Indicator (EHSI) is designed for the general aviation market. It combines traditional heading and navigation functionality with mapping and an optional traffic display in true 3 ATI format. The display uses colour active-matrix LCD technology to improve viewing at night and in bright daylight.

The KI 825 is designed to interface with the most common systems in general aviation. It uses analogue, ARINC-429 and -407 and RS232 standards to connect with VHF navigation, GPS, heading, traffic and weather systems.

Announced during 2004, an enhanced, NVIS-compatible version of the KI 825, incorporating Ferris wheel compass card headings and auto course slew, where the KI 825 automatically positions the course pointer to the flight plan way point. In addition, during instrument approaches, sensor displayed sensor data automatically changes colour to alert the pilot to the change of flight mode.

Specifications
Dimensions: 3.26 × 3.26 × 7.67 in (3 ATI)
Weight: approx 3.0 lb
Power supply: 11 to 33 V DC
Temperature range: –20 to +70°C
Certification: FAA-TSOs C113, C6d and C9c

Status
In production and in service in a wide variety of helicopters, piston and turbine aircraft. During the second quarter of 2004, Honeywell announced that it had gained US Federal Aviation Administration (FAA) certification for the NVIS-compatible KI 825 EHSI.

Contractor
Bendix/King.

Bendix/King KI 825 EHSI (Bendix/King)   0589144

KMD 550/KMD 850 MultiFunction Displays (MFDs)

Type
MultiFunction Display (MFD).

Description
The KMD 550 is a 5 in diagonal, full-colour, AMLCD designed for piston and light turbine-engined aircraft. The KMD 850 is the same apart from the inclusion of a digital and analogue weather radar interface.

Both units provide a moving map function that displays airports, special-use airspace, DMEs, VORs, and other features such as terrain elevation.

The KMD 550 and 850 can also form part of the Bendix King Integrated Hazard Avoidance System (IHAS). The KMD 550 is included in the IHAS 5000 designed for non-radar-equipped aircraft, while the KMD 850 forms part of the IHAS 8000 for aircraft with radar. IHAS capability allows datalink to the display of weather information, terrain information deriving from EGPWS and air traffic

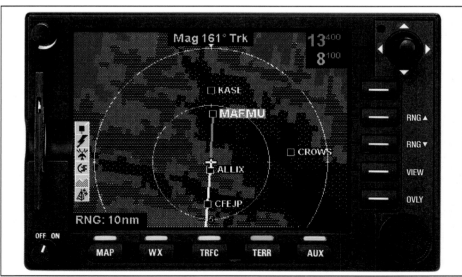

Bendix/King KMD 850 showing terrain and route information (Bendix/King)   0589143

information deriving from Mode S sources, to provide comprehensive situational awareness.

Specifications
Dimensions: 158 × 101 × 254 mm
Weight: 2.8 kg max with rack
Temperature: –20 to +70°C
Altitude: up to 35,000 ft

Status
In production and in service.

Contractor
Bendix/King.

KNS 660 flight management system

Type
Flight Management System (FMS).

Description
Aimed at the turbine-powered corporate and commuter sector of general aviation, KNS 660 is the designation for a family of great-circle navigation management systems to work in conjunction with Gold Crown III avionics and other compatible units.

The KNS 660 has its own database, ARINC 429 input/output formats, analogue processing, air data function, vertical navigation computation, frequency management and optional Omega/VLF sensor. It has a choice of two control/display units, the 152 mm (6 in) KCU 568 or the 114 mm (4.5 in) KCU 567 for installations with limited space. The KNS 660 will handle signals from VOR/DME, Omega/VLF, Loran C, INS, GPS, AHRS (Attitude Heading and Reference System) and compass. The KTU 709 Tacan system can replace the DME if required.

An optional receiver/processor providing a choice between Omega/VLF and Loran C is now available. A potential growth area is MLS.

Status
Production began in July 1984. The KLN 670 GPS Navstar sensor was developed and certified as part of the KNS 660 system in 1987.

Contractor
Bendix/King.

KRA 10A radar altimeter

Type
Flight instrumentation, radar altimeter.

Description
The KRA 10A radar altimeter is a low-cost system suitable for independent use or in combination with King Silver Crown avionics equipment and tailored to general aviation requirements. The system features a KI 250 indicator which displays altitude from 20 to 2,500 ft and meets published accuracy specifications between 50 and 2,000 ft. There is a standard facility for presetting decision height which produces a visual and aural warning on reaching the set altitude. The KRA 10A uses the KA 131 antenna which is suitable for mounting on surfaces either parallel to the ground or sloped at between 6 and 20°. The KRA 10A is an all-solid-state system with short warm-up time and can be fitted with an auxiliary output to interface with flight director and autopilot installations.

Specifications
Dimensions:
(indicator) 100 × 83 × 83 mm
(transmitter/receiver unit) 79 × 89 × 203 mm
(antenna) approx 100 × 100 mm aperture
Weight:
(indicator) 0.4 kg
(transmitter/receiver unit) 0.9 kg
(antenna) 0.4 kg
Power supply: 28 V DC, 6 VA

Altitude: up to 2,500 ft
Accuracy:
(0 to 100 ft) 5 ft
(100 to 500 ft) 5%
(>500 ft) 7%
Temperature:
(indicator) –15 to +71°C
(transmitter/receiver unit) –54 to +71°C

Status
In service.

Contractor
Bendix/King.

KRA 405 radar altimeter

Type
Flight instrumentation, radar altimeter.

Description
Part of the King Gold Crown avionics range, the KRA 405 is an all-solid-state radar altimeter suitable for twin-engine general aviation and regional airliner types. The KRA 405 interfaces with King KPI 553A HSI and KFC 300 autopilot to give smooth tracking of the glideslope beam. It can provide indications from 2,000 ft above ground level and a usable output is available from 2,500 ft above ground level for ground-proximity warning system operation. Separate transmit and receive horn antennas are used.

Specifications
Dimensions:
(indicator) 83 × 83 × 170 mm
(transmitter/receiver unit) 83 × 133 × 296 mm
(antennas) each 178 mm diameter
Weight:
(indicator) 0.8 kg
(transmitter/receiver unit) 2.9 kg
(antennas) 1.2 kg total
Power supply: 28 V DC, 24 VA
Frequency: 4,300 MHz
Altitude: up to 2,500 ft
Accuracy:
(0 to 500 ft) 5%
(>500 ft) 7%

The KRA 405B updates the KRA 405 by reducing the number of primary circuit boards from seven to two, reducing overall weight of the receiver/transmitter by 50 per cent, and adding updated software. System elements are KRA 405B receiver/transmitter; KNI 415 (fixed-wing) or KNI 416 (rotary-wing) indicator; two KA 54A antennas; optional CM2000 configuration module. Altitude information above ground level is output as analogue voltages and in ARINC 429 digital format allowing the KRA 405B to interface with components such as GPWS, TCAS and autopilot enunciators, and so on.

Specifications
KRA 405B
Dimensions:
KRA 405B: 279 × 76 × 90 mm
KNI 415/416: 170 × 83 × 83 mm
KA 54A: 93 × 89 × 19 mm
Weight:
KRA 405B: 1.36 kg
KNI 415/416: 0.77 kg
KA 54A: 0.09 kg (each)
Altitude: up to 2,500 ft
Accuracy:
(0 to 500 ft) ±5 %
(>500 ft) ±7%
Temperature:
KRA 405B: –55 to +70°C
KNI 415/416: –20 to +71°C
KA 54A: –54 to +85°C

Status
The KRA 405B is in production.

Contractor
Bendix/King.

Lasernav II navigation management system

Type
Flight Management System (FMS).

Description
Introduced in 1984, Lasernav II is an inertial system for airlines and general aviation which combines self-contained, strapdown laser inertial position and aircraft motion sensors with externally sensed radio signals to provide a single efficient integrated guidance package. The notable advantages of previous Honeywell laser inertial systems were maintained in Lasernav II: 2½ to 10 minutes alignment time dependent upon latitude, three to four times the reliability of earlier systems and reduced size, weight and power consumption. The system can also be mounted in an unpressurised environment.

Position data obtained from the Honeywell GPS sensor unit and VOR/DME and Omega/VLF stations is blended with inertial position information using high-speed digital computing techniques. DME/DME updating is obtained by way of an auto-tuning function that requires no pilot inputs. VOR/DME updating is obtained by manually tuning the radios. Triple inertial navigation system mixing combines inertial data from two other Lasernav II systems to calculate a composite inertial position.

In addition to normal navigation functions, Lasernav II can replace all conventional attitude and heading sensors including vertical and directional gyros, flux valves and compass controllers. In a typical dual installation, the total box count can be reduced by up to 14 separate boxes, resulting in weight savings of up to 60 kg and a greatly simplified installation.

Lasernav II comprises a navigation management unit and a control/display unit. An internal non-volatile memory can store 20 flight plans of up to 20 waypoints each. A Global Wulfsberg NDB-2 worldwide database (required for DME auto-tuning) facilitates the automatic flight planning feature that requires the pilot to input only the departure and destination points. A great circle route is computed, intermediate waypoints are selected and range and bearing to nearest VOR/DME is displayed, all automatically.

Specifications
Dimensions:
(navigation management unit)
    322 × 324 × 193 mm
(control/display unit) ARINC 562
Weight:
(navigation management unit) 22.1 kg
(control/display unit) 3.2 kg
Power supply:
(total) 115 V AC, 400 Hz, 160 W
    28 V DC, 19 W
Accuracy (95% probability):
(position) 2 n miles/h
(velocity) 8 kt
(heading) 0.4°
(pitch and roll) 0.1°

Status
In service. Installations approved include Cessna Citation III, Gulfstream II and III, Dassault Falcon 50 and Canadair Challenger CL-600 and CL-601 business jets, and Boeing 737-200 airliners.

Contractor
Honeywell Inc, Commercial Electronic Systems.

Lasernav laser inertial navigation system

Type
Flight Management System (FMS).

Description
Introduced during early 1983, Lasernav® exploits the strapdown inertial reference system developed for the Boeing 767 and 757, but also has facilities that make it suitable for long-range business and corporate jet aircraft.

In addition to normal navigation functions, Lasernav can replace all customary attitude and heading sensors, including compass system components such as the flux detector, resulting in a reduction of up to 14 separate boxes. This is said to result in a weight saving of up to 60.3 kg, and a volume reduction of up to 50 per cent by comparison with equivalent dual installations in other systems.

Lasernav comprises two units; an inertial navigation unit and a control/display unit. The memory can store the co-ordinates of up to 255 waypoints in 20 routes and up to 20 waypoints per flight plan, for immediate recall. In dual installations each system can store a different flight plan, but can share it with the other system if required. For international routes or long overwater sectors, the system can use a Global Wulfsberg NDB-2 database and only the departure point and destination co-ordinates need to be inserted. Lasernav then computes a great circle route, selects and identifies air traffic reporting points and lists bearing and distance to the nearest VOR/DME on the eight-line by 14-character control/display unit.

A notable advantage is the sharp reduction in alignment time; the interval from switch-on to ready is as little as 2½ minutes at the equator and 10 minutes at 60° latitude.

Specifications

Dimensions:
(inertial navigation unit) 322 × 324 × 193 mm
(control/display unit) ARINC 561
Weight:
(inertial navigation location) 21.1 kg
(control/display unit) 3.2 kg
Power supply:
(total) 115 V AC, 400 Hz, 146 W
 28 V DC, 28 W

Status
In service.

Contractor
Honeywell Inc, Commercial Electronic Systems.

Lasertrak navigation display unit

Type
Multi Function Display (MFD).

Description
A companion device to the Honeywell Laseref, Laseref II and Laseref III inertial reference systems, the Lasertrak navigation display unit provides back-up waypoint navigation using positions from up to three IRS. In the event of a flight management system failure the Lasertrak allows continued navigation along the planned route. Accepting up to 10 waypoints, Lasertrak computes and displays desired track and cross-track error for the intended course.

The Lasertrak can be used in flight management and inertial reference systems architecture when flight dispatch with a failed FMS is desired.

Specifications
Dimensions: 146 × 114 × 152 mm
Weight: 2.3 kg
Power supply: 27 V DC, 10 W

Status
In service in Dassault Falcon 50 and 900, Gulfstream IV and Bombardier Challenger aircraft.

Contractor
Honeywell Inc, Commercial Electronic Systems.

Liquid Crystal Control Display Units (LCCDU)

Type
Control and Display Unit (CDU).

Description
The LCCDU is a microprocessor-based communications control and display unit. It

Liquid Crystal Crew Display Unit (LCCDU)
0018162

provides monaural or binaural monitoring capability of up to 30 channels for each ear as required to each crew position equipped with an LCCDU, and it provides the microphone interface to the Communications Control Unit (CCU).

Through the full function keyboard, the LCCDU controls transmit and receive functions including the control and reassignment of all communication assets, crypto assignments, and Built-In Test (BIT) readouts. Complete status of the communications assets for each operator is available through the eight-colour liquid crystal display. Each LCCDU is identical and completely interchangeable with any other position location. The CCU may be programmed in real time if necessary to allow or disallow access of any given communications asset by any given crew position via the LCCDU.

The integral display intensity control is capable of adjusting the display brightness from off to full intensity without external controls or potentiometers. The LCCDU interfaces with the CCU via a serial RS-422 control data twisted-shielded wire pair.

Options include: tone generation crypto/modem switching; dual crew binaural audio; cross-band or in-band radio relay.

Specifications
Dimensions: 165.1 × 146 × 152.4 mm
Weight: 2.73 kg
Power: 28 V DC, 28 W

Contractor
Palomar Products Inc.

Liquid Crystal Display indicators

Type
Cockpit display.

Description
Poly-Scientific solid-state Liquid Crystal Display (LCD) indicators employ advanced microprocessor circuitry with liquid crystal technology. Interface can be accomplished with analogue and digital signals from inputs of thermocouples, strain gauges, potentiometers and LVDTs. The indicators can be adapted for direct replacement of electromechanical indicators.

The LCDs function over a wide temperature range and work well under harsh environmental conditions. They can be grouped together in one display or installed as separate indicators and can be supplied in round scale formats or vertical bar graphs in various sizes. Digital augmentation is available to provide greater accuracy when required. Displays can be back-lit with electroluminescent, incandescent lighting or LEDs, and can be made NVG-compatible.

Formats available include a compact space-saving design as small as 1.5 in (38 mm), vertical scales with or without digital display for torque, tachometer, pressure or temperature display and large formats available as a single-pointer, dual-pointer or single-bar display augmented with a digital readout.

Status
In use in military, commercial and business fixed- and rotary-wing aircraft.

Contractor
Northrop Grumman Corporation, Component Technologies, Poly-Scientific.

Liquid Crystal Displays (LCDs)

Type
Cockpit display.

Description
Aerospace Display Systems, Inc produces a range of customised LCDs for civil and military (NVG compatible) use, including the Autopilot Readout Device (ARD) used on A310, A320 and A321 aircraft; the Tension Skew Indicator (TSI) used in the MH-53E Super Stallion helicopter during minesweeping operations; and the Vertical Torque Indicator (VTI) used in commercial helicopters. Products can have specialised display formats with high-contrast options, backlighting, night vision goggle compatibility and a temperature range of −55 to +85°C.

Contractor
Aerospace Display Systems, Inc.

Aerospace Display Systems tension skew indicator
0051690

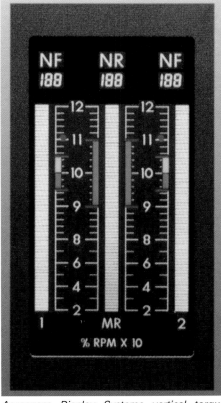

Aerospace Display Systems vertical torque indicator
0051691

LM Series compact indicators

Type
Flight instrumentation.

Description
The LM Series indicators are a low-cost response to the FAA FAR 121.343 requirements for in-flight recording of engine parameters. Standard LM Series features include back-lighted dial and pointer, 360° dial arc, BIT, smooth pointer movements, modular construction for easy servicing, compact 3.5 in (88.9 mm) long design and 40,000 h MTBF.

The simple mechanical design incorporates only one moving part – the pointer – so there are no brushes or wipers to wear out. Surface-mounted PCB design helps to ensure long, trouble-free service life for the electronic components. The indicators are designed for display of primary and secondary engine parameters in retrofit as well as new production aircraft installations.

Optional features available include flight recorder inputs, high-accuracy redundant digital display, RS-232 interface, output warning signals and exceedance and trend monitoring.

Specifications
Power:
(Indicator): 7 to 32 V DC, 3 W max
(Lighting): 0 to 5 V DC, 2 W max
Accuracy:
(Dial): ±1.0% of full scale
(Digital display): ±0.2% of full scale
Operating temperature: −40 to +70°C

Status
LM Series indicators are installed in Fokker F28, Mitsubishi YS-11, Saab SK60 and AeroVodochody L139 and L159 aircraft.

Contractor
Ametek Aerospace Products.

LRA-900 low-range radio altimeter

Type
Flight instrumentation.

Description
The LRA-900 low-range radio altimeter includes enhancements to a digitally controlled variable bandwidth filter which provides improved noise rejection and leading-edge tracking. Signal microprocessors perform automatic calibration that continuously compares the received ground return signal frequency with the frequency produced by separate precise delay lines in each channel. This results in fine resolution, providing accuracy to support autoland flare and touchdown computations. Resolution at touchdown is quoted as better than 1.2 in.

Low-component count and low-stress level circuits give the LRA-900 high reliability and quick, easy and effective onboard fault isolation. A comprehensive self-test monitor in the system is capable of determining operational status to a 99 per cent confidence level.

The LRA-900 is designed to fulfil all new environmental requirements. The system is protected for Cat K lightning and Cat Y High-Intensity Radiated Fields (HIRF). In addition, it is fully functional over 200 ms of power interruptions. It meets ARINC 707, 429 and 600, TSO-C87, DO-160C, DO-155 and DO-178A.

Specifications
Dimensions: 3 MCU
Weight: 4.3 kg
Power supply: 115 V AC, 380–420 Hz, 24 W nominal
Frequency: 4,300 ±25 MHz
Altitude: −20 to 5,000 ft
Accuracy: ±1 ft or 2% of indicated altitude
Temperature range: −40 to +70°C

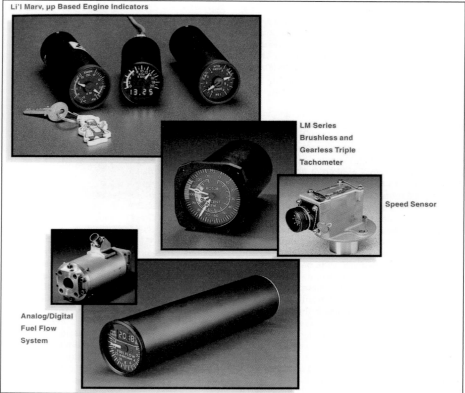

Li'l Marv, μp Based Engine Indicators

LM Series Brushless and Gearless Triple Tachometer

Speed Sensor

Analog/Digital Fuel Flow System

Ametek's LM Series indicators　　　0051692

Status
In production and in service.

Contractor
Rockwell Collins.

MARS-II data recording system

Type
Flight/mission recording system.

Description
MARS-II provides a systems approach to data recording. An SCSI computer enables an electronic module to utilise tape, disk or solid-state memory, with one electronic module supporting up to seven storage modules. Combined in this system is the capability to record or reproduce eight completely asynchronous, electronically configurable channels of PCM or dual-redundant MIL-STD-1553B data in addition to one channel of IRIG time code data and one channel for voice annotation. In addition, MARS-II accepts NRZL, R-NRZL, biphase and analogue data formats.

Specifications
Dimensions:
(electronic module) 127 × 304.8 × 342.9 mm
(storage module) 127 × 238.1 × 342.9 mm
Weight:
(electronic module) 15.88 kg
(storage module) 6.8 kg

Power supply: 28 V DC
115/220 V AC, 47–400 Hz
Temperature range: −54 to +50°C
Altitude: up to 50,000 ft

Status
Variants of the MARS-II available include: Enhanced MARS-IIe (40 Mbits/s) version; MARS-II-SB (Single Box); MARS-II-L (Laboratory), and MARS-II DAS (Data Analysis System).

MARS-II-SB is specifically designed for rugged or rough-ride applications with limited or confined space; it provides two channels of standard PCM rates and one channel of high rate PCM, together with one channel of IRIG time code data and one channel of voice annotation.

Distributed in the UK and Benelux countries by Metrum Information Storage Ltd (UK).

Contractor
Metrum-Datatape Inc.

MDR-80™ Mission Data Recorder

Type
Flight/mission recording system.

Description
The TEAC MDR-80™ is a new digital mission data recorder, featuring digital video capture, in excess of 14 Gbytes of memory, MIL-STD-1553B

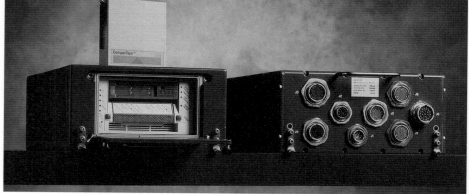

The Metrum-Datatape MARS-II consists of (left) the storage module and (right) the electronics module　　　0504172

TEAC's MDR-80 has been specified for the Eurocopter Tiger HAD. (TEAC America) 1128661

interface and the capability to transmit and receive digital video signal. Other features and benefits of the system include:

- Plug-in application cards
- Form, fit and functional replacement for TEAC V-80 series AVTR
- Support for the TEAC V-80 serial I/O communication protocol
- Simultaneous record and playback
- MPEG-2 video compression
- Up to two channels of video recording via user application cards
- Dynamic modification of video and audio encoding bit rates
- User and/or aircraft controls via high-speed or low-speed databus
- Software configuration control enables upgrade and expansion without hardware modification
- Solid-state Removable Memory Module (RMM) interfaces directly to ground debriefing station via industry standard interface
- Multiple RMM configurations available
- Industry-standard processor architecture and operating system assures supportability
- Optional time-synchronised ACMI data recording via application-specific ACMI card
- Expansion cards to add functional growth.

The basic unit provides for single channel video. Options include additional video channels, MIL-STD-1553B interface, ACMI recording, digital video receive/transmit and 'smart' control functionality. The addition of user application cards further enhances the system, including MPEG-2 video, digital video receive and transmit, fibre channel and 10BaseT/100BaseT interfaces.

Specifications
Video:
Signal standard: EIA RS-170, NTSC, CCIR (PAL)
Resolution: 500 lines
Linearity: 10 grey scale
Signal-to-noise ratio: Better than 43 dB
Audio:
Channels: One (two optional)
Bandwidth: 50 Hz to 15 kHz
Signal-to-noise ratio: Better than 40 dB
Dimensions: 149 × 122 × 162 mm
Weight: 3.0 kg
Power: 28 V DC (+4, –8), 35 W (45 W with heater); MIL-STD-704A, -704D, -704E

Status
During July 2006, TEAC announced receipt of a multi-year contract award from Eurocopter España, S.A, an EADS Company, for MDR-80 solid state digital Mission Data Recorders to support over 60 Tiger HAD helicopters. The Eurocopter Tiger HAD aircraft will be deployed by both Spain and France. TEAC contract delivery will commence with qualification units to be delivered in the Autumn of 2006 followed by production units in 2007.

Contractor
TEAC America Inc.

MDR-87™ Mission Data Recorder

Type
Flight/mission recording system.

Description
The TEAC MDR-87™ is a new solid-state, digital mission data recorder, providing a form, fit and

functional replacement for the TEAC V-83 and V-1000 series AVTRs. Features and benefits of the system include:

- Plug-in application cards
- Support for the TEAC V-83 discrete and serial I/O communication protocols
- Simultaneous record and playback
- MPEG-2 video compression, selectable hit rates
- Up to six channels of video recording via user application cards
- Dynamic modification of video and audio encoding bit rates
- User and/or aircraft controls via high-speed or low-speed databus
- Software configuration control enables upgrade and expansion without hardware modification
- Solid-state Removable Memory Module (RMM) interfaces directly to ground debriefing station via industry-standard interface
- Multiple RMM memory capacity configurations available
- Industry-standard processor architecture and operating system assures supportability
- Optional time-synchronised ACMI data recording via application-specific ACMI card
- Seven expansion slots for growth.

The basic unit provides for single-channel video. Options include additional video channels, MIL-STD-1553B interface, ACMI recording, digital video receive/transmit and 'smart' control functionality. The addition of User Application Cards (UACs) further enhances the system, including MPEG-2 video, PCM recording, fibre channel and 10BaseT/100BaseT interfaces.

Specifications
Video:
Signal standard: EIA RS-170, NTSC, CCIR (PAL)
Resolution: 500 lines
Linearity: 10 grey scale
Signal-to-noise ratio: Better than 43 dB
Audio:
Channels: One (two optional)
Bandwidth: 50 Hz to 15 kHz
Signal-to-noise ratio: Better than 40 dB
Dimensions: 243 × 134 × 330.2 mm
Weight: 7.7 kg
Power: 28 V DC (+4, –8), 61 W (3 UACs); MIL-STD-704A, -704D, -704E

Contractor
TEAC America Inc.

MFD 5200 colour AMLCD MultiFunction Display

Type
MultiFunction Display (MFD).

Description
Designed for the general aviation market, the MFD 5200 colour Active Matrix Liquid Crystal Display (AMLCD) is a lightweight, low-cost version of those installed in jet transport aircraft. The system consists of a low-profile, 1.15 kg remote-mounted symbol generator and 1.3 kg colour AMLCD control display unit. The sunlight-readable 5 in diagonal display provides Situational Awareness (SA) from onboard systems, including navigation, air data, engine and airframe, as well as environmental elements such as the weather. The basic function of the MFD 5200 is moving map navigation.

Position data input from any GPS or LORAN shows as an aeroplane or helicopter icon over a digitised aviation chart. The flight plan automatically uploads from the Arnav STAR 5000 or FMS 5000 GPS navigators or Bendix/King KLN 89-89B and KLN 90-90B GPS navigators, thereby reducing pilot workload. It will interface to the next-generation Goodrich WX-500 Stormscope®, to display thunderstorms, lightning strikes, and building storms in relation to the aircraft position and flight plan. It is FAA approved for depiction of broadcast NEXRAD precipitation radar and METAR weather reports. An optional VIPER (Video InPut EncodeR) board converts the

MFD 5200 from its digital moving map role, to a composite NTSC host for video or infra-red camera display.

Moving map display
The MFD 5200 uses a combination window and paging system which provides a main screen and menus, submenus and pop-up windows. The moving map is the largest window and is the main page or screen. All system displays/functionality can be accessed from the main menu:

- Geographic waypoint database
- VFR and IFR moving map SA display
- Graphic display of DIR, user defined arc, bearing, distance and ETA at destination
- Picture graphics of airports, VORs, vector airways, intersections, man-made obstructions, latitude/longitude grid lines, and other user-defined waypoints
- MORA (Minimum Off-Route Altitude) and obstruction proximity display system data
- Audible alarm annunciator for obstruction and special use airspace warnings
- Eight scale levels, ranging from a 1 n mile view to national or hemispherical view
- Track, ground speed, and altitude displays based on GPS or LORAN C input
- Complete flight planning capability with in-flight updates or integration of flight plans from any Arnav 5000 or 7000 series LORAN or GPS navigator
- Eight user-defined electronic checklists, each containing 10 items of up to two lines of 30 characters each
- Display of rivers, state lines, coastlines, lakes, islands and other major geographical features
- Optional flight recorder for trip or mission track and altitude recording.

WxLink™ weather datalink option
WxLink™ is a one-way datalink that uplinks weather products to the aircraft for display on the MFD in graphic or text formats. WxLink™ requires the addition of a radio receiver that is used to receive the digital weather data.

Terminal weather display
The terminal weather display feature includes the following information:

- Ceiling and visibility converted to a graphical reporting icon, placed near airport identifier
- Easy pilot identification of airports that are at or below minimums
- Four levels for ceiling and visibility – VFR, marginal VFR, IFR, low IFR
- Full METAR text available for selection by pilot.

In-cockpit NEXRAD radar display
The NEXRAD radar display includes the following features:

- Graphical display of NEXRAD 8 km base reflectivity
- 16 levels of NEXRAD base reflectivity converted to 4 levels for airborne display
- Information displayed on moving map using own-ship position reference to NEXRAD image
- Each NEXRAD transmission image is 400 × 300 n miles.

Engine Instrument Caution Advisory System (EICAS)
The addition of an Engine Monitoring Module will add EICAS functionality to the MFD 5200. The

MFD 5200 installed in a Bell 206 helicopter
0526163

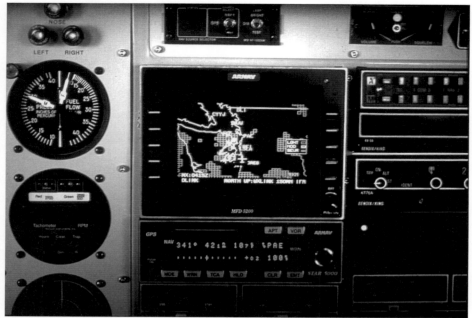

MFD 5200 showing moving map and Nexrad weather radar via WxLink™   0018179

EICAS displays sensed engine data and alerts the crew to out-of-range conditions. The EICAS image can be redirected to an alternate MFD screen in the event the primary display becomes inoperative. The image can also be 'windowed' onto the MFD in addition to the moving map information.

Status
Certified under a multiple STC, the MFD 5200 is now operating in Sikorsky, Bell, Mooney, Cessna, Piper, and Raytheon aircraft.

Contractor
Arnav Systems Inc.

MFD 5200s colour AMLCD MultiFunction Display

Type
MultiFunction Display (MFD).

Description
The (MFD) 5200s is the newest member of the MFD 5000 family. This new MultiFunction Display (MFD) is designed for in-panel or out-of-panel mounting. Its 2 in depth permits the MFD 5200s to be mounted in any aircraft, while retaining all of the basic features of the MFD 5200 (see separate entry).

The MFD 5200s uses an improved processor to decrease drawing times and increase the amount of information that can be displayed. Map features are as for the MFD 5200.

Terrain Obstruction Proximity System (TOPS)
The MFD 5200s offers a new terrain awareness feature called the Terrain Obstruction Proximity System (TOPS). TOPS provides constant real-time altitude awareness based on aircraft position, ground speed, track, altitude and an extensive onboard terrain database. Look-Around and

MFD 5200s   0526166

Look-Ahead features provide advanced warning of any possible terrain or obstruction hazard.

In the Look-Around window, the MFD 5200s calculates the distance that the aircraft may travel in any direction within five minutes. This information is compared with information obtained from the onboard database to provide a five-minute 'protection zone' around the aircraft. At any stage, if the protection zone is threatened by a high obstruction or high terrain, visual and audio warnings are provided and the hazard is plotted on the moving map display.

The Look-Ahead window extends the protection of the Look-Around window by projecting a 60 n mile swathe, 4 n miles wide, along the instantaneous track of the aircraft. If any ground obstruction or high terrain threatens the projected line, at current aircraft altitude, the Look-Ahead window provides an indication of how far ahead the potential threat is. As the aircraft turns or changes altitude, the Look-Ahead profile is dynamically updated to show a new projection of all terrain and obstruction hazards.

The Look-Around and Look-Ahead windows provide valuable information in high terrain and around unfamiliar airports. In times of deteriorating weather, TOPS can simplify terrain avoidance in reduced visibility and at night.

Improved data entry
The MFD 5200s includes a new data entry knob, similar to the company's GPS and LORAN navigation systems. The knob facilitates faster and easier entry of waypoint data and movement of the screen cursor during on-screen selections. The smart ID option completes many entries automatically and eliminates superfluous data entry.

Specifications
Display
Display: 5.0 in (diagonal) colour AMLCD
Dimensions: 121 × 159 × 51 mm (H × W × D)
Weight: 1.0 kg
Power: 25 W
Operating temperature: –20 to +70°C

Remote Processing Unit (RPU)
Dimensions: 51 × 159 × 235 mm (H × W × D)
Weight: 1.1 kg
Power: 4 W
Operating temperature: –20 to +70°C

Status
The MFD 5200s is certified for a wide variety of aircraft.

Contractor
Arnav Systems Inc.

MFD-68S Multifunction Display

Type
MultiFunction Display (MFD).

Description
The MFD-68S offers user-defined graphics for primary flight displays, engine instruments, synoptics, text, fuel management, targeting reticles and automated checklists. The 6 × 8 in (152 × 203.2 mm) active matrix liquid crystal display presents video, graphics, video with graphics overlay, split-screen video/graphics and split-screen graphics/graphics.

The MFD-268 series is offered in either a 'smart' or 'video' display to support a wide variety of architectures. The display contains internal graphics generation, analogue and digital input/output interfaces and a MIL-STD-1553B interface. The operational software can be modified over the MIL-STD-1553B bus without removal of the display from the installed position.

The MFD is programmable in the Ada language for local display format generation. A high-level graphics language simplifies tailoring of display formats.

Specifications
Dimensions:
254 (depth) × 264 (width) × 208 mm (length)
Weight: 7.71 kg nominal
Display: 152 × 203.2 mm active area

Status
Selected for the avionics upgrade of 58 Black Hawk helicopters of the Turkish Armed Forces, together with a dual Flight Management System (FMS).

Rockwell Collins has developed a common architecture configuration applicable to a variety of platforms, including the CH-53 and the S-92.

Contractor
Rockwell Collins.

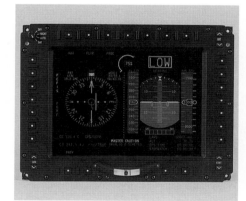

MFD-68S multifunction display   0504211

MiniARMOR-700 multiplex/demultiplex system for digital data recording

Type
Flight/mission recording system.

Description
The L-3 Telemetry-East MiniARMOR-700 multiplex/demultiplex system for digital data recording is based on the design concepts of the Asynchronous Real-time Multiplex and Output Reconstructor (ARMOR) developed by Calculex, Inc.

MiniARMOR-700 systems combine a wide variety of analogue and digital signals into a single high-speed (up to 240 Mbps) composite digital data stream for digital recording. A system configured to reconstruct the composite signals does so while maintaining inter-channel data coherency.

The MiniARMOR-700 supports two operational formats to maintain full backward compatibility

MiniARMOR-700 miniature asynchronous real-time multiplexer and output reconstructor
0002445

with Calculex ARMOR 1 products while affording the full flexibility and performance of the advanced MiniARMOR-700 design.

The MiniARMOR-700 is a modular system with a full compliment of modules and options to support virtually any mix of analogue and digital signal types to support specific user applications. Modules may be changed in the field for user ease.

The MiniARMOR-7000 is available in a 5, 7, 9 and 11 slot configuration, with a maximum measurement of 136 × 190.5 × 261.6 mm and a maximum weight of 6.16 kg. It operates from 28 V DC, 115/220 V AC, 46-63 Hz and has an operating temperature range of –30 to +70°C.

Contractor
L-3 Telemetry-East.

Mission computer and display system for the C-17

Type
Aircraft mission computer.

Description
In September 1986, the then Delco Electronics was awarded a contract for the development of a prototype mission computer and electronic display system for the C-17.

Although the computer is similar to a flight management system, it features a number of other functions including parachute cargo extraction guidance and automatic flight control system coupling for station-keeping in formation flying.

Status
In service on US Air Force C-17 transport.

Contractor
General Dynamics.

MME-64 Miniature Multiplexing Encoder

Type
Flight/mission recording system.

Description
The MME-64 miniature multiplexing PCM encoder is a low-cost data system designed for airborne data acquisition and telemetry applications to accommodate up to 64 single-ended analogue inputs or up to 24 discrete bilevel inputs. Unfiltered PCM output is NRZ-L 0 to 5 V TTL, CMOS-compatible. Filtered PCM output is six pole Bessel response, factory-adjustable from 0 ± 50 mV to 0 ± 2.5 V. Bit rate can be programmed up to 1 Mbit. The frame pattern is controlled by EEPROM and is factory-programmed.

Specifications
Dimensions: 82.6 × 76.2 × 30.4 mm
Temperature range: –30 to +70°C

Contractor
L-3 Telemetry-East.

Model 32HE/64HE variable speed digital recorder

Type
Flight/mission recording system.

Description
The first in series of Harsh Environment (HE) recorders by Metrum-Datatape, the Model 32HE is designed, built and qualified specifically for harsh conditions. Both the Model 32HE and Model 64HE are sealed units; once the tape door is closed, the media, electronics and transport mechanism are protected from the environment. Suitable for use in fixed- and rotary-wing aircraft operations, including ASW, flight testing and other situations demanding large storage volumes.

Using the latest 100 kbpi version of Metrum-Datatape technology, the two models capture data at variable streaming rates from 0 to 32 Mbps (and 0 to 64 Mbps) and burst rates from 0 to 160 Mbpi. They record data at a density of 100,000 bpi on to broadcast-quality, high – energy, Metrum-Datatape-certified ST-160 S-VHS cassettes, that provide 13.8 Gbytes of data storage, giving 57 minutes minimum (linearly increasing as data rate reduces).

Both the Model 32HE and Model 64HE include three interfaces: a single-board mux – eight channel PCM and two channel analogue; series RS-422, TTL, ECL; parallel RS-422, TTL, ECL.

Specifications
Dimensions: 381 × 165.1 × 386.1 mm
Weight: 15.9 kg
Power: 28 V DC, 100 W (300 W with heaters operating)

Status
The Model 32HE was introduced in 1997. The Model 64HE, was introduced in October 1999. It provides a variable data rate up to 64 Mbits/s and stores up to 27.5 Gbytes on a single, low cost, S-VHS tape. Model 32HE recorder/reproducers can be upgraded to 64HE capability.

Distributed in the UK and Benelux countries by Metrum Information Storage Ltd (UK).

Contractor
Metrum-Datatape Inc.

Model 32HE variable speed digital recorder
0015332

Model 960 extended environment COTS computer

Type
Aircraft mission computer.

Description
The model 960 extended environment COTS computer is a repackaged implementation of the Digital Equipment Corporation 500 MHz AS/500 Alpha-based workstation.

Status
Raytheon is contracted to provide Boeing with 174 workstation processors and 55 input/output processors for incorporation into the Boeing-supplied Tactical Command System for the UK Royal Air Force Nimrod MRA4 aircraft.

The Tactical Command System integrates and displays the output from the primary sensors, the defensive aids subsystem, armament control system and communications suite. Each aircraft carries seven workstations located in the aircraft's rear cabin which are capable of being reconfigured, control and display mission system elements. Two workstations are dedicated to the acoustic system, and the remaining five can be reconfigured into any sensor display format to meet mission requirements.

Contractor
Raytheon Company, Space and Airborne Systems.

Model 1044 altimeter

Type
Flight instrumentation.

Description
The Model 1044 radar altimeter was designed and developed to be a low-cost, lightweight, very accurate, high-rate production unit. The altimeter transmits and receives a very low-RF signal that is Phase Shift-Key (PSK) modulated using a pseudo-random code. The PSK technique was selected because it provides the equipment with the capability to measure over a wide range of altitudes from 0 to 10,000 ft with low power and fewer parts than other types of radar altimeters. A commanded built-in test verifies that the unit is operational by testing 98 per cent of all the altimeter functions.

The altimeter comprises two subassemblies. The microwave subassembly contains a filter board assembly and four microwave hybrids: the transmitter, receiver, dielectric resonant oscillator and video amplifier. The microwave hybrids provide increased reliability with reduced parts count. The electronic assembly contains an analogue processor, digital processor and power supply.

Contractor
Northrop Grumman Corporation, Navigation Systems Division.

Model 3044 radar altimeter

Type
Flight instrumentation, radar altimeter.

Description
The Model 3044 radar altimeter was designed and developed for manned aircraft applications and operates from 0 to 10,000 ft. It employs hybrids, application-specific integrated circuits and MIL-STD-1553B interface to provide a highly reliable lightweight production unit. The altimeter transmits and receives a very low-RF signal that is phase shift-key modulated using a pseudo-random code. The phase shift-key technique provides the equipment with the capability to measure over the full altitude range using a very low-level RF output.

The altimeter comprises two subassemblies. The microwave subassembly contains a filter board assembly and four microwave hybrids: the transmitter, receiver, dielectric resonant oscillator and video amplifier. The microwave hybrids provide increased reliability at reduced part count. The electronic subassembly contains an MIL-STD-1553B interface, signal processor and power supply. The application-specific integrated circuit incorporated into the signal processor enables digital signal processing at a greatly reduced part count.

Contractor
Northrop Grumman Corporation, Navigation Systems Division.

Model 3255B Integrated Data Acquisition and Recorder System (IDARS)

Type
Flight/mission recording system.

Description
Smiths' Integrated Data Acquisition and Recorder System (IDARS) is versatile and easily adaptable to a wide variety of requirements that go beyond flight data and voice recording and into information management. Its modular design allows the IDARS to perform all the functions of a Cockpit Voice Recorder (CVR), Flight Data Recorder (FDR) and a Flight Data Acquisition Unit (FDAU) using two cards in one unit. There are three additional card slots to accommodate growth features.

IDARS supports custom tailoring of hardware and software configurations to meet specific vehicle acquisition and processing requirements.

The IDARS computes, compresses, and stores data from a wide range of analogue, digital and discrete sources. Configurable software controls the entire system and is easily modified for specific applications.

The system is water recoverable, with an optional acoustic beacon to 20,000 ft depth.

The airborne software's operational flight program characteristics can be reprogrammed and data download accomplished via a front panel connector or a readily accessible separate maintenance connector. Common ground support software tools are available to permit complete and easy access to the recorded voice and data.

Stored flight data and voice are synchronised and offer an integrated data analysis database for optional animated flight replay capability. This is supported by the Graphical Replay Animation System (GRAS) to enhance effective aircrew flight analysis and crew training. The GRAS can be hosted on a desktop or laptop to permit total or segmented flight replay.

Specifications
Dimensions: ½-ATR Short chassis per ARINC 404; 320.5 (L) × 123.95 (W) × 193.5 (H) mm
Weight: 7.2 kg, maximum without acoustic beacon
Impact resistance: >3,400 g
Power: +28 V DC, 40 W (maximum)
Compliant: EUROCAE ED-55 and ED-56a; FAA TSO-C123a and -C124a

Status
- In production. Selected for the US Air Force B-1B, KC-135, T-6A II, U-2S, UH-1N;
- UK MOD CH-47, Lynx, Sea King, Super Puma;
- Israeli Air Force C-130, CH-53, UH-60;
- Eurocopter EC 135, BK 117, BO 105.

Contractor
Smiths Aerospace.

Smiths' Model 3255B IDARS   0051332

Model 8000 Horizontal Situation Indicator (HSI)

Type
Flight instrumentation.

Description
Aeronetics® Model 8000 Horizontal Situation Indicator (HSI) is a slaved compass system specifically designed for rotary-wing and high-performance fixed-wing aircraft.

The Model 8000 HSI system consists of the Model 8130 HSI, the Model 8100 remote gyro and the Model 8140 flux detector/transmitter. The gyro employs electromagnetic erection, eliminating the performance and reliability degradation frequently experienced with air-erected gyros.

Simultaneous indications are provided for selected course, course deviation and selected heading. Glideslope deviation is displayed when an active ILS frequency has been selected. Other system features include:
- Warning flags to indicate validity of NAV, glideslope and compass information
- An Automatic Emergency Mode (AEM) which provides for continuous heading information in the event of gyro failure. An annunciator indicates when the AEM is in operation
- Outputs for RMI or autopilot and flight director systems.

Specifications
(Model 8130 HSI)
Dimensions: 3 ATI
Weight: 1.4 kg
Power supply: 28 V DC, 1.5 A (max)
Lighting: Internal, 5 or 28 V, white
Temperature range: –30 to +55°C
Altitude: –1,000 to 40,000 ft
TSO: C6d, C34e, C36e, C40c
Environmental: DO-160A
System accuracy: ±2°

Status
In production.

Contractor
Northrop Grumman Corporation, Component Technologies, Poly-Scientific.

Model 8000 Multifunction Display (MFD)

Type
MultiFunction Display (MFD).

Description
The Model 8000 MFD is fully compatible with AN/AAQ-16 night vision systems for either new

Model 8000 Multifunction sunlight readable display   0001393

installation or for retrofit purposes. It is also compatible with Generation II and Generation III night vision goggles.

The Model 8000 is sunlight readable, with a display output greater than 200 fL (up to 400 fL) and a contrast ratio of 6:1 @ 10,000 fc ambient illumination.

The front control panel features 14 push-buttons with 14-bit encoded outputs, day/night/off switch, contrast and brightness adjustment, display size, centre and symbol brightness adjustment, Video 1 or 2 select. Front panel functions can be customised for specific requirements.

Specifications
Weight: <6.4 kg
Largest viewable display area: 122 × 162 mm

Contractor
Palomar Products Inc.

Modular Mission and Display Processor MDP

Type
Aircraft mission computer.

Description
Astronautics' modular MDP is a high-performance computer which provides two main functions – interface with other aircraft avionic systems to perform the Operational Flight Programme (OFP) and generation of the symbology for the various displays in response to key aircraft parameters, such as pitch, roll and heading and encompassing weapon delivery and navigation functions. The processing power of the MDP enables it to perform at the heart of display, navigation and weapon delivery computation systems.

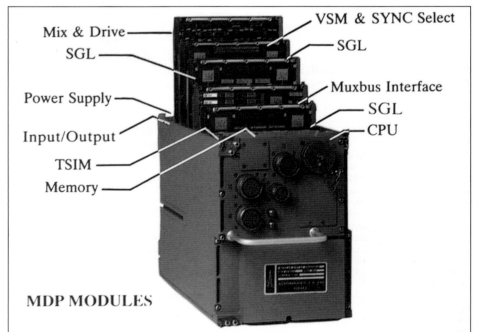

Astronautics modular mission and display processor modules   0015279

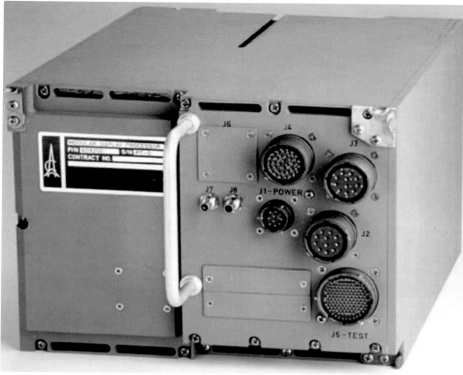

Astronautics modular mission and display processor (Astronautics)   0568535

micro-programmed processor capable of driving stroke and raster displays in various combinations and formats; the video front end which mixes selected inputs with synthetic symbols; the input/output modules; power supply.

Other key features of the MDP include:
- Can be used as a display and/or Mission Computer (MC)
- Multiple processors for advanced mission and graphics processing
- Adaptable to any display system configuration
- Provides fast and efficient symbol generator and imagery control
- Supplied with software display package which represents full interface between the OFP and the MDP hardware
- Software Graphic Editor is available for efficient development of display templates.

Specifications
CPU: RISC, Pentium™ and/or Power PC
Memory: 2 Mbyte standard (4 Mbyte by upgrade)
Graphics engine: 5 million pixels/s/channel (3 graphics engines)
Graphics capability: up to 6 independent graphics channels; up to 512 graphics characters/symbols; 8 colours
Interface: ARINC-429, MIL-STD-1553B, RS-232, RS-422, discretes, analogues, synchro
Number of card slots: 12
Power: 115 V AC, 400 Hz at 220 VA or 28 V DC
Dimensions: 279.4 × 261.6 × 193 mm
Weight: 15.9 kg
MTBF: 1,200 h

Status
In production and in service in A-4, AC-130, F-5, Kfir, MB-339, T-38 and other aircraft.

Contractor
Astronautics Corporation of America.

The modular MDP is based on modern processor technology and distributed architecture. By selecting the appropriate number of functional modules for the MDP, the aircraft displays to be driven can be stroke, raster, stroke on raster, monochrome, or colour formats, per the requirements of HUD/HMDs and MFDs. An integrated map module is also available.

In systems where video inputs are supplied, the symbology can be superimposed on the displayed video scene or displayed as symbology only on each of the displays (under system or pilot control).

The MDP comprises five main items: the processor, which includes the CPU and the main memory; the graphics engine, a high-speed

---

MultiFunctional Display (MFD)

Type
MultiFunction Display (MFD).

Description
The MultiFunctional Display (MFD) is the latest addition to the Miligraphic product line. Using the Miligraphic operating system, the 14 in (355.6 mm) flat tension mask CRT display with the 0.28 mm pitch shadow-mask is capable of displaying high-contrast high-visibility red/green/blue colour images. Three multifunctional displays are used on the ES-3A aircraft: one of these is in the cockpit and two are at crew stations. Each MFD is supported by an individual controller housed in a separate common chassis rack. This enables installation of the MFD monitors in limited access areas, with remote installation of the control electronics.

The MFD features contrast of 4:1 with a visor in 8,000 ft-candles ambient light, 91 lines/in resolution and compact construction. The monitor weighs 41.28 kg and the control unit 42.64 kg. The MFD meets MIL-E-5400F, MIL-T-5422F and MIL-STD-461.

Status
The MFD is used in the Lockheed Martin ES-3A.

Contractor
BAE Systems North America, Information and Electronic Warfare Systems.

Multi-input Interactive Display Unit (MIDU)

Type
Control and Display Unit (CDU).

Description
The MIDU is designed in accordance with ARINC 739 Multipurpose Control Display Unit (MCDU) screen formats and keyboard operation. It enables flight crew to manage the input, output and display of up to 11 separate systems. Examples of MIDU subsystem control include Aircraft Communications Addressing and Reporting System (ACARS) management unit; Aircraft Condition Monitoring Systems (ACMS) data management unit; satellite communications system; central maintenance computer.

The MIDU achieves a full data transfer capability using a credit card-sized removable memory (1 – 120 Mbyte capacity units available) for mass storage and data transfer in association with appropriate Airborne Data Loader (ADL) and Quick-Access Recorder (QAR) systems. MIDU supports the following standard aircraft interfaces: ARINC 429, ARINC 573 (QAR) and RS-422.

The VHF Radio Management Panel (RMP) application enables the MIDU to act as a VHF radio controller in accordance with ARINC 716, in which mode it supports conventional frequency (and new European) frequency spacing, displays active and standby frequencies, and provides certifiable voice-mode switching between VHF voice and ACARS data.

The MIDU display is a colour Active Matrix Liquid Crystal Display (AMLCD) with 14-line × 24-character format in accordance with ARINC 739.

Specifications
Dimensions: 114.3 (H) × 146.1 (W) × 165.1 (D) mm
Display area: 61.0 (H) × 73.7 (W) mm
Weight: 2.95 kg
Power: 40 W, 115 V AC or 28 V DC

BAE Systems' control and display system for the Lockheed Martin ES-3A   0581519

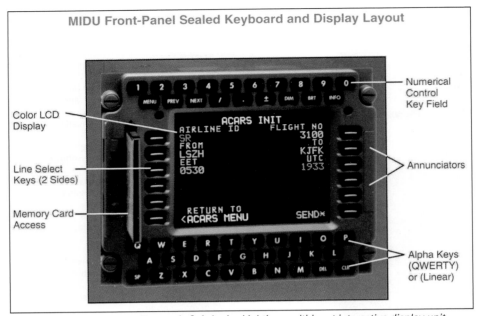

MIDU Front-Panel Sealed Keyboard and Display Layout

Numerical Control Key Field

Color LCD Display

Line Select Keys (2 Sides)

Memory Card Access

Annunciators

Alpha Keys (QWERTY) or (Linear)

The Honeywell Aerospace, Electronic & Avionics Lighting multi-input interactive display unit showing typical display formats
0062200

Contractor
Honeywell Aerospace, Electronic & Avionics Lighting.

MultiMissions Management System (MMMS)

Type
Mission Management System (MMS).

Description
The MMMS incorporates the advanced technology, system design, features and capabilities included in the UNS line of Flight Management Systems (FMS). In addition, the MMMS includes special interfaces, and the ability to fly the following six distinct types

Examples of MMMS patterns displayed on the UNS-1D, UNS-1C and UNS-1K systems
0051643

of pattern: rising ladder, expanding square, racetrack, sector search, orbit and border patrol. From the Control Display Unit (CDU), the operator is able to select the pattern and define the specific parameters appropriate to the mission. The type of pattern is graphically displayed, together with active/interrupted status and leg sequencing. Track, bearing, time and distance are numerically displayed as referenced to appropriate pattern waypoints throughout mission operations. Patterns can be activated, cancelled, or interrupted at any time. The pattern being flown can be interrupted and a new pattern selected, as required.

The MMMS systems interface with many of the aircraft systems, sensors and cockpit displays, including TACAN, radar, Doppler, EFIS and other systems. Additionally, the integral 12-channel GPS receiver meets the requirements for primary means of navigation remote/oceanic and provides seamless navigation throughout all phases of flight including non-precision approaches and special mission patterns worldwide. Flight management features include full SID, STAR and approach procedures, airways, advanced VNAV, holding patterns, heading mode, 3-D approach mode, fuel management and other facilities.

The MMMS specialised software is available for the UNS-1C, UNS-1D and UNS-1K FMS.

Status
Believed in production and in service.

Contractor
Universal Avionics Systems Corporation.

MultiPurpose Colour Display (MPCD) for the F-15C/D MSIP and F-15E

Type
MultiFunction Display (MFD).

Description
The MultiPurpose Colour Display (MPCD), developed for the F-15 fighter as part of a continuing Multinational Staged Improvement Programme (MSIP) enhancement for the C and D versions, is a very high-resolution 5 × 5 in (127 by 127 mm) shadow-mask system with stroke, raster, or combined stroke/raster writing in any of 16 colours as determined by software in the symbol generator. Raster symbology is a full-colour representation of a colour sensor output. Functions include built-in test for a number of aircraft systems, graphic representation of aircraft stores configuration (replacing the electromechanical armament control panel on earlier F-15s), display of video from electro-

optic sensors and weapons and interface for secure tactical information from a JTIDS communications system.

Honeywell has been contracted for a second version of the colour display for the F-15E programme. The F-15E fleet, which incorporates air-to-ground and air-to-air mission capabilities, is equipped with three Honeywell MPCDs on each aircraft. Honeywell is also under contract for the display processor for F-15E aircraft. The equipment not only powers the aircraft's three colour displays, but all the displays on the aircraft.

The display's delta gun shadow-mask tube uses dynamic and static convergence for maximum symbol fidelity. A bandpass filter and 17,000 in/s writing speed facilitates viewing in high ambient light levels, or even direct sunlight. Two CRT versions are available, one having 0.008 in phosphor dot spacing said, by Honeywell, to be only two-thirds the dot size and spacing of other shadow-mask CRTs designed for airborne applications.

The Honeywell programmable signal data processor interfaces with other equipment and sensors, generating monochrome and colour symbology for other displays. It comprises two elements: a three-channel symbol generator and a general purpose processor. Self-test circuits monitor all critical signals continuously and failures are flagged on a built-in test indicator.

Specifications
Dimensions: 172 × 180 × 387 mm
Weight: 11.8 kg
Screen size: 127 × 127 mm
Linewidth: 0.016 in typical
Video bandwidth: DC 10 MHz
Television scan resolution: 70 line pairs/in

Status
In service with F-15C/D MSIP and F-15E fighters. A version of this display is in the F-117A aircraft.

Contractor
Honeywell Defense Avionics Systems.

NeoAV IIDS Integrated Instrument Display Systems

Type
Cockpit display.

Description
The Rogerson Kratos NeoAV Integrated Instrument Display Systems comprise a family of cockpit displays using advanced Active Matrix Liquid Crystal Display (AMLCD) technology. IIDS can efficiently replace up to 38 conventional instruments, as well as the caution advisory system. Standard features include engine instrument indication, fuel quantity measurement indications, hydraulic systems indications, electrical system indications, caution, warning, and advisory messages, and outside air temperatures. IIDS also has FADEC or EEC compatibility, trend monitoring and weight and balance synoptic indications.

Special helicopter capabilities include mast torque indications and chip detector warning outputs.

Functions are monitored and managed with colour AMLCDs, featuring redundancy and maintenance capabilities. Multiple reversionary display modes operated by bezel-mounted buttons enable the system to be 'fail operational' in case of in-flight malfunctions. Built-In Test Equipment (BITE) is available. Metric or English displays are available, as is built-in non-volatile memory for engine history and maintenance logs. Engine performance instruments are typically presented on one display, while aircraft systems instruments are on a second display. Solid-state components, redundant architecture and derating assure high reliability. IIDS also offers significant reductions in weight, heat and power consumption.

Status

IIDS displays have been certified for the Bell 430, Sikorsky S76 and Canadair CL415. Certification is in process for the Casa 212, 235 and 295, Bell 412 and 427, as well as for Zeppelin Airship. Rogerson Kratos IIDS installations will be factory standard on all Bell twin-engined helicopters (Models 212, 412, 427 and 430).

Contractor

Rogerson Kratos, a Rogerson Aircraft Corporation subsidiary.

The Rogerson Kratos NeoAV IIDS display system was fitted to the Bell 427 helicopters as it entered flight test
0022211

Bell 430 cockpit showing both the IIDS Integrated Instrument Display System and the Model 550 EFIS Electronic Flight Instrument System
0018163

NeoAV Model 500 EFIS Electronic Flight Instrument Display

Type

Electronic Flight Instrumentation System (EFIS).

Description

The NEOAV™ 500 EFIS series is the foundation of a family of AMLCD displays with options to tailor the display to an application's needs. The 5 ATI self-contained, plug-replaceable unit is designed to be interchangeable with most existing aircraft flight director systems.

The series features multiple display formats, analogue and digital interfaces, display reversions and cross cockpit monitors. It has a map mode and can display ARINC 453 weather radar information. It is also capable of displaying TCAS II, EGPWS/TAWS and GPS information.

Status

Certified by the US FAA on the B 727-100 in 1994 and on the Bell Model 430 in 1996. The NeoAV Model 500 replaced existing HSI/ADI instruments in Federal Express Boeing 727 and DC-10 aircraft. Also selected for the Boeing 707

NEOAV 500 displays are shown here in a Boeing 727 cockpit   0504215

and 737, Gulfstream GII and Bell Helicopter Model 430 helicopter. Selected for the Bell Model 427.

Contractor

Rogerson Kratos, a Rogerson Aircraft Corporation subsidiary.

NeoAV Model 550 EFIS Electronic Flight Instrument System

Type
Electronic Flight Instrumentation System (EFIS).

Description
The NeoAV Model 550 is an Active Matrix Liquid Crystal Display (AMLCD) Electronic Flight Instrumentation System (EFIS). The system is a microprocessor-based, self-contained unit that includes the symbol generator, Colour Active Matrix Liquid Crystal Display (CAMLCD) and interface hardware/software. The system design also incorporates full-time, Built-In Test (BIT) operated by a high-speed 32-bit microprocessor. The NeoAV 550 EFIS also includes a manual, self-test feature for the pilots and maintenance personnel. Faults are continuously recorded and stored in non-volatile memory for review after the aircraft has landed. Each NeoAV 550 component in the system is interchangeable (single part number) and is networked for maximum redundancy. Both analogue and digital interfaces are available, offering efficient, 'plug in' display interchangeability. Self-contained (bezel-mounted) EFIS control or remote controllers are available as options.

Status
NeoAV was certified on the Boeing 727 in 1994. The Bell Helicopter 430 was certified in 1996, follow-on certifications were for the Bell 412 and Bell 427. Rogerson Kratos EFIS installations is factory standard on all Bell twin-engined helicopters (Models 212, 412, 427, and 430). Other certifications include the Agusta 109 as a standard factory fit in 1998, and the US Air Force C18B EFIS upgrade in 1998.

Contractor
Rogerson Kratos, a Rogerson Aircraft Corporation subsidiary.

Night Hawk Head-Up Display (HUD)

Type
Head-Up Display (HUD).

Description
Night Hawk, developed as CMC Electronics' most advanced Head-Up Display (HUD), is a dual-combiner system with a wide 30° total Field of View (FoV) coupled with a 6 in

CMC Electronics' Night Hawk HUD is fitted to the Lockheed Martin X-35 concept demonstrator. The aircraft has been selected as the new F-35 Joint Strike Fighter (JSF) 0098625

CMC Electronics' Nighthawk HUD (CMC Electronics) 1030040

aperture, providing for enhanced weapon aiming and greater pilot situational awareness. An automatic control adjusts brightness in stroke mode and brightness and contrast in raster mode. Options include an integral Up-Front Control Panel (UFCP) and a colour video camera/recorder

Other features of the Night Hawk system include:
- Stroke only or stroke-on-raster display
- 30-minute operational level test/remove and replace
- Integrated colour HUD camera and video recorder option (NTSC or PAL format)

- High brightness and clarity
- Flexible design for retrofit applications
- May be driven by up to two MMDPs (see separate entry)
- Electronic boresight.

Night Hawk is compatible with all current types of NVIS.

Specifications
Weight: 11.4 kg
Combiner:
Single or dual combiner
P-53 Phosphor
80% transmissivity
135,000°/sec max draw rate
<1% image distortion
Accuracy 0.0 to 2.5 mr
Contrast Ratio 1.3 to 1 against 10,000 ftL ambient background
Power supply: 28 V DC per MIL-STD-704D
Power consumption: 60 W
MTBF: 3,500 h
Environmental: MIL-STD-810E per MIL-5400 Class 1, MIL-STD-461D/462D

Status
Night Hawk, together with CMC Electronics' FV-3000 MMDP was originally installed in the Lockheed Martin X-35 concept demonstrator which was selected during the fourth quarter of 2001 as the new F-35 Joint Strike Fighter (JSF) for the USAF and other participating airforces. While the X-35 was fitted with the Nighthawk HUD, it is highly unlikely that the F-35 will incorporate the system.

Contractor
CMC Electronics Inc.

NSD Series Horizontal Situation Indicators (HSI)

Type
Flight instrumentation.

Description
Century offers four NSD Series HSI models:
- **NSD360A slaved HSI** has all the features available in an HSI, matched with the simplicity of air gyro operation.
- **NSD360A slaved HSI with RMI bootstrap** provides accurate heading information to a variety of other flight instrument displays.
- **NSD360A non-slaved HSI** provides effective HSI performance.
- **NSD1000 HSI** does not require a remote gyro, but combines the power of a remote gyro with the light weight, reliability and simple installation of a self-contained one-box instrument. Slaving is a standard feature. RMI bootstrap is optional.

The NSD HSIs feature 360° heading presentation, rectilinear course deviation indicator, full-view glide slope indicator, masking glide slope warning flag, 45° tick marks, referencing heading bug, failed gyro

The Century NSD1000 HSI (left) and NSD360A HSI (right) 0504210

warning flag, free gyro mode, gyro caging knob, lost power warning flag, discrete nav warning flag and Rnav, GPS and Loran compatibility. They also feature autopilot outputs for heading and course, continuously caged heading and course selection knobs, reference aircraft and heading lubber line, diffused incandescent perimeter lighting and course arrow with reciprocal indicators. The slaved models have built-in slaving indicator and automatic magnetic gyro slaving. The NSD1000 includes a built-in electric gyro.

Specifications
Dimensions: 85.6 × 85.6 × 220.7 mm
Panel cutout: 3-ATI
Weight: 2.09 kg
Power supply: 14 or 28 V DC
Compliance: TSOs C5e, C6d, C9c, C52a

Status
In production.

Contractor
Century Flight Systems Inc.

NuHums™ health and usage monitoring system

Type
Flight/mission recording system.

Description
The NuHums is a modular, lightweight design that performs all signal processing in a single 1/2 ATR unit. The NuHums can be upgraded to add an optional flight data/cockpit voice recorder or third party diagnostic system.

The NuHums performs real-time onboard diagnostic processing for in-flight analysis of helicopter rotors, drivetrains, gearboxes and engines. It utilises the RADs-AT rotor track and balance system, and provides immediate feedback on any parameter exceedance.

NuHums modular architecture permits the customer to select features to match the requirement, and to add new features when required. The only essential element is the Data Acquisition Unit (DAU); sensors, cockpit displays, recorders, and other elements are options.

Specifications
DAU
Channels: 48 analogue and 48 digital channels
Internal memory: 64 Mbytes
Dimensions: 1/2ATR
Power: 28 V DC, 30 W
Weight: 4.55 kg

Status
In service on a number of helicopter types.

Contractor
SPS Signal Processing Systems, a division of Smiths Aerospace.

NuHums health and usage monitoring system   0015335

PACER CRAG C/KC-135 avionics upgrade

Type
Integrated avionics system.

Description
Rockwell Collins is the prime contractor for the C/KC-135 PACER CRAG upgrade. This programme is an avionics-driven upgrade for the aircraft fleet and includes the integration and installation of the FMS-800 flight management system, FDS 255 colour flat-panel flight display system, a WXR-700X forward-looking weather radar replacement system and

an embedded INS/GPS navigation system to replace the existing compass. An open systems architecture has been utilised with installed growth and functional capabilities to ensure continued interoperability with current and planned Global Air Traffic Management (GATM) requirements.

The PACER CRAG integration and upgrade is aimed at modernising the C/KC-135 cockpits and, through human factors techniques, reducing crew workload and automating many routine cockpit functions.

Rockwell Collins is prime contractor for the programme and has responsibility for design, test and integration of the flight management system, displays, weather radar and embedded INS/GPS for compass replacement.

The radar provides weather, windshear and skin paint functions. The skin paint capability allows radar detection, identification and separation maintenance of aircraft during refuelling operations.

Additionally, as part of the PACER CRAG modification, an integrated Traffic alert Collision Avoidance System (TCAS) and Enhanced Ground Proximity Warning System (EGPWS) are being installed.

Rockwell Collins is also installing a digital interphone system to replace the existing AN/AIC-10 system.

Status
In July 1996, Rockwell Collins delivered the first C/KC-135 PACER CRAG (Compass, Radar And GPS) aircraft with avionics upgrade to the US Air Force, signifying the formal release of the aircraft into the Qualification Test and Evaluation (QT&E) phase of the programme. Ultimately, more than 550 C/KC-135 aircraft will undergo the upgrade.

The PACER CRAG system was declared operationally suitable in October 1997. In October 1998, the US Air Force certified that the PACER CRAG system provided adequate reduction in workload to permit safe operation without a navigator. The first production aircraft was delivered in June 1998. By 30 November 1999, the fifth production option had been awarded, bringing the total shipset quantity to 505.

During June 2001, delivery of the 547th and final production kit completed PACER CRAG deliveries. Installation was then planned to be completed by the USAF by September 2002.

Contractor
Rockwell Collins.

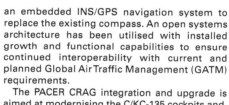

PACER CRAG flight deck   0051642

Panoramic Cockpit Display (PCD) for the F-35

Type
Avionic display system.

Description
L-3 Display Systems' Panoramic Cockpit Display (PCD) for the F-35 Lightning II is a full colour 20 × 8 in landscape AMLCD incorporating (zoned) touch screen capability. As part of the International Integrated Product Development Team (I-IPDT), Barco will produce parts of the display, drawing on experience with ruggedised touch screen technology and leveraging display technology currently integrated into the Multi Role Tanker Transport (MRTT) aircraft. The display will feature a high-brightness panel for daylight readability, combined with full NVIS compatibility for night operations with Night Vision Equipment (NVE). The display computer is described as highly redundant, as is the power supply; this last point is important when the majority of the tactical capability of the cockpit is vested in a single display.

The PCD will significantly increase the total effective screen area available to the pilot. The display is divided into four main zones, providing any combination of tactical, sensor, systems and weapons information as selected by the pilot, who will have the ability to 'customise' zones according to personal preference.

LynuxWorks, LynxOS-178 RealTime Operating System (RTOS) will power the PCD, delivering information for all the major functions of the F-35, including flight and sensor displays, communication, radio and navigation and IFF systems to provide for enhanced pilot Situational Awareness over legacy designs. The DO-178B-certified RTOS adheres to open standards, is Linux compatible, interoperable with a POSIX (Portable Operating System for UNIX) API, and supports the ARINC-653 specification. DO-178B is a safety critical standard for developing avionics software systems developed by the Radio Technical Commission for Aeronautics (RTCA) and the European Organization for Civil Aviation Equipment. POSIX is a set of programming interface standards for application development controlled by the Open Group,

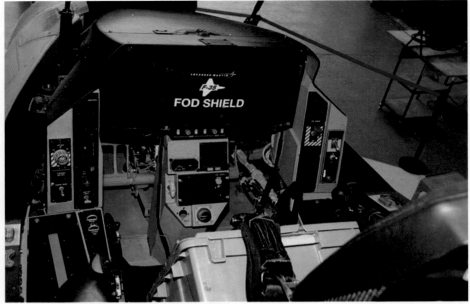

The cockpict of an early production-representative F-35 Lighting II, showing the panoramaic cockpit dispaly (PCD - covered with a FOD shield) and absence of a fixed HUD. This image also cleary shows the advantages of the sidestick configuration -the pilot has a clear view of his standby horizon unobstructed to the CDU (Lockheed Martin)   1184934

a vendor- and technology-neutral consortium promoting integrated information based on open standards and global interoperability. The ARINC-653 spec and associated APEX interface are designed for the system partitioning and scheduling required in safety-critical systems.

POSIX and ARINC are open industry standards that protect software applications for incurring rewrites or re-certification when migrating from one generation of hardware to the next. Based on open standards, the LynxOS-178 will provide security through virtual machine brick-wall partitions that make it impossible for system events in one partition of the RTOS to interfere with events in another.

These software interfaces also simplify the integration of other POSIX- or ARINC-based applications, saving development time and eliminating software flaws that would

otherwise be introduced when software module interfaces are rewritten to proprietary operating systems.

Specifications
Display area: 20 × 8 in
Compliance: DO-178B; MIL-L-85762A Class A

Status
The contract for the PCD was originally awarded to Rockwell Collins, but it was reported that problems with Collins' projection display technology resulted in a subsequent award during November 2005, to L-3 Communications Display Systems for the display and Display Management Computer (DMC) for the F-35 which is due to enter service in 2012.

Contractor
L-3 Communications, Displays.

Pegasus Flight Management System (FMS)

Type
Flight Management System (FMS).

Description
Honeywell's new-generation Pegasus FMS had its first flight on a Boeing MD-90 in May 1997 and was certified on the Boeing 757/767 and MD-90 in March 1998.

Pegasus is designed to bring the benefits of the emerging Communications, Navigation, Surveillance/Air Traffic Management (CNS/ATM) environment to the operators of air transport aircraft.

Pegasus is designed to be used across multiple aircraft hardware platforms, including: Airbus A320, A330, A340; Boeing 717, 757, 767, MD-11 and MD-90.

Honeywell claim that Pegasus offers more than 25 times the throughput capacity and 16 times more total memory than the previous-generation FMS, making it ideal for powering the functionality that will be required in the coming Free Flight era.

Pegasus is being marketed by Honeywell as one of the major components of its WorldNav™ package to meet the CNS/ATM environment. The other main elements of the WorldNav™ package are: the Honeywell Traffic alert and Collision Avoidance System TCAS 2000 and the Honeywell Air Data/Inertial Reference System

Honeywell's new-generation Pegasus flight management system   0051636

(ADIRS), or the ANSIR 2000 Air Data/Inertial Reference Unit.

Status
First flight May 1997; first certified March 1998. Selected by a large number of airlines for both

Airbus and Boeing fleets. US FAA certified on the A330/A340 during 2000, and for the A320 family in 2001.

Contractor
Honeywell Inc, Commercial Electronic Systems.

Primus 1000 integrated avionics system

Type
Integrated avionics system.

Description
The Primus 1000 integrated avionics system is designed for mid-sized business jets and regional turboprops. Advanced processing technologies and the integration of key functions result in increased capability, reliability and flexibility while achieving significant reductions in size, weight, power requirements and installation costs.

The Primus 1000 is based on the IC-600 integrated avionics computer which combines a display processor, flight director, Cat II fail-passive autopilot and EICAS processor in a single ½ ATR box. The Primus 1000 utilises an ARINC 429 architecture and is composed of an EFIS electronic display system incorporating from two to five 8 × 7 in (203.2 × 177.8 mm) large format displays, single or dual flight director and an optional single or dual fail-passive autopilot. The other components of the system include AZ-840 micro air data computer, separate vertical and directional gyros or the AH-800 fibre optic AHRS, Primus II radio system and the Primus 650 weather radar system. Standard options are Honeywell's TCAS, MLS, lightning sensor system, Primus 870 turbulence detection weather radar, Primus 700 or Primus 450 weather radar and the Laseref III inertial reference system.

Status
The Primus 1000 was selected for the Lear 45 in September 1992 and later for the Citation Bravo, Citation V Ultra and Embraer ERJ-135 and ERJ-145. A Primus 1000 system selected for the Sino Swearingen SJ30-2 in April 1996 specified dual IC-600 computers, AZ-850 all-digital Micro Air Data Computers (MADC), Primus II digital integrated radios and Primus 650 weather radar.

Contractor
Honeywell Inc, Commercial Aviation Systems.

The Primus 1000 integrated avionics system installed in the Embraer-145 aircraft, showing a five display EFIS/Engine Instrument and Crew Advisory System (EICAS) using DU-870 8 × 7 in colour displays   0131547

Primus 1000/2000 EFIS 8 × 7 in Navigation Display (ND), showing MAP and TCAS information   0099082

Primus 1000/2000 EFIS 8 × 7 in Navigation Display (ND), showing MAP and TCAS information   0099081

Primus 2000 advanced avionics system

Type
Integrated avionics system.

Description
The Primus 2000 advanced avionics system is designed for twin-turbine aircraft in the business and regional airliner markets. Its small size and weight are the result of using the most advanced technology in components, packaging techniques and systems design. The system incorporates surface-mount technology, very large-scale integrated circuits, application specific integrated circuits and high-density multilayer circuit boards. The Primus 2000 also offers flexibility and maximum growth potential through the use of an advanced architecture built around Honeywell's Avionics Standard Communications Bus (ASCB).

Interconnection between all major systems is accomplished by the ASCB, which provides both the total data handling capacity and requires a lower wire count than the one-way ARINC 429 standard. The bidirectional databus has critical level capability that eliminates the need for numerous dedicated lines between avionics systems. The ASCB architecture enhances system level availability, allowing the ASCB to be the sole means of interconnect for most avionics subsystems.

The Primus 2000 is composed of an electronic display system incorporating from two to six large screen 8 × 7 in (203 × 177.8 mm) display units and associated control panels, an integrated

The Honeywell Primus 2000 integrated avionics system on the Citation III aircraft 0131544

avionics computer containing electronic display processors, a fault warning computer, a fail-operational/fail-passive automatic flight control system and an optional flight management system, micro air data computers, attitude and heading reference system or inertial reference system, data acquisition units, Primus II radio and Primus 650 weather radar systems and the avionics standard communications bus. Honeywell's flight management system, Traffic alert and Collision Avoidance System (TCAS), Global Positioning System (GPS), Microwave Landing System (MLS), lightning sensor system, Primus 870 turbulence detection radar and laser inertial reference system are offered as standard options.

Status
PRIMUS 2000 was first certified on the Fairchild Dornier 328 in 1998, with initial deliveries in 1999. Also now certified on the Cessna Citation X and on the Dassault Falcon 900C/EX, in a five, 8 × 7 in, full colour EFIS configuration, including two MFDs, two primary flight displays and an EICAS display.

Contractor
Honeywell Inc, Commercial Aviation Systems.

Primus 2000XP integrated avionics cockpit

Type
Flight management and cockpit display system.

Description
Honeywell's Integrated Avionics Computer IC-800 is standard on the Global Express in a triple configuration. It combines the function of five major subsystems into a compact, lightweight package that is 75 per cent smaller than the combined volume of the units it has replaced.

The Primus® 2000XP integrated cockpit was designed specifically for the Global Express. Its high-accuracy air data system precisely measures speed, altitude, temperature and other critical flight information, while six 8 × 7 in displays keep the pilot fully informed about all aspects of the flight.

Primus 2000XP standard features include:
1. six-tube Electronic Flight Instrument System (EFIS)/Engine Instrument and Crew Advisory System (EICAS) using DU-870 8 × 7 in colour displays;
2. dual fail-operational autopilot;

3. dual integrated NZ-2000 Flight Management Systems (FMS);
4. triple long-range Nav source (two FMS and LaserTrak);
5. Global Navigation Satellite Sensor Unit (GNSSU) Global Positioning System (GPS);
6. dual full flight regime autothrottle triple Integrated Avionics Computers (IAC);
7. triple AZ840 Micro Air Data Computers (MADC);
8. triple LASEREF III Inertial Reference Systems (IRS);
9. quad Ametek DA-810 Data Acquisition Units (DAU);

10. Primus 880 colour weather radar;
11. Central Aircraft Information Maintenance System (CAIMS);
12. dual Primus II Integrated Nav/Com/Ident radios;
13. TCAS II Traffic Alert and Collision Avoidance System.

Optional avionics for the Global Express include: a third integrated FMS; second integrated GPS; third VHF com; lightning sensor system; three or six channel SATCOM, featuring worldwide phone, fax and data satellite communication.

The IC-800 features include: two symbol generators; fault warning computer; tone and voice generator; Radio System Bus (RSB) interface; Avionics Standard Communication Bus (ASCB) controller and interface; flight guidance computer; flight management computer; autothrottle control and power supply.

The exceptional system integration within the IAC is made possible by the lastest technologies, including: Very Large Scale Integration (VLSI); Surface Mount Technology (SMT) and advanced manufacturing techniques to reduce circuit card size. Single cards within two of the IACs contain entire FMS computers. The IACs also house the electronics for automatic flight control, autothrottle and systems monitoring.

Status
In service.

Contractor
Honeywell Inc, Commercial Electronic Systems.

Primus Epic™ avionics system

Type
Flight management and cockpit display system.

Description
The heart of the Primus Epic™ avionics system is the Virtual Backplane Network™ developed exclusively for this system. This architectural concept blends the cabinet-based modular

PlaneView™ display system in the Gulfstream G550 (P Allen) 0569530

The PlaneView™ Cockpit Display System (CDS) is designed to integrate with Honeywell's 2020 HUD and EVS (P Allen) 0569531

PlaneView™ display system has been certified in the Gulfstream G550   0530194

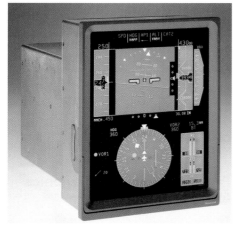

The DU-1080 flat-panel display selected for the Hawker Horizon 5-display Primus Epic™ installation   0051695

capabilities of the Honeywell 777 AIMS system with the aircraft-wide network capabilities of the Primus 2000 system. The architecture allows a very high degree of system integration and scalability by allowing all data generated by any function to be globally available within the system.

A key component of the Virtual Backplane Network is the open architecture afforded with the bidirectional Avionics Standard Communication Bus (ASCB). The ASCB continues the evolution of the ASCB used in many business and regional aircraft. The Primus Epic™ version of the ASCB provides the throughput equivalent of 100 high-speed ARINC 429 databusses. The Primus Epic™ bus is claimed to significantly reduce wire weight, power use and installation cost, while greatly increasing the capacity to support aircraft utility systems control and other future aviation requirements.

The Primus Epic™ system hardware is built on the Modular Avionics Unit (MAU) derived from Honeywell technology, developed for the Boeing 777. Computing modules within the MAUs and flat panel displays utilise Honeywell's Digital Engine Operating System (DEOS™), which allows the different aircraft and functions to run simultaneously and independently. The full size 8 × 10 in colour flat panel liquid crystal displays, in a two- to six-display configuration, feature new functionality within a point-and-click Graphical User Interface (GUI) environment, which supports moving maps, ground-based weather, real-time video and electronic pilot manuals.

Pilots may choose traditional interfaces or new cursor control devices, including touchpad, joystick, light pen or tracker ball to interact with on-screen 'soft key' controls. In future, Primus Epic™ will offer voice command as a control option for some functions.

The initial utilities integrated into the Primus Epic™ system via the MAU were the landing gear control and anti-skid braking systems, but the intention is to integrate a wide range of utility systems including: air conditioning and environmental control; electrical power distribution; fire protection; fuel; hydraulics; ice protection; lights; oxygen; APU; engine vibration and others.

A variant of the system, designated Primus Epic™ CDS Retrofit (CDS/R), became available in June 1999. This system includes an upgraded Control Display System (CDS), using 8 × 10 inch LCD Primary Flight Displays (PFDs) and MultiFunction Displays (MFDs). It is designed around the Primus 1000 and Primus 2000 systems and utilises the IC-1080 integrated computer, with optional internal FMZ-2000 Flight Management System (FMS), DU-1080 LCD display units and display controllers. It includes built-in growth capabilities to support future CNS/ATM requirements and to further enhance

system features with products such as Flight Management Systems (FMS), Global Positioning Systems (GPS), Satcom. GPS Landing Systems (GLS) and Traffic alert and Collision Avoidance Systems (TCAS).

The latest development of the Primus Epic™ display system, designed to integrate with Honeywell's HUD and Enhanced Vision System (EVS – see separate entry), is designated PlaneView™. As installed in the Gulfstream G550, the PlaneView™ cockpit features four 360 mm (14 in) AMLCD displays in a tight side-by-side arrangement across the width of the main instrument panel. Each display consists of multiple 'panes' which are user-configurable to show primary flight, navigation, system and sensor information, with on-screen navigation and menu selection via cursor control.

Status

During the second quarter of 2001, Honeywell announced successful first flights of Primus Epic™ CDS/R on the Gulfstream II, Gulfstream III, L-382 and Cessna Citation V, with certification expected by the end of 2001. Primus Epic™ is to be standard equipment on the new Raytheon Hawker Horizon business jet, where the installation will comprise a five-display integrated system with two flight situational displays, two multifunction displays, two cursor control devices and glareshield controllers; the display unit chosen for the system is the DU-1080 flat panel colour liquid crystal display unit (8 × 10 in). The aircraft made its first flight in August 2001 and is expected to complete certification testing in 2003, with first deliveries scheduled for 2004.

Primus Epic™ was also selected by Fairchild Aerospace for the 728/928 JET family, by Embraer for the ERJ170/190 aircraft, and by Agusta-Bell for the AB139 helicopter. The Primus Epic™ configuration for the 728/928 JET aircraft family comprises three MAUs and five 8 × 10 in colour flat-panel displays. In the case of the AB 139 helicopter, four configurations are available: a basic VFR system; a 3-axis AFCS IFR installation; a 4-axis AFCS IFR installation; and a Search And Rescue (SAR) version.

PlaneView™ has been certified on the Gulfstream V/G500 and V-SP/G550.

Contractor

Honeywell Inc, Commercial Electronic Systems.

Pro Line II integrated avionics system

Type

Integrated avionics system.

Description

Collins' Pro Line II integrates many individual units into major subsystems that are tied together via a digital data bus. Reliability issues, weight and maintenance complexity of analogue systems are replaced with a more simple and efficient overall package. Electronic Flight Instrumentation System (EFIS), autopilot, air data, attitude heading reference and communications/navigation subsystems are incorporated into an integrated avionics system which can be retrofitted into business jet or turboprop aircraft.

Electronic flight instrumentation

Collins EFIS-85 provides airspeed information in a digital readout. The EFIS-86 incorporates airspeed with two vertical scales. Both systems also provide airspeed trend information to facilitate smooth manual power applications. Additional data displayed on the Electronic Attitude Director Indicator (EADI) includes inputs from the digital autopilot, such as mode annunciation and climb and cruise speeds. The EFIS displays are optimised for ease of interpretation and clarity. The pilot may select a variety of modes for different flight regimes and can add or delete information from displays when desired.

Collins EFIS utilises high-resolution Cathode Ray Tube (CRT) techniques that combine raster

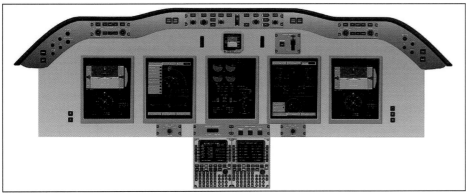

The Primus Epic™ integrated avionics display system for the Hawker Horizon aircraft   0081856

Collins Pro Line II installed in a Falcon 50   0105283

and stroke displays. Air data information is integrated on both the five- and six-inch displays. EFIS-85 includes a proven full sky/ground display that adds airspeed and acceleration/deceleration information. Deviation from preselected altitude is also shown. The larger size of the EFIS-86 display allows the addition of a vertical airspeed scale on the left side of the ADI. A high or low scale is selected automatically, depending on airspeed; acceleration/deceleration is also displayed.

Collins WXR-350 weather radar system is fully integrated with Collins EFIS to provide a picture of significant weather combined with navigation data on the Horizontal Situation Indicator (HSI) and multifunction display (MFD). Features of the radar include the display of a fourth level of precipitation to indicate rainfall rates greater than 50 mm per hour. Another feature of the radar is path attenuation compensation. The possible danger of cells hidden or reduced in intensity behind heavy rainfall areas is identified to the pilot for increased safety. The radar also provides ground mapping with different colours used to clearly identify the mode selected. The radar annunciates a target alert if a contourable cell appears from 60 to 150 nm ahead.

System architecture of the EFIS includes an MFD as an integral part of the system, providing large-screen presentation of any two navigation systems, checklists, weather or navigation data from four long-range navigation systems. This design allows for reversion and a high level of fault tolerance. In the event of a CRT failure, composite ADI/HSI information can be displayed on any of the remaining tubes. In addition, the HSI display can be moved to the MFD and full ADI information can be displayed on the HSI. Hence, no critical flight information will be lost due to a single-tube failure. Full reversionary capability of EFIS remote units is also assured.

Flight Control System (FCS)
Core to the performance and efficiency of the Collins digital FCS is the strapdown AHS-85 Attitude Heading Reference System (AHRS). The system provides all attitude and heading measurements as well as angular rates and linear accelerations in the three aircraft axes. Multiple piezoelectric sensors detect aircraft rates and acceleration. This information is processed by the AHRS and transmitted to the autopilot as well as the EFIS and weather radar.

Supporting the FCS is the ADS-85 Air Data System. The ADS-85 combines into one digital system the various air data sensors and instruments previously distributed throughout the aircraft. The system includes encoding

altimeter, altitude preselector/alerter, Mach/airspeed indicator, vertical speed indicator, true airspeed/temperature indicator and vertical NAV computer/indicator displays.

The APS-85 autopilot system is a fully digital, fail-passive autopilot with dual independent flight guidance computers. The system is capable of Category II operation with growth to Category IIIA automatic landing. The APS-85 utilises digital inputs from the AHS-85 that provide immediate acceleration, rate and position information to the autopilot. These inputs are used by the autopilot to provide accurate flight path tracking in all flight regimes and fast response in windshear situations. Attitude and heading inputs from the Collins AHRS combine with air data to allow the new autopilot to sense minute changes and react immediately, whereas legacy autopilot systems responded to changes in aircraft position after they have occurred and were significant enough to be sensed.

The APS-85 receives inputs from the AHRS and air data when force is exerted on the aircraft before major displacement occurs, reacting quickly and positively to avoid undesired movement. This capability provides smoother, more positive control over aircraft guidance.

Autopilot modes are designed to enhance aircraft performance, save fuel and reduce pilot workload. The CLIMB mode, when selected, allows the aircraft to automatically fly a preprogrammed Indicated Airspeed (IAS)/Mach profile that is optimised for each aircraft type. The autopilot is internally programmed for optimum airspeed at light, medium and heavy weights, based on static air temperature and altitude. SPEED mode combines the IAS/Mach functions in a single button selection. The autopilot computer determines whether IAS or Mach is selected depending on altitude. DESCEND mode provides a smooth transition from cruise to an aircraft-optimised rate of descent. Standard autopilot modes, including HEADING, 1/2 BANK, NAV, APPROACH, ALTITUDE and VNAV are also provided.

Communications/navigation
Through extensive application of microprocessor technology, Pro Line II radios offer a wide variety of capabilities in communication, navigation and other critical flight functions for a variety of aircraft types. Engineered with the Commercial Standard Data Bus (CSDB), Pro Line II radios are fully interchangeable with previous Pro Line systems and are compatible with most digital and analogue applications. Pro Line II Series 400 radios, compatible with the ARINC 429 data bus, interface with current digital avionics systems.

Five basic functions are included in the system: Voice and data communication, navigation, Distance Measuring Equipment (DME), Automatic Direction Finding (ADF) and transponder.

Options include an individual control head for each radio and an integrated radio tuning unit or flight management system control display unit. Installation with remote units that can be located together or distributed to accommodate space constraints enhances the versatility of the system. Self-diagnostic and flight deck display capabilities test the system and isolate a malfunction down to the line replaceable unit or individual circuit. As a result, corrective action is conveniently expedited and accurately performed. Moreover, specially developed troubleshooting programmes allow technicians to quickly return radios and controls to full service, effectively minimising downtime.

Options
A large number of options can be added and integrated to the system, including Traffic Alert and Collision Avoidance System (TCAS), satellite communications, Airborne Communications Addressing and Reporting System (ACARS), HF COMM and a third VHF COMM.

Typical configuration
A typical 3-tube EFIS installation includes 2 × EFD-85 Electronic Flight Displays, 1 × MFD-85B Multifunction Display, 1 × CHP-86B Course Heading Panel, 1 × DCP-85G Display Control Panel, 1 × 332D-11T vertical gyro, 1 × MPU-85N Multifunction Processor Unit, 1 × DPU-85N Display Processor Unit, 1 × LAC-80 Longitudinal Accelerometer and 1 × INCLIN inclinometer.

Status
In service in a wide range of regional and business aircraft.

Contractor
Rockwell Collins.

Pro Line 4 integrated avionics system

Type
Integrated avionics system.

Description
Rockwell Collins' Pro Line 4 is a fully integrated avionics system designed for business and regional aircraft. The system is inherently flexible, facilitating the control of all additional subsystems demanded by the mission requirements of the host aircraft. These subsystems include the Electronic Flight Instruments System (EFIS), Engine Indication and Crew Alerting System (EICAS), automatic Flight Controls System (FCS), Maintenance Diagnostics System (MDS), Flight Management System (FMS) and aircraft sensors.

The system is composed of the following components:

Integrated Avionics Processing System (IAPS)
The Pro Line 4 system is controlled via the Integrated Avionics Processing System (IAPS). The subsystems housed within the IAPS are the flight control computer, flight management computer and maintenance diagnostic computer.

Electronic Flight Instrument System (EFIS)
A wide variety of EFIS are offered, featuring 6 × 7 and 7¼ in square displays. All data processing and symbol generation accomplished in the control head. The basic systems consists of Primary Flight Displays (PFDs) and Multifunction Displays (MFDs).

The PFD displays information such as attitude, airspeed, altitude, vertical speed, heading, active navigation source, autopilot/flight director modes, weather radar information as well as other selected information. Display formats vary according to aircraft manufacturer and customer options, although core philosophy remains consistent in all systems.

The main function of the MFD is to display map data. This data contains a graphical depiction of

Pro Line 4 as installed in the Bombardier Challenger 604 – a six-tube configuration, showing primary flight (×2), navigation, weather radar, and engine and systems information   0105285

the aircraft flight plan as well as other orientation/situational awareness data, including optional weather and traffic information. The MFD also displays data windows for such information as sensor information and navaid selection. Other information can be displayed if required. Either MFD can display PFD information in reversionary mode.

Engine Indication and Crew Alerting System (EICAS)

The EICAS is designed to provide for all engine instrumentation and crew annunciation in an integrated format. As part of the EICAS, graphical depiction of aircraft systems can be selected, including electrical, hydraulic, anti-icing, environmental and flight controls. System monitor information and crew awareness messages are also displayed on the EICAS.

Flight Control System (FCS)

The avionic FCS is a fully digital, fail-safe autopilot, certified to Category II operation. The system interfaces with a wide range of sensors, including Attitude and Heading Reference Systems (AHRS), Inertial Reference Systems (IRS) and Air Data Systems (ADS).

Pilot interface with the FCS is via the Flight Control Panel (FCP). The FCP provides for selection of flight mode, airspeed, reference and trim, with hardware configuration tailored to aircraft manufacturer requirements.

Maintenance Diagnostics System (MDS)

Maintenance diagnostics are enhanced by the Pro Line 4 architecture. Unlike more traditional diagnostics systems, the MDS provides fault analysis with constant monitoring, immediate identification, recording and display of maintenance information.

A complete fault history of each Line Replaceable Unit (LRU) can be shown on the aircraft MFD or downloaded to disk. The system also provides the ability to perform a trend analysis. These combined capabilities are extremely valuable in troubleshooting intermittent or transient failures and aid flight and/or maintenance crew identify the appropriate preventative or corrective action.

In addition to the basic features, EICAS-equipped aircraft are capable of recording engine parameter and trend analysis data. This data can

be effectively used for determining engine/aircraft maintenance scheduling. A long-term history of engine exceedances is also provided.

Flight Management System (FMS)

There are four series in the Collins FMS family applicable to Pro Line 4. The 6000 series is

designed for heavy business operations, the 5000 series for medium/light business operations, the 4000 series for regional aircraft and the 3000 series for light business operations. The systems provide for full flight planning, multiwaypoint lateral and full profile vertical navigation, complete Standard Instrument Departures (SIDs) and Standard Arrivals (STARS), airways, holding patterns and full performance envelope data.

The FMS can be tailored specifically for the mission requirements of the aircraft. Computer interface is via Control Display Units (CDUs), ranging from a $6^3/_8$ in LCD CDU to a 9-in ARINC 702 CDU, according to aircraft/application.

The FMS uses a complete compliment of sensors: VHF Omnidirectional Radio Range (VOR), Distance Measuring Equipment (DME), Global Positioning System (GPS), Inertial Reference System (IRS) and Instrument Landing System (ILS). These sensors are blended into a single position using Kalman Filtering to yield the optimum navigational solution for all phases of flight.

Navigation sensors

Navigation sensors associated with Pro Line 4 include:

- ADF-60/462 (coverage up to 2182 kHz).
- DME-42/442 (three-channel scanning; displays current ground speed, distance from the station and the ETA at the station).
- VIR-32/432 (VOR/ILS functions, including localizer, glideslope and marker beacon compliant with ICAO Annex 10 FM immunity requirements).
- ALT-4000 (provides accurate measurement of terrain from minus 20 to 2500 ft; supports Cat III and single installation Cat II).

Rockwell Collins Pro Line 4 EICAS display   0105284

For details of the latest updates to *Jane's Avionics* online and to discover the additional information available exclusively to online subscribers please visit
jav.janes.com

Other sensors associated with the system include AHRS and ADS.

Collins' AHRS represents a new approach in sensing aircraft attitude and heading information which eliminates complex and vulnerable electromechanical gimbaled platforms. Attitude and heading information is obtained by electronically processing three-axis rate and acceleration information sensed by Rockwell Collins' new rotating piezoelectric multisensor. This new concept eliminates many issues associated with conventional gyroscopic devices, resulting in improved accuracy, increased reliability and reduced installation costs. The system consists of an attitude heading computer, flux detector unit, mount and internal compensation unit, which combine to provide three-axis position, rate and acceleration data.

The ADS is an all-digital solid-state system that senses, computes and displays all parameters associated with aircraft movement through the atmosphere. This new design features a solid-state piezoresistive sensor, digital computation and display, fault isolation and diagnostics. In addition, programmable outputs and interfaces are available, facilitating installation of the system in a wide range of aircraft.

Communication equipment
Collins' VHF radios include extended frequency ranges to 152 MHz and are upgradeable to the new 8.33 kHz channel spacing requirement, as well as providing data capability. HF Comm is available as a standalone unit or integrated into any Pro Line avionics system, providing communication and datalink capabilities on up to 280,000 channels.

The Collins SAT-906 satcom system (see separate entry) provides digital voice, data or fax communications. The SAT-906 provides up to six channels of simultaneous communications that can be used either in the cabin or in the flight deck. The system utilises so-called 'smart' digital handsets and data ports for fax/modem transmissions, each with its own separate on-board address, enabling callers to dial a specific seat or fax/modem location within the aircraft.

Weather Radar
The Collins TWR-850 Turbulence-detection Weather Radar is associated with Pro Line 4. Features of the system include: Capability to interface with various antenna sizes (12-, 14- and 18-in), solid-state design (no magnetron), turbulence detection, sector scan, split sweep display, ground clutter suppression, Path Attenuation Compensation (PAC) alert and antenna autotilt, all combined in a single unit.

TCAS
Collins' TCAS-94 Traffic Alert and Collision Avoidance System (TCAS II) (see separate entry) can detect and track up to 64 aircraft simultaneously. Each 'target' is filed and prioritised by level of threat and displayed on the traffic display. The system also uses a diversity Mode S transponder to communicate aircraft to aircraft, as well as to provide ground communication and surveillance.

The system consists of a TCAS Receiver/Transmitter (R/T) computer, a single (or dual, if required) directional TCAS antenna, a single (or dual) Mode S transponder and controls/displays. The display can either be standalone or integrated into the Collins Pro Line EFIS.

Bombardier PrecisionPlus™ avionics upgrade
During the 4th quarter of 2001, Bombardier Aerospace announced the new Bombardier PrecisionPlus™ upgrade for its Challenger 604 business jet. Developed by Rockwell Collins as an enhancement to the aircraft's current Pro Line 4 avionics suite, the upgrade offers several new features, including automation of both V-speed calculation and thrust setting as primary information, and three-dimensional display of the aircraft's flight plan.

After receiving certification from both Transport Canada and the United States Federal Aviation Administration, the avionics upgrade was integrated into all new production aircraft manufactured after June 2001 and is available to current Challenger 604 operators as a retrofit.

The basic PrecisionPlus™ upgrade package includes the following features:
- Automatic look-up and display of takeoff, approach, landing and missed-approach speeds.
- Automatic look-up and display of thrust setting (N1) for takeoff, climb, cruise and go-around.
- Blending of actual observed wind and entered wind to improve the prediction of flight time and fuel requirements and enhance mission planning.
- Position reporting in non-radar environments such as the North Atlantic.
- Improved polar navigation, enabling the pilot to navigate and steer the aircraft at latitudes over 89 degrees.
- Full-time Distance Measuring Equipment (DME) reporting on the pilot's Multifunction Display (MFD).
- Engine Indication and Crew Alerting System (EICAS) improvements, including the addition of metric fuel indication capabilities, logic enhancements and Flight Management System (FMS) performance enhancements.
- Full integration with the Flight Dynamics HGS Head-Up Display (see separate entry), and with Safe Flight's Mark II Auto Throttle System.

In addition, the PrecisionPlus™ upgrade offers four optional features:
- A 3-D flight plan map, providing a three-dimensional graphical representation of the programmed flight plan and predicted flight path on the MFD.
- A long-range cruise feature, allowing pilots to select a cruise speed computed by the FMS for either maximum range or maximum speed.
- A search pattern feature offers automatic generation of waypoints which enables pilots to fly fixed search patterns.
- An expanded Flight Data Recorder (FDR) will provide operators with the ability to record additional FDR parameters as required by FAR 135.152.

Status
In production and in service. Pro Line 4 is currently installed in the Beechjet 400A, Canadair Regional Jet and Challenger 604, Dassault Falcon 20, 50 and 2000, Gulfstream 100® and 200®, Learjet 60 and the Saab 2000 regional airliner.

Contractor
Rockwell Collins.

Pro Line 21 integrated avionics system

Type
Integrated avionics system.

Description
A major emphasis of Pro Line 21 is the design of the flight deck. A major effort in terms of the effective application of technology together with system integration has been made to provide major advancements in three key areas: Human Machine Interface (HMI), Information Management (IM) and Situational Awareness (SA). Pro Line 21 was introduced in 1995 for corporate and regional aircraft.

Pro Line 21 flight decks are custom-configured with two to five adaptive flight displays that utilise a mix of AMLCD formats, including a 6 × 8 in and 8 × 11 in active area display. In addition to primary flight and navigation information, this LCD technology allows clear presentation of approach plates, terrain maps, real-time video and other highly detailed – and even three-dimensional – graphics that deliver flight operations information to pilots in innovative formats developed and refined by pilots.

Pro Line 21's key subsystems include the Collins family of Flight Management System (FMS) satellite-based precision navigation and communication systems with the GPS-4000 Global Positioning System (GPS) sensor, an advanced-technology Attitude Heading Reference System (AHRS) and an advanced flight control system with fail-passive autopilot. Also standard are solid-state weather radar and Pro Line radio sensors, including transponder, TCAS and DME. An optional maintenance system displays current LRU status, fault history and diagnostic data on the multifunction display.

Pro Line 21 provides a flexible user interface that can be tailored to the needs of individual flight decks. Collins' goal with the system is to provide the required information, exactly when it is needed. Pro Line 21 shows information in the most effective location with a simple, natural method for the crew to assimilate and manage the display of data. Pro Line 21 integrates Collins 's Head-up Guidance System™ (HGS), which enables the pilot to keep his focus outside/forward to increase SA.

The system supports growth through field-loadable software, enabling changes to be loaded onto the aircraft without removal of LRU/LRMs. Pro Line 21 system architecture is flexible and can be configured in a variety of ways to fit a particular mission. For example, a system

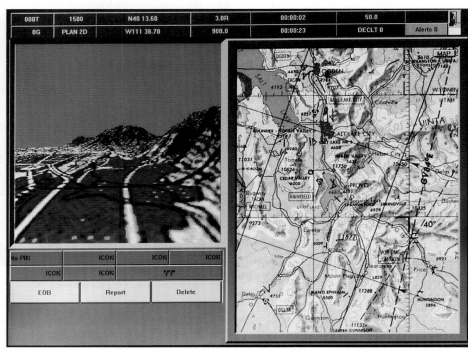

Pro Line 21 real-time Situational Awareness (SA) 3-D information on surrounding airspace and terrain

0001386

Pro Line 21 Retrofit in the Hawker 800XP (Rockwell Collins)   0534153

for a light aircraft may have as few as two or three Adaptive Flight Displays (AFDs), while a long-range aircraft may opt for five AFDs. This flexibility facilitates the optimum cost/capability trade-off for a given set of mission requirements. The system architecture supports a high degree of interface flexibility through standard ARINC 429 interfaces that enable various types of equipment from various manufacturers to be interfaced with Pro Line 21.

Pro Line 21 Retrofit

Collins' Pro Line 21 Retrofit integrates new avionic technologies into legacy flightdeck environments. It takes advantage of new capabilities, including advanced display technologies, solid-state geophysical sensors, all digital CNS, integrated FMS and a flexible architecture to provide the bandwidth required to meet current and future operational and performance requirements. Pro Line 21 Retrofit allows operators to select the level of capabilities desired while maximising life cycle cost improvements. The system may be applied in a number of flight deck configurations, allowing operators to choose between integrated systems or individual products.

Collins' Pro Line 21 Retrofit offers advanced capabilities providing proactive decision tools, including:

• Data communications
• Up-linked graphical weather
• 3-D flight plan maps
• Overlays to normal navigation maps/ geopolitial airways and airspace
• Approach and airport charts and associated textual charts.

The system also supports future functionalities such as enhanced electronic charting, mapping and improved terrain awareness. New capabilities are managed through user-friendly on-screen menus that simplify human-machine interface and reduce installation complexity. Possible system applications include Adaptive Flight Display (AFD), Flight Management System (FMS), Automatic Flight Control Systems (AFCS), Maintenance/ Diagnostics System (MDS) and advanced processors. AFDs incorporate Active Matrix Liquid Crystal Display (AMLCD) technology, available in 8×10 in format. Installations range from one to four displays, depending on aircraft configuration and operational requirements. Available products include the AFD-3010, FMS-3000/5000/6000, AHS-3000 Attitude Heading Reference System (AHRS), ADS-87A Air Data System, TCAS II Traffic Alert and Collision Avoidance System, Doppler™ Weather Radar and Collins Pro Line CNS sensors.

Pro Line 21 Continuum

During the last quarter of 2000, Rockwell Collins introduced Pro Line 21 Continuum to provide advanced avionics capabilities for current-technology flight decks.

Pro Line 21 Continuum takes advantage of the new technologies of the Pro Line 21 system, including both high-performance partitioned processing and Ethernet interfaces, to provide the bandwidth required to meet operational and performance requirements. These technologies are applied to key processing modules in the aircraft's avionics and display systems, resulting in a flexible, cost-effective and expandable architecture.

The system provides advanced capabilities that increase performance and ease pilot workload, including data communications, uplinked graphical weather and 3-D FMS planning maps. The system also supports such future functionality as enhanced mapping, electronic charting and improved terrain awareness, further increasing crew efficiency and situational awareness. New capabilities are managed through user-friendly, on-screen menus that simplify human-machine interfaces and reduce installation complexity.

Continuum allows operators to select the level of capabilities desired while maximising life-cycle cost improvements. The system may be applied in a number of flight deck configurations, allowing operators to choose between integrated systems or individual products.

System applications include Adaptive Flight Displays (AFDs), FMSs and Automatic Flight Control Systems (AFCSs), a maintenance/ diagnostics system, and advanced processors. The AFDs incorporate AMLCD technology, available in 8×10, 7×7 and 7×6 in sizes. Installation ranges from two to five displays,

depending on aircraft configuration and operational requirements. Available products include the FDS-2000 Flight Display System, AHS-3000 Attitude Heading Reference System, ADS-87A Air Data System (ADS), TCAS II Traffic alert and Collision Avoidance System, Doppler weather radar and Collins Pro Line radio/ navigation sensors.

The system has also been designed for growth. As future CNS/ATM requirements are implemented, the system will support new operational capabilities, including:

• Required Time of Arrival (RTA)
• Automatic Dependent Surveillance (ADS-B)
• Controller-Pilot Data Link Communications (CPDLC)
• Aeronautical Telecommunications Network (ATN)
• Wide Area Augmentation System (WAAS)
• Local Area Augmentation System (LAAS).

To simplify the addition of new capabilities, the system is designed to accommodate future changes through on-aircraft software downloads, reducing the need for major flight deck or architecture modification. Operators of Collins Pro Line 4-equipped aircraft will also be able to upgrade to these new capabilities.

Pro Line 21 Integrated Flight Information System (IFIS)

The Pro Line 21 IFIS is designed to increase the capability of the current Pro Line 4/21 architecture by integrating the latest information management and decision-making tools to further improve SA and increase the operational effectiveness of the total avionics system.

The Pro Line 21 IFIS provides the following enhancements:

• Electronic charting
• Graphical weather
• Enhanced FMS mapping with multiple overlays
• Increased software upload capability.

A key characteristic of the IFIS is its ability to gather data (from host aircraft systems or via datalink/uplink) and provide it to the flight crew in an efficient and timely manner. The Pro Line 21 IFIS takes advantage of new technologies, including high performance, partitioned architecture and Ethernet interfaces that enhance functional processing. This technology is applied to key processing modules (File Server Unit) and in the display system (AFD).

A typical IFIS configuration includes:

File Server Unit (FSU)

The File Server Unit (FSU-5010) is the heart of the Pro Line 21 IFIS. It consists of a microprocessor

Pro Line 21 Continuum installed in a Bombardier Challenger 601 (Rockwell Collins)   0534154

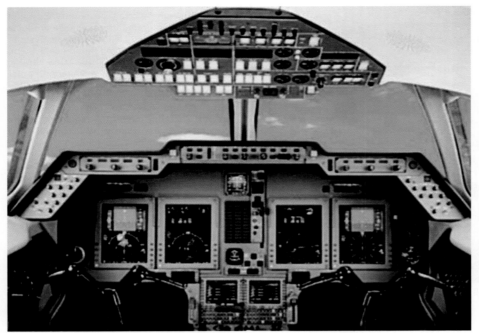

Pro Line 21 Continuum installed in a Cessna Citation III (Rockwell Collins)   0534156

card, Ethernet node card and a Flash memory card (2.0 Gbytes for FSU-5010, upgradeable to 16 Gbytes). The FSU-5010 is packaged in a 2 MCU-like package with an external fan and external compensation unit mounted on the tray. The FSU-5010 has an additional slot for an extra processor card and an I/O card for future growth.

Slot Upgrade
A spare slot in Pro Line 21 Adaptive Flight Displays (AFD) is designed with growth in mind. The already fielded AFD-3010 is upgradeable to IFIS specification by populating an Ethernet card and upgrading the graphics processor card. This new card contains 1 channel of Ethernet (full-duplexed) with an additional channel for video (growth). The previous graphic processor card is upgraded to dual-graphics processors if the existing display does not already have this functionality.

Cursor Control Panel (CCP)
A Cursor Control Panel (CCP) is required to provide necessary crew interface to the new enhanced features.

Collins Portable Access Software
Collins Portable Access Software (CPAS) application software facilitates loading of large databases to the FSU, recurring on a bi-weekly or monthly basis. In addition, the CPAS application software will be capable of providing navigation database, flight plans and company routes to the FMS, MDC database download, RIU software, VHF software, FSU software and AFD software download capability.

Status
Pro Line 21 is in production and in service in Hawker 800XP, Bombardier Challenger 300, Gulfstream 150 and the Raytheon Premier I, which also includes Collins' AHS-3000 AHRS, IAPS-3000 Integrated Avionics Processing System, AFC-3000 Automatic Flight Control system, and a complete radio and radar package. The system has also been selected for the Bell Agusta 609 civil tilt-rotor, which will feature three 6 × 8 in AMLCDs, including two Primary Flight Displays (PFDs) and one MultiFunction Display (MFD), and the Bombardier Continental.

The system is also installed in the Cessna Citation CJ1 and CJ2 and is selected for the new CJ3. This information is displayed on the CJ3 flight deck's 10 × 8 in high-resolution, AMLCD adaptive flight displays. Bezel-mounted menu selectors allow easy access to display formats adapted for each phase of flight. The Citation CJ3 will also be the launch aircraft for Collins'

new onboard File Server Unit (FSU). The FSU integrates with the Pro Line 21 system's high-performance partitioned processing capabilities, Ethernet interfaces and AMLCD flight displays to provide enhanced maps, optional electronic charts and graphical weather information.

Cessna has targeted the first Citation CJ3 customer delivery for late 2004.

Pro Line 21 Continuum has been fitted to a wide range of business jets, including the Bombardier Challenger 600/601, Cessna Citation III, Falcon 20/50, Hawker 700/800 and Gulfstream II, IIb and III.

The IFIS is a applicable to existing AFD-3010 Pro Line 21 Systems, installed in Bombardier Challenger 601 retrofit (10 × 8 AMLCD), Falcon 20/50 retrofit (10 × 8 AMLCD), Falcon 900 retrofit (10 × 8 AMLCD), Hawker 800 retrofit (10 × 8 AMLCD), Raytheon Premier, Hawker 800XP, Cessna CJ1/2, Cessna CJ3, Gulfstream G150, Astra retrofit (10 × 8 AMLCD), Piaggio P180 and Citation 501 retrofit aircraft.

Contractor
Rockwell Collins.

Pro Line 21 Integrated Display System (IDS)

Type
Avionic display system.

Description
Collins' Pro Line 21 IDS interfaces with existing Collins autopilots, providing a flexible and cost-effective retrofit solution for legacy aircraft flightdecks. Integrating the latest in display technologies, including the Pro Line 21 8 × 10 in AMLCD, with onboard sensors, radios, flight management systems and autopilots, Pro Line 21 IDS offers upgrade without the expense normally associated with major avionic retrofits.

Key user benefits and major features of the Pro Line 21 IDS include:
- High-resolution 8 × 10 in AMLCD Adaptive Flight Displays (AFD)
- 2-, 3- or 4-display systems
- Utilises existing Communication, Navigation, and Surveillance (CNS) sensors
- Displays support EGPWS, Collins, weather radar, TCAS II and other advanced features
- Designed to Interface with Collins, autopilots, AP-105, APS-65, 80 and 85 autopilots, Honeywell SPZ-500 (via the DIU-3010)
- Higher reliability
- Reduced installation costs

- Supports growth for future CNS/ATM requirements.
  Options include:
- Integrated Flight Information System (IFIS)
- Electronic Charts – SIDs, STARs, Approaches & Airport Diagrams
- Graphical weather – XM Satellite
- Enhanced maps
- Collins' Air Data Computers
- Collins' Attitude Heading and Reference System (AHRS)
- Collins' Radio Tuning Units (RTU)
- Collins' Traffic Alert Collision Avoidance System (TCAS II)
- Pro Line 21 Digital Radios.

The major system components include 8 × 10 in AMLCD displays, the DIU-3010 Display Interface Unit (DIU), a Display Control Panel (DCP) and Course Heading Panel (CHP). The 8 × 10 in AFD-3010 display provides air data, Attitude Direction Indicator (ADI), Horizontal Situation Indicator (HSI), TCAS, Collins' weather radar and EGPWS on the Primary Flight Display (PFD). The companion AFD-3010E is employed as a partner MultiFunction Display (MFD). The DIU-3010, the backbone of the system, provides the interface between existing aircraft sensors, radios and Collins' Auto Pilot (AP). The DIU-3010 converts CSDB digital and analogue data from existing air data, AHRS, radio, FMS and autopilot sensors into ARINC 429 for interface with Pro Line 21 AMLCD Displays. The DCP and CHP provide the primary interface and integration management between the pilot and displays. The DCP and CHP permit the pilot to select course and heading, sequence through the active navigation source data, choose display formats and select range.

Specifications
Dimensions:
AFD-3010 208 (H) × 254 (W) × 246 (L) mm
CHP-3010 38 (H) × 146 (W) × 131 (L) mm
DCP-3030 215 (H) × 48 (W) × 204 (L) mm
DIU-3000 200 (H) × 57 (W) × 368 (L) mm
AFD-3010E 208 (H) × 277 (W) × 246 (L) mm
CCP-3000 48 (H) × 146 (W) × 131 (L) mm
Weight:
AFD_3010: 5.9 kg
CHP-3010: 0.64 kg
DCP-3030: 1.13 kg
DIU-3000: 2.49 kg
AFD_3010E: 6.35 kg
CCP-3000: 0.77 kg
Certification:
AFD-3010: FAA TSOs C2d, C3d, C4c, C6d, C8D, C0c, C10b, C34e, C35d, C36e, C40c, C41d, C43c, C46a, C52b, C63c, C66c, C87, C92c, C95, C101, C104, C105, C110a, C113, C115b, C117a, C119b, C129a; DO-160D, DO-178B Level A
Others: FAA TSO C113; DO-160D

Status
In production and in widespread service.

Contractor
Rockwell Collins.

Projection display systems

Type
MultiFunction Display (MFD).

Description
Kaiser Electronics is applying Commercial Off-The-Shelf (COTS) projection technology, and employing microdisplay components, to produce aircraft Projection Display Systems (PDSs) which provide superior performance and viewing qualities when compared with current Active Matrix Liquid Crystal Displays (AMLCDs). The suite of displays features scalable design, which facilitates customer-driven display size and resolution, according to application. The key to the technology is an open architecture design, coupled with a single 'optical engine' which can be utilised for a wide variety of display sizes, from 5 in square to

37 in diagonal, with only slight modification to the folded optics.

Other benefits and features of the system include:

- Panoramic size, conformal and multiple aspect ratio available from common core design
- True colour display at pixel level
- High reliability
- Affordable touchscreen capability
- Rapid cold start without heaters
- Low-cost, multiple-source components. Current PDS contracts include:
- Boeing, to provide a 6 × 6 in (152 × 152 mm) PDS for the F/A-18E/F
- Lockheed Martin, to provide an 8 × 8 in (203 × 203 mm) PDS for the F/A-22 Raptor
- Lockheed Martin, to provide a 20 × 8 in (508 × 203 mm) PDS for the F-35 JSF.

Status
During the second quarter of 2002, Kaiser successfully flew its first PDS on a Boeing F/A-18E/F at the Naval Air Warfare Centre Weapons Division (NAWCD) at China Lake, California. The Digital Expandable Colour Display (DECD), is a 152 × 152 mm (6 × 6 in) reflective micro Liquid Crystal Display (LCD) projection-based smart display, which is part of the Boeing F/A-18E/F Cost Reduction Initiative (CRI) aimed at reducing costs and improving performance and reliability of the Super Hornet.

Contractor
Rockwell Collins Kaiser Electronics.

Radar Control Display Unit (RCDU) for UK E-3D

Type
Control and Display Unit (CDU).

Description
Honeywell Defense Avionics Systems will develop the colour Radar Control Display Unit (RCDU) for the United Kingdom's fleet of seven E-3D Airborne Warning And Control System (AWACS) aircraft under a subcontract to Northrop Grumman Electronic Sensors and Systems Division of Baltimore, Maryland as part of the Boeing Company's United Kingdom AWACS Radar System Improvement Programme (RSIP).

The RCDU adapts proven Commercial-Off-The-Shelf (COTS) hardware Honeywell Air Transport Systems originally developed and integrated on the Boeing 777, incorporating Honeywell's flat panel, active matrix liquid crystal display technology.

Status
In service in the UK E-3D AWACS aircraft.

Contractor
Honeywell Defense Avionics Systems.

Radio Magnetic Indicators (RMIs)

Type
Flight instrumentation.

Description
The Model 3100 is a dual-switched Radio Magnetic Indicator (RMI) in a 3 in (76.2 mm) form factor. The VOR pointers will accept AC sine/cos bearing information and the ADF pointer will accept ARINC synchro information. The slaved heading card accepts ARINC synchro data bearing information for selected ADF or VOR stations. The unit is TSO'd and is applicable to any fixed-wing or helicopter application.

The Model 3337 is a dual-switched Radio Magnetic Indicator (RMI) in a 3 in (76.2 mm) form factor. The pointers in VOR position will accept either AC sine/cos or ARINC synchro information. In the ADF position, the pointers will accept either DC sine/cos or ARINC synchro

The Model 3100 RMI is fitted in the Jetstream 61 and de Havilland Dash 8   0504204

signals. The slaved heading card will accept ARINC synchro signals.

Specifications
Weight: 1.13 kg
Power supply: 28 V DC
   (Model 3100) 0.55 A (max)
   (Model 3337) 0.75 A (max)
Accuracy:
   (card) ±1°
   (pointers) ± 2°
Temperature range:
   (Model 3100) –30 to +55°C
   (Model 3337) –30 to +70°C
Altitude:
   (Model 3100) –1,000 to 20,000 ft
   (Model 3337) –1,000 to 40,000 ft
Environmental: DO-138

Status
The Model 3100 is fitted in the Jetstream 61, de Havilland Dash 8 and other aircraft.

The Model 3337 is fitted on the ATR-42 and ATR-72, Saab 340, UK Royal Air Force Tucano, Jetstream 41 and corporate and commuter aircraft, as well as helicopters.

Contractor
Northrop Grumman Corporation, Component Technologies, Poly-Scientific.

RD-800 series Horizontal Situation Indicators (HSI)

Type
Flight Instrumentation.

Description
The RD-800 Horizontal Situation Indicator (HSI) features three digitally driven servoed displays in conjunction with two four-digit gas tube displays showing time and distance to waypoints. Microprocessor control gives improved versatility in navigational data processing.

The Honeywell RD-850 horizontal situation indicator   0503953

Specifications
Dimensions: 5 ATI
Weight: 4 kg
Power supply: 115 V AC, 400 Hz or 26 V AC, 400 Hz

RD-800J horizontal situation indicator
In the RD-800J HSI the readout of true airspeed is provided by conventional counter displays and for ease of interpretation the command bars are colour identified.

Specifications
Dimensions: 5 ATI
Weight: 4 kg
Power supply: 115 V AC, 400 Hz or 26 V AC, 400 Hz

RD-850 horizontal situation indicator
The RD-850 HSI features the most up-to-date applications of instrument technology including microprocessor control. Coloured display elements are included together with distance to go and groundspeed counters. Automatic direction-finder annunciators are fitted in the lower instrument area.

Specifications
Dimensions: 5 ATI
Weight: 4.7 kg
Power supply: 115 V AC, 400 Hz or 26 V AC, 400 Hz

Contractor
Honeywell Inc, Commercial Electronic Systems.

Reconnaissance Management Systems (RMSs)

Type
Mission Management System (MMS).

Description
Reconnaissance Management Systems (RMSs) are Smiths products which link aircraft avionics to unique sensor interfaces to control, monitor

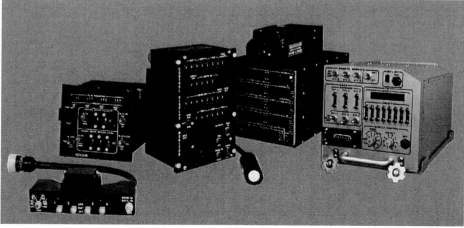

Smiths' Reconnaissance Management Systems (RMSs)   0051647

and annotate multisensor film and/or EO reconnaissance sensor suites.

RMSs provide a single interface between aircraft and sensor suite. Evolving from early Airborne Data Annotation Systems (ADASs), today's Control Processor and Annotation (CPA) products include mux bus interfaces and micro processor-controlled, low-power, high-density electronics which control and report sensor operation while retaining the film annotation function. Image identification and management are greatly enhanced with machine readable, MIL-STD-782 code matrix or alphanumeric data blocks. Datalink output of this reconnaissance information is also available for correlation with images on the ground by image interpreters.

Recording Head Assemblies (RHAs) provide the means to annotate film with all mission parameters at the instant a sensor collects image data.

With the transition from film sensors to Electro-Optical (EO) sensors (analogue and digital) for usable images in near-realtime, the airborne video processing function has been added to the basic control and annotation requirements. This addition converts analogue or digital imagery data from EO sensors and incorporates annotation data into the reformatted video.

Smiths' Code Matrix Reader (CMR) is used to read the annotated films' MIL-STD-782 code block for efficient management of film results and automated search for archival retrieval of data. CMR includes a film gate through which the film is passed at a high rate. The code block data is then transmitted to an electronics unit for processing and I/O control by a host computer.

RMS combines four functions:

- Aircraft interface: The aircraft interface provides primary systems and displays interface via an existing standard digital bus: MIL-STD-1553, ARINC standards, or through analogue or discrete interface as required
- Sensor control interface: Interfaces to sensors are implemented with the latest analogue and discrete techniques providing operator confidence and reliability, that is, synchronising, level shifting, impedance matching, on-off-ready, special interest, angular velocity stabilisation, BIT initiate, preflight testing. Special temperature sensing data, control of doors, or Environmental Control System (ECS) have been incorporated
- Annotation: High- or low-speed (for LOROP) data annotation provides the capability to annotate data in cameras, infra-red sets and radar systems via CRT, LED or other special devices during flight. Aircraft parametric data can also be output in a serial digital form for flight recording or datalinking so that all information is synchronised with images on the ground with the same time and position data. EO video is annotated by recording data along with video. This is done in an auxiliary audio track on videotape or per new JSIPS format
- Video processing: Charge Coupled Device (CCD) output data of EO sensors and/or IRLS output data are converted for compatibility with airborne monitors, videotape recorders and/or datalink formats in Video Processor Units (VPUs). Further image processing functions: video compression, enhancements, are optional. These video functions are usually provided in a unit separate from the control unit to isolate high-frequency video signals in the RMS.

Status
RECCE aircraft RMS systems
RF-4 AN/ASQ-90
OV-1D AN/AYA-10
F-14 TARPS AN/ASQ-172 or AN/ASQ-197
CP-140 RDAS
RF-5E PSCS, A/A24Q-1 (V)
F/A-18D (RC) AN/ASQ-197 (X)
RF-111C AN/ASQ-197A
AV-8B AN/ASQ-197 (X)

Contractor
Smiths Aerospace.

RGC350 Radar Graphics Computer (RGC)

Type
Avionic display processor.

Description
The L-3 RGC350 Radar Graphics Computer (RGC) provides simultaneous overlays of traffic, terrain and weather avoidance data on existing radar indicators. Interfacing with the many current radar systems (including Allied Signal RDS-81, -82, -84, -86, Collins' WXR-220, -270, -300 and Honeywell Primus 200, 300, 400, 660 and 800), the RGC350 enables a combination of L-3 SkyWatch®, SkyWatchHP® or TCAS I Traffic Avoidance information and L-3 LandMark TAWS information, with L-3 Stormscope® lightning detection data from a WX1000E (429 EFIS) or WX500 Stormscope® information to be displayed all on the same indicator. The RGC350 also allows information generated by another graphics computer, such as navigation data, to be displayed.

Key features of the system include:

- Overlay of traffic, terrain and lightning on existing weather radar displays
- Multiple display modes with easy mode selection
- Small 'all-in-one panel' or console-mounted unit
- RS-232 serial interface
- Fault tolerant pass-through circuitry for other graphics computers
- Connections for remote yoke-mount switches.

Specifications
Dimensions: 37.6 (H) × 145.5 (W) × 200.9 (D) mm
Weight: 0.68 kg

Cooling: Internal fan
Input voltage: 18–32 V DC, 12 W
Operating temperature: –20 to +55°C
Altitude: Up to 55,000 ft
Compliance: FAA TSO-C105, RTCA DO-160D, DO-178B Level D

Status
In production and in service.

Contractor
L-3 Communications, Avionics Systems.

Rotorcraft Pilot's Associate (RPA)

Type
Avionic display system.

Description
The RPA is a display system concept, developed by a team led by Boeing, under contract to the US Army's Aviation Applied Technology Directorate (AATD). The objective is to use artificial intelligence to help pilots exploit the full potential of onboard and offboard sensors and other sources of data input to optimise mission effectiveness and flexibility.

The RPA programme aims to establish revolutionary improvements in combat helicopter effectiveness. This will be achieved through the application of knowledge-based systems for cognitive decision-aiding and the integration of advanced pilot facilities, acquisition, armament and fire control, communications, controls and displays, navigation, survivability and flight-control equipment.

The rotorcraft pilot's associate concept demonstrator configuration   0048663

RPA builds on advanced avionics technologies being developed by the US Army for the updated AH-64D Longbow Apache and the RAH-66 Comanche. These systems enhance the automation of such functions as flight control, information processing and weapons management. In addition to voice recognition, the RPA features an advanced helicopter pilotage system and advanced data fusion. RPA technology has applications in the civilian marketplace and offers an opportunity to aid human performance in a variety of fields.

RPA moves into the cognitive realm of data interpretation, hypothesis formulation, planning and decision making. The result is an intelligent associate that will assist the pilot in understanding the vast array of battlefield information, planning the mission and managing the complex systems in modern military aircraft. Development and evaluation will occur in three stages. Initial design and assessment will be accomplished in a rapid prototyping laboratory environment. As the design matures, it will move into full mission simulation, where more rigorous and in-depth evaluations will be conducted. The third stage will install RPA in an AH-64D Longbow Apache attack helicopter, incorporating the Boeing Company's advanced digital flight control system as well as the US Army's advanced helicopter pilotage sensor system.

The Boeing Company will be responsible for system integration, architecture design, prototype development, full mission simulation, offensive systems, vehicle management and flight test. Lockheed Martin Systems Integration will provide computing technology, including the Massively Parallel Processor, and has responsibility for data distribution, mission planning, defensive systems management and external situation awareness.

Status
The RPA concept is being developed by a team led by Boeing (Mesa) and including Associate Systems Inc (Atlanta), Honeywell (Minneapolis and Teterboro), Kaiser Electronics (San Jose), Lockheed Martin Advanced Technology Laboratories (Camden), Lockheed Martin Systems Integration (Owego) and Raytheon Electronic Systems (Dallas). First flight of a specially adapted AH-64D Longbow Apache occurred in October 1998. Testing is continuing. The current configuration is understood to present data to a pilot helmet-mounted display for use in the heads-up/eyes-out-of-the-cockpit role, and to a second crew member both on helmet-mounted display and three 10 × 8 inch head-down displays for mission and weapon system management.

Data sources used in the current testing are understood to include onboard navigation, radar and electro-optic surveillance sensors, as well as aircraft survivability sensors. Offboard data sources include the Tactical Receiver Intelligence eXchange System (TRIXS), Joint Surveillance and Target Attack Radar System (JSTARS), Joint Tactical Information Distribution System (JTIDS), Battlefield Combat Identification System (BCIS) and tactical command centre data, integrated via the Improved Data Modem (IDM) and JTIDS.

RPA presents the pilot with a number of planning systems optimised for attack planning and sensor planning. The pilot can accept, reject, modify or override RPA suggestions.

Concept planning includes the use of RPA in other attack helicopters, in Unmanned Combat Air Vehicles (UCAVs) and possibly in fixed-wing aircraft in the Joint Strike Fighter (JSF) timeframe.

Contractor
The Boeing Company, leading a consortium.

Ruggedised Optical Disk System (RODS)

Type
Flight/mission recording system.

Description
The RODS is a digital data storage and retrieval device utilising a fully erasable and rewritable 5.25 in diameter optical disk as the storage medium. RODS is designed for use in special mission aircraft and for other military applications requiring highly reliable data storage and transfer in harsh operating environments.

Typical areas of application include digital map terrain data storage, electronic document storage, signal and mission data recording, maintenance and structural data recording and general purpose data loading and data transfer.

Data capacity is 300 Mbytes user data per side and sustained transfer rate is 518 kbytes/s. Access is 100 ms maximum.

Specifications
Dimensions: 127 × 190.5 × 317.5 mm
Weight: 6.8 kg
Power supply: 115 V AC, 400 Hz or 28 V DC
Reliability: 5,000 h MTBF

Contractor
Honeywell Aerospace, Electronic & Avionics Lighting.

SAMSON system

Type
Multimission system.

Description
SAMSON (Special Avionics Mission Strap-On-Now) is a quick turn around kit facilitating reconfiguration of C-130 aircraft for special missions such as search and rescue, photo reconnaissance, drug interdiction, electronic surveillance/countermeasures, air sampling and radio relay. The kit comprises an equipment pod which replaces the C-130E/H external fuel tank, a cargo pallet-mounted operators' console, interface wiring, and any necessary additional external components. The kit may be installed or removed within one work shift by a crew of four to six persons and is accomplished with no permanent modification to the aircraft.

A variant of the SAMSON pod, designed for the Open Skies Treaty, carries three Recon-Optical KS-87B framing film cameras, one Recon-Optical KS-116A panoramic film camera and two video cameras. Space has been reserved for future additions of a synthetic aperture radar and an infra-red linescanner.

Status
The system prototype has undergone numerous field tests by US and foreign C-130 users. The first Open Skies Pod variant was delivered to the Belgian Air Force in October 1996. It is currently being used by the Pod Group consisting of Belgium, Canada, France, Greece, Italy, Luxembourg, the Netherlands, Norway, Portugal and Spain.

Contractor
Lockheed Martin Aeronautics Company.

Sentinel® Data Transfer Systems (DTS)

Type
Flight/mission recording system.

Description
The Raymond Engineering Sentinel® Data Transfer Systems (DTS) comprise a family of small, lightweight, solid-state, data storage systems that have been specifically designed to provide reliable data storage in severe environment military applications.

Sentinel® Model 9410 SCSI DTS
The Sentinel Model 9410 SCSI DTS is a growth version of the Model 9510 SCSI DTS and features an industry standard SCSI-2 system interface with differential termination and dual slots which can accommodate either two type III or any combination of Sentinel PCMCIA (Personal Computer Memory Card International Association) Flash memory cards. The SCSI DTS allows for rapid insertion and removal of the Sentinel cards and features a rugged retention and ejection mechanism with a sealed field access cover.

The Sentinel SCSI DTS includes a 28 V DC power supply, dual PCMIA controller, control microprocessor and SCSI-2 interface electronics.

The system operates as a peripheral device or LRU under control of a host computer. The Sentinel SCSI DTS is designed to accept two Sentinel PC Cards which incorporate the latest in Flash memory technology.

Sentinel cards utilise the PC-Card ATA standard command set and are compatible with PCMCIA equipped computers and laptops for programming and downloading purposes.

Specifications
Data capacity: system accepts one or two PC cards.
Interface: SCSI-2 differential.
System performance:
Data transfer rate
to/from card 6.0 Mbytes/s burst
to/from system 1.1 Mbytes/s burst

Lockheed Martin Special Avionics Mission Strap-On-Now (SAMSON) pod   0505259

Sentinel Model 9410 SCSI Data Transfer System (DTS) 0051291

Average access time 5.5 ms
Power: 28 V DC, <10 W
Dimensions: 163 (L) × 125 (W) × 63.5 (H) mm
Weight: 2.32 kg
MTBF: 40,000 h

Sentinel® Model 9415 SCSI DTS
The Sentinel Model 9415 is also a growth version of the Sentinel Model 9510 SCSI DTS, designed specifically for aircraft panel-mount installations.

Sentinel® Model 9422 DTS
The Sentinel Model 9422 is designed for use with an RS-422 system interface DTS, and specifically for aircraft panel-mount installations. It can accept either Type II or Type III Sentinel PCMCIA Flash memory cards, and contains a 28 V DC power supply, PCMCIA controller, control microprocessor and RS-422 interface electronics that includes automatic baud rate selection.

Sentinel® Model 9450 DTS
The Sentinel Model 9450 DTS is designed for use with MIL-STD-1553A/B system interfaces, and aircraft panel-,mount installations. The system accepts two separately addressable Type II Sentinel PCMCIA Flash memory cards, and contains memory control electronics, data buffer, MIL-STD-1553 interface and power supply.

Sentinel® RS-232/-422 DTS
The Sentinel RS-232/-422 has been designed for use in severe military environments. It contains a PCMCIA controller, microprocessor, RS-232/-422 interface electronics and automatic baud rate selection.

Status
The Sentinel Model 9410 SCSI DTS has been selected for the C-130J of the Royal Australian Air Force and also for the SH-2G helicopter upgrade.

Contractor
Kaman Aerospace Corporation.

Sentinel® instrument system

Type
Avionic display system.

Description
The Sentinel® instrument system is designed to meet the need for fully integrated aircraft

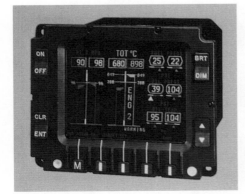

Sentinel® instrument system is installed in the Agusta A 109 helicopter 0504207

systems providing a complete solution from signal interface to cockpit display. The system consists of multiple cockpit Electronic Display Units (EDUs) and a dual-redundant Data Acquisition Unit (DAU), which consolidates large amounts of analogue, digital and discrete data to ARINC 429 format. All signals are acquired and transmitted to the displays in digital format by both halves of the DAU which can be mounted in any convenient location in the aircraft.

Typical system applications include primary and secondary engine instruments and caution/advisory panels. Auxiliary functions such as fuel and electrical power management can be incorporated on additional selectable screens or can be displayed on a separate display.

The 117.8 × 83.8 mm (4.4 × 3.3 in) active matrix liquid crystal display provides crisp full-colour graphics in a presentation which can be custom-designed for any application. Graphics and fonts can be created and changed without impact on hardware. Individual symbols and fonts can be selected from a large variety of existing styles or can be tailored to meet specific needs.

Bezel-mounted soft keys provide the pilot with access to lower-level functions and allow input of information such as barometric pressure or target set points. Menu options direct the pilot to the desired information. Solid-state design provides high reliability to meet demanding requirements. The Sentinel® can display up to 256 colours simultaneously, selectable from a palette of 4,096 colours. Colours can be easily selected and matched to a standard chromatic co-ordinate. Easily replaceable fluorescent lamps are used to provide a uniformly lit 125 ft-lambert average white brightness, suitable for viewing in direct sunlight.

Specifications
Weight: 2.49 kg
Power supply: 28 V DC or 115 V AC, 25 W without heater
Temperature range: –40 to +70°C
Display: 111.8 × 83.8 mm, 960 × 234 pixels

Status
Installed on the Agusta A 109 helicopter, and the Canadair Global Express and MD-10 (FedEx DC-10 conversion) aircraft.

Contractor
Ametek Aerospace Products.

Series 900 avionics system

Type
Avionics system.

Description
Series 900 avionics include HF, VHF, ADF, VOR, MMR, DME, LRA, TPR, weather radar, TCAS and satellite communications systems. They meet industry and FAA requirements for equipment certified to Criticality Level 2. This requires

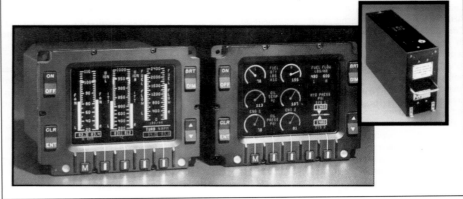

The Sentinel ® electronic display units and a data acquisition unit 0081843

tolerance to aircraft environments including 200 ms extended power interrupts, immunity to high-intensity radiated fields and conformity to the stringent environmental requirements of DO-160C. Conformity to the strict software documentation standards of DO-178C and the higher VHF FM interference levels required by ICAO Annex 10 are also included.

Series 900 is packaged in ARINC standard LRUs. Only the high-power systems require forced-air cooling, due to improvements in heat dissipation techniques. All other systems can be passively cooled.

All software is in Ada. External software loading capability has been enhanced and expanded to allow easy shop or on-aircraft modification.

Series 900 units are the next generation to the Series 700 products. This allows fleet commonality and use of the same top-level test equipment.

Status
In production and in service for Airbus A300-600, A310/319/320/321/330/340 and Boeing 737NG, 747-400, 757, 767, 777 aircraft.

Contractor
Rockwell Collins.

SH-2G(A) Super Seasprite Integrated Tactical Avionics System (ITAS)

Type
Avionic display system.

Description
Working with Northrop Grumman, Kaman has created an advanced glass cockpit which contains a highly automated Integrated Tactical Avionics System (ITAS), enabling a crew of two to fly the aircraft and manage its multimission equipment suite. The ITAS is a low-risk avionics system designed specifically to meet the requirements of the Royal Australian Navy.

The Kaman-Northrop ITAS, driven by two mission data processors, integrates the input of radar, thermal imager and electronic protection measures for manageable cockpit presentations. It enables the Super Seasprite crew to attack targets with a variety of anti-ship missiles. Electronic Flight Instrumentation System (EFIS), engine and transmission data, tactical plots, and sensor imagery are posted on any of the SH-2G(A)'s four-colour multifunction displays.

The new glass cockpit retains the under-glare shield caution/advisory panels introduced originally on the SH-2G, but eliminates all electromechanical instruments except for back-up airspeed and altitude gauges and a standby compass.

Status
In service in SH-2G(A) helicopters to the Royal Australian Navy.

Contractor
Kaman Aerospace Corporation.
Northrop Grumman Corporation, Navigation Systems.

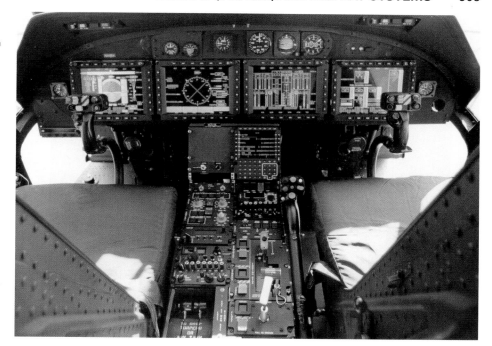

The SH-2G Super Seasprite advanced glass cockpit developed by Kaman Aerospace and Northrop Grumman
(Kaman)
0534152

Skymap II™ and Tracker II™

Type
Avionic Digital Map System (DMS).

Description
The Skymap II™/Tracker II™ system is a monochrome variant of the Skymap/Tracker product line, which can be knee, yoke, swivel, gimbal, panel or rack mounted. The monochrome reflective supertwist liquid crystal display providing sunlight readability and wide-angle viewing.

Skymap II™ provides GPS moving map precision navigation worldwide. Tracker II™ is the repeater equivalent of Skymap II™, and has been engineered to interface with existing GPS and Loran receivers.

Skymap II™ and Tracker II™ include interchangeable data modules that contain regional Jeppesen data, Skyforce geographic data, and the entire operating software. This architecture permits Skyforce to offer periodic operating system enhancements within the normal Jeppesen updating process. A worldwide navigational capability is available in three cassettes to cover the Americas, Atlantic International and Pacific International. Operator facilities are similar to those listed for Colour Skymap/Tracker.

Specifications
Screen: 127 mm diagonal high-contrast reflective supertwist back-lit LCD (128 × 240 pixels)

Dimensions: 158 × 115 × 35 mm
Weight: 0.65 kg
Temperature: –20 to +70°C
Power: 10 to 33 V DC, 3 W max; 6 AA cells dry/rechargeable
Inputs: RTCM-104 format (with differential option)
Outputs: RS-232, NMEA 0183, annunciator
User database:
(waypoints) 500 user-defined waypoints and 25 user-configurable airfields that are non-IACO listed
(routes) 99 user-defined routes with up to 99 turning points

GPS Receiver (Skymap II™ only)
Receiver: 8 channel parallel, simultaneous tracking
Acquisition: 12 s (almanac, position, time and ephemeris known); 43 s (almanac, position and time known)
Reacquire: 1.5 s
Accuracy: 15 m (without S/A); 1 to 5 m with differential option

Skymap II™ Version 2.00+
The following additional features were added in summer 1997: Aeronautical data – Jeppesen Plus+; cartographic data version 2.00+; operating system version 2.00 Plus+ to offer a choice of three languages: German, French and English selectable from the set-up menu.

Status
Amongst many military and civil installations, Skymap II™ is fitted to the Hawk aircraft of the Royal Air Force Red Arrows Display Team. At the beginning of 1997, new features added included: HSI, E6B calculator and flight logging capabilities, and map detail was improved. In summer 1997, Skymap II™ was enhanced to the Skymap II™ Version 2.00+ standard.

Contractor
Bendix/King.

Skymap IIIC™ and Tracker IIIC™

Type
Avionic Digital Map System (DMS).

Description
Skymap IIIC™ and its Tracker IIIC™ companion are full-screen colour updates of Skymap IIC™ and Tracker IIC™, which again can be knee, yoke, swivel, gimbal, panel or rack mounted.

Skymap IIIC™ utilises an integral eight-channel parallel GPS receiver, while Tracker IIIC™ requires an input from an external GPS or Loran receiver.

Standard language options for the text now include English, French, German and Spanish.

Specifications
Screen: 127 mm (5 in) diagonal
Dimensions: 158 × 115 × 65 mm
Weight: 0.9 kg
GPS receiver (Skymap IIIC™ only):
channels: 8 parallel
acquisition: 12 s, almanac, position, time and ephemeris known
43 s, almanac, position and time known
Reacquire: 1.5 s
Accuracy: 15 m (without S/A); 1 to 5 m with differential option
Internal database: geographical; Jeppesen®; updates (1/3/6 or 12 months); English, French, German and Spanish
Waypoints: 500 user-defined waypoints and 25 user-configurable airfields for non-ICAO listed airfields
Routes: 99 reversible routes, with up to 99 turning points

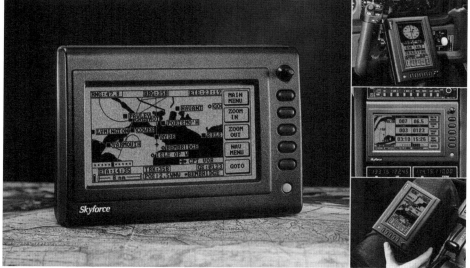

Skymap II™ and Tracker II™ display and optional mounting configurations
0018190

Skymap IIIC™ moving map colour display   0051684

Contractor
Bendix/King.

Smartdeck™ integrated flight displays and control system

Type
Avionic display system.

Description
L-3 Communications' SmartDeck™ integrated flight displays and control system is an integrated avionics suite, designed to enhance flight safety through the application of leading-edge technology, human factors engineering, and 'smart' systems integration. SmartDeck™ economically provides general aviation pilots with an electronic situational display with primary flight cues, together with moving map, weather, traffic and terrain information on 10 in diagonal flat panel displays. SmartDeck™ technology fuses data from all available aircraft sources to enhance pilots' situational awareness with 'Highway-In-The-Sky' and moving map presentations. SmartDeck™ will also monitor engine and aircraft systems for early detection of potentially hazardous situations.

Status
Available.

Contractor
L-3 Communications, Avionics Systems.

SparrowHawk™ Head-Up Display (HUD)

Type
Head-Up Display (HUD).

Description
CMC Electronics' SparrowHawk™ is a single- or dual-combiner, raster-capable Head-Up Display (HUD) designed for new fighter aircraft and for upgrading existing aircraft, easily customised to provide fleet commonality or to ease pilot training and/or transition across aircraft types.

SparrowHawk™ provides a stoke or stroke-on-raster display with an electronic boresight and 25° Field of View (FoV). An automatic control adjusts brightness in stroke raster mode and brightness and contrast in raster mode. An optional Up-Front Control Panel (UFCP) allows the pilot to select the system mode and to enter numerical data manually and includes a Light Emitting Diode (LED) character display presenting alphanumeric entries used by the HUD computer. An optional colour HUD camera and video recording system can be integrated with the system. Two different symbol generators can drive SparrowHawk™, facilitating system redundancy in the event of failure.

Air-to-air functions include LCOS guns and IR missile; air-to-ground functions include Continuously Computed Impact Position (CCIP), Air-Ground Gun (AGG) and rocket attacks, with both manual and automatic release.

The system has been designed to provide a fully customised instrument interface, with symbology and up-front controls. At a total system weight of 7.8 kg, SparrowHawk™ is one of the lightest weapon aiming systems available. It can provide aircraft with weapon aiming combined with a wide range of sensors, from platforms having only basic attitude heading reference systems and no radar, to those with full inertial reference and radar. A wide range of interfaces is available, from analogue and synchro to ARINC 429 to MIL-STD-1553B. SparrowHawk™ features extensive BIT, with most faults diagnosed on-aircraft. Internal data logging and continuous in-flight testing keeps track of system health.

Specifications
Display:
  Single or dual combiner
  Transmissivity 80% (dual-combiner)
  P-53 phosphor
  5 in aperture
  Stroke only or stroke-on-raster
  Symbology brightness adjustable 0–3,000 fl
  25° circular FoV
  135,000°/sec maximum draw rate
  <1% image distortion
  Accuracy 0.0 to 2.0 mr
  Contrast Ratio 1.3 to 1 against 10,000 fl ambient
Weight: 7.8 kg (17.25 lb)
Power supply: 28 V DC per MIL-STD-704D
Power consumption: 60 W
MTBF: 3,500 h
Environmental: MIL-STD-810E per MIL-5400 Class 1, MIL-STD-461D/462D

Status
In production and in service. Sparrow Hawk™ is fitted to a number of light attack/training aircraft, including Pilatus' PC-7 Mk II, PC-9 and PC-21 and Korean Aerospace Industries (KAI) KT-1C (in the PC-21 and KT-1C, it is combined with the FV-4000 mission computer), Aero L 139 Albatros and L 159; also specified for the IAI KFIR fighter

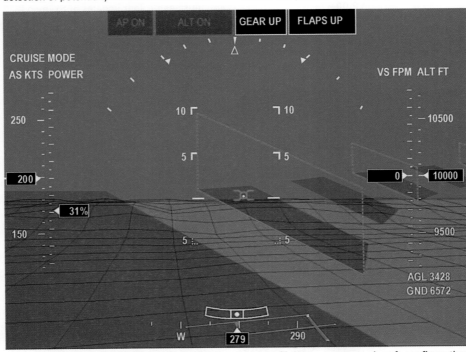

The Smartdeck™ pilots' primary flight display, showing flight parameters, aircraft configuration, flight guidance and terrain information   0103881

The SparrowHawk™ HUD fitted to US Navy F-14B aircraft   0101840

SparrowHawk ™ HUD (CMC Electronics) 1034684

and for the US Navy F-14A/B upgrade, where it is combined with the FV-3000 MMDP (see separate entry).

In its original configuration, the F-14A/B HUD symbology is projected directly onto the windscreen; the integration of the Sparrow Hawk™ HUD, while almost doubling the instantaneous FoV for the pilot, necessitated fitment of a new windscreen assembly in the F-14D.

Contractor
CMC Electronics Inc.

SPZ-7600 integrated SAR avionics

Type
Avionics system.

Description
The SPZ-7600 is a system designed to fill the civil search and rescue role for operations in all weather conditions worldwide. It includes dual FZ-706 flight control computers, helicopter optimised EDZ-705 EFIS displays with special SAR symbology, the Primus 700 Series surface mapping/weather/beacon radar and solid-state AA-300 radio altimeter systems with circuitry and displays specifically for rotary-wing applications.

The SPZ-7600 is built around the single pilot IFR technology of the SPZ-7000 (see item above), proven in service on Sikorsky S-76 helicopters in global conditions.

During critical operations, the SPZ-7600 can be programmed to execute an automatic hands-off approach to hover and auto-hover with velocity hold at an electronically pinpointed datum.

Status
In production. Certified for Bell 412 and Sikorsky S-76 helicopters.

Contractor
Honeywell Inc, Commercial Electronic Systems.

SPZ-8500 integrated avionics system

Type
Integrated avionics system.

Description
Honeywell's SPZ-8500 integrated avionics system for the Gulfstream V aircraft includes six 8 × 8 in EFIS/EICAS CRT displays with system synoptic pages; dual IC-800 integrated avionics computers providing dual, fail operational autopilot/flight directors with coupled go-around and Flight Path Angle (FPA) modes; FMZ-2000 Flight Management System (FMS) with worldwide navigation database, including airfield departure

Honeywell SPZ-8500 EFIS 8 × 8 in Primary Flight Display (PFD) 0099081

and approach procedures and coupled vertical guidance. Other system features include triple micro air data computers, triple Laseref III inertial reference systems and Primus 870 Doppler turbulence detecting weather radar.

Status
Certified on Gulfstream V in April 1997.

Contractor
Honeywell Inc, Commercial Aviation Systems.

ST-180 HSI slaved compass system

Type
Flight instrumentation.

Description
The ST-180 system combines a magnetically slaved gyroscopic compass with a VOR/Localiser and glide slope display. The ST-180

Honeywell SPZ-8500 integrated avionics system on Gulfstream V flight deck 0529624

was designed specifically for rotary-wing and high performance fixed-wing aircraft. This HSI slaved compass system offers a single convenient display which provides the pilot with all necessary information about the aircraft's position relative to ground-based navigational aids. Simultaneous indications are provided for selected course, course deviation and selected heading. Glide slope deviation is displayed when an active ILS frequency has been selected. The ST-180 HSI Slaved Compass System was developed and approved for helicopter operations.

The ST-180 consists of the Horizontal Situation Indicator (HSI), the remote electric gyro, the magnetic flux sensor and the model slaving panel. The gyro employs electromechanical erection thereby eliminating the performance and reliability degradation frequently experienced with air-erected gyros.

Critical circuits within the HSI are continuously monitored to minimise potential erroneous information. Should the flux sensor fail, the system may be switched to a free-gyro mode. Should a gyro failure be detected, the system may be transitioned into the 'Automatic Emergency Mode' (AEM). In AEM, the compass card is controlled by the Flux Sensor and behaves similarly to a normal wet compass. The pilot can continue to use the HSI.

Specifications
FAA TSO: C6e; C34e; C36e; C40c; C6d; C9c; C52a
Input power: 27.5 V DC (supplied by remote gyro)
Compass card accuracy: ±1°
Dimensions: ARINC specification 408 ATI-3
Weight: 1.4 kg

Status
In production.

Contractor
Meggitt Avionics/S-TEC.

Standard Helicopter Electronic Flight Instrument (EFI)

Type
Electronic Flight Instrumentation System (EFIS).

Description
Astronautics' 5 ATI Electronic Flight Instrument (EFI) is certified as a direct replacement for electromechanical HSIs and ADIs in helicopters and fixed-wing aircraft.

The EFI can function as an ADI or HSI with several functional overlays. Similarly to the other Astronautics EFIs (see separate entries), the Standard Helicopter EFI is designed around an AMLCD and is capable of receiving and processing ARINC-429/702 data from RNAV/FMS/GPS and TCAS/CDTI systems, analogue/ARINC-429 data from radios (VOR, DME, ADF) and ARINC-453/708 from Colour Weather Radar (CWR) and Enhanced Ground Proximity Warning Systems. The EFI also

Astronautics' standard helicopter EFI, as installed in the S-61 (Astronautics)   0568544

has RS-422 (cross-talk), analogue, synchro, discrete, and differential resolver interface capabilities.

The EFI has been certified to for use on civilian and military helicopters and fixed-wing aircraft. The unit is certified to function as an ADI, HSI, combined ADI/HSI, CWR, FMS/Map, EGPWS/TAWS (optional) display, TCAS display, CDTI/ADS-B display (optional) and storm scope (optional).

Other features of the system are as for the Helicopter EFI (see separate entry).

Specifications
Useable screen area: 4.24 × 4.24 in (107.7 × 107.7 mm)
Resolution: 524 × 524
Display luminance: 150 fl (200 fl option)
Viewing angle: ±60°
Processor: Pentium™
Interfaces: ARINC-429, -453, -561, -407 (synchro), RS-422, analogue, discrete, video (optional)
Dimensions: 5 ATI
Weight: <3.64 kg
Power: 28 V DC
Control panels: Multiple configuration options

Status
In production and in service. Applications include the A109, Super Puma and S-61 helicopters.

Contractor
Astronautics Corporation of America.

Standby engine indicator

Type
Cockpit display.

Description
The standby indicator provides a display of four engine parameters for each engine. The digital displays are part of an LCD, having white characters on a black background. These are easily readable in direct sunlight, as well as at dusk or night, when a separate lighting circuit provides a high-brightness, high-contrast presentation.

A special electronics design eliminates digit toggling during static conditions and provides fast response and sequential counting during dynamic conditions. The display update rate is variable, causing the display to simulate a mechanical counter and giving the flight crew a sense of the rate change of the parameter. When a parameter reaches a programmable limit, the respective display will flash to communicate the warning. If more than one parameter is over limit, they will flash synchronously to avoid confusion.

Loss of signal for any of the inputs is detectable and results in a display of three dashes. Loss of power results in a blank display. BITE permits mechanics to determine if a fault resides outside the indicator. Thus a high mean time between unscheduled removals is achieved.

Specifications
Dimensions: 82.8 × 103.1 × 188.7 mm
Weight: 1.59 kg
Power supply: 10–32 V DC, 2.8 W per channel

Status
In service on Cessna Citation X, Gulfstream GIV and Fokker 100 aircraft.

Contractor
Ametek Aerospace Products.

Standby Instruments for the Boeing 777

Type
Electronic flight instrument.

Description
Rockwell Collins is supplying flat-panel colour LCD standby indicators for the Boeing 777.

These 3 × 3 in instruments include the attitude indicator, airspeed indicator and altimeter. They are passively cooled.

Specifications
Dimensions: 76.2 × 76.2 × 215.9 mm
Reliability: >15,000 h MTBF

Status
Certified on Boeing 777. In production.

Contractor
Rockwell Collins.

Super SVCR-V301 high-resolution airborne video recorder

Type
Flight/mission recording system.

Description
The high-resolution Super SVCR-V301 recorder is lightweight and compact. It is designed to record video camera, infra-red sensor and multifunction displays in the stringent environment of fighter aircraft. It is designed and tested to meet MIL-STD-810C/D, including rain, sand and dust, and EMI-tested to MIL-STD-461C and -462. It is designed specifically to meet the stringent electrical, mechanical and environmental requirements encountered in modern flight test applications.

The V301 incorporates Super VHS format, rewind and playback, over 2 hours of recording, high-speed forward and reverse search, a visual event marker, comprehensive BIT, electronic frame indexing, serial and parallel interface, three audio channels and 525-, 875- and 1,023-line scan rates.

The V301's Super VHS format is not just an improvement to standard VHS. It is a distinct new format providing significantly higher picture clarity with a full 400 lines of horizontal resolution in both colour and black and white recording, providing significant improvement in line picture detail. The Super SVCR-V301 provides higher luminance signal frequency and wider frequency deviation and separates luminance and chrominance signals to minimise the degradation of image quality from cross colour and dot interference. The signal-to-noise ratio in the V301 has been significantly increased by broadening the frequency deviation from 1 to 1.6 MHz. Raising the carrier frequency also reduces interference with chrominance signal and substantially increases contrast range.

The V301 will record in both standard and Super VHS formats. This allows the use of existing VHS ground playback equipment until it is replaced with higher resolution Super VHS equipment. A host of full-function commercial ground playback equipment is currently available, all capable of playing cassettes recorded on the V301. Super VHS format tapes cannot be played back on standard VHS systems. However, they can be transferred or edited down to ¾ in Umatic or the standard VHS format.

Specifications
Dimensions: 111 × 212 × 290 mm
Weight: 7.2 kg
Power supply:
115 V AC, 400 Hz, 120 W (heater only)
22–30 V DC, 33 W

Status
Flight test aircraft on which the V301 has been installed include the F-15, E-2C, F/A-18, P-3 and RF-4C. Operational programmes include the Tornado GR. Mk 4, AH-1W Upgrade and F/A-18 export military sales. The V301 is under consideration for a number of other advanced operational programmes.

Contractor
Photo-Sonics Inc.

Synthetic Vision System

Type
Avionic Synthetic Vision System (SVS).

Description
Collins' Synthetic VISion (SVIS) system provides a three-dimensional 'virtual view' of terrain, intended to increase pilot Situational Awareness (SA) in both normal- and Low-Visibility Operations (LVO). Collins' Advanced Technology Centre, in partnership with the Langley Research Centre at NASA and Jeppesen Sanderson, is developing the terrain database that supports the SVIS system. When the database is combined with the GPS location of the aircraft, the pilot can clearly 'see' the flight environment from three different views: from the cockpit, from outside the aircraft (external view), and from above the aircraft (map view).

A major step forward for the system came in February 2000 with the launch of the Shuttle Radar Topography Mission (SRTM). Analysts at the National Imagery and Mapping Agency (NIMA) will use the SRTM data to generate 3-D topographic maps of the earth.

SVIS technology offers enhanced SA, real-time guidance, predictive alerting and improved flight planning capabilities; the system also shows promise for increasing the efficiency of surface operations and enhancing safety during approach and landing.

The SVIS system provides intuitive guidance cues to reduce pilot workload, and enhances the safety of military fixed- and rotary-wing missions by supporting all-weather, day/night operations into both prepared and austere landing zones.

SVIS is an integrated solution and will rely on traditional subsystems similar to Traffic alert and Collision Avoidance Systems (TCASs), Ground Proximity Warning Systems (GPWSs), Flight Management Systems (FMSs) and displays to provide warning functions.

Status
Initial flight testing of Collins' SVS began at Dallas-Fort Worth in 2000. The following year the testing was transferred to Eagle, Colorado, using NASA's B757. Terrain around Eagle normally results in high-minimums, although American Airlines, one of a number of operators utilising the airport, had received FAA approval to lower its minimums to 1,400 ft Above Ground Level (AGL). During the Eagle tests, NASA simulated

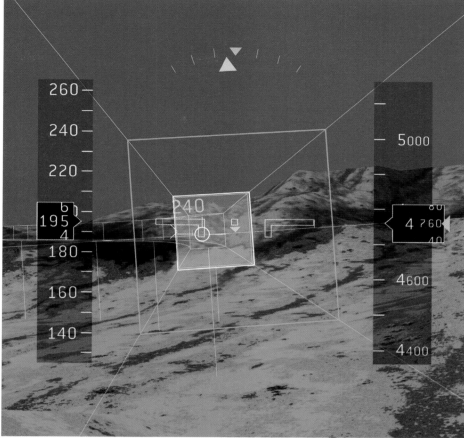

Collins' Synthetic Vision (SVIS) system (Rockwell Collins)   0528581

'engine out' departures and made approaches down to 200 ft AGL with SVS.

During early 2003, Rockwell Collins announced the inclusion of SVS onboard the Boeing Technology Demonstrator. The Next-Generation 737-900 was equipped with a suite of new and emerging flight deck technologies to assess the value for enhancing safety and operational efficiency across the Boeing fleet of airplanes. Rockwell Collins' systems included the Head-up Guidance System (HGS), Surface Guidance System (SGS), Synthetic Vision System (SVS) and a Multimode Receiver (MMR) with GPS Landing System capabilities (GLS).

During 2004, test pilots from NASA and Gulfstream flew a GV equipped with Collins' SVS and a combination of other sister technologies, including head-up displays, multi-scan weather radar, a voice-recognition system and cockpit displays including 3-D computer-generated views of the terrain, obstacles, runways and flight path.

Gulfstream has stated that SVS will be incorporated into their aircraft, as the technology matures.

Contractor
Rockwell Collins.

System-1 integrated avionics

Type
Integrated avionics system.

Description
System-1 is a new generation of integrated avionics; it includes Universal's UNS-1 Flight Management System (FMS); Terrain Awareness and Warning System (TAWS); Flat Panel Integrated Displays (FPIDs); Universal Cockpit Display (UCD); Data Transfer Unit (DTU); and UniVision cabin information system.

UNS-1 Flight Management System (FMS)
At the heart of the system is Universal's UNS-1 FMS (see separate entry). Worldwide, it provides a flight management capability from take-off to landing, with both lateral and vertical guidance, using position data from integral 12-channel GPS receivers that complement external navigation sensors. The UNS-1 database can include full procedures for Standard Instrument Departures (SIDs), Standard Terminal Arrival Routes (STARs) and three-dimensional approaches. Both lateral and pseudo glideslope outputs provide the pilot with ILS look-alike, IFR certified and autopilot coupled, guidance on the FMS Control Display Unit (CDU) for every approach.

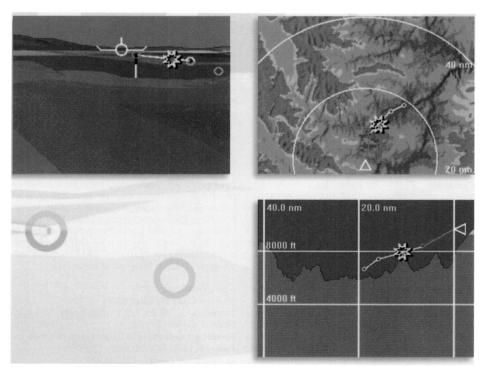

The Universal Avionics Systems' cockpit display and representative data formats top left: 3-D perspective view; top right: plan view; bottom right: profile view   0062213

Terrain Awareness and Warning System (TAWS)

Universal's TAWS system (see separate entry) provides terrain situational awareness relative to current and predicted aircraft position, together with an advanced Ground Proximity Warning System (GPWS). The system provides alert information to the flight crew both visually and aurally. Using a worldwide terrain map database, TAWS provides outputs for display of terrain in several views including profile and perspective views on video-capable displays, such as the FPIDs, FMS CDU and UCDs that form part of System-1. ARINC 453 outputs are also available for use with weather radar displays.

Flat Panel Integrated Displays (FPIDs)

Universal offers a range of colour FPIDs to present System-1 data. They provide anti-aliased graphics capabilities and ±60° horizontal viewing angle with vertical electronic steering for best performance. Standard ARINC sizes are available for Primary Flight Data (PFD) and Navigation Data (ND); MultiFunction Data (MFD) including selectable moving maps, weather radar, weather map, TAWS, Enhanced Ground Proximity Warning System (EGPWS) data, and Terrain alert and Collision Avoidance System (TCAS) data; engine performance and other data. Current models include EFI-340 (3ATI), EFI-450 (4ATI), EFI-500 (4 × 5ATI), EFI-550, EFI-600 (5ATI), EFI-640 (5 × 6ATI), MFD-640, EFI-710 (6ATI), PFD-840 and PFD-890. See separate entries for details of Universal's FPIDs and FPID control panels.

Universal Cockpit Display (UCD)

Universal's UCD comprises a supplemental, 10 in, touchscreen, display terminal for the cockpit, together with a remote-mounted computer unit. The UCD provides the pilot with access to electronic displays, including JeppView™ terminal area charts as well as procedural checklists. It also accommodates video inputs to provide full TAWS display capabilities. See separate entry.

Data Transfer Unit (DTU-100)

The DTU-100 utilises 100 Megabyte Zip^R disks. It is utilised for updating the FMS navigation database and the TAWS terrain database via a 10 MHz ethernet connection.

UniVision multimedia cabin information system

The UniVision system provides a communication and multimedia access and presentation capability.

Contractor

Universal Avionics Systems Corporation.

Tactical Aircraft Moving Map Capability (TAMMAC) Digital Map Computer (DMC)

Type

Avionic Digital Map System (DMS).

Description

Harris Corporation provides the Digital Map Computer (DMC) for the Tactical Aircraft Moving MAp Capability (TAMMAC – see separate entry).

The DMC is described as an affordable, high-performance system utilising readily available, Commercial Off-The-Shelf (COTS) components in a ruggedised, easily programmable unit.

The system provides the host platform with two independent map channels, 3.1 Gb of internal mass memory, full raster graphics and a full range of real-time terrain data displays. The DMC is optimised for Real-Time In the Cockpit (RTIC) display of mission critical data, especially terrain-referenced information, such as ground threats, targets, waypoints and no-fly zones. The real-time display of threat intervisibility facilitates extremely high levels of Situational Awareness (SA).

Status

In production for US Navy F/A-18E/F Super Hornet and other Navy and international aircraft.

Contractor

Harris Corporation.

Tactical Disk Recording System–LANTIRN (TDRS-L)

Type

Flight/mission recording system.

Description

The TDRS-L provides the capability to compress and record digital video from the LANTIRN FLIR simultaneously with other aircraft navigation and sensor data. It is configured for installation in the standard LANTIRN targeting pod. The unit consists of a chassis with liquid coolant cold plate, a Recorder Electronics Unit (REU), a Removable Memory System (RMS) and a Power Converter Unit (PCU). Interfaces to the TDRS-L include the FLIR video, one RS-232 port for control and status, a second RS-232 port for maintenance, an Ethernet port and an external SCSI port.

The digital video is compressed by a jpeg module through the selection of one to five predefined quantisation tables and formatted by the processor into an industry-standard graphics file format (AVI). All of the data is recorded on a removable disk cartridge, model RMS-1910 (> 90 minutes recording time, typical). The digital video files can be replayed on a PC or workstation with a standard CODEC and standard multimedia software. The aircraft data is recorded in separate files on the removable disk. This data is synchronised to the digital video by frame number.

Specifications

Dimensions: 255.3 × 262.9 × 152.4 mm
Weight: 11.35 kg
Formatted capacity: 9 Gbytes
Power: 115 V AC, 60–400 Hz
MTBF: 5,000 h

Contractor

General Dynamics Information Systems.

Tactical Disk Recording System– RAID (TDRS-R)

Type

Flight/mission recording system.

Description

The Tactical Disk Recording System-RAID (TDRS-R) provides a high-performance data storage system suitable for use in harsh environments. Minimum sustained transfer rates of over 90 Mbytes/second and burst transfers of 100 Mbytes/second are possible. The TDRS-R is housed in two ½ ATR-style chassis; the Storage Array Module (SAM) and the electronics unit. The SAM is mounted in a SAM Isolation Frame (SIF) that provides shock and vibration isolators for the disk array and can be easily inserted and removed without the use of tools. The electronics unit, which can be configured with various I/O interfaces, contains the RAID controller and an optional general processor. The standard I/O interface for TDRS-R is Fibre Channel Arbitrated Loop (FC-AL). The SAM currently has a total formatted capacity of 288 Gbytes when used in a RAID configuration, or 360 Gbytes raw capacity. SAM modules can be daisy-chained for greater capacities.

Specifications

Storage Array Module (SAM)
Dimensions: 238.8 × 132.1 × 208.3 mm
Weight: 6.4 kg
Formatted capacity: 288 Gbytes with redundancy
Power supply: +5 V DC and +12 V DC
Temperature range: –40°C to +71°C
Shock: 15 g 11 ms
Linear acceleration: ±9 g all axes

Altitude: 50,000 ft
Reliability: 26,000 hours

Electronics Unit
Dimensions: 355.6 × 124.5 × 198.8 mm
Weight: 5.9 kg without cards installed
Power supply: + 115 V AC single phase or +28 V DC
Temperature range: –40°C to +71°C
Shock: 15 g 11 ms
Linear acceleration: ±12 g all axes
Altitude: 50,000 ft
Reliability: 18,000 hours (includes electronics)

Contractor

General Dynamics Information Systems.

Tactical Optical Disk System (TODS)

Type

Flight/mission recording system.

Description

The Tactical Optical Disk System (TODS) is a severe environment digital data storage and retrieval device utilising a fully erasable and rewritable 5.25 in diameter optical disk as the storage medium. TODS is designed for unrestricted operation in high-performance military aircraft and other platforms requiring highly reliable data storage and transfer in very severe operating environments.

Areas of application include digital map terrain data storage, signal and mission data recording, maintenance and structural data recording and electronic document storage.

Data capacity is 300 Mbytes user data and sustained transfer rate is 518 kbytes/s. Access time is 120 ms maximum.

The tactical optical disk system   0503887

Specifications

Dimensions: 127 × 165.1 × 254 mm
Weight: 8.16 kg
Power supply: 115 V AC, 400 Hz or 28 V DC
Reliability: 5,000 h MTBF

Contractor

Honeywell Aerospace, Electronic & Avionics Lighting.

TAMMAC Tactical Aircraft Moving MAp Capability

Type

Avionic Digital Map System (DMS).

Description

The new US Navy Tactical Aircraft Moving MAp Capability (TAMMAC) is fitted with the Smiths Industries Aerospace Advanced Memory Unit (AMU) which will replace earlier US Navy data storage and mission data loader equipment.

Status

The TAMMAC system is applicable to US Navy F/A-18, AV-8B, AH-1W, UH-1N, CH-60, F-14, S-3, CH-53, SH-60 and P-3 aircraft.

Contractor

The Boeing Company, Information and Electronic Systems Division.

TCAS RA Vertical Speed Indicators

Type
Flight instrumentation.

Description
Smiths Aerospace has developed special vertical speed indicators for use with TCAS I and TCAS II Traffic alert and Collision Avoidance Systems. The units are form, fit and function interchangeable with most existing VSIs. Designed for ARINC 735 compatibility, the units may be used with TCAS II systems now entering service.

The 2074 Series instruments are outwardly similar to the Smiths Aerospace 2070 Series VSIs already in service with many airlines. The 2074 Series is intended for those applications where traffic advisories will be presented through some other medium such as a weather radar. Resolution advisories are presented through red and green LED 'eyebrows'. Green segments indicate those vertical speeds which will maintain safe separation from other aircraft, while red depicts potentially dangerous vertical speeds.

The 2074 flat-panel vertical speed/TCAS I instrument provides TCAS I data in addition to acting as a standard VSI. This instrument has successfully interfaced with both the Goodrich and AlliedSignal TCAS I systems. An ARINC 429 channel provides good readability with a full-colour display.

Status
The 2074 RA VSI is certified for use with AlliedSignal, Honeywell and Rockwell Collins TCAS systems. Over 800 units are currently in operation. The 2074 provides a dedicated RA display capability with TCAS Change 7.0 compliance.

Contractor
Smiths Aerospace.

Smiths Aerospace's Model 2074 RA VSI

0503960

TILS II Landing System receiver for the JAS 39 Gripen

Type
Aircraft landing system, Instrument Landing System (ILS).

Description
Telephonics is producing the Tactical Instrument Landing System receiver (TILS II) for the JAS 39 Gripen. The airborne unit weighs less than 5 kg and is approximately 120 × 280 × 125 mm in size. It provides outputs to a MIL-STD-1553B databus.

The system works in conjunction with the TILS ground equipment developed by Telephonics and currently deployed throughout Sweden and Finland. This receiver and ground system are currently in production.

A variation of the receiver with DME has also been supplied for the Space Shuttle programme.

A multimode version incorporating ILS, GPS and J-band MLS is in development. This version

The TILS II receiver for the JAS 39 Gripen

0504087

will provide users with the flexibility to land at airfields equipped with ILS as well as Tactical Instrument Landing System (TILS) equipped roadways. The GPS feature can be used for en route navigation. All three functions will be contained in the same form factor as the current TILS II version.

Status
In service on JAS 39 Gripen.

Contractor
Telephonics Corporation, Command Systems Division.

TRA 3000 and TRA 3500 radar altimeter

Type
Flight instrumentation, radar altimeter.

Description
The TRA 3000 radar altimeter provides precise height Above Ground Level (AGL) information from 2,500 to 40 ft. The transmitter/receiver and antenna fit in a single, lightweight aerodynamic unit that can be installed on the fuselage or under the wing. The TRA 3000 altimeter is used with the TRI 40 radar indicator on 28 V aircraft.

The TRA 3500 is designed for aircraft, helicopters and seaplane operations and provides accurate height AGL from 2,500 ft down to ground level. It has a patented 'anti-hover' capability that prevents altitude wandering during the hover. The TRA 3500 consists of a remote unit and dual-antennas for greater accuracy and ARINC analogue outputs allow integration with other on-board equipment such as flight director systems. The TRA 3500 is used with the TRI 40 indicator.

Specifications
Dimensions:
(indicator) 35 × 88 × 189 mm
(transmitter/receiver/antenna) 192 × 126 × 25 mm

Weight:
(indicator) 0.27 kg
(transmitter/receiver/antenna) 0.68 kg
Frequency:
(TRA 3000): 100 MHz sweep, within 4.2 to 4.4 GHz
(TRA 3500): 4.3 GHz centre frequency sweep, 4.25 to 4.35 GHz
Accuracy:
(TRA 3000): 40–100 ft ±5 ft, 100–500 ft ±5%, 500–2,500 ft ±7%
(TRA 3500): 0–100 ft ±5 ft, 100–500 ft ±5%, 500–2,500 ft ±7%

Status
In production and in service.

Contractor
FreeFlight Systems.

Universal Cockpit Display (UCD)

Type
Cockpit display.

Description
The Universal Cockpit Display (UCD) is an Electronic Flight Bag system that conforms to TSO C-113 and incorporates DO-178B Level C software. The UCD features a slimline 10 in touchscreen cockpit display terminal which provides the pilot access to electronic terminal area charts, checklists, Terrain Awareness and Warning System (TAWS) displays and other data.

JeppView™ charts and associated NOTAMs and airport information can be loaded onto the remote-mounted UCD Computer (UCDC). Departure and arrival airport information supplied from the Flight Management System (FMS) will prompt the display of the associated charts. Manual searches are accomplished through entry of airport name or ICAO identifier and the FMS aircraft position is depicted by a green donut overlay on the geo-referenced approach and airport charts. In addition to normal zoom functions, a touch zoom feature allows the pilot to view a selected area on a chart at a higher magnification.

Aircraft specific procedural checklists can be downloaded or pilot defined. Colours indicate the checked status of each item. Notes can also be incorporated. Checklists are presented in index-chapter format yielding quick access for both normal, abnormal and emergency situations.

The 780 × 1,024 pixel, high-resolution UCD Terminal (UCDT), also accommodates video inputs and supports display of two simultaneous views from the TAWS including selection of plan, profile and 3-D perspective displays.

The Universal Cockpit Display (UCD)

0062215

Growth potential includes access to maintenance manuals and other reference materials, UniLink and UniVision interfaces, and display for external video cameras.

Specifications
UCDC
Dimensions: 2 MCU
Weight: 3.4 kg

UCDT
Dimensions: 213.4 × 304.8 × 22.4 mm
Weight: 1.8 kg
Display: 6.1 × 8.1 in, 10 in diagonal, 780 × 1,024 pixels

Status
In service.

Contractor
Universal Avionics Systems Corporation.

UNS-1B Flight Management System (FMS)

Type
Flight Management System (FMS).

Description
The UNS-1B FMS comprises a Control/Display Unit (CDU), Navigation Computer Unit (NCU) and Data Transfer Unit (DTU) for uploading and downloading navigation database information. The CDU is available with a 127 mm colour display, 10 line-select keys, 10 function keys and a full alphanumeric keyboard.

The UNS-1B uses position data from long- and short-range navigation sensors to determine the best computed position. Automatic scanning DME/DME/DME positioning with slant range error correction, as well as en route Rho/Theta, is computed. Vertical navigation, three-dimensional approach mode – including GPS approaches – and holding pattern are also included. All 20 leg types can be flown, including heading to altitude, radial intercept, DME arcs and procedure turns – all manoeuvres required to fly complete SIDs, STARs and holding pattern procedure accurately. The UNS-1B FMS accepts fuel flow data from up to four engines.

Analogue and digital outputs are provided to flight directors, autopilots, EFIS, multifunction displays and radar navigation displays. Digital communications are provided using the ARINC 429/571/561/575 formats.

The UNS-1B has a 3.1 Mbyte database capacity, equivalent to 200,000 waypoints, and can include SIDs, STARs and approaches. It also includes airports, navaids, en route and terminal waypoints and airways in the database. The Jeppesen database is formatted on 3.5 in diskettes and is updated on a 28-day cycle. Pilot-defined data can include 200 routes, with up to 98 waypoints in each for a total of 3,000 waypoints, 100 arrivals/departures, 100 approaches and 100 runways. Built-in batteries prevent memory loss when the unit is removed from the aircraft.

Additional system features include a heading mode for direct control of aircraft heading through the CDU, bank angle commands correlated to altitude and turn anticipation to eliminate overshoots. Options include frequency management for centralised control of aircraft navigation and communication radios, AFIS interface capabilities for airborne ground link information and aircraft specific performance data functions.

The latest variant is UNS-1B plus. It utilises a colour flat-panel control/display. The 2MCU-sized navigation computer unit houses the new ASCB-interface printed circuit board. This board can also be installed in the UNS-1C and UNS-1D models as retrofits for the Challenger 601-3A and Falcon 900 aircraft.

The UNS-1B plus with GPS-1200 sensors is certified for GPS operation in en route, terminal and approach phases of flight (TSO C1156 and C129 Class B1/C1), and meets the requirements for primary means of navigation in oceanic/remote airspace.

UNS-1B with GPS-1000 12-channel receiver, RAIM and Fault Detection and Exclusion (FDE) has received certification by the Civil Aviation Authority of New Zealand for GPS primary means of navigation in remote/oceanic airspace.

Specifications
Dimensions:
(CDU) 169 × 146 × 200.2 mm
(NCU) 194 × 57 × 388 mm
(DTU) 53.8 × 146 × 205 mm
Weight:
(CDU) 3.45 kg
(NCU) 3.45 kg
(DTU) 1.47 kg
Power supply: 27.5 V DC, 60 W (max)
26 V AC, 400 Hz, 1 VA

Status
In production and in service.

Contractor
Universal Avionics Systems Corporation.

UNS-1E, UNS-1F, UNS-1L 'super' Flight Management Systems (FMS)

Type
Flight Management System (FMS).

Description
UNS-1E, UNS-1F and UNS-1L FMS supersede the UNS-1C, UNS-1D and UNS-1K respectively. The new systems are described as 'super' FMSs by Universal Avionics. Each system contains an integral 12-channel combined GPS/GLONASS receiver providing increased accuracy, integrity and availability for navigation en route and on non-precision approach worldwide. To allow for future software enhancements, each 'super' FMS incorporates a new 32-bit processor, which provides 25 times the computational performance of the current systems. The program memory has been increased fourfold, and the navigation database capacity has been increased to 32 Mbytes. The new FMSs also feature an Ethernet communication port for interfacing to Universal's new line of System-1 products.

The UNS-1E features an integrated 5 in display. The UNS-1F comprises a remotely mounted 2 MCU navigation computer, and separate 5 in control display unit. The UNS-1L includes a 2 MCU navigation computer unit and a separate 4 in control display unit. The 5 in displays are graphics- and video-capable and support weather and TAWS displays.

Status
In production and in service.

Contractor
Universal Avionics Systems Corporation.

UNS-1K Flight Management System (FMS)

Type
Flight Management System (FMS).

Description
The UNS-1K is positioned between the higher end UNS-1C and UNS-1D systems and the lower placed UNS-1M navigation management system produced by Universal Avionics Systems Corporation. The UNS-1K utilises the same software, SCN 602, as the UNS-1C and UNS-1D systems and features 10 line select keys and large 4 in colour flat-panel display. The control display unit is housed in a standard 4.5 in tall × 5.75 in wide Dzus-mounted unit 3.25 in deep. The navigation computer unit is 2 MCU in size, and includes an integral 12-channel GPS receiver. It also includes the Multi-Missions Management System (MMMS).

The system will meet emerging Required Navigation Performance/Actual Navigation Performance requirements around the world, including new European B-RNAV requirements.

System features include flight planning with full SID/STAR procedures, airways and approaches. A best computed position is based upon inputs from the integral GPS receiver, auto-scanning DME, and the operator's complement of external navigation sensors. The system will fly all ARINC 424 leg types. En route manoeuvre capabilities include a dedicated direct-to function, FMS heading commands, PVOR tracking, coupled VNAV with computed top of descent and vertical direct-to commands. Procedural holding patterns and approaches along with their transitions and missed approach procedures are contained in the database. The Approach Mode provides IFR-approved, pseudo-localiser, pseudo-glide slope guidance to any airport making all approaches look like an ILS. Fuel Management and optional Frequency Management functions are available

UNS-1K flight management system

0018224

and the MultiMission Management System (MMMS) is available on the UNS-1C/-1D models. A standby power-off mode retains flight plan and fuel initialisation data for up to 8 hours. A comprehensive test mode is incorporated to facilitate installation check out and return to service.

The integral 12-channel GPS receiver provides real-time and predictive Receiver Autonomous Integrity Monitoring (RAIM), automatic Fault Detection and Exclusion (FDE), step detection, and manual satellite deselection capabilities. The UNS-1K is certified for GPS operation en route, terminal and approach phases of flight and meets the GPS navigation operational approvals and Minimum Navigation Performance Specifications (MNPS) for navigation in the North Atlantic Track (NAT) airspace. The system can also be approved under FAA Notice 8110.60 for primary means of navigation in remote/oceanic airspace using GPS alone in conjunction with Universal Avionics' off-line PC-based Flight Planning and RAIM Fault Detection and Exclusion programme. The UNS-1K will also be compatible with Universal Avionics' GLS-1250 GPS landing system (currently in flight test) which will provide future growth by adding GPS precision approach capability.

The UNS-1K will also interface with Universal Avionics' new UniLink air-to-ground datalink. Adding UniLink provides the operator with such capabilities as predeparture clearances, flight plan up/downlinking, oceanic clearances, position reporting, digital ATIS, messages and text weather, through the UNS-1K control display unit. The UniLink UL-600 is housed in a 1 MCU size unit which can provide datalink information through several communications media including VHF, telephony and satcom systems. The UNS-1K with UniLink combine to provide the operator with full capability Communication/Navigation/Surveillance (CNS) avionics suite.

Status

Both the UNS-1K and UniLink are available. UNS-1K is certified TSO C129a Class A1/B1/C1 and C115b, with Supplemental Type Certification (STC) on Boeing 737-200. Selected for a number of regional aircraft types.

The UNS-1K was replaced by the UNS-1L so-called Super FMS from February 2000. (See separate entry).

Contractor

Universal Avionics Systems Corporation.

UNS-1M navigation management system

Type
Flight Management System (FMS).

Description
The all-in-one UNS-1M is a full navigation management system, self-contained in a control/display unit, which meets the standards of TSO C-129 A1/B1/C1. It features a flat-panel display, integral 12-channel GPS receiver with Receiver Autonomous Integrity Monitor (RAIM) and full alphanumeric keyboard. Smart auto-scanning DME/DME/DME provides continuous DME updating. Three external long-range sensor inputs accommodate combinations of Omega/VLF, Loran C, inertial and GPS sensors. The UNS-1M provides digital and analogue outputs for autopilot, mechanical HSI and EFIS systems. The unit's three-dimensional approach mode provides guidance on approach and is IFR certified for non-precision GPS, RNav, VOR and VOR/DME approaches. The fuel management system uses DC analogue sensors to provide real-time data on fuel weight, gross weight and landing weight. A worldwide Jeppesen navigation database is stored on, and updated via, a non-volatile flash memory card. The database includes airports, VORs, DMEs, VOR/DMEs, ILS/DMEs, VORTacs, Tacans, NDBs and en route and terminal waypoints.

Specifications
Dimensions: 114 × 146 × 241 mm
Weight: 2.81 kg
Power supply: 19–32 V DC, 35 W (max) at 27.5 V 26 V AC, 400 Hz

Status
In production. Selected by Executive Airlines for ATR-42 and ATR-72 aircraft.

Contractor
Universal Avionics Systems Corporation.

V-80AB-F, V-82AB-F and V-83AB-F Hi-8 mm Airborne Video Tape Recorders (AVTR)®

Type
Flight/mission recording system.

Description
The TEAC V-80AB-F, V-82AB-F and V-83AB-F Hi-8 mm AVTRs offer single-, dual- or triple-deck configurations respectively, that provide fully qualified, off-the-shelf, ruggedised capability, designed for HUD, FLIR, RWR, MFD recording, based on 15 years actual combat aircraft operational service.

Performance features include a 4 MHz bandwidth, 400 TV lines of resolution, full remote control and event mark generation, using standard commercial cassettes.

An optional plug-in board can be added to directly time stamp GPS, ZULU, or IRIG clocks onto the video tape. This option allows for exact synchronisation of video from multiple aircraft involved in tactical, test, or training missions. Flight crews can also select whether to write the time on the visible picture or use Vertical Interval Time Code (VITC) stamping that does not encroach on the picture space.

The V-83AB-F triple-deck requires no change to existing V-1000AB-R aircraft mounting trays or control panels.

The TEAC Integrated Debriefing System (TIDS) ground stations can synchronise and control up to four commercial video recorders when using Hi-8 mm tape cassettes, and each TIDS can be daisy-chained to create a larger system comprising up to 32 synchronised playback decks.

Specifications
V-80AB-F:
Recording decks: one
Tape format: Hi-8 mm and 8 mm (NTSC/PAL/S-Video)
Recording time: 120 min (NTSC); 90 min (PAL)
Horizontal resolution: 400 TV lines (nominal)
Control interface: manual, discrete, RS-422
Dimensions: 148 (W) × 120 (H) × 161 (D) mm
Weight: <3 kg
Power: 28 V DC, 15 W (45 W with heater)

V-83AB-F
Recording decks: three independent (two for V-82AB-F)
Tape format: Hi-8 mm and 8 mm (NTSC/PAL/S-Video)

The TEAC V-80AB-F Hi-8 mm ruggedised single-deck airborne video tape recorder   0051334

The TEAC V-83AB-F Hi-8 mm ruggedised triple-deck airborne video tape recorder   0051335

Recording time: 120 min (NTSC), or 360 min in series
Horizontal resolution: 400 TV lines (nominal)
Control interface: discrete, RS-422
Dimensions: 243 (W) × 151 (H) × 330 (D) mm
Weight: <8 kg
Power: 28 V DC, 45 W (135 W with heater)

Status
TEAC claims that its single-, dual-, and triple-deck units are flying on over 75 per cent of the Western world's military aircraft.

In November 1999, TEAC was awarded a contract to supply the V-83AB-F triple-deck AVTR® system to the Royal Australian Air Force for the F/A-18 Cockpit Video Recording System (CVRS) programme. The system's proprietary time code insertion capability will allow each tape to be coded during recording with a time code synchronised to the Global Positioning System (GPS) universal time reference, thus permitting the synchronisation of the recordings of several aircraft on the same mission. Deliveries began in 2000.

Contractor
TEAC America Inc.

Very large format AMLCD

Type
Electronic Flight Instrumentation System (EFIS).

Description
The Innovative Solutions & Support Inc (ISS) very large format (14.6 × 11.1 in viewing area), flat screen, Active Matrix Liquid Crystal Display (AMLCD) replicates the traditional 'T-shaped' airspeed, attitude, altitude and HSI data presentation, but without the cluttered vertical tapes of typical EFIS systems. Angle Of Attack (AOA) and Vertical Speed Indication (VSI) are also presented, and notable features are the size, brightness, resolution (1,280 × 1,024 pixels), contrast (200:1), off-boresight (±80°) viewing, and anti-aliasing qualities.

These capabilities have been obtained by bringing together 'enabling technologies' from the commercial electronics market, and utilising them in the context of aviation without the constraints imposed by use of conventional

The ISS 14.6 × 11.1 in viewing area AMLCD   0051697

single-purpose black-box displays. As well as use of the latest commercial glass capabilities, modern very high throughput (1.2 billion instructions per second) commercial processors are utilised in dual redundant, dual channel, configuration to provide the computing power required to support the display and symbology sets envisaged. The display is linked to a remote 3MCU electronics unit.

Additional data such as VOR, TCAS, engine data, flight path data, and even threat warning data could be added in the space available, to meet customer requirements.

Status

Launch customer for the ISS AMLCD was the Pilatus PC-12. However, ISS believes the real opportunity for this display is the military training-aircraft market, because the display can be redefined to display any combination of conventional formats, simply by software change, and thus to represent any operational aircraft required, resulting in much reduced operational conversion flying requirements.

Contractor

ISS Innovative Solutions & Support Inc.

VSC-80 video systems

Type

Flight/mission recording system.

Description

The VSC-80 series comprises the VSC-80A and the updated VSC-80B. The VSC-80B comprises the TEAC V-80AB-F Hi-8 mm 525-line airborne video recorder with the Merlin Engineering precision scan converter, to produce a 60 per cent improvement in image quality and resolution over the existing TEAC V-1000AB-F U-matic

The TEAC VSC-80B video system   0081863

Airborne Video Tape Recorder (AVTR) system. Both the VSC-80A and VSC-80B are direct form/fit replacements for the V-1000AB-F.

FLIR or day-TV images from video sensors can be captured at a rate of four per second and viewed immediately on the cockpit display. If desired, captured images can be selected for transmission, cropped, deleted, or saved for later transmission. Images received from ground stations, or other aircraft, can also be immediately viewed on the onboard display and retransmitted when required. Using image compression, transmission time is less than 15 seconds.

Specifications

Tape format: Hi-8 mm and 8 mm
Recording time: 120 min (NTSC) min
Function: record and playback; burst transmission
Dimensions: 271.8 (W) × 165.5 (H) × 369.5 (D) mm
Weight: <8.4 kg
Power: 115 V AC, 400 Hz, 75 V A (115 V A with heaters)

Status

Selected for US Army and Netherlands Army AH-64D Longbow Apache helicopters.

Contractor

TEAC America Inc.

WorldNav™ CNS/ATM avionics

Type

Avionics system.

Description

Honeywell has developed its WorldNav™ CNS/ATM avionics product line to meet the requirements of the emerging Communications, Navigation, Surveillance/Air Traffic Management (CNS/ATM) operating environment. Central to Honeywell's WorldNav™ concept is the Pegasus flight management system; other items of equipment included in the WorldNav™ concept were initially the TCAS 2000 Collision Avoidance System, the ADIRS Air Data Inertial Reference System.

Most recent addition to the WorldNav™ system is the new CM-950 Communications Management Unit (CMU) to provide new technology needed to implement the Aeronautical Telecommunications Network (ATN).

Status

The three initial elements of the WorldNav™ CNS/ATM system are in production, and this suite of WorldNav CNS/ATM avionics has been selected by a consortium of Latin American airlines to equip their fleet of A319 and A320 aircraft for future CNS/ATM operations.

More recently in mid-1999, Honeywell signed EVA Air as the launch customer for the CMU. The initial installation was for new Boeing 747-400 freighter aircraft. Follow-on installations included EVA Air's complete fleet of existing Boeing 747-400, 767 and MD-11 aircraft.

Contractor

Honeywell Inc, Commercial Electronic Systems.

AIRBORNE ELECTRO OPTIC (EO) SYSTEMS

Australia

Long-Range Tactical Surveillance (LRTS) sensor

Type
Airborne Electro-Optic (EO) surveillance system.

Description
BAE Systems Australia's LRTS thermal imaging sensor is part of a family of thermal imaging sensors for land, sea and air use. The LRTS is optimised for the maritime environment. Notable features include a high-definition 640 × 486 pixel staring array detector, with high-performance optics and two or three fields of view, providing for enhanced performance in humid conditions.

Automated functions such as scan, track and flexible digital interfaces permit integration with a range of tactical platforms and systems including fixed- and rotary-wing aircraft, military vehicles and maritime forces. Interface and display facilities are provided for cueing the thermal imaging system to contacts detected by other sensors.

Integration options include sharing of display and controls with other sensors and provision of a video cassette recorder for mission reconstruction.

Specifications
Detector: Platinum Silicide (PtSi): 640 × 486 pixels
Spectral response: 3–5 µm
Field of Regard:
  Azimuth: 360° continuous
  Elevation: +35 to –120°
Field of View:
  Narrow: 2.9 × 2.2°
  Mid: 11.3 × 8.65°
  Wide: 37.7 × 28.5°
Dimensions:
  Turret (d × h): 406 × 550 mm
  Control electronics: ½ ATR 350 × 124 × 194 mm
Weight:
  Turret: 42 kg
  Control electronics: 9 kg
Power: 28 V DC, MIL-STD-704
Control/Data: dual-redundant 1553B, remote terminal unit; RS-232 and RS-422; asynchronous serial 12-bit digital video
Interfaces: RS422/232
Video format (frame rate): CCIR (25 Hz) or RS-170 (30 Hz); programmable 640 × 486 composite monochrome or RGB graphic overlay

Status
In service with the Royal Australian Navy.

Contractor
BAE Systems Australia Ltd.

Canada

12DS/TS200 imaging turret

Type
Airborne Electro-Optic (EO) surveillance system.

Description
WESCAM's original Model 12DS was designed specifically to meet airborne law enforcement requirements. Weighing less than 23 kg with the optional Smartlink Interface Unit (SIU), the 12 in diameter dual-sensor camera features a high-resolution two Field of View (FoV), Indium Antimonide (InSb) staring array, Thermal Imager (TI) and a colour CCD Daylight TV camera with ×14 zoom lens. Active gyrostabilisation and vibration isolation enables the Model 12DS to operate with less than 35 micro-radians Line-of-Sight (LOS) jitter.

Developed from the 12DS and incorporating all of its core capabilities, the model 12DS200 is a longer range variant of basic system, designed for Unmanned Aerial Vehicle (UAV) rotary- and fixed-wing aircraft. The 12DS200 provides for long-range detection, recognition, identification and tracking of vehicles or personnel by day or night and in poor weather conditions. Most recently, with the inclusion of a third payload in the form of an IR illuminator, the system becomes the 12TS200 and facilitates operations at extremely low light levels, down to total darkness.

Specific features and benefits of the 12DS/TS200 include:
- High image quality
- 4-axis active gyro-stabilisation
- 6-axis passive gyro-stabilisation
- Compact and lightweight gimbal
- 3–5µm InSb FLIR TI and 3 FoV optics
- Daylight colour CCD camera with ×20 zoom lens
- Plug and Play SmartLink Interface Unit (SIU)
- Optional laser illuminator
- Seamless integration with WESCAM microwave downlink equipment
- High reliability and ease of use

SmartLink Interface Unit (SIU)
Smartlink is a single unit that provides all the interfaces and options needed to control a stabilised airborne camera and associated peripheral equipment. Through the SIU, the 12DS/TS200 system is compatible with related mission equipment such as microwave downlink equipment, searchlights, auto trackers, moving maps, GPS, radar and intercom systems.

Specifications
Gimbal
Two-axis inner (pitch/yaw) and two-axis outer (azimuth/elevation) active gyro stabilisation
Weight: 18.2 kg
Dimensions: 305 mm (diameter) × 370 mm (height)
Power supplies: 28 V DC
Azimuth/elevation slew rate: Max >90°/s
Azimuth range: Continuous 360°
Elevation range: +90 to –120°
Line of sight jitter: <35 microradians RMS

Thermal Imager (TI)
Mid-Wavelength Infra-Red (MWIR) InSb staring Focal Plane Array (FPA) (DS and TS)
Spectral range: 3–5 µm
Resolution: 256 × 256
Cooling: Stirling-cycle cooler
FoV:
25 (H) × 25° (V) at 17 mm
7.3° (H) × 7.3° (V) at 60 mm
2.2° (H) × 2.2° (V) at 200 mm (12DS200)

Colour Daylight CCD Camera
1-CCD Sony XC-999 in conjunction with ×20 continuous zoom lens
Format: NTSC or PAL
Resolution: 470 TV lines/460 TV lines (PAL)
Focal lengths: 16 to 160 mm
FoV:
23° (H) × 17° (V) at 16 mm
2.3° (H) × 1.7° (V) at 160 mm

Status
In production and in service in a wide variety of news, maritime surveillance and reconnaissance applications onboard fixed- and rotary-wing aircraft and UAVs.

Contractor
L-3 Communications WESCAM.

14TS/QS gyrostabilised camera system

Type
Airborne Electro-Optic (EO) surveillance system.

Description
Extensively deployed on UAVs, the Model 14TS can accommodate three sensors and the Model 14QS four sensors. The range of sensor payloads available for the Model 14 include: a 3 to 5 µm 6 FoV TI, which is available in Platinum Silicide (PtSi) or InSb detector formats; a colour daylight CCD TV camera (DLTV) with 955 mm long-range spotter lens; a colour daylight CCD TV camera with 10X zoom lens, and an 'eye-safe' Laser Range-Finder (LRF).

Specifications
Gimbal
Two-axis inner (pitch/yaw) and two-axis outer (azimuth/elevation) active gyrostabilisation
Weight: 34 kg
Dimensions: 360 mm (D) × 420 mm (H)
Power supplies: 28 V DC
Azimuth/elevation slew rate: Max >90°/s
Line-of-Sight (LOS) range: Continuous 360°
LOS tilt range: +90 to –120°
LOS tilt: TI/DLTV zoom: +10 to –120°
DLTV spotter/LRF: +30 to –120°
Line of sight jitter: <35 microradians RMS

WESCAM 12DS/TS200 imaging turret   1113764

For details of the latest updates to *Jane's Avionics* online and to discover the additional information available exclusively to online subscribers please visit

jav.janes.com

Model 14TS 0002205

Model 14TS mounted on a RQ-IK AV UAV
0089256

Thermal Imager (TI) – InSb
Mid-Wavelength Infra-Red (MWIR) InSb staring
  Focal Plane Array (FPA)
Spectral range: 3–5 μm
Resolution: 256 × 256
FoV: 1×, 2× (optical)
40.9 × 40.9° at 11 mm
20.1 × 20.1° at 22 mm
6.3 × 6.3° at 70 mm
3.1 × 3.1° at 140 mm
1.6 × 1.6° at 280 mm
0.8 × 0.8° at 560 mm

Thermal Imager (TI) – PtSi
Mid-Wavelength Infra-Red (MWIR) PtSi staring
  Focal Plane Array (FPA)
Spectral range: 3–5 μm
Resolution: 512 × 512
FoV: 1× 2× (optical)
40.9 × 31.3° at 19 mm
20.2 × 15.5° at 38 mm
10.9 × 8.4° at 70 mm
5.4 × 4.2° at 140 mm
2.7 × 2.1° at 280 mm
1.4 × 1.0° at 560 mm

Colour Daylight CCD Camera with Spotter
  1-CCD Sony XC-999
Format: NTSC or PAL
Resolution: 470 TV Lines/460 TV Lines (PAL)
Focal length: 955 mm
FoV: 0.38° (H) × 0.29° (V)

Colour Daylight CCD Camera with Zoom Lens
  1-CCD Sony XC-999 in conjunction with
  10× continuous zoom lens
Focal lengths: 16 to 160 mm
FoV: 23° (H) × 17° (V) at 16 mm
2.3° (H) × 1.7° (V) at 160 mm

Eye-safe LRF (14QS only)
Laser: Erbium glass, 1.54 μm wavelength
Pulse rate: 1 Hz
Range/accuracy: 49,995 m (claimed)/±5 m

Status
No longer in production. In widespread service
on fixed- and rotary-wing platforms, including
UAVs.

Contractor
L-3 Communications WESCAM.

16SS725/1000 gyrostabilised broadcast camera

Type
Airborne Electro-Optic (EO) surveillance system.

Description
The WESCAM™ 16SS725 and 16SS1000
gyrostabilised broadcast camera systems offer
a variety of camera and lens options to suit the
specific requirements of the broadcast news and
sports industries. Combined with the SmartLink
Interface Unit, SIU-800, any of the 16SS systems
can be integrated into a complete airborne package
with the addition of Wescam's line of Omnipod and
Skypod microwave transmission systems.

With a range of broadcast camera and zoom
lens options, the Wescam 16SS systems can be
custom configured to meet the preferences of
each application. An optional second camera and
wide Field of View (FoV) lens package may be
installed inside the turret along with the primary
camera to allow both wide angle and continuous
zoom imagery. Using the picture-in-picture
capabilities of the SIU-800, the operator can view
the wide angle for situation awareness while
zooming in on the subject for close-up shots.

The 16SS series systems are actively gyrostabilised
in five axes and vibration isolated in six axes for
jitter-free images under extreme flight conditions.

Specifications
Gimbal
Two-axis inner (pitch/yaw) and two-axis outer
  (azimuth/elevation) active gyrostabilisation.
Weight: 34 kg (16DS-A), 39 kg (16DS-W), 44.1 kg
  (16DS-M)
Dimensions: 400 (d) × 510 mm (h)
Power supplies: 28 V DC
Azimuth/elevation slew rate: 60°/s (typical)
Azimuth range: Continuous 360°
Elevation range: +90 to –120° (–180° 16DS-M)
Line of sight jitter: <35 microradians RMS

Thermal Imager (TI) – 16DS-A
AGEMA THV 1000 TI, SPRITE 5-bar focal plane
Spectral range: 8–12 μm
Cooling: Stirling cycle cooler
FoV: 20° (h) × 13° (v) (wide)
5.0° (h) × 3.3° (v) (narrow)

Thermal Imager (TI) – 16DS-W
Northrop Grumman Micro-FLIR TI, SPRITE 8-bar
  focal plane
Spectral range: 8–12 μm
Cooling: Closed cycle Stirling cooler
FoV: 10.0° (h) × 7.5° (v) (wide)
3.0° (h) × 2.3° (v) (narrow)

Thermal Imager (TI) – 16DS-M
Mid-Wavelength Infra-Red (MWIR) PtSi 640 × 480
  pixel Focal Plane Array (FPA) TI
Spectral range: 3–5 μm
Cooling: Stirling cycle cooler
FoV: 36.9° (h) × 28.6° (v) (wide)
11.15° (h) × 8.5° (v) (intermediate)
2.85° (h) × 2.2° (v) (narrow)
Features: Automatic gain control, polarity
  switching, electronic zoom (2×/4×), image freeze

Colour Daylight CCD Camera – 16DS-A
1-CCD Sony XC-999
Format: NTSC or PAL
Resolution: 470 TV Lines/460 TV Lines (PAL)

Colour Daylight CCD Camera – 16DS-M, 16DS-W
3-CCD Sony XC-003 in conjunction with 14 × 10.5
  Fujinon zoom lens
Format: NTSC or PAL
Resolution: 570 TV Lines (NTSC/PAL)
Lens FoVs: 3.0° (h) × 9.8° (v) at 21 mm
0.93° (h) × 0.7° (v) at 294 mm

WESCAM™ 16SS320/16SS750/16SS725
The Model 16SS320 single-sensor daylight
broadcast TV system delivers high-resolution
TV broadcast-quality images for electronic
newsgathering applications, surveillance and
observation, suitable for nosemounting on small
fixed and rotary wing aircraft. The sensor is a

daylight broadcast 8–320 mm colour camera with
a × 20 zoom lens with imaging to 4,000 ft (typically).
The Model 16SS750 is also a single-sensor
daylight broadcast TV system, suited to long
stand-off range applications. The sensor is a
three-CCD colour video camera coupled to a
10.5–750 mm lens. This model is available in two
configurations, the 16SS725 or 16SS750, with
×33 or ×36 magnification.

Status
No longer in production. In service on a variety
of fixed- and rotary-wing platforms, including
the Cessna 182, Bell 206, 407, AS350 and MD900
aircraft.

Contractor
L-3 Communications WESCAM.

MX-12 imaging turret

Type
Airborne Electro Optic (EO) navigation and
targeting system.

Description
The MX-12 is the latest compact imaging system
in WESCAM's MX-series of modular multispectral
turrets (which includes the MX-15, -15D and -20;
see separate entries), utilising the same design
approach in a more compact and lightweight
(<25 kg) airborne package. Drawing on the
advanced optics, high-resolution sensors and
advanced stabilisation of the larger members of
the MX-series, the MX-12 is claimed to deliver
range performance superior to competitive mid-
sized turrets, in a package which is approximately
half the weight of competitive systems.

The modular design, using common components
within the MX-series turrets and a common set of
interior avionics, minimises life cycle costs and
facilitates a high degree of common technology
insertion. All MX-series products are qualified to
military environmental and EMI/EMC requirements
(MIL-STD-461 and MIL-STD-810).

With its highly efficient packaging, the MX-12 can
be configured to meet mission-specific objectives
at a lower price point compared to larger systems.
The MX-12 features a high-magnification Thermal
Imager (TI), incorporating a three-step zoom, with
a choice of other sensors:
• Colour daylight camera with zoom lens
• Colour daylight spotter
• Laser range-finder
• Laser illuminator to provide for a highly
  customisable and application-tailored
  surveillance/targeting system.
• The turret comes with an integral Inertial
  Measurement Unit (IMU) that delivers
  maximum location accuracy, with geo-pointing,
  geo-location and geo-focus functions.

Status
In production.

Contractor
L-3 Communications WESCAM.

MX-15D advanced multispectral turret

Type
Airborne Electro Optic (EO) navigation and
targeting system.

Description
The MX-15D (Designation) is the result of
continued development of the MX-15 system (see
separate entry). Featuring key improvements in
the areas of range, resolution and magnification,
the MX-15D is suitable for fixed- and rotary-wing
applications, including Situational Awareness
(SA), Armed Reconnaissance (AR) and targeting.
The modular design of the MX-15D supports
from four to six high-performance sensors:
• Colour daylight camera with zoom lens
• Monochrome daylight camera with spotter lens

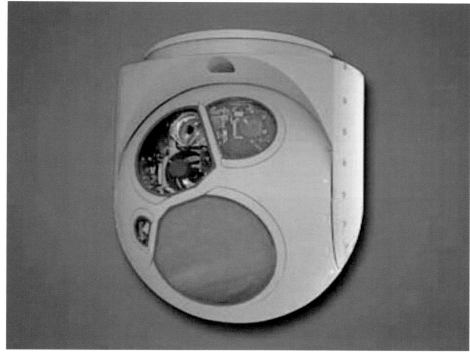

MX-15D multispectral imaging turret (L-3 Communications, WESCAM) 1113769

Wide FoV IR image taken from WESCAM's MX-15D (L-3 Communications, WESCAM) 1113773

Mid FoV IR image of the same scene taken from WESCAM's MX-15D (L-3 Communications, WESCAM) 1113770

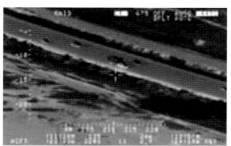

Narrow FoV IR image of the same scene taken from WESCAM's MX-15D (L-3 Communications, WESCAM) 1113771

Very Narrow FoV (Spotter) IR image of the same scene taken from WESCAM's MX-15D (L-3 Communications, WESCAM) 1113772

- IR with high magnification 4-step zoom
- High MTBF Laser Designator (LD)
- Eyesafe Laser Range Finder (LRF)
- Laser target illuminator

The MX-15D features tight boresighting specifications across the different sensors and between the Fields of View (FoV) for each sensor, facilitating seamless sensor transition during mission execution.

By focusing on the three factors that drive the maximum range of a sensor, namely magnification, resolution and stabilisation, WESCAM claims class-leading performance for mid-size Electro Optic (EO) turrets, facilitating longer-range target identification and thus greater stand-off. To achieve these goals, the MX-15D features custom-designed large aperture lenses arranged in a high-density configuration serving high-quality sensors.

In addition to high quality sensors and optics, the MX-15D incorporates enhancement algorithms to further refine output imagery. The system performs IR Local Area Processing (ILAP) which greatly improves image detail and scene contrast by increasing dynamic range, noise reduction and optimising pixel performance. Operator selectable optics filters plus excellent near IR sensitivity in the EO imagers ensures effective haze penetration.

Integrated into the MX-15D payload assembly is a compact, efficient and reliable diode-pumped LD. This technology provides for superior performance and increased MTBF relative to more traditional flashlamp technology.

In common with all MX-series turrets, the MX-15D features an integrated Inertial Measurement Unit (IMU) mounted directly to the optical bench. This high-performance unit incorporates compact, solid-state Fibre-Optic Gyro (FOG) for high sensitivity and reliability. Mounting the IMU on the inner gimbal with the payload allows the system to determine the absolute Line of Sight (LoS) of the systems cameras, without needing to collect relative information from other sensors, mounting alignment calibration data, or derive LoS from the host platform's Inertial Navigation System (INS). In concert with these IMU-derived accuracies, WESCAM's MX-GEO Features Package provides for a number of user-oriented enhancements to the system:
- GEO-Pointing allows hands-off operator tracking. The system locks all sensors on a precise latitude/longitude position and aims the turret at that fixed point, compensating for aircraft movement or obstructions impinging LoS.

- GEO-Steering eliminates over-controlling of the LoS by smoothing the motion of the turret and facilitating faster acquisition and lock-on.
- GEO-Focus ensures correct focus at all times, including alternating between sensors, and automatically refocuses onto the Point Of Interest (POI) subject if the LoS is broken by obstructions, such as broken cloud cover. The operator's input is reduced to occasional fine-tuning.

As a civilian system, the MX-15D Meets MIL-STD-810, 704 and 461 standards for military applications in a number of fixed- and rotary-wing platforms.

Status
In production.

Contractor
L-3 Communications WESCAM.

MX-15 multispectral turret

Type
Airborne Electro-Optic (EO) navigation and surveillance system.

Description
The MX-15 combines advanced gyro-stabilisation and high-quality long-range optics in a multispectral sensor suite. The system has been military-qualified as the AN/AAQ-35. The MX-15 embraces a variety of applications, depending on sensor configuration. Since the introduction of the system in 2001, the MX-15 has been the subject of significant and ongoing R&D:

Payload
Up to 6 optional sensors, including IR imaging, laser illuminator, Laser Range Finder/Designator (LRF/D)
Multiwaveband coverage – day/night CCD with laser illuminator

Performance
50% improvement in maximum IR detection range
5% increase in day spotter camera detection range (dual channel)
70% increase in stabilisation performance
Increased day CCD TV sensitivity
Advanced Local Area Processing (LAP), for cleaner images

Logistical
Integrated Inertial Measurement Unit (IMU)
¾ and ½ ATR MCU available
Increased Mean Time Between Failures (MTBF)
Expanded range of integration kits available

The modular nature of the MX-15 facilitates a very high level of customisation, with support

MX-15 multispectral turret (L-3 WESCAM) 1113776

This series of four images demonstrates the Fields of View (wide, mid, narrow and very narrow) of the MX-15 multispectral turret (L-3WESCAM)   1113777

This series of four images demonstrates the Fields of View (wide, mid, narrow and very narrow) of the MX-15 multispectral turret (L-3WESCAM)   1113774

This series of four images demonstrates the Fields of View (wide, mid, narrow and very narrow) of the MX-15 multispectral turret (L-3WESCAM)   1113775

This series of four images demonstrates the Fields of View (wide, mid, narrow and very narrow) of the MX-15 multispectral turret (L-3WESCAM)   1113768

A close-up of a MX-20 turret installed aboard a CASPER-configured USCG HC-130H aircraft (Jane's/Patrick Allen)   0543449

for any desired combination of up to six sensors, including:

- High magnification MWIR (3 to 5 μm) IR Thermal Imager (TI) with 4-step zoom
- Colour 1-CCD daylight camera with zoom lens
- Colour 3-CCD daylight camera with spotter lens
- Monochromatic charge multiplying CCD night camera with spotter lens
- Diode-pumped laser illuminator
- Eye-safe laser LRF

The MX-15 incorporates a number of features which enhance operational capability and ease of use:

Geo-pointing enables the turret to follow a target 'hands-free'. It locks all sensors on a precise latitude/longitude and aims the turret Line of Sight (LoS) while compensating for airframe movement, target obscuration and sensor drift.

Geo-steering smooths the motion of the turret and relieves the operator of continuous compensation.

Geo-focus ensures correct focus at all times, even when alternating between sensors. Operator input is reduced to occasional fine-tuning.

Autotracking registers a digital point of reference and follows that pattern. This operates on any of the sensors and tracks either a subject in the FoV or the entire video frame.

In addition, an optional moving map, embedded in the IMU, provides for enhanced operator Situational Awareness (SA).

Further enhancements to the MX-15, introduced from 2005, include:

Night Spotter
Augmenting the 3-CCD day spotter camera, a charge-multiplied CCD facilitates detection and identification of non IR-significant Points of Interest (POI) in low-light conditions.

Laser-illuminated Night Spotter
Under extreme low-light conditions, down to total darkness, charge-multiplying devices do not receive enough energy to form a useable image. Under such circumstances, illumination of the target area with energy of a compatible wavelength will produce reflected energy which can then be utilised by the sensor to create an image. Accordingly, the MX-15 incorporates an optional long-range laser illuminator to support the Night Spotter camera. The entire FoV of the Night Spotter is illuminated with energy in the 0.8 μm wavelength (invisible to the unaided eye), enabling long-range covert

identification of vehicle license plates, ship names etc under conditions of total darkness.

The entire MX Series is based upon a common architecture, including system interfaces, user interface and internal components, resulting in logistic and service benefits on fleet-wide installations, combined with enhanced ease of use for integrated applications that use a number of different airframes.

Status
In production and in service on a wide variety of fixed- and rotary-wing platforms. As of 2005, most recent MX-15 installations include Bell 412 helicopters of the New York Police Department (NYPD), Spanish Navy SH-3D Sea King helicopters. In addition, the system has been selected for the Swedish Coast Guard Aircraft Replacement Programme and the US Navy COP Programme.

Contractor
L-3 Communications WESCAM.

MX-20 multispectral turret

Type
Airborne Electro-Optic (EO) navigation and surveillance system.

Description
The MX-20 was developed specifically for long-range/stand-off surveillance and identification applications, combining highly accurate gyrostabilisation with multiple, high-magnification, day and night vision sensors. The MX-20 turret supports up to six high-performance sensors.

The MX-20 system comprises a 20 in sensor turret, operator and master control units, a joystick controller, a MIL-STD-704A power conditioning unit and interfaces for up to two image recorders and three display monitors.

MX-20 turret on a US Marine Corps AH-1Z attack helicopter   0554718

The MX-20 turret features a 5-axis gyrostabilisation and 6-axis vibration isolation system which achieves an extremely low line-of-sight jitter in both fixed- and rotary-wing aircraft. The Mid-Wave IR (MWIR) Thermal Imager (TI) incorporates both large (640 × 512) and small (320 × 240) formats, together with a 4-step zoom capability. Multiple focal lengths, combined with an elite 1.5× optical extender, and fast step switching between focal lengths enable operators to rapidly detect, acquire and identify Pints Of Interest (POI) and maintain tracking, all at long stand-off ranges. A high performance EO sensor compliments the optics, with two different camera/lens options for advanced target recognition/identification purposes:

Colour daylight camera/zoom lens
This option comprises a continuous zoom colour camera, referred to as EO Wide. This combination produces low to medium image magnification, to provide scene context, target detection and classification during surveillance operations.

Colour or monochromatic daylight camera/ spotter lens
This option comprises a four-position step-zoom 3 CCD colour or monochromatic camera with high magnification spotter lens, referred to as EO Narrow. This combination produces high to very-high magnification, to provide for accurate target recognition and identification. The colour camera is typically selected for cluttered urban environments, where colour provides essential extra information to aid target identification. The monochromatic camera is typically selected for long-range maritime

MX-20 multispectral turret (L-3 Communications WESCAM)   1113779

MX-20 Wide FoV (L-3 Communications WESCAM)
1113782

MX-20 Mid FoV (L-3 Communications WESCAM)
1113780

MX-20 Narrow FoV
(L-3 Communications WESCAM)
1113781

MX-20 Very Narrow FoV
(L-3 Communications WESCAM)
1113778

applications where colour images are less useful and every display pixel can be utilised to produce the highest possible resolution image.

In addition to the above capabilities, the MX-20 incorporates further optional features to enhance the night-time mission:

Night camera/spotter lens
Augmenting the 3 CCD day spotter, the charge-multiplied CCD night spotter enables the operator to discern scene detail which would defeat the day spotter due to insufficient scene illumination. This optional enhancement is particularly suited to covert surveillance operations at night. WESCAM claims the employment of a charge-multiplied system provides up to 4× the effective identification range of active illuminators, such as Active Laser Illuminator (ALI) systems.

Laser illuminated night spotter
During extremely dark nights where there is insufficient ambient light to drive even a charge-multiplied CCD effectively, some form of active illumination is required to facilitate surveillance operations. Accordingly, WESCAM has intergated a long-range laser illuminator with the night spotter camera. The illuminator covers the entire FoV of the night spotter with a been of 0.8 μm energy which is invisible to the unaided human eye.

The MX-20 also features a geo-location capability which generates accurate ground target geographical co-ordinates.

The MX-20 is manufactured to meet demanding environmental, quality, reliability and maintainability requirements, and has been qualified to military, aeronautical and commercial aviation standards.

The entire MX series of products is based upon a common architecture. The common system interfaces, user interface and internal components enable logistic and service benefits on fleet-wide installations, and ease of use for integrated applications that use a number of different airframes.

Specifications
Turret
Azimuth: 360° (continuous)
Elevation: −120 to +30°
Gimbal Active gyro-stabilisation: 5-axis
Vibration stabilisation: 6-axis passive isolation
Gimbal line of sight jitter: <5 μrad RMS
Azimuth/elevation slew rate: 60°/s
Dimensions: 175 × 356 × 152 mm (master control unit); 200 × 100 × 80 mm (joystick controller); 530 (Ø) × 670 mm (turret)

Weight: 0.7 kg (joystick controller); 3.6 kg (operator control unit); 22.7 kg (master control unit); 84.1 kg (turret)
Power: 500 W (nominal); 1,000 W (max)

Thermal Imager
Detector: MWIR (3–5 μm) InSb staring array
Resolution: 640 × 512 (large format) and 320 × 240 (small format)
Focal lengths: 40, 200, 1,000 mm, plus 1.5× optical extender (0.5 sec step time)

Colour daylight TV with zoom lens
Type: 3CCD
Resolution: 800 TV lines
Field of View: 30.3° to 1.64°

Monochrome TV with step-zoom spotter lens
Type: 1CCD
Resolution: 570 TV lines
Field of View (h): 0.61° to 0.11° in four stages

Colour TV with step-zoom spotter lens
Type: 3CCD
Resolution: 800 TV lines
Field of View: 0.47° to 0.09° in four stages

Status
The WESCAM MX-20 belongs to the MX family of modular, turreted, EO surveillance and targeting systems, the others being the MX-12 and the MX-15/15D (see separate entries). WESCAM adapted its MX-20 (designated AN/ASX-4) for the US Navy's P-3C Advanced Imaging Multi-spectral System (AIMS) in the mid-1990s. AIMS provides for long-range maritime surveillance and identification as part of the P-3C Orion's Anti-surface warfare Improvement Programme (AIP). It replaces the existing AN/AAS-36 Infra Red Detection Set (IRDS) and Electro Optical Sensor (EOS). In 1997, the company received a US Navy contract to provide two AN/ASX-4 turrets and, as of April 2004, 64 systems had been delivered. In June 1998, the US Coast Guard ordered the AN/ASX-4 AIMS as part of a sensor upgrade for its HC-130H. As of April 2004, 17 units had been delivered.

During the first quarter of 2003, 21 MX-20 systems were ordered for the Canadian CP-140 Aurora Incremental Modernisation Programme, with Lockheed Martin Canada providing the fuselage retractors and acting as prime contractor for the CUSD40 million programme.

Similarly, the MX-20 was also selected as part of a Royal New Zealand Air Force (RNZAF) maritime patrol aircraft upgrade programme for its P3K Orions.

More recently, the MX-20 has been adopted as the turret assembly for the AN/AAQ-30 Hawkeye target sight system (see separate entry) for the US Marine Corps' AH-1Z SuperCobra upgrade. It is also part of Galileo Avionica's Airborne Tactical Observation and Surveillance (ATOS) system on Italian Ministry of Finance ATR-42 MP Surveyor maritime patrol aircraft.

Contractor
L-3 Communications WESCAM.

MX-series modular Electro Optic turrets

Type
Airborne Electro Optic (EO) navigation and targeting system.

the MX-series of multispectral turrets includes the MX-12, MX-15 (AN/AAQ-35) and MX-20 (AN/ASX-4) (L-3 Communications, WESCAM)
1113783

Description

WESCAM claim that the MX-series is the industry's only modular family of turreted EO/ IR/laser surveillance and targeting systems. The family features four high performance multispectral turrets which incorporate maximum commonality of components and avionics. The products in this series are the MX-12, MX-15, MX-15D and MX-20. The modular approach of the MX-series conveys the following benefits:

- Improved usability through standard operator interfaces
- Common systems training
- Simplified product support
- Common test equipment
- Common maintenance training
- Reduced product and life cycle support costs
- Leveraged IRAD technology insertion across entire product family

All MX-series products are qualified to military environmental and EMI/EMC requirements.

Status

In production.

Contractor

L-3 Communications WESCAM.

Denmark

Modular Reconnaissance Pod (MRP)

Type

Airborne reconnaissance system.

Description

In 1994, the Royal Danish Air Force (RDAF) contracted TERMA to develop an all-new reconnaissance pod for the F-16 that would be sufficiently flexible to accommodate current and future sensor systems. The MRP I comprises three elements: a common pod structure, a sensor-particular element, and a platform-particular element.

The common pod structure includes the pod body, strongback, and electronic control system. The sensor-particular element is fitted with fixings for LRUs and sensors. The platform-particular element includes pylon attachments and electrical/mechanical interfaces.

The RDAF selected the TERMA EWMS to control the MRP on its F-16MLU (Mid-Life Update) aircraft.

The MRP I can be delivered 'empty', to be configured by the customer. Payloads which have been successfully integrated into the pod include the Recon/Optical CA-261, CA-260/25A, AN/ZSD-1(V), VIGIL and LAEO (Low-Altitude EO)/ MAEO (Medium-Altitude EO) sensors.

MRP II

As part of a Swedish Air Force (SwAF) reconnaissance pod project, TERMA is developing a new-design modular system known as MRP-II (the SwAF designation is SPK 39, Spaningskapsel 39). The preliminary design review for MRP-II was passed in August 2002. Proven concepts from the original MRP-I (such as the strongback design which allows for rapid configuration upgrades without the need for extensive flight testing) have been retained. However, the MRP-II has a new shape, a new environmental system, a new rotating window section and the capability to be jettisoned. The last is not possible with the MRP-I which is designed to make maximum use of the available space underneath the F-16 centreline station and which is therefore conformally shaped.

The MRP-II is also different in that it will be possible to mount equipment not only on the strongback, but also on the fairings.

SPK 39 reconnaissance pod on a JAS 39 Gripen   1041511

The sensor layout of the Modular Reconnaissance Pod I   0044998

For the SwAF configuration, the rotating camera position will be occupied by a Recon Optical CA 270 (see separate entry), initially only in the electro-optical variant but with the option to later upgrade to dual-band operation. Data will be captured with an L-3 Communications RM-8000 solid-state recorder.

It is understood that the second sensor will be the legacy wet-film SKA 24C (Spaningskamera 24C) camera. Both the RM-8000 recorder and the SKA 24 film cassette can be accessed via a large service door. The MRP-II will have radomes forward and aft to allow for later installation of a datalink system.

Specifications

MRP I
Dimensions: 4,496 × 762 × 610 mm
Weight:
(empty) 227 kg
(loaded) 544 kg

Flight envelope: –2 to +9 g
Data interfaces: tape recorders and datalink

MRP II (SPK 39)
Dimensions: 4,572 × 685 × 635 mm
Weight: (loaded) 680 kg

Status

More than 50 MRP I have been delivered to the following air forces: Royal Danish Air Force, Belgian Air Force, Royal Netherlands Air Force, and US Air Force Air National Guard.

TERMA has confirmed a contract for nine SPK 39/MRP-II pods for the Swedish Air Force; introduction to service is planned for JAS 39C/D Gripens of the Swedish Air Force Rapid reaction unit (SWAFRAP) from 2006.

Contractor

TERMA AS.

France

ATLIS II Laser Designator Pod (LDP)

Type

Airborne Electro-Optic (EO) targeting system.

Description

The Automatic Tracking Laser Illuminating System (ATLIS) II is a pod-mounted Laser Designation Pod (LDP) which features an automatic, dual-mode (visible and near IR) TV tracker, together with a laser stabilisation system intended to reduce pilot workload and facilitate tracking/target designation by a single-seat aircraft while manoeuvring at low level. As the system is pod mounted it may be installed on a wide variety of fast-jet aircraft.

The roll section, at the front of the pod, houses a steerable, stabilised mirror which provides the optical line of sight, while a pitch/yaw rate stabilised inertial platform provides rejection of high-frequency dynamic motions. The roll turret drive unit is used in conjunction with the pitch/yaw stabilisation system to provide line of sight steering. The dual-mode tracker provides both area correlation and contrast/ point tracking. The area correlator mode is used to stabilise the scene and provide designation for area or low-contrast targets.

The acquired image is reflected from the stabilised mirror into a fixed optical assembly which folds and focuses the image into the TV camera. The output from the camera provides two optical fields of view to the display in the cockpit

and to the automatic tracker, and a further six for target acquisition/identification. Laser energy is reflected to the dichroic portion of a combining glass located within the combined image path. This combining glass passes shorter wavelength image data (0.5–0.9 µm) but reflects the laser energy (1.06 µm) so that it leaves the combiner collinear with the scene data.

The system has a claimed pointing accuracy of 1 m on a target at an average firing range of 10 km. High image quality, coupled with the tactical benefits of a dual-mode tracker and a high degree of magnification (up to ×20), ensure optimum target acquisition and identification. The system also facilitates Battle Damage Assessment (BDA) with video recording and a reconnaissance mode.

An F-16 with an ATLIS II LDP mounted on the starboard chin station 0005520

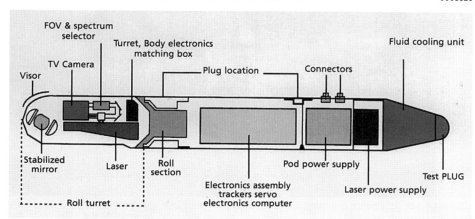

A schematic of the ATLIS II pod 0005518

ATLIS II LDP mounted on a French Aéronavale Dassault Super Etendard Modernisé (SEM) 0094064

Specifications
Dimensions: 2,520 (length) × 305 mm (diameter)
Weight: 170 kg
Power supply: 115 V AC, 400 Hz, 2.3 kW
Laser: 1.06 μm NdYAG
Laser spot accuracy: ±1 m at 10 km range
Field of Regard (FoR):
　Roll: unlimited
　Pitch: −160 to +15°
TV tracker wavebands: 0.5–0.7 μm (visible) and
　0.7–0.9 μm (near IR)
Carriage: standard 760 mm (30 in) NATO bomb rack

Status
In production and in service. The ATLIS II LDP equips French Air Force Jaguars, where it was used operationally during Operation Desert Storm, and has been ordered for export to equip Mirage 2000 and F-16 aircraft. Integration of the ATLIS II targeting pod and Aerospatiale AS-30L laser-guided missile forms the third part (Standard 3) of a four-part programme of work for the French Aéronavale Dassault Super Etendard Modernisé (SEM) that is due to be completed in 2004, for service until 2010.

The system is also reported as having been delivered to the air forces of Pakistan and Thailand.

Contractor
Thales Optronics.

BIRD'CAST airborne surveillance system

Type
Airborne Electro Optic (EO) surveillance system.

Description
BIRD'CAST is described as an airborne surveillance system designed for use in the co-ordination of ground operations from airborne platforms. The system enables the acquisition, tracking and recording of images, including transmission and processing to efficiently provide near-realtime data to commanders to facilitate timely and effective decision making in fluid tactical situations.

The system comprises a TV camera, an infra-red camera and a video console for onboard image control. Image transmission is performed through digital radio link down to fixed or mobile ground stations.

Acquired data are then transmitted to a central station that has all the required tools to co-ordinate ground operations. The operation centre receives in near-realtime the position of the observation aircraft, via digital mapping, in addition to recorded video.

BIRD'CAST is designed to meet the requirements of border surveillance, urban surveillance and policing operations, VIP protection, Search And Rescue (SAR) operations, environmental monitoring and the protection of industrial assets (surveillance of remote installations and pipelines).

Status
In production.

Contractor
SAGEM Défense Sécurité, Optronics and AirLand Systems Division.

Chlio-WS multisensor airborne FLIR

Type
Airborne Electro-Optic (EO) surveillance system.

Description
Thales has developed a version of the Chlio thermal imaging system, Chlio-WS, which incorporates a second-generation detector. Chlio-WS is intended for search- and rescue-operations and is installed on helicopters.

The system uses the Wescam Synergi thermal imaging modules. The detector is a 288 × 4 cadmium mercury telluride focal plane array, operating in the spectral band 8–12 μm.

The system has four fields of view and is mounted on a gyrostabilised turret which is stabilised in two axes electro-mechanically and two axes electro-optically.

Specifications
Thermal imager
Spectral range: 8–12 μm
Fields Of View (FOV) (H×V):
(wide) 24.0 × 18.0°
(intermediate) 12.0 × 9.0°
(narrow) 3.0 × 2.2°
(very narrow) 1.5 × 1.1°
FOV switch time: 0.5 s
Features: remote switching of FOV; ×2 electronic zoom; polarity switching; automatic and manual gain; edge enhancement; symbology; NVG compatible

Colour camera
Camera type: 3CCD Sony XC-003
Lens: 14 × 10.5 Fujinon zoom lens
FOV (H×V):
(wide) 13.0 × 9.8° 21 mm
(narrow) 0.93 × 0.70° 294 mm

Gimbal
Line Of Sight (LOS) pan range: 360° continuous
LOS tilt range: −180 to +90°
Slew rate: 60°/s typical

Weights
(gimbal) 40 kg
(IR electronics unit) 5.6 kg

The Chlio-WS multisensor airborne sensor system 0079245

(system control unit) 1.5 kg
(hand controller) 0.7 kg

Dimensions
(gimbal) 400 mm (diameter) × 510 mm (H)
(IR electronics unit) 245 × 181 × 160 mm

Status
Chlio-WS is in production for C160, Alizé and
French Navy Falcon 50 Surmar aircraft. It is
in service on French Air Force Super Puma
helicopters and with export customers. The
Wescam Synergi thermal imager is also used
in the latest variant of Thales' Convertible Laser
Designator Pod (CLDP).

Contractor
Thales Optronics.

CN2H Night Vision Goggles (NVGs)

Type
Aviator's Night Vision Imaging System (ANVIS).

Description
The CN2H NVGs are part of a range of night
vision systems produced for military applications;
the goggles are specifically designed for use in
helicopters and fixed-wing aircraft for night piloting
in tactical situations. They are fixed to the helmet,
with a power pack on the back and can be quickly
discarded in the event of an emergency. Focusing
and positional adjustments are available to suit the
wearer. The goggles are compatible with a wide
range of French, British and American helmets
and they incorporate second- or third-generation
image intensifiers according to requirements.

Specifications
Weight: 0.95 kg, including power module, battery
   and helmet adapter; 0.55 kg (goggle only)
Power supply: 28 V DC (off-board); 3.5 V PS 31
   battery; 3 V BA 5567U battery; option 2 × sets
   of 1.5 V AA switchable batteries
Battery life: approximately 20 h (PS 31 battery)
Field of view: 40°
Eye relief: 20 mm nominal
Eyepiece adjustment: −5 to +2 dioptres
Magnification: ×1
Focus capability: 40 cm to nominal infinity
System gain: II tube dependant
Resolution: 0.86 cy/mr (standard II tubes);
   1.00 cy/mr (high resolution option)
Compatibility: French Guenau 459 and Elno FPH-
   500, British Helmets Ltd Mark 4 and Alpha and
   US Gentex SPH-4 and -5

Status
No longer in production. In service with French
Army Aviation (ALAT), Air Force and Navy and
with export customers.

Contractor
SAGEM Défense Securité, Optronics and AirLand
   Division.

CN2H-AA Mk II Night Vision Goggles (NVGs)

Type
Aviator's Night Vision Imaging System (ANVIS).

Description
The CN2H-AA Mk II NVG is the latest development
of the CN2H (see separate entry), incorporating
new, high-resolution optics. They are designed
for rotary- and fixed-wing aircraft of all types,
including high-performance fighter aircraft, where
the goggle can be rapidly and easily discarded with
either hand before ejection. The major difference
between the CN2H-AA and the CN2H is that the
power supply is relocated, since the battery pack
arrangement on the rear of the helmet, as in the
CN2H, is unsuitable for ejection seat applications.
The CN2H-AA can be equipped with European Gen

CN2H-AA Mk II goggles mounted on a FPH-500 helmet. Note the integrated visor assembly facilitates simple mounting of the goggle and allows the pilot to employ the visor with the NVG – birds do fly at night! (Patrick Allen)   0568559

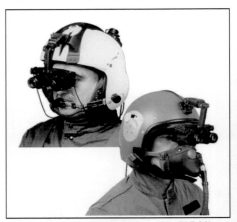

CN2H-AA NVGs are in service with FAF Mirage 2000, Mirage F1 and French Navy Super Etendard pilots   0568557

II Image Intensifier Tubes (IIT), including SuperGen,
HyperGen or XH-72, XD-4/XR-5 or US Gen III tubes
(a resolution of >72 lp/mm is recommended).
A standard Minus Blue (645 nm) filter is available,
or other specification per customer requirement.
Autogated power supplies are available,
depending on IIT specification. The goggle
can be mounted on a wide variety of aircrew
helmets, including the French OS 458/459 and
FPH-500/600, Gallet LH250 and LA100, the British
Alpha and the American HGU-55, SPH5 and
Russian ZSh-5.

Specifications
Weight:
   Goggle: 590 g
   Power pack: 150 g
   Helmet mount: 100 g
Power supply: 28 V DC, PS31 or 2 × 1.5 AA
   battery
Duration: 20 hours (PS31)
Field of view: 40°
Focus: 40 cm to nominal infinity
Eye relief: Nominal 20 mm
Interpupillary adjustment: 52 to 72 mm
Eyepiece adjustment: −5 to +2 dioptres
Magnification: ×1
System gain: >3,000
Resolution: >1.33 cy/mr

Status
In production and in service with the French
Air Force (FAF) and Navy. The CN2H-AA Mk II
equips Mirage 2000 units of the FAF. Available for

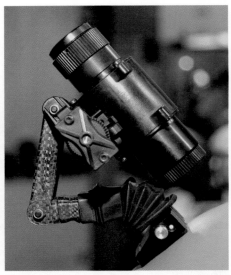

CN2H-AA Mk II goggles in stowed position (Patrick Allen)   0568558

export subject to clearance of French Ministry of
Defence (MoD).

Contractor
SAGEM Défense Securité, Optronics and AirLand
Systems Division.

Convertible Laser Designator Pod (CLDP)

Type
Airborne Electro-Optic (EO) targeting system.

Description
The Convertible Laser Designator Pod (CLDP) has
been developed from Thales Optronics' ATLIS
(Automatic Tracking Laser Illuminating System)
pod (see separate entry).

In the CLDP, the laser designator may be
supported by either a TV (CLDP/TV) (TeleVision), or
a Thermal Camera (CLDP/CT) ('Camera Thermal' –
note that in the accompanying schematic,
'Camera Thermal' has been anglicised to 'Thermal
Camera' (TC)). The pod features a common body,
with laser transceiver, electronic assembly and
environmental control system, and two separate
nose sections, which can be changed in 2 hours.

A Mirage 2000D with CLDP/CT on the right shoulder pylon   0010610

Mirage 2000Ds. The upper aircraft is carrying an ATLIS II LDP, while the lower aircraft is carrying a CLDP/CT pod on the right shoulder pylon   0569625

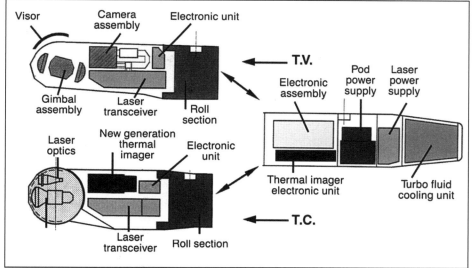

Schematic of the CLDP/TV and TC pods   0005521

The TV head features a camera, gimbaled mirror and a roll-stabilisation device. It also has a Field of View (FoV) selector and a visible or near infra-red spectrum selector. This has four magnifications and corresponding FoVs.

The Thermal Imaging (TI) head features a gimbaled optical head with a laser and thermal imager optics and roll-stabilisation device. There are four FoVs: 12/6° for navigation and 4/2° for target acquisition and tracking.

The latest version of the system, CLDP/CT-S, based on a Wescam Synergi TI using a 288 × 4 Cadmium Mercury Telluride (CMT) Focal Plane Array (FPA), is in production for French forces' CLDP installations.

Specifications
Physical
Weight: 290 kg
Length: 2.85 m

TV
Spectral band: 0.5–0.9 µm
Magnification: ×2.5, ×5, ×10, ×20
Fields of view: 0.75, 1.5, 3 and 6°

Thermal Camera
   Spectral band: 8–12 µm

Laser
   Wavelength: 1.06 µm

Status
In production and in service. Flight tests of the CLDP/TV began in mid-1986. Flight tests of the CLDP/CT began in January 1988 using a Jaguar aircraft.

In September 1988 at the Landes flight test centre, a Jaguar successfully launched an AS 30L missile at a speed of 470 kt from an altitude of 213 m at a range of 8 km. The CLDP was reported to have acquired the target at a range of 13 km.

In service with the French Air Force, fitted to the Mirage 2000D where the system saw service during air operations over Iraq in 2003, and with the Italian Air Force on their Tornado IDS aircraft. CLDP was also selected by Abu Dhabi for their Mirage 2000 and Saudi Arabia for their Tornado IDS. In addition, the Italian Air Force signed a USD23 million contract in mid-2001 to equip at least 36 AMX aircraft with the system as part of a wider upgrade programme, due to be completed by 2006.

Elsewhere, CLDP is being offered as part of the Sukhoi Su-25M5 and MiG-29 SMT upgrade programmes.

CLDP is now co-produced by Thales Optronics and Galileo Avionica.

Contractor
Thales Optronics.
Galileo Avionica SpA.

Cyclope 2000 infra-red linescan sensor

Type
Airborne reconnaissance system.

Description
The Cyclope 2000 infra-red linescan sensor is designed for airborne applications including: reconnaissance, and battlefield observation and surveillance (on aircraft, helicopters or UAVs); as a navigation aid; the monitoring and surveillance of sensitive areas; forest fire detection, pollution detection (oil on the sea's surface) and mine detection. It provides day and night infra-red images of the 8 to 12 µm spectral bandwidth to a thermal sensitivity of 0.1°C, with an angular resolution of 1 mrad. The field of view can be adapted from 60 to 120°.

The basic modular configuration can be adapted to various aircraft. Improved versions are offered with a different spectral bandwidth, multispectral detection or stereoscopy.

Specifications
Dimensions: 170 × 170 × 200 mm
Weight: <6 kg
Power supply: 28 V DC, 70 W

Status
In production.

Contractor
SAGEM Défense Securité, Optronics and AirLand Division.

Damocles multifunction Laser Designator Pod (LDP)

Type
Airborne Electro Optic (EO) targeting system.

Description
Damocles is a multimode, multifunction Laser Designator Pod (LDP) that incorporates a staring Focal Plane Array (FPA) third-generation Thermal Imager (TI), operating in the 3 to 5 µm waveband. The pod specification also includes a Laser Spot Tracker (LST), Laser Range Finder/Target Designator (LRF/TD) and an optional CCD TV camera. A Navigation FLIR (NAVFLIR), also operating in the MWIR band and providing a 24 × 18° Field of View (FoV), can be installed in the pod or integrated into the host pylon. NAVFLIR imagery is projected on to the Head-Up Display (HUD), with targeting information being shown

Dassault Rafale with Damocles pod mounted on the right shoulder (Patrick Allen)   1024653

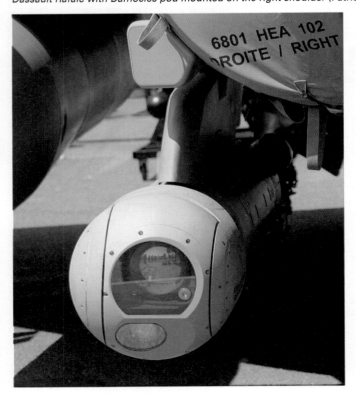

Close-up of Damocles pod mounted on a Dassault Rafale multirole fighter (Patrick Allen)
0554711

on one of the multifunction cockpit displays, allowing the pilot to view both simultaneously.

Damocles is designed primarily for laser designation of ground targets; Thales claims that the LRF/TD is sufficiently powerful for use at a slant range of 45 km from an altitude of 45,000 ft. The pod can also be used for supplementary navigation, air-to-air identification and reconnaissance.

Specifications
Thermal imager
Spectral band: 3–5 μm
Fields of view:
  1.0 × 0.75°
  4.0 × 3.0°
  24 × 18° (navigation field)
Electronic magnification: ×2

Laser spot tracker/Laser range-finder
Wavelength: 1.06–1.54 μm (eye-safe)
Compliance: STANAG 3733
Dimensions: (l × Ø) 2,500 mm × 370 mm
Weight: 265 kg

Status
In production and in service. Damocles was developed for the Mirage 2000 and Rafale multirole fighters. The system entered flight test during 2000. Damocles is marketed as a stand-alone targeting pod, or in combination with the NAVFLIR system (see separate entry), which facilitates integration of EO navigation and targeting systems into a single weapons station and/or dedicated pylon. The combined system was ordered by the United Arab Emirates (UAE) to equip approximately 63 Mirage 2000-9 fighters, with first delivery during the latter part of 2005. In UAE configuration the combined system is known as SHEHAB (Damocles) and NAHAR (NAVFLIR). Another customer for the combined system is reported to be the Greek Air Force, having specified the system as part of its Mirage 2000-5 Mk2 order.

The French Navy selected Damocles for installation on its Super Etendard carrier-based strike aircraft, with delivery of 15 pods during 2002–03.

Contractor
Thales Optronics.

Gerfaut Helmet-Mounted Display System (HMDS)

Type
Helmet-Mounted Display System (HMDS).

Description
The Gerfaut HMDS, under development for the Dassault Rafale multirole fighter, will facilitate rapid aiming of air-to-air weapons, particularly in the close combat environment with agile missiles. The system employs Electro Optic (EO) Helmet Tracking (HT), utilising miniature cameras around the cockpit and diodes on the outer shell of the helmet, to accurately determine pilot Line of Sight (LoS). System electronics are calibrated once during manufacture of the airframe to produce a highly accurate 'cockpit map', which is used to determine compensation factors in HT algorithms. Each pilot will have a specially fitted helmet, which will also be subject to calibration within the cockpit map, allowing full and unrestricted head movement within the designated 'head box'.

The system can be adapted to fit any fighter cockpit.

Status
The Gerfaut HMDS was chosen for French Air Force and Navy Rafale fighters in 2003. A total of 336 Gerfaut HMDs have been ordered for all French Air Force Rafale pilots, to be fielded with the F3 build standard of the aircraft. The contract for the system was signed in 2004, with integration ongoing for initial delivery for test and evaluation in early 2006. Production is anticipated to begin in 2007.

Contractor
SAGEM Défence Securité, Navigation and Aeronautic Systems Division.

HESIS airborne surveillance system

Type
Airborne Electro Optic (EO) surveillance system.

Description
HESIS is a day/night airborne surveillance system, featuring near-realtime image processing via digital datalink to a ground control station.

High magnification camera optics facilitate long stand-off range for helicopters or patrol aircraft to enable discrete point/area surveillance by day and night. The data can be either recorded on board, or directly transmitted to the ground station or field forces.

HESIS meets a wide range of mission requirements such as maritime surveillance, border patrol and paramilitary operations in urban areas.

The optical system, comprising TV and IR cameras, is supported by digital microwave technology which provides for broadcast of recorded images to remote stations via standard interfaces.

Status
In production and in service.

Contractor
SAGEM Défense Securité, Optronics and AirLand Division.

IRIS new generation FLIR

Type
Airborne Electro Optic (EO) Forward-Looking Infra-Red (FLIR).

Description
IRIS is a high-sensitivity modular thermal imager with high resolution and image quality. IRIS's main features include: up to three switchable fields of view, automatic gain and offset control, polarity selection; ×2 zoom, athermalised focusing, extended BITE, boresight alignment; digital image enhancement, hot point detection and tracking and low power consumption.

The latest enhancement given to SAGEM's nav/attack system (MAESTRO) has been the night operation capability, based on integration of the IRIS internal FLIR. The IRIS second-generation IRCMOS camera provides a one-to-one infra-red image, super-imposed on the external world in the head-up display, in combination with the normal symbology. A narrow field of view image is also available to be displayed in the head-down display, for easy and precise target designation.

This new configuration has recently been qualified in-flight onboard a Mirage III and Mirage 2000.

The compact design of IRIS allows for internal installation leaving all hardpoints free for weaponry.

Specifications
Wavelength: 8–12 µm
Detection module: integrated detector/dewar microcooler device
288 × 4 elements IRCCD focal plane array Cd Hg Te
closed-cycle Stirling microcooler
NETD: T < 0.02°C
Video output: CCIR

Status
In production for Mirage 2000, Mirage III and UAV programmes. IRIS is also fitted into the main sight of several attack helicopters (French and export), including the sighting system for the Rooivalk attack helicopter, the Strix sight in French, Australian and Spanish Tiger attack helicopters, the OLOSP gimballed Electro Optic (EO) pod in the Eurocopter N90 and the OSIRIS sight in German Tiger attack helicopters.

Contractor
SAGEM Défense Securité, Optronics and Airland Systems.

IRIS thermal imager (SAGEM)   1146545

MDS 610 MultiDistance Sensor

Type
Airborne reconnaissance system.

Description
The MDS 610 is a passive electro-optical airborne reconnaissance sensor designed for daytime intelligence-gathering missions, including low-, medium- and high-altitude tactical reconnaissance, stand-off oblique and vertical reconnaissance, fixed or moving target localisation and identification.

The system is gyrostabilised in two axes and can be pod- or fuselage-mounted on combat, reconnaissance and surveillance aircraft. The sensor has both 'pushbroom' (narrow field, along aircraft track) and panoramic scanning modes. It can be manually or automatically controlled, with real-time image display in the cockpit and transmission via datalink. Automatic control is achieved by pre-programming according to the mission.

Thales' MultiDistance Sensor (MDS) 610   0505267

MDS 610 is of modular design with an electro-optical CCD detector, giving unlimited mission duration (normal film is optional), stereo viewing of small surface areas, and in-flight recording and replay capability.

Specifications
Lens focal length: 610 mm – f/4
CCD detector: 10,000 pixels (0.4–1.1 µm)
Field of view: 11°
Lateral coverage: ±110°
Longitudinal coverage: ±20°
Weight: 120 kg
Dimensions: (d × l) 350 × 1,500 mm

Status
The MDS 610 was installed in Thales' Desire reconnaissance pod (see separate entry) demonstrator which was tested on a Mirage F1 CR aircraft. It is also installed in the follow-up Presto pod, based on the Desire demonstrator. Seven Presto pods were ordered by the French Air Force to equip Mirage F1 CR and Mirage 2000D aircraft.

Contractor
Thales Optronics.

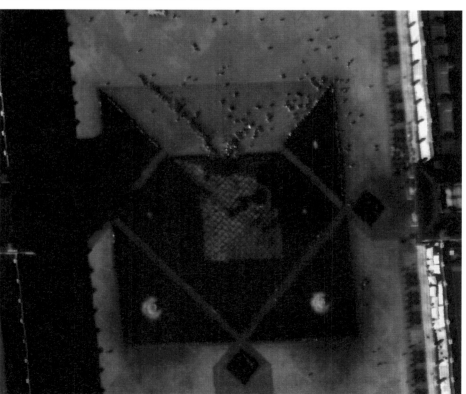
Reconnaissance imagery taken with an MDS 610, taken at 31,000 feet at × 4 and × 8 zoom   0058728

A French Air Force Mirage F1 CR carrying an MDS 610 electro-optical sensor in an early Presto demonstration pod　　　0505458

Reconnaissance imagery taken with an MDS 610, taken at 31,000 feet at × 4 and × 8 zoom　　0058729

NAVFLIR navigation and attack system

Type
Airborne Electro Optic (EO) Forward-Looking Infra-Red (FLIR).

Description
Thales Optronics has developed the NAVFLIR airborne Forward-Looking Infra-Red (FLIR) navigation and attack system. This provides assistance for low-altitude flight at night and medium-range targeting in day or night-time conditions by projecting an infra-red image onto the HUD sighting device. This image reproduces the terrain in the path of the aircraft. Aircraft installation can be on the nose of a standard pylon or on a chin pod. The FLIR has a ×2 electronic zoom.

NAVFLIR can detect a target out to 20 km and perform reconnaissance at ranges of up to 10 km.

Specifications
Detector: 3rd-generation starring array
Spectral band: 3–5 μm
Field of view
　Wide: 18 × 24°, or matched to the HUD FOV
　Narrow: 6°
Electronic zoom: ×2
Performance:
　(for a 20 × 20 m target) 10 km
　(for a 100 × 100 m target) 50 km
Video standard: STANAG 3350
Databus: 1553 Digibus

Weight: 20 kg (onboard version, no cooling system)
Cooling: independent system or provided by aircraft air conditioning system

Status
In production and in service. NAVFLIR is marketed as a combined installation, utilising a single weapon station, with the Damocles multimode pod (see separate entry). The combination is in service in Mirage 2000-9 multirole fighter aircraft of the United Arab Emirates, where it is designated SHEHAB (Damocles) and NAVHAR (NAVFLIR).

Contractor
Thales Optronics.

OLOSP 400 stabilised optronic platform

Type
Airborne Electro Optic (EO) surveillance system.

Description
OLOSP 400 is a stabilised optronic platform designed to be integrated in ground or airborne observation systems, covering missions such as maritime patrol, Search And Rescue (SAR) as well as border surveillance and police operations in urban areas.

According to mission and application, the system can be chin- or roof-mounted. OLOSP 400 comprises MATIS thermal imagers, TV cameras, an eye-safe Laser Range Finder (LRF) and a Laser Target Designator (LTD). Acquired images can be transmitted in real time to a ground station by STERN (see separate entry), a broadcast-quality digital image transmission system.

Status
In production.

Contractor
SAGEM Défense Securité, Navigation and Aeronautic Systems Division.

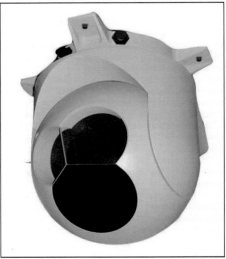

OLOSP 400 stabilised optronic platform (SAGEM)　　　1146547

Optronique Secteur Frontal (OSF) for the Rafale aircraft

Type
Airborne Electro Optic (EO) Infra-Red Search and Track System (IRSTS).

Description
Thales Optronics and SAGEM SA are co-operating for the development and manufacture of the TV and Infra-Red Search and Track System (IRSTS) for the Dassault Aviation Rafale ACT and ACM.

The OSF (Optronique Secteur Frontal) is designed to aid covert missions, firing under jamming,

Thales' pylon-mounted NAVFLIR navigation and attack system partners the Damocles targeting pod　　　0079244

Optronique Secteur Frontal (OSF) IRST (left) and TV (right) sensor heads fitted to a French Air Force Rafale multirole fighter (Patrick Allen)   0554710

visual identification and damage assessment in air-to-air, air-to-ground and air-to-sea operations and to provide navigation/piloting assistance. Key features include long-range Infra-Red (IR) passive detection, very low false alarm rates, high-definition CCD imagery, an eye-safe laser range-finder and very large Field of Regard (FoR), which may be supplemented by employing the seeker heads of Mica missiles fitted to the wingtip stations. The utilisation of separate optical assemblies for the IR and TV sensors facilitates multiple simultaneous search/identification/telemetry operation.

The system is fully integrated with RBE2 radar and weapons system, with target hand-off from radar to OSF facilitating passive approach and engagement. Future improvements to the OSF includes enhancing the video system to a day/night camera.

Specifications
Sensor coverage
IRSTS: 180° horizontal FoR
CCD TV: 60° horizontal total FoV
IRST operational range: estimated 150 km (maximum)

Status
In continuing development for the Rafale aircraft. OSF is part of the Rafale F2 build standard.

Contractor
SAGEM Défence Securité, Navigation and Aeronautic Systems Division.
Thales Optronics.

Presto/Desire reconnaissance pod

Type
Airborne reconnaissance system.

Description
The Desire reconnaissance pod was a demonstrator built by Thales Optronics under a French Ministry of Defence programme.

Thales' Presto electro-optical reconnaissance pod   0079241

This led to the development of the Presto pod for standoff or high-speed penetration missions. Presto provides real-time imaging, digital recording and transmission to ground processing stations.

The system incorporates Thales' MDS 610 multidistance sensor (see separate entry), a 610 mm focal length camera with a range of up to 50 km, a Reconnaissance Management System (RMS), a high-speed digital recorder and an associated ground station. Presto is capable of operating in both narrow and panoramic modes, and can function either under direct pilot control or autonomously in flight under pre-programmed control.

Status
In service. The French Air Force ordered seven Presto pods for Mirage F1 CR and Mirage 2000D aircraft. Presto entered the in-flight development stage at the Cazaux flight test establishment in December 1999, with operational evaluation carried out at the French Air Force's Mont-de-Marsan flight test centre. By 2001, pods were upgraded with a digital 10,000 pixel CCD (Charge Coupled Device) Electro Optical (EO) sensor and a 240 Mbps high-speed recorder.

Contractor
Thales Optronics.

Reco-NG reconnaissance pod

Type
Airborne reconnaissance system.

Description
Reco-NG (Reconnaissance, New Generation) is being developed to equip the Dassault Rafale (F3 standard) and upgraded versions of the Mirage 2000 (-5F, -D and -N). The podded system is an all-new design and integrates high-resolution dual-band sensors with a high-rate imagery downlink for near real-time transmission compatible with STANAG 7085 for greatly reduced sensor-to-shooter times. In addition, the system offers a NATO interoperable (STANAG 7023) ground exploitation segment including an open link to C4ISR systems to provide a key component in strategic/tactical reconnaissance data exchange. In support of the airborne segment, Reco-NG features advanced mission planning software for long-range stand-off manned reconnaissance, aircraft and/or pod tasking.

Reco-NG is capable of performing strategic reconnaissance tasks, collecting imagery from 50,000–60,000 ft altitude and from a standoff range of several hundreds of kilometres, and tactical reconnaissance at altitudes down to 300 ft and high speed (500–600 kt).

The high- to medium-altitude sensor used in Reco-NG is the Thales-developed DB-STARS digital,

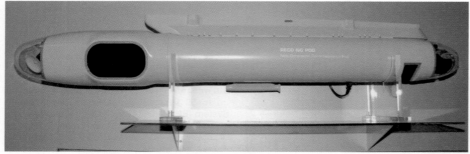

A mock-up of Thales' Reco-NG pod was shown at the 2005 Paris Air Show. In this image, the forward datalink antenna can be clearly seen   1144917

dual-band (NIR/MWIR) camera, which combines two large-format, high sensitivity Focal Plane Arrays (FPAs), one in the visible/NIR (0.4–1.5 μm) and one in the Mid Wave IR (3–5 μm) band. Each detector utilises the camera's main high-resolution optical telescope, made by Carl Zeiss Optronics, which offers both a wide and a narrow Field of View (FoV). The standoff camera will feature a 'walk in video mode', enabling the sensor to follow a winding road. The DB-STARS sensor is housed in a rotating section of the 1,000-kg Reco-NG pod, allowing it to stay locked on to a specific Point Of Interest (POI) during aircraft manoeuvring.

The tactical low-altitude sensor in Reco-NG is the high-speed SuperVIGIL all-digital 8–14 μm Infra Red Line Scanner (IRLS), supplied by Thales Optronics.

A key element of the Reco-NG system, in a network-enabled operational context, is the ability to be permanently connected to the overall C4ISR network by means of a new-generation high-speed datalink. The Thales-developed TDH 6000-family (multipayload datalink) high-rate (>100 Mbps) secure digital datalink will send imagery down to a ground station over distances of up to 350 km. The airborne equipment of this NATO-interoperable, STANAG 7085-compliant system includes a TMA 6011 Travelling Wave Tube (TWT) based transmitter, a high-power amplifier, a crypto set and two Ku-band datalink antenna units. The antennas are mounted on the front and rear of the pod to guarantee Line of Sight (LoS) at all times. The ground data terminal features a TMG 6011 receiver unit, a crypto set and a Ku-band turret equipped with a 1.5 m antenna dish for automatic tracking of the airborne datalink system.

In addition to the high-rate datalink, the system also features a low-speed UHF-band STANAG 4372-compliant omnidirectional service link. The system is bidirectional, facilitating the uplink of mission data as well as the downlink of imagery.

Status
In development. Reco-NG flight testing to date has focused on datalink, sensor performance, data processing and the ground station. As of early 2005, the focus of the test programme has shifted to overall system performance. Testing is expected to be complete by the end of 2005.

As of August 2005, Thales Land & Joint Systems was reported as 'expecting' a contract worth more than EUR100 million (USD121 million) from the French DGA for Reco-NG airborne reconnaissance systems. The contract, expected to be for 20 (plus four on option) reconnaissance pods, follows extensive flight demonstrations (50–70 sorties) and trials with two Reco-NG prototype pods that began at the end of 2004 at Cazaux Air Base using a specially instrumented Mirage 2000.

Reco-NG is to equip Dassault Rafale F3-standard multirole fighters in both the French Air Force and Navy (with an initial operational capability planned for 2008), as well as Mirage 2000 (particularly the 2000N standard K3) aircraft in the French Air Force. In both cases, the Reco-NG pod will be carried on the centreline station.

Contractor
Thales Optronics.

SDS 250 electro-optical reconnaissance sensor

Type
Airborne reconnaissance system.

Description
The SDS 250 is a compact passive electro-optical airborne reconnaissance sensor designed to perform the following daytime intelligence gathering missions: low- and medium-altitude tactical reconnaissance, vertical reconnaissance and up to 20 km standoff oblique reconnaissance.

The sensor has 'pushbroom' (narrow field, along aircraft track/heading) scanning mode, electronic roll stabilisation and selectable operating spectral band – either visible or near-infra-red, or both. The system provides real-time image display in the cockpit and line of sight position is selected from the cockpit. The system is compatible with real-time image transmission via datalink. The compactness of the sensor allows fuselage installation.

Specifications
Lens focal length: 250 mm – f/5.6
CCD detector: 6,000 pixels (0.4–1.1 μm)
Field of view: 14°
Lateral coverage: ±85°
Weight: 30 kg
Dimensions: 410 × 400 × 230 mm

Thales' SDS 250 medium-/low-level electro-optic sensor   0079242

Status
The SDS 250 was selected by the French Air Force and the French Aeronavale for the upgraded Super Etendard aircraft.

Contractor
Thales Optronics.

Strix day/night sights

Type
Airborne Electro-Optic (EO) targeting system.

Description
SAGEM has developed a family of day/night nose- and roof-mounted sights for helicopters. The Strix system has been designed for terrain observation, target acquisition and tracking and weapon aiming for guns, rockets, air-to-air and laser-guided missiles.

The Strix sight incorporates an IRCMOS Thermal Imager (TI), direct view channel, TV channel, laser range-finder and micromonitor, with interface via MIL-STD-1553B databus. Two-axis stabilisation and automatic in-flight bore-sighting throughout the entire operating range of the helicopter ensure the accuracy of the targeting information. In addition, several automatic target tracking functions (single or multiple targets) are available. TI and TV video images, combined with aiming symbology, are projected into the sight eyepiece.

Status
Strix is in production for the French ALAT HAP (multirole support) and the Australian ARH (Armed Reconnaissance) Tiger helicopter. Nightowl, a nose-mounted version of Strix, is in production for the South African Rooivalk attack helicopter programme.

Contractor
SAGEM Défense Securité, Optronics and Air Land Systems Division.

Strix roof-mounted sight on a Tiger HAP attack helicopter   0096419

Tango thermal imager

Type

Airborne Electro-Optic (EO) surveillance system.

Description

The Tango thermal imager is a modular thermal imaging system for long-range maritime patrol aircraft. It is an 8 to 12 µm CMT thermal imager with three fields of view. The system was developed for the Dassault Aviation Atlantique 2 maritime surveillance aircraft. Key features include: a large aperture for very long range imaging; high resolution; fine image stabilisation; automatic aiming towards designated targets; aircraft databus coupling; and line-to-line integration. It is incorporated in a Sere-Bezu gyrostabilised platform, fixed under the nose of the Dassault Aviation Atlantique 2.

Day and night missions include passive detection of ships and snorkels, long-range identification of surface vessels, reconnaissance and search and rescue.

Tango 2G is a second-generation thermal imager based on the Synergi thermal imaging modules developed by Thales Optronics and Zeiss-Eltro Optronic. Synergi uses a 288 × 4 IRCCD detector developed by Sofradir. Tango 2G has four fields of view. It is a multisensor integration and incorporates a CCD detector for the visible and near-infra-red channel and has an auto-search mode. It is being developed for the Atlantique third-generation, the aircraft ATL3G.

Specifications

Tango modular thermal imaging system
Dimensions: 600 mm turret diameter
Turret weight: 85 kg
System weight: 120 kg
Gyrostabilised field of view
   (Azimuth) ±110°
   (Elevation) +15 to −60°
Tracking speed with speed/accuracy optimisation:
   1 rad/s
Infra-red channel
   Detector: CMT
   Cooling: Stirling engine
   Spectral band: 8–12 µm
Field of view
   (Wide) 6.45 × 4.30°
   (Medium) 2.15 × 1.43°
   (Narrow) 1.07 × 0.7°

Tango 2G
Dimensions: 600 mm turret diameter
Turret weight: 75 kg
System weight: 98 kg
Gyrostabilised field of view
   Azimuth ±360°
   Elevation +15 to −93°
**Tracking speed with speed/accuracy optimisation,
auto-search mode:** 1 rad/s

Tango 2G modular thermal imaging system for long-range maritime patrol aircraft   0505447

Thales' Tango thermal imager mounted on the nose of the Dassault Aviation Atlantique 2 aircraft
0504240

Infra-red channel
   Detector: 288 × 4 IRCCD CMT focal plane array
   Spectral band: 8–12 µm
Fields of view: 15 × 11.2°
   7.5 × 5.6°
   1.5 × 1.1°
   0.75 × 0.6°
Visible and near infra-red channel
   Detector: CCD
Field of view: 1.5 × 1.1°

Status

Tango is in production and in service in French Navy Atlantique 2 aircraft.

Contractor

Thales Optronics.

TIM laser range-finders

Type

Airborne Laser Range-Finder/Target Designator (LRFTD) system.

Description

TIM laser range-finders are high-repetition rate eye-safe laser range-finders designed to be integrated into an airborne fire control system to measure the distance to a ground-based, aerial or naval target.

They consist of a 1.54 µm transmitter integrating a 1.06 µm laser, a Raman conversion cell, a receiver, the power supplies, an interface chronometry/control and serial link RS-422 with the host system.

Specifications

Dimensions: 300 × 180 × 150 mm
Weight: 8.5 kg
Power supply: 220 V AC, 3 phase or 115 V AC, 400 Hz

Status

The TIM laser range-finder is in production and in service in Dassault Rafale aircraft and Eurocopter Tiger helicopters.

Contractor

Compagnie Industrielle des Lasers (CILAS).

TMS303 laser range-finder

Type

Airborne Laser Range-Finder (LRF).

Description

The TMS303 Laser Range-Finder (LRF) is intended to be integrated into short- and medium-range weapon systems for range measurement of land and airborne targets, such as the SFIM Viviane sight for HOT missiles on the Gazelle helicopter.

The sight consists of a 1.54 µm transmitter integrating a 1.06 µm laser, a Raman conversion cell and a power supply; an interface/control/range processing card; a receiver and a low-voltage converter.

To remain compatible with existing systems, the TMY303, a 1.06 µm version has been developed. This could easily be upgraded to the 1.54 µm eye-safe configuration.

Specifications

Dimensions: 190 × 119 × 78.5 mm
Weight: 2.2 kg
Power supply: 28 V DC

Status

TMS303 and TMY303 are in service. Production status unknown.

Contractor

Compagnie Industrielle des Lasers (CILAS).

TMV 630 airborne laser range-finder

Type

Airborne Laser Range-Finder/Target Designator (LRFTD) system.

Description

The TMV 630 airborne laser range-finder equipment has been designed to meet single unit, small installation requirements and can be fitted easily to a wide range of aircraft. It provides high-precision aircraft-to-target range measurement and is claimed to increase considerably the performance of conventional weapon-aiming systems. The large field of view

Thales' TMV 630 airborne laser range-finder   0503750

provided is compatible with all head-up displays and the electrical interfaces are compatible with almost all aircraft types. The high-speed and accurate laser beam-steering is specifically adapted for continuously computed impact point attacks irrespective of terrain or the nature of the weapons or aircraft altitude.

Specifications
Dimensions: 190 ×190 × 520 mm
Weight: 15 kg
Power supply: 28 V DC, 12 A
Wavelength: 1.06 µm
Range: up to 19 km
Accuracy: better than 1 mrad

Status
In production as part of nav/attack systems on Dassault Mirage aircraft.

Contractor
Thales Optronics.

TMV 632 airborne laser spot tracker and range-finder

Type
Airborne Laser Ranger and Marked-Target Seeker (LRMTS) system.

Description
The TMV 632 ground attack laser range-finder was developed at the request of the French DGA (Délégation Générale de l'Armament) and combines eye-safe laser ranging and angular tracking functions into a single compact monobloc system, for use on tactical ground support and training aircraft. It provides the weapon systems with extremely accurate fire control data, for both laser-guided and conventional munitions. Several variants are available, with different digital interfaces. The dual-function TMV 632 offers the same level of performance and accuracy as single-function systems.

The TMV 632 is mounted in the airframe (embedded) or fitted in a mini-pod, or inside a store-carrying pylon. The TMV 632 detects and identifies the laser spot on illuminated ground-based targets. Acquisition and tracking are automatic. A sighting reticle in the head-up display allows the pilot to aim at the target. The laser beam of the range-finder is locked to the position of the tracker, making accurate measurement of the aircraft-to-target distance possible. There are two independent safety interlocks in the TMV 632. If required, cooling air is supplied by the aircraft.

The TMV 632 airborne laser spot tracker and range-finder   0018355

The TMV 632 laser spot tracker and range-finder in underbelly housing   0005531

Specifications
Overall Dimensions (L × W × H): 530 × 170 × 190 mm
Total weight: 18 kg
Wavelength
  Laser: 1.54 µm
  Spot tracker: 1.06 µm
Range: up to 20 km in range-finding mode and 15 km in tracking mode
Field of regard: 40° azimuth, 20° elevation
Interfaces: TM632: digibus, TM632A: ARINC, TM632B: bus 1553
Power: 28 V DC, 8 A

Status
In service in Mirage F1CT aircraft.

Contractor
Thales Optronics.

Topowl® day/night Helmet-Mounted Sight Display (HMS/D)

Type
Helmet-Mounted Display System (HMDS).

Description
Thales Avionics' Topowl® binocular HMS/D for helicopters is based on a modular design, with optimised head-supported mass, for day or night missions. The basic helmet provides conventional physiological protection and communications, while the display module projects imagery on the visor, using integrated night vision sensors (image intensifier tubes) or Forward-Looking Infra-Red (FLIR) video and/or synthetic symbology.

The HMS/D is a purpose-built, rotary-wing optimised system, utilising proven European technology. Technical and operational features include the following:
- Enhanced situational awareness and safety due to visor-projection technology
- Binocular, wide field of view of 40°
- Lightweight (2.2 kg), well balanced helmet with optimised centre of gravity
- Modular design, with a basic helmet, specially adapted to the needs of each pilot and a display module that remains in the helicopter
- Accurate head-position sensing (tracking) technology, which is fully qualified and flight proven in the helicopter environment
- NBC compatibility, head-in displays and aircrew equipment kits
- Image intensifier tubes optically integrated in the display module, giving safe dual-sensor night-mission capability
- Dual-power source for the image intensifier tubes with a backup battery installed in the quick release connector

Topowl® shown demonstrated in a Ka-52   0006798

Topowl® is proving to be the success story in helicopter HMDs, with its recent selection by the US Marine Corps for its AH-1Z SuperCobra   0137430

- Integrated cursive symbol generator providing safe and clearly readable symbology, even in full daylight.
- Compatible with head-in display, NBC kit and opthalmic glasses

Status

In production and in service, with over 75 Topowl® HMS/D systems delivered to date and significant orders from helicopter operators around the world, including South Africa (for the Rooivalk), France, Germany and Australia (for the Eurocopter Tiger) and France, Germany, Italy, Portugal, Sweden, Norway, Finland and Netherlands (for the NH 90). During June 2002, Topowl® was selected by the US Marine Corps (USMC) for its Bell AH-1Z SuperCobra and UH-1Y helicopters, with delivery of the first two HMS/D units to Bell Helicopters five months later during November. On December 20th, two more pre-production units were delivered to Fort Worth. The remaining TopOwl units for the EMD phase were delivered through the end of June 2003 at a rate of two per month. A total of 18 of these units were ordered, with production of 560 production TopOwl systems (two each for 180 UH-1Z and 100 UH-1Y aircraft) to begin after satisfactory completion of flight testing.

As part of the USMC operational acceptance effort, during September 2004, Topowl® was tested with and without the Communications Ear Plug (CEP) for sound attenuation. Insertion loss measurements were made for the HMS/D without the CEP in accordance with ANSI guidelines. It was found that the HMS/D worn without the CEP does not meet the Navy/Marine Corps sound attenuation requirement but exceeds the requirement when worn with the CEP. Accordingly, the US Army Aeromedical Research Laboratory (AARL) recommended that Topowl should not be used without the additional hearing protection and enhanced communications afforded by insert hearing protection and communication devices such as the CEP.

During Mid-2004, it was reported that Topowl® had successfully completed operational flight testing with the French Army for the Tiger and the USMC for the AH-1Z. Thales predicts that over 1,500 Topowl® systems will be produced for 11 countries over the next 10 years.

Contractor

Thales Avionics SA.

Topsight® Helmet-Mounted Sight/Displays HMS/D

Type

Helmet-Mounted Display System (HMDS).

Description

Thales Avionics developed the Topsight® family of Helmet-Mounted Sight/Displays (HMS/D) to meet the requirements of modern combat aircraft pilots.

The first two products in this family are a fully integrated Topsight® HMS/D, designed specifically for the Rafale multirole fighter, together with the modular Topsight® E, designed for export markets.

The Topsight® E comprises a basic helmet with a removable display module that projects symbology on the visor. Depending on the assigned mission, the display module can be replaced with a double visor module (like a conventional helmet), or an ejection-compatible night vision module. In all cases, the pilot keeps his own oxygen mask and inner helmet.

Featuring an optimised centre of gravity and head-supported weight of less than 1.5 kg, Topsight® E ensures effective target designation and acquisition for all missions. A camera integrated in the display module also provides full mission replay capability.

The Topsight® E is designed from the outset for excellent physiological protection against shock, punctures, fire and facial injury, as well as safe ejection at speeds up to 625 kt.

The Topsight® E HMS/D has been associated with an Indian Air Force Procurement of MiG-29K fighter aircraft
0051713

Using a highly accurate electromagnetic head tracking system, the pilot can perform off-boresight target designation and weapon firing, enabling him to maintain optimum tactical position and/or aspect relative to any engagement/attack. Once the target is acquired, the pilot's line of sight is then transmitted to the aircraft's other systems, including sensors, avionics and weapon systems, for lock-on.

The Topsight® helmet-mounted sight/display family also allows reverse cueing from the aircraft systems to Topsight® itself, which means the pilot's eyes are guided to the target tracked by aircraft sensors.

The Topsight® family maximises the pilot's Situational Awareness, by displaying flight and navigation data, weapons system status, warnings and other vital information directly on the visor. Information displays can be customised to meet specific needs, ensuring superiority in both aerial combat and ground attack missions.

Specifications

Field of view: 20° (monocular)
Interfaces: MIL-STD-1553 bus, RS-422
Eye relief: 60 mm

Status

Flight evaluation of Topsight® commenced in May 1997, as part of the Rafale development programme for the French Air Force and Navy. Despite being regarded as the front runner for selection as the HMD for Rafale, Topsight® eventually lost out to Sagem's Gerfaut HMD in 2004.

Topsight® E flight evaluations were completed by the French Air Force on a Mirage 2000 during 2000, with reports of an Indian Air Force order for the system to equip its MiG-29K fighter aircraft during 2004.

Contractor

Thales Avionics SA.

Victor thermal camera

Type

Airborne Electro Optic (EO) targeting system.

Description

The Victor thermal camera is designed to be connected to Viviane and Strix gyrostabilised sights, for SA342 Gazelle HOT and HAP Tiger escort helicopters. It can also be fitted on AS 365M Panther, BO 105 or other types of helicopters.

Victor converts the thermal radiation of landscape and objects into a visible image at TV standard and is suitable for air-to-ground and air-to-air gun or missile firing, rocket firing and as a flying aid by day and night and under adverse weather conditions.

The system displays symbols in the eyepiece of the sight or on a TV monitor and has a magnifying function of ×2 to enlarge the image. There is a processing board option for improved performance. Initial sight stabilisation is by the platform and electronic fine stabilisation is by the imager. There is a specific video output for tracking.

Thales' Victor camera on a French Army Gazelle helicopter   0504241

Specifications
Weight:
(total) 25 kg
(on roof) 17 kg
Power supply: 20–32 V DC, 140 W
Wavelength: 8–13 μm
Trifocal lens: 30 × 20°, 6 × 4°, 2.4 × 1.6°
Range: up to 4,000 m

Status
In service.

Contractor
Thales Optronics.

Germany

HELLAS helicopter Obstacle Warning System (OWS)

Type
Airborne Laser Obstacle Warning System (OWS).

Description
HELLAS (HELicopter LASer) is an active laser radar (ladar) system designed to provide warning of wires and cables and other similar obstacles to helicopters flying Nap-Of-the-Earth (NOE) and other low-altitude missions, at up to 1,000 m range, both for military and civil operations.

Obstacle detection is performed with an imaging eye-safe ladar, which generates images of the scene in front of the helicopter, while range data processing is performed in the processing unit. Processing results can be configured for display as warning information for the pilot.

Depending on customer-specific requirements, the warning information can be indicated by an optical warning indicator (day, night, NVG-compatible), or a helicopter display unit (MFD, HUD, HMD). Additionally, the system will provide a warning signal (an acoustic alarm) if the pilot exceeds a specified safety threshold. HELLAS gives a timely warning to the pilot with a high probability of detection of >99.5 per cent per second.

Standard ARINC, STANAG and MIL-STD-1553B interface capability facilitates installation in all types of helicopter.

Specifications
Sensor: 3-D imaging laser radar
Wavelength: solid-state Erbium fibre, Class III, NOHD = 0 m, eye-safe at 1.54 μm
Laser pulse power: 4 kW (10 kW option)
Receiver: InGaAs APD hybrid (Avalance Photo Diode)
Scanning: 2 axes. horizontal: fibre optic; vertical: oscillating mirror
Image repetition frequency: 2 to 4 Hz
Field of view: 32 × 32° (32 × 40° option)
Range:
>1,000 m (extended area objects, in good visibility)
>500 m (wires >10 mm, good visibility)
>400 m (extended area objects, adverse weather)
>300 m (wires >10 mm, poor weather, oblique incidence)
Range resolution: <1 m
Weight: <27 kg

Cockpit displays and controls
Display: Two optical warning indicators (3 zones warning), standard MFD
Control panel: ON/OFF, NORMAL (for cruise), APPROACH (takeoff and landing)

Status
Since 1995, HELLAS has accumulated hundreds of flight hours on BK-117, CH-53, EC 145 and UH-1D helicopters. The German Federal Border Guard has acquired 25 systems for its EC 135 and EC 155 helicopters, with deliveries completed during 2002/2003. Elsewhere, the Canadian

The HELLAS OWS has been fully integrated into a Bk 117 of the German All-Weather Rescue Helicopter (AWRH) programme.   0054090

HELLAS on a Federal German Border Guard EC135 helicopter (Eurocopter)   1128601

Defence Research Establishment (DREV) ordered HELLAS for its Bell 406. HELLAS is part of the French All-Weather Helicopter Programme HTT, flying on an EC 155. The German Federal Office of Defence Technology and Procurement selected the military version of the system for the NH90 TTH helicopter and the system will also be employed by Finland on its NH90s for the NSHP. Dornier is partnered with the USSOCOM in a Foreign Comparative Test (FCT) Programme, with HELLAS undergoing flight tests on a UH-60. Further systems will be delivered for the MH47, MH53 and MH60 helicopters. In August 2003, Hellas achieved type certification from the German airworthiness authority, LBA, with further certification planned for the UK, USA, France, Austria and Switzerland.

Contractor
Dornier GmbH (an EADS company).

KRb 8/24 F reconnaissance camera

Type
Airborne reconnaissance system.

Description
The KRb 8/24 F camera achieves wide-angle panoramic coverage with undistorted framing camera geometry on a single 9.5 in (240 mm) wide film. The camera performs at highly survivable parameters for low- to medium-altitude reconnaissance missions. Each exposure covers 143° across track, with true angle forward motion compensation across the entire format. This format affords sequential along-track stereoscopic coverage. Images are without the cylindrical distortion inherent in panoramic cameras. Special Zeiss optics provide performance into the near infra-red, allowing the use of all aerial film types, including colour and camouflage detection, without refocusing. Small size and low weight permit easy installation in RPVs, pods and aircraft.

Specifications
Dimensions: 356 × 374 × 311 mm
Weight: 22 kg
Power supply: 28 V DC, 250 VA
Focal length: 80 mm
Aperture range: f/2.56-f/16

Contractor
Z/I Imaging GmbH.

KS-153 modular camera system

Type
Airborne reconnaissance system.

Description
The KS-153 is a modular camera system consisting of three different focal length configurations: The Pentalens 57, Trilens 80 and Telelens 80. The Pentalens configuration is the latest design development in the system. These three configurations have a common camera body, film cassettes and film cassette holder. The desired focal length lens and shutter assembly with format mask can easily be attached to the camera body to accommodate mission requirements. The high parts commonality provides a benefit to multiple-type camera users.

The modular KS-153 is a fully electronic, microprocessor-controlled design and uses Carl Zeiss optics optimised for high resolution. The camera system is built and tested to meet and exceed reliability and maintainability criteria for use in modern military or commercial aircraft, pods and UAVs.

The KS-153 Pentalens 57 is a high cycle rate, pulse operated, sequential frame camera designed for low- to medium-altitude, wide-angle, photography from high-performance aircraft.

The camera combines wide-angle panoramic camera coverage with framing camera geometry on a single 240 mm (9.5 in) wide film. The wide-angle coverage is provided by an optical assembly of five TOPAR A2 2/57 lenses having an effective maximum aperture of f/2.7. The 182.7° lateral coverage is displayed by five across-track images per frame, each frame covering an angular field of view of 18.27° across-track by 47.4° along-track. The field view of the side lenses are deflected by front-mounted prisms.

The KS-153 Trilens 80 is a high cycle rate, pulse operated, sequential frame camera designed for low- to medium-altitude, wide-angle photography from high-performance reconnaissance aircraft. The camera accommodates both 4 mm standard base and 2.5 mm thin base roll film in any panchromatic, infra-red or colour emulsion. Each frame covers an angular field of view of 143.5° across-track and 48.5°

Z/I Imaging KS-153 modular camera system   0044999

Z/I Imaging KS-153 Pentalens 57 camera   0051001

Z/I imaging KS-153 Trilens 80 camera   0051002

Z/I Imaging KS-153 Telelens 610 camera   0051003

along-track. The wide lateral coverage is provided by the optical assembly of three S-TOPAR A1 2/80 lenses having an effective maximum aperture of f/2.56 and the field of view of the side lenses deflected by front-mounted prisms. Major camera features are true angle corrected Forward Motion Compensation (FMC), constant velocity focal plane shutter, integral intervalometer, Automatic Exposure Control (AEC), easily interchangeable interface card and continuously monitoring BITE. This BITE feature is further enhanced by an integral non-volatile BITE memory to enable post-flight analysis of in-flight transient failures.

The KS-153 Telelens 610 is a pulse-operated, sequential frame camera designed for medium- to high-altitude oblique photography from high-performance reconnaissance aircraft. The camera accommodates both 4 mm standard base and 2.5 mm thin base roll film in any panchromatic, infra-red or colour emulsion. Each frame covers an angular field of view of 21.4° across-track by 10.7° along-track. The camera mounts in an integral ring bearing, which allows in-flight rotation under electronic control to any desired oblique angle. The Telelens 610 has the same major camera features as the Trilens version of the KS-153 camera.

Stereoscopic viewing greatly increases intelligence gained from aerial photography and the KS-153 images appear side-by-side to provide maximum convenience for direct stereoscopic viewing without cutting the film.

Specifications
Dimensions:
(Pentalens 57) 439 × 467 × 502 mm
(Trilens 80) 439 × 467 × 506 mm
(Telelens 610) 423 × 809 × 470 mm
Weight:
(Pentalens 57) 59 kg
(Trilens 80) 57 kg with 500 ft of film
(Telelens 610) 110 kg with 200 ft of film
Power supply: 115 V AC, 400 Hz, 3 phase
Focal length:
(Pentalens 57) 57 mm
(Trilens 80) 3.5 in (80 mm)
(Telelens 610) 24 in (610 mm)
Cycle rate:
(Pentalens 57) 10/s (max)
(Trilens 80) 10/s (max)
(Telelens 610) 4/s (max)
Shutter speed: 1/2,000 to 1/150 s
Aperture range:
(Pentalens 57) f/2.56 to f/16
(Trilens 80) f/2.56 to f/16
(Telelens 610) f/4 to f/16
Format:
(Pentalens 57) 50 × 212 mm
(Trilens 80) 72 × 222 mm
(Telelens 610) 230 × 115 mm
Film length:
(Pentalens 57) 500 ft
(Trilens 80) 200 or 500 ft
(Telelens 610) 200 ft

Contractor
Z/I Imaging GmbH.

RMK TOP aerial survey camera system

Type
Airborne reconnaissance system.

Description
The RMK TOP is an aerial photography system used for survey and cartographic purposes. It provides enhanced image quality with minimum distortion and extensive image motion compensation by

Z/I Imaging RMK TOP15 configuration showing the TOP15 camera (centre), T-TL central control unit (left) and T-CU central computer unit (right) 0051004

FMC and a gyrostabilised suspension mount. System controls and functions are monitored by a compact computer and microprocessor and a pulsed rotating disk shutter with a constant access time of 50 ms. It is particularly suitable for use with GPS-controlled navigation systems.

The basic components of an RMK TOP camera are the RMK TOP15 camera body with wide-angle PLEOGON A3 4/153 lens (or RMK TOP30 camera body with standard TOPAR A3 5.6/305 lens), T-TL central control unit, T-CU central computer unit, T-MC film magazine with FMC and T-AS suspension mount gyrostabilised in three axes.

The following options are available for navigation and system control: T-FLIGHT GPS-supported photoflight management system for flight planning and mission documentation, T-NT visual navigation telescope, T-NA automatic navigation meter for automatic V/H measurement and interfaces for aircraft specific navigation systems.

Specifications
Weight:
(RMK TOP15) 176.3 kg total
(RMK TOP30) 169.6 kg
Focal length:
(RMK TOP15) 6 in (153 mm)
(RMK TOP30) 12 in (305 mm)
Aperture:
(RMK TOP15) f/4 to f/22
(RMK TOP30) f/5.6 to f/22
Exposure time: 1/50 to 1/500 s

Contractor
Z/I Imaging GmbH.

VOS 40/270 digital EO reconnaissance camera

Type
Airborne reconnaissance system.

Description
A member of Zeiss' VOS family of digital cameras, the VOS 42/270 digital EO camera was specifically designed to meet the requirements of the RecceLite tactical reconnaissance pod (see separate entry). The camera fits into the stabilised front section of the RecceLite pod, facilitating a very large total Field of Regard (FoR). The sensor can also be integrated into UAVs and other aerial platforms.

Major features and benefits of the VOS 42/270 camera include:
• High resolution via large pixel count
• Automatic exposure control
• Automatic focus control

• Auto balancing (compensates for zoom movement)
• No mechanical shutter for increased reliability
• High-angular resolution
• Cockpit image display capability.

Specifications
Detector: FPA CCD 2,048 × 2,048 pixels
Pixel size: 7.4 × 7.4 μm²
Lens: Carl Zeiss, f = 42 – 270 mm
FoV:
  Narrow FoV: 3.2 × 3.2°
  Medium FoV: 7.1 × 7.1°
  Wide FoV: 20.5 × 20.5°
Relative aperture: 1:3.2
Frame rate: 3 frames/sec maximum
Serial control interface: RS-422
Output: 12/8-bit digital video
Dimensions: 290 × 270 × 110 mm
Weight: <8 kg
Power: 28 V DC, ≤45 W (average)

Status
In production and in service in RecceLite pods fitted to Spanish Air Force EF-18A+/B+ aircraft.

Contractor
Carl Zeiss Optronics GmbH.

VOS digital video colour camera system

Type
Airborne reconnaissance system.

Description
The VOS digital video colour camera system is designed to provide high-resolution airborne photographic surveillance.

VOS uses a high-resolution, multi-spectral detector. It consists of three parallel photodiode line arrays, each with 6,000 pixels for the colours red, green and blue, covering a spectral band from 350 to 1,050 nm. A cut-off filter in front of the lens limits the spectral response to 650 nm, generating a true colour image.

The line array of the camera system operates in a push-broom mode and a continuous image is generated by forward motion of the aircraft. The resolution in the forward direction is directly dependent on the aircraft speed. In order to obtain square image pixels, the control unit automatically calculates the correct line rate dependent on the altitude, speed and focal length. Post-processing is used to optimise the output of the camera, recording and processing functions.

The SMV-1 stabilised platform is used as a mount for the VOS camera system. It employs gyro technology, active control components and passive vibration damping to stabilise the camera body.

Specifications
VOS electro-optical sensor
Detector: Kodak, colour, 3 × 6,000 pixels (red, green, blue)
Pixel size: 12 × 12 μm², 12 μm pitch
Line space: 96 μm
Line rate: 1.6 kHz, maximum
Spectral response: 350–1,050 nm; colour with 650 nm filter
Data rate: 230 Mbits/s, maximum
Lens: Carl Zeiss Planar®, FL 80 mm, f/2.8
Field of View: 48.5° across flight path
Dimensions: 154 × 154 × 210 mm
Weight: 3.5 kg
Power: 28 V DC, 28 W

Sensor control unit
Dimensions: 483 × 133 × 390 mm
Weight: approximately 15 kg
Power: 28 V DC, 125 W

Sensor control panel
Dimensions: 150 × 200 × 105 mm
Weight: approximately 1.5 kg
Power: supplied by control unit

SMV-1 stabilised platform
Gimbals: pitch and roll axes stabilised
Gimbal freedom: ±10° each axis
Stabilisation: <70 μrad rms
Positioning accuracy: <0.5° (1σ)
Weight: approximately 10 kg (excluding camera)
Dimensions: 460 × 310 × 220 mm
Power: 28 V DC, 50 W

Status
The VOS system is operated by the German 'Open Skies' force, in a three-camera fit. One system is vertically installed in the observation aircraft, the other two systems in oblique positions on the left and right side, with overlapping FoVs for wide area coverage.

Contractor
Carl Zeiss Optronics GmbH.

The Carl Zeiss Optronics VOS digital video colour camera system 0079246

International

AN/AVS-7 ANVIS/HUD system

Type
Helmet-Mounted Display System (HMDS).

Description
The AN/AVS-7 ANVIS/HUD system projects flight data into the views of the pilots NVGs. By eliminating the need to shift attention and focus to the cockpit interior, the AN/AVS-7 Aviator's Night Vision System/Head-Up Display (ANVIS/HUD) enhances flight safety and operational effectiveness. Pilots can fly head-up, viewing the situation outside the cockpit, while at the same time receiving all essential flight data. The AN/AVS-7 features a real-time, high resolution, lightweight display unit which is easily mounted on NVGs. It is adaptable to any aircraft platform and has four independent display modes, each with declutter facilities. The pilot and co-pilot can independently select symbols and display modes from controls on the centre console and collectives, using ANVIS compatible control panel and display facilities. BIT provides 95 per cent failure detection and fault isolation.

Day-HUD optics interface directly with the AVS-7 harness to provide day symbology, including expanded symbology for day/night pilot operations, plus maintenance and training.

Specifications
Field of View (FoV): 34°
Display type: stroke (highly stable)
Resolution: 512 × 512 pixels
Interfaces: MIL-STD-1553B, ARINC 429, RS-422, analogue, discrete and synchro
Reliability: >1,000 h
Power: 28 V DC, <5A
Avionics B kit
Dimensions:
(converter control (CC)): 63.5 × 139.7 × 76.2 mm
(display unit (DU)): 88.9 × 38.1 × 39.1 mm
(signal data converter (SDC)): 274.3 × 190.5 × 198.1 mm
Aircraft installation A kit: sensors and harness, as required

Status
AN/AVS-7 is built by Elbit Systems Ltd and supported in the US by EFW, Fort Worth, Texas. AN/AVS-7 is installed on many Israeli Air Force, US Army, US Marine Corps and US Navy rotary- and fixed-wing aircraft. AN/AVS-7 adapts to the following aircraft types: CH-46E, CH-47D, CH-53E, HH-60H, KC-130T, MH-47E, MH-60K, MV-22, UH-1N and UH60A/L. A total of 3,500 systems are in use worldwide.

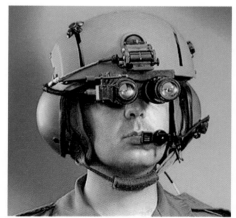

AN/AVS-7 ANVIS/HUD system (Elbit Systems)
0569535

A derivative system (Gideon) has been ordered by the US Marine Corps, and by unspecified other forces for Bell AH-1W helicopters.

EFW and Elbit Systems Ltd have upgraded the system for the US Army and other customers; the upgraded system carries the designation ANVIS/

The AN/AVS-7 ANVIS/HUD is installed in US Boeing MV-22 Osprey helicopters   0093859

ANVIS/HUD 24 day display (Elbit Systems)   0569534

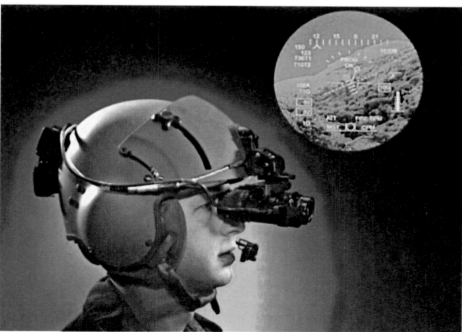

ANVIS/HUD 24 AN/AVS-7 component; night clip-on display (Elbit Systems)   0569533

HUD 24 (see separate entry). The upgrade doubles the number of inputs and outputs to transfer data to the Crash Survival Memory Unit (CSMU), supports HUMS, increases CPU speed, solves obsolescence issues, facilitates system programming via a front panel connector, increases reliability and reduces hard-manoeuvre display latency.

Contractor
Elbit Systems Ltd.
EFW.

ANVIS/HUD 24 system

Type
Helmet-Mounted Display System (HMDS).

Description
Elbit Systems' ANVIS/HUD 24 provides for helmet-mounted display of flight data across the entire 24-hour spectrum of helicopter operations. The

---

system combines an updated version of the AN/AVS-7 ANVIS/HUD system (also designated AN/AVS-503 and DNVG) with the Day HUD add-on module to facilitate the display of the same symbology set to the pilot under all flight conditions.

By eliminating the need to shift attention and focus to the cockpit interior, ANVIS/HUD 24 enhances flight safety and operational effectiveness.

The ANVIS/HUD 24 system features multiple interfaces which enables installation on virtually any type of helicopter and integration with most avionic systems, both western and non-western in origin. Further features and benefits of the system include:

- Dual-cockpit HOCAS-driven operation
- Modular system
- Two spare card slots to support target cueing (head tracking), digital map, VADR, HUMS, and so on
- Single system Signal Data Converter (SDC)
- Eight selectable display modes
- Symbology set compatible with existing cockpit instrumentation and displays
- Independently-controlled display to each crew member
- Enhanced pilot Situational Awareness (SA)
- Increased survivability
- Enhanced crew co-ordination
- Integrates with eyeglasses and NBC mask
- Quick disconnect facility for safe rapid emergency egress

Specifications
Field of View (FoV): 32°
Display type: stroke
Contrast ratio: 2:1 min (at 0.03 fc)
Distortion: <3%
Interfaces: MIL-STD-1553B, ARINC 429, RS-422, analogue, discrete and synchro
MTBF: >3,000 h
Power: 28 V DC, <48 W (total system); <35 W (SDC)
Weight: <70 g (clip-on module); <6 kg (SDC)
Compliance: MIL-E-5400 Class 1A; MIL-STD-461C Part 2

Status
In production and in service. The ANVIS/HUD 24 system components are qualified on over 25 platforms.

Contractor
Elbit Systems Ltd.
EFW.

Archer helmet display system

Type
Helmet-Mounted Display System (HMDS).

Description
Derived from the Guardian Helmet-Mounted Display System (HMDS), the Archer helmet display system provides visor-projected displays and optical Line of Sight (LoS) measurements. This enables the user to aim weapons and cue sensors by simply looking at the target.

The Archer Electro-Optic Helmet Tracking System (EOHTS) measures the orientation of the pilot's Line of Sight (LoS); the HTS employs an array of LEDs embedded in the outer helmet which are observed by cockpit-mounted cameras. Compared to AC or DC electromagnetic tracking systems, the EOHTS is relatively immune to many interference effects, such as canopy rail/windscreen interference and sunlight reflections. The EOHTS is common to all Guardian and Archer helmets, measuring all six degrees of freedom of the pilot's head and providing a digital output to the associated electronics unit.

The Archer display helmet is a two piece (inner and outer helmets) design, with the family comprising both fixed-wing and rotary-wing variants:

The fixed-wing variant consists of a single common inner helmet which may be matched with one of two outer helmets, which provide a Helmet Mounted Sight (HMS), providing a fixed-legend display, or a Helmet Mounted Display and Sight

(HMDS), providing a dynamic display of weapon, navigation, aircraft and flight information.

The rotary-wing variants provide for visor-projected display as well as Night Vision Enhancement (NVE) by using Night Vision Goggles (NVG) or NVE cameras to provide day and night display capability. Turreted weapons and sensors are slaved to the user's LoS, while the display enables the user to maintain look-out and enhanced Situational Awareness (SA).

The Archer rotary-wing HMDS comprises similar electronics and HTS to the fixed-wing version, combined with a number of possible helmet variations, all based on the common inner helmet:

- HMS, providing a fixed-legend display
- HMDS, providing a dynamic display of weapon, navigation, aircraft and flight information
- Night Vision Display System (NVDS), providing a clip-on display module for NVG and full symbology display
- Full binocular HMDS providing a fully integrated day-night helmet with full dynamic display capability and detachable NVE cameras.

The inner helmet is individually matched as a personal-issue item to each aircrew user. The inner helmet carries all life support elements, including communications and oxygen mask.

The basic HMS display medium is a red LED, which provides the symbols for off-boresight aiming and, in conjunction with other sensors, cueing arrows to allow fast acquisition of a target or Point Of Interest (POI). The red symbology is reflected off the visor to the pilot's eye.

The HMDS is available as a monocular or binocular display. The monocular display is similar to the HMS, with the LED replaced with a miniature high-brightness CRT. Full symbology is available, giving the wearer full information on aircraft parameters, navigation data, weapon status and fuel. The symbology is fully programmable to meet user requirements. A feature of the monocular display is the ability to provide the pilot with raster imagery from an external sensor such as a targeting pod or Infra Red Search and Track System (IRSTS).

The binocular system provides a Wide FoV display which projects full symbology as in the monocular display, but with the addition of helmet-mounted NVE camera imagery. The night vision cameras form a key component of the system and are designed to replace the night-only clip-on NVG module. The use of NVE cameras rather than a traditional NVG/Clip-on module system reduces overall helmet weight.

Status
It has been reported that SAGEM's Gerfaut HMD for the Rafale F3 is based on the Archer design, using an optical tracking system.

Contractor
Thales Optronics Ltd.
Cumulus, a business unit of Denel (Pty) Ltd.

Eurofighter Typhoon Integrated Helmet

Type
Helmet-Mounted Display System (HMDS).

Description
A development of the Striker HMDS, the Eurofighter Typhoon Integrated Helmet has been developed to meet the specific requirements of the Eurofighter application for pilot protection and life support, combined with full Helmet-Mounted Display (HMD) functions. The system is an integral part of the Eurofighter avionics suite, providing Night Vision Equipment (NVE) and Forward-Looking Infra-Red (FLIR) sensor display, combined with full navigation and weapon-aiming symbology. The helmet also provides for interface between the pilot and the Direct Voice Input (DVI) system.

The helmet is a two-part modular and re-configurable design, featuring low head-supported mass, balance and comfort. The requirement for ejection safety at up to 600 kts was found to be the major contributor to the total mass of the helmet, which, nonetheless, has been restricted to 1.9 kg for the day helmet (without NVE cameras) and 2.3 kg for the night helmet (NVE cameras fitted). This compares with approximately 1.8 kg for the day-only Guardian and JHMCS systems (see separate entries) and approximately 1.4 kg for a standard fast-jet aircrew helmet.

The inner helmet is designed to provide for the high stability required to maintain the pilot within the design eye position for the optics, while remaining comfortable and providing sufficient adjustment range to fit all head sizes (one size fits all). The inner helmet includes a new lightweight oxygen mask, based on the MBU-20/P, which provides for pilot pressure breathing. Other features of the inner helmet are:

- Advanced suspension system
- Forced air ventilation

BAE Systems has produced an HMD system for the F-35 JSF under contract from Lockheed Martin, known as the Alternative HMD (AHMD). Note the similarity with the Eurofighter helmet, which is no disadvantage in this case, as the AHMD benefits from the extensive and lengthy development of the Typhoon Tranche 2 system. This image also clearly shows the intended large-format displays of the F-35 and one of the Electro Optical (EO) helmet-tracking cameras (on the seat frame) (BAE Systems)

1127373

The Eurofighter Integrated Helmet with NVE cameras fitted on to each side of the pilot's head. These are detachable if not required by the mission, thus saving head-mounted weight (Eurofighter)
0528491

The Eurofighter Integrated Helmet. Note the diodes on the outer shell for the optical helmet tracking system (BAE Systems)
0121549

- Brow pad moulded to fit individual aircrew
- Can be integrated with Nuclear Biological and Chemical (NBC) hood.

The outer helmet attaches to the inner helmet, combining primary aircrew protection with a lightweight and rigid platform for the optical components of the system. The outer helmet incorporates diodes for the optical helmet tracking system and all of the equipment for the projection system: twin CRTs, optical relays, a brow mirror, twin detachable NVE cameras and a blast/display visor. At present, although much lighter than traditional Night Vision Goggles, all solid-state NVE cameras cannot rival current NVGs in terms of display resolution and brightness; however, new hybrid II/CMOS camera technology is under development, employing considerable front-end image processing, which has demonstrated perceived optical performance up to that of in-service Generation III NVGs. Resolutions of up to 1280 × 1024 pixels are available, with work ongoing to improve sensor resolution even further. The optical helmet tracking system, via a helmet-mounted Light Emitting Diode

(LED) array and optical trackers mounted in the cockpit, provides for high accuracy and low latency operation within the aircraft's performance envelope, particularly in the areas of instantaneous turn and roll rate capability. The projection system includes a fully overlapped 40° FoV (which can also be monocular if required) and features a high degree of modularity for cameras, NBC, laser protection and so on. High-performance display processing is VME-based, utilising dual-processors.

Specifications

Display: Dual high resolution CRT; stroke/raster/mixed modes
FoV: 40° binocular, fully overlapped
Eye relief: >50 mm
Exit pupil: >15 mm
Head supported mass: 2.3 kg (including NVE cameras), 1.9 kg (NVE cameras detached); inlcusive of oxygen mask (MBU-20/P) and twin visors
NVE cameras: Solid-state, effective down to 0.5 mlux; detachable

Wavelength: 0.715 to 0.910 μm
Resolution: 1280 × 1024 pixels

Status
Introduction of the Typhoon Integrated Helmet is planned as part of the Tranche 2 Typhoon standard. The helmet forms part of the Striker family of HMDS. During 2003, BAE Systems announced a collaboration effort with Cumulus, a business unit of Denel (Pty) Ltd, to integrate their HMD expertise to capture a major portion of the worldwide market for military helmet-mounted systems.

During September 2004, BAE Systems announced a contract to perform the initial design of an Alternative HMD (AHMD) for the F-35 Joint Strike Fighter (JSF) under a contract from Lockheed Martin. The design, benefiting from the extensive development of, and based heavily on, the Eurofighter Typhoon system, will be utilised for UK F-35s.

During March 2005, BAE Systems announced that the helmet had begun formal flight testing on Eurofighter Typhoon aircraft, marking the first flights of a binocular, visor-projected, night vision capable HMD on a fighter aircraft. Addition flights were carried out through 2005, with production-ready helmets being readied for flight at the end of 2005.

Contractor
BAE Systems leading a consortium comprising:
  Galileo Avionica SpA, Avionic Systems and Equipment Division.

Guardian Helmet Display System (HDS)

Type
Helmet-Mounted Display System (HMDS).

Description
The Guardian family of display helmets is a range of systems that have evolved from the strategic teaming of Thales Optronics and Cumulus. The result is a fully integrated, lightweight, balanced, comfortable, high-performance range of display helmets for fixed- and rotary-wing applications.

The Guardian concept is based on use of an inner helmet that is individually matched as a personal-issue item to each aircrew user. The inner helmet carries all life support elements, including communications and oxygen mask. Active noise reduction and chemical defence are offered as options.

The inner helmet can support a variety of outer helmets, which carry the Electro-Optic (EO) platform's visors, optics, display media and other electronics.

Sight Helmet
The Guardian Sight Helmet is designed to allow fighter and light attack aircraft operators to slew off-boresight missiles, thereby greatly increasing the operational effectiveness of the aircraft. The display medium is a red LED, which provides the symbols for off-boresight aiming and, in conjunction with a radar, cueing arrows to allow fast acquisition of the target. The red symbology is reflected off the visor to the pilot's eye.

With the Guardian optical head tracker constantly providing head angles and position to the weapon system, all the pilot needs do is align the LED cross with the target and press the missile designate button. The missile head then moves to the target angles and locks-on. On receipt of the missile tone, the pilot fires the missile.

There is no dynamic symbology and minimal software in the system, minimising cockpit and aircraft integration time and cost. The Guardian Sight Helmet is in production.

Monocular display
The monocular display is similar to the Sight Helmet design, but the LED is replaced with a miniature high-brightness CRT and the drive electronics are increased in capability to match the potential of the CRT. Full symbology is available, giving the wearer full information on aircraft

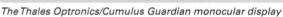

The Thales Optronics/Cumulus Guardian monocular display   0079250

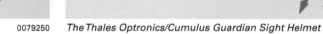

The Thales Optronics/Cumulus Guardian Sight Helmet   0079249

parameters, navigation data, weapon status and fuel. The symbology is fully programmable to meet user requirements. A feature of the monocular display is the ability to provide the pilot with raster imagery from an external sensor such as a targeting pod or infra-red search and track system. The Guardian monocular display is in production.

Binocular display

The first Guardian binocular helmet developed is aimed at the helicopter market and has a wide field of view display (up to 60°), which can project night vision, or helmet-mounted night vision camera imagery. The night vision cameras form a key component of the system and are designed to replace the Night Vision Goggle (NVG), or folded NVG system. Pilkington claims that the use of night vision cameras, rather than NVGs, is advantageous in reducing helmet bulk and providing a better growth path to high-definition TV-standard imagery and improved brightness control. The first binocular display system flew in April 1997.

Head tracking system

The optical head tracker is an innovative, high-performance, proven system, which is now in production. Infra-red LEDs are positioned on the helmet and are tracked by cockpit-mounted miniature cameras. The system is immune to cockpit metalwork, 100 per cent sunlight compatible and highly accurate. Performance accuracies better than 4 mrad have been demonstrated. The head tracking system is in production.

Drive electronics

The display helmet drive box provides aircraft interfaces (MIL-STD-1553B, ARINC 422/429, analogues, discretes and synchros), processing, symbol generation, display processing, deflection amplifiers and power to the system. It is able to drive all the Guardian family of sights and displays. The drive electronics are in production.

Specifications

Accuracy: better than 5 mrad
Measuring envelope
Azimuth: −180 to +180°
Elevation: −90 to +90°
Roll: −90 to +90°

HMD imagery

Angular dimensions: 20° circular (day); 40° circular (night)
Exit pupil: 16 mm
Colour: green
Focused at infinity
Weight
Helmet (without NVGs): <1.6 kg
Total system weight: <15 kg
Power consumption: <120 W

Status

Progenitors to the current Guardian HDS have been in service in South African Air Force Mirage/Cheetah fighter aircraft since the early 1980s.

In June 1999, Thales Optronics delivered the first of two Guardian monocular display helmets to the Aircraft Systems Directorate of Defence Matériel Administration in Sweden for fitting to the Gripen simulator. The first flight of the system was conducted during 2000.

Thales Optronics has worked closely with Guardian team member Cumulus in optimising the system for the Gripen aircraft. The optical head tracker installation has been adjusted to match the Gripen's cockpit layout.

In a parallel effort to Guardian HMD development, during 2003, Denel secured two contracts from BAE Systems-Saab worth USD10.5 million to produce advanced tracking helmet systems for the Eurofighter Typhoon and for components for the Gripen HDS. Denel has developed its HTS for installation on the Typhoon.

BAE Systems and Denel have a strategic teaming agreement to use their respective strengths in the advanced helmet technology sector to capture a share of the worldwide market for military aircraft helmet systems. The team has competed for a number of additional new research and development contracts:

Eurofighter Typhoon

Denel has developed its high-speed HTS for the Eurofighter Typhoon. This system comprises three cockpit sensors (video cameras) that detect a series of LEDs built into the pilot's outer helmet. A head-tracker processor takes in data from these sensors and rapidly calculates the angle and position of the pilot's head. The information is used to correctly position the

display of vital symbology on the pilot's helmet-mounted display. The helmet-mounted display projects vital flight, instrumentation, navigation and mission data together with weapons and countermeasures status, directly onto the pilot's visor. The processor also drives external sensors and missile seekers, keeping them aligned with the pilot's line of sight. The latest version of the HTS was reported as tracking and processing data three times faster than its predecessor.

Export Gripen

BAE Systems Avionics Systems Division was awarded a contract during 2003 for the Guardian HDS for the export variant of Gripen. The system comprised a binocular display helmet, head tracker sensors and an Electronics Unit (EU) to be developed by Denel's Cumulus business unit.

The EU contains all the interfaces, processing and display drives necessary for the generation of symbology for display on the helmet visor. Development models were delivered during 2004. Subsequently, it is understood that the Guardian HDS was superseded by a variant of BAE Systems' Eurofighter Typhoon Integrated Helmet, designated Cobra (see separate entry).

Contractor

Thales Optronics Ltd.
Cumulus, a business unit of Denel (Pty) Ltd.

Helmet-Mounted Display (HMD) for the F-35 JSF

Type

Helmet-Mounted Display System (HMDS).

Description

The HMD for the F-35 JSF will replace the traditional Head-Up Display (HUD) as the Primary Flight Display (PFD) and provide for enhanced weapons cueing and the display of Electro Optic sensor information from the F-35's suite of sensors, including target imagery from the Electro Optic Targeting System (EOTS) and a 3 to 5 µm IR scene from the Distributed Aperture System (DAS) for night pilotage. Considering the HMD must also fulfil the more mundane responsibilities of impact protection,

This early configuration for the F-35 HMD features a monocular night vision camera mounted in the brow area of the helmet 1127385

noise attenuation and meet stringent weight and balance criteria, the task is not without risk, but the benefits in terms of overall cost savings and weight reduction are significant.

The VSI advanced HMDS employs a combination of electro-optics and head position and orientation tracking software algorithms to present critical flight, mission, threat and safety symbology on the pilot's visor. The system allows the pilot to direct aircraft weapons and sensors to an area of interest or issues visual cues to direct the pilot's attention. The HMDS comprises the helmet-mounted display, a Display Management Computer (DMC-H) and helmet-tracking system.

Fundamental requirements for the F-35 HMDS include visor-projected, binocular, Wide Field of View (WFoV), high-resolution, low latency imagery and symbology, equivalent accuracy to current HUDs (to qualify as a PFD), night vision for 24-hour operations, together with fit, comfort and safety during ejection. Proper weight and balance are crucial in minimising pilot fatigue resulting from high-*g* manoeuvres and reducing head and neck loads during ejection.

The display system employs a flat panel, Active Matrix LCD (AMLCD), combined with a high-intensity back light, as its image source. The fully

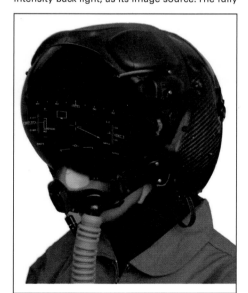

A significant change to the current configuration for the F-35 HMD involves the attached Night Vision Equipment (NVE). The earlier single centrally-mounted Night Vision (NV) camera (providing only biocular vision for the pilot) has been replaced by twin brow-mounted cameras (providing a fully overlapped binocular image for the pilot) in a similar arrangement the BAE Systems Integrated Helmet, which was eliminated as the Alternate HMD (AHMD) for the F-35 during the latter part of 2005. (Vision Systems International)   1180316

overlapped display provides a binocular 40 × 30° image (W × H). The digital image source provides both symbol writing and video capability with sensor overlap. The system includes a clear, optically coated visor for night operations and a shaded visor for daylight operations. Imagery is provided via the Distributed Aperture System (DAS) or a helmet-mounted day/night camera. F-35 pilots can select imagery and symbology via HOTAS commands. As is the custom with current high-performance HMDS, the two-part helmet is composed of an inner helmet, which provides a custom fit for maximum comfort and stability, and an outer helmet which incorporates the display electronics.

Specifications
Field of view: 40 × 30° binocular with full overlap
Eye relief: 50 mm on axis
HMD/DU weight: 1.86 kg (helmet, optics, oxygen mask and 3 inches of hosing)

Status
During August 2003, VSI announced that it had been awarded a USD84.65 million contract from Lockheed Martin for the development of its advanced Helmet Mounted Display (HMD) system for the F-35 JSF. The contract covers the Joint Strike Fighter (JSF) System Development and Demonstration (SDD) phase and planning for Low-Rate Initial Production (LRIP) and subsequent full-rate production phases. Under the contract, VSI will develop and deliver HMD units to support the JSF development and integration laboratories, JSF simulators and JSF flight test platforms. Additionally, the contract includes options for Technology Refreshment (TR) and Operational Test and Evaluation (OT&E) spares.

After an assessment phase, a production contract for more than 3,000 HMD systems, in support of JSF production, is expected to follow.

Contractor
Vision Systems International.

Joint Airborne Navigation and Attack (JOANNA)/Airborne System for Target Recognition, Identification and Designation (ASTRID)

Type
Airborne Electro-Optic (EO) navigation and targeting system.

Description
BAE Systems (Edinburgh) and Thales Optronique collaborated on the JOint Airborne NavigatioN and Attack (JOANNA) Technology Demonstration Programme (TDP) under a GBP20 million contract awarded by Defence Procurement Agency (DPA) in March 2001. France's Délégation Générale pour l'Armement (DGA) provided half of the government money and the two companies – which conducted the TDP as an integrated project team – also contributed to the funding. It was reported at the time that Italy and Spain were negotiating to join the effort as equal partners at both government and industry levels.

JOANNA was designed to act as a risk-reduction effort for the fourth-generation Airborne System for Target Recognition, Identification and Designation (ASTRID). Development of ASTRID was planned to take place during 2005-07, with production beginning in 2008. The pod-mounted version is planned to enter service in 2009-10 to equip aircraft such as the Eurofighter Typhoon and Rafale, in their proposed post-2010 versions (Eurofighter Tranche 3 and Rafale Standard F4). Elements of the system may also be installed internally aboard future Unmanned Combat Air Vehicles, possibly in support of the The UK Ministry of Defence's (MoD's) Strategic Unmanned Aerial Vehicle Experiment (SUAVE).

The emphasis of the JOANNA TDP is to research new technologies to achieve the aforementioned aims of greater stand-off and automation of tasks, leaving pilots with greater

capacity to perform support tasks. Facilities are likely to include fully automated tracking of multiple moving vehicle-sized targets. Other new technologies to be assessed will include active laser target illumination to provide very high-resolution images, automatic C2 (a feature that will autonomously generate 'assumptions' regarding the tactical situation, based on the platform types and movements of multiple targets tracked by JOANNA) and Automatic Target Recognition/Identification (ATR/I).

BAE Systems is responsible for the design of the electro-optical sensing section, with Thales providing the thermal imagers, in addition to handling image processing and overall management functions within the pod.

The design and manufacturing effort was due to begin during 2003, with its first flight during 2005 aboard a BAC 111 testbed aircraft. As a podded design, the sensor ball protrudes from the belly of the aircraft, with its electronics being mounted in the cargo hold, while the processing and supporting avionics are installed in the cabin.

The JOANNA TDP concluded with the submission of a final report at the end of 2005, which provided for recommendations to adjust the requirements for the ASTRID programme. The BAC 111 will continue to act as a testbed for complementary follow-on TDPs, in line with the philosophy that ASTRID will form one component of an integrated Situational Awareness (SA) suite that includes additional sensors – such as a radar, an infra-red search-and-track system, and electronic support measures – together with datalinks and navigation packages.

ASTRID will allow a varied target set to be automatically detected, tracked, identified and engaged from greater stand-off ranges than is possible with the current generation of pods and reduce crew workload. The system is expected to perform air-to-air and air-to-ground tasks with equal capability.

The follow-on TDPs are expected to take place in 2006-9/10. They will involve AESA radar technologies and hyperspectral sensors, in addition to ASTRID.

Status
ASTRID is believed to be ongoing.

Contractor
BAE Systems.
Thales Optronics.

Joint Helmet-Mounted Cueing System (JHMCS)

Type
Helmet-Mounted Display System (HMDS).

Description
Vision Systems International (VSI) was selected by Boeing and Lockheed Martin Tactical Aircraft Systems (jointly) to develop the JHMCS to equip US Air Force and US Navy fighters (F-15, F-16, F/A-18 and F/A-22) that the two companies produce.

The JHMCS is primarily designed to provide first shot, high off-boresight air-to-air weapons engagement under high-*g* conditions, allowing pilots to lock on and fire at enemy aircraft without having to manoeuvre their own aircraft into position; in combination with the AIM-9X, the JHMCS is known as HOBS – the High Off-Boresight Seeker, with the pilot utilising the helmet system to 'look and lock' to exploit the high angle-off capability of the missile.

The JHMCS, combining a magnetic head tracker with a display projected onto the pilot's visor, allows the pilot to aim sensors, air-to-air and air-to-ground weapons and view flight parameters without recourse to aircraft instruments or HUD. The helmet portion (either HGU-55/P or HGU-68/P) is common, with the removable display module replaced by the Night Vision Cueing and Display (NVCD) for night flying duties. While this should obviate the need for two separate helmets and facilitate pilot transition from day into night during a mission, suitable in-cockpit storage facilities for the fragile day/night modules in aircraft such as the F-16 could prove problematic. The overall system is designed to have low-HMD weight and optimised CG for the demands of fast-jet operations.

While primarily designed for the employment of highly agile air-to-air missiles, the system has been found to bring some interesting and highly valuable benefits to pilots currently using it. US Air Force F-15 pilots have reported significantly increased Situational Awareness (SA) while utilising JHMCS during all phases of combat. The ability to rapidly gain a three-dimensional air picture by scanning with the JHMCS and noting target cues (from onboard sensors) displayed, even though the bogeys and/or bandits are Beyond Visual Range (BVR), has been found to greatly enhance co-ordination between elements and flexibility of response to enemy manoeuvre. This feature has been found to be so effective, many pilots engaged in air-to-air operations elect to retain JHMCS at night, in preference to conventional NVGs, although this shortfall in JHMCS capability will be addressed with the introduction of the NVCD system.

To date, JHMCS has been deployed exclusively in single-seat or for the pilot only in dual-seat fighter configurations, which significantly limits full exploitation of the system's capabilities. In recognition of this situation, full integration of the helmet into the rear seat of the Boeing F/A-18D began during the early part of 2005. This will undoubtedly bring greater SA to both crew, facilitated by almost instantaneous target hand-off between cockpits and seamless integration of the pilot's and Weapon System Operator's Mental Air Picture (MAP).

QuadEye™ Modular NVG

A VSI Wide Field Of View Night Vision Goggle (WFOVNVG), designated QuadEye™, has been selected for the NVCD system, the night vision 'partner' for JHMCS. The NVCD will dramatically expand the capability and effectiveness of JHMCS by providing image intensified night vision integrated with standard HMD symbology and Line of Sight (LoS) cueing. An interesting feature of QuadEye™, in contrast to the similar Integrated Panoramic Night Vision Goggle (IPNVG – see separate entry) is that it is modular, providing a twin-tube circular 40° FoV in standard configuration, which can be expanded, as/if required, to 100 × 40° FoV by the addition of a further two outboard tubes.

The US Navy's prerequisite for integrated night vision included the requirement to leverage existing technology without modification to the aircraft's installed JHMCS hardware. As JHMCS is a modular 'day-only' system, the JHMCS Display Unit (DU) can be quickly exchanged with the NVCD QuadEye™ DU to support round-the-clock missions.

Specifications

Field of regard: unlimited
Field of view: 20° monocular, right eye
Exit pupil: 18 mm on-axis; 16 mm off-axis
Enhanced cueing: 2 LED reticles
Eye relief: 50 mm on axis
Compatible helmets:
   US Air Force: HGU-55/P
   US Navy: HGU-68/P
HMD/DU weight: 1.82 kg (with mask)
MTBF: 1,000 h
Electronics dimensions: 228.6 × 177.8 × 127.0 mm
Electronics weight: 6.82 kg

Status

Vision Systems International LLC (VSI) was established by Elbit Systems Ltd of Israel, through its US subsidiary EFW Inc of Fort Worth, Texas and Kaiser Aerospace & Electronics Corporation of Foster City, California, specifically to pursue Helmet-Mounted Display (HMD) systems opportunities worldwide. Kaiser was subsequently acquired by Rockwell Collins and EFW Inc is now an Elbit Systems of America (ESA) company.

In production and in service in US Navy (USN) F/A-18E/F Super Hornets and US Air Force (USAF) F-15C Eagle aircraft.

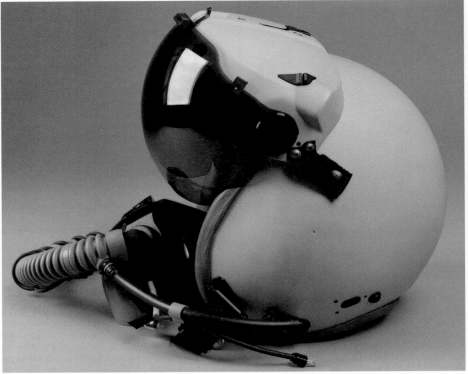

The VSI joint helmet-mounted cueing system, shown fitted with the removable display module
0099195

A Royal Danish Air Force (RDAF) pilot wearing JHMCS during start-up in a Block 10 F-16B. The helmet is part of the current M3 update for RDAF F-16s undertaken by a consortium of the RDAF, TERMA A/S and EADS Defence Electronics, Ulm   1039339

First flights of JHMCS were performed on USAF F-15 and USN F/A-18 aircraft in January 1998. Boeing received approval for Low-Rate Initial Production (LRIP) for the system from the USN during July 2000, and subsequently issued orders to VSI, in a contract worth approximately USD15 million, to begin production of JHMCS for the F/A-18E/F Super Hornet aircraft. The Navy received an initial batch of approximately 40 systems during 2002.

At this time, Boeing was also contracted by the USAF to commence LRIP of the system for its F-15 and F-16 aircraft. The USD33 million contract was for 9 units for F-15s and 28 units for F-16s. The Hellenic Air Force also ordered 55 units for its F-16s, with options for a further 10.

During the first quarter of 2003, Boeing awarded a third LRIP contract to VSI, worth USD60.1 million, for more than 300 JHMCS to fill domestic production commitments and retrofit obligations for the F-15, F-16 and F/A-18.

At the beginning of May 2003, VSI was awarded a USD3.3 million contract from the USN for the development of a Night Vision Cueing and Display system (NVCD), based on their QuadEye™ WFOVNVG.

During the first quarter of 2004, VSI announced a fourth LRIP contract (Batch 4) for a further 400+ JHMCS under a USD62.6 million contract from Boeing. These systems will further equip USAF F-15 and F-16 fighters, together with USN F/A-18s. They will also be supplied under the US Foreign Military Sales (FMS) programme to Australia (F/A-18), Chile (F-16), Finland (F/A-18) and Poland (F-16). In addition, under a commercial sale, JHMCS will be supplied to the Republic of Korea Air Force for its F-15Ks. The initial value of the contract, USD37.3 million from US Air Force Materiel Command is expected to increase to at least USD54.9 million. Deliveries are due to be completed by December 2005.

Integration of the system into the F-15K will most likely be conducted in parallel with the F-15E, as operated by the USAF.

During August 2004, VSI was awarded a USD75.6 million contract for more than 300 JHMCS systems from Boeing. This award was the first Full Rate Production (FRP) contract for the JHMCS. This contract was to fill additional US government domestic requirements for the USAF and Air National Guard (ANG) F-15, F-16 and USN F/A-18 platforms as well as FMS production commitments for Australia (F/A-18), Finland (F/A-18), Greece (F-16), Poland (F-16) and Switzerland (F/A-18). Deliveries under FRP 1 commenced in March of 2005 and will continue through December 2006.

During March 2005, VSI announced a contract from Hellenic Air Force for JHMCS for its new fleet of F-16 block 52 aircraft.

Elsewhere, JHMCS was also integrated into the F-35 Joint Strike Fighter (JSF) aircraft demonstrator and Lockheed Martin Tactical Aircraft Systems (LMTAS) has been conducting further development work on an aircraft modification kit to incorporate the JHMCS and the Link 16 datalink into the F-16 aircraft, to support FMS to Belgium, Denmark, the Netherlands and Norway as part of the F-16 MLU.

During May 2005, the USN awarded VSI a USD3.3 million contract for the development of the NVCD. Later, during July, the NVCD made its first flight on board an F/A-18F. The mission profile included two sorties in an F/A-18 two-ship formation. As part of the familiarisation process, the pilots performed various air-to-ground, air-to-air, low level and formation manoeuvres. During September, the NVCD made its first dual-seat flight in a Naval Air Warfare Centre (NAWCAD) F/A-18F. For the demonstration, the pilot's and weapons systems officer's independent Lines of Sight (LoS) were integrated with the aircraft weapons and sensors.

Continuing a highly successful 2005, in October, VSI announced the receipt of several new contracts with a total value of more than USD100 million from Boeing for the delivery of more than 500 additional JHMCS as part of a second FRP (FRP-2) for US government domestic requirements for USAF F-15 and F-16, ANG F-15, USN F/A-18 single- and dual-seat platforms as well as FMS production and spares commitments.

During November 2005, VSI announced the delivery of the 1,000th JHMCS; at the same time, the US Air Force announced its intention to integrate Laser Eye Protection (LEP) into the JHMCS day module.

As of mid-2006, VSI had been awarded three FRP lots totalling 1,300 systems over and above LRIP awards.

Contractor
Vision Systems International.

LITENING airborne laser target designator and navigation pod

Type
Airborne Electro-Optic (EO) targeting system.

Description
LITENING is a multipurpose, day/night, precision laser targeting and navigation pod originally conceived and developed for fighter aircraft by the Rafael Armament Development Authority of Israel during the early 1990s. After Rafael conduced an agreement with Northrop Grumman for further co-development and production of the system in 1995, the first US variant was designated LITENING II, or AN/AAQ-28(V)2, with later variants following this nomenclature. While the original LITENING design, fielded by the Israeli Air Force during the mid-1990s, incorporated a mechanically scanned Forward Looking Infra-Red (FLIR), all subsequent variants incorporate a staring array MWIR FLIR, Charge-Coupled Device (CCD) TV, Laser Spot Tracker/Range Finder (LST/RF), laser marker and multifunction tracker (area, inertial, correlation). An on-gimbal Inertial Navigation Sensor (INS) has a stabilised Line of Sight (LoS) and automatic boresight capability.

All sensors are fixed relative to each other, providing for good operational boresight retention. The INS stabilises the line-of-sight and automatically aligns the pod boresight to the aircraft INS, making mechanical boresighting

unnecessary. The pod front section has an integral compressor that maintains a minimum atmospheric pressure, allowing the laser designator to operate at altitudes up to 50,000 ft. Modular design of software facilitates aircraft integration.

These components of the LITENING system provide the following capabilities:
- The high-resolution FLIR provides a day/night attack capability
- The CCD TV is designed to enhance the stand-off capability of the system during daylight operations
- The FLIR's wide field of view provides a HUD-compatible display for low-altitude navigation and enhanced Situational Awareness (SA) during air-to-ground attacks
- The laser spot tracker provides the system with the capability to track laser spots from a second laser source (either ground-based or 'buddy lasing'), while the laser marker allows the pod to illuminate targets (at approximately 0.8 µm) for co-operative attack by forces equipped with Night Vision Goggles (NVGs), which are sensitive in the 0.6 to 0.9 µm waveband.

The INS and software design is optimised to permit easy integration in a wide variety of US combat aircraft, such as the B-52, A-10, AV-8B, F-4, F-5, F-15, F-16 and F/A-18, as well as international combat types, including the Jaguar, JAS 39 Gripen, MiG-21, Mirage 2000, Tornado, Eurofighter Typhoon and others.

Northrop Grumman and Rafael have integrated many improvements to the baseline LITENING II system since initial contract award for the US Marine Corps, Air National Guard and Air Force Reserve Command in 1998, resulting in the LITENING ER and AT variants (as shown below).

LITENING ER – AN/AAQ-28(V)3
LITENING ER incorporates a 640 × 512 pixel FLIR camera, which is claimed to improve target tracking and recognition ranges by 30 per cent, enabling the target acquisition portion of the system to more closely match the standoff capability of the high-power laser and extended release ranges of modern laser-guided weaponry.

Further extending the capabilities of the system, during January 2003, Northrop announced the successful modification of a LITENING ER pod to enable a datalink between a US Marine Corps AV-8B aircraft and a ground station for a Pioneer UAV. Modifications involved mounting a Government-Furnished Equipment (GFE) data/video link transmitter, compatible with the Pioneer UAV system in the pod, along with a Commercial Off-The-Shelf (COTS) transmit antenna. Operational evaluation flights of the AV-8B aircraft equipped with the modified LITENING pod were conducted at the US Marine Corps Air Station in Yuma, Arizona, where cockpit imagery from the pod was successfully relayed to a Pioneer ground station via the GFE transmitter.

LITENING AT – AN/AAQ-28(V)4
The latest variant of the system, designated LITENING AT (Advanced Targeting), a further development of LITENING ER, incorporates an improved 640 × 512 pixel FLIR, new circuitry and algorithms designed to extend target detection range and improve Target Co-ordinate Generation (TCG) accuracy to support J-Class GPS-guided munitions, together with enhanced image processing capabilities, multitarget cueing and improved air-to-air performance.

LITENING AT also includes provision for a diode-pumped NdYag dual-wavelength (operational /eyesafe) laser to increase laser energy and improve designation range, while also providing safe training.

LITENING AT+
Continuing the Spiral Development of the LITENING system, Northrop Grumman and Rafael have developed further enhancements to the now baseline LITENING AT. The first enhancements involved upgrade of the onboard datalink, previously taken from the Pioneer UAV, to that from the Predator UAV. Additionally, the introduction of Electronic Stabilisation (E-Stab), microscan and an improved Automatic Gain Control (AGC+) conveys an increased identification range of the IR sensor of up to 30 per cent. A 5-target multitrack capability will enable more weapons to be employed on a single pass. All of these upgrades are defined by the Company as 'software only'. LITENING AT+ was introduced during 2005.

LITENING AT++
Somewhat imaginatively, the next stage of development for the LITENING system is described as LITENING AT++; this is a hardware and software modification that introduces a high-resolution CCD TV sensor (1,000 × 1,000 pixels) for improved identification range during daytime. Further into the future (LITENING AT+++), Northrop and Rafael plan to upgrade the IR sensor to a 1,000 × 1,000 array to match the CCDTV, possibly to coincide with the introduction to service of the F-35 Joint Strike Fighter.

Specifications
(LITENING AT)
Dimensions: 2,200 (L) × 406 mm (D)
Weight: 200 kg
FLIR FPA
picture elements: 640 × 512
FoV:
(narrow) 1.0 × 1.0°
(medium) 4.0 × 4.0°
(wide) 24 × 24°
CCD camera
picture elements: 768 × 494
FoV:
(narrow) 1.0 × 1.0° (with zoom down to 0.5 × 0.5°)
(wide) 3.5 × 3.5°

The LITENING II pod saw significant action with the US ANG over Afghanistan in 2002   0079248

A-10A Thunderbolt with a LITENING ER pod
(Northrop Grumman) 0566135

Laser designator and range-finder
Energy: 100 mJ per pulse (135 mJ in LITENING AT+)

Non-firing training mode incorporated (32 mJ training laser in LITENING AT+)

Trackers: advanced correlator/inertial laser spot search and track

Gimbals:

(fields of regard) +45 to −150° pitch; ±400° roll (stabilisation) 10 μrad

Flight envelope: (at low altitude) M1.2; (manoeuvre) 9 *g*

Cooling: Self-contained environmental control unit

Status
AN/AAQ-28(V)1 LITENING
No longer in production, in service. AN/AAQ-28(V)2 LITENING II
No longer in production, in service.

AN/AAQ-28(V)3 LITENEING ER
No longer in production, in service.

AN/AAQ-28(V)4 LITENING AT
In production and in service.

AN/AAQ-28(V) LITENING AT+/LITENING III
In production and in service.

AN/AAQ-28(V) LITENING AT++
Under development.

LITENING has been integrated with the following aircraft:
- German IDS Tornados, as part of the Mid Life Update (MLU) programme carried out by EADS Deutschland for the German Air Force and Navy.
- Indian Jaguar and Mirage 2000.
- Israeli F-16C/D Block 30 and Block 40 aircraft.
- Romanian MiG-21, as part of the Lancer upgrade programme.
- Venezuelan F-16A/B Block 15 aircraft.
- Greek F-4 aircraft, as part of the EADS MLU.
- Italian Navy AV-8B Harrier II (LITENING II).
- Spanish Navy AV-8B Harrier II (LITENING II).
- Spanish Air Force F-18.
- Swedish Air Force JAS 39 Gripen.
- Singaporean F-16 (LITENING III).
- Chilean Air Force F-16 (LITENING III).

In the case of the German Tornado MLU, the 36 pods (20 for the Air Force and 16 for the Navy) were fitted with electro-optical systems provided by Zeiss Optronik GmbH.

Regarding US activity, in August 1998, Northrop Grumman and Rafael were awarded an initial contract worth nearly USD18 million to supply LITENING II targeting pods to the US Air Force Reserve Command (AFRC) and the US Air National Guard (ANG) for use on F-16 aircraft, where it was designated the Precision Attack Targeting System (PATS). The initial contract was for eight pods to achieve initial operating capability by March 2000. Qualification tests were flown at Edwards AFB, California, in late 1999, with operational test and evaluation occurring in early 2000. The initial order involved a total of 168 LITENING II pods for AFRC/ANG F-16 C/D aircraft, with first deliveries made to the 457th Fighter Squadron, AFRC in February 2000.

LITENING II saw operational service with the ANG over Afghanistan in 2002, where accuracy and, importantly, reliability were reported to have

A LITENING ER pod being loaded onto a B-52H (Northrop Grumman) 0576355

been impressive, particularly when compared with older LANTIRN systems (see separate entry).

Integration of the LITENING ER pod was carried out for the US Marine Corps' AV-8B Harrier II in August 2001 and for the US Air Force's F-16 in January 2002. During July 2002, a LITENING ER-equipped A-10 made eight successful flights at Davis-Monthan Air Force Base, Tucson, Arizona, in support of a precision engagement risk reduction operational utility evaluation. The integration required no changes to the aircraft's existing software. A 'smart' cable, connecting the pod to the aircraft, allowed full functionality with only minor updates to the LITENING software. Following the flights, the A-10 deployed to Nellis Air Force Base in Las Vegas, Nevada, where it used laser-guided weapons and participated in the Joint Experimental Force Experiment.

Integration of LITENING ER with more platforms followed in November 2002 (F-15E), with an order for 24 pods, worth USD32.6 million, following in March 2003. Integration of LITENING ER on the B-52H Stratofortress was announced in April 2003, with an order for 12 LITENING AT systems, made in April 2004, awaiting funding approval.

During January 2003, initial demonstration flights of LITENING ER onboard a US Navy F/A-18D Hornet aircraft were carried out at the Patuxent Naval Air Station in Maryland, building on techniques employed by the Spanish Air Force on their EF-18s. During the demonstrations, the pod successfully tracked targets using both CCD TV and FLIR sensors. During March 2004, the USMC replicated this effort for their F/A-18Ds.

LITENING AT successfully completed initial evaluation at the end of 2002, with additional testing scheduled for March 2003 as part of an incremental update/test process. During March 2003, Northrop Grumman announced the first US Air Force contract to upgrade ANG LITENING pods to AT configuration. The USD19.7 million contract for ANG F-16 aircraft provided for the retrofit of 45 LITENING II and 19 LITENING ER pods to AT status, with options for additional upgrades. By March 2004, the first four production LITENING AT systems and upgrade kits, together with 14 additional systems had been delivered. Subsequently, the ANG awarded Northrop Grumman a separate USD10.3 million contract for a further eight LITENING AT pods. At the same time, Northrop announced that, as part of a USMC contract worth some USD40.3 million, the first ten LITENING AT systems of a total order for 20 pods, spares and support, combined with the retrofit of 47 ER pods, had been delivered.

Also during March 2004, Northrop announced its support for an initiative by the US Navy's F/A-18

Programme Office at the Patuxent Naval Air Station to integrate LITENING AT onboard USMC F/A-18 Hornets. Later that year, on 27 August the USMC successfully completed integration testing of pod. The tests culminated with a pod on a F/A-18D guiding a laser-guided bomb to a direct hit on a target. The USMC was reported as intending to procure 60 pods to support its fleet of 72 F/A-18Ds.

During September 2004, The Royal Australian Air Force (RAAF) announced that it intended to equip its F/A-18s with an advanced target-acquisition system under a program costing more than AUD100 million (USD71 million). The three contenders were the Raytheon ATFLIR, the Lockheed Martin PANTERA and the Northrop Grumman/Rafael LITENING AT. LITENING AT emerged as the victor, announced in 2005, with the first squadron to be equipped by early 2007.

During the summer of 2005, the US Marine Corps fleet of F/A-18D Hornet aircraft passed 10,000 combat flight hours with LITENING AT.

During 2006, F/A-18C/D Hornet aircraft received a new clearance to carry the LITENING pod on Station 4 (left chin station), in addition to the extant clearance for Station 5 (centreline). The US Marine Corps and Royal Australian Air Force (RAAF) both intend to operate with the LITENING AT pod aboard their F/A-18s on station 4. The Hornet typically carries its targeting system on station 4 to preserve the centreline station for a fuel tank or other stores.

At the end of August 2006, Northrop Grumman announced more detailed plans for the development and fielding of its fourth development (AN/AAQ-28(V)5) of the LITENING system. The latest variant will amalgamate the planned upgrades of the aforementioned LITENING AT++ and AT+++, with an advanced 1,024 × 1,024 pixels FLIR sensor for improved target detection and recognition ranges under day/night conditions, new two-way data links and other networking capabilities to enable improved communications between ground-based and airborne forces, new sensors for improved target identification and other advanced target recognition and identification features. Integration of the 1k × 1k CCD sensor, also planned as part of the new variant, have been completed ahead of schedule.

As of August 2006, over 360 LITENING pods of all variants had been delivered out of a total order of 400 covering six different platforms.

Contractor
Rafael Armament Development Authority Ltd, Missiles Division.
Northrop Grumman, Electronic Systems Sector, Defensive Systems Division.

Noctua Pilot Night Vision System (PNVS)

Type
Pilot's Night Vision System (PNVS).

Description
The Noctua Pilot Night Vision System (PNVS) comprises the Guardian binocular Helmet Display System (HDS – see separate entry) and head-steered piloting FLIR, enabling the pilot to fly low-level nap-of-the-earth missions in total darkness.

The high slew rate of the piloting FLIR has been designed specifically to keep up with the pilot's head movement. Using a second-generation 8 to 12 µm FLIR, it provides a fully enhanced night vision image for display.

The system is designed to optimise conventional utility helicopter operations such as combat search and rescue, reconnaissance, scout and Special Forces insertion missions. The pilot can select the optimum sensor from either the turret (thermal camera), or helmet (integrated image-intensified cameras).

Specifications
HMD
Positional accuracy: better than 5 mrad
Tracking envelope
Azimuth: –180 to +180°
Elevation: –90 to +90°
Roll: –90 to +90°
Exit pupil: 16 mm
Weight
Helmet (without NVGs): <1.6 kg
PNVS
Field of view (h × v): 40 × 30°
Spectral band: 8–12 µm
Detector: 2nd-generation (288 × 4 IRCC)
Interface: RS422
Video output: PAL CCIR 625 lines
Elevation: –50 to +45°
Azimuth: –120 to +120°
Positional accuracy: ≤5 mrad
Angular slew rates: >150°/s
Angular accelerations: >1,000°/s²
Platform diameter: 280 mm
PNVS total weight: 24.5 kg

Status
In production and in service in South African Air Force (SAAF) Rooivalk attack helicopters.

Contractor
Thales Optronics Ltd.
Cumulus, a business unit of Denel (Pty) Ltd.

SAAF Rooivalks are equipped with a nose-mounted turret for the Target Detection and Tracking System (TDATS), which accommodates thermal and television cameras, a laser-rangefinder, a laser designator and an autotracker. The upper (smaller) turret is the Noctua PNVS
0083238

OSIRIS: Tiger anti-tank helicopter mast-mounted sight

Type
Airborne Electro-Optic (EO) targeting system.

Description
OSIRIS is a day/night Mast-Mounted Sight (MMS) which detects, acquires and transfers potential target information to its associated missile system. The gyro-stabilised OSIRIS sight head includes a CONDOR 1 Thermal Imager (TI), low light TV camera and an optional laser range-finder. The system operates in three wavebands: 0.5 to 0.7 µm, 0.7 to 1.0 µm (NIR) and 8 to 12 µm (LWIR). The system is mast-mounted, which, coupled with a completely passive detection capability, enables the host platform to remain concealed and undetected during the reconnaissance and target acquisition phases of an attack.

OSIRIS was developed as part of the third-generation European LR-TRIGAT anti-armour missile system for the Tiger attack helicopter.

Status
OSIRIS production started in June 1999 for the German Army UHT version of the Tiger helicopter.

Contractor
SAGEM Défense Securité, Navigation and Aeronautic Systems Division.

SAGEM OSIRIS Mast-Mounted Sight (MMS)
0085465

PIRATE Infra-Red Search and Track System (IRSTS)

Type
Airborne Electro Optic (EO) Infra-Red Search and Track System (IRSTS).

Description
Galileo Avionica, part of Selex Sensors and Airborne Systems, is the leader of the Eurofirst Consortium, also comprising Thales Optronics Ltd and Grupo Tecnobit, which is developing the Passive Infra-Red Airborne Track Equipment (PIRATE) for the Eurofighter Typhoon.

PIRATE is a combined Infra-Red Search And Track (IRST) and Forward-Looking Infra-Red (FLIR) system, capable of passive target detection at extreme range (IRST mode, depending on target signature), over a wide Field of View (FoV) (mode dependent) and under certain conditions of poor visibility (subject to atmospheric attenuation effects). As a totally passive sensor, it enables the aircraft to gather early intelligence of threats and to manoeuvre into a tactically advantageous position without being detected by hostile ECM systems. As part of the Eurofighter integrated avionics suite, PIRATE provides sensor information and imagery for cockpit Multifunction Head-Down Displays (MHDDs) and Helmet-Mounted Display (HMD) (see separate entries), enhancing pilot Situational Awareness (SA) in both air-to-air and air-to-ground missions.

In IRST Multiple Target Track (MTT) mode, the PIRATE system performs automatic detection, prioritisation of multiple targets over the entire Field of Regard (FoR), performed on a Track-While-Scan (TWS) basis. Single Target Track (STT) can be selected for any target acquired within this or any other selected scan volume. The system Line-of-Sight (LoS) can also be slaved to other aircraft sensors or the pilot's HMD to enable high angle-off weapons employment. In Steerable Infra-Red Picture onto Helmet (SIRPH) mode, the IRST is slaved to the pilot's LoS, displaying the sensor picture on the visor.

For the Eurofighter Typhoon installation, the PIRATE sensor is mounted above the radome (optimum positioning for IRST air-air operations) and slightly to the left of the aircraft centreline (allowing for limited look-down in FLIR mode). Thus, FLIR-mode air-ground applications (particularly at low level) will be inevitably compromised by this arrangement, although it is understood that much work has been undertaken to minimise interference with

Eurocopter Tiger prototype with OSIRIS Mast-Mounted Sight (MMS)
1043989

Eurofighter DA 1 on take-off for an Advanced Medium-Range Air-to-Air Missile (AMRAAM) firing trial. Note the PIRATE IRSTS on the upper left side of the nose and the inward facing high-speed cameras mounted in dummy drop tanks   0121565

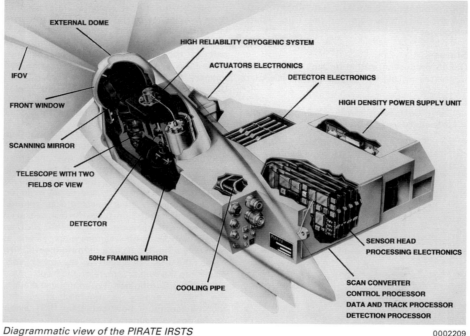

Diagrammatic view of the PIRATE IRSTS   0002209

look-down modes due to energy absorption by, and re-radiation from, the radome. With these limitations to the Eurofighter installation in mind, the PIRATE system provides the following FLIR modes:

- Navigation (fixed forward view overlaid on the HUD, or slaved to the pilot's HMD)
- Thermal cueing (hot spots, according to spot size, thermal contrast, position relative to flightpath/HMD sightline)
- Identification (in STT mode, the IR image can be presented on the MHDD and frozen to allow for visual identification).

The system utilises proven signal-processing technology derived from the Thales (formerly Pilkington) Optronics Air Defence Alerting Device (ADAD), which demonstrates a very high suppression rate of false alarms.

Specifications
Spectral band: 3–5 and 8–11 μm
Detection range: excess of 50 nm

Status
A development contract to supply equipment for the Eurofighter Typhoon aircraft was awarded in 1992. Flight trials were conducted during 1999 using pre-production hardware, with full system testing and development of production-standard equipment during 2000/2001. As of 2004, development of the system was continuing towards introduction to operational service.

Contractor
The Eurofirst Consortium, comprising:
Selex Sensors and Airborne Systems, Radar Systems Business Unit.(prime).
Thales Optronics Ltd.
Grupo Tecnobit.

Scout helmet-mounted cueing system

Type
Helmet-Mounted Sight System (HMSS).

Description
The Scout helmet-mounted cueing system, derived from the military Guardian family (see separate entry) of HMDS, is a simple, monocular system designed for civilian/paramilitary helicopter applications, including Border Patrol, Search and Rescue and police observation. The system cues a helicopter observation system directly towards an area of interest, providing immediate acquisition and tracking of any moving or stationary item, thereby significantly reducing the observer's workload and acquisition time.

Scout employs a similar EO Helmet-Tracking System (EOHTS) to the Guardian/Archer HMDS, together with a unique monocular cueing module.

Status
In production.

Contractor
Thales Optronics Ltd.
Cumulus, a business unit of Denel (Pty) Ltd.

Terminator infra-red targeting system

Type
Airborne Electro-Optic (EO) targeting system.

Description
Raytheon Electronic Systems leads a private venture team that is developing the Terminator infra-red targeting system for the high-performance standoff targeting role. Derived from the Advanced Targeting Forward-Looking Infra-Red (ATFLIR) system (see separate entry), Terminator is a new, claimed 'third-generation' design, incorporating advanced technology with better performance and reliability than existing first-generation targeting pods, and at lower cost.

BAE Systems, Avionics has contributed its Mid-Wave Infra-Red (MWIR) navigation FLIR, Laser Spot Tracker (LST), roll drive unit and

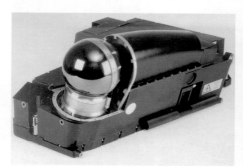

PIRATE IRSTS sensor   0121566

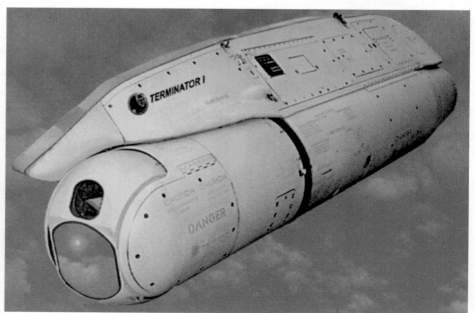

The Terminator infra-red targeting pod   0079253

aircraft-specific adapters to the design, while Raytheon has contributed its experience of mid-wave, large-format, staring array technology.

Specifications
Focal plane: 640 × 480 InSb
Spectral band: 3.7–5.0 μm
Fields of view: 5.0 and 0.83°
Recognition range: 10 n miles
MTBF: >300 h

Status
In production and in service.

Contractor
Raytheon Company, Space and Airborne Systems.
BAE Systems, Avionics.
Fairchild Controls.

Tiger helicopter Helmet-Mounted Display (HMD)

Type
Helmet-Mounted Display System (HMDS).

Description
Developed by Selex (who also call the system Knighthelm) and TELDIX, this binocular 40° field of view system incorporates a one-piece display module for both day and night mission requirements. For low-visibility and night missions, the system displays both intensified images from third-generation image intensifiers and imagery from a Forward-Looking Infra-Red (FLIR) sensor. This, together with flight and weapon symbology information, is projected onto clear combiners placed in front of each of the pilots' eyes. The combiners may be rapidly flipped up out of the line of sight if required. Pilot comfort and mission capability requirements have been fully integrated, and all helmet controls for tactical sensors and weapons may be controlled by Hands-On Collective and Stick (HOCAS) or via a control panel. Sensors and weapons are interfaced to the helmet via Selex' advanced Helmet Tracking System (HTS – see separate entry).

Specifications
Weight: 2.2 kg
Field of view: 40°
Eye relief: 30 mm

Status
Selected for the development phase of the Integrated Helmet System for the German Tiger helicopter. Development of the HMD is finalised with qualification complete. Integration into Tiger was conducted during 1998/9. Subsequently selected for German Tiger variants, although French Tigers will utilise the Topowl HMD (see separate entry).

Contractor
Selex Sensors and Airborne Systems.
TELDIX GmbH.

Viper Helmet-Mounted Display (HMD) systems

Type
Helmet-Mounted Display System (HMDS).

Description
The Viper family of helmet mounted displays provides a low-cost, lightweight visor-projected

Viper I HMD　　　　　　0001461

display capability by addition of a display module to almost any type of flying helmet. The Viper family comprises four variants; Viper I, Viper II and Viper III (now redesignated Night Viper) and Viper IV.

Delft Sensor Systems supplies the optical modules for Viper I and II to BAE Systems. Viper III (Night Viper) was developed by Delft Sensor Systems in co-operation with BAE Systems. Viper IV is a BAE Systems development.

Viper I
The Viper I helmet mounted display is a monocular system providing a field of view suitable for off-boresight missile aiming.

The visor reflects dynamic flight and weapon aiming data to the pilot from a high-efficiency miniature CRT display projected via an optical relay assembly. The optical design allows the use of a standard aircrew visor with the addition only of a neutral density reflection coating. This technique enables a high outside world transmission without colouration. The design also ensures that the displayed image is stable and accurate even when the visor is raised.

Although primarily configured for daytime use, the Viper I system is capable of displaying video from a sensor, providing the pilot with an enhanced cueing system after dark or in bad weather.

Specifications
Viper I
Weight: 1.72 kg (on a US standard flying helmet)
Eye relief: >80 mm
Eye relief: >70 mm
Exit pupil: 15 mm
Real world transmission: >70%

The Tiger integrated helmet-mounted display　　　　0001459

Viper II HMD　　　　　　0001460

Night Viper pilot's day and night helmet 0051715

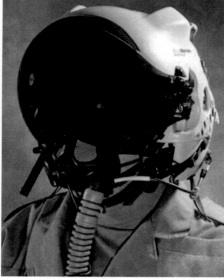

Viper IV HMD 0080272

Viper II

The Viper II helmet mounted display provides a binocular, fully overlapped, 40° field of view capability. The system is able to receive sensor video (for example, from a Forward Looking Infra-Red (FLIR) sensor), and display this together with overlaid flight and weapon aiming symbology.

The Viper II has a similar optical concept to the Viper I, and by use of a spherical visor, maintains a stable image within the large exit pupil regardless of visor position. Full interpupillary adjustment is also provided.

Specifications

Viper II
Weight: 1.54 kg (on a US standard flying helmet)
Field of view: 40°
Eye relief: 81 mm to visor
Exit pupil: >15 mm
Real world transmission: 88%

Other common Viper I and II features

1. both models fit neatly onto all sizes of standard UK, US Air Force and US Navy fixed-wing aircraft helmets;
2. the HMD incorporates provision for a lightweight video camera head. When the output of this camera is mixed with scan converted helmet-mounted display symbology, a video output is available for recording purposes and subsequent use on the ground for mission review and training;
3. they provide full space stabilised symbology when coupled to a suitable Head Tracking System (HTS), such as the advanced DC system developed by BAE Systems;
4. the Viper helmet mounted display system includes an Electronics Unit (EU) and a Cockpit Unit (CU). These units can be configured to provide a variety of functions, depending on user requirements and which Viper variant

is required. These functions include: display processor, video processing, MIL-STD-1553 bus interface, symbol generation, CRT display drive, high voltage power supply, display control and head tracker interface; further improvements to the Viper family of HMDs are currently in development.

Night Viper (Viper III)

The Night Viper is a day and night helmet with integrated image intensifier tubes and a binocular visor projected helmet mounted display. The lightweight concept is based on the successfully flight proven Viper I and Viper II concepts and was formerly designated Viper III (The designation was changed to Night Viper in November 1998). Delft Sensor Systems developed the pilot night vision system integrated into the helmet. The visor combines and reflects night vision images to the pilot, which are collected by the helmet-mounted objectives. Use of a spherical visor means display accuracy is insensitive to visor rotation. The product provides both see-through vision and an extremely high degree of ejection safety. Applications include both helicopter and fixed wing aircraft without any update or modification of any electronic system. The Night Viper module can be fitted to all sizes of HGU55/P helmets. The optical relay system can be adjusted for individual interpupillary settings with no changes or modifications to the visor. Since there is no interface to the aircraft, an electronics box is not part of the system.

Specifications

Field of View: full binocular 45°
night image: 40°
overlapping L + R: >30° horizontal
see-through image: 180 × 110°
Exit pupil: 15 mm on axis
Image intensifier tubes: standard ANVIS 18 mm inverting SuperGen or Gen3

Resolution at 30 mlux with Omnibus IV Gen3 tubes: better than 1 mrad
Dioptre adjustment: not applicable
Eyerelief: >75 mm, compatible with eye glasses
Interpupillary distance: 55 to 74 mm with centred exit pupil
Transmission: real world 35% – other values upon request
Weight on HGU55/P:
(incl oxygen mask, earphones, liners, and so on): <2.3 kg
Centre of gravity: in 'Knox box' (COG displacement <0.89 inches)

Viper IV

Viper IV is a visor projected HMD system, offering a 40° field of view, using monocular or binocular display on a spherical visor. Viper IV provides a large (>20 mm) exit pupil. The system is claimed to provide excellent stability with no loss of display at high *g* and has been windblast tested to 550 kt. Viper IV was the HMD employed in the Variable Stability In-Flight Simulation Test Aircraft (VISTA) F-16 aircraft during research into a virtual HUD concept, itself part of risk-reduction activities concerned with future fighter cockpit environments.

Viper IV provides off-boresight missile designation with reverse cueing and multiple target designation and tracking in the air-to-air scenario. For air-to-ground operations, it provides target designation with CCIP/CCRP symbology. Head Line Of Sight (LOS) is computed by an electromagnetic head tracker system, which is also used to space-stabilise the symbology, and to slew missile seeker heads or other sensors.

The Viper IV display system uses either a single (monocular) or dual (binocular) CRTs, mounted in the spherical 'bumps' in the upper part of the outer shell, to provide stroke symbology over sensor video. It uses a high-brightness daytime mode; at night, video from head-steered sensors, such as FLIR or night vision cameras, can be overlaid.

Status

The operational status of the various Viper HMD products is understood to be as follows:

Viper I has been trialled as part of the European Partner Air Forces (EPAF) F-16 Mid-Life Update (MLU) programme and was also involved in the X-31 concept programme;

Viper II was evaluated by DASA in the German Air Force Tornado turreted FLIR programme and employed by QinetiQ as the HMD for research into a Visually Coupled System (VCS).

Night Viper was trialled by the Royal Netherlands Air Force.

Viper IV was employed as part of a Lockheed Martin/BAE Systems trial into a Virtual HUD (VHUD) concept during 2000. It is believed that a derivative of this helmet will form the HMD for the EPAF F-16 MLU.

Contractor

Selex Sensors and Airborne Systems.
Delft Sensor Systems.

Israel

COMPASS: COmpact MultiPurpose Advanced Stabilised System

Type

Airborne Electro Optic (EO) navigation and targeting system.

Description

COMPASS is designed to provide day/night search-and-track of land and sea targets, onboard weapon control and day/night navigation. It is designed for land, sea, or airborne application and is an adaptable, integrated concept, featuring modular components, add-on modules and

interface/plug-in capability. It can be installed as a stand-alone system or as part of a larger weapons system. It provides multi-operator control capability and, when installed in a helicopter, can be operated from the pilot and/or navigator/weapons operator positions.

COMPASS incorporates three sensors: A Long-Wave (LWIR) or Mid-Wave (MWIR) Forward-Looking Infra-Red (FLIR), colour or black and white (interchangeable) zoom CCD camera, and Laser Range-Finder/Designator (LRFD). The system is capable of manual or automatic search and track operation against naval, aerial and ground targets, as well as weapons guidance and day/night navigation. All of these operations may

be carried out concurrently, using the combined sensor system.

Modular in design, the COMPASS concept system can be readily customised to individual user requirements. System integration is facilitated by provision of a MIL-STD-1553B databus interface and RS422 avionics interface.

Specifications

FLIR: LWIR (8–12 μm) or MWIR (3–5 μm)
Angular coverage: azimuth: 360°; elevation: +35 to −85° (−35° to +85°)
Stabilisation: <20 μrad
Maximum flight speed: operating: 300 kt; endurance: 400 kt

COMPASS: COmpact MultiPurpose Advanced Stabilised System for helicopters 0092869

Temperature: operating: –40° to +50°C; endurance: –45° to +71°C
Weight: 31 kg (without laser); 34 kg (with laser)

Status
In production and in service.

Contractor
Elop Electro-Optics Industries, an Elbit Systems Ltd company.

CRYSTAL high resolution thermal imaging sensor

Type
Airborne Thermal Imager (TI).

Description
CRYSTAL is a high-resolution, 3 to 5 μm (MWIR) Focal Plane Array (FPA), 3 FoV Thermal Imaging (TI) sensor coupled with a dual-axis microscan mechanism and providing a high 512 × 512 pixel sampling resolution. The system is designed for integration into gimballed, stabilised payloads for fixed-wing, helicopter and UAV applications, stabilised and non-stabilised mast mounted EO sensor clusters, target acquisition and night sight systems.

The main features of the system include:
- Lightweight, small volume, single unit configuration
- Easily integrated with higher level systems
- High sampling resolution
- Microscan
- High image quality from proprietary ASIC
- Miniature closed cycle cooler, integral with dewar

Specifications
Detector: InSb 256 × 256 element FPA
Cooler: Closed-cycle, integral dewar assembly
Spectral band: 3–5 μm
Microscan: dual axis, optpo-mechanical
Sampling resolution: 512 × 512 pixels
Aperture: 100 mm

CRYSTAL thermal imager 0109664

Fields of view:
(narrow) 1.5 × 1.5°
(medium) 5.25 × 5.25°
(wide) 21.0 × 21.0°
Video format: CCIR, 625-line, 50 Hz
Interface: RS-422, serial, asynchronous
Weight: 5.0 kg
Power: 30 W, 24 V DC
MTBF: >2,000 hours

Status
In production and in service.

Contractor
Elop Electro-Optics Industries, an Elbit Systems Ltd company.

Day-HUD

Type
Helmet-Mounted Display System (HMDS).

Description
Safe, daytime nap-of-the-earth flight demands that both pilots pay full attention to the external world. The ability to present comprehensive cockpit and navigational information directly in the pilot's field of view dramatically increases Situational Awareness and flight safety during complicated manoeuvres in degraded weather conditions (hover, landing, rescue, shipborne operations, over water, snowfields and in dust, mist and during dusk hours), thus reducing crew workload. The Day-HUD enables both pilots to fly head-out and receive all vital data, including altitude, height, speed and aircraft condition, flight and platform warnings at eye level. The Day-HUD head assembly adapts to ANVIS-AVS-6 or similar NVG-mountings. Day HUD is a component of Elbit Systems' ANVIS/HUD 24 (see separate entry), displaying the same symbology set as the partner upgraded AN/AVS-7 ANVIS/HUD utilised for night operations as part of a full 24-hour flight solution. Day HUD employs the same aircraft system interface as the ANVIS/HUD 24 and fits onto any standard NVG helmet mount. The system comprises a miniature display source and an optical combiner; symbology brightness can be adjusted according to background lighting conditions.

Addition benefits and features of the system include:
- AMLCD technology
- Facilitates training in the employment of ANVIS/HUD in more benign daytime operations
- Enhanced tactical survivability during ultra low-altitude manoeuvring
- Quick disconnect for rapid egress.

Elbit Systems' Day HUD (Elbit Systems) 0569541

Specifications
Field of view: 25°
Exit pupil: >12 mm
Eye relief: >50 mm (compatible with eyeglasses and NBC equipment)
Contrast ratio: 1.4:1 min (at 10,000 fc)
Brightness: 1,400 fl min
Distortion: <2%
Transmission: >42%
Weight: <0.23 kg (helmet-mounted element)

Status
In production for the US Army.

Contractor
Elbit Systems Ltd.

Display And Sight Helmet (DASH)

Type
Helmet-Mounted Display System (HMDS).

Description
The Elbit Systems' Display And Sight Helmet (DASH) allows the crew of a combat aircraft to direct missiles or sensors on to targets or points of interest by simply looking at that point. Head

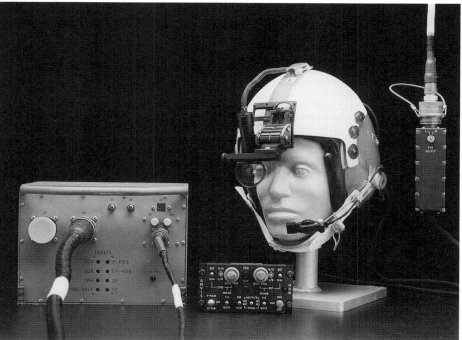

The Elbit Systems/Elop Day-HUD System 0504275

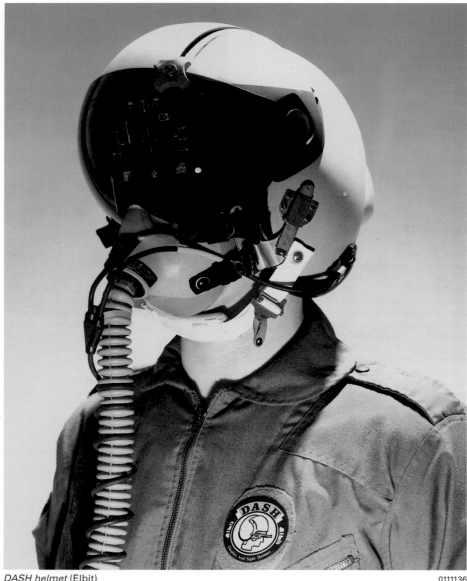

DASH helmet (Elbit)   0111136

DSP-1 dual sensor payload   0099742

position, and hence sightline, is computed. DASH slaves all armament systems to the pilot's line of sight. The pilot directs missile, radar, EO targeting pod and INS Line of Sight (LoS) to specific targets by looking at them and receives feedback on his visor on the target that has been acquired. The pilot can also point out a target to a second crew member by looking at it.

DASH presents the head-up display information directly on the pilot's visor so that he is always aware of flight conditions. The position of a target seen by one crew member can be electronically cued in the visor of any other person linked into the system.

The LoS tracker provides active measurement of pilot LoS and helmet position with high accuracy in all dynamic environments. With pilot LoS slaved to aircraft sensors and vice versa, the result is improved air-to-ground target detection and improved air-to-air interception capability.

Specifications
Weight: 1.9 kg head supported
Field of view: 20° monocular, right eye
Eye relief: >50 mm
Exit pupil: 15 mm
Head tracker coverage:
(azimuth) ±180°
(elevation) ±90°
(roll) ±60°
Accuracy: 6 mrad
Reliability: 1,000 h MTBF

Status
DASH is in service on Israeli Air Force F-15 and F-16 aircraft, and with foreign customers on the F-5, F-16, Mirage 2000 and MiG-21.

Contractor
Elbit Systems Ltd.

DSP-1 Dual Sensor Payload

Type
Airborne Electro Optic (EO) surveillance system.

Description
DSP-1 is a compact day/night observation system designed for use on light reconnaissance aircraft and helicopters (as well as UAVs and patrol boats). It is a four-gimbal system, gyrostabilised in azimuth and elevation. It uses two channels: a third-generation focal plane array InSb FLIR night sensor with a continuous (×22.5) zoom lens; and a high-resolution colour Charge Coupled Device (CCD) daylight channel equipped with a (×20) zoom lens.

Options include an Intensified Charge Coupled Device (ICCD), laser pointer; 8 to 12 µm FLIR (first or second generation), video tracker; radar designated pointing, MIL-STD-1553B databus and GPS interface.

Specifications
FLIR sensor
Spectral range: 3–5 µm
Detector: InSb FPA 256 × 256 InSb
Lens: × 22.5 continuous optical zoom
Field of view:
(narrow) 0.98 × 0.92°
(wide) 21.7 × 20.6°
IFoV: 67 µrad
NEDT: 0.02 K
Video output: PAL or NTSC
Cooler: closed cycle

Daylight camera
Type: high-resolution colour CCD 768 (H) × 494 (V) pixels
Lens: × 20 zoom

Field of view:
(narrow) 0.92 × 0.7°
(wide) 18.6 × 13.9°
Acquisition range:
Truck detection 25 km (daylight camera), 25 km (FLIR)
Truck recognition: 10 km (daylight), 7.5 km (FLIR)

Electromechanical
Type: 4-gimbal
Field of regard:
(elevation) +10° to −110°
(azimuth) 360° continuous
Stabilisation: better than 25 micro radians RMS
Pointing accuracy: 0.7°
Power: 28 V DC, 110 W
Dimensions: (height × diameter) 500 × 320 mm
Weight: 26 kg

Contractor
Controp Precision Technologies Ltd.

EO/IR-LOROP Electro-Optic/ Infra Red Long-Range Oblique Photography system

Type
Airborne reconnaissance system.

Description
The EO/IR-LOROP Electro-Optic Long-Range Oblique Photography system provides high-performance aircraft with a day/night stand-off reconnaissance capability incorporating high-rate data transmission or data record/transmit options, in pod- or nose-mounted configurations.

Elop claims that EO/IR-LOROP is unique in its combination of high performance (in terms of resolution and coverage rate) with lightweight and small size. A highly stable scanning system and athermal telescope design contribute to the system's accuracy.

The system is incorporated into a Multi Mission Modularity pod, which is based on a standard F-16 300 gallon centreline fuel tank. Employing a simple interface to the host aircraft avionic system, the camera is located in an interchangeable centre section for tactical and strategic reconnaissance missions. It uses a Cassegrain Ritchey-Chretien mirror dual/multispectrum telescope with a linear array of butted CCD detectors and IR detectors in the focal planes. It has three-axis gimballed stabilisation and can take positional data from the aircraft inertial navigation system and GPS via MIL-STD-1553B databus (RS-422 optional) and provide RS-170 video feed for internal display on a cockpit multifunction display.

The system can operate at medium to high altitudes (10,000 to 50,000 ft) using a side oblique sector panoramic scan with LoS stabilised in two axes. The system can provide imagery in

EO/IR LOROP (Elop) 0547527

the visible and IR spectrums simultaneously via a X-/Ku-band airborne datalink, allowing near real-time data transmission or data record/transmit options. System control modes include preloaded mission file, remote control from the ground station and cockpit control. The ground station incorporates the tracking antenna, datalink transmitter and receiver, advanced intelligence capabilities, image enhancement, archiving capability and hard copy and soft copy displays.

The EO/IR LOROP system provides a full reconnaissance capability, from mission planning to data dissemination, featuring COTS computer hardware and proprietary software, including Windows NT™ servers and workstations. Among the capabilities of the ground station are multisensor image manipulation (visible, IR and SAR), near-realtime and offline data reception, reformatting and decompression, near-realtime and offline image interpretation, template-driven report generation and an interface to command, control, communications, computers and intelligence systems for online report dissemination and the entry of interpretation requests.

A typical operational flight envelope for the pod is 10,000 to 50,000 ft at ground speeds between 350 to 600 kt up to M1.6.

The system can undertake stand-off missions by day or night with a pilot override for targets of opportunity.

Specifications
Camera type: E-O/IR LOROP
Focal plane array: Visible and IR FPAs
Spectral regions: Visible 0.55–0.9 μm; IR 3-5 μm (MWIR)
Resolution:
Visible NIIRS-6 from 50 km, 40,000 ft
Visible NIIRS-5 from 90 km, 40,000 ft
Camera weight: 120 kg
Power requirements: 115 V AC, 400 Hz 3Ø, 28 V DC

Status
In service in F-16 and RF-4 aircraft. Elop has been selected by several air forces to replace legacy film cameras with podded systems based on the EO-LOROP system.

Contractor
Elop Electro-Optics Industries, an Elbit Systems Ltd company.

ESI-1 stabilised camera system

Type
Airborne Electro Optic (EO) surveillance system.

Description
The ESI-1 is a three-gimbal, gyrostabilised camera system, designed for electronic news gathering, law enforcement, industrial and

Controp Precision Technologies' ESI-1 stabilised camera system 1031810

agricultural surveillance applications. Low mass and compact dimensions make the ESI-1 suitable for installation into all types of air vehicle.

The system is integrated with an automatic video tracker and can be supplied with customer-specified video graphics.

Specifications
Camera
Type: ⅓ in 3-CCD
Lens: ×16 zoom (9.5 to 152 mm)
Field of view:
(narrow) 1.8 × 1.35°
(wide) 28 × 21°
Resolution: 800 TV lines
Controls: Zoom, focus, iris
Electronic shutter: Automatic
AGC: Automatic
White balance: Automatic
Electromechanical
Type: 3-gimbal gyrostabilisation system
Field of regard:
(azimuth) 360° continuous
(elevation) +6 to –96°
Gimbal angular rate:
(azimuth) 0 to 40°/s
Stabilisation: better than 10 μrad rms
Pointing accuracy: 0.7°
Power: 21 to 32 V DC, 50 W
Altitude: up to 6,100 m (20,000 ft)
Acceleration: up to 10 *g*
Temperature:
(operating) 0 to +40°C
(non-operating) –20 to +55°C
Dimensions: 300 (Ø) × 435 mm (H)
Weight
 Turret: 13.1 kg
 Electronics unit: 2.3 kg

Status
In production.

Contractor
Controp Precision Technologies Ltd.

ESP-1H Light Weight Observation System

Type
Airborne Electro Optic (EO) surveillance system.

Description
The ESP-1H is a lightweight observation system designed for light aircraft and UAVs, capable of detecting vehicular-sized targets at ranges up to 20 km and identifying them at up to 7 km.

The system features a three-gimbal mount, which is gyrostabilised in azimuth and elevation, and is equipped with two daylight TV cameras; a wide-angle camera with a 16 mm fixed focal length and a high-resolution black and white or colour Charge Coupled Device (CCD) camera with a 6× (50 to 300 mm) zoom lens.

Specifications
Camera #1
Type: high-resolution black/white or colour CCD
Field of view: 1.5 × 26°
Lens: 6× zoom
Camera #2
Type: wide-angle
Lens: 16 mm fixed focal length
Gimbal
Field of regard:
(azimuth) 360° continuous
(elevation) +10 to –110°
Stabilisation: > 50 mrad
Pointing accuracy: 0.7°
Angular velocity: 0–50°/s
Dimensions: 283 (D) × 415 mm (H)
Weight: 8 kg

Status
Operational on various (unspecified) platforms.

Contractor
Controp Precision Technologies Ltd.

ESP-600C high-resolution colour observation payload

Type
Airborne Electro Optic (EO) surveillance system.

Description
ESP-600C is a very high-resolution, lightweight, stabilised daylight observation system, designed primarily for scout helicopters, light reconnaissance aircraft and UAVs (including IAI Scout and Searcher) on daylight operations. It is a three-gimbal system, gyrostabilised in azimuth and elevation. ESP-600C carries two high-resolution colour CCD cameras: one with a very wide fixed field of view for target detection and one with a ×15 zoom lens for target recognition and identification. The system can be integrated with an automatic video tracker and when mounted on a helicopter or light aircraft, the payload is controlled by an on-board control and display unit. Options include 3-CCD cameras, extended focal length for longer acquisition ranges.

Specifications
Electro-Optical
Zoom camera: high-resolution. Colour CCD (PAL or NTSC)
Lens: 15 × zoom
FoV: 0.7 to 22.6°
WFOV camera: high-resolution. Colour CCD (PAL or NTSC)
Field of view: 22.6 × 17°
Acquisition ranges:
Vehicle recognition: 12 km
Vehicle detection: 30 km

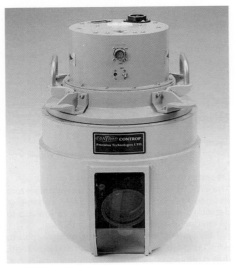

The ESP-600C stabilised payload　0099739

Electromechanical
Field of regard:
(azimuth) 360° (xn) continuous
(elevation) +10 to −110°
Angular velocity: 0 to 40°/s
Stabilisation accuracy: 10 μrad
Fully stabilised in the lower hemisphere
(including Nadir)
Pointing accuracy: 0.7°
Power: 28 V DC, 40 W
Dimensions: 300 (D) × 435 mm (H)
Weight: 12.3 kg

Contractor
Controp Precision Technologies Ltd.

EVS system

Type
Airborne Electro-Optic (EO) Enhanced Vision
System (EVS).

Description
The EVS camera is designed to provide day/night
improved orientation for ground taxiing and
pilotage. It allows visual landing in reduced visibility
conditions, such as fog, haze, dust, smog, and so
on. The system provides a fused, visual and NIR
picture and displays a video image, superimposed
on the pilot's Field of View (FoV). The EVS video
image displayed on the HUD coincides with the
regular view observed by the pilot through the
aircraft window. During approach and landing in
poor weather conditions, the EVS camera enables
the pilot to gain visual landing cues (landing area
or runway approach lights) which would otherwise
be impossible, thus forcing a go-around.
    The EVS camera comprises the following
components:

Focusing lens
The focusing lens is diffraction-limited. Its
geometrical spot diameter on the detector
faceplate is less than half detector pitch for the
entire detector faceplate. The lens focal length is
17 mm and the lens transmittace 89 per cent,
with new coating materials developed. The lens
useful spectral band ranges from 1.2 to 5.5 μm.

Detector dewar cooler assembly
The camera is based on an InSb photovoltaic
staring array, cooled to 77 K. The snapshot detector
contains 320 × 240 elements and its 8 readout video
lines allow collection of all array information within
2 msec. The upper limit of the system f number is
defined by the detector cold shield and is equal to
1.5. A detector cycle contains light collection and
information readout takes about 4 msec to complete.
The system usually runs at 240 frames per second,
with a maximum of 300 frames per second.

Exposure time control
Sophisticated electronics control the detector
exposure time in a range of 100,000:1 in order

to achieve a total dynamic range of 10,000,000:1.
The detector integration time is updated 30 times
per second. Integration time varies according
to the amount of light collected. The integration
time per frame defines the number of frames per
second. For high radiation scenery, the camera
runs at 300 frames per second, while for very low
radiation backgrounds the frame rate will drop to
30 frames per second.

Non-uniformity correction
Non-uniformity correction corrects each pixel and
brings the whole picture to a uniformity dictated
by the required SNR value. The non-uniformity
correction process is implemented by a high-
order polynomial expansion of a few variables
implemented in 32 bits floating point.

Atmospheric transmittance estimation
A mathematical function utilises the input
raw video signal to estimate a relative value
that describes the atmospheric transmittance.
Atmospheric transmittance is usually measured
by using a calibrated laser beam travelling
through the atmosphere for a known distance.
Since the EVS camera does not incorporate a laser
beam and distance to the scenery is unknown, the
EVS camera can estimate only a relatively vague
value of atmospheric transmittance. The relative
atmospheric transmittance is measured and
estimated 30 times per second. A sophisticated
time filter is used to limit the fluctuations of this
value and to increase estimation accuracy.

Demodulator
The demodulation level is controlled by the
estimated atmospheric transmittance value.
For high-transmittance atmospheric conditions
the demodulation level is very low and for low
atmospheric transmittance, the demodulation
level is very high. The demodulator operates
constantly, even in perfect atmospheric conditions.
By operating the demodulator at maximum
capacity on high contrast, the contrasts between
small objects are enhanced without prejudicing
the ability to correctly interpret the image.

Amplification function
Enables the display of very low contrast (0.01%)
information.

Compression Block
Translates any signal collected into a limited
number of grey levels at the FLIR output.

Status
The EVS contains a custom camera, a Head Up
Display (HUD), an external window mounted in
the aircraft and an electronic control box built by
Kollsman Inc. The EVS passed the FAA proof of
concept during August and September 2000, with
first certification flight in the summer of 2001.

Contractor
Opgal Optronic Industry Ltd.

FSP-1 FLIR Stabilised Payload

Type
Airborne Electro Optic (EO) surveillance system.

Description
FSP-1 is a high-resolution, stabilised, 8 to 12 μm
FLIR sensor with a three Field of View (FOV)
telescope, designed primarily for scout helicopters,
light reconnaissance aircraft, UAVs and marine
patrol boats. FSP-1 features a four-gimbal system,
gyrostabilised in azimuth and elevation and
protected against weather effects and contamination
by a rotating dome with built-in optical window.
    Options include third-generation 3 to 5 μm
Focal Plane Array (FPA) or second-generation 8
to 12 μm FLIR camera, video tracker, extended
environmental conditions and laser pointer.

Specifications
FLIR sensor
Spectral range: 8–12 μm
Detector: Cadmium Mercury Telluride (CMT)

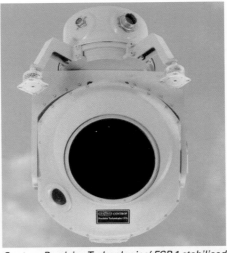

*Controp Precision Technologies' FSP-1 stabilised
payload*　0099740

Field of view:
(narrow) 2 × 1.5°
(medium) 7 × 5.3°
(wide) 24 × 18°
Resolution: 0.07 × 0.11 mrad
Sensitivity: 0.8° C (MRTD at 7 cy/mrad)
Cooler: closed-cycle split-Stirling cycle
Acquisition ranges:
Vehicle recognition: 4 to 5 km
Vehicle detection: 12 to 20 km

Electromechanical
Type: 4-gimbal gyrostabilisation system
Field of regard:
(azimuth) 360° continuous
(elevation) +10 to −105°
Gimbal angular rate:
(azimuth) up to 45°/s
(elevation) up to 32°/s
Stabilisation: better than 25 μrad rms
Pointing accuracy: 0.7°
Power: 28 V DC, 125 W
Serial Connection: RS-422A (bi-directional)
Altitude: up to 6,100 m (20,000 ft)
Acceleration: up to 5 *g*
Temperature:
(in operation) −20 to +50°C
(in storage) −32 to +71°C
Dimensions: 320 (D) × 500 mm (H)
Weight: 28 kg

Status
The FSP-1 is operational on various platforms
including the IAI Searcher UAV.

Contractor
Controp Precision Technologies Ltd.

Laser Range-Finder Designator (LRFD) systems

Type
Airborne Laser Range-Finder/Target Designator
(LRFTD) system.

Description
Elop manufactures a range of Laser Range-
Finder Designator (LRFD) systems, which have
been selected for a number of modern aircraft
and helicopter weapon systems, including:
1. the Laser Range-Finder Designator (LRFD)
    for the now cancelled RAH-66 Comanche
    reconnaissance attack helicopter of the US
    Army;
2. the Range-Finder Target Designator Laser
    (RFTDL) of the Night Targeting System A
    (NTS A), integrated by TAMAM Division,
    Electronics Group, Israel Aircraft Industries
    and Kollsman, is fitted to AH-1W Cobra
    helicopters of the US Marine Corps;
3. the laser designator element of the F/A-18
    NITE Hawk pod requirement for the US Navy,
    integrated by Lockheed Martin Missiles and
    Fire Control Division;

Elop's Laser Range-Finder Designator (LRFD) for the RAH-66 Comanche helicopter Electro-Optic Sensor System (EOSS)   0018349

Elop's Laser Designator for the F/A-18 NITE Hawk pod   0018348

4. the Laser Tracker Receiver (LTR) widely fitted to Apache attack helicopters;
5. the Laser Designator of the Litening pod, manufactured by Rafael Electronic Systems Division, for the air forces of Israel, Germany, Romania, Venezuela and the US Air National Guard;
6. the Laser Range-Finder Designator (SELRD) for the Kiowa Warrior (OH-58) helicopter of the US Army, where it will be incorporated into the mast-mounted sight for aiming laser-guided weapons, replacing the present laser system, with improvements in operational performance, reliability and training flexibility;
7. the laser Range-Finder Designator (dual mode) with diode pumped and OPO technology for the Apache (AH-64D) helicopter of the US Army, where it will be incorporated into the TADS sight for aiming laser-guided weapons, replacing the present laser system, with improvements in operational performance, reliability and training flexibility.

Laser Range-Finder Designator (LRFD) for RAH-66 Comanche helicopter
A new LRFD developed by Elop has been selected to form the LRFD element of the Electro-Optic Sensor System (EOSS), being integrated by Lockheed Martin Missiles and Fire Control Division for the US Army's RAH-66 Comanche reconnaissance attack helicopter.

The new Elop LRFD operates on two different spectral wavelengths, within the single system, to provide optimum performance for both the laser target designation/guidance function (1.06 μm),and the laser ranging function (1.57 μm) simultaneously.

The system is based on use of innovative diode-pumped technology, which improves both operation and reduces power consumption.

RFTDL for the Cobra helicopter   0001226

Elop's Laser Range-Finder Designator (SELRD) for the Kiowa Warrior (OH-58) helicopter (MMS)   0051006

A critical aspect of the design is that the basic wavelength is converted, through use of solid-state devices and advanced optics, to produce an eye-safe system, thus enhancing realistic crew training and training flexibility.

Range-Finder Target Designator Laser (RFTDL)
The RFTDL is a high-repetition rate laser designator which forms part of the NightTargeting system A (NTS A) of the US Marine Corps' Cobra AH-1W helicopter. Integration of the RFTDL into the NTS A system is byTAMAM Division of Israel Aircraft Industries and Kollsman.

Contractor
Elop Electro-Optics Industries, an Elbit Systems Ltd company.

LOTUS thermal imaging sensor

Type
AirborneThermal Imager (TI).

Description
LOTUS is a long-range, high resolution, 3–5 μm (MWIR) two-dimensional Focal Plane Array (FPA), 3 FoVThermal Imaging (TI) sensor with dual-axis microscan capability. The system is designed for integration into gimballed, stabilised payloads for fixed-wing, helicopter and UAV applications.
The main features of the system include:
• Lightweight, small volume, single unit configuration
• Easily integrated with higher level systems
• High sampling resolution
• 4:1 microscan
• Advanced detector technology
• High image quality from proprietary ASIC
• Low power consumption.

Specifications
Detector: InSb 320 × 240 element FPA
Cooler: Closed-cycle, integral dewar assembly
Spectral band: 3–5 μm

LOTUS thermal imager   0122600

Microscan: 4:1
Sampling resolution: 640 × 480 pixels
Aperture: 150 mm
Fields of view:
(narrow) 0.6 × 0.5°
(medium) 2.0 × 1.6°
(wide) 6.6 × 5.3°
Video format: CCIR or RS-170
Interface: RS-422
Dimensions: 250 × 160 × 240 mm (L × W × H)
Weight: 4.8 kg
Power: 25 W, 24 V DC compliant with MIL-STD-704
MTBF: >2,000 hours

Status
In production and in service.

Contractor
Elop Electro-Optics Industries, an Elbit Systems Ltd company.

MIDASH Modular Integrated Display And Sight Helmet

Type
Helmet-Mounted Display System (HMDS).

Description
MIDASH is Elbit's latest addition to its line of head-mounted systems, developed specifically for attack and reconnaissance helicopters.

MIDASH allows pilots to fly head out day and night and integrate weapon and targeting systems using the pilot's helmet line of sight and dedicated flight and targeting symbology.

MIDASH draws on Elbit's experience in production of head-mounted systems, among them, DASH, JHMCS, AN/AVS-7 and ANVIS/HUD 24, designed for both fixed- and rotary-wing platforms.

The Battle Hawk targeting configuration   0080291

Elbit's MIDASH HUD   0051714

Controp Precision Technologies' Mini-Eye   1031816

MIDASH provides attack and reconnaissance helicopter pilots with wide Field of View (FoV), see-through binocular night imagery, symbology and Line-of-Sight (LoS) cueing for both day and night operation. It comprises a standard shell helmet and a dark visor with personal fitting device, Helicopter Retained Units (HRU) and a small quick disconnect Vest-Mounted Unit (VMU). The HRU and VMU are standard aircraft units attached to any pilot by snap-connectors. In addition, a stand-alone mode enables the pilot to operate the image intensified channels while outside the helicopter.

The Image Intensified Display (IID) is based on a pair of XD4 (Gen III) tubes used for night operation and a single CRT overlaid symbology channel. Both of these are projected on transparent-designated combiners. MIDASH integrates fully with any aircraft-installed AN/AVS-7 or ANVIS/HUD 24 systems without further aircraft modification.

Specifications
Night image FoV: 50° (H) × 40° (V)
Day/night symbology FoV: 30° circular
Day/night operation transition: Instantaneous, by single switch operation
Total weight night operation: 2.2 kg
Eye relief: >50 mm
Adjustments: performed once on personal fitting device; no pre-flight adjustments necessary
Symbology contrast for day: >1:1.4 @ full sunlight (10,000 fc ambient)
Transmission: >50%

Status
MIDASH is in service in armed Black Hawk (Battle Hawk) and Puma helicopters.

Contractor
Elbit Systems Ltd.

Mini-Eye stabilised payload

Type
Airborne Electro Optic (EO) surveillance system.

Description
Controp Precision Technologies' Mini-Eye is a small, lightweight, daylight observation system especially designed for small UAVs. Mini-Eye features an airframe-conformal design which minimises drag and thus maximises platform endurance. The high-resolution colour CCD camera is gyrostabilised in pitch and roll.

Applications include daytime observation, damage assessment, Search and Rescue, traffic surveillance, border patrol, anti-terrorist operations, anti-smuggling and power line inspection. Options include the substitution of a Thermal Imager (TI) for the CCD camera.

Specifications
Physical
Weight: 4.5 kg
Dimensions: 418 × 182 × 196 mm (L × H × W)

Electromechanical
Field of regard:
(roll) +55 to −55° of Nadir
(elevation) +65 to −25° of Nadir
Stabilisation: better than 50 µrad RMS per axis
Angular velocity: Up to 35°/sec

Electro Optical
Camera: High-resolution colour CCD
Lens: ×10 zoom
Field of View:
(narrow) 1.8 × 1.3°
(wide) 17.6 × 13.3°
Video standard: PAL/NTSC
Controls: Zoom, focus, iris
AGC: Automatic
Electronic shutter: Automatic
White balance: Automatic

Environmental
Temperature: −20 to +50°C
Vibration: 2.5 g, 5 to 2,000 Hz
Shock: 12 g/6 msec half-sine
Altitude: Up to 15,000 ft

Electrical
Power: 28 V DC, +15 V DC
Consumption: 30 W (nominal)
Video: NTSC or PAL
Interface: RS-422

Contractor
Controp Precision Technologies Ltd.

MSSP-1 Multi-Sensor Stabilised Payload

Type
Airborne Electro Optic (EO) surveillance system.

Description
Controp Precision Technologies' MSSP-1 is a rugged day/night surveillance system especially configured for use on attack helicopters (as well as multiwheeled terrestrial vehicles and marine patrol boats). It is a four-gimbal system, gyrostabilised in azimuth and elevation, and equipped with three sensors – a high-resolution 8–12 µm FLIR sensor, a high-resolution Charge Coupled Device (CCD) daylight camera and a laser range-finder.

Options include an Intensified Charge Coupled Device (ICCD), 3–5 µm InSb Focal Plane Array (FPA) FLIR camera, eyesafe Laser Range-Finder (LRF), extended environmental conditions capability, video tracker, radar designated pointing, MIL-STD-1553 and a GPS interface.

Specifications
FLIR sensor
Spectral range: 8–12 µm
Detector: Cadmium Mercury Telluride (CMT)

Controp Precision Technologies' MSSP-1 multisensor stabilised payload   0099741

Field of view:
(narrow) 2 × 1.5°
(medium) 7 × 5.3°
(wide) 24 × 18°
Cooler: closed cycle

Daylight camera
Type: high-resolution black/white CCD
Lens: 15 × zoom
Field of view:
(narrow) 1.2 × 0.9°
(wide) 18 × 13.7°

Laser range-finder
Wavelength: 1.06 μm
Range resolution: 5 m
Repetition rate: 1 pps

Electromechanical
Type: 4-gimbal
Field of regard:
(azimuth) 360° continuous
(elevation) +25 to −110°
Stabilisation: better than 25 μrad
Power: 28 V DC, 300 W
Dimensions: 400 (diameter) × 650 mm (height)
Weight: 41 kg

Contractor
Controp Precision Technologies Ltd.

MSSP-3 MultiSensor Stabilised Payload

Type
Airborne Electro Optic (EO) surveillance system.

Description
Controp Precision Technologies' MSSP-3 is a day/night observation system especially designed for Maritime Patrol applications on board aircraft, helicopters and patrol boats. In common with the MSSP-1, it is a four-gimbal system, gyrostabilised in azimuth and elevation and equipped with three sensors: a high-resolution, third-generation 3 to 5 μm InSb Focal Plane Array FLIR camera with a dual Field of View (FoV) lens; a high-performance, black and white/colour CCD camera with a ×15 zoom lens; and an optional eyesafe Laser Range-Finder (LRF).

Options include an Intensified Charge Coupled Device (ICCD), colour CCD, a ×22.5 continuous optical FLIR zoom lens for the InSb thermal imager, a second-generation 8 to 12 μm thermal imager, interface to GPS and MIL-STD-1553.

Specifications
Physical Weight:
Turret: 38 kg
Control: unit 4.5 kg
Joystick: 1.5 kg
Turret Dimensions: 400 × 570 mm (D × H)

Electromechanical
Field of regard:
(azimuth) 360° continuous
(elevation) +35° to −110°
Stabilisation: better than 25 μrad RMS
Angular velocity:
(azimuth) 60°/s (max)
(elevation) 50°/s (max)

FLIR Sensor
Wavelength: 3 to 5μm
Detector: InSb 320 × 240 FPA
Field of view:
(narrow) 2.2 × 1.65
(wide) 11.0 × 8.2
Cooler: closed cycle cooler
Gain Control: automatic/manual

Daylight Camera
Camera: high-resolution black/white CCD
Resolution: 550 TV lines
Lens: ×15 zoom
Field of view:
(narrow) 1.2° × 0.9°
(wide) 18.0° × 13.7°

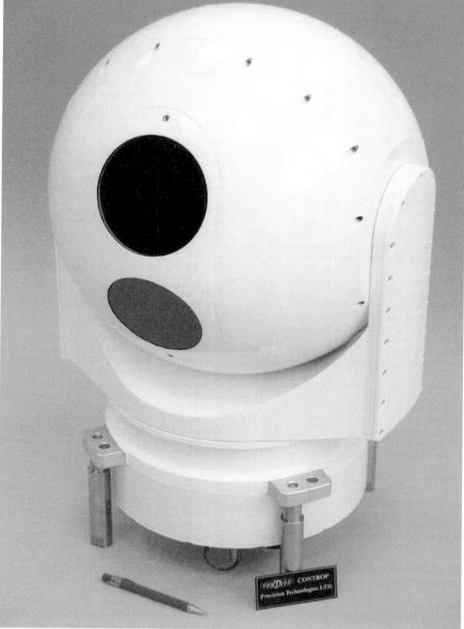

Controp Precision Technologies' MSSP-3 MultiSensor Stabilised Payload   0077827

Gain Control: automatic/manual
LRF (optional)
Wavelength: 1.54 μm
Range: 20,000 m (limited by atmospheric attenuation)
Accuracy: ±5 m
Repetition: 10 ppm (1 pps burst)

Environmental
Temperature: −20 to +50°C
Vibration: 2.5 g, 5 to 2,000 Hz
Shock: 20 g/11 ms duration
Speed: Up to 300 kt

Electrical
Power: 28 V DC, 200 W
Video: CCIR or RS-170
Interface: RS-422

Contractor
Controp Precision Technologies Ltd.

Multimission Optronic Stabilised Payload (MOSP)

Type
Airborne Electro Optic (EO) targeting system.

Description
Multimission Optronic Stabilised Payload (MOSP) is a lightweight dual or triple sensor, target acquisition and range-finder/pointing payload.

Tamam helicopter Multimission Optronic Stabilised Payload   0018347

MOSP has the capability to provide a stabilised view throughout the lower hemisphere, its coverage including the nadir point.

Sensor package options include: single monochrome (or colour CCD or triple colour CCD) day channel; 8–12 μm first-/second-generation

Tamam MOSP on Eurocopter SA 342 Gazelle helicopter   0018346

and a 3 to 5 μm third-generation Focal Plane Array (FPA) FLIRs; laser range-finder of up to six pulses/s, or laser target illuminator.

Specifications
Dimensions: 354 (diameter) × 548 mm (height)
Weight: 26–36 kg (varies with sensors carried)
Spatial coverage:
(elevation) +15 to −105°
(azimuth) n × 360° (unlimited)
Fields of view:
TV: channel monochrome, or one- or three-colour CCD:
(narrow) 0.37° (for the long-range MOSP)
(zoom) 1.3 to 18.2°
FLIR channel: 3–5 or 8–12 μm
(narrow) 2.4°
(medium) 8.2°
(wide) 29.2°
Power consumption: 280 W (day/night), 28 V DC

Status
Over 300 MOSP systems have been delivered for varying types of air, sea and land-based platforms.

Contractor
Israel Aircraft Industries Ltd, Tamam Division.

Multisensor Stabilised Integrated System (MSIS)

Type
Airborne Electro Optic (EO) surveillance system.

Description
The Multisensor Stabilised Integrated System (MSIS) is a lightweight fully stabilised electro-optical system, for day and night passive surveillance and tracking of surface and airborne targets. The system is designed to detect, recognise and track the complete range of manoeuvring targets from a rubber dinghy to a supertanker, as well as all types of helicopters, fixed-wing aircraft and sea-skimming missiles.

MSIS integrates three sensors: a CCDTV camera, a 6 in Thermal Imager (TI) and a laser-range finder. It has a built-in automatic target-tracking computer. Optionally, it can also be equipped with a laser pointer and CCD camera. The stabilised ball turret provides a low wind resistance and the system obtains a picture with good resolution, high tracking accuracy and very good recognition ranges.

Specifications
Weight: 60 kg
Turret dimensions: 780 mm (h) × 600 mm (d)
Line of sight stabilisation: <20 μrad
Angular coverage:
Elevation: −35 to +85°
Azimuth: n × 360°
Slew rate: 60°/s

Thermal imager
Detector: second-generation 8–12 μm CMT, 480 × 4 elements or 3–5 μm FPA
Cooling: integral Stirling engine

Laser range-finder
Wavelength: 1.064 μm or 1.54 μm eye-safe

TV camera
Type: B/W CCD

Status
In production. The system is operational on Israeli Navy Eurocopter SA 366G Panther shipborne helicopters. Some 180 MSIS systems have been ordered or delivered both for the Israeli Navy and for unidentified foreign customers.

During 2003, Elop announced an order for the upgrade of Israeli Navy systems expected to exceed USD4 million. The upgrade covers replacement of the current TI camera with a new so-called 'third-generation' FLIR camera, while system electronics will also be upgraded. In addition to the third-generation FLIR, the new-generation MSIS employs a daylight TV camera and a rapid LRF.

Contractor
Elop Electro-Optics Industries, an Elbit Systems Ltd company.

Night Targeting System (NTS)

Type
Airborne Electro Optic (EO) targeting system.

Description
Since 1982, Tamam has been involved in upgrade programmes to provide Cobra attack helicopters with laser ranging, designation and night attack capabilities. Developed by Tamam under sole-source contract from the US Marine Corps and Israeli Air Force, NTS has proven its effectiveness during extensive operational employment.

There are three variants of the NTS – the basic version (NTS), an enhanced version (NTS-A) and the latest version of the system, designated Super NTS.

NTS-A
NTS-A was designed to provide enhanced performance over the basic system for installation in newer attack helicopters, including the AH-1 W/S/P Cobra. For NTS-A, the original optical tube is removed and flat-panel displays are introduced, thus freeing space in the cockpit facilitating greater compatibility with different types of attack helicopter which employ 'glass cockpit' architectures, such as the US Marine Corps AH-1W 4BW upgrade programme.

NTS-A has the following capabilities:
2nd Generation Time Delay Integration (TDI) MCT FLIR operating in the 8–12 μm waveband
Compatibility with TOW I, TOW II, Hellfire, and other weapons
Target acquisition during day, night and limited visibility conditions
Laser Designator and Range-finder System (LDRS) for laser-guided weapons and for measuring target range

NTS-A night targeting system-A   0015363

Super NTS mounted on an AH-1S attack helicopter (Israel Aerospace Industries)   1128681

TV Tracker (TVT), providing target auto-tracking during day and night
Fully-automatic in-flight boresight capability
Navigation and tactical data display
Display of NTS-A standard video signal (both the FLIR and TV camera pictures) on a multifunction display in both the gunner and pilot cockpits
Built-in growth potential for future integration with other systems onboard, via (2) MIL-STD-1553B databus

Super NTS
Super NTS is the latest version of the NTS, carrying over the improvements of NTS-A, while upgrading core sensors and tracking capability to provide greater detection, recognition and identification ranges.

Enhancements over NTS-A include:
2nd Generation TDI MCT operating in the 8–12 μm waveband, or Gen 2½ InSb Focal Plane Array (FPA) FLIR operating in the 3–5 μm waveband
Laser Spot Tracker (LST)
Advanced video tracker, providing enhanced tracking performance over extended ranges providing for reduced crew workload
TV cameras replaced by high-resolution colour CCD cameras
Extensive Built-In Test (BIT)
Greater MTBF and reduced MTTR

Specifications
Super NTS
FLIR Sensor:
240 × 4 TDI MCT, 8–12 μm; 256 × 256 FPA InSb, 3–5 μm
Field of View (FoV) (H × V)
Wide: 18.0 × 24.0° × 2.0
Option 30.0 × 40.0° (pilotage)
Medium: 5.2 × 3.9° × 9.0
Narrow: 1.5 × 1.1° × 32.1
Zoom: 0.73 × 0.55° × 64.3
TV sensor:
1/2 in Charge Coupled Device (CCD), 780 × 576 array
Fields of View (FoV) (H × V)
Wide: 14.4 × 10.8°
Medium: 2.8 × 2.1°
Narrow: 0.72 × 0.54°

Scanning: 20° × 10° (H × V)
Turret:
Field of regard
(azimuth) 90° to the left, 95° to the right
(elevation) up 30°, down 60°
Slew rate:
(low-angular velocity) 2°/s
(high-angular velocity) 90°/s
(acceleration) 60°/s³

Super NTS FLIR targeting imagery (Israel Aerospace Industries)   1128682

Missile interfaces available:
Hellfire missile; Rafael NT-D missile
all types of TOW missiles (TOW, TOW-1, TOW 2A, TOW 2B)
Weight: 129 kg (excluding aircraft installation kit)
Power: 400 W (average); 650 W (peak, during laser operation)

Status
More than 450 NTS and NTS-A systems are in service on AH-1 variants worldwide. Super NTS is suitable for a wide range of helicopters including the AH-1F, -1P, -1S and -1W, and the A129 Mangusta.

Contractor
Israel Aerospace Industries Ltd, Tamam Division.

Plug-in Optronic Payload (POP)

Type
Airborne Electro Optic (EO) surveillance system.

Description
The Plug-in Optronic Payload (POP) is a modular, compact, lightweight electro-optical payload, designed for day/night surveillance, target acquisition, identification and location. POP is designed for light aircraft, helicopters and UAVs. Installed on unmanned platforms, POP is operated via datalink from a remote ground station. In manned platform applications an onboard observer uses a hand-control grip and video monitor. POP is based on a plug-in sensor module – or 'Slice' – which can easily be replaced in the field within minutes. The 'Slice' exists in several configurations incorporating sensors such as a focal plane FLIR operating in the 3 to 5 µm range, long-range colour CCD TV, Laser Range-Finder (LRF) and laser pointer. Automatic video tracking is an option.

Specifications
Dimensions: 260 (d) × 380 mm (h)
Weight (typical): payload 15 kg; control unit 0.7 kg
Sensor combinations: FLIR and colour TV; FLIR, colour TV and laser pointer; FLIR and LRF
FLIR FOV:
(Wide) 22 × 16°
(Medium) 6.9 × 5.2°
(Narrow) 1.7 × 1.3°
(Super narrow) 0.85 to 0.65°

Tamam POP-200 Plug-in Optronic Payload
0018345

POP installed on helicopter   0018344

FLIR detector: InSb 320 × 240 FPA, 3–5 µm
TV zoom: 1:16 optical, ×2 electronic
TV FOV: 27 to 0.85° continuous zoom

Status
In production and in service.

Contractor
Israel Aerospace Industries Ltd, Tamam Division.

RecceLite reconnaissance pod

Type
Airborne reconnaissance system.

Description
The Reccelite modular reconnaissance pod is based on the Litening airborne laser target designator and navigation pod (see separate entry). The pod shares 75 per cent commonality with Litening, using the same structure and support equipment.

Where the Litening pod includes a 3 FoV Thermal Imaging (TI) camera, a 2 FoV CCD camera and a Laser Range Finder and Target Designator (LRFTD), the RecceLite pod replaces the LRFTD with a MWIR (3 to 5 µm) Focal Plane Array (FPA), 3 FoV Infra-Red (IR) sensor and incorporates a new, 3 FoV FPA CCD manufactured by Zeiss Optronik GmbH. In addition, RecceLite incorporates an Imager Handling Unit (IHU) and an integral Inertial Reference System (IRS), to enhance image stability during aircraft manoeuvres and annotate imagery with exact positional information.

With all sensors, including the IRS, mounted within the same four-axis gimbal arrangement as the Litening pod, conversion of a Litening pod to the RecceLite standard can be done as a field upgrade. This facility will be of particular interest

to operators of F-16 and F-15E aircraft, which utilise a non-weapon intake pylon to carry Litening, whereas current podded reconnaissance sensors usually utilise a weapon-capable centreline or shoulder pylon. This level of integration also enables role-change of the front section without the need for specialist alignment procedures; no pre-flight alignment is required.

A digital solid-state flight recorder, using flash memory, is located in the non-stabilised section of the pod, giving up to 2½ hours of recording time, while a datalink can be used for near real-time reconnaissance applications.

While a conventional IR Line Scanner (IRLS) Field of Regard (FoR) is tied to the aircraft flightpath, RecceLite allows for some aircraft manoeuvre while maintaining tracking, by virtue of its gimballed sensor arrangement. Operating modes include wide-area search, either below or to either side of the flight path, sideways path scanning, and spot (Point of Interest, POI) collection. The pod gathers visual and IR imagery simultaneously, and may be used for target detection, recognition and identification. Currently, the pod is intended for medium- and low-altitude use, but the system has the growth potential to deal with the high-altitude role. Potential improvements include improved optics, and an integrated IRLS.

The Reccelite system includes a portable ground station for mission planning and image exploitation.

Specifications
Stabilisation: Four-axis gimbal, stabilised to 25 mrads
Sensors
  Visual CCD sensor: VOS 40/270 Digital EO Camera (see separate entry): 2,048 × 2,048 pixel, f = 42 to 270 mm, 3 FoV
  IR sensor: 640 × 512 pixel FPA, 3–5 µm waveband, 3 FoV
FoR: +45 to −150° (pitch), ±360° (roll)
Cockpit display: 512 × 512 pixel
Output: 12/8-bit digital video
Dimensions: 2,200 (L) × 400 (Ø) mm
Weight: 210 kg
Cooling: Self-contained Environmental Control Unit (ECU)

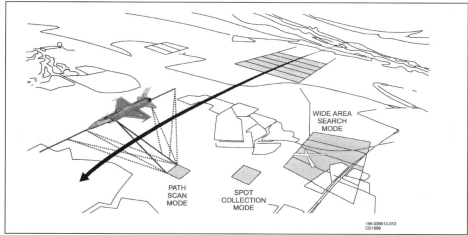

RecceLite modes of operation   0034872

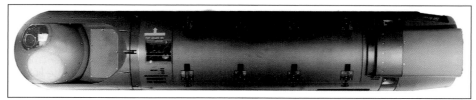

The RecceLite reconnaissance pod (above). Compare the front end of the pod with a standard LITENING II targeting pod (below)
0058709

For details of the latest updates to *Jane's Avionics* online and to discover the additional information available exclusively to online subscribers please visit
jav.janes.com

The RecceLite reconnaissance pod (above). Compare the front end of the pod with a standard LITENING II targeting pod (below)   0034873

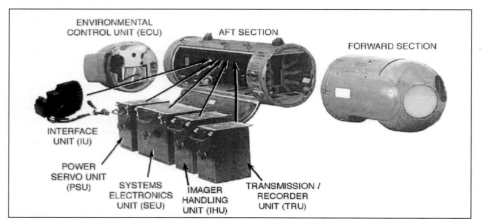

RecceLite schematic, illustrating the modular concept of the system   0034869

Status

In production and in service. The Spanish Air Force has been operating RecceLite on its EF-18A+/B+ Hornet aircraft since 2001. Under deal agreed in mid-2000, 24 pods were acquired at a cost of USD38 million. For the Spanish Air Force configuration, the system includes a Calculex MONSSTR(R) 5000Z solid-state non-volatile recorder (see separate entry).

Reccelite meets all NATO standards for tactical reconnaissance and is compatible with all Litening-capable aircraft, including the F-4, F-5, F-15, F-16, AV-8B, A-10, JAS 39 Gripen, Jaguar, Mirage 2000 and others.

More recently, Rafael supplied 10 Reccelite pods (and 6 Litening pods) to the Brazilian Air Force for its AMX aircraft in a deal worth around USD50 million which also included 3 Reccelite ground exploitation stations.

The company claims that RecceLite is approximately 30 per cent of the cost of a LOROP-class system.

Contractor

Rafael Armament Development Authority Ltd, Missile Division.

TADIR thermal imaging sensor

Type

Airborne Thermal Imager (TI).

Description

TADIR is an advanced second-generation, light weight thermal imager which utilises a 480 × 4 element Time Delay and Integration (TDI) detector.

TADIR (Elop)   0109667

The system features a closed-cycle cooler and continuous optical zoom with automatic athermalisation. Powerful signal processing is based on proprietary ASIC with real-time algorithms for image enhancement. The image output uses CCIR/RS170 composite or 8/12 bit digital video output.

System applications include stabilised airborne payloads and helicopter night pilotage systems.

Specifications

Detector: MCT-PV, 480 × 4 element TDI array
Cooler: Closed-cycle
Spectral band: 8–12 µm
Aperture: 150 mm
Fields of view: Continuous zoom,
  2.0 × 1.5° to 40 × 30°
Video format: CCIR, RS170, 8/12 digital
Interface: RS-422
Dimensions: 250 × 250 × 150 mm
  (L × W × H)
Weight: 7 kg
Power: 70 W, 28 V DC to MIL-STD-704

Status

In production and in service.

Contractor

Elop Electro-Optics Industries, an Elbit Systems Ltd company.

Toplite multi-sensor optronic payload

Type

Airborne Electro-Optic (EO) targeting system.

Description

Toplite is a multisensor stabilised EO surveillance and targeting system, incorporating a Second Generation 240 × 2 scanning, or a Generation 2½ 3-5 µm 320 × 240 (Toplite II) or 640 × 480 (Toplite III) pixel array Thermal Imager (TI). The turret also houses a 1× or 3× CCD, low-light TV camera, a 1.06/1.57 µm, operational/eye-safe Laser Range Finder (LRF) and an optional Night Vision Goggle (NVG) compatible, 0.808 µm laser target illuminator.

Other system features include:
- Auto tracking
- Single LRU
- Interfaces for fire control system, digital moving map, GPS, INS and radar
- Goniometer for CLOS (TOW) missile automatic homing
- Built-in test

Specifications

Dimensions: 662 × 406 mm (H × Ø)
Weight: 59 kg (system)
Azimuth coverage: 360° continuous/±165°
Elevation coverage: +35 to –85°
Slew rate: Up to 90°/sec
LoS acceleration: Up to 100°/sec²
LoS stabilisation: <20 µrad (manoeuvring helicopter)

System	Toplite-II	Toplite-III
TI:	320 × 240	640 × 480
	3–5 µm array	3–5 µm array
Wide FoV:	24 × 18°	24 × 18°
Medium FoV:	4.6 × 3.5°	3.73 × 2.8°
Narrow FoV:	1.3 × 1.0°	0.99 × 0.77°

Video interface:
12-bit digital video, CCIR and PAL, RS-170 and NTSC
System interface:
  Ethernet, MIL-STD-1553, C and RS-422
Power: 24–28 V DC
Consumption: 550/300 W maximum (with/without laser designator)

Status

In production and in service in a wide range of airborne and maritime applications.

Contractor

Rafael Armament Development Authority Ltd, Systems Division.

Very Light Laser Range-finder (VLLR)

Type

Airborne Laser Range-Finder (LRF).

Description

The VLLR is designed to respond directly to the need for greater accuracy, particularly when aiming at moving targets. It upgrades existing platforms and pods and enhances the capabilities of new systems. It is suitable for a variety of airborne systems, including UAVs, as well as seaborne and ground platforms. The VLLR is compact, integrating easily into a variety of systems; modular, being adaptable to different systems, high-speed, sending up to twenty laser pulses per second; ruggedised, to meet severe environmental conditions, and has trouble-free maintenance, with malfunction indicator, external test pins and plug-in PC boards. The housing is customised according to customer requirements.

The VLLR features high-efficiency, low-heat dissipation to the pod and low divergence of the laser beam.

Specifications

Weight: <3.8 kg
Power consumption: 220 W (max)
Wavelength: 1.064 µm
Output energy: 80 mJ
PRF: single shot or up to 20 pps
Beam divergence: less than 0.4 mrad
Pulsewidth: 15 ns (nominal)
Range: 200–9,995 m
Range accuracy: ±5 m

Status

In production.

Contractor

Elop Electro-Optics Industries, an Elbit Systems Ltd company.

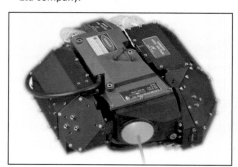

Elop VLLR (Elop)   0101627

Italy

ATOS tactical observation and surveillance system

Type
Airborne Electro-Optic (EO) surveillance system.

Description
The ATOS tactical observation and surveillance system is designed to fulfil the requirements of offshore patrol, search and rescue, anti-pollution surveillance and fishery protection. With additional specialised sensors, the system is also capable of long-range observation and detection for Anti-Submarine Warfare (ASW).

ATOS is based on a modular avionic architecture, facilitating the integration of a variety of different sensor packages and enabling simple tailoring of the system to customer requirements. The 4-axis stabilised turret provides 360° azimuth coverage and incorporates a number of high-resolution sensors, including a multiple FoV Thermal Imager (TI), a colour video camera featuring ×38 magnification and a B + W, narrow FoV (0.35°) TV camera for long-range observation. A search light, with remotely controlled IR filter is slaved to the Electro-Optic (EO) sightline.

ATOS is managed by a proprietary computer system which is compliant with current commercial standards, including ATX, PCI, Intel ×86 processors and the Windows XP(™) operating system. The Mission Management System (MMS) provides:
- Full mission management
- Tactical situation display over digital maps
- Data correlation
- Control of mission sensors
- Real-time collection of data and storage of in a relational database.

The MMS also includes a full-range of V/U/HF transceivers and wideband secure voice and data communications.

The configuration and structure of the operator console may be changed according to requirement and includes the following equipments:
- Control Electronics Unit (CEU)
- Sensors monitor
- System processor
- Operator display, keyboard and hand controller
- GPS interface
- Video recorder.

An optional datalink is also available for the transmission of sensor data/video over short- and medium-ranges to ground stations.

Status
Two variants of the ATOS system, designated ATOS-ES (Environmental Surveillance) and -LW (Light Weight) are available for installation on fixed- and rotary-wing platforms.

Contractor
Selex Sensors and Airborne Systems, Radar Systems Business Unit.

Electro Optical Surveillance and Tracking System (EOST)

Type
Airborne Electro Optic (EO) surveillance system.

Description
The Electro Optical Surveillance and Tracking System (EOST) family of gyrostabilised turrets are designed for long-range, day/night observation and surveillance operations, including maritime patrol, search and rescue, border patrol, coastguard and law enforcement applications.

EOST-45
The EOST-45/M is a 14 in (350 mm) diameter, lightweight 4-axis gyrostabilised turret

incorporating a 3-sensor payload. In addition to the turret/payload combination, the system includes a separate Platform Control Unit (PCU), which provides for turret control and interface with the host platform avionic system. The EOST-45 is designed to support maritime patrol, search and rescue and police operations in small manned platforms and UAVs.

Specifications
Gimbal
Four-axis active gyrostabilisation
Weight: 32 kg (turret plus payload); 5 kg (PCU)
Dimensions:
  Turret: 350 mm (diameter) × 500 mm (height)
  PCU: 129 × 233 × 231 mm (H × W × D)
Power Supply: 28 V DC; 400 W (max)
Vertical Field of Regard (FoR): +20 to −120°
Line of Sight (LoS) range: Continuous 360°
LoS stability: 20 μrad
Pointing accuracy: 3 mrad (3σ)

Thermal Imager (TI)
Mid Wave Infra Red (MWIR) staring Focal Plane Array (FPA)
Spectral Range: 3–5 μm
FoV: 20 × 15°, 2.4 × 1.8° (remotely switchable)

Colour daylight TV
Format: NTSC or PAL
Zoom: ×10
FoV: 2.7 to 27° Horizontal FoV (HFoV)

B+W TV camera with high-magnification telescope
Horizontal FoV: 0.4°
Interface: Dedicated serial link, PAL and CCIR/625 video
Options: Dedicated video monitor, video tracker, hand controller, video recorder, integrated GPS, NTSC and RS-170 interfaces

EOST-42/M3
The EOST-42/M3 is a 16 in (406 mm) diameter, lightweight 4-axis gyrostabilised turret incorporating a 3-sensor payload. In addition to the turret/payload combination, the system includes a Electronics Unit (EU), which provides for turret and payload control, interface with the host platform mission avionics, IR and TV automatic video tracking, colour symbology generation and electronic zoom and freeze frame. The EOST-42 is designed to support maritime patrol, search and rescue and law enforcement operations in fixed- and rotary-wing aircraft.

Specifications
Gimbal
  Four-axis active gyrostabilisation
Weight: 44 kg (turret plus payload); 12 kg (EU)
Dimensions:
  Turret: 406 mm (D) × 495 mm (H)
  EU: 132 × 400 × 210 mm (H × W × D)
Power Supply: 28 V DC; 550 W (max)
Vertical Field of Regard (FoR): +30 to −120°
Line of Sight (LoS) range: Continuous 360°
LoS stability: 20 μrad
Pointing accuracy: 1.5 mrad (3σ)

Thermal Imager (TI)
Mid-Wave Infra-Red (MWIR) staring Focal Plane Array (FPA)
Spectral Range: 3–5 μm
FoV: 22 × 16°, 5 × 3.8°, 1.2 × 0.9°

Colour daylight TV
Format: NTSC or PAL
Zoom: ×10
FoV: 2.7 to 27° HFoV

B+W TV camera with high-magnification telescope
Horizontal FoV: 0.4°
Interface: Dedicated serial link, PAL and CCIR/625 video, MIL-STD-1553B dual-redundant databus

Options: Dedicated video monitor, video tracker, hand controller, video recorder, integrated GPS, NTSC and RS-170 interfaces, UV camera (replacing B + W camera), Laser Range Finder (LRF – replacing B + W camera)

EOST-23
The EOST-23 is a 5-axis gyrostabilised turret incorporating a 3-sensor payload. The EOST-23 is designed to support maritime patrol, search and rescue and coast guard/police operations in larger fixed-wing Maritime Patrol Aircraft (MPA).

Specifications
Gimbal
Four-axis active gyro stabilisation
Weight: 84 kg (turret plus payload); 12 kg (EU)
Dimensions:
  Turret: 534 mm (diameter) × 665 mm (height)
  EU: 132 × 400 × 210 mm (H × W × D)
Power Supply: 28 V DC
Vertical Field of Regard (FoR): +30 to −120°
Line of Sight (LoS) range: Continuous 360°
LoS stability: 5 μrad
Pointing accuracy: 3 mrad
Airspeed: Up to 400 kt

Thermal Imager (TI)
Second-generation Long-Wave Infra-Red (LWIR)
Spectral Range: 8–12 μm
FoV: 28 × 21°, 8.2 × 6.1°, 2.5 × 1.9° (remotely switchable)
Colour daylight TV
Zoom: 2.4 to 39° continuous

B+W TV camera with high magnification telescope
FoV: 0.1, 0.2, 0.3 and 0.6° (remotely switchable)
Interface: MIL-STD-1553B dual-redundant databus, dedicated serial link
Options: Dedicated video monitor, video tracker, hand controller, video recorder, integrated GPS, NTSC and RS-170 interfaces, UV camera (replacing B+W camera), Laser Range Finder (LRF – replacing B+W camera), MWIR TI (replacing LWIR TI), power conditioner for 115 V AC/400 Hz power supply.

The EOST range is completed by the EOST-53, a 14 in (350 mm) diameter, lightweight 4-axis gyrostabilised turret incorporating a 3-sensor payload, designed for civil helicopter applications.

Status
All EOST variants are in production and in service on a variety of fixed- and rotary-wing platforms, including the ATR 42MP, Piaggio P166 Albatross and P68 aircraft and AB412 and A109 helicopters.

Contractor
Selex Sensors and Airborne Systems, Radar Systems Business Unit.

FLIR 111 helicopter navigation FLIR

Type
Airborne Electro Optic (EO) Forward-Looking Infra-Red (FLIR).

Description
The FLIR 111 is a high-performance compact Forward Looking Infra-Red (FLIR) navigation system, designed for helicopters. The system features a second-generation Thermal Imager (TI) operating in the 7.5- to 10.5 μm waveband mounted in a twin-axis steered platform. The video image generated by the FLIR 111 TI is presented on the pilot's Helmet-Mounted Display (HMD), facilitating ultra low-level Nap Of the Earth (NOE) pilotage and navigation.

Specifications

Gimbal
Weight: 20 kg
Power consumption: 150 W
Dimensions:
  Turret: 282 × 210 mm (diameter)
  Mount: 165 × 250 (H × W)
Vertical Field of Regard (FoR)
  Elevation: –70 to +45°
  Azimuth: ±130°
Angular speed
  Elevation: 150°/sec
  Azimuth: 140°/sec
Angular acceleration
  Elevation: 1,140°/sec²
  Azimuth: 1,000°/sec²
Pointing accuracy: 0.2 mrad

Thermal Imager (TI)
Long-Wave Infra-Red (LWIR) 288 × 4 Time Delay
  Integration (TDI) linear array
Spectral range: 7.5–10.5 μm
FoV: 30 × 40°
Cooling: Linear split-stirling cooler
Noise Equivalent Temperature Difference (NETD):
  <0.1 K

Status

The FLIR 111 has been qualified for the Eurocopter
NH90TTH and Tiger hleicopters.

Contractor

AEG-Infrarot-Module.
Galileo Avionica.
Hensoldt Systemtechnik GmbH.

GaliFLIR ASTRO electro-optic multisensor system

Type

Airborne Electro Optic (EO) navigation and
targeting system.

Description

GaliFLIR ASTRO is a multisensor electro-optic
system which is available in a number of
configurations and is mainly intended for avionic
applications such as navigation, surveillance,
observation, reconnaissance, aiming and targeting.

GaliFLIR comprises a stabilised sensor platform
and sensor pack. The sensor pack can include
FLIR, laser designator, day/night TV cameras
to suit customer requirements. The FLIR has
been designed on a modular basis to improve
its flexibility. The modules for series parallel
scanning have been designed for applications in
fixed-wing aircraft, helicopters and RPVs.

The FLIR with dual field of view optics is
mounted on a high-accuracy stabilised platform.
A day and night TV camera can be mounted as an
option on the same platform.

*The Alenia Difesa GaliFLIR ASTRO electro-optic
multisensor system*   0080282

Specifications

Volume: 10 litres
Weight: 32 kg
Power supply: 28 V DC, <100 W
Wavelength: 8–12 μm
Field of view:
(scanner) 40 × 27°
(wide) 16 × 10.8°
(narrow) 4 × 2.7°
Resolution: 0.15 mrad (narrow FOV)
Detector: Sprite, CMT 8 elements
IR lines: 512
Display: TV monitor (standard CCIR 625/50)

Status

In production. GaliFLIR ASTRO has been selected
by the Italian Navy and Coast Guard. It is installed
on Agusta AB-212 and AB-412 helicopters and on
EH 101 and SH-3D helicopters.

Contractor

Selex Sensors and Airborne Systems, Radar
  Systems Business Unit.

HIRNS Plus

Type

Pilot's Night Vision System (PNVS).

Description

The Helicopter Infra-Red Navigation System
(HIRNS) Plus has been developed to enhance
the night flying capabilities of the Agusta A129
Mangusta light attack helicopter, replacing the
previous HIRNS system (see separate entry).

HIRNS Plus features a second-generation
LWIR (8–12 μm) detector manufactured by AEG
Infrarot Module GmbH, which will provide a

*HIRNS Plus will significantly increase the night
flying capability of the A129 Mangusta light
attack helicopter*   0022506

significantly improved IR image to the pilot via
a MultiFunction Display (MFD) or the Integrated
Helmet And Display Sighting System (IHADSS –
see separate entry), facilitating Nap Of the Earth
(NOE) missions.

The system is integrated with a TEAC video
recorder, which also receives inputs from the
helicopter Target Sighting Unit (TSU) and cockpit
intercom systems. The sensor is integrated into
BAE Systems' Type 239 two-axis unstabilised
platform (see separate entry).

Status

Developed for and in service in the Agusta A129
helicopter. The system has also been adopted for
the NH 90 tactical transport helicopter.

Contractor

Selex Sensors and Airborne Systems, Radar
  Systems Business Unit.

Pilot Aid and Close-In Surveillance (PACIS) FLIR

Type

Pilot's Night Vision System (PNVS).

Description

The PACIS FLIR is a thermal imaging system in
the 8 to 12 μm range, designed to be installed
on helicopters and fixed-wing aircraft in order

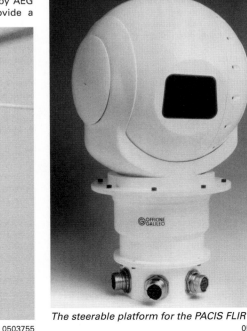

The steerable platform for the PACIS FLIR   0503756

GaliFLIR undergoing operational testing on an Agusta A109 helicopter   0503755

to provide them with increased capability by day, night and in adverse weather operations. It creates a TV-compatible video signal for viewing in the cockpit on a standard display.

PACIS is provided with a telescope with two switchable fields of view and is steerable by a position control grip. It can be used for navigation, day/night surveillance, border patrol, search and rescue, remote sensing and monitoring or as a take-off and landing aid. PACIS can be interfaced with the aircraft avionic system.

The system is composed of a steerable platform, electronic unit and FLIR control grip connected by a cable to the electronic unit. The platform aims the FLIR optical axis in azimuth and elevation. The FLIR is equipped with a two field of view telescope: the wide field of view is used for navigation and surveillance, while the narrow is used to identify and track targets.

Specifications
Dimensions:
(steerable platform) 300 × 511 × 300 mm
(electronic control unit) 170 × 225 × 390 mm
(control panel) 146 × 124 × 165 mm
Weight:
(steerable platform) 23 kg
(electronic control unit) 8.5 kg
(control panel) 1.5 kg
Power supply: 28 V DC, 140 W (average), 450 W (peak)
Wavelength: 8–12 μm

Field of view:
(wide) 40 × 26.7°
(narrow ×4 magnification) 10 × 6.6°
Field of regard:
±170° azimuth,
+45 to –70° elevation
Video format: CCIR 625 lines at 50 Hz
Interface: RS-422

Status
In production and in service.

Contractor
Selex Sensors and Airborne Systems, Radar Systems Business Unit.

Poland

PNL-3 Night Vision Goggle (NVG)

Type
Aviator's Night Vision Imaging System (ANVIS).

Description
The PNL-3 NVG, developed by PCO (Przemyslowe Centrum Optyki) of Warsaw, has been designed for helicopter pilots and crew. The Type 1, fully overlapped, 40° circular Field of View (FoV) design maintains scene perspective to the pilot during night pilotage. The PNL-3 can be adapted to fit the THL-5NV or Alpha helmet with full adjustment of eye relief, height, tilt and Inter-Pupillary Distance (IPD) to facilitate an accurate fit for a wide range of anthropometric sizes. In common with other helicopter NVGs, the power supply is mounted onto the back of the helmet to counterbalance the NVG and to reduce fatigue.

Interestingly, the PNL-3 features a special design feature which automatically detaches and extends the binocular unit away from the face by the action of a g-force safety interlock, triggered at a nominal 5 g, in the event of an emergency crash landing.

Specifications
II Tube: Generation III
Resolution: 56 lp/mm
Magnification: ×1
Field of View: 40° circular, fully overlapped
Exit pupil: 22 mm
Focus: 25 cm to infinity
Interpupillary distance adjustment: 60 to 72 mm
Dioptre adjustment: –6 to +2 Dioptre
Power: 2.7 V DC (battery); 3.6 V DC (onboard power supply)
1 × AA (3.6 V) or 2 × AA (1.5 V)
Weights:
Total system: 885 g
Goggle: 483 g
PSU: 356 g

Status
During 2006, the PNL-3 was selected as Poland's standard NVG system, and recommended for use on all Polish military aircraft. On 21 February 2006, following final qualification and verification tests, the Armament Council of the Ministry of National Defence (MND) officially approved the system, whereupon it became 'referred equipment' and therefore could be ordered on a sole source basis in the future.

Production on the PNL-3 began in 2005, when the MND's procurement department awarded PCO (part of Bumar Group) a contract worth up to PLN17.8 million (USD5.6 million) to equip the Polish Land Forces Aviation with the PNL-3 NVG. A test batch of 20 sets of the PNL-3 had been previously ordered by the MND for qualification purposes.

Deliveries are scheduled to begin during 2006 and will run through 2010. It is understood the contract covers up to 350 PNL-3 sets. Equipped with Gen III (or substitute) Image Intensifier Tubes (IIT), the PNL-3 was developed to meet all current military requirements and standards used by NATO and the Polish MND, although the relatively low breakaway limit and stow position for the binocular section might cause problems in some high-performance applications.

Test pilots at PZL Świdnik have confirmed the capabilities of the PNL-3 during tests conducted during 2005 with NVGs installed on an Alpha helmet.

PCO is conducting further development of the system to incorporate Head-Up Display (HUD) capability, similar to that offered by Elbit Systems' AN/AVS-7 ANVIS/HUD-7. A prototype PNL-3 with HUD capability is to be used on the Gluszec combat support helicopter, a modified PZL W-3WA Sokol, now being developed by PZL-Świdnik in co-operation with the Polish Air Force Institute of Technology (ITWL) in Warsaw.

Later in 2006, the PNL-3 will be delivered to PZL Mielec to equip a pair of PZL M28B/PT Bryza

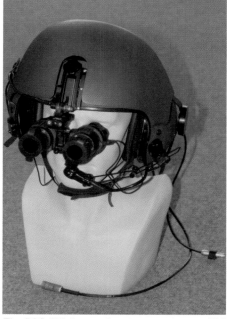

The PCO PNL-3, seen here mounted on an FAS THL-5NV flying helmet, has been selected as the standard issue night-vision goggle for Polish military aircrew (Grzegorz Holdanowicz) 1130652

patrol aircraft and to Poland's naval helicopter fleet, to equip PZL W-3RM Anakonda Search And Rescue (SAR) helicopters, followed later by the Mil Mi-14PS/PL 'Haze' SAR and anti-submarine warfare helicopters.

Contractor
PCO Warszawa.

Russian Federation

A-84 panoramic aerial camera

Type
Airborne reconnaissance system.

Description
The A-84 panoramic aerial camera is designed to provide wide-area photographic coverage of the earth's surface during daylight conditions, from medium and high altitudes. It can be set to operate automatically by onboard control system, or be controlled manually from its control panel.

The A-84 camera is equipped with image motion compensation and automatic exposure control. The camera film records navigation data from the aircraft navigation system.

Specifications
Focal length: 300 mm
Aperture: f/4.5
Frame size: 118 × 748 mm
Film size: 480 m (length) × 130 mm (width)
Nominal overlap in centre frame: 25%
Along track filming distance: 160 × aircraft altitude
Linear resolution: 0.4 m at range = 2 × altitude; 0.8 m at range = 6 × altitude
Power: 27 V DC, less than 300 W; 115 V AC, 400 Hz, less than 900 V A
Weight: 160 kg

Status
Fitted to Tu-22 medium bomber (presumably the Tu-22MR variant), and to the Tu-154 medium transport aircraft for 'Open Skies' operations. Also reportedly fitted to the M-17 high-altitude reconnaissance and research aircraft.

Contractor
Zenit Foreign Trade Firm, State Enterprise P/C S.A. Zverev Krasnogorsky Zavod.

AK-108Ph vertical and oblique aerial camera

Type
Airborne reconnaissance system.

Description
The AK-108Ph aerial camera is designed for simultaneous vertical and oblique photography using three across-track width options (described as Routes 1, 2 and 3), together with three vertical/oblique angular options (defined as 0° (vertical), 75° and 80°). The camera head mirror and magazine can be rotated to provide the required coverage angle.

Camera features include: automatic exposure control; linear compensation of image shift during exposure; mirror stabilisation; temperature and pressure focus compensation and aircraft navigation recording.

Specifications
Focal length: 1.8 m
Aperture: f/5
Frame size: 180 × 180 mm
Angular field of view: 6°
Film size: 240 m (length) × 190 mm (width)
Route options: 1, 2 and 3
Photographic angle options: 0°, 75° and 80° from vertical

Angle of Photography,	Across track coverage relative to height		
	Route 1	Route 2	Route 3
0°	0.1	0.18	0.26
75°	1.7	3.2	5.3
80°	3.8	8.6	21.4

Overlap, nominal: 20%
Power: 27 V DC, less than 300 W; 115 V AC, 400 Hz, less than 900 VA; 36 V AC, 400 Hz, less than 100 VA
Dimensions: 3.1 m × 600 mm diameter
Weight: 600 kg

Status
Fitted to Su-24MR reconnaissance aircraft.

Contractor
Zenit Foreign Trade Firm, State Enterprise P/C S.A. Zverev Krasnogorsky Zavod.

The Zenit AK-108Ph vertical and oblique aerial camera   0018337

The AK-108Ph camera is fitted to SU-24MR reconnaissance aircraft   0048910

GEO-NVG-III Night Vision Goggles

Type
Aviator's Night Vision Imaging System (ANVIS).

Description
Geophizika claims that its GEO-NVG-III (Russian designation reported to be GEO-ONV (Ochki Nochnogo Videniya)) night vision goggles employ third-generation GaAs photocathode technology to provide high responsivity in starlight/overcast conditions. The goggles are ruggedised to meet Russian military requirements for low-altitude helicopter combat, reconnaissance, and search and rescue operations.

Features include: full 40° field of view, F/1.1 at 43 mm eye relief and 10 mm exit pupil; full

binocular night vision; full peripheral vision; automatic brightness control; internal power supply; quick disconnect.

Specifications
Illumination: 10^{-5} to 10 lx
Magnification: ×1
Field of view: 40°
Exit pupil and eye relief: 10 and 43 mm
Objective lens: fixed focus 25 mm, F/1.1
Focus range: 300 mm to 00
Eyepiece lens: 25 mm
Weight: 0.78 kg
Voltage required: 3 V DC, 50 mA (2 × AA batteries)
Mechanical adjustment:
(vertical) 20 mm
(fore and aft) 24 mm
(tilt) 15°
(interpupillary) 56-73 mm
(eyepiece dioptre) +4 to −4 dioptres
Photocathode: GaAs
Sensitivity:
luminous 2,856 K 1,200 uA/1m
radiant (830 nm) 120 mA/W
equivalent brightness input (at 10^{-4} lx)
2.5×10^{-7} lx
S/N (at 10^{-4} lx) 15
centre resolution: 32 mm
useful cathode diameter: 17.5 mm

Status
Claimed to be in widespread use on Russian military helicopters.

Contractor
Geophizika-NV.

GEO-NV-III-TV day/night tracking system

Type
Airborne Electro-Optic (EO) targeting system.

Description
The GEO-NV-III-TV image-intensified, solid-state, charge-coupled device (CCD) camera is a versatile day/night tracking system, mounted on a gyrostabilised platform, that incorporates three separate channels: a CCD day sensor; an image intensifier; and a CCD night channel.

Specifications
Intensified CCD night channel
Scene illumination: 10^{-5} to 10^{-1} lux
Sensor instantaneous field of view: 10°
Combined sensor/platform field of view:
+30 to −50°
Focus range: 0.3 m to ∞
Objective lens: f = 75 mm; F/1.5
Camera slew rate: 20°/s
System accuracy: 0.5°
Dimensions: 200 (length) × 60 mm (diameter)
Weight: 0.65 kg
CCD day channel
Pixels: 512 (H) × 582 (V)
Pixel size: 7.6 × 6.3 mm
Active imaging cell size: 4.6 (H) × 3.5 mm (V)
Resolution: 480 TV lines
Field of view: 13.5°
Objective lens: f = 25 mm; F/1.8
Grey scales: 10
Dimensions: 50 (height) × 60 mm (diameter)
Weight: 0.1 kg
Image intensifier
Photocathode: GaAs

Status
Geophizika states that the GEO-NV-III-TV system is fitted to Kamov Ka-50 helicopters and to a wide range of Mil helicopters, including Mi-17, Mi-24, Mi-26 and Mi-28. Geophizika has also proposed the system for the Ka-52 helicopter.

Contractor
Geophizika-NV.

Gyrostabilised Optical Electronic System (GOES)

Type
Airborne Electro Optic (EO) navigation and targeting system.

Description
GOES platforms are designed to carry Thermal Imagers (TI), day- and low-light-level TV cameras, cine and video cameras, Laser Range-Finders (LRFs), and similar equipment. The first four models produced were designated GOES-1/2/3/4. They were all designed to carry electro-optical payloads varying from 16 to 100 kg. The systems were suitable for both civil and military use, with payload specification by customer choice.

These early models were superseded by the GOES-310/320/330 platforms, with designations representing single-, double- and triple-channel systems respectively. All are designed for detection and recognition of objects in a broad range of vision angles, in severe rolling and vibration conditions for both civil and military environments.

Sensors available for fitment to the GOES-310/320/330 series include TI, cine and video cameras, LRF, and Infra-Red (IR) sensors in any combination of fits. System options include compatibility with GPS, RS 232 interface, video monitor, VHS/SVHS videotape recording and air-ground microwave datalink.

GOES-520 turrets have been shown with vertically arranged and side-by-side aperture configurations. It has been reported that the side-by-side arrangement, possibly intended for installation in the Ka-50 attack helicopter, is also known as TOES-520. In both cases, the turret is utilised for pilotage, employing low light TV and TI sensors   0089875

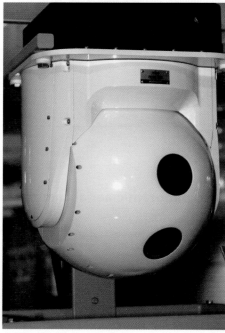

GOES-520 Electro Optic turret, shown at the Farnborough Airshow in 2002 (Patrick Allen)   1109106

GOES-1/2/3/4

Platform	GOES-1	GOES-2	GOES-3	GOES-4
Payload weight	100 kg	16 kg	30 kg	65 kg
Payload volume	85 dm³	13.5 dm³	20 dm³	55 dm³
Weight of optical turret	140 kg	25.5 kg	20 kg	85 kg
Dimensions of optical turret	Ø720 × 980 mm	Ø340 × 552 mm	Ø460 × 613 mm	Ø640 × 850 mm
Look angles				
(azimuth)	±170°	±170°	±235°	±135°
(elevation)	+80 to −40°	+10 to −30°	+45 to −115°	+10 to −30°
Stabilisation (micro radians):	50	70	50	50

GOES-321/342/344/346

Platform	GOES-321	GOES-342	GOES-344	GOES-346
System components	TI, LRF	TI, TV, LRF	TI, TV, LRF	TI, TV, LRF
System weight	85 kg	185 kg	90 kg	105 kg
Dimensions of optical turret	Ø460 × 613 mm	Ø460 × 613 mm	Ø460 × 613 mm	Ø640 × 850 mm
Look angles				
(azimuth)	±230°	±230°	±150°	±230°
(elevation)	+40 to −30°	+25 to −115°	+85 to −20°	+30 to −115°

The GOES-321, -342 and -346 are in production. The -321 includes a sighting function for unguided missiles and guns. The -342 and -346 include laser weapons guidance.

The newest model in the range is the GOES-520 platform, intended for day/night surveillance, search and detection, employing TI and TV sensors.

Specifications
GOES-310/320/330
Platform
Stabilisation: 5-axis

GOES-346   0089893

Pointing error: <50 μrad RMS
Field of regard:
(azimuth) ±230°
(elevation) + 30 to −110°
Maximum angular rate: 60°/s
Dimensions:
(opto-mechanical unit): Ø460 × 613 mm
(electronic unit): 330 × 485 × 225 mm
(control unit): 225 × 50 × 57 mm
Weight:
(opto-mechanical unit): 55 kg
(electronic unit): 20 kg
(control unit): 0.43 kg

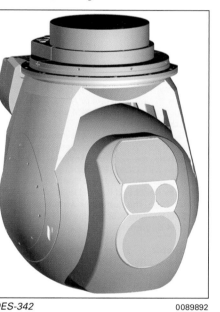

GOES-342   0089892

Power: 27 V DC, 500 W; 115/200 V AC, 3-phase, 400 Hz, 250 VA
Thermal imager
Type: AGEMA THV-1000
Receiver: 5-bar SPRITE focal plane
Spectral range: 8–12 μm
Cooling: integral Stirling cooler
Pixels: 580 × 386
Field of view (H&V):
(narrow) 5.0 × 3.3°
(wide) 20.0 × 13.3°
NETD: 0.18°C
TV system
Type: CCD Sony EVI-331 colour
Pixels: 752 × 582
TV lines: 480
Focal range: f = 5.4 to 64.8 mm with ×12 optical zoom
Field of view: 48.8 × 37.6° to 4.4 × 3.3°
Laser range-finder
(made by Production Association Urals Optical and Mechanical Plant (PA UOMZ)
Wavelength: 1.54 μm
Energy: 0.01-0.07 J
Beam divergence: 2–3 minutes of angle
PRF: 1 Hz
Maximum range: 10 km
Measurement error: <5 m

GOES-520
Dimensions: Ø350 × 500 mm
Weight: 45 kg
Look angles: ±180° (azimuth), +35 to −85°

The GOES-310/320/330 gyrostabilised optical electronic systems (from top to bottom)   0079247

GOES-342 turret beneath the nose of an upgraded Mi-24M in place of the original Raduga-Sh daylight optical sight. The large, circular window is for the TI; the two smaller windows above it are for the TV camera (right) and Ataka missile tracker; the large upper aperture is for the Laser-RangeFinder/Designator (LRF/D). The housing to the right of the turret contains antennas for the datalink used to control Shturm or Ataka anti-tank missiles   0578042

Status

The fire-control system of the Ka-52 Alligator combat helicopter consists of the Myech-U (Sword) radar system and the Samshit-BM (Boxwood) Electro Optical (EO) sight. Myech-U, developed by Phazotron-NIIR, carries the official designation FH01 Arbalet-52 and consists of two sensors – a nose-mounted Ka-band radar for ground observation and homing of air-to-surface weapons and an L-band radar for self-protection, with its antenna installed on the top of the rotor mast. Samshit-BM incorporates the GOES-451 gyro-stabilised turret, which accommodates TI and TV cameras, a Laser-RangeFinder/Designator (LRF/D), a laser spot tracker and a laser generator for the beam-riding Vikhr (AT-16) anti-tank missile.

The single-seat Ka-50K and the two-seat Ka-50-2 export variant carry a reduced fit, comprising the Arbalet-50 radar and Samshit-50 EO sight. The former consists of the mast-mounted L-band segment of Arbalet-52, while the latter is a version of Samshit-BM using the smaller GOES-346 turret carrying TI and TV cameras, an LRF and the laser beam-riding system for Vikhr.

The mid-life upgrade for the Russian Mi-8 armed utility helicopters includes the installation of a GOES-321 turret, containing a TI camera and LRF. This supports observation and reconnaissance, together with aiming of guns and unguided rockets. The first turret was installed in January 2000, and about 25 upgraded Mi-8MTKOs are now in service. A similar installation of the GOES-321VMI variant aboard Mi-24s was proposed in 1999 as a response to an Urgent Operational Requirment (UOR) to enhance the performance of Mi-24s operating in Chechnya. An experimental installation began flight trials in May of that year, but the plan was abandoned because the GOES-321 can be used only with unguided armament.

Turning to the Mi-24 Hind, there have been a number of EO upgrade proposals. UOMZ stated that orders were placed for a GOES-320-based system at the MAKS-97 Airshow. The GOES-320 unit was further shown on a Hind 'night vision' proposal (Mi-24VN, also designated Mi-35O) mockup shown at MAKS-99. At the Paris Airshow in June 2001, a night attack Mi-35M (Mi-24VK-2) helicopter was displayed fitted with a GOES-342 turret under the nose. At the MAKS-2001 Airshow, the Mi-35M/GOES-342 combination was shown again, together with a Mi-24VK-1 proposal fitted with a GOES-321 turret. Of these, it seems only two proposals survive – the Mi-24PN (Mi-35PN, which features a fairing for a Zarevo FLIR and LRF in the nose lip; this upgrade configuration is only possible for Mi-24P which has side-mounted guns) and the Mi-24M (designated Mi-35M for export customers), which includes a GOES turret. The core of the Mi-24M program involves the installation of an OPS-24N day/night targeting system, based on the GOES-342 turret, which is mounted beneath the starboard side of the forward fuselage (the Raduga-Sh daylight optical sight is removed). The new turret houses TI and TV cameras (each with wide

Three members of UOMZ Electro Optic (EO) turrets. The GOES-451 (centre) equips the Ka-52; the GOES-321M (left) is employed in upgraded Mi-8s; the GOES-520 is employed for pilotage and navigation in a number of different platforms. The GOES-321 and -520 accommodate Agema THV-1000 sensors (Piotr Butowski)   0523845

A mock-up of a Mi-35M was displayed at MAKS-97 equipped with a GOES-320 turret   0099638

and narrow FOVs), an LRF and an IR goniometer for use with Shturm (AT-6) or Ataka (AT-9) anti-tank missiles. Testing of the Mi-24M with the GOES-342 began during 2002.

In September 2003, it was reported that the three Augur Au-17 Bars tethered aerostats ordered by the City of Moscow will each be equipped with GOES-520 turrets.

GOES systems have also been observed fitted to Kamov Ka-29 and Ka-50N helicopters.

As of 2006, the company were promoting the GOES-321 for Mi-8 upgrades and the GOES-342 for Mi-24 upgrades.

Contractor

Production Association Urals Optical and Mechanical Plant (PA UOMZ).

Helmet-mounted aiming system (HSTs-T)

Type
Helmet-Mounted Sighting System (HMSS).

Description
The Electroautomatika HSTs-T helmet-mounted aiming system is designed to compute and display line of sight angular co-ordinates of the visual target as cued by the pilot's head turn; generate collimated images of the sighting marker and launching marker in the field-of-view of the pilot's right eye.

System components comprise an easily removable helmet-mounted optical assembly with reference points device and sighting parameters display system; two pilot's head position sensors (trackers) installed in the cockpit and, as a rule, installed with a collimator-type indicator used as a mounting base; control and interface unit.

The helmet-mounted sight system provides for data exchange over MIL-STD-1553B or ARINC-429 data standards; autonomous computation of sighting angles; high-accuracy target designation for effective employment of modern weapons. Data is presented to the pilot in the form of sighting and launch markers. These markers are presented at a maximum luminance of 35,000 cd/m² to provide for excellent daylight readability, automatically adjusting the level down as background scene levels decrease.

Specifications
Angular coverage:
(azimuth:) ±60°
(elevation) –15 to +60°

GOES-321   0089890

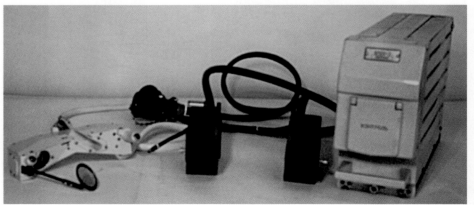

The Electroautomatika helmet-mounted aiming system   0062242

Max aiming error:
in the central aiming area lying between ±35°
azimuth and −15 to +60° elevation 20′ of arc
outside the central aiming area: 35′ of arc
Data refresh rate: 100 Hz
System weight:
(helmet-mounted assembly) 0.3 kg
(head position trackers) 2 × 0.5 kg
(control and interface unit) 6 kg
MTBF: 2,000 flight hours
Operating temperature range: −40 to +55°C
Power: 80 VA maximum

Status
In service.

Contractor
Electroautomatika OKB.

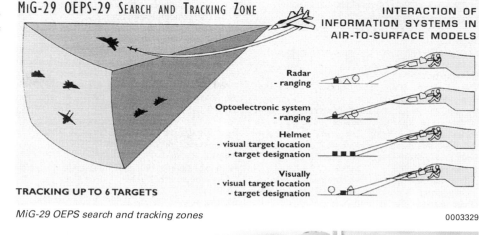

MiG-29 OEPS-29 SEARCH AND TRACKING ZONE

INTERACTION OF INFORMATION SYSTEMS IN AIR-TO-SURFACE MODELS

Radar
- ranging

Optoelectronic system
- ranging

Helmet
- visual target location
- target designation

Visually
- visual target location
- target designation

TRACKING UP TO 6 TARGETS

MiG-29 OEPS search and tracking zones   0003329

OEPS Opto-electronic sighting system

Type
Airborne Electro Optic (EO) Infra-Red Search and Track System (IRSTS).

Description
The OEPS opto-electronic sighting system provides for search, detection, tracking and ranging of airborne and ground targets. Two versions of the system are in service: the OEPS-29, as installed in MiG-29 aircraft, and the OEPS-27, in the Su-27. The OEPS-27 is larger and heavier, offering greater detection range and Field of Regard (FoR). Functionality of the system is similar for both variants, with full integration with the SURA and earlier SHCH-3UM Helmet Mounted Target Designation Systems (HMTDSs – see separate entry).

As part of the Su-27's SUV-27 weapons targeting complex, the OEPS-27 sensor has a range of up to 27 n miles (50 km), depending on the IR signature of the target. The collimated laser range-finder has a reported range of 4.3 n miles (8 km), functioning through common optics in a transparent housing forward of windscreen.

The Mig-29's OEPrNK-29 weapon aiming and navigation system includes the OEPS-29 electro-optical sight and Laser Range Finder (LRF). The sensor head is protected by removable fairing for non-operational flights.   1130230

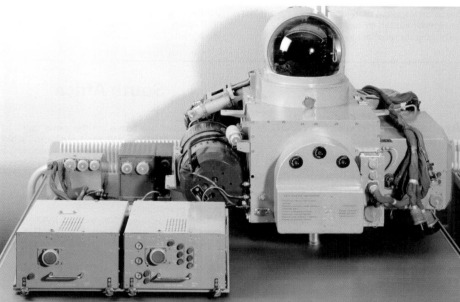

OEPS-27 opto-electronic sight system   0089874

An integral part of the MiG-29's OEPrNK-29 weapon aiming and navigation system, the OEPS-29 (with a smaller IRST sensor) has a reported range of approximately 8 n miles (15 km).

Specifications
OEPS-29
Weight: 78 kg
Field of regard: ±60° (azimuth), +30 to −15° (elevation)
OEPS-27
Weight: 174 kg
Field of regard: ±60° (azimuth), +60 to −15° (elevation)

Technical specifications of laser emitters associated with the OEPS-27 and OEPS-29 systems are given in a previous entry, headed 'Laser range-finder/target illuminators'.

Status
In production and in service on MiG-29 and Su-27 aircraft and derivatives.

It has been reported that an improved variant of the system, fitted to the Su-30MKK, is designated OEPS-31E-MK.

Contractor
Production Association Urals Optical and Mechanical Plant (PA UOMZ).

Shkval sighting system

Type
Airborne Electro-Optic (EO) targeting system.

Description
The Shkval sighting system is designed as a comprehensive anti-tank electro-optical fire-control system, incorporating: TV sighting sensor, laser range-finder and target designator, and laser beam-rider. It is also fitted with a three-axis field of view stabilisation system and an automatic image correlator to ensure tracking against ground, sea or sky backgrounds. There is ×23 magnification to extend detection/tracking ranges. Tracking angles are quoted as +15 to −80° in elevation and ±35° in azimuth. Laser guidance system accuracy is quoted as 0.6 m.

Status
Shkval is part of the weapon system on the Ka-50 Black Shark and Ka-52 Alligator helicopters, and on the Su-24T and Su-39 multimission attack aircraft.

Contractor
Zenit Foreign Trade Firm, State Enterprise P/C S.A. Zverev Krasnogorsky Zavod.

OEPS-29 opto-electronic sight system   0089876

Sapsan Electro Optic (EO) targeting pod

Type
Airborne Electro Optic (EO) targeting system.

Description
The Sapsan (Peregrine) target acquisition/designator targeting pod, being developed by the Urals Optical and Mechanical Plant, is designed for the MiG-29 and Su-27/30-series combat aircraft. The company is also looking at the requirements of potential customers with a view to integrating the pod on Su-35 aircraft.

The Sapsan pod is an integrated system, incorporating a stabilised Thermal Imager (TI), Laser Range Finder and Designator (LRFD), together with associated electronics units and a thermal control system. It is designed to enable precision targeting of guided weapons from fast-jet combat aircraft by day or at night.

Other features of the system include:

- Automatic lock-on and tracking
- 3-axis FoV stabilisation
- Advanced tracking algorithms
- Thermal control system

Two variants are under development – Sapsan for the Russian Air Force (with applications including the Su-34 and Yak-130) and Sapsan-E for export. Target markets are seen as China (Su-30MK/MKK) and India (MiG-29K and Su-30MKI).

Specifications
Dimensions: 360 × 3,000 mm
Weight: 250 kg
Field of regard: ±10° (azimuth); +10 to −150° (elevation); ±180° (roll)

Status
In development.

Contractor
Production Association Urals Optical and Mechanical Plant (PA UOMZ).

Sapsan-E pod mounted on the starboard intake station of a Su-27SM at MAKS 2005 (E L Downs)   1128677

South Africa

Argos airborne observation system

Type
Airborne Electro-Optic (EO) surveillance system.

Description
The Argos airborne observation system, a military derivative of the LEO family (see separate entry), provide target sighting solutions for airborne, land system and maritime applications. Argos helicopter-mounted targeting sights are designed for aerial surveillance, search and detection, target recognition and target engagement. The system provides control logic and interfaces to slave-turreted cannons, integrates with rocket launchers and missiles (target designation or missile guidance) and assists in bombing missions.

Various ARINC and 1553B interface options are available.

Specifications
Thermal imagers: 3–5 μm or 8–12 μm TI
Colour TV cameras: 1-CCD or 3-CCD colour camera
Laser range-finders: various
Options: GPS position display; auto-tracker, microwave downlink
Turret diameter: 400 mm

Status
In production and in service.

Contractor
Cumulus, a business unit of Denel (Pty) Ltd.

LEO-II airborne observation system

Type
Airborne Electro-Optic (EO) surveillance system.

Description
LEO-II is a range of gyrostabilised airborne observation systems specifically developed for the paramilitary security market. Suitable for both rotary- and fixed-wing aircraft operating by day or night, LEO II is a powerful force multiplier in applications ranging from law enforcement through search and rescue to border patrolling.

Headed by the latest model combining Forward-Looking Infra-Red (FLIR) with broadcast-quality colour TV, the range comprises various single-, dual- and triple-sensor models fitted with a selection of Thermal Imagers (TIs), colour TV cameras and laser range-finders.

All models are based on the same compact, high stabilisation, 410 mm platform, which utilises an open modular architecture to guarantee future upgrade capability. The systems carry appropriate civil aviation approvals and are built to ISO 9002 and MIL-STD-10C/C/E standards. A full suite of options is available to enhance system operation.

Announced during 2000, the LEO-II-A3, merges the Kentron-developed Kenis 3 to 5 μm TI, a high-performance broadcast quality TV camera with 54× zoom lens and a fixed, long-range 'spotter' TV camera. The Kenis TI is described as particularly suited to long-range imaging in hot and humid conditions.

Specifications
(LEO-II-A3)
TI: 3–5 μm or 8–12 μm TI camera
Kenis TI: 3–5 μm, 384 × 288 CMT focal plane array
Colour TV camera: 3-CCD, 54× zoom

The operator's console and monitor for the LEO observation system mounted in a Eurocopter BO 105 helicopter   0581491

LEO-II airborne observation system   0051007

Wide spectrum spotter TV: Dual-CCD, 780 mm focal length, 0.35° FoV
Laser range-finders: various
Stabilisation: 4-axis; <2μrad
FoR: 360° azimuth; +20 to −120° elevation
Platform size: 410 mm diameter
Basic system weight: <40 kg
Options: GPS position display; searchlight slaving; embedded video auto-tracker; microwave downlinks; ARINC 429 interface

Status
As of Decmber 2003, LEO airborne observation systems were reported as in service in South Africa and 34 other countries, including 90 per cent of UK police air support units.

Contractor
Denel Optronics.

LEO-II-A2 with 8 to 12 μm FLIR and broadcast-quality TV camera
0065938

Sweden

Airborne laser range-finder

Type
Airborne Laser Range-Finder (LRF).

Description
The airborne Laser Range-Finder (LRF) is a very compact unit, which is easy to integrate into existing navigation and weapon delivery systems.

This high repetition-rate laser range-finder uses a simple, modular design consisting of transmitter, receiver, range counter and a deflection unit for the optical axis.

The laser is aimed at the target by slaving the deflection unit to the aircraft sighting system.

Specifications
Transmitter:
(laser type) Nd YAG
(wavelength) 1.06 μm
(pulse energy) 20 mJ
(pulse length) approx 10 ns
(pulse repetition frequency) 1–10 Hz in bursts
(beam divergence) 0.7 mrad
Receiver:
(detector) Silicon avalanche diode
(field of view) 0.5 mrad
Range Counter:
(range, max) 20,000 m
(range, min) 200 m
(range resolution) 5 m
Range:
(typical range) ≥10 km at optical visibility >20 km
Deflection Unit:
(azimuth travel) ±10°
(elevation travel) ±10°
(accuracy) <1 mrad
(slow rate) >60°/s
Interface:
(digital bus) ARINC 429
Weight:
(total weight) approx 14 kg
Dimensions: approx 500 × 160 × 160 mm
Power:
(power supply) 28 V DC
(power consumption) 225 W

Status
In service.

Contractor
Saab Bofors Dynamics AB.

The Saab Bofors Dynamics airborne laser range-finder 0002210

HELIOS Helicopter Observation System

Type
Airborne Electro Optic (EO) targeting system.

Description
The HELIOS Helicopter Observation System is a helicopter-mounted anti-tank system, or, if ordered without weapons provision, a scout observation system. The system includes high resolution, direct view, wide and narrow FoV optics, TV and FLIR sensors, together with an optional Laser Range Finder/Designator (LRF/D) for battlefield surveillance and targeting for self-employed or co-operatively employed weapons.

Agusta A 109 with roof-mounted HELIOS and HELiTOW carrying eight TOW missiles 0125220

The system can be configured as a day-only system or with FLIR added, a day/night system, either in a stand-alone configuration, or interfaced with the host platform via MIL-STD-1553B databus.

HELIOS employs rate-integrating gyros for stabilisation. Both roof- and nose-mounted configurations are available, making the system compatible with almost any helicopter. High-resolution optics provide both wide and narrow fields of view.

Status

In production and in service on the Eurocopter BO 105, Bell B406Cs Combat Scout, Agusta A 129 Mangusta, Eurocopter AS550 Fennec and the Agusta A 109. Over 200 systems have been produced.

Contractor

Saab Bofors Dynamics AB.

IR-OTIS Infra-red Optronic Tracking and Identification System

Type

Airborne Electro Optic (EO) Infra-Red Search and Track System (IRSTS).

Description

Saab Bofors Dynamics' IR-OTIS is a multifunctional Infra-Red Search and Track (IRST) system, intended to provide passive situation awareness for the JAS 39 Gripen aircraft at long range, during day and night operations against air and ground targets. The system can operate both as an IRST and a traditional FLIR, although it should be noted that performance in air-ground applications will be limited by aircraft installation constraints (traditionally, IRST systems are mounted above the radome to enhance air-air look-up), severely curtailing look-down for weapon aiming.

In IRST mode, the system scans a designated section of airspace, either in support of other aircraft sensors or autonomously, detects targets automatically, and tracks them while continuing to search the volume, effecting a passive Track-While-Scan (TWS) similar to active TWS in AI radars. In this mode, the system can scan with a narrow Field of View (FoV) to give long-range detection or with a FoV to cover a larger sector in a shorter time.

In the FLIR mode, the system generates a stabilsed image of the Field of Regard (FoR) covering a defined sector (according to flight regime and tactical requirements) ahead of the aircraft.

In air-to-air applications, IRST mode is used for target search, with automatic lock-on and TWS to facilitate passive identification and engagement of hostile aircraft.

In air-to-ground applications, the FLIR mode is used for passive navigation and target detection, acquisition and identification for weapons employment.

Specifications

Sensor wavelength: 8–12 μm
Sensor elements: 1,100–1,200 elements
Field of view: several, selectable
Field of regard: >one hemisphere, limited by aircraft installation
Sensor unit: 30 kg and 30 litres volume
Signal processing unit: 10 kg and 10 litres volume

Status

In development for the Swedish Defence Material Administration for JAS 39 Gripen; the first development model was flight tested on a JA 37 Viggen aircraft.

Since the sensor system is designed to be fitted on the Gripen and requires relatively little space, integration in other aircraft should be possible

IR-OTIS installed on a Swedish Airforce JAS 39 Gripen. Note the slightly offset installation above the radome to allow for some limited look-down in air-ground applications   0116567

IR-OTIS assemblies: sensor unit (top), electronics unit (lower left) and aircraft HUD and large format head-down displays (lower right) for the JAS 39 Gripen   0116566

Development of IR-OTIS was carried out on the JA 37 Viggen   0116568

either in a new build arrangement or as retrofit equipment.

Contractor

Saab Bofors Dynamics AB.

Turkey

ASELFLIR-200

Type
Airborne Electro Optic (EO) navigation and targeting system.

Description
The Aselsan ASELFLIR-200 forward-looking infrared (FLIR) system is a light-weight, multipurpose, thermal imaging sensor for pilotage/navigation, surveillance, search and rescue, automatic tracking, target classification and targeting. The ASELFLIR-200 is an open architecture and hardware/software flexible unit which can be adapted to various air platforms, including rotary-wing, fixed-wing and unmanned air vehicles.

Key features of ASELFLIR-200 include Electronic Image Stabilisation (EIS), Local Area Processing (LAP) for image enhancement, MultiMode Tracking (MMT), analogue and digital video outputs for transmission and/or recording, MIL-STD 1553/ARINC and other discrete databusses to interface with onboard avionics, such as radar, navigation and weapon systems. The ASELFLIR-200 has three fields of view: Narrow Field of View (NFoV) for recognition and identification, Medium Field of View (MFoV) for detection and a unity Field of View (FoV) for navigation and pilotage.

The system is available in single-, dual- and triple-sensor configurations:
- FLIR only
- FLIR+colour CCD or FLIR + eye-safe Laser Range-Finder (LRF)
- FLIR+colour CCD + eye-safe LRF.

ASELFLIR-200 is in full production and installed on various rotary-wing and fixed-wing platforms. It incorporates a second-generation 4 × 240 Focal-Plane Array (FPA) detector that operates in the 8 to 12μm band. The key features of the system provide greatly improved range performance over conventional first-generation linear array detectors and improve mission capability.

There are two weapon replaceable assemblies: a turret unit, WRA-1 and electronics unit, WRA-2. Options include a laser range-finder and/or a CCD day TV camera.

Specifications
Field of View (FoV):
(wide) 22.5 × 30°
(medium) 5 × 6.67°
(narrow) 1.3 × 1.7°
Parallel detector channels: 240 × 4 FPA
Spectral Band: 8 to 12 μm
Electronic zoom: 2:1 and 4:1
Gimbal angular coverage:
(azimuth) 360° continuous
(elevation) 40° up; 105° down
Gimbal acceleration: head steering compatible at aircraft speeds
Gimbal slew rate: 3 radians/sec
Laser range-finder: optional
Day TV: optional
Compliance: MIL-STD-E-5400, MIL-STD-810
Video outputs: analogue and digital video outputs are provided
Cooling: self-contained
Weight:
(turret unit) <31.8 kg
(electronics unit) <22.73 kg
Dimensions:
(turret unit) 323.85 (diameter) × 372.87 mm (height)
(electronics unit) 306.3 (width) × 413.5 (length) × 199.1 mm (height)
Power: standard aircraft power

Status
As of November 2005, in production and in service on various rotary- and fixed-wing platforms.

Contractor
Aselsan Inc, Microelectronics, Guidance and Electro-Optics Division.

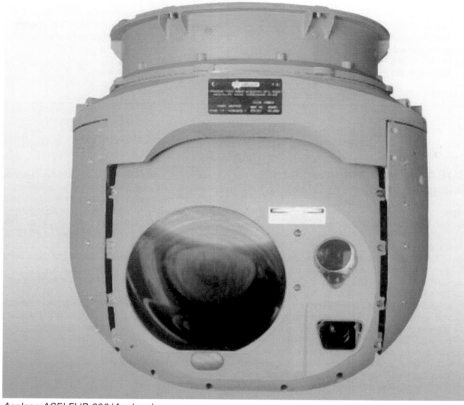

Aselsan ASELFLIR-200 (Aselsan)   1128602

M929/M930 aviator's Night Vision Goggles (NVG)

Type
Aviator's Night Vision Imaging System (ANVIS).

Description
The M929/930 and M929A/930A aviator's Night Vision Goggles (NVG) are designed for both fixed- and rotary-wing applications. They can be fitted with a variety of standard form 18 mm Image Intensifier Tubes (IIT). The latest model offers improved flash response over earlier variants of the system. The M929 and M930 are identical except for their respective mounts – the M929/929A features a standard mount assembly, while the M930/930A features an offset mount assembly, for application in the Cobra attack helicopter.

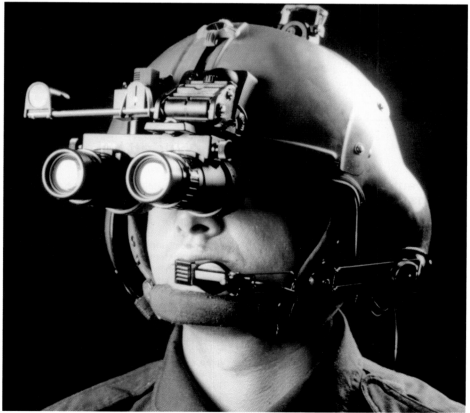

Aselsan M930 aviators' Night Vision Goggles, note offset mount for Cobra attack helicopter application (Aselsan)   0103875

Specifications
Magnification: ×1 ±5%
FoV: 40°, circular (+1, −2°)
Image Intensifier Tube (IIT) Gen III (M929/930)
   Advanced Gen III (M929A/930A)
System gain:
   2,500 (M929/930)
   5,000 (M929A/930A)
Resolution (on axis):
   1.01 lp/mR (M929/930)
   1.28 lp/mR (M929A/930A)

F-number: F/1.2
Eye relief: 25 mm nominal
Weight:
(Binocular assembly) 590 g
(Visor-mount assembly) 200 g
Operating temperature range: −35 to +49°C
Maximum operating altitude: 15,000 ft
Power supply: 2 × AA size battery (rotary-wing configuration)
Status
In production and in service.

Contractor
Aselsan Inc, Microelectronics, Guidance and Electro-Optics Division.

Ukraine

SURA Helmet-Mounted Target Designation System (HMTDS)

Type
Helmet-Mounted Sight System (HMSS).

Description
The SURA Helmet-Mounted Target Designation System (HMTDS), developed as a successor to the SHCH-3UM HMTDSs used on MiG-29 and Su-27 aircraft, is designed to be used either independently, or in combination with other aircraft systems, for airborne observation and the aiming of guided weapons and gun systems.

The HMTDS generates target designation signals for weapons in proportion to the angles of turn of an operator's (pilot's) head, as well as collimating the image of an aiming mark and initiating one-time commands to his field of view.

The HMTDS helmet tracking system is implemented using small IR emitting diodes integrated with an optical sight in a detachable unit, mounted on the pilot's helmet, together with two scanning units (designated scanning unit A and scanning unit B) mounted one each side of the Head-Up Display (HUD). This concept facilitates use of the system in a variety of

SURA helmet-mounted, scanning and electronic units (Arsenal)   0097082

aircraft with minimum requirement for aircraft modification. It also meets safety requirements for pilot ejection and emergency evacuation and makes it possible to install the helmet-mounted unit on current pilot's helmets, such as the Russian ZSH-7 and French OS 600. The HMTDS comprises:
- A helmet-mounted unit, attached to the pilot's helmet
- Scanning units A and B, mounted each side of the HUD
- An electronics unit that processes sensor information, and interfaces it to other aircraft weapon and communications systems.

Specifications
Target designation angles:
(horizontal) +70 to −70°
(vertical) +60 to −30°
Target designation accuracy (1σ): <3 mrad
Output data format: MIL-STD-1553B or ARINC-429
Power: 115 V AC, 400 Hz, <150 VA
Weights:
(helmet-mounted sighting device) 0.36 kg
(scanning units) 0.8 kg
(electronics unit) 3.5 kg
Temperature limits: −54 to +60°C
Ground warm up time of HMTDS: <1 min
Continuous operation: limited to 5 h, followed by 25 min cycle time

Status
SURA is fully ground and flight tested, in serial production for and in service on a number of fixed-wing aircraft, including the Su-27 and Su-30. SURA has been offered as part of a weapon-aiming upgrade for the Mi-24 'Hind' helicopter.

Contractor
Arsenal Central Design Office.

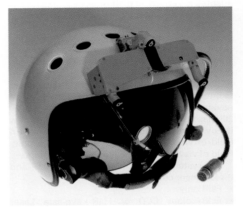

SURA helmet-mounted unit on a Russian ZSH-7AP helmet (Arsenal)   0097080

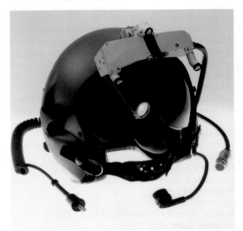

SURA helmet-mounted unit on a French OS 600 helmet (Arsenal)   0097081

United Kingdom

ADEPT automatic video tracker systems

Type
Airborne Electro Optic (EO) tracking system.

Description
ADEPT 30
The ADEPT 30 tracker is the core tracker used in both the CAATS and CAATS 2 systems. It comprises a well integrated software design on a single VME card designed for use in the Octec airborne CAATS systems, and easy integration into other manufacturers' avionics VME electronics.

The ADEPT 30 unit has a great deal of flexibility in its design, allowing closed loop control of a platform and easy integration to control panels.

A range of daughter boards gives even greater flexibility and performance, allowing input of high-speed digital data, further signal processing, picture stabilisation, improved detection

capability or additional tracking channels – this configuration is known as ADEPT 30+ (see side heading below).

A range of enclosures is available from a commercial 19 in rack unit to fully ruggedised avionic enclosures. Build standards are available to meet different standards from commercial up to full MIL Spec.

Octec has integrated its trackers with over 100 different platforms, including most widely used helicopter sensor platforms.

Specifications
Dimensions: 233.4 × 160 mm Double Euro
Power: +5 V 3 A, +12 V 0.2 A, −12 V 0.2 A
Video input: Composite video 625/525 line CCIR or RS-170
No of video inputs: 2
Track modes: Centroid, Correlation, Edge, Multiple target track, Scenelock
Video output: 1 with symbology overlay
Interfaces available: VME, RS-232/422, Analogue, Discretes

Automatic video detection: variable from 2–90% of FoV

ADEPT 30+
A range of daughter boards is available which allow the ADEPT 30+ to carry a wide range of additional functions, including:
1. input of high-speed digital data from modern sensors;
2. provision of a second tracking channel (see below);
3. additional preprocessing;
4. electronic picture stabilisation;
5. improved detection capability;
6. multimode tracking;
7. enhanced filtering;
8. colour processing.

ADEPT 33 automatic video tracker
The ADEPT 33 tracker is the latest in the range of Octec video trackers and is a new tracker with smaller form factor for requirements where size and weight are more critical. It comprises an integrated

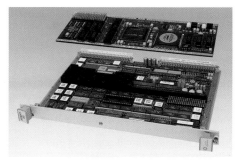

ADEPT 30+ VME-based automatic video tracking system   0051010

software design on a single Eurocard for easy integration into other manufacturers electronics.

The ADEPT 33 unit has a great deal of flexibility in its design, allowing closed loop control of a platform and easy integration to control panels. Built-in interfaces include RS-232/422 and the video output includes video with symbology overlay. Build standards are available to meet different standards from commercial up to full MIL specification.

Specifications

Mechanical: single height VME board; forced air or conduction cooled variants available
Dimensions: 100 × 160 mm single Euro (3U) format
Power: +5 V 2.5A, +12 V 0.2 A, −12 V 0.2 A
Video input: composite video 625/525 line CCIR or RS-170
No of Video Inputs: 2
Track modes: centroid, correlation, edge, scenelock, multiple target track (optional)
Video output: 1 with symbology overlay
Interfaces available: RS-232/422
Automatic video detection: variable from 2 to 90% of FoV

ADEPT 34

The ADEPT 34 is a PCI format single-board automatic video tracker. It is software-based and ideally suited to simulator and training systems. No external hardware is required, beyond a standard motherboard, and it can be controlled via the PCI bus or on-board RS 232/RS 422D serial links.

Multiple channel automatic video trackers

The ADEPT family of video trackers are available in multiple channel configurations. These can range from two PECs dual-gate trackers to complex systems with multiple tracking channels.

The dual-gate tracker provides dual-channel functionality and is optimised for the acquisition, tracking, and transition between high-velocity targets such as a host aircraft and released bombs or missiles.

The multiple channel tracker utilises ADEPT family video tracker PECs in conjunction with a Single Board Computer (SBC). The SBC provides tracker/target sequencing, subsystem management and target trajectory prediction.

Status

Some 500 ADEPT 30 trackers have been shipped to customers worldwide and it continues in full production. ADEPT 33 is in full production for a NATO UAV.

VSG 30 image processing system

The VSG 30 image processing system comprises a board, using the VMEbus format, which provides a range of complementary facilities to the ADEPT tracker range. The board provides for three video inputs in either monochrome or colour. One or more of these can be passed through the image processing modules, which provide a range of facilities including electronic zoom, freeze-frame, frame-to-frame integration and filtering. Some of these processing facilities may be combined to provide more complex filtering.

A full colour symbology generator is supplied, operating in either monochrome, or in up to 15 selectable colours.

VSG 30 may be supplied with a separate single-board computer to provide additional facilities

ADEPT 33 single height VME bus automatic video tracker   0051011

such as frame grabbing and storage in a modular system. Selection of inputs, outputs and filtering can all be remotely controlled over datalinks.

The VSG 30 is built in a range of environmental specifications to match the ADEPT 30.

Specifications

Video Inputs: 3
Format: Composite Video 1.0 V p-p
Lines: 625/525 Line
Fields: 50/60 Hz
Standards: PAL, NTSC, Y/C
Video Output: 6 max
Format: Composite (3), Y/C (3), RGB (1)
Level: 1.0 V p-p
Mechanical: VMEbus
Dimensions: 233.4 × 160 mm double Euro

Status

VSG 30 entered production in the third quarter of 1998.

Contractor

Octec Ltd.

AF500 Series roof observation sights

Type

Airborne Electro-Optic (EO) targeting system.

Description

The AF532 roof-mounted helicopter sight superseded the AF120 sight introduced in 1970. The new sight is half the weight of its predecessor, but confers greatly improved optical performance.

This gyrostabilised, monocular, periscopic telescope is designed for the gunner/observer in reconnaissance helicopters, particularly when scouting targets for anti-tank helicopters.

The device has a built-in interface for a laser designator and range-finder. It can also be adapted for night vision equipment, helmet sights and weapons, and there are facilities for attaching a television recording camera for training or intelligence gathering. The design of the optical system is such that the varying eye positions in different helicopter installations can be easily accommodated.

The gyrostabilised head protrudes above the roof forward of the rotor mast, while the down-tube and eyepiece extend downwards from the roof so that the eyepiece falls into a comfortable viewing position. The down-tube is adjustable in height and retracts sideways, locking close to the roof when not in use. A control handle is extended by the operator and adjusted in tilt so that it can be used with the right forearm resting on the knee. A horizontal thumbstick is used to steer the sightline and a direction indicator, to show its direction relative to aircraft heading, is mounted on the glareshield in front of the pilot.

The sight provides a stabilised image of the chosen field of ×2.5 magnification for search and ×10 for identification and laser operation. The sightline may be steered through ±30° and ±120° in pitch and yaw planes respectively.

AF580 systems, fitted with BAE Systems, Avionics laser ranger and target designators and integrated with a Rockwell Collins Automatic Target Handover System have been evaluated by the US Army in Bell OH-58C Kiowa helicopters. This combination of systems permits the range and bearing of the target to be determined by the observation helicopter and transmitted by datalink to an attack helicopter. In a further development, a thermal image from a separate FLIR was injected into the sight to permit night observations. The FLIR and the sight were steered via the same controller.

Status

In service in British Army Air Corps' Westland Gazelle helicopters. Laser Target Designation and Range Finder (LTDRF) equipment has been fitted to a number of systems and is now in service.

Contractor

Ferranti Technologies Limited.

ATLANTIC podded FLIR system

Type

Airborne Electro Optic (EO) Forward-Looking Infra-Red (FLIR).

Description

The Airborne Targeting Low-Altitude Navigation Thermal Imaging and Cueing (ATLANTIC) pod-mounted FLIR system is designed to give ground attack aircraft night and poor weather capability on high-speed low-level missions.

The AF532 sight unit installed on the roof of a British Army Westland Gazelle helicopter   0503970

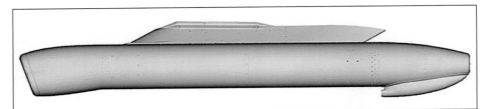

The ATLANTIC II FLIR pod 0533856

ATLANTIC II FLIR combined sensor head 0533857

The Atlantic pod employs the BAE Systems modular FLIR system as used in the Harrier GR Mk 7 and it can also include an advanced thermal cueing system for early target detection and a laser spot tracker. The system, with multitarget compatibility, has a MIL-STD-1553B databus interface and can be integrated with existing weapons and avionics systems.

The telescope is selected to match the FLIR image to the aircraft HUD field of view. Pod cooling is by means of a self-contained environmental conditioning unit and the detector is cooled by a closed-cycle cooling engine. Time to readiness for the system is 31/2 minutes typically from 20°C.

ATLANTIC II LSL
The latest variant is the ATLANTIC II, designed specifically for the European Partners Air Forces (EPAF) nations' F-16 Mid-Life Update (MLU), which integrates a Laser Spot Locator (LSL) for target identification, location and target tracking and a Thermal Cuer (TC) to provide early target detection and prioritisation for weapon delivery. The system is available in both internal and podded configurations with optional 8 to 10 μm or 3 to 5 μm staring Focal Plane Array (FPA) FLIRs.

ATLANTIC II sensors and advanced signal processing are claimed to provide outstanding image quality on the host aircraft Head-Up Display (HUD).

The pod features high-reliability, extensive Built-In Test (BIT) equipment and ease of maintenance to ensure maximum equipment availability and low life cycle costs.

Specifications
ATLANTIC II
Dimensions: 2,410 × 254 mm diameter
Weight: 94 kg (excluding LSL); 129 kg (including LSL)
Mechanical interface: F-16 nacelle mount, NATO standard pylon, custom pylon
Cooling: Self contained, passive, environmental conditioning unit
Power supply: 28 V DC nominal, 210 W typical; 115/200 V AC, 400 Hz, 146 W typical
Flight envelope: M1.6 at 72,000 ft, 9 g
Temperature range: –40 to +70°C
FoV: Selectable for aircraft HUD
28 × 21° for F-16/F-5
25 × 16° for Mirage/Tornado
20 × 13° for A-4/AV-8B
Video output: 525-line, 60 Hz RS170 or 625-line, 50 Hz CCIR
Resolution: 790 × 512 pixels (H × V) (CCIR System 1)
LSL
Wavelength: 1.064 μm
Pulse width: 15 ηs (–5, +10)
Pulse rates/interval deviation: Per STANAG 3733 Bands 1 and 2
Pulse rise/fall times: 3 to 10 ηs

A computer generated image of the ATLANTIC II LSL on an F-16B 0533858

Status
Standard equipment on a wide variety of strike aircraft, including F-16, A-4, Hawk 100 and the AV-8B/Harrier II+.

Contractor
Selex Sensors and Airborne Systems.

CAATS 2 – Compact Airborne Automatic video Tracker

Type
Airborne Electro-Optic (EO) tracking system.

Description
The CAATS 2 unit provides the same, or even better, tracking performance than the earlier CAATS unit, but is lighter, less than half the size, and uses less power than the original design. It is a sealed unit using conduction cooling. Several versions are available to meet different environmental scenarios. CAATS 2 also has the capability, using daughter boards, to enhance performance including electronic image stabilisation and input of digital video.

The CAATS 'Scene Lock' tracking feature is retained and performance improved. This is a particularly important feature for the helicopter fit. Centroid tracking is also provided together with a range of submodes, including automatic cueing, which allows optimisation for particular target scenarios. A video symbology generator is integrated to allow on-screen data to be shown and recorded.

Specifications
Weight: 2.8 kg
Dimensions: 250 (W) × 210 (D) × 50 mm (H)
Power: 28 V DC, 1.0 A mean or 115/220V AC using adaptor
Video input: RS-170 or CCIR video input 50/60 Hz
No of video inputs: 2
Tracking modes: centroid, scenelock (other modes optional)
Video output: 1 with symbology overlay.
Automatic video detection: variable from 2–90% of FoV

Status
First shipments took place during March 1997.

Contractor
Octec Ltd.

Electro-Optical (EO) Helmet-Tracking System (HTS)

Type
Airborne Electro-Optic (EO) Helmet-Tracking System (HTS).

Description
Selex' electro-optical Helmet-Tracking System (HTS) is designed to satisfy the requirement for a tracking system to interface with helmet-mounted displays and sights that is fast, accurate and remains uninfluenced by metal structure within the aircraft cockpit.

One or more optical sensors located on the aircraft structure detects the position of several helmet-mounted light emitting diodes (LEDs). A multiplexer energises each LED in turn and the position of the helmet in the cockpit is determined from the sensed relative position of the LED from each sensor. The aircraft-mounted electronics unit can be provided in stand-alone form or as a card set for integration into an existing unit. The system does not need to be aligned or boresighted after initial installation.

Status
In production for various fixed- and rotary-wing applications.

Contractor
Selex Sensors and Airborne Systems.

FIN 1010 FLIR

Type
Electro Optic (EO) navigation system.

Description
The FIN 1010 is a modular FLIR which provides a totally passive solution to the requirements of night navigation and, to a limited extent, target

CAATS 2 automatic video tracker 0001239

The BAE Systems Harrier GR7 is fitted with the FIN 1010 modular FLIR; the sensor is accommodated in the blister on top of the nose, thereby aligning the sensor window close to the pilot's LoS and facilitating use of the FLIR to aid in takeoff and landing in poor weather (L J Cartwright)   0589480

acquisition. The FIN 1010 may be installed either in a pod or integrated directly into the aircraft, with the optics looking forward through a small window. The system incorporates the Thermal Imaging Common Module II (TICM II) which features a lightweight miniaturised scanner and advanced signal processing to satisfy the demanding space and performance requirements of the airborne role. Hands-off fully automatic operation in gain and offset minimises aircrew workload. The modular design of the FLIR permits simple reconfiguration to meet the differing space constraints of a diverse range of aircraft, including the Tornado GR4, Harrier GR7 and Hawk 100. The production equipment for the Hawk 100 comprises two LRUs: the sensor head and the electronics unit. The electronics unit includes space provision for future growth in performance and capability. The configuration for the Harrier and AV-8B can be fitted within a 254 mm diameter pod.

By projecting a high-resolution image of the terrain ahead on the HUD, the FLIR permits the pilot to carry out aggressive manoeuvres at low altitude. In addition, an integrated thermal cuer detects hot objects, which may be potential targets, within the scene and marks them on the HUD.

The telescope is selected to match the FLIR image to the aircraft HUD field of view. The detector consists of eight parallel CMT TEDs, cooled by a closed-cycle cooling engine. Time to readiness is typically 3½ minutes at 20°C.

Other versions are available for tactical transport aircraft, such as the C-130 Hercules, or as enhanced vision systems on civil aircraft.

For further information regarding the operation of the FIN 1010 modular FLIR, please refer to the Electro Optics Analysis section.

Specifications
Waveband: 8–12 μm
Sensor: TICM II
Temperature range: –40 to +70°C
FoV: 16 × 25° (V × H)
Power supply: 28 V DC nominal, 200 W typical 115/200 V AC, 400 Hz, 146 W typical
Video output: 625-line, 50 Hz or 525-line, 60 Hz
Interface: MIL-STD-1553 and/or discrete hardwired

Status
In production for the UK Royal Air Force Tornado GR Mk 4 and Harrier GR Mk 7, AV-8B for the US Marine Corps, Spanish and Italian navies and the Hawk 100.

Contractor
Selex Sensors and Airborne Systems.

Heli-Tele television system for helicopters

Type
Airborne Electro-Optic (EO) surveillance system.

Description
The Heli-Tele is a broadcast-standard television surveillance system designed for mounting on helicopters. The system provides long-range real-time airborne surveillance to meet the requirements of police, military, paramilitary, civil and other security forces.

Heli-Tele consists of a colour camera, with a high-magnification zoom lens, mounted on a gyrostabilised, steerable, platform. The operator can steer the camera and adjust the magnification to display any selected area of the ground scene. Video information is transmitted to any number of ground stations via an air-to-ground microwave link with a range of over 90 km. For night or poor visibility surveillance, a standard, modular, thermal imaging sensor package may be installed in place of the TV camera payload, with a turnaround time of only 30 minutes.

Status
In service. The Heli-Tele has been certified by the UK's Civil Aviation Authority and is fitted to many types of helicopter.

Contractor
Selex Sensors and Airborne Systems.

Helmet-Mounted Sighting System (HMSS)

Type
Helmet-Mounted Sight System (HMSS).

Description
The HMSS is a visor-projected sighting system which comprises an optical subassembly mounted to a largely unmodified helmet (in the case of the UK Helmet-Integrated Systems Limited Mk 4D or Mk 10B, the sight is attached using the standard Night Vision Goggle mounting holes), with the associated electronics remotely located in the aircraft's avionics bay or similar. Sighting and cueing information is presented to the pilot by means of a high-brightness LED reticle, relayed by a prism and reflected into the pilot's eye by a dichroic patch coating on the inner surface of the clear visor.

The HMSS, originally developed in conjunction with the UK Defence Evaluation Research Agency (DERA, now QinetiQ), has undergone various stages of upgrade to provide maximum optical performance with minimum obscuration. The system comprises a BAE Systems Helmet-Mounted Sight (HMS) in conjunction with Honeywell's Advanced Metal Tolerant Tracking (AMTT) technology.

The sight enables the pilot to perform off-boresight missile target acquisition, and to fully exploit the advantages of modern missile seeker head capabilities, including a proven ASRAAM interface. Aircraft sensors (TIALD, EO-LOROP) can also be pointed using the sight, and, with the reticle slaved to the sensor, to cue the pilot. In a typical air-to-air situation, the HMSS can be

The FIN 1010 FLIR picture projected onto the HUD of a UK RAF Harrier GR7; HUD symbology is overlaid onto the FLIR scene to enable the pilot to maintain 'heads out' during low-level flight at night (E L Downs)   0589481

Selex Helmet-Mounted Sighting System (HMSS)   0080274

The HMSS can be used to slew aircraft sensors, such as TIALD (Crown Copyright)   0533845

employed to effect an IR lock onto a target at up to 30° off boresight.

Specifications
Weight: 0.15 kg
Field of view (circular): 40°
Exit pupil: 16 mm at centre field of view
Eye relief: Compatible with prescription spectacles

Status
A contract was awarded for pre-production units for use on the UK Royal Air Force Jaguar GR. Mk 1B/T2B as part of the Jaguar '97 upgrade. The production-standard HMSS was flown for the first time on a UK Royal Air Force Jaguar GR Mk.3A aircraft at Boscombe Down in early 1999. The system is in service in Jaguar aircraft of the UK Royal Air Force and the Omani Air Force.

Contractor
Selex Sensors and Airborne Systems.

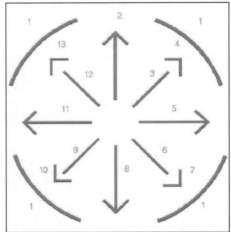

HMSS symbology set   0533846

Laser Ranger and Marked Target Seeker (LRMTS)

Type
Airborne Laser Ranger and Marked Target Seeker (LRMTS).

Description
The LRMTS is a dual-purpose unit which can be used as a self-contained laser ranger or as a target seeker with simultaneous range-finding. In the target seeking role it can be used to detect and attack any target designated by ground troops with a compatible laser, enhancing the effectiveness of battlefield close air support.

The LRMTS is an Nd:YAG laser mounted in a stabilised cage, which allows beam-pointing and stabilisation against aircraft movement. The seeker can detect marked targets outside the head movement limits. It operates at a relatively high pulse repetition frequency, thus allowing continuous updating of range information during ground attack. As range is a crucially important

parameter for accurate weapon delivery, and yet virtually unobtainable on a non-laser-equipped aircraft, the LRMTS is a vital additional sensor.

In a typical operation, a forward air controller with pulsed laser target-designation equipment directs the aircraft to a location within laser detection range before switching on the ground marker equipment. Radio communications between the forward air controller and aircraft crew are minimised and positive identification of even small, hidden or camouflaged targets is assured.

Once the LRMTS has detected the laser energy reflected from the target it provides steering commands to the pilot on the head-up display. Ranging data is also shown and fed directly into weapon aiming computations for the accurate and automatic release of weapons.

The LRMTS is easy to install and harmonise with other aircraft systems and is said to be more effective than any alternative sensor during operations at the grazing angles used in low-level ground attack. By improving weapon delivery accuracy, the sensor ensures a high probability of success in single-pass, high-speed attacks.

Associated with the LRMTS head is an electronics unit which contains power supplies and ranging and seeker processing. The laser needs a transparent window and in the Jaguar is mounted behind a chisel-shaped nose, with two sloping panels. For the Harrier, where there is more chance of debris accumulation during VTOL operations, the optics are protected by retractable eyelid shutters.

A specially designed installation for the UK Royal Air Force Panavia Tornado has incorporated the LRMTS into an underbelly blister. In addition

to the systems in service with the UK Royal Air Force, the equipment is also used by three other air forces.

An eye-safe version of LRMTS has been developed and successful trials have been completed. Eye-safe operation has been achieved by altering the output of the laser with a Raman cell from 1.06 μm to 1.54 μm which is in the eye-safe region of the spectrum. Existing LRMTS systems can be retrofitted with this modification which will significantly reduce safety restrictions for training.

Specifications
Dimensions:
(LRMTS head) 300 × 269 × 607 mm
(electronics unit) 330 × 127 × 432 mm
Weight:
(LRMTS head) 21.5 kg
(electronics unit) 14.5 kg
Power supply: 200 V AC, 400 Hz, 3 phase, 700 VA 28 V DC, 1 A
Wavelength: 1.06 or 1.54 μm
PRF: 10 pps
Angular coverage:
(elevation) +3 to −20°
(azimuth) ±12°
Roll stabilisation: ±90°
Detection angle: ±18° from aircraft heading
Max range: >9 km

Status
In service. Development of the LRMTS began in 1968 under a government contract and prototype units were first flown in 1974. Deliveries to the UK Royal Air Force, accounting for over 200 units, for installation in nose housings of the BAE Systems Harrier and Sepecat Jaguar aircraft were completed in 1984. Over 800 LRMTS have been delivered to the Tornado, Jaguar and Harrier programmes, in the UK and overseas.

Contractor
BAE Systems.

MultiSensor Turret System (MSTS)

Type
Airborne Electro Optic (EO) surveillance system.

Description
The MSTS is a versatile compact lightweight thermal imaging system suitable for helicopter and subsonic fixed-wing aircraft operations.

The LRMTS mounted beside the nosewheel bay of a fully loaded UK RAF Panavia Tornado GR Mk.1 (I Black)   0533855

The gyrostabilised turret platform provides 360° steerable Field of Regard (FoR) in both azimuth and elevation and is fully operational at airspeeds up to 300 kt.

The standard thermal imaging payload utilises UK TICM II modules, a closed-cycle cooling engine and a continuous zoom telescope with magnification of ×2.5 to ×10. Alternative payloads with a variety of IR and TV sensors and telescopes are available.

The basic system consists of two LRUs: the turret and a control switch joystick unit. This can be expanded by the addition of a third unit to cater for the range of options which provides full integration with other aircraft systems. The options currently available include MIL-STD-1553B, ARINC 429 and RS-422 databus interfaces, video tracker, electronic magnification, radar-designated target hand-over and a variety of control units.

Specifications
Weight:
(turret) 42 kg
(control switch joystick unit) 2 kg
(optional control electronics unit) 12–20 kg
Detector: 8-parallel CMT SPRITE, InSb array (option)
Wavelength: 8–13 μm, 3–5 μm (option)
Power supply: 28 V DC nominal, 340 W typical
Video output: 625-line, 50 Hz or
525-line, 60 Hz

Status
In production for S-61, S-76, Lynx and Sea King helicopters and Fokker F50, Dornier 228 and CN-235 maritime patrol aircraft. A quantity of MSTs have also been fitted to Nimrod MR Mk.2 aircraft. Ordered for UK Royal Air Force EH 101 Support Helicopters and as part of a surveillance system being installed on Agusta-Bell 412 EP helicopters by an export customer.

Contractor
Selex Sensors and Airborne Systems.

SiGMA Thermal Imager (TI)

Type
Airborne Thermal Imager (TI).

Description
SiGMA (Sensor Integrated Modular Architecture) is the latest Staring Focal Plane Array (SFPA) Thermal Imager (TI) from BAE Systems. It provides high-performance passive infra-red imaging for land, sea and airborne military operations worldwide. SiGMA has been designed as a stand-alone imager, or as part of an integrated system, with application specific enclosures and optics as required by the systems integrator. At the heart of SiGMA imagers are processing electronics designed to readily accommodate a range of detectors including ½- and full-TV formats, plus the capability to accept future High-Definition TV (HDTV) formats.

The incorporation of an FPA, eliminating the requirement for a complex scanning mechanism, provides for significantly higher reliability with lower acquisition and through-life ownership costs for the system. Other features and benefits of SiGMA include:
- Programmable configuration
- Automatic or manual control of gain and offset
- Programmable gating region
- User-defined text and graphic symbols
- Freeze frame
- Up to ×16 electronic zoom and pan
- Four programmable correction tables
- Auto calibration mode
- Military specification
- Simple system integration
- Flexible video output and control interface
- Lightweight compact design
- Low through-life cost of ownership.

MSTS fitted under the starboard wing of a Nimrod MR Mk.2   0005551

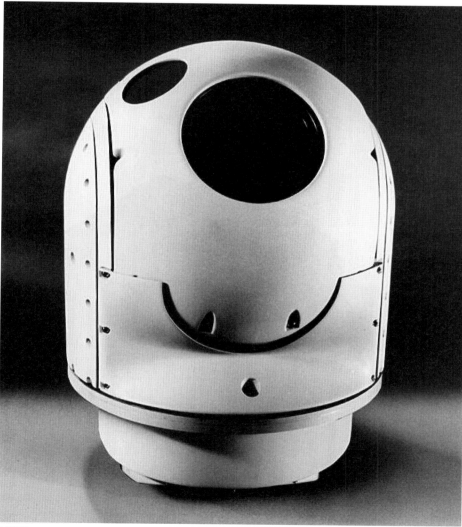

MSTS   0005549

Specifications
Dimensions: 175 × 130 × 100 (L × W × H) mm (excluding lens and enclosure)
Weight: <4 kg
Power: <35 W (at 24 V DC) (low power standby mode available)
Reliability: >5000 h
Detector: Staring Focal Plane Array (FPA)
Array: up to 640 × 512 (no microscan)
Wavelength: 3–5 μm (MWIR) or 8–10 μm (LWIR)
Horizontal FoV: Selectable by lens
Non uniformity correction: Quadratic

Video: 625 line 50 Hz STANAG 3350 Class B, or 525 line 60 Hz STANAG 3350 Class C, or 640 × 512 VESA (single output)
Digital output: 14-bit uniformity corrected, full dynamic range, or 8-bit video
Interface option: RS422
Operating temperature: −33 to +55°C
Environmental: DEF STAN 00-35, MIL-STD 810E

Status
In production as the TI camera for the Titan 385 MultiSensor Turret System (MSTS) (see separate entry).

SiGMA Thermal Imager (TI) (BAE Systems Avionics)   0523595

The STAIRS C thermal imager   0053836

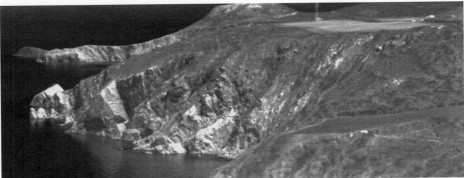

Image taken using a prototype STAIRS C thermal imager   0111402

Contractor
Selex Sensors and Airborne Systems.

STAIRS C Thermal Imager (TI)

Type
Airborne Thermal Imager (TI).

Description
The STAIRS C thermal imaging system is a set of ruggedised modules developed from the original technology demonstrators sponsored by QinetiQ (formerly the UK Defence Evaluation and Research Agency, DERA) and built by Thales Optronics (formerly Pilkington Optronics). The system will provide high-performance, second-generation Thermal Imaging (TI) for all land, sea and air platforms, intended to replace the UK TICM II (Thermal Imaging Common Module).

The STAIRS C detector is a cooled Cadmium Mercury Telluride (CMT) device similar in format to that used for the Eurofighter Typhoon IRSTS, but further developed for advanced imaging. STAIRS C offers greatly improved target recognition and identification ranges, up to double that of many first-generation systems, due to its extended spatial resolution and thermal sensitivity.

STAIRS C uses high-resolution SXGA and standard TV video formats and features EOPM compatibility, automatic control and comprehensive graphics, zoom and freeze-frame. The module interfaces have been defined to ease system integration, maintenance and testing.

Specifications
Telescope options:
Dual 3.6 × 2.2° and 20 × 12°, 180 mm aperture
Dual 5 × 3° and 17.5 × 10.5°, 110 mm aperture
Sightline stability: ≤1 mrad
Spectral band: 8 to 9.4 μm (FIR)
Detector: 768 × 8 CMT diode array with 6 element TDI
Cooling: Integrated detector-cooler assembly
Image formats: 1280 × 768 (25/50 Hz); 1024 × 768 (50 Hz); 768 × 576 (50 Hz); 576 × 768 (50 Hz)
Display: SXGA: 625 line, 50 Hz (CCIR 1); 525 line, 60 Hz (RS-170)
Control: Automatic with manual override
Power: 120 W
Weight: 20 kg
Dimensions:
Sensor Head Module (SHM): 392 × 200 × 142 mm
Electronics Unit (EU): 200 × 200 × 142 mm
Environment: specified for rotary and fixed-wing aircraft and tracked vehicles
MTBF: 5,000 h

Status
Thales Optronics and QinetiQ started work on the STAIRS C programme in 1995. The first demonstration systems have been under intensive trials evaluation by QinetiQ (Malvern) since late 1997. During 2000, the UK MoD invited Thales Optronics to tender, single source, for the STAIRS C Productionisation (SCP) Technology Demonstrator Programme (TDP). The aim is to develop a production-standard, high-performance, second-generation TI suitable for land, sea and air platforms. The system features a very long linear CMT detector of 768 × 8 diodes, derived from the detector being developed for the PIRATE IRST of the Eurofighter Typhoon.

Contractor
Thales Optronics Ltd.

Striker Helmet-Mounted Display (HMD)

Type
Helmet-Mounted Display System (HMDS).

Description
The Striker HMDS is a high-performance, visor-projected Helmet-Mounted Display (HMD), intended for both fixed- and rotary-wing

applications. The imagery displayed on the visor is relayed to both eyes from onboard systems such as head-steered Forward Looking Infra-Red (FLIR), or head-mounted sensors, such as integrated detachable Night Vision (NV) cameras; the system is therefore a fully 24-hour-capable system (under Visual Meteorological Conditions). The basic Striker system combines an advanced HMD with a high level of pilot protection, low head-supported weight and proven high-accuracy electromagnetic helmet tracking technology. The Striker system provides for a net reduction in operator workload, increased pilot awareness, combined with the capability to transfer data from the aircraft platform into the electronic battlefield through sensor fusion and network-centric operation.

System description

The HMD comprises a display module, fitted onto a lightweight, tailored-for-fit basic helmet; the system provides a fully overlapped binocular, visor-projected display of high-resolution raster imagery. When displaying imagery from the two night vision cameras (which are detachable for daylight missions), the HMD provides for night vision in the NIR waveband. In addition to the two-part helmet, the system comprises:

- Helmet Electronics Unit (HEU) – this unit provides all display and head tracker processing electronics. These functions can also be provided as modules to be housed in a mission computer
- Helmet Cockpit Unit (HCU) – located within the cockpit and providing those electronic functions that need to be physically located close to the HMD
- Helmet tracker transmitter – located at a defined position and orientation within the cockpit, the unit outputs very precise reference magnetic fields within the cockpit
- Quick Release Connector (QRC) – provides a single-point connection between the aircraft and the HMD. The QRC is specially designed to provide for a safe and rapid egress from the cockpit. In addition to the primary quick release, it incorporates secondary failsafe connector separation features
- Boresight Reticle Units (BRU) – two per cockpit, providing Line-of-Sight (LoS) and roll boresighting references for the helmet tracker.

The HEU houses the display processing, symbol generator, NV camera processing and interface electronics and also acts as a video distribution system, interfacing with video from other sources, such as an onboard targeting FLIR. It also facilitates the output of NV camera video for mission recording or onward transmission.

The HMD system meets the requirements for low head supported mass. The total head supported mass of the basic HMD is 1.68 kg (3.7 lb) rising to 2.18 kg (4.8 lb) with both NV cameras attached; the centre of gravity is quoted as well within safety guidelines specified for day and 24-hour HMD configurations. The display field of view is 40° with a 100 per cent binocular overlap, providing stroke symbology for daytime viewing and overlay on camera video for night-time operations. The display source is a high-resolution CRT that has been developed to meet the requirements for the display of the NV camera video, while also providing high-brightness symbology against the outside world for daytime missions.

The high-accuracy helmet LoS tracker uses a proven DC magnetic tracker developed by Selex primarily for rotary-wing applications; the system is currently in production for the Knighthelm HMD for the Eurocopter Tiger (see separate entry). This provides a level of accuracy sufficient for applications such as weapon aiming. A key feature of the DC tracker is its enhanced tolerance to cockpit metal compared with many similar AC

electromagnetic systems. The presence of cockpit metal distorts the reference magnetic fields emitted by the helmet tracker transmitter. The helmet-mounted sensor samples the reference magnetic fields to provide the information necessary to compute helmet orientation and position. The distortion due to cockpit metal may therefore induce errors in this measurement process. This error is removed by characterising the cockpit metal in a process know as cockpit mapping. As the DC tracker is relatively tolerant of cockpit metal, it is only necessary to map a representative sample of cockpits to generate a generic cockpit map, which can then be used for all aircraft with that cockpit configuration. This characteristic of the Striker system obviates the requirement of some AC systems to map every cockpit or to re-map if cockpit instruments are changed.

Night Vision Devices (NVDs)

There are two main approaches to integrating the function of a Night Vision Goggle (NVG) and HMD: optical image combination of Image Intensifer (II) tube and CRT images is now well proven. However, it involves providing multiple optical channels on the helmet for the II and CRT display and can result in a bulkier design. Also, it does not readily enable the export of the intensified night image as a video signal. The approach adopted for Striker is to exploit electronic image combination using NV cameras. This approach will be implemented in the Eurofighter HMD (see separate entry), with claimed 'comparable performance to NVGs', while facilitating the sharing of information with other airborne or ground forces. For this approach, two miniature helmet-mounted NV cameras provide enhanced night vision. The cameras employ a specially developed image intensifying Charge Coupled Device (CCD). The NV camera video signal is fed back to the remote display drive electronics within the HEU, where it is electronically combined with the symbology and displayed on the helmet-mounted CRTs.

This approach maximises the performance of the CRT display by removing the need to optically mix II and CRT imagery. It also minimises head-supported mass and bulk, which is a critical parameter in the development of HMD systems. The relatively small size and weight of the NV cameras has facilitated a configuration which allows them to be easily removed and replaced on the HMD by the user, in flight, with gloved hands.

Although still utilising II technology, the intensified imagery is now presented in a high-definition raster display format, allowing the user more control of the brightness and contrast of the NV picture. The NV image is displayed on the helmet visor such that it exactly overlays the outside world scene viewed through the visor.

Selex is developing the NV cameras with the goal of at least equalling the resolution of current NVGs when viewed through the HMD. The company claims the design exceeds the minimum requirements with either of the clip-on cameras providing for see-through, visor-projected night pilotage superior to Generation III Omni III direct view NVGs.

In the urban environment, the shortcomings of conventional II tubes used in NVGs result in halos around bright point light sources which obscure the detail within the scene (see Analysis section). The NV camera applies real-time processing technology to counteract this effect. This is called anti-blooming technology (ABT) and produces a dramatic improvement in the performance of the NV system, as it allows the camera to see through typical blooming. The camera also uses Enhanced Automatic Gain Control (E-AGC), which dynamically compensates images to improve dark scene imagery while preserving highlight detail. Both ABT and E-AGC are fully

automatic, responding in real time without any loss in sensitivity. The intrascene dynamic range is one million to one.

The NV camera provides a claimed resolution better than 1cy/mrad. The interscene dynamic range is 100 million to one, and the camera operates in light levels from overcast starlight up to overcast day. This is achieved by autogating of the intensifier power supply and by controlling the microchannel plate. The camera is a very small unit weighing only 260 g. The mount design allows the user to fit or remove the camera, automatically aligning it with the HMD optical axis and providing the electrical connection with the HMD umbilical cable.

Helmet

Pilot protection and life support remain paramount requirements, but many current in-service flying helmets are proving to be a barrier to adding new functionality. Lessons learnt from many HMD applications show that adapting current aviator helmets by addition of modules has not been the optimum solution and often compromises safety. Key characteristics such as weight, stability, accuracy of fit, comfort, centre of gravity, life support equipment compatibility and crash safety tend to be compromised by the addition of new functions on the standard helmet shell. Most current aircrew helmets were not designed for the tasks they are now being asked to perform. The designs have to some extent been adapted to meet increased demands, but HMDs need a better and more accurate fit, stability, comfort and improved helmet retention.

The HMD visor is also part of the optical system and introduces a further set of integration issues. The protection properties of helmet and display module combined must also be considered. The combined weight and centre of gravity of the HMD must meet stringent crash safety requirements.

The Striker family of visor-projected, two-part helmets has been developed to provide aircrew with a helmet which maintains all of the protection and life support functions of standard helmets while providing the display function within a lightweight, integrated modular assembly. The basic inner helmet is custom fitted to the head. It provides a comfortable, stable platform which is accurately fitted. It also houses the communications subsystem and provides interfaces for the UK AR5 NBC respirator system. The inner helmet has been designed to accommodate a very wide range of head sizes and shapes using a single-size shell, and employs a unique helmet fitting system to ensure an accurate, comfortable and stable fit.

The mission-specific outer module is driven by the platform and/or mission requirements. It contains the visor-projected display, helmet tracker, display and glare visors. It is a single-size design so that any outer can be fitted to any inner. The modular design allows the outer to be tailored to the application and allows for ease of maintenance and reconfiguration. The two NV cameras clip onto the outer display module. The display module is secured to the basic helmet using three quick-release latches that may be operated with a gloved hand. The display module can also be fitted to the basic helmet while it is being worn, allowing the pilot to walk out to the aircraft wearing the inner helmet shell and then to don the display module in the cockpit.

Specifications
Striker
FoV:
    Optical system: 40°, circular
    NV cameras: 40 × 30°
Binocular overlap: 100%
Exit pupil: >15 mm, on axis
Eye relief: >70 mm

For details of the latest updates to *Jane's Avionics* online and to discover the additional information available exclusively to online subscribers please visit
jav.janes.com

Early work on the Striker display helmet provided the basis for the Eurofighter Typhoon Head Equipment Assembly (HEA)   0080273

HMD adjustment: Interpupillary 60 to 75 mm
Video format: 1080 line 60 Hz 2:1 interlaced
Display contrast ratio (day): >1.2:1 (10,000 fl ambient condition)
Helmet tracker:
Type: DC electro-magnetic (LoS and roll)
Angular coverage: ±180° azimuth, ±90° elevation, ±180° roll
Update rate: 100 Hz (interleaved, two-cockpit operation)
Weights:
    Basic: <1.68 kg
    With NV cameras: <2.18 kg

System growth and technology insertion
The Striker HMD has been designed from the outset with upgrade paths to allow insertion of new technologies as they become available. The miniature CRT remains the display device of choice for high-performance HMD applications, but significant research and development activity has been expended on the replacement of CRTs with alternative miniature display technologies. Prototype HMDs based on the Striker design have been produced to demonstrate the feasibility of various display technologies. One of these is a full-colour binocular display using reflective Active Matrix Liquid Crystal Display (AMLCD) technology with a three-colour laser illumination system. It is claimed that this display is the first full-colour binocular visor-projected HMD. Monochrome displays using non-laser sources of illumination have also been produced.

For some applications, a larger HMD FoV may be required. The current design could be adapted to provide a larger horizontal FoV by reducing

The Striker family of helmet-mounted displays includes early work for the Helmet-Mounted Sighting System (HMSS) as used in Jaguar aircraft   0106288

the binocular overlap in a similar manner to the MIDASH HMD (see separate entry). However, evidence suggests that partially overlapped systems which reduce the binocular overlap significantly lower than 40 per cent are less than ideal. Therefore, a new visor-projected optical configuration that allows a 50 per cent growth in the FoV has been designed. The availability of display technologies that support such a FoV is limited, but solutions providing the required pixel density are in development. The major technical issue to be solved is the provision of a non-CRT display technology, which provides the necessary display contrast against bright sunlight without resorting to a glare visor.

While DC or AC electromagnetic head tracking is provided in the basic striker system, BAE Systems also offer a high-accuracy Electro-Optical head tracking system for applications where cockpit configuration may interfere with the operation of electromagnetic systems.

Cobra
The latest variant of the Striker design, selected for integration into South African Air Force (SAAF) Gripen multirole combat aircraft, is the Cobra HMDS. In common with Striker, the helmet is a two-shell design, although head tracking is via infra-red Light Emitting Diodes (LEDs) on the outer shell and the detachable NV cameras are omitted (although they remain an option). Helmet position is tracked using a Denel Optronics EO tracking system comprising three CMOS sensors mounted on the canopy rail and either side of the rear cockpit. The inner shell, personally tailored

to the pilot, contains communications equipment and oxygen mask mounting points. Pilots' heads are laser-scanned to ensure a perfect fit.

The fully overlapped 40° FoV, like Striker, can display raster- or vector-generated graphics. The Gripen HMDS features a further LoS camera for pilot training purposes and mission recording and also employs a completely new symbology set, developed by the Gripen team.

Specifications
Cobra
FoV:
    Optical system: 40°, circular
    Binocular overlap: 100%
    Exit pupil: >15 mm, on axis
    Eye relief: >50 mm
LoS camera: 525 line daylight CCD camera
Helmet tracker:
    Type: LED Electro Optical (LoS and roll)
    Weight: 1.975 kg
Options: Active noise reduction; NV cameras

Status
In May 1999, the Striker helmet was successfully tested in a 600 kt ejection from a specially modified YF-4J Phantom aircraft at China Lake US Naval Air Warfare Center, as part of an assessment for a revised parachute canopy for the F/A-18 aircraft.

In 2004, Cobra development was ongoing with four major simulation programmes, each involving five pilots. Trials of the new helmet began with centrifuge tests in the UK, with relatively gradual g onsets of approximately 1 g per second, before moving on to similar trials in Sweden with a faster onset rate of up to 6 g per second up to 9 g. These tests are intended to prove the helmet's stability and comfort before flying trials.

System integration is being carried out during the second quarter of 2004, with flight trails planned for the fourth quarter. Pre-production units are scheduled for the fourth quarter of 2005 with full production expected from 2007.

Contractor
Selex Sensors and Airborne Systems.

Super VIGIL Infra-Red Linescan Sensor (IRLS)

Type
Airborne reconnaissance system.

Description
Thales' Super VIGIL digital linescan sensor is packaged as a single line replaceable unit, interfacing via MIL-STD-1553B with Reconnaissance

Super VIGIL IR Linescan imagery   1114882

Super VIGIL IR Linescan imagery   1114884

The Cobra HMD has been tested in the Saab Gripen multirole fighter. Note the LED array on the outer helmet, which has been modified from that employed in the Eurofighter Integrated Helmet to allow for the differing cockpit environment of the Gripen, and the bracket for mounting the optional NV camera which aligns it with the pilots unaided eyeline (Gripen International)   0589477

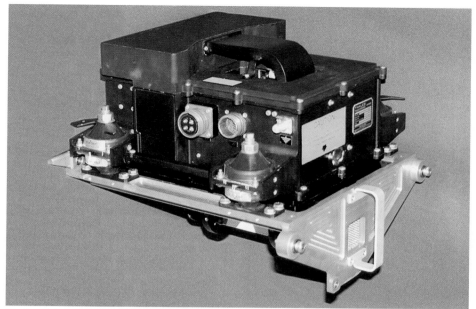

Thales Optronics Super VIGIL IR Linescan Sensor (IRLS)   1114886

Super VIGIL IR Linescan imagery   1114885

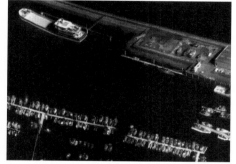

Super VIGIL IR Linescan imagery   1114883

Management Systems (RMS) and providing a 24-hour Infra Red (IR) reconnaissance capability. The system outputs a 1,200 line/s Hotlink II digital video output of continuous, along-track imagery. The system converts scans into analogue signals which contain both line synchronisation and video. The signal generated can be directly interfaced with a video cassette recorder, a datalink transmitter and/or a ground imagery exploitation station for display on a conventional video monitor.

Super VIGIL is equipped with two SPRITE CMT detector elements and is configured for optimal ground coverage at high Velocity/Height (V/H) ratios (employing Bowtie effect compensation, with typical performance being achieved with a V/H value that equals five).

Integrated with the host platform's navigation system, internal accelerometers compensate for image defects at roll values of up to 60° port and starboard and at roll rates of up to 50°/s. Super VIGIL also compensates for lower IR emittance levels in the scene at large across-track scan angles in order to maintain full dynamic range across the scan.

Specifications

Spectral band; 8–14 μm
Detector: dual-element SPRITE CMT
Transverse field-of-view
Scanned: 190°
Displayed: 180°
Angular resolution: 0.53 mrad (nominal 8,192 pixel/line sampling rate, 1,200 line/s line rate)
Thermal sensitivity: >0.16° NETD
Velocity/Height range: up to 5 rads/s
Line rate: 1,200 lines/s
Pixels: 8,192 per line
Cooling system: closed cycle Stirling cooling engine
Power: 28 V DC 160 W (running)
Dimensions: 322 × 293 × 274 mm
Weight: <20 kg
Environmental: MIL-STD-810 and DEF STAN 07-55

Status

Super VIGIL is in production for the Reco NG (See separate entry) reconnaissance pod used by French Air Force Mirage 2000 and Rafale aircraft.

Contractor

Thales Optronics (Vinten) Ltd.

Thermal Imaging/Airborne Laser Designator (TIALD) pod

Type

Airborne Electro Optic (EO) targeting system.

Description

The Thermal Imaging/Airborne Laser Designator (TIALD) pod provides military aircraft with automatic tracking and laser designation of targets, day or night, for the delivery of laser guided and ballistic weapons. The pod carries three Electro-Optic (EO) sensors – a thermal imager, TV and a laser designator, which all share a common optical aperture. This unique packaging solution results in the smallest diameter pod of this type and facilitates the maintenance of an accurate common boresight for all sensors. The large sightline Field of Regard (FoR) afforded by the pod roll and gimbal arrangement facilitates designation without flying directly towards the target. In many cases, the full FoR of the pod is restricted in part by the airframe of the carriage aircraft, therefore, in order to prevent the laser impacting the airframe during manoeuvres, specific obscuration profiles are stored within the pod. The Infra-Red (IR) and TV sensors operate simultaneously with either image displayed by single switch selection. This allows the operator to quickly compare IR and TV performance in the target area and choose the optimum sensor for the attack. The IR sensor has two Fields of View (FoV), wide and narrow, while the TV sensor operates only in the Narrow Field of View (NFoV). In NFoV, an electronic zoom facility facilitates maximum target standoff for avoidance of co-located defences.

Apart from the cockpit display and controls the pod is self-contained, taking its power from the aircraft primary supplies. Interface with the avionics is via a MIL-STD-1553B databus.

The forward section of the pod contains the thermal imager and TV sensors, the telescope and the laser designator transceiver unit. The laser, TV and thermal imager optical paths are combined within the telescope and steered over a wide angle of regard by the pod roll and gimbal arrangement. The combined optical path is stabilised against aircraft movement and pod vibration by a stabilised mirror.

The static rear sections of the pod contain the ram-air cooler, electronics units and power supplies. The automatic video tracker, produced by BAE Systems, Plymouth, is included within the electronics units.

In operation (current -400 Series), the sensor sightline would be directed onto the target area, primarily by commands from the aircraft nav/attack system, or, in reversionary modes or during unplanned attacks, by the crew. After assessment of the target area in Wide Field of View (WFoV), the crew would acquire and identify the target and specific Desired Mean Point of Impact (DMPI) in Narrow Field of View (NFoV). After confirmation of target identification on the video display, the video tracker may be engaged and locked to the target. Once the video tracker has locked onto the target, the aircraft may be manoeuvred within pod gimbal limits, with the system automatically keeping the sensor sightline locked onto the target. Depending on weapon type and delivery platform, the laser would be fired at a predetermined time to enable terminal guidance and ensure optimum weapon impact parameters.

Although used primarily to designate targets for laser-guided weapons, TIALD may also be used as an integral aircraft sensor, used to update the nav/attack system before release of ballistic weapons, thus ensuring the most accurate release solution possible. TIALD has also been used extensively in a reconnaissance/surveillance mode at medium level, tracking and pinpointing the positions of points of interest and recording video footage for post-mission analysis.

The latest development of TIALD is the -500 Series, which incorporates enhancements over the in-service -400 Series in three major areas.

The present Thermal Imaging Common Module (TICM) II scanning camera, which operates in the LWIR (8–12 μm) band, is replaced by a MWIR (3–5 μm) design that packages a 384 × 288 element staring array developed by BAE Systems into a camera provided by Kentron.

The second upgrade involves the installation of a Northrop Grumman LN-200 Inertial Measurement Unit (IMU), which provides the pod with a local immediate source of accurate position information and permits target hand-off to or from other systems. For the -400 Series, this information is provided by the host aircraft Inertial Navigation System (INS), which introduces data latency problems with increasingly overworked aircraft databus systems (such as the MIL-STD-1553B), while also (in some cases) limiting the ultimate initial pointing accuracy of the pod sightline. Depending on aircraft type/era, this accuracy/latency problem varies from one aircraft to another (in UK RAF application, TIALD pods are supplied to squadrons as required), and is reported to present particular problems after violent manoeuvring, leading to delays in establishing the pod's new sightline while the aircraft system settles down.

The third improvement comes from the addition of a transfer alignment function that automatically harmonises the pod sightline with the aircraft nav/attack system boresight (radar/laser/FLIR/HUD and so on). At present, the alignment process is carried out manually by the pilot, by placing the aircraft on a known fixed point, then aligning both the pod sightline and the HUD sightline (and hence, assuming the aircraft system is perfectly harmonised, the nav/attack system boresight) with a distant feature. While this may be merely problematic at remote operating locations, it is, in practice, impossible aboard an aircraft carrier.

A close-up of TIALD, fitted to the centre fuselage pylon of a UK RAF Jaguar GR3 (unstowed, in Boresight position)   0111719

A TIALD 500 Series pod mounted on the left shoulder station of a UK RAF Tornado GR4 (Jamie Hunter)   1184361

The availability of modern digital memory will also allow the -500 Series to be transferred from one aircraft type to another without requiring new software.

Specifications
Dimensions: (length) 2,900 × (diameter) 305 mm
Weight: 210 kg (-400); 230 kg (-500)
Sensors:
TV 0.7–1.0 μm
Laser 1.06 μm
Thermal imager 8–12 μm
Field of regard: (Tornado GR1)
Ball + 30 to –155°
Roll continuous
Field of view:
(wide) 10 × 6.7°
(narrow) 3.6 × 2.4°
Electronic zoom: ×2, ×4
Power supply: 200 V AC, 400 Hz, 3 phase, 3.3 kW (max), 1.2 kW (average)

Status
The -400 and -500 Series TIALD pod is in service with the RAF on Tornado GR. Mk 1/4, Jaguar GR. Mk 1B/3 and Harrier GR. Mk 7 aircraft. Over the life of the system (-100, -200, -300 and -400 Series), TIALD pod development has been continuous with a series of enhancements, including more powerful head motors and the integration of VITS automatic tracking systems (see separate entry). The combination of TIALD and the Enhanced Paveway (EPW – GPS and laser guided) proved to be highly successful during RAF air operations over Iraq in support of Operation TELIC in 2003.

Flight trials of latest variant, the -500 Series pod, were completed during 2003, with the pod entering service in the UK RAF Tornado GR4 during 2005. The upgraded system is reported to deliver an increase in performance (acquisition, identification and target tracking) from medium level of approximately 30 per cent over the -400 Series.

Contractor
BAE Systems.

Titan 385 MultiSensor Turret System (MSTS)

Type
Airborne Electro-Optic (EO) navigation system.

Description
Drawing on extensive experience in the area of turret sensors, Selex has developed the Titan 385 MultiSensor Turret System (MSTS), which combines a number of (in-house) high-resolution sensors into a single Line-Replaceable Unit (LRU). The system features an advanced, large-format 3 to 5 or 8 to 10 μm waveband IR surveillance camera which employs the SiGMA (Sensor InteGrated Modular Architecture) Thermal Imager (see separate entry), combined with optional additional sensors, including a navigation camera, low-light and colour TV cameras, a Laser Range-Finder (LRF) and laser designator. Standard features of the Titan 385 MSTS include an autotracker and a proven navigation system interface. The turret may be mounted upright or underslung below the host platform.

Specifications
System
Dimensions: 385 (Ø) × 525 (H) mm
Weight: 45 kg
Slew rate: up to 100°/s
Airspeed: 300 kt (operational); 400 kt (carriage)
IR Sensor
Detector: Staring Focal Plane Array (FPA)
Array: 320 × 256; 640 × 512
Wavelength: 3–5 μm (MWIR); 8–10 μm (LWIR)
FoV
   (320 × 256): 11.25 × 9° (WFoV); 2.25 × 1.8° (NFoV)
   (640 × 512): 18.0 × 14.4° (WFoV); 3.6 × 2.8° (NFoV)
All plus ×2 electronic magnification
Colour TV camera
Detector: 3 CCD
Optics: 10:1 continuous optical zoom
FoV: Matched to IR sensor
Low-light TV camera:
Dynamic range: Starlight to full sunlight
Optics: 26°, 13° or 8.8° fixed FoV
Features: Anti-blooming and over-exposure immunity
Navigation IR camera
Detector: Uncooled staring array
Array: 256 × 128
Wavelength: 8–12 μm (LWIR)
Optics: 40° fixed FoV

A UK RAF C-130 equipped with the Titan 385 multisensor turret package (Jane's/Patrick Allen)   1209359

Laser Range-Finder (LRF)
Type: Erbium glass (1.54 µm)
Accuracy: ±10 m
Range: 80 m to 20 km

Status
In production and in service.

During the third quarter of 2002, the UK Royal Air Force announced a GBP20 million upgrade of part of its fleet of 25 Lockheed C-130K Hercules C.1 tactical transport aircraft with an enhanced vision package for night operations, which includes Selex' Titan 385 MSTS and SiGMA TI. The system entered service during 2004.

Contractor
Selex Sensors and Airborne Systems.

Tornado Infra-Red Reconnaissance System (TIRRS)

Type
Airborne reconnaissance system.

Description
The TIRRS for the Tornado GR.Mk1A comprises two Side-Looking Infra-Red (SLIR) sensors and an IRLS 4000 Infra-Red Linescanner manufactured by Thales Optronics. The 8 to 14 µm IR sensors embody elements of the UK TICM II SPRITE detector programme.

TIRRS is the first system of its type was first employed operationally during the 1990–91 Gulf War. The system is mounted internally in the fuselage and provides horizon-to-horizon across-track coverage with roll stabilisation and gives a real-time display in the cockpit.

The output from the sensors is recorded on videotape and the operator in the rear cockpit can monitor the imagery while the sortie is under way, directly from the sensors or by replaying from the videotape recorders. The system offers a high-definition thermal picture which can be magnified or enhanced in flight.

Status
In service in UK Royal Air Force Tornado GR. Mk4A and Royal Saudi Air Force Tornado IDS reconnaissance aircraft.

Contractor
Thales Optronics (Vinten) Ltd.

Tornado GR1A Infra-Red Line Scanner (IRLS) with protective shroud open, showing the rotating sensor which provides for horizon-to-horizon coverage (E L Downs)   1047552

Type 105 Laser Range-Finder

Type
Airborne Laser Range-Finder (LRF).

Description
The Type 105 is a high-repetition rate, Nd:YAG steerable laser range-finder developed privately by BAE Systems to provide compact low-cost accurate target ranging sensors for ground attack aircraft.

The 105 Series includes several variants to suit different aircraft installation and avionics requirements. The latest variant has been specifically designed as a two-box system for ease of installation in a wide range of aircraft. The 105 is now MIL-STD-1553B databus-compatible

The BAE Systems Type 105 laser ranger is fitted in the Hawk 100   0581493

and is particularly suitable for aircraft updates such as the F-5E, A-4, A-10, Hawk and AMX, as well as new ground attack aircraft.

Target range is measured to an accuracy of 3.5 m, standard deviation to a range of 10 km, effectively removing the largest source of error in air-to-surface weapon delivery.

Low power consumption, ease of integration with aircraft avionic systems, flexible configuration, small size and frontal area are seen to be important factors in the suitability of the Type 105 for fit or retrofit in ground attack aircraft.

Specifications
Dimensions: 190 × 200 × 370 mm
Weight:
(laser) 12.5 kg
(electronics unit) 4.5 kg
Power supply: 28 V DC, 300 W
Wavelength: 1.06 µm
PRF: 10 pps
Angular coverage: within 10° semi-apex angle cone
Range: 10 km
Accuracy: 3.5 m standard deviation
Reliability: >1,000 h MTBF

Status
In service.

Contractor
BAE Systems.

Type 126 laser designator/ranger

Type
Airborne Laser Range-Finder/Target Designator (LRFTD) system.

Description
The Type 126 is a high-energy Nd:YAG range-finder and designator system developed for the UK MoD Thermal Imaging Airborne Laser Designation (TIALD) pod. The Type 126, comprising separate transmitter unit and power supply/control unit for ease of pod or aircraft installation, provides very high-output energies at pulse repetition rates of up to 20 Hz in a temperate environment ranging from –54 to +71° C.

Status
In production and in service in BAE Systems' Thermal Imaging Airborne Laser Designation (TIALD) Series 400 pod.

Contractor
BAE Systems.

Type 221 thermal imaging surveillance system

Type
Airborne Electro Optic (EO) surveillance system.

Description
The Type 221 thermal imaging surveillance system is designed for service with military helicopters.

Developed by BAE Systems Avionics (now Selex) in conjunction with Pilkington Optronics, the system incorporates an IR18 thermal imager and telescope by the latter company, with sightline stabilisation steering provided by a stabilised mirror. The assembly is contained in a pod which either can be mounted beneath the nose of a helicopter or can project through an aperture in the aircraft floor.

The IR18 imager unit provides a normal field of view of 38° in azimuth and 25.5° in elevation. In the Type 221 application, users can choose a telescope magnification of either ×2.5 or ×9, with corresponding wide or narrow fields of view. The wider field of view (15.2° azimuth by 10.2° elevation) would be used for general surveillance, target acquisition or navigation; the narrow field of view (4.2° azimuth by 2.6° elevation) permits detailed observation for target identification or engagement of targets detected in the wide field of view mode. In the pod installation the system has fields of regard of +15 to –30° in elevation and ±178° in azimuth. The entire sensor system is vertically mounted above the mirror which is angled, periscope fashion, at 45° to the horizontal to provide views in the horizontal plane.

The sensor system employs SPRITE detector units cooled by a Joule-Thompson minicooler supplied with high-pressure compressed air. The air source is a bottle, mounted on the equipment and charged immediately before flight. This has a capacity of 1 litre and provides a system operation time of approximately 2½ hours. If greater endurance is required other cooling options, involving the use of mini-compressors permanently connected to the equipment, are available. The system operates in the 8 to 13 µm band and has a sensitivity of between 0.17 and 0.35°C to target background and surroundings.

The display may be presented on either 525- or 625-line television monitors. The output is either in CCIR or EIA composite video formats, as required. This television-compatible output may be displayed on one or more monitors throughout the aircraft to aid winching operations.

The Type 221 system is intended for use in medium to large helicopters and roles envisaged include maritime reconnaissance, search and rescue and integration with onboard weapon systems to improve all-weather capability.

Specifications
Dimensions:
(pod unit) 865 long × 420 mm (max) diameter
(electronics unit) 127 × 432 × 330 mm (max)
Weight:
(pod unit) 75 kg
(electronics unit) 8 kg

Status
In service. The Type 221 thermal imager has been fitted to a number of Aerospatiale/Westland Puma helicopters operated by the UK Royal Air Force.

Contractor
Selex Sensors and Airborne Systems.
Thales Optronics Ltd.

Type 239 Electro-Optic (EO) turret

Type
Pilot's Night Vision System (PNVS).

Description
The Type 239 is a compact, steerable platform, designed to house Electro-Optic (EO) Pilot's Night Vision Systems (PNVSs) such as the Helicopter Infra-Red Navigation System (HIRNS – see separate entry) on the Agusta A129 Mangusta helicopter.

The platform incorporates two axes of movement, with ±130° range in azimuth and +75 to –60° in elevation. The platform's acceleration and slew rates for both axes are compatible with helmet tracking systems to enable the platform to function with a Helmet Mounted Display (HMD).

The platform is suitable for mounting on almost any helicopter, due to its small size and low weight, and can be fitted with a range of compact imaging systems to meet customer requirements.

Status
In production and in service on the Agusta A129 and Italian HH3 helicopters.

Contractor
Selex Sensors and Airborne Systems.

Type 8010 electro-optical sensor

Type
Airborne reconnaissance system.

Description
The Type 8010 electro-optical sensor is a form and fit replacement for the F95 and other 70 mm film format framing cameras such as the Types 360, 518 and 544 manufactured by W Vinten Ltd, and can therefore be retrofitted into existing reconnaissance systems as well as integrated into new installations. The primary role for the sensor is day low- and medium-altitude tactical reconnaissance and surveillance. A secondary role is day low- and medium-altitude surveillance to aid various government agencies, including the police, coastguard, drug enforcement and fishery protection agencies.

The Type 8010 sensor may be fitted internally in a wide range of aircraft, RPVs, drones and UMA or in podded systems.

The sensor is a solid-state push broom device employing a 12 μ 4096 pixel linear Charge Coupled Device (CCD) array. It scans at 1,800 lines per second in the visible to Near Infra-Red (NIR) spectral range (500 to 950 nm). It may be fitted with a range of lenses and may be mounted in oblique, vertical, split pair or fan configurations depending on operational requirements. Imagery is recorded on an airborne digital recorder and can be displayed on a Ground Imagery Exploitation System (GIES). Imagery may also be transmitted to the ground when suitable datalinking equipment is fitted to the aircraft.

Specifications
Dimensions:
226 × 181 × 264 mm (1.5 and 3 in lenses)
226 × 181 × 258 mm (6 in lens)
Resolution:
(A version) 12 μm, 4,096 elements
(B version) 8 μm, 6,144 elements
Weight: 7.8 kg plus lens
Lenses: 152 mm, 76 mm, 38 mm
Field of view: 18.3°, 35.7°, 65.6°

Status
In service with the Royal Air Force and other air forces for tactical reconnaissance operations.

Contractor
Thales Optronics (Vinten) Ltd.

An A129 Mangusta light attack helicopter employs the Type 239 turret to house the HIRNS, located just above the gun
0137478

Type 8010 E-O sensor compatible with 152 mm lens
0001243

Type 8040B electro-optical sensor

Type
Airborne reconnaissance system.

Description
The Type 8040B utilises the same technology as the Type 8010, but provides a replacement for the 126 mm film format Type 690 camera.

The Type 8040B sensor has an 8 μm 12,288 element charge-coupled device mounted in the focal plane. The sensor electronics record the imagery onto either a digital or analogue output.

The pixel size within the focal plane array gives 'photographic quality' resolution in the imagery. When associated with the advanced technology 450 mm lens, which offers a corresponding high-modulation transfer function, the 8040B can offer long-range resolution previously only associated with longer focal length systems.

The Type 8040B EO sensor is designed to operate between 200 and 40,000 ft at slant ranges of 300 m to 40 km. Although the spatial resolution is comparable with that of film from the Type 690 film camera, the Type 8040B provides a significant advantage over film against

Type 8040B EO sensor mounted in standoff module
0062748

Type 8040B E-O sensor image   0001246

low-contrast targets, especially when imaged through a hazy atmosphere.

Specifications
Dimensions: depending on podded or internal installation
Weight: depending on podded or internal installation
Lenses: 450 mm
Fields of view: 12.4°, 6.2°

Status
In production and in service with the UK Royal Air Force on the Jaguar and Belgian Air Force on the F-16.

Contractor
Thales Optronics (Vinten) Ltd.

VICON 18 Series 601 reconnaissance system

Type
Airborne reconnaissance system.

Description
The VICON 18 Series 601 pod is one of the wide range of VICON 18 reconnaissance pods designed for a variety of operational roles. Specifically, the Series 601 provides day time reconnaissance from a low-cost, lightweight pod for use at low, medium and high altitudes. Implicit in the pod design is the ability to integrate the pod with a variety of airframes. The pod is currently fully flight cleared on the BAE Systems Tornado, Jaguar, Harrier and Hawk aircraft. VICON 18 Series pods are also operational on F-5, MiG, Learjet and many other aircraft types.

Film variant
The VICON 18 Series 601 pod contains two sensors:
Type 690 (F144) 126 mm film format framing camera with a 450 mm focal length high-resolution lens which is mounted in a rotatable nose cone;
Type 900B 70 mm film format panoramic camera of 76 mm focal length which is mounted in the rear centre section.

The sensors are driven by a VICON 2000 Reconnaissance Management System under the control of the pilot or systems operator. The VICON 2000 System can be interfaced to a wide range of aircraft avionics systems including the MIL-STD-1553B databus. Navigational data can be imprinted on the imagery of both sensors.

The pod provides capabilities for the following operational roles: tactical standoff photography; long-range oblique photography; low-level tactical reconnaissance; area coverage.

Specifications
Dimensions:
(overall length) 2,250 mm
(standard diameter) 457.2 mm
(depth over saddle) 508 mm
(weight (estimated)) 254 kg
(centre of gravity) midway between the 14 in or 30 in attachments
Performance:
(sea level) up to 730 kt (M1.1) at 36,000 ft (11,000m): up to 1,033 kt (M1.8)
Symmetrical normal (accelerations): +ve 9.0 *g* −ve 2.0 *g*
Asymmetrical normal (acceleration): +ve 4.0 *g* −ve 2.0 *g*
Rate of roll: 150°/s

Status
In service.

VICON 18 Series 601 Electro-Optical/Infra-Red Reconnaissance Pod (Joint Reconnaissance Pod)
The original Jaguar Replacement Pod (JRP) programme, began with a conversion/upgrade of the Vicon 18 GP-1 shell, with the two wet-film cameras replaced by a pair of Vinten Type 8010 EO sensors. The Type 8010 EO sensor features a 12 μm, 4,096 pixel CCD linear array and may be fitted with a range of lenses (38, 76 or 152 mm focal length). For the JRP, the Type 8010 sensors are mounted side-by-side in oblique positions, together with a vertically-mounted Vinten VIGIL IRLS and video recording facilities. The Type 8010 sensors are typically set at an 18° angle and use a 3in lens, although other lenses and angles can be used. The Type 8010 produces NATO Standard 7023 images on SVHS(C) video cassette on Airborne Reconnaissance Integrated

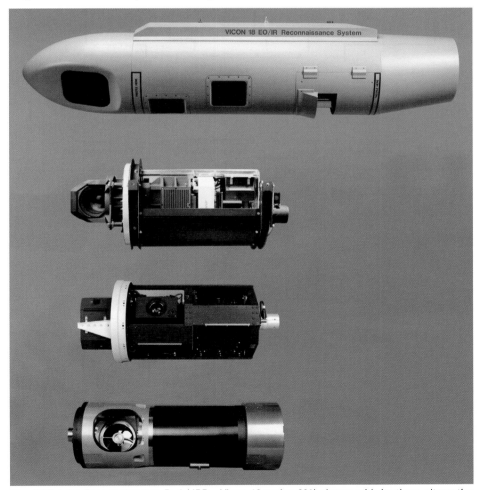

The RAF Joint Reconnaissance Pod (JRP – Vicon 18 series 601) shown with its three alternative payloads (from above): the medium-level module, the low-level module and the dual-band sensor (Thales)   1121078

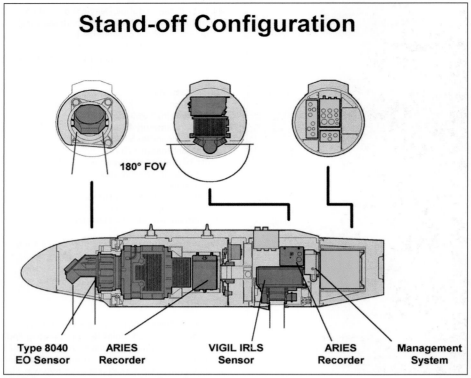

VICON 18 electro-optical standoff configuration   0062750

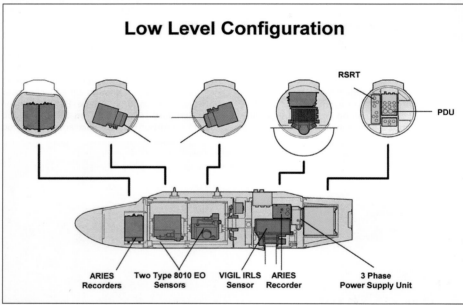

VICON 18 electro-optical low-level configuration   0062751

VICON 18 GP (1) pod installed on the centre pylon of a UK RAF Jaguar aircraft (forward rotating sensor window in the stowed/parked position)   0034608

Electronics Suite (ARIES) recorders, offering 35 minutes of recce imagery plus five minutes at either end for data. Each sensor has its own ARIES recorder and the pod is completed with Vinten's Recce Management System (RMS). Four of the original GP-1 pods were converted in this way. This configuration is termed the Low Level Configuration.

Full JRP

The Full JRP configuration was developed from the EO GP-1 (re-designated Interim JRP). The full JRP has a MIL/STD-1553 databus connection to the aircraft systems, allowing GPS-derived positional data to be inserted onto the recce imagery.

A medium-level JRP variant has been developed with the two Type 8010 EO sensors being replaced by a single Type 8040B LOROP (Long Range Oblique Photography) EO sensor with a 12.4° FoV. This uses the same mirror-rotating head and optics of the F144 framing camera of the GP-1 pod, allowing the camera to be positioned anywhere from horizon-to-horizon. The electro-optical sensor incorporates an 8 μm 12,288 pixel CCD linear array and an ultra-high-resolution 450 mm focal length lens. The lens focuses automatically from 150 m to infinity, providing an excellent stand-off capability, and FoV of 12.4°. The rear portion of the pod retains the VIGIL IRLS, ARIES recorder and RMS electronics.

For a medium-altitude JRP mission, typical altitude would be 20,000 ft with a look angle of 60°. During Operation Resinate North, a typical recce mission with two photo runs of 25 minutes (each with up to 10 targets) could last an average of 3.5 hours.

If there is a mission change due to weather or threat, since target and aircraft location (in three dimensions) are known, the rotating head can be re-aligned at the press of a switch. For the Jaguar pilots, the introduction of the Helmet-Mounted Sight System (see separate entry) allows targets of opportunities to be marked on the aircraft navigation system and then, at the press of a switch, transferred as an extra target to the pod. Once on the ground, the cassettes are rapidly passed to the Ground Imagery Exploitation System (GIES).

New capabilities

Thales Optronics (Vinten) have been tasked with the replacement of the existing video-based recording system with a solid-state digital recorder for the Type 8040-equipped medium-altitude JRP, to be implemented in 2004. This will offer either an equivalent resolution at 50 per cent greater stand-off distances or an increased volume of recorded data with improved image quality (from operating at higher scanning rates).

A near-realtime image transmission capability is made possible for Jaguar GR3As employing Symmetrics' Improved Data Modem (IDM) 3.02 with Phototelesis PRISM (Photo Reconnaissance Intelligence Strike Module) in conjunction with a secure radio facility such as the AN/ARC-210 (see separate entries). These systems have been evaluated under the Extendor communications relay operational concept demonstration.

The Harrier GR9 upgrade should see the installation of a single-card, VME-based, four-channel version of the IDM from Innovative Concepts Inc, although initially, capability would be restricted to data transmission only.

Status

In production and in service. The VICON 18 Series 601 GP-1 EO/IR pod and Ground Imagery Exploitation System (GIES) were selected by the UK Royal Air Force to replace older reconnaissance equipment carried by Jaguar GR1A aircraft. The pod includes a role change capability, providing either day EO short- to medium-range standoff, or low-level reconnaissance, together with day/night Infra-Red (IR) capability.

Integration of the pod onto the Jaguar was carried out under the Jaguar Replacement Pod (JRP) programme, completing low-level sensor performance trials, designed to evaluate the performance of the EO and IR linescan sensors in both day and night conditions.

Avionics integration of the JRP to the Jaguar GR3A (also known as Jaguar '97) was conducted by QinetiQ (formerly DERA). Introduction to service on Jaguar aircraft of RAF 41 Sqn was achieved during the latter part of 2000. Although considered an interim solution, pending the development of a Full JRP, they were used by 41(F) Squadron Jaguar GR3As during Operation Resinate North.

Four new-build Full JRP pods were contracted for at the same time as the Interim JRP GP-1s. A further six were ordered in March 1999.

Of the 12 VICON 18 GP-1 pods procured, four were converted to Interim JRP standard and 10 new Full JRP pods were procured from Vinten. Around this time, the acronym JRP changed to Joint Reconnaissance Pod, due to the system's integration onto the Harrier GR.7 and Tornado GR4A. The RAF declared a limited acceptance of the JRP on land-based Harriers in April 2002 and then brought it into use just ahead of Operation Telic in March 2003. The JRP went into operational service on Tornado GR4A aircraft of II(AC) Squadron at RAF Marham in June 2003.

The RAF is reported to be seeking to acquire a further six pods.

Contractor
Thales Optronics (Vinten) Ltd.

VICON 70 general purpose modular pod

Type
Airborne reconnaissance system.

Description
Since the introduction of the VICON 70 Series of general purpose modular pods in the early 1980s over 50 different variants have been designed or built to meet a wide range of reconnaissance, surveillance, night illumination, Forward Looking Infra-Red (FLIR) and Instrumentation requirements.

Variants have been supplied for operations on helicopters, light aircraft, civil aircraft and a range of fighter bombers flying at speeds up to M2.2.

Most variants have been designed to meet Vinten requirements but many have been built for other Original Equipment Manufacturers (OEM) particularly for research and development purposes.

Reconnaissance and Surveillance
The VICON 70 pod will accept a wide range of reconnaissance and surveillance sensors including film framing and panoramic cameras recording on 70 mm film, Infra-red Linescan (IRLS) and Electro-Optical (EO) sensors recording on either 70 mm film or on video-tape and night illumination systems operating with either white light or IR flash.

Systems are designed to meet customers' operational requirements including the integration of CCD TV video sighting devices with onboard real-time display.

VICON 70 pod   0002223

Combinations of day and night sensors have been supplied in VICON 70 pods from low V/H applications on helicopters to V/H 5 on fighter-bomber aircraft.

FLIR
A number of different FLIRs including GEC TICM II, HUGHES, and IR-18 have been integrated into VICON 70 pods for various OEMs and customers for a number of applications including navigation and attack.

Vinten-supplied FLIR pods are designed for reconnaissance and surveillance operations and are fitted with the GEC TICM II FLIR sensor, the associated processing electronics, and detector cooling system.

The imagery is recorded on videotape and in some applications the imagery is displayed in real time on board the aircraft on a video monitor.

Instrumentation pods
Variants of the VICON 70 pod have been designed to accept Photo-Sonics high-speed instrumentation cameras for recording release of weapons and other stores from combat aircraft. The pods will accept the 16 mm IVN or 16 mm IPL cameras operated from either aircraft or battery power.

Specifications
Performance: Variants cleared up to M 2.2
Attachment lugs:
  Twin suspension. NATO standard 14 in or to customer specification
Connectors: Location and type to customer specification
Standard diameter: 355 mm (14 in)
Length: from 1,447 mm (57 in) minimum up to 2,300 mm (100 in) maximum approx.

Status
In service and in production.

Saab and BAE Systems selected a variant of the VICON 70, designated VICON 70 Series 72c,

for the Gripen aircraft. The pod is fitted on the underfuselage pylon of the aircraft. A variety of electro-optic and infra-red sensors may be installed which, when combined with onboard sensors such as the radar, will enable Gripen to undertake the full spectrum of reconnaissance missions.

Contractor
Thales Optronics (Vinten) Ltd.

VICON 2000 digital reconnaissance management system

Type
Airborne reconnaissance system.

Description
The VICON 2000 modular digital reconnaissance management system entered service in 1986. It is now in service on a wide variety of combat aircraft types, interfacing with 64 K NAVHARS, MIL/STD-1553B databus and ARINC 407A, 429/10, 561 and 571. Some podded reconnaissance systems operate on multiple aircraft types, with minimal changes required to move the system from one aircraft type to another.

The VICON 2000 system is designed for use with all types of reconnaissance sensors fitted either internally or in podded systems. The system normally comprises a navigational interface unit, systems management unit and Sensor Interface Units (SIU). New sensors incorporate the SIU, reducing the number of LRUs in the system. The VICON 2000 may be controlled by a conventional cockpit control unit or via the aircraft computer system.

Components of the VICON 2000 are linked by RS-422A serial datalinks over two twisted pairs.

Status
In service in a wide variety of combat aircraft types.

Contractor
Thales Optronics (Vinten) Ltd.

VIGIL Infra-Red Linescan Sensor (IRLS)

Type
Airborne reconnaissance system.

Description
The VIGIL IRLS is a single LRU sensor which incorporates the electronics and the control interface which may be either single discrete

Cessna Citation fitted with VICON 70 pod   0001248

VIGIL IRLS flown at 1,000 ft AGL-day 0001242

VIGIL IRLS flown at 1,000 ft AGL-night 0001241

signals or RS-422 serial command link for interfacing to advanced avionic databusses.

VIGIL is designed for installation in RPVs, drones, UMA, helicopters and medium-performance fixed-wing aircraft up to a velocity/height ratio of 2.5.

It is a high-performance day and night airborne sensor with horizon-to-horizon coverage operating in the 8–14 μm waveband. The linescanner detector is a single element SPRITE CMT which is cooled by a continuous-rated Stirling closed-cycle cooling engine.

The infra-red radiation from the terrain is scanned by the sensor, producing continuous along-track imagery. VIGIL produces an analogue signal containing both line sync and video. This signal is suitable for direct interfacing with either video cassette recorder for airborne or ground recording, datalink transmitter for transmission to a ground station or an imagery display processor for scan conversion and image manipulation for display on a conventional video monitor.

Specifications

Spectral band; 8–14 μm
Detector: single-element SPRITE CMT
Transverse field-of-view:
  Scanned: 190°
  Displayed: 180°
Angular resolution: >0.67 mrad
Thermal sensitivity: >0.16° NET
Velocity/height range: up to 2.5 rads/s
Roll stabilisation:
  Mode 1: ±30° with signal input from roll gyro of 0.5 V DC
  Mode 2: locked to airframe
  Line rate: 600 lines/s
Pixels: 8,192 per line at 4.92 MPixel/s
Cooling system: closed cycle Stirling cooling engine
Outputs:
(analogue) video/sync composite 2.0 MHz bandwidth, frame sync (50 Hz)
(digital (optional) image depth, compression, Bit rate
Power: 28 V DC 165 W (running)
Dimensions: 316 × 254 × 247 mm
Weight: 11.5 kg
Environmental: MIL-STD-810 and DEF STAN 07-55

Status

In production and service in the Modular Reconnaissance Pod (MRP) in Belgian Air Force (BAF) F-16AM and in Thales' Vicon Series 601 in the Joint Reconnaissance Pod (JRP) for UK Royal Air Force (RAF) Harrier GR7/9, Jaguar GR3 and Tornado GR4 aircraft.

Contractor

Thales Optronics (Vinten) Ltd.

VIGIL IRLS including recorder
0001240

Visually Coupled System (VCS)

Type
Helmet-Mounted Display System (HMDS).

Description
A Visually Coupled System (VCS) has been developed at QinetiQ (formerly the Defence Evaluation and Research Agency), Boscombe Down, which allows research into Helmet-Mounted Display (HMD) symbology and investigation of Human Factors (HF) implications of VCS operation in fast jet aircraft.

A Tornado F2 aircraft has been modified to include a nose-mounted, steerable FLIR sensor that provides imagery for a binocular 40° Field of View (FoV) Helmet Mounted Display (HMD), worn by the pilot in the front seat. An AC electromagnetic Head Tracking System (HTS) allows the Line of Sight (LOS) of the FLIR sensor head to be slaved to the HMD. Latency between the two is minimised by the use of fast response optics in the FLIR and a high speed serial datalink from the HTS. A dedicated symbol generator provides a variety of symbology in the HMD in addition to the FLIR imagery, to allow aircraft state and attitude information to be presented to the pilot outside the FoV of the fixed Head Up Display (HUD).

Trials of the system have assessed its utility for a variety of operational day and night flying tasks, including low-level flying, producing interesting results, particularly in the area of HF. Several elements of the system are being upgraded to evaluate improved HMD symbology and imagery from miniature sensors mounted directly onto the pilot's helmet.

Contractor
QinetiQ.

VITS automatic tracking systems

Type
Airborne Electro Optic (EO) tracking system.

Description
The Video Image Tracking System (VITS) is a family of high-performance automatic tracking systems that employ an expanding architecture. They provide superior performance to more traditional systems, employing faster processing and reduced operator workload.

VITS is fully compliant with the complete range of military specifications and provides both VME and RS-422/RS-232 serial interfaces. The family embodies Ada software, custom ASICs and extensive BIT facilities.

The VITS family consists of VITS 1000, VITS 1500, VITS 2000 and VITS 3000.

The current, in-service VITS 1000 is incorporated in both ground and airborne systems, such as the

The head-steerable FLIR sensor mounted close to pilot's eyeline on a Tornado F2 interceptor

0005542

BASE VERDI-2 ATDT (Automatic Target Detection and Tracking) system and BAE Systems' Thermal Imaging Airborne Laser Designator (TIALD) pod (see separate entry). VITS 1000 features dual-standard video input, high-precision centroid and correlation tracking, low data latency for high system bandwidth, automatic selection of optimum tracking mode, track quality measures to aid system performance, an aim-point refinement, and a manual boresight alignment facility.

The VITS 2000 is intended to enhance TIALD performance against mobile targets. While the VITS 1000 automatic target-tracking facility allows the TIALD operator to manually cue an acquisition box over a target on his display, with the tracker maintaining and facilitating precise pointing of the laser designator, the VITS 2000 automates the acquisition process by cueing potential targets. This enhancement is achieved through recent advances in digital-processing technology: early analogue cuers, such as those developed to meet the UK Ministry of Defence (MoD) Staff Requirement (Air) 1010, operate by simply detecting hotspots on the battlefield. VITS 2000 employs data from the TIALD sensors defining target size and (slant) range (generated from the number of pixels that the target subtends, together with range inputs) to reject objects that are significantly smaller or larger than the intended target.

This automatic cueing facility could potentially be further enhanced to include an analysis of parameters such as target outline, and to

compare the resulting shapes with those stored in an electronic library. Other features of VITS 2000 include multiple object tracking, target cueing and prioritisation, automatic target acquisition, and robust tracking despite decoys and obscurants.

The most capable variant of the family, the VITS 3000, will include all the features of the other VITS models, plus automatic target classification using neural networks and enhanced rejection of false alarms.

The VITS systems are available both as off-the-shelf products and as application-tailored derivatives. One example of a tailored product has resulted in the VITS 1500 system – a multiple target detection and tracking system but without the sophisticated correlation tracking subsystem.

Status
VITS 1000 is in production and in service. The VITS 1000 tracking system has been integrated with the BAE Systems TIALD pod and, following extensive user evaluation, all TIALD pods will be retrofitted with the system. Trials programmes are now continuing with the integration of VITS 2000 into the pod. VITS 1000 has also been supplied to a variety of UK and offshore prime contractors and defence establishments for both ground and airborne use. VITS 2000 and 3000 are under evaluation.

Contractor
Selex Sensors and Airborne Systems.

United States

Advanced Airborne Reconnaissance System (AARS)

Type
Airborne reconnaissance system.

Description
BAE Systems' Advanced Airborne Reconnaissance System (AARS) is based around the F-9120 electro-optical, dual-band (EO/IR) Long-Range Oblique Photographic System (LOROPS). The F-9120 sensor provides simultaneous imagery in the visible and infra-red spectra, operating in the 0.45 μm to 0.95 μm (visible to NIR) and 3.4 to 5 μm (MWIR) wavebands. The sensor incorporates three Fields of View (FoV), each with focal lengths matched to produce the same size/scale of image in the ground station. The low-level FoV provides

contiguous imagery at 550 kt up to approximately 1,000 ft above ground level, and stereo imagery from about 2,500 ft up to a slant range of approximately 11,000 ft, whereupon automatic switchover to the medium FoV takes place. The medium FoV operates out to a slant range of approximately 15 nm, when automatic switching to the long-range/high-level FoV takes place. The high-level FoV (with a focal length of 120 in) gives a NIRRS rating of 5 out to over 75 nm in the visible mode.

Aside from the F-9120 sensor, the AARS comprises three Line Replaceable Units, the Reconnaissance Management Unit and Solid State Recorder (RMU/SSR), Imager Control Electronics (ICE) and Imaging Sensor Electronics (ISE). The AARS is enclosed in a BAE Systems-developed Advanced Recce Pod (ARP) suitable for carriage on many types of tactical fighter. In podded configuration, the AARS

LRUs are integrated with an environmental control system at the rear of the pod and a NATO-standard Common Data Link (CDL), operating in the X- or Ku-bands with an Line of Sight (LoS) range out to 200 nm, in the nose of the pod. The F-9120 sensor is enclosed in a rotating centre section for horizon-to-horizon coverage with roll, pitch, yaw and forward motion compensation.

Specifications
Focal length: 13 in
Operating spectrum: 0.45 to 0.95 and 3.4 to 5.0 μm
Infra-red detector: Indium Antimonide (InSb)
Autofocus: continuous
Weight
  AARS pod including datalink: <727 kg
  F-9120: <136.4 kg
Dimensions
  AARS pod: Length 4,200 mm; diameter 635 mm

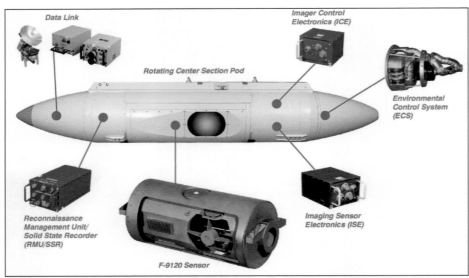

Components of the Advanced Airborne Reconnaissance System (AARS)   0079783

The AARS pod mounted on a US Air Force F-16 Block 50 aircraft at the Eglin Air Force Base Test Centre in Florida   0563492

Status

The F-9120 was initially designed and developed as an Internal Research and Development (IR&D) programme but work continued under a US Naval Research Laboratory contract, partly as a risk reduction effort for the US Navy Shared Reconnaissance Pod (SHARP). A prototype was flown on a P-3 Orion in 2001.

The AARS is in production. Oman has ordered two AARS pods, one ground station and integrated logistics support for use by its F-16C fighters in February 2004 for a reported contract value of USD27.5 million.

Contractor

BAE Systems, Reconnaissance and Surveillance Systems.

Advanced Remote Ground Unattended Sensor (ARGUS) programme

Type

Air-delivered tactical target location system.

Description

The Advanced Remote Ground Unattended Sensor (ARGUS) programme is intended to address the rapid detection, location and engagement of mobile surface targets on the battlefield. As demonstrated during the recent operations in Iraq, despite the advances made in Unmanned

Air Vehicles (UAVs), manned intelligence aircraft and sophisticated surveillance satellites, there is still a significant delay in the detection and tracking of fleeing vehicles such as tanks and surface-to-surface missile launchers.

To address this shortfall, the US Air Force (USAF) began an Advanced Remote Ground Unattended Sensor (ARGUS) programme. The USAF intends to field a family of autonomously operating sensors, which could be positioned in places such as road/rail chokepoints and bridges, to monitor time-sensitive targets and send the data collected back to an air operations centre in near-realtime.

ARGUS is intended to augment less persistent sensors by acting as a 'tip-off' sensor, informing field commanders of initial movment and providing guidance as to where to concentrate reconnaissance effort.

ARGUS units will be powered internally and capable of self-testing, two-way communication and data storage. They will pass information via the Distributed Common Ground Station (DCGS) for dissemination to theatre commanders.

Proposals from industry were requested for mid-2003, with a winning design to be selected later that year, with the first production units available in FY04. The programme contains an option to pursue an Advanced Air-Delivered Sensor (AADS) for the US Marine Corps.

Both soldiers on the ground and aircraft will be able to deploy the ARGUS sensors. Initially, the USAF will integrate them with their F-16 multirole fighter aircraft, but eventually the A-10

The ARGUS system is based on an earlier 'Steel Eagle' demonstration effort. The sensor is visible, mounted to the left forward intake station of the aircraft   0530630

A close-up of the 'Steel Eagle' sensor   0530631

ground-attack aircraft, F-15E multirole fighter and General Atomics MQ-9 Predator B UAV will deploy them.

Block upgrades could include modules that sense seismic, acoustic, magnetic, thermal, electromagnetic, electro-optical, radio, radiological, chemical and biological signatures.

The ARGUS programme builds on the Defense Intelligence Agency's Unattended Ground Sensor (UGS) advanced concept technology demonstration; during this initiative, the agency demonstrated a 'Steel Rattler' hand-emplaced unit and a 'Steel Eagle' airdropped variant. The USAF and US Special Operations Command assisted in the project.

The US Army also has an interest in affordable, expendable, unattended ground sensors as part of its Future Combat Systems programme.

Status

Textron Systems has won a USD19.2 million contract from the US Air Force to produce the Advanced Remote Ground Unattended Sensor (ARGUS) system. The company will also develop a derivative system known as the Advanced Air-Delivered Sensors (AADS) system for the US Marine Corps under this contract. The work is expected to be complete by September 2009.

Contractor

Textron Systems.

AeroVision™ HMD (Helmet Mounted Display)

Type

Helmet-Mounted Display System (HMDS).

Description

The AeroVision™ HMD is designed for law enforcement, paramilitary and military applications, providing for the display of external sensor information to the wearer via two colour 17.80 mm Liquid Crystal Displays (LCDs). The display unit is attached to a NVG helmet mount, enabling the AeroVision™ HMD to utilise the same helmet mounting and adjustment mechanisms as the Aviators' Night Vision Imaging System (ANVIS) AN/AVS-6/9 and is therefore compatible with all helmets which utilise that system.

The HMD is a fully qualified system that incorporates a 28 V DC power converter and interconnect harness assembly. An optional independent battery pack is also available. The LCDs and associated optics provide full colour, high resolution, high brightness, wide field of view and 100% stereo overlap in a small lightweight configuration – total weight less than 0.9kg (2 lb).

Optional features include a Video Source Selector (VSS), allowing selection of up to four video sources for display, and a Multiple Display System (MDS), enabling the addition of up to four HMDs to the system.

This innovative design allows flight crew to use one standard helmet for both day and night missions, without recourse to customised

A German Federal Border Police EC 155B, fitted with a thermal imaging camera providing sensor data for the rear crewmember's Aerovision HMD   0114858

mountings, as required more complex helmet-mounted displays such as the JHMCS (see separate entry); AN/AVS-6/9 NVGs may be rapidly interchanged with the AeroVision™ HMD.

Specifications
Field of view: 30°
Focus: Fixed at 3.4 m (11 ft)
Stereo overlap: 100%
Eye relief: may be worn with eyeglasses
Display: two full colour 17.80 mm LCDs
Resolution: 180,000 pixels per LCD
Input: multi-NTSC channel, field sequential
Video interface: multi channel BNC input
Power: 28 V DC input; 6 V DC output; 2.5 W

Status
Evaluated by the German Federal Border Police on their Eurocopter EC 155B during 2000.

Contractor
AeroSolutions.

Airborne Surveillance Testbed (AST)

Type
Airborne Electro Optic (EO) surveillance system.

Description
The AST project is a technology demonstration programme that supports development and evaluation of defensive systems to counter InterContinental and Theatre Ballistic Missiles (ICBMs and TBMs) and their warheads.

As an airborne platform, the AST also supports risk reduction testing and the evaluation of developing technologies.

Formerly known as the Airborne Optical Adjunct (AOA), AST is a key element of the US Department of Defense's Ballistic Missile Defense Organisation (BMDO).

Initial research was aimed at evaluating whether an airborne infra-red (IR) sensor could reliably provide early warning using detection, tracking and target discrimination methods, as well as provide ICBM tracking information to ground radar. The programme was later renamed AST, and the original support mission was changed to data-gathering for ballistic missile defence development and resolving issues associated with target characteristics.

Under contract with the US Army's Space & Strategic Defense Command, Boeing has modified a 767 commercial jet aircraft to carry a large multicolour IR sensor housed in a cupola atop the 767's fuselage.

IR sensors detect the comparative heat of objects. The AST sensor, comprising more than 30,000 detector elements, is sensitive enough to detect the heat of a human body at a distance of more than 1,000 miles against the cold background of space. The AST sensor has demonstrated the capability to detect, track and discriminate warheads from missile components, debris and decoys, both from the ambient temperature of an object before launch and from the heat generated during re-entry into Earth's atmosphere.

Boeing was awarded the original AOA contract in July 1984. As the integration contractor, Boeing's responsibilities included procuring the data processor, as well as all recording,

communications and support equipment. Hughes Aircraft Electro-Optical Data Systems Group designed, built and maintains the IR sensor.

The AST and its mission system equipment were initially used to gather data that confirmed the system's ability to acquire, track, discriminate warheads from decoys, and provide track information to ground units. Originally, the emphasis was on ICBMs. At the time of the 1990–91 Gulf War national emphasis shifted to defending military installations and troops from TBMs. To support the US Army's studies on TMD, the AST system was used to gather data applicable to TBMs, as well as ICBMs.

AST has successfully participated in 65 missions, including eight operational exercises with real-time tactical communication links to US Army, Navy and Air Force elements to demonstrate the utility of an airborne IR platform in a TMD role.

Information acquired during tests at national missile ranges is added to a database that is used in computer-simulation models, and to confirm the system's capability to discriminate between objects. Data acquired during flights has validated onboard software, computers, and communication equipment necessary for real-time processing of tracking data and transmission of that data to ground-based defensive units.

The AST frequently operates at the Western Test Range, off the coast of California, and observes missiles launched from: Vandenburg Air Force Base, California; the US Army's Kwajalein Missile Range in the central Pacific Ocean; Pacific Missile Range Facility, Hawaii; White Sands Missile Range, New Mexico; Eastern Test Range in Florida; and Wallops Flight Facility, Virginia.

During a typical strategic test mission, an ICBM launched from Vandenberg Air Force Base, for example, sends one or more ballistic missile re-entry vehicles into the Kwajalein Missile Range. Depending on the objectives assigned to AST, the aircraft will be positioned 150 to 300 miles to the side of the missile's trajectory or down range of the impact point to acquire data during boost, exoatmospheric flight, or re-entry.

For short-range TBMs, AST can observe the entire trajectory. The data acquired by the sensor is processed in real time to generate a track that may be transmitted to the ground after each target-sighting update – about every 1.5 seconds. The computers on board AST extrapolate the track forward to predict an impact point, and can backtrack to estimate the launch point.

All sensor and tracking data generated by the computers is recorded on board AST for post-test analysis and system evaluation, which takes place at a laboratory in Kent, Washington. This information is also provided to the US Army's data facility in Huntsville, Alabama. It is available to all Department of Defense services and defence contractors engaged in ICBM and TBM defence projects to enhance their knowledge of missile threats and warheads.

The AST aircraft is based in Seattle, Washington, and can deploy to any national test range within one day. It carries a flight crew of 15 for typical missions, with room for observers from various government agencies and contractors. Missions last approximately 6 to 8 hours, and the aircraft usually flies at an altitude above 42,000 feet, somewhat higher than commercial aircraft.

The technology developed for AST is applicable to many sectors of the defence community. AST

personnel currently are supporting the US Navy Theatre Wide Captive Carry (NTWCC) Standard Missile (SM)-3 and AirBorne Laser (ABL) programmes.

For the NTWCC SM-3 programme, Boeing has completed aircraft, integration and two successful flight tests of the sensor for system risk reduction.

Status
Under contract to the US Army Space and Missile Defence Command, the AST has successfully completed over 65 data collection missions. In November 1998 the AST successfully tracked two US Navy missiles with the captive SM-3 seeker, built by Raytheon Missile Systems in Tucson Arizona. Technology developed for the AST has been utilised extensively in the Boeing AirBorne Laser (ABL), installed in a heavily modified B747-400F airframe (see separate entry).

Contractor
The Boeing Company, Integrated Defense Systems, Space and Intelligence.

All Weather Window™ Enhanced Vision System (EVS)

Type
Airborne Electro-Optic (EO) Enhanced Vision System (EVS).

Description
The Kollsman All Weather Window™ is an Enhanced Vision System (EVS) based on a specialised Infra-Red (IR) sensor suite. The system includes a damage-resistant IR window, mounted in the host aircraft's radome area and an electronics processor, to provide interface between the system IR sensor and aircraft Head-Up Display (HUD). The sensor image is projected onto the HUD, providing the pilot with a conformal IR scene overlaid onto his direct view of the outside scene. Thus, the system enhances the ability of the pilot to detect ground features (runway surfaces, buildings and so on) at night and/or in conditions of poor visibility, thereby improving situational awareness and, ultimately, the safety of aircraft approaches and low-level flight in poor weather conditions. The system may also enable pilots to carry out approaches in <Cat I conditions, subject to the necessary approvals.

Specifications
Sensor waveband: 1–5 µm
Field of view: 30 × 22.5°

Night Window™
Kollsman's Night Window™ EVS is a compact system which consists of an 8 to 12 µm FLIR sensor with an integrated IR window and an optional electronics processing/power supply box for added interface flexibility. The system uses proprietary video processing software algorithms proven by prior EVS certifications to improve and display an IR image of the surrounding airport structures and terrain on a Head Down Display (HDD) and/or HUD.

Night Window™ facilitates installation into very small aircraft and helicopter platforms, which would be too small to accept the full EVS. All Weather Window™.

Boeing Airborne Surveillance Testbed   0131127

Kollsamn's All Weather Window™ EVS recently obtained FAA STC on the Gulfstream V   0100518

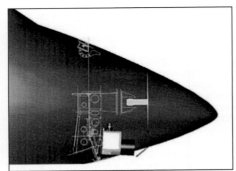

Kollsamn's All Weather Window™ EVS arrangement in Gulfstream V   0121194

Enhanced Flight Vision System (EFVS)

Kollsman recently launched its Enhanced Flight Vision System (EFVS), developed from the basic EVS but incorporating a number of significant enhancements, including a new infra-red detector integrated with a new optical system. These enhancements, combined with new image processing hardware and software, provide superior capabilities in marginal weather conditions. EFVS is designed to interface with standard digital busses and features an enlarged environmental envelope to cope with high-altitude commercial and corporate jet operations.

Status

A Gulfstream EVS was certified by the FAA on 14 September 2001, following 7½ years of research, development and testing. As of early 2004, more than 100 EVS had been installed on six models of Gulfstream aircraft. EVS is standard equipment on the large cabin, ultra long-range G550, an option on the G400, G300 and a retrofit on the GV, GIV and GIV-SP models. The EVS is a standard option on the Gulfstream long-range, large-cabin G450.

During October 2003, Kollsman announced the first installation of EVS on large Air Transport class aircraft which will be operated by the FedEx Express. EVS will be standard equipment on FedEx Express Boeing MD-10, and MD-11, and Airbus A300, and A310 aircraft. The EVS is scheduled for certification by end of 2006 with installations beginning in 2007 in conjunction with HUD installations.

NASA has acquired a Gulfstream GV business jet as the platform for its evaluation of the experimental Synthetic Vision System (SVS) and Runway Incursion Prevention System (RIPS). Both SVS and RIPS are part of NASA's Aviation Safety and Security Programme, which researches and evaluates new onboard systems that improve a pilot's Situational Awareness (SA), resulting in reduced incidents of Controlled Flight Into Terrain (CFIT) and runway incursions. NASA and Gulfstream began the seven-week flight-test programme in June 2004.

As of mid-2005, Kollsman reported over 400 orders for its new EFVS.

Night Window™ is available for installation in commercial, military, business aircraft and helicopters.

Contractor

Kollsman Inc.

AN/AAD-5 Infra-red reconnaissance set

Type

Airborne reconnaissance system.

Description

The AN/AAD-5 is a dual field of view infra-red reconnaissance system which scans the terrain beneath an aircraft's flight path. The system consists of seven LRUs: the receiver, recorder, film magazine, control indicator, infra-red performance analyser, cooler and power supply.

The receiver includes the scanning optics, cooler cryostat, two 12-element detector arrays, 24 preamplifiers and the associated buffer electronics. The detector arrays are Mercury

Cadmium Telluride (MCT) photoconductors which are sensitive to infra-red radiation in the 8 to 14 μm waveband. One array is used for the wide field of view and the other for narrow field of view.

The recorder consists of a CRT, recording optical components, video and sweep electronic cards, digital timing circuits, a high-voltage power supply, film speed control circuits and an auxiliary data annotation set.

An improved version, the AN/AAD-5(RC), is derived from the AN/AAD-5. The AN/AAD-5(RC) imaging process is identical to that of the AN/AAD-5. The receiver is two-thirds the size of the AN/AAD-5 receiver and uses an integral 0.25 W split-Stirling miniature cryogenic refrigerator.

Various electro-optical output options were available for both AN/AAD-5 and AN/AAD-5(RC) equipped systems. Upgrades using the Lockheed Martin IR linescanner real-time display system can add a cockpit real-time display with onboard video recording, with an optional datalink to ground stations, while maintaining the current film system operation.

Specifications

Dimensions:
(receiver) 460 × 380 × 330 mm (AN/AAD-5)
    430 × 345 × 279 mm (AN/AAD-5(RC))
(recorder) 410 × 580 × 230 mm
(magazine) 380 × 180 × 250 mm
(control indicator) 150 × 100 × 80 mm
(cooler) 410 × 20 × 300 mm
(analyser) 150 × 50 × 300 mm
(power supply) 250 × 430 × 280 mm
Weight:
(AN/AAD-5) 130 kg
(AN/AAD-5(RC)) 63.2 kg
Power supply: 115 V AC, 400 Hz, 725 VA nominal
    28 V DC, 118 W nominal
Magazine capacity:
    106 m (conventional film)
    213 m (thin-based film)

Status

In service with the US Air National Guard in the RF-4C and the US Navy in the F-14 TARPS (see separate entry). Also used by a number of air forces including Australia for the RF-111 and Germany, Greece, Korea, Spain and Turkey for the RF-4. Over 600 AN/AAD-5 systems and 168 AN/AAD-5(RC) systems have been delivered.

Contractor

BAE Systems, North America.

AN/AAQ-16 night vision system

Type

Airborne Electro-Optic (EO) navigation and surveillance system.

Description

The AN/AAQ-16 infra-red imaging system entered initial production for the US Department of Defense in 1984. It has been selected by the US Army, Navy, Marine Corps and Air Force and international customers for a variety of helicopters.

The system provides 24-hour mission capability during night or degraded weather conditions in support of special operations, air assault, search and rescue, and anti-surface warfare.

An advanced thermal imaging system, AN/AAQ-16's high-resolution, TV-like imagery provides for low-level pilotage and navigation in the wide field of view, and in the narrow field of view, provides long-range target detection and identification. The AN/AAQ-16 combines automatic FLIR performance optimisation, and DC restoration to respond to various environmental conditions, with hands-off features of automatic gain, level, and focus for a high-quality image and eased operator workload. Either a black-hot or white-hot image can be selected by the operator. Additional features which enhance mission success include an autotracker and operator-controlled autoscan.

Variants of the AN/AAQ-16 system include: the AN/AAQ-16B which utilises the 8 to 12 μm

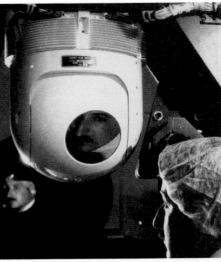

Close-up of the FLIR turret of the AN/AAQ-16 night vision system   0503770

waveband and two fields of view (30 × 40° and 5 × 6.7°); the AN/AAQ-16C which also utilises the 8 to 12 μm waveband and three fields of view (30 × 40°, 5 × 6.7° and 1.9 × 2.5°) and a dual-mode tracker.

Specifications

Dimensions:
(FLIR turret) 356 (length) × 305 mm (diameter)
(electronics unit) 305 × 200 × 412 mm
(multifunction control unit) >6.2 × 76.2 × 157.5 mm
(system control unit) 38.1 × 146 × 115 mm
Weight:
(FLIR turret) 24.49 kg
(electronics unit) 20.87 kg
(multifunction control unit) 0.59 kg
(system control unit) 0.45 kg
Wavelength: 8–12 μm
Field of view:
(×1 magnification) 30 × 40°
(×6 magnification) 5 × 6.7°
(×16 magnification option) 1.9 × 2.5°
Reliability: >300 h MTBF

Status

More than 400 systems delivered or on order for the US Army UH-60L, CH-47D, MH-47E, MH-60K; US Air Force HH/MH-60G, US Marine Corps CH-53E, Australian S-70B. Co-produced with NEC for the Japanese Defence Force.

Contractor

Raytheon Company, Space and Airborne Systems.

AN/AAQ-17 infra-red detecting set

Type

Airborne Electro-Optic (EO) navigation and surveillance system.

Description

The AN/AAQ-17 infra-red detecting set is a multipurpose thermal imaging system. Typical missions are navigation, search and rescue, surveillance and fire control. AN/AAQ-17 systems are fitted on US Air Force AC-130A, AC-130H and AC-130U gunships, HC-130P and HC-130 tankers and C-141 special operations transports.

In the US Air Force gunship programme the AN/AAQ-17 FLIR replaces the AN/AAD-7 FLIR on the AC-130H, providing improved performance and reliability at lower cost. The AN/AAQ-17 is a derivative of the AN/AAQ-15. Both of these systems are lightweight and easily adapted to new aircraft. Originally, the AN/AAQ-15 FLIR was developed for the US Air Force HH-60 Nighthawk helicopter and has also been produced for the MC-130H Combat Talon II aircraft.

The AN/AAQ-17 consists of four LRUs: the infra-red receiver (LRU1), control-converter (LRU2), gimbal position control (LRU3) and infra-red set control (LRU4). Environmental qualification, reliability demonstrations and maintainability demonstrations are complete. The AN/AAQ-17 uses standard DoD FLIR common modules to convert long wavelength radiation into a composite TV video.

A 13.7 × 18.3° wide field of view allows navigation, area search and detection of larger targets. The 3 × 4° narrow field of view allows small target detection and target recognition. A precision gimbal provides accurate line of sight angular measurement. An adaptive gate video tracker reduces operator workload by providing hands-off automatic line of sight control. Operation of the FLIR is through manual controls or over a MIL-STD-1553B databus. This produces a flexible system adaptable to many aircraft.

Specifications
Weight: (infra-red receiver) 43.09 kg
(control converter) 21.77 kg
(gimbal position control) 2.27 kg
(infra-red set control) 1.81 kg
Field of view: (wide) 13.7 × 18.3°
(narrow) 3 × 4°
Field of regard: (azimuth) ±200°
(elevation) +15 to −105°

The AN/AAQ-17 infra-red detecting set is installed in US Air Force AC-130 gunships   0521571

Status
In service in AC-130A, AC-130H, AC-130U, HC-130N and HC-130P and C-141 aircraft.

Contractor
Raytheon Company, Space and Airborne Systems.

AN/AAQ-18 Forward Looking Infra-Red (FLIR) system

Type
Airborne Electro-Optic (EO) navigation system.

Description
The AN/AAQ-18 FLIR system is a common module update programme. The system improves reliability and readiness of the US Air Force Special Forces by replacing the AN/AAQ-10 system. System installations include the MH-53J helicopter and the MC-130E Combat Talon aircraft.

The primary mission of the AN/AAQ-18 is navigation. It is also ideal for search and rescue missions. The AN/AAQ-18 consists of 5 LRUs.

Specifications
Power supply: 115 V AC, 400 Hz, 3 phase, 3 kVA
28 V DC, 2 A
Field of view:
(wide) 18.2 × 13.7°
(narrow) 4.1 × 3.1°
Field of regard:
(azimuth) ±190°
(elevation) +15 to −105°

The AN/AAQ-18 is installed in US Air Force Special Forces MC-130E Combat Talon aircraft   0008493

Status
Operational in the MH-53J helicopter and US Special Forces MC-130E Combat Talon aircraft.

Contractor
Raytheon Company, Space and Airborne Systems.

AN/AAQ-27 (3 FoV) Mid-Wave Infra-Red (MWIR) imaging system

Type
Airborne Electro Optic (EO) navigation and targeting system.

Description
AN/AAQ-27 (3 FoV) is a third-generation, three-Field of View (FoV), Mid-Wave Infra-Red (MWIR) system for helicopter navigation, surveillance, and targeting applications. With three FoVs available, pilots can fly and navigate on low-level missions or detect and identify long-range targets from higher altitudes. Raytheon states that the system uses non-developmental production components to provide good reliability, higher-resolution imagery and better range performance than current long-wavelength infra-red systems.

The AN/AAQ-27 (3 FoV) is a modified version of the AN/AAQ-27 IR system which Raytheon produces for the US Marine Corps MV-22 Osprey (see separate entry), which features substantial commonality with the combat-proven AN/AAQ-16 system. The AN/AAQ-27 (3 FoV) is compatible with existing AN/AAQ-16 system mountings and is easily interchangeable for upgrade programmes.

The above images show an unmodified Royal Australian Navy S-70B-2 (left) and an aircraft modified under Project Sea 1405, with an AN/AAQ-27 (3 FoV) MWIR imaging system under the nose (right). Note also the AN/AAR-54(V) passive MAWS flanking the IR turret sensor   0094208

The above images show an unmodified Royal Australian Navy S-70B-2 (left) and an aircraft modified under Project Sea 1405, with an AN/AAQ-27 (3 FoV) MWIR imaging system under the nose (right). Note also the AN/AAR-54(V) passive MAWS flanking the IR turret sensor 0111788

Specifications
Detector: staring array (640 × 480 InSb)
Spectral band: 3–5 µm
FoV:
  Wide: 30 × 40°, ×1
  Medium: 5.0 × 6.7°, ×6
  Narrow: 1.3 × 1.73°, ×23
Dimensions:
  Turret FLIR unit (TFU): 360.4 (height) × 303.5 mm (diameter)
  Systems electronics unit (SEU): 199.1 (height) × 413.5 (length) × 306.3 mm (width)
System weight: 42.27 kg
Tracker: dual-mode (centroid/correlation)
Options: eye-safe Laser Range-Finder (LRF); Laser Range-finder/Designator (LRFD)

Status
Selected by Kaman Aerospace International Corporation for the Royal Australian Navy SH-2G Super Seasprite helicopter programme. In addition, Tenix Defence was prime contractor for the Project Sea 1405 programme, which involved the upgrade of Australian Navy S-70B-2 Seahawks, which included installation of AN/AAQ-27 (3 FoV) systems. During 2003, 16 aircraft were reportedly upgraded.

Contractor
Raytheon Company, Space and Airborne Systems.

The AN/AAQ-27 (3 FoV) MWIR imaging system 0023396

AN/AAQ-27 MWIR staring sensor

Type
Airborne Electro Optic (EO) navigation and targeting system.

Description
The AN/AAQ-27 Mid-Wave Infra-Red (MWIR) staring sensor is an advanced IR system that features substantial commonality with the combat-proven AN/AAQ-16B IR system. It uses non-developmental, in-production components to provide higher-resolution imagery than current Long-Wave Infra-Red (LWIR) systems, and incorporates an MWIR indium antimonide (InSb) staring focal plane array with 480 × 640 detector elements.

The system features a turreted FLIR weighing 22.7 kg, with a total system weight under 42.3 kg. Options include a video autotracker, a third Field of View (FoV) and eye-safe Laser Range-Finder (LRF) (see separate entry).

Specifications
Wavelength: 3–5 µm (MWIR)
Field of view: 30 × 40° and 5 × 6.7°

Status
The system is in production for the US Marine Corps MV-22 Osprey.

Contractor
Raytheon Company, Space and Airborne Systems.

AN/AAQ-30 Hawkeye Target Sight System (TSS)

Type
Airborne Electro Optic (EO) navigation and targeting system.

Description
The AN/AAQ-30 Hawkeye Target Sight System (TSS), part of the USMC's Bell AH-1 Upgrade Programme, facilitates target detection, recognition and identification by day or night and during adverse weather. The wide Field of View (FoV) optics provide a secondary navigation capability when light levels are low and Night Vision Goggles (NVGs) are ineffective.

Key factors which affect imaging performance include aperture size (for enhanced detection and identification ranges and poor weather performance), optics/sensitivity (for good resolution, which directly affects recognition range and thus stand-off distance according to selected magnification factor) and stabilisation (which affects resolution, designation/ engagement range and third-party targeting ability).

The TSS, developed by Lockheed Martin Missiles and Fire Control, packages a variety of sensors into the same 52 cm-diameter WESCAM Model 20 turret/gimbal assembly (see separate entry) that is already operational aboard US Navy P-3C and US Coast Guard C-130 maritime patrol aircraft. In order to accommodate the large turret assembly, Lockheed Martin fabricates an entirely new nose section for the AH-1Z. The AAQ-30 also shares many key elements with Lockheed Martin's Sniper XR fixed-wing targeting system (see separate entry), recently adopted by the USAF as its Advanced Targeting Pod (ATP), and

The AN/AAQ-27 Mid-Wave Infra-Red (MWIR) sensor is in production for the US Marine Corps MV-22 Osprey (Boeing) 0524706

the Electro-Optical (EO) targeting system for the F-35 Joint Strike Fighter (JSF).

The TSS employs proven hardware from fielded products, with 73 per cent of its content made up of Commercial Off-The-Shelf (COTS)/non-developmental items. The system is of modular design, to permit ease of maintenance and future growth. An important part of the design is a five-axis 'soft-mount' gimbal, which provides sightline stabilisation of better than 15 mrads. All sensors are auto-boresighted to each other, and to the helicopter's dual-EGI (Embedded GPS/INS) systems. This minimises the target-location error, which is typically 7.5 m when the helicopter's current position is known precisely but would increase to 20 m with an aircraft position error of 14 m.

The TSS accommodates a Forward Looking Infra-Red (FLIR), colour television (TV) camera, Laser Range Finder/Designator (LRFD), Inertial Measurement Unit (IMU), boresight module and Electronics Unit (EU). All units built so far also include a laser spot tracker, although the USMC has not yet decided whether this will form part of the operational fit.

The FLIR, which has a comparatively large (21.7 cm) aperture, accommodates a staring array of 640 × 512 indium antimonide (InSb) detectors, operating in the Mid-Wave Infra-Red (MWIR, 3 to 5 μm) waveband. The Sony DXC-950 TV camera (to be replaced by the -390 model in production units), which functions in the visible and Near IR (NIR) wavebands (using a high-pass filter for operation at low light levels or in haze), includes three Charge Coupled Device (CCD) detector arrays. The camera is fitted with a Canon ×2.5 extender lens and provides continuous zoom up to a magnification of ×18. Two FoVs are matched to two of the four offered by the FLIR, permitting easy switching between the sensors.

The LRFD is of the same type as that in the LANTIRN targeting pod (see separate entry), of which approximately 1,000 units are in service, with a selectable eyesafe function.

The TSS supports autonomous target acquisition and re-engagement. Automatic tracking of up to three targets simultaneously, using correlation, contrast or centroid tracking modes, is available with both the FLIR and the TV camera. Each target can vary in size from a single pixel up to 80 per cent of the total FoV. The system can store track files for up to 10 additional ground targets, even after they exit the FoV, by the use of an inertially-derived 'coast' function (similar to that of the TIALD LDP – see separate entry); this function also allows track to be maintained through short-duration obscuration due to weather.

The commercially derived, open architecture of the TSS and its use of PowerPC processors, together with the comparatively large volume available in the turret, supports growth to include additional facilities. These could include sensor fusion, an NVG-compatible laser pointer, a low-light-level electron-bombarded CCD colour television camera, integration of a navigation FLIR (NAVFLIR), the adoption of a long-wave FLIR (LWIR, 8 to 12 μm) based on quantum-well devices, and other sensors. The aircraft IMU supports growth to electronic stabilisation.

Although it is not part of the TSS baseline fit, Hawkeye can also take advantage of Lockheed Martin's XR (eXtended Range) image-processing technique. This electronically enhances imagery in part of the FOV to provide a 60 per cent improvement in recognition and identification ranges. Lockheed Martin has incorporated prototype XR electronics and their associated algorithms in TSS units that it has built under the engineering and manufacturing development (EMD) programme.

Specifications
FoR: ±120° (azimuth), +45 to –120° (elevation)
Detector: staring array (640 × 512 InSb); spectral band 3 to 5 μm
FoV: 21.7 × 16.3° (wide), 4.4 × 3.3° (medium), 0.88 × 0.66° (narrow), 0.59 × 0.44° (very narrow)
System weight: 116 kg (turret 83 kg, EU 33 kg)
Tracker modes: contrast, centroid, correlation
Options: eye-safe LRFD; Laser Spot Tracker; XR image processing; 8 to 12 μm NAVFLIR

The AN/AAQ-30 TSS integrated electro-optic sighting and fire-control system will be fitted to USMC AH-1Z Cobra attack helicopter (Bell Helicopter)   0111619

Real-time processed TSS FLIR imagery of Orlando, at a range of 13.6 km, utilising all four fields of view. From left to right, these are wide (21.7 × 16.3°), medium (4.4 × 3.3°), narrow (0.88 × 0.66°) and very narrow (0.59 × 0.44°). In the very narrow field of view, performance has been enhanced by the XR image processing system. The images were acquired in 16 km visibility at a temperature of 23.8°C (75°F), although no figures for absolute humidity are known (Lockheed Martin)   0111627

The AN/AAQ-30 TSS mounted on a AH-1Z during flight trials. Note the extensively modified nose and non-standard flight test recording equipment mounted on the side of the nose to monitor the operation of the turret (Lockheed Martin)   0568344

AN/AAQ-30(V) CATSeye

Lockheed Martin Missiles and Fire Control is developing a more compact version of the AN/AAQ-30 – the AAQ-30 (V) CATSeye, which employs a Wescam Model 16 (see separate entry) turret of 40 cm diameter rather than the 51 cm of the TSS, providing approximately 70 per cent of the performance available from the larger system at 70 per cent of the cost.

Potential applications include attack helicopters that are too small to accommodate the full TSS, such as the Agusta Westland A 129, together with unmanned aerial vehicles and other compact platforms. The unit can incorporate a tracker for the TOW anti-tank missile.

CATSeye weighs 61 kg, compared with 116 kg for Hawkeye. It accommodates: a MWIR TI based on a staring 640 × 480 indium antimonide (InSb) detector derived from the TSS, with a 15.2 cm aperture and three fields of view; a Sony ICX058AL intensified Charge Coupled Device (CCD) television camera; and a Laser Range Finder/Target Designator (LRF/TD) with eye-safe mode. Stabilisation of the four-axis gimbal is better than 20 μrad.

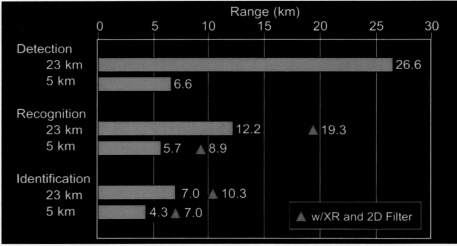

Detection, recognition and identification ranges for the TSS FLIR (upper) and TV camera (lower) against a 2.5 m diameter target. The figures 23 km and 5 km refer to ambient visibility. The asterisks (upper diagram) and triangles (lower diagram) indicate performance with the XR image-processing facility and a 2-D filter. The latter performs edge enhancement to improve the sharpness of the picture (Lockheed Martin) 0111625

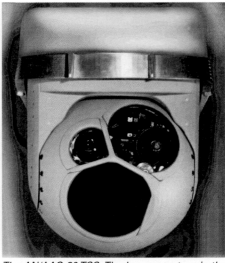

The AN/AAQ-30 TSS. The lower aperture is the FLIR sensor, while the upper left is the TV sensor and the upper right is shared by the LRFD and optional Laser Spot Tracker (LST) (Lockheed Martin) 0111629

A Northrop Grumman LN-200 Inertial Measurement Unit (IMU) is mounted on the gimbal for stabilisation and aimpoint geoposition data supplied to onboard weapons employing Global Positioning System (GPS) guidance.

Status
Forms part of the USMC upgrade for the AH-1Z attack helicopter. Five units have been manufactured under Engineering and Manufacturing Development (EMD) contract, with options covering another 21 units for Low-Rate Initial Production (LRIP). The USMC requires approximately 201 AH-1Z helicopters.

Lockheed Martin completed a prototype of the AN/AAQ-30(V) CATSeye during the first quarter of 2002, and the system has subsequently successfully supported several field demonstrations.

The XR image processing algorithm is also available for the CATSeye variant of the system.

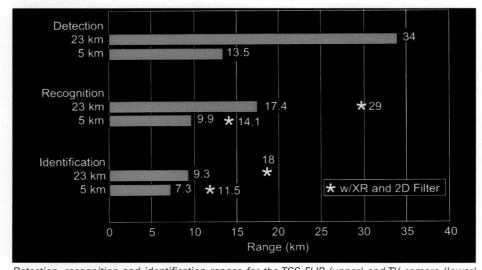

Detection, recognition and identification ranges for the TSS FLIR (upper) and TV camera (lower) against a 2.5 m diameter target. The figures 23 km and 5 km refer to ambient visibility. The asterisks (upper diagram) and triangles (lower diagram) indicate performance with the XR image-processing facility and a 2-D filter. The latter performs edge enhancement to improve the sharpness of the picture (Lockheed Martin) 0111624

Contractor
Lockheed Martin Missiles and Fire Control.

AN/AAQ-32 Internal FLIR Targeting System (IFTS)

Type
Airborne Electro Optic (EO) navigation and targeting system.

Description
The AN/AAQ-32 IFTS, together with the AN/APG-80 Agile Beam Radar (ABR) and Falcon Edge Integrated Electronic Warfare Suite (IEWS) (see separate entries) together make up the primary sensor suite of the Lockheed Martin F-16E/F Block 60 aircraft ordered by the United Arab Emirates (UAE).

The IFTS incorporates two similar 3–5 μm Mid-Wave Infra-Red (MWIR) sensors; the navigation FLIR is installed in a small turret mounted above the radome, offset slightly to afford some measure of look-down. The targeting FLIR is pod-mounted on the left chin station and also incorporates a Laser Range-Finder/Designator (LRF/D). Performance for the system is said to be comparable to the latest systems, such as Raytheon's AN/ASQ-228 ATFLIR and Lockheed Martin's Sniper XR (see separate entries).

Status
The first UAE Block 60 F-16F aircraft, equipped with the IFTS, was rolled out in the latter part of 2003.

Contractor
Northrop Grumman Corporation, Electronic Systems Sector.

AN/AAR-50 Navigation FLIR (NavFLIR)

Type
Airborne Electro-Optic (EO) navigation system.

Description
The AN/AAR-50 Navigation FLIR (NavFLIR) is a derivative of the AN/AAQ-16 night vision system installed in US Army helicopters. It uses a thermal imaging sensor to provide pilots of fixed-wing aircraft on low-level missions at night or in bad weather with a TV-like image of the terrain ahead projected on to a head-up display. The system as configured for the F/A-18 is pod-mounted in a fixed forward-staring position but could be configured in different pods for a variety of aircraft.

The AN/AAR-50 NavFLIR mounted on the starboard intake of a US Navy F/A-18 0503771

NavFLIR consists of four major weapon-replaceable assemblies: FLIR sensor unit, pod electronics unit, thermal control unit and pod adaptor.

Specifications
Dimensions: 1,981 × 254 mm diameter
Weight:
(pod) 73.48 kg
(adaptor) 23.13 kg
Wavelength: 8–12µm
Field of view: 19.5 × 19.5° displayed
Reliability: >410 h MTBF

Status
In service.

Contractor
Raytheon Company, Space and Airborne Systems.

AN/AAS-32 laser tracker

Type
Airborne Laser Range Finder/Target Designator (LRFTD).

Description
The Boeing AN/AAS-32 is a laser tracker produced for the Bell AH-1S Cobra light attack helicopter. A derivative of the system is part of the Target Acquisition Designation System/Pilot Night

The AN/AAS-32 is in service in the AH-1S Cobra light attack helicopter   0022471

Viewing System (TADS/PNVS) in the Boeing AH-64 Apache attack helicopter.

The system features a Wide Field of View (WFoV) sensor, with targets illuminated by the host helicopter's laser designator or by ground-based systems.

Specifications
Weight:
  receiver: 9 kg
  electronics unit: 3.4 kg
Dimensions:
  receiver: 214 × 188 mm
  electronics unit: 152 × 152 × 188 mm

Field of regard:
(azimuth) –90 to +90°
(elevation) –60 to +30°
Field of view:
(azimuth) 20°
(elevation) 10°
Aperture: 127 mm

Status
In service.

Contractor
The Boeing Company, Integrated Defense Systems.

AN/AAS-38A F/A-18 targeting pod

Type
Airborne Electro-Optic (EO) targeting system.

Description
The AN/AAS-38A Forward-Looking Infra-red (FLIR) targeting pod enables pilots of US Navy and Marine Corps F/A-18 Hornet aircraft to attack ground targets day or night with a precision strike capability. The system provides Infra-Red (IR) imagery on a cockpit panel display and accommodates a laser range-finder/designator that can pinpoint targets for both laser-guided and conventional weapon delivery.

The pod is integrated with the aircraft's avionics system through a MIL-STD-1553 databus, allowing the pod to receive command and cue signals from the onboard mission computer and provide status and targeting information to the cockpit display and weapon delivery system.

The AN/AAS-38A consists of 12 weapon replaceable assemblies (WRAs) that can be readily accessed and replaced without the need for calibration, alignment, special tools or handling equipment.

The AN/AAS-38A configuration accepts the two laser subsystem WRAs, a laser transceiver and laser power supply to provide the aircrew with the capability for laser target designation and ranging (LTD/R).

Design qualification and flight test were successfully completed in August 1994 thus providing the US Navy a second source for AAS-38 pods and spares (see also Lockheed Martin Electronics & Missiles entry entitled NITE Hawk).

The Targeting FLIR, when integrated with the AN/AAR-50 Navigation FLIR and Night Vision Goggles, provides the pilot/aircrew with the capability to maintain situational awareness, navigate/avoid terrain, acquire/designate targets and assess battle damage for deployment of the Pave Way/GBU-24 precision-guided weapon series.

Specifications
Dimensions: 1,830 (length) × 330 mm (diameter)
Weight: 172.7 kg
NFOV: 3 × 3°
WFOV: 12 × 12°
Field of regard:
(pitch) +30° to –150°
(roll) ±540°
Video: RS-343 875 lines

Status
In service.

Contractor
Raytheon Company, Space and Airborne Systems.

AN/AAS-40 Seehawk FLIR

Type
Airborne Electro-Optic (EO) navigation system.

Description
The AN/AAS-40 Seehawk is a thermal imaging system using US Department of Defense common module FLIR components to provide high-resolution imagery. Designed for an aircraft or surface vessel, the system's current principal application is aboard a US Coast Guard Sikorsky HH-52A helicopter serving the primary role of search and rescue, law enforcement, maritime environmental control, marine and border control and navigational assistance. It is also installed on the Northrop Grumman S-2(T) ASW aircraft.

The AN/AAS-40 Seehawk FLIR on the US Coast Guard HH-52A helicopter   0503781

The Northrop Grumman AN/AAS-40 Seehawk lightweight FLIR system consists of a turret assembly, the control electronics unit and the power supply unit. FLIR imagery is fed to the cockpit-mounted display. An automatic scan capability provides constant search coverage in elevation and azimuth. The autosearch mode is enhanced by inclusion of automatic lock on which reacts to either large or small targets, as selected by the operator.

Status
The Seehawk system successfully completed two years' flight demonstration aboard a Beech 200T aircraft. The system was fully operational in conjunction with the other onboard avionics systems, demonstrating maritime patrol, surveillance and reconnaissance applications. The latest-generation AN/AAS-40 Seehawk has been installed aboard a modified Sikorsky S-76 helicopter which was used as a demonstrator in support of the now cancelled US Army RAH-66 Comanche programme. The AN/AAS-40 is in production for the US Coast Guard HH-65A Dolphin helicopter and the Northrop Grumman S-2(T) ASW aircraft.

Contractor
Northrop Grumman Corporation, Electronic Systems Sector.

AN/AAS-42 Infra-Red Search and Track System (IRSTS)

Type
Airborne Electro Optic (EO) targeting system.

Description
The AN/AAS-42 Infra-red Search and Track System (IRSTS) is designed to permit the multiple tracking of thermal energy emitting targets at extremely long range to augment information supplied by conventional tactical radars. The system enhances performance against low radar cross-section targets while providing immunity to electronic detection and RF countermeasures. High-resolution IRST provides dramatically improved raid cell count at maximum declaration ranges – information that can stand alone or be fused with other sensor data to enhance situational awareness.

The AN/AAS-42 is fitted to the F-14 Tomcat in US Navy service. Note the AAS-42 sensor head (right) alongside the magnifying optical identification sensor (right) – the two are utilised in parallel to provide for entirely passive long-range identification of bogeys (Lockheed Martin) 1044579

The sensor head (WRA-1) is part of the hardware configuration of the AN/AAS-42 IRSTS (Lockheed Martin) 1044578

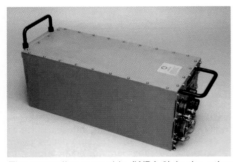

The controller assembly (WRA-2) is the other main hardware component of the AN/AAS-42 IRSTS (Lockheed Martin) 1044580

The IRSTS consists of a sensor head mounted beneath the nose of the F-14D and an electronics unit just aft of the cockpit. The system is integrated with the F-14's central computer system and complements the AN/APG-71 radar providing the aircrew both target track data and infra-red imagery displays. The AN/AAS-42 operates in six discrete modes, with selectable and individually controlled scan volumes in azimuth (±80°) and elevation (±70°).

Specifications
Dimensions:
(pod) 1,366.6 × 248 mm diameter
(sensor head) 914.4 × 228.6 mm diameter
(electronics unit) 190.5 × 190.5 × 482.6 mm

Weight:
(pod) 86.5 kg
(sensor head) 41.28 kg
(electronics unit) 16.78 kg

Status
Production is complete. In service in US Navy F-14D aircraft.

Contractor
Lockheed Martin Missiles and Fire Control.

AN/AAS-44(V) thermal imaging/laser designation system

Type
Airborne Electro Optic (EO) targeting system.

Description
The AN/AAS-44(V) is a multipurpose thermal imaging and laser designation system. It provides

long-range surveillance, target acquisition, tracking, range-finding and laser designation for the Hellfire missile and for all US tri-service and NATO laser-guided munitions.

Status
The AN/AAS-44(V) is deployed on US Navy HH-60H helicopters and on the US Navy Light Airborne MultiPurpose System (LAMPS) MK III SH-60B armed helicopters as part of the Block I upgrade programme. Production continues for both programmes. By mid-1999, 69 systems were in US Navy service with a further 35 on order.

Contractor
Raytheon Company, Space and Airborne Systems.

AN/AAS-52 Multispectral Targeting System (MTS)

Type
Airborne Electro Optic (EO) navigation and targeting system.

Description
The AN/AAS-52 MTS is a combined Electro-Optic (EO), Infra-Red (IR) and laser ranging system. Using state-of-the-art digital architecture, this advanced EO/IR system provides long-range surveillance, target acquisition, tracking, range-finding and laser designation for the HELLFIRE missile and for all tri-service and NATO laser-guided munitions. As installed in the MQ-1 Predator, the system provides for target detection, identification and engagement.

The MTS uses Raytheon's Local Area Processing (LAP), an automatic, 'hands-off' image optimisation technique that maximises displayed image information, enhancing both Situational Awareness (SA) and long-range surveillance capability.

The AN/AAS-52 comprises a turret unit (Weapon Replaceable Assembly (WRA) – 1), which incorporates an integrated Inertial Measurment Unit (IMU) and an electronics unit (WRA-2).

To ensure continuous operational advantage, the MTS has been designed for growth options such as multiple-wavelength sensors, TV cameras (near-IR and colour), illuminators, eyesafe range-finders, spot trackers, and other avionics. The advanced electronic and optical design provide a clear growth path for image fusion and other performance enhancements through add-in circuitry.

Specifications
Spectral band: 0.4–0.7 and 0.6–0.8 µm (TV and NIR); 3–5 µm (MWIR)
Focal plane array: 640 × 480 InSb
IRFOV: 0.6 × 0.8° (ultra-narrow); 1.2 × 1.6° (narrow); 5.7 × 7.6° (medium); 15 × 20° (medium-wide); 33 × 44° (wide)

The AN/AAS-52 MTS turret (Raytheon) 0521308

TV FOV: 0.21 × 0.28° and 0.6 × 0.8° (ultra-narrow); 1.2 × 1.6° (narrow); 5.7 × 7.6° (medium); 15 × 20° (medium-wide); 33 × 44° (wide)
Electronic zoom: 2:1 (0.11 × 0.14°TV and 0.3 × 0.4° IR); 4:1 (0.06 × 0.07°TV and 0.15 × 0.2° IR)
Power: 28 V DC (30 A, nominal)
Interfaces: DVI, Firewire, NTSC and RS-170, (video); Ethernet, MIL-STD-1553 and RS-422 (system)
Temperature: –54°C to +55°C
Altitude: up to 9,144 m
Dimensions: 475 (H) × 443 (Ø) mm (WRA-1); 343 (W) × 318 (L) × 235 (H) mm (WRA-2)
Weight: 56.3 kg (WRA-1); 21.8 kg (WRA-2)

Status
The US Air Force is planning to upgrade the sensor suite in its fleet of General Atomics RQ-1A/B UAVs with the replacement of the existing AN/AAS-44(V) stabilised chin-turret with the AN/AAS-52 MTS. New production Predators will be delivered with the MTS. While not currently funded, the per-unit cost of retrofitting the MTS to the Predators currently in service is estimated at approximately USD1.5 million. (The basic Predator system costs USD2.5 million per vehicle.)

Three prototype MTS systems were fitted to Predators before the system was committed to action over Afghanistan during 2003. All of these systems were reportedly used in action, including self-designation for onboard Lockheed Martin AGM-114 Hellfire missiles. In addition, the system has undergone operational testing at Nellis Air Force Base, Nevada.

During 2004, Raytheon announced the integration and successful demonstration of MTS, in concert with the SeaVue radar (see separate entry), onboard a General Atomics Aeronautical Systems Mariner UAV, a derivative of Predator B.

The US Air Force Special Operations Command (AFSOC) has selected the MTS to replace the All-Light-Level Television (ALLTV) installation aboard its AC-130U Spectre gunships. AFSOC has so far ordered five installations.

The US Air Force plans to replace the original EO turret onboard its Predators with the AN/AAS-52 MTS (General Atomics) 0521309

Elsewhere, the US Naval Air Warfare Center Aircraft Division (NAWCAD) awarded Optics 1 a USD12.3 million Small Business Innovation Research (SBIR) Phase III contract to design, build, install, support and provide training for a variant of the MTS to equip special-projects aircraft.

Contractor
Raytheon Company, Space and Airborne Systems.

AN/ASD-12(V) Shared Reconnaissance Pod (SHARP)

Type
Airborne reconnaissance system.

Description
The AN/ASD-12(V) Shared Reconnaissance Pod (SHARP pod) is intended to replace the film-based Tactical Air Reconnaissance Pod System (TARPS – see separate entry) that equips US Navy F-14s. SHARP, which is the size of a 330 gallons (US) fuel tank, can mount on any standard weapons pylon with a MIL-STD-1760 stores-management interface. The pod can accommodate Recon/Optical's CA-279/M medium-altitude and CA-279/h high-altitude dual-band cameras, which can operate at altitudes ranging from 2,000 to 50,000 ft and at ranges of up to 50 n miles. The interchangeable sensors provide stereo or mono imagery, captured in a framing format, at a coverage rate of up to 10,000 nm²/h. The sensors are installed in a rotating mid-section to optimise coverage and protect the 45 × 28 cm optical window (rotated to face the strongback). Sensors Line of Sight (LoS) can be pointed automatically by the aircraft mission management system, or automatically by the pilot's HMD, aircraft radar, HUD or targeting system (such as ATFLIR – see separate entry). Other system elements include a digital data storage system with a capacity of 64 Mb, a Ku-band Common DataLink (CDL) terminal and an environmental control unit.

Specific features and benefits of SHARP include:
- Strongback and bulkhead design
- Non-load-bearing midsection
  Rotating Midsection:
- Optimises total Field of Regard (FoR)
- Reduces size of optical window
- Protects window on takeoff and landing
- Self-contained environmental control system
- High- and medium-altitude EO/IR sensors for day/night operation
- Adaptable to SAR, hyperspectral and other sensors
- High-capacity digital storage system
- NITF JPEG format compliant
- Real-time imagery compression
- Programmable compression rate and selectable image size
- Ku-band Common DataLink (CDL) compatible with Distributed Common Ground Station (DCGS)
  Onboard inertial navigation system:
- Optimises sensor pointing accuracy
- Tactical Airborne Mission Planning System (TAMPS) compatible.

Developed by Raytheon in conjunction with the US Navy, other contributors to the programme include Honeywell, L-3 Communications, Recon/Optical and Smiths Aerospace. SHARP will provide the US Navy with a day/night air reconnaissance capability for fixed-wing aircraft such as the F/A-18E/F.

Physically, the SHARP pod is composed of (from nose to tail) the CDL antenna/radome assembly and five successive equipment bays. Bay 1 houses the CDL and sensor electronics, a heat exchanger and a power supply. Bay 2 houses a data recorder, a second power supply, a sensor electronics unit, an auxiliary data processor and an interface reception unit. Bay 3, the rotating sensor bay, houses the suite of EO/IR sensors. The open-architecture design of the system facilitates the installation of a wide variety of alternative sensors if required. Bay 4 accommodates inertial navigation and pod management units. At the rear of the pod, bay 5 houses the system's environmental control unit.

Specifications
EO/IR performance (specification):
System resolution: >1 m
Data storage: 60 min recording time (minimum)
Transmitted imagery: Annotated with mission number, time, date, target and platform location, platform heading, altitude and groundspeed
Forward avionics bay
Cooling: Air cooled
Cooling: Conduction
Operating temperature: −40 to 71°C
Temperature stability: ±10°C per hour
Sensor bay
   Cooling: Air cooled
   Operating temperature: 0 to 40°C

Requirement	Threshold	Objective
Resolution, medium-altitude overflight, visible	VIS NIIRS 6	VIS NIIRS 7
Resolution, medium-altitude overflight, IR	IR NIIRS 5	IR NIIRS 6
Resolution, medium-altitude stand-off, visible	VIS NIIRS 4	VIS NIIRS 5
Resolution, medium-altitude stand-off, IR	IR NIIRS 3	IR NIIRS 4
Resolution, high-altitude stand-off, visible	VIS NIIRS 5	VIS NIIRS 6
Resolution, high-altitude stand-off, IR	IR NIIRS 4	IR NIIRS 5
Operational availability (complete SHARP system)	0.70	0.85

Temperature stability: ±10°C per hour
Datalink range: >150 nm
Datalink transmission rate: >137 Mbps
Suspension: NATO standard 30 in lugs
Weight: 950 kg
MTBF: >1,000 h

Status
During 2002, Naval Air Systems Command tested a Raytheon SAR (a development of the Falcon SAR system) installed in a SHARP pod on a Northrop Grumman F-14, with the intention that the radar could be implemented as a pre-planned product improvement programme for SHARP pods equipping F/A-18s. The addition of a SAR will provide imagery from ranges of up to 50 n miles in poor weather conditions, enhancing all-weather reconnaissance and facilitating precision air-to-ground targeting. Potential export customers for SHARP include the RAAF under Project Air 5421.

The USN accepted its first prototype SHARP pod to be integrated on the F/A 18E/F Super Hornet in June 2002. A further four more prototypes followed in the Engineering and

Onboard the USS Nimitz , technicians access Bay 2 of a SHARP pod, mounted on the centreline station of an F/A-18F, which houses the Digital Data Storage (DSS) system   0557211

AN/ASD-12(V) SHARP mounted on the centre fuselage station of an F/A-18E Super Hornet (US Navy)   1041687

For details of the latest updates to *Jane's Avionics* online and to discover the additional information available exclusively to online subscribers please visit
jav.janes.com

Manufacturing Development (EMD) phase. In March 2003, Raytheon was awarded a contract for eight full-rate production SHARP pods, with the first 10 Low-Rate Production (LRP) units (out of an expected total order of 34) delivered in April 2003. During November 2004, a follow-on order for eight additional pods was received and, in July 2005, a further six pods (plus six Common Datalink suites and four data storage cartridges) for F/A 18E/F aircraft were requested. Delivery of this last batch is expected by March 2007.

SHARP achieved early operational capability with the F/A-18Fs operated by Carrier Air Wing 11 aboard the USS Nimitz, where it was used in combat during Operation Iraqi Freedom during 2003. Initial Operational Capability (IOC) was reached in 2004 on F/A-18Fs flying from the USS *Kitty Hawk*.

The introduction of SHARP has been accompanied by the deployment of complementary airborne and shipboard systems that can exploit its imagery. The USN implemented the Fast Tactical Imagery-II (FTI-II) photo-reconnaissance intelligence strike module, an enhanced variant of that which equips F-14s, aboard eight F/A-18Fs operating from the aircraft carriers USS Abraham Lincoln and Nimitz during Operation Iraqi Freedom.

The company has also offered the sensor in another pod suitable for carriage by German Tornado reconnaissance aircraft and a modular variant of the SAR (Mod SAR) system has been demonstrated for carriage by aircraft such as the Lockheed C-130 transport aircraft. A roll-on/roll-off version of the Mod SAR has been flown on a C-130, installed on a pallet and positioned in the door plug in place of the aircraft's rear left paradrop door. The sensor, which features a 24 in antenna and has a search range of 20 to 30 n miles in Ku-band and 40 to 60 n miles in X-band, can be installed or removed in less than 60 minutes. Described by company officials as a 'poor man's reconnaissance capability', the design could also be scaled down for carriage by platforms including business jets or a small aircraft such as the Beechcraft 200. The Asia-Pacific region has been identified as a potential market place for such an all-weather reconnaissance and surveillance concept, according to Lockheed Martin officials. The Mod SAR system could also be carried by UAVs, such as the General Atomics RQ-1 Predator and Northrop Grumman RQ-4A Global Hawk. The SAR payload is expected to have a unit cost of around USD5 million.

Contractor
Raytheon Technical Services Company.

AN/ASQ-228 Advanced Targeting Forward-Looking Infra-Red (ATFLIR) pod

Type
Airborne Electro Optic (EO) navigation and targeting system.

Description
The AN/ASQ-228 ATFLIR pod, a member of Raytheon's Terminator family of IR targeting pods (see separate entry), includes both infra-red targeting and navigation systems. One of the primary design aims for the ATFLIR system was to achieve sufficiently accurate long-range performance for F/A-18E/F crews to be able to deliver their air-to-ground weapons from beyond the range of defensive anti-aircraft artillery and many surface-to-air missile systems.

The AN/ASQ-228 is intended to replace three existing systems: the AN/AAS-38A/B targeting FLIR, AN/AAR-50 navigation FLIR and AN/ASQ-173 laser designator tracker/strike camera (see separate entries). The pod, approximately the same size as the AN/AAS-38 at 1.83 m long by 33 cm in diameter and weighing 191 kg (compared with over 350 kg for the AAS-38/AAR-50/ASQ-173 combination), is designed to function effectively in high temperatures and humidities.

The targeting FLIR uses the same third-generation Mid-Wave Infra-Red (MWIR) 640 × 480 staring focal plane array technology that has been used in the

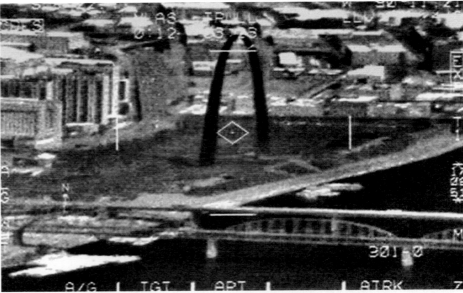

FLIR image of St Louis, taken from a AN/ASQ-228 ATFLIR pod, at a slant range of approximately 22 n miles. The increased performance of the MWIR sensor allows crews to identify targets at ranges beyond the capability of earlier systems such as BAE Systems' TIALD   0111413

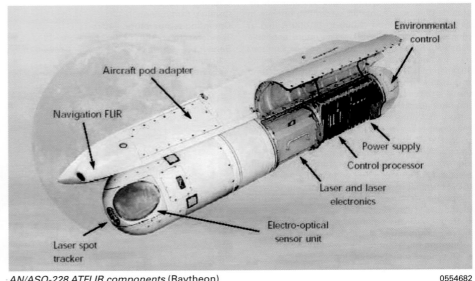

AN/ASQ-228 ATFLIR components (Raytheon)   0554682

US Marine Corps MV-22 Osprey (the AN/AAQ-16-27 system), providing up to ×30 zoom capability. In addition to the targeting FLIR, the system includes a laser range-finder and target designator (which incorporates a next-generation diode-pumped laser that has been demonstrated at altitudes of up to 50,000 ft), a laser spot tracker, a Charge Coupled Device (CCD) television camera and a NAVigation FLIR (NAVFLIR – provided by BAE Systems). The CCD camera incorporates a ×2 electronic zoom and provides approximately twice the resolution of the targeting FLIR. The sensors and laser share a common optical path with continuous autoboresight alignment. The three targeting sensors are mounted on an optical bench that also incorporates an Inertial Measurement Unit (IMU), with inputs from the host aircraft's Global Positioning System (GPS) providing pinpoint accuracy.

In defining the major difference between ATFLIR and other advanced targeting systems (such as Sniper XR and Litening AT+) – the key point is integration – ATFLIR communicates threat co-ordinates to other onboard sensors and works seamlessly with the host aircraft's radar, including the new APG-79 AESA of the F/A-18E/F Super Hornet. Using MIDS, it also supports communication with off-board sensors on both manned and unmanned platforms, expanding the systems scope into network-centric operations.

Specifications
Targeting FLIR
Focal plane: 640 × 480 InSb array
Spectral band: 3.7–5.0 μm (MWIR)
FoV: 6.0° (wide), 2.8° (medium), 0.7° (narrow)
Detection range: Up to 40 nm

Laser
Type: Diode pumped, high power
Modes: Operational, eyesafe
Altitude: Demonstrated up to 50,000 ft

Pod
Interface: MIL-STD-1553B
Reliability: >600 hr MTBF
Dimensions: 1,828.8 × 330.2 mm (L × Ø)
Weight: 191 kg

Status
In production and in service.

Raytheon began engineering and manufacturing development of the ATFLIR system in late 1997, with eight systems undergoing qualification and flight testing in 1999. These completed more than 250 flights during developmental and operational testing, including carrier-suitability trials and weapons testing.

The operational requirements document for the system calls for target identification and weapon release from at least 20,000 ft with 'sufficient accuracy' to deliver weapons such as the Joint Direct Attack Munition (JDAM) and Joint StandOff Weapon (JSOW). The ability to deliver laser-guided weapons from altitudes between 20 and 40,000 ft implies long standoff, which, in turn, demands target 'identification' (not to be confused with target detection) prior to weapon release (Rules of Engagement, RoE) and a high-power laser to deliver sufficient energy for the terminal seeker of the weapon to effectively 'see' the target.

Naval Air Systems Command awarded Raytheon a USD62.3 million contract in March 2001 for the

The AN/ASQ-228 ATFLIR pod, on the port chin station of an F/A-18C/D (BAE Systems) 1127368

first Low-Rate Initial Production (LRIP) batch, totaling 15 pods plus spares. In July 2002, a second LRIP contract, for 24 pods and worth around USD95 million, was awarded, with deliveries due to begin in December 2003. In August 2002, Raytheon was awarded a USD47 million contract for system components and spares, with work to be completed by the end of 2004.

The AN/ASQ-228 achieved early operational capability during 2002 with the first F/A-18E/F squadron, VFA-115, which deployed 12 aircraft aboard the aircraft carrier USS *Abraham Lincoln* on 24 July. This was described as an 'early operational capability', since it preceded the formal operational evaluation which began at the Naval Air Warfare Center Weapons Division, China Lake, California, later that year.

The system achieved Initial Operating Capability (IOC) with F/A-18E/F Super Hornets of US Navy squadron VFA-102, based at Lemore naval air station, California, in September 2003.

During Oct 2003, the pod achieved two significant program milestones – the successful completion of Operational Evaluation (OPEVAL)

tests and Initial Operational Capability (IOC) for US Navy and Marine Corps F/A-18 squadrons. During OPEVAL, ATFLIR was tested in a variety of tactical scenarios that simulated the operational environment in order to assess its operational effectiveness as a weapon system and its suitability to be maintained and operated by the Navy and Marine Corps. Testing included characteristics such as reliability, maintainability, interoperability and training. To achieve IOC, Raytheon delivered 10 Block 1 ATFLIR pods to the fleet and put in place training of aircrew, support personnel, maintenance support and logistical support to sustain operations throughout an extended deployment.

In December 2003, Raytheon announced it had been awarded a USD298.2 million contract from the Naval Air Systems Command (NAVAIR) for Full-Rate Production (FRP) of the AN/ASQ-228 pod for the US Navy and Marine Corps' F/A- 18 fleet. The contract, for FRP lots 1 and 2, calls for the production of 88 units, associated spares, special tooling and test equipment, and sustaining engineering support. Contract

deliveries are currently scheduled to be complete by November 2006.

As of January 2006, 100 pods had been delivered to the US Navy.

Raytheon is continuing to integrate new and advanced technologies into the ATFLIR pod. During 2006, a new laser marker system was installed together with a high-speed secure Ku-band datalink system. Current datalinks used in targeting pods rely on the C-band frequency, which is slow (600 Kbyte/s compared with the Ku-band capability of 10Mbyte/s. Growth provision for full duplex (two way communications) functionality along with a sophisticated encryption system to protect the flow of information has been planned. This will allow near instantaneous communications between aircrew and ground forces, as well as providing enhanced communications with shipboard command and control systems.

During March 2006, Raytheon announced that the US Navy had ordered a further batch of ATFLIR pods. Worth just under USD92 million, the contract will see the company produce a further 50 full-rate production ATFLIR pods for the navy's F/A-18C/D by November 2008.

Future planned enhancements include sensor fusion, automatic target recognition, an improved EO (CCD) camera and laser spot tracker, plus increased detection ranges.

Total production will reach 574 units (240 for F/A-18E/F Super Hornets and 334 for retrofit to F/A-18C/D aircraft), worth approximately USD1 billion, to equip the US Navy and Marine Corps. Traditionally, the USN has bought targeting pods in the ratio of one for every three operational aircraft; ATFLIR procurement is based on an allocation of 10 pods per 12 aircraft squadron and 12 for those with 14 aircraft. The plan includes pods for all three fleet-replacement squadrons, plus notional numbers for reserve units, together with 20 to form a pool of spares. ATFLIR will be authorised for international sales as it enters FRP.

Other potential customers include Canada, which has a requirement for a multifunction infra-red sensor to equip its CF-18s.

Raytheon is also developing a self-cooled version of the ATFLIR for possible use on the F-15, F-16, and Eurofighter; this version will remain 85 per cent common with the original version.

Contractor

Raytheon Company, Space and Airborne Systems.

AN/AVS-6(V)2 Night Vision Goggle (NVG)

Type

Aviators' Night Vision Imaging System (ANVIS).

Description

ITT's Aviator's Night Vision Imaging System (ANVIS) AN/AVS-6(V)2 is a Gen III Night Vision Goggle (NVG) with new features and improved performance compared to the previous (V)1 version. These include high-quality 25 mm eye-relief eyepieces (similar to the AN/AVS-9, see separate entry), independent eye-span adjustment for each monocular, smoother focusing, more stable mounting and increased fore and aft adjustment.

The AN/AVS-6 incorporates two high-resolution Gen III Image Intensifiers (II) with higher gain and increased photoresponse, yielding increased performance across the range over earlier models. Exact tube specification may vary, from OMNIBUS III-Plus (OMNI III+) to the latest OMNI V specification, according to customer requirements and US export restrictions.

The nominal 25-mm eye-relief and independent eye-span adjustment enables the AN/AVS-6 to accommodate a wide range of individual physical characteristics as well as eyeglasses. The system is made of Ultem engineering plastic with improved resistance to chemicals.

The two-piece (mount and binocular assemblies) NVG can be mounted to a variety

ITT's AN/AVS-6(V)2, distinguished from earlier models by 25 mm eyepieces and independent eyespan adjustment controls (either side of the bridge) 0131123

of helmets and is powered by a dual-battery power pack, integrated into the mount for fast-jet applications or in a separate unit attached to the rear of the helmet (facilitating some degree of counter balancing to alleviate fatigue) for helicopters. The NVG can be fully powered by either battery, providing for enhanced safety of the system via redundant supplies. Power packs use either universally available AA alkaline batteries or half-size military batteries.

Specifications
(OMNI IV specification)
Spectral response: 0.6 to 0.9 μm (NIR)
Field of view: 40°
Magnification: 1:1
Resolution: 1.3 cy/mrad
Gain: 5,500
Dioptre adjustment: +2 to –6
Interpupillary adjustment: 52 to 72 mm
Fore and aft adjustment: 27 mm, range
Tilt adjustment: 10°
Objective lens: EFL 27 mm F/1.23, T/1.35
Eyepiece lens: EFL 27 mm
Filter: Cut-off at 625 ηm
Exit pupil eye relief: on-axis: 14 mm at 25 mm distance; full-field: 6 mm at 25 mm distance
Battery type: AA size alkaline
Battery life: nominally 30 h under standard conditions
Weight of binocular: 590 g (mount not included)
Automatic breakaway: 11 to 15 g
Operating temperature range: –32 to +52°C

Status
In production and in service.

Contractor
ITT Night Vision.

AN/AVS-6(V)3 Night Vision Goggle (NVG)

Type
Aviator's Night Vision Imaging System (ANVIS).

Description
Northrop Grumman Electronic Systems (formerly Litton) produce the AN/AVS-6(V)3 Night Vision Goggle (NVG). While the physical characteristics of the two-piece design are identical to the ITT Night Vision AN/AVS-9 (F4949 – see separate entry), Northrop Grumman can supply the AVS-6(V)3 with its own Image Intensifier (II) tubes.

Aside from the installation of higher performance II tubes, Northrop Grumman has made significant design improvements to the AN/AVS-6(V)3 over the (V)2:
• 25mm eyepieces[1]
• More ergonomic interface controls[2]
• Dual-independent interpupillary adjustment[1]
• Enhanced mounting options ensuring AN/AVS-6(V)3 compatibility with virtually all flight helmets.
[1]included in later AN/AVS-6(V)2 goggles
[2]as AN/AVS-9

Northrop Grumman's new contoured design, Low-Profile Power Pack (LPPP) securely attaches to flight helmets using 'hook and pile' fasteners. Twin compartments accept two AA batteries each, for primary and alternate power. Additionally, the LPPP incorporates an aircraft power interface port, enabling the use of aircraft-supplied power for improved life-cycle cost savings and serves as an additional power source.

For customer requirements from Gen II High Definition (HD) and above, the design of the AN/AVS-6(V)3 allows for direct drop-in interface with all 18 mm MX-10160 series image intensifiers, facilitating upgrade to higher performing image intensifiers without requiring system replacement.

Specifications
(OMNI IV specification)
Spectral response: 0.6 to 0.9 μm (NIR)

The Editor's HGU-55/P fitted with a banana clip mount, AN/AVS-6(V)2 and LPPP attached to the rear of the helmet with a Velcro patch (E L Downs)   0580571

Field of view: 40°
Magnification: 1:1
Objective focus range: 41 cm to infinity
Resolution: 1.3 cy/mrad (typically 1.36)
Gain: 5,500 (minimum)
System distortion: <3%
Dioptre adjustment: +2 to –6
Interpupillary adjustment: 52 to 72 mm
Vertical adjustment: 25 mm min
Fore and aft adjustment: 27 mm min
Tilt adjustment: 10° min
Objective lens: EFL 27 mm F/1.23, T/1.35
Objective adjustment: 3/4 turn (338°)
Eyepiece lens: EFL 27 mm
Filter: Cut-off at 625 ηm (Class B)
Power: 2.0–3.0 V DC (100 mA, steady)
Low battery warning: 2.3 V
Weight of binocular (M949 assembly): 590 g (max)
Automatic breakaway of binocular assembly: 11–15 g (per specification)
Temperature range: –40 to +52°C

Status
In production and in service.

Contractor
Northrop Grumman Corporation, Electronic Systems Sector.

AN/AVS-9 (F4949) Night Vision Goggle (NVG)

Type
Aviator's Night Vision Imaging System (ANVIS).

Description
Developed in 1992 to meet specific requirements of the USN, ITT's Aviator's Night Vision Imaging System (ANVIS) AN/AVS-9 (USN designator F4949) is a Gen III Night Vision Goggle (NVG) offering improved performance and lighter weight than the earlier AN/AVS-6(V)2 (see separate entry).

Since the advent of the AN/AVS-6(V)2, performance margins between the two NVGs have narrowed considerably, with both goggles accepting a wide variety of Image Intensifiers (II) up to the latest OMNIBUS VI (OMNI VI) specification, according to customer requirements and US export restrictions.

Physical characteristics of the AN/AVS-9 are similar to those for the AN/AVS-6, with identical mount/power supply options for fast-jet and helicopter applications.

Two versions of the F4949 are available, according to application. The F4949G is designed for fighter and other applications where a Head-Up Display (HUD) is used – this version incorporates a 'notch filter', which enables a portion of the HUD energy to be 'seen' by the NVG. The F4949R is intended for helicopter or other applications which do not employ a HUD, therefore this NVG does not include the notch filter and offers marginally better performance.

Physically, rotary-wing versions of the AN/AVS-9 goggles feature a rear-mounted, low-profile battery pack allowing operation for more than 50 hours. Power is provided by a cable extending from the battery pack, over the helmet and into a connector in the mount. Fixed-wing versions feature a front-mounted battery pack that allows operation for more than 16 hours. An optional battery pack adapter enables connection of the fixed-wing version to the rear-mounted battery pack, providing the same operational time as the rotary version, but only when ejection is not a consideration.

Specifications
(OMNI IV specification)
Spectral response: 0.6 to 0.9 μm (NIR)
Field of view: 40°
Magnification: 1:1
Resolution: 1.3 cy/mrad (typically 1.36)
Gain: 5,500 (minimum)
Dioptre adjustment: +2 to –6
Interpupillary adjustment: 52 to 72 mm
Fore and aft adjustment: 27 mm, range
Tilt adjustment: 10°
Objective lens: EFL 27 mm F/1.23, T/1.35
Eyepiece lens: EFL 27 mm
Filter: Cut-off at 625 ηm (Class B)
Exit pupil eye relief: on-axis: 14 mm at 25 mm distance; full-field: 6 mm at 25 mm distance
Power: 3.0 V DC (100 mA, steady)
Cell Life: 10 to 22 hours (at 70°F, AA batteries, both cells used)
Weight of binocular: 550 g (max)
Mount weight: 250 gms (fast jet), 330 gms (helicopter)
Temperature range: –32 to +52°C

Status
In production and in service with all US air arms and numerous operators worldwide.

ITT's AN/AVS-9 is physically very similar to the AN/AVS-6(V)2, distinguished by the more pronounced objective focus adjustment rings on the front of each monocular, and identical to the AN/AVS-6(V)3

0131124

Contractor
ITT Night Vision.

AN/AVS-10 Wide Field Of View Night Vision Goggle (WFOVNVG)

Type
Aviator's Night Vision Imaging System (ANVIS).

Description
The Wide Field Of View Night Vision Goggle (WFOVNVG), also known as the Panoramic NVG (PNVG), employs a second pair of II tubes to increase the horizontal FoV of the WFOVNVG over the standard issue AV/AVS-9 item (see separate entry). In order to minimise weight and bulk, the system utilises the latest 16 mm II tubes and 'folds' the optical path of the outer II tubes. The WFOVNVG features a partially overlapping 100° × 40° (H × V) intensified FoV, where the central 30° of the horizontal field is binocular, while the flanking 35° portions are only seen by the corresponding eye. When viewing the outside

scene through the PNVG I, a thin demarcation line separating the binocular image from the flanking monocular peripheral images is seen. The WFOVNVG combination eyepieces are fixed and non-focussing (fixed dioptre), necessitated by the non-circular nature of the left and right FoV. Ongoing studies in the US have identified a 'standard' dioptre setting of −1.0 which is projected to be suitable for the vast majority of serving US airmen without need for additional corrective optics.

WFOV Variants
PNVG I
PNVG I, which is configured to fit onto the US HGU-55/P helmet, is a Type 2 (folded optics) design which features newly-developed 16 mm II tubes rather than the currently fielded 18 mm format items. Along with the goal of offering comparable performance to current baseline OMNI IV tubes, the II tube weight is reduced by almost 50 per cent. The 16 mm tubes have longer fibre optics on the outside optical channel than the inner optical channel. Dual-fixed eyepieces (tilted and fused) and four objective lenses (the inner two adjustable and the outer two fixed)

make up part of the folded optical approach. The inner optical channels include very fast f/1.17 objective lenses as compared with the f/1.25 objective lenses of the currently fielded AN/AVS-6 and AN/AVS-9 goggles. The outboard channels, due to size and weight constraints, incorporate f/1.30 objective lenses. The objective lenses incorporate Class B (Leaky Green) filters, ensuring compatibility with the latest colour cockpits and aircraft head-up displays. Eyepiece effective focal length is 24 mm while the design eye clearance has been optimised at 20 mm. A specially designed single left side and single right side power supply is remotely located but allows each side inner and outer optical channels to be controlled independently. The multiple adjustments of the standard AN/AVS series NVGs has been carried over to the PNVG, including tilt, independent Inter-Pupilary Distance (IPD), up/down, and fore/aft. Customised visors will also be incorporated into the overall design, ensuring cockpit compatibility, mechanical stability, and escape protection, with holes cut for the NVG objective lenses to protrude. A new latching mechanism affords one-handed on/off capability.

In the event of aircraft ejection, two 'AAA' alkaline batteries located in the Remote Electronics Module (REM) provide power (up to 16 hours) to the PNVG during the ejection sequence and later for ground-based Escape and Evasion (E and E) and Combat Survival And Rescue (CSAR).

Two configurations of the PNVG I will be built. The first, PNVG I-Configuration 1, features an REM, which attaches to a 'universal connector' (as used in the Visually Coupled Acquisition and Targeting System (VCATS) daytime helmet module). The universal connector provides aircraft data and power to the PNVG. This configuration also features a 640 × 480 Active Matrix Electro Luminescent display (AMEL) for symbology overlay, a magnetic head-tracker and an electronics package. PNVG I-Configuration 2, a stripped down version of Configuration 1, does not include an AMEL display, magnetic head-tracker, or electronics package. Since the majority of the HGU-55/P helmets are not equipped with the VCATS universal connector, a special banana clip mount has been designed that will accept the PNVG module on any HGU-55/P helmet. PNVG II

Another variant of the PNVG has been developed, designated PNVG II, which is a Type 1 (straight-through optics) design, suitable for transport aircraft and helicopters, where ejection load constraints on the NVG have not influenced the configuration. The partial overlap 100° × 40° (H × V) intensified FoV is maintained, but the system resembles currently fielded aviator NVGs. Whereas PNVG I currently mates only to the HGU-55/P helmet, PNVG II is compatible with any helmet that incorporates the standard ANVIS mounting bracket. Similarly to PNVG I, the central 30° × 40° (H × V) FoV is completely binocular, with no overlap outside this area. The dual fixed eyepieces, tilted and fused together, and four objective lenses (the inner two adjustable and the outer two fixed) are as for PNVG I. The non-folded inner optical channels are designed with extremely fast f/1.05 objective lenses, while the folded outer channels use PNVG I inner channel optics with f/1.17 objective lenses. Eyepiece effective focal length remains at 24 mm while the physical eye clearance has been increased to 27 mm. All of the mechanical adjustments currently available on the AN/AVS-6 and AN/AVS-9 remain (for example, tilt, independent IPD adjustment, up/down, fore/aft). Power for the PNVG II will be provided via the batteries that are currently integrated with the AN/AVS-6/9 mounting systems. Various configurations of PNVG II were developed, including a 640 × 480 AMEL display and Class A/B filters.

IPNVG
In April of 2000, a follow-on development effort was initiated, designated the Integrated Panoramic Night Vision Goggle (IPNVG). Lessons learned from the PNVG effort were incorporated into the IPNVG programme, with additional specific objectives, including wide field-of-view,

A pilot of the 422nd Test and Evaluation Squadron at Nellis Air Force Base testing a set of IPNVG/PNVG II in a A-10 Thunderbolt II. The A-10 was the first aircraft type to receive the AN/AVS-10, employing it in recent night operations (US Air Force)

1127358

integrated laser protection, fit/comfort, image quality, integrated symbology/imagery display, field support, ejection/crash/ground egress safety, compatibility with existing systems, affordability, supportability, maintainability, production and reliability.

The prototype IPNVG has a 95° horizontal FoV and a 38° vertical FoV. In common with the PNVG II, the eyepieces are fixed, therefore, in order to compensate for the loss of dioptre adjustment, the IPNVG eyepieces are fixed to −1.0 D, with two other dioptre settings (−0.25 D, and −2.0 D) available via snap-on lenses. The snap-on lenses clip over each ocular of the IPNVG.

The IPNVG also has an integral wraparound power supply housed in the image intensifier tube itself that is insulated for protection against environmental elements. The IPNVG design affords interchangeability of the image intensifier tubes and facilitates easy maintenance of the goggles.

In a parallel effort, in September 2001, eMagin Corporation was selected by the US Air Force to supply its high-resolution active matrix Organic Light Emitting Diode (OLED) microdisplays for a number of military display applications, including the Strike Helmet 21 system that uses the Integrated Panoramic Night Vision Goggle (IPNVG), slated for F-15E operations in the 2003–2004 timeframe. Programme aims included energy efficient brightness levels of over 30,000 candelas per square meter (cd/m2). This see-through capability facilitates superimposition of data and information over a real-world scene, enhancing navigation, targeting, and Situational Awareness (SA) for the pilot, even in conditions of bright sunlight.

Specifications
PNVG II Baseline
II Tube diameter: 16 mm
Spectral response: 0.6 to 0.9 µm (NIR)
Field of view: 100° × 40° (H × V)
Magnification: 1:1
Resolution: 1.3 cy/mrad or better
Gain: 5,500 or better
Dioptre adjustment: Fixed dioptre (−1.0)
Inter Pupillary Distance (IPD) adjustment: 52 to 72 mm
Fore and aft adjustment: 27 mm, range
Tilt adjustment: 10°
Objective lens: EFL 27 mm F/1.23, T/1.35
Eyepiece lens: EFL 27 mm
Filter: Cut-off at 625/665 ηm, according to application
Exit pupil eye relief: on-axis: 14 mm at 25 mm distance; full-field: 6 mm at 25 mm distance
Battery type: AA size alkaline
Battery life: nominally 30 h under standard conditions
Weight of binocular: 590 g (mount not included)
Automatic breakaway: 11 to 15 g
Operating temperature range: −32 to +52°C

Status
In July of 1999 the Air Force Research Laboratory (AFRL) took delivery of the last of twelve Panoramic Night Vision Goggle (PNVG) systems, which were evaluated on a wide variety of fixed- and rotary-wing platforms, including F-15C, F-15E, F-16, A-10, F-117, C-17, C-5, C-130, and HH-60.

During 2003, a production contract was awarded by the US Air Force to Insight Technology, Inc for a WFOVNVG based on the IPNVG.

The long-term goal of the programme is to refine the baseline WFOVNVG/PNVG with additional capability at defined increments, designated 'spiral' developments, thus creating an Integrated Panoramic Night Vision Goggle (IPNVG) with expanded capability. The levels of spiral development were originally defined as:
- A baseline (non-ejection mount) PNVG (II) that mounts to the current helmet with a standard AN/AVS-9 mount. This commenced qualification testing during 2002, with initial production deliveries from the end of 2004.
- The first spiral development is an ejection-capable PNVG that follows the baseline PNVG by approximately six months. It is believed that this Type 2 development may have been postponed.

- The second spiral development incorporates data display capability.
- The third spiral development incorporates weapon and sensor cueing ability.

The spirals are scheduled to be ready for qualification testing by the first quarter of FY05 and are for PNVG development only, not aircraft or helmet upgrades.

As of early 2007, (I)PNVG spiral development was ongoing, with US A-10 pilots first to receive the baseline AN/AVS-10.

Contractor
Aeronautical Systems Centre, Wright-Patterson AFB (Programme Manager).
Insight Technology Inc (WFOVNVG).
ITT Night Vision (II tubes).

Arrowhead M-TADS/PNVS

Type
Airborne Electro Optic (EO) navigation and targeting system.

Description
During the last quarter of 2000, the US Army selected a new sensor system intended to give the AH-64D significantly greater night vision and targeting capabilities over its standard Target Acquisition Designation Sight/Pilot Night Vision Sensor (TADS/PNVS). A USD78.5 million development contract was awarded to Team Apache Systems (TAS), a limited liability company comprising Boeing and Lockheed Martin, for the Arrowhead advanced targeting and navigation system. Drawing extensively on technology integrated into the RAH-66 Comanche Electro-Optic Sensor System (EOSS), the new sensor suite is a combination of improvements to the basic TADS/PNVS utilising elements of the EOSS, providing improved performance and more effective integration with the Apache weapon system. The new system's official designation is Modernised TADS/PNVS, or M-TADS/PNVS.

While functionally similar to the basic system, M-TADS/PNVS offers a performance improvement of nearly 100 per cent, with life cycle costs halved.

The core of the Arrowhead upgrade is a DRS Technologies 8 to 12 µm detector, which uses the Standard Advanced Dewar Assembly II (SADA II) detector with a 480 × 4 Cadmium Mercury Telluride (CMT) Focal Plane Array (FPA). This new Second Generation (see analysis section for definitions) LWIR (8–12 µm) scanning sensor, as a result of

M-TADS/PNVS on a US AH-64 Longbow Apache. Compare this image with that for the standard system – the main external difference between the two is the different configuration of the PNVS sensor on the upper gimbal.
(Lockheed Martin)   1128658

The TADS Electronic Display and Control (TEDAC) provides a marked improvement in the operator interface for the TADS
(Lockheed Martin)   0569784

more information gathered during each scan, will enable crews to detect, acquire and identify targets at greater ranges. Arrowhead's digital video output enhances recording capability and facilitates still-frame video imagery transmission to the ground commander or to other aircraft during normal operations.

The M-TADS/PNVS sensors are designed to increase the lethality and survivability of the AH-64 Apache, with enhanced capabilities to detect, identify and engage targets at greater ranges. It will allow target cueing from the Longbow fire-control radar or radar frequency interferometer for rapid target identification. The new sensor suite also provides for improved image quality and increased effectiveness in day/night and adverse weather operations. Coupled with an improved Integrated Helmet and Display Sighting System (IHADSS – see separate entry), the pilot will have greatly enhanced nap-of-the-earth capability with built-in growth for a wide (30 × 52°) FoV helmet display, although, at present, the IHADSS FoV is limited by the FLIR FoV (30 × 40°).

With the upper portion of the rotating turret assembly housing the PNVS, the lower turret mounts the Arrowhead Targeting System FLIR capable of a 40 per cent increase in targeting range over the standard system and an improved CCD TV camera for daylight viewing. Improvements to the laser ranging/target marking system are derived from the Comanche project. Direct-view optics are eliminated, thus enhancing pilot survivability, by the replacement of the standard optical relay tube with the new TADS Electronic Display and Control (TEDAC), which integrates a 5 × 5 in Active Matrix Liquid Crystal Display (AMLCD), increasing useful space in the cockpit and reducing maintenance action by 90 per cent.

Status
The Arrowhead MWIR sensor was proposed to the US Army as an option for a TADS/PNVS modernisation competition conducted during 2000, but the option was not exercised at contract award due to funding constraints. The full Arrowhead system engineering, manufacturing and development phase began during the fourth quarter of 2000. In January 2004, Boeing announced that an Apache upgraded with the Arrowhead M-TADS/PNVS had logged its first flight in November 2003 and was conducting further flight testing.

The test programme followed a government-industry agreement leading up to a USD260 million US government contract for Lot 1 production contract in November 2003. This covered 19 Arrowhead units to be field-retrofitted on to US Army Apache Longbows, with first units operating from June 2005 installed on the final Apache Longbows being built under a multi-year contract with the US Army. The Lot 1 agreement also covered integration of the Arrowhead on Apache Longbows already in service in both the Block I and Block II configurations, together with units for 36 Apache Longbows being built for international customers.

The Lot 2 contract, for 97 systems for the US Army and foreign customers, was awarded in January 2005 with deliveries beginning in July 2006. The Lot 3 contract was expected in the June/July 2006 timeframe. Formal delivery of the first Arrowhead system to US Army was on 10 May 2005. The US Army plans to equip all of its 704 AH-64Ds with Arrowhead by 2011.

The UK Ministry of Defence ordered the Arrowhead M-TADS/PNVS for the British Army Air Corps Apache AH.1 fleet in May 2005. Flight trials will begin in 2007, with first delivery in January 2009 and completion of the the retrofit to all 67 Apaches by the end of 2010.

The Royal Netherlands Air Force has also indicated its intention to retrofit Arrowhead M-TADS/PNVS to its AH-64D fleet.

A variant of the PNVS, designated Pathfinder, is offered as an upgrade to cargo/utility helicopters to enhance flight operations in 'brown-out' and 'white-out' conditions and provide greater pilot Situational Awareness (SA) during night and/or adverse weather conditions.

Contractor
Lockheed Martin Missiles and Fire Control.

ATARS Advanced Tactical Airborne Reconnaissance System

Type
Airborne reconnaissance system.

Description
The Advanced Tactical Airborne Reconnaissance System (ATARS), developed for US Navy and US Marine Corps F/A-18D Hornet aircraft, provides an integral tactical reconnaissance capability for the aircraft without recourse to an external role-fit pod, as for the F-14 TARPS system (see separate entry). The ATARS system is mounted in the nose of the aircraft, with the 20 mm cannon removed in order to accommodate the sensor package. The gun openings in the nose are blocked off with a blank plate and different front nose undercarriage doors are installed, with a wiring kit to accommodate the new electronics. Aerodynamic limits and flying qualities are said to be unchanged from a standard F/A-18D. Interestingly, the signature 'false canopy' on the underside of the cockpit is omitted due to the sensor suite.

The ATARS has three sensors, the LAEO (Low-Altitude Electro-Optical), the MAEO (Medium-Altitude Electro-Optical) and the IRLS (Infra-Red Line Scanner). The LAEO and IRLS are used primarily for altitudes of between 200 and 3,000 ft, while the MAEO is used between 3,000 and 20,000 ft. The MAEO sensor allows the aircrew to obtain high-resolution imagery from 3 to 5 miles stand-off, without direct overflight of targets in high-threat areas.

The Reconnaissance Management System (RMS) provides control of the sensors, recorders and datalink, and manages the flow of data from the sensors to the digital recorder, and from the datalink to the ground station and cockpit displays. The system can store 12 preplanned point, strip or area targets and 20 targets of opportunity.

Two 19 mm tapes are loaded into the aircraft externally for the recording of SAR (Synthetic Aperture Radar) information, EO or IRLS data, and are set to record either SAR or ATARS (LAEO, MAEO, IRLS) information prior to take-off. Tape capacity is reported as 30 min (SAR) and 45 min (ATARS). The ATARS F/A-18D can also carry a DataLink (DL) pod, on the centreline station, which facilitates transmission of reconnaissance imagery to airborne or ground-based Command and Control (C²) for

ATARS Sensor Suite

BAE Systems' ATARS sensor suite   0079784

A USMC pilot pre-flights the ATARS system on a F/A-18D test aircraft (US Navy)   0083496

A F/A-18D aircraft equipped with ATARS – note the covers on the sensor windows under the nose of the aircraft   0137818

near-realtime sensor-to-shooter applications or bomb damage assessment. In addition to sending the information using the DL, the recording tapes are used concurrently to ensure that no data is lost.

Specifications

Low-Altitude Electro-Optical (LAEO) sensor:
200–3,000 ft above ground level
Field of view: 138°
Vertical or forward oblique
High-resolution, dawn-to-dusk below the weather, high-speed sensor
Medium-Altitude Electro-Optical (MAEO) sensor:
3,000–20,000 ft above ground level
Field of view: 22°
Horizon-to-horizon field of regard
High-resolution, medium-range standoff sensor for daylight operations in high-threat environments
Infra-Red LineScanner (IRLS) sensor:
200–10,000 ft above ground level
140° or 70° FoV
8–12 µm waveband
High-resolution, low- to medium-altitude day/ night reconnaissance capability

Status

Designed for internal fitment in US Navy and US Marine Corps (USMC) F/A-18D aircraft, ATARS can be configured for internal or pod fit; current production is for internal fit to US Marine Corps F/A-18D aircraft, and pod fit for US Air Force F-16 aircraft (when it is designated TARS – Theatre Airborne Reconnaissance System).

Delivery of the first sensor-equipped aircraft was during 1992, to MCAS El Toro for trials and evaluation. A USD50 million contract was awarded to Boeing in early 1997 for the first four Lockheed Martin Fairchild Systems AN/ASD-10(V) ATARS units comprising the initial phase of low-rate production. A second phase followed, with four more units ordered in FY99 at total cost of USD23 million and five in FY00 at a cost of USD35.3 million. ATARS became operational on F/A-18D aircraft of VMFA(AW)-332 of the US Marine Corps and deployed during Operation Allied Force. The final unit to receive the upgrade was VMFA(AW)-242 at MCAS Miramar in 2002. Each of the USMC's six night-attack VMFA(AW) F/A-18D units now have three ATARS-equipped aircraft.

During the first quarter of 2004, BAE Systems' Information and Electronic Systems Integration Division announced that it would be integrating a solid-state digital recorder into ATARS for US Navy F/A-18 aircraft under a USD9.5 million contract from US Naval Air Systems Command. The project was due for completion in June 2005.

Contractor

BAE Systems North America, C⁴ISR Systems.

BRITEStar SAFIRE™ imaging sensor/target designator

Type

Airborne Electro-Optic (EO) surveillance system.

Description

Developed from the Star SAFIRE™, The BRITEStar SAFIRE™ combines a high-resolution 3–5 µm indium antimonide (InSb) Focal Plane Array (FPA) IR imager, a CCD TV camera and an eye safe laser designator/range finder. The laser designator is fully compatible with all Band 1 and tri-service PRF and PIM codes. The integral boresight module automatically aligns the Thermal Imager (TI) and TV sensors to the centroid of the laser spot.

Options include an autotracker, laser spot tracker, MIL-STD-1553B, ARINC and RS-232/422 interfaces, digital video output and a Night Vision Goggle compatible laser illuminator.

BRITEStar SAFIRE™ applications include Maritime Patrol, Search and Rescue, Anti-Surface Warfare (AsuW), Anti-Submarine Warfare (ASW), surveillance and reconnaissance.

BRITEStar SAFIRE™ mounted on a Bell UH-1Y   0111622

BRITEStar II turret on an ARH-70A helicopter at the 2006 Farnborough International Air Show (Patrick Allen)   1132881

The system is engineered to be easily configured for inverted or upright mounting on aircraft or surface platforms.

Developed from the (renamed) BRITEStar I, BRITE Star II is a derivative of the original system, incorporating the latest advances in sensor and processing technology.

Specifications
BRITEStar I
Turret
Dimensions: 490 × Ø411 mm
Weight: 51.2 kg
Stabilisation: 4 axis
Azimuth coverage: 360° continuous
Elevation: +32 to –100°
Control: HCU, serial digital
Environmental: MIL-STD-810E, MIL-STD-461D Class A1B

Thermal Imager
Spectral band: 3–5 µm
Detector: 320 × 240 InSb FPA
Resolution: microscanned to 640 × 480
Field of view:
(narrow) 0.8 × 0.6°
(medium) 3.4 × 2.6°
(wide) 24.9 × 18.7°

TV sensor
Type: Monochrome CCD
Field of view:
(narrow) 0.8 × 0.6°
(medium) 3.4 × 2.6°
(wide) 24.9 × 18.7°
Resolution: 768 × 494
Laser designator/range finder
Designator type: NdYAG, 1.06 µm, Class 4
Range finder type: Class 1 eye safe, 1.57 µm
Interfaces: MIL-STD-1553B, RS-232, RS-422, ARINC 429

BRITEStar SAFIRE™   0089857

Status

In production and in service. BRITEStar I systems are installed onboard US Marine Corps UH-1N helicopters, which were deployed in-theatre during Operation 'Iraqi Freedom' in 2003. UH-1N helicopters fitted with the system were operational in Afghanistan and Iraq during 2006.

During July 2006, Bell Helicopter and FLIR Systems jointly announced that Bell Helicopter had selected BRITEStar II as the Target Acquisition Sensor Suite (TASS) for the US Army's Armed Reconnaissance Helicopter (ARH-70A) Programme.

Contractor

FLIR Systems Inc.

CA-260 digital framing reconnaissance camera

Type
Airborne reconnaissance system.

Description
The CA-260 digital framing camera is designed specifically to provide near-photographic quality images while enhancing the survivability of the tactical reconnaissance platform at low to medium altitudes. It is configured for external pod or internal aircraft mounting on a wide variety of reconnaissance platforms.

Advanced digital framing technology reduces the amount of time required to cover a target area compared with conventional E-O linescan sensors. Wafer-scale processing has been used to develop the 4 M pixel or 25 M pixel array CCDs which are used in the CA-260 camera to give a wide field of view E-O image with continuous stereo coverage of targets, at 56 per cent overlap. An on-chip motion compensation architecture eliminates the image blur normally associated with a wide field of view framing camera.

Because the mounting configuration and physical envelope are identical to that of the KS-87 film camera, the CA-260/4 or CA 260/25 can provide an interim E-O framing capability to users of existing KS-87 cameras.

Specifications
Lens: 1.5 in fl, f/4.5; 3.0 in fl, f/4.5; 6.0 in fl, f/4.0; 12.0 in fl, f/4.0
Field of view:
(CA-260/4)
35.8° with 1.5 in fl lens; 18.3° with 3.0 in fl lens; 9.2° with 6.0 in fl lens; 4.6° with 12.0 in fl lens
(CA-260/25)
76.9° with 1.5 in fl lens; 43.3° with 3.0 in fl lens; 22.4° with 6.0 in fl lens; 11.3° with 12.0 in fl lens

CA-260/25 25 Mega pixel imagery using 7:1 data compression and ×16 magnification   0018321

Recon/Optical, Inc CA-260 E-O framing reconnaissance camera   0018322

An image taken from a CA-236 digital framing camera 11 minutes after sunset from a TARPS F-14 aircraft   0005562

Pixel size (pitch): 0.012 × 0.012 mm
Frame rate: 2.5 frames/s max
Dimensions:
(camera) 175.3 × 261.6 × 401.3 mm
(power supply) 134.6 × 274.3 × 464.8 mm
Weight:
(camera) 27.27 kg (without lens)
(power supply) 8.18 kg
Power: 115 V AC, 400 Hz, 210 VA; 28 V DC, 140 W

Status
In service. Successful engineering tests were completed in July 1993. RF-4C and F-14 TARPS (see separate entry) demonstrations were completed in 1994. Subsequently, the CA-260 supplanted the KS-87 framing camera in the TARPS-DI pod, used extensively by the US Navy on F-14 Tomcats in support of operations in the Persian Gulf and over Bosnia and, most recently, Afghanistan. The CA-260/4 is in service on US Air Force/Air National Guard F-16s. The CA-260/25 is in service in the US Air Force Theatre Airborne Reconnaissance System (TARS) Programme, and for Royal Danish Air Force F-16 aircraft.

Contractor
Recon/Optical, Inc.

This clear × 30 enlargement illustrates the effect of on-chip motion compensation to reduce blurring in WFoV applications   0014359

CA-261 digital step framing camera

Type
Airborne reconnaissance system.

Description
The CA-261 digital step framing camera is designed specifically to provide photographic-quality images while enhancing the survivability of the tactical reconnaissance platform at medium altitudes. It is configured for external pod or internal aircraft mounting on a wide variety of reconnaissance platforms.

The camera combines the proven performance of the CA-260 25-Mpixel digital framing camera with the stepping capability of a proven two-axis stabilised step head with de-rotation prism. This combination produces a system that captures a series of 25-Mpixel images through a 12-in focal length lens in the cross track direction. These images allow for wide area coverage of up to 180° in a digital framing format.

The captured imagery can be displayed in mosaic form to provide a 'birds eye' view of the entire area of interest. This display retains

CA-261 flight test image: 12 in lens in f1 at an altitude of 20,245 ft and 44,990 ft slant range   0023181

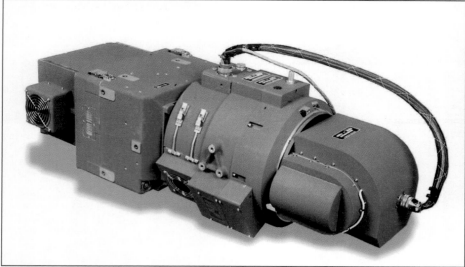

CA-261 digital step framing camera   0125120

Enlargement of a section of the left image using 7:1 data compression   0023180

the capability to manipulate individual images in the same manner as the CA-260 (see separate entry). De-rotation optics are employed to eliminate image rotation. The high resolution of the CA-260 is maintained by incorporating stabilisation electronics and software developed by Recon/Optical, Inc. This combination yields residual rates <.001 rad/s even at disturbance input rates in excess of 10°/s.

Advanced digital step framing technology reduces the amount of time required to cover a target area compared with conventional Electro-Optic (EO) linescan sensors. Recon/Optical, Inc employed wafer-scale processing to develop the 25-Mpixel array CCDs used

in the CA-260 to give a wide field of view image. An on-chip motion compensation architecture eliminates the image blur normally associated with a wide field of view framing camera.

Four modes of operation are available:
- Mode 1: Automatic. Maximum across-track coverage centred on a selected depression angle
- Mode 2: N-Select. Operator-selected number of steps and depression angle
- Mode 3: Stereo. Maximum across-track coverage with at least 56 per cent forward overlap
- Mode 4: Spot. Multi-aspect imaging of a specific target.

Specifications
Dimensions: 356 × 406 × 965 mm
Weight: 76.4 kg
Power supply: 28 V DC, 115 V AC, 400 Hz
Lens: 12 in (304.8 mm) f/6 or 18 in (457.2 mm) f/8
Field of view: 11.3 × 11.3° (12 in), 7.6 × 7.6° (18 in), each frame
Field of regard: 180 (in cross track) × 11.3° (in line of flight)
Frame rate: 2.5 frames/s
Angular resolution: 39.4 µrad/pixel (12 in), 26.2 µrad/pixel (18 in)
Max V/R: >0.44 rad/s (12 in), >0.30 rad/s (18 in)

Status
As of 2006, in production and in service with the air forces of several countries.

Contractor
Recon/Optical, Inc.

CA-265 IR digital framing camera

Type
Airborne reconnaissance system.

Description
The CA-265 IR digital framing camera is designed for low- to medium-altitude operation in high-speed tactical aircraft, as a replacement camera for earlier conventional, downward-looking Infra-Red Line Scanners (IRLS). It features vertical, forward-oblique, or side-oblique mounting. Design characteristics include: a very large frame of view; 1,968 × 1,968 Platinum Silicide (PtSi) (3–5 micron) focal plane array with patented on-chip Forward Motion Compensation (FMC); near-realtime data availability with digital output; high-resolution, long-range IR imaging; electronic exposure control for wide dynamic range and imaging during manoeuvre.

Specifications
Aircraft speed: 100–580 kt
Aircraft altitude: 500–50,000 ft (200 ft with optional lens)
Lens focal length/f number: 12 in (304.8 mm), f/2; option 6 in (152.4 mm), f/2
FoV/frame: 11.6 × 11.6° (12 in, f/2), 21.9 × 21.9° (6 in f/2)
Operating waveband: 3–5 µm
Focal plane array: 1,968 × 1,968 PtSi
Frame rate: 2.5 frames/s
Exposure time: variable
Recorder interface: Ampex DCRsi-240 or solid-state recorder
Recorder output format: ROI standard, non-proprietary with annotation and digital sub-stamped image support
Video output (optional): RS-170
Control interface: RS-232 serial port
Dimensions:
  ISU without lens: 312.4 × 325.1 × 368.3 mm (L × W × D)

Recon/Optical CA-265 IR digital framing camera   0051039

VDPU: 558.8 × 259.1 × 292.1 mm (L × W × D)
power supply: 345.5 × 302.3 × 91.5 mm (L × W × D)

Status
In production and in service. Flight tested by the US Navy on a P-3 aircraft in October 1998.

Contractor
Recon/Optical, Inc.

CA-270 dual-band digital framing camera

Type
Airborne reconnaissance system.

Description
The CA-270 dual-band (visible and IR) digital framing camera is designed to meet the demands of modern tactical reconnaissance missions.

The CA-270 combines digital framing technology with patented electronic, on-chip graded image motion compensation to produce

high-resolution reconnaissance imagery. Two imaging modules, one for the visible and one for the infra-red spectrum, provide the dual-band imaging capability. The images can be recorded on-board the aircraft on a digital data recorder, or transmitted directly to a ground station via a datalink.

Specifications
Operating spectra:
(visible/near infra-red) IR 515–900 µm
(infra-red) 3.0–5.0 µm
Lens type:
(both spectra) refractive
(obscuration) none
Lens focal length and/number:
(visible spectrum): 12.0 in (304.8 mm), f/6 or 18 in (457.2 mm), f/8
(infra-red spectrum) 12.0 in (304.8 mm), f/3.5 or 18 in (457.2 mm), f/8
Detector arrays:
(visible spectrum): 25.4 megapixel Silicon CCD, squared format, 10 µm pixel pitch
(infra-red spectrum): 4.1 megapixel array, MWIR, Indium Antimonide (InSb), square format, 25 µm pixel pitch
Operating modes: visible alone, infra-red alone, visible/infra-red simultaneously
Field of view/frame:
(visible spectrum): 6.3 × 6.3°
(infra-red spectrum): 6.3 × 6.3°
Frame rates: both spectra: up to 4 frames/s
Scan coverage rates:
(visible alone): 24°/s XLOF
(infra-red spectrum): 24°/s XLOF
Maximum V/H rates : 0.40 rad/s (0.226 kt/ft)
FMC: both spectra: on-chip, graded
Stabilisation: two-axis stabilised, roll and azimuth, roll axis <1.0 mr/s residual against a 10°/s roll disturbance

Status
In production and in service. The CA-270 has been designed to meet the requirements of the US Navy's Super Hornet SHAred

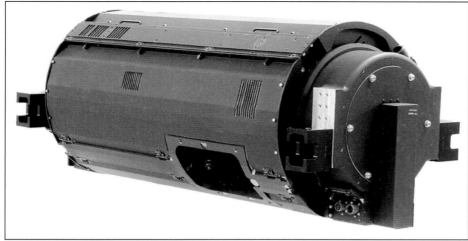

CA-270 dual-band framing camera (Recon/Optical)   1176154

Reconnaissance Pod (SHARP) programme (where it is designated CA-279/M). Also installed in the Swedish SPK 39 pod on the JAS 39 Gripen.

Contractor
Recon/Optical, Inc.

CA-295 dual-band digital framing camera

Type
Airborne reconnaissance system.

Description
The CA-295 is a dual-band (visible and IR), long-range oblique step framing, digital reconnaissance camera, designed for medium- and high-altitude application. It provides high-resolution, single- or simultaneous dual-band coverage and stereoscopic imagery for advanced targeting systems and enhanced photo interpretation.

The camera features an actively stabilised optical system and a common aperture, allowing both the visible and IR channels to use the same primary optical element, facilitating precise harmonisation between spectral bands. A choice of relay optics in the visible channel provides a selection of camera focal lengths, giving the CA-295 the ability to meet a variety of mission requirements within a standard system design.

The CA-295 comprises three major subassemblies. The Stabilised Imaging Unit (SIU) contains both visible and IR imaging modules, long-range precision optics, common camera and stabilisation electronics and power supplies. The Image Processing Unit (IPU) contains the system controller electronics, I/O interfaces and a scalable set of image processing boards for formatting and processing both IR and visible image data. The Power Conversion Unit (PCU) connects to the aircraft power system and supplies filtered, switched power to the SIU and IPU.

The CA-295 can provide precision pointing and target location using an integrated Inertial Navigation System/Global Positioning System (INS/GPS). This capability enables highly accurate georegistration of images and enhances the system's ability to generate three-dimensional data.

Specifications
FoR: horizon-to-horizon (with suitable window)
Coverage modes: selectable single- or simultaneous dual-band wide area, stereo, area, multi-aspect spot
Wavebands: 0.5 to 0.9 μm (visible/NIR), 3.0 to 5.0 μm (MWIR)
Detector focal plane array/format/pitch: 25 Mpixel/5,040 × 5,040/10 μm (visible/NIR), 4 Mpixel/2,016 × 2,016/25 μm (MWIR)
Stabilisation: Active two-axis Fibre Optic Gyro (FOG); passive isolation
Frame rate: 4.0 frames/s
Maximum V/H rate: 0.084 rad/s (0.048 kt/ft), 84-in;
0.14 rad/s (0.081 kt/ft), 50-in
Lens:
Visible/NIR: 50 in (1,270 mm), f/4, 72 in (1,829 mm) f/5.8, 84 in (2,134 mm) f/6.7
MWIR: 50 in (1,270 mm), f/4
FoV (frame): 2.27 × 2.27° (50 in), 1.58 × 1.58° (72 in), 1.35 × 1.35° (84 in)
FMC: On-chip, electronic, graded on both channels

Status
In production and in service. Selected for the US Navy SHAred Reconnaissance Pod (SHARED, see separate entry), where it is designated CA279/H.

Contractor
Recon/Optical, Inc.

CA-880 reconnaissance pod

Type
Airborne reconnaissance system.

Description
The CA-880 is designed for tactical and strategic LOng-Range Oblique Photographic (LOROP) reconnaissance missions. An oblique KS- 146B camera is installed in a modified fuel tank which has been certified for centreline carriage on the RF-4, F-4E, Mirage III, and Mirage V. The system also includes cockpit control and status panels, left and right oblique sights for camera pointing, and a master power distribution unit.

The camera features a 1,676 mm (66 in) focal length, f/5.6 lens, passive and active stabilisation, forward motion compensation, automatic exposure control, autocollimation for focus optimisation, manual and automatic pointing control, pod and camera thermal control systems and built-in test functionality. Either Electro-Optical (EO framing or scanning Image Sensor Units (ISUs) can be used.

Specifications
Weight: 666 kg
Power: 115 V AC, 400 Hz 3 phase, 28 V DC
FoV: 3.9° in line of flight, 3.9, 7.5, 11.0, 14.5 or 21.6° across line of flight
Film length: 305 m
Cycle rate: 0.45 cycles/s max
Exposure time: 1/30 to 1/1,500 s
Overlap: 12 to 56%

Status
As of early 2006, in production and in service.

Contractor
Recon/Optical, Inc.

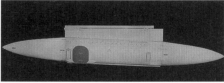

Recon/Optical, Inc CA-880 LOng-Range Oblique Photographic system (LOROP)   0505262

CA-890 tactical reconnaissance pod

Type
Airborne reconnaissance system.

Description
The CA-890 tactical reconnaissance pod is designed for operation at altitudes of 60 to 9,144 m, employing one to three sensors singly or simultaneously. The nose section of the pod houses a KS- 153A tri-lens camera on a forward-oblique or vertical in-flight rotatable mount. The centre section contains a KA-95B panoramic camera configured for 40,90 and 140° vertical scans and 40° left and right oblique scans. The tail section houses an infra-red linescanner, incorporating multiple selectable vertical fields of view.

The pod is a modified 1,400 litre (370 gallon) fuel tank that has been flight certified for centreline carriage on the F-4E. The environmental control system is integral to the pod and controls the nose and centre section. The tail section is open to the atmosphere for the IRLS receiver. The system also includes cockpit control and status panels and a master power distribution unit.

The Recon/Optical, Inc CA-890 tactical reconnaissance pod   0505263

CA-295 dual-band digital framing camera (Recon/Optical)   1176155

Specifications
Weight: 544 kg
Power: 115 V AC, 400 Hz 3 phase, 28 V DC

Status
In service.

Contractor
Recon/Optical, Inc.

Dark Star laser designator is in service on F-117A stealth fighters   0051017

Dark Star laser designator

Type
Airborne Laser Range Finder/Target Designator (LRFTD).

Description
The Dark Star laser designator is designed to provide accurate, first-strike capability with all types of laser-guided weapons. The two-assembly system has been optimised for low volume and weight, and includes transmitter and high-energy converter line-replaceable modules. Incorporating a newly-developed high-brightness resonator, this Nd:YAG (1.064 µm) air-cooled equipment provides state-of-the-art beam divergence at high efficiency.

Specifications
Wavelength: 1.064 µm
Power: 115 V AC, 3-phase, 400 Hz, 250 W; ±15 V DC, 20 W
Operating temperature: –10 to +47.5°C
Cooling: forced ambient air
Dimensions:
  transmitter: 292 × 292 × 61 mm
  HEC: 216 × 145 × 112 mm

Weight:
  transmitter: 5.0 kg
  HEC: 2.0 kg

Status
In service on the F-117A stealth fighter.

Contractor
Northrop Grumman Corporation, Electronic Systems, Laser Systems.

DB-110 Dual-Band reconnaissance system

Type
Airborne reconnaissance system.

Description
The DB-110 Dual-Band reconnaissance system is based on a two-axis gimballed design derived from the Senior Year Electro-optical Reconnaissance System (SYERS) II. An 11 in diameter direct viewing Cassegrain telescope, which can be articulated in the pitch and roll axes, is used to collect long-range standoff imagery. A second set of optics is deployed by rotating the optical bench through 180°, providing Wide Field of View (WFoV) imaging at shorter ranges.

Both optical subsystems utilise a common set of visible and Infra-Red (IR) focal planes. Either simultaneous dual-band or single spectral band imagery can be collected.

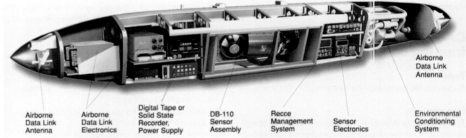

A schematic of the Reconnaissance Airborne Pod for Tornado (RAPTOR) (Goodrich)   0533866

The EO/IR sensor is of a compact lightweight modular design that is available as a dual-band, day-only or night-only sensor. The day sensor has a silicon CCD array and the IR sensor is Mid-Wave IR (MWIR) with a choice of indium antimonide arrays. The imagery obtained by the visible and IR sensor exceeds level 5 on the National Image Interoperability Rating Scale (NIIRS 5). The DB-110 can be used at stand-off target-to-sensor ranges from 5 km out to 100 km.

The modular design provides flexibility to tailor the DB-110 to individual customer requirements, such as focal length modification, and to accommodate technology upgrades throughout the service life of the system.

Multiple operational modes provide mission planning flexibility to interleave wide area search, spot collection, and target tracking/stereo modes. Both pre-programmed and manual control modes are provided. Control is via a MIL-STD-1553B databus. The system is compatible with digital tape recorders and digital datalinks.

Reconnaissance Airborne Pod for TORnado (RAPTOR)
The DB-110 sensor forms the core of the UK RAF's Reconnaissance Airborne Pod for TORnado (RAPTOR) pod. The contract for RAPTOR specified the provision of an end-to-end stand-off tactical reconnaissance system for the RAF Tornado GR4A; the original requirement specified a day/night long-range system able to operate simultaneously in the visual and IR wavebands, with target recognition at 72 km (visual) and 36 km (IR).

RAPTOR capabilities include interleaved wide area search, spot collection and target tracking/stereo modes so as to allow imaging of widely dispersed targets. Depending on the mission-planning scenario, in excess of 200 targets can be acquired when in spot collection mode. In target-tracking mode, up to 10 scans of the same sector can be performed. In wide area search mode, ground coverage can be up to 90,000 nm²/h. Continuous operation is possible using a datalink, limited only by the range of data transmission and aircraft endurance. Recording time is determined by data rate and tape capacity when using an onboard recorder.

In addition to the centrally mounted DB-110 sensor and its associated electronics, the RAPTOR pod comprises an airborne datalink antenna at

IR Wide Field-of-View (14 in. FL)

Visible Wide Field-of-View (16 in. FL)

Long range telescope (110 in. Vis FL) (55 in. IR FL)

Dual IRFPA

Visible CCD

The DB-110 dual-band reconnaissance system hardware (Goodrich)   0533867

RAPTOR is mounted on the port shoulder station of the Tornado GR4   0557219

the nose and tail, datalink electronics, a digital tape (or solid-state) recorder, power supply, Reconnaissance Management System (RMS) and an environmental conditioning system.

Weighing just under a tonne, the RAPTOR pod is mounted on the port shoulder pylon of the Tornado GR4A and has three window positions (port, vertical and starboard). Imagery can be displayed on the left TV tab display in the rear cockpit. The pod is not yet fully integrated with the GR4 avionics system (this will be addressed in the forthcoming Stage 2 software) and operates autonomously. However, the system receives positional data from the aircraft main computer and is integrated to the Thermal Imaging Airborne Laser Designator (TIALD – see separate entry) interface (which is used to trigger datalink transmissions). Work is under way to optimise procedures for using the L-3 Communications Common Data Link (CDL) on RAPTOR, which was not employed during operations over Iraq in 2003.

For each target, a variety of factors are considered in order to create what the pilots refer to as the target 'doughnut'. These include the required NIIRS standard, the aircraft speed and height, the position of the sun, the wind direction, the terrain and the atmospherics (haze, mist, fog, rain, snow, and so on). During operations over Iraq in 2003, two or three pairs of aircraft were assigned to each air tasking order and targets included troops, installations and equipment formations.

Specifications
Sensor
resolution:
(EO NIIRS) 5 at 60 km
(IR NIIRS) 5 at 30 km
focal lengths:
(EO) 110 in

(IR) 55 in
dimensions: 1,270 × 470 mm (diameter)
weight: 159 kg
power: 115 V AC, 400 Hz, 28 V DC

Optical
Type: Cassegrain reflector
Focal length: to fit application
(visible) 2,800 mm nominal
(infra-red) 1,400 mm nominal
Aperture: 280 mm
(visible) f/10
(infra-red) f/5

Focal planes
Visible (0.6 to 1.0 µm): Silicon CCD array; 5,120 × 64 TDI line array
Infra-red (3 to 5 µm): Indium antimonide (InSb) array; 512 × 484 area array
Data output: Max data rate: 260 Mbps; data compression available to meet recorder or datalink requirments
Digital tape-recorder: Data rate up to 240 Mbps continuously variable; 48 Gbytes on tape (equivalent to 20,000 nm²)
Interface: Compatible with Common DataLink (CDL)

Operation
Field of Regard (FoR):
(across line of flight) 180°
(along line of flight) ±20°
Geometry: panoramic/sector scan (4-28°)
Overlap: variable from 0–100%
Performance:
(visible) up to NIIRS 6
(infra-red) up to NIIRS 5

RAPTOR
Electronics (5 LRUs)

Dimensions
PPDV: 457 × 406 × 102 mm
SCU: 165 × 198 × 418 mm
RMS: 257 × 410 × 236 mm
PAA: 72 × 255 × 175 mm
INS: 178 × 119 × 303 mm
Weight: 50 kg (total)

Performance
Max data rate (digital datalink): 150 Mbits/s, J-band, line of sight
Altitude: 10,000–80,000 ft
Ground speed: M0.1-1.6
Field of regard: 180° across line of flight, ±20° along line of flight
Panoramic/sector scan: 4 to 28°
Overlap: variable from 1 to 100%

Status
The DB-110 is a compact, lighter weight derivative of the SYERS II (see separate entry) system in the Lockheed U-2. The system has been flight tested on the F-15E, F-111 and RF-4 and is also marketed as part of the HISAR® Integrated Surveillance And Reconnaissance system.

It has been reported that several systems have been built for the Japanese Maritime Self-Defence Force for use on their P-3 Orion MPA. In September 2003, the Polish Ministry of National Defence announced its selection of the DB-110 for integration into a reconnaissance pod for integration with F-16C/D Block 52M+ aircraft on order from the USA. It is believed the pod will be a variant of the BAE Systems Theatre Airborne Reconnaissance System (TARS) pod, which is already qualified on the F-16.

The DB-110 was selected by the UK Royal Air Force for Tornado GR Mk.1A and GR Mk.4/4A aircraft as part of the RAPTOR sensor suite. The complete RAPTOR system also encompasses ground exploitation stations. Eight airborne pods, two Data Link Ground Stations (DLGS) and four mission planners were ordered, of which two pods and one DLGS have still to be delivered.

The original in-service date for RAPTOR was May 2001, but a series of developmental and integration problems saw this slip to September 2002, when four pods and a DLGS system were deployed to Kuwait in time to prepare imagery ahead of Operation TELIC. According to Goodrich, more than 46 per cent of TELIC missions were carried out at night using the IR channel. Overall, the performance of RAPTOR was considered very good, but the results are being checked, assessed and refined after the action. The latest 8.2.14 operational software for the system (now being evaluated) should bestow full capability with the use of the datalink, event replay and the full IR mode.

Contractor
Goodrich Surveillance and Reconnaissance Systems.

Distributed Aperture System (DAS)

Type
Pilot's Night Vision System (PNVS).

Description
In late 1999, Northrop Grumman was awarded a USD9.9 million contract by the US Navy for research and development of the Multifunction Infra-red Distributed Aperture System (MIDAS) Advanced Technology Demonstrator (ATD) programme.

MIDAS continued directly from the Distributed Aperture Infra-Red Sensor (DAIRS) technology development programme.

The aim of the MIDAS ATD programme was to reduce the risk associated with the then DAIRS concept for the Joint Strike Fighter (JSF), and to allow an acceptable entry to an Engineering & Manufacturing Development (E&MD) implementation of DAIRS.

Renamed more simply the Electro-Optical (EO) Distributed Aperture System (DAS), the system consists of multiple IR cameras providing 360°

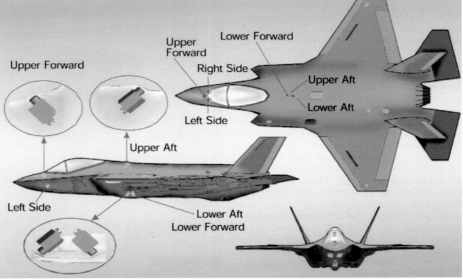

Distributed Aperture System sensor locations on the F-35 Joint Strike Fighter (JSF)   0568342

spherical coverage for Missile Approach Warning (MAW), short-range Infra-Red Search and Track (IRST), NAVFLIR, pilotage and Situational Awareness (SA).

For the JSF project, Northrop Grumman Electronic Sensors and Systems Sector (ES[3]) teamed with Lockheed Martin Electronics and Missiles to provide EO systems for the F-35 JSF. Northrop Grumman has the lead for the DAS and Lockheed Martin has the lead for the Electro-Optical Targeting System (EOTS – see separate entry).

While few details concerning the configuration of the system have been released, it is understood that the system will incorporate MWIR (3 to 5 µm) IR sensors for short-range pilotage requirements. Resolution is maintained across the entire Field of Regard (FoR) via optical tiling, where the individual airframe sensors' Fields of View (FoVs) video channels are electronically 'stitched' together to create a WFoV/R for the pilot's Helmet-Mounted Display (HMD). The effect can be likened to viewing a video wall, made up of a large number of smaller video monitors. At the system design stage, DAS was defined as the only enhanced vision sensor for pilotage; accordingly, significant research effort is ongoing into providing the pilot with high resolution and high brightness/contrast imagery to rival current direct optical path systems such as Night Vision Goggles (NVG). The Dutch TNO Physics and Electronics Laboratory (TNO-FEL) is contributing its newly developed Infra-Red Sensor Conditioning (IRSC) algorithms to the F-35 DAS. For the DAS application, the IRSC algorithms are expected to enhance the clarity of the IR digital image before it is fed to the pilot's HMD.

Status

In development. During 2004, Northrop announced that it had configured two DAS sensors on its BAC-111 test bed aircraft, with a third sensor to be added later in the test programme.

Contractor

Northrop Grumman Corporation, Electronic Systems Sector.

The F-35 JSF will include a six-sensor DAS
0521685

DAS Aperture (Northrop Grumman)
1127357

Electro-Optical Targeting System (EOTS) for the F-35

Type

Airborne Electro-Optic (EO) targeting system.

Description

The Electro-Optical Targeting System (EOTS), which, along with the Distributed Aperture System (DAS – see separate entry), forms the Electro Optical Sensor System (EOSS) for the F-35 Joint Strike Fighter (JSF), is under development by Lockheed Martin Missiles and Fire Control. The internally mounted EOTS will provide extended range detection and precision targeting against ground targets, plus long-range detection of air-to-air threats. The system is lightweight (expected to be under 91 kg) and also features passive and active ranging, plus the capability to generate highly accurate geo co-ordinates for precision attack requirements.

Described as 'highly common' with Sniper XR (see separate entry), EOTS features an advanced staring MWIR (3 to 5 µm) Focal Plane Array (FPA) for long-range target recognition in a highly reliable, easily maintained system. The system is integrated into the lower forward nose of the F-35 in a rugged, low-profile sapphire-faceted window to give some measure of stealth. It is linked to the aircraft's integrated central computer through a high-speed fibre-optic interface.

A mockup of the F-35 JSF at the 2003 Paris Air Show with the multifaceted EOTS under the nose
0568460

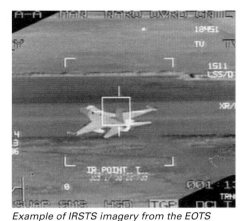

Example of IRSTS imagery from the EOTS   0568461

Example of FLIR imagery from the EOTS. Note the inset zoom facility allowing for extended range identification of targets   0568462

Other features of the system include:
- Compact, single-aperture design
- Advanced, large format, MWIR FPA
- FLIR
- IRSTS
- Automatic boresighting
- Tactical/eyesafe diode-pumped Laser Range-Finder/Designator (LRF/D)
- Laser spot tracker.

The modular design of EOTS allows for two-level maintenance to reduce life-cycle costs. It shares a common integrated detector assembly with the F-35's Distributed Aperture System (DAS – see separate entry), being developed by Northrop Grumman Electronic Systems, plus COTS-based electronic modules. BAE Systems (UK) will be supplying the laser.

The system is now being further developed and refined, with Lockheed Martin now actively seeking other industry suppliers from within the JSF partnership to supply components for the system.

Specifications
As a derivative of the Sniper XR system, detailed specifications for the EOTS are believed to be as for Sniper XR.

Status
Work on the EOTS was started at the beginning of 2002, following a USD171 million subcontract award from BAE Systems. In support of the EOTS effort, Northrop Grumman's BAe One-Eleven Co-operative Avionics Test Bed (CATB) has been fitted with EOTS sensors, processors and software and has been flying since the first quarter of 2000. Successful demonstrations included automatic target cueing and handoff to the AN/APG-81 Active Electronically Scanned Array (AESA) radar, and the electronic warfare suite.

During July 2004, BEI Technologies' Precision Systems and Space Division was awarded an up to USD2 million contract by Lockheed Martin for the design, test and manufacture of EO sensor components for EOTS through System Development and Demonstration (SDD).

The first EOTS avionics test bed flight trials with a refined system are scheduled to begin in August 2005, with the first EOTS-equipped F-35 flight due in April 2007.

Contractor
Lockheed Martin Missiles and Fire Control. BAE Systems, UK.

F/A-18 C/D Laser Target Designator/Range-finder (LTD/R)

Type
Airborne Laser Range Finder/Target Designator (LRFTD).

Description
The LTD/R is integrated into the AAS-38 Night Hawk FLIR pod and provides the F/A-18 C/D Hornet, flown by the US Navy and Marine Corps, with advanced fighting capabilities through improved target ranging and designation. It consists of a laser transceiver Weapons Replaceable Assembly (WRA) and a laser power supply WRA.

Specifications
Wavelength: 1.064 μm
Range limits: 315–24,966 m
Beam diameter: 0.28 in
Beam alignment: 0.3 milliradians
Pulse repetition rate: Band I and Band II PRF codes
Weight: LTR: 5.9 kg
LPS: 4.9 kg

Status
In production and in service in US Navy and US Marine Corps F/A-18C/D aircraft.

Contractor
Northrop Grumman Corporation, Electronic Systems, Laser Systems.

FLIR common modules

Type
Airborne Thermal Imager (TI).

Description
The FLIR common module programme has been in a state of continuous production since April 1991. The family of modules includes two different 30 Hz Scan-Interlace and one 60 Hz Scan-Interlace modules. Associated equipment includes a 3 and 5 V bias regulators, an auxiliary control, pre-amplifier and a post-amplifier.

In avionic application, FLIR common modules are integrated into the FLIR systems for the OH-58 Kiowa Helicopter and the AH-64 Apache Helicopter.

Customers for the modules have included the US Army CECOM, Hughes, Raytheon and Aeroflex International.

Status
In production and in service in wide variety of ground and airborne platforms.

Contractor
Symetrics Industries Inc.

Helicopter Infra-Red Navigation System (HIRNS)

Type
Airborne Electro Optic (EO) navigation system.

Description
The Helicopter Infra-Red Navigation System (HIRNS) is a low-weight television-compatible FLIR designed to aid navigation by day and night. The sensor head is integrated with BAE Systems' Type 239 two-axis unstabilised platform (see separate entry).

An A129 pilot equipped with IHADSS, which displays flight, weapon system information and the HIRNS image   0111830

The LTD/R is in service on F/A-18 C/D aircraft   0051018

The HIRNS sensor is located below the Helitow sighting system (Patrick Allen)   0569543

A serial scanning thermal imager is used because balancing of amplifiers is not required to provide a high-resolution display. It also has the advantage of simpler processing electronics, while DC restoration ensures a bloom-resistant image and clear horizon definition. Energy from the scene is scanned horizontally with a continuously rotating eight-faceted mirror, while a flat nodding mirror carries out vertical scanning. The output signal from the SPRITE detectors is amplified and processed to provide gain and level control and is then displayed in a standard TV format.

The system may be linked with an Integrated Helmet And Display Sighting System (IHADSS – see separate entry) to allow the crew to perform Nap Of the Earth (NOE) missions.

Specifications
Weight:
(FLIR) 6.35 kg,
Field of regard:
(bearing) ±130°
(elevation) +20 to –60°

Status
In service on the Agusta A129 helicopter.

Contractor
Honeywell Sensor and Guidance Products.

Helicopter Night Vision System (HNVS)

Type
Pilot's Night Vision System (PNVS).

Description
The Helicopter Night Vision System (HNVS) is installed in CH-53 Super Stallion transport helicopters for use on low-level tactical missions in adverse weather conditions. The system is based on equipment similar to that installed by Northrop Grumman on 10 AH-1S helicopters which were used as surrogate trainers for AH-64 pilots.

The system includes a data entry panel, master control assembly, the Honeywell Integrated Helmet And Display SubSystem (IHADSS) and control panel, video monitor/recorder and control panel, power distribution unit, heater/filter assembly, system control electronics, vapour cycle unit, symbology generator, multiplexing remote terminal, control grips and the Lockheed Martin Pilot Night Vision System (PNVS).

The PNVS is a thermal imaging system which optically senses the heat emitted by objects and converts it into video image for display to the pilot and co-pilot. The PNVS, mounted as a turret on the chin of the CH-53E, is controlled

by the IHADSS worn by the pilot or co-pilot. The IHADSS monitors head position and converts head movement into azimuth and elevation commands which are sent to the PNVS. Video from the turret is displayed and flight symbology is overlaid on the cockpit displays, allowing either the pilot or the co-pilot to fly the aircraft in a head-up attitude.

Status
In service in the CH-53 Super Stallion transport helicopter.

Contractor
Northrop Grumman Corporation, Electronic Systems Sector.

Helmet-Mounted Display System (HMDS)

Type
Helmet-Mounted Display System (HMDS).

Description
CMC Electronics' Helmet Mounted Display System (HMDS) consists of a two-piece helmet (inner and outer), an Electro-Optical (EO) Helmet Tracking System (HTS) and a miniature Cathode Ray Tube (CRT) monocular visor display system. The HMDS is designed to operate in conjunction with the FV-4000 mission computer in both rotary- and fixed-wing applications.

The inner helmet provides for accurate and comfortable fitting to a wide range of head sizes, impact protection and communications; the outer helmet incorporates the HTS diodes, CRT optics, visor system and on-helmet power supplies (similar to the Guardian HMD).

For daylight operations, the visor display system provides flight information (attitude, height, heading speed and navigational steering) to the pilot's right eye in a circular 20° FoV. For night operations, a Night Vision Display System (NVDS), comprising a binocular Night Vision Goggle and CRT optics module, can be fitted to the helmet in place of the day optics.

Specifications
Display source: miniature CRT (right eye)
Display type: cursive (stroke) or raster
FoV: 20° circular
Helmet tracker: Electro-Optical via LED diode array
Interface: MIL-STD-1553B, RS-422, ARINC, synchro, analogue

Status
In production. As of 2002, a binocular HMDS was reported in development.

Contractor
CMC Electronics Inc.

Infra-Red Acquisition and Designation System (IRADS)

Type
Airborne Electro Optic (EO) targeting system.

Description
Due to the demanded stealth characteristics of the F-117A Nighthawk, a new and innovative navigation and attack sensor system was required which could deliver laser-guided weapons accurately, without recourse to a traditional (non-stealthy) multimode radar system.

The Infra-Red Acquisition and Designation System (IRADS), reportedly derived from the AN/AAS-38A Forward Looking Infra-Red (FLIR) system carried by US Navy and Marine Corps F/A-18 aircraft, is a totally passive target acquisition and weapon delivery system. IRADS utilises two targeting units, designated FLIR and DLIR (Downward Looking Infra-Red), with the DLIR incorporating a laser designator as well as the IR sensor. This arrangement enables the aircraft to maintain sensor Line-of-Sight (LOS) to

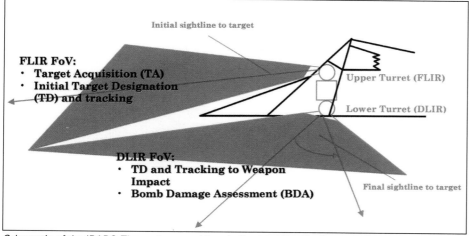

Schematic of the IRADS. The system provides similar total FoR (in the vertical plane) to externally-mounted pods, such as TIALD, while maintaining the stealth characteristics of the airframe
0524723

The upper FLIR turret of the IRADS, with the sensor unstowed (US Air Force) 1127374

the target during the entire Time-of-Flight (TOF) of a Laser-Guided Bomb (LGB – the primary weapon of the aircraft), while operating at medium altitude, without recourse to aircraft manoeuvre – an important consideration when aircraft radar signature is also affected by aspect.

The FLIR is mounted in a recess just ahead of the cockpit front windshield, located in a steerable turret assembly which incorporates the dual Field of View (FoV) sensor. When the FLIR is not in use, it is rotated 180° towards the aircraft fuselage to protect the sensor from damage/erosion. The DLIR is mounted in a recess underneath the forward fuselage and to the right of the nose gear. Both the FLIR and the DLIR recesses are covered by a RAM mesh screen and the edges of the recesses are serrated, with fasteners covered with RAM putty to prevent unwanted radar reflections.

The DLIR includes a bore-sighted laser designator, slaved to the IR sensors, for attack with LGBs. The spot size of the laser on the ground is reported to be of the order of (30.5–45.7 cm (12–18 in)), stabilised by the turret/Inertial Navigation System (INS).

Without an active sensor for navigation/attack system update and a requirement for highly accurate pointing of the FLIR/DLIR for successful first-pass target acquisition, a highly accurate INS was required for the F-117. Initially, the AN/ASN-131 Standard Precision Navigator/Gimballed Electrostatic Aircraft Navigation System (SPN/GEANS – see separate entry), developed originally for the B-52 bomber, was chosen for its high level of accuracy and reliability.

However, by 1991, despite its undoubted accuracy for a mechanical system, the SPN-GEANS took more than 40 minutes to align and was becoming increasingly difficult to support, so the navigation sensor for the F-117A became the subject of the Operational Capabilities Improvement Programme III (OCIP III). As part of this programme, the Ring Laser Gyro Navigational Improvement Program (RNIP), replaced the SPN-GEANS with a new Honeywell H-423/E Laser INS (LINS), which aligns rapidly, is more accurate and offers a much improved Mean Time Between Failures (MTBF). The subsequent

integration of a Rockwell-Collins GPS (with low-observable antennas) brought about a programme re-designation from RNIP to RNIP+, together with a further significant improvement in navigational accuracy.

The F-117A cockpit is equipped with an F/A-18-derived Head-Up Display (HUD) and two 12.7 cm (5 in) CRTs, while the main FLIR/DLIR CRT features a bespoke 30.5 cm (12 in) screen. For night operations in clear weather at low altitudes, the main pilot aid is the FLIR/DLIR CRT.

In operation, the INS is used to guide the aircraft to the target area and initially point the FLIR (wide FoV) towards the target/Point of Interest (POI). As the aircraft approaches the target, the pilot monitors the view presented by the FLIR and, when the target is identified, switches to the narrow FoV and locks onto the target. As the target passes under the aircraft, the FLIR sensor will be obstructed by the aircraft nose, so targeting is taken over by the DLIR, which acquires and continues tracking of the target. After weapon release, the DLIR laser is turned on and illuminates the target, enabling the LGB to home in on the reflected IR energy.

Specifications
NFOV: 3 × 3°
WFOV: 12 × 12°

Status
Over its service life, the F-117A has been under a near constant state of upgrade and modification to maintain a high level of operational effectiveness. The baseline IRADS, manufactured by Texas Instruments, was the subject of a Block 1 modification (essentially a wiring upgrade) between 1986 and 1988. A new turret, incorporating upgraded TI sensors, completed flight test during February 1993 with production turrets retrofitted by late 1996.

As part of a connected aircraft upgrade, part of the F-117A OCIP II (OCIP II – a follow-on to OCIP I which integrated the GBU-27), the F-117A cockpit was modified. The original, centrally mounted

FLIR/DLIR display was replaced by a Harris Corporation digital moving map display, with IR imagery displayed on one of two Honeywell colour multifunction LCDs, which replaced the original monochrome, Texas Instruments MFDs on the left and right hand sides of the main panel. OCIP I also included a new Flight Management System (FMS) linked to an updated Flight Control System (FCS) with autothrust and full authority autopilot. OCIP II was completed in 1995.

The upgrade of the core navigation system, OCIP III, was completed in 2000.

More recently, a second-generation IRADS, manufactured by Raytheon, was integrated into the F-117A from 2000. The system adds a new video tracker and system controller to facilitate more accurate target designation. In addition, an imaging IR sensor has been incorporated into the system, which provides high-quality imagery which can be transmitted to other aircraft or ground stations using a Low Probability of Intercept (LPI), encrypted datalink. The reported USD10 million contract for 50 sets (43 new production units and seven refurbished development sets) was placed by the USAF Aeronautical Systems Center (ASC) at Wright-Patterson AFB during 1998.

Noting the previous comments concerning continued upgrade of all of the F-117A systems, the Combat Capability Sustainment Programme (CCSP) aims to replace obsolete avionics systems, establish new vendors and improve reliability and maintainability to 'keep the F-117 operational through its service life'. The CCSP began Concept & Technology Development (CTD) in FY00 with Congressional funding. As of February 2005, IRADS continues to be the focus of upgrade,

The upper FLIR turret of the IRADS, in the stowed position, can be seen below the front coaming in this head-on view of an F-117A
0524724

The IRADS automatically switches from the upper FLIR to the lower DLIR during LGB fly-out, ensuring continuous designation throughout weapon Time-of-Flight (ToF)
0014027

with further modifications, as yet unspecified, intended to eliminate known obsolescence and supportability issues and to facilitate continued IRADS operational capability.

Contractor
Raytheon Company, Space and Airborne Systems.

Infra-Red Imaging Subsystem (IRIS)

Type
Airborne Electro-Optic (EO) targeting system.

Description
The Infra-Red Imaging Subsystem (IRIS) is part of the AN/AAS-38 FLIR imaging system. This is a self-contained pod designed for use on the US Navy F/A-18 aircraft for target acquisition and recognition and weapon delivery. It also allows reconnaissance under day and night and adverse weather conditions.

The IRIS includes dual fields of view of 12 × 12° and 3 × 3° with automatic thermal focus. Built-in image stabilisation provides a natural horizon display to the pilot.

A key feature of IRIS is the automatic video tracker contained in the controller-processor. This tracker provides automatic target acquisition, line of sight control and offset designation for accurate weapon delivery.

The controller-processor provides video processing necessary to produce 875-line RS-343 television video. The processor adds track and field of view reticles for cockpit displays. A MIL-STD-1553 digital databus provides communications with the aircraft AN/AYK-14 mission computer. The infra-red receiver uses advanced switching regulator designs and high-density packaging to provide efficient primary to secondary power distribution, control and regulation for the IRIS and pod system.

Status
In service on the US Navy F/A-18 aircraft.

Contractor
Raytheon Company, Space and Airborne Systems.

Integrated Helmet And Display Sighting System (IHADSS)

Type
Helmet-Mounted Display System (HMDS).

Description
Both the pilot and co-pilot/gunner are provided with helmet units and controls to allow independent and co-operative use of the system. The IHADSS sight component provides off-boresight line of sight information to the fire-control computer for slaving weapon and sensor to the pilot's head movements. Real-world sized

Honeywell IHADSS helmet sight and display

0503973

video imagery from the slaved and gimballed infra-red sensor is overlaid with targeting as well as flight information symbology and projected on a combiner glass immediately in front of the pilot's eye. The IHADSS allows night nap-of-the-earth flight at below treetop altitudes, without reference to cockpit instruments, and rapid target engagement.

Specifications
Dimensions:
(sensor surveying unit × 4) 111.76 × 129.54 × 78.74 mm
(sight electronics unit) 177.8 × 177.8 × 289.56 mm
(display electronics unit) 139.7 × 177.8 × 289.56 mm
(display adjust panel × 2) 152.4 × 182.88 × 76.2 mm
Weight:
(helmet × 2) 1.4 kg each
(sensor surveying unit × 4) 0.57 kg each
(sight electronics unit) 7.26 kg
(boresight reticle unit × 2) 0.23 kg each
(helmet display unit × 2) 0.57 kg each
(display electronics unit) 6.35 kg
(display adjust panel × 2) 1.58 kg each
(total weight) 23.45 kg
Power supply: 115 V AC, 400 Hz, 3 phase, 460 W
Field of view: (horizontal) 40° × (vertical) 30°
Coverage:
(azimuth) ±180°
(elevation) ±90°
Accuracy: 3 to 10 mrad RMS
Slew rate: 120°/s

Status
In production and in service on the Boeing AH-64A Apache and Agusta A 129 helicopters.

Contractor
Honeywell Sensor and Guidance Products.

KA-95 panoramic camera

Type
Airborne reconnaissance system.

Description
The KA-95, Recon/Optical's most compact panoramic camera using 5 in (127 mm) film, is ideal for medium-altitude reconnaissance. Mounted internally in manned reconnaissance aircraft or externally in a pod, the camera utilises a 12 in lens which produces large-scale wide-angle photographs of up to 190° coverage, each being equivalent to the coverage provided by six framing cameras. Using electronic synchronisation of all components, the camera is of modular design to simplify maintenance.

An important advantage of the KA-95 is sector scan. The camera can be set up for six different

Recon/Optical Inc KA-95 panoramic camera

0018317

across-track scan angles. The quality of the photography produced is enhanced by built-in roll stabilisation. To eliminate image blur arising from motion of the platform, the KA-95 uses translating lens forward motion compensation.

The KH-95B is designed for the RF-5E.

Specifications
Weight: 63.5 kg including film
Power supply: 115 V AC, 400 Hz, 3 phase; 28 V DC
Stabilisation: up to 10°/s roll
Lens: 12 in (304.8 mm) focal length, f/4
Film: 5 in (127 mm) aerial roll film, 1,000 ft 2.5 mil film

Status
As of early 2006, in production and in service. The KA-95 is currently being used by a participant in the verification of the Open Skies treaty.

Contractor
Recon/Optical, Inc.

KS-127B long-range oblique film camera

Type
Airborne reconnaissance system.

Description
The KS-127B is an optical Long-Range Oblique Photographic (LOROP) camera, designed to fit into the nose sensor compartment of the RF-4 tactical reconnaissance aircraft. The camera has a 1,676 mm (66 in) focal length, f/8.0 lens system, active and passive image stabilisation, autofocus and thermal stabilisation. Typically, the KS-127B can take detailed photographs

Recon/Optical's reconnaissance pod solutions

0051040

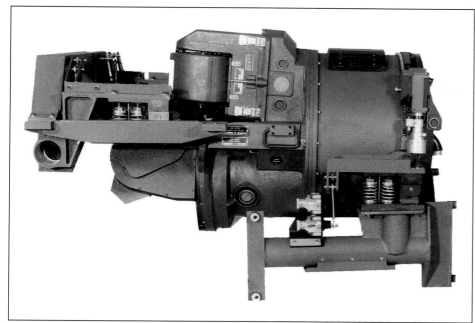

KS-127B photographic camera (Recon/Optical)   1176157

KS-127B image taken at 26,000 ft and 15.5 n miles slant range   0023184

of sites over 35 km distant from an aircraft patrolling at 35,000 ft. This coverage permits missions to be conducted from safe, long-range, standoff distances.

The KS-127B is a stepping frame camera that includes both manual and automatic operating modes. Manual mode provides 3.9 × 3.9° coverage in single-step operation. Automatic mode offers two-step vertical, or three- or four-step oblique coverage. The two-step mode provides 3.9 × 7.5° of coverage, while the three-step mode provides 3.9 × 11° and the four-step mode 3.9 × 14.5°. Under manual operation, sights in the rear cockpit are used to point the camera at selected targets.

Specifications
Power: 115/208 V RMS, 400 Hz, 28 V DC
Lens: 1,676 mm (66 in), f/8.0
Film: 127 mm (5 in) aerial roll film; 1,000 ft of 2.5 mm film; 700 ft of 4 mm; 500 ft of 5.2 mm
Shutter speed: 1/300 to 1/1,500 s
Overlap: 12 to 56%

A section of the left image, enlarged ×21
0023185

Status
In service in the RF-4.

Contractor
Recon/Optical, Inc.

KS-146A long-range oblique film camera

Type
Airborne reconnaissance system.

Description
The KS-146A reconnaissance camera uses a 1,676 mm (66 in) focal length, f/5.6 lens to provide LOng-Range Oblique Photography (LOROP), designed for podded applications.

The system is divided into two subsystems – the camera assembly and the electronics unit. The camera assembly contains the scan head, lens, autofocus drive, thermal system, roll drive, shutter, optical filter and film magazine. The electronics unit contains power supplies, microprocessor-based control electronics and servo controls for camera operation, including built-in test, cycle rate, exposure, focus, stabilisation, roll drive and thermal system controllers.

The scan head is a two-axis gimbal, providing scan mirror mounting, pointing, stepping and active stabilisation for roll and azimuth. The optical system uses a nearly-diffraction-limited lens yielding high modulation even at low contrast levels.

Specifications
Weight:
(camera) 317 kg
(total system) 385 kg
Power: 115 V AC, 400 Hz, 28 V DC
Field of view: 3.9 to 14.5°
Film: 127 mm (5 in) roll film; 1,000 ft of 0.0025 in film
Shutter speed: 1/30 to 1/1,500 s

Status
In service in reconnaissance applications in Europe and Asia.

Contractor
Recon/Optical, Inc.

KS-146A LOROP camera   0505264

KS-147A long-range oblique film camera

Type
Airborne reconnaissance system.

Description
The KS-147A reconnaissance camera uses a 66 in focal length f5.6 lens to provide LOng-Range Oblique Photography (LOROP). Target coverage, on either side of the aircraft, can be selected from the cockpit. The camera incorporates automatic exposure control, passive and active stabilisation, auto-focus and thermal stabilisation. Typically, the KS-147A takes detailed pictures of sites to either the left or right side of the aircraft, from 18.5 km to over 92.6 km distant from an aircraft flying at 30,000 ft or higher.

To meet space and weight constraints, the camera is subdivided into two parts – the camera assembly and the electronics assembly. The camera assembly contains the scan head, lens, fold mirrors, autofocus drive, roll drive, shutter, optical filter and film magazine. The electronics assembly includes the power supply, control electronics and servo-controls. For installation in the RF-5E, the camera assembly is on a removable pallet that becomes part of the aircraft's nose structure.

Specifications
Weight:
(camera) 245 kg
(total system) 270 kg
Power supply: 115/208 V AC, 400 Hz, 700 VA; 28 V DC, 300 W
Lens: 1,676 mm (66 in) focal length, f/5.6
Film: 127 mm (5 in) roll film; 1,000 ft of 0.0025 in film
Depression: 4–39°
Cycle rate: 0.375 to 1.5 cycles/s
Overlap: 12 to 56%

Status
As of early 2006, the KS-147A was in service in the RF-5E.

Contractor
Recon/Optical, Inc.

The KS-147A LOROP camera   0505265

The KS-147A LOROP camera installed in the nose section of an RF-5E Tigereye   0503372

KS-157A long-range oblique film camera

Type
Airborne reconnaissance system.

Description
The KS-157A is a compact, folded optics LOng-Range Oblique Photographic (LOROP) camera, designed for both tactical and strategic reconnaissance installed in business jet aircraft. Capable of standoff distances of 55.5 km or more, the camera uses a 1,676 mm (66 in) focal length, f/5.6 optical system to provide high-resolution frame imagery suitable for both detection and, more importantly, identification of military targets. The system offers in-flight selection of

The Recon/Optical KS-157A LOROP camera
0505266

photographing either the right or left oblique positions and can record targets at depression angles from the horizon to 30° below.

The KS-157A provides high-resolution performance even under low-light conditions. Controls in the cabin include selection of variable depression angles and frame coverages from a wide range of altitudes, velocities and standoff distances, and repeat frames for specified targets.

The camera incorporates automatic exposure control, passive and active stabilisation, auto-focus and thermal stabilisation.

Specifications
Weight:
(camera) 317 kg
(total system) 385 kg

Power supply: 115/200 V AC, 400 Hz
28 V DC
Lens: 66 in focal length, f/5.6
Film: 127 mm (5 in), 305 m (1,000 ft) capacity
Format: 114 × 114 mm

Status
As of early 2006, in production and in service.

Contractor
Recon/Optical, Inc.

LAMPS Mk III laser designator

Type
Airborne Laser Range Finder/Target Designator (LRFTD) system.

Description
The LAMPS Mk III laser designator is designed for use with all types of laser-guided weapons.

The laser designator system has been optimised for low volume and weight. Built-in test verifies critical aspects of system performance. LAMPS Mk III is a fully qualified military laser designator.

Specifications
Wavelength: 1.064 μm
Weight: 5.0 kg
Dimensions: 292 × 292 × 61 mm
Power: 115 V AC, 400 Hz, 3-phase, 250 W; ±15 V DC, 20 W
Operating temperature: −10 to +47.5°C
Cooling: Forced ambient air

Status
In production and in service in US Navy SH-60 helicopters.

Contractor
Northrop Grumman Corporation, Electronic Systems, Laser Systems.

LAMPS Mk III laser designator in service on US Navy SH-60 helicopters
0051019

LANTIRN ER

Type
Airborne Electro Optic (EO) navigation and targeting system.

Description
During 1999, Lockheed Martin began a series of modifications to the basic LANTIRN system, intended to provide improved performance in terms of detection/recognition ranges and laser effectiveness. These modifications addressed the necessities of more effectively exploiting the latest variants of long-range laser-guided weapons and successfully keep pace with the US Air Arms' evolving high-medium level LGB operations. The enhanced system, previously designated Modernised LANTIRN, is now known as LANTIRN ER (Extended Range).

The major enhancements to the core system, drawn from Lockheed's Sniper XR system (see separate entry) include:

1. A MWIR (3 to 5 μm) InSb staring array Forward Looking Infra-Red (FLIR) sensor providing for dramatic improvements in detectivity and, thus, detection range, enabling target acquisition at the extended ranges demanded by weapons such as the GBU-24 series LGB. Furthermore, the new staring array is largely solid state compared to previous generation detectors, facilitating a claimed 23 per cent increase in reliability.

2. A solid-state diode-pumped laser, which, as a result of lower beam divergence and greater pointing accuracy, enables operation of the targeting pod at greater range and altitude (up to 40,000 ft is claimed).

3. An eye-safe training laser with tactical performance and range is integrated into the system, which is claimed to be 17 per cent more reliable, thanks to an improved power supply, fewer parts and a cooler operating temperature.

4. An enhanced computer system, which is smaller, weighs half as much, and uses half as much

power as the computer it replaces. Throughput, memory and reliability are optimised; software, cabling and interfaces remain the same.

In addition to the above core enhancements which make up the LANTIRN ER system, further customer-specified optional features include:

1. An Automatic Target Recognition (ATR) system to reduce pilot workload in the acquisition and identification of targets.

2. A Laser Spot Tracker (LST), to further aid in target acquisition and identification via third party marking.

3. A digital disk recorder for use in Battle Damage Assessment (BDA) of target strikes and reconnaissance mission support in surveillance modes.

4. A TV sensor, which has been successfully tested and flown, will provide added capability in the event of poor IR sensor performance due to atmospheric conditions.

While significantly enhancing LANTIRN's traditional mission – day/night strike/interdiction – LANTIRN ER will facilitate a wider range of aircraft mission capabilities to include air-to-air tracking, theatre missile defence and battle damage assessment. The optional further refinements to LANTIRN ER provide for greater targeting flexibility and enhanced reconnaissance capability.

Tiger Eyes
Tiger Eyes is an evolution of LANTIRN ER, combined with an Infra-Red Search and Track System (IRSTS) utilising technology derived from the AN/AAS-42 in US Navy service, Tiger Eyes integrates a true multirole capability into the core LANTIRN system. Tiger Eyes is a combined navigation/targeting system and IRSTS. The sensor suite includes the navigation pod and targeting pods of the LANTIRN ER system with a pylon-mounted Long-Wave IR (LWIR) IRSTS of similar capability to the AN/AAS-42, giving the total system a true multispectral third-generation IR capability. The passive LWIR sensor searches for and detects heat sources within its field of view, significantly enhancing air-to-air engagement range.

Status
All elements of the LANTIRN ER system may be applied as upgrades to the over 700 LANTIRN systems delivered to date. LANTIRN ER is in production.

While externally identical, the LANTIRN ER system offers greatly increased detection and identification range, coupled with a high-power laser to facilitate long-range standoff during targeting of heavily defended targets (Lockheed Martin)
1044584

During August 2002, Lockheed Martin received a USD163.7 million order from Boeing to produce the Tiger Eyes sensor suite for F-15K fighter aircraft ordered in 2002 by the Republic of Korea Air Force.

During April 2006, Lockheed Martin was selected by Boeing as the preferred supplier of the EO sensor suite for Republic of Singapore Air Force (RSAF) F-15SGs. The sensor suite includes the Sniper Advanced Targeting Pod (ATP) and the Tiger Eyes navigation pod/IRSTS.

Contractor
Lockheed Martin Missiles and Fire Control.

Lite Eye™ Helmet-Mounted Display (HMD)

Type
Helmet-Mounted Display System (HMDS).

Description
The Lite Eye™ Helmet-Mounted Display (HMD) is a compact, lightweight, low-cost monocular display system designed for helicopters and combat support aircraft. Specifically designed to attach to the standard ANVIS Night Vision Goggle (NVG) system, Lite Eye™ HMD provides crew members basic HUD capability (aircraft flight, engine performance and weapons symbology) in both day and night operations. Designed to accommodate a wide range of aircraft configurations and mission requirements, the core Lite Eye™ HMD system is easily expanded to include capabilities such as head tracking (magnetic or non-magnetic) and symbol generation. Also designed to accommodate a wide variety of users, Lite Eye™ HMD can be used with all military aviator helmets and provides full ranges of InterPupillary Distance (IPD), fore/aft and vertical adjustments.

Specifications
Field Of View: 20 × 15°
Exit pupil: 15 mm
Eye relief: 22 mm
Contrast ratio:
  1.4 at 10,000 fL
  1.7 at 5,000 fL
  4.5 at 1,000 fL
Resolution: 0.9 cy/mr
Transmission: 30%; 448 to 650 nm
Adjustments: IPD; fore/aft; vertical
Optical focus: Infinity
Display:
Type: Active Matrix ElectroLuminescence (AMEL)

VGA: 640 × 480
Colour: monochromatic yellow
Dimming range: 100:1
Interface: VESA FPDI-1 (display controller input interface)
Average power: 5 W (5 V DC)

Status
In production and in service.

Contractor
Rockwell Collins Kaiser Electronics.

Littoral Airborne Sensor – Hyperspectral Programme (LASH)

Type
Airborne Electro Optic (EO) mine detection system.

Description
In December 1999, following an initial Proof Of Concept (POC) demonstration, the US Office of Naval Research (ONR) awarded Science and Technology International (STI) a contract worth USD50 million over five years to develop a family of imagers under the Littoral Airborne Sensor – Hyperspectral (LASH) programme. This was initially focused on Anti-Submarine Warfare (ASW), but has since expanded to include Mine CounterMeasures (MCMs) and other applications.

LASH operates in the Pushbroom mode, using three-axis active stabilisation to maintain the optics pointing directly at the target, regardless of platform motion. Two Charge-Coupled Device (CCD) cameras, with a combined 40° Field of View (FoV) build up an image, one line at a time, in the direction of flight. A spectrometer splits incoming reflected light into different colours (wavelengths) containing spectral information in the visible range of the electromagnetic spectrum (approximately 430–840 ηm). These are stored as a 'data cube' with up to 288 wavelengths for the entire imaged scene. Each layer holds reflectance values for a particular wavelength. These values are used to discriminate and classify objects of interest, exploiting characteristics not discernible by the human eye.

The spatial and spectral resolutions can be changed in software, depending on the specific needs of the application. Altitude, platform speed and frame rate all play a part in determining the Ground Sampling Distance (GSD) (pixel size). Inputs from an inertial navigator and a GPS

receiver are used to georegister each pixel in the image to an accuracy of <1 m.

Extensive testing in the ASW role has taken place aboard the P-3 Orion Maritime Patrol Aircraft (MPA) and Seahawk helicopter. The imager has demonstrated its ability to detect, geolocate and classify submerged submarines at tactical depths in littoral waters, where sonar performance is often limited. In September 2002, a LASH-equipped P-3 operating in the vicinity of the Yellow Sea was successfully cued by the US Navy's Distant Thunder acoustic surveillance system.

A variant of LASH, known as the Littoral Mine Countermeasures Rapid Reconnaissance System (LMCRRS) and installed in a pod carried by a Bell 206 helicopter, participated in Fleet Battle Experiment Hotel in August 2000 and in Exercise 'Kernel Blitz' at Camp Pendleton, California, during March 2001. The sensor demonstrated its ability to find mines in the surf zone and to detect kelp, which can substantially affect amphibious missions.

Passive optical remote sensing of shallow water and the surf zone poses several challenges. The signal received at the sensor is a combination of that from the water itself, surface irradiance reflected off bottom features, glint, and inherently complex in-water optical effects. One of the largest sources of non-target signal is surface clutter – glint from foam, detritus and wave action.

In an effort to solve this problem, STI has developed two clutter-removal methods. The first is a strictly statistical approach, which is computationally efficient and mathematically simple. The second employs hyperspectral algorithms that were developed for the detection of submerged targets in deep water, in which the glint is subtracted from the scene before image segmentation and anomaly detection. The latter 'linear unmixing' method does not appreciably increase the computation time and provides what STI describes as 'startlingly better' results.

Under an extension of the LASH programme, initiated in August 2001, STI has recently installed its hyperspectral imager and other sensors for testing aboard a Skyship 600 airship at several locations on the US eastern seaboard. The trials are examining the suitability of such a combination for applications including force protection, harbour security patrol and homeland security. In addition to LASH, the airship carries a WESCAM MX-20 gimbal fitted with a broadcast-quality colour video camera, thermal imager and laser range-finder,

STI has integrated its hyperspectral imager and several other sensors into the 9 m long gondola of a Skyship 600 airship under an extension of the LASH programme (STI)   0532375

Lite Eye™ HMD   0054204

The US Navy has conducted extensive trials with STI's LASH hyperspectral pod on board several platforms, including the P-3 Orion MPA (STI)   0535248

Raytheon SeaVue and Furuno commercial radars, the VIPER sniper-detection system developed by the National Research Laboratory (NRL), and navigation equipment and military radios.

Further expansions of the fit could include signals-intelligence sensors, together with the use of data fusion. STI is also developing a hyperspectral package known as Pueo (named after a Hawaiian owl), for potential use aboard expendable UAVs to direct munitions engagements, which it hopes to test aboard the airship in February or March 2003.

Status
In development.

Contractor
Science and Technology International.

Low-Light-Level TeleVision system (LLLTV)

Type
Airborne Electro Optic (EO) surveillance system.

Description
The Lockheed Martin Low-Light-Level TeleVision (LLLTV) is a general purpose multirole system. It is designed for maritime surveillance missions such as search and rescue, monitoring of territorial waters and policing offshore environmental laws. Principal features are high resolution and sensitivity at low-light levels and a small (16 mm diagonal) format. It is claimed to be capable of resolution densities in

excess of 30 lines/mm and to have a wide dynamic range, which provides useful imagery around brightly lit parts of the area surveyed.

The camera head can be mounted in any attitude and is designed for hands-off operation. It uses a 16 mm vidicon tube and an 18 mm second-generation hybrid photocathode intensifier, which has extended sensitivity at the red end of the spectrum.

The system's electronic unit has a switchable line rate of either 525 or 875 lines a frame as standard but other line/frame rates are optionally available. Normal aspect ratio is 4:3 but a version with an aspect ratio of 1:1 is also available. Frame rate is 30 Hz. Resolution is 600 television lines horizontal and the dynamic range permits operation at face illumination levels from 0.1 ft candle down to starlight conditions.

The LLLTV has been chosen by the Spanish Navy for a shipboard fire-control application. The US Coast Guard has also selected an active gated version for possible use with its Dassault Aviation HU-25 Guardian aircraft, the application being to monitor and police territorial waters. In this configuration the equipment is designated AN/ASQ-174 Active Gated TeleVision (AGTV) and was flight-tested during 1984/85.

Specifications
Dimensions:
(camera head) 76 × 228 mm
(electronics unit) 241 × 184 × 165 mm
Weight:
(camera head) 1.36 kg
(electronics unit) 4.54 kg
Power supply: 28 V DC, 30 W

Status
In service. The LLLTV system also forms part of the US Air Force AC-130H Spectre Gunship, in which role it enables the crew to covertly illuminate targets with the aid of a separate flashlight laser. In January 1999, the US Air Force Special Operations Command contracted Lockheed Martin to produce an upgraded capability with a larger aperture lens giving improved light-gathering capability and increased magnification, an intensified CCD camera, an ambient cooled laser illumination and improved servo system.

Contractor
Lockheed Martin Missiles and Fire Control.

Magic Lantern for SH-2G helicopter

Type
Airborne Electro-Optic (EO) mine detection system.

Description
Magic Lantern uses a blue-green laser and camera array to scan the water below the helicopter from surface level down to keel depth. The system correlates multiple scans to identify mines. It is able to sweep the entire 'upper column' at and below the ocean surface, and is cleared for day and night operations. Accurate navigation data from the GPS enables the SH-2G crew with Magic Lantern to locate mines precisely.

The podded Magic Lantern system replaces the Magnetic Anomaly Detector (MAD) on a strengthened hardpoint on the right side of the helicopter. The Magic Lantern system display in the aft cabin of the SH-2G shows the Sensor Operator (SENSO) mine detection symbology and real-time video imagery of suspected mine contacts. The ASN-150 Tactical Navigation system (TACNAV) provides signals to the HSI in the cockpit enabling the pilots to fly predetermined search patterns. A Tactical Decision Aid (TDA) has been developed to aid aircrews in determining search patterns and analysing post-mission data. A miniaturised airborne GPS is part of the SH-2G modifications to provide EOD crews with precise position data.

Status
In service.

Contractor
Kaman Aerospace Corporation.

Magnetrak Helmet-Mounted Sight (HMS)

Type
Helmet-Mounted Sight (HMS).

Description
The line of sight data acquired by the sensor/source arrangement is passed into the aircraft weapon aiming system using a MIL-STD-1553B databus link. Conversely, target data detected by one of the aircraft's sensors such as the radar or thermal imager, can be relayed to the Magnetrak system, to give target cueing information on the pilot's visor. Thus the pilot's own line of sight can be directed to a possible target for visual identification before attack.

The components of the Magnetrak system include a small three-axis sensor and source, system electronics unit and memory unit, the latter supplying mapping data to the former. There is also a visor display consisting of a parabolic visor, (and cover) and an integral LED reticle generator which projects a collimated cross-hair image on to the visor, allowing the pilot's eye to focus on target and reticle simultaneously. Discrete dots around the cross-hair give a cueing facility. The visor fits any

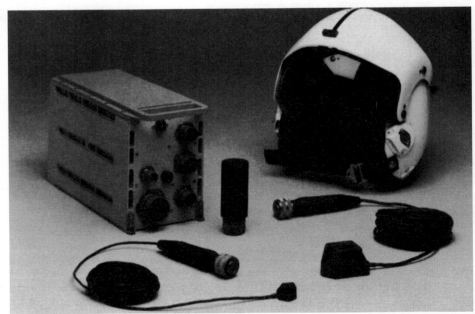

The Polhemus Magnetrak helmet sight. Left to right: system electronics unit and connecting cables, aircraft memory unit, magnetic field sensor, helmet with visor display and source unit   0581562

standard military helmet without modification to the shell. The electronics unit is 1/2 ATR in format.

The Magnetrak offers a resolution of 0.1° and covers 360° movement in azimuth and roll axes and ±90° in elevation. The motion box, inside which the head must remain for accurate system performance, is 410 × 254 × 150 mm either side of a central point.

Specifications
Dimensions:
(source) 61 × 35 × 35 mm
(sensor) 28 × 23 × 18 mm
(electronics unit) 124 × 174 × 283 mm
(memory unit) 109 × 33 mm
Weight:
(source) 0.16 kg
(sensor) 19 g
(electronics unit) 5.13 kg
(memory unit) 0.17 kg
Power supply: 115 V AC, 400 Hz, single phase, 0.7 A

Status
In production and in service.

Contractor
Polhemus Inc.

Mast-Mounted Sight (MMS)

Type
Airborne Electro-Optic (EO) targeting system.

Description
In October 1984, the US Army ordered the MMS for its Army Helicopter Improvement Programme (AHIP) Bell OH-58D helicopters. The sight is mounted over the main rotor drive shaft of the host helicopter main rotor, thus allowing the aircraft to remain behind cover (foliage/terrain), while the sight *only* is unmasked to the target.

The MMS integrates a stabilisation platform and electrical systems with a sensor suite purchased from Northrop Grumman. The sensors include a low-light television, a thermal imaging sensor and laser range-finder/designator. The sensors contained within a 650 mm carbon-epoxy sphere, with a sightline 810 mm above the plane of rotor rotation. Data is channelled into the cockpit area via cables, through a 23 mm tube running inside the drive shaft. A feature of AHIP is the Automatic Target Handover System, by which targeting information from one aircraft can be transmitted to another or to ground-based weapons. AHIP is the precursor to the much more ambitious RAH-66 light battlefield helicopter system.

In order to minimise the vibration levels associated with the rotor shaft mounting of the system, Boeing devised a 'soft mount' which provides a high degree of isolation. Performance of the anti-vibration system is such that target bearing can be measured to within 20 mrad. The pivotal requirement to minimise total system weight above the main rotor resulted in a sensor package weighing only 73 kg. Equipment bay systems add another 41 kg.

The television camera has a silicon-vidicon dawn to dusk capability with an 8° field of view for target acquisition and a 2° field for recognition. The FLIR sensor has a 120 element common module detector array and two fields of view of 10 and 3°. Television and laser systems share a common optical path to minimise the number of components. A video tracker and digital scan converter in the data processor together permit the incorporation of other features such as autotrack, frame-freeze and point-track.

In the cockpit are two multifunction displays for the presentation of video from the MMS, together with flight guidance and com/nav information.

The MMS is applicable to a variety of helicopter platforms, including the Boeing 500 Series, the

The mast-mounted sight on an OH-58D helicopter   0503977

TV image of a Kiowa Warrior seen through the mast-mounted optics. Senosr images can be captured and saved for onward transmission   0523337

The mast-mounted sight display is on the left of this Bell OH-58D helicopter cockpit. The view is adjusted by controls on the stick on the pilot's right-hand side   0503978

Sikorsky H-76, the Bell 406, the Agusta A 129 and the Eurocopter BK 117. The MMS can be used to provide these platforms with both AHIP scout features and light attack capability, using Hellfire and/or TOW missiles.

Specifications
System weight: 113.4 kg
Turret
　Dimensions: 647.7 × 1193.8 mm (Ø × H)
　Weight: 72.57 kg

Stabilisation: 2-axis, <20 μrad
Azimuth: ±190°
Elevation: ±30°
CCD television camera
   Type: Low-light silicon vidicon
   Spectral range: 0.65–0.9 μm
   Field of view: 2° NFOV; 8° WFOV
   Line format: 875
Thermal imaging sensor
   Detector: 120-element HgCdTe scanning
   Spectral band: 8–12 μm
   Aperture: 16.76 cm
   Field of view: 2.8° NFOV; 10° WFOV
   Line format: 875
Laser range-finder/designator
   Type: flashlamp NdYAG
   Wavelength: 1.06 μm

Status

In production and in service. Over 400 systems delivered to the US Army for the OH-58D Kiowa helicopter. MMS was used extensively in Operation Desert Storm, designating the first tank kill to weapons fired from a US Army AH-64 Apache. It is reported that Taiwan has purchased 51 MMS systems for its OH-58D helicopters. In total, more than 500 units in total have been purchased by the US and international customers.

Contractor

DRS Technologies, Inc.

Mast-Mounted Sight Laser Range-Finder/Designator (MMS-LRF/D)

Type

Airborne Laser Range Finder/Target Designator (LRFTD) system.

Description

Northrop Grumman Laser Systems produces the MMS-LRF/D, a NdYAG laser system, which equips the US Army's OH-58D helicopters (see separate entry). Use of advanced packaging techniques enables the electronics, high-voltage power supplies, cooling system, range receiver and laser transmitter to be housed in a single LRU. The unit includes an asynchronous digital interface for precise communication with the MMS computer internally. All subassemblies interface with a common bus under control of a main processor, which maintains system timing through software control. Built-in test circuitry periodically monitors system operation, allowing the processor to compensate automatically for component degradation.

The MMS suite is comprised of a thermal imaging sensor for night operations, a low-light level television camera, a multimode tracker and a laser range-finder/designator. Later variants of the system incorporate an eye-safe mode, which shifts the laser output to 1.54 μm.

Specifications

Wavelength: 1.064 μm (tactical), shifted to 1.54 μm for eye-safe training applications
Pulse Repetition Frequency: (PRF): Band I, Band II, tri-service codes
Beam diameter: 0.52 in
Weight: 6.4 kg
Dimensions: 143 × 210 × 271 mm
Power: 28 V DC, 15 V DC, 8.5 V DC

The MMS-LRF/D is in service on the OH-58D Aeroscout helicopter   0051022

Status

In production and in service. A modified unit has been integrated into the all-light-level television system in the Lockheed Martin AC130U.

Contractor

Northrop Grumman Corporation, Electronic Systems, Laser Systems.

MicroStar™ compact dual imager system

Type

Airborne Electro-Optic (EO) surveillance system.

Description

FLIR Systems' MicroStar™ compact dual imager is designed to be fitted to weight- and/or size-restricted airframes, including UAVs and employed in the tactical surveillance, force protection and Search and Rescue (SAR) roles. The system features a lightweight, low-profile, fully sealed 9 in gimbal designed to minimise drag, thus decreasing fuel consumption and thereby increasing on-station time for the host platform.

The MicroStar™ incorporates dual imaging sensors – high-resolution infra-red and boresighted CCD-TV with low-light capability. Continuous zoom optics allows the operator to customise the Field of View (FoV) as required without losing sight of the target; with fixed FoV systems, the operator may lose sight of the target during magnification changeover. FLIR Systems claims that no other IR system combines continuous zoom optics with an IR sensor.

MicroStar™ control functions are configured in FLIR Systems' 'Terse Binary Protocol' software, designed to work with the Company's Hand Control Unit (HCU). The HCU is a simple, intuitive and non-fatiguing control device created to help operators during extended periods of use and intense action.

An Autotracker reduces workload and fatigue by keeping the selected scene or target within the FoV without constant operator inputs. Pushbutton access is provided for the three distinct target tracking modes:
- Centroid – used to track moving targets
- Scenelock – for tracking larger stationary scenes
- Correlation – used to track small, slow moving targets.

Specifications
System
   Gyrostabilisation: 2-axis, fibre optic gyros
   Azimuth/Elevation slew rate: 0–50°/s
   Power: 18-32 V DC at 15 A

MicroStar™   0137882

Weight: 11.8 kg
Dimensions: 229 × 343 mm

Thermal imager
   Sensor: 320 × 240 InSb Focal Plane Array
   Wavelength: 3–5 μm
   FoV: 2.2–22° horizontal, 10:1 continuous zoom

Daylight camera
   Pixel arrangement:
   768H × 494V lines (NTSC)
   752H × 582V lines (PAL)
   Zoom: 18:1 continuous zoom (×4 digital zoom)
   Resolution: 470+ television lines
   FoV: 2.7–48° horizontal, continuous zoom
   Sensitivity: 3 lux at f/1.4 or 0.2 lux at f/1.4

Electronic Control Unit
   Power: 18-32 V DC at 15 A
   Weight: 4.13 kg
   Dimensions: 242 × 100 × 203 mm
   Interface: RS 232/422
   Video output: PAL/NTSC
   Additional components: RS 232 serial GPS interface; Stow mode; On-screen display of geoposition, date/time and tracker mode
   Options: Monochrome CCD daylight camera; laser pointer; ARINC interface; searchlight slave interface; 30 MW laser illuminator

Status

In production and in service. MicroStar™ has been selected for the US Army's RQ-7A Shadow UAV.

Contractor

FLIR Systems Inc.

NITE Hawk targeting FLIR

Type

Airborne Electro Optic (EO) targeting system.

Description

Development of NITE Hawk began in March 1978 with the requirement to provide the F/A-18 Hornet with a day and night strike capability. The first systems, designated AN/AAS-38, were delivered to the US Navy in 1983.

The NITE Hawk presents the pilot with real-time passive thermal imagery in a television formatted display to assist in the location, identification and tracking of targets. The system provides the aircraft mission computer with accurate target line of sight pointing angles and angle rates. Automatic in-flight boresight compensation is used to correct for dynamic flex and thus maintain accurate pointing angles throughout the aircraft's full performance envelope. Increased weapon delivery accuracy is provided through

a Laser Target Designator/Ranger (LTD/R) feature (AN/AAS-38A). The LTD/R provides precise target range information and designation capability for precision-guided munitions.

Additional operational capabilities were incorporated into the NITE Hawk AN/AAS-38B, which has been in production since 1992. A laser spot tracker has been included which allows the system to search for, acquire and track targets, which have been laser designated by ground or other airborne sources. A multifunction

In some Lockheed Martin literature, the latest variant of NITE Hawk is also referred to as NITE Hawk Block III. The Block III capability includes a new third-generation FLIR to replace the first-generation FLIR in the original variant. The modification much improves detection and identification ranges and reduces maintenance costs. Installation of the new FLIR is said to involve very minor modifications to the NITE Hawk system.

NITE Hawk carried on the port shoulder pylon of a F/A-18D Hornet of the US Marine Corps (Lockheed Martin)
1044581

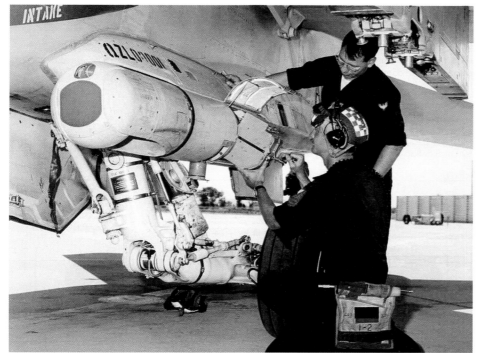

The NITE Hawk FLIR is a modular system; in this image, the control processor LRU is being replaced (Lockheed Martin)
1044582

The gimballed head of the NITE Hawk FLIR provides for full freedom in roll (Lockheed Martin)
1044583

Specifications

Dimensions:
(AN/AAS-38/38A/38B) 1,840 × 330 mm diameter
(NITE Hawk (SC)) 2,440 × 330 mm diameter
Weight:
(AN/AAS-38) 158 kg
(AN/AAS-38A/38B) 168 kg
(NITE Hawk (SC)) 195 kg
Field of view:
(wide) 12 × 12°
(narrow) 3 × 3°
Field of regard:
(pitch) +30 to −150°
(roll) ±540°
Stabilisation: 35 µrad
Tracking: 230 µrad
Pointing: 400 µrad
Tracking rate: 75°/s

Status

The first production NITE Hawk system was delivered to the US Navy in December 1983. Systems have been delivered to, or are on order for: Australia, Canada, Kuwait, Malaysia, Spain, Switzerland and Thailand.

NITE Hawk SC has been flight tested on the F-14, F-15, F-16 and AV-8B. The NITE Hawk SC system is the baseline laser/designator pod for the Spanish Eurofighter Typhoon.

Lockheed Martin completed development of the NITE Hawk AN/AAS-46 configuration for the F/A-18E/F at the end of 1996.

With over 500 NITE Hawk systems delivered, it is envisaged that the AN/ASQ-228 Advanced Targeting FLIR (ATFLIR – see separate entry) will gradually replace the AN/AAS-38/46 in US Navy service on the F/A-18E/F, while the Northrop Grumman Litening AT (see separate entry) has been successfully integrated in a number of US Marine Corps platforms.

Contractor

Lockheed Martin Missiles and Fire Control.

Pathfinder navigation/attack system

Type

Airborne navigation and laser designation system.

Description

Pathfinder is an international derivative of the LANTIRN navigation system (see separate entry).

The Pathfinder navigation/attack system uses hardware derived from the LANTIRN night navigation system. It consists of three LRUs which can be integrated into a pod or embedded in the aircraft fuselage or pylon. These LRUs are an infra-red sensor, power supply and environmental control unit. The first two of these LRUs are derived directly from LANTIRN but the environmental control unit is smaller and lighter, with reduced power requirements.

As with the LANTIRN navigation pod, Pathfinder imagery may be presented on a HUD or any other cockpit video display. Unlike the navigation pod, in addition to a ×1 wide field of view it also contains a ×3 magnified slewable field of view for standoff target acquisition.

Pathfinder has been demonstrated in a pod on the F-16, integrated into a pylon on the A-7 and embedded in the B-1B. It is suitable for installation on a wide variety of additional aircraft including the F-5, A-10, C-130, Dassault/Dornier Alpha Jet, Dassault Rafale, British Aerospace Hawk and Panavia Tornado.

---

autotracker also provides scene track, intensity centroid and geometric tracker algorithms for improved acquisition and maintenance of target lock on during attack of air-to-ground targets. Enhancements were also made to the air-to-air detection and tracking capabilities of the system.

Two further versions of the system have been developed: the AN/AAS-46 configuration for the F/A-18 E/F aircraft, and the NITE Hawk SC (self-cooled) system.

The NITE Hawk SC is designed to perform in a supersonic flight environment, providing single or multiseat aircraft with 24-hour strike capability against land- or sea-based targets. NITE Hawk SC interfaces with other aircraft avionics over a MIL-STD-1553B multiplex databus.

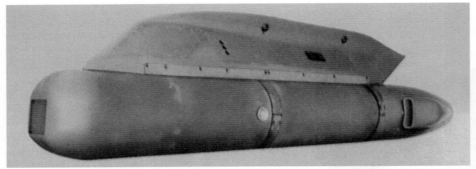

The Lockheed Martin Pathfinder is a FLIR sensor pod derived from the LANTIRN navigation pod

0504135

Specifications

Dimensions: 1,950 (length) × 248 mm (diameter)
Weight: 90.5 kg
Power supply: 115 V AC, 400 Hz, 3 phase, 2.8 kW
  28 V DC, 2 A
Field of regard: 77 × 84°
Field of view:
(wide) 21 × 28°
(narrow) 7 × 9°
Temperature range: –40 to +90°C
Reliability: 539 h MTBF

Status

In production and in service with a number of air forces worldwide, including those of Egypt, the US and Taiwan.

Contractor

Lockheed Martin Missiles and Fire Control.

RAYFLIR-49 and -49(LG) Day/Night Airborne Thermal Sensor (DNATS)

Type

Airborne Electro Optic (EO) navigation and surveillance system.

Description

The Raytheon Electronic Systems Forward-Looking Infra-Red (FLIR) RAYFLIR-49 is a low-weight, multiple-purpose, Focal Plane Array (FPA) thermal imaging sensor for navigation, surveillance, maritime, search and rescue and troop transport missions.

The RAYFLIR-49 includes two Weapon Replaceable Assemblies (WRA): Turret Unit WRA-49 (TU) WRA-1 and Electronics Unit WRA-49 (EU). The RAYFLIR-49(LG) combines a three-chip, colour TV, with continuous zoom and an eye-safe laser range-finder with the FPA FLIR.

Features of the RAYFLIR-49 system include multiple field of view; electronic image stabilisation; Local Area Processing (LAP); an adaptable interface and multimode video tracker; MIL-STD-1553, ARINC 429 and RS-422 controls; 525 or 625 line video standard.

Specifications

Fields of view:
(wide) 22.5 × 30°
(medium) 5 × 6.67°
(narrow) 1.3 × 1.7°
Electronic zoom: 2:1 and 4:1
Gimbal angular coverage:
(azimuth) 360° continuous
(elevation) +40 to –150°
Gimbal slew rate: 3 rads/s
Gimbal stability: 4 axis <25µrad
Gimbal angular resolution: <100 µrad
Interface slaving: radar, INS, mission or weapon
  computers
Control interfaces: MIL-STD-1553, ARINC 429,
  RS-422
Video outputs: 525 RS-170, 625 CCIR
Max airspeed: >300 kt
Dimensions:
(turret unit) 323.85 diameter × 375.92 height mm
(electronics unit) 306.32 (W) × 413.52 (L) ×
  199.14 (H) mm
Power: standard aircraft power
Options: cameras, lasers, illuminators,
  displays, recorders, spot trackers and
  other avionics

Status

In service.

Contractor

Raytheon Company, Space and Airborne Systems.

RAYFLIR-49 and -49(LG) day/night airborne thermal sensor

0062231

RISTA Reconnaissance, Infra-red, Surveillance, Target Acquisition

Type

Airborne reconnaissance system.

Description

Northrop Grumman, under contract to the US Army's Joint Precision Strike Demonstration Office, has developed an advanced multimode/multimission electro-optical reconnaissance system that detects, identifies and pinpoints threat targets from an airborne platform.

The system consists of a second-generation IR sensor located in the airborne platform, a ground-based processor for Aided Target Recognition (AiTR) and an Image Processing Facility (IPF) which houses the AiTR and processes detected target images for dissemination to using organisations. This advanced concept provides the army with the capability to search over a wide area and locate threats in real time to provide critical intelligence to battlefield commanders.

The approach to RISTA builds upon advanced image processing algorithms developed for the (now cancelled) Comanche programme and represents the culmination of more than two decades of advanced image processing research.

The RISTA sensor is derived from the army's Airborne Standoff Minefield Detection System (ASTAMIDS) programme. The ASTAMIDS second-generation sensor has been upgraded to include a Forward-Looking Infra-Red (FLIR),

RISTA sensors are also part of the Pilatus PC-12M/MRA proposal

0106533

while retaining the line scanning capability of the original. Similarly, the AiTR processor uses a design employed on the army's MSAT-AIR programme.

The IPF is derived from the army's ETRAC programme. Configured for use in a HMMWV, the workstations provide mission planning and sensor control, real-time image display and exploitation and reporting capability.

The system has two modes of operation: spotlight and line scanning with either mode selectable during flight from the IPF. Each mode is designed to support a particular portion of the reconnaissance mission.

The spotlight mode (FLIR) is used to perform battle damage assessment or provide longer viewing of a suspected target to enhance recognition. In this mode, the line-of-sight of the RISTA sensor is geographically stabilised (geo track) to a set of UTM co-ordinates commanded by the sensor control function within the IPF. Imagery is transmitted to the IPF at a 30 Hz rate.

The line scanning mode is used to perform reconnaissance over a broad area. In this mode,

the sensor swathe covers an eight-mile wide track on the ground and the AiTR screens the incoming data for likely targets. Image target chips centred about AiTR target detections are presented to the image analyst for confirmation and dissemination.

Status

In 1998, Northrop Grumman was selected by the Danish companies Per Udsen and Terma Elektronik to outfit the Royal Danish Air Force's fleet of F-16 fighters with RISTA sensors. The award is the first international sale of the system. To date, it is understood that the RDAF has procured two RISTA medium altitude infra-red sensors, which are part of the sensor payload of the Tactical Reconnaissance System (TRS) pod, carried in the forward section. RISTA is also one of the sensor options of the Pilatus PC-12M/MRA.

Contractor

Northrop Grumman Corporation, Electronic Systems Sector.

SAFIRE™ thermal imaging system AN/AAQ-22

Type
Airborne Electro-Optic (EO) surveillance system.

Description
Military-qualified as the AN/AAQ-22 thermal imaging system, the SAFIRE™ (Shipborne/Airborne Forward Looking Infra-Red Equipment) is a digital, high-resolution 8 to 12 µm wavelength system with three-axis gyrostabilisation and full 360° turret rotation. The SAFIRE™ system, designed for fixed- or rotary-wing applications, features dual fields of view: Wide Field of View (WFoV – 28°) is designed for navigation and area searches, while Narrow Field of View (NFoV – 5°), designed for target tracking, is supplemented by electronic zoom and freeze-frame facilities. System modularity allows for a variety of payload configurations and provides many automated features, such as auto image adjustment and autoscan.

The system includes a hand controller and flight-hardened video display. Options include an autotracker, Laser Target Designator (LTD), Laser Range Finder (LRF), searchlight slaving, and radar and navigation interfaces. The hermetically sealed SAFIRE™ turret is flight qualified for airspeeds over 400 kt. SAFIRE™ systems are certified on more than 20 fixed- and rotary-wing aircraft.

Specifications
Detector: 4 × 4 CMT array
Spectral Band: 8–14 µm
Scanning: serial/parallel
Cooling: closed cycle
Cool downs time: <6 mins
Dimensions:
   Gimbal: 446 × Ø 384 mm
   Central Electronics Unit (CEU): 320 × 260 × 196 mm
   Controller: 229 × 114 × 117 mm
   Monitor: 220 × 256 × 268 mm
Weights:
   Gimbal: 42.5 kg
   CEU: 10.2 kg
   Controller: 1 kg
   Monitor: 4.76 kg
Field of regard:
   Azimuth: 360° continuous
   Elevation: +30 to −120°
Field of view:
   (narrow) 5 × 3°
   (wide): 28 × 16.8°
Zoom: ×1.85, ×10.5, ×21 (electronic zoom)
Pixels per line: 525 × 392 active pixels
Stabilisation: 3-axis
Interfaces: MIL-STD-1553B, RS-170, RS-232, RS-422, ARINC 429, CCIR

Status
In service worldwide in 50 nations. Fielded by all branches of the US military. Other users include the Royal Danish Navy, the Japanese Maritime Safety Agency, the Royal Netherlands Navy, and the Royal Saudi Naval Forces.

SAFIRE™ hand controller, with optional LRF/ LTD controls mounted on top   0132674

SAFIRE™ with optional Laser Range Finder (LRF)/Laser Target Designator (LTD)   0015295

The basic SAFIRE™ turret-mounted system, showing narrow- (upper) and wide-(lower) FoV TI imagers   0005547

Contractor
FLIR Systems Inc.

Sharpshooter targeting pod

Type
Airborne Electro Optic (EO) targeting system.

Description
Sharpshooter is an international derivative of the targeting pod used in LANTIRN. Sharpshooter carries the designator AAQ-14(V)1 to differentiate it from the AN/AAQ-14 pod of the LANTIRN system. It can be integrated with the Pathfinder navigation pod to permit pilots to fly at low altitudes in total darkness to the target area.

Sharpshooter provides round-the-clock targeting capabilities through the use of a 200 mm aperture infra-red sensor that has a 1.7° narrow field of view and a 6° wide field of view. A stabilised line of sight includes a 150° look-back angle and continuous roll tracker, which can track either stationary or moving targets or track a scene using area correlation. The laser designator and ranger is boresighted to the centre of the field of view of the targeting pod and is programmable for coding. Sharpshooter includes an eye-safe laser for training.

Status
In production and in service with a number of air forces worldwide, including those of Bahrain, Egypt, Greece, Israel and Saudi Arabia. Historically, the US Air Force ordered 20 Sharpshooter (and 20 Pathfinder) pods in 1998 for F-16 aircraft for Taiwan. The USAF ordered a further 39 Pathfinder/Sharpshooter pods for Taiwan in June 2000, also for fitment to F-16 aircraft.

Contractor
Lockheed Martin Missiles and Fire Control.

Sniper® Advanced Targeting Pod (ATP)

Type
Airborne Electro-Optic (EO) targeting system.

Description
Sniper, designated AN/AAQ-33 by the US Air Force, is an advanced, long-range precision targeting system developed by Lockheed Martin for application on current and future US fighter aircraft. PANTERA (Precision Attack Navigation and Targeting with Extended Range and Acquisition) is the export variant of Sniper.

Sniper is modular design, allowing for podded, semi-conformal and internal configurations. The current system is in the form of a lightweight pod, which differs from current targeting pod designs, such as the Low Altitude Navigation and Targeting Infra-Red for Night (LANTIRN), Thermal Imaging And Laser Designation (TIALD) and

LITENING (see separate entries), in employing a dual-faceted sapphire fixed-shroud window nose section, which provides for roll only; the internal optical equipment provides for freedom in pitch. This approach avoids the requirement for a ball-type head and attendant cavities which are reported to be susceptible to airflow-induced acoustic vibrations. The point faceted front window should also provide some degree of stealth for external carriage on current fighter aircraft such as the F-16. Further, Sniper's small diameter compared to rival systems results in reduced drag in external carriage.

To achieve the required high performance from the sensor package, Sniper incorporates a highly stabilised platform, with the optical bed stabilised with six mechanical isolators and smoothing algorithms employed to eliminate any residual jitter in the image. In common with current advanced targeting pods, Sniper XR features a 5 in common aperture, eliminating the need for continuous auto-boresighting.

A dual-CCD TV/IR sensor package, including a Generation 2½ staring array FLIR operating in the 3 to 5 μm band (MWIR), is integrated into the single-aperture design, providing for long-range identification of targets and reduced overall system weight. The FLIR employs an Environmental Control Unit (ECU) for cooling located at the rear of the pod. FLIR electronics include microscan, 2-D resolution enhancement algorithms and automatic boresighting and stabilisation techniques (including an electronic 'de-rolling' facility to elevate system performance beyond that offered by current systems).

Other Sniper features include an autotracker, a diode-pumped laser with eye-safe training mode, passive air-to-air target detection and tracking, a Night Vision Goggle (NVG) compatible laser marker and a laser spot tracker. Pilot interface is via Hands-On Throttle And Stick (HOTAS) and high-resolution MultiFunction Display (MFD).

The pod was designed from the outset to be compatible with a variety of different aircraft types without needing to address the host platform's Operational Flight Programme (OFP), thereby facilitating rapid switching of pods in the field. This is achieved via an integration module – a 5,000 line software package which tailors the pod to a specific aircraft type. If a pod switch is demanded from one aircraft to a different type, then the only required change is to this module.

The Sniper pod has been demonstrated at supersonic speed, with Lockheed Martin claiming significant performance improvements over current Targeting Pod (TGP) designs, coupled with far lower acquisition and ownership costs.

In operational testing, Sniper has demonstrated target acquisition and identification range performance over twice that of current Target Designation Pod (TDP) designs. While much of this improvement in performance is undoubtedly due to the high resolution sensors, outstanding area/target track, autofocus and FLIR calibration features facilitate reduced workload for the operator, particularly for pilots of single-seat aircraft such as the Lockheed Martin F-16.

PANTERA is an advanced navigation and targeting system developed from Sniper for application on current and future international aircraft. Offering the same benefits as Sniper, PANTERA is aimed at aircraft such as the F-16

This front view of an F-16 carrying Sniper shows the slightly non-symmetrical rotating head mount and dual-faceted sapphire window (Lockheed Martin) 0568202

A Sniper pod mounted on the port intake pylon of an F-15E. Note the ECU arrangement at the rear of the pod (Lockheed Martin) 0568248

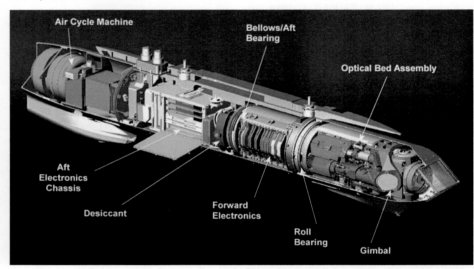

A schematic of Sniper, which is comprised of LRUs with a median weight of less than 2.7 kg. Circuit cards and laser cards are mounted back-to-back to save space; the design uses commercial technologies such as PowerPC processors (Lockheed Martin) 0111721

Block 30/40/50 and AM/BM MLU, F/A-18 Hornet, Eurofighter Typhoon and JAS 39 Gripen. While PANTERA hardware is the same as for Sniper, software is customised and platform dependent.

Specifications
FLIR
IR detector: 512 × 640 pixel staring focal plane array
IR waveband: 3–5 μm (MWIR)
Field of view
wide: 3.6 × 3.6°
narrow: 1.0 × 1.0°
digital expansion: 0.5 × 0.5°
Electronic zoom: 4× continuous

TV camera
CCD: 659 × 494 pixels
Field of view: wide/narrow
Electronic zoom: 4× continuous

Pod
Field of regard:
pitch: +35 to −155°
roll: continuous
Stabilisation: Better than 20 μrad (6 g RMS) (demonstrated 10-11 μrad)
Boresight error: 115 μrad
Laser Range Finder (LRF): 1.064 μm (operational) and 1.57 μm (eyesafe)
Dimensions: 302 mm diameter; 2,300 mm long (without ECU); 2,490 mm long (with ECU)

Weight: 177 kg (without ECU); 181 kg (with ECU)
MTTR: 18 min
MTBF: >600 h

Status
In production and in service. The US Air Force (USAF) selected the Lockheed Martin Missiles and Fire-Control Sniper XR (eXtended Range) system as the winner of its Advanced Targeting Pod (ATP) competition in August 2001. The initial contract provided pods, plus associated equipment, spares and support.

First deliveries of the system began in January 2003, to equip US Air Force F-16CJ Block 50 and Air National Guard (ANG) F-16C Block 30 aircraft. As of January 2005, the Sniper has been deployed operationally with USAF F-16C/D Block 30/40/50 and F-15E units.

The pod is also destined for the A-10C and B-1B and a variant of the core Sniper system will be carried internally by the F-35 Joint Strike Fighter (JSF). Integration into US Marine Corps and Navy F/A-18 aircraft has been proven, although with Northrop Grumman's LITENING ER/AT equipping US Marine Corps F/A-18C/D and Raytheon's Advanced Targeting FLIR (ATFLIR) in service in US Navy F/A-18E/F aircraft, confirmed sales may be harder to achieve.

The 174th Fighter Wing of the New York ANG attained Initial Operating Capability (IOC) with Sniper on their F-16C/D on 21 April 2006.

Since entering USAF service in January 2005, Sniper has accumulated over 20,000 flying hours in more than 3,000 sorties in support in Operation 'Iraqi Freedom'. Sniper has proven effective in providing stand-off detection of improvised explosive devices (IEDs) along convoy routes in Iraq. As a result of this experience, Lockheed Martin is to fit a video down-link capability to the Sniper under the terms of a USD9 million cost-plus contract from the USAF, announced on 23 May 2006. This will allow troops on the ground to simultaneously view the same display as the pilot in his cockpit, via a Rover III ISR (Intelligence, Surveillance and Reconnaissance) datalink man-pack receiver manufactured by L-3 Communications.

During September 2005, Singapore ordered Sniper for its procurement of F-15SG multi-role fighter aircraft.

As of 25 May 2006, there were 135 Sniper ATPs on order for the USAF and ANG, 100 of which had been delivered. The planned procurement currently stands at 522 Sniper pods, valued at USD843 million, for US active-duty and reserve units by 2008, but this is expected to rise due to A-10 and other bomber requirements.

PANTERA export sales include Oman, Poland and Norway, with Sniper/PANTERA in competition to equip Australian and Canadian F-18s. In September 2003, PANTERA was delivered to the Royal Norwegian Air Force (RNoAF) with first deployment of the system onboard an F-16AM aircraft during exercise Maple Flag in May 2004. In October 2004, a RNoAF F-16AM carrying a PANTERA pod equipped with a data downlink capability successfully demonstrated one of the new capabilities that the advanced Sniper design brings to the targeting pod – that of near-realtime reconnaissance and surveillance, enhancing Situational Awareness (SA) for both air and ground forces and providing a powerful tool in modern network-centric operations. Immediately after the demonstration, the RNoAF exercised its option to acquire additional PANTERA pods.

Contractor
Lockheed Martin Missiles and Fire Control.

Star Q™ airborne observation system

Type
Airborne Electro-Optic (EO) surveillance system.

Description
Star Q™ is a four-axis gyrostabilised multisensor platform fully ruggedised to military standards (MIL STD 810E). Star Q™ features CCD TV and a three FoV Thermal Imager (TI) which employs an advanced FPA detector operating in a narrow sub-band (8 to 9 μm) of the LWIR region (8 to 14 μm).

The QWIP TI provides significantly enhanced platform capability, combining high sensitivity and an optical magnification ratio of 25:1 to facilitate target detection, recognition and identification at large stand-off ranges.

The Star Q™ gimbal design features open modular architecture, allowing the addition of a variety of further payload options including: eye-safe Laser Range-Finder (LRF) and very narrow FoV CCD TV spotter scope.

Specifications
Turret
Gyrostabilisation: 4-axis

Star Q™ turret   0547501

FoR:
Azimuth: full 360° coverage
Elevation: +20 to −105°
  Weight: 42 kg
  Dimensions: 490 × 410 mm

Thermal Imager
Sensor: GaAs Quantum Well Infra-red Photodetector
(QWIP)
Resolution: 320 × 240 or 640 × 480 FPA
Cooling: Integrated Stirling cooler
Wavelength: 8-9 μm
FoV (320 × 240 array):
Narrow: 0.98 × 0.74° (h × v)
Medium: 6 × 4.5°
Wide: 25 × 18.75°
Electronic Zoom: ×4 with digital interpolation

TV camera
Camera: 1/3 in 3-CCD sensor
Resolution: 800 TV lines (PAL)
FoV: 0.67 to 35° zoom
Focal length: 9.5 to 256.5 mm f3.5 to 512 mm f7.0
Zoom: × 54 (× 27 with × 2 switchable extender)
Options: Wide spectrum spotter scope, LRF (LP-15)

Status
In production and in service. During November 2001, the UK Ministry of Defence (MoD) awarded an GBP8 million (USD11.6 million) contract to FLIR Systems International Ltd, the UK subsidiary of FLIR Systems Inc to supply Star Q™ EO imaging systems to equip Royal Air Force and Royal Navy Sea King Search and Rescue (SAR) helicopters. The contract covered 16 systems for RAF Sea Kings, with options for a further six for Royal Navy Sea Kings. The helicopter upgrade work was completed during 2004.

Contractor
FLIR Systems Inc.

Star SAFIRE II™ thermal imaging system

Type
Airborne Electro-Optic (EO) surveillance system.

Description
Developed from the Star SAFIRE™, the Star SAFIRE II™ is a military qualified COTS multisensor gyrostabilised airborne platform which can incorporate up to five boresighted payload options, including an advanced Thermal Imager (TI), near-IR sensitive low-light CCD, high-magnification spotter scope, laser illuminators or pointers, and a laser range-finder. The advanced 5-axis stabilisation system provides precise motion correction in the fourth and fifth axes, enabling increased magnification to be utilised to increase effective detection and, more importantly, identification ranges.

In addition, Star SAFIRE II™ incorporates a new, all-digital, Central Electronics Unit (CEU),

Star SAFIRE II™   0137861

designated SPECTRUM™, which incorporates a simplified interface, facilitating aircraft integration and enhanced growth capability. The system is available with embedded moving map/ mission planning systems.

Optional features include a multimode autotracker, high magnification spotter, laser illuminator/pointer/range-finder, haze/low-light image processing, geopointing/target positioning, MIL-STD-1553B interface, radar, navigation and weapon/FCS interfaces, handheld/panel mount and custom/remote system controllers, quick disconnect mounts, flat-panel displays, RF downlinks and VTR accessories.

Specifications
Platform
Dimensions: 446 × Ø383.5 mm
Weight: 44.5 kg (including all options)
Stabilisation: 5-axis
Azimuth coverage: 360° continuous
Elevation: +30 to −120°
Slew rate: variable up to 60°/sec
Airspeed: Up to 405 kt (flight tested)
Environmental: MIL-STD-810, -461

Thermal Imager
Spectral band: 3–5 μm
Detector: 320 × 240 InSb FPA
Resolution: microscanned to 640 × 480
Field of view:
  (narrow) 0.8 × 0.6°
  (medium) 3.4 × 2.6°
  (wide) 25.2 × 18.8° (optional 33.3 × 25°)
Cooling: Stirling linear cooler

TV sensor
Type: Colour CCD
Sensitivity: 0.2 lux
Output: 450/460 lines (PAL/NTSC)
Field of view: 24 to 2.1°
Zoom: ×18, continuous, plus × 4 electronic

Laser Range Finder (LRF)
Type: Class 1 (eye safe), 1.54 μm
Divergence: <1 mrad
Accuracy: ±5 m (≤10 km)

Laser illuminator
Wavelength: 0.82–0.85 μm

Long-range spotter scope
Field of view: 0.4 × 0.3°
Interfaces: MIL-STD-1553B, RS-232, RS-422, ARINC 429, RS-170, CCIR

Status
In production and in service. During the latter part of 2002, FLIR Systems received an order for eight Star SAFIRE II™ systems worth USD5.5 million for upgrading Mexican Navy CASA C-212-200 aircraft. The turret will be mounted under the forward fuselage and will form part of CASA's Fully Integrated Tactical System (FITS) which provides combined radar/FLIR mission capability. Deliveries began in 2003 and will run until 2005. Initial integration will be handled by CASA in Seville. In addition, a USD3.5 million contract from the Modern Technologies Corporation, in conjunction with the US Air Force Warner Robins Air Logistics Centre was placed to supply the system for installation in US Air National Guard (ANG) Lockheed HC-130H Hercules Search And Rescue (SAR) aircraft. The programme covers the procurement and installation of eight SAFIRE II™ systems up to 2005. Deliveries were expected to begin in 2003. The contract also provides for future delivery of up to 37 additional systems, through to 2006, for the entire fleet of HC-130 aircraft. If fully funded and exercised, this future delivery option is valued at approximately USD16 million.

In November 2002, Agusta SpA awarded FLIR Systems a USD10 million contract to equip Royal Danish and Portuguese Air Force EH101 helicopters with 20 Star SAFIRE II™ systems. The two-year delivery period commenced in February 2003.

Contractor
FLIR Systems Inc.

Star SAFIRE III™ thermal imaging system

Type
Airborne Electro-Optic (EO) surveillance system.

Description
The AN/AAQ-36 Star SAFIRE III™ is a military qualified COTS multisensor, gyrostabilised airborne platform which incorporates up to six boresighted payload options, including a high-resolution advanced Thermal Imager (TI) with extended range performance, an Intensified CCD (ICCD) camera, a high magnification, multi-FoV spotter scope, narrow and wide area laser illuminators/pointers and laser range-finder. Star SAFIRE III™ is FLIR Systems' first imaging system to offer >640-pixel resolution, autofocus for all sensors, digital video and optional image fusion. Available with an integrated Mission Management System (MMS – moving map/tactical mission planner).

Specifications
TI
Type: 640 × 480 InSb FPA with enhanced fidelity
Spectral waveband: 3–5 μm
FoV: 4 FoV from 25° to 0.4° (optional wide field of view >33°)

LLTV
Type: Colour LLTV with near-IR sensitivity
Sensitivity: 0.2 lux
ICCD: Multi-FoV with matched optics for image fusion
Zoom: × 25 continuous zoom plus × 4 electronic

CCD
Type: Colour CCD with >800 TV lines
FoV: Multi-FoV with matched optics for image fusion (NFoV 0.29°)

System
Weight: 44.5 kg (including all options)
Standard: 5-axis stabilisation; autofocus; ARINC-419/-429, single-/dual-RS-232/-422/-485 interfaces
Options: 320 × 240 InSb FPA (enhanced); NTSC or PAL; ICCD; high-magnification spotter; multimode autotracker, laser illuminators/pointers/range-finder; haze/low-light image processing, geo-pointing/target positioning; MIL-STD-1553B; radar, navigation and weapon/FCS interfaces; handheld/panel mount and custom/remote system controllers; quick disconnect mounts; flat panel displays; RF downlinks and VTR/accessories

Status
In production and in service.

In January 2004, the US Army Space and Missile Defense Command ordered 27 Star SAFIRE III systems (plus two Star SAFIRE IIs) in a USD16.8 million contract. In April 2004, the Royal Australian Air Force placed an AUD10 million order for Star SAFIRE III systems to equip its fleet

Star SAFIRE III™   0547504

of AP-3C Orion MPAs. In June 2005, FLIR Systems announced that it had received an order for an unspecified number of Star SAFIRE III from the US Army's Space and Missile Defense Command. At the same time a contract was received from Canada to equip 28 Sikorsky H-92 (CH-148) Cyclone maritime helicopters.

Star SAFIRE III is in US Coast Guard (USCG) service, equipping the new EADS CASA HC-235A maritime patrol aircraft being procured for the Deepwater Recapitalisation Progamme, as well as being retrofitted to the service's HC-130J from January 2007.

Contractor
FLIR Systems Inc.

Star SAFIRE™ airborne thermal imaging system AN/AAQ-22

Type
Airborne Electro-Optic (EO) surveillance system.

Description
Originally conceived for the maritime environment, the AN/AAQ-22 Star SAFIRE™ is a digital, high-resolution system enhanced for imaging over water and in humid atmospheric conditions. Star SAFIRE™ is a military-qualified COTS airborne infra-red system equipped with a third-generation 3–5 μm indium antimonide (InSb) 320 X 240 Focal Plane Array (FPA) detector.

The military qualified, Commercial-Off-The-Shelf (COTS), multisensor gyrostabilised airborne platform incorporates advanced detector technology and three Thermal Imager (TI) fields of view to provide for long-range identification and image clarity in the most adverse climates. Although compact in size, the rugged turret and modular design can incorporate up to four imager payloads for day or night operations, with options including a CCD TV camera, an eye-safe laser range-finder, NVIS-compatible

FLIR Systems' Star SAFIRE™ (FLIR Systems)
0089852

laser illuminator, long-range spotter scope and interface to navigation and radar systems.

Specifications
Dimensions: 446 × Ø383.5 mm
Weight: 43.6 kg
Stabilisation: 3-axis
Azimuth coverage: 360° continuous
Airspeed: >400 kts
Elevation: +30 to −120°

Thermal Imager
Spectral band: 3–5 μm
Field of view:
(narrow) 1.4 × 1.03°
(medium 5.7 × 4.3°)
(wide) 30 × 22.5°

TV sensor
Type: 0.33 in 1-CCD colour
Resolution: 450/460 lines (PAL/NTSC)
Minimum illuminance: 4.5 lx
Field of view:
(minimum) 2.1 × 1.6°
(maximum) 21 × 15.6°
Zoom: ×10

Long-range spotter scope
Field of view: 0.4° with 688 mm focal length lens
Interfaces: MIL-STD-1553B, RS-232, RS-422, ARINC 429, RS-170, CCIR

Status
No longer in production, having been superseded over time by Star SAFIRE II and III (see separate entries); many current users of the system have upgraded components to enhance performance in line with more current variants.

Star SAFIRE™ systems are certified on more than 20 fixed- and rotary-wing aircraft. Operators include the US Marine Corps for its fleet of UH-1N Huey helicopters in the SAR and surveillance roles, the US Navy and Customs Service for P-3 Orion aircraft for maritime patrol and surveillance missions, the New Zealand Navy for its SH-2G(NZ) helicopters and the New South Wales police for law enforcement.

Contractor
FLIR Systems Inc.

SU-172/ZSD-1(V) Medium-Altitude Electro-Optical Sensor (LAEO)

Type
Airborne reconnaissance system.

Description
The SU-172/ZSD-1(V) Medium-Altitude Electro-Optical (LAEO) sensor operates from 2,000 to 25,000 ft and features a 22° narrow Field of View (FoV), 304.8 mm (12 in) focal length, f/5.6 lens for daylight operations. High-resolution imagery

Star SAFIRE III has been procured for EADS CASA HC-235A maritime patrol aircraft being procured by the USCG for the Deepwater Recapitalisation Progamme (Jane's/Peter Felstead)   1132432

is captured on a 10 × 10 μm, 12,000-pixel CCD. The SU-172/ZSD-1(V) has been designed with emphasis on an integrated system approach to reduce life-cycle costs. The sensor is automated to reduce aircrew workload and for use in UAVs.

Specifications
Dimensions:
(imaging unit) 511 × 381 × 330.2 mm
Weight: <62.14 kg
Power supply: 28 V DC, 440 W
Altitude: 2,000–25,000 ft

Status
In production and in service as part of the AN/ASD-10(V) Advanced Tactical Airborne Reconnaissance System (ATARS, see separate entry) equipping 18 US Navy/Marine Corps F/A-18D Hornet aircraft and as part of the AN/ASD-11(V) Theatre Airborne Reconnaissance System (TARS, see separate entry) equipping US Air National Guard F-16C aircraft.

Contractor
BAE Systems, Reconnaissance and Surveillance Systems.

SU-173/ZSD-1(V) Low-Altitude Electro-Optical Sensor (LAEO)

Type
Airborne reconnaissance system.

Description
The SU-173/ZSD-1(V) Low-Altitude Electro-Optical (LAEO) sensor operates from 200 to 3,000 ft with a low distortion 140° wide field of view and vertical or forward oblique fixed focus. The sensor provides high-resolution visible spectrum imagery for high-speed, low-altitude area coverage on tactical reconnaissance missions.

The SU-173/ZSD-1(V) consists of a sensor and electronics unit that collect imagery, provide roll correction of imagery, perform preliminary image processing and output digital image data. The low-altitude sensor system has been designed to be flexible and responsive to both mission and platform tailoring. The system may be internally mounted in manned aircraft such as the RF-4C/E or F/A-18(RC) or pod-mounted in the F-14, F-16, Tornado, Mirage, Jaguar and JAS 39 Gripen, in addition to being configured for unmanned aerial vehicles.

Specifications
Dimensions:
(imaging unit) 284.5 × 121.9 × 363.2 mm
Weight: <24.5 kg
Power supply: 28 V DC, 231 W
Altitude: 200–3,000 ft

Status
In production and in service as part of the AN/ASD-10(V) ATARS (see separate entry) equipping 18 US Navy/Marine Corps F/A-18D Hornet aircraft.

Contractor
BAE Systems, Reconnaissance and Surveillance Systems.

SureSight™ Enhanced Vision System (EVS)

Type
Airborne Electro-Optic (EO) Enhanced Vision System (EVS).

Description
CMC Electronics has developed the SureSight™ Enhanced Vision System (EVS) family of products to increase flight crew situational awareness by helping them see through fog, haze, precipitation and at night to improve overall safety and aircraft economic efficiency. This is accomplished by using Infra Red (IR) and Millimetric Wave Radar (MMWR) sensor systems.

EVS provides an image on the Head-Up Display (HUD), a Head-Down Display (HDD), or both, to enable the pilot(s) to see the terrain/airport environment in low visibility situations. EVS significantly improves situational awareness, not only during take-off, approach and landing, but also during ground manoeuvring.

SureSight™ is a modular design, facilitating simple integration into aircraft equipped with raster-capable displays in the general aviation, air transport and military markets. The first modules to be produced were the infra-red sensors, to be followed by a MMWR sensor system.

The EVS-IR and optional EVS-MMWR sensors are designed to interface and operate with all raster HUD systems. The infra-red sensor is installed inside of the weather radar radome, and requires an elliptical forward looking, infra-red-transparent, 'window', either built into the radome or located immediately below it. Since the EVS-MMWR has no requirement for a dedicated 'window', it is planned that the EVS-MMWR will share space within the radome, sited below the weather radar antenna. In order to accomplish this, the MMWR package must be made small and light enough to fit into the radome space and not compromise the aircraft mounts. The radar antenna will be electronically stabilised about the aircraft flight path, thereby assuring that the radar scan will always be centred on the aircraft's trajectory. The infra-red sensor and the EVS-MMWR sensor can operate in the unpressurised environment of the radome compartment. However, the image fusion processor, requiring a pressurised environment, would need to be located below the cabin floor or in an avionics rack located within the pressurised areas of the host aircraft.

Two versions of SureSight™ are available, the I-series and the M-series. They have the following capabilities and applications:

CMA-2600 SureSight-I™
The CMA-2600 is a high-performance EVS, designed for the most demanding applications. The main features of the system comprise:
- Advanced, second-generation EVS-IR sensor, providing superior sensitivity and image quality
- High sensitivity for maximum weather penetration
- Single dual-band focal plane array to support certified approach and landing systems without the need for fusion hardware/software
- Single LRU for maximum ease of installation
- Compact, lightweight design minimises impact on aircraft
- Proven second-generation technology for superior performance and long life
- Sealed unit to simplify maintenance
- Compatible with any video capable HDD such as MCDU, EFIS and EFB.

The total installed weight of the CMA-2600 is expected to be less than 8.2 kg, including wiring harness, brackets, connectors and radome modification.

CMA-2610 SureSight-M™
The CMA-2610 is a low-cost variant of the EVS, providing enhanced situational awareness improves situational awareness with a single-band LWIR uncooled sensor:

Specifications
(CMA-2610)
Weight: <1.8 kg
Sensor: Uncooled IR
Wavelength: 8.0 – 14.0 μm LWIR
Resolution: 320 × 240
Certification: RTCA DO-178B
Dimensions: 71mm (Ø) × 177.8 mm (L)
Installation: Tail fin or nose

Status
Flight trials of the EVS infra-red sensor system were held in Everett, Washington, during May 2000 and in the Northeast US during July 2001.

CMC Electronics gained Canadian certification for its I-series sensor during the first quarter of 2005. The sensor is utilised in the EVS developed by Bombardier for the Global business jet family, where it is integrated with a Thales HUD.

Rockwell Collins is also reported to have adopted the SureSight sensor for its integration activities on several business jet types.

Work is ongoing to combine the outputs of the IR and MMWR sensors, using sensor fusion techniques, to blend the data from the two sensors, on a pixel-by -pixel basis, to produce an accurately matched image, which is then projected onto the basic HUD combiner, at the identical scale and orientation of both the combiner guidance information and the outside scene. The EVS-MMWR sensor, together with sensor fusion electronics will be retrofittable to the basic HUD/EVS-IR package at a later date.

Contractor
CMC Electronics Inc.

SYERS 2 reconnaissance system

Type
Airborne reconnaissance system.

Description
The principal imaging intelligence sensors of the U-2R high-altitude reconnaissance aircraft are the Senior Year Electro-optical Reconnaissance System (SYERS 2) and the Advanced Synthetic Aperture Radar System (ASARS 2 – see separate entry), both manufactured by Goodrich Surveillance and Reconnaissance Systems (formerly Itek Optical Systems). One or the other is mounted in the nose of the U-2, according to mission requirements.

The original SYERS programme was born out of a requirement to replace the film camera on the U-2 with a long-range, high-resolution Electro-Optical (EO) sensor whose images could be recorded on tape or data linked to a ground station. SYERS became operational on the U-2 fleet in the late 1980s. A SYERS upgrade programme was initiated in the early 1990s, resulting in SYERS 2.

SYERS 2 is a long focal length multiband sensor with three visible and four infra-red imaging channels, two Mid-Wave IR (MWIR) and two Short-Wave IR (SWIR), for day and night imaging capability. The system fits into a rotating nose section of the U-2R, facilitating a horizon-to-horizon Field of Regard (FoR) from a stand-off position or directly overhead. SYERS 2 offers improved range, resolution and area coverage over the original system, and features multispectral imaging capability with seven imaging bands (as opposed to two, one visible and one MWIR in the original SYERS system). The inclusion of two SWIR imaging bands in SYERS 2 improves the capability of the sensor in poor visibility during daylight hours while bolstering the system's ability to defeat many forms of camouflage through multi-spectral imaging.

When SYERS 2 is powered up, the nose section of the U-2R rotates into the operating position and automatically scans large rectangular areas along the flight path of the aircraft. The system transmits continuous Push broom still-frame

The SYERS II system is housed in the rotating nose section of the U-2R high-altitude reconnaissance aircraft   0055084

images of each rectangular area in all seven bands simultaneously to a ground station.

Status
Successful Engineering and Manufacturing Development (EMD) tests were completed in 2000. Now in production. Goodrich delivered the first production systems to the US Air Force U-2 fleet during 2002.

Contractor
Goodrich Surveillance and Reconnaissance Systems.

Tactical Airborne Reconnaissance Pod System (TARPS)

Type
Airborne Electro-Optic (EO) reconnaissance system.

Description
The Tactical Airborne Reconnaissance Pod System (TARPS) is carried on the F-14 Tomcat. The 5.2 metre, 17-foot, 841 kg pod houses three camera sensors mounted in equipment racks. The front of the pod houses a two-position (vertical and forward oblique) Recon/Optical KS-87 frame. In the central portion of the pod is a KA-99 low-altitude panoramic camera, with the rear position filled by an AN/AAD-5 Imaging Infrared (II) sensor camera (see separate entries for details of the camera systems). TARPS is currently the US Navy's primary organic reconnaissance system.

An upgrade to the original system, known as TARPS-DI (Digital Imagery), replaces the KS-87 frame camera with a Recon/Optical CA-260 digital frame camera, enabling the F-14 host aircraft to downlink digital imagery via Common Data Link (CDL) in near real-time. TARPS-DI is used primarily in the location and identification of high-priority

A US Navy F-14 showing off its centreline TARPS pod   0034302

threats to the Battlegroup, with each downlinked image carrying a latitude and longitude 'stamp', facilitating rapid reaction by strike aircraft.

The TARPS-DI upgrade is viewed as an interim step to the acquisition of a true real-time large FoV, high-resolution tactical reconnaissance capability, possibly by the acquisition of the Advanced Tactical Airborne Reconnaissance System (ATARS).

TARPS is carried by F-14A and D model aircraft, having been active in operations such as Desert Storm, Deny Flight and, most recently, over Afghanistan, where TARPS-DI has been utilised to great effect, gathering timely and accurate information on enemy movements, enabling fast reaction by ground forces.

Specifications
See relevant entries for sensors.

Status
A contract was awarded in 1996 to supply five sets of digital cameras for testing in the TARPS pod on the F-14 Tomcat. The system, known as TARPS-DI, replaces the KS-87 framing camera with a digital frame camera (Recon/Optical CA-260/261) which is able to store up to 200 images in digital format for onward transmission via datalink. The US Navy is reported to have ordered 24 TARPS-DI pods, to be delivered by 2003 at a cost of USD8.6 million.

TARPS is due to be replaced by the Shared Reconnaissance Pod (SHARP) in US Navy service.

Contractor
Northrop Grumman Corporation, Electronic Systems Sector.

TADS laser range-finder/target designator

Type
Airborne Laser Range Finder/Target Designator (LRFTD) system.

Description
Northrop Grumman is the designer and manufacturer of the laser range-finder/target designator, integrated into the Target Acquisition Designation Sight/Pilot Night Vision Sensor

(TADS/PNVS – see separate entry) and fitted to the Boeing AH-64 Apache attack helicopter. The system enables laser ranging and designation of targets for the helicopter's Hellfire weapon system and other laser-guided munitions.

Specifications
Wavelength: 1.064 µm
Weight:
  laser targeting unit: 6.6 kg
  laser electronics unit: 8.2 kg

Power: 115 V AC, 400 Hz, 3-phase
Operating temperature: +11 to +55°C

Status
In production and in service on AH-64 Apache helicopters of many nations, including the UK WAH-64 programme.

Contractor
Northrop Grumman Corporation, Electronic Systems, Laser Systems.

Target Acquisition Designation Sight/Pilot Night Vision Sensor (TADS/PNVS)

Type
Airborne Electro Optic (EO) navigation and targeting system.

Description
Developed for the Boeing AH-64A Apache attack helicopter, Lockheed Martin's Target Acquisition Designation Sight/Pilot Night Vision Sensor (TADS/PNVS) is designed to provide day, night and limited adverse weather target information and navigation capability. The TADS/PNVS system comprises two independently functioning subsystems known as TADS (AN/ASQ-170) and PNVS (AN/AAQ-11).

The AN/ASQ-170 TADS subsystem provides the co-pilot/gunner with search, detection and recognition capability by means of direct-view optics, TV or FLIR sighting systems, which may be used singly or in combination, according to tactical, weather or visibility conditions.

TADS consists of: a rotating turret, mounted on the nose of the helicopter and containing the

Litton's TADS laser range-finder/designator   0051021

TADS/PNVS on a UK Army WAH-64 Apache Longbow (E L Downs)   0126991

TADS/PNVS on a US Army AH-64A. Note the TADS (lower gimbal), which is slaved to the front gunner's eyeline, while the PNVS (upper gimbal) is not in use (the pilot is looking directly ahead, while the PNVS is stowed) (Boeing)   0126994

sensor subsystems; an optical relay tube, located at the co-pilot/gunner station; three electronic units in the avionics bay; and cockpit controls and displays. TADS turret sensors have a Field of Regard (FoR) covering ±120° in azimuth and from +30 to −60° in elevation.

By day, either direct vision or television viewing may be used. The direct vision system has a Narrow Field of View (NFoV), 3.5° at ×18.2 magnification, and a Wide Field of View (WFoV), 18° at ×3.5 magnification. The television system provides a NFoV of 0.9° and a WFoV of 4°. For night operations, the FLIR sensor has three FoVs: narrow (3.1°), medium (10.2°) and wide (50°).

Once acquired, targets can be tracked manually or automatically for attack with 30 mm gun, 70 mm rockets or Hellfire anti-tank missiles. The associated Litton laser (see separate entry) may be used to designate targets for attack by other helicopters or by artillery units firing the laser-guided anti-armour Copperhead shell.

The AN/AAQ-11 PNVS subsystem provides the pilot with flight information symbology, which permits ultra-low level/nap-of-the-earth flight

to enhance survivability in dense ground-air environments.

PNVS consists of a FLIR sensor system packaged in a rotating turret mounted above the TADS, an electronics unit located in the avionics bay and the pilot's display and controls. The system covers a FoR of ±90° in azimuth and from −45 to +20° in elevation. Instantaneous FoV is 30 × 40°.

TADS is designed to provide a back-up PNVS capability for the pilot in the event of the latter system failing. The pilot or co-pilot/gunner can view, on his own display, the video output from either TADS or PNVS. Although designed primarily for combat helicopters flying nap-of-the-earth missions, PNVS may also be used individually in tactical transport and cargo helicopters.

Status

TADS/PNVS has been fielded with the AH-64 since 1981. More than 1,200 systems have been delivered to date. In addition to the US Army, international customers include Egypt, Greece, Israel, Netherlands, Saudi Arabia, the United Arab Emirates and the United Kingdom.

The first TADS/PNVS for the UK WAH-64 Apache was delivered in May 1998 by Lockheed Martin, working with a team of British companies to build an internationally manufactured TADS/PNVS system.

During the last quarter of 2000, the US Army selected a new sensor system intended to give the AH-64D significantly greater night vision and targeting capabilities. A USD 78.5 million development contract was awarded to Team Apache Systems (TAS), a limited liability company comprising Boeing and Lockheed Martin, for the Arrowhead advanced targeting and navigation system. Drawing extensively on technology integrated into the Comanche Electro-Optic Sensor System (EOSS), the Modernised TADS/PNVS (M-TADS/PNVS) provides improved performance and more effective integration with the Apache weapon system (see separate entry).

Contractor

Lockheed Martin Missiles and Fire Control.

Theatre Airborne Reconnaissance System (TARS)

Type

Airborne reconnaissance system.

Description

The Theatre Airborne Reconnaissance System (TARS) consists of a forward facing Recon/Optical CA-260 electro-optical framing sensor (see separate entry) and a BAE Systems AN/ZSD-1(V) Medium Altitude Electro Optical (MAEO) sensor (see separate entry) installed in a TERMA Modular Reconnaissance Pod (MRP). Control of the TARS is exercised through the F-16 mission computer utilising the AN-ALQ-213 countermeasures system control panel and MultiFunction Display (MFD).

A BAE Systems ground exploitation system comprises five transit boxes, offering field-deployable viewing, screening and exploitation capability to give squadron commanders near-realtime reconnaissance data.

Specifications – per individual sensors.

Status

Production complete. The AN/ASD-11(V) Theatre Airborne Reconnaissance System (TARS) is in service with the US Air National Guard, which has operated 20 systems on F-16C aircraft since mid-1998.

Components of BAE Systems' Theatre Airborne Reconnaissance System (TARS)   0079767

FLIR Systems' Ultra 7500™, 8000™ and 8500™
0137867

In 2003, a Pre-Planned Product Improvements (P³I) to the TARS saw a STANAG 4575-compliant Solid State Recorder (SSR) and an Airborne Information Transmission (ABIT) common datalink installed.

Contractor
BAE Systems, Reconnaissance and Surveillance Systems.

TIFLIR-49

Type
Airborne Electro-Optic (EO) navigation and surveillance system.

Description
TIFLIR-49 is a low weight, multiple purpose, thermal imaging sensor for navigation, surveillance, maritime search and rescue and troop-transport missions.

TIFLIR-49 includes two weapon replaceable assemblies: turret unit WRA-1 and electronics unit WRA-2. These units, together with off-the-shelf displays, controls and recorders provide a flexible, low-cost FLIR system for fixed- or rotary-wing aircraft.

Features include: a second generation focal plane array; electronic image stabilisation; dual-mode video tracker; multiple fields of view; local area processing; MIL-STD-1553B interface.

Specifications
Fields of view
(wide) 22.5 × 30°
(medium) 5 × 6.67°
(narrow) 1.3 × 1.7°
Electronic zoom: 2:1 and 4:1
Gimbal angular coverage

(azimuth) 360°
(elevation) +40 to −105°
Gimbal slew rate: 3 rad/s
Gimbal acceleration: 30 rad/s² at 300 kt
Gimbal angular resolution: <100 µrad
Gimbal angular position: <2.0 mrad RMS, dynamic
Cooling: self-contained
Weight
  Turret unit: 25.9 kg
Electronic unit: 191.5 kg
Dimensions
  Turret unit (Ø × H): 323.9 × 372.9 mm
Electronic unit (W × L × H): 306.3 × 311.9 × 199.1 mm
Maximum airspeed: 300 kt

Status
Available. TIFLIR is understood to be essentially the same as the AN/AAS-44, but stripped of the laser, for international sales. The system completed a successful installation and test on board a US Coast Guard HC-130 aircraft.

Contractor
Raytheon Company, Space and Airborne Systems.

Ultra 7500™ thermal imaging system

Type
Airborne Electro-Optic (EO) surveillance system.

Description
FLIR Systems' Ultra 7500™ features a lightweight, dual-sensing gimbal and optional laser illuminator to facilitate 24-hour, high-altitude, long-range search and surveillance from airborne platforms. Its triple payload capability features a 320 × 240 pixel InSb Infra-Red (IR) imager and Low-Light TV (LLTV) camera.

With a 23 cm diameter and weighing approximately 12 kg, the fully sealed gimbal

unit is designed to minimise drag and provide enhanced ground clearance for smaller airborne platforms. The LLTV camera, featuring a ×18 optical zoom, delivers clear images in daylight, while also enhancing operations during dusk and dawn. The infra-red optics includes continuous zoom, allowing the operator to customise the Field of View (FoV). An autotracker feature keeps designated targets or scenes within the FoV without the need for constant operator input.

Specifications
Turret dimensions (h × d): 343 × 229 mm
Turret weight: 11.8 kg
Field of Regard: 360° continuous in azimuth and elevation
Slew rates: 0.02 to 50° per second in both axes
Stabilisation: Fibre-optic gyro

Thermal imager
Detector: 320 × 240 InSb FPA
Waveband: 3–5 µm
Fields of view: 2.2 × 1.65°, 22 × 16.5°
Zoom: 10:1, continuous
Calibration: Internal Nonuniformity Correction (NUC)

Daylight imager
Pixel arrangement: 768 (h) × 494 (v) (NTSC); 752 (h) × 582 (v) (PAL)
Resolution: 460 (PAL), 470 (NTSC) television lines
Telescope: 18:1 continuous zoom, 4× electronic zoom
Fields of view: 0.7° (e-zoom) to 48° horizontal continuous zoom, NFOV 2.7°
CCD sensitivity: 0.2 lux at f/1.4 w/out filter, 3 lux at f/1.4 w/filter

Electronic control unit
Power requirements: 18 V DC to 32 V DC input
Optional Features: ARINC, RS-232 serial GPS, radar, SLASS interfaces, laser pointer/illuminator, IRCCD and CCD optical extenders.

Status
Launched September 2001. In production and in service in a variety of fixed- and rotary-wing platforms.

Contractor
FLIR Systems Inc.

TIFLIR-49 turret unit WRA-1   0018315

During 2002, ENAER completed a flight test programme of its T-35TD Turbo Pillán trainer with FLIR Systems' Ultra 7500 sensor (ENAER)   0562817

Ultra 8000™ thermal imaging system

Type
Airborne Electro-Optic (EO) surveillance system.

Description
The lightweight, triple-payload-capable FLIR Systems' Ultra 8000™ is designed for pursuit and patrol missions. It includes a high-definition digital Indium Antimonide (InSb) thermal imager and 18×, 0.2 lux low-light TV camera. The user can select either continuous zoom or three preset fields of view. The TV and IR imagery zoom can lock together. An optional Class III3b CW diode laser pointer allows ground forces equipped with NVGs to see the 30 mw laser's beam and spot. Icon-based on-screen colour graphics and a simplified hand control are used to control the system. The Ultra 8000™ supports moving maps and searchlights, and downlinks and displays GPS data with an optional input kit.

Specifications
TI: 320 × 240 InSb FPA
Spectral waveband: 3–5 µm
Zoom: Continuous zoom telescope with × 10 magnification
FoV: continuous zoom and 3 preset: 21.7° × 16.5° (WFoV), 8° × 6° (MfoV), 2.2° × 1.65° (NFoV)
Video output: PAL or NTSC
TV: 768 (H) × 494 (V) (NTSC); 752 (H) × 582 (V) (PAL)
TV Lines: 460 (PAL), 470 (NTSC)
Zoom: 18:1 continuous zoom, 4× electronic zoom
FoV: 48° × 32° to 2.2° × 1.65° continuous or matched to 3 FOV IR settings
Weight: 13 kg
Optional features: ARINC, RS-232 serial GPS, radar and SLASS interfaces, PIP monitor

Status
In production and in service.

Contractor
FLIR Systems Inc.

Ultra 8500™ thermal imaging system

Type
Airborne Electro-Optic (EO) surveillance system.

Description
The Ultra 8500™ is a compact 9 in (230 mm) military-qualified, stabilised turret with up to three boresighted sensors. Optimised for light and medium fixed- and rotary-wing aircraft operation, the Ultra 8500™ includes an advanced Thermal Imager (TI) with auto focus and continuous zoom IR lens, a colour CCD TV with auto focus and low-light TV mode, and an optional Class IIIb long-range laser illuminator to pinpoint ground targets. In addition to continuous zoom, the TI has three preset FOVs and also a 'focus memory', which almost eliminates the need to re-focus between zoom settings. The low-light TV camera has an 18× optical and digital zoom and can be locked to the IR sensor FOV. The system provides real-time automatic image optimisation, an imbedded multimode autotracker, and an intuitive, icon-based colour graphical overlay. Designed for law enforcement, search and rescue, and general surveillance missions, the Ultra 8500™ provides powerful imaging capabilities in a small and lightweight (13 kg) gimbal package.

Specifications
TI: 320 × 240 InSb FPA
Spectral waveband: 3–5 µm
Zoom: Continuous zoom telescope with × 10 magnification
FoV: continuous zoom or 3 presets: 21.7° × 16.9° (WFoV), 8 × 6° (MfoV), 2.2 × 1.65° (NFoV)
Video output: PAL or NTSC

TV: 752 (H) × 582 (V) (NTSC); 768 (H) × 494 (V) (PAL)
TV Lines: 460 (PAL), 470 (NTSC)
Zoom: 18:1 continuous zoom, 4× electronic zoom (maximum magnification × 72)
FoV: 48° × 32° to 2.2 × 1.65° continuous or matched to 3 FOV IR settings
Weight: 13 kg
Optional features: ARINC, RS-232 serial GPS, radar and SLASS interfaces, PIP monitor, Terse Binary Protocol (TBP) for remote and UAV functionality, compact electronic control box for UAV applications

Status
In production and in service.

Contractor
FLIR Systems Inc.

Ultra 8500FW™ thermal imaging system

Type
Airborne Electro-Optic (EO) surveillance system.

Description
Derived from the Ultra 8500™, the Ultra 8500FW™ is a 9 in (230 mm) military-qualified, stabilised turret with up to three boresighted sensors. The Thermal Imager (TI) in the Ultra 8500FW™ features 450 mm optics, focus memory and a continuous zoom IR lens with three preset Fields of View, a colour CCD camera with auto focus and low-light mode and an optional Class IIIb long-range laser illuminator to pinpoint ground targets. Other characteristics of the system are as for the Ultra 8500™. Designed for law enforcement, search and rescue and general surveillance missions, the Ultra 8500™ provides for extended range operations in the same small and lightweight (13 kg) gimbal package. The system also includes an NVIS-compatible laptop computer or simplified hand controller.

Specifications
TI: 320 × 240 InSb FPA
Spectral waveband: 3–5 µm
Zoom: Continuous zoom telescope with × 10 magnification and × 1.8 extender
FoV: continuous zoom or 3 presets: 21.7 × 16.9° (WFoV), 8 × 6°/4 × 3° (MFoV), 2.2 × 1.65°/1.22 × 0.915° (NFoV)
Video output: PAL or NTSC
TV: 752 (H) × 582 (V) (NTSC); 768 (H) × 494 (V) (PAL)
TV Lines: 460 (PAL), 470 (NTSC)
Zoom: 18:1 continuous zoom, 4× electronic zoom (maximum magnification × 72)
FoV: 48 × 32° to 2.2 × 1.65° continuous or matched to 3 FoV IR settings
Weight: 13 kg
Optional features: ARINC, RS-232 serial GPS, radar and SLASS interfaces, PIP monitor, Terse Binary Protocol (TBP) for remote and UAV functionality, compact electronic control box for UAV applications

Status
In production and in service.

Contractor
FLIR Systems Inc.

UltraFORCE II™ airborne observation system

Type
Airborne Electro-Optic (EO) surveillance system.

Description
The UltraFORCE II™ is a high-performance multisensor airborne observation system. It features three payloads – a high resolution, advanced LWIR QWIP TI, a 3-CCD broadcast quality high-magnification TV camera and a dual-camera high-magnification spotter scope.

FLIR Systems' UltraFORCE II™   0137866

The IR imager features autofocus, auto image optimisation and three-field, high-magnification lenses. The optional dual-CCD spotter scope provides extreme long-range capability and the ability to have colour imaging in daylight and low-light monochrome imaging at night. Optimised for airborne law enforcement, the system features numerous aircraft mounts and interfaces.

Specifications
Gyrostabilised platform
Type: 4-axis gyrostabilised gimbal
Turret dimensions (h × d): 490 × 410 mm
Weight: 39.5 kg

Thermal imager
Camera type: LWIR QWIP FPA; three FoV optical system w/auto focus
Detector: GaAs QWIP 320 (h) × 240 (v) FPA
Spectral band: 8–9 µm

Fields of view
Wide: 25 (h) × 18.75° (v)
Medium: 6 (h) × 4.5° (v)
Narrow: 0.98° (h) × 0.74° (v)
Electronic zoom: ×4 with digital interpolation
Format: PAL or NTSC
Cooling: Integrated long-life Stirling cooler

Colour zoom TV camera
Sensor format: 1/3 in 3-CCD sensor
Resolution: 800 TV lines (PAL)
Active pixels: 752 (h) × 582 (v)
Field of view: 0.67° to 36° HFoV zoom
Zoom ratio: 54 × (27 × with 2× switchable extender)
Format: PAL or NTSC
Optional features: Dual-sensor spotter scope, eye-safe Laser Range Finder (LRF), laser pointer, digital autotracker, real time video and audio downlink, and SLASS capability.

Status
In production and in service.

Contractor
FLIR Systems Inc.

WF-360 surveillance and tracking infra-red system

Type
Airborne Electro-Optic (EO) targeting system.

Description
The WF-360 system features a Forward-Looking Infra-Red (FLIR) sensor operating in the 8 to 12 µm long-wave spectrum, with options for adding a high-resolution day TV camera and an eye-safe laser range-finder. These sensors are boresighted together on an optical bed housed in a stabilised gimbal designed for aircraft applications. The system provides high-resolution day and night detection and tracking of airborne,

The Northrop Grumman WF-360 surveillance and tracking infra-red system   0581494

maritime and land-based targets. System capabilities include passive search and track in both air-to-air and air-to-surface modes. ARINC and MIL-STD-1553 bus interfaces are provided to communicate with other aircraft sensors such as the AN/APG-66/68 radar, flight management systems and inertial navigation systems. A high-throughput processor provides cueing of the system to radar targets or navigation waypoints, real-time display of the track point co-ordinates and fire-control solutions.

The WF-360 infra-red sensor marries US Army common modules with electronics providing DC restoration, automatic gain and level control and digital scan conversion for a standard US RS-170 525 line video output. These features, together with an advanced digital noise reducer and image enhancer, provide superior video imagery. Virtually all FLIR systems in the US DoD inventory use common module parallel scan long-wave technology because of its capability for widely varying operational scenarios throughout the world.

The aerodynamically streamlined turret utilises a two-axis stabilised platform to stabilise the optical lines of sight and point them throughout the entire lower hemisphere with look-up limited only by the aircraft structure. A broadband servo system featuring solid-state rate sensors and an inner acceleration loop provides stabilisation better than 35 μrad in typical aircraft environments ranging from helicopters to the US Air Force A-10. The turret is environmentally self-contained, employing an internal liquid-to-air heat exchanger to allow proper operation of the turret without the need for bleed or cabin air.

The WF-360's astronomical telescope provides two fields of view: a 4.5° narrow field of view and an 18.5° wide field of view. The TV sensor uses a ×6 zoom lens which can match either FLIR field of view or zoom continuously.

In addition to the TV and laser range-finder, other optional plug-in modules provide capabilities including FLIR/TV image fusion, wide area correlation video tracking and covert laser illumination for night TV imagery. The computer-aided track mode, in conjunction with the automatic video trackers, provides an excellent coast mode in the event of short-term target obscuration.

Status
Used extensively for surveillance and drug interdiction by the US Coast Guard in the HU-25 Falcon jet and RU-38, the US Air Force in the C-26 and the US Army in the DH-7.

An improved system, featuring the second-generation common module 480 × 4 scanned focal plane array is available.

Contractor
Northrop Grumman Corporation, Electronic Systems Sector.

YAL-1A Airborne Laser (ABL)

Type
Airborne tactical laser system.

Description
Boeing is leading a team selected by the US Air Force to develop and demonstrate the Airborne Laser (ABL). The high-power laser is coupled with a revolutionary optical system capable of focusing a 'basketball-sized' spot of heat that can destroy a boosting missile from hundreds of miles away. The laser and optical systems are controlled by a sophisticated computer system that can simultaneously track and prioritise potential targets.

The team includes the USAF, Northrop Grumman and Lockheed Martin. Boeing is responsible for developing the ABL surveillance, Battle Management, Command, Control, Communications, Computers and Intelligence (BMC⁴I) architecture, integrating the weapon system, and also supplies the 747-400F aircraft. Northrop Grumman will provide the Chemical Oxygen Iodine Laser (COIL) and ground support. Lockheed Martin is developing the Beam Control/Fire Control (BC/FC) system.

ABL is one part of a 'Family of Systems' (FOS), designed to counter Theatre Ballistic Missiles (TBMs). ABL will destroy hostile TBMs while they are still in the highly vulnerable boost phase of flight before separation of the warheads. Operating above the tropopause, the system will autonomously detect and track missiles as they are launched, using an onboard surveillance system. The BC/FC system will acquire the target,

then accurately point and fire the laser with sufficient energy to weaken the missile casing to the point where the violent aerodynamic loads experienced by the missile during this phase of flight will be sufficient to cause structural failure.

The megawatt-class offensive laser is some of the oldest technology on the venerable 747-400F. Developed by the USAF during the 1970s, the laser functions by a chemical reaction between chlorine, hydrogen peroxide and iodine to create what is termed as 'an explosion of light'. This light travels down a mirrored tube and flexible hose to the rotating nose turret (see diagram). It is estimated that the aircraft will fire between 20 and 40 kill shots before a landing for replenishment is necessary. In addition to missile defence, the ABL will have inherent capabilities to perform other activities, such as engaging threat aircraft, temporarily blinding enemy satellites, performing imaging surveillance and providing cruise missile defence.

Main features of the ABL are:

Active Ranger System (ARS)
Mounted above the upper deck of the 747-400F behind the cockpit area, the Active Ranger System (ARS) pod utilises a C-130 wing fuel pylon and an F-16 centreline fuel tank (thus reducing design and fabrication costs) to house the Active Laser Ranger (ALR). The ALR consists of a modified LANTIRN 2000 system (see separate entry), with a high-power CO_2 laser. The ARL will receive vector information from the Infra-Red Search and Track (IRST) system to enable it to track the target and point the CO_2 laser to acquire range information.

Infra-Red Search and Track System (IRSTS)
Six IRST sensors, sourced as Military-Off-The-Shelf (MOTS) components from the F-14 programme, are positioned on the tail, fuselage and on the chin pod, which also houses the relocated weather radar. Full 360° coverage is provided for surveillance, initial detection and tracking of TBMs during boost phase.

BMC⁴I
The BMC⁴I segment provides surveillance, communication, planning, and the central command and control of the ABL weapon system. It performs the following functions:

- Infra-red surveillance, detection, and tracking of multiple targets
- Target typing and prioritisation
- Distributed predictive avoidance (deconfliction)
- Mission planning (orbit selection and management)
- Military communications
- Crew/system Interface
- Theatre interoperability.

The BMC⁴I suite provides capability to deliver missile launch and impact point prediction information to the rest of the FOS; a mission

The ABL concept is based on a heavily modified 747-400F airframe. Note the Active Ranging System (ARS), mounted above the cockpit, and the relocated weather radar, mounted together with the forward Infra-Red Search and Track (IRST) sensors on a chin pod below the main laser turret
(Boeing)   0126996

The ABL is designed to integrate into the 'Family of Systems' (FOS) for space-, air- and ground-based TBM defence (Boeing)   0126997

data processing subsystem, utilising ruggedised commercial hardware, hosts the BMC⁴I software and supports the distribution of information between the weapon system segments over the Local Area Network (LAN).

Illuminator laser

Lockheed Martin is responsible for the illuminator laser, which consists of the Tracking ILluminator Laser (TILL) and the Beacon ILluminator Laser (BILL); both are solid-state, diode-pumped 2nd-generation devices.

The TILL is the heart of the beam control/fire control system, projecting rapid, powerful pulses of light on a small section of a boosting target missile. The light will be reflected back to an extremely sensitive camera. The reflected light data is interpreted as information about the target's speed, elevation and probable point of impact.

Nose-mounted turret

Lockheed Martin has fabricated the nose-mounted turret, which will house a 1.7 m conformal window intended to focus laser energy onto the target and collect return signals and image data. The lens took five years to manufacture and is described as one of the most complex optics ever developed by the US Missile Defense Agency. Azimuth Field of Regard (FoR) is 120°, with the window protected against bird strike and harmful atmospheric constituents (lightning strike, dust, and weather effects) by inward rotation, where the outer optical surface is protected by a gasket and shield. During an operational mission, three of the four lasers carried by the aircraft will be fired through this aperture.

High-Energy Laser (HEL)

Northrop Grumman is responsible for the high-energy COIL, which boasts record chemical efficiency and utilises advanced materials (plastics, composites and titanium) to reduce weight for airborne application. The laser features a closed chemical system, designed for aircraft safety and field maintainability, and incorporates modularity to allow for graceful and controlled degradation in case of failure. The COIL operates with liquid and gaseous chemicals, a mixture of hydrogen peroxide and potassium hydroxide, a salt that enhances and sustains the chemical reaction inside the laser. The beam-generating process within the COIL begins when chlorine gas is injected into a spray of hydrogen peroxide and chemical salts, producing excited oxygen. Iodine gas is then mixed with the excited oxygen to produce excited iodine. When the iodine returns in a

normal or ground state it emits photons, which are collected and amplified to create a beam capable of destroying a target several hundred miles away.

Beam control system

Lockheed Martin is responsible for the Beam Control (BC) system, which is intended to point, focus and fire the laser which provides sufficient energy to destroy an enemy missile during the highly vulnerable boost phase of flight and before separation of its warheads. The BC performs the following functions:

- Target acquisition and tracking
- Fire control engagement sequencing, aim point-and-kill assessment
- HEL beam wavefront control and atmospheric compensation
- Jitter control, alignment/beam-walk control and beam containment for HEL and illuminator lasers
- Calibration and diagnostics providing autonomous real-time operations and post-mission analysis.

Boeing engineers at the company's Laser & Electro-Optical Systems organisation in West Hills, California, under the direction of Lockheed Martin Space Systems, were responsible for the development and delivery of steering mirrors which are an important part of the overall ABL BC/FC system.

Three types of steering mirrors have been developed to meet various requirements throughout the ABL aircraft. A 31-cm 'slow mirror' is a lower-bandwidth mirror that maintains high-energy laser alignment as the 747-400F structure flexes in flight. The mirror is

responsible for ensuring the megawatt-class laser beam stays aligned within the aircraft. A 31-cm 'fast mirror' is a high-bandwidth design responsible for targeting the laser. It must correct high-frequency tilt errors caused by atmospheric turbulence. A 13-cm mirror meets both low- and high-bandwidth applications in the illuminator laser path.

The mirror substrates will be coated to protect against heating from the high-energy laser and to reflect all other illuminator, infra-red and alignment wavelengths in the beam control system.

One of the major functions of the BC/FC system is to compensate for the vibrations associated with flight and the distortion of light due to the Earth's atmosphere in order to successfully concentrate the laser's power on the targeted missile. This task requires a sophisticated network of lasers, mirrors and precision optics controlled by real-time software.

Communications subsystem

The communications subsystem has been developed from MOTS hardware and software, providing HF, VHF, UHF and AFSATCOM connectivity for both voice and data (Link 16 and TIBS), fully supporting the TMD FOS.

Display and control

The ABL Proof-of-Concept (POC) aircraft will feature eight operator consoles, while production aircraft will include four, facilitating selection and display of all parameters controlling the automatic sequencing of the weapon system.

Major engineering work to modify the basic Boeing 747-400F was required to accommodate the ABL systems, including relocation of the aircraft weather radar to a chin pod and installation of the so-called '1,000 bulkhead', which lies 1,000 in aft of the forward airframe datum and isolates the two pilots and four (for the production system) weapons systems operators from the potentially lethal chemical reaction which generates the laser discharge.

Specifications

Laser: Chemical Oxygen Iodine Laser (COIL)
Wavelength: 1.315 μm
Laser power: Megawatt-class (1–5 MW)
Range: ≥320 km (200 miles) required; approximately 450 km estimated

Programme history

The idea of utilising a laser for military operations was first proposed in the late 1960s, with work beginning on a project to field a laser-equipped aircraft in the 1970s. Initially, a KC-135A was chosen as the platform for a carbon dioxide gas dynamic laser. Designated the Airborne Laser Laboratory (ALL), the specially modified aircraft shot down its first target, a towed drone, over the White Sands Missile Range in New Mexico on 2 May 1981. The event marked the first time a high-energy laser beam had ever been fired from an airborne aircraft. On 26 July 1983, the USAF announced that the ALL had been used to shoot down five Sidewinder air-to-air missiles. It

Cutaway of the ABL 747-400F aircraft. Note the station 1,000 bulkhead, located behind the crew stations, which isolates all crewmembers from the potentially lethal chemical reactions which generate the laser discharge (Boeing)   0126998

During a flight test on the 17 May 2005, the nose turret of the YAL-1A was rotated to the operating (unstowed) position for the first time in flight (US Air Force)   0585606

The YAL-1A during flight testing over the Edwards Air Force Base range complex during December 2004   1041524

marked the apogee of the programme, although tests would not end until the ALL shot down yet another drone four months later. The aircraft was retired in 1984 and four years later was flown to Wright-Patterson Air Force Base in Dayton, Ohio, where it is now on display at the Air Force Museum.

Although the ALL had shown that a laser mounted on an aircraft could be a formidable defensive weapon, it was generally viewed as impractical. Its carbon dioxide laser was too bulky, it was dependent on an external power source, and it did not generate enough power to be effective at extended ranges. However, as a result of the SCUD missile threat encountered during the Persian Gulf War in 1991, the concept of an anti-missile laser was revived.

By this time, technological advances had dictated the replacement of the gas dynamic laser in the ALL with a vastly superior chemically-operated device, invented at the Air Force Weapons Laboratory at Kirtland Air Force Base, New Mexico, called a Chemical Oxygen Iodine Laser (COIL). The COIL resolved many of the doubts planners had about the ALL system. A number of times more powerful than the gas dynamic laser, the COIL had an internal power source, it was much more compact, and it was capable of producing a lethal beam over long distances.

As a result, rather than reviving the ALL, the USAF built an entirely new system, changing not only the laser but also the type of aircraft that would carry it. The Airborne Laser (ABL) was designed to incorporate multiple COIL modules (six in the prototype version; 14 in the production model) installed in pairs in the rear of a Boeing 747-400 freighter. In addition, a sophisticated optical system, capable of projecting a beam over hundreds of kilometres and compensating for any atmospheric disturbances that might exist between the aircraft and its target, would ensure best use of this increased laser power.

Status

In November 1996, the USAF awarded Boeing, TRW (Northrop Grumman has subsequently inherited TRW's involvement) and Lockheed Martin a USD1.3 billion Programme Definition and Risk Reduction (PDRR) contract to develop an ABL system, also known as the YAL-1A Attack Laser, that would detect, track, and destroy theatre ballistic missiles during their boost phase.

In addition, four adjunct missions for the ABL were studied during this phase: self-protection, protection of other High-Value Airborne Assets (HVAA), the role of the ABL in cruise missile defence, and the role ABL can play in airborne surveillance.

In April 1998, the ABL programme passed a key laboratory test, when a scaled beam control system demonstrated the laser pointing and focusing performance required for the ABL mission. A 'First Light' test of the flight-weighted module, a kW class COIL, was conducted at TRW's test centre in June 1998, with further progressively higher power tests scheduled to follow. The configuration of the ABL was fixed during a final design review in April 2000, initiating PDRR Phase II continuing into 2002. Heavy engineering work to modify the Boeing 747-400F platform, serial 00-00001, was completed in early 2002, including replacing the aircraft's nose with a turret for the laser and beam-control optics, adding steel struts and titanium supports to reinforce the fuselage and installing the '1,000 Bulkhead'.

Final ground tests began during the second quarter of 2002, with the first flight on 18 July, circling above McConnell Air Force Base, Kansas for 82 minutes, during which the crew evaluated the flight characteristics of the platform. The aircraft flew without the COIL and beam-control systems, both of which were still undergoing developmental work at the time. Flight-test activities began with airworthiness tasks, such as validating air-to-air refuelling capability.

During February 2004, engineers at the ABL facility at Edwards Air Force Base mixed the first batch of the chemicals needed by the COIL modules installed on the ABL aircraft. Shortly after a shipment of 4,400 gallons of hydrogen peroxide was delivered to the laser's chemical mixing facility, engineers mixed a 1,200 gallon batch of chemicals for the COIL.

Also during the early part of 2004, two large optical control assemblies for the Beam Control/Fire Control (BC/FC) system were installed on the YAL-1A prototype aircraft – the Beam Transfer Assembly (BTA) and the Multi-Beam Illuminator (MBIL) Bench. The BTA contains the beam control sensors and deformable mirrors that compensate for atmospheric distortion and fix the ABL's laser on targets. The sensors facilitate automatic target detection and tracking, and detect the atmospheric distortion information provided by the BILL. The MBIL bench, installed in the aircraft's mid-section, includes low-energy lasers and their alignment optics used to illuminate and track missiles and point the high-energy laser. The BTA contains the sensors, steering mirrors and deformable mirrors used to focus the high-energy laser on the target missile.

Lockheed Martin performed extensive testing to verify that the system accurately controls every mirror at operational data rates. The tests validated that the BC/FC system was capable of acquiring a target, initiating tracking of the target, initiating atmospheric compensation, firing the high-energy laser and shutting down the system, while maintaining beam quality and accuracy. To accomplish the tasks at the required speeds, the BC/FC system executes over 600,000 lines of C and Ada high-order software using the computer processing power of more than 80 Power PCs, capable of executing over 72 billion instructions per second.

During the second quarter of 2004, the US Missile Defense Agency (MDA) admitted that the problems of mastering the engineering and integration of a large number of 'revolutionary' technologies had been challenging and therefore keeping the ABL on its original schedule and within budget had proven difficult. Accordingly, the agency stated that it did not anticipate having the ABL available as part of the initial portfolio of ballistic missile defence capabilities, dubbed the 'Block 2004' system, which was envisaged for later that year. The Block 2004 system was rescheduled to operate to FY2006, with key enhancements incorporated thereafter.

Continuing the 'First Light' test programme, on 10 November 2004, the MDA successfully made a further test firing of the megawatt-class laser on the ABL laser testbed at the Systems Integration Laboratory (SIL). SIL is a special building at Edwards Air Force Base, which houses a modified Boeing 747 freighter fuselage, where all elements of the laser system are being assembled and tested. The test involved the simultaneous firing of all six laser modules, including the associated optics that comprise the COIL. The laser systems were reported as producing an amount of infra-red laser energy that was within pre-test expectations.

At the beginning of December 2004, the YAL-1A flew for the first time in approximately two years, after engineers had installed the system's laser-beam control system. Further flight testing concentrated on air worthiness and the functioning of the system without its megawatt-class kill laser on board.

During May 2005, the ABL's 1.7 m wide conformal window was unstowed for the first time during flight, with further demonstration of the BMC[4]I suite's ability to autonomously detect and hand off targets using Link 16 secure communications.

At the beginning of August 2005, Lockheed Martin announced that it had completed initial flight testing of the BC/FC elements, with more than 20 hours of tests validating the performance of the low-power passive beam-control system aboard the YAL-1A aircraft, including the system's ability to align the megawatt-class kill laser (not yet fitted). During October, low power systems integration passive testing, including ground and flight tests of the BMC[4] and further testing of the BC/FC segment, were completed. The tests demonstrated the stability and alignment of the two BC/FC optical benches with the turret, proving the system's pointing and vibration control functions as well as its ability to acquire targets as directed by the battle-management segment. A series of more than 20 flights verified the BC/FC system's target-tracking capability and its ability to align the HEL's full optical path in the dynamic environment of flight. Other accomplishments included the collection of data in the flight environment to verify the jitter performance of the BC/FC system and the in-flight exposure of the flight turret assembly's conformal optical glass. Any effects of the reconfigured nose on either the aircraft handling or pilot air parameter measurements were calibrated. The team has also demonstrated the ability of the BMC[4] system to autonomously detect and hand off targets using Link 16 secure communications.

During December 2005, Boeing announced the successful completion of a series of tests involving its high energy laser at the SIL. During this test series, lasing duration and power were demonstrated at more than 10 seconds and approximately 83 per cent of its design power. While the original plan had been that the laser would not be installed on the aircraft until it produced 100 per cent of its specified power, on December 9, 2005, the Director of the MDA gave the programme permission to disassemble the SIL and begin installation of the laser on the aircraft. Although the ABL had not reached its design power, the designers believe that the 83 per cent power achieved in ground tests was sufficient to achieve 95 per cent of the planned maximum lethal range against all classes of ballistic missiles.

Also reported during late 2005, a laser phenomenon known as jitter was considered a major technical risk to the ABL system's overall performance. Jitter control is crucial to the successful operation of the ABL because the beam projected by the COIL must be stable enough to focus sufficient energy on a fixed location on the target missile to rupture its fuel or oxidiser tank.

With the completion of this phase of testing, the Boeing 747-400 aircraft was moved to Boeing's Wichita facility to undergo final modification to accommodate installation of the HEL and begin low power system integration active ground testing.

During June 2006, the ABL team made a major step towards demonstrating the capability of the system by successfully firing surrogate lasers from inside the aircraft. During ground tests at Boeing facilities in Wichita, Kansas, the team placed the lasers in the ABL aircraft and fired them repeatedly into a range simulator. The tests verified that the optical beam train, a series of optical components, steering and deformable mirrors and sensors, were correctly aligned. The equipment exercised is part of the BC/FC system. The lasers used were low-power surrogates for the ABL's HEL and BILL/TILL. The program planned to install actual illuminators in the aircraft for ground and flight tests later in 2006.

On October 27th, 2006, the ABL team rolled out the YAL-1A after its ground modification work was completed in Wichita. The roll-out marked the completion of a number of key modifications to the aircraft:

- Full integration of the BC/FC system inside the ABL aircraft.
- The addition of floor reinforcements and chemical-fuel tanks to the back of the aircraft to prepare it for installation of the HEL in 2007.
- Northrop Grumman finished ground-testing the optics that will shape the high-energy laser beam and direct it from the laser to the beam control/fire control system. The optics underwent inspection and refurbishment after the laser achieved lethal power and run-times in a ground laboratory in December 2005.

As of late-2006, the ABL team were preparing for another major activity later in the year. The TILL was planned to be fired in flight at an instrumented target board located on a missile-shaped image painted on a NKC-135 test aircraft. This activity will verify ABL's active tracking and atmospheric compensation capabilities.

The ABL team are planning for a first missile shoot-down test in 2008.

As of late-2005, full-scale production was planned for six aircraft (including serial 00–00001, modified to production standard), with the USAF planning to deploy a total fleet of seven AL-1 aircraft around the 2010 to 2015 timeframe.

The USAF requires the ABL to have a lethal range of at least 320 km (200 miles). Programme officials say that, under the Bush administration, the ability of the system to engage missiles of intermediate and even strategic ranges will be determined only by its technical capabilities and not limited by policy restrictions.

The cost of the programme since its inception and up through the initial missile shootdown exercises is expected to total under USD2 billion, with the MDA requesting USD598 million for the project in FY 2003, including USD85 million to initiate the purchase of the second ABL aircraft, known as the Block 2008 variant. It will incorporate improvements over the YAL-1A, but will still be a test aircraft and not a fully operational system.

Airborne Tactical Laser (ATL)

During 1999, Boeing successfully completed POC testing of a smaller version of the ABL designed specifically for tactical weapons applications. The new device utilises the same COIL technology, optimised for power levels of 100 to 500 kW, operating at ground level and emitting no exhaust.

Tactical COIL technology facilitates a more mobile, self-contained laser weapon with claimed significant lethality at engagement ranges up to 10 km for ground-to-air defensive systems, and over 20 km for air-to-ground or air-to-air systems. Packaging concept studies have indicated that a complete weapons system can be accommodated in rotorcraft (V-22, CH-47), fixed-wing aircraft (AC-130) and ground vehicles.

The ATL is expected to destroy, damage or disable targets with little to no collateral damage, supporting missions on the battlefield and in urban operations, producing scaleable effects, meaning the weapon operator will be able to select the degree and nature of the damage done to a target by choosing a specific aimpoint and laser shot duration.

Application studies have concentrated on installation of a 300 kW laser into a V-22 Osprey platform with an onboard optical sensor suite. Operating below cloud ceiling, ATL can provide a fast-response defensive screen against low-altitude anti-ship or overland cruise missiles in high-threat environments. A ground-based Tactical COIL, sized to counter short-range tactical rockets, can be fully contained in one or two vehicles. With modified sensors and fire control, the ATL also offers a unique ultra-precise strike

capability for operations such as peacekeeping or enforcement, where pinpoint accuracy, tactical stand-off and no collateral damage are predominant considerations.

Status

During the second quarter of 1999, a POC demonstration laser operated routinely at approximately 20 kW during the test. With reliability and demonstrated repeatability, these tests explored performance over a wide range of operating conditions. For several of the tests, the laser exhaust gases were completely captured in a small sealed exhaust system. Results confirmed overall laser efficiency and the sealed exhaust system's ability to meet the requirements for a scaled-up tactical COIL weapon system.

During October 2006, flight testing for the ATL Advanced Concept Technology Demonstration (ACTD) programme began, having achieved 'first light' from the ATL's high-energy chemical laser in ground tests during September. During the low-power flight tests, the ATL ACTD system located and tracked ground targets at White Sands Missile Range, New Mexico. The low-power, solid-state laser served as a surrogate for ATL's HEL. To prepare for the tests, the ATL aircraft, a C-130H from the US Air Force 46th Test Wing, was fitted with flight demonstration hardware, which included a beam director and optical control bench, which directs the laser beam to its target and weapon system consoles, which displays high-resolution imagery and enables the tracking of targets.

It is planned that, by 2007, the full system will be installed on the test aircraft and fired in-flight at mission-representative ground targets to demonstrate the military utility of high energy-lasers. The laser will fire through a rotating turret that extends through an existing 50 in diameter hole in the aircraft's belly.

Following the 2007 tests, it is anticipated that the US DoD will approve starting ATL's full-scale development, with the system expected to enter service at the same time as the full-scale ABL system.

Contractor (ABL Programme)

Boeing Integrated Defense Systems.
Lockheed Martin Missiles & Space.
Northrop Grumman, Electronic Systems, Laser Systems Division.
US Air Force Phillips Laboratories.

Contractor (ATL Programme)

Boeing Integrated Defense Systems.
L-3 Communications.
HYTEC, Inc.

For details of the latest updates to *Jane's Avionics* online and to discover the additional information available exclusively to online subscribers please visit
jav.janes.com

AIRBORNE RADAR SYSTEMS

AIRBORNE RADAR SYSTEMS

Canada

APS-504 (V) series radar

Type
Airborne surveillance radar.

Description
The APS-504(V) series of airborne search radars are designed primarily for maritime patrol applications. They can be installed in either fixed- or rotary-wing aircraft. In addition to coastal and offshore surveillance missions, these radars can also be used for weather avoidance, low-resolution land mapping and navigation.

The APS-504(V)2 is a commercial version of the AN/APS-504 airborne search radar, which was developed specifically for Canadian Forces' Tracker aircraft. It uses the same 100 kW (peak power) I-band magnetron and transmitter pulse widths of 0.5 and 2–4 µs. The system consists of a two-axis antenna unit with parabolic antenna, transmitter/receiver unit, analogue PPI display and radar control unit. Fifty-three APS-504(V)2 radars have been produced.

Following the success of the APS-504(V)2, the APS-504(V)3 was developed to include an improved transmitter/receiver. It also includes a digital signal processor and scan converter that produces a ground-stabilised PPI display in a high-resolution 875-line video format. Navigation and cursor data are overlayed on the non-fading radar video, which can also be recorded and played back. Several sizes of high-performance flatplate antennas are available and are mounted on a two-axis pedestal. Twenty-five APS-504(V)3 radars have been produced.

The APS-504(V)5 radar is the most advanced of the Litton APS-504 family of airborne search radars. It employs a TWT-based transmitter with wideband frequency agility, high-ratio pulse compression, scan-to-scan integration and digital signal processing to enhance the detection of sea-surface targets, including targets with radar cross-sections as small as 1 m² in Sea State 3. The APS-504(V)5 can be configured to meet various installation and performance requirements. It has been installed in aircraft ranging from small twin-engined turboprops, such as the Beech 200 to larger jet aircraft such as the Boeing 737. More than sixty-seven APS-504(V)5 radars have been produced.

The APS-504(V)5 is available with a choice of two-axis or three-axis pedestals. The two-axis pedestal provides the smallest swept volume. The three-axis pedestal provides pitch-and-roll stabilisation of the antenna at scan rates up to 120 rpm, but requires a larger radome in order to accommodate the larger swept volume. A broadband antenna offers the best performance for maximum range and clutter reduction, but a selection of offset-horn parabolic antennas is also available.

Northrop Grumman (formerly Litton) also offers the Tactical Data Management System, which interfaces to the APS-504(V)5 and other avionics and provides advanced tactical capabilities.

Specifications
Power: per MIL-STD-704C; 115 V AC, 400 Hz 3-phase; and 28 V DC
Control interface: MIL-STD-1553B or radar control unit
External interfaces: MIL-STD-1553B, RS-422, ARINC 407, ARINC 429, discretes
Transmitter: Travelling Wave Tube Amplifier (TWTA)
Transmit frequency: 8.9–9.4 GHz, 16 frequencies, with selectable agility patterns
Transmit power: 6.6 kW (peak)
Compressed pulsewidths: 200 ns and 32 ns
Scan rates: 7.5–120 rpm
Track-while-scan: 20 targets
Display type: RS-343 875-line video
Formats: PPI/multi,: full scan, sector scan, range delay
Resolution: 800 × 800 pixels; alphanumeric overlay: 400 × 495 pixels

Range scales: 3–200 n miles
Weight: 180 kg (with three-axis pedestal); varies with configuration

Status
The APS-504(V)2 and APS-504(V)3 are no longer in production. The APS-504(V)5 is still in production.

Contractor
Northrop Grumman Navigation Systems Canada.

OASys Obstacle Awareness Radar System

Type
Airborne obstacle avoidance radar.

Description
Designed specifically for helicopters, the OASys radarairborne Obstacle Awareness Radar System (OARS) alerts pilots to unseen obstacles such as aircraft, towers, terrain and power lines, in difficult conditions such as smoke or fog. Amphitech claims that OASys is the only all-weather obstacle awareness system available. Benefits of the system include:
- Significantly reduced probability of CFIT
- Detection of small and dangerous obstacles
- Detection of nearby aircraft
- Provides timely warning that allows for evasive manoeuvres
- Provides increased Situational Awareness (SA) during take-off and landing.

The system is based on innovative 35 GHz radar technology utilising proprietary components, providing 100 per cent coverage of the helicopter flight path while also acting as an optimised weather radar with a 40 nm range. The system is claimed to be capable of detecting stationary and moving obstacles and power lines at up to 1 nm in low visibility conditions.

Key components of the system include:
- **Very low sidelobe antenna**
  Essential to system performance, especially at low altitude where it keeps undesired ground reflections at a level below the minimum detectable signal. Also features a high gain for increased system sensitivity
- **High-power pulsed RF source**
  Allows detection of small obstacles and provides best range resolution
- **High-accuracy pitch and roll stabilised antenna platform**
  Stabilisation within 0.1° guarantees that the antenna will radiate exactly along the required direction, thereby maximising detection sensitivity at low altitudes while minimising 'false alarms'
- **Advanced signal processing**
  Features an optimised combination of hardware, firmware and software: hardware for fast and broadband signal processing, configurable firmware for optimal processing in a given mode of operation, and system software for final video and alarm processing

OASys radar fitted to a Bell 212 helicopter (Amphitech) 0134964

The OASys radar was employed during 2002 for a US Navy UAV advanced navigation concepts study. The system is seen here mounted on the nose of the Scaled Composites Proteus (Amphitech) 0533862

- **Integrated system software**
  Provides for take-off mode, cruise mode, landing mode, and a special helicopter weather mode. Highly integrated CPU platform.

OASys incorporates Digital Signal Processing (DSP) for high performance and features audible tone/voice and visual warning signals combined with an adaptive safety zone, which adjusts to the aircraft flight profile automatically via a high-precision dual-axis gyrostabilisation system.

Specifications
Antenna: 28 cm, very low sidelobe, Ka-band
Scan rate: up to 150°/sec at 90° coverage (horizontal)
Stabilisation: better than 0.1° in a ±30° azimuth/ elevation window
Scan volume:
(typical) ±30° azimuth, ±11° elevation centred about aircraft flight vector
(maximum) ±90° azimuth, –85° to +25° elevation
Display unit
  Radar range: up to 4 nm (40 nm for weather)
  Range accuracy: 2 m
  Safety zone automode: Adaptive up to 1 nm
Power consumption: 28 V DC; 75 W standby, 250 W operating
Power transmitted RF: 2.5 W average maximum
Temperature: –40 to +60°C
Weight: 20 kg
Interface: ARINC 429, RS-422, RS-232
Software: DO-170B level C; extensive BIT
Environmental: DO-160D
Software: D0-178B Level C

Status
In production and in service. During 2001, the US Navy evaluated the system aboard a commercial helicopter, where the system was able to track a Cessna 172-sized aircraft target at over 6 miles. In September 2002, Amphitech announced its first sale of OASys to Canadian Helicopters Limited (CHL), for installation in a Bell 212 in service supporting a CHL customer in Labrador. In March 2002, the OASys radar was employed during US Navy testing of advanced UAV navigation concepts.

Contractor
Amphitech International.

The OASys obstacle awareness radar (Amphitech) 0134965

SLAR 100 Side-Looking Airborne Radar

Type
Airborne surveillance radar.

Description
The first SLAR 100 system was installed in a modified de Havilland Dash 7 operated by the Canadian Atmospheric Environment Service, which uses the aircraft to monitor ice floes and other reconnaissance duties. The equipment entered service at the end of 1985 and has been joined by a second set, fitted to a Lockheed L188 Electra.

The basic SLAR consists of a control unit, hard-copy film recorder, transmitter/receiver, central processor unit and 5.18 m antennas. The basic SLAR can be upgraded to a synthetic aperture fixed focus radar by integrating an options processor containing digital downlink interface and/or tape drive interface, Doppler beam-sharpening processor, moving target indicator, 2.44 m antennas and constant false alarm rate.

The SLAR 100 has a maximum range of 100 km on either side of the aircraft, producing map-like displays in swaths of 25, 50 or 100 km, at scales of 1:1,000,000, 1:500,000, 1:250,000 or 1:125,000 overlaid with a latitude and longitude grid.

The system has two antennas positioned along the underside of the fuselage for the Dash 7 application, each 5.28 m long and 40 cm high; equally the antennas could be mounted in an external pod. A magnetron transmitter is used. The radar imagery is combined in a central

The SLAR 100 is fitted in the Boeing Canada DHC-7 Dash 7 for use over the Northern Territory in all weather conditions   0503697

processor with aircraft attitude and navigation data before being recorded on a roll of thermally developed black and white film. The radar data can also be displayed in the aircraft or datalinked to a ground station.

The SLAR 100 radar in the Dash 7 would normally operate at between 5,000 and 10,000 ft. Aircraft roll angle must be maintained within ±4° and yaw to within ±15°.

The transmitter operates with a peak power of 200 kW, a PRF of 800 Hz, a pulse-width of 0.23 μs and at a frequency of 9,250 MHz. In the SLAR 100 the along-track range resolution is proportional to target range, being 7.8 m/km. Across track the range resolution is constant at 37.5 m.

Specifications
Dimensions:
(transmitter) 444 × 482 × 584 mm
(central processor) 265 × 482 × 559 mm

(control unit) 265 × 482 × 406 mm
(antenna) 404 × 5,285 mm
Weight:
(transmitter) 60 kg
(central processor) 29 kg
(control unit) 22 kg
(antenna (each) 36.5 kg
(total system) 267 kg

Status
In service in Boeing Canada DHC-7 Dash 7 aircraft operated by the Canadian government.

Contractor
EMS Technologies Inc.

SLAR 300 Side-Looking Airborne Radar

Type
Airborne surveillance radar.

Description
The SLAR 300 is a real aperture side-looking non-coherent airborne imaging radar with a 100 km range on either side of the aircraft which produces a 200 km swath. The radar's high sensitivity is due to the use of a 250 kW peak power magnetron transmitter operating at I/J-band, although a G-band version can be provided. The SLAR 300 operates with a vertical or horizontal polarisation, which is selectable if a dual-polarisation antenna is used. The system works with either fixed or gimballed antennas mounted on both sides of the aircraft.

EMS Technologies' own lightweight, low-cost, high-efficiency modular dual-polarised microstrip antenna can be used with this system, where size and weight are primary considerations. Output devices include a film recorder, downlink, digital tape recorder and video display.

SLAR 300 is the basic model of an upgradeable family of radars. Options such as synthetic aperture processing, polarimetric processing and moving target indicator processing may be added after delivery to upgrade this system.

Specifications
Range: 100 km each side
Range resolution: 37.5 m
Azimuth resolution: 7.5 m/km of range
Swath width: 25, 50 or 100 km
Swath offset: 0, 25, 50 or 75 km
Aircraft groundspeed: 150–330 kt
Altitude: 5,000–20,000 ft
Max squint angle: ±15°

Contractor
EMS Technologies Inc.

Tri-mode Synthetic Aperture Radar (TriSAR)

Type
Airborne surveillance radar.

Description
The TriSAR produced by Array Systems Computing is a real-time, high-resolution, airborne image processing system. It produces detailed radar cross-section images in three modes: strip map; Range Doppler Profiling/Inverse SAR (RDP/ISAR); and spotlight.

The ability to provide detailed radar cross-section data enables TriSAR to fulfil a number of specialist maritime patrol functions, including iceberg classification; tactical ice surveillance; ocean mapping; search and rescue; harbour

surveillance; battlefield surveillance; land mapping.

In the spotlight mode, the antenna is directed at the target for a predetermined length of time, data is collected and processed to produce a single, static, high-resolution image of the target. The spotlight mode is able to image both stationary and moving targets. For moving targets, an adaptive sub-aperture focusing algorithm automatically constructs a finely focused image of the moving target. Aircraft motion is fully compensated. The radar illumination time for a target is substantially reduced. Spotlight SAR is the operator's preferred mode for imaging a potentially threatening maritime target. The spotlight technique is also capable of producing fine resolution images of non-moving targets

In the strip map mode, TriSAR provides an endless strip of imagery parallel to the aircraft's flight; the strip map mode delivers high-resolution imagery in real time.

The RDP/ISAR mode is similar to the traditional ISAR, in that it produces a continuous series of snapshots of moving targets, but it can also continue taking snapshots, while analysis of earlier snapshots is in progress.

Array Systems Computing provides data recording and ground processing facilities to support operational use of the TriSAR system.

Contractor
Array Systems Computing.

China

204 Airborne interceptor radar

Type
Airborne Fire-Control Radar (FCR).

Description
It is the first I/J-band monopulse airborne interceptor radar, successfully developed by

Leihua for all-weather F-8 fighters. It has search, acquisition and tracking capabilities and can be used for attack on flying targets with gun, rockets and missiles in association with an onboard fire-control computer and optical gunsight.

It provides good anti-interference, high reliability and easy maintenance.

Specifications
Detection range: 29 km
Scan range:
(azimuth) ±38°
(elevation) –12 to +24°
Volume: 0.145 m³
Weight: 145 kg
Frequency: I/J-band

China Leihua Electronic Technology Research Institute 204 airborne interceptor radar 0504232

Status
In service.

Contractor
China Leihua Electronic Technology Research Institute.

698 side-looking radar

Type
Airborne surveillance radar.

Description
The I/J-band 698 side-looking radar is designed specifically for detection of periscopes and ships. The radar features coherent moving target detection, slotted feed double parabolic reflector antenna, parametric amplifier, high-stability local oscillator, coherent receiver, IF log amplifier and digital filter. Detection ranges are quoted as 60 km against ships and 17 km against a periscope.

Specifications
Detection range: periscope 17 km; ship 60 km
Display ranges:
(transversal) 60 km (normal)
(longitudinal) 30 km (searching)
High resolution:
(transversal) 300 m (searching)
(longitudinal) 50 m
Operational altitude: 50–500 m (searching)
Volume: 0.8 m³
Weight: 230 kg

Status
In service.

Contractor
China Leihua Electronic Technology Research Institute.

CWI illuminator

Type
Airborne target illumination/tracking radar.

Description
The continuous wave illuminator, now used for F-8 fighters, is designed to perform semi-active radar guidance of air-to-air medium-range interception missiles when operated in combination with airborne radar.

Status
In production. There is a series of CWIs, including CWI-A, CWI-B, CWI-C and CWI-D, available for different types of aircraft.

Specifications
Frequency: I/J-band
Radiation power: 200 W
FM noise: LFM <−99 dB/Hz/10 KHz
AM noise: LAM <115 dB/KHz/10 KHz
Weight: 40 kg
Volume: 0.035 m³

Contractor
China Leihua Electronic Technology Research Institute.

JL-7 fire-control radar

Type
Airborne Fire-Control Radar (FCR).

Description
The multifunction JL-7 fire-control radar for the F-7C aircraft is designed to search, detect and track airborne targets and carry out air-to-ground ranging. It can be used for attack on air or ground targets using missiles, guns or bombs in association with a gunsight or HUD.

Specifications
Volume: 0.12 m³
Weight: 115 kg
Frequency: J-band
Range:
(detection) 27.8 km
(track) 18.5 km
Coverage:
(azimuth) ±35°
(elevation) −13 to +17°
Altitude: 2,300 to 65,000 ft

Status
In service with Chinese Air Force F-7C aircraft.

Contractor
China Leihua Electronic Technology Research Institute.

China Leihua Electronic Technology Research Institute JL-7 fire-control radar 0001964

China Leihua Electronic Technology Research Institute JL-7 fire-control radar installed on the F-7C aircraft 0001965

JL-10A airborne interceptor radar

Type
Airborne Fire-Control Radar (FCR).

Description
The JL-10A airborne radar is designed for the fighter requirements of medium-range omnidirectional attack, close-range dogfight, look-up and look-down and surface moving target attack over land and sea. It is the first airborne full-wave pulse Doppler radar, with high, medium and low PRF, produced in China. The JL-10A is a highly digitised system which uses a slotted array antenna, signal exciter and sophisticated signal processor.

Specifications
Range: 59.3 km look-up, 53.7 km look-down (5 m² target)
Tracking range: 29.6 km
Range resolution: 150 m
Range accuracy: 15 m
Reliability: >70 h MTBF

Status

Development and flight trials completed. Further status unknown.

Contractor

China Leihua Electronic Technology Research Institute.

China Leihua Electronic Technology Research Institute JL-10A airborne interceptor radar
0001191

PL-7 fire-control radar

Type

Airborne Fire-Control Radar (FCR).

Description

The PL-7 is a lightweight, monopulse, J-band fire-control radar developed for the Chinese Air Force and designed for use in fighter aircraft. It has air-to-air and air-to-ground modes and can operate in conjunction with a fire-control computer, IFF, head-up display or aiming sight.

The radar consists of 18 LRUs located in the nose and cockpit. It has five modes in air-to-air: search (from 400 m to 30 km and through ±45°); manual acquisition; boresight; attack/track (up to 15 km, through ±45° and down to 2,300 ft altitude) and transponder. The three air-to-ground modes are: slant range, attack and acquisition.

A horizontally polarised antenna is used, with a 3.4° beamwidth in azimuth, 5.6° in elevation and a 30 dB gain.

Specifications

Volume: 0.23 m³
Weight: 115 kg

Peak power: 75 kW
Reliability: 50 h MTBF

Contractor

China National AeroTechnology Import & Export Corporation.

Shenying multimode airborne radar

Type

Airborne Fire-Control Radar.

Description

The Shenying multimode airborne radar is a coherent I/J-band pulse Doppler system which will provide fighters with capabilities of medium-range dogfight, look-up and look-down weapon delivery and ground or sea moving target attack.

The antenna is a flat plate slotted array which features low sidelobes and full azimuth and elevation monopulse operation. A gridded TWT transmitter is employed, which operates at low, medium and high PRFs. The receiver, with low-noise front end, consists of two channels which provide the monopulse sum and multiplexed difference channels (azimuth and elevation). Shenying incorporates digital signal/data processors to handle radar mode control, to conduct the built-in test and to perform radar signal and data processing.

Specifications

Detection range:
(search) 80 km (max); 54 km (look-down)
(tracking) 40 km (look-up); 32 km (look-down)
Search angle:
(azimuth) ±60°
(elevation) ±60°
Frequency: I/J-band

Status

Status uncertain.

Contractor

China Leihua Electronic Technology Research Institute.

France

Agrion maritime surveillance

Type

Airborne surveillance radar.

Description

Agrion is a member of the Iguane family of maritime surveillance radar systems and exists in several versions. It is designed primarily for use aboard helicopters or light aircraft forming a part of task forces, employed for support at sea or for coastal protection. Several types of antenna are available to meet the requirements of various aircraft. The Agrion 15 version allows the guidance of the AS 15TT Aerospatiale air-to-surface missile.

Agrion operates in the I/J-band, using pulse compression and frequency agility to ensure high performance on maritime targets in all combinations of weather, sea state and operating altitude. These same techniques also provide maximum protection against electronic countermeasures.

The system provides operational missions such as surface and anti-submarine warfare, over-the-horizon targeting for shipborne surface-to-surface missiles, search and rescue, marine

Chin-mounted Agrion 15 radar on an AS 565SA Panther anti-submarine/anti-ship warfare helicopter
0514666

environmental protection, navigation and weather avoidance.

Status

Agrion 15 radars are reported to be in service aboard Eurocopter AS 565SA Panther anti-submarine/anti-ship warfare helicopters operated by the Saudi Navy. Also reported to be in service with the People's Republic of China's Naval Air Force in their Z-9C helicopters.

Contractor

Thales Systemes Aeroportes SA.

AMASCOS multisensor system

Type

Airborne surveillance radar.

Description

AMASCOS (Airborne Maritime Situation Control System) is designed for building up and updating tactical situations in real time and as a decision aid for operators. It is a family of maritime systems, which uses a modular approach to system design and can be integrated on any type of fixed-wing aircraft or helicopter.

The three versions of AMASCOS – AMASCOS 100, AMASCOS 200 and AMASCOS 300 – correspond to the broad categories of mission requirement ranging from simple maritime surveillance to anti-surface and anti-submarine warfare. The typical AMASCOS configuration integrates Thales equipment such as radar, FLIR, sonics, MAD and communications; the modular nature of the core system makes it possible to tailor each system to a specific requirement.

AMASCOS 100

AMASCOS 100 is a lightweight configuration, weighing less than 250 kg. It includes radar and FLIR plus a tactical computer and is suited for a wide range of missions, such as EEZ surveillance, search and rescue and law enforcement. AMASCOS 100 is suitable for fitment to light turboprop aircraft or carrier-based helicopters with an operating crew of one to two.

AMASCOS 200

AMASCOS 200 adds ESM equipment to the AMASCOS 100. It is suitable for anti-surface warfare and can be extended to provide an anti-surface warfare capability. AMASCOS 200 is suitable for fixed- and rotary-wing aircraft of the 8 ton class, with two or three operators.

AMASCOS 300

The AMASCOS 300 is the most versatile version and is likely to comprise the Ocean Master radar, Nadir Mk II inertial GPS, Chlio FLIR, DR 3000 ESM, Link W datalink, Sadang 1000 sonobuoys, HS 312S dipping sonar and MAD Mk III. This version is suitable for both anti-surface and anti-submarine warfare missions and is also suitable for command and control assignments. The heart of the system is a dedicated tactical computer, which collates and processes data from different sensors and other onboard equipment. AMASCOS 300 is suitable for installation on any maritime patrol aircraft of over 10 tons with a crew of three or more operators.

Status

In production and in service.

In 1996, the Indonesian Navy chose AMASCOS for its six NC-212 maritime patrol aircraft (Ocean Master radar and Chlio FLIR) and its three NBO 105 helicopters (Ocean Master radar). System fits also include the Gemini navigation computer.

In 1998, AMASCOS was selected by the United Arab Emirates for its IPTN CN235-220 Maritime Patrol Aircraft (MPA). This installation includes the Ocean Master radar, Thales' Chlio FLIR and Gemini navigation system. It is also likely to include Thales' DR 3000 ESM system.

An AMASCOS configuration that incorporates Thales' DR 3000 ESM system, an Ocean Master

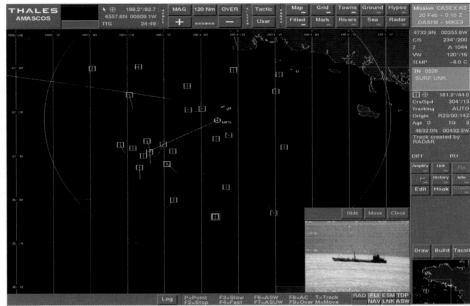

AMASCOS display format
0102014

Pakistani Atlantic 1 maritime patrol aircraft are understood to have been upgraded with an AMASCOS 300 system
0102012

radar, a Sadang C1 sonobuoy signal processing system and a Thales Avionics navigation suite was reported to have been supplied to Pakistan, for use in a Dassault Breguet Atlantic 1 maritime patrol aircraft upgrade programme, that was completed during 1998. During 2002, it was further reported that Turkey had signed a USD400 million contract covering the supply of AMASCOS suites for retrofit to Turkish Navy/Coast Guard CN-235M Maritime Patrol Aircraft (MPA).

During 2004, AMASCOS was reported to have been procured for installation aboard two Japanese Coast Guard Gulfstream GV aircraft.

During 2005, the application was selected for installation aboard 10 ATR 72 ASW maritime patrol/anti-submarine warfare aircraft being procured by Turkey.

During April 2006, Thales and Airod concluded an agreement for the upgrade of two Royal Malaysian Air Force Beech King Air 200TB maritime patrol aircraft with an AMASCOS configuration that included a tactical command system, an Ocean Master search radar and a FLIR.

Contractor

Thales Systemes Aeroportes SA.

Antilope V radar

Type

Airborne multimode radar.

Description

The Antilope V is a J-band (10 to 20 GHz) airborne radar designed for the Mirage 2000N and 2000D. Its basic functions are terrain-following, ground-mapping, interlace (terrain-following and ground-mapping), air-to-air and air-to-surface. Essential characteristics of Antilope V are:

- J-band transmission providing high ground reflectivity

- high-speed vertical scanning of the antenna
- asymmetric antenna, providing accurate localisation of obstacles in the path of the aircraft
- an antenna of the flat slotted-array type producing a weighted polar diagram with very low level diffuse sidelobes
- receiver with a wide dynamic range and image-frequency suppression
- image sharpening of the monopulse type with compression in the elevation plane for highly accurate determination of the height of obstacles
- real-time radar data processing
- continuous and automatic test system
- protection against reception by antenna sidelobes

Radar information is displayed on a head-up display and on a three-colour multimode CRT head-down display, as well as being sent to the navigation and weapon systems. The system can provide terrain-following commands at 300 ft (91 m) and 600 kt, computing a preset obstacle clearance height and with a preselected g level.

Antilope V terrain-following and navigation radar
0503702

Status

The Mirage 2000N radar has been in service since 1987. First deliveries of an upgraded version of Antilope V for the Mirage 2000D began in late 1992. Development work on a synthetic aperture mode for the Antilope V began in 1996.

Contractor

Thales Systemes Aeroportes SA.

BATTLESCAN radar

Type

Airborne surveillance radar.

Description

The BATTLESCAN airborne stand-off battlefield surveillance radar has been developed using experience gained during work on the Horizon battlefield surveillance system (see separate entry) and takes the form of a family of ground surveillance radars mounted onboard various types of air vehicles including light aircraft, helicopters, airships and drones. The radar operates in the I/J band and has a range of 150 km in all weathers. It is designed to detect and locate, in a single scan of a few seconds, columns of vehicles, ships and formations of helicopters, complementing surveillance performed by ground-based radars, and to cover areas hidden by ground contours. The scanning sector width can be up to 90° within a scanning sector axis of 360°. The scanning speed can be 2, 4 or 8° per second. BATTLESCAN is designed to provide raw data, which is first processed on board the aircraft and interpreted to provide alert and tactical information for battlefield commanders. The information is then transferred to a mobile ground station via secure datalink which has a range of 120 km. The radar is fully interoperable with the US Joint Surveillance Target Attack Radar System (Joint STARS – see separate entry). Features of the BATTLESCAN radar include:

* a carrier aircraft motion self-compensating system
* moving target indication
* a flat, very low sidelobe modular antenna
* a wideband travelling-wave tube frequency-agile transmitter
* a digital signal processor
* high-performance electronic counter-countermeasures
* operator console with colour display showing target location and speed sorting, on a map with suitable symbols
* operational modes including MTI, Sea Mode and SAR/ISAR

The operator console includes automatic operator guiding functions.

Status

In production. BATTLESCAN is reported as having been used in the French Horizon battlefield surveillance system.

Contractor

Thales Systemes Aeroportes SA.

HORIZON™ battlefield surveillance system

Type

Airborne surveillance radar.

Description

The HORIZON™ system (Hélicoptère d'Observation Radar et d'Investigation sur Zone) has been developed and manufactured for the French Army Light Aviation (Aviation Légère de l'Armée de Terre, ALAT) for tactical intelligence data gathering. It is a development of the earlier Orchidée concept evaluation system.

Each HORIZON™ system comprises one fully equipped ground station with secure Agatha datalink and two AS 532 UL Cougar helicopters, each equipped with Target MTI radar unit and operator console, navigation and communication equipment and Agatha datalink.

Horizon radar on AS 532 UL Cougar helicopter   0001192

The radar consists of three modular units (transmitter/receiver, processor and control units), which are fitted inside the helicopter, and a wide span flat antenna, mechanically positioned and mounted outside the helicopter. In flight, the complete antenna mechanism is set vertically under the helicopter and the antenna scans over 360°. On landing the antenna is locked crosswise and is raised mechanically under the helicopter tail.

The I-band (8–10 GHz) ground surveillance radar can detect moving objects over large areas and up to 150 km standoff, including vehicles (wheeled and tracked), helicopters (moving and hovering), aircraft and ships. Each moving object is detected, localised and automatically analysed and classified (as a vehicle, helicopter or other target). Up to 20,000 km² can be surveyed every 10 seconds. The targeting radar element is an all-digitial, frequency agile, I/low J-band (8–12 GHz) Doppler MTI radar. It combines mechanical and electronic scanning and provides instantaneous panoramic surveillance of either 360° or a sector bounded by 60 and 90° sectors at scanning rates of 2, 4 or 8° per second, independent of helicopter course/heading and speed. Airborne activities are controlled from an operator panel in the Cougar helicopter. Radar data is processed both aboard the aircraft (one workstation) and on the ground (two workstations). The hardware and software used in each of these positions is optimised to operator workload, responsibility level and user organisation. Collected data is transmitted to a ground station over a secure, all-digital, frequency-hopping, Agatha datalink, up to a distance of 150 km.

Each HORIZON™ system is capable of autonomous operation, co-ordinated operation with other HORIZON™ systems, or as part of a larger C³I system.

The system has also been proposed to NATO, within the framework of an Alliance Ground Surveillance acquisition programme.

System elements are provided by the following contractors: Eurocopter International: Cougar helicopter; Thales Systemes Aeroportes SA: Target MTI radar, Agatha datalink and ground station.

Specifications

Target radar
Range: 150 km (all weather)
Resolution:
(range) 40 m
(velocity) 2 m/s
Scan sector: 360°
Scan rate: 2,4 or 8°/s

Agatha datalink
RF: J-band
Data rate: up to 0.5 Mbyte/s
Range: up to 200 km

Status

In production and in service.

Orchidée technology is reported to have been under test since 1986. In June 1990, the French government cancelled the Orchidée programme because of funding difficulties. Despite this, the Orchidée demonstrator was deployed operationally during the 1991 Gulf War where it completed 24 surveillance sorties (under the codename 'Horus') in support of French ground forces and the American XVIII Airborne Corps. In addition to its surveillance work, the Orchidée/Horus aircraft was used to locate hostile radar jammers and provide targeting data for US AH-64 attack helicopters. In October 1992, the HORIZON programme began and the French army went on to order a total of four operational systems, the first of which was delivered during the summer of 1996. A second system was delivered during July 1997. During 1999, HORIZON™ was used operationally in support of NATO's Operation 'Allied Force' air interdiction campaign against Serbia and by mid-2000, all four Horizon systems had been delivered and declared operational.

An upgraded variant of the system, designated HORIZON NG, is under development. Enhancements to the system are reported to include lower weight and improved tracking capability.

Contractor

Thales Systemes Aeroportes SA.
Eurocopter International.

ORB 32 radar systems

Type

Airborne surveillance radar.

Description

ORB 32 is a range of I-band (8 to 10 GHz) airborne radars. It features very low weight and power consumption and can be installed on a wide range of helicopters and aircraft. ORB 32 is built from modular subassemblies, enabling easy extension and is designed for exclusive economic zone control, anti-surface warfare, anti-submarine warfare, active missile fire control, search and rescue, radar navigation and weather avoidance.

It features pitch and roll stabilisation, 360° azimuth and 30° elevation scanning and 60, 120, 180 and 240° sector scans, and azimuth, bearing or true motion stabilisation. The peak power is typically 70 kW.

The ORB 32 is available in a number of versions:

The ORB 3201 and ORB 3211 are simple, compact and lightweight systems suitable for small aircraft or helicopters, specially designed for surface reconnaissance, exclusive economic zone control and search and rescue. They provide navigation information and weather avoidance.

The ORB 3202 and ORB 3212 are airborne reconnaissance and target designation radars. When integrated into a weapon system, their purpose is to detect, designate and accurately track two sea targets. Target co-ordinates may be automatically transmitted to active missiles carried by aircraft or helicopters or to a launch vessel for over-the-horizon targeting.

The ORB 3203 and ORB 3214 are one element of an anti-submarine warfare weapon system for helicopters or aircraft. They enable helicopter station holding in ASW, tactical situation information, guidance and aircraft attack on a designated target, navigation, weather and mapping. The ORB 3203 and 3214 perform both primary and secondary radar functions. Use of transponders makes identification of helicopters flying at low altitude possible, even if the primary echo is in sea clutter.

Status
In service with the French Navy and the armed forces of a number of other countries. Installed in Dauphin, Super Frelon, Nord 262, Super Puma and Boeing Vertol 107 helicopters.

Contractor
Thales Systemes Aeroportes SA.

Thales' ORB 32 radar is used for surveillance and fire-control applications in maritime aircraft

0503707

ORB 37 radar system

Type
Airborne weather radar.

Description
ORB 37 has been designed to meet the French Air Force navigation requirements for the C-160 Transall transport aircraft. It carries out weather avoidance and accurate ground-mapping functions and is fitted with an interrogation facility for beacon homing.

The system consists of seven units; a slotted array flat-plate antenna, a transmitter/receiver, a power supply, a Plan Position Indicator (PPI) high-definition circular display for ground-mapping at the navigator's station, a digital PPI display on the flight deck and two control units, one for each station.

For maximum efficiency, the antenna scans at low rate for the weather mode and at high rate for the ground-mapping mode. The corresponding pulsewidths are 2.5 and 0.4 μs.

Specifications
Frequency: I-band (9,375 MHz)
Power output: 10 kW

Status
ORB 37 is in service with the French Air Force.

Contractor
Thales Systemes Aeroportes SA.

Thales' ORB 37 weather and navigation radar

0514669

Over-The-Horizon Target (OTHT) designation system

Type
Airborne surveillance radar.

Description
The Over-The-Horizon Target (OTHT) designation system is designed to detect and locate targets beyond the horizon and to transmit the co-ordinates to a coastal battery or ship. The complete system consists of equipment for acquisition and transmission.

The ORB-32 panoramic radar detects, locates, identifies through the use of IFF data and automatically tracks the targets. The ORB-32–03 version for helicopters contains an antenna with IFF capability in the radome and, in the fuselage,

Thales' Raphael-TH side-looking, synthetic aperture radar mounted on a French Air Force Mirage F1CR reconnaissance aircraft

1150265

a frequency-agile transceiver, junction box, radar control unit, scan converter, 9 in TV scope and a track-while-scan processing unit. The IFF option includes an IFF adaptation unit and IFF control unit.

The transmission of data takes place by conventional or jam-protected UHF radio over the frequency range 225 to 400 MHz. The jam-resistant version consists of a TDP-500 Series unit which allows data transmission in the frequency-hopping mode at 64 hops/s and generates system synchronisation, and an ERM-9000 transceiver equipped with a frequency-hopping unit. The conventional version consists of an MSA-300 modulator which converts the message into two LF tones for the UHF transceiver, and a CDM-6000 which receives information in the form of LF tones and converts them back to digital form.

Contractor
Thales Communications.

Raphael-TH surveillance radar

Type
Airborne surveillance radar.

Description
The Raphael-TH, also known as the SLAR-2000, is an all-weather side-looking airborne radar employing synthetic aperture and pulse compression techniques to provide high-quality mapping. The airborne part of the system is pod-mounted on a combat aircraft. It can also be installed in the cargo bay of a commuter or transport aircraft. Radar information is transmitted via datalink to a ground station where it is displayed in real time.

Operating in the I/J-band (8–20 GHz) and featuring an effective beamwidth of a few mrads, the radar is highly directional, providing sharp and accurate ground mapping. The system also features a Moving Target Indicator (MTI) capability. Ground returns are processed on the aircraft, then transmitted to a ground station for automatic exploitation. This ground station is air- or ground-transportable.

Status
In service with the French Air Force and the air forces of several other countries. It was used operationally during the 1991 Gulf War and in support of NATO operations in Kosovo. As of 2003, the Raphael-TH system was slated for upgrade with a high-resolution imagery capability.

Contractor
Thales Systemes Aeroportes SA.

RBE2 airborne radar

Type
Airborne multimode radar.

Description
The RBE2 is the first of a new class of airborne radars using the Thales Systemes Aeroportes' electronic scanning process. The RBE2 is a passive Electronically Scanned Array (ESA) multirole radar with air-to-air and air-to-surface capabilities designed to meet the operational requirements of the French Air Force's Rafale B/C and French Navy Rafale m. The radar is made up of four Line Replaceable Units (LRUs). The antenna features passive electronic scanning and high-beam agility. The Exciter/Receiver is used for coherent X-band frequency generation in four independent channels. The transmitter is matched to high, low and medium PRFs and the programmable Signal/Data Processor carries out target detection and ECCM processing.

For the air defence role the RBE2 is able to carry out all air defence functions including search in look-up and look-down modes, identification, automatic multi-target tracking and dogfight. Detection is optimised by means of automatic waveform selection of high, medium or low pulse recurrence frequencies. Multitarget tracking capability is improved by the use of a two-plane ESA which can track existing targets and simultaneously search for new targets to provide the RBE2 radar with its simultaneous multimode capability. The radar can cope with up to 40 targets during the early stages of an engagement, prioritise eight and then fire at four simultaneously from Beyond Visual Range (BVR).

The RBE2 generates simultaneous radar-to-missile datalinks for the air-to-ground role and provides Rafale with all-weather deep strike capability. It also provides close support and battlefield interdiction, by means of automatic Terrain-Following/Terrain-Avoidance (TF/TA), high-resolution mapping and ground moving-target search and tracking. As a result of the electronic scanning antenna, the simultaneous operation of TF/TA and air-to-air modes brings vital self-defence capabilities while flying penetration missions and ground clearance capabilities in air-to-air missions.

Thales' RBE2 radar is installed in the Dassault Rafale aircraft   0001193

In addition to air defence and air-to-ground roles, the RBE2 is optimised for shipping strike missions with long-range detection, multitarget tracking and target recognition capability.

The RBE2 radar is fully integrated with the EW optronics suites to provide a real-time, multisensor, weapon system.

Status
When Thales first introduced the RBE2 radar for the Rafale, it became one of the very first ESAs in a modern fighter aircraft and, as such, one of the most capable multimode radars of the period. The first production standard RBE2 radar was delivered during 1997, with operational qualification (to F1 standard) following in 1998. As of 2004, the radar was operational aboard French Navy Rafale m aircraft and is scheduled to enter service (in F2 configuration) with the French Air Force during 2005.

Since the late 1990s, Thales has been working on an updated version of the RBE2 that utilises an Active Electronically Scanned Array (AESA) in place of the passive array originally developed for the aircraft. In April 2002, the French Procurement Agency (DGA) contracted Thales to develop a first AESA radar demonstrator based on the RBE2. This radar successfully achieved its objectives, flying for the first time on the Mystère 20 testbed aircraft at the Cazaux flight test centre in December 2002 and for the first time on the Rafale in April 2003. During July 2004, the DGA awarded Thales a new contract, worth EUR85 million, to develop a second demonstrator for the RBE2 AESA. It was reported that the original RBE2 AESA demonstrator was built using US-made Transmit/Receive (T/R) modules, whereas the new contract is intended to demonstrate the feasibility and benefits of equipping Rafale with an AESA based on European technology. It also covers risk assessment associated with radar integration (thermal, mechanical, electrical environment) and in-flight operational testing. The new AESA variant of the RBE2 is intended to enter service with the French Air Force in 2012 as part of the next-generation F4 configuration.

Thales has stated that the upgrade from passive ESA to AESA (available to customers from 2008) is relatively straightforward – the move to an AESA was part of the core design of the RBE2 from the outset.

One notable improvement that an AESA will bring will be a significant enhancement to Rafale's ECM and Non Co-operative Target Recognition (NCTR) functions. With information from the radar track concerning target aspect, NCTR works by matching a library of threat aircraft signatures to that 'seen' by the radar to effect identification; the technique works particularly well against frontal aspect targets which cannot shroud the engine compressor disk.

The RBE2-AA formed part of the Rafale specification for the Republic of Singapore Air Force (RSAF) air superiority fighter competition, eventually won by Boeing's F-15.

Contractor
Thales Systemes Aeroportes SA.

RC 400 compact multimission, multitarget radar

Type
Airborne multimode radar.

Description
The Radar Compact (RC) RC 400 is a modular, lightweight, compact, multimission radar. The radar has been designed for a wide range of fighters and advanced training aircraft. RC 400 technology is derived from the RDY-2 radar (see separate entry), which equips the Mirage 2000-5. The baseline architecture comprises four Line-Replaceable Units (LRUs): the antenna unit; the exciter/receiver; the transmitter; and the processor unit. This configuration can be reconfigured as required to fit the nose of most aircraft types and only requires non-filtered air-cooling. Several antenna options are available, based on a low inertia, flat slotted-plate design, with very low side lobes.

Thales' RBE2 Active Electronically Scanned Array (AESA) radar (Thales)   1150268

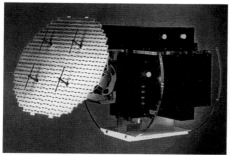

Thales' RC 400 compact multimission multitarget radar   0011895

The Mirage 2000B is fitted with Thales' RDI radar   0110916

The RC 400 design features track-while-scan, smart automatic management of scanning and target prioritisation. This provides accurate, multiple fire-and-forget capability, together with datalink requirements for missile control. The radar utilises high-, medium- and low-PRFs to optimise performance in all Beyond Visual Range (BVR) interception missions, as well as combat phases of operation. Within a given mode, the PRF is automatically managed from bar to bar, with respect to attack geometry.

The RC 400 offers the following operational capabilities:

- Air-to-air: all-aspect look up/look down detection; automatic management of low-, medium- and high-PRFs;TWS automatic lock-on; simultaneous multitarget engagements; IFF
- Air-to-ground: improved ground mapping with azimuth beam compression; Doppler beam sharpening; ranging; optional SAR high-resolution mapping; contour mapping for low-altitude penetration; surface MTI andTWS track
- Air-to-sea: track-while-scan on two targets; sea target calibration

Specifications

Transmitter unit: I/low J-band (8–12 GHz sub–band); air cooled
Receiver unit: 2 channels; wide dynamic range
Processor unit: fully programmable signal and data processing
Antenna unit: monopulse flat slotted array; elliptical or circular; IFF dipoles
Weight: 115 kg
Average transmitted power: 400 W
Power consumption: 3.6 kVA

Status

Flight tested during 1999. Marketed for multirole fighter upgrade, new-build lightweight fighter and role flexible advanced trainer applications. Specifically, the MiG-29 SMT, Mirage F.1CH/EH fighter and MAKO advanced trainer have all been linked with this radar.

Contractor

Thales Systemes Aeroportes SA.

RDI (Radar Doppler à Impulsions)

Type

Airborne Fire-Control Radar (FCR).

Description

The RDI is one of two pulse Doppler radars (the other being the RDM) developed for France's Dassault Mirage 2000. It is intended for the all-altitude air superiority and interception version and is based on a travelling wave tube, I/J-band (8 to 12 GHz) transmitter radiating from a flat, slotted plate antenna. The radar range capability is designed for high-performance Beyond Visual Range (BVR) interception and is commensurate with employment of the 40 km range Matra Super 530D semi-active homing air-to-air missile. The performance of the radar, and of other systems on the aircraft, benefits from the digital signal handling and transmission of information by databus. Considerable electronic countermeasures resistance is built into the equipment, which can operate in air-to-air search, long-range tracking and missile guidance, and

automatic short-range tracking and identification modes. Although designed for air-to-air operation, the system incorporates ground-mapping and air-to-ground ranging modes.

The high PRF available (100 kHz+°) guarantees accurate target speed assessment, combined with a Thales patented process which derives range data at maximum range in search mode as well as in tracking mode.

Air-to-air modes include air-to-air search; long-range TWS or continuous target tracking and missile guidance; automatic short-range tracking for missiles or guns. Air-to-ground modes include ground mapping; contour mapping; air-to-ground ranging. Total weight of the 11 line-replaceable units is 255 kg.

Status

In service with the French Air Force. RDI has been installed on late model Mirage 2000B and early model Mirage 2000C aircraft.

Contractor

Thales Systemes Aeroportes SA.

RDM (Radar Doppler Multifunction)

Type

Airborne multimode radar.

Description

The RDM monopulse Doppler I/J-band radar is in service with the French Dassault Mirage 2000. Whereas the RDI is designed for interception and air combat, the coherent, multimode, all-digital, frequency-agile RDM is intended largely for the multirole export version. It operates in air defence/air superiority, strike and air-to-sea modes.

In the air-to-air role, the system can look up or down, range while searching, track-while-scan, provide continuous tracking, generate aiming signals for air combat and compute attack and firing envelopes. For the strike role it provides real-beam ground-mapping, navigation updating, contour-mapping, terrain-avoidance, blind let-down, air-to-ground ranging and Ground Moving Target Indication (GMTI). In the maritime role it provides long-range search, track-while-scan and continuous tracking and can designate targets for active missiles.

For air-to-air combat, the RDM provides a 120° cone of coverage, the antenna scanning at either

RDI pulse Doppler radar for the air superiority versions of the Mirage 2000   0514662

50 or 100°/s, with ±60, ±30 or ±15° scan. A 5 m² radar cross section target can be detected at up to 111 km range. For air-to-air gun attacks, the 3.5° beam can be locked to the target at up to 19 km range, with automatic tracking within the head-up display field of view, or in a 'super-search' area, or in a vertical search mode. In look-down, air-to-air scenarios, a 5 m² radar cross section target can be detected at up to 46 km range.

Options include a Continuous Wave Illuminator (CWI) and Doppler Beam Sharpening (DBS). Comprehensive Electronic Counter-Countermeasures (ECCM) are incorporated.

Significant improvements to the radar, particularly to the look-down function, including hardware and signal processing, have been incorporated.

Status

In service with more than 250 radars delivered to date. The RDM radar is installed in Mirage 2000B, 2000C, 2000E single-seat and Mirage 2000ED two-seat export aircraft. The RDM radar equips Mirage 2000 squadrons of the French Air Force, as well as the air forces of Egypt, Greece, India, Peru and the UAE.

Contractor

Thales Systemes Aeroportes SA.

RDM Doppler radar for the multirole version of the Mirage 2000   0514661

RDY multifunction radar

Type

Airborne multimode radar.

Description

RDY is the I/J–band (8–12 GHz) multifunction Doppler radar designed for the Mirage 2000-5 aircraft. The three main modes of operation are air-to-air, air-to-ground and air-to-sea. The system features a 60 cm diameter, flat-plate, phased-array antenna, incorporating four integral high-frequency dipoles for IFF interrogation. The antenna is capable of a 60° (semi-angle) conical scan pattern.

In air-to-air mode, the RDY detects very low- or high-altitude targets at long range, irrespective of their angle of approach. It presents the pilot with tactical situation analysis, offers multitarget tracking, Track-While-Scan (TWS) track and IFF interrogation.

In air-to-ground mode, the RDY employs Doppler Beam Sharpening (DBS) to provide all-weather, low-altitude penetration capability, including terrain avoidance and ground-mapping functions.

Thales' RDY multitarget airborne fire-control radar   0504234

In air-to-sea mode, the RDY is able to detect targets, even in high sea states at up to 150 km, perform multitarget tracking and target allocation for ASMs such as AM 39 or Kormoran II.

The radar uses high-, medium- and low-PRF waveforms. High PRF is used for long-range detection of low-flying targets; low PRF for high-flying targets and a medium PRF for all-aspect medium-range detection in ground clutter.

The target handling capability is quoted as the ability to track up to 24 targets, TWS for eight of those and firing solutions for the four which pose the greatest threat by position/vector.

RDY-2

The RDY-2 is the latest development of the radar, featuring a number of notable enhancements to the basic system. Installed in Mirage 2000-5 Mk2 and later variants, the X-band RDY-2 enhances the air-to-ground capabilities of the latest Mirage 2000 multirole fighter aircraft. Enhanced functionality includes Air-to-Ground Ranging (AGR), Ground Moving Target Indicator (GMTI) and SAR high-resolution ground mapping.

Signal processing is carried out by a Programmable Signal Processor (PSP) with a very large computational capability – a speed of 100 Mcops is claimed. This high computational throughput facilitates GMTI, DBS and SAR, while upgrading system capability in other data-intensive modes, such as TWS and air-to-sea.

The RDY-2 is designed to reduce maintenance and support costs, with extensive Built-In Test Equipment (BITE), which performs automatic monitoring/diagnosis functions, indicating results in the cockpit. Further, the modular design of the radar (exciter/receiver, transmitter, PSP, data processor and antenna) facilitates line level servicing; within the Line Replaceable Units (LRUs), two automatic test benches identify faulty components at an intermediate level.

Specifications

(RDY-2)

Exciter: X-band (8–12.5 GHz)
Transmitter: Travelling Wave Tube (TWT); liquid cooled; dual-peak power matched to H/M/L PRF
Antenna: 60 cm diameter, high gain, flat slotted array; IFF dipoles (option)
Scan: ±60° azimuth and elevation
Receiver: Monopulse; 3 independent channels (Σ, Az, El)
Target handling: Up to 24 targets (track); up to eight targets (firing)

Thales' RDY-2 offers enhanced air-to-ground capabilities   1036652

Mirage 2000-5, with radome open, showing the antenna array of its RDY radar   0559969

Radar Modes

Air-to-Air	**Air-to-Ground**	**Air-to-Sea**
Velocity Search	Ground Mapping	Specific Search
Range-While-Search	(RWS) Doppler Beam Sharpening (DBS)	MultiTarget Track (MTT)
Track-While-Scan (TWS)	Synthetic Aperture Radar (SAR)	Target Radar Cross Section (RCS) assessment
		Freeze Frame
MultiTarget Track (MTT)	Air-to-Ground Ranging (AGR)	
Single-Target Track (STT)	Ground Moving Target Indicator (GMTI)	
Combat/Dogfight	Ground Moving Target Tracking (GMTT)	
Raid Assessment		
IFF Interrogator ECCM		

Signal processor: Programmable, incorporating in-built ECCM capabilities
Detection range (estimated): 140 km (maximum detection); 80 km (look-down, shoot-down)

Status

In production and in service. Development of the RDY radar began in 1984 with the first prototype radar tested in a Falcon 20 during 1987. The first production radar was delivered in 1995. Full operational capability for the radar was achieved in 1997.

At the 1999 Paris Air Show, Thales unveiled a new SAR high-resolution imaging function for the RDY that provided the radar with an all-weather reconnaissance capability.

The RDY-2 is installed in Mirage 2000-5F aircraft of the French Air Force, Greek Mirage 2000-5 Mk 2, Mirage 2000-5EDA/-5DDA of the Qatar Air Force and Taiwanese Mirage 2000-5Ei/Di.

Contractor

Thales Systemes Aeroportes SA.

Varan sea surveillance radar

Type

Airborne surveillance radar.

Description

Varan is essentially an Iguane radar (see earlier item) with a smaller antenna which makes it suitable for virtually all the present and planned lightweight maritime patrol aircraft and helicopters. It has been fitted to the Dassault Falcon Gardian of the French Navy and selected for the naval version of the ATR 42 transport aircraft and Eurocopter SA 365F Dauphin 2 helicopter.

The system provides real-time pollution and ice detection. Key features are I/J-band (8 to 20 GHz) operation, pulse compression and

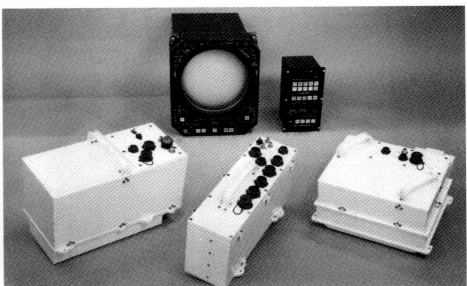

Thales Systemes Aeroportes SA' Varan surveillance radar LRUs   0503708

frequency agility for optimum performance in all combinations of weather, sea state and operating altitude, coupled with enhanced electronic counter-countermeasures capability. The unspecified but low-peak power level, associated with high receiver sensitivity, increases the difficulty of detection by hostile radars. Typical detection ranges, in Sea State 3 to 4 are snorkel 55 km, fast patrol boat 110 km and freighter 240 km. Overall system weight (six LRUs and antenna) is 111 kg.

Status
In service on the French Navy's Falcon 20H Guardian aircraft. In service on Super Puma helicopters and Falcon 20 aircraft with foreign customers.

Contractor
Thales Systemes Aeroportes SA.

The Varan airborne radar
0514664

International

Captor Radar

Type
Airborne multimode radar.

Description
The multimode radar for the Eurofighter Typhoon is currently in development by the Euroradar Consortium, led by BAE Systems. The consortium was awarded the contract after a competitive evaluation with other systems.

Captor is a third-generation coherent I/J-band (8–20 GHz) multimode radar, based on the technology of Blue Vixen. It incorporates a significant increase in processing power, which will exploit fully the high information content of the advanced transmission waveform, with wideband spread-spectrum, allowing operations to be maintained in a hostile EW environment. The 193 kg radar comprises 61 shop-replaceable items packaged into six line-replaceable units (scanner, waveguide unit, two-module receiver and a two-module transmitter). System cooling is by means of air and liquid and the radar is compatible with STANAG 3910C and MIL-STD-1553B databusses for integration into an overall avionics system.

System features include:
Transducer-computer architecture (using software to achieve complexity);
Employment of multiple waveforms (high, medium and low pulse repetition frequencies), optimally selected for specific roles;
a very high-power aperture with minimal loss;
Flexible ECM provision with inherent growth;
Supportable design.

Air-to-air features of the Captor radar include: all-aspect detection (look-up and look-down); automated, multitarget Track-While-Scan (TWS); automatic track initiation and transition; RAID assessment, target identification and prioritisation; target adaptive scan pattern adjustment; Non-Co-operative Target Recognition (NCTR) and four selectable combat functions. Electronic Counter Measures (ECM) resistance is incorporated to classify and counter jamming. The radar is compatible with all modern air-to-air missiles, including the AIM-120 AMRAAM and the recently announced MBDA Meteor Beyond Visual Range Air-to-Air Missile (BVRAAM).

Air-to-surface features include Ground Moving Target Indicator (GMTI), ground mapping/ranging, high-resolution ground mapping, air-surface tracking, terrain avoidance, weapon release computation and sea surface search.

Captor radar for the Eurofighter Typhoon (BAE Systems)
1127361

Development Eurofighter Typhoon DA7 over Sardinia, displaying a weapon load of 6 × AIM 120, 2 × AIM 9 and 2 × 1,500 litre drop tanks (Eurofighter)
1127362

While the baseline Captor-M is a mechanically scanned radar, work is ongoing to further develop the system to keep pace with the latest electronically scanned arrays. The Captor Active Electronically Scanned Array Radar (CAESAR), also known as Captor-E, integrates an electronically scanned array into the existing Captor radar. The concept for CAESAR was to take the existing Captor radar, replace its mechanically scanned antenna with the AESA array and to leave the whole of the back-end (processor, and so on) unchanged (with the exception of the antenna control unit and the antenna power supply unit). This approach, in common with recent efforts by Raytheon with the AN/APG-63(V)2/3, facilitates rapid physical integration and lower technological risk.

Status

In production and in service in Eurofighter Typhoon multirole fighters. Initial flight trials were conducted using a BAC 1–11 testbed aircraft, proving radar performance and system reliability. Prototype DA4 and DA5 Eurofighter Typhoon aircraft both flew with operational Captor radars on their maiden flights early in 1997, carrying on to conduct full operational evaluation of radar properties for the detection and tracking of airborne targets. During early radar trials, the radar was reported to have detected and tracked fighter targets at over 160 km and large aircraft at up to 320 km.

By mid-2001, a total of 16 preproduction C model Captor radars had been delivered to the Eurofighter construction consortium and the sensor was reported as having completed over 2,000 test hours at the EADS Defence and Security Systems facility. On 19 June 2001, the Euroradar Consortium announced that it had formerly delivered the software package required to facilitate the radar's full range of air-to-air and air-to-ground operating modes.

BAE Systems Avionics was awarded a serial production contract for Captor radars to equip the Eurofighter Typhoon for all partner nations. The contract covers the establishment of processes and facilities across the Euroradar Consortium to facilitate the production of radars through to 2015. The initial production order authorised an initial batch of 147 radars and spares to meet the production requirements for the first tranche of aircraft.

The go-ahead for Tranche 2 production was finally given on 14 December 2004, bringing to an end protracted negotiations and moving the Eurofighter fleet towards a credible ground attack role. Tranche 2 is worth GBP13 billion (USD25 billion) and involves 68 aircraft for Germany, 46 for Italy, 33 for Spain and 89 for the UK. Deliveries are expected to run until 2010. As far as UK Eurofighters are concerned, the baseline precision attack fit for Batch 5, Tranche 1 aircraft will be the LITENING laser designation

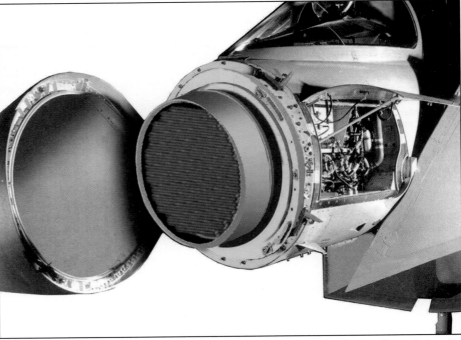

Computer-generated image showing the CAESAR prototype installation. Thus configured, the radar is also known as Captor-E (EADS)   1132643

and targeting pod (see separate entry), combined with Paveway Laser-Guided Bombs (LGBs). This configuration has been confirmed as the minimum initial capability for the UK's Tranche 2 aircraft. RAF aircraft will be operational in this configuration by the end of 2007.

CAESAR has been in development since April 2002 under the auspices of the four-nation Euroradar consortium, which is led by Selex of the UK and includes EADS Defence & Security of Germany, Galileo Avionica of Italy and INDRA in Spain. While the new array is heavier than the mechanical version, weight and centre of gravity issues have been dealt with. Captor-E is expected to be more expensive than the standard Captor-M, although the evolutionary approach of utilising the existing processor back-end will undoubtedly mitigate this for existing Eurofighter operators.

Flight tests in 2006 involved seven missions, during which a total of 50 test runs were made, with the AESA array operating (for over 20 hours with only a single T/R module failure) against an Alpha Jet trainer and a Hawker Siddeley 748 twin-turboprop aircraft flying co-operative target profiles. These tests may lead to full-scale development of the Captor-E radar for Typhoon Tranche 3, deliveries of which are scheduled to start in 2012. The next step will be to perform CAESAR flight tests onboard a Typhoon, the main goal being

to check out t he radar's features in relation to the overall weapons system, possibly during 2006.

During February 2006, the Captor Active Electronically Scanned Array Radar (CAESAR) radar demonstrator completed its first airborne operational test aboard Selexs BAC111 testbed aircraft. During the three-hour test sortie, the radar was reported to have successfully engaged air targets and to have demonstrated significant operational and performance improvement as a result of incorporating E-scan. Further flight trials of the CAESAR demonstrator were scheduled for later in the year. The CAESAR upgrade is understood to be part of the Typhoon package offered to export customers during recent competitions.

Contractor

Selex Sensors and Airborne Systems.
EADS, Defence and Security Systems Division.
Galileo Avionica, Avionic Systems and Equipment Division (part of the Selex Sensors and Airborne Systems).
INDRA Sistemas SA.

ENR European Navy Radar for the NH 90-NFH

Type

Airborne surveillance radar.

Description

EADS Deutschland GmbH, Selex Sensors (formerly Galileo Avionica) and Thales are co-operating on the design, manufacture and marketing of an airborne sea-surveillance radar – the ENR European Navy Radar – for the naval version of the NH 90 helicopter (NH 90-NFH).

The ENR is directly derived from the APS 784 produced by the then Eliradar and the Ocean Master radar produced by Thales Systemes Aeroportes SA and EADS Deutschland GmbH. It features state-of-the-art developments from these radars, including Inverse-Synthetic Aperture Radar (ISAR) processing.

Status

Based on the proposal submitted by the ENR partners, the radar was selected by the participating governments. The contract was placed in July 1998.

Contractor

Selex Sensors and Airborne Systems, Radar Systems Business Unit.
EADS Deutschland GmbH, Defense and Civil Systems.
Thales Systemes Aeroportes SA.

During February 2006, the Captor Active Electronically Scanned Array Radar (CAESAR) radar demonstrator completed its first airborne operational test aboard Selex's BAC111 (Selex)   1046241

RDR-1500B multimode surveillance radar

Type
Airborne surveillance radar.

Description
The RDR-1500B is a lightweight airborne digital colour display 360° multimode radar designed specifically for helicopters and fixed-wing aircraft in a multitude of low- and medium-altitude missions including anti-surface vessel operations, surveillance and patrol, search and rescue, customs and fishery protection. The system is available in single- and dual-display configurations. The dual-display configuration is provided when an operator console is available on the aircraft. It consists of eight units: receiver/transmitter, interface unit, two digital colour displays, cockpit control unit, cabin/console control unit, antenna assembly with antenna drive and flat plate array and switch unit.

The transmitter/receiver operates as a short-range pulse radar for high-resolution sea search and terrain-mapping, and also as a long-range pulse radar for long-range sea search, terrain-mapping and weather avoidance. Standard radar modes include weather detection, ground-mapping, search, beacon detection and identification. Ground-stabilised, aircraft heading and north-oriented display modes are available. The RDR-1500B has the facility to offset the sweep centre to any location on the display and provide target marker capability. Information from a variety of onboard navigation sensors such as INS, Omega, VOR, DME and FLIR can be displayed independently or as overlays (except for FLIR).

Other capabilities include target position transmission via datalink. An optional video processor allows use of CCP/PAL European standard colour display and videotape recording of the images on a VHS standard video recorder.

The video processor allows the radar operator to superimpose FLIR or TV colour images automatically over the radar picture. The video processor is also used to point the stabilised gimbal of FLIR/TV sensors to the selected target. Automatic target tracking is provided by the video processor unit. The modular design of the system allows for additional growth.

Specifications
Dimensions:
(transmitter/receiver)	194.1 × 123.2 × 320.5 mm
(colour indicator)	152.4, 228.6 or 254 mm
(control panel)	133.4 × 115.5 × 168.7 mm
(interface unit)	194.1 × 189.2 × 323.9 mm

Weight: 34.4 kg (complete system)
Frequency:
(transmitter/receiver for weather and search) 9,375 ±5 MHz
(receiver for beacon) 9,310 ±5 MHz
Power output: 10 kW (nominal)
PRF: 1,600, 800 or 200 Hz
Pulsewidth: 0.1, 0.5 or 2.35 µs

The RDR-1500B is fitted to Agusta AB 412 helicopters   0504113

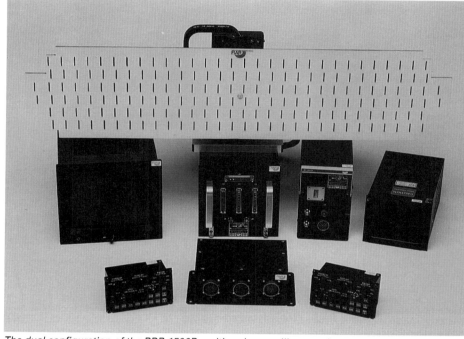

The dual configuration of the RDR-1500B multimode surveillance radar   0504114

Beamwidth:
(azimuth) 2.6°
(elevation) 10.5°
Range: 300 km
Selectable ranges: 1.16 km; 2.32 km; 4.63 km; 9.26 km; 18.52 km; 37.04 km; 74.08 km; 148.16 km; 296.32 km
Antenna gain: 31 dBi (AA-1504A)
Scan angle: 120° (sector scan); 360°
Scan rate: 28°/s (120° sector scan); 45–90°/s
Interfaces: ARINC-407, –419 and -429; SIN COS; XYZ

Status
In service. The RDR-1500B is installed in a wide variety of aircraft and helicopters, including the AgustaWestland 109, Lynx, Sea King, Antonov An-32, Bell 230 and 412, BAE Systems Jetstream 31, CASA 212, Convair T–29, deHavilland Canada DHC-7, Dornier Do 228, Eurocopter BO-105, Dauphin, Fennec and Super Puma, Fokker F27, Heli-Dyne Sentinel, Piaggio P166, Raytheon Beech King Air C90 and 200, Reims Caravan 2 and Shorts Sherpa.

Contractor
Telephonics Corporation, Command Systems Division.

Israel

EL/M-2001B radar

Type
Airborne multimode radar.

Description
The EL/M-2001B is a range-only I/J-band radar for single-seat tactical aircraft operating in air-to-air and air-to-ground modes. The target is detected visually while acquisition and tracking is accomplished automatically by the radar. The system can operate in heavy ground clutter. Information from the radar can be displayed on the head-up display or fed into the weapon control computer for weapon delivery computation. The six LRUs are based on solid-state technology, with the exception of the travelling wave tube, and have considerable reserves for future growth.

Specifications
Frequency: I/J-band (8–20 GHz)
Dimensions:
(diameter) 450 mm
(length) 490 mm
Antenna aperture: 195 mm
Weight: <50 kg
Power supply: 115 V AC, 400 Hz, 3 phase, 1 kVA, DC 30 W

The Elta EL/M-2001B radar on an Israeli Air Force Kfir aircraft   0504051

The EL/M-2001B is fitted to Israeli Air Force (IAI) Kfir fighters   0079632

Status

In production and in service. The radar is described as suitable for new-build or retrofit applications.

The EL/M-2001B is operational in Israeli Air Force (IAI) Kfir fighters. The EL/M-2001B has also been installed in upgraded Chilean Mirage 50C/CN aircraft and the 75 close air support and 10 two-seat training MiG-21 aircraft included in Romania's Lancer upgrade programme. The radar is also understood to be the core of the FIAR Pointer ranging radar, installed in Italian Air Force AMX aircraft.

Contractor

Elta Electronics Industries Ltd.

EL/M-2022 maritime surveillance radar

Type

Airborne surveillance radar.

Description

The EL/M-2022 multimode maritime surveillance radar is designed for fixed- (EL/M-2022A) or rotary-wing aircraft (EL/M-2022H) and UAVs (EL/M-2022U). The main missions for the radar are maritime target surveillance, ASW, search and rescue and economic zone control. Modular hardware design, software/remote control and a flexible avionic interface ensure easy installation in a variety of airborne platforms. The radar features multiple Track-While-Scan (TWS) for up to 256 targets, expand and freeze capabilities and sector or full-scan coverage. It features high MTBF and extensive BIT. Optional features include integral IFF compatibility and air-to-air operational modes due to a high degree of commonality with the EL/M-2032 Fire-Control Radar (FCR – see separate entry).

EL/M-2022A

Operational modes consist of long-range sea surveillance, small target detection for periscopes based on a high-resolution waveform, high-resolution mapping based on Doppler beam-sharpening, moving target indication, navigation and weather capabilities, and SAR/ISAR/range signature classification.

The EL/M-2022A consists of an ultra-low sidelobe, planar-array, vertically polarised antenna, which is stabilised in azimuth and elevation. It also features automatic and manual tilt control and a coherent TWT-based transmitter. The receiver/processor features a wide dynamic range receiver and programmable signal processor.

Radar azimuth and elevation is stabilised, with manual and automatic tilt control. The EL/M-2022A receiver/processor incorporates a programmable digital signal processor providing a wide dynamic range. The radar's waveforms are optimised for each of its operating modes and its transmission chain incorporates coherent, Travelling Wave Tube (TWT) amplification. The antenna electronic enclosure houses a low noise amplifier and the antenna drive control unit, while system cooling is by means of a mixture of forced and ambient air.

The sensor's display features include full and multiple sector scan control, freeze, image expansion, classification tools (target feature), weather video and support for all major video standards. Interfaces include ARINC 429 (navigation data option), discrete, MIL-STD-1553B (data), RS-170 (radar/weather radar video), RS-343A (CCIR) and RS-422 (optional radar slew control of an associated forward-looking infra-red sensor). The system also features continuous in-flight self-test and automatic failure alert, power-up operational readiness test and Built In Test (BIT).

Three basic configurations have been identified: EL/M-2022A(V)1, EL/M-2022A(V)2 and EL/M-2022A(V)3.

EL/M-2022A(V)1 is a lightweight radar for small airborne platforms such as helicopters and UAVs, with medium-range detection and a 50 target TWS capability. Typical detection ranges in Sea State 3 are 130 km against a small ship. Weight of the EL/M-2022A(V)1 is 65 kg.

EL/M-2022A(V)2 is a version with a long-range periscope detection capability. Typical detection ranges in Sea State 3 are 55 km on a periscope and 130 km on a small ship. The EL/M-2022A(V)2 weighs 86–95 kg.

EL/M-2022A(V)3 is a version with long-range classification, periscope detection and a 100 target TWS capability. Detection ranges in Sea State 3 are 55 km on a periscope and 130 km on a small ship. The EL/M-2022A(V)3 weighs 95–103 kg.

EL/M-2022H

The EL/M-2022H is optimised for helicopter applications in surface/littoral warfare, maritime surveillance, economic exclusion zone patrol, drug/illegal immigration interdiction, coast guard, fisheries patrol and search and rescue operations.

EL/M-2022H operating modes include:
- air (airborne target detection and tracking)
- classification (range signature and ISAR imaging with automatic ISAR library target classification)
- imaging (spot SAR mode for littoral surveillance, performance dependent on platform velocity)

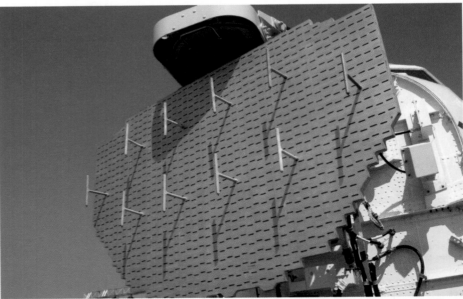

Elta's EL/M-2022A airborne maritime surveillance radar is associated with a number of different antenna configurations (Jane's/Patrick Allen)   1024375

- navigation/weather (real beam mapping and a four colour weather avoidance display)
- sea (surface target detection with automatic tracking of all detected targets)

Specifications
EL/M-2022A
Range: Up to 370 km (detection, large target)
Tracking capability: Up to 256 targets (including IFF tracking)
Altitude: 0–30,000 ft
Temperature: –20 to +55° C
Weight: 90–100 kg (approximately) excluding operator console
Power: 115 V, 400 Hz, 3 phase, 2 kVA maximum (MIL-STD-704)
Interfaces: MIL-STD-1553B data; ARINC 429 (nav data option); RS-343A, CCIR, or RS-170 radar video; optional RS-422 for radar slew command to an external FLIR turret system.

Status
The EL/M-2022A(V)3 is in production and in service. The radar has been selected for installation aboard export Antonov AN-72P/-76 MPA and Australian AP-3C ASW aircraft. The radar has also been reported as an option for EADS-CASA's Fully Integrated Tactical System (FITS) package for maritime patrol aircraft. During February 2006, Elta announced that it was to supply the EL/M-2022A radar into two P-3 modernisation programmes, with the work being valued at then year USD37 million.

Contractor
Elta Electronics Industries Ltd.

EL/M-2032 Multimode Airborne Fire Control Radar

Type
Airborne Fire-Control Radar (FCR).

Description
The EL/M-2032 is an advanced pulse-Doppler multimode fire-control radar designed for multimission fighters, for air-to-air, air-to-ground and air-to-sea missions.

Modular hardware design, all-software control and flexible MIL-STD-1553B avionic interface ensure that the radar can be installed in various fighter aircraft, such as the F-4, F-5, F-16, Mirage and MiG-21 and customised to meet specific requirements. Antenna size can be adapted to the space available in the aircraft nose.

In air-to-air operation the radar offers the following modes of operation: range while search, single- and dual-target track, Track While Scan (TWS), Situational Awareness (SA), raid assessment, vertical scan, slewable Air Combat (S-AC), Head Up Display Air Combat (HUD-AC), boresight and adaptive boresight.

In air-to-ground missions the modes are: high-resolution mapping (SAR) with image tracking, Surface Moving Target Indicator (SMTI) and tracking, Real Beam Map (RBM), Doppler Beam Sharpening (DBS), air-to-ground ranging, beacon and weather avoidance.

The air-to-sea modes are: sea search, sea target (TWS), sea target continuous track, inverse SAR sea target classification and range signature sea target classification.

The radar consists of the antenna, transmitter and receiver/processor.

The planar-array antenna features ultra-low sidelobes and two-axis monopulse operation.

The Elta EL/M-2032 installed in MiG-21 0504235

Turkish Air Force F-4E 2020 fighters are fitted with the EL/M- 2032 FCR 0569494

The transmitter is a TWT coherent transmitter.

The receiver/processor includes a programmable signal processor which provides full software control of the system.

Specifications
Weight: 72–100 kg depending on antenna size
Power consumption: 2–3 kVA
Performance:
(Air-to-Air) detection and tracking up to 80 n miles
(Air-to-Ground) ground mapping, high-resolution mapping and surface target detection up to 80 n miles
(Air-to-Sea) detection, tracking and classification up to 160 n miles

Status
More than 250 EL/M-2032 radars have been installed or contracted as part of eight upgrade programmes, including retrofit to Chilean F-5E, Romanian MiG-21 Lancer C, Indian Air Force Jaguar-M (replacing the Thales Agave radar) and Navy Sea Harrier FTS Mk 51 (replacing the Blue Fox) and Turkish F-4 Phantom 2020 aircraft. In addition, a configuration has been developed for the Northrop Grumman F-5 Plus package and the radar is part of the Israel Aerospace Industries (IAI) F-16 ACE upgrade package (see separate entry).

Contractor
Elta Electronics Industries Ltd.

EL/M-2060P airborne Synthetic Aperture Radar (SAR) pod

Type
Airborne Ground Mapping Radar (GMR).

Description
The EL/M-2060P is an innovative, field-proven, high-performance, pod-mounted SAR reconnaissance system, that can be fitted to a standard stores position on the centreline station of modern fast jet aircraft such as F-16, F-18, Gripen and Tornado.

The EL/M-2060P is completely autonomous, self-contained, all-weather, day and night high-resolution reconnaissance radar system. ELTA claims that it produces images that approach

photographic resolution. It is able to maintain capability in poor visibility conditions, under smoke and cloud cover and against a wide variety of man-made camouflage.

The system consists of a detachable, pod-mounted, externally carried, SAR; a Ground Exploitation System (GES); and an in-built bidirectional datalink.

The collected SAR imagery and data undergoes on-board, on-line, real-time processing and is transmitted to the GES for further automated interpretation. As an alternative or parallel mode, the collected data is recorded on board for retransmission or later interpretation on the ground.

Operation of the pod is highly automated, minimising the load on the pilot. In a typical fast jet installation, the EL/M-2060P system provides real-time collection and interpretation of intelligence data, with the ability to provide surveillance out to 120 km range, giving total coverage of more than 50,000 km^2/h.

Modes of operation include:
1. Strip mode: strip mode provides fast coverage of large areas at standoff ranges, and mapping in sufficient detail for target detection and overall assessment of an entire region;
2. Spot mode: spot mode provides detailed examination of designated areas of interest and high resolution for target classification;
3. Strip/Ground Moving Target Indication (GMTI) mode: strip/GMTI mode highlights moving targets within the SAR strip.

The integral datalink belongs to Elta's EL/K-1850 microwave datalink family and operates in C- (4 to 8 GHz), X- (8 to 12.5 GHz) or Ku- (12.5 to 18 GHz) band; it provides a full duplex capability and has more than 250 n miles range depending on line of sight.

The GES integrates Commercial-Off-The-Shelf (COTS) technology with advanced data processing software to perform computer-aided data exploitation and intelligence reporting.

Specifications
Pod weight: 590 kg
Aircraft interfaces:
(power) 4.3 kW maximum (3.0 kW typical)
(cooling) self sufficient
(video output) RS-170 (05 CCIR)
(control) MIL-STD-1553

The EL/M-2060P synthetic aperture radar system concept 0064379

The Elta EL/M-2060P synthetic aperture radar pod exploded view   0064380

The EL/M 2060P synthetic aperture radar pod installed on an F-16 aircraft   0067727

Status

In production.

ELTA and Lockheed Martin Tactical Aircraft Systems signed an agreement to co-operate in the sale of the EL/M-2060P SAR in the United States and the rest of the world. The US Air Force has performed demonstration trials aboard an F-16 aircraft.

An upgraded version of the EL/M-2060P SAR reconnaissance pod was offered to the US Navy for its SHAred Reconnaissance Pod (SHARP) requirement. The upgraded pod, known as ELMSAR, was intended to increase F/A-18E/F reconnaissance capability in the 2003 timeframe as the F-14 equipped with the TARPS reconnaissance pod (see separate entry) was retired from service. The ELMSAR differs from the original EL/M-2060P in replacing the Israeli-designed datalink with multiple alternative datalinks of US origin.

Contractor

Elta Electronics Industries Ltd.

EL/M-2060T SAR/GMTI system

Type

Airborne Ground Mapping Radar (GMR).

Description

The EL/L-2060T is described as being a SAR (with GMTI capability) that is suitable for transport aircraft installations, comprising a radar sensor (with an integral online signal processor), airborne and ground exploitation segments and a datalink. Of the subsystems, the radar sensor is made up of a low sidelobe planar array antenna, a transmitter and a processor that incorporates a coherent exciter, a high-speed signal processor. The transmitter incorporates coherent Travelling Wave Tube (TWT) amplification. The ground exploitation segment (designated as the GES) makes use of Commercial-Off-The-Shelf (COTS) technology and is capable of computer-aided

mission planning/management, data exploitation and intelligence reporting. Within the host aircraft, the system incorporates up to two exploitation consoles, with the GES being configured with up to two (tactical shelter installation), five (mobile S-280 shelter installation) or 10 (fixed-site installation) workstations. The datalink utilised in this application is part of the EL/K-1850 series. It is a full duplex system that operates in the C, X or Ku-band and has a range of over 250 n miles depending on line of sight. The radar operates in three modes. Strip mode gives fast coverage of large areas at stand-off ranges. This provides overall assessment of an entire region with sufficient mapping detail for target detection. Spot mode gives a detailed examination of a designated area of interest providing high resolution for target classification. Strip/GMTI mode highlights moving targets on top of the Strip SAR image. There is also a radar squint capability to facilitate flexible area coverage geometry. System options include maritime patrol, inverse SAR and air-to-air modes together with integration with electro-optic and signals intelligence sensors.

Specifications

Power supply: Peak 4.3 kW; (3.5 kW typical)
Power consumption: 300 W (air data terminal); 750 W (airborne workstation); 3,000 W (M-2050T – radar sensor)
Weight: 10 kg (air data terminal); 100 kg (airborne workstation); 170 kg (M-2050T – radar sensor)

Status

As of 2006, in production.

Contractor

Elta Electronics Industries Ltd.

EL/M-2075 Phalcon AEW radar

Type

Airborne Early Warning (AEW) radar.

Description

The EL/M-2075 is a solid-state D-band conformal array radar system for use on a Boeing 707 and other aircraft. Phalcon, as the complete AEW mission suite is referred to, is intended for airborne early warning, tactical surveillance of airborne and surface targets and intelligence gathering. It also integrates the command and control capabilities needed to employ this information.

The system uses six panels of phased-array elements: two on each side of the fuselage, one in an enlarged nosecone and one under the tail. Each array consists of 768 liquid-cooled, solid-state transmitting and receiving elements, each of which is weighted in phase and amplitude. These elements are driven by individual modules and every eight modules are connected to a transmit/receive group. Groups of 16 of these eight module batches are linked back to what is described as a prereceive/transmit unit, and a central six-way control is used to switch the pretransmit/receive units of the different arrays on a time division basis.

As used in its Chilean Boeing 707-based application, the lateral fairings measured approximately 12 × 2 m and were mounted on floating beds to prevent airframe flexing degrading the radar accuracy.

Each array scans a given azimuth sector, providing a total coverage of 360°. Scanning is carried out electronically in both azimuth and elevation. Radar modes include high PRF search and full track, track-while-scan, a slow scan detection mode for hovering and low-speed helicopters (using rotor blade returns) and a low PRF ship detection mode. These modes can be interleaved to provide multimode operation in any scanning sector. Typically, 2 to 4 second scan rates are used in high-priority sectors and 10 to 12 second rates in low-priority sectors.

Radar range was quoted as 200–215 n miles nominal (according to target characteristics and operating mode), with simultaneous track capacity up to 100 targets. For the Chilean application, the EL/M-2075 is believed to have been integrated with a monopulse Identification Friend-or-Foe (IFF) capability, command and control and electronic/communications intelligence subsystems. Subsequent to the EL/M-2075 configuration, at least two succeeding generations of Phalcon AEW radars have been developed.

The only known operational application of Elta's putative EL/M-2075 AEW radar is aboard Chile's Boeing 707-385C Cóndor AEW aircraft   0504053

Status

The Chilean Air Force operates a Boeing 707-based Phalcon command and control aircraft equipped with a three-antenna EL/M-2075 radar. Additionally, the radar has been specified for three Indian A-50EI AEW and Control (AEW & C) and four Israeli Gulfstream G550 'Etam' Compact AEW (CAEW) aircraft. For the CAEW application, the complete Phalcon mission suite is understood to have been designated as the EL/W-2085.

The Phalcon AEW mission suite was selected for installation aboard a Chinese AEW platform based on the A-50 airframe. Work on installing a Phalcon-type radar in a prototype aircraft was well advanced when, in July 1999, the project was cancelled due to intense US pressure on the Israeli government to stop the sale of advanced early warning technology to China.

Variants of the system have been offered to Australia (Airbus A310), China (Beriev A-50), India (Ilyushin Il-76) and South Korea.

Contractor

Elta Electronics Industries Ltd.

A close-up of the nose and starboard side antenna housings fitted to the Chilean Boeing 707-385C Cóndor AEW platform
0507418

EL/M-2057 airborne surveillance radar

Type

Airborne surveillance radar.

Description

The EL/M-2057 is a lightweight GMTI/SAR radar, designed for light fixed-wing aircraft (including the King Air B200, Socata TB 700 and Piper Cheyenne) and medium helicopter applications. The system comprises an antenna assembly, Airborne Exploitation Station (AES) and a radar processor unit. There is a complimentary Ground Exploitation Station (GES) segment.

The two elements are connected by a bi-directional data link (drawn from Elta's 370 km plus range L- (1 to 2 GHz), S- (2 to 4 GHz), C- (4 to 8 GHz), X- (8 to 12.5 GHz) or Ku- (12.5 to 18 GHz) EL/K-1850 equipment family).

In SAR mode, EL/M-2057 produces strip and spot imagery and has a squint capability for increased area coverage. Initial SAR imagery and GMTI plot interpretation takes place on board the host platform, while in-depth exploitation is carried out in the associated GES that receives real-time data via the data link and uses previous in-flight recorded information for additional interpretation.

The AES can accommodate up to two operator stations; the GES is a modular configuration and can be fixed or shelter mounted and features automatic and/or computer-aided tools for data interpretation and reporting.

Status

As of late 2006, the EL/M-2057 was reported as available.

Contractor

Elta Electronics Industries Ltd.

Italy

APS-717 search and rescue adar

Type

Airborne multimode radar.

Description

The APS-717 family consists of two search and navigation radar systems which are tailored to the requirements of individual customers. It is suitable for many roles, including search and rescue, surveillance, navigation and target designation.

The APS-717(V)1 is a lightweight radar which is suitable for both fixed-wing aircraft and helicopters. It operates in the I/J-band, providing detection over 180° in azimuth with automatic stabilisation. It can be integrated with the navigation system and a FLIR sensor. Other features include Constant False Alarm Rate (CFAR), scan-to-scan integration, pulse-to-pulse integration, a freeze mode and colour display with graphics.

APS-717 (V)1 search and rescue radar   0011884

The APS-717(V)2 is a high-performance upgrade of the APS-717(V)1 and offers a number of additional features. These include 360° azimuth coverage, integration and automatic initialisation of FLIR and LLTV, a video recorder output and an optional track-while-scan capability covering 32 targets.

Status

The APS-717(V)1 is in service on HH-3F helicopters and G222 aircraft of the Italian Air Force. The APS-717(V)2 is in service on Italian Harbour Authority AB 412 helicopters.

Contractor

Selex Sensors and Airborne Systems, Radar Systems Business Unit.

Creso airborne battlefield surveillance radar

Type

Airborne surveillance radar.

Description

Creso is one of the sensor systems under development as part of the Italian Army's Surveillance and Target Identification Subsystem (SORAO) CATRIN command, control, communications and intelligence system. It comprises both air and ground elements. The airborne element is installed aboard an Agusta AB 412 helicopter, and comprises the Creso battlefield surveillance radar, with ESM and FLIR

Creso airborne battlefield surveillance radar
0011889

sensors, together with a datalink system for air/ground data transfer.

Creso's operational roles are detection of ground moving targets beyond the FEBA (Forward Edge of the Battle Area); production of a target count in battlefield areas designated to it; high-precision localisation of designated targets.

Creso is an I-band, pulse Doppler radar; it utilises a coherent TWT transmitter and features wideband frequency agility; high-resolution pulse compression; programmable FFT processor; zoom capability; high ECCM resistance; and growth capability for air-to-air surveillance.

The system is made up of six Line-Replaceable Units (LRUs). Key elements are a 0.4 × 2 m flat planar slotted-array antenna, a wideband helical Travelling Wave Tube (TWT) transmitter, a microwave assembly, a double conversion receiver exciter, a signal processor and a MIL-STD-1553B data bus compatible data processor. Signal processing utilises Fast Fourier Transform and constant false alarm rate techniques, while the data processor generates display formats, controls antenna function and manages the radar's frequency agility.

As installed in the AB 412, the system is installed in two avionics bays built into the rear of the helicopter's main cabin and an operations console mounted centrally behind the flight crew positions.

Status
Trials of the system began in 1997. It is understood that the prototype system, fitted to an Augusta AB 412 helicopter, is still undergoing tests with final production on hold until the NATO multinational Alliance Ground Surveillance (AGS) system has been defined.

Contractor
Galileo Avionica.
FIAR.

Grifo multimode radar family

Type
Airborne multimode radar.

Description
The Grifo multimode pulse Doppler I/J-band radar is a compact radar with a high degree of modularity and low life cycle costs designed for air superiority aircraft. It has look-up and look-down capabilities.

The radar features a monopulse flat plate antenna array, coherent TWT transmitter, pulse compression, wideband frequency agility and a programmable waveform generator and FFT processor. It is fully compatible with semi-active and active missiles and has a high immunity to ECM.

The Grifo is suitable for fitting to a number of aircraft such as the Mirage, A-4, F-5, MiG-21, Super 7 and several trainer/light attack aircraft.

Grifo 7 radar
The Grifo 7 is designed to be installed in the nose of CAC F-7 aircraft. It has full look-up and look-down air-to-air capabilities through the use of pulse Doppler and medium PRF waveform, plus an air-to-ground ranging mode to support CCIP/CCRP. The system also incorporates a dual-channel receiver and extensive Electronic Counter-Counter Measures (ECCM) provisions.

Two modes are selectable in air-to-air: Super-Search (SS) is used for the acquisition and tracking of the highest priority target in the HUD field of view. The radar allows the missile seeker to be slewed to the target line of sight for offset delivery. In Boresight (BST), fixed antenna pointing is used for automatic acquisition and tracking of the nearest target. Both of these modes feature automatic transition to Single Target Track (STT).

Most of the hardware is common with the other versions of the Grifo family. Grifo 7 has compatibility with IR missiles, rockets, guns and free-fall bombs.

Specifications
Frequency: X-band (8–12.5 Ghz, NATO I/J Band)
Power: 850 W
MTBF: >200 hrs
Weight: 55 kg

Status
In production for the CAC F-7. 100 ordered by Pakistan for retrofit to the PAF F-7 fleet.

Grifo F radar
The Grifo F has been developed for retrofit in the Northrop F-5E/F. It is an X-band (8–12.5 Ghz, NATO I/J Band) multimode Pulse Doppler (PD) radar which offers eight air-to-air, four air combat and nine air-to-surface modes. It uses a TWT transmitter with both pulse compression and wideband frequency agility, a programmable FFT processor and a monopulse flat plate array antenna, of which two sizes are available.

Agusta AB 412 helicopter showing Creso radar antenna beneath the nose; also shown is the four-port ESM system above the radar, and the datalink below the tailboom   0515155

Elements of the Grifo 7 radar for F-7 aircraft   0077470

Operational modes for air-to-air employment include range-while-search, velocity search, Track-While-Scan (TWS), Single Target Track (STT), Situation Awareness (SA) and air combat. For air-to-ground operation the modes are Real Beam Mode (RBM), Doppler Beam Sharpening (DBS), Air-to-Ground Ranging (AGR), sea map, Ground Moving Target Indicator (GMTI) and Ground/Sea Moving Target Track (GMTT/SMTT). Navigation modes are Beacon (BCN), Weather Avoidance (WA) and Terrain Avoidance (TA).

Specifications
Frequency: X-band (NATO I/J)
MTBF: >200 hrs
Power: <1.5 kW
Weight: 85 kg

Status
In production for upgrade of Singapore Air Force Northrop F-5E/F and RF-5E aircraft.

Grifo L radar
Grifo L is a variant of Grifo F selected by the Czech Air Force for integration into its Aero Vodochody L-159 fighter trainer.

Grifo M3 radar
Similar to the Grifo F, the 87 kg Grifo M3 has been developed for the Mirage III. It makes use of a different antenna array to fit the nose of the Mirage III. Modes are as for the Grifo F.

The Grifo M3 is compatible with a variety of weapon systems including semi-active missiles. The Grifo M21 is similar to the M3 and is under development for the MiG-21.

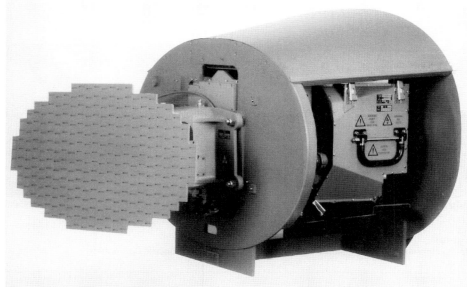

Alenia Difesa Grifo F radar for the F-5 E/F 0077469

Elements of the Grifo m radar for Mirage III aircraft 0077468

Status
In production for the Mirage III, 35 ordered by Pakistan.

Contractor
Galileo Avionica SpA, Avionic Systems and Equipment Division.

HEW-784 Early Warning Radar (EWR)

Type
Airborne Early Warning (AEW) radar.

Description
The HEW-784 is a Heliborne Early Warning Radar (EWR) for the Airborne Early Warning (AEW) version of Italian Navy's EH101 helicopter. The system is capable of all-weather surveillance against intruding aircraft and maritime surveillance and patrol. Features of the HEW-784 include:

Roles:
Primary: Detection and tracking of multiple high- and low-level airborne targets
Detection of sea-skimming missiles
Secondary: Detection and tracking of multiple sea-surface targets
Ground mapping

Radar Modes:
Air-to-air: Surveillance Look-down (MTI) Detection/Tracking of high radial speed targets (up to 128) in presence of ground clutter (with Doppler filtering)
Air-to-surface: Detection and tracking of sea-surface targets (up to 64)
Ground mapping at short/medium/long range

Specifications
Frequency: X-band (8-12.5 GHz, NATO I/J-band)
Antenna: parabolic with 360° scan and sector scan capability; the antenna requires 3 m diameter radome

Transmitter: coherent travelling wave tube with very high average power
Features: pulse compression; pulse-to-pulse frequency agility; adaptive, multiple target look-up, look down, air-to-air, track-while-scan strategies; scan-to-scan integration; datalink to task force surface ships; Inverse Synthetic Aperture Radar (ISAR)
Display: digital multifunction displays, with two separate scan converters for video displays; different formats and orientations can be presented on each; integrated Identification Friend or Foe (IFF)
Modules: eight separate Line Replaceable Units (LRUs)
Interfaces: MIL-STD-1553B, STANAG 3350 Class B (two independent outputs), Mark XII Identification Friend-or-Foe (IFF) interface
Weight: 270 kg (including mounting trays)

Status
In production for the Italian Navy EH 101 helicopter

Contractor
Selex Sensors and Airborne Systems, Radar Systems Business Unit.

Scipio radar family

Type
Airborne multimode radar.

Description
The Scipio family is a series of coherent, lightweight, compact frequency-agile radars. The range is designed to suit one-man operations on fighter aircraft with multi-role capabilities. It has a total weight of less than 75 kg, with a high MTBF and low MTTR. An extensive BIT system provides fault detection and location. Scipio offers an air-to-air mode with look-down capabilities, air combat mode with automatic detection, designation and tracking, sea mode for target detection and tracking of surface vessels and ground ranging and mapping.

The SCP-01 is the first member of the Scipio family. Its main features include I-band frequency, pulse compression and pulse Doppler techniques, frequency-agile TWT transmitter, high throughput, fully coherent, software reconfigurable signal processing, monopulse tracking, track-while-scan, MIL-STD-1553B interface and colour video output with graphics.

Status
The SCP-01 is in production and in service in Brazilian Air Force AMX light fighter aircraft.

Contractor
Selex Sensors and Airborne Systems, Radar Systems Business Unit.

The SCP-01 airborne radar has been designed for Brazilian Air Force AMX aircraft 0503710

Japan

Multifunction Radar for the F-2

Type
Airborne multimode radar.

Description
A new airborne radar for the Mitsubishi F-2 (formerly the FS-X), selected as Japan's replacement for the Mitsubishi F-1, has been developed by Mitsubishi Electric Corporation under the management of the Japanese Defence Agency's Technical Research and Development Institute. The radar has a 66 cm diameter active phased-array antenna made up of 750 modules. The radar is reported to have full air-to-air and air-to-ground functionality, including Track-While-Scan (TWS) and real beam ground mapping, with rapid re-moding and interleaving capability.

Mitsubishi XF-2B two-seat operational trainer   0093615

Status
In development for the Mitsubishi F-2 fighter.

Contractor
Mitsubishi Electric Corporation.

Russian Federation

Epolet phased array

Type
Airborne Active Electronically Scanned Array (AESA).

Description
Associated with the Adyutant radar system, the Epolet phased array is designed to support and/or replace standard legacy radar antennae on aircraft such as the MiG-23 Flogger, although the basic array can be applied to various applications where a small and lightweight (conformal) phased array is required.

In common with most modern Russian radar systems, the Adyutant/Epolet combination has been designed to operate as part of a fully integrated airborne opto-electronic system, able to provide control for missiles such as the R-27.

Applied to aircraft such as the MiG-29 Fulcrum, after acquisition and targeting with its main array, Epolet conformal arrays mounted on either side of the nose facilitate missile support while allowing the aircraft to turn up 90° away from the target, enabling the Fulcrum to extend its range-to-target and ensure 'first kill' before the bogey's own missiles arrive.

Specifications
Tracking range: Up to 65 km
Coverage: ±45° azimuth scan; ±85° missile support
Simultaneous airborne target handling: 2–4 targets (tracking)
Weight: 5 kg
Power consumption: 15 W

Status
In production.

Contractor
State Enterprise V Tikhomirov Scientific Research Institute of Instrument Design (NIIP).

FH01 Arbalet multifunction radar for the Kamov-52 Alligator

Type
Airborne multimode radar.

Description
Arbalet is a multifunction air-to-surface and air-to-air radar designed for the Kamov-52 Alligator attack helicopter, providing day/night/all-weather combat capability. The Ka-52 fire-control system consists of the Myech-U (Sword) radar system and the Samshit-BM (Boxwood) electro-optical sight. Myech-U developed by Phazotron-NIIR. The suffix U in the radar's designation stands for 'Udarnyi' (strike), indicating that its main task is to detect and track ground targets and to support

weapon delivery against them. The radar also carries the official designation FH01 Arbalet-52.

Air-to-surface modes include the following over-land and over-water capabilities:
target detection and localisation;
moving target indication and data track;
ground mapping;
terrain-following/avoidance;
air-to-surface missile and gun control.
Air-to-air modes include:
air target detection and tracking;
air-to-air missile and gun control.

Arbalet combat helicopter multifunction radar
0062856

The system consists of two sensors. A nose-mounted Ka-band radar is used for ground observation and guidance of air-to-surface weapons; it is reported to be capable of detecting a tank-sized object from a range of 8–12 km. In addition, an L-band radar, intended mainly for self-protection, with its antenna installed on the top of the rotor mast, is reported to be able to detect a fighter aircraft from a distance of 15 km, and an anti-aircraft missile from 5 km.

Status
Trials of the Myech-U aboard the Ka-52 began during 2003. A further variant of the FH01, designated Arbalet-50, is part of the Ka-50K sensor suite.

Contractor
Phazotron-NIIR, Joint Stock Company.

Gukol weather/navigation radars

Type
Airborne weather radar.

Description
Four variants of the Gukol radar have been marketed by Phazotron as I/J-band (8–12 GHz)

Ka-52 Alligator helicopter showing nose-mounted radar and mast-mounted radome   0062854

weather/navigation radars for light strike/attack aircraft, helicopters and military/civil transports. Operating modes are claimed to include weather and obstacle detection/avoidance; real beam and synthetic aperture mapping; navigation and beacon tracking modes.

Gukol has been designed to meet relevant ARINC interface requirements.

The four variants have been offered with antenna sizes varying from 370 to 670 mm and equipment weights from 15 to 65 kg.

The largest variant, marketed for larger transport/tanker types, is said to have an L/M-band blind landing capability.

Status

In production. It is understood that orders have been received for Il-96 and Tu-204 aircraft.

Contractor

Phazotron-NIIR, Joint Stock Company.

Kopyo airborne radars

Type

Airborne multimode radar.

Description

Four variants of the Kopyo radar are believed to be available: Kopyo, Super Kopyo, Kopyo-25 and Kopyo-Ph.

Kopyo airborne radar

The Kopyo radar is an all-weather, coherent, multimode, multiwaveform search-and-track radar that uses digital processing to provide the features and flexibility needed for both air-to-air and air-to-surface missions. It was derived from technology developed for the Zhuk radar.

Air-to-air modes include Range-While-Search (RWS) in look-up and look-down mode; Single Target Track (STT); Track-While-Scan (TWS) of four targets and simultaneous engagement of two targets; air combat modes (vertical scan, HUD search, wide angle, boresight).

Air-to-surface modes include real beam ground map; Doppler Beam Sharpening (DBS) to 0.45° (1:10); synthetic aperture beam sharpening to give 30 × 30 m resolution; enlargement, freezing capability; TWS two targets; Ground Moving Target Indicator/Track (GMTI/GMTT); Air-to-Ground Ranging (AGR).

The Kopyo radar is compatible with many weapons including: Kh31A, R27R1, R27T1, R-73E, R60MK, RVV-AE (R-77), other precision weapons and iron bombs.

Specifications

Frequency: I/low J-band (8–12.5 GHz)
Detection range (5 m² target):
Look-down: 57 km approaching targets
   25 km receding targets
Look-up: 57 km approaching targets
   30 km receding targets
Lock-on range: 40 km
Simultaneous target capability: 10 (tracking);
   2 (engaged)
Antenna: Mechanically scanned, 0.5 m diameter
   slotted array
Angular coverage: ±10°, ±20°, ±40° azimuth, ±40°
   in elevation
Peak power: 5 kW
Average power: 1 kW
Input power: 8.5 kVA, 400 Hz, 1 kW DC
Weight: 105 kg
Volume: 185 dm³
Reliability: 120 h MTBF
Cooling: air, liquid

Status

Kopyo was part of the Indian Air Force upgrade of 125 MiG-21bis aircraft to MiG-21–93 configuration.

Proposed to China for the F-7II and A-5 aircraft as Komar and for the FC-1 aircraft as Super Komar. It has been reported that the Komar variant was subsequently abandoned.

Kopyo-25 system units   0001202

Kopyo-25 radar fitted to the centreline station of the SU-25TM demonstrator '20' (Paul Jackson)
                                                           0507531

Marketed for MiG-21, MiG-23, F-5, Mirage 1 and Hawk 200 upgrades.

Super Kopyo airborne radar

Super Kopyo is described as a modernised Kopyo system, featuring new signal and data processors and improved detection ranges, combined with a reduction in overall weight to 90 kg.

Status

It is understood that the Super Kopyo radar was proposed as a replacement for the Sapfir-23 radar in the MiG-23 Flogger fighter aircraft, although MiG preferred a solution based on the more advanced Moskit-23.

Kopyo-A

Kopyo-A is a multifunction radar that is designed for naval helicopter applications. Operating on a centre frequency of 10 GHz (3 cm wavelength), Kopyo-A incorporates a mechanically scanned, Glass Reinforced Plastic (GRP) antenna weighing 100 kg. The radar is able to detect a 1 m² Radar Cross Section (RCS) surface target at a range of 30 km, a 5 m² RCS air target at 70 km, a 150 m² RCS surface target at up to 130 km and a `large naval vessel' at up to the radar horizon.

Status

Testing of Kopyo-A computer system and software algorithms is understood to have been completed by the end of 2002. The system has

been offered for retrofit applications onboard Kamov Ka-27 naval helicopters.

Kopyo-F

Kopyo-F, incorporating a phased-array antenna, is designed for retrofit applications and as a tail-warning radar for installation aboard Sukhoi Su-27IB/Su-35 combat aircraft. As such, the radar is reported as incorporating programmable data and signal processors, enhanced counter-countermeasures proofing and a coherent travelling wave tube transmitter that is rated at 150 W, 400 W or 1,000 W depending on platform installation.

Specifications

Frequency: 8-12.5 GHz
Detection range (5 m² target):
Look-down: 70 km approaching targets
   40 km receding targets
Look-up: 75 km approaching targets
   45 km receding targets
Air-to-ground modes: Ground mapping, Doppler
   Beam Sharpening (DBS), Synthetic Aperture
   Radar (SAR) (medium and high resolution)
Simultaneous target capability: 10 (tracking);
   2 (engaged)
Antenna: Mechanically scanned, 0.6 m diameter
   slotted array
Antenna gain: 30 dB
Angular coverage: ±70° azimuth and elevation
Peak power: 3 kW

Average power: 0.3 kW
Input power: 1.8 kVA, AC; 0.25 kVA, DC
Weight: 125 kg
Volume: 165 dm³
Reliability: 200 h MTBF

Kopyo-M

Kopyo-M incorporates a reprogrammable TS501F data processor and a TS181F digital computer and offers additional operating modes including raid assessment, wide-angle close combat, ground moving target tracking (up to four targets simultaneously) and an enhanced TWS capability that tracks up to 12 targets and engages four. Basic detection range is 75 km, or 50 km for a fighter sized target in the enhanced TWS mode. Basic tracking range is 55 km. Weighing 90 kg, Kopyo-M occupies a 230 dm³ volume with a MTBF of 200 hours.

Status

Kopyo-M was displayed at the Aero India 2003 tradeshow in February 2003.

Kopyo-29

Kopyo-29 is a Kopyo-M radar adapted for the MiG-29B fighter aircraft.

N027 Kopyo-25 airborne radar

The N027 Kopyo-25 is a derivative version of the Kopyo radar, designed specifically for carriage on the Su-39 attack aircraft. It is carried in an under-fuselage pod. Addition of the Kopyo-25 radar to the Su-25/39 gives the aircraft day/night, all-weather capabilities as well as considerably enhanced efficiency in air-to-air combat.

Operational capabilities include detect and track air targets in the automatic mode, (including targets flying at low altitude over land or sea); designate targets and engage them with radar- and IR-guided air-to-air missiles or guns; high-speed vertical search and automatic lock-on of visible targets in close combat, in association with the use of high-manoeuvrability dogfight missiles; ground mapping with real beam and Doppler sharpening and scaling-up of the chosen sector of the map.

Specifications

As for Kopyo.

Status

Designed for Su-25TM, also shown at Zhukovsky 97 Air Show on the Su-39 Strike Shield aircraft. Offered as a competitor to the N012 radar for the Su-32 fighter-bomber.

Kopyo-Ph (Kopyo-50)

Kopyo-Ph (also designated Kopyo-50) is a derivative of the Kopyo radar that employs a phased-array antenna using technology derived from the N011M programme tested on Su-35. The system is also reported as featuring digital signal and data processors and a 12-bit analogue-to-digital converter. Air-to-air and air-to-surface modes include look up/down Range-While-Search (RWS), Track-While-Scan (TWS), Single Target Track (STT), air combat, real beam ground mapping, DD and SAR.

Status

Believed development complete and offered for integration into new build and legacy aircraft.

Contractor

Phazotron-NIIR, Joint Stock Company.

Moskit radar

Type

Airborne multimode radar.

Description

The Moskit radar is a coherent, multimode, digital fire-control radar that provides weapon delivery and dogfight capabilities. It is smaller, lighter and less expensive than fighter radars of its class. Technology employed in the Moskit radar is reported to derive from the Kopyo programme. Moskit is claimed to detect and track targets at

all aspects and altitudes, and to provide the following capabilities:

Air-to-air modes: Range While Search (RWS) in look-up and look-down; eight target Track-While-Scan (TWS) and two target simultaneous engagement; air combat: HUD search, slewable scan, boresight, vertical scan.

Air-to-ground modes: real beam ground map; Doppler Beam Sharpening (DBS); Synthetic Aperture Radar (SAR); enlargement; freeze; beacon; two target TWS; Air-to-Ground Ranging (AGR); Ground Moving Target Track (GMTT).

Moskit comprises five LRUs: a flat slot array antenna; air-cooled TWT transmitter; monopulse four-channel coherent receiver; 280 Mflops programmable signal processor; 1 Mflops effective speed, 512K static RAM, 1.5M ROM data processor. The Moskit radar is compatible with such weapons as Kh-29L, Kh-29TD, Kh-31A, Kh-31PE, Kh-38, RVV-AE, as well as KAB500KR iron bombs.

Moskit-23

The Moskit-23 variant was proposed as part of the MiG-23-98-1 upgrade, offering similar capabilities as the basic Moskit radar, but with increased power and operating range in a heavier package suitable for installation in the MiG-23 airframe.

Specifications

As for Moskit, except:
Detection range: 90 km (approaching targets)
Tracking range: 60 km
Weight: 110 kg
Cooling: air

Mosquito

A smaller and lighter variant of Moskit, designated Mosquito, has been developed for a proposed Indian Air Force (IAF) upgrade programme for its Jaguar interdictor aircraft. The system provides the following capabilities:

Air-to-sea modes: detection of sea targets to 100 km in Sea State 4 to 6, co-ordinates measurement accuracy of 300 m², and engagement using Sea Eagle air-to-surface missiles.

Air-to-surface modes: detection and co-ordinates measurement of sea ports and fleet anchorage and engagement of them using unguided missiles.

Air combat modes: detection, lock-on and tracking of air targets and engagement of them using Western and Russian-guided air-to-air missiles; HUD screen; slewable scan; boresight; and vertical scan.

This very small radar has a 25 km range and can track only one target at a time. It has a +30° detection/track angle in azimuth and two or four bars in elevation. The radar in IAF application will weigh 70 kg and will be pod mounted.

Specifications

Frequency: X-band (8–12 GHz)
Detection range: 45 km (approaching targets)
Lock-on range: 30 km
Angular coverage: ±10°, ±30° azimuth, 2 or 4 bars in elevation
Peak power: 4 kW
Average power: 0.3 kW
Input power: 2.1 kVA, 200 V, 400 Hz; 0.2 kW 27 V DC
Weight: 70 kg
Volume: 300 dm³
Cooling: air

The Moskit radar was proposed as part of the MiG-23-98 upgrade (Yefim Gordon)   0524599

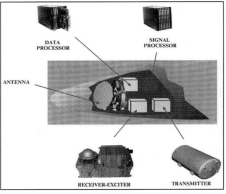

Moskit airborne radar for advanced combat trainers   0011877

Status

Moskit has been marketed for MiG-23, MiG-23BN and MiG-27 upgrades.

Mosquito was offered to the Indian Air Force as a maritime radar upgrade for its Jaguar aircraft.

Moskit-21 has been marketed by Phazotron as an upgrade for the MiG-21M/MF (MiG-21–98) and possible candidate for the MiG-ATC (Advanced Trainer Combat) concept variant of the MiG-AT. As yet, no orders have been placed for the combat capability. The Moskit-23 was MiG's preferred upgrade for the MiG-23 aircraft (MiG-23-98-1 and -98-2).

Contractor

Phazotron-NIIR, Joint Stock Company.

N001 Mech Fire-Control Radar (FCR)

Type

Airborne Fire-Control Radar (FCR).

Description

During 1974, the Soyuz Programme was instigated to develop advanced radars for the MiG-29 Fulcrum and Su-27 Flanker fighter aircraft. Both radar sets were to have maximum commonality, with the only main difference being antenna size and power output. They were to use flat, slotted-array antennas, digital signal processing and employ Pulse-Doppler (PD) processing and Track-While-Scan (TWS) for enhanced look-down, shoot-down performance against multiple targets. The maximum ranges were to be 100 and 200 km, for the MiG-29 and Su-27 respectively. The MiG-29 radar was to be developed by NIIR, while NIIP would tackle the Su-27 system.

However, both programmes failed to meet their performance requirements and in the late 1970s, both companies produced less capable radars – NIIR designed and built the N019 Rubin (see separate entry) radar for the MiG-29, while NIIP developed the N001 Mech radar for the Su-27. Both radars carry the NATO designation Slot Back and feature mechanically scanned planar arrays, combined with advanced solid-state technology and digital signal processing. Initial performance of the N001 was rather disappointing – head-on maximum detection range was 140 km against a large aircraft and 100 km against a fighter-size

The Su-27SK is fitted with the N001VE radar

1047026

target. At low aspect, the ranges decreased by 50 per cent. Compared to earlier Russian radars, however, look-down, shoot-down ranges were relatively good. After much work, the N001 radar was accepted into service in the early 1990s.

In the Su-27, the N001 radar forms part of the RLPK-27 integrated radar sighting system.

The radar is capable of maintaining track information on up to 10 targets simultaneously, with a single target selected for full tracking and engagement (STT).

The N001 system incorporates a 1.0 m diameter twist-cassegrain antenna and features medium- and high-pulse repetition frequency modes to optimise look-down. Other system features include a datalink capability, the ability to track jamming sources and the ability to hand-off data to the associated OEPS-27 Infra-Red Search and Track System (IRSTS – see separate entry).

N001VE for the Su-27SK

The N001VE radar has been developed from the baseline N001 Mech radar for the Su-27SK aircraft.

The N001VE radar is a multimode, PD radar capable of target detection and tracking against both approaching and receding targets in look-up and look-down modes.

The radar can detect an approaching aircraft at 100 km and lock-on at 80 km. Other characteristics are as for the baseline radar, including STT.

The N001VE radar provides all target illumination and guidance commands needed for operation of the complete range of R-27 medium-range missiles (R-27R1 radar semi active missile, R-27T1 heat-seeking missile, including the R-27ER1 and R-27ET1 extended range derivatives), and R-73 short-range air-to-air missiles (R-73E heat-seeking), as well as the GSh-301 30 mm cannon.

The RLPK-27 integrated radar sighting system represents one element of the overall SUV-27 weapon control system. Other elements of the SUV-27 system include the OEPS-27 optical electronic sighting system, the SEI-31 integrated cockpit display system, the ILS-31 head-up display, the Shchel-3UM helmet-mounted display, together with IFF and integrated computing system.

Status

The N001VE radar is in production, and in service on Su-27SK aircraft.

NIIP has further upgraded the N001VE to improve ECCM characteristics, upgrade the air-to-air multitarget capability and to provide air-to-ground capability; this radar is designated N001VEP.

The improved air-to-air capability enables employment of RV-AE medium range active radar-homing air-to-air missiles and enable the Su-27SK to engage two air-to-air targets simultaneously.

A new computer has been added to the radar to add ground mapping and the selection of moving targets, providing for surface target detection and anti-ship operations, using Kh31A

active radar-homing anti-ship missiles and Kh-31P or Kh25MP anti radar missiles, together with smart bombs.

The Su-30MK2 version has been fitted with the N00VEP radar.

Specifications
Frequency: X-band
Antenna: 1.0 m diameter flat planar array, mechanical scanning
Power output: 1 kW (average)
Search range: 80–100 km (high aspect, 3 m² target)
Track range: 65 km (high aspect, 3 m² target)
Coverage: ±60° azimuth scan
Simultaneous target handling: 10 targets (detection); Single Target Track (STT)
MTBF: 100 h

Status
In service in Su-27 series combat aircraft.

Contractor
State Enterprise V Tikhomirov Scientific Research Institute of Instrument Design (NIIP).

N007 Zaslon Fire-Control Radar (FCR)

Type
Airborne Fire-Control Radar (FCR).

Description
The N007 (also known as the RP-31 Zaslon) is the airborne fire-control radar fitted to MiG-31 aircraft. The designator C-800 is also used in association

with this radar and its NATO designation is Flash Dance.

Zaslon is an I-band (9–9.5 GHz sub band) Pulse-Doppler (PD) radar, with a look-down/shoot-down capability. It uses phased array radar techniques to scan a near 140° conical sector, and to track up to 10 targets simultaneously. Up to four engagements can be handled simultaneously, with a maximum detection range of 200 km. Radar data can be transferred between MiG-31 aircraft by a datalink to aid raid management.

An upgraded variant of Zaslon, designated Zaslon-A, was developed for the MiG-31B and -31BS aircraft, which went into production in 1990. The new system featured a new data processor, an enhanced range capability, improved countermeasures resistance and was integrated with new modes of operation for the aircraft datalink and upgraded navigation computers.

During 2004, a further upgrade of the Zaslon-A was reported. The Zaslon-AM retains Zaslon-A's existing antenna array, but replaces existing Argon-15A processors with Leninetz 'Bagiet' units.

Meanwhile, a further derivative of the Zaslon radar, designated Zaslon-M, was developed for the MiG-31M development aircraft. Equipped with a 1.4 m diameter antenna assembly and a range of 400 km against a 20 m² Radar Cross Section (RCS) target, Zaslon-M is quoted as capable of tracking up to 24 targets simultaneously and engage six of them with semi-active R-37 missiles. However, the MiG-31M did not go into series production.

By 1995, a multirole export derivative of the MiG-31, designated MiG-31FE, was being marketed, which included an anti-ship mode, and the ability to use antiship missiles. This aircraft did not attract any sales, but in January 1999, a Russian Air Force variant of this aircraft, designated MiG-31BM, was demonstrated at MAKS '99 and is reported to form part of the Russian Air Force's fourth generation aircraft upgrade programme, with the MiG-29SMT. This aircraft is said to carry an upgraded variant of Zaslon with increased target detection range and surface target detection capability.

Specifications
Frequency: 9–9.5 GHz
Antenna diameter: 1,100 mm
Range: 65 km (detection, maximum, 0.3 m² RCS target); 70 km (track, low aspect, bomber-sized target); 90 km (search, low-aspect, bomber-sized target); 120–150 km (tracking, high-aspect, bomber-sized target); 180–200 km (search, high-aspect, bomber-sized target); 300 km (detection, maximum, large target)
Simultaneous targeting capacity: up to 10 targets (detection); up to 4 targets (tracked)
Power output: 2.5 kW (average)
Angular coverage: –60 to +70° (elevation); ±70° (azimuth)

Zaslon on MiG-31

0514693

The Zaslon radar equips Russian Air Force MiG-31 interceptors   0096659

MTBF: 55 h
Weight: 300 kg (antenna assembly); 1,000 kg
(complete system)

Status

The Zaslon is in service with MiG-31 and Zaslon-A with the MiG-31B aircraft of the Russian Air Force and the Kazakhstan Air Force. First tests of the radar were reported to have taken place during 1973, with its first test flight being undertaken during 1976. On 15 February 1978, a Zaslon radar detected and tracked 10 simultaneous targets for the first time and the first MiG-31 equipped with the sensor entered service during 1981. Zaslon was declared fully operational during 1983.

Zaslon-M is associated with the MiG-31M development programme, and is not believed to have been put into active service. MiG-31BM carries the latest variant of Zaslon, which includes antiship capabilities.

Contractor

State Enterprise V Tikhomirov Scientific Research Institute of Instrument Design NIIP.

N010 Zhuk Fire-Control Radar (FCR)

Type

Airborne multimode radar.

Description

The N010 Zhuk (Beetle) airborne radar was designed by the NIIR Moscow (Scientific Research Institute for Radio Engineeering, Moscow), now part of Phazotron-NIIR. The N010 (together with the N011 radar that equips Su-27M) was the first multifunction Russian radar capable of tracking air as well as surface targets. The radar uses a slotted flat-plate antenna and digital computer. It was introduced in 1988 on the MiG-29M. Development has continued under Phazotron, producing a large number of variants:

Zhuk is the definitive type. It is a coherent, multimode, multimission, digital fire-control radar intended for MiG-29 and its upgrades.

Zhuk-8-II is reported as being a flat-plate antenna variant of the N010 radar that is designed for installation aboard upgraded Chinese J-8-II interceptors.

Zhuk-27 is a variant marketed for Su-27 upgrades and Su-30 variants.

Zhuk-Ph (also known as Zhuk-F) features a phased-array antenna system and advanced intrerleaved air-to-air and air-to-ground capability. This radar is believed to form the basis of the design of both

the Zhuk-MFE and -MSFE airborne Fire Control Radars (FCRs).

Zhuk-M appears to be the most advanced mechanically-scanned Zhuk so far developed. It is marketed for MiG-29 upgrades.

Zhuk-ME (also known as RP-35) was designed for the MiG-29SMT (see separate entry).

Zhuk-MFE is a phased array variant of the Zhuk-ME radar.

Zhuk-MS is described as a 'Sukhoi' oriented variant that is equipped with a 960 mm diameter flat-plate antenna and offers 3 m resolution synthetic aperture and terrain following modes. Believed to have been installed in late production PLAAF Su-30MKK fighters.

Zhuk-MSE (also known as Pharaon) is a lightweight variant designed for use as a light fighter nose radar or heavy fighter tail radar (see separate entry).

Zhuk-MSFE (also known as Sokol) is a phased array antenna variant designed for next-generation multirole aircraft (see separate entry).

A key feature of the Zhuk radar is its programmability. The radar can be programmed to respond to new threats, to incorporate improved operating modes, to integrate new weapons and to respond to new electronic countermeasures by software change.

Zhuk is compatible with the datalink requirements of a wide range of air-to-air and air-to-surface weapons, including: Kh-31A, R-27T1, R-73E, RVV-AE and other precision-guided weapons and iron bombs.

Zhuk is claimed to have the following capabilities:

Air-to-air modes: 10 to 12 target Track-While-Scan (TWS) and simultaneous engagement of two to four targets; Range-While-Search (RWS) look-up/look-down; Air Combat Manoeuvring (ACM) modes – vertical scan, HUD search, boresight, wide angle.

Air-to-surface modes include two target track-while-scan, Ground Moving Target Indicator (GMTI), Air-to-Ground Ranging (AGR), real beam ground mapping, Doppler Beam Sharpening (DBS), aircraft velocity measurement for navigation system updating, Synthetic Aperture Radar (SAR), enlargement and freezing capability and automatic terrain avoidance.

Zhuk-Ph combines 17 air-to-air and air-to-ground operating modes, similarly to the basic Zhuk, but including 8-target TWS with engagement of up to eight simultaneously.

Zhuk-ME appears to be a very major redesign of the basic system. Size and weight are significantly reduced, while power and capability have been increased; antenna design is also significantly changed. Performance figures quoted are for 20 track-while-scan targets, and 4 simultaneous engagements, at up to 120 km detection range. Zhuk-ME is also claimed to be able to perform automatic terrain avoidance and hovering helicopter detection.

Specifications

Zhuk (baseline)
Frequency range: 8–12.5 GHz (X-band)
Antenna: 680 mm dish, mechanical scanning
Peak power: 5 kW
Average power: 1 kW

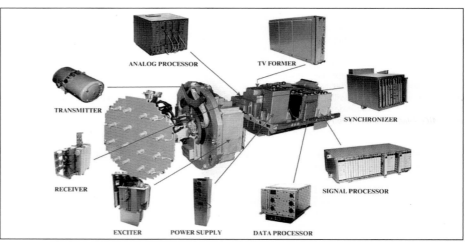

Zhuk airborne radar   0011873

The Su-27/30 Flanker employs the Zhuk-27 Airborne FCR   0118166

Detection range: 80 km (forward hemisphere); 40 km (rear hemisphere) (5 m² RCS target)
Simultaneous targeting: up to 12 (detection); up to 4 (tracked); up to 2 (engaged)
Angular coverage: two/four bar scan in elevation (–40/+60°); ±20, ±60 or up to ±90° in azimuth (depending on range)
Cooling: air/liquid
Weight: 220 kg
Reliability: 120 h MTBF

Zhuk-8-II
Detection range: 70 km (forward hemisphere, 5 m² RCS target)

Simultaneous targeting: up to 10 (detection); up to 2 (tracked); up to 1 (engaged)
Angular coverage: –40/+55° (elevation); ±85° (azimuth)
Weight: 240 kg

Zhuk-27
Detection range: 130 km (forward hemisphere, 5 m² RCS target)
Simultaneous targeting: up to 10 (detection); up to 4 (tracked); up to 2 (engaged)
Angular coverage: –40°/+55° (elevation); ± 90° (azimuth)
Weight: 260 kg

Zhuk-Ph (Zhuk-F)
Detection range: up to 200 km (forward hemisphere, 5 m² RCS target)
Simultaneous targeting: up to 24 (detection); up to 8 (tracked); up to 4 (engaged)
Angular coverage: ±70° (azimuth and elevation)
Weight: 300 kg

Zhuk-M
Detection range: 120 km (forward hemisphere, 5 m² RCS target)
Simultaneous targeting: up to 10 (detection); up to 4 (tracked); up to 2 (engaged)
Peak power: 6 kW
Average power: 1.5 kW
Angular coverage: –40/+55° (elevation); ±85° (azimuth)
Weight: 180 kg
Reliability: 200 h

Zhuk-MFE
Power output: 2 kW (average); 8 kW (peak)
Detection range: 80 km (rear hemisphere); 180 km (forward hemisphere, 3 m² RCS target)
Simultaneous targeting: up to 30 (detection); up to 6 (tracked); up to 4 (engaged)
Angular coverage: ±70° (azimuth and elevation)
MTBF: >150 h
Cooling: air and liquid
Weight: 250–270 kg class
Volume: 600 dm³

Zhuk-MS
Antenna diameter: 960 mm
Detection range: 170 km (forward hemisphere); 60 km (rear aspect) (5m² target)
Peak power: 6 kW
Average power: 1.5 kW
Simultaneous targeting: up to 20 (detection); up to 4 (tracked); up to 2 (engaged)
Weight: 230 kg

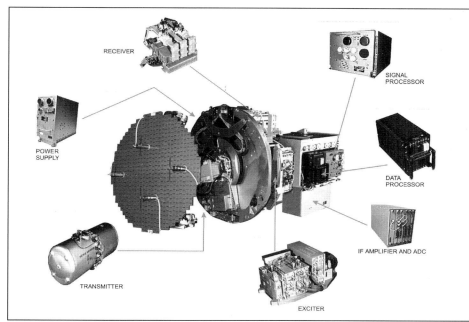

Zhuk-M airborne radar 0062857

The MiG-29M employs the N010 Zhuk-M airborne radar 0064899

A variant of the Zhuk-MFE radar is reported to have been offered to the PLAAF for its J-10 multirole fighter 0573725

Status
Variants of the Zhuk are in service installed in MiG-29SMT and Su-27/30 aircraft. Advanced versions are being marketed for Russian Air Force and export aircraft. When fitted to the Chinese J-8IIM aircraft, the radar is designated Zhuk-8-II.

A variant of the Zhuk-MFE radar is reported to have been offered to the PLAAF for its J-10 multirole fighter.

Contractor
Phazotron-NIIR Corporation and Joint Stock Company.

N011 Fire-Control Radar (FCR)

Type
Airborne Fire-Control Radar (FCR).

Description
During the late 70s and early 80s, when NIIR was developing the N010 Zhuk (Beetle) radar for the MiG-29 (see separate entry), NIIP was simultaneously working on the N011 Bars radar for the Su-27M (Su-35). Both radars had similar capabilities, but the N010 was more powerful, with a larger antenna and therefore significantly outperformed the N011, leading to that radar's eventual demise in its baseline form.

A new variant of the radar was subsequently developed and designated N011M. The N011M Bars is described as operating in the L- (1 to 2 GHz) and X- (8 to 12.5 GHz) bands, with a 960 mm diameter antenna with 36 dB gain and main and average sidelobe values of –25 dB and –48 dB respectively.

The antenna arrangement is noteworthy as it employs a combination of mechanical and electronic scanning. The main advantage of electronic scanning is that the radar beam can be moved to another point in space within milliseconds, bestowing a true multitarget capability on the host platform. However, in the case of the N011M, passive Electronically Steered Array (ESA) incorporates a relatively small scan angle (40° semi-angle), which would, in isolation, be insufficient for a modern fighter. Accordingly, the antenna of the N011M has been mounted on a hydraulically driven mount that can be rotated to right and left by 30°, giving a total scan of 140°. The antenna's weight is 100 to 110 kg, but, due to the electronic scanning contribution, the mechanical rotation of the antenna need not be as dynamic as in a typical fighter radar with purely mechanical scanning. Primarily, electronic scanning is employed and mechanical rotation is necessary only if demanded by the tactical situation. When fitted with the Chelnok Travelling Wave Tube (TWT), the radar has an average power output of 5 to 7 kW and its operating modes include air-to-air velocity search, Range While Search (RWS), Track While Scan (TWS), target identification and various close-combat options. Air-to-surface modes include real beam mapping, Doppler Beam Sharpening (DBS), Synthetic Aperture Radar (SAR), Ground-Moving Target Indication (GMTI), Ground Target Tracking (GTT), ground-mapping, terrain-following and terrain-avoidance. Anti-shipping modes include sea surface search, moving target selection, target tracking and target identification. The radar is also said to use engine compressor face returns for Non Co-operative Target Recognition (NCTR), with five such returns being identified every second.

The radar that can simultaneously track 15 airborne targets and engage four to six of them is also reported to be immune to clutter, with an effective 'zero' velocity gate. Search range for a target with a 2 m² Radar Cross Section (RCS) is reported as 80 to 100 km in a head-on position, or 30 to 40 km in a tail-on position. A fighter-size target can be detected at a maximum range of 140 km. The maximum search range for large air targets, such as an E-3A AWACS aircraft, is 400 km. In air-to-ground mode, the N011M can acquire surface targets at ranges of up to 200 km and small ground targets, such as Main Battle Tanks (MBTs), out to 40 to 50 km. The radar can also detect naval targets; destroyer-size ships are detected at a range of 120 to 150 km.

Specifications

Frequency: X-band (L-band IFF interrogator)
Antenna: 960 mm diameter passive ESA, combined electronic and mechanical scanning
Power output: 4–5 kW
Search range: 80–100 km (high aspect, 2 m² target); 30–40 km (low aspect, 2 m² target)
Coverage: ±70° azimuth scan (±40° electronic scan plus ±30° mechanical scan)
Simultaneous airborne target handling:
15 targets (detection); 4 targets (tracking, within 40° semi angle)
Simultaneous ground target handling: 2 targets (tracked)

Status

During 1991, the Tikhomirov NIIP institute, Zhukovsky, built two N-011M experimental radars, with flight testing beginning in 1996 using a Su-27M.

More recently, the N011M has been further developed for application to Indian Air Force (IAF) Su-30MKI multirole fighter aircraft. It has been reported that, due to the delayed work on the radar, Su-30MKI aircraft in the various batches delivered to the IAF are equipped with different radars. Su-30MKIs of the first batch (2002) have the N011M Bars Mk 1 radar, an almost exact copy of the N011M of the 1990s with a limited scope of tasks. The radar uses the Russian TS101 computer, the same as in series-produced MiG-29s and Su-27s. The radar is air-to-air only. The aircraft of the second batch (2003) are equipped with the Bars Mk 2 radar, using the previous computer but with a wider scope of tasks. This radar enables simultaneous attack against four targets with RVV-AE (R-77) air-to-air missiles and can perform limited ground mapping. The aircraft is also capable of launching Kh-31A anti-ship missiles and Kh-59ME long-range TV-guided missiles. Fighters of the final delivery (December 2004) have been equipped with the Mk 3 radar, with an Indian computer and i960 processor manufactured by Hindustan Aeronautics Limited (HAL) Hyderabad Division, securing performance

The lightweight Bars-29 radar was unveiled at the beginning of 2005. The radar is being marketed for new versions of the MiG-29   0590593

The N011M Bars radar in its most recent Mk3 variant is capable of simultaneous engagement of four airborne targets or one surface and one airborne target   1039081

This Su-30MKI prototype has been fitted with the N011 Bars radar and new avionics   1039076

of the full scope of the navigation and combat functions, including the use of the weapons in manoeuvring flight. This version of the radar enables simultaneous attack against one surface and one air target while continuing full scanning functions. The radar can also designate targets for all types of armament. Radar-acquired target hand-off to an LDP, such as the Rafael Litening targeting pod (see separate entry), the fighter can launch laser-guided bombs such as the KAB-500L and KAB-1500L.

N011M Bars-29

Tikhomirov NIIP unveiled their new lightweight version of the N011M Bars radar at Aero India in 2005. The mock-up, designated Bars-29, is being marketed for new versions of the MiG-29 multirole fighter aircraft. The radar is aimed at the MiG-29M2 Multi-Role Combat Aircraft (MRCA), which is being offered for India's requirement for 126 lightweight multirole fighters, as well as for mid-life upgrades of about 50 MiG-29s currently in service with the IAF.

NIIP officials said that the Bars-29 radar can be installed in other fighters including the Dassault Mirage 2000.

Two variants of Bars-29 are under consideration. The first will retain most of the components of N011M Bars. A smaller system will feature a 600 mm antenna and weigh 50 to 60 kg, about half the weight of the original Bars. Other new subassemblies of the Bars-29 will include a super high-frequency receiver and driving oscillator made using the same technology as N011M Bars. The overall weight of the Bars-29 radar will be 350 to 400 kg, about 100 kg less than the N011M Bars. Other more ambitious variants of the Bars-29 radar will have new equipment modules made with more recent technology. In such cases, the radar weight may be reduced to between 250 and 300 kg.

Status

NIIP have stated that the Bars-29 radar could be ready by 2008.

Contractor

State Enterprise V Tikhomirov Scientific Research Institute of Instrument Design (NIIP).

N019M Topaz multifunction Fire Control Radar (FCR)

Type

Airborne Fire-Control Radar (FCR).

Description

The N019M Topaz is a multifunction, multimode, coherent, pulse Doppler, air-to-air modernisation of the N019 Sapfir 29 radar (see separate entry). N019M is intended for installation in MiG-29 aircraft. The modernisation is claimed to increase operational effectiveness, particularly in ECM conditions. Topaz is compatible with a wide range of air-to-air weapons, including the R27ER1, R27T1, R27ET1, RVV-AE and R73E.

N019M1

Phazotron has developed a complete modernisation package for the N019M, which is also applicable to the similar N001 radar which equips the Su-27/Su-30 Flanker. The upgrade is achieved by the replacement of existing Line Replaceable Units (LRUs) with new ones on a form, fit and function basis, offering the following air-to-air and air-to-ground features:

- Air-to-air range increased by 30 to 50 per cent
- 10 target Track-While-Scan (TWS)
- Simultaneous engagement of four targets (non-supported missiles)
- Four advanced close combat modes

Topaz multifunction fire-control radar　　0011874

- Target group/class identification
- Weather modes
- Collision alarm
- Ground mapping:
- Low resolution real beam
- Medium resolution Doppler Beam Sharpening (DBS)
- High resolution Synthetic Aperture (SAR) (up to 5×5 m)
- Moving Target Indicator (MTI) against ground targets
- Kh31A/Kh35A air-to-surface missile integration
- Target hand-off to TV channel

The upgrade also provides for greater reliability and lower operating costs over earlier variants, although as of late 2005, this variant of the radar has not received any orders.

N019MP/MF

Phazotron also developed two versions of the N019M optimised for air-to-air (M019MF) and air-to-ground (N019MP) operations. This upgrade is also achieved by the replacement of existing Line Replaceable Units (LRUs) with new ones on a form fit and function basis.

Specifications

Detection range (against 3 m² target):
(approaching target) 90 km
(receding target) 40 km
Lock-on range: 60 km (TWS)
Scan angle:
(azimuth)±15°, ±25°, ±70°
(elevation) 4, 6 lines
MTBF: 100 h
Weight: 390 kg

The N019ME is fitted in Malaysian Air Force MiG-29ME aircraft (I Black)　　0528521

The rear cockpit of a Malaysian MiG-29, showing the N019ME radar display (I Black)　　0528522

Status

N019M is installed in later model production MiG-29 aircraft. The 'E' designator indicates that the radar is an export variant.

Contractor

Phazotron-NIIR, Joint Stock Company.

N035 Irbis Fire Control Radar (FCR)

Type

Airborne Fire-Control Radar (FCR).

Description

The N035 Irbis Passive Electronically Scanned Array (PESA) radar, developed NIIP for the Su-27SM2 (N035) and Su-35 (N035E) Flanker, is a development of the N011 Bars radar (see separate entry) of the Su-30MKI, utilising some legacy Bars modules, including the control system, low and ultra-high frequency receivers and master oscillator.

The amplifier of the N035 has been developed from the N011M radar and the two-plane hydraulic antenna mount is an advanced version of the single-plane system on Bars. The Irbis processing system is new, comprising two computers – the Solo-35.01, which performs initial signal processing, and the Solo-35.02, which performs data processing and exercises radar control. Irbis software employs elements from the Bars system as well as from the N001V/VEP radar (see separate entries).

The N035 Irbis radar is to be tested in a Su-30MK2 in late 2006 (NIIP)   1155804

The X-band Irbis features a number of important improvements over the Bars radar:

- Wider range of operational frequencies
- Greater resolution, facilitating longer range discrimination of formation elements during target identification/sorting/allocation
- A more powerful transmitter, providing longer range and better resistance to jamming
- An improved antenna and two-dimensional hydraulic mount providing for a larger instantaneous search volume and increased angular capability, facilitating greater freedom of manoeuvre for the aircraft while maintaining fly-out missile support during an engagement

The 900 mm antenna is installed on a hydraulically driven stabiliser providing two planes of movement (azimuth and roll); the combination of electronic and mechanical scanning brings some highly important operational advantages to the Irbis radar. In the electronic scanning mode, the antenna provides a 60° semi-angle in azimuth; the mechanical mount facilitates a further 60° in azimuth, providing a total angular capability of 120°. Against a bogey armed with a similar active or semi-active missiles, but only the current average 60-70° semi-angle radar capability, assuming launch parameters (platform speed at launch, no look up/down advantage) equal, this extended angular capability at longer range allows a Flanker equipped with Irbis to turn further away from the bogey after launch (during the 'missile support' phase of fly-out), thus greatly extending the time to impact of the bogey's missiles to enable a 'kill' before the bogey's inbound missile(s) reach their own autonomous phase to become an immediate threat. Such an advantage would make head-on engagement of Irbis-equipped Flankers a highly risky proposition unless countering with stealth to deny acquisition.

Additional antenna rotation by up to 120° in roll facilitates a change from vertical wave polarisation to horizontal for better performance in the acquisition of sea targets. Apart from typical targets such as aircraft, helicopters and cruise missiles, Irbis is also designed to detect and engage hypersonic targets, air-to-air and surface-to-air missiles, as well as short- and medium-range ballistic missiles.

The N035E is capable of handling up to 30 airborne targets in Track While Scan (TWS) mode, eight of which can be tracked and simultaneously engaged by eight medium-range active-radar AAMs (RVV-AE) or by four long-range active-radar AAMs (K-100). Two targets can be engaged simultaneously with semi-active radar missiles. When deployed against surface targets, Irbis is capable of ground-mapping while simultaneously engaging up to four surface targets.

Average power for the air-to-air search mode is 5 kW, with a peak power of 20 kW, enabling the radar to detect a high-aspect (near head-on) airborne target with a radar cross-section of 3 m² at between 350-400 km at high level, decreasing to up to 150 km at low aspect (tail-on).

Advanced capabilities for the new radar include automatic identification of the target by class (large, medium or small; helicopter or cruise missile, etc) and type by comparison with up to 20 platforms in the system's library. Claimed resolution is sufficient to distinguish individual returns spaced 100 m apart at a range of 50 km, with velocity threshold of 5 m/sec.

Specifications

Frequency: X-band
Antenna: 900 mm diameter PESA, combined electronic and mechanical scanning
Power output: 5 kW (average; 20 kW (peak)
Search range: 350–400 km (high aspect, 3 m² target); 150 km (low aspect, 3 m² target)
Coverage: ±120° azimuth scan (±60° electronic scan plus ±60° mechanical scan)
Simultaneous airborne target handling: 30 targets (detection); 8 targets (tracking)
Simultaneous ground target handling: 4 targets (tracked)

Status

In development; the export derivative of the system is known as Irbis-E.

An experimental version of the Irbis with a Pero antenna (see separate entry) began airborne trials on 24 June 2005 on a Su-30MK2. The tests confirmed search ranges as well as the possibility of the simultaneous handling of up to nine targets – two real and seven simulated. The radar also proved dual-mode operation (interlaced air-to-air and air-to-surface modes).

As of September 2006, NIIP had made two more Irbis radars – the first of which is to be tested in the air on Su-30MK2 (side number 503) in late 2006.

Russian press reports revealed that a plasma screen, proposed by the Institute of Theoretical and Applied Electrodynamics (ITPE) and tested on a N011M radar installed in a T-10M developmental aircraft, would be employed to reduce the cross-section of the radar's antenna.

Contractor

State Enterprise V Tikhomirov Scientific Research Institute of Instrument Design (NIIP).

Osa light fighter radar

Type

Airborne multimode radar.

Description

Osa (Wasp) is a multifunctional, electronically scanned, light fighter radar, designed to perform interleaved air-to-air and air-to-surface operations, using adaptive beam shape and sidelobe levels. Osa is claimed to be able to simultaneously track four targets, while engaging two of them, whilst maintaining acquisition search.

The associated antenna system is the SKAT-m phased array, which features two-dimensional electronic scanning in the X- and L-bands. It provides simultaneous shaping of a sum and two tracking patterns in a combination antenna of two arrays within a common aperture.

Osa is compatible with a wide range of Russian active and semi-active radar missiles.

Specifications

Radar frequency: X-band and L-band
Transmitter power: 700 W (average); 3.5 kW (peak)
Coverage: Conical 120° scan
Detection range: 85 km/40 km for 6 m² target
Tracking range: 65 km
Simultaneous target capacity: 10 (detection); 4 (tracking); 2 (engaged)
Volume: 420 dm³
Weight: 120 kg (system)
Power consumption: 3.6 kW at 400 Hz; 500 W at +27 V DC
SKAT-m antenna
Aperture: 460 mm
Coverage
Two-dimensional scanning ±45°

Osa light fighter radar (Jane's/Patrick Allen)   1109275

The Osa light fighter radar has been proposed for combat versions of the Yak-130 advanced training aircraft   0590975

Pero airborne radar antenna   0118164

Beam setting: 150–300 msec (ferrite phase shifters); 50 msec (pin diode phase shifters)
Side lobes: <–23 dB
Weight: 22 kg

Status
Reported as having been tested aboard a MiG-29UBT aircraft and proposed as part of a modernisation package for the MiG-29UB for Russia and foreign customers. The system is also marketed for MiG-21 and other MiG-29 upgrades and proposed for combat versions of the MiG-AT and Yak-130 training aircraft.

The SKAT-m antenna has been proposed for Su-27 and MiG-31 upgrades.

Contractor
State Enterprise V Tikhomirov Scientific Research Institute of Instrument Design (NIIP).

Pero dual-band antenna system

Type
Aircraft antenna, radar.

Description
Tikhomirov NIIP has developed a phased-array antenna assembly designed for both retrofit and new-build applications (including later MiG-29 and Su-27 combat aircraft). The Pero antenna can be integrated with the N001 radar (see separate entry) for increased performance. Typical radar range increases to approximately 190 km for the Su-30 upgrade, with tracking capability increased to 12 targets and engaging four of them with R-77 missiles.

The antenna system comprises an X-band reflective phased array with optical power supply and an L-band transmissive phased array.

Specifications
Frequency: 1–2 and 8–12.5 GHz (L- and X-band)
Scanning sector: ±55° (electronic scanning)
Gain: 31.5 dB (X-band, MiG-29); 34 dB (X-band, Su-27)
Phase level (mean): 35 dB (X-band, MiG-29); 38 dB (X-band, Su-27)
Aperture: 750 mm (MiG-29-based applications); 1,050 mm (Su-27-based applications)
Weight: 40 kg (MiG-29); 82 kg (Su-27)

Status
According to Russian sources, two experimental Pero systems were assembled during 2001, with one being delivered to China for testing. The antenna has been utilised to upgrade a number of in-service Russian radars to date.

The Pero antenna has been employed during development of the N035 Irbis multimode radar (see separate entry) during 2005 on a Su-30MK2.

Contractor
State Enterprise V Tikhomirov Scientific Research Institute of Instrument Design (NIIP).

Russian radars

Type
Airborne radar.

Description
The table outlines Russian Federation radar systems and programmes. Full details are not available for all, but data is provided in the subsequent equipment entries, where current Russian source information is available.

China has procured 100 examples of the Zhuk-8-II radar for retrofit to its J-8-II interceptor   0116938

Name	Contractor	Aircraft	Type	Remarks
SPPZ	AeroPribor	Civil/military aircraft	Ground proximity warning system	
IFF 60P	All-Russian JSC	Civil/military aircraft	IFF system	
SRZO-KR	Kazan Scientific Research Institute	Military aircraft	IFF system	
Berkut	Leninetz Holding Company	IL-38 (May)	ASW radar	
Kinzhal-V	Leninetz Holding Company	Mi-28N, Ka-50	Attack radar	Possibly pod-mounted under stub-wing
Khishnik (B004)	Leninetz Holding Company	Su-27IB/Su-34	Multifunction fire-control radar	Phased array antenna
Korshin	Leninetz Holding Company	Tu-142 (Bear-F), Ka-25 (Hormone)	ASW radar	
NIT	Leninetz Holding Company	–	Side-looking airborne radar	
Obzor	Leninetz Holding Company	Tu-95MS (Bear-H)/Tu-160 (Blackjack)A	–	
PN	Leninetz Holding Company	Tu-22K, Tu-22M	–	
PNA-D	Leninetz Holding Company	Tu-22M3 (Backfire-C)	–	
Aisberg-Razrez	Leninetz Holding Company	Military/civil survey	Dual-band side-looking radar, ground/water surface surveying	
Duet	Leninetz Holding Company	Civil aircraft	Dual-band weather/navigation radar	
High-resolution airborne radar	Leninetz Holding Company	Proposal	Obstacle avoidance and poor visibility aid	
Neva	Leninetz Holding Company	Civil aircraft. Mapping radar	Multifunction weather/obstacle/ground	
Sea Dragon	Leninetz Holding Company	Tu-142MK, Il-38, Ka-28	Maritime surveillance	

Name	Contractor	Aircraft	Type	Remarks
VID-95	Leninetz Holding Company	Civil aircraft and landing radar	Short-range, high-resolution, approach	
Kvant	NIIP, Vega	An-71 (Madcap), YAK-44	Airborne early warning radar	
Sabla	NIIP, Vega	MiG-25RB (Foxbat-D)	Side-looking airborne reconnaissance radar	
Shmel	NIIP, Vega	A-50 (Mainstay)	Airborne early warning radar	
Shomol	NIIP, Vega	MiG-25RB (Foxbat-D)	Side-looking airborne reconnaissance radar	
Shtyk	NIIP, Vega	SU-24MR (Fencer-E)	Side-looking airborne reconnaissance radar	
N001 Mech	NIIP, Zhukovsky	Su-27	X-band fire-control radar	NATO designation Slot Back. Modes include RWS, Threat Assessment, Combat Manoeuvre. Max detection range (fighter target): 100 km (high aspect); 50 km (low aspect). Max detected targets: 10. Max tracked/engaged targets: 2/1
N001E	NIIP, Zhukovsky	Su-27SK	X-band fire-control radar	Radar for export version of Su-27
N001V	NIIP, Zhukovsky	Su-27SM	X-band fire-control radar	Upgrade of the N001, featuring a new phased array antenna incorporating air-to-ground modes, slightly increased range and superior processing capabilities. Air-to-air modes include identification of target type, selection of a single target within a group and indication of hovering helicopters. Air-to-ground modes include ground mapping and MTI Max detection range (fighter target): 150 km (high aspect). Max tracked/engaged targets: 2/2
N001VE/VEP	NIIP, Zhukovsky	Su-27SKM, Su-30 MKK/MK2	X-band fire-control radar	Export version of N001V. Reported to exclude upgraded processor of N001V, which limits maximum detection range and denies identification of target type
N007 Zaslon	NIIP, Zhukovsky	MiG-31	Passive Electronically Scanned Array (PESA) fire-control radar	Also designated RP-31; NATO designation Flash Dance
N007 Zaslon-M	NIIP, Zhukovsky	MiG-31M	Passive Electronically Scanned Array (PESA) fire-control radar	Design brief included interception of short-range surface-to-surface missiles Reported abandoned by NIIP
N010 Zhuk	Phazotron-NIIR	MiG-29M; marketed for MiG-25, MiG-29, MiG-33 upgrades	X-band fire-control radar. Basic radar is air-to-air only; later Models multimode multifunction coherent PD, AA/AG	680 mm flat plate, slotted array antenna. Modes include TWS, RWS, Combat, Manoeuvre. Max range: 80 km, Max detected targets: 10. Max tracked/engaged targets: 4/2
N010 Zhuk-8-II	Phazotron-NIIR	Installed in upgraded Chinese J-8-II interceptors	Multimode, multifunction, coherent PD, AA/AG	Variant of the N010 radar. Modes include TWS, RWS, Combat Manoeuvre. Max range: 70 km (fighter sized target). Max detected targets: 10. Max tracked/engaged targets: 2/1
N010 Zhuk-27	Phazotron-NIIR	Marketed for SU-27, SU-30MK	X-band multimode, multifunction, coherent PD, AA/AG	960 mm flat plate, slotted array antenna. Derivative of Zhuk aimed at Su-27 upgrade market. Large diameter antenna and high peak power output variant
N010 Zhuk-F	Phazotron-NIIR	Originally intended for installation aboard the MiG-29M3	X-band multimode, multifunction, coherent PD, AA/AG	Mates the N010 radar with a phased-array antenna. Also designated Zhuk-PH
N010 Zhuk-M	Phazotron-NIIR	MiG-29SMT-2	X-band multimode, multifunction, coherent PD, AA/AG	Upgraded Zhuk, replacing C.90 processor to enable air-to-ground capability. Also reported as RP-35. Export version designated Zhuk-ME. Modes include TWS, RWS, Combat Manoeuvre, Threat Assessment, DBS, SAR, Picture Freeze, Terrain Avoidance, GMTI, AGR. Weight: 220 kg. Max detection range (3m² target): 140 km (high aspect); 65 km (low aspect) Max detected targets: 24 (TWS). Max tracked/engaged targets: 4/2
N010 Zhuk-MSF	Phazotron-NIIR	Proposed for Su-30/Su-35 and PAK-FA	X-band multimode, multifunction, coherent PD, AA/AG	Mates Zhuk-MS radar with 980 mm phased array antenna of Zhuk-F; also known as Sokol (see below). 620 mm antenna. Export version designated Zhuk-MFE. Modes include TWS, RWS, Combat Manoeuvre, Threat Assessment, DBS, SAR, Picture Freeze, Terrain Avoidance, Terrain Following, GTD, GTT. Weight: 270 kg. Max detection range (3m² target): 190 km (high aspect); 80 km (low aspect) Max detected targets: 24 (TWS). Max tracked/engaged targets: 6/6
N010 Zhuk-MS	Phazotron-NIIR	Su-30MKK3	X-band multimode, multifunction, coherent PD, AA/AG	960 mm diameter flat-plate antenna and offers 3 m resolution synthetic aperture and terrain-following modes. Export version designated Zhuk-MSE. Weight: 230 kg

Name	Contractor	Aircraft	Type	Remarks
N011 Bars	NIIP, Zhukovsky	Su-27M	X-band fire control radar	Max detection range (5m^2 target): 170 km (high aspect); 60 km (low aspect) Max detected targets: 20 (TWS) Max tracked targets: 4 Phased array radar with combined mechanical/electronic scanning Modes include TWS, RWS, Combat Manoeuvre, Real Beam Mapping, DBS, SAR, GMTI, GTT
N011M	NIIP, Zhukovsky	Su-30MKI, Su-35/Su-37	L- and X-band multimode, multifunction radar	Max detection range; 100 km (high aspect); 60 km (low aspect) Max detected targets: 12 Max tracked/engaged targets: 4/2 Phased array radar with combined mechanical/electronic scanning; Non-Cooperative Target Recognition (NCTR) capability Modes include TWS, RWS, Combat Manoeuvre, Real Beam Mapping, DBS, SAR, GMTI, GTT
N012	NIIP, Zhukovsky	Su-37	Rearward facing 'sting' radar to facilitate employment of highly agile (or rearward facing) AAMs	Max detection range; 140 km (fighter sized target), 400 km (large aircraft); for 2 m^2 target 100 km (high aspect)/40 km (low aspect) Max detected targets: 15 Max tracked/engaged targets: 6/4 Max range 4 km
N014	NIIP, Zhukovsky	Sukhoi T-50 PAKFA	Multimode, multifunction, coherent PD, AA/AG	Max detected targets: 20 (TWS) Max tracked targets: 4 Reported cancelled
N035	NIIP, Zhukovsky	Su-35	L- and X-band multimode, multifunction radar	Also known as Irbis Passive Electronically Scanned Array (PESA) high-power radar with combined mechanical/electronic scanning Max detection range; 350–400 km (high aspect) and 150 km (low aspect) for 3 m^2 sized targe Max detected targets: 15 Max tracked/engaged targets: 30/8 (medium-range active missiles); 30/4 (long-range active missiles); 30/2 (semi-active missiles)
VEGA-M	NIIP, Zhukovsky	Tu-154M	Airborne surveillance radar for open skies operations	
M002	Phazotron-NIIR	Yak-41M	Multifunction air-to-air/ air-to-ground/map/terrain follow-avoid	N010 Zhuk development
Arbalet	Phazotron-NIIR	Kamov helicopters: Ka-52 Alligator	Multifunction, air-to-surface, air-to-air	
Gukol	Phazotron-NIIR	Marketed for light strike/ attack aircraft, transports	Weather/navigation radar	
Kopyo (or Komar)	Phazotron-NIIR	MiG-21–93	Multimode, multifunction, coherent PD air-to-air/ air-to-ground	
Kopyo-25	Phazotron-NIIR	Su-25TM (Frogfoot-B), Mi-28N	Multimode, multifunction, coherent PD air-to-ground	
Kopyo-Ph	Phazotron-NIIR		Multimode, multifunction, coherent PD air-to-air/ air-to-ground	Phased-array version of Kopyo
Moskit	Phazotron-NIIR	MiG-ATC advanced combat trainer aircraft	Multimode, multifunction, coherent PD air-to-air/air-to-ground	
Mosquito	Phazotron-NIIR	Marketed for Jaguar upgrade	Multimode, multifunction, coherent PD air-to-air/air-to-ground/air-to-sea maritime radar	
Phathom	Phazotron-NIIR/ Thomson-CSF	Marketed for SU-22 upgrade	Multimode, multifunction, coherent PD air-to-air/ air-to-ground	Co-operative derivative of Kopyo
RP-21 Sapfir	Phazotron-NIIR	Many versions of MiG-21	Basic air-to-air radar	
RP-22 Sapfir-21	Phazotron-NIIR	Many later versions of MiG-21	Basic air-to-air radar	
S-23 Sapfir-23	Phazotron- NIIR	Many versions of MiG-23	PD air-to-air radar	
RP-25 Sapfir-25	Phazotron-NIIR	MiG-25 variants	look-up air-to-air radar	Nato designation High Lark 4 Enlarged and more powerful variant of the Saphir-23. Additional ECCM capabilities Integrated with modernised AA-6 missiles Max detection range; 100 km (detection), 75 km (tracking) (high aspect); Max detected targets: 4 Max tracked/engaged targets: 1/1
N019 Sapfir-29	Phazotron-NIIR	MiG-29	X-band PD look-down shoot-down air-to-air radar	NATO designation Slotback Mechanically scanned planar array Single Target Track (STT) only Nato designation High Lark 4 Max detection range: 100 km (detection), 70 km (tracking) (high aspect, look-up); look-down ranges decrease by 20% max detected targets: 10 max tracked/engaged targets: 4/1

Name	Contractor	Aircraft	Type	Remarks
N019M Topaz	Phazotron-NIIR	MiG-29S, ME. Marketed for MiG-23, MiG-29 upgrades	X-band PD look-down shoot-down air-to-air radar	Modified version of N019 Sapfir-29 More powerful digital processor More advanced ECCM capabilities Max detection range; 90 km (detection)
N019M1	Phazotron-NIIR	MiG-29SE proposal	X-band PD look-down shoot-down air-to-air radar	Alternative fit to N019MP
N019ME Topaz	Phazotron-NIIR	Export version of N019M; reported fitted to Malaysian MiG-29	X-band PD look-down shoot-down air-to-air radar	
N019MF	Phazotron-NIIR	MiG-29 upgrades	X-band PD look-down shoot-down	Modified version of N019M, optimised for air to air- Variant of N019M
N019MP Topaz	State Enterprise V Tikhomirov Scientific Research Institute of Instrument Design NIIP	MiG-29 SMT	air-to-air radar X-band PD look-down shoot-down air-to-air- radar with basic air-to ground functionality	More powerful processor Modes include RWS, Combat Manoeuvre, Threat Assessment, Terrain Mapping, DBS, GTD, GTT Max detection range: 90 km Max detected targets: 20 Max tracked/engaged targets: 4/2
Osa	State Enterprise V Tikhomirov Scientific Research Institute of Instrument Design – NIIP	MiG-29UBT	X-band multimode, multifunction, coherent, digital PD	Phased array radar Modes include TWS, RWS, Combat Manoeuvre, Threat Assessment, DBS, SAR, Picture Freeze, Terrain Avoidance, Terrain Following, GTD, GTT Max detection range; 85 km Max detected targets: 8 Max tracked/engaged targets: 4/2
Pharaon	Phazotron-NIIR	Light fighter nose radar, or heavy fighter tail radar	Multifunction fire-control, air-to-air and air-to-ground	Derivative of Zhuk-MSFE (see above)
Sokol	Phazotron-NIIR	Marketed for latest Sukhoi aircraft, ,	Multimode, multifunction coherent PD	Also known as N010 Zhuk-MSFE (see above)
Zhemchug	Phazotron-NIIR	Marketed for MiG-29. Possibly proposed for the J-10	X-band multimode, multifunction, coherent, digital PD	Phased array radar Modes include TWS, RWS, Combat Manoeuvre, Threat Assessment, GTD, GTT Max detection range; 80 km (fighter sized target) Max detected targets: 20 Max tracked/engaged targets: 4/4

The Zhuk-MS is reported to be installed in PLAAF Su-30MKK fighters   0100801

Leninetz' B004 phased-array antenna in the nose of an early Su-27IB prototype   0070065

A Russian Air Force Su-27SM, a 'mid-life update' to the Flanker which inlcuded the N001V radar (KnAAPO)   0560831

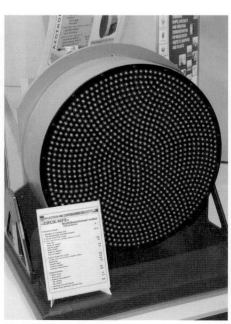

Zhuk-MFE radar mockup at the 2004 Farnborough Airshow   1097639

Zhemchug multifunction Fire Control Radar (FCR)

Type
Airborne Fire-Control Radar (FCR).

Description
The Zhemchug multifunction Pulse Doppler (PD) radar is designed for medium-sized fighter aircraft such as the MiG-29. The radar is claimed to be capable of tracking up to 20 targets, prioritising and engaging the four most threatening. Zhemchug is compatible with a wide range of Russian air-to-air weapons, including the R27ER1, R27T1, R27ET1, RVV-AE and R73E, and air-to-ground weapons, including surface-to-air missiles and smart bombs.

Specifications
Detection range:
(approaching target)   80 km
(receding target)   40 km
Lock-on range: 60 km
Weight: 180 kg

Status
Available.

Contractor
Phazotron-NIIR, Joint Stock Company.

Zhuk-M multimode airborne radar

Type
Airborne multimode radar.

Description
The Zhuk-M (Zhuk-ME in its export form) is a coherent, multimode, digital fire-control radar that provides a comprehensive set of all-weather air-to-air and air-to-surface modes, with superior dogfight and weapons delivery capabilities. The air-to-air modes provide the capability to detect, track and engage targets at all aspects, even in the presence of ground clutter. Air-to-surface modes provide extensive ground mapping, target detection, location and tracking capabilities, as well as navigation features.

The Zhuk-M/ME is designed for use with the MiG-29SMT aircraft, and is compatible with a wide range of air-to-air and air-to-surface weapons, including: Kh-29T, Kh-31A, Kh-35U, R-27ER1, R-27ET1, R-27R1, R-27T1, R-73E, RVV-AE, KAB-500KR.

The system was originally designed with a phased-array antenna, to provide electronic scanning and high gain/low sidelobe operation at extreme scan angles – in this configuration the system is designated Zhuk-MF(E). However, during 2000, the phased array antenna was deleted, with a cheaper mechanically scanned slotted array replacing it to drive the cost of the system down. This was mated to a liquid-cooled Travelling Wave Tube (TWT) and three-channel receiver. Despite this reduction in specification, the stated basic operational requirement for the system, the simultaneous attack of 4 targets, could still be met. However, in order for the mechanically scanned array to adequately support all four missiles in flight, all targets were required to be relatively close to boresight (±40°); widely spaced targets (±70°) were restricted to dual-simultaneous engagement only.

The system that comprises a high gain, low sidelobe, electronically scanning phased-array antenna, a liquid-cooled Travelling Wave Tube (TWT) transmitter, an exciter, a three-channel microwave receiver and programmable signal and data processors.

The Zhuk-ME is claimed to have the following capabilities:

Air-to-air modes: 24 Track-While-Scan (TWS), Range While Search (RWS); air combat – vertical scan, HUD search, boresight, wide angle, velocity

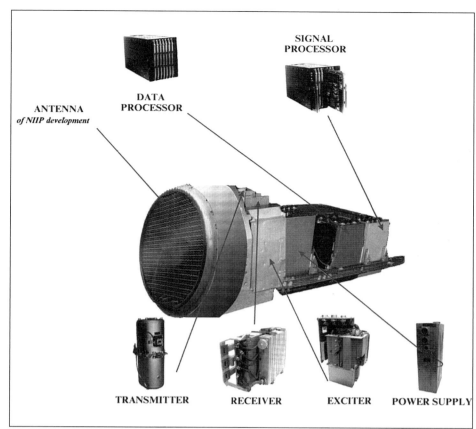

Zhuk-MF/MFE multimode airborne radar in its original configuration with phased-array antenna
0011876

A variant of the Zhuk-ME, combined with a phased array antenna, has been proposed for the Chinese Air Force (PLAAF) J-10 multirole fighter
1043741

search, raid cluster resolution, automatic terrain avoidance.

Air-to-surface modes: two target TWS, Ground Moving Target Indication/Track (GMTI/GMTT), Air-to-Ground Ranging (AGR), Real Beam ground mapping, Doppler Beam Sharpening (DBS), Synthetic Aperture Radar (SAR), Enlargement, Picture Freeze, Beacon Mode.

Specifications
(Zhuk-ME)
Detection range: 140 km (high aspect); 65 km (low aspect)
Performance: Simultaneous detection of up to 24 targets; Track-While-Scan of 4 and engagement of 2
Angular coverage: ±20°, ±40°, ±60° in azimuth, 2 or 4 bars in elevation

Radar frequency: X-band (8–12.5 GHz)
Peak power: 8 kW
Average power: 2 kW
Input power: 12 kVA, 200 V, 400 Hz; 2 kW 27 V DC
Weight: 220 kg
Volume: 350 dm³
Cooling: air/liquid
Reliability: >120 h MTBF

Status
Designed for the MiG-29SMT.

Phazotron stated at the Dubai Air Show in 2000, that the phased array antenna had been deleted and a cheaper slotted array with mechanical scanning had been substituted to reduce price. The company went on to state that the cheaper configuration could still achieve the basic operational requirement for engagement of two to four targets – albeit less satisfactorily – in that engagement of 4 targets is still possible if all are close to the boresight (+/–40°), but only 2 targets for widely separated angles (+/–70°).

A variant of the radar has also been proposed for the Chinese J-10, although it is probable that this proposal includes the phased array antenna.

Contractor
Phazotron-NIIR, Joint Stock Company.

Zhuk-MSFE multimode airborne Fire Control Radar (FCR)

Type
Airborne multimode radar.

Description
Zhuk-MSFE
The Zhuk-MSFE (also reported as Sokol (Falcon)) is a coherent, multimission, digital fire-control radar that provides a comprehensive set of all-weather air-to-air and air-to-surface modes. The radar essentially combines the Zhuk-MS radar with the 980 mm phased array antenna of Zhuk-MFE (see separate entry).

The diverse operating modes required to meet the multimission design concept are achieved by employing a variety of complex and flexible waveforms involving low-, medium- and high-pulse repetition frequencies.

The Zhuk-MSFE is intended to be installed in next-generation multirole aircraft; it is compatible with the datalink requirements of a wide range of air-to-air and air-to-surface weapons, including: Kh-31A, R-27R1, R-27T1, R-73E, RVV-AE and other precision-guided weapons.

The antenna is a passive Electronically Scanned Array (ESA) design which provides high gain and low sidelobes at all scan angles. Originally it was to a Tikhomirov-NIIP design, but a Phazotron-NIIR design has been used to replace it. The transmitter is a liquid-cooled TWT. The receiver is a multiple-channel system, with low noise figure. The signal and data processors use flexible high-order language programming.

Zhuk-MSFE is claimed to have the following capabilities:

Air-to-air modes: 24 target Track While Scan (TWS) and simultaneous engagement of six targets; Range While Search (RWS) look-up/look-down; Air Combat Manoeuvring (ACM) – vertical scan, HUD search, boresight, wide angle; velocity search; raid cluster resolution, automatic terrain avoidance.

Air-to-surface modes: two target TWS; Ground Moving Target Indication/Track (GMTI/GMTT); Air-to-Ground Ranging (AGR); real beam ground map; Doppler Beam Sharpening (DBS); precision velocity update; Synthetic Aperture Radar (SAR); enlargement; Freeze Picture; Beacon Mode.

Pharaon
Pharaon, derived from the Zhuk-MSFE, is a lightweight (75 kg) radar, designed to offer similarly comprehensive air-to-air and air-to-ground capabilities to its larger sibling and marketed as a nose fire-control radar for light fighters or as the tail cone radar for heavy fighter aircraft such as the Su-37.

Pharaon employs a new method of implementing a phase-controlled antenna that radically reduces beam-switching time and hence increases target handling capability. The antenna employs a novel, radiating element concept, in which the launching aperture is not a conventional series of regularly spaced slots in a horizontal waveguide, but a series of non-regularly spaced, circularly polarised ferrite phase shifters, placed in the face of feeders that trace the path of a circular arc, fed from the outside of the antenna. The result of this implementation is claimed to be beam-switching times reduced by a factor of 10 when compared with the conventional slotted waveguide design, and a consequent increase in simultaneous target handling capability from 12 to 30.

This antenna technology can be retrofitted into other radars, with Zhuk-ME cited as a prime candidate.

The N010 Zhuk-MSF/Sokol radar is designed for high-performance heavy fighters such as the Su-37. The radar would be displayed on either of the two front-panel MFI-10 MultiFunction Displays (MFDs)
0121096

A further variant, designated Pharaon-M, is reported as an all solid-state version of Pharaon, weighing 45 kg.

Specifications
Zhuk-MSFE
Frequency: X-band, NATO I/low J-band (8–12.5 GHz)
Antenna diameter: 980 mm
Detection range:
(approaching targets)
190 km (3 m² target)
(receding targets) 80 km
Lock-on range: 150 km
Angular coverage: ±70° spatial coverage
Simultaneous target capacity: 24 (TWS); 6 (engaged)
Peak power: 8 kW
Average power: 2.5 kW

Input power: 12 kVA, 200 V 400 Hz; 2 kW 27 V DC
Weight: 275 kg
Volume: 600 dm³
Cooling: liquid
Reliability: >120 h MTBF
Pharaon
Frequency: X-band, NATO I/low J-band (8–12.5 GHz)
Detection range: 70 km against air targets
Angular coverage: +/–70° in azimuth and elevation
Peak power: 4 kW
Average power: 0.3 kW
Weight: 75 kg

Status
The Zhuk-MSFE has been designed to meet the requirements of the Su-27KUB. In late 1999, the first prototype underwent laboratory testing,

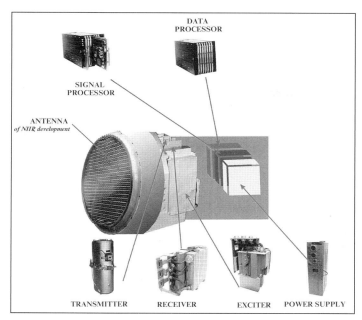

While the Sokol radar is suitable for nose radar duties, its smaller sibling, Pharaon, is designed to fitted as a tail 'sting' radar   0103527

Zhuk-MSFE multimission airborne fire-control radar
0062855

with subsequent fight testing carried out in a Su-27KUB airframe.

Pharaon is designed for use in the Sukhoi Su-27IB and Su-33KUB.

Contractor
Phazotron-NIIR, Joint Stock Company.

Pharaon multifunction airborne fighter radar
0067189

Zhuk-MA multimode airborne radar

Type
Airborne multimode radar.

Description
Phazotron-NIIR has designed the Zhuk-MA (Zhuk-MAE in its export form) Active Electronically Scanned Array (AESA) radar for the MiG-35.

The company began researching an AESA radar circa 2000 as part of a Russian Air Force contest for a fifth-generation fighter. The resulting Zhuk-A was originally destined for a light new-generation MiG fighter and for upgrades for the MiG-29. A mock-up of Zhuk-A, shown at the MAKS exhibition in Zhukovsky in August 2005, had a 700 mm diameter antenna, tilted 20° upwards, with 1,088 Transmit Receive (T/R) modules divided into 272 blocks with four modules in each. In that form, the radar weighed in at 450 kg, deemed unacceptably heavy for light fighter applications. A redesign, involving perforating the body of the radar and substituting magnesium alloy in parts of the structure, reduced weight to 300 kg; this was still deemed too heavy. In order to reduce weight further, more drastic action was required – the antenna diameter was reduced to 600 mm, with 680 modules (170 blocks with four modules each) and set vertically. These changes facilitated a more

The Active Electronically Scanned Array (AESA) Zhuk-A has been developed by Phazotron-NIIR into the Zhuk-MAE; this radar features the same configuration as the Zhuk-ME, but employs a 600 mm AESA (Piotr Butowski)   1184541

forward mounting in the aircraft, which gave extra space in the fuselage for the associated electronics. Reduction of the antenna diameter also reduced power consumption to 5–6 kW and this in turn enabled the radar to be cooled by the existing systems within the MiG-29. This is an important point, considering problems encountered with the development of the AN/APG-80 radar for the F-16E/F – cooling requirements for the AESA make it a difficult retrofit for earlier marks of the aircraft, severely restricting its potential. As a result of the decrease in antenna diameter, range for the Zhuk-A decreased from 200 km to 130 km, although this was considered sufficient for light fighter applications.

While the baseline Zhuk-A had been thoroughly defined, production imperatives forced the utilisation of comparatively heavier components from the production Zhuk-ME radar for the MiG-35, resulting in a hybrid radar, designated Zhuk-MA.

The Zhuk-MA/MAE is a multifunction X-band radar that can track and engage air, ground and sea targets. Claimed performance includes a maximum beam deflection of 70° without parasitic sidelobes. System weight is 220 kg to 240 kg, of which the antenna is 105 kg.

Phazotron has reported that, to date, it is able to design and manufacture all components of the radar, except for the T/R modules, which require Monolithic Microwave Integrated Circuits (MMIC). After attempting to first manufacture its own T/R modules, then subcontracting the activity, with limited success, Phazotron has utilised US-manufactured MMICs for its developmental radars, although production of the Russian-made MMICs is currently underway.

Further development of the Zhuk-MA could see integration of a Vivaldi tapered-slot antenna (AN/APG-79/RBE2A).

The Zhuk-MAE AESA radar will be installed in the MiG-29M2 (Piotr Butowski)   1184540

Specifications
Detection range: 130 km (high aspect)
Performance: Simultaneous tracking of up to 30 targets and engagement of 8 of them
Angular coverage: ±70° in azimuth
Radar frequency: X-band (8–12.5 GHz)
Power: 5–6 kW
Weight: 220–240 kg

Status
As of November 2006, RSK MiG was promoting the MiG-35 equipped with the Zhuk-MAE radar for India's Multirole Combat Aircraft (MRCA) requirement for 126 fighters.

At the same time, Phazotron was completing assembly of two radar prototypes, with first flight of a MiG-29M2 with a developmental Zhuk-MAE during the first half of November, with the intention of displaying the aircraft/radar combination at the Aero India 2007 exhibition in Bangalore, in February 2007.

It has been reported that the Indian Air Force intends to employ the active antenna

as an upgrade to the N011 Bars radar (see separate entry) installed in its fleet of Su-30MKI fighters.

When it flys, the Zhuk-MAE will be the first Russian airborne-tested AESA fighter radar.

Phazotron have stated that the present design of the Zhuk-MAE radar is not definitive; the 700 mm diameter antenna may be re-introduced to increase range to 200 km, while keeping weight down to 280 kg. The +20° rigging angle for the antenna would also broaden the azimuth scan with the aircraft banked and provide an extended scan in the vertical with wings level. An increase in the number of T/R modules per block is also planned; however it has been reported that, as of November 2006, factors including a lack of processing capacity in the control system and limitations of the individual modules have hampered this process.

Contractor
Phazotron-NIIR, Joint Stock Company.

Sweden

Erieye AEW&C Airborne Early Warning & Control mission system radar

Type
Airborne Early Warning (AEW) radar.

Description
The Erieye AEW&C mission system radar features active, phased-array technology. The antenna is fixed, and the beam is electronically scanned, which Ericsson claims provides improved detection and significantly enhanced tracking performance compared with radar-dome antenna systems.

Erieye detects and tracks air and sea targets out to the horizon (and beyond due to anomalous propagation); instrumented range is 450 km. Typical detection range against fighter-sized targets is approximately 350 km, in a 150° broadside sector, both sides of the aircraft. Outside these sectors, performance is reduced in forward and aft directions.

Erieye is understood to operate as a medium- to high-PRF pulse Doppler, solid-state radar, in E/F-band (3 GHz), and to comprise 192 two-way transmit/receive modules, that produce a 1° pencil beam, steered as required within the operating 150° sector each side of the aircraft (one side at a time). It is understood that Erieye has some ability to detect aircraft in the 30° sectors fore and aft of the aircraft heading, but has no track capability in this sector. The aircraft could be manoeuvred to permit tracking of targets in these sectors.

The electronically scanned antenna is controlled by an automatic intelligent energy management system, developed to utilise the phased-array technology implemented, pulse-by-pulse, to illuminate any desired azimuth. The ability instantaneously to direct the radar energy in any wanted direction is used by the

operator to optimise power management for any particular scenario by assigning priorities to areas of interest, thus optimising probabilities of detection and overall system performance.

Status
Erieye is in series production for the Swedish Air Force, where it is implemented in the Saab 340B aircraft. This system is understood to carry the Swedish Air Force's system designation FSR 890, and the Erieye radar to be designated PS-890. Four aircraft have been handed over to the Swedish Air Force, with two more systems due to have been delivered during 1999.

The Erieye system has been selected by the Brazilian government as the airborne element

of the SIVAM system for surveillance of the Amazonas. The contract is for five Erieye systems. Delivery began in 1999 for installation in the Brazilian Embraer EMB-145 aircraft, designated EMB-145SA when converted. The EMB-145SA first flew in May 1999 with planned in-service dates for the first three systems in 2001 and the final two in 2002. Equipment includes the Erieye radar, with integrated IFF system, and command, control and communication system, that will be integrated into the SIVAM command and control system.

In December 1998, Greece selected Erieye as supplier of its AEW&C requirement, based on four systems integrated on EMB 145 aircraft. A final contract was signed on 1 July 1999 with deliveries beginning in 2002. The system is to be

The Swedish Air Force has selected the Saab 340B as the carrier for the Erieye mission system   0010932

delivered by a consortium formed by Ericsson Microwave Systems and Thales Airborne Systems (called Ericsson Thomson-CSF AEW Systems AB for the discharge of this contract). Thales Airborne Systems will provide the NATO-interoperable TSX2500 IFF identification and communication systems and the self-protection systems.

Marketing discussions are in progress with a number of other countries in Asia, Europe and the Americas.

Contractor
Ericsson Microwave Systems AB.

PS-05/A multimode multimission radar for the JAS 39 Gripen

Type
Airborne multimode radar.

Description
Ericsson Microwave Systems AB produces the multimode, multimission, I-band (9 to 10 GHz), pulse Doppler radar for the JAS 39 Gripen. Ericsson collaborated with GEC-Marconi for the development of the PS-05/A, which, in original configuration, shared some elements with the Blue Vixen radar installed in the UK Royal Navy Sea Harrier FA2 aircraft.

The PS-05/A radar provides the following multimode capabilities: air combat, ground attack, high-resolution mapping and reconnaissance to meet the multimission operational requirements of the JAS-39 Gripen aircraft. All principal modes are software driven, and include:

- Air-to-air operation, using high-PRF and medium-PRF pulse Doppler waveforms, to provide long-range search; auto-acquisition; multiple-target track-while-scan; multiple-priority target tracking; short-range, wide-angle search and track for air combat; high-resolution single-target tracking; raid assessment; missile mid-course update
- Air-to-ground operation, using low-PRF pulse Doppler and frequency agility waveforms, to provide search; tracking; high-resolution mapping; air-to-surface ranging.

The radar matches the datalink requirements of the AMRAAM and MICA air-to-air missiles, and is claimed to have excellent ECCM capabilities.

The main waveform modes are:
- HPD – a high-PRF, pulse Doppler mode for clutter rejection, primarily designed for use against airborne approaching targets
- MPD – a medium-PRF, pulse Doppler mode for clutter rejection, primarily designed for use against approaching and receding targets; a special high-resolution sub-mode is designed for target tracking
- LPD – a mode using Doppler processing for clutter rejection designed for use against surface targets
- LPRF – a low-PRF mode with pulse-to-pulse frequency agility for use against surface targets and for real-beam mapping
- AGR – a mode exclusively designed for ground target ranging
- DBS (Doppler Beam Sharpening) – a Synthetic Aperture Radar (SAR) mode utilising Doppler processing for high-resolution mapping, with high-angular coverage obtained by continuous antenna scanning
- SLM (SpotLight Mode) – an SAR mode utilising Doppler processing for very high-resolution mapping.

The PS-05/A radar comprises four Line-Replaceable Units (LRUs):
- Antenna Unit: a lightweight, slotted waveguide, monopulse, planar array, featuring guard antenna, IFF dipoles and all-digital servo control, weighing 25 kg
- Power Amplifier Unit and Transmitter Auxiliary Unit: 1,000 W average power, flexible waveform, liquid cooled, TWT system, with two LRUs together weighing 73 kg
- High-Frequency Unit: multiple-channel receiver, monopulse, pulse-to-pulse frequency-agile, microwave integrated circuit design, internally software controlled, weighing 32 kg

The Ericsson Erieye Airborne Early Warning & Control (AEW&C) system on the Embraer EMB 145 aircraft   0079238

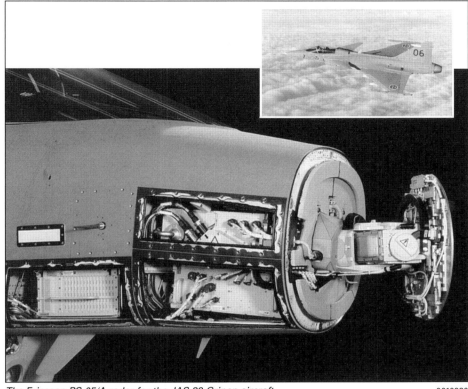
The Ericsson PS-05/A radar for the JAS 39 Gripen aircraft   0010930

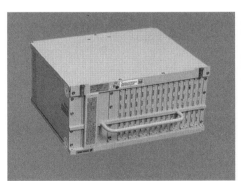
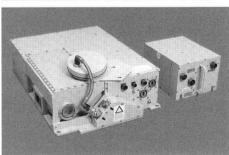

PS-05/A radar antenna unit, (bottom right) power amplifier and transmitter auxiliary units (bottom left), high-frequency unit (top right) and signal and data processor unit (top left)   0010931

- Signal Data Processor: D80 multiprocessor concept fully programmable unit, employing ASIC technology, with programming in High Order Language, Pascal, weighing 23 kg.

Specifications
Weight: 156 kg
Power: 115/200 V AC, 400 Hz, 8.2 kW; 28 V DC, 250 W
Antenna: 600 mm planar-array
Power output: 1 kW (average)
MTBF: 250 h (predicted)
Interface: MIL-STD-1553B

Status
In excess of 100 units of the PS-05/A radar for the Gripen aircraft have been delivered to the Czech, Hungarian, South African and Swedish Air Forces.

The radar has been the subject of a continuous upgrade effort:

Mk 3
Also known as the JAS Upgrade Radar (JURA) programme, the PS-05/A Mk 3 is the baseline sensor for Batch 3 JAS 39C/D aircraft and incorporates the Modular Airborne Computer System (MACS), replacing the D80 computer. The Mk3 radar brings improvements to the Gripen's air-to-air and air-to-ground missions. Radar system software was also upgraded to facilitate future upgrades.

Mk 4
Also known as the Gripen Reconnaissance and Enhanced Target Acquisition (GRETA) programme, the PS-05/A Mk 4 builds on the Mk 3 configuration and adds a high-resolution synthetic aperture mode together with an enhanced passive surveillance mode. As of mid-2004, the PS-05/A Mk 4 was described as being at the demonstration stage.

Mk 5
Also known as the Not Only a RAdar (NORA) programme, the PS-05/A Mk 5 is aimed at future Gripen Mid Life Upgrade (MLU) efforts, incorporating an Active Electronically Scanned Array (AESA). A NORA system also figures in the Ericsson Integrated Radio-frequency Avionics (EIRA) concept, in which an AESA radar would form part of the proposed Multifunction Integrated Defensive Avionics System (MIDAS), which fuses radar and electronic

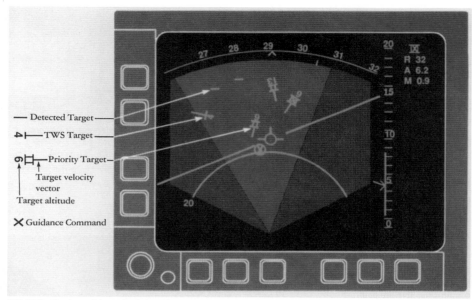

PS-05/A display with typical data format   0001209

warfare sensors into a unified whole. Raytheon has supplied Ericsson with an AESA antenna assembly with which to test a NORA-type system architecture and its associated signal processing. The array was scheduled to begin anechoic chamber tests during 2004, with flight testing on a JA-37 Viggen testbed aircraft following in 2005.

Contractor
Ericsson Microwave Systems AB.

United Kingdom

ARI 5980 Searchwater radar

Type
Airborne surveillance radar.

Description
Searchwater is the commercial name for the ARI 5980 radar which is standard equipment on UK Royal Air Force Nimrod MR. Mk 2 maritime reconnaissance aircraft; it was designed to replace the ASV Mark 21 radar. The system formed part of a major mid-life refit and 31 Nimrod MR. Mk 1s were converted to Nimrod MR. Mk 2 configuration by the addition of Searchwater and other improvements.

Searchwater is designed for all-weather, day or night operation outside the defensive range of potential targets.

The system comprises a frequency-agile radar which uses pulse compression techniques and a pitch and roll stabilised scanning antenna with controllable tilt and automatic sector scan. IFF equipment is included to interrogate surface vessels and aircraft.

The signal processor enhances the detection of surface targets (including submarine periscopes) in high sea states. An integrating digital scan converter permits plan-corrected presentation and classification of target and transponder returns. The single radar observer in the aircraft is presented with bright, flicker-free television-type PPI, B and A scope displays in a variety of interactive operating modes. Weather radar and navigation facilities are provided within the system.

A real-time dedicated digital computer relieves the radar operator of many routine tasks, while continuously and automatically tracking, storing and analysing data to provide position information and automatic classification for a number of ship targets at the same time. Built-in test facilities provide for automatic detection and diagnosis of faults.

The facilities offered by Searchwater reduce the vulnerability of the host aircraft by permitting operation in a standoff mode, avoiding the need to fly over the target for visual identification. Over-the-horizon targeting is also provided for such missiles as Sea Eagle and Harpoon.

The system is entirely modular, with the interfaces and mechanical construction designed for ease of fault location and replacement. Major

Thales' Searchwater radar display in a UK Royal Navy Sea King Mk 2 AEW helicopter   0503723

Thales' Searchwater antenna on a UK Royal Navy Sea King Mk 2 AEW helicopter. Note the 90° rotated stowed position, with the radome in a semi-deflated condition   0059616

units are functionally self-contained as far as possible with a minimum of interconnections. Extensive use is made of hybrid and integrated circuit techniques. The transmitter uses solid-state frequency generators and mixers, followed by two cascaded travelling wave tubes. A fluorocarbon liquid cooling system is employed. The scanner, which both transmits and receives radar and IFF signals, uses a reflector of lightweight construction based on resin-bonded carbon fibre.

Status

Searchwater entered service aboard UK Royal Air Force Nimrod MR. Mk 2s in 1979.

Searchwater in the Sea King

As a result of demands from the UK Royal Navy for more effective radar sensors for organic fleet defensive surveillance following operations in the Falklands, Searchwater was modified for use aboard converted Westland Sea King anti-submarine helicopters, redesignated Sea King AEW Mk 2. Ten Searchwater systems were acquired to permit 24 hour AEW coverage for the UK Royal Navy.

The system was also supplied to the Spanish Navy for AEW helicopters under a GBP13 million deal announced in September 1984. Deliveries began in October 1986; the system entered service in mid-1987 and is operational on three Spanish Navy Sikorsky SH-3D helicopters.

Status

The first deliveries of Sea King AEW Mk 2s, fitted with modified Searchwater radars, to the UK Royal Navy were in 1985. Deliveries to the Spanish Navy began in 1986. The UK Royal Navy bought two more radars in mid-1991.

Contractor

Thales Defence Ltd.

ASTOR airborne standoff radar

Type

Airborne surveillance radar.

Description

The Airborne STandOff Radar (ASTOR) is a UK MoD sponsored project designed to provide detailed over-the-border surveillance of the land battle and major hostile ground forces. The system is designed to fulfil UK MoD Staff Requirement (Land/Air) 925 (SR 925), valued at GBP800 million and scheduled to achieve Initial Operating Capability (IOC) during 2005.

The core requirement is for a system capable of providing high-resolution static imagery with the ability to detect moving targets. The tactical objective is to provide 24-hour observation and targeting intelligence of enemy first and second echelon forces and to support peacekeeping operations. To that end, the core capabilities of ASTOR inlcude:

- Combined Ground Moving Target Indicator/ Synthetic Aperture Radar (GMTI/SAR)
- Interleaved (not concurrent) radar operating modes
- SAR swathe and spotlight (0.5 m resolution) operating modes
- Satellite communications
- Air-to-air datalink (Link 16)
- 8 – 12.5 GHz and 12.5 – 18 GHz band primary imagery datalinks
- Secure radios for air-to-ground and air-to-air voice traffic
- Capability to detect and track helicopters and low-speed moving ground targets
- Interoperability with other (non-national) surveillance systems
- The contract award is for five aircraft, plus eight supporting ground stations.

Raytheon Systems Limited (RSL) was awarded the prime contract, covering development, production and support, in December 1999. The Raytheon Systems bid is based on the Bombardier Global Express airframe, with 25 per cent input from Bombardier's subsidiary, Short Brothers in Belfast. Ultra Electronics, with Cubic Defence Systems, is to provide the secure broadcast

Thales' Searchwater antenna on a UK Royal Navy Sea King Mk 2 AEW helicopter. When in use the radome is inflated. The system rotates clockwise through 90° to give clearance for take-off and landing　　　0001211

Viewed from below, the Sentinel R Mk 1's lower fuselage is dominated by the radar antenna; note the towed decoy launcher located just aft of the trailing edge of its starboard wing (Raytheon)　　　0583745

This in-flight image of UK RAF Sentinel R Mk 1 ZJ690 clearly shows the dorsal satellite communications antenna radome, ventral radar antenna fairing, starboard side CMDS mounting and starboard side missile warning sensor on the tail fairing (Raytheon)　　　0583746

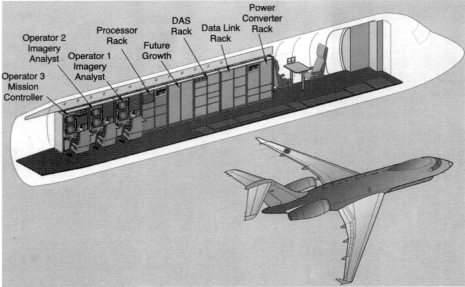

The interior design of the Raytheon Systems Limited ASTOR airborne system   0062746

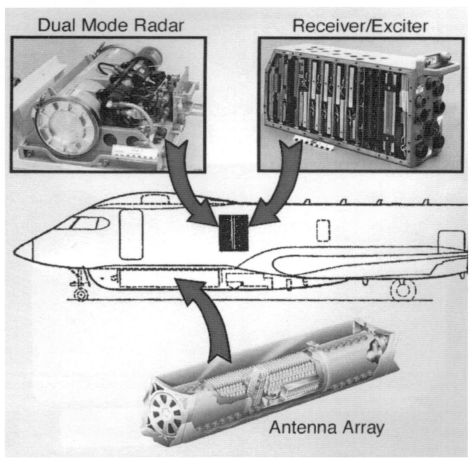

The ASTOR radar arrangement. In this case, the ASARS-2 Plus is depicted as the 'Dual Mode Radar'
(MS/Raytheon Systems)   0054853

datalink. Ground stations will be provided by Motorola, in conjunction with RSL, Marshal Specialist Vehicles and QinetiQ. Six of the ground stations will be mobile variants, mounted on Steyr Pinzgauer 6 × 6 trucks; the remaining two will be built as deployable shelter-based versions.

ASARS-2 Plus Radar

The ASARS-2 Plus is a combined SAR/MTI surveillance radar, developed from Raytheon Systems' ASAR-2 radar which is the primary sensor aboard the US Air Force U-2R/S high-altitude reconnaissance aircraft. During 1999, Raytheon conducted flight trials of an enhanced ASARS-2 sensor (designated ASARS-2 Improvement Programme – AIP) in which a Commercial-Off-The-Shelf (COTS) Power PC-based onboard processor was used to perform tasks previously undertaken at the ground station. Overall improvement in system performance as a result of the AIP included real-time precision targeting,

onboard processing, broad area synoptic coverage, enhanced ground MTI, a four-fold increase in search coverage, a nine-fold increase in spot coverage and a new, very high-resolution, search mode.

During 1990, the ASARS-2 Plus variant was introduced, described as a development of the base system that presented an ASTOR-compliant, dual-mode SAR/MTI radar. The ASARS-2 Plus was reported to incorporate the latest USAF-standard receiver/exciter, upgraded processing (probably derived from the AIP variant), a new 4.6 m antenna (to be provided by BAE Systems, who will also provide the defensive aids suite) and design consultancy in motion compensation and autofocus provided by QinetiQ, Malvern.

In December 1999, it was reported that ASARS-2 Plus complied with all UK MoD requirements in terms of performance, risk, cost and supply of sensitive technologies. A predicted detection range 'in excess of' 296 km (184 miles), from

an altitude of 15,545 m (50,988 ft) and a grazing angle of 2°, has been claimed.

Status

The UK MoD and RSL signed the UK ASTOR contract in December 1999. A development Bombardier Express in ASTOR configuration made its initial flight on 3 August 2001. The aircraft had been fitted with the various external shapes and aerodynamic modifications planned for the production ASTOR platform. Launching from Wichita, Kansas, the flight lasted 3 hours and 18 minutes with the aircraft reaching 463 km/h and 7,620 m. Flight trials continued, with the next major milestone being the delivery of the first production airframe on 7 February 2002. Bombardier delivered the first production Sentinel R Mk 1 airframe to Greenville, Texas for modification and systems integration.

During May 2003, the UK RAF reported that the service had designated the ASTOR aircraft as the Sentinel R Mk 1 and that its operating unit was to be No 5 (AC) Squadron, to be based at RAF Waddington in Lincolnshire, England.

On 4 February 2004, RSL announced that it had taken delivery of three Global Express airframes for conversion to Sentinel R Mk 1 standard at its Broughton, North Wales facility. Sentinel R Mk 1 ZJ690 made its maiden flight from Greenville, Texas on 26 May 2004. During the 4-hour 24-minute flight, the aircraft reached an altitude of 4,572 m.

Anticipated in-service date for the Sentinel R Mk1 is late 2005, with two airframes and three crews, Initial Operating Capability (IOC) by mid-2007 and Full Operating Capability (FOC) by April 2008, with a minimum of four airframes.

To smooth entry into service, a Joint Test Team has been established to run operational tests, with an Operational Evaluation Unit (OEU) to develop tactics and procedures. The JTT brings together representatives of QinetiQ, the Defence Science and Technology Laboratory (DSTL), No. 5 Squadron and the OEU, who will all be involved in various phases of the test and evaluation process. The first phase of the test and evaluation process involves conducting tests to see if radar data from the aircraft can be downloaded to a ground station and interoperability tests with the US Joint Surveillance Target Attack Radar System (JSTARS) and the UK RAF Reconnaissance Airborne Pod for Tornado (RAPTOR) (see separate entries).

Following these trials, Operational Test (OT) will aim to demonstrate that ASTOR can meet its required performance goals in a series of operationally representative tests. The tests will run in two phases that will aim to accurately replicate a cycle of tasking – mission planning – deployment – connectivity with tactical and operational headquarters – data collection – image and data processing/exploitation – reporting.

Upon successful completion of OT, the UK RAF will formally accept ASTOR into service and Operational Evaluation (OE) will begin. This phase will continue until full operational capability is declared in April 2008, when all five Sentinel aircraft and their eight supporting ground stations are planned to be delivered to the RAF and British Army. This will culminate in three major evaluation exercises, which will allow all stakeholders to fully understand the system's capabilities and limitations.

Contractor

Raytheon Systems Limited.

Blue Kestrel maritime surveillance radars

Type

Airborne surveillance radar.

Description

The Blue Kestrel surveillance radar family consists of the Blue Kestrel 5000, 6000 and 7000. They are allied to the Seaspray family of radars, using common modules throughout the range. This minimises cost and provides natural growth paths for the customer.

Blue Kestrel 5000 maritime surveillance radar

Blue Kestrel 5000 is a pulse compression radar sensor developed for the UK Royal Navy EH

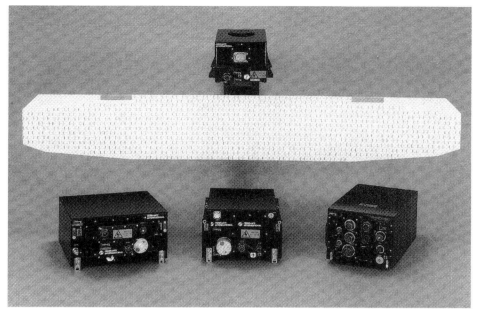

The main units of the BAE Systems Blue Kestrel 5000 radar　　0503716

The EH 101 Merlin helicopter is being fitted with the BAE Systems Blue Kestrel 5000 radar, and may receive the Blue Kestrel 7000 in future capability upgrade　　0503717

101 Merlin helicopter. It consists of a large flat plate antenna and Travelling Wave Tube (TWT) transmitter to provide the Merlin with the optimum power aperture product, combined with a high-gain receiver and digital processor. The radar sensor is interfaced with and controlled by the platform's mission avionics management system via the MIL-STD-1553B databus.

This highly integrated sensor approach to the radar results in a high-performance pulse compression surveillance radar in a particularly compact and lightweight four-line replacement unit configuration. It is suitable for application in a range of sophisticated naval helicopters or maritime patrol aircraft.

Specifications
Scanner type: planar-array
Frequency: I-band
Transmitter: low-peak power, high-mean power TWT
Pulsewidths: selectable
PRF: selectable
Coverage: 360°
System weight: 102 kg
Features: pulse compression, CFAR, multiple TWS and operator-selectable scan-to-scan integration

Status
Blue Kestrel 5000 has completed development and production deliveries have commenced. The development phase included the building

of 10 systems, all to full flying standard, and the completion of several thousand hours running time in ground-based rigs, two specially converted Sea King flying testbeds, and the EH 101 Merlin prototype. Lockheed Martin ASIC, the Merlin programme prime contractor, placed the production contract for Blue Kestrel 5000 in October 1992.

Blue Kestrel 6000 maritime surveillance radar
Blue Kestrel 6000 is a coherent pulse Doppler radar sensor. It is conceived as a modular upgrade to the Blue Kestrel 5000 sensor, where the only major change required is to insert the pulse Doppler upgrade module. It brings together BAE Systems' experience with the pulse compression Blue Kestrel 5000, the multimode pulse Doppler Blue Vixen, and Inverse Synthetic Aperture Radar (ISAR) work, which has been jointly funded by QinetiQ and BAE Systems.

Flight trials of a modified Blue Kestrel 5000 in a QinetiQ Sea King testbed have been successfully completed, producing real-time images of co-operative and opportunity surface contacts from a helicopter. The resulting radar sensor provides enhanced air-to-surface and air-to-air detection, standoff classification capability, and a superior anti-submarine warfare capability with sub-clutter target detection. Additional options include moving target detection and high-resolution synthetic aperture radar ground mapping.

Specifications
Frequency: I-band
Type: coherent multimode
Transmitter: TWT fixed frequency/frequency agile
Scanner type: planar-array
Coverage: 360°
LRUs: 4
System weight: 125 kg
Features: CFAR, enhanced TWT, ISAR classification, ASW and air-to-air modes
Options: MTI and SAR ground mapping

Status
Blue Kestrel 6000 development is complete.

Blue Kestrel 7000 maritime surveillance radar
Meanwhile, a more comprehensive upgrade, designated Blue Kestrel 7000, is nearing development completion. Blue Kestrel 7000 will have a full pulse Doppler, multifunction, COTS-based, open architecture, reprogrammable capability (the same capability can also be accommodated within the smaller Seaspray radar, where it will be designated Seaspray 7000). Blue Kestrel 7000 is planned for a future EH 101 capability upgrade programme.

Specifications
Frequency: I-band (8–10 GHz)
Pulse-width: selectable (optimised for mode)
PRF: selectable (optimised for mode)
Transmission modes: fixed frequency; frequency agile
Dimensions: 220 × 340 × 150 mm (scanner); 390 × 390 × 195 mm (transmitter); 500 × 390 × 195 mm (processor/receiver)
Weight: 27.4 kg (scanner/antenna); 28.5 kg (transmitter); 30.1 kg (processor/receiver)

Contractor
Selex Sensors and Airborne Systems.

Foxhunter airborne interception radar

Type
Airborne Fire-Control Radar (FCR).

Description
The Air Intercept (AI) 24 Foxhunter radar is the multimode airborne interception radar for Tornado F3. A substantial part of the signal processing is performed digitally in addition to digital radar data handling. The equipment design anticipated trends in offensive tactics, such as low-level penetration and use of ECM, and the latest improvements give growth potential into supporting active missiles. Additionally, Foxhunter has the flexibility to operate as part of ground- or AEW-based control environments while retaining the ability to perform autonomously.

Foxhunter is designed to detect and track subsonic and supersonic targets at ranges in excess of 185 km at both low and high level.

Foxhunter operates in I-band (3 cm), and in its primary mode uses the pulse Doppler Frequency Modulated Interrupted Continuous Wave (FMICW) technique. At the heart of the radar is a master timing and synchronising unit, employing phase-locked loop techniques, which generates the complex waveforms for amplification by the transmitter, and the accurately related reference signals and precise timing pulses employed within the receiver and signal processing circuits.

Compact and lightweight surface acoustic wave devices, provide signal waveforms for the pulse compression modes of the radar. The antenna uses the 'Elliot' twist-reflecting cassegrain principle which combines rigidity and light weight with extremely low levels of spurious radiation lobes, a key feature in the rejection of ground clutter and jamming signals. This type of antenna has been developed to a very advanced level of performance. The scanner employs a hydraulic drive mechanism and very high-grade servos to achieve the speed and precision of beam pointing and stabilisation demanded especially during manoeuvre.

The Stage 3 radar update, part of the AMRAAM Optimisation Programme (AOP), facilitated the introduction of AMRAAM and ASRAAM to replace the Tornado F3's legacy weapons suit of Skyflash and AIM-9L, shown here (BAE Systems)   1127382

BAE Systems Foxhunter radar installed in the nose of a UK Royal Air Force Tornado F. Mk 3 aircraft   0503722

The TWT transmitter has the inherent flexibility to handle the differing waveforms used in the various modes of operation. The heart of the signal processing system is a digital processor employing Fast Fourier Transform techniques of frequency analysis to filter the signal returns into narrow frequency channels. Target echoes are segregated from clutter returns and a particularly uniform detection threshold is achieved against all target velocities.

Foxhunter also provides target illumination for the Sky Flash medium-range semi-active radar air-to-air missiles and, since a significant ECM operating environment can be expected, the radar system incorporates many EPM features.

Product improvement programme

Since Foxhunter was introduced into service, a continuous product improvement programme has been implemented. Stage 1 introduced numerous modifications and permitted radar functions that were closely tied to HOTAS modes. The TWS and EPM algorithms were also refined. The Stage 2 standard introduced a number of major improvements, including a completely redesigned data processor and supporting

software, as well as significant modification to both the transmitter and receiver paths. The resultant effect is a much higher quality of TWS performance in a high RF noise environment. The processor also enables sophisticated automatic and manual scan management techniques to be employed in TWS; it is also capable of handling a significantly greater number of tracks. Additionally, it provides scope for significant growth potential beyond Stage 2, including a limited ability to support active radar missiles.

The next improvement to British and Italian Tornado F3 aircraft came with Stage 2G+. Stage 2G+ was developed as a result of lessons learned in the 1991 Gulf War, and included much more automation of search techniques. Royal Saudi Air Force aircraft remained at Stage 1 standard.

Stage 2H incorporated maintenance and support improvements and mechanical interfaces for further pre-planned hardware product improvements.

The latest update to the Foxhunter has been undertaken as part of the AMRAAM Optimisation Programme (AOP). AOP includes the modification of the Foxhunter radar to Stage 3 standard, to allow the radar to pass updated target positions

via datalink to the missile. This ensures that during missile time-of-flight the weapon can manoeuvre appropriately to position the seeker correctly whereupon the missile's own seeker can lock-on and autonomous guidance can take over. At that stage, the firing aircraft's role in the engagement is complete and it can break lock and turn away.

Status

Foxhunter is in operational service with the UK Royal Air Force, the Royal Saudi Air Force and the Italian Air Force. Stage 1 was introduced into service in 1989. The latest development standard, Stage 2H, has entered service with the UK Royal Air Force and the Italian Air Force.

The advanced weapons integration effort began in May 1999 with the GBP125 million Tornado F3 Capability Sustainment Programme (CSP), led by BAE Systems and the Defence Aviation Repair Agency (DARA), which eventually led to the introduction of the MBDA Advanced Short Range Air-to-Air Missile (ASRAAM) and AMRAAM onto the aircraft. As discussed, the AOP effort, launched to specifically exploit the AIM-120 AMRAAM to its full potential, facilitating employment of the missile at its maximum kinematic potential, was launched in June 2001. In parallel with the Stage 3 radar update, the missile launchers on the underside of the aircraft and the cockpit displays have been modified to allow carriage and correct employment of the missile. Raytheon's Successor Identification Friend or Foe (SIFF) interrogator has been installed to give the F3 Mode 4 interrogator ability, which adds a critical element of RoE compliance in BVR attacks.

The AOP has bestowed the ability to engage multiple targets simultaneously, thereby increasing crews' Situational Awareness (SA) and delivering greater firepower. It also allows the aircraft to employ weapons at range with sufficient average speed to ensure that crews can engage targets while employing tactics that ensure they stay outside the target's own effective killing zone. AOP was introduced to a cadre of RAF aircrew at the end of 2003. The cadre participated in a successful AMRAAM operational evaluation (OPEVAL) in January 2004 during a Tactical Leadership Training (TLT) exercise. On completion of the OPEVAL, the capability was rolled out at the RAF Leuchars Wing in February 2004 when both 43(F) and 111(F) squadrons commenced training.

Contractor
BAE Systems.

Pod SAR

Type
Airborne surveillance radar.

Description
QinetiQ (formerly the UK Defence Evaluation Research Agency) is leading a UK MoD applied research programme for a podded Synthetic Aperture Radar (SAR) and Ground-Moving Target Indicator- (G-MTI) mode radar, to provide detailed battlefield all-weather day and night penetrating reconnaissance capabilities for Tornado and Eurofighter Typhoon aircraft in areas where standoff systems and other more vulnerable systems cannot see or be employed.

QinetiQ has awarded Thales Defence Limited the single concept study to design the technology demonstrator. The demonstration radar will have a high-resolution SAR output of both spot and strip modes, together with G-MTI capabilities. Pod SAR will complement other assets, such as ASTOR. The radar will be installed in the Smart Pod, which can be configured for a variety of sensor fits.

Status
Technology demonstrator programme. Intended to enter service during 2004/05.

Contractor
QinetiQ.
Thales Defence Ltd.

Searchwater 2000AEW radar

Type
Airborne Early Warning (AEW) radar.

Description
In February 1997, the UK MoD appointed the then Racal Defence Electronics Limited (now Thales Defence Limited) as prime contractor for the radar and mission system upgrade of the UK Royal Navy's Sea King AEW Mk 2 helicopters. Acting as prime contractor, Thales Defence Limited is responsible for all aspects of the upgrade, including equipment supply, installation, aircraft modification and certification, together with provision of logistic support.

Thales is supplying its new-generation Searchwater 2000 AEW radar for the upgrade. This radar variant retains the high-performance maritime surveillance features of the existing in-service Searchwater radar, but offers considerably improved AEW performance using new high-power transmitters and advanced pulse Doppler signal processing techniques. The new radar was under development as a private venture for five years, and provides an advanced high-technology hardware and software solution with considerable weight saving and improved reliability. Searchwater 2000MR (for Nimrod MRA4) and Searchwater 2000 AEW (for Sea King AEW) have a high degree of commonality, which will provide operational and logistic standardisation benefits to both the UK Royal Navy and the UK Royal Air Force.

The updated helicopter is designated the Sea King Mk 7. The 10 in-service Sea King Mk 2 helicopters are being updated over seven years (1997-2003). The contract includes an extensive logistic support package covering mission and maintenance trainers, and workshop support down to second line, including the provision of support on board aircraft carriers at sea. Since contract award, Thales has received an order for three additional systems to enable expansion of the AEW fleet to 13 aircraft. The additional three helicopters will be conversions from Sea King Mk 5/6 AEW.

Radar modes include air-air (look-up and look-down), Moving Target Indicator (MTI), maritime surveillance (ASW/ASuW), littoral, open waters, navigation/ground mapping, weather, beacon and target identification/classification.

Status
Trials of the system were conducted at QinetiQ Boscombe Down during 2001, with entry into service during 2002.

Contractor
Thales Defence Ltd.

Thales' Searchwater 2000AEW radar incorporates a comprehensive Man-Machine Interface (MMI), including large high-resolution flat panel colour displays and Interactive Control Panels (ICPs)
0098709

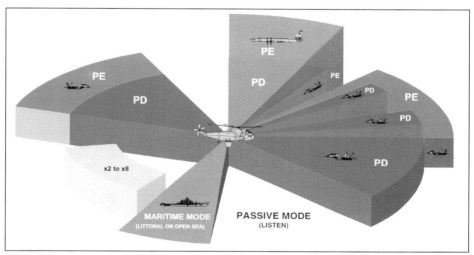

Thales' Searchwater 2000AEW radar incorporates multi-bar scanning for multiple-level raid detection. Pulse Doppler/Pulse Envelope interlacing allows targets to be detected at maximum range and successfully tracked through background clutter
0098707

Searchwater 2000MR maritime radar

Type
Airborne surveillance radar.

Description
As part of the contract for BAE Systems' Nimrod MR4A mission system, Thales Defence Limited was in turn contracted by Boeing Space and Defense for the supply of over 20 Searchwater 2000MR radars.

Searchwater 2000MR is a high-performance Maritime Patrol Aircraft (MPA) radar specifically designed for all-weather Anti-Submarine Warfare (ASW) and Anti-Surface Warfare (ASuW) operations in littoral waters or open seas. The radar is optimised to have the highest probability of detecting fleeting submarine masts in poor sea conditions using the latest detection algorithms, together with an extensive range of proven techniques to reduce the effects of sea clutter. The broad antenna beam shape permits continuous coverage at all ranges. Automatic tracking of over 100 targets provides accurate bearing, range and speed information for maintaining the surface plot for employment of air-surface weapons or for third party targeting.

A full suite of classification tools ensures that surface threats can be identified quickly and safely at long stand-off ranges. Array Systems Computing Inc supplies radar imaging software from its TriSAR programme.

A Pulse Doppler (PD) mode provides for Situational Awareness (SA) during ASW and ASuW operations, detecting and localising aircraft within the search volume. The radar incorporates an IFF interrogator and includes a search and rescue transponder display function. A navigation/weather avoidance mode display

Thales' Searchwater 2000MR radar will enter service on the UK RAF Nimrod MR4A
0098708

Thales' Searchwater 2000MR radar includes SAR and ISAR classification modes   0098710

can be provided as a separate video channel to the cockpit.

Specifications

Type: Multimode high-mean power TWT, coherent surveillance radar with all-digital processing and adaptive threshold control
Frequency: X-band (8–12 GHz), selectable
Elevation coverage: Single sweep (cosec2 beam shape)
Polarisation: V & H
Modes: ASW, AsuW, swathe SAR, air-air PD, search and rescue
Tracking: automatic >100 targets
Power: 115 V AC, 3 phase, <4 kVA
Cooling: air
Interface: MIL-STD-1553B, 2 RGB video outputs

Status

Thales Defence is contracted to supply 21 systems for the UK RAF Nimrod MRA4 up to 2004. The radar is also suitable for P3 and ATL MPA platforms.

Contractor

Thales Defence Ltd.

Seaspray 3000 airborne radar

Type

Airborne surveillance radar.

Description

Seaspray 3000 (previously known as Seaspray Mk 3) introduces significant advances over Seaspray Mk 1, most notably through the provision of a new digital processor and full 360° scan. The system is designed primarily for operation in light, agile, shipborne naval helicopters to provide detection and tracking of small targets in adverse conditions for the Sea Skua missile.

BAE Systems Seaspray 3000 radar units   0503720

For this purpose it retains the combat proven monopulse lock-follow target illumination of its predecessor.

Consisting of six LRUs, Seaspray 3000 is configured to operate on a MIL-STD-1553B databus, with multiple additional standard interfaces being provided. The resultant lightweight, compact and flexible system is applicable to both the retrofit and new aircraft. Operating in the I-band, Seaspray 3000 uses a magnetron to provide high transmitted power and very high-speed agility to provide these platforms with high-detection performance in sea and weather clutter and in ECM conditions. The digital processor provides advanced features, including constant false alarm rate and track-while-scan facilities to optimise target detection and reduce operator work load.

The operator is provided with a comprehensive tactical situation display of scan-converted television format, in monochrome or colour. In addition, the output of other sensors (FLIR, ESM, Datalink and so on) may be displayed. The control unit employs a menu structure with soft key options.

Specifications

Frequency: I-band
Peak power: 90 kW
Transmitter: high-speed, spin-tuned magnetron
Pulsewidths: 2 selectable
PRF: 4 selectable
Coverage: 360°
LRUs: 6
System weight: 90 kg

Target illumination: monopulse lock-follow
Features: CFAR and operator-selectable scan-to-scan integration
Options: operator-selectable circular polarisation, a range of antenna sizes (up to 1 m diameter), and a range of display options

Status

Seaspray 3000 is in service with the Republic of Korea Navy and Brazilian Navy in GKN Westland Super Lynx, in the Turkish Navy in Agusta Bell 212, and in the German Navy in GKN Westland Sea King helicopters. The German Navy has also ordered Seaspray 3000 for its new Super Lynx Mk 88A helicopters and the Royal Navy is upgrading its Seaspray Mk 1 radars to Seaspray 3000 standard in its Lynx Mk 1 helicopters. Seaspray 3000 has also seen service in a Fokker F27 maritime patrol aircraft and has been proven with Sea Skua in land-based coastal battery and fast patrol boat applications.

Contractor

Selex Sensors and Airborne Systems.

Seaspray 7000 airborne radar

Type

Airborne surveillance radar.

Description

Seaspray 7000 is a pulse Doppler, pulse compression, multimode, COTS-based, open architecture, programmable airborne maritime surveillance radar. It can be operated as a stand-alone radar or as the heart of a fully integrated avionics suite. It is the resultant system from combining the proven man/machine interface and processing of Seaspray 3000 with BAE Systems' Travelling Wave Tube-based (TWT-based) pulse compression technology, developed in conjunction with the Blue Kestrel 7000 radar.

Comprising six LRUs, which are form/fit compatible with the Seaspray 3000, Seaspray 7000 provides the optimum performance for medium-size maritime patrol aircraft and naval helicopters. Processing capabilities include: pulse compression, Synthetic Aperture Radar (SAR), Inverse SAR (ISAR), and Moving Target Indication (MTI).

Specifications

Frequency: I-band
Transmitter: Ring bar TWT power amplifier
Pulsewidths: selectable
PRF: selectable
Coverage: 360°
LRUs: 6
System weights: total 62.2 kg

Westland Super Lynx helicopters of the South Korean Navy are equipped with the Seaspray 3000   0503721

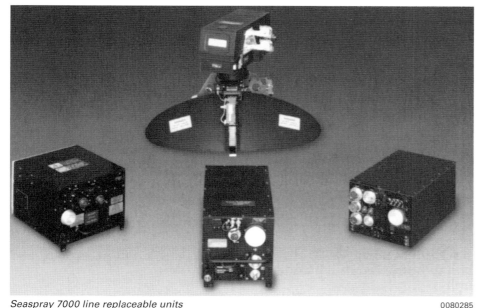

Seaspray 7000 line replaceable units 0080285

The Seaspray 7000 is installed in the chin radome of the Super Lynx 300 helicopter 0528526

(scanner): 10.2 kg
(transmitter): 22.5 kg
(receiver): 12.0 kg
(processor): 17.5 kg

Dimensions:
(scanner): 737 mm swept dimension
(transmitter): 145 (W) × 381 (D) × 243 (H) mm
(receiver): 207 (W) × 381 (D) × 225 (H) mm
(processor): 260 (W) × 475 (D) × 200 (H) mm

Features: pulse compression, CFAR, multiple TWS, operator-selectable scan-to-scan integration, SAR and ISAR

Options: operator-selectable circular polarisation, range of antennas and displays

Status
The Seaspray 7000 airborne radar was designed for, and is in service in, the Super Lynx 300. The numbering gaps in the Seaspray range (4000, 5000 and 6000) are explained as follows: Seaspray 4000 was a development project that has matured into the Seaspray 7000; the designators 5000 and 6000 have been reserved for the Blue Kestrel range of radars.

Contractor
Selex Sensors and Airborne Systems.

Skyranger airborne radar

Type
Airborne Fire-Control Radar (FCR).

Description
Skyranger is a lightweight airborne weapon control radar developed for light fighter and light attack aircraft in the retrofit market. It consists of three main units; antenna, transmitter/receiver and signal processor/power supply, and since the amount of space available for retrofit programmes can often be limited and irregular in shape, the modularity of Skyranger has been established at printed circuit card level. The individual cards can, therefore, be packaged into housings designed for the space available. The equipment has been designed as part of an integrated avionics suite, the other system being an air data computer, radar altimeter, head-up display, weapon aiming computer and secure communications.

Skyranger accepts discrete digital commands from a cockpit-mounted control panel and provides output data in the form of a digital serial link (ARINC 429) to a HUD and other weapon aiming systems. It has two main modes, guns and missiles, the former having a shorter range wide-angle beam. In the missile mode the radar energy is fed from the feed horn and reflected back from the parabolic antenna in a 6° beam with a maximum range of 15 km. For gun attacks, the radar energy is fed out directly from the antenna through a polarised window. This results in an 18° beamwidth and a range of 5 km. Minimum range is 300 m for guns and 150 m for missiles, with a ranging accuracy in the order of ±15 m below 3 km and ±30 m above. Target relative velocities from –500 to +1,000 m/s may be handled.

The equipment operates in I-band and has a 5 per cent pulse-to-pulse agility. MTBF is given as 200 hours and the equipment contains built-in test systems. The current version has a fixed antenna.

Specifications
Frequency: I-band
Range: 15 km for missiles, 5 km for gun operation
Range resolution: 150 m
Pulse-to-pulse agility: 5%
Power supply: 27 V DC, <50 W; 115 V 400 Hz, single phase, <400 VA
Weight:

(antenna)	4 kg
(transmitter/receiver)	25 kg
(signal processor/power supply)	8 kg
(total installed weight)	40 kg

Status
In service in export F-7M fighter aircraft supplied by China to various customers. More than 300 systems have been produced.

Contractor
Selex Sensors and Airborne Systems.

Super MAREC radar

Type
Airborne surveillance radar.

Description
Super MAREC (MAritime REConnaissance) is an upgrade of the MAREC II involving improved software and the replacement of the MAREC II's 430 mm plotting display with the 356 mm colour television-type display developed for Super Searcher. This saves weight and space and increases the tactical navigation facilities available to the operator. Additional facilities include multiple-target Track-While-Scan (TWS) and navigation overlay data.

Enhancements to Super MAREC include the addition of a full ISO contoured weather avoidance mode, a ground-mapping mode and an extra radar mode to give improved close-range target detection. Digital map overlays are also provided. A smaller display and joystick option is offered as a replacement for the large display and keyboard to permit radar operation from the cockpit by the second pilot.

Status
In service with the Indian coastguard.

Contractor
Thales Defence Ltd.

Super Searcher airborne radar

Type
Airborne surveillance radar.

Description
Super Searcher is a development of the Sea Searcher and Super MAREC radars and is fitted to UK Royal Navy Sea King helicopters. Designed for multithreat maritime operations, it is a lightweight I-band command and control radar with a high-definition display on a range of colour raster scan CRTs that can show true motion or centre PPI with variable sector scan.

By comparison with Sea Searcher and Super MAREC, the Super Searcher has a greater detection, target tracking and guidance performance. The system has an inbuilt guidance capability, which can be adapted for all fire-and-forget, anti-surface vessel, sea-skimming missiles.

The system incorporates three selectable pulsewidths, one of which provides high definition of small targets in bad weather. Contact recognition is also improved by the use of the latest microprocessor techniques and signal processing algorithms.

The colour CRT facilities include freeze-frame and memory storage with graphical overlays of tactical and navigational symbology. The display can show either true or relative motion with or without offsets. Data is displayed either in latitude and longitude or as a grid reference. The system has multiple track-while-scan capability and provides a range of

Thales' Super Searcher radar   0503724

navigational facilities including waypoint markers. As an option, digitised maps can be loaded prior to the mission, allowing the use of map overlays on the radar image. A comprehensive library of symbols, that can be vectored as required, eases the operator's task, particularly in intelligence storage and extraction. The system is compatible with IFF/SSR interrogators and can be interfaced with a wide range of other sensors such as FLIR, ESM and sonar.

Super Searcher Ground-Mapping Radar (GMR) is a development of the Super Searcher radar and is a further addition to Thales' family of lightweight airborne radars. The radar uses a three-axis cosecant^2 antenna combined with high-resolution displays and upgraded processing to enhance ground features. The low-cost system can be used for high-altitude radar fixing, ground-mapping and ground target attack training.

Status
In production. Super Searcher is the principal sensor on board the 20 Sea King Mk 42 ASW helicopters ordered by the Indian Navy. It has also been installed on the eight Sikorsky S-70B helicopters of the Royal Australian Navy. Brazil has been supplied with 20 Super Searchers, for fitting to Embraer EMB-111 maritime patrol aircraft. Further GMR systems have been fitted to UK Royal Air Force Dominie aircraft for navigation training. Over 100 Super Searchers have been sold worldwide.

Contractor
Thales Defence Ltd.

Track-while-scan system processor (left),
digital solid-state radar indicator unit (centre)
and joystick control unit (right)
0018195

Track While Scan (TWS) system and airborne radar indicator unit

Type
Airborne surveillance radar.

Description
Track-While-Scan (TWS) System
The Caledonian Airborne Systems Ltd TWS system interfaces with the AlliedSignal RDR-1500B radar and navigation systems to enable on-screen target track and marker placement. TWS, radar and Scorpio 2000 datalink system control is by point-and-click using the joystick control. The system is capable of tracking up to 20 independent targets, providing information on present position (PP), ground speed and heading. Aircraft, target and marker data are overlaid on the radar indicator unit; operator prompts and messages are presented on the lower screen. In addition, the TWS system has the capability to insert 50 user-defined geostationary markers and store 100 screen images.

Airborne Radar Indicator Unit
The Caledonian Airborne Systems Ltd airborne radar indicator unit is a direct solid-state replacement for existing CRT-based indicators. It handles a variety of radar, composite and VGA video standards and provides a high-resolution flicker-free display. The 10.4 in multifunction display depicted serves as a direct replacement for the AlliedSignal IN-1502A radar indicator, and as the display for the Caledonian Airborne Systems Ltd Track-While-Scan (TWS) system. The design can, however, be tailored to suit displays from 4 to 16 in sizes and a number of video standards. The radar display can also be used as a multifunction display for other sensor data, in which application it is able to provide map overlay and to operate in night vision goggles conditions.

Specifications
Display resolution:
800 × 600 pixels (SVGA) in a 264 mm diagonal
upgrade path to: 1024 × 768 pixels (XVGA)
Video standards:
RGB computer graphics (up to SVGA) with
   separate H & V syncs
   composite and S-video (Y/C)
   CCIR/PAL & RS-170/NTSC scaled or native
   formats
upgrade path to: computer graphics (up to
   XVGA) and RDR-1500B radar video
Colours: 262,144
Luminance:
1000 Cd/m^2
upgrade path to: 1500 Cd/m^2
Keypad: 8 keys including 5 programmable
Interface: RS422
Power: 28 V DC at 105 W; 5V DC at 1 W
Weight: 5 kg
Dimensions: 268 × 221 × 131 mm

Status
In production and in service.

Contractor
Caledonian Airborne Systems Ltd.

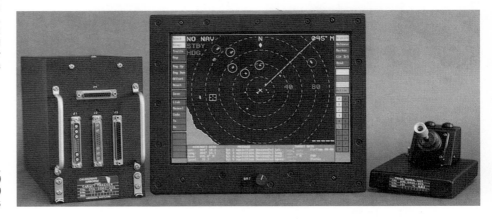

United States

AN/APG-63(V) fire-control radar

Type
Airborne Fire-Control Radar (FCR).

Description
The AN/APG-63 is the principal sensor for the F-15 Eagle. The system was designed around three main objectives: capability, reliability and maintainability. Operationally, the system was designed to provide single pilot control for the tracking of hostile aircraft at long and short range, in a look-down situation, with a clutter-free display incorporating all appropriate information and including guidance and steering information on a Head-Up Display (HUD), with simplified controls weapons co-ordination to include a secondary air-to-ground facility.

The requirement for the AN/APG-63 called for an MTBF of 60 hours, a level never before approached for this class of equipment. As a corollary, maintenance time would be reduced to about a quarter of that of previous systems. High reliability and relatively simple maintenance schedules reduce the turnaround time. This was reflected by the smaller numbers of flight line and maintenance personnel needed. This was largely due to the use of hybrid and integrated circuits to reduce the number of components and to the use of only four basic module sizes in the entire system.

The APG-63 system comprises nine LRUs: exciter, transmitter, antenna, receiver, analogue processor, digital processor, power supply, radar data processor and control unit. The system has a wide look-angle and its antenna is gimballed

in all three axes to maintain target lock during extreme manoeuvring. The clutter-free head-down radar display gives a clear look-down view of target aircraft silhouetted against the ground, even in the presence of heavy clutter from ground returns. This look-down, shoot-down ability is achieved by using both high and medium PRFs, by digital data processing and by using Kalman filtering in the tracking loops. The system's gridded Travelling Wave Tube (TWT) permits variation of the waveform to suit the tactical situation.

False alarms are eliminated, regardless of aircraft altitude and antenna look-angle, by a low-sidelobe antenna and frequency rejection of both ground clutter and vehicles moving on the ground, so that only real targets are displayed.

An F-15C Eagle fighter with nose radome swung aside to show the AN/APG-63(V)2 AESA (Boeing)
0103897

An F-15C Eagle escorts a F/A-22 Raptor; upgrades to the F-15 APG-63 radar, including incorporation of an AESA will ensure the type's continued effectiveness until replaced the F/A-22/APG-77 combination
0583426

Primary controls for the multimode, pulse Doppler I-/J-band radar are located on the control column, allowing the pilot to maintain 'heads-up' during dogfighting. Three special modes are provided for close in combat: supersearch, vertical scan and boresight. These enable automatic acquisition of, and lock on to, targets within 18.5 km. In the supersearch mode the radar locks on to the first target entering the HUD Field of View (FoV). In the vertical scan mode the radar locks on to the first target that enters an elevation scan pattern at right-angles to the aircraft lateral axis. In the boresight mode the antenna is directed straight ahead and the radar locks on to the nearest target within the beam. In all tracking modes the target position is displayed on the HUD if the target is within the FoV, thus greatly increasing the range of visual detection.

The use of interleaved high- and medium-PRF waveforms provides look-down, shoot-down capability for both approaching and retreating targets. The technology utilises digital signal processor. Incoming signals are sampled and their frequency content analysed by Fourier transforms on individual samples.

Although designed specifically for the air-to-air role, the F-15 incorporates a ground attack capability, with the APG-63 providing target ranging for automatic bomb release for a visual attack, a mapping mode for navigation and a velocity update for the inertial navigation system.

The APG-63 is compatible with the AIM-7 Sparrow, AIM-120 AMRAAM and AIM-9 Sidewinder air-to-air missiles. Antenna search patterns and radar display presentations are selected automatically by means of a three-position switch on the throttle for the type of weapon to be used (medium-range radar homing, short-range infra-red missiles or M61 cannon).

While the eventual reliability goal of 60 hours specified by MIL-STD-1781 was met consistently on bench tests, in the field, in-service equipment yielded about 30 to 35 hours MTBF.

In 1979, the AN/APG-63 was the first airborne radar to incorporate a software Programmable Signal Processor (PSP). The PSP allows the system to quickly respond to new tactics or accommodate improved modes and weapons through software reprogramming rather than by extensive hardware retrofit. This change was introduced with the improved F-15C and F-15D models. Aircraft produced before the PSP was introduced have received a retrofit item.

In December 1983, the US Air Force awarded a USD274.4 million contract to upgrade the control computer and armament control system as part of the MultiStaged Improvement Programme (MSIP) launched in February of that year. MSIP arose out of a McDonnell Douglas study, beginning in June 1982, which showed the need to improve the avionics system, and specifically the radar, in order to maintain reasonable combat superiority over likely adversaries. Radar improvements include a memory increase to 1 M words, a radar data processor speed tripled to 1.4 Mops and improvements to both transmitter and receiver. The outcome of this activity was the AN/APG-70, equipping F-15E and late-model F-15C and D aircraft in US service, together with F-15I and F-15S aircraft of the Israeli and Saudi Arabian air forces respectively.

Meanwhile, Raytheon also developed an upgrade to the basic AN/APG-63, designated APG-63(V)1. The APG-63(V)1 was designed from the outset as field retrofit for APG-63s in the F-15, focussing on enhanced system performance, reliability (120 hours MTBF), maintainability and supportability. The upgrade utilised software from the APG-70 to provide a substantial increase in computing power and Electronic Counter-CounterMeasures (ECCM) capability, while improving basic air-to-ground modes. The APG-63(V)1 features include improved Built-In Test (BIT), compatibility with both mechanically and electronically scanned antennas, increased receiver/exciter Signal-to-Noise Ratio (SNR), increased transmitter bandwidth and increased processing power and memory. System operating modes include boresight, supersearch, vertical scan, auto guns, Track-While-Scan (TWS), Range-While-Scan (RWS), precision velocity update (navigation), Ground Moving Target Indicator (GMTI) and high-resolution ground mapping.

During the last quarter of 1999, the US Air Force placed a contract to equip 18 of its F-15C fleet with a further upgrade of the AN/APG-63 radar, designated AN/APG-63(V)2. The Boeing Phantom Works unit led a team that received a USD250 million contract to install Raytheon's Active Electronically Scanned Array (AESA) radar. The stated aim of the upgrade was to improve Situational Awareness (SA) and to make better use of the AIM-120 missile capability by increasing air-to-air detection and tracking range. However, there has been speculation that the APG-63(V)2 radar was developed as part of a larger scale effort to plug a perceived gap in the US' capability against low-altitude cruise missiles – the characteristics of an AESA provide a greatly enhanced capability to detect and target small, low-flying and sometimes relatively slow-moving targets. The upgrade also included an Advanced Identification Friend or Foe (AIFF) system from BAE Systems North America. The program was used to provide experience of the technology that underpins the US Air Force F/A-22 and F-35 Joint Strike Fighter (JSF) radar systems. The Air Force F-15 System Program Office's Projects Team at Wright-Patterson Air Force Base, Ohio, managed the programme on behalf of the US government. The mechanically scanned antenna in the APG-63(V)1 is one of the highest maintenance items on the aircraft, demanding a high-pressure hydraulic circuit and mounting system capable of withstanding the rigours of the extremely rapid scan of the antenna. The AESA eliminates several hydraulic and electrical systems required to support a mechanically scanned array.

While the production run of 18 APG-63(V)2 radars is complete, much of the technology of this radar is reported to be present in Raytheon's current variant of the radar, designated AN/APG-63(V)3. Similarly to the upgrade from APG-63 to APG-63(V)1, the (V)2 to (V)3 upgrade is reported as an antenna replacement; the 'back end' of the radar remains much the same. The design of the (V)3 variant makes use of 'next-generation' Transmit/Receive (T/R) modules, a benefit of crossover with the AN/APG-79 AESA installation in the F/A-18E/F Hornet, to replace the earlier 'brick' arrangement in the (V)2. This facilitates increased resolution and performance, together with a weight reduction reported to be as much as 100 kg.

The AN/APG-63(V)1 is the current standard equipping the majority of US Air Force F-15 aircraft
0589036

The newest variant of the radar is designated the AN/APG-63(V)4, which mates the antenna from the (V)3 with a more advanced processing unit derived from that in the AN/APG-79 installed in the F/A-18E/F Super Hornet. This arrangement should enable the radar to more adequately exploit the enormous data gathering capabilities inherent to AESA technology, although similar systems have found an urgent requirement to upgrade datalink communications in parallel in order to offload data such as radar imagery and Electronic Support Measures (ESM) information.

Specifications
AN/APG–63 baseline
Volume: 0.25 m³
Weight: 221 kg total
Number of LRUs: 9
Frequency: I/J-band (8–20 GHz)
Transmit power: 12.975 kW
Reliability: 60 h MTBF (APG-63); 120 h MTBF (APG-63(V)1)

Status
The baseline AN/APG-63 is no longer in production but remains in service in the F-15 Eagle. The last system was delivered in September 1986, by which time some 1,000 sets had been built, including co-production in Japan. Aside from the F-15 application, four P-3A Orions, on loan from the US Navy, have been modified to include an AN/APG-63 radar mounted in the aircraft nose, for anti-drug smuggling operations with the US Customs Service. About 700 radars are still operational in F-15As, Bs, and early model Cs and Ds operated by the US Air Force and the air forces of Israel, Japan, and Saudi Arabia.

The AN/APG-63(V)1 programme began in October 1994 with the award of an Engineering and Manufacturing Development (EMD) contract valued at USD175 million. This phase was essentially complete by the end of 1999. Meanwhile, performance testing began in 1997 under Low-Rate Initial Production (LRIP), funded under contracts totalling USD96 million. In July 1999, a contract for the AN/APG-63(V)1 Production 1 lot of 22 radars was awarded for aircraft retrofit by November 2001. A further production order for 25 sets was placed in June 2000 for delivery by December 2002. To date, Raytheon has received a total of five production orders to deliver 161 AN/APG-63(V)1 radar systems, plus spares, to the US Air Force. Raytheon is also on contract to deliver 40 radar systems and spares to Korea for the F-15K, with production deliveries scheduled through November 2005 at a maximum rate of five systems per month.

The final three F-15C aircraft upgraded with the AN/APG-63(V)2 were delivered during December 2000 to complete unit strength at Elmendorf Air Force Base, Alaska. The Air Force has transferred its AESA-equipped F-15C aircraft to the Air National Guard (ANG) for homeland defence of the continental US. While other F-15C, D and E models are likely to receive an AESA upgrade, it is possible that budgetary constraints may conspire to deny the F-15C/APG-63(V)2 the same. However, much of this discussion depends on a future force of AESA-equipped aircraft, such as the F/A-22 and F-35, are fielded in sufficient numbers to cover all of the emerging missions for such aircraft, including expeditionary, anti-cruise missile, ECCM and ESM.

The AN/APG-63(V)3 began airborne testing during May 2006, with the radar installed in Raytheon's modified Boeing 727 testbed aircraft. During July the radar is expected to be transferred onto an F-15C for further flight testing at Eglin Air Force Base in Florida from September. The APG-63(V)3 has been ordered by the US ANG for its F-15C aircraft. The MINDEF officials at the Asian Aerospace 2006 air show, finally cleared the manufacturer to acknowledge widespread speculation that the APG-63(V)3 system was also included in the mission system package for Boeing's successful bid in 2005, to sell an initial round of 12 F-15SGs to Singapore. Singapore requires up to 24 new fighters, with an initial tranche to be delivered in 2008–9.

Should budgetary constraints in the US force further reductions in F-22A procurement and delay fielding of the F-35 Joint Strike Fighter (JSF), then US Air Force contingency plans to retain a number of F-15C/D squadrons in the interim may provide further opportunities for upgrading of radars to (V)3 standard.

Staying with US procurement, the AN/APG-70 radar installed in the F-15E is slated to be replaced by an AESA-equipped radar during the 2010 timeframe. While Raytheon is developing the AN/APG-63(V)4 for this purpose, it is highly likely that a competition will take place, involving all of the new US AESA radars, including Raytheon's own APG-79. A decision is expected during 2007; the F-15E order will be for around 224 radars.

Contractor
Raytheon Company, Space and Airborne Systems.

AN/APG-65 multimission radar

Type
Airborne multimode radar.

Description
In the air-to-air role the APG-65 radar incorporates the complete range of search, track and combat modes, including several previously unavailable in an operational radar. Specifically these modes are gun acquisition (the system scans the head-up display field of view and locks on to the first target it sees within a given range); vertical acquisition (the radar scans a vertical slice of airspace and automatically acquires the first target it sees, again in a given range); and boresight (the radar acquires the target after the pilot has pointed the aircraft at it). In the raid assessment mode the pilot can expand the region around a single target that is being tracked, giving increased resolution around it and permitting separation of closely spaced targets. As a gunsight, the radar operates as a short-range tracking and lead-computation device using frequency agility to reduce errors due to target scintillation. All the pilot has to do is to put the gunsight pipper on the target and press the firing button.

Other modes are long-range velocity search (using a high-PRF waveform to detect oncoming aircraft at high relative velocities); range while search (high- and medium-PRF waveforms interleaved to detect all-aspect targets, not only head-on but at any line of sight crossing angle), and track-while-scan which can track up to 10 targets simultaneously and display eight. When combined with autonomous missiles such as AMRAAM, this mode confers a launch-and-leave capability and the simultaneous engagement of multiple targets.

In the F/A-18's air-to-ground role, the APG-65 has six modes: terrain-avoidance, for low-level penetration of hostile airspace; precision velocity update, when the radar provides Doppler signals to update or align the aircraft's inertial navigation system; tracking of fixed and moving ground targets; surface vessel detection, in which the system suppresses sea clutter by a sampling technique; air-to-surface ranging on designated targets; and ground-mapping. Two Doppler beam-sharpening modes are provided for these air-to-ground modes.

During tests in early 1983, the system demonstrated (a year ahead of schedule) the 106 laboratory test hours MTBF required by contract. Two systems chosen at random for a reliability

A technician removes an LRU from the AN/APG-65 radar of a F/A-18A Hornet (Raytheon) 0554684

The AV-8B Harrier II Plus is equipped with the AN/APG-65 radar (US Navy) 1127378

US Navy F/A-18A Hornets were originally equipped with the AN/APG-65 radar (US Navy) 1127377

German Air Force F-4F ICE programme

In May 1985, the German Ministry of Defence chose the AN/APG-65 as a major element in the Improved Combat Efficiency (ICE) programme for the German Air Force F-4F Phantom. This radar is also designated AN/APG-65GY.

EADS (formerly DaimlerChrysler Aerospace AG) built the radars under licence and ICE includes a Honeywell laser inertial navigation system, Marconi Electronic Systems air data computer, EADS radar display and Raytheon Electronic Systems cockpit displays. EADS was also contracted to provide the ICE upgrade to the Hellenic Air Force as part of its F-4E upgrade.

Radar for AV-8B

A variant of the APG-65 entered service in the AV-8B Harrier II Plus with the US Marine Corps and Italian and Spanish navies during 1993.

Specifications

(APG-65 in F/A-18)
Volume: <0.126 m^3 excluding antenna
Weight: 154.6 kg
Number of LRUs: 5
Frequency: 8–12.5 GHz (X-band)
Antenna: low-sidelobe planar-array with fully balanced direct electric drive replacing hydraulics and mechanical locks of previous systems
Transmitter: liquid-cooled, contains software-programmable gridded travelling wave tube amplifier
Receiver: contains A-D converter
Radar-data processor: general purpose with 50K 16-bit word bulk-storage disk memory
Signal processor: fully software-programmable, runs at 7.2 Mops
Reliability: 120 h MTBF. Built-in test equipment detects 98% of faults and isolates them to single replaceable assemblies that can be changed in 12 min without adjustments or setting up

Status

In service with US Navy, US Marine Corps, Canadian Forces, Royal Australian Air Force, Spanish Air Force and Kuwaiti Air Force F/A-18 aircraft; US Marine Corps, Spanish Navy and Italian Navy AV-8B Harrier II Plus aircraft. The AN/APG-65, manufactured under license by EADS Deutschland, was also a core component of the German Air Force F-4F Improved Combat Efficiency (ICE) programme and the Hellenic Air Force F-4E Avionics Upgrade Programme (AUP).

The AN/APG-65 continues in production for the AV-8B Harrier to AV-8B Harrier II Plus upgrade for the US Marine Corps, Italian Navy and Spanish Navy.

Contractor

Raytheon Company, Space and Airborne Systems.

AN/APG-66(V) radar

Type

Airborne multimode radar.

Description

The baseline AN/APG-66 is a multimode radar system installed on 16 different airborne platforms operating in 20 countries. The APG-66(V) provides improved performance, functionality, reliability and maintainability over its predecessor, the AN/APG-66, which was originally designed for the Lockheed F-16. Over 2,300 of the original version of the radar have been produced and deployed worldwide.

Technology originally developed for the F-16 Mid-Life Update (MLU) has been carried over to the APG-66(V) family, which includes the following major in-service variants:

APG-66(V)2 – F-16 MLU build standard.
APG-66(V)3 – Believed to incorporate a CW illuminator for Taiwanese F-16s.
APG-66H – Variant optimised for BAE Systems Hawk 200 aircraft. The APG-66H features a slightly smaller antenna than the standard equipment and

demonstration operated under test for a total of 149 hours without a failure.

All fault-finding is conducted with Built-In Test (BIT) which currently operates at module LRU level. The biggest improvement in BIT technology with the AN/APG-65 has been the reduction in false alarm rate. The US Navy requirement is to be able to detect and locate 98 per cent of all radar faults by BIT. The US Navy also specifies 12 minutes MTTR, which involves running a BIT test, locating the fault, removing and replacing the faulty line-replaceable unit and running a BIT check on the new unit.

In its first six months of operation, the initial Marine Air Group II at El Toro in California had only six radar engineers to support a 12 aircraft unit flying 30 hours a month.

Under the terms of an Australian Industrial Participation (AIP) programme signed in December 1981, Philips Electronics Systems in Australia is engaged in final assembly and test of the radar and co-produces the data processor. In February 1984, it was announced that Marconi Española SA had been licensed to build low-voltage power supply modules for the radars being produced for Spain's F/A-18s.

AN/APG-66(V)2 for participating European governments' F-16A/B mid-life upgrade programme
0001217

The AN/APG-66(V)2 radar display is shown on the left MultiFunction Display (MFD) in this F-16 MLU cockpit
0105081

a new signal data processor, which is reported to incorporate most of the modes available in the APG-68 radar (see separate entry).

APG-66J – Variant for the upgrade of Japanese Air Self-Defence Force F-4EJ aircraft.

Improvements over the baseline radar incorporate the latest technology include a newly developed signal data processor, higher power transmitter, low-power RF speed and sensitivity and faster antenna phase shifting.

Consolidating the functions of the radar computer and digital signal processor reduces system size, weight, power and cooling requirements, while providing seven times greater processing speed and 20 times greater non-volatile memory than the AN/APG-66 system. A Doppler correlator reduces the false alarm rate by classifying radar returns as either ground vehicles, weather, mutual interference or sidelobe returns. The processing and fully programmable graphics capability in the signal data processor provides numerous options for radar operational mode growth, colour display support and growth capabilities to perform combined multisensor processing functions such as for FLIR or missile warning.

Significant advantages over the original AN/APG-66 have been realised, including twice the operational range, 40 per cent higher reliability and 16 per cent reduction in system weight. A

four to one improved ground-map resolution has also been achieved. Greater situation awareness is provided by Track-While-Scan (TWS), Situation Awareness Mode (SAM), multitarget track and improved protection against electromagnetic interference. Support for multiple employment of up to six missiles is provided and the radar is compatible with a variety of Beyond Visual Range (BVR) air-to-air missiles and anti-ship missiles.

The radar provides air-to-air and air-to-surface modes in all weathers against all-aspect targets in look-up/shoot-up and look-down/shoot-down engagements, including high-clutter environments. The flexible multimode fire-control sensor is designed to provide integrated target and navigation data. Air-to-air modes provide the capability to detect and track multiple targets in the presence of ground clutter and electromagnetic interference. Air-to-surface modes provide extensive mapping for navigation, target detection, tracking and ranging.

The upgrade of the air-to-air capability of both individual aircraft and the F-16 units provides a significant air superiority margin that will also include a degree of air space management, incorporating both control aircraft and ground station datalinks providing real-time target information. Operating modes include 10 target TWS, Range-While-Search (RWS) look-up and look-down, six shot AMRAAM against six

targets, medium-resolution DBS (Doppler Beam-Sharpening) and enhanced ECCM capabilities.

Key radar features in air-to-air are search and track mode incorporating TWS, RWS and SAM functions. SAM offers a better track quality than TWS on multitarget track, plus the capability to track one or two targets while maintaining RWS scan volume. The TWS mode can, in addition to providing information on up to 10 targets, provide search information on up to 64 other targets.

There is a multitargeting capability for up to 10 target tracks, two target SAM tracks, single target track and BVR engagements. Inclusion of a BVR capability is a key point in order to provide MLU-modified aircraft with the ability to use medium-range radar-guided missiles such as Sparrow, AMRAAM and MICA.

Air Combat Manoeuvre (ACM) features automatic acquisition of targets at short range in four predetermined scan patterns. Other information available includes the search altitude display which provides a reference to the altitude of all radar search returns and an indication of target relative groundspeed as well as an indication of the target aspect.

In the air-to-surface mode the APG-66(V)2 radar provides a significant improvement in both ground attack and anti-ship roles, especially in conditions of extreme clutter. Functions exist for the provision of a radar map video for target detection and navigation as well as providing an improved capability against shadowed or hidden targets. Using the ground map as a base, the APG-66(V)2 is also able to provide an overlay of navigation and reference points especially for rendezvous locations, weapon release points and waypoints. Operating modes include real-beam map, enhanced ground map, medium and coarse Doppler beam-sharpening, fixed target track, ground moving target indication and SAM in ground map interleaved mode.

Continuous updating of the navigation system is also enhanced through air-to-ground ranging which, combined with mission data information, provides confirmation of reference points along a proposed mission route and information regarding terrain-avoidance for obstacles that can be preprogrammed into the mission data package.

Additional features include moving ground target tracking that can also be superimposed on an improved definition ground map and, in both ground and sea modes, the ability to detect and fix on targets in open country or at sea. The sea mode provides a capability for the accurate detection and tracking of targets in both low and high sea states. The long-range sea mode provides a detection range reported to be as much as 148 km.

The radar may be readily adapted to a platform through expeditious software modifications supported by the fast Express software development environment and by selecting one of the nine antennas available or by resizing the array as necessary.

Features providing for integration flexibility include NTSC and PAL/CCIR video formats, 480 × 480 colour video resolution with 64 grey shades or 256 colours, fully programmable multifunction displays, MIL-STD-1553B remote terminal or bus controller, discrete input/output for HOTAS, CW illumination LRU and RAM or ECS cooling air capability.

Possible growth capabilities include weather awareness, blind let-down, wide swath map at medium altitude, terrain-following/terrain-avoidance, synthetic aperture radar, internal FLIR, missile warning and ESM processing.

Specifications
Volume:
(APG-66(V)1) 0.102 m³
(APG-66(V)2) 0.097 m³
(APG-66H) 0.082 m³
(APG-66T) 0.08 m³
Weight:
(APG-66(V)1) 134.3 kg
(APG-66(V)2) 115.9 kg
(APG-66H) 107.7 kg
(APG-66T) 98.4 kg

Power supply: 115 V AC, 3 phase, 400 Hz,
3,209 VA
28 V DC, 115 W
Frequency:
(APG-66(V)1) 6.2–10.9 GHz
(APG-66(V)2) 8–10 GHz
Search angle: 120° (azimuth and elevation)
Azimuth scan: ±10°, ±30°, ±60°
Elevation coverage: 1, 2 or 4 bar
Range scale: 10, 20, 40, 80 n miles (19, 37, 74,
148 km)
Electronic protection: multiple features, EMI
pulse editor, fast phase shifting, frequency
agility.
Maintainability: 5 min MTTR
Reliability: 206 h MTBF

Status

The initial variants in the AN/APG-66 Series, the APG-66T and APG-66H, have been deployed in the AT-3 trainer and BAE Systems Hawk 200 aircraft respectively. The APG-66H is currently in production.

The APG-66(V)1 was developed for the US Air Force. The contract for 270 systems was completed in 1991.

Development of the APG-66(V)2 for the European F-16A/B Mid-Life Update (MLU) programme was initiated in 1992. The programme will update 48 Belgian, 61 Danish, 136 Netherlands and 56 Norwegian air force AN/APG-66 radars to AN/APG-66(V)2 standard. DT&E was conducted at Edwards AFB in May 1995. Test missions were flown by the Netherlands Air Force in Autumn 1996. Production radar software was completed in January 1997.

Flight trials for both the APG-66(V)2 and (V)3 radars were conducted in 1994. These successfully demonstrated doubled operational detection range; false alarms reduced by a factor of 10 even in the presence of multiple radars and severe ECM; simulated six-shot AMRAAM capability and ground-mapping range improved to 148 km.

The APG-66(V)3 is closely related to APG-66(V)2, configured for use aboard Block 20 F-16A/B aircraft. It has been reported that that APG-66(V)3 incorporates a Continuous Wave Illumination (CWI) capability, developed for the Taiwanese Air Force for its 150 AIM-7 Sparrow-equipped F-16A/Bs.

Northrop Grumman has also delivered 30 ARG-1 radars, a customised APG-66(V)2 for installation in 32 Argentinean Air Force A-4AR aircraft.

In May 1999, US FMS ordered a further 49 AN/APG-66(V)2 upgrade kits, 24 for Belgium and 25 for Portugal.

As of June 2001, in excess of 550 APG-66(V)2 and (V)3 radars were reported as being operational with more than 150 being new-build equipments.

Northrop Grumman is offering a derivative of the APG-66(V)2/(V)3, under the name APG-66(V)X (or APG-66(V)10), for upgrade of Block 15/20 and MLU aircraft. This radar is believed to utilise the antenna, modular receiver/exciter and common radar processor from the AN/APG-68(V)9 and to offer similar capabilities to the AN/APG-68(V)10 (see separate entry) upgrade kit for F-16C/D aircraft.

During 2002, Northrop Grumman announced a 23-year engineering services technical support contract for spares, repair and systems modification for APG-66 and APG-68 radars in service with the US and foreign air forces.

Contractor

Northrop Grumman Corporation, Electronic Systems Sector.

AN/APG-67(F) attack radar

Type

Airborne Fire-Control Radar (FCR).

Description

The AN/APG-67(F) has been developed for retrofitting into Northrop F-5 aircraft and other fighter aircraft.

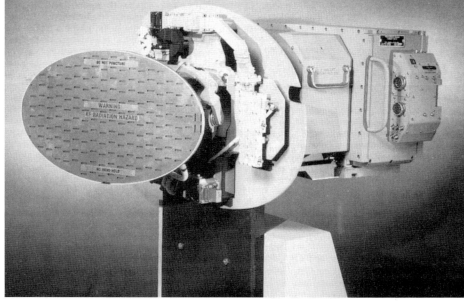

The AN/APG-67(F) radar for the F-5 retrofit market   0503729

A significant characteristic of the APG-67, stemming from its Modular Survivable Radar (MSR) ancestry, is the choice and layout of circuitry to enable the system to be updated for new technologies as they appear, such as very large-scale integration, without having to redesign the LRUs. Growth features can also be added in this way.

In air-to-air missions the system searches for, automatically acquires and tracks targets in both look-up and look-down situations. In its air-to-surface role the radar provides real-beam mapping, high-resolution Doppler beam-sharpened ground-mapping (40:1), and air-to-ground ranging. Variants of the radar can include synthetic aperture imaging.

The heart of the processor contains 16 Shark 2106D processors. Twelve perform signal processing functions and the other four perform Hardware I/O, radar mode control and aircraft mission computer interface functions, together with self-test radar processing associated with motion compensation, target tracking, antenna motion control and raster scan output to the cockpit display.

The processor integrates raw target data from the antenna so that reliable reports can be established. The unit incorporates a programmable signal processor and its use is considered to be the key to what is claimed to be this unit's exceptional performance. The processor performs a variety of functions, including fast Fourier transformation, moving target indication, pulse compression and motion compensation, and can operate with variable waveforms and bandwidth. The transmitter produces coherent I/J-band radiation from a travelling wave tube and can operate at low, medium and high PRFs, with variable power outputs and pulsewidths. The antenna is a flat plate slotted array with low-sidelobe sensitivity and with ±60° scan in azimuth and elevation. Scan width is selectable and elevation search can be accomplished in 1, 2 or 4 bar modes. The receiver has a low-noise front-end to maximise detection range.

The AN/APG-67(F) version for the new aircraft and retrofit market offers a full range of air-to-air capabilities including range-while-search out to 148 km, three air combat modes and single target track. It also provides a complementary set of air-to-ground modes including map, expand, Doppler beam-sharpening, freeze and surface moving target indication and track over ground or rough seas.

It consists of three LRUs and incorporates a MIL-STD-1553B database interface and MIL-STD-1750A computer architecture. There is extensive built-in test.

Lockheed Martin has collaborated with Geophysical and Environmental Research Corp to develop GSAR, a variant of the AN/APG-67 radar to equip business aircraft for the SAR role. GSAR has a range of 150 km, with swath widths up to 40 km wide. Resolution is reported to be 3 m, improving to 1 m in spotlight mode.

Specifications

APG-67(F)
Volume: 0.049 ms
Weight: 74 kg
Frequency: I/J-band
Power: 2,600 VA
Transmit power: 350 W
Antenna size: 279 × 711 mm
Detection range (fighter-size targets):
(look-down) 39 km
(look-up) 72 km
(sea targets, 50 m²) 57 km Sea State 1
(max range, all modes) 148 km
Accuracy (air-to-ground): 15 m or 0.5% of range
Reliability: design 235 h MTBF

Status

In production. Taiwan has selected the APG-67 for its KTX-II Indigenous Defence Fighter project.

Contractor

Lockheed Martin Naval Electronics & Surveillance Systems.

AN/APG-68(V) multimode radar

Type

Airborne multimode radar.

Description

The AN/APG-68(V) is a coherent, digital, multimode radar, derived from the AN/APG-66(V)2 (see separate entry), introduced into service in the Lockheed F-16 from Block 30/40 onwards. The radar features an electrically driven antenna, a Modular Low-Power Radar Frequency (MLPRF) unit that features plug-in modules for growth and ease of maintenance, and an air-cooled Dual Mode Transmitter (DMT) that permits the radar to operate using low-, medium- and/or high-pulse repetition frequencies (PRFs). This hardware capability and flexibility allow the radar to be optimised easily for any air-to-air or air-to-ground search, track or mapping scenario.

The radar has been subject to an almost constant upgrade process, enhancing performance and, notably, reliability. Since its introduction, early

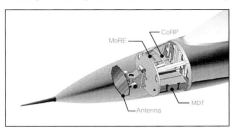

The core upgrade components for the APG-68 (V) are form, fit and functional replacements for legacy systems   0533864

variants of the radar (APG-68(V)1 to (V)4 on Block 30/40s) demonstrated a Mean Time Between Failures (MTBF) of 160 hours. This figure was increased to 264 hours in the Block 50 aircraft installation (APG-68(V)5 for USAF F-16s, (V)7 and (V)8 on export aircraft) and 390 hours for the Block 50/52+ Baseline installation (AN/APG-68(V)9). The radar is claimed to be the most reliable in the US Air Force inventory.

The AN/APG-68(V) incorporates a total of 25 air-to-air and air-to-surface modes of operation. Air-to-air modes provide the capability to detect, track and engage targets at all aspects and at all altitudes even in the presence of ground clutter. Air-to-surface modes provide extensive mapping, target detection, location and tracking, as well as navigation capability. The radar employs low PRF for air-to-surface engagements and medium and high PRF modes for long-range air interception. As the nomenclature suggests, a high PRF (and hence high data rate) velocity search mode increases detection range for high-speed targets, while a medium PRF Range-While-Search (RWS) mode is used, after acquisition, to enhance range and aspect information. In the Track-While-Scan (TWS) mode, the AN/APG-68(V) can track up to 10 targets simultaneously, assess relative threat and prioritise targets accordingly for engagement in threat order. The system employs high-resolution Doppler techniques to resolve individual targets at long range. Automatic acquisition of targets within the search volume enhances reaction times in air combat engagements. The system employs a velocity threshold in look-down, which enables the system to distinguish airborne targets from ground vehicles.

In assessing the performance and features of variants of the AN/APG-68(V), it should be noted that the vast majority of F-16s in current operational service are Block 30 (approximately 700), with smaller numbers of Block 40 and 50 aircraft (400 and 250 respectively in late 2001). The current 'baseline' radar for the US Air Force is the AN/APG-68(V)5, which features dual target Situational Awareness (SA), enhanced monopulse ground-map, electronic boresight, enhanced long-range target tracking and enhanced Air Combat Manoeuvring (ACM) acquisition modes.

In May 2000, the AN/APG-68(V)9 (formerly the AN/APG-68(V)XM) was introduced for new build Block 50 aircraft. The new radar incorporates many enhancements and advanced features over the (V)5, as well as the previously mentioned increased MTBF:

- Advanced RF architecture
- Increased transmitter power/increased air-to-air range
- A high-resolution Synthetic Aperture Radar (SAR) mode, including a modified antenna assembly, incorporating a Strapdown Inertial Measurement Unit (IMU) 50 per cent growth capability in software for enhanced upgradeability
- A modular, receiver/exciter unit which is a form-fit replacement for earlier low-power radio frequency units
- Simple integration with Block 50 software
- A programmable, commercial-off-the-shelf, open architecture, common radar processor that replaces the existing Programmable Signal Processor (PSP)/Advanced Airborne Signal Processor (APSP), offering a 3:1 improvement in programming efficiency
- Compatible with the AIM-120, AIM-9X, Joint Helmet Mounted Cueing System (JHMCS), 'smart' munitions, Advanced Targeting Pods (ATP) and Electronic Warfare (EW) systems.

Alongside APG-68(V)9, Northrop Grumman has developed the APG-68(V)10, derived from the (V)9, which is described as an 'upgrade kit' for F-16C and D fighters. The APG-68(V)10 makes use of the (V)9 antenna, receiver/exciter and common radar processor and a modified (V)9 transmitter (for F-16A/B aircraft). Installed in F-16C/D aircraft the APG-68(V)10 provides identical performance and reliability to the (V)9. A smart interface within a new Common Radar Processor (CoRP – see below) allows the (V)10

architecture to automatically sense and adapt to the host aircraft's avionics configuration. Installed in F-16A/B aircraft, the (V)10 upgrade represents a baseline capability that is equivalent to the AN/APG-66(V)2–3 (see separate entry) together with increases in detection range/tracking accuracy and MTBF.

The APG-68(V)10 is also reported to be available for the upgrade of fighters other than the F-16.

Upgrade path
In order to facilitate upgrade from earlier variants of the radar, Northrop Grumman has conducted extensive research and development work to produce a range of retrofit components that offer form, fit and upgraded functional replacement for legacy items:

Modified APG-68 Antenna: By adding a strapdown Inertial Measurement Unit (IMU), this upgrade enables Synthetic Aperture Radar (SAR).
Modular Receiver/Exciter (MoRE): Replaces the Modular Low-Power Radio Frequency (MLPRF). The MoRE features the latest open architecture design and provides wideband waveform guides for SAR mode and high-bandwidth A/Ds.
Common Radar Processor (CoRP): Replaces the Advanced Programmable Signal Processor (APSP). The CoRP is a COTS-based, open architecture design which provides enhanced processing power and facilitates growth.
Medium-Duty Transmitter (MDT): The MDT provides higher average and peak power.

By utilising these upgrade components, the latest APG-68(V)9 can be compatible with any core F-16 Fire Control Computer (FCC) from Block 25 through to Block 50 via software modification.

The AN/APG-68 radar fitted to US Air Force F-16C/D aircraft   0018074

AN/APG-68(V) functionality includes the AIM-120 AMRAAM   0103990

In common with all AN/APG-68(V) variants, the (V)9 is a form and fit replacement for earlier models
0532345

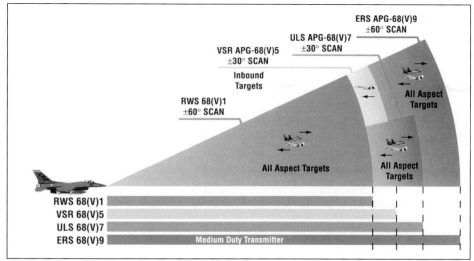

Each subsequent variant of the APG-68 (V) radar offers increased performance and greater functionality

0533865

Israeli F-16I Block 52+ aircraft are equipped with the AN/APG-68(V)9 radar (Lockheed Martin)

0083057

Specifications

Volume: 0.13 m³
Weight: 172 kg (AN/APG-68(V)), 164 kg (AN/APG-68(V)9)
Frequency: 8–12.5 GHz
Transmitter: gridded, multiple peak power travelling wave tube
Power: 5,600 VA ((V)9)
Antenna: planar array, 740 × 480 mm
Search range:
(air-to-air) 160 nm (296 km)
(air-to-ground) 80 nm (148 km)
Range scales: 10, 20, 40, 80, 160 n miles
Azimuth scan: ±10°, ±25°, ±30°, ±60 °
Elevation coverage: 1, 2, 3 or 4 bar
Number of units: 4 LRUs, all in the nose of the aircraft
Maintenance: 30 min MTTR
Reliability: 264/390 h MTBF ((V)5/(V)9)
Cooling: 9.9 kg/min at 27°C ((V)9)

Status

More than 6,500 copies of the AN/APG-68 and its predecessor, the AN/APG-66, have been produced. In addition to US air arms, the AN/APG-68(V) is installed in F-16C/D aircraft of the air forces of Bahrain, Egypt, Greece, Israel, Korea and Turkey.

The most recent orders and deliveries for AN/APG-68(V) radars are as follows:

24 new radars provided for Egyptian F-16s were designated AN/APG-68(V)8.

During 2001, Northrop Grumman announced orders for 125 APG-68(V)9 radars. It was reported that these radars were for new F-16C/D Block 50+ aircraft ordered by the Hellenic Air Force, and Israeli F-16I Block 50/52 aircraft.

During 2002, Northrop Grumman announced a 23-year engineering services technical support contract for spares, repair and systems modification for APG-66 and APG-68 radars in service with the US and foreign air forces.

During the early part of 2003, the US Air Force Combined Test Force (CTF) began testing the APG-68(V)9 high resolution ground mapping modes, with full operational capability achieved at the end of the year.

During August 2004, the US Air Force awarded Northrop Grumman a USD22 million initial contract to develop a radar upgrade for its F-16 fighters equipped with the APG-68(V)5 radar.

As of mid-2005, Northrop Grumman had delivered 200 APG-68(V)9 radars to prime contractor Lockheed Martin Corporation.

Most recently, in July 2005, Northrop Grumman announced a further, three-year contract from the US Air Force to complete development of the aforementioned radar upgrade, now designated the AN/APG-68(V)10, for approximately 240 F-16C and D aircraft.

Contractor

Northrop Grumman Corporation, Electronic Systems Sector.

AN/APG-70 radar for the F-15E

Type

Airborne multimode radar.

Description

The radar for the two-seat F-15E combat aircraft is a substantially improved version of the APG-63, designated AN/APG-70. It is also installed in a limited number of F-15C/D aircraft.

By comparison with the earlier system, the APG-70 has a far greater RF bandwidth, a larger look-down target detection range, a one-third increase in MTBF and is packaged in eight units instead of nine. Four of the LRUs (radar data processor, programmable signal processor, analogue signal converter and receiver/exciter) are completely new, while the transmitter and control unit have been modified; only the power supply and the antenna remain unchanged. The radar uses VLSI technology, making the programmable signal processor five times faster than previous designs while having three times the memory.

Expanded built-in test is provided, giving unambiguous fault detection and isolation and the ECCM capability is improved to combat new and more advanced threats. The system is compatible with existing and new missiles such as AIM-7F/M Sparrow, AIM-9 Sidewinder and AIM-120 Advanced Medium-Range Air-to-Air Missile (AMRAAM), and with 20 mm cannon.

In air-to-air modes, the radar has five search modes: range while scan with high PRF, medium PRF or interleaved PRF, range-gated high PRF and velocity search. The radar also has single-target track, track-while-scan and raid assessment track modes and vertical search, super search, boresight and auto-guns target acquisition modes.

In the air-to-ground mode the radar can produce a high-resolution or a real-beam ground map; the specification calls for 2.6 m resolution at 37 km range. There is also a precision velocity update mode and air-to-ground ranging. Terrain-following/terrain-avoidance, ground moving-target track indicator and fixed-target track modes have also been incorporated during a continuous upgrade and enhancement programme for the radar.

In December 1987, Raytheon announced a USD58 million contract to supply a variant of the APG-70 radar as the fire-control system in the US Air Force Special Operations Force's AC-130U aircraft. This aircraft employs a 105 mm howitzer and 25 mm and 40 mm cannon. Five new air-to-ground modes were developed for this application: fixed-target track, ground moving target indication and track, projectile impact point position, beacon track and a weather mode. The existing antenna and signal processor were modified and a digital scan converter added to the system. The first system was delivered at the end of 1988.

Specifications

Mechanical: 3 gimbal axes, mechanical scan
Volume: 0.25 m³
Weight: 251 kg
Frequency: 8–20 GHz selectable
PRF: multiple
Number of LRUs: 8
Range:
(air-to-air) 185 km
(automatic acquisition) 150 m to 18.5 km
(ground map) >80 km
Resolution:
(ground map) 2.6 m at 37 km
Reliability: 80 h MTBF

Status

In service in US Air Force F-15E and late model F-15C/D and on the F-15I (Israel) and F-15S (Saudi Arabia) aircraft. It has been reported that APG-70S radars fitted to Saudi aircraft feature an application-specific processor, antenna drive rate and processor memory. The variant installed in the AC-130U gunship aircraft is designated the AN/APQ-180. During the second quarter of 2002, trials of an APG-70 processor upgrade were completed and shortly afterwards, Raytheon was awarded USD45 million Indefinite Delivery/Indefinite Quantity (ID/IQ) contract for the provision of engineering and technical services support for the development of Operational Flight Programme (OFP) upgrades and maintenance/reliability work for the entire family of radars. Work is to be undertaken at Warner Robins Air Force Base, to be complete by 2007.

During the third quarter of 2002, Raytheon was awarded a USD14 million contract for the provision of APG-70 spares for the Israeli Air Force, scheduled for completion by the end of 2004.

During October 2002, Raytheon announced that it had been awarded a firm, fixed-price, ID/IQ follow-on contract for the repair of APG-70 radars worth USD23 million.

Contractor
Raytheon Company, Space and Airborne Systems.

AN/APG-71 fire-control radar for the F-14D

Type
Airborne Fire-Control Radar (FCR).

Description
AN/APG-71 is an enhancement of the AN/AWG-9 radar originally fitted to F-14 aircraft.

Compared with the AWG-9, the APG-71 offers better overland performance, expanded velocity search capability, a larger target engagement zone, a raid assessment mode and programmable electronic countermeasures and clutter control features.

The APG-71 is essentially a digital version of the radar section of the AWG-9, with greatly improved ECM performance acknowledging the new and vastly more sophisticated jamming technologies that have appeared since design of the F-14 was frozen. New modes include medium-PRF all-aspect capability, monopulse angle tracking, digital scan control, target identification and raid assessment, but the number of boxes comes down from 26 to 14. The system also employs some of the elements developed by Hughes for the new F-15 radar, the AN/APG-70; for example the APG-71 has a signal processor which is 86 per cent common in modules with that in the APG-70. The AWG-9's transmitter, power supply and aft cockpit tactical information display are retained for the APG-71.

The APG-71 also incorporates Non-Co-operative Target Recognition (NCTR), by which radar contacts may be identified as friendly or hostile, at beyond visual range, through close examination of the raw radar returns at high resolution; this technique alleviates problems with deficiencies and ambiguities in IFF equipment.

The APG-71's antenna retains the gimbal system of the AWG-9 and adds a new array with low sidelobes and a guard channel to eliminate sidelobe penetration of ground clutter and electronic warfare interference. The APG-71 also provides an improved radar master oscillator which significantly increases the number of radar channels and provides frequency-agile operation, with low sidelobes.

The signal processor has four processing elements as against the three found in the APG-70, giving an operating speed of 40 Mcops (million complex operations per second). The radar data processor also has a large degree of commonality with that in the APG-70, differing only in the interface cards, and operates at 3.2 Mips (million instructions per second).

The aft cockpit digital display for the APG-71 was originally developed for the AWG-9 but never put into production, while the tactical information display, originally designed for the F-106 aircraft, remains largely unchanged from the version in the F-14A. Significantly, the APG-71's software is written in Jovial.

Specifications
Frequency: X-Band (NATO I-/J-Band, 8–12.5 GHz)
Antenna: Slotted planar array
Power: 500 W (pulse), 7 kW (Pulse Doppler) average; 10 kW peak
Range: 210 km
Scan rate: 80°/s (horizontal), 2 scans/s (vertical)
Tracking: Up to 24 targets simultaneously
Volume: 0.78 m^3
Weight: 590 kg

Status
In service in US Navy F-14D fleet defence aircraft. Testing of the AN/APG-71 began at the end of 1986, with flight trials during 1988. Operational evaluation of the radar was completed during

The AN/APG-71 radar in an F-14D   0503734

1990, with delivery of production units continuing until 1993 for a total of 45 F-14D aircraft.

Support of the system is ongoing; Raytheon was awarded a USD 5 million contract in 2002 for the supply of ten antenna subassemblies for use in the repair of in-service units.

Contractor
Raytheon Company, Space and Airborne Systems.

AN/APG-73 radar for the F/A-18

Type
Airborne multimode radar.

Description
The AN/APG-73 radar is an all-weather, coherent, multimode, multiwaveform search-and-track radar for upgraded versions of the F/A-18A, later versions of the F/A-18C/D and the F/A-18E/F aircraft. It is based on the AN/APG-65 fitted to earlier versions of the F/A-18, but uses new signal and data processors and a revised receiver/exciter to perform both air-to-air and air-to-surface missions. Raytheon Electronic Systems claims this will give the new radar more than three times the speed and memory of the AN/APG-65 and will make it compatible with an Active Electronically Scanned Antenna (AESA).

The new data processor unit is a multifunction processor consisting of a signal processor function and a data processor function. The signal processor has four million words of bulk memory capacity and its throughput has been increased to 60 million complex operations per second by use of multichip modules. The data processor function is a general purpose, dual-1750A computer that

provides software loads to the signal processor and controls to the rest of the radar.

Phase II of the AN/APG-73 upgrade incorporates a Motion-Sensing Subsystem (MSS) unit; a stretch waveform generator module; and a Special test equipment, Instrumentation and Reconnaissance (SIR) module, as well as reconnaissance software. These enhancements give the F/A-18 the hardware capability to produce high-resolution ground maps (comparable with those of the F-15E and U-2). It also allows precision strike missions using advanced image-correlation algorithms, further improving weapon designation accuracy.

Air-to-air modes include: high PRF; velocity search; high/medium PRF range-while-search; four short range automatic acquisition modes; track-while-scan; gun director; raid assessment/situation awareness; single target track.

Air-to-surface modes include: Doppler beam-sharpened sector and patch mapping; medium resolution SAR; radar navigation ground mapping; real beam ground mapping; fixed and moving ground target indication/track; air-to-surface ranging; terrain avoidance; precision velocity update; inverse range angle; and sea surface search with clutter suppression.

Reconnaissance modes are strip map and spotlight map.

Specifications
Volume: (excluding antenna) 0.126 m^3
Weight: 154 kg
Frequency: 8–12 GHz
Number of LRUs: 5 (transmitter, MSS, power supply, data processor, receiver) plus antenna

Status
First operational units of the AN/APG-73 were delivered in mid-1994.

Raytheon's AN/APG-73 equips F/A-18C/D fighter aircraft (Raytheon)   1127376

AN/APG-76 multimode radar   0514948

ballistic missile launcher detection and ground surveillance sensor aboard a USAF F-16 and a US Navy S-3B.

Contractor
Northrop Grumman Corporation, Electronic Systems Sector, Norden Systems.

AN/APG-77 multimode radar

Type
Airborne multi mode radar.

Description
A joint venture of Northrop Grumman's Electronic Systems and Sensors Sector and Raytheon Electronic Systems, the AN/APG-77 multimode radar employs a low observable active aperture, electronically scanned array which incorporates approximately 2,000 individual transceiver modules. The antenna provides the agility, low radar cross-section and wide bandwidth necessary to support the F/A-22 Raptor's multirole mission, providing sophisticated interleaved air-to-air and air-to-ground, long-range, multiple-launch/multiple-target, all-weather capability. In addition to the Active Electronically Scanned Array (AESA), the radar sensor subassemblies include a radar support electronics unit (similar to a traditional receiver/exciter), a radio frequency receiver unit, an array power supply, an aircraft installation kit and a coolant distribution manifold. The system employs Monolithic Microwave Integrated Circuit (MMIC) and Very High-Speed Integrated Circuit (VHSIC) technologies in order to enhance reliability and achieve a claimed Mean Time Between Failures (MTBF) of more than 400 hours.

The AN/APG-73 is in production for the US Navy F/A-18E/F aircraft. It is also in production for the US Marine Corps F/A-18A upgrade, designated ECP-583.

The APG-73 was also specified for F/A-18C/Ds ordered by Finland, Malaysia, Switzerland and Thailand.

In January 2000, Raytheon was awarded a contract USD200 million to supply 71 AN/APG-73 radar units for the Royal Australian Air Force F/A-18 A/B aircraft upgrade (presently fitted with the APG-65), together with 40 radar kits for the US Navy (19) and US Marine Corps (21). A further award, valued at USD8.8 million, was made in May 2000 for nine updated receivers to support the Royal Australian Air Force requirement for completion in May 2002. The APG-73 radar, similarly, forms part of the projected Canadian F/A-18 upgrade programme.

Latest development models for the F/A-18E/F Super Hornet have been reported to use the designator AN/APG-73(V).

It is intended to replace the AN/APG-73 in the F/A-18E/F with the AN/APG-79 Active Electronically Scanned Array (AESA) radar, designed to improve the air-to-air detection and tracking range, add higher resolution air-to-ground mapping modes at longer ranges, improve situational awareness and reduce support costs. In November 1999, under an Advance Agreement between Boeing and the US Navy, Raytheon was tasked to develop an integrated AESA radar prototype. Drawing on experience gained with the AN/APG-63(V)2 AESA system fitted as an upgrade package to USAF F-15C aircraft, Raytheon demonstrated the APG-79 in November 2002, with first flight in 2003 and an Initial Operating Capability (IOC) expected during 2006.

Contractor
Raytheon Company, Space and Airborne Systems.

AN/APG-76 MultiMode Radar System (MMRS)

Type
Airborne multimode radar.

Description
APG-76 MMRS is designed to provide the F-4 Phantom with enhanced air-to-air and air-to-ground capabilities. Air-to-air capabilities include look-up, look-down and beacon mode, as well as air track and air combat modes. Air-to-ground performance is enhanced through the inclusion of real-beam ground map, high-resolution Synthetic Aperture Radar (SAR) and Doppler beam-sharpening, ground mapping with simultaneous ground moving target indication, and beacon modes. A powerful clutter suppression interferometer provides a clutter-free resolution SAR map, as well as multitarget tracking capability. The radar's air-to-ground capabilities have also been used in the Gray Wolf technology demonstration (see operational status). Here, the radar is mounted in a pod and is described as offering a number of operating modes including real beam, Doppler sharpening and spotlight. In real beam, the Gray Wolf application is noted as having a maximum detection range in excess of 185 km, while the Doppler beam-sharpening capability allows 26 × 26 km sectors to be scanned with high-resolution values. The equipment's spotlight mode is thought to incorporate three submodes, the most sensitive of which provides 0.3 m resolution on targets at ranges of up to 130 km. The Gray Wolf application is also noted as being able to track up to 75 moving surface targets simultaneously.

Status
A podded APG-76 codenamed Gray Wolf is noted as having been tested as a theatre

The AN/APG-77 multimode radar   0073156

The F-22 Raptor carries the AN/APG-77 multimode radar   0085268

The AH-64D Longbow Apache helicopter with mast-mounted Longbow radar   0018072

AN/APG-78 Longbow mast-mounted radar dome   0044987

Designed from the outset for operation in dense RF environments, the APG-77 is reported as having a maximum range in excess of 180 km. In addition to the normal Search, Track-While Scan (TWS), Multiple Target Track and Dogfight modes, the radar incorporates an ultra-high resolution target recognition mode, offering centimetric resolution at the extremes of its range envelope. Returns generated by the ultra-high resolution mode are matched to an onboard library to facilitate Non-Co-operative Target Recognition (NCTR). Considering potential fighter/attack duties for the F/A-22, the radar, in common with other AESA designs, provides inherent Electronic Counter Measures (ECM) support and is capable of jamming ground defence radars, thus enhancing the survivability of the aircraft and strike package.

Status
Major elements of the first radar, including active electronically scanned antenna array, supporting electronics and search/track software were delivered in April 1998 to the Boeing Company's F/A-22 Avionics Integration Laboratory in Seattle, Washington, where engineers integrated the radar with other F/A-22 avionics. Additional radar systems were flight-tested on a Boeing 757 Flying TestBed (FTB) aircraft and on F/A-22 flight test aircraft. Teamed with Lockheed Martin to design and build the F/A-22 Raptor, Boeing is responsible for integrating the radar with all avionics systems to produce an integrated sensor suite featuring fused sensor target detection and tracking.

At the end of 2000, the first flight of the AESA, on board Raptor 4004, took place. This was the first F/A-22 fighter aircraft to fly with advanced avionics hardware and software. Initial reports from the test pilot and instrumented data indicated that the radar was successfully tracking multiple targets almost immediately after the aircraft lifted off from Lockheed Martin's Marietta Georgia facility.

In January 2001, Northrop Grumman reported that all 11 Engineering and Manufacturing Development (EMD) APG-77 radars had been delivered.

Low-rate initial production of the radar began during 2001.

Full radar functionality was planned for the Block 3.0 avionics software integrated into Raptor 4005. Further airborne tests of the radar system will focus on those capabilities and scenarios that could not be duplicated using the FTB.

AN/APG-77 functionality is currently somewhat limited; the radar incorporates no air-to-ground modes, with in-flight retargeting and/or system updating accomplished by manual entry of data points/co-ordinates received via a Link 16 text message or radio communication. The datalink terminal operates in receive-only mode.

During the second quarter of 2002, the US Air Force proposed the expansion of the role of the Raptor to encompass air-to-ground as well as air-to-air missions, heralding the redesignation of the aircraft as the F/A-22. Taking advantage of some of the inherent design features embodied in the baseline air-to-air version, such as the ability to supercruise on military (non-afterburning) power at speeds above M1.5, high-altitude capability, high manoeuvrability and an extremely stealthy airframe, the F/A-22 will be able to attack heavily defended targets while providing for its own defence against enemy fighters.

The proposed enhancements, which involve both sensors and weapons, could be implemented by the Block 5 software that is scheduled to be incorporated into the fleet by the end of 2007.

The upgrades would be incorporated in Lot 6 aircraft, production of which is due to begin in early FY06. Block 5 adds air-to-ground modes for the APG-77, interfaces to the Small-Diameter Bomb (SDB) and other weapons, and has an electronic attack capability and the ability to transmit over Link 16.

New software, planned for Block 5, will provide the radar with automatic target detection and classification, geo-registration of imagery with Digital Terrain Elevation Data (DTED) to provide a three-dimensional perspective view and reduce the target-location error, a Moving-Target Indication (MTI) mode and a 'far look' capability in Synthetic Aperture Radar (SAR) modes, with 3 m resolution and an air-track update facility.

In June 2004, Northrop announced the successful first flight test of a new so-called 'fourth generation' variant of the APG-77, which adapts elements of the AN/APG-80 and -81 radars for the F-16 Block 60 and F-35 Joint Strike Fighter (see separate entries). The new design significantly reduces the required number of parts for the radar, with commensurate reductions in production and maintenance costs.

While the current order for F/A-22 aircraft remains somewhat uncertain in terms of numbers, as of 2004, Northrop expects to deliver over 200 AN/APG-77 FCR for the aircraft.

Specifications
Frequency: 8–12 GHz
Power per antenna module: 10 W
Reliability: 2,000 h MTBF estimated

Contractor
Northrop Grumman Corporation, Electronic Systems Sector.
Raytheon Electronic Systems.

AN/APG-78 Longbow radar

Type
Airborne multimode radar.

Description
A Lockheed Martin/Northrop Grumman joint venture company (Longbow Limited Liability Company LLLC) is working with The Boeing Company to supply the Longbow radar and RF Hellfire fire-and-forget missile to the AH-64D Apache helicopter for the US Army. The joint venture is also working with GKN Westland Helicopter Ltd to supply the radar and missile for the WAH-64 Apache for the UK. A lightweight variant is expected to be integrated into the US Army's RAH-66 Comanche reconnaissance and light attack helicopter.

The AN/APG-78 Longbow radar system comprises a mast-mounted, millimetric-wavelength (35 GHz), fire-control radar which allows the crew to search, detect, locate, classify and prioritise ground targets (tanks, air defence units, trucks and so on) in a 55 km² area. Up to 16 of the highest priority targets can be displayed simultaneously, at ranges up to 8 km for moving targets. An integrated Radar Frequency Interferometer (RFI) is located below the radar. It performs passive seeking to provide directional cues for the fire-and-forget capability of the Hellfire missile. The RFI is designated AN/APR-48A and is supplied by Lockheed Martin. The Longbow system provides exceptional performance in adverse weather, and in the presence of obscurants and countermeasures.

Status
Following a successful proof-of-principle programme, the joint venture received a USD314.9 million contract in 1991 for full-scale development of the radar. The Longbow system completed Initial Operational Test and Evaluation (IOT&E) in 1995, demonstrating a claimed 28-fold improvement in combat effectiveness over the AH-64A. Contracts for the first two lots comprising a total of 20 fire-control radars were awarded in March 1996 and February 1997.

Meanwhile, in April 1996, contracts were awarded to Longbow International (LBI), the joint venture's international marketing unit, to supply Longbow systems to the British Army for all 67 of its Westland Attack Helicopter WAH-64s. The first Apache Longbow radar was delivered to GKN Westland in June 1998, two months ahead of schedule.

In November 1997, the US Army awarded the joint venture a five-year, USD565 million contract to build 207 Longbow radars. The US Army plans that half of its fleet of helicopters will carry the Longbow radar/RFI. Similarly, the Royal Netherlands Army procurement for AH-64D includes only a limited fit of the Longbow radar/RFI systems. Other nations that have selected the Apache/Longbow system include the air forces of Israel, Japan and the Republic of Singapore, which received the first of 20 AH-64D during May 2002. Shortly afterwards, the UAE requested remanufacture of its AH-64A helicopters to -64D standard, including 32 AN/APG-78 FCRs, in a total order of up to USD1.5 billion. During the third quarter of 2002, and after previous objections which led to cancellation of a USD800 million order for AH-64Ds, the US approved the sale of eight Longbow FCRs to Kuwait, as part of a sixteen-aircraft package worth up to USD2.1 billion. Deliveries are due to begin in 2005.

Contractor
Lockheed Martin Missiles and Fire Control.
Northrop Grumman Electronic Systems.

AN/APG-79 Active Electronically Scanned Array (AESA) radar

Type
Airborne Active Electronically Scanned Array (AESA).

Description
Raytheon is developing the AN/APG-79 Active Electronically Scanned Array (AESA) radar for the Boeing F/A-18E/F Super Hornet.

One of the major obstacles to overall performance in current multifunction radars is the utilisation of a mechanically scanned antenna array; in order to service the demands of near simultaneous interlaced air-to-air and air-to-ground functionality, the rate of scanning of the antenna must be extremely high, thereby placing great loads on the aircraft hydraulic system and antenna gimbal arrangement. AESA radars like the APG-79 incorporate a 'tile array' of solid-state transmit/receive modules with electronic antenna steering, thus facilitating near instantaneous radar axis shifts. This rapid beam scan feature provides a foundation for vastly superior performance and capabilities over current systems.

The APG-79 system comprises the AESA, Motion Sensor (MS), a Radar Power System (RPS), a Common Integrated Sensor Processor (CISP) and a Receiver/Exciter (R/E). The MS compensates for aperture motion, while the RPS is described as a fault tolerant system that supports modular transceiver architecture. The CISP is a commercial-off-the-shelf, PowerPC™-based, open architecture, and modular system, while the R/E features wide bandwidth, fast frequency agility, low noise and programmable waveform generation.

Raytheon describes the APG-79 as a 'wideband' device – it is expected that the radar is capable of operating well beyond the X Band (8 to 12.5 GHz) capability of the AN/AG-73 equipping F/A-18C/D aircraft of the US Navy and Marine Corps.

Radar modes include air-to-air search/track/ Track While Search (TWS), passive ESM, sea surface search, Real Beam Mapping (RBM), Synthetic Aperture Radar (SAR), Ground Moving Target Indication (GMTI) and SAR on GMTI. In

A modified pre-series F/A-18F Super Hornet, the Airborne Electronic Attack (AEA) demonstrator, unofficially known as EA-18G Growler, first flew during November 2001. The aircraft was modified to carry three Northrop Grumman AN/ALQ-99 jamming pods and two fuel tanks (Boeing)   0525965

two-seat applications, the APG-79 provides for independent, dual-cockpit operation.

When compared with the current AN/APG-73, in air-to-air modes the APG-79 provides for two to three times the detection range, combined with far greater discrimination against closely spaced targets. In air-to-ground modes, a threefold increase in resolution facilitates high-resolution SAR imaging (with four steps in SAR resolution). As alluded to earlier, all air-to-ground radar modes can be performed concurrently with air-to-air modes. Inherent in the AESA design, the APG-79 incorporates a number of coherent jamming and electronic support features, which will undoubtedly further enhance the capability of the EA-18G Airborne Electronic Attack (AEA) aircraft currently under development for the US Navy (USN).

Benefiting from the entirely solid-state antenna construction, the APG-79 features dramatically improved reliability and lower cost – indeed, such is the confidence that the antenna will need no attention other than for battle damage, a modified radome, which slides forward instead of hinging to the right, has been introduced for the F/A-18E/F, saving valuable space in aircraft carrier hangars, where the radome is often damaged during routine maintenance.

Specifications
Weight: 43.1 kg (AESA)
MTBF: >15,000 h

Status
Under development.

The AN/APG-79 entered Engineering and Manufacturing Development (EMD) in February 2001, with full demonstration of the system carried out in November 2002, when an integrated system successfully transmitted, received, and collected real beam ground mapping radar data of the California coastline.

Following the laboratory development phase, an extensive flight test programme at Naval Air Warfare Centre at China Lake commenced in 2003 with the radar installed in an F/A-18E/F Super Hornet.

The system is planned to be integrated into US Navy fleet service in the F/A-18E/F, with an Initial Operating Capability (IOC) in 2006.

The Navy plans to buy 415 systems to equip its F/A-18 aircraft, including newly acquired F/A-18E/F Super Hornets, EA-18G AEA and upgrades of F/A-18C/D aircraft.

Concurrent Research and Development (R&D), commenced during mid-2004 at the Naval Air Warfare Centre, is intended to explore the potential for the APG-79 (and other AESA radars) to radically enhance the overall combat power of offensive formations by developing the inherent jamming and ESM capabilities of the radar, as part of a wider net-centric effort.

The lifetime value of the APG-79 AESA program is estimated to be at least USD1 billion. Internationally, the APG-79 will be available for upgrade of any of the eight operating countries' F/A-18s.

Contractor
Raytheon Company, Space and Airborne Systems.

AN/APG-80 Agile Beam Radar (ABR)

Type
Airborne Active Electronically Scanned Array (AESA).

Description
The AN/APG-80 ABR is the designation of the Foreign Military Sales (FMS) radar offered with the Lockheed Martin F-16 Block 60+ aircraft. Featuring a wideband Active Electronically Scanned Array (AESA), the APG-80 is the first production system to benefit from Northrop Grumman's fourth-generation transmitter/receiver module technologies which provide significantly enhanced performance over the latest AN/APG-68(V)9 (see separate entry) which equips current Block 50/52 F-16s of the USAF, while maintaining installation commonality with that system.

Considering the multirole mission of the F-16, the AN/APG-80 baseline configuration represents a significant upgrade in capability, with advanced interleaving of modes enabling the pilot to maintain Situational Awareness (SA) and weapons-quality tracking of air-air targets while prosecuting air-ground attacks. Air-air features and improvements over current versions of the AN/APG-68 include:

- Expanded bandwidth
- Greater detection range
- 140° track volume
- 20 target multitrack (with growth potential for up to 50 for greater SA)
- 6 targets tracked simultaneously at Single-Target-Track (STT) accuracy

The AN/APG-79 AESA radar for the F/A-18E/F Super Hornet   0118181

The first flight of F-16F RF-01 for the UAE took place on 6 December 2003   0563220

The UAE took delivery of its first batch of F-16E/F Desert Falcon aircraft on 3 May 2004. This formation consisted of three F-16E single-seat and two F-16F twin-seat aircraft, with the lead F-16F trailing from its wingtip-mounted 'Smokewinders' (Lockheed Martin)   1116063

- Maintains tracking at greater range and target Line-of-Sight (LOS) rates
- Low RCS providing Low Probability of Intercept (LPI) by target.

These features combine seamlessly with improvements in air-ground capability:
- Automatic Terrain Following (ATF)
- Ultra-High-Resolution Synthetic Aperture Radar (UHRSAR)
- Extensive Electronic Counter-Counter Measures (ECCM).

In addition to the baseline features of the system, optional features include:
- Enhanced SAR/Automatic Target Cueing (ATC)
- Moving Target Indicator (MTI)-on-SAR.

The AN/APG-80 incorporates extensive growth potential with a Mean Time Between Failures (MTBF) of 500 hours, double that of the current baseline radar for Block 50 aircraft, the AN/APG-68(V)5 (see separate entry).

Status
The AN/APG-80 radar is part of the F-16 Block 60 Desert Falcon which has been ordered by the United Arab Emirates (UAE). While the radar is also offered as an upgrade to current operators, it is reported that existing airframes would require extensive upgrades to power supplies and cooling systems to facilitate fitment of the radar.

During August 2003, Northrop Grumman delivered the first AN/APG-80 to Lockheed Martin for the F-16 Block 60 aircraft intended for the UAE order. As Northrop Grumman commences delivery of production radars, testing of additional software modes will continue into 2004 using test radars on board the company's BAC 1–11 test bed aircraft in Baltimore. APG-80 radar deliveries are scheduled to continue through to late 2005 for the fleet of 80 F-16 Block 60 aircraft destined for the UAE.

The first flight of the Block 60 F-16, as part of the ongoing Lockheed Martin-led weapons systems development effort, took place on the 6 December 2003, with delivery of the first batch of UAE aircraft taking place on 3 May 2004. UAE F-16Fs will carry an identical avionics fit to the single-seat fighter, this being achieved through the use of a dorsal avionics compartment. Suitable for use as training platforms, the two-seat aircraft will also be fully capable of conducting dedicated strike missions, with the rear crew station configured for use by a weapon

systems operator. Once in squadron service, the Block 60 aircraft are expected to be based at Al Dhafra and Liwa. Initial deliveries of the Block 60 aircraft will be made in a Lot 1 configuration, which will introduce the new radar, EW suite and advanced cockpit systems. A subsequent Lot 2 standard will introduce advanced EW modes, an expanded weapons capability and a terrain-following mode for the APG-80. A final Lot 3 configuration may introduce the Joint Helmet-Mounted Cueing System (JHMCS – see separate entry).

In addition to the radar, Northrop Grumman is providing the Integrated Forward-Looking Infra-red and Targeting System (IFTS) and the Integrated Electronic Warfare System (IEWS) for the F-16 Block 60 aircraft (see separate entries). Included in the Block 60 contract is the Combined Intermediate Automatic Test Equipment (CIATE) program. The CIATE is capable of automatically testing all three Northrop Grumman sensor systems, the APG-80, IFTS and IEWS, and will detect faults and allow subsystem repair down to the component level.

Contractor
Northrop Grumman Corporation, Electronic Systems Sector.

AN/APG-81 multimode radar for the F-35 Joint Strike Fighter

Type
Airborne Active Electronically Scanned Array (AESA).

Description
Drawing on experience with the AN/APG-77 radar for the Lockheed Martin F/A-22 Raptor, Northrop is developing the AN/APG-81 multimode Active Electronically Scanned Array for the F-35 Joint Strike Fighter (JSF). The radar will be an integrated part of the aircraft sensor suite, which includes the Northrop/Lockheed Electro-Optic Targeting System (EOTS) and Distributed Aperture System (DAS) (see separate entries).

The radar will fulfil air-to-air, air-to-ground and Electronic Warfare (EW) duties for the multirole JSF.

Status
In development. The radar system has been undergoing trials on board Northrop Grumman's BAC-111 Co-operative Avionics Test Bed (CATB) aircraft. During the second quarter of 2005, Northrop Grumman delivered the first APG-81 to prime contractor Lockheed Martin Aeronautics Company. The radar was destined for radome integration testing at Lockheed Martin's facility in Palmdale, California. Shortly afterwards,

The AN/APG-81 radar installed in the nose of Northrop's modified BAC-111 CATB testbed aircraft (Northrop Grumman)   1151472

The F-35's multimode AN/APG-81 will integrate with its Electro-Optical Distributed Aperture and Search and Track systems   0521685

the radar passed a major milestone in system integration testing by detecting airborne targets at Northrop's integration laboratory.

Contractor
Northrop Grumman Corporation, Electronic Systems Sector.

AN/APN-215(V) radar

Type
Airborne Ground Mapping Radar (GMR).

Description
The AN/APN-215(V) colour radar is a weather, surface search and precision terrain-mapping system derived from the successful and widely used RDR-1300 commercial system. It is designed for heavy twins, turboprops and transport helicopters. Low weight and a 445 km range suit it to utility and reconnaissance aircraft and it was chosen for the US Army versions of the Beech King Air, the U-21 and the RU-21.

In conjunction with other equipment the APN-215 can display navigation pictorial information overlaid on the weather map, together with pilot-programmable pages of checklist information such as en route navigation data and emergency procedures.

The system comprises three units: a 305 mm pitch and roll-stabilised antenna, transmitter/receiver and colour control/display unit.

Status
In service with the US Army and Coast Guard.

Contractor
Honeywell Aerospace, Sensor and Guidance Products.

AN/APN-234 multimode radar

Type
Airborne surveillance radar.

Description
The AN/APN-234 is a lightweight airborne digital colour display multimode radar designed to provide sea search weather detection and terrain-mapping for a variety of military aircraft including rotary- and fixed-wing types ranging from light to heavy twins. The system consists of a receiver/transmitter, combined colour display/control unit, stabilised antenna and optional interface unit.

The AN/APN-234 is identical to the AN/APN-215 except for the addition of the sea search function.

Status
In service with the US Navy on the Northrop Grumman C-2A Greyhound and the Lockheed Martin EP-3E aircraft.

Contractor
Honeywell Aerospace, Sensor and Guidance Products.

AN/APN-241 airborne radar

Type
Airborne multi mode radar.

Description
The AN/APG-241 is a lightweight, fully coherent Pulse Doppler (PD) radar that is based on the Northrop Grumman AN/APG-66/68 series of fire-control radars. It was developed to provide precision airdrop and navigation radar capabilities for military tanker and transport aircraft and offers the following operating modes and features:

Weather (WX): WX mode uses an ARINC 708A six-colour display format and is able to detect and display weather phenomena, through intervening weather, at ranges of up to 320 nm (593 km). Turbulence can be detected and displayed at ranges of up to 50 nm (93 km).

WindShear (WS): In WS mode, APN-241 provides up to 90 seconds' warning of a microburst, has a claimed <10^4/flight hour false alarm rate and can be configured with an audio alert if required.

Ground Mapping: In Ground Mapping mode, the radar uses Monopulse Ground Mapping (MGM) for all ranges and azimuths (claimed to offer a 2.5 to 10× improvement Real Beam mapping) together with Doppler Beam Sharpening (DBS) for high-resolution, off-the-nose coverage. The Ground Mapping mode is claimed to offer a 31 m Circular Error of Probability (CEP).

Skin Paint (SP): In SP mode, the radar can detect a C-130-sized target at ranges of up to 20 nm (37 km), showing variable target sizes based on radar cross-section. Doppler vectors are used to establish head-on/tail-on aspect.

Beacon Mode (BCN): The APN-241's BCN mode is able to receive and interrogate both airborne and ground-based beacons.

Station Keeping (SKE): The SKE function provides APN-169C data in offset or centred Plan Position Indicator (PPI) formats.

Flight Plan (FP): FP mode makes use of navigation data from a Self-Contained Navigation System (SCNS) or other form of SNU 84 reference.

Traffic Collision Avoidance System (TCAS): TCAS mode provides for enhanced Situational Awareness (SA).

BCN, SKE, FP and TCAS modes can be displayed autonomously or be overlaid on any other mode's display.

The system can interleave radar modes, allowing the crew to view and control separate modes simultaneously while, at the same time, overlaying any of the three display modes. It is capable of accommodating a crew of two or three. The open systems architecture of the AN/APN-241 will allow advanced functions such as Synthetic Aperture Radar (SAR), terrain-following/terrain-avoidance, maritime surveillance and autonomous landing capability to be added without significant development costs. A SAR upgrade was successfully flight tested during 1999.

APN-241NT

Northrop Grumman was contracted by BAE Systems Australia to develop and supply seven examples of an updated variant of the APN-241, known as APN-241NT, using an antenna that allows it to equip medium-sized transport aircraft as part of the upgrade of Royal Australian Air Force HS 748 navigator trainers.

AN/APN-241 Radar Antenna   0092870

Specifications

Type: Dual-channel monopulse
Frequency: 9.3–9.4 GHz (coherent PD)
Range scales: 1.5, 3, 5, 10, 20, 40, 80, 160, 320 nm
Azimuth sectors: ±15°, ±30°, ±60° (centred about cursor) and ±135°
Elevation coverage: −25° to +10°
Stabilisation: Three-axis
Power input: 950 W (single-phase, 400 Hz, 115 V)
Power output: 9.5 W (average); 116 W (peak)
MTBF: greater than 1,000 h
MTTR: 17.5 min
Maintenance: two-level
Cooling: unpressurised air
Display formats: centred, offset or expanded 2:1; 4:1 PPI
Display orientation: heading, north, track or drop zone up (with SCNS)
Interfaces: APN-169C (SKE mode); ARINC 429; LTN-72; LTN-92; MIL-STD-1553B; RA (AL-101 or CARA); RS-170; SCNS; TCAS II
Dimensions: 66 × 81 cm (antenna assembly); 66 × 25 × 20 cm (TP unit)
Weight: 28 kg (TP unit); 31 kg (antenna assembly)

Status

By the first quarter of 2001, in excess of 230 AN/APN-241 radars had been procured worldwide. Current applications include C-130H transport aircraft flown by the air forces of the USA, Australia (12 aircraft) and Portugal (six), together with seven HS.748 navigation trainers operated by the Royal Australian Air Force (APN-241NT variant).

The APN-241 is the baseline radar for the C-130J transport aircraft and the C-27J Spartan, Lockheed Martin Alenia Tactical Transport Systems' medium tactical airlifter. In 1999, Northrop proposed the APN-241B radar to replace the existing AN/APS-133(V) systems on the C-17A Globemaster III transport aircraft. The most significant difference between the two variants was the replacement of the 36-inch aperture radar used on the basic model with one of 22-inch aperture to fit into the C-17. During 2000, Northrop Grumman and the European Aeronautic, Defence and Space (EADS) Company announced their joint promotion of a variant for use on the A400M next-generation transport aircraft. Also during 2000, Northrop Grumman announced that it had been awarded a USD7.7 million contract covering the supply of 24 APN-241 radars for installation aboard USAF C-130H aircraft, with delivery completed during 2002.

Currently, APN-241 growth potential includes a software-driven synthetic aperture radar mode, together with terrain following, drop zone wind measurement and autonomous landing guidance capabilities.

Contractor

Northrop Grumman Corporation, Electronic Systems Sector.

AN/APN-242 search/navigation radar

Type

Airborne Ground Mapping Radar (GMR).

Description

The AN/APN-242, formerly the AN/APN-59(X), is a long short-range colour weather and navigation radar with integrated navigation and air data overlays via the AN/ASN-165 Display Group. Principal features include weather detection and display in colour, black/white, or green for night vision goggle compatibility; enhanced terrain-mapping with navigation using a latitude/longitude-stabilised electronic cursor; aircraft detection/skin painting concurrent with other operating modes; beacon interrogation and display; and optionally, IFF display.

The APN-242 was designed by the then Litton Marine Systems as a follow-on to the AN/APN-59 now in use worldwide on over 1,500 C-130 and C-135 aircraft. The radar is suitable both for

new installations and as a form, fit and function replacement for the APN-59. Performance has been improved in every operating mode by increasing range, resolution and accuracy. By redesigning high failure rate components, the APN-242 achieves a system Mean Time Between Failure (MTBF) rate greater than 1,000 h.

The APN-242 antenna subsystem is a flat plate array which is stabilised to aircraft attitude reference systems and which, through the elimination of all gears, achieves reliabilities approaching 7,000 h. The array rotates 360° or can sector-scan. The antenna beam can be tilted vertically and the pattern can be instantaneously switched between pencil and fan to achieve the desired illumination. Pulsewidths and PRFs are operator selectable. The low-noise receiver and lower-power transmitter improve APN-59 performance and the long life 10,000 h magnetron has the power to skin paint fighter aircraft at extended ranges through intervening rain showers. The AN/ASN-165 is the standard display group.

When replacing the APN-59, the APN-242 does not require an aircraft modification for installation and either the antenna or receiver/transmitter (R/T) subsystems can be dropped in as direct replacements for APN-59 subsystems. On these aircraft, the system can be installed on the flight line in a day by unit-level maintenance using existing radar cabling, connections and mounting brackets.

Specifications

Frequencies:
(radar operation) I-band; 9,375 ±10 MHz
(beacon reception) 9,310 MHz
Transmitted power: 25 kW nominal peak (new high-reliability magnetron)
Ranges: 2.5 to 20 (2.5 n mile increments), 25, 30, 50, 100, and 240 n miles
Pulse length: multiple lengths (0.20, 0.8, 2.35, and 4.5 ms) automatically selected for different ranges and functions
Scanning features: 360°
Scan rates 12 rpm on long-range functions; 45 rpm on short-range functions
Sector scan (basic) approx 80°, centred about forward position, variable sector with alternate control
Antenna beam selection: pencil or equal energy (fan) beam; both with 3° azimuth beamwidth, instantaneous electronic switching
Antenna stabilisation: stabilised to existing aircraft reference throughout a range of ±15° pitch and ±30° roll
Controls: independent navigator and pilot controls
System components/weight:
(basic) 7 components/70 kg
Power:
(basic) 115 V ±5% (380 to 420 Hz), 800 VA average

Status

The AN/APN-242 is in service in US Air Force C-130 and KC-135R aircraft.

Contractor

Northrop Grumman Corporation, Sperry Marine.

AN/APQ-122(V) radar

Type

Airborne Ground Mapping Radar (GMR).

Description

The AN/APQ-122(V) is a dual-frequency nose radar developed for use in the US Air Force Adverse Weather Aerial Delivery System (AWADS) programme for installation in C-130E transport aircraft. This long-range navigation sensor is used for weather avoidance and navigation in supply dropping missions. The equipment provides ground-mapping out to more than 385 km, weather information up to 278 km and beacon interrogation up to 444 km when using the I-band frequency radar. J-band frequencies are used when short-range

high-resolution performance and target location are required. In the J-band mode the radar provides a high-resolution ground map display to permit target identification and location for position fixing and aerial delivery missions. In this mode the radar will detect and display targets with a radar cross-section of 50 m^2 while operating in rainfall of 4 mm/h.

In addition to the dual-frequency system designed for AWADS, designated AN/APQ-122(V)1, three other configurations have been developed. The AN/APQ-122(V)5 is a single-frequency I-band radar which has been developed as a direct replacement for the AN/APQ-59 radar used in C-130 and E-4B aircraft. Facilities include long-range mapping, weather evaluation and avoidance and rendezvous. A navigation training version of the AN/APQ-122(V)5, the AN/APQ-122(V)7, has been designed for use in the T-43A aircraft. Another dual-frequency radar, the AN/APQ-122(V)8, incorporates a terrain-following capability and is used on Combat Talon 1 MC-130 aircraft.

Status

The AN/APQ-122(V) has been supplied to the US Air Force and the air forces of Argentina, Australia, Bolivia, Cameroon, Congo, Denmark, Ecuador, Egypt, Gabon, Greece, Indonesia, Iran, Israel, Italy, Jordan, Libya, Malaysia, Morocco, Nassau, New Zealand, Niger, Nigeria, Oman, Philippines, Portugal, Saudi Arabia, Singapore, Spain, Sudan, Thailand, Venezuela and Zaïre. The radar has been installed in C-130H, RC-130A, KC-135A, RC-135A, RC-135C and E-4B aircraft.

Contractor

Raytheon Company, Space and Airborne Systems.

AN/APQ-158 radar

Type

Airborne Ground Mapping and Terrain-Following Radar (GMR/TFR).

Description

The AN/APQ-158 is a multimode forward-looking radar used primarily for terrain-following/terrain-avoidance at low altitudes in the Pave Low III night/adverse search and rescue helicopter, the Sikorsky MH-53J. The equipment is similar to the AN/APQ-126 but is modified for compatibility with the unique helicopter characteristics and the Pave Low III mission requirements. The radar contains 15 LRUs which provide the same basic modes of operation as the AN/APQ-126.

System upgrades have provided this radar with the ability to supply updates in all modes except terrain-following, and to perform terrain-following missions over very high clutter areas such as cities.

Status

In service in MH-53J helicopters.

Contractor

Raytheon Company, Space and Airborne Systems.

AN/APQ-159(V)5 Fire Control Radar (FCR)

Type

Airborne Fire-Control Radar (FCR).

Description

The AN/APQ-159(V)5 is the latest member of the APQ-159 family of Fire Control Radars fitted to the F-5 family of fighter aircraft. Designed as an upgrade to the APQ-159(V)3 radar as employed by US Navy Aggressor F-5E aircraft, the APQ-159(V)5 features improved performance, reliability and maintainability over earlier variants. Compatible with other avionic upgrades to the F-5E, aircraft wiring and structure require no modification during installation of the APQ-159(V)5.

F-5E of the ROKAF   0576472

The main components of the upgrade are form/ fit/function replacements for receiver/transmitter and radar processor. Minor modifications are required for the antenna and a new video amplifier and erase-pulse generator is incorporated into the DVST indicator.

Claimed enhancements over previous APQ-159 installations include:

- 100 per cent greater acquisition and tracking ranges
- 14 per cent improvement in typical detection range (at 20,000 ft)
- 17 per cent improvement in typical lock-on range (at 20,000 ft)
- 200 per cent improvement in MTBF.

Specifications
Coverage: ±45° azimuth; +45/–40° elevation
Receiver: GaAs FET low noise
Range: 40 nm (search); 20 nm (acquisition/track)
Typical detection range: 16 nm (at 20,000 ft)
Typical lock-on range: 11.5 nm (at 20,000 ft)
MTBF: 150 hours

Status
The AN/APQ-159(V)5 is in production and in service with numerous F-5E operators.

Contractor
Systems & Electronics Inc.

AN/APQ-164 multimode radar

Type
Airborne Ground Mapping Radar (GMR).

Description
The AN/APQ-164 is the radar installed in the US Air Force B-1B aircraft. This radar combines technology from the F-16 AN/APG-68 radar and the Electronically Agile Radar (EAR) programme of the US Air Force.

The B-1B radar generates data for navigation, penetration, weapon delivery, and for certain other functions such as air refuelling. There are four modes in the AN/APQ-164 system that provide the navigation capability. The primary mode is a high-resolution synthetic aperture radar mapping mode, backed up by a monopulse-enhanced real beam ground-mapping mode. The system also detects weather ahead and can display ground beacon returns over a real beam image. The penetration functions of the radar include automatic terrain-following and terrain-avoidance. For weapon delivery the radar provides four different functions. The first is a velocity update mode, similar to a Doppler navigator, which generates velocity information for the inertial navigation system. Coupled with an accurate Global Positioning System receiver in the avionics system, velocity update produces a dynamic, precision antenna calibration correction. Second, there is a ground moving-target detection and tracking capability for both fast- and slow-moving vehicles. Third is a high-altitude altimeter function that provides a very accurate measure of local height above the ground. Fourth is a monopulse targeting mode that provides accurate height to the on-scene selected fixed target.

The synthetic aperture mode provides the operator with a high-resolution image of an area of ground that can be chosen by the avionics system or the operator. Long-range maps can be made and five different map scales displayed. The synthetic aperture mapping mode accepts the co-ordinates of a waypoint from the avionics system and makes a map centered on that point. To make an image, the antenna is electronically scanned to the waypoint location. The radar transmits a train of pulses, gathers data for the image, and then switches itself off. At the same time, the image is stored in the radar and presented on the display in a rectangular, ground co-ordinate display.

The radar provides the basic data required for automatic terrain-following. It scans the ground in front of the aircraft and measures the terrain in a range versus height profile out to 19 km and stores that data in the computer. The profile data is sent across the multiplex bus to the terrain-following control unit where the data is used to generate climb/dive commands. This flight profile is then automatically fed into the pilot's flight control system. Since the radar is not continuously scanning in terrain-following, a very low update is used, helping to reduce the risk of detection. This rate is variable and depends on aircraft altitude, manoeuvres, groundspeed and terrain roughness. Under normal conditions updates are made at 3 to 6 second intervals. However, if the terrain demands it, data can be gathered continuously.

The AN/APQ-164 in the B-1B is a dual-redundant system, with two complete and independent sets of Line-Replaceable Units (LRUs), except for the phased-array antenna. This was the first airborne application of this technology for combat aircraft. Only one set of LRUs is used at a time, the other being maintained on standby.

The phased-array is an outgrowth of the antenna developed on the EAR programme. It contains 1,526 phase control modules and allows virtually instantaneous beam movement to any point in the antenna field of regard. When the radar mission requires a forward, right or left region of regard, the antenna is physically movable to three different positions on a roll detent mount. The radar can, therefore, look off to either side of the aircraft or forward by rolling the antenna about an axis. The normal antenna position is looking forward. However, when the antenna is rolled to one side, the field of view extends from the aircraft nose back to about 115°, permitting a look off to the side of interest without having to change aircraft heading. Once physically moved to one of the three available positions, the antenna is locked into a detent. From the fixed spot, it can be scanned electronically ±60° in azimuth and elevation by means of a unit on the antenna called the beam-steering controller, which controls all 1,526 phase control modules.

Specifications
Frequency: I-band
Transmitter: gridded, multiple, peak-power TWT (similar to the AN/APG-68 transmitter)
Antenna: phased-array electronically scanned, 1,118 × 559 mm
Operating modes: (air-to-ground) high-resolution mapping, monopulse enhanced real beam mapping, automatic terrain-following, manual terrain-avoidance, velocity update, ground moving target detection and track, high-altitude calibrate, ground beacon; (air-to-air) weather mapping, air-to-air beacon, rendezvous mode. Growth for full conventional standoff capability and a full air-to-air mode complement is provided.
Weight: 570 kg

Status
AN/APQ-164 is installed aboard US Air Force B-1B Lancer bomber aircraft.

Contractor
Northrop Grumman Corporation, Electronic Systems Sector.

AN/APQ-168 multimode radar

Type
Airborne multimode radar.

Description
The Sikorsky HH-60D Night Hawk helicopter is designed to penetrate hostile territory during darkness to rescue downed aircrews or deliver and retrieve special operations teams. The task calls for long-distance nap of the earth flying and accurate navigation.

The radar can operate in terrain-clearance, terrain-avoidance, air-to-air ranging and cross-scan modes, the latter combining ground-mapping or terrain-avoidance with terrain-following. A terrain storage facility permits the radar to have a reduced duty cycle thereby reducing the probability of detection by enemy ESM equipment.

The system has increased electronic countermeasures resistance, improved weather penetration, better guidance in turning flight, a power management function for semi-covert operation and low-beam reflectivity. Extensive BITE provides a high degree of fault isolation and

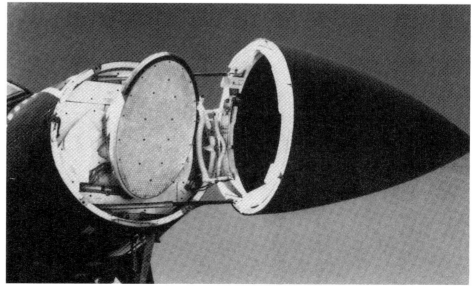

AN/APQ-164 airborne radar   0514682

detection. The system is carried in a pod in the nose of the aircraft.

Specifications
Dimensions: 1,420 mm long × 330 mm diameter
Weight: 113 kg
Power supply: 115 V AC, 3 phase, 200 VA
28 V DC
Reliability: 144 h specified MTBF

Status
In service with the US Air Force Sikorsky HH-60D Night Hawk helicopter.

Contractor
Raytheon Company, Space and Airborne Systems.

The AN/APQ-168 radar for the Sikorsky HH-60D Night Hawk helicopter 0503739

AN/APQ-170(V)1 dual-radar

Type
Airborne Ground Mapping and Terrain-Following Radar (GMR/TFR).

Description
Systems and Electronics Inc developed the AN/APQ-170 radar to equip 28 MC-130H Combat Talon II aircraft procured for the US Air Force Special Operations Forces. The radar has terrain-follow/avoidance capabilities, as well as ground map, weather and beacon modes. The AN/APQ-170(V)1 is a dual radar for redundancy and best performance in each mode. The system has high reliability and, therefore, a high probability of mission success. Delivery of the first AN/APQ-170-equipped aircraft took place in 1989.

The Combat Talon II aircraft incorporate the avionics suite developed by Lockheed Martin, including the Raytheon Systems Company AN/AAQ-45 infra-red system and a four-CRT cockpit display system.

Status
The current version of the radar in service on the 28 MC-130H Combat Talon II aircraft is the AN/APQ-170(V)1. A programme to improve mission effectiveness and reliability and to reduce maintenance costs was awarded to Lockheed Martin Federal Systems (Owego) and Systems and Electronics Inc (SEI) as its subcontractor, in January 1997 for development, testing and installation for all 28 aircraft by November 1999. The latest upgraded/modified configuration is designated the AN/APQ-170(V)1–425 Mod.

Contractor
Systems & Electronics Inc.

AN/APQ-174 MultiMode Radar (MMR)

Type
Airborne multimode radar.

Description
The AN/APQ-174 multimode radar has been developed for US Army, Navy and Air Force combat rescue and special operations missions, for use on aircraft such as the HH/MH-60, CH/MH-47, HH-53 and the V-22. The radar is a derivative of the LANTIRN terrain-following radar and the AN/APQ-168 multimode radar and maintains commonality with five of the six LANTIRN LRUs. The system will enable an aircraft to perform special operations and search and rescue missions at night, in adverse weather conditions and in a high-threat environment.

AN/APQ-174 modes include normal, power management and weather, terrain-following, terrain-avoidance, ground-ranging, beacon and weather. Set clearances are 100, 150, 200, 300 and 500 ft. Weather performance is enhanced by the use of selectable circular polarisation and operation in 10 mm/h of rain is claimed. The system includes extensive internal monitoring,

The AN/APQ-174 MultiMode Radar (MMR) provides the ability to operate at altitudes down to 100 ft in adverse weather by day or night 0503738

periodic and manually initiated BIT and end-to-end test.

The AN/APQ-174 allows operations at low altitudes, down to 100 ft above the ground by day or night. MFR improvements include the addition of weather detection and beacon interrogation modes to help in navigation and rendezvous. Other upgrades to the radar include expansion of software memory, conversion to electrically erasable memory and addition of an obstacle warning signal to existing video displays. Another upgrade, adding a new set clearance altitude to the terrain-following mode, will allow the aircrew to train at a safer altitude.

Specifications
Dimensions:
(pod) 330 mm diameter × 1,090 mm
(radar interface unit) 760 × 330 × 480 mm
Weight: 114 kg
Reliability: 144 h MTBF specified

Status
The AN/APQ-174 MMR is deployed on the US Army Special Operations Aircraft (SOA) MH-60K and MH-47E.

Contractor
Raytheon Company, Space and Airborne Systems.

AN/APQ-175 radar

Type
Airborne Ground Mapping Radar (GMR).

Description
The AN/APQ-175 radar was developed to replace the AN/APQ-122(V) for the US Air Force in Adverse Weather Aerial Delivery System (AWADS) C-130E aircraft. It is designed to enable the C-130 to airdrop and air land personnel and equipment

during poor weather conditions. Features of the equipment will include long- and short-range precision ground-mapping, weather detection, station-keeping, and beacon integration and reception. Independent multi-purpose displays are used for both pilots and the navigator so that each can independently observe certain different radar functions. Interface to the aircraft is via MIL-STD-1553 bus to facilitate radar data distribution and interaction with other aircraft systems.

The AN/APQ-175(V)X provides the C-130 with a modern navigation and weather radar. It is a dual-frequency system which, together with frequency agility, selectable beam shaping and selectable polarisation, provides optimum performance for the various modes and conditions. Very high resolution and antenna stabilisation are provided in order to achieve the needed accuracies at all ranges for autonomous air-drop and landing. Other features include echo contouring, rainfall rate determination, false return rejection, aim point offset, and built-in test.

Status
Delivery of production units began in 1990. The current version is in service with the US Air Force. The AN/APM-464 depot-level system test equipment, provided by Systems & Electronics Inc, is also in service with the US Air Force.

Contractor
Systems & Electronics Inc.

AN/APQ-181 radar for the B-2

Type
Airborne Ground Mapping and Terrain-Following Radar (GMR/TFR).

Description
The AN/APQ-181, designed specifically for the Northrop Grumman B-2 Spirit stealth bomber,

operates in the J-band (12.5 to 18 GHz) using 21 separate modes for terrain-following and terrain-avoidance; navigation system updates; target search, location, identification and acquisition; and weapon delivery.

The radar is a completely redundant modular system which employs two electronically scanned antennas, sophisticated software modes and advanced low probability of intercept techniques that match the aircraft's overall stealth qualities.

To meet reliability specifications and provide for operational redundancy, the AN/APQ-181 provides two separate radar sets, each consisting of five Line-Replaceable Units (LRUs): antenna, transmitter, Radar Signal Processor (RSP), Radar Data Processor (RDP) and receiver/exciter, with all but the antennas able to function for either or both radar units. These LRUs weigh 955 kg and have a volume of 1.485 m³.

The LRUs are installed in the aircraft in three zones. Each 260 kg antenna is mounted behind a large radome some 8 ft outboard of the aircraft centreline, just below the flying wing's leading edges. Antenna locations are marked by large, slightly darker, rectangular patches visible on the underside of the aircraft. Each antenna has a large unobstructed field of view forward and to the side of the aircraft fuselage reference line. Six units (two each of the transmitter, receiver and RSP units) are located symmetrically in openings in each sidewall of the nosewheel well; the two RDPs are located one above the other in an opening in the aft wall of the nosewheel well.

The antenna is electronically steered in two dimensions and features a monopulse feed design to enable fractional beamwidth angular resolution. It includes a beam steering computer that determines and commands the phase settings of the beam steering phase shifters in response to a pointing direction command from the RDP. The antenna is fitted with a Smiths Industries' motion sensor subsystem, which is a modified strapdown inertial platform used to measure antenna motion to enable compensation during SAR mode operation. The antenna is equipped with its own power supply and is liquid cooled. It is designed to have carefully controlled and very low scattering performance (low radar cross-section) with respect to both in- and out-of-band Radio Frequency (RF) illumination.

The radar transmitter is a single unit, including dedicated high-voltage power supplies and utilising a liquid-cooled, high-power gridded travelling-wave tube RF amplifier, similar to other Hughes radars.

The receiver/exciter LRU performs several functions usually requiring more than one LRU in contemporary radars. These include generating RF waveforms for amplification by the transmitter (exciter) and amplification, detection and frequency down-conversion to baseband (receiver) of signals received from the antenna. The receiver/exciter also digitises the received signal stream and performs pulse compression to enhance range resolution.

The radar signal processor extracts target images and measurement information from the digitised signal stream and converts this information into a format usable by interfacing avionics or displays. In addition to a digital data output bus, it also has a video output bus that enables direct drive of cockpit displays. The RSP is fully programmable.

The RSP is a dual-central processor unit, general purpose-type computer. It is the command controller for all radar units and serves as the radar terminal on the B-2 avionics databus.

All radar system components communicate over a dual-redundant MIL-STD-1553 databus, and are hardened to withstand transient radiation and electromagnetic pulse effects. The radar was designed for very stringent environmental requirements exceeding those of other radars, due to the extreme Noise, Vibration and Harshness (NVH) environment expected during low-altitude flight in the rigid B-2 airframe.

Specifications

Frequency band: 12.5–18 GHz
Modes: 21 modes including terrain following and terrain avoidance
Line Replaceable Units (LRUs): 10
Weight: 955 kg
Volume: 1.485 m³
Power: 22 kVA AC; 500 W DC (peak)

Status

The AN/APQ-181 is in service in US Air Force B-2 Spirit strategic bombers. The USAF is embarking on a five-year spending plan from Fiscal Year 2004 (FY04) to address a potential frequency-interference issue, which concerns the possibility that the radar system will interfere with commercial satellite communications after 2007. When the USAF acquired the B-2 in the early 1980s, the issue of frequency interference was not widely foreseen. However, in recent years, more and more of the frequency spectrum has been made available to commercial users. At the end of 2007, the frequency in which the radar operates will be turned over to commercial users. Unmodified, the B-2 radar will disrupt their transmissions and could damage commercial communication satellites, for which, it is likely, the USAF would be liable, according to industry sources.

To address this, during October 2002, Northrop Grumman Corporation's Integrated Systems sector was awarded a USD34.2 million contract by the US Air Force for the first phase of the B-2 radar Pathfinder programme, a multiyear effort to design and integrate a new radar antenna on the Stealth Bomber. The total programme value for Northrop Grumman, prime contractor for the B-2 programme, is estimated at more than USD900 million to the company through to 2011.

Installation of the new radar antenna on the B-2 fleet is scheduled to be completed by the end of this decade. The modification consists of an Active Electronically Scanned Array (AESA) radar system that will resolve the aforementioned conflicts in radio frequency usage between the B-2 and commercial systems and, importantly, facilitate future upgrades to improve radar performance.

Initial phase effort, completed during 2003, consisted of system engineering leading to the establishment of performance requirements. The Raytheon Company's Space and Airborne Systems in El Segundo, California, which provided the original B-2 radar, is the principal subcontractor to Northrop Grumman on the radar programme.

During the second quarter of 2003, Northrop Grumman and Raytheon were awarded further contracts for radar upgrade work. Northrop was awarded a USD85.9 million contract for the establishment of specifications, the design of new and modified components and risk-reducing technical demonstrations, with work scheduled for completion by the end of August 2004. Raytheon was awarded a USD63 million contract for work with phase 2 of the APQ-181 upgrade's Component Advanced Development (CAD) effort, to include design and prototyping activities with regard to the proposed AESA antenna. This work was also due to be complete by the end of August 2004.

During September 2004, Northrop Grumman was awarded a USD388 million contract by the US Air Force for the next phase of the programme. During this System Development and Demonstration (SDD) phase, Northrop Grumman will develop and test the antenna and integrate six new radar systems on B-2 aircraft for initial demonstration and operational training. The subsequent production and deployment phase will include Low-Rate Initial Production (LRIP) and Full-Rate Production (FRP) to field the upgraded radar. Installation of the new antenna

The Northrop Grumman B-2 bomber carries the Raytheon Electronic Systems AN/APQ-181 radar
0103893

---

into the B-2 fleet will take several years to complete. In addition to replacing the antenna, Northrop Grumman will modify the B-2 defensive management system and the radar transponder to support the change in operating frequency. In addition to Raytheon, subcontractors include Lockheed Martin Corporation in Owego, New York, for defensive management system modifications and BAE Systems Information and Electronic Systems Integration in Greenlawn, New York, for radar transponder modifications. The radar modernisation and other efforts to upgrade the B-2 will improve its ability to communicate and exchange data with joint force commanders and share updated target information during a mission.

Under another programme, Northrop Grumman recently delivered the first B-2 with a specially formulated coating developed to improve the aircraft's combat readiness. Another significant improvement in the aircraft's capability is a bomb rack assembly that enables the B-2 to deliver up to 80 GPS-guided weapons on a single pass, five times as many as its current capacity.

Contractor
Raytheon Company, Space and Airborne Systems.
Northrop Grumman Corporation, Integrated Systems Sector.

AN/APS-124 search radar

Type
Airborne surveillance radar.

Description
The AN/APS-124 search radar was specially designed to be part of the comprehensive avionics suite for the US Navy Sikorsky SH-60B Seahawk ASW helicopter built to satisfy the Light Airborne MultiPurpose System (LAMPS) Mk III requirement. One of the problems associated with the operation of these medium-size helicopters from the 'Spruance' class destroyers on which they serve is that of stowage, particularly the height limitation. The APS-124 is therefore designed around a low-profile antenna and radome and consists of six LRUs.

Optimum detection of surface targets in rough sea is accomplished by several unique features including a fast-scan antenna and an interface with the companion OU-103/A digital scan converter to achieve scan-to-scan integration. The system is associated with a multipurpose display and with the LAMPS datalink so that radar video signals generated aboard the aircraft can be displayed on LAMPS-equipped ships.

The system operates in three modes covering long- and medium-range search and navigation and fast scan surveillance. The display ranges are selectable up to 74 km and the false alarm rate is adjustable to suit conditions.

The system is designed around the MIL-STD-1553 digital databus to communicate with other aircraft equipment and the modular design facilitates installation on other aircraft.

Specifications
Weight: 95 kg
Coverage: 360° azimuth
Display range: out to 160 n miles (selectable)
Pulse length:
(long range) 2 µs
(medium range) 1 µs
(short range) 0.5 µs
PRF:
(long range) 470 pps
(medium range) 940 pps
(short range) 1,880 pps
Scan rate:
(long range) 6 rpm
(medium range) 12 rpm
(short range) 120 rpm

Status
Service deployment of the LAMPS III Seahawk helicopter began in 1983. More than 300 systems are in service.

Contractor
Raytheon Company, Space and Airborne Systems.

AN/APS-125/-138/-139/-145 airborne early warning radars

Type
Airborne Early Warning (AEW) radar.

Description
The AN/APS-125/-138/-139/-145 series of airborne early warning radars have all been developed for the Northrop Grumman E-2C Hawkeye aircraft.

AN/APS-125
The AN/APS-125 radar was designed for the airborne early warning role, and operates in the UHF band. It was reported to have been the original fit in US Navy E-2C Group 0 aircraft, and in the E-2C Group 0 aircraft exported to Egypt, Israel and Japan, during the late 1970s and early 1980s.

AN/APS-138
The AN/APS-138 was an update of the AN/APS-125 during the mid-1980s. It reportedly included minor circuit changes and a low-sidelobe antenna, designated the Total Radiation Aperture Control Antenna (TRAC-A). AN/APS-138 radars were reportedly in production between 1983 and 1987, and the US Navy is understood to have updated AN/APS-125 radars to the AN/APS-138 standard. AN/APS-138 radars were also reportedly sold to Singapore on its E-2C aircraft.

AN/APS-139
The AN/APS-139 radar is a development of the AN/APS-138 that improves ECCM performance and surface surveillance capabilities, reportedly in production from 1987 to 1989. Reportedly no radars have been updated from AN/APS-138 to AN/APS-139 standard.

AN/APS-145
Latest in the series is the AN/APS-145. It is an update of the AN/APS-138 and AN/APS-139 radars, and has been retrofitted to all US Navy E-2C aircraft from 1990. The AN/APS-145 development reportedly specifically addresses the problem of overland clutter and provides fully automatic overland targeting and tracking capability, an improved IFF system, expanded processing and new colour displays. The present radar is said to perform very well over sea and desert, but to degrade rapidly when terrain becomes more rugged. To reduce the false alarm rate, a feature known as 'environmental processing' is being developed. This adjusts the sensitivity of the radar cell by cell, according to the clutter and traffic in each cell. To enable large aircraft to be detected at long range (up to 400 n miles), a new lower PRF is used and, to match this development, the E-2C's rotordome rotation rate is slowed from 6 to 5 rpm. A third PRF is also introduced allowing the radar to operate with different PRFs during scanning, to eliminate blind speed problems caused by single PRF operation. The AN/APS-145 entered service in 1992. It is fitted to US Navy E-2C Group II aircraft and to Taiwanese E-2C and to French E-2C aircraft.

E-2C Group II radar display   0018073

Lockheed Martin and Northrop Grumman have signed an MoU to collaborate in supplying airborne early warning and control (AEW&C) systems to overseas customers, based on Lockheed Martin supplying the AN/APS-145 radar and Northrop Grumman being responsible for integration of the mission system. This agreement applies to ADE&C variants of the Lockheed Martin C-130J and to other platforms, including the Northrop Grumman E-2C, but excluding the Boeing E-3 AWACS.

Status
Aircraft fits are reported to be as follows: E-2C Group 0 aircraft fitted with AN/APS-125/-138; E-2C Group I aircraft fitted with AN/APS-139; E-2C Group II aircraft fitted with AN/APS-145; all US Navy aircraft upgraded to AN/APS-145 configuration. Four US Coast Guard P-3 AEW&C aircraft are reportedly fitted with the AN/APS-138 radar.

Because the E-2C will serve the US Navy for many years, more capable Hawkeyes are on order. Northrop Grumman restarted its Hawkeye production line in 1994, after the US Navy ordered the first four of an expected 36 new Group II E-2Cs. In April 1996, the eighth and ninth new Hawkeyes were initiated under an advance procurement contract. In addition, Northrop Grumman is updating Group I Hawkeyes to Group II configuration for the US Navy. Modified aircraft are fitted with the AN/APS-145 radar system with fully automatic overland targeting and tracking capability, an improved IFF system, a 40 per cent increase in radar and IFF ranges, expanded processing capacity, new high-target-capacity colour displays, JTIDS for improved secure, anti-jam voice and data communications and GPS-based navigation capability.

From 2001, all US Navy production deliveries have been Hawkeye 2000s; in addition, all international customer deliveries are based on this configuration. Japan has conducted a phased upgrade of its E-2Cs to incorporate the Hawkeye 2000 technology standard.

France has ordered one and the Republic of China two. Meanwhile, Northrop Grumman is marketing the concept of an 'Advanced Hawkeye' for the period after 2010 based on use of an electronically scanned radar. The development radar will be incorporated into a Lockheed Martin C-130 AEW aircraft for flight testing.

Contractor
Lockheed Martin Naval Electronics & Surveillance Systems.

The AN/APS-145 surveillance radar has been retrofitted to US Navy E-2C Hawkeye AEW aircraft
0504237

AN/APS-130 radar

Type
Airborne Ground Mapping Radar (GMR).

Description
The 12.5–18 GHz AN/APS-130 radar was developed for the US Navy's EA-6B electronic warfare aircraft. Functions of the radar include:

- Search
- Ground mapping
- Tracking and ranging of fixed or moving targets
- Terrain-avoidance or terrain-following
- Beacon detection and tracking.

A track-while-scan capability provides simultaneous range, azimuth and elevation data for weapon delivery. As in other systems, range and azimuth markers must be placed on the target, but elevation data are available on a continuous basis and are derived from a separate phase interferometer array carried below the main scanner dish. The latter has a width of about 1 m and is illuminated by a conventional horn feed to produce a very narrow beam in azimuth.

The AN/APS-130 was reported operational on some US Navy EA-6B electronic warfare aircraft in 2001   0106205

The beam has a cosec^2 profile in elevation and this, with the interferometer elevation data provided, eliminates the need for mechanical scanning in the elevation plane. The interferometer array consists of two adjacent rows of 32 horns and moves with the main dish. Energy reflected from ground targets arrives at the upper and lower rows with a time difference, which is measured by phase comparison techniques and translated into angular information. Terrain profile data from the radar system is also presented on a vertical display for the pilot.

The system incorporates comprehensive built-in test facilities. System weight is approximately 227 kg.

Status
The radar was reported as operational during 2001, installed in subsequent new-build Prowlers and retrofitted to some of the existing inventory.

Contractor
Northrop Grumman Corporation, Electronic Systems Sector, Norden Systems.

AN/APS-131/APS-135 side-looking radars

Type
Airborne surveillance radar.

Description
The AN/APS-131 is a Side-Looking Airborne Radar (SLAR) used for the detection of ships and boats, for search and rescue, and for the detection of oil pollution. The aircraft installation consists of six main subassemblies: antenna, receiver/transmitter, synchroniser, amplifier, recorder/processor/viewer and control unit.

An area of up to 200 km on either side of the aircraft can be mapped when both arrays are in use and one can be selected if mapping of only one side is required. A range control determines the width of the target area to be mapped and presented on the photo-radar map. This control has four settings corresponding to 25, 50, 100 and 200 km wide scans by each antenna. When used in conjunction with the antenna switch to select either left, right or both arrays, maps corresponding to four standard scales can be presented on the display at 1:250,000, 1:500,000, 1:1 million and 1:2 million. Radar and aircraft operational data is annotated on the film. This data, together with the latitude and longitude printed on the film, helps the measurement of map co-ordinates for any feature observed on the radar image.

Status
The AN/APS-131 was developed under contract to the US Coast Guard and is in service on HU-25A aircraft. A similar radar, the AN/APS-135 is used on the US Coast Guard HC-130.

Contractor
Motorola Inc, Government & Systems Technology Group.

AN/APS-133 radar

Type
Airborne multi mode radar.

Description
The AN/APS-133 digital colour radar is a high-performance weather, beacon-homing and terrain-mapping system designed for large commercial and military transports. The RDR-1FB (Type 1) was originally launched for the retrofit of the US Air Force's C-141 Starlifter fleet and has since been installed on the C-5, E-3 and KC-10 aircraft among others.

The multicolour display can be used in conjunction with other equipment to show

programmable checklists or to superimpose navigation or other information on the weather map. The system employs digital processing and microcomputer techniques, as well as a solid-state modulator.

In November 1984, the company delivered to the US Navy and Marine Corps the first units of the RDR-1FB(M) (Type 2) improved land-mapping version for its fleet of KC-130 tankers and C-130 transports; the unit now equips the entire fleet of 70 aircraft. It was specially suited to US Marine Corps requirements, with a high PRF, short pulsewidth, enhanced digital processor and selectable sector-scan antenna to improve radar navigation at low level.

The US Air Force has selected the Type 2 for its E-4 NEACP and VC-25A (Air Force 1 Boeing 747) aircraft and the US Navy uses the unit in its EA-6A Intruder aircraft fleet.

Significant landmarks and continental shorelines up to 555 km away can be portrayed in the ground-mapping mode by using the high-power output concentrated into a pencil beam. At the same time, discrete details such as lakes, rivers, bridges, runways and runway approach reflectors readily show up on the colour display. To improve range resolution at short ranges the system operates with 0.4 μs (RDR-1FB(M)) or 0.5 μs (RDR-1FB) pulses in contrast to the 5 μs pulses used for long-range ground- and weather-mapping.

In the air-to-air mode, the APS-133(V) detects and tracks other aircraft during rendezvous, formation and air refuelling. Aircraft of C-130/C-141 size can be tracked to 56 km, but may still be resolvable at ranges as little as 550 m depending on relative bearing, aspect and altitude.

To provide long-range homing to remote ground destinations or tanker aircraft, the APS-133(V) operates at I-band frequencies – 9,375 MHz for beacon interrogation and 9,310 MHz for beacon reception. The identification of closely spaced pulse reply codes at long ranges is made possible by the marker and delay modes of the radar indicator. In the marker mode a variable marker is positioned on the screen just in front of the beacon reply. When switched to the delay mode the display presentation starts at the marker range. The range switch can then be moved to select a shorter range scale, yielding an expanded view of the area containing the beacon reply.

Derived from the Bendix/King RDR-1F used on many hundreds of airliners, the AN/APS-133(V) comprises five LRUs: a 762 mm fully stabilised split-axis parabolic antenna that provides specially shaped search or fan beams for terrain-mapping and skin painting, a transmitter/receiver, a colour display, a radar control unit and an antenna sector-scan control unit (RDR-1FB(M) only).

Specifications
Weight:
(antenna) 15.8 kg
(transmitter/receiver) 22.2 kg
(sector-scan control unit) 1.2 kg
(colour indicator) 6.3 kg
Frequency: I-band (9,375 MHz transmit, 9,310 MHz receive)
Power output: 65 kW
PRF:
200 pps (Type 1 system)
200 and 800 pps (Type 2 system)
Pulsewidth:
(weather) 5 μs
(beacon) 2.35 μs
(mapping) 0.5 μs (or 0.4 μs Type 2, selectable)

Status
In service with US Air Force transport aircraft, notably C-5A, C-17 and C-141 and KC-10 Extenders, also E-4A, E-3A, VC-25A and in US Navy/USMC aircraft such as C/KC-130, EA-6A, E-6A, A-3 and YP-3C.

Contractor
Honeywell Aerospace, Sensor and Guidance Products.

AN/APS-133(TTR-SS) multimode radar system

Type
Airborne multi mode radar.

Description
The AN/APS-133(TTR-SS) multimode radar is a coherent pulse Doppler system. It is designed for military tanker and transport aircraft.

The APS-133's advanced digital signal processing architecture provides operational advantages like frequency agility, pulse compression and Doppler beam-sharpening with monopulse resolution enhancement. The system has a precision ground-mapping capability and the capability to detect and display weather and turbulence. To aid in differentiating between mountain shadows and low reflectors such as lakes, and to allow detection of ridge lines, the system utilises a selectable fast time constant. When selected, this gives the appearance of a three-dimensional picture. Both the receiver/transmitter and the digital processor provide fault isolation to the LRM level, and fault isolation and storage to the LRU level.

The APS-133(TTR-SS) features calibrated turbulence detection and display on all range settings, signal attenuation compensation for more accurate weather display, solid-state digital design and independent roll axis stabilisation.

The APS-133 (TTR-SS) multimode radar is designed for military tanker and transport aircraft   0503725

There is automatic beacon decoding for APX-78 and APN-69 beacons, monopulse operation for separating closely spaced targets, and a freeze-frame facility. The system has a MIL-STD-1553B dual-digital bus for FMS and INS interface, antenna stabilisation amplifiers for pitch, roll and tilt and multishade monochrome display capability with colour enhancement. It can interface with and display station-keeping data from the AN/APN-169C and IFF data from the AN/APX-76.

Specifications
Weight:
(total system) 62 kg
Power supply: 115 V AC, 275 W
5 V AC, 10 W (panel lighting)
28 V DC, 100 W

Contractor
Honeywell Aerospace, Sensor and Guidance Products.

AN/APS-134(V) radars

Type
Airborne surveillance radar.

Description
The APS-134(V) anti-submarine warfare and maritime surveillance radar is the international successor to the US Navy's AN/APS-116 periscope detection radar. The APS-134(V) incorporates all the features of the former system while improving performance and adding capabilities, including a new surveillance mode.

The heart of the radar is a fast-scan antenna and associated digital signal processing which, says Raytheon Systems Company, form the only proven and effective means of eliminating sea clutter. This technique is used in two of the three operating modes, the third being a conventional slow scan for long-range mapping and navigation. The transmitter power is 500 kW.

In Mode 1, periscope detection in sea clutter, high-resolution pulse compression is employed with a high PRF and a fast-scan antenna, actual values being 0.46 m, 2,000 pps and 150 rpm. Display ranges are selectable to 59 km. There is an adjustable false alarm rate to set the prevailing sea conditions and scan-to-scan processing is employed.

Mode 2, long-range search and navigation, operates at medium resolution and with a low PRF, low scan and display ranges selectable to 278 km. Actual values are 500 pps and 6 rpm.

Mode 3 operates, again at high resolution, for maritime surveillance. A low PRF (500 pps) is used in conjunction with an intermediate scan speed of 40 rpm. Display ranges are selectable to 278 km and an adjustable false alarm rate is used together with scan-to-scan processing.

The system is also available in an offline configuration, with its own 10 × 10 in (254 × 254 mm) CRT control/display unit. Online operation linked in with other aircraft systems is accomplished via a MIL-STD-1553 digital databus, with the digital scan converter providing raster scan video for other aircraft displays. The weight of the entire APS-134(V), including the waveguide pressurisation unit, is 237 kg. The equipment is compatible with the inverse synthetic aperture radar techniques developed by Raytheon Systems Company for long-range ship classification.

Status
In service with Dornier/Dassault Aviation Atlantique ASW aircraft of the German Navy, Republic of Korea Navy and Pakistan Navy P-3C aircraft, Portuguese Air Force P-3P aircraft and as part of the Royal New Zealand Air Force Lockheed Martin P-3B Orion update programme.

The AN/APS-134(V)6 radars on the Republic of Korea Navy P-3C aircraft are to be updated by the addition of ISAR (Inverse Synthetic Aperture Radar) capability. Some AN/APS-134(V) radars are also being upgraded to a so-called AN/APS-137(V)6 standard.

AN/APS-134(V)7 radar
The AN/APS-134(V)7 offers technological improvements over its predecessor, the AN/APS-134(V). It features the developments of the US Navy's AN/APS-116 family of periscope detecting radar systems.

The AN/APS-134(V)7 radar system includes the periscope detection, long-range maritime surveillance and navigation modes from the AN/APS-134 radar. To that are added capabilities for improved periscope detection, advanced digital signal processing, multiple track-while-scan, dual-channel digital scan conversion, ESM countermeasures and 0-level built-in radar system diagnostics. There is also an optional capability to record radar and FLIR video.

The system is designed to detect small targets in high sea states at long range. Long-range performance is achieved by using a 500 kW high-power transmitter, 35 dB high-gain antenna and custom-developed low-noise preamplifier of less than 3 dB noise figure. These features provide the signal-to-noise ratio necessary to achieve long-range capability. Detecting small targets in the sea clutter environment is an inherent problem of maritime surveillance radar. To overcome this limitation, pulse compression and scan-to-scan processing are employed in the AN/APS-134(V)7.

The AN/APS-134(V)7 features simultaneous 32-target high-resolution tracking, 360° capable PPI coverage with sector scan, picture within picture PPI and B scan display format, dual-channel multilevel digital scan conversion and advanced digital signal processing. Multiple radar configurations are available for offline, MIL-STD-1553B and ANEW databusses. The system is adaptable to multiple radar video display configurations. The AN/APS-134(V)7 interfaces with other aircraft systems such as IFF, FLIR and ESM.

Status
In production for the Fokker Maritime Enforcer Mk 2 patrol aircraft.

Contractor
Raytheon Company, Space and Airborne Systems.

AN/APS-137(V) inverse synthetic aperture radar

Type
Airborne surveillance radar.

Description
The APS-137(V) is an improved version of the AN/APS-116 periscope detection radar which is standard on the US Navy's S-3 aircraft. Over 200 units of the AN/APS-116 I/J-band radar have been produced and the APS-137(V) introduces an Inverse Synthetic Aperture Radar (ISAR) mode. Funding for this development, which increases radar processing and introduces a standard surveillance and automatic classification capability, started in 1982 and was scheduled to continue into 1994 as part of the avionics upgrade which denotes the S-3B version.

The APS-137(V) offers long-range detection and classification of ships, the radar producing a recognisable image of the target vessel. The image is derived from the Doppler shifts of the returns, compared with the reference level. Using pulse compression and fast scan processing to eliminate sea clutter, the APS-137(V) provides improved periscope detection and high-altitude maritime surveillance. Multiple target tracking (track-while-scan) is also available.

The APS-137(V) is compatible with the seekers in Harpoon, Tomahawk and other missiles, and can interface directly into weapons system computers. By comparing images before and after an attack, target battle damage can be assessed.

Status
Full production for the S-3B retrofit programme commenced in 1987. In January 1987, Raytheon Electronic Systems received a contract to supply AN/APS-137(V) radars to equip US Navy S-3B and P-3C and US Coast Guard C-130 aircraft, with deliveries extending to 1992. This was extended at the end of 1987 to include a further nine sets and long lead items for 16, to equip the S-3B. A further contract was awarded in August 1991 for 24 AN/APS-137(V) systems.

The AN/APS-137B(V)5 radar forms part of the P-3C Anti-surface warfare Improvement Programme (AIP) for 146 US Navy P-3C Update III-equipped aircraft.

Contractor
Raytheon Company, Space and Airborne Systems.

AN/APS-143(V)1 and (V)2 radars

Type
Airborne surveillance radar.

Description
The AN/APS-143 radar is designed to provide aircraft and aerostats with good detection capability against small targets in very high sea states. It is a lightweight, travelling wave tube radar with pulse compression. This provides better detection, resolution and clutter rejection, higher average power and greater receiver/transmitter reliability. The system includes a signal processor incorporating track-while-scan, scan conversion and databus interfaces.

Specifications
Weight:
(inc display) <110 kg
Frequency: 9.3 to 9.5 GHz
Agility: fixed, or can operate with 11 agility steps of around 20 MHz over the band
Peak power: 8 kW min, 10 kW nominal
Effective peak power: 700 kW up to 93 km range; 2.4 MW above
Compression ratio: 70:1 at ranges up to 93 km; 240:1 above
Pulsewidth: 5.0 and 17.0 µs uncompressed; 0.1 µs compressed (weighted)
PRF: 2,500, 1,510, 750 or 390 Hz (range dependent)

Status
In production. In operation with US government aerostats and US Air Force de Havilland DHC-8 aircraft used for test range surveillance. The Malaysian Air Force has taken delivery of the APS-143 for Beech 200T aircraft, as has the Japanese Maritime Safety Agency on the Saab 340B.

Contractor
Telephonics Corporation, Command Systems Division.

AN/APS-143(V)3 and AN/APS-143B(V)3 sea surveillance radars

Type
Airborne surveillance radar.

Description
The AN/APS-143(V)3 is a maritime surveillance and tracking radar designed for installation in a variety of fixed-wing aircraft and helicopters. It is also known as OceanEye. The system uses frequency agility and pulse compression

techniques and consists of three units: an antenna, receiver/transmitter and signal processor. Radar control is via a dedicated control panel with on-screen controls, or by a central universal keyset via MIL-STD-1553B databus. Features include TWS for 30, 100 or 200 targets, air search with MTI, integrated electronic support and IFF system interfaces and electronic ECCM provision (including sector blanking and staggered pulse repetition frequencies).

The flat-plate planar antenna array, which can be fitted into any radome, is stabilised for ±30° in pitch and roll. The transmitter is a TWT type with a peak power output of 8 kW.

The latest variant, APS-143B(V)3, can be upgraded with a complete imaging capability: range profiling, ISAR, spotlight SAR, strip-map SAR. The system can also incorporate software interfaces, via an embedded Tactical Data Management System (TDMS), for external systems such as FLIR, ESM, IFF and TDL. The TDMS capability also includes overlay of worldwide Database II or vector shoreline maps onto the radar display.

Specifications
APS-143B(V)3
Antenna: Flat plate planar array with V or H polarisation
Antenna stabilisation: –40 to +10° (roll and pitch)
Sector scan: 45–300° or continuous 360° (operator selectable)
Transmitter: Helix type TWT
Frequency: I-band (9.2–9.7 GHz)
Frequency agility: 450 MHz
Peak power: 8.5 kW (min)
Pulsewidth: 5 or 17 μs to 40 μs for B(V)3, compressed width 100 ns (0.1 μs); and 23.4 μs compressed to 6.5 ns for B(V)3 imaging.
PRF: 2,500, 1,500, 800 or 400 Hz
ECCM: PRF, jitter, sector blanking
Max range: 370 km
Compression ratio: 50:1/170:1 (1:1 to 3,600:1 for B(V)3)
Range resolution: 15 m
Azimuth accuracy: 0.5°
Image resolution: 1 m
Video output: monochrome or colour (user selectable); RS-170, RS-343, CCIR601, SVGA-SXGA
Weight: 81.8 kg

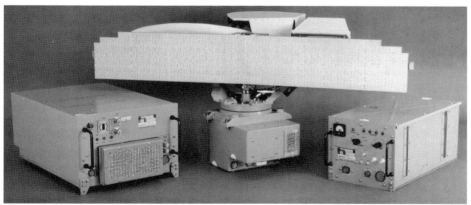

The AN/APS-143(V)3 is in service in S-70 helicopters, shown with integrated ESM antennas 0504055

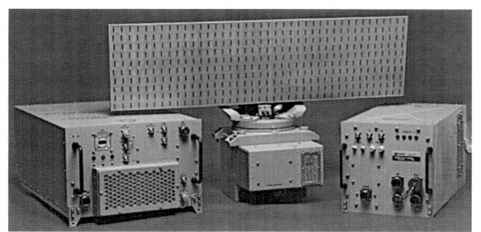

The APS-143B(V)3 radar system 0044997

Status
The APS-143(V)3 and APS-143B(V)3 are the latest variants of the APS-143 family. Worldwide, APS-143(V)- radars are installed or have been selected for a large number of platforms:

APS-143(V)3: Two CATPASS 250 aircraft of the Ecuadorean Navy, five SH-2G helicopters the New Zealand Navy and twenty S-70C(M)-1 helicopters of the Taiwanese Navy.

APS-143B(V)3: Royal Australian Navy SH-2G helicopters (with imaging capability), South African Super Lynx 300 helicopters (selected); orders placed by the Danish Air Force (three systems for CL-604 Challenger aircraft), US Coast Guard (six systems for HU-25A aircraft) and US Government (two systems for the C-26).

Contractor
Telephonics Corporation, Command Systems Division.

AN/APS-147 multimode airborne radar

Type
Airborne multimode radar.

Description
The AN/APS-147 multimode radar, designed for the US Navy SH-60R helicopter, is an Inverse Synthetic Aperture Radar (ISAR) which uses the latest in high-throughput signal and data processing. Flexibility through programmability provides a product optimised for the maritime surveillance mission.

Advanced processing allows the AN/APS-147 to use a variety of waveforms to perform its mission at an output power substantially lower than traditional counterparts in maritime surveillance radars. This results in a radar with an extremely Low Probability of Intercept (LPI). Using a low peak power waveform with frequency agility, the radar can detect medium- to long-range targets without the threat of ESM interception.

Radar modes include target imaging, small target and periscope detection, long-range surveillance, weather detection and avoidance, all-weather navigation, short-range search and rescue, enhanced LPI search and target designation.

The AN/APS-147 features a flexible modular design which can be tailored to meet specific requirements and can be easily upgraded, LPI, high-resolution images for rapid classification, lightweight construction through the use of composite materials, low-input power, simple design for high reliability and maintainability, fully programmable signal processor with multiple waveform exciter and high-throughput rates and integrated IFF and SAR option.

Status
Telephonics Corporation delivered its first AN/APS-147 radar to Lockheed Martin Federal Systems, Owego for integration and test aboard a US Navy SH-60R helicopter in June 1999.

Contractor
Telephonics Corporation, Command Systems Division.

AN/APY-2 radar for the E-3 AWACS

Type
Airborne Early Warning (AEW) radar.

Description
The Boeing Company is prime contractor to the US Air Force for the E-3 programme. Northrop Grumman's Electronic Sensors and Systems Sector, based in Baltimore, Maryland, builds the production AN/APY-2 surveillance radar, which is installed in a Boeing 707-320 aircraft modified with a large radome for the radar and other antennas. The AN/APY-2 surveillance radar is also installed on the Boeing 767 AWACS aircraft. The radar transmitter and receiver, communications gear, and other command and control systems are located inside the aircraft.

The E-3 radar provides full, long-range surveillance of high or low-flying aircraft. A maritime capability provides surveillance of moving or stationary ships. The E-3 aircraft operates during all kinds of weather and above all types of terrain.

The E-3 surveillance system detects and tracks both enemy and friendly aircraft in a large volume of air space. Low-flying aircraft, which can escape detection by ground-based radar, are detected by the E-3 aircraft.

The AN/APY-2 high-PRF pulse Doppler radar with its digital signal processing provides the look-down surveillance capability that is the key to the E-3 airborne warning and control system.

The AN/APY-2 radar looks down at the ground, distinguishing between ground reflections (clutter) and radar returns from aircraft hugging the ground to escape detection. In addition to detecting either high- or low-flying aircraft and ships out to the aircraft horizon, the E-3 radar also offers long-range aircraft surveillance above the horizon. For an E-3 aircraft altitude of 9,000 m (30,000 ft), the radar look-down range to the horizon is about 400 km (245 miles).

The E-3 aircraft provides a real-time assessment of both enemy action and friendly resources. With an E-3 aircraft, the airborne commander has available the information that is needed to detect, assess and counter an enemy threat.

High-PRF pulse Doppler radar – concept
The basic advantage of high-PRF Doppler radar is that it provides better Doppler separation of moving targets from ground clutter. Additional radar techniques minimise ground clutter and maintain high sensitivity. Some of these are an extremely low-sidelobe antenna, ultrastable frequency generation and digital processing techniques.

An inherent feature of the E-3 radar design is its flexibility to accommodate future growth through software control. Many signal processing functions are accomplished with programmed instructions rather than the specific hard-wired arrangements of circuitry so that functions or processes can be updated by altering programme instructions. These are programmable in flight; thus tactical programmes can be adjusted to respond to changes in the tactical situation.

Radar operating modes

The target-handling capability of the E-3 radar is enhanced by its operation in various modes, depending on the nature of the tactical situation. Each 360° radar scan can be divided into as many as 24 azimuth sectors and each sector can be operated with its own set of operation modes. Available radar modes are as follows:

1. pulse Doppler non-elevation mode scans with high-PRF Doppler to provide look-down surveillance of aircraft out to the radar's horizon, but it does not measure target elevation.
2. pulse Doppler elevation scan is similar to the previous mode, but includes an electronic vertical scan of the radar beam to provide target elevation.
3. beyond-the-horizon mode of operation is used for long-range surveillance of medium- and high-altitude aircraft. Since the radar beam is above the horizon, there is no ground clutter and a low-PRF radar pulse is used to obtain the range and azimuth of target aircraft.
4. passive scanning mode operates with the radar transmitter 'off' and the receiver 'on' to obtain ECM information, such as the locations of enemy jammers.
5. maritime mode uses a very short radar pulse to provide the high resolution required to detect moving and anchored surface ships.

Multimode operation and sectoring permits maximum potential of the radar to be concentrated in sectors where the need is greatest. The pulse Doppler mode and beyond-the-horizon mode can be used simultaneously in an interleaved manner, as can the maritime and pulse Doppler non-elevation modes.

Technological features

To permit cancellation of the main-beam clutter in the radar receiver, extremely stable signal generation and advanced signal processing are used. The improvements in signal generation capability necessary for the E-3 radar were made through advances in circuit components, including oscillator crystals and transmitter tubes.

The signal processing advances were made possible by the application of digital signal processing and by analogue-to-digital converter developments for the E-3 application. Radar control is accomplished by operating commands from the central computer.

Much of the high performance achieved by the E-3 pulse Doppler radar derives from the antenna and rotodome design. Low antenna sidelobes are necessary to minimise sidelobe clutter return and avoid reduction in detection performance for target returns that do fall within the sidelobe clutter region. The low sidelobe patterns that have been achieved in the E-3 radar antenna and rotodome represent a significant aspect in antenna design. The advance was made possible by the application of digital computing techniques to the design of the antenna, and by the use of high-precision digital techniques in manufacturing the antenna. The Boeing rotodome design enabled the low-sidelobe characteristics of the antenna to be maintained when radiating through the rotodome.

In addition to minimising sidelobe clutter, the low-sidelobe antenna design is also a major contributor to the radar's resistance to jamming. The highly directional nature of the antenna when receiving, rejects a jamming signal except when the antenna is pointed directly at the jamming sources. Hence, jamming signals can be easily identified and located.

E-3 radar Electronic Counter-CounterMeasures (ECCM) capabilities

The E-3 airborne surveillance radar system, with its inherent ECCM capabilities, can provide effective command and control under jamming conditions.

The E-3 ECCM features, such as low antenna sidelobes and inherent chaff rejection, are an integral part of the radar design, rather than add-on 'fixes'. This results from the pulse Doppler radar design to eliminate the effect of ground

A UK RAF E-3D escorted by two Tornado F3 interceptors   0131118

The rotodome of the Boeing E-3A Sentry houses the AN/APY-2 surveillance radar scanner   0503742

returns (clutter) and the requirement for the E-3 to operate effectively against a variety of electronic countermeasures.

Filtering is used in the pulse Doppler radar receiver to remove mainbeam clutter. This filter also eliminates the radar signals received from stationary or slowly moving targets, such as chaff.

Another form of ECCM available is the ability to switch to a radar frequency which is not being jammed. This negates the jammer effect even though the jammer remains within the radar line of sight.

The primary E-3 radar ECCM capability is derived from the very low antenna sidelobes. Flight tests of the E-3 radar, integrated with other E-3 subsystems such as a computer and tracking display consoles, have shown that the E-3 can operate successfully against powerful airborne and ground jammers.

The E-3 radar also has the ability to determine the relative bearing of a jammer with the radar in either an active or passive mode. The mobility of the E-3 enables it to perform self-triangulation. The system computer indicates the position data on an operator's console and computes the intercept path of the fighter aircraft designated to attack the jammer.

E-3 mobility also permits the use of jammer avoidance tactics to reduce or eliminate the jammer effects. The E-3 aircraft can drop below the radar horizon in relation to the jammers, thereby eliminating reception of the jamming energy. This tactic would result in some loss of low-altitude coverage but will enable target detection above the horizon line of sight to continue unaffected by the jammers.

Simultaneous operation of two or more E-3 aircraft enhances the overall system effectiveness against jammers and non-jamming targets. The E-3 has the ability to transmit information via a datalink system to other E-3 aircraft or to ground stations, producing a synergistic effect and improving the surveillance and battle management functions.

E-3 radar design

Radars for E-3 aircraft consist of three major subsystems: the slotted planar-array antenna located in the rotodome on top of the aircraft, the radar receivers and processors located in the centre of the aircraft cabin, and the radar

transmitter located in the lower cargo bay. Total weight of the radar system is 3,742 kg.

The slotted planar-array antenna rotates with the aircraft rotodome at six rpm to provide horizontal radar scanning. The antenna face consists of 30 slotted waveguide sticks and measures 7.3 m × 1.5 m.

Vertical scanning and height-finding are performed by electronic scanning techniques using ferrite phase shifters. The phase shifters, phase-control electronics, receiver protectors and receiver paramplifiers are mounted on the back of the antenna. The phase shifters are located on one side of the antenna and the electronics on the other side for weight balance and easy access during maintenance. Northrop Grumman-developed receiver protectors use a radioactive igniter power source for long life and fail-safe operation even during radar shutdown.

The radar transmitter consists of eight pressurised vessels located in the lower cargo bay. An overhead rail system permits easy removal of the transmitter units through a lower hatch.

The high-power transmitter chain is completely redundant with an inflight switchover capability should a malfunction occur. This transmitter redundancy, along with extensive redundancy in other parts of the E-3 system, assures high radar reliability and high probability of E-3 mission success.

Only two tubes are used in the transmitter chain: a high-power Klystron and a travelling wave tube driver. All other elements of the transmitter are solid state.

The radar receiver and digital signal processor are located in a single cabinet in the centre of the E-3 aircraft. Digital printed circuit boards and receiver electronics are accessed through hinged doors. Critical circuitry is backed up by redundant circuit boards.

A radar-dedicated digital computer and high-speed digital data processor control and monitor radar operation, reject ground clutter from radar returns, perform frequency analysis on signal returns, correlate radar returns to determine presence of legitimate targets, and digitally format the output target reports in range, velocity, azimuth and elevation.

Digitised radar information is passed in near-real-time to the E-3 central processor and

The slotted planar-array antenna of the AN/APY-1/2 radar   0018075

displays. The central processor correlates radar reports over successive scans to form target tracks. Navigation inputs are supplied to the radar computer to adjust for aircraft motion and altitude.

Digital radar signal processing provides significant advantages over conventional analogue processing in terms of cost, weight, complexity, reliability, maintainability and operability. Since the data processor is software programmable, system flexibility and room for potential system growth to meet changing environments also are advantages.

Use of high-reliability components throughout the radar, digital technology, use of integrated circuits and functional groupings of circuits, redundancy of critical circuitry with automatic switchover, built-in-test and fault isolation to an individual circuit board, and inflight maintenance capability all add up to make the E-3 radar highly reliable and easy to maintain.

Built-in test constantly monitors radar operation under control of radar computer software. Normal system operation is interspersed with fault detection tests. A 98.5 per cent probability of detecting online faults was demonstrated by the radar. In the event of a malfunction, the radar computer automatically reconfigures the system using available redundant circuitry. After correction, the results are displayed to the radar maintenance technician.

Spares for most non-redundant portions of the E-3 radar are carried on the aircraft and substituted while in the air. This attention to radar reliability and easy maintenance assures a high probability of mission success for the E-3 aircraft. Operational availability has consistently been greater than 95 per cent in the AWACS worldwide operation.

AWACS Radar System Improvement Programme (RSIP)

Northrop Grumman's Electronic Sensors and Systems Sector (ESSS), under a contract worth more than USD350 million from the US Air Force Materiel Command's Electronic Systems Center, undertook the Engineering Manufacturing Development (EMD) phase of an E-3 AWACS Radar System Improvement Programme (RSIP). RSIP is a joint Air Force/NATO programme.

The RSIP contract includes the design, development and flight test of improvements to the AWACS AN/APY-1 and -2 radars to maintain operational capability against the growing threat from smaller radar cross-section targets, cruise missiles and electronic countermeasures. The contract also includes significant improvements in the man-machine interface and reliability and maintainability. RSIP represents the most significant upgrade to the E-3 radar since its development in the early 1970s.

The major portion of RSIP is devoted to increasing radar sensitivity against small targets through replacement of the digital Doppler

processor and radar data correlator with a state-of-the-art Surveillance Radar Computer (SRC), and translation of the associated software into Ada language. In addition to handling the upgraded radar processing load, the SRC will contain adequate growth reserves to accommodate further radar upgrades.

Improvement in the man-machine interface will result from modification of the radar control and maintenance panel by incorporation of a spectrum analyser, special test equipment and new displays for monitoring the surveillance environment as well as the maintenance status of the radar system.

The RSIP EMD programme included fabrication of the first five modification kits for the US Air Force and one modification kit for NATO.

Under the initial production contract, Boeing and Northrop Grumman (ESSS) were to build 18 modification kits for NATO, four for the US Air Force, seven for the UK. Follow-on options have been exercised so that RSIP production for 43 of the total production of 70 aircraft has been approved. All 43 systems are scheduled to be operational by 2002.

RSIP kit installation is by US Air Force at Tinker AFB for US Air Force; DaimlerChrysler Aerospace AG for NATO; BAE Systems (Operations) Limited for UK.

Status

The E-3 AWACS became operational with the US Air Force in early 1978. To date, 34 E-3s have been delivered to the US Air Force, 18 to NATO, and five to the Royal Saudi Air Force. Production of seven E-3 radars for the United Kingdom and four E-3 radars for France has also been completed. Additionally, the first 4 radars for the Japanese 767 AWACS have been delivered, with 2 of the additional Japanese 767 systems becoming operational in early 1999.

The RSIP has been installed on all 18 NATO aircraft, all seven UK Royal Air Force aircraft and three US Air Force aircraft.

In May 2000, the US Air Force awarded Raytheon Electronic Systems a contract to incorporate 435 Common Large Area Display Sets (CLADS) in the US Air Force AWACS aircraft.

Contractor

Northrop Grumman Corporation, Electronic Systems Sector.

AN/APY-6 airborne surveillance radar

Type

Airborne surveillance radar.

Description

The AN/APY-6 is a high-performance airborne Synthetic Aperture Radar (SAR) that provides

Ground Moving Target Indication (GMTI) and sea surveillance capability. It has been developed as a technology demonstration system for the US Office of Naval Research. It had its first flight in the first quarter of 1999.

The AN-APY-6 radar uses a Travelling Wave Tube (TWT) power amplifier and a flexible planar-array antenna that can be configured for forward looking, side looking or 360° cover in nose or belly-mounting locations. The receiver is a 4-channel unit that provides three MTI channels and one SAR channel.

Precision location of fixed and moving targets is achieved, with slow speed, clutter suppression MTI, to provide single pass two- or three-dimensional strip map data at 0.3 m resolution.

System architecture is based on an Ethernet with fibre channel interconnectivity and industry-standard programming in 'C'.

The AN/APY-6 system is air-cooled and weighs less than 205 kg.

The technology applied to this project is reported to complement Northrop Grumman work on the Joint Strike Fighter (JSF) Multifunction Integrated Radio Frequency System (MIRFS) technology demonstrator.

The project name Gray Wolf has been linked to this technology demonstrator programme.

Contractor

Northrop Grumman Corporation, Electronic Systems Sector, Norden Systems.

AN/UPD-8/-9 and AN/APD-14 side-looking reconnaissance systems

Type

Airborne surveillance radar.

Description

AN/UPD-8/-9

The AN/UPD-8 and AN/UPD-9 are updated and improved versions of the AN/UPD-4, with much greater range. The AN/UPD-8 is a side-looking, synthetic aperture airborne reconnaissance radar system for the US Air Force. The system provides an all-weather, day/night, standoff, tactical and strategic reconnaissance capability and, with its extended-range antenna pods and airborne datalink electronics, is the latest in a sequence of Lockheed Martin airborne reconnaissance radars.

Operating in the I/J-band, the system records with equal effectiveness during daylight and darkness and through periods of adverse weather that would render other types of reconnaissance systems ineffective. The airborne radar system's standoff capability, which is increased by 60 per cent in the new equipment, allows coverage of border regions, harbours and coastal areas without violating another nation's airspace or national waters. Maximum range, limited by power, is about 130 km. At such distances, resolution is 6.1 m in azimuth and 4.6 m in range but resolution increases at shorter ranges. At 50 to 90 km the radar resolution is 3 m, allowing large aircraft to be identified and showing clear images of vehicle formations.

Lockheed Martin has also developed a pod-mounted version of the UPD-8. Similar in size to a 1,250 litre fuel tank and weighing less than 682 kg, the pod includes the main antenna, support electronics and a datalink. A version of this pod has been flight tested on an RF-4B and is to be supplied for US Marine Corps F/A-18(RC) reconnaissance aircraft. The podded system can also be installed on an executive jet aircraft.

Another upgrade of the UPD-4 is the AN/UPD-9 developed for the US Marine Corps. This is similar to the UPD-8, with a datalink, but does not have the long-range features. Basic standoff range is about 80 km.

The airborne sensor part of the overall system, the AN/APD-12, operates in eight modes selectable by the operator according to aircraft altitude and the distance to the target area. The system can record to left or right of the aircraft track with the normal antennas or with the extended-range antennas which are mounted in a centreline pod. Target information can be transmitted to the ground in real time through a datalink or can be recorded in flight for later processing on the ground. The former facility enables a ground commander, perhaps hundreds of kilometres away, to evaluate targets while the aircraft continues its mission.

Specifications
Volume:
(UPD-8) 1.305 m³
(UPD-9) 0.58 m³
Weight:
(UPD-8) 93.5 kg
(UPD-9) 321 kg
Frequency: I/J-band
Swath width: 18.5 km
Range: (UPD-8) 93 km, (UPD-9) 55 km
Resolution: 3 m

Status
In operational service with Israeli Air Force RF-4B aircraft.

AN/APD-14
Sandia National Laboratory has modified the AN/UPD-8 design to form the AN/APD-14 SAR for Open Skies (SAROS) set. The treaty on Open Skies specifies SAR resolution no better than 3 m resolution. SAROS, installed aboard an OC-135 aircraft, has a centre frequency of 9.6 GHz and a mapping swath of 10 n miles.

Contractor
Lockheed Martin Tactical Defense Systems.

ASARS-2 Advanced Synthetic Aperture Radar System

Type
Airborne surveillance radar.

Description
The high-resolution ASARS-2 side-looking radar was designed for the US Air Force U-2 high-altitude battlefield surveillance aircraft, and satisfies a US Air Force/Army requirement for a standoff intelligence gathering system. ASARS-2 was launched in 1977 under the designation AN/UPD-X and first test flown on board a U-2R reconnaissance aircraft in 1981. It is likely that the resolution approaches limits set by altitude, residual airframe vibration and atmospheric turbulence. The antenna is V-shaped to enable the ground on either side of the aircraft's track to be surveyed without the aircraft having to manoeuvre. Guidance for the ASARS-2 antennas is provided in part through several other sensors and information is transmitted via a datalink to a special ASARS-2 deployable processing station on the ground. Here, the signals are converted into strip-maps and spotlights and are available within minutes for air force and army commanders.

ASARS-2 consists of a search mode for wide area ground coverage, featuring both a Moving Target Indicator (MTI) and a fixed target indicator and spotlight mode, for high-resolution coverage of smaller areas (fixed targets only).

In 1988, Raytheon Electronic Systems was awarded USD20 million to develop an Enhanced Moving Target Indicator (EMTI) which, in part, was designed to add MTI to the previously mentioned spotlight mode. This entailed software modifications to the ground-based processing component and the addition of components in the airborne ASARS-2 receiver/exciter and processor control unit.

The radar has been gradually upgraded as part of the ASARS-2 Improvement Programme (AIP), with the current production ASARS-2 radar, designated ASARS-2A, providing coverage over an area four times greater and supporting spot imaging over an area nine times greater than the baseline design. In search mode, ASARS-2A has a resolution of 1 m over a swathe 2.8 n miles wide, 2 m over a 11 n mile swathe and 3 m over a 21 n mile swathe, compared to a fixed resolution of 3 m over a 100 n mile swathe for the baseline sensor. Spot mode resolution has been improved from 1 m to 30 cm over a 1 × 1 n mile area.

The latest U-2 ASARS-2A radar provides improvements to functionality and resolution over the baseline sensor   0036973

The ASARS-2A enhancements facilitate broad area synoptic coverage and the supply of complex imagery for Measurement and Signals Intelligence (MASINT) applications and supports real-time targeting of Precision Guided Munitions (PGMs) through the supply of highly accurate GPS coordinates.

Status
In service with US Air Force U-2 high-altitude reconnaissance aircraft.

The AIP, which began in June 1996, is divided into several phases. The initial phase (completed at the end of 1998), known as the contingency phase, consisted of integrating the legacy ASARS-2 electronic scanning antenna, transmitter and low-voltage power supply with two new units based on COTS hardware – the Receiver/Exciter/Controller (REC) and the onboard processor, based on a PowerPC commercial architecture. Flight testing, completed at the end of 1998, tested and evaluated the new Asynchronous Transfer Mode (ATM) interfaces and the hardware/software of the onboard signal/data processor and the REC. The ATM interfaces connect the radar to the air/ground datalink and provide COTS architecture foundation.

The next phase of the AIP, commenced in 2002, consists of enhancements to image quality, including the addition of an Inverse Synthetic Aperture Radar (ISAR) mode for three-dimensional imaging and coherent change detection.

Although not yet adopted as part of the AIP, further upgrades could include the adoption of an Active Electronically Scanned Array (AESA) antenna.

Contractor
Raytheon Company, Space and Airborne Systems.

ASSR-1000 surveillance radar sensor

Type
Airborne surveillance radar.

Description
The ASSR-1000 is a radar sensor designed for the long-range detection of both air and maritime targets. It provides detection over rough terrain and high sea states. It is lightweight and adaptable for installation on a variety of platforms, including the Sentinel 1000 and 1200 airships.

The ASSR-1000 utilises technology from the Northrop Grumman Low-Altitude Surveillance System (LASS), with a solid-state transmitter, signal processor and post-processor. The lightweight radar has been carefully adapted for the airship airborne application and is planned to fly at altitudes up to 15,000 ft, providing a line of sight of 222 km. It is a D-band fully solid-state coherent pulse compression system, featuring an ultra-lightweight antenna, fully solid-state

The antenna for the ASSR-1000 surveillance radar is 7 m wide and just under 4 m high   0503743

transmitter, frequency diversity and automatic target detection and tracking. It is capable of interfacing with operational displays, navigation equipment and datalinks. The ASSR-1000 includes an integrated AN/TPX-54 beacon interrogator.

It has a low probability of false alarm with a high probability of detection on small fluctuating targets, with a field-proven 60 dB moving target improvement factor over terrain and sea conditions. The antenna is a circular reflector, stabilised in both pitch and roll, with a wideband feed horn used for both the search and the beacon radars. The signal processor has special features that allow it to detect small targets in the surrounding environments. These features include full range I and Q processing, pulse compression, six pulse

canceller with variable time periods between pulses, constant false alarm rate processing and dual-frequency transmission selectable from 93 frequencies.

Specifications
Dimensions:
(antenna) 4,572 mm diameter
Weight: 850 kg nominal
Power: 30 kW
Frequency: 1,215 to 1,401 MHz
Pulsewidth: pulse pairs of 48 and 51 μs
  separated by 60 MHz
PRF: 375 pps average
Range:
(instrumented) 300 km
(2 m² air target) 278 km in clutter

(4 m² marine target) out to horizon in Sea
  State 3
Range resolution: 0.3 km
Azimuth coverage: 360°
Azimuth accuracy: 0.25°
Data refresh rate: 12 s
Reliability: >1,500 h MTBCF

Status
In production and in service with the US Customs Service and several overseas customers.

Contractor
Northrop Grumman Corporation, Electronic Systems Sector.

FMR-200X multimode weather radar

Type
Airborne weather radar.

Description
The Rockwell Collins FMR-200X Flight Multimode Weather Radar System is a standard Non-Developmental Item/Commercial Off-The-Shelf (NDI/COTS) ARINC 708A X-band coherent Doppler colour weather radar system. This system provides full precipitation detection, turbulence detection, forward-looking windshear detection and an additional active skin paint mode, capable of detecting tanker-size aircraft at ranges up to 15 n miles.

New multimode weather radar software has been designed and certified to RTCA DO-178B level D standards. The system has been qualified to meet environmental standards described in RTCA DO-160C (with a few exceptions). MIL-STD-461D is used for radiated emissions, antenna spurious and harmonics. All operations other than the skin paint mode, sixteen-level mapping and minor display and control bus modifications are defined in accordance with ARINC 708A

Specifications
WRT-701X
Receiver/transmitter
Size	8 MCU per ARINC 600
Weight	14.1 kg (maximum) including R/T mount
Cooling	Mount based fan
Power	200 W nominal at 115 V ac 400 Hz

WCP-701
Mode control panel
Size	10.3 H × 22.6 W × 23.6 D mm
Weight	0.77 kg (maximum)
Cooling	Natural convection
Power	5 W (maximum) (powered from R/T)

(airborne weather radar with forward-looking windshear detection capability). This system is based on the Collins WXR-700X Radar System that has over 6,600 installations worldwide.

The WXR-700X originally entered service in 1980 and was the first Air Transport airborne radar to incorporate a Doppler turbulence detection feature. In 1986 Rockwell Collins began a co-operative programme with NASA to expand the system's capability to detect

WMA-701X
Antenna pedestal
Size	Per ARINC 708
Weight	12.7 kg (maximum) (including flatplate)
Cooling	Natural convection
Power	100 W nominal at 115 V ac 400 Hz

WFA-701X
Flatplate antenna
Size	110 × 134 mm
Weight	3.1 kg (maximum)
Beamwidth	2.5° elevation × 3.5° azimuth
Sidelobe Performance	>30 dB

windshear/microburst events ahead of the host aircraft.

Status
The FMR-200X is in service in US Air Force KC-135 aircraft.

Contractor
Rockwell Collins.

Ground-mapping and terrain-following radar for the Tornado

Type
Airborne Ground Mapping and Terrain-Following Radar (GMR/TFR).

Description
The Tornado IDS radar has no official designation, but is often referred to as the Tornado Nose Radar (TNR). The TNR is also used in the ECR version of the Tornado.

The all-weather day or night radar comprises two essentially separate systems that share a common mounting, power supply and computer/processor. They are the Terrain-Following Radar (TFR) and the Ground-Mapping Radar (GMR). The first is used for automatic high-speed, low-level approach to the target and escape after an attack. The second is the primary attack sensor for the IDS Tornado and operates in air-to-ground and air-to-air modes to provide high-resolution mapping for navigation updating, target identification and fire control.

The radar enables the crew to fix the aircraft position by updating the Doppler-monitored inertial navigation system, provides range and tracking information for offensive or defensive weapon delivery, and commands, via the autopilot, a contour-hugging flight profile that reduces the chance of detection by hostile air-defence radars. ECCM, dramatically enhanced with the introduction of the Phase 1 upgrade improvement, is used to provide relative immunity from interference in severe ECM environments.

The three units comprising the system are the radar sensor (transmitter/receiver package for TFR and GMR), a digital scan converter and a radar display unit in which a moving-map image

can be superimposed on a radar picture for navigation updating and target identification.

The GMR operates in the J-band frequency band with nine modes: readiness, test, ground-mapping, boresight contour mapping, height-finding, air-to-ground ranging, air-to-air tracking, land/sea target lock on and beacon homing.

The TFR operates at J-band frequencies and has four modes: standby, ground standby, test and terrain-following. In the latter mode the aircraft can be flown automatically or manually through

The attack radar for the Panavia Tornado IDS showing the ground-mapping antenna (above) and terrain-following antenna (below)   0503740

head-up display steering information. The pilot can also select ride comfort (for a given speed, the closer the allowable ground clearance, the less comfortable the ride owing to the greater g levels needed to stay on the commanded flight profile).

Both systems have extensive built-in test features to ensure a high degree of fault isolation and comprehensive reversionary modes.

Status
In service on Tornado IDS and ECR aircraft.

Contractor
Raytheon Company, Space and Airborne Systems.

HISAR® Integrated Surveillance And Reconnaissance system

Type
Airborne surveillance radar.

Description
HISAR is an I/J-band Synthetic Aperture Radar (SAR) mapping and surveillance system. It is designed for border surveillance, remote sensing, and specialist monitoring of all types (including oil pollution, deforestation and economic zones).

HISAR's development is based on experience with ASARS-2 and the AN/APQ-181 radar that equips the B-2 bomber.

HISAR features three imaging modes:
1. Wide-area search with a resolution of 20 m.
2. A combined SAR/MTI Strip mode, with 6 m resolution at a range of 20 to 110 km.
3. A SAR Spot mode with 1.8 m resolution.

HISAR® Integrated Surveillance and Reconnaissance system operators console   0001216

HISAR® synthetic aperture radar installed aboard the Beech King Air 200T.   0018071

Moving Target Indication (MTI) is a key feature of the HISAR radar that detects and tracks moving land and sea targets.

HISAR is supported by a self-contained, compact mission-planning system, that is used for pre-flight planning, and post-mission analysis.

Specifications
Dimensions: 0.83 m³
Weight: 245 kg (full system)
Power consumption: 4,930 W

Status
Raytheon states that HISAR is installed on five different platforms, including the US Army Airborne Reconnaissance Low-Multifunction (ARL-M), the Beechcraft King Air 200T and the DARPA Global Hawk high-altitude endurance unmanned aerial vehicle.

HISAR is also being marketed internationally for civil applications. These include environmental monitoring, border surveillance and maritime patrol. The system is marketed as a complete integrated reconnaissance and surveillance suite, in which form additional sensors are added, typically, the Raytheon Electronic Systems DB

110 electro-optical sensor and the AN/AAQ-16 infra-red sensor.

Contractor
Raytheon Company, Space and Airborne Systems.

Integrated MultiSensor System (IMSS)

Type
Airborne Fire-Control Radar (FCR)/Electro-Optic (EO) navigation and surveillance system.

Description
The Integrated MultiSensor System (IMSS) was initially configured specifically to perform day and night all-weather air interdiction and maritime patrol as part of the anti-drug efforts of the US Customs Service, a role in which it has been exceptionally successful. By combining and integrating a high-performance multimode radar and an infra-red imaging system, IMSS is also effective in performing reconnaissance, surveillance and search and rescue missions.

The IMSS consists of the AN/APG-66 radar fully integrated with an infra-red detection set such as the Northrop Grumman WF-360, controls and displays and an inertial navigation system.

The AN/APG-66 is a digital, coherent, multimode radar system developed to serve both the strike and fighter demands of the F-16 aircraft. It provides long-range detection and acquisition of targets at all altitudes and aspect angles in the presence of heavy background clutter. The current version of the APG-66 incorporates a new signal data processor which replaces several radar units and eliminates the need for special interface hardware when used in IMSS. The result is significant improvement in system weight, cooling, power, capability and reliability.

The primary function of the infra-red system is short-range tracking and observation of airborne,

maritime and ground targets. As integrated in the IMSS, the infra-red capability has, as a passive system, proved exceptionally effective in the covert tracking of aircraft. The infra-red system can be applied independently to track a designated target by locking on to the heat differential generated by the target; it can be slaved to the radar so that the infra-red line of sight follows radar-tracked airborne targets and it can be directed to acquire and track ground targets automatically, using data from the INS. Additional capabilities include a TV camera, laser range-finder and video recorder.

The IMSS controls and displays are integrated with the radar, infra-red and INS through the radar signal data processor. A hand control unit provides the sensor operator with single-hand slew control of both sensors, including antenna elevation, radar cursor in azimuth and range, target designation and infra-red line of sight. Separate displays are provided for the radar and the infra-red, and these displays can be duplicated at several stations within the aircraft.

The INS provides the inertial references required by the radar and infra-red systems, such as roll, pitch, yaw, heading and velocities in three axes.

The IMSS incorporates continuous self-testing and BIT to isolate a fault down to an easily replaceable unit.

Contractor
Northrop Grumman Corporation, Electronic Systems Sector.

Joint STARS AN/APY-3 Joint Surveillance Target Attack Radar System

Type
Airborne surveillance radar and battle management system.

Description
The Joint Surveillance Target Attack Radar System (Joint STARS) designated AN/APY-3, is a long-range air-to-ground surveillance and battle management system. It is capable of looking deep behind hostile borders to detect and track ground movements, in both forward and rear echelon areas and to detect helicopter and fixed-wing aircraft. Joint STARS provides air and ground commanders with the intelligence and targeting data for management of their war-fighting assets.

Joint STARS is a complex of systems. It comprises an airborne platform, four major subsystems and a ground station module that receives, in near-realtime, radar data processed in the aircraft. The four major subsystems, all integrated on the airborne platform, consist of an advanced radar; internal and external communications including UHF, VHF and HF voice links, the Joint Tactical Information Distribution System (JTIDS) and a newly developed surveillance and control datalink; operations and control including advanced computers and 18 operator display stations which perform the data processing and display functions for tens of thousands of targets and C³I operation; and a self-defence suite which is in the process of being specified. Joint STARS is a joint US Air Force and US Army development, with the air force being responsible for the airborne segment and the army for the ground segment.

Joint STARS detects, locates, classifies, tracks and targets potentially hostile ground movements in virtually all weather. It operates in near-realtime and in constant communication through a secure datalink with army mobile ground stations that, in turn, can use TACFIRE and the advanced field artillery tactical data system to talk to artillery for fire support or to the all-source analysis system using US message format. Joint STARS will also maintain constant communication with air force tactical command posts via JTIDS. The platform for Joint STARS is a modified and militarised version of the Boeing 707-300 series aircraft.

The JSTARS aircraft, showing the JSTARS radar under the forward fuselage 0018076

The technology employed in the radar was initially demonstrated as part of the Air Force/DARPA Pave Mower programme in the late 1970s. Major technological achievements of Joint STARS include the software-intensive displaced phased-centre slotted array antenna radar with several concurrent operating modes; the unique 8 m antenna mounted under the fuselage of the aircraft developed by Norden Division of United Technologies; very high-speed processors, each capable of over 600 Mops; high-resolution colour graphic and touchscreen tabular displays; the wideband surveillance and control datalink and over one million lines of integrated software code. The radar operates as either a side-looking synthetic aperture radar for the detection of fixed or stationary targets, or a Doppler radar to track slow-moving targets such as tanks or troop platforms.

In September 1987, the air force, following the August Joint Requirements Oversight Committee review, increased the number of E-8 aircraft from 10 to 22. In April 1988, the Conventional Systems Committee, after reviewing the Joint STARS airborne segment, concluded that the airborne platform should be changed from used aircraft to new 707s. However, in October 1989, the air force returned to the plan to continue the airborne portion of the programme with used 707 aircraft, on the grounds of the higher cost of new aircraft, to meet an initial operational capability in 1997.

In September and October 1990, operational field demonstrations in a dense electromagnetic environment and poor weather were completed in Europe. After six weeks of flying demonstrations, consisting of 25 missions and 110 flying hours, Joint STARS concluded the demonstration of the system's capabilities and data gathering for the continuing development of the programme. US Air Force and Army officials assessed the demonstration as a complete success.

In November 1990, Northrop Grumman received a contract for the development of the third Joint STARS full-scale development aircraft and follow-on full-scale development work. Meanwhile, the two E-8A prototype aircraft (T-1 and T-2) achieved outstanding success in the 1991 Gulf War, and provided strong impetus for full development of the E-8C production configuration.

Specifications

Antenna: 7.3 m long, side-looking, phased-array, housed in canoe-shaped radome under forward fuselage aft of nose landing gear; scanned electronically in azimuth, steered mechanically in elevation from either side of aircraft.

Operating modes: Wide area surveillance; fixed target indication; Synthetic Aperture Radar (SAR); moving target indicator; target classification.

Workstations: 17 identical workstations for system operators; one navigation/self-defence workstation; each operator can perform: flight path planning and monitoring; generation and display of cartographic and hydrographic map data; radar management, surveillance and threat analysis; radar; radar data review; time of arrival calculation; jammer location; pairing of weapons and targets, and other functions.

Communications: Surveillance and Control DataLink (SCDL) for transmission to ground stations; JTIDS; TActical Data Information Link-J (TADIL-J); Satcom; two encrypted HF radios; three encrypted VHF radios with provision for SINgle Channel Ground and Airborne Radio System (SINCGARS).

Status

Work on the Joint STARS programme began as a joint service programme during 1983. With the aim of providing an airborne surveillance and anti-tank radar system, three industrial teams bid for the programme, with that led by the then Grumman (now Northrop Grumman) being awarded a USD657 million contract for the airborne element of the programme during September 1985. Development of the APY-3 radar subsystem was in the hands of the then Norden Systems (now Northrop Grumman Norden Systems), while Motorola began work on the Joint STARS GSM under a separate US Army contract. Two Boeing 707-328C airliners were acquired from American Airlines and Qantas to act as Joint STARS test beds. After modification by Boeing, the two airframes were delivered to Grumman for fitting out during 1986–1987. The first fully configured prototype (known as the E-8A) made its maiden flight on 22 December 1988. The Joint STARS development programme was interrupted by the 1991 Gulf War, which saw both development aircraft brought up to an interim operational capability and deployed to Saudi Arabia under the control of the USAF 4411th Joint STARS Squadron. The first operational E-8A sortie was flown on 14 January 1991 and, by the end of the conflict, the two aircraft were reported to have flown a total of 534 hours in the course of 49 surveillance missions.

In April 1992, Grumman was awarded a USD125 million low-rate E-8 initial production advanced procurement contract. Initially, the production standard aircraft was to have been a new-build E-8B model but this was abandoned after the delivery of the first airframe in favour of reworking existing Boeing 707 airframes. Such aircraft are designated as E-8Cs, the first example of which was completed in March 1994. After a favourable programme review in May 1993, construction of six E-8Cs was commenced at a rate of two per year. It is understood that a total of 16 E-8C aircraft will be procured by the USAF.

Outside the US, the Joint STARS system has been proposed to meet both the UK's Airborne STand Off Radar (ASTOR – see separate entry) and NATO's Alliance Ground Surveillance (AGS) system requirements. New higher-resolution SAR modes offered to NATO include Enhanced SAR (ESAR), Inverse SAR (ISAR) and SAR Swath.

Alongside the continuing production effort, the E-8 is also to be the subject of a Multi-Stage Improvement Programme (MSIP). The first phase of the MSIP, introduction of the TActical Digital Intelligence Link-Joint (TADIL-J) datalink and infra-red countermeasures, has been funded. Phase 2 of the MSIP is understood to involve the integration of satellite communications, an improved data modem and an automatic target recognition facility. Trials of the countermeasures capability took place at Eglin Air Force Base, Florida during late 1995 using an E-8A fitted with five AN/ALE-47 IR CMDS and an AN/AAR-47 MAWS.

In November 1998, the US Air Force awarded Northrop Grumman a Pre-Planned Product Improvement (P3I) contract to develop the next-generation advances to JSTARS. The P3I contract, which could have a value of USD776 million, is the first element of the Radar Technology Insertion Program (RTIP). Including production, the total value of RTIP is expected to exceed USD1.3 billion. More recently, Northrop Grumman received a separate Indefinite Delivery/Indefinite Quantity (ID/IQ) contract with a potential value of USD1.2 billion for upgrade work on JSTARS. The contract extends to March 2005. The first project in the ID/IQ contract is development of an integrated satellite communication system for JSTARS, valued at approximately USD40 million.

The main elements of the RTIP include a new 2-D electronically scanned array radar, a Commercial-Off-The-Shelf (COTS) technology signal processor and improved software for operation and control of the workstations inside the JSTARS platform. The new array is reported to be approximately half the size of the APY-3 unit.

Northrop Grumman and Raytheon Electronic Systems have agreed a 50/50 workshare on the radar portion of the RTIP.

To date, the USAF has ordered its 16th E-8 aircraft, with the intention of requesting funding for a 17th.

Contractor

Northrop Grumman Corporation, Norden Systems.

LAIRS Lockheed Martin Advanced Imaging Radar System

Type

Airborne surveillance radar.

Description

The current LAIRS is the fourth generation of upgraded Synthetic Aperture Radar (SAR) electronics based upon the original LAIRS that was developed from the Advanced Synthetic Aperture Radar System (ASARS-1). Although ASARS-1 was developed for the SR-71 Mach 3 environment, LAIRS has been specifically built and tailored to operate in the medium altitude,

The JSTARS radar antenna unit 0018077

subsonic through transonic speed regimes. LAIRS uses the latest in SAR algorithms and is composed of both an airborne segment and a ground segment.

The airborne segment is built in a modular design and can be easily modified to accommodate specific customer requirements. The system incorporates commercial-off-the-shelf (COTS) technology in many of the components, including the airborne or ground processing system. The collected data from the airborne sensor can be transmitted directly to the ground via a datalink as phase history data, recorded on the aircraft for later processing on the ground or processed in real time on the aircraft and transmitted to the ground as imagery data. The datalink also has a feature for both downlink and uplink of data.

LAIRS is capable of providing both fixed target imagery (FTI) and moving target imagery (MTI). The FTI modes are swath mode and spotlight modes. The swath mode provides a wide area

(normally 19 km wide) image of the earth or sea for surveillance and provides medium-resolution imagery (normally 3 m resolution). The swath can be continuous in either a parallel path or an oblique path to the flight vector. Wider swaths and finer resolution swaths are available, and are easily incorporated into the modular design of LAIRS. The second FTI mode is spotlight mode. LAIRS has two spotlight modes, one for classification and a higher-resolution mode for identification of targets. Spotlight modes can image targets either forward or aft of the aircraft and within the squint limits of the antenna. LAIRS also has an MTI mode for imaging moving targets. The mode typically uses a sector scan image out to 185 km and can image moving targets within a 50 km sector that can be positioned anywhere from 35 to 185 km away from the aircraft.

The LAIRS ground station uses state-of-the-art computer equipment and software to store, retrieve, combine, exploit and report.

Typically, LAIRS is displayed on soft copy so that the user can optimise data analysis from imagery produced by the LAIRS system. The number of workstations, the management of imagery software, and hard copy production and display equipment can be tailored to the specific requirements of the user.

Status
LAIRS radar equipment is reported to be operational aboard the US Customs Service P-3 aircraft.

Specifications

	Swath mode	Spotlight mode	MTI mode
Min slant range	<19 km	<19 km	35 km
Max slant range	>185 km	>185 km	185 km
Resolution	<3 m	fine and ultra-fine	<3 m

Contractor
Lockheed Martin Tactical Defense Systems.

Multirole Electronically Scanning Aircraft (MESA) system

Type
Airborne Active Electronically Scanned Array (AESA).

Description
The MESA system is a low-cost, high-performance modular airborne surveillance radar system, installed on a slightly modified Boeing 737-700 aircraft, as the main sensor of the Boeing 737 Airborne Early Warning & Control (AEW&C) system.

Northrop Grumman Electronic Sensors and Systems Sector (ESSS) is responsible for the MESA radar; Boeing Information & Surveillance Systems is performing all aircraft modification and systems integration work.

The MESA radar supports a variety of air surveillance missions including airborne, ground, maritime, environmental, drug interdiction and border patrol missions. The system provides for the detection of low-flying aircraft, helicopters, tactical ballistic missiles and both stationary and moving targets in a clutter environment. The MESA antenna is an innovative aperture that provides a 360° azimuth scan, with no mechanical rotation. MESA employs Pulse Doppler (PD) radar forms for air search and pulse forms for maritime surface search. It will also provide in the same aperture an integrated civil and military Identification Friend-or-Foe (IFF) capability. Other sensor avionics that can be integrated with MESA to create a comprehensive AEW&C capability include IFF, ESM, FLIR and LLTV, all coupled to a suite of workstations and voice and data communications facilities.

The radar utilises a static, electronically scanned antenna on the top of the aft main fuselage, constructed from an advanced composite structure that supports side-emitting electronic manifold arrays and a 'top hat' end-fire array atop the dorsal fin. This ultra-light foam composite sandwich construction facilitates integration of the system into the airframe at a fraction of comparable system weights, thereby minimising the fuel penalty to the 737-700 platform.

The dorsal fin arrays provide coverage +/–60° from the broadside axis, the 'top hat' array covers +/–30° from the fore and aft axis (with overlaps). The operating frequency is 1,200 to 1,400 MHz, and the antenna arrays comprise 'hundreds' of silicon carbide solid-state transmit/receive elements.

The MESA radar provides 360° horizontal coverage to 190 nm. It functions in three different modes: low Pulse Repetition Frequency (PRF) for ground and maritime detection; high PRF for air-to-air detection; and spot mode for detailed tracking of high-priority targets and enhanced long-range performance. 360° scan time can be varied by the operator, between 3 and 40

seconds, and it can be optimised between the three available modes during every rotation. Sector width is constrained by beamwidth, which varies between 2 and 8°. Increased range and higher probability of detection is provided in the spot mode by use of an increased beam dwell period to increase average power on target.

The three antenna arrays provide Identification Friend or Foe (IFF) capability, as well as active radar. The arrays cannot function simultaneously in both modes, the 'top hat' section may function in IFF-mode, while the dorsal fin arrays are in active radar mode, and vice versa.

The mission system, implemented by Boeing Information & Surveillance Systems, is an open system architecture derived from experience gained in the AirBorne Laser and Nimrod MRA 4 mission system programmes. Up to 14 workstations can be fitted. All workstations are similarly configured. All positions can be used for the full airborne early warning and control function, including management of communications. Radar and communications management functions have been automated to the point where no specialist operators are required for these activities, and no specialist airborne maintenance position is planned. The workstations are configured against the skin of the aircraft to maximise space and minimise weight.

The open systems architecture permits integration of other onboard sensors, including infra-red, television and electronic support measures systems. It can also interface datalink systems to co-operating air or ground platforms for data transfer. Boeing says that the overall Human Machine Interface (HMI) will be compatible with the NATO AWACS mid-term upgrade programme.

The 737-700 aircraft is fitted with additional fuel tanks, and the strengthened wing and landing gear of the 737-800. An additional distinguishing feature is a fin below the rear fuselage added for aerodynamic stability. At present it does not contain any electronic components, but its location appears to be suitable for self defence equipment.

The flight deck is standard 737-700, but with the addition of a sixth flat screen display, between the pilots, for presentation of mission tactical data and threat scenario information. The Marconi Electronic Systems HUD 2020 Visual Guidance System (VGS), certified on the 737-800, is also available, and forms a part of the Wedgetail installation.

Mission ceiling is 41,000 ft, with normal operations between 29,000 and 40,000 ft. Time on task at 300 nm is 8 to 8.5 hours, without flight refuelling. A flight refuelling capability is an option.

Status
A radar proof-of-concept flight demonstration was completed in 1994, and flight testing continued until 1996.

Boeing Information & Surveillance Systems incorporated the MESA radar in its Boeing 737 Airborne Warning and Control (AEW&C) proposal for the Royal Australian Air Force (RAAF) Australian Project Air 5077 Wedgetail requirement. An initial contract award for seven aircraft was made in 1999, although by 2002, the requirement had been revised down to a purchase of four aircraft, with first aircraft delivery planned for 2004. Northrop Grumman Corporation announced completion of the first MESA radar antenna during November 2002, which was subsequently installed on a test range

The Boeing 737 AEW&C aircraft, showing the MESA radar antenna on the top of the fuselage

0079022

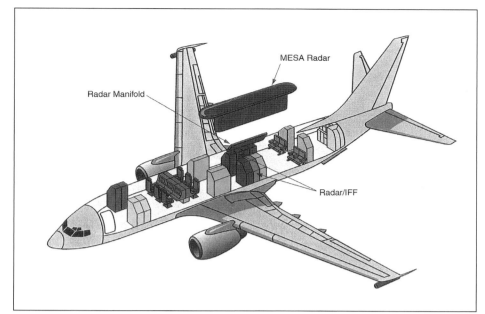

The Boeing 737 AEW&C equipment configuration   0079023

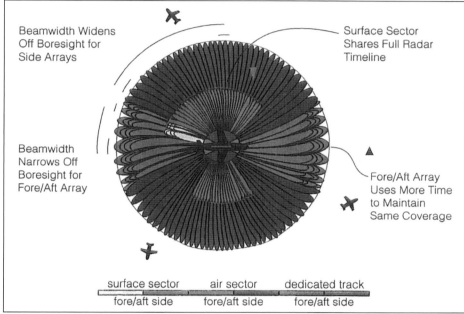

Boeing 737 AEW&C typical radar scan pattern   0079024

The radar indicator interfaces with Honeywell's LSZ-860 lightning sensor system and Data Nav for complete severe weather avoidance, navigation and checklist capability.

High-resolution and high-sensitivity mapping modes are available, designated GMAP 1, which is used to locate small targets at sea, and GMAP 2, which provides ground mapping over land or coastline. Display modes include three display colours, which differ from those used for the weather display and pilot-selectable sea clutter reduction. Range scales of 0.5, 1, 2.5, 5, 10, 25, 50, 100, 200 and 300 n miles are available and the shortest range scale provides a resolution of 55 ft.

The Primus 701 includes a beacon capability which makes low-visibility approaches possible where standard navaids may not be available. It also allows air-to-air rendezvous. The Primus 701 operates in radar only, beacon only or both beacon and radar modes. Beacon targets are shown in contrasting colours in weather- and ground-mapping modes.

Primus 700/701 BITE includes comprehensive and continuous internal fault monitoring and a menu function which allows access to monitor pages. A non-volatile memory records internal fault data for later retrieval by maintenance technicians.

Status
In production. Primus 700 has been selected by the US Army as a retrofit for 55 MH-60 and CH-47 helicopters, and by Boeing as the standard factory-installed radar in the Chinook CH-47SD.

Primus 701, which includes beacon capabilities, has been selected by Cougar Helicopters for its Eurocopter AS 332 Super Puma helicopters.

Contractor
Honeywell Inc, Commercial Electronic Systems.

Primus 880, 660 and 440 weather radars

Type
Airborne weather radar.

Description
The Primus 660 and 880 weather radars are high-power (10 kW) successors to the Primus 650 and 870 systems, respectively. The Primus 440 is designed as a powerful, reliable weather radar for light-class business aircraft.

All three systems have a stabilised antenna, up to 24 in for the Primus 880, and are packaged in a Transmitter/Receiver/Antenna (TRA) architecture that weighs 6.36 kg. Each system is compatible with Honeywell's LSZ-860 Lightning Sensor System and may be displayed on either the Electronic Flight Instrument System (EFIS) or on a dedicated weather radar indicator.

The Primus 880 features Doppler turbulence detection pulse pair processing that detects spectrum spreading caused by turbulence within any storm cell, regardless of rainfall rates. Once detected, turbulent areas are displayed in white on all ranges up to 50 n miles, allowing pilots to safely manoeuvre around potentially hazardous weather.

For the first time on a Primus radar system, Primus 880 also features Built-In Test Equipment (BITE) on two of the most important components of the system, the transmitter and receiver, providing a complete RF loop-back which continuously tests the transmitter power and receiver sensitivity and reports any faults to the pilot.

Other features of the Primus 880 include Honeywell's exclusive Rain Echo Attenuation Compensation Technique (REACT) which alerts pilots to storms hidden behind other storms; Target Alert (TA) which notifies pilots of potentially hazardous weather directly in front of the aircraft; Altitude Compensated Tilt (ACT), which allows detection of weather that may affect the aircraft en route and reduces the amount of tilt management performed by the pilot; Ground Mapping (GM), which serves as a navigation aid

---

for integration and pattern testing. The first Wedgetail 737-700 rolled off Boeing's production line at Renton, Washington, during the previous month.

Contractor
Northrop Grumman Corporation, Electronic Systems Sector.

Primus 700/701 Series surface mapping, beacon and colour weather radar

Type
Airborne weather radar.

Description
A three-box system including receiver/transmitter, indicator and antenna, the Primus 700/701 is compatible with certain Honeywell EFIS. Its powerful 10 kW magnetron transmitter has six pulse-widths, seven bandwidths and four PRFs for maximum performance on all ranges in all modes. Ten selectable range scales from 0.5 to 300 n miles (1 to 556 km) provide optimal range scales in every condition.

Five antenna sizes from 10 to 24 in make the Primus 700/701 system suitable for any airframe.

The Primus 700/701 dual-EFIS interface capability and antenna sweep time-sharing in effect make two radars available to the crew throughout the

flight. Each crew member can select their own range, mode, gain and tilt display on their EHSI.

With the radar indicator, pilots will have the Honeywell features of a variable range mark and azimuth cursor with digital distance and bearing readouts. A new menu function allows the indicator switches to do double duty, controlling infrequently used features such as heading display on/off. Without a radar indicator installed, the system includes one or more WC-700 radar controllers.

Primus 700/701 weather radar features include a four-colour display of rainfall intensity. On ranges of 50 n miles (93 km) or less, turbulence detection shows areas where there is moderate or stronger levels of turbulence.

The Honeywell Rain Echo Attenuation Compensation Technique (REACT) safety feature performs three distinct functions. First, it maintains target calibration by compensating for attenuation caused by intervening rainfall. Returns remain properly calibrated for the storm behind the storm. Second, REACT advises pilots of areas where target calibration cannot be maintained even with maximum compensation. For those areas, REACT changes the screen background to blue, warning that calibration is no longer possible and attenuation may be hiding areas of severe weather. Third, any target displayed in the blue field will appear in magenta to alert the pilot of its probable severity.

The system also includes Honeywell weather radar features of ground clutter reduction and target alert.

Primus 880 weather radar   0001215

by depicting terrain features not available in this clarity and detail with lower-power radars.

Honeywell's optional LSZ-860 Lightning Sensor System overlays lightning information on to the precipitation/turbulence display to provide a very powerful severe weather detection capability. The LSS accurately displays the position and lightning rate of up to 50 storm cells at the same time.

In July 1999, an interface became available to integrate the Primus 440, 660 and 880 radars to the Honeywell Aerospace, Electronic & Avionics Lighting Mk VII Enhanced Ground Proximity Warning System (EGPWS).

Status
In service.

Contractor
Honeywell Inc, Aerospace Electronic Systems.

RDR-4A/RDR-4B Forward Looking WindShear radar (FLWS)

Type
Airborne weather radar.

Description
Chosen by Boeing as standard equipment for the 767 and 757 transports, the RDR-4A was designed to meet ARINC 708 requirements. The I/J-band system features a solid-state transmitter and line of sight antenna with split-axis performance and is compatible with the EFIS flight decks of the Boeing 767, 757, MD-80, DC-10, Airbus A310, and Lockheed Martin L-1011 transports and other designs. The range is 592 km.

The current RDR-4B operates in the X-band and incorporates forward-looking windshear detection and avoidance capabilities. The windshear detection capability is easily incorporated into existing RDR-4A radars, without form or fit changes to the installation.

The RDR-4B is a Doppler weather radar that measures actual horizontal windspeed, using reflections from the moisture that is always

present in the atmosphere, and penetrates weather systems and detects microbursts embedded in rain. It provides specific windshear locations on a radar PPI presentation, giving 10 to 60 seconds of advance warning, on a display free from interference such as ground clutter. The RDR-4B can display turbulence up to 46 miles ahead and 90° left or right. It can display windshear up to 5¾ miles ahead up to 40° left or right.

The system operates automatically any time the aircraft is below 2,300 ft AGL, although the mode is selectable at any time.

The RDR-4B radar display is a map-like presentation that shows areas of weather and their locations relative to the aircraft. Light rain is shown in green, moderate rain in yellow, and heavy rain in red. A function known as penetration compensation allows more accurate presentation of storms even when viewed through intervening rainfall. Turbulence is magenta and windshear is a symbol called a windshear icon – a pattern of red and white arc-shaped stripes.

The RDR-4B can generate audible windshear warnings and alerts whenever the aircraft is less than 1,500 ft AGL. During approach, windshear ahead within 1.7 miles causes a 'go around' warning; in the landing phase, this warning is reduced to 0.6 miles. During take-off, windshear within 3.4 miles generates a warning.

The latest variant of the RDR-4B, designated the RDR-4B Predictive Wind Shear Radar (PWSR), uses a computer processor to identify windshear through moisture movement/mixing in the air mass ahead of the aircraft, providing up to a full minute's warning to the crew.

From late 1999, a further development of the RDR-4B radar became available. Automatic Tilt (AT) setting of the antenna angle (also known as auto-tilt) is provided by interfacing the RDR-4B with the Enhanced Ground Proximity Warning System (EGPWS) to predict the optimum setting for the antenna tilt. The ability to integrate the RDR-4B with the EGPWS is being designated Terrain Data Correlation (TDC). The system determines the optimum tilt setting for the radar antenna by measuring the average terrain elevation over five defined sectors of the total scan; as the radar

scans the volume ahead of the aircraft, the tilt is automatically adjusted according to the scanned sector. The advantage to the pilot is that the combined capability supplements EGPWS data where database coverage is poor, may lower minima at airports with precipitous terrain and alerts pilots to obstacles not in the EGPWS database.

From 2000, an enhanced high-altitude turbulence detection capability was integrated into the system, based on using advanced windshear processing technology with a higher pulse transmission rate to measure turbulence ahead of the aircraft.

Specifications
RDR-4B receiver/transmitter
Transmitter frequency: 9,345 MHz
Transmitter power: 125 W nominal
Maximum detection ranges:
(weather and map modes): 320 n miles
(turbulence mode): 40 n miles
(windshear): 5 n miles
Pulse-width:
(radar modes): 6 and 18 μs interlaced
(windshear mode): 1.5 μs
Pulse repetition rate:
(weather and map modes): 380 Hz
(turbulence mode): 1,600 Hz
(windshear mode): 6,000 Hz
Dimensions: 8 MCU
Weight: 13.2 kg

PPI-4B colour indicator
Display size: 3.3 × 4.3 in
Display ranges and marks: 5/2, 10/5, 20/5, 40/10, 80/20, 160/40, 320/80 n miles
Display colours:
(radar mode):
green: level 2 rainfall
yellow: level 3 rainfall
red: level 4 rainfall and above
magenta: turbulence
(windshear mode):
red and black icon overlays standard weather display
yellow flashlight beams mark end of icon
(TCAS overlay):
red: resolution advisory
yellow: traffic advisory
white: proximate and non-traffic threat
(EGPWS terrain data):
green: no terrain threat
yellow: caution terrain
red: warning terrain
Weight: 6.7 kg

Antenna and drive
Antenna:
(REA-4B/DAA-4A): 30 in flat plate, gain 35 dB, beamwidth 2.9°, weight 13.7 kg, scan 180° in azimuth and ± 15° in elevation
(REA-4A/DAA-4B): 24 in flat plate, gain 33 dB, beamwidth 3.6°, weight 5.5 kg, scan 160° in azimuth and ± 15° in elevation
Stabilisation limits:
(REA-4B/DAA-4A): ±25° in pitch and ±43° in roll with a combined pitch/roll limit of ±43°
(REA-4A/DAA-4B): ±15° in pitch and ±30° in roll with a combined pitch/roll limit of ±35°

Status
More than 6,000 RDR-4A radars are in service with numerous operators.

The RDR-4B was first certified in September 1994, and first use of the RDR-4B in commercial service was on a Continental Airlines Boeing 737-300 in November 1994. Over 1,500 RDR-4B radars are now in service.

The RDR-4B Predictive WindShear Radar (PWSR) was granted a supplemental type certification by the US FAA for the Boeing 737-300 and 737-500 aircraft in September 1999. United Airlines has ordered 495 for its existing fleet of late-generation airplanes including Airbus A319 and A320, Boeing 737-300, 737-500, 747-400, 757, 767, and 777 models. United Airlines has specified that all its new aircraft are to be delivered with factory-installed RDR-4B radars.

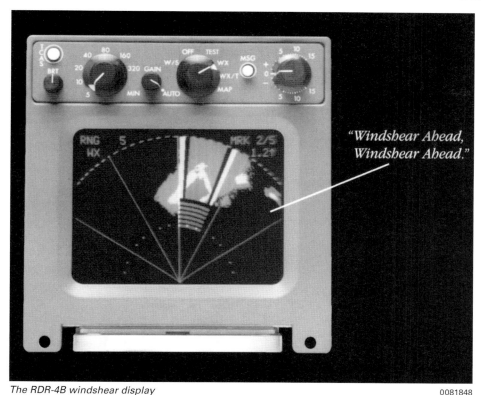

The RDR-4B windshear display   0081848

During the third quarter of 2001, American Airlines signed a 10-year agreement with Honeywell enabling purchase of predictive windshear radar systems for installation on current and future fleet aircraft. During the initial phase of the programme, Honeywell will supply RDR-4B systems, although the agreement allows for optional upgrade to Honeywell's HDR-1 next-generation radar system, which became available in 2004.

The RDR-4B PWSR/AT is in service in the Airbus A340-600.

Contractor
Honeywell Aerospace, Electronic & Avionics Lighting.

RDR-1400C colour weather and search and rescue radar

Type
Airborne weather and surveillance radar.

Description
The RDR-1400C is designed for helicopters and fixed-wing aircraft as a weather avoidance and search and rescue radar. Its main features are:

- Weather avoidance
- Search and Rescue
- Detection of oil slicks
- Beacon Modes
- Special sea clutter rejection circuitry to detect small boats and buoys down to a minimum range of 300 m
- Precision ground mapping in situations where high target resolution is important.

The RDR-1400C is a weather detection system. It has a 240 nm maximum display range, giving the user time to plan weather avoidance manoeuvres. For clear, detailed close-ups, two modes permit selection of full-scale ranges of either 1 n mile or 0.5 n miles, enhancing safety and precision of movement.

Different surveillance missions require different capabilities, so the RDR-1400C provides three specialised search modes. Search 1 incorporates special sea clutter rejection circuitry to help detect small boats or buoys down to a minimum range of 300 m.

Search 2 is designed for precision ground-mapping, where high target resolution is important. Search 3 mode, which includes normal ground-mapping, can also be used to

detect and track sea-surface phenomena such as oil slicks.

The RDR-1400C complies with TSO C-102, enabling land or sea approaches in 200 ft ceiling, half-mile visibility minima. Its beacon tracking mode permits operation with either current beacon codes or the newer DO-172/16 format, changed via push-button.

Specifications
Dimensions:
(receiver/transmitter) 127 × 158.8 × 352.4 mm
(indicator) 158.8 × 158.8 × 276.2 mm
Weight:
(receiver/transmitter) 6.58 kg
(indicator) 5.22 kg
Power requirements:
4.2 A, 28 V DC
3 A, 115 V AC, 400 Hz
TSO compliance: C-102 and C-63
Frequency: I-band (9.375 GHz)
RF power output: 10 kW
Scan angle: 120 or 60°
Scan rate: 28°/s
Display range/marks: 0.5/0.125; 1/0.25; 2/0.5; 5/1.25; 10/2.5; 20/5; 40/10; 80/20; 160/40; 240/60 nm
Min tracking range: 300 m
Beacon range: line of sight to 160 nm
Temperature:
(receiver/transmitter) –50 to +55°C
(Indicator): –40 to +55°C

Status
Following the acquisition of the former AlliedSignal/Bendix Search and Weather Radar products by the Telephonics Corporation, the RDR-1400C has been integrated with Telephonics' APS-143 OceanEye imaging radar and AN/APS-147 multimode radar product lines.

To date, more than 8,000 of the RDR-1400 family are in use worldwide. RDR-1400C radar systems are installed on the following civil and military platforms:

Helicopters	Aircraft
Agusta A109	Beech King Air
AB 212/412	Dornier Do 228
Bell 214	CASA C-212
Eurocopter BK 117	CASA CN-235/C-295
Eurocopter MBB 105	C-130
Eurocopter 145/155	C-141
Eurocopter AS 332 Super Puma	Fokker F-27
Eurocopter 365 Dauphin	L-410
Eurocopter Ecureuil/Fennec	

Sikorsky S-76/HH-60
Agusta/Westland Cormorant EH-101
Westland Super Lynx/Sea King

Contractor
Telephonics Corporation, Command Systems Division.

RDR-1500B multimode surveillance radar

Type
Airborne multi mode radar.

Description
The RDR-1500B is a development of the RDR-1300C radar fitted to US Coast Guard HH-65A helicopters. It is designed to provide multimode radar capability for helicopters and fixed-wing aircraft engaged in low- and medium-altitude maritime missions. Primary modes are surveillance and search; secondary modes include terrain mapping, weather avoidance, beacon navigation and the display navigation information from the aircraft navigation system.

The RDR-1500B is a lightweight, X-band, 360° digital colour radar system consisting of six line-replaceable units: the RT-1501A transceiver, the IN-1502A or -1502B multifunction colour indicator, a CN-1506A control panel, an IU-1507A interface unit, an AA-1504A antenna array and a DA-1503A or DA-1203A antenna drive.

RT -1501A receiver/transmitter
The RT-1501A receiver-transmitter functions as a short-range pulse radar for high-resolution sea search and terrain mapping, and also as a long-range pulse radar for long-range sea search, terrain mapping and conventional weather avoidance.

IN-1502A/1502B multifunction colour indicator
The 216 × 241 mm (8.5 × 9.5 in) IN-1502A multifunction colour indicator provides a continuous, three-colour display of sea and ground targets, weather returns, beacon, VOR, and navigation information within the area scanned by the antenna. The indicator is NVIS-compatible and has a raster format auxiliary input to display FLIR or TV in monochrome.

The IN-1502B is a 127 × 152 mm (5 × 6 in) cockpit display which can be used in place of, or in addition to, the IN-1502A indicator. With similar capabilities to the IN-1502A, this indicator is intended for installation where space is limited.

CN-1506A control panel
The NVIS-compatible CN-1506A control panel contains all the controls for the radar system.

IU-1507A interface unit
The IU-1507A interface unit processes the digitised video output from the receiver-transmitter and supplies red, green and blue data in digital form to the multifunction indicators. It also supplies the line blank and scan blank pulses to the indicators, drive signals to the antenna unit, and PRF and mode selection signals to the receiver-transmitter.

AA-1504A antenna array and DA-1503A antenna drive
The basic AA-1504A antenna array is a 991 × 227 mm flat-plate phased-array unit with 2.6 × 10.5° (azimuth × elevation) beamwidth. The antenna is remotely controlled in the elevation axis by a tilt control on the CN-1506A control panel.

The DA-1503A antenna drive positions the antenna array in azimuth and elevation. The antenna is motor driven, with combined pitch, roll, and tilt Line Of Sight (LOS) stabilisation of up to ±25° of true vertical. It scans through 360° with selectable speeds of 45°/s and 90°/s. Stabilisation is achieved via pitch and roll signals from the aircraft vertical gyro and the CN-1506A control unit tilt control, selectable ±15° from the horizontal.

The DA-1203A antenna drive unit provides for a 120° scan if required.

System options include track-while-scan (TWS), a FLIR sensor pointing output, a target datalink and a video link interface.

Specifications
Dimensions:
(RT-1501A) 194 × 123 × 321 mm
(IN-1502A) 216 × 241 × 325 mm
(IN-1502B) 127 × 152 × 284 mm
(CN-1506A) 133 × 116 × 169 mm
(IU-1507A) 194 × 189 × 324 mm
(AA-1504A) 991 × 227 mm
(DA-1503A) 211 × 249 × 241 mm
Weight:
(RT-1501A) 6.8 kg
(IN-1502A) 8.8 kg
(IN-1502B) 4.7 kg
(CN-1506A) 1.0 kg
(IU-1507A) 7.25 kg
(AA-1504A) 1.8 kg
(DA-1503A) 8.8 kg
Transmitter frequency: 9,375 MHz
Receiver frequency:
(Search/weather modes) 9,375 MHz
(Beacon mode) 9,310 MHz
PRF: 200, 800, 1,600 Hz
Pulse width: 0.1, 0.5, 2.35 µs
Antenna gain: 31 dBi (AA-1504A)
TSO compliance: C-63
Frequency: X-band
RF power output: 10 kW peak
Scan angle: 360°
Scan rate: 28°/s (120° sector scan); 45–90°/s
Ranges: 0.625, 1.25, 2.5, 5.0, 10.0, 20.0, 40.0, 80.0, 160.0 n miles

Status
Co-developed with FIAR of Italy, the RDR-1500B is installed in a wide variety of helicopters and fixed-wing aircraft, including the Agusta 109, BAE Systems Jetstream 31, CASA 212, Eurocopter Super Puma, Fokker F-27, Westland Lynx, Beech King Air 200, Antonov AN-32, de Havilland DHC-7 and Bell 412.

Contractor
Telephonics Corporation, Command Systems Division.

RDR-1600 search and rescue/ weather radar

Type
Airborne multi mode radar.

Description
Telephonics Corporation has developed the RDR-1600 search and rescue and weather avoidance radar to be fully compatible with modern integrated flight decks, including the latest glass cockpit designs. The system is claimed to be 25 per cent lighter and consume 25 per cent less power than the industry standard radar. In addition, the short-range performance of the RDR-1600 allows a minimum detection range of 450 ft for offshore oil platform and other precision landing applications.

The RDR-1600 is designed for helicopters and fixed-wing aircraft. Its main features are:
- Five primary operational modes
- 10 kW peak power
- Built-in test circuitry
- Weather detection/target alert
- Search and rescue
- Surveillance
- Beacon Trac tracking mode
- Normal and precise ground mapping
- Oil slick detection and mapping
- Narrow pulse precision approach mode (450 ft minimum detection range)
- Improved clutter detection
- ARINC 429 and 453 interfaces.

Specifications
Dimensions: (W × D × H)
(Receiver/transmitter) 127 × 352 × 159 mm
(Antenna) 254–168 mm (10 in), 305–194 mm (12 in), 457–270 mm (18 in) × 195 mm (depth)
(Control panel) 159 × 276 × 159 mm

Weight:
(Receiver/transmitter) 7.2 kg
(Antenna) 3.4 kg (10 in), 3.5 kg (12 in), 4.0 kg (18 in)
(Control panel) 0.77 kg
Minimum detection range:
(Search mode) 137 m (450 ft)
(Weather mode) 457 m (1500 ft)
Beacon range: LOS or up to 160 n miles
TSO compliance: C-102
Frequency: X-band
RF power output: 10 kW
Scan angle: 120 or 60°
Scan rate: 28°/s
Ranges: 0.5, 1.0, 2.0, 5.0, 20.0, 40.0, 80.0, 160.0, 240.0 n miles

Status
In production and in service.

Contractor
Telephonics Corporation, Command Systems Division.

RDR-1700 multimode surveillance radar

Type
Airborne multi mode radar.

Description
An updated version of the RDR-1500B (see separate entry), the RDR-1700 is an I-band multimode radar originally designed for integration with glass cockpit configurations. Its flexible design allows installation in helicopters as well as fixed-wing aircraft primarily for airborne search and surveillance but with secondary roles of terrain mapping, weather avoidance, beacon navigation and oil slick detection.

The RDR-1700 is a three box, lightweight, 360° (belly-mounted) or 120° (nose-mounted) colour system. Antenna options include 737 × 229 mm, 838 × 229 mm and 991 × 229 mm 360° scan units, together with 254, 305, 457 mm (diameter) and 457 × 305 mm 120° scan assemblies.

The LRUs are the antenna/pedestal unit, the receiver-transmitter unit and the interface unit. Control is provided via glass cockpit multifunction displays using an ARINC 429 interface. In addition to the operational features of the RDR-1500B, the RDR-1700 has an integral 20-target Track-While-Scan (TWS) capability.

System options include interfaces for Inertial Navigation (IN) and Global Positioning System (GPS), Forward Looking Infra-Red (FLIR) and Low-Light (LL) TV systems.

Specifications
Frequency: 9.310 GHz (beacon mode receive); 9.375 GHz (transmit and search/weather mode receive)
Transmitter power: 10 kW (nominal)
Pulse Repetition Frequency (PRF): 200, 800 or 1,600 Hz
Pulse width: 0.1, 0.5 or 2.35 msec
Antenna gain: 26–31 dBi (antenna dependent)
Scan angle: 120° sector or full 360°
Scan rate: 28° per sec (120°); 45° per sec (360°); 90° per sec (360°)
Range scales: 0.5, 1.0, 2, 5, 10, 20, 40, 80, 160 nm
Operating modes: Surface search, terrain mapping, beacon navigation, weather avoidance
Interface: ARINC 429
Weight: 23 kg (120° scan configuration); 27 kg (360° scan configuration)

Status
In production and in service. During 2001, Telephonics announced a contract from Agusta Westland for up to 23 RDR-1700 radars for installation in naval helicopters.

Contractor
Telephonics Corporation, Command Systems Division.

RDR 2000 Vertical Profile® weather radar

Type
Airborne weather radar.

Description
The RDR 2000 is a Vertical Profile® weather radar. In addition to normal weather radar features, it adds a vertical display of weather along an azimuth selected by the pilot using the track line. This enables the pilot to monitor storm development by observing the rate of growth of the tops of storm cells. In addition, it is possible to check the angle of a storm cell's leading edge to see which way it is heading and to check the 'radar tops'. Four levels of colour are used to display information. The system is available with a 12 or 10 in antenna, is stabilised to ±30° combined pitch and roll and is TSO'd to 55,000 ft. It is EFIS-compatible using the ARINC 429 and 453 databus structure.

Typically, an RDR 2000 radar would be paired with the GC 360A radar graphics unit and a moving map GPS unit, such as the KLN 90B or KLN 900 or CDU-XIs Flight Management System (FMS) to provide a comprehensive display capability. This includes fault annunciation, antenna tilt readout and independent dual-indicator operation.

Specifications
Dimensions:
254 or 309 mm antenna
Power output: 3.5 kW nominal (rated 4.0 kW)
Horizontal scan: 100° at 25°/s
Vertical scan: 60°
Weather Avoidance: 400 km typical for 309 mm antenna system
Range scales: 10, 20, 40, 80, 160 and 240 n miles
Weight: 12.02 kg

Status
In service.

Contractor
Bendix/King.

The RDR 2000 Vertical Profile® weather radar system showing a storm cell in conventional radar display in the vertical profile mode

0081850

RDR 2100 Vertical Profile® weather radar

Type
Airborne weather radar.

Description
The RDR 2100 Vertical Profile® four-colour, digital, weather radar includes sophisticated features such as Automatic Range Limiting (ARL), which alerts pilots to the possibility of a foreground storm cell blocking the picture of a more distant cell.

For aircraft equipped with a MultiFunction Display (MFD), the RDR 2100 can provide split-screen view to show both vertical and horizontal views simultaneously, sector scan to provide more frequent updates, and a logarithmic scale to show nearby storms in large detail while still allowing the pilot to see distant storms.

The Vertical Profile system scans vertically at the azimuth selected by the pilot using a track line. This enables the pilot to examine the angle of a storm cell's leading edge to determine the direction of movement, to check the radar tops and to distinguish between ground and weather returns. The radar is fully stabilised to ±30° combined pitch and roll. Four levels of colour are available with switchable range scales of 5, 10, 20, 40, 80, 160, 240 and 320 n miles.

The system is fully EFIS compatible using ARINC 429 and ARINC 453 databus structures.

Specifications
Transmit power: 6 kW
Weight: 12.02 kg
Antenna: 305mm (standard); 254 mm (optional)
Horizontal scan: 120°
Vertical scan: 60°

Contractor
Bendix/King.

RDR 2100VP Vertical Profile® weather radar system

Type
Airborne weather radar.

Description
The RDR 2100VP vertical profile radar functions equally well as a standalone unit, or in conjunction with an electronic display system, either flat panel or EFIS.

In addition to offering the standard features of a four-colour, solid-state, digital radar, it includes a vertical display of weather information, selected by a push-button selection (Vertical Profile). The pilot can thus monitor the rate at which individual storm cells are developing, and by checking the angle of the cell's leading edge, determine the way the weather is moving. If the aircraft is fitted with multifunction displays, both the horizontal and vertical modes can be displayed simultaneously, by choosing a split-screen display.

The radar's full-width scan covers 120°, but a faster update 60° scan is available, and also an auto-step feature, in which horizontal scans are made at 4° intervals across a ±10° vertical span. A logarithmic weather display permits the pilot to watch close-in weather activity, while monitoring more distant weather patterns.

Configuration options for the Horizontal Situation Indicator (HSI) include standard weather returns in a 120° sector scan overlaid with navigation courses, navaids, waypoints, airports and other key data.

Honeywell Aerospace claims a three-fold improvement in magnetron life by comparison with current systems in service, and suggests that the RDR 2100VP radar is ideal for regional airline use.

Specifications
Antenna/receiver/transmitter
Performance index:
(10-in antenna) 213.9 ±2.5 dB
(12-in antenna) 216.7 ±2.5 dB
Displayed weather ranges: 5, 10, 20, 40, 80, 160, 240, 320 n miles
Weather colours: 5, including black
Vertical profile scan angle: ±30°
Ground map variable gain: 0 to −20dB (configurable at installation)
Peak output power: 6.0 kW, nominal
Pulse-width: 4 μs
Pulse repetition frequency: 106.5 ±5 Hz
Antenna scan angle: 100 or 120° (configurable at installation)
Antenna scan rate: 25°/s
Tilt angle, manual: −15 to +15°
Stabilisation: ±30° combined pitch, roll and tilt

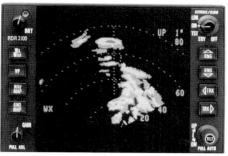 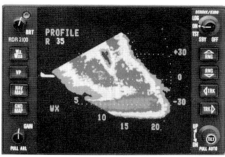

The RDR 2100VP displays a storm cell in traditional form in the left hand display and in vertical profile mode on the right hand display, as would be seen by the pilot on a split-screen display   0062202

Stabilisation adjustments (non-volatile)
stored in configuration module (part of aircraft installation)
Power: 28 V DC, 3.0 A
Weight: 4.5 kg
Altitude: 55,000 ft
Temperature range: −55 to +70°C
TSO: C63c Class 7

Indicator IN-862A
Display ranges:
(weather) 5, 10, 20, 40, 80, 160, 240, 320 n miles
(navigation only) up to 1,000 n miles
Power: 28 V DC, 2.0 A
Weight: 4.43 kg
Altitude: 35,000 ft unpressurised
Temperature range: −20 to +55°C
Dimensions: 114.3 (H) × 162.8 (W) × 344.7 (L) mm
TSO: C63c

Contractor
Bendix/King.

RDR-4000 Vertical Profile (VP) weather radar system

Type
Airborne weather radar.

Description
RDR-4000
The RDR-4000, Honeywell's latest airborne weather radar, incorporates a variety of design features intended to provide a far better weather picture ahead of the aircraft. Approximately half the size of previous radars and nearly 20 lb lighter, the new radar employs new scanning and computer-modelling techniques to provide

a greatly enhanced 3-D picture of significant weather. The RDR-4000 introduces a number of new technologies to civilian weather radars:

3-D volumetric scanning The radar automatically scans the airspace 90° either side of the nose, from the ground to 60,000 ft, out as far as 320 n miles ahead of the aircraft (>1.5 million cubic miles). Current systems search only a single (pilot-selected or automatically referenced to flightpath) flat plane. This 'volumetric' (lateral and vertical) information provides pilots with increased information with which to decide on avoidance options.

Pulse compression Pulse compression technology and a very low-noise, coherent-frequency receiver provides far greater weather detection range, increased resolution, predicted windshear detection, Doppler turbulence detection and cloud temperature/moisture content. Previously, pulse compression has been exclusively employed by military radar systems.

Ground clutter extraction Limited to only 2 scans of data, multi-scanning radars cannot easily separate weather returns from ground returns, so extraneous ground clutter must be suppressed with a simple filter, which only works well at short ranges on boresight and can significantly reduce weather returns. The RDR-4000 uses an internal high-resolution terrain database, adapted and scaled-down from Honeywell's Enhanced GPWS (EGPWS), to extract ground clutter from the display, without the associated reduction of weather returns.

Advanced antenna design The system integrates an advanced antenna. A direct drive motor system, coupled with dual-redundant azimuth and elevation motors, removes all gearing, improving reliability and producing a quieter antenna system that provides faster scanning with greater pointing accuracy.

Honeywell's RDR 4000 airborne weather radar has been selected for the Airbus Industrie A380 ultra-large transport aircraft (Honeywell)   1044236

For details of the latest updates to *Jane's Avionics* online and to discover the additional information available exclusively to online subscribers please visit
jav.janes.com

Higher system sensitivity The RDR-4000 eliminates the installation and maintenance costs associated with waveguide runs and wave-guide switches, at the same time increasing performance.

Advanced operational modes The RDR-4000 provides advanced weather modes for automatic weather detection and storm cell threat analysis. The automatic mode uses the FMS flight plan to automatically determine threats along the aircraft's flight path. Analysis modes include vertical profile modes along the aircraft's track, a selected azimuth of interest, and along an un-wound flight plan. This data is combined with terrain and flight plan information to provide a complete vertical analysis of the terrain and weather ahead of the aircraft. Another mode, Constant Altitude, provides a full 180-degree view of weather at selected altitudes.

While not a new technology, and employed by radar systems in business aircraft since the early 1990s, the RDR-4000 introduces the first Vertical Profile (VP) display for an airliner; this comprises a side view of the weather that can be aligned along the direction of flight or compared with the flight plan (both lateral and vertical portions) stored in the Flight Management System (FMS). The side view is shown on the radar display in a separate window below the top-down/god's eye view. Also, rather than adjusting antenna tilt angle to have the radar look-up or -down, pilots can select a 'constant altitude' mode (which takes earth curvature into account), providing weather information for a given altitude.

As an entirely new design, in addition to providing greatly increased reliability over legacy systems, the RDR-4000 takes advantage of the latest technologies to reduce in-service maintenance costs:

- Reduced size, weight and power – 3MCU package

- Elimination of waveguide runs, which can develop cracks, leading to loss of performance
- Elimination of waveguide switches
- Elimination of gears in the antenna drive system
- On-board high-speed Ethernet software data loading
- Enhanced BITE and fault logging by flight leg
- Digital radio techniques eliminates tuning points providing improved repeatability, reliability and lower drift
- Removable compact flash memory for data recording and download
- No forced air cooling required

RDR-4000M

The RDR-4000M is a military derivative of the commercial RDR-4000 system offering the following modes:

Terrain mapping The RDR-4000M offers both an automated mapping mode and real-beam ground mapping capabilities. Automated ground map is a highly sensitive, "hands-off", long-range mode in which the terrain display presentation is developed from the collected 3D volumetric buffer data. This mode is suitable for detecting prominent terrain features, such as coastlines and large bodies of water at long ranges and can be used as a check on navigation.

High-resolution terrain map This mode provides a depiction of the ground illuminated by the active antenna beam at the tilt angle selected by the pilot. Using pulse compression techniques, high-resolution terrain map provides 50 ft resolution out to 5 nm. Resolution at longer ranges is commensurate with display range.

Skin-paint Skin-paint mode takes advantage of pulse compression to optimise radar parameters for detecting C-130 (30 m²) size targets at ranges of up to 10 miles (in up to 4 mm/hr rainfall).

EMCON The RDR-4000M offers an Emissions Control mode where transmissions can be interrupted with a master control signal for silent operations.

The RDR-4000M is also designated AN/APS-150 for US military applications.

Specifications
Antenna/receiver/transmitter
Dimensions: 3MCU
Weight: 18 lb lighter than a traditional radar unit in a single installation, 28 lb lighter in a dual-installation
Antenna diameter: 12 in
Displayed weather ranges: 5, 10, 20, 40, 80, 160, 240, 320 n miles
Altitude: 55,000 ft
Temperature range: –55 to +70°C

Status
In production and in service. Airbus Industrie has selected the radar for the A380, as part of that aircraft's integrated Aircraft Environment Surveillance System (AESS), which Honeywell is developing. It has also been selected by American Airlines for retrofit across its fleet of Boeing airliners. The total estimated value of the program should top USD1 billion, according to the Company, which also seeks customers from business jet and regional airliner operators.

The US DoD has selected the RDR 4000M for the C-17 Globemaster heavy airlifter. The USD30 million contract covers initial development and equipment cost for 180 aircraft. The radar has also been selected for the Japanese C-X transport.

Contractor
Bendix/King.

SeaVue (SV) surveillance radar

Type
Airborne surveillance radar.

Description
Develped from Raytheon's AN/APS-134(V) and AN/APS-137(V), the SeaVue (SV) family of X-band radars are modern, lightweight, high-performance systems. They employ proven long-range surface detection techniques that support the entire range of surveillance missions. Inverse Synthetic Aperture Radar (ISAR) and SAR modes provide real-time imaging of maritime, land-based and coastline targets. The AN/APS-134(LW) was the initial product of the SV radar family.

Radar features include Sea Vizion™, Raytheon's unique combination of system parameters and target signal processing; fully coherent ISAR, SAR and MTI processing; pulse compression, pulse-to-pulse frequency agility and PRF jitter.

Radar modes available include surface surveillance and small target detection; navigation and mapping; weather avoidance; ISAR and SAR processing. Available enhancements include moving target discriminator; ISAR classification aids; Doppler Beam Sharpening (DBS); coherent look-down air target detection and tracking; control and display options.

STEP upgrade
The latest variant of the SV radar features an open-architecture arrangement, facilitating future upgrades and utilising Commercial-Off-The-Shelf (COTS) technology to keep costs down. During April 2001, a processor upgrade for the SV radar was announced, Dy 4 Systems' System Technology Enhancement Programme (STEP). The upgrade comprises a 400 MHz G4 PowerPC dual-processor board, replacing chips believed to be in the range of 10–20 MHz. While the core processor upgrade significantly enhances all of the baseline capabilities of the radar, of particular

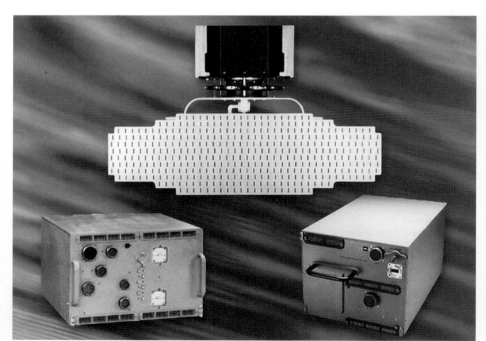

SeaVue surveillance radar system, showing antenna, receiver/exciter/synchroniser/processor, and transmitter units
0521913

significance is the increased performance in littoral waters and against small targets in high sea states, critical areas for Search And Rescue (SAR) platforms.

Raytheon claims this upgrade path has allowed the company to save time and money in the development of the system. STEP will be able to accept still faster chips and boards in the future.

The Dy 4-enhanced variant is now the baseline for new-build SV systems, which are suitable for both fixed- and rotary-wing aircraft; Raytheon

is planning to build 15 to 20 Dy 4-enhanced SV systems a year.

Specifications
Detection performance (claimed):
(tanker) 230 n miles
(patrol boat) 95 n miles
(life raft) 30 n miles
Transmitter: coherent, solid-state, TWT; 8, 15 or 50 kW peak power options; 9.5–10 GHz; linear FM, fixed, frequency agile, or biphase coded waveforms

The JASDF's U-125A SAR aircraft are equipped with the SV-1011 variant of the SeaVue radar (Kanematsu/Raytheon)
0011044

Antenna: parabolic or flat plate; stabilised, 360° scan; 6, 60, 120 rpm; sector and searchlight search, IFF capability

Receiver/Exciter/Synchroniser/Processor (RESP): linear FM pulse compression; digital pulse compression; Sensitivity Time Control (STC); Automatic Gain Contol (AGC)

MTBF: >500 h

Power: 28 V DC; 115 V AC, 400 Hz

Dimensions	Width	Height	Depth	Weight
antenna	1,240 mm	650 mm	n/a	23 kg
transmitter	330 mm	279 mm	498 mm	30 kg
RESP	391 mm	257 mm	498 mm	37 kg

Status

Demonstrated on PC-12 and IPTN CN-235 aircraft. Operational on Royal Australian Air Force Coastalwatch Dash-8, Japanese Air Self-Defense Force (JASDF) U-125A SAR and Italian Customs ATR-42 aircraft. SeaVue is currently used in six countries, including the United States.

In 2001, SV was selected to equip the two EMB-145 Maritime Patrol Aircraft on order from Embraer for the Mexican Air Force. Later that year, the Mexican Navy announced that it would acquire eight systems, to be retrofitted to CASA C-212-200s as part of a wider upgrade of the aircraft as a maritime patrol asset, including an order for Star Safire II multisensor infra-red surveillance systems (see separate entry).

During July 2002, Raytheon was awarded a USD5.7 million follow-on contract for five SV component kits for the JASDF (which currently employs the SV-1011 variant, assembled by Toshiba). Under the contract, Raytheon assembled and delivered five primary SV component kits, configured for small target detection, search and rescue and weather information. The five systems were installed aboard Hawker 800 executive-class aircraft. Raytheon has delivered more than 40 SeaVue kits to the JASDF to date.

During May 2004, Raytheon announced the successful integration of SeaVue and the AAS-52 Multispectral Targeting System (MTS-A, see separate entry) aboard a General Atomics Aeronautical Systems Mariner Unmanned Aerial Vehicle (UAV), a derivative of the Predator B.

Contractor

Raytheon Company, Space and Airborne Systems.

Stormscope® WX-500 weather mapping system

Type

Airborne weather mapping system.

Description

The Stormscope® WX-500 weather mapping system has been designed to interface with the new generation of multifunction displays currently manufactured by Archangel, ARNAV, Avidyne, Eventivde, Garmin, L-3, Honeywell, Sandel Avionics and UPS Aviation Technologies. All sensor functions are controlled through the multifunction display.

The WX-500 with its advanced digital ranging algorithms combines the most popular features associated with the Series I and Series II systems to provide precision mapping of electrical discharges which are associated with thunderstorm activity.

There are two modes of operation – cell mode and strike mode. It operates in 25, 50, 100 and 200 n mile ranges, and displays a 360° view of electrical activity and a forward 120° view.

Heading stabilisation, a strike rate indicator, and continuous (and operator initiated) self-tests are other features of the unit.

Specifications

Dimensions:
antenna 25.4 × 87.6 × 174 mm
WX-500 processor w/o tray
130.8 × 44.6 × 228.6 mm

Weight:
antenna 0.42 kg
WX-500 processor 1.12 kg

Power supply: 11 to 32 V DC
Temperature range: −55 to +70°C
Altitude: 55,000 ft MSL

Status

In production and in service in a wide variety of aircraft.

Contractor

L–3 Communications, Avionics Systems.

Stormscope® WX-900 weather mapping system

Type

Airborne weather mapping system.

Description

The Stormscope WX-900 weather mapping system maps electrical discharges and thunderstorm activity, clearly alerting crews to storms containing lightning.

Unlike radar, the WX-900 displays the electrical discharges associated with cumulus and mature and dissipating thunderstorms. It provides a full 360° view of weather or 180° during monitor modes, with pilot-selectable ranges of 25, 50 and 100 n miles (46, 92 and 185 km). The Supertwist LCD is a self-contained panel-mount unit with electroluminescent backlighting.

The WX-900 performs a self-test after turn-on, then performs tests continuously during system operation. The system features push-button selection of view, pilot initiated self-test programme, brightness adjustment, time mode which provides elapsed flight time and approach timer functions, battery monitor mode which monitors the aircraft electrical system and a noise analyser mode. An integral service menu includes strike test, spectrum analyser and board test modes.

Specifications

Dimensions:
(antenna) 254 × 87.6 × 167 mm
(display/processor) 86 × 86 × 192 mm

Weight:
(antenna) 0.42 kg
(display/processor) 0.71 kg

Power supply: 10.5 to 32 V DC, 8 W

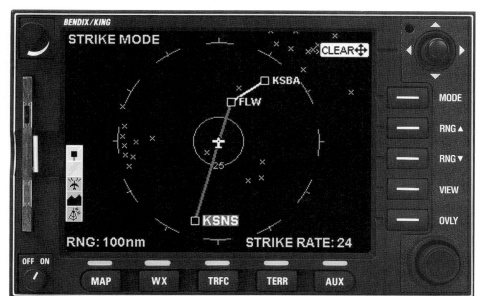

Stormscope® WX-500 radar display on Bendix/King KMD 850 MultiFunction Display (MFD) (Bendix/King)
0589142

Temperature range:
(display) 0 to +55°C
(antenna) −55 to +70°C
Altitude: up to 20,000 ft

Status
In production and in service.

Contractor
L-3 Communications, Avionics Systems.

Stormscope® WX-900 weather mapping system display
0098571

Stormscope® WX-950 weather mapping system

Type
Airborne weather mapping system.

Description
The Stormscope® WX-950 weather mapping system provides two modes of operation – cell mode and strike mode. In cell mode, the WX-950 detects lightning up to 200 n miles out and displays storm cell areas of greatest intensity. During periods of heavy electrical activity, cell mode can be used to identify a path that avoids thunderstorms. In Strike mode the system detects and maps individual strikes aiding the identification of the beginnings of storms. It uses a high-resolution 3 in ATI display/processor for optimum image precision.

The WX-950 operates in 25, 50, 100 and 200 n mile ranges, displays a 360° view of electrical activity in all directions and a forward 120° view that doubles the resolution for analysing activity ahead.

When configured with a compatible heading system, the WX-950 provides heading stabilisation. A discharge rate indicator, integrity indicator and built-in self tests are other features of the unit.

Specifications
Dimensions:
(antenna) 25.4 × 87.6 × 174 mm
(display/processor) 3 ATI
Weight:
(antenna) 0.38 kg
(display/processor) 1.3 kg
Power supply: 11 to 32 V DC
Temperature range: −20 to +55°C
Altitude: 55,000 ft

Status
In production and in service.

Contractor
L-3 Communications, Avionics Systems.

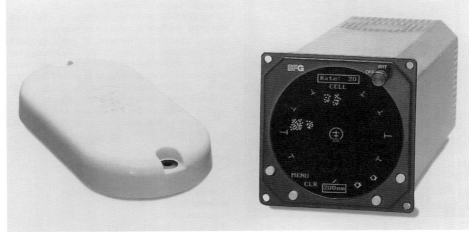

Stormscope® WX-950 weather mapping system
0001378

Stormscope® WX-1000/WX-1000+/WX-1000E weather mapping systems

Type
Airborne weather mapping system.

Description
The Stormscope WX-1000 and WX-1000+ series are weather mapping systems that map electrical discharges. Time and date information, stopwatch and timing information and checklists can be selected from the main menu. All units feature a continuous self-test with error messages that assure accurate weather detection.

The WX-1000 has pilot-selectable ranges of 25, 50, 100 and 200 n miles (46, 92, 185 and 370 km), and provides either 360° or 120° viewing options. The high-resolution CRT display provides a sunlight-readable storm activity picture that can be used in tandem with conventional radar systems to determine the severity and stages of storms. Electrical discharge information is acquired and stored on all ranges simultaneously. Other features provide six programmable checklists, each containing a maximum of 30 lines with up to 20 characters per line. The WX-1000 can be upgraded to the WX-1000+.

The WX-1000+ is a heading-stabilised WX-1000 which displays digital heading information in degrees when operating in the weather-only modes and a flag advisory in the event of heading source malfunction. The WX-1000+ is also know as the military AN/AMS-2. The WX-1000+ can be upgraded to the 'E' version with navaid option, to display navigational information from a Loran or GPS.

The WX-1000E uses an RS232 or RS422 input from a compatible Loran/GPS to display thunderstorm information overlaid with a course line to 10 Loran or GPS-generated waypoints. Other

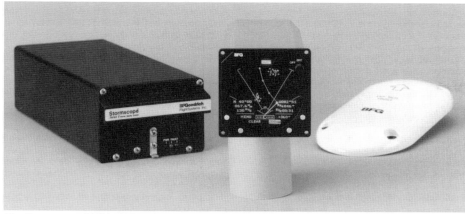

The Stormscope® WX-1000 series consists of (left to right) the processor, display and antenna

0504095

Specifications

Dimensions:
(processor) 86 × 124 × 322 mm
(display) 86 × 86 × 210 mm
(antenna) 29 × 114 × 256 mm
Weight:
(processor) 3.02 kg
(display) 1.03 kg
(antenna) 0.91 kg
Power supply: 10.5 to 32 V DC, 28 W
Temperature range:
(display) −20 to +70°C
(processor/antenna) −55 to +70°C

Status

In production and in service.

Contractor

L-3 Communications, Avionics Systems.

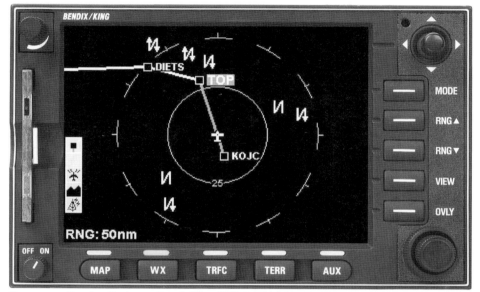

Stormscope® radar display on Bendix/King KMD 550 MultiFunction Display (MFD) (Bendix/King)

0589685

features include a Loran/GPS-generated Course Deviation Indicator (CDI) and the display of 6 of 14 user-selectable Loran or GPS-generated flight parameters. Full flight plan within the selected range can be displayed when used with receivers that transmit flight plan information. The system also gives warning of weak or missing Loran/GPS signals.

The WX-1000E EFIS Interface allows thunderstorm activity to be displayed on a standard EFIS display. Thunderstorm data is transmitted via an ARINC 429 standard databus.

Stormscope Series II The WX-1000E with Navaid and WX-1000E EFIS Interface are also incorporated into the Stormscope Series II.

Target Locating Radar

Type

Airborne surveillance radar.

Description

The Target Locating Radar (TLR) is a self-contained radar detection, location, classification and tracking system. It can be used in aircraft or missiles or UAVs and is totally reconfigurable externally. TLR is available with a variety of wideband antennas. It searches for targets of interest and detected pulses of interest are processed to determine signal characteristics and angle of arrival. These data are reported over the bus interface for multiple targets. TLR can provide data for use by other sensors, the crew or as a direct input to a guidance system or can be used as a mid-course or terminal sensor. Other features of the TLR include light weight, small size and high accuracy.

Specifications

Volume: 0.0142 m³
Weight: 8.2 kg
Frequency: 2–18 GHz
Pulsewidth detection: 100 ns without DF, 600 ns with DF
Target tracks: 30 simultaneously
Accuracy: 1° RMS

Contractor

TRW Avionics Systems Division.

Terrain-Following Radar (TFR) for LANTIRN

Type

Airborne Terrain-Following Radar (TFR).

Description

The LANTIRN TFR retains the basic facilities available in the Tornado, including the ability to operate under enemy jamming in bad weather and at very low level. In addition, system modifications permit the host aircraft to maintain terrain-following capability while executing turns at a maximum rate of 5.5°/s compared with the Tornado's 2°/s, and at angles of bank up to 60°. Whereas the Tornado TFR, and previous terrain-following radars, employed magnetrons, the LANTIRN TFR has a travelling wave tube, giving considerably better performance, particularly in an ECM environment, and higher reliability. The pilot can control the aircraft's flight path by selecting terrain-clearance heights of from 100 to 1,000 ft.

The LANTIRN TFR subsystem began testing in a Lockheed Martin F-16 during 1983 and quickly demonstrated the ability to fly down to 100 ft above the terrain.

LANTIRN under F-16

0503741

Status

In service on US Air Force F-15E and F-16D, Israeli Air Force F-15I and Royal Saudi Air Force F-15S aircraft.

Contractor

Raytheon Company, Space and Airborne Systems.

TWR-850/WXR-840 weather radar

Type
Airborne weather radar.

Description
Collins' TWR-850/WXR-840/WXR-800 weather radar systems employ solid-state electronic and computer technologies to provide operational convenience and radar features more normally found on large air transport aircraft. The compact size and light weight of the TWR-850/WXR-840/WXR-800 makes these radars suited to turboprop commuter aircraft. The solid-state transmitter improves reliability and eliminates one of the most troublesome components of conventional weather radar – the magnetron vacuum tube. In addition to detecting precipitation, the TWR-850 also detects accompanying turbulence by using Doppler frequency shift.

The sensor portion of the TWR-850/WXR-840/WXR-800 system is a single unit comprising an integrated receiver, transmitter, antenna and processing circuitry. This entire unit weighs less than 11 kg and mounts in the aircraft radome. Weather radar system control is via the WXP-850/WXP-840/WXP-800 Weather Radar Control Panel (WRCP) or the DCP-2000 Display Control Panel (DCP). As an option on the WXP-840/WXP-850, two such panels can be installed, providing each pilot independent control of the radar picture. The weather radar picture can be displayed on the Electronic Flight Instrument System (EFIS), Multifunction or Adaptive Flight Display (MFD/AFD) in a four-colour (green, yellow, red, magenta) format. Magenta indicates areas of very heavy rainfall rates of two inches per hour or greater. In addition, the Path Attenuation Compensation (PAC) alert indicates areas of radar shadow due to signal attenuation caused by intervening areas of precipitation.

The radar system is fully digital and requires a minimum of interconnecting wires. Standard ARINC 429 interfaces are used for control information and digital altitude inputs while radar data is transmitted from the sensor unit via digital ARINC 453 buses. The system will also accept conventional analogue altitude signals for antenna stabilisation.

The maximum display range of 300 nm provides long-range viewing capability. The minimum full-scale display range of five nm provides an expanded view useful in the terminal area. Picture interpretation and pilot workload are reduced with the selectable ground clutter suppression mode. A feature of Collins solid-state radar is the automatic rotation of the picture while the aircraft is turned, maintaining radar picture detail. Auto tilt (TWR-850/WXR-840 only) is an additional feature designed to decrease workload. During altitude or range changes, it automatically maintains the antenna tilt angle to the approximate angle desired by the pilot based on the previous selected tilt settings. A sector scan (TWR-850/WXR-840 only) function of ±30° is provided to facilitate rapid picture update. The TWR-850/WXR-840/WXR-800 also provides gain settings both above and below the calibrated position. There are three steps in each direction with each step providing approximately six dB of gain change that result in a total range of ±18 dB (two colour levels) from the calibrated position.

The all solid-state RTA-8XX receiver/transmitter/antenna/processor assembly is the central item of the TWR-850/WXR-840/WXR-800 system. This unit mounts on the forward bulkhead where normally only the radar antenna is situated. Mounted on the forward portion of this unit is the flat plate antenna. Directly behind the antenna is the RF assembly, which consists of a solid-state transmitter and the receiver, which facilitates the deletion of a traditional waveguide. The receiver/transmitter/antenna moves as a single unit as the system scans. This assembly is mounted on a drive mechanism that contains the motors and gears for the scan and tilt process. The drive assembly is attached to the base, a cylindrical unit 15 in in diameter and less than two in depth. The base contains power supplies and signal processing circuitry.

Control of all weather radar functions is performed through the WXP-850/840/800 WRCP, associated with the TWR-850/WXR-840/WXR-800 systems, or the DCP-2000 DCP. All provide rotary, push button and push-pull switches for the pilot and also contain the circuitry required to process the radar data from the sensor unit and output it for use by the display system.

Maximum weather avoidance range is 420 nm.

Specifications
Frequency:
I-band, 9,345 MHz
Scan rate: 27° per second
Transmitter power: 35 W nominal
Tilt: ±15°
Stabilisation: ±30° combined pitch, roll and tilt
Weather avoidance range: 340–420 nm (depending on variant)

Dimensions
(RTA-852/842/800) 220 × 380 mm (L × Ø)
(RTA-854/844) 220 × 380 mm (L × Ø)

Weight
(RTA-852/842/800) 8.8 kg
(RTA-854/844) 9.1 kg
Temperature: −55 to +70° C
Altitude: 55,000 ft

Status
In production and in service.

Contractor
Rockwell Collins.

WXR-700( ) Forward-Looking Windshear radar (FLW)

Type
Airborne weather radar.

Description
Rockwell Collins' WXR-700( ) colour weather radar system is an all solid-state weather radar system suitable for new or retrofit applications with dedicated radar indicators or with electronic flight instruments. More recently, Rockwell Collins has added a Forward-Looking Windshear (FLW) detection capability to the WXR-700( ) radar system. This capability, per ARINC 708A characteristic, provides forward looking windshear detection out to 5 nm and turbulence detection to 40 nm. In addition to commercial airline applications, Rockwell Collins

was selected to provide the US Air Force with the first Commercial-Off-The-Shelf (COTS) FLW radar with skin paint function in conjunction with the PACER CRAG programme.

The WXR-700( ) comprises four units: a slotted array flat-plate antenna (for good sidelobe reduction), a microprocessor-controlled transmitter/receiver, a display unit and a control unit. The microprocessor control system in the transmitter/receiver supervises all control and data transfers, programmes and controls the RF processes such as pulsewidth, bandwidth and PRF selection and directs antenna scan and stabilisation. The unit also contains circuits to reduce ground clutter suppression when operating in the weather mode. An optional feature is pulse-pair Doppler processing, whereby, with the addition of a single circuit board to the transmitter/receiver, the horizontal velocity of rainfall can be sampled. This technique is recommended by the NSSL as being particularly suitable for the analysis of storm cells.

The CRT indicator uses a high-resolution shadow-mask tube with a multicolour display scheme. The CRT provides alphanumeric identification of radar modes and incorporates annunciators and controls. The receiver has a sufficiently wide dynamic range to detect the Z-5 and Z-6 levels of rainfall that indicate a high probability of hail.

It provides both visual and aural alerts of windshear events occurring up to 90 seconds ahead of the aircraft flight path. The radar is automatically activated below 1,200 ft AGL and scans a detection path of 30° either side of the aircraft centreline.

The WXR-700( ) is available at 9.3 GHz X-band frequency as the WXR-700X.

The WRT-701X Receiver/Transmitter (R/T) features a crystal controlled multipulse width transmitter and a truly coherent receiver with programmable bandwidth signal processing. These features provide pulse-pair processing to allow ground clutter suppression, path attenuation compensation, Doppler turbulence detection and provide a high-resolution picture when used with an EHSI or 5ATI LCD. Microprocessor control achieves superior stabilisation performance (with both digital and analogue attitude inputs), flexible installation configuration and programmable growth for future requirements.

Quality enhancements have been implemented in the WRT-701X R/T. Improvements include a single-channel multiplier assembly, redesigned power supply and amplifier and 150 W reduction in power consumption, while maintaining 150 W peak power output.

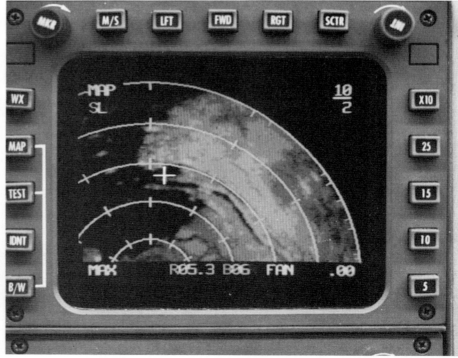

A Rockwell Collins WXR-700 ground-mapping display shown offset to give maximum coverage to the right of the aircraft's track

0503728

In support of OEM decisions to include FLW radar on most production aircraft, Collins has introduced the -612, -623, and -633 versions of the WRT-701X FLW radar containing the newly required EGPWS interface, enhanced BITE and enhanced fault logging memory.

Key features of the FLW radar enhancements include:

- Added EGPWS interface for display alert prioritisation
- Revised Boeing FLW parameters:
- Elimination of 'advisory' icon displays from 3 to 5 nm
- Expansion of takeoff caution alert inhibits to 80 kt to 400 ft AGL
- Maintain takeoff warning alert inhibits from 100 kt to 50 ft AGL
- Expanded 'tilt code' fault message reporting
- Data loadable software via RS-232
- Backwards compatible with existing FLW installations.

Specifications

for C- and X-band systems (equivalent to NATO G- and I-band)
Dimensions:
(antenna) ARINC 708
(transmitter/receiver) 8 MCU per ARINC 600
(indicator) ARINC 708 Mk II
(control unit) 146 × 67 × 152 mm
Weight:
(antenna) 12.25 kg
(transmitter/receiver) 12.25 kg
(indicator) 8.16 kg
(control unit) 1.04 kg
Frequency:
(C-band) 5,440 MHz
(X-band) 9,330 MHz
Power output:
(C-band) 200 W
(X-band) 100 W
PRF:
(C-band) 180–1,440
(X-band) 180–1,440
Pulsewidths:
(C-band) 2–20 µs
(X-band) 1–20 µs
Range:
(C-band) 445 km
(X-band) 593 km

Status

In production and service. The WXR-700 FLW radar has received final type certification approvals for installation in Airbus A-319, -320, -321, -330 and -340. It has also been certified on Boeing aircraft including Boeing -737, -747, -757, -767 and -777.

The US Navy has selected the WXR-700 FLW and Rockwell Collins' FDS-255 flight display system for its C-9 aircraft upgrade.

A special 'missionised' version of the commercial WXR-700 radar, designated the FMR-200X (see separate entry), is being fitted to US Air National Guard (ANG) C/KC-135 PACER CRAG upgrade aircraft as a replacement for the AN/APN-59 radar. The FMR-200X radar will provide a colour weather system incorporating FLW, skin paint detection and limited ground map capability.

A derivative version of the WXR-700, designated WXR-701X, upgraded with Doppler turbulence detection has been selected by the UK Ministry of Defence for the upgrade of its C-130K aircraft.

Contractor

Rockwell Collins.

WXR-2100 windshear radar system

Type

Airborne weather radar.

Description

Collins' WXR-2100 MultiScan™ is a fully automatic synthetic X-band (9.3 GHz) radar that provides advanced weather detection and turbulence avoidance features. The WXR-2100 automatically selects tilt settings, which optimise all weather returns from 0 to 320 nm. A second radar beam provides OverFlight™ protection, which is claimed to virtually eliminate inadvertent thunderstorm top penetrations and significantly reduces the threat of turbulence associated with convective thunderstorms. Growth features such as enhanced OverFlight™ protection and E-TURBn (enhanced turbulence protection) ensure that the WXR-2100 will be able to grow with the development of future safety enhancements.

The WXR-2100 is suitable for new or retrofit applications with dedicated radar indicators or with electronic flight instruments.

The WRT-701X receiver-transmitter (see separate entry) features a crystal-controlled multipulsewidth transmitter and a truly coherent receiver with programmable bandwidth signal processing. These features provide pulse-pair processing to allow ground clutter suppression, path attenuation compensation, Doppler turbulence detection, and excellent image resolution when used with an EHSI or SATI LCD. Microprocessor control is utilised to ensure superior stabilisation performance (with both digital and analogue attitude inputs), flexible installation configuration and programmable growth for future requirements.

In support of OEM fitment requirements for many production aircraft, Collins has introduced –612, -623, and -633 versions of the WRT-701X FLW radar, which include EGPWS interface, enhanced BITE and enhanced fault logging memory.

In detail, the main features and benefits of the system include:

Fully automatic operation:
The WXR-2100 is designed to work in the fully automatic mode. Pilots select only the desired range; tilt and gain inputs are not required.

Clutter free display
Collins' third-generation ground clutter suppression algorithms are employed to eliminate approximately 98 per cent of ground clutter, providing for an essentially 'clutter free' display of threat weather.

Optimised weather detection
Weather data from multiple scans at varying tilt angles is stored in the system memory. When the flight crew selects a desired range, information from the various scans is extracted from memory and merged on the display. Since both long- and short-range weather information is available due to the use of multiple tilt angles, the display presentation presents an optimised weather picture regardless of the aircraft altitude or the range scale selected.

GainPlus™
GainPlus™ incorporates the following functions:

Conventional increase and decrease of gain: The system enables the flight crew to increase and decrease gain during both manual and automatic operations.

Variable temperature-based gain: Variable temperature-based gain automatically compensates for low thunderstorm reflectivity during high-altitude cruise.

Path Attenuation Compensation and Alert (PAC Alert): Compensation for attenuation due to intervening weather is provided within 80 nm of the aircraft. When compensation limits are exceeded, a yellow PAC Alert bar is displayed to warn the flight crew of an area of radar shadow.

OverFlight™ protection: OverFlight protection reduces the possibility of inadvertent thunderstorm top penetration at high cruise altitudes. The system's lower beam information and memory capability are utilised to prevent thunderstorms that are a threat to the aircraft from disappearing from the display until they pass behind the aircraft.

Geographic weather correlation: The WXR-2100 automatically compensates for the changes in reflectivity in different geographic environments to provide a more accurate weather presentation during over water operations.

Comprehensive low altitude weather
The use of multiple tilt angles at low altitude allows the radar to protect against vaulted thunderstorm energy by scanning along the aircraft flight path, scan for growing thunderstorms beneath the aircraft, and view weather at extended ranges.

Windshear detection
Automatic forward looking windshear detection is provided in the landing and take off environment.

Ground mapping
Ground mapping mode enables the detection of major geographical features; such as cities, lakes and coast lines.

Split function control
For Boeing aircraft, the split function control provides the captain and first officer with independent control of range, gain and mode of operation. When operating in manual mode independent tilt control is also available.

Simultaneous display updates
The captain's and first officer's displays update simultaneously during automatic operation, even when different ranges and modes are selected.

Other system features include:

- Added EGPWS Interface for display alert prioritisation
- Revised Boeing Forward Looking Windshear (FLW) parameters
- Elimination of 'Advisory' Icon displays from 3 to 5 nm
- Expansion of take-off caution alert inhibits up to 80 kt/400 ft AGL
- Maintenance of take-off warning alert inhibits from 100 kt/50 ft AGL

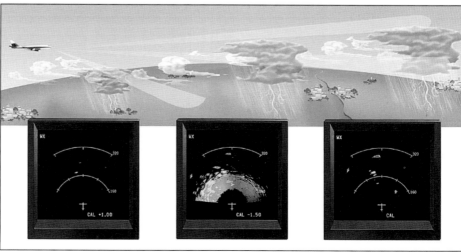

Collins' WXR-2100 MultiScan™ system (Rockwell Collins) 0534159

- Expanded 'tilt code' fault message reporting to facilitate system troubleshooting
- Data-loadable software via RS-232
- FLW detection to 5 nm
- Turbulence detection to 40 nm
- Enhanced ground clutter suppression
- Backwards compatible with existing FLW installations.

Specifications

General
Certification: ARINC 708, 708A, 429, 453, 600; TSOs C63b, C63c; RTCA DO-160D, DO-178B Level C
Interfaces: TCAS, ACARS, EFIS, CMC/CFDS, radio altimeter, EGPWS
Performance index (avoidance): 236 dB (580 nm range)

Receiver-transmitter
Dimensions: 8 MCU ARINC 600 form factor
Weight: 12.07 kg
Input power: 115 V AC, 400 Hz, single phase
Power dissipated: 145 W nominal

Transmitter
PRF: 180 (up to 9,000) pp/s
Pulse widths: 1 to 20 microseconds
Frequency: 9.33 GHz
Peak power: 150 W nominal

Receiver
Noise figure: 4 dB
Bandwidth: optimised to selected pulse width
Minimum discernible signal: −125 dBm typical

Antenna
First side lobe: −31 dB
Size: per ARINC 708
Weight (with flat plate): 12.25 kg, dual; 10.89 kg single
Power dissipated: 100 W nominal
Beam width: 3.5°
Gain: 34.4 dB nominal

Status
In production and in service in a wide variety of commercial aircraft. The system is applicable to Airbus 318, 319, 320, 321, 330, 340, Boeing 737, 747-400, 757, 767 and 777 aircraft.

Contractor
Rockwell Collins.

AIRBORNE ELECTRONIC WARFARE (EW) SYSTEMS

AIRBORNE ELECTRONIC WARFARE (EW) SYSTEMS

Australia

ALR-2002 Radar Warning Receiver (RWR)

Type
Airborne Electronic Counter Measures (ECM), Radar Warning Receiver (RWR).

Description
BAE Systems Australia's ALR-2002 was the first indigenously designed and manufactured Radar Warning Receiver (RWR) to meet the operational requirements of the Royal Australian Air Force (RAAF) F/RF-111C/6 strike, F/A-18 tactical fighter, S-70A Black Hawk and air transport aircraft.

The modular design of the ALR-2002 has enabled development of a family of RWRs, with common use of hardware and software models.

The system utilises a dual receiver architecture with parallel processing to maximise probability of intercept and Situation Awareness (SA).

Key features of the ALR-2002 system include: near 100 per cent probability of intercept against specified emitters; full situation awareness to aircrew during high-density emitter environments; high angle of arrival accuracy over a wide RF bandwidth; the ability to detect a variety of continuous wave and pulsed emitters; software developed in Ada.

ALR-2002 hardware comprises four Quadrant Receivers (QRs), a Low-Band Receiver Unit (LBRU), a Data Processor (DP), a Track and Interface Processor (TIP) and an Azimuth Display Indicator (AZDI). Functionally, ALR-2002 detects, direction-finds, analyses and classifies acquired radar emissions and warns (aurally and visually) its host platform's crew as to which signals pose a threat.

The ALR-2002 RWR has been designed to operate in conjunction with other electronic warfare systems fitted to the host platform. The system can co-ordinate the responses to specific threats from a variety of sensors in addition to its own receivers. It has also been designed to act as the bus controller in a fully integrated EW suite.

The system software is a key element of the design and will be 100 per cent developed by British Aerospace Australia. The signal processing algorithms have been produced using knowledge

BAE Systems Australia ALR-2002 forms part of Project Echidna, a wide-ranging EW upgrade to all of Australia's military platforms (BAE Systems Australia) 0590757

and experience gained through the ALR-2002 Concept Demonstrator programme, which was successfully demonstrated to the Royal Australian Air Force in 1993.

ALR-2002 is an important component of the Royal Australian Air Force's Project Air 5416 (Project Echidna), an integrated EW upgrade to a range of Royal Australian Air Force, Army and Navy fixed- and rotary-wing aircraft, including F-111C, C-130J, S-70A-9, CH-47D and Sea King Mk 50A.

As such, the ALR-2002 system is designated ALR-2002A for strike aircraft (F/RF-111), ALR-2002B for fighter (F/A-18) and ALR-2002D for helicopter (S-70A-9 and CH-47) applications.

Specifications
Emitters: Complex, CW, pulse-Doppler and low/high PRF
Interface: dual redundant MIL-STD-1553B
Power: 28 V DC (MIL-STD-704) and 115 V, 400 Hz
Altitude: 70,000 ft
Temperature: −40 to +71°C (operating)
Dimensions/weight:
AZDI: 95 × 108 × 46 mm; 1.4 kg
QR: 179 × 43 × 194 mm; 2.2 kg (each)
LBRU: 308 × 172 × 81 mm; 5.2 kg
DP: 330 × 287 × 96 mm; 11.8 kg
TIP: 335 × 123 × 193 mm; 9.3 kg

Status
The Australian Department of Defence selected BAE Systems as a 'preferred supplier' for its AUD250 million (USD163 million) project Echidna, to provide the ALR-2002 RWR as part of the RAAF Electronic Warfare Self-Protection (EWSP) system. The type-specific Black Hawk EWSP system is scheduled to achieve Initial Operating Capability (IOC) by the end of 2005.

On 3 February 2003, BAE Systems announced that initial flight trials of the ComBat RWR aboard an RAAF Hornet UpGrade (HUG) standard F/A-18 had been successfully completed. The trials demonstrated equipment performance in a range of operational scenarios and followed-on from a similar exercise carried out with the equipment installed aboard a RAAF F-111 during 2000.

During February 2005, BAE Systems announced that it had been awarded a then year USD120 million for the design, integration and installation of EWSP suites aboard Australian Army CH-47D and S-70A-9 helicopters. Forming Phase 2A of Project Air 5416, the suite proposed included the company's ALR-2002 RWR and its Sensor Independent Integrated Defensive Aids Suite (SIIDAS) controller. The work is scheduled for completion by the end of 2009.

Contractor
BAE Systems Australia Ltd.

Canada

AN/ASA-64 Magnetic Anomaly Detector (MAD)

Type
Airborne Magnetic Anomaly Detection (MAD) system.

Description
The AN/ASA-64 MAD identifies and marks local distortions in the earth's magnetic field induced by the presence of submarines. The operator is alerted by visual and aural alarms, thereby reducing the level of experience needed to operate the system. As the system does not require constant monitoring, the operator can devote more time to other sensors.

CAE completed a product improvement programme in support of the AN/ASA-64 submarine anomaly detector originally built for the US Navy P-3C Orion maritime patrol aircraft. The improved version has increased processing power. A variant for helicopters has also been developed.

Specifications
Dimensions:
(control unit) 102 × 146 × 90 mm
(ID-1559 processor) 229 × 150 × 153 mm
Weight:
(control unit) 0.68 kg
(ID-1559 processor) 3 kg

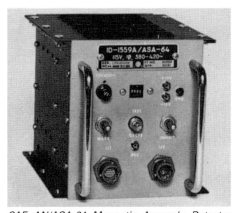

CAE AN/ASA-64 Magnetic Anomaly Detector (MAD) 0503683

Power supply: 115 V AC, 20 W
Environmental: MIL-E-5400

Status
In service.

Contractor
CAE Inc.

AN/ASA-65(V) nine-term compensator

Type
Airborne Magnetic Anomaly Detection (MAD) system.

Description
CAE developed the AN/ASA-65(V) semi-automatic Magnetic Anomaly Detector (MAD) compensator to improve the effectiveness of MAD on aircraft with only manual compensation for aircraft interference with the earth's magnetic field. Previously, fixed-permalloy strips and copper coils were mounted in the MAD boom to create induced and eddy-current fields equal and opposite to those caused by the aircraft. These compensators had to be custom-designed for each individual aircraft, took a long time to adjust on flight test and did not cater for the changes which take place during the aircraft's life. CAE also says that new, more sensitive MAD equipment needs greater precision than fixed compensators can provide.

The AN/ASA-65(V) compensates for permanent interference after only five minutes' flying, compared with about an hour needed for manual compensators. The system allows for manoeuvre interferences after 30 to 45 minutes, improving MAD detection range, especially when frequent manoeuvres are performed and conditions are turbulent. This compares favourably with manual compensation procedures which traditionally took 90 minutes.

The all-solid-state AN/ASA-65(V) is compatible with all current MAD systems. Internal patch connectors are used to adjust the system for the aircraft concerned.

Specifications
Dimensions:
(control indicator) 229 × 146 × 165 mm
(electronic control amplifier) 197 × 149 × 346 mm
(magnetometer assembly) 152 mm cube
(coil assembly) 89 mm cube
Total weight: 13.4 kg
Power supply:
115 V AC, 100 W
28 V DC or AC, 10 W for panel lamps
Figure of merit: <1 gamma
Max compensation field: 50 gamma on each side of aircraft
Reliability: >1,800 h MTBF

Status
In service. The nine-term compensator is used by the US Navy P-3C Orion and S-3A Viking ASW aircraft.

Contractor
CAE Inc.

AN/ASQ-504(V) Advanced Integrated MAD System

Type
Airborne Magnetic Anomaly Detection (MAD) system.

Description
The AN/ASQ-504(V) Advanced Integrated MAD System (AIMS) is an inboard system for helicopters, fixed-wing aircraft and lighter-than-air platforms. This fully automatic system improves detection efficiency while reducing significantly the operator's workload. For helicopter installation, the detecting head is mounted inboard the aircraft, thus providing 'on-top' contact when over a target by eliminating the time delay inherent in a towed detecting head system.

The AN/ASQ-504(V) system combines sensitivity and accuracy with ease of operation. Its advanced signal processing capabilities eliminate aircraft-generated interference, reduce geological and solar interference and provide automatic contact alert, both visually and audibly. Detection data, via the control indicator, (an avionics bus interface or RS-343 video output), allows the operator to determine if the aircraft is within target acquisition range.

AIMS eliminates the hazard associated with towed systems in a helicopter application. It also allows surveillance and manoeuvrability at higher speed, thereby increasing patrol range and detectability and reducing the incidence of false alarms. When used with dipping sonar, transition between the systems can be performed quickly and effectively.

AIMS comprises a very high sensitivity optically pumped magnetometer, vector magnetometer, amplifier computer and control indicator. AIMS can operate independently, or it can accept and execute commands from common control/display units via a MIL-STD-1553 digital databus, while displaying latency free magnetic traces through video output.

Specifications
Dimensions:
(control indicator) 181 × 146 × 145 mm
(amplifier computer) 193 × 262 × 552 mm
(vector magnetometer) 165 × 165 × 165 mm
(detecting head) 178 × 801 mm
Weight: 30 kg
Power supply: 108/118 V AC, 380/420 Hz, single-phase, 200 VA
Sensitivity: 0.01 gamma (in flight)
Feature recognition: automatic target detection; visual and audible operator alert
Environmental: MIL-E-5400

Status
In service. In July 1987, CAE announced a CAD38 million contract to supply 242 ASQ-504(V) AIMS systems to equip Sea King helicopters of the UK Royal Navy and Nimrod MR. Mk 2 aircraft of the UK Royal Air Force. Deliveries started in early 1989 and continued until late 1991. The system has also been ordered by several other countries for various helicopter and fixed-wing applications.

Contractor
CAE Inc.

AN/UYS-503 sonobuoy processor

Type
Airborne Electronic Counter Measures (ECM), sonar system.

Description
The AN/UYS-503 is a small, lightweight acoustic processing system designed for use in a variety of airborne, surface and subsurface surveillance platforms. The system is hosted within a highly-ruggedised and custom-designed ½-ATR CCAs designed to reduce overall system weight and volume and provides very high failure-mode robustness and redundancy. The UYS-503 operates as a complete system by providing all of the required input signal conditioning, signal processing and analysis, post-detection processing, control and display processing and support for digital raw acoustic data recording and replay via a ground-based Fast Time Analysis System (FTAS).

The modular and scaleable architecture of the system facilitates concurrent processing of 8 or 16 sonobuoys. The AN/UYS-503 provides operationally-proven sonobuoy signal processing, target detection, classification, localisation and tracking capability for ASW applications:

- Demultiplexing and processing of all sonobuoy types in current NATO inventory: DIFAR, LOFAR, VLAD, BARRA, DICASS
- Passive processing: narrowband, swathe, broadband and transient signal detection, classification and localisation. Colour GRAMs with bearing encoded as hue
- Monostatic Active: DICASS echo detection and localisation
- Fully-integrable with OEM sonobuoy receivers, data management systems and acoustic and/or mission data recorders
- Tactical situation display showing buoy positions with contact and target position data overlays
- Fully-automated passive/active contact-feature autodetection and target tracking

Specifications
Dimensions: 262 × 396 × 246 mm
Weight: 27 kg (16 DIFAR system)
Frequency: full-band DIFAR
Input channels: Any standard sonobuoy receiver
Control input: MIL-STD-1553B, RS-232C or RS-422
Tactical data output: \MIL-STD-1553B, RS-232C, RS-422 or other as specified
Video output: RS-343 composite video colour or monochrome

Status
The AN/UYS-503 sonobuoy processing system is installed in Australian AP-3C Maritime Patrol Aircraft (MPA). Previous versions of the system are installed in Canadian Seaking (CH-124), USN Seasprite (SH-2) and Australian Seahawk (SH-60B) ASW helicopters as well as ASW/MPA aircraft of Sweden and Japan.

The AN/AQS-970 & 971 sonobuoy processing systems in service with the UK RAF's Nimrod MPA fleet are 32/64 buoy processing evolutions of the previous AN/UYS-503 system. General Dynamics Canada is currently under contract to upgrade the UYS-503 systems onboard the Australian AP-3C to provide a capability broadly equivalent to the AQS-971 (UK Nimrod) acoustic system. The AN/UYS-503 acoustic processor has now been superseded by the AN/UYS-504 sonobuoy processor (see separate entry).

Contractor
General Dynamics Canada.

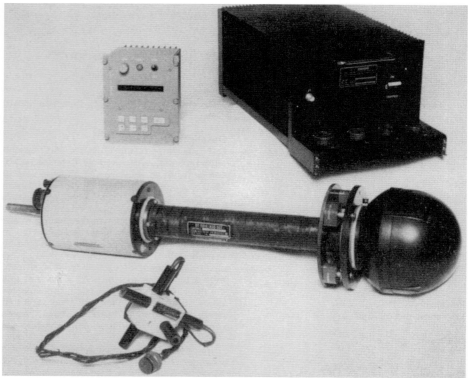

The CAE AN/ASQ-504(V) Magnetic Anomaly Detector (MAD)   0503684

AN/UYS-504 sonobuoy processor

Type

Airborne Electronic Counter Measures (ECM), sonar system.

Description

The AN/UYS-504 is General Dynamics (GD) Canada's latest sonobuoy acoustic processing system. The system has been developed from the UYS-503 (see separate entry) and the AN/AQS-970/971 sonobuoy processors.

The UYS-504 is an Open-Architecture 6U-VME hardware system comprising advanced COTS DSPs, microprocessors, Operating System (OS) and a Windows(™) operator interface. The UYS-504 functions as a complete system by providing all of the required input signal conditioning, signal processing and analysis, post-detection processing, control and display processing and includes an option for a removable disk that allows full mission digital recording and replay via a ground-based Fast Time Analysis System (FTAS).

The UYS-504 features modular and scaleable architecture that facilitates concurrent processing of 8, 16, 32 or 64 sonobuoys. The UYS-504 processing and display functionality is hosted in two distinct hardware configurations – the UYS-504(FW) uses an air-cooled enclosure and is intended for 16/32/64 buoy fixed-wing MPA applications, while the UYS-504(RW) is packaged in a conduction-cooled enclosure in is intended for 8/16 buoy ASW helo applications.

Both the UYS-504(FW) and UYS-504(RW) satisfy all relevant and applicable military environmental qualification standards.

The UYS-504(RW) processing system also features packaging of sonobuoy VHF receivers, sonobuoy processing and processing for the L3 Communications, Ocean Systems HELRAS dipping Sonar in a single conduction-cooled enclosure.

Other features and benefits of the AN/UYS-504 include:

- Demultiplexes and processes all sonobuoy types in the NATO and ABCA inventory: DIFAR, HIDAR, LOFAR, VLAD, BARRA, DICASS, CAMBS, SSQ-110 with sonobuoy Command Function Select capability
- Passive processing: Narrowband, Swathe, Broadband, Diesel Engine, Transient and Wideband Intercept detection, classification and localisation. Colour GRAMs with bearing encoded as hue
- Monostatic Active: DICASS and CAMBS echo detection and localisation, featuring multi-ping history 'Chevron' display formats
- Multistatic Active: Impulsive (SQQ-110) and Coherent (FM & CW) source waveform and echo detection and localisation processing

- Fully integrating with OEM sonobuoy receivers, data management systems and acoustic and/or mission data recorders
- Tactical situation display showing buoy positions, coastlines, bathymetry contours with contact and target position data overlays
- Fully-automated passive/active (tonal, signature and echo-based) contact-feature autodetection and target tracking
- Fully-integrated and automated energy-map buoy-field data fusion: Passive, Monostatic and Multistatic Active localisation/fusion capability
- Comprehensive, hi-fidelity embedded operator training simulation capability for all types of buoys processed with external tactical training scenario controlled by an instructor using a remotely connected laptop PC.

Specifications

System components: VME Acoustic Signal Processor (VASP), Dual-Hard Disk Drive (DHDD)
Dimensions: (W × H × D)
VASP: 390.7 × 269.7 × 498.3 mm
DHDD: 19.1 × 269.7 × 320.5 mm
Weight:
   VASP: 29.5 kg
   DHDD: 12.3 kg

Status

In production and in service.

The AN/UYS-504(FW) (designated MVASP for this application) is installed in Canadian CP-140 (Aurora) P-3 MPA aircraft. The AN/UYS-504(RW) (designated CVAR for this application) has been selected for fleet-wide installation in Canadian CP-148 Cyclone maritime helicopters to be deployed on all City and Tribal Class warships.

Contractor

General Dynamics Canada.

AN/AQS-970/971 sonobuoy processor

Type

Airborne Electronic Counter Measures (ECM), sonar system.

Description

The AN/AQS-971 is a direct evolution of the widely deployed AN/UYS-503 sonobuoy processing system. The AQS-971 system is hosted within highly-ruggedised and custom-designed ½-ATR CCAs, designed to reduce overall system weight and volume combined with high failure-mode

robustness and redundancy. The AQS-971 functions as a complete system by providing all of the required input signal conditioning, signal processing and analysis, post-detection processing, control and display processing and supports digital raw acoustic data recording and replay via a ground-based Fast Time Analysis System (FTAS).

The AQS-971 features a modular and scaleable architecture that provides concurrent processing of up to 32 sonobuoys. The AQS-970 is a 64-buoy processing version of the AQS-971.

Other features and benefits of the AN/AQS-970/971 include:

- Demultiplexes and processes all sonobuoy types in the NATO and ABCA inventory: DIFAR, HIDAR, LOFAR, VLAD, BARRA, CAMBS, SSQ-110 and ALFEA (electro-acoustic active source capability) with sonobuoy Command Function Select capability
- Passive processing: Narrowband, Swathe, Broadband, Diesel Engine, Transient and Wideband Intercept detection, classification and localisation. Colour GRAMs with bearing encoded as hue
- Monostatic Active: CAMBS echo detection and localisation, featuring multi-ping history 'Chevron' display formats
- Multistatic Active: Impulsive (SQQ-110) and Coherent (FM & CW) source waveform and echo detection and localisation processing
- Fully integrating with OEM sonobuoy receivers, data management systems and acoustic and/or mission data recorders
- Tactical situation display showing buoy positions, coastlines, bathymetry contours with contact and target position data overlays
- Fully-automated passive/active (tonal, signature and echo-based) contact-feature autodetection and target tracking
- Fully-integrated and automated energy-map buoy-field data fusion: Passive, Monostatic and Multistatic Active localisation/fusion capability
- Comprehensive, hi-fidelity embedded operator training simulation capability for all types of buoys processed with external tactical training scenario controlled by an instructor using a remotely connected laptop PC.

Status

In production and in service.

The AN/UYS-971 (32-buoy) sonobuoy processing system is installed onboard UK RAF Nimrod MR2 maritime patrol aircraft.

The AN/AQS-970 (64-buoy) will be deployed on the UK RAF Nimrod MRA4 aircraft.

Contractor

General Dynamics Canada.

Chile

DM/A-104 radar warning receiver

Type

Airborne Electronic Counter Measures (ECM), Radar Warning Receiver (RWR).

Description

The DM/A-104 is a wideband radar warning receiver for helicopters and combat aircraft that provides instantaneous detection of threat radar emitters. The system consists of four orthogonal spiral antennas, each connected to wideband crystal video receiving channels that provide 360° coverage in the 2–18 GHz frequency range for the detection of most search, acquisition and fire-control radars.

A powerful digital processor and advanced de-interleaving software ensure real-time automatic radar sorting and evaluation. A cockpit display presents information on the most dangerous threats. In addition, an audio warning system is sent to the intercom system. The threat itself

is classified and the pilot is informed whether his aircraft is being targeted by a surveillance, acquisition or fire-control radar in lock on mode and whether the transmission is from a continuous wave radar.

The DM/A-104 can be interfaced with the DM/A-202 chaff/flare dispensing system, to provide an automatic self-protection capability, and onboard systems for blanking of self-initiated signals. It features a compact and modular design that does not demand much aircraft space and can be retrofitted easily on any combat aircraft. It is designed for a high level of reliability and maintainability, with a complete BIT capability for online diagnosis.

Specifications

Frequency: 2–18 GHz in 4 sub-bands
C/D-band 0.7–1.3 GHz
Sensitivity: –50 dBm
Accuracy: >10° RMS

Contractor

DTS Ltda.

DM/A-202 chaff and flare dispensing system

Type

Airborne Electronic Counter Measures (ECM), Counter Measures Dispensing System (CMDS).

Description

The DM/A-202 is a self-protection system for helicopters and combat aircraft that provides chaff and/or infra-red-flares to break the lock of radar and IR-guided missiles. It consists of five LRUs: the cockpit control unit and four launching units containing the chaff and flare cartridges. The control unit includes a bright display which shows the amount of chaff and flare cartridges remaining and a mode selector for selecting one of four different launching sequences.

Each launching unit comprises an easily removable magazine, allowing quick reloading of the cartridges, and the associated firing circuits. A typical configuration consists of 108 chaff cartridges and 54 flare cartridges contained in the four launching units. The system can be expanded to handle up to eight launching units.

Flexibility and ease of operation have been achieved by the use of a fast reprogrammable microprocessor that ensures adaptability of the system to changing tactical situations. Simplicity of operation has also been enhanced by the installation of control switches on the aircraft throttle and/or stick, allowing the pilot to operate the system without removing his hands from the controls. Modes of operation are manual, semi-automatic and automatic. A safety switch is incorporated in the system so that all the stores can be jettisoned in an emergency.

To avoid reducing the aircraft operational load carrying capacity, the launching units are normally attached externally to the rear fuselage or to the sides of the ventral and wing pylons. Internal or semi-recessed installation can also be adopted, depending on the aircraft configuration and available space.

The DM/A-202 also provides an interface with the DM/A-104 radar warning receiver for full aircraft self-protection.

Status
In service.

Contractor
DTS Ltda.

EWPS-100 EW system

Type
Airborne Electronic Counter Measures (ECM), integrated suite.

Description
The EWPS-100 has been developed to protect helicopters and combat aircraft from present and future radar-controlled weapon systems. It operates over the 0.7 to 18 GHz frequency band and provides a low-cost and effective answer to operational requirements in the air-to-air and air-to-ground roles. An integrated EW system architecture has been developed using modular techniques, proven software and hardware building blocks. The EWPS-100 integrates the DM/A-104 radar warning receiver, DM/A-202 chaff/flare dispenser and DM/A-401 self-protection jammer.

The main features of the EWPS-100 are a high probability of threat interception, high sensitivity, initiation of countermeasures, power management in time and frequency and control of chaff/flare cartridge dispensing.

Specifications
Frequency: 0.7–18 GHz
Detection: pulse, pulse Doppler, CW
Sensitivity: –50 dBm
Maximum pulse density: 500,000 pps
Display: 16 threats simultaneously

Status
In service.

Contractor
DTS Ltda.

Itata ELINT system

Type
Airborne Electronic Counter Measures (ECM), Electronic Intelligence (ELINT) system.

Description
The Itata ELINT system is a high-sensitivity electronic intelligence-gathering system that can detect, locate and measure the parameters of emissions from search, acquisition and fire-control radars. Itata consists of a fully programmable superheterodyne receiver, a digital pulse analyser and a high-gain wideband rotating parabolic antenna which provides 360° coverage and bearing information to an accuracy of within a few degrees. Although intended primarily for light transport aircraft, Itata can also be installed in ships or ground vehicles.

The receiver operates over a frequency range of 3 MHz to 18 GHz in six bands. It can be used either in a wide open mode over the complete frequency range or in a selective mode over a single band. After detection of a transmission, the receiver locks on automatically and measures the frequency and other parameters. Digitised data of each intercepted signal can be recorded automatically for subsequent analysis.

Specifications
Frequency: 3 MHz–18 GHz in 6 bands
Azimuth coverage: 360°
Azimuth beamwidth: (E/F-band) 8°, (J-band) 1.8°
Polarisation: circular

Status
In service with the Chilean Air Force on Beech 99A aircraft.

Contractor
DTS Ltda.

China

BM/KG 8601 repeater jammer

Type
Airborne Electronic Counter Measures (ECM), radar jamming system.

Description
The BM/KG 8601 repeater jammer operates in the E/F- and G/H-bands and is available for installation in strike and fighter/bomber aircraft to counter airborne tracking, anti-aircraft fire control and SAM guidance radars. The jammer has a high power output, minimal repeater delay time, threat management through RF channelling and wide antenna coverage. It provides multijamming techniques.

Status
In service. Production status unknown.

Contractor
Southwest China Research Institute of Electronic Equipment.

BM/KG 8605/8606 smart noise jammers

Type
Airborne Electronic Counter Measures (ECM), radar jamming system.

Description
The BM/KG-8605 operates in the I/J-band as a smart noise jammer performing a hybrid type of jamming which incorporates some of the features of both noise and deception jamming.

The BM/KG 8606 operates in the I-band and features both orthogonal and dual circularly polarised jamming techniques.

Both the BM/KG 8605 and the BM/KG 8606 operate in conjunction with chaff/flare dispensers providing cross polarisation jamming and fast and accurate set on. They are small, lightweight and have a low power consumption.

Status
In service. Production status unknown.

Contractor
Southwest China Research Institute of Electronic Equipment.

BM/KJ 8602 airborne radar warning system

Type
Airborne Electronic Counter Measures (ECM), Radar Warning Receiver (RWR).

Description
The BM/KJ 8602 is a radar warning receiver designed for tactical and other combat aircraft. It consists of a digital signal analyser, a CRT display unit, control box, several receivers and a number

The BM/KJ 8602 airborne radar warning system
0503798

of antenna units. It features wide frequency coverage in two bands, 0.7 to 1.4 GHz and 2 to 18 GHz, and is capable of dealing with multiple threats. Automatic sorting and identification of threat emissions are provided. The system can operate in conjunction with ECM units and chaff/flare dispensers.

The BM/KJ 8602 system is compact and lightweight, making it suitable for tactical aircraft where space and weight are strictly limited.

Specifications
Weight: 20 kg
Frequency: 0.7–14 GHz and 2–18 GHz
Response time: 1 s
Capacity: 16 threats simultaneously
Coverage:
(azimuth) 360°
(elevation –30 to +30°
Accuracy: 15° RMS

Status
In service. Production status unknown.

Contractor
Southwest China Research Institute of Electronic Equipment.

BM/KJ 8608 airborne ELINT system

Type
Airborne Electronic Counter Measures (ECM), Electronic Intelligence (ELINT) system.

Description
The BM/KJ 8608 ELINT system detects, locates, identifies and analyses radar emitters deployed on the ground and at sea with high probability of intercept, high sensitivity and the accurate

measurement of parameters. It features wide frequency coverage, high sensitivity and long operational range, automatic signal identification, emitter position fixing capability, operations in a dense RF environment and BITE.

Specifications
Power supply: 115 V AC, 400 Hz; 28 V DC
Frequency: 1–18 GHz
Frequency accuracy: 5 MHz
Coverage: (azimuth) 360°
Accuracy: (bearing) (1–8 GHz) 5°, (8–18 GHz) 3°

Status
In service. Production status unknown.

Contractor
Southwest China Research Institute of Electronic Equipment.

GT-1 chaff and infra-red flare dispensing set

Type
Airborne Electronic Counter Measures (ECM), Counter Measures Dispensing System (CMDS).

Description
The GT-1 is a chaff and flare dispensing set which has been developed to provide self-protection to fixed-wing aircraft and helicopters. Chaff provides radar countermeasures in the 2 to 18 GHz frequency range and flares provide IR countermeasures in the 1 to 3 μm and 3 to 5 μm wavelengths.

The GT-1 dispensing chaff and flares from a Nanchang A-5C aircraft   0503797

The complete system consists of a programme controller, an operations control, dispensers and cartridges and can be interfaced with a threat warning system to form a self-protection system. The standard configuration is 36 chaff and 18 flare cartridges. Dispensing can be manual or automatic and cartridges can be launched singly or in pairs.

Specifications
Weight: 40 kg (without cartridges and cables)
Power supply: 27 V DC, <0.7 A (static), <3 A (ignition)
380 V AC, 50 Hz, 3 phase, 2 kW

Coverage:
(azimuth) 360°
(elevation) 0 to +85°

Status
In service. Production status unknown.

Contractor
China National Electronics Import and Export Corporation.

Denmark

Apache Modular Aircraft Survivability Equipment (AMASE)

Type
Airborne Electronic Counter Measures (ECM), integrated suite.

Description
The AMASE DAS, designed to enhance the survivability of the AH-64D battlefield attack helicopter against Infra-Red (IR) Surface-to-Air Missiles (SAMs), comprises the AN/ALQ-213(V) Electronic Warfare Management System (EWMS), two-colour threat displays and two Stub Wing Pods (SWPs), each of which houses Missile Approach Warning (MAW) sensors and IR decoy flare launchers.

The EWMS control panel is located on the cockpit pedestal in the pilot's cockpit, with the threat display units installed in both pilot and the gunner cockpits. Threat display units are driven by a tactical data unit that can record data for post-flight analysis.

Each of the SWPs is equipped with three MAW sensor heads and two decoy launcher magazines. Two of the sensor heads provide 45° coverage fore and aft, with the third looking vertically downwards on the starboard pod and vertically upwards on the port unit to provide full 360° spherical coverage.

Decoy launcher magazines are arranged to fire outwards and forward.

Status
The AMASE was originally developed to meet a Royal Netherlands Air Force (RNLAF) requirement in support of peace-keeping operations in Afghanistan. TERMA was awarded an initial production contract for eight systems during February 2000. The first AMASE modified Apaches entered operational service in Iraq on 28 May 2000.

The pilot's cockpit of a Royal Netherlands Air Force (RNLAF) AH-64D Apache helicopter, showing the EWMS control panel mounted in the centre pedestal, below the two MultiFunction Displays (MFDs)
1097345

A RNLAF Apache AH64D, equipped with the Apache Modular Aircraft Survivability Equipment (AMASE), at Basra Air Station during operations in Iraq during 2000 1139975

Subsequently, a second contract was awarded for the remaining RNLAF Apaches, with deliveries completed by August 2005. Future systems may include the integration of a Directed Infra-Red Counter Measures (DIRCM) system, a Radar Warning Receiver (RWR), Laser Warning Receiver (LWR) and an active RF jamming capability.

Contractor
TERMA A/S.

C-130 self-protection concept

Type
Airborne Electronic Counter Measures (ECM), integrated suite.

Description
The Royal Danish Air Force (RDAF), the Royal Netherlands Air Force (RNLAF) and the Portuguese Air Force (POAF) decided to implement a self-protection suite for their C-130s consisting of a Radar Warning Receiver (RWR) and an Advanced CounterMeasure Dispenser System (ACMDS) of the ALE-40 type/format, together with different types of active ECM systems and Missile Approach Warning Systems (MAWS).

The concept was developed by TERMA and the RDAF. After extensive testing both in Denmark and in international trials like MACE and EMBOW the RDAF, RNLAF, and the POAF awarded TERMA production contracts. The system is operational on the three air forces' C-130 aircraft.

The installation is a symmetrical installation with eight standard ALE-40 type dispenser magazines on each side and 2×2 magazines in the nose gear bay (for flares only). Four magazines on each side are tilted downwards primarily for ejection of flares, but the system can be equipped with chaff and/or flares in any of the dispensers as required for the mission. The dispensers are operated by four sequencers and two EMI-filters on each side, and two sequencers and one EMI-filter in the nose gear bay.

The RWR antennas and preamplifiers are installed in wingtip pods providing perfect coverage with no shadowing and almost no cable losses. The installation can be performed by the customer or a customer-selected company with supervision by TERMA.

The ACMDS and wingtip pods can be delivered as a kit consisting of wingtip pods, magazine canisters, magazines, sequencer housings including sequencers and EMI-filters, fairing

cable kits and the Electronic Warfare Management System (EWMS). Alternative control systems, as specified by the customer, can be used. Customer Furnished Equipment (CFE) would be the RWR, MAWS, ECM system to be installed.

One of the major benefits of the system is that there is no penetration of the pressure hull. The structural impact and requirements for reinforcements are thus minimised. The impact on other systems is limited to re-routeing of a few tubes, wires and cables.

The EWMS controls the system and provides advanced dispensing techniques for both chaff and flares. The EWMS provides control of all on-board EW-systems – and links them together for automatic response. Test flights have proved that the wingtip pods act as tip fences and reduce stalling speed by a few knots as well as making the aircraft more stable at slow speeds. When the dispensers are not in use, a set of cover plates can be installed to protect magazines and breech-plates from the outside environments.

Status
In service. The system has accumulated 30,000+ flight hours, on C-130 aircraft of the Royal Danish Air Force (three aircraft), Royal Netherlands Air Force (two aircraft) and Portuguese Air Forces (six aircraft).

Contractor
TERMA AS.

Electronic Warfare Management System (EWMS)

Type
Airborne Electronic Counter Measures (ECM), control system.

Description
The EWMS provides a control system for the co-ordination and operation of electronic warfare self-protection subsystems for fighter aircraft, transports and helicopters. In US service the EWMS is designated AN/ALQ-213(V).

The EWMS was originally developed for the F-16 in close co-operation with the Royal Danish Air Force. The objectives were to reduce workload and to ensure prompt and effective use of onboard EW subsystems. It was subsequently developed into a generic EWMS to fulfil the requirements of EW control for a large number of aircraft types, including: fighters, helicopters and transport aircraft.

The EWMS replaces all existing discrete EW control panels and indicators, except the Radar Warning Receiver (RWR) azimuth indicator.

EWMS functions
EWMU and ECAP. All subsystems are controlled from the Electronic Warfare Management Unit (EWMU), either by operating the software driven menus or through dynamic processing and automatic activation by the Electronic Combat Adaptive Processor (ECAP).

Modes of operation include:
- Manual: the pilot/operator selects and activates the countermeasure programme for the threat
- Semi-automatic: the ECAP processor analyses the threat signals, computes the most effective combination of countermeasures and cues the pilot. He then initiates the computer script programme by activating a consent switch located on the control stick. The pilot keeps his hand on throttle and control stick
- Automatic: analysis and selection of the threat adapted countermeasure responses take place as in semi-automatic mode, but the system will automatically initiate the response without pilot intervention. The pilot will be notified via synthetic voice message, display message/graphical symbols or multidimensional audio.
EWPI. Up-front control and display is effected through the Electronic Warfare Prime Indicators (EWPI) (one type for the F-16, and one type for other aircraft). The EWPI presents the pilot with prioritised up-front information about the status of the EW subsystems, as follows:

The TERMA C-130 self-protection concept 0051277

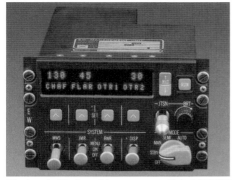

Electronic Warfare Management Unit (EWMU)
for all aircraft types   0051247

- Dispenser: indication of usable decoy payloads remaining, including low level cues and indication of dispenser system activity
- Jammer: operating mode/technique selection and jammer status
- Radar warning: dedicated function buttons and status lights
- Missile warning: subsystem activation and indication of declared missile threat approach angle and time to impact, data displayed depends on the capability of the missile warning system concerned.

TERMA has developed a new Tactical Threat Display (TTD), which will replace the EWPI and display information on a new NVIS compatible colour CRT. As for the EWPI, a dedicated version for the F-16 has been developed, along with a generic version for other aircraft.

EWAP. System software – the Operational Flight Program (OFP) and mission data can be loaded through the Electronic Warfare Aircraft ground equipment Panel (EWAP), or via the MIL-STD-1553B mux bus. The EWMS provides for multiple in-flight selections of pre-flight loaded jamming and dispensing programmes via menus.

EWMS interfaces

In addition to integrating all EW self-protection subsystems into one co-ordinated electronic warfare self-protection suite, the EWMS also interfaces the EW suite with the aircraft core avionics system.

The EWMS provides multiple electrical interface signals, including MIL-STD-1553B avionics and EW buses, RS-232, RS-422, RS-485, and PPD serial digital buses, plus a number of software programmable discrete I/O lines.

The EWMS is designed with a flexible systems architecture, which enables it to integrate and control a variety of EW subsystems. The EWMS functions are not limited to EW system control, the EWMS has also been chosen by the Danish, Belgian and US air forces to operate their F-16 Tactical Reconnaissance Systems. The EWMS provides a multiprocessor architecture, providing growth to meet new EW, and other operational requirements. Examples of EW subsystems controlled by the EWMS are:

- Dispensers: ACMDS, PIDS +, ALE-47, -50
- RWRs: ALR-56M, -67, -68, -69, -69IV, and SPS-1000
- Jammers: ALQ-119, -131, -162, -165, -176, -184

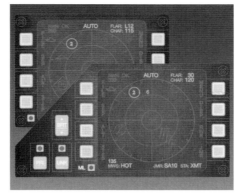

Tactical Threat Display (TTD)   0051250

- Missile Warning Systems (MWS): AAR-47, -54, -57, -60, and EL/M-2160.

EWMS upgrades

TTD. TERMA offers a Tactical Threat Display (TTD), based on multifunction display technology, for aircraft that have inadequate threat display capability. The TTD incorporates a three-dimensional audio warning for the pilot's earphones. The TTD is available in two versions: a semi-triangular version for the F-16, as a replacement for the EWPI and the RWR CRT, and a rectangular version for other aircraft types. The TTD is a graphics colour display and it provides a number of soft keys to operate EW subsystems.

TDU. The Tactical Data Unit (TDU) is a Commercial-Off-The-Shelf (COTS)-derived item, which associates with the TTD to identified operational requirements for integrated EW system mission load, organic threat data recording, post-mission database reporting, and effective analysis by means of the detachable PC-compatible PCMIA Tactical Data Cartridge (TDC), plugged into the unit.

EWMS integrated self-protection suite. The EWMS integrated self-protection suite typically comprises a fully integrated EW self-protection suite, controlled by the EWMS and upgraded with the TTD and TDU. The suite contains an RWR, MWS, advanced countermeasures dispensing system, active ECM jammer, Towed Radar Decoy (TRD) and also in some examples, a reconnaissance pod.

Status

In production and in service. EWMS has been selected by Australia, Belgium, Denmark, Germany, The Netherlands, Norway, Portugal and the US. It has also been selected for the F-16 MLU by Belgium, Denmark, The Netherlands and Norway and is fitted to a number of transport types, including C-130 aircraft of the Royal Danish Air Force, Royal Netherlands Air Force and the Portuguese Air Force. EWMS is installed in US Air Force HH-60Gs, US Air Force Reserve and US Air National Guard A-10s and F-16s. In the US Air Force, the system is also utilised to control the F-16 Tactical Reconnaissance System (TRS).

The Royal Australian Air Force has contracted TERMA to fit the EWMS for an interim upgrade programme to the F-111C, before Project Echnida is completed.

Contractor

TERMA AS.

Modular Countermeasures Pod (MCP)

Type

Airborne Electronic Counter Measures (ECM), Counter Measures Dispensing System (CMDS).

Description

There are three variants of the Modular Countermeasures Pod (MCP) – MCP-7 (seven magazines), MCP-10 (ten magazines) and MCP-F (eight magazines, supersonic).

The basic building block of the MCP system is a barrel module containing two ALE-40 or ALE-47 chaff and flare dispenser magazines and a sequencer switch. Barrels may be indexed at 15° radial intervals around the circumference of the pod in order to provide flexibility in dispensing direction, according to application.

The baseline MCP-7 system is composed of two rotating barrels in the main pod, with a nosecone containing a further dispenser magazine and a tailcone accommodating a safety switch. The strongback is fitted with standard NATO 14 in lugs.

The MCP is controlled from a TERMA EW Management System (EWMS) or similar installation. Options include the addition of RWR, DIRCM, MAW and towed decoys.

The MCP is able to carry various combinations of chaff and flares (except the forward magazine/s of the MCP-7/10, which accommodate only flares), including:

Flares: M206, MJO-7B
Chaff: RR-170, RR-180.

The MCP-F fighter aircraft countermeasures pod, which externally bears a strong resemblance to the Saabtech BOZ 100 series chaff/flare pod widely employed on Panavia Tornado aircraft, features eight magazines for a total of up to 240 standard chaff/flare expendables (typically 180 flares plus 60 chaff rounds). The MCP-F reapplies the technology and design principles applied in the earlier MCP-7 and -10 variants, incorporating two downward firing fixed magazines (typically used for chaff) and six magazines placed in three barrel modules mounted in rotating housings which can be positioned to fire sideways or downward at any angle.

Derived from the 10-magazine MCP-10 pod that is certified on German Air Force C-160 transport aircraft, the MCP-F incorporates new nose and aft sections to make the pod suitable for carriage on supersonic high-performance fighter aircraft. The mechanical and aerodynamic characteristics of the pod permit installation on both port and starboard outboard wing stations of Tornado fighter aircraft and the electrical installation includes a set of pod interfaces to adapt and distribute control inputs from the aircraft EW control system. In addition, the MCP-F pod will include an onboard 3 × 115 V AC-driven power management system to provide 28 V DC power for the pod-mounted subsystems. For its intended Tornado application, there would be no requirement to install the AN/ALQ-213(V) EWMS, since the existing EW controller would be able to support the MCP-F.

Specifications

MCP-7
Length: 2,270 mm
Diameter: 381 mm
Height: 434 mm

Electronic Warfare Prime Indicators (EWPI) for F-16 (left) and other aircraft (right). These indicators will be replaced by two variants of the new TTD   0051248

Forward section of the MCP-10, showing the downward-firing magazines; these are designed for flares only   0566810

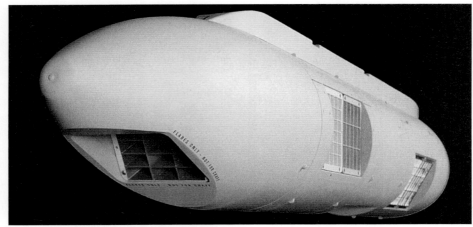

The MCP-7 Modular Countermeasures Pod   0051245

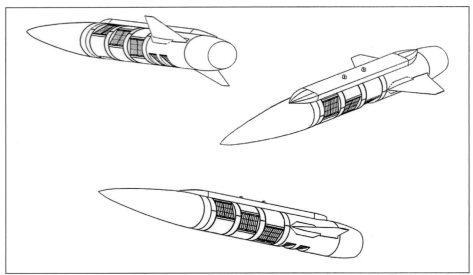

Line drawing of the MCP-F showing the three rotating barrel housings, which facilitate precise ejection of expendables, and two fixed downward-firing magazines. The form resemblance to the Saabtech BOZ 100 series chaff/flare dispenser illustrates the intended recipient of the MCP-F, namely the Panavia Tornado, although the system is compatible with most fast jet types   0566806

Weight:
(empty) 65 kg
(with typical load) 132 kg

MCP-F
Length: 4,000 mm
Diameter: 381 mm
Height: 434 mm
Weight:
(empty) 75 kg
(with typical load) 152 kg (estimated)

Status
In production and in service. The MCP-7 is installed in a twin-pod underwing configuration on Royal Netherlands Air Force Fokker 60U aircraft, typically carrying a load of 240 Infra-Red (IR) flares and 360 chaff cartridges.

The MCP-F, under development since 2002, is aimed at the German Air Force Tornado ECR aircraft based at Lechfeld Air Base. TERMA is offering the pod as a enhancement to the ECR's chaff/flare capability in support of its SEAD role.

Also under contract to the German Air Force, TERMA is supplying 24 CMDS upgrade kits for C-160 Transall tactical transport aircraft. Each kit includes one MCP-10 pod under each wing, one SCD-4 scab-on pod with four magazines each on both sides of the fuselage (aft of the cockpit), an AN/ALQ-213(V) EWMS and colour threat display in the cockpit, combined with an AN/ALR-68 Radar Warning Receiver (RWR) and an AN/AAR-54 Missile Warning System (MWS). In addition to the existing chaff dispensers mounted on the wheel wells (four magazines on each side), a fully fitted Transall will thus mount a total of 36 countermeasures dispenser magazines for a total of 1,080 chaff/flare rounds.

Contractor
TERMA AS.

Pylon Integrated Dispenser System (PIDS)

Type
Airborne Electronic Counter Measures (ECM), Counter Measures Dispensing System (CMDS).

Description
The PIDS system, in the baseline configuration, integrates an F-16 underwing stores pylon with a Counter Measures Dispensing System (CMDS) (AN/ALE-47 or Chemring Chaffblock), thus facilitating no loss of weapons carriage capability. In parallel, a variant of PIDS, designated the Electronic Combat Integrated Pylon System (ECIPS) integrates a Radar Warning Receiver (RWR) to the pylon in place of the CMDS. An enhanced version of PIDS, designated PIDS +, is the latest standard of the system designed for use on F-16 stations 3 and/or 7. PIDS + integrates a RWR, CMDS and a passive Ultra-Violet Missile Warning Systems (UV MWS) into the F-16. The system is integrated with, and operated from, the existing onboard control system (AN/ALE-47 control panel or TERMA Electronic Warfare Management System (EWMS)). Standard F-16 A/B/C/D pylons can be modified and the PIDS can be used on standard F-16s as standard weapon pylons. Existing PIDS can be retrofitted to PIDS + configuration. Compared with standard fuselage dispensers, the greater turbulent airflow produced aft of the PIDS provides for better chaff blooming, especially at high-tactical airspeed.

On F-16 MLU aircraft, only a cable kit has to be installed from the electronics bay to the wheel well. On non-MLU F-16 aircraft an additional harness kit is required, installed in the wheel well. The nose and aft sections of the pylons are rebuilt to accommodate the dispenser system and the MWS. The pylons are modified at the manufacturer's facility inclusive of repaint and a complete systems check, thus ready for immediate installation when returned.

The PIDS + dispenser comprises two standard AN/ALE-40/47 ACMDS-type dispensers installed in the aft fairing, interchangeable for inboard/outboard dispensing. The two dispensers are located identically to the two forward dispensers on the baseline PIDS. The MWS installation comprises six sensors (three in each pylon), compatible with AN/AAR-54, AAR-57, AAR-60 and the Rafael Guitar MWS. Up to 90 + per cent sensor coverage is possible with this MWS configuration.

A further variant of the PIDS system, designated ECIPS/PIDS Universal (ECIPSU/PIDSU), supports MIL-STD-1760 weapons operation, with field upgrade from the PIDS + configuration.

The Royal Danish Air Force (RDAF) has 50 + pylons operational with CMDS and the AN/ALQ-162 and associated antennas. With the pylons permanently installed on stations 3 and 7 it is possible to install an adapter on the missile launcher which makes it possible to convert the F-16 from the ground-attack role to air-to-air role in a few minutes. The adapter simply fits to the MAU-12 rack on the pylon. The adapter fits both the AIM-9 Sidewinder and the AIM-120 AMRAAM. The advantages are thus that the conversion time from ground-attack to air-to-air is greatly reduced and that the wing loading is reduced compared with installation on stations 2/8.

Specifications
	Standard pylon	PIDS	PIDS +
Weight	125.5 kg	160.9 kg	170.9 kg
Length	2,515 mm	2,566 mm	2,802 mm
Width	114.3 mm	228.5 mm	256.5 mm
Height	381 mm	381 mm	381 mm

Status
In production and in service.

In the mid-1980s, the Royal Danish Air Force tasked TERMA to develop a modification for the weapon pylon of the F-16 to accommodate additional chaff dispensers and new dispenser

TERMA F-16 PIDS +   0051276

electronics. The F-16 PIDS was developed and is now operational in the Belgian Air Force, Royal Danish Air Force, Royal Netherlands Air Force, US Air Force Reserve and US Air National Guard with more than 600 units having been delivered.

The development of PIDS + and PIDSU was conducted under a multinational programme sponsored by the airforces of Belgium, Denmark, Netherlands and Norway.

At the end of 2004, TERMA announced a collaborative project, in partnership with EADS Defence Electronics, to enhance the operational effectiveness and survivability of Danish Air Force F-16AM/BM fighters. TERMA will integrate an EADS-supplied Missile Warning System (MWS) based on the AN/AAR-60(V)2 MILDS-F. The MWS will comprise six MILDS-F sensors, three on each of two new integrated pylon units on underwing stations 3 and 7. An Electronic

Combat Integrated Pylon System Universal Plus (ECIPSU +) will be mounted on station 3, together with an integrated Northrop Grumman AN/ALQ-162(V)6 radar jamming system. A PIDSU + will be mounted on station 7, plus the MILDS-F computer processor and two chaff/flare dispenser magazines.

Contractor
TERMA AS.

France

5000 series chaff/flare countermeasure dispensers

Type
Airborne Electronic Counter Measures (ECM), Counter Measures Dispensing System (CMDS).

Description
Alkan has developed countermeasure dispensers that can accommodate either chaff or infra-red flare cartridges or a combination of the two. The cartridges are arranged in interchangeable, easily handled magazines loaded in the modular dispenser. Typically, a single dispenser will accommodate five to seven modules. Each module houses a magazine containing, for example, 18 × 40 mm diameter chaff cartridges.

The dispenser electronic management system performs the firing sequences created by software whether it is connected to an RWR or not. It permanently manages the inventory of available cartridges and provides the necessary information to the cockpit control unit which displays the status of the complete equipment.

The CADMIR dispenser pylon using the same technologies is specifically designed for semi-conformal installation on the Mirage 2000 and 2000–5.

Status
The system is in service in French Air Force Jaguar aircraft. Two dispensers are fitted under

the wing in a conformal installation near to the aircraft fuselage. Each dispenser contains seven modules (Alkan Type 5020).

The same system is also in service on the Dassault Mirage F1, on which it is installed on a special wing hard point.

The Alkan Type 5081 pod is designed to fit either to the JATO point of the MiG-21 or to any 14 in (356 mm) standard armament hard point. The system was in service on the French Navy Super Etendard until the type's retirement.

The same concept is used in an internal configuration on the Mirage 5. The Type 5013 dispenser which houses four standard magazines is in series production for an export customer.

Contractor
Alkan SA.

ABD detector-jammer

Type
Airborne Electronic Counter Measures (ECM), radar jamming system.

Description
ABD is a self-protection detector/jammer designed for the export version of the Mirage 2000. This multimission aircraft is intended to operate in a wide range of theatres and its self-protection system is able to counter all types of threats.

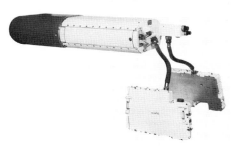

ABD 2000   0009386

ABD is installed internally, thereby avoiding using up ordnance pylon points and/or limiting the aircraft flight envelope. The complete system weighs 80 kg and consists of four LRUs: a main unit containing the receiver, jamming channels, transmitter and aft antennas; a left-hand conformal unit containing the electrical power supplies for the system; a right-hand conformal unit containing the computers controlling the system operation; an antenna providing forward coverage, located on the vertical fin of the aircraft.

The ABD system is capable of detecting and identifying RF emitters, selecting the more dangerous threats, alerting the pilot on the countermeasures display unit and automatically initiating jamming appropriate to the type of threat. It has a very wide frequency coverage, is entirely controlled by microprocessors, and can be programmed in accordance with user's operational scenarios. The system is operated by the countermeasures control panel fitted in the cockpit. Using this panel the pilot can select one of four positions: Off, SIL (the system detects a threat), Jam, Test.

Specifications
Frequency: H-, I- and J-bands (6–20 GHz)
Coverage: all sectors
Reaction time: 0.6 s
Detector sensitivity: up to –60 dBm
Measured parameters: RF frequency, frequency agility, pulsewidth, PRI, duty cycle, sector, polarisation, level, type
Multijamming modes: continuous and spot noise, barrage noise, cover pulse jamming, count down, deception, blinking
Weight: 80 kg

Status
In service on Mirage 2000H aircraft of several air forces. The latest variant of the system, designated ABD 2000, can also be integrated with the Thales Systems Aeroportes/MBDA Integrated Counter Measures System (ICMS – see separate entry).

Contractor
Thales Systemes Aeroportes SA.

The Alkan 5020 decoy dispenser on a Jaguar aircraft   0503799

For details of the latest updates to *Jane's Avionics* online and to discover the additional information available exclusively to online subscribers please visit
jav.janes.com

ASTAC airborne ESM/ELINT system

Type
Airborne Electronic Counter Measures (ECM), Electronic Support Measures (ESM) system.

Description
The Analyseur de Signaux TACtiques (ASTAC) electronic reconnaissance system consists of an internally or pod-mounted airborne sensor package and an associated ground processing station. It is intended to perform detection, identification and location of any radar type in a very dense environment. A datalink between the pod application and the ground station enables a very rapid build-up of the electronic order of battle of the observed area.

The main characteristics of the system are a very wide frequency coverage, wide instantaneous bandwidth, high sensitivity, high discriminating power and high direction measurement accuracy by interferometer. The system is fully automatic, fully reprogrammable and possesses a very high-speed processing capability of up to 20 radars/s. It can process pulse modulated radar with pulse repetition internal diversity or agility, radio frequency diversity or agility, as well as pulse compression, Continuous Wave (CW) and interrupted CW systems.

ASTAC uses two wideband compressive surface acoustic wave receivers. One receiver is used to obtain a very precise measurement of the radar frequency and the two together can handle frequency-agile emitters. The system uses interferometer phase-measuring antenna arrays to determine the azimuth of any threat emitter operating within its 0.5 to 18 GHz (with 18 to 40 GHz or 0.1 to 0.5 GHz as an option) frequency range. When packaged as a pod, ASTAC can store acquired data in an onboard recording subsystem as well as transmit it to its associated ground station using an Ultra High Frequency (UHF 300 MHz to 1 GHz) datalink. When installed on a two-

The ASTAC ESM/ELINT pod mounted, under fuselage, on a French Air Force Mirage F1-CR reconnaissance aircraft
0515211

Parameter measurement:

	Range	Resolution	Accuracy
Frequency	0.5–18 GHz	0.125 MHz	0.2 MHz RMS
	18–40 GHz	0.125 MHz	0.4 MHz RMS
PRI	3–32 µs	1 µs	±1.5 µs
	32 µs–16 ms	0.125 µs	±0.25 µs
			<10 ns (technical analysis mode)
Pulsewidth	0.1 µs–2 ms	62.5 ns	60 ns +5% RMS
Pulse Amplitude	50 dB	0.33 dB	
Spectrum	0.125–10 MHz	125 kHz	200 MHz
Antenna rotation period	1–20 s	0.1 s	0.1 s RMS
TOA	0–120 s	0.125 µs	

seat aircraft, the ASTAC system can be configured to display to the rear crew member in tabular and liquid crystal formats. In such an installation, a keyboard is provided for the operator to interface with the equipment.

Specifications
Frequency coverage: 0.5–18 GHz (18–40 or 0.1–0.5 GHz option)
Location accuracy: <1% of the distance between platform and emitter (aircraft at 40,000 ft and Mach 0.9)
Instantaneous area coverage: 164 km2 (aircraft at 40,000 ft and Mach 0.9)
Emitter location time: 2-3 min (111 km range, aircraft at 40,000 ft and Mach 0.9)
Antenna coverage: ±20° (elevation); 120° (azimuth, lateral antenna, 0.5–4 GHz sub-band); 240° (azimuth, forward antenna, 2/4–18 GHz sub-band)
Dimensions (pod): 4,100 long × 406 mm diameter
Weight: 400 kg
Power: <1.7 kVA

Status
The ASTAC system was first demonstrated on a Japanese Air Self-Defence Force (JASDF) RF-4 during 1992.

The ASTAC ELINT system has been installed aboard French Air Force C160 Gabriel aircraft (internal installation), where it was combat proven during the Gulf War. Other installations are reported as the DC-8 SARIGUE NG (internal) and Mirage F1-CR tactical reconnaissance (pod) aircraft together with JASDF RF-4EJ tactical reconnaissance platforms (pod). For the Japanese programme, the ASTAC pod is apparently designated as the TACtical Electronic Reconnaissance (TACER) system, with national contractor Mitsubishi acting as prime.

The French Air Force is known to have used ASTAC operationally over Bosnia Herzegovina and combat missions during the war in Kosovo in 1999.

The ASTAC pod is currently integrated on the Mirage 2000-5 aircraft.

During late 2002, Thales announced the development of an ASTAC Mk 2 configuration that incorporates a new computer unit, frequency converter and high-band radio frequency preamplifier/switching unit.

Contractor
Thales Systemes Aeroportes SA.

ASTAC airborne ESM/ELINT system   0504251

Barem/Barax jamming pod

Type
Airborne Electronic Counter Measures (ECM), radar jamming system.

Description
The Barem/Barax system is a pod-mounted radar band detector-jammer designed for tactical/strike aircraft applications. It incorporates a Travelling Wave Tube (TWT) and provides instantaneous coverage over the H-, I- and J-bands. It can detect, identify and simultaneously jam multiple threats from a variety of ground-to-air and air-to-air Pulse Doppler (PD) and Continuous Wave (CW) radars. The equipment has an extensive memory capacity and features a modular software design that is reprogrammable to cater for future threat developments. The system can also be equipped with a techniques generator which deals with the most modern coherent radar systems.

The M2.0 capable pod includes reception and transmission antennas, a receiver and a transmitter. Threats are detected by the antennas and analysed by the receiver against a wide range of radar parameters. The ultra-wideband TWT amplified transmitter uses noise and deception modes to jam the radar in less than 1 second.

Thales' Barem/Barax jamming pod, shown here installed on a French Air Force Mirage F1, can be flown on tactical aircraft at speeds above M2.0
0504064

Operation is fully automatic. Two threats can be countered at the same time at the aircraft front or rear. Barem/Barax incorporates an integrated superheterodyne receiver that automatically adjusts to the pervading RF environment, can be easily reprogrammed on the flightline and provides flight reports for situation monitoring in quasi real time. For internal installation, the equipment can be packaged into two units. Housed in a pod cleared for flight at speeds above M2.0, the system has front and rear receive and transmit antennas coupled to the superheterodyne receiver and TWT transmitter which work under automatic microprocessor control over threats detected by the threat identification library and the jamming techniques generator are easily reprogrammable on the flight line. Received signals are recorded in flight for subsequent analysis and storage in the library.

Specifications
Dimensions: 3,450 (length) × 160 mm (diameter)
Weight: 85 kg (pod); 65 kg (internal fit)
Power required: 700 VA

Status
In service on French Air Force and export Mirage and Jaguar and French Navy Super Etendard Modernisé aircraft. The system is combat proven and in its modernised form is designated PAJ-FA.

Contractor
Thales Systemes Aeroportes SA.

BEL expendable radar jammer

Type
Airborne Electronic Counter Measures (ECM), Expendable Radar Decoy (ERD).

Description
BEL is an expendable, wideband, coherent jammer for self-protection of aircraft against missiles and tracking radars. It presents a credible, aircraft-like electromagnetic signature with regard to velocity, direction and range. BEL is able to counter Continuous Wave (CW), pulse and Pulse-Doppler (PD) radar threats, in particular those used in active and semi-active missile seeker heads, either air-to-air or ground-to-air and including radars with monopulse direction-finding or other counter-countermeasures. Dimensions of the BEL cartridge are compatible with current-generation decoy dispensers.

Status
BEL was reported to have entered Limited Rate Initial Production (LRIP) during the early part of 2001.

Contractor
Thales Systemes Aeroportes SA.

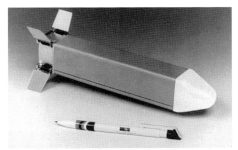

BEL expendable radar jammer   0009389

Carapace threat warning system

Type
Airborne Electronic Counter Measures (ECM), Radar Warning Receiver (RWR).

Description
Carapace is a version of the passive part of the EWS-16 system for the F-16 aircraft (see later item).

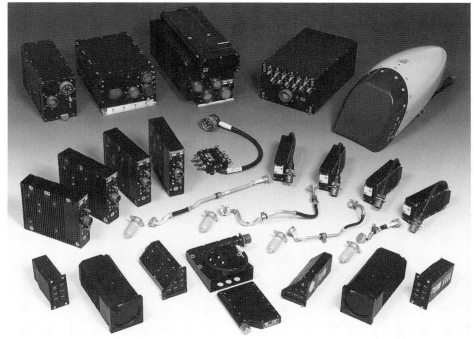

The elements making up the Carapace threat warning system as applied to the two-seat F-16B
0515227

The main part of the system is a hybrid receiver, including IFM, crystal video and superheterodyne receivers, plus an interferometric direction-finding array.

Carapace is believed to cover the C to K frequency bands (0.5–40 GHz) and is designed to detect, identify and localise all modern threats with great accuracy. It is able to analyse accurately, on a pulse-to-pulse basis, emitters at long range in severe ECM environments and offers an ESM capability. Accuracy of direction-finding is classified but is probably of the order of 1°.

Status
Carapace is in service with Belgian Air Force F-16 aircraft and was employed successfully in offensive missions during Operation Allied Force in 1999.

Contractor
Thales Systemes Aeroportes SA.

Corail countermeasures equipment

Type
Airborne Electronic Counter Measures (ECM), Counter Measures Dispensing System (CMDS).

Description
Corail is a radar and optronic countermeasures equipment designed for the Mirage F1 fighter. The system is housed in a conformal pod; two pods can carry up to 256 decoy cartridges according to type – EM, IR and EO cartridges can be used.

Status
In service with Mirage F1 aircraft of the French Air Force.

Contractor
MBDA.

Corail self-protection countermeasures system under the French Air Force Mirage F1 CT wings
0504247

DDM-SAMIR missile launch detector in Rafale configuration   0001258

DDM-SAMIR missile launch detector

Type
Airborne Electronic Counter Measures (ECM), Missile Approach Warner (MAW).

Description
The DDM-SAMIR (Système d'Alerte Missile Infra Rouge) missile launch detector provides automatic detection of ground-to-air and air-to-air missile threats by detecting the missile launch plume, locating the missile in flight and instantaneously transferring threat data to the aircraft ECM system. The equipment makes use of passive IR detection techniques and is reported to be able to function in severe countermeasures environments, detecting all types of missile and optimised against very short-range air defence weapon systems. The system features advanced algorithms which facilitate classification of threats and rejection of parasitic sources in order to generate an extremely high probability of detection, combined with low false alarm rates. Performance is further enhanced by the use of a multispectral mosaic detector array to provide IR signature discrimination.

DDM-SAMIR is made up of two component modules (an electro-optical head and a signal processing unit) that can be packaged to meet specific platform size and weight requirements. Applications include combat, transport, VIP transport and rotary-winged aircraft types. Complete angular coverage is achieved via the use of multiple electro-optical heads, with each head providing 180°-sector coverage. DDM-SAMIR has been produced in two configurations:

Mirage 2000 variant
The DDM-SAMIR Mirage 2000 variant features a combined electro-optical head and signal processing unit integrated into a single assembly mounted at the aft end of the Magic AAM launchers on either side of the aircraft. In this application, DDM-SAMIR is understood to be integrated with the aircraft's SPIRALE Counter Measures Dispensing System (CMDS – see separate entry).

Rafale variant
The DDM-SAMIR Rafale variant forms part of the aircraft's SPECTRA Defensive Aids Sub System (DASS). In this application, the electro-optical head and signal processing unit form separate modules.

Specifications
Weight:
  (electro-optical head) 5.6 kg
  (signal processing unit) 3.6 kg
Coverage: 180° (per sensor)
Accuracy: ±2°

Status
In service with French Air Force Mirage 2000N, 2000D and Rafale aircraft. Full-scale development of the DDM-SAMIR airborne missile detector is understood to have begun during 1985, with flight testing commencing in 1988 and aircraft qualification achieved during 1990. An initial DDM-SAMIR (for the Mirage 2000) production contract was awarded during October 1994, with a second award (relating to production of the system's Rafale variant) being received during 1995. DDM-SAMIR is also understood to have been evaluated by the Japanese Air Self-Defence Force (JASDF).

DDM-SAMIR launch detector as installed as part of the countermeasure equipment of a Mirage 2000 combat aircraft   0504250

Contractor
MBDA (prime).
SAGEM SA, Defence and Security Division (subcontractor).

DUAV-4 helicopter sonar

Type
Airborne Electronic Counter Measures (ECM), sonar system.

Description
The DUAV-4 is an active/passive directive sonar designed for submarine surveillance and location (azimuth, distance and radial speed). It is specially designed for use on board light ship-based helicopters such as the Lynx. It may also be fitted on small surface vessels.

The DUAV-4 differs from conventional sonars in its signal processing system, which is designed to give improved detection, especially in severe reverberation conditions such as in shallow waters. The sonar can be operated in either the active or passive mode. True bearing, range and radial speed are measured in the active mode; true bearing only in the passive mode.

A combined display unit permits surveillance display, for initial detection, or plotting display, for precise azimuth determination. Total weight, including the electronic rack, cables and dome, is 250 kg.

Status
The DUAV-4 is in service with the French Navy, Royal Netherlands Navy and several other navies. Over 90 systems have been produced. In some French Navy Lynx helicopters, the DUAV-4 will be replaced by the HS12.

Contractor
Thales Underwater Systems SAS.

ÉCLAIR-M Counter Measures Dispensing System (CMDS)

Type
Airborne Electronic Counter Measures (ECM), Counter Measures Dispensing System (CMDS).

Description
The ÉCLAIR-M CMDS has been designed as a complement to the SPRIALE system (see separate entry) employed on the Mirage 2000 by increasing overall decoy capacity. The ÉCLAIR-M system comprises a mechanical framework, six dispenser racks (each capable of holding eight

Underside of a Mirage 2000, showing SPIRALE (top left) and ÉCLAIR-M (centre) CMDS (MBDA)
0558777

flares or 18 chaff cartridges) and an electronics unit. With ÉCLAIR-M supplementing SPIRALE, flare capacity is quadrupled. The ÉCLAIR-M electronics unit is the same as that employed in the Rafale CMDS, mounted in the Mirage 2000 drag-chute bay. A serial datalink connects ÉCLAIR-M to the SPIRALE digital system unit in order to facilitate system integration and enable the two systems to operate as a fully functional Defensive Aids Sub System (DASS) under control of SPIRALE.

Status
In production and in service for all versions of the Mirage 2000. To date, 260 systems have been ordered.

Contractor
MBDA.

The ELIPS chaff/flare dispenser for helicopters and light aircraft   0504137

ELIPS helicopter self-protection system

Type
Airborne Electronic Counter Measures (ECM), Counter Measures Dispensing System (CMDS).

Description
The Electronic Integrated Protection Shield (ELIPS) has been designed for helicopters and light aircraft. It is a multidecoy launching system adaptable to any kind of chaff or flares, operating in automatic, semi-automatic and manual modes. Direct coupling to any type of Radar Warning Receiver (RWR) or Missile Approach Warner (MAW) for automatic sequence selection and launching is available.

The type and quantity of ammunition available, selected firing sequence and functioning status of the system are displayed either on the cockpit control unit or on a centralised multifunction ECM display. Loading of the various decoys is identified through codes allocated to magazines or through an integrated recognition system.

Two, four or more magazines may be controlled without any modifications to the system.

Specifications
Weight:
(2 chassis configuration) <12 kg empty, <23 kg loaded with IR flares
Firing interval: 25 ms-1 min
System response: <50 ms

Contractor
Alkan SA.

EWR-99/FRUIT radar warning receiver

Type
Airborne Electronic Counter Measures (ECM), Radar Warning Receiver (RWR).

Description
The EWR-99 FRUIT radar warning receiver is designed for helicopters. It is a user programmable, database oriented system covering a very wide frequency range and making use of full-band Instantaneous Frequency Measurement (IFM) technology. The system is able to detect all RF signals including CW, pulse, and pulse Doppler emitters in high-density electromagnetic environments. More than a simple radar warning receiver, the EWR-99 offers smart management of onboard dispensers and can interface with a missile warning system.

Status
In service. Installed in French Army Aviation Eurocopter AS 341/342 Gazelle, AS 330 Puma, AS 332 Super Puma and AS 532 Cougar helicopters; also selected by PZL – Swidnik to equip its Sokol multirole helicopters. Also installed in

EWR-99/FRUIT Radar Warning Receiver (RWR) on a Gazelle helicopter (Thales)   1150267

other countries' Cougar, Puma and Fennec helicopters.

French Army Aviation units employed the EWR-99 in combat operations during Operation Allied Force in 1999.

Contractor
Thales Systemes Aeroportes SA.

EWS-16 self-protection system

Type
Airborne Electronic Counter Measures (ECM), integrated suite.

Description
The EWS-16 is an integrated system which consists of a threat warning system and an active jammer. The threat warning system can detect, identify and localise all modern threats with great accuracy. All threats can be identified in less than 1 second without any ambiguity, even in dense electromagnetic environments. The active jammer features high radiated power and can counter pulsed and CW radars. Analysis of threats is on a pulse-by-pulse basis, with very accurate identification and priority assessment, a clear pilot interface, and full integration with the weapon system. The EWS-16 is fully programmable and makes use of a separate threat library. The ESM subsystem can accurately locate ground-based radars and can record data during flight.

Features of the EWS-16 system include: a management and compatibility unit based on a 32-bit processor; a crystal video receiver and a high-speed wideband superheterodyne receiver; an instantaneous wideband direction-finding interferometer; an IFM receiver; real-time spectral analysis processing; a plug-in Emitter Identification (EID) and mission report module; a multiple threat jammer employing a high-power transmitter; and synergy between jammer and decoy dispenser.

Status
A version of the passive part of the EWS-16, known as Carapace, is in service in the Belgian Air Force for its entire fleet of F-16 aircraft.

Contractor
Thales Systemes Aeroportes SA.

EWS-A radar warning system (AIGLE)

Type
Airborne Electronic Counter Measures (ECM), Radar Warning Receiver (RWR).

Description
The EWS-A (Electronic Warning System for Aircraft) is a compact radar warning system (10 kg) designed for modernisation of aircraft self-protection systems. The system is lightweight and can detect all radar threats including CW, pulsed and pulsed Doppler emitters, even in the densest environments. EWS-A is also sometimes designated as AIGLE.

The key features of the EWS-A are a very short reaction time, high confidence of emitter identification due to Instantaneous Frequency Measurement (IFM) and a 100 per cent probability of interception. The EWS-A will also control jammers and/or chaff launchers (if fitted to the aircraft), and can be integrated to the aircraft weapon system through serial link and/or multiplex bus interface. The EWS-A is fully user programmable.

Status
In service in French Air Force Mirage F1 aircraft. Three variants of EWS-A were developed to meet the following aircraft upgrade programmes and have been validated during flight tests: one for Mirage III, Mirage 5 and Mirage 50 aircraft with Dassault Aviation, one for F-5 aircraft with Northrop Grumman and one for MiG-21 aircraft with Mikoyan.

Contractor
Thales Systemes Aeroportes SA.

Gabriel SIGINT system

Type
Airborne Electronic Counter Measures (ECM), Signals Intelligence (SIGINT) system.

Description
Thales has developed SIGINT electronic intelligence systems for integration onboard aircraft such as the DC-8, C-160 Transall and C-130. One of these, Gabriel, configures Thales' own Analyseur

This airborne picture of the Gabriel SIGINT aircraft shows many of the type's distinctive identification features including the ventral and aft fuselage radomes, the wingtip antenna pod configuration and the dorsal blade aerial array 0008504

de Signaux TACtiques (ASTAC) technology for detection, analysis and localisation of radar emissions and a COMINT subsystem, provided by Thales Communications, for the detection, interception, classification, listen-in, analysis and localisation of radio communications. The system offers a high degree of automation to assist the operators to accomplish all types of mission.

The Gabriel tactical SIGINT platform utilises a modified Transall C-160NG airframe that is externally characterised by the installation of radomes beneath its forward fuselage (retractable) and on both sides of its rear fuselage and vertical fin; wingtip antenna pods; a short-base, interferometric direction-finding blade array above its forward fuselage and modified main undercarriage sponsons. The aircraft accommodates a flight/mission crew of 14, nine of whom are system operators. The aircraft baseline ELectronic INTelligence (ELINT) capability is understood to have been built around a first-generation application of the ASTAC system (see separate entry).

Status
Two C-160 Transall aircraft equipped with the Gabriel system are in service with the French Air Force. Gabriel was used operationally during the Gulf War in 1990-91 and in support of NATO air operations in Bosnia and Kosovo.

During 2002, it was reported that a C-160 Gabriel mission system upgrade was in development and was scheduled to be retrofitted to French Air Force aircraft during 2007.

Contractor
Thales Systemes Aeroportes SA.

HS 12 helicopter sonar

Type
Airborne Electronic Counter Measures (ECM), sonar system.

Description
The HS 12 is an active/passive panoramic helicopter version of the SS 12 small ship sonar and uses the same electronics as that version. It has similar capabilities for operation in shallow or noisy waters and has a system weight of 230 kg, making it suitable for installation on light helicopters such as the Lynx. The HS 12 transducer is lowered and raised by a hydraulic winch at high speed.

Operation in CW and FM modes is possible and digital signal processing is employed by the system's microprocessor. Automatic tracking of two targets and transmission of elements to an external equipment, such as a plotting table, are provided.

The system operates on 13 kHz in the active mode and in the 7 to 12 kHz range passively. A total of 12 preformed beams is employed, giving 30° sectors, with a maximum range of about 10 km. The display consists of four quadrants, these being obtained by processing adjacent beams. The operator can select a CW mode which provides target range and Doppler. FM processing can also be selected and a sector mode is provided.

Specifications
Frequencies: 11.5, 13, 14.5 kHz
Transmission level: >=212 dB/μPa/m
Weight: 230 kg
Transducer cable length: 302 m

Status
In service in the Indian Navy where it equips Sea King helicopters and in the Finnish Navy on Cougar helicopters. The system also equips Super Frelon and Dauphin helicopters of the People's Republic of China Navy.

Contractor
Thales Underwater Systems SAS.

HS 312 ASW system

Type
Airborne Electronic Counter Measures (ECM), sonar system.

Description
The HS 312 is an acoustic system for helicopters, incorporating the facilities of the HS 12 system and the SADANG acoustic processor. The equipment functions in both passive and active CW and MF modes and has a longer range than the HS 12. The acoustic subsystem can be fitted to process sonobuoys simultaneously with the dipping sonar.

Only a single operator is required and the light weight and compactness of the system mean that the HS 312 can be fitted to any type of light- or medium-size helicopter. Performance of the basic components has been improved by integration of the processing units, integration of a standardised keyboard and the use of only one display screen.

Status
In production and in service in Chilean and UAE Navy Cougar helicopters.

Contractor
Thales Underwater Systems SAS.

ICMS Integrated CounterMeasures Suite

Type
Airborne Electronic Counter Measures (ECM), integrated suite.

Description
Thales Airborne Systems and MBDA have developed an integrated internally mounted EW suite for the Mirage 2000 aircraft. This is a highly sophisticated system, known as the Integrated Counter Measures Suite (ICMS), where all parts are linked to a central interface and management unit.

The system incorporates three warning receivers designed by Thales Airborne Systems. These receivers consist of a version of the Serval equipment (see later entry), a superheterodyne receiver to detect CW radar, pulse compression signals and low-power pulse Doppler signals, and a receiver/processor mounted in the aircraft nose to detect missile command links. The missile detector function can also incorporate the DDM-SAMIR (see separate entry) infra-red warning receiver.

Two detector-jammers are included, each with its own receiver which allows it to operate should the basic radar warning receiver be out of action. These detector-jammers consist of a high frequency sub-unit to counteract airborne and surface-to-air threats, and a low-frequency sub-unit to operate against surface-to-air threats in the lower part of the spectrum. The MBDA Spirale chaff/IR flare dispenser (see separate entry) is also included in the overall system to provide passive countermeasures. Spirale is an internally mounted equipment which dispenses stores through openings in the aircraft structure.

The latest version of the system is designated ICMS Mk 3. It includes an improved Radar Warning Receiver (RWR), incorporating

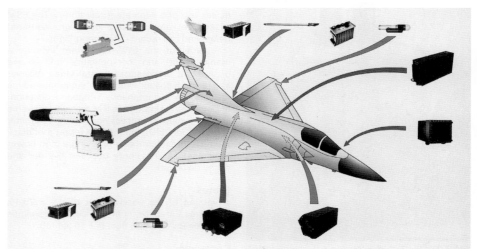

ICMS integrated self-protection system for the Mirage 2000 0051252

instantaneous frequency measurement and a high sensitivity receiver utilising the latest digital technologies. Self protection is provided by improved electromagnetic jamming featuring a Digital RF Memory (DRFM). To increase decoy payload capability, the Éclair-M chaff and flare dispenser (see separate entry) has been added to the Spirale system. The ICMS Mk 3 also performs ESM functions as well as high accuracy threat localisation and tactical situation display.

Status
ICMS Mk 1 and Mk 2 are in service on aircraft operated by the air forces of Greece, Qatar and Taiwan. A derivative is being developed for the UAE Mirage 2000 (Mirage 2000-9) upgrade, designated IMEWS (Integrated Mission Electronic Warfare Suite), by Thales Airborne Systems in partnership with Elettronica and MBDA. IMEWS is reported to incorporate electronically-steered jamming antennas and highly accurate DF.

ICMS Mk 3 is in production for the Mirage 2000-5 Mk2 and Mirage 2000-9.

Contractor
Thales Systemes Aeroportes SA.
MBDA.

LEA (Leurre Electromagnétique Actif)/Spider active expendable jammer

Type
Airborne Electronic Counter Measures (ECM), radar jamming system.

Description
MBDA, in association with Thales Airborne Systems, has developed the LEA/Spider expendable self-protection jammer which can be fitted in current chaff and flare containers. This decoy is made of an electronic payload, using the latest technologies such as MMIC, a GaAs amplifier with a high degree of integration to fit in the limited volume, a vehicle housing the payload and a battery.

LEA/Spider is designed to counter most modern threats such as active coherent missile homing heads. It has the basic capability to defeat monopulse tracking.

Artist's rendition of the LEA Spider jammer deployed by a Mirage 2000 (MBDA)   1027208

Status
During 2001, the LEA active radar was reported as having demonstrated a jamming capability against Pulse-Doppler (PD) during flight trials. As of 2003, LEA was reported available to export customers.

Contractor
Thales Systemes Aeroportes SA.
MBDA.

Modular Self-Protection System (MSPS)

Type
Airborne Electronic Counter Measures (ECM), integrated suite.

Description
MSPS is a lightweight EW suite designed for retrofit applications. It combines the Sherloc

Radar Warning Receiver (RWR) with the Barem/Barax radar band detector-jammer and an Alkan Type 5081 podded decoy system. System capabilities include wide frequency coverage and rapid re-programming. The system features a combined crystal video/superheterodyne receiver subsystem and fast data transmission between subsystems to facilitate automatic responses to detected threats. MSPS-based systems can be installed on a variety of fixed- and rotary-wing aircraft and the suite's jamming subsystem can be installed internally if a sufficiently large space envelope is available within the host airframe. The system can be augmented with add-ons such as laser and missile warners and support jamming equipment.

System weight is 80 kg when mounted internally, rising to 100 kg if the radar jamming subsystem is pod-mounted.

Status
MSPS is reported to be fitted to French Navy Super Etendard Modernisé strike aircraft.

Contractor
Thales Systemes Aeroportes SA.

MSPS is reported as installed in French Navy Super Etendard aircraft   0017853

MWS-20 Missile approach Warning System

Type
Airborne Electronic Counter Measures (ECM), Missile Approach Warner (MAW).

Description
MWS-20 is an active missile approach warning system designed for self-protection of helicopters (MWS-20H), and transport aircraft (MWS-20TA).

The system utilises a Pulse Doppler (PD) missile approach warner, and comprises a

Components of the MWS-20 Missile Warning System (MWS)   1036728

transceiver/processor unit, four antennas and a cockpit display/control box. MWS-20 performs direction-of-arrival, time-to-impact and missile range, speed and bearing calculations on passive, active and semi-active surface-to-air, air-to-air, anti-shipping and anti-radiation missiles that use radar, infrared, laser, fibre optic or wire guidance. The equipment makes use of active technology in order to minimise false alarm rates while providing an all-weather capability and insensitivity to decoys. MWS-20 can also exercise automatic or semi-automatic control over a decoy dispenser subsystem.

The system incorporates built-in test and makes use of miniaturisation to reduce weight and volume. The use of application specific integrated circuitry and solid-state transmitter technology is claimed to enhance the system's reliability.

Specifications
MWS-20 (H and TA)
Angular coverage: 360°
Power Consumption: 600 VA
Weight: <20 kg

Status
In production and in service. MWS-20 has been reported as installed aboard (or specified for) helicopters operated by French Army Special Forces, French Army Horizon radar surveillance and French Air Force Cougar combat search and rescue helicopters, and C-130 transport aircraft. In French Air Force service, MWS-20 is understood to be designated as the Damien system. As of 2004, MWS-20 variants formed part of Thales' SPS-H and SPS-TA integrated self-protection systems for helicopter and transport/wide-body aircraft applications respectively.

MWS-20 has also been employed for self-protection of an Airbus A340 ultra-large civilian transport aircraft as part of the L3-COM Avisys Wipps system.

Contractor
Thales Systemes Aeroportes SA.

PAJ-FA detector/jammer

Type
Airborne Electronic Counter Measures (ECM), radar jamming system.

Description
PAJ-FA is a pod-mounted detector/jammer system, operating in the H- through J-band (6 to 20 GHz), designed to counter acquisition, target tracking and missile seeker radars. PAJ-FA incorporates Digital Radio Frequency Memory (DRFM) technology and offers power-managed amplitude modulation, barrage and spot noise, clutter, combination, false target and range/velocity gate pull-off jamming modes. Other system features include:

- a twin travelling wave tube transmission chain
- full software control
- a user programmable mission library
- prioritisation of detected threats
- the ability to track and jam multiple threats on separate channels
- fore and aft transmission arrays
- output matched to threat polarisation
- built-in test
- generation of flight reports for post-mission analysis
- Mach 2 flight speed compatibility

PAJ-FA can be used as a stand-alone system or as part of an airborne self-protection suite, where it can control associated countermeasures dispensing systems.

Specifications
Frequency coverage: 6–20 GHz
Dimensions: 160 × 3420 mm
(diameter × length)
Weight: 85 kg
Power consumption: 1 kW
MTBF: >250 h

Status
PAJ series detector/jammers were reported as having been procured by France and Spain. Of these, French PAJ systems (also known as Barax or Barem equipment prior to 'modernisation') have, over time, been applied to Jaguar, Mirage III and F1 and Super Etendard aircraft. The Spanish application appears to be flown on Mirage F1 aircraft and is described as being a country-specific variant, manufactured in co-operation with Spain's defence electronics industry. According to Jane's sources, the system received a new techniques generator during the 1990s and in the French Mirage F1 application, is 'fully' integrated with the type's TDS-FA radar warning receiver.

As of mid-2005, the PAJ-FA detector/jammer was confirmed as a fully supported system.

Contractor
Thales Systemes Aeroportes SA.

The PAJ-FA detector-jammer   0009384

SADANG

Type
Airborne Electronic Counter Measures (ECM), sonar system.

Description
The two-operator SADANG system is adaptable to all types of ASW aircraft. It is capable of processing LOFAR (up to 64 simultaneously), DIFAR, DICASS and Barra sonobuoys. The basic module, which can be fitted on ASW fixed- or rotary-wing aircraft, handles mission tactics and processes sonobuoy data on one console manned by a single operator. Automated functions relieve operators from repetitive tasks, particularly during the search phase. Operators can choose to display sonobuoy detection or classification images.

As a processing system, SADANG can be integrated with new-generation data management systems, or can operate in a stand-alone mode, handling complex information, including tactical data and fire control.

The system is designed both to accommodate a large number of configurations and to offer a large potential for evolution, which will enable it to process data from advanced digital sensors such as sonobuoy networks and new active arrays. Additionally, a general purpose beamformer has been developed and tested in co-operation with Australian affiliate Thales Underwater Systems.

Using the latest processing technologies, coupled with current graphics displays and data fusion techniques, the systems are designed for the sanitising and survey of both deep and shallow waters. Up to 16 DICASS can be processed simultaneously, enabling a typical coverage of about 200 n miles.

Advanced processing covers the full spectrum of noise including narrowband frequency lines, broadband signals, swaths, transients and Demon signals.

SADANG 2000 offers an improvement in performance using modern, proven technology. This system features an impressive range of operator aids, including colour-coded software tools for improved classification and data fusion. It also provides lightweight, optimised operations, designed to meet a wide range of threats and tactical situations.

Status
SADANG 2000 1C equips 25 French Navy Dassault Atlantique 2 Maritime Patrol Aircraft. SADANG 2000 1C is also operational aboard the three Pakistani Atlantic 1 MPA. The system has also been acquired by the UAE Navy for its CN235 MPA, where it operates in conjunction with the HS 312 (see separate entry) dipping sonar.

Contractor
Thales Underwater Systems SAS.

SAPHIR chaff and flare dispenser

Type
Airborne Electronic Counter Measures (ECM), Counter Measures Dispensing System (CMDS).

Description
The SAPHIR chaff and flare dispenser system was originally designed for the protection

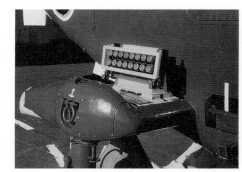

SAPHIR-B self-protection countermeasure system equipping a Lynx helicopter of the French Navy   0504249

of helicopters. The system can be deployed manually, fully or semi-automatically or in a survival mode. The system comprises two to eight cartridge dispensers (each containing 24 to 72 cartridges) and a control unit that facilitates 12 different dispensing programmes which define the number and time interval between cartridges and/or salvos. In addition to standalone operation, SAPHIR can be incorporated into an integrated countermeasures system.

Status
SAPHIR-A was designed to equip the Super Puma helicopter but has also been installed on Puma, Cougar, Lynx and Ecureuil helicopters. During the 1991 Gulf War, French Army Gazelles and Pumas, and French Air Force Pumas were equipped with Saphir A.

SAPHIR-B was designed for light helicopters and is in service on the Gazelle and Lynx.

SAPHIR-M is in production for the NH 90 TTH (Tactical Transport Helicopter), NH 90 NFH (NATO Frigate Helicopter) and Tiger helicopters for both European and export customers. A Saphir derivative is also being proposed for the A400M European transport aircraft.

Contractor
MBDA.

Sarigue SIGINT system

Type
Airborne Electronic Counter Measures (ECM), Signals Intelligence (SIGINT) system.

Description
Thales has developed SIGINT electronic intelligence systems for integration onboard aircraft such as the DC-8, C-130, C-160 and business jets. One such system, known as Sarigue, configures two subsystems. The first is a Thales Airborne Systems ELINT subsystem for detection, analysis, identification and localisation of radar emissions. The other is a COMINT subsystem for detection, interception, classification, listen-in, analysis and localisation of radio communications stations provided by Thales Communications.

The latest version of the overall system is known as Sarigue NG (New Generation). It is based on the ASTAC equipment (see separate entry) and the TRC 290/600 series communications equipment. The TRC 290 receivers operate in the VHF/UHF band. The TRC 600 series is a range of receivers,

SAPHIR-A chaff and flare system operating on a Cougar helicopter   0051251

analysers and direction-finding equipment covering the frequency range 0.1 to 1,350 MHz. The COMINT function can be extended as far as the 20 GHz region via the use of the ELINT subsystem.

Status
A single DC-8-55F Sarigue SIGINT platform was reported to be in service with the French Air Force's 51e Escadron Electronique based at Evreux. Sarigue was used operationally during the 1990-91 Gulf War and has been active in support of UN/NATO operations in Bosnia and Kosovo. In May 1993, Thales Airborne Systems was awarded prime contractorship on the Sarigue NG programme. Sarigue NG utilises an existing French Air Force DC-8 airframe and is in service.

Contractor
Thales Systemes Aeroportes SA.

Sherloc Radar Warning Receiver (RWR)

Type
Airborne Electronic Counter Measures (ECM), Radar Warning Receiver (RWR).

Description
The Sherloc Radar Warning Receiver (RWR) is a system designed for fixed-wing aircraft or helicopters. The system is capable of detecting, classifying and localising all types of pulse radar as well as Continuous Wave (CW) emitters, is able to integrate data from laser and missile warning receivers and to control a decoy dispenser. Sherloc incorporates a crystal video receiver, high-speed digital processor and a radar signals library which is easily reprogrammable on the flight line. The system also delivers operational flight reports. Of modular design, there are many ways to build Sherloc units into a system to meet specific requirements for extensive self-protection.

Threat data is presented to the pilot in the form of alphanumeric symbology on a colour liquid crystal display unit; symbols denoting the identity of the threat are displayed according to relative bearing and signal strength, with up to eight emitters presented simultaneously. In addition, the system incorporates an aural warning facility (including optional synthetic voice).

Alternatively, a simple light-emitting diode display can be used, indicating threat classification and signal strength.

Sherloc operates in the D- through J-bands (1 to 20 GHz), with extensions available at both ends of the range to cover the C- through K-band (0.5 to 40 GHz). System weight is 13 kg for aircraft and 9.4 kg for helicopters.

The latest version of the system, designated Sherloc F, incorporates an Instantaneous Frequency Measurement (IFM) receiver.

Specifications
Weight: 10 kg
Frequency: D- through J-band (1 to 20 GHz); optional extension to C- through K-band (0.5 to 40 GHz)
Receiver: crystal video and IFM
DF accuracy: better than 10°

The Sherloc RWR, in combat aircraft configuration. The display on the extreme right is a smaller unit for helicopters   0514856

Weight: 9.4 kg (helicopter application); 13 kg (fixed-wing application)

Status
In production and in service. Over 200 Sherloc RWRs have been sold. In service in French Air Force C-135F, C-160 and DC-8 transport aircraft, and Mirage 50 and F1 fighters and Super Puma and Dauphin helicopters of export customers.

Sherloc F has been installed on C-160, C-130 (Sherloc-SF) transport aircraft, French Navy Super Etendard naval fighters and Panther, Cougar and Super Puma helicopters.

Contractor
Thales Systemes Aeroportes SA.

SPECTRA EW system

Type
Airborne Electronic Counter Measures (ECM), integrated suite.

Description
Thales Systemes Aeroportes and MBDA have developed the SPECTRA self-protection system for the Rafale ACT/ACM aircraft. The system is the first ever in France to cover electromagnetic, laser and infra-red domains. It makes use of sophisticated techniques, such as interferometry, digital frequency memory, electronic scanning, multispectral infra-red detection, artificial intelligence, image processing and substrate technologies (MMICS on GaAs substrates and VHS INS). SPECTRA includes an active phased-array transmitter.

The system is fitted internally in the Rafale, with over 10 locations distributed throughout the aircraft, and integrated through a specific EW databus and a central processor. SPECTRA may also be installed on the outside of the aircraft.

Thales is responsible for EW system integration, electromagnetic detection, jamming and laser warning functions. The laser warning function is provided by the DAL (Détecteur d'Alerte Laser) system. MBDA provides the DDM-SAMIR (Missile Launch Detector) (two-sensor type) and the chaff and flare dispenser (with four cartridge dispenser modules and two chaff dispenser dual-tubes). Dassault Aviation is responsible for aircraft integration.

Status
The SPECTRA programme was launched during 1990, with the first prototype system being delivered during 1993. Flight trials (aboard a Mystère 20 testbed and Rafale aircraft M02) started during 1994. As of 2004, Jane's sources were reporting the 'F1' SPECTRA suite

configuration as having been in service with the French Navy since December 2000 and as having been used operationally during the French contribution to Operation 'Enduring Freedom'.

More recently, the F2 SPECTRA suite configuration has been delivered to the French Air Force and began operational evaluation during April 2005; the F3 configuration is due to be delivered in 2008. SPECTRA deliveries are expected to total some 185 (F2/F3/F4 configurations) equipments for both the French Air Force and Navy.

Contractor
Thales Systemes Aeroportes SA.
MBDA.

SPIRALE chaff and flare system

Type
Airborne Electronic Counter Measures (ECM), Counter Measures Dispensing System (CMDS).

Description
The SPIRALE chaff and flare system was developed for the Dassault Mirage 2000 fighter and entered service at the end of 1987. The system comprises two cartridge dispensers located under the rear fuselage, two chaff dispensers located at each wingroot and two electronics units. DDM-SAMIR missile warning functionality (see separate entry) can be added, via two infra-red missile detectors mounted in Magic missile launchers. DDM-SAMIR signals are sent to the SPIRALE electronics for final processing. SPIRALE is totally integrated into the Mirage 2000 Integrated CounterMeasures System (ICMS) via the Digibus datalink, with decoy types and sequences related to the threats, but it can also be operated as a standalone system if required. The operating modes are: automatic, semi-automatic, manual, survival and navigation. Different programmes can be selected for each mode.

The SPIRALE ammunition payload consists of 16 IR cartridges and 112 chaff packs. The ÉCLAIR-M cartridge dispenser (see separate entry) can be added to the SPRIALE system to increase decoy payload.

Status
SPIRALE entered service on the Mirage 2000 at the end of 1987. The system is in service with the French Air Force and has been ordered by Greece (Mirage 2000-5 Mk 2 aircraft) and the United Arab Emirates (Mirage 2000-9 aircraft). The system has been employed operationally by French Air Force Mirage 2000 aircraft during multinational operations over the Balkans and Afghanistan.

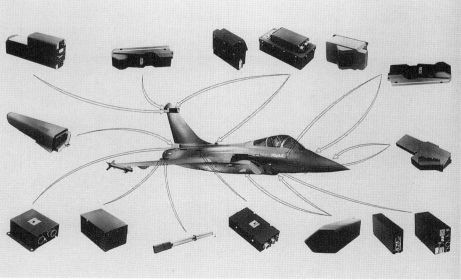

Layout of SPECTRA components within the Rafale airframe   0504246

SPIRALE dispenser mounted on a Mirage 2000 aircraft   0503801

The Sycomor ECM system on a Mirage F1 fighter   0504063

Contractor
MBDA.

SPIRIT electronic warfare system

Type
Airborne Electronic Counter Measures (ECM), Counter Measures Dispensing System (CMDS).

Description
SPIRIT is an electronic warfare system for transport aircraft. It comprises a high-capacity dispenser providing up to 84 cartridges, containing up to 376 flares, on the aircraft. It consists of a Type 5160 decoy dispenser, scabbed on to each side of the fuselage, and an NVG-compatible control unit which offers automatic, semi-automatic or manual operation.

Status
Fitted to French Air Force C-160/C-160NG aircraft interfaced with Thales' Sherloc radar warning receiver and an Elta missile approach warning system.

Contractor
Alkan SA.

SPS-H and SPS-TA Self-Protection Systems

Type
Airborne Electronic Counter Measures (ECM), integrated suite.

Description
Thales has developed the SPS family of integrated DAS systems, designed for helicopter (SPS-H) and fixed-wing transport aircraft (SPS-TA).

SPS-H is available in V1 and V2 configurations. The SPS-H V1 includes the MWS-20 Missile Approach Warner (MAW) and the EWR-99 Radar Warning Receiver (RWR) (see separate entries), combined with a Counter Measures Dispensing System (CMDS). The SPS-H V2 configuration incorporates the TDS-H RWR (see separate entry) replacing the EWR-99. Optional equipment includes various available Laser Warning Receiver (LWR) systems and interface software for additional infra-red and/or radio frequency jammers to complement the CMDS element of the core system.

A central computer exercises threat data fusion, smart control of countermeasures (automatic, semi-automatic and manual modes), decoy control, electromagnetic compatibility functions and interface control via MIL-STD-1553B, ARINC or RS databus.

User self-sufficiency is provided under software control of the Operational Flight Program (OFP), which is clearly separated from the threat library.

As with SPS-H, SPS-TA is available in V1 and V2 configurations, utilising similar components to those described for the SPS-H variants (the SPS-TA V2 utilises the TDS-TA RWR – see separate entry) and offers similar capability packages that are tailored to the needs of fixed-wing transport aircraft such as the C-130 and the C-160.

Status
In production and in service. SPS-H is installed in French Army Eurocopter AS 532 Cougar helicopters and French Air Force C-130 aircraft.

Contractor
Thales Systemes Aeroportes SA.

Sycomor chaff and flare system

Type
Airborne Electronic Counter Measures (ECM), Counter Measures Dispensing System (CMDS).

Description
The Sycomor chaff and flare system is intended for the various export versions of the Dassault Mirage F1 fighter and can be packaged either in a 2.95 m long externally mounted pod or in a 2.5 m conformal pack. Each pack has three chaff dispensing tubes and seven cartridge magazines. Each pod has the capacity for two packs.

Status
Sycomor entered service in 1984 and is operational with Mirage F1s flown by several air forces.

Contractor
MBDA.

TDS radar warning receivers

Type
Airborne Electronic Counter Measures (ECM), Radar Warning Receiver (RWR).

Description
The Threat Detection System (TDS) family of Radar Warning Receivers (RWR) are designed for employment on fighter aircraft (TDS-FA), transport aircraft (TDS-TA) and helicopters (TDS-H).

The Threat Detection System for Fighter Aircraft (TDS-FA) is a compact (10 kg), user-reprogrammable RWR that is designed for the upgrading of existing self-protection suites. The system detects Continuous Wave (CW), pulse and Pulse-Doppler (PD) emitters operating in dense environments. Emitter identification is based on the use of instantaneous frequency measurement technology with a claimed 100% probability-of-intercept. In addition to threat warning, TDS-FA is capable of controlling onboard jammers and/or countermeasures dispensing systems; the system can also be integrated with the aircraft system via serial link and/or multiplex bus interfaces.

Similar to TDS-FA, the Threat Detection System for Transport Aircraft (TDS-TA) includes the same instantaneous frequency measurement technology in a system designed for the requirements of transport aircraft.

The Threat Detection System for Helicopters (TDS-H) RWR completes the TDS family of RWRs, incorporating similar features to TDS-FA and -TA. TDS-H weighs only 4 kg.

Status
In production and service. TDS-FA is installed on Mirage F1 aircraft of the French Air Force, combined with the PAJ-FA jamming pod (see separate entry). Other variants of TDS-FA are reported to have been developed for the Mirage III/V, the F-5 and the MiG-2l.

TDS-TA is in production to equip French Air Force C-130 and C-160 transport aircraft.

TDS-H is operational on French Army helicopters, including Gazelle, Puma and Super Puma, and has been selected for the entire French Army Light Aviation (Aviation Légère de l'Armée de Terre, ALAT) combat helicopter force, and by the Polish armed forces for the Sokol helicopter.

TDS-H is part of Thales' SPS-H V2 integrated helicopter self-protection system.

Contractor
Thales Systemes Aeroportes SA.

French Army Eurocopter Cougar helicopters are fitted with SPS-H   0526397

Upgraded DUAV-4 helicopter sonar

Type
Airborne Electronic Counter Measures (ECM), sonar system.

Description
The upgraded DUAV-4 (DUAV-4 UPG) differs from the DUAV-4 in its acoustic processor which is now based on the SADANG new-generation acoustic processor. The system offers more powerful signal and data processing, as well as an improved man/machine interface.

The DUAV-4 UPG also offers specific software tools and operator aids designed to meet the requirement for detection in shallow, rocky bottom waters.

Status
The upgraded DUAV-4 sonar is in service in Swedish Navy Chinook helicopters.

Contractor
Thales Underwater Systems SAS.

Germany

ALTAS Laser Warning Receiver (LWR)

Type
Airborne Electronic Counter Measures (ECM), Laser Warning Receiver (LWR).

Description
The Advanced Laser Threat Alerting System (ALTAS) is EADS' latest laser threat alerting system. ALTAS is a passive design which detects, tracks and provides warning of hostile laser sources aimed at the host platform. ATLAS detects all sources within the laser waveband, including all Laser-RangeFinders (LRF), Laser Target Designators (LTD) and, optionally, Laser Beam Riders (LBR).

The system provides all necessary laser source parameters, including precise Angle of Arrival (AoA), necessary to facilitate effective countermeasures.

The modular architecture of the LWR provides for easy integration and simple installation in helicopters, large transport aircraft or fighters. All processing activities are performed within the (identical) sensor heads – the system therefore requires no dedicated processor unit. Other key capabilities of the system include a low false alarm rate, short activation time and high AoA resolution, combined with high Mean Time Between Failures (MTBF).

ALTAS can be easily integrated into higher level Electronic Warfare (EW) suites utilising standard interfaces (RS-422 or MIL-STD-1553B databus). Featuring a high MTBF and single LRU architecture, logistic support for the system is greatly reduced over more traditional configurations.

Specifications
FoV: 4 sensors provide 90° elevation × 360° azimuth; six sensors provide 180° elevation × 360° azimuth
Wavelength: 0.5–1.65 μm
Threat capability: 4 simultaneous
Parameters: AoA (2° RMS), Pulse Repetition Frequency (PRF), pulse width, waveband, identification of laser type (via library), laser power level
Dimensions: (L × W × H) 85 × 85 × 95 mm (excluding flange/connector)
Interface: RS-422; MIL-STD-1553B optional
Power consumption: <2 W per sensor
Weight: <1.5 kg per sensor

Status
In production. ALTAS is designed to readily integrate with Electronic Warfare (EW) suites via standard interfaces.

Contractor
EADS Deutschland GmbH, Systems and Defence Electronics, Airborne Systems.

LWR Laser Warning Receiever

Type
Airborne Electronic Counter Measures (ECM), Laser Warning Receiver (LWR).

Description
The EADS LWR is designed to be integrated with the Threat Warning Equipment (TWE) (see entry) to provide multiple threat warning capability for tactical helicopter systems.

EADS' Laser Warning Receiver   0018292

The LWR utilises high-sensitivity near infra-red detectors to provide precise measurement of the angle of arrival of laser pulses in azimuth, with single pulse detection close to 100 per cent, and high immunity against sunlight and explosion flashes.

The LWR provides detection, analysis, identification, bearing, measurement and transmission for visual display and audio alarm of threat laser illumination from laser range-finders and illuminators.

Specifications
Sensor coverage: 360° in azimuth, adjustable to customer requirements in elevation (up to 90°)
Azimuth accuracy: 10° rms
Reaction time: audio alarm within fractions of a second
False alarm rate (after processing): <1.5 per hour
Power consumption: 51 W (2 × 25.5 W)
Dimensions:
  (height) 110 mm
  (width) 129 mm
  (length) 240 mm
Weight: 3.6 kg (2 × 1.8 kg)

Status
Designed to be part of the TWE Threat Warning Equipment, selected for the NH90 TTH (Tactical Transport Helicopter).

Contractor
EADS, Systems and Defence Electronics, Airborne Systems.

PIMAWS

Type
Airborne Electronic Counter Measures (ECM), Missile Approach Warner (MAW).

Description
During the last quarter of 2001, Bodenseewerk Gerätetechnik GmbH (BGT) began flight trials of a new passive infra-red Missile Approach Warning System (MAWS) for airborne platforms. The system, known as PIMAWS, aimed at the Eurofighter Typhoon, has been under development since 1997, covered by a German Defense Ministry (MoD) contract but with significant company funding. Development of the system was reported complete in 2003.

PIMAWS is described as a 'step-and-stare' Imaging Infra-red (IIR) system, which would require only two sensor units to achieve full spherical coverage for a fighter aircraft-sized platform. It employs intelligent real-time image processing algorithms and would be able to track threat missiles from launch up to, and including, the post-burnout phase. This last point is important because when medium-range missiles are part of the threat envelope, an effective MAWS must be able to detect and track hostile missiles even after their rocket motor has burnt out.

BGT has is proposed PIMAWS for incorporation into the company's proposed Defensive Aids Sub System (DASS) for the Airbus A400M transport aircraft. This would involve a three-year industrialisation and engineering effort. BGT has also proposed the system for later batches of Eurofighter Typhoon aircraft. Currently, the EuroDASS self-protection suite incorporates a (Selex-supplied) active radar missile warning system, but BGT hopes that some of the partner nations may prefer a passive system.

Flight trials were performed with an EADS/Dornier Do-228 twin-turboprop aircraft, provided

PIMAWS on Eurofighter   0111343

by the German MoD. Live firing trials involved ground-based scenarios, including simulated missile launches, with emphasis on typical false alert sources.

The PIMAWS sensor unit features coverage of 360° in azimuth and 105° in elevation. Not a traditional staring array, the system uses the principle of 'stop and stare', obviating the need for a large number of sensors per aircraft to achieve full coverage. It is envisaged that a PIMAWS-equipped aircraft will require just two sensors, rather than as many as six conventional types.

The PIMAWS sensor employs a 256 × 256 CMT detector array which operates in the 3 to 5 μm (MWIR) waveband. The automatic scanning rate is such that the sensor can cover the full hemisphere six times per second. It does this by taking 4 × 12 images (each counting 256 × 256 = 65,536 pixels) every one-sixth of a second. This is claimed to provide sufficient reaction time to deploy countermeasures such as chaff, flares or a directed energy system, even against a shoulder-fired missile from short range. The image processing of the sensor data is handled by the same systolic array processor technology that is used in the IIR seeker of the IRIS-T short-range air-to-air missile. The systolic array processor technology features 1,024 individual processors (down to 512 or 256 processors) grouped within a single ASIC-chip.

PIMAWS will also feature a manual scanning mode, allowing the pilot to take a detailed look at a specific sector of interest. However, resolution would not be sufficient to produce reconnaissance-type imagery.

Status
Technology demonstrator. Proposed for the Eurofighter Tyhoon and Airbus A400M.

Contractor
Bodenseewerk Gerätetechnik GmbH (BGT).

Sky Buzzer Towed Radar Decoy (TRD)

Type
Airborne Electronic Counter Measures (ECM), Towed Radar Decoy (TRD).

Description
The EADS Sky Buzzer is a high-power fibre optic Towed Radar Decoy (TRD) system designed for use on fighter and large body aircraft, available in several configurations. When fitted with an integrated Techniques Generator (TG), Sky Buzzer can be operated as a stand-alone offboard jammer or supplemented with an onboard transmitter to extend RF coverage down to lower frequencies. A new generation, wideband DRFM, incorporating special techniques to defeat all modern monopulse radars, forms the heart of the Sky Buzzer ECM TG.

The onboard electronic package and the launch/retrieval mechanism can be housed in several different pods. Alternatively, the towed transmitter may be driven by an existing jammer system and the launcher integrated into the aircraft structure or a wing pylon. Depending on space available, either a parachute-based recovery system or an active winch system for in-flight retrieval can be specified.

Status
Operational tests on F-4 and Tornado IDS aircraft have demonstrated the reliability of the Sky

EADS' Sky Buzzer towed decoy   0018295

Tornado pod installation of the Sky Buzzer towed decoy system   0018294

Buzzer TRD and the effectiveness of the system against the latest generation of radar threats. Highly stable flight behaviour ensures there is virtually no impact on the host aircraft in terms of performance, with no restrictions to the normal flight envelope.

Sky Buzzer is available to order.

Contractor
EADS Deutschland, Systems and Defence Electronics, Airborne Systems.

Tornado ECR system

Type
Airborne Electronic Counter Measures (ECM), Emitter Locator System (ELS).

Description
The Tornado ECR is based on the sixth Tornado production batch build standard and incorporates a MIL-STD-1553B databus, upgraded radar warning and active electronic countermeasures equipment and an improved missile control unit.

Specific Tornado ECR features include an Emitter Location System (ELS) to pinpoint, identify and display hostile radar emitters, a FLIR to enable covert low-level flight in adverse weather conditions and at night, advanced displays combined with more powerful computers to give the crew more time for tactical decision making and last, but not least, the HARM anti-radiation missile.

The Raytheon Systems Company ELS allows the passive autonomous acquisition, identification and precise location in range and angle of radiating threats. The information is displayed to both members of the crew for target selection. After selection, the co-ordinates are used for cueing the HARM missiles as well as for handover to follow-on forces or for onboard storage.

The ELS detects, identifies and locates radar emissions through the use of a high probability of intercept receiver system. It features multi-octave RF coverage, phase interferometric antenna arrays for precision direction-finding, passive ranging channelised receivers and a multiple MIL-STD-1750A digital processor. The system operates across the RF spectrum for all primary surface-to-air and airborne threats. Data acquired by the system is transferred to the tactical displays of both crew members for threat assessment. The ELS is interconnected with the aircraft avionic and defensive aids databusses,

and the emitter library is loaded from the mission data transfer system. The ELS provides for HARM targeting, with some ECR aircraft also equipped with a special data recorder, which records all raw data from the ELS receivers for post-flight ground evaluation.

To further improve the ranging performance of the ELS, the Cross Platform Sensor Correlation technique was developed, utilising data from two or more ECR aircraft to refine the overall ranging solution; an experimental version of this system was flown by the German Air Force (GAF) during the second quarter of 2003, proving its effectiveness and enhanced performance over single aircraft sensing.

The internally mounted FLIR, developed by Zeiss, provides both aircrew with navigation, reconnaissance and attack information in adverse weather and night conditions. FLIR enhances the covert penetration and attack capabilities of the Tornado ECR.

Status
In service with the German and Italian air forces. German Airforce Tornado ECR aircraft flew operationally over Bosnia and Kosovo in support of NATO peacekeeping missions.

Contractor
EADS Deutschland.

Tornado Self-Protection Jammer (TSPJ)

Type
Airborne Electronic Counter Measures (ECM), radar jamming system.

Description
The TSPJ is a third-generation self-protection jamming pod, designed by EADS for German Air Force and German Navy Tornado aircraft.

Main design features are: it is a generic, modular, jamming system with no threat-specific hardware; has an integrated receiver/processor; has a large repertoire of jamming techniques; all functions are user-programmable via Mil-Bus; there are autonomous and radar warning controlled modes; the modular design and intelligent interface permits adaptation to other platforms.

The system includes a Digital Radio Frequency Memory (DRFM) for the production of coherent jamming signals to defeat coherent

The TSPJ self-protection jammer is carried on the Tornado outer wing station   0051253

German Tornado ECRs flew SEAD missions in support of NATO operations over Bosnia and Kosovo   0069282

threat radars; the DRFM developed by EADS not only contains the RF and memory section to digitise and store signals precisely, but also all techniques generators needed for generation of the required jamming signals.

Operationally, the TSPJ provides a fully software programmable ECM system, capable of operation against both coherent and non-coherent threat systems.

Status

The first TSPJ systems were delivered to the German Air Force in early 1998. In production.

Contractor

EADS Deutschland, Systems and Defence Electronics, Airborne Systems.

International

Advanced Self-Protection Integrated Suite (ASPIS)

Type

Airborne Electronic Counter Measures (ECM), integrated suite.

Description

Northrop Grumman Advanced Systems Division, Raytheon Electronic Systems and BAE Systems North America have teamed to produce the Advanced Self-Protection Integrated Suites (ASPISs) which automatically detect and counter hostile threats.

ASPIS uses interchangeable subsystems, so the configuration is flexible. The suite can grow by upgrading existing subsystems, by adding subsystems or by reprogramming to counter the changing threat environment. This flexibility also facilitates installation in a wide variety of aircraft.

ASPIS consists of the Northrop Grumman AN/ALR-93(V) RWR/EWSC, the Raytheon ALQ-187 active jammers and the BAE Systems North America AN/ALE-47 Counter Measures Dispensing System (CMDS).

Interfaced with the AN/ALR-93(V), the Raytheon AN/ALQ-187 is a fully integrated, power-managed electronic countermeasures system. It can counter multiple pulse, pulse Doppler and CW threats, including SAM, AAA and air-to-air weapons. The ALQ-187 features user-programmable threat data and ECM techniques files for automatic detection of single or multiple threat radars.

The AN/ALE-47 TACDS (Threat Adaptive Countermeasures Dispenser System), is a threat adaptive dispensing system that can prioritise and automatically dispense the correct type and amount of chaff and/or flares to counter single or multiple threats.

Status

During February 1993, the Hellenic Air Force selected ASPIS for its F-16s. The specific configuration comprised the AN/ALR-93(V)1 RWR, Raytheon's DIAS radar jammer (a country specific variant of the ALQ-187 system) and the AN/ALE-47 CMDS. Deliveries to Greece began during 1994 with a prototype suite starting flight trials in March 1996. Delivery of production systems to Greece began later in 1996. By 1999, Raytheon reported having delivered 80 systems to the Hellenic Air Force. Subsequent to this first programme, Greece selected the ASPIS II suite (incorporating Digital Radio Frequency Memory technology, the ALR-93(V) RWR, the ALQ-187 jammer and the ALE-47 CMDS) for installation aboard its F-16 aircraft. The Greek order involved the supply of 60 ASPIS II suites for installation aboard Block 52 + aircraft, 29 ALQ-187 radar jammers for installation aboard selected Block 30/50 aircraft and the upgrading of the existing ASPIS DIAS jammers with Digital Radio Frequency Memory (DRFM) technology.

Contractor

Northrop Grumman Corporation, Electronic Systems Sector.
Raytheon Electronic Systems.
BAE Systems North America, Information and Electronic Warfare Systems.

Airborne Active Dipping Sonar

Type

Airborne Electronic Counter Measures (ECM), sonar system.

Description

The Airborne Active Dipping Sonar (AADS) system consists of the sonar array, reeling machine, sonar transmitter/receiver and signal and display processor. It is designed to be fully compatible with all types of maritime helicopters in service today.

The sonar transducer contains the acoustic projector for sonic pulse generation and an expandable array of receiving hydrophones. The projector resonates at the frequency and duration of the pulse generated by the sonar transmitter. It is capable of producing maximum acoustic source levels at any depth below 6 m. Centre frequencies are selectable for optimal acoustic performance in different environments. The expandable receiving array provides a large acoustic aperture for increased sensitivity and direction. The expanding mechanism is designed to fail-safe standards to ensure safe retrieval of the array.

Rapid deployment and retrieval of the sonar transducer is provided by the lightweight reeling machine. Layering and stowage of the required length of the extra strong cable has been tested with an MTBF of 4,440 cycles. Replacement of cable and assembly can be accomplished during the refuelling cycle.

The Sonar Transmitter/Receiver (STR) generates the transmit waveforms and provides the output power to drive the acoustic projector. It also processes the signal returns from the transducer receive array. The STR is located in the helicopter cabin and provides local control functions for the dipping sonar subsystem, including the reeling machine, the acoustic transducer and the interfaces between the dipping sonar system and the signal processor.

Status

In service.

Contractor

Thomson Marconi Sonar SAS.
Raytheon Electronic Systems.

AN/AAQ-24(V) Nemesis Directional Infra-Red Counter Measures (DIRCM) system

Type

Airborne Electronic Counter Measures (ECM), Infra-Red Counter Measures (IRCM) system.

Description

Nemesis is a Directional Infra-Red Counter Measures (DIRCM) system that detects, acquires and tracks IR missiles, then defeats the threat by accurately focusing an intense beam of modulated IR energy onto the seeker head of the incoming missile.

Northrop Grumman is the team leader and overall system integrator. The company also provides the passive warning element, based on its AN/AAR-54 PMAWS (Passive Missile Approach Warning System), together with the jamming lamp and techniques generator. Boeing provides the Fine Track Sensor (FTS) for the transmitter azimuth axis. Selex Sensors and Airborne Systems (S&AS) provides the turret hardware, pointing and tracking subsystem, operator controls, power supplies, threat modelling and the transmitter unit.

The passive Missile Warning System (MWS) element of the system detects missile plume energy, tracks multiple energy sources and classifies each source as a lethal missile, a non-lethal missile (not intercepting the aircraft), or clutter. Its very fine Angle-of-Arrival (AoA) capability delivers rapid and accurate hand-off to the IRCM pointing/tracking subsystem. Fine AoA processing provides detection ranges nearly double that of existing passive systems and greatly reduces false alarm rates. This provides poor weather and all-altitude operation

ALE-47 TACDS Chaff/Flare Dispenser System

ALR-93(V)1 Threat Warning System and Suite Controller

MILDS III Missile Launch Detection System

ALQ-144(V) Infrared Countermeasures System

LWS-250 Laser Warning System

Advanced Self-Protection Integrated Suite (ASPIS)

ALQ-187H Pulse and CW Jammer

Coordinated Countermeasures Control and Feedback

LITTON Threat Warning System and System Controller

Data Fusion and Countermeasures Optimization Algorithms

ELECTRONIC COUNTERMEASURES SUBSYSTEM

COUNTERMEASURES DISPENSING SUBSYSTEM

Advanced Self-Protection Integrated Suite (ASPIS)   0018268

The current DIRCM configuration with the Small Laser Transmitter Assembly (SLTA) located on a Terma AMASE countermeasures pod. The arrangement was part of a trial of the combined system on an AH-64D Apache of the Royal Netherlands Air Force, displayed at Farnborough in July 2006 (Jane's/Patrick Allen)
1183836

while protecting against multiple simultaneous engagements in dense clutter environments.

The system utilises a wide Field-of-View (FoV) sensor and compact processor, with each sensor covering a 120° Field-of-View (FoV). From one to six sensors can be employed, providing up to full spherical coverage.

The Selex S&AS transmitter consists of a pointing turret with a payload of a Fine Track Sensor (FTS) and an IR jammer. The transmitter unit, when cued by the MWS, acquires the incoming missile, tracks it and projects a high-intensity IR beam at the target. Each Nemesis transmitter is equipped with a laser path to facilitate future upgrade of in-service systems.

The agile pointing system ensures that the DIRCM transmitter unit can rapidly acquire the approaching missile; the four-axis tracking system ensures that maximum jamming energy is maintained on the target throughout the engagement.

The AAQ-24 employs a high-resolution, high-sensitivity, large area Focal Plane Array (FPA) imaging system. The FPA employs a Cadmium Mercury Telluride (CMT) detector, particularly sensitive in the 3–5 μm waveband, making the system highly sensitive to IR missile plumes and benefiting from the increased atmospheric transmission, higher contrast and increased discrimination characteristics within the Mid-Wave IR (MWIR) window.

The FTS, located on the azimuth axis, features fast cool-down time and high sensitivity, which affords post-burn out tracking over an extended temperature range. During a threat engagement, the image of the incoming missile is electronically processed by the FTS. The resulting electronic image is used by the system to lock the transmitter onto and maintain an IR beam on the incoming missile until it is defeated.

The latest Full Rate Production (FRP) systems employ a Small Laser Transmitter Assembly (SLTA). The SLTA uses the Viper™ compact solid-state, diode-pumped laser system developed by Northrop Grumman in collaboration with Fibertek. The latest transmitter configuration, which entered Low Rate Initial Production (LRIP) in 2006, is the Mini Pointer-Tracker Assembly (MPTA), which provides comparatively higher performance in a smaller, lightweight low-drag package, that can form-fit to the existing STLA mounting, facilitating simple upgrade. This arrangement will become standard for AN/AAQ-24 systems and the associated Guardian™ civil DIRCM mounting for commercial applications.

Northrop developed two versions of the baseline system. The standard design, for small aircraft and helicopters, employed a single lamp and uses DC power. A larger variant, designated the Large Aircraft Infra-Red Counter Measures (LAIRCM) system, incorporated two flashlamps

and employs AC power. In the larger installation, each of the two flashlamps had more than twice the power of the single lamp in the smaller variant. The Nemesis system was designed from the outset to permit an easy upgrade to laser jamming, since suitable laser technology was not available for DIRCM at the design stage. The laser beam passes through the same optical path as the tracking sensor. The maturing of laser technology allowed the introduction of an air-cooled laser, which weighs less than 4.5 kg, is lighter than the flashlamp and measures only 5 cm high. It provides high levels of energy, with a peak power of 380 W and an average of 150 W (Bands 1, 3 and 4), allowing large aircraft to be protected by a smaller number of installations.

The jamming head is a stabilised four-axis system, allowing it to be installed anywhere on the aircraft. It employs two methods of jamming: guidance degradation, which generates false targets in the seeker of an attacking missile; and optical break-lock, following which the missile often corkscrews out of control. Break-lock generally occurs within 1 s. Testing has included operation against multiple threats – either sequentially, or by assigning different transmitters to each – and body tracking following motor burn-out.

LAIRCM

The Large Aircraft IR Counter Measures (LAIRCM) system is an evolution of Nemesis, retaining

the AN/AAQ-24 designator as AN/AAQ-24(V)13, employing the Viper™ laser rather than a pulsed krypton lamp.

System evolution

As of mid-2006, a new generation of missile approach warner was in development. Known as the Multi-Image Multi Spectral (MIMS), it is a two-colour IR sensor which uses the distinctive features of missile IR signatures to detect missile launches at significantly greater ranges than the current UV-based passive warning technology allows. The use of a Personal Computer Memory Card Interface Adapter (PCMCIA) or 'smart card' enables the user to tailor each AN/AAQ-24 system installation for techniques and jamming codes, re-programming, maintenance and aircraft configuration.

Specifications
Dimensions
SLTA: 356 × 420 mm (Ø × L)
MPTA: 356 × 354 mm (Ø × L)
Standard MAW: 121 × 171 mm (Ø × L)
MIMS MAW: 107 × 156 mm (Ø × L)
CIU: 208 × 146 × 48 mm
Processor: 515 × 195 mm × 194 mm
Weight
SLTA: 31.9 kg
MPTA: 25.8 kg
Standard MAW: 1.9 kg
MIMS MAW: 2.8 kg
CIU: 1.2 kg
Processor: 16.6 kg
Power
SLTA: 580/1,905 W (standby/maximum)
MPTA: 300/1,200 W (standby/maximum)
Standard MAW: 17/55 W (standby/maximum)
MIMS MAW: 100/140 W (standby/maximum)
CIU: 25 W
Processor: 399 W

Status

The AN/AAQ-24 Nemesis (UK designation, ARI 18246) is in production and in service. Northrop Grumman Electronic Systems Integration International Inc, a subsidiary of Northrop Grumman Corporation, leads an international team consisting of Northrop Grumman Electronic Sensor and Systems Sector, Selex Sensors and Airborne Systems and the Boeing Company.

The DIRCM was developed under Operational Emergency Requirement (OER) 3/89, a joint UK/US programme initiated by the UK MoD in 1989 and joined by the US Special Operations Command (SOCOM) in 1993. In March 1995, the UK and the US Special Operations Command awarded a contract worth approximately USD450 million, if all options were exercised, to develop and produce the Nemesis system.

Initial UK flight trials of the system began aboard a Sea King helicopter in October 1997 at

The AN/AAQ-24(V) Nemesis installation on an RAF TriStar C2 transport (Jane's/Patrick Allen) 1179865

Mini Pointer Tracker Assembly (MPTA)   1159092

GKN Westland Helicopters facility in Yeovil, UK and were completed in July 1998.

Nemesis successfully completed its first live-fire test at the US Army's White Sands Missile Range in June 1998, when the system correctly detected, declared, acquired, tracked and jammed the modern Band IV IR missile threat in a series of live-fire test engagements. Testing at White Sands continued up to November 1999, when the target used was a UH-1 helicopter suspended below a cable car.

LAIRCM was cleared into low-rate initial production in August 2002. The initial goal was to equip 79 C-17s (which will use three defensive laser turrets) and C-130s by 2007, with the first installations in 2004. However, about a dozen C-17s were equipped with an interim single-jammer system known as BOSS. The configuration was reported as a bespoke processor (believed to be the LAIRCM processor less two interface techniques generators and two video processors), a CIU, an SLTA, six ultra-violet missile warning sensors and two repeaters.

LAIRCM was also ordered by the Royal Australian Air Force (RAAF) in 2002 for installation on its Boeing 737-based Wedgetail airborne early warning and control aircraft. As of January 2006, all four Royal Air Force C-17A Globemaster C1 airlifters had been equipped with LAIRCM. In April 2006, the US Defense Security Co-operation Agency (DSCA) revealed a requirement for LAIRCM systems for the four C-17s on order from the RAAF and, in September 2006, the DSCA indicated that the four C-17s to be procured by Canada will each be equipped with the AN/AAQ-24(V)13 LAIRCM system.

In July 2006, the Royal Netherlands Air Force (RNLAF) trialled a variant of its Apache Modular Aircraft Survivability Equipment (AMASE) pod fitted with a centrally-mounted AN/AAQ-24 SLTA head. The AMASE was originally developed for the RNLAF by Terma to equip AH-64D Apache helicopters deployed to Afghanistan.

In July 2006, Northrop Grumman announced its first delivery order for LAIRCM under a five-year Indefinite Delivery/Indefinite Quantity (IDIQ) contract from the US Air Force Aeronautical Systems Centre at Wright-Patterson Air Force Base. With a potential value of USD49.5 million, the company will deliver LAIRCM system hardware and support for the C-17 and C-130 transport aircraft. The total value of the IDIQ contract has an expected ceiling of USD3.2 billion and will further allow for the purchase of LAIRCM hardware for Foreign Military Sales and other government agencies until December 2010.

Also in July 2006, Selex and Northrop Grumman signed a Strategic Alliance Agreement, which will see the former continue to manufacture and support the transmitters, as well as setting up a production process to co-produce the MPTA for future applications, including Guardian™.

In November 2006, Northrop Grumman received a USD104.6 million contract for the retrofit of LAIRCM defensive systems on NATO's fleet of 17 E-3A Sentry airborne early warning aircraft. The programme will be complete by December 2009. The first prototype installation will be performed on a NATO E-3A (NL-1) at L-3 Communications in Greenville, Texas, with the second installation being conducted at EADS' Manching facility in Germany on aircraft NL-2.

Contractor

Northrop Grumman Corporation, Electronic Systems Integration International.
Northrop Grumman Corporation, Electronic Sensors and Systems Sector.
Selex Sensors and Airborne Systems.
The Boeing Company.

AN/AQS-22 Airborne Low-Frequency Sonar

Type

Airborne Electronic Counter Measures (ECM), sonar system.

Description

The AN/AQS-22 Airborne Low-Frequency Sonar (ALFS) is a US Navy project for a dipping sonar which will largely replace the AN/AQS-13F. It is scheduled to be fitted to the SH-60R, as part of the update from SH-60B, SH-60F and SH-60H to SH-60R standard.

The sonar operates in one of five frequency bands within the 1.2 to 5.6 kHz frequency range, with up to 24 beams formed at a fixed angle of depression. The sonar generates high power waveforms – CW, LFM and HFM. FM and CW pulses may be combined within a single pulse. The projector uses piezo-electric ceramic

rings and cardioid beamforming is performed. A system to control the angle of depression is under development. The associated acoustic processor will be the UYS-2A. The UYS-2A has operator selectable full-band, eighth-band and quarter-band verniers, and constant resolution modes are available for up to eight omni-directional or four directional sonobuoys over their full bandwidth. These processing options are available concurrently with the active or passive modes of the dipping sonar. The processor can handle up to 16 sonobuoys. The AQS-22 utilises the expandable sonar array and reeling machine subsystem of Thales Underwater Systems' (TUS) FLASH (Folding Light Acoustic System for Helicopter) system.

The design integrates via a MIL-STD-1553B databus with existing display subsystems. The sonar is controlled by the operator using menus on the existing aircraft control display unit. A dedicated control panel provides fail-safe control of the reeling machine.

Status

In service. During January 1992, the consortium of the then Hughes Aircraft Co. and Thomson Marconi Sonar SAS was awarded a USD31.3 million contract to develop the ALFS system. The five-year contract, awarded by Naval Air Systems Command, provided for full-scale development and a follow-on production option of up to 50 systems.

The AQS-22 successfully completed its operational test phase in early 2000; initial installations included SH-60B LAMPS and SH-60F helicopters.

At the end of 2002, Raytheon was awarded a USD16.5 million contract by the Naval Air Systems Command for the low-rate initial production of four ASQ-22 sonars for the MH-60R helicopter. Raytheon was responsible for systems integration, plus the sonar transmitter and receiver. In turn, TUS was awarded a contract by Raytheon to supply outboard arrays and reeling machines for the four sonars.

Contractor

Thales Underwater Systems.
Raytheon Integrated Defense Systems.

The AN/AQS-22 ALFS undergoing trials on a SH-60 Seahawk   0006973

Eurodass Defensive Aids SubSystem (DASS)

Type

Airborne Electronic Counter Measures (ECM), integrated suite.

Description

Eurodass DASS is designed to provide Eurofighter Typhoon aircraft with spherical ESM coverage, forward and rearward missile warning, fibre optic towed decoys, active phased

array jamming, and options for future growth to meet the aircraft's mission as an air superiority fighter.

The Eurodass Consortium comprises BAE Systems (Operations) Limited in the UK, Elettronica SpA in Italy and INDRA in Spain; together they will provide the Eurodass DASS for the Eurofighter Typhoon aircraft being procured for UK, Italy and Spain. German participation in DASS is subject to negotiation. Design activity for the system is led from BAE Systems (Operations) Limited at Stanmore, UK.

Status

In production and in service in Eurofighter aircraft. The Eurodass Consortium was awarded a GBP200 million contract in March 1992 for the development of DASS. The work-share was initially split approximately 60 per cent to the UK, and 40 per cent to Italy. In May 1998, a GBP170 million contract was awarded for the Production Investment (PI) contract for the Eurodass DASS. The contract covers the establishment of processes and facilities across the Eurodass Consortium to enable the production of DASS through to 2015.

Cutaway of Eurofighter wingtip pod showing DASS LRUs 0083808

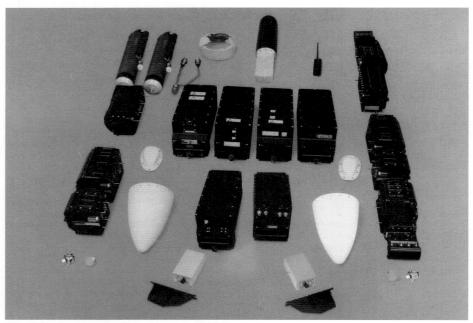

Eurofighter Typhoon DASS components (BAE Systems) 1127367

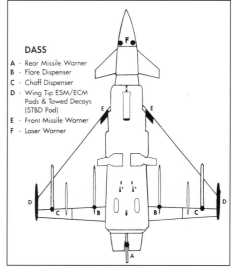

DASS

A - Rear Missile Warner
B - Flare Dispenser
C - Chaff Dispenser
D - Wing Tip ESM/ECM
    Pods & Towed Decoys
    (STBD Pod)
E - Front Missile Warner
F - Laser Warner

Diagrammatic representation of Eurofighter DASS elements 0044917

Contractor

Eurodass Consortium members:
BAE Systems (Operations) Limited,
Elettronica SpA,
INDRA Sistemas.

MILDS® AN/AAR-60 missile detection system

Type
Airborne Electronic Counter Measures (ECM),
Missile Approach Warner (MAW).

Description
The Missile Launch Detection System (MILDS),
AN/AAR-60 is a passive imaging sensing device.

It is optimised to sense the solar blind spectral
band of radiation that is emitted from an
approaching hostile missile exhaust plume.

MILDS® AN/AAR-60 is designed to detect
potential missile threats and provide maximum
warning time.

The MILDS® AN/AAR-60 system comprises
one to six identical sensor heads including optics,
filters and imaging processing. The imaging
sensor allows the system to track and classify UV
sources, determine angle of attack and priority.
Then the system can accurately initiate the
proper countermeasures for the threat and warn
the pilot, so he can begin evasive manoeuvres.
The inherently high spatial resolution of MILDS®
combined with advanced temporal processing
significantly increases the probability of detection
while virtually eliminating false alarms.

The sensor head is a self-contained LRU requiring
no additional hardware. One of the sensor heads
operates as a master while the rest are slaves. Only
the master sensor head handles the communication
with the Electronic Warfare System Computer
(EWSC) and countermeasure dispenser; this being
a purely software function, the hardware of the
different sensor heads is identical.

Each MILDS® AN/AAR-60 sensor outputs a
fully processed signal. The system can be fully
integrated into existing threat warning systems
or integrated with directed IR countermeasures

systems, reporting through existing display and
communication paths, or it can be installed as a
stand-alone system capable of driving a display
and activating countermeasures.

MILDS® F AN/AAR-60(V)2
MILDS® F, designated AN/AAR-60(V)2 is based on
the MILDS® design. Several sensors linked to a
Countermeasures Signal Processor provide full
spherical coverage and rapid reaction. The system
is optimised for the installation either in pylons
or in the fuselage of fighter aircraft. The first
application for MILDS® F is for installation onboard
Danish Air Force F-16 fighter aircraft. For the Danish
programme, MILDS® F will be integrated into
the Pylon Integrated Dispenser System (PIDS+)
and Electronic Combat Integrated Pylon System
(ECIPS+) and linked to the AN/ALQ-213 EW
Management System (see separate entries).

Specifications
Field of view: 4 sensors provide 95°
    elevation × 360° azimuth
    6 sensors provide 180° elevation × 360°
    azimuth
Power consumption: <14 W per sensor
Weight: <2.0 kg/sensor
Dimensions: (L × W × H) 108 × 107 × 120 mm
    excluding flange/connector
Interfaces: RS-422 optional MIL-STD-1553B,
    discrete lines

Status
MILDS® AN/AAR-60 is in production for many
customers worldwide. Platforms include the
NH90, Tiger, AS532 Cougar, CH-47, UH-60, SH-60,
SH-3D, UH-1, S-70, AH-1, A129 Mangusta, AB205,
AB212, AB412, Mi-17 helicopters and CN-235,
C-130 Hercules and Airbus fixed-wing aircraft.

Announced during the 2004 Farnborough
Airshow, MILDS® F will first be installed onboard
Danish Air Force F-16 aircraft. The Danish MILDS®
F-based EW suite will be marketed worldwide by
EADS Deutschland and TERMA A/S.

Contractor
EADS Deutschland.

TWE Threat Warning Equipment

Type
Airborne Electronic Counter Measures (ECM),
integrated suite.

Description
TWE was designed initially for the Eurocopter
Tiger helicopter. It is designed to provide warning
of hostile surveillance and fire-control radars,
tracking radars, CW illuminators, and missile
seekers, as well as laser range-finders, laser
illuminators and laser missile guidance systems.

TWE comprises radar and laser warning
receivers, a Central Processing Unit (CPU), a library
module, a symbol generator, a multifunction
display, radar antennas, laser heads and a control
box. Coverage is E- through K-band (2 to 40 GHz)
in the radar domain and bands I and II for lasers.
The radar warning subsystem is capable of
instantaneous frequency measurement and can
handle pulse, Pulse Doppler (PD) and Continuous
Wave (CW) emitters. The CPU processes received
radar and laser signals, interfaces with the library
module for threat identification and maintains
electromagnetic compatibility.

MILDS® six sensor LRUs functioning in master-slave configuration 0018288

Thales' Threat Warning Equipment (TWE)   1150266

TWE is designed to be the core of a Defensive Aids Suite (DAS); the CPU is capable of managing any onboard self-protection suite (CMDS/MAW), interfacing through a MIL-STD 1553B databus.

The system generates a colour threat display which shows information on threat type, bearing and danger level. The visual format is backed up by an audio warning which sounds each time a radar or laser threat is detected.

TWE also allows crew members to call up data lists from the library module during flight in order to change priorities for different phases of a mission. The system incorporates a built-in test routine and is claimed not to require a test bench for routine maintenance.

Specifications
Frequency:
(radar) E- through K-bands (2 to 40 GHz)
(laser) 0.4–1.1 μm (option: 0.4–1.7 μm)
Coverage: 360° azimuth, ±45° elevation
Accuracy:
(angular) better than 10° RMS
(frequency) less than 20 MHz
Library size: 2,000 radar/laser threats
Dimensions:
  (CPU) 199 × 194 × 350 mm
  (antenna) Ø100 × 88 mm
Weight: <15 kg
Power: <200 W

Status
TWE is standard fit for Eurocopter Tiger attack and NH90 tactical transport helicopters currently operated by or to be procured by the armed forces of Australia, France, Finland, Germany, Italy and Portugal. For the NH90, TWE is reported as being combined with the MILDS® AN/AAR-60 passive missile warner and possibly the SAPHIR-M chaff and flare dispenser. As installed in the Tiger, TWE is a component of the helicopter's Electronic Warfare System (EWS).

The then DaimlerChrysler Aerospace (now a part of EADS) received the first Tiger EWS production contract (covering 160 systems) during March 2000, with first deliveries of the system in 2002. Most recently, during March 2004, Thales announced that a contract award from EADS, worth almost EUR100 million, for the supply of more than 500 TWE systems for installation aboard both Tiger and NH90 helicopters. At the time of the announcement, deliveries were scheduled to take place through to 2012.

Contractor
Thales Systemes Aeroportes SA.
EADS Deutschland, Defence Electronics.

Israel

Advanced Countermeasures Dispensing System (ACDS)

Type
Airborne Electronic Counter Measures (ECM), Counter Measures Dispensing System (CMDS).

Description
The Advanced Countermeasures Dispensing System (ACDS) is a computer-controlled airborne self-defence system with enhanced capabilities for chaff, flare and decoy dispensing, applicable to fixed- and rotary-wing aircraft. The ACDS is capable of performing both as a stand-alone system, where the crew selects and initiates prestored dispensing programmes, and as an integrated part of a full EW suite, executing dispensing programmes according to instructions from an EW controller and/or aircraft avionics command via digital data links.

Programme execution in the semi-automatic mode requires operator consent; in fully automatic mode, the dispensing programme is executed immediately on threat detection.

The system includes an NVIS-compatible Mini-Control and Display Unit (MCDU), which provides visual display of threat direction, as detected by associated EW sensors, and generates an audio warning. Threat Matrix Generator (TMG) software allows the customer to modify dispensing programmes according to

operational requirement. A computer notebook-based Memory Loader & Verifier (MLV) facilitates efficient downloading of Threat Matrix™ data to the ACDS during O-Level Maintenance.

The ACDS family includes a wide range of 'smart' dispensers, including internal, dual and scab-on types.

Specifications
Payloads: 1 × 1, 1 × 2, 2 × 2.5 in
Dispensing:
Automatic, semi-automatic and manual modes
Controls up to 24 separate dispensers (30-pin or 60-pin)
Simultaneous dispensing of up to 4 payloads from single dispenser
16 manual programmes for single payload (chaff, flare, RF)
10 cascaded programmes
Cascaded escape programme
500 for chaff
500 for flares
Automatic programmes
Fast jettison programme
Mixed payload dispensing within single programme
Weights:
MCDU: 1.2 kg
Smart dispenser assembly: 2.8 kg

Magazine: 3.0 kg
Payload: 0.18 kg (1 × 1 in)
Interface: 3 × RS-422 serial links; MIL-STD-1553B databus
Power:
MCDU: 28 V DC; 1.3 A (power up), 0.6 A (steady state)
Smart dispenser assembly: 28 V DC; 0.2 A (power up), 0.1 A (steady state)
Payload: 28 V DC; 5 A (20 ms spacing)

Status
In service, with many units delivered worldwide.

Contractor
BAE Systems ROKAR International Ltd.

Advanced Digital Dispensing System (ADDS)

Type
Airborne Electronic Counter Measures (ECM), Counter Measures Dispensing System (CMDS).

Description
The Advanced Digital Dispensing System (ADDS) is a computer-controlled, threat-adaptive

Schematic of the ACDS (BAE Systems ROKAR)   0561210

The Advanced Digital Dispensing System (ADDS)   0122562

countermeasures dispensing system. It is designed to protect aircraft from both ground and air threats by dispensing decoy payloads. The system is configured for high-performance aircraft, helicopters, transport aircraft and maritime patrol aircraft.

ADDS can dispense chaff, flares and RF expendables in single shot or multiple simultaneous dispense programmes to provide multispectral responses that best counter the threat.

ADDS incorporates improvements derived from combat experience and hundreds of installations on a wide variety of aircraft, including the A-4, AH-64, F-4, F-5, F-15 and F-16.

Status
In service; several hundred systems have been delivered worldwide.

Contractor
BAE Systems ROKAR International Ltd.

AES-210/E ESM/ELINT RWR system

Type
Airborne Electronic Counter Measures (ECM), Electronic Support Measures (ESM) system.

Description
The AES-210/E is a sophisticated system, which performs ESM, ELINT and radar warning system tasks in very dense RF environments. It can be installed in small- or medium-size aircraft or ships for applications such as maritime and overland surveillance, ELINT information gathering and platform self-protection. As a programmable system designed with a modular and flexible architecture for future growth, the AES-210/E is optimised to receive all known and foreseeable radar signals, identifying highly complex signals and threats and employing an advanced combination of receiving techniques.

The AES-210/E offers very high intercept probability, sensitivity and reliability, providing very accurate direction-finding. A number of different modern DF techniques can be provided to meet operational requirements, including advanced interferometric methods to achieve the most accurate data. The system has five fast CPUs and can be operated automatically or by a single operator.

System features include:
1. Automatic detection, measurement and identification of radars.
2. Long-range detection together with high probability of intercept, using a combination of wide- and narrow-band receivers.
3. Accurate direction finding implemented by amplitude comparison.
4. DTOA and interfometric techniques.
5. Onboard emitter library updating and mission planning.
6. Integrated self-protection capability, including RWR display in cockpit.
7. In-flight recording of data for further analysis.
8. MIL-STD-1553 and ARINC 429 interfaces.
9. Datalink communications to remote users.
10. Windows NT® operation.

Specifications
Weight: 40 kg typical
Frequency: 0.5–18 GHz (optional extension up to 40 GHz available)
Coverage: 360° azimuth
DF accuracy: 3° (typical fine DF); 7° (typical coarse DF)
Identification library: over 1,000 emitter modes (can be updated in flight)
Interface: RS-232, RS-422, MIL-STD-1553B, ARINC
Environmental: MIL-E-5400T
Power: 500 W

Status
The AES-210/E has been installed on several types of aircraft, including: Beechcraft B200/B200T, Dornier 228, Falcon 20, Learjet 36A, M117, Sea Hawk and Sea Sprite.

The AES-210/E has been selected for the Royal Australian Navy for installation aboard its 16 S-70B-2 and 11 SH-2G(A) helicopters for its Maritime Patrol requirement, together with Elisra's LWS-20 laser warning system and chaff and flare dispenser system. Elisra is the main EW system integrator.

Contractor
Elisra Electronic Systems Ltd.

ALR-2001 ESM

Type
Airborne Electronic Counter Measures (ECM), Electronic Support Measures (ESM) system.

Description
The ALR-2001 is a sophisticated airborne ESM able to measure radar-operating parameters with sufficient precision to 'fingerprint' individual radars from others of the same type.

ALR-2001 also has COMINT capabilities from V/UHF to microwave datalink.

It provides direction-finding capability, reportedly to better than 1°.

Status
Selected by the Royal Australian Air Force for its P-3C aircraft.

Contractor
Elta Electronics Industries Ltd.

ASPS EW suite for the F-15I and F-16I

Type
Airborne Electronic Counter Measures (ECM), integrated suite.

Description
The ASPS self-protection suite for fighter aircraft comprises active (SPJ) and passive SPS subsystems, employing a combination of proven and new technologies that include monolithic microwave integrated circuitry, application specific integrated circuitry and multiminiature Travelling Wave Tubes (TWT). The SPS subsystem provides wide- and narrow-band reception and a signal parameter measurement

capability, with selectable operating bandwidth to enhance sensitivity and signal separation. Data processing and analysis is performed using multiple processors and gate array modules. The SPJ component is described as offering a high effective radiated power value, wideband transmission and reception coverage, wide angular coverage, the ability to cope with a multithreat emitter environment.

Countermeasures techniques include Doppler, range, amplitude modulation and noise modes, with the system providing a pilot's threat display (threat azimuth and type), missile launch alerting and the ability to display via a Multifunction Display Unit (via MIL-STD-1553B) if required. System control is by means of front panel switches on a cockpit control unit.

Specifications
Signal reception: CW, high PRF, pulse and pulse-Doppler
Coverage: full arc in azimuth
Interface: MIL-STD-1553B; RS-422
Data processing: multiple autonomous processors
Data loading: portable field loader (emitter table and operational software)
Recording: flight data (downloaded via the field loader)
Audio warnings: composite, missile launch and new threat
Display: CRT (graphic representations of emitter type, reception angle, relative lethality and threat status)
Interfaces: radar and CMDS
Environmental: MIL-E-5400 Class II qualified

Status
The ASPS EW system is Elisra's most advanced EW suite. It is based on a combination of proven and new-generation technologies. It is in production and operational aboard Israel Air Force F-15I aircraft. The system was selected for 50 new F-16I aircraft which Israel agreed to purchase in July 1999.

Contractor
Elisra Electronic Systems Ltd, a member of the Elisra Group.

BRITENING Directional Infra-Red Counter Measures (DIRCM) system

Type
Airborne Electronic Counter Measures (ECM), Infra Red Counter Measures (IRCM) system.

Description
BRITENING is a fully autonomous DIRCM system designed for installation in air transport and passenger aircraft and provides protection against shoulder-launched, heat-seeking, surface-to-air and anti-tank missiles fired at the host platform during taxying, take-off and/or landing. The system comprises a directional tracking and jamming turret, a mission computer and an array of IR or ultra-violet missile warning sensors. A requirement of civil certification of such systems, BRITENING requires no flight crew intervention during operation and offers wide

Units of the Elisra AES-210/E ESM/ELINT RWR system 0062851

The DIRCM turret of BRITENING; note the missile warning sensors around the mounting ring 0567272

spatial coverage, multiple threat engagement capability, storage of all events and real-time reporting to ground facilities. The equipment is designed for retrofit to legacy aircraft, with system configurations available for a wide range of civilian air transport and tactical airlifters. BRITENING can act as a standalone DIRCM or as part of a fully integrated Defensive Aids Sub System (DASS).

Specifications
MTBF: 6,000 h (calculated)
Weight: <150 kg (including Group A provision)

Status
BRITENING is reported to be a derivative of Rafael's AERO-GEM military helicopter/fixed-wing aircraft EO self-protection suite. AERO-GEM (see separate entry) incorporates Rafael's GUITAR-350 passive missile warning system, JAM-AIR DIRCM and a chaff and IR decoy flare dispenser. During March 2003, BRITENING was successfully tested on an Arkiav Airlines Boeing 757.

While DIRCM-based systems are reported to be the more expensive solution (compared with flare-based systems), they are 'clean' systems, with no expendables ejected from the host aircraft, a factor which may sway agencies such as the US DHS in its choice of system for

deployment on US airliners which frequently operate in and out of airports which border significant population centres.

Contractor
Rafael Armament Development Authority Ltd, Systems Division.

CDF-3001 airborne V/UHF COMINT/DF system

Type
Airborne Electronic Counter Measures (ECM), Communications Intelligence (COMINT) system.

Description
CDF-3001 is an airborne self-contained COMINT/DF system designed to acquire, monitor and measure the Direction Of Arrival (DOA) of tactical radio transmissions. The operational system is driven by a Windows-based man/machine interface, and advanced hardware implementation provides fast scan and DF capability. The system is linked to the host aircraft navigation and command and control system via an RS-232 interface. The CDF-3001 system is suitable for both fixed- and rotary-wing application.

CDF-3001 airborne V/UHF COMINT/DF system installed under helicopter   0018285

Specifications
Frequency coverage: 20–500 MHz (optionally 1.5–1,000 MHz)
Monitored signals: AM and FM

Contractor
Tadiran Electronic Systems Ltd, a member of the Elisra Group.

COMINT/ELINT system for the Arava

Type
Airborne Electronic Counter Measures (ECM), Communications Intelligence (COMINT) system.

Description
The COMINT/ELINT system for the Arava twin-engined light transport can be rolled on or off a multipurpose aircraft or be permanently installed. The COMINT operator uses sensors such as the Elta EL/K-1250 with omnidirectional antennas, direction-finding and signal recording. The ELINT operator has 0.5 to 18 GHz bandwidth coverage with omnidirectional antennas, signal analyser, data processor and recorder facilities. Jamming can be accomplished using antenna arrays in the aircraft tail, giving approximately 240° azimuth coverage in the lower hemisphere. Spot and band jammers of 20 to 400 W power are used.

Specifications
In production.

Contractor
Elta Electronics Industries Ltd.EL/K-7010 tactical communications jammer system

EL/K-7010 tactical communications jammer system

Type
Airborne Electronic Counter Measures (ECM), communications jamming system.

Description
The EL/K-7010 is a modular family of communication jamming systems designed for stand-alone operation and providing for signal search and acquisition, preset channel monitoring and automatic jamming modes. It can also be integrated within a larger electronic warfare system. Computer control of power management, fast reaction jamming and multikilowatt effective radiated power provide simultaneous multiple target capability. New interception tasks are accomplished rapidly by a fast scanning receiver and automatic signal sorting in the system computer. In addition to airborne installations, ground-based systems suitable for armoured personnel carriers and air conditioned shelters are available.

Status
In production.

Contractor
Elta Electronics Industries Ltd.

EL/K-7032 airborne COMINT system

Type
Airborne Electronic Counter Measures (ECM), Communications Intelligence (COMINT) system.

Description
The EL/K-7032 is designed for the surveillance and interception of radio signals in the 20 to 500 MHz frequency range, operation being largely computer-controlled.

A typical airborne system would include a supervisor's station having a computer controller, two VHF/UHF radios, a display and a data recorder. This station can work with up to four traffic collection stations, each having up to four radios and data recorders.

The EL/K-7032 has a frequency resolution of 1 kHz (with 10 Hz an option) and can intercept AM, FM, CW and HF-SSB transmissions as required.

Specifications
Frequency: 20–500 MHz
Resolution: ±1 kHz (10 Hz optional)
Modes: AM and FM (CW and SSB optional)

Status
In production and in service. Fitted to the Phalcon early warning aircraft.

Contractor
Elta Electronics Industries Ltd.

The Elta EL/K-7032 airborne COMINT system
0503810

EL/K-7035 all-platform COMINT system

Type
Airborne Electronic Counter Measures (ECM), Communications Intelligence (COMINT) system.

Description
The EL/K-7035 COMINT system intercepts, monitors, locates, analyses and reports on radio communications in the 20 to 500 MHz range. It is suitable for air, sea or ground applications.

The standard EL/K-7035 system includes a supervisor console, two to five operator consoles, a system controller and mass storage, RF distribution and antenna, a DF system, plotter position and datalink. The supervisor and operator consoles would each be equipped with two to four COMINT receivers, two to four controllers, two to four dual-channel tape recorders and a ruggedised computer. As an option these consoles could also have an IF panoramic display and time code generator/reader.

Contractor
Elta Electronics Industries Ltd.EL/L-8222 self-protection jamming pod.

EL/L-8222 self-protection jamming pod

Type
Airborne Electronic Counter Measures (ECM), radar jamming system.

Description
The EL/L-8222 is a power-managed, podded, Airborne Self-Protection Jammer (ASPJ) with an integrated ESM receiver. Using Pre-Flight Message (PFM) intelligence data programmed into the equipment, the EL/L-8222 performs electronic countermeasures against detected RF emissions emanating from hostile surface and airborne weapon systems that have been classified as threats.

Main features of the system are:
- Autonomous threat environment analysis and identification by an integrated ESM receiver, supporting threat acquisition, sorting and initiation of appropriate jamming techniques
- Power-managed jamming in the time, frequency and amplitude domains
- Flight-line programming capability, facilitating rapid adaptation to changing threat parameters
- Supersonic capability
- Internal BIT and modular construction, facilitating low level maintenance.

In operation, threat RF signals are received via antennas which are mounted in the forward and aft sections of the pod. After amplification and filtering in the input section, the signal is analysed to identify the threat. The appropriate electronic countermeasure responses, associated with each identified threat, are initiated by the logic section. Corresponding modulations and final amplification take place in the RF high power section. The RF jamming signals are then transmitted via the forward and aft antennas.

The PFM is loaded into the system by a Data Field Loader (DFL). The PFM contains identification data of the threats to be countered and the corresponding ECM responses. The PFMs are prepared before flight using the PFM Generator (PFMG).

The system is equipped with a built-in cooling system, which consists of a closed-loop cooling circuit and a heat exchanger. A ram air intake is incorporated in the pod housing.

The EL/L-8222 can operate autonomously or can be integrated with other EW and avionics systems. Communication between the system and host aircraft may be accomplished via MIL-STD-1553B databus, RS-422 or discrete line.

Status
Operational on Kfir, F-16, F-5, F-4, Jaguar and MiG-21 aircraft operated by a number of air forces

The Elta EL/L-8222 self-protection jammer pod mounted under the wing of an Israel Air Force F-16
0515277

worldwide. During 2000, the RAAF selected the EL/L-8222 for its F-111C/G strike aircraft. The pods will be supplied with comprehensive ground support equipment including automatic test equipment, software support facility and operational support equipment.

Contractor
Elta Electronics Industries Ltd.

EL/L-8233 Integrated Self-Defence System (ISDS)

Type
Airborne Electronic Counter Measures (ECM), integrated suite.

Description
Elta's EL/L-8233 ISDS was specifically designed to meet the operational requirements of small, light fighter aircraft. Featuring a modular architecture, advanced technology, proven software and hardware, ISDS as developed for F-5E and F-5F aircraft includes: a Radar Warning Receiver (RWR); mini CW repeater jammer; and chaff/flare dispenser, all integrated through a MIL-STD-1553B databus.

The RWR provides warning of: pulse, pulse Doppler, CW and other exotic radar threats. It provides high probability of detection beyond lethal threat envelopes, in a very short cycle time, with positive identification based on threat frequency and time domain characteristics. It offers high integration with the active jamming system and automatic release of chaff and flares, together with recording of unidentified threats, and field programmability.

The CW repeater jammer provides self defence deception against enemy weapons that use pulse Doppler radar and semi-active radar missile seekers in conjunction with CW illumination. Both Velocity Gate Pull Off (VGPO), and angle deception techniques are exploited in the CW repeater. The system is programmable using ELINT data and Pre-Flight Message (PFM).

The chaff and flare dispenser is a smart, threat-adaptive design that can be tailored in capacity to meet customer requirements.

Specifications
RF coverage: 7–17 GHz (radar jammer)
ERP: 200 W (radar jammer)
Target coverage: 4 simultaneously (radar jammer)
Weight: 50 kg (radar jammer)

EL/L-8233 Integrated Self-Defence System (ISDS)   0018251

Status
In service.

Contractor
Elta Electronics Industries Ltd.

EL/L-8240 self-protection system

Type
Airborne Electronic Counter Measures (ECM), integrated suite.

Description
The EL/L-8240 is an internally mounted self-protection system suitable for modern combat aircraft. It offers radar warning, jamming and deception over a large frequency range and wide angular coverage. It can be integrated with other avionics systems. A distributed processing system is used and the unit can be readily reprogrammed to meet changing threats.

Specifications
Weight: 136 kg
Power required: 7 kVA

Status
The EL/L-8240 is reported to be installed in Batch 3 F-16C and D aircraft of the Israeli Air Force.

Contractor
Elta Electronics Industries Ltd.

EL/L-8300 MPA ESM/ELINT system for MPA

Type
Airborne Electronic Counter Measures (ECM), Electronic Support Measures (ESM) system.

Description
ELTA's EL/L-8300 is an ESM/ELINT system for tactical and strategic missions. It is designed for Maritime Patrol Aircraft (MPA) and has simultaneous ESM and ELINT capabilities. The system performs surveillance, tactical support, anti-surface warfare, anti-submarine warfare, tracking and targeting missions.

The system features complementary receiving sources, including a 4 GHz Channelised Receiver and a wideband SHR. It measures instantaneous 360° Direction Of Arrival (DOA) based on

The Elta EL/L-8300 SIGINT system in operation in an Israeli Air Force Boeing 707 aircraft. Also installed on Boeing 707 ECM aircraft of the Spanish Air Force   0503812

The Elta EL/L-8300 SIGINT system in operation in an Israeli Air Force Boeing 707 aircraft. Also installed on Boeing 707 ECM aircraft of the Spanish Air Force   0503813

Differential Time Of Arrival (DTOA) technology. The system measures and automatically processes signals received and provides classification, identification and location of targets. Multimode operations allow a single operator to conduct ELINT measurements while monitoring other signals of interest. There may also be an optional RWR display in the cockpit.

Specifications
Weight: <175 kg
Frequency coverage: 0.5 – 18 GHz (MMW optional)
Instantaneous bandwidth: 4 GHz
Bandwidth selectivity: 250 MHz, 50 MHz, 10 MHz (others optional)
System sensitivity: better than –70 dBm, better than –80 dBm for ELINT
Accuracy: better than 2 RMS for P-3C

Status
Installed in Israeli Air Force Boeing 707 aircraft, and incorporated into the SIGMA mission suite installed in the Spanish Air Force's Boeing 707 SIGINT aircraft. It is also reported that the South African Air Force operates at least one Boeing 707 with a version of the EL/L-8300. The system has also been selected for Australian Lockheed Martin P-3C Orion and Singaporean Fokker Enforcer maritime patrol aircraft.

The EL/L-8300 is being provided to Lockheed Martin Fairchild Systems for integration into the DASS for the UK Nimrod MRA.4 (designated EL/L-8300(UK); in this role it will function as an ESM system.

Contractor
Elta Electronics Industries Ltd.

EL/L-8312A ELINT/ESM system

Type
Airborne Electronic Counter Measures (ECM), Electronic Support Measures (ESM) system.

Description
The EL/L-8312A system forms a part of the EL/L-8300 SIGINT set (see earlier entry) or can operate alone, covering the 0.5 to 18 GHz frequency band to detect, analyse and identify signals out to the radar horizon. Direction-finding is fast and accurate and bearings are stored in the associated computer and correlated with aircraft navigational data to give a display of the actual location of selected transmissions on a colour graphic console. The EL/L-8312A incorporates the EL/L-8312R receiver, EL/L-8320 signal parameters measurement equipment and the EL/L-8610 computer.

Status
In production and service with the Phalcon and a variety of other aircraft. In June 1991, it was announced that Singapore had awarded a USD20 million contract for EL/L-8312A systems for four Enforcer Mk 2 maritime patrol aircraft. The EL/L-8312A has also been supplied for retrofit to Royal Australian Air Force P-3C Orions.

An enhanced version of the EL/L-8312A ESM is being offered, together with an enhanced variant of the Phalcon Airborne Early Warning (AEW) radar, by the joint Raytheon Electronic Systems/Elta Electronics Industries Ltd consortium that is marketing worldwide its Airborne Early Warning and Command and Control (AEW&C) systems.

Contractor
Elta Electronics Industries Ltd.

EL/M-2160 missile approach warning system

Type
Airborne Electronic Counter Measures (ECM), Missile Approach Warner (MAW).

Description
The EL/M-2160 missile approach warning system provides warnings of missile attack and automatically activates chaff and flare dispensers to protect the aircraft. The EL/M-2160 is an all-solid-state pulse Doppler radar. It provides time-to-impact and direction information to enable timely and effective response.

The equipment consists of the Transceiver Processing Unit (TPU), RF head, four to six antennas (according to platform and coverage requirements) and an optional cockpit control unit. The system has an all-weather capability at detecting any approaching missile type in boost, sustain and post burnout phases. It has 360° coverage and can be used throughout the flight envelope. It has a low false alarm rate, short reaction time and an extensive BIT and fault isolation capability.

Specifications
Dimensions: (TPU) 508 × 254 × 216 mm
Weight: 32 kg (TPU and RF head)
Power consumption: 400 W
Interfaces: 1553B, PPD, RS-422, RS-232

Status
The system has been designed for F-16 aircraft and adapted for transport aircraft and helicopters. The system is currently in service with several European and Asian air forces on a variety of aircraft and helicopters, including C-130, C-160, Fokker, Antonov-32, Bell-212 and Mi-17. The system has been successfully tested against live missile firing and has proved effective in protecting the platforms in several missile firing engagements.

Contractor
Elta Electronics Industries Ltd.

Elta's EL/M-2160 missile approach warning system
0054165

EL/M-2160F Flight Guard civil aviation protection system

Type
Airborne Electronic Counter Measures (ECM), integrated suite.

Description
EL/M-2160F Flight Guard self-protection system has been designed to protect military, para-military and civil aircraft. It is in service and has been installed on over 150 aircraft to date. The detection sub-system is based on the EL/M-2160 lightweight Pulse-Doppler (PD) radar system with four to six antennas to provide up to 360° coverage of the host aircraft. Elta claim the employment of PD radar, as opposed to either UV or IR detectors, provides a near zero false alarm rate, an important point considering the necessary fully autonomous operation of civilian protection systems. The system is fully automatic in operation and response with a very short reaction time, accurate time to impact measurement and optimal and timely countermeasures activation.

Flight Guard employs an IMI Counter Measures Dispenser System (CMDS), consisting of two flare dispenser magazines positioned on the aircraft fuselage, firing behind the engines. Flare ejection speed can be varied to provide adjustable coverage depending on engine position. The flares are described by the company as 'environmentally friendly', featuring rapid burn-out to permit low-altitude ejection during the critical takeoff and landing phases of flight.

While the deployment of expendable flares from civilian aircraft may prove problematic around some of the world's busy metropolitan

The IAI/Elta test aircraft at the 2003 Paris Airshow, fitted with two FlightGuard Pulse Doppler (PD) antennas, each covering a 60° sector (for a total of 120°) of the stern area of the aircraft (Patrick Allen)
0554708

The IAI/Elta test aircraft at the 2003 Paris Air Show, nose radome removed, showing the FlightGuard Pulse Doppler (PD) radar antennas, covering the front and beam sectors (Patrick Allen)
0554706

airports, Elta claim far greater effectiveness of their system compared to DIRCM-based offerings, particularly against multiple attacking missiles. Further, with an estimated fly-away cost of around USD500,000 per aircraft, FlightGuard is approximately half the cost of its competitors.

Specifications
Dimensions: 306 × 361 × 207 mm
Weight: 60 kg
Power consumption: 500 W

Status
Flight Guard is in service on aircraft and helicopters of the Israeli Air Force. The system was displayed fitted to an Israel Aircraft Industries Boeing 737 flight test aircraft at the 2003 Paris Air Show as a showcase for the system's capabilities for the protection of civilian aircraft. In the wake

of terrorist attacks on civilian transport aircraft in Kenya and Afghanistan during 2002–2003, the Israeli Ministry of Transport and Ministry of Defence embarked on a full-scale civil aviation protection plan at the end of 2003. During the second quarter of 2004, Israel was reported to be seeking US approval to operate Flight Guard-equipped El Al flights into US airports. Initial tests of the Flight Guard system on El Al aircraft were scheduled for June 2004, after which an initial installation onto six El Al aircraft was planned for the end of 2004. During July 2004, Flight Guard passed an airborne test against a SA-7, successfully acquiring the missile and deploying flares to create a significant miss distance.

As of early 2005, certification of the system was still pending, although when achieved, certification with Israel's Civil Aviation Authority is expected to suffice for the US Federal Aviation Authority (FAA), which has a mutual certification

IMI is testing Flight Guard, with visible flares, on IAI's 737 multimission aircraft   0564399

amendment with its Israeli counterpart. IAI has teamed with a US company that will market and install Flight Guard systems in the United States.

Contractor
Elta.
IMI.

Guitar-350 passive missile warning system

Type
Airborne Electronic Counter Measures (ECM), Missile Approach Warner (MAW).

Description
The Guitar-350 is a lightweight totally passive, autonomous warning system which detects missiles by sensing the electro-optical emissions from their engines. The system provides audio and visual alarms, allowing time for the aircraft to activate its defensive systems.

Guitar-350 is based on a patented sensor and detection algorithm, which enables it to identify the type of missile, track its trajectory and transfer data on time-to-impact for automatic control of the infra-red countermeasures subsystem. It emits no electromagnetic or electro-optical signals, which might be detected. The system is designed to detect ground-to-air and air-to-air missiles and has an extremely low false alarm rate.

Guitar-350 features coverage of all possible attack angles and has high discrimination against background interference. It withstands adverse environmental conditions and is compatible with cockpit displays.

Specifications
Azimuth coverage: 360°
Elevation coverage: 120°
Warning time: 4–6 s
Weight: less than 15 kg
Power: less than 200 W
Interface: MIL-STD-1553B

Status
Suitable for helicopters, fighter and transport aircraft. Guitar 350 supersedes the earlier Guitar 300 system.

The FlightGuard PD radar control LRU (Patrick Allen)   0554707

The IAI/Elta test aircraft was fitted with two IMI CMDS one on each side of the rear portion of the fuselage, positioned to cover the rear of the aircraft's two podded turbojet engines (Patrick Allen)   0554709

Guitar 350 passive missile warning system detector   0051256

For details of the latest updates to *Jane's Avionics* online and to discover the additional information available exclusively to online subscribers please visit
jav.janes.com

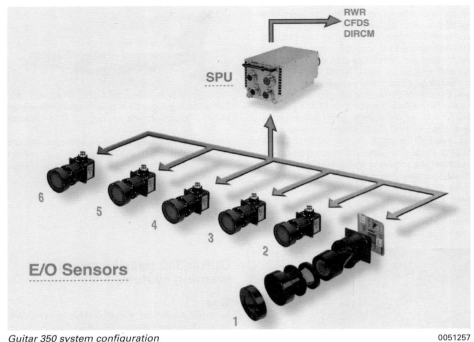

Guitar 350 system configuration   0051257

and the system as a whole incorporates built-in test of its front end, preprocessor, input/output interface and central processor unit memory.

Specifications
Power source: 28 V DC (MIL-STD-704)
Power consumption: 120 W
Interfaces: active electronic countermeasures; MUX bus (option); onboard laser source blanking; RS-422
Environmental: to MIL-E-5400 Class IB
Weight: 0.2 kg (control panel); 0.6 kg (sensor head – each); 1.4 kg (display unit); 2.5 kg (LWA); 6.5 kg (complete system – stand-alone configuration)

Status
It is understood that LWS-20 variants have been selected or procured for installation aboard S-70 and SH-2G helicopters of the Royal Australian Navy, Canadian Forces CH-146 Griffon helicopters and German Army Aviation CH-53G helicopters (both as part of an SPS-65(V) application).

Contractor
Elisra Electronic Systems Ltd.

Contractor
Rafael Armament Development Authority Ltd, Systems Division.

JAM-AIR Electro-Optical (EO) jamming system

Type
Airborne Electronic Counter Measures (ECM), Infra-Red Counter Measures (IRCM) system.

Description
The JAM-AIR EO jamming system is a DIRCM designed for helicopters and fixed-wing applications and is claimed to be effective against IR-guided surface-to-air, IR-guided air-to-air and anti-tank missiles. The system can be operated as a standalone equipment or as part of a Defensive Aids Sub System (DASS) such as Rafael's AERO-GEM and BRITENING systems, where JAM-AIR is combined the GUITAR-350 Missile Warning System (MWS) and a Counter Measures Dispenser System (CMDS). System features include:

- Dual-axis, remotely operated gimbal
- High power, pulsed, arc lamp source
- Automatic activation of associated CMDS
- Interfaces for a range of missile warning systems
- Optional tracking unit to provide real-time assessment of countermeasures effectiveness.

Status
In production.

Contractor
Rafael Armament Development Authority Ltd, Systems Division.

Display model of the JAM-AIR head (Jane's/ Patrick Allen)   1137020

LWS-20 laser warning system

Type
Airborne Electronic Counter Measures (ECM), Laser Warning Receiver (LWR).

Description
The LWS-20V-2 is the latest identified variant of Elisra Electronic System's LWS-20 laser warning system. Designed for stand-alone operations or integration into a multithreat warning system (such as the SPS-65V – see separate entry), the stand-alone LWS-20V-2 configuration comprises four sensor heads, a Laser Warning Analyser (LWA), a system control panel and a pilot's display. Functionally, the four sensor heads receive and detect pulsed laser signals and send the detected pulses to the LWA. Here, each detected pulse is characterised by its angle of arrival (for direction-finding purposes), relative time of arrival and amplitude. Threat identification is by means of library comparison. The LWA incorporates a powerful 32-bit microprocessor with complementary metal oxide semiconductor-based Random Access Memory (RAM), for temporary data storage, and electronically erasable programmable Read-Only Memory (ROM), for combat software and programmable tables. Identified threats are presented in alphanumeric and symbolic formats; threat types are identified by azimuths and lethalities. Audio warnings are also generated within the aircraft intercommunications system. Software is based on 'proven self-protection' code

PAWS Passive Airborne Warning System

Type
Airborne Electronic Counter Measures (ECM), Missile Approach Warner (MAW).

Description
The Passive Airborne Warning System (PAWS) is a lightweight infrared missile launch and approach warning system designed for fighter aircraft and attack helicopters. It was specified, characterised and integrated by Elisra. It consists of four to six infra-red staring sensors and one processor. The system may operate independently or may be integrated with an EW suite.

PAWS detects the missile heat and tracks it even in high-clutter environments. It determines when a missile threatens the aircraft, provides an accurate readout of approach direction and an estimate of the time to intercept. It can also select the appropriate narrowbeam, or flare dispenser, countermeasures and activate them automatically. The system has a very low false alarm rate even in violent manoeuvres and when operating in a high-clutter environment.

PAWS consists of a high-power processor, built by Elisra, and an IR sensor built by Elop Electronic Industries. It features missile launch and approach warning, discrimination between threatening and non-threatening missiles, multithreat warning capability, accurate approach direction and time-to-go estimation and automatic activation of countermeasures.

LWS-20 laser warning receiver   0002225

PAWS Passive Airborne Warning System 0131094

PAWS integrates with the SPS-65 radar warning receiver, the LWS-20 laser warning receiver, and a chaff and flare dispenser. It is also able to enslave a directional infra-red countermeasures system.

Specifications
Power source: 28 V DC (MIL-STD-704D)
115 V, 400 Hz, 1Ø
Power consumption: 200 W
Dimensions:
(sensor) 120 × 120 × 230 mm
(processor) short 1/2 ATR (203 × 389 × 127 mm)
Weight:
(sensor) 4 × 2.5 kg
(processor) 10 kg
MIL specification: MIL-STD-5400, Class I B
Communication interfaces: RS-422, MIL-STD-1553B MUX BUS, transputerlink

Status
In production.

Contractor
Elisra Electronic Systems Ltd.

SPJ-20 self-protection jammer

Type
Airborne Electronic Counter Measures (ECM), radar jamming system.

Description
The SPJ-20 is a powerful, yet compact, self-protection jamming system, designed to provide protection against ground and airborne radar-guided fire control systems. It is suitable for both fixed- and rotary-wing aircraft.

The main features include:
1. An acquisition receiver and ECM exciter all in a single LRU – no separate RWR is required.
2. Powerful, multiport, pulse and CW transmitters (up to two transmitters supported).
3. Wideband operation.
4. ECM response against pulse, CW and pulse Doppler threats.
5. Interfaces to other onboard avionics and chaff and flare dispensers.
6. In-flight data recording and a field programmable threat library.

The acquisition receiver is a wideband, VCO-based, fast IFM/superheterodyne receiver. It performs amplitude monopulse direction of arrival measurement threat identification and classification.

The ECM resources include: a noise transponder based on a fast VCO; pulse and CW repeater channels; multi-resource range, Doppler, noise and amplitude ECM techniques, generators and trackers.

The transmitter is a 4-tube, multimode (pulse/CW) miniTWT cluster with TWT outputs providing combined in phase and amplitude. The combined output signal is steerable to any combination of output ports on a pulse-by-pulse basis.

Mission support equipment includes: a Mission Loader/Verifier (MLV) on a laptop PC; a Pre-Flight Message Generator (PFMG); and a Post-Mission Data Analyser (PMDA).

Specifications
Frequency: 6–17.5 GHz (transmit); 2–18 GHz (receive)
Interfaces: MIL-STD-1553B, RS 422, RS 232
Environmental qualification: MIL-E-5400T Class 2X
Weight: receiver/jammer: 12 kg; each transmitter 30 kg

Power: receiver/exciter: 28 V DC, 240 W; each transmitter: 115 V AC (3-phase), 3,000 W

Status
In service.

Contractor
Elisra Electronic Systems Ltd.

SPS-20V airborne self-protection system

Type
Airborne Electronic Counter Measures (ECM), Radar Warning Receiver (RWR).

Description
The wideband SPS-20V2 is the latest configuration of Elisra Electronic Systems' SPS-20V airborne radar warning system. As such, it is designed to integrate with existing airborne installations and to detect and display pulsed radar threats operating within the 0.7 to 18 GHz frequency range. Continuous Wave (CW) radar threats are detected by a dedicated high-sensitivity receiver. An 8 cm (3 in) Display Unit (DU) provides an alphanumeric presentation of the type, coarse angle of arrival, relative lethality and status of the analysed radar threats.

The SPS-20V2 features a digital signal analyser that carries out the data processing and interfacing tasks and contains a reprogrammable emitter library file. Functionally, threats are displayed on the DU screen together with audio warnings via the aircraft intercommunication system. Up to 200 different symbols can be programmed to represent specific threat types. Displayed threats on the screen allow the pilot to establish the azimuth and emitter type of the

The SPJ-20 self-protection jammer 0062852

The Elisra SPS-20V warning system 0504066

16 most lethal threats. The displayed symbol and two additional arrowheads indicate the relative threat lethality. Up to 30 different symbols can be programmed to represent specific threat types. Additional alerting capability for missile launch status is also provided. As well as in stand-alone applications, the SPS-20V forms part of Elisra's SPS-65V Multisensor Self-Protection System (MSPS – see separate entry).

Specifications
Frequency coverage: 1–20 GHz
Sensitivity: –38 dBm (DJ pulse); –40 dBm (CD pulse)
Threat sorting parameters: PRI, stagger, jitter, PW, conical scan modulation, angle of arrival
Weights: 0.2 kg (control unit); 0.4 kg (single-channel receiver – 2 used); 1.1 kg (dual-channel receiver – 2 used); 1.2 kg (display unit); 1.4 kg (analyser); 1.63 kg (optional CW receiver)
Power supply: 28 V DC (MIL-STD-704)
Environmental: MIL-STD-5400

Status
It is understood that SPS-20V variants have been selected or procured for use aboard Canadian CH-146 Griffon helicopters (SPS-65V application), German CH-53G helicopters (SPS-65V application) and upgraded Romanian MiG 21 (Lancer programme) and MiG 29 (Sniper programme) aircraft.

Contractor
Elisra Electronic Systems Ltd.

SPS-45V integrated airborne self-protection system

Type
Airborne Electronic Counter Measures (ECM), Radar Warning Receiver (RWR).

Description
The SPS-45V is an airborne self-protection system which integrates two subsystems: the SPS-20V which detects pulse radars from the low band to 18 GHz and the SRS-25 which is a superheterodyne receiver detecting CW, high PRF and low ERP radars. The LWS-20 may also be incorporated as an option to detect laser pulses. Identified threats detected by the subsystem are displayed to the pilot on a 3 in unit which provides a graphical representation of type, angle of arrival, relative lethality and status.

The SPS-45V offers complete threat reprogrammability for new and changing environments through an extensive, easily updated emitter library. A portable memory loader/verifier enables the loading or unloading of the operational software and emitter tables in the field.

Four orthogonally dispersed cavity-backed spiral antennas receive pulsed signals in the 2 to 18 GHz frequency range and send them to the two dual receivers. The video pulse output from both receivers is fed to the analyser which tags every pulse with its angle of arrival, amplitude, pulsewidth and time of arrival.

The analyser CPU processes these characterised pulses, de-interleaves individual pulse trains and identifies the threatening emitter by comparing the pulse parameters with preprogrammed library information. Pulse signals in the low band are also detected and characterised to provide additional information and threat status identification.

The superheterodyne receiver is fed via couplers from the dual receiver.

The input circuitry enables selection of one of the four antennas. The signal detected from the selected antenna passes a preliminary filter, mixes with the output signal of a voltage-controlled oscillator and goes to an IF section. The signal is then amplified, filtered and detected. The video signal is transferred to the video circuit for conversion to digital values. The final analysis and identification of the threat is done by a microprocessor and the results are transferred to the analyser via a serial datalink.

The analyser is responsible for integrating the data from the subsystems into one common threat file.

The analysed threats are displayed on the display unit screen. In addition, audio warnings are dispatched to the pilot through the aircraft intercom system. The threat representations on the display screen allow the pilot to establish the azimuth, lethality and emitter type for up to 16 of the most lethal threats. Up to 250 different symbols can be programmed to represent specific threat types. Additional alert capability for missile launch status is provided.

The system operating mode is controlled by front panel push-buttons on a single control unit. This unit enables the pilot to power the system, activate the automatic self-test procedure, delete preprogrammed threats from the display and define modes of operation.

Each of the two subsystems incorporates its own microprocessor with EEPROM and RAM memory. Each microprocessor works autonomously using its own emitter and mission table.

Specifications
Frequency: low band –18 GHz
Received signals: CW, low EPR, high PRF and pulsed
Coverage: 360° azimuth (depending on antenna installation)
Communications channels: RS-232/-422; TTL in/out
Threat sorting: angle of arrival, conical scan modulation, frequency, jitter, PRI, PW and stagger
Audio: composite, missile launch and new threat
Display: alphanumeric and special symbology

Interfaces: MIL-STD-1553B bus (optional), CMDS, missile warner, radar blanking, radar jammer and simplified blanking centre
Weight:
(SPS-20) 7.5 kg
(SRS-25) 3.8 kg
Power supply: 28 V DC (MIL-STD-704)
Environmental: MIL-STD-5400, Class II

Status
In production.

Contractor
Elisra Electronic Systems Ltd.

SPS-65(V) Self-Protection System

Type
Airborne Electronic Counter Measures (ECM), integrated suite.

Description
The SPS-65(V) is a sophisticated lightweight system with outstanding sensitivity which is capable of intercepting and analysing a wide range of known and anticipated emissions from the battlefield, including laser radiation sources. The design of the SPS-65(V) has been enhanced by technically implementing combat experience.

The SPS-65(V) comprises three subsystems:
- The SPS-20(V) low-volume lightweight radar warning system detects, processes and displays radar threats operating within the 0.7 to 18 GHz frequency range
- The SRS-25 superheterodyne receiver detects modern CW, high PRF and low ERP radars
- The LWS-20 laser warning receiver detects, identifies and locates hostile laser-guided weapons.

A three-inch display unit provides an alphanumeric representation of the type, relative bearing, relative lethality and status of up to 16 analysed threats. In addition, audio warnings are dispatched to the pilot through the intercom system. A flare/chaff dispensing system, which can be automatic or crew-activated, is available for countermeasures capabilities.

Software and emitter tables can be loaded in the field into the emitter library in a matter of seconds. This provides complete and immediate adaptability to changing threat environments. The analyser records flight events and unloads them into a tape cassette for playback after flight.

SPS-65(V)2 integrated RWR and LWS
The SPS-65(V)2 is a sophisticated low-weight, low-volume system that integrates three subsystems: SPS-20(V)2, SRS-25 and LWS-20. The SPS-20(V)2

The SPS-45V consists of SPS-20V and SRS-25 subsystems   0504067

All SPS-65(V)3 electronics and software are contained in a single Line Replaceable Unit (LRU)   1039907

The Elisra SPS-65(V) self-protection system 0089536

detects classical pulse radars within the low-band up to 18 GHz. The SRS-25 uses a superheterodyne receiver to detect modern continuous wave radars, high-PRF (Pulse Repetition Frequency) radars and low-ERP (Effective Radiated Power) radars. The LWS-20 detects and analyses laser signals incident on the aircraft.

SPS-65(V)3 integrated DAS and ESM

The latest iteration of the SPS-65(V) is the SPS-65(V)3, designed to be integrated onto helicopters, fixed-wing aircraft and Unmanned Aerial Vehicles (UAVs). The system combines a range of advanced Defensive Aids Subsystem (DAS) and Electronic Support Measures (ESM) functionalities into a single Line Replaceable Unit (LRU). According to the company, functionalities that can be incorporated into the SPS-65(V)3 LRU include ESM, threat warning (radar, laser and missile approach warning), Counter Measures Dispenser System (CMDS) control and overall Electronic Warfare System (EWS) control.

ESM and radar threat warning functionality is provided by the SPS-20 wideband digital receiver subsystem (which detects pulse and continuous wave radars within the low band up to 18 GHz) and the NBDR-25 narrowband digital receiver subsystem (for detecting continuous-wave, high pulse-repetition frequency and low ERP radars). Laser warning capability is provided by feeding the data from four LWS-20 laser warning sensors into the SPS-20 analyser, where measurements of received laser pulses aimed at the aircraft are processed. Missile approach warning can be achieved by feeding the data from a passive missile warning system, such as Elisra's infra-red sensor-based PAWS system, into the SPS-65(V)3 LRU. This would not only provide missile launch warning and bearing to the threat, but also other data including Hostile Fire Indicator (HFI) and weapons direction. This last capability results from the ability of the PAWS to localise the target position accurately.

Another possible capability of the SPS-65(V)3 combined with PAWS would be to provide the pilot with panoramic infra-red night vision. This could be achieved by integrating the pilot's Helmet-Mounted Display (HMD) with the SPS-65(V)3 so that PAWS imagery could be projected onto the helmet display, in a similar manner to the Distributed Aperture System (DAS) for the F-35 Joint Strike Fighter (JSF). For helicopter applications, the panoramic view would cover 360° in azimuth and 70° in elevation (four PAWS sensors, mounted one each in the nose, tail and on both sides of the helicopter). For fighter applications a fifth and sixth sensor would be added, looking up and down to provide 360° spherical coverage.

The system would also be able to provide collision warning to enhance Situational Awareness (SA) during Night Vision Goggle (NVG) helicopter formation flying.

The company claims a key advantage of integrating all capabilities into a single system is that the SPS-65(V)3 will be capable of

detecting and identifying a threat, correlating the data from multiple sensors, automatically initiating self-defence measures and helping the pilot to direct weapons to engage the source of the threat.

Specifications

Frequency: 0.7–18 GHz (SPS-20(V)); 6.5–18 GHz (SRS-25)

Signal types received: CW, low ERP, high PRF and pulse

Interfaces: CMDS, MIL-STD-1553B, missile warner, radar blanking, radar jammer, RS-232/-422, simplified blanking centre and TTL in/out

Field loading: via portable memory loader/verifier

Threat sorting: angle-of-arrival, conical scan modulation, frequency, jitter, PRI, PW and stagger

Audio: composite, new threat and missile launch

Display: alpha-numeric and special symbols

Weight:
  (SPS-20V) 7.5 kg
  (SRS-25) 4.0 kg
  (LWS-20) 2.5 kg

Power supply: 28 V DC

Frequency: 0.7–18 GHz

Environmental: MIL-STD-5400 Class II

Status

Already in use by major defence forces worldwide, the SPS-65(V) has also been fully qualified by the German military authorities and is in service in the CH-53G helicopter.

The SPS-65(V)2 is in service.

The SPS-65(V)3 has been offered to European and Asian operators; one of these is the UK, where the system is being offered in a competition to

equip the AgustaWestland Future Lynx in army and naval variants as well as the AgustaWestland EH-101 Merlin helicopter. Elisra claim that the use of a single LRU brings considerable advantages in terms of acquisition cost, logistic support, spare parts and life-cycle cost.

Elisra have been working with partner Elbit Systems on the provision of HMD-displayed night vision, although integration is not, as yet, complete. Elisra demonstrated the SPS-65(V)3 infra-red vision capability at the 2005 Paris Air Show, using a virtual reality helmet display and a video recorded during helicopter operations.

Contractor

Elisra Electronic Systems Ltd.

SPS-1000V-5 self-protection system

Type

Airborne Electronic Counter Measures (ECM), Radar Warning Receiver (RWR).

Description

The SPS-1000V-5 is a compact, lightweight, state-of-the-art, airborne, self-protection system, which was specifically designed to be a Radar Warning Receiver (RWR) form-fit replacement for a wide range of front line fighter aircraft.

The system is field proven and is operational on a variety of airborne platforms. The system detects, identifies and displays modern emitters over the C- to J-bands. It can be extended to higher frequencies if required.

The SPS-1000V-5 features two integrated, independent dual receivers, which are controlled by three microprocessors interconnected by a fast internal bus, allowing for a large throughput and fast response time. The time domain receiving subsystem provides continuous and instantaneous coverage of the full frequency range and near 100 per cent probability of intercept. A multichannel superheterodyne/IFM receiver subsystem provides higher sensitivity/detection range and selectivity for CW, high-PRF, low-ERP and pulse Doppler radars. An additional microprocessor controls the displays human and aircraft interfaces.

The Mission Support System (MSS) was designed to provide immediate response for any operationally required update. It allows for complete programmability and analysis of multiple PRF threats as well as the human machine interface of the system, such as

The SPS-1000V-5 self-protection system 0062853

displayed symbols (selectable from a built-in library), audio tones, missile launch warning, fade in/fade out times and other data.

The system has growth potential to meet the requirements of different platforms and operational scenarios. The SPS-1000V-5 system contains a large array of interfaces including: discretes, MIL-STD-1553B, RS 422, RS 232, and blanking inputs and outputs.

Specifications
Weight:
  (central receiver processor) 14 kg
  (pilot display and control) 2.4 kg
  (antennas and front end receivers) 6 kg
  (RF switch) 0.8 kg
Power supply: 115 V AC, 400 Hz
  28 V DC (MIL-STD-704)
Environmental: MIL-STD-5400, Class II

Status
Selected for the Royal Australian Air Force C-130H interim upgrade. The SPS-1000V-5 also forms part of the F-16 Avionics Capability Enhancement (ACE) upgrade package (see separate entry).

Contractor
Elisra Electronic Systems Ltd.

SRS-25 airborne receiver

Type
Airborne Electronic Counter Measures (ECM), Radar Warning Receiver (RWR).

Description
The SRS-25 detects and analyses CW, high-PRF and medium-ERP radar signals and determines their direction of arrival. It forms part of the SPS-45V and SPS-65V integrated helicopter self-protection systems. Continuous wave or high-PRF signals in the 6.5 to 18 GHz frequency range are analysed and signals can be recorded for later playback, analysis or training. An extensive threat library, which can be updated easily between flights, is incorporated in the unit. Four cavity-backed spiral antennas are used for accurate direction-finding.

Specifications
Frequency: 6.5–18 GHz
Received signals: CW, high PRF, medium ERP

Reception coverage: ±30° (elevation); 360° (azimuth)
Frequency accuracy: ±10 MHz
Frequency resolution: 2 MHz
Dimensions: 101 × 122 × 250 mm
Weight: 4 kg

Status
The SRS-25 is part of the SPS-65V multisensor threat warning systems which have been procured for installation aboard Canadian Forces CH-146 Griffon and German Army CH-53G helicopters.

Contractor
Elisra Electronic Systems Ltd.

TACDES SIGINT system

Type
Airborne Electronic Counter Measures (ECM), Signals Intelligence (SIGINT) system.

Description
TACDES is an airborne SIGINT system consisting of the RAS-1B or RAS-2A ELINT systems. TACDES is installed on large transport aircraft such as the Boeing 707 or Lockheed Martin C-130 and performs real-time acquisition, location, processing and reporting of communications and radar signals between 20 MHz and 18 GHz.

Accurate instantaneous wide-angle coverage is provided by interferometric COMINT and ELINT DF systems. The system features an enhanced command and control capability and real-time secure voice and data communications.

RAS-1B ELINT and ESM system
The RAS-1B airborne interferometric ELINT system, installed on large transport aircraft, covers the 0.7 to 18 GHz frequency band (0.5 to 40 GHz optional) to obtain the enemy Electronic Order of Battle (EOB). It can measure frequency, pulse repetition frequency, amplitude and Direction Of Arrival (DOA) data on an average of two emitters per second, with exceptionally high accuracy. Operation has been optimised for weak emitter recognition in high-density electromagnetic environments and an activity file is maintained which contains identification, classification and updated data on recognised threats. A

combination of wide bandwidth acquisition IFM receiver and a narrowband superheterodyne analysis receiver is used to ensure 100 per cent probability of detection and high system throughput. Operation can be fully automatic or man/machine oriented in simple manual mode to accommodate up to two operators.

Specifications
Frequency: 0.5–18 GHz (0.5–40 GHz optional)
Frequency resolution: 1.25 MHz
Frequency accuracy: 2.5 MHz
DOA accuracy:
  (from 2–18 GHz) 0.7° (typical), 0.3° (optional)
  (from 0.5–2 GHz) 1.3° (typical), 0.5° (optional)
PRI resolution: 100 ns
PRI accuracy: 200 ns
Instantaneous bandwidth:
  (analysis) 40 MHz
  (acquisition) 8 GHz
Sensitivity: –70 dBm typical
De-interleaving capability: Up to 8 pulse trains in the same frequency/DOA cell

RAS-2A ELINT and ESM system
The RAS-2A airborne, wingtip-mounted, interferometric tactical ESM system is deployed on small- and medium-size transport aircraft. In the 0.5 to 18 GHz frequency band it can measure frequency, PRI, amplitude and direction of arrival data with excellent accuracy, with 360° azimuth coverage. Operation has been optimised for a single operator who can both display an EOB scene and passively target an anti-radiation or anti-ship missile. The high sensitivity of the system ensures weak signal processing even at long ranges. Automatic file activity is maintained for quick identification and classification of threats.

The RAS-2MT is the configuration for maritime surveillance and targeting.

Specifications
Frequency: 0.5–18 GHz
Frequency resolution: 0.2 MHz typical
Frequency accuracy: 0.4 MHz typical
DOA accuracy:
  (left/right) 0.8°
  (fore/aft) 5°
  Coverage:
  (azimuth) 360°
  (elevation) ±10°
Sensitivity: –75 dBm
Instantaneous bandwidth: 1, 5, 15, 40 MHz

Status
In production for use on large transport aircraft that are required for special surveillance tasks.

Contractor
Tadiran Electronic Systems Ltd, a member of the Elisra Group.

Top Scan/A

Type
Airborne Electronic Counter Measures (ECM), Electronic Support Measures (ESM) system.

Description
Suitable for all types of airborne platform, Rafael's Top Scan/A ES/ELINT system is designed to detect, identify and locate ground-based radar emitters. The system comprises two modules for sensors and digital/electronics:

The sensor module includes the system's Direction-Finding (DF) antenna array, warning antennas, Radio Frequency (RF) front-end and a frequency source unit. The sensor module can be accommodated in either an external pod or a conformal mounting.

The digital/electronics module, housed within the host platform, comprises the power supply, Intermediate Frequency (IF) receiver, Analogue-to-Digital (A/D) converter, signal processor and system controllers.

The RAS-1B ELINT and ESM system operators' consoles   0503814

Other system features and benefits include:
- High Probability of Intercept – POI
- Suitable for manned and unmanned vehicle applications
- Automatic detection, collection, measurement, identification and localisation of target emitters
- Autonomous or controlled operating modes
- Option for control by a remote ground station in controlled mode
- Datalink communication with a remote airborne or ground-based station
- Integral reprogrammable (updating and mission planning) emitter library
- high resolution dual-axis interferometer DF and localisation

- utilises a wide-aperture 'sparse' (few elements) antenna array
- Pulse Repetition Interval (PRI) tracking mode for simultaneous multiple target tracking
- Proprietary ambiguity, multipath and outlier elimination algorithms
- Covers modern/complex pulse types in dense RF environments

Specifications
Frequency coverage: 0.5–18 GHz
Azimuth coverage: 360°
Elevation coverage: >90°
Dual-axis DF accuracy: 2° RMS
Sensitivity: –70 dBm (at bandwidth of 500 MHz)
Bandwidth: 20, 100 or 500 MHz (selectable)

Dynamic range: 40/60 dB (sampling, immediate); 55 dB (detection)
Pulse width: 0.10–120 s and continuous wave
Receiver recovery time: 250 ns
Frequency measurement accuracy: 1 MHz RMS
Frequency measurement time: 100 ns
Power consumption: <150 W
Weight: <15 kg

Status
In production and in service.

Contractor
Rafael Armament Development Authority Ltd, Systems Division.

Italy

Apex jamming pod

Type
Airborne Electronic Counter Measures (ECM), radar jamming system.

Description
The Apex deception jamming pod uses the existing ELT 555 pod shell and is claimed to be effective against all types of fire-control radars. Frequency coverage against both pulse and CW emitters is quoted as being extended. The pod weighs 140 kg.

Apex uses fast digital RF memory technology to facilitate coherent response techniques and the system is front-line programmable.

Status
Apex is suitable for aircraft such as the Hawk 100/200, MB-339, F-5 and Mirage F1.

Contractor
Elettronica SpA.

ELT/156X radar warning receiver

Type
Airborne Electronic Counter Measures (ECM), Radar Warning Receiver (RWR).

Description
The Elettronica ELT/156X radar warning receiver features very wideband miniature crystal video receiving systems and provides full azimuth and frequency coverage up to millimetric-waves and including distributed multiprocessing. The use of bit-slice microprocessing and EEPROM technology allows flexible and high-speed processing, giving fast reaction times and adaptability to threat evolution. Threat parameter software is reprogrammable at flight line level. The dual cockpit display incorporates both a raw mode, to give an immediate picture of the surrounding environment, and a synthetic mode, to give a clear indication of threat identity. The system also includes an aural alarm.

The ELT/156X may be integrated with other onboard EW systems and sensors via two RS-422A serial links.

The system consists of two antennas/RF heads, signal processor, display and control panel. The version for helicopters, the ELT/156X(V2), employs four single-channel RF heads.

Specifications
Weight:
(antenna/RF head × 2) 2.2 kg
(signal processor) 10.5 kg
(display) 2.1 kg

The ELT/156X radar warning receiver showing (from left to right) antennas, processor and (front) the control panel, (rear) the display   0503819

(control panel) 0.7 kg
(total weight aircraft version) 17.7 kg
(total weight helicopter version) 20.1 kg

Status
In production and in service.

Selected by the Italian Navy for its AS/ASVW and AEW EH 101 helicopters, also selected for the Brazilian AMX light fighter aircraft.

Contractor
Elettronica SpA.

ELT/263 ESM system

Type
Airborne Electronic Counter Measures (ECM), Electronic Support Measures (ESM) system.

Description
The ELT/263 airborne ESM is designed principally for use in maritime aircraft such as the Beech 200T but has been configured for other types such as the Guardian, CASA C212, Learjet 35A, F27 Maritime, Bandeirante, Piaggio P166 and Britten-Norman Maritime Defender. The

surveillance of coastal waters is its main function. Detection, analysis and identification of emissions in the E- to J-bands is performed, in addition to bearing measurement; this permits emitter location by using a series of successive bearing measurements.

The ELT/263 equipment, in addition to its surveillance tasks, provides for emitter location by the triangulation method. Emission analysis, identification and bearing information can be transferred to external users such as ground stations, naval units and the aircraft's radar operator and tactical navigator and can also be recorded on an optional printer.

The main items of the ELT/263 are a set of four DF antennas, an omnidirectional antenna set, DF receiver, IFM receiver, ESM display and control console, radar warning processor and radar warning display unit.

Status
In service. The equipment is in current production for several customers.

Contractor
Elettronica SpA.

ELT/457-460 supersonic noise jamming pods

Type
Airborne Electronic Counter Measures (ECM), radar jamming system.

Description
A set of four related noise jamming pods are available for use on high-performance strike/fighter or light attack aircraft, mounted on an available underwing pylon. All units are self-contained with a ram-air turbine generator in the nose section and each is dedicated to a particular waveband. A heat-exchanger is situated behind the turbine and there are fore and aft antennas in all pods. The ELT/459 and ELT/460 versions have additional antennas beneath the body of the pod. Processing within the system allocates threat jamming power on a proportional basis against any type of pulsed radar threat and includes built-in test equipment and control of blanking with other aircraft systems. All units are claimed to be highly resistant to ECCM, including any type of frequency agility, and incorporate a large threat library. Maximum operating speeds are M1.1 at sea level and M1.5 at 40,000 ft.

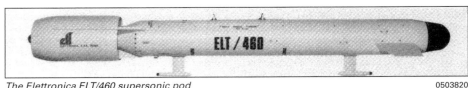
The Elettronica ELT/460 supersonic pod   0503820

Specifications
Dimensions: 3,120 (length) × 340 mm (diameter)
Weight: 145 kg
Interface: STANAG 3726 and MIL-A-8591D

Status
In service on a number of fighter platforms.

Contractor
Elettronica SpA.

ELT/553(V)-2 airborne pulse and CW jammer

Type
Airborne Electronic Counter Measures (ECM), radar jamming system.

Description
The ELT/553(V)-2 airborne deception jammer features fully automatic operation and is programmable at flight line level. It is designed to ensure maximum flexibility, operating either as a stand-alone system, or in co-operation with on board RWRs.

ELT/553(V)-2 comprises:
- A low-band pulse module (E- to H-band)
- A high-band pulse module (H- to J-band)
- A CW module dedicated to counter semi-active missile threats.

ELT/553(V)-2 utilises:
- A channelised receiving section architecture for quick and reliable lock-on and recognition, even in very dense environments
- Repeater jammer philosophy to increase effective jamming power and hence jamming effectiveness in the presence of ECCM techniques
- Dedicated CW mode to counter semi-active missile threats
- Effective operation against staggered and jittered PRI emitters, as well as against frequency diversity and frequency agility threats
- Automatic deception of locked-on radars
- Multithreat capability
- RS422 interfaces.

Optional capabilities include:
- Automatic deception of pulse and CW radars, TWS included

- Pulse-by-pulse power management and optimisation across 10 channels of operation
- Very high ERP
- Flight line programmability
- Complete BITE.

Specifications
LRUs	quantity	weight (kg)
High-band pulse module	1	40
Low-band pulse module	1	37
CW module	1	20.4
Control panel	1	1.4
High-band antenna	2	0.2 × 2
Low-band antenna	2	0.4 × 2

Power consumption: 2.5 kVA

Status
In production and in service in Brazilian AMX and Italian Tornado aircraft.

Contractor
Elettronica SpA.

ELT/554 deception jammer

Type
Airborne Electronic Counter Measures (ECM), radar jamming system.

Description
The ELT/554 is specifically designed for installation in attack helicopters and operates in the J-band. It is lightweight and compact and reacts rapidly to incoming signals. The jamming programmes are intelligent and several threats can be jammed simultaneously.

A self-protection suite for modern attack helicopters comprising the ELT/554 lightweight deception jammer (left) and the ELT/156 radar warning receiver   0503821

Contractor
Elettronica SpA.

ELT/555 supersonic self-protection pod

Type
Airborne Electronic Counter Measures (ECM), radar jamming system.

Description
The ELT/555 has a similar configuration to the ELT/457 system, with the same ram-air turbine on the nose of the pod. Internal equipment includes fore and aft facing antennas which receive pulsed and continuous wave signals and transmit pulsed responses, plus separate fore and aft facing continuous wave transmitter antennas on the undersurface. The system is sensitive to H- to J-band (6 to 20 GHz) threats and is designed to operate in dense electromagnetic environments. It has multiple target contrast capability and features BITE. Operating envelope and mechanical interface specifications are the same as those of the ELT/457. The cockpit display can be tailored to customers' requirements.

Specifications
Dimensions: 3,000 (length) × 340 mm (diameter)
Weight: 150 kg

Status
In service.

Contractor
Elettronica SpA.

ELT/553(V)-2 airborne pulse and CW jammer   0018283

The Elettronica ELT/555 deception jammer pod on a Learjet 35A multimission trainer   0503822

ELT/558 self-protection jammer

Type
Airborne Electronic Counter Measures (ECM), radar jamming system.

Description
The ELT/558 self-protection jammer operates at the lower end of the frequency spectrum and features wide-angle coverage, fully automatic responsive jamming operation, high ERP, low false alarm rate, high immunity to ECCM, very short reaction time and effective reaction against many threats. The system can be integrated with the aircraft's avionics.

Status
In production and in service on the Mirage 2000.

Contractor
Elettronica SpA.

The ELT/558 receiver unit (left) and ELT/558 transmitter unit (right)   0018282

ELT/562 and ELT/566 deception jammers

Type
Airborne Electronic Counter Measures (ECM), radar jamming system.

Description
The ELT/562 and ELT/566 are repeater jammer systems for internal installation on aircraft and helicopters. ELT/562 is designed to counter pulse threats, while ELT/566 combats continuous wave threats. One or both jammers may be associated with an airborne ESM system such as the Elettronica Colibri. The jammers are effective against H- to J-band threats and are intended to protect aircraft and helicopters in battlefield environments.

Status
In production and in service.

Contractor
Elettronica SpA.

Fast Jam ECM system

Type
Airborne Electronic Counter Measures (ECM), communication jamming system.

Description
Fast Jam is designed for operations over the VHF and UHF communications frequencies. It is intended mainly for airborne applications, where maximum advantage is obtained by the platform elevation and speed, although the compact and lightweight construction of the system also allows installation on land vehicles and ships. The main features of the system include modularity, coverage of the VHF and UHF bands with optional coverage of the HF band, high speed and intercept capability and multiple frequency jamming capability.

In the Fast Jam configuration the system consists of a search and analysis receiver subsystem, monitoring subsystem and jammer subsystem. The first two subsystems perform the same functions as the basic ESM system described in the later entry for Smart Guard. The addition of the jammer subsystem allows the jamming of intercepted communications channels, to disrupt enemy communications.

The jammer subsystem operates under the control of the system computer and is fully automated. By time-sharing, it is able to jam up to six channels simultaneously without performance degradation. The channels to be jammed are either preset or can be selected by the operator according to the specific operational requirements. A look-through function allows an adaptive jam/receive time management so that simultaneous search and jamming functions can be performed.

Manual control of the jammer subsystem is available as a back-up to fully automatic operation. A high-power wideband all-solid-state amplifier is used in the jammer subsystem and VSWR and thermal automatic switch off circuits are employed to avoid permanent damage to the amplifier in case of overheat or antenna failure. Various types of modulation are available from external modulation sources.

Status
In service. The system has been installed in a number of different types of aircraft.

Contractor
Elettronica SpA.

RALM-01/V2 Laser Warning Receiver (LWR)

Type
Airborne Electronic Counter Measures (ECM), Laser Warning Receiver (LWR).

Description
The RALM-01/V2 LWR, an improved version of the RALM-01, provides laser threat detection and classification capability. Specifically designed for aircraft and helicopters, RALM-01/V2 can either be integrated with other EW equipment or work as a stand-alone equipment and directly drive countermeasures systems.

The use of fibre optic technology and an optical external head provides field verified performance in terms of false alarm rate and probability of detection.

RALM-01/V2 detects pulsed wave laser radiation, discriminates it from background clutter, identifies the laser threat type, measures its characteristics and compares it with an internal database. The database is stored in a non-volatile, erasable memory and can be easily modified and reloaded whenever it is necessary to match different threat scenarios. The system then identifies and refines the direction of arrival of the threat with sufficient accuracy to effectively deploy countermeasures. RALM-01/V2 deals with multiple simultaneous threats and operates at an extremely low false alarm rate, largely (it is claimed) due to the unique fibre optic architecture of the system, which offers outstanding immunity to EMC and adverse environmental conditions.

RALM-01/V2 has a wide interface capability that facilitates its integration with on-board systems for control and information exchange purposes.

The standard interface is based on RS-422 serial link and/or MIL-STD-1553B databus.

Specifications
Azimuth coverage: 360°
Elevation coverage: 80°
Angular accuracy: ±22.5° (16° RMS)
Sensor band: 0.5 to 1.8 μm
False alarm rate: Less than 1 in 16
Reaction time: 100 msec RMS
Weight:
  (electronic unit) <4.5 kg
  (sensor) <0.3 kg
  Power: 28 V DC; <35 W
Dimensions:
  (sensors) 100 × 70 × 60 mm (nominally × 2)
  (processor) 3/8 ATR short
Environmental: MIL-STD-810E; MIL-STD-461C
Interface: MIL-STD-1553B, RS-422, discrete signals

Status
RALM-01/V2 LWRs are now installed in attack helicopters, Combat Search And Rescue (CSAR) helicopters, combat support and utility helicopters and large transport aircraft, including the Mangusta A 129, HH-3F and C-130J.

Contractor
Marconi Selenia SpA.

The RALM-01/V2 LWR is fitted to the A 129 Mangusta light attack helicopter   0521344

Sea Petrel RQH-5(V) airborne ESM/ELINT systems

Type
Airborne Electronic Counter Measures (ECM), Electronic Support Measures (ESM) system.

Description
The Sea Petrel RQH-5(V) is a family of systems which can meet EW requirements ranging from threat detection and analysis to electronic intelligence. The various configurations and options allow the system to be tailored to a specific requirement. All the RQH-5(V) Series of equipments have capabilities for integration with other systems and provide target parameters and direction of arrival information for weapon systems. The small size and low weight of RQH-5(V) components make the system readily adaptable to current airborne, naval and ground installations with a minimum of effort. The system can be operated after minimal instruction and is designed for maximum operating time with minimum servicing.

The Elettronica ALR-733(V)2 ESM system for maritime patrol aircraft   0080283

The Sea Petrel RQH-5(V) ESM/ELINT system   0504138

The system covers frequencies from 0.65 to 18 GHz, with an optional extension to 40 GHz, and is entirely automatic. It provides real-time automatic extraction, analysis and tracking of all incoming radar signals. It also provides pulse, intrapulse and fine analysis for ELINT, including frequency fine measurement, measurement of jitter and stagger, frequency or PRI agility analysis, histogram preparation, detection and recording of antenna pattern and related amplitude histogram. The operator has only to view the system display which provides data on up to 200 emitters. The RQH-5(V) can operate without any prior knowledge of the electromagnetic scenario with no significant reduction in performance. On the other hand, the ELINT capability in terms of emitter parameter statistical analysis and pulse and intrapulse analysis, allows complete characterisation and analysis of the radar signals and keeps records of them for post-flight data collection.

The basic components of the RQH-5(V) are the antennas, direction-finder receiver, IFM receiver and data extractor.

The antenna group includes one omnidirectional antenna and four or eight DF antennas. Various DF and fine DF antenna types are available to cover different frequency ranges. The modular approach of each antenna module allows different installation configurations, to cater for any platform constraints.

Each antenna unit incorporates the associated electronic circuitry. Direction-finding is performed by a wide open omnidirectional and instantaneous receiver using amplitude comparison monopulse techniques. An eight port configuration is normally adopted, but it can also use a four element DF antenna subsystem for radar warning receiver applications.

The IFM receiver features high sensitivity and high probability of detection, wide open operations, fast response in order to operate in a multimillion pps environment and very high instantaneous dynamic range. Two versions are available – the FR-6 and FR-7 – providing different RF measurement accuracy.

The data extractor consists of a very powerful multiprocessor structure, specially adapted for real-time applications as a derivative of the AYK-204 airborne computer. The resulting automatic data extraction process has a very high acquisition speed of up to 60 new emitters every 20 ms, even in a completely unknown environment. This unit includes some standard I/O interfaces, such as serial lines, video graphic, memory expansion and MIL-STD-1553 bus developed for the Alenia Difesa AYK-204 airborne computer and its derivative versions.

Several graphic and alphanumeric display modes are available to the operator. This includes frequency and direction of arrival of the signal, tabular lists, emitter characteristics, true or relative bearing, frequency-agile deviation, PRF, jitter and stagger values, emitter name and threat level and emitter scan period and type. Tactical, panoramic and geographic modes for ESM, radar and navigation are also available.

The identification library handles up to 3,000 modes of emitter parameters. The operator can store previous known data together with pre-assigned threat level and confidence level. The system compares extracted emitter data with the library and provides an immediate alert on high-interest emitters.

Several optional components are available. A high-accuracy direction-finder uses a multiple beam antenna system and a crystal video amplitude sectoral monopulse receiver.

Several multibeam flat arrays, covering different frequency ranges, are available to provide different instantaneous fields of view and very high DF accuracy. A K-band high-gain steerable antenna and a formatter unit provide high-sensitivity detection and direction of the emitter signals. A fine analysis ELINT receiver gives supplementary information on the active extracted emitter, including the presence of simultaneous multiple RF or intrapulse modulations.

The following configurations are available:

The SL/ALR-730 Series
offers electronic support measures and ELINT for all types of platforms including large and medium maritime patrol aircraft and large, medium and small helicopters. It is based on superheterodyne receiver technology.

The SL/ALR-740 Series
offers RWR functions combined with automatic signal analysis for post-flight intelligence, and is designed for installation on small aircraft or helicopters. Average DF accuracy of 10° RMS is provided, with automatic warning and emitter parameter measurements. The ALR-741-R uses multiple-IFM receivers.

The SL/ALR-780 Series
allows the integration of ECM modules in the above ESM equipment.

Specifications
Frequency: 0.6–18 GHz
Sensitivity: –60 dBm
Accuracy: ±2.5° RMS

Status
In service. Current and future applications are quoted to include the AB412, EH 101 and NFH-90 helicopters.

A version of the SL/ALR-730, the ALR-735(V)3, has been selected by the Italian Navy for the EH 101 helicopter in its ASW/ASVW (Anti-Submarine Warfare/Anti-Surface Vessel Warfare) and AEW (Airborne Early Warning) roles.

The ALR-733(V)2, an ESM system for maritime patrol aircraft, is reported to have been selected by the Italian Guardia di Finanza for use on an ATR-42 aircraft. The line replaceable units appear to be different, but nonetheless the ALR-733(V)2 is believed to be part of the overall RQH-5(V) family.

The system is also said to represent the Italian contribution to a joint Italian/German EADS programme for future maritime patrol aircraft requirements.

Contractor
Elettronica SpA.

SL/ALQ-234 self-defence pod

Type
Airborne Electronic Counter Measures (ECM), radar jamming system.

Description
The SL/ALQ-234 self-defence pod fulfils the requirements for aircraft self-defence with a minimum of penalty to flight and combat capabilities. The SL/ALQ-234 is designed for standard fuselage or wing pylon installation. A small cockpit control panel provides the pilot with threat warning and jamming reaction information. Self-contained ram-air turbine power generation and cooling, as well as the modular pylon attachment technique, make the SL/ALQ-234 suitable for immediate installation on a variety of supersonic aircraft.

Real-time processing and power management are the main factors leading to the multiple threat self-defence capability of the SL/ALQ-234. For pulse threats the system covers the I- and J-bands and provides noise and deception jamming. CW coverage is provided in the H- to J-bands and the threat is countered by deception jamming.

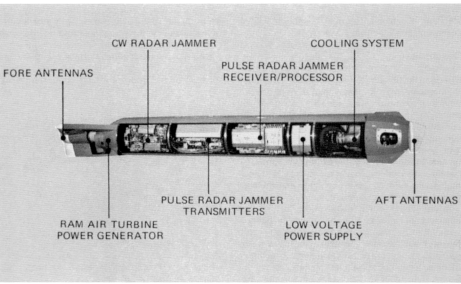

The internal arrangement of the SL/ALQ-234 self-defence pod   0503816

SL/ALQ-234 self-defence pod 0503815

The system can operate even in the complete absence of prestored threat data, through software generated self-adaptive modes. Against known threats prestored in the system memory it can perform more complex and specific reactions. Positive self and mutual screening of the aircraft is achieved against the most advanced ground-based air defence systems.

Specifications
Dimensions: 3,825 (length) × 414 mm (diameter)
Weight: 270 kg
Frequency:
(pulse) I/J-bands
(CW) H/J-bands
Power output: 7.5 kVA
Altitude: sea level to 30,000 ft
Speed: M1.1 at sea level, M1.5 at 30,000 ft

Status
In service with various air forces.

Contractor
Selex Sensors and Airborne Systems, Radar Systems Business Unit.

Smart Guard COMINT system

Type
Airborne Electronic Counter Measures (ECM), Electronic Support Measures (ESM) system.

Description
The Smart Guard COMINT system is designed for intelligence monitoring of the VHF/UHF communications band. Although it is mainly

The Smart Guard DF and emitter fixing subsystem 0503824

intended for airborne applications, where maximum advantage is obtained by the platform height and speed, the compact and lightweight construction of the system allows installation on ground vehicles and in ships.

The basic ESM system consists of a search and analysis receiver and a monitoring subsystem. In this configuration the system allows the search and intercept of communications through a computer-controlled operation. The search is carried out over the complete bandwidth, or on specified sub-bands or channels. On operator control the receiver stops on a specific channel, so allowing the demodulation and analysis of the particular channel. A continuous monitoring of up to eight channels is provided by up to eight remotely controlled receivers and associated recorders. The frequency tuning of each monitoring receiver is set automatically by the system computer. Voice and associated data are recorded for subsequent analysis. In this configuration the system is manned by a single operator.

The Smart Guard monitoring and recording subsystem 0503825

In the Smart Guard configuration, the basic ESM system is augmented by a DF and fixing subsystem. The latter aids the capability to measure the DOA of the communications transmissions, and to determine their location. The DOA measurements are performed by a specific dual-channel superheterodyne receiver controlled by the system computer, and are obtained through an interferometric measurement. In addition to the DOA, the frequency value and signal strength are measured for each specific channel.

An interface with the navigation system provides the actual position of the platform in such a way that for each specific emission a set of data is stored. This contains DOA, frequency value, signal strength and platform position. The fixing of specific communication emission is performed by a dedicated computer and associated software algorithms by using the DOA and platform position data of a specific channel. At least two significant DOA measurements are required for a fixing computation.

In the Smart Guard configuration, two operators are required: one to control the basic ESM system and one dedicated to the fixing operation. A ground-based retrieval and analysis system is available as an option.

Integration of the Smart Guard COMINT system with the ELT/888 ELINT system makes it possible to monitor effectively the full electromagnetic environment and to identify and locate all enemy weapon systems.

Status
In service.

Contractor
Elettronica SpA.

The Smart Guard basic ESM subsystem 0503823

Japan

J/APQ-1 rear warning receiver

Type
Airborne Electronic Counter Measures (ECM), Radar Warning Receiver (RWR).

Description
The J/APQ-1 Radar Warning Receiver (RWR) has been developed for JASDF F-15J aircraft. It is designed to cope with both radar and infra-red threats and will automatically activate chaff and flare countermeasures. Installed in a fairing on the aircraft's starboard fin, the system is said to utilise a 20 cm antenna and is reported to generate both visual and audio threat alerts and to be able to automatically activate an onboard Counter Measures Dispensing System (CMDS) when a threat is confirmed.

Status
Developed for the F-15J. Believed to have entered service in 1992.

Contractor
Mitsubishi Electric Corporation.

J/APR-4/4A radar warning system

Type
Airborne Electronic Counter Measures (ECM), Radar Warning Receiver (RWR).

Description
The J/APR-4 was designed for the F-15J/DJ aircraft. It is able to process multiple inputs simultaneously in a dense electromagnetic environment, and has a digital computer with a reprogrammable software package to allow reconfiguration for future requirements. The indicator provides for daylight viewing and multithreat data presentation in alphanumeric and graphic format. The system is also designed to interface with other EW equipment such as the J/ALQ-8.

The J/APR-4A is the advanced model of the J/APR-4 radar warning receiver. Its specification calls for the ability to process multiple inputs simultaneously in a dense electromagnetic environment. The system incorporates a digital processor with reprogrammable software which permits reconfiguration to meet developing threats. A tactical situation CRT display presents multiple-threat data in alphanumeric and graphic form. Interfaces with other onboard electronic countermeasures systems, such as J/ALQ-8 jammers, can be accommodated.

Status
In production for the F-15J/DJ.

Contractor
Tokimec Inc.

J/APR-5 and J/APR-6 radar warning systems

Type
Airborne Electronic Counter Measures (ECM), Radar Warning Receiver (RWR).

Description
The J/APR-5 and J/APR-6 are developments of the J/APR-4 system, with additional capability designed to cope with current threats. Actual sizes and weights vary according to application.

Status
All RF-4E reconnaissance aircraft are fitted with the J/APR-5. The J/APR-6 equips the F-4EJ Kai aircraft.

Contractor
Tokimec Inc.

The Tokimec J/APR-5 RWR equips Japan Air Self-Defence Force RF-4E aircraft   0503826

The J/APR-6 is installed in Japan Air Self-Defence Force F-4EJ Kai aircraft
0504069

Russian Federation

20 SP M-01 airborne flare dispenser system

Type
Airborne Electronic Counter Measures (ECM), Infra-Red Counter Measures (IRCM) system.

Description
The 20 SP M-01 flare dispenser system provides protection against infra-red homing missiles. It has the following modes of operation:
1. automatic: the system automatically selects the optimum flare dispense programme, responds to the commands of the EW suite and allows the aircrew to concentrate on their primary mission;
2. manual: pre-programmed dispense programmes are available for manual selection and activation by the aircrew;

3. accelerated: employment takes place at the maximum dispense rate, on aircrew command—emergency ejection.

The system provides multiple firing pulses to enable single, double, triple and quadruple payload dispensing when required. Emergency ejection is also available.

System architecture comprises: a program loading unit; a control unit; two switching units; and four dispenser magazines.

Specifications
Maximum payload: 120 flare cartridges
Salvo length: 1–4 flares
Salvo spacing: 0.01–10.0 s
Power: 27 V DC, 100 W

Status
Fitted to MiG-29 and Su-27 aircraft.

Contractor
Avia Avtomatika.
Joint Stock Company PRIBOR.

Airborne Laser Jamming System (ALJS)

Type
Airborne Electronic Counter Measures (ECM), Infra Red Counter Measures (IRCM) system.

Description
The Automatic System Corporation has developed a new, retrofittable Airborne Laser Jamming System (ALJS) for both the commercial and military aircraft, designed to detect and jam the seeker heads used in a range of IR-guided threats that includes the Advanced Short Range Air-to-Air Missile (ASRAAM), AIM-9L/M/R/X, Magic, R-73E and R-74E air-to-air, and Stinger, Mistral and Strela-2/-2M surface-to-air missiles.

ALJS comprises two 150 kg packages, each of which incorporates a control unit, a power supply, a front end signal processor, a laser source, a cooling system, an optical/mechanical head unit and a processing unit (see illustration) that can be pod mounted or scabbed onto the port and starboard sides of the host aircraft fuselage. For fighter applications, the system is designed to be integrated with the host aircraft's Electronic Counter Measures (ECM) suite, navigation system and other IR sensors, to aid in target cueing, to provide:
• RWR cued hostile aircraft tracking and missile launch warning

20 SP M-01 airborne flare dispenser system   0018280

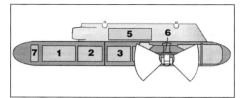

A schematic showing the layout of the ALJS pod (Rosoboronexport) *1 control unit, 2 power supply, 3 front-end signal processor, 5 cooling system, 6 optical unit and 7 computer*   0522986

- Sectored IR missile threat detection and tracking
- Angular threat co-ordinates for use by other elements of the host platform's defensive aids suite
- Pre-launch IR lock-on jamming.

Operating modes encompass target tracking and jamming which is triggered (in suitably equipped aircraft) by the host platform's avionic system. In civilian aircraft applications, operating mode initiation is understood to be autonomous. The company claims that the ALJS functions effectively in both natural and manmade interference environments.

In operation, once an incoming missile is detected, the nearest pod to the incoming missile tracks it with the laser and dazzling the seeker head, creating loss of guidance and increasing miss distance to beyond the effective lethal radius of the missile warhead. The pod provides the pilot with confirmation that the laser is hitting the target as the laser registers positive reflection back at the emitter. The company claims that the system is highly effective in a close-range dogfight, capable of acquiring and disabling two missiles simultaneously, virtually as they are launched.

Specifications
Probability of target detect/tracking: At least 90%
Probability of target suppression: At least 80% (sector coverage after 1.5 s of laser exposure)
Jamming coverage: –45 to +30° (elevation); 360° (azimuth)
Power supply: 27 V DC; 115–200 V AC (400 Hz)
Power consumption: ≤1 kW (stand-by, 27 V DC supply); ≤4 kW (stand-by, 115–200 V supply); 10 kW (operating, 115–200 V supply)
Dimensions (L × Ø): 1.6 m × 38 cm (each pod shell)
Weight: 300 kg (complete system, 2 × 150 kg equipment packages)

Status
According to the company, the system has been in development for more than 10 years and was authorised in 2002 for foreign sale under the auspices of Rosoboronexport. Customer demonstrations were conducted later in the year. To date, no export orders have been reported.

Contractor
Automatic System Corporation.

Airborne radar jammers – Gardeniya, Schmalta and Sorbtsiya

Type
Airborne Electronic Counter Measures (ECM), radar jamming system.

Description
Gardeniya, Schmalta and Sorbtsiya are all airborne jammers, that are believed to belong to the same family.

The Gardeniya family includes both pod-mounted and internal radar jammers, that can operate autonomously and automatically. Family capabilities are reported to include: self-protection; noise jamming in the 10, 20 and 70 cm

Sorbtsiya on an Su-34   0504265

wavebands; and communications jamming (Gardeniya IFUE – fitted to Mi-17P). Internally mounted versions of the Gardeniya family are reported to be fitted to MiG-29 aircraft.

Sorbtsiya (also reported as Sorbtsiya-S) wide spectrum jamming pod has been shown fitted in wing tip pods on Su-271B, naval Su-27 'Flanker', Su-34 and the Su-35 advanced 'Flanker' derivative.

Status
Many systems are in service; it is probable that development continues. The Su-27SK is reported to carry an optional Gardeniya installation, using two wing-tip pods.

Contractor
Pleshakov Scientific & Industrial Corp (GosCNRTI), Design Bureau.
Moscow.

Irtysh EW system on Su-39 Strike Shield

Type
Airborne Electronic Counter Measures (ECM), integrated suite.

Description
The Central Scientific Institute for Radiotechnical Measurements TSNIITI, Omsk is the system designer for the complete EW system – known as Irtysh – on the Su-39 Strike Shield aircraft (also known as Su-25TM). Data from the following elements of the system is displayed on both the Head-Down Displays (HDDs) and the Head-Up Display (HUD).

Su-39 tail view, showing the Shokogruz active infra-red countermeasures system, and above it the rear-facing antennas of the Pastil RWR   0018335

Pastil radar warning receiver
The Pastil RWR covers radio frequencies from 1.2 to 18 GHz, with the ability to intercept pulse, pulse Doppler, and continuous wave signals. It can operate in a stand-alone mode or be integrated with the electronic countermeasures system. It

Su-39 starboard wing view, showing the Omul ECM pod on the outer weapon station, and the Pastil RWR antennas on the forward end and side of the wing tip fairing. The port wing carries an identical installation   0018334

Su-39 Strike Shield aircraft, showing: Shkval EO sighting system in the nose, Kopyo radar pod under belly, Omul ECM pod under each wing, Pastil RWR on each wingtip   0018336

YB-3A emergency ejection of all flares   0018299

has antennas in the front and side of both wing tip fairings, and in the tail-sting of the aircraft.

An upgraded version, described as Pastel-K, has been reported (the difference in spelling presumably being due to transliteration and the designator 'K' possibly indicating a K-band capability).

Omul MSP-25 Electronic CounterMeasures ECM pod

The ECM system is located in the two pods on the outer weapon stations of both wings. It is said to 'cover the necessary radio frequency bands to counter expected threats', and to provide essentially 360° angular coverage, except for a 15° half angle cone each side of the normal to the aircraft centreline. The configuration comprises two identical pods, both of which have receive and transmit capability fore and aft; the operational configuration being to receive on one pod and to transmit on the other to overcome isolation problems. The system is reported to provide both noise and deception countermeasures, including range, angle and velocity gate pull-off.

Shokogruz Infra-Red CounterMeasures IRCM

The main IRCM system, known as Shokogruz, is an active jammer mounted in the tail-sting. It is a modulated IR power source, which is claimed to protect the engines at all thrust levels up to 95 per cent. Above 95 per cent thrust, flares are used to augment the system; however, a new active jammer called SNOP is being developed to cater for the higher thrust level requirement. The IR flare dispenser system, known as UV-26, holds a total of 192 flares.

Status
Installed in the Su-39.

Contractor
Central Scientific Institute for Radiotechnical Measurements TSNIITI, (possibly also known as Omskavtomatiki) Omsk.

YB-3A flare dispenser system

Type
Airborne Electronic Counter Measures (ECM), Infra-Red Counter Measures (IRCM) system.

Description
The YB-3A flare dispenser is a modular design, that offers a large number of hardware

configuration options, and considerable programming flexibility.

The basic configuration comprises four electronic line replaceable units: the control panel, a programme selector (double unit), a computer unit, and a safety unit. Two types of dispenser units are available, capable of dispensing 8 × 50 mm calibre cartridges and 32 × 26 mm calibre cartridges respectively. The control system is capable of addressing up to 512 flare locations, contained in multiple dispenser unit configurations.

Control capabilities include:
1. 50,000 + flexible random interval programming options;
2. eight preliminary installed programs, with flexible selection in flight;
3. five variable program parameters;
4. 1–8 cartridges in salvo;
5. 0.025–16 seconds interval between salvos;
6. 18 minutes maximum program duration, or until 'all gone';
7. 8 – 512 flare payload;
8. one-year programmed life;
9. emergency ejection of all flares.

The control panel provides the following capabilities: program loading, fire/stop control, payload remaining indication, rapid fire and built-in-test control, variable brightness display.

The program selector provides the following features: automatic operation, random intervalometer capability, one year programmed life.

Specifications
Dimensions and weights:
(cockpit control panel):164 × 64 × 104 mm; 0.5 kg
(program selector): 186 × 221 × 212 mm; 5.5 kg
(computer unit):172 × 373 × 214 mm; 9.8 kg
(safety unit):94 × 90 × 66 mm; 0.35 kg

YB-3A flare dispenser double program selector unit (Paul Jackson)   0018303

YB-3A flare dispenser 32 × 26 mm cartridge dispenser unit (Paul Jackson)   0018301

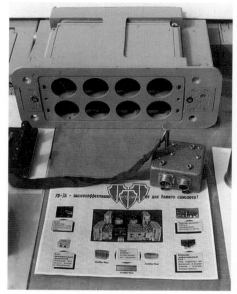

YB-3A flare dispenser 8 × 50 mm cartridge dispenser unit (Paul Jackson)   0018302

L-218-1 and L221 flare cartridges for YB-3A flare dispenser (Paul Jackson)   0018300

YB-3A flare dispenser control panel (Paul Jackson)   0018304

YB-3A flare dispenser – typical configuration   0018305

(dispenser units) 130 × 384 × 172 mm for the 32 × 26 mm cartridge dispenser; 7.4 kg 130 × 384 × 285 mm for the 8 × 50 mm cartridge dispenser; 8.5 kg
Power: 24–30 V DC, 150 W; 115 V AC, 400 Hz, 200 VA
Cartridges: Developed by the Scientific Research Institute of Applied Chemistry.

	L218-1	L-221
Calibre	26.6 mm	50.2 mm
Length	80 mm	202 mm
Weight	0.11 kg	1.0 kg
Aircraft speed (max)	1,350 km/h	1,350 km/h
Dispenser barrel life	700 bursts	250 bursts

Status
Designed for both aircraft and helicopter application; widely deployed.

Contractor
Vympel NPO.

South Africa

Airborne Laser Warning System (LWS)

Type
Airborne Electronic Counter Measures (ECM), Laser Warning Receiver (LWR).

Description
The airborne Laser Warning System (LWS) provides threat identification and Direction-Finding (DF) indication of laser range-finders, designators and lasers used for missile guidance purposes. The system is designed to interface with an existing onboard RWR/ESM host system and is available with one of three LWS sensor models (LWS-200, LWS-300 and LWS-400). All the sensor models use the same Laser Warning Controller (LWC) for data processing and interfacing to the host EW system. On detection of a threat, an audio and visual (display) alarm is generated via the host EW system, or if configured as a stand-alone system, via a dedicated Threat Display and Control Unit (TDCU) developed by Avitronics. The LWS-300 and LWS-400 are form-fit compatible, and all three sensor models are electrically compatible allowing for easy and cost effective upgrading.

Broad coverage of the laser spectrum ensures detection of all known current threats. The threat library is user programmable and field loaded into the LWS via the host EW system.

The system is able to record threat parameters encountered during the mission for post analysis.

Specifications
LWS-200
Wavelength coverage: 0.6–1.8 μm
Threat coverage: Ruby, GaAs, NdYAG, Raman Shifted NdYAG and Erbium Glass lasers
AOA accuracy: azimuth 11° rms
Spatial coverage:
azimuth 360° (99° per sensor)
elevation 60°
Probability of intercept: >99% for a single pulse
Dimensions: 103 × 86 × 64 mm
Weight: 0.8 kg per sensor

LWS-300
Wavelength coverage: 0.5–1.8 μm
Threat coverage: the same as LWS-200 plus doubled NdYAG lasers
AOA accuracy: azimuth 15° rms
Spatial coverage:
azimuth 360° (90° per sensor)
elevation 60°
Probability of intercept: >99% for a single pulse
Dimensions: 115 × 90 × 76 mm
Weight: 1.2 kg per sensor

LWS-400
Wavelength coverage: 0.5–1.8 μm and 2–12 μm
Threat coverage: the same as LWS-300 plus CO_2 lasers
AOA accuracy: azimuth 15° rms
Spatial coverage:
azimuth 360° (90° per sensor)
elevation 60° (0.5–1.8 μm)
elevation 40° (2–12 μm)
Probability of intercept: >99% for a single pulse
Dimensions: 107 × 90 × 76 mm
Weight: 1.2 kg per sensor

Laser warning controller
Dimensions: 188 × 89 × 131 mm
Weight: 2.5 kg

Airborne Laser Warning Controller with sensor models (from left to right) LWS-400, LWS-200 and LWS-300   0051259

Status
In service.

Contractor
Avitronics.

Electronic Surveillance Payload (ESP)

Type
Airborne Electronic Counter Measures (ECM), Electronic Support Measures (ESM) system.

Description
The Electronic Surveillance Payload (ESP) is a derivative of Avitronics' Emitter Location System (ELS) tailored for installation on UAVs. The system was originally designed to operate as a stand-alone ESM system integrated with the Kentron Seeker II UAV, providing information on an enemy's electronic order of battle through emitter identification and location.

The ESP consists of an acquisition and analysis receiver and controller integrated into a single unit, coupled with a nose-mounted interferometric antenna array. Emitter data is transferred via the UAV data link to a ground based REmote Terminal (RET) for threat display and control of the system.

The ESP provides the following functions:
- acquisition, analysis and precision DF of emissions from search, tracking and fire-control radars
- accurate bearings and signal parameter measurement of emitters
- gathering and recording of detailed emitter data for ESM/ELINT analysis
- high Probability Of Intercept (POI) for search radars using the wide-open acquisition radar
- onboard mini-flash data recorder
- frequency measurement for designated emitters

- an autonomous mode which provides for the detection of low POI emitters.

Specifications
Dimensions:
controller: 193 × 127 × 343 mm
Mass:
controller: 10 kg
antenna array: 6 kg
Antennae: phase amplitude matched
Frequency coverage: 0.5 to 18 GHz
Frequency resolution: 1 MHz
Instantaneous bandwidth: 1 GHz or 80 MHz narrow band
Direction-finding: 1° RMS (>2 GHz); 3.5° RMS (700 MHz)
Field of view (fully calibrated): 240° in azimuth (in 3 sectors); 70° in elevation

Status
In service.

Contractor
Avitronics.

Emitter Location System (ELS)

Type
Airborne Electronic Counter Measures (ECM), Emitter Locator System (ELS).

Description
Accurate direction-finding is important to ESM systems for the geo-location of emitters and for directing jammers and weapons. Integrated with the MSWS multisensor warning system, the ELS provides the high accuracy DF required for these tasks.

The ELS consists of an integrated receiver and controller and a number of interferometric antenna arrays, dependent on the system requirements. The main features of the system are: low mass and volume by using a single channel switched receiver; intra-pulse channel switching for a single pulse DF capability; high DF accuracy using a combination of phase and amplitude comparison technique; pulse Doppler handling capacity; high sensitivity.

The ELS is designed as an integral part of MSWS to enhance the DF capability of the radar warning function. All display and control is via the host system. The ELS functions include: acquisition and analysis of search, track and fire control radars; provision of accurate bearings for emitters designated from the host system; frequency measurement for designated emitters; detection of low probability of intercept emitters in an autonomous mode; gathering of emitter data for ESM/ELINT analysis.

The ELS is designed for installation on fixed- and rotary-wing aircraft and remotely piloted vehicles.

Emitter Location System (ELS) showing ELS controller (left) and 2–18 GHz antenna array   0051258

Specifications

Dimensions:
    (controller) 193 × 127 × 343 mm
    (2–18 GHz antenna array) 110 × 90 × 250 mm
    (0.5–18 GHz antenna array) 180 × 80 × 500 mm
Mass:
    (controller) 14 kg
    (2–18 GHz antenna array) 5 kg
    (0.5–18 GHz antenna array) 7 kg

Frequency coverage: 2–18 GHz (0.5-18 GHz optional)
Frequency resolution: 2 MHz in narrowband mode
Instantaneous bandwidth: 1 GHz/80 MHz
Direction-finding: 1° RMS
Field of view: 120° in azimuth per antenna array; 70° in elevation per antenna array

Status
In service.

Contractor
Avitronics.

MAW-200 Missile Approach Warning System

Type
Airborne Electronic Counter Measures (ECM), Missile Approach Warner (MAW).

Description
The MAW-200 Missile Approach Warning system operates as stand-alone equipment, or as part of an integrated Defensive Aids Sub-System (DASS). It is a totally passive ultraviolet system which provides detection and timely warning of the approach of surface-to-air and air-to-air missiles. Upon positive detection of the approaching missile, a priority interface to the chaff and flare dispensing system is activated for immediate and automatic dispensing of countermeasures against the threat. A visual Direction-Finding (DF) warning is provided to the aircrew via the display unit of an EW suite (or in the case of a stand-alone system via the Threat Display and Control Unit – TDCU) accompanied by the appropriate audio alarms.

The MAW subsystem consists of four sensors and a processing card, which resides in the Electronic Warfare Controller (EWC) of the EW suite. On the stand-alone version, processing is carried out in a dedicated processor unit. Each sensor is responsible for processing its detection algorithms. The processing card inside the EWC is responsible for the further processing of data received from the various sensors and for the built-in test control and management of the MAW sensors. The MAW-200 has a multi-threat capability and can track at least eight targets simultaneously.

Main features include: totally passive ultra-violet detection; low false alarm rate; no in-flight recalibration required; instantly online, no cooling required; comprehensive self-test routines.

MAW-200 sensors with Electronic Warfare Controller (EWC) and Threat Display And Control Unit (TDCU)   0103886

Specifications
EWC dimensions: 343 × 127 × 193 mm
EWC weight: 14 kg
TDCU dimensions: 128 × 127 × 120 mm
TDCU weight: 2.2 kg
Sensor dimensions: 230 × 130 × 130 mm
Sensor weight: 3.1 kg
Detection method: passive ultra-violet
Detection range: >5 km for shoulder-launched missiles
Spatial coverage: 360° azimuth with 4 sensors
DF resolution (azimuth): better than 5°
Multithreat capability: at least 8 targets simultaneously

Status
In production.

Contractor
Avitronics.

Multi-Sensor Warning System (MSWS)

Type
Airborne Electronic Counter Measures (ECM), integrated suite.

Description
The Multi-Sensor Warning System (MSWS) provides tactical aircraft with a complete warning capability for self-protection. The capability includes radar warning, laser warning, and missile approach warning. The modular architecture provides for a variety of

sensors to be integrated into and managed by the system, allowing the user to upgrade the system.

Generic design and low unit count allow easy installation in aircraft ranging from helicopters to fighters. Complete spherical coverage is available and the system provides full threat identification. Threat identification parameters are user definable. The MSWS is flight line programmable and includes extensive BIT facilities.

The system includes a radar warning function for pulse Doppler and CW radars in high pulse density environments, a man/machine interface via a multifunction display and interface to and control of automatic chaff and flare dispensing systems.

The RWR features an Instantaneous Frequency Measurement (IFM) receiver, covering the 2 to 18 GHz band in 4 GHz steps. It is reported to be able to cope with pulse densities up to 2.5 Mpps and to display worst situation threats within 500 ms, using 32-bit parallel processors.

The MultiSensor Warning System (MSWS)   0051261

Malaysia ordered 18 Su-30MKM fighters in August 2003, to be equipped with Avitronics' MSWS
0014635

The laser warning system is reported to cover 0.5 to 12 μm wavelengths, offering detection capability against laser range-finders, designators and missile guidance lasers, providing both threat classification and bearing.

The solar-blind MAW uses ultra-violet detection techniques and is said to typically provide 5,000 m warning of shoulder-launched missiles.

The standard configuration comprises four sensor heads for each of the RWR, LWR and MAW functions. The display shows the nature and status of received signals, together with relative bearing, lethality and tabulated parametric data. Audio warning is provided to alert the crew to display data.

The system can be integrated with recording facilities for use in the Intelligence gathering role.

Growth options include an interface with an active ECM system activated automatically on threat detection, avionic system interface via a MIL-STD-1553B bus.

The system is supplied with pre-flight data compiler and flight data analyser software tools that enable the operator to tailor the system to suit the developing threat environment. A memory-loading unit allows software to be loaded and unloaded on the flight line and various pre-flight data sets can be loaded and selected in flight as a mission proceeds. All relevant data is recorded during the mission for later downloading and analysis.

Specifications
Dimensions:
(EW controller) 343 × 127 × 193 mm
(threat display and control unit)
    128 × 127 × 120 mm
(front end receiver ×4) 176 × 45 × 158 mm
(spiral antenna ×4) 110 × 110 × 67.5 mm
(LWS-400 sensor ×4) 107 × 90 × 76 mm
(MAW-200 sensor) 230 × 130 × 130 mm
Weight:
(EW controller) 14 kg
(threat display and control unit) 2.2 kg

(front end receiver ×4) 3 kg (per unit)
(spiral antenna ×4) 0.7 kg
(LWS-400 sensor) 1.2 kg (per sensor)
(MAW-200 sensor) 3.1 kg (per sensor)

Radar Warning System -50 (RWS-50)
Frequency coverage:
    0.7–40 GHz (pulsed signals)
    0.7–18 GHz (CW signals)
Direction-finding:
    10–12° RMS for pulsed signals in the 2–40 GHz range
Spatial coverage:
    360° (azimuth or spherical)
    90° (elevation or spherical)
Pulse density capability: >2.5 million pulses per second
Frequency resolution: 10 MHz
Laser Warning System-400 (LWS-400)
Wavelength coverage:
    0.5–1.8 μm and 2–12 μm
Direction-finding:
    15° rms (azimuth)
Spatial coverage (per sensor):
    azimuth 90°
    elevation 60° (0.5–1.8 μm)
    elevation 40° (2–12 μm)
Laser threat coverage: doubled NdYAG, Ruby, GaAs, NdYAG, Raman Shifted NdYAG, Erbium Glass and CO_2 lasers
Laser threat types: range-finders, designators and lasers used for missile guidance (beam riders)
Probability of intercept: >99% for a single pulse

Missile Approach Warning-200 (MAW-200)
Operating frequency:
    solar blind UV band
Direction-finding:
    better than 5°
Spatial coverage: 94° conical field of view per sensor
Multithreat capability:
capable of tracking at least 8 targets simultaneously

Status
In production and in service. During 2002, Avitronics received an order from Agusta Westland for the MSWS), reported to be worth more than R150 million (USD15 million), for Super Lynx helicopters ordered by Oman. In addition, Switzerland and the United Arab Emirates also selected the MSWS for their Super Puma fleets.

Most recently, during April 2004, Avitronics announced that the Malaysian government had placed an order with them for the MSWS to be fitted to the Royal Malaysian Air Force's 18 new Sukhoi-30MKM fighters. Work on the first two phases of the order, valued at USD20 million, was scheduled to commence immediately. The first phase is concerned with system engineering to fully integrate the MSWS with the aircraft. The second phase of about 18 months will see a prototype developed and the production systems manufactured.

Contractor
Avitronics.

RWS-50 Radar Warning System

Type
Airborne Electronic Counter Measures (ECM), Radar Warning Receiver (RWR).

Description
The RWS-50 system provides tactical aircraft with a comprehensive radar warning capability for self-protection. This capability can be extended to include laser and missile approach warning.

In its basic configuration, the RWS-50 consists of four 2 to 18 GHz spiral antennas, two-dual detector amplifiers, an analyser unit and a colour

multifunction display and control unit. This configuration includes an interface for automatic or manual control of a chaff and flare dispensing system.

The RWS-50 features a versatile threat library, flexible architecture, parallel processing, high sensitivity, high probability of intercept/low cycle time and low power consumption.

Upgrade options include; 0.7 to 1.4 GHz detection capability (omni or full DF); CW detection capability (omni or full DF); extended RF range 0.7 to 40 GHz (in one antenna); frequency measurements (via external superheterodyne/IFM subsystem); increased sensitivity/dynamic range; MIL-STD-1553B

interface; LWR capability; MAW interface; spherical coverage.

Specifications
Dimensions:
(EW controller) 343 × 127 × 193 mm
(threat display and control unit)
    128 × 127 × 120 mm
(dual-detector amplifier) 176 × 45 × 158 mm
(spiral antenna) 110 × 70 × 70 mm
Weight:
(EW controller)14 kg
(threat display and control unit) 2.2 kg
(dual-detector amplifier ×2) = 3 kg
(spiral antenna ×4) 0.4 kg

Power supply: 28 V DC, 140 W
Frequency: 2–18 GHz
Spatial coverage:
(azimuth) 360°
(elevation) 90°
Direction-finding: 10° rms

Status
In production and in service.

Contractor
Avitronics.

*Avitronics' RWS-50
Radar Warning
System*
0103887

Spain

EN/ALR-300(V)1 Radar Warning Receiver (RWR)

Type
Airborne Electronic Counter Measures (ECM), Radar Warning Receiver (RWR).

Description
The EN/ALR-300(V)1 RWR has been designed to detect, locate, identify and display to the pilot the existing threat environment form pulse and continuous wave emissions, regardless of illumination mode. The two systems in the series are the ALR-300(V)1 and the ALR-300(V)2 (see separate entry). The main difference between them is that the (V)2 incorporates a superheterodyne receiver/DIFM.

Operation is controlled by software based on real-time multiprocess, employing four 68000 microprocessors.

The equipment consists of: four E/J-band DF spiral antennas; four E/J-band channelised crystal video receivers; a processor control unit and an azimuth indicator display with synthetic audio warning, together with a C/D-band monopole antenna and channelised receiver.

The system provides identification of all detected pulse and CW radar emitters and warns the pilot of threats by means of both an alphanumeric graphical CRT display and voice synthesised messages in accordance with the lethality of the detected emitters. The system includes the ability to record up to 100 emitters during flight.

The EN/ALR-300(V)1 includes full mission and maintenance hardware and software support facilities. Reprogramming on the flight line can be accomplished by use of EEPROM mission loading equipment that forms part of this support capability.

Specifications
Frequency range: C-J in 5 sub-bands (0.5 to 20 GHz)
Signals detected: pulse and CW
Antennas:
C/D-band: one monopole (0.5 to 2.0 GHz)
E/J-band: 4 flat spirals (2 to 20 GHz)
Wide band receivers:
C/D-band: one channelised receiver
E/J-band: 4 quadrantal, channelised receivers
Azimuth coverage: 360°
Elevation coverage: ±30°
DF accuracy: 12° RMS
Simultaneous emitter tracking: 30

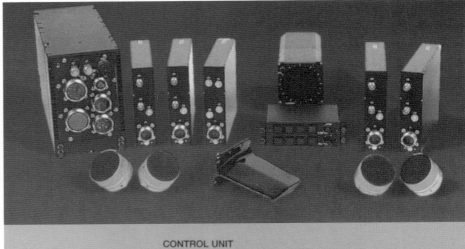

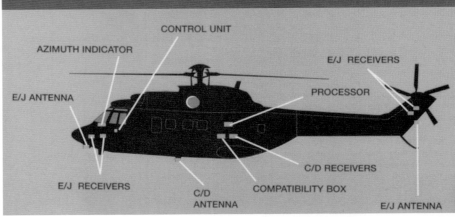

The ALR-300 radar warning receiver system and aircraft installation locations   0080278

Simultaneous emitter display: 15
Emitter library: 100
Standard interfaces: MIL-STD-1553B, RS422
Dedicated interfaces: chaff/flare dispensers
Weight: 20 kg
Power consumption: 200 W

Status
In operation on Mirage F-1 fighter aircraft, CASA C-101 trainer aircraft, and on Chinook and Super Puma helicopters. In production for Beech aircraft.

Contractor
INDRA Sistemas SA.

EN/ALR-300(V)2 Radar Warning Receiver (RWR)

Type
Airborne Electronic Counter Measures (ECM), Radar Warning Receiver (RWR).

Description
The EN/ALR-300(V)2 is a computer controlled RWR adopted for use by the Spanish Air Force for all of its combat aircraft. The EN/ALR-300(V)2 upgrades the EN/ALR-300(V)1 (see separate entry) by the addition of a superheterodyne/DFIM receiver, a new power supply and new software to improve the equipment performance.

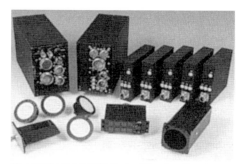

The EN/ALR-300(V)2 Radar Warning Receiver
0554701

Radar data are presented to the operator through alphanumeric symbols on a high-brightness display. The position of the symbols on the display indicates lethality and bearing of the detected signal and identification of the radar. Blinking of the symbols, together with sounding of an audio alarm indicate high-priority radar threats.

Specifications
Weight: 38 kg
Power consumption:400 W
Frequency: C-J (0.5-20 GHz) in four sub-bands
Coverage: 360° azimuth
Accuracy: >12° RMS
Options: DF capability in C/D wavebands (0.5–2 GHz)

Status
The EN/ALR-300(V)2 is the standard equipment for all Spanish combat aircraft. It is in operation on the Mirage F1 fighter aircraft, and in production for Super Puma helicopters.

Contractor
INDRA Sistemas SA.

EN/ALR-310 Radar Warning Receiver (RWR)

Type
Airborne Electronic Counter Measures (ECM), Radar Warning Receiver (RWR).

Description
The EN/ALR-310 is a fully programmable crystal video RWR that incorporates a digital computer to provide emitter identification in complex signal environments. The system incorporates a readily programmable emitter library and provides unique identification of all detected pulsed and CW emitters by alphanumeric symbols on the NVIS-compatible, high-brightness CRT display. Up to 15 radars may be displayed simultaneously.

Specifications
Weight: 20 kg
Power consumption: 180 W
Frequency: 0.7–1.5 GHz and E/J-band, in 2 bands
Coverage: 360° azimuth
Accuracy: >15° RMS
Radars detected: pulsed and CW
Number of radars simultaneously displayed: 15
Symbology: programmable
Receiver: crystal video
Processor memory: EEPROM

Status
The EN/ALR-310 is suitable for fighter, light strike/attack and helicopter applications. Selected for Spanish Army helicopters.

Contractor
INDRA Sistemas SA.

Receiver

D/F antenna

Omndirectional Blade antenna

Control unit

Display

Processor

The EN/ALR-310 is fitted in Spanish Army helicopters
0503827

SIGMA SIGINT system

Type
Airborne Electronic Counter Measures (ECM), Signals Intelligence (SIGINT) system.

Description
The SIGMA airborne Signals Intelligence (SIGINT) system incorporates multi-spectral radar and communications sensors and is designed to monitor and analyse dense electromagnetic environments and detect, locate and identify emitters and signals traffic. SIGMA employs a modular and configurable architecture that is built around a range of standard elements that includes an operator processor console, use of the Ada programming language, a UNIX-based operating system, an Ethernet Local Area Network (LAN) and standard interfaces. The system is also able to accommodate both INDRA developed radar and communications band receiver equipment and other electromagnetic and Electro Optic (EO) sensors. The SIGMA system also incorporates a ground-based operational support facility.

Status
In service onboard a Spanish Air Force Boeing 707 SIGINT aircraft.

Contractor
INDRA Sistemas SA.

The SIGMA SIGINT system is installed on a Spanish Air Force Boeing 707
0063866

Signal Identification Mobile System (SIMS)

Type
Airborne Electronic Counter Measures (ECM), Signals Intelligence (SIGINT) system.

Description
The Signal Identification Mobile System (SIMS) is configured for airborne, shipborne and land-based applications. It allows detection, direction-finding, analysis and library storage over the 1 to 18 GHz frequency range and transfers all data to remotely located sites through a built-in datalink.

SIMS consists of an antenna unit, RF unit, direction-finding and panoramic signal displays, a computerised system controller, video analyser and peripherals.

Specifications
Power supply: 115 V AC, 400 Hz
Frequency: 1–18 GHz pulse or CW signals
Accuracy: ±4° (1–2 GHz)
  ±3° (2–4 GHz)
  ±2° (4–18 GHz)

Status
In service.

Contractor
INDRA Sistemas SA.

SOCCAM COMINT system

Type
Airborne Electronic Counter Measures (ECM), Communications Intelligence (COMINT) system.

Description
The modular communication observation and control system (SOCCAM) is configured for airborne, shipborne and land-based applications. It is an Electronic Support Measures (ESM) and Direction Finding (DF) COMINT system for tactical and strategic missions, operating over the Very/Ultra High Frequency (V/UHF – 30 MHz to 1 GHz) wavebands.

SOCCAM provides functions for scanning, searching and detecting active transmissions. The system detects activity in a series of discrete bands and analyses intercepted signals to allow the operator to determine transmission characteristics and store them for subsequent analysis.

Further system features and benefits include digital processing, multiple receivers, programmable (by threat and mission type) system-specific frequency search sub-bands, standard cartographic applications, digital audio recording, Ethernet Local Area Network (LAN) and Ada/UNIX-based ruggedised workstation capabilities.

SOCCAM has two different operating modes: operation in an unknown scenario when the system searches for active transmissions in the mission area, and operation in a known scenario, where the operator has prior knowledge of threats in the area and seeks to locate and monitor them.

Specifications
Frequency: 20–500 MHz
Sensitivity: –100 dBm

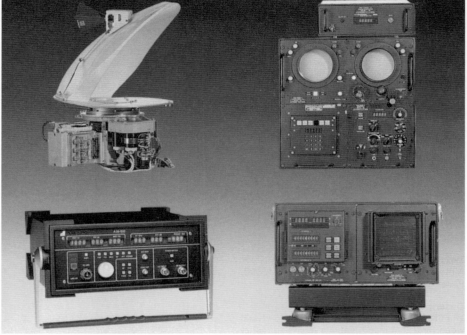

The Signal Identification Mobile System   0504070

Modes: AM, FM, CW and PLS
Accuracy: 4° RMS
Interfaces: RS-422, IEEE-488

Status
In service onboard CASA C-212 and Fokker F-27 aircraft.

Contractor
INDRA Sistemas SA.

TARAN airborne ESM/ECM system

Type
Airborne Electronic Counter Measures (ECM), Electronic Support Measures (ESM) system.

Description
TARAN is a modular, multi-operator, computer-controlled airborne EW system for both ESM and ECM in the communications and tactical navigation/identification bands (C/D-band, 0.5 to 2 GHz). The system is intended primarily for airborne applications and can be installed on a wide range of fixed-wing aircraft and helicopters.

Two different bands can be operated: low band for communication systems and high band for navigation/identification systems. The low-band subsystem consists of a search and analysis receiver, monitoring console and jamming set. The high-band subsystem comprises a search and analysis receiver, monitoring console and a deception jammer.

The main features of TARAN include the ability to operate in a dense electromagnetic environment and resist multiple threats simultaneously. There is a high degree of automation in order to reduce reaction time, and growth potential to cater for the evolving threat. The jamming function is automatic with target frequency selection being made in-flight or prior to the start of the mission. Environmental data is displayed to the operators in the form of tables that can be manipulated using keyboard-driven screen menus. Recorded data can be downloaded for post-mission analysis and there is a computer-based system remote control facility. TARAN has total communication capacity with the command and control system with a simple man/machine interface and ergonomic design. The system is modular, able to adapt to the needs of different platforms and has high reliability and ease of maintenance.

The system is installed with a suitable set of consoles with interconnection routing depending on the type of airborne platform.

Specifications
Power supply:
115 V AC, 400 Hz, 1.36 kVA (2.6 kVA with jammer operating)
28 V DC, 1.5 kW (15.2 kW with jammer operating)

Status
In service with the Spanish Air Force on the Falcon 20.

Contractor
INDRA Sistemas SA.

A TARAN workstation   0514867

Sweden

BO series Counter Measures Dispensing System (CMDS)

Type
Airborne Electronic Counter Measures (ECM), Counter Measures Dispensing System (CMDS).

Description
The SaabTech BO Counter Measures Dispensing System (CMDS) consists of BOA, BOL and BOP CounterMeasures Dispensing Systems (CMDS) and a BOC controller module. The BO system has been developed to be applicable to a wide range of aircraft types/configurations, with particular emphasis on high reliability and safety. Dependent on the host aircraft Electronic Countermeasures (ECM) equipment standard, the BO system can operate in fully automatic, semi-automatic or manual modes.

BOA
The BOA CMDS is an electro-mechanical system designed to protect civil airliners, VIP-jets and Head-of-State jets and turboprops from surface-to-air IR-guided missiles. The design is an adaptation of the BOL CMDS, employing an electro-mechanical dispenser using pyrophoric IR flares. The choice of an electro-mechanical dispensing system is driven by civilian requirements which demand minimal probability of inadvertent release of

flares – traditional squib-based pyrotechnic dispensers are more susceptible in this respect. In addition, the use of pyrophoric flares means that any residue impacting the ground after a low-level release (which may be in the immediate area of an airport and/or metropolitan area) is also minimised.

BOA is integrated into the engine pylons of the host aircraft, which maximises the effects of IR decoys, which are ejected directly aft of the engine into the area of the exhaust plume.

BOL CMDS
The BOL CMDS was originally developed for insertion into the LAU-7 Sidewinder launch rail (the combined launcher/dispenser is known as LAU-138 in US service), thus facilitating the installation or enhancement of passive EW capability in an aircraft without reduction in available air-air weapon stations. In this configuration, the LAU-7 launcher has been redesigned to incorporate a BOL module in the rear half and a revised and rearranged missile seeker cooling system in the nose. The BOL system is currently compatible with a wide range of air-to-air missile launch rails which includes the LAU-127/LAU-128/LAU-129 family of AIM-120 AMRAAM and AIM-9X missile launchers, the Frazer-Nash Common Rail Launcher (CRL) and the ACMA MPRL.

The system comprises an electromechanical dispenser assembly which can accommodate up to 160 standard packages of either chaff (US nomenclature RR-184) or IR countermeasures (MSU-52B), mounted in plastic holding frames. With spacing controlled by an onboard countermeasures computer, the individual payloads are ejected into the turbulent airflow aft of the launcher, which disperses the packages and aids in rapid 'blooming' of the resultant chaff cloud. If required, the BOL dispensing mechanism can be installed in a conformal housing where no suitable launch rail is available. Saab Avionics claims that the BOL electromechanical dispensing method is more precise than pyrotechnic ejection and offers advantages in terms of safety on the ground and maintainability. BOL operates with standard dispensers such as the ALE-39, -40, -45 and -47. Loading and downloading of CM packages is achieved quickly (less than one minute) and easily while the dispenser is fitted to the aircraft; this can be done either by hand or with a special BOL loader. The BOL CM dispenser can be function-tested with the aid of the BOL test set.

BOL Customisation
SaabTech has developed customised BOL applications for specific fighter types. In each case, the shape of the dispenser allows it to be sited in the most desirable locations on an aircraft. Located on the wings, vortices can be utilised to improve dispersion. With dispensers on each wing, the spatial separation of payload clouds significantly increases radar cross-section (if used with chaff) or the extension of the IR radiating source (if used with IR payload). BOL systems are thus usually mounted in a symmetrical twin or quadruple configuration on the wings. BOL's high payload-to-volume ratio, non-pyrotechnical release mechanism and effective dispersion techniques provides for highly effective dispensing for both chaff and IR payloads. The latter allows covert dispensing of a special material that has proved very effective against advanced IR missiles.

A complete countermeasures system typically combines BOL dispensers with conventional pyrotechnical dispensers for spot flares. For aircraft already equipped with pyrotechnic dispensers, retrofit of BOL significantly enhances the protection of the aircraft by increasing the total payload capacity.

BOL chaff dispenser in LAU-7 launch rail configuration for the Royal Air Force   0018276

BOL/LAU-138 shown with AIM-9L loaded to a US Navy F-14A Tomcat   0017863

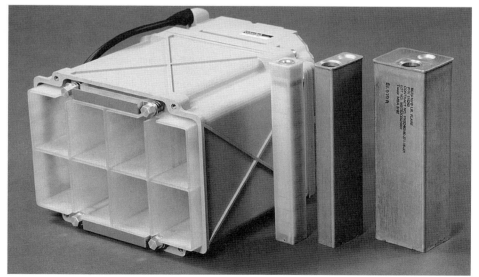

BOP/C countermeasures dispenser, shown with 1 × 1 in, 2 × 1 in and 2 × 2.5 in flare cartridges
0081490

For details of the latest updates to *Jane's Avionics* online and to discover the additional information available exclusively to online subscribers please visit
jav.janes.com

BOL/LAU-128 shown with AIM-9L   0017864

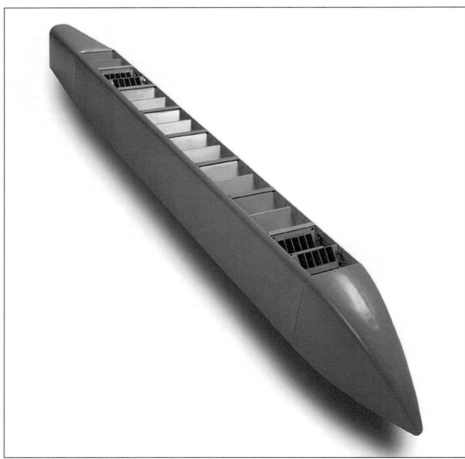

BOP/AT/AX countermeasures dispenser   0081493

payload downwards and towards the rear of the host platform and is loaded from below. BOP/A is designed for ease and rapidity of installation; a typical fighter fit incorporates two launch assemblies. For larger airframes (such as transport aircraft), a four launch assembly installation is used. In the case of fighter applications, BOP/A can also be installed conformally to minimise drag. Maximum reload time is given as 60 seconds and individual BOP/A launch assemblies communicate with their host aircraft's Electronic Warfare (EW) system via an integral electronics unit and an RS-422 serial datalink. The system can be fully integrated into an onboard countermeasures suite and an optional IR sensor can be mounted in the rear of the launch assemblies to provide confirmation of IR decoy flare ignition.

BOP/AT, BOP/AX CMDS

The BOP/AT and BOP/AX CMDS make up a family of countermeasure dispensers for fighter and transport aircraft. BOP/AT is designed to be scab mounted onto the top of the fuselage of a fighter aircraft. So installed, the equipment ejects expendables upwards and forwards, an arrangement that allows IR decoy flares to develop their full spectrum characteristics in close proximity to the aircraft, thereby maximising their effectiveness. BOP/AX is designed to be scab mounted onto the side of the fuselages of a transport aircraft. The equipment can eject payloads upwards, downwards and forwards. By fitting double or quadruple installations, BOP/AX can be configured to dispense payloads in almost any direction required, thereby allowing the equipment's control circuitry to select the most effective ejection sequence for any detected threat.

BOP/AT and BOP/AX dispensers can accept all types of expendables with lengths up to 210 mm. These include standard 1×1 in, 2×1 in and 2×2.5 in and 36 or 40 mm diameter cartridges, in a single-load or mixed-load configuration (for example, one type of chaff cartridge and two different types of flare cartridge). The dispenser automatically identifies each payload and the quantity of each type is reported to the control system. Ignition pulses are automatically adjusted by the dispenser software to match the particular cartridge ignition characteristics. The dispenser can carry up to nine magazines, each holding up to 24 cartridges, depending on type. The electronic control system is derived from the control system of the BOP/B and BOP/C dispensers and offers very high dispense rates and other advanced features such as automatic load identification. The electronics are fully integrated in the dispenser.

Dispensing is normally controlled via a databus and can be performed in almost any sequence. Nested sequencing and random dispensing are supported. The large capacity of the dispenser and the sophisticated control system allow preventive serial dispensing. If an expendable should fail to eject, the dispenser detects this and a new attempt is immediately initiated with another cartridge of the same type.

If an integrated MAW is used, the system can be set to automatic, manual or preprogrammed response.

BOP/B CMDS

The BOP/B CMDS is designed for fighter aircraft applications and, in common with the other BOP CMDS, can be loaded with chaff and/or expendable jammers in addition to flares. BOP/B is scab mounted or as a rear extension of a pylon and can accommodate six 55 mm calibre cartridges. The system's dispenser module comprises a magazine, a breech plate and an electronics unit, and features breech plate coding pins to identify the type of payload inside a particular cartridge. The ejection of ECM payloads is controlled by the BOC control unit via databus or discrete signal.

BOP/C CMDS

BOP/C, based on a dispenser developed for Sweden's JAS-39 Gripen, is a high-performance countermeasures dispenser designed for fixed-wing and helicopter applications. BOP/C may be fitted either internally or externally and features interchangeable magazines, which allows for

In the F-16, BOL dispensers are installed on stations 2 and 8. The original Under Wing Adapter (UWA) is replaced by a new adapter which is designed to allow fitment of two BOL dispensers, constituting a BOL 'Twin Pack', without interfering with any original weapons load or flight control surfaces. All functionality of the original UWA (including carriage of LAU-129/16S200/16S210) is retained. The BOL dispenser is fitted inside a conformal housing which acts as an aerodynamic fairing as well as mechanical attachment for the BOL to the new UWA. An aerodynamic cover is used if no BOL is installed on either side of the Twin Pack UWA. With a BOL Twin Pack installed on stations 2 and 8 respectively, up to 640 countermeasure packages can be loaded, resulting in a substantial pre-emptive as well as reactive CMDS capability. The Twin Pack UWA can also be installed on stations 3a and 7a.

In the F-18 BOL is fitted inside a conformal housing which acts as an aerodynamic housing as well as mechanical attachment for the BOL to the SUU-63 pylon. The pylon is fitted with a new side panel designed by Aerostructures in Melbourne, Australia, to accept the BOL. The design of the new side panel ensures maintained structural integrity of the original pylon. This installation has also been achieved without interfering with weapons load or flight performance.

BOP/A CMDS

The 31 kg (empty) BOP/A CMDS is made up of beam-shaped launch assemblies that accommodate six, two round capacity magazine modules and are compatible with 55 mm diameter chaff and Infra-Red (IR) decoy flare (up to three per round) cartridges. The system ejects its

BOP/B countermeasures dispenser on Gripen aircraft   0018277

BOL Twin Pack for F-16 applications   1039736

different expendable cartridge sizes, or a mix of differing cartridge sizes in the same payload. This feature enables the system to accept different types of expendables with lengths up to 210 mm, including standard 1 × 1 in, 2 × 1 in and 2 × 2.5 in cartridges. In common with the other BOP dispensers, BOP/C features include payload type and quantity identification, databus/serial link ejection control, preventative serial dispensing, nested sequencing and random dispensing and automatic adjustment of ignition pulses to suit payload characteristics. In addition, BOP/C offers high dispensing rates coupled with extremely short reaction times, control of dual squib cartridges and capability to communicate with advanced expendables (allows for programming of advanced RF decoys prior to dispensing or while under tow. BOP/C is designed to be integrated with the host aircraft threat sensors and controlled via Defensive Aids Sub System (DASS) computer, or the complimentary BOC Control Unit.

BOP/D
The BOP/D CMDS is based on a dispenser developed for the JAS 39 Gripen. It is designed for both internal and external installations and can be configured to accept single type or mixed loads in a range of formats, including 25 × 25 mm (1 × 1 in), 50 × 25 mm (2 × 1 in) and 50 × 64 mm (2 × 2.5 in) and 36 and 40 mm diameter cartridges with lengths of up to 210 mm. BOP/D allows interchange of magazines for different cartridge sizes and it automatically identifies payloads by dispenser, type and number available. Dispensing is controlled by datalink, with software-controlled ignition pulses facilitating the use of cartridges with different ignition characteristics. The system detects failed ejections, immediately following up with a new attempt to fire another cartridge of the same type. BOP/D is also capable of supplying power to and communicating with advanced expendables such as active Towed Radar Decoys (TRD).

The BOP/D architecture can accommodate up to eight individual dispensers (with each dispenser being individually controlled); system control is by means of host platform Defensive

Aids Subsystem (DAS) or SaabTech's dedicated BOC control unit.

BOP-L
The BOP-L series is described as a family of lightweight, intelligent and high performance CMDS designed for helicopter and fixed-wing transport aircraft applications. The modular BOP-L incorporates the following features:
Integral functional/personal safety features
Flight-line programmable dispensing sequences
Full integration with onboard Electronic Warfare Controller (EWC)
Payload misfire detection/compensation
Payload/magazine identification
Automatic, semi-automatic, manual, standby and back-up modes of operation
Emergency payload jettison (all modes)
Smart dispenser selection (based on selected DAS mode, host aircraft dynamics and threat direction)
Up to 32 dispensers within a single system
Sixteen payload types

BOC Control Unit
The BOC is a combined cockpit control unit and system programmer. The controller can easily be connected to different types of warners and aircraft systems by means of databusses and discrete signals. Reprogrammable non-volatile memories are used for storage of mission-specific data. BOC can control up to eight BOL or BOP dispensers simultaneously and can also control the most current pyrotechnic dispensers. The BOC controller can provide adaptive processing, to optimise response in a specific threat situation, and stores management, to keep an accurate record of the remaining load and give full flexibility in load configuration and optimum use of expendables.

Specifications
BOA CMDS
Payload capacity: 80 packs
Control signals: ARINC/MIL/-STD-1553B; discretes; RS-485 serial datalink
Power supply: 115 V AC (400 Hz, 28 VDC)

BOL CMDS
Weight (fully loaded):
  (LAU-138): 56.5 kg
  (scab-on fit): 24.0 kg
  (dispenser only): 19.0 kg
Capacity: 160 packs
Reloading time: <1 min
Control signals: discrete input – one dispense signal MIL-STD-1553B or RS-485 serial link
Power: 115 V AC, 400 Hz

BOP/A CMDS
Chaff capacity: twelve standard 55 mm diameter cartridges
Flare capacity: twelve standard 55 mm diameter cartridges, each cartridge holds two or three flares
Reloading time: 1 min (max)
Control signals: discrete input – one dispense signal MIL/STD-1553B or RS-422 serial link
Safety signal: MASS (+28 V)
Weight: 31 kg (without cartridges)

BOP/AT, BOP/AX CMDS
Cartridge capacity: nine magazines of: twenty-four 1 × 1 in, twelve 2 × 1 in, six 2 × 2.5 in; sixteen 36 mm or ten 40 mm diameter payloads
Reloading time: 3 min (max)
Control signals: one dispense signal (discrete input, +28 V) MIL/STD-1553B or RS-485 (databus)
Safety signals: MASS (+28 V); safety pin (option)
Weight: 2.5 kg (single empty magazine); 40 kg (dispenser structure)
Dimensions: (L × W × H): 2,650 × 240 × 355 mm

BOP/B CMDS
Chaff capacity: six standard 55 mm diameter cartridges
Flare capacity: six standard 55 mm diameter cartridges, each cartridge holds two or three flares
Reloading time: 1 min (max)
Control signals: discrete input – one dispense signal MIL-STD-1553B or RS-422 serial link
Safety signal: MASS (+28 V), safety pin
Weight: 11.2 kg (with empty cartridges)

BOP/C CMDS
Capacity: forty 1 × 1 in, or twenty 2 × 1 in, or eight 2 × 2.5 in cartridges
Repetition rate: 10 ms
Power supply: 115 V single phase, 400 Hz
Control signals: one dispense signal (discrete input, +28 V) MIL-STD-1553B or RS-485
Safety signal: MASS (+28 V), safety pin (option)
Weight: 7 kg (including magazine)
Dimensions: 179 (L) × 236 (W) × 281.2 (H) mm

BOP/D CMDS
Capacity: 40 × 1 × 1 in, 20 × 2 × 1 in or 8 × 2 × 2.25 in cartridges
Repetition rate: 10 ms
Power supply: 115 V (three-phase, 400 Hz) or 28 V DC
Control signals: 28 V dispense enable, single 28 V dispenser, RS-485 serial datalink and safety pin (optional)
Weight: 3.9 kg (excluding magazine)
Dimensions: 236 (L) × 199 (W) × 106 (H) mm

Status
In production. The BOL dispenser is currently in service with the UK Royal Air Force on the Harrier GR Mk 7 and the Tornado F3, the US Navy on the F-14 Tomcat and the Swedish Air Force JA-37 Viggen aircraft. It will also be integrated into the CRL for Eurofighter Typhoon and scab mounted to a weapons pylon on the JAS-39 Gripen. Flight tests of BOL integrated into the LAU-128 MRL on the F-15 began in 1997 as part of the US Department of Defense's Foreign Comparative Test (FCT) programme; additional trials leading to complete certification of BOL on the F-15 were conducted during 2000. The US Navy has successfully trialled the scab mounted variant of BOL on the F/A-18C/D Hornet and S-3 Viking, with evaluation and certification work on the LAU-128/BOL variant for wingtip mounting on the F/A-18C/D ongoing.

BOP-L

	BOP-L23	BOP-L29	BOP-L31
Capacity:	23 × 1 × 1 in (including dual mode) or 11 × 2 × 1 in cartridges or Chemring Modular Expendable Block (MEB)	29 × 1 × 1 in (including dual mode) or 14 × 2 × 1 in cartridges or Chemring Modular Expendable Block (MEB)	31 × 1 × 1 in (including dual mode) or 15 × 2 × 1 in cartridges
Repetition rate:	10 ms	10 ms	10 ms
Power supply:	28 V DC	28 V DC	28 V DC or 115 V AC (400 Hz)
Control signals:	dispense, dispense enable, safety interlock and serial datalink	dispense, dispense enable, safety interlock and serial datalink	dispense, dispense enable, safety interlock and serial datalink
Weight:	<2 kg (magazine/payload excluded)	<2 kg (magazine/payload excluded)	<2 kg (magazine/payload excluded)
Dimensions (L × W × H):	177 × 122 × 60 mm	177 × 150 × 60 mm	231 × 122 × 60 mm

BOC control unit
Dimensions: 76 × 147 × 165 mm

BOP/A CMDS is installed aboard Viggen aircraft of the Swedish Air Force with BOP/B systems being fitted to J 35J Draken and JAS 39 Gripen aircraft of the same service. BOP/AT has been tested aboard a Tornado strike aircraft.

Contractor
SaabTech AB.

BO2D RF expendable decoy

Type
Airborne Electronic Counter Measures (ECM), Expendable Radar Decoy (ERD).

Description
BO2D is designed to provide fighter aircraft with self-protection against active and semi-active radarseeker missiles. It is a broadband repeater pyrotechnically ejected from a chaff/flare dispenser. It utilises internal battery power and communicates with the host aircraft through a towline. RF transmissions can be switched on or off and different ECM modes can be chosen while the decoy is being towed. The pilot is thus able to launch the BO2D without transmitting when closing in on a possible threat. After use, the towline is cut with a pyrotechnic cutter.

The decoy is a broadband, high-gain repeater with FM capability. It utilises all solid state advanced MIC technology. It comprises the following units: receiving antenna, signal amplifier, modulator, power amplifier and transmitting antenna.

The decoy is housed in a standard 55 mm diameter dispenser magazine, such as the BOZ/BOP types of pyrotechnic dispenser, to provide convenient and cheap installation. The cartridge, which also contains the towline and a brake mechanism, is about 100 mm longer than the standard flare cartridge. The loading and operation of the decoy are similar to those of a flare cartridge. For aircraft without a 55 mm dispenser, a small dedicated launcher is also available.

Specifications
RF coverage: H-, I- and J-bands
Weight: <2 kg (including cartridge)
Cartridge size: 55 mm diameter

Status
The BO2D is being developed for the Swedish Air Force. Qualification testing was conducted in 1998, with prototype and initial aircraft testing conducted during 1999. The system is reported as being production-ready.

Contractor
SaabTech AB.

BOQ A110 jammer pod

Type
Airborne Electronic Counter Measures (ECM), radar jamming system.

Description
The BOQ A110 is based on the proven REWTS (Responsive Electronic Warfare Training System) concept, which traces its heritage to the JAMMER A100 (ALQ-503) system, operational with the Swedish, Swiss and Canadian air forces.

The BOQ A110 is a complete dual-use system that may be used for training without revealing or compromising the system's tactical capability. The switch between training and tactical roles is accomplished in seconds by downloading software suitable for the planned mission via the system's interface. With the ability to operate in both autonomous and manually-controlled modes, the BOQ A110 system provides both an effective ECM training system and also a capable tactical jammer, which in addition incorporates an ESM capability.

The BOQ A110 comprises four main subsystems: A110 pod; Control Unit (CU); Display Unit (DU); and Mission Planning Workstation (MPW).

The BOQ A110 pod is designed for standard NATO 14 and 30 in hard-point installations. The pod is ram-air cooled. The pod has 360° receiver and transmitter coverage, provided by three 120° antennas. The system incorporates three receivers: a Continuous Wave/High Pulse Doppler (CW/HPD) receiver; a Direction-Finding Receiver (DFR); and a Set-On Receiver (SOR). The CW/HPD receiver alerts the system to high-duty cycle emitters. The DFR identifies the sector of operation. The SOR is a narrowband receiver used in some jamming modes and in ESM modes to analyse emitters.

The jamming transmitter can be directed through any one of four antennas: three evenly spaced 120° antennas covering 360° and a high-gain antenna in the forward sector. The pod can generate spot and barrage noise, noise deception, amplitude modulation, velocity deception, coherent and FML-based range deception, radar simulation and combination modes.

Specifications
Frequency range: 6.8–10.5 GHz
Sensitivity:
(low-duty cycle) –45 dBm
(high-duty cycle) –60 dBm
ERP:
(wide-lobe antennas) 1 kW (typical at 9.5 GHz)
(high-gain antenna) 10 kW (typical at 9.5 GHz)

BOQ A110 jammer pod under the starboard wing of a Swedish Air Force JA37 Viggen (SaabTech)
1127523

Antenna coverage:
(wide-lobe antennas) 3 antennas, each 120 × 40°
(high-gain antenna) 20 × 20°
(Doppler receiver) 2 antennas, each 70 × 40°, and 2 antennas, each 120 × 40°
Input power: 115 V AC, 2.5 kVA; 28 V DC, 6A
Dimensions:
(pod length) 3,140 mm
(pod diameter) 424 mm
(control unit) 250 × 145 × 103 mm
(display unit) 178 × 134 × 134 mm
Weights: (pod) 235 kg
(control unit) <4 kg
(display unit) <4 kg
MTBF: >200 h

Status
In service with the Swedish Air Force (SwAF), where the pod carries the designation U95.

Contractor
SaabTech AB.

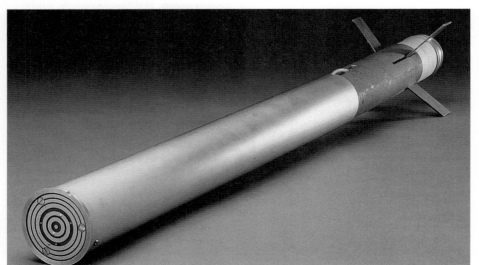

SaabTech BO2D expendable decoy   0081491

BOQ X300 jammer and Electronic Warfare (EW) system

Type
Airborne Electronic CounterMeasures (ECM), radar jamming system.

Description
The BOQ X300 high-performance jamming pod is the latest pod developed and manufactured by Saab. The pod is a modular system that integrates a sophisticated jammer, supported by a RWR and ESM system. As an option, the pod can be configured with a dual fibre optic towed decoy to provide effective countermeasures against monopulse threats.

The BOQ-X300 provides self-protection for high value assets such as fighter, attack and reconnaissance aircraft. The pod is designed to suppress legacy threats, surface based as well as airborne. A secondary role for the BOQ X300 is to provide jamming for training of radar operators in airborne as well as ground- or sea-based environments. As a training jammer, the BOQ-X300 is a part of the Saab REWTS (Responsive Electronic Warfare Training System) concept.

Specifications
Jamming: Full spectrum jamming, DRFM, Doppler techniques, programmable mission data sets, RWR and ESM functionality
Frequency coverage: 0.5 to 40 GHz
Dimensions: 4,070 × 430 mm (L × Ø)
Weight: 270–330 kg (role dependent)
Power: 115/200 V, 400 Hz, 4–6 kVA.

Status
In production.

Contractor
SaabTech AB.

BOW-21 Radar Warning Receiver (RWR)

Type
Airborne Electronic Counter Measures (ECM), Radar Warning Receiver (RWR).

Description
The SaabTech BOW-21 is an E- through J-band (2-20 GHz) Radar Warning Receiver (RWR) designed to cope with extremely dense RF environments. The system features high sensitivity, high selectivity and a claimed 100 per cent Probability of Intercept (POI).

The basic BOW-21 employs a wideband IFM in combination with a narrowband superheterodyne receiver system. The wideband receiver comprises a four-channel amplitude monopulse design with high dynamic range and a high-resolution Digital Frequency Discriminator (DFD). Switchable filters are used to cope with interoperability effects.

In contrast to traditional RWR designs, the BOW-21 receiver includes a four-channel, narrowband device with full monopulse capability. This means that the narrowband receiver can be used to detect not only CW and high PRF signals, but any type of signal that falls within its bandwidth. The local oscillator is a high-speed synthesiser which permits optimised search based on library information. The precise synthesiser and a narrowband DFD are claimed to yield excellent frequency accuracy and resolution. Both receivers have their own video processors, which independently characterise every pulse. Good direction-finding accuracy is achieved by continuous calibration of the RF chains. The narrowband receiver is normally searching but can also be cued by the wideband receiver.

The basic system has four antennas yielding toroidal coverage. To obtain full spherical coverage, the number of antennas and receiver channels, wideband as well as narrowband, can optionally be increased to six. Other options are C/D-band (0.5-2 GHz) coverage, K band (20-40 GHz) coverage and interferometric receivers for increased direction-finding accuracy.

The BOW-21 can also be equipped with a digital receiver which will significantly increase performance with respect to selectivity and passive ranging.

Signal processing duties are shared between the pulse processor and the RWR computer. Within the pulse processor, all processing is done in real time to ensure a rapid response and that no pulses are lost. The correlation is done in two steps, pulse to burst and burst to emitter, and all primary and derived parameters are used in an optimised mix in the correlation process. The primary functions of the RWR computer are emitter identification, passive ranging and threat evaluation. Secondary functions are BIT control, recording of mission data and emitter simulation for training.

The RWR computer features a single board design incorporating several serial channels, flash memory, Ethernet and two PCI interfaces for standard or customised I/O. High-level software executes under the industry-standard VxWorks real-time operating system. The RWR computer has spare capacity and can optionally also perform EW computer functions such as the control of jammers and chaff/flare dispensers.

The system is designed to interface via MIL-STD-1553B databus with existing cockpit displays, although an optional dedicated radar warning display is available.

The BOW-21 system comprises the following LRUs:
- Four (optionally six) Receiver Front-End Units (RFUs). In the basic version the RFU includes antenna, RF preamplifier, filters and a microcontroller. Optionally, the RFU may be equipped with a K-band antenna and front-end and/or interferometer antennas and phase discriminator. The mechanical design of the unit must be tailored to the specific aircraft installation

- A Low Band Antenna (LBA). This is a passive blade antenna required for C/D-band operation
- A Receiver Processor Unit (RPU). The RPU includes receivers, pulse processor, RWR/EW computer and aircraft interfaces. In the basic version the RPU has several empty slots which can be used for the spherical coverage option, the C/D-band option and/or the digital receiver option. The RPU subunits follow the VME standard; unit size is 1-ATR Short.

Specifications
Frequency range: 2–20 GHz (0.5–2 GHz and 20–40 GHz bands as options)
Coverage: ±45° (elevation, -5 dB, baseline configuration); ±90° (elevation, option); 360° (azimuth)
DF accuracy: 1° RMS (interferometric option); 7° RMS (baseline configuration, narrow and wideband)
RF accuracy: 1 MHz (narrowband); 5 MHz (wideband)
Dynamic range: 75 dB
Pulse density capability: 2 Mpps
Tracked emitters: 500
Emitter modes in library: 10,000
Reaction time: 1 s (max)
A/c interfaces: MIL-STD-1553B; RS-422
Power: 3 × 115 V (50 V A – RFU); 3 × 115 V (500 V A – RPU)
Cooling: conduction (RFU) and forced air (RPU)
Dimensions (w × h × d):
44 × 114 × 113 mm (LBA)
125 × 125 × 150 mm (RFU – excl installation specific casing)
256 × 194 × 387 mm (RPU)
Weights:
0.4 kg (LBA)
2.5 kg (RFU)
15 kg (RPU)

Status
In 1985, the Swedish Air Force ordered the AR830 system for the first versions of the JAS 39 Gripen. By the mid-1990s, a requirement for a more capable RWR for the Gripen led to the development of the BOW-21 range of systems. BOW-21 utilises the same basic principles as the AR830 but its performance with respect to range, selectivity and processing power is considerably improved. Extensive flight testing was performed with various system configurations, and in late 1999, the then CelsiusTech Electronics (now SaabTech) received a contract for full-scale development and production. More recently, a version of the BOW-21 system was selected for the mid-life upgrade of German Luftwaffe Tornado PA-200 aircraft.

Contractor
SaabTech AB.

BOZ 100 series ECM dispenser

Type
Airborne Electronic Counter Measures (ECM), Counter Measures Dispensing System (CMDS).

Description
The SaabTech BOZ 100 series is an advanced ECM chaff and IR flare dispenser originally developed for the Swedish Air Force, but also supplied to overseas customers. It is capable of sophisticated break-lock and corridor operation at subsonic and supersonic speeds. The unit is characterised by a relatively large chaff capacity and has a comprehensive EW system interface; the system is suitable for use in deep penetration, strike, reconnaissance and electronic warfare roles. It is microprocessor-

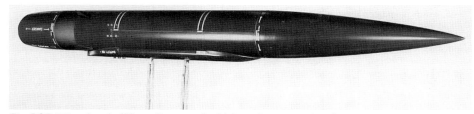

The BOZ 100 series chaff/flare dispenser for high-performance aircraft 0001266

controlled and has a reprogrammable program memory.

Specifications
Dimensions: 4,000 (length) × 380 mm (diameter)
Weight: 325 kg

Attachments: 356 or 762 mm (14 or 30 in) Minimum Area Crutchless Ejector (MACE) or NATO standard lugs

Status
In service with the Swedish Air Force under the designation BOX 9. About 700 pods have

been supplied for the Panavia Tornado IDS programme. The German Air Force and Navy use the BOZ 101, the Italian Air Force uses the BOZ 102 and the Royal Air Force uses the BOZ 107. The UK also employs the BOZ 107 on the Canberra PR9 reconnaissance aircraft. As a result of lessons learned during the first Gulf War, installation of outboard pylons has enabled the BOZ 107 (coupled with the Sky Shadow ECM pod) to be fitted to UK Royal Air Force Tornado F3 interceptors as part of an active and passive upgrade to the aircraft ECM capability.

Contractor
SaabTech AB.

Close-up of a Tornado GR4 armed with eight Brimstone, four ALARM missiles and a BOZ 107 CMDS pod on the port outer pylon
1121452

BOZ 3 training chaff dispenser

Type
Airborne Electronic Counter Measures (ECM), Counter Measures Dispensing System (CMDS).

Description
BOZ 3 is a high-capacity chaff dispenser originally developed for a Swedish Air Force requirement and since adapted for training use. It can be manually or automatically initiated and has both break-lock and corridor chaff modes.

Status
In service with several countries for training purposes.

The BOZ 3 training chaff dispenser   0503829

Contractor
SaabTech AB.

BOZ ED ECM dispenser

Type
Airborne Electronic Counter Measures (ECM), Counter Measures Dispensing System (CMDS).

Description
The latest development of the BOZ 100 series advanced ECM pod is designated BOZ ED (Extended Duration). Physically a modification of the original BOZ pod, the system is form, fit and functional replacement for all -100 series Counter Measures Dispensing Systems (CMDS). BOZ-ED employs a dispensing technique similar to the BOL dispenser, where the payload is released electro-mechanically into the wing vortices of the host aircraft for rapid dispersion of the chaff or pyrophoric IR material.

BOZ ED provides the host aircraft with high capacity and extended duration pre-emptive and reactive CM protection, dispensing chaff and standard pyrotechnic flares. The system is also capable of dispensing non-pyrotechnic chaff/dark flares, which have been proven onboard US F-14 Tomcat and F-15 Eagle fighter aircraft.

Dispensing is by pre-defined or datalink-controlled sequences with claimed BOL-effectiveness. Discrete chaff packs are ejected into the wing vortex with no 'bird-nesting'.

BOZ ED can also incorporate an aft missile approach warning system. The system features a sophisticated and comprehensive EW system interface and is suitable for use in deep penetration, strike, reconnaissance and electronic warfare roles. It is microprocessor-controlled and has a reprogrammable program memory.

Specifications
Dispenser: Chaff or Special Material (SM)
Capacity: 1,160 SM packs; 420 chaff packs; 20 2 × 1 in flare cartridges

IR protection time: 20 min (continuous, pre-emptive)
Flare cartridge dispensing direction: 45°
Dimensions: 4,000 (L) × 380 mm (D)
Weight: 325 kg
Attachments: 356 or 762 mm (14 or 30 in) Minimum Area Crutchless Ejector (MACE) or NATO-standard lugs

Status
In production.

Contractor
SaabTech AB.

Electronic Warfare Core System (EWCS) for Gripen

Type
Airborne Electronic Counter Measures (ECM), integrated suite.

Description
The Electronic Warfare Core System (EWCS) is Saab's integrated EW system for the Gripen multirole fighter aircraft. The system is described as a VME-based, modular, open architecture system, with a built-in flexibility allowing the system to be configured to specific customer needs. This modular approach is also claimed to facilitate subsequent upgrades to incorporate technological advances or added functionality, thus enabling the system to cope with future EW environments.

The basic system consists of an RWR/ESM system and an internal jammer, making a total of seven Line Replaceable Units (LRUs): four wingtip units, one fin pod unit, one forward transmitter unit and an Electronic Warfare Central Unit (EWCU). Basic frequency coverage for the RWR is E- to J-band (2–20 GHz), while the internal jammer covers the H- to J-bands (6–20 GHz); however, RF coverage may be increased to suit specific customer needs.

The EWCS employs an open architecture based on the VME bus standard. Application software includes functions for emitter identification, estimating emitter location, performing dynamic threat analysis and managing countermeasures. The system is capable of controlling a full onboard EW suite, including Passive Missile Warner (PMW), Laser Warning Receiver (LWR), Towed Radar Decoy (TRD), Countermeasures Dispensers (CMDS) and an additional external jamming pod. Communication with the host aircraft Stores Management System (SMS) and other aircraft avionics is via a MIL STD 1553B databus, and the system is designed to integrate with, or be controlled by, the host aircraft Main Computer (MC). In the current (Gripen) configuration, the host MC provides the EWCS with mission data such as threat parameter and EW countermeasures library information. In return, the EWCS provides the MC with actual threat detection, location and identification, together with deployed countermeasures information, for display, audio and mission recording purposes.

EWCS is designed to integrate with Saab Avionics' range of podded jammers, BOL/BOP CMDS and the BO 2D TRD. With a full suite of available options, EWCS will not only provide a powerful self-protection capability, but also enable the host aircraft to fulfil a limited escort jamming role in both air-to-air and air-to-ground missions.

Status
The first generation of the EWCS (formerly known as EWS 39) is in service with the Swedish

The basic EWCS may be complemented by the addition of further external jamming pods, as seen on the starboard shoulder pylon of this Swedish Air Force JAS 39 Gripen   0051265

The A100 ECM training pod carried under a Pilatus PC-9   0503808

The pod is entirely self-contained and requires only power from the carrier aircraft. The system is controlled by an ECM operator, through a cockpit control box, and programmable EEPROM; it is 100 per cent reprogrammable in the air. The analysis and subsequent jamming of incoming signals over 360° with coverage for the selectable high- and low-gain antennas give the system high flexibility in tactical flying and training.

Specifications
Dimensions: 3,235 (length) × 426 mm (diameter)
Weight: 210 kg
Mounting: NATO standard 14 or 30 in lugs
Power supply: 115 V AC, 400 Hz, 3 phase, 3 kVA
   28 V DC, 5 A
Frequency: H- through to I/J-bands
   (6.8–10.5 GHz)
Output power: 350 W, ERP 1–2 or 10 kW
Coverage:
(horizontal) 360°
(vertical) ±30°
Speed: M0.2 to M1.0+

Status
In service in the Canadian, Swedish and Swiss air forces.

Contractor
SaabTech AB.

Air Force. The second generation (Gen 2) of the system is under development, with narrowband receivers added to the wideband receivers used in the first generation system, for increased sensitivity and selectivity. As an option, the system can be supplied with interferometers for accurate angle-of-arrival measurements, as well as several antennas for increased spatial coverage.

The Gen 2 RWR design also forms the basis of the Saab Avionics AB RWR selected for the German Air Force and Navy Tornado IDS and ECR update programme.

Contractor
SaabTech AB.

JAMMER A100 (ALQ-503)

Type
Airborne Electronic Counter Measures (ECM), radar jamming system.

Description
The JAMMER A100 system is a manually or automatically computer-controlled jammer pod for tactical use and ECCM training of air defence fighters and AAA operators. The system

provides Responsive Electronic Warfare Training and Support (REWTS) by giving the operator situational awareness with built-in RWR, look-through capability and a set on receiver. Over 50 smart noise, advanced range, velocity and angle deception modes and combinations of these modes are available. The pod is also capable of providing simulation of missile seeker radars. Single or multithreat capability is provided by an advanced frequency memory loop, a set on receiver and selectable bandwidths.

The A100/AN/ALQ-503 pod   0503807

Turkey

AN/ALQ-178(V)3 and (V)5 integrated self-protection systems

Type
Airborne Electronic Counter Measures (ECM), integrated suite.

Description
The AN/ALQ-178(V)3 and (V)5 integrated self-protection systems, are manufactured by MiKES Microwave Electronic Systems Inc, for installation in Turkish Air Force aircraft. The AN/ALQ-178 is an advanced, internally mounted, self-protection system specifically designed for high-performance fighter aircraft, including the F-16 and F/A-18. It has been fully operational since 1986.

The AN/ALQ-178 receiver utilises a superheterodyne receiver to provide precision frequency measurement and required sensitivity. The integrated architecture of the AN/ALQ-178, utilising a common receiver system, results in rapid threat identification and counter responses (both RF ECM and chaff and flare) for an

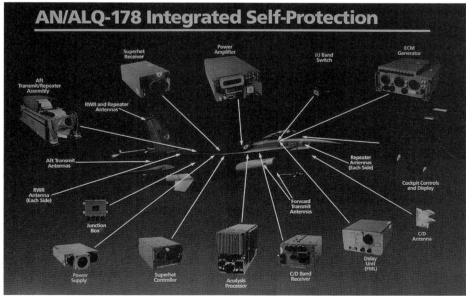

MiKES Microwave Electronics Systems Inc AN/ALQ-178 integrated self-protection system   0002228

exceptional level of aircraft survivability in dense environments.

Status
The AN/ALQ-178(V)3 (also known as SPEWS-1) is operational on the Turkish Air Force's 160 F-16 Block 30/40 aircraft. The AN/ALQ-178(V)5 has been selected for the Turkish Air Force's 80 F-16 Block 50 aircraft.

Contractor
MiKES Microwave Electronic Systems Inc.

Aselsan EW Self-Protection System (ASES)

Type
Airborne Electronic Counter Measures (ECM), integrated suite.

Description
ASES has been designed to protect large, special mission aircraft and may be modified for helicopters. Integrated with an EW Central Management Unit (CMU), ASES provides for radar and missile warning, Situational Awareness (SA) and chaff/flare dispensing. The system provides a high level of protection, coupled with minimum workload for the crew.

ASES, as an integrated EW Self-Protection System (SPS), receives information from associated sub-systems, makes a threat assessment (detection, classification and identification) and decides on the most suitable countermeasures to defeat the radar and/or missile threat. ASES provides an advanced Human Machine Interface (HMI) with multifunction display support, indicative audio warnings and alerts, in-flight event recording, single-point Mission Data File (MDF) download and mission report upload, NVIS compatibility and GPS/INS integration.

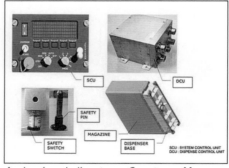

Aselsan's helicopter Counter Measures Dispensing System (CMDS) (Aselsan)   1128606

ASES is supported on the ground by an ASES SW Support Station, which facilitates MDF preparation before the mission and post-mission data analysis via playback of the recorded mission events on a digital geographical map.

ASES incorporates the following sub-systems:
- Central Management Unit (CMU)
- Cockpit Control and Display Unit (CCDU)
- Memory Loading Unit (MLU)
- Missile Warning System (MWS)
- Radar Warning Receiver (RWR)
- Counter Measures Dispensing System (CMDS).

ASES is integrated around a MIL-STD-1553 architecture, with CCDU interface via ARINC-429. Currently, ASES is integrated with the AN/AAR-60 MILDS (MWS), AN/ALE-47 (CMDS), SPS-45 (RWR).

The inherent flexibility of the system architecture facilitates customising according to requirement.

Status
ASES is in production and has been in service since 2002. The system was selected as the EW Self-Protection System for Turkish Navy CN-235 MELTEM Maritime Patrol Aircraft.

Contractor
Aselsan Inc, MST Division.

Helicopter Counter Measures Dispensing System (CMDS)

Type
Airborne Electronic Counter Measures (ECM), Counter Measures Dispensing System (CMDS).

Description
Aselsan's CMDS is designed for helicopter applications, dispensing Chaff/Flare decoys against RF-/IR-guided missiles. The system comprises five LRUs:
- System Control Unit (SCU)
- Dispense Control Unit (DCU)
- Dispenser base
- Magazine
- Safety switch.

The CMDS dispenses 1 × 1 inch IR or RF decoys. Up to four magazines can be connected to a DCU, with each magazine containing 30 decoys. The system is also able to dispense each half of RR-180 type dual-chaff decoys independently, increasing the total number of RF decoys by a factor of two. Each magazine can be loaded with mixed types of decoys.

The onboard Mission Data File (MDF) contains up to 2,400 individual dispensing programmes.

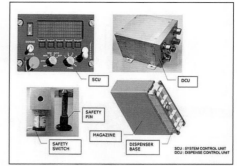

Aselsan's helicopter Counter Measures Dispensing System (CMDS) (Aselsan)   1128606

The dispense programmes can be initiated manually, semi-automatically or automatically, with two payloads dispensed simultaneously if required. In semi-automatic (dispensing with pilot consent) and automatic (dispensing initiated by system) modes, the system receives warnings from external Missile Warning Systems (MWS), Radar Warning Receivers (RWR) or Laser Warning Receivers (LWR).

In-flight loading of MDF is available via MIL-STD-1553B databus and the system provides for automatic warning in the event of critical (low) expendable level and misfire.

The system can be integrated with a centralised EW suite controller.

Specifications
Performance
Dispensing programmes: Up to 2,400
   Minimum programmable delay between dispensing programmes: 15 ms
Modes
   Automatic: Dispensing initiated by system; ESCAPE, JETTISON, DISPENSE commands available
   Semi-automatic: Dispensing initiated by pilot consent; ESCAPE, JETTISON, DISPENSE commands available
   Manual: Dispense initiated by pilot via ESCAPE, JETTISON or DISPENSE commands
   Bypass: Bypass mode is utilised with critical failure in SCU. Pilot dispense commands routed direct to DCU(s) via discrete
   Interface: MIL-STD-1553B; RS-422; RS-232; discrete
Compliance: MIL-STD-810F; MIL-STD-461C; MIL-L-85762A

Status
In development.

Contractor
Aselsan Inc, MST Division.

United Kingdom

700 Series sonobuoy and VHF homing receivers

Type
Airborne Electronic Counter Measures (ECM), sonar system.

Description
700 Series receivers are frequency synthesised, dual-bandwidth VHF receivers designed to form part of a system to provide port and starboard and fore and aft VHF homing with automatic on-top indication. The dual bandwidth permits reception of 25 kHz spaced signals when in the narrowband mode, or 375 MHz spaced sonobuoy signals in the wideband mode. An audio output is available to provide aural port-starboard homing and recognition of amplitude-modulated signals.

The 700 Series was developed from and embodies technology and subsystems used in the Chelton VHF/UHF homing systems, of which more than 700 are in service for worldwide

military and civil operations. A complete system comprises four LRUs plus antennas, and features frequency-synthesised dual-bandwidth operation, up to 99 preset channels with full OTPI interface, low weight and small size.

Specifications
Type 700 receiver
Weight: 1.325 kg
Power supply: 28 V DC, 500 mA
Frequency:
(Type 700-1) 100.00–173.500 MHz,
(Type 700-2) 117.975–173.975 MHz

Type 710 control unit
Weight: 0.98 kg
Power supply: 22–31.5 V DC, < 1.0 A
Frequency:
(Type 710-1) 100.00–159.975 MHz,
(Type 710-2) 117.975–173.975 MHz
Channel mode frequency:
(Type 710-1) 136.00–173.500 MHz,
(Type 710-2) 136.00–173.500 MHz

Status
In production and in service. Chelton 700 Series sonobuoy and VHF homing receivers have been ordered by Bristow Helicopters, the Helicopter Services (Navy), the UK Royal Navy, the Royal Norwegian Navy and the Royal Swedish Navy.

Contractor
Chelton (Electrostatics) Ltd.

730 series sonobuoy homing and channel occupancy system

Type
Airborne Electronic Counter Measures (ECM), sonar system.

Description
The 730 series 99-channel sonobuoy homing and channel occupancy system operates over

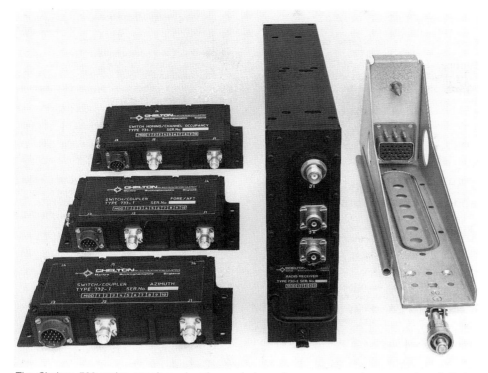

The Chelton 730 series sonobuoy homing and channel occupancy system comprises (left, top to bottom) the Type 731-1 homing/channel occupancy switch, the Type 733-1 fore and aft switch coupler, the Type 732-1 azimuth switch coupler, (centre) the Type 730-1 radio receiver and (right) the Type 734-1 receiver mounting tray   0503686

The 1220 Series laser warning receiver consists of electronic unit, control unit and sensors   0503830

the frequency range 136.000 to 173.500 MHz and has homing or channel occupancy independent modes of operation.

In the homing mode, the system supplies azimuthal information so that the aircraft may be flown overhead a sonobuoy. Additional information indicates whether the sonobuoy is ahead or astern of the homing aircraft. This information is further decoded to give an indication of the instant the aircraft passes overhead the sonobuoy.

In the channel occupancy mode, the receiver measures the received signal strength to show when a selected channel is clear for the reception of sonobuoy signals.

The system comprises eight LRUs: the Type 730-1 radio receiver, Type 731-1 homing/channel occupancy switch, Type 732-1 azimuth switch coupler, Type 733-1 fore and aft coupler, Type 734-1 receiver mounting tray, azimuth antenna assembly, fore and aft antenna assembly and channel occupancy antenna assembly.

The heart of the system is the radio receiver which interfaces the system to the aircraft via a dual-redundant MIL-STD-1553B databus.

In the channel occupancy mode, the radio receiver provides the necessary control signals to disable the azimuth and fore and aft switch couplers. The receiver cycles through channels as designated by the mission computer and provides information on the databus channel status, together with received signal strength.

In the homing mode, the radio receiver provides the necessary control signals of the azimuth and fore and aft switch couplers to switch the RF input of the radio receiver between the antenna assemblies.

BITE is capable of diagnosing 95 per cent of all defects to LRU level.

Specifications
Weight: (all LRUs) 8 kg

Status
In production and in service. The 730 series sonobuoy homing and channel occupancy system has been selected for the Westland/Agusta EH 101 Merlin ASW helicopter.

Contractor
Chelton (Electrostatics) Ltd.

1220/1223 Series Laser Warning Receiver (LWR)

Type
Airborne Electronic Counter Measures (ECM), Laser Warning Receiver (LWR).

Description
The 1220 Series laser warning receiver is a modular expandable system with centralised processing and miniaturised remote sensor heads. It may be operated as a standalone LWR with its own display or as part of an integrated system with, for example, the Selex Sensors and Airborne Systems Sky Guardian 200 RWR.

A pre-mission programmable threat library provides the capability for selective threat identification. The panel-mounted display is NVIS-compatible and provides sector indication and other threat data.

An optional module provides a record of up to several thousand date and time tagged laser detection.

Specifications
Dimensions:
(sensor) 25 (length) × 50 mm (diameter)
(electronics unit) 127 × 194 × 323 mm
(display) 80 × 80 mm
Weight:
(sensor) 0.4 kg
(electronics unit) 5 kg
(control panel) 0.4 kg
(display panel) 0.4 kg
Power supply: 24–28 V DC or 115 V AC, 400 Hz
Field of view: 360° azimuth × 45° elevation
Angular resolution:
(azimuth) ±22.5° (15 or 10° option)
Spectral range:
(basic system) 0.35–1.1 μm
    1–1.8 μm and 8–11 μm options
Spectral resolution: 15–100 nm
Interfaces: RS-422 or MIL-STD-1553B

Status
In production for high-performance aircraft applications. The system has been sold to unspecified customers in Europe and Asia.

An updated variant, the 1223 LWR, forms part of the HIDAS system (see separate entry) selected for the British Army WAH-64D Apache helicopter. The model 1223 has very high sensitivity and extended waveband coverage, to

provide detection of laser beam-riding threats at the point of launch, as well as rangefinders and target designators.

Contractor
Selex Sensors and Airborne Systems.

AA34030 sonobuoy command transmitter

Type
Airborne Electronic Counter Measures (ECM), sonar system.

Description
The AA34030 sonobuoy command transmitter is a variant of the AD 3400 VHF/UHF transmitter/receiver, which has been in full production for a number of years. This variant has been developed in conjunction with the AQS-903 acoustic processing system and is currently in service with several air forces around the world.

The equipment is a UHF transmitter which has been specifically developed to provide the RF downlink between an airborne acoustic processing system and active sonobuoys. The transmitter provides AM or FM analogue or digital communications over the frequency range 282 to 292 MHz with channel increments of 25 kHz, although for normal sonobuoy operations 50 kHz is used. Control of the transmitter is via an ARINC 429 digital data highway and a number of discrete inputs. Control information is normally provided by the acoustic processing system, although a discrete control unit can be provided.

Specifications
Frequency range: 282–292 MHz
Channel spacing: 50 kHz
No of channels: 200
Frequency accuracy: 2 ppm
Transmitter power: 50 W FM, 15 W AM
Dimensions: 194 × 125 × 256 mm
Weight: 6.5 kg

Status
In production and in service in UK Royal Navy EH 101 Merlin helicopters.

Contractor
Selex Sensors and Airborne Systems.

AA34030 sonobuoy command transmitter   0518509

Apollo radar jammer

Type

Airborne Electronic Counter Measures (ECM), radar jamming system.

Description

Apollo is an advanced radar jammer for aircraft self-protection. It is designed to ensure survival in hostile environments by countering a wide range of threats and is available as a pod or installed within the airframe.

Apollo comprises modules developed from in-service Selex Sensors equipment. These include warning receivers, jammers and digital processors combined in a lightweight compact system. Apollo operates automatically, so that jamming management does not add to the pilot's workload.

The podded Apollo fits on a standard hardpoint and can be quickly installed and removed, giving commanders the freedom to redeploy it quickly on other aircraft.

Different performance specifications are available: from a simple repeater jammer, to a more capable system that incorporates a frequency set on receiver and noise deception capability, through to advanced single or dual DRFM options.

Apollo's jamming responses are software driven. As well as producing noise, it generates carefully modulated response patterns to confuse and deceive fire-control radars. By using various jamming techniques, Apollo is effective against pulse and CW radar signals. It is capable of dealing with multiple threats simultaneously.

Where the aircraft is fitted with a radar warning receiver such as the BAE Systems Sky Guardian, data can be passed between Apollo and the RWR to optimise intelligent interaction. This combination offers the advantages of an integrated defensive aids system and controls the decoy dispensers to fire chaff and flares at the best possible moment to co-ordinate with the jamming.

Apollo is designed to be as self-sufficient as possible. If the aircraft's own electrical system cannot generate the power needed, a ram-air turbine is fitted to the podded variant to supply the unit.

Apollo's digital processors identify radars by comparing their parameters with the built-in EW library. The jamming technique most likely to defeat that radar is selected and used. The control system organises jamming responses in order of priority by assessing various factors. These include whether a signal is known to be a threat, whether it is locked on and its direction relative to the aircraft. Apollo can be programmed on the flight line with the latest threat intelligence.

In an aircraft fitted with a radar warning receiver, Apollo momentarily blanks out the jamming signal so that the RWR can sample the threat environment and this updated threat information is then passed back to the jammer. The operation is synchronised to maintain peak jamming performance.

Specifications

Dimensions: 2,650 (length) × 280 mm (diameter) (length with RAT) 2,985 mm

Weight:
130 kg, 160 kg (with RAT)
40 kg (inboard)

Power supply: 115 V AC, 400 Hz, 3 phase
28 V DC
Frequency: H- to J-bands (other bands optional)

Status

Fully developed and flight tested. A variant of Apollo forms part of the Helicopter Integrated Defensive Aids System (HIDAS), installed in British Army WAH-64D Apache attack helicopters.

Contractor

Selex Sensors and Airborne Systems.

AQS 902/AQS 920 series acoustic processors

Type

Airborne Electronic Counter Measures (ECM), sonar system.

Description

The AQS 902 and its export variants, the AQS 920 series, have been designed for ASW helicopters, maritime patrol aircraft and small ships. This range of lightweight and versatile systems will handle data from all current and projected NATO inventory sonobuoys and will interface with a wide range of 31 or 99 RF channel sonobuoy receivers. An AQS 902/920 series can process and display up to eight buoys or a combination of sonobuoys and dipping sonar simultaneously. The modular structure of the AQS 902/920 series allows systems to be tailored to specific customer requirements. The signal processor unit converts received sonobuoy signals into digital form, carries out the necessary filtering and analysis and processes the data into a suitable form for display. It is packaged in a 1 ATR(S) box. Acoustic data is displayed on a CRT or electrographic paper chart recorder. Either or both may be specified, but the CRT display requires a post-processor unit to prepare data for presentation and provides many additional operator aiding facilities. The operator uses a key panel and rollerball or stiff stick cursor control device to specify processing modes and display formats, measure and extract information and access the wide range of automatic and semi-automatic system facilities.

The AQS 902/920 offers numerous processing and display facilities including simultaneous passive and active processing, wide and narrowband spectral analysis, broadband correlation, CW and FM active modes, passive auto alerts, Doppler fixing, calculation of target range and speed, bathythermal processing and ambient noise measurement. Additionally, the system provides automatic tracking and calculation of line frequency, target bearing and range and signal-to-noise ratio.

A unique feature of these systems is the acoustic localisation plot which presents the operator with a geographical plot of sonobuoy bearing and range information.

Status

The AQS 902G-DS acoustic processor was withdrawn from operational service with UK Royal Navy Sea King HAS Mk6 helicopters early in 2004 when the last Sea King Mk6 was withdrawn from ASW operations.

The ASQ 903 acoustic processor (see separate entry) has superseded the AQS 902/920 series processors. Some export variants of the AQS 902/920 remain in service.

Contractor

Thales Underwater Systems Ltd.

AQS 903/AQS 903A acoustic processing system

Type

Airborne Electronic Counter Measures (ECM), sonar system.

Description

The AQS 903/903A acoustic processors are powerful, lightweight systems designed for both helicopter and fixed-wing applications. Capable of processing data from between 8 and 32 sonobuoys in any combination, the system's flexibility is derived from a design based around a number of concurrent processing 'pipelines', each able to handle a number of passive and/or active sonobuoys. Four of these pipelines provide total processing power equivalent to eight times that of the earlier AQS 901, all at a fraction of that system's size and weight.

The AQS 903/903A make extensive use of distributed processing and include the latest component technology to achieve optimum performance. The system provides for a wide range of processing options to enable the operator to smoothly progress from initial contact/detection through to classification, localisation and tracking. Analysis and display cues are in plain language, requiring minimal operator training.

The AQS 903/903A can process data from LOFAR, DIFAR, HIDAR, Barra, DICASS and CAMBS sonobuoys.

The Apollo jamming pod on a UK Royal Navy Sea Harrier   0009916

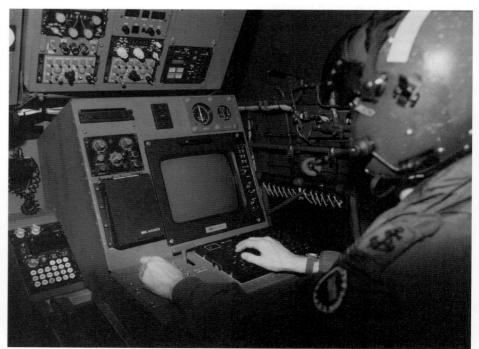

The AQS 902G/DS installation in a UK Royal Navy Sea King Mk 6   0503687

The EH 101 Merlin ASW helicopter   0504047

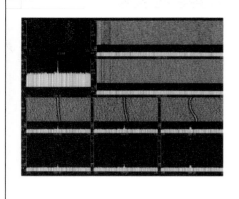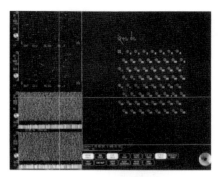

Ultra Electronics Limited, Sonar and Communication Systems AQS 970 acoustic processor hardware and two display formats   0044961

Status

To date, 44 AQS 903 systems have been delivered to the UK Royal Navy for their EH 101 Merlin Helicopters. The AQS 903A upgrade is due to be completed and accepted into service in 2004. The basic configuration will process data from the complete range of NATO inventory sonobuoys.

Contractor

Thales Underwater Systems Ltd.

AQS 940 acoustic processor

Type

Airborne Electronic Counter Measures (ECM), sonar system.

Description

AQS 940 is an advanced, powerful airborne acoustic processor which uses a cost-effective and totally open architecture based on rugged state-of-the-art, off-the-shelf hardware. The existing and proven algorithms used in the AQS 903/930 series have been re-hosted and enhanced with new colour broadband and data fusion techniques. The system can also be provided with the latest processing and display techniques for multistatics, providing increased area coverage in littoral waters.

The system is designed for use by one to four operators, with each operator having access to all the processed data as graphical data fusion plots, localisation plots, acoustic formats or tabular information.

All current and projected NATO sonobuoy types can be processed; sonobuoy data may be received from either traditional analogue receivers or from new digital receivers.

Operator workload is reduced by the user-friendly interface and, more importantly, the wealth of operator aids. In particular, the data fusion display focuses the operator's attention on areas of acoustic interest.

The system can be easily interfaced to other avionics systems via any industry standard.

Status

In service.

Contractor

Thomson Marconi Sonar Limited.

AQS 970 acoustic processor

Type

Airborne Electronic Counter Measures (ECM), sonar system.

Description

The Ultra AQS 970 airborne acoustic processor has been selected for the Nimrod MRA4

programme and is being developed through collaboration between Ultra Electronics and Computing Devices of Canada. The CDC UYS-503 system is already in service with the defence forces of Australia, Canada, Sweden and the US. The AQS 970 interfaces with the Tactical Command System (TCS) and the Sonobuoy Receiving System (SRS) in fixed-wing Maritime Patrol Aircraft (MPA) and Anti-Submarine Warfare (ASW) helicopters, providing signal conditioning, signal and data processing, display generation and control functions.

The Ultra AQS 970 is designed to meet the detection, localisation and attack requirements posed by the latest-generation nuclear and diesel submarines operating in open ocean and littoral water environments. The modular architecture of the AQS 970 and its ample processing power and memory capacity enable it to accommodate readily new buoy types and processing techniques without hardware modification.

The system employs a scalable hardware and software architecture, configurable as 16, 32 or 64 identical processing channels, providing full 64-buoy processing capability (subject to buoy RF channel limitations) for all NATO variants of the following existing or planned sonobuoy types:

Passive search and localisation
DIFAR
LOFAR
VLAD
CODAR (DIFAR/omnipairs)
HIDAR (digital DIFAR)
BARRA.

Active search and localisation
CAMBS
DICASS
RO
SR(SA)903 Active sonobuoy search system.

Special purpose
Ambient Noise Monitoring (ANM)
BATHY
Built-in Acoustic Test Signal Generator
(ATSG) and Tactical Acoustic Trainer (TAT).

Processing and display capabilities include bearing coherent processing and effective colour-coded displays; energy map and acoustic situation displays; narrowband, broadband/swath, diesel, DEMON, and transient detection processes; variable band CODAR (using DIFAR buoys); adaptive interference rejection/anti-jam processing (DIFAR, BARRA, CAMBS)

Operator aids include Gram markers and harmonic dividers, CPA analysis, Lloyd's mirror analysis, display magnify/freeze/zoom/pan, cross-hair and hook, adjustable thresholds and integration times, concurrent multiresolution

processing, rapidly selectable operator-defined display formats and combinations.

General aids provided include frequency/bearing/range trackers, synchronised audio, range prediction, data fusion and contact tracking, auto alerts.

Status

Ultra Electronics is providing the integrated acoustic system for Nimrod MRA4.

Contractor

Ultra Electronics, Sonar and Communication
 Systems.

Ariel airborne Towed Radar Decoy (TRD)

Type

Airborne Electronic Counter Measures (ECM), Towed Radar Decoy (TRD).

Description

Ariel is a compact, lightweight and cost-effective radar jamming system which is recoverable during or after flight for repeated operational employment on subsequent missions.

Ariel can be mounted within the airframe of the host aircraft, in a pod on a standard underwing pylon or as a scabbed-on fit to the fuselage skin. Three variants of the system have been developed:

Ariel 1, the first-generation decoy, was a relatively high-power RF linked repeater decoy, utilising the host platform's Radar Warning Receiver (RWR) to cue an onboard techniques generator, control/modulation data from which is passed to the TRD via a decoy interface module and an optical fibre link incorporated in the tow cable. The TRD then transmits the counter measures signal.

Ariel 2, the current second-generation version in UK Royal Air Force service, is a high-power datalinked decoy that can function as both a towed radar repeater and deception jammer. Ariel 2, as configured for the Eurofighter Typhoon, is a double installation deployed from the rear of the aircraft's starboard wingtip pod fairing.

Both of these variants depend on equipment aboard the host aircraft, including the host aircraft's RWR, the EHT power supply unit and the winch control unit. The tow cable is of multi-element fibre optic construction and includes electrical conductors and strength members.

The towed body includes the preamplifiers and final TWT amplifiers, transmit antennas and the power conditioning unit. The RF-linked decoy uses the aircraft RWR and an onboard ECM

Tornado F3 aircraft adapted to the defence suppression role carried the Ariel TRD for self defence. The decoy was carried in a modified BOZ 107 chaff pod seen on the port wing of the aircraft   0578815

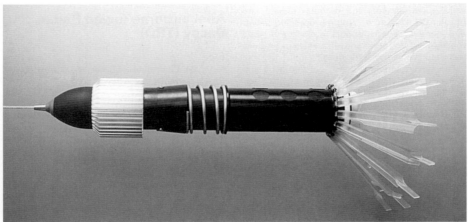

Ariel 2 Towed Radar Decoy (TRD) (Selex)   0528523

A Nimrod MR. Mk2 on takeoff. Note the BOZ 107 pod under the left wing   0130703

techniques generator, with the decoy interface module passing signals down an optical fibre to a photo detector in the decoy. The decoy transmits all types of ECM signals, as generated by the onboard techniques generator or the existing onboard ECM systems.

The latest variant of the system, Ariel 3, is understood to be an all solid-state decoy.

For large aircraft, the decoy is deployed and recovered using an associated onboard winch mechanism. After deployment on smaller combat aircraft, it is jettisoned before landing and may be recovered by optional parachute for reuse.

Regardless of its deployment method, the Ariel fibre optic tow cable is constructed to allow the decoy to be deployed throughout a full sortie, rather than just when a threat is detected.

Ariel is effective against monopulse, semi-active radars and home-on-jam systems. It has fully programmable countermeasures techniques and can operate either in a standalone mode or be integrated with an existing onboard EW suite.

Specifications
(Ariel 1)
Frequency: 2–18 GHz
Coverage: spherical
Techniques: noise, repeater and advanced countermeasures
Speed range: Flight tested at speeds between 150 kt and M1.2

Status
First fitted on UK Royal Air Force Nimrod MR. Mk 2 aircraft under an Urgent Operational Requirement (UOR) for the 1991 Gulf War; this configuration has been reported as no longer in service. Ariel has also been installed on Tornado F. Mk 3 aircraft. In both of the aforementioned applications, the system was housed in a modified BOZ 107 pod (see separate entry). Flight testing, on a MC-130 aircraft fitted with an ALQ-131 pod, has also been carried out by the US Air Force.

The current production standard, Ariel 2, is in service in Eurofighter Typhoon aircraft. Ariel 3 is under development.

Contractor
Selex Sensors and Airborne Systems.

ARR 970 sonobuoy telemetry receiver

Type
Airborne Electronic Counter Measures (ECM), sonar system.

Description
The ARR 970 sonobuoy receiver system has been designed to satisfy the requirements of current Anti-Submarine Warfare (ASW) aircraft, while providing additional facilities to meet future requirements. It amplifies and demodulates FM sonobuoy transmissions and provides the output to an acoustic processor for analysis and display. The ARR 970 provides excellent RF and acoustic performance as well as lightweight and compact size.

Two models are available. A multichannel version offers 16-, 20- and 32-channel options, while a single channel version provides only the 32-channel capability. Both models include options for an On-Top Position Indicator (OTPI) function, channel scanning and a sonobuoy positioning system.

Specifications
16-channel system
Dimensions: 545 × 215 × 345 mm
Weight: 33 kg
Power supply: 115 V, 3 phase, 400 Hz, MIL-STD-704A: 300 VA
RF channels:
99 + test (375 kHz spacing, standard)
505 (75 kHz spacing, optional)
Audio interfaces: balanced, 2 Vrms for 75 kHz peak deviation
IF bandwidth: 240 kHz, standard
Control interfaces: MIL-STD-1553B, RS-422

32-channel system
Dimensions: 545 × 328 × 345 mm
Weight: 39 kg
Power supply: 115 V, 3 phase, 400 Hz, MIL-STD-704A: 450 VA
RF channels:
99 + test (375 kHz spacing, standard)
505 (75 kHz spacing, optional)
Audio interfaces: balanced, 2 Vrms for 75 kHz peak deviation
IF bandwidth: 240 kHz, standard
Control interfaces: MIL-STD-1553B, RS-422

Ultra Electronics ARR 970 sonobuoy telemetry receiver multichannel (16-, 20 and 32-channels) (left) and single-channel (32- channels) (right) configurations 0011858

Status
Selected for use on Nimrod MRA4.

Contractor
Ultra Electronics, Sonar and Communication Systems.

AWARE ESM system

Type
Airborne Electronic Counter Measures (ECM), Electronic Support Measures (ESM) system.

Description
The Advanced Warning of Active Radar Emissions (AWARE) range of equipments provides advanced warning of radar emissions for use in light strike and transport aircraft and helicopters or in marine craft and ground vehicles.

The system is modular in concept and design, providing variants of the basic system by simple changes of subassembly or LRUs. The expansion facilities offered provide an ESM suite and integration with ECM systems and will cover from C-band up to the millimetric region.

The flexibility of design of AWARE allows systems to be configured to match user requirements, from a basic RWR to a full ESM suite, and it is suitable for operation in a wide range of airframes, from light battlefield helicopters to high-performance aircraft. There are currently four variants in the AWARE range: AWARE-3, AWARE-4, AWARE-5 and AWARE-6. All have the same basic modules, the only difference being in the man/machine interface.

AWARE-3
AWARE-3 is the basic ESM system, covering the E- to J-bands and intended for battlefield helicopters, light tactical and logistic aircraft. The system detects and identifies pulse, pulse Doppler, CW and ICW signals. An immediate identification of the threat is provided and the particular threat type is identified.

Identification is achieved by comparison with a stored threat library. The library, threat priorities and ECM interfaces are under software control which is loaded before a mission, allowing rapid updating of threats and priorities on a mission by mission basis. Additionally, security of system and intelligence data is enhanced by the ability to clear the library and software. The system operates in peak pulse densities of several hundred thousand pulses/s, radiated by many simultaneous emitters.

The AWARE display is heading-compensated with computer-controlled symbology. An extensive symbol library can be programmed with symbols of the user's choice.

AWARE-3 consists of four planar spiral antennas, a hand-portable programme loading unit, two dual-channel crystal video receivers, a fast Instantaneous Frequency Measurement (IFM) receiver and a powerful signal processor. Control of the system can be achieved either as an autonomous system, where the control unit and a 3 in display are employed, or as an integral part of the mission management system, with control being exercised via a MIL-STD-1553B standard interface. All LRUs are convection cooled.

The four antennas are mounted in mutually orthogonal azimuth directions. The signals received are passed to the two dual-channel crystal video receivers. These receivers provide the preamplification, detection and compression of the video signals for input to the signal processor. They also provide controlled RF outputs to the IFM receiver. The output of the IFM receiver is presented to the signal processor as a digital number.

The signal processor determines the direction of arrival and time of arrival of the signals. The results of these operations, when combined with output of the IFM receiver, allows de-interleaving of the incoming signals to be performed. Once de-interleaved and characterised, the signals are classified by comparing them with a stored library of known emitters. Following classification, the source of emission and its range and bearing may be displayed in simplified plan form on the monitor or passed to the aircraft mission management system via a monitor or via a MIL-STD-1553 databus.

AWARE-4
AWARE-4 is an ESM system for maritime helicopter applications. The right-hand screen retains the PPI presentation of emitters to maintain the radar warning role, while the left-hand screen is a tote for the display of emitter parameters.

AWARE-5
AWARE-5 retains the AWARE-3 control unit but has no display. In this variant, the data are displayed on either the aircraft HUD or the AI radar display, via the MIL-STD-1553B or ARINC 429 databusses. AWARE-5 is for use in light strike aircraft.

AWARE-6
AWARE-6 is similar to AWARE-4, but the display unit is replaced by two 10 in colour monitors. AWARE-6 is for use in maritime patrol aircraft operations where more space is available and a larger tote that can display the whole track table is required.

Specifications
System weight:
(AWARE-3)13 kg
(AWARE-4)15 kg
Power supply: 28 V DC
150 W (AWARE-3)
180 W (AWARE-4)
Frequency: 2–18 GHz
Accuracy: better than 10° RMS

Status
In service UK forces and other operators. AWARE-3, under the designation ARI 23491 (Rewarder), equips Lynx and Gazelle helicopters of the British Army. AWARE-3 also equips Lynx helicopters of the Royal Netherlands Navy. Systems are being used for both airborne and shipborne applications.

Contractor
Selex Sensors and Airborne Systems.

The AWARE ESM system is in service in Lynx and Gazelle helicopters 0504140

Corvus ELINT system

Type
Airborne Electronic Counter Measures (ECM), Electronic Intelligence (ELINT) system.

Description
Corvus is a self-contained ELINT receiving system which includes signals analysis and magnetic recording. The equipment provides a complete turnkey system requiring only an antenna and turntable system.

A major feature of the system is its ability to record and analyse intrapulse modulation on wideband frequency-agile emitters. This modulation is measured at up to 200 MHz sampling rate and the measurements are correlated with the standard pulse descriptors.

Analysis of the live or recorded data can be performed by the integral pulse analyser. Alternatively, the EP4240 separate off-line analysis system is available. This allows a more detailed analysis to be made and ELINT libraries to be compiled, with rapid retrieval of data.

The system covers the frequency range of 0.5 to 18 GHz and options are available for wider ranges. The instantaneous bandwidth of the channelised receiver is 1 GHz.

Modern radars employ modulation during each pulse in order to improve range resolution and produce profiles of targets. Some radars have unintentional modulation during each pulse and this can be used to provide unique identification.

Corvus can record every pulse of a radar over many antenna scans, complete with intrapulse data. This is measured by sampling every 5 ns and measuring both instantaneous amplitude and instantaneous frequency. These samples are then digitised and recorded directly on to a magnetic tape recorder.

The sampling and digitising technique produces a recording bandwidth of up to 50 MHz and this is essential for observing the intrapulse modulation of modern complex military radars without distortion.

Corvus links the recording technique to the channelised receiver to enable every pulse of a frequency-hopping radar to be reproduced, even if the hop range is as wide as 1 GHz.

A 2 GHz bandwidth channelised receiver is now in development.

Contractor
Thales Communications.

CRISP Compact Reconfigurable Interactive Signal Processor

Type
Airborne Electronic Counter Measures (ECM), sonar system.

Description
The CRISP (Compact, Reconfigurable Interactive Signal Processor) system has been developed using Thales' Series 5 technology to provide a compact, portable mission support system.

A high-resolution (1,024 × 1,024 pixels) colour graphics display (available as a flat screen) with pop-up menus and mouse roll and click selection allows interactive and user-friendly operation, utilising a Windows™ style environment.

Using modular boards already in operational naval service, CRISP offers the following capabilities:

- Simultaneous broadband and narrowband analysis
- Digital DIFAR processing (16 channels – near-realtime in omni and steered cardoid with bearing extraction)
- Frequency range from infrasonic to intercept
- Very fast time replay (×8)
- Short-term event capture and analysis
- Rapid software reconfiguration

Spectral processing analysis bands cover 2,400, 1,200, 800, 400, 200, 100, 80, 40, 20 and 10 Hz. Demon processing covers input bands of 300, 600 and 1,200 Hz with analysis bands of

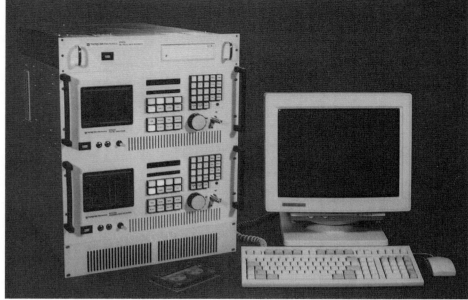

The Corvus ELINT system   0504145

40, 80 and 200 Hz. In LOFAR, the CRISP system processes eight channels from 0 to 19 kHz.

The CRISP architecture allows it to be an onboard data gatherer or a mission support, faster than real-time analyser.

Status
In service with the UK Royal Navy, Royal Air Force and Canadian Air Force.

Contractor
Thales Underwater Systems Ltd.

EP4000 Series of EW modules

Type
Airborne Electronic Counter Measures (ECM), Electronic Support Measures (ESM) system.

Description
A range of advanced modules for electronic warfare has been developed by Thales. The range includes the EP4410 pulse analyser and the EP4220 channelised receiver.

The EP4410 is designed to interface with the EP4220 or conventional swept superheterodyne receivers. It is an advanced pulse analyser and uses qualification on descriptor parameters to enable signals to be selected for analysis and displayed on the high resolution colour graphics display. In addition to pulse descriptor information, the VME processor-based analyser measures intrapulse frequency and amplitude information by frequent sampling of the signals through each pulse. The EP4410 has the capacity to perform an analysis on up to 64,000 pulses, with intrapulse information automatically correlated with pulse descriptor data.

A digital receiver option is available for users wishing to analyse amplitude, frequency and phase intrapulse characteristics with minimum receiver colouration effects.

The EP4220 is a wideband channelised receiver and has been specifically designed for ELINT, ESM and signal monitoring applications which require high sensitivity and wide instantaneous bandwidth. Operating in the 0.4 to 18 GHz frequency band, it offers very high performance, detecting complex signals in a dense electromagnetic environment. Its wide 1 GHz bandwidth enables full examination of frequency-agile transmitters. Within each channel, the detection bandwidth is sufficiently narrow to give measurements comparable with narrowband ELINT receivers.

A 2 GHz bandwidth channelised receiver is now in development.

Specifications
EP4410 pulse analyser
Dimensions: 483 × 178 × 575 mm
Weight: 20 kg
Power supply: 115/240 V AC, 40–60 Hz, 300 W

EP4220 channelised receiver
Dimensions: 483 × 310 × 580 mm
Weight: 50 kg
Power supply: 115/240 V AC, 40–60 Hz, 400 W
Frequency: 0.5–18 GHz

Contractor
Thales Defence Ltd.

Hermes ESM system

Type
Airborne Electronic Counter Measures (ECM), Electronic Support Measures (ESM) system.

Description
Hermes is an EW surveillance system for the interception and analysis of radar signals, providing data for planning operations and for defence. It can form part of an air defence

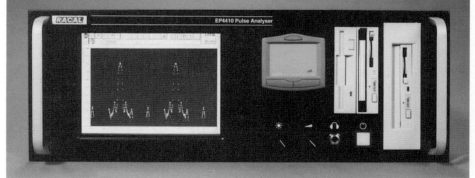

The EP4410 Pulse Analyser   0001269

King Mk 42B of the Indian Navy   1046308

network, both in the maritime surveillance role and from land-based sites.

In the airborne mode, Hermes can be configured from a basic radar warning receiver to a complete ESM system for the AEW role and reconnaissance. It can also be fitted into UAVs.

Hermes intercepts radar signals at long range using wideband superheterodyne receivers that give unrivalled handling of continuous wave and high-duty cycle signals. The modular design means that Hermes can be installed in a wide variety of platforms and can be fitted inboard on helicopters and medium-size aircraft or in a pod on fast jets.

Hermes is fitted with interfaces for navigation, data exchange and the automatic triggering of electronic countermeasures. These interfaces can be tailored to specific platform requirements.

Specifications

Weight: 110 kg approx
Frequency: C- to J-bands (0.5–20 GHz)
Signal types: pulse, CW, ICW, RF agile, PRI agile, jammers
Coverage:
(azimuth) 360°
(elevation) ±45°
Emitter library: >1,000 modes

Status

The system is in service with the Indian Navy on its Westland Mk 42B Sea King helicopters.

Contractor

Selex Sensors and Airborne Systems.

HIDAS Helicopter Integrated Defensive Aids System

Type

Airborne Electronic Counter Measures (ECM), integrated suite.

Description

Founded on the needs of the demanding attack helicopter role – where short engagement ranges demand rapid reaction times – HIDAS components are designed to operate together to provide radar, laser and missile threat warning, together with advanced radar and IR countermeasures and off-board decoys.

Component systems of HIDAS are as follows:
1. Radar Warning Receiver (RWR): Sky Guardian 2000 RWR from Selex Sensors and Airborne Systems which covers the E-J band, with optional frequency extensions downwards to C/D bands, and upwards to K band.
2. Laser Warning Receiver (LWR): Series 1223 LWR from Selex Sensors and Airborne Systems, with full beam-rider detection capability.

3. Common Missile Warning System (CMWS): AN/AAR-57(V) from Sanders, a Lockheed Martin Company; the AN/AAR-57(V) is the CMWS component of the ATIRCM/CMWS; each CMWS set provided for the HIDAS system will include an Electronic Control Unit (ECU), which processes threat data, and four passive Electro Optic (EO) missile sensors, being produced by Lockheed Martin Infra-Red Imaging Systems.
4. Radar Frequency (RF) jammer: Apollo from Selex Sensors and Airborne Systems, covering the H-J band, with optional extensions down to the E-G bands, and upwards to the K band.
5. Counter Measures Dispensing System (CMDS): Vicon 78 series 455 from Thales Optronics, comprising three dispensers, one for chaff and two for flares. The chaff will be fired upwards towards the tail rotor. Flares will be fired from dispensers on the rear fuselage. It is understood that HIDAS will employ the latest 'smart'/'special materials' IR decoy flares to improve IRCM capability.
6. Directional Infra-Red Counter Measures (DIRCM) from Selex (Nemesis programme) to provide IR jamming in bands I-IV.

To achieve the high degree of integration essential to timely reaction, the DAS management module is incorporated within the Sky Guardian 2000 RWR. This enables data from all the DAS sensors (RWR, LWR, CMWS) to be correlated against the large threat library. The DAS manager then determines and automatically responds with the optimum mix of radar, IR and off-board countermeasures. The DAS communicates with the aircraft mission system over a MIL-STD-1553B

databus and can display data or system status reports on the aircraft multifunction displays or on dedicated EW displays.

By combining and correlating data from all its sensors, HIDAS is able to present to the crew a comprehensive picture of the surrounding environment. Thus, crews have the option to take pre-emptive action; for instance, the HIDAS detection can be used to cue the on-board weapon system.

HIDAS is configured to ensure radar, laser and missile threats are detected, identified, declared to the crew and, where appropriate, countermeasures automatically instigated, in under two seconds – typically one second from threat detection.

HIDAS automatically selects the appropriate form or combination of radar jamming, IR jamming, chaff or flare dispensing. Where the crew require it, the countermeasures can be restricted to semi-automatic operations and a manual override is available at all times.

HIDAS is fully user reprogrammable at the flight line and provision is made for a mission library of up to 4,000 threat modes. Individual modes can be linked readily to enable the positive identification of weapon systems.

A 'smartcard' PCM port can be provided on the control unit to facilitate the loading of pre-flight messages. The large capacity of these cards allows the same device to be used for recording in-flight mission data. Software loading and recording functions can also be carried out by aircraft systems over the MIL-STD-1553B Bus.

The BAE Systems MERLIN is a PC-based software support system that enables the user to quickly create their own mission libraries, specify the required countermeasures and carry out post flight analysis of mission data.

Control of HIDAS is exercised either through a simple dedicated control unit with associated display, or through 'soft' controls on multifunction displays via the aircraft mission computer.

HIDAS supports both MIL-STD-1553B and RS-422 interfaces, and provides a combined situation picture and status report for display to the crew on the aircraft multipurpose display.

Status

In production and in service in the British Army's 67 WAH-64D Apache attack helicopters. Release To Service (RTS) for the British Army was granted in January 2001, following a successful series of HIDAS trials at the US Army's Proving Ground at Yuma, Arizona during 2000.

Elsewhere, HIDAS has been selected for 12 AH-64D helicopters on order for the Greek Army, with delivery scheduled for 2007, and 16 AH-64Ds on order for the Kuwaiti Air Force.

Contractor

Selex Sensors and Airborne Systems.

HIDAS helicopter integrated defensive aids system   0528524

A HIDAS-equipped WAH-64 Apache successfully completed trials of the system at the US Army Proving Grounds at Yuma, Arizona, during 2000 (GKN Westland Helicopters)   0073978

Kestrel ESM system

Type

Airborne Electronic Counter Measures (ECM), Electronic Support Measures (ESM) system.

Description

An ESM system in production for ELINT, surveillance and reconnaissance tasks, Kestrel has a near 100 per cent probability of intercept of any radar emission within its frequency range. Kestrel also utilises the latest processing techniques to give the operator virtually real-time information. In service with UK forces, it is known as Orange Reaper.

The system receives and processes radar emissions over 0.6 to 18 GHz. A six-port amplitude comparison system measures the bearing of the emissions and provides instantaneous digital information over 360° in azimuth. At the same time, a frequency measurement receiver provides an instantaneous frequency indication.

Digitised information on all threats, as accumulated pulse by pulse, is passed to the processor. In this equipment overlapping pulse trains from different radars are derived and their pulse repetition frequency and frequency agility characteristics determined. The digitised data is then passed into the main processor where long-term information is extracted. Radar identification is made by comparing measured and derived radiation parameters with those stored in a library of known emitters. Full information about the radar signal environment is then presented to the system operator on an ordered tabular or tactical display or, alternatively, the data can be transferred to remote displays using a standard data highway.

Orange Reaper is installed in the Westland/Agusta Royal Navy EH 101 Merlin helicopter and

has been supplied to Denmark for use in Lynx helicopters.

Specifications
Frequency: 0.6–18 GHz (optionally up to 40 GHz)
Frequency accuracy: 5 MHz RMS
Accuracy: ±3.5° RMS
Processing time: 1 s (max)
Receiver types: log video amplifier; digital log video amplifier; instantaneous frequency measuring; continuous wave superheterodyne
Coverage: 45° (elevation); 360° (azimuth)
Warning/identification: all pulse types including continuous wave, interrupted continuous wave, pulse-Doppler, 3-D low probability of intercept, jitter/stagger/agile, pulse compression; frequency-agile radars and unknown emitters
Emitter library storage: at least 2,000 modes (main library)
Display: high brightness, colour cathode ray tube with computer controlled, software programmable symbology
Emitter displays: up to 400 pages of 20 emitters per page, each showing tote format parametric data and emitter identity
Range bearing displays: 25 highest priority emitters showing lethality, type and track identification
Expanded track information: showing full parametric data and ranges for a single track
BITE: continuous monitoring of key system parameters; provision of `on request' fault isolation for more than 95 per cent of the system; provision of DOA quality check
Additional features: true or relative bearing display presentation; processor controlled audio alarms; operator initiated emitter hand-off; operator controlled 'specialised' computer modes; an ELINT data recording interface
Weight: 55 kg (display configuration dependent)
Power consumption: 1 kVA
Interfaces: serial and parallel interfaces for jammers, chaff/flare systems, displays, telemetry links and recorders
MTBF: over 800 h (demonstrated to MIL-STD-781C)
Environment: MIL-E-5400 Class II

Status
In production and operational on Lynx helicopters of the Royal Danish Navy, with Orange Reaper versions ordered for 44 EH 101 for the Royal Navy under a GBP30 million contract.

An export version of Kestrel, known as Kestrel II, is being evaluated by a number of countries.

Contractor
Thales Defence Ltd.

Thales' Kestrel ESM system is fitted to the Royal Navy EH 101 Merlin helicopter   0503832

Elements of the Kestrel ESM system   0009390

MIR-2 ESM system

Type

Airborne Electronic Counter Measures (ECM), Electronic Support Measures (ESM) system.

Description

The MIR-2 airborne ESM system, also known as Orange Crop, combines airborne warning system and search receiver functions using an advanced digital receiver, served by a fully solid-state wideband antenna system covering the 0.5 to 18 GHz radar frequencies. Six antennas monitor a full 360° in azimuth and provide a very high intercept probability, improving the crew's location and identification of friendly and enemy radars. The control-indicator unit, on the cockpit coaming, is a compact light-emitting diode display which indicates frequency band, amplitude and relative bearing on the main display. On an associated fine bearing unit the true bearing and radar PRF are indicated to a high degree of accuracy. The complete system of eight units comprises six antennas, a processor and the control-indicator unit.

A UK Royal Navy Westland Lynx is equipped with Thales' MIR-2 warning receiver   0503834

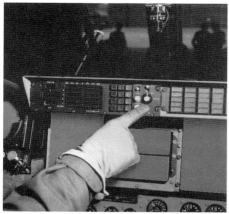

Thales' MIR-2 ESM cockpit control-indicator unit 0503833

Specifications
Weight: 47.7 kg
Frequency: 0.5–18 GHz
Pulsewidth: 0.15–10 μs
PRF: 0.1–10 kHz
Coverage: 360°

Status
MIR-2 is in service in helicopters of the Brazilian Navy (Lynx) and the UK Royal Navy (Lynx and Sea King). Over 300 systems are reported to have been supplied to customers worldwide.

Contractor
Thales Defence Ltd.

Operator Aiding SYStem (OASYS)

Type
Airborne Electronic Counter Measures (ECM), sonar system.

Description
The performance of the current generation of ASW sonar systems depends upon the successful fusion of the skills of a highly trained operator and the use of sophisticated computer equipment. A balance must be maintained between the need to present the operator with the widest choice of detailed information and the ability to view, comprehend and evaluate such large volumes of data.

The OASYS provides the acoustic system operator with a toolset of functions and graphical displays which enable the operator to make the most effective use of the acoustic processor and available sonobuoy stores. The OASYS includes a set of database functions for providing environmental, target, sonobuoy and processor information. By making use of all this information, OASYS can provide tools for sonobuoy stores management, tactical advice and processing recommendations. Although OASYS has been designed primarily as an operational system management aid, its portability and extensive toolset allows it be used as a mission planning tool within a ground-based PC environment.

Contractor
Thomson Marconi Sonar Limited.

PA7030 laser warning equipment

Type
Airborne Electronic Counter Measures (ECM), Laser Warning Receiver (LWR).

Description
The PA7030 laser warning equipment offers both instantaneous warning and bearing information on radiation by ruby and neodymium lasers. The system consists of a direct detector head unit and up to four indirect detectors mounted externally on the aircraft. A compact display and control unit is mounted in the cockpit.

The indirect detectors are sensitive to pulsed laser radiation scattered by the aircraft and so can detect designators and range-finders even when these do not directly illuminate the direct detector. Electronic filtering discriminates against glints and flashes to give an extremely low false alarm rate. The direct sensor gives instantaneous bearing on a radiation source to within 15°.

Specifications
Dimensions:
(PA7031 direct detector) 152 × 48 mm
(PA7032 scatter detector) 105 × 80 × 130 mm
(PA7034 signal conditioner) 260 × 160 × 87 mm
(PA7033 display) ¼ ATR short
Weight:
(direct detector) 2 kg
(scatter detector) 1.5 kg
(signal conditioner) 2 kg
(display) 2 kg
Power supply: 28 V DC, <12 W
Coverage:
(direct detector) 360° (azimuth), –45 to +10° (elevation)
(scatter detector) as required

Status
Development equipment has completed trials in helicopters and fixed-wing aircraft in the UK, USA and Canada.

Contractor
BAE Systems.

Prophet radar warning receiver

Type
Airborne Electronic Counter Measures (ECM), Radar Warning Receiver (RWR).

Description
Prophet is an operationally proven radar warning system designed to reduce combat aircraft vulnerability to radar associated threats on the battlefield. It provides the crew with timely and unambiguous threat warning, permitting appropriate countermeasures, such as manoeuvres or chaff discharge, to be employed. System design has taken account of the likely need for frequency extensions in order to increase the survivability of the battlefield helicopter.

The receiver detects signals over 2 to 18 GHz and, by means of a processor with a programmable threat library, alerts the pilot to imminent hostile action. The system is designed to operate in a dense RF environment. A four-port antenna system provides direction-finding, using signal amplitude comparison to derive bearing. The system has an extremely low false alarm rate and threats are recognisable with a high degree of confidence. An audible warning is provided by a tone generator and fed to the aircraft internal communications system.

The Prophet display uses light-emitting diodes and has three lines, each dedicated to a threat. Each line can show an arrow indicating threat direction and a three-character alphanumeric identifier. Display colour and brightness is such that it can be viewed in bright sunlight and through night vision goggles. Also provided, as an option for fixed-wing aircraft, is a full threat display on the HUD and multifunction displays.

The system can be readily reprogrammed and has high in-service reliability.

Specifications
Weight: 10 kg
Power supply: 28 V DC, 80 W
Frequency: 2–18 GHz (pulse and CW); options: millimetric wave and laser threat warning
Coverage: more than 100 radar modes
(azimuth) 360°
(elevation) ±30°
DF accuracy: 10° RMS
Detection time: <1 s

Thales' Prophet radar warning receiver 0503835

Status
In service with UK Royal Navy Sea King HC4s and British Army Gazelles and other air forces. Prophet is in production for the BAE Systems Hawk 100 and 200.

Contractor
Thales Defence Ltd.

PVS2000 Missile Approach Warner (MAW)

Type
Airborne Electronic Counter Measures (ECM), Missile Approach Warner (MAW).

Description
The PVS2000 is a radar-based Missile Approach Warning (MAW) equipment originally designed to equip the UK Royal Air Force Harrier aircraft. The MAW uses a low-powered pulse Doppler to detect approach of missiles, and automatically initiates countermeasures. The system is optimised to provide protection against air-to-air and ground-to-air heat-seeking missiles.

The MAW comprises: a transmitter/receiver unit, lightweight antenna, signal processing unit and cockpit control unit.

The Transmitter/Receiver Unit (TRU) contains a low-power solid-state transmitter and receiver, analogue processing and digital conversion circuitry. It is designed for installation in an unconditioned area of the aircraft tail within 1 m of the antenna.

The remote lightweight antenna is connected to the TRU, using low-loss Gore cable, or may be attached directly to the TRU. The antenna must have a clear view aft.

The Signal Processing Unit (SPU) houses the digital signal processing necessary for detection and assessment of signal returns. All interfaces with other aircraft avionics systems, BITE monitoring and module level diagnostics are handled by this unit. The SPU is installed in a conditioned avionics bay.

The optional Cockpit Control Unit (CCU) provides on/off control and incorporates a BITE status indicator lamp. A small auxiliary blade antenna is mounted on any suitable site with a relatively clear view forward.

The radar is mounted in the tail of the aircraft and should be linked to an expendables dispenser for the automatic deployment of flares and chaff.

Specifications
Dimensions:
(transmitter/receiver unit) 156 (length) × 250 mm (diameter)
(signal processing unit) ½ ATR short
(optional cockpit control unit) 38 × 146 × 75 mm
(main antenna/radome) 100 (length) × 260 mm (diameter)
(auxiliary antenna) 150 × 100 × 260 mm
Weight:
(transmitter/receiver unit) 5.5 kg
(signal processing unit) 7 kg
(cockpit control unit) 1 kg
(main antenna/radome) 0.7 kg
(auxiliary antenna) 0.25 kg

Status
In service. The Eurofighter Typhoon Defensive Aids Sub System (DASS) incorporates Missile Approach Warners derived from the PVS2000, one each in the port and starboard wing roots

PVS2000 family of equipment　　　0504255

(near the cockpit) and one in the rear fuselage (near the tail). In this application, the MAW is directly linked to the flare launchers allowing an instantaneous response to a local launch.

Contractor
Selex Sensors and Airborne Systems.

R605 sonobuoy receiving set

Type
Airborne Electronic Counter Measures (ECM), sonar system.

Description
The R605 sonobuoy receiver is in use on the Royal Navy EH 101 Merlin ASW helicopter. The receiver amplifies and demodulates four simultaneous FM sonobuoy transmissions and provides the output to the acoustic processor for analysis and display. It provides coverage of 99 RF channels per receiver module and is mechanically interchangeable and connector compatible as a direct replacement for the AN/AAR-75.

For applications where the receiving sets must be mounted at a distance from the receiving antenna, a remote preamplifier is provided. MIL-STD-1553B control is available. Control functions include RF channel selection, sonobuoy type for optimal bandwidth control, RF level reporting and self-test initiation and reporting.

The R605 receiver features a large RF dynamic range and high sensitivity over a broad base bandwidth. In addition, the receiver provides a wide, linear phase, audio output with close uniformity

The Ultra R605 sonobuoy receiver　　　0504110

between each of the four audio channels. Digital sonobuoy data undergoes matched filter detection and clock regeneration in the receiver. Differentially driven clock and data signals are available for each of the four receiver outputs at any of the data rates and code functions.

Specifications
Dimensions: $159 \times 351 \times 232$ mm
Weight: 10 kg
Power supply: 115 V AC, 400 Hz, 3 phase, 100 VA max
Frequency: 136–173.5 MHz
Channels: 99 RF, 4 acoustic
Interfaces: MIL-STD-1553, RS-422, ARINC 429
Reliability: >2,000 h MTBF

Status
In serial production. In service on UK Royal Air Force Nimrod MR.2 Mk 2 and Royal Australian Air Force P-3C Orion aircraft, and selected for the UK Royal Navy EH 101 Merlin helicopter.

Contractor
Ultra Electronics, Sonar and Communication Systems.

Radar Homing and Warning Receiver (RHWR)

Type
Airborne Electronic Counter Measures (ECM), Radar Warning Receiver (RWR).

Description
The Radar Homing and Warning Receiver (RHWR) is produced in two versions: an amplitude comparison version for strike aircraft and a high-accuracy version for air defence aircraft. Both versions are effective against radar threats likely to be encountered and are combat proven. Frequency coverage is available from C- to J-bands. RHWR can detect and identify threats up to and beyond the radar horizon. The information obtained is correlated with the aircraft interception radar tracks to allow for the classification and allocation of target priorities. High-accuracy DF is obtained by using interferometers. RHWR uses a powerful digital processor, which accepts the parametric data from the receivers, including frequency, PRI, pulsewidth and scan characteristics. The information is compared with a comprehensive emitter library. The identified emitters are passed to the cockpit displays which provide the crew with both tabular and graphical presentation of threats, as well as the overall radar scenario.

The RHWR was upgraded by the then BAE Systems (now Selex Sensors and Airborne Systems) to meet the requirements of UK MoD Air Staff Requirement 907. Increased processing power and additional memory was added and

Front and rear hemisphere RHWR aerials mounted on the fin of a UK RAF Tornado GR4 (E L Downs)　　　0109929

capability for the RHWR to exchange data with the Sky Shadow ECM pod was provided. The Sky Shadow system was also upgraded under the same programme. While each system retains the ability to operate independently, under normal conditions the RHWR acts as the defensive aids system controller of an integrated system.

Status
In service in the Tornado GR4 strike attack aircraft of the UK Royal Air Force, Tornado F Mk 3 fighter aircraft of the UK Royal Air Force and Italian Air Force and with the Tornado ADV of the Royal Saudi Air Force.

Contractor
Selex Sensors and Airborne Systems.

Shrike TG Digital Radio Frequency Memory (DRFM) Techniques Generator (TG)

Type
Airborne Electronic Counter Measures (ECM), radar jamming system.

Description
Thales Defence Ltd has been contracted by Lockheed Martin Fairchild Systems to provide its Shrike DRFM TG for integration into the Nimrod MRA4 Defensive Aids SubSystem (DASS).

System architecture provides for the Lockheed Martin Fairchild Systems AN/ALR-56M(V) RWR to cue the DRFM TG, which is then used to drive the Fibre-Optic Towed Decoy (FOTD) version of the Raytheon Systems Company AN/ALE-50 towed decoy system.

The DRFM TG has been developed by Thales in its work on advanced jamming system technology in association with QinetiQ (formerly the Defence and Evaluation Research Agency). It is understood that a twin-DRFM installation is to be used to meet the high data throughput specification for Nimrod MRA4 and ASTOR (see separate entry) systems.

Status
All 21 Nimrod MRA4 aircraft are due to be fitted with the system. The Shrike DRFM TG has also been selected for the UK RAF ASTOR aircraft.

Contractor
Thales Defence Ltd.

Sky Guardian 2000 Radar Warning Receiver (RWR)

Type
Airborne Electronic Counter Measures (ECM), Radar Warning Receiver (RWR).

Description
Sky Guardian 2000 is configured for rapid response in a high-density electromagnetic environment and has the processing capability to display multiple threat signals in less than a second. The system operates either as a stand-alone radar warner or integrated with an ECM system such as the Apollo radar jammer or other countermeasures. Sky Guardian 2000 will accept data from the Type 1220 laser warning receiver and display the threats on a common indicator bearing unit.

Sky Guardian will provide long-range detection of airborne and surface threat radars, emitter identification and built-in recording for post-flight analysis. Other key features are display persistence for fleeting intercepts, low-band targeting options and the ability to integrate into a multifunction cockpit control and display. The system is extendable to meet future threats.

Sky Guardian 2000 employs advanced signal processing algorithms and provides high sensitivity as well as accurate RF measurement. An emitter library of 4,000 emitter descriptions can be loaded by a PCMCIA smart card

The Sky Guardian 2000 radar warning receiver   0504143

incorporated into the control unit. Sky Guardian 2000 has been designed from the outset as the core of a Defensive Aids System (DAS) and includes DAS control functions.

When configured for the Electronic Support Measures (ESM) mission, Sky Guardian 2000 is designated as Sky Guardian 2500.

Specifications
Weight: 13 kg
Frequency: E- to J-bands (C/D- and K-bands optional)
Frequency measurement: pulse, pulse Doppler, CW and ICW
Coverage: 360° instantaneous
Accuracy: better than 10°
Pulse density: >1 Mpps
Polarisation: dual-polarisation option
Emitter library: over 4,000 emitter modes

Status
Selected for UK Royal Air Force Merlin HC.3 support helicopters. Sky Guardian 2000 also forms part of the Helicopter Integrated Defensive Aids Suite (HIDAS) for the British Army WAH-64D Apache Longbow helicopter.

Sky Guardian 2500 is installed aboard the Malaysian Navy's six Super Lynx 300 anti-surface warfare helicopters.

Contractor
Selex Sensors and Airborne Systems.

Sky Shadow ECM pod

Type
Airborne Electronic Counter Measures (ECM), radar jamming system.

Description
The ARI 23246/1 Sky Shadow ECM pod was developed for use with the UK Royal Air Force Panavia Tornado.

A high-power travelling wave tube amplifier is used with a dual-mode capability for deceptive and continuous wave jamming. A voltage-controlled oscillator uses a varactor-tuned Gunn diode to cover the full frequency band. The set-on receiver is an original Racal (now Thales) design, and signal processing uses Selex Sensors and Airborne Systems hardware.

The pod incorporates both active and passive electronic warfare systems and includes an integral transmitter/receiver, processor and cooling system. It has radomes at both ends and is stated to be capable of countering multiple ground and air threats, including surveillance, missile and airborne radars. Power management is automatic and modular construction allows

the system to be adapted to differing operational missions.

An enhanced version of Sky Shadow has been developed. This features a high-performance data processor running on Ada software. The range of jamming activities has been extended, with greater flexibility for front line reprogramming.

Sky Shadow has been upgraded, with the Tornado Radar Homing and Warning Receiver (RHWR), as part of the UK MoD Air Staff Requirement 907. The improved Sky Shadow is able to exchange data with the RHWR. When operating in the linked mode, RHWR acts as a Defensive Aids System (DAS) controller and both equipments offer the synergistic benefits of an integrated system.

Specifications
Dimensions: 3,350 (length) × 380 mm (diameter)

Status
BAE Systems (now Selex) has upgraded the Sky Shadow ECM pod for the UK Tornado GR4 aircraft. The main feature of the upgrade was the intercommunication between the Skyshadow pod and the Tornado ARI 18241/2 RHWR. Thus modified, the two systems can operate as a federated suite, thereby improving their overall operational effectiveness.

The upgraded Sky Shadow system is in service in UK RAF Tornado GR4 aircraft.

Contractor
Selex Sensors and Airborne Systems.

A Royal Air Force Tornado carrying the ARI 23246/1 Sky Shadow ECM pod on each outer pylon   0503831

TMS 2000 series acoustic processing system

Type
Airborne Electronic Counter Measures (ECM), sonar system.

Description
The TMS 2000 is an advanced and powerful airborne acoustic processor. It uses a cost-effective and totally open architecture based on rugged, state-of-the-art, off-the-shelf hardware. The TMS 2000 uses existing and proven algorithms which have been developed as part of the AQS 900 and SADANG series of processors. They have been rehosted and enhanced with new colour broadband, data fusion, and multistatic processing techniques.

The system is designed for use by one to four operators, with each operator having access to all the processed data as graphical data fusion plots, localisation plots, acoustic formats or tabular information. The system can support multiple high-resolution colour monitors.

The MMI makes extensive use of the Geographic Energy Plot (GEP) to provide results from data fusion. This GEP can be superimposed on an acoustic localisation plot (sonobuoy position, coastlines and sea bottoms) facilitating operator efficiency.

All current and projected NATO sonobuoy types are processed and sonobuoy data may be received from either traditional analogue receivers or from new digital receivers.

System control uses the latest techniques of pull-down or pop-up menus, augmented by on-screen editing and fast dedicated keys.

The system can be easily interfaced to other avionics systems via any industry standard or MIL-STD-1553B interface.

Passive functions
The TMS 2000 provides a range of advanced processing detection modes, which cover a full range of target noise characteristics including narrowband frequency lines, broadband signals, swaths, transients and DEMON signals. Each channel has two main processes, plus up to six auxiliary processes which are tuned to the target characteristics and environmental conditions. Computer assistance is provided to link database information to observed signal data to aid target classification. DOUBLE DEMON processing ensures full coverage of both the carrier noise bandwidth and modulating spectrum to achieve permanent detection of transient cavitation phenomena.

Active functions
The TMS 2000 has the capability to process up to 16 DICASS sonobuoys simultaneously in the multistatic mode as well as the monostatic mode, and so satisfies typical brown shallow water operational requirements. Typical coverage is 200 square nautical miles with 16 DICASS sonobuoys.

Specifications
Sonobuoy capability: 8, 16, or 32 sonobuoy channels; compatible with all NATO sonobuoys
Passive processing: up to 64 DIFAR per unit; two main and up to six auxiliary processes for each channel; manual/automatic noise nulling; cardioid processing; computer-assisted classification
Active processing: 16 DICASS multistatic/monostatic modes; ping-to-ping integration; auto-handling of pulse sequences; reverberation suppression and stationary artefact removal; new-generation LF projectors (LFAS, LODICASS)
Displays: passive and active, with colour enhancement; passive and active geographic energy plots

Status
In production. TMS 2000 has been proposed the MPA-R and French Navy Atlantique 2 mid-life upgrade in a 64-channel version. A 16-channel version has been selected for the French Navy NH 90. Ordered by the Spanish Air Force and UAE Navy.

Contractor
Thales Underwater Systems Ltd.

Type 453 laser warning receiver

Type
Airborne Electronic Counter Measures (ECM), Laser Warning Receiver (LWR).

Description
The Type 453 laser warning receiver is designed for use in fixed-wing aircraft, helicopters and armoured fighting vehicles. It provides a countermeasure to laser-guided weapon systems and provides an audible alarm and visual display showing the direction from which the threat has originated. Output from the receiver can be combined with radar warning equipment to give an integrated battlefield threat warning system or with a defensive aids system.

The system uses a number of dispersed sensors to detect incident laser radiation. This counters the problem of detecting a very narrow beam which at any instant would be illuminating only a small part of the airframe. The sensor head configuration can be tailored to provide spherical or hemispherical cover and presents the laser threat bearing to the crew as sectoral information. The number of sensors required is tailored to the aircraft type. The sensor protrudes through the aircraft skin as a 25 mm hemisphere and occupies a space of 50 mm depth behind the skin.

The sensor heads are completely passive. Laser radiation is routed to the central processing unit by fibre optic cables. This feature eliminates risks of false alarms being generated by radio frequency interference.

Status
In production and in service in British and Spanish Eurofighter Typhoons, where a single sensor is mounted below the windscreen.

Contractor
Selex Sensors and Airborne Systems.

Vicon 70 countermeasures pod

Type
Airborne Electronic Counter Measures (ECM), Counter Measures Dispensing System (CMDS).

Description
Vicon 70 is a modular, lightweight dispensing pod which uses components of the Vicon 78 Series 455 system to provide self-protection against radar threats and heat-seeking missiles. It is designed for helicopters and fixed-wing aircraft where structural or other constraints preclude the use of airframe-mounted dispensers.

Vicon 70 can be configured for two dispensers, or for up to eight dispensers by adding modules to the pod. Modules can also be used to house radar warning receiver and missile approach warner LRUs. It can be fitted to any fixed-wing aircraft or helicopter type equipped with weapons racks or fuselage or wing pylons with 14 in NATO attachments. The firing trajectory of the dispenser modules can be preset to optimise the effectiveness of the chaff and flares. Manual or fully automatic versions of Vicon 70 are available. The automatic version interfaces directly with the radar warning receiver via a serial datalink or databus.

Unlike the internal CMDS installation, the podded solution requires no major airframe modification and provides a greatly enhanced chaff dispensing capability. The additional benefit of the pod is that countermeasures dispensing may be carried as role equipment and can be downloaded should self-protection not be required.

Specifications
Dimensions: 2,705 length × 356 mm diameter
Weight: 210 kg fully loaded with 8 dispensers

Status
In production and in service.

Contractor
Thales Optronics (Vinten) Ltd.

Vicon 78 airborne decoy dispensing systems

Type
Airborne Electronic Counter Measures (ECM), Counter Measures Dispensing System (CMDS).

Description
Vicon 78 is a family of advanced lightweight chaff and IR decoy dispensing equipment in production for high-performance combat, tactical transport and maritime patrol aircraft,

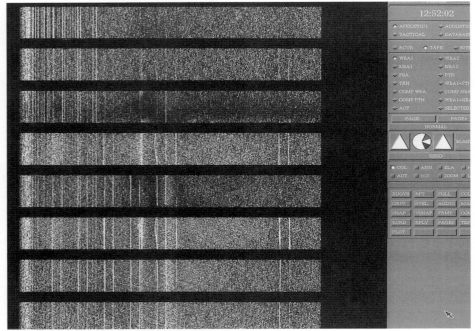

TMS 2000 series acoustic processing system (typical passive display)   0001187

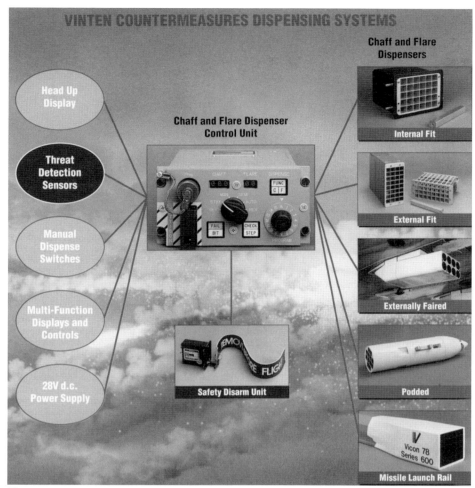

Vinten countermeasures dispensing systems   0001268

helicopters and UAVs. Vicon 78 systems can be fitted as original equipment or upgrade existing systems. It is a defensive aids system which provides self-protection by passive ECM against radar-guided and IR-seeking air- and ground-launched missiles and radar-directed anti-aircraft artillery fire.

Vicon 78 systems provide manual or fully automatic threat adaptive capability. Fully automatic operation, greatly reducing aircrew workload, is achieved by interfacing the dispensing system to a threat detection sensor. In automatic mode the programme to be dispensed is downloaded from the Radar Warning Receiver (RWR) or Missile Approach Warner (MAW). In semi-automatic mode the programme to be dispensed is downloaded from the RWR and initiated by the crew. In manual mode the crew can dispense preset chaff or IR flare programmes.

Systems can be offered with multidispenser configurations to cater for large transport and maritime patrol aircraft down to a single lightweight dispenser for smaller aircraft, each of which can accept interchangeable chaff or flare magazines containing up to 64 cartridges. Dispensers can also be configured to dispense

Rear view of a Hawk showing the Vicon 78 Series 300 countermeasures dispenser   0504146

expendable jammers, towed decoys and other format expendables as required. The dispensers may be internally mounted, semi-recessed with fairings or carried in a lightweight pod.

Vicon 78 Series 200

Vicon 78 Series 200 equipment has been supplied for the Sea Harrier programme. The Series 200 CounterMeasures Dispensing System (CMDS) is a two-dispenser system, controlled by a salvo control processor unit and manually operated by the pilot. Vicon 78 Series 200 equipment has successfully undergone EMC testing at Boscombe Down and has been selected by the UK MoD. The Series 205 CMDS has the ability to dispense ALE-39 and ALE-40 format rounds from a common dispenser.

Vicon 78 Series 203 equipment was developed and supplied to CASA of Spain for an export CASA 101 programme. It is a four-dispenser system controlled by a two-box processor with a fully automatic mode of operation via an interface to the aircraft Litton AN/ALR-80(V) RWR. The Series 203 system also equips the CASA CN-235 tactical transport aircraft, interfacing to the Litton AN/ALR-85 RWR.

The Vicon 78 Series 210 was selected by the Royal Air Force for operations by the Tornado F3 in the Gulf War. It utilises a new format dispenser that fires a 55 mm two shot infra-red decoy flare. After the 1990-91 Gulf War, the Series 210 was incorporated as a fully approved modification on the Tornado F3 aircraft.

Vicon 78 Series 300

The Vicon 78 Series 300 equipment was developed for the BAE Systems Hawk 100 and 200 Series aircraft. In the Hawk, the Series 300 CMDS is a digitally controlled, fully automatic two-dispenser system that interfaces via an

RS-422 link to the Racal Prophet or Marconi Sky Guardian RWRs. The Series 300 system has been designed with the capability to control up to 16 dispensers, with each dispenser having the capability to be configured to hold up to 64 chaff or flare cartridges. Each dispenser can accept interchangeable chaff and flare magazines. When power is applied, the system will identify the number and type of expendables in each dispenser and display the count of chaff and flare cartridges on the cockpit control unit.

A continuous BIT facility is executed by the system; in the event of failure, fail indications are displayed on the cockpit control unit. The BIT is also designed as a first line maintenance facility to identify faults and so speed rectification.

The Vicon 78 Series 300 CMDS has been designed to interface with any RWR or ESM system that has a serial datalink interface capability. The RWR/CMDS interface has been successfully achieved with Litton, BAE Systems, Avionics and Racal. Agreements are also in place with other RWR manufacturers. Vicon 78 CMDSs have the ability to interface to MAWs and missile launch detectors.

Vicon 78 Series 400

The Vicon Series 400 incorporates the advanced electronics of the Series 300, which have been improved and miniaturised, with the eight cartridge 16 shot 55 mm dispenser format of the Series 210. The system utilises a single miniature LRU Chaff and Flare Dispenser Control Unit (CFDCU), which can be either cockpit- or remote-mounted, and provides the system control, processing and programme loading functions.

The Vicon 78 Series 400 has been selected for a fleetwide upgrade of Royal Air Force Tornado F Mk 3 aircraft.

Vicon 78 Series 420

The Series 420 is based on the 400 system electronics, with the CounterMeasures Dispensers (CMD) being reconfigured to a 1 × 1 in payload, 8 × format suitable for helicopter platforms. Each 8 × 4 CMD can carry 32 expendables and the system can be configured for up to four dispensers. The 420 system is currently equipping British Army Air Corps Lynx and Gazelle helicopters.

Vicon 78 Series 455

The Series 455 CMDS is the latest advanced lightweight version of the VICON 78 series suitable for helicopters, fast jet, Maritime Patrol Aircraft (MPA) and transport aircraft. Due to its low mass and reduced number of LRUs, the Series 455 CMDS is ideal for retrofit and upgrade programmes, while allowing ease of integration into new build aircraft. The system can control up to 24 countermeasures dispensers and up to two BOL chaff dispensers. Several dispenser formats are available, including variants that are interchangeable with ALE-40/-47 LRUs. A wide selection of payload options is supported including standard, dual-shot and modular expendable block decoys. In addition, all Series 455 dispensers are able to accommodate up to six different types of payload simultaneously. The system can be expanded to suit large,

The Vicon 78 Series 400 is fitted to UK and Italian Tornado F Mk3 aircraft   0504147

For details of the latest updates to *Jane's Avionics* online and to discover the additional information available exclusively to online subscribers please visit
jav.janes.com

The VICON 78 Series 455 countermeasures dispensing system selected for the UK WAH-64 Apache HIDAS
0062752

A UK Royal Air Force Canberra PR9 reconnaissance aircraft dispenses flares from its VICON 78 CMDS
0521403

multidispenser installations by the addition of further dispensers without the need for sequencer LRUs. The VICON 78 Series 455 has been selected as the CounterMeasures Dispensing System (CMDS) element of the UK WAH-64 Apache attack Helicopter Integrated Defensive Aids System (HIDAS), and forms part of the Nimrod MRA4 defensive system. It has also been configured for Hawk 100/200 upgrades.

Vicon 78 Series 500

The Vicon 78 Series 500 lightweight stand-alone dispenser utilises the electronics of the Vicon 78 Series 300 system reconfigured to provide a simple and cost-effective dispensing solution for light aircraft, helicopters and RPVs. The 12-shot dispenser has a simple mechanical and electrical interface and can be operated from the cockpit or, in the case of an RPV, via an RF link. The system can be configured to interface to an MAW.

Vicon 78 Series 600

The Vicon 78 Series 600 is a chaff and flare dispenser which can be mounted on the end of a missile launch rail. The system allows additional countermeasures to be carried without requiring major airframe modifications and with only a small weight and drag penalty.

Status

Vicon 78 dispenser applications are as follows: Vicon 78 Series 200 UK Royal Navy Sea Harrier aircraft; Vicon 78 Series 203 export-CASA 101 (interfacing with AN/ALR-80(V) RWR) and CN235 (interfacing with the AN/ALR-85 RWR) programmes; Vicon 78 Series 210 fully approved modification for Royal Air Force Tornado interceptors post 1990/1991; Vicon 78 Series 300 all export Hawk training/light attack aircraft; Vicon 78 Series 400 UK and Italian Tornado F Mk 3 aircraft; Vicon 78 Series 420 British Army Air Corps Gazelle and Lynx helicopters; Vicon 78 Series 455 Royal Air Force Canberra PR9, Nimrod MRA4, ASTOR, export Hawk training/light attack aircraft, Aerovodochody L159 light multirole combat aircraft, export C-130 aircraft, Wedgetail AEW&C aircraft, CN-235 and several rotary-wing aircraft including the UK WAH-64 Apache, Agusta A109, Super Lynx, Oryx, Cougar and Mi-24.

Contractor

Thales Optronics (Vinten) Ltd.

Zeus ECM system

Type

Airborne Electronic Counter Measures (ECM), integrated suite.

Description

The ARI 23333/1 Zeus is a compact multipurpose EW integrated defensive aids system for combat aircraft. It provides a complete defensive system consisting of a radar warning receiver and a jammer. The RWR is the latest in a series of passive intercept systems and is based on IFM and fast superheterodyne techniques. It is able to intercept and measure the characteristics of all radar-controlled systems which may threaten the aircraft. The RWR will measure parameters including direction of arrival, time of arrival, frequency, PRI, pulsewidth, amplitude, scan interval and scan rate. Signals are passed to the Zeus digital processor which identifies radar type, displays it and gives audio warning. The processor also controls the jammer and other means of countermeasures such as chaff, flares and decoys. The transmitter will jam both pulse and CW radars, while control features ensure that home-on-jam radars are not given sufficient time to lock on to the aircraft. Different jamming modes are available with Zeus for the operational requirements of individual customers.

The Zeus internal ECM system
0504144

Specifications

Weight: 118 kg
Frequency: C- to J-bands (0.5–20 GHz)
Coverage:
(azimuth) 360°
(elevation) ±45°
Response: 1 s typically
Jamming: noise, VGPO, deception, co-operative

Status

Zeus is in service with the UK Royal Air Force on the Harrier GR. Mk 7 aircraft and was employed during that type's deployment to Bahrain during Operation TELIC in 2003. Installations have been designed for other aircraft, including the JAS 39 Gripen, AV-8B, F-16 and F/A-18. Zeus has also undertaken technical and operational evaluations in France and the US.

As of 2005, Selex was supporting Zeus and is understood to have tested new hardware (including a techniques generator and a digital radio frequency memory) as potential upgrades for the system.

Contractor

Selex Sensors and Airborne Systems.

United States

AAR-58 Missile Approach Warning System (MAWS)

Type
Airborne Electronic Counter Measures (ECM), Missile Approach Warner (MAW).

Description
The AN/AAR-58 passive airborne Missile Approach Warning System (MAWS) is a multicolour, scanning, infra-red system, designed to overcome the deficiencies of ultra-violet warning systems in sunlight at high altitudes. The system features a part-hemispherical search volume, automatically warning the aircrew of missile approach and vector, combined with the capability to command countermeasures deployment.

Major system features include:
- Multicolour infra-red detection
- Track-while-search processing
- Multiple threat processing
- Countermeasures discrimination (target colour, intensity, size, trajectory)
- False-alarm discrimination (terrain reflection, solar radiation).

The system is able to detect all current air-air missiles, with fast reaction and accurate angular resolution, enabling cueing of laser Infra-Red CounterMeasures (IRCM) systems, such as the Northrop Grumman AN/ AAQ-24(V) DIRCM Nemesis system (see separate entry).

The AN/AAR-58 may be internally or pod mounted and fully integrated as part of a Defensive Aids SubSystem (DASS). It also can be installed into the lower gondola of Raytheon's ALQ-184 ECM pod.

Specifications
Dimensions
  Baseline: 101.6 × 152.4 × 330 mm plus 152.4 mm dome
  Alternative: 101.6 × 152.4 × 406.4 mm
Weight: 9.1 kg
Sensor coverage:
  Azimuth: 360°
  Elevation: ±135°
Angular resolution: <1°
Power supply: 115 V AC, 400 Hz, 1 A; 28 V DC, 1.1 A
Temperature range: –54 to +71°C
Altitude: up to 45,000 ft
Interface: MIL-STD-1553B, RS-422, RS-232
MTBF: >500 h

Contractor
Raytheon Company, Space and Airborne Systems.
CMC Electronics Cincinnati.

Advanced Air Delivered Sensor (AADS) programme

Type
Airborne Electronic Counter Measures (ECM), Electronic Intelligence (ELINT) system.

Description
The Advanced Air Delivered Sensor (AADS) programme, derived from the Advanced Remote Ground Unattended Sensor (ARGUS) programme, is intended to address the rapid detection, location and engagement of mobile surface targets on the battlefield. As demonstrated during the recent operations in Iraq, despite the advances made in Unmanned Air Vehicles (UAVs), manned intelligence aircraft and sophisticated surveillance satellites, there is still a significant delay in the detection and tracking of fleeing vehicles such as tanks and surface-to-surface missile launchers.

To address this shortfall, the US Marine Corps (USMC) began the AADS programme. The USMC intends to field a family of autonomously operating sensors, which could be positioned in places such as road/rail chokepoints and bridges, to monitor time-sensitive targets and send the data collected back to an air operations centre in near-realtime.

AADS is intended to augment less persistent sensors by acting as a 'tip-off' sensor, informing field commanders of initial movement and providing guidance as to where to concentrate reconnaissance effort.

AADS units will be powered internally and capable of self-testing, two-way communication and data storage.

The USMC will deploy AADS on their AV-8B ground support aircraft. Block upgrades could include modules that sense seismic, acoustic, magnetic, thermal, electromagnetic, electro-optical, radio, radiological, chemical and biological signatures.

The AADS programme builds on the Defense Intelligence Agency's Unattended Ground Sensor (UGS) advanced concept technology demonstration; during this initiative, the agency demonstrated a 'Steel Rattler' hand-emplaced unit and a 'Steel Eagle' airdropped variant. The USAF and US Special Operations Command assisted in the project.

The US Army also has an interest in affordable, expendable, unattended ground sensors as part of its Future Combat Systems (FCS) programme.

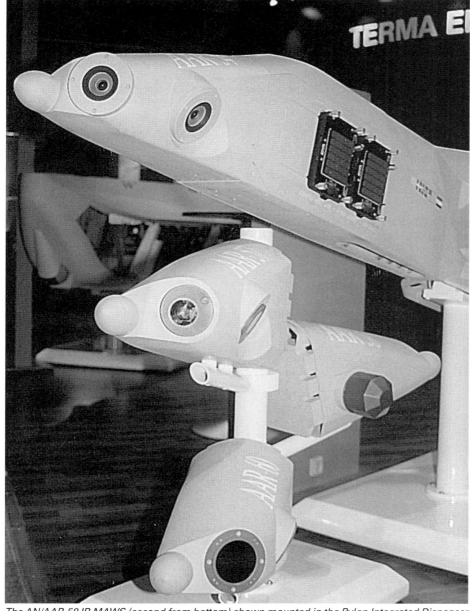

The AN/AAR-58 IR MAWS (second from bottom) shown mounted in the Pylon Integrated Dispenser System (PIDS)   0083807

The AADS system is based on an earlier 'Steel Eagle' demonstration effort. The Steel Eagle sensor is visible, mounted to the left forward intake station of the aircraft   0530630

A close-up of the 'Steel Eagle' sensor   0530631

Status

Textron Systems has won a USD19.2 million contract from the USMC to develop the AADS system. Low-Rate Initial Production (LRIP) will begin in 2006.

Contractor

Textron Systems.

Advanced Digital Dispenser System

Type

Airborne Electronic Counter Measures (ECM), Counter Measures Dispensing System (CMDS).

Description

The Advanced Digital Dispenser System (ADDS) is a microprocessor, computer-controlled, threat adaptive countermeasures dispensing system providing aircraft self-protection against current and future ground and airborne threats. ADDS utilises user programmable dispensing programmes, adapted for the specific threat and optimum engagement parameters, to control and dispense chaff, including dual chaff cartridges, flares, RF and future types of offboard expendable decoys.

With simple software modifications, aircraft configurations can be tailored for high-performance aircraft, helicopters and transport and maritime patrol aircraft. ADDS incorporates improvements derived from combat experience and hundreds of installations on a wide variety of aircraft including the A-4, F-4, F-5, F-15, F-16 and others.

ADDS features include automatic, semi-automatic and manual operational modes; payload inventory; discrete inputs from aircraft MIL-STD-1553B and RS-422 databusses to interface with onboard RWR, MLD and TWS warning sensors; solid-state sequencer switch providing multiple firing pulses with low burst intervals and automatic and crew initiated BIT.

Status

In production for many air forces. Installed as standard equipment on a number of aircraft types.

Contractor

BAE Systems North America.

AN/AAR-44 advanced Missile Warning System

Type

Airborne Electronic Counter Measures (ECM), Missile Approach Warner (MAW).

Description

The AN/AAR-44A is the latest version of the passive AAR-44 airborne IR Missile Warning System (MWS), with a claimed detection range in excess of any current MWS, able to determine an incoming missile's angle of arrival to less than 1°, sufficient for threat hand-off to a Directional Infra-Red CounterMeasures (DIRCM) system. The AAR-44 detects and tracks the threat missile's IR energy to provide warning. The system continuously searches the Field of View (FoV) to detect, declare and track incoming missiles. It warns the aircrew of missile position and automatically controls countermeasures to neutralise threats and enhance survivability.

The system provides long-range, multithreat detection while continuously tracking IR clutter. It is effective over the complete flight envelope and provides hemispheric coverage with one sensor unit or spherical coverage with two sensors.

Other system features include visual and audio threat warning to aircrew, countermeasures command, multispectral discrimination, full system operation with or without databus integration, a DIRCM interface, robust scanning technology and the ability to be reprogrammed on the flight line.

Design features include automatic warning, track-while-search processing and the ability to

track multiple missile threats simultaneously. Multidiscrimination modes against solar radiation, terrain and water reflection eliminate false alarms. MIL-STD-1553B bus interface capability is standard.

Previously, L-3 Cincinnati Electronics had teamed with Raytheon to develop the AN/AAR-44(V), which was later re-designated AN/AAR-58. Several prototypes were made but the system never entered production.

Specifications
(AN/AAR-44A)

Type: Scanning IR MWS
Warning: Visual, audio and external countermeasures command
Dimensions
  Control and display: 102 × 142 × 150 mm
  Processor: 200 × 212 × 252 mm
  Sensor: 400 × 369 mm diameter
Weight
  Overall system: 27.97 kg
  Control and display: 1.13 kg
  Processor: 8.27 kg
  Sensor: 18.57 kg
Altitude: to 50,000 ft
Coverage (per sensor head): ±135 × 360° (elevation × azimuth)
Cueing accuracy (AOA): better than 1°
Multiple threats: unlimited
Reliability: over 500 h
Interfaces: MIL-STD-1553B
Temperature: –54 to +90°C (storage); –54 to +71°C (operating)
Power: 115 V AC, 400 Hz

Status

The AN/AAR-44A completed USAF operational testing in 1997. Now in production and in US Air Force service in AC-130H/U and MC-130E/H versions of the Hercules and the HH-53 'Super Jolly' helicopter.

As of August 2003, the US Air Force Materiel Command was looking at the use of Artificial Intelligence (AI) algorithms to reduce false alarm performance of the AN/AAR-44A, which can then be incorporated into the latest operational flight programme.

Contractor

L-3 Cincinnati Electronics.
Raytheon Electronic Systems.

AN/AAR-44 missile warning receiver

Type

Airborne Electronic Counter Measures (ECM), Missile Approach Warner (MAW).

Description

The AAR-44 infra-red passive airborne warning receiver continually searches a hemisphere while tracking and verifying missile launches. It warns the crew of missile position and automatically controls countermeasures to neutralise threats and enhance survival.

Design features of the equipment include: automatic pilot warning and countermeasures command, continuous track-while-search processing, multiple missile threat capability and countermeasures discrimination capability. Multidiscrimination modes against solar radiation and terrain and waer reflection, minimise false alarms. MIL-STD-1553B bus interface capability is incorporated. A number of sensor unit configurations are available to fulfil specific field of view requirements.

An improved version, designated AN/AAR-44A, which is reported to incorporate upgraded componentry to enhance performance and reliability, offers similar performance to the AAR-44, but with a lesser power requirement (155 W). The AAR-44A also includes a DIRCM interface.

Specifications
System
Temperature: –54 to +71°C
Altitude: Up to 50,000 ft

Control and display unit
Dimensions: 102 × 142 × 150 mm
Weight: 1.12 kg

Processor
Dimensions: 200 × 212 × 252 mm
Weight: 8.82 kg

Sensor
Dimensions: 400 × 369 mm (Ø) (each sensor)
Weight: 17.91 kg
Displays: sector threat indicators; external countermeasures command and audio tone.

Status

The AN/AAR-44 is in service onboard US Air Force C-130 aircraft. AAR-44 equipment operated by the US Air Force's Special Operations Command (SOCOM) is understood to have been the subject of a processor upgrade programme. The AAR-44 is also reported to be in Belgian Air Force service.

The AN/AAR-44A is understood to have completed flight test and entered service in 2000/2001.

Contractor

BAE Systems North America, Mason.

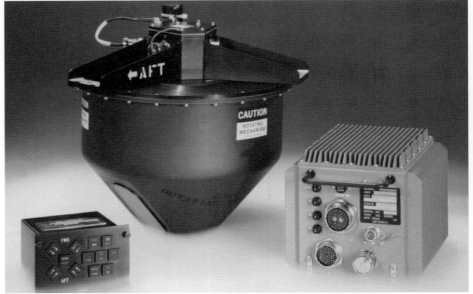

AN/AAR-44 missile warning system with (left to right) control display unit, conical detector head and processor

0503839

AN/AAR-47 Missile Warning Set (MWS)

Type
Airborne Electronic Counter Measures (ECM), Missile Approach Warner.

Description
The AN/AAR-47 Missile Warning Set (MWS) detects the plume of approaching surface-to-air missiles and gives the pilot an indication of range and bearing. Decoy systems, such as the AN/ALE-39 flare/chaff dispenser, can be automatically triggered and the system is also compatible with the AN/APR-39A radar warning receiver.

The feasibility of such a warning system was first demonstrated by Loral in December 1977 and, following development of the AAR-46 system in 1979, full-scale development of the AAR-47 started in March 1983. The AAR-47 is small, needs no cryogenic cooling and has an MTBF of 1,500 hours.

The system consists of six sensors, a central processor and a control indicator. The sensors are based on ultra-violet technology. They are hard-mounted on the skin of the aircraft and provide full coverage, with overlap to protect against blanking. The complete system weighs only 18.2 kg. Within 30 seconds of electrical power a self-initiated BIT programme is completed and the system is operational with no in-flight down time for recalibration. The processor analyses the data from each sensor independently and as a group, and automatically deploys the appropriate countermeasures. In the case of a flare failure, the AAR-47 automatically commands a second flare.

Multidirectional threats are automatically analysed and prioritised for countermeasures sequencing. The control indicator displays the incoming direction of the highest priority threat for tactical manoeuvres.

A shorter sensor has been developed for use in locations, such as the F-16 fuselage, where depth is limited.

AN/AAR-47 technology is used as the basis of the Common Missile Warning System (CMWS) element of the Advanced Threat Infra-Red CounterMeasures (ATIRCM) and AN/AAR-57 Common Missile Warning System (CMWS – see separate entry) being developed by BAE Systems North America, Information and Electronic Warfare Systems (IEWS) (formerly Sanders).

Specifications
Dimensions:
(sensor) 120 × 200 mm
(processor) 203 × 257 × 204 mm

Weight:
(4-sensor system) 14 kg
(sensor) each 1.5 kg
(processor) 7.9 kg
Power supply:
(4-sensor system) 28 V DC, 75 W
(4 W per sensor, 59 W for processor)
Coverage: 360° azimuth (given by 6 sensors)

Status
Many companies have been responsible for the production and further development of the AN/AAR-47 MWS. Both Alliant Techsystems and Sanders are manufacturing the core system, with both sources utilised by the US DoD. At present, after the acquisition of Sanders by BAE Systems North America, that company has now become BAE Systems North America, Information and Electronic Warfare Systems. To date, over 1,500 AAR-47 systems have been manufactured and installed on helicopters and fixed-wing aircraft operated by Australia, Canada, UK and USA.

Alliant also continues to supply the baseline AN/AAR-47 system and has developed upgraded variants, designated AN/AAR-47(V)1 for the basic upgrade, and AN/AAR-47(V)2 for a Laser Warning (LW)-capable system (see separate entry).

Contractor
BAE Systems North America, Information and Electronic Warfare Systems (AN/AAR-47)

AN/AAR-47 Missile Warning System (MWS)

Type
Airborne Electronic Counter Measures (ECM), Missile Approach Warner (MAW).

Description
The AN/AAR-47 missile warning system is a small lightweight passive electro-optic threat warning device used to detect surface-to-air missiles fired at helicopters and low-flying fixed wing aircraft, provide audio and visual-sector warning messages to the aircrew and automatically allocate countermeasures.

The basic system consists of multiple Optical Sensor Converter (OSC) units, a Computer Processor (CP) and a Control Indicator (CI). A set of four to six OSC units are mounted on the aircraft exterior to provide omnidirectional protection. The OSC units detect the rocket plume of missiles and send appropriate signals to the CP for processing. The CP analyses data from each OSC and automatically deploys the

The AN/AAR-47 missile warning system 0514977

appropriate countermeasures. The CP also contains comprehensive BIT circuitry. The CI displays the incoming direction of the threat, so that the pilot can take appropriate action.

The basic AN/AAR-47 system was first fielded in the late 1980s and has undergone extensive upgrade to ensure that supportability is maintained out to the 2010-2015 timeframe. The upgrade improves the original system performance by increasing sensor sensitivity, reducing recovery time, and extending the overall temperature range. The upgrade involves an approximately 80 percent overhaul of the system: upgraded sensors, CI and CP.

A Laser Warning (LW)-capable variant of the system provides detection and warning of laser guided and laser aided threats. This version integrates the LW into the whole system, rather than simply grafting it onto the basic arrangement.

The basic upgrade is designated the AN/AAR-47(V)1, while the LW-capable units are designated the AN/AAR-47(V)2.

Specifications
Dimensions:
  OSC: 120 × 200 mm
  CP: 203 × 257 × 204 mm
Weight:
  OSC: 7.9 kg
  CP: 1.5 kg (each)
Power supply: 28 V DC; 4 W (per OSC), 59 W (CP)
Angular coverage: 360° (six OSC array)

Status
In excess of 2,500 sets of the AN/AAR-47 Basic system have been ordered for the US Navy, US Air Force, US Army and several foreign customers. Alliant Integrated Defense Company (formerly Hercules) was appointed as second source for AN/AAR-47 basic system and the prime contractor for the AN/AAR-47(V)1 and AN/AAR-47(V)2 systems.

The AN/AAR-47 missile warning set   0503855

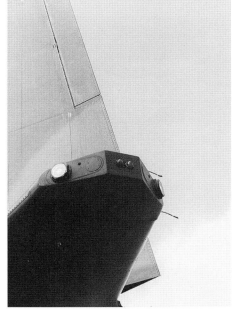
Twin aft-facing AN/AAR-47 OSCs mounted on the tail of a C-130 tactical transport aircraft (Tim Ripley)   0515242

BAE Systems North America (formerly Sanders) also supplies a version of the AN/AAR-47 to the US military and foreign customers (see separate entry).

Contractor
Alliant Integrated Defense Company.

AN/AAR-54(V) Missile Warning System (MWS)

Type
Airborne Electronic Counter Measures (ECM), Missile Approach Warner (MAW).

Description
The AN/AAR-54(V) Missile Warning System (MWS) can be applied to most airborne platforms, including helicopters, fighter and transport aircraft. The system can also be used to provide identification, missile tracking information and target cueing to Directed Infra-Red Counter Measures (DIRCM) systems. A major benefit of the inherent adaptive design of the AN/AAR-54(V) is that all applications are able to employ common hardware and software.

The fine 1° angle of arrival discrimination capability of the AAR-54(V) contributes to a greatly reduced false alarm rate as well as detection ranges nearly double that of existing ultra-violet systems. The system features outstanding passive time-to-intercept, clutter rejection, long range and short shot missile detection, rapid automatic cueing for Counter Measures Dispensing Systems (CMDS) and enhanced Situational Awareness (SA) via dedicated warning display, aircraft MultiFunction Display (MFD) or HUD.

The system consists of Wide Field of View (WFoV), high-resolution ultra-violet sensors and a modular electronics unit. One to six sensors can be utilised, providing up to full spherical coverage. Full in-flight Built-In Test (BIT) and fault isolation to a single sensor or electronics Line-Replaceable Unit (LRU) provides level-2 maintenance.

Specifications
Sensor FoV: 120° (full spherical coverage achieved with six sensors)
Dimensions (L × W × H)
  Sensor: 112 × 87 × 87 mm
  Electronics unit: 216 × 160 × 96.5 mm
Weight
  Sensor: 1.75 kg
  Electronics unit: 4.8 kg
  Power
  Sensor: 8 W (approx 76 W per sensor during anti-ice)

The AN/AAR-54(V) Passive Missile Approach Warning System (PMAWS) consists of up to six sensors and an electronic unit   0504264

Electronics unit: 34 W
Temperature: –54 to +71°C
MTBF
  Sensor: 57,804 h
  Electronics unit: 7,128 h
MTTR: 30 min
Interface: MIL-STD-1553B, RS-422, discrete

Status
During 2001, Northrop Grumman, the US DoD and the UK Ministry of Defence completed AN/AAR-54(V) design verification testing. The trials programme included over 300 live fire demonstrations with the warner installed on QF-4 drones, a cable car test rig and ground vehicles, and over 100 hours of clutter flight testing and analysis. The system has also been integrated into and demonstrated with a DIRCM system and the AN/ALQ-131 electronic measures system (see separate entries). In April 1995, the UK Ministry of Defence and the US Special Operations Command awarded Northrop Grumman a USD35 million Engineering, Manufacturing and Development (EMD) contract for the AN/AAQ-24(V)/ARI 18246 DIRCM system (see separate entry), which includes the AAR-54(V). In this application, the equipment provides missile detection, lethal missile declarations and fine angle of arrival hand-off to the DIRCM. The AN/AAQ-24(V)/ARI 18246 is being installed on over 15 types of UK and US fixed-wing aircraft and helicopters.

Outside the AN/AAQ-24 application, the AN/AAR-54(V) has been procured by the Australian Navy (SH-2G(A)) and the air forces of Australia (Boeing 737 'Wedgetail' AEW +C aircraft), Denmark (F-16 Mid Life Update (MLU)), Germany (C-160 transport aircraft), the Netherlands (AS 532U2, CH-47D and F-16 MLU), Portugal (C-130H Hercules), Norway (F-16 MLU) and Japan (H-60). Of these, the Netherlands' helicopter application involves 13 CH-47D and 17 AS 532U2 aircraft, while

the Portuguese installation combines a six-sensor AN/AAR-54(V) configuration with the AN/ALE-40 Countermeasures Dispensing System (CMDS – see separate entry) and TERMA's Electronic Warfare Management System (EWMS) to create a full Defensive Aids Sub System (DASS). The F-16 MLU application integrates six AN/AAR-54(V) sensors into TERMA's Pylon Integrated Dispenser System Plus (PIDS +) (see separate entry).

The AN/AAR-54(V) is in full-rate production at greater than 70 sensors per month for CMDS, DIRCM and avionics suite integration programmes.

Contractor
Northrop Grumman Corporation, Electronic Systems Sector, Defensive Systems Division.

AN/AAR-57 Common Missile Warning System (CMWS)

Type
Airborne Electronic Counter Measures (ECM), Missile Approach Warner.

Description
The AN/AAR-57 CMWS provides missile warning for rotary-wing, transport and tactical aircraft in all military services. It is claimed to have a very low false alarm rate and to exceed NATO staff requirements for a missile warning system in this respect. It is also able to detect and declare missile identity before missile burnout. A very high detection-processing speed and self-contained image-stabilisation assembly ensure high performance on manoeuvrable fighter aircraft.

The CMWS consists of an Electronics Control Unit (ECU) developed by Sanders (now BAE Systems, following acquisition in 1999) and up to six Electro-Optical Multiple Sensors (EOMS), designed and developed by Lockheed Martin Infra-red Imaging Systems (LMIRIS) of Lexington, Massachusetts. The ECU processes threat data and provides cueing for expendables and directable infra-red jammer elements of the Advanced Threat Infrared Counter Measures (ATIRCM) system. The EOMS employs ultra-violet detector technology, to detect radiation from the plume of the attacking missile.

Specifications
Weight:
(AN/AAR-57, including one ECU and six EOMS) 14.44 kg
(EOMS): 1.18 kg
(ECU): 7.36 kg
Power:
(AN/AAR-57, including one ECU and six EOMS) 449 W (353 W without anti-icing)
(EOMS): 12 W without anti-icing, 28 W with anti-icing
(ECU): 281 W

Status
The US Army, PEO Aviation, is managing the tri-service CMWS programme as part of the ATIRCM programme, in Redstone Arsenal, Alabama.

In August and September 1999, the US Army and Sanders conducted 12 flight tests of the CMWS system (with a single EOMS) aboard a US Army UH-60 Blackhawk helicopter. The system was exposed to both simulated and actual industrial and military ultra-violet sources of radiation at varying altitudes in varying weather conditions.

Customers include the US Army, Navy, Marine Corps and Air Force for A-10, Apache, Blackhawk and C-5 Galaxy. Prospective US military customers for the CMWS include the C17, C-130, C-141, Chinook, CV-22, F-15, F-16, Kiowa and RC-135. Other national armed forces service programmes, include the UK WAH-64 Apache and UK Nimrod MRA4.

Contractor
BAE Systems North America, Information and Electronic Warfare Systems.

AN/AAR-54(V) installation on the Pylon Integrated Dispenser System (PIDS +) of an F-16 MLU aircraft   0051266

AN/ALE-43 chaff cutter/ dispenser pod

Type
Airborne Electronic Counter Measures (ECM), Counter Measures Dispensing System (CMDS).

Description
The AN/ALE-43 is a high-capacity chaff system which holds rolls of chaff material and cuts it to the appropriate dipole length during operation. Lundy claims to have overcome the difficulties associated with past chaff cutters by developing a unique chaff roving supply system.

The system has a chaff roving hopper behind which is the chaff cutter. A remote processor controls operation of the cutter.

Chaff is drawn simultaneously from up to nine chaff roving packages in the hopper. Each roving package passes through a guide tube which terminates at draw rollers and a cutting roller. As dipole lengths are cut, they are discharged into a turbulent airflow and distributed efficiently behind the pod. Each cutter assembly consists of a drive motor, clutch/brake unit and three indexing cutting rollers, the latter embodying blades which yield specific combinations of dipole lengths.

The system can be podded or mounted internally. It is frequently used as a training aid, but its primary wartime functions would be for anti-ship missile defence, area saturation operations over battlefields, corridor seeding and aircraft self-protection.

Specifications
Podded version
Dimensions: 3,370 (length) × 480 mm (diameter)
Weight:
(empty) 139 kg
(loaded) 284 kg
Chaff payload: 8 × RR-179 roving packages
Max dispensing rate: 7.2×10^6 dipole in/s

Internal version
Weight:
(basic) 37 kg
(max loaded) 191 kg
Chaff payload: 9 × RR-179 roving packages
Max dispensing rate: 8.1×10^6 dipole in/s

Common characteristics
Power supply: 115 V AC, 400 Hz, 1.7 kVA
28 V DC, 2.5 A
Max continuous dispense time: 11 min
On-time: select 1–9 s in 1 s steps or continuous
Off-time: select 1–9 s in 1 s steps
Altitude: 0–50,000 ft

Status
In service. The podded version can fit a variety of standard pylon attachments and is used on the B-52, F-4 Phantom, EA-4A Skyhawk and EA-6B Prowler. Internally mounted versions have been used on the ERA-3B Skywarrior and NKC-135.

Contractor
Alliant Defense Electronics Systems Company.

AN/ALE-43(V)1/(V)3 chaff countermeasures dispenser set

Type
Airborne Electronic Counter Measures (ECM), Counter Measures Dispensing System (CMDS).

Description
The AN/ALE-43(V)1/(V)3 chaff countermeasures dispensing set is designed for two types of installation: as a pod (mounting per MIL-A-8591 both 14 and 30 in mounting), or internal mount. In the pod installation, the chaff cutter assembly is mounted on a structural bulkhead at the rear of the pod centre section. This assembly consists of a drive motor, clutch/brake unit, cutting mechanism and supports. The cutting mechanism comprises a rubber platen roller and three cutter rollers.

Specifications
AN/ALE-43(V)1 Pod Version
Dimensions: 3,370 (length) × 482 mm (diameter)
Weight (empty): 154 kg
Power requirements: 115/208 V, 400 Hz, 3-phase 1,700 VA and 28 V DC, 2.5 A
Chaff supply: standard RR-179/AL roving bundle 18 kg
Chaff payload: 144 kg (8 bundles)
Dispensing rate: 216 g/s per 8 roving bundles
Modes (2):
(pulse) 1-9 s in 1 s intervals
(continuous) 660 s

AN/ALE-43(V)3 Internal Version
Dimensions: 408 × 338 × 310 mm
Weight (empty): 36 kg
Power requirements: 115/208 V, 400 Hz, 3-phase 1700 VA and 28 V DC, 2.5 A
Chaff supply: standard RR-179/AL or special roving bundle
Chaff payload: up to 225 kg (up to 9 roving bundles)
Dispensing rate: 27 g/s per roving bundle
Modes (2):
(pulse) 1-9 s in 1 s intervals
(continuous) 880 s

Status
The AN/ALE-43(V) pod version has been proven on SH-3 helicopter, EA-6A, EA-7A, F-4, F-16, F-18, P-3, Learjet 36 and other aircraft. Internal installation versions are operational on such aircraft as: EW Falcon, EW Challenger, ERA-3B, NKC-135 and LearJet 35.

Contractor
Alliant Defense Electronics Systems Company.

AN/ALE-47 Threat Adaptive Countermeasures Dispenser System (TACDS)

Type
Airborne Electronic Counter Measures (ECM), Counter Measures Dispensing System (CMDS).

Description
The AN/ALE-47 TACDS provides an integrated, threat-adaptive, reprogrammable, computer-controlled capability for dispensing expendable decoys (such as the ALE-39, ALE-40 or M-130) to enhance aircraft survivability in sophisticated threat environments. The threat-adaptive features of the TACDS are provided by the customer-defined Table-Look-Up Mission Data File (MDF). The TACDS Operational Flight Program (OFP) derives all (fixed and adaptive) dispensing programmes based entirely upon the customer-developed MDF. The system interfaces with other onboard EW and avionics systems through two MIL-STD-1553 databusses, two RS-422 busses or discrete lines.

The AN/ALE-47 comprises the following major elements:

A Cockpit Control Unit (CCU), which provides aircrew/system interface, inventory/built-in test indication and system mode/inhibit functions

An F-16-compatible dispenser/magazine assembly provides payload housing/firing, magazine-to-sequencer identification and 'smart' payload interface functions. The dispenser accepts a total of five different magazine types and up to 32 such assemblies can be accommodated within a single system/host platform

An F-18-compatible dispenser/magazine assembly, essentially a different form/fit variant of the F-16 dispenser/magazine

A helicopter dispenser/magazine assembly. As above

An optional programmer unit that provides threat adaptive processing, threat evaluation, dispensing strategy, built-in test and payload management and CMDS-Electronic Warfare (EW) system/air data computer link functions. Can be supplied as a VME card.

The following operating modes are available: automatic, semi-automatic, manual and bypass. Mode selection is performed by the crew using a Cockpit Display Unit (CDU). An emergency jettison mode is available to dispense all expendables on command. In automatic, the CMDS evaluates the threat data from on-board EW sensors and integrates it with stored threat information to determine the optimum countermeasures dispense program; the system dispanses expendables without operator action. In semi-automatic mode, the CMDS determines the dispense program as in the automatic mode, but signals the operator to manually activate the dispense. In manual mode, the operator manually selects and depresses one of six pre-determined programs. In bypass mode, the operator manually dispenses expendables in degraded mode after a partial system failure by reconfiguring the system in flight.

The AN/ALE-47 can accept customised dispenser/magazine assemblies (external conformal, special payload form factor and/or special internal shape) and is claimed to be compatible with next-generation expendable payloads. The system can be integrated with a wide range of EW systems, including the AN/ALR-47, -56M, -66, -67, -69 and -93 RWRs, AN/AAR-47, -54, -57 and -60 MWR and AN/ALQ-126, -136, -165 and -167 airborne jammers.

Other features of the system include: Built-In Test (BIT) to the Shop Repairable Unit (SRU);

AN/ALE-47(V) TACDS   0051271

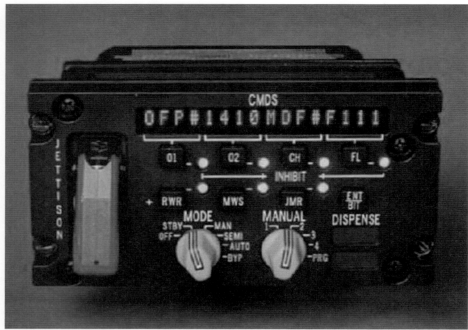

AN/ALE-47 helicopter Cockpit Control Unit (CCU)   0051272

MDFs customised to the customer's threat environments; Memory Loader Verifier (MLV) for loading MDFs into the system; and a Maintenance Test Station (MTS) to provide depot level repair capability.

Specifications
Dimensions:
CCU 953 × 146 × 173 mm
F-16 dispenser/magazine 216 × 170 × 256 mm
F-18 dispenser/magazine 250 × 258 × 167 mm
Helicopter dispenser/magazine 328 × 227 × 118 mm
Programmer unit 953 × 146 × 159 mm
Weight
CCU <1.82 kg
F-16 dispenser/magazine <2.27 kg (empty); 10.16
    12.34 kg (full)
F-18 dispenser/magazine <2.27 kg (empty); 10.85
    12.78 kg (full)
Helicopter dispenser/magazine <2.27 kg (empty);
    10.16–12.34 (full)
Programmer unit <2.27 kg
Power
CCU 23 W
Programmer unit 20 W

Status
Currently in production and in service in a wide variety of fighter, transport and rotary-wing aircraft of US and international air forces. Well over 2,000 systems have been delivered to date.

Contractor
BAE Systems North America.

a Shop Replaceable Assembly (SRA); hardware/ software growth provisions; reprogrammable manual dispense routines; ALE-40 form-fit compatibility and ALE-39 Sequencer form-fit compatibility. Support equipment available for the TACDS includes the Tracor Mission Planning System/Threat Matrix Generator (TRACOMPS/ TMG) that permits efficient development of

---

AN/ALE-50 Towed Decoy System (TDS)

Type
Airborne Electronic Counter Measures (ECM), Towed Radar Decoy (TRD).

Description
The ALE-50 TDS provides protection against RF threats. When deployed, the decoy seduces RF guided missiles away from the host aircraft. This stand-alone system requires no threat-specific software and communicates health and status to its host aircraft over a standard databus.

The ALE-50 TDS consists of a launch controller, launcher and towed decoy. The decoy control/ monitor electronics and power supply are contained in the launch controller. The launcher, containing a magazine that holds, typically, three decoys, can be customised to fit any candidate aircraft. Decoys have a 10-year shelf life and are packaged in a sealed canister, which also contains the pay-out reel.

The AN/ALE-50 launcher is also compatible with BAE Systems' AN/ALE-55 Fibre-Optic Towed Decoy (FOTD).

Status
The ALE-50 programme was originally a joint US Navy/US Air Force/Raytheon Electronic Systems development. The MultiPlatform Launch Controller (MPLC) is the standard launch controller for all installations.

US Air Force F-16 aircraft were the first operational users, for which decoy production began in early 1997. The F-16 installation employs a specially adapted AMRAAM pylon, manufactured by Lockheed Martin Tactical Aircraft Systems. Numerous contracts placed between 1998 and 2002 have covered expanded F-16 fleet usage (US Air Force and Air National Guard units) and operational employment of the system in tactical air operations over Serbia and Kosovo.

In 1998, Raytheon was awarded a contract for 50 launchers and 150 magazines for the Rockwell B-1B bomber for operations over Kosovo in 1999. In 2002, after problems encountered with the AN/ALE-55 FOTD, part of the Defensive Systems Upgrade Programme (DSUP), the USAF announced a sole-source contract to the company

to upgrade the ALE-50 to an FOTD equivalent and to demonstrate integration on both the B-1B and the F-15 in place of the ALE-55.

Elsewhere, the ALE-50 is operational on US Navy F-18C/D and the service has acquired a small number of systems for test and contingency use on the F/A-18E/F.

An installation has been developed for integration with the AN/ALQ-184 ECM pod, designated AN/ALQ-184(V)9. The US Air Force

ALE-50 pylon on Block 50 F-16 (small underwing pylon outboard of larger wing weapons pylon)
0051267

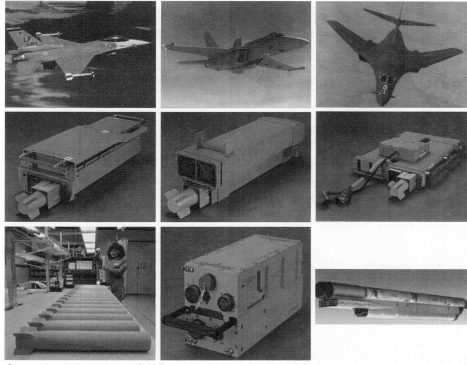

Composite 9-view photo of ALE-50 installation, showing: F-16 Launch/Launch Controller (left); F/A-18E/FT-3 Launcher (centre); B-1B 1 × 4 Launcher (right); MultiPlatform Launch Controller (bottom centre); ALQ-184(V)9 pod with 1 × 4 launcher (bottom right); and ALE-50 production configuration (bottom left)

0018259

approved full-rate production of the modified AN/ALR-184(V)9 version in October 1999.

A fibre-optic derivative of the ALE-50 is also being developed for the UK Royal Air Force Nimrod MRA. Mk 4 maritime patrol aircraft, where it will be linked to the AN/ALR-56M RWR for threat cueing and to a techniques generator being developed by Thales, based on DRFM technology.

Most recently, the US Air Force awarded Raytheon a USD32.5 million contract to provide 1,249 decoys for Air Force and Navy aircraft. The completion of the Lot 9 programme will bring the total number of AN/ALE-50 towed decoys produced by the company to 22,495.

Contractor
Raytheon Company, Space and Airborne Systems.

AN/ALE-55 Fibre-Optic Towed Decoy (FOTD)

Type
Airborne Electronic Counter Measures (ECM), Towed Radar Decoy (TRD).

Description
The ALE-55 FOTD, a component of the joint US Air Force/US Navy Integrated Defensive Electronic CounterMeasures (IDECM) Radio Frequency CounterMeasures (RFCM) system, is a new 'smart' Towed Radar Decoy (TRD).

After deployment, the FOTD is towed behind the host aircraft on a length of fibre-optic cable; its fins adjust to changes in the local airflow to minimise the stress on the fibre optic line during flight and aircraft manoeuvre. The system comprises a number of Electronic Countermeasures (ECM) systems and electronic emitters. Data for these systems is transmitted down the cable from the host aircraft's organic ECM systems to 'lure away' homing missiles, thereby increasing miss distance to a value grater than the lethal radius of the missile warhead.

The AN/ALE-50 Multiple Platform Launch Controller (MPLC) and launcher magazine is compatible with the ALE-55 FOTD.

Status
The AN/ALE-55 FOTD is under development as part of the IDECM RFCM programme.

During flight tests at the US Naval Air Station, Patuxent River, on a Boeing F/A-18E/F fighter, the decoy had suffered from problems caused by multiple exposures of the fibre optic towline to the afterburner plume from the aircraft engines, which resulted in damage to or destruction of the line. After modification, however, the redesigned line operated successfully for more than three times the required duration in the course of combat-representative flight manoeuvres.

Further flight tests on board a US Air Force B-1B bomber, including the first use of the decoy at supersonic speeds, were successfully completed at Edwards Air Force Base late in 2002. The tests demonstrated decoy deployment, safe separation and in-tow performance, while exploring the extremes of the decoy's flight envelope, which included towing decoys at supersonic speeds while performing multiple wing sweeps. It was reported that decoys were successfully deployed from the aircraft on four different occasions and all maintained towline integrity until commanded to sever. Signal line continuity results varied, with maximum sustained signal continuity for approximately 45 minutes. Preliminary results indicated good decoy stability in the supersonic region.

Flight trials continuing.

BAE Systems' AN/ALE-55 Fibre-Optic Towed Decoy (FOTD)   0515382

Contractor
BAE Systems North America.

AN/ALQ-99 Tactical Jamming System (TJS)

Type
Airborne Electronic Counter Measures (ECM), radar jamming system.

Description
The AN/ALQ-99 is a large and sophisticated ECM jammer designed for the SEAD Suppression of Enemy Air Defences role for the US Navy EA-6B and US Air Force EF-111A aircraft.

The system has been operational for over 20 years and has undergone numerous upgrades to meet changing operational requirements and to take advantage of technical advances. During this time, it has progressed through nomenclature changes up to the current configuration, which is believed to be known as AN/ALQ-99F(V).

The system was originally designed in two configurations – as a full internal fit for the EF-111A and as up to five external pods (4 underwing and one underfuselage) for the EA-6B. Since the EF-111A has now been withdrawn from US Air Force service, the EA-6B/AN/ALQ-99 combination is being further upgraded to meet current and near-term US Navy and the US Air Force SEAD requirements.

The EA-6B configuration comprises either two or three operator positions, from which specialist crew members exercise automatic/semi-automatic/manual control over the System Integrated Receiver (SIR) situated in the aircraft tail fin, and the underwing/underfuselage pods. Each pod includes transmitter elements and high-gain electronically steerable antennas (in both nose and tail), together with a ram-air turbine to provide aircraft-independent electrical power.

The ALQ-99 was designed from the outset to meet specific threat requirements with very high effective radiated power. To meet this design aim, the pod is fitted with high-gain, directional antennas, and specific limited bandwidth transmitters. The original version only covered four bands, but additional bands have been added through the years to meet new operational requirements, with coverage in the latest versions from VHF up to NATO I/J-band (approximately 30 MHz to 10 GHz) in separate frequency-specific bands, designated from band 1 up to band 9/10.

Recent upgrades to the ALQ-99 pod have been integrated with EA-6B aircraft upgrades. In 1976, ICAP 1 (Increased CAPability) improvements were introduced to reduce processing time, add new displays, and improve communication, navigation and IFF functions. Then, in 1982, ADVCAP (ADVanced CAPability) was implemented to update the OR-262 receiver/processor and AN/ALR-73 passive detection system (see separate entry).

In the 1980s, ICAP 2 configurations, designated -82, -86 and -89 for the years of their introduction, were implemented. Current ICAP 2 Prowlers carry five integrally powered pods with 10 very high-power jamming transmitters. Each pod contains an exciter that optimises the jamming signals for transmission by two powerful transmitters. The ICAP 2 Prowler carries exciters that generate signals in any of seven frequency bands. Each pod can jam in different bands simultaneously. In addition, the Prowler can carry any mix of pods, depending on mission requirements.

The latest defined configuration is designated ICAP 2 Block 89A, flown for the first time in June 1997. ICAP 2 Block 89A is understood to include the following upgrades to the ALQ-99 system: an upgrade to the universal exciter, low- and high-band transmitters; and the AN/USQ-113(V)2 Phase III communications jammer. It is also understood to include the following avionics upgrades:

A US Navy EA-6B aircraft fitted with AN/ALQ-99 pods on both outer wing stations and under the belly. Note the ram air turbines mounted on the nose of each pod   0085270

new radios (AN/ARC-210); an embedded inertial navigation system; a global positioning system, as well as hardware and software enhancements to the navy's standard mission computer AN/AYK-14; new instrument landing system and COTS electronic flight instrumentation system.

In January 1998, Tracor Aerospace Electronic Systems Inc (now part of BAE Systems North America) was awarded a USD60 million contract to manufacture 120 AN/ALQ-99 Band 9/10 transmitters for the US Navy Air Systems Command (with options for 61 additional transmitters and spares). A further contract for the fabrication of 84 Band 9/10 transmitters was issued in 1999, valued at USD42 million.

Having achieved fleet commonality with the ICAP Block 89A configuration in 2001, this standard will serve as the baseline for the new ICAP 3. Under development by a consortium of Northrop Grumman (prime), Litton (a wholly owned Northrop Grumman subsidiary), PRB Associates and BAE Systems North America, the ICAP 3 configuration incorporates new band 1/2/3 and band 9/10 transmitters, an enhanced universal exciter, a 'reactive' SIR group (the Litton LR-700 equipment with a top-end frequency of 18 to 20 GHz), full integration of the TJS with the AN/USQ-113(V) communications jammer (see separate entry), new operator graphical interfaces, improved connectivity and situational awareness (displays and tactical datalink) and a 'selective reactive' narrow band jamming capability for use against frequency-agile threats. ICAP 3 is also understood to include the requirement for the wiring and interface for the MIDS (Multifunction Information Distribution System – see separate entry) to enable the EA-6B to transmit data to, and receive from, other US military airborne and ground locators, as well as NATO platforms. The first flight of one of the two prototype ICAP 3 aircraft was made in November 2001. ICAP 3 Initial Operational Capability (IOC) is scheduled

for 2004. EA-6B aircraft are expected to remain in operational service until at least 2015.

A potential replacement for the EA-6B Prowler, the Boeing EA-18 ('Growler'), also equipped with 3 AN/ALQ-99 jamming pods, made its first flight in November 2001 as well. Boeing claims that, with 99 per cent airframe commonality with the F-18E/F Super Hornet and integrating Prowler ICAP 3 electronics, the EA-18 could begin a System Development and Demonstration phase during 2003.

Contractor
Northrop Grumman Corporation, Integrated Systems Sector

While the major contractor for the AN/ALQ-99 has been Northrop Grumman, many others have been involved through the years. Participants include:

AIL Systems Inc.
BAE Systems, North America.
COMPTEK Amherst Systems Inc.
Lockheed Martin Systems Integration – Owego.
Northrop Grumman Corporation, Advanced Systems Division.
PRB Associates.

AN/ALQ-126B Deception Electronic Counter Measures (DECM) system

Type
Airborne Electronic Counter Measures (ECM), radar jamming system.

Description
The AN/ALQ-126 is a DECM system developed by the US Naval Air Systems Command for its Tactical Air Electronic Warfare Programme. It provides wider frequency coverage than the preceding AN/ALQ-100 system and was initiated in response to new threats from anti-aircraft, surface-to-air missile and airborne interception systems. The initial production model (ALQ-126A) was widely used by the US Navy aboard aircraft such as the A-4, A-6, A-7 and F-4 aircraft.

An improved version, the ALQ-126B, offers increased frequency coverage, and incorporates a digital instantaneous frequency measurement receiver, improved deception techniques and more modem construction, packaging and cooling arrangements. The ALQ-126B includes a distributed microprocessor control system to enable the system to be reprogrammed on the flight line and improve signal processing. Covering the 2 to 18 GHz band frequency band, the system is capable of generating a variety of jamming modulations including inverse conical

scanning, range gate pull off, swept square wave and main lobe blanking.

The ALQ-126B operates either autonomously or as part of an integrated weapon system comprising the AN/ALR-45F, APR-43 or ALR-67 radar warning equipment, the AN/ALQ-162 continuous wave radar jamming system, and the HARM, Sparrow, Phoenix and AIM-120 missiles. The ALQ-126B is compatible with the A-4, A-6, A-7, F-4, F-14 and F/A-18 aircraft. The ALQ-126B also forms part of the ALQ-164 ECM pod (see separate entry).

Specifications
ALQ-126B
Dimensions: 411 × 270 × 609 mm
Weight: 86.3 kg

Status
The US Navy awarded BAE Systems North America contracts to the value of nearly USD500 million for the production of the ALQ-126B for its F/A-18 aircraft. The equipment is also installed on Australian F/A-18, Kuwaiti F/A-18, Malaysian F/A-18, Canadian CF-18 and Spanish EF-18 aircraft. Deliveries to these countries are complete.

Contractor
BAE Systems North America, Information and Electronic Warfare Systems.

AN/ALQ-128 threat warning receiver

Type
Airborne Electronic CounterMeasures (ECM), Radar Warning Receiver (RWR).

Description
The AN/ALQ-128 is the standard threat warning receiver on the US Air Force's F-15 aircraft, forming part of the Tactical Electronic Warning System (TEWS) with the ALR-56, ALE-45 and ALQ-135 systems. The ALQ-128 has been in production since 1980. Little is known about this system's performance; it possibly provides coverage of the higher-frequency bands above the J-band limit of the ALR-56.

Status
In service on the US Air Force F-15.

Contractor
Raytheon Company, Space and Airborne Systems.

AN/ALQ-131 noise/deception jamming pod

Type
Airborne Electronic Counter Measures (ECM), radar jamming system.

Description
The AN/ALQ-131 is an automatic, highly reliable modular self-protection system. It was designed to provide advanced broadband coverage against all types of modern radar-guided weapons. The ALQ-131 is carried externally on a variety of front-line, high-performance aircraft. It is certified on front-line aircraft such as the A-7, A-10, C-130, F-4, F-15, F-16, and F-111.

The modular design of the pod structure and electronic assemblies, plus its central computer software architecture, enable the ALQ-131 system to adapt quickly to a broad spectrum of EW applications. This feature proved valuable in the Gulf War, where ALQ-131s provided over 48 per cent of the US Air Force tactical aircraft EW self-protection. The Block II version experienced no combat losses in over 12,000 combat sorties. AN/ALQ-131 pods also provided protection during Bosnian operations, on US Air Force F-16, A-10, C-130 and Royal Netherlands Air Force F-16.

Increased effectiveness can be achieved by incorporating various mission modules,

In ICAP 2 aircraft, TJS SIR group receivers and antennas are located in the large tail fairing (Martin Streetly)   0121191

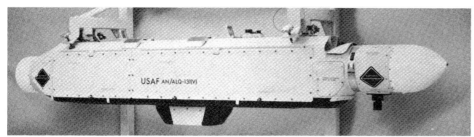

The Northrop Grumman AN/ALQ-131(V) ECM pod   0503862

including missile warning systems, offboard and towed countermeasures dispensers and advanced technique generators. This capability allows maximum flexibility in countering all threat types.

The basic structure elements of the pod are modular canisters that provide structural support, cooling and environmental protection. Each canister is an I-beam structure which also serves as a cold plate. Both sides of the I-beam form equipment bays into which functional equipment modules are mounted. The modules can also be mounted in lower equipment bays on the bottom surface of the I-beam. Using common mounting techniques, each canister can accommodate several equipment modules which can be removed directly from the bays without disassembly of the pod. The system is 2.82 m long and its weight ranges from 260 to 306 kg.

The functional organisation of the system is centred around the Interface and Control (I/C) module which contains a programmable digital computer as the system controller. The modules required for a given configuration are connected to the I/C by a digibus that carries all sensor and control data. A MIL-STD-1553B databus is currently being added to the system to enhance both internal and external capabilities. A memory loader/verifier allows operational flight and mission-specific program software to be loaded into the pod on the flight line in less than 15 minutes.

The I/C module also contains a digital waveform generator that can permit up to 48 simultaneous waveforms for deception modulation. When any ECM technique requires a deception waveform, the latter's values are transmitted to the onboard equipment via a waveform distribution bus.

The ALQ-131 is compatible with both C-9492 and AN/ALQ-213 Cockpit Control Units (CCUs).

Maintenance of the ALQ-131 is based on the pod's Centrally Integrated Test System (CITS) which provides a comprehensive functional check of system operation, both in flight and on the ground. During flight, the CITS continuously monitors the operational status of the equipment, including repeater channel modulation, high voltage of the TWTs, noise power output, primary bus voltage and the integrity of the computer memory.

The pod is a software reprogrammable system which allows a tactical commander to tailor the ALQ-131's responses for the mission requirements. Utilising the ALQ-131's ability to be flight line reprogrammed, mission-specific data can be created in response to threat changes and, by using a memory loader verifier, changes loaded into the pod's digital computer.

Another capability of the system is its self-contained power management feature. This is included in a receiver/processor module which detects radar threats, measures their key parameters and performs weapon type and operational mode identification. This information is then used by the ALQ-131's I/C module to select optimum jamming techniques automatically and tailor their parameters for countering all detected threats. Numerous threats can be countered simultaneously, each with independent techniques, using the receiver/processor's PRI tracking capability.

Specifications
Dimensions:
(one band shallow) 2,210 × 297.2 × 533.4 mm
(two band shallow) 2,819.4 × 297.2 × 533.4 mm
(three band deep) 2,819.4 × 297.2 × 635 mm
Weight:
(one band shallow) 175.54 kg
(two band shallow) 260.36 kg
(three band deep) 305.72 kg
Power supply: 115 V AC, 400 Hz
Reliability: 125 h MTBF (200 h MTBF achieved during Operation Desert Storm with an operational readiness rate of 95%)

Status
More than 560 Block I and 460 Block II pods have been delivered to the US Air Force. A production programme to update the original Block I configuration to the latest Block II version is

under way. Over 370 pods have been procured by the air forces of Bahrain, Belgium, Egypt, Israel, Japan, Netherlands, Norway, Pakistan, Portugal, Singapore and Thailand. International deliveries are continuing.

In October 1993, the US Air Force completed EMI/EMC testing of the ALQ-131 modified with an AN/ALQ-153 active missile warning system and AN/ALE-47 countermeasures dispenser on an F-16 aircraft. In April 1994, the ALQ-131 was demonstrated with a fully integrated AN/AAR-54 passive missile warning system and AN/ALE-47. In August 1998, the ALQ-131 was demonstrated with a high-power fibre optic Towed Radar Decoy (TRD) on a C-130 aircraft.

A self-contained external ECM suite configuration of ECM pod, missile warning system and countermeasures dispenser is being considered, to provide tactical aircraft with protection against IR-guided missiles while eliminating Group A aircraft modifications.

The pod has been combat proven in Bosnian operations on US Air Force A-10, C-130 and F-16 aircraft and Royal Netherlands Air Force F-16s.

New contracts for the AN/ALQ-131 system continue to be issued by the US Air Force, both for continued customer support and for system upgrades for both US Air Force and Foreign Military Sales (FMS) requirements. Recent awards for Block II conversion kits have been made on behalf of the Royal Norwegian Air Force (16 kits) in June 1999, and the Egyptian Air Force (39 kits) in January 2000.

In 1999, Northrop Grumman was contracted by Mitsubishi Electric Corporation to provide ALQ-131 production kits for local assembly, test and delivery to the Japanese Air Self-Defence Force.

The ALQ-131(V) is currently undergoing a mid-life upgrade to improve the transmitter, ECM techniques and unit sustainment and availability. The upgrade kits will be available for installation in 2004.

Contractor
Northrop Grumman Corporation, Electronic Systems Sector.

AN/ALQ-135 jamming system

Type
Airborne Electronic Counter Measures (ECM), radar jamming system.

Description
The AN/ALQ-135 Internal Countermeasures Set (ICS) is a component of the Tactical Electronic Warfare System (TEWS) for the US Air Force F-15; it is installed in various configurations in the F-15A, C, and E variants and operates with the AN/ALR-56 radar warning system and the AN/ALE-45 countermeasures dispenser.

The AN/ALQ-135 is an advanced jamming system which uses high-powered transmitters. All equipment is mounted internally and jamming system management is self-contained in the F-15A, C and E.

The basic ALQ-135 (Band 1.5 and Band 3 variants) consists of seven LRUs plus appropriate waveguides and antennas. Two LRUs are receivers, four are amplifiers and one is a preamplifier. Over the years the system has continued to evolve along with the capabilities of the aircraft and changes in the threat. While maintaining commonality with the original system and support electronics, the AN/ALQ-135 has been updated to include full band coverage and extremely effective jamming techniques flexibility.

For installation in the two-seat F-15E, the frequency range of the system has been expanded by the addition of a Band 3 transmitter/receiver/processor and power amplifier for the aft radiating antenna. Bands 1 and 2 have individual jammers in older F-15s, but in the F-15E variant they are being combined into a Band 1.5 low-band jammer that is half the size of the Bands 1 and 2 jammer. From a logistical

The F-16 carries the AN/ALQ-131 pod under the fuselage, or, in this case on an outboard wing station   0504155

The Northrop Grumman AN/ALQ-135 jamming system is carried internally in the F-15　　0503856

standpoint, the Band 1.5 system is more than 70 per cent common with the existing Band 3 system, thereby maximising commonality and minimising support costs. Major features of the AN/ALQ-135 system are:

Receiver: channelised, multimode for fast response and activity detection throughout the band, variable bandwidth superheterodyne with IFM and sophisticated blanking/lookthrough.

Control: 20 microprocessors in a federated architecture with parallel processing for fast system response and flexibility/ reprogrammability.

Techniques: demonstrated capability for emerging/developing radar threats with advanced techniques and coherent response.

Status
In service with F-15. Northrop Grumman has produced more than 1,750 AN/ALR-135 systems.

The Band 1.5 subsystem was flight tested during 1998, with approval for initial upgrade of 170 AN/ ALQ-135 systems to Band 1.5 configuration granted by the US Congress in November of that year. The first operational installation of the Band 1.5 ECM subsystem in an F-15E aircraft was carried out in December 2000. With the system having passed Initial Operational Test and Evaluation (IOTE) in January 2001, Northrop Grumman received a USD30 million contract (Lot II) from the US Air Force to provide 17 AN/ALQ-135 Band 1.5 ECM subsystems for the F-15E, with a requirement for work to be completed by December 2001. This marked the transition to full production for the system. In March 2002, the US Air Force awarded the third (Lot III) of the five-lot production programme, worth USD 65 million, for a further 28 Band 1.5 subsystems, to be completed by September 2003. By March 2002, eleven of the systems had been delivered to Elmendorf Air Force Base, Alaska, for installation.

To date, the AN/ALQ-135 has operated in more than 6,600 hours of combat, beginning with Operation Desert Storm and continuing through recent operations in the Balkans, Kosovo and Afghanistan, without a single aircraft loss due to threats covered by the ECM system.

Contractor
Northrop Grumman Corporation, Electronic Systems Sector, Defensive Systems Division.

AN/ALQ-136 radar jamming system

Type
Airborne Electronic Counter Measures (ECM), radar jamming system.

Description
The AN/ALQ-136 family of airborne jamming systems is designed to protect rotary- and fixed-wing aircraft from radar-guided weapons. The AN/ALQ-136 enables the host aircraft to carry out a mission from a low-level approach, pop up and complete its mission before the hostile radar can initiate a fire sequence.

The system consists of three main units: a control unit, two spiral antennas – one each for transmit and receive – and a transmitter/receiver. When the aircraft is illuminated by a hostile radar, the jammer automatically analyses the received pulses, compares them with its threat library, assigns a priority and then provides the most appropriate countermeasure response.

The AN/ALQ-136 is software reprogrammable to allow adaptation to the swiftly changing threat scenarios and can simultaneously handle multiple threats. The system is internally mounted and integrated on numerous US Army helicopters.

Status
In service. More than 1,400 systems have been delivered to the US Army and are deployed on AH-64 Apache and AH-1 Cobra attack helicopters. Other platform applications include the MH-53J Pave Low, MH-60 Black Hawk, EH-60 Quick Fix, MH-47E Chinook and AH-1W SuperCobra helicopters and the A-10 and RC-12/RU-21 fixed-wing aircraft.

AN/ALQ-136(V)2
The AN/ALQ-136(V)2 is an enhanced version with increased receiver sensitivity, frequency range and threat-handling capability. The system design employs extensive use of Thick Film Hybrid technology (TFH) and multilayered PCBs. A MIL-STD-1553B databus is used to facilitate system integration.

Specifications
Dimensions:
(receiver/transmitter) 177.8 × 330.2 × 408 mm
Weight: 31.8 kg

Status
200 systems have been delivered to the US Army.

Contractor
ITT Industries, Avionics Division.

AN/ALQ-144A(V)

Type
Airborne Electronic Counter Measures (ECM), radar jamming system.

Description
BAE Systems' ALQ-144A(V) countermeasures system provides protection against a wide spectrum of infra-red threats. Its small size, light weight and low power consumption make the AN/ALQ-144A(V) ideal for helicopter installation. It is an omnidirectional system consisting of a cylindrical source surrounded by a modulation system to confuse the seeker of the incoming missile. Phase Lock (PL) installations of two, or more, systems permit protection of larger helicopters or fixed-wing aircraft. The units may be mounted on either the top or bottom of the airframe or, in the case of the PL configuration, both. The new phase lock configuration (V)5 is a pair of ALQ-144A transmitters that are electrically phased through a junction box. They are controlled by a single Operator Control Unit (OCU), which has the capability to select from nine different jamming options.

Specifications
Weight: 12 kg (transmitter); 0.45 kg (control unit)
Dimensions: transmitters: 320 mm (diameter) × 340 mm (height)
control units 28 × 147 × 134 mm
Field of view: 360° azimuth
MTBF: 300 h

Status
More than 5,000 systems have been delivered to the US Army CECOM for fielding in all US armed services and to a dozen export customers, for the following applications:
ALQ-144A(V)1 – Utility, fitted to: HH-60, MH-60, Puma, UH-1H and UH-60B helicopters;
ALQ-144A(V)1 – Observation, fitted to: Lynx, OH-58D helicopters and OV-10 aircraft;
ALQ-144A(V)3 – Attack, fitted to: A-109, A-129, AH-1F, AH-1W, AH-64 and Tiger helicopters;

BAE Systems AN/ALQ-144A(V) transmitter and control unit　　0051269

The AN/ALQ-136(V)1/5 jamming system is carried by US Army AH-64 Apache attack helicopters
0504076

ALQ-144A(V)5 – Phased Lock Special Mission, fitted to: SH-2F/G, SH-3D, SH-60B/R, Super Puma and VH-60 helicopters.

In February 2000 an order for the Turkish government was completed – for a large number of AN/ALQ-144A(V)1 units and a smaller number of AN/ALQ-144A(V)5. The system continues in production.

Contractor

BAE Systems North America, Information and Electronic Warfare Systems.

AN/ALQ-156 Missile Approach Warner (MAW)

Type

Airborne Electronic Counter Measures (ECM), Missile Approach Warner (MAW).

Description

The AN/ALQ-156 is an airborne, internally mounted, pulse Doppler missile-warning system, originally developed for slow- and low-flying helicopters and fixed-wing aircraft, but later modified for use in other types of aircraft. The system is solid-state, incorporating four doughnut-shaped antennas, a transceiver and a control unit. The system recognises missile threats by comparing closure rates and other ballistic parameters, warns the pilot aurally and triggers the countermeasures command to dispense flares, chaff or other expendables. When coupled with radar and laser warning sensors, the ALQ-156 can initiate automatic selection of the appropriate countermeasure.

The AN/ALQ-156(V) is a 360° pulse Doppler radar missile detector that illuminates an incoming missile, detects the RF reflection and measures the missile's range and velocity. It then accurately determines time-to-go and provides the optimum triggering of an expendable to protect the host platform. The system can be integrated with a range of countermeasures dispensers such as the ALE-40, ALE-47 and M-130 systems to provide infra-red protection and with a range of RWRs, such as the APR-39, ALQ-56, ALR-56 and MK-20. Integrated it can also improve situational awareness.

A later model, the AN/ALQ-156A, was developed for the US Navy and was designed for use in tactical aircraft. It integrates an RWR and deceptive ECM data to allow smart control of decoys and flares and has a much longer detection range.

Status

The AN/ALQ-156 is fitted to a variety of aircraft, including the US Army CH-47, EH-1H, EH-60B, BV-10, REC-12D/H/K, RU-21A/B/C/D and RV-1D platforms. The US Air Force uses the system on its C-130 transports and the Royal Air Force on its Chinook helicopters. Several other nations use the system on C-130 aircraft and Chinook helicopters. Nearly 1,000 systems have been delivered. An external pod-mounted configuration has been flight tested successfully.

In January 2000, the US Army Communications and Electronics Command (CECOM) contracted BAE Systems North America to supply 17 systems (designated AN/ALQI-156(V)2/(V)3T) for the Hellenic Army's CH-47D helicopters. This version of the AN/ALQ-156(V) was developed as part of the NATO C-130 Apollo programme.

The AN/ALQ-156 (V) detects the approach of IR missiles   0504154

It incorporates improved false alarm rejection capability.

Contractor

BAE Systems North America, Information and Electronic Warfare Systems.

AN/ALQ-161 ECM system for tthe B-1B

Type

Airborne Electronic Counter Measures (ECM), integrated suite.

Description

AIL Systems (now a business unit of the EDO Corporation) is the prime contractor for the AN/ALQ-161 defensive countermeasures suite for the US Air Force Rockwell B-1B Lancer. This was launched in 1972 under AIL Systems leadership for the Rockwell B-1A bomber, but the effort came to a halt when the aircraft was cancelled in June 1977. The programme was, however, reinstated in October 1981. AIL Systems accordingly resumed its task as ALQ-161 project leader, having participated in limited flight trials with one aircraft since 1979. The current equipment was developed under restart and initial production contracts, worth more than USD1,700 million, which were awarded to AIL Systems and its subcontractors. Each aircraft set is understood to have cost around USD20 million, approximately one-tenth the cost of a complete B-1B, and comprises no fewer than 108 LRUs, with a total weight of over 2,300 kg exclusive of cabling, displays and controls.

The operating frequency range is approximately 0.5 to 10 GHz, to cover early warning, ground-controlled interception, surface-to-air missile and interceptor radar frequencies. Jamming signals in the higher regions of the ECM spectrum are emitted from three electronically steerable, phased-array antennas: one in each wing-glove leading-edge and the third in the fuselage tailcone. Each antenna provides 120° azimuth and 90° elevation coverage. Lower frequency signals are emitted from quadrantal horn antennas mounted alongside the high-frequency equipment.

Major subcontractors in the AN/ALQ-161 programme were Northrop Grumman, Litton Industries and Sedco Systems. All companies served as subsystem managers with responsibilities for receivers, data processing

and jamming techniques. Northrop Grumman provided the low-frequency jamming antennas and Sedco supplied the phased-array antennas.

AIL Systems was responsible for system integration and for the LRUs, including 51 unique designs. Most of the LRUs are about 0.03 to 0.06 m3 in volume and weigh between 18 and 36 kg. The majority are readily accessible and can be removed easily or installed by one or two people. Total power required is about 120 kW.

Improvements introduced during the programme included a frequency extension into the K-band to increase the warning and jamming performance, extension into the low-frequency domain below 200 MHz to improve the response of the warning system, introduction of a digital radio frequency memory to permit the deception jamming of more advanced radars, notably those employing pulse Doppler techniques, and introduction of a new Tail Warning Function (TWF).

An IBM 101D computer functions as the central processor. The computer, employing Jovial-based software, identifies the function of every hostile radar, assesses its potential threat and assigns a jamming priority. Three sets of phased-array antennas mounted in the wing leading-edges and tail provide full 360° coverage in Bands 6, 7 and 8, antennas for lower frequencies being located fore and aft in the airframe.

Two complete systems were delivered for continuing trials with a modified Rockwell B-1A in the Summer of 1984 and for the first production B-1B; the latter made its initial flight in October 1984. Production rate was built up to four systems a month by May 1986. In August 1985, AIL Systems was awarded a USD1,800 million contract for the final 92 ALQ-161 shipsets.

Problems in development were reported in mid-1986, leading to the first 22 B-1Bs not being equipped with the TWF which had been scheduled to enter service in mid-1987. Problems with the TWF and with repeatability of results from the system as a whole were part of a list of deficiencies in the B-1B's performance made public at the end of 1986. Of a USD600 million request made by the US Department of Defense at that time to extend the capabilities of the B-1B, almost USD100 million was allocated to bring the ALQ-161 up to its original specification and solve other problems (according to the DoD), while the bulk of a USD131 million element within the USD600 million total requested for expenditure in 1988/89 was to upgrade the defensive avionics suite to meet capabilities which have emerged since the original baseline was set in 1982.

The AIL AN/ALQ-161 system for the B-1B   0503837

The AN/ALQ-161 system has undergone continuous upgrade to maintain the combat effectiveness of the B-1B, including work to integrate the AN/ALE-50 Towed Radar Decoy (TRD) system (US Air Force)   0524707

System operation

Operation of the AN/ALQ-161A is primarily in an automatic mode, where the system receives, identifies, and jams threat radars instantaneously. However, the Weapons Systems Operator (WSO) can manually intervene if desired. Complete Situational Awareness (SA) is provided on two threat display formats to the WSO to include constant information of threat type, mode, location jamming status, frequency and relative amplitude. Other additional system and/or threat information is available on call to the operator through a correlated joystick that allows him to designate specific threats and request data from either display format. Additional manual functions such as jamming modifications, receiver attenuation changes and threat characteristic changes are available to the WSO.

The system is highly reprogrammable, on the ground and in the air, which allows it to adapt to any scenario in minimum time. The receiver continually observes the entire environment to detect most conventional threats at a moderate sensitivity. For those emitters that require higher sensitivity for detection, the receiver employs a directed, high sensitivity search routine that is programmed by the user for maximum effectiveness.

The TWF provides a Pulsed Doppler (PD) radar function to detect any missile threatening the bomber from the aft sector. If a missile is detected, the system provides immediate warning to the crew for evasive action, and can automatically dispense RF and IR decoys to counter the threat.

Status

Production of 100 systems is complete and the system is installed in the B-1B. In February 1991, AIL was awarded a USD16.5 million contract for AN/ALQ-161A systems for the Northrop B-2 and an additional USD5.491 million to upgrade systems for the B-1B.

The B-1B Defensive Systems Upgrade Program (DSUP), reportedly being applied to the whole B-1B fleet, includes Block B, C, D and E enhancements to the core ALQ-161A system. The details of Block B upgrades are unknown. Upgrades designated Block C to E are understood to involve: Block C improved direction finding and sector blanking (initiated in 1996); Block D improved jamming capability (to begin in 1999); Block E improved emitter detection (later). During the second quarter of 2002, AIL Systems, through its parent company EDO, was awarded contracts totalling USD11.4 million from Warner Robins Air Logistics Centre, to provide continuing support and performance improvement for the AN/ALQ-161A for the B-1B fleet, which included integration work for the AN/ALE-50 Towed Radar Decoy (TRD) system.

The AN/ALQ-161A was reported to have been 'reliable and effective' during operations in Afghanistan and the most recent Gulf War over Iraq.

Perhaps as a result of this performance, on the 22 May 2003 EDO announced it had received two further B-1B programme awards in support of the AN/ALQ-161, totalling some USD11.2 million, enabling the US Air Force to maintain and upgrade the bomber's defensive suite. The first award was for 12 months of engineering support under a new 10-year ID/IQ (Indefinite Delivery/Indefinite Quantity) contract. The task covers sustaining engineering services, which include software-test facility support, mission-data optimisation analysis, technical studies and analyses, software testing and diminishing manufacturing sources studies. The second award covers continued support for subassembly repairs for 6 to 24 months.

Contractor

AIL Systems, Inc.

AN/ALQ-162(V) radar countermeasures set

Type

Airborne Electronic Counter Measures (ECM), radar jamming system.

Description

Development of the ALQ-162 was started in 1979 under contract to the US Navy; the US Army later joined the programme. It is a small, low-cost, lightweight reprogrammable Radar Frequency (RF) jamming system which can be supplied with its own receiver/ESM management processor, or made compatible with many existing types of radar warning receiver processor systems. The system is software-programmable to meet new threats, and can be installed in pod, pylon or internal fit configurations.

The ALQ-162(V) is capable of defeating Pulse Doppler (PD) and Continuous Wave (CW) threats and is fully integrated with the following systems: radar warning receivers (AN/APR-39, -43, -45, -46, -66, -67 and -69); pulse jammers (AN/ALQ-126 (A and B) and AN/ALQ-136); and CM dispensers (AN/ALE-36, -40 and M-130). The system provides self-protection against radar threats by continuously scanning the threat signal environment, identifying and prioritising emitters and then generating specific countermeasures against them.

The antennas provide nominal coverage of ±60° in azimuth and ±30° in elevation. The ALQ-162(V) will also interface with additional antennas if required.

The system has been subject to continuous upgrade, with the latest variant designated AN/ALQ-162(V)6, also known as Shadowbox II.

AN/ALQ-162 (V) installed in the outboard pylon on F/A-18   0018248

This upgrade incorporates advanced Pulse capabilities, to counter emerging threats and provide increased aircraft survivability, and advanced Microwave Power Module (MPM) technology that will more than double the output of the system over earlier variants. Digital Radio Frequency Memory (DRFM) technology enables the system to defeat PD threats while continuing to counter CW threats. In common with all previous versions, no changes to form, fit, or aircraft wiring are required to implement this upgrade. All of the system upgrades incorporated into the ALQ-162(V)6 are available for retrofit into earlier systems.

Potential growth areas for the system include integration with Towed Radar Decoys (TRDs), Directional Infra-Red CounterMeasures (DIRCM) systems and the RF component of the Suite of Integrated Sensors and CounterMeasures (SISCM) for helicopters. Additionally, an increased threat-handling capability is under development to enable the system to counter multiple threats in different bands simultaneously and to improve PD capability. The AN/ALQ-162(V) can be installed internally or in a pod.

Specifications
Dimensions: 161 × 184 × 420 mm
Weight: 19 kg
Power supply: 115 V AC, 400 Hz, 3 phase, 650 W
Heat dissipated: 480 W
Reliability: >260 h MTBF demonstrated

Status
ALQ-162(V) installations include: US Navy – A-4M, A-7E, AV-8B, F-4S and RF-4B; US Army – EH-1, EH-60, OV-1D, RC-12D, RU-21, RV-1D; US Army Special Operations Command (ASOC) – MH-47D/E and MH-60K/L; NATO – C-130, F-16 and Saab F-35 Draken, more than 650 installations.

Following test flights of Shadowbox II aboard a US Navy F/A-18, the system was acquired by the Royal Norwegian Air Force and Danish Air Force for their F-16 aircraft during 2000. During the first quarter of 2001, Northrop Grumman received a USD15 million Foreign Military Sales (FMS) contract from the US Navy for 44 AN/ALQ-162(V)6 systems to equip Egyptian Air Force AH-64D helicopters, the first time the system has been applied to a rotary-wing platform.

Contractor
Northrop Grumman Corporation, Electronic Systems Sector, Defensive Systems Division.

AN/ALQ-164 ECM pod

Type
Airborne Electronic Counter Measures (ECM), radar jamming system.

Description
The AN/ALQ-164 is an airborne pod-mounted jammer for both pulse and CW modes. The pod incorporates unmodified ALQ-126B and ALQ-162 jammers. It can operate fully autonomously or integrated with an RWR and other avionics on board the aircraft. It is reprogrammable to provide for threat changes. The ALQ-164 is intended for fitting on the centreline pylon of the AV-8B aircraft and has applications for a wide variety of tactical and transport aircraft.

Specifications
Dimensions: 2,160 × 445 × 406 mm
Weight: 188 kg
Power supply: 115 V AC, 400 Hz, 3 kVA
Frequency: G- to J-bands

Status
In service with AV-8B Harrier aircraft of the US Marine Corps and Spanish and Italian Navies.

Contractor
BAE Systems North America, Information and Electronic Warfare Systems.

AN/ALQ-165 Airborne Self-Protection Jammer (ASPJ)

Type
Airborne Electronic Counter Measures (ECM), radar jamming system.

Description
The AN/ALQ-165 Airborne Self-Protection Jammer (ASPJ) is designed for F-14D, F-16C, F-18C/D, A-6E and AV-8B aircraft and can be installed internally or in a pod. The equipment incorporates the latest technology in TWTs, microwave components and packaging. It can be electrically reprogrammed on the flight line and the BIT and modular plug-in design permits rapid replacement of assemblies at the operational level without external test equipment. Weight of the system is between 91 and 150 kg, depending on configuration; it occupies a space of 0.6443 m³.

The AN/ALQ-165 has the ability to select automatically the best jamming techniques to use against any given threat, based on the jamming system's own computer data and real-time data of the threat signal from the receiver/processor. The computer software can be modified to accommodate new threats as they arise. The equipment covers the frequency range in two bands and is technically expansible to cover a greater frequency range if required.

The transmitters within the system can jam a large number of threats simultaneously over various ranges and in different modes. The computer selects the power and duty cycle criteria based on threat parameters detected and processed by an advanced receiver subsystem. An augmented version with an additional transmitter power booster is also available.

AN/ALQ-165(V)
Following renewed enthusiasm for the ASPJ system in the US, the US Naval Air Systems Command in December 1997 issued a USD45 million firm fixed price contract to procure: 36 AN/ALQ-165(V) ASPJ systems for US Navy F/A-

ASPJ AN/ALQ-165 internal fit on F-14D 0018271

The ASPJ has been offered as an upgrade to Malaysian F-18D aircraft 0118194

18 C/D and F-14D aircraft; 15 AN/ALQ-165(V) ASPJ systems as US Navy spares; 10 AN/ALQ-165(V) ASPJ spares for the US Air Force; and data services support for the US Navy ASPJ programme. Contract fulfilment to be by December 1999. The nature of the upgrade represented by the addition of (V) to the nomenclature is not known.

Specifications
Dimensions:
(processor) 141.7 × 199.4 × 403.9 mm
(low-band receiver) 141.7 × 199.4 × 400.6 mm
(high-band receiver) 141.7 × 199.4 × 400.6 mm
(low-band transmitter) 121.4 × 208 × 644.7 mm
(high-band transmitter) 121.4 × 208 × 644.7 mm
Weight:
(processor) 17.69 kg
(low-band receiver) 16.78 kg
(high-band receiver) 16.78 kg
(low-band transmitter) 30.84 kg
(high-band transmitter) 29.48 kg

Status
In September 1994, Finland procured the system for its F/A-18 aircraft. Three months later, the Swiss Air Force decided to fit the ASPJ to its F/A-18 aircraft. In July 1995, deployed on US Navy and US Marine Corps F/A-18 C/D aircraft and in May 1996 approved for unrestricted fleet use on the F-14D. Since its initial deployment, the SPJ has logged more than 40,000 operational flight hours on those aircraft flying over Bosnia and the Persian Gulf.

In January 1997, the SSPJ announced a third export sale to the Republic of Korea for its F-16 fleet, with the first modified aircraft delivered to ASPJ standard in Feb 1999.

In December 1997, a contract for 30 new ASPJ systems was announced for US Navy and Marine Corps aircraft.

Most recently, the US offered Malaysia a USD45 million EW upgrade package for its eight F/A-18D aircraft, based on the provision of five AN/ALQ-165(V) ASPJ kits.

Contractor
Consolidated Electronics Countermeasures. (An ITT Industries, Avionics Division/Northrop Grumman Corporation, Electronic Systems joint venture.)

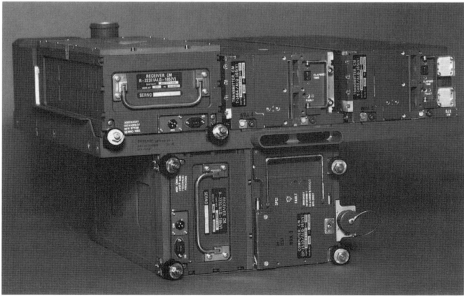

The ASPJ configuration for Korean F-16 aircraft 0515376

AN/ALQ-167 jamming pod

Type
Airborne Electronic Counter Measures (ECM), radar jamming system.

Description
The AN/ALQ-167 pod generates a wide range of jamming techniques in multiple bands extending from B-band through K-band (250 MHz to 40 GHz). The system is a modular design which facilitates tailoring of frequency range and performance characteristics to customer requirements.

AN/ALQ-167 jamming scenarios are selectable using a cockpit-mounted control box (single or dual) or laptop (IBM PC) computer communicating with the jamming system via RS-422 serial bus. Interface with aircraft Multi Function Display (MFD) and/or Electronic Support Measures (ESM) equipment via RS-422 or MIL-STD-1553B data bus is also available. The parameters and features that can be selected include operating bandwidth, jamming scenario, jamming parameters, set-on receiver, forward/aft radiation and power level. Up to 16 pre-programmed jamming scenarios are available.

Recent upgrades to the system have increased Effective Radiated Power (ERP) and added both DRFM and Set-On Receiver (SOR) capabilities.

Options available include expansion provisions, UAV environment, alternative MIL-STD-1553B serial data bus and internal fitment.

A radar simulator pod, based on the core AN/ALQ-167 design, is also available.

Specifications
Physical
Pod dimensions: 3,500 × 260 mm (L × Ø) (excluding antennas, fins and hardback)
Weight: 175 kg
Power: 115/200 V AC (±10%), 3-phase 400 Hz; 7.5 A per phase
Cooling: Ram air

Cockpit control panel
Dimensions: 67 × 150 × 170 mm (H × W × D)
Weight: 0.9 kg

Frequency coverage and midband ERP
AN/ALQ-167 (CD/NM): 0.85 to 1.4 GHz (C/D-band); 4.0 kW (66 dBm); circular polarisation
AN/ALQ-167 (EF/NM): 2.5 to 3.8 GHz (E/F-band); 8.5 kW (69.3 dBm); linear polarisation
AN/ALQ-167 (GIJ/NM): 5.25 to 10.5 GHz (G/I/J-band); 7.5 kW (68.8 dBm); circular polarisation (9–18 GHz, high J-band option)

Programmable jamming techniques
The following jamming technique categories, including various overlaid combinations, can be called up in-flight without need to re-configure the system:

Cockpit control panel functions
- Jammer scenario switch (16 programmable scenarios)
- Transmit command switch (OFF/STBY/OPER)
- FWD/AFT switch (antenna select)
- HORZ/VERT/TEST switch (antenna polarisation select).

Status
AN/ALQ-167 jamming pods are in widespread service in many types of military and civilian fixed-wing aircraft and helicopters.

Contractor
Rodale Electronics Inc.

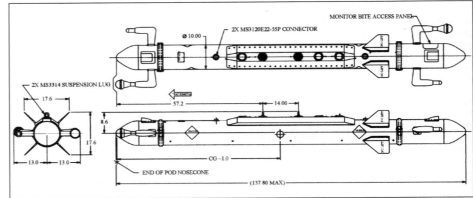

The AN/ALQ-167 airborne jamming pod (Rodale Electronics)   0580572

AN/ALQ-172 Electronic Counter Measures system

Type
Airborne Electronic Counter Measures (ECM), integrated suite.

Description
The AN/ALQ-172 is an advanced ECM system which was combat proven during Operation Desert Storm. It is installed in US Air Force B-52 strategic bombers and Special Operations MC-130E/H Combat Talon I and II and AC-130U/H Gunships.

System design features of the AN/ALQ-172 include full automation, multiband coverage, simultaneous multiple threat recognition and jamming, digital computer control, advanced jamming techniques, high effective radiated power, threat reprogrammability, high-gain array antenna, threat warning display, dual MIL-STD-1553B databus interface and extensive BIT.

Reportedly there are three upgraded versions of the system: the AN/ALQ-172(V)1, AN/ALQ-172(V)2 (which incorporates a phased-array antenna) and the AN/ALQ-172(V)3.

Status
In service in US Air Force B-52 and Special Operations MC-130E/H Combat Talon I and II, AC-130H and AC-130U/H gunship aircraft. System upgrades under way include expansion of AN/ALQ-172 system memory and extended frequency coverage designated AN/ALQ-172(V)3 for the AC-130H upgrade programme.

In 1998, the US Special Operations Command contracted ITT to upgrade its C-130 fleet to the AN/ALQ-172(V)1 (ECP-93) standard at a cost of USD18 million. Procurement options permit increase to USD44.6 million. This upgrade provides expanded system ECM and OFP memory capacity.

Contractor
ITT Industries, Avionics Division.

AN/ALQ-176(V) ECM pod

Type
Airborne Electronic Counter Measures (ECM), radar jamming system.

Description
The AN/ALQ-176(V) is a pod-mounted jamming system designed, developed and tested to fulfil aircraft mission requirements for ECM support, standoff jamming and combat evaluation and training. The pod system comprises transmitter modules and a complementary antenna housed in canisters to form a slim pod with a 10 in (25 cm) diameter. The AN/ALQ-176(V) is mounted on standard aircraft wing or fuselage stores/munitions stations. The pod offers the option of using aircraft internal power or a ram-air turbine.

The modular design of the AN/ALQ-176(V) pod allows flexibility in ECM configurations. A two- or three-canister arrangement is available. The AN/ALQ-176(V)1 two-canister configuration houses up to three voltage-tuned magnetron transmitters. The AN/ALQ-176(V)2 three-canister version provides housing for five transmitters. Dependent on frequency band and tube selection, each transmitter produces 150 to 400 W CW (up to 30 per cent bandwidth) at efficiencies of greater than 50 per cent. The transmitters provide ECM across the specified threat frequencies. Transmitter protection circuits safeguard the system and provide fault status to the cockpit control panel.

The AN/ALQ-176(V) has been specifically engineered to allow rapid turnround in meeting the demands of new threat frequencies. It can accommodate new demands through selection of different stored parameters or through easy replacement of the transmitter tube or antenna. Transmitter spares, maintenance costs, times and training skills required are greatly reduced by the use of standard transmitter design. In addition the design incorporates built-in test and provides for parameter/mode selection at the cockpit control panel.

The flexibility in ECM pod configuration and standard aircraft stores/munitions stations installation enables the AN/ALQ-176(V) ECM pod to be used on a wide variety of tactical and support aircraft.

Specifications
Dimensions: 250 mm diameter
(AN/ALQ-176(V)1) 1,990 mm length
(AN/ALQ-176(V)2) 2,590 mm length
Weight:
(transmitter module) 17 kg
(AN/ALQ-176(V)1) pod) 112 kg
(AN/ALQ-176(V)2) pod) 196 kg
Input/output power:
(AN/ALQ-176(V)1) 2.6 kVA/1.2 kW
(AN/ALQ-176(V)2) 4.5 kVA/2 kW
Frequency: 0.8–10.5 GHz
Transmitter power output: 150–400 W per tube CW
(dependent on frequency and tube selection)
Modulation: noise (various)
Control: ALQ-10725 and C-9492 aircraft ECM cockpit control panel

Status
The US Air Force, the Royal Norwegian Air Force, the Royal Thai Air Force, and the Canadian Ministry of Defence have the AN/ALQ-176(V) in their inventories.

Contractor
Alliant Integrated Defense Company.

CD/NM		EF/NM		GIJ/NM	
Noise	**Deception**	**Noise**	**Deception**	**Noise**	**Deception**
SWPT	-	SWPT	MFR	SWPT	MFR
SPT	-	SPT	RSAM	SPT	RSAM
NCDB	-	NCDB	VGPO	NCDB	VGPO
BCDB	-	BCDB	RGPO	BCDB	RGPO
BAR	-	BAR	RD	BAR	RD
NSAM	-	NSAM	NBRN	NSAM	NBRN
BSAM	-	BSAM		BSAM	
DRFM (option)	-	DRFM (option)		DRFM	
SOR (option)	-	SOR (option)		SOR	

AN/ALQ-178(V) radar warning and Electronic CounterMeasures (ECM) suite

Type
Airborne Electronic Counter Measures (ECM), integrated suite.

Description
AN/ALQ-178(V) is an integrated radar warning and active countermeasures suite for tactical fighter aircraft, such as the F-16. An earlier version was known as RAPPORT III. The system utilises a central programmable computer for data analysis and system control, with independent microprocessors to direct Radar Warning Receiver (RWR), display, jamming and countermeasures dispensing functions. The wideband RWR continuously scans the threat radar environment. Detected signals are de-interleaved and identified by radar type and displayed to the pilot on a cathode ray tube in legible, unambiguous format. The power management algorithm optionally matches the countermeasures to the RWR's constantly changing threat picture. Separate forward and aft jammers are used for maximum spatial coverage. The jammers are accurately set for maximum power output at each victim radar's frequency to maximise effective radiated power and jammer effectiveness. The jammer frequency range provides coverage of the entire threat band while the jammer power is sufficient to counter these radars throughout their lethal area. The countermeasures dispenser is automatically controlled, including the pilot's display and cues, to provide optimum response to the threat between active jamming and chaff dispensing.

Although the ALQ-178(V) consists of an RWR, an active jammer and countermeasures control, the RWR can be installed as an independent threat warning system (controlling a countermeasures dispensing system), with the jammer added later.

The latest derivative of the ALQ-178(V) family incorporates technological advancements and packaging to increase capabilities significantly.

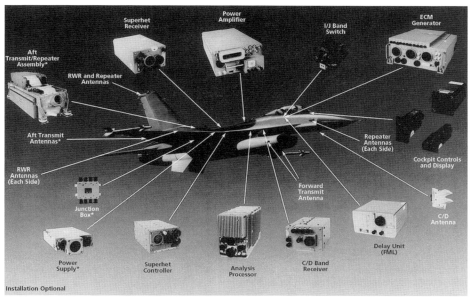

Elements which make up the ALQ-178(V) ECM system   0515250

Enhanced performance is provided by Digital RF Memory (DRFM) technology, agile jamming channels, distributed high speed/capacity processors and precision direction-finding capabilities.

A follow-on to ALQ-178(V), combined with the AN/ALR-56M or other RWR, is known as the AN/ALQ-202(V) autonomous jammer.

Specifications
RF coverage: E-J band
Targets: 32 targets simultaneously
Weight: 203 kg
Power: 2.5 kW
Interfaces: MIL-STD-1553B and RS-232

Status
ALQ-178(V)3 was selected by Turkey in January 1989 for its F-16 aircraft and, in October 1989, it was announced that Turkey had awarded a contract worth USD325 million for supply of systems, test and support, and ancillary equipment. Deliveries ran from October 1991 to early 1996. The then Lockheed Martin and the Turkish company Kavala formed a joint venture known as MiKES to produce the ALQ-178(V)3 in Turkey after initial deliveries from the USA.

ALQ-178(V)1 is fitted to F-16s of the Israeli Air Force with the threat warning subsystem. A portion of the system was built under licence in Israel by Elisra. The Israeli application may cover aircraft up to batch 3 F-16C/Ds but excludes the country's F-16D-30 aircraft. ALQ-178(V)1 incorporates a set of crystal video (DF) receivers which have been eliminated from the ALQ-178(V)3.

Contractor
BAE Systems, North America, Information and Electronic Warfare Systems.

AN/ALQ-184(V) self-protection jamming pod

Type
Airborne Electronic CounterMeasures (ECM), radar jamming system.

Description
The AN/ALQ-184(V) ECM pod is designed to provide effective countermeasures against surface-to-air missiles, radar-directed gun systems and airborne interceptors by selectively directing high-power jamming against multiple emitters. An upgrade of the AN/ALQ-119(V) pod retaining the same dimensions is the AN/ALQ-184(V), which functions as a repeater, transponder or noise jammer.

Multibeam architecture based on proven Rotman lens antenna technology provides substantially reduced countermeasures response time with a tenfold increase in ERP at 100 per cent duty factor (a figure of 9 kW is quoted in some sources). High reliability is inherent in the architecture due to redundant mini-TWTs, reduced operating voltages and lower operating temperatures. Maintainability has been improved by the use of digital circuitry, automatic gain optimisation, elimination of manual adjustments and extensive computer-driven built-in test.

Other features of the AN/ALQ-184(V) include:
- Near instantaneous RF signal processing
- DF on every received signal, independent of frequency
- Selective directional high-power effective radiated power against multiple emitters
- Full 100 per cent duty cycle transmit capability
- Customised ECM techniques for maximum effectiveness against engaged emitters

AN/ALQ-184 (V) jamming pod being prepared for installation on a US Air Force F-16   0515251

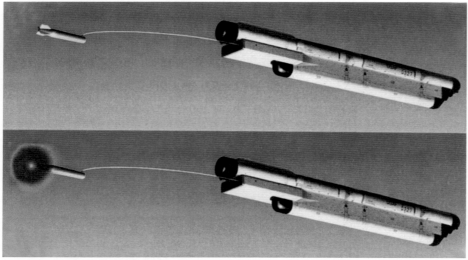

AN/ALQ-184 (V)9 combined ALQ-184 ECM pod and ALE-50 towed decoy system, indicating the design concept of jammer and decoy operation   0018256

AN/ALQ-184(V)9 configuration, showing 1 × 4 ALE-50 launcher unit at centre rear   0018257

In operation, the system compares incident signal parameters to those in a threat the library, determines the correct response and automatically initiates the necessary countermeasures. ECM coverage is accomplished through forward and aft transmitters, providing for combined mid-/high-band channels.

The AN/ALQ-184(V) can be configured as two- ('mid/high') or three-band ('low' and 'mid/high') systems. Of these, the two-band configuration weighs 210 kg while the three-band variant weighs 289 kg. Mean time between failure values for the two configurations are 150 hours and 80 hours respectively.

Specifications
Length: 3,957 mm
Weight: 289 kg

Status
Raytheon Electronic Systems has delivered more than 880 pods to the US Air Force since 1989. The ALQ-184 operates on US Air Force A-10, F-4, F-15 and F-16 aircraft. AN/ALQ-184(V) has also been selected by the Taiwanese government for its F-16 aircraft, and deliveries began in 1997. Sale of an additional 48 pods to Taiwan was authorised by the US government in June 2000.

In 1996, the US Air Force awarded Raytheon a contract to upgrade the ALQ-184. Under the contract, Raytheon Electronic Systems built 225 reprogrammable low-band kits, providing the pods with rapid flight line programming, improved life cycle costs and additional modulation jamming techniques. The contract also included options for upgrade kits to all ALQ-184 ECM pods produced to date.

On 10 April 2000, Raytheon was awarded a USD5 million modification to the aforementioned contract, covering the supply of 298 modification kits designed to upgrade pods used on USAF A-10, F-15 and F-16 aircraft to ALQ-184(V)11 standard. Work was completed during 2001.

AN/ALQ-184(V)9
The AN/ALQ-184(V)9 is a development of the AN/ALQ-184 ECM pod, in which a scab-fit unit of four ALE-50 towed decoy system units is added to the aft end of the pod.

The ALE-50 system consists of a launch controller, launcher and towed decoy. The decoy protects the host aircraft against radar-guided missiles by presenting a more significant target to the missile, seducing it away and thus creating a miss distance greater than the lethal radius of the missile warhead.

ECM techniques co-ordination between the two systems is managed by an Advanced Correlation Processor (ACP). The ACP has been tested by the US Air Force Air Warfare Center.

The ACP will make the decision between ALQ-184 and ALE-50 threat responses in order to employ the most effective counter to the threat.

Integration of ALE-50 into the ALQ-184 also results in significant improvement of basic ALQ-184 capability. To provide space for the ALE-50 launch controller and four-decoy launcher, the low-band controller was modernised and made field programmable by upgrade of componentry. This so-called 'two-card' (denoting the replacement of 12 original circuit cards with just two modern items) low-band modification improves MTBF to the degree that the addition of the ALE-50 LRUs is completely offset and the resultant ALQ-184(V)9 MTBF is better than that of the basic system.

The ALQ-184(V)9 incorporates improved ability to communicate with its host aircraft and other onboard systems, such as missile warning receivers and radar warning systems, via dual-redundant MIL-STD-1553B interfaces.

As with the basic ALE-50, ALQ-184(V)9 will offer growth to Fibre-Optic Towed Decoy (FOTD) capability using proven techniques generation in the ALQ-184 pod.

Status
In service with the US Air Force and Air National Guard.

Contractor
Raytheon Company, Space and Airborne Systems.

AN/ALQ-187 internal countermeasures system

Type
Airborne Electronic CounterMeasures (ECM), radar jamming system.

Description
The AN/ALQ-187 is an internally housed, fully automatic jammer system integrated with radar warning and flare/chaff systems for tactical aircraft self-protection. The only fully integrated jammer on the F-16, it is also readily adaptable to installation in the F-4, F-18, A-7 and Mirage 2000 aircraft. The system detects and defends against surface-to-air missiles, anti-aircraft artillery and air-to-air interceptor weapon systems. It can interface with the AN/ALE-39 or AN/ALE-40 chaff/flare dispensers and the AN/ALR-66, AN/ALR-69 or AN/ALR-74 radar warning receivers.

The fully software programmable system automatically detects single and multiple-threat radars of the same or different technologies and selects the most effective of 13 Electronic Counter Measures (ECM) techniques to counter pulse, Pulse Doppler (PD) or Continuous Wave (CW) ground-based, shipborne or airborne emitters. ECM techniques include Range Gate Pull Off (RGPO), Range on False Targets, Inverse Gain, Velocity Gate Pull-Off (VGPO), Doppler False

The AN/ALQ-187 internal jamming system consists of four LRUs   0504080

Target, Swept Spot Noise and False Scan/False Lobe. The system is flight line reprogrammable, allowing immediate incorporation of the latest intelligence threat data and ECM techniques. Advanced power management ensures maximum jamming effectiveness. The AN/ALQ-187 jammer also forms part of the Raytheon Advanced Self-Protection Integrated Suites (ASPIS) system.

Specifications
RF coverage: H-J bands (6.5–18 GHz)

Dimensions:
(control unit) 86 × 146 × 83 mm
(forward transmitter unit) 222 × 336 × 533 mm
(aft transmitter) 256 × 222 × 533 mm
(processor) 203 × 188 × 458 mm
Weight:
(control unit) 1.35 kg
(forward transmitter) 43.1 kg
(aft transmitter) 29.5 kg
(processor) 14.1 kg

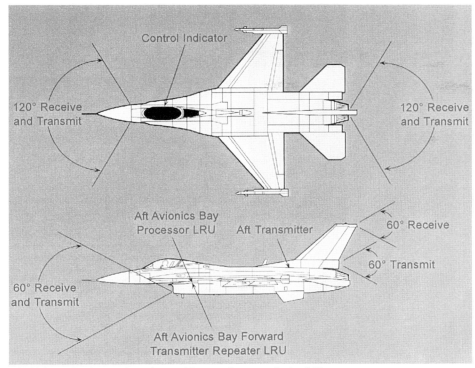

AN/ALQ-187 Equipment Location and System Coverage in the F-16   0092871

Status
The AN/ALQ-187 is in service in Greek Air Force F-16 aircraft.

Contractor
Raytheon Company, Space and Airborne Systems.

AN/ALQ-196 low band jammer

Type
Airborne Electronic Counter Measures (ECM), radar jamming system.

Description
The AN/ALQ-196 low band jammer is designed to provide effective defence against current and future Radio Frequency (RF) missile threats. The AN/ALQ-196 is an internally mounted Electronic Warfare (EW) system which has been in operation on US Air Force Special Operations Command (AFSOC) MC-130E Talon I aircraft since 1994; a similar system has also been in service aboard US Air Force U-2 Dragon Lady aircraft for more than two decades.

Status
During the second quarter of 2003, BAE Systems was selected by The Boeing Company to provide the AN/ALQ-196 low band jammer for the US Air Force Special Operations Command (AFSOC). The value of the contract was estimated at up to USD80 million over the next five years. BAE Systems Information & Electronic Warfare Systems (IEWS) will work with the prime contractor, Boeing, to integrate the ALQ-196 onboard AFSOC MC-130H Talon II and AC-130U gunship aircraft.

Contractor
BAE Systems North America.

AN/ALQ-202 autonomous jammer

Type
Airborne Electronic Counter Measures (ECM), radar jamming system.

Description
The AN/ALQ-202 autonomous jammer is an advanced capability ECM jamming system for internal installation on F-16, F/A-18, and other

aircraft. The ALQ-202 provides automatic and prioritised ECM responses to the full spectrum of radar threats, including advanced airborne interceptors and surface-to-air missiles. The ALQ-202 is fully integrated with onboard avionic systems including Radar Warning Receiver (RWR), Fire-Control Radar (FCR), chaff/flare dispenser and mission computer. System interfaces are designed for non-interference, allowing fully compatible operational performance of all onboard avionic systems. The ALQ-202 integration with the CounterMeasures Dispenser (CMD) system enables fully automated/co-ordinated electronic and dispensed countermeasures operations. The ALQ-202 is designed to operate independently, or interfaced with any RWR, including the ALR-56M Advanced Radar Warning Receiver (ARWR), and earlier ALR-67 and ALR-69 systems.

The ALQ-202 incorporates an internal Jammer Support Receiver (JSR). The JSR automatically scans through a programmed frequency range of interest, providing a rapid and independent threat detection capability against both pulsed and CW threat signals. The ALQ-202 further determines the direction of arrival of each identified threat radar, and selects from a prioritised multilevel list of ECM techniques to automatically counter

threat radars with an optimised directional response.

The ALQ-202 continually searches for new threats and updates system information on those previously detected. Jamming is both transponder and repeater type, and includes continuous and time-gated responses. The system utilises an extensive library of complex deception and denial jamming techniques, with proven effectiveness.

The ALQ-202 system comprises an ECM generator, multitransmitter unit(s), Digital RF Memory (DRFM) and system control unit. The ECM generator contains the jammer support receiver and ECM technique generating functions. The DRFM is provided for additional jamming techniques capability. Multitransmitter units utilise new technology travelling wave tubes for improved efficiency and reliability, with critical low (non-interfering) thermal noise level, to assure compatible operation with other aircraft systems.

Contractor
BAE Systems, North America, Information and Electronic Warfare Systems.

AN/ALQ-204 Matador IR countermeasures system

Type
Airborne Electronic Counter Measures (ECM), Infra-Red Counter Measures (IRCM) system.

Description
The AN/ALQ-204 Matador is a family of infra-red countermeasures systems which has 11 different configurations. The system is suitable for all types of large transport aircraft with unsuppressed engines, one transmitter per engine being recommended for maximum protection. The basic system consists of transmitters, a Controller Unit (CU), Power Supply Unit (PSU) and an operator's controller. Transmitters are electronically synchronised by the CU which controls and monitors one or two transmitters. The operator's controller is common to all configurations, controls from one to seven transmitters and incorporates a system status display.

Each transmitter contains an IR source which emits pulsed radiation to combat an IR missile. Preprogrammed multithreat jamming codes are selectable on the operator's control unit and all new codes can be entered as required to cope with new threats. Each transmitter weighs 22 kg nominal and can be pod mounted for aircraft attachment.

Status
In service on the Boeing 747 and 707 (derivatives), Lockheed Martin L-1011 and BAE Systems VC10,

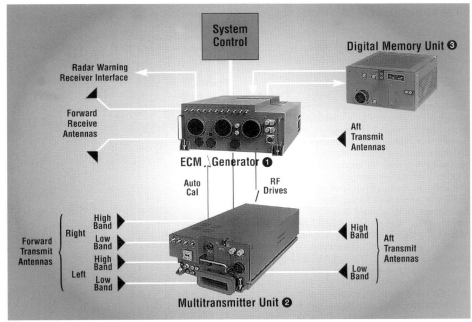

AN/ALQ-202 autonomous jammer units   0009915

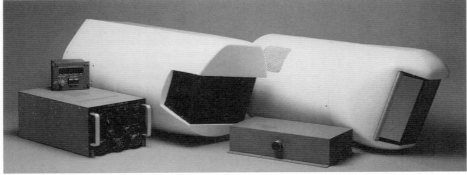

The AN/ALQ-204 infra-red countermeasures system is installed in a number of large fixed-wing aircraft   0503853

The AN/ALQ-204 has been integrated into a number of platforms, not least that of a famous head of state (BAE Systems)   0533860

146, Andover, Airbus -340 and Gulfstream G-IV aircraft.

All current applications are FAA certified for operation on Head of State, commercial and civilian aircraft.

Contractor
BAE Systems North America, Information and Electronic Warfare Systems.

AN/ALQ-210 Electronic Support Measures (ESM) system

Type
Airborne Electronic Counter Measures (ECM), Electronic Support Measures (ESM) system.

Description
The AN/ALQ-210 Electronic Support Measures (ESM) system is a multiple-bandwidth phase, frequency and amplitude measuring receiver. Incorporating digital receiver technology, the system features wide instantaneous bandwidth to provide enhanced probability of intercept as well as Radio Frequency (RF)-agile emitter acquisition.

Utilising these technologies, it is claimed that the AN/ALQ-210 locates both moving and stationary targets to an order of magnitude far exceeding the capability of classic Direction-Finding (DF) equipments. Enhanced threat identification and mode determination is accomplished via operationally proven algorithms and digital parametric accuracy. The AN/ALQ-210 provides autonomous processing to reduce operator workload, combined with manual intervention to allow for tailoring of parameters.

The design features a single-receiver processor unit combined with four identical antenna modules, Open Systems Architecture (VME) and Commercial-Off-The-Shelf (COTS) processing to ensure long-term supportability and growth.

AN/ALQ-210 ESM system is designed for a wide variety of aerospace applications, including ELectronic INTelligence (ELINT), Anti-Submarine Warfare (ASW) and Airborne Early Warning (AEW) duties. The simplified antenna requirements associated with the advanced location techniques are claimed to provide more installation flexibility than traditional interferometer solutions.

Specifications
RF Bandwidth: Up to 1 GHz, instantaneous
Processor: DSP plus 200 MHz PowerPC™
Dimensions:
  Receiver/Processor: 1,512 cu in
  Antenna modules: 1,317 cu in (4 modules)
MTBF: >3,000 h

Status
In production and in service on US Navy Sikorsky SH-60R multimission helicopters.

Contractor
Lockheed Martin Systems Integration – Owego.

AN/ALQ-211(V) Suite of Integrated RF Countermeasures (SIRFC)

Type
Airborne Electronic Counter Measures (ECM), integrated suite.

Description
The AN/ALQ-211(V) SIRFC system is a fully integrated airborne electronic combat system that provides real-time battlefield situational awareness, radar warning and self-protection capabilities to a multitude of aircraft.

SIRFC counters multiple, simultaneous pulse, pulse Doppler and continuous wave threats. It intercepts, analyses, passively ranges and geolocates multiple air- and ground-based RF threat signals and co-ordinates appropriate countermeasures or weapons cueing.

High-power pulse and continuous wave ECM transmitters provide protection against coherent and non-coherent radars. Through its advanced countermeasures module, SIRFC is claimed to provide robust countermeasure techniques against monopulse radars.

The system can act as an overall Electronic Warfare (EW) suite manager with the capability to automatically co-ordinate the dispensing of appropriate combinations of RF and/or IR countermeasures.

The AN/ALQ-211(V) SIRFC system provides an embedded capability which directly supports the battlefield digitisation initiatives of the US Army. SIRFC's real-time situational awareness reports and updates can be disseminated horizontally and vertically to aircrews and battlefield commanders. The location of radar-directed threats can be transmitted from the aircraft to the ground commander over the tactical datalink and displayed on appliqué hardware. The Situational Awareness (SA) and onboard countermeasures capabilities of the AN/ALQ-211(V) SIRFC system allows aircrews to evade or defeat a diversity of airborne and ground-based threats, significantly enhancing survivability.

The hardware elements of the SIRFC include a receiver-processor (Line Replaceable

Line Replaceable Unit 1 receiver/processor as installed in the Longbow Apache. SIRFC is in the lower right hand corner of the avionics equipment bay   0083138

The AN/ALQ-210 equips US Navy SH-60R helicopters   0034479

The AN/ALQ-211(V)2 configuration is optimised for the CV-22 Tiltrotor   0049445

AN/ALQ-217 Electronic Support Measures (ESM) system

Type
Airborne Electronic Counter Measures (ECM), Electronic Support Measures (ESM) system.

Description
Drawing on technology utilised in the Lockheed Martin AN/ALQ-210 (see separate entry), the AN/ALQ-217 Electronic Support Measures (ESM) system is an integral element of the E-2C Hawkeye 2000 mission system. The aircraft supports air and surface surveillance, intercept control, strike control, tanker coordination, search-and-rescue and drug interdiction missions. The main features of the system are:

- High Probability-of-Intercept (POI) in open ocean and dense littoral waters
- Rapid targeting solutions against all radar threats
- Emitter identification for high confidence under stringent Rules-of-Engagement (RoE)
- Fast reaction time and threat mode variance detection
- High commonality with the AN/ALQ-210 ESM system
- NDI-based active front ends for additional sensitivity
- Open-systems architecture (VME) and Commercial Off-The-Shelf (COTS) processing
- High Mean Time Between Failures (MTBF).

The system interfaces with the E-2C's central mission computer and tactical displays to provide the US Navy with enhanced capability to passively detect radar systems and disseminate the tactical picture to other airborne, shipborne or land-based tactical command centres.

The system upgrade incorporates Lockheed Martin's proven digital receiver technology with a militarised commercial architecture to provide improved system performance at lower weight and with increased MTBF compared with older systems. Weight reduction is achieved by an advanced antenna design, Radio Frequency (RF) component miniaturisation and increased individual module functionality. This weight reduction has allowed integration of other key sensors in the multirole mission of the E-2C without major airframe modification.

The AN/ALQ-217 consists of front end amplifiers, four antenna arrays and one receiver/processor, which utilises the Lockheed Martin SP-I03A Power PC 603e/704e/740™ Single Board Computer (SBC). Utilising technology inherent in the AN/ALQ-210 ESM system, installed in US Navy SH-60R multimission helicopters, has enabled Lockheed Martin to maintain a relatively low unit cost for the AN/ALQ-217.

Unit 1), a transmitter (LRU-2), an 'advanced countermeasures' module (LRU-3), four antenna quadrant electronic assemblies, a low-band detection antenna and associated converter unit, four quadrant receiver antennas, four transmit antennas and a transmit switch. Total weight of the system is approximately 45 kg.

The SIRFC is compatible with ITT Avionics' High Power Remote Transmitter (HPRT) radio frequency countermeasures unit. The HPRT is installed aboard MH-47E and MH-60K special forces helicopters.

The AN/ALQ-211(V)'s flexible design provides for potential installation on a variety of platforms. Known configurations and associated platforms include:

- **AN/ALQ-211(V)1:**
AH-64 Apache. Precision radar warning, threat geolocation and advanced jamming.
- **AN/ALQ-211(V)2:**
CV-22 Tiltorotor. Precision radar warning, threat geolocation, advanced jamming/SA and integration with the host platform's avionics.
- **AN/ALQ-211(V)3:**
RAH-66 Comanche (cancelled programme).
- **AN/ALQ-211(V)4:**
F-16 Falcon. Digital radar warning, enhanced SA, advanced high-/low-band jamming (optimised for air-to-air functionality) and Counter Measures Dispenser System (CMDS) control.
- **AN/ALQ-211(V)5:**
Eurocopter NH90. Precision digital radar warning, enhanced SA, Electronic Support (ES) capability, integration with the host platform's avionics and a defined growth path towards the implementation of a jamming capability.
- **AN/ALQ-211(V)6:**
MH-47E. Precision radar warning, threat geolocation, enhanced SA, high-power jamming and integration with the host platform's avionics.
- **AN/ALQ-211(V)7:**
MH-60K. As for ALQ-211(V)6.
- **AN/ALQ-211(V)8:**

The SIRFC makes extensive use of SEM-E modules and application specific and monolithic microwave integrated circuitry (ITT Avionics)   0515257

ERJ-145 ACS. Digital radar warning, enhanced SA, integration with the host platform's avionics and a defined growth path towards the implementation of a high-power jamming capability.

Status
In production and in service.

As of December 2005, the AN/ALQ-211(V)1 was reported as being at the Programme Objective Memorandum stage, with the ALQ-211(V)2, (V)4, (V)5, (V)6, (V)7 and (V)8 configurations all under contract. However, the termination of the US Army's USD879 million Aerial Common Sensor (ACS) programme during the early part of 2006 calls into question the status of the associated ALQ-211(V)8.

Contractor
ITT Industries, Avionics Division.

The AN/ALQ-217 ESM system is an integral element of the E-2C Hawkeye 2000 upgrade (Northrop Grumman)   0121193

AN/ALQ-211(V) hardware (ITT Avionics)   0017869

Specifications

Dimensions:
Receiver/Processor: 1.7 cu ft
Active front ends/Antennas: 3.5 cu ft
Weight:
Receiver/Processor: 42.3 kg (93 lb)
Active front ends/Antennas: 44.1 kg (97 lb)
MTBF: >1,400 h

Status

In production and in service on US Navy E-2C
Hawkeye 2000 aircraft.

Contractor

Lockheed Martin Systems Integration – Owego.

The AN/ALQ-504 airborne VHF/UHF intercept, DF, jamming and deception systems   0062921

AN/ALQ-504 airborne VHF/UHF intercept, DF, jamming and deception systems

Type

Airborne Electronic Counter Measures (ECM), communications jamming system.

Description

The AN/ALQ-504 is a small, lightweight airborne integrated intercept, DF, jamming and deception system, that can be installed in a wide range of fixed- and rotary-wing aircraft. The system can also be adapted for installation in wheeled or tracked vehicles, or on board ships.

The system performs manual or fully automatic search and detection of communications signals in the VHF and UHF bands between 20 and 400 MHz. Frequency extensions up to 2,500 MHz are available.

The AN/ALQ-504 system measures DF lines of bearing, and computes the location of intercepted signals. It enables the operator to jam selected signals or to transmit false messages for deception or training purposes. Signal jamming can also be linked to DF bearing, providing jamming of signals within a selected azimuth sector.

Jamming power output is 400 W to allow operation at a considerable standoff distance. Up to four active signals from a prioritised list of up to 10 frequencies can be jammed simultaneously using various strategies. The operator can seek and identify hostile signals, and can either jam each selected signal with one of a variety of computer-generated modulations, or implement the system's deception capabilities. The operator can use pre-recorded or processed messages as deception tools.

Contractor

ZETA, an Integrated Defense Technologies Company.

AN/ALR-52 ECM receiver

Type

Airborne Electronic Counter Measures (ECM), Electronic Support Measures (ESM) system.

Description

The AN/ALR-52 is a multiband Instantaneous Frequency Measuring (IFM) receiver for airborne or land applications. A naval variant is also produced, as the AN/WLR-11. Typical systems cover the microwave frequency band from 0.5 to 18 GHz, using receiver modules that cover octave bandwidths. Additional capabilities include: provision for selection of a particular signal or frequency band; blanking of signals which are of no further interest; separation of interleaved pulse trains; measurement of radar parameters; analysis of CW in addition to pulse trains; and direction of arrival measurement. The complete equipment provides for two operator positions, each equipped with a control unit and a display. The two IFM control units are capable of complete control over any band selected. If the same band is selected by both units, the video processing unit gives priority control to one position only. All intercepted signals, however, are presented on the displays at both positions.

A modified version of the ALR-52 has been developed and deployed, with one of the operator positions replaced by an interface unit to link the IFM receiver with a digital computer. This system employs a 1 to 18 GHz DF antenna system. The computer uses the data to tag every received pulse with its frequency, pulsewidth, time of arrival and azimuth. This information is stored and used to develop a library of emitter parameters.

Status

In operational service. The AN/ALR-52 is fitted to the EP-3E ELINT aircraft operated by the US Navy.

Contractor

ARGOSystems (a Condor Systems business unit).

AN/ALR-56A/C Radar Warning Receiver (RWR)

Type

Airborne Electronic Counter Measures (ECM), Radar Warning Receiver (RWR).

Description

The ALR-56 RWR is used with the AN/ALQ-135 Internal Countermeasures Set jamming equipment to form the Tactical Electronic Warfare System (TEWS) of the F-15 fighter aircraft.

The basic ALR-56 system incorporates the R-1867 processor/low-band receiver, the R-1866 high-band receiver, the IP-1164 display, the C-9429 immediate action control unit, the PP-6968 power supply, a TEWS controller and an antenna array. In more detail, the processor and low-band receiver unit contains three major sections: a single-channel low-band superheterodyne receiver, a dual-channel Intermediate Frequency (IF) section, and a processor.

The low-band receiver is electronically tuned under control of the processor. The dual-channel IF section operates with either the high-band dual-channel receiver or the low-band single-channel receiver. Selection of the receiver to operate with the dual-channel IF section is under the control of the processor.

The processor contains a preprocessor and a general purpose digital computer. The preprocessor contains all the video circuits for analysing intercepted signals. It also provides digital outputs to the computer which represent the measured signal parameters. On the basis of these measurements, a digital output is provided to the display and an audio signal generated. The computer also controls all ALR-56 system functions and is software programmable for the expected threat environment.

The solid-state, digitally controlled, dual-channel high-band receiver is capable of scanning an extremely large portion of the electromagnetic spectrum. Its single conversion superheterodyne design provides high sensitivity and the use of dual yttrium indium garnet Radio Frequency (RF) preselection is claimed to afford excellent selectivity and spurious rejection. High frequency accuracy is obtained through the use of a frequency synthesised local oscillator.

The RF input ports are configured to accept two main antenna inputs to each channel and two additional RF inputs which may be designated to either channel. All switching functions are provided internally.

The receiver is capable of receiving signals over a large dynamic range for analysis and Direction-Finding (DF) measurements. A self-contained precision signal source is provided for calibration and/or built-in test over the entire tuning range. Provision is also made for multiplexing out important receiver functions.

All receiver functions are controlled by serially generated NRZ Manchester coded data. The precise data rate also serves as the reference clock for the receiver frequency synthesiser.

On the countermeasures display, rapid threat evaluation for aircraft defensive manoeuvres is accomplished through automatic intensity control utilising ambient light sensing techniques. Sharp, high-contrast, unambiguous alphanumerics and special symbology allow immediate assessment of the overall tactical threat, without requiring pilot display adjustments as external ambient light fluctuates. Phosphor screen-optical filter

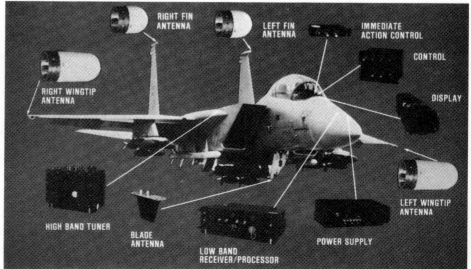

AN/ALR-56C radar warning equipment on the US Air Force F-15 Eagle   0503848

matching provides contrast enhancement to overcome direct sunlight contrast degradation.

A range-bearing data presentation is provided with special clutter elimination programmes as well as special threat status. Built-in test circuits determine display system malfunction and automatically indicate GO conditions. An integral lighting panel provides control illumination for night-time viewing.

The antenna system consists of four circularly polarised spiral antenna assemblies (each within its own radome) and a blade antenna. Collectively, this antenna system provides omnidirectional acquisition and direction-finding over the operating frequency range of the RWR. The four spiral antennas, which cover the high-band frequency range, are mounted in a manner that provides 360° azimuth coverage. DF operation is accomplished under computer control. The blade antenna provides omnidirectional coverage of the low-band frequency range.

The TEWS control unit provides for control of the RWR as well as other countermeasures subsystems. A single switch controls the entire RWR on/off function, and separate switches and relays are provided for the subsystems under RWR control. Audio tone cues are controlled from this unit also.

The other control panel is for activating in-flight functions that require instant pilot reaction. In addition to countermeasures subsystem mode control, interrogated data readout on this unit can be called up from the subsystems by means of push-button operation.

A major update known as the ALR-56C has now been completed with improvements in the processor to handle new threats, greater signal densities and other changes. ALR-56C is a digitally controlled, dual-channel superheterodyne receiver covering the E- to J-bands, and is capable of sorting and identifying all current and projected threats to the aircraft. RF input ports are configured to accept two main antenna inputs to each channel and two additional RF inputs that may be designated to either channel, with all switching functions provided internally.

ALR-56C was designed to improve on the ALR-56A signal acquisition capability to include the latest threat parameters and to improve on situational awareness and terminal radar direction. The system maintains the ALR-56A capabilities in the areas of analysis, establishment of priorities, jammer management, passive countermeasures management, and visible and audio displays and warnings.

Status
ALR-56 is in service aboard US Air Force and Saudi Arabian F-15 aircraft. ALR-56C equipments are reported to be installed on F-15E strike aircraft and as being retrofitted to F-15Cs and Ds as part of the two types' Multinational Staged Improvement Programme. ALR-56C is reported to be replacing ALR-56A throughout the F-15 inventory as well as being the standard installation on all new build aircraft.

Contractor
BAE Systems, North America, Information and Electronic Warfare Systems.

AN/ALR-56M(V) Radar Warning Receiver (RWR) system

Type
Airborne Electronic Counter Measures (ECM), Radar Warning Receiver (RWR).

Description
The AN/ALR-56M(V) has become the US Air Force's standard RWR and is designed to meet the 21st century threat environment and operational requirements. ALR-56M(V) was selected as the replacement for the AN/ALR-69(V) RWR in a US Air Force competitive fly-off programme in 1988. ALR-56M(V) has successfully completed all required US government testing, including US DoD operational test and evaluation and has been

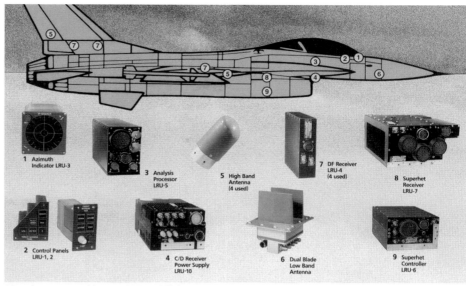

The elements making up ALR-56M(V) as applied to the F-16   0503849

1 Azimuth Indicator LRU-3
2 Control Panels LRU-1, 2
3 Analysis Processor LRU-5
4 C/D Receiver Power Supply LRU-10
5 High Band Antenna (4 used)
6 Dual Blade Low Band Antenna
7 DF Receiver LRU-4 (4 used)
8 Superhet Receiver LRU-7
9 Superhet Controller LRU-6

in production since 1989. Over 900 ALR-56M(V) systems are known to be in service with the US Air Force or export customers.

ALR-56M(V) is stated to have an internal jammer hardware interface; sensitivity greater than maximum threat engagement ranges; adaptive selectivity options to provide threat sorting in dense environments; frequency agility to allow rapid adaptive RF scanning; measurement ability to sort and identify time of arrival, pulse repetition frequency, signal strength, pulsewidth, angle of arrival and RF frequency on a monopulse basis; throughput to handle high pulse density within the response time requirement; and an interface capability to all aircraft avionics via dual redundant MIL-STD-1553B busses – one for aircraft avionics (inertial navigation system, fire-control radar and control/display systems) and one for electronic warfare system integration (AN/ALE-47 countermeasures dispenser, internal or external jammer, flight test instrumentation and memory loader/verifier unit). ALR-56M(V) also provides the ability to store multiple, in-flight selectable mission data tables. The operational flight programme and the mission data tables are completely separate entities, allowing for easy reprogramming of parametric data. ALR-56M(V) supports two-level maintenance, enhanced by extensive built-in test.

The system consists of four direction-finding receivers, each connected to one of the four high-band quadrant antennas, a C/D-band receiver/power supply, connected to a low-band antenna assembly, a superheterodyne receiver, a superheterodyne controller and an analysis processor which communicates with the control panel and azimuth indicator and provides two MIL-STD-1553B busses. ALR-56M(V) is effectively a time-shared, rapid, agile single-channel superheterodyne receiver for processing the signals intercepted by the four high-band antennas and the low-band antenna assembly. Wideband, fast, frequency-agile, sensitive receivers provide intra-pulse parametric measurement capability. A computer-controlled tunable radio frequency notch filter, adaptive digital signal processing and parametric screening provide the ability to control the input to the high-speed control processor unit resulting in an unambiguous rapid threat warning, even when operating in the densest, most sophisticated threat environment.

Status
ALR-56M(V) entered worldwide operation in the US Air Force in 1992. ALR-56M(V) has been selected for overseas application on F-16 and C-130 aircraft, including Foreign Military Sales contracts for South Korea and Taiwan.

ALR-56M(V) is reported to be mandated for all US Air Force AC-130U, CV-22 and Block 50/52 F-16C/D aircraft. The system is also a retrofit item for US Air Force Block 40/42 F-16 aircraft and

under consideration for use in B-1B strategic bombers. In addition, it is the recommended replacement for the AN/ALR-69(V) on the F-4, MH-53J, C-130, Joint STARS, A-10, MC-130E, EC-130 and C-141. It is the baseline offer for C-130J aircraft.

A January 1998 US Air Force contract provided for delivery of 490 replacement computer kits. More than 1,000 units have been ordered.

Contractor
BAE Systems, North America, Information and Electronic Warfare Systems.

AN/ALR-66B(V)3 surveillance and targeting system

Type
Airborne Electronic Counter Measures (ECM), Electronic Support Measures (ESM) system.

Description
The AN/ALR-66B(V)3 was developed for the US Navy as an enhancement in capability for its multimission aircraft. The ALR-66B(V)3 surveillance and targeting system is the successor to the US Navy's AN/ALR-66A(V)3 system.

Integrated with the aircraft's radar antenna, the system provides ultra-high system sensitivity and precision DF accuracy. Interfaced with other aircraft sensors, the AN/ALR-66B(V)3 provides operation on a non-interfering basis, as well as interfacing with aircraft navigation systems, central computer and display.

Radar antenna modification techniques provide C- to J-band signal reception and precision direction-finding capabilities required for over-the-horizon targeting. Simultaneous operation of the radar and ESM surveillance and targeting functions is allowed. Alternatively, a dedicated spinning antenna may be used for 360°C- to J-band precision measurements.

Advanced signal processing techniques for instantaneous, positive emitter identification in high-density environments are used. A multimode emitter library, which uses EEPROM technology, permits rapid reprogramming of total library scenarios and individual emitters. Emitters not found in the library are displayed by their generic characteristics with precise parameter measurements. The use of EEPROM technology for the AN/ALR-66B(V)3 eliminates the need for hardware modification when changes to the memory are made.

The basic modes of operation are:

Surveillance mode: the highest priority emitters are presented on the plasma display in positions corresponding to their range and bearing, using unique symbology which is either emitter or platform related.

The AN/ALR-66B(V)3 surveillance and targeting system   0503843

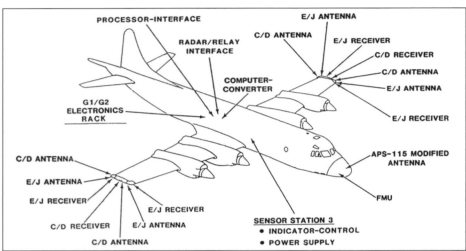

The AN/ALR-66B(V)3 layout in the Lockheed P-3C Orion ASW aircraft   0503844

Targeting mode: precise targeting data on any emitter is presented on the plasma display. This operating mode provides the OTH targeting data required for weapon activation.

Emitter waveform analysis: video is accepted from the computer-converter and data is presented directly on the plasma display, allowing the operator to analyse emitter waveforms.

Specifications
Weight: 87.3 kg
Receiver: crystal video
Frequency: contiguous over the C- to J-bands
Warning and identification: all pulsed radars including pulse Doppler, CW, ICW, LPI, 3-D, jitter/stagger, pulse compression and agile
Radar storage: >2,000 emitter modes in removable library storage module
Symbology: 1, 2 or 3 symbols per emitter as desired, programmable

Status
The AN/ALR-66(V)3 has been procured by the US Navy and other international customers for Lockheed Martin P-3B and P-3C aircraft.

The AN/ALR-66(V)3 has been upgraded to the AN/ALR-66B(V)3 for the US Navy.

Contractor
Northrop Grumman Corporation, Electronic Systems Sector, Defensive Systems Division.

AN/ALR-67(V) countermeasures warning and control system

Type
Airborne Electronic Counter Measures (ECM), Radar Warning Receiver (RWR).

Description
The AN/ALR-67 countermeasures warning and control system is the successor to the US

Navy's AN/ALR-45 system installed in A-6E and F-14A tactical aircraft. It is the standard threat warning system for tactical aircraft and was specifically designed for the A-6E/SWIP, AV-8B, F-14B, F-14D and F/A-18. The system detects, identifies and displays radars and radar-guided weapon systems in the C to J frequency range. The system also co-ordinates its operation with onboard fire-control radars, datalinks, jammers, missile detection systems and anti-radiation missiles. The dispensing of expendables, such as chaff, flares and decoys, is controlled by the AN/ALR-67(V)2.

The AN/ALR-67(V)2 consists of broadband crystal video receivers, a superheterodyne receiver, an integrated low-band receiver, an

antenna array and an alphanumeric azimuth indicator. The system is field-programmable and includes provision for software updates at squadron level. The system is fully compatible with MIL-STD-1553B databus requirements and features interface control of such systems as HARM, ALQ-126A/B, ALE-47, ALQ-162 and ALQ-165.

The AN/ALR-67(V)2 comprises the following units:
1. four small spiral high-band antennas to provide 360° azimuth RF coverage
2. four wideband, high-band quadrant receivers
3. a low-band array plus receiver to provide 360° azimuth low-band coverage
4. a narrowband superheterodyne receiver for signal analysis functions
5. twin CPU
6. threat display
7. control unit.

The AN/ALR-67(V)2 in turn has been given a significant enhancement in capability, through Engineering Change Procedure ECP-510 to the AN/ALR-67E(V)2 standard.

AN/ALR-67E(V)2
ECP-510 provided a card-for-card upgrade of the AN/ALR-67(V)2 to the AN/ALR-67E(V)2 standard. It provides a significant increase in system sensitivity in the presence of strong signals and offers a significant increase in computer pulse processing capability. The upgrade also features the capability to detect and exploit unique signals for improved tactical awareness. To better manage high density signal environments, the AN/ALR-67E(V)2 incorporates Application-Specific Integrated Circuit (ASIC) chips that increase the system's processing power five-fold. The AN/ALR-67E(V)2 provides additional enhancements including a 10-fold improvement in detection ranges when in the presence of a wingman's radar signals; it also incorporates INS stabilisation for accurate display in high 'G' manoeuvres.

The AN/ALR-67E(V)2 modular architecture permits further enhancement, without change to the aircraft system, to provide yet more computer power, increased detection range, and improved target discrimination.

AN/ALR-67(V)3/4
For clarity, note that the AN/ALR-67(V)3/4 developments of the AN/ALR-67(V) are manufactured by the Raytheon Systems Company.

Status
In production. Over 1,600 AN/ALR-67(V) and AN/ALR-67(V)2 systems have been sold. The AN/ALR-67(V) has been supplied for the AV-8B, F-14 and F/A-18A/B/C/D aircraft of the US Navy and US Marine Corps and the air forces of Australia,

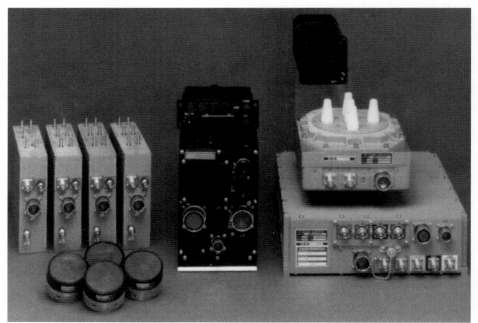

Northrop Grumman AN/ALR-67(V)2 countermeasures warning and control system   0018269

Canada, Finland, Kuwait, Malaysia, Spain and Switzerland.

Contractor

Northrop Grumman Corporation, Electronic Systems Sector, Defensive Systems Division.

AN/ALR-67(V)3 countermeasures receiving set

Type

Airborne Electronic Counter Measures (ECM), Radar Warning Receiver (RWR).

Description

The AN/ALR-67(V)3 is commonly known as the Advanced Special Receiver (ASR) set. It has been designed to be the standard US Navy Radar Warning Receiver (RWR) for carrier-based tactical aircraft such as the F/A-18C/D, F/A-18E/F, F-14C/D, A-6E and AV-8B.

The system comprises seven Weapon Replaceable Assembly (WRA) types to give a total of 13 WRAs. These consist of four integrated antenna detectors, four quadrant receivers, countermeasures receiver, countermeasures computer, low-band integrated antenna, control status unit and azimuth display indicator.

Each Integrated Antenna Detector (IAD) includes a dual-polarised microwave antenna, MMW antenna and supporting electronics. For ease of installation in various aircraft types, the IAD is available in three different housings.

Each IAD feeds signals into an associated Quadrant Receiver (QR) which conditions the received energy for processing. Conditioning involves filtering, amplification and frequency conversion to IF. Band structure has been optimised to minimise interference from onboard jammers.

The Countermeasures Receiver (CR) accepts preconditioned signals from the QR and low-band WRAs and generates digital words describing the parameters of the pulsed and CW radar waveforms detected. Parameters include amplitude, angle of arrival, time of arrival, frequency, pulsewidth and modulation.

The Countermeasures Computer (CC) incorporates an Ada programmable 32-bit JIAWG-compatible computer. Software in the CC processes the pulse and CW data to characterise, identify and prioritise intercepted threats based on their potential lethality. The CC manages all external interfaces, including two MIL-STD-1553B busses and special interfaces to various jammers and missile launch computers.

The low-band integrated antenna receives and conditions signals in the lower frequency region and is intended to counter those elements of the SAM threat that utilise lower frequencies.

The Control Status Unit (CSU) is a push-button WRA (carried over from the AN/ALR-67(V)2), fitted in the cockpit that enables the pilot or maintenance technician to issue commands to the system.

The Azimuth Display Indicator (ADI) is a 3 in (76.2 mm) diameter CRT cockpit display, also carried over from the AN/ALR-67(V)2, used to show intercepted threats. The ADI may not be required in the F/A-18E/F, AV-8B or F-14D programmes.

The AN/ALR-67(V)3 also forms part of the IDECM programme, including an interface to the AN/ALE-50 towed decoy system.

Status

In production and in service in F/A-18E/F aircraft. In August 1999, Raytheon was awarded an initial contract for full-rate production of the AN/ALR-63(V)3 for the US Navy F/A-18E/F, totalling 34 complete installations, together with 40 spare quadrant receivers and five countermeasures receivers. These were delivered during 2001/02. Further production contracts followed, with the latest in April 2005, where Raytheon received its seventh production contract for 42 systems totalling USD44 million.

Contractor

Raytheon Company, Space and Airborne Systems.

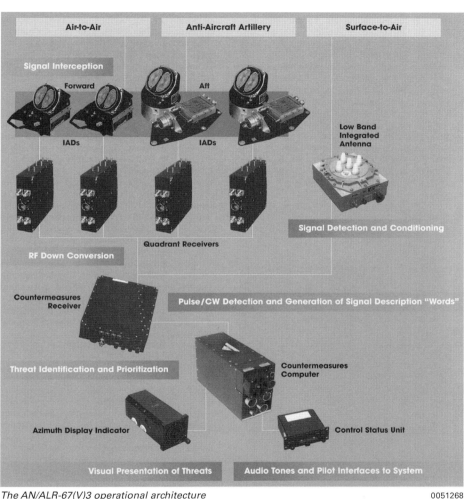

The AN/ALR-67(V)3 operational architecture   0051268

Components of the AN/ALR-67(V)3 (Raytheon)   1127371

AN/ALR-68A(V)3 advanced radar warning system

Type

Airborne Electronic Counter Measures (ECM), Radar Warning Receiver (RWR).

Description

The AN/ALR-68A(V)3 advanced radar warning system is now deployed operationally by the German Air Force. It is a digital, radio frequency, threat warning receiver system that replaced the AN/ALR-46.

The ALR-68A(V)3 is a broadband crystal video, field-programmable system which provides in-cockpit threat parameter programming and hand-off to tactical ECM systems. It uses a digital threat processor containing software provisions unique to the German threat scenario.

Status

Manufactured in association with EADS Deutschland, the AN/ALR-68A(V)3 is installed in C-160, F-4F and RF-4E aircraft of the German Air Force.

Contractor

Northrop Grumman Corporation, Electronic Systems Sector, Defensive Systems Division.

Northrop Grumman Advanced Systems' AN/ALR-68A(V)3 advanced radar warning receiver   0062862

AN/ALR-69A(V) Radar Warning Receiver (RWR)

Type
Airborne Electronic Counter Measures (ECM), Radar Warning Receiver (RWR).

Description
Raytheon claims that the AN/ALR-69A(V) is the world's first all-digital RWR. It uses broadband digital receiver technology integrated into an open architecture design to provide improved detection ranges and together with accurate and unambiguous emitter identification in dense RF environments. 360° coverage of the host platform is facilitated by four independent quadrant receivers.

Status
In production and in service. During the third quarter of 2001, Raytheon was selected for the US Air Force AN/ALR-69 upgrade programme and awarded a USD26 million contract to design, develop, test, manufacture and integrate a radar warning receiver subsystem to be installed, initially, in US Air Force (USAF) C-130 aircraft. The company was tasked to develop a receiver subsystem which provides for threat detection, identification and situational awareness to air force pilots. The programme required three prototype units and five pre-production units to be used for initial flight testing on C-130 and F-16 aircraft. The programme also included optional enhancements incorporating geolocation, improved azimuth accuracy and determination of specific emitter identification of detected threats. The new ALR-69A(V) system requires less space than the existing AN/ALR-69 RWR, while providing greatly enhanced situational awareness to the aircrew.

Contractor
Raytheon Company, Space and Airborne Systems.

AN/ALR-73 detection system

Type
Airborne Electronic Counter Measures (ECM), Electronic Support Measures (ESM) system.

Description
The AN/ALR-73 passive detection system is an airborne ESM equipment developed for the US Navy's E-2C airborne early warning aircraft. It is an improved version and successor to the AN/ALR-59 and, because of the extensive update, has been given its own designation. The ALR-73 is intended to augment the AEW, surface, subsurface and command and control functions of the E-2C by enhancing the threat detection and identification capabilities of the aircraft. It is a completely automatic, computer-controlled, superheterodyne receiver processing system that communicates directly with the E-2C command and control central processor. The design of the system was motivated by four major considerations: very high probability of intercept in dense environments; automatic system operation; high reliability; and ease of maintenance.

Features of the ALR-73, which are related to its intercept probability performance, include 360° antenna coverage, four independently controlled receivers, dual-processor channels and a digital closed-loop rapid tuned antenna. Features concerned with automatic operation are low false alarm rate, automatic overload logic, an AYK-14 computer that adaptively controls hardware and degraded mode operation.

The system uses 52 grouped antennas in four sets, one for each of several wavebands. Each complete set is positioned to look at a 90° sector. The forward and aft antennas are in the fuselage extremities and the sideways-looking aerials are in the tailplane tips. All receiver sets are under separate control, so wavebands are scanned independently and simultaneously in all sectors. Each antenna has dual-processing channels and uses digital closed-loop rapid tuned local oscillators. The latter provide instantaneous frequency measurement with fast time response and high-accuracy frequency determination.

Receiver outputs are collected at a signal preprocessor unit which performs pulse train separation, direction-finding correlation, band tuning and timing and built-in test equipment tasks. Data is then in a form suitable for the general purpose digital computer, which has overall control of electronic surveillance measures operations and will vary frequency coverage, dwell time and processing time according to prescribed procedures. Control of these parameters is aimed at maximising the probability of intercepting signals on particular missions. Other onboard sensor data and crew inputs will determine the technique adopted. Data such as signal direction of arrival, frequency, pulsewidth, pulse repetition frequency, pulse amplitude and special tags are sorted by the computer and transmitted to the E-2C central processor.

The ALR-73 can measure direction of arrival, frequency, pulsewidth and amplitude and PRI simultaneously. Scan rate information is also available if called for by the central processor. Special emitter tags can be provided. The ALR-73 detects and analyses electromagnetic radiation within the microwave portion of the spectrum and sends emitter reports of pulsewidth, PRI, direction of arrival, frequency, pulse amplitude and special tag to the E-2C's central processor via its own data processor. The ALR-73 immediately reports new emitters to the central processor, which performs the identification function. It eliminates redundant data on emitters for a programmable period of time, thus significantly reducing the data rate to the central processor. The ALR-73, a multiband, parallel scan, mission programmable system covers the frequency range in four bands through step sweeping. Programmable frequency bands and dwell time permit very rapid surveillance of priority threat bands. Non-priority bands are also monitored,

Raytheon's digital AN/ALR-69A(V) Radar Warning Receiver (RWR) (Raytheon)   1127370

The US Navy E-2C Hawkeye carries the Litton AN/ALR-73 ESM system   0503842

but at a reduced rate. Probability of intercept is increased without sacrificing sensitivity through the detection of both real and image sidebands.

Status
In production, and in service on the US Navy Northrop Grumman E-2C and with the forces of France, Japan, Singapore and Taiwan.

The AN/ALR-73 has also been successfully installed on C-130 aircraft.

Contractor
Northrop Grumman Corporation, Electronic Systems Sector, Defensive Systems Division.

AN/ALR-76 ESM system

Type
Airborne Electronic Counter Measures (ECM), Electronic Support Measures (ESM) system.

Description
The AN/ALR-76 is a form, fit and function replacement for the ALR-47 (see separate entry) and is part of the Weapons Systems Improvement Program (WSIP) for the S-3B ASW aircraft. Compared with the ALR-47, the ALR-76 has an extended frequency range, which includes the microwave frequency region, with particular emphasis on radar transmissions of short duration. Capabilities include detection, tracking, classification, high-accuracy location and identification of emitters in dense RF environments.

The AN/ALR-76 combines Electronic Support (ES) and radar warning functions in a single system, consisting of two sets of spiral antennas (four in each), two multiband receivers and a signal comparator. Audio alerts are provided, as are outputs for a Counter Measures Dispenser

The wingtip ALR-76 antenna assembly as applied to a US Navy EP-3E Aries II land-based SIGINT aircraft (Martin Streetly)   0515217

System (CMDS). Dual-channel operation provides for full 360° instantaneous azimuth coverage. The signal comparator measures the characteristics of the outputs from the four receiver channels simultaneously and provides an instantaneous 360° azimuth Field of View (FoV) and a monopulse direction-finding capability. All emitter parametric data is digitally encoded. Computer programmes, contained in a pulse processor and a general purpose Control and Correlation Processor (CCP), both of which are contained within the signal comparator unit, perform signal processing and system control functions.

Specifications
Weight: 61 kg

Status
The AN/ALR-76 is in full production for the US Navy S-3B WSIP, as well as the EP-3E and ES-3A electronic reconnaissance aircraft. It is also installed on the Canadian EST Challenger aircraft.

Contractor
Lockheed Martin Systems Integration – Owego.

AN/ALR-93(V)1 Radar Warning Receiver/Electronic Warfare Suite Controller (RWR/EWSC)

Type
Airborne Electronic Counter Measures (ECM), Radar Warning Receiver (RWR).

Description
The AN/ALR-93(V)1 is a computer-controlled RWR that provides automatic detection and display of RF signals in the C/D and E- to J-bands.

The AN/ALR-93(V)1 is designed to operate in dense, complex RF threat emitter environments with near 100 per cent Probability of Intercept (POI). By combining the wideband acquisition capability of an amplified Crystal Video Receiver (CVR) with the fast frequency measurement of an Instantaneous Frequency Measurement (IFM) receiver and the selectivity of a SuperHeterodyne Receiver (SHR), the AN/ALR-93(V)1 is able to reduce threat ambiguities, increase detection range and still achieve a high POI.

The AN/ALR-93(V)1 provides threat warning to the aircrew both visually and aurally; display of warnings can be integrated with other aircraft warning systems.

The AN/ALR-93(V)1 also provides a flexible mission recording and software support system, providing for post-mission analysis and mission optimisation.

Specifications
Weight: <27.2 kg
Frequency coverage: C/D, E to J-band
Receiver types: amplified CVR, IFM, SHR
Radar types: all pulsed radars including pulse Doppler, CW, ICW, LPI, jitter/stagger, pulse compression, frequency-agile, PRI agile and agile-agile
DF accuracy: 15° rms (E- to J-bands), omni directional (C/D-band)
Emitter library storage: 1,800 modes
Reliability: 525 h MTBF

Northrop Grumman AN/ALR-93(V)1 radar warning receiver and electronic warfare suite controller
0062867

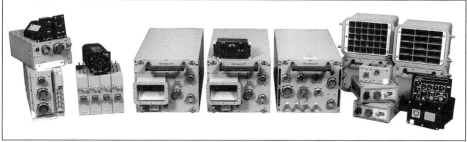

Aspis EW system   0062861

Interface options: 2 dual redundant MIL-STD-1553B, RS-232C, RS-422, discretes
Warnings: visual and aural
Programming: both OFP and emitter library are contained in EEPROM
Software: 'C' higher order language.

Status

Operational on 80 F-16C/D aircraft for a NATO customer. Believed to be part of the Greek ASPIS EW system. Also provided for fit to CN-253 and A-4M aircraft for international customers.

Contractor

Northrop Grumman Corporation, Electronic Systems Sector, Defensive Systems Division.

AN/ALR-94 electronic warfare system

Type

Airborne Electronic Counter Measures (ECM), integrated suite.

Description

The AN/ALR-94 electronic warfare system is an integral element of the F/A-22 Raptor sensor suite. It is reported to provide RF warning and countermeasure functions, together with missile-launch warning, and to be integrated with the AN/ALE-52 flare dispenser.

The F/A-22 Raptor aircraft   0079236

Status

In February 1999, BAE Systems North America completed production and delivery of the first 11 AN/ALR-94 suites for the F/A-22 EW Engineering and Manufacturing Development (EMD) programme.

Units shipped from BAE Systems to the F/A-22 Avionics Integration Laboratory (AIL) in Seattle, include: advanced apertures and associated electronics, a Remote Aperture Interface Unit (RAIU) and RF unit, as well as aircraft power supplies, test stands and workstations. BAE Systems is also providing, among other items, the overall Communications, Navigation and Identification (CNI) antennas for the F/A-22.

Contractor

BAE Systems North America, Information and Electronic Warfare Systems.

AN/ALR-95(V) maritime patrol ESM system

Type

Airborne Electronic Counter Measures (ECM), Electronic Support Measures (ESM) system.

Description

The AN/ALR-95(V) Electronic Support (ES)/Radar Warning Receiver (RWR) system is designed to provide automatic signal intercept, identification and Direction-Finding (DF). The system incorporates a signal processor, tuner, multiple radio frequency distribution units, a digital instantaneous frequency measuring receiver, a Windows-based workstation, an Omni-directional antenna, a narrowband precision DF antenna and an array of wideband antennas. Other system features include:

- High-sensitivity set-on receiver channel;
- Special signal processing;
- Real-time signal analysis displays;
- Amplitude monopulse DF;
- Integrated ES and threat warning functions;
- Frequency/azimuth blanking and interference suppression;
- Precision parameter measurement;
- Emitter identification (>10,000 mode emitter library);
- Ability to utilise existing AN/ALR-66 ES system wingtip antennas/receivers;
- Optional AN/ALE-47 CMDS interface.

Specifications

Power: 30 W (wideband receiver); 70 W (narrowband DF antenna); 85 W (RF distribution unit); 140 W (tuner); 220 W (DIFM); 500 W (processor)
Dimensions (H × W × D): 91 × 137 × 213 mm (wideband receiver); 91 × 165 × 357 mm (RF distribution unit); 102(D) × 114(Ø) mm (wideband antenna); 140(H) × 147(Ø) mm (omni antenna); 198 × 147 × 533 mm (tuner); 216 × 389 × 584 mm (DIFM); 267 × 274 × 483 mm (processor); 584(H) × 610(Ø) mm (narrowband DF antenna)
Weight: 0.45 kg (wideband antenna); 1.8 kg (omni antenna); 4.1 kg (wideband receiver); 7.7 kg (RF distribution unit); 16.3 kg (tuner); 18.1 kg (processor); 22.7 kg (narrow band DF antenna); 26.3 kg (DIFM)

Status

In production and in service. Installed in US Navy Lockheed P-3C Orion Maritime Patrol Aircraft (MPA) as part of the Avionics Improvement Programme (AIP) for the aircraft.

Contractor

EDO Reconnaissance and Surveillance Systems.

AN/ALR-606(V)2 series surveillance and direction-finding systems

Type

Airborne Electronic Counter Measures (ECM), Electronic Support Measures (ESM) system.

Description

The AN/ALR-606(V)2 surveillance and direction-finding system is specifically designed for use in maritime patrol aircraft, offering coverage in the C- to J-bands and over-the-horizon emitter location. It was derived as an export version of the AN/ALR-66A(V)3 system. The equipment provides advanced capabilities in such areas as precision DF accuracy, high sensitivity for over-the-horizon detection, precise frequency measurement, advanced signal processing coupled with expanded data memory, multimode operator interactive display and controls, precision emitter parameter measurements, integration with other aircraft primary sensors and EEPROM flight line reprogramming. Upgrades to the ALR-606A(V)2 can be accomplished with minimal or no aircraft wiring change. Upgrade includes higher sensitivity E- to J-band receivers, expanded emitter library capacity, improved processing and in-flight signal-of-interest programming. Additional upgrades to ALR-606B(V)2 have been developed which include incorporation of a dedicated, controllable DF antenna and the addition of a data recorder to support ground analysis of ESM intercepts.

Status

The AN/ALR-606(V)2 has been provided for the Northrop Grumman S-2 maritime patrol aircraft, Sikorsky S-70 helicopter and for numerous international customers. The ALR-606(V) as well as the upgraded ALR-606A(V)2 and ALR-606B(V)2 systems are in production.

Contractor

Northrop Grumman Corporation, Electronic Systems Sector, Defensive Systems Division.

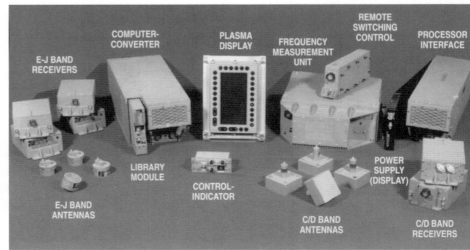

Units of the AN/ALR-606(V)2 surveillance and direction-finding system   0504077

AN/APR-39(V) Radar Warning Receiver (RWR)

Type
Airborne Electronic Counter Measures (ECM), Radar Warning Receiver (RWR).

Description
The APR-39(V)1 RWR equipment provides automatic warning of emitters in the E, F, G, H, I and most of J radar bands, as well as the appropriate portions of C- and D-bands. It is intended for use on either fixed-wing aircraft or helicopters. The equipment provides indications of bearing, identity and the mode of operation of detected signals with the acquired data being displayed on a cockpit indicator. Proportional pulse repetition frequency of displayed signals and alarm tones are presented to the crew via an integral audio warning subsystem.

The basic APR-39(V)1 system comprises two dual-video receivers, four spiral cavity-backed antennas, one blade antenna, an indicator unit, a comparator and a control unit. Without cables and brackets, this weighs 3.63 kg.

An updated version, APR-39(V)2 differs in that the comparator is replaced by a CM-480/APR-39(V) digital processor. This performs signal sorting, identification of emitters, bearing computation and character generation for the presentation of threat details in alphanumeric form on the cockpit display. The unit weighs 6.5 kg and incorporates an adaptive noise threshold and angle-gate, programmable pulse repetition interval filters, and coded emitter outputs. A 19,000-word programmable read-only memory/ random access memory is provided.

Status
APR-39(V) series RWRs are reported to have been installed in 15 types of aircraft and patrol/ fast attack craft. Several thousand APR-39(V) systems are reported to have been produced, a figure which includes 230 for installation aboard German PAH-1 anti-tank helicopters.

Contractor
BAE Systems, North America, Information and Electronic Warfare Systems.

AN/APR-39A(V) threat warning systems

Type
Airborne Electronic Counter Measures (ECM), Radar Warning Receiver (RWR).

Description
AN/APR-39A(V)1
The AN/APR-39A(V)1 is an upgrade of the earlier analogue APR-39(V)1 radar warning system. It is a lightweight system designed for helicopters and light fixed-wing aircraft. It provides visual and aural warning of hostile radar energy incident on the host aircraft over a very wide frequency range. The system displays multiple threats, (with the highest priority indications highlighted) while keeping track of other emitters in the environment.

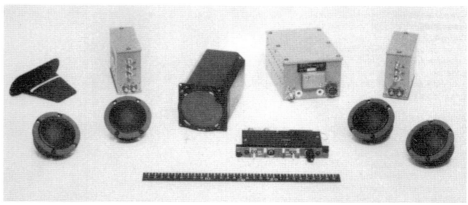

Units of AN/APR-39(V)1 RWR system   0514897

A digital display clearly identifies the threat type and azimuth from the aircraft. It also indicates if the threat is searching, locked and tracking, or when the radar lock is broken. The alphanumeric display will automatically separate symbols that are grouped too closely together. The aural warning is by synthetic voice.

AN/APR-39A(V)3
AN/APR-39A(V)3 uses APR-39(V)1 technology and provides continuous coverage through dual-channel crystal video receivers. This system is deployed on US Army and NATO aircraft. It is specifically designed for helicopters and other light aircraft operating at very low levels. It consists of 10 line-replaceable units which are form-fit compatible with the analogue APR-39(V)1. Frequency range covered is E/J and C/D.

Status
In service and, in the case of the AN/APR-39A(V)3, also in production. (See also Northrop Grumman Advanced Systems Division entry for AN/APR-39A(V)2 and 'AN/APR-39A(V)3 with CW' and threat warning systems).

Contractor
BAE Systems, North America, Information and Electronic Warfare Systems.

AN/APR-39A(V)1, AN/APR-39A(V)2, AN/APR-39A(V)3 and AN/APR-39B(V)1/3 threat warning systems

Type
Airborne Electronic Counter Measures (ECM), Radar Warning Receiver (RWR).

Description
Note: The original AN/APR-39(V) was an analogue system designed and developed by E-Systems (now part of Raytheon Electronic Systems). The AN/APR-39A(V) was a digital form-fit replacement. The AN/APR-39A(V)1 and AN/APR-39A(V)3 are made by BAE Systems North America, with Northrop Grumman as a second source. The AN/APR-39A(V)2 and AN/APR-39B(V)1/3 are made by Northrop

Grumman Electronic Systems Sector, Defensive Systems Division.

AN/APR-39A(V)1
Designed for helicopters and other light aircraft operating at low level and in Nap-Of-the-Earth (NOE) mission environments, the digital AN/APR-39A(V)1 provides audio and visual (NVG compatible) warning. LRUs include: one digital signal processor; 2 H/M band crystal video receivers; four H/M band spiral antennas; one C/D band receiver; one C/D band blade antenna; one display unit and one control unit.

Specifications
Frequency range: H/M and C/D bands
Weight: 7.05 kg
Power supply: 28 V DC, 58 W
Interfaces: RS-422 databus, MIL-STD-1553B databus optional, E-O warning systems, missile launch detectors, radar jammers, CW warning receivers

AN/APR-39A(V)2
The AN/APR-39A(V)2 is designed for use on Special Electronic Mission Aircraft (SEMA), helicopters and non-high-performance fixed-wing aircraft such as the C-130 and CV-22. In addition to threat warning, the system acts as a controller in an integrated, digitally controlled, electronic warfare survivability suite. The AN/APR-39A(V)2 is integrated with laser warning, missile warning systems and countermeasures dispenser systems. Major LRUs include: four quadrant antennas; two dual channel receivers; one central processor/receiver; one display unit (NVG compatible); one control unit; and one blade antenna.

AN/APR-39A(V)2 and -39B(V)2
The current configurations of threat warning systems are the APR-39A(V)2 and APR-39B(V)2. Both cover the frequency range from C- to K-band and provide conventional radar warning. The -39A(V)2 is used by US Army and Marine Corps helicopter and fixed-wing aircraft. In addition to threat warning, it acts as a control for the EW survivability suite. It is integrated with the laser warning receiver and missile warner.

Specifications
Frequency range: E-K and C/D bands
Weight: 15.8 kg
Power supply: 28 V DC, 200 W
Interfaces: RS-422 databus, MIL-STD-1553B databus optional, E-O warning systems, missile launch detectors, radar jammers, CW warning receivers, and mission data recorders

AN/APR-39A(V)3
Designed for helicopters and other small aircraft operating at low level and NOE missions, the AN/APR-39A(V)3 provides C/D and E to J band coverage with audio and visual warning (NVG compatible). It comprises 10 LRUs, which are form/fit compatible with the analogue APR-39(V)1: one digital signal processor; 2 crystal video receivers; 4 E/J band antennas; one C/D band antenna; one display unit; one control unit.

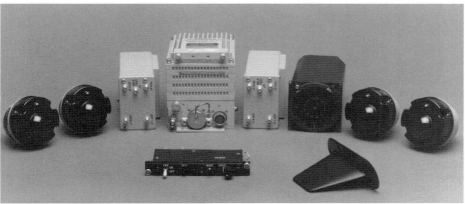

AN/APR-39A(V)1 threat warning system   0514898

Specifications

Frequency range: E-J and C/D bands
Weight: 8 kg
Power supply: 28 V DC, 58 W
Interfaces: RS-422 databus, MIL-STD-1553B databus optional, E-O warning systems, missile launch detectors, radar jammers, CW warning receivers

AN/APR-39B(V)1/3

The AN/APR-39B(V)1/3 utilises APR-39A(V)2 technology and includes dual-channel crystal video receivers, plus a Tuned RF (TRF) receiver, to provide enhanced direction-finding of CW and pulse Doppler radars, enhanced frequency measurement for improved ambiguity resolution, and better high-pulse density performance. The system is fully programmable. AN/APR-39A(V)1 and AN/APR-39A(V)3 can be upgraded to this new RF threat warning configuration. The upgrade adds one LRU to the AN/APR-39A(V)1 configuration at a weight penalty of 4.9 kg, and with no change within the cockpit.

AN/APR-39A(VE) threat warning system

The AN/APR-39A(VE) is a version of the AN/APR-39A(V)1/3 used by the Royal Norwegian Air Force. These systems have now been upgraded using the AN/APR-39B(V)1/3 upgrade kits and are now known as the Viking threat warning system.

Status

The AN/APR-39A(V)1, AN/APR-39(V)2, and AN/APR-39A(V)3 are operational in US and foreign military service. The AN/APR-39B(V)1/3, or Viking, is operational in Norwegian service. Sales continue, and in June 1999, the then Litton Advanced Systems Division announced a USD70 million order for 350 AN/APR-39A(V)2 systems for US combat helicopter and fixed-wing platforms.

Contractor

Northrop Grumman Corporation, Electronic Systems Sector, Defensive Systems Division.
BAE Systems North America, Information and Electronic Warfare Systems.

AN/APR-39A(V)1,3,4 family of Radar Warning Receivers (RWRs)

Type

Airborne Electronic Counter Measures (ECM), Radar Warning Receiver (RWR).

Description

Northrop Grumman Electronic Systems' AN/APR-39A(V)1/3/4 are a family of low-cost RWRs, combining lightweight, small size and low power consumption. The A(V)1/3/4 provides rapid 360 degree detection and identification of RF threats. The APR-39A(V)1 is the standard US Army RWR and is also widely utilised in USAF, USN and USMC aircraft. The APR-39A(V)3 is generally for export and is identical to the A(V)1 except for frequency coverage. The APR39A(V)4 is identical to the A(V)1 with the addition of a 1553B interface card and is used specifically on the AH-64D.

Status

There are in excess of 5,000 APR-39A(V)1/3/4 systems fielded worldwide; this includes all 4 US services and approximately 15 international users. Current installations for the system include the AH-1G/S/W, AH-64A/D, CH-47D, MH-4, UH-1H and A-109 helicopters, the F-5A/B, L-35, L-100 and DH-7/8 fixed-wing aircraft, together with various shipboard applications. The APR-39(V)1 and APR-39A(V)1 systems in the USMC/USN are being replace by the new APR-39A(V)2 and APR-39B(V)2.

Contractor

Northrop Grumman Corporation, Electronic Systems Sector, Defensive Systems Division.

The AN/APR-39A(V)2 threat warning system   0504263

The AN/APR-39A(V)1 threat warning system   0062865

The AN/APR-39A(V)3 threat warning system   0062864

The APR-39B(V)1/3 threat warning system   0062866

Specifications

	AN/APR-39A(V)1	AN/APR-39A(V)3	AN/APR-39A(V)4
Frequency Range	E-M and C/D bands	E-K and C/D bands	E-M and C/D bands
Weight	6.8 kg (15 lb)	6.8 kg (15 lb)	7.0 kg (15.5 lb)
Power	100 W, 28 V DC	100 W, 28 V DC	110 W, 28 V DC
Display	NVG Compatible CRT	NVG Compatible CRT	MFD/Mission Computer
Interfaces	RS-422	RS-422	Dual 1553B (RT) RS-422
Integration	AAR-47, AVR-2, AVR-2A, PC-M/LV	AAR-47, AVR-2, AVR-2A, PC-M/LV	AAR-47, AVR-2, AVR-2A, PC-M/LV

AN/APR-39A/B(V)2 Radar Warning Receiver (RWR)/ Electronic Warfare Management System (EWMS)

Type
Airborne Electronic Counter Measures (ECM), Radar Warning Receiver (RWR).

Description
Northrop Grumman Advanced Systems' AN/APR-39A(V)2 and AN/APR-39B(V)2 provide state-of-the-art technology, combining low acquisition and running costs, light weight, small size and low power consumption. The APR-39A/B(V)2's highly sensitive multireceiver architecture provides rapid 360° detection and identification of RF threats in the E-K Bands and is the EWMS for the USMC/USN Suite of Integrated Sensors and Countermeasures (SISCM). SISCM is comprised of an EWMS, RWR, Missile Warning System (MWS), Laser Warning System (LWS) and Countermeasures Dispensing System (CMDS), providing multispectral threat warning and semi-automated and fully automated countermeasures deployment. An RF Jamming capability is currently under consideration.

The AN/APR-39B(V)2 is the newer variant and differs from the A(V)2 in several significant areas. The primary area is an Electronic Warfare Management System (EWMS) card, which is added to the system. The EWMS has its own stand alone 603R processor specifically for EW management, integration and situational awareness applications. The EWMS card was originally designed to integrate with new glass cockpit aircraft such as the AH-1Z, UH-1Y, KC-130J and MV-22. The system has a Duplex, Dual 15538 capability, allowing it to be a bus controller on the EW bus and an RT on the avionics bus. This serves to improve situational awareness by allowing the sharing of EW sensor information, EOB data and onboard navigation data that can all be overlaid on a cockpit digital map. The EWMS card also provides for mission data recording and playback.

Status
The system entered service in November 2000 and is being installed in a wide range of USMC and USN rotary- and fixed-wing aircraft, including the AH-1W/Z, UH-1Y/N, CH/MH-53, HH-60, MV-22 and KC-130. Additional customers for the system include Greece, New Zealand and Taiwan.

Contractor
Northrop Grumman Corporation, Electronic Systems Sector, Defensive Systems Division.

AN/APR-39B(V)4 Radar Warning Receiver (RWR)

Type
Airborne Electronic Counter Measures (ECM), Radar Warning Receiver (RWR).

Description
Northrop Grumman Advanced Systems' AN/APR-39B(V)4 is a low-cost system that

Northrop Grumman AN/APR-39B(V)2 Suite of Integrated Sensors and Countermeasures (SISCM) (Northrop Grumman)　　1127366

Specifications

	AN/APR-39A(V)2	AN/APR-39B(V)2
Frequency Range	E-K and C/D bands	E-K and C/D bands
Weight	15.8 kg (35 lb)	15.8 kg (35 lb)
Volume	12,710 cc (775 cu in)	12,710 cc (775 cu in)
Power	200 W, 28 V DC	200 W, 28 V DC
Display	NVG Compatible CRT	MFD/Mission Computer
Mission Data Recorder	N/A	2–8 h recording time
Interfaces	BC Dual 1553B	(BC) Dual 1553B (RT)
	RS-422 Dual 1553B	RS-422 RS-232
Integration	AAR-47, AVR-2, AVR-2A,	AAR-47, AAR-47(V)1,
	AAR-60, ALE-47, USQ-131	AAR-47(V)2, AVR-2,
		AVR-2A, AAR-60,
		ALE-47, USQ-131,
		PC-M/LV, AH-1Z/UH-1Y
		Mission Computers

provides state-of-the-art technology in RWRs, combining light weight, small size and low power consumption. The APR39B(V)4's highly sensitive multireceiver architecture provides rapid 360 degree detection and identification of RF threats in the E-K Bands. The -39B(V)4 can be purchased as a new stand-alone system or as an upgrade kit to the current AN/APR39A(V)1/3 RWR. The AN/APR-39B(V)4 upgrade kit consists of one new small LRU (RF Deck), four new plug and play CCAs and new quadrant receivers and antennae. In most cases, the old A(V)1/3 wiring harness can be retained and only RF cables would need to be added.

Improved capabilities over the old AN/APR-39A(V)1/3 system include:
- Greatly improved ambiguity resolution
- Much higher sensitivity
- Higher pulse density handling capability
- AI/PD handling
- Complex/modern threat detection
- Directional CW detection
- Increased processing capability

- Dual 1553B databus
- Built-in growth capability.

Specifications
Frequency range: E-K and C/D bands
Weight: 11.8 kg (26 lb)
Power: 150 W, 28 V DC
Display: NVG compatible CRT or MFD interface
Interfaces: Dual 1553B (RT), RS-422
Integration: AAR-47, AVR-2, AVR-2A, PC-M/LV

Contractor
Northrop Grumman Corporation, Electronic Systems Sector, Defensive Systems Division.

AN/APR-46A, AN/APR-46A(V)1 ESM systems

Type
Airborne Electronic Counter Measures (ECM), Electronic Support Measures (ESM) system.

Description
The AN/APR-46A receiving system is a rapid scanning, highly sensitive heterodyne receiver, able to detect threats and provide warning of DF radar signals at very long range. The receiver is also equipped for emitter bearing determination, as well as signal parameter extraction and identification. It can be interfaced to onboard avionics through a dual-redundant MIL-STD-1553B interface.

EDO RSS' AN/APR-46 receiving system　　0504259

The intercept portion of the AN/APR-46 provides frequency coverage between 30 MHz and 18 GHz. Direction-finding capability is available between 0.5 and 18 GHz. Antenna installations containing the omnidirectional and DF antennas can be mounted on both the top and bottom of the aircraft. A single operator can select either or both and configure the receiver scan strategy to optimise coverage against expected threats. Fully automated search, identification and reporting functions are possible for stand-alone systems.

Threat identification is provided to assist the operator via analysis of the radio frequency, pulse repetition frequency and pulse-width of the intercepted signal. These values are then compared with parameters contained in an emitter library. It is reported that the system also includes monopulse DF capability.

Specifications

Frequency: 0.5–18 GHz
Emitter PRF: CW to 100 kHz
Emitter pulsewidth: 200 ns to CW
Probability of intercept: 100% within bandwidth of receiver
Accuracy:
   10° RMS (0.5–2 GHz)
   5° RMS (2–18 GHz)

Status

In production for the US Air Force.

Contractor

EDO Reconnaissance and Surveillance Systems.

AN/APR-48A Radar Frequency Interferometer (RFI)

Type

Airborne Electronic Counter Measures (ECM), Emitter Locator System (ELS).

Description

The AN/APR-48A provides 360° detection, identification and target azimuth information for a range of emitters, including early warning, ground targeting, counter battery and airborne radars, when installed in fixed-wing and aircraft and helicopters, including the AH-64D and WAH-64D Longbow Apache and the OH-58D Kiowa Warrior. For these helicopter applications, the system is mast-mounted and includes a four-element interferometer. A four-element coarse Direction-Finding (DF) array with a 360° Field of View (FoV) is used for initial signal acquisition, while the specific AH/WAH-64D application features a four-element, long baseline interferometer (with a rotating 90° FoV) to enable it to provide fine DF cuts (better than 1°). The system is also noted for its ability to look through the host aircraft's rotor disc to enable target detection at low elevation. Total system weight is given as 13.4 kg.

The AN/APR-48A has been designed for both helicopter and fixed-wing applications (including unmanned aerial vehicles). It can operate as a stand-alone system, where it may be employed as the primary sensor for the host aircraft's Defensive Aids Sub System (DASS), or as part of an integrated target acquisition system, where it is used to cue Electro-Optic (EO) or Radio Frequency (RF) sensors. In the AH-64D application, the equipment is claimed to significantly reduce the aircraft's exposure time to threats, thereby increasing survivability. The equipment also has application in Suppression of Enemy Air Defence (SEAD) and reconnaissance tasks, as was demonstrated during the US Army's AN/APR-48A initial Operational Test and Evaluation (OT&E) programme during 1995. During this effort, the system proved capable of detecting radars and providing continuous situational awareness in extremely dense air defence emitter/Electronic Counter Measures (ECM) environments.

The AN/APR-48A is reported to make use of advanced packaging techniques which include the use of an ultra-lightweight, second-generation wideband antenna design, integrated gallium arsenide RF components and Very Large Scale Integrated (VLSI) monopulse parameter measurement technology. The receiver used is a four-channel, wide instantaneous bandwidth, superheterodyne design which employs delay line discriminators at a high intermediate frequency, typical of instantaneous frequency measuring systems. The manufacturer is also understood to have developed system upgrades that reduce AN/APR-48A's weight, volume and power requirements for use in fixed-wing air defence applications while maintaining current system performance.

Status

The AN/APR-48A has been developed for the US Army AH-64D Longbow Apache and OH-58D Kiowa helicopters. 222 systems are in production for the US Army AH-64D Longbow Apache. Deliveries began in November 1998 and are projected to continue until February 2003. In addition, 62 sets are on order for UK WAH-64 Longbow Apache helicopters. Deliveries were scheduled from December 1998 to February 2003.

Contractor

Lockheed Martin Systems Integration – Owego.

AN/APR-50 Defensive Management Suite (DMS)

Type

Airborne Electronic Counter Measures (ECM), integrated suite.

Description

AN/APR-50 Defensive Management System (DMS) is installed in USAF B-2 aircraft. The system

is classified and very few technical details have been released. It is believed that the system is designed to present aircrew with intercepted emitter information overlaid on a pre-programmed display of known emitter locations. The installation features a number of antennas, distributed all around the airframe, feeding nine radio frequency front ends, which detect and analyse a wide variety of signals. Each of these front ends may be tuned to a different part of the frequency spectrum. Five receivers take the outputs of the front ends and pass them to the processor.

There is considerable speculation that the AN/APR-50 utilises an ECM technique known as 'active cancellation' – this stealth technique employs an array of antennas to transmit a signal which is out of phase with incoming radar emissions, thus effectively reducing the intensity of the reflected returns through interference. If the emitted interference signal, travelling in the same direction, is exactly matched in terms of amplitude, period and phase, to the reflected radar signal, then the threat radar would not be able to detect any return signal, thus failing to 'see' the aircraft. This is called destructive interference. In terms of applying such ECM techniques to an airborne platform, incoming signals will have many different characteristics of amplitude, period and phase, which, combined with the many different directions of reflection, resulting in phase/amplitude shift, will make true 'cancellation' extremely difficult to achieve in the real world. It is more likely that the characteristics of the strongest incident signal would be selected by the system processor for destructive interference.

Status

The Northrop Grumman Company designation of the system is ZSR-63. In January 1993, Northrop Grumman was awarded a USD117 million contract to continue development of the AN/APR-50. Northrop Grumman was also awarded USD53.9 million to carry on with ESM development, including extension of the frequency range. It is believed that this was originally for Band 2 and the extension was to cover Band 4 from 500 MHz to 1 GHz. While it is believed that baseline versions of the system failed to meet operational expectations, software changes completed during the second quarter of 1998 were incorporated to address these problems. AN/APR-50 systems installed in Block 30 aircraft are described as 'fully capable in Bands 1 to 4'. Additional software upgrades were implemented during 2001–2002.

Contractor

Northrop Grumman Corporation, Integrated Systems Sector.
Lockheed Martin Systems Integration – Owego.

AN/AQS-13 sonar for helicopters

Type

Airborne Electronic Counter Measures (ECM), sonar system.

Description

The AN/AQS-13 is a helicopter dunking system, the -13B and -13F models being in production. It is one of a series of equipments which began with the AN/AQS-10 in 1985. The AN/AQS-13B is a long-range active scanning sonar which detects and maintains contact with underwater targets through a transducer lowered into the water from a hovering helicopter. Opening or closing rates can be accurately determined and the system also provides target classification information.

The AN/AQS-13B has significant advantages in operation and maintenance over earlier systems. To aid the operator, some electronic functions have been automated. Maintenance has been simplified by eliminating all internal adjustments and adding BITE circuits. These advantages were brought about by the use of the latest electronic

The AN/APR-48A radar frequency interferometer   0081867

circuits and packaging techniques, which also reduced system size and weight.

An Adaptive Processor Sonar (APS) has been developed for the system, to enhance detection capability in shallow water and reverberation-limited conditions, while eliminating false alarms from the video display. The APS is a completely digital processor employing fast Fourier transform techniques to provide narrowband analysis of the uniquely shaped CW pulse transmitted in the APS mode. The display retains the familiar PPI readout of target range and bearing, but APS adds precise digital readout of the radial component of target Doppler. With APS, processing gains of greater than 20 dB with zero false alarm rates have been measured for target Dopplers under 0.5 kt.

The AN/AQS-13E was the first system to integrate APS and sonobuoy processing in a common processor, the sonar data computer. Improvements to this system led to the AN/AQS-13F.

The higher energy transmitted with the longer pulse APS mode, combined with the narrowband analysis, also substantially improves the figure of merit in the non-reverberant conditions typical of deep water operations. Measured processing gains for APS under ambient wideband noise limited conditions exceed 7 dB.

The AN/AQS-13F has been designed to provide rapid tactical response against the most advanced submarine threats. It is a sister equipment to the AN/AQS-18 (see next item) and is identical in many respects. A new transducer, when lowered to depths down to 450 m, permits instantaneous range improvements of over 100 per cent compared to previous systems. Very high-speed reeling allows a dip to maximum depth to be completed in approximately 3 minutes. The powerful omnidirectional transducer providing 216±1 dB source level is integrated with a sensitive directional receiver array providing azimuth resolution in a small rugged unit. The sonar data computer offers digital matched filter processing for 200 and 700 ms sonar pulses, as well as sonobuoy control and processing. The azimuth and range indicator and receiver provides a video display for the operator.

Specifications
Weight:
(13A) 373 kg
(13B) 282 kg
(13F) 280 kg
Frequency: 9.25 to 10.75 kHz
Sound pressure level:
(13B) 113 dB
(13F) 216 dB
Range scales: 1, 3, 5, 8, 12, 20 n miles (0.9, 2.7, 4.6, 7.3, 11, 18.3 km)
Operational modes:
(13A) active 3.5 or 35 ms, MTI, APS, passive, voice communications, key communications
(13F) active 3.5 or 35 ms rectangular pulse, 200 or 700 ms shaped pulse, MTI, passive 500 Hz (bandwidth 9 to 11 kHz), SSB voice communications on 8 kHz
Visual outputs:
(13A) range and bearing
(13B) range, range rate, bearing and operator verification
Audio output:
(13A) single channel with gain control
(13B) dual channel with gain control plus constant level to aircraft intercom
Recorder operation: bathythermograph, range, aspect, MAD self-test
Operating depth:
(13F) 1,450 ft at 50 ft hover

Status
The AN/AQS-13 is widely used by US forces and 1,000 sets have been ordered or supplied for the helicopters of 15 foreign navies in Asia, Europe, the Middle East and South America. The AN/AQS-13F was selected by the US Navy and is now in operation on the SH-60F carrier-based ASW helicopter.

Contractor
L-3 Communications, Ocean Systems.

AN/AQS-14/14A/14A(V1) sonar

Type
Airborne Electronic Counter Measures (ECM), sonar system.

Description
The AN/AQS-14 sonar equips the US Navy Sikorsky MH-53E helicopter; first deliveries took place in 1984 and the system has been in service since June 1986. To date, 32 systems have been delivered. Northrop Grumman has delivered additional AN/AQS-14A systems with a new airborne electronics console which greatly improved system performance. The AQS-14 saw extensive use in the Red Sea during 1984 and in the Gulf during 1987/88 and 1990/91 during Operations Desert Shield and Desert Storm.

Used for minehunting duties, towed from helicopters, hovercraft or small surface vessels, the AQS-14 is a side-looking multibeam sonar with electronic beam-forming, all-range focusing and an adaptive processor. The underwater vehicle is 3 m long and has an active control system which allows it to be towed at a selected distance above the seabed or under the surface. The vehicle is controlled by an operator in the helicopter and it is connected by a non-magnetic cable. The operator, assisted by computer-aided detection and classification, has a real-time sonar display on which he can mark targets of interest. A tape recorder allows the recording, classification, position logging and review of data concerning mines and similar objects.

The latest variant of the system is the AN/AQS-14(V1). In May 1999, Northrop Grumman was awarded a one year contract, valued at USD3.2 million, for development, integration and test of an engineering design model to incorporate laser line scanning technology. The addition of this, laser-driven, electro-optical capability will enable operators to actually identify mines. The upgrade will also provide an increase in side-scan sonar resolution of ×4 the current resolution at close range and ×2 the current resolution at long range. Testing of the system was completed in November 2000.

Status
In production and in service with the US Navy on MH-53E helicopters, small craft, remotely controlled vehicles and multimission air cushion vehicles.

Contractor
Northrop Grumman Corporation, Electronic Systems Sector.

AN/AQS-18 sonar for helicopters

Type
Airborne Electronic Counter Measures (ECM), sonar system.

Description
The AN/AQS-18 is a helicopter long-range active scanning sonar. The system detects and maintains contact with underwater targets through a transducer lowered into the water from a hovering helicopter. Active echo-ranging determines a target's range and bearing and opening or closing rate relative to the aircraft. Target identification information is also provided.

The AN/AQS-18 is an advanced version of earlier dunking sonars made by AlliedSignal and includes digital technology, improved signal processing and improved operator displays. The system consists of a small high-density transducer with a high sink and retrieval rate, a built-in multiplex system to permit use of a single conductor cable, a 330 m cable and compatible reeling machine and a lightweight transmitter built into the transducer package. The Adaptive Processor Sonar (APS), which provides enhanced performance in shallow water areas, is an integral part of the system.

The AN/AQS-18 offers a number of improvements over earlier dipping sonars. These include increased transmitter power output to give longer range, high-speed dip cycle time and reductions in weight of all units.

The APS increases detection capability in shallow water and limited reverberation conditions, while eliminating false alarms from the video display. The APS is a digital processor which uses fast Fourier transform techniques to provide narrowband analysis of the uniquely shaped CW pulse transmitted in the APS mode. The PPI display retains the normal readout of target range and bearing.

The APS processing gain improvement over the normal AN/AQS-18 analogue processing is 20 dB for a 2 kt target and 15 dB for a 5 kt Doppler target. The higher energy transmitted with the longer pulse APS mode, combined with the narrowband analysis, also improves operation in the non-reverberant conditions more typical of deep water. In general, the gain improvement above a speed of 10 kt exceeds 7 dB.

The latest version is the AN/AQS-18(V) which is available with both 300 and 450 m length cables.

Specifications
Weight: 252 kg plus 13.3 kg for APS
Frequency: 9.23, 10, 10.77 kHz
Sound pressure level: 217 dB/µPa/yd (0.9 m)
Range scales: 1,000, 3,000, 5,000, 8,000, 12,000, 20,000 yds
Modes: 3.5 or 35 ms pulse (energy detection) and 200 or 700 ms pulse (narrowband nalysis)
Visual outputs: range, range rate, bearing, operator verification
Audio output: dual channel with gain control plus constant level to aircraft intercom
Recorder operation: bathythermograph, range, ASPECT, MAD, BITE
Operating depth: 330 m

Status
In production and in service with the German, Greek, Italian, Japanese, Portuguese, Spanish, Taiwanese and US navies on Lynx, SH-3, SH-60J and S-70C(M)-1 helicopters. Selected for the Egyptian SH-2G(E).

Contractor
L-3 Communications, Ocean Systems.

AN/AQS-18(V) dipping sonar system

Type
Airborne Electronic Counter Measures (ECM), sonar system.

Description
The AN/AQS-18(V) is the export version of the AN/AQS-13F helicopter dipping sonar. These systems employ a transducer which is lowered into the water from a hovering helicopter to detect and maintain contact with underwater targets. Active echo-ranging determines target range and bearing and opening and closing rate relative to the helicopter. Target classification indications are also provided.

The AN/AQS-13F/18(V) series of dipping sonars is specifically designed for ASW helicopters where great mobility is required for fast reaction. AN/AQS-13F/18(V) equipped helicopters are well-suited for redetection of contacts, target localisation and weapon delivery against shallow and deep water threats. ASW helicopters are often required to search areas which are difficult for other sensor platforms, including shallow water with high noise areas, coastal regions, constrained passages, high-density shipping lanes and areas of concentrated naval activity.

Features of the series include high source level to provide long-range, shallow water signal processing for high reverberation areas, capability to control, process and display sonobuoys in a single integrated system, high-speed reeling machine to achieve maximum depth and retrieval within 3 minutes and adaptable reeling

The AN/AQS-18(V) deployed from a German Navy Westland Lynx Mk 88 helicopter   0504107

machine designs compatible with a wide range of helicopters.

AN/AQS-13F/18(V) systems include digital technology, improved signal processing, extensive use of hybrid integrated circuits and improved operator displays.

AN/AQS-13G/18A

The AN/AQS-13G/18A is the latest stage in evolutionary development of the system. The improved 'dry end' of the system provides 14 dB improvement over earlier versions, thus providing search rates which are more than four times greater.

Specifications
Weight:
Frequency: 9.23, 10.003 and 10.744 kHz
Operating depth: 440 m
Sound pressure level: 217 ±1 dB/µPa/yd (0.9 m)
Range scales: 0.9, 2.7, 4.6, 7.3, 11, 18 km
Modes:
(active) 3.5, 35 ms pulses, 200, 700 ms shaped pulses (passive), (communicate) SSB at 8 kHz
Raise speed: 6.7 m/s average
Lower speed: 4.9 m/s
Water exit speed: 1.5 m/s

Status
Users include Greece, South Korea, Spain and Taiwan.

Contractor
L-3 Communications, Ocean Systems.

AN/ARN-146 On-Top Position Indicator (OTPI)

Type
Airborne Electronic Counter Measures (ECM), sonar system.

Description
The AN/ARN-146 radio receiving set is a 99-RF channel On-Top Position Indicator (OTPI) which is used on board ASW rotary- and fixed-wing aircraft to provide bearing and on-top position indication of deployed sonobuoys. When used in conjunction with a suitable ADF system, the AN/ARN-146 enables an operator to locate and verify the position of sonobuoys which are operating on any of 99 RF channels.

The AN/ARN-146 has a modular solid-state design for use with computer control, either via RS-422 directly or when connected with the AN/ARR-84 sonobuoy receiver. The AN/ARN-146 is form and fit interchangeable with the R-1651/ARA and R-1047 A/A OTPI receivers. It is compatible with ARA-25, ARA-50 and OA-8697/ARD ADF antenna systems.

The AN/ARN-146 set consists of the R-2330/ARN-146 receiver and the C-11699/ARN-146 radio set control. The R-2330/ARN-146 is a VHF AM receiver, which receives signals from sonobuoys operating on any of 99 RF channels in the frequency range from 136 to 174 MHz. This receiver houses all the electronics, including internal transfer circuits for the switching of RF, baseband, power and phase compensation circuits for sharing the DF system between the OTPI and an associated UHF receiver system for the ADF. The R-2330/ARN-146 also contains a PLL synthesised local oscillator, digital address decoder, voltage tuned BP filters, AGC and BIT circuitry.

The C-11699/ARN-146 radio set control is an optional manual control box which is used in place of the RS-422 bus or the MIL-STD-1553B dual bus of the AN/ARR-84. This control box provides the capability to apply power to the AN/ARN-146 system to select any one of 99 RF channels, to activate the BIT circuitry and to display adequate signal strength and results of the BIT via the adequate signal strength indicator.

Specifications
Dimensions:
(radio receiver) 125.8 × 134.6 × 77.47 mm
(radio set control) 57.15 × 127 × 146 mm
(mounting plate) 15.2 × 144.8 × 77.47 mm
Weight:
(radio receiver) 2.15 kg
(radio set control) 0.57 kg
(mounting plate) 0.27 kg
Power supply: 115 V AC, 400 Hz, single phase, 15 VA
28 V DC, 20 W
Frequency: 136-173.5 MHz (99 channels)
Sonobuoy compatibility: DIFAR, LOFAR, Ranger, BT, CASS, DICASS, VLAD, CAMBS, Barra, ERAPS, HLA, ATAC, SAR
Reliability: 3,000 h MTBF

Status
In production and in service with the US Navy. Fitted to SH-60B helicopters.

Contractor
Flightline Systems.

AN/ARR-502-type sonobuoy receivers

Type
Airborne Electronic Counter Measures (ECM), sonar system.

Description
AN/ARR-502-type sonobuoy receivers are used with a number of acoustic processors and are installed on various platforms, providing for reception and demodulation of sonobuoy signals in the 136 to 174 MHz frequency range. Different configurations of the system are possible through the use of standardised cards that support each available function. For example, additional acoustic channels use a common receiver module configuration supported by additional backplane circuitry and/or enclosure. Current installations utilise an external preamp to account for transmission losses between the antenna and receiver, although the unit is capable of being configured with an internal preamp and antenna filter.

An Acoustic Test Signal Generator (ATSG) function is provided with all units, and supports a Built-In Test (BIT) function to provide for a complete end-to-end system test and fault isolation to the module level.

The Sonobuoy Position System (SPS) function provides real-time geolocation of up to 99 deployed buoys, using interferometer techniques with the supply of aircraft navigation and metrics. SPS is also capable of providing On Top Position

Flightline Systems' AN/ARR-502B(V)X for the Aurora CP-140   0044964

BAE Systems' Nimrod MR2 MPA employs Flightline Systems' R624 sonobuoy receiver   0130705

The AN/ARR-502B(V) is fitted to Canadian CP-140 Aurora MPA   0002283

or RS-422 bidirectional serial communications dependent on the platform.

Other key features of the system include:

- Extremely low acoustic receiver baseband noise floor
- 99 RF Channels with tuneable resolution from 5 kHz to 375 kHz between each channel
- Up to 64 (for the BAE Systems Nimrod configuration which utilises two R624 receivers) simultaneous acoustic channel reception with any mix of analogue or digital signals
- 5 dB noise figure
- 80 dB dynamic range
- 12 dB signal-to-noise ratio at 0.5 µV and 20 dB at 1.0 µV input
- AM immunity: Progressive AGC with programmable time constants for improved multipath immunity
- >140 dB out-of-band interference rejection
- >80 dB in-band interference rejection
- MTTR <16 minutes for module replacement
- Excellent third-order intermodulation characteristics with 'active' mixer technology
- MIL-STD-704A compliant 115V AC 400 Hz 3-phase power supplies.

Specifications
Acoustic channels
ARR-502B(V)1: 16
AAR-502B(V)3: 16
ARR-502B(V)4: 16
R624(V)1: 32
R624(V)2: 32
Weight
ARR-502B(V)1: 31.4 kg
AAR-502B(V)3: 33.6 kg
ARR-502B(V)4: 33.2 kg
R624(V)1: 49.5 kg
R624(V)2: 47.7 kg
Reliability: >850 h MTBF predicted
Power: 115 V AC, 400 Hz, 3-phase MIL-STD-704A
Frequency: 136–174 MHz
Sonobuoy compatibility: All standard NATO types

Status
The AN/ARR-502 sonobuoy receiver is in service in the CP-140 Aurora and Nimrod Maritime Patrol Aircraft, and selected for the NH 90 helicopter.

Contractor
Flightline Systems.

(OTP), RF scanning and RF levels. The system supports various current antenna configurations, from three antennas on the NH 90 up to 10 antennas on the Aurora CP-140.

Since the standard receiver modules are both FM and AM capable, the OTP indicator function can be obtained through the addition of a standard receiver module, connected to a DF antenna system (for example, DF-301, ARA-50 or ARA-25), or is available with the Sonobuoy Positioning System (SPS) through software.

All receivers are compatible with most analogue and digital sonobuoys, including DIFAR, LOFAR, RANGER, BT, CASS, DICASS, VLAD, CAMBS, BARRA, ERAPS, HLA, ATAC and SAR.

Control and monitoring of the receiver is accomplished through either MIL-STD-1553B bus

---

AN/ARR-72 sonobuoy receiver system

Type
Airborne Electronic Counter Measures (ECM), sonar system.

Description
The AN/ARR-72 sonobuoy receiver system is used on the P-3C patrol aircraft, in conjunction with acoustic signal processors and a digital computer. The AN/ARR-72 system receives, amplifies and demodulates FM signals transmitted by deployed sonobuoys in the 162.25 to 173.5 MHz VHF band. The AN/ARR-72 is compatible with the AN/SSQ-36, 41, 50, 53 and 62 sonobuoys. The receiver is in five parts: AM-4966 preamplifier, CH-169 receiver, SA-1065 audio assembly, C-7617 control indicator and SG-791 Acoustic Sensor Signal Generator (ASSG) which performs diagnostic functions. The AN/ARR-72 system is compatible with LOFAR, CODAR, BT, RO, CASS and DICASS equipment.

The AN/ARR-72 is a dual-conversion superheterodyne VHF receiver system. Radio frequency signals are received at the dual-aircraft VHF blade antennas, amplified by the system's AM-4966 dual RF amplifiers and passed on to the CH-169 31-channel receiver assembly. Within this assembly, a multicoupler distributes the preamplified RF to the 31 fixed tuned receivers, where it is further amplified and demodulated to provide baseband audio and RF level signals.

US Navy SH-60 LAMPS Mk III helicopters are fitted with Flightline Systems' AN/ARR-72 sonobuoy receivers   0503691

Modules of the Flightline Systems' AN/ARR-72 sonobuoy receiver (from left): preamplifier, acoustic sensor signal generator, receiver, audio assembly and control indicator   0503692

Each of the 31 receiver channels contains a channelised first converter, an IF filter, a second converter and a discriminator/amplifier. The first converter contains a crystal-controlled local oscillator and mixed/IF circuit. The plug-in discriminator/amplifier provides RF level and FM signal detection. The optional phaselock discriminator is directly interchangeable with the discriminator/amplifier module. The first and second converter and discriminator/amplifier assemblies are plug-in units and, excepting the local oscillator crystals, are identical for the 31 channels.

The SA-1605 audio assembly includes 19 audio switching and amplifier cards and two audio power supply regulators. The audio assembly accepts the baseband and RF level outputs of the 31 receivers and outputs them to the computer and processing equipment. Each of the 19 audio channels contains a 31 by 1 switching matrix to select the output of a given receiver. This receiver signal is then amplified and provided through two individually buffered outputs to the processing equipment. Selection of a particular receiver channel may be accomplished by the digital computer or the C-7617 dual-channel control indicator. RF level for the selected receiver is displayed on the corresponding control-indicator meter.

The SG-791 ASSG is the BIT for the receiver system. RF test signals are generated for application to the preamplifier, the multicoupler or to a radiating antenna. Internal circuits generate simulated signals for the testing of normal LOFAR, extended LOFAR, range only and BT processing equipment. External modulation inputs are provided to accept modulation from devices such as the target generator in the demultiplexer of the AN/AQA-7(V) DIFAR equipment, thus allowing end-to-end checks of sophisticated ASW equipment.

The ASSG contains a redesigned multifrequency oscillator to provide for generation of the 31 individual RF frequencies. This replaces a previously utilised crystal turret. This synthesiser reduces the channelling time from 10 seconds to less than a second.

Specifications
Weight:
(preamplifier) 0.9 kg
(receiver) 26.53 kg
(audio) 17.69 kg
(ASSG) 8.71 kg
(8 control indicators) 14.51 kg
Total weight: 68.36 kg
Power supply: 115 V AC, 400 Hz, single phase, 300 W
28 V DC, 280 mA for panel lighting
18 V DC, 250 mA for ASSG annunciators
Frequency: (31 channels) 162.25–173.5 MHz at 0.375 MHz spacing
Noise figure: 5 dB max (3.5 dB is typical)
IF rejection: 66 dB min (>100 dB typical)
Image rejection: 66 dB min (>100 dB typical)
High audio level: 16 VRMS at ±75 kHz deviation
Standard audio level: 2 VRMS at ±75 kHz deviation
Crosstalk: >54 dB min
Output isolation: 60 dB min
Audio frequency response: ±1 dB from –20 Hz to 20 kHz; ±6 dB from 5 Hz to 40 kHz
Audio distortion: –20 Hz to 5 Hz <3%
5–18 kHz <5%
Controlled audio phase characteristics: –100 Hz to 3 kHz ±5°
Audio noise: 10-300 Hz 2 mV
301–2,400 Hz 3 mV
Operational stability: 500 h
Operating life: 20,000 h
Reliability: 500 h MTBF including ASSG and 10 dual-channel control indicators

Status
In service with P-3C aircraft and Sikorsky SH-60 LAMPS Mk III helicopters.

Contractor
Flightline Systems.

AN/ARR-75 sonobuoy receiving set

Type
Airborne Electronic Counter Measures (ECM), sonar system.

Description
The AN/ARR-75 sonobuoy receiving set is a 31-RF channel FM receiver designed for ASW fixed-wing aircraft and shipboard applications. Independent receiver modules provide four simultaneous demodulated audio outputs, each capable of selecting one of 31 RF input channels.

The AN/ARR-75 receiving set is composed of two units. The first of these units is the OR-75/ARR-75 receiver group assembly, which consists of the PP-6551/ARR-75 power supply, four R-1717/ARR-75 receiver modules and the CH-670/ARR-75 chassis. The receiver group assembly contains the majority of the system's electronics.

The second receiving set unit is the C-8658/ARR-75 or C-10429/ARR-75 radio set control. This control unit provides independent RF channel selection and signal strength monitoring of any one of the 31 RF channels for each of the four receiver modules.

Specifications
Dimensions:
(receiver) 203 × 186 × 305 mm
(controller) 122 × 145 × 71 mm
Weight:
(receiver) 9.8 kg
(controller) 1.1 kg
Power supply: 115 V AC, 400 Hz, 3 phase, 75 VA max 27 V (C-8658) or 5 V (C-10429), 5.6 W for lighting at either 27 V (C-8658) or 5 V (C-10429)
Frequency: 162.25–173.5 MHz (31 channels)
Noise figure: 5 dB at 50 ohms
Specifications: MIL-STD-461, 462, 463, 781, MIL-E- 5400, MIL-R-81681, AR-5, 8, 10, 34
Operating life: 20,000 h
Reliability: 1,500 h MTBF

Status
In service on the LAMPS Mk I, LAMPS Mk III and SH-3H helicopters.

Contractor
Flightline Systems.

Flightline Systems' AN/ARR-75 sonobuoy receiver 0503693

AN/ARR-78 (V) Advanced Sonobuoy Communication Link (ASCL) 0011864

AN/ARR-78(V) Advanced Sonobuoy Communication Link (ASCL)

Type
Airborne Electronic Counter Measures (ECM), sonar system.

Description
ASCL is the US Navy standard sonobuoy receiver and is deployed on the Orion P-3C Update III and Viking S-3B ASW aircraft. The P-3C Update III employs two ASCLs in a 32-acoustic-channel expansion configuration (CHEX). ASCL operates on both types of aircraft in conjunction with standard VHF antennas. The following units comprise a typical ASCL configuration.

RF preamplifier
The AM-6875 is an optimised high-performance, low-noise VHF preamplifier that provides amplification and prefiltering of received RF signals.

Radio receiver
The R-2033 (P-3C, Update III) or the R-2066 (S-3B) contains 20 fully synthesised receiver modules (16 acoustic and four auxiliary) and one each of the following modules: RF/ADF amplifier multicoupler, reference oscillator, I/O Proteus digital channel (P-3C), I/O Manchester digital channel (S-3B), I/O processor, clock generator, BITE and DC power supply module. Each single-conversion module includes mixer conversion, frequency-synthesised local oscillator, demodulator, and output interface circuits.

Each of the acoustic receiver modules processes FM/analogue signals at any of the 99 channels in the extended VHF band. Each of the four auxiliary receiver modules processes FM/analogue signals at any of the 99 channels and provides one channel for selection and processing of the On-Top Position Indicator (OTPI) signals, two channels for the operator to monitor acoustic information and one channel to monitor the RF signal level in any of the RF channels. Common receiver modules are interchangeable.

BITE circuits provides comprehensive end-to-end evaluation of each receiver from the VHF preamplifiers to the receiver output interface circuits. BITE is initiated automatically by the computer (such as Proteus) and/or by the operator through the Indicator Control Unit (ICU). Performance status is displayed on the ICU and routed to the computer.

Indicator Control Unit (ICU)
The C-10126 ICU provides the operator with a means for manual control of each receiver channel

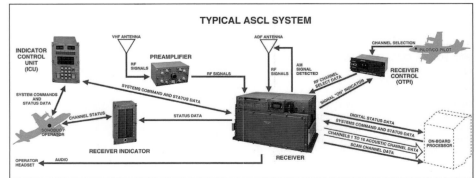

Typical ASCL system 0011865

frequency assignment, receiving mode and self-test. It also displays status of the receiving set and operator entry information.

Receiver Indicator (Receiver Status)
The ID-2086 Indicator continuously displays the control mode setting, the RF channel number, and the received signal level for each receiver.

Receiver Control (OTPI)
The C-10127 Receiver Control Unit provides the operator with control over the OTPI.

Specifications
Frequency: extended VHF
Receiver: 20 (16 acoustic/4 auxiliary)
Channels: 99 per receiver
Audio Output:
(analogue) 2 V rms balanced
(monitor) 50 mW, 300 ohms
Power: 115 V AC ±10%, 380 to 440 Hz, 3-phase, 500 W; 18 to 32 V DC, 7 W; 26.5 V AC ±10%, 400 Hz, 50 W

Status
Installed in US Navy P-3C Orion and S-3B Viking aircraft.

Contractor
BAE Systems North America, Greenlawn.

AN/ARR-84 sonobuoy receiver

Type
Airborne Electronic Counter Measures (ECM), sonar system.

Description
The AN/ARR-84 is designed to receive signals from all current and planned future US and allied sonobuoys. This receiver has four acoustic channels capable of receiving signals from up to four deployed sonobuoys simultaneously on any of 99 RF channels. The AN/ARR-84 is installed on a variety of ASW platforms, including rotary- and fixed-wing aircraft as well as large surface combatants and fast patrol craft. This receiver is a form and fit replacement for its predecessor, the AN/ARR-75 31-channel sonobuoy receiver.

The AN/ARR-84's major assemblies include a power supply, four identical receiver modules and the electrical equipment chassis. The power supply contains the input/output assembly for MIL-STD-1553B and the FSK printed circuit board, allowing digital sonobuoy data extraction.

Additional features of the AN/ARR-84 include dual-selectable IF bandwidths for optimum performance with a wide variety of sonobuoy types. Outstanding AM rejection and mechanical vibration immunity severely reduce spurious returns due to propeller/rotor multipath and platform vibration. Low susceptibility to conducted and radiated energy prevents desensitisation and allows the AN/ARR-84 to be utilised near strong onboard emitters and shipboard search radars.

Flightline also provides an optional radio set control, which is intended for use when an online computer is not available and which provides power control and BIT operation for the receiver group. RF channel selection, sonobuoy type selection and RF level readout is provided independently for each of the four receivers in the group. Operator input is effected through a multifunction keypad, with signal strength provided as a histogram display and entry readback/failure data provided by a 16-character message display.

Specifications
Dimensions: 190.5 × 381 × 254 mm
Weight:
(receiver) 11.34 kg
(optional control box) 1.81 kg
Power supply: 115 V AC, 400 Hz, 3 phase, 100 VA
Frequency: 136-173.5 Hz (99 channels)
Sonobuoy compatibility: DIFAR, LOFAR, Ranger, BT, DICASS, VLAD, CAMBS, Barra, ERAPS, HLA, ATAC, SAR
Environmental: MIL-E-5400 Class 1B
Reliability: 1,500 h MTBF
Interfaces:
(control) MIL-STD-1553 dual-bus, RS-422, manual control box (optional), discrete channelisation out-puts to tactical computer
(RF) single RF antenna input, auxilliary preamplified output from single antenna
(acoustic) high- and low-level analogue data and regenerated data and clock for digital sonobuoy
(RF level) analogue discrete or digital via control bus

Status
In service with aircraft of the US Navy. Fitted to SH-60B helicopters.

Contractor
Flightline Systems.

AN/ASQ-81(V) Magnetic Anomaly Detection system (MAD)

Type
Airborne Magnetic Anomaly Detection (MAD) system.

Description
The AN/ASQ-81(V) Magnetic Anomaly Detection (MAD) system was developed for US Navy use in the detection of submarines from an airborne platform. The system operates on the atomic properties of optically pumped metastable helium atoms to detect variations of intensity in the local magnetic field. The Larmor frequency of the sensing elements is converted to an analogue voltage which is processed by bandpass filters before it is displayed to the operator.

Four configurations of the AN/ASQ-81(V) are available: two for use within an airframe and two for towing behind an aircraft. The US Navy uses the AN/ASQ-81(V)1 in the land-based P-3C Orion, where it is housed in a tail sting. The AN/ASQ-81(V)3 is installed in the carrier-based S-3 Viking aircraft, where it is extended on a boom.

The AN/ASQ-81(V)2 is a towed version employed by the US Navy on Sikorsky SH-3H and Kaman SH-2D helicopters. It is also in service with other countries, including the Netherlands for use on the Westland Lynx, Japan for use on the Mitsubishi HSS-2, and with forces employing the Hughes 500D helicopter.

The second towed version, the AN/ASQ-81(V)4, is used by the US Navy on the Sikorsky SH-60B LAMPS III helicopter.

All versions of the AN/ASQ-81(V) have the same C-6983 detecting set control, AM-4535 amplifier and power supply unit. The AN/ASQ-81(V)1 and 3 use a DT-323 magnetic detector, while the AN/ASQ-81(V)2 and 4 have a TB-623 magnetic detecting towed body. The towed version is controlled by the C-6984 reel control, which works the RL-305 magnetic detector launching and reeling machine.

Status
In production and in service.

Contractor
Raytheon Company, Space and Airborne Systems.

AN/ASQ-208 Magnetic Anomaly Detection (MAD) system

Type
Airborne Magnetic Anomaly Detection (MAD) system.

Description
The AN/ASQ-208 is a derivative of the AN/ASQ-81 magnetometer that operates on the atomic properties of optically-pumped metastable helium atoms to detect variations in total magnetic field intensity.

The AN/ASQ-208 is a digital system which incorporates microprocessor technology to achieve aircraft compensation, multiple channel filtered display and threshold processing. Two system configurations are available: one for inboard installations and one for towed installations.

The system features a high-sensitivity helium sensor and choice of inboard or towed configuration. Control implementation is by MIL-STD-1553B or Manchester databus for online operation or dedicated offline control unit.

Enhanced signal recognition is provided by three-channel filtered data for optimum operator display and automatic detection with range/confidence estimate. Automatic aircraft compensation is provided by electronic compensation of aircraft interfering terms and an automatic figure of merit estimator.

The AN/ASQ-208(V) is configured for minimum impact on existing aircraft or new aircraft installations and utilises existing AN/ASQ-81(V) aircraft wiring. It requires a single cable change for the vector sensor.

System performance enhancements include shallow water detection, elimination of dedicated Magnetic Anomaly Detection (MAD) compensation flights and automatic detection.

Status
In production and in service with Sikorsky S-76N.

Contractor
Raytheon Company, Space and Airborne Systems.

Flightline Systems' AN/ARR-84 99-RF channel sonobuoy receiver shown with the AN/ARN-146 OTPI (on the right)   0503694

The Royal Netherlands Navy P-3C Orion uses the AN/ASQ-81 (V) 1   0503696

AN/ASQ-213 HARM Targeting System (HTS)

Type
Airborne Emitter Location System (ELS).

Description
The AN/ASQ-213 HTS was designed to be fitted to US Air Force F-16C/D Block 50D aircraft and provide targeting information for the AN/AGM-88A HARM missile in the manned SEAD role. US Air Force aircraft in the HTS role are designated F-16CJ.

The 200 × 1,420 mm (Ø × L), 41 kg HTS pod, mounted on the starboard chin station of the F-16CJ, detects, identifies, and locates hostile radars and provides data needed by the HARM system for calculation of targeting and launch parameters. In its most effective mode ('range known'), the HTS displays the target location to the pilot for HARM designation and firing.

HTS/HARM-equipped F-16 aircraft often operate with RC-135 Rivet Joint or EA-6B Prowler aircraft to optimise effectiveness.

HTS
The baseline configuration achieved Initial Operating Capability (IOC) during 1994 – this is often reported as the Lot I configuration. A 1996 Pre-Planned Product Improvement (P³I) programme resulted in an improved specification, reported as incorporating increased processing power (to facilitate a greater number of discrete targets), greater capability against low frequency emitters and increased target resolution; while described at the time as an 'upgrade', the configuration is also known as the Lot II configuration.

Additional improvements fielded during mid-2000 enhanced the equipment's search speed, memory size and target identification/handling capabilities. In parallel with the aforementioned hardware upgrades, a series of software revisions have been undertaken, designated R5 and R6. Software release R5 is understood to be the partner to the baseline (Lot I) configuration, with R6 implemented on 132 US Air Force HTS pods by the end of December 2001.

Pilot interface is via a cockpit Control and Display Unit (CDU), facilitating threat display, tactical data and target selection; the system can handle up to nine classes of threat emitter (library via data transfer cartridge), with the pilot selecting or de-selecting threat classes using option select buttons on the CDU.

HTS(E)
Housed in an identical pod to the baseline system, Raytheon markets a variant designated HTS(E) (believed to stand for export). This configuration facilitates employment of the missile in its most effective 'range known' mode,

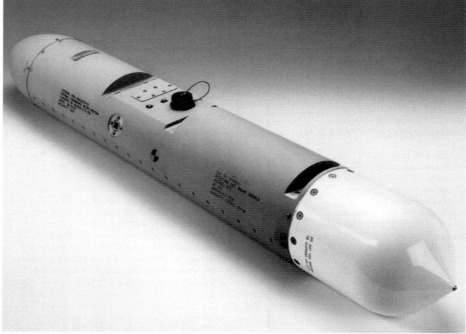

The HTS pod   0104006

with the system able to autonomously detect, identify and locate radar guided threats at extended ranges; modes include search, single target track and multiple target track. HTS(E) is fully field reprogrammable, designed for high reliability/maintainability.

HTS R7
The latest variant of the HTS is designated HTS R7, also known as the Smart Targeting and Identification via Networked Geolocation (STING). HTS R7 incorporates Global Positioning System (GPS) and digital RF receivers, an enhanced power supply and a new software standard (R7). Major enhancements to the system include:
- Precision emitter geolocation capability, including multi-platform co-operative ranging with weapons system update to facilitate the use of precision weapons
- Precision emitter identification capability
- Virtual emitter training
- Employment of external targeting data acquired via its host platform's Link 16 capability

For US Air Force F-16CJ aircraft, HTS R7 is carried on the aircraft's port chin station, to allow for the fitment of an Advanced Targeting Pod (ATP), such as the AN/AAQ-33 Sniper (see separate entry) on the starboard chin station.

Status
The HTS is in production and in service. An HTS R7 programme definition and risk reduction study was launched during 2000, with a System Development and Demonstration (SDD) contract being awarded in February 2001. Raytheon announced that it had delivered the first such pod to the US Air Force on 28 September 2006.

As of 2006, the US Air Force has procured a total of 209 ASQ-213 pods (including 77 built as new to the R6 standard), with 200 funded for upgrade to R7/STING standard, with implementation scheduled for 2006 ((83 pods) and 2007 (117 pods).

Elsewhere, during 1999, Greece requested the supply of 20 HTS(E) systems for use on its F-16C/D Block 52 + (also known as the Advanced Block 52) aircraft, although no delivery has been reported.

Contractor
Raytheon Company, Space and Airborne Systems.

AN/AVR-2A(V) laser detecting set

Type
Airborne Electronic Counter Measures (ECM), Laser Warning Receiver (LWR).

Description
The AN/AVR-2A(V) laser detecting set detects, identifies and characterises optical signals and provides audible and visible warning of laser threats to the aircrew. It consists of four staring array SU-130A(V) sensor units and a CM-493A interface unit comparator. The AN/AVR-2A(V) interfaces with all variants of the AN/APR-39(V) Series radar signal detecting set to function as an integrated radar and laser warning receiver system.

The AN/AVR-2A(V) system provides 360° coverage and can identify laser range-finders, designators and beamriders. It contains a reprogrammable, removable, user data module. P³I expansion includes a MIL-STD-1553B interface.

Status
In production for the US Army, Navy, Marine Corps and Special Forces for the AH-1F Cobra, AH-64 Apache, AH-64D Longbow Apache, AH-1W SuperCobra, MH-60K Black Hawk, H-47E Chinook, OH-58D Kiowa Warrior, HH-60H Combat Rescue, UH-1N Huey, and V-22 Osprey helicopters. Also selected for the UK EH 101SH and WAH-64 helicopters. Under development for: RAH-66

US Air Force F-16CJ carrying an HTS pod on the starboard chin station (just below the drop tank) and an AGM-88 HARM missile under the wing   0111359

The AN/AVR-2A laser detecting set consists of four sensor units and an interface unit comparator

0504153

Comanche, C-130 Hercules, SH-60B LAMPS, and SH-60F CV helo aircraft.

Contractor
Raytheon Company, Space and Airborne Systems.

AN/AYR-1 ESM system

Type
Airborne Electronic Counter Measures (ECM), Electronic Support Measures (ESM) system.

Description
The AN/AYR-1 is an advanced ESM system which is part of an upgrade to US Air Force and NATO E-3 AWACS aircraft. The system undertakes passive interception, identification and analysis of radar and radio signals. The antennas are installed in a blister on either side of the aircraft, just aft of the cockpit, and in nose and tail arrays, to provide 360° azimuth coverage. The extra processing units required to analyse signals quickly and classify radar returns, plus associated equipment, add approximately 855 kg to the overall aircraft weight.

Status
In production for the US Air Force and NATO. In January 1997, a contract was awarded to Boeing Defense and Space Group provided for integration of four ESM systems to E-3F AWACS aircraft at a then value of USD32.4 million.

Contractor
ARGOSystems (a Condor Systems business unit).

AN/TLQ-17A countermeasures set

Type
Airborne Electronic Counter Measures (ECM), communications jamming system.

Description
The modular AN/TLQ-17A is an advanced tactical communications countermeasures set developed for the US Army. It is configurable to meet user requirements for a number of host platforms including helicopters.

The AN/TLQ-17A incorporates computer-controlled search, surveillance and jamming in the HF and VHF ranges, microprocessor

technology for advanced EW applications, BITE and modularity for ease of maintenance.

Specifications
Frequency:
(band 1) 1.5–20 MHz
(band 2) 20–80 MHz
Operating modes: search/lockout, priority search/lock on, monitor/automatic and scan band sectors
Effective radiated power: 10–550 W
Receiver tuning time: 1 ms
Preselected frequencies: 255
Reliability: 400 h MTBF

Contractor
BAE Systems, North America, Information and Electronic Warfare Systems

AN/USQ-113 communications jammer

Type
Airborne Electronic Counter Measures (ECM), communications jamming system.

Description
The AN/USQ-113 provides the EA-6B Prowler aircraft with its communications jamming capability. To meet the expanded role of the aircraft now that it has subsumed the US Air Force support jamming tasks (after retirement of the EF-111A) as well as those of the US Navy, an upgrade programme has been initiated.

Status
In September 1996, BAE Systems North America was contracted to develop three preproduction systems during a non-recurring Engineering and Manufacturing Development (EMD) phase. The contract also called for the upgrade of 30 existing USQ-113 systems, beginning in July 1997, scheduled for completion before the end of 1998. Under another contract issued in October 1998, Sanders were contracted to provide 33 AN/USQ-113(V)2 Phase III systems and two improved operator panels. The Phase III systems consist of: new receivers, power amplifiers and transmitters developed to extend the frequency range of the system and to replace the operator display. The new display has an improved Liquid Crystal Display (LCD) that runs Windows(tm)-based software. Additionally, a signal recognition device is being developed for insertion in a new system controller. With the device, aircrews will be able to analyse the signals they receive. This upgrade

forms part of the ICAP III (Improved Capabilities III Warfighter Upgrade System) upgrade to the EA-6B Prowler aircraft managed for the US Navy by Northrop Grumman Corporation.

Contractor
BAE Systems North America, Information and Electronic Warfare Systems.

AQS-18A dipping sonar system

Type
Airborne Electronic Counter Measures (ECM), sonar system.

Description
The AQS-18A dipping sonar system represents a development of the medium frequency AN/AQS-18(V) dipping sonar. The dome control, reeling machine and transducer of the AN/AQS-18(V) have been interfaced with a powerful digital processor, control unit and colour display. The advanced processing brings greater performance through high-resolution digital processing, greater contact memory space and the flexibility to increase the number of sonar beams, type and length of pulses and menus of operator displays.

The AQS-18A has additional pulse lengths of 1.6, 3.2 and 4 seconds. The longer pulses put more energy on the target and provide higher Doppler resolution for maximum performance in high reverberation shallow water conditions. An FM mode is available for extremely low Doppler target detection and maximum range resolution. The total ASW system improvement of the AQS-18A is 14 dB over current systems and can provide more than four times the area search rate of the AN/AQS-18(V).

The AQS-18A has spare processing power and space for additional processing features such as computer-aided detection and classification, multisensor target fusion, embedded training and performance prediction, based on environmental data collected during past or current missions.

MIL-STD-1553 databus protocol facilitates integration with other aircraft subsystems and components. BIT eases support and boosts availability. The new weapon-replaceable assemblies have a significantly higher MTBF than current systems.

Specifications
Weight: 265 kg total
Frequency: 9.23, 10.003, 10.774 kHz
Sound pressure level: 217 ±1 dB/μPa/yd (0.9 m)
Range scales: 0.9, 2.7, 4.6, 7.3, 11, 18, 29 km
Operating depth: 440 m
Modes:
(active) 3.5, 35 ms pulses, 0.2, 0.7, 1.6, 3.2, 4 s shaped; 0.625 s FMs (passive), (obstacle avoidance), (communicate) SSB at 8 kHz
Raise speed: 6.7 m/s average
Lower speed: 4.9 m/s
Water exit speed: 1.5 m/s

Status
The AQS-18A is in production and has been delivered to the Egyptian and Italian Navies.

Contractor
L-3 Communications, Ocean Systems.

AR-730 ESM/DF system

Type
Airborne Electronic Counter Measures (ECM), Electronic Support Measures (ESM) system.

Description
The AR-730 ESM/DF system is a modular equipment designed for maritime patrol aircraft and anti-submarine warfare applications. It provides surveillance and threat warning over the frequency range 0.5 to 18 GHz with automatic signal and data processing and precision DF.

The system consists of antenna assemblies, an RF processing assembly and a control/processing assembly.

The antenna assemblies are designed for simple installation on any aircraft and consist of an omnidirectional antenna channel for signal acquisition, rotating DF antenna for direction-finding and optional coarse DF antennas and crystal video receivers. For the AR-730, maximum weight savings are realised by mounting the high-gain DF antenna on the back of the radar antenna. Alternative configurations include separate antenna/radome mountings on the underside of the aircraft.

The RF processing assembly contains the receivers. A DIFM receiver provides high probability of intercept. A superheterodyne receiver provides added sensitivity for finding low-power emitters, allowing searching of dense portions of the frequency spectrum and the 0.5 to 18 GHz band.

The control/processor assembly includes a signal processor and optional cassette recorder, printer and operator display/keyboard. Automated acquisition and identification of intercept is provided, with prompt notification to the avionics computer system. Information provided to the avionic system includes intercept reports, intercept updates, threat alerts and systems alarms. The only information required for processing is a threat/identification database.

Contractor
ARGOSystems (a Condor Systems business unit).

AR-900 ESM system

Type
Airborne Electronic Counter Measures (ECM), Electronic Support Measures (ESM) system.

Description
The AR-900 is a high-performance, fully automatic, passive surveillance system that has been designed for both Electronic Warfare (EW) and ELINT applications. It provides both 100 per cent Probability Of Intercept (POI) and full 2 to 18 GHz frequency coverage. The system comprises three major assemblies: the antenna assembly, the receiver/processor unit, and the operator's workstation. The antenna assembly uses a wideband omnidirectional antenna and a wide open amplitude monopulse DF subsystem that provides better than 3° RMS bearing accuracy.

The receiver/processor unit comprises: two Digital Instantaneous Frequency Measurement (DIFM) receivers, an amplitude-monopulse bearing processor, and a signal processor.

This equipment can process 1,000,000 pps. The signal processor receives digitised high-resolution frequency measurements for each received Radio Frequency (RF) pulse and Continuous-Wave (CW) signal from the DIFM receivers, while the amplitude-monopulse bearing processor provides direction of arrival and amplitude of each pulse. The signal processor processes this information in parallel, compares the pattern with those in a signal library to identify the emitter and its platform, and alerts the operator to a high-threat signal within one second of acquiring the signal. Emitter libraries can be programmed with up to 10,000 emitter modes. The operator's workstation is made up of a keyboard with integrated trackball; a high-resolution colour display; a 3.5 in floppy disk drive; CD-ROM; printer; and an embedded computer.

AR-900 ESM system   0018250

The AR-900's operator is provided tactical and intelligence information through simple, comprehensive displays. The activity, tactical graphics, and tactical summary pages give the operator threat-warning information, while the frequency × azimuth, frequency × PRI, and frequency × amplitude pages are provided for analysis. Various displays aid in system setup, and an intercept report generator is provided to prepare intelligence reports.

Specifications

	2 to 6 GHz	6 to 18 GHz
Frequency range	2 to 6 GHz	6 to 18 GHz
System sensitivity	–65 dBm	–65 dBm
Azimuth coverage, instantaneous	360°	360°
DF measurement accuracy, rms	3.5°	2°
Displayed resolution	1.0°	1.0°
Dynamic range	>70 dB	
Signal types received	Conventional pulse trains, agile frequency (±10%), staggered PRI (2-16 positions), jittered PRI (±10%), frequency/ phase/amplitude or pulse, CW, FMCW, pulse Doppler	
Frequency measurement		
(displayed resolution)	1 MHz	1 MHz
(accuracy, rms)	2 MHz	3 MHz
Pulsewidth measurement		
(range)	0.1–230 µs (0.05–230 µs with 2 dB reduction in sensitivity)	
(resolution)	0.05 µs	
Amplitude measurement		
(range)	>60 dB	
(resolution)	1 dB	
PRI measurement		
(range)	2–20,000 µs	
(resolution)	0.1 µs	
(display)	PRI (µs) or PRF (Hz)	
Scan types	Circular, conical, bidirectional, unidirectional	
Polarisations received	Horizontal, vertical, slant linear, circular	
System alarms	Threat, steady illumination, CW, amplitude	
Threat library capacity	10,000 emitter modes	
Number of signals tracked	500	
System reaction time	1 s max	
Pulse density capacity	1,000,000 pps	
Options		
(RF extensions)	0.5–2.0 and 18–40 GHz	
(superheterodyne receiver)	0.5–18 GHz	
(fine DF)	1.0°	

Contractor
ARGOSystems (a Condor Systems business unit).

AR-7000 airborne SIGINT system

Type
Airborne Electronic Counter Measures (ECM), Signals Intelligence (SIGINT) system.

Description
The AR-7000 is designed to provide airborne reconnaissance, direction-finding and geolocation of emitters in all bands from 20 MHz to 18 GHz. Three subsystems are used to provide this full band coverage of ELINT, COMINT and ESM.

The ELINT subsystem provides for operator controlled signal search, analysis and recording of all pulsed signals from 0.1 to 18 GHz. Three superheterodyne tuners are provided for the 0.5 to 18 GHz tuning range; a fourth heterodyne tuner covers the 0.1 to 1 GHz range. A pulse analyser

performs detailed numeric emitter parameter measurements.

The ESM subsystem provides instantaneous wideband intercept, automatic detection, identification and DF on all signals from 0.5 to 18 GHz. The wideband instantaneous coverage in a dense complex environment is provided by three DIFM receivers and by crystal video receivers which furnish monopulse DF data. Additional DF capability is given by a fine DF subsystem which uses phase interferometers for high-accuracy DF measurement on a tasked basis. An emitter library of up to 5,000 emitters may be tracked in the active emitter file.

The COMINT subsystem provides up to 20 receivers from as many as five operators to use in monitoring the 20 MHz to 1 GHz band. In addition, a DF subsystem furnishes DF on communication signals from 20 to 500 MHz. The DF subsystem uses two phase-matched receivers and two antenna arrays to perform highly accurate multiple differential phase measurements on signals from 20 to 500 MHz. The multiple differential phase measurements are converted to DF estimates by the subsystem processor.

Status
The AR-7000 forms the basis of the Fokker 50 SIGINT Black Crow aircraft.

Contractor
ARGOSystems (a Condor Systems business unit).

ATIRCM/CMWS AN/ALQ-212(V)

Type
Airborne Electronic Counter Measures (ECM), integrated suite.

Description
The AN/ALQ-212(V) Advanced Threat IR CounterMeasures (ATIRCM) system, combined with the AN/AAR-57 Common Missile Warning System (CMWS), is in development by BAE Systems, North America Information and Electronic Warfare Systems (IEWS). The system combines the missile warning (CMWS), with an advanced jammer that integrates laser and xenon lamp technology, and an enhanced dispenser to counter current threats. The design also includes pre-planned product improvement plans for countering future threat systems. When the missile warning sensors detect a threat, the ATIRCM system will slew one or two jamming heads in the direction of the missile. A missile tracker on the jammer head will further refine the direction of the jamming transmission, ensuring that the main beam remains on the target. The system will also cue the release of flares from the dispenser.

BAE Systems' directable IRCM suite provides comprehensive protection against a multi-tier array of infra-red threats. The system detects, acquires, tracks and then jams these missiles with a focused beam of infra-red energy. The AN/ALQ-212(V) incorporates both a robust laser and an arc lamp to provide protection in all IR threat bands and has proven its ability to defeat modern IR missiles in live-fire tests.

The ATIRCM/CMWS suite functions as a stand-alone warning system (the AN/ALQ-57(V)), using common missile warning sensors and electronics coupled to expendables, or as a complete jamming suite (the AN/ALQ212(V)), incorporating one or more jam heads.

The ATIRCM system is claimed to provide over 100 times the jamming power, at less weight, than traditional IRCM systems.

Status
The US Army is the main customer for the AN/ALQ-212(V) and has issued BAE Systems North America with a contract to integrate the system onto the AH-64D Longbow Apache. The system is also slated for other US services and other countries are considering application to their platforms, or have already selected it. It had been expected that the system would be fitted to about 25 different types of US Army,

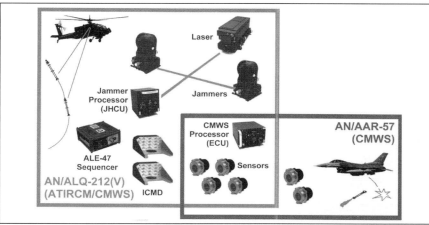

The AN/ALQ-212(V) ATIRCM/CMWS configuration   0051270

The CMA-2082F EWCDU   1030035

ATIRCM/CMWS counters advanced threats using the Common Missile Warning System to cue directable jammers and dispensers 0125563

Navy, Marine Corps and Air Force rotary-wing, transport and fighter aircraft. However, in April 2000, the US Air Force announced a reduction in its projected requirement from 853 systems to 362 systems, and at the same time the US Navy reduced its projected requirement from 665 to 264 systems.

Contractor
BAE Systems, North America, Information and Electronic Warfare Systems.

Challenger IR jammer

Type
Airborne Electronic Counter Measures (ECM), Infra-Red Counter Measures (IRCM) system.

Description
A small omnidirectional IR jammer, especially suitable for helicopters and light aircraft, Challenger can be installed in fixed aircraft

apertures or deployed on retractable platforms. The jammer head weighs about 9 kg and power consumption is within the range 1 to 4 kW. The jammer assembly is driven by a compact controller in the aircraft. Single or dual transmitter configuration can be selected to provide 360° azimuth protection. Challenger, an upgraded AN/ALQ-157, was designed to be flexible and to offer a wide variety of installation options.

Status
In production and in service on Lynx and UH-1 helicopters.

Contractor
Lockheed Martin IR Imaging Systems.

CMA-2082F Electronic Warfare Control and Display Unit (EWCDU)

Type
Airborne Electronic Counter Measures (ECM), control system.

Description
The CMA-2082F Electronic Warfare Control and Display Unit (EWCDU) is a high-performance Multifunction Control and Display Unit (MCDU), with onboard EW processing capability. The unit may be mounted in either a landscape or portrait orientation within the cockpit.

The NVIS compatible full-colour 4 × 4 in (101.6 × 101.6 mm) 480 × 480 RGB AMLCD is fully sunlight readable. Viewing angle is up to ±45°. The system interfaces through two dual-redundant MIL-STD-1553B databusses. The MCDU incorporates a powerful Central Processing Unit (CPU), based on a 25 MHz Intel 80486DX processor core, with extensive onboard memory.

The CMA-2082F incorporates a comprehensive keyboard comprising 18 soft keys and 32 data keys.

Specifications
Physical
Dimensions: 210 × 148 × 208 mm
Weight: 5.0 kg
Power: 28 V DC; 75 W (115 W warm-up)
Cooling: Forced air with built-in fan

Signal interface
Standard: two dual-redundant MIL-STD-1553B BC/RT; RS-422 ×2
Optional: ARINC-429, analogue or discrete I/O, colour video output, additional processor (such as Power PC)
Spare card slots: 2

Operator interface (keyboard)
Data keys: 32
Soft keys: 18
Integral lighting: 5 V AC/DC, externally supplied

Display
Type: Full colour AMLCD
Pixel array: 480 × 480 RGB × 64 grey shades (6-bit colour)
Viewing angle: ±45 horizontal, +20 to −10° vertical at CR =10:1
Contrast ratio: Min 5:1 at 10,000 fc (at optimum viewing angle)
Luminance: 0.10 to 200 fl
NVIS compatibility: MIL-L-85762A Type 1 Class B

Processor/Software
Applications: Intel 80486DX operating at 25 MHz
Graphics/symbol generation: Texas Instruments TMS 34020
Serial interface: Intel 386EX
Applications memory: 4 Mb flash; 2 Mb static RAM
Programme language: ANSI C/Ada

Environmental/EMI
Temperature: −40 to +55°C continuous, +70°C for 30 minutes
Altitude: Up to 50,000 ft
Compliance: MIL-STD-810, MIL-STD-461C Class A1b, DO-160C Category C2

Status
In production. Selected for a NATO fighter platform.

Contractor
CMC Electronics Inc.

Comet Infra-Red Counter Measures (IRCM) system

Type
Airborne Electronic Counter Measures (ECM), Infra-Red Counter Measures (IRCM) system.

Description
Comet is an advanced Infra-Red CounterMeasures (IRCM) system suitable for front-line attack and transport aircraft, including the A-10 and C-130. An alternative to more expensive missile warning

The Lockheed Martin Challenger infra-red jammer   0503854

For details of the latest updates to *Jane's Avionics* online and to discover the additional information available exclusively to online subscribers please visit
jav.janes.com

Raytheon's Comet Infra-red Countermeasures dispenser installed on the port outboard pylon of an A-10 aircraft (Raytheon)   0554683

systems, Comet facilitates combat identification, engagement of multiple targets, and operations below cloud cover. The system is completely autonomous, with no cueing required.

Using technology derived from special-material flares, Comet dispenses an advanced pyrophoric material directly into the airstream behind the aircraft. The material is fully covert in the visible spectrum, creating a continuous false target that seduces IR threats away from the host platform. The Comet pod consists of six canisters containing enough infra-red decoy material to provide aircraft protection for up to 30 minutes. A precision dispenser system meters the material directly into the airstream behind the aircraft. A simple drive system allows each canister to be driven individually or simultaneously. Thus, the dispense rate of the infra-red material can be varied to fit the situational environment and the plume tailored to mimic the engine signature of virtually any aircraft.

The Comet system's novel approach and inherent design simplicity minimises maintenance and servicing. The payload cartridge is sealed and requires no maintenance. All six canisters can be loaded and ready for use in less than one minute. In addition, all components in the multiplatform dispenser controller can be serviced while the pod is on the aircraft.

Status
Tested by the US Air Force on the A-10. It has been reported that the US Air Force FY03 Unfunded Priority List (UPL) recommends the acquisition of Comet infra-red countermeasure pods for US Air Force A-10 aircraft. Proposed to the US Department of Homeland Security.

Contractor
Raytheon EW Systems.

CS-2010 HAWK Receiver/DF system

Type
Airborne Electronic Counter Measures (ECM), Electronic Intelligence (ELINT) system.

Description
The CS-2010 is a scalable, single- or multiple-operator, pooled resource manual ELINT/ESM system covering the frequency range 20 MHz to 40 GHz, designed for airborne, shipborne or land-based applications. The family of equipments includes microwave tuners, demodulators, VHF/UHF up converters, 18 to 40 GHz down converters and spinning DF antenna subsystems. System elements, or resources, are maintained by the system as a pool of resources and are shared by the operators. Operators acquire resources from the pool as required and release them back to the pool when no longer needed. Operators

interface to the CS-2010 through a Resources Control Console (RCC), consisting of a C-144 Resource Control Unit (RCU), a KY-102 keyboard and a pair of high-resolution IP-109 raster scan RGB display units. The RCC provides the operator with control of the pooled resource elements and provides the operator with real-time displays of both antenna and receiver functional outputs.

The CS-2010 features a variety of operational modes, including Band Scan, Receiver Analysis and DF Analysis. Band Scan is a display synthesised from the RF scan data produced by all unused tuners in the system resource pool – that is, the display is produced by parallel scanning all tuners which have not been acquired by an operator for use in analysis of DF operations. The normal and expanded analysis displays provide the operator with RF, IF pan displays and LIFM for receiver channels acquired by the operator from the resource pool. The operator has control over five independent traces representing up to five independent digitally generated polar, inverse polar and rising raster rectilinear, and supports the manual DEF capabilities of the system.

The system is built around VersaNet™, a Time Domain, Multiple Access (TDMA) bus, developed by EDO RSS for ELINT system integration. VersaNet™ distributes control information between system elements and carries digitised display and audio information

gathered in the system elements to the RCC. VersaNet™ enables additional resources such as tuners, demodulators, resource control units and pedestal electronic units to be added to the system as required. The principle tuner in the family is the TN-618 microwave tuner which features very low phase, 500 MHz wide IF and 10 kHz step size over 0.5 to 18 GHz. Block up converters and down converters extend the frequency coverage. The tuner also includes a built-in single bandwidth demodulator. Other CS-2010 HAWK system units are:

MD-128/CCY-128 analysis demodulator/ demodulator mainframe. The MD-128 analysis demodulator is a full-function demodulator providing dedicated LOG, LIN and FM video outputs, as well as audio. Eight selectable IF filters and a 0–60 dB stepped attenuator are also provided. The MD-128 is package in a 'slice' configuration package designed to be mounted in the CY-128 demodulator mainframe. The CY-128 houses up to 4 MD-128 analysis demodulators.

CV-131 down converter. The CV-131 down converter is used to extend the range of the CS-2010 from 18 to 40 GHz. The CV-131 features phase locked oscillators and low noise amplifiers.

PE-137 Pedestal Electronics Unit (PEU). The PE-137 PEU supplies and controls the motor drive current for a single DF antenna pedestal assembly. The PE-137 operates in spin, point and sector modes. Pedestal position vs. time is relayed to the system RCUs via VersaNet™ for use in DF display generation.

SA-118 switch matrix. The Sa-118 is an 8 × 8 non-blocking switch matrix designed to provide up to 8 tuners access to up to 8 inputs. Each input/output covers the entire 0.5–18 GHz frequency range and multiple SA-118s can be combined to increase the number of input ports. Switch control is provided automatically through the auxilliary control port of the tuner allowing various functional capabilities, including antenna port scanning.

A typical CS-2010 HAWK system configuration consists of 4 operator positions (4 RCCs), 8 TN-618 tuners, 8 MD-128/CY-128 demodulator channels, a SA-118, a CV-131 and 2 PE-137 PEUs.

Specifications
Frequency range: 0.5 to 18 GHz basic; 20 MHz to 40 GHz (with CV-131 down converter and CV-137 up converter)
Noise: 16 dB max (0.5 to 18 GHz, 23°C))
Phase noise: 1.0° RMS max (100 Hz to 20 MHz)
Tuning resolution: 10 kHz steps (1 kHz option)
Dynamic range: 90 dB min (1 dB comp); 60 dB min (2 tone) (reference 1 MHz BW)

EDO RSS' CS-2010 HAWK receiver/DF system   0504151

IF outputs: 1 GHz (500 MHz BW), 160 MHz WB (75 MHz BW), 160 MHz NB (0.1 to 50 MHz BW selectable)

IF bandwidths: 0.1, 0.5, 1.0, 2.5, 5.0, 10, 25, 50 MHz

IF attenuation: 0 to 60 dB, 1 dB steps

Remote interface: RS-232, Ethernet, IEEE-488, Opt

Temperature: –54 to +55°C (TN-618)

Altitude: Up to 15,000 ft

Power: 115/230 V AC, 47 to 440 Hz

Status

Maintained by service, supported by contractor.

Contractor

EDO Reconnaissance and Surveillance Systems.

CS-2040 Mini-HAWK Receiver/DF system

Type

Airborne Electronic Counter Measures (ECM), Electronic Intelligence (ELINT) system.

Description

The CS-2040 is a single-operator ELINT receiver system based on EDO Systems' HAWK architecture and utilising elements of the CS-2010 HAWK system (see separate entry). The system features a wideband superheterodyne tuner (TN-618), and analysis demodulator (MD-128) and an operator control and display unit (C-244).

The CS-2040 features a modular design, which allows from 1 to 16 receiving channels and 1 to 4 operators; the system can also (as an option) control and display one or two spinning DF antennas.

Other major features of the CS-2040 include:

- 0.5 to 18 GHz frequency coverage with an instantaneous bandwidth of 500 MHz
- Analysis demodulator with 8 IF filters, IF attenuator, LOG/LIN/FM video
- Built-in IF PAN and LIFM modules
- 6.5 in colour LCD display with 640 × 480 resolution
- Illuminated 64-key keyboard, tuning knob and four parameter encoders
- Extensive BIT
- VersaNet™ LAN distributes control, status and digitised trace data

Specifications

Frequency range: 0.5 to 18 GHz

Noise: 16 dB max

Phase noise: 1.0° RMS max (100 Hz to 20 MHz)

Tuning resolution: 10 kHz steps

Dynamic range: 90 dB min (1 dB comp); 60 dB min (2 tone 3rd order)

IF outputs: 1 GHz (500 MHz BW), 160 MHz WB (75 MHz BW), 160 MHz NB (0.1 to 50 MHz BW selectable), 21.4 MHz (0.1 to 2.5 MHz BW selectable)

IF bandwidths: 0.1, 0.5, 1.0, 2.5, 5.0, 10, 25, 50 MHz (other bandwidths available)

IF attenuation: 0 to 55 dB, 1 dB steps

Remote interface: RS-232

Temperature: –54 to +55°C (TN-618); 0 to 50°C (all other units)

Altitude: Up to 15,000 ft

Power: 115/230 V AC, 47 to 440 Hz

Dimensions (H × W × D):
  TN-618: 195.6 × 147.3 × 523.2 mm
  C-244: 177.8 × 444.5 × 508 mm

Weight:
  TN-618: 15 kg
  C-244: 18.2 kg

Options: 0.5° phase noise, 1 kHz tuning steps, Ethernet, IEEE-488 remote interfaces, frequency extension to 40 GHz with down converter, frequency extension to 20 MHz with up converter, DF antenna control and displays, AGC LIN video, LIN IF PAN video

Status

In service.

Contractor

EDO Reconnaissance and Surveillance Systems.

CS-3001 pulse analyser

Type

Airborne Electronic Counter Measures (ECM), Signals Intelligence (SIGINT) system.

Description

The CS-3001 is a VME-based pulse analyser that accepts 160 MHz and 1 GHz IF inputs and provides de-interleaved pulse descriptor words for analysis. Combined with a fully featured Java/Network Friendly GUI control interface, the system provides excellent visualisation and statistical display of collected emitter intercept data. It uses a mixture of analogue hardware, high-speed digitisers, and FPGA/DSP-based processing.

The CS-3001 pulse analyser base configuration consists of three single slot 6U VME cards (demodulator, digitiser and BITE) along with a pair of two slot, high-performance control and analysis CPU 6U VME cards for a total consumption of seven VME slots.

The system is available as a VME board set or as a complete, self-contained pulse analyser mounted in a general use VME rack, a ruggedised VME rack or other form factor enclosure. The CS-3001-1 (DEMOD) VME card provides a wide 500 MHz bandwidth for rapid signal acquisition and to accommodate the capture of complex modulations on modern signals of interest. In addition, this card provides higher sensitivity narrower bandwidths (75, 25, 10 MHz) that can be used for signal isolation and to perform fine-grained pulse analysis. The 160 MHz channel can process bandwidths up to 75 MHz, or as low as 10 MHz. High-dynamic range LOG video amplifiers, and discriminators along with carefully designed filters provide the basis of a high-performance demodulator that is packaged as a single slot VME card.

The heart of the system, the CS-3001-2 (DIGITISER) VME board, contains multiple 10-bit high-speed digitisers as well as a series of FPGA devices. This board provides the precision time tags used to derive pulse width, TOI, and PRI, signal characterisation of amplitude and frequency, video intrapulse collection, and detection flags for noting the presence of various intrapulse MOP features. This digital signal characterisation is fused together to form 160-bit Pulse Descriptor Words (PDWs) for subsequent storage, display and analysis. The last card in the pulse analysis engine, VME card CS-3001-3 (BITE), contains the system timing reference(s), interfaces, and BITE functions. It accepts 1-pps pulses from GPS receivers to implement system synchronisation. This board has an internal reference oscillator that can be used as a free running reference with buffered outputs, or can be locked automatically to an external precision 10 MHz reference source.

Using the on-board oscillator as a single reference for other system elements, such as the tuner, provides a coherent system with excellent precision and tracking. The CS-3001-4 A/B (CPUs) devices are the control and analysis CPU's provided for the operation of the system. One CPU acts as the external interface providing control access, data distribution, post-processing analysis, and insulation for the pulse analysis engine. The second CPU provides precision real-time device control and data collection. The two controllers execute the embedded CS-3001-SP software package that collects the pulse information, performs analysis, co-ordinates control, and interfaces with system real time devices (such as tuners and antenna controllers). The system is designed with 6 serial ports to support interfacing with various external equipment such as GPS, platform navigation, ComSol ELINT tuners, and antenna controllers.

An external workstation processor runs the CS-3001-DP software that operates under a Windows NT/2000 platform and provides fully functioned display and analysis capabilities to support the mission critical needs of ELINT operators throughout the world. This software is designed in a Network Friendly Java-based format that supports control of the pulse analyser assets, visualisation, and analysis of collected data via an internet browser.

The CS-3001 is packaged in either a general use (CS-3001-5) or ruggedised (CS-3001-6), custom VME case with additional slots available for the CS-5111 VME microwave tuner (see separate entry), real-time devices (GPS, IRIG, and so on), or special purpose user functions.

The CS-3001 is designed to be upgradeable in the future to meet specialised signal identification requirements and expanding mission roles.

Status

In production.

Contractor

Rockwell Collins.

CS-5020C microwave receiver

Type

Airborne Electronic Counter Measures (ECM), Electronic Intelligence (ELINT) system.

Description

The CS-5020C microwave receiver is a high-performance, super-heterodyne, set-on and sweep receiver that converts signals in the frequency range of 0.5 to 18 GHz into IF outputs of 1.0 GHz (standard) and optional 70, 140, and 160 MHz (selectable) outputs. Video (AM/FM/LOG) and audio demodulated outputs are also available. The receiver can be extended in frequency coverage to 0.1- to 22 GHz without the need for external extenders. A millimetre-wave downconverter interface also can be included internally, which allows the operator to attach a remote downconverter to provide further coverage in that frequency range.

The CS-5020C can be configured for a wide variety of applications in COMINT or ELINT missions. The 1.0 GHz/500 MHz BW wideband IF output allows processing of various high data rate/high complexity signals. Combined with available external millimetre-wave downconverters, the system provides for the search and collection of signals of interest in the microwave and millimetre-wave frequency bands from 100 MHz to 110 GHz.

The CS-5020C combines high performance and flexibility in multimission applications. A high dynamic range with a typical noise figure of 13 dB, third-order intercept of +2 dBm, and 100 Hz tuning resolution makes the system extremely capable in demanding missions. The 70/140/160 MHz IF outputs can be provided with fixed gain or optional AGC/MGC control and are normally upright, but can be inverted upon command. The receiver's RF/IF gain is microprocessor normalised to provide excellent flatness across the entire 0.1 to 22 GHz frequency range. The CS-5020C is fully synthesised with phase noise less than 0.65 RMS (0.5 typical), and is phase coherent to either an internal crystal oscillator or an external 10 MHz reference.

The system is contained in a half-width, two unit high, rack-mount package that uses forced air cooling for high reliability. The AC power input accepts 90 to 260 V AC, 47 to 400 Hz. The internal power supply has sufficient output for the aforementioned optional external millimetre-wave downconverter and preamplifier. It is designed to meet the requirements of MIL-STD-461 and includes mechanical shielding techniques such as EMI gasketing and waveguide-beyond-cutoff hole patterns for cooling vents.

A front panel is now a standard feature with keypad plus optical encoder tuning. All control and monitoring functions are available locally via the front panel or remotely via either the standard RS-232/RS-422 or other optional digital interfaces. Significant functions include 100 Hz display resolution, front panel 'EXT REF' LED, ability to store and recall one hundred settings, BITE, sweep functions, frequency reference auto-update, and appropriate IF and demodulation functions.

Status

In production.

Contractor

Rockwell Collins.

CS-5060 ESM/ELINT system

Type
Airborne Electronic Counter Measures (ECM), Electronic Support Measures (ESM) system.

Description
The CS-5060 is a fully automatic ESM/ELINT system which combines long-range signal detection, high probability of intercept, fine grain parameter measurement and complex signal handling in the microwave frequency range. The system is designed to address both operational and technical ESM/ELINT requirements, covering the frequency range of 0.5 to 18 GHz (expandable to 40 GHz). The CS-5060 system consists of a high-gain antenna used for spinning DF, a microwave tuner, a wideband signal processor, and an operator workstation for control and display. CS-5060 units are all Non-Development Items (NDI) that are currently deployed in a variety of shipboard, airborne and land-based applications and installations.

The spinning DF antenna provides for high-gain, long-range signal detection and bearing measurement over the entire frequency range. The TN-618 tuner is a fully synthesised microprocessor-controlled 0.5 to 18 GHz superheterodyne tuner. The SP-103 signal processor (see separate entry) accepts a 1 GHz, 500 MHz bandwidth input and performs automatic pulse parameter measurement, deinterleaving, emitter characterisation and identification. The SP-103 measures and characterises emitter frequency, pulse repetition interval and agility, pulsewidth, line of bearing, amplitude, time first seen and time last seen. The processor also compares the measured parameters of the target emitter to an internal library of known emitters to identify the target.

The combination of a high-gain antenna, tuner with 500 MHz instantaneous bandwidth and the intelligent search strategies resident in the signal processor facilitates rapid signal acquisition and a high-confidence emitter identification.

Specifications
Frequency range: 0.5 to 18 GHz (with optional increased coverage)
Frequency accuracy: 1 MHz RMS
Instantaneous bandwidth: 500 MHz
IF bandwidths: 20, 50, 100, 250, 500 MHz
Video bandwidths: 5 and 20 MHz
System sensitivity: −65 to −90 dBm
Auto DF accuracy: 2–5° RMS
PRI range: 2 msec to 10 msec
PRI accuracy: 50 ηsec
Scan rate range: 0.05–100 Hz
Scan rate accuracy: 2% RMS
Scan types: Circular, bi-directional, conical, steady, complex
System dimensions:
  Antenna: 482.6 × 736.6 mm (Ø × H)
  Processor: 269.2 × 218.4 × 457.2 mm W × H × D)
  Tuner: 147.3 × 198.1 × 558.8 mm (W × H × D)
System weights:
  Antenna: 31.8 kg
  Processor: 16.4 kg
  Tuner: 15 kg
Power:
  Antenna: 200 W
  Processor: 250 W
  Tuner: 125 W

Status
Designed for airborne, shipborne and land-based applications. In production and in service.

Contractor
EDO Reconnaissance and Surveillance Systems.

CS-5106C millimetre wave receiver

Type
Airborne Electronic Counter Measures (ECM), Electronic Intelligence (ELINT) system.

Description
The CS-5106C is a low-cost millimetre wave receiver that is tuneable, but designed for set-on applications. A triple conversion frequency scheme provides low phase noise, high-dynamic range, and low cost in a self-contained instrument.

The internal phase locked LOs and synthesiser are referenced to a single internal crystal oscillator which can in turn be locked to an external 10 MHz reference. The unit switches automatically according to the external reference presence.

The CS-5106C is designed to allow the user in the field to tune the receiver over a limited band (approximately 3 GHz) using a remote interface and notebook computer (not supplied) for control. It is also possible to change the first LO to provide overall range changes (such as change from 23.6 GHz to 36.4 GHz). In this manner, a CS-5106C can cover any frequency in the 18 to 40, 40 to 60, 60 to 80, 80 to 100, or 100 to 110 GHz frequency range by selecting or changing the LO and mixer assemblies. Different plug-in tuning heads can be installed within the case or remoted to the antenna via an external cable, providing the best possible sensitivity while maintaining flexibility and ease of use. The tuning head tends to be very wideband in nature, often requiring only the LO to be changed to cover a very wide range. The LO can be a fixed frequency module or a variable frequency module with provides a 3 GHz minimum range. The second LO provides 1 kHz step size fine tune capability.

Optionally, the CS-5106C can be provided to cover frequencies below 18 GHz (down to 6 GHz), or with up to three bandwidths at either 70, 140, or 160 MHz IF centre frequencies. An optional AGC section can also be provided and a FM video section added. A command is also accepted to switch between inverted and non-inverted IF output sense.

The frequency control of the CS-5106C is handled via an RS-232 interface and the internal microprocessor has battery backed up memory. Thus, a notebook computer can be used to tune the instrument over a 3 GHz range with 1 kHz tuning resolution. Once programmed, the unit will remain at the frequency until changed via the RS-232 interface. Should power be lost, the CS-5106C will return to the last set frequency when power is restored. Windows NT™ and Window 95™ compatible computer programs are provided along with the unit to control the CS-5106C on a notebook computer.

Typically, the CS-5106C will have a coaxial input up to 40 GHz and waveguide inputs above 40 GHz. Companion, low noise, external preamplifiers can be supplied with either coaxial or waveguide connectors as appropriate for the band of interest.

The CS-5106C has a front panel power switch, cooling fans and BITE LED which indicates a loss of lock or other problem should this occur. The CS-5106C is AC powered using a universal 90–260 V AC input, with circuit breaker protection. The plug-in tuning head slides into a rear panel opening via a blind mated interface connector. All power, RF input, IF input, video/LOG outputs and external reference connections are on the unit rear panel.

Status
In production.

Contractor
Rockwell Collins.

CS-5111 VME microwave tuner

Type
Airborne Electronic Counter Measures (ECM), Electronic Intelligence (ELINT) system.

Description
The CS-5111 is a high-performance microwave tuner that is housed in a 2 slot VME chassis.

The CS-5111 converts RF/microwave signals in the 0.4–18 GHz frequency range into a standard IF output, centred at 1 GHz, with 500 MHz of bandwidth. An optional IF output, centred at 160 MHz, with 85 MHz of bandwidth is also available. Frequency coverage can be optionally expanded to cover 0.15 to 20 GHz without requiring external extenders.

The system has been configured to optimise front-end signal capture in demanding ELINT missions. The CS-5111 tuner uses switched filters for pre-selection to allow for fast tuning speeds and an input limiter protects the tuner from damage from signal power levels up to +28 dBm. TTL controlled, high-speed input blanking is provided to accommodate high-density radar environments. With typical tuning speeds of <1 ms ensure a high confidence in probability of intercept in complex radar applications. The 1 GHz IF output's bandwidth of 500 MHz supports fast rise time/narrow pulse width signals of interest.

The double conversion architecture and a low-noise figure provide for a very high, tuner dynamic range. An input third-order intercept of +2 dBm typical permits processing of dense environments with very good intermodulation performance. An optional internal input attenuator is available that can provide 0, 10, or 20 dB of input attenuation to extend the dynamic range and provide extended intermodulation performance for processing higher power level signals of interest. The CS-5111 features fully synthesised operation with 1 MHz tuning resolution at the 1 GHz IF output and 10 kHz tuning resolution at the optional 160 MHz IF Output. In addition, the tuner operates phase coherent to either an internal 100 MHz low-phase noise crystal oscillator or external system 10 MHz reference.

The CS-5111 unit is housed in a small and rugged 2 slot module package that allows for ease of operation in various VME racks. VME power and control connections are located on the rear panel, while all other connections are established at the front panel.

The package is designed to meet the requirements of MIL-STD-461 for RF interference suppression and includes mechanical shielding techniques such as EMI gasketing and waveguide-beyond-cutoff hole patterns for cooling vents. BITE status indicators are located on the front panel to permit easy visibility of unit status.

The unit is controlled through a VME standard control bus, which is connected to at the rear panel and all tuner function parameters can be controlled through the VME bus. All functions and BITE status can also be monitored through the VME bus. DC power is also provided through the standard VME rack connections, as part of the VME bus connector.

Specifications
Dimensions: 2 slot VME, 6U
Weight: 3.6 kg max
Input Frequency: 0.4–18 GHz range (standard); 0.15 GHz lower extension, 20 GHz upper extension (option)
Input VSWR: 2.5:1 max (50 Ω)
Maximum CW input signal level: +28 dBm (no damage)
LO re-radiation: −90 dBm max (0.4–18 GHz)
Noise Figure (at 25 C): 13 dB typical, 15 dB max; 17–20 dB max (option)
Input third-order intercept (at 25 C): +2 dBm typical, 0 dBm min (no input attenuation)
Spurious free dynamic range: Single tone 65 dB min, two tone 57 dB min (50 MHz BW)
Linear dynamic range: 70 dB min (50 MHz BW)
Image rejection: 60 dB min
IF rejection: 65 dB min
Internal frequency stability (over temperature range): ±8 × 10-6 max after 1 hour warm-up, ageing at 1 × 10-6/year
External reference: Frequency 10 MHz, amplitude 0 ±3 dBm, automatically switched, SMA female connector
Tuning modes: CW and step tune
Settling time (after last command bit): 500 μsec typical, 2 msec max

Tuning resolution: 10 kHz (at 160 MHz IF output), 1 MHz (at 1GHz IF output)
LO spurious rejection: −55 dBc min (>±1 kHz from Fc)
Integrated phase noise (100 Hz-25 MHz): 0.8° RMS typical, 1.0° RMS max
IF output frequency/bandwidth
1 GHz IF output: 500 MHz (when Fc ≥ 0.4 GHz)
160 MHz IF output: 85 MHz (when Fc ≥ 0.4 GHz)
RF/IF gain: 15 dB min (no input attenuation)
RF/IF gain flatness: ±1 dB max
Input attenuation settings: 0, 10 or 20 dB
IF gain control (manual): 60 dB typical, 50 dB max
Group Delay
1 GHz IF: 2 ηsec typical over 80% of the IF bandwidth
160 MHz IF: 3.5 ηsec typical over 80% of the IF bandwidth
IF Output VSWR: 2.0:1 max (50 Ω)
Temperature
Operating: 0 to +50°C
 Storage −40 to +85°C
Cooling: VME rack forced air

Status
In production.

Contractor
Rockwell Collins.

CS-5510 electronic surveillance and collection system

Type
Airborne Electronic Counter Measures (ECM), Electronic Intelligence (ELINT) system.

Description
The CS-5510 is a wideband, superheterodyne, multi-channel acquisition, receiving and ELINT collection and analysis system. The system supports multiple operator positions and includes Windows-based HMI with collection control, signal analysis and signal recording functions.

Due to the particular requirements of an ELINT system to acquire signals at significant standoff ranges in the sidelobes or backlobes of target radar systems, the CS-5510 is designed to use high-gain antennas and narrowband receivers to maximise signal reception.

The system produces real-time RF spectrum displays and IF PAN displays from set-on receivers. Due to the 'scan-while-locked' nature of the tuner control, the displays show the signals at the precise RF positions, without need for re-tuning.

The system is designed to cover 0.5 to 18 GHz, with optional frequency extension capability to include 20 to 500 MHz and 18 to 40 GHz. At the core of the CS-5510 is a signal processor, which controls a wideband tuner and characterises signals in up to 500 MHz bandwidths. Up to 8 TN-618 tuners can be controlled by this processor. Parallel-receiver architecture

is employed, which allows simultaneous automatic and manual collection capabilities. The processor automatically creates pulse descriptor words from the entire suite of tuners, deinterleaves them and compares the results with a user-furnished signals library for emitter identification.

Specifications
Frequency range: 0.5 to 18 GHz (with optional increased coverage)
Frequency step size: 100 kHz
Processing channels: Up to 16 simultaneous 50 MHz wide channels, one 500 MHz wide channel
Demodualtion: LOG, LIN, FM
Signal types: Conventional stagger (32 positions), jitter (FMOP flag) and others
Pulse density: >4 MPPS
Phase noise: 1.0° RMS max
Tuning speed/resolution: <1 MHz steps at 100 μsec; >1 MHz steps at1 msec (all frequencies)
Spur free dynamic range: 60 dB min
IF outputs: 160 MHz (50 MHz BW), 1 GHz (500 MHz BW)
IF PAN: Real-time display
Sensitivity: −80 dBm (antenna assets dependant)
Response time: 1.0 sec (with 10,000 mode library)
Workstation operating system: Windows-NT or UNIX

Status
Designed for airborne, shipborne and land-based ELINT applications.

Contractor
EDO Reconnaissance and Surveillance Systems.

DASS 2000 Electronic Warfare (EW) suite

Type
Airborne Electronic Counter Measures (ECM), integrated suite.

Description
The DASS 2000 electronic warfare suite is a turnkey, fully integrated defensive aids suite which can be configured to meet specific customer requirements and is applicable to a wide range of platform types. The suite's long-range, multispectral threat detection and identification capabilities provide the aircrew with full situation awareness through sensor fusion, and ensure the application of optimum self-protection countermeasures. These co-ordinated capabilities ensure platform survivability in all phases of mission prosecution.

DASS 2000 resources typically include a radar warning receiver, missile warning system, laser warning system, chaff/flare dispensers, electronic

countermeasures jammer, towed radar decoy and infra-red countermeasures. BAE Systems will adapt the configuration to user requirements, integrate it into the platform ensuring full compatibility, and provide a full range of training and logistics support.

DASS 2000 features interactive functions and a full colour display that provides the EW operator with tactical awareness. The display enables threat assessment as well as automatic countermeasures response by using the defensive aids resources.

A DASS 2000 based system is on order by the UK MOD, via British Aerospace, as the Nimrod MRA.4 DASS.

Status
A system based on DASS 2000 has been selected for the UK Nimrod MRA.4.

Nimrod MRA.4 Defensive-Aids SubSystem (DASS)
The Nimrod MRA.4 DASS is not yet fully defined, but the following elements are understood to be included: the BAE Systems AN/ALE-56M radar warning receiver; the Raytheon Electronic Systems ALE-50 towed radar decoy; the Racal Defence Electronics Digital Radar Frequency Memory Techniques Generator (DRFMTG).

System architecture is based on receipt of threat warning via the AN/ALE-56M and/or an as yet unspecified missile approach warning receiver; production of a suitable threat countermeasure in the techniques generator, and its transfer via EO cable to the AN/ALE-50 towed radar decoy.

The Racal techniques generator is based on advanced twin Digital Radio Frequency Memory (DRFM) units, which are used to classify the threat and customise the most coherent jamming response.

Status
In development.

Contractor
BAE Systems, North America, Information and Electronic Warfare Systems.

EC-130E airborne TV and radio broadcasting station

Type
Airborne Electronic Counter Measures (ECM), Signals Intelligence (SIGINT) system.

Description
Variants of the psychological operations airborne TV and radio broadcasting EC-130E aircraft are capable of broadcasting TV in full colour, in any format, worldwide. Aircraft modification kits

EDO RSS' CS-5510 electronic surveillance and collection system   0547232

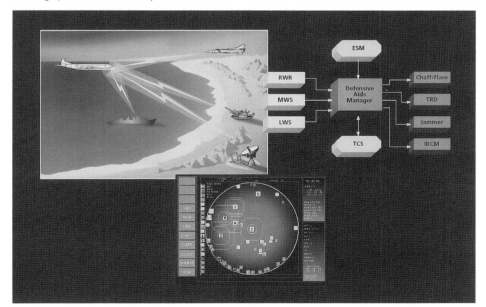

DASS 2000 provides maritime patrol with an effective multispectral, integrated EW capability   0515253

and installations include VHF and UHF antennas housed in 6 ft diameter, 23 ft long (1.829 × 7.01 m) pods under each wing dedicated to higher frequency TV and rotatable to alter polarities, VHF low-frequency TV antennas mounted on each side of the tailplane, upgraded transmitters, equipment for formatting TV signal to worldwide standards and increased output power from 1 to 10 kW. Two retractable trailing-wire antennas provide both HF and VHF AM omnidirectional radio broadcast coverage.

The aircraft are upgraded versions of EC-130E Volant Solo aircraft.

Status

In service on US Air National Guard EC-130E aircraft.

Contractor

Lockheed Martin Aircraft and Logistics Centers.

Emitter Location System (ELS)

Type

Airborne Emitter Location System (ELS).

Description

Raytheon Electronic Systems, under contract to DaimlerChrysler Aerospace AG on behalf of Panavia, developed an Emitter Location System (ELS) for use on the Tornado ECR aircraft. The ELS detects, identifies and locates radar emissions through the use of a high probability of intercept system. It features multi-octave frequency coverage, phase interferometric antenna arrays for precision direction-finding and passive ranging, channelised receivers and multiple 1750A digital processors for operation in dense signal environments.

The ELS operates across the frequency spectrum for all primary surface-to-air and air-to-air threats. Data acquired by the system is transmitted, via a MIL-STD-1553B databus, to the tactical displays of both crew members, who can then make threat assessments and activate the appropriate countermeasures. Threats that have been located by the ELS can also be communicated to follow-on forces through the Operational Data INterface (ODIN).

This data can then be used for threat avoidance or suppression operations.

The design uses a SAW channeliser in a cued analysis receiver configuration. In this configuration the delay lines allow the channeliser to measure the frequency of the incoming signal and then, using a fast-settling local oscillator, tune and cue up the narrowband receivers for subsequent analysis and pulse report generation. The channeliser provides the cued analysis contiguous high signal-selectivity across a wide instantaneous bandwidth. The basic channeliser design consists of multiple parallel SAW filter banks, each bank having multiple signal channels.

High-speed real-time emitter analysis is accurately processed for positive threat identification. In addition to this, accurate DF and passive ranging is computed within the processor, correctly locating and displaying large numbers of emitters. The processed information can then be used to cue weapons, and could be configured to steer jamming pods.

The configuration of the ELS in the Tornado ECR consists of 10 individual LRUs. The two antenna/RF converter units are mounted conformally with the aircraft's left and right wing nibs near the base. The remaining units are distributed within the wing shoulder and gun bays. An operator control is provided in the ELS control panel and the Tornado TV tab displays.

Status

In service in German and Italian Air Force Tornado ECR aircraft.

Contractor

Raytheon Company, Space and Airborne Systems.

ES5000 COMINT/ELINT system

Type

Airborne Electronic Counter Measures (ECM), Electronic Support Measures (ESM) system.

Description

The ES5000 is a combined communication and electronic intelligence system which can be installed in most types of aircraft. It performs signal interception, direction-finding and analysis, data collection and location of communication emitters from 2 to 1,200 MHz and non-communication emitters from 0.02 to 18 GHz. Information is transmitted by microwave link to a ground station. The airborne SIGINT equipment can be controlled either by operators on board the aircraft or remotely from the ground analysis centre.

On a typical mission, one or two airborne collection centres communicate by microwave link with the ground centre. One ES5000 can perform signal intercept, direction-finding and emitter location, producing emitter bearing data. A two-aircraft configuration can perform instantaneous emitter location on short-burst transmissions. Both systems pass data to the ground analysis centre.

Specifications
COMINT
Frequency: 2–1,200 MHz
Instantaneous coverage: 500 MHz
DF rate: 12/s coarse, 4/s fine
Field of view:
    120° each side (primary)
    ±30° front and rear (secondary)
Elevation: –10 to +2°
Accuracy:
    <2° RMS (primary)
    <3° RMS (secondary)

ELINT
Frequency: 0.02–18 GHz
Coverage: 360° azimuth
Accuracy: 1–3° RMS
Measurements: PRI/PRF, pulsewidth, amplitude, scan time, DF, time of arrival, intrapulse
PRI range: 2–10,000 µs
Radar types: pulsed, 16 level stagger, CW, interval pulse, complex
Emitter library: 10,000 modes
Active emitters: 1,024

Microwave link
Frequency: E/F-band
Range: 350 km at 10,000 ft altitude

Status

In production and in service.

Contractor

TRW Systems & Information Technology Group.

Escort ESM system

Type

Airborne Electronic Support Measures (ESM) system.

Description

Escort is a family of electronic support measures for maritime patrol aircraft, small ships and ground-based vehicles. It is generally applicable to adapted versions of business aircraft. Escort provides signal detection, sorting, analysis, identification and reporting of radar frequency emitters. Wide instantaneous spatial and frequency coverage, combined with sophisticated processing techniques, provides high probability of intercept. All mission data can be recorded on tape cartridge for post-mission analysis. The ELINT configuration provides precision direction-finding by using phase interferometer antennas in conjunction with the basic receiver processor and display system.

The basic system operates over the 2 to 18 GHz frequency band with growth options available. It uses two antenna units, each consisting of a four-element antenna array, a high-speed RF switch and various control and amplification circuitry. The receiver unit accepts RF from port and starboard antenna units and performs frequency, time of arrival, amplitude and angle of arrival measurements on each received pulse. The processor consists of a 16-bit general purpose computer with 128,000

Layout of the emitter location system in the ECR Tornado   0503859

16-bit words of memory and a variety of interface circuits. The search and dwell process used by the receiver as it steps between sub-bands to intercept and collect emitter data is controlled by the processor software. Because of the wide bandwidth of the receiver, the search cycle over the entire frequency/azimuth domains is typically accomplished within one main beam illumination of a radar, ensuring high probability of intercept and a very low intercept time for new emitters.

The ESM operational software program maintains an extensive emitter data file, which includes the measured parameters as well as time of initial and most recent intercepts. Formats of CRT displayed outputs vary according to user requirements.

Specifications
Volume: 0.112 m³
Weight: 91 kg
Frequency: 2–18 GHz
Processing: 100 emitters
Accuracy: 8° DF, 5 MHz frequency, 1 µs PRI

Status
The system is believed to form part of the RC-350 Guardian airborne reconnaissance system.

Contractor
Raytheon Company, Space and Airborne Systems.

EW-1017 surveillance system

Type
Airborne Electronic Counter Measures (ECM), Electronic Support Measures (ESM) system.

Description
As an airborne electronic surveillance system, EW-1017 automatically acquires and identifies emissions within the C- to J-bands. The system is also designed to receive and identify all those emissions illuminating the aircraft (including short bursts) in particular when it is operating in very dense signal environments. The warning of possible danger is given both visually and aurally on a display unit while preferential scan is used to ensure immediate recognition of possible lethal threats.

EW-1017 consists of antenna arrays, receiver, a processor system, a pilot display, and ESM operator interactive display subsystem. Broadband spiral antennas are used to provide omnidirectional coverage which, together with their separate multiband receivers, are mounted in pods on each wingtip. This location drastically limits aircraft 'shadowing' and the proximity of the receiver cuts signal losses to a minimum. Angular bearing of the emissions is determined by using selected pairs of antennas.

The hybrid superheterodyne receiver combines high acquisition probability, high sensitivity, frequency accuracy and a high degree of frequency selection and selectivity. A broad bandwidth is used in the acquisition mode to obtain the initial intercept with a narrow bandwidth then being used for accurate bearing measurement and analysis. To ensure processing capability in highly dense signal conditions a high-speed digital computer performs the data processing functions, supplemented by microprocessors. This enables the receivers to scan the frequency band continuously on a reprogrammable basis, so that conventional, continuous wave and agile signals are processed for identification. For special applications, the receiver can be interfaced via a smart post processor unit to a centralised tactical display and control system.

An interactive display subsystem provides a full range of operator facilities to manage and optimise collection. It also provides a readily accessible real-time emitter and platform library storage and analysis capability, facilitated by a

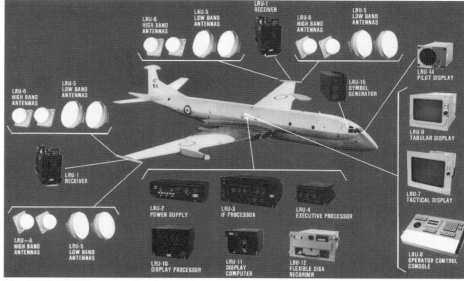

The EW-1017 ESM system showing the location of units on the Nimrod MR2. Mk2 aircraft   0504079

modern data management system. A control/display unit allows the operator to monitor and control the automatic surveillance function to resolve possible ambiguities and evaluate and use to the best advantage the data displayed.

Status
EW-1017 is in service with the German Naval Air Arm (Atlantique maritime patrol aircraft) and the UK Royal Air Force (Nimrod MR2 Mk 2 maritime patrol and E-3D Sentry airborne warning and control system aircraft). In UK service, the system is designated as ARI-18240 Yellowgate.

In April 1999, Thales Defence was contracted by the UK Royal Air Force to upgrade the processing system and threat management software for the UK E-3D aircraft using Commercial-Off-The-Shelf (COTS) processors.

Contractor
BAE Systems, North America, Information and Electronic Warfare Systems.

Falcon Edge Integrated Electronic Warfare Suite (IEWS)

Type
Airborne Electronic Counter Measures (ECM), integrated suite.

Description
The Falcon Edge IEWS, together with the AN/AAQ-32 IFTS and the AN/APG-80 Agile Beam Radar (ABR) (see separate entries) together make up the primary sensor suite of the Lockheed Martin F-16 Block 60 aircraft ordered by the United Arab Emirates.

Falcon Edge provides complete angular coverage for the aircraft, employing a number of Transmit (Tx)/Receive (Rx) antenna arrays:
- Forward- and aft-facing E-J Band (2–20 GHz) passive Rx/active Rx (outer portion of each wing)
- C-D Band (0.5–2 GHz) Rx (Compass Sail Array – nose section)
- C-J Band (0.5–20 GHz) Rx (both sides of forward fuselage)

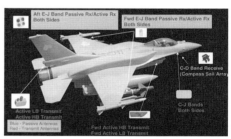

Falcon Edge transmit/receive antenna array
0524712

- High- and Low-Band active Tx (chin station and rear fin root).

Onboard electronics comprise a digital receiver Line Replaceable Unit (LRU), digital techniques generator LRU and two forward-transmitter LRUs housed in the aft equipment bay, together with the aft-transmitter LRU housed in the tail root.

Status
Falcon Edge is part of the F-16E/F Desert Falcon ordered by the United Arab Emirates (UAE). The first batch of UAE aircraft was delivered on 6 May 2004.

During 2002, BAE Systems' Advanced Systems Division was awarded a contract by Northrop Grumman for the development of the Precision Direction Finding (PDF) antenna subsystem for the Falcon Edge Programme. The initial contract for the design, development, prototype and manufacture of two Engineering Development Model (EDM) shipsets was valued at USD1.98 million. A follow-on contract was awarded for more antenna assemblies, which, together with anticipated production of the antennas and associated radomes for 80 aircraft and spares (for the UAE Desert Falcon order), amounted to a contract value of just over USD16 million with a performance period starting in 2002 and ending in 2004.

Contractor
Northrop Grumman Corporation, Electronic Systems Sector.

Falcon Edge LRUs   0524711

GuardRail/Common Sensor (GR/CS)

Type
Airborne Electronic Counter Measures (ECM), Signals Intelligence (SIGINT) system.

Description
The US Army's GR/CS is an airborne SIGINT system that intercepts, locates and classifies target systems and transmits data to ground processors to provide real-time intelligence

information. The GR/CS suite consists of a transportable ground-based system and RC-12 aircraft carrying remotely controlled mission equipment.

The Guardrail system is primarily for COMINT, but an ELINT capability has been added to later aircraft. The concept behind Guardrail is that data from multiple receivers are microwave downlinked to a ground-based Integrated Processing Facility (IPF).

The airborne relay facilities part of the Guardrail system is the airborne communications intelligence payload installed in electronic mission RC-12K, N and P aircraft. They contain intelligence intercept, signal classification and direction-finding equipment, and include provisions and interfaces for other signal intelligence subsystems, mission datalinks and aircraft navigation and communications systems. Collected data is datalinked directly to the IPF, or via an RC-12Q mothership.

Through a series of demonstrations, known as Precision SIGINT Targeting System (PSTS), Guardrail has proven its ability to work with national resources to provide targeting data for cross-programs operations. Guardrail has participated in the US Office of Naval Research (ONR)-sponsored Advanced Concept Technology Demonstration (ACTD), and is claimed to be already compliant with the US Joint Avionics SIGINT Architecture (JASA), with Guardrail/Common Sensor System 2 being the first operational implementation.

For the 21st century, GR/CS will be part of the US Army's Aerial Common Sensor system, which will also use the Direct Air-to-Satellite Relay (DASR) operational with Guardrail.

Status

Guardrail has been in operational service with the US Army for a number of years on RU-21H and RC-12D aircraft. The original Guardrail was updated to IIA and was later updated to Guardrail IV. Further updating provided the Guardrail V under the nomenclature AN/USD-9(V)2. The US Army has developed a modification to Guardrail V that includes the Advanced Quick Look ELINT system produced by Electronics & Space Corporation and the Communications High-Accuracy Airborne Location System (CHAALS) provided by Lockheed Martin. This integrated system, known as GuardRail/Common Sensor (GR/CS), was deployed in the RC-12K.

Latest versions RC-12K, N, P and Q are fitted with upgraded CHAALS equipment and have access to the Defense Satellite Communications System (DSCS).

The US Army Communications Electronics Command (CECOM) began fielding the GR/CS Interoperability Subsystem (GRIS) in October 1998.

The GRIS is a real-time sensor management system that will provide the GR/CS with the ability to collect and analyse SIGINT data collected by the US Air Force within line of sight, using the Interoperable DataLink (IDL).

GRIS-capable variants of the aircraft include: RC-12 K, N, P and Q.

The IDL design is based on, and interoperable with, the US Air Force Contingency Airborne Reconnaissance System/Deployable Ground Station (CARS/DGS).

Contractor

TRW Systems & Information Technology Group.

HELRAS Active Dipping Sonar

Type

Airborne Electronic Counter Measures (ECM), sonar system.

Description

HELRAS, previously known as LFADS (Low Frequency Active Dipping Sonar) is similar in size and weight to the mid-frequency AN/AQS-18(V). HELRAS has been demonstrated on SH-3 Sea King, EH 101 and SH-60 helicopters. The cable interface, sonar processor, sonar control unit,

display, reeling machine and control devices are common between the 10 kHz AQS-18 and the 1.38 kHz HELRAS systems.

HELRAS is capable of depths down to 440 m and has figure-of-merit sufficient to achieve convergence zone detections in deep water, and transmission/receive characteristics optimised for extremely long ranges in shallow water. The low-frequency capability designed into HELRAS using proprietary transducer and beam-forming technology allows multiple boundary interactions and reduced reverberation contamination of the received signals. Use of high-resolution Doppler processing and shaped pulses achieves detection of targets even at speeds as low as 1 kt. Extended duration FM pulses are available to detect the near zero Doppler target as well.

Specifications

Operating depth: 440 m
Operating frequencies: 1.33–1.43 kHz
Source level: 220 dB/µPa/yd
Range scales: 1, 3, 5, 10, 15, 20, 30, 40, 60, 80, 100, 120 kYd (thousands of yards); 0.9, 2.7, 4.6, 9, 14, 18, 27, 37, 55, 73, 91, 110 km
Operational modes: Active-CW to 10 s pulses (FM to 5 s pulses); passive; UQC; raytrace
Raise speed (avg): 4.3 m/s
Lower speed (max): 4.9 m/s
Water exit speed: 1.5 m/s
Seating speed: 0.6 m/s
System weight: (stand-alone)
(Processor and Display) 59 kg
(R/M and Cable) 95 kg
(Transducer) 141 kg
(Total weight) 295 kg

Status

HELRAS has been proven in a variety of differing water conditions and demonstrated in the Timor and Mediterranean Seas, Vestfjorden Fjord against varying targets, including diesel-electric submarines.

HELRAS is in service in Italian Navy EH 101 helicopters and selected for the NFH 90 helicopter.

Contractor

L-3 Communications, Ocean Systems.

Integrated Defensive Electronic Counter Measures (IDECM) Radio Frequency Counter Measures (RFCM) system AN/ALQ-214(V)

Type

Airborne Electronic Counter Measures (ECM), integrated suite.

Description

The AN/ALQ-214(V) IDECM RFCM system is a joint US Navy/US Air Force programme, lead by the US Navy. It is intended to provide a range of aircraft types with next-generation protection from RF threats. The system's primary application is the F/A-18E/F carrier-borne, multirole combat aircraft where it is integrated with the AN/ALR-67(V)3 radar warning receiver, the AN/AAR-57 Common Missile Warning System (CMWS) and the AN/ALE-55 Fibre Optic Towed Decoy (FOTD).

System design objectives include improved situational awareness through use of:
1. the Common Missile Warning System (CMWS);
2. real-time information fusion;

3. IR threat countermeasures based on the kinetic ASTE round and 'smart' dispensing routines;
4. Radio Frequency CounterMeasures (RFCM) to defeat radar threats, that incorporates features such as monopulse angle tracking, signal coherency and manual tracking techniques.

The IDECM RFCM system comprises a Techniques Generator (TG), an Independent WideBand Repeater (IWBR) and the ALE-55 Fibre Optic Towed Decoys (FOTD). Both the TG and the IWBR interface with the FOTDs through the decoy dispensing system. Electronic CounterMeasures (ECM) techniques are synthesised in the TG and transduced to optical frequencies for transmission to the FOTD via its tow line. Within the FOTD, the optical data is converted back to RF format for amplification and transmission. The IWBR provides an alternate source of ECM by passing the threat signal (as received on the host aircraft) to the FOTD.

The AN/ALQ-214(V) is understood to be capable of both independent repeater mode, and coherent digital RF memory mode operation, based on use of a coherent Digital Radio Frequency Memory (DRFM), and interchangeable RF transmitters.

The IDECM RFCM system can be tailored to customer applications by use of any combination of the elements described above. It is mandated for the US Navy F/A-18E/F and US Air Force B-1B, F-15C/E and U-2 aircraft, but could meet the future needs of a wide variety of other aircraft.

Specifications

(predicted)
Total weight: 71.9 kg
Total power consumption: 1,397 W
MTBF: 600 h

Status

In November 1995, the contractor team of Sanders and ITT Avionics announced that Sanders (as prime) had been awarded a USD26.8 million, five-year duration, IDECM RFCM system Engineering and Manufacturing Development contract. The contractor team is required to deliver and integrate five TGs, 150 FOTDs and 100 FOTD mass models for use in development flight testing. Initial flight testing started in June 1999, with testing of the IDECM RFCM system with an operational FOTD at the Naval Air Warfare Center – Weapons Division at China Lake, California. This was part of the Engineering and Manufacturing Development (EMD) phase of the IDECM RFCM programme. The FOTD was successfully deployed and the IDECM system correctly detected, identified and radiated – via the FOTD – against prioritised threats.

The system's primary application is the US Navy F/A-18E/F aircraft. Under a contract option, work is also being done on a common, high-powered Towed Decoy System design, development and test effort for the US Air Force F-15 and B-1B aircraft, together with a B-1B architecture study to determine how the IDECM RFCM system can support the B-1B's Defensive System Upgrade Program (DSUP). Acceptance testing of the techniques generator (receiver, modulator and processor), the signal conditioning assembly and equipment rack were completed at Sanders plant in October 1998. Electronic countermeasures data generated by the on-board system is carried via fibre optic cable to the towed decoy, which converts and amplifies the signal.

Following further integration and testing at Boeing's systems integration laboratory in St Louis, the US Navy's Air Warfare Center at Point Mugu and at ITT and Sanders factories,

IDECM RFCM LRUs, showing, from left: the signal conditioning assembly; the receiver/processor/techniques generator; the onboard transmitters (optional); the fibre optic towed decoy 0001274

the US Naval Air (NAVAIR) Systems Command announced its intention, in early 2000, to award Sanders a contract for the low-rate initial production of the IDECM RFCM. The requirement is expected to be for up to 80 RFCM systems and up to 340 FOTDs to support US Navy, US Air Force and foreign customer requirements. Initial deliveries for the F/A-18E/F began during 2003.

Most recently, BAE Systems was awarded a USD18 million modification to a previous contract for the fabrication, assembly, test and delivery of seven ALQ-214(V)2 IDECM RFCM systems. Work is scheduled for completion by the end of 2004.

Contractor
BAE Systems, North America, Information and Electronic Warfare Systems.

Joint Strike Fighter (JSF) electronic warfare equipment

Type
Airborne Electronic Counter Measures (ECM), integrated suite.

Description
In March 1999, Lockheed Martin Tactical Aircraft Systems selected the team of Sanders

The Lockheed Martin F-35C JSF (naval variant)   0096382

(now BAE Systems North America) and Litton Advanced Systems (now Northrop Grumman Advanced Systems) to be the Electronic Warfare (EW) equipment supplier for the then Lockheed Martin X-35 proposal, now the F-35 Joint Strike Fighter (JSF). BAE Systems will provide the infra-red and Radio Frequency Counter Measures (RFCM) elements, while Northrop Grumman and BAE Systems will be jointly responsible for the Electronic Support Measures (ESM) aspects.

Contractor
BAE Systems North America.
Northrop Grumman Advanced Systems.

Lightweight 16-channel sonobuoy VHF receiver

Type
Airborne Electronic Counter Measures (ECM), sonar system.

Description
The lightweight 16-channel sonobuoy VHF receiver is the newest addition to Flightline Systems' range of sonobuoy receivers. This lightweight receiver is intended for a wide variety of Maritime Patrol Aircraft (MPA) and ASW helicopters. The basic configuration provides for the simultaneous reception of 16 separate acoustic channels. With the addition of Sonobuoy Location System (SLS) modules, the receiver is capable of providing bearing data for deployed sonobuoys.

The sonobuoy VHF receiver consists of specialised plug-in modules which afford quick and reliable maintenance. The receiver module contains advanced circuitry and software-selectable tuning, allowing for the reception of the standard 99 RF channel frequencies (505 subchannels). The receiver is controlled via RS-422 bus and includes a self-contained pre-amplifier. A self-contained, forced air fan assembly provides receiver cooling.

In common with other Flightline receivers, the system gives an extremely low acoustic receiver baseband noise floor, exceptional third-order intermodulation and spurious response performance, threshold-extending FM detector facilitating sonobuoy-type optimisation,

progressive AGC with programmable time constants to improve multipath, propeller artifact, and countermeasure rejection. The VHF receiver also incorporates surface acoustic wave IF filtering for improved error rate for digital buoys, and signal strength indication and high-resolution local oscillator tuning, which can be used to optimise performance with off-tuned sonobuoys and for operating in a countermeasures environment.

The lightweight 16-channel sonobuoy VHF receiver is fully solid-state with no mechanical relays. Comprehensive Built-In Test (BIT) capability provides initial system Go/No Go status, as well as fault isolation to increase maintainability. No special test equipment is required.

Other features of the system include:
- 99 RF Channels supported
- Simultaneous reception of up to 16 acoustic channels with any mix of analogue or digital signals
- 5dB noise figure
- 80dB dynamic range
- 12dB SNR at 0.5 µV input and 20dB SNR at 1.0 µV input
- AM immunity: progressive AGC with programmable time constants for improved multipath immunity
- >80dB in-band interference rejection
- MTTR <16 minutes for module replacement
- Excellent third order intermodulation characteristics with 'active' mixer technology
- Highly immune to out-of-band interference (125 dB minimum rejection)
- Accommodates RF signals over the range of 0.5 µV to 100,000 µV (–113 dBm to –7 dBm)
- Extended acoustic base band frequency response: 0.1 Hz to 125 kHz
- Audio Distortion ≤1% from 2 Hz to 5 kHz, ≤3% from 5 kHz to 50 kHz for input levels of 10 to 100,000 µV
- Voltage Standing Wave Ratio (VSWR) does not exceed 2.0 to 1 over the 136 to 174 MHz range
- Acoustic RF signal strength reporting ability.

Specifications
Dimensions: 485.8 × 247.9 × 383.5 mm (including rear connectors and vibration isolators)
Weight: <25 kg
Reliability: >1,500 hrs MTBF predicted
Power: 115 V AC, 400 Hz, 3-phase Wye
Frequency: 136–173.5 MHz
Sonobuoy compatibility: All standard NATO types

Status
The lightweight 16-channel sonobuoy VHF receiver has been qualified on the NH 90 helicopter.

Contractor
Flightline Systems.

LR-100 Warning and Surveillance Receiver

Type
Airborne Electronic Counter Measures (ECM), integrated suite.

Description
The LR-100 is a lightweight radar signal receiver covering 2 to 18 GHz. Options are available for 70 to 200 MHz and 18 to 40 GHz. This receiver has been built to provide precision RWR, ESM and ELINT measurements with a total installed weight of less than 23 kg, including receiver, antennas, cables and brackets. The LR-100 system is a complete, two-channel interferometer receiver system with a 500 MHz bandwidth and VME-based processor.

Phase and amplitude calibration signals are injected at the antenna to achieve precision angle and location measurements.

Emitter identification and location data are passed digitally for warning, display and analysis and recording. The only support equipment needed for the LR-100 is a Windows™-compatible PC. Help and maintenance manuals are built into Windows™-based software. User-defined receiver function and identification parameters are programmed with the same software tool, which can also graphically display this information on the vehicle or via datalink. The LR-100 can be modified in real time for special receiver modes or directed tuning.

Specifications
Dimensions:
(receiver) 254 × 177.8 × 127 mm
(array) 304.8 × 304.8 × 90 mm
Weight:
(receiver) 11 kg
(array) 1.8 kg
Field of view: ±65° (azimuth and elevation)
Accuracy:
amplitude DF: <15°
interferometer DF: 0.78° rms

Flightline Systems' lightweight 16-channel sonobuoy VHF receiver   0134487

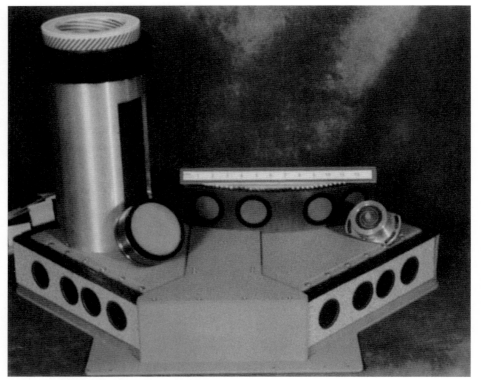

Northrop Grumman LR-100 warning and surveillance receiver　　　0062860

LR-100 array open　　　0062870

The MIRTS installed on a UK Royal Air Force Raytheon Hawker 800 (Paul Jackson)　　　0503858

Status

The LR-100 was initially designed with the Unmanned Aerial Vehicle (UAV) mission in mind and has since been purchased for the Kaman SH-2G Seasprite naval helicopter, the Sikorsky S-70 Naval hawk and other DoD applications.

Contractor

Northrop Grumman Corporation, Electronic Systems Sector, Defensive Systems Division.

LT-500 Emitter Targeting System

Type

Airborne Electronic Counter Measures (ECM), Emitter Locator System (ELS).

Description

LT-500 is a passive precision interferometric RF Emitter Targeting System (ETS). The fully militarised LT-500 ETS is a joint development of Northrop Grumman Electronic Systems Sector and TRW Avionics. Designed to achieve the goal of balancing performance, cost, and reliability in a single receiver/processor electronics package, its primary application is passive precision RF targeting of modern air-ground weapons via retrofit and upgrade of existing tactical fighter aircraft.

This cost-effective implementation consists of a totally modular 'building block' architecture (SEM-E size electronics modules weighing 34 kg total without antenna) that may be easily tailored in capabilities to match specific emitter tracking/targeting mission requirements. The LT-500 ETS may be installed with a variety of interferometer antenna array configurations and is operated via a MIL-STD-1553 control/display interfaces. An embedded, 32-bit JIAWG compliant, RISC processor hosts the Ada language 'Operational Flight Program'.

Employing Northrop Grumman proprietary interferometric direction-finding and ranging techniques, the LT-500 achieves very rapid situation awareness, emitter identification and precision geolocation in dense electromagnetic environments. The LT-500 ETS is compatible with either internal or pod installation on board the F-15, F-16, F/A-18 or similar tactical aircraft. Because of its modularity, it may be easily adapted to a wide variety of other airborne, shipboard, or ground-based platforms.

Contractor

Northrop Grumman Corporation, Electronic Systems Sector, Defensive Systems Division.

Modularised Infra-Red Transmitting System (MIRTS)

Type

Airborne Electronic Counter Measures (ECM), Infra-Red Counter Measures (IRCM) system.

Description

The Modularised Infra-Red Transmitting System (MIRTS) is a derivative of the AN/AAQ-8(V). It is an advanced subsonic, infra-red countermeasures system for deployment in a wide range of aircraft, including helicopters, which can be carried internally or pod mounted. MIRTS utilises advanced jammer technologies including a variable optics/reflector design to provide optimum aircraft infra-red signature coverage, combined with advanced digital electronics and mode-switching power supplies to enhance system reliability, maintainability and versatility.

Specifications

Transmitter/receiver
Dimensions: 228 × 240 × 635 mm
Weight: 23.6 kg
Power supply: 115 V AC, 400 Hz, 3 phase, 3.3 kVA
　28 V DC, 5.6 W
Operator Control Unit
Dimensions: 146 × 57 × 127 mm

Weight: 0.68 kg
Power supply: 5 V AC, 400 Hz, 3 phase, 28 V DC, 10 VA
Electronics control unit
Dimensions: 190 × 259 × 318 mm
Weight: 6.6 kg
Power supply: 115 V AC, 400 Hz, 3 phase, 308 VA 28 V DC, 22.4 W

Status

In service. The system has been tested on the rear fuselage of a UK Royal Air Force VC10. Aircraft installations include the Boeing 707, 747 and DC-8, Eurocopter Puma; Falcon 20, Fokker 27, and Raytheon Hawker 800.

Contractor

Northrop Grumman Corporation, Electronic Systems Sector, Defensive Systems Division.

Multifunction Integrated Radio Frequency System (MIRFS) for Joint Strike Fighter (JSF)

Type

Airborne Electronic Counter Measures (ECM), integrated suite.

Description

A team led by Northrop Grumman's Electronic Sensors and Systems Sector (ESSS) is working under a USD48.5 million contract from the Joint Strike Fighter (JSF) programme office to design, build and flight test the Multifunction Integrated Radio Frequency System (MIRFS).

Under the contract, the ESSS-led team is designing a new active electronically scanned array, a Multi Function Array (MFA), which will help to reduce the future JSF aircraft's avionics system cost by 30 per cent and its weight by 50 per cent. Other team members are Litton Advanced Systems Division, Raytheon Electronic Systems, BAE Systems North America and Harris.

The Northrop Grumman antenna will perform radar, high-gain electronic support measures, and Communication/Navigation/Identification (CNI) functions. The antenna will be flight-tested using existing radar support electronics from Northrop Grumman's APG-77 radar being developed for the F-22 aircraft, and a Commercial-Off-The-Shelf (COTS) emulator of the Integrated Core Processor (ICP). Software will be reused from the F-22 radar and from Northrop Grumman Norden Systems' AN/APG-76 radar. Northrop Grumman will flight-test the new MFA in the company's BAC 1–11 testbed aircraft against various airborne and ground-based targets, such as the SCUD missiles used during the Gulf War. The radar is reported to use pulse-to-pulse MTI and SAR technology to provide two-dimensional imaging of the target.

In parallel, Northrop Grumman is developing a Common Integrated RF (CIRF) concept that will combine radar and electronic warfare equipment into an advanced open architecture RF environment. The CIRF will replace the radar support electronics used in the initial flight tests and along with the advanced antenna and ICP will form a complete flying prototype of the MIRFS.

Status

In development.

Contractor

Northrop Grumman Corporation, Electronic Systems Sector.

Passive Ranging Sub System (PRSS)

Type

Airborne Electronic Counter Measures (ECM), Emitter Locator System (ELS).

Description

The Passive Ranging Sub System (PRSS) is a digital receiver designed to provide accurate, real-time emitter location and autonomous Electronic Support Measures (ESM) cueing (detection, identification and location) in dense Radio Frequency (RF) and Electronic Counter Measures (ECM) environments. The system utilises self-scaling Doppler and self-resolving, long-baseline interferometer techniques for single aircraft emitter location. The PRSS is an open system, modular VME architecture design, based on commercial standards, facilitating upgrade and enhancement via additional plug-in modules.

Status

In production and in service. The PRSS is used as the basis for the AN/ALQ-210 and AN/ALQ-217 ESM systems.

Contractor

Lockheed Martin Systems Integration – Owego.

R-624(V) sonobuoy receiver

Type

Airborne Electronic Counter Measures (ECM), sonar system.

Description

The R-624(V) sonobuoy receiver is a variant of the AN/ARR-502 receiver (see separate entry), intended for Maritime Patrol Aircraft (MPA) applications. The R-624(V) receiver is capable of simultaneously tuning and demodulating 32 NATO standard sonobuoy transmissions. Two receivers may be used in an acoustic suite to allow for processing of 64 simultaneous sonobuoy transmissions. The standard 99 RF frequencies for Sonobuoys (505 subfrequencies for special sonobuoys) are supported.

The R-624(V) is controlled via a dual-redundant MIL-STD-1553B databus. If a two-receiver configuration is employed, intra-receiver communication is achieved via a RS-422 link. The sonobuoy receiver can be commanded to determine the signal strength of all or selected channels and execute Built-In Test (BIT) for end-to-end determination of system health, control receiver module parameters for particular buoy-type optimisation, and perform calibration functions of the Sonobuoy Position System (SPS) (if so equipped). With SPS, the sonobuoy receiver is capable of directional location of sonobuoys and Search And Rescue (SAR) emergency beacons through interferometer techniques. In this configuration, On-Top Position Indicator (OTPI) information is also available.

In common with other Flightline sonobuoy receivers, the R-624(V) provides an extremely low acoustic receiver baseband noise floor, exceptional third-order intermodulation and spurious response performance, threshold-extending FM detector facilitating sonobuoy-type optimisation, progressive AGC with programmable time constants to improve multipath, propeller artifact and countermeasure rejection. Also provided are surface acoustic wave IF filtering (for improved error rate for digital buoys), signal strength indication and high-resolution local oscillator tuning.

The R-624(V) receiver is fully solid-state with no mechanical relays. Comprehensive BIT capability provides initial system Go/No Go status, as well as fault isolation to increase maintainability. No special test equipment is required.

Other features of the system include:
- 99 RF channels (505 sub-channels) supported
- Simultaneous reception of up to 64 acoustic channels with any mix of analogue or digital signals (using two 32-channel sonobuoy receivers)
- 5dB noise figure
- 80dB dynamic range
- 12dB SNR at 0.5 µV input and 20dB SNR at 1.0 µV input
- AM immunity: progressive AGC with programmable time constants for improved multipath immunity
- >140 dB out-of-band interference rejection
- >80dB in-band interference rejection
- MTTR <16 minutes for module replacement
- Excellent third-order intermodulation characteristics with 'active' mixer technology
- Accommodates RF signals over the range of 0.5 µV to 100,000 µV (–113 dBm to –7 dBm)
- Extended acoustic base band frequency response: 0.1 Hz to 125 kHz
- Audio Distortion ≤1% from 2 Hz to 5 kHz, ≤3% from 5 kHz to 50 kHz for input levels of 10 to 100,000 µV
- Voltage Standing Wave Ratio (VSWR) does not exceed 2.0 to 1 over the 136 to 174 MHz range
- Acoustic RF signal strength reporting ability.

Specifications

Dimensions: 604.5 × 308.9 × 374.6 mm (including rear connectors and vibration isolators)
Weight: <120 kg
Reliability: >1,200 hrs MTBF predicted
Power: 115 V AC, 400 Hz, 3-phase Wye
Frequency: 136–173.5 MHz
Sonobuoy compatibility: All standard NATO types

Status

The R-624(V) receiver is in service on the Nimrod MRA-4 MPA.

Contractor

Flightline Systems.

R-1651/ARA On-Top Position Indicator (OTPI)

Type

Airborne Electronic Counter Measures (ECM), sonar system.

Description

The solid-state modular construction R-1651 radio receiving set is a 31-channel On-Top Position Indicator (OTPI) which is being used on board ASW rotary- and fixed-wing aircraft to provide bearing information and on-top position indication of deployed sonobuoys. This OTPI, when used in conjunction with a suitable ADF system, enables the operator to locate and verify the position of deployed sonobuoys operating on any of 31 VHF RF channels.

The R-1651/ARA is form and fit interchangeable with the R-1047 A/A OTPI. It is compatible with the ARA-25 and ARA-50 antennas and also available in a modified configuration which is compatible with the DF-3010E antenna system. The R-1651/ARA is controlled by the optional C-3840/A control box.

The R-1651/ARA is a single-conversion, 31-channel superheterodyne VHF receiver which is designed to receive signals from deployed sonobuoys over the frequency range of 162.25 to 173.5 MHz.

The antenna input is switched by a relay to the receiver or the UHF receiver, on external command. The RF signals pass through bandpass filters, are amplified and converted to an IF of 25 MHz. After IF amplification and detection, the resultant audio signal is amplified to the level required by the ADF system. The AGC level is developed at the detector, amplified and used for gain control of the IF and RF stages and to operate the adequate signal strength circuits.

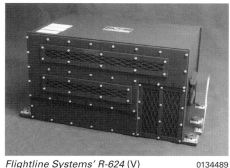

Flightline Systems' R-624 (V) 0134489

The local oscillator provides one of 31 possible LO signals. The same number of crystals is appropriately selected using external controlled code lines containing a 6-bit binary code. Line receivers, decoding gates and matrix drivers connect one of the 31 crystals to the local oscillator circuits.

The optional C-3840/A radio receiver control is used to switch the antenna system, by energising an external coaxial relay, from the UHF/ADF mode of operation to the sonobuoy OTPI mode of operation to select any one of the 31 RF channels and to indicate adequate signal strength when the R-1651/ARA receives an RF signal of –86 dBm or greater.

Specifications
Dimensions: 76.2 × 124.46 × 134.6 mm
Weight: 1.59 kg
Power supply: 115 V AC, single phase, 15 VA
   28 V DC, 20 W
Frequency: 162.25–173.5 MHz (31 channels)
Altitude: up to 35,000 ft
Temperature range: –54 to +55°C
Operating life: 5,000 h
Reliability: 1,000 h MTBF

Status
In service in the US Navy P-3C Updates I and III and the Sikorsky SH-60 LAMPS Mk III.

Contractor
Flightline Systems.

SATIN EW system

Type
Airborne Electronic Counter Measures (ECM), integrated suite.

Description
SATIN (Survivability Augmentation for Transport INstallation) is a minimum modification approach to equip C-130 aircraft with self-protection sensors and countermeasures. The installations are designed to be installed by a small team in a few days. All external or cockpit components are provided with replacement cover panels, so that the systems can be removed when not required, or used on other aircraft.

SATIN components are selected from the standard military inventories of the particular user to minimise logistics problems. Systems which have been installed for various users include the AN/ALR-69 Radar Warning Receiver, APR-39A(V)1 Radar Signals Receiving Set, AN/ALQ-156 ESM system, AN/AAR-47 Missile Warning Set, AN/ALQ-157 Self-Protection System and the AN/ALE-39, ALE-40 and ALE-47 Countermeasures Dispenser Systems. Other systems can be incorporated, as dictated by user requirements.

Status
SATIN has been installed on various versions of C-130 aircraft for several users and the basic installation approach has been incorporated into production C-130 aircraft.

Contractor
Lockheed Martin Aeronautics Company.

SP-103 signal processor

Type
Airborne Electronic Counter Measures (ECM), Electronic Support Measures (ESM) system.

Description
The SP-103 wideband signal processor provides real-time analysis and identification of radar signals for ESM/ELINT applications. The unit also includes built-in receiver and antenna controls for directed search utilising compatible receivers and antennas. Windows NT/UNIX-based operator interface software provides all required controls and displays.

The **SP-103** can be integrated with new or existing superheterodyne receivers and accepts a 1 GHz (500 MHz bandwidth) input. System output consists of parametric emitter characterisation (frequency, PRI, pulse-width and amplitude, time-first-seen, time-last-seen, scan period and scan type) that can be displayed to the operator and/or transmitted to an external user. The processor also compares the measured and characterised parameters of the target emitter with a customer-furnished internal library of known emitters to determine target identity.

When used in an ES/ELINT system with a rotating DF capability, the **SP-103** performs DF processing using automatic AoA measurement software. To support detailed emitter analysis and/or emitter identification ambiguity resolution, the system provides 'interactive' analysis capabilities.

The system incorporates automated BIT and diagnostics.

Specifications
Type: 1 GHz, 500 MHz BW ELINT signal
   processor
IF bandwidths: 20, 50, 100, 250 or 500 MHz
   (selectable)
Video bandwidth: 5 or 20 MHz (selectable)
Max input burst rate: 2 Mpps
Frequency resolution: 125 kHz
Frequency accuracy: 1 MHz RMS
PRF range: 100–500,000 pps
PRI range: 2 μs–10 ms
PRI types: constant, continuous wave, jitter,
   pulse groups and stagger (up to 32 levels)
Pulse width
Resolution: 40 ηs
Range: 50 ηs–2.62 ms
Accuracy: 40 ηs RMS
Scan
Rate range: 0.05–100 Hz
Accuracy: 2% RMS
Types: bidirectional, circular, complex, conical,
   sector and steady
Active emitters: up to 1,024
Emitter library: 10,000 mode, programmable
Temperature: –30 to +55°C (operating); –51
   to +71°C (storage)
Power consumption: 250 W
Operating voltage: 115/230 V AC (47-440 Hz)f
Standard interface: IEEE 402.3 Ethernet
Dimensions: 215.9 × 269.2 × 457.2 mm
   (H × W × D)
Weight: 16.4 kg

Status
Designed for airborne, shipborne and land-based applications. In production and in service.

Contractor
EDO Reconnaissance and Surveillance Systems.

SP-160 pulse processor

Type
Airborne Electronic Counter Measures (ECM), Electronic Support Measures (ESM) system.

Description
The SP-160 signal processor is designed to digitise an intercepted radar signal into a series of Pulse Descriptor Words (PDWs) that characterise the RF frequency, time of arrival (for PRI determination), pulsewidth, amplitude and bearing. These PDWs are automatically processed (de-interleaved) into active emitter reports for each radar. Parametric measurements (pulse modulation type and characteristics) of the active emitter reports are then compared against a library for emitter identification.

Operation is automatic, giving the operator a display of signal activity with identification of specific radar types. A continuous log of activity is stored in memory for post-mission analysis. Requiring only a single video input, the SP-160 is compatible with any type of radar receiver. The associated EI-1400 control/display, featuring a 5 in (127 mm) display and built-in floppy disk

drive, is designed for aircraft installations where space and weight are limited.

The SP-160 processes normal, stagger, jitter, CW and complex radar signals and provides de-interleaving, precision pulse measurement and signal identification of radar emitters in a complex environment.

Specifications
IF input: 160 MHz (70 MHz BW), 1,000 MHz (500
   MHz BW)
Video input: Detected video
Options: Integrated Windows NT colour
   workstation; internal Windows NT HMI;
   External Windows NT HMI via 100 base T Lan

Status
Designed as a direct replacement for the AN/ULQ-16 and SP-2060 signal processors. In production and in service.

Contractor
EDO Reconnaissance and Surveillance Systems.

Technique Control Modulator

Type
Airborne Electronic Counter Measures (ECM), control system.

Description
Symetrics' Technique Control Modulator (TCM) is an electronic chassis developed for the Electronic Counter Measures (ECM)/Electronic Attack (EA) Technique Generator (TG) for US Navy aircraft. The TCM includes a chassis, power supply and motherboard and serves as the housing for seven circuit card assemblies.

TCMs are used in the AN/ULQ-21 ECM set, which provides simulation of present and projected ECM threats, utilised to evaluate ECM capabilities of weapon systems and to train US Navy Weapon Systems Operators (WSOs).

Status
In production and in service in US Navy aircraft.

Contractor
Symetrics Industries Inc.

ZS-1915 airborne communication signals intercept and DF system

Type
Airborne Electronic Counter Measures (ECM), Electronic Support Measures (ESM) system.

Description
The ZS-1915 system provides for the intercept, detection, monitoring, DF and location of communication signals in the 20 to 500 MHz frequency range, with optional extensions to 1,000 or 2,700 MHz. The system is designed for installation in a wide variety of aircraft, from small single-engined turboprop aircraft upwards. The ZS-1915 DF antenna array comprises a set of standard blade antennas or a radome assembly mounted on the aircraft underside. DF accuracy depends on the aircraft installation, but it is normally better than 3° RMS. The ZS-1915 system provides for continuous, or discrete, high-speed frequency search with simultaneous DF measurement on active signals. Active signals can be handed off to one or more monitor receivers, whose audio outputs can be recorded on an integrated Digital Audio Record & Playback (DARP) subsystem.

An interface to read the aircraft navigation/GPS system is provided to automatically establish emitter location from signal intercepts. An integrated Geographic Information System (GIS) accepts digital maps in a variety of formats. The location of an emitter is automatically calculated from a series of lines of bearing taken over a period of time as the aircraft flies past the emitter. Both lines of bearing and the calculated emitter location are displayed as overlays on the digital map.

Emitters operating in a network can be independently identified and located. All data obtained during a flight mission are automatically entered and stored in the system database, including flight path, intercepts, emitter locations and audio recordings. This data can be transferred by removable disk to a ground analysis station, where the complete mission can be 're-flown' and analysed.

The ZS-1915 system operator uses a rugged PC-based workstation with colour monitor and keyboard/trackerball together with the Microsoft Windows NT® operating system to provide a familiar, easy to use operating environment. A wide variety of display modes and operating screens is available to suit different operational missions. Additional monitor receivers, an increased channel capacity DARP, and a signal analysis capability are available as options.

RC 232C remote-control and network (LAN) interfaces to other airborne or ground-based systems are available as options. The client/server-based software design, using standard TCP/IP protocols, permits easy expansion to a more extensive COMINT/DF system.

Contractor

ZETA, an Integrated Defense Technologies Company.

ZS-3015/4015 ultrascan intercept and direction finding systems

Type

Airborne Electronic Counter Measures (ECM), Electronic Support Measures (ESM) system.

Description

The ZS-3015/4015 ultrascan intercept and direction finding systems are designed for tactical and strategic signal collection requirements demanding high performance in local or remote fixed-site, vehicle, airborne or shipboard applications.

The ZS-3015 (2-channel) and ZS-4015 (5-channel) systems use wideband digital

The ZS-3015/4015 ultrascan intercept and direction finding systems   0062919

receivers and Digital Signal Processing (DSP) techniques to achieve high-performance signal intercept and direction finding over the 20 to 3,000 MHz frequency range.

Precision DF measurements result from the use of a proprietary, proven, correlation interferometry DF technique. Very high spectrum scan speeds are achieved through use of a 6.4 MHz wideband multichannel receiver with Fast Fourier Transform (FFT) processing using parallel Digital Signal Processors (DSPs). A comprehensive Windows NT® Graphic User Interface (GUI) enables the operator to select a wide variety of automatic and manual operating modes.

SQL database management supports building Signals Of Interest (SOI) files with all pertinent signal parameters and mission intercept files. An integrated Geographic Information System (GIS) accepts a variety of digital maps, and it displays lines of bearing and emitter locations as overlays on the map.

Other system features include Digital Audio Record and Playback (DARP) of received audio, with correlated operator comments of signal intercept events, local or remote control, built-in networking capabilities between systems and open client-server architectures via LAN/WAN using RF, fibre-optic or wire links.

Contractor

ZETA, an Integrated Defense Technologies Company.

AIRCRAFT CONTROL AND MONITORING SYSTEMS

AIRCRAFT CONTROL AND MONITORING SYSTEMS

Canada

E1553™

Type
Avionic interface system.

Description
Edgewater is the developer of Extended 1553™, or E1553™, a new version of the existing MIL-STD-1553B communications standard for military avionics. In early 1990s, the US DoD recognised that the internal communications network capabilities of its aircraft needed to be improved in order to meet the advanced requirements of the next generation of processors and sensor technologies being developed to enable network centric operations.

The current MIL-STD-1553 was developed during the 1970s and has evolved to become the predominant LAN technology used in western military aircraft. Currently MIL-STD-1553B delivers a data throughput rate of 0.7 – 1.0 Mbps, which, despite the introduction of the MIL-STD-1760 weapons databus, installed in many fighter applications, still leaves a significant shortfall in data handling capabilities for aircraft employing advanced communication and sensor suites, including Active Electronically Scanned Array (AESA) radar systems. With many legacy third- and fourth-generation fighter aircraft using 1553 having over 20 years of serviceable life left, the US DoD recognised that the cost of rewiring all its aircraft with new LAN technology (such as a fibre-optic backplane for high-speed data transfer) would be prohibitive. Hence, the US Air Force began investigating methods by which extant MIL-STD-1553B technology could be extended to achieve bandwidth rates of 100 Mbps, without affecting existing communications.

In 2000, Edgewater demonstrated its E1553™ technology to the US Air Force and collaborated with them to to produce the MIL-STD-1553B Notice 5 standard, which defines a vastly increased data throughput capability over the existing standard.

To achieve the increase in throughput, Edgewater took an approach similar to that of Digital Subscriber Line (DSL) technology widely used by Internet Service Providers (ISPs) to transmit data over phone lines without interfering with voice services. In this case, however, the

Extended 1553 technology transmits data over the 1553 data bus cables without interfering with the legacy 1553 connectivity. The Company conducted a detailed study that determined commercial off-the-shelf DSL technologies, such as Carrier-less Amplitude and Phase (CAP) and Digital Multi-Tone (DMT), designed for point-to-point, non-real-time Internet-type communications would not be suitable for the development of E1553™, therefore electing to design a bespoke system in order to accommodate predictable and deterministic operation, immunity to Electro Magnetic Interference (EMI) and electromagnetic compatibility.

Edgewater's final design is a multi-carrier waveform system that can readily adapt to the unpredictable channel variations of the 1553 data bus media. The waveform physical-layer protocol includes efficient line encoding and forward error correction to achieve high throughput with low bit error rates in the face of narrowband jammers and wideband transients.

Edgewater's E1553™ data bus increases data throughput from 1 Mbps to an uncompressed 200 Mbps over existing aircraft wiring while maintaining a 10^{-12} bit error ratio and without impact to existing 1553 communications.

EHS-SV601 VME module
In conjunction with Hill Air Force Base, Edgewater has developed an evaluation circuit card for a cockpit display and data-processing unit in the F-16 Fighting Falcon aircraft. The EHS-SV601 VME module provides for high-speed 1553 bus connectivity into military and civilian embedded computing systems, offering MIL-STD-1553B Notice 5 compatibility in a single 6U VME card format using Edgewater's Extended 1553™ technology.

Features include VME64, transformer coupling, VxWorks driver (other OS by application) and Built-In Test (BIT).

The same capabilities are available in a PCI card hardware format, as the EHS-SC601 compact PCI module.

Specifications
EHS-SC/SV601
Type: One MIL-STD-1553B Notice 5 MCW Interface at 100 Mbps
Dimensions: 6U VME card

Cooling: Conduction
Temperature: –40 to +71°C operating
Shock and vibration: MIL-STD-810E
EMC: MIL-STD-461E
Interface: Support for a single width PMC interface

Status
Edgewater has successfully demonstrated E1553™ to the US Air Force and Navy. In December 2005, the US Air Force awarded Edgewater a USD75 million cost-plus-fixed-fee Indefinite Delivery/Indefinite Quantity (ID/IQ) contract to help the US Air Force to prepare for the deployment of E1553™ and to support additional avionics and electronics capabilities that require high bandwidth data bus networking.

In March 2006, the US Air Force announced an updated military standard for data bus networking in avionics. MIL-STD-1553B Notice 5, developed by Edgewater in support of the US Government, increases the data throughput on the 1553 bus by a factor of 200 over the current 1553B standard without the need for rewiring or changing the infrastructure of the host platform.

In June 2006, Rockwell Collins and Edgewater entered into an agreement to validate the functionality of Edgewater's E1553™ hardware in a simulated operational environment. Rockwell Collins and Edgewater chose a C-130 System Integration Lab (SIL), which replicates the aircraft's avionic and databus architecture, to perform the testing. A successful test was carried out in August 2006, where Rockwell Collins integrated E1553™ hardware with an avionics cockpit display and a Tactical Targeting Network Technology (TTNT) radio. The TTNT waveform, which has been selected as the Airborne Networking Waveform by the US Air Force, provides a high-bandwidth, reliable, ad-hoc network required to connect forces. A data throughput in excess of 50 Mbps was required to support the demonstrated applications. Edgewater's E1553™ successfully supported this requirement with no degradation to the legacy 1553, which ran concurrently at its maximum rate of approximately 700 kbps.

Contractor
Edgewater Computer Systems, Inc.

Warning and caution annunciator display

Type
Avionic crew alerting system.

Description
The warning and caution annunciator display consists of three LED dot matrix annunciation display zones, an LED dot matrix resettable cue display, an edge-lit information panel with key switches and a dual-multiplex databus receiver/transmitter.

The left and right LED modules and the cue display are ANVIS yellow for the display of cautionary messages. The upper zone of the centre LED module is ANVIS green B to display advisory messages and the lower zone is ANVIS red for the presentation of warning messages. The edge-lit information panel has seven push-button control switches for scrolling, display brightness, cue reset and self-test.

Specifications
Dimensions: 115 × 396 × 224 mm
Weight: 6.58 kg
Power supply: 28 V DC, 180 W, 5 V AC

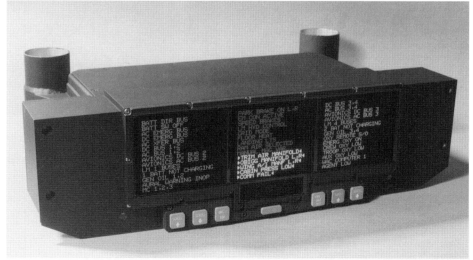

The Northrop Grumman warning and caution annunciator is in production for the C-17   0504091

Status
In production for US Air Force C-17 aircraft.

Contractor
Northrop Grumman – Canada Ltd.

China

NG-2 altitude preselector

Type
Avionic crew alerting system.

Description
The NG-2 altitude preselector receives ARINC 429 serial digital inputs to provide alerts and aural warnings to the pilot. The system incorporates a self-test function.

Specifications
Range: 0–25,000 m
Inner level: −60 to +60 m
Outer level: −150 to +150 m
Accuracy: ±10 m
Weight: 0.7 kg
Dimensions: 197 × 86 × 12 mm
Power: 27 V DC; <10 W
Environmental: GJB 150-86

Status
In service.

Contractor
Chengdu Aero-Instrument Corporation (CAIC).

Czech Republic

AMOS Aircraft Monitoring System

Type
Onboard system monitor.

Description
AMOS is designed to provide acquisition, processing and storage of information for the L159 light attack aircraft; it monitors airframe fatigue data; flight data; accident data; pilot response data. There is an optional facility for in-flight transmission of the acquired data.

AMOS comprises a Flight Data Acquisition Unit (FDAU); crash-protected Flight Data Recorder (FDR-159); Function Signalling Panel (FSP); Flight Data Entry Panel (FDEP); Ground Evaluation Equipment (GEE); and Ground Support Unit (GSU).

Using the GEE/GSU, it is possible to replay the mission in 3-D, as well as to analyse and evaluate aircraft system data.

Contractor
SPEEL PRAHA Ltd.

PMS 39 Aircraft Monitoring System (AMS)

Type
Onboard system monitor.

Description
The PMS 39 Aircraft Monitoring System, derived from the AMOS monitoring system (see separate entry) is intended for acquisition, processing and storage of flight parameters for the Aero Vodochody L-39 ZA/C light attack/training aircraft.

Differences between PMS 39 and AMOS are mainly in the sensor input conditioning circuitry and number of parameters processed, which is dependent on the number of processor modules in the system.

The PMS 39 system consists of the MU 39 Flight Data Acquisition Unit (FDAU) and the ZJ 39 *recording unit. The* MU 39 gathers and processes flight parametric data from the host L-39 and the ZJ 39 unit stores the information in a solid-state memory.

The MU 39 also acquires engine vibration data and stores it in an internal memory for subsequent harmonic analysis. MU 39 status is indicated to the pilot/engineer by an indicator light located in the cockpit. The ZJ 39 receives information from the MU 39 via a RS232 serial interface. The ZJ 39 solid-state memory is divided into two parts, operation and standby (backup) memory. Data is stored into both memory parts simultaneously.

The system utilises Windows™ based PANDA software which combining with corresponding GEE/GSU configuration, provides for data downloading and evaluation.

Data download from the ZJ 39 solid-state memory is via a PMU-F RS422 interface, with selected records or the complete memory downloaded by operator selection. When the PMU-F is transferred to GEE ground equipment, data is saved via RS232 interface directly to the hard disk file on GEE. Data download can also

Specifications

	FDR-159	FDAU	FSP
Dimensions	185 × 134 × 330 mm	240 × 260 × 390 mm	145 × 32 × 100 mm
Weight	<8 kg	<10 kg	<0.35 kg
Power	28 V DC, <8 W	28 V DC, <55 W	28 V DC, <10 W
Memory capacity	16 Mbyte (custom)	4 Mbyte (custom)	
MTBF	>5,000 flight h	>3,000 flight h	>20,000 flight h
Communications interfaces	RS-232; RS-422; ARINC 429; MIL-STD-1553B		
Compliances	MIL-STD-810E; EUROCAE ED-55; TSO-C124		

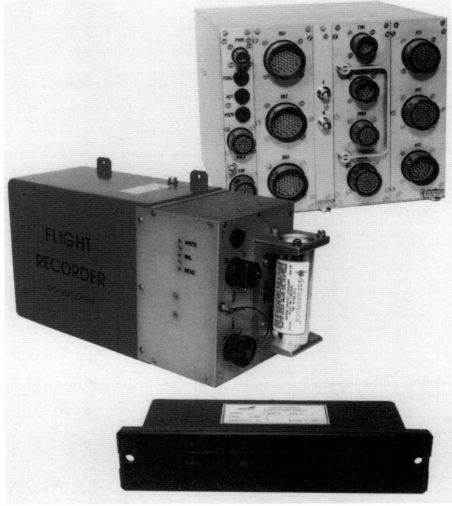

SPEEL PRAHA's AMOS aircraft monitoring system, showing FDAU, FDR-159, and FSP   0015299

be achieved directly into a GSU device via RS232 interface.

Specifications
MU 39
Dimensions: 267 × 201 × 236 mm
Weight: 6 kg (MU); 0.6 kg (frame)
Power: 16 to 32 V DC, 20 W (max)
Memory capacity: 16 Mb (engine parameter recording – approx 1 hr)
MTBF: 3,000 flight hr/min
Accuracy: Analogue signals ±0.6% ±2 bits; aircraft power (28 V, 3 × 36 V, 115 V) ±1% ±2 bits; frequency ±0.2%

Operating temperature: −54 to +60°C
ZJ 39
Dimensions: 286 × 120 × 187 mm
Weight: 8 kg max
Power: 16 to 32 V DC, 7 W (max)
Memory capacity: 8 Mb ×2 (operational and standby memories)
MTBF: 10,000 flight hours
Operating temperature: −54 to +60°C

Contractor
SPEEL PRAHA Ltd.

Denmark

Airborne pollution surveillance system

Type
Flight Inspection/calibration system.

Description
The TERMA surveillance system provides integration of a wide variety of different surveillance sensors, navigation equipment, and video systems to monitor oil and other polluting agents at sea; the system can also be used to monitor fishing violations, to map ice and for search and rescue missions.

The system can include any, or all, of the following elements, according to customer requirement:

Side-Looking Airborne Radar (SLAR)
The I-band (8-10 GHz) maritime surveillance SLAR has become the primary long-range sensor for oil pollution surveillance, typically covering a 20 n miles swathe to each side of the aircraft from preferred search altitudes. An oil slick is detected by the variation in the reflected radar signals between oil-covered water and normal seawater. In applications like ice mapping and ship surveillance, the SLAR system covers a 40 n miles swathe to each side.

Infra-Red/Ultra-Violet (IR/UV) scanner
The IR/UV scanner is provided for close-range imagery and allows a general assessment of the oil slick to be made, as the aircraft passes overhead of the slick first detected by the SLAR.

The infra-red system can be operated both day and night. It provides information on the spreading of oil and indicates the variations of thickness within the slick. The ultra-violet sensor is only used during daylight. It maps the complete area covered with oil, irrespective of thickness.

Video and photo cameras
Video cameras are used to secure evidence of oil pollution, fishery violation and other illegal actions. The scenes taken with the camera can be recorded on a videotape or stored as still photographs in the computer system. Real-time navigation data is integrated into the picture. The video camera can be normal colour, a Low Light LevelTeleVision (LLLTV), and/or an infra-red sensor.

For photographic documentation, a hand-held camera with a real-time annotation can be integrated into the system.

Data downlink
Data downlink equipment is used for transmission of real-time or stored data to a ground station or ship. Mission control aboard a ship is often more efficient when images of the actual situation are made by an aircraft flying at high altitudes.

Data storage
The information from microwave and optical sensors can be recorded either on a standard videotape recorder or on a high-resolution digital tape recorder. The digitally stored sensor information can be further analysed on the ground in an image processing system.

Display console
The TERMA concept is to utilise only one common control. Typically, it will comprise a 20

OBSERVER CONSOLE

The TERMA airborne pollution surveillance system operator console and observer console (inset)
0062179

in colour Sensor Image Display (SID). The SID is used to display data from the sensor currently in use; stored data can also be displayed. Real-time navigation data can be integrated into the bottom of the image display. The operator can annotate real-time data via a trackball. The SID provides the operator with an outline map of the area under surveillance.

All sensor control functions are performed via a Windows™ operating environment.

Specifications
SLAR
Ranges: 20 and 40 n miles
Navigation interfaces: ARINC, RS 422, RS 232
Power: 28 V DC, 15 Amp
Antenna: 2 single or 1 double
(horizontal beamwidth) <0.6°
(vertical beamwidth) 19 ± 3°
(gain) >33 dB
(polarisation) VV
(dimensions) 3,658 mm wide

Transceiver:
(frequency) 9,375 ± 30MHz
(PRF) 0 to 2,000 Hz
(pulse-width) 0.1-1.0 μsec
(peak power) 25 kW ±1 dB
Consoles:
(operator console dimensions) 1,300 (L) × 550 (W) × 1,300 (H) mm
(operator console weight) 130 kg

Status
More than 40 systems have been delivered to customers worldwide.

Contractor
TERMA A/S.

France

AC 68 multipurpose disk drive unit/airborne data loader

Type
Data processing system.

Description
The multipurpose disk drive unit/airborne data loader enables the uploading and downloading of onboard computers and allows remote loading of the computer program database or program initialisation for ACMS, FMC, ACARS and TCAS, memory computer downloading and recording data reports, QAR function and other functions, from aircraft computers. The AC 68 data loader is also compatible with ARINC-615-3 and, in portable configuration, with the ARINC-603 standard.

The storage medium used for transfer is a standard 3.5 in (89 mm) double-side high-density magnetic diskette. The formatted capacity is 1.44 Mbytes.

Status
In production for Airbus and Boeing aircraft. In portable configuration, available to fit into any platform.

Contractor
SAGEM Défense Securité, Navigation and Aeronautic Systems Division.

SAGEM AC 68 multipurpose disk drive unit/ airborne data loader   0503864

AFCS 155 autopilot/flight director system

Type
Autopilot system.

Description
The AFCS 155 is a series duplex, fail-passive autopilot. Optimised for both single- and dual-pilot IFR operations, covering the whole flight envelope from hover to VNE, it is IFR-certified on the Eurocopter Super Puma and Dauphin helicopters and can be integrated with SAGEM couplers. An STC is also available for the Erickson S-6A Air Crane helicopter. These couplers include the FDC 85 three- or four-axis and FDC 155 four-axis digital coupler. In the duplex configuration each channel has its own power supplies, sensors and interconnections.

Basic functions include long-term attitude and heading hold, turbulence compensation, collective link mode and autotrim. All autopilot configurations satisfy single pilot IFR requirements and there are three uppermodes: heading select, altitude hold and airspeed hold.

Fly-through or transparent handling characteristics allow the pilot to make quick attitude and heading changes while benefiting from dynamic damping. On releasing the controls the autopilot returns to long-term stabilised flight at the previously set attitudes. If new attitude settings are required the pilot can either use the stick release button on the cyclic pitch grip or can change the settings at a slow rate by using the stick-top four-way 'beep' trim button. The 'beep' trim button is also used to alter the reference airspeed slowly when the automatic airspeed hold mode has been selected. All versions of the system include automatic trim, which keeps the

series actuators centred so that the autopilot has full authority.

Additional flight director coupler facilities provide IFR automatic navigation, radio navigation and approach capabilities, including steep approach MLS beam capture and track. There are also additional modes for anti-submarine warfare and offshore or search and rescue operations. The latter includes automatic pattern following, automatic up and down transitions to and from selected radio altimeter height, Doppler/radio height or sonar cable hover and low Doppler speed automatic hold. Compatible couplers include FDC 85 three- and four-axis navigation and approach couplers (the latter is Cat II certified), and the FDC 155 all mission digital coupler.

Specifications
Weight:
(autopilot computer) 8 kg
(servo amplifier) 4.2 kg
(autopilot control box) 1.6 kg
(4 actuators) 1 kg each
(3 trim servos) 1.3 kg each
(barosensor) 1.5 kg
(FDC 85) 2 kg
(FDC 155) 8 kg
(flight director coupler box) 2 kg
(collective pitch motor) 2 kg

Status
In production and installed on Eurocopter AS 332 Super Puma, AS 365N Dauphin and S-6A Erickson Air Crane helicopters.

Contractor
SAGEM Défense Securité, Navigation and Aeronautic Systems Division.

AFCS 165/166 digital Automatic Flight Control Systems (AFCS)

Type
Flight Control System (FCS).

Description
The AFCS 165/166 and variants are four-axis flight control and flight director systems intended for medium and heavy-capacity helicopters. The power components are either electromechanical or electro-hydraulic, the sensors being digital or analogue. A dual/dual microprocessor system provides fail-passive and fail-operational capabilities without performance loss.

The basic AFCS 165/166 comprises Category I navigation and approach modes and allows single pilot IFR operations. A modular system approach is also possible. Modes relating to specific missions such as ASW, SAR or anti-tank are offered as an option, software integrated.

AFCS 166 control unit as fitted in the naval version of the Hindustan Advanced Light Helicopter (ALH)   0581499

In addition to the monitoring and safety functions, the AFCS is fitted with preflight test, a computer self-test and built-in first and second line maintenance test facilities.

Status
Both AFCS are in production. The AFCS 165 was selected by Eurocopter for the Super Puma Mk II, and the AFCS 166 was selected for the Advanced Light Helicopter developed by Hindustan Aeronautics for both civil and military applications.

Contractor
SAGEM Défense Securité, Navigation and Aeronautic Systems Division.

Aircraft Condition Monitoring System (ACMS)

Type
Onboard system monitor.

Description
The ACMS has been designed for the acquisition of aircraft parameters in accordance with the latest airworthiness regulatory agency requirements. It processes engine, APU and aircraft monitoring, airline incident and specific investigation report data and transmits this data to a digital flightdata recorder or quick access recorder. Provision is also made for customised reports. The customisation is realised through Ground Support Equipment (GSE); software is PC-compatible. The system interfaces with ACARS, multifunction Control Display Unit or CDU, onboard printer, data loader and centralised fault display system.

A new option that utilises a removable PCMCIA drive offers flight data storage equivalent to use of external recorders.

Status
In service in the Airbus A320, A330, A340 and Boeing 737, 747-400, 757, 767, MD-87, -88 and -90, ATR 42 and 72, Embraer 120 and ERJ 135/145.

Contractor
SAGEM Défense Securité, Navigation and Aeronautic Systems Division.

APM 2000 autopilot module

Type
Flight Control Computer (FCC).

Description
The APM 2000 is the core of the Automatic Flight Control Subsystem for the new generation of helicopters, that permit single-pilot IFR operations.

The functions of the APM 2000 are: stability and control augmentation; long-term attitude hold; and AP modes of flight. Depending on the configuration, the system can operate either in simplex APM operation or in duplex APMs. The APM 2000 is a fail-passive digital computer that uses highly integrated technology and is the basis of a computer family that can meet requirements from two to four-axis autopilot system implementation. Two identical APM 2000 modules can be integrated to build a fail-operational configuration, to improve the mission reliability.

The APM 2000 can interface either with the Electro-Hydraulic Actuators (EHA) through a SAS 2000 computer or directly with Smart Electro-Mechanical Actuators (SEMA).

A typical APM 2000 installation could include one APM 2000, two to four trim actuators, three to four SEMA actuators, one APMS and two AHRS (digital). According to version, the APM 2000 could be connected to flight display systems, radio navigation systems, a Doppler system, a MFDAU/HUMS, or one APM 2000 (duplex configuration).

SAGEM APM 2000 autopilot module   0001398

Specifications
Dimensions: 337 × 183 × 38 mm
Power supply: 16 to 32.2 V, 25 W
Weight: 2 kg
Environmental: DO 160 C

Status
In production. Eurocopter Deutschland has selected the APM 2000 system for the EC 135, and the EC 145; it has also been selected by Eurocopter for the EC 155 and EC 225.

Contractor
SAGEM Défense Securité, Navigation and Aeronautic Systems Division.

CARNAC 21 Flight Inspection System (FIS)

Type
Flight inspection/calibration system.

Description
CARNAC 21 is a fully automatic, small and lightweight system designed for navaids calibration. The multitasking system offers five main functions: mission planning, flight inspection, reporting, training and system maintenance. CARNAC 21 is a modular system, removed or installed within 30 minutes and easily adaptable to any aircraft.

The trajectory system used in CARNAC 21, designated VP D-GPS, offers certified real-time positioning (10 cm), with long-range coverage. The system provides for flight inspection capability up to Category III and covers all navigation and landing aids, including GNSS. It is also adaptable to new technologies and accepts hardware update to the avionics or data processing equipments without requirement to modify software. The Man-Machine Interface (MMI) incorporates multilanguage pull-down menus.

Options include REPLICATOR, a real-time board-to-ground curves transmission system. This unit has been developed to improve the dialogue between airport maintenance and flight inspectors, with the aim of reducing flight inspection time. The transmission of all acquired parameters and aircraft position status enables maintenance engineers to monitor, in near real-time, the effects of beacon adjustment and measurements performed by the flight inspection system.

Specifications
(basic)
Aircraft position fixing accuracy: VP (Very Precise) D-GPS – 10 cm
Dimensions: 500 × 1,150 × 950 mm
Weight: 135 kg
Power: 28 V DC, 25 A
Compliance: ICAO Doc 8071 (Vol 1 and 2), ICAO Annex 10, Annex 14

Status
Operational in over 40 countries. CARNAC 21 with VP D-GPS has been operational for more than 8 years.

Contractor
SAGEM Défense Securité, Navigation and Aeronautic Systems Division.

Centralised Fault Display Interface Unit (CFDIU) for the A320

Type
Avionic interface system.

Description
The concept of a Centralised Fault Display Interface Unit (CFDIU), pioneered for the A320, suits the need for effective maintenance tools that enable troubleshooting down to the faulty LRU. The CFDIU collects and processes all built-in test equipment messages from the main aircraft avionics subsystems. Maintenance information can be displayed for diagnosis either by the flight crew or by ground personnel.

The CFDIU receives all failure message output and stores it for subsequent use. It also performs a correlation check between messages, separated by a short time interval, that are assumed to have been generated by the same fault. It is possible for the CFDIU to transmit messages to the ground during flight by a datalink, or printout all the in-flight messages after the aircraft has landed.

The CFDIU is packaged in a 4 MCU ARINC 600 box and consumes 25 W at 28 V DC.

Status
In service in the Airbus A320.

Contractor
Thales Avionics SA.

DataLink Interface Processor (DLIP)

Type
Avionic interface system.

Description
The DLIP is a key component for managing L1, L11, L14, L16 and L22 tactical datalinks. The DLIP is built around a modular, scalable, architecture based on COTS modules. The DLIP is able to simultaneously process messages and protocols from multiple datalinks. This allows it to integrate datalinks within command and control systems or platforms.

Status
The DLIP is available in air, land and seaborne versions. It is being proposed for NATO's ACCS LOC 1 air defence programme, and for the Erieye AEW&C aircraft selected by the Hellenic Air Force.

Contractor
Thales Communications.

Digital Automatic Flight Control System (AFCS) for the A300, A310 and A300-600

Type
Flight Control System (FCS).

Description
Digital processors replace the largely analogue elements of the original automatic flight control system in the Airbus Industrie A300 wide-body airliner, bringing the standard of this AFCS up to that of an almost wholly digital system. This is available in all A300 production with the forward-facing crew compartment. The system was first flown on the A300 in December 1980 and entered service with Garuda Indonesian Airways in January 1982. It is certified for Cat IIIB operation. The newer A310, the first of which entered service in April 1983, has the new digital automatic control system as standard.

The digital automatic control system provides the flight augmentation functions of pitch trim in all modes of flight, yaw damping, including automatic engine failure compensation when the autopilot is engaged, and flight envelope protection. It has a comprehensive complement of autopilot and flight director modes that permit automatic operations from take-off to landing and roll-out, a thrust control system which operates throughout the flight envelope, and a derate capability. It contains protection features against excessive angle of attack and has a fault isolation and detection system for line maintenance.

The AFCS also integrates windshear detection as a combination of vertical and horizontal shear components, with annunciation and flight guidance functions.

Design has been in accordance with ARINC 600 and 700 characteristics and has led to the adoption of ARINC 429 databusses between the automatic control system processors and sensors. Four to six processors are used, comprising two flight augmentation computers, one or two flight control computers (the second unit being necessary only if Cat III automatic landing capability is required), together with a thrust control computer. A second thrust control option is available, and the system also includes a flight control unit providing pilot interface with the autopilot/flight director and autothrottle functions, and a thrust rating panel which allows crew access to thrust limit computations.

A further new item of equipment is an engagement unit, with pitch trim and yaw damper engage levers, autothrottle arm and engine trim controls, which is mounted in the flight deck roof panel. Two pitch and roll dynamometric rods are also used as control wheel steering sensors. There are two trim actuators, an autothrottle actuator, a coupling unit on each engine and two further dynamometric rods connected to the throttle control linkage.

Flight deck controller equipment has been revised and a new autopilot/flight director and autothrottle mode selector is installed in the centre glareshield. Variable data can be entered by rotating selector knobs and shown by liquid crystal display readouts. The various modes are engaged by push-buttons. Modes available are altitude capture and hold, heading select, profile to capture and maintain vertical profiles and thrust commands from the flight management system, localiser, landing and speed reference. Autothrottle modes include delayed flap approach, speed/Mach number select, and engine N1 or engine pressure ratio selection.

The thrust rating panel is mounted above the centre pedestal. This has comprehensive controls permitting thrust levels to be selected, depending on operating mode and providing for selection of such facilities as derated thrust take-off.

The fault isolation and detection system has a dedicated maintenance/test panel which, on a two-line by 16-character display, provides written alert messages based on fault information from automatic testing activities. This is conducted in all LRUs and includes fault isolation, tests to

Airbus A310 uses the digital automatic flight control system   0503908

Units of the FCS 60B autopilot and automatic flight control system for commuter aircraft   0581500

check for correct operation after maintenance action and, on the ground, automatic landing availability checks. Up to 30 faults from six flights can be stored and retrieved.

The system is jointly produced by Thales Avionics (formerly Sextant) as prime contractor, Smiths Industries Aerospace (UK) and Bodenseewerk Gerätetechnik GmbH (Germany).

Computer units
1 or 2 flight control computer(s) (10 MCU size)
1 or 2 thrust control computer(s) (8 MCU size)
2 flight augmentation computers (8 MCU size)

Control units
flight control unit (glareshield)
thrust rating panel (centre panel)
FAC/ATS engagement unit (roof panel)
Maintenance/test panel

Other units
2 pitch dynamometric rods
2 roll dynamometric rods
2 trim actuators
autothrottle actuator
2 engine coupling units
2 engine dynamometric rods

Status
In service in the Airbus A300-600ST and A310.

Contractor
Thales Avionics SA.

DMU–ACMS Data Management Unit – Aircraft Condition Monitoring System

Type
Data processing system.

Description
The SAGEM DMU-ACMS is a common unit for use on the A319/A320/A321 series of commercial aircraft. The unit is designed to acquire engine and aircraft data reports, and data required for airline operational management, as may be programmed by the users.

A new option to the system has been introduced which utilises a removable PCMCIA card and offers flight data storage capability equivalent to that of external recorders such as QAR/DAR.

A319/A320/A321 DMU-ACMS common unit
    0015271

Specifications
Interfaces:
(MCDU) ARINC 739
(on-board printer) ARINC 740 or 744
(DAR recorder) ARINC 591 (fully programmable frame)
(ACARS MU) ARINC 724B
(MDDU/PDL) ARINC 615
(CFDS (Centralised Fault Data System)) ABD 0048
Acquisition:
(discretes) 20
(ARINC 429 buses) 55
Dimensions: 3 MCU
Weight: 4.5 kg
Power: 115 V AC, 400 Hz, 40 VA

Status
In production and in service.

Contractor
SAGEM Défense Securité, Navigation and Aeronautic Systems Division.

EMTI data processing modular electronics

Type
Aircraft utilities computer.

Description
For the future generation of avionics core systems, Thales is developing and marketing the EMTI (Ensemble Modulaire de Traitement de l'Information), or MDPU (Modular Data Processing Unit), in co-operation with Dassault.

The main characteristics are modular data processing assembly; open-ended system; adaptable to existing interfaces; simplified maintenance.

The system ensures independence between the basic software and the application programs. The programs are being developed with the ODILE software engineering environment and are defined in co-operation with Dassault Aviation. New technologies (object-oriented open environment, automatic code production) are used so that system object libraries, developed under past programs, can be re-used and the simulations and proofs can be incorporated into the environment.

Status
This new generation of products will be installed on the Mirage 2000-9 aircraft. The products are evaluated for use under the Rafale programme and for the retrofitting of different weapon systems.

Contractor
Thales Avionics SA.

Dassault Aviation. FCS 60B automatic Flight Control System

Type
Flight Control System (FCS).

Description
The FCS 60B is designed for commuter aircraft to provide ease of operation for flight profiles

ranging from initial climb out to Cat II ILS or MLS approach.

In the BAE Systems Advanced Turboprop, the system consists of a pair of three-axis autopilot/flight director fully digital computers, which drive dual primary and trim servos. The FCS 60B also acts as a back-up fly-by-wire system for the three axes. This mechanisation allows independent and redundant en route operation of the pilot and co-pilot flight directors, as well as cross-monitoring for Cat II approaches.

Status
Certified on the BAE Systems Advanced Turboprop aircraft.

Contractor
SAGEM Défense Securité, Navigation and Aeronautic Systems Division.

Fly-by-wire systems for the A330 and A340

Type
Flight Control System (FCS).

Description
As on the A320, the A330 and A340 flight controls are hydraulically actuated and electrically or mechanically controlled. Pilot controls in the cockpit consist of two sidesticks, conventional rudder pedals, mechanical pitch trim and electrical rudder trim.

Electrical flight control is achieved by seven computers of three different types: three Flight Control Primary Computers (FCPC), which are in charge of generating control laws and controlling surfaces; two Flight Control Secondary Computers (FCSC), which are also in charge of controlling surfaces; and two Flight Control Data Concentrators (FCDC), which interface the flight control system with other aircraft systems to provide an isolation function.

Each of the two FCSCs can control the power elements used to activate the aircraft control surfaces. In normal operation, the FCSC achieves spoiler control, rudder trim control and rudder travel limiting. In back-up mode, as in the case of a failure of the FCPC, the FCSC achieves aileron control, elevator control and yaw damping.

The FCDC performs data concentration, warning and maintenance functions. In data concentration, the FCDC transmits information, such as control surface positions, to the display management computers and to the flight data interface unit. In warning, the FCDC indicates flight control failure status to the flight warning computers and to the display management computer. For maintenance the FCDC isolates and memorises the flight control system failures and interfaces with the centralised maintenance computer.

Status
In production for the Airbus A330 and A340.

Contractor
Thales Avionics SA.

The A330 shares a common flight control architecture with other types in the Airbus fleet 0085606

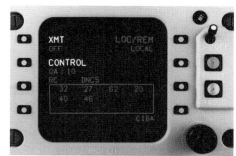

The Rockwell-Collins France MCU-2202F data terminal set control 0504104

Fuel Control and Monitoring Computer (FCMC) for the A330 and A340

Type
Aircraft system control, fuel.

Description
The A330 and A340 have two fuel tanks in each wing, one in the tailplane and one in the fuselage.

The A330 and A340 fuel control and monitoring system is composed of about 96 fuel height probes and other sensors for temperature, densitometers and level detectors installed in the fuel tanks, one to three refuelling panels and two identical Fuel Control and Monitoring Computers (FCMC).

The FCMC in the A330 and A340 concentrates functions previously scattered in distinct computers in the A310 and A320 such as fuel quantity measurements, Hi-Lo measurement and centre of gravity control.

The role of the FCMC is to measure the fuel quantity that is available on the aircraft and to indicate it to the aircrew. It also controls automatic refuelling of tanks to preselected levels and monitors associated pumps and valves, together with monitoring and controlling the fuel temperature in the tanks and fuel tank utilisation sequence. The FCMC also detects too low or too high levels in tanks and manages the associated warnings, controls fuel distribution for wing load alleviation and controls the centre of gravity of the aircraft by transferring fuel from one tank to another.

Status
In production for the Airbus A330 and A340.

Contractor
Thales Avionics SA.

IMA data processing equipment

Type
Data processing system.

Description
Thales has developed data processing equipment for new avionics core systems which is designed to provide military aircraft with scalable computing power depending on the aircraft mission and other electronics equipment complexity. It can be used for mission control (including navigation, weapons management, electronic warfare and radar control).

The main characteristics are modular data processing; open architecture; easy adaptation to existing interfaces; independent of basic software and application programs.

The modules in production are the central processing module; graphics data processing module; mass memory module; bus management module; input/output module.

This data processing equipment is provided with a module development kit; an object oriented software environment that handles system object libraries and makes software reuse easy for other applications or simulations and automatic code production.

The data processing equipment is intended for the new military aircraft as well as for the retrofit of military aircraft already in operation.

Status
As of April 2004, IMA was reported as in production and in service.

Contractor
Thales Systemes Aeroportes SA.

Low-airspeed system

Type
Aircraft flight computer.

Description
Thales Avionics has acquired the exclusive VIMI licence and has developed the CLASS low-airspeed system as an answer to the problem of low-airspeed measurements in helicopters for flight management, navigation and weapon firing requirements.

Using the helicopter's flight characteristics and, in particular, the measurement of cyclic pitch controls and attitudes, CLASS allows the combination of low-airspeed-computed data with conventional air data measurements to provide airspeed information. This information is valid throughout the flight range and requires no additional external mobile probes.

Status
In production for the Eurocopter Tiger helicopter.

Contractor
Thales Avionics SA.

MCU-2202F data terminal set control

Type
Data processing system.

Description
The MCU-2202F data terminal set control has been designed for the control of the Collins

MDM-2202 Link 11 modem. It is a processor-controlled unit which provides configuration and control instructions to the data terminal set and provides address storage and sequencing instructions for control of the network roll call operation.

Specifications
Dimensions: 146 × 105 × 242 mm
Weight: 3.6 kg
Power supply: 28 V DC
Temperature range: −20 to +70°C
Altitude: up to 55,000 ft

Status
In production for aircraft of the French Navy.

Contractor
Rockwell-Collins France, Blagnac.

NH 90 Flight Control Computer (FCC)

Type
Flight Control Computer (FCC).

Description
SAGEM manufactures the Flight Control Computer (FCC) for the Eurocopter NH 90 helicopter.

The FCC is an essential component of the fly-by-wire system. It comprises the primary flight-control system, which provides basic control capabilities to the helicopter and its stabilisation, together with the Automatic Flight Control System (AFCS), which provides the 'hands-off' high-level Auto Pilot (AP) control modes that are required by the helicopter to fulfil its mission.

Status
In production for the NH 90 helicopter.

Contractor
SAGEM Défense Securité, Navigation and Aeronautic Systems Division.

SAS 2000 Stability Augmentation System

Type
Flight Control System (FCS).

Description
The SAS 2000 computer (Stability Augmentation System) is used to control the stability of helicopters. It is configured as two separate optional equipment packages in order to offer the maximum customer flexibility and choice: yaw SAS and/or pitch and roll SAS.

The yaw SAS consists of an integrated yaw rate gyro and control law computer, which provides command signals to an electromechanical series actuator driving the input lever of the tail rotor hydraulic boost.

The equipment consists of: a Fibre Optic Gyro (FOG) and Smart Electro-Mechanical Actuator (SEMA). The FOG measures the yaw angular rate and is used to provide damping about the vertical axis. In addition to the rate signal, the unit is also used to implement the control laws and output signals to feed the SEMA.

SAGEM pitch and roll SAS computer   0001396

The SEMA is a high performance electromechanical actuator using modern brushless motors with rear earth permanent magnets. It incorporates a complete position servo feedback loop using Hall effect sensors within the same housing.

The pitch and roll SAS computer is designed to be interfaced with a conventional attitude sensor, the SAS analogue pitch and roll computer controls the slaving loop of an electro hydraulic actuator. The pitch and roll SAS computer interconnects a full digital auto pilot module to the EHAs. In the VFR configuration the pitch and roll SAS computer is connected to helicopter generation, EHAs, trim actuator units (pitch and roll), cyclic stick grip and instrument panel switches, a remote vertical gyroscope or panel-mounted gyro horizon and pitch/roll axes solenoids.

Specifications
FOG
Mass: 0.75 kg

Software: RTCA DO 178B
Power supply: 16 V DC to 32.2 V DC
Environmental: DO 160 C

SEMA
Input: ARINC 429
 Adjustable working stroke (± 2.5 and 8 mm)
 Working speed up to 18 mm/s
Stall load: >50 N
Limit load: >4000 N
Mass: <0.8 kg

Pitch and roll SAS computer
Dimensions: 200 × 122 × 112 mm
Weight: 1.3 kg
Power supply: 28 V, 20 W
Environmental: DO 160 C

Status
In production, and installed on Eurocopter EC 135.

Contractor
SAGEM Défense Securité, Navigation and Aeronautic Systems Division.

Warning and maintenance system for the A330 and A340

Type
Aircraft utilities computer.

Description
The A330 and A340 warning and maintenance system is composed of two major subsystems: the flight warning system (FWS) and the central maintenance system (CMS).

The FWS presents the aircrew with visual and warning messages related to failures. It also indicates the seriousness of each failure and the corrective actions to be taken.

The system consists of two system data acquisition concentrators (SDACs), which acquire data then generate signals which go to three display management computers (DMCs), and/or two flight warning computers (FWCs). The DMCs use the SDAC signals to generate displays of systems pages and engine parameters on two Electronic Centralised Aircraft Monitor (ECAM) display units. The FWCs generate alert messages, memos, aural alerts and synthetic voice messages for ECAM display, together with radio altitude callouts, decision height callouts and landing speed increments. Routine control of ECAM displays is exercised via an ECAM control panel.

The onboard maintenance system generates and displays maintenance data on the multipurpose control and display units (MCDUs) for use by aircrew and ground maintenance personnel.

The system consists of two centralised maintenance computers (CMCs), three MCDUs and one onboard printer.

The CMS operates in two main modes: In flight, the CMS operates in Normal (Reporting) mode, and records and displays failure messages transmitted by each system built-in test equipment (BITE). On the ground, the CMS operates in Interactive (Menu) mode, and allows the connection of any BITE directly with the MCDU in order to display maintenance data or to initiate a test.

Development and production is conducted jointly by Thales Avionics and Airbus Industrie.

Status
In service on the Airbus A330 and A340.

Contractor
Airbus Industrie.
Thales Avionics SA.

Germany

AD-FIS Flight Inspection System

Type
Flight inspection/calibration system.

Description
AD-FIS is a highly integrated, digital flight inspection system which utilises advanced methods of real-time data processing and satellite navigation technologies. The system is compliant with ICAO recommendations and performs calibration of all ground-based navigation aids including ILS, MLS, TACAN, NDB, DME, PAR, SSR, VHF/UHF Comms, VDF and UDF.

The system is composed of the following subsystems: flight inspection receivers; position reference sensors; data processing equipment; operator console and aircraft interface.

Data from the system's sensors, which include airborne receivers and ground-based position reference equipment, is interfaced to the computer via a high speed databus. Data from ground-based sensors is transmitted to the aircraft via a UHF datalink. Sensor information received by the computer is stored in raw, unprocessed format on a magneto optical disk, allowing future access to all sensor data, and is simultaneously processed by the computer. Processed data is displayed in real time to the operator in graphical format via the operator's LCD screens and finally printed on the system's colour printer.

The system hardware is housed in a compact rack design, tailored to the host aircraft. The rack houses one real-time and at least one semi-real-time computer, flight inspection receivers, colour printer, operator station and additional items such as a spectrum analyser. Ground-based reference equipment includes a highly accurate laser tracking system and Differential GPS (DGPS) station.

Specifications
Dimensions: according to customer specifications
Weight: approx 150 kg; varies according to system design requirements

Status
In production and in service.

Contractor
Aerodata Flugmesstechnik GmbH.

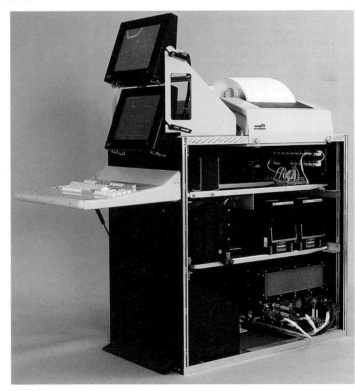

AD-FIS Flight Inspection System
0005428

Cabin lighting products

Type
Aircraft system control, internal lighting system.

Description
Diehl produces range of aircraft cabin lighting products for a variety of different civilian airliners and regional jets.

Airbus product family
Advanced Integrated Ballast Unit (AIBU)
15, 18 and 30 W T8 standard fluorescent lamps operating at 115 V AC/400 Hz. Continuously dimmable down to 5 per cent.

AIBU LED add-on
LED add-on for 15 and 30 W AIBU, employed as low-level lighting or coloured feature lighting (option). Continuously dimmable.

AIBU reflector
Reflector panel for 15 and 30 W AIBU.

Duo-AIBU
2 × 15 W and 2 × 30 W T8 standard fluorescent lamps operating at 115 V AC/400 Hz. Continuously dimmable down to 5 per cent.

Electronic ballast unit
15 and 30 W T8 standard fluorescent lamps operating at 115 V AC/400 Hz. Two-step dimmable (50 per cent and 10 per cent).

Boeing product family
Bin light
13, 17, 19, 32, 39 and 45 W T8 fluorescent Octron lamps operating at 115 V AC/400 Hz. Two-step dimmable (65 per cent and 35 per cent).

Sidewall light
13, 23 and 29 W T8 fluorescent Octron lamps operating at 115 V AC/400 Hz. One-step dimmable (25 per cent).

Ceiling light
13, 20, 25 and 30 W T8 fluorescent Octron lamps operating at 115 V AC/400 Hz. One-step dimmable (25 per cent).

Regional jet product family
Electronic duo ballast unit T5
2 × (6–18 W) T5 fluorescent rapid-start lamps, featuring remote operation and one-step dimming (25 per cent). Operating voltage 28 V DC.

Electronic ballast unit
13 to 40 W T8 fluorescent Octron lamps, featuring remote operation. Available in different lengths for all power ratings.

LED-based products
LED spot light
The LED spot light produces a round or rectangular homogeneously illuminated footprint covering distances 40 to 75 in (1,010 to 1,900 mm). The units combine high reliability and low power consumption and are mechanically compatible with standard reading light modules. Operating voltage is 12 or 28 V DC.

LED strip light
With application in areas such as lavatories or stairs, the LED strip light is available in various lengths from 90 to 540 mm. Operating voltage is 28 V DC.

LED mirror light
The LED mirror light is available with half-oval or round illuminating elements and in two lengths (430 and 510 mm). Operating voltage is 28 V DC.

LED dome light
The LED dome light is used as an emergency light for main and cross-aisle applications. With an operating voltage of 6 V DC, the dome light produces a homogeneously illuminated footprint.

LED ceiling emergency light
Applicable as an emergency and/or night light, this unit features redundant circuits for high system reliability. Operating voltage is 6 V DC.

Floor Proximity Emergency Escape Path Marking (FPEEPM) system
The FPEEPM system consists of LED exit sign and escape path marking, featuring redundant circuits for high reliability. Operating voltage is 6 V DC and the system is available in different lighting lengths.

Power supplies
Emergency Power Supply Unit (EPSU)
DLE produces a range of Emergency Power Supply Units (EPSUs), designed to provide power for onboard emergency lighting units in the event of main system failure:

Slide Release Power Supply Unit (SRPSU)
The SRPSU, based on EPSU P/N 3214-62-30, provides power to the aircraft escape slide release system. The SRPSU provides outputs for emergency exterior slide lights (2 × 6 V DC/1.0 A max), proximity switches (1 × 6 V DC/150 mA max), release circuits (2 × 24 V DC/0.6 A max) and squib circuits (1 × 5 A/>50 ms). The unit features a RS-232 serial data interface and charging time is 40 minutes. Test functions include monitoring of external loads, self-test, load test and battery capacity.

Autonomous Standby Power Supply Unit (ASPSU)
The ASPSU provides power to the aircraft passenger door warning lights and proximity switch control unit. The unit provides a maximum output of 28 V DC/2.3 A, with signal inputs of 2 × onboard power 28 V DC and 1 × logic input (open/ground). Charging time is 50 min and the system features a RS-232 serial data interface. Test functions include monitoring of external loads, self-test and battery capacity.

Status
In production and in service.

Diehl Aerospace (formerly Diehl Luftfahrt Elektronik) produces interior lighting systems for:

Airbus: all commercial aircraft programs (including the A380).

Boeing: all commercial aircraft programs (including the 787).

Embraer: ERJ 135/140/145, EMB 170/190.

In recent years, Diehl has tracked trends towards 'mood' lighting, replacing traditional functional lighting and the gradual replacement of fluorescent and incandescent sources with solid-state illuminants, such as LEDs. Moodlighting systems are installed in A330/340, A380, Boeing 777 and 787 aircraft.

Contractor
Diehl Aerospace GmbH.

Cabin management and security systems

Type
Aircraft system control, integrated.

Description
DLE's cabin management product range comprises control units and Human Machine Interfaces (HMI). HMI comprise Flight Attendant Panels (FAPs) and handsets; control units comprise Area Lighting Controllers (ALC) and Lighting Interface Standardisation Adapters (LISA).

Safety and security products includes emergency illumination, with complete systems compliant with the requirements of FAR 25.812. These systems include:
- Emergency illumination
- Floor path marking
- Exit marking
- Emergency Power Supply Units (PSU)
- Lighting components are primarily designed with LEDs and Emergency PSUs feature environmentally friendly battery technology.

Status
All products are in production and in service in a wide variety of commercial transport aircraft.

Contractor
Diehl Aerospace GmbH.

Flight Control Data Concentrator (FCDC)

Type
Aircraft flight computer.

Description
The Flight Control Data Concentrator (FCDC) is part of the flight control system of the Airbus Industrie A319, A320 and A321 aircraft. Its main functions are to validate, concentrate and store in-flight status and failure data of the flight control system. The electrical interfaces are a number of ARINC 429 and discrete inputs and outputs. Outline and mounting are in accordance with ARINC 600.

Specifications
Dimensions: 2 MCU
Weight: 3.4 kg
Power supply: 28 V DC, 16 W
Interfaces:
(input/output)
ARINC 429, SGS and SAV discrete
(input)
DC analogue
Reliability: >18,000 h MTBF

Status
In production. The FCDC entered operational service in 1988 on board the A320, which is equipped with a dual FCDC installation. It is also in service on the A319 and A321 aircraft.

Contractor
LITEF GmbH.

	P/N 3214-54-10/20	P/N 3214-54-50/60	P/N 3400-20/30/40/50
Output	6 V DC/10 A (9 A when operating simultaneously with AC output) 6 V DC output: 6 × 3.6 A each (max) 115 V AC output: 3 × for 70 square inch EL lamps (-20 only)	6 V DC/5 A (4.3 A when operating simultaneously with AC output) 6 V DC output: 6 × 3.6 A each (max) 115 V AC output: 2 × for 70 square inch EL lamps (-60 only)	6 V DC/5 A (-20); 4.4 a (-30); 3.8 A (-40); 3.2 A (-50) 6 V DC output: 4 × 1.5 A each (max) 115 V AC output: 150 mA
Signal input	1 × onboard power 28 V DC 1 × onboard power 115 V AC 6 × logic (open/ground)	1 × onboard power 28 V DC 1 × onboard power 115 V AC 6 × logic (open/ground)	3 × onboard power 28 V DC
Charging time	40 min (quick charge)	40 min (quick charge)	90 min
Interface	RS-232 serial data interface	RS-232 serial data interface	
Test functions	Monitoring of external loads, self-test, load test and battery capacity test	Monitoring of external loads, self-test, load test and battery capacity test	Monitoring of external loads

Modular avionic computers

Type
Avionic computer.

Description
A range of modular avionic computers are designed by Teldix for a variety of airborne roles.

The Missile Control Unit (MCU) is designed to control intelligent missiles fitted to aircraft such as the Panavia Tornado. The MCU serves as a computing and interface system between the aircraft's main digital computer and the relevant input/output circuits of the missiles.

The MCU is a 16-bit multimicroprocessor containing the interface circuits to convert, process and pass internally fed data to the missiles via the MIL-STD-1553B databus. The Launcher Decoder Unit (LDU) provides the interface between the MCU and individual missiles.

Status
In service.

Contractor
TELDIX GmbH.

RM 3300 series RMI converters

Type
Avionic interface system.

Description
The RM 3300 series converters enable Sine/Cosine or composite video signals from navigation receivers to be used to drive RMI indicators and other display devices which require XYZ synchro inputs.

By varying the circuitry of the internal modules in the RM 3300 family, the converters can be used to provide inputs for XYZ synchros per ARINC 407, Sin/Cos DC or Sin/Cos AC indicators. There is even a model which simply functions as a synchro amplifier, to enable low-power synchro signals to drive corresponding high-power synchros.

These converters are approved for installation in unpressurised areas of aircraft operating up to 50,000 ft. There are no altitude restrictions for installations in pressurised areas.

Specifications
Dimensions: 54 × 139 × 214 mm (excluding mating connector)
Weight: 0.75 kg

Status
In production and in service.

Contractor
Becker Avionic Systems.

Static Inverters

Type
Aircraft system control, electrical.

Description
Diehl produces single-phase airborne static power units, rated between 1,000 and 2,500 VA, which utilise aircraft 28 V DC battery power to deliver a 115 V AC, 400 Hz output. Design features include a crystal-controlled oscillator for output frequency stability and a Pulse Width Modulated (PWM) square-to-sine wave power conversion scheme for efficient and reliable operation. An integrated cooling fan forces ambient air through the unit, exhausting through the rear panel.

The unit employs a two-stage power conversion technique that is claimed to provide for superior regulation of the output voltage over wide input line variations. One DC/DC converter scales and isolates the 28 V DC input to a high internal DC voltage, which is then transferred to a DC/AC inverter to generate the 115 V/400 Hz output. The switching frequency for the DC/DC converter is 50 kHz and 25 kHz for the DC/AC inverter.

Specifications
Input
Rated voltage: 28 V DC nominal
Operating voltage: 22 to 32.5 V DC (full performance); 17 to 32.5 V DC (reduced performance)
Current: 44 A (at 28 V DC rated power)

Output
Voltage: 115 ±5 V AC
Frequency: 400 ±2 Hz
Rated power: 1.0–2.5 kVA
Efficiency: >82%

General
Fault indication: overheat, fan failure, output over voltage, frequency failure
Dimensions: 216 × 109 × 305 mm
Weight: <5.3 kg

Status
In production and in widespread service.

The following programs are equipped with Diehl Aerospace's (formerly Diehl Luftfahrt Elektronik) Static Inverters or other secondary power equipment:

Commercial transports, including A318- A321 and A330/340, A380.

Military helicopters, inlcuding CH53G.

Contractor
Diehl Aerospace GmbH.

International

Automatic Flight Control System (AFCS) for the Eurocopter Tiger helicopter

Type
Flight Control System (FCS).

Description
The Automatic Flight Control System (AFCS) is a digital autopilot, which is superimposed on the mechanical flight control system of the helicopter. The duplex digital AFCS provides the following functions: command and stability augmentation/attitude hold, to stabilise the aircraft as a weapon platform and reduce gust sensitivity; tactical modes, including three-axis hover, a line-of-sight mode for weapon aiming/weapon delivery and gunfire compensation; standard cruise modes providing four-axis control.

Each AFCS consists of two identical Flight Control Computers (FCC) and two identical Control Panels (CPs). The FCC contains five modules with the following elements and functions: a CPU Board with two microprocessors, one for system management functions and the other one for running the control laws; an Input/Output Controller (IOC) for input/output control, analogue I/O, cross-channel datalink, communication and hardware synchronisation; an input/output board for discrete I/O requirements; a Power Supply Unit (PSU) for the generation of internal voltages and LVDT/RVDT excitation; and a rear module for EMC filtering and lightning strike protection.

The AFCS can be controlled from either of the two control panels, which are installed in the front and rear cockpits. Apart from the ARINC link, communication between avionics equipment and the AFCS is established with a MIL-STD-1553B bus.

The AFCS software is programmed in Ada using state-of-the-art software development techniques.

The Tiger AFCS is a joint development by Nord-Micro (acting as prime contractor) and Thales Avionics.

The Tiger PAH-2 automatic flight control system   0079239

Status
The first flight of an AFCS-equipped Tiger helicopter was in 1993. Tiger, in various configurations has been ordered by Australia, France and Germany; interest has been shown by Spain.

Contractor
Nord-Micro AG & Co OHG.
Thales Avionics SA.

Avionics computers for Eurofighter Typhoon

Type
Avionic computer.

Description
Air data transducer for Eurofighter Typhoon
The Eurofighter Typhoon air data transducer, currently being developed by an international consortium consisting of Selex Sensors and Airborne Systems (formerly BAE Systems Avionics), Bavaria Avionik Technologie and Tecnobit, provides high integrity and fast dynamic response air data information.

The air data transducer features a combined multifunction pitot-static and flow angle mobile vane developed by Thales. It provides local air data parameters, such as pitot, static and differential pressures, via a dedicated MIL-STD-1553B databus to the flight control computer. Calculations of airspeed, altitude, Mach number, angle of attack and angle of sideslip are generated to support the Eurofighter Typhoon's artificial aerodynamic stabilisation.

Specifications
Dimensions: 138 × 125 × 155 mm
Weight: 3.2 kg
Power supply: (heater) 115 V AC, 450 W
±20 V DC, 15 W

Contractor
Selex Sensors and Airborne Systems, UK.
Bavaria Avionik Technologie GmbH, Germany.
Tecnobit SA, Spain.

Attack Computer/Navigation Computer (AC/NC) for Eurofighter Typhoon
The main functions of the Avionic Computer (AC) and Navigation Computer (NC) are to provide data for navigation, attack and identification. The two computers, which are identical in their hardware configuration, are based on a modular multiprocessor architecture. The central processing units are equipped with Eurofighter-selected standard microprocessors MC 68020 CPU and MC 68882 floating point co-processor. In order to meet the stringent performance requirements for floating point arithmetic, the central processing units are equipped with additional customised, floating point, accelerators, which can perform arithmetic operations extremely quickly. This special purpose co-processor is based on a high performance RISC technology.

Each computer is equipped with interfaces to two STANAG 3910 high-speed fibre-optic digital data buses.

Avionic Computer (AC) for Eurofighter Typhoon
0051280

Eurocopter Tiger   0523875

The software provided is developed mainly in Ada, applying standardised software development tools and methods. It includes built-in-test software, an adapted Ada run-time system and a set of Ada packages to provide a defined interface between target specific input/output interfaces and the application software developed by Eurofighter GmbH. (Main contract from DASA).

Contractor
TELDIX GmbH, Germany (prime).
Selex Sensors and Airborne Systems, Italy (workshare partner).
General Dynamics United Kingdom Ltd, UK (workshare partner).

Cockpit Interface Unit (CIU) for Eurofighter Typhoon
The Cockpit Interface Unit (CIU) handles and controls all cockpit-relevant data. The extensive data through-put from a STANAG 3910 high-speed fibre optic digital databus to a STANAG 3838 electrical databus and vice versa is handled by a modular multiprocessor system.

The modular multiprocessor system of the CIU is based on the Eurofighter-selected MC 68020 CPU supported by MC 68882 floating point co-processor. This high data throughput is realised by a complex hardware structure and TELDIX developed ASIC's. Integrated built-in-test functions (BIT) are used.

Cockpit Interface Unit (CIU) for Eurofighter Typhoon   0051281

The software provided is developed mainly in Ada, applying standardised software development tools and methods. It includes built-in-test software, an adapted Ada run-time system and a set of Ada packages to provide a defined interface between target specific input/output interfaces and the application software developed by Eurofighter GmbH. Each of two CIU's necessary per single seat aircraft has the capability to control the STANAG 3838 cockpit bus. (Main contract from BAE Systems).

Contractor
TELDIX GmbH, Germany.

Defensive Aids Computer (DAC) for Eurofighter Typhoon
The Defensive Aids Computer (DAC) performs defensive aids management functions and calculations to assist the pilot in achieving mission success.

The integrated modular microprocessor architecture system based on the Eurofighter-selected MC 68020 CPU handles the interfaces for activity request, suppression and blanking discretes and chaff/flare dispenser control.

Information is transferred and received via one STANAG 3838 electrical databus and one STANAG 3910 high-speed fibre optic digital databus.

The software provided is developed mainly in Ada, applying standardised software development tools and methods. It includes built-in-test software, an adapted Ada run-time system and a set of Ada packages to provide a defined interface between target-specific input/output interfaces and the application software developed by Eurofighter GmbH. (Main contract from DASA).

Contractor
TELDIX GmbH, Germany (prime).
Selex Sensors and Airborne Systems, Italy (workshare partner).

Eurofighter Typhoon front computer   0001276

Interface   Processor   Unit   for   Eurofighter
Typhoon   0051282

Front computer for Eurofighter Typhoon

ENOSA, leading a consortium with Selex Sensors and Airborne Systems (formerly Galileo Avionica), Smiths Aerospace and VDO, has been awarded the development contract for the front computer for the Eurofighter Typhoon. The front computer forms part of the Utilities Control System (UCS) of the aircraft; it is of compact design and is fully integrated with other primary aircraft systems.

The front computer provides all the functions for control and monitoring of: the environmental and temperature control system; the life support system; the crew escape system.

Design features of the front computer are that it is fully compliant with Eurofighter Typhoon requirements and that it is databus MIL-STD-1553B compatible, allowing it to communicate with the rest of the UCS. It is based on a Motorola 68020 microprocessor.

Contractor

ENOSA, Spain.
Selex Sensors and Airborne Systems, Italy.
Smiths Aerospace, UK.
VDO, Germany.

Interface Processor Unit (IPU) for Eurofighter Typhoon

The Interface Processor Unit (IPU) is the data acquisition, management and processing unit of the integrated monitoring, test and recording system for the Eurofighter. It is based on a modular multiprocessor architecture and includes the Eurofighter-selected MC 68020 CPU standard microprocessors and MC 68882 floating point co-processor.

The IPU is equipped with interfaces to two STANAG 3910 high-speed fibre optic digital databusses and one STANAG 3838 electrical databus. Intelligent controllers are provided for the interfaces, which perform time critical functions for data transfer on the high-speed bus, the addressing and formatting of data and error checking.

In addition, the IPU also contains both electrical and fibre optic links to various aircraft subsystems. A digital signal processor handles sensor data and compresses digitised audio data, using special compression algorithms. (Main contract from BAE Systems).

Contractor

TELDIX GmbH, Germany (prime).
Selex Sensors and Airborne Systems, Italy (workshare partner).

Digital fly-by-wire flight control system for the Eurofighter Typhoon

Type

Flight Control System (FCS).

Description

BAE Systems (Operations) Limited leads a consortium comprising BGT of Germany, Galileo Avionica, Avionic Systems and Equipment Division of Italy and INDRA Sistemas of Spain, which is under contract to DaimlerChrysler Aerospace (DASA) to develop the Flight Control Computer (FCC) and Stick Sensor and Interface Control Assembly (SSICA) for the Eurofighter Typhoon aircraft.

The Flight Control System (FCS) provides the aircraft with a full-time fly-by-wire control system which controls the aircraft via its 11 primary and secondary flying control surfaces to give the aircraft carefree handling characteristics together with outstanding agility and manoeuvrability throughout the flight envelope. The Eurofighter Typhoon has no mechanical back-up system and is therefore totally dependent on this digital system. Redundancy of the system is ensured through replication of both the hardware and software, and should a failure occur in any one computer, this can be isolated and the aircraft will continue to fly normally.

To provide the necessary integrity and safety, the system uses a quadruplex design with each of the four FCCs containing identical hardware and software. Each computer contains Motorola 68020 32-bit microprocessors and software compiled in Ada. Comprehensive built-in test provides a

Eurofighter Typhoon displaying the carefree handling characteristics of its quadruplex FBW system
0103894

continuous system monitoring for both ease of system maintenance and assurance of the continued high integrity performance. ASICs have been widely used to ensure a low component count to maximise performance and reduce system size and weight. The FCC also contains an interface to the aircraft's utility bus to provide integrated vehicle management and to a STANAG 3910 fibre optic avionics bus to allow integration with the mission avionics.

Status
In production and in service.

Contractor
BAE Systems (Operations) Ltd.
Bodenseewerk Gerätetechnik GmbH (BGT).
Galileo Avionica, Avionic Systems and Equipment Division.
INDRA Sistemas.

Flight control system for the AMX

Type
Flight Control System (FCS).

Description
BAE Systems (Operations) Limited is co-operating with Selex Sensors and Airborne Systems (formerly Galileo Avionica) in the Fly-By-Wire (FBW) flight-control system for the Italian/Brazilian AMX strike aircraft. The system comprises two dual redundant flight-control computers each based on 16-bit microprocessor hardware and incorporating fail-safe software. The system commands the movement of seven control surfaces and incorporates a recently developed autopilot facility and automatic pitch, roll and yaw stabilisation. Analogue computing is used for the actuator control loops, the pilot command paths and rate damping computation. Digital computing handles gain scheduling, electronic trim and integration of the airbrake. Redundant microprocessors in the flight control computer units monitor system performance and are associated with built-in test facilities designed to provide a high confidence and rapid comprehensive system check. Testing is initiated by the pilot before flight and is conducted automatically thereafter. In addition to the flight control computers, the FBW system also includes pilot control position sensors, three-axes rate gyros and air data components.

Status
In production and in service.

Contractor
BAE Systems (Operations) Limited.
Selex Sensors and Airborne Systems.

AMX flight control computer   0001411

Fly-by-wire system for the Tornado

Type
Flight Control System (FCS).

Description
The Panavia Tornado flight control system uses triplex electronic signalling and processing, with quadruplex actuation, to provide a high degree of manoeuvrability throughout the entire flight envelope and has a mechanical back-up system for emergency control though with degraded handling qualities. The fly-by-wire processing is incorporated in the Command and Stability Augmentation System (CSAS) and the automatic flight functions, which include safety critical capability such as automatic terrain-following, are integrated with the automatic flight control system. These are separately configured but very closely allied systems. The triplex CSAS components are associated with a duplex Spin Prevention and Incidence Limiting System (SPILS) which ensures that the crew can fly the aircraft to its structural and aerodynamic limits without the risk of loss of control.

Design, development and production of the systems was a trinational venture between BAE Systems (Operations) Limited in the UK, Bodenseewerk Gerätetechnik (BGT) in Germany and Galileo Avionica (now Selex Sensors and Airborne Systems) in Italy.

Command Stability Augmentation System (CSAS)
The CSAS is an analogue FBW manoeuvre demand system. It provides electrically signalled pitch, roll and yaw control and automatic stabilisation of aircraft response to pilot command or turbulence. Gain scheduling improves handling qualities and control stability over the entire flight envelope and operates in conjunction with the spin prevention and incidence limiting system. 'Carefree' manoeuvring allows the exploitation of the aircraft's full lift capability under all flight conditions without the risk of structural damage resulting from a pilot's control demand for control

Panavia Tornado SPILS computer opened up to show the four-board front connector layout   0503924

surface movement that would exceed the design strength of the airframe.

Individual LRUs are the CSAS pitch computer, CSAS lateral computer, CSAS control unit, pitch, roll and yaw rate gyros and pitch, roll and yaw position transmitters.

Spin Prevention and Incidence Limiting System (SPILS)
SPILS is a duplex analogue system which operates in conjunction with the CSAS to achieve maximum aircraft lift in low-level flight. It limits aircraft incidence, irrespective of the pilot's

Panavia Tornado CSAS equipment includes computer units, control panels, triplex position transmitters, triple gyro packs and quadruplex first stage actuators   0503923

The Editor demonstrates the carefree handling of the FBW of the Tornado GR. Mk 1 at ultra-low level during a training exercise in Northern England   0105291

demands, when maximum safe angles of attack are reached. Individual LRUs are the SPILS computer and SPILS control unit.

Automatic Flight Director System (AFDS)
The digital Autopilot and Flight Director System (AFDS) automatically controls the flight path in all modes, including terrain-following, and sends signals to the director instruments enabling the crew to monitor autopilot performance or to fly the aircraft manually. Pitch autotrim is also incorporated. The duplex self-monitoring

processor configuration provides high-integrity automatic control, permitting low-altitude cruise with appropriate safety margins. The flight director remains available after most single failures. Autopilot manoeuvre demand signals are routed to the control actuators through the command and stability augmentation system. A 12-bit processor is used with 6 kbits words of stored program and 1 kbits words of data store. Typical computing speed is around 160 Kops and program cycle time is 32 ms. Individual LRUs are the two AFDS computers, AFDS control unit,

autothrottle actuator and pitch and roll stick-force sensors.

Status
In service in all variants of the Panavia Tornado.

Contractor
Selex Sensors and Airborne Systems, Radar Systems Business Unit.
Bodenseewerk Gerätetechnik GmbH.
BAE Systems (Operations) Limited.

Interface unit for the Tiger helicopter

Type
Avionic interface system.

Description
The interface unit is a data acquisition, management and preprocessing computer for the basic avionics of the French/German Tiger helicopter. It is based on a modular multiprocessor

architecture and includes the MC 68020 processor. It is equipped with input/output modules for data transfers via interfaces of different types, such as MIL-STD-1553B, ARINC 429, analogue, frequency measurement and discrete.

The equipment and the application software is developed in Ada. It includes BIT capabilities, an adapted Ada run-time kernel and flexible data transmission, handling, formatting and processing so that interface parameters are configurable via a parameter table which allows

changes in acquisition rate or selection of preprocessing without any modification of the software itself.

Status
In service in development aircraft.

Contractor
TELDIX GmbH.
Sextant (co-op partner).
VDO Luftfahrtgeräte Werk (co-op partner).

Maintenance Data Panel (MDP) for Eurofighter Typhoon

Type
Data processing system.

Description
A consortium consisting of INDRA Sistemas and EADS produces the Maintenance Data Panel (MDP) and Portable Maintenance Data Store (PMDS) for the Eurofighter Typhoon. The MDP and PMDS apply the most up-to-date technology for monitoring and recording information from the aircraft for maintenance operations.

The MDP is the primary panel for ground crew maintenance; it collects and records fatigue data and status information from aircraft systems and engines. Using an electroluminescent touchscreen display, the MDP facilitates user-guided operation and handling.

The following types of data are displayed: refuelling/defuelling; weapon stores; data loading; failure data (actual and last flight status); limit exceedance; status of consumables; life usage data; and aircraft tail number.

All in-flight event monitoring and status, as well as fatigue data, for avionics, aircraft and engine systems will be stored in the non-volatile solid-state Portable Maintenance Data Store (PMDS).

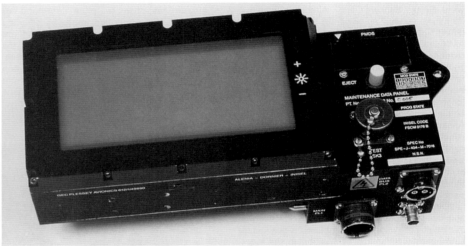

Maintenance data panel for Eurofighter Typhoon   0001277

The PMDS is used to transport maintenance data for ground analysis, and to load maintenance-related data into the aircraft.

Status
In production and in service.

Contractor
INDRA Sistemas SA. (prime contractor)
EADS Deutschland GmbH.

Utility control system for the Eurofighter Typhoon

Type
Aircraft utilities computer.

Description
The Eurofighter Typhoon utility control system is based on the utilities management system which was successfully demonstrated on the Experimental Aircraft Programme (EAP). The

system integrates, via a MIL-STD-1553B databus, the control and monitoring of a number of aircraft functions, including environment, cabin temperature, life support, crew escape, fuel management, fuel gauging, secondary power, hydraulics and other miscellaneous systems. The main benefits derived from the system are in weight saving and reliability. In addition, there is a greater fault tolerance and better damage resistance, system health is monitored and trends can be identified, aiding planned maintenance.

Status
In limited production.

Contractor
Selex Sensors and Airborne Systems, Radar Systems Business Unit.
INDRA EWS.
Smiths Industries Aerospace.
VDO Luftfahrtgeräte Werk Adolf Schindling GmbH.

YAK/AEM-130 quadruplex digital flight control system

Type
Flight Control System (FCS).

Description
The YAK/AEM-130 multirole lightweight jet trainer is specified with a quadruplex, digital, fly-by-wire, flight control system designed by Teleavio srl of Italy and BAE Systems North America. Design features of the control system include fault-tolerant redundancy management providing electrical two-fail-operate capability; a quad-channel redundant primary actuation

Equipment items of the YAK/AEM-130 quadruplex digital flight control system   0062222

system; multifunction air data computation and sensing; and extensive in-flight built-in testing.

The system provides for stability and flying qualities in accordance with MIL-STD-1797A and controllability to at least 35° Angle-of-Attack (AoA), together with 'carefree' handling (*g* limiting, stall/spin prevention) and automatic reversion in the event of damage or failure.

Contractor

BAE Systems North America, Aircraft Controls
  Teleavio srl.

The Yak-130 (Yefim Gordon)
0525061

Israel

ACE-5 computer

Type
Avionic computer.

Description
The ACE-5 is a modular high-performance airborne general purpose computer which is fully compatible with MIL-STD-1750A and MIL-STD-1553A and B. It is designed for multirole fighter mission management.

Computation performance is 1.7 Mips, with growth potential to 2.5 Mips. Some 256 kwords of memory with battery back-up are programmable at flight line level, with growth capacity to 1 Mwords. BITE provides continuous monitoring of functional integrity.

Specifications
Dimensions: 137 × 198 × 472 mm
Weight: 10 kg
Power supply: 115 V AC, 400 Hz, 3 phase
  28 V DC
Environmental: MIL-E-5400 Class II
Reliability: >500 h MTBF

Status
In service on Israeli F-16C/D aircraft.

Contractor
Elbit Systems Ltd.

Advanced Brake Controller (ABC)

Type
Aircraft system control, undercarriage/wheelbrakes.

Description
The Advanced Brake Controller (ABC) is designed for the Lockheed Martin F-16 fighter aircraft, providing improved braking performance, especially on contaminated runways, coupled with fully redundant operation. System benefits include reduced pilot workload during the landing phase, improved availability and reduced in-operation costs through increased braking system MBTF and reduced MTTR.

The main features of the ABC include:
- Improved braking algorithms (normal and backup)
- No aircraft modification required (retrofit flightline executable)
- Dual-redundant electronics (two independent controllers).

Advanced self BIT and diagnostics covering entire braking system:
- Automatic fault isolation
- Automatic recording of all landings and rejected takeoffs
- Detected fault will automatically trigger braking channel selection
- Failure condition 'trapping'
- MTBF >25,000 h
- Re-programmable via communication channel. Standard USAF GSE hand-held Test and Data Extractor Unit (TDEU):
- Dedicated Expert System software
- Enables diagnosis of braking system sensors serviceability
- Flightline test of ABC does not require aircraft power
- NSN 4920-01-424-8886DQ
- Optional laptop configuration
- PC-based workstation for ground evaluation and diagnostics of braking system.

Specifications
Dimensions: 102 × 130 × 163 mm (H × W × L)
Weight: 1.8 kg
Processors: Intel 80C196
Power consumption: 17 W
Qualification: MIL-STD-810C, -461B, -704A

Status
The ABC is qualified for the F-16A/B/C/D/I. In production and in service.

Contractor
RSL Electronics Ltd.

Advanced Fuel System Controller (AFSC)

Type
Aircraft system control, fuel.

Description
The Advanced Fuel System Controller (AFSC) is designed for the AH-64 attack helicopter, providing the pilot with fuel system data and management functions. The controller features a colour display that provides both analogue and digital displays coupled with user-friendly menu driven commands.

The main features of the AFSC include high MTBF, embedded BIT and simplified fuel system diagnostics for improved mission effectiveness and streamlined fuel system maintenance.

Specifications
Processor: Intel 80196 kc
Dimensions: 159 × 126 × 127 mm (H × W × L)

Weight: 2.1 kg
MTBF: >10,000 h
Power consumption: 11.9 W
Qualification: MIL-STD-810C, -461B, -704A
Display
  Type: 5 in AMLCD, day and night compatible
  Luminance: 320 cd/m²
  Contrast ratio: 60:1
  Resolution: 234 × 320
  Viewing angles: ±65° (H), +40/–65° (V at CR > 5)
Functional keys: 7

Fuel management functions
Fuel transfer sequence
Maintenance of longitudinal centre of gravity within limits
Fuel jettison
Selective fuel transfer (priority override)
Pilot advisories
  Available fuel
  Fuel flow
  Remaining endurance
  Leak detection
  Fuel transfer faults
  BINGO fuel

Status
The AFSC is in production and in service in AH-64 helicopters of the Israeli Air Force.

Contractor
RSL Electronics Ltd.

Armament Control System (ACS)

Type
Aircraft system control, Stores Management System (SMS).

Description
The Armament Control System (ACS) provides, in conjunction with the host aircraft avionic and other systems, all the necessary means for

The Weapon Control Panel for the Armament Control System (ACS)
0137242

operation, control, release, fire and jettison of the various weapons and other stores carried by fixed- and rotary-wing aircraft. The operational system consists of two main subsystems: Armament Interface Unit (AIU) and Weapon Control Panel (WCP). For training purposes, a third unit is available – an Aft Weapon Display Panel (AWDP).

Specifications
Weapon Release Modes:
  Bombs:
    Manual or computed release
    Single or ripple release
    Quantity and interval selection
    Nose, nose/tail and tail fusing options
  Rockets:
    Manual release only
    Single or ripple release
    Quantity and interval selection
  Gun:
    Preloading of rounds amount for each gun
    Burst limit setting
    Automatic and manual recocking
  Missiles:
    Automatic missile selection
    Control of electrical power, cooling and uncage functions
    Command seeker head position and audio tone routeing
Dimensions:
  WCP: 146 × 140 × 94 mm
  AWDP: 146 × 140 × 94 mm
  AIU: 155 × 325 × 191 mm
Weights:
  WCP: 1.6 kg
  AWDP: 0.8 kg
  AIU: 6.6 kg
Power: 28 V DC per MIL-STD-704A
Operating temperature: –25 to +70°C (WCP, AWDP), –40 to +70°C (AIU)

Status
In production and in service.

Contractor
BAE Systems ROKAR International Ltd.

Autonomous Combat manoeuvres Evaluation (ACE) system

Type
Onboard flight monitoring system.

Description
The Autonomous Combat manoeuvres Evaluation (ACE) system is a flight training debriefing system aimed mainly at multiparticipant flight training. The development of ACE is the direct outcome of an operational requirement for a squadron-level replacement for the current Air Combat Manoeuvring Instrumentation (ACMI) ranges.

ACE, unlike ACMI, is not ground range dependent and this enables debriefing of any kind of flight executed anywhere, without the need for carriage of external pods or telemetry. It is installed internally on each aircraft. Installation is simple and no software or cockpit changes are required for aircraft equipped with a MIL-STD-1553 databus. The ground debriefing station is based on commercial hardware and is designed to be used at squadron level on a daily basis.

Rada Autonomous Combat manoeuvres Evaluation system onboard unit   0001295

Aircraft flight data required to reconstruct manoeuvres and the operation of the avionics system is recorded independently on board each aircraft by the ACE flight monitor unit on the aircraft's VTR. The use of a C/A or P code GPS as an integral part of the ACE airborne system provides position accuracy and synchronisation, without being limited to a specific antenna range and telemetry. After landing, the cassettes of all the participants, containing the digital data and video and audio playbacks, are processed on a ground debriefing station from which a unified display file is produced and displayed graphically. Synchronised HUD video playback and audio complement and complete the debriefing material.

Specifications
Dimensions: 70 × 165 × 320 mm
Weight: 4 kg
Power supply: 28 V DC

Status
In service on Israeli Air Force F-16 aircraft with an option for F-15 and F-4 2000 aircraft, to become the standard fit on all Israeli Air Force fighters.
  In service on Chilean F-5E aircraft.

Contractor
Rada Electronic Industries Ltd.

Diagnostic Engine Starting System Controller (DESSC)

Type
Aircraft system control, power plant.

Description
The Diagnostic Engine Starting System Controller (DESSC) is an all-digital direct form-fit replacement for the original Engine Starting System (ESS) controller installed in the Lockheed Martin F-16. The DESSC, characterised by improved Jet Fuel Starter (JFS) control logic and full diagnostic capabilities, is claimed to significantly improve engine starting system performance and reliability, while streamlining maintenance and reducing costs.
  The main features of the DESSC include:
- Improved engine starting algorithms (successful engine starting at first attempt over the entire flight envelope)
- Embedded HUMS
- Dual-redundant electronics (two independent controllers)
  Advanced self-BIT and diagnostics:
- Automatic fault isolation
- Automatic recording of all engine starts
- Monitoring and recording of JFS exceedances
- Automatic cancellation of unsafe starts
- Pilot alert on unsafe starting condition
- Immediate fault detection at system power-up
- MTBF >15,000 h
- Soft clutch engagement.
  Standard USAF GSE hand-held Test and Data Extractor Unit (TDEU):
- Dedicated Expert System software
- Enables individual testing of JFS system components
- Flightline test of DESSC does not require aircraft power
- NSN 4920-01-424-8886DQ
- Optional laptop configuration
- Reprogrammable via communication channel
- PC-based workstation for ground evaluation and diagnostics of braking system.

Specifications
Dimensions: 128 × 103 × 268 mm (H × W × L)
Weight: 2.2 kg
Processors: Intel 80C88-2
Power consumption: 7 W
Qualification: MIL-STD-810C, -461B, -704A

Status
The DESSC is fully qualified by Lockheed Martin and the US Air Force (NSN 2925-01-436-7325YP) for the F-16A/B/C/D/I for F-100 and F-101 engines. In production and in service.

Contractor
RSL Electronics Ltd.

Digital Generator Control Unit (DGCU)

Type
Aircraft system control, electrical.

Description
The Digital Generator Control Unit (DGCU) is designed to replace the existing F-4 Phantom voltage, frequency and load sharing system to improve electrical power system reliability and maintainability via increased system MTBF and decreased MTTR. The Digital Generator Control and Monitoring system consists of two identical DGCUs which together control the aircraft electrical power, perform continuous system protection and automatically record in-flight failures. The two DGCUs replace the existing analogue system of three LRUs.

The main features of the AFSC include power control (voltage/frequency regulation, electrical load control, generator load sharing and protection and system synchronisation), continuous in-flight BIT, on-board recording of failures and maintenance related data.

Specifications
Processor: Intel 80C88-2
Dimensions: 211 × 100 × 260 mm (H × W × L)
Weight: 4.1 kg
Power consumption: 10 W
Qualification: MIL-STD-810C, -461B

Status
The DGCU is in production and in service in Israeli F-4 aircraft.

Contractor
RSL Electronics Ltd.

Digital Temperature Control Amplifier (DTCA)

Type
Aircraft system control, power plant.

Description
RSL produces a Digital Temperature Control Amplifier (DTCA) as a form-fit replacement for the analogue exhaust temperature (T5) controller coupled with the engine Health and Usage Monitoring System (HUMS) installed in GE J-79 (DTCA J-79) and J-85 (DTCA J-85) turbojet engines. The DTCA is fully qualified by GE.
  The main features of the DTCA include:
- Engine environment qualified
- Compatible with J-79 17/15/19/J1E and J-85-5/13/21 engines
- Dual-redundant electronics (two independent controllers); failure triggered channel switching
- Embedded HUMS (monitors engine and afterburner operating time, ignitions counting and exceedance)
- Simple T5 calibration procedure
- Acceleration reset switch by software.
  Advanced self-BIT and diagnostics:
- Tests all DTCA functions, modules and engine sensors
- Automatic fault isolation
- Optional pilot-initiated recording (DTCA J-85)
- Optional extended monitoring capabilities – up to 6 new inputs (DTCA J-85)
- Automatic monitoring and recording of all engine starts and exceedances.
  GSE package (DTCA J-79):
- Hand-held Test and Data Extractor Unit (TDEU)

- Flight-Line Tester (FLT)
- D Level Test Station (DLTS), with dedicated Expert System software
- PC-based workstation for ground evaluation and diagnostics of braking system. GSE package (DTCA J-85):
- Standard USAF GSE TDEU (NSN 4920-01-424-8886DQ)
- Optional laptop configuration of the same TDEU
- Dedicated Expert System software
- Enables individual testing of engine components (alternator, thermocouples, and so on)
- PC-based workstation for ground evaluation and diagnostics of braking system
- Flightline test of DTCA does not require aircraft power.

Specifications
DTCA J-79
Dimensions: 125 × 73 × 320 mm (H × W × L)
Weight: 2.7 kg
MTBF: >10,000 h
Power consumption: 10 W
Qualification: MIL-STD-810C, -461B, -704A

DTCA J-85
Processor: Intel 80C196
Dimensions: 64 × 115 × 292 mm (H × W × L)
Weight: 1.6 kg
MTBF: >17,000 h
Power consumption: 7 W
Qualification: MIL-STD-810C, -461B, -704A

Status
The DTCA is in production and in service. The DTCA J-79 is installed in F-4, F-104 and Kfir aircraft operated by the Israeli, German, Turkish and Japanese airforces. The DTCA J-85 is installed in T-38 and F-5 aircraft.

Contractor
RSL Electronics Ltd.

Mini Total Health and Usage Management System (MT-HUMS)

Type
Onboard system monitor.

Description
Mini T-HUMS (MT-HUMS), a derivative of T-HUMS for helicopters (see separate entry), is designed engine monitoring in Unmanned Air Vehicles (UAVs) and utilises the same core technology as its larger cousin.

MT-HUMS features a smart array of sensors coupled with powerful processing to conduct in-flight diagnostics and prognostics of the power plant, including gearbox, accessories and propeller. MT-HUMS provides the UAV ground station with mission-related advisories while airborne and post-landing diagnostics to facilitate cost-effective maintenance.

The system also facilitates condition-based maintenance, which reduces the overall maintenance effort and thus cost by enabling repair while damage is still minor, thereby increasing Mean Time Between Failures (MTBF) and reducing Mean Time To Repair (MTTR).

The open architecture of the system facilitates customer-defined configuration options, including propeller balancing post-mission debrief.

Direct operational/maintenance benefits of the system include:
- Improved flight safety
- Improved mission reliability and effectiveness
- Fewer In-Flight Shutdowns (IFSD)
- Improved fleet availability
- Lower life cycle costs
- System supports Flight Operation Quality Assurance (FOQA) programmes
- Simplified logistics
Other features of the system include:

- In-flight monitoring, recording and diagnostics of power plant(s)
- Vibration analysis (engines, gearbox, drive chain/propeller); facilitates on-the-wing balancing of engine/fan/propeller
- On-the-wing balancing
- Fatigue cycles and tracking of engine starts
- Flight regime recognition
- Optional airframe load and stress monitoring
- Optional flight controls monitoring
- Optional video and voice recording
- Optional integral GPS module
- Configurable data extraction – LAN, PCMCIA card, IR, wireless communications
- Wide interface spectrum – MIL-STD-1553B, RS-232/422, ARINC 429
- Diagnostics/prognostics using Expert System, Neural Network and Fuzzy Logic solutions
- User-friendly tools for effective data mining
- Vibration analysis tools
- Data archiving
- Web-based ground system
- MT-HUMS is fully MIL-STD-810E and -461B compliant.

Status
In production. MT-HUMS has been specified by the Israeli Air Force (IAF) for its UAVs.

Contractor
RSL Electronics Ltd.

Modular MultiRole Computer (MMRC)

Type
Avionic computer.

Description
The MMRC performs tasks for a full avionic suite. The MMRC is a single-multimodule computer designed to handle a broad spectrum of tasks previously performed by several individual units.

It is easily upgradeable to accept new systems and capabilities, and has built-in growth potential in processing capability, memory and interfaces.

Each module of the MMRC performs a different function: fire control; stores management; display processor (MFDs, MFCDs, HUD, HMD); display and sight helmet; integrated communication radio navigation and identification system; digital image processing and communication system; and video and VTR controller.

The MMRC is also the central element of the HALO advanced helicopter avionics suite offered as a flexible system upgrade for a wide range of helicopters.

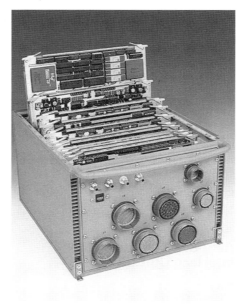

The Modular MultiRole Computer (MMRC)
0504266

Status
In service in AMX advanced trainer aircraft.

Contractor
Elbit Systems Ltd.

Total Health and Usage Management System (T-HUMS)

Type
Onboard system monitor.

Description
T-HUMS is designed for turbofan, turbojet and turboshaft engine monitoring in commercial and military fixed- and rotary-wing aircraft. Two further variants of the system offer extended capabilities specifically for helicopters, tilt-rotors and Unmanned Air Vehicles (UAVs). T-HUMS features a smart array of sensors coupled with powerful processing capabilities, to conduct in-flight tracking, trend monitoring and exceedance evaluation, as well as diagnostics and prognostics of power plant, rotors system and other critical aircraft systems. T-HUMS provides the pilot with mission-related advisories while airborne, and maintenance crews with accurate diagnostics for cost-effective maintenance.

The open architecture of the system facilitates customer-defined configuration options, including propeller track and balancing, an optional crash survival memory module, video and voice recording for post-mission debrief and a pilot control panel to form the basis of a flight data recording system.

Direct operational/maintenance benefits of the system include:
- Improved flight safety
- Improved mission reliability and effectiveness
- Fewer In-Flight Shutdowns (IFSD)
- Improved fleet availability
- System supports Flight Operation Quality Assurance (FOQA) programmes
- Simplified logistics
The system also facilitates condition-based maintenance, which reduces the overall maintenance effort and cost by enabling repair while damage is still minor in nature, increased aircraft MTBF and reduced MTTR.

Other features of the system include:
- In-flight monitoring, recording and diagnostics of power plant(s)
- Vibration analysis (engines, gearbox, drive chain/propeller); facilitates on-the-wing balancing of engine/fan/propeller
- Fatigue cycles and tracking of engine starts
- Flight regime recognition
- Optional hot start rejection
- Optional airframe load and stress monitoring
- Optional flight controls monitoring
- Optional video and voice recording
- Optional integral GPS module
- Configurable data extraction – LAN, PCMCIA card, IR, wireless communications
- Wide interface spectrum – MIL-STD-1553B, RS-232/422, ARINC 429
- Diagnostics/prognostics using Expert System, Neural Network and Fuzzy Logic solutions
- User-friendly tools for effective data mining
- Vibration analysis tools
- Data archiving
- Web-linked RSL support
- Optional mission debriefing workstation applications
- T-HUMS is fully MIL-STD-810E and -461B compliant

Status
As of mid-2006, in production and in service with the Israeli Air Force (IAF) AH-64 Apache and CH-53 helicopters, with the UH-60 to follow. T-HUMS is also offered for CH-47, Lynx, Puma and other rotary-wing platforms.

Contractor
RSL Electronics Ltd.

Italy

Advisory panel for the NH90

Type
Avionic crew alerting system.

Description
Logic produces an Advisory Panel (APN) for NH Industries NH90 helicopters. The panel has controls for managing visual test indicators, top recorder, landing gear position and acoustic alarms shutdown. It is an NVG-compatible unit that uses latest generation LEDs for the panel backlight to give improved reliability.

Status
In service with NH Industries NH90 helicopters.

Contractor
Logic SpA.

Aermacchi M346 Engine Fuel Panel (EFP)

Type
Aircraft sensor, power plant/Fuel Quantity Indication (FQI).

Description
The Logic M346 Engine Fuel Panel (EFP) displays high- and low-pressure compressor speeds, engine gas temperature and fuel quantity. It receives its data from a MIL-STD-1553 databus interface and also has an RS422 serial interface. Data acquisition supervision and coherency checks are performed by the microprocessor. The display is compatible with NVGs.

Status
In production and in service in development Aermacchi M346 aircraft.

Contractor
Logic SpA.

Aermacchi M346 Fuel Control System

Type
Aircraft system control, fuel.

Description
Logic manufactures the fuel control system used in the Aermacchi M346. It has passive capacitive probes and thermistor-based low-level sensors both fed by the control unit. The system offers on ground and in-flight partial and total fuel quantity calculation with flight attitude correction. It manages valves and pumps for automatic fuel balancing and manages on land and in-flight refuelling, allowing selection of the fuel quantity to be supplied. Tank pressure is controlled by dedicated valve management.

Status
In production and in service on development Aermacchi M346 aircraft.

Contractor
Logic SpA.

Aermacchi M346 Generator Control Unit (GCU)

Type
Aircraft system control, electrical.

Description
The Logic Generator Control Unit (GCU) is fully digital with a TI DSP TMS320 core. The control law implemented by the software ensures repeatable performance over time and gives RMS voltage regulation independent from waveform distortion. Protection is provided from short circuit, over-current, over-voltage, under-voltage, differential overload and under-frequency. The GCU also offers ground power unit management with dedicated over-voltage, under-voltage and over/under-frequency protection, and generator field control and protection. A dedicated I/O link co-ordinates busbar crossfeed with a redundant GCU. There is also an RS422 interface with the MISCO vehicle computer (see separate entry) for the transmission of electric equipment status.

Status
In production and in service on development Aermacchi M346 aircraft.

Contractor
Logic SpA.

Airborne Strain Counter (ASC)

Type
Onboard system monitor.

Description
The Airborne Strain Counter (ASC) unit, is designed to determine the different load spectra and/or local stresses that arise during operation in all aircraft structures submitted to variable loads. The unit is suitable for monitoring the fatigue life of complex structural elements, where the local stresses cannot be easily correlated to the standard usage parameters.

The ASC automatically accomplishes acquisition of a maximum of eight parameters, received from strain gauges or accelerometers positioned on the most significant points of the aircraft structure; conversion of the acquired data into digital form; processing of the data and outputting of a stress matrix; storage of the data in a non-volatile memory; interfacing with the dedicated ground support unit to allow data transfer and subsequent processing; supply of power to the sensors; and performance of self-test to check serviceability of the equipment.

The ASC features eight fully independent channels for acquisition, preprocessing and storage of data. Each channel is capable of managing the interface signals of the associated strain gauge or accelerometer. Each channel amplifies the analogue signal received from the sensor and converts it into a digital signal through an A/D converter, providing information on number of stress cycles experienced by the structure, mean values of the stress for every cycle and peak-to-peak value of the stress for every cycle.

Information is stored in a non-volatile memory on aircraft serial number, card serial number, ASC serial number, total number of stress cycles, status word, stress matrix, total number of minutes of aircraft motion and total number of minutes of operation in the presence of failures.

An RS-422 databus is used to transfer the data stored from the measurement channels to the ground support unit.

At system power on, each channel performs a self-test to check hardware and software integrity.

Status
In service on the Aermacchi MB-339 and the Aermacchi/Alenia/Embraer AMX.

Contractor
Logic SpA.

ANV-803 Computer and Interface Unit (CIU)

Type
Avionic computer.

Description
The ANV-803 is designed to handle real-time environments in modern avionic systems where

The Marconi ANV-803 computer and interface unit 0504159

powerful computation capability, small size and low weight are required. The powerful RISC engine and modular design of the ANV-803 meet a wide range of applications while keeping a simple monoprocessor architecture, resulting in a good price/performance ratio and maintenance cost.

System options include a complete range of input/output support modules and a comprehensive set of software development tools for Ada programming. Different interface cards are available, including MIL-STD-1553B BC/RT/BM, ARINC 429, discrete, serial and analogue.

Specifications
Processor: RISC R3000 family of 32-bit architecture; maths co-processor
Software: Ada
Memory provision: Cache, flash EPROM, static RAM, non-volatile RAM, RS-232 lines, timers, interrupt handling, clock generation, reset, watchdog, bus timeout and power fail logic
Dimensions: 124 × 194 × 380 mm
Weight: 7.0 kg
Power supply: 28 V DC or 115 V AC, 400 Hz
Interface: High-performance 32-bit bus with CPU module

Contractor
Marconi Selenia SpA.

Armament computers for the Eurofighter Typhoon

Type
Aircraft system control, Stores Management System (SMS).

Description
Selex Sensors and Airborne Systems (formerly Galileo Avionica) is responsible for the armament computers for the Eurofighter Typhoon, including the following elements:

Safety-critical armament controller
Dual-channel, fully redundant digital computer, dedicated to the management of all safety-critical functions of the armament system, such as release, firing and jettison of the external stores.

Specifications
Weight: 10 kg
Power Consumption: 90 W
Size: ¾ ATR Long

Non safety-critical armament controller
Dual-channel, fully-redundant digital computer, dedicated to the management of all non-safety-critical functions of the armament system, such as pre-selection of missiles, computing of weapon aiming co-ordinates etc.

Specifications
Weight: 10 kg
Power Consumption: 125 W
Size: ⅝ ATR Long

Distribution unit

This unit is dedicated to the distribution of all the range of RF signals to the external stores, e.g. air-to-air and air-to-surface missiles. It is fully compliant with MIL STD 1760, Class 2.

Specifications

Weight: 5.4 kg
Power Consumption: 40 W
Size: $^5/_8$ ATR Long

Status

In production for Eurofighter Typhoon.

Contractor

Selex Sensors and Airborne Systems, Radar Systems Business Unit.

CM 115 E voice/data encryption equipment

Type

Data processing system.

Description

The CM 115 E equipment is a digital ciphering device that can be configured for voice or data applications. It can be used in conjunction with HF, VHF and UHF radios providing either narrow- or wide-band channels. The voice analogue signal is internally converted to a digital signal, encrypted through a high-grade security algorithm, modulated and sent to the external radios for transmission. The crypto algorithm is a proprietary Marconi Selenia Communications algorithm that can be customised by the end user. Plain mode is also available. The operator can select the working mode as well as the key variable. Rapid erasure of all keys is possible. The equipment can store up to 60 key variables. Key loading is performed by means of portable electronic transfer devices (tape reader, fill gun). The chassis is a lightweight aluminium alloy casting composed of three sub-assemblies, a front panel, main chassis and rear panel. The equipment is based on modular assembly allowing easy module substitution and maintenance. The CM 115 E can be provided with the following Ancillary Units:

- CP1144/D Remote Control Unit (RCU);
- MT1133/D mounting slide.
  The equipment operates in the following modes:

Narrow band

Voice mode:

- Headphone interface (Voice in/out, 0 dBm/ 600 Ω, PTT);
- Analogue-to-Digital Voice conversion (LPC10 Vocoder at 2,400 bps);
- Digital Encryption;
- Analogue modulation (modem set in voice mode);
- Radio interface.

Data mode:

- DTE interface (electrical interface V.10);
- Synchronous encryption and transmission at 300, 600, 1,200, 2,400 bps;
- Analogue modulation (modem set in data mode);
- Radio interface.

Wide band

Voice mode:

- Headphone interface (Voice in/out, 0 dBm/ 600 Ω, PTT);
- Analogue-to-Digital Voice conversion (CVSDM 16 kbps);
- Digital Encryption;
- Radio interface (BB/Diphase modulation).

Data mode:

- DTE interface (electrical interface V.10);
- Synchronous encryption and transmission at 8 or 16 kbps;
- Half-duplex/full-duplex transmission mode.
  The CM 115 E also incorporates a Crypto Ignition Key (CIK), anti-tampering mechanisms, a battery for key holding, crypto alarms, RCU and BITE functions.

Specifications

Applications: Secure voice and data communications over HF/VHF/UHF radio channels

Narrow-band mode (HF)
Analogue interface (plain and secure): 0 dBm ±3 dB
Data interface (plain and secure): V.10/V.11 selectable
Baud rate: 300, 600, 1,200, 2,400 bps in synchronous mode
Voice coding: LPC10 at 2,400 bps

Wide-band mode (UHF/VHF)
Analogue interface (plain): 0 dBm ±3 dB
Data interface: V.10/V.11 selectable
Baud Rate (plain): 300, 600, 1,200, 2,400 bps, 8, 12, 16 kbps in synchronous mode
Baud Rate (secure): 16 kbps in synchronous mode
Voice Coding: CVSDM at 8, 12, 16 kbps

Environmental
Temperature
　Operating: −40 to +55°C
　Transport/storage: −55 to +70°C
Humidity: 95% max between 25 and 55°C, non-condensing (MIL-STD-2036 paragraph 5.1.2.7)
Altitude
　Operating: up to 4,270 m
　Transport/storage: up to 10,700 m

Electrical
EMI/EMC: According to MIL-STD-461B part 4 class 3
Back-up battery: BA1372/U 6.5V or equivalent, lasting 1 year in normal condition (replaceable from the front panel)

Physical
Dimensions: 146 × 123 × 136.5 mm (W × H × D)
Weight: ≤5.3 kg
Power: 28 V DC; 30W

Status

In production and in service.

Contractor

Marconi Selenia SpA.

De-ice Controller Board (DCB) for NH90

Type

Aircraft system control, ice detection and anti-ice.

Description

Logic, in partnership with Artus, has developed a De-ice Controller Board (DCB) for the NH Industries NH90 helicopter de-icing system. The system has redundant architecture based on two Power De-icing Units (PDUs) and two DCBs and is designed to prevent icing on the rotor blade and horizontal stabiliser. The DCB manages each PDU section power and implements over-current and over-voltage protection. There is also redundant over-temperature protection against the mats overheating. The DCB BITE includes the PDU BITE. The hardware platform is based on the Motorola MC68332 and the de-icing sequences are easily programmable using software certified to DO-178B Level B.

Status

In service on NH Industries NH90 helicopters.

Contractor

Logic SpA.

Emergency Flotation Unit (EFU) for NH90

Type

Aircraft system control, emergency/survival.

Description

The Logic Emergency Flotation Unit (EFU), built for the NH Industries NH90 helicopter, activates the inflation cartridges in case of emergency ditching. The inflation sequence is automatically

activated by means of immersion sensors; there are double input commands required for manual inflation so that no single malfunction can cause unwanted activation of the inflation sequence. The system has hot twin architecture, which is fully redundant with independent power supplies.

Status

In service on NH Industries NH90 helicopters.

Contractor

Logic SpA.

Engine Condition and Control System – Boeing H46

Type

Aircraft system control, power plant.

Description

Logic produces an Engine Condition and Control System (ECCS) which is installed in US Marine Corp and US Navy CH46 helicopters. The system consists of two control units and one quadrant to adjust the fuel control unit in accordance with the power demand. An analogue control system keeps the Nr constant at 100 per cent and keeps the torque of the engines matched throughout the different power demands. There is an independent and redundant manual control system in case the normal system fails.

Status

In service on US Marine Corps and US Navy CH46 helicopters.

Contractor

Logic SpA.

External Fuel Control (EFC) system for the NH90

Type

Aircraft system control, fuel.

Description

The Logic External Fuel Control (EFC) system for the NH90 helicopter is composed of one control unit and one turbine flow meter. There are ARINC 429 and discrete I/O interfaces with the main fuel system. The system acquires data from the flow meter and calculates instantaneous fuel flow and total fuel mass used. There is fuel control valve management and fuel pump activation and protection such that no single failure can lead to a pump overheating. The same unit operates on 115 V AC or 28 V DC pumps.

Status

In service on NH Industries NH90 helicopters.

Contractor

Logic SpA.

Folding Automatic Controller (FAC) for NH90

Type

Aircraft system control, power plant.

Description

Logic produces a Folding Automatic Controller (FAC) that is installed into the naval version of NH Industries NH90 helicopter. The unit has fully authority management of main and tail rotor folding/unfolding operations. The system has triple redundancy position BITE for safety and mission performances and in-system re-programmable hardware sequencers (no software) that are validated as per DO-254. There are solid-state drivers for the DC actuators and MIL relays for the AC power drivers. There is independent over-current and short circuit protection for each DC and AC actuator and a fully independent power inhibition circuit.

Status

In service on naval versions of the NH Industries NH90 helicopter.

Contractor

Logic SpA.

Landing gear control for Aermacchi M346

Type

Aircraft system control, undercarriage/wheelbrakes.

Description

Logic produces a landing gear control lever subsystem to control the landing gear extension and closure on the Aermacchi M346. It is an electromechanical unit with NVIS-compatible front panel back lighting.

Status

In production and in service on development Aermacchi M346 aircraft.

Contractor

Logic SpA.

Stores management system for NH90

Type

Aircraft system control, Stores Management System (SMS).

Description

Logic, in partnership with Marconi and Smiths, has designed and developed a Stores Management System (SMS) for use on the NH Industries NH90 helicopter. The Release Control Panel (RCP) is designed fully by Logic, while the Storage Management Unit chassis and backpane, EMI protection board and lighting, and external load control board are also produced by the company.

Status

In service on NH Industries NH90 helicopters.

Contractor

Logic SpA.

Japan

Altitude computer

Type

Aircraft flight computer.

Description

The altitude computer receives pitot and static pressure and total temperature inputs for the calculation of aircraft altitude. It has a microprocessor and performs digital calculations. The altitude computer includes BIT and performance monitoring.

Status

In production for the T-2 trainer.

Contractor

Shimadzu Corporation.

AP-120 DFCS

Type

Flight Control System (FCS).

Description

AP-120 Digital Flight Control System (DFCS) is a high-grade comprehensive flight control system which consists of Digital Flight Control Computer (DFCC), two altitude controllers, two accelerometers, main and sub control panels. The DFCC, as a core unit, has both autopilot and flight director functions. It features dual-redundant fail-operative/fail-passive capability and is inertially smoothed to give high-accuracy stable altitude-hold capability. High maintainability is ensured through extensive use of an In-Flight Performance Monitor (IFPM), Built-In Tester (BIT) and non-volatile maintenance memories, and the use of selected high-reliability parts and advanced thermal design techniques. The system can be expanded to provide a MIL-STD-1553B databus interface and other enhancements.

Status

As of 2004, the AP-120 was in production and service in the Lockheed/Kawasaki P-3C aircraft and its derivatives.

Contractor

Tokyo Aircraft Instrument Co Ltd.

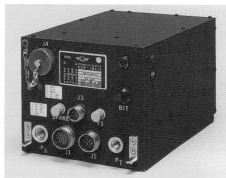

The altitude computer is in production for the T-2 trainer   0504160

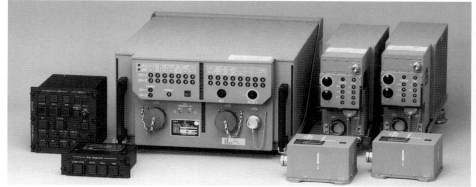

The Tokyo Aircraft Instrument Co AP-120 DFCS   0504089

Russian Federation

AGB-96, AGB-98 and AGR-100 horizon gyros

Type

Aircraft sensor, attitude/Angle of Attack (AoA).

Description

The AGB-96, AGB-98 and AGR-100 horizon gyros are designed for aircraft pitch/roll indication.

Aviapribor AGR-81 (left) and AGR-29 (right) standby horizon gyros   0018206

Aviapribor AGR-100, AGB-98, AGB-96 horizon gyros (left to right)   0018207

Specifications
Compliance: ENLGS P 8.1.2

	AGB-96	AGB-98	AGR-100
Readiness time	2.0 min	2.0 min	2.0 min
Angular range			
(roll)	±360°	±360°	±360°
(pitch)	±85°	±85°	±85°
Power			
(gyro) all systems:	27 V DC		
(sensors) all systems:	36 V AC, 400 Hz, 0.8 A		
Temperature range all systems:	−20 to +70°C		
Dimensions	105 × 105 × 250 mm	85 × 85 × 250 mm	61 × 61 × 220 mm
Weight	2.2 to 2.5 kg	1.8 to 2.0 kg	1.5 kg

Contractor
Aviapribor Corporation.

AGR-29 and AGR-81 standby gyros

Type
Aircraft sensor, attitude/Angle of Attack (AoA).

Description
The AGR-29 and AGR-81 standby horizon gyros are designed for aircraft pitch/roll indication. The AGR-29M can be used as a remote indicator of pitch and roll from another vertical sensor.

Contractor
Aviapribor Corporation.

Specifications
Compliance: ENLGS P 8.1.2

	AGR-29	AGR-81
Readiness time	2.0 min	2.0 min
Angular range		
(roll)	±360°	±360°
(pitch)	±360°	±360°
Power		
(gyro)	27 V DC	
(sensors)	36 V AC, 400 Hz, 1.0 A	
(back lighting)	6 V DC	
Temperature range	−30 to +70°	
Dimensions	105 × 105 × 250 mm	85 × 85 × 250 mm
Weight	up to 3.1 kg	2.2 to 2.5 kg

Aircraft Systems Control System (ASCS) – Electronic Flight Engineer

Type
Aircraft system control, integrated.

Description
GosNIIAS has developed the Electronic Flight Engineer: Aircraft Systems Control System (ASCS), as an outgrowth of its work with Rockwell Collins and Smiths Industries on the following elements of the Il-96 M/T avionics development programme: the Integrated Display System (IDS), Flight Management Computer System (FMCS), Automatic Flight Control System (AFCS), Thrust Management Computer (TMC) and Central Maintenance Computer System (CMCS); and on experience gained in production of the Collins Traffic alert and Collision Avoidance System (TCAS).

The Electronic Flight Engineer is intended for aircraft systems automation and serviceability/maintenance enhancement. The automation element permits reduction of the flight crew to two.

The Electronic Flight Engineer automates control of over 20 aircraft systems, including: hydraulics, electrical power, air conditioning, pressure regulation and other systems. The system comprises three computer units (ASCC) and eight MFUs.

Status
Certified by the US FAA in early 1999.

Contractor
GosNIIAS State Research Institute of Aviation Systems.

ASShU-204 fly-by-wire flight control system

Type
Flight Control System (FCS).

Description
The ASShU-204, together with the SUR-204 surfaces control system, provides for full-authority aircraft control, processing input electrical signals for aerodynamic control surface deflection in accordance with the specified control laws, and generating output signals to the relevant aircraft control systems, to provide:
- Aircraft control and pitch trim, in concert with VSUP-85 parameters (see separate entry)
- Static and dynamic characteristics, together with lateral and longitudinal stability computations in accordance with ENLGS
- Automatic warning or prevention of aircraft limit exceedence depending on flight mode.

The ASSHU-204 generates electrical signals to provide full control in pitch, roll and yaw (together with trim functions), spoiler control for roll and lift-dump during the landing roll and autobraking. The system provides continuous monitoring of control signal status, including threshold, and outputs these signals to the SUR-204. The ASSHU-204 also includes full internal BIT of its own units as well as the associated SDU-6 and PAB-204 systems, with results (pass/fail) output to the PPO-5-1 maintenance panel. The ASSHU-204 comprises the following subsystems:
- BVUU-1 stability and control computers: 4 units
- BUC-8-1 analogue units: 3 units
- PPO-5-1 pre-flight maintenance panel: 1 unit
- DLUC-3 linear acceleration: 6 units
- DUSU-M angular rate sensors: 15 units
- DSP-6 position sensors: 15 units
- BT-5 transformer units: 3 units.

Specifications
Reliability: ≥4,000 h
Design life: 45,000 h in 20 years

Power:
4 independent AC power supplies, 115/200 V, 3-phase, 400 Hz
4 independent DC power supplies, 27 V
Total system weight: 117.4 kg

Status
Fitted to Tu-204, Tu-204-100 and Tu-204-120 aircraft.

Contractor
Aviapribor Corporation.

ASShU-334 fly-by-wire flight control system

Type
Flight Control System (FCS).

Description
The ASShU-334 fly-by-wire flight control system is a digital and analogue system designed to provide stability and control for regional aircraft

ASShU-334 fly-by-wire flight control system

0018242

during manual, automatic and override manual flight conditions. System components are as follows:

- Stability and control computers: 4 units
- Analogue units: 2 units
- Pre-flight maintenance panel: 1 unit
- Linear acceleration: 6 units
- Angular rate sensors: 18 units
- Position sensors: 7 units
- Transformer units: 3 units.

Specifications
No of redundant channels: 6
Weight: 90 kg
Power: 500 W

Status
Fitted to the An-334 twin turbofan medium-range airlifter aircraft.

Contractor
Aviapribor Corporation.

ASUU-96 automatic control and stability augmentation system   0018241

ASUU-96 automatic control and stability augmentation system

Type
Flight Control System (FCS).

Description
The ASUU-96 automatic control and stability augmentation system is a digital-analogue system, designed for the Il-96 aircraft; it comprises the following units:

- BVUU-1 control and stability computer unit: 5 units (4 for ASUU-96M)
- BUK-6 and BUK-7 control monitor unit: 4 units (2 each)
- RA-98 actuator unit: 8 units
- PPO-5-1 pre-flight maintenance panel: 1 unit
- DLUC-3 linear acceleration transducer: 4 units
- DPS-6 triplex position pick-off: 8 units (4 for ASUU-96M)

- DUSU-M angular rate sensor: 12 units
- RGM-18, 19, RM-50 Mounting racks: 4 units (2 RGM-19)
- BEP-2 electrical drive unit: 1 unit
- MVE-25D5 electrical actuator: 1 unit
- Transformer: 2 units
- RGM-18 mounting rack: 1 unit
- RGM-19 mounting rack: 2 units
- RM-50 mounting rack: 1 unit.

The system provides the following functions and protections:

- Fulfills stability and controllability requirements
- Restricts rudder deviation to reduce loads on the vertical tail
- Operates in parallel with the VSUT system (see separate entry)
- Adjusts airframe loading when aircraft approaches limiting values of Angle-of-Attack (AoA) or loading (n).

The ASSU-96 includes automatic and manual BITE, controlled through BVUU-1-1 computer and PPO-5-1 maintenance panel.

Specifications
No of redundant channels: 4
Interface: Serial bipolar code GOST 18977-79 and RTM 1495-75 Mod3 (ARINC 429 equivalent); analogue discrete
Total system weight (all components): 220 kg (204 kg ASUU-96M)
Power: 200/115 V AC, 400 Hz; 27 V DC, 14 A
MTBF: >300 h

Status
Fitted to all variants of the Il-96 wide-bodied airliner.

Contractor
Aviapribor Corporation.

BTsVM-386 airborne digital computer

Type
Avionic computer.

Description
The BTsVM-386 airborne digital computer is a 32-bit, PC-compatible, multitask, modular, open architecture system, which can be extended by peripheral and graphics coprocessors. Three processor options are available. Program languages: C++; Modula-2; Assembler.

Status
In production and in service.

Contractor
Ramenskoye Design Company AO RPKB.

BTsVM-386 airborne digital computer   0015276

Specifications
Memory:
(ROM)   1.5 Mbytes (extended to 65 Mbytes)
(static RAM)   0.5 Mbytes (extended to 64 Mbytes)
(flash memory)   extended to 256 Mbytes
Input/output:
(ARINC-429)   32 independent input; 16 independent output
(multiplex MIL-STD-1553B)   3 channels
(event signals)   32 input; 16 output
MTBF: 10,000 h
Power: 115 V AC; 26 V DC, 90 W
Weight: 8 kg

	i386/387-20	i486DX-50	i860-25
Word length	32-bit	32-bit	64-bit
Speed			
(fixed point)	10 Mops	50 Mops	50 Mops
(floating point)	0.7 Mops	3.2 Mops	50 Mops
(graphics)			25 k polygons/s

DAP 3-1 air data sensor

Type
Aircraft sensor, pressure.

Description
The DAP 3-1 air data sensor is designed to measure local angle of attack, pitot and static pressures.

Specifications
Airspeed range: 150 to 1,600 km/h
Altitude: up to 30,000 m (98,400 ft)
Vane deflection range: ±90°
Temperature: ±60° C
Power: 115 V AC, 400 Hz (vane heating); 27 V DC (sensor)
Weight: 5 kg

Contractor
Aviapribor Corporation.

DAP 3-1 air data sensor   0103864

DAU-19 airflow direction transducer

Type
Aircraft sensor, Angle of Attack (AoA).

Description
The DAU-19 airflow direction transducer measures local angle of attack and outputs values via a proportional electrical signal. The unit is designed to be fitted to amphibians.

Specifications
Airspeed range: up to 900 km/h
Altitude: 1,000 m (3,280 ft)
Vane deflection range: ±30°
Error: ±0.25°
Power: 27 V DC, 20 W
Weight: 0.65 kg

Contractor
Aviapribor Corporation.

DAU-19 airflow direction transducer   0103865

EDSU-77 fly-by-wire flight control system

Type
Flight Control System (FCS).

Description
The EDSU-77 is designed to provide aircraft stability and control in pitch, yaw and roll for the An-70 tactical transport aircraft in manual, automatic and override control modes of flight operation; it comprises the following units:
- Stability and control computers: 4 units
- Analogue interfaces: 2 units
- Maintenance panel: 1 unit
- Linear acceleration sensors: 12 units

- Angular rate sensors (unified): 18 units
- Position sensor (triple): 20 units
- Emergency trim control: 1 unit.
  Electromechanical actuators and sensors as required

Specifications
No of redundant channels: 6
Weight: 250 kg
Power: 115 V AC, 400 Hz, 1,000 VA

Status
Fitted to An-70 propfan medium range, wide bodied, tactical transport aircraft.

Contractor
Aviapribor Corporation.

FCS-85 flight control system

Type
Flight Control System (FCS).

Description
FCS-85 is part of a standard digital avionics system designed for aircraft automatic control during flight and for director control during take-off and approach.

In conjunction with other avionics control systems FCS-85 provides control functions including:
- in conjunction with TCS-85 thrust control system, automatic control in altitude and heading, automatic hold of barometric altitude;
- with the control column, control of pitch, heading and roll;
- automatic approach and automatic landing in compliance with ICAO Cat IIIA requirements using ILS and MLS radio beacons;
- automatic approach to decision height in compliance with ICAO Cat II requirements using ILS, MLS, Landing System-50 (LS-50) radio beacons;
- automatic director approach to decision height in compliance with ICAO Cat I requirements using ILS, MLS, LS-50, SRNS radio beacons;
- automatic monitoring of FCS operation;
- prevention of exceedance conditions.

FCS-85 comprises the following LRUs: flight control computers (FCC-1) 3 linked units, each 8 MCU, ARINC 429 compatible; control panel (CP-56) 1 unit.

Specifications
FCC-1
Power: 115 V AC, 400 Hz, 150 VA; 36 V AC, 400 Hz, 1.5 VA; 27 V DC, 30 W
Weight: 12 kg
MTBF: 5,000 h
CP-56
Power: 115 V AC, 400 Hz, 75 VA; 6 V AC, 400 Hz, 12 VA; 27 V DC, 1 W
Weight: 8 kg
MTBF: 6,000 h

Status
In service in Il-96-300 and Tu-204 airliners.

Contractor
NIIAO Institute of Aircraft Equipment.
Moscow Institute of Electromechanics and Automatics.

FILS Fault Isolation and Localisation System

Type
Avionic Fault Warning System (FWS).

Description
FILS is the top level, of three levels, of onboard BITE and fault isolation system. It integrates the BITE activities and data management and storage of all major flight management systems, such as EFIS, FCS, TCS.

FILS is designed to: process fault diagnosis and isolation; output fault data on the databus and displays and ground test facilities; store data on faults for up to 10 flights and 30 faults per flight in non-volatile memory; input data on faults into aircraft utility systems.

FILS is a single LRU, with alphanumeric display, mounted in the flight engineer position, that complies with ARINC 604 and 734.

Specifications
ROM capacity: 40 kwords
Non-volatile memory: 16 kwords
Inputs; up to 25 channels (ARINC 429)
Outputs: up to 6 channels (ARINC 429); up to 12 discrete instructions
Weight: 10 kg
Power: 115 V DC, 400 Hz, 60 VA
MTBF: 5,000 h

Status
In service in Il-96-300 and Tu-204 airliners.

Contractor
NIIAO Institute of Aircraft Equipment.

Fly-By-Wire (FBW) Container

Type
Flight Control Computer (FCC).

Description
The FBW Container represents the latest generation of integrated modular avionics. It contains flight control, autopilot and air data modules. The Container interacts with sensors and servos, performing all tasks involved with manual and automatic flight control and augmentation. The FBW Container is an open system which and can be expanded to include indication systems,

Fly-By-Wire Container   0103866

EDSU-77 fly-by-wire control system for the An-70 aircraft   0018228

GPWS or other modules, depending on customer requirements. The system is compatible with a wide variety of aircraft types/missions.

Contractor
Aviapribor Corporation.

MVD-D1 air data module

Type
Aircraft sensor, pressure.

Description
The MVD-D1 air data module is designed to measure pitot and static pressures, calculate total air temperature (using data from a temperature sensor) and output electrical signals to onboard systems of the host aircraft utilising an ARINC-429 interface. The MVD-D1 module facilitates the calculation of altitude/airspeed parameters directly, without utilising conventional pressure systems, thereby introducing redundancy into the aircraft instrumentation system.

Specifications
Pitot pressure: 12 to 2,830 (±0.36) Hpa
Static pressure: 12 to 1,080 (±0.36) Hpa
Dimensions: 188 × 132 × 95 mm (with connector)
Temperature: –60° to +240°C
Power: 27 V DC; 13 W
Weight: 0.9 kg

Contractor
Aviapribor Corporation.

MVD-D1 air data module   0103868

PVD-2S/PVD-K air pressure probes

Type
Aircraft sensor, pressure.

Description
The PVD-2S provides pitot and static pressure information, while the PVD-2K probe provides full pitot pressure information to air data systems.

Specifications

	PVD-2S	PVD-K
Airspeed range	Up to 1,100 km/h	Up to 450 km/h
Altitude	Up to 30,000 m	Up to 7,000 m
Pitot error	±0.02 to ±0.07 q	±0.02 q
Static error	±0.025 to ±0.058 q	±0.016 q
Electrical	27 V DC	27 V DC
Power	140 W	120 W
Weight	0.4 kg	0.25 kg

Contractor
Aviapribor Corporation.

PVD-2S and PVD-K pressure sensors   0103870

Tachometer Indicators

Type
Aircraft sensor, speed.

Description
ElectroPribor Kazan Plant manufactures 14 types of tachometer for measuring engine shaft speed, calibrations can be either in revolutions per minute (rpm) or per cent.

Specifications
ITE-2TB
Operating temperature range: –60 to 60°C
Reading range: 10-110%
Error: +0.5%
Weight: 1.30 kg

ITE-2
Operating temperature range: –60 to +60°C
Reading range: 10-110%
Error: +0.5%
Weight: 0.95 kg

2TE15-1M
Operating temperature range: –60 to +60°C
Reading range: 1,000–15,000 rpm
Error: +75 rpm
Weight: 0.90 kg

Status
In service and in production for a wide variety of military and civilian fixed- and rotary-wing aircraft throughout the RFAS.

Contractor
ElectroPribor Kazan Plant.

ElectroPribor Kazan Plant tachometers   0018200

Temperature gauges

Type
Aircraft sensor, temperature.

Description
ElectroPribor Kazan Plant manufactures 10 types of temperature gauges and thermocouples for measurement of gas flow temperatures in turbojet and turboprop engines, and in the cylinder-head of piston engines.

Specifications
ITG-1
Measuring temperature range: +200 to +1,100°C
Error: +12°C
Weight: 0.76 kg

TBG-1
Measuring temperature range: +300 to +900°C
Error: +7°C
Weight: 0.76 kg

ElectroPribor Kazan Plant temperature gauges
0018199

TUT-9
Measuring temperature range: –40 to +300°C
Error: +14°C
Weight: 0.25 kg

Status
In service and in production for a wide variety of military and civilian fixed- and rotary-wing aircraft throughout the RFAS.

Contractor
ElectroPribor Kazan Plant.

VSUP-85 flight control computer system

Type
Flight Control System (FCS).

Description
The VSUP-85 flight control computer system is designed to generate control data for all phases of flight of modern transport aircraft types; it comprises two types of line-replaceable units: flight control computer (three units) and cockpit control unit (one unit).

Specifications
Compliance: NLGS, DO-160
Weight:
(Tu-204) 51.5 kg
(Il-96) 36 kg
Power: 115 V AC, 400 Hz, 525 VA
27 V DC, 30 W
36 V AC, 400 Hz, 4.5 VA
6 V AC, 400 Hz, 25 VA
MTBF:
(computer) 5,000 h
(control unit) 6,000 h

Status
Fitted to the Tu-204 twin turbofan medium-range airliner, and the Il-96 wide-bodied airliner. The -85 element of the designator indicates that the original development contract was awarded in 1985.

Contractor
Aviapribor Corporation.

VSUP-85 flight control computer system
0018239

VSUPT-334 flight and thrust control computer system for Tu-334

Type
Aircraft system control, power plant.

Description
The VSUPT-334 flight and thrust control computer system performs both flight control and thrust control functions for the Tu-334 medium-range airlifter and its twin turbofan ZMKB Progress D-436T1 engines; it comprises the following line-replaceable units: flight/thrust computer (three units), thrust unit, cockpit control unit, thrust actuator.

Specifications
Compliance: NLGS, DO-160
Weight: 67 kg
Power: 115 V AC, 400 Hz, 525 VA
27 V DC, 400 W
36 V AC, 400 Hz 4.5 VA
6 V AC, 25 VA

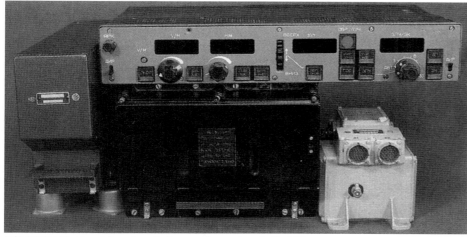

VSUPT-334 flight and thrust control computer system for Tu-334   0018238

VSUT-85 thrust control computer system
0018237

MTBF:
(computer) 5,000 h
(cockpit control unit) 6,000 h
(thrust actuator) 5,000 h

Status
Fitted to the Tu-334 twin turbofan medium-range airlifter.

Contractor
Aviapribor Corporation.

VSUT-85 thrust control computer system

Type
Aircraft system control, power plant.

Description
The VSUT-85 thrust control computer system is designed to provide automatic control/hold of indicated airspeed or Mach number by controlling power plant thrust on Tu-204 and Il-96 aircraft. The system operates in the following modes/flight phases:
• Take-off (derate/non-derate)
• IAS/Mach speed hold
• Vertical speed hold
• Flight level change
• Approach
• Go-Around (GA).
   The system comprises the following line-replaceable units: thrust control computer (two units), thrust control, thrust actuator.

Specifications
Compliance: NLGS-3, DO-160
Dimensions:
BVUT-1 190 × 387 × 194 mm
BUPRT-96 194 × 368 × 90 mm
MRT-96 234 × 264 × 205 mm
MRT-204 175 × 138 × 128 mm
PUT-3 146 × 80 × 176 mm
Weight:
VSUT-85-1 (Il-96-300) 38 kg
VSUT-85-2 (Tu-204-120) 32 kg
VSUT-85-3 (Tu-204-100, Tu-214) 32 kg
Power:
Two channels, 115 V AC, 400 Hz, 120 VA
Two channels, 27 V DC, 60 W
Two channels, 36 V AC, 400 Hz, 1 VA
One channel, 6 V AC, 400 Hz, 4 VA
MTBF:
(computer BVUT-1) 5,000 h
(thrust actuator MRT-204) 5,000 h
(thrust control panel PUT-3) 8,000 h
Interface: RTM 1495 Mod3 (ARINC 429), analogue, discrete

Status
Fitted to the Tu-204-100, -120 and Tu-214 twin-turbofan medium-range airlifters and the Il-96 four-turbofan, wide-bodied airliner. The -85 element of the designator indicates that initial development contract award was in 1985.

Contractor
Aviapribor Corporation.

ZBM Baget-53 integrated airborne digital computer

Type
Avionic computer.

Description
The Ramenskoye Design Company's ZBM Baget-53 integrated airborne digital computer employs the VME bus system and Vx Works software. The central processor module uses R3000/R3010 25 MHz microprocessors. Static RAM is not less than 1 Mbyte and 2 Mbyte of ROM is included, together with a 256 Mbyte volatile memory module and graphics processor module. The system offers 32 signal inputs and 32 signal outputs. It also complies with GOST standards for control of multifunctional control/display modules and fibre optic communication systems.
   Overall dimensions are as specified by 2.5K, 3K and 4K GOST 6765.16-87 standards.

Status
In service. Production status unknown.

Contractor
Ramenskoye Design Company AO RPKB.

Sweden

JAS 39 Gripen aircraft interface unit

Type
Avionic interface system.

Description
The JAS 39 aircraft interface unit, which is based on a Pentium® computer system, is used to interface aircraft sensors in the JAS 39 Gripen to the main electronic system. The unit contains logic to handle warning information and operate the control panel.
   Power supply control of the electronic system, normal control of the undercarriage (extension and retraction) and operation of a crash recorder are other functions of the unit. The main part of the logic in the unit is provided by software. Critical functions are implemented in dedicated logic.
   The aircraft interface unit features modular design, LRU concept to facilitate maintenance, chassis integrated power supply for optimal heat dissipation, SMT for increased packing density and built-in redundancy.

Status
In production and in service in the JAS 39 Gripen.

Contractor
SaabTech AB.

JAS 39 Gripen Data Entry Panel (DEP)

Type
Data processing system.

Description
The Data Entry Panel (DEP) facilitates inputting data into the avionics system and, to a limited extent, it also enables the output of certain data that is displayed to the pilot. The front panel of the DEP has integrated push-button switches, indicators, a rotary switch and a display. All switches have backlit legends for easy identification. The fully sunlight readable display and indicators are of the 'dead face' variety and thus hidden when not 'activated'. The panel is NVIS-compatible according to MIL-L-85762A. The DEP is connected to the central avionics system via a serial multiplex data bus compliant with MIL-STD-1553B.

Status
In production and in service in the JAS 39 Gripen.

Contractor
SaabTech AB.

JAS 39 Gripen Environmental Control System Controller (ECSC)

Type
Aircraft system control, environmental.

Description
The ECSC is a computer that controls cockpit temperature, cockpit airflow, cooling pack temperature and avionics airflow. The controller receives input signals from the cockpit control panel, with various sensors and switches producing output signals for the control valves and indicators. Communication to and from the aircraft avionics systems is made via a redundant MIL-STD-1553B serial bus. The ECSC features a modular design, LRU concept to facilitate maintenance, chassis-integrated power supply for optimal heat dissipation, and SMT for increased packing density and built-in redundancy.

Status
In production and in service in the JAS 39 Gripen.

Contractor
SaabTech AB.

JAS 39 Gripen Environmental Control Unit (GECU)

Type
Aircraft system control, environmental.

Description
The Gripen Environmental Control Unit (GECU) is a Utility Management Unit (UMU) equipped with a general purpose computer, redundant power supply (28 VDC and 115 V AC, 400 Hz) and integrated I/O functionality. The main tasks for the system are measurement, monitoring and control of the fuel system, hydraulics system and the Environmental Control System (ECS). The GECU is housed in a 4 MCU box and the computer is compatible with the ADA software development environment. Communication to and from the aircraft avionics systems is made via a redundant interface MIL-STD-1553B serial bus. The GECU features include a modular design, LRU concept to facilitate maintenance, chassis-integrated power supply for optimal heat dissipation, COTS industrial components, and SMT for increased packing density and built-in redundancy.

Status
In production and in service in the JAS 39 Gripen.

Contractor
SaabTech AB.

JAS 39 Gripen stores management unit

Type
Aircraft system control, Stores Management System (SMS).

Description
The stores management unit distributes and interfaces signals between the various weapons and tactical systems in the aircraft. It is based on a Pentium® computer system. The unit is partly operated from the main mode selector, status panel and target acquisition panel. The unit also handles the preparation of stores, functions for firing and releasing stores in co-operation with the system computer, stores monitoring, control of the gun and control of countermeasures.

The stores management unit features modular design, LRU concept to facilitate maintenance, chassis integrated power supply for optimal heat dissipation, SMT for increased packing density and built-in redundancy.

Status
The MIL-STD-1553B variant of the JAS 39 Gripen stores management unit is now in production for Lot 1 and Lot 2 aircraft. The system is being

The stores management unit (left) and aircraft interface unit (right) for the JAS 39 Gripen
0504222

updated to MIL-STD-1760 for Lot 3 aircraft, with pre-production units delivered.

Contractor
SaabTech AB.

Modular Airborne Computing System (MACS)

Type
Avionic computer.

Description
MACS is a standardised computer concept developed for the JAS 39 Gripen aircraft programme. It replaces the previous standard computer system – the SDS80. The system comprises the MACS computer and its software engineering environment SEEMACS.

At present, the D80E computer is used for data processing in the Gripen mission computer, radar, display and ECM systems; it uses Pascal/D80 software language. Future Gripen processing requirements will need increased processing power and use of the Ada software language. MACS is designed to fulfil these requirements, with 10 times greater processing power within similar power and volume limits. SEEMACS replaces the PUS80 support environment.

A MACS computer consists of three functional units which communicate using VME. The processor unit handles all calculations and each processor uses its own program and data memory. The communication unit facilitates communication with external devices via MIL-STD-1553B databus. The mass memory unit supports the other units.

Status
In production.

Contractor
Ericsson Microwave Systems AB.

SDS80 standardised computing system

Type
Avionic processor.

Description
Designed for airborne, land and naval applications, the SDS80 is intended as a general purpose processor with cost savings extending over software development as well as hardware. The current design consists of three modules: the D80E computer, PUS80 programme development system and Pascal/D80 high-order language. By adding D80E modules, the system can be expanded from a relatively simple unit to an extremely high-performance embedded computer system. Using an intermodule bus, up to 15 processing units, input/output modules and bulk storage facilities can be connected.

The D80E computer is the system computer for mission data, radar, display and ECM systems in the Saab JAS 39 Gripen combat aircraft. It was also a component of the mid-life upgrade programme for the UK Royal Navy's Sea Harriers.

Specifications
Dimensions: ARINC 600
Weight: 6.5 kg
Interface: MIL-STD-1553B
Cooling: forced air or fan

Status
In service in the JAS 39 Gripen. The latest, improved version of the D80 computer has been retrofitted into the first and second batches of Gripen aircraft, which originally carried an earlier version of the system.

Contractor
Ericsson Microwave Systems AB.

The latest, improved version of the D80 computer, retrofitted into the first batch of Gripen aircraft
0015278

United Kingdom

Active noise control

Type
Aircraft system control, Noise, Vibration and Harshness (NVH).

Description
Selex (formerly BAE Systems Avionics) has developed an active noise control system which lowers the noise within the aircraft cabin to provide a reduction in the fatigue level for the cabin staff and an improvement in the comfort level for the passengers. Primarily targeted at turboprop aircraft, it can be fitted to any aircraft where tonal noise is considered to be intrusive.

All major tonal noise sources present within a particular spectrum are identified and monitored via an array of microphones discretely placed within the aircraft trim and connected to a central controller. Antiphase signals are

then calculated using powerful Digital Signal Processors (DSPs), which then control the tonal noise present in the cabin via a set of loudspeakers also discretely placed in the aircraft trim.

Flight trials have taken place on a number of aircraft including the Jetstream 41, for which the system has now been selected. The system can either be specified as an initial fit or retrofitted.

Trials have also been conducted on the Lockheed Martin C-130 military transport where significant reductions in noise level were demonstrated in both the cockpit and cargo bay areas.

Status
In service in the Jetstream 41.

Contractor
Selex Sensors and Airborne Systems.

Selex Sensors and Airborne Systems' active noise control
0001409

Advanced Data Transfer Equipment (ADTE)

Type
Data processing system.

Description
Smiths Aerospace offers a system solution for memory intensive avionic requirements. The Advanced Data Transfer Equipment (ADTE) allows aircraft integrators the flexibility to include advanced capabilities into their system architecture, including Digital Terrain Systems (DTSs), digital moving maps, solid-state video recording, network-accessible solid-state mass storage and high-speed interfaces.

The ADTE is available in two basic configurations. The first consists of two components; an intelligent cockpit-mounted receptacle called the Advanced Data Transfer Unit (ADTU), and the removable Advanced Data Transfer Cartridge (ADTC). The convection-cooled ADTU uses the same form factor as Smiths' Data Transfer Unit (DTU), providing all aircraft interfaces and serving as the receptacle for the ADTC. The second configuration has three components; a small form-factor, non-processor-based Data Transfer Cartridge Receptacle (DTCR) that is typically mounted in the cockpit, an air-cooled Data Management Unit (DMU) that is mounted in an equipment bay, and the ADTC. The DTCR serves as the receptacle for the ADTC and interfaces to the DMU via a high-speed bus. The DMU provides all aircraft interfaces.

Both the ADTU and DMU share common, compact PCI form-factor circuit cards. The DMU was designed, however, for those applications that required more functionality equating to more circuit card assemblies and the resulting higher heat dissipation. The ADTE is available with numerous interfaces designed to facilitate data movement between the ADTE and other aircraft avionics. These interfaces include MIL-STD-1553B, RS- 422, both copper and fibre optic fibre channel, Digibus, ARINC 429, and Fast Ethernet. The system can be configured to customer requirements.

Specifications
ADTU
Dimensions: 178 × 127 × 113 cm
Weight: 2.9 kg
Input power: 115 V AC, 400 Hz, single phase
Operating temperature: −40 to +85°C

ADTC
Dimensions: 191 × 119 × 41 cm
Weight: 1.1 kg
Input power: 115 V AC, 400 Hz, single phase
Operating temperature: −40 to + 85°C

DTCR
Dimensions: 197 × 146 × 114 cm
Weight: 2.4 kg
Power: ±12 V DC (from DMU)
Operating temperature: −40 to +71°C

DMU
Dimensions: 210 × 130 × 147 cm
Weight: 5.5 kg
Power: 115 V AC, 400 Hz
Operating temperature: −40 to +71°C

Status
In service in Lockheed Martin F-16, BAE Systems Jaguar and Dassault Mirage 2000 aircraft.

Contractor
Smiths Aerospace.

Automatic Voice Alert Device (AVAD)

Type
Avionic crew alerting system.

Description
The Automatic Voice Alert Device (AVAD) stores prerecorded human speech in digitised form and operates under microprocessor control to assemble messages from words or phrases held in a vocabulary store. Since the voice is not synthesised, it can be either male or female in any language to provide the appropriate degree of stress and urgency for any situation. Racal claims that high-quality reproduction ensures that the voice remains recognisable and intelligible under all conditions.

Message priority order, repetition rate and volume are programmable to individual requirements to provide the maximum information to aircrew without disrupting their primary tasks. Racal claims that existing discrete audio warning systems can be replaced by AVAD with a minimum of installation cost and complexity.

The units in the current AVAD range are the V694, V695 and V697.

V694
The AVAD V694 is designed to integrate and control audio alerts keyed by signals from system sensors and push-buttons. The message format is completely flexible and the system conforms to ARINC 577 audible warning systems and ARINC 726 flight warning computer systems for civil aviation applications.

Up to 4 minutes of prerecorded speech can be encoded and stored as a vocabulary of messages, phrases, words and tones. Messages can be constructed using words and phrases drawn from the unit memory under software control. This allows the total duration of all messages to exceed the vocabulary by a considerable amount. Up to 15 message channels are based on positive or zero volt keying. Four channels can respond to variable DC voltage inputs, offering a sensitivity of up to 256 logic switching steps across the input range. A stabilised 5 V DC output is available for use as a reference for an external potentiometer. Typical variable readouts might be aircraft altitude, electrical system voltages or engine pressure ratios.

Unit operation and vocabulary store may be defined by the user and program software allows control of each message cycle. Each message has normal, regrade and test priority values assigned to it. The output sequence is then controlled by the regrade and test inputs.

There are two audio outputs, one for a telephone or headset and the other for driving a loudspeaker direct. A variant can be provided where the loudspeaker drive output is removed and replaced by an auxiliary audio input. This auxiliary input facilitates the summing of existing aircraft audio warnings with the single AVAD output for routeing into the aircraft audio integration/intercommunication control system. This AVAD configuration is particularly suitable for helicopter use. The nominal level of output can be adjusted by individual potentiometers offering a 20 dB range, with the relative volume of each message or message format under software control. The complete message format, including repetition rate, pauses, message inhibit/de-inhibit and tones, may be similarly controlled. A key input is provided, which inhibits any message in progress without affecting other keyed messages. A message will remain inhibited until its sensor input is removed and reapplied. Alternatively, the inhibit may be removed after a specified time if

Thales' AVAD units V694, V695 and V697 (right to left) 0044841

the warning message keyline remains activated during this period.

Full test/reset facilities provide fast checking of unit operation. Correct operation of the microprocessor and amplifier circuitry and control program status are continuously self-monitored. In the event of a fault condition, both audio outputs are automatically inhibited and a fail output is provided to activate an indicator to warn of unit failure.

V695
The V695 is a smaller seven-channel unit with microprocessor control. It offers all the programming flexibility of the V694, but has a storage capacity of 13.5 seconds. An auxiliary audio input is provided.

V697
The V697 provides a single output for applications in which only one warning, such as 'fire' or 'pull up' is required. A second keyline may be used to provide a system enable/disable input. The V697 can be enabled by either positive or zero volt keying and the resultant warning output can be either a continuous or one message cycle per key event.

All the above units provide a bandwidth of 100 Hz to 4 kHz. Audio distortion is less than 4 per cent. Volume is preset and adjustable by 20 dB. The V694 is also programmable to different levels for each message and variation during messages.

Specifications
Dimensions:
(V694) 165 × 115 × 51 mm
(V695) 134 × 61 × 32 mm
(V697) 110 × 50 × 28 mm
Weight:
(V694) 0.95 kg
(V695) 0.2 kg
(V697) 0.14 kg
Power supply: 28 V DC
Temperature range: −40 to +70°C

Status
In widespread service in a variety of fixed- and rotary-wing aircraft including UK Royal Air Force Jaguar fixed-wing aircraft, together with Puma and Sea King helicopters.

Contractor
Thales Defence Ltd.

Central warning unit

Type
Avionic Fault Warning System (FWS).

Description
The central warning unit has been designed and manufactured for use on modern fighter aircraft. The unit houses all the control circuitry and displays to provide 70 illuminated captions, along with audio tone and synthetic voice generation.

Each unit incorporates 28 primary red channels and 42 secondary amber channels. Additional

Page Aerospace central warning unit   0581507

features include day/night mode dimming of caption illumination, test facilities for active and non-active sensor channels, ground activation of certain channels with override facility for test purposes and generation of audio horn tones and synthetic voice.

Various combinations of caption are available in the range, while all variants incorporate a self-illuminating Betalight panel over the switches. Electrical connections are made via a pair of multipin connectors. The unit is capable of total NVG-compatibility.

Specifications
Dimensions: 134 × 165 × 165 mm
Power supply: 22.5–30.5 V DC normal (16–30.5 emergency)
Temperature range: –40 to +70°C
Altitude: up to 40,000 ft

Contractor
Page Aerospace Ltd.

Common Cockpit Keyset (CCK)

Type
Aircraft system control, integrated.

Description
The Common Cockpit Keyset is a highly versatile control panel, capable of displaying major functions and their subsets with up to 1,000 stored switch legends. The panel is adaptable to any environment or vehicle where a large number of systems must be controlled from limited available operator real estate.

The CCK was developed for the LAMPS MK III CH-60 and SH-60 helicopters in order to interface with the aircraft mission processor to select and control the multiple systems available on the platform. The keyset also allows for future expansion and changes via software modifications rather than a rework of the entire hardware/software package. The system is PowerPC based and supports multiple interfaces. It is designed to work in conjunction with the CH/SH-60 flight mission computer, but may be adapted to work with any fire-control or mission computer. The interface is through redundant RS-422 Asynchronous Serial Interface (ASI) channels.

The switch panel consists of highly reliable stainless steel dome switches with a millions-of-depressions life cycle and solid-state electronics package for a combined MTBF of over 5,500 hours. The displays are compatible with Type 1, Class B, Gen III Night Vision Goggles (NVGs).

Specifications
Dimensions: 127 × 419 × 146 mm
Weight: 6.8 kg
Input power: 100 W maximum
Interfaces: dual RS-422 Serial; Ethernet for Power PC programming
Environmental: RTCA/DO-160C

Status
The Smiths Aerospace Common Cockpit Keyset was developed for the CH-60 and SH-60 helicopters.

Contractor
Smiths Aerospace.

Data Storage and Transfer Set (DSTS)

Type
Data processing system.

Description
The solid-state data storage and transfer set is currently available with 1 Mbyte of memory capacity in a fully qualified format for installation in military aircraft.

The data transfer module has a quick-release handle mechanism for ease of operation and handling. The system is configured to be compatible with aircraft standard console mountings and includes the power supply and a databus interface for control and data transmission. The system has full read/write capability for upload and download of data.

The data transfer system can be used for upload of mission, navigation and other operational data or for the download of maintenance and mission data. Higher capacity variants (up to 400 Mb) have been developed.

Status
In service on the UK Royal Air Force Harrier GR Mk.7 and T. Mk.10.

Contractor
Honeywell (Normalair-Garrett Ltd).

The solid-state data transfer system has been selected for UK Royal Air Force Harrier GR. Mk 7 and T. Mk 10 aircraft   0504163

Digital Fuel Quantity Indicating System (FQIS) for Airbus

Type
Aircraft sensor, Fuel Quantity Indication (FQI).

Description
Smiths' digital Fuel Quantity Indicating System (FQIS) combines optimum safety with maximum operational economy and high accuracy, enabling operators to avoid carrying excess fuel and, hence, maximise payload.

The system computer, a two-channel, dual-redundant, digital data processing unit, calculates the fuel mass and performs control, monitoring and output functions. Fuel quantity is gauged with DC capacitance fuel probes mounted in each tank. Selected sensors also provide fuel temperature and permittivity measurements. In addition, densitometers and compensators mounted in the bottom of each tank provide fuel density and further permittivity measurements required for accurate gauging.

Refuelling is controlled from a preselector unit on the aircraft refuelling panel. The system automatically fills the tanks simultaneously to a quantity selected by the operator. This ensures minimum refuel duration and maintains the physical balance of the aircraft during refuelling. Refuelling and defuelling can also be controlled manually via switches on the refuel panel.

The FQIS provides 1 per cent fuel gauging accuracy at normal ground and cruise attitudes and this can be improved further by the installation of additional probes. Fuel quantity indications for each tank and total fuel indications are sent via ARINC 429 databuses for display by aircraft systems. Ground operations (refuel/defuel control and display panel) are also linked by the databuses to the Fuel Quantity Processor Unit. Overall power consumption can be 20 per cent below that of AC systems.

Status
The digital FQIS is in production and in service in Airbus A318, A319 and A320 aircraft.

Contractor
Smiths Aerospace.

Electrical Load Management System (ELMS)

Type
Aircraft system control, electrical.

Description
Smiths' ELMS provides a comprehensive range of functions, including distribution and protection of primary and secondary electrical power, and control of aircraft utilities subsystems. The system also interfaces with the cockpit controls and displays via an ARINC 629 databus for control switching and system status reporting.

For application to the Boeing 777, the ELMS has three primary panels and four secondary panels: left, right and auxiliary power and ground handling/service distribution, left power, right power, and standby power.

The design of the ELMS features compact primary panels with line-replaceable contactors. Three of the secondary power management panels incorporate dual-ARINC 629 bus interfacing, dual-processing elements and line-replaceable electronic and switching relay modules. The design makes the maximum use of multiple-redundancy techniques and immunisation against high-intensity radio frequencies.

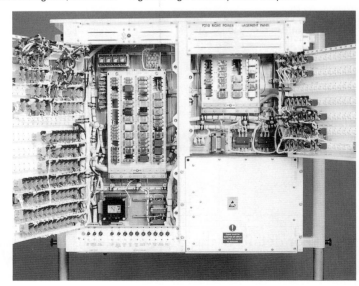

Smiths Aerospace ELMS for the Boeing 777
0003120

The introduction of new technologies, particularly Solid State Power Control (SSPC) and Smart Battery Charging (SBC) has resulted in further applications on the F/A-22 Raptor battery charger control system, the C-130J electronic circuit breaker unit (using SSPC technology), the F-35 JSF Electrical Power Management System (EPMS) and the AH-64D Longbow Apache Improved Electrical Power Management System (IEPMS).

The EPMS installation in the Longbow Apache comprises four LRUs: two high-power switching modules contain the primary power buses and two electrical load centres provide the monitoring and control of the system and also contain the secondary power distribution buses.

Status
In production for, and in service on, the Boeing 777, 767, AH-64D (EPMS) and UCAV X-45A, and under development for the Lockheed Martin C-130J and F-35 JSF.

Contractor
Smiths Aerospace.

Electrical Power Management System (EPMS) for the Apache

Type
Aircraft system control, electrical.

Description
Smiths' EPMS for the AH-64D Apache comprises four line-replaceable units. Two high-power switching modules, located close to the generators, contain the primary power buses, high-power contactors and some circuit breakers. Two electrical load centres, located in the avionics bay, provide the monitoring and control of the system. They also contain the secondary power distribution buses, circuit breakers and remote power switches up to 25 A. Smiths Aerospace also supplies the 350 A Regulated Transformer Rectifier Unit (RTRU).

Systems features and benefits include:
1. Reduced life cycle cost
2. 35% weight saving
3. Fault tolerance
4. Automatic load switching
5. Improved diagnostics and maintainability
6. Reduced aircraft build time
7. Electronic fault protection.

System functions include:
1. Managing electrical power (90 kVA)
2. Switching and protection of aircraft primary AC, DC and battery power
3. Use of smart I^2t contactors to improve trip co-ordination

4. Monitoring of primary power system over MIL-STD-1553B databus
5. Pilot selections made via MIL-STD-1553B databus
6. Switching and protection of secondary loads in dedicated load centres.

Status
In production and in service in AH-64D attack helicopters.

Contractor
Smiths Aerospace.

Eurofighter Typhoon Direct Voice Input (DVI) system

Type
Aircraft system control, avionic.

Description
The Man-Machine Interface (MMI) of the Eurofighter Typhoon is designated the Voice, Throttle and Stick (VTAS) control system, which incorporates Hands On Throttle And Stick (HOTAS), Head-Up Display (HUD), Glareshield Control Panel (GCP), MultiFunction Head Down Displays (MFHDDs) and Direct Voice Input (DVI).

The Eurofighter DVI System utilises a Speech Recognition Module (SRM), developed by Smiths Aerospace and the then Computing Devices (now General Dynamics UK), which was the first production DVI system utilised in a military cockpit. DVI will provide the Eurofighter pilot with an additional natural mode of command and control over approximately 26 non-critical cockpit functions, to reduce pilot workload, improve aircraft safety and expand mission capabilities.

The Eurofighter SRM is a double-sided surface mounted card which is integrated within the Communications and Audio Management Unit (CAMU). General Dynamics (UK) Hastings Ltd manufactures the CAMU in the UK. An important technological breakthrough during the development of DVI occurred in 1987 when Texas Instruments produced the TMS-320-C30 Digital Signal Processor (DSP). This greatly advanced the packaging of DVI from large complex systems to a single card module. This early advance in semiconductors provided the hardware to produce a practical and viable system. This advance enabled Smiths Aerospace and General Dynamics (UK) to develop the SRM for the Eurofighter Typhoon. The project was given the go ahead in July 1997, with DVI/SRM development and pilot assessment carried out on the Eurofighter Active Cockpit Simulator at Warton.

The SRM is a speaker-dependent, connected word, voice recognition system with a maximum vocabulary of 200 words, of which 107 are implemented in the operational unit. Recognition performance is a priority, required to be in excess of 95 per cent in the cockpit environment with further development towards stress-tolerant algorithms to 98 per cent. The module accepts continuous natural speech at 100 to 140 words/min with an average 120 ms response time. The voice generation provides high-quality digitised speech encoding at the rate of 120 to 160 words/min utilising a 200-word vocabulary. The SRM has been designed to provide robust speech recognition performance with flight-proven compensation to overcome the environmental factors of noise, G loading and pilot stress. Potential speech recognition algorithms include artificial neural nets and the classical Markov Model, both with implementations specifically designed for the airborne environment. The SRM is packaged onto a single card module (120×100 mm and 320 g).

The SRM recognises continuous speech and is speaker dependent, requiring the pilot to register his/her voice on the PC-based Ground Support Station (GSS). This 'enrolment' takes less than one hour and the resulting voice template is transferred to the aircraft via the Smiths Aerospace Mission Data Loader/Recorder. SRM commands range from simple display control commands to the more complex commands providing for management of the Eurofighter avionics system. The system will handle routine assignments such as selecting radio frequencies and head-down display moding, although safety-of-flight and weapons release related functions will not be included into the DVI system.

Specifications
- High recognition performance, >95% in flight
- Fast response time, >200 ms
- DVI vocabulary of 200 words
- Flight proven countermeasures to overcome the environmental factors of noise, G loading and stress
- Speaker dependant, language independent with full multilingual capability
- Flexible command syntax for easy reconfiguration to meet the Eurofighter requirements
- Comprehensive PC-based GSS.

Status
In development for the Eurofighter Typhoon. Memory growth to 400 words is available. The SRM has been fully flight qualified and passed the initial ground performance acceptance testing, in which 12 pilots from the four partner nations tested the SRM under simulated cockpit noise conditions. The first 15 SRM pre-production units have been delivered with flight trials commencing during 2002.

Contractor
Smiths Aerospace.

Fibre Channel – Data Transfer Equipment (FC-DTE)

Type
Data processing system.

Description
Today's avionics architectures demand continually increasing amounts of mass storage. This mass storage must be extremely rugged, operate at very high access rates with low latency and be cost effective. Smiths' Fibre Channel Data Transfer Equipment (FC-DTE) is designed to operate on two 1.062 Gbyte Fibre Channel Arbitrated Loop (FC-AL) compliant nodes. The FC-DTE supports sustained read rates of 16 Mbps and write rates of 8 Mbps, and also provides a dual-redundant MIL-STD-1553 bus for use with older avionics not yet possessing a high-speed FC interface.

The FC-DTE consists of an aircraft-mounted Data Transfer Receptacle (DTR) and a removable

Electrical Power Management System for the AH-64D Longbow Apache   0001395

Data Transfer Cartridge (DTC). The DTR manages and controls the data exchanged between the DTC and other avionics subsystems via the aircraft's MIL-STD-1553 and Fibre Channel buses. Avionics using each type of bus can use the FC-DTE concurrently as a mass storage/file server device. The DTR can also pre-and post-process the data which is being stored/accessed from the DTC. Current configurations allow for memory densities in the DTC of between 100 Mbytes and 2 Gbytes. Memory upgrades to in excess of 30 Gbytes should be available in the future.

Status
In production.

Contractor
Smiths Aerospace.

Fly-by-wire control stick assemblies

Type
Flight control inceptor.

Description
BAE Systems has produced high-integrity position sensors for a wide range of aircraft including the pilot's stick sensors and the rudder pedal position sensors for the Tornado aircraft and pilot position sensors for the AMX. This range of sensors has been extended to include fly-by-wire passive control sticks for the Eurofighter Typhoon and F/A-22 in centre and side console configurations respectively.

Eurofighter Typhoon centre stick controller
The 'passive' unit utilises springs and dampers to provide feel to the pilot and contains high integrity electrical position sensors. They are configured as two axis displacement units utilising multiredundant position sensing to provide high integrity input demands to the flight control system. No mechanical outputs are provided from these passive units and the force and damping characteristics are fixed. Mass balancing can also be specified which minimises stick movement caused by aircraft acceleration (see also International Section for details of consortium partners).

F/A-22 side stick controller
The modular construction provides easy maintenance and also enables the force/displacement characteristics of the units to be tailored to meet the particular aircraft requirements. All of the units are high integrity, flight safety critical items manufactured to the highest standards using class 1 components and manufacturing techniques. Careful choice of materials ensures the strength-to-weight ratio is optimised to meet aircraft weight constraints, provide a robustness capable of withstanding an applied pilot load in excess of 200 lb and still give reliable precision manoeuvring throughout the flight envelope and life of the aircraft.

Further developments
A design currently under development is an electrically back-driven active pilot controller which will allow adjustable spring rates, breakout forces, detents and damping ratios to provide situational awareness to the pilots at all times. With feel characteristics capable of being tailored to suit individual applications, it will also permit electrical connection between two sticks such that if one stick is moved, the other will follow.

Active throttle units using the same technology and possessing a high degree of commonality with the active stick unit are also under development.

In aircraft with a requirement for dual sticks, such as a civil aircraft, helicopter or military trainer, and fitted with a full-time fly-by-wire system, linked active control sticks can be provided, packaged to suit the individual requirements.

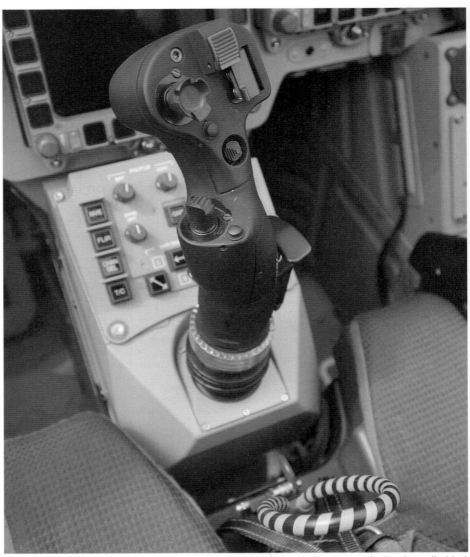

The Eurofighter Typhoon centre stick assembly. The stick incorporates a gun/missile trigger (behind, forefinger actuation), Late Arm (top centre), weapon release/commit (top right), radar/IRST/air combat rocker switch (top left), HMD split radar/FLIR select (below rocker), trim control (centre), air-to-air weapon selector (low toggle), IFF/NCI interrogate (in front of low toggle), autopilot engage/disengage (behind, middle finger), and Instinctive Cut Out (ICO)/Nose Wheel Steering (NWS) disconnect (Eurofighter)
1128621

F-22 side-stick controller pedestal
0001413

Status

Stick assemblies are being manufactured and supplied as part of the Eurofighter Typhoon and the F/A-22 Raptor programmes.

More recently, BAE Systems supplied active stick units for both of the US Joint Strike Fighter (JSF) contenders and now supplies units for the Lockheed Martin F-35 JSF.

Contractor

BAE Systems.

Fuel quantity gauging and indication systems

Type

Aircraft sensor, Fuel Quantity Indication (FQI).

Description

Smiths Aerospace designs and manufactures electronic equipment and systems for the measurement, management and indication of fuel in civil and military aircraft. Current fuel system applications include large civil transport aircraft such as the Boeing 777 and Airbus family; commuter aircraft and business jets such as the Raytheon Hawker 800/1000, BAE Systems 146 and Jetstream 41 and 61 aircraft; and military aircraft such as the BAE Systems Hawk, AMX and Eurofighter Typhoon. Smiths is also a partner in collaborative programmes, including those for the Airbus A300, A310, A319 and A320.

Analogue and digital displays

Fuel contents indicators range from simple moving coil analogue types to servo-pointer and digital multitank displays driven from an ARINC 429 databus. Internal illumination is optional and dial presentation and colour are displayed to specification. Solid-state LED or LCD indicators can be used to provide numeric, analogue or graphical presentations of fuel quantity.

Analogue and digital signal processors

The analogue output from capacitance probes can be processed by entirely analogue means and used to drive either analogue or digital indicators. Where higher accuracy or additional facilities are required, probe signals can be digitised and processed digitally to provide outputs of fuel mass in ARINC 429 or MIL-STD-1553 formats. All analogue and digital processors incorporate BITE, which performs levels of self-test varying from basic confidence checks to comprehensive system testing and calibration.

Fuel level sensing systems

Fuel level sensing systems provide an accurate and safe means to detect fuel levels. They feature a low-cost, solid-state fluid level sensor connected to a separate switch unit which can either stand alone or be incorporated into another unit within the fuel system. The lightweight sensor unit is small enough to be mounted on a tank wall or fuel gauging probe and is immune to temperature effects over a wide operating range. Applications include high- and low-level warning indication or control, automatic shut-off switching for refuelling, automatic control of liquid transfer and sequential draining and filling of tanks. Fuel level sensors may also be applied to other fluids such as oil or hydraulic fluid.

Ultra-sonic and capacitance tank probes

Fuel height within tanks can be measured using either capacitance or ultra-sonic probes. Capacitance probes can be used in analogue or digital fuel systems. Systems using ultra-sonic probes are digital throughout. It is usual to install several probes in each tank so that fuel levels can be gauged accurately over a wide range of aircraft attitudes and fuel contents. Non-linear height and volume characteristics of fuel tanks can be accommodated either by mechanically profiling the linear electrodes of the probes or, in digital systems, by incorporating appropriate software in the processor. The ultra-sonic fuel probe incorporates a piezo-electric crystal that transmits sound at a frequency imperceptible to the human ear. This gives a time signal related to fuel height that is drift free over the long term. This technology has a lower susceptibility to electro-magnetic interference than traditional capacitance systems.

Hawker Horizon Fuel Management System

Smiths Aerospace is responsible for designing and delivering the complete fuel system for the Raytheon Hawker Horizon aircraft.

Status

In production and in service in a large number of aircraft types.

Contractor

Smiths Aerospace.

Ground roll director system

Type

Flight Director (FD) system.

Description

The Smiths Aerospace ground roll director system provides pilots with head-free guidance information. It meets the requirements for ground roll guidance in Cat IIIB weather conditions. It has been designed so that pilots can easily revert to para-visual guidance when forward visual references disappear in conditions of deteriorating visibility such as drifting fog, RVR reporting failure, or differing fog densities.

The system is based on the Para-Visual Director (PVD) concept developed during the early 1960s. A display unit is positioned in the glareshield directly in front of each pilot. During head-free operation, while concentrating on external visual cues, the pilot immediately registers any movement of the black and white bands in his peripheral vision and makes instinctive corrections to the azimuth steering controls without looking directly at the display unit. For the Boeing 777, an Active Matrix Liquid Crystal Display (AMLCD) replaces the original electromechanical 'barber's pole' indicator, providing a 100 × 25 mm usable area that offers the additional use as a multifunction display for messages such as those associated with an ATC datalink or FMC.

Specifications

Dimensions:
(display) 204 × 63.6 × 33.8 mm
(computer) 321 × 193.5 × 61.5 mm
Weight:
(display) 0.55 kg
(computer) 2 kg

Status

In production and in service in Boeing 747-400, 757, 767, 777 and MD-11 aircraft.

Contractor

Smiths Aerospace.

Groundspeed/drift meter Type 9308

Type

Aircraft flight computer.

Description

The Type 9308 standard aircraft indicator size unit shows groundspeed within 3.5 kt at 100 kt or 5 kt at 300 kt. Drift is registered within 0.5° of true value. If the signal is lost, the last measured groundspeed and drift are displayed. The Type 9308 is compatible with Doppler 71 and 72 sensors.

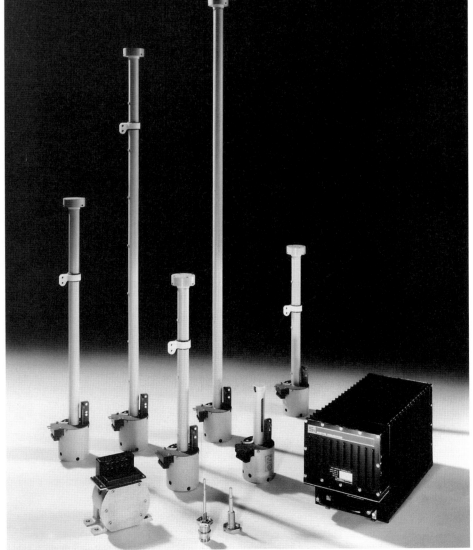

Ultra-sonic fuel quantity gauging system   0001423

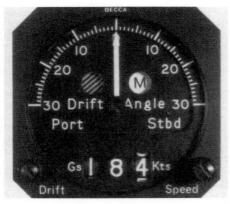

Thales Avionics' Type 9308 groundspeed/drift presentation 0503935

Status
In production.

Contractor
Thales Avionics SA.

Ice and Snow Detection System (ISDS)

Type
Aircraft system control, ice detection and anti-ice.

Description
The Ice and Snow Detection System (ISDS) measures liquid water content of the atmosphere with low dependency on air velocity and is especially suited to helicopters. It can take measurements down to the hover and eliminates the need for air velocity signal inputs to the ice protection controller. The electronic unit accepts a computed airspeed input to enhance the liquid water content calculation accuracy.

In addition to ice detection capability, the ISDS can also detect snow and is able to categorise and display the severity of both conditions. The electronic unit requires an input of outside air temperature to categorise icing as light, moderate or severe and snow conditions as trace, light, moderate, heavy and severe.

Specifications
Weight:
(detector) 0.9 kg
(meter) 0.34 kg
Airspeed: 0–250 kt
Altitude: sea level to 15,000 ft
Reliability: 10,850 h MTBF

Status
A version incorporating snow detection and digital outputs has been selected for the EH 101 helicopter where automatic control of the anti-de-ice systems is provided.

Contractor
Penny & Giles Aerospace Ltd.

Penny & Giles ice detection system 0581509

Landing Gear Control and Interface Unit (LGCIU) for Airbus

Type
Aircraft system control, undercarriage/wheelbrakes.

Description
The LGCIU is in service on the Airbus A319, A320, A321, A330 and A340. The LGCIU controls the extension and retraction of landing gear and doors, provides an instantaneous on-wheels indication and monitors the position of cargo door locks and flap position. It interfaces with the fault warning and centralised aircraft maintenance system, providing status signals, built-in test and data to the maintenance recording systems.

Landing gear system status is monitored using a two-wire proximity sensor system. Interface with other aircraft systems is through an ARINC 429 databus.

Status
In service on the Airbus A319, A320, A321, A330 and A340.

Contractor
Ultra Electronics Ltd, Controls Division.

Micro Cartridge Interface Device (MicroCID™)

Type
Data processing system.

Description
Smiths' Micro Cartridge Interface Device (MicroCID™) was designed to support current and future Data Transfer Cartridge (DTC) interface requirements. The MicroCID™ can be ordered separately or as a part of a Ground Support Equipment (GSE) package. The standard MicroCID™ is connected to a host computer via a SCSI-2 interface. The product has been designed to support other types of standard interfaces by installation of various interface cards into the internal mezzanine slot.

The MicroCID™ supports the interface between a PC and the DTC from a variety of current products such as the UDTE, ADTE, DMVR and AVSR. The MicroCID™ enables loading of the aircraft's mission data and recovery of mission results and maintenance information recorded on the aircraft's Data Transfer Cartridge (DTC) during flight.

The MicroCID™ is a microprocessor controlled unit that is available as an external peripheral to standard laptop and desktop PCs. The system provides a modular interface and logic to interface with and play data back from Smiths Aerospace DTCs. These DTCs range in memory capacity from 16 Kbytes to 30 Gbytes. This feature provides interface and upgrade capability to the existing equipment currently in use.

Specifications
Dimensions: 130 × 50 × 230 mm
Weight: 1.1 kg
Power requirement: 12 V DC, 4 A, 48 W maximum
Temperature: 0 to +50°C operating
Reliability: 56,000 h MTBF

Status
In service with the F/A-22, F-16, A-10, B-1B, B-52, Mirage 2000, Jaguar and IDF.

Contractor
Smiths Aerospace.

MicroInterface Unit (MIU)

Type
Avionic interface system.

Description
Smiths' MIU-30E is the first in a new range of microinterface units, providing affordable 'off-the-shelf' high-density interfacing to sensitive and specialised sensors.

MIUs are cost-effective, line-replaceable modules which are designed to be located in remote locations, away from the main aircraft processing resources but in the vicinity of the subsystem itself. Their prime role is to provide a serial digital interface between subsystem sensors and a higher-level computing resource, such as Smiths' Remote Interface Unit (RIU). In addition to operating as a data-gathering slave to a RIU or other computing resource, some variants of MIU can provide basic autonomous local closed-loop control.

The modular software and hardware characteristic of the MIU provides significant flexibility to the user. This enables common MIUs to be used throughout the aircraft, with units taking on a different interfacing role in each location. This flexibility both reduces the cost of initial acquisition and future upgrades.

Specifications
Temperature: –40 to +85°C (operating)
Start-up: From –55°C cold soak
Dimensions: 76 × 76 × 19 mm
Power: 28 V DC, MIL-STD-810A to E
Weight: 0.15 kg

Status
In production.

Contractor
Smiths Aerospace.

Onboard Maintenance System (OMS)

Type
Onboard system monitor.

Description
Smiths' engine life computer and standard flight data recorder have been combined into a single Onboard Maintenance System (OMS). The engine life computer records running hours and engine starts, and provides real-time calculations of low-cycle fatigue and snapshots of data to highlight deviations, as well as recording excursions

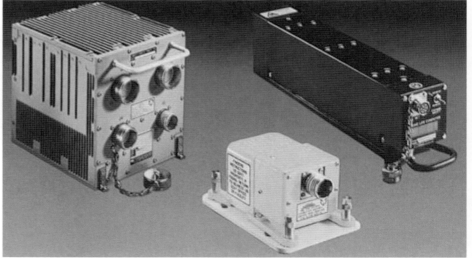

The Onboard Maintenance System consists of (left to right) the signal acquisition unit, crash survivable memory unit and engine life computer 0504171

outside defined limits. The standard flight data recorder records structural, engine, tracking, maintenance, mishap and training or mission replay data.

The benefits of the OMS are crash survivability, crash position location and water recovery, fast accurate data transfer, management of major engine components, improved flight safety and longer life and lower cost of ownership. The system is suitable for both new build aircraft and retrofit. It consists of a Signal Acquisition Unit (SAU), Crash Survivable Memory Unit (CSMU) and engine life computer. The SAU serves as the interface to aircraft data signals and power inputs. It acquires, computes, compresses and stores data. Choosing which data to record is controlled by software and can be tailored to any application. The SAU is entirely solid state.

The CSMU can be adapted to a variety of requirements. Retention of the last 15 minutes or a full 25 hours of mishap data is available and the capability to store the last 2 hours of voice is in development.

The engine life computer processes data to compute engine life usage, thermal creep, exceedances and performance information. This is transferred into internal non-volatile memory for retrieval either manually via the integral LED display or semi-automatically via a data transfer device. The computer uses standard hardware modules and the software architecture is structured for multiple applications. External programming enables general engine algorithms, such as low-cycle fatigue and exceedance levels, to be factory programmed to suit specific applications and customer requirements. The computer includes 64 kbytes of non-volatile

memory to store cumulative life usage totals, performance data and incident records. Extensive BIT is incorporated to monitor the system for correct functionality. The computer can be configured to drive annunciators for predefined input conditions, exceedances and BIT. A Deployable Flotation Unit (DFU) houses the CSMU and is equipped with a radio beacon and strobe. The buoyant DFU is released at impact or at a specified depth. The beacon and strobe provide visual and radio position location and will operate for at least 72 hours continuously.

Status
In production and in service.

Contractor
Smiths Aerospace.

PA7010 Expendable System Programmer (ESP)

Type
Avionic interface system.

Description
The PA7010 Expendable System Programmer (ESP) is designed to interface with radar warning receivers, jamming systems and chaff and flare dispensers, and to integrate the various EW sensors for the maximum operational effectiveness. The interface is produced via digital datalinks and discrete signals; crew monitoring and intervention is embodied in the design.

On detection of a threat, ESP is alerted to command the release of expendables. An easily reconfigurable stores library determines the appropriate release pattern and sequence. ESP then commands the response from one or more dispensers.

ESP has been designed for military combat aircraft and has a predicted reliability in excess of 1,000 hours MTBF. Modular construction and comprehensive BIT simplify maintenance.

Specifications
Dimensions: 127 × 127 × 190.5 mm
Weight: 2.7 kg
Power supply: 28 V DC, 10 W

Status
In production. The system is fitted in the UK Royal Air Force Harrier GR. Mk 7, in which it interfaces with the Zeus ECM system and onboard countermeasures dispensers. It has also been selected for the Eurofighter Typhoon.

Contractor
BAE Systems.

PLU 2000 program loader

Type
Data processing system.

Description
Penny & Giles' computer program loaders are based on industry standard DC300A data cartridges and expanded capacity DC300XL units. They provide a range of features appropriate to military or commercial fixed-wing aircraft and helicopters, and land- and sea-based vehicles and vessels, in severe environments where low error rates are essential.

The PLU 2000 program loader can accommodate up to three optional modular interfaces, has a capacity of 3.2 Mbits expansible to 6.4 Mbits, and can load up to 10 preselected programs. It is operated by a remote-control unit.

Status
In service on UK Royal Air Force Nimrod aircraft and UK Royal Navy Sea King helicopters.

Penny & Giles airborne and ground program loaders   0503880

Contractor
Penny & Giles Aerospace Ltd.

Primary Flight Computers (PFCs) for the Boeing 777

Type
Flight Control Computer (FCC).

Description
The three Primary Flight Computers (PFCs) form the core element of the Fly-By-Wire (FBW) system for the Boeing 777. The complete Primary Flight Control System (PFCS) provides control in all three axes together with protection and compensation facilities.

The FBW system provides full time control in pitch, roll and yaw axes with pilot commands being input to the PFCs via a triplex ARINC 629 bus and Actuator Control Electronics (ACE) units. Pilot commands are provided from conventional control column and pedal inputs via multiple position transducers to the ACEs which convert the signals into digital form for transmission to the PFCs.

Primary flight computer for Boeing 777   0001415

The PFCs compute the pilot commands for control of a single tabbed rudder surface, two elevators, a single all moving tailplane, an outboard aileron, a flaperon and seven spoiler surfaces on each wing. These computed commands are then routed via the ACEs to the appropriate electrohydraulic actuator. The computation also implements the C* control laws, together with the protection and compensation facilities for ease of crew workload.

All three primary flight computers are identical and contain three dissimilar computing lanes. The secondary redundancy management of these lanes allows fault tolerant operation in the event of failures, Within each primary flight computer, each lane uses a different 32-bit microprocessor. Three different compilers are used to convert the Ada high order software design into the microprocessor's object code. This level of dissimilarity provides protection against residual compiler and processor errors.

Extensive use has been made of ASICs ensuring a low component count and high reliability. This high reliability, coupled with the fault tolerant architecture, provides benefits to the operator through reduced life cycle costs, improved maintainability and increased system availability.

Status
In service on the Boeing 777 aircraft.

Contractor
BAE Systems.

Propeller Electronic Control (PEC)

Type
Aircraft system control, power plant.

Description
The PEC is being developed for the de Havilland Dash 8–400. The PEC carries out safety-critical propeller control functions, including pitch control, synchrophasing and speed control. Additionally, the PEC controls routine feather and auto-feather operations, and the Automatic Take-off Thrust Control System (ATTCS). In the event of an engine failure at take-off, ATTCS minimises crew workload by auto-feathering the propeller on the failed power plant and up-trimming the propeller on the healthy engine; swapping roles at each power-up dual-channel active/standby control lane ensures redundancy. Control lanes are microcontroller based.

Fault monitoring and detection indicated to the FADEC through an RS-422 link. The PEC also signals the FADEC to prevent dispatch in the event of certain fault conditions.

Status
In service on the Dash 8–400.

Contractor
Ultra Electronics Ltd, Controls Division.

RIU-100/429 general purpose Remote Interface Unit

Type
Avionic interface system.

Description
Smiths' RIU-100/429 is the latest in a new range of general purpose remote interface units, which were originally developed for Vehicle Management Systems (VMSs) applications, where interface flexibility and physical robustness are essential.

In a compact package measuring $6 \times 6 \times 1.1$ in $(152 \times 152 \times 28$ mm), the RIU-100/429 provides 100 interface channels, combined with a an ARINC 429 interface. A microcontroller-based I/O core provides the highly flexible generic interfaces,

which have been designed to suit most common vehicle system applications.

Specifications
Temperature: DO-160C section 4.5 (–40 to +85°C operating, –65 to +90°C storage)
Humidity: DO-160C section 6.0
Vibration: DO-160C section 8.0, random
Shock: DO-160C section 7.0 (30 g peak, 11 ms half-sine wave)
Altitude: DO-160C section 4.6 (–1,000 to +70,000 ft)
EMC: DO-160C section 20 Cat Y, DO-160C section 21 Cat Z
Power: DO-160C Cat Z (28 V DC); 5 W
Weight: 0.7 kg

Status
In production.

Contractor
Smiths Aerospace.

RIU-100 general purpose Remote Interface Unit

Type
Avionic interface system.

Description
Smiths' RIU-100 is the first in a new range of general purpose remote interface units, giving a true 'off-the-shelf' solution to the challenge of providing low-cost, high-density interfacing within an integrated avionics architecture. The RIU concept was originally developed for Vehicle Management Systems (VMSs) applications, where interface flexibility and physical robustness are essential.

In a compact package measuring $6 \times 6 \times 1.1$ in $(152 \times 152 \times 28$ mm), the RIU-100 provides 100 interface channels, combined with a dual-redundant MIL-STD-1553B remote terminal or an ARINC 429 databus interface. A microcontroller-based I/O core provides the highly flexible generic interfaces, which have been designed to suit most common vehicle system applications. Each interface is configured by a user-defined data table. The interface unit can be utilised in a number of ways, including remote data concentrator, a controller for power distribution panels and a subsystem controller.

As a remote data concentrator, the RIU-100 provides digital signal conditioning for system processors. When used as a controller for power distribution panels, it provides remote power switching combined with software logic interlocking, status monitoring and data bus interfacing. As a subsystem controller, the RIU-100 can perform closed loop control, execute external commands and provide health and status information to other systems via its databus interface.

The RIU-100 incorporates an adaptable modular architecture, with a predefined set of generic interfaces chosen to suit a wide range of common aircraft sensors and systems.

Specifications
Temperature: MIL-STD-810E, method 501.2/502.3 (–40 to +85°C operating, –65 to +90°C storage)
Humidity: MIL-STD-810E, method 507.3
Vibration: MIL-STD-810E, method 514
Shock: MIL-STD-810E, method 516.4 (30 g peak, 11 ms half-sine wave)
Altitude: MIL-STD-810E method 500.3 (–1,000 to +70,000 ft)
EMC: MIL-STD-461D/462D and lightning/nuclear ElectroMagnetic Pulse (EMP)
Power: MIL-STD-704A/E (28V DC); 5W
Weight: 0.7 kg

Status
In production and selected for a number of applications by Boeing and Lockheed Martin.

Contractor
Smiths Aerospace.

SEP10 Automatic Flight Control System (AFCS)

Type
Flight Control System (FCS).

Description
The SEP10 automatic flight control system is fitted to the British Aerospace 146. It provides three-axis control or stabilisation and incorporates a pitch and roll two-axis autopilot, elevator trim, flight director and yaw damping facilities. It uses simple, well-proved control laws and the minimum of sensors. There is also a synchronise control facility which allows the pilot to disengage the autopilot clutches and sensor chasers temporarily and to manoeuvre the aircraft manually, so adjusting the data of the basic and manometric autopilot modes.

The autopilot is based on rate control laws. Pitch and roll rate signals are derived from ARINC three-wire attitude references, thus eliminating the need for rate gyros. Other ARINC standard interfaces accept a wide range of sensor inputs, including those from barometric and radio navaid sensors, and allow systems to be tailored to suit operators' needs. The autopilot computer uses digital computing techniques to provide outer loop control and to organise the mode logic, and has capacity to accommodate optional facilities. Analogue computing is used for the inner loop stabilisation computing, servo-drive amplifiers and safety monitors.

The system can be supplied with either a parallel acting yaw damper, which uses a rotary servomotor to drive the rudder and rudder pedals, or a duplicated series yaw damper which drives linear actuators in series with the rudder control run. In each case, the yaw damping system is self-contained and consists of an analogue yaw computer, sensor and the relevant actuator or servo motor.

Flight director computations are performed within the digital section of the autopilot computer, which can supply commands to V-bar or split-axis flight directors. The flight director and autopilot share common mode selection and outer loop guidance but, if desired, they can be operated independently. Emphasis has been placed on maintainability and ease of testing, both for the installed system and for individual units in the workshop. Routine testing is designed to confirm correct functioning of safety devices, the tests being performed by operating a test-button in conjunction with buttons on the mode selector. Modular construction has been used extensively to ensure that faulty equipment can be corrected and recertified easily and quickly.

The following descriptions of individual LRUs outline the operation of a full SEP10 system.

Autopilot controller
The autopilot controller, in addition to providing autopilot and yaw damper engage or disengage controls, also includes pitch rate and roll angle selectors, and the elevator and rudder trim indicators. Engagement of the autopilot and yaw damper is confirmed by the illumination of a legend within each selector. Pitch control uses a spring-centred lever which has a non-linear feel so that minor adjustments can be made instinctively. Roll control is accomplished by rotation of the control knob, which remains offset by a displacement proportional to the roll angle demanded in the basic mode, but returns automatically to the central position on selection of an alternative mode.

Mode selector
There are 11 push-button switches, each illuminating as mode indicators for the selection of both autopilot and flight director functions; control mode engagement is confirmed by the illumination of a white triangle on the appropriate button.

A turbulence facility is included to soften flight disturbances in turbulent air. This reverts the autopilot to the basic stabilisation mode and at the same time reduces the overall gain of the system. Autopilot and engagement lights are provided so that the engagement state of the system can be seen on the mode selector. There is also provision for remote mode indication.

Autopilot computer

The autopilot computer receives both analogue and logic information from sensors, controllers and selectors and processes it to formulate the pitch and roll axis demands and the flight director commands. The majority of autopilot computing is performed digitally, although analogue techniques are used to provide pitch and roll stabilisation and authority limitations. Correct functioning of the computer safety circuits is verified by a test facility at a convenient remote station.

Yaw computer

This unit takes short-term damping information from a yaw rate gyro, a lateral accelerometer and the roll VRU; it drives the series rudder actuator to provide yaw damping and turn co-ordination. For aircraft types requiring a parallel damper, such a system is available.

Altitude selector

The altitude selector provides facilities for altitude preselect mode as well as the normal altitude alerting functions. Altitude information is obtained from either a servo altimeter or an air data source. Selected altitude is presented on a counter display. A warning flag obscures this display in the event of a power failure or absence of altitude valid signal, and a test facility allows checking of the associated audio-visual signals and altitude preselect function.

Air data unit

Where there is a requirement for a Mach hold facility to secure better fuel economy, the basic airspeed sensor can be replaced with an air data unit providing the necessary extra outputs.

Monitor computer

For operation to Cat II weather minima, this unit computes the performance monitor functions necessary to provide a fail-safe pitch channel. It independently monitors autopilot pitch rate, localiser and glide slope deviation and provides outputs that can be used to disconnect the autopilot and provide warnings to the pilot. The computer is completely independent of the autopilot, and a self-test facility allows a check to be made on the correct operation of all the monitoring functions.

Status

In service in the BAE Systems 146.

Contractor

Smiths Aerospace.

SEP20 Automatic Flight Control System (AFCS)

Type

Flight Control System (FCS).

Description

Both the autostabiliser and the autopilot in the SEP20 AFCS are fully digital and achieve levels of reliability and repeatability higher than was possible with earlier analogue systems. In addition, digital technology allows the mode of operation to be modified according to varying flight conditions or aircraft configurations. If required, the pilot can uncouple the autopilot and fly the aircraft using flight director commands provided by the system.

A comprehensive built-in test facility is incorporated, providing output data on the status of the system. Maintenance and release test functions are included, together with preflight safety checks, in-flight monitoring and the ability to identify a faulty LRU quickly.

The full AFCS comprises two identical digital Flight Control Computers (FCC), a Pilot's Control Unit (PCU), a Dynamic Sensor Unit (DSU) and a Hover Trim Control (HTC) unit for helicopter SAR applications. Extensive use is made of

ARINC 429 both for external communications and for communications within the system. The equipment is designed to suit widely differing primary roles, such as anti-submarine operations or civil passenger transport, necessitating exacting safety standards.

The two identical FCCs are packaged in a 6 MCU configuration. Each comprises nine printed circuit cards incorporating four microprocessors of two widely used types – the Intel 80286 and Motorola 68000. Each microprocessor has been programmed independently to minimise the possibility of common mode faults.

The PCU enables the pilot to select the required AFCS control mode and displays the state of mode engagement. The unit is divided into two segregated sections in order to maintain integrity and fault survivability.

To achieve the level of sensor signal redundancy for failure survival in the yaw axis, the DSU incorporates a yaw rate gyro, lateral accelerometer and normal accelerometer. Output signals are provided to ARINC 429 digital format.

For helicopter applications the HTC is integral with the winch controller and provides the winchman with limited authority control of the aircraft through the AFCS hover trim mode.

Status

In production and in service in the Westland/Agusta EH 101 helicopter.

Contractor

Smiths Aerospace.

SLAved Searchlight System for helicopters (SLASS)

Type

Aircraft system control, Line of Sight (LoS).

Description

The SLASS SLAved Searchlight System converts a standard remotely controlled searchlight to a servo-controlled system. When installed, the system provides the means to slave the searchlight to an E-O gimbal system, so that it automatically follows and points in the same direction as the gimbal. It consists of a servo control module and a remote-control unit which is used in conjunction with an E-O gimbal system. It operates with Spectrolab and other searchlight systems.

SLASS increases the performance of existing searchlight/E-O gimbal systems by reducing operator fatigue and minimising operator disorientation when his view changes to and from the video monitor scene and the outside illuminated scene.

Specifications

Dimensions (W × D × H): 127 × 165 × 38 mm
Weight: 1.0 kg
Power: 28 V DC 0.5 A mean
115/230 V with AC power converter
Environmental: temperature operating: –40 to +70-C
humidity 10 to 80% non-condensing

Status

SLASS is in production for US and European markets.

Contractor

Octec Ltd.

Smart fuel probes

Type

Aircraft sensor, Fuel Quantity Indication (FQI).

Description

Smiths' 'Smart' fuel probes are particularly suitable for operation in military aircraft, where their simple interface allows connection to

multiple processing units, thus improving redundancy and overall system availability.

The probe is formed from two concentric aluminium tubes, which constitute the plates (electrodes) of a capacitor. The two electrodes are spatially separated and located with high-quality, fuel-resistant polymer support pins. The probe can be mounted within the airframe using a number of different techniques, the exact design of which will be dependent on installation and insulation/bonding requirements. These techniques include integral metallic flange for attachment to the tank wall, attachment points for clamps, and brackets fitted to the probe.

Key features of the system include:
- Uses known law where capacitance is proportional to fuel height
- Not affected by aircraft harness stability
- Virtually immune to EMI
- Simple connection through circular 3-wire connector
- Easy-to-achieve multiple sensing channels for redundancy
- Simple computer interface
- 20 per cent less power consumption compared with AC probes
- Better than 0.5 per cent probe height can be achieved.

Status

In service in BAE Systems Nimrod MRA4, Boeing UCAV X-45, GKN Westland EH101 and Eurofighter Typhoon.

Contractor

Smiths Aerospace.

SN500 Series Automatic Flight Control Systems (AFCS)

Type

Flight-Control System (FCS).

Description

The SN500 series of Automatic Flight Control System (AFCS) are high-integrity designs for the most demanding applications. The system can be integrated both with existing and future facilities, such as MLS, and is suited for all Sea King roles including ASW, OTHT, SAR, AEW and utility. It features duplex stabilisation, attitude and heading hold, fly-through manoeuvring, automatic trimming, barometric height hold, airspeed hold, heading acquire, RNav, radar altitude hold, transition up and down, hover with hover trim and overfly.

The analogue autostabiliser provides tight hands-off control combined with flexible hands-on manoeuvring. The duplex lanes of computing are integrated with existing simplex series actuators and, in the cyclic axes, with existing hydraulic beepers. The system is designed to provide a level of safety suited to demanding environmental conditions. The computing circuits are modular and use separate plug-in boards in each lane. Each board is self-contained with its own stabilised power supply. Lane 1 and Lane 2 circuits are separated by a partition and use separate wiring. BITE facilitates ground testing and preflight check-out provides continuous interlane monitoring.

The digital autopilot uses the latest technology to provide coupled control of a wide range of modes. The use of digital techniques allows the specific requirements of each mode, both standard and special to role, to be taken into account. The use of extensive self-testing facilities and performance monitoring within the simplex microprocessor give unambiguous indication to the crew of any detected failures and, where safety requires it, initiates a fly-up procedure. In all axes but collective, autopilot outputs are summed into the duplex lanes of the analogue autostabiliser. In the collective axis, functions are monitored using analogue techniques.

The pilot's controller for the SN500 flight control system has engagement buttons for the duplex autostabiliser and autopilot modes   0504185

Specifications

Dimensions:
(sensor unit) 65 × 130 × 212 mm
(autostabiliser computer unit) 194 × 125 × 398 mm
(pilot's controller) 245 × 146 × 213 mm
(autopilot computer unit) 194 × 125 × 398 mm
(hover trim controller) 86 × 146 × 277 mm

Weight:
(sensor unit) 1.9 kg
(autostabiliser computer unit) 5 kg
(pilot's controller) 5 kg
(autopilot computer unit) 5 kg
(hover trim controller) 1 kg

Power supply: 115 V AC, 400 Hz
28 V DC

Status

The SN500 is in production and service with the SH-3 Sea King helicopter. The SN501 has been certified for the Sikorsky S61 helicopter.

Contractor

Smiths Aerospace.

Solid-state power distribution system for F/A-22

Type

Aircraft system control, electrical.

Description

The Smiths family of power management and distribution products includes the F/A-22 solid-state power distribution system, which controls and distributes secondary 28 V DC and 270 V DC power to aircraft loads. This advanced technology solid-state switching system is housed in four aircooled line replaceable Power Distribution Centres (PDCs), which also control

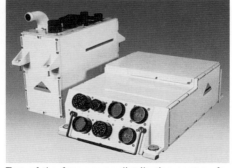

Two of the four power distribution centres for the F-22   0051646

the distribution of 115 V AC over airframe located circuit breakers and relays.

The modular design of the F/A-22 power distribution system allows greater flexibility of load management than conventional power distribution systems. Each PDC uses a common footprint Solid-State Power Controller (SSPC) design approach. The 28 V and 270 V SSPCs are the same size and are available in four amperage ratings: 5, 7.5, 12.5 and 22.5 A.

The SSPC modules are double-sided, surface mount boards which incorporate Smiths' patented integral buried bus bar. This common design for both 28 V and 270 V SSPC modules further enhances design flexibility and cost-effectiveness.

Status

In production and in service in F/A-22 Raptor aircraft.

Contractor

Smiths Aerospace.

Speech Recognition Module (SRM)

Type

Aircraft system control, avionic.

Description

Smiths Aerospace offers voice recognition and speech generation for use in both fixed- and rotary-wing aircraft and land applications. The system has been designed to provide robust speech recognition performance with flight-proven techniques to overcome the environmental factors of noise, g loading and pilot stress. The Speech Recognition Module (SRM) recognises continuous speech and is speaker dependent, requiring the pilot to register their voice on a PC-based Ground Support Station (GSS). The module provides an additional natural mode of command and control over many cockpit functions which reduces pilot workload, improves aircraft safety and expands mission capabilities.

Smiths speech recognition technology has been packaged onto various Mil-spec and commercial modules – Eurofighter SRM, AYK14 IVM, DVI 104 and PCI.

The main features of the system include:

- High recognition performance >95% in flight
- Fast response time <200 ms
- DVI vocabulary of 200 words
- Flight-proven counter measures to overcome the environmental factors of noise, g loading and stress
- Speaker dependent, language independent with full multilingual capability
- Flexible command syntax for easy reconfiguration
- Comprehensive PC-based ground support equipment
- Memory growth to 500 words available.

Status

A variant of the SRM is in development for the Eurofighter Typhoon (see separate entry).

Contractor

Smiths Aerospace.

STS 10 full flight regime autothrottle

Type

Aircraft system control, power plant.

Description

Designed for the Boeing 737-300 and now installed as standard equipment, the Smiths Aerospace autothrottle has been developed from the highly successful system supplied to Boeing for the 727-200 and 737-200 aircraft. In 1990, the STS 10 was fitted to the 737-500 and is capable of operating with intermixed engine situations.

A single unit will fit either 737-300, 737-400 or 737-500. It interfaces with flight management systems, digital air data systems, inertial reference systems and digital autopilots and uses advanced digital techniques for higher reliability, easier maintenance and lower cost of ownership.

The system comprises a digital computer with independent electromechanical drive to each throttle lever. The computer, which is housed in a single ½ ATR long box, accepts analogue and digital information from sensors and systems on board the aircraft. After processing this data the computer generates outputs to drive servo-actuators which adjust the position of each throttle lever independently, so achieving optimum engine performance. A further output from the same computer drives the fast/slow indicators on the ADIs or EFIS displays.

The autothrottle includes a number of unique features designed to enhance performance and promote flight safety. A particular feature of the system is the ability to override the actuator drive and adjust the throttle levers manually, without the pilot applying more force than he would normally use in manual operation.

To achieve precise control throughout the full flight regime, the autothrottle computer continuously monitors all the necessary engine and aircraft parameters and adjusts the thrust in accordance with the prevailing flight conditions. Protection is included to prevent exceeding predetermined N1 engine limits and maximum aircraft incidence.

The system includes damping controls which are designed to minimise throttle activity during normal flight conditions.

If a large change in vertical windspeed occurs during the approach a command is inserted which enables the system to achieve the required level of thrust more quickly.

With the launch of the Boeing 737-600/700/800 the autothrottle computer is being repackaged into a smaller lighter case. By the use of ASIC technology the autothrottle computer is being re-engineered into a ⅜ ATR Short case. The software embodied in the new unit will remain largely unchanged but is adapted for the differences in airframe performance and changed engine characteristics. The full flight regime features of the current B737-300/400/500 autothrottle computer will be embodied in the new autothrottle computer for the 737-600/700/800.

The single largest change is that the 737-600/700/800 aircraft will use FADEC controlled engines in the same way as the 777. This similarity has lead to a change in the servo drive for the autothrottle system. In the 737-300/400/500 it was necessary to drive the cables which routed the throttle level commands to the engine. In FADEC controlled aircraft it is only necessary to drive the pilots' throttle levers and therefore the technology used in Autothrottle Servo Motor (ASM) developed for the 777 has been adapted for use in the 737-600/700/800. The main change for Smiths Aerospace was the need to adapt the 777 unit into a smaller envelope for the new 737 variants.

Status

In production. The unit is standard fit on 737–300, 737–400, 737–500 and 737–700 for which a new ⅜ ATR unit will be produced.

Contractor

Smiths Aerospace.

UltraQuiet Active Noise and Vibration Control (ANVC) systems

Type

Aircraft system control, Noise, Vibration and Harshness (NVH).

Description

The UltraQuiet Active Noise and Vibration Control systems for passenger aircraft cabins

UltraQuiet Active Noise and Vibration Control System 1041440

reduce noise by duplicating the primary sound field with an additional secondary sound field in anti-phase. The UltraQuiet system can operate over the complete flight envelope including take-off and landing, reduce the harmonics of the blade passing frequency, track the variation in the blade passing frequency which occurs during air turbulence and when the aircraft banks, and reduce beat caused by poor propeller synchrophasing.

UltraQuiet Active Tuned Vibration Attenuators (ATVAs) are used in place of loudspeakers, and produce the anti-phase sound-field by vibrating the fuselage of the aircraft. This reduces both cabin noise and vibration. A typical system comprises 24 loudspeakers, 48 microphones and an electronic controller. The loudspeakers are lightweight devices specially designed for optimum performance at low frequencies. They are positioned in the trim beside the seats and in the roof. The exact positions are determined

by measurement and practicalities of the trim design. The loudspeakers each have a low-power amplifier which is mounted on, or adjacent to, the loudspeaker. This gives maximum flexibility when deciding the number of loudspeakers required. The microphones are positioned in the rear of the trim. The electronic controller consists of printed circuit boards inside a 1/2 ATR short ARINC 600 case. The unit has a comprehensive internal built-in test system.

The performance of the system is optimised for cabin wide noise reduction, with no areas where noise is unacceptable. The system compensates automatically for the changes in sound field caused by staff and passengers moving around the cabin. The system typically achieves attenuations of 8 to 12 dBA throughout the flight envelope.

The UltraQuiet seat integrates broadband Active Noise Control (ANC) technology into a single passenger seat creating a localised zone of quiet around the head of a seated passenger. This addresses the noise found on wide-bodied jet aircraft, a noise that is less dominated by the tones found on turboprops and rear-engined jet aircraft. The seat uses microphones in the headrest to monitor the original noise field. A digital signal processor located in the seat drives loudspeakers in the headrest in opposition to the original noise field. This leaves a residual noise field that is up to 10 dBA lower than the original.

Status
The launch customer for the ATVA system was de Havilland. The system is standard on the

Beech 1900D, Dash-8Q Series 100, 200, 300 and 400; Saab 340B+; Saab 2000 and King Air 350. It is optional on the Bombardier Challenger and other King Air aircraft.

Contractor
Ultra Electronics Ltd, Noise & Vibration Systems.

Versatile electronic engine controller

Type
Aircraft system control, power plant.

Description
Developed for the Pratt & Whitney PW305 turbofan engine, this is a versatile digital engine control unit which can be integrated with minimal redesign on a wide range of aero engines and airframes. Reduced unit cost from volume production, together with minimal non-recurring costs and benefits from reliability growth in related applications, combine to make this an attractive and cost-effective unit for both military and civil applications.

Status
In service in the Raytheon Hawker 1000, Learjet 60 and Astra Galaxy aircraft.

Contractor
Goodrich Corporation, Fuel and Utility Systems.

United States

253M-4A emergency panel

Type
Aircraft system control, emergency/survival.

Description
The 253M-4A emergency panel is a four-switch panel that provides the crew with rapid, automated, single-switch action control over several emergency communication radio and IFF functions. It also provides single-switch cancellation of mission sensitive data.

The 253M-4A incorporates four dedicated locking switches for communication emergency, IFF emergency, or 'zeroize' (guarded) switch requirements. The communication switch also disables both the auxiliary guard receiver and all receiver squelches. The IFF emergency switch causes the transponder to respond with emergency code 7700, independent of the status of the MIL-STD-1553B multiplex data bus. The 'zeroize' switch clears all codes, presets, and data bases.

The unit requires 5 V DC at 0.9 amps for edge lighting requirements and emits NVIS green in accordance with MIL-L-85762A for Night Vision Goggle (NVG) compatibility. The appropriate output discretes are grounded when the operator activates the panel switches, but each is diode isolated to prevent an external ground signal on the panel's output from inadvertently cancelling mission data.

Specifications
Dimensions: 28 × 146 × 132 mm (H × W × D)
Weight: 0.36 kg
Power: 5 V AC or DC; 0.94 Amp
Cooling: Convection
Qualification: MIL-STD-810C, MIL-STD-461B

Status
In production and in service.

Contractor
Rockwell Collins.

Advanced Lighting Control System (ALCS)

Type
Aircraft system control, external lighting system.

Description
Sanmina-SCI's Advanced Lighting Control System (ALCS) provides conditioned power for non-NVIS and NVIS-compatible avionics lighting. The system generates and controls up to 501 channels of 5 V DC, 28 V DC, and 115 V AC lighting power. The conditioned DC power eliminates audio frequency interference problems common in other light dimming systems. Features designed into the system facilitate the trimming and balancing of crew station lighting to accommodate wiring and optical loses. This balancing results in uniform lighting intensity, continuously controllable from the off condition to maximum intensity.

ALCS provides all functionality and interfaces required to implement a fully NVIS compatible lighting system.

The Lighting Control Units (LCUs) provide the lighting power and interface directly to aircraft interfaces such as dimmer controls and aircraft discretes. The Trim and Status Panel (TSP) is used to balance the lighting and perform diagnostics on the crew station lighting system. The status of all lighting channels

Sanmina-SCI ALCS 0098792

and all ALCS aircraft interfaces is monitored and displayed by the TSP. The TSP provides an RS-232 interface for use with an external computer to perform maintenance and to reconfigure the characteristics of the crew station lighting.
System features include:
• >1 kW lighting power per LCU
• Up to 12 lighting zones per LCU
• Over temperature shutdown
• Over current/short circuit shutdown
• BIT
• Reconfigurable on-aircraft.

Specifications
Dimensions:
LCU 279.4 × 228.6 × 304.8 mm (H × W × L)
TSP 101.6 × 136.5 mm (DZUS rail mount)
Weight:
LCU 26.8 kg (fully populated)
TSP 2.3 kg
Power: 28 V DC, 40 A; 115 V AC, 400 Hz, 1 A
Temperature: −54 to +55°C
Environmental: MIL-STD-810

Status
In production and in service on the C-130H.

Contractor
Sanmina-SCI Corporation.

Advanced Memory Unit (AMU) Model 3266N

Type
Data processing system.

Description
The Smiths Aerospace Advanced Memory Unit (AMU), Model 3266N, provides reliable in-flight mass memory data transfer and recording under extreme environmental conditions. The AMU is a high-performance memory management subsystem, featuring a PowerPC processor and dual PCMCIA PC Card receptacles. Communication

The advanced memory unit model 3266N and representative-fit aircraft 0051283

capabilities include MIL-STD-1553A/B remote terminal and bus monitor, High-Speed Interface (HSI, fibre channel) and RS-422/RS-485 interface buses. The cockpit receptacle accepts a Type I or II PC Card for mission data and a Type I, II or III card for maintenance and map data.

The AMU is part of the US Navy Tactical Aircraft Moving MAp Capability (TAMMAC) programme, developed by the Naval Air Systems Command, PMA-209 and The Boeing Company. TAMMAC includes the AMU, a Digital Map Computer (DMC) and a HSI cable assembly. The AMU and DMC communicate over the HSI bus. Mission and maintenance data are transferred over a MIL-STD-1553A/B bus. The AMU can be used within the TAMMAC system or as a stand-alone data loader/recorder. Inherent AMU processor and storage capabilities allow for optimised functional growth, including PC-Card-loadable, aircraft-unique user programmes. The AMU is externally software updatable. The TAMMAC AMU is designed to replace earlier US Navy Data Storage Set (DSS or MU) and Mission Data Loader (MDL) equipment, in addition to providing for increased functionality.

Specifications
Dimensions: 146 × 76 × 210 mm
Weight: 2.0 kg
Power: 28 V DC, 25 W maximum
Reliability: 18.000 h predicted MTBF

Status
Baseline TAMMAC aircraft include the F/A-18C/D, F/A–18E/F, AV-8B, AH-1Z, UH-1Y and MH-60S CSAR. Additional Navy TAMMAC aircraft are planned.

Contractor
Smiths Aerospace.

AH-64 caution and warning system

Type
Avionic Fault Warning System (FWS).

Description
This system provides caution and warning annunciation for the AH-64 Apache helicopter.

It consists of a 60 station pilot panel, 24 station co-pilot panel and master caution warning panel. Subsystem inputs are monitored and fault information displayed as applicable on any or all three units. The legend modules are fully sunlight-readable and incorporate front lamp replacement.

Specifications
Dimensions:
(pilot panel) 146 × 206.3 × 177.8 mm
(co-pilot panel) 146 × 95.2 × 177.8 mm
(master caution) 193.5 × 25.4 × 69.8 mm
Weight:
(pilot panel) 4.4 kg
(co-pilot panel) 2.36 kg
(master caution) 0.73 kg

Status
In production and in service in the AH-64 Apache helicopter.

Contractor
Smiths Aerospace Electronic Systems, Long Island.

Airborne File Server (AFS)

Type
Data processing system.

Description
The Airborne File Server (AFS) is the next generation of the Astronautics Onboard Maintenance Terminal Computer Unit (OCU), built for the use as a maintenance unit and network server on aircraft. The AFS family of computers can be used for applications that include maintenance, Internet connections, telecommunications, flight log and other flight activities. The basic design has been TSO-approved by the Federal Aviation Administration.

Specifications
Processor: Pentium™ III/K6 266 MHz
Memory: 128 Mb RAM, 512 K L2 cache, 8.4 Gb non-removable, PCMCIA Type II/III removable, 48 Mb Flash
Dimensions: ARINC-600 4 MCU (124 × 324 × 194 mm)
Weight: 6.36 kg
Interface: ARINC-429, -717 (FDR), RS-422, -232, Discrete, IEEE 1284 (Parallel), universal serial bus, Ethernet, modem, E1/T1

Status
Produced in support of Penny and Giles Aerospace and Aeronautical Information Solutions Group.

Contractor
Astronautics Corporation of America.

Airborne vibration monitoring system

Type
Aircraft system control, Noise, Vibration and Harshness (NVH).

Description
The airborne vibration monitoring system is designed for engine rotor imbalance monitoring and onboard two-plane engine trim balancing. The system consists of two major components:

Airborne vibration monitoring system for the Boeing 777 1041437

a remote charge converter and a signal conditioner.

Vibration data is transmitted to the cockpit for display via an ARINC 429 or 629 databus. Advanced digital signal processing techniques provide accurate vibration magnitude and phase management. Utilising in-flight data provides superior trim balance information, allowing accurate one shot engine balancing and eliminating requirements for engine ground runs for balance verification. The system simplifies engine balance operations by providing specific maintenance instructions. Extensive BIT capabilities isolate faults between accelerometers, the two LRUs and the interconnection cabling.

For system retrofit applications, an integrated signal conditioner, incorporating the functions of a remote charge converter, is available in the ARINC 429 configuration.

Specifications
Dimensions:
(signal conditioner) ARINC 600 3 MCU
(remote charge converter) 56 × 61 × 114 mm
Weight:
(signal conditioner) 2.8 kg
(remote charge converter) 0.45 kg

Status
In service in the Boeing 777, Canadair Global Express and Nimrod.

Contractor
Ametek Aerospace Products.

Aircraft Condition Analysis and Monitoring System (ACAMS)

Type
Data processing system.

Description
This Aircraft Condition Analysis and Monitoring System (ACAMS) monitors multiple data sources and stores a history in its 20 Gb flash array. A second processor, using Artificial Intelligence (AI) techniques, processes this array and makes decisions regarding potential problems with the host aircraft. ACAMS also monitors the health of the aircraft power system. Standard Compact PCI 3U cards and a 6-slot backplane allow for customised configurations of the system to suit customer requirements.

The system is convection cooled (external forced air will aid cooling), with low power consumption. Ethernet and RS-232 connections are available on the front panel, along with push buttons and indicators. There is an ARINC 600 connector at the rear for signals.

Specifications
Processor: PowerPC (bus controller or slave)
Dual processors can be expanded to five
Memory: 4 Mb Flash; 128 Mb RAM; dual-DMA
Main memory module: 10 Gb flash array per module (80 Gb module optional);
non-volatile FERAM bad block and frequently accessed variable storage;
up to 3 modules per system
Interface: Ethernet; dual RS-232;
CPCI bus (I/O transfers)
ARINC 429 I/O: 16 transmit channels;
32 receive channels;
Global status register for fast scanning
Power control: 200 msec system hold up using storage capacitors;
Early fail interrupt;
Generates common reset for processors;
Power: 28 V DC or 120 V AC, 400 Hz
Environmental: DO-160 Cat B
Dimensions: 330.2 (D) × 190.5 (H) × 184.2 (W) mm
Weight: 3.18 kg

Status
In production.

Contractor
Symetrics Industries Inc.

Aircraft Monitoring Unit (AMU)

Type
Onboard system monitor.

Description
DRS Technologies' Aircraft Monitoring Unit (AMU) acquires host platform flight parameters in addition to acquiring and compressing audio signals.

Specifications
Data acquisition
Analogue inputs
60 channels used to monitor all sensor inputs
Channel usage per sensor:
   1 channel per potentiometer
   1 channel per temperature probe
   1 channel per frequency (tachometer)
   2 channels per 144 discretes
   3 channels per synchro
   3 channels per magnesym (fuel flow)
Expandable to 120 channels.

Data acquisition
Audio
3 input channels, 50 to 3,500 Hz (pilot, co-pilot, third crew member)
1 input channel, 50 to 6,000 Hz (area mic)
1 replay channel, 50 to 6,000 Hz (for monitor)

Data acquisition
ARINC-429
8 ARINC-429 input channels (fast/slow)
2 ARINC-429 output channels (fast/slow)

Data acquisition
MIL-STD-1553
2 redundant MIL-STD-1553 input bus
Bus monitor/remote terminal.

Data acquisition
Real-time output
Flight data output in real-time in ARINC-717 or RS-232 format.

Data acquisition
Output to data recorder
RS-422 at 1 Mbits/sec
4 frames at every second
Frame synchronisation pattern
4 sub-frames per frame
Each sub-frame contains up to 90 words of 12 bits each.

Power: 28 V DC or 115 V AC/400 Hz
Dimensions: 317.5 × 124.5 × 193.0 mm
Weight: <4.55 kg
Compliance: ED-55, ED-56A

Status
In production.

Contractor
DRS Technologies, Inc.

Aircraft Propulsion Data Management Computer (APDMC)

Type
Data processing system.

Description
The APDMC is primarily an engine monitoring, aircraft warning and display and data transfer unit. It acquires engine data from electronic engine controls and provides engine health, status, maintenance and limit functions.

Avionics systems are connected to the warning and caution systems MIL-STD-1553B bus and ARINC 429 bus through the APDMC with 37 other aircraft LRU analogue and discrete input/output parameters. The APDMC provides aircraft stall, overspeed, take-off and horizontal stabiliser warning and display; records engine maintenance data and system flight data; and provides cockpit display data for system LRUs, fault history, propulsion and exhaust gas temperature.

The signal input/output complement includes dual-redundant MIL-STD-1553B multiplex databusses; eight receive and seven transmit ARINC 429 serial databusses; two ARINC 573-7 serial databus outputs of which one is used for an AIMS recorder output; 35 analogue inputs including a mix of RVDTs and LVDTs plus excitation; a mix of AC and DC absolute and ratiometric signals; 26 discrete inputs including series and shunt; and six discrete outputs.

The MIL-STD-1750 microprocessor and memory management unit controls an expandable 128 kbyte onboard ROM, 32 kbyte RAM and 32 kbyte NV fault storage. System BIT allows fault isolation to the offending LRU with 97.5 per cent confidence. Internal BIT provides onboard unit status. Non-volatile fault storage allows ground-based interrogation of LRU faults.

Specifications
Dimensions: 123.9 × 193 × 320 mm
Weight: 5.55 kg
Power supply: 28 V DC, 40 W
   115 V AC

Status
In production for the US Air Force C-17.

Contractor
Hamilton Sundstrand Corporation.

Altitude Hold and Hover Stabilisation system (AHHS)

Type
Flight Control Computer (FCC).

Description
The AHHS is designed to reduce pilot workload and increase safety during low altitude and low-speed aircraft operations by providing the H-60 pilot with a variety of altitude hold and stabilised hover/low-speed control modes. The main features of the system include:
- Interfaces to existing sensors
- Fail safe design
- Pilot maintains full override authority
- High reliability (4,500 h MTBF)
- Utilises proven system interfaces
- Utilises GFE/MIL parts

AHHS functions include radar altitude hold, barometric altitude hold, hover stabilisation with position hold, hover stabilisation with radar altitude hold, programmed velocity drift and collective trim.

System Components
System Control Unit (SCU)
The SCU is a DZUS-mounted, NVIS-compatible control panel that permits mode selection and sets the radar altitude hold height. The SCU also displays system failure information.

System Electronics Unit (SEU)
The SEU inputs data from sensors and aircraft systems and converts data to commands for the collective, pitch, and roll trim servos. The SEU also provides system health status.

Forward Relay Unit (FRU)
The FRU is an optional switching unit for displaying AHHS cues on existing displays. It is not required if a hover display is present.

System Interface Unit (SIU)
The SIU is an optional unit for power distribution and switching relays in some applications.

Collective Trim Services (CTS)
CTS provides the electromechanical interface to the collective for altitude control. It also provides electrical trim for the collective control when the AHHS is not engaged.

The AHHS system can easily be adapted to any H-60 installation. The software can be modified to accept any ARINC-429 or -575 data configuration from the navigation system. The other interfaces used by the AHHS are standard to most H-60 aircraft configurations. Unique design requirements can be accommodated through minor changes to the SEU interface circuitry. The FRU switches data to existing instruments such as the VSI when AHHS is active, in the event a dedicated hover display is not present.

Specifications
System Control Unit (SCU)
  Weight: 1.14 kg
  Dimensions: 146 × 76 × 178 mm

System Electronics Unit (SEU)
  Weight: 5.45 kg
  Dimensions: 406 × 76 × 178 mm
  Power requirements: 115 V AC, 400 Hz, 250 mA;
    28 V DC, 1 A

Forward Relay Unit (FRU)
  Weight: 0.68 kg
  Dimensions: 165 × 137 × 51 mm

System Interface Unit (SIU)
  Weight: 1.6 kg
  Dimensions: 95 × 73 × 194 mm

Collective Trim Services (CTS)
  Weight: 5.11 kg
  Dimensions: 133 × 267 × 127 mm

Status
In production and in service in the H-60 helicopter.

Contractor
DRS Technologies, Inc.

Altitude Preselect/Alerter 1D960

Type
Avionic crew alerting system.

Description
The 1D960 Altitude Preselect/Alerter provides precise altitude preselecting and/or alerting capabilities, in addition to ATC (100 ft resolution) or RS-232 (1 ft resolution) altitude encoding.

During climb or descent, the desired capture altitude is entered on the 1D960 with the rotary switch. The 1D960 is armed by pressing the 'AP Arm' switch on the face of the unit. The Altitude Preselect/Alerter compares the selected altitude with the information provided by the encoding altimeter. As the altitude approaches to within 1,000 ft of selected altitude the display will flash and give two audible tones. Upon closure of 200 ft from selected altitude the display stops flashing and gives one audible tone. At selected altitude the autopilot will smoothly transition to altitude mode and the 1D960 will change from Arm Mode to display current encoder altitude. If the aircraft deviates more than 200 ft from selected altitude the display will flash and give one audible tone.

The 1D960 Altitude/Preselect/Alerter can be installed with the two-axis Century 2000 or Century 41 Flight Control Systems, or used as a stand-alone altitude alerter. It has a five position switch to enter the following modes: altitude readout; radar altimeter setting; adjustable pixel dimming; barometric/millibar setting; and altitude select/arm setting.

1D960 altitude preselect/alerter   0001429

Specifications
Dimensions: 37 × 81 × 229 mm
Panel cutout: 3 in ATI
Weight: 0.61 kg
Power: 14 or 28 V DC

Status
In service.

Contractor
Century Flight Systems Inc.

AMS-2000 Altitude Management and alert System

Type
Avionic crew alerting system.

Description
The Shadin AMS-2000 Altitude Management and alert System uses information from Mode-C altitude encoders to deliver: time-based altitude alerting, rather than fixed altitude buffers, to notify the pilot 15 seconds before an altitude target or limit is reached – regardless of rate of climb or descent; automatic calculation and display of density altitude; real-time display of instantaneous vertical speed without the inherent lag of the aircraft's static system; calculation and display of aircraft and engine performance percentage.

Specifications
Dimensions: 38 × 80 × 146 mm
Power: 10 to 28 V DC, at 300 max
Weight: 0.25 kg

Status
In service.

Contractor
Shadin Co Inc.

AMS-2000 Altitude Management and alert System   0001445

AN/AKT-22(V)4 telemetry data transmitting set

Type
Data processing system.

Description
The AN/AKT-22(V)4 system relays up to eight channels of sonobuoy data from an ASW helicopter to a ship. It comprises the T-1220B transmitter-multiplexer, the C-8988A control indicator, an AS-3033 antenna and a TG-229 actuator. The sonobuoy signals are received by dual AN/ARR-75 radio receiving sets and passed into the transmitter-multiplexer. The control indicator has four trigger switches, each of which disables two data channels. Composite trigger tones are brought into the multiplexer separately via the control indicator and combined with the sonic data channels and the single voice channel. The resulting FM signal is used to modulate the transmitter.

The data transmitting set has a DIFAR operating mode in which the extra voice channel is inoperative and the composite FM modulating signal is disconnected from the transmitter input. Two DIFAR sonobuoy transmissions enter the transmitter-multiplexer on dedicated channels. After conditioning in an amplifier-adaptor, the DIFAR signal is split into two components: DIFAR

A and DIFAR B. The A signal is conditioned in a low-pass filter, while the B signal drives a variable-cycle oscillator centred at 70 kHz and then passes through a bandpass filter. The two filter outputs are combined linearly and the resulting composite modulation signal drives the transmitter. A switch on the controller indicator controls whether the normal sonic or composite DIFAR signals are transmitted.

The AS-3033 antenna has two sections: a VHF element for receiving the sonobuoy signals and a UHF part which sends the multiplexed data down to the ship. The ship receives the information on an AN/SKR telemetric data receiving set.

Specifications
Power supply: 115 V AC, 3-phase, at 0.85, 1.5, and 1 A respectively
Warm-up time: <1 min in standby mode < 15 min under environmental extremes
Operating stability: >100 h for continuous or intermittent operation
Frequency: 2,200–2,290 MHz, 1 of 20 switch-selectable E-L-band channels
Multiplexer inputs:
(a) 8 sonar data channels (7 with 10–2,000 Hz bandwidth; 1 with 10–2,800 Hz), at 0.16–16 V
(b) 4 sonar trigger channels, 26–38 kHz, at 1–3 V
(c) 1 voice channel, 300–2,000 Hz bandwidth, at 0–0.25 V
(d) 2 composite DIFAR channels, 10-2,000 Hz bandwidth at 3–6 V or 10.6 V
Channel phase correlation: difference in phase delay between any 2 passive data channels <1° (10–500 Hz)

Status
In service with the US Navy.

Contractor
Flightline Systems.

AN/APN-171 radar altimeter

Type
Flight instrumentation, radar altimeter.

Description
For over 25 years, US Navy helicopters have been flying with the APN-171 (HG9000) as their standard altimeter. Many functions are available with this system, including low-altitude warning, radar altitude warning set input, aircraft rate, aircraft altitude errors and landing gear warnings.

Three basic Honeywell AN/APN-171 radar altimeter systems are currently available: the HG9010 is a 0 to 1,000 ft system, the HG9025 is a 0 to 2,500 ft system and the HG9050 is a 0 to 5,000 ft system. All AN/APN-171 systems are available with standard or special output signals to represent a particular altitude range.

Specifications
Dimensions: 337 × 194 × 125 mm
Weight: 6.1 kg
Power supply: 115 V AC, 400 Hz, 10 VA
Altitude:
(HG9010) 0–1,000 ft
(HG9025) 0–2,500 ft
(HG9050) 0–5,000 ft
Accuracy: ±5 ft ±3%

Status
In service as standard on all US Navy helicopters. A modification kit is available to upgrade existing systems.

Contractor
Honeywell Sensor and Guidance Products.

AN/APN-194 radar altimeter

Type
Flight instrumentation, radar altimeter.

Description
The AN/APN-194 radar altimeter is standard on all navy fixed-wing and high-performance aircraft

including the F-14 and F/A-18. Functions include low-altitude warning, radar altitude warning set input, aircraft rate, aircraft altitude errors, landing gear warnings and both analogue and digital outputs.

Several height indicators interface with the APN-194. The AN/APN-194 was introduced in the early 1970s as a form, fit and function replacement for the AN/APN-141 and Honeywell has now produced over 8,000 of these altimeters. Since then, the AN/APN-194 has been upgraded twice, incorporating a solid-state transmitter and production/installation enhancements.

Options include transmitter power management. The AN/APN-194 can interface with up to four height indicators.

Specifications
Dimensions: 185 × 97 × 82 mm
Weight:
(radar altimeter) 2 kg
(height indicator) 0.7 kg
Power supply: 115 V AC, 400 Hz, 25 W
Frequency: FCC approved 4.3 GHz
Transmit power: 5 W
PRF: 20 kHz
Pulsewidth: 0.02 or 0.20 μs
Altitude: 0–5,000 ft
Accuracy: ±3 ft ±4%

Status
In service as standard on all US Navy fixed-wing high-performance aircraft.

Contractor
Honeywell Sensor and Guidance Products.

AN/APN-209 radar altimeter

Type
Flight instrumentation, radar altimeter.

Description
The AN/APN-209 radar altimeter is standard on all US Army helicopters. Functions include transmitter power management, low- and high-altitude warnings, NVG compatibility, analogue and digital outputs and integration of indicator, receiver and transmitter. For more installation flexibility, a version of the AN/APN-209 with the transmitter/receiver separate from the indicator is available.

Honeywell has produced over 9,000 AN/APN-209 altimeters. It was introduced in the early 1970s and, since then, has been upgraded twice, incorporating a solid-state transmitter and producibility enhancements.

Options include MIL-STD-1553B databus, voice warning and LPI enhancements.

Specifications
Dimensions:
(receiver/transmitter) 145 × 83 × 83 mm
(indicator/receiver/transmitter) 199 × 83 × 83 mm
Weight:
(receiver/transmitter) 1.4 kg
(indicator/receiver/transmitter) 1.9 kg
Power supply: 28 V DC, 25 W
Altitude: 0–1,500 ft
Accuracy: ±3 ft ±3%

Status
In production and in service as standard on all US Army helicopters.

Contractor
Honeywell Sensor and Guidance Products.

AN/APN-224 radar altimeter

Type
Flight instrumentation, radar altimeter.

Description
The APN-224 was developed specifically for the Boeing B-52 and meets the nuclear hardening and high-reliability specifications of that aircraft. The system's performance and ability to withstand severe environments led to its selection by the US Air Force for the B-1B. The APN-224 has also been selected for the US Air Force's A-10 and the US Air National Guard's F-16.

Functions of the AN/APN-224 include nuclear hardness, aircraft rate, blanking pulse and both digital and analogue outputs.

A modified version of the AN/APN-224, which has a MIL-STD-1553B databus, has been developed for the F-16 LANTIRN aircraft.

Specifications
Dimensions:
(B-52) 213 × 127 × 86 mm
(F-16) 213 × 160 × 102 mm
Weight:
(B-52) 2.1 kg
(F-16) 3.6 kg
Power supply: 115 V AC, 400 Hz, 45 VA
Altitude:
(B-52) 0–5,000 ft
(F-16) 0–10,000 ft
Accuracy: ±5 ft ±4%

Status
In service on US Air Force/US Air National Guard B-52, B-1B and F-16 LANTIRN aircraft.

Contractor
Honeywell Sensor and Guidance Products.

AN/APN-232 Combined Altitude Radar Altimeter

Type
Flight instrumentation, radar altimeter.

Description
The Combined Altitude Radar Altimeter (CARA) is an all-solid-state 0 to 50,000 ft FM/CW radar altimeter system operating at a nominal frequency of 4.3 GHz. It consists of a receiver/transmitter, antennas and indicators and offers inherent low probability of intercept and anti-jam capability as well as conventional analogue and digital outputs for aircraft avionic systems.

Low Probability of Intercept (LPI) performance is achieved by control of the power output so that the transmitted power is the least amount required for signal acquisition and tracking. The system mechanisation automatically adjusts the required transmitter power to maintain normal system operation over varying terrain, aircraft altitude and attitudes. The FM/CW technology, operating over a 100 MHz bandwidth, provides inherent spread-spectrum capability which further reduces detectability.

The AN/APN-232 is easily maintained and offers low life cycle cost. The equipment incorporates highly modular packaging. The assemblies, subassemblies and components are 100 per cent screened and tested to stringent military standards. Ease of maintenance is achieved by using proven microprocessor-based digital display fault monitoring. These elements combine to produce a predicted system MTBF of greater than 2,000 hours. Low life cycle cost is obtained from high system MTBF, automatic self-test with fault isolation to both the line- and shop-replaceable units and a two echelon maintenance concept which maximises the use of shop-replaceable units that may be discarded rather than repaired.

Specifications
Dimensions:
(receiver/transmitter) 88.9 × 160 × 222.2 mm
(indicator) 82.55 × 82.55 × 104.1 mm
(fixed-wing antenna) 101.6 × 101.6 × 5.3 mm
(rotary-wing antenna) 142.2 × 158.7 × 5.3 mm
Weight:
(receiver/transmitter) 4.76 kg
(indicator) 1.13 kg
(fixed-wing antenna) 0.23 kg
(rotary-wing antenna) 0.34 kg
Power supply: 28 V DC, 100 W max
Accuracy:
(analogue) ±2 ft ±2%
(digital) ±2 ft (0–100 ft), ±2% (100–5,000 ft), ±100 ft (5,000–10,000 ft), ±1% (>10,000 ft)

Status
In production, with over 10,000 APN-232s delivered. CARA is the US Air Force's standard altimeter.

Contractor
NavCom Defense Electronics Inc.

AN/ASK-7 data transfer unit

Type
Data processing system.

Description
The AN/ASK-7 data transfer device, originally designed for use with the Boeing AGM-86 air-launched cruise missile, has now been adapted for the US Air Force B-52 and B-1B aircraft.

The unit transfers large quantities of digital information from ground-based master systems to aircraft equipment. It will also monitor performance parameters as part of the central integrated test system on the B-1B. In the B-52, it forms a small part of the current Offensive Avionics System (OAS) update programme to improve the effectiveness of the aircraft.

Major features of the AN/ASK-7 include full record/replay capability, MIL-STD-1553B interface, error correction code, remote electronics, extensive built-in test and nuclear hardening. The system comprises a control unit, containing most of the recorder electronics, and an instrument-mounted cassette drive unit accommodating two cassettes at a time.

Specifications
Weight:
(cassettes) 3.1 kg
(cassette drive unit) 3.6 kg
(control unit) 6.8 kg
Power supply: 115 V AC, 400 Hz, 20 W

Status
In service.

Contractor
Honeywell Aerospace, Sensor and Guidance Products.

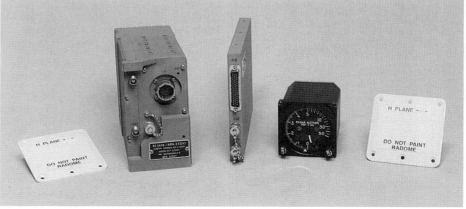

The Navcom AN/APN-232(V) CARA system

0504119

AN/ASQ-195 signal data converter set

Type
Data processing system.

Description
The AN/ASQ-195 signal data converter set is a multifunctional digital processor that provides control and data interchange for radios, instrument landing systems, Tacan, automatic direction-finders, sensors such as LANTIRN and radar altimeters, air-to-air interrogation, IFF and avionics BIT. The system has been designed for the F-15E but can be adapted to other aircraft. Each of the two units serves as back-up or redundant data processor for the other unit for many of the functions.

The AN/ASQ-195 features three dual-redundant MIL-STD-1553B databus interfaces, three RS-422 databus interfaces, three MIL-STD-1750A microprocessors, six serial microcontrollers, 120 kbyte EEPROM memory, 240 programmable discrete output interfaces, 355 programmable discrete input interfaces, four synchro interfaces and 28 analogue interfaces.

Display of the INS parameters, frequencies and channel selection is provided by the up-front control panel, communicating with the AN/ASQ-195 via the RS-422 databus.

Contractor
The Boeing Company.

AN/ASQ-197 Sensor Control-Data Display Set (SC-DDS)

Type
Data processing system.

Description
Smiths Aerospace's AN/ASQ-197 Sensor Control-Data Display Set (SC-DDS) provides a link between the reconnaissance sensor suite, the reconnaissance operator and the aircraft systems providing function control, data annotation, and navigational data distribution. Originally designed for use in F/A-18D(RC)'s film system, the SC-DDS is expandable to accommodate an E-O sensor suite with an interface to an external image processing unit. This also provides an SC-DDS hybrid film and E-O sensor capability.

As presently configured, the AN/ASQ-197 controls three to five sensors. Command messages, via the MIL-STD-1553 mux bus or serial interfaces to the aircraft and cockpit, determine sensor ready, run, extra picture and so on. Analogue control signals for Image Motion Compensation (IMC), altitude, and cycle rate pulses are generated for each station based on aircraft Vg/H. Sensor suite

status is transmitted over the mux bus or output via discretes, providing periodic update of current system conditions.

Annotation data is formatted in both alphanumeric and MIL-STD-782 BCD formats. These data blocks are then output to appropriate recording head assembly, CRT or LED, for film annotation or for E-O video annotation to the separate image processing unit.

AN/ASQ-197 front panel controls and displays allow the operator to enter operational data, initiate system self-tests and monitor system status. Remote operate and internal BIT sections allow the operator to initiate built-in-test on any of the connected sensors, on the SC-DDS itself, or to run any of the sensors with fixed control parameters (FMC, cycle rate). Integrated AN/ASQ-197 configurations are designed for F/A-18D (RC), F-14A/D TARPS, RF-111C AUP and AV-8B ETARS aircraft.

Specifications
Data inputs: SSI, mux bus – A3818, A5232, or MIL-STD-1553B.
Sensor suite: system/sensor feedback, data demands and reference voltages.
Manual preflight: ADAS modes, fixed data film remaining, bit initiated.
Data outputs:
Aircraft mux bus: time, running, film remaining, status.
Sensor suite: Operate commands.
Filming annotation CRT, LED matrix or strip RHA.
Power: 115 V AC, 1 phase, 400 Hz at 100 VA max 28 V DC at 56 W max.
Weight: (SC-DDS and tray) – 12.3 kg.
Dimensions: 225.6 (W) × 222.3 (H) × 406.4 (L) mm.

Status
The AN/ASQ-197 is in service in F/A-18D (RC), F-14 A/D TARPS, RF-111C (AUP), AV-8B ETARS.

Contractor
Smiths Aerospace, Germantown.

AN/ASQ-215 Digital Data Set (DDS)

Type
Data processing system.

Description
Smiths Aerospace's Digital Data Set (DDS) is a high-capacity, solid-state military airborne data storage and retrieval system. It provides rapid mission initialisation of the aircraft's avionics suite, as well as recording pertinent mission and maintenance data during flight; the DDS also supports in-flight processing. The AN/ASQ-215

The mission data loader consists of (left) the data transfer module and (right) the interface receptacle unit   0504164

has been adopted as the US Navy standard Digital Data Set (DDS). The DDS will support Global Positioning System data entry requirements for all MIL-STD-1553 multiplex databus-equipped US Navy aircraft. However, because many added features have been incorporated into this data management system, the DDS is suitable for a variety of tri-service military applications.

The Digital Data Set consists of two elements, a hand-carried, non-volatile solid-state data storage device known as the Data Transfer Module (DTM), and the cockpit-mounted, intelligent receptacle for the DTM, the Interface Receptacle Unit (IRU). The IRU manages and controls the data exchange between the DTM and other avionic subsystems via the aircraft's serial digital MIL-STD-1553 multiplex databus. The DTM serves as the transportable storage medium for both pre- and post-mission information exchange between ground computer stations and the airborne system. An upgraded DTM is now available to meet increased mission memory requirements. All fielded AN/ASQ-215 receptacles can have their software upgraded to accept the PCMCIA DTM (PDTM). In addition, the PCMCIA card may be accessed from laptop computers.

Operating primarily as a remote terminal on the MIL-STD-1553 multiplex bus, the IRU also has the provisions to operate as a bus controller. The IRU provides an exhaustive self-test/Built-In-Test function for all circuitry, including the DTM while performing real-time bulk memory error detection and correction.

On a single circuit board, the DTM provides 2 Mbytes, expandable to 40 Mbytes with an additional card, of high-speed random access non-volatile data storage. The DTM contains only bulk memory and address decode logic to reduce overall operating expense and simplify future growth. A high degree of data integrity is assured through the use of a proven error detection and correction method.

Specifications
Dimensions:
(IRU) 203 × 127 × 53 mm
(DTM & PDTM) 152 × 81 × 32 mm
Weight:
(IRU) 1.50 kg
(DTM) 0.31 kg
(PDTM) 0.47 kg
Input power: 28 V DC at 7.5 W (max)
Reliability:
(System) 16,239 h MTBF
(IRU) 43,000 h MTBF
(PDTM) 25,970 h MTBF with one PCMCIA card
PDTM capacity: 2 Type-II or one Type-III PCMCIA ATA memory cards

Status
In service.

Contractor
Smiths Aerospace, Germantown.

AN/AYA-8C data processing system

Type
Data processing system.

Description
The AN/AYA-8C data processing system for the US Navy P-3C Orion anti-submarine warfare

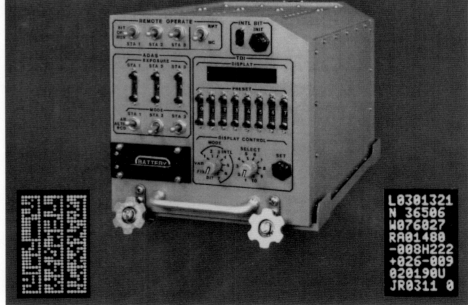

The AN/ASQ-197 Sensor Control-Data Display Set (SC-DDS)   0051289

aircraft provides the interface between the AN/ASQ-212 general purpose digital computer and the aircraft systems and so constitutes a major part of the P-3C's mission avionics.

The system is connected to all the crew stations on the aircraft: tactical co-ordinator, non-acoustic sensor, navigation/communications, acoustic sensor and flight deck. In addition, it communicates directly with the radar interface unit, armament/ordnance system, navigation systems, sonar receiver, magnetic anomaly detector, infra-red detection system, electronic support measures, sonobuoy reference system and Omega.

The P-3C's data processing activities are divided into four logic units, forming separate boxes in which all electronic operations are conducted in a combination of digital and analogue formats. Various keysets and panels complete the system hardware.

Logic Unit 1 interfaces between the central computer and four types of peripheral information system: manual data entry, system status, sonobuoy receiver and auxiliary readout display.

The manual entry subsystem provides the communication between the various operator stations and the central computer. Each operator has a panel of illuminated switches and indicators by which he communicates with the central computer. System status for the navigation and magnetic anomaly detector is received and stored by part of Logic Unit 1, then transmitted to the central computer. The computer receives the status information on demand, or when any status changes. Finally, the auxiliary readout display logic interfaces between the central computer and the auxiliary displays at the tactical co-ordinator and nav/com stations. The radar, sonar antenna, infra-red detection system and electronic support measures interfaces are also achieved by Logic Unit 1.

Logic Unit 2 is concerned with communicating between the central computer and the navigation and armament/ordnance systems. This logic unit transmits Doppler and inertial navigation data to the computer, and instructions to launch search-and-kill stores.

Logic Unit 3 controls the CRT displays provided for the tactical co-ordinator, sensor operators and pilot. The tactical co-ordinator and pilot displays can generate characters, vectors and conics, while the sensor displays use characters and vectors only.

Logic Unit 4 is mainly an expansion unit comprising two items: the Data Multiplexer Sub-unit (DMS) and the Drum Auxiliary Memory Sub-unit (DAMS). These provide extra input/output capacity and memory capacity respectively.

The DMS can service four input and output peripherals, as selected by the central computer. One output channel presents characters, vectors and conics for the auxiliary sensor display and one input/output channel is used for the aircraft's Omega navigation system, a command launch system for the Harpoon anti-ship missile and sonobuoy reference system.

The DAMS was incorporated to give an additional 393,216 words of memory to the computer, so that the operational program could be expanded to accommodate extra functions and equipment.

Various keysets and control panels allow access to the central computer via the data processing system. A universal keyset allows the transfer of information between the computer and the nav/com operator. The pilot uses his own keyset for controlling the information presented on his CRT display for entering navigation stabilisation data, dropping weapons and flares and entering information on visual contacts. The ordnance panel displays the commands which the computer has given to the ordnance operator concerning status and position of the search stores, such as sonobuoys, which are available for deployment. Finally, there is an armament/ordnance test panel which monitors the output from the data processing system Logic Unit 2 to those systems.

Status
In service on P-3C Orion aircraft.

Contractor
Lockheed Martin Naval Electronics & Surveillance Systems.

Angle Of Attack (AOA) system

Type
Aircraft sensor, attitude/Angle of Attack (AoA).

Description
The Angle Of Attack (AOA) system is designed to assist pilots to obtain optimum aircraft performance on descent, approach and landing. It also provides useful data during instrument flying, in-flight turbulence, navigation and low-speed flight. The system consists of the AOA transmitter, AOA control panel indicator and AOA approach indexer.

The AOA transmitter is the foundation of the system. It senses the airflow direction at the side of the aircraft's fuselage. Specially equipped with heaters for de-icing and moisture control, the transmitter has a unique drainage feature to prevent water intake in the air or on the ground.

The AOA indicator is an easy-to-read indicator which displays aircraft lift information on a graded scale from 0 to 1.0. 0 represents zero lift and 1.0 represents 100 per cent lift, or stall. Combined with the system's flap position information, the AOA control panel indicator's display is valid for all flap configurations. The indicator also operates the indexer and the fast-slow pointer on flight director systems. The display shows the angle of attack for maximum lift/drag ratio to allow the aircraft to be flown at the best endurance speed or best glide angle. It also shows an approach reference angle of attack to be flown when in the final approach configuration. Finally, there is an adjustable V/Vs reference which is set using the REF SET knob.

Mounted on the aircraft's glareshield, the AOA approach indexer is a three-light/three-colour unit that instructs the pilot on the best speed of approach, based on the angle of attack.

Status
The system is fitted on US Air Force fighter aircraft, US Navy and US Marine Corps carrier-based aircraft, and on other civilian aircraft such as the Astra, Avanti, Beechjet, Citation, Falcon, Gulfstream, Jetstar, Sabreliner, Starship and Westwind.

Contractor
Avionics Specialties Inc.

AN/USQ-131A Memory Loader/Verifier Set (MLVS)

Type
Data processing system.

Description
The AN/USQ-131A Memory Loader/Verifier Set (MLVS), is a small portable unit that is capable of loading/verifying multiplex bus single point or direct loading of Operational Flight Programmes (OFPs), and User Data Files (UDFs) into each aircraft's Weapon Replaceable Assemblies (WRAs). The system comprising, a MLVS unit, transit case and an extended BIT adapter, is capable of uploading and verifying various avionic and Electronic Warfare (EW) reprogramming systems, via MIL-STD 1553 data bus and utilising MIL-STD-2217 protocols B-H.

Various reprogrammable systems onboard a wide variety of aircraft, including the F/A-18 A-F, F-14A/B and D, AV-8B, V-22, HH-60H, SH-60B, VH-60N and P-3C AIP, can be addressed by the USQ-131A.

The MLVS is powered via a 28 V DC utility connector on the host aircraft.

Further features of the MLVS include:
- Supports RS-232 and RS-422 interfaces
- Direct UUT loading and single point loading
- Utilises PCMCIA Flash cards (2)
- LCD display
- No calibration required.

Specifications
MTBF: 2,750 h (required); 4,900 hours (demonstrated)
Dimensions: 228.6 (D) × 101.6 (H) × 139.7 (W) mm
Weight: 2.5 kg

Status
In production. Compatible with a wide range of aircraft.

Contractor
Symetrics Industries Inc.

APS-65 autopilot

Type
Autopilot system.

Description
Introduced in 1982, the APS-65 was claimed to be the first digital autopilot offered for turboprop types with dual microprocessor computation. The basic operating modes are: roll hold, pitch hold, heading hold, navigation mode, approach mode, indicated airspeed hold, vertical speed hold, climb and descent, altitude select and go-around. Additionally, when climbing, the autopilot can be programmed to fly the aircraft at optimum efficiency, thus providing fuel savings.

A typical APS-65 consists of four servos, a dual-channel air data sensor, slip/skid sensor, one yaw rate sensor, one flight control panel, one autopilot control panel and one autopilot computer. The heart of the system is the APC-65 Autopilot Computer that utilises a command channel and a servo channel to provide fail-safe operation. The FCP-65 Flight Control Panel provides mode selection and annunciation for the system and the APP-65A Autopilot Panel provides for crew selection of the yaw damper functions and adjustment of the vertical mode reference and roll angle.

A new control technique for the pilot to use when introducing pitch, altitude, airspeed and vertical speed hold changes has been incorporated. This employs a rocker switch that can be operated to select precise alterations. Altitude can be adjusted in increments of 25 ft, pitch in increments of 1°, airspeed in increments of 1 kt and vertical speed in increments of 200 ft/min. The new technique has been incorporated to reduce pilot workload and permit smoother flying. Operational safety is enhanced through the dual processor configuration, which ensures that no single equipment or software fault can result in exceeding preset limits or a malfunction in more than one axis of control. Built-in monitoring ensures automatic disengagement when a fault is detected and appropriate diagnostics assist in fault detection in any element. The system is lighter and has a lower parts count than comparable analogue systems; higher reliability is attributed to these features.

The autopilot is suitable for the heavier business turboprop types such as the King Air, and can be integrated with electromechanical or electronic flight director systems and navigation systems produced by Rockwell Collins.

Specifications
Dimensions:
(APC-65) 85 (H) × 129 (W) × 354 (L) mm
(FCP-65) 48 (H) × 146 (W) × 87 (L) mm
(APP-65A) 48 (H) × 146 (W) × 87 (L) mm
Weight:
(APC-65) 2.68 kg
(FCP-65) 0.45 kg
(APP-65A) 0.36 kg
Power:
(APC-65) 28 V DC, 1.2 A
(FCP-65) 28V DC, 0.55A
(APP-65A) 28 V DC, 30 mA
Temperature: –55 to +70°C
Altitude: 55,000 ft
FAA TSO: C52a (Flight director); C9c (Autopilot)

Status
In production and in service. First certified on the Beech King Air in April 1983. The APS-65 has now been chosen for 28 types of turboprop business aircraft. The King Air was noteworthy as the first general aviation all-digital turboprop to be certified. Its Collins avionics suite included APS-65 autopilot, EFIS-85 electronic flight instrument

system, ADS-80/85 air data system and nav/com equipment.

Contractor
Rockwell Collins.

APS-85 digital autopilot

Type
Autopilot system.

Description
The APS-85 autopilot completes the Pro Line III family of digital avionics for general and business aviation. It uses either an FCC-85 or -86 Flight Control Computer: the FCC-86 is a dual-channel computer that provides two independent flight director functions; the FCC-85 provides a single flight director function. The system incorporates an MSP-85/85A Mode Select Panel to provide crew interface with the flight director function. The panel contains a push button and annunciator light for each mode selection available. The APP-85 Autopilot Panel provides crew interface with the autopilot and yaw damper. It contains the autopilot and yaw damper engage levers as well as push buttons for flight director transfer and turbulence mode. The system has been certified for Cat II operation; Cat IIIA landings will be possible with some growth.

Apart from its digital nature, the APS-85 has a number of features not found on earlier Rockwell Collins general aviation autopilots, including the APS-80. For example it includes climb and descent modes. When selected in the climb mode, the system flies the aircraft according to an airspeed or Mach number schedule appropriate to that type of aircraft and its weight at take-off. In the descent mode the system sets up a rate of descent tailored to the aircraft manual. Three diagnostic modes – report, input and output – are incorporated. In the first, faults in the flight control computers are displayed to the crew. In the second, the system reads out information from outside entering the system, for example from the air data system or control surface position sensors. In the final mode, the system provides particular flight control computer outputs for examination. The diagnostics on the APS-85 are self-contained and do not require additional test equipment.

Specifications
Dimensions:
(FCC-86) 197 (H) × 129 (W) × 408 (L) mm
(MSP-85) 38 (H) × 146 (W) × 140 (L) mm
(APP-85) 48 × 146 × 144 mm

Weight:
(FCC-86) 6.0 kg
(MSP-85) 0.45 kg
(APP-85) 0.64 kg
Temperature: –55 to +70°C
Altitude: 55,000 ft
Power:
(FCC-86) 28 V DC, 67W
(MSP-85) 28 V DC, 9 W
(APP-85) 28 V DC 10.5 W
FAA TSO: C52a (Flight director), C9c (Autopilot)
Environmental: DO-160A
EUROCAE: ED-14A

Status
In production and in service. The APS-85 has been certified as the autopilot on the Saab 340, Canadair CL-600B and -601, the Raytheon Hawker 800, Falcon 50B, Falcon 20 C, D, E and F, Learjet 55B/C, Gulfstream 100, DC-8, Xian Y7200A and C-130.

Contractor
Rockwell Collins.

ASPRO ASsociative PROcessor

Type
Avionic processor.

Description
The ASPRO Associative Processor is a small but powerful fully military-qualified 0.44 cu ft parallel processor embedded in the Northrop Grumman E-2C Hawkeye's avionics suite. It performs millions of operations per second to track potentially thousands of targets. It also performs additional display processing functions on the aircraft.

Since it was originally developed, ASPRO's processing throughput has been increased and its memory doubled, with no increase in volume. In addition, a system to meet MIL-E-16400 has been built, for use in 'Los Angeles' class submarines. Lockheed Martin has developed the ASPRO-VME, which features increased memory, system throughput, and expansion capability to accommodate new applications.

Specifications
Dimensions: 203 × 229 × 254 mm
Weight: 14.5 kg
Power: 200 W
Throughput: 50 Mops
Reliability: 4,500 h MTBF

Status
In service in the E-2C Hawkeye aircraft.

Contractor
Lockheed Martin Naval Electronics Surveillance Systems.

ASPRO-VME parallel/ associative computer

Type
Avionic processor.

Description
The ASPRO-VME, Lockheed Martin's fourth-generation parallel processing computer, is a modular open architecture VME-compatible card set. It is capable of performing between 150 Mflops and 2.4 Gflops. The basic three-module 512 processor configuration can be expanded from 512 to 8,192 processor elements. Each of the parallel processors, called Processing Elements (PEs), contains a full 32-bit IEEE floating-point processor and a bit-serial processor.

A module occupies a single-VME card slot and consists of two printed circuit boards attached to a cold plate. The entire ASPRO-VME, in its basic configuration, requires three VME slots.

Programmable in Ada, ASPRO-VME is supported by a powerful software development tool set which allows application programmes to be easily developed. Because of ASPRO's single instruction multiple data stream architecture, modular expansion from 512 PEs to 8,192 PEs can be accomplished without rewriting software.

The parallel architecture and associative search capability produce significant gains over conventional processors for sophisticated tracking, correlation, data fusion, and situational awareness algorithms.

ASPRO-VME is available in a rugged and full Mil-Spec configuration. It can operate in a stand-alone mode or be directly embedded in commercial and rugged workstations with 6U VME slots. Applications for ASPRO-VME include command and control, correlation and tracking, data fusion, database management, signal processing, expert systems, neural networks and image processing.

Specifications
Weight: 2.59 kg
Power: 80 W

Status
Production units available.

Contractor
Lockheed Martin Naval Electronics & Surveillance Systems.

Automatic Flight Control System (AFCS) for the A 109

Type
Flight Control System (FCS).

Description
The Agusta A 109 flight control system has been designed to reduce pilot workload and improve reliability and safety at a realistic cost. A typical installation includes two autopilot (helipilot) systems for redundancy, which may be used with or without a flight director system.

This duplex system consists of one directional and two vertical gyros, controller panel, two computer units and five series actuators, two each for pitch and roll and one for yaw control. A trim computer and two trim actuators for pitch and roll adjustments are also part of the system.

A flight director facility can be incorporated but requires the addition of a mode controller and navigation receiver inputs, plus ADI and HSI. Flight director functions include heading, vertical speed, barometric altitude and airspeed select, go-around procedure selection, and VOR, ILS, MLS, Omega/VLF, RNav system and Loran or Tacan coupling.

The main flight panels of the Raytheon Hawker 800. The Rockwell Collins APS-85 autopilot mode select panels are fitted on the coaming in front of the pilot and co-pilot   0503915

Status
In service in the Agusta A 109 helicopter.

Contractor
Honeywell Inc, Commercial Electronic Systems.

AutoPower® automatic throttle system

Type
Aircraft system control, power plant.

Description
The role of the Safe Flight Instrument Corporation AutoPower® system is to provide aircrew with precise control of N1 for take-off, climb and go-around, and speed (IAS/Mach) throughout the remainder of the flight. The system monitors N1 thrust limits from the FMC to provide protection from engine exceedance and integrates with the vertical modes of the flight control system.

The system comprises four main components: a glareshield-mounted digital Indicated Air Speed (IAS) display; an AutoPower computer; a clutch pack; and a set of yoke-mounted increase/ decrease switches. An AutoPower engage switch, cockpit annunciators and circuit breakers are also required. The AutoPower computer contains all the electronics necessary for system operation. The clutch pack contains one clutch per throttle and a servo drive motor.

AutoPower can be disengaged by pressing the Go-Around switch. In addition, the pilot can override the AutoPower system at any time, by moving the throttles with normal throttle control force.

Status
In production and in service. The AutoPower® system is installed on variety of civil aircraft, including the Boeing 767, Bombardier CL850, CRJ200, 604 and 605 and Gulfstream G200, as well as military types, including the Lockheed F-117A.

Contractor
Safe Flight Instrument Corporation.

Avionic common module systems

Type
Avionic processor.

Description
Lockheed Martin Management & Data Systems is the technology leader for development, application, implementation and integration of JIAWG/MASA/SHARP avionics common module processing clusters and systems. These capabilities and resources are aimed at allowing module users complete integration control from subsystem development to system platform implementation.

The foundation of common module information processing systems is extensive work in VHSIC design. A limited number of these standard VHSIC chips enables creation of a family of modules that can be used in different avionics systems. Designed in the Standard Electronic Module (SEM) E format, each module measures 149.3 × 162.6 × 14.7 mm.

The Lockheed Martin Tactical Defense Systems family of avionics common modules consists of seven processing and Input/Output (I/O) SEM-E module types, a power supply SEM-E module, backpanels, liquid or air-cooled racks and active and passive star coupler technology. Reusability and standardisation have been stressed within the design philosophy for the avionic equipment family. Lockheed Martin Tactical Defense Systems has created a generic modular, module functional design to promote reuse of ASIC devices and to support integrated diagnostics.

The Advanced General Purpose Processor Element (AGPPE) is a 32-bit RISC-based data and signal processor module ideally suited for avionic systems applications. Equipped with a Mips R3000 RISC processor, its pipelined architecture yields very high throughput while its standard SEM-E format and three input/output interfaces allow easy integration in various systems.

Advanced ceramic circuit boards and surface-mount components contribute to AGPPE's light weight, low power consumption and dense packaging. Full 32-bit operation is enhanced by a five-stage pipeline, on-chip cache control and an on-chip memory management unit. Block refilling of both instruction and data caches is supported.

The Mips R3000 processor executes instructions up to 20 times faster than the VAX 11/780. In addition the R3010 floating point co-processor chip handles floating point arithmetic compliant with the ANSI/IEEE standard.

Onboard memory resources include a 1 Mbyte SRAM which can be accessed synchronously in two CPU clock cycles. Bootstrap and debugging code can be stored in a 128 kbyte EEPROM. Separate 16 kbyte data and instruction caches, which effectively double the available cache memory bandwidth, provide instructions and operands at the CPU clock rate. Other features that enhance performance include a four-word buffer for block refill of each cache and a one-word write buffer for writes to main memory.

The module features TM bus and a Lockheed Martin designed maintenance controller ASIC with an IEEE-488 channel for console operations. A network interface controller provides a one-chip interface from the AGPPE to a data flow network, a 32-bit parallel bus.

MIL-STD-1750A processor
The MIL-STD-1750A processor module is a 3.85+ Mips VHSIC processor with 512 kwords of local SRAM, plus 8 kwords of start-up ROM and PI-bus, TM-bus and IEEE-488 standard I/O interfaces on a single-width double-sided

¾ ATR size SEM-E module. The module's ceramic printed circuit boards and surface-mount components provide a dense, lightweight, low-power, general purpose data processor module. The highly parallel pipelined architecture enables high throughput. Advanced built-in test techniques reduce life cycle costs by supporting two-level maintenance. In addition, the standard SEM-E size and three standard I/O interfaces allow easy integration into a variety of systems.

The processor module has an onboard MIL-STD-1750A maintenance controller to automate built-in test on the module, communicate with the other modules via the TM-bus and handle console operations via the IEEE-488 bus.

The MIL-STD-1750A data processor's three CMOS gate arrays, the CMOS semi-custom maintenance controller common to the Unisys common module family and the ECL clock chip are equivalent to over 160,000 gates.

High-Speed DataBus interface (HSDB)
The High-Speed DataBus (HSDB) interface module integrates the high-performance processor with a linear token-passing high-speed fibre optic databus on a single-width double-sided ¾ ATR size SEM-E module.

The HSDB module provides a dual-redundant 50 Mbit/s fibre optic system interface, processor, 256 k words of local SRAM, a subsystem interface, maintenance via the TM-bus and IEEE-488 bus I/O interfaces.

The module functions are highly integrated, using surface-mount components. The HSDB can be configured with processor/bus combinations consisting of either the high-performance 1750A processor or the Mips-based RISC processor and with either a PI or N bus subsystem interface. The HSDB interface module has a CMOS gate array, an ECL gate array and hybridised fibre optic transmitters and receivers for the HSDB interface.

DC-DC converter
The Lockheed Martin Tactical Defense Systems DC-DC converter is a single-width ¾ ATR SEM-E size power supply that provides over 200 W of regulated power from an unregulated 270 V DC bus. The converter uses a standard power supply topology with hybrids and high-frequency techniques to attain high efficiency, high reliability and small size.

Input power to the converter is +135 V DC and −135 V DC (270 V DC line-to-line) MIL-STD-704D. Output voltages are +5 V DC, −15 V DC, −5.2 V DC and +80 V DC. The efficiency of the DC-DC converter is 80 per cent at full load. Up to six modules may be paralleled to increase current capacity and/or provide redundancy.

The converter has a patented digital controller to enhance stability, improve testability and provide fault detection. The module has a slave TM-bus interface to communicate with other modules in the subsystem.

Contractor
Lockheed Martin Management & Data Systems.

C-130J Cockpit control panels

Type
Aircraft system control, avionic.

Description
These cockpit control panels provide the functional interface between the operator and

the applicable subsystems. They provide data, system status, controls and information displays.

Interfaces are available through MIL-STD-1553B databus, ARINC 429 databus or via discrete wire signals. The system is available with Night Vision Imaging Systems (NVIS)-compatible lighting, and with or without edge lighting panels.

Examples of current control panels designs include: aerial delivery; air conditioning; automatic flight control system; bleed air; caution and warning; cursor control; fire handles; flight select; fuel management; heading/course select; hoist and winch control; ice protection; landing gear; lighting control; pressurisation; radar control; and reference set/mode select.

For details of the latest updates to *Jane's Avionics* online and to discover the additional information available exclusively to online subscribers please visit
jav.janes.com

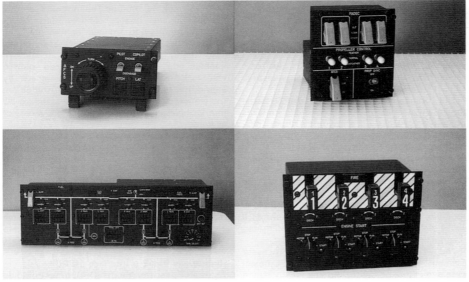

Cockpit control panels (C-130J)   0581511

Status
In production and in service in the C-130J.

Contractor
Smiths Aerospace Electronic Systems, Long Island.

C-130J Digital Autopilot/Flight Director (DA/FD)

Type
Autopilot system.

Description
The system consists of dual, Automatic Flight Control Processors (AFCP), each with two central processor units for redundant control law computation and mode logic implementation for: autopilot, flight director and autothrottle performance. System inputs to the control surfaces (ailerons, rudder, elevators) are made through three, individually housed, servo drive units containing: servo motor, tachometer, synchro, solenoid engage clutch, drive spline and electrical connector.

Honeywell's digital autopilot/flight director provides a proven solution for the C-130J programme. Early versions of the system flew on C-130E/Hs with the US (RAMTIP and PRAM), Royal Australian and Royal New Zealand Air Forces. An identical hardware version is flying on the Belgian Air Force C-130Hs. Aerodynamic control laws and flight director functions have been thoroughly tested in a series of development programmes for the US Air Force and for foreign governments.

The C-130J version of this autopilot is qualified to both MIL-STD and Federal Aviation Administration requirements in accordance with MIL-STD-810, MIL-STD-454, MIL-STD-461, AC 25.1309-1A, DO-178A and DO-160C. The autopilot interfaces with C-130J sensors and avionics via a MIL-STD-1553B databus. The interface to the three servo drive units and the autothrottle is via direct wire, to ensure adequate response time.

Key design features include: digital electronics programmed in Ada; fail-passive operation; dual flight control computers, each with two central

Automatic flight control processor   0051638

processor units for redundant control law and mode logic computation (internal comparison of LRU outputs verify each LRU operation); flight director system internal to the autopilot computer; comprehensive built-in test covering 98 per cent of circuitry (power-up, continuous, maintenance diagnostics); parameter modification permitting selected control law gains to be adjusted under flight test-only conditions; autothrottle control for precise speed control; each LRU provides all of the main computing and power amplification functions for the digital autopilot/flight director system; comparison of servo outputs for validity and tolerance (disagreement results in system disengagement).

Operational modes for the autopilot include: heading and altitude select; coupled navigation (global positioning system, inertial navigation, Doppler); radio navigation (VOR/LOC, back-course, TACAN, GS); altitude hold; turn/pitch knob adjust; coupled flight management system operation; pitch/roll go-around; autothrottle; heading-hold; speed-hold; vertical navigation; yaw damper; turn co-ordination; and speed select.

Specifications
Dimensions: 292 × 203 × 505 mm
Weight: 15.8 kg

Status
In production.

Contractor
Honeywell Defense Avionics Systems.

CC-2E Data Processing System (DPS)

Type
Data processing system.

Description
The CC-2E Data Processing System (DPS) is the central processor for the Boeing E-3 AWACS aircraft which provides early warning of threats and airborne control of friendly aircraft. It receives and processes data from onboard sensors, sends the completed airborne view to mission operators and links critical messages to ground commanders. The CC-2E DPS is the fourth-generation multipurpose airborne processor for the E-3.

A recent value engineering change proposal effort resulted in the development of increased memory in the CC-2E DPS, reducing the number of monolithic memory units from five to three per system.

Status
In service.

Contractor
Lockheed Martin Systems Integration – Owego.

Century 41 autopilot/flight director

Type
Autopilot system.

Description
The first Century autopilot to feature digital processing, the Century 41 is a two-axis system with built-in VOR/Loc/GS couplers and outputs to drive a single cue V-bar flight director. It employs both position and rate signals to command the flight control servos. Other features include synchronised pitch attitude and altitude hold modes, a pitch modifier, automatic and manual electric trim and automatic preflight check schedule. VOR/Loc/GS capture intercepts are tailored to groundspeed, intercept angle, wind direction and distance from the ground station. The VOR tracking circuitry incorporates gain reduction to ensure 'soft' passage around the station.

When using the go-around mode, the pilot merely has to press a single button, clean up the aircraft and add power; the autopilot flies to a calibrated pitch-up attitude appropriate to the single engine safety speed for that particular type of aircraft, and turns on to a new heading preset by the pilot. The NSD-360A slaved or unslaved HSI may be substituted for the standard directional gyro. Both vacuum and electric variants are available.

The Digital Altitude Preselector/alerter (DAP) model ID960 is available as an option for all Century 41 installations.

Specifications
Dimensions: 3 ATI panel displays
Weight (installed): 16.2–17.5 kg

Status
In service.

Contractor
Century Flight Systems Inc.

Century 2000 autopilot/flight director

Type
Autopilot system.

Description
The Century 2000 is a panel-mounted, one-, two- or three-axis modular design autopilot which provides economical upgrading from any of several levels of capability.

Any Century 2000 can be expanded to include fully coupled roll and pitch plus yaw damper three-axis autopilot with flight director and altitude preselect by installing expansion kits. The basic system is roll axis with heading and built-in VOR/Loc coupler. Adding pitch axis expansion to the system makes available functions such as altitude command and hold, glideslope capture and track, control wheel steering, proportional pitch trim and automatic pitch synchronisation. Flight director expansion computes roll and pitch attitudes needed to intercept and maintain headings, courses, attitudes and altitudes. The solutions are then displayed through the flight director horizon as steering commands. The addition of a yaw damper completes the Century 2000 package by enhancing directional stability and maintaining co-ordinated flight during turns.

The Century 2000 has a number of automatic performance and safety features such as soft track and internal circuit checks. The system will interface with the NSD-360A HSI and any other ARINC HSI. The flight director steering horizon is the single-cue delta/vee presentation. Non-flight director and heading only directional gyros are also compatible.

The Digital Altitude Preselector/alerter (DAP) model ID960 is available as an option for all Century 2000 installations.

Specifications
Dimensions:
(panel space) 158.3 × 57 mm
Weight: 8.3 kg

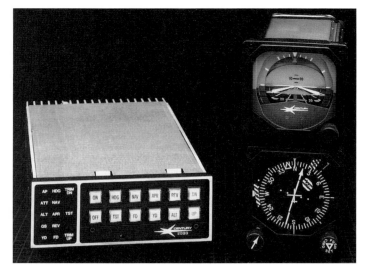

Century 2000
autopilot/flight
director
0001432

Power supply: 14 or 28 V DC, 4.5 A
Compliance: TSOs C4c, C6c, C52a

Status
In production and in service.

Contractor
Century Flight Systems Inc.

Century I autopilot

Type
Autopilot system.

Description
The Century I autopilot is an all-electric, rate-based lightweight single-axis wings level/heading system. An electric actuator in the aileron circuit provides the control power for attitude stabilisation and pilot commanded, knob controlled turn rates of up to 200°/min. A tilted rate gyro inside a standard 3 in (76 mm) case senses roll rate and rate of turn, for both instruments and servo, and the system can be slaved to VOR/Loc or panel-mounted GPS. Century I can also be used as a back-up to the Century IIB, III or IV vacuum/electric systems (see later items), sharing the same roll servo.

Specifications
Dimensions: 89 × 89 × 194 mm
Weight: 3.17 kg
Power supply: 14 or 28 V DC, 1.25 A

Status
In service.

Contractor
Century Flight Systems Inc.

Cockpit control panels

Type
Aircraft control system, avionic.

Description
G6990 series TCAS panels
Gables TCAS panels are available for all Boeing commercial aircraft types in both keyboard and rotary layouts. The G6990-51 is certified for Boeing New Generation aircraft. The G6990-40 has been selected by British Airways for installation on its A320 aircraft.

G6993-08 and G6992-22 have been selected by United Parcels Service, Airborne Express

Gables' G7130-03 ATC/TCAS general aviation
control panel   0022248

and Federal Express for use in the Cargo Airline Association ADS-B programme, installed on DC-9 and B-727 aircraft.

Gables has teamed with Honeywell Inc, Commercial Aviation Systems, to develop an ATC/TCAS (Mode-S/ATCRBS) control panel that is compatible with the new Honeywell TCAS 1500 system.

G7130-03 ATC/TCAS control panel
The G7130-03 control panel has been certified for the Falcon-900 aircraft, compatible with the Rockwell Collins 822-0078-005 radio.

G7185 series cargo smoke detection and G7183/4 fire suppression control panel
Gables' series of smoke detection and suppression control panels are designed to alert crew members to a fire event in ample time to avoid a catastrophic situation.

A specific design criterion of these units is the avoidance of false alarms, which has been a problem with smoke detection systems. The control panels include LED annunciators.

The G7185, integrated cargo smoke detection/suppression control panel, automatically switches from dual- to single-loop operation. This switch is made without affecting the dual-loop mode of additional detectors that have not suffered a fault. Depending upon where the smoke is detected, the control panel installed in a two-bay aircraft will automatically arm the extinguishing system in either the forward or aft bay of the aircraft. The panels in aircraft with three cargo bays will automatically arm the system in the forward, mid or aft bays of the aircraft.

AirTran Airways has selected the G7183 smoke detection control panel and the G7184 cargo fire-extinguishing control panel for use on its fleet of DC9 aircraft. Continental Airlines has chosen the G7185 smoke detection control panel for use on its MD80 aircraft.

G7400 series VHF control panels
Continuing VHF control panel programmes include six variants of the G7400 series for Boeing

Gables' G7409 radio control panel   0022249

Gables' G7404 radio tuning panel   0022250

737, 757 and 767 aircraft. Included in this series is the G7406, which is capable of tuning both ARINC 566A and 716/750 series radios. The G7404 radio tuning panel is now available in three variants for use on Boeing 747-400 aircraft. Gable's radio tuning panels have also been certified on Boeing 737 New Generation aircraft. The G7409 control radio panel has been selected for the Boeing MD-10 aircraft.

G7406-25 has been selected to provide VHF comm control panels with 8.33 kHz channel spacing for the US Air Force KC10 fleet; installation began in April 1999.

G7400 series control panels are also certified on the Russian Aviation Register.

G7500-03 VHF navigation control panel
The G7500-03 panel is standard on Boeing 737 new-generation aircraft. It can select VHF navigation frequencies in the range from 108.00 to 117.95 MHz using 50 kHz spacing. It can also select MLS channels from 500 to 699 in unit increments. It is compliant with ARINC standards 709, 710, 711, 720 and 727 and is certified to TSOs C36, C40c, C66c and C104. The G7500-03 has an MTBF of 22,000 operating hours.

Contractor
Gables Engineering Inc.

Combo Data Transfer Systems (DTS)

Type
Data processing system.

Description
The Combo DTS is a military Data Transfer System produced in two versions. It loads and records data to a 3.5 in floppy disk or to dual PCMCIA flash memory cards to and from the avionics system, via the MIL-STD-1553B databus or the Ethernet databus. It can support two separate MIL-STD-1553B channels and function as a bus controller. Additionally, it also provides two independent Ethernet interfaces.

The Combo DTS acts as a MIL-STD-1553B remote terminal to load navigation databases and as a bus controller in the memory/loader/verifier mode to load operational programs into aircraft systems.

The floppy disk is MS-DOS compatible, readable by IBM-compatible PCs and supported by the AFMSS and PFPS mission planning systems.

65000 Combo DTS
The 65000 Combo DTS is a derivative of the current military production system and produces two MIL-STD-1553B interfaces.

67000 Combo DTS
The 67000 Combo DTS is a modification to the 65000 Combo DTS to add two Ethernet interfaces. The DTS 67000 performs all of the functions of the DTS 65000 in addition to the Ethernet data transfer functions.

ARINC 615 portable data loaders
Demo Systems' ARINC 615 portable data loaders are used to upload navigational databases and operational programs to commercial DTS, using 3.5 in computer disks and to download maintenance and other information.

Status
The 65000 Combo DTS is in use on MH-53J and HH-60 helicopters, AC-130H and AC-130U gunships, a variety of C-130 programmes including the MC-130H Combat Talon II aircraft and on RC-135 and WC-135 aircraft.

The 67000 Combo DTS is in use on MH-53M helicopters.

Demo Systems claim that their ARINC 615 portable data loaders support about 80 per cent of the world's airlines and are used on a wide range of systems and aircraft, including Airbus, Boeing, Lockheed Martin and Raytheon aircraft.

Contractor
Demo Systems LLC.

Common Boresight System (CBS)

Type
Flight inspection/calibration system.

Description
DRS's Joint Service Common Boresight System (CBS) draws on extensive experience gained with other boresighting systems, such as the Captive Boresight Harmonisation Kit (CBHK) for the AH-64 Apache and AH-1F Cobra, the A/U36M-V1 Rapid Armament Boresight System (RABS) for the AH-1W Cobra, the APQ-180 Radar for the AC-130U Gunship and the CBS for the Hawk T Mk1. The CBS features the proven Triaxial Measurement System (TMS), a single-beam instrument that simultaneously measures azimuth (yaw), elevation (pitch) and roll with laser accuracy.

Through the use of airframe-specific mounting adapters, the CBS can be tailored to boresight virtually any aircraft, as well as most naval vessels, ground vehicles and radar systems. It is compatible with all current applicable aircraft types and requires no platform modifications. In operation, the small, lightweight TMS transmitter is positioned relative to the Aircraft Armament Data Line (AADL). It transmits a collimated, linearly polarised laser beam to the TMS receiver mounted on the aircraft system under evaluation. The TMS Control Display Unit (CDU) processes the azimuth, elevation and roll misalignment information from the TMS receiver and displays these data at a high refresh rate in all 3 axes simultaneously. The display presents measurements in both analogue and digital formats. Built-In-Test (BIT) of the TMS is controlled from CDU, which also supplies DC power and synchronisation signals to the transmitter and receiver. The CDU contains a data port to allow interface with RS-232 and MIL-STD-1553B architecture, PCs and other data processing formats.

TMS calibration is via a dedicated calibration/set-up fixture, which mounts the TMS transmitter and receiver in known on-axis and off-axis facing positions. The CDU is used to adjust the displays to coincide with the known positions.

With the TMS Transmitter and Receiver mounted on the aircraft, boresighting can be completed at ground level. This eliminates the need to work with optical, inertial or other alignment equipment while on ladders or raised platforms when making accurate boresight measurements. The 3 TMS operational components (transmitter, receiver and Control and Display Unit) are common to all applications, substantially reducing inventory and training requirements, with commensurate life cycle cost savings.

The main features of the CBS include:
- System accuracy is 0.50 (3 Sigma) and 0.17 (1 Sigma) or better in all axes, depending on specific aircraft requirements
- Analogue and digital readout to permit tighter tolerances than possible with the human eye and ensuring repeatability of results
- Simple on-site calibration – TMS accuracy can be quickly checked and recalibrated in the field
- Continuous measurement of all 3 axes at the same time by the same instrument thereby significantly reducing alignment time
- The CBS routinely harmonises all fire control, weapons and navigation systems in less than 1 hour
- No aircraft hangar required
- Small and lightweight – requires only two operators
- Interchangeable parts for all platforms
- The CBS can be used in all weather conditions – high wind, snow, rain, blowing sand, full sunlight and on carrier flightdecks
- EMI Protected – complies with MIL-STD-461C
- Operation of the system is the same for all platforms and consistent with existing training and procedures
- The CBS approach is combat proven, and the TMS technology is mature
- Ruggedised, with all components field tested and supplied in sealed cases

Specifications
Compliance: MIL-T-28800, MIL-STD-810, MIL-L-85762, MIL-H-46855, MIL-STD-454, MIL-STD-5400, MIL-STD-461C, MIL-STD-704, MIL-STD-167, MIL-STD-882
Operating range: 0.5 to 100 ft
Laser safety: Exceeds ANSI Z136.1
Display: 3-axes display simultaneously and continuously; digital readout accuracy to 0.025 mrad
MBTF: Demonstrated in excess of 1,700 h
Interface: RS-232 and MIL-STD-1553B
Other applications: Land and shipboard radar systems, RF antenna/sensor systems, land- and sea-based gun mounts, alignment of inertial reference, electro-optic and propulsion systems

Status
The CBS is compatible with the F-16 Falcon, F-15 Eagle, F/A-18 Hornet/Super Hornet, AV-8B Harrier, F-22 Raptor, AH-64A and AH-64D Apache, AH-1S/F, AH-1W and AH-1Z Cobras, OH-58D Kiowa Warrior, Panther, Tiger, A 129 Mangusta, Hawk 100/200, Tornado GR4 and Eurofighter Typhoon aircraft.

Contractor
DRS Technologies, Inc.

CS-1095 IF/baseband converter

Type
Avionic interface system.

Description
The CS-1095 IF/baseband Converter is a small, lightweight converter that translates an IF input signal into analogue and digital baseband outputs. This converter accepts a standard 70 MHz IF input or can be optionally configured to accept multiple IF centre frequencies (21.4, 70, 140, 160 MHz or 1 GHz) that are user selectable. The optional frequency IF inputs are first converted to a 70 MHz IF frequency and then processed into baseband outputs using a combination of RF and Digital Signal Processing (DSP) techniques. The unit can feature both programmable narrowband DSP based bandwidths, as well as analogue wideband bandwidths. Pre-selection is available in the CS-1095 for improved out-of-band rejection of unwanted signals. A multichannel capability (option) provides for DSP clock and timing synchronisation between multiple converters. An external 10 MHz reference oscillator can be used during multichannel operation to provide for RF LO phase coherency for all channels.

The CS-1095 is described as well suited for demanding applications in signal collection and analysis, with the optional selectable IF input frequencies providing flexibility and compatibility for diverse, multimission roles. The DSP technology permits 'change-on-the-fly' filtering selections that provide optimum conversion characteristics for a multitude of signal types.

The CS-1095 unique RF/DSP architecture delivers extremely low differential group delay and very low ripple over a wide percentage of bandwidth. Existing all-analogue systems have much higher differential group delay and higher ripple over a narrower bandwidth percentage. The CS-1095 converter offers a superior level of performance that is stable over time and temperature with none of the drift effects associated with all-analogue systems. A unique low-cost, snapshot digitiser and replay capability is available, which utilises an Ethernet interface for storage and replay loading. The signal data, over a specific timeframe, is stored, along with bandwidth and clock rate configuration in RAM for downloading to external storage devices such as the CS-711B (see separate entry). This data can then be later reloaded for replay including endless loop replay for extended analysis of the digital or analogue outputs from signals of interest.

The CS-1095 is housed in a full rack width, 1U tall (1.75 in), rack-mount chassis. The chassis utilises forced air cooling for high reliability and long life. The unit is designed to meet the requirements of MIL-STD-461 for the RF portion, and includes mechanical shielding techniques such as EMI gasketing and waveguide-beyond-cutoff hole patterns for cooling vents.

The standard unit is remotely controlled via an RS-232/RS-422 interface. The unit may also be locally controlled via the front panel with keypad, encoder and display. Optional IEEE-488, Ethernet and RS-232 PC compatible remote control interfaces are also available. The optional Ethernet interface is required when using the optional snapshot feature. All converter functions can be controlled and/or monitored either locally or remotely, including BITE status. The universal AC power input accepts 90 to 260 V at 47 to 63 Hz.

Status
In production.

Contractor
Rockwell Collins.

D7000 airborne data acquisition and recording system

Type
Onboard flight monitoring system.

Description
The D7000 airborne data acquisition and recording system is designed for a wide range of operational and flight test applications. Featuring adaptable signal interfacing and interchangeable media cartridges, the D7000 provides a highly flexible avionics data acquisition and recording platform in a very compact package. With a rugged

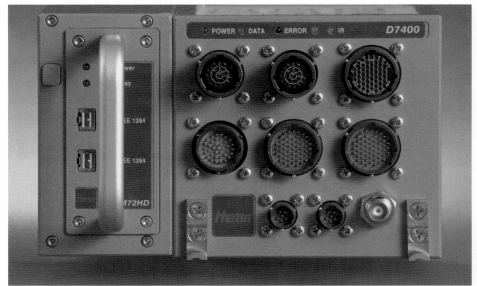

The D7000 offers a variety interfaces to provide for a form/fit replacement of a large number of legacy systems (Heim Data Systems)   1127613

mechanical design to cope with harsh airborne environments, the compact external dimensions and configurable connector interfaces of the D7000 are designed to allow direct replacement of legacy video and data recorders.

With up to 256 Mbps total system data rate, the system employs media contained in interchangeable cartridges, allowing the user to quickly choose and load either rugged hard drive or solid-state media at any time during the service life of the equipment. The packet-recording format ensures accurate data time tagging and consistent channel-to-channel phase relationships. These systems are compatible with both Heim Data format and the IRIG106 Chapter 10 standard, including a compliant IEEE-1394b data download interface on the media cartridge. The D7000 mainframe also has GigE data download capability and the media SCSI interface provides a high-speed output to other Heim data download and analogue replay systems that are fitted with media cartridge receivers.

Complete with an integral IRIG time code generator, the D7000 Series can also synchronise with external IRIG and GPS time sources to provide sustained high-accuracy time tagging of recorded data. The D7000 can optionally be fitted with an integral GPS receiver, providing time reception, distribution and data time tagging into a single package. The media cartridge can also be fitted with an integrated clock that allows laboratory based time references to be automatically transferred to the recorder each time a cartridge is loaded.

The mainframe holds up to 4 multichannel signal interface cards in any combination, providing a wide selection of interfacing, including concurrent recording of multiple video, PCM and MIL-STD-1553 bus sources in a single unit. The user has access to configuration and control functions through integral serial, network and discrete contact interfaces.

The D7000 also features an additional power connector, to allow the unit to be powered from ground support equipment without disturbing flight power connectors. The D7000 military style connectors are all front-mounted and can be supplied in a generic configuration or with specific custom connector types and pin-out connector, so the unit can be powered from ground. This enables direct form-fit replacement of both legacy video recorders and avionics data acquisition systems with a single box solution.

Status
In production.

Contractor
Heim Data Systems, Inc.

Data Acquisition Unit (DAU)

Type
Data processing system.

Description
The Data Acquisition Unit (DAU) is designed for analogue or digital processing applications where large amounts of data need to be processed or consolidated. The system can be configured as a single- or dual-channel unit. A dual-channel configuration provides complete hardware redundancy for all parameters.

The DAU is designed using a modular approach with motherboard and plug-in function boards, which can be added as required to accommodate any aircraft sensor or communication bus. Custom boards can be designed to handle specialised sensors or communications. The system utilises advanced technology to process analogue, digital and discrete engine and airframe signals. Typical signals are filtered, converted to digital data, scaled and formatted for cockpit display or use by other aircraft signals.

In addition to analogue inputs, each DAU channel includes two ARINC 429 inputs and an RS-232 interface for communications with Sentinel displays or other aircraft systems such as the flight management system and FADECs.

All processed or stored information is transmitted via the standard ARINC 429 output bus. Optional databusses or buffered analogue and discrete outputs can be incorporated into the DAU. The built-in RS-422 port can be used to output real-time data or as an access point for maintenance interrogation.

In addition to the normal tasks of conditioning and processing signals, the DAU can perform a variety of maintenance functions. Health monitoring, exceedance recording and trending algorithms can be run in a background mode while the processor would normally be idling.

Extensive BIT isolates problems to the faulty sensor or circuit. Self-test is run continuously, with results stored in non-volatile memory for evaluation. Cross-channel communications are used to verify channel integrity.

Specifications
Weight: 4.08 kg
Power supply: 10–32 V DC or 115 V AC, 400 Hz
Temperature: –40 to +70°C

Status
Fitted to Bombardier Global Express aircraft.

Contractor
Ametek Aerospace Products.

Data Loader Recorder (DLR) RD-664/ASH

Type
Data processing system.

Description
The Data Loader Recorder (DLR) provides the capability to store and retrieve digital data on a non-volatile, transportable media. The Removable Media Cartridge (RMC) is based on field-proven, high-capacity 3.5-inch (9 Gigabytes) Winchester disk products currently used on many shipboard, airborne, ground-mobile, and ground-fixed platforms. The DLR contains one 9 Gigabyte RMC, a Power Converter Assembly (PCA), and a SCSI Differential Converter (SDC). The system is designed to facilitate operational use and maintenance; it does not require any periodic maintenance. The DLR has a computer system interface that meets SCSI-2 and SCSI-3 differential. The DLR operates from 28 VDC input through an EMI filter and the PCA converts to +5 VDC and +12 VDC to power the RMC and the SDC. When the hinged access door to the RMC is opened, power is removed from the RMC. The DLR can operate through significant shock and vibration levels through the use of a shock/vibration isolation frame for the RMC. The operating temperature is –20°C to +55°C with an MTBF of 15,000 hours. The DLR performs self-diagnostics on power up and when commanded over the SCSI bus. A fault indicator and a disk activity indicator are provided on the front panel.

Specifications
Dimensions: 146.1 × 228.6 × 279.4 mm
Weight: 7.26 kg
Formatted capacity: 9 Gbytes
Power supply: 28 V DC
Temperature range: –20°C to +55°C
Shock: 20 g 5.5 ms
Altitude: 35,000 ft
Reliability: 15,000 h

Contractor
General Dynamics Information Systems.

Data Management System (DMS)

Type
Data processing system.

Description
The Data Management System (DMS) is a step towards achieving paperless operations, with electronic control of information. In the cockpit, approach and departure information, taxi and ground diagrams are electronically maintained and can be narrowed to the definition required by the pilot with the pan and zoom feature. Maintenance, cabin and operations data can be accessed at a single terminal, which also serves as the data load and retrieval centre. The Collins DMS is packaged in a single cabinet, using line-replaceable modules. These allow significant reduction in space and weight requirements and need no forced-air cooling.

The DMS consists of a multichannel computer, with information contained in mass storage modules for quick and efficient data access. Information is displayed on high-resolution colour flat-panel displays. The entire system is networked with a fibre optic distributed data interface.

The Rockwell Collins DMS is adaptable to virtually all aircraft types.

Status
Available as an option for the Boeing 777. First deliveries were made in 1995.

Contractor
Rockwell Collins.

Data Transfer Equipment (DTE)

Type
Data processing system.

Description
The Data Transfer Equipment (DTE), also known as the Data Insertion Device (DID), fulfils five main functions. These considerably reduce the workload imposed on crews and eliminate the risk of human error. The aircraft is automatically initialised and mission start procedures are launched simply by inserting a cartridge programmed during the preflight preparation. The DTE takes full control of navigation during the automatic ground-hugging phases of the flight and operates the weapons system. It also has electronic mapping and mission playback and debriefing on the various phases. The DTE reduces ground maintenance since it records and analyses the main aircraft parameters throughout the mission.

The DTE is capable of interfacing with all ground mission preparation and debriefing systems. It is operable from any allied or friendly logistic platform. The DTE also allows training in an intensive electronic warfare environment without requiring any special infrastructure.

Specifications
Dimensions:
(data transfer unit) 178 × 127 × 113 mm
(data transfer cartridge) 191 × 119 × 41 mm
Weight:
(data transfer unit) 3 kg
(data transfer cartridge) 0.7 kg
Power supply: 115 V AC, 400 Hz, single phase
Cartridge capacity: 8 kwords expandable to 2 Mwords

Status
In service with the US Air Force, Army and Navy aboard more than 35 different aircraft, including A-6, A-10, B-1B, C-17, F-14, F-16, and SH-60. Also in service with NATO, Japan, the Republic of South Korea, and the Republic of China (Taiwan).

In late 1999, Orbital Fairchild Defense (now Smiths Aerospace) was contracted to supply an updated version of the DTE to the United Arab Emirates for use on Mirage 2000-9 aircraft.

Contractor
Smiths Aerospace, Germantown.

The data transfer unit (right) and data transfer cartridge (left) 0051286

Data Transfer Equipment/Mass Memory (DTE/MM)

Type
Data processing system.

Description
Smiths Aerospace developed the F/A-22 Data Transfer Equipment/Mass Memory system (DTE/MM). This system is composed of an aircraft cockpit resident Data Transfer Unit/Mass Memory (DTU/MM) and a portable Data Transfer Cartridge (DTC).

The DTC performs all data transfer functions such as mission data load, loading of operational flight programs, and in-flight data recording.

This cockpit resident DTU/MM is an intelligent mass memory system that contains up to 576 Mbytes of user non-volatile memory that provides modular growth to 2 Gbytes, The growth is obtained by the insertion of next-generation memory modules. The 100 per cent Ada operational flight program in the DTU/MM contains all functions necessary to support memory expansion including memory management up to 2 Gbytes of storage in the DTU/MM as well as the DTC.

The DTC is a portable non-volatile memory device that is used to transfer mission planning data to the aircraft for pre-mission initialisation, and is carried, typically by the pilot, from the aircraft to the mission debriefing area for unloading of in-flight recorder mission and maintenance data from the cartridge. The DTC uses a modular memory approach similar to the DTU/MM. Four memory module slots provide growth to 2 Gbytes. Present configuration provides 84 Mbytes of high-speed non-volatile random access memory. Also, this next-generation cartridge has maintained the common cartridge interface to ensure compatibility with existing ground support equipment and mission planning systems in the field.

The DTE/MM contains two memory volumes, the DTC and the Mass Memory (MM) for storage of user data files. The system maintains separate directories for each volume containing status and management data for each file present. Each volume also contains a header file used to control special functions such as automated file download for system initialisation ('boot-up' of other computers connected to the DTE/MM) and to specify bus parametrics.

File management commands allow the user to create, delete and erase files, read and write file data, copy files from DTC to MM, report file and system status information, format the volumes and perform built-in-test.

Data written into files is appended with a Reed Solomon error detection and correction code to provide robust data protection. Furthermore, if an error is detected during a data write cycle, the system provides recovery through a memory re-mapping capability, effectively enhancing system reliability.

The DTU/MM may be configured with up to nine Flash memory modules providing user memory capacity up to 576 Mbytes. This capacity may be doubled (or more) through insertion of new memory modules using next-generation memory components. Addressing capability of the DTU/MM supports up to 2 Gbytes of user memory.

The DTC may be configured with up to four SRAM memory modules providing user memory capacity up to 84 Mbytes and may also be expanded through new-generation module insertion. The DTC also supports addressing growth to 2 Gbytes. System features include full implementation of the JIAWG J88-N2 Linear Token-Passing High-Speed DataBus (HSDB) operating at 50 Mbits/s; automated download of specified system files in support of system initialisation; file management command structure; full Reed Solomon error detection and correction capability on all user memory; built-in fault log memory separate from DTU/MM and DTC memories; automatic over temperature detection with (override) system shutdown; extensive built-in-test (BIT) capability: startup BIT,

Data transfer equipment/mass memory for the F/A-22 Raptor   0051288

initiated BIT, periodic BIT; DTU/MM operational software may be reprogrammed (updated) with system installed.

Specifications
Dimensions:
DTU/MM: 209.6 (L) × 146.1 (H) × 195.6 (W) mm
DTC: 190.5 (L) × 119.4 (W) × 40.6 (H) mm
Weight:
DTE/MM: 5.5 kg
DTC: 1.6 kg
Input power: 28 V DC, 45 W (typical)
Mass Memory data capacity: up to 576 Mbytes, growth to 2 Gbytes
Cartridge data capacity: up to 84 Mbytes, growth to 2 Gbytes
Software: 100% coded in Ada
System interface: Fibre-optics high-speed databus (JIAWG J88-N2)
Reliability (MTBF): DTE/MM system – 11,000 h

Status
In production and in service in the F/A-22 Raptor.

Contractor
Smiths Aerospace, Germantown.

Data Transfer Systems (DTSs)

Type
Data processing system.

Description
Data Transfer Systems (DTSs) are a selectable family of automated digital data loading, recording and post-mission downloading subsystems for aircraft and surface vehicles using digital avionics. The first solid-state DTS was initiated by the then Smiths Industries in the 1970s to obviate digital information transfer problems encountered during the preflight loading of mission-related information. Errors and delays in aircraft readiness were being caused by the manual

insertion of preflight information such as target co-ordinates and mission waypoints via aircraft system keyboards. The introduction of the DTS has basically eliminated preflight data entry errors while, at the same time, reducing cockpit flight deck initialisation time from about 30 minutes to a few seconds.

In addition to the loading and retrieval of normal mission data, DTS equipment is being used to initialise Joint Tactical Information Distribution System (JTIDS), Navstar GPS (Global Positioning System), missile guidance control units, digital mapping system and voice control interactive devices. DTS equipment can also provide immediate post-mission printout and analysis of operational and flight test information. It is also used within Flight Data Recorder systems as an information recording and download mechanism.

A typical DTS includes a small, portable, solid-state data transfer memory media, a receptacle into which the media is inserted and a ground-based mission data computer terminal containing an application-specific software database.

Available memory media includes a small shirt pocket-sized Data Transfer Module (DTM, or cartridge) which is offered with SRAM, Flash EPROM and combined SRAM/Flash EPROM, depending on customer preference/need. DTM memory capacities range from 16 kbytes to 150 Mbytes, with planned growth to 600 Mbytes achievable by 2000. Smiths Industries has introduced seven backward-compatible DTM improvement configurations since its initial DTM design in the 1970s. Optional PCMCIA (PC Card) portable memory media solutions are also offered for selected Smiths Industries DTS applications. PC card memory capacities currently exceed 1 Gbyte, with 6 Gbyte PC cards forecast for 2000.

The data transfer memory media receptacle, mounted in a convenient location within the aircraft, accepts the DTM and/or PC Card(s) via a safety spring-loaded door. Depending on user

Data Transfer System   0001307

needs, Smiths Industries receptacles are modularly expandable to accommodate standard electronic interfaces, such as RS-422, MIL-STD-1553A/B, SCSI-2, Fibre Channel, Ethernet and combinations of these interfaces. Newer DTM/PC Card receptacles also offer a media data management microprocessor, additional memory and/or other electronics to increase system functionality. The receptacle electronics can be provided either within the receptacle or packaged remotely to conserve cockpit space. Smiths' cockpit receptacles can be configured to accommodate the DTM, PC Cards (PCMCIA), or a combination of DTM and PC Card memory media.

DTM/PC Card ground interface and mission planning computer terminals incorporate user-friendly software to accommodate efficient digital data loading and retrieval functions. Mission planning terminals have progressed from large expensive machines to smaller, more capable and less expensive ground computer systems. Smiths offers its own tactical Mission Data Ground Terminal (MDGT) based on Personal Computer (PC) technology. Also offered is a family of DTM ground interface and test equipment for use within customer designated mission planning systems. Selectable DTM ground interface and test unit/kit solutions include DTM to PC-ISA/EISA, RS-422, IEE-488 and SCSI-2 interfaces. Included within these standard mission planning systems are the Computer Aided Mission Planning System (CAMPS), Mini-CAMPS, Mission Support System-1 (MSS-1), MSS-2, MSS-2+, Air Force Mission Support System (AFMSS), Portable Mission Planning System (PMPS) and Army Aviation Mission Planning System (AMPS).

Status
In production. Over 80 types of fighters, multi-engine aircraft, helicopters and surface vehicles have been equipped with the Smiths DTS, including over 8,500 systems to date.

Contractor
Smiths Aerospace.

Digital Flight Control System (DFCS) for the F-15E

Type
Flight Control System (FCS).

Description
Whereas the standard F-15 has a Lockheed Martin analogue flight control system, the additional requirements of the F-15E for deep penetration with emphasis on terrain-following and terrain-avoidance, allied with new sensors, called for a digital triplex automatic system of greater performance and reliability. Digital technology permits control laws to be optimised both for air combat and for high-speed low-level flight. Lear Astronics is also responsible for the digital pressure sensors used to provide speed and aerodynamic pressure sensing for the system. The system is based on MIL-STD-1750 instruction set architecture and uses Lockheed Martin Fairchild Systems 9450 microprocessors.

Status
In service on F-15E.

Contractor
BAE Systems North America.

Digital flight control system for the F/A-18

Type
Flight Control System (FCS).

Description
The F/A-18 flight control system is a digital four-channel Fly-By-Wire (FBW) system operating the aileron, stabilator and rudder primary flying controls, leading-edge and trailing-edge flaps and nosewheel steering. The system incorporates

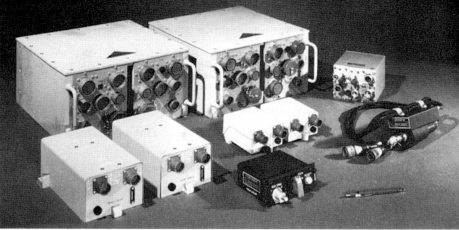
The flight control electronics set for the F/A-18 Hornet 0503925

32 servo loops to drive and control these functions. At the same time the conventionally mounted control column, in contrast to the F-16's sidestick controller, has a mechanical link to the ailerons for reversionary pitch and roll control. All FBW computations are accomplished by four digital computers operating in parallel, accepting inputs from linear variable differential transformers sensing control column and rudder pedal movement and analogue motion sensors, and formulating commands to the redundant electrohydraulic servo-actuators driving the control surfaces. The system therefore remains operational after two failures. It has a high degree of integration with other aircraft equipment, communication being accomplished by means of a MIL-STD-1553 databus.

Two special display modes are used in conjunction with the system. The first is a flight control failure matrix, the second provides recovery guidance during spins and both use the same CRT unit. In the first mode, the display is selected to show details of the fault, after the pilot's attention has been drawn to its presence by an indication on the central warning system. The display signals 'X' at the position of the fault on a schematic diagram of the system painted on the screen. In the second mode, the display shows the position of the control column needed to recover from a spin, assumed to exist when yaw rates greater than 15°/s occur simultaneously with speeds of 125 kt or below. This mode is selected automatically, having absolute priority over other displays when this combination of speed and yaw rate occurs. The two modes were originally provided for flight test purposes, but have been retained in production aircraft following recommendations from the Naval Air Test Center and test pilots.

The system comprises two flight control computers, each incorporating two microprogrammable digital processors specially designed for flight control applications, two rate sensor boxes, two acceleration sensor boxes, an air data sensor box, rudder pedal force sensor and pilot's control panel.

Status
In service on the F/A-18.

Contractor
BAE Systems Controls.

Digital fuel gauging systems

Type
Onboard system monitor.

Description
Smiths Aerospace has had its 2300 Series digital fuel quantity indicators in service for some years on an increasing variety of civil and military aircraft. The system is standard fit on the Boeing 737, 747-200/300 and 777.

Smiths Aerospace has developed an integrated fuel quantity system in a single LRU. This device performs automatic refuelling, valve control,

Smiths Aerospace's digital fuel gauging repeater and master indicators for the Boeing 747 0581521

preselection of fuel quantity and digital data output for EFIS displays.

Status
In addition to supplying Boeing, for the 727, 737 and older 747s, Smiths Aerospace has certified the system for retrofit to a variety of aircraft including Boeing DC-8, DC-9 and DC-10 and the Fokker F28.

Military retrofits have been completed on the A-4 Skyhawk, C-130 Hercules, C-141 and F-4 Phantom.

Contractor
Smiths Aerospace.

Digital fully fly-by-wire system for the F-16C/D

Type
Flight Control System (FCS).

Description
The system has evolved from the equipment developed by Honeywell Aerospace for the US Air Force Advanced Fighter Technology Integration (AFTI) programme based on an F-16 airframe. A fundamental change from the flight control system in the AFTI aircraft is, however, the progression from three digital plus one analogue channels per axis to four digital channels, the configuration favoured for operational reliability (the system has to remain operational after any two failures). The four digital channels use MIL-STD-1750A architecture utilising Jovial high-order language and each has a processor, memory, input/output functions, discrete failure logic, MIL-STD-1553B digital databus and serial link to other channels. The computer is designed to remain fully operational following any two consecutive failures within the quadruplex part of the system. The single unit, weighing 22.7 kg and housed in a 1 ATR long box, has the equivalent function of the four separate units on the AFTI aircraft and occupies about half the space. In all other respects the system meets the form, fit and function requirements that permit it to replace directly the earlier analogue system.

Computing and processing technology is based on large-scale and very large-scale integration and gate arrays. It is designed to accommodate further growth, incorporating VHSIC components, without the need for new software. Flight critical functions are hard-wired into the system for the highest integrity, but less critical signals communicate with other equipment via the digital databus. The system has 48 kwords of PROM memory, 2 kwords of RAM scratchpad memory, 8 kwords of input/output scratchpad memory and a 2 kword memory to record faults.

Benefits of the all-digital system over the previous analogue one are given as better reliability, lower power density, smaller number of components, greater ease of tuning to meet changes and lower life cycle costs.

Honeywell Aerospace considers this flight control system to be an important step forward, since it can be seen as the basis for future integrated flight-critical full authority flight control systems incorporating thrust/side-force vectoring, terrain-following and fire control.

Specifications
Dimensions: 514 × 273 × 222 mm
Weight: 22.7 kg
Power: 200 W
Reliability: >2,150 h demonstrated MTBF

Status
Installed in later F-16C and F-16D single-seat and two-seat fighters.

Contractor
Honeywell Aerospace, Sensor and Guidance Products.

DL-900 data loader

Type
Data processing system.

Description
Honeywell's DL-900 data loader is compatible with the NZ-600, NZ-800, NZ-900 and NZ-2000 navigation computers and the IC-615, IC-800 and IC-1080 integrated avionics computers. The system features an RS-422 interface and accepts high- and low-density diskettes. With suitable software (version 5.2 and greater) the DL-900 is capable of simultaneous NDB uploads.

Specifications
Dimensions: 146 × 57 × 210 mm
Weight: 1.27 kg
Power: 28 V DC; 2.5 W

Status
In production and in service with associated navigation systems.

Contractor
Honeywell Aerospace, Electronic & Avionics Lighting.

DMU 100 and 101 Data Management Units

Type
Data processing system.

Description
The DMU 100/101 Data Management Units are the brain for the ACMS system, analysing the real-time information from the FDAU, the ADAU or the DFDAU. They control the digital ACMS recorder, having program logic that determines what information to record and when to do so. They also provide data to displays, including the FDEP and airborne printer, and contain extensive built-in test equipment.

Specifications

	DMU 100	DMU 101
Dimensions	1 ATR long	6 MCU
Weight	16.36 kg	5.7 kg
Power	180 W	55 W

Status
The DMU is part of the company's Mk II and expanded Mk III ACMS system in the following aircraft:
- DMU 100: Airbus A300, Boeing 747, and DC-10
- DMU 101: Airbus A310 and Boeing 767.

Contractor
Hamilton Sundstrand Corporation.

DMU 120 Data Management Unit

Type
Data processing system.

Description
The DMU 120 is a powerful microprocessor-based data acquisition and processing tool utilised on the A320 aircraft for aircraft, engine and APU condition monitoring and for intensive aircraft systems troubleshooting. It collects comprehensive aircraft and engine data via ARINC 429 serial input ports. Real-time data processing and analysis is performed. Information in the cockpit is made available in convenient format via the MCDU and printer. Automatic, cockpit and uplink requested information supplied by the DMU is transmitted to the ground by ACARS. The DMU provides for data reduction recording to the DAR which permits economical ground-processing of airborne data.

Specifications
Dimensions: 195 × 95.5 × 287.6 mm
Weight: 4.4 kg
Power: 30 VA

Status
In production for the Airbus A320.

Contractor
Hamilton Sundstrand Corporation.

The DMU 120 is installed in the Airbus A320
0503883

DR-200 PCMCIA data transfer system

Type
Data processing system.

Description
The DR-200 PCMCIA (PC Card) receptacle is an intelligent microprocessor-based unit that permits access to aeronautical worldwide databases as well as mission specific data. It provides two PC Card sockets and communications over the MIL-STD-1553B databus. The unit will accept both 3.3 and 5 volt commercially available Type I, II, or III memory cards.

The DR-200 was selected competitively by the US Army and provides connector and software compatibility with the previous US Army standard DR-902 data receptacles and cartridges.

When used in conjunction with a navigational database loaded onto a PC Card by MPGS or the SOF DAFID Conversion utility, the DR-200 provides rapid retrieval of all stored airport, heliport, communication and navigation information.

The main capabilities and functionality of the DR-200 are:
- Host aircraft bus-controller controls all DR-200 operations using MIL-STD-1553B messages
- DR-200 Operational Flight Programme (OFP) is loaded directly from one of the PC Card sockets
- All information stored on mission data PC card is zeroed through both discrete and software commands at commencement
- Rockwell Collins designed DR-200 microprocessor operating at 24 MHz
- Module contains 64K RAM, 64K Program Memory, 64K non-volatile memory
- Hardware interfaces include MIL-STD-1553B, discrete I/O, RS-422 and ARINC-429
- To provide backward capability; the unit performs read and write services in the same manner as the Rockwell Collins DR-902 data receptacle
- DR-200 strictly observes the use of sub-addresses and bus timing to ensure true compatibility
- Status, type and directory information for each inserted card
- Responding to ICAO lookup command, the unit performs data retrieval and display formatting for the requested worldwide navigation database including airport, heliport, navaid and intersection data
- DR-200 correctly handles data provided by the Army SOF ground station software executing on a desktop or laptop computer. It is compatible with the SOF mission aircraft platforms
- Worldwide database on the PC Card for the five nearest airports
- User is able to copy the contents of similar fixed memory cards between sockets
- DR-200 correctly handles data provided by the MPGS ground station software, operating on desktop or laptop computer, provides flight planning, transfer and filtering of Jeppesen ARINC-424 or NIMA DAFIF information to PC Cards
- Retrieval of waypoints, airports, approaches, navaids and flight plan data from the PC Card on command from the FMS-800 (see separate entry).

Specifications
Configuration: Dual Type-I, II, or III socket, microprocessor I/O board, modular power supply/filter
Dimensions: 57.2 × 146.1 × 152.4 mm (2.25 (H) × 5.75 (W) × 6 (D) in)
Weight: 0.91 kg (max)
Cooling: Convection
Power: 5 W (max)
Mounting: Dzus rail
Qualification: PCMCIA 2.1, MIL-STD-1553B, MIL-STD-810E, MIL-STD-461D, MS25212/MS25213

Status
In production.

Contractor
Rockwell Collins.

DSS-100 Data Storage Set

Type
Data processing system.

Description
The DSS is a major element of the US Navy Flight Incident Recorder and Aircraft Monitoring System (FIRAMS) on the F/A-18 aircraft. It also performs data storage functions on the A-6, F-14, AV-8B and V-22 aircraft. The DSS consists of a Data Storage Unit (DSU) and a Data Storage Unit Receptacle (DSUR). The DSUR is mounted in the cockpit. The DSU is a modular electronic memory unit which slides into the DSUR for flight and is removable for analysis off the aircraft. The DSU employs a microprocessor to extract data from a MIL-STD-1553B bus for storage in memory and vice versa. The microprocessor also expands the effective

amount of internal memory by executing a data compression algorithm. The memory medium is non-volatile EEPROM installed in blocks of from 2 to 8 Mbytes. The DSUR contains no electronics or active components.

Specifications
Dimensions:
(DSU) 43 × 119 × 205 mm
(DSUR) 53 × 127 × 229 mm
(faceplate) 102 × 165 mm
Weight: 1.9 kg
Power: 12 W

Status
In production and service; installed in the A6, AV-8B, F-14, F/A-18.

Contractor
Hamilton Sundstrand Corporation.

Electronic Flight Control Set (EFCS) for the B-2

Type
Flight-Control System (FCS).

Description
The digital fly-by-wire Electronic Flight Control Set (EFCS) consists of two parts: a quadruple redundant set of Flight Control Computers (FCCs) and quadruple redundant pilot sensor assemblies. Each FCC is a single-channel, real-time digital computer, connected via a MIL-STD-1553 bus controller to the avionics sytems, remote terminals, pilot sensor assemblies, cockpit panels, nosewheel steering and hydraulic system. The computer includes a PACE 1750A processor with a throughput of 2 Mips.

There are three pilot sensor assemblies: yaw pedal, pitch stick and roll stick. The sensor assemblies transform pilot input to an electronic signal which communicates with the FCC.

Specifications
Dimensions:
(FCC) 388.4 × 345.9 × 193.5 mm
(pilot sensor assembly) 215.9 × 66 × 91.4 mm
Weight:
(FCC) 18.6 kg
(pilot sensor assembly) 1.81 kg

Status
In service in the Northrop Grumman B-2.

Contractor
Lockheed Martin Control Systems.

Electronic Flight Control System for the C-17

Type
Flight Control System (FCS).

Description
The C-17 transport is designed to carry men and equipment into austere airfields and drop zones near the front line. Operation of a large aircraft at low speed needs a flight control system that provides stable yet responsive control, augments the basic stability of the aircraft at low speeds and is extremely reliable. These requirements are being met through development of a quadruple redundant digital fly-by-wire control system for both the primary and secondary flight controls of the C-17. The Electronic Flight Control System (EFCS) controls the actuators that drive the movable ailerons, flaps, leading-edge slats, spoilers, horizontal stabilisers, elevators and rudders.

All flight controls are electrically commanded in their normal mode of operation. A mechanical system provides back-up control between pilot controls and the elevator, aileron and lower rudder.

The EFCS is a full-time, full-authority fly-by-wire control system with stability augmentation in all axes. Electronic flight control functions

Components for the C-17 electronic flight control system 0503926

are provided by quadruple redundant sensor, computation and actuation channels. In the fly-by-wire control mode, the sensors relay pilot force signals to the four Flight Control Computers (FCC). The FCCs are programmed, using other sensor information and inputs, to generate flight control signals that control flight path and attitude. In addition, the EFCS provides automatic aileron and stabiliser trim, autopilot, flight director and autothrottle functions. An angle-of-attack limiting system is provided to prevent stall during high-lift configuration.

The EFCS is designed to provide continued operation following both failures and battle damage. This capability is achieved through four-channel architecture which can continue full-function operation on two channels. The channels are separated on the aircraft and the components are located so as to minimise the risk of losing multiple channels through battle damage.

Status
In service on the C-17.

Contractor
BAE Systems Controls.

Engine Data Multiplexer (EMUX)

Type
Data processing system.

Description
The Engine Data Multiplexer (EMUX) is used for reduction of large amounts of analogue, digital and discrete data to ARINC 429 format. Some typical sources of data are the engine and transmission sensors, prop speed sensors and all the various aircraft system discretes.

The EMUX features extremely flexible architecture, permitting maximum cost effectiveness in any application. Included on individual, easily serviced plug-in cards are all the signal conditioning and data conversion circuitry needed to interface the engine sensors with the digital databus. For use in systems where reliability is of the utmost importance, two totally independent channels are provided in each engine data converter.

For use with aircraft systems, the engine data converter offers additional optional features such as internal storage and periodic reporting of engine limit parameters, engine serial number encoding and non-volatile storage of limit exceedances, durations and engine cycle. It can also be structured to interface with other databusses, such as MIL-STD-1553B.

The design is sufficiently flexible so that it can be used with any type of engine. The software, as well as the analogue to digital circuit cards, is modular, so that changes to the engine sensors and engine performance characteristics can be readily accommodated.

One EMUX is assigned to each engine. It is a dual-redundant device. Each half is a totally independent functioning unit having its own power supply, input converter cards, discrete units, microprocessor and digital output circuits. It shares only a common interconnect between its two halves. Isolation between the two halves is such that a failure of one half will not affect the other half's operation.

Specifications
Weight: 5.22 kg
Power consumption: 7 W per channel typical, 10 W per channel (max)
Reliability: 5,000 h MTBF

Status
The EMUX has applications on the Beach Starship and Fokker 100.

Contractor
Ametek Aerospace Products.

Engine Diagnostic Unit (EDU)

Type
Onboard system monitor.

Description
Pratt and Whitney's F100-PW-220 and 229 high-performance engines are fitted with Hamilton Sundstrand's advanced engine monitoring system which includes both engine-mounted equipment and the associated ground-based diagnostic units.

An EDU is the on-engine module which acquires engine data from controls and sensors, records operating time and cycles, detects critical events and stores selected event parameters. The EDU also performs extensive self-health and data validity checks and stores this data, along with Digital Electronic Engine Control (DEEC) diagnostic data, for further analysis.

The hand-held Data Collection Unit (DCU) is used to gather data from the EDU, clear the EDU's memory and perform EDU/DEEC system diagnostics. Data from multiple aircraft may be stored in the DCU's removable memory module, providing easy data transfer between the flight line and a ground-based computer for logistics data recording.

The ground-based portable Engine Analyser Unit (EAU) interfaces with the EDU or the DEEC for data acquisition, memory examination or modification and data monitoring of the EDU/DEEC output data streams during engine operation. The EAU is also capable of performing diagnostic testing of the DEEC or EDU and exercising combined EDU/DEEC fault logic and event codes, providing a fast, accurate determination of engine problems.

Specifications
Dimensions:
(EDU) 241 × 279 × 127 mm
(DCU) 475 × 424 × 412 mm

Weight:
(EDU) 4.13 kg
(DCU) 23.6 kg

Status
In production.

Contractor
Hamilton Sundstrand Corporation.

Engine Monitoring System Computer (EMSC)

Type
Aircraft system control, power plant.

Description
Ametek manufactures the Engine Monitoring System Computer (EMSC) for the General Electric F110 engine which equips the US Air Force F-15 and F-16C/D aircraft. The computer provides in-flight monitoring of engine exceedance, faults or trends. Engine-related signals are acquired from the engine monitoring system processor via the MIL-STD-1553B engine signal databus. Diagnostic data is retained in non-volatile memory which annunciates this data visually to the cockpit.

A secondary function downloads the data via the RS-232C series communication link to the data display and transfer unit for ground support evaluation. This link also uploads information to the EMSC such as the aircraft engine diagnostic information, time data, aircraft serialisation and life usage data.

Status
In service on the General Electric F110 engine on the F-15 and F-16C/D. Most of this system's modules are also used in the engine analyser unit for the Northrop Grumman E-2C's Allison T56-A-427 engine.

Contractor
Ametek Aerospace Products.

F/A-22A – VMS/IVSC: Vehicle Management System/ Integrated Vehicle Subsystem Control

Type
Flight Control System (FCS).

Description
The VMS combines flight and propulsion controls, while the IVSC controls the aircraft utilities. The VMS utilises a triplex digital flight control system, with no electrical or mechanical back-up, to provide full carefree handling and enhanced manoeuvrability, with a sidestick fly-by-wire controller.

The IVSC modules control the following services: electric power, hydraulics, fuel systems, integrated warning/caution/advisory functions, diagnostics and health monitoring, auxiliary power, environmental and life support functions.

Equipment modules are included in the avionics common module racks. Raytheon Electronic Systems provides common 1750A processor modules for these systems.

Status
Installed in F/A-22A.

Contractor
BAE Systems North America, Aircraft Controls.

F-16 general avionics computer

Type
Avionic processor.

Description
The F-16 general avionics computer acts as the mission computer for the F-16C/D aircraft, providing avionics and weapons control solutions, IFF processing and navigation functions. Designed for MIL-STD-1750 ISA performance, the

computer offers as standard features throughout of up to 6 Mips DIAS, 512 kbytes or greater RAM or EPROM, battery back-up MIL-STD-1553B input/output with up to four channels per module, with bus controller or remote terminal. Options include single module and custom packaging, custom input/output and BIT functions.

Specifications
Dimensions: 135 × 312 × 333 mm
Weight: 9.09 kg
Power supply: 115 V AC, 400 Hz, 70 W
Reliability: >4,000 h MTBF

Status
In production for the F-16C/D in service with the US Air Force and other air forces.

Contractor
Northrop Grumman Electronic Systems, Navigation Systems Division.

F-16 Voice Message Unit (VMU)

Type
Avionic Voice Messaging System (VMS).

Description
The Voice Message Unit (VMU) is a compact, ruggedised voice warning system currently in service as an integral part of the F-16 avionics. The VMU is capable of monitoring twelve 28 V DC discrete and four 5 V DC differential inputs for message activation. The VMU can be arranged to react to aircraft systems' sensor activity, delays, priority and a number of occurrences under software control of the VMU microprocessor. The VMU has the capacity for 12 seconds of verbal or other messages which can be broadcast over an intercom or other facilities. As the VMU is microprocessor controlled it can be adapted to monitor and provide voice/tone warnings for most aircraft and vehicle systems.

Specifications
Dimensions: 63.5 × 98.6 × 82.6 mm
Weight: 0.63 kg
Power: (max) 4 W

Status
The VMU is currently in service on the F-16 and interfaces with the avionic systems to provide the pilot with an audible indication of aircraft system status which has exceeded preset parameters.

Contractor
Sanmina-SCI Corporation.

FCC100 automatic flight control system

Type
Flight Control System (FCS).

Description
The FCC100 system was designed specifically for the US Army Sikorsky UH-60 Black Hawk helicopter. It meets requirements calling for quick and easy field maintenance and repair and is claimed to have been the first application of a dual-digital processor-based stability augmentation system for a production helicopter. Features include three-axis stability augmentation with turn co-ordination, pitch, roll and yaw attitude and altitude/airspeed hold modes.

Specifications
Dimensions: 356 × 305 × 217 mm
Weight: 8 kg
Power: 75 W

Status
In service on the Sikorsky UH-60 Black Hawk helicopter.

Contractor
Hamilton Sundstrand Corporation.

FCC105 automatic flight control system

Type
Flight Control System (FCS).

Description
The FCC105 automatic flight control system introduced modern digital technology features in Sikorsky HH-60, CH-53E, SH-60 and MH-53E helicopters. It is a dual-redundant system which provides three-axis stability augmentation, turn co-ordination and hands-off flight. It can also include a stick-force feel system and has four degrees of freedom facilities and hover augmentation. The most recent version, on the Sikorsky MH-53E helicopter, includes MIL-STD-1553B digital data interfaces. A non-redundant derivative is utilised on the SH-60B. A further derivative of this, with the nomenclature 'general computer', has been developed for use on H-60 helicopters other than the SH-60B.

The baseline 53E computer used in the Sikorsky SH-60B includes automatic approach to hover, three-axis Doppler and radio altitude coupling, with provision for tactical navigation computer coupling.

Specifications
Dimensions: 190 × 432 × 266 mm
Weight: 11.5–13.5 kg
Power: up to 83 W

Status
In service on Sikorsky CH–53E, MH-53E, HH-60 and SH-60 helicopters.

Contractor
Hamilton Sundstrand Corporation.

FCC-105-1 automatic flight control system

Type
Flight Control System (FCS).

Description
Based on the FCC-105 system, the FCC-105-1 automatic flight control system is applicable to the Sikorsky S-76 and H-76 helicopters and has been in production since 1977. It offers a wide range of features up to full three-axis stability augmentation with turn co-ordination, pitch, roll and yaw attitude and altitude/airspeed hold modes. Each system amplifier has independent pitch, roll and yaw channels contained in individual modules which can be selected separately. This building block principle permits customers to select as many features as necessary to meet their particular requirements.

The system comprises a stability augmentation system amplifier, yaw switch, heading hold amplifier, airspeed switches, control panels, linear electromechanical actuators, indicator panel, rate gyros, cyclic switches and airspeed hold amplifier.

Specifications
Dimensions
(main processor) 190 × 318 × 113 mm
Weight
(main processor) 3.4 kg
Power: 35 W

Status
In production for the Sikorsky S-76 and H-76 helicopters.

Contractor
Rockwell Collins.

FCS-700A autopilot/flight director system for the 747-400

Type
Autopilot system.

Description
The FCS-700A is an enhanced derivative of the autopilot on the Boeing 757 and 767 and

is designed to provide LRU interchangeability between those two aircraft types and the 747-400. It is a Cat IIIB fail-operative system tailored to the control requirements of the 747-400 and is usable within the complete flight envelope for autopilot and autoland functions including automatic landings with a decision height of zero feet and a runway visual range of 600 m.

The FCS-700A consists of three flight control computers and a mode control panel. The current computer is the FCC-703, which replaces the FCC-702 and FCC-701. Each computer has dual-Rockwell Collins Adaptive Processor System (APS) microprocessors for increasing processor capacity to allow for future growth in autopilot capability for MLS. The FCC-703 enables easier system upgrades due to the incorporation of dataload capability via either front or rear connector.

The system replaces 14 dedicated LRUs on previous 747 aircraft, saving considerable space and weight.

Status
In service in Boeing 747-400 aircraft.

Contractor
Rockwell Collins.

FCS-4000 Digital Automatic Flight Control System (DAFCS)

Type
Flight Control System (FCS).

Description
The FCS-4000 is an evolutionary digital automatic flight control system, with hardware and software architecture derived from the APS-85/850 autopilot systems (see separate entry). The FCS-4000 shares power supply, input/output, signal conditioning and environmental control with other functions residing in the Integrated Avionics Processing System Card Cage (IAPS CC). This resource sharing reduces wiring and installation complexity and improves reliability.

The FCS-4000 Flight Control System (FCS) is an integrated autopilot and flight director system. It provides dual-redundant flight guidance computations and a three-axis autopilot with automatic pitch trim control signals. The yaw damper may be a series or parallel configuration, and is dual and independent in some versions. Additional channels of automatic trim can be provided with optional trim computer modules. The FCS-4000 may also be installed in a dual-autopilot configuration.

The FCS-4000 FCS requires two FCC-40XX Flight Control Computers, which are located in the integrated avionics processing system. Each FCC provides an independent flight director function by performing the guidance computations that result in commands to the crew displays. The FCCs operate together to provide the autopilot function. Each FCC computes its own servo command, which passes through a mid-value voter, resulting in a signal to drive the servos. The FCCs also operate together to provide the pitch trim function.

A typical FCS-4000 system consists of two FCC-40XX, one FCP-400X, three SVO-85/SMT-87 servo motors/mounts and a pitch trim servo motor. One or two linear actuators are used for yaw damping in place of the SVO-85 servo motor in some applications. The heart of the system is a digital FCC that is implemented via two Line Replaceable Modules (LRM), each with two microprocessors. This computational power provides for precise flight control and operational flexibility for two independent flight director channels. The configuration of the FCP offers a single point for either crew member to make lateral and vertical mode selections, as well as reference adjustments. The servo interface with the airplane FCS is through lightweight, high reliability servo motors. The primary servo motors have dual monitoring tachometers to ensure that faults at the servo will be detected and quickly isolated. The servo is connected to the flight control system through a slip clutch mechanism, which provides an over-ride capability in the event of a servo motor jam.

The Flight Control Panel (FCP-400X) offers a single point for either crew member to make lateral and vertical mode selections, as well as reference adjustments. The panel allows display of either flight guidance computation, on both primary flight director displays and coupled to the autopilot. Autopilot engagement control is also on the FCP, with standard lateral and vertical modes provided. A Flight Level Change mode holds the airspeed reference, while executing the change of altitude. A VNAV mode is available that allows integrated coupling of the FMS vertical flight plan to the autopilot.

System operation
FCC-40XX Flight Control Computer
The FCC-40XX is a LRM that resides in the integrated avionics processing system. The FCS-4000 requires two of these modules. Each FCC provides an independent flight director function. The flight director operates in the mode selected by the crew and performs the guidance computations that result in commands provided to the flight crew displays via V-bars or crosshairs and to the autopilot function. The FCCs operate together to provide the autopilot function. Each FCC computes its own servo command to drive the servo motors. These commands are routed through a midvalue voter. These analogue commands are also torque limited, with servo rate used as feedback of servo movement to close the servo loop. The FCCs also operate together to provide the automatic pitch trim function. The FCCs sense torque on the primary pitch servo and generate arm and command signals to drive the pitch trim servo or external trim control system. Discrete signals are received from the trim system for feedback of trim system response to commands, as well as for monitoring of unexpected trim system operation.

FCP-400CX Flight Control Panel
The FCP-400X provides crew interface with most flight director and autopilot functions. It contains a pushbutton and, typically, a Light Emitting Diode (LED) annunciator light for each available mode. The FCP transmits button presses to the FCC for mode selection requests and illuminates LEDs which confirm that the FCC has accepted the mode request. Various versions of the FCP are possible to accommodate the application. Variations include available mode selections, panel colour and lighting power. The FCP-400X also provides the crew interface with the autopilot and yaw damper. It contains the autopilot and yaw damper, flight director transfer and turbulence mode pushbuttons. The FCP system also provides clutch engage power to the servos when the FCC acknowledges the engagement.

MSP-85/APP-85
The FCS-4000 is compatible with the MSP-85/APP-85 control panels used in the APS-85 system.

SVL-4000
The SVL-4000 is a direct current motor and linear actuator with position feedback outputs. For these actuators, the FCC provides low-level command signals, as the amplifier circuits are located in the actuator.

SVO-85( ) servo motor
The SVO-85( ) is a DC servo motor with dual-rate feedback tachometers. The motor is mechanically coupled to the servo capstan through a magnetic engagement clutch. Three versions of servo may be applied to meet the torque requirements of the control system(s). Variations of each type exist to accommodate physical installation needs. Each servo type provides a choice of electrical harness interface, such as standard unit connector or pendant cable.

SMT-85/87 servo mount
The SMT-85/87( ) provides the mechanical interface with the aircraft control system. The capstan accommodates the bridal cable or other mechanism for coupling to the aircraft control system. The capstan couples to the servo motor through a magnetic engagement clutch. A slip clutch is also incorporated to allow for crew override in the event of a servo motor jam. Several versions of the SMT exist to accommodate the servo motor torque rating and the type of aircraft control system.

Specifications
Dimensions:
(FCC-40XX) 222 (H) × 151 (W) × 52 (L) mm
(FCP-400X) varies with application
Weight:
(FCC-40XX)1.04 kg
(FCP-400X) varies with application
Temperature:
(FCC-40XX) −55 to +70°C
(FCP-400X) −20 to +70°C
Altitude: 55,000 ft
Power:
(FCC-40XX) 28 V DC, 25 W
(FCP-400X) 28 V DC, 4 W
Certification: FAA TSOs C52a (Flight Director), C52b and C9c (Autopilot); DOs-160C and -160D; EUROCAE ED-14C

Status
In production and in service in FAR Part 25 certifications, including Category II approach. The system has been certified on the Beech TTTS/TCX, Beechjet 400A, IAI Astra/Galaxy, Bombardier CRJ-100, -200, -700, -900, Bombardier Challenger 604, Dassault Falcon 20/50EX/2000, Saab 2000 and the Gulfstream 100 and 200.

Contractor
Rockwell Collins.

FGS-3000 digital flight guidance system

Type
Autopilot system.

Description
The FGS-3000 is the most recent evolution of digital automatic flight control systems at Rockwell Collins. This system was developed from the FCS-4000 and incorporates a hardware and software architecture suitable for Fail Passive certifications. The system is offered for integration with avionics systems oriented towards the smaller Part 25 and Part 23 aircraft market. It shares power supply, input/output, signal conditioning and environmental control with other functions residing in the Integrated Avionics Processing System Card Cage (IAPS CC). This resource sharing reduces wiring and installation complexity and improves reliability.

The FGS-3000 Flight Guidance System (FGS) is an integrated autopilot and flight director system that typically consists of two FGC-3000 Flight Guidance Computers, one FGP-300X Flight Guidance Panel and three SVO-3000/SMT-65 servo motor/mounts. It provides dual independent flight guidance computations and a three-axis autopilot with automatic pitch trim control signals. It may be installed for single- or dual-pilot cockpits. The yaw damper may be a series or parallel configuration.

The FGP offers a single point for either crew member to make lateral and vertical mode selections as well as reference adjustments. Optional Mode Select Panels (MSPs) may be installed when there are space limitations. If two MSPs are installed, they are automatically synchronised. The panel allows display of flight guidance commands from either computer to be displayed on both primary flight director displays and coupled to the autopilot. Autopilot engagement control is on the FGP or an Autopilot Panel (APP). Standard lateral and vertical modes are provided. A Flight Level Change mode holds the airspeed reference while executing the selected change in altitude. A VNAV mode is available which facilitates coupling of the FMS vertical flight plan to the autopilot.

Specifications

Dimensions:
(FGC-3000) 222 (H) × 151 (W) × 53 (L) mm
(FGP-3000) varies with application
Weight:
(FGC-3000) 1.04 kg
(FGP-3000) varies with application
Temperature:
(FGC-3000) –55 to +70°C
(FGP-3000) –20 to +70°C
Altitude: 55,000 ft
Power:
(FGC-3000) 28 V DC, 25 W
(FGP-3000) 28 V DC, 4 W
Certification:
(FGC-3000) FAA TSOs C52a (flight director), C52b,
   C9c (autopilot); DO-160C
(FGP-3000) FAA TSOs C20 (flight director), C52b,
   C9c (autopilot); DO-160D

Status

In production and service in FAR Part 23 and Part
25 certifications including Category II Approach.
The system has been certified on the CJ 1, CJ 2,
Premier I and Hawker 800XP.

Contractor

Rockwell Collins.

Fibre Channel – Data Transfer Equipment (FC-DTE)

Type

Data processing system.

Description

The FC-DTE consists of an intelligent, aircraft-
mounted receptacle called a Data Transfer
Receptacle (DTR) and a removable Data Transfer
Cartridge (DTC).

The DTR manages and controls the data
exchanged between the DTC and other avionic
subsystems via the aircraft's MIL-STD-1553 and
Fibre Channel busses. These busses may be used
simultaneously thus allowing MIL-STD-1553 and
Fibre Channel-based avionics to concurrently use
the FC-DTE as a mass storage/file server device. In
addition to providing the interface to the aircraft,
the DTR contains significant processing capability
which may be used to pre- and post-process the
data which is being stored/accessed from the DTC.
The DTR also contains a spare card slot which is fully
wired to accept other Smiths Aerospace and other
third-party circuit cards. With this capability, single
card functions such as a digital moving map can be
added to the FC-DTE. These circuit cards have direct
access to the DTC's mass storage and can interface
directly to other Fibre Channel and MIL-STD-1553
devices. Thus, future aircraft upgrades can take full
advantage of the FC-DTE's capabilities.

The Data Transfer Cartridge has been optimised
to provide extremely high-speed access to the
solid-state mass memory which it contains. Current
DTC configurations allow memory densities of
between 100 Mbytes and more than 2 Gbytes. The
design of the DTC allows future memory upgrades
and memory densities in excess of 30 Gbytes will
be available in the foreseeable future.

FC-DTE features support for SCSI-3 upper level
protocol with sustained write rates of more than 20
Mbytes/s and sustained read rates of more than 30
Mbytes/s; significant growth is provided via DTR
spare processing throughput, DTR spare card slot,
and DTC modular memory design; DTC is Air Force
Mission Support System (AFMSS) compatible; the
DTC is removable in explosive atmospheres; it
supports NSA-approved declassification of stored
data; it is qualified for multi-aircraft use including
those aircraft with radiation hardness requirements;
it provides 95 per cent fault detection and 95 per
cent fault isolation.

Specifications

Dimensions: DTR – 208.3 (L) × 127 (W) × 152.4
   (H) mm
DTC – 190.5 (L) × 119.4 (W) × 40.6 (H) mm
Weight: DTR – 2.95 kg, DTC – 1.42 kg
Power: 115 V AC, 400 Hz, single phase, 43 W
   maximum

Software: Ada
DTR I/F:
1 dual-redundant MIL-STD-1553B
2 independent FC-AL Nodes
DTC I/F: AFMSS compatible
MTBF: 12,900 h
Environmental: MIL-D-5400 Class 2

Status

In production and in service.

Contractor

Smiths Aerospace, Germantown.

Fire Detection Suppression system (FiDS)

Type

Aircraft fire-detection system.

Description

The Goodrich Aerospace FiDS is an integrated
system which includes smoke detectors, flight
deck panel, maintenance bay panel, wiring,
plumbing, and halon bottles. It is designed to
satisfy the requirements of FAR Pt 28.855.

The system provides 60 second detection
and 60 minute suppression for narrow-body
aircraft, and 60 second detection and 180 minute
suppression for widebody models. Should smoke
be detected the existing master caution warning
light illuminates both on the glare shield and on
the flight deck panel, and an audible warning
is provided to the flight deck crew. Despatch
reliability is assured using redundant dual-loop
smoke detectors operating independently in each
zone and redundant electronic circuitry.

Status

FiDS is certified for Boeing 727, MD-82/83 and DC
10 aircraft. It is adaptable to several models of
aircraft including Boeing 737, DC-9, Fokker 100
and BAC 146.

Contractor

Goodrich Corporation, Fuel and Utility Systems.

Fire/Overheat Detection System (F/ODS)

Type

Aircraft fire-detection system.

Description

The Fire and Overheat Detection System (F/ODS)
provides continuous fire and overheat detection
in engine nacelles and the auxiliary power
unit compartment. The system also detects
hot air leakage from engine bleed air ducting
and identifies the location of the overheat. On
detection of an overheat, the system sends
commands to the bleed air isolation valves. The
F/ODS reports this information to the mission
computer via the MIL-STD-1553 databus.

The system consists of a dual channel controller
that monitors 44 sensor loops and operates as a
master and slave with cross-channel fault checking
to achieve high-integrity fire detection. There are
dual-loop continuous thermal sensors to provide
fire detection in six zones and overheat detection in
sixteen zones. Fire detection occurs within 5 seconds.
The system can locate overheat conditions in eleven
zones to an accuracy of 5 per cent of loop length.

Based on this technology, a system can be
designed and customised to meet the fire
detection needs of most aircraft. The system may
also be incorporated into an integrated utility
system where several aircraft subsystems are
combined in a single package.

Specifications

Dimensions: 318 × 191 × 194 mm, ARINC 404
   {¾} ATR short
Weight: 6.9 kg

Contractor

Goodrich Corporation, Fuel and Utility Systems.

Flexcomm C-1000/C-1000S control head

Type

Aircraft system control, avionic.

Description

The C-1000/C-1000S control head is designed for
use with Wulfsberg's Flexcomm communication
system. It is capable of controlling the RT-30, RT-
118, RT-138, RT-138F, RT-406F and RT-450 model
radios. Flexcomm provides FM communications
in low-band VHF, high-band VHF, UHF and
VHF AM.

The C-1000/C-1000S is a fully frequency-agile
control unit which provides thumbwheel control
of all available Flexcomm channels and provides
for the storage of up to 30 preset channels for
simplex or semi-duplex operation. Any of the
channels may be changed by the operator.
However, this capability may be disabled if the
operator so desires. It also provides control
of CTCSS tones in both receive or transmit.
The C-1000 has edge lighting, using either 5
or 28 V. When used with a complete Flexcomm
system it automatically selects the appropriate
radiotelephone unit depending on the desired
frequency.

The system also contains full discrete switches
for use with external devices such as antennas
and DTMF coders.

The 'S' model has been modified to enable/
disable the Motorola DVP encryption system.

Specifications

Dimensions: 146 × 76 × 191 mm
Weight: 1.2 kg
Preset channel memory: 30 channels

Status

In production and in service.

Contractor

Chelton Avionics Inc, Wulfsberg Electronics
   Division.

Flexcomm II C-5000 control head

Type

Aircraft system control, avionic.

Description

The C-5000 is the newest generation control head
from Wulfsberg Electronics. This unit can control
and independently monitor any combination up
to three of the following transceivers: RT-30, RT-
118, RT-138F, RT-406F, RT-450, RT-5000, RT-7200
and RT-9600F.

The C-5000 includes channel identification via
alphanumeric or frequency display, simultaneous
control of up to three RT systems, built-in dual-
tone multifrequency, built-in dual-independent
microphone inputs which allow independent
operation of two RTs, multiple control head
installation capability, simulcast, relay and
repeater modes, programme scan and a 350-
channel memory. Each channel has individual
transmit and receive frequencies and selection
of CTCSS and DCS tones is possible with the RT-
138F, RT-406F and RT-9600F models.

The operator can monitor main and
Guard receiver audio from all transceivers
simultaneously and the C-5000 has independent
volume control of the active or selected RT
system and secondary RT systems permitting the
systems to perform relays, simulcasts, repeater
and full-duplex operations. The communications
management controller is easy to operate, with
vacuum fluorescent display, which is easily
readable in bright sunlight, and optional NVG-
compatibility.

Status

In production.

Contractor

Chelton Avionics Inc, Wulfsberg Electronics
   Division.

Flight Control Electronics Assembly-Upgraded for JAS 39 Gripen

Type
Aircraft flight computer.

Description
The JAS 39 Flight Control Electronics Assembly-Upgraded (FCEA-U) by Lockheed Martin is a digital, high performance triplex flight control computer which provides full authority, fly-by-wire control of seven primary control surfaces, four secondary systems, and interfaces with a variety of sensors and I/O devices. The FCEA-U system is provided to Saab Scania AB, Saab Military Aircraft for the Swedish Air Force.

Within each channel of electronics, the FCEA-U contains two high performance 32-bit processors. A Motorola 68040 performs primary flight control law processing and is interfaced via dual port RAM to a Texas Instruments TMS320C30 processor which acts as the I/O processor in addition to computing back-up control laws. A 10 MHz serial cross-channel datalink with built-in error checking is closely coupled with a data acquisition system that can exchange data between channels with less than 150 μs latency.

The electronics are housed in a single, air-cooled, lightweight chassis which is designed with an integrated flexible harness/mother-board and removeable circuit cards for easy maintenance. Reliability of the FCEA-U exceeds 1,500 h.

Specifications
Dimensions: 358 × 454 × 192 mm.

Status
In production for and in service in the JAS 39 Gripen aircraft.

Contractor
BAE Systems Controls.

Flight Control Electronics Assembly-Upgraded (FCEA-U) for JAS 39 Gripen 0001443

Flight Control Electronics Assembly-Upgraded (FCEA-U) for JAS 39 Gripen 0001442

Flight Control System for the V-22 Osprey

Type
Flight Control System (FCS).

Description
The V-22 Osprey Flight Control System (FCS) is a fly-by-wire system composed of triple, dual Primary Flight Control System (PFCS) processors and triple Automatic Flight Control System (AFCS) processors. The Flight Control Computers (FCCs) provide interfaces for the swashplate, elevator, rudder, flap and pylon primary actuators. The FCCs also provide interfaces for the cockpit control force/driver actuator, cockpit control thrust driver actuator and nosewheel steering secondary actuators. The FCC interfaces with the MIL-STD-1553B avionics multiplex bus and also incorporates a

dedicated flight control system multiplex bus in each channel. The system is two fail-operational with respect to PFCS functions and one fail-operational with respect to AFCS functions.

Each FCC incorporates dual 1750A processors for the PFCS control functions and one 1750A processor for the AFCS control function. An input/output processor is used for data management and built-in test functions. The PFCS and AFCS processors interface with the I/O bus via a dual-port memory to enhance computerised throughput.

The PFCS electronically connects the pilot's controls to the various control surface actuators for safe operation of the aircraft. The automatic flight control system provides the necessary level of control augmentation required for reliable mission performance.

The PFCS provides basic control of the aircraft by connecting and mixing pilot control inputs during the helicopter, transition and aeroplane modes of operation. All actuators are commanded from the PFCS processors and each dual PFCS processor has failure detection and shutdown logic so that either processor in each channel can independently shutdown the actuator channel controlled by that channel.

The automatic flight control system provides stability and control augmentation and mission related selectable modes of the flight control system. It is a triplex cross-channel monitored system which provides operation after first failure in most triplex sensor paths. On a second failure, the function is shut down in a fail-safe manner.

Automatic flight control system inputs to the primary flight control system are voted at the input interface of the PFCS. The selected inputs are further restricted by rate and authority limiting in the control laws of the primary flight control system.

A non-redundant, dual-computation analogue back-up computer, in conjunction with analogue elements of the PFCS, provides for continued control of the aircraft in the event of a total loss of the digital processing functions. It also directly controls the back-up modes of the pylon actuator and electronic engine controls.

The flight control computer processor monitors the back-up computers continuously in normal operation. Built-in test for the back-up is controlled by the I/O processor during preflight and maintenance modes of operation.

The system consists of:

Flight control panel: this provides pilot/co-pilot control for the automatic flight control system.

Engine control panel: lever type engine condition controls are provided for the V-22's two engines. Illuminated push-button switches are used to select and indicate which engine controller is active. A third lever controls the rotor brakes.

Cockpit interface unit: the Cockpit Interface Unit (CIU) passes cockpit discrete data to the dedicated flight control system multiplexer bus. The CIU also contains a roll rate sensor.

Flight control computers: one flight control computer using 1750A architecture is provided for each channel. Primary and automatic flight control system functions each incorporate separate processing. Within the computer, the systems communicate via a common dual-port memory designed so that the hardware failures on the AFCS bus or memory side do not cause failures of the PFCS functions.

Analogue back-up computer: a non-redundant, dual computation channel analogue back-up computer provides for continued control of the aircraft in the event of a system failure of the digital processing functions.

Conversion actuator electronic interface unit: this provides control of the electric back-up pylon actuator motor.

Flap panel: this allows the pilot/co-pilot to select the operation of the flaps.

Status
In production for the V-22 Osprey.

Contractor
BAE Systems Controls.

Flight director computers

Type
Flight Director (FD) system.

Description
Astronautics manufactures both analogue and digital flight director computers for most military fixed-wing aircraft and helicopters. The flight director computer provides both two- and three-cue steering information to the pilot which may be presented on either cross-pointers or a combined single indicator.

A flight director computer which is compatible with GPS is currently being manufactured for the US Army Sikorsky UH-60A helicopter. The flight director computer accepts digital signals directly from the GPS receiver and other radio navigation aids and converts them to analogue signals for display on the flight instrument. The navigation

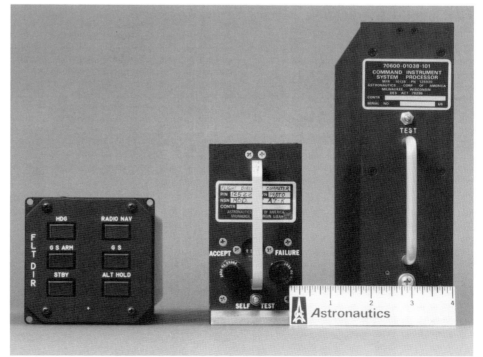

The Astronautics Macchi MB-339 flight director computer (left), A-10 flight director computer (centre) and Black Hawk command instrument system processor (right) 0503914

information received from the GPS receiver provides horizontal deviation, vertical deviation and glide slope angle.

Status
In production and in service.

Contractor
Astronautics Corporation of America.

Flight director systems

Type
Flight Director (FD) system.

Description
Astronautics produces a wide range of analogue and digital Flight Director Systems (FDS) for helicopters and fighter, trainer and transport aircraft. The FDS will accept inputs from various aircraft sensors, instruments, mode selectors and navigation equipments to perform signal switching and computations of pitch and roll steering commands, which in turn drive cockpit instruments/displays.

The main functions of the systems include:
- Flight director cues
- Attitude director cues
- Horizontal Situation Indicator (HSI)
- Search patterns
- 4D intercept/approach
- GPS interface.

The FDS systems all include a flight director computer (or steering computer), dual-Attitude Director Indicators (ADIs), dual-Horizontal Situation Indicators (HSIs) and pilot control unit/mode selector panel.

Specifications
Physical characteristics are dependent on installation
Power: 115 V AC
Indicator sizes: 3, 4 or 5 in
ADI: 2 or 3 axis
Interface: Analogue, discrete, synchro and optional ARINC-429
Qualification: MIL-qualified

Status
In production. Astronautics' FDS are in service in UH-60 helicopters and P-3 Maritime Patrol Aircraft (MPA).

Contractor
Astronautics Corporation of America.

Fuel Quantity Indicating Systems (FQIS)

Type
Aircraft sensor, Fuel Quantity Indication (FQI).

Description
Goodrich Fuel & Utility Systems provides fuel quantity indicating systems for both commercial and military, fixed- and rotary-wing aircraft,

Retrofit fuel quantity indicating system for Lockheed Martin C-130 aircraft   0018232

Boeing 737-700 FQIS tank unit hardware   0018231

including: Boeing, Airbus, Saab, Embraer, de Havilland, Lockheed Martin, Sikorsky, Bell, and other foreign OEMs. These fuel quantity indicating systems provide high reliability and typically provide a standard accuracy of ±2.2 per cent at full and ±1.2 per cent with a densitometer. Increased accuracy is possible for custom applications.

Retrofit FQIS have been chosen by numerous airlines for their Boeing 747-200, -300, and 757/767 cargo and passenger aircraft.

The in-tank hardware for the fuel quantity indicating systems on these aircraft include all new harnesses, fuel quantity sensors, compensators for providing signal variation in dielectric constant of the fuel, plus densitometers which provide direct fuel density measurement information to the processor. Flight deck and refuelling panel indicators provide visual verification of total fuel and individual tank quantities. The processor unit controls the entire fuel system by calculating fuel quantity, controlling the refuelling sequence and performing extensive health monitoring on both wiring and electronics. The processor provides messages on operational health and history. In addition to measuring the fuel quantity, these

systems control pumps and valves to transfer the fuel between tanks thus maintaining centre of gravity for optimum performance.

Status
Goodrich, formerly Simmonds Precision, has provided dedicated and interged system solutions for many aircraft manufacturers including Boeing, Airbus Industrie, Northrop Grumman Corporation, Lockheed Martin, Saab, Embraer, Bombardier and EADS.

Contractor
Goodrich Corporation, Fuel & Utility Systems.

Hard Disk Subsystem (HDS)

Type
Data processing system.

Description
The HDS provides a highly reliable form and fit replacement for Miltope and Ampex magnetic tape transports in an airborne environment. Based on field-proven 3.5 in magnetic disk technology, the HDS provides a ×65 increase in storage capacity and reductions in weight and power. Each HDS enclosure emulates up to three tape transports with electrically isolated and redundant hardware for each magnetic tape transport equivalent set. An operator control panel provides a user-friendly interface, while extensive BIT capability allows for easy field maintenance and eliminates the need for special test equipment. The disk media is in a removable cartridge which currently uses a drive with 1 Gbyte of formatted storage.

The HDS is designed to support incorporation of future higher-capacity disk technology by simply replacing the drive in the removable cartridge with new 3.5 in disks. Capacities of 8 Gbytes are planned.

Specifications
Dimensions:
(enclosure) 469.9 × 787.4 × 314.2 mm
(1,054 Mbyte cartridge) 58.4 × 124.5 × 177.8 mm
Weight:
(enclosure) 43.1 kg
(1,054 Mbyte cartridge) 2.27 kg
Power supply: 115/200 V AC, 400 Hz, 3 phase, 165 W
Reliability: 14,500 h MTBF

Status
In service.

Contractor
General Dynamics Information Systems.

Health and Usage Monitoring System (HUMS)

Type
Onboard system monitor.

Description
Smiths Aerospace combines the proven Cockpit Voice Recorder/Flight Data Recorder aircraft monitoring technology with proprietary Health and Usage Monitoring System (HUMS) technology to provide an advanced multi-aircraft-capable, 'single-box' real-time helicopter diagnostic system.

HUMS provides critical flexibility for cross-platform and future growth requirements. The HUMS represents the logical evolution of multibox systems to a single box utilising the latest monitoring, recording, and diagnostic systems technology.

Design features include: automatic rotor balancing; onboard rotor track and balance; aircraft usage monitoring; rotor, engine, transmission and gearbox health monitoring; powertrain, gearbox and bearing diagnostics; engine and structures life usage monitoring; flight regime recognition; crash-protected voice and flight data recording; wide interface capability MIL-STD 1553B/RS-429/ARINC 422/717;

Boeing 757/767 fuel quantity indicating system   0018233

Collins' HF-9000 is certified on the Hawker 800
0583448

correspondence network) are preprogrammed into nonvolatile memory permitting phone patch capabilities over thousands of miles.

Built-In-Test Equipment (BITE) provides diagnostic testing and monitoring to determine if the system is capable of providing the specified performance. Should a fault occur, BITE will identify the specific unit, module/circuit card and the failed circuit.

Several pressurised and non-pressurised automatic antenna couplers are available in versions for tuning grounded wire, grounded tube and shunt antennas. These provide full operational capability and reliability at high altitudes and extreme temperatures. The antenna coupler quickly matches the power amplifier in the receiver/transmitter to the antenna to maximise the radiated power. Each antenna coupler has the capability to learn and store tuning data for the 99 preset channels as well as storing the last 50 manually tuned frequencies. This learning is accomplished during the first tune cycle and is then stored; upon recalling the preset channel or a previously used manual frequency, the antenna coupler tuning cycle time is nominally 40 milliseconds.

The HF-9000 utilises fibre optic cables to transfer all control and status information between the system units. A fibre optic cable is much lighter than a normal data cable while providing two additional advantages: Firstly, fibre optics eliminate the electromagnetic interference to which copper cable is susceptible. Secondly, it is capable of handling large amounts of data within a very short period of time.

The system is available in various configurations. The receiver/transmitter may be remotely mounted in the aircraft avionics bay and controlled by the control unit or via a MIL-STD-1553B data bus, ARINC 429 data bus or other control scheme. This flexibility of interface makes the HF-9000 system compatible with advanced flight deck management systems. Optional data bus configurations can also be implemented. For military applications, the system can be provided with NVG-green panel lighting (HF-9010).

The latest variants of the system (HF-9087D/HF-9087F) incorporate a new lightweight Digital Signal Processor (DSP)-based communications system that provides a single integrated solution to HF voice and data communications, offering significantly improved reliability over earlier systems. The integrated multimode system provides data communication capability over HF modems, video imaging system, secure voice devices and data encryption devices, while continuing to provide voice HF communications. Embedded system functionality includes MIL-STD-188-141A Automatic Link Establishment (ALE),

The Smiths Industries Aerospace HUMS
0080287

and extensive recording/monitoring/playback/analysis capability.

Contractor
Smiths Aerospace.

HF-9000 series HF radios

Type
Avionic communications system, HF/data.

Description
Collins HF-9000 is a series of lightweight HF radios developed to meet the requirements of civilian and military platforms, ranging from helicopters to fighter aircraft.

Initial emphasis during the development of the system was to satisfy the needs of light-fixed and rotary-wing tactical aircraft. That system, designated the HF-9000 and the AN/ARC-217(V) in its military form, employs a modular design utilising fibre optics, microprocessor technology, digital synthesisers and couplers, combined with MIL-STD-1553B or ARINC 429 control.

The main features of the HF-9000 are:
- Full-frequency HF system permits direct tuning of any of 280,000 frequencies between 2.0 and 29.9999 MHz in 100 Hz steps when operated in the discrete frequency mode
- Rapid-tune antenna coupler with up to 40 ms computer tuning facilitates quick channel changes
- Fibre optic technology is used to interconnect all system elements for reduced EMI and installation weight
- 99 user-programmed half-duplex or simplex channels, 249 half-duplex ITU channels and six emergency channels may be called up by their channel numbers when operated in the channel mode
- Discrete frequency mode provides USB, UD, AM equivalent, CW, LSB and LD operation (the 99 user-programmed channels may be programmed to operate in any of these modes)
- Selectable power outputs of 175, 50, or 10 W PEP (optional 200 W receiver/transmitter available)

- BITE is utilised for diagnostic testing and monitoring
- Control utilizes diachronic LCD for optimum readability in all lighting conditions

The system can be operated in simplex or duplex modes over the frequency range from 2 to 29.9999 MHz in 100 Hz increments in both Upper SideBands and Lower SideBands (USB, LSB) voice and data, Amplitude Modulation Equivalent (AME) and Continuous Wave (CW) modes, with growth to HF datalink operations. Up to 99 programmable pre-set channels can be stored in a non-volatile memory and each memory channel can store separate receive and transmit modes and frequencies. The transceiver uses a direct digital frequency synthesiser for rapid frequency changes with microprocessor control to improve stability. The antenna coupler is designed to permit rapid tuning of a wide range of antennas in a variety of aircraft.

To minimize user workload, the HF-9000 system has the capability to store 99 user-programmed preset channels. The receiver/transmitter will store frequency data and RF emission mode for either simplex or half-duplex communication in nonvolatile memory. In addition, two emergency channel frequencies and all 249 half-duplex ITU maritime radiotelephone channels (public

Collins HF-9000
0079231

MIL-STD-188-110A data modem functionally, MIL-STD-188-148 ECCM capabilities, Independent SideBand (ISB) data operation and ARINC 714-6 SELCAL decoding with growth capability for future HF waveforms. The HF-9000 is compatible with the requirements of MIL-STD-203-1A for LINK 11 (TADIL A) data communications. Built-In Test Equipment (BITE) is utilised for diagnostic testing and monitoring.

Specifications

Dimensions (typical):
(HF-9010 control unit) 67 × 146 × 149 mm
(HF-9087D transmitter/receiver) 194 × 141 × 320 mm
(HF-9040 coupler) 192 × 97 × 320 mm
Weight (typical):
(HF-9010 control unit) 1.2 kg
(HF-9087D transmitter/receiver) 8.6 kg
(HF-9040 coupler) 3.6 kg
(HF_9041 coupler) 4.9 kg
Certification: FAA TSO-C31d, -C32d

Status

Introduced in 1986. In service in Gulfstream IV and Hawker 800 business aircraft and with the Royal Australian Air Force as a replacement for the 618T HF radio on aircraft such as the C-130 Hercules.

By mid-2003, over 7,000 systems had been installed worldwide in more than 60 types of rotary and fixed-wing aircraft. The system has also been selected as the standard HF airborne system on most large corporate aircraft and has received commercial airworthiness approval in over 20 countries.

Contractor

Rockwell Collins.

IAPS-3000

Type

Avionic processor.

Description

The Integrated Avionics Processing System (IAPS) provides the infrastructure required to interconnect and manage the various avionic subsystems in the aircraft. The IAPS cabinet may be considered to be a part of the aircraft wiring that physically houses resident Line Replaceable Modules (LRMs).

The IAPS consists of a fully wired card cage containing the following:
- ICC – IAPS Card Cage (ICC-3001)
- IEC – IAPS Environmental Controller (IEC-3001)
- PWR – Power Supplies (PWR-3000)
- IOC – Input/Output Concentrators (IOC-3100)
- CSU – Configuration Strapping Units (CSU-3100)
- OCM – Options Configuration Module (OCM-3100)
- FGC – Flight Guidance Computers
- FMC – Flight Management Computers
- MDC – Maintenance Diagnostic Computers.

The ICC contains hardware to mount and interconnect all resident LRMs. An internal motherboard provides interconnection between all units that reside in the card cage. The IEC monitors temperature transducers in both halves of the ICC and operates heaters or cooling fans to control the IAPS environment. The PWRs provide independent power to the LRMs and isolated power to each stabiliser pitch-trim potentiometer. The IOCs provide data management functions by acting as a central data collection and distribution point. These concentrators receive databus inputs from each major Line Replaceable Unit (LRU) on the aircraft, process the data words and then transmit only the words of interest to each receiving LRU. The CSUs, along with the OCMs, provide a matrix of configuration shunts that configure the IAPS with optional features.

The FGCs are part of the autopilot system that provides three-axis autopilot/yaw damper, dual-flight director, and automatic pitch trim functions. The FMCs are part of the flight management system that provides flight management functions. The MDCs provide computation and storage of maintenance parameters for the avionics LRUs.

The IAPS contains several subsystems in a common cabinet (card cage) that is used in place of discrete LRUs and mounting trays. The basic enclosed subsystems are:
- Dual-ARINC 429 serial data concentrators
- Dual-flight guidance systems providing three-axis autopilot/yaw damper operation
- Single- or dual-flight management system
- Optional maintenance and diagnostic computers for system troubleshooting.

The ICC functions as the mount for several IAPS subsystems and provides interconnections between external (aircraft) subsystems and the internally-mounted IAPS modules. The ICC provides lightning and High-Intensity Radiated Fields (HIRF) protections, as well as EMI emission filters, for all subsystems mounted within the ICC. The ICC contains a single circuit board with two functionally identical sides, with the only physical difference being that the right side is configured with the MDC module interconnection. The two-sided design provides for adequate segregation of data. The ICC printed circuit boards contribute to aircraft wiring reduction as LRM-to-LRM wiring is performed internal to the cabinet. Aircraft interconnection is provided by the cabinet.

The IEC is a fan module that provides increased subsystem reliability by controlling the temperature for the IAPS cage. The IEC monitors temperature transducers in both halves of the cage and operates heaters or cooling fans. The IOCs are the ARINC 429 serial data processors, utilised to reduce aircraft wiring by having a common input/output data collection point that re-transmits multiple sensor input buses to various internal and external subsystems. The FMS and FGS subsystems share these common I/O LRMs resulting in a reduction of volume, weight, and power compared to each having their own dedicated I/O functions. This also greatly reduces the installation cost, connector volume, and wiring weight within the aircraft.

The PWR modules are used in association with the IOCs; they provide two separate power sources that independently power the left FGC, left IOC, right FGC, and right IOC. One PWR powers the left side LRMs and associated portions of the ICC. The other PWR powers the right side LRMs and associated portions of the ICC. The PWRs also reduce the weight, volume, and system power through multisystem application.

The FGCs provide independent flight guidance computation and operate together to provide flight guidance, 3-axis autopilot, automatic pitch trim, and yaw damper functions. The FGC receives critical AHRS data directly from the AHRS computer and receives air data, navigation sensor, and FMS data from the IAPS concentrators. The FMCs provide coupled vertical navigation (VNAV) and NAV-to-NAV capture, navigation aid data base storage, burst radio tune capabilities, and several control/planning functions. The MDCs receive diagnostic words from the aircraft LRUs through the IOCs.

The MDC processes and records data for fault messages and LRU fault history. Additional storage is available for engine trend/exceedance data and maintenance data from other airplane systems. The stored maintenance data can be downloaded with the Collins Portable Access System (CPAS).

The CSUs contain a strapping matrix that defines the system configuration for each side of the IAPS. The CSU contains DIP switches that are set to program system operation for the aircraft. They provide centralised strapping for various internal and external subsystems as well as customer strapping of optional equipment (for example, third inertial reference system or third COMM). The OCMs mount to the CSUs and provide protected strapping for configuration strapping word-0.

Specifications

Dimensions:
(ICC-3000/3010) 260 (H) × 422 (W) × 187 (L) mm
(ICC-3001) 296 (H) × 422 (W) × 202 (L) mm
Typical system weight: 7.78 kg (ICC-3001)
Temperature (with IEC-3000):
(ICC-3000/3010) –40 to +70°C

(ICC-3001) –40 to +70°C
Temperature (with IEC-3001):
(ICC-3000/3010) –55 to +70°C
(ICC-3001) –55 to +70°C
Altitude: 55,000 ft
Power: 28 V DC (supplied by host aircraft)
Certification:
(ICC-3000/3010) FAA TSO C9c, C52a, CII5b
(ICC-3001) FAA TSO C9c, C52b, CII5b

Status

In production and in service.

Contractor

Rockwell Collins.

IAPS-4000

Type

Avionic processor.

Description

The Integrated Avionics Processing System (IAPS) provides the infrastructure required to interconnect and manage the various avionic subsystems in the aircraft. The IAPS cabinet may be considered to be a part of the aircraft wiring that physically houses resident Line Replaceable Modules (LRMs).

The IAPS consists of a fully wired card cage containing the following:
- IEC – IAPS Environmental Controller (IEC-4001)
- PWR – Power Supplies (PWR-4000)
- IOC – Input/Output Concentrators (IOC-4100)
- CSU – Configuration Strapping Units (CSU-4100)
- OCM – Options Configuration Module (OCM-4100)
- LHP – Lightning/High Intensity Radiated Field Protectors (LHP-4000/4001)
- FCC/FGC – Flight Control/Guidance Computers
- FMC – Flight Management Computers (optional)
- MDC – Maintenance Diagnostic Computers (optional).

The IAPS Card Cage (ICC-4009) contains hardware to mount and interconnect all resident LRMs. Temperature sensors are installed on each quadrant of the ICC motherboard. The IEC monitors the temperature sensors and operate cooling fans to automatically regulate the IAPS environment. The LHP provides an isolation interface between the resident LRMs and external units. The PWR provides separate sources that independently power the FCC and the FMC functions. The IOC provides a data management function by acting as a central data collection and distribution point. These concentrators receive databus inputs from each major Line Replaceable Unit (LRU) on the aircraft, process the data words and then transmit only the words of interest to each receiving LRU.

The CSU provides a matrix of configuration shunts that configure the IAPS with optional features.

The MDCs provide computation and storage of maintenance parameters for the avionics LRUs. As an option, the MDC processes and records the data for fault messages, engine trend and exceedance data, life cycle information and LRU fault history. The stored maintenance data can be downloaded to a diskette in the Data Base Unit (DBU).

Note that the empty ICC is an LRU. Each environmental controller, lightning/HIRF protector, power supply, concentrator, strapping unit and computer installed in the card cage is an individual LRM. If any of these modules fail, the individual module may be replaced versus the entire IAPS assembly.

The IAPS contains several subsystems in a common cabinet (card cage) that is used in place of discrete LRUs and mounting trays. The basic enclosed subsystems are:
- Dual-ARINC 429 serial data concentrators
- Fail passive three-axis Flight Control System (FCS) (optional dual)
- Single-, dual- or triple-flight management system(s).

Additionally, an optional maintenance and diagnostic computer is available for system troubleshooting.

The IAPS cabinet is a LRU that contains LRMs, with several versions and options available. The ICC printed circuit boards contribute to aircraft wiring reduction as LRM-to-LRM wiring is performed internal to the cabinet. Aircraft interconnections are provided by Lightning HIRF Protector (LHP) modules.

The IEC is a fan module that provides increased subsystem reliability by initiating forced-air circulation. For systems installed outside the pressurised cabin, optional IECs with heaters are available for cold-temperature operation. The IOC are the ARINC 429 serial data processors, utilised to reduce aircraft wiring by having a common input/output data collection point that re-transmits multiple sensor input buses to various internal and external subsystems. For example, the FMS and FCS subsystems share these common I/O LRMs resulting in a reduction of volume, weight and power compared to these subsystems each having their own dedicated I/O functions. Additionally, the installation cost, connector volume and wiring weight within the aircraft are greatly reduced. These IOCs are dual on each side of the card cage providing for high availability.

The PWR modules are used in association with the IOC; they provide low-volt power to the IOCs and FCS/FMS subsystems. External power is available for use by external IAPS-related subsystems. The PWRs also reduce the weight, volume and system power by having a multisubsystem use, as one PWR may power an IOC, FCC or FMC/MDC. The FCC/FGC is part of the autopilot system that provides three primary axis and pitch trim control. The system is fully integrated into the flight deck. A dual autopilot system, Auxiliary Trim Coupler (ATC) roll or yaw trim computers are optional.

The MDC is an optional module used for system troubleshooting as well as engine data storage. Faults and exceedance data may be stored for later retrieval by maintenance personnel. Part of the IOC subsystem are the CSU's; these provide centralised strapping for various internal and external subsystems as well as customer strapping of option equipment (for example, third inertial reference system, third COMM).

Specifications
Dimensions:
(ICC-400X) 515 (H) × 178 (W) × 410 (L) mm
(IOC-4100) 17 (H) × 149 (W) × 222 (L) mm
(PWR-4000) 33 (H) × 151 (W) × 222 (L) mm
(CSU-4100) 14 (H) × 149 (W) × 222 (L) mm
Weight:
(ICC-400X) 5.45 kg
(IOC-4100) 0.45 kg
(PWR-4000) 0.68 kg
(CSU-4100) 0.36 kg
Temperature (with IEC-4000): –40 to +70°C
Temperature (with IEC-4001): –55 to +70°C
Altitude: 55,000 ft
Certification:
(ICC-400X) FAA TSO C9c, C52a, CII5b
(ICC-4100) FAA TSO C9c, C52b, CII5b

Status
In production and in service.
Contractor
Rockwell Collins.

IC-800 Integrated Avionics Computer (IAC)

Type
Avionic processor.

Description
Honeywell has developed an Integrated Avionics Computer (IAC) which incorporates the functions of four major avionics subsystems into a compact lightweight package that is 75 per cent smaller than the units it replaces.

The IC-800 is a key component of Honeywell's Primus 2000 advanced avionics system. The units incorporated in the IAC include a digital automatic flight control computer which has all the features and functions of the SPZ-8000 digital autopilot, dual-display processors, a fault

The IC-800 is in production for the Bombardier Global Express (Bombardier)   0105288

warning computer and a full Flight Management System (FMS) computer. The IAC, which is a single LRU, has hardware expansion capability for incorporating additional functions.

The IAC provides at least 50 per cent more processor capacity and 100 per cent more memory for each function than was previously available. Its FMS computer has a 1.9 Mbyte database. This additional memory and processor power will allow substantial growth as each function is enhanced in the future.

The IAC's reduced size and weight result from the extensive use of advanced circuit and packaging technologies. These technologies include Very Large Scale Integration (VLSI), Surface Mount Technology (SMT), Application Specific Integrated Circuits (ASIC), high-density memories, rigid flex motherboards, multilayer circuit boards and advanced thermal management. The IAC is housed in an ARINC {1/2} ATR short package with an integral cooling fan.

Status
In production for the Dornier 328 and the Bombardier Global Express business jet.

Contractor
Honeywell Inc, Commercial Electronic Systems.

Ice detection system

Type
Aircraft system control, ice detection and anti-ice.

Description
DNE Technologies has developed an ice detection probe which employs what is claimed to be the most sensitive method of ice accretion detection currently available. The probe operates in a cyclic fashion using the thermal characteristics of the ice which forms on it combined with the heat of fusion effect as the ice is formed. The signal produced may be used to initiate a warning signal or to activate the aircraft's de-icing system automatically. Between detection cycles, the probe is cleared of ice by an integral heater circuit. It is claimed that an ice accretion thickness of 0.12 mm can be detected within 5 seconds. A test facility for airborne confidence checking and for ground servicing purposes is incorporated. The standard ice detector is designed to sense ice formation on fuselage and nacelle air intakes for turbine engines, but other versions are available for carburettor ice detection on piston-engined aircraft and for other ice sensitive areas.

Status
DNE Technologies ice detectors are in service on the B-1B, B-2, the F-16C/D, F-117 and F/A-18E/F.

Contractor
Ultra Electronics – DNE Technologies.

Integrated Processing Centre (IPC)

Type
Avionic processor.

Description
The Rockwell Collins IPC provides cost-effective, adaptable core processing capability and is the primary integrating component of the Flight2 architecture. The architecture and modular

components of the IPC are shared across all three Rockwell Collins businesses (Air Transport, Business and Regional, and Government).

The IPC is a direct derivative of the Collins Integrated Avionics Processing Systems (IAPS – see separate entries), currently flying in over 1,000 commercially certified aircraft, including the US Air Force's T-1A. It uses modular architectural elements that are compliant with AFR part 25 civil certification standard.

The IPC configuration for each specific aircraft platform is achieved through a building-block approach, which uses a suite of modules that are standard components across the Rockwell Collins business units. These modules perform centralised application processing as well as input/output data concentration from peripheral subsystems and devices. An avionics-qualified dual-redundant, full-duplex 100 base-Tx Ethernet Local Area Network (LAN) configured in a hub-and-spoke topology forms the next-generation data transfer system for all inter- and intra-IPC data transfers.

The IPC modular hardware is designed to allow LTM replacement at the operator level even under power-on (hot swap) conditions. This LRM concept lowers mean repair time and reduces spare requirements, resulting in significantly lower life-cycle costs. The hardware design is full dual path (dual power supplies, dual LAN, dual I/O) to provide increased levels of aircraft availability. The LRM is based on commercially available standards that easily support the integration of third party designs. In addition, the LRM is sized to accommodate standard VME form factor cards. A number of cabinet sizes are available to accommodate a wide variety of module configurations. Growth slots are provided to support embedded functions such as Terrain Avoidance Warning System (TAWS), digital map, autopilot, communication management or Head-Up Display (HUD) processing.

The IPC provides a software architecture comprised of three software protocol layers for device drivers, system services and applications. Each layer of the architecture strictly enforces open commercial Application Program Interfaces (APIs). The layered protocol design insulates software applications from changes to the underlying processing resources or network devices. The architecture's adherence to open commercial standards also allows for the introduction of third-party developed applications.

Execution of multiple software applications on the same hardware resource is also supported in the architecture. This feature allows a processing resource to host multiple applications of different critical levels and functions, thereby allowing future changes and capability upgrades to be certified on an incremental basis, without having to perform regression testing on unrelated applications.

Status
In production and in service.

Contractor
Rockwell Collins.

Interference Blanker Unit (IBU)

Type
Aircraft system control, avionic.

Description
The Interface Blanker Unit co-ordinates the operation of the aircraft's onboard transmitters and receivers

Strategic Technology Systems interference blanker unit 0051287

to prevent mutual interference and/or receiver damage when operating on multiple overlapping frequencies. Each aircraft transmitter issues a 'blanking request' discrete before activation of its RF output stage. The IBU detects these 'blanking request' signals and issues a discrete 'blanking output' signal to any victim receivers which are operating on overlapping frequency bands. The victim receiver uses the 'blanking output' signal to either disable its front end receiver circuitry or to filter out any input signals for the duration of the 'blanking output' signal. The IBU automatically handles any combination of simultaneous overlapping 'blanking request' signals to ensure that any victim receiver is disabled as long as any offending transmitter is active. The mapping of 'blanking request' channel to 'blanking output' channel, the 'blanking request' trigger threshold and detection characteristics, the 'blanking output' hold-off delay, pulse-width, pulse extension and pulse truncation characteristics, and the 'blanking output' voltage level are all fully programmable on a channel by channel basis.

Specifications
Dimensions: ARINC ½ ATR short
Weight: 4.55 kg
Power: 115 V AC, 400 Hz, single phase, 100 V A
MTBF: 10,000 h

Status
Contracted for F/A-18E/F; backward compatible to F/A-18C/D, AV-8B, AH-64D and other aircraft.

Contractor
Strategic Technology Systems Inc.

KFC 275 flight control system

Type
Flight Control System (FCS).

Description
The KFC 275 digital flight control system is designed primarily for piston twins. This system uses the same KCP 220 autopilot computer with four microprocessors and the same KDC 222 air data sensor as the KFC 325. However, the KFC 275 uses a different flight instrument system and (76 mm) electromechanical instruments, including the KI 256 V-bar flight command indicator and the KCS 55A slaved pictorial navigation indicator system.

The KMC 221 mode controller provides mode selection and annunciation for most modes, along with the KAP 185A mode annunciator. Annunciations are provided for both armed and coupled modes when appropriate.

Like the KFC 325, the KFC 275 can be configured with optional altitude/vertical speed preselect. There is, however, a choice of either the KEA 130A three-pointer encoding altimeter or the KEA 346 counter-drum pointer servoed altimeter to go with two different versions of the KAS 297 altitude/vertical speed preselector.

Status
In service.

Contractor
Bendix/King.

KFC 325 digital flight control system

Type
Flight Control System (FCS).

Description
The KFC 325 is a three-axis digital flight control system designed to suit high-performance turboprop aircraft. An extensive preflight test of full-time monitors ensures system integrity while airspeed-compensated control maintains proper response to changing aircraft configurations. Manual electric trim speed is also adjusted to aircraft speed.

The remote-mounted flight computer contains four microprocessors, with one dedicated to each of the following functions: roll, pitch, yaw damp and logic. In addition to providing these computations, the microprocessors provide extensive preflight test of pitch, roll and accelerometer monitors which ensures system integrity during flight operations.

The KFC 325 is configured with a 102 mm (4 in) electromechanical ADI and HSI with growth provisions for interface with the EFS-10, EFS-40 and EFS-50 electronic flight instrument system EADI and EHSI displays. The electromechanical instruments include the KCI 310A rotating sphere flight command indicator and the KPI 553A pictorial navigation indicator with radiate of DME distance, groundspeed and TTS plus radar altitude and distance from 1,000 ft AGL to touchdown.

In addition to such standard modes as altitude hold (ALT), heading select (HDG), nav (VOR/RNav), approach (APR), glideslope (GS), reverse localiser (BC), control wheel steering (CWS), indicated airspeed (IAS), hold and yaw damp (YD), the KFC 325 also has the standard comfort modes of soft ride and half bank. Altitude/vertical speed preselect is optional. The optional KAS 297C altitude and vertical speed selector is a panel-mounted unit which

can interface with a KEA 346 counter-drum pointer servoed altimeter. It provides altitude alerting 1000 ft prior to reaching, and on reaching, a preselected altitude and also anytime the aircraft deviates from that altitude by more than 300 ft. Altitudes may be selected in 100 ft increments up to 35,000 ft with vertical speed in 100 ft per minute increments to ±3,000 ft per minute (strappable to ±5,000 ft per minute depending on aircraft certification).

The KFC 325 also includes as standard equipment the KDC 222 air data sensor which provides altitude and airspeed as well as normal and sideslip inputs to the flight control system.

The servos use capstan assemblies which may be left in the aircraft should servo repair be required, thus allowing the aircraft to remain rigged.

Specifications
TSO compliance: TSO C9c and C52a; RTCA DO-168B
Power: 28 V DC

Status
In production and in service.

Contractor
Bendix/King.

KFC 400 digital flight control system

Type
Flight Control System (FCS).

Description
Intended for turboprop and turbine-powered general aviation aircraft, the KFC 400 is a dual-channel system designed to fail-passive. The system is microprocessor driven and when used in conjunction with the KNS 660 navigation and frequency management system can provide full three-dimensional flight guidance with automatically scheduled climb and descent profiles.

The system includes a digital air data computer providing altitude, airspeed, vertical speed and

KFC 325 digital flight system showing (top left) the EFS 40 EADI, (bottom left) the EFS 40 EHSI and (right) the KMC 321 digital mode selector 0131553

KFC 400 digital flight control system   0503912

Mach number via an ARINC 429 digital databus, and it can also drive a range of electromechanical flight instruments. Continuous fault monitoring is incorporated.

Status
In service on the Beechjet 400.

Contractor
Bendix/King.

LSZ-860 lightning sensor system

Type
Avionic crew alerting system.

Description
Honeywell's newest lightning sensor, the LSZ-860, is an upgraded version of the LSZ-850. The LSZ-860 senses both visible and high-energy invisible electrostatic and electromagnetic disturbances caused by electrical discharge activity within a 200 n mile radius around the aircraft. When lightning occurs, the system carefully analyses the discharge and creates the proper symbol for display on most Primus colour radar indicators or on most Honeywell EFIS/MFD displays. The system's computer rapidly and accurately determines the rate of vertical lightning in a fixed geographical area and then displays the centre of that area with the lightning rate symbol. Wide bandwidth, extensive signal processing and lightning stroke recognition algorithms ensure a more accurate display. Extraneous signal filtering minimises noise which would otherwise clutter and confuse interpretation.

The LSZ-860 system computes the location and lightning rate for up to 50 thunderstorm areas. The computer tracks the location of each of these areas. To ensure accurate tracking of lightning areas, all displays are both heading and velocity stabilised to keep the symbol over the same ground position regardless of aircraft manoeuvring.

The system gathers information in a full 360° pattern around the aircraft, even in the standby mode. Thus, the lightning sensor system is always ready to present the weather picture in either the full 360° display mode without a radar overlay or in the sector display mode that can include a radar and navigation data overlay.

Three distinct levels of lightning rate are computed for display, each depicted by a unique lightning rate symbol. The symbols represent the vertical lightning rate-of-occurrence. The symbol location is the average position of the lightning that occurred in the previous two minutes, and is displayed inside the radius of the range selected. Each lightning symbol represents the centre of a circular area with a radius dependent on the range selected (8 n mile radius at a range of 25 n mile, 18 n mile radius at a range of 100 n miles and 30 n mile radius at a range of 200 n miles). Whenever any lightning activity is detected at any range, the lightning sensor computer will place a magenta lightning alert symbol at the proper bearing and at the end of the selected range for five seconds.

Specifications
Dimensions: (LP-860) 375 × 194 × 61.5 mm; (AT-850) 294 × 31.8 × 154 mm
Weight: (LP-860) 3.06 kg; (AT-850) 1.14 kg
Power supply: 28 V DC, 28 W

Status
In service.

Contractor
Honeywell Inc, Commercial Electronic Systems.

M362F general purpose processor

Type
Avionic processor.

Description
The M362F is a general purpose unit which uses parallel, binary, floating point, two's complement 16-bit processing. A typical instruction mix yields an operating speed of about 340 kips.

The M362F can be tailored to particular operations by specifying from a wide choice of memory types and standard input/output circuit modules. It is mechanised on two operating modules. The instruction repertoire can be varied by adding microprogramme memory or changing the existing microprogramme memory which consists of 10 integrated circuits. The M362F will operate with core and/or semiconductor memory. The processor provides 72 basic machine instructions, including nine special purpose types that are mechanised using the basic input/output instructions. Micro-instructions are held in a 512-word control memory, but this can be expanded to 2,048 words, providing for such operations as byte (8-bit) control jumps, skips and transfers, register variable shifts, register/register floating point arithmetic, logical immediate and register/register, and macro instructions. The last provides square root and trigonometric functions much faster than by using subroutine executions.

Development of M362F software can be accomplished on support equipment such as mini-computer directed systems, IBM 360/370 or similar facilities. Mini-computer equipment enables software development on a stand-alone basis in the laboratory.

The M362F is used as the fire-control computer in the Lockheed Martin F-16 and has a comprehensive set of Jovial or assembly coded software facilities.

Specifications
Dimensions: ½ ATR case
Weight: 6.5 kg
Power: 28 V DC
Computer type: parallel, binary, fixed and floating point, two's complement
Word length: 16-bit
Typical speed: 340 kips
Max address range: 65,536 (64 k) memory locations
Instruction set: 72 (expansible)
Input/output options
  analogue/digital/analogue converter

analogue input/output multiplexer
discrete input/output (28 V)
MIL-STD-1553 bus
multipurpose serial digital processor

Status
In service as the fire-control computer in the F-16.

Contractor
General Dynamics.

M362S general purpose processor

Type
Avionic processor.

Description
Related to the M362F, the M362S is a 32-bit, high-speed general purpose processor. It uses microprogrammed, parallel, binary, fixed and floating point, two's complement operations and has a typical operating speed of 750 kips.

The processor provides 96 basic machine instructions, microprogrammed in a 512-word control memory, with possible expansion to 1,024 words. A RAM system is available and comprises semiconductor CMOS memory modules, memory controller and an error detection and correction unit. A total of 65,536 (64 k) 16-bit memory locations can be accommodated.

Packaging is either for forced-air cooling, as in aircraft bays, or a radiant cooling frame for space applications. The system has been selected for NASA's Inertial Upper Stage programme. A comprehensive set of Jovial and assembly coded software facilities is available.

Specifications
Dimensions: 365 × 365 × 152 mm
Weight: 24.5 kg
Power supply: 28 V DC, 220 W
Computer type: binary, fixed and floating point, 2's complement
Word length: 32-bit
Typical speed: 750 kips
Max address range: 65,536 (64 k) memory locations
Instruction set: 96 (expansible)
Input/output options
  analogue/digital/analogue converter
  analogue input/output multiplexer
  discrete input/output (5 or 28 V DC)
MIL-STD-1553 bus
  multipurpose serial digital processors

Status
In production.

Contractor
General Dynamics.

M372 general purpose processor

Type
Avionic processor.

Description
The M372 is a development of the M362F used as the fire-control computer for the Lockheed Martin F-16. It has been designed for applications requiring extended performance in memory, input/output, throughput and computational speed. Operating speed is typically between 570 and 720 kips, and non-volatile, electrically alterable memory of 32, 64, 128 or 256 kwords capacity can be accommodated and addressed. In addition to analogue and discrete input/output channels, up to three MIL-STD-1553 dual-redundant digital databus channels can be accommodated and the computer executes the US Air Force MIL-STD-1750 standard instruction set architecture.

The system is supported by Jovial software, and packaging dimensions can be between {1/2} and 1 ATR, depending on the features incorporated.

Specifications

Dimensions: 112 × 170 × 190 mm
Weight: 7.3 kg
Power supply: 28 V DC, 100 W
Computer type: binary, parallel, microprogrammed,
  fixed and floating point, 2's complement
Word length: 16-, 32-, or 48-bit
Typical speed: 570–720 kips
Max address range: 1,576,058 (1,024 k) words
Instruction set: MIL-STD-1750A, Notice 1, 262 instructions

Status

In production. The M372 computer is used as the F-16's enhanced fire-control computer, as the Lockheed Martin LANTIRN pod control computer and as the C-5B MADAR multiplexer/processor.

Contractor

General Dynamics.

Magic IV general purpose processor

Type

Avionic processor.

Description

The hardware used in this processor is unrelated to earlier Delco processors. Magic IV is a high-performance, all-large-scale integrated microcomputer system, which promises to provide lower cost, reduced power, weight and size and greater modularity and reliability advantages over existing machines. Typical operating speed is 250 kops. There are three basic machines: the M4116, M4124 and M4132 available with 16-, 24- and 32-bit word architectures respectively. They are seen as suitable for remote terminal or sensor orientated processor applications.

Almost exclusive use of large-scale integrated circuits has resulted in a great reduction in the number of components compared to a conventional processor. The large-scale integrated units are:

	M4116 (16-bit)	M4124 (24-bit)	M4132 (32-bit)
CPU control unit	1	1	1
CPU arithmetic unit	4	6	8
Input/output control unit	2	3	4
Memory controller	2	3	4
Total	9	13	17

Additional large-scale integrated circuits may be needed to meet programmable communication interface and digital input requirements. MTBF estimates range from 27,000 hours for a simplex configuration to 150,000 hours for a dual-redundant system. NMOS large-scale integrated circuits are used with typical component densities of 1,000 gates and 5,000 transistors per 250 mil chip, with pair gate delays below 5 ns. Delco claims better nuclear radiation tolerance than contemporary dynamic NMOS circuits due to the static logic design, substrate bias design and exclusive use of NOR gate. Digital inputs compatible with ARINC 561, 575 and 583 can be provided.

Specifications

Dimensions: 167 × 69 × 127 mm
Weight: 1.4 kg
Power: 25.3 W
Computer type: binary, parallel, fixed point, 2's complement
Word length: 16-, 24- or 32-bit
Typical speed: 250 kops
Max address range: 32,768 (32 k) words
Instruction set: 89 (expansible)

Status

In production. The computer is used as the Fuel Savings Advisory and Cockpit Avionics System (FSA/CAS) computer for the US Air Force's C-135 and KC-135 aircraft. It is also used in the performance management system in the Boeing 747 and DC-10 and MD-80 series.

Contractor

General Dynamics.

Magic V general purpose processor

Type

Avionic processor.

Description

The Magic V processor is an all VLSI implementation of a CPU and memory system that executes the MIL-STD-1750A, Notice 1 instruction set architecture.

The Magic V processor utilises 3 µm bulk CMOS technology to configure a complete MIL-STD-1750A central processor in just 10 VLSI chips. An eleventh VLSI chip provides extended memory management to address up to 1 Mwords and a twelfth chip provides an IEEE-488 bus interface which enables external communication with the CPU and memory for monitoring performance and for developing software. These 12 VLSI chips are mounted on one side of a single 1/2 ATR size circuit card assembly. The back side of the single circuit card assembly mounts the VLSI memory controller that interfaces the CPU with the memory devices mounted on this side. Typical configurations of the M572 include up to 192 kwords of CMOS RAM plus a start-up ROM, or various combinations of RAM, EEPROM and UV PROM devices, all capable of being addressed by the programmable VLSI memory controller.

The M572 typically operates at 850,000 to 1 million operations running the DAIS instruction mix. A built-in feature of the M572 is the ability to couple multiple CPU/memory cards in a multiprocessor configuration.

Specifications

Dimensions:
(CPU and memory) 109 × 163 × 13 mm
Weight: 0.45 kg with memory side fully populated
Power:
(CPU) 2.2 W
(CPU and memory) 7.3 W
Computer type: general purpose, microprogrammed, fixed and floating point, custom CMOS VLSI, TTL compatible interfaces
Word length: 16-, 32- and 48-bit; 48-bit logic unit
Instruction set: MIL-STD-1750A, Notice 1; all specified options
Memory size: up to 192 kwords CMOS RAM on 1/2 ATR size circuit card; up to 256 kwords CMOS RAM on 3/4 ATR size circuit card
Max address range: 1 Mwords
Addressing modes: direct, indirect, immediate, indexed, non-indexed, relative, base relative, BIT, byte
Test interface: built-in IEEE-488 interface for DMA operations, test communications, software development, and real-time performance monitoring
Input/output: all mandatory features and non-application dependent MIL-STD-1750A options implemented, discrete and digital outputs/inputs, analogue outputs/inputs, and MIL-STD-1553B bus interfaces implemented in various system level applications

Status

In service on C-17A and F-14D in the display systems. The system was selected for engine monitor and control systems for the now cancelled RAH-66 Comanche helicopter.

Contractor

General Dynamics.

MDC-3000/4000 Maintenance Diagnostic Computer

Type

Onboard system monitor.

Description

The Collins MDC-3000 / MDC-4000 Maintenance Diagnostic Computer (MDC) performs an integral function within Pro Line 4 avionics systems by monitoring Line Replaceable Units (LRUs) to detect failures, identifying those requiring replacement, and logging fault data to facilitate fault diagnosis. Along with the system's ability to provide avionics maintenance diagnostics, the MDC can also display aircraft systems data, engine maintenance information and extensive airframe/user customised checklists, depending on aircraft configuration. The MDC information is accessed through the cockpit multifunction displays.

The maintenance diagnostic computer is packaged in the Pro Line 4 Integrated Avionics Processing System (IAPS), reducing avionics weight and size, and simplifying avionics interfacing. This integration allows the MDC to gain extensive access to available aircraft system diagnostic information.

Status

In production and in service. The MDC-3000 and MDC-4000 are available for business and regional aircraft utilising the Pro Line 4 avionics system.

Contractor

Rockwell Collins.

MDC-3100/4100 Maintenance Diagnostic Computer

Type

Onboard system monitor.

Description

The Collins MDC-3100/MDC-4100 Maintenance Diagnostic Computer (MDC) performs an integral function within Pro Line 4 and Pro Line 21 avionics systems by monitoring Line Replaceable Units (LRUs) to detect failures, identifying those requiring replacement, and logging fault data to facilitate fault diagnosis. Along with the system's ability to provide avionics maintenance diagnostics, the MDC can also display aircraft systems data, engine maintenance information and extensive airframe/user customised checklists, depending on aircraft configuration. In addition, the MDC provides a single access point for test and rigging functions for LRUs.

Magic V has been selected for engine monitor and control systems for the Boeing RAH-66 Comanche helicopter (Boeing)

0105287

The MDC information is accessed through the cockpit multifunction displays, remotely through a Portable Maintenance Access Terminal (PMAT), or through reports downlinked via ACARS.

The maintenance diagnostic computer is packaged in the Pro Line 4 or Pro Line 21 Integrated Avionics Processing System (IAPS), reducing avionics weight and size, and simplifying avionics interfacing. This integration allows the MDC to gain extensive access to available aircraft system diagnostic information.

Specifications

(MDC-4100)
Dimensions: 17 × 222 × 154 mm (H × W × L)
Weight: 0.36 kg
Temperature: −40° to +70°C
Altitude: 55,000 ft
Certification: FAA TSO C113; RCTA DO-160D; EUROCAE ED-14D

Status

In service. The MDC-3100 and MDC-4100 are available for business and regional aircraft utilising the Pro Line 4 or Pro Line 21 avionics systems.

Contractor

Rockwell Collins.

Mega Data Transfer Cartridge with Processor (MDTC/P)

Type

Data processing system.

Description

The Mega Data Transfer Cartridge with Processor (MDTC/P) stores Digital Terrain Elevation Data (DTED) and the Digital Vertical Obstruction File (DVOF) used in Terrain Referenced Navigation (TRN) and Terrain Awareness and Warning Systems (TAWS). The first application of the MDTC/P is for a Digital Terrain System (DTS) provided by BAE Systems (Operations) Limited TERPROM software for the F-16.

INS, barometric and radar altimeter data are transmitted over the aircraft databus to the MDTC/P and INS update and pilot cueing and warnings are transmitted by the MDTC/P over the bus to the INS, multifunction display, HUD and voice warning system.

Other applications for the MDTC/P include the digital moving map database for the F-22, EW simulation and the Tactical Air Combat Mission Analysis and Training System.

Status

Ordered by the US Air Force Reserves for use with the DTS (TERPROM) in the F-16.

Contractor

Smiths Aerospace, Germantown.

Micro-Aircraft Integrated Data System (Micro-AIDS)

Type

Data processing system.

Description

The Micro-Aircraft Integrated Data System (Micro-AIDS) has been developed to automate the various data logging functions on an aircraft, including the gathering and processing of data concerning engine condition and operational performance monitoring, exceedances and fuel usage and so on. A single unit is used to gather the data which is stored in solid-state memory.

The Micro-AIDS operates on the ARINC 573 data stream and recorded data can be recovered by ground crew using a portable data retrieval unit.

Specifications

Dimensions: ARINC 600 2 MCU (3 MCU in expanded version)
Weight: 2.3 kg (3.6 kg expanded)
Power supply: 115 V AC, 400 Hz, 15 W
Capacity: 25 flights

Status

In production.

Contractor

Honeywell Aerospace, Electronic & Avionics Lighting.

Mini-Flight Data Acquisition Unit (Mini-FDAU)

Type

Data processing system.

Description

The Mini-FDAU is available in either the ARINC 404 or 600 form factor.

Output to any ARINC digital flight data recorder is generated in standard ARINC 573 format at 768 bits/s, Harvard biphase. Provisions are included to receive the DFDR BITE signal and individually display both the FDAU and DFDR status. Provisions are also included for interfacing with a Sundstrand flight data entry panel.

Input flexibility is available to permit interfacing with any aircraft configuration and to comply with diverse regulatory requirements. Data acquisition modules are available to accommodate use with analogue or digital ARINC 429 systems or with a combination of both analogue and serial databus inputs.

Input acquisition capacity is provided for recording of the FAA 17 mandatory parameters requirement. Capacity can be expanded to comply with the proposed ICAO requirements by the addition of one module.

The analogue acquisition design achieves a universal capability to acquire synchro, AC ratio, low-level DC and potentiometer signals with a single input circuit. The unit also allows automatic determination of the input signal based only on the aircraft wiring connections. This allows the Mini-FDAU to be interchanged among different aircraft without the need for reprogramming of a configuration PROM.

The Mini-FDAU is expandable to accommodate future additional input to include parameters such as those required for engine condition monitoring, a separate CPU and memory capable of screening data for storage in onboard solid-state memory. An ACARS interface expansion allows access to stored information and downlinking automatically or on demand.

Specifications

Dimensions: ARINC 600, 3 MCU
Weight: 3.63 kg (max)

Contractor

Honeywell Aerospace, Electronic & Avionics Lighting.

MMSC-800 MicroMiniature Signal Conditioner

Type

Avionic processor.

Description

The MMSC-800 microminiature signal conditioner/PCM encoder combines L-3 Telemetry-East's experience of analogue and digital conditioning into a 12-bit PCM encoder. A complete family of modules provides signal conditioning for various sensors. EEPROM programmable gain/offset and sample rates provide the user with hands-off control of measurement characteristics and output data formatting. The MMSC-800 utilises thick film hybrid modular assemblies and has been qualified for missile and aerospace environments. It is available in stand-alone or master remote configurations.

The MMSC-800 features 12-bit PCM encoder with integral signal conditioning and multiplexing, EEPROM programmable gain, offset, sample rate and PCM format, ruggedised modular construction, extremely small size and RS-232 programming capability. It is military and

airborne qualified and has high reliability. The system accuracy is quoted as 0.5 per cent.

Contractor

L-3 Telemetry-East.

MMSC-800-RPM/E remote pressure multiplexer/encoder

Type

Avionic processor.

Description

The MMSC-800-RPM/E remote pressure multiplexer/encoder accepts pressure inputs directly, electronically scans the pressure inputs and conditions and encodes the resulting information. It is based on the MMSC-800 signal conditioner and PCM encoder, which has been used on many commercial test programmes and applications. Operating as a remote unit, the RPM/E can be placed in areas of the aircraft well away from the Model PCU-700 Series III central controller. Communication between the master and remote is via a 10-wire command response interface. The channel sequence and gain/offset combinations are fully programmable through the central controller. The temperature reference of the PSI scanner is electronically monitored and transmitted to the central controller to allow the user to correct for errors due to temperature changes in the scanner. Features are controlled EEPROM, programmable via an RS-232 from any personal computer terminal.

Specifications

Dimensions: 135.9 × 45.7 × 96.5 mm
Weight: 3.63 kg
Input channels: 32
Temperature range: −20 to +80°C

Contractor

L-3 Telemetry-East.

Modular Airborne Processor (MAP)

Type

Avionic processor.

Description

The MIL-STD-1750A Modular Airborne Processor (MAP) is a 1.5 Mips VHSIC, CMOS processor for use in UAVs such as smart weapons, RPVs, targets, and land, sea and air autonomous vehicles.

It has up to 256 kwords each of SRAM and EEPROM with 8,000 start-up memory. The basic MAP consists of two 6 × 9 in (152.4 × 228.6 mm) PCs: a processor board and an Input/Output (I/O) board. The processor board contains the 1750A chip set and resource controller ASIC along with system SRAM. The I/O board contains a dual-redundant MIL-STD-1553B bus controller interface, a MIL-STD-1553B remote terminal interface compliant with the requirements of MIL-STD-1760A, MIL-STD-1760A discretes and EEPROM memory. Two additional cards can be provided to accommodate user-defined I/O requirements.

All MAP power is provided by a pluggable PC card DC/DC converter requiring MIL-STD-704 28 V DC input. A spare slot is reserved for the optional Programme Development Card/Performance Monitor Module (PDC/PMM).

The PDC provides a standard IEEE-488 control interface for applications development and test. The PDC also provides a plug-in PMM interface to a standard VAX DRQ3B I/O channel providing enhanced Ada software development capability, including real-time interactive data collection and debugging.

The MAP is housed in a lightweight EMI compliant enclosure utilising MIL-C-38999 connectors. It is designed for conduction cooling requiring no fans or environmental controls. The MAP utilises low-cost glass-epoxy boards with pin-grid array ASICs and DIP integrated circuits.

The MAP design utilises a 3.8 Mip 1750A chip set with the RC coupled to a modified system information transfer and execution bus open architecture. This permits modular expansion on the I/O bus as required by the application. The MAP contains BIT coverage of 98 per cent as per MIL-STD-2084.

Contractor
Lockheed Martin Management & Data Systems.

MPC-800 miniature signal conditioner and encoder

Type
Avionic processor.

Description
The L-3 Telemetry-East miniature signal conditioner and encoder (MPC-800) is a data acquisition system component that provides all the most common required signal conditioning and encoding functions in a small modular package. The design minimises size, weight and cost, and improves survivability in extremely hostile environments. The unit is constructed from stackable conditioning and overhead modules which contain discrete, thick film hybrid and custom ASIC circuit technologies.

By utilising modular architecture, the MPC-800 can be configured to meet the exact needs of any flight test program and still allow the user the ability to reconfigure the unit in the field to adapt to changing mission requirements. Full programmability of gains, offsets, filter cutoffs, and channel sampling rates is offered through a Graphical User Interface (GUI). The MPC-800 may be operated as a master, remote, or stand-alone system.

A wide variety of standard analogue and digital signal conditioning modules are available for virtually any sensor. Custom modules can be created for unique measurement requirements.

Contractor
L-3 Telemetry-East.

MPC-800 miniature signal conditioner and encoder   0002350

Multi-Application Control Computer (MACC)

Type
Avionic computer.

Description
The Multi-Application Control Computer (MACC) is a flexible computer that can be applied to a wide range of flight and subsystem control fixed-wing aircraft and helicopter applications. A key feature of the MACC architecture is that any number of MACCs can be interconnected to form a simplex, dual, triplex or quad configuration without hardware or software changes. For example, the quad configuration can be reconfigured as two separate, but linked, dual-channel computers performing different tasks. This flexibility is made possible by the Input Output Computer (IOC) which performs all Input/Output (I/O) processing in firmware, leaving the application specific processing to the general purpose processors.

The hard separation between I/O management and application processing is key to the success of the MACC architecture. Both the IOC firmware and application software can be loaded and monitored over an RS-422 link from a portable PC. The same interface can also be used for system testing and integration.

The IOC contains redundancy management, BIT and fault detection and isolation algorithms which are performed at significantly faster speeds compared to general purpose processors. Many of these algorithms are embedded in silicon microcircuits, giving the MACC system significant I/O processing bandwidth. Specifically, the IOC can perform all the I/O processing needed for 50,000 sensor sets/s. The result is a system that is robust, fault tolerant and highly flexible for meeting the requirements of most embedded control systems.

Specifications
Dimensions: 257.8 × 193.8 × 254 mm
Weight: 7.76 kg
Power supply: 115 V AC, 28 V DC, 55.5 W

Status
In production. Present applications include flight control, vehicle management system control, actuator and subsystem controllers.

Contractor
Hamilton Sundstrand Corporation.

N1 computer and display

Type
Aircraft system control, power plant.

Description
The Safe Flight Instrument Corporation N1 computer system continuously monitors altitude and air temperature from the aircraft's air data computer, together with anti-ice and environmental control system mode logic from the applicable aircraft systems. The N1 computer interpolates and displays the N1 thrust setting in real time for takeoff, climb, cruise and go-around, each based on the airplane flight manual performance charts. The panel-mounted display instantly displays the appropriate target N1 thrust setting for each flight mode selected by the crew.

Selecting takeoff/go-around will display the flight manual target N1 thrust setting for takeoff. Once airborne, select N1 thrust settings for climb, cruise, or go-around with the mode switch, and the system continuously displays the appropriate N1 thrust setting schedule for the selected mode and given conditions.

Status
As of March 2006, in production and in service.

Contractor
Safe Flight Instrument Corporation.

Safe Flight Instrument Corporation N1 computer displays   0081842

PCU-800 series signal conditioner and PCM encoder

Type
Avionic processor.

Description
The PCU-800 series signal conditioner and PCM encoder provides a ±0.5 per cent system accuracy over its operating temperature range. Available in

2, 4, 8 or 16 card slot configuration, the PCU-800 features include: EEPROM programmable format and channel parameters via RS-232; plug-in system overhead and signal conditioning cards, both digital and analogue types; universal bridge conditioning with completion, calibration and substitution; programmable balance via software control and 12-bit resolution.

The PCU-800 will operate as a stand-alone or master controller. Channel capacities vary depending on the type of signal inputs, but they can range up to more than several hundred channels per unit. Up to 40 PCUs can be clustered in a distributed data acquisition system to provide a total channel capacity of more than several thousand.

The PCU-800 is directly compatible with many other ADAS-7000 and ADAS-8000 airborne products such as the PMU-700 Series III programmable master controller, Albus-1553 all-bus monitor, MPBM-1553 microbus monitor and ATD-800 Series II ruggedised tape deck.

Contractor
L-3 Telemetry-East.

Processor Interface Controller and Communication (PICC) module for the F/A-22

Type
Avionic processor.

Description
Raytheon Electronic Systems produces the MIL-STD-1750A instruction set architecture-based Processor Interface Controller and Communication (PICC) module for the F/A-22 aircraft. Each module provides data processing, MIL-STD-1553B bus communication, intermode communications via serial interchannel datalink and analogue and digital input/output for the vehicle management system.

Status
In production and in service.

Contractor
Raytheon Company, Space and Airborne Systems.

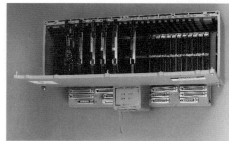

The processor interface controller and communication module for the F-22   0503872

PTA-45B airborne data printer

Type
Data processing system.

Description
The PTA-45B data printer uses a high-efficiency switching power supply, which provides sufficient power to continuously print 'all-black' (all 600+ dots printing at the same time). Since normal operation runs at only 15 per cent of the machine capacity, the printer is claimed to be capable of providing up to 10 times the reliability of previously available models.

The printer permits flight crew access to hard copies of uplinked flight plans, messages and graphics, and can also print a copy of datalink information displayed on the weather radar display. The printer may be driven by a datalink system or a maintenance computer. User features include paper guides to help prevent.

Units of the PTA-45B data printer family 0503882

paper jamming, a hinged spool to ease paper changing and a paper level indicator to tell how much paper is left. The unit has a MTBF in excess of 5,000 hours.

Status

In service on BAE Systems 146 aircraft. The PTA-45B is standard equipment on the Airbus A320 and has been selected for the Boeing 737, 747-400, 757 and 767, MD-11, MD-80 and MD-88 and Airbus A310-300, -400 and -500 airliners.

Contractor

Honeywell Aerospace, Electronic & Avionics Lighting.

Removable Memory Systems

Type

Data processing system.

Description

General Dynamics Information Systems produces a range of Removable Memory Systems (RMS) intended for applications requiring removable ruggedised data storage. The series provides compact, cost effective, and upgradable memory solutions.

RMS-1018

The RMS-1018 is based on field-proven, high-performance 2.5 in Winchester disk technology. The RMS-I 018 consists of an environmentally sealed disk with over 18 Gbytes of formatted data storage. It has internal mechanical isolation for shock and vibration and a standard SCSI-2 interface. An internal heater and heater control for extended temperature operation is optional. Compatibility with previous RMS series products is maintained.

Specifications

Dimensions: 58.4 × 124.5 × 177.8 mm
Weight: 1.54 kg
Formatted capacity: 18 Gbytes
Power supply: +5, +12 or +28 V DC
Temperature range: –51 to +71°C
Shock: 30 g 11 ms and 125 g 2 ms
Altitude: 50,000 ft
Reliability: 130,000 h

RMS-1072

The RMS-1072 is based on 3.5 in Winchester magnetic disk technology. The system consists of an environmentally sealed disk with 72 Gbytes of formatted data storage and either a standard Ultra SCSI or a standard Fibre Channel Arbitrated Loop (FC-AL) interface. An internal heater and controls for extended temperature operation is optional. The heater requires 30 watts of either +12 V DC or +28 V DC.

Specifications

Dimensions: 58.4 × 124.5 × 177.8 mm
Weight: 2 kg
Formatted capacity: 72 Gbytes
Power supply: 115 V AC, 60–400 Hz
Temperature range: –40 to +71°C
Shock: 15 g 11 ms
Altitude: 60,000 ft
Reliability: 130,000 h

RMS-1910

Similarly to the RMS-1072, the RMS-1910 is based on 3.5 in Winchester magnetic disk technology. The RMS-1910 consists of an environmentally sealed disk with 9.1 Gbytes of formatted data storage vibration and a standard SCSI-2

interface. An internal heater and controls for extended temperature operation is also optional. The heater requires 30 watts of either +12 V DC or +28 V DC.

Specifications

Dimensions: 58.4 × 124.5 × 177.8 mm
Weight: 2 kg
Formatted capacity: 9.1 Gbytes
Power supply: 115 V AC, 60–400 Hz
Temperature range: –40 to +71°C
Shock: 15 g 11 ms
Altitude: 60,000 ft
Reliability: 130,000 h

RMS-1648

The RMS-1648 is based on 2.5 in Winchester disk technology. The RMS-1648 consists of an environmentally sealed disk with over 6.4 Gbytes of formatted data storage. It has internal mechanical isolation for shock and vibration and a standard SCSI-2 interface. An internal heater and heater control for extended temperature operation is optional. Compatibility with previous RMS series products is maintained.

Specifications

Dimensions: 58.4 × 124.5 × 177.8 mm
Weight: 1.54 kg
Formatted capacity: 6.4 Gbytes
Power supply: +5, +12 or +28 V DC
Temperature range: –51 to +71°C
Shock: 30 g 11 ms and 125 g 2 ms
Altitude: 50,000 ft
Reliability: 130,000 h

Contractor

General Dynamics Information Systems.

SBS-500 single-rate bit synchroniser

Type

Avionic processor.

Description

Processing PCM data in the SBS-500 single-rate bit synchroniser suppresses receiver noise that may accompany the digital input signal and converts the input to a clean digital output. Available as a stand-alone 275 cm³ package, the SBS-500 is suitable for missiles, aircraft and space applications.

It has a clock synthesis oscillator and a data detection match filter determined by the specific bit rate. The SBS-500 can also perform code conversion from digital transmission codes and provide biphase or NRZ outputs. The SBS-500 performs well under low-input signal-to-noise conditions and is typically within 2 dB of theoretical bit detection accuracy. Acquisition time is 10 clock cycles typical and the SBS-500 has output options of TTL, HC, RS-232 and RS-422.

Specifications

Power supply: 28 V DC
Bit rate: 1–250 kHz
Temperature range: –40 to +85°C

Contractor

L-3 Telemetry-East.

The SBS-500 single-rate bit synchroniser
0503868

Sentinel™ instrument system Data Acquisition Unit (DAU)

Type

Onboard system monitor.

Description

The DAU is used to acquire and transmit critical information from a variety of aircraft systems. It is a dual-redundant 3 MCU unit for monitoring systems such as the electronic engine controller, electrical system, air data computer, hydraulic system and a variety of other aircraft subsystems.

The DAU transmits data through both ASCB and ARINC 429 busses. Information will then be processed by the IC-800 integrated avionics computer and may be shown on the EICAS.

Specifications

Inputs: 30 analogue, 8 digital buses and 120 discrete (base configuration)
Power: 28 V DC (10 to 32 V)
Power consumption: 7 W typical per channel
Weight: 4.1 kg
Temperature range: –40 to +70°C operating
Interfaces: ARINC 429, RS422, RS232

Status

The DAU has been integrated with the Honeywell Primus 2000 XP integrated avionics system for the Bombardier Global Express business jet.

Contractor

Ametek Aerospace Products.

SPZ-600 automatic flight control system

Type

Flight Control System (FCS).

Description

The SPZ-600 automatic flight control system is designed specifically for long-range high-performance business jets and is a complete dual-channel system from the sensors to the servos. All roll, pitch and (optional) yaw channels are fully operational at all times and the performance of each is continuously monitored and compared. If a failure that resulted in a hardover manoeuvre occurs in any channel, it is immediately shut down, resulting in single channel operation in that axis only. Performance status is continuously displayed on the autopilot status panel, as well as on the autopilot master warning annunciator. Honeywell says that the unique monitoring system and duplex servo design have been thoroughly tested in transport aircraft. Each of the duplex servos incorporates two independent servo motors that operate from signals applied by their own autopilot channels. The common tie to a single control surface is accomplished through a mechanical differential gear mechanism.

In the event of a system fault, the master warning annunciator flashes amber and may be cancelled by pressing the annunciator. The status panel then indicates the system has automatically disconnected the malfunctioning channel and is a single channel in that axis only. The status panel also enables the manual selection of single channel operation in any axis and has provisions for testing the dual-autopilot channels and monitoring circuits before flight.

Control of the autopilot for basic stabilisation and attitude command is provided through the autopilot controller. Engaging the system with no flight director mode selected causes the aircraft to maintain the existing pitch attitude, roll to wings level and hold the existing heading. With a navigation or vertical path mode selected, engaging the autopilot automatically couples the selected mode. When the autopilot is engaged the yaw damper is automatically in use, when not

HSI and ADI flight director displays, computers and controllers are part of the Honeywell SPZ-600 automatic flight control system   0503918

engaged the yaw damper may be selected separately to assist the pilot during manual flight. The autopilot controller also has a turn knob and pitch wheel, which allow the pilot to insert pitch and roll commands manually. The amount of bank or pitch change is proportional to the command selected. A soft ride engage button reduces autopilot gains for smoother operation in turbulence and a couple button selects which flight director is driving the autopilot.

Touch control steering allows the pilot to take control of the aircraft momentarily without disengaging the system. He pushes the touch control steering button on the control wheel and manually changes the flight path as desired. While the button is pressed, the autopilot synchronises with the existing aircraft attitude and on releasing the button the system holds the new attitude or resumes the coupled flight mode.

The flight director system uses standard 5 in (127 mm) instruments and there is a choice of displays, with either split cue or V-bar directors and vertical scale differences. The ADZ-E 242 air data system includes a full set of appropriate instruments, including altitude alert controller and true airspeed/temperature indicator.

Specifications
Autopilot
Dimensions:
(controller) 69 × 146 × 114 mm
(status/switching panel) 48 × 146 × 114 mm
(2 duplex servos and brackets) 106 × 205 × 298 mm
(computer) 194 × 71 × 321 mm
(yaw actuator) 54 diameter × 232 mm
(normal accelerometer) 31 × 25 × 61 mm
Weight: 14.17 kg total

Flight director system
Dimensions:
(ADI (AD-650B unit) 129 × 129 × 224 mm
(HSI (RD-650B unit) 103 × 129 × 203 mm
(remote controller) 38 × 146 × 66 mm
(computer) 194 × 71 × 321 mm
(mode select) 47 × 146 × 114 mm
(rate gyro) 46 × 52 × 65 mm
Weight: 10.16 kg total

Air data system
Dimensions:
(computer) 193 × 125 × 361 mm
(altimeter) 83 × 83 × 159 mm
(vertical speed indicator) 83 × 83 × 146 mm
(Mach/airspeed indicator) 83 × 83 × 186 mm
(VNav computer/controller) 38 × 87 × 500 mm
Weight: 9.44 kg total

Gyro references
Dimensions:
(vertical gyro) 157 × 165 × 238 mm
(directional gyro) 191 × 154 × 229 mm
(flux valve and compensator) 121 diameter × 73 mm
Weight: 6.08 kg total

Status
In service. SPZ-600 variants are available for the Fokker F27, Raytheon Hawker 125–700,

Bombardier Challenger, Gulfstream III and Cessna Citation aircraft.

Contractor
Honeywell Inc, Commercial Electronic Systems.

SPZ-7000 series digital flight control system

Type
Flight Control System (FCS).

Description
The SPZ-7000 is Honeywell's fourth-generation helicopter autopilot, and is claimed to be the first digital pitch, roll, yaw and collective four-axis system to receive civil certification. The system is microprocessor-based and has two flight computers providing full autopilot and stability augmentation, dual-flight director mode selectors, automatic preflight and en route testing features and comprehensive diagnostic circuits.

A version of the system designated SPZ-7300 was chosen by Agusta for search and rescue versions of the A 109, AB 212 (a licence-built version of the Bell 212) and AB 412. The SPZ-7300 couples a helipilot system with a digital flight path computer added to provide important new functions for the search and rescue mission. In addition to the standard flight director modes, it generates a two-stage decelerating approach to the hover, hover augmentation and the ability to hover while coupled to a Doppler navigation reference for better hover performance.

Status
The system was certified aboard the Sikorsky S-76 Mk II in November 1983 (now also certified on the S-76B model) and in December 1984 was specified by the airframe company as the factory standard option for this upgraded version of the corporate helicopter; in the S-76 Mk II the dual-channel system permits Instrument Meteorological Conditions (IMC) operation with one pilot. The system has also been certified on the Eurocopter AS 365N helicopter and for the Bell 222UT utility transport helicopter. All versions in production.

Contractor
Honeywell Inc, Commercial Electronic Systems.

SPZ-8000 digital flight control system

Type
Flight Control System (FCS).

Description
The SPZ-8000 digital flight control system was designed with emphasis on corporate aircraft such as the Raytheon Hawker 800. The system is unusual among its kind in having a bidirectional databus – the Honeywell ASCB avionics standard communications bus– and includes a flight management system accommodating both lateral and vertical guidance. Honeywell believes that its ASCB system provides both the higher refresh rates needed for flight control applications and a greater flexibility in use than the one way ARINC 429 standard. The system is built around two FZ-800 flight computers and is fail-operational.

Status
In production. The first aircraft to be certified was the de Havilland Canada Dash 8. This was followed by the Raytheon Hawker 800 and 1000, Aerospatiale/Alenia ATR 42, Cessna Citation III and the Dassault Falcon 900. The Gulfstream G-IVSP and Canadair Challenger CL-601-3A followed in 1987. Latest standard installations include the ATR-72 and Citation VII and Gulfstream IV.

In the Gulfstream IV installation, the SPZ-8000 system integrates: the Electronic Flight Instrument System (EFIS); the Engine Instrument and Crew Alerting System (EICAS); Digital Flight Control System (DFCS); Flight

The Honeywell SPZ-7000 digital automatic flight control system is installed in the Eurocopter AS-365N helicopter   0503919

SPZ-8000 all-digital autopilot for fixed-wing aircraft 0503920

CPU throughput: 429 kips, DAIS mix
Interfaces: RS-232, MIL-STD-1553B
I/O: 96 analogue inputs, 12 analogue inputs, 128 discrete I/O, 1553 and cross-channel datalink.(Above I/O is not a limitation, but current capability.)
Programs: Jovial and Assembly

Status
In service in the EA-6B Prowler. The system has also been adapted for US Air Force A-10 and TR-1 aircraft and US Army AH-64 helicopters.

Contractor
BAE Systems Controls.

Management System (FMS); Laseref II Inertial Reference System (IRS); and Primus 870 Weather Radio System.

Contractor
Honeywell Inc, Commercial Electronic Systems.

ST-360 LCD Altitude selector/alerter

Type
Avionic crew alerting system.

Description
The new S-TEC LCD altitude selector/alerter is a compact, lightweight unit combining the computer and programmer in a single package. It is fully TSOd to C9c standards.

The ST-360 requires input from the aircraft's altitude encoder to function. It offers digital selection of altitude and vertical speed for climb and descent with the vertical speed automatically reducing prior to altitude capture. The system also has a selectable decision height and minimum descent altitude function. Visual and aural alerts are used with these functions and also provide warning if the aircraft departs the selected altitude.

Specifications
Power required: 14/28 V DC
Weight: 0.6 kg
Dimensions: 41 × 87 × 172 mm

Status
In production.

Contractor
Meggitt Avionics/S-TEC.

LCD altitude selecter/alerter 0001449

Stability Augmentation/Attitude Hold System (SAAHS) for the AV-8B

Type
Flight Control Computer (FCC).

Description
The SAAHS was the first digital system to be applied to any of the Harrier series of aircraft. It comprises a limited authority stability augmentation system with some conventional autopilot functions to reduce pilot workload. The single-channel system operates on all three axes and is accompanied by a mechanical back-up for reversionary operation. Extensive self-monitoring circuitry switches out the system in the event of a fault, so that it fails passive. Monitoring is accomplished in three ways: equipment, software monitoring of equipment

and software monitoring of performance. In addition a comprehensive preflight schedule is provided.

Stability augmentation modes comprise three-axis rate damping in both vertical and cruise flight and in transition between these phases. It also includes rudder/aileron and stabiliser/aileron interconnects to improve turn co-ordination. Autopilot modes include altitude, attitude and heading hold, automatic trim, control column steering and airspeed limit schedules. A rudder pedal shaker alerts the pilot to the onset of potentially dangerous sideslip during the hover.

Principal element of SAAHS is a 6.1 kg digital flight control computer with 2901-bit slice, 16-bit processor working at 470 Kips, with UVEPROM. The flight control programme occupies 20 kwords of memory of which 9 kwords are taken by the control law code and 11 kwords by the built-in test schedule. The other units comprise a three-axis rate sensor, lateral accelerometer, stick sensor and forward pitch amplifier. In December 1982, the system demonstrated a completely hands-off automatic vertical landing.

Specifications
Weight: 12.3 kg total

Status
In service on the US Marine Corps AV-8B.

Contractor
Honeywell Defense Avionics Systems.

Standard Automatic Flight Control System (SAFCS) for the EA-6B

Type
Flight Control System (FCS).

Description
The Standard Automatic Flight Control System (SAFCS) for the EA-6B Prowler is a standard computer design suitable for application to a variety of Navy aircraft. This capability is made possible by the inclusion of additional input/output circuitry and multiple memory options. Growth in redundancy up to triplex is possible through the inclusion of a cross-channel databus for exchanging information between computers.

The EA-6B SAFCS currently interfaces with analogue sensors and a MIL-STD-1553B avionics databus and drives electrohydraulic and trim actuators. Command augmentation, attitude and directional control servo-loops provide aircraft stability and control.

An extensive built-in test is provided together with non-volatile RAM for in-flight fault data storage. Maintenance information displayed in the front panel simplifies testing and logistic support. It also provides a quick turnaround resulting in higher availability. Use of hybrid circuits and low-power electronics minimises space, power and weight considerations while maximising reliability.

Specifications
Dimensions: 178 × 145 × 267 mm
Weight: 9.1 kg
Power supply: 115 V AC and 28 V DC
Power dissipation: 42 W
Cooling: convection, no forced cooling air

Supplemental Flight Data Acquisition Unit (FDAU)

Type
Data processing system.

Description
The FDAU has been designed to provide a low-cost convenient means for increasing the number of flight parameters recorded by Honeywell Aerospace's standard five-parameter Universal Flight Data Recorder (UFDR). Without replacing the UFDR, the Supplemental FDAU and recorder can together accommodate up to 11 parameters. This is accomplished by merging new parameters into the ARINC 573 data stream generated by the UFDR's internal FDAU and returning the data to the recorder section of the UFDR. The Supplemental FDAU provides a status output that is compatible with the ARINC 542 status-reporting system, causing minimal impact to existing wiring. If the Supplemental FDAU is not installed or its status is invalid, the UFDR automatically reverts to its normal five-parameter operation.

The Supplemental FDAU can accept a variety of sensor and signal types. The programmable analogue inputs accept synchro, AC ratio No 1, AC ratio No 2, 0-5 V low-level DC and potentiometer signals. Dedicated inputs include shunt and series discretes, ARINC 573 serial data from the UFDR and UFDR validity status.

The BIT capabilities of the Supplemental FDAU determine whether the recording is functioning and enunciate system status. Additionally, the Supplemental FDAU will provide validity indication that will allow faults to be isolated to the LRU.

Specifications
Dimensions: ARINC 600, 2 MCU
Weight: 2.27 kg

Contractor
Honeywell Aerospace, Electronic & Avionics Lighting.

System 20/30/30ALT/40/50/55/60/65 Flight Control Systems

Type
Flight Control System (FCS).

Description
The basic System 20 Flight Control System consists of a combined programmer/computer/annunciator in the turn co-ordinator and an electrically operated roll servo. It permits both straight and turning flight and offers dual-gain NAV/LOC/GPS tracking. Low gain is used for en-route tracking with smooth station passage, while high gain is for flying approaches. Heading Pre-select and Hold functions are operational if a heading system is installed, such as a directional gyro with heading bug.

The System 30 has all the functions and options of the System 20 but includes an altitude hold mode. It also includes a compact, remote pitch computer that provides pitch trim annunciation. An accelerometer confers pitch stability and automatically disconnects the system in the event of a pitch axis fault. In addition, the system incorporates a preflight test feature that tests internal limiter circuitry.

The System 40 Flight Control System consists of a 3 ATI-size, panel-mounted, mode selector/programmer/annunciator, a turn co-ordinator and an electrically operated roll servo. The roll axis computer is contained within the 3 ATI case. The system permits both straight and turning flight, while inputs from a radio beacon provide the means for VOR tracking for en route navigation, localiser tracking for approach and reverse or back course tracking. The system can be upgraded to include a heading mode by adding the optional directional gyro.

The System 50 has all the foregoing functions and options but includes an altitude hold mode and elevator trim indicators from a pitch axis computer contained in the 3 ATI case. An accelerometer confers pitch stability and automatically disconnects the system in the event of a pitch axis fault. In addition the system incorporates a preflight test feature that tests internal limiter circuitry.

The System 55X is the first of the 'full function' two-axis autopilots and it is a pure rate-based autopilot. It combines programmer, computer, annunciator and servo-amplifier functions into one panel and is the first S-TEC autopilot designed specifically to be integrated into the aircraft radio panel. It offers course intercept capability in NAV, HDG/NAV or HDG/APR modes along with VOR/LOC/GS/REV/GPS coupling, and course deviation and NAV flag warning. The System 55 has GPS Steering (GPSS) which responds directly to left/right steering commands output from the newer GPS navigators. This gives very accurate hands-off GPS navigation. It also has altitude trim capability and vertical speed command.

The System 60 is a full-function flight control system using rate of change of aircraft motion instead of attitude to generate demands to the flying controls. The System 60 is a single- or two-axis system (60-1 and 60-2) comprising an electric turn co-ordinator, 3 in (76 mm) air-driven directional gyro, mode programmer/annunciator, pitch and/or roll guidance computers and servos, master switch and control wheel disengage switch and altitude transducer.

The System 65 flight control system has the same performance standards as the System 60 but includes automatic elevator pitch trim, liquid crystal display programmer and a remote annunciator.

A liquid crystal display altitude selector/alerter combines the computer and programmer in a single package.

System 60 PSS provides a cost-effective way of adding vertical flight control to virtually any single axis autopilot system. It provides altitude hold, glide slope coupling and vertical speed capabilities. The system does not interface with existing roll axis autopilots, but complements them. Being self-contained the System 60 PSS does not connect to the aircraft pneumatic or vacuum systems.

Status
In production.

Contractor
Meggitt Avionics/S-TEC.

System Processors (SPs) and Weapon Processors (WPs)

Type
Avionic processor.

Description
Sanmina-SCI's SPs and WPs are avionics processors designed for use in aircraft where low weight, small size and high performance are important considerations. The SP provides navigation, communications, electronic warfare and engine monitoring functions, while the WP provides integrated sighting systems and weapons control functions.

Three different versions of the processors have been produced with varying I/O makeup and

Two different versions of the ASP are being produced for the AH-64D Longbow Apache   0503871

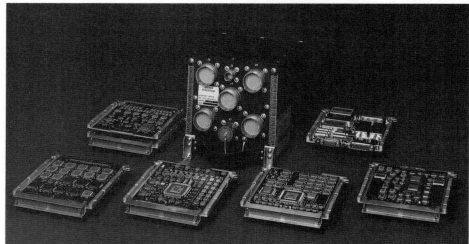

The SCI aircraft systems processor can house up to six modules   0503870

processor performance. The processors make extensive use of SMT and ASICs to yield modular units in a form-factor similar to ½ ATR short chassis weighing less than 7 kg. An aluminium thermal core in each of the six modules conducts component heat to the forced air plenum built into the sidewall of the processor housing.

The highest performance version is based on a 300 MHz Power PC 755 CPU with 1 Mb of backdoor L2 cache. Memory consists of 32 Mb of SDRAM, 16 Mb of User FLASH, 2 Mb of Boot FLASH and 256 Kb of NVRAM. A high-speed PCI bus connects the Power PC to the Fibre Channel, Ethernet and MIL-STD-1553 interfaces.

The processors provide up to 330 digital and analogue I/Os. Serial interfaces include dual 1 GBaud Fibre Channel interfaces, Ethernet 10/100, dual MIL-STD-1553B interfaces with BC/RT/MON capabilities and ARINC 429 receivers. Other I/Os include up to 201 discrete input channels, up to 64 discrete output channels, and seven RS-422 serial data channels capable of up to 230 kbaud. I/O also includes up to 42 analogue inputs, up to 22 analogue outputs, one synchro input, and two resolver outputs. The I/O channels are designed to allow parallel connection of multiple processors for redundant operation, incorporating extensive provisions for BIT and internal loop back of I/O signals.

VxWorks Board Support and BIT code packages are also available.

The processors are designed for MIL-E-5400, Class 1A operation and tested for EMI operation to MIL-STD-461, RS03 200V/M environments. BIT provides 98 per cent fault detection and isolation to the replaceable module level.

Status
Processors with the Power PC 755 CPU entered full production in 2002. Over 1000 processors with R3000 MIPS CPUs have been delivered for use in the US Army AH-64D and UK Army WAH-64D Longbow Apache attack helicopters.

Contractor
Sanmina-SCI Corporation.

Tactical Data System (TDS)

Type
Data processing system.

Description
The Rockwell Collins Tactical Data System (TDS) is installed on the Royal Australian Navy's Role Adaptable Weapons System (RAN/RAWS), the Sikorsky S-70B-2 Seahawk helicopter. With this system, the Seahawk can perform ASW missions with three crew members, do all sensor processing on board the aircraft, locate, classify and prosecute targets autonomously and transfer target data automatically via digital datalink.

The TDS consists of a tactical display unit, a horizontal situation video display, a multifunction keyboard for communication navigation and identification management, an advanced mission adaptable computer, a data loader that simplifies

The Rockwell Collins Tactical Data System (TDS) is in service on the Royal Australian Navy's Sikorsky S-70B-2 Seahawk helicopters   0504108

the preflight mission preparation and a datalink to transfer target and mission data.

The S-70B-2 with the TDS has the capability to assist the crew in making tactical decisions autonomously and each crew member has access to virtually all tactical data. That implies greater situational awareness, so that the entire mission can be flown by three crewmen. System functions include navigation, flight plan management, datalink, tactical display and databus controller.

The tactical display unit is a multifunction display which provides an integrated pictorial representation of processed sensor information to the tactical

co-ordinator and sensor operator. As it is raster driven, there is no limit to the amount of information that can be displayed. Two identical, yet independent, 8 × 8 in (203 × 203 mm) CRTs display standard naval combat data system symbology and various special display icons in the presentation of the tactical plot. Crew members use multifunction slew controllers to select symbols to provide access to the various display functions and to designate displayed data to the system. There are no complicated or extensive sets of buttons and switches to master and during a mission, the crew can concentrate on the tactical display, not on managing a host of individual subsystems.

A powerful mission adaptive computer links all the aircraft sensors and displays via a MIL-STD-1553B multiplex databus management system. A very efficient functionally redundant integrated design places specific emphasis on maintaining a consistently high mission success rate through graceful degradation.

Status
In service on Royal Australian Navy Sikorsky S-70B-2 Seahawk helicopters.

Contractor
Rockwell Collins.

Triden series autopilot

Type
Autopilot system.

Description
Century Flight Systems latest product is the Triden series autopilot. The advanced computer aided design of the system, employing state of the art Surface Mount Technology (SMT) digital electronics, bestows superior performance and reliability over current general aviation autopilots, rivaling the performance of systems used in commercial airliners.

The Triden autopilot is a two-axis system which uses both rate and position data to smoothly control the aircraft, including flight in turbulent conditions. The system also features Century

Flight Systems' expandable system architecture, which facilitates installation of an autopilot with later addition of features as they become available.

Features of the system include:
- LCD mode annunciator
- Adjustable display contrast
- Backlit mode selectors
- Trim prompting
- Automatic Heading select NAV and approach gain select
- Heading hold
- GPS/VOR/LOC coupling
- Back course
- Automated approach mode
- Automatic Glideslope (GS) capture and coupling from above or below GS
- Automatic intercepts up to 179° when using an HSI or 45° with a DG
- Altitude hold
- Vertical speed select adjustable from 100 to 1000 fpm
- Control wheel steering
- Yoke mounted emergency disconnect switch
- Voice or tone annunciator (installer selectable)
- Attitude command
- Attitude hold
- Automatic self test with continuous gyro system status monitoring.

Optional features, which may be retrofitted to the Triden, include automatic and manual electric trim control, a single cue flight director steering horizon, digital altitude preselect with visual and aural altitude alerting and a yaw damper with dual mode sensing (for a complete three-axis system).

Specifications
Panel cutout: 3 in ATI standard
Weight: 3.95 kg (including computer and two servos)
Power: 14 or 28 V DC

Century Flight Systems' Triden series autopilot   0134051

Status
In service.

Contractor
Century Flight Systems Inc.

U1638A MIL-STD-1750A computer

Type
Avionic computer.

Description
The U1638A computer is a militarised radiation-hardened general purpose avionic computer that meets all the requirements of MIL-STD-1750A Notice 1. The U1638A computer uses the U1635 high-speed MIL-STD-1750A central processing unit. The U1638A combines this high-density high-speed microprogrammed CPU with memory, input/output interfaces and power conditioning to provide the high-speed data computation, data storage and data transfers required for avionic systems.

The U1638A computer is packaged as a MIL-STD-1788 type 10 chassis which includes 12 shop-replaceable units. One is a power supply/power conditioner and 11 are processor and Input/Output (I/O) printed circuit cards.

Using the U1635 CPU in conjunction with a high-performance cache memory, the U1638A achieves a DAIS mix performance in excess of 1.5 Mips. The main memory consists of one million words of non-volatile core memory organised as two 512,000 modules. The primary I/O interfaces are the four MIL-STD-1553B dual-redundant and multiplex databusses.

The U1638A includes a programmed I/O channel and 16 input and 16 output discretes which are programme-controlled and can be defined to meet the user's needs. Four external interrupts and five identification bits are also provided. The unit is radiation-hardened with a radiation detector, power dump circuitry, a microcontrolled main memory radiation event recovery system independent of macro software and EMP protection.

Specifications
Dimensions: 193.5 × 322.3 × 319 mm
Weight: <23 kg
Power supply:
(dual AC input) 115 V AC, 400 Hz, 3 phase, 350 W

Status
In service in US Air Force B-2 Spirit stealth bombers.

Contractor
Lockheed Martin Management & Data Systems.

Ultrasonic Fuel Quantity Indicating System (UFQIS)

Type
Aircraft sensor, Fuel Quantity Indication (FQI).

Description
The use of ultrasonic sound wave transmission is a recent innovation in fuel gauging. This technology offers a number of benefits at a comparable acquisition cost to that of a capacitance system. The heart of the Ultrasonic Fuel Quantity Indicating System (UFQIS) is the fuel quantity processor unit, which contains the interface circuitry for the in-tank fuel probes, the temperature sensor, the density measuring devices, the refuel panel and the external databuses. The architecture provides an effective 'brick wall' between fuel tanks, ensuring that each tank is treated entirely separately. By building redundancy into each circuit block, no single component failure will prevent gauging or degrade performance of the system, and no two failures will cause total loss of gauging.

The ultrasonic fuel probe incorporates a piezoelectric crystal that transmits sound at a defined frequency above the range perceived by the human ear. This provides a time domain signal related to fuel height which is drift-free over the long term. Since it utilises a time domain signal, this technology also has a lower susceptibility to ElectroMagnetic Interference (EMI) than traditional capacitance systems. The body of the ultrasonic probe is electrically inactive, and so can be bonded to aircraft structures. This reduces the means by which electrical transients could be transmitted into the fuel tank, thus improving safety. Unlike capacitance systems, the ultrasonic system does not depend on harness screen integrity.

Status
In service in the Boeing 777.

Contractor
Smiths Aerospace.

Unmanned Combat Air Vehicle (UCAV) gauging system

Type
Aircraft sensor, Fuel Quantity Indication (FQI).

Description
Smiths Aerospace UCAV Gauging System features high accuracy and Electro Magnetic Interference (EMI) immunity. It is based on Smiths' 'Smart Probe' and Remote Interface Unit (RIU) technologies.

The fuel probe is a capacitance device which integrates electronic signal conditioning. This produces a time domain signal which responds linearly with fuel height in the tank. The probe is temperature-compensated; it provides an output signal which is compatible with digital systems, and has a very low susceptibility to EMI. Careful circuit design and packaging, together with robust standard parts, has resulted in a high-accuracy, high-reliability, temperature-compensated sensor.

The probe interfaces with a Smiths RIU via an Intrinsic Safety Barrier Unit (ISBU). The ISBU provides a safe DC supply to the probe and ensures that no hot shorts or induced voltages on the signal lines can enter the tank, thus improving the safety of the system. The time domain signals are interfaced to an RIU, which measures the signal and converts the value into a probe height which is transmitted over a digital databus for further processing by other aircraft systems. If required, the RIU can produce a fuel volume or fuel mass.

The RIU provides additional facilities for connection to up to 100 analogue and discrete inputs and outputs which can be used for sensor interface.

Status
Installed in the Boeing Unmanned Combat Air Vehicle (UCAV) X-45A demonstrator.

Contractor
Smiths Aerospace.

Upgraded Data Transfer Equipment (UDTE)

Type
Data processing system.

Description
Smiths Aerospace's (formerly Orbital Fairchild Defense) Upgraded Data Transfer Equipment (UDTE) comprises an aircraft resident receptacle called an Upgraded Data Transfer Unit (UDTU) and a removable data transfer cartridge called a Mega Data Transfer Cartridge with Processor (MDTC/P); UDTE is a form, fit, and function replacement for Smiths Aerospace's Data Transfer Equipment. In addition, UDTE provides an expanded array of functions.

Data transfer capabilities include a Digital Terrain System (DTS); an Embedded Data Modem (EDM); an Embedded GPS Receiver; an Embedded Display Processor; an Embedded

Training System; an Integrated Electronic Combat System; a high-performance processor to back up flight critical systems; up to 2 Gbytes of solid-state mass storage.

UDTU
Using state-of-the-art circuit design and packaging techniques, the three circuit cards in the original DTU have been reduced to a single circuit card. This card uses an industry standard 32-bit microprocessor which has enough processing throughput to retain all the original DTU's functionality while also hosting advanced software algorithms such as those required for Embedded Training and an Integrated Electronic Combat System. In addition, new capabilities may be added to the UDTU via the two unused card slots. The list of circuit cards which can populate these spare slots is continually growing and includes a ruggedised GPS receiver, an EDM and a Display Processor.

MDTC/P
The MDTC/P is a form, fit and function replacement for Orbital Fairchild Defense's (now Smith's Aerospace's) standard Data Transfer Cartridges and remains fully AFMSS compatible. The MDTC/P contains up to 280 Mbytes of non-volatile, solid-state mass memory and a high-performance Digital Signal Processor (DSP) ideally suited for mathematically complex algorithms. Typical uses for the MDTC/P include mission planning, storage of Digital Map Data, storage of avionics data gathered during flight, in-flight execution of BAE Systems (Operations) Limited TERrain PROfile Matching (TERPROM) algorithm which is used as part of the Digital Terrain System and post-mission playback associated with training.

Real-Time Intelligence in the Cockpit (RTIC)
RTIC becomes a reality with the inclusion of the Embedded Data Modem (EDM) within the UDTU. Located in a spare card slot within the UDTU, the EDM takes full advantage of the UDTU's prewired backplane and connectors to bring real-time data into the aircraft. The EDM was first flown at Edwards Air Force Base in 1995 and has been proven effective with ARC-164 and ARC-210 radios, aircraft intercoms, and KY-58 crypto sets. Using standard protocols, the EDM allows real-time data, including reconnaissance photos and imagery, to be sent to and from the aircraft.

Digital moving map
The UDTU's backplane and connectors are also wired to accept a single board digital moving map. Smiths Aerospace's fully featured moving map provides anti-aliased RS-170 RGB outputs and has capabilities normally associated with much more expensive, stand-alone map systems. These capabilities include sun angle shading, height above threshold, threat intervisibility, continuous zoom, graphics overlays, and user-defined annotation. The moving map may also be used to display two and three-dimensional imagery.

The mass memory utilised by the moving map resides within the removable data transfer cartridge. Accordingly, the moving map electronics and mass memory are co-located within the data transfer equipment and external high-speed busses are not required at the aircraft level. As well as the obvious aircraft wiring advantages associated with this approach, the use of the MDTC/P for the map storage allows standard mission planners to be used to load and prepare the map database.

Digital Terrain System (DTS)
In the Digital Terrain System (DTS), BAE Systems' TERPROM® algorithm is hosted on the digital signal processor within the MDTC/P. DTS allows aircraft integrators to incorporate significant aircraft safety of flight and lethality improvements without compromising the throughput of other aircraft systems. The numbers of aircraft using DTS continues to grow and includes platforms such as the F-16, Mirage 2000, and Jaguar.

By correlating inertial navigation and altimeter data with Digital Terrain Elevation Data (DTED)

stored within the MDTC/P, precision Terrain Referenced Navigation (TRN) is accomplished. Once the aircraft is accurately located with respect to the terrain, DTS provides numerous aircraft enhancements including predictive ground collision avoidance, obstruction warning and cueing, database terrain following, and weapons ranging.

Video recording
In recent years, the amount of solid-state memory available within the MDTC/P has seen a dramatic increase. Functions previously considered to be not feasible due to the large amount of required storage are now well within the capabilities of the equipment. One such function is the recording of video data. By populating one of the UDTU's spare card slots with a video compression card, between 1.5 and 2 hours of imagery can be recorded. This video can be synchronised with other aircraft data stored in the cartridge thus improving the effectiveness of mission debriefs by eliminating the separate, non-synchronised recorders which are typically used today.

Embedded training
The advanced capabilities of the UDTE allow embedded training to be accomplished without the need to add external aircraft pods or the need to fly on an ACMI range. The use of the UDTE for embedded training was first demonstrated at Edwards AFB in 1995.

For this function, the UDTU serves as a bus monitor recording aircraft data pertinent to mission debrief, and as a bus controller able to inject synthetic threat data into the aircraft's electronic warfare systems. The data recorded during the training session is retrieved from the MDTC/P via standard mission planning ground equipment. Time tagging using GPS time can be used to synchronise several aircraft thus allowing multi-aircraft playback.

Synthetic threat data can be preplanned and stored in the MDTC/P or real-time data can be sent to the aircraft via the embedded data modem or other aircraft datalinks. By using the DTS's TRN capability, these synthetic threats are accurately positioned on the ground. In addition, DTED information stored in the MDTC/P is used to accurately simulate terrain masking. The end result is a highly accurate, life-like training scenario without the need for very expensive ranges or external equipment.

Embedded GPS
The UDTU's backplane and rear panel contain a location for the addition of an RF connector thus supporting a third party GPS card housed within one of the UDTU's spare card slots. Including the GPS card within the UDTU co-locates the GPS function with GPS almanac data stored on the data transfer cartridge. Functioning as a bus controller or remote terminal on the MIL-STD-1553 interface, the UDTU can supply the GPS information to other avionic systems.

Bus Monitor and Analysis Software (BMAS)
Avionic configurations are increasingly complex and the isolation and debug of a problem can cause extended downtime. BMAS allows the UDTE to function as MIL-STD-1553 test equipment and record user-specified MIL-STD-1553 data. All pertinent MIL-STD-1553 data can be replayed on a standard personal computer using Smith's Aerospace data logger software. At the completion of debug, the UDTE can resume normal operation.

Ground support peripherals
The MDTC/P is fully AFMSS compatible. For those applications where an AFMSS is not available, Smiths Industries' SCSI Cartridge Interface Device (ScsiCID™) allows any personal computer or workstation to communicate with the MDTC/P over a standard SCSI-2 interface. Data rates in excess of 5 Mbytes/s allow data to be loaded rapidly into the MDTC/P's mass memory.

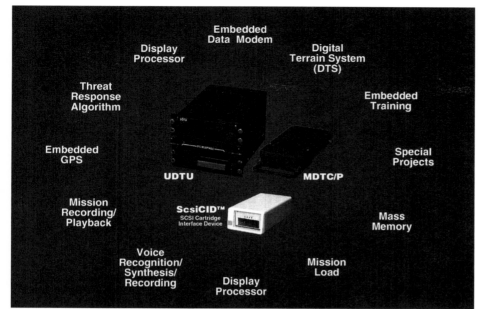

Smiths Aerospace's Upgraded Data Transfer Equipment (UDTE)   0002320

VIA equipment configuration   0051696

Specifications

Dimensions:
UDTU: 178 × 127 × 113 mm
MDTC/P: 191 × 119 × 41 mm
Weight:
UDTU: 3.0 kg
MDTC/P: 1.6 kg
Input power: 115 V AC, 400 Hz, single phase
32 W max
Software: Ada

Status

In production and in service.

Contractor

Smiths Aerospace, Germantown.

Versatile Integrated Avionics (VIA)

Type

Avionic processor.

Description

VIA is a mature, flexible, general purpose processor developed for the commercial airline market. VIA has direct application for military aircraft and offers a true Commercial-Off-The-Shelf (COTS) solution.

The VIA design lends itself to both new production (for example the Boeing 717) and retrofit (Federal Express MD-10) applications. Since the beginning of 1999, Honeywell has been selected by several customers, including the US military, to put the VIA system on several military aircraft, including the C-5, KC-10 and E-6.

The baseline VIA system consists of a common chassis and power supply hosted in an ARINC standard 8 MCU Line Replaceable Unit (LRU). Internal cards provide generic computing and I/O communication. The innovative ARINC 659 backplane and construction of each of the cards provides high integrity, high-availability processing and communications. For the ARINC 659 backplane, I/O data is transmitted up to four times, once on each of the Ax, Ay, Bx and By busses. The ARINC 659 data rate is approximately 60 Mbits/s.

The baseline VIA contains one Core Processing Module (CPM) with dual AMD29050 RISC microprocessors executing identical code on a cycle-for-cycle basis; two I/O controller Modules (IOM) providing redundant I/O interfaces; and one high-integrity commercial power supply.

For military applications, the VIA contains a data transfer module for MIL-STD-1553B bussing with provisions for VIA operations either as a bus controller, back-up bus controller or remote terminal.

The baseline VIA has provisions for two additional core processing modules, graphics generators to drive AMLCDs, and three additional I/O modules should increased computational and/or I/O capability be necessary for specific applications.

Solutions involving two or more VIAs use a Cross Channel Link (CCL) providing a communication tie between VIA processors.

The VIA design uses an open system architecture to support customer or third party participation in software development and follow-on support. ARINC standards and FAA-approved tools allow VIA customers to develop their own software to provide unique customer solutions.

The overall VIA design is based on ARINC 651 design guidance for IMA. The VIA external interfaces are commercial and military standards such as AEINC 429 and MIL-STD-1553B. The VIA backplane is designed to ARINC 659.

A single VIA, operating in a partitioned mode, provides solutions to multiple avionics functions previously provided by separate LRUs. Robust partitioning is designed into the VIA hardware and software to ensure the VIA will operate as a virtual LRU for a selected time frame. Virtual LRU operations are possible for flight management, integrated displays, COMM/NAV/Surveillance (CNS) management, flight controls, central caution and warning, central maintenance diagnostics, built-in test and other functions. The partitioning scheme guarantees virtual LRU operation by ensuring that a functional partition is not allowed to affect another partition's code, I/O or data storage area; a partition is not allowed to consume shared processor or I/O resources to the exclusion of another partition; and a single failure of common hardware is not allowed to prevent safe flight and landing.

The VIA is designed to easily accommodate future Global Air Traffic Management (GATM) and FADEC military upgrades by using the partitioning concept to add the desired performance in a separate partition or to an existing partition.

Specifications

Dimensions: 203.2 (H) × 304.8 (W) ×
406.4 (D) mm
Weight: 20.45 kg

Status

VIA is a second-generation Honeywell product based on an Integrated Modular Avionics (IMA) design for the Boeing 777 aircraft. The IMA was first certified by the FAA in April 1996. The Boeing 777 IMA design was repackaged into a discrete LRU called VIA. VIA certification by the FAA was achieved in October 1997 on the Boeing 737-700 series aircraft.

The latest application of the system is designated VIA 2000. In the Boeing 717, VIA 2000 supplies the following systems: the Category IIIa Auto Flight System, including windshear detection and stall warning (the system can be upgraded to Cat IIIb capability); dual Air Data/Inertial Reference Systems (ADIRS); and the central fault display system.

VIA is in production. Applications include both new production aircraft (Boeing 717, 737-700 and MD-90) and retrofit applications (Federal Express MD-10). VIA has been selected for the C-5 Avionics Modernisation Programme (AMP) to meet Global Air Traffic Management (GATM) requirements for operating worldwide missions in the 21st century; other applications include the KC-10 and E-6.

Contractor

Honeywell Inc, Commercial Aviation Systems.
Honeywell Inc, Defense Avionics Systems.

AIRCRAFT COCKPITS

AIRCRAFT COCKPITS

Introduction

Long gone are the days of flying an aircraft 'by the seat of the pants'. While early aircraft could indeed be flown satisfactorily by 'feel', the necessity for accurate information concerning aircraft performance and systems status came with increasing airframe and propulsion system sophistication and capability. Most pilots of the current era will be familiar with the 'Standard T', an arrangement of aircraft performance instruments which placed the main Attitude Indicator (AI) at the head of a 'T' shape, flanked by altitude and airspeed indicators, with the Horizontal Situation Indicator (HSI) below the AI. Other instruments, such as the Turn and Slip,

Vertical Speed Indicator (VSI) and Angle-of-Attack (AoA) indicator, were added around this basic arrangement. This standardised configuration of aircraft instrumentation enabled pilots to rapidly assimilate the necessary performance information through 'instrument scan' – centred on the main AI, the pilot selectively reads each performance instrument as required, always coming back to the AI. While a highly developed instrument scan may not be required for flying during clear conditions, it is certainly a critical requirement for operations in Instrument Meteorological Conditions (IMC). The 'Standard T' has served pilots very well for many years, but these analogue instruments have been steadily giving way to modern Cathode Ray Tube (CRT),

or, more latterly, Active Matrix Liquid Crystal Displays (AMLCDs), which offer greater reliability (MTBF), lighter weight and less bulk – an entire 'T' can be replaced by a single display.

While modern large and military aircraft are almost exclusively designed with 'Glass Cockpits', many legacy aircraft are also being retrofitted with CRT or AMLCDs. In common with the previously mentioned 'Standard T', current approaches employ broadly similar display configurations, conforming to particular size/bulk constraints, according to host aircraft and/or type/role. The following aircraft cockpit images illustrate current design philosophies for various types of military and civilian aircraft, including both fixed- and rotary-wing platforms.

Civilian aircraft cockpits

Airbus A300/310

The flightdeck of an Airbus Industrie A300/310 wide-body airliner. As the company's first wide-body airliner, the cockpit of the A300/310 followed current tradition, with control yokes and a mixture of analogue and Cathode Ray Tube (CRT) displays. Of note in this image is the selection of colours in the pilots' Primary Flight Displays (PFDs) and the Flight Mode Annunciator (FMA) demarcation at the upper edge of the PFDs, which were carried over to the new generation of cockpits which Airbus introduced with its Fly-By-Wire A320. It is perhaps unsurprising that, in the absence of an established customer base, Airbus should first offer a cockpit familiar to pilots of Boeing 757/767 aircraft. However, compare this with the radical departure to side sticks and an all-glass flightdeck in the A320 and the reader will begin to appreciate the valiant efforts of the Airbus designers in challenging the then established flightdeck layout for large commercial aircraft (Airbus Industrie)
1044245

Airbus A320

The flightdeck of an Airbus Industrie A320 narrow body airliner. As the first of the current generation of the highly successful Airbus Fly By Wire (FBW) designs, the cockpit exemplifies the Airbus 'common cockpit' design philosophy (see other Airbus entries) – six Cathode Ray tube (CRT) displays for Electronic Flight Information System (EFIS – in front of each pilot) and Electronic Centralised Aircraft Monitor (ECAM – vertically stacked on the centre panel), together with side-stick controllers provide for an extremely 'clean' environment for the pilots. Below the EFIS displays, a pull-out map table capitalises on the side-stick design, facilitating the completion of flight logs and map planning or, indeed, eating during long sectors! This commonality between Airbus cockpits also extends to the positioning of all major controls (thrust levers, flap lever, airbrake/spoiler, communications panels and Flight Management System controls), enabling pilots to rapidly and efficiently transition between Airbus types with minimal cross-training required (Airbus Industrie)   0569003

Airbus A330

The flightdeck of an Airbus Industrie A330 wide-body airliner. Following Airbus 'common cockpit' design philosophy established with the introduction of the narrow body A320, the side sticks, the six Cathode Ray tube (CRT) displays for Electronic Flight Information System (EFIS – in front of each pilot) and Electronic Centralised Aircraft Monitor (ECAM – vertically stacked on the centre panel), are instantly recognisable to any pilot 'moving up' from the A320. Such commonality undoubtedly brings great savings in cross training; indeed, our pilot will see little change as he or she makes their next 'move' to the A340 – the editor has observed a simulator configuration change between A330 and A340 which took 30 minutes and involved only the removal and replacement of the overhead panel and thrust levers. Airbus' new long-range twin, the A350, originally adopted the A330 cockpit configuration, but, perhaps in response to the advanced cockpit layout of the rival Boeing B787, adopted the A380 cockpit avionics system during early 2006 (Airbus Industrie)
1044244

Airbus A340-600

The flightdeck of a development (note the test equipment to the rear left of the centre pedestal) widebody Airbus A340-600. While retaining the same basic layout as the A340-300 and other current generation Airbus FBW designs, the -600 instrumentation includes slightly larger AMLCD displays, replacing the earlier CRT items, offering greater brightness and enhanced readability under high light conditions. In addition, to the left of the upper ECAM display, a LCD secondary Attitude Indicator, incorporating ILS display functionality, replaces the analogue display found in -300 series aircraft (Airbus Industrie)   0569001

Airbus A350

The flightdeck of Airbus Industrie's new A350 is the same variation of Thales' Integrated Modular avionics System (IMS – see separate entry) as found in the A380 ultra large airliner. Having originally adopted the then near 20-year old configuration found in the A330, after a redesign of the entire nose section, space was found to accommodate the extra depth required to install the portrait displays of the new system. This decision to introduce a more advanced cockpit was perhaps spurred on by the competing Boeing B787's standard large format display/HUD layout. Of note is the employment of QWERTY keyboards for data input via the Keyboard Cursor Control Units (KCCUs); Airbus have made this departure from the traditional ABCD keyboard found in current Multifunction Control and Display Units (MCDUs) in response to the proliferation of personal computers and hence a greater familiarity with the QWERTY keyboard format. As a current operator, the editor is in broad agreement with this move, although Boeing has notably resisted following this departure with the B787 – it retains a more 'traditional' MCDU format with ABCD keyboard. Boeing has cited 'operator preference' as its main driver in retaining the old layout; noting that a majority of pilots, certainly in long-haul operations, carry a laptop computer in their flight bag and the keyboard of most of the new Electronic Flight Bags (EFBs), increasingly integrated in new cockpits and retrofitted into legacy cockpits, feature QWERTY keyboards, it would seem to be a more logical decision by Airbus in this case (Airbus Industrie)   1128630

AvCraft Dornier 328

The cockpit of an AvCraft Dornier 328 which incorporates a Honeywell Primus 2000 avionic display suite. Note the portrait AMLCDs and dual-Primus II integrated radio system and Mode S transponder. The narrower cockpit of this relatively small twinjet airliner means that a single central AMLCD performs all EICAS duties (Patrick Allen)

0589486

BAE Systems Jetstream 41

The cockpit of a BAE Systems Jetstream 41 turboprop regional transport aircraft. The Jetstream, in various guises, has been in service for many years so it is unsurprising that the flightdeck shows a distinct 'evolution' – note the mix of analogue and digital instrumentation. Small-format EFIS displays in front of each pilot are supported by analogue secondary instruments. A large Central Warning Panel (CWP – upper centre), combined with engine instrumentation, rather than a full EICAS, is another indicator as to the origins of this cockpit (BAE Systems)

1129575

Boeing B747-400

The flightdeck of a widebody Boeing B747-400. Note the considerable commonality with the following B777-200, which extends well beyond the corporate choice of flightdeck colour. While 6 CRTs certainly give the impression that the 747-400 is a thoroughly modern design, the aircraft certainly lacks the sophistication of the later A340, with many of the control systems carried over from the -200. Of note, although not easy to make out from this image, is that Boeing produced a more 'transatlantic-friendly' cockpit in the -400, with a few important features which make operations that bit easier than in its French counterpart. Firstly, the standby altimeter incorporates both mb and inches of mercury scales side-by-side, which makes setting equivalent pressures very simple (this facility was introduced in the standby system of A340-600). Also, the navigation system displays individual components (IRS 1,2,3) positions, providing an easy to interpret comparison of positions for estimation of navigational error prior to Minimum Navigation Performance Specification (MNPS) airspace. Display area is also greater than in the A340, although it is the editor's opinion that the European company's use of colour is superior, again, primarily as a result of later design. One aspect of the Boeing cockpit which produces some discussion amongst pilots is the tiller (shown just above the seat backrests on the upper left and right bulkheads); the 'push/pull' action for 'left/right' (although the lack of intuitiveness makes this statement not necessarily exactly true!) is liked by some and equally hated by others!   0131797

Boeing B777-200

The flightdeck of a widebody Boeing B777-200. In contrast to the previous Airbus cockpits, Boeing steadfastly retains traditional control yokes in its aircraft – while sympathetic with the views of many pilots who argue that this arrangement allows the non-handling pilot a much greater awareness of the actions of his colleague and 'feel' for the aircraft's responses to control inputs, in these days of true Fly-By-Wire (FBW), 'feel', I'm afraid, is a misnomer, and current training reflects the new requirements of interpretation of Flight Mode Annunciators (FMAs) and far less 'hands-on' flying. To that end, it is the editor's considered opinion that the ergonomic advantages and gains in useable cockpit area offered by side stick controls indicates the way that all modern air transport cockpits will be designed in the future. Of further note are the two Electronic Flight Bags (EFBs), shown just outboard of each of pilot position. These provide aircraft technical systems information (rather than having to reference paper manuals), together with ground navigation (linked to the aircraft navigation system to provide positional information) and electronic display of approach plates, area maps and so on. The use of EFBs and other related Onboard Information Systems (OIS) is becoming widespread, with many in-service commercial transport aircraft receiving retrofit Class 3 EFBs and the majority of new designs featuring integrated systems. Final word for the B777 concerns the choice of flightdeck colour – beige, while undoubtedly easy on the eye, evokes mixed emotions to those who make this their 'office' for long periods of time! Pilots who have exclusively flown Boeing aircraft (unsurprisingly) seem not to notice the colour scheme, while those with experience of other makes are moved to immediate (negative) comment. While it is not the intention of the editor to indulge in some 'mud-slinging' (sic!) concerning Boeing colour schemes, merely to perhaps promote some discussion as to what the optimum might be (Patrick Allen)   1144360

Bombardier Challenger 601

The flightdeck of A Bombardier Challenger 601 aircraft, fitted with Rockwell Collins' ProLine 21 Continuum avionics system which comprises four 10 × 8 in AMLCD Adaptive Flight Displays, the FMS-6000 FMS, AHS-3000 Attitude Heading Reference System (AHRS), TCAS-4000 Traffic Alert and Collision Avoidance System, TWR-850 turbulence weather radar and ProLine radios (Rockwell Collins)   0569547

Bombardier CRJ-700

The flightdeck of a Bombardier CRJ-700 fitted with Rockwell Collins' ProLine 4 avionics. Note the six large-format displays, featuring two EFIS displays for each pilot and two Engine Indication and Crew Alerting System (EICAS) displays on the centre panel. The Captain's position also includes a Head-Up Display (HUD) to facilitate aircraft operations in less than Category I (CAT 1) weather conditions (Bombardier)   0569004

For details of the latest updates to *Jane's Avionics* online and to discover the additional information available exclusively to online subscribers please visit
jav.janes.com

Bombardier Dash 8 Q400

The flightdeck of a Bombardier Dash 8 Q400 large turboprop airliner. Note the five large-format portrait AMLCDs and HUD for the Captain's position
(Bombardier)
0569006

Bombardier Global Express

The impressively uncluttered flightdeck of A Bombardier Global Express large business jet, which also forms the basic platform for the UK's Airborne Stand Off Radar (ASTOR) system. Note the six large portrait-format AMLCDs, for EFIS and EICAS information, flanked to the left and right quarters by aircraft systems information systems, enabling the pilots to retrieve vital aircraft technical information without recourse to paper manuals (Bombardier)

0569005

Learjet 31a

The 'snug' cockpit environment of the Learjet 31a including a Bendix/King avionics suite featuring individual AMLCD instrumentation, including the pilots' Artificial Horizons (AH – upper) and Horizontal Situation Indicators (HSI – lower and centre panel). Note the plethora of circuit breakers ranged on outer bulkheads – opinions vary as to whether the incorporation of large circuit breaker panels within the cockpit constitutes a vital tool in the rectification of onboard faults by the pilots, or a dangerous distraction when airborne – as a practising pilot, I am a firm believer in having circuit breakers available to me while in the air (Bombardier)
0569007

Learjet 45

The cockpit of a Learjet 45 business jet. Despite the limited space available, particularly in terms of cabin width, the pilots have large-format portrait AMLCDs, providing flight and systems information, combined with two smaller, centrally-positioned communication/navigation control/displays. Note the standby instrumentation is centrally positioned to be equally readable by both pilots and the forward edges of the outboard circuit breaker panels, just visible at the edges of the photograph (Bombardier)

0569008

Learjet 60

The cockpit of a Learjet 60, in this case fitted with Rockwell Collins' ProLine 4 avionics suite. The cockpit features a mixture of analogue and digital display information, combined with two Multi Function Control and Display Units (MCDUs – centre pedestal behind the thrust levers) for individual control of the aircraft Flight Management System (FMS). Note also the outboard circuit breaker panels (Bombardier)   0569009

Tupolev Tu-204-300

The flightdeck of a Tupolev Tu-204-300 narrow body airliner, which first flew in August 2003. In common with earlier Tu-204 variants, the cockpit features the current arrangement of six LCDs for EFIS (in front of each pilot) and ECAM (vertically stacked on the centre panel). In this image, while the Inertial Reference Systems (IRS) may be switched off, the Display Units (DUs) formats can be clearly seen to echo that of current Airbus/Boeing configurations. The standby instruments are biased towards the Commander's position (the co-pilot would have great difficulty interpreting the standby airspeed indicator, which is directly in front of the Captain) and purely analogue in this aircraft. In common with other Russian aircraft, the flightdeck is finished in the familiar light blue colour scheme – perhaps some research regarding operational imperatives (reflections under high brightness conditions and/or low light/night conditions with/without visual aids) and the physiological effect of cockpit colour schemes on pilots during extended range operations could determine an optimum solution, rather than some of the seemingly arbitrary and sometimes outright distressing schemes in service today (Yefim Gordon)   0587978

Tupolev Tu-334

The flightdeck of a Tupolev Tu-334 airliner, which shows great commonality with the previous Tu-204 cockpits, including the arrangement of six LCDs for EFIS and ECAM. With inertial systems aligned, this image shows the Primary Flight Display (PFD) and Navigational Display (ND) formats in front of each pilot, which would be familiar to current Airbus pilots both in terms of layout and colour scheme. Further commonality is shown in the MCDU and NAV/COMM panel arrangement of the centre pedestal. It is assumed some maintenance work is ongoing in this image, as an inspection panel has been removed in the Commanders (left) footwell, exposing two equipment connectors! (Yefim Gordon) 0587979

Helicopters

Agusta Bell 412

This glass cockpit is available as an option for the Agusta Bell 412/Griffon. While physical characteristics of the cockpit are very much as for the Bell 212/412, the four large colour Active Matrix Liquid Crystal Displays (AMLCDs) and single Multifunction Control and Display Unit (MCDU) dominate the panel. Ancillary equipment (radios, nav/comm, IFF and so on) and standby instrumentation are as for the standard 412. Contrast this arrangement with that of the Bell Super Huey – while the Huey cockpit displays cost a fraction of those shown here, if mission priorities demand the clearest and most logical cockpit layout, then the very clean glass cockpit layout shown here should be high on the shopping list! 0589298

Agusta Westland EH-101

The cockpit of an Agusta Westland EH-101 military transport helicopter, which exemplifies the current side-by-side cross-platform configuration of six multifunction AMLCDs, dual FMS mounted on the centre pedestal, with back-up instrumentation via three smaller AMLCDs. The Six primary AMLCDs comprise one PFD directly in front of each pilot, one ND/tactical display each inboard, with the central pair reserved for systems/EICAS. The main panel shows sufficient room for larger format portrait displays which may form part of a future upgrade to the aircraft (P Allen)   1096872

Bell Super Huey

The cockpit of a Helipro upgraded Bell Super Huey helicopter. Compared to the previous Huey II cockpit, the Helipro upgrade includes upgrade to CRT display for both the pilot's PFDs as well as tactical/NDs. In contrast, the centre console remains largely as per the original (Helipro) 0536748

Eurocopter NH90

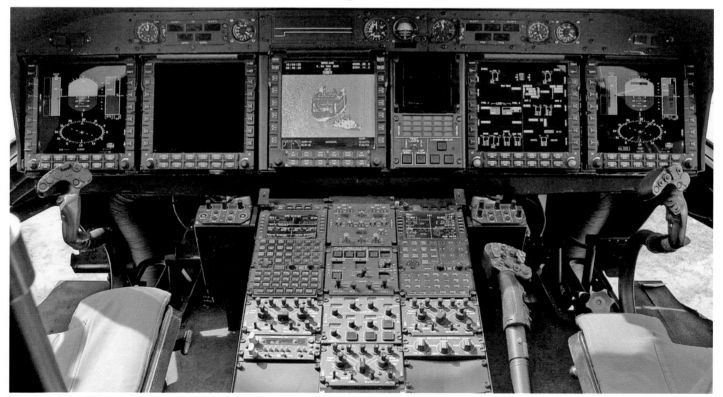

The Eurocopter NH90 military transport helicopter shares a similar configuration to that of its rival, the Agusta Westland EH-101, although employs purely analogue standby instrumentation (top row, main panel). The central tactical display shows an illustrative IR image, while the multifunction display to the left of the commander's PFD shows EICAS information. All FMS and communications systems are controlled from the centre pedestal (Patrick Allen) 0589484

Eurocopter Puma HC Mk1

During 1996, Thales Avionics completed the final airframe upgrade of the UK Royal Air Force Eurocopter Puma HC Mk1 fleet. The upgrade included an updated hover facility to EHSI (Thales Avionics)

0051617

Sikorsky S-92

The cockpit of a Sikorsky S-92 twin turboshaft medium-lift helicopter, featuring a Rockwell Collins dual FMS and display suite which includes five 8 × 6 portrait-format AMLCDs; the four main displays in front of the pilots show PFD, EICAS, navigational and weather radar information. The centre-mounted display is intended to be utilised for optional sensor data, such as FLIR or moving map data (Rockwell Collins)   0569010

Military aircraft cockpits

Aermacchi M-311

The cockpit of the Aermacchi M-311 features CMC Electronics' Cockpit 4000 (also to be found in the Raytheon T-6B – see later), including three 5 × 7 in MFDs in each cockpit and full HUD and UFCP in the front cockpit. In the rear cockpit, the centre MFD is in an elevated position, displaying the picture of the front-seat HUD. The M-311 avionics system is based on two mission computers working in parallel and providing full redundancy. The inertial/ GPS navigation platform is similar to that in the RQ-4 Global Hawk Unmanned Aerial Vehicle (UAV). The avionics system allows the introduction of flight-plan data, visualisation of analogue-type instrument displays, and supports air-to-air and air-to-ground combat simulation with guns, missiles and various types of bombs. Remaining avionics include a battery powered emergency instrument system and radio transceivers 1039724

Airbus A400M

The cockpit of the Airbus Military A400M epitomises current practice with the employment of civilian derived and/or Commercial Off The Shelf (COTS) avionics and displays (in this case drawing heavily on the Airbus A380 cockpit). The cockpit features a development of Thales' Integrated Modular avionics System (IMS) with large portrait format 6 × 8 in AMLCDs, combined with Primary Flight Display-qualified Digital Head-Up Display Systems (D-HUDS). In common with Airbus' civilian transports, onboard systems are monitored with an Electronic Centralised Aircraft Monitor (ECAM – centre panel displays) and data input is accomplished via a Keyboard Cursor Control Unit (KCCU) – as found in the A380. It is hoped that the civilian seat configuration shown in this Airbus image will accommodate pilots wearing Night Vision Goggles (NVGs), together with battery packs/counterweights on the back of their helmets (Airbus Military SL)

1129143

BAE Systems Hawk Mk127

The Hawk Mk127, as ordered by the Royal Australian Air Force (RAAF), is designed to provide effective mission system management training to pilots progressing to the Boeing F/A-18 multirole fighter. To that end, a dual-combiner HUD, Up Front Control Panel (UFCP) and three MultiFunction Displays (MFDs) and Hands-On-Throttle-And-Stick (HOTAS) inceptors are pivotal elements of any advanced trainer, where it can be argued that training pilots to manage a multirole avionics system is just as important as teaching the fundamentals of high-performance jet handling   0549850

Embraer Super Tucano

Embraer's Super Tucano, while comparable with the Pilatus PC-21 in terms of intended roles, displays a more compromised approach to the redesign of an existing aircraft. Restricted cockpit width has meant that only two 8 × 6 in portrait displays could be incorporated, with a mixture of LCD and analogue primary flight instrumentation occupying the central portion of the main panel. A dual-combiner HUD and multiline UFCP, combined with full HOTAS inceptors complete the upgrade to facilitate advanced training and light attack duties

0131837

Lockheed Martin C-130K Hercules

For those C-130 operators who are not looking to replace their older variants with the new C-130J, but desire the benefits of a 'glass cockpit', there are a number of upgrade packages available. This is the cockpit of a modified Spanish Air Force KC/C-130. As well as the cockpit displays, a full avionics upgrade was carried out to standardise and modernise the fleet. The new system integrates the flight control, flight management, communications, navigation and self-protection systems. The integrated control and flight management system is designed around a MIL-STD-1553 databus, 2 mission computers with five 6 × 8 in AMLCD screens controlled by 3 CDUs (EADS-CASA)   0051299

Lockheed Martin U-2S

In 1998, the Engineering and Manufacturing Development (EMD) phase of the U-2 Reconnaissance Avionics Maintainability Program (RAMP) cockpit upgrade commenced, with the purpose of replacing legacy cockpit avionic systems to increase pilot Situational Awareness (SA) and reduce workload. The cockpit portion of the RAMP upgrade effectively replaces the aircraft's original analogue cockpit instrumentation with commercial-off-the-shelf items – three L-3 Communications 15 × 20 cm (6 × 8 in) colour AMLCDs, a new UFCP and CDU and a Meggitt Avionics Secondary Flight Display (SFD) (USAF) 0583060

Northrop Grumman T-38C Talon

US Air Force T-38 Talon jet trainers were upgrade to T-38C standard under a Boeing-run T-38 Avionics Upgrade Programme (AUP), part of the 'Pacer Classic' initiative that aims to extend the life of the supersonic trainer until 2020. On the 5 May 2005, the 300th T-38C was delivered to the US Air Force. Upgrades include a Wide Field of View (WFoV) Head-Up Display (HUD) in the front cockpit, multifunction head-down displays, electronic engine instruments and Up-Front Control Panels (UFCPs) in both cockpits. In addition, there is an integrated Global Positioning System/Inertial Navigation System (GPS/INS) suite and a Traffic Collision Avoidance System (TCAS). Honeywell Military Avionics was selected by Boeing to supply the avionics for the programme, which include embedded Inertial Navigation/Global Positioning System (IN/GPS) based on the H-764G, embedded radar altimeter module, embedded yaw damper stability augmentation system and a standalone air data computer. For the weapons training task, a digital, air-to-ground, simulated ('no drop') weapons scoring system is installed. This cockpit, in terms of layout and implementation, is representative of many upgrades, like the Saab JA37 Viggen, and, to a lesser extent, the Embraer Super Tucano, where mission electronics and displays are 'squeezed' into existing space, necessitating compromise in screen size, and/or limited analogue instrumentation retained to constrain costs. Contrast this cockpit with the Pilatus PC-21 and Raytheon T-6B, whose cockpits were very much part of the core design, facilitating a much more coherent layout of major flight and mission displays (US Air Force)

1129577

Pilatus PC-21

The front cockpit of the first development Pilatus PC-21 basic/advanced military training aircraft. The cockpit features full HOTAS controls and three 8 × 6 in Barco AMLCDs – the central display acts as a Primary Flight Display (PFD), while the left and right displays are configurable to customer specification to show navigation/attack, stores inventory, datalink/pseudo-radar, digital map and sensor information. The CMC Electronics HUD is integrated with a proprietary UFCP, which includes a single-line scratch-pad for data input. The UFCP is flanked by a pair of Meggitt Secondary Display System (SDS) LCDs (Pilatus)
0569002

For details of the latest updates to *Jane's Avionics* online and to discover the additional information available exclusively to online subscribers please visit
jav.janes.com

Raytheon T-6B

The front cockpit of the Raytheon T-6B basic/advanced military training aircraft. The aircraft is a derivative of the Pilatus PC-9 and, thus shares the width of cockpit which allows the T-6B and PC-21 to install three side-by-side large-format portrait AMLCDs. For the T-6B, a variant of CMC Electronics' Cockpit 4000 is employed, which also features a SparrowHawk™ HUD system. HOTAS controls are another common feature between the T-6B and PC-21, although, by inspection, it would appear that the T-6B biases more functionality towards the throttle inceptor (Patrick Allen)

0589485

Saab JA 37 Viggen

The cockpit of the Saab JA37 Viggen shows the results of a continuous upgrade from an all-analogue arrangement to a hybrid part analogue (primary flight instrumentation)/part digital (radar, tactical communications and digital map display) configuration. While this upgrade process has enabled the JA37 to remain operationally effective during its service life, the inevitable compromises in ergonomics and limited integration of such arrangements will always be inferior to a (more costly) brand new design (Saab)

0569545

Saab JAS 39C Gripen

The cockpit of the Saab/BAE Systems JAS 39C Gripen exemplifies the working environment for the pilot of a modern fourth-generation fighter – a wide angle holographic HUD, UFCP with multiline scratchpad, 3 large format 8 × 6 in portrait AMLCDs and datalink control and display unit (lower left) provide a veritable deluge of Situational Awareness (SA) information. Note also the arrangement of the left and right consoles; generally, since all controls in these areas are outside of the pilot's normal FoV, switches are configured to be differentiated by feel (external formation/anti collision and intensity controls) and/or over-centre guarded to avoid inadvertent selection (IFF emergency). Also, since the pilot should not need to remove his hand from the control column, the left console houses switches which may need to be manipulated during flight; on the right console, the environmental controls and internal lights dimmers can be set prior to takeoff and, under normal operations, should not be required to be adjusted whilst airborne (Saab)

0589478

Sukhoi Su-27 SMK

The cockpit of the Sukhoi Su-27 SMK single-seat multi-role fighter. While the aircraft offers near state-of-the-art aerodynamic performance and is capable of carrying a vast array of both air-to-air and air-to-ground ordnance on its 12 hardpoints, the cockpit environment is very much a 'work in progress', with little of the coherence of display types and integration of western rivals. Two large format and one central small format AMLCDs dominate the cockpit, showing engine/systems, radar warning and radar information. For such a modern fighter, the number of analogue instruments, arrayed below the AMLCDs, is noteworthy – the reliability of current CRT/LCD designs has seen substitution of such arrays with a single secondary electronic display, driven by essential services power and/or batteries; space saved enables the fitment of even larger format (portrait) displays, facilitating more rapid assimilation of information by the pilot. The configuration of this cockpit is also reminiscent of legacy Soviet designs, with only vestigial Up-Front controls – any pilot of similar western fighter aircraft would immediately miss the absence of comm/nav functionality in this area – when the Su-27 was first introduced, the communications fit was reported as two hard-wired frequencies to facilitate take-off and landing and instructions from ground-based fighter controllers! Another area of note is the standard colour scheme in the cockpit, which follows a long-standing tradition, again with roots in the Soviet era. While the colour easily marks this cockpit as a Russian design, it is this pilot's opinion that cockpit colour should not merely be dictated in this way, with a more analytical approach taken to reflect physiological and operational imperatives. This seemingly arbitrary selection of cockpit colour is not unique to Russian designs, particularly with respect to civilian cockpits – note the use of colour in Boeing (beige) and Airbus (grey/blue) commercial aircraft (Yefim Gordon) 1044056

CONTRACTORS

CONTRACTORS

A

AAC (Accurate Automation Corporation)
7001 Shallowford Road
Chattanooga
Tennessee 37421
United States
Tel: (+1 423) 894 46 46
Fax: (+1 423) 894 46 45
e-mail: marketing@accurate-automation.com
Web: www.accurate-automation.com

AAVD
Parc de Gavanière Bât 8
F-38120 Saint Egrève
France
Tel: (+33 4) 38 02 05 81
Fax: (+33 4) 38 02 08 53
Web: www.aavd.fr

ABE Elettronica SpA
Via Leonardo da Vinci 92
Loc Settecamini
I-24043 Caravaggio BG
Italy
Tel: (+39 0363) 35 10 07
Fax: (+39 0363) 507 56
e-mail: webmaster@abe.it
Web: www.abe.it

ACSS (Aviation Communication & Surveillance
    Systems)
Primary Service Center
19810 North 7th Avenue
Phoenix
Arizona 85027
United States
Tel: (+1 623) 445 70 00
Fax: (+1 623) 445 70 01
Web: www.l-3com.com/acss

AD Aerospace Ltd
Abbots Park
Preston Brook
Cheshire WA7 3GH
United Kingdom
Tel: (+44 870) 442 45 20
Fax: (+44 870) 442 45 24
e-mail: adenquiry@ad-aero.com
Web: www.ad-aero.com

Aerodata Flight Inspection GmbH
Hermann-Blenk-Strasse 34
D-38108 Braunschweig
Germany
Tel: (+49 531) 235 27 74
Fax: (+49 531) 235 27 99
e-mail: info@afi-bs.de
Web: www.afi.aero

Aerofina Avionics Enterprise
5 Fabrica de Glucoza
District 2
R-72322 Bucharest
Romania
Tel: (+40 21) 242 07 72
Fax: (+40 21) 242 04 44
e-mail: serv@aerofina.ro
Web: www.aerofina.ro

Aeropribor-Voskhod JS
19 Tkatskaya Street
105318 Moscow
Russian Federation
Tel: (+7 095) 363 23 01
Fax: (+7 095) 363 23 43

AeroSolutions
Manassas Regional Airport
10681 Frank Marshall Lane
Manassas
Virginia 20110
United States
e-mail: sales@aerosolutions.com
Web: www.aerosolutions.com

Aerospace Display Systems, Inc
2321 Topaz Drive
Hatfield
Pennsylvania 19440-1936
United States
Tel: (+1 215) 822 60 90
Fax: (+1 215) 822 79 74
Web: www.aerospacedisplay.com

Aerosystems Consultants
Rowan House
White Hart Street
Foulden
Norfolk IP26 5AW
United Kingdom
Tel: (+44 1366) 32 88 48
Fax: (+44 1366) 32 88 49
e-mail: enquiries@aerosystems.uk.com

Aerosystems International Ltd
Alvington
Yeovil
Somerset BA22 8UZ
United Kingdom
Tel: (+44 1935) 44 30 00
Fax: (+44 1935) 44 31 11
e-mail: enquiries@aeroint.com
Web: www.aeroint.com

AFMC (US Air Force Materiel Command)
Aeronautical Systems Center
Wright-Patterson Air Force Base
Ohio
United States
Tel: (+1 937) 255 27 25
Fax: (+1 937) 656 40 22
e-mail: ascpa@wpafb.af.mil
Web: www.wpafb.af.mil/ascpa

AgustaWestland UK
Yeovil Plant
Lysander Road
Yeovil
Somerset BA20 2YB
United Kingdom
Tel: (+44 1935) 47 52 22
Fax: (+44 1935) 70 21 61
e-mail: info@whl.co.uk
Web: www.agustawestland.com

AIM (Aerosonic Corporation)
1212 North Hercules Avenue
Clearwater
Florida 33765
United States
Tel: (+1 727) 461 30 00
Fax: (+1 727) 447 59 26
e-mail: mindy@aerosonic.com
Web: www.aerosonic.com

Airbus Military SL
15 Avenue Didier Daurat
F-31707 Blagnac Cedex
France
Tel: (+33 5) 61 93 30 51
Fax: (+33 5) 61 93 34 70
Web: www.airbusmilitary.com

Alkan SA
Rue du 8 Mai 1945
BP 23
F-94460 Valenton
France
Tel: (+33 1) 45 10 86 39
Telex: f 266126
e-mail: info@alkan.mgn.fr
Web: www.alkan.fr

All-Russia JSC Nizkegorodshaya Yamaska
PO Box 648
420032
Kazan Republic of Tartastan
Russian Federation
Tel: (+7 8432) 55 71 63
Fax: (+7 8432) 55 34 85

Ametek Aerospace
50 Fordham Road
Wilmington
Massachusetts 01887
United States
Tel: (+1 978) 988 41 01
Fax: (+1 978) 988 49 44

Ametek Aerospace
Sellersville Division
900 Clymer Avenue
Sellersville
Pennsylvania 18960
United States
Tel: (+1 215) 257 65 31
Fax: (+1 215) 257 18 88
Web: www.ametekaerospace.com

Ampex Data Systems Corporation
Global Headquarters
1228 Douglas
Redwood City
California 94063-3199
United States
Tel: (+1 650) 367 41 11
Fax: (+1 650) 367 46 69
Telex: 348464
e-mail: sales@ampex.com
Web: www.ampex.com

Amphitech International Inc
3440 Francis-Hughes
Suite 120
Laval
Québec H7L 5A9
Canada
Tel: (+1 450) 663 45 54
Fax: (+1 450) 663 71 34
Web: www.amphitech.com

AMS (Aerospace Monitoring and
    Systems (Pty) Ltd)
PO Box 1980
1685 Halfway House
South Africa
Tel: (+27 11) 695 76 04
Fax: (+27 11) 315 16 45
e-mail: marketing@amshums.com
Web: www.amshums.com

Andrew Corporation
Richardson Office
3 Westbrook Corporate Center
Suite 900
Westchester
Illinois 60154
United States
Tel: (+1 972) 952 97 00
Fax: (+1 972) 952 00 00

Arc Industries Inc
PO Box 867
Destin
Florida 32541
United States
Tel: (+1 850) 77 74
Fax: (+1 850) 77 79

ArgonST
Imaging Group
300 Parkland Plaza
PO Box 1869
Ann Arbor
Michigan 48106
United States
Tel: (+1 734) 769 56 49
Fax: (+1 734) 769 04 29
Web: www.argonst.com

ARGOSystems Inc
324 N Mary Avenue
Sunnyvale
California 94088-3452
United States
Tel: (+1 408) 737 20 00
Fax: (+1 408) 524 20 23
Telex: 6717064
Web: www.argosystems.com

ARINC Inc
2551 Riva Road
Annapolis
Maryland 21401-7465
United States
Tel: (+1 410) 266 40 71
Fax: (+1 410) 573 22 00
e-mail: info@arinc.com
Web: www.arinc.com

Array Systems Computing Inc
1120 Finch Avenue West
7th Floor
Toronto
Ontario M3J 3H7
Canada
Tel: (+1 416) 736 09 00
Fax: (+1 416) 736 47 15
Web: www.array.ca

Arsenal CDB (Arsenal Central Design Bureau)
8 Moskovska Street
01010 Kyiv
Ukraine
Tel: (+380 44) 254 59 02
Fax: (+380 44) 25 93 15 09
e-mail: info@arsenalcdb.com.ua

Artex Aircraft Supplies Inc
14405 Keil Road Northeast
Aurora
Oregon 97002
United States
Tel: (+1 503) 678 79 29
Fax: (+1 503) 678 79 30
Telex: 4942194
e-mail: sales@artex.net
Web: www.artex.net

Aselsan Inc
Microelectronics, Guidance and Electro-Optics
P.K. 30
TR-06011 Etlik Ankara
Turkey
Tel: (+90 312) 847 53 00
Fax: (+90 312) 847 53 20
e-mail: marketing@mgeo.aselsan.com.tr
Web: www.aselsan.com.tr

ASTA Components
226 Lorimer Street
Port Melbourne
Victoria 3207
Australia
Tel: (+61 3) 96 47 31 11
Fax: (+61 3) 96 46 22 53

Astro (Astronautics CA Ltd)
16 Martin Gehl Street
PO Box 3351
IL-49130 Petah Tikvah
Israel
Tel: (+972 3) 925 15 55
Fax: (+972 3) 925 15 50
Web: www.astronautics.co.il

Astronautics Corporation of America
4115 North Teutonia Avenue
PO Box 523
Milwaukee
Wisconsin 53201-0523
United States
Tel: (+1 414) 449 40 00
Fax: (+1 414) 447 82 31
Web: www.astronautics.com

ATE (Advanced Technologies and
    Engineering Co (Pty) Ltd)
PO Box 632
1685 Halfway House
South Africa
Tel: (+27 11) 266 76 00
Fax: (+27 11) 314 53 79
e-mail: ategen@iafrica.com
Web: www.ate-aerospace.com

ATK Missile Systems Company
Integrated Systems Division
13133 34th Street North
PO Box 4648
Clearwater
Florida 33758
United States
Tel: (+1 727) 572 19 00
Fax: (+1 813) 572 21 80
Web: www.atk.com

Avalon Electronics Ltd
Langhorne Park House
High Street
Shepton Mallet BA4 5AQ
United Kingdom
Tel: (+44 1749) 34 52 66
Fax: (+44 1749) 34 52 67
e-mail: info@avalon-electronics.com
Web: www.avalon-electronics.com

Avia Avtomatika
Design Bureau of the Pribor Joint Stock Company
47 Zapolnaya Street
305000 Kursk
Russian Federation
Tel: (+7 122) 264 68
Fax: (+7 122) 244 15

Aviapribor Corp
5 Aviatzionny Pereulok
125319 Moskva
Russian Federation
Tel: (+7 095) 152 48 74
Fax: (+7 095) 152 26 31
Web: www.aviapribor.ru

Avidyne Corporation
55 Old Bedford Road
Lincoln
Massachusetts 01773
United States
Tel: (+1 781) 402 75 85
Fax: (+1 781) 402 75 99
e-mail: info@avidyne.com
Web: www.avidyne.com

Avionics Specialties Inc
Route 743
PO Box 6400
Charlottesville
Virginia 22906-6400
United States
Tel: (+1 434) 973 33 11
Fax: (+1 434) 973 29 76
Telex: cvl 5105875469 tdyavio
e-mail: marketing@avionics-specialties.com
Web: www.avionics-specialties.com

Avtech Corporation
3400 Wallingford Avenue North
Seattle
Washington 98103
United States
Tel: (+1 206) 695 80 00
Fax: (+1 206) 695 80 11
Web: www.avtcorp.com

Aydin Displays, Inc
1 Riga Lane
Birdsboro
Pennsylvania 19508
United States
Tel: (+1 610) 404 53 53
Fax: (+1 610) 404 81 90
e-mail: sales@aydindisplays.com
Web: www.aydindisplays.com

B

BAE Systems Australia
North Ryde Office
40-52 Talavera Road
North Ryde
New South Wales 2113
Australia
Tel: (+61 2) 98 55 80 00
Fax: (+61 2) 98 55 88 84

BAE Systems Australia
Sydney Office
Level 3, Westfield Towers
100 William Street
Sydney
New South Wales 2011
Australia
Tel: (+61 2) 93 58 29 00
Fax: (+61 2) 93 58 48 16

BAE Systems Electronic & Integrated Solutions
Communication Nav Identification & Reconnaisance
Greenlawn
New York 11740-1600
United States
Tel: (+1 516) 262 70 00
Fax: (+1 516) 262 80 02

BAE Systems Inc
Headquarters
1601 Research Boulevard
Rockville
Maryland 20850-3173
United States
Tel: (+1 301) 838 60 00
Fax: (+1 301) 838 69 25
Web: www.na.baesystems.com

BAE Systems Rokar International Ltd
PO Box 45049
IL-91450 Jerusalem
Israel
Tel: (+972 2) 532 98 88
Fax: (+972 2) 582 25 22
Web: www.rokar.com

BAE Systems, North America Flight Systems Inc.
16880 Flight Systems Drive
Mojave
California 93501-1666
United States
Tel: (+1 661) 824 46 01
Fax: (+1 661) 824 09 43
e-mail: fs.marketing@baesystems.com
Web: www.na.baesystems.com

BAE Systems, Regional Aircraft
Woodford Aerodrome, Chester Road
Woodford
Cheshire SK7 1QR
United Kingdom
Tel: (+44 161) 957 46 30
Fax: (+44 161) 955 30 08
e-mail: raebusiness@baesystems.com
Web: www.regional-services.com

BAE Systems
Corporate Headquarters
Warwick House
Farnborough Aerospace Centre
PO Box 87
Farnborough
Hampshire GU14 6YU
United Kingdom
Tel: (+44 1252) 37 32 32
Fax: (+44 1252) 38 30 00
Web: www.baesystems.com

Ball Aerospace & Technologies Corp
10 Long Peaks Drive
Broomfield
Colorado 80021
United States
Tel: (+1 303) 939 40 00
e-mail: jjmeyer@ball.com

Barco
President Kennedypark 35
B-8500 Kortrijk
Belgium
Tel: (+32 56) 23 35 79
Fax: (+32 56) 23 30 13
e-mail: sales.aerospace@barco.com
Web: www.barcodefense.com

Becker Flugfunkwerk GmbH
Baden Airpark B 108
D-77836 Rheinmünster
Germany
Tel: (+49 7229) 30 50
Fax: (+49 7229) 30 52 17
e-mail: info@becker-avionics.de
Web: www.becker-avionics.com

BKT (Bavaria Keytronic Technologie GmbH)
Boschstrasse 23
PO Box 500272
D-22761 Hamburg
Germany
Tel: (+49 40) 89 69 90
Fax: (+49 40) 890 30 14

C

CAE Inc
Headquarters
8585 Côte de Liesse
PO Box 1800
Saint-Laurent
Québec H4T 1G6
Canada
Tel: (+1 514) 341 67 80
Fax: (+1 514) 734 56 17
e-mail: marine@cae.com
Web: www.cae.com

CAIC (Chengdu Aero-Instrument Corp)
PO Box 229
610091 Chengdu
Sichaun
China
Tel: (+86 28) 87 40 90 18
Fax: (+86 28) 87 40 01 42
e-mail: trade@caia-china.com
Web: www.caic-china.com

Carl Zeiss Optronics GmbH
Carl Zeiss Strasse 22
PO Box 1129
D-73446 Oberkochen
Germany
Tel: (+49 7364) 20 65 30
Fax: (+49 7364) 20 36 97
e-mail: optronics@zeiss.de
Web: www.zeiss.de/optronics

CAS (Caledonian Airborne Systems Ltd)
Caledonian House
6 Ninian Road
Aberdeen Airport
Aberdeen AB2 0PD
United Kingdom
Tel: (+44 1224) 72 22 74
Fax: (+44 1224) 72 28 96

CEIEC (China National Electronics Import &
   Export Corp)
Electronics Building A23
Fuxing Road
PO Box 140
100036 Beijing
China
Tel: (+86 10) 68 29 65 10
Fax: (+86 10) 68 21 23 61
e-mail: ceiec@ceiec.com.cn
Web: www.ceiec.com.cn

Century Flight Systems Inc
Municipal Airport
PO Box 610
Mineral Wells
Texas 76068
United States
Tel: (+1 940) 325 25 17
Fax: (+1 940) 325 25 46
Web: www.centuryflight.com

China Leihua Electronic Technology
   Research Institute
108 Liangxi Road
PO Box 607
214063 Wuxi
Jiangsu
China
Tel: (+86 510) 551 16 07
Fax: (+86 510) 551 13 87

China National Aero-Technology Import &
   Export Corp
5 Liang Gua Chang Road
Dongcheng Qu
East City District
PO Box 647
100010 Beijing
China
Tel: (+86 1) 44 58 31

CILAS (Compagnie Industrielle des Lasers)
8 Avenue Buffon
ZI La Source
Box Postale 6319
F-45063
Orléans Cedex
France
Tel: (+33 2) 38 64 15 55
Fax: (+33 2) 38 64 40 72
Web: www.cilas.com

CMC (CMC Electronics Inc)
600 Dr Frederik Philips Boulevard
Ville St-Laurent
Québec H4M 2S9
Canada
Tel: (+1 514) 748 31 48
Fax: (+1 514) 748 31 00
Web: www.cmcelectronics.ca

CMC (Ultra Electronics Inc)
600 Dr Frederik Philips Boulevard
Québec H4M 2S9
Canada
Tel: (+1 514) 748 31 48
Fax: (+1 514) 748 31 00
Web: www.cmcelectronics.ca

Controp Precision Technologies Ltd
PO Box 611
IL-45105 Hod Hasharon
Israel
Tel: (+972 9) 760 49 71
Fax: (+972 9) 744 16 86
e-mail: cntrpnir@netvision.net.il
Web: www.controp.com

Cubic Corporation
P.O. Box 85587
San Diego
California 92186-5587
United States
Tel: (+1 858) 277 67 80
Fax: (+1 858) 277 18 78

Cubic Defense Applications
Headquarters
PO Box 85587
San Diego
California 92186-5587
United States
Tel: (+1 858) 277 67 80
Fax: (+1 858) 505 15 80
e-mail: service@cubic.com
Web: www.cubic.com

D

David Clark Company Incorporated
360 Franklin Street
Worcester
Massachusetts 01604
United States
Tel: (+1 508) 751 58 00
Fax: (+1 508) 753 58 27
Web: www.davidclark.com

Denel Optronics
Centurion
PO Box 8859
0046 Centurion
South Africa
Tel: (+27 12) 674 01 46
Fax: (+27 12) 674 01 99
e-mail: marketing@deneloptronics.co.za

DI (Dynamic Instruments Inc)
3860 Calle Fortunada
San Diego
California 92123
United States
Tel: (+1 858) 278 49 00
Fax: (+1 858) 278 97 00
e-mail: pwhitte@dynamicinst.com

Diamond J Inc
PO Box 9526
Wichita
Kansas 67277
United States
Tel: (+1 316) 264 06 00
Fax: (+1 316) 264 06 00
e-mail: info@diamondjinc.com
Web: www.diamondjinc.net

Diehl BGT Defence GmbH & Co
PO Box 10 11 55
D-88641 Ueberlingen
Germany
Tel: (+49 7551) 89 01
Fax: (+49 7551) 89 28 22
Web: www.bgt.de

DLE (Diehl Luftfahrt Elektronik GmbH)
Donaustrasse 120
D-90451 Nürnberg
Germany
Tel: (+49 911) 949 44 10
Fax: (+49 911) 949 44 09
e-mail: marketing@dle.diehl.com
Web: www.diehl-dle.com

DNE Technologies
50 Barnes Park North
30 Wallingford
Connecticut 06492-0030
United States
Tel: (+1 203) 265 71 51
Fax: (+1 203) 265 91 01
e-mail: info@ultra-dne.com
Web: www.ultra-dne.com

DRS Communications Company, LLC
1200 East Mermaid Lane
Wyndmoor
Pennsylvania 19038-7695
United States
Tel: (+1 215) 242 72 53
Fax: (+1 215) 233 99 47
e-mail: info@drs-c3.com

DRS Signal Solutions (West)
17680 Butterfield Boulevard
Morgan Hill
California 95037
United States
Tel: (+1 408) 852 08 00
Fax: (+1 408) 852 08 01
e-mail: info@drs-ssw.com

DRS Systems & Electronics
201 Evans Lane
St Louis
Missouri 63121-1126
United States
Tel: (+1 314) 553 49 60
Fax: (+1 315) 553 49 49
Web: www.seistl.com

DRS Technologies, Inc
Corporate Headquarters
5 Sylvan Way
Parsippany
New Jersey 07054
United States
Tel: (+1 973) 898 15 00
Fax: (+1 973) 898 47 30
e-mail: info@drs.com
Web: www.drs.com

DTS (Desarrollo de Tecnologias y Sistemas Ltda)
Rosas 1444
8340259 Santiago
Chile
Tel: (+56 2) 397 10 00
Fax: (+56 2) 397 10 01
e-mail: info@dts.cl
Web: www.dts.cl

E

E&IS (BAE Systems Electronic & Integrated
  Solutions)
305 Richardson Road
Lansdale
Pennsylvania 19446-1485
United States
Tel: (+1 603) 885 28 17
Fax: (+1 603) 886 28 13
Web: www.ae.na.baesystems.com

E&IS (BAE Systems Electronic & Integrated
  Solutions)
65 Spit Brook Road
PO Box 868
Nashua
New Hampshire 03061-0868
United States
Tel: (+1 603) 885 43 21
Fax: (+1 603) 885 28 13
e-mail: stephanie.diehl@baesystems.com
Web: www.eis.na.baesystems.com

E&IS (BAE Systems Electronic & Integrated
  Solutions)
Communications, Navigation & Identification
164 Totowa Road
Wayne
New Jersey 07474-0975
United States
Tel: (+1 973) 633 61 40

E&IS (BAE Systems Electronic & Integrated
  Solutions)
Infrared Imaging Systems
2 Forbes Road
Lexington
Massachusetts 02421-7306
United States
Tel: (+1 781) 863 48 11
Fax: (+1 781) 863 41 93
Web: www.eis.na.baesystems.com

E&IS (BAE Systems Electronic & Integrated
  Solutions)
Integrated Defense Solutions
6500 Tracor Lane
Austin
Texas 78725-2070
United States
Tel: (+1 512) 929 28 24

E&IS (BAE Systems Electronic & Integrated
  Solutions)
Platform Solutions Sector
600 Main Street
Johnson City
New York 13790
United States
Tel: (+1 607) 770 20 00
Fax: (+1 607) 770 35 24
Web: www.ps.na.baesystems.com

EADS CASA
Headquarters
Avenida de Aragon 404
E-28022 Madrid
Spain
Tel: (+34 915) 85 70 00
Fax: (+34 915) 85 76 66
e-mail: sales@casa.eads.net
Web: www.casa.eads.net

EADS Deutschland GmbH
Military Aircraft
D-81663
München 80
Germany
Tel: (+49 89) 60 70
Fax: (+49 89) 60 72 64 81
e-mail: pressmilitaryaircraft@eads.com
Web: www.eads.net

EADS France
Headquarters
37 Boulevard de Montmorency
F-75781
Paris Cedex 16
France
Tel: (+33 1) 42 24 24 24
Fax: (+33 1) 42 24 21 32
Web: www.eads.net

Edgewater Computer Systems, Inc
1125 Innovation Drive
Ottawa
Ontario K2K 3G6
Canada
Tel: (+1 613) 271 11 01
Fax: (+1 613) 271 11 52
e-mail: info@edgewater.ca
Web: www.edgewater.ca

EDO Corporation
Defense Programs & Technologies
455 Commack Road
Deer Park Long Island
New York 11729-4591
United States
Tel: (+1 631) 595 38 07
Fax: (+1 631) 595 30 68

EDO Reconnaissance & Surveillance Systems
18705 Madrone Parkway
Morgan Hill
California 95037
United States
Tel: (+1 408) 201 80 00
Fax: (+1 408) 201 80 10
e-mail: busdev@condorsys.com
Web: www.condorsys.com

Eidel (Eidsvoll Electronics A/S)
Nedre Vilberg vei 8
N-2080 Eidsvoll
Norway
Tel: (+47) 63 95 97 00
Fax: (+47) 63 95 97 10
e-mail: eidel@eidel.no
Web: www.eidel.no

Elan GmbH
Fritz-Ullmann Straße 2
D-55252 Mainz-Kastel
Germany
Tel: (+49 6134) 719 50
Fax: (+49 6134) 17 92

Elara Cheboksary Apparatus and Instrument
Plant Joint Stock Company
40 Moskovsky Prospect
428015 Cheboksary Chuvash Republic
Russian Federation
Tel: (+7 8352) 49 95 32
Fax: (+7 8352) 45 20 83
e-mail: elara@elara.ru
Web: www.elara.ru

Elbit Systems Electro-Optics Elop Ltd
Advanced Technology Park
Kiryat Weizmann
PO Box 1165
IL-76111 Rehovot
Israel
Tel: (+972 8) 938 66 06
Fax: (+972 8) 938 60 10
e-mail: webmaster@el-op.com
Web: www.el-op.com

Electroautomatika Experimental Design Bureau
Ulitsa Marshala Govorova 40
198095 St Petersburg
Russian Federation
Tel: (+7 812) 252 13 98
Fax: (+7 812) 252 38 17

Electropribor Voronezhski Plant
20-letiya Oktyabrya, 59
394006 Voronezh
Russian Federation
Tel: (+7 732) 57 55 25

Elettronica GmbH
Industriepark Kottenforst
Am Hambuch 10
D-53340 Meckenheim
Germany
Tel: (+49 2225) 88 06 24
Fax: (+49 2225) 88 06 47

Elisra
Corporate Headquarters
48 Mitzva Kadesh Street
IL-51203 Bene Beraq
Israel
Tel: (+972 3) 617 55 22
Fax: (+972 3) 617 58 50
e-mail: marketing@elisra.com
Web: www.elisra.com

ELTA
Head Office
14, place Marcel Dassault
F-31702 Blagnac Cedex
France
Tel: (+33 5) 34 36 10 00
Fax: (+33 5) 34 36 10 01
Web: www.elta.fr

Endevco Corporation
30700 Rancho Viejo Road
San Juan Capistrano
California 92675
United States
Tel: (+1 949) 493 81 81
Fax: (+1 949) 661 72 31
Telex: 685608
e-mail: applications@endevco.com
Web: www.endevco.com

Enertec SA
185 Avenue du Général de Gaulle
BP 316
F-92143 Clamart Cedex
France
Tel: (+33 1) 41 28 87 87
Fax: (+33 1) 41 28 87 00
e-mail: sales@enertecgroup.com
Web: www.enertecgroup.com

Enterprise Control Systems Ltd
31 The High Street
Wappenham
Northamptonshire NN12 8SN
United Kingdom
Tel: (+44 1327) 86 00 50
Fax: (+44 1327) 86 00 58
e-mail: sales@enterprisecontrol.co.uk
Web: www.enterprisecontrol.co.uk

ESW – Extel Systems Wedel GmbH
Industriestrasse 33
D-22876 Wedel
Germany
Tel: (+49 4103) 60 35 15
Fax: (+49 4103) 60 40 05
e-mail: info@esw-wedel.de
Web: www.esw-wedel.de

Eventide Inc
One Alsan Way
Little Ferry
New Jersey 07643-1001
United States
Tel: (+1 201) 641 12 00
Fax: (+1 201) 641 16 40
e-mail: audio@eventide.com
Web: www.eventide.com

F

Flight Components AG
Bitzibergstrasse 5
CH-8184 Bachenbülach
Switzerland
Tel: (+41 1) 861 12 00
Fax: (+41 1) 861 17 15
e-mail: sales@flightcomponents.com
Web: www.flightcomponents.com

FLIR Systems Inc
Worldwide Headquarters
16505 South West 72nd Avenue
Portland
Oregon 97224
United States
Tel: (+1 503) 684 37 31
Fax: (+1 503) 684 32 07
Web: www.flirthermography.com

FN Herstal SA
33 Voie de Liège
B-4040 Herstal
Belgium
Tel: (+32 4) 240 82 97
Fax: (+32 4) 240 88 99
e-mail: info@fnherstal.com
Web: www.fnmfg.com

France Aerospace SARL
Regina-2
Place de Pyramides
F-75001 Paris
France
Tel: (+33 1) 42 60 31 10
Fax: (+33 1) 40 15 95 16

FTL (Ferranti Technologies Ltd)
Cairo House
Waterhead
Oldham
Lancashire OL4 3JA
United Kingdom
Tel: (+44 161) 624 02 81
Fax: (+44 161) 624 52 44
Web: www.ferranti-technologies.co.uk

FUS (Goodrich Fuel & Utility Systems)
Group Headquarters
100 Panton Road
89305 Vergennes
Vermont 05491
United States
Tel: (+1 802) 877 40 00
Fax: (+1 802) 877 41 13
Web: www.fus.goodrich.com

G

Galileo Avionica SpA
Avionic Systems and Equipment Division
Via S Alessandro 10
I-00131 Roma
Italy
Tel: (+39 06) 41 88 31
Fax: (+39 06) 41 88 38 00

Galileo Avionica SpA
Radar Systems Business Unit
Via G.B. Grassi 93
I-20157 Milan
Italy

Garmin International Inc
1200 East 151 Street
Olathe
Kansas 66062
United States
Tel: (+1 913) 397 82 00
Fax: (+1 913) 397 82 82
Web: www.garmin.com

GEI (Gables Engineering Inc)
247 Greco Avenue
Coral Gables
Florida 33146
United States
Tel: (+1 305) 774 44 00
Fax: (+1 305) 774 44 65
e-mail: sales@gableseng.com
Web: www.gableseng.com

General Dynamics C4 Systems
Scottsdale
8201 East McDowell Road
Scottsdale
Arizona 85252
United States
Tel: (+1 480) 441 30 33
Fax: (+1 480) 726 29 71
e-mail: info@gdds.com
Web: www.gdc4s.com

General Dynamics Canada
Head Office
3785 Richmond Road
Ottawa
Ontario K2H 5B7
Canada
Tel: (+1 613) 896 77 86
Fax: (+1 613) 596 73 96
e-mail: info@gdcanada.com
Web: www.gdcanada.com

General Dynamics UK Limited
East Sussex Office
Castleham Road
St Leonards-on-Sea
East Sussex TN38 9NJ
United Kingdom
Tel: (+44 1424) 85 34 81
Fax: (+44 1424) 85 15 20
e-mail: sales@generaldynamics.uk.com
Web: www.generaldynamics.uk.com

General Dynamics United Kingdom Limited
Bryn Brithdir
Units 3 & 4
Oakdale Business Park
South Wales NP12 4AA
United Kingdom
Tel: (+44 1424) 79 83 92
Fax: (+44 1424) 79 80 09
e-mail: info@generaldynamics.uk.com
Web: www.generaldynamics.uk.com

General Dynamics United Kingdom Limited
London Office
11-12 Buckingham Gate
London SW1E 6LB
United Kingdom
Tel: (+44 20) 79 32 34 00
Fax: (+44 20) 79 32 34 01
Web: www.generaldynamics.uk.com

General Dynamics United Kingdom Limited
Tewkesbury Office
Building 61
DSDC Ashchurch
Tewkesbury GL20 8LZ
United Kingdom
Tel: (+44 1684) 85 54 00
Fax: (+44 1684) 85 54 50
Web: www.generaldynamics.uk.com

Geophizika-NV
Matrosskaia Tishina Street
House 23, Building 2
107016 Moskva
Russian Federation
Tel: (+7 095) 269 27 42
Fax: (+7 095) 268 01 42

GST (Grintek Ewation)
PO Box 912561
0127 Silverton
South Africa
Tel: (+27 12) 421 62 12
Fax: (+27 12) 421 62 16
e-mail: marketing@ewation.co.za

H

H R Smith Group
Street Court
Kingsland
Leominster
Hereford and Worcestershire HR6 9QA
United Kingdom
Tel: (+44 1568) 70 87 44
Fax: (+44 1568) 70 87 13
e-mail: sales@hr-smith.com
Web: www.hr-smith.com

HAL (Hindustan Aeronautics Ltd)
Corporate Headquarters
15/1 Cubbon Road
PO Box 5150
Bangalore 560 001
India
Tel: (+91 80) 22 86 69 07
Fax: (+91 80) 22 86 87 58
e-mail: hal_engine_mktg@vsnl.net
Web: www.hal-india.com

Hamilton Sundstrand
Headquarters
One Hamilton Road
Windsor Locks
Connecticut 06096-1010
United States
Tel: (+1 860) 654 60 00
e-mail: hs.general@hsd.utc.com
Web: www.hamiltonsundstrand.com

Harris Corporation
Government Communications Systems Division
PO Box 37
Melbourne
Florida 32902-0037
United States
Tel: (+1 321) 727 40 00
Web: www.harris.com

Heim Data Systems, Inc
PO Box N
Belmar
New Jersey 07719
United States
Tel: (+1 732) 556 23 18
Fax: (+1 732) 556 23 19
e-mail: marketing@heimdata.com
Web: www.heimdata.com

Heim Systems GmbH
Technologiepark Bergisch Gladbach
Friedrich-Ebert-Strasse
D-51429 Bergisch Gladbach
Germany
Tel: (+49 2204) 84 41 00
Fax: (+49 2204) 84 41 99
e-mail: info@heim.zodiac.com
Web: www.heim-systems.com

Honeywell Aerospace, Engines & Systems
Engine Systems and Accessories, South Bend
717 North Bendix Drive
South Bend
Indiana 46620-1000
United States
Tel: (+1 219) 231 20 00
Fax: (+1 219) 231 33 35

Honeywell AG
Kaiserleistrasse 39
D-63067 Offenbach am Main
Germany
Tel: (+49 69) 806 40
Fax: (+49 69) 81 86 20
Web: www.honeywell.de

Honeywell Bendix/King Avionics
One Technology Centre
23500 West 105th Street
M/D #19 Olathe
Kansas 66061-1950
United States
Tel: (+1 913) 712 31 45
Fax: (+1 913) 712 56 97
Web: www3.bendixking.com

Honeywell Inc
Defense and Space Electronic Systems
9201 San Mateo Boulevard North East
Albuquerque
New Mexico 87113-2227
United States
Tel: (+1 505) 828 52 45

Honeywell Inc
Sensor and Guidance Products Headquarters
2600 Ridgway Parkway
MN17-3725
Minneapolis
Minnesota 55413
United States
Tel: (+1 612) 951 10 00

Honeywell
Headquarters
3333 Unity Drive
Mississauga
Ontario L5L 3S6
Canada
Tel: (+1 905) 608 60 35
Fax: (+1 905) 608 60 61
Web: www.honeywellaerospace.com

I

IAI (Israel Aerospace Industries Ltd)
Tamam Division
Yehud Industrial Zone
PO Box 75
IL-56100 Yehud
Israel
Tel: (+972 3) 531 50 03
Fax: (+972 3) 531 51 40
Web: www.iai.co.il

IAI/ELTA (ELTA Systems Ltd)
PO Box 330
IL-77102 Ashdod
Israel
Tel: (+972 8) 857 25 43
Fax: (+972 8) 856 45 68
e-mail: breshef@elta.co.il

Indra
Aranjuez Division
Joaquin Rodrigo 11
E-28300 Aranjuez
Madrid
Spain
Tel: (+34 91) 894 88 00
Fax: (+34 91) 894 89 17

INS (Interscan Navigation Systems)
5418-11 Street North East
Calgary
Alberta T2E 7E9
Canada
Tel: (+1 403) 730 55 55
Fax: (+1 403) 730 55 11
Web: www.pelorus.com

Insight Technology, Inc
3 Technology Drive
Londonderry
New Hampshire 03053
United States
Tel: (+1 603) 626 48 00
Fax: (+1 603) 647 72 34
e-mail: info-mil@insightlights.com
Web: www.insightlights.com

INSYTE (BAE Systems, Integrated System
  Technologies)
Basildon
Christopher Martin Road
Basildon
Essex SS14 3EL
United Kingdom
Tel: (+44 1268) 52 28 22
Fax: (+44 1268) 88 31 40

INSYTE (BAE Systems, Integrated System
  Technologies)
Head Office
Eastwood House
Glebe Road
Chelmsford
Essex CM1 1QW
United Kingdom
Tel: (+44 1245) 70 27 02
Fax: (+44 1245) 70 27 00
e-mail: insyte@baesystems.com
Web: www.amsjv.com

Intergraph Deutschland GmbH
Ulmer Strasse 124
D-73431 Aalen
Germany
Tel: (+49 7361) 889 50
Fax: (+49 7364) 88 95 29
e-mail: info-germany@intergraph.com
Web: www.intergraph.de

International Aerospace Inc
234 Garibaldi Avenue
PO Box 340
Lodi
New Jersey 07644
United States
Tel: (+1 973) 473 00 34
Fax: (+1 973) 473 87 48

IRI (Ideal Research Inc)
1810 Parklawn Drive
Rockville
Maryland 20852
United States
Tel: (+1 301) 984 56 94
Telex: 904059 wsh

Iridium Satellite LLC
Corporate Headquarters
6701 Democracy Boulevard
Suite 500
Bethesda
Maryland 20817
United States
Tel: (+1 301) 571 62 00
Fax: (+1 301) 571 62 50
Web: www.iridium.com

IS&S (Innovative Solutions and Support Inc)
720 Pennsylvania Drive
Exton
Pennsylvania 19341
United States
Tel: (+1 610) 646 98 00
Fax: (+1 610) 646 01 46
Web: www.innovative-ss.com

Italtel SpA
Headquarters
Via A di Tocqueville 13
I-20154 Milan
Italy
Tel: (+39 02) 438 81
Fax: (+39 02) 43 88 52 20
Telex: i 314840 sitele
Web: www.italtel.it

ITT Corp A/CD
Aerospace/Communications
756 Lakefield Road
Building J Westlake Village
Westlake Village
California 91361-2624
United States
Tel: (+1 260) 451 60 42
Fax: (+1 260) 451 61 26
Web: www.acd.itt.com

ITT Corp Night Vision
7635 Plantation Road
Roanoke
Virginia 24019
United States
Tel: (+1 540) 362 80 00
Fax: (+1 540) 362 45 74
e-mail: nvsales@itt.com
Web: www.nightvision.com

ITT Corp
Avionics Division
100 Kingsland Road
Clifton
New Jersey 07014
United States
Tel: (+1 973) 284 38 30
Fax: (+1 973) 284 28 32
Web: www.acd.itt.com

J

JAE (Japan Aviationt Electronics Industry Ltd)
1-21-2 Dogenzaka
Shibuya-ku
Tokyo 150-0043
Japan
Tel: (+81 3) 37 80 27 52
Fax: (+81 3) 37 80 29 45
Web: www.jae.co.jp

JDA (Japan Defense Agency)
5-1 Ichigaya
Honmura-cho
Shinjuku-ku
Tokyo 162-8801
Japan
Tel: (+81 3) 32 68 31 11
Fax: (+81 3) 34 08 64 80
e-mail: info@jda.go.jp
Web: www.jda.go.jp

JP Instruments
PO Box 7033
Huntington Beach
California 92615-7033
United States
Tel: (+1 714) 557 38 05
Fax: (+1 714) 557 98 40
Web: www.jpinstruments.com

K

KAC (Kaman Aerospace Corporation)
Bloomfield Office
Old Windsor Road
PO Box 2
Bloomfield
Connecticut 06002-0002
United States
Tel: (+1 860) 243 73 36
Fax: (+1 860) 243 70 43
Telex: us 99326
e-mail: frenchm-kac@kacman.com
Web: www.kamanaero.com

KAC (Kaman Aerospace Corporation)
Raymond Engineering Operations
217 Smith Street
Middletown
Connecticut 06457-9990
United States
Tel: (+1 860) 632 45 84
Fax: (+1 860) 632 46 66
Web: www.raymond-engrg.com

Kazan Electropribor Plant
20 N Ershov Street
420045 Kazan
Russian Federation
Tel: (+7 8432) 76 40 01
Fax: (+7 095) 420 20 12

KDA (Kongsberg Defence & Aerospace AS)
Headquarters
PO Box 1003
N-3601 Kongsberg
Norway
Tel: (+47) 32 28 82 00
Fax: (+47) 32 28 79 50
e-mail: kda.office@kongsberg.com
Web: www.kongsberg.com

KGN (Kearfott Guidance & Navigation Corp)
1150 McBride Avenue
Little Falls
New Jersey 07424
United States
Tel: (+1 973) 785 61 83
Fax: (+1 973) 785 59 05
Telex: us 133440
e-mail: marketing@kearfott.com
Web: www.kearfott.com

Kollsman Inc
220 Daniel Webster Highway
Merrimack
New Hampshire 03054-4844
United States
Tel: (+1 603) 886 22 73
Fax: (+1 603) 886 20 06
Web: www.kollsman.com

L

L-3 Cincinnati Electronics
7500 Innovation Way
Mason
Ohio 45040-9699
United States
Tel: (+1 513) 573 61 00
Fax: (+1 513) 573 67 41
e-mail: sales@cmccinci.com
Web: www.l-3com.com/ce

L-3 Communications Systems – West
640 North 2200 West
PO Box 16850
Salt Lake City
Utah 84116-0850
United States
Tel: (+1 801) 594 20 00
Web: www.l-3com.com/csw

L-3 Communications, Electronic Systems
Toronto Facility
25 City View Drive
Toronto
Ontario M9W 5A7
Canada
Tel: (+1 416) 249 12 31
Fax: (+1 416) 246 20 01

L-3 Wescam Inc
649 North Service Road West
Burlington
Ontario L7P 5B9
Canada
Tel: (+1 905) 633 40 00
Fax: (+1 905) 633 41 00
e-mail: sales@wescam.com
Web: www.wescam.com

Leica Geosystems Geospatial Imaging
5051 Peachtree Corners Circle
Suite 100
Norcross
Georgia 30092
United States
Tel: (+1 770) 776 34 00
Fax: (+1 770) 776 35 00
Web: gi.leica-geosystems.com

Leninetz Holding Company
212 Moskovskiy Prospekt
196066 St Petersburg
Russian Federation
Tel: (+7 812) 378 32 19
Fax: (+7 812) 448 81 38

LITAL (Litton Italia SpA)
Via Pontina km 27,000
I-00040 Pomezia
Roma
Italy
Tel: (+39 06) 91 19 22 39
Fax: (+39 06) 91 19 22 85
e-mail: info@lital.it
Web: www.lital.it

Litef GmbH
Loerracherstrasse 18
D-79115 Freiburg im Breisgau
Germany
Tel: (+49 761) 490 10
Fax: (+49 761) 490 14 80
e-mail: info@litef.de
Web: www.litef.de

LLI (Liebherr-Aerospace Lindenberg GmbH)
Pfaenderstrasse 50-52
PO Box 1363
D-88153
Lindenberg Allgäu
Germany
Tel: (+49 8381) 460
Fax: (+49 8381) 46 43 77
e-mail: info@lli.liebherr.com
Web: www.liebherr.com

LM Aero (Lockheed Martin Aeronautics
    Company)
PO Box 748
Fort Worth
Texas 76101-0748
United States
Tel: (+1 817) 763 40 85
Fax: (+1 817) 763 47 97
Web: www.lockheedmartin.com

LMALC (Lockheed Martin Aircraft & Logistics
    Centers)
107 Frederick Street
Greenville
South Carolina 29607
United States
Tel: (+1 864) 422 66 26
Fax: (+1 864) 422 62 76
Web: lmalc.external.lmco.com

Lockheed Martin Canada Inc
Head Office
3001 Solandt Road
Kanata
Ontario K2K 2M8
Canada
Tel: (+1 613) 599 32 70
Fax: (+1 613) 599 32 82
Web: www.lockheedmartin.com/canada

Lockheed Martin Canada Inc
Montréal Office
6111 Royalmount Avenue
Montréal
Québec H4P 1K6
Canada
Tel: (+1 514) 340 83 10
Fax: (+1 514) 340 83 18

Lockheed Martin Corporation
Systems and Sensors – Owego
1801 State Route 17C
Owego
New York 13827-3998
United States
Tel: (+1 607) 751 20 00
Web: www.lockheedmartin.com

Lockheed Martin Integrated Systems and
    Solutions
700 North Frederick Avenue
Gaithersburg
Maryland 20879
United States
Tel: (+1 301) 240 79 00
Web: www.lockheedmartin.com/iss

Lockheed Martin Missiles and Fire Control
Dallas Office
PO Box 650003 M/S PT-42
Dallas
Texas 75265-0003
United States
Tel: (+1 972) 603 10 00
Fax: (+1 972) 603 10 09
Web: www.missilesandfirecontrol.com

Lockheed Martin Missiles and Fire Control
Orlando
5600 Sand Lake Road
PO Box 5837
Orlando
Florida 32819-8907
United States
Tel: (+1 407) 356 95 24
Fax: (+1 407) 356 03 08

Lockheed Martin UK
6801 Manning House
22 Carlisle Place
London SW1P 1JA
United Kingdom
Tel: (+44 20) 77 98 28 88
Fax: (+44 20) 77 98 28 51
e-mail: chris.trippick@imco.com
Web: www.lockheedmartin.co.uk

Logic SpA
Via Galileo N.5
I-20060 Cassina dé Pecchi
Milano
Italy
Tel: (+39 02) 92 44 01
Fax: (+39 02) 92 10 25 28
e-mail: info@logic-spa.com
Web: www.logic-spa.com

M

M&DS (Lockheed Martin Management &
    Data Systems)
PO Box 8048
Philadelphia
Pennsylvania 19101
United States
Tel: (+1 610) 531 74 00
Fax: (+1 610) 962 20 96
Web: mds.external.lmco.com

M&DS (Lockheed Martin Management &
    Data Systems)
ISR Systems
PO Box 85
Litchfield Park
Arizona 85340
United States
Tel: (+1 623) 925 64 46
Fax: (+1 623) 925 71 59

M/A-Com
1011 Pawtucket Boulevard
PO Box 3295
Lowell
Massachusetts 01853-3295
United States
Tel: (+1 978) 442 50 00
Fax: (+1 978) 442 53 50
Web: www.macom.com

Manufacturing Stock Company AVECS
17 I-ya UI
Yamskova Polya
Moskva
Russian Federation
Tel: (+7 812) 257 08 46
Fax: (+7 812) 257 77 32

Martec Serpe-Iesm
Zone Industriel des cinq Chemins
F-56520 Guidel
France
Tel: (+33 2) 97 02 49 49
Fax: (+33 2) 97 65 01 91
Telex: 950535
e-mail: contact@saraviation@martec.fr
Web: www.martec.fr

MBDA
20-22 rue Grange Dame Rose
PO Box 150
F-78141
Vélizy-Villacoublay Cedex
France
Tel: (+33 1) 34 88 14 46
Fax: (+33 1) 24 88 22 88
Telex: f 69 52 38
Web: www.mbda.net

MBDA
Via Tuburtina Km. 12,400
I-00131 Roma
Italy
Tel: (+39 6) 419 71
Fax: (+39 6) 413 10 91
Web:www.baesystems.com/facts/awards/
    locations8.htm#italy

MBDA
Headquarters
11 Strand
London
WC2N 5RJ
United Kingdom
Tel: (+44 20) 74 51 60 59
Fax: (+44 20) 74 51 60 89
Web: www.mbda.net

MBDA
Stevenage Office
Six Hills Way
Stevenage
Hertfordshire SG1 2DA
United Kingdom
Tel: (+44 1438) 75 28 11
Fax: (+44 1438) 75 46 88

MDA (MacDonald, Dettwiler & Associates Ltd)
Head Office
13800 Commerce Parkway
Richmond
British Columbia V6V 2J3
Canada
Tel: (+1 604) 231 27 43
Fax: (+1 604) 278 29 36
e-mail: info@mdacorporation.com
Web: www.mdacorporation.com

Meggitt Avionics
7 Whittle Avenue
Segensworth West
Fareham
Hampshire
PO15 5SH
United Kingdom
Tel: (+44 1489) 48 33 00
Fax: (+44 1489) 48 33 40
Web: www.meggitt.com

Merlin Engineering Works, Inc
1888 Embarcadero Road
Palo Alto
California 94303-3308
United States
Tel: (+1 650) 856 09 00
Fax: (+1 650) 858 23 02
e-mail: sales@merlineng.com
Web: www.merlineng.com

Mesit Prístroje spol sro
Sokolovská 573
CZ-686 01 Uherské Hradiste
Czech Republic
Tel: (+420 572) 80 12 00
Fax: (+420 572) 80 16 02
e-mail: mo@msp.mesit.cz
Web: www.msp.mesit.cz

MID SpA Marconi Italian Defence
Via Le Castello della Magliana 75
I-00148 Roma
Italy
Tel: (+39 6) 65 96 04 51
Fax: (+39 6) 65 96 04 65

Mid-Continent Instruments
9400 East 34th Street North
Wichita
Kansas 67226
United States
Tel: (+1 316) 630 01 01
Fax: (+1 316) 630 07 23
e-mail: info@mcico.com
Web: www.mcico.com

Mikes Microwave Electronic Systems Inc
Cankiri Yolu 5km
TR-06750 Akyurt
Ankara
Turkey
Tel: (+90 312) 847 51 00
Fax: (+90 312) 847 51 14
e-mail: ayca.aktas@mikes.com.tr

Mikrotechna Praha as
Barrandova 409
CZ-143 00
Praha 4
Czech Republic
Tel: (+420 2) 25 27 35 12
Fax: (+420 2) 25 27 35 14
e-mail: pacakova@mikrotechna.cz
Web: www.mikrotechna.cz

Mitsubishi Corporation
Aerospace Division
2-6-3 Marunouchi
Chiyoda-ku
Tokyo 100-8086
Japan
Tel: (+81 3) 32 10 21 21
Fax: (+81 3) 32 10 47 83
Web: www.mitsubishi.co.jp

Monit'air
ZAC des Pres Rouz
81 rue Alain-Fournier
F-38920 Crolles
France
Tel: (+33 4) 76 08 14 39
Fax: (+33 4) 76 08 89 04

Moscow Scientific Research Institute of Instrument Engineering
34 Kutuzovsky Prospect
121170 Moskva
Russian Federation
Tel: (+7 95) 247 07 04
Fax: (+7 95) 148 79 96
Telex: su 412268 vega

Motorola Inc
Missile and Fuze Systems Office
8220 East Roosevelt
PO Box 9040
Scottsdale
Arizona 85252
United States
Tel: (+1 602) 441 29 58

MRCM GmbH
Woerthstrasse 85
D-89077 Ulm
Germany
Tel: (+49 731) 392 52 33
Fax: (+49 731) 392 20 49 24
e-mail: ewation.marketing@sysde.eads.net
Web: www.mrcm.net

MRPC Avionica (Moscow Research and Production Complex Avionica)
Ulitsa Obraztsova 13
127055 Moskva
Russian Federation
Tel: (+7 095) 684 27 55
Fax: (+7 095) 681 38 46
e-mail: avionika@orc.ru
Web: www.mnpk.ru

MS Instruments plc
Bridge House
Rye Lane
Dunton Green
Sevenoaks
Kent TN14 5JD
United Kingdom
Tel: (+44 1732) 74 95 80
Fax: (+44 1732) 46 39 37
Web: www.msinstruments.co.uk

MS2 (Lockheed Martin Maritime Systems and Sensors)
Moorestown
199 Borton Landing Road
PO Box 1027
Moorestown
New Jersey 08057-0927
United States
Tel: (+1 856) 722 50 00
Fax: (+1 856) 231 93 63
Web: www.lockheedmartin.com/ms2

N

NAAP (National Association of Avionics Producers)
16 Bolshaya Monetnaya Street
197101 St Petersburg
Russian Federation
Tel: (+7 812) 233 74 59
Fax: (+7 812) 233 83 06

NAT (Northern Airborne Technology Ltd)
1925 Kirschner Road
Kelowna
British Columbia V1Y 4N7
Canada
Tel: (+1 250) 763 22 32
Fax: (+1 250) 762 33 74
e-mail: sales@natech.com
Web: www.northernairborne.com

Navcom Defense Electronics Inc
4323 Arden Drive
El Monte
California 91731-1997
United States
Tel: (+1 626) 579 86 89
Fax: (+1 626) 444 76 19
e-mail: nde@navcom.com
Web: www.navcom.com

NEC Corporation
Headquarters
7-1 Shiba 5-chome
Minato-ku
Tokyo 108-8001
Japan
Tel: (+81 3) 34 54 11 11
Fax: (+81 3) 37 98 15 12
e-mail: webmaster@nec.co.jp
Web: www.nec.co.jp

NIIAO Institute of Aircraft Equipment
18 Tupolev Street
Zhukovsky 140182
Moskva
Russian Federation
Tel: (+7 095) 556 23 22
Fax: (+7 095) 556 54 42
e-mail: info@niiao.ru
Web: www.niiao.ru

NIIP (V Tikhomirov Scientific Research Institute of Instrument Design)
3 Gagarin Street
140180 Zhukovsky Moscow Region
Russian Federation
Tel: (+7 095) 721 37 85
Fax: (+7 095) 556 88 87
e-mail: niip@niip.ru
Web: www.niip.info

Nord-Micro AG & Co OHG
Victor-Slotosch-Strasse 20
D-60388 Frankfurt am Main
Germany
Tel: (+49 6109) 30 30
Fax: (+49 6109) 30 32 33
e-mail: mail@nord-micro.de
Web: www.nord-micro.de

Northrop Grumman Corporation
Corporate Headquarters
1840 Century Park East
Los Angeles
California 90067-2199
United States
Tel: (+1 310) 553 62 62
Web: www.northropgrumman.com

Northrop Grumman Corporation
Electronic Systems
Mail Stop A255
PO Box 17319
Baltimore
Maryland 21090
United States
Tel: (+1 410) 993 68 48
e-mail: d.mccallam@ngc.com
Web: www.es.northropgrumman.com

Northrop Grumman Corporation
Government Relations
1000 Wilson Boulevard
Suite 2300
Arlington
Virginia 22209-2278
United States
Tel: (+1 703) 875 84 00
Fax: (+1 703) 875 85 21

Northrop Grumman Corporation
Integrated Systems Sector
One Northrop Grumman Avenue
El Segundo
California 90245
United States
Tel: (+1 310) 332 10 00
Fax: (+1 469) 420 81 80
e-mail: calawge@mail.northgrum.com
Web: www.northgrumman.com

Northrop Grumman Electronic Systems
Navigation Systems Division
21240 Burbank Boulevard
Woodland Hills
California 91367-6675
United States
Tel: (+1 818) 715 24 70
e-mail: customerservice.nsd@northropgrumman.
   com
Web: www.nd.es.northropgrumman.com

O

Octec Ltd
The Western Centre
Western Road
Bracknell
Berkshire RG12 1RW
United Kingdom
Tel: (+44 1344) 46 52 00
Fax: (+44 1344) 46 52 01
e-mail: sales@octec.co.uk
Web: www.octec.com

OIP Sensor Systems
Westerring 21
B-9700 Oudenaarde
Belgium
Tel: (+32 55) 33 38 11
Fax: (+32 55) 33 38 02
e-mail: sales@oip.be
Web: www.oip.be

OIS (Optical Imaging Systems Inc)
47050 Five Mile Road
Northville
Michigan 48167
United States
Tel: (+1 734) 454 55 60
Fax: (+1 734) 207 13 50

Opgal Optronic Industry Ltd
Industrial Zone
PO Box 462
IL-20100
Karmiel
Israel
Tel: (+972 4) 995 39 07
Fax: (+972 4) 995 39 00
Web: www.opgal.com

OSSD (Goodrich Optical & Space Systems)
5 Omni Way
Chelmsford
Massachusetts 01824-4142
United States
Tel: (+1 978) 967 22 56
Fax: (+1 978) 967 22 16
Web: www.oss.goodrich.com

P

Page Aerospace Ltd
Forge Lane
Sunbury-on-Thames
TW16 6EQ
United Kingdom
Tel: (+44 1932) 78 76 61
Fax: (+44 1932) 78 03 49
e-mail: sales@pageaerospace.co.uk
Web: www.pageaerospace.com

Palomar Display Products, Inc
1945 Kellogg Avenue
Carlsbad
California 92008
United States
Tel: (+1 760) 931 33 43
Fax: (+1 760) 931 32 98
e-mail: displays@palomardisplays.com
Web: www.palomardisplays.com

PCO (Przemyslowe Centrum Optyki SA)
Ostrobramska 75
PL-04-175 Warszawa
Poland
Tel: (+48 22) 613 77 44
Fax: (+48 22) 613 78 44
Telex: pl 813877 pco
e-mail: pcodt@ikp.atm.com.pl
Web: www.pcosa.com.pl

Penny + Giles Aerospace Ltd
1 Airfield Road
Christchurch
Dorset BH23 3TH
United Kingdom
Tel: (+44 1202) 48 17 71
Fax: (+44 1202) 48 48 46
e-mail: sales@pennyandgiles.com
Web: www.pennyandgiles.com

Peschges Variometer GmbH
Zieglerstraße 11
D-52078 Aachen
Germany
Tel: (+49 241) 18 05 94 00
Fax: (+49 241) 160 19 51
e-mail: infovp@peschges-variometer.de
Web: www.peschges.com

Phazotron – NIIR Corporation and Joint Stock
   Company
1 Electricheskiy Pereulok
123557 Moskva
Russian Federation
Tel: (+7 095) 253 69 88
Fax: (+7 095) 253 04 95
e-mail: phaza@aha.ru
Web: www.phazotron.com

Photo-Sonics Inc
820 South Mariposa Street
Burbank
California 91506-3196
United States
Tel: (+1 818) 531 32 19
Fax: (+1 818) 842 26 10
e-mail: info@photosonics.com
Web: www.photosonics.com

Phototelesis Corporation
7800 IH 10 West
Suite 700
San Antonio
Texas 78230
United States
Tel: (+1 512) 349 20 20
Fax: (+1 512) 341 10 20
Web: www.photot.com

PhotoTelesis
700 Lincoln Center
7800 IH 10 West
San Antonio
Texas 78230
United States
Tel: (+1 210) 349 20 20
Fax: (+1 210) 349 20 70
Web: www.photot.com

Pirometr St Petersburg Joint Stock Company
Bolshaya Monetnaya 16
197101 St Petersburg
Russian Federation
Tel: (+7 812) 238 72 45
Fax: (+7 812) 233 83 06

Planar Systems Inc
Corporate Headquarters
1195 NW Compton Drive
Beaverton
Oregon 97006-1992
United States
Tel: (+1 503) 748 11 00
Fax: (+1 503) 748 14 93
e-mail: sales@planar.com
Web: www.planar.com

Polhemus Inc
40 Hercules Drive
PO Box 560
Colchester
Vermont 05446-0560
United States
Tel: (+1 802) 655 31 59
Fax: (+1 802) 655 14 39
e-mail: sales@polhemus.com
Web: www.polhemus.com

Polyot Research and Production Company
1 Komsomolskaya Place
GSP 462
603950 Nizhnyi Novgorod
Russian Federation
Tel: (+7 8312) 42 21 04
Fax: (+7 8312) 42 31 57
e-mail: polyot@nnov.rfnet.ru
Web: www.polyot.nnov.rfnet.ru

Polytech AB
PO Box 20
SE-64032 Malmkoping
Sweden
Tel: (+46 157) 217 42
Fax: (+46 157) 213 48
e-mail: info@polytech.se
Web: www.polytech.se

PPI (Esterline Palomar Products)
23042 Arroyo Vista
Rancho Santa Margarita
California 92688-2604
United States
Tel: (+1 949) 766 53 00
Fax: (+1 949) 766 53 53
e-mail: sales@palpro.com
Web: www.palpro.com

Pribor JSC
47 Zapolnaya Street
305040 Kursk
Russian Federation
Tel: (+7 0712) 51 31 25
Fax: (+7 0712) 51 31 37
e-mail: obk@aviavtom.kursk.ru
Web: www.aviaavtomatika.ru

Production Association Urals
Optical Mechanical Plant
33B Vostochnay Street
620100 Yekaterinburg
Russian Federation
Tel: (+7 3432) 24 81 09
Fax: (+7 3432) 24 18 44
e-mail: infouomz@uomz.com
Web: www.uomz.ru

Q

QinetiQ
Company Headquarters
Cody Technology Park
Ively Road
Farnborough
Hampshire GU14 0LX
United Kingdom
Tel: (+44 1252) 39 33 00
Fax: (+44 1252) 39 33 99
Web: www.qinetiq.com

R

R&S (Rohde & Schwarz GmbH & Co KG)
PO Box 801469
D-81614 München
Germany
Tel: (+49 89) 412 91 17 47
Fax: (+49 89) 412 91 30 06
e-mail: air.operations@rohde-schwarz.com
Web: www.rsd.de

Rada Electronic Industries Ltd
Company Headquarters
7 Giborei Israel Street
Sapir Industrial Park
PO Box 8606
IL-42504 Netanya
Israel
Tel: (+972 9) 892 11 11
Fax: (+972 9) 885 58 85
e-mail: mrkt@rada.com
Web: www.rada.com

Radiopribor Joint Stock Company
2 Fatkullin Str.
420022 Kazan
Russian Federation
Tel: (+7 8432) 37 79 64
Fax: (+7 8432) 37 40 93

Rafael Armament Development Authority Ltd
Missile Division
PO Box 2250/30
IL-31021 Haifa
Israel

Ramsensky Instrument Engineering Plant
39 Mikhalevich Street
140100 Ramenskoye
Russian Federation
Tel: (+7 095) 501 41 11
Fax: (+7 095) 556 43 28
Telex: 9117725 roza

Raytheon Company
Corporate Headquarters
870 Winter Street
Waltham
Massachusetts 02451-1449
United States
Tel: (+1 781) 862 66 00
Fax: (+1 781) 860 25 20
e-mail: corpcom@raytheon.com
Web: www.raytheon.com

Raytheon Systems Company
Sensors and Electronic Systems
PO Box 902
El Segundo
California 90245
United States
Tel: (+1 310) 616 10 22
Fax: (+1 310) 616 63 30

Raytheon Technical Services Company LLC
Engineering and Production Support
6125 East 21st Street
Indianapolis
Indiana 46219-2058
United States
Tel: (+1 317) 306 79 92

RCF (Rockwell Collins France SA)
6 Avenue Didier Daurat
BP 20008
F-31701 Blagnac Cedex
France
Tel: (+33 5) 61 71 77 00
Fax: (+33 5) 61 71 78 34
e-mail: rcf@rockwellcollins.com
Web: www.rockwellcollins.com

RDI (Reutech Defence Industries)
New Germany
PO Box 118
3620 KwaZulu Natal
South Africa
Tel: (+27 31) 719 57 11
Fax: (+27 31) 719 58 75
e-mail: info@rdi.co.za
Web: www.rdi.co.za

Revue Thommen AG
Haupstrasse 85
CH-4437 Waldenburg
Switzerland
Tel: (+41 61) 965 22 22
Fax: (+41 61) 961 81 71
e-mail: info@thommenag.ch
Web: www.thommenag.ch

RIIE (State Research Institute of Instrument
  Engineering)
125 Prospect Mira
129226 Moskva
Russian Federation
Tel: (+7 095) 181 16 38
Fax: (+7 095) 181 33 70

Ring Sights Defence Group Ltd
PO Box 22
Bordon
Hampshire GU35 OJR
United Kingdom
Tel: (+44 1420) 47 22 60
Fax: (+44 1420) 47 83 59
e-mail: defence@ringsights.com
Web: www.ringsights.com

Rodale Electronics Inc
Company Headquarters
20 Oser Avenue
Hauppauge
New York 11788
United States
Tel: (+1 631) 231 00 44
Fax: (+1 631) 231 13 45
e-mail: rodaleel@rodaleelectronics.com
Web: www.rodaleelectronics.com

Rogerson Kratos Corporation
403 South Raymond Avenue
Pasadena
California 91009
United States
Tel: (+1 626) 449 30 90
Fax: (+1 626) 449 48 05
e-mail: rksales@rogerson.com
Web: www.rogersonkratos.com

ROI (Recon/Optical, Inc)
Corporate Headquarters
550 West Northwest Highway
Barrington
Illinois 60010-3094
United States
Tel: (+1 847) 381 24 00
Web: www.reconoptical.com/default.asp

RPKB (Ramenskoye Design Bureau JSC)
2 Gurjev Street
140103 Ramenskoye
Russian Federation
Tel: (+7 096) 463 39 32
Fax: (+7 096) 24 63 19 72
e-mail: rpkb@space.ru

RSL (Raytheon Systems Ltd)
Essex Office
The Pinnacles
Elizabeth Way
Harlow
Essex CM19 5BB
United Kingdom
Tel: (+44 1279) 42 68 62
Fax: (+44 1279) 40 75 46
Web: www.raytheon.co.uk

RSL (Raytheon Systems Ltd)
Head Office
80 Park Lane
London W1K 7TR
United Kingdom
Tel: (+44 20) 75 69 55 00
Fax: (+44 20) 75 69 55 96
e-mail: corporatecommunications@raytheon.co.uk
Web: www.raytheon.co.uk

RSL Electronics Ltd
Ramat Gabriel Industrial Zone
PO Box 21
IL-10550 Migdal Ha'emek
Israel
Tel: (+972 4) 654 75 10
Fax: (+972 4) 654 75 20
Web: www.rsl.co.il

Russ Research Center
2800 Indian Ripple Road
Dayton
Ohio 45440-3696
United States
Tel: (+1 877) 836 43 20
Web: www.russresearchcenter.com

Russkaya Avionika Joint Design Bureau
M.M. Gromov L11
140160 Zhukovsky 2
Russian Federation
Tel: (+7 095) 556 52 91
Fax: (+7 095) 556 50 83

S

Saab Avitronics
Pretoria
PO Box 8492
0046 Centurion
South Africa
Tel: (+27 12) 672 60 29
Fax: (+27 12) 672 62 22
e-mail: avitronics@grintek.com
Web: www.avitronics.co.za

Saab Grintek Communications
PO Box 10252
0046 Centurion
South Africa
Tel: (+27 12) 671 45 00
Fax: (+27 12) 671 45 11
e-mail: gcs@grintek.com

Saab Microwave Systems AB
Solhusgatan (Göteborg)
Flöjelnergsgatan 2A
SE-431 84 Mölndal
Sweden
Tel: (+46 31) 747 00 00
Fax: (+46 31) 747 29 02

Saab Transpondertech AB
Låsblecksgatan 3
SE-589 41 Linköping
Sweden
Tel: (+46 13) 18 80 00
Fax: (+46 13) 18 23 77
Web: www.transpondertech.se

SaabTech AB
Communications Division
Vattenkraftsvagen 8
SE-135 70 Stockholm
Sweden
Tel: (+46 8) 798 09 00
Fax: (+46 8) 798 84 33
Web: www.saabcom.se

Safe Flight Instrument Corporation
20 King Street
White Plains
NewYork 10604-1206
United States
Tel: (+1 914) 946 95 00
Fax: (+1 914) 946 78 82
e-mail: mail@safeflight.com
Web: www.safeflight.com

SAFRAN
Group Headquarters
2, Boulevard du general Martial Valin
F-75724 Paris Cedex
France
Tel: (+33 1) 40 60 80 80
Fax: (+33 1) 40 60 81 02
Web: www.safran-group.com

Sagem Avionics Inc
Pierce County Airport
16923 Meridian East
Puyallup
Washington 98375
United States
Tel: (+1 253) 848 60 60
Fax: (+1 253) 848 35 55
e-mail: general.delivery@arnav.com

Sagem Avionics Inc
2701 Forum Drive
Grand Prairie
Texas 75052-7099
United States
Tel: (+1 972) 641 35 58
Web: www.sagemavionics.com

Sagem Défense Sécurité
Montluçon Plant
Usine de la Côte Rouge
PO Box 3247
F-03106 Montluçon Cedex
France
Tel: (+33 4) 70 28 75 00
Fax: (+33 4) 70 28 28 01
Telex: f 990451
Web: www.sagem.com

Sagem Défense Sécurité
Navigation and Aeronautic Systems Division
Le Ponant de Paris
27 rue Leblanc
F-75512 Paris
France
Tel: (+33 1) 40 70 63 63
Fax: (+33 1) 40 70 66 40
Web: www.sagem.com

Sagem Défense Sécurité
Optronics and AirLand Systems Division
Le Ponant de Paris
27 rue Leblanc
F-75512 Paris Cedex 15
France
Tel: (+33 1) 40 70 63 63
Fax: (+33 1) 40 70 84 36
Web: www.sagem.com

Sanmina-SCI Corporation
2700 North First Street
San Jose
California 95134
United States
Tel: (+1 860) 434 45 73
Fax:t (+1 860) 434 45 99
e-mail: products@sanmina-sci.corp
Web: www.sanmina-sci.com

Satori SA
Batiment 66
Aeroport du Bourget
BP 151
F-93352 Le Bourget Cedex
France
Tel: (+33 1) 48 62 73 00
Fax: (+33 1) 48 64 98 56

Scientific Research Institute of Radio Equipment
Shkipersky Pzotok 19
RF
199106 St Petersburg
Russian Federation
Tel: (+7 812) 356 18 34
Fax: (+7 812) 352 37 51

Selex Communications SpA
Via A. Negrone, 1a
I-16153 Genova Cornigliano
Italy
Tel: (+39 010) 600 21
Fax: (+39 010) 650 18 97
e-mail: genova.pmpp@marconicomms.com
Web: www.marconi.com

SELEX S&AS (Selex Sensors and Airborne
   Systems)
Edinburgh 1
South Gayle Crescent
Edinburgh EH12 9EA
United Kingdom
Tel: (+44 131) 332 24 11
Fax: (+44 131) 314 28 39

SELEX S&AS (Selex Sensors and Airborne
   Systems)
Edinburgh 2
Building 19, CreweToll
Ferry Road
Edinburgh EH5 2XS
United Kingdom
Tel: (+44 131) 332 24 11
Fax: (+44 131) 332 06 90

SELEX S&AS (Selex Sensors and Airborne
   Systems)
Portsmouth
Browns Lane
The Airport
Portsmouth
Hampshire PO34 5PH
United Kingdom
Tel: (+44 23) 80 70 23 00
Fax: (+44 23) 80 31 67 77

Selex Sistemi Integrati
Head Office
ViaTiburtina Km 12,400
I-00131 Roma
Italy
Tel: (+39 06) 415 01
Fax: (+39 06) 413 11 33
Web: www.selex-si.com

Shimadzu Corporation
1, Nishinokyo-Kuwabaracho
Nakagyo-ku
Kyoto 604-8511
Japan
Tel: (+81 75) 823 11 11
Web: www.shimadzu.co.jp

Sifam Instruments Ltd
Woodland Road
Torquay
DevonTQ2 7AY
United Kingdom
Tel: (+44 1803) 40 77 34
Fax: (+44 1803) 40 77 40
e-mail: info@sifam.com
Web: www.sifam.com

Signature Industries Ltd
SARBE Search & Rescue
Tom Cribb Road
Thamesmead
London SE28 0BH
United Kingdom
Tel: (+44 20) 83 16 44 77
Fax: (+44 20) 83 16 62 18
e-mail: sales@sarbe.com
Web: www.sarbe.com

Sigtek
9075 Guildford Road
Suite C-1
Columbia
Maryland 21046
United States
Tel: (+1 410) 290 39 18
Fax: (+1 410) 290 81 46
e-mail: info@sigtek.com
Web: www.sigtek.com

Sigtronics Corporation
949 North Cataract Avenue
Suite D
San Dimas
California 91773
United States
Tel: (+1 909) 305 93 99
Fax: (+1 909) 305 94 99
e-mail: sales@sigtronics.com
Web: www.sigtronics.com

Simrad Optronics Ltd
Hever, 7 Amberley Court
County Oak Way
Crawley
West Sussex RH11 7XL
United Kingdom
Tel: (+44 1293) 56 04 13
Fax: (+44 1293) 56 04 18
Web: www.simrad-optronics.com

Skyquest Aviation Ltd
High Walls
Pinkneys Drive
Maidenhead
Berkshire SL6 6QD
United Kingdom
Tel: (+44 1628) 78 51 43
Fax: (+44 1628) 63 74 46
e-mail: info@skyquest.co.uk
Web: www.skyquest.co.uk

Speel Praha Ltd
Beranovych 130
CZ-199 05 Praha 9 Letnany
Czech Republic
Tel: (+420 2) 86 92 36 19
Fax: (+420 2) 86 92 37 21
e-mail: info@speel.cz
Web: www.speel.cz

Sperry Marine UK
Burlington House
118 Burlington Road
New Malden
Surrey KT3 4NR
United Kingdom
Tel: (+44 20) 83 29 20 00
Fax: (+44 20) 83 29 24 58
e-mail: marketing@sperry-marine.com
Web: www.sperry-marine.com

Spirit AeroSystems
Public Relations Office
M/C K12-12
PO Box 780008
Wichita
Kansas 67278-0008
United States
Tel: (+1 316) 526 31 53
e-mail:communications@spiritaerosystems.com

State Research Institute of Aviation Systems
7 Victorenko Street
125319 Moskva
Russian Federation
Tel: (+7 095) 157 70 47
Fax: (+7 095) 943 86 05
e-mail: info@gosniias.ru
Web: www.gosniias.msk.ru

Strategic Technology Systems Inc
1 Electronics Drive
PO Box 3198
Trenton
New Jersey 08619
United States
Tel: (+1 609) 584 02 02
Fax: (+1 609) 584 05 05

Sunair Electronics Inc
3005 SW 3rd Avenue
Fort Lauderdale
Florida 33315-3312
United States
Tel: (+1 954) 525 15 05
Fax: (+1 954) 765 13 22
e-mail: sunair@sunairhf.com
Web: www.sunairhf.com

Symetrics Industries, LLC
1615 West NASA Boulevard
Melbourne
Florida 32901
United States
Tel: (+1 321) 254 15 00
Fax: (+1 321) 259 41 22
e-mail: symetrics@symetrics.com
Web: www.symetrics.com

Symmetricom Ltd
4 Wellington Business Park
Dukes Ride
Crowthorne
Berkshire RG45 6LS
United Kingdom
Tel: (+44 1344) 75 34 80
Fax: (+44 1344) 75 34 99
Web: www.symmetricom.com

Symmetricom
3750 Westwind Boulevard
Santa Rosa
California 95403
United States
Tel: (+1 978) 232 14 77
Fax: (+1 707) 527 66 40
e-mail: ttm_sales@symmetricom.com
Web: www.symmetricom.com

Sypris Data Systems
160 East Via Verde
San Dimas
California 91773-5120
United States
Tel: (+1 877) 797 74 78
Fax: (+1 909) 962 94 01
Web: www.syprisdatasystems.com

T

T&T (Thrane & Thrane A/S)
Lundtoftegaardsvej 93 D
DK-2800 Lyngby
Denmark
Tel: (+45) 39 55 88 00
Fax: (+45) 39 55 88 88
e-mail: info@tt.dk
Web: www.thrane.com

Tadiran Spectralink Ltd
29 Hamerkava Street
PO Box 150
IL-58101 Holon
Israel
Tel: (+972 3) 557 31 02
Fax: (+972 3) 557 75 79
e-mail: info@tadspec.com
Web: www.tadspec.com

TAV (Thales Aerospace Division)
Head Office
45 rue de Villiers
BP 49
F-92 526 Neuilly sur Seine Cedex
France
Tel: (+33 1) 57 77 80 00
e-mail: contact.info@fr.thalesgroup.com

Teac America Inc
7733 Telegraph Road
Montebello
California 90640
United States
Tel: (+1 323) 727 48 66
Fax: (+1 323) 727 48 77
e-mail: airborne@teac.com
Web: www.teac.com

Teac Corporation
3-7-3 Naka-cho
Musashino-shi
Tokyo 180-8550
Japan
Tel: (+81 422) 52 50 00
Fax: (+81 422) 55 89 59
Web: www.teac.co.jp

Team
10 Place Vauban
Silic 127
F-94523 Rungis Cedex
France
Tel: (+33 1) 49 78 66 00
Fax: (+33 1) 49 78 66 29
e-mail: team.france@team-avionics.com
Web: www.team-avionics.com

Tech Comm Inc
3650 Coral Ridge Drive
Coral Springs
Florida 33065-2558
United States

Technisonic Industries Limited
240 Traders Boulevard
Mississauga
Ontario L4Z 1W7
Canada
Tel: (+1 905) 890 21 13
Fax: (+1 905) 890 53 38
Web: www.til.ca

Tecnobit SL
Avenue de Europa 21
Parque Empresarial La Moraleja
E-28108 Alcobendas
Madrid
Spain
Tel: (+34 91) 661 71 61
Fax: (+34 91) 661 98 40
e-mail: tecnobit@tecnobit.es
Web: www.tecnobit.es

Tekhpribor
Korpusnoj Proezd 1a
196084 St Petersburg
Russian Federation
Tel: (+7 812) 296 73 83
Fax: (+7 812) 296 88 89
Web: www.airshow.ru/firms/e225.htm

Teldix GmbH
PO Box 105608
D-69046 Heidelberg
Germany
Tel: (+49 6221) 51 20
Fax: (+49 6221) 51 25 58
e-mail: pr@teldix.com
Web: www.teldix.de

Teleavio Srl
Via A Negrone 1A
I-161 53 Genoa Cornigliano
Italy
Tel: (+39 010) 600 37 25
Fax: (+39 010) 600 36 99

Teledyne Controls
12333 West Olympic Boulevard
Los Angeles
California 90064
United States
Tel: (+1 310) 820 46 16
Fax: (+1 310) 442 43 24
e-mail: info@teledyne.com
Web: www.teledyne-controls.com

Teledyne Electronic Technologies
Microelectronics
12964 Panama Street
Los Angeles
California 90066
United States
Tel: (+1 310) 822 82 29
Fax: (+1 310) 574 20 15
e-mail: microelectronics@teledyne.com
Web: www.teledynemicro.com

Teledyne Electronic Technologies
Microwave Production
1274 Terra Bella Avenue
Mountain View
California 94043-1885
United States
Tel: (+1 415) 962 69 44
Fax: (+1 415) 966 15 21
e-mail: microwave@teledyne.com
Web: www.teledynemicrowave.com

Telefunken Radio Communication Systems
GmbH & Co KG
Eberhard-Finckh-Strasse 55
D-89075 Ulm Boefingen
Germany
Tel: (+49 731) 155 32 61
Fax: (+49 731) 155 31 12
e-mail: avionics@sysde.eads.net
Web: www.tfk-racoms.com

Telephonics Corporation
Corporate Headquarters
815 Broad Hollow Road
Farmingdale
New York 11735
United States
Tel: (+1 631) 755 75 83
Fax: (+1 631) 549 60 18
Web: www.telephonics.com

Telex Communications Inc
Corporate Headquarters
12000 Portland Avenue South
Burnsville
Minnesota 55337
United States
Tel: (+1 952) 884 40 51
Fax: (+1 952) 884 00 43
e-mail: info@telex.com
Web: www.telex.com

Tellumat (Pty) Ltd
PO Box 30451
7966 Tokai
South Africa
Tel: (+27 21) 710 29 11
Fax: (+27 21) 712 12 78
e-mail: avionics@tellumat.com
Web: www.tellumat.com

Terma A/S
Aerospace and Defense
Fabrikvej 1
DK-8500 Grenaa
Denmark
Tel: (+45 8779) 52 00
Fax: (+45 8779) 52 01
e-mail: ihr@terma.com
Web: www.terma.com

Thales Air Systems
Headquarters
Aerospace Division
45 Rue de Villiers
F-92526 Neuilly sur Seine
France
Tel: (+33 1) 57 77 88 10
Fax: (+33 1) 57 77 80 24
e-mail: info@fr.thalesgroup.com
Web: www.thales-airborne.com

Thales Land & Joint Systems
Bury
Vicon House
Western Way
Bury St Edmunds
Suffolk IP33 3SP
United Kingdom
Tel: (+44 1284) 75 05 99
Fax: (+44 1284) 75 05 98
Web: www.thalesgroup-optronics.com

Thales Land & Joint Systems
Glasgow
1 Linthouse Road
Govan Glasgow G51 4BZ
United Kingdom
Tel: (+44 141) 440 40 00
Fax: (+44 141) 440 40 01
Web: www.thalesgroup-optronics.com

Thales Land & Joint Systems
Taunton
Lisieux Way
Taunton
Somerset TA1 2JZ
United Kingdom
Tel: (+44 1823) 33 10 71
Fax: (+44 1823) 27 44 13
e-mail: laser.capabilities@uk.thalesgroup.com
Web: www.thalesgroup-optronics.com

Thales Navigation
Consumer Products
960 Overland Court
San Dimas
California 91773
United States
Tel: (+1 909) 394 50 00
Web: www.thalesnavigation.com

Thales Nederland BV
PO Box 42
NL-7550 GD Hengelo
Netherlands
Tel: (+31 74) 248 81 11
Fax: (+31 74) 242 59 36
e-mail: info@nl.thalesgroup.com
Web: www.thales-nederland.com

The Boeing Company
Integrated Defense Systems, Headquarters
PO Box 516
St Louis
Missouri 63166-0516
United States
Tel: (+1 253) 773 28 16
Fax: (+1 314) 234 82 96
Web: www.boeing.com/ids

The Titan Corp
Communications and Electronic Warfare Division
3033 Science Park Road
San Diego
California 92121
United States
Tel: (+1 858) 552 99 08
Fax: (+1 858) 552 99 09
Web: www.titan.com

TOKIMEC Inc
Electronics Systems Division
2-16-46 Minami-Kamata
Ohta-ku
Tokyo 144-8551
Japan
Tel: (+81 3) 37 37 86 41
Fax: (+81 3) 37 37 86 68
e-mail: corp-comm@tokimec.co.jp
Web: www.tokimec.co.jp

Tokyo Aircraft Instrument Co Ltd
1-35-1 Izumi-honcho
Komae-shi
Tokyo 201-8555
Japan
Tel: (+81 3) 34 89 11 25
Fax: (+81 3) 34 88 55 21
Telex: j 2422246 tkkwad
e-mail: tkksales@mars.plala.or.jp

Toshiba Corp
Headquarters
1-1-1 Shibaura
Minato-ku
Tokyo 105-8001
Japan
Tel: (+81 3) 34 57 49 24
Fax: (+81 3) 54 44 93 33
Telex: j 22587 toshiba
Web: www.toshiba.co.jp

Trackair Ltd
PO Box 1916
Marlow Bottom
Buckinghamshire SL7 3UT
United Kingdom
Tel: (+44 1628) 48 25 00
Fax: (+44 1628) 48 58 18
e-mail: sales@aerialsurvey.com
Web: www.aerialsurvey.com

Transas Aviation Ltd
54-4 Maly Prospect, V.O.
199178 St Petersburg
Russian Federation
Tel: (+7 812) 325 31 31
Fax: (+7 812) 567 19 01
e-mail: info@transas.com
Web: www.avia.transas.com

Trimble Navigation Limited
Corporate Headquarters
935 Stewart Drive
Sunnyvale
California 94085
United States
Fax: (+1 408) 481 77 81
Web: www.trimble.com

TUS (Thales Underwater Systems Ltd)
Ocean House
Templecombe
Somerset BA8 0DH
United Kingdom
Tel: (+44 1963) 37 05 51
Fax: (+44 1963) 37 22 00
e-mail: marketing.tus@uk.thalesgroup.com
Web: www.thales-naval.nl

TUS (Thales Underwater Systems)
Headquarters
525 route des Dolines
BP 157 Valbonne
F-06903 Sophia-Antipolis Cedex
France
Tel: (+33 4) 92 96 45 35
Fax: (+33 4) 92 96 40 19
Web: www.thales-underwater.com

Tyco Electronics UK Ltd
M/A-Com
Featherstone Road
Wolverton Mill
Milton Keynes
Buckinghamshire MK12 5EW
United Kingdom
Tel: (+44 1908) 57 42 00
Fax: (+44 1908) 57 43 00
Web: www.macom.com

U

Ultra Electronics Controls Division
417 Bridport Road
Greenford Middlesex UB6 8UE
United Kingdom
Tel: (+44 20) 88 13 44 44
Fax: (+44 20) 88 13 43 51
e-mail: marketing@uecd.co.uk
Web: www.uecd.co.uk

Ultra Electronics Holdings plc
Sonar and Communication Systems
419 Bridport Road
Greenford Middlesex UB6 8UA
United Kingdom
Tel: (+44 20) 88 13 45 67
Fax: (+44 20) 88 13 45 68
e-mail: information@ultra-electronics.com
Web: www.ultra-electronics.com

Ulyanovsk Instrument Design Office
10a Krylova Street
432001 Ulyanovsk
Russian Federation

Universal Avionics Systems Corporation
Corporate Headquarters
3260 East Universal Way
Tucson
Arizona 85706
United States
Tel: (+1 520) 295 23 00
Fax: (+1 520) 295 23 90
e-mail: marketing@uasc.com
Web: www.uasc.com

Universal Instrument Corporation
6090A Northbelt Parkway
Norcross
Georgia 30071
United States
Tel: (+1 770) 242 74 66

V

Viasat Inc
6155 El Camino Real
Carlsbad
California 92009-1699
United States
Tel: (+1 760) 476 22 00
Fax: (+1 760) 929 39 41
Web: www.viasat.com

For details of the latest updates to *Jane's Avionics* online and to discover the additional
information available exclusively to online subscribers please visit
jav.janes.com

Vibro-Meter SA
Route de Moncor 4
CH-1701 Fribourg
Switzerland
Tel: (+41 26) 407 15 22
Fax: (+41 26) 407 15 55
Telex: ch zrhvm8x
e-mail: aerospace@vibro-meter.ch
Web: www.meggitt.com

VSI (Vision Systems
   International LLC)
641 River Oaks Parkway
San Jose
California 95134
United States
Tel: (+1 408) 232 53 00
Fax: (+1 408) 432 84 49
e-mail: sales@vsi-hmcs.com
Web: www.vsi-hmcs.com

Vympel NPO
90 Volokolamskoje Street
123424 Moskva
Russian Federation
Tel: (+7 095) 491 03 68
Fax: (+7 095) 491 02 22

W

W J Communications, Inc
Corporate Headquarters (Components and
   Thin Film)
401 River Oaks Parkway
San Jose
California 95134-1916
United States
Tel: (+1 408) 577 64 11
Fax: (+1 408) 557 66 21
e-mail: sales@wj.com
Web: www.wj.com

Walter Dittel GmbH
Erpftinger Strasse 36
Postfach 101261
D-86899 Landsberg Lech
Germany
Tel: (+49 8191) 335 10
Fax: (+49 8191) 33 51 49
e-mail: info@dittel.com
Web: www.dittel.com

Weston Aerospace Limited
124 Victoria Road
Farnborough
Hampshire GU14 7PW
United Kingdom
Tel: (+44 1252) 54 44 33
Fax: (+44 1252) 37 12 16
e-mail: websales@westonaero.com
Web: www.westonaero.com

Wind River
Corporate Headquarters
500 Wind River Way
Alameda
California 94501
United States
Tel: (+1 510) 748 41 00
Fax: (+1 510) 749 20 10
Web: www.windriver.com

Z

Zenit Foreign Trade Firm
8 Rechnaya Street
Krasnogorsk 143400
Moskva
Russian Federation
Tel: (+7 095) 561 33 77

INDEXES

Alphabetical Index

W

X

Y

Z

Manufacturer's Index

For details of the latest updates to *Jane's Avionics* online and to discover the additional information available exclusively to online subscribers please visit
jav.janes.com

NOTES

NOTES